Aids to Calculation

Conversion Factors

A conversion factor is a fraction in which the numerator (top) and the denominator (bottom) express the same quantity in different units. For example, 2.2 pounds (lb) and 1 kilogram (kg) are equivalent; they express the same weight. The conversion factors used to change pounds to kilograms and vice versa are:

$$\frac{1 \text{ kg}}{2.2 \text{ lb}} \text{ and } \frac{2.2 \text{ lb}}{1 \text{ kg}}.$$

Because a conversion factor equals 1, measurements can be multiplied by the factor to change the *unit* of measure without changing the *value* of the measurement. To change one unit of measurement to another, use the factor with the unit you are seeking in the numerator (top) of the fraction.

Example 1 Convert the weight of 130 pounds to kilograms.

- Choose the conversion factor in which the kilograms are on top and multiply by 130 pounds:

$$\frac{1 \text{ kg}}{2.2 \text{ lb}} \times 130 \text{ lb} = \frac{130 \text{ kg}}{2.2} = 59 \text{ kg}.$$

Example 2 Consider a 4-ounce (oz) hamburger that contains 7 grams (g) of saturated fat. How many grams of saturated fat are contained in a 3-ounce hamburger?

- Because you are seeking grams of saturated fat, the conversion factor is:

$$\frac{7 \text{ g saturated fat}}{4 \text{ oz hamburger}}.$$

- Multiply 3 ounces of hamburger by the conversion factor:

$$3 \text{ oz hamburger} \times \frac{7 \text{ g saturated fat}}{4 \text{ oz hamburger}} = \frac{3 \times 7}{4} = \frac{21}{4} =$$

5 g saturated fat (rounded off).

Percentages

A percentage is a comparison between a number of items (perhaps the number of kcalories in your daily energy intake) and a standard number (perhaps the number of kcalories used for Daily Values on food labels). To find a percentage, first divide by the standard number and then multiply by 100 to state the answer as a percentage (*percent* means "per 100").

Example 3 Suppose your energy intake for the day is 1500 kcalories (kcal): What percentage of the Daily Value (DV) for energy does your intake represent? (Use the Daily Value of 2000 kcalories as the standard.)

- Divide your kcalorie intake by the Daily Value:

1500 kcal (your intake) ÷ 2000 kcal (DV) = 0.75.

- Multiply your answer by 100 to state it as a percentage:

0.75 × 100 = 75% of the Daily Value.

Example 4 Sometimes the percentage is more than 100. Suppose your daily intake of vitamin C is 120 milligrams (mg) and your RDA (male) is 90 milligrams. What percentage of the RDA for vitamin C is your intake?

120 mg (your intake) ÷ 90 mg (RDA) = 1.33.

1.33 × 100 = 133% of the RDA.

Example 5 Sometimes the comparison is between a part of a whole (for example, your kcalories from protein) and the total amount (your total kcalories). In this case, the total is the number you divide by. If you consume 60 grams (g) protein, 80 grams fat, and 310 grams carbohydrate, what percentages of your total kcalories for the day come from protein, fat, and carbohydrate?

- Multiply the number of grams by the number of kcalories from 1 gram of each energy nutrient (conversion factors):

$$60 \text{ g protein} \times \frac{4 \text{ kcal}}{1 \text{ g protein}} = 240 \text{ kcal}.$$

$$80 \text{ g fat} \times \frac{9 \text{ kcal}}{1 \text{ g fat}} = 720 \text{ kcal}.$$

$$310 \text{ g carbohydrate} \times \frac{4 \text{ kcal}}{1 \text{ g carbohydrate}} = 1240 \text{ kcal.}$$

- Find the total kcalories:

$$240 + 720 + 1240 = 2200 \text{ kcal.}$$

- Find the percentage of total kcalories from each energy nutrient (see Example 3):

    Protein: $240 \div 2200 = 0.109 \times 100 = 10.9 = $ 11% of kcal.

    Fat: $720 \div 2200 = 0.327 \times 100 = 32.7 = $ 33% of kcal.

    Carbohydrate: $1240 \div 2200 = 0.563 \times 100 = 56.3 = $ 56% of kcal.

    Total: $11\% + 33\% + 56\% = 100\%$ of kcal.

In this case, the percentages total 100 percent, but sometimes they total 99 or 101 because of rounding—a reasonable estimate.

Ratios

A ratio is a comparison of two (or three) values in which one of the values is reduced to 1. A ratio compares identical units and so is expressed without units.

Example 6 Suppose your daily intakes of potassium and sodium are 3000 milligrams (mg) and 2500 milligrams, respectively. What is the potassium-to-sodium ratio?

- Divide the potassium milligrams by the sodium milligrams:

    3000 mg potassium ÷ 2500 mg sodium = 1.2.

The potassium-to-sodium ratio is 1.2:1 (read as "one point two to one" or simply "one point two"), which means there are 1.2 milligrams of potassium for every 1 milligram of sodium. A ratio greater than 1 means that the first value (in this case, potassium) is greater than the second (sodium). When the ratio is less than 1, the second value is larger.

Weights and Measures

LENGTH
1 meter (m) = 39 in.
1 centimeter (cm) = 0.4 in.
1 inch (in) = 2.5 cm.
1 foot (ft) = 30 cm.

TEMPERATURE

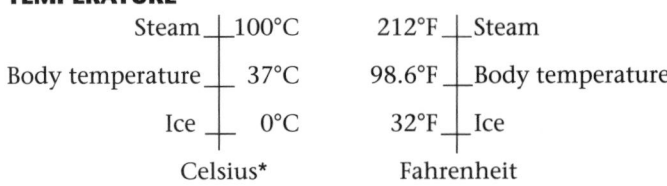

- To find degrees Fahrenheit (°F) when you know degrees Celsius (°C), multiply by 9/5 and then add 32.
- To find degrees Celsius (°C) when you know degrees Fahrenheit (°F), subtract 32 and then multiply by 5/9.

VOLUME
1 liter (L) = 1000 mL, 0.26 gal, 1.06 qt, or 2.1 pt.
1 milliliter (mL) = 1/1000 L or 0.03 fluid oz.
1 gallon (gal) = 128 oz, 8 c, or 3.8 L.
1 quart (qt) = 32 oz, 4 c, or 0.95 L.
1 pint (pt) = 16 oz, 2 c, or 0.47 L.
1 cup (c) = 8 oz, 16 tbs, about 250 mL, or 0.25 L.
1 ounce (oz) = 30 mL.
1 tablespoon (tbs) = 3 tsp or 15 mL.
1 teaspoon (tsp) = 5 mL.

WEIGHT
1 kilogram (kg) = 1000 g or 2.2 lb.
1 gram (g) = 1/1000 kg, 1000 mg, or 0.035 oz.
1 milligram (mg) = 1/1000 g or 1000 µg.
1 microgram (µg) = 1/1000 mg.
1 pound (lb) = 16 oz, 454 g, or 0.45 kg.
1 ounce (oz) = about 28 g.

ENERGY
1 kilojoule (kJ) = 0.24 kcal.
1 millijoule (mJ) = 240 kcal.
1 kcalorie (kcal) = 4.2 kJ.
1 g carbohydrate = 4 kcal = 17 kJ.
1 g fat = 9 kcal = 37 kJ.
1 g protein = 4 kcal = 17 kJ.
1 g alcohol = 7 kcal = 29 kJ.

*Also known as *centigrade*.

Nutrition Therapy and Pathophysiology

Second Edition

Nutrition Therapy and Pathophysiology

Second Edition

Marcia Nahikian Nelms
The Ohio State University

Kathryn Sucher
San Jose State University

Karen Lacey
University of Wisconsin, Green Bay

Sara Long Roth
Southern Illinois University Carbondale

Diane Habash
The Ohio State University

R. Gerald Nelms
University Center for the Advancement of Teaching at Ohio State University

Christina Lee Frazier
Southeast Missouri State University

Melissa Hansen-Petrik
The University of Tennessee–Knoxville

Robert D. Lee
Central Michigan University

Thomas J. Pujol
Southeast Missouri State University

Joshua E. Tucker
Michigan State University

Jeremy T. Barnes
Southeast Missouri State University

Mildred Mattfeldt-Beman
Saint Louis University

Roschelle A. Heuberger
Central Michigan University

Kathy Jones Irwin
Mercy Health Systems

Ethan A. Bergman
Central Washington University

Susan N. Hawk
Central Washington University

Deborah A. Cohen
University of New Mexico

Cade Fields-Gardner
The Cutting Edge

Elaina Jurecki
BioMarin Pharmaceutical Inc.

Joyce Wong
Kaiser Permanente Medical Center, Northern California

BROOKS/COLE
CENGAGE Learning™

Australia • Brazil • Japan • Korea • Mexico • Singapore • Spain • United Kingdom • United States

BROOKS/COLE
CENGAGE Learning™

Nutrition Therapy and Pathophysiology, 2e

Authors: Marcia Nahikian Nelms,
Kathryn Sucher, Karen Lacey, Sara Long Roth

Publisher/Executive Editor: Yolanda Cossio

Acquisitions Editor: Peggy Williams

Developmental Editor: Elesha Feldman

Assistant Editor: Elesha Feldman

Editorial Assistant: Alexis Glubka

Media Editor: Miriam Myers

Marketing Manager: Laura McGinn

Marketing Assistant: Elizabeth Wong

Marketing Communications Manager: Linda Yip

Senior Content Project Manager: Carol Samet

Creative Director: Rob Hugel

Art Director: Yvo Riezebos/Riezebos Holzbaur

Print Buyer: Paula Vang

Rights Acquisitions Account Manager, Text:
    Roberta Broyer

Rights Acquisitions Account Manager, Image:
    Dean Dauphinais

Production Service: S4Carlisle Publishing
    Services

Text Designer: Riezebos Holzbaur Design Group

Photo Researcher: Scott Rosen, Bill Smith Group

Cover Designer: John Walker

Cover Image: Tetra Images/Corbis

Compositor: S4Carlisle Publishing Services

Design Photos: Part 1—Marcia Nelms; Part 2—
    © Imagebroker/Alamy; Part 3—© Custom
    Medical Stock Photo/Alamy; © Ace Stock
    Limited/Alamy FM/BM—© Tomo
    Jesenicnik 2009. Used under license from
    Shutterstock.com

For product information and technology assistance, contact us at
Cengage Learning Customer & Sales Support, 1-800-354-9706
For permission to use material from this text or product,
submit all requests online at **www.cengage.com/permissions**
Further permissions questions can be e-mailed to
permissionrequest@cengage.com

Library of Congress Control Number: 2010920875

Student Edition:

ISBN-13: 978-1-4390-4962-4

ISBN-10: 1-4390-4962-9

Brooks/Cole Cengage Learning
20 Davis Drive
Belmont, CA 94002-3098
USA

Cengage Learning is a leading provider of customized learning solutions with
office locations around the globe, including Singapore, the United Kingdom,
Australia, Mexico, Brazil, and Japan. Locate your local office at
www.cengage.com/global

Cengage Learning products are represented in Canada by
Nelson Education, Ltd.

To learn more about Brooks/Cole, visit **www.cengage.com/brookscole**

Purchase any of our products at your local college store or at our preferred
online store **www.cengagebrain.com**

Printed in the United States of America
3 4 5 6 15 14 13 12

Dedication

For our colleagues in
nutrition and dietetics

For our students: past,
present and future.

For Jerry, Taylor, and Emory
Marcia Nahikian-Nelms

For my supportive and loving husband
Peter, and my son Alexander
Kathryn Sucher

For my husband Jim; your
encouragement and support
mean the world to me
Karen Lacey

JKR . . . your love and friendship make
life a joyous experience
Sara Long Roth

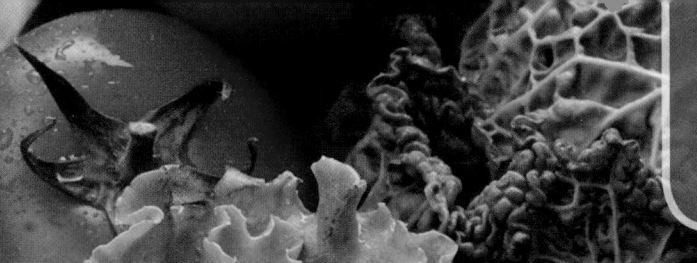

Brief Table of Contents

Table of Contents

9

Cellular and Physiological Response to Injury: The Role of the Immune System 149

10

Nutritional Genomics 183

14

Diseases of the Upper Gastrointestinal Tract 340

15

Diseases of the Lower Gastrointestinal Tract 376

16

Diseases of the Liver, Gallbladder, and Exocrine Pancreas 437

17

Diseases of the Endocrine System 471

20

Diseases and Disorders of the Neurological System 609

21

Diseases of the Respiratory System 648

22

Metabolic Stress and the Critically Ill 682

26

Metabolic Disorders 805

Preface

The authors of this text are educators, clinicians, and researchers. Therefore, our purpose in the second edition of this text is to continue to provide the most up-to-date research and application of evidence-based nutritional care for students, clinicians, and researchers as they seek to understand and treat nutrition-related disease. Most of us look to primary reference texts as the cornerstone of our practice. Many names come to mind—Zeman's *Clinical Nutrition*; Harrison's *Book of Internal Medicine*; and *The Merck Manual of Diagnosis and Therapy*. We continue to strive for this text to not only provide the reference material necessary to understand clinical nutrition practice, but provide it in such a way that the learning environment will support students' development of critical thinking, clinical reasoning, and decision-making skills.

What continues to make *this* text different from other clinical nutrition texts? The clinical environment evolves as a result of the impacting forces of research, health care funding, evidence-based nutrition practice, and development of the nutrition care process, standardized language, and standardized nutrition diagnoses. To meet the demands of these evolving forces, this text includes not only the most current research and integration of evidence-based practice within the context of the nutrition care process, but also an overview of health care systems and the dietitian's role within these systems as a member of the health care team; guidelines for documentation and other professional writings; and coverage of emerging fields such as nutrigenomics. Furthermore, as the framework for the nutrition care process has progressed over the previous three years, the structure for our text has organized its pedagogy to be consistent with each step of the nutrition care process. This text incorporates standardized language, the Evidence Analysis Library, the American Dietetic Association's Nutrition Care Manual, and Standards of Practice.

The text begins with a discussion of the dietitian's role as a nutrition expert, and then proceeds through a guided tour of the nutrition care process, introducing the basics of assessment, diagnosis, intervention, and monitoring/evaluation. Next, a comprehensive review of physiological concepts required to integrate nutrition therapy as a component of medical care is presented. These foundational chapters cover physiological response to injury, the immune system, fluid and electrolyte balance, pharmacology, and genetics—focusing specifically on the application of each of these topics to clinical nutrition practice.

The final section of this text is organized using a systems approach consistent with other medical texts. Each nutrition therapy chapter discusses normal structure and function of a body system, explains how the disease process interrupts normal functioning, and then describes appropriate medical and nutrition interventions. This second edition has retained the pedagogical features students and educators found especially helpful—Clinical Applications boxes, case studies, overviews of common medical care and drug-nutrient interactions, and interviews with current clinical practitioners. New features for this edition include Life Cycle Perspectives boxes, nutrition assessment summary tables, sample documentation, and PES statements. This approach allows any health care professional to benefit from this text.

Though every effort has been made to address the most recent research and the most common clinical and medical practices, this text has the same limitation any medical textbook will have: new diagnoses, new drugs, new treatments, and a new understanding of the relationship between nutrition and disease will inevitably continue to be cultivated after publication. Thus, this book strives to educate students about not only facts and theories that comprise current medical knowledge, but also the process of skill development that empowers students to grow in expertise within their field. As practitioners of the future utilize the nutrition care process, it will be refined even as their knowledge of disease and its treatment evolves.

As clinical practitioners and current dietetic educators, we have experienced a need for not only a different approach to a clinical nutrition text, but also a reference for clinical practitioners. We believe this text continues to fill both voids.

New to This Edition

The second edition of *Nutrition Therapy and Pathophysiology* has built upon the strengths of the first edition: a comprehensive focus on pathophysiology and medical treatment with a thorough review of the most current research. While this text continues to provide the all-in-one reference for the study of nutrition therapy, a number of content and organizational changes have been made to more closely integrate the nutrition care process and illustrate practical applications of clinical guidelines. Specific diets and food recommendations are now covered within each chapter, and new research and life-cycle perspectives are integrated throughout. The second edition's new chapter organization will allow the student and practitioner to follow the steps of the nutrition care process. Nutrition therapy within each systems chapter has been augmented with an emphasis on real-life application of the standards in patient care and has been updated with the latest evidence-based practice. Figures and tables have been modified to provide visual explanations of concepts within the text. New photos of real clinical settings have been added to both enhance chapters pedagogically and add visual appeal.

Specific changes for the second edition include the following:

Part 1 The Role of Nutrition Therapy in Health Care

- *Chapter 1 Role of the Dietitian in the Health Care System* has been streamlined to focus specifically on the RD's role as a member of the health care team and how nutritional care fits within the overall treatment approach. This chapter introduces the nutrition care process, evidenced-based practice, and the Standards of Professional Practice.

Part 2 The Nutrition Care Process

- *Chapter 2 The Nutrition Care Process* has been both updated and expanded to include the new terminology for all steps of the NCP. It offers a simple, straightforward explanation of how the nutrition care process frames nutrition practice. Many new examples help the student and practitioner to connect the theoretical discussion to the practical application.
- *Chapter 3 Nutrition Assessment: Foundation of the Nutrition Care Process* has undergone a major revision with a focus on the tools needed for successful completion and interpretation of nutrition assessments. New photographs and tables provide for easy references.
- *Chapter 4 Nutrition Intervention*—a new chapter—builds on the intervention terminology from Chapter 2 to explain the process of developing interventions, beginning with oral diets as examples of interventions within the acute care setting. This chapter includes new coverage of both counseling and nutrition education, with a focus on their use in acute care.
- *Chapter 5 Enteral and Parenteral Nutrition Support* has been extensively revised to incorporate ASPEN and EAL guidelines for prescribing nutrition support. The chapter is supported by a significant review of the literature and features new step-by-step, practical examples of the development of both enteral and parenteral prescriptions.
- *Chapter 6 Documentation of the Nutrition Care Process* This chapter focuses on documentation of the nutrition care process and the development of PES statements within a variety of chart formats. New to this edition are examples of electronic charting.

Part 3 Introduction to Pathophysiology

- *Chapter 7 Fluid and Electrolyte Balance* and *Chapter 8 Acid-Base Balance* have been updated with the latest research to provide a thorough review for the student and comprehensive reference for the practitioner.
- *Chapter 9 Cellular and Physiological Response to Injury: The Role of the Immune System* integrates basic concepts of cellular injury, the physiological response to it, and immunology—formerly in two separate chapters—to illustrate how the immune response occurs alongside this physiological response to injury and disease. The basics of organ transplantation are also covered within the context of immunology. A newly expanded section on food allergy emphasizes nutrition therapy and elimination diets.

- *Chapter 10 Nutritional Genomics* presents extensive new research on the influence of genetics in disease risk, the clinical applications of genetic testing, and the impact of these findings on dietetics practice.
- *Chapter 11 Pharmacology* includes excellent references for comprehension of drug-nutrient interactions, and a new box and table on complementary/alternative therapies.

Part 4 Nutrition Therapy

The nutrition therapy sections within these chapters have been reorganized to more closely mirror the steps of the nutrition care process. The redesigned Application of the NCP features now walk the student through the process using a case study format. Furthermore, each chapter provides updated coverage of the common diagnostic procedures and medications and a reference table for complementary/alternative treatments that are often encountered by the registered dietitian.

- *Chapter 12 Energy Balance and Body Weight* features many new figures, including graphs of updated obesity statistics and a comparison of four bariatric surgical procedures. It presents new findings on nonalcoholic fatty liver disease, "screen time" and obesity in children, and the effectiveness of various weight-loss plans.
- *Chapter 13 Diseases of the Cardiovascular System* includes new findings about cardiac cachexia, guidelines for prevention of adverse nutritional effect of medications, up-to-date CVD statistics, and current WHO criteria for metabolic syndrome.
- *Chapter 14 Diseases of the Upper Gastrointestinal Tract* has been revised to provide extensive nutrition therapy for numerous diagnoses with excellent coverage of pathophysiology and medical care.
- *Chapter 15 Diseases of the Lower Gastrointestinal Tract* is infused with extensive new research on irritable bowel syndrome and inflammatory bowel disease. Also new to Chapter 15 are excellent guidelines for the use of probiotics within numerous diagnoses and a figure outlining the multifactorial etiology of irritable bowel syndrome.
- *Chapter 16 Diseases of the Liver, Gallbladder, and Exocrine Pancreas* has been reorganized to focus on chronic liver disease and its nutrition consequences. It includes new boxes on diagnostic procedures, acute liver failure, and hepatitis B testing; and new tables on staging scales for hepatic encephalopathy and cirrhosis, supplementation for alcoholics, and pancreatic enzymes.
- *Chapter 17 Diseases of the Endocrine System* has been revised to include more practical application of nutrition therapy in diabetes, including carbohydrate counting and carbohydrate-to-insulin ratios. The discussion of the pathophysiology of diabetes complications has been enhanced with greater detail and a new figure.
- *Chapter 18 Diseases of the Renal System* features the latest National Kidney Foundation standards for nutrition therapy, and has been reorganized to focus on nutrition therapy for chronic kidney disease and its complications.
- *Chapter 19 Diseases of the Hematological System* meticulously covers the pathophysiology of hematologi-

cal disorders, with extensive nutrition therapy for anemias. New boxes focus on iron nutrition: its supplementation in pregnancy, and its possible connections with restless leg syndrome.

- *Chapter 20 Diseases and Disorders of the Neurological System* includes an extensive discussion of the ketogenic diet as a treatment for seizure disorders as well as up-to-date nutrition therapy interventions for numerous other neurological disorders.

- *Chapter 21 Diseases of the Respiratory System* includes new data on the role of supplementation in the treatment of asthma and other respiratory disorders, and a new table on using the EAL to plan nutrition therapy for chronic obstructive pulmonary disease.

- *Chapter 22 Metabolic Stress and the Critically Ill* now thoroughly integrates ASPEN nutrition support recommendations for the critically ill patient. Highlights include the latest diagnostic criteria for SIRS and MSOF and research on permissive underfeeding.

- *Chapter 23 Neoplastic Disease* presents new dietary recommendations for cancer prevention; new research on the etiology, diagnosis, and treatment of cancer cachexia; and the latest advances in medical treatments for cancer.

- *Chapter 24 HIV and AIDS* has been updated with current medical treatments and their impact on nutritional status, as well as a new box on diagnostic tests for HIV.

- *Chapter 25 Diseases of the Musculoskeletal System* features a new section on systemic lupus erythematosus.

- *Chapter 26 Metabolic Disorders* has been thoroughly updated with new information on screening and diagnosis, acute medical treatments, and evidence-based nutrition therapy. It includes a new section on organic acidemias, focusing on nutrition therapy for propionic acidemia.

Acknowledgements

We have had significant assistance in development and writing of this text.

We are grateful to our contributing authors for their expertise, persistence, and dedication in the development of each of their chapters:

Jeremy T. Barnes
Southeast Missouri State University

Ethan A. Bergman, PhD, RD, CD, FADA
Central Washington University

Deborah A. Cohen, DCN, RD
University of New Mexico

Christina Frazier, PhD
Southeast Missouri State University

Cade Fields Gardner, MS, RD, LD, CD
The Cutting Edge

Diane Habash, PhD, RD, LD
The Ohio State University

Melissa Hansen-Petrik, PhD, RD, LDN
The University of Tennessee–Knoxville

Susan N. Hawk, PhD, RD
Central Washington University

Roschelle A. Heuberger, PhD, RD
Central Michigan University

Kathy Jones Irwin, MS, RD
Mercy Health Systems, Knoxville, Tennessee

Elaina Jurecki, MS, RD
BioMarin Pharmaceutical Inc.

Robert D. Lee, DrPH, RD
Central Michigan University

Mildred Mattfeldt-Beman, PhD, RD, LD
Saint Louis University

Thomas J. Pujol, EdD, FACSM
Southeast Missouri State University

Joshua E. Tucker, MS
Michigan State University College of Osteopathic Medicine

Joyce Wong, MS, RD
Kaiser Permanente Medical Center, Northern California

We believe the practitioner interviews will assist students in understanding the role of the clinical dietitian within the health care practice and serve as significant role models. We would like to thank the following individuals for their gracious consent for interviews within this text:

Mary Ellen Beindorff, RD, LD
Barnes-Jewish Hospital, St. Louis, Missouri

Shelly Case, BSc, RD
nutrition counselor and author of the book *Gluten-Free Diet*

Nancy Duhaime, MS, RD, LD
Bartlett Regional Hospital, Juneau, Alaska

Robin Gaff, RD, LD, CSO
The Ohio State University Medical Center

Linda White Gray, RD, CDE
University of California, San Francisco Medical Center

Kathleen Huntington, MS, RD, LD
Oregon Health & Science University, Portland

Marianne Hutton, RD, CSR, CDE
advertising editor for the *Renal Nutrition Forum* and Program Chair, National Kidney Foundation Council on Renal Nutrition 2010 Spring Clinical Meeting

Betty Kovacs, MS, RD
New York Obesity Research Center, St. Luke's-Roosevelt Hospital, New York

Eileen MacKusick, MS, RD
Watsonville Community Hospital, Watsonville, California

Julie Matel, MS, RD
Lucille Packard Children's Hospital

Melody O'Donnell, RD
University of California, San Francisco Medical Center

La Paula Sakai, RD, MS, CNSD
Retired

Lynne Schonder, RD, MS, CNSC
Valley Care Medical Center, Pleasanton, California

Kathryn Sikorski, RD, CDE
Santa Clara Valley Medical Center, San Jose, California

Valerie Simler, MS, RD
Valley Care Medical Center, Pleasanton, California

Gretchen Vannice, MS, RD
Omega-3 RD™ Nutrition Consulting

Jill Weisenberger, MS, RD, CDE
National Clinical Research, Inc. – Norfolk in Norfolk, Virginia

Beth Zupec-Kania, RD, CD
Children's Hospital of Wisconsin

We would like to thank the many reviewers whose insights and suggestions proved invaluable during the writing process:

Mallory Boylan
Texas Tech University

Tina Crook
University of Central Arkansas

Jamie M. Erskine
University of Northern Colorado

Bahram Faraji
University of Texas Pan American

Bj Friedman
Texas State University

Rubina S. Haque
Eastern Michigan University

Molly M. Michelman
University of Nevada, Las Vegas

Patricia Hughes Petitt
University of Montevallo

Maria T. Spicer
Florida State University

Julie Poh Thurlow
University of Wisconsin, Madison

Melinda Wells Valliant
The University of Mississippi

Shahla Wunderlich
Montclair State University

Linda O. Young
University of Nebraska, Lincoln

Finally, while many researchers, dietetic educators, clinicians, and practitioners have authored and reviewed this text, it would not be the document it is without the dedication and incalculable input from our developmental editor, Elesha Feldman. We would also like to acknowledge the significant assistance of acquisitions editor Peggy Williams in keeping us on schedule and having faith in the contributions that this book can make. This book could not have been accomplished without them.

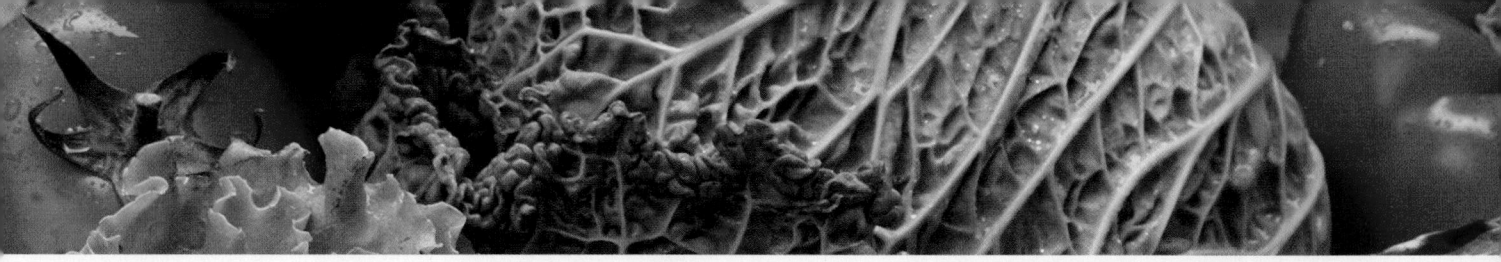

Nutrition Therapy
and Pathophysiology

Second Edition

Part 1

The Role of Nutrition Therapy in Health Care

1

Role of the Dietitian in the Health Care System

Kathryn Sucher, ScD, RD
San Jose State University

Introduction

The connection between diet and health has long been recognized. However, the profession of dietetics was first defined in 1899 by the American Home Economics Association as "individuals with knowledge of food who provide diet therapy for the medical profession." After 1917, dietitians were affiliated with the American Dietetic Association (ADA).[1] Dietitians who were employed in hospitals became known as *clinical dietitians*. Over time, the clinical dietitian's role became the provision of specialized care and modification of diets to treat various medical conditions.

In the early 1970s, after high levels of malnutrition in hospitalized patients were reported[2] and new and improved procedures for delivering enteral and parenteral nutrition were developed, clinical dietitians began to take a leadership role in screening patients and monitoring their needs for adequate nutrition support. In addition, as research pointed to the role of diet in the development of chronic disease, clinical dietitians began to provide primary and secondary disease prevention for such diseases as atherosclerosis, cancer, and type 2 diabetes mellitus.[3] The information provided in this chapter is meant to help you understand where you might find potential sources of employment, your contribution to the nutrition care of a patient as part of the heath care team, reimbursement issues that you might encounter, and your professional responsibilities, and to help you develop critical thinking skills that are necessary for the nutrition care process.

The Registered Dietitian in Clinical Practice

The Role of the Clinical Dietitian

The practice of clinical nutrition is called *nutrition therapy*. Clinical dietitians are the educated and trained professionals who can best deliver nutrition therapy by using the nutrition care process (NCP). The nutrition care process consists of four major components: (1) nutrition assessment, (2) nutrition diagnosis, (3) nutrition intervention, and (4) nutrition monitoring and evaluation.[4]

Scope of Practice Framework

The American Dietetic Association has developed a framework (Figure 1.1) to help guide dietitians and to better define dietetics practice. The Scope of Practice Framework defines the roles, functions, responsibilities, and activities that dietetics practitioners are educated and authorized/proficient to

evidence-based dietetics practice—dietetics practice in which systematically reviewed scientific evidence is used to make food and nutrition practice decisions

health insurance—financial protection against health care costs associated with treatment of disease or accidental injury

medical doctor—a health professional who has earned a post-bachelor degree of doctor of medicine or doctor of osteopathy and who has passed a licensing examination

nurse—a health care professional who has earned at least an associate's degree in nursing, has been licensed by the state, and assists patients in activities related to maintaining or recovering health

occupational therapist—a health professional who has obtained a master's degree and passed a national registration exam, who helps individuals with mentally, physically, developmentally, or emotionally disabling conditions improve their ability to perform tasks in their daily living and working environments

outcomes research—evaluation of care that focuses on the status of participants after receiving care

pharmacist—a licensed health professional with a doctorate of pharmacy who compounds and dispenses medications, checks laboratory results for therapeutic drug levels, and reviews risk for drug interactions

social worker—a professional with at least a bachelor's degree in social work who provides persons, families, or vulnerable populations with psychosocial support, advises family caregivers, counsels patients, and helps plan for patients' needs after discharge

speech-language pathologist—a health professional who has earned a master's degree and passed a national examination, who assesses, diagnoses, treats, and helps to prevent speech, language, cognitive, communication, voice, swallowing, fluency, and other related disorders

---

perform within the boundaries of federal, state, and facility regulations. You will notice that the framework is divided into three broad categories. The first area is Foundation Knowledge, which you are now in the process of obtaining. The second one is Evaluation Resources, which are meant to help evaluate a dietitian's performance to determine if the standards of the job are being met. The last area, Decision Aids, includes resources to help further define the scope of practice. The framework is meant to be flexible so that, as the profession changes or as an individual specializes or advances in her or his practice, evaluation resources and decision aids will also be modified.

The Clinical Nutrition Team

Health care is defined as the prevention, treatment, and/or management of illness. Clinical dietitians are employed in a number of acute and chronic health care facilities, as listed in Table 1.1. Depending on the facility, nutrition therapy services may be organized along different lines. The manager of the services may have the title of chief clinical manager or clinical nutrition manager. This person often reports to the director of nutrition service, who commonly supervises the clinical nutrition manager and food service manager/directors. In turn, inpatient and outpatient clinical dietitians usually report to the clinical manager. Other important personnel in nutrition therapy services are registered dietetic technicians (DTR), who assist dietitians in the nutritional screening of patients and provision of nutrition education in addition to other duties, and dietary assistants/diet clerks who are often responsible for the documentating and processing diet orders and assuring accuracy of the meals that are provided for patients. Table 1.2 provides common job specifications for clinical nutrition team members.

Clinical dietitians' services may be provided to general patient care units, such as those on a general medical or surgical floor, or may be based on a medical specialization, such as treatment of patients in intensive care units (e.g., burn/trauma unit or pediatric/neonatal intensive care units).

Table 1.1 Types of Acute and Chronic Health Care Facilities in the United States

Acute Care Facilities	
Hospitals	
Public Not for Profit	Often owned and managed by the county or state government
Private Not for Profit	Owned or managed by the community, a religious organization, district health councils, or their own hospital board
Private Profit	Investor-owned (for-profit) health care organizations
Veterans and Military	Government-run health care facilities for veterans of the U.S. military service and active-duty enlisted men and women
Clinics	
Outpatient	For preventative, primary health care (e.g., treatment for ear infection); and secondary health care (e.g., treatment of type 2 diabetes)
Urgent Care	Primary care
Longer-Care Facilities	
Skilled Nursing Facilities	Custodial services for the chronically ill and disabled*
Residential/Assisted Living	Custodial services for activities of daily living (e.g., bathing)*
Rehabilitation/Restorative	Provide assistance with recovery from acute or chronic illness and/or surgical procedures (e.g., stroke)*
Hospice	Comfort care for those who are not expected to live more than six months*

*http://ucare.utah.gov/what_is_ltc.html

In addition, clinical dietitians may be certified in a medical specialty and become, for example, diabetes educators, lactation consultants, or nutrition support specialists. Nutrition therapy practice certifications and their requirements are listed in Table 1.3.

Figure 1.1 The ADA Scope of Practice Framework

Block One: Foundation Knowledge

Definition of Dietetics as a Profession: "The integration and application of principles derived from the sciences of food, nutrition, management, communication, and biological, physiological, behavioral, and social sciences to achieve and maintain optimal human health" within flexible scope of practice boundaries to capture the breadth of the profession.

Five Characteristics of the Profession	Professionals Who Demonstrate This Characteristic...	Core Professional Resources	
Code of Ethics	Follow a Code of Ethics for practice	Code of Ethics	Ethics Opinions
Body of Knowledge	Possess a unique body of theoretical and science-based knowledge that leads to defined skills, abilities, and norms	Philosophy and Mission: • Research Philosophy and Diagram	Research, Position Papers, Practice Papers, Published Literature
Education	Demonstrate competency at selected level by meeting set criteria and passing credentialing exams	CADE® Core Competencies and Emphasis Areas	CDR® Certification (RD®, DTR®)
Autonomy	• Are reasonably independent and self-governing in decision-making and practice • Demonstrate critical thinking skills • Take on roles that require greater responsibility and accountability both professionally and legally • Stay abreast of new knowledge and technical skills	The CDR Professional Development Portfolio Process offers a framework for credentialed professionals to develop specific goals, identify learning needs and pursue continuing education opportunities. This may encompass certificates (such as weight management), specialty certificates (such as CSR® advanced practice certification), or advanced degrees.	
Service	Provide food and nutrition care services for individuals and population groups and other stakeholders. Additional functions may include: • Managing food and other material resources • Marketing services and products • Teaching dietitians and other professionals or students • Conducting research • Managing human resources • Managing facilities	Nutrition Care Process and Model	Nationally developed guidelines ADA Evidence-Based Guides for Practice
		Practice-Based Evidence • Dietetics Practice Outcomes Research • Dietetics Practice Audit	

The Framework consists of three building blocks with flexible boundaries. The blocks describe the full range of responsibilities, roles, and activities that dietetics professionals are educated and authorized to perform. The flexible boundaries allow for new roles to emerge. Because of the complexity of our profession, it is impossible to present this information as a list of isolated activities that are parceled out at different levels. Rather, a stepped algorithmic approach is needed to capture the breadth of the profession, allow individual practitioners to draw from the full range of resources, and lend our scope of practice the flexibility it needs to evolve as new research in dietetics and practice emerge.

From an individual perspective, whether an activity is within *your* scope of practice is influenced by every level of the Framework—our Foundation Knowledge, Code of Ethics, Standards of Practice, and Standards of Professional Performance—as well as by licensure and certification laws, research, guides for practice and expert opinion, and new research.

Block Two: Evaluation Resources

The evaluation resources listed here are intended for use in conjunction with relevant state, federal and licensure laws. Together with the laws, they serve as a guide for ensuring safe and effective dietetics practice. Practitioners can use them **to evaluate performance, make hiring decisions, determine whether a particular activity falls within the legitimate scope of the practice, or to initiate regulatory reform.** The core standards are based on the Nutrition Care Process and Model (NCPM) and Commission on Accreditation for Dietetics Education (CADE) educational core competencies. Specialty and advanced standards can evolve for specific practice areas.

Code of Ethics	DTR Standards of Practice in Nutrition Care RD Standards of Practice in Nutrition Care ↓ RD Specialty or RD Advanced Standards of Practice	Standards of Professional Performance for Dietetics Professionals ↓ RD Specialty or RD Advanced Standards of Professional Performance

Block Three: Decision Aids

The healthcare environment in which we work is highly diverse and always evolving. The resources listed here are intended to help dietetics professionals respond to new demands. By using the Decision Tree and Decision Analysis Tool, professionals can fully consider whether a new role or activity falls within the legitimate scope of their practices, and thereby grow their practices to encompass new areas. This is particularly helpful when state, federal, organizational and educational guidelines have not yet expanded to address a need. The other resources can be applied to seek guidance when making such decisions, or when effecting change at the local or national level to reflect emerging trends and needs.

Decision Analysis Tool	Decision Tree	Definition of Terms

Supporting Documentation for use within Decision Tree and Decision Analysis Tool

Licensure/Certification/Credentials
Examples include: State Licensure, CDR Credentials, Specialty Certificates, Advanced Practice Certification, or Advanced Degrees.

Organizational Privileging

Individual CDR Professional Development Portfolio
Portfolio Learning Plan and Learning Activities Log

Evidence Based Practice
• Existing Research and Literature
• ADA Position and Practice Papers, Ethics Opinions
• Nationally-Developed Guidelines and ADA Guides for Practice

Practice Based Evidence
• Dietetics Practice Outcomes Research

The arrows reflect the flexible, dynamic nature of the Framework. At both the individual practitioner level and our collective professional level, developments in one area of the Framework influence others. For example, as new trends in dietetics practice emerge, education, certification, and standards of practice and professional performance will change to address them. Likewise, as a practitioner tailors his or her individual scope of practice through experience and training, this will influence the resources utilized at every level.

Source: Reprinted from Journal of the American Dietetic Association 105:4, J O'Sullivan Maileet, J Skates and E Pritchett; ADA: Scope of Practice Framework, © Copyright 2005, with permission from Elsevier.

Table 1.2 Responsibilities and Tasks of Clinical Nutrition Team Members

Clinical Nutrition Team Member	Responsibilities	Major Tasks
Clinical Nutrition Manager	Directs the activities of clinical dietitians, dietetic technicians, and dietetic assistants	Hiring, evaluating, and training employees; reviewing productivity reports, writing job descriptions, scheduling employees, developing policies and procedures, designing performance standards, and developing and implementing goals and objectives of the department*
Registered (Clinical) Dietitian (RD)	Provides nutritional care for patients	Nutritional screening/assessment of patients to determine the presence or risks of developing a nutrition-related problem, development of nutritional diagnosis, nutrition intervention, and monitoring and evaluation of the nutrition care plan
Dietetic Technician (DTR)	Assists the clinical dietitian	Gathering data for nutritional screening; assigning a level of risk for malnutrition according to predetermined criteria; administering nourishment and dietary supplements for patients and monitoring tolerance; and providing information to help patients select menus and giving simple diet instructions
Dietetic Assistant/Diet Clerk	Assists the clinical dietitian and/or dietetic technician in some routine aspects of nutritional care	Processing diet orders, checking menus against standards, setting up standard nourishment, tallying special food requests; distributing and collecting patient menus and trays; may be involved in evaluating patient food satisfaction and helping to gather food records used to evaluate nutrient intake.

*Digh EW, Dowdy RP. A survey of management tasks, completed by clinical dietitians in the practice setting. J Am Diet Assoc. 1994;94:1381–84.

Table 1.3 Dietetic Practice Certifications Requirements

Specialty	Certifying Organization (webpage)	Requirements
Board Certified Specialist in Pediatric Nutrition (CSP)	American Dietetic Association / Commission on Dietetic Registration (www.eatright.org)	Current RD, and two years minimum length of RD status, and 2,000 hours of pediatric practice within the last five years, and successful completion of the Board Certification as a Specialist in Dietetics examination.
Board Certified Specialist in Renal Nutrition (CSR)	American Dietetic Association / Commission on Dietetic Registration (www.eatright.org)	Current RD, and two years minimum length of RD status, and 2,000 hours of renal practice within the last five years, and successful completion of the Board Certification as a Specialist in Dietetics examination.
Board Certified Specialist in Gerontological Nutrition (CSG)	American Dietetic Association / Commission on Dietetic Registration (www.eatright.org)	Current RD, and two years minimum length of RD status, and 2,000 hours of gerontological practice within the last five years, and successful completion of the Board Certification as a Specialist in Dietetics examination.
Board Certified Specialist in Sports Dietetics (CSSD)	American Dietetic Association / Commission on Dietetic Registration (www.eatright.org)	Current RD, and two years minimum length of RD status, and 1,500 hours of sports dietetics practice within the last five years, and successful completion of the Board Certification as a Specialist in Dietetics examination.
Board Certified Specialist in Oncology Nutrition (CSO)	American Dietetic Association / Commission on Dietetic Registration (www.eatright.org)	Current RD, and two years minimum length of RD status, and 2,000 hours of oncology dietetics practice within the last five years, and successful completion of the Board Certification as a Specialist in Dietetics examination.
Certified Diabetes Educator (CDE)	National Certification Board for Diabetes Education (www.ncbde.org)	Minimum of two years experience working as a registered dietitian is required. Minimum of 1,000 hours of professional practice experience in diabetes self-management education with a minimum of 40% (400 hours) accrued in the most recent year preceding application. Minimum of 15 clock hours of continuing education activities applicable to diabetes within the two years prior to applying for certification. Successful completion of the Certified Diabetes Educator Examination.
Certified Nutrition Support Dietitian (CNSD)	National Board of Nutrition Support Certification (www.ptcny.com/clients/NBNSC)	It is recommended that candidates have at least two years of experience in specialized nutrition support (parenteral and enteral nutrition). Successful completion of the Certification Examination for Nutrition Support Dietitians.
Lactation Consultant (IBCLC)	The International Board of Lactation Consultant Examiners (www.iblce.org/index.php)	Completed at least 45 hours of continuing education in lactation, must have 1,000 hours of breastfeeding counseling experience within five years, and pass the certification examination.

Table 1.4 Education and Certification Requirements of Selected Members of the Health Care Team

Health Profession	Education	Degree Initials	Credentialing
Medical Doctor and Osteopathic Doctor	Four-year post-bachelor degree plus internship and residency	MD; DO	State licensure exam
Nurse	Two- or four-year degree	AA (two-year); BSN (four-year)	State licensure exam RN
Pharmacist	Six-year postsecondary education	PharmD	State licensure exam
Occupational Therapist	Master's degree	MOT, MS, or MA	National exam for registration (OTR)
Speech-Language Pathologist	Master's degree plus a clinical fellowship	MS or MA	National exam for Certificate of Clinical Competence (CCC)
Social Worker	Bachelor's degree or master's degree	BSW or MSW	State licensing, certification, or registration

Source: Occupational Outlook Handbook (OOH), 2008-09 Ed. [Internet] Washington, DC. U.S. Bureau of Labor Statistics. Available at http://www.bls.gov/oco/

Other Health Professionals— Interdisciplinary Teams

In the health care setting, individuals from different disciplines communicate with each other regularly in order to best care for their patients.[5] Dietitians are integral members of the patient's health care team and collaborate with physicians, pharmacists, nurses, speech pathologists, occupational therapists, social workers, and many others when providing nutritional treatment. Dietitians must know the roles of the other team members in order to be effective and to ensure optimal patient care. Table 1.4 covers the education and training requirements for health professionals and the job roles with which a dietetic student should be familiar when first starting to practice dietetics.

The practice of medicine by **medical doctors** includes the diagnosis, treatment, correction, advisement, or prescription for any human disease, ailment, injury, infirmity, deformity, pain, or other condition, physical or mental. All physicians in the United States have advanced training and certification in a specialized area of medicine or surgery.[6] Table 1.5 lists the recognized board specialties and subspecialties. Nutritionally, doctors are responsible for prescribing nutrition support and nutrition prescriptions for their patients.

The largest group of health care workers in the United States is **nurses**. Registered nurses (RNs) assist in the treatment of patients, administer medications and intravenous solutions, educate patients on various medical conditions, and provide advice, follow-up care, and emotional support to patients' family members.[6] Since they provide care 24 hours a day, 7 days a week, nurses are commonly responsible for doing the initial nutrition screening of patients and then documenting a patient's food intake while hospitalized as well as notifying the dietitian if a patient has inadequate intake.

A licensed **pharmacist** dispenses medications and advises the medical staff on the selection and effects of drugs. In addition, pharmacists monitor laboratory results for therapeutic drug levels as well as electrolyte levels for patients receiving parenteral nutrition, and review risks for drug-drug and drug-nutrient interactions. Pharmacists are commonly responsible for compounding sterile solutions including parenteral nutrition support solutions.[6]

Occupational therapists assist patients to improve their ability to perform tasks in living and working environments. Many of their clients suffer from disabling mental, physical, developmental, and/or emotional conditions. Occupational therapists help clients to perform all types of activities, from using a computer to caring for daily needs (dressing, cooking, and eating).[6]

They often work with patients with swallowing disorders and clients with physical disabilities to provide special instructions on eating and use of adaptive feeding devices.

Speech-language pathologists—sometimes called *speech therapists*—assess, diagnose, treat, and help to prevent speech, language, cognitive, communication, voice, swallowing, fluency, and other related disorders. Speech-language pathologists working in a health center provide clinical services to individuals with swallowing disorders, and they work closely with physicians, nurses, and dietitians to help assess the need for and to provide nutrition support.[6]

Table 1.5 American Boards of Medical Specialties

Specialty Board	
Allergy & Immunology	Orthopedic Surgery
Anesthesiology	Otolaryngology
Colon & Rectal Surgery	Pathology
Dermatology	Pediatrics
Emergency Medicine	Physical Medicine & Rehabilitation
Family Practice	Plastic Surgery
Internal Medicine*	Preventive Medicine
Medical Genetics	Psychiatry & Neurology
Neurological Surgery	Radiology
Nuclear Medicine	Surgery
Obstetrics & Gynecology	Thoracic Surgery
Ophthalmology	Urology

*Subspecialties of Internal Medicine include Adolescent Medicine, Cardiovascular Disease, Clinical & Laboratory Immunology, Critical Care Medicine, Endocrinology, Diabetes & Metabolism, Gastroenterology, Geriatric Medicine, Hematology, Infectious Disease, Interventional Cardiology, Medical Oncology, Nephrology, Pulmonary Disease, Rheumatology, Sleep Medicine, Sports Medicine, and Transplant Hepatology.

(This list was effective March 2009. American Board of Medical Specialties®; www.abms.org/) The subspecialties are noted only for Internal Medicine.

Medical **social workers** work with individuals and families to provide the psychosocial support needed to cope with chronic, acute, or terminal illnesses. They also educate family caregivers, counsel patients, and help plan for patients' needs after discharge by arranging community and financial resources to cover medical needs, food-related services, and costs.[6]

Health Care Services and Reimbursement for Medical Nutrition Therapy (MNT)

Where do nutrition services fit within our current health care picture? Nutrition therapy remains an essential component of medical treatments, and research indicates its importance will continue to be recognized. The provision of nutrition therapy and its reimbursement are affected by health care financing.

The pluralistic system of health care in the United States includes many components: private insurance, group insurance, Medicare, Medicaid, workers' compensation, the Veterans Health Administration medical care system, Department of Defense hospitals and clinics, the Public Health Service's Indian Health Service, state and local public health programs, and the Department of Justice's Federal Bureau of Prisons. Currently, the system is structured around the provision of **health insurance**. In 2007, 84.4% of the U.S. population was insured, and 15.6% was not.[7] Many Americans are forced to live without health insurance because they cannot afford it. In the United States currently, there are two general categories of health insurance: private and public. Approximately 68% of the U.S. population has private insurance, and 27% is covered by public health insurance provided by the government. Table 1.6 lists the types of private and public health insurance and typical reimbursement for medical nutrition therapy.

Developing Critical Thinking Skills and Professional Performance

The 2001 Medicare benefit legislation defined medical nutrition therapy as "nutritional diagnostic, therapy, and counseling services for the purpose of disease management, which are furnished by a registered dietitian or nutrition professional."[9] As mentioned previously, clinical dietitians are educated and trained professionals, and are the members of the health care team best able to deliver accurate and appropriate clinical judgments in order to provide appropriate nutritional care. Completing the didactic program in dietetics is your first step to becoming a registered dietitian (RD). The next step, supervised practice, provides the opportunity for a dietetic intern to apply his or her education in the clinical setting.

As a dietetic student, you will acquire a great deal of knowledge during your didactic education. However, students usually do not have the opportunity to apply their knowledge other than through hypothetical disease case assignments. When a student enters the clinical environment, usually as a dietetic intern, he or she quickly finds that providing nutrition care requires more than mastery of a textbook. This textbook provides you with information on the nutrition care process, or a medical condition, its diagnosis, and dietary treatment, but it does not integrate the diagnosis or treatment with the patient's own experiences, symptoms, behaviors, values, social perspectives, and other medical problems.

To provide optimal nutritional care, all of the aspects of a patient's life must be considered. To do this, the practitioner must be able to think critically in order to solve problems and develop the best solution for a client's needs. Dietetic educators know that the dietitians' problem-solving skills, along with their critical thinking skills, evolve with experience and practice. Thus, the path to becoming an RD requires both education and practice.

Table 1.6 Reimbursement for Nutrition Therapy Services

Private Insurance	
Private Health Insurance*	*Alliance Healthcare Initiative:* Children 3–18 years old and their families will be able to see an RD for nutrition counseling (a minimum of four visits).[8]
Group Contract (e.g., Managed Care Org.)	RDs can obtain provider numbers that allow for reimbursement of services.
Public Insurance	
Medicare	Since 2002, Medicare pays RDs who enroll in the Medicare program as providers. Providers are able to bill Medicare for MNT services provided to Medicare beneficiaries with type 1 diabetes, type 2 diabetes, gestational diabetes, nondialysis kidney disease, and post-kidney transplant status.[9]
Medicaid	Some states include benefits for nutrition services, but there is significant variability among states.

*Differences in coverage for nutrition services vary by region and insurance.

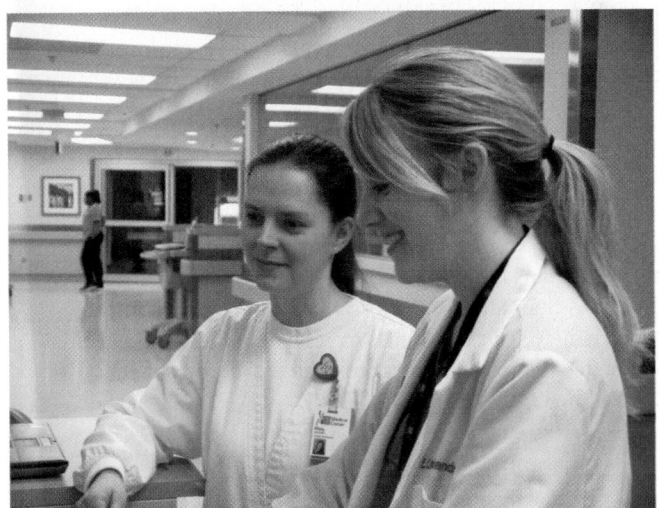

Nutritional care requires collaboration among members of the entire health care team. Dietetic technicians and registered dietitians work together to provide optimal patient care.

Source: Courtesy of Marcia Nelms.

Definition of Critical Thinking

The act of thinking involves using the mind to organize and integrate information, identify relationships, make inferences, form conclusions, and make decisions. When thinking critically, one also challenges assumptions, creates alternatives, and makes informed decisions. In 1990, a group of 46 expert critical thinkers convened the Delphi Consensus Group and defined critical thinking as:

> . . . purposeful, self-regulatory judgment which results in interpretation, analysis, evaluation, and inference as well as explanation of the evidential, conceptual, methodological, criteriological or contextual considerations upon which that judgment is based.[10]

Critical thinking skills are very important, but few students or practitioners understand their application in the nutrition care process (see Chapter 2). The following section will outline the general components of critical thinking and their applications as well as the broader implications of critical thinking for the American Dietetic Association's Standards of Practice for RDs in Nutrition Care; and Standards of Professional Practice for RDs.[11]

Components of Critical Thinking

The dietitian who effectively uses critical thinking skills will make clinical judgments that result in effective nutritional care. Five components have been identified as essential in critical thinking: specific knowledge base, experience, competence, attitudes, and standards.[12]

Specific Knowledge Base The first component of critical thinking is the dietitian's knowledge about nutrition and its role in health and disease. RDs will all have a minimum level of knowledge based on the Standards of Education of the American Dietetic Association set by the Commission on Accreditation for Dietetic Education.[13] Most dietitians exceed the minimum standards, depending on the programs from which they have graduated, the continuing education choices they make, and the advanced degrees they pursue. The dietitian's knowledge base includes information and theories related to communications, physical and biological sciences, social sciences, integration of information, research, food science, nutrition, counseling, management, health care systems, and governance of dietetics practice.

The dietitian's knowledge base is also continually changing and expanding. Learning is a lifelong process, and dietitians engage in continuing education throughout their careers. Nutrition is a developing science, and as new information becomes available, dietitians must apply the new developments to practice. The ADA Standards of Professional Performance for RDs require dietetic professionals to reflect on their practice in order to anticipate and react to change and remain effective practitioners.[11]

Experience The second component of critical thinking is experience in dietetics practice. Dietetic students in nutrition therapy courses often feel overwhelmed by all the information they are expected to know. Though they can effectively apply the information to "mock patients" in case studies, students do not think themselves capable of applying what they have learned to patients in the "real" clinical setting. This occurs in part because dietitians do not learn from textbooks alone; they also learn by observing, listening to patients, interacting with other health care professionals, and reflecting on the situations that arise.

Dietitians are required to complete a supervised practice experience (dietetic internship or coordinated program) before they can be eligible to write the exam for registration. This period of supervised practice allows the dietetic intern to gain experience in all areas of dietetic practice. Real patients with real problems provide the most effective learning experiences by stimulating the dietetic intern's intellectual curiosity and promoting retention of the information.

As an illustration, consider a skill that you developed in the past and now may take for granted, the skill of driving a car. Sometime around the age of 15 or 16, you had the opportunity to drive a car after years of observing someone else drive a car. You most likely had to complete a driver education course that included learning about all the legal aspects of driving. Then you probably got behind the wheel with the driver training instructor or one of your parents. No doubt that first ride was a little frightening, and you may have made some decisions that could be improved upon. Now compare your performance that first time behind the wheel with the way you drive today. The difference is that now you have had experience driving and making related decisions.

Competencies The third component of critical thinking involves the cognitive processes that a dietitian goes through to make clinical judgments. These processes are essentially the same as those used by physicians and other health care professionals and are referred to as *medical problem solving*.[14]

In addition to having knowledge and skills related to nutritional care, you must also have the ability to identify nutrition-related problems and make decisions regarding the most appropriate nutrition-related solutions. These competencies or abilities include the scientific method, problem solving, decision making, and diagnostic reasoning.

SCIENTIFIC METHOD The basic steps in the scientific method are as follows:

- Identify the phenomenon.
- Collect data about the phenomenon.
- Formulate a hypothesis to explain the phenomenon.
- Test the hypothesis through experimentation.
- Evaluate the hypothesis.

Consider the following application of the scientific method in a clinical setting. The dietitian is alerted that an elderly patient is not eating most of the food on his trays at mealtime (identification of phenomenon). The dietitian checks the medical record and sees that the patient lost his wife six months ago and has no family to visit him. The dietitian visits the patient, and she or he learns that the patient wears dentures. By observing and asking questions, the dietitian determines that the patient is experiencing a great deal of discomfort with them (collection of data about the phenomenon). The dietitian suspects that the cause of the patient's poor intake might be that he cannot chew the foods due to pain (formulation of a hypothesis to explain the phenomenon); hence, the dietitian using the nutrition care process recommends that soft, easily chewed foods be served to the patient (test of the hypothesis through experimentation). The next day, the nurse reports that the patient's intake has been 100% for the last three meals (evaluation of the hypothesis).

EVIDENCE-BASED DIETETICS PRACTICE As defined by the American Dietetic Association, "**evidence-based dietetics practice** is the incorporation of systematically reviewed scientific evidence into food and nutrition practice decisions. It integrates professional expertise and judgment with client, customer and community values and evaluates outcomes."[15]

Changes in nutrition therapy recommendations are inevitable because of new developments in science and medicine, including ongoing research in nutrition therapy. Some of what you learn in school today will be outdated by the time you finish your internship and become a registered dietitian. As a dietitian, you must be able to critically review research findings by utilizing the research methodology skills you learned during your dietetic education. Dietetics practice should not be based on tradition but on evidenced-based research. For example, imagine you are an RD working in a facility where the standard of practice is that the initial diet order for all postoperative patients is clear liquids. However, you have just attended the local dietetic district meeting, and the speaker mentioned that most abdominal postoperative patients can tolerate a regular house diet and do not necessarily need to be on clear liquids. What do you do? A search of the medical literature reveals several research articles that support the use of regular diets postoperatively.[16–18] You should critically analyze each article and summarize the evidence. If the evidence supports use of regular diets postoperatively, this information should be presented to the appropriate staff members at your facility for discussion.

The ADA has been instrumental in the development of the Evidence Analysis Library and ADA Evidence-Based Guidelines, which are posted online for its members. As defined by ADA, a *guideline* is a "systematically developed statement based on scientific evidence to assist practitioner and patient decisions about appropriate health care for specific clinical circumstances."[15] Guidelines are tied to the evidence-based library, which is updated as new research is published. This process is not unique to dietetics, and it is rapidly becoming the standard for all health care professions.

PROBLEM SOLVING The process of problem solving involves obtaining information about the problem and then using the information to effectively solve the problem. For instance, suppose you walk into a room and flip the light switch, but the light does not go on. To determine the source of the problem, you would probably check the lightbulb first. If the lightbulb is not burned out, then other possible sources of the problem need to be checked. A similar process is used to determine a patient's nutritional problem. The practitioner can assume that there is a problem when the patient's nutritional status is not optimal; the patient's nutrition-related information is then collected in order to find "clues" that point to the nutritional diagnosis and solution. Consider the client who presents with abdominal cramps, diarrhea, and flatulence throughout the day. Assessment information includes: 50-year-old Asian-American female, postmenopausal; family medical

history includes a 75-year-old mother who developed osteoporosis resulting in a broken hip; and her 24-hour diet recall indicates she recently started drinking an 8-oz glass of 1% fat milk at every meal and sometimes before going to bed, because her doctor had told her to consume more calcium in her diet. Although additional information remains to be collected to rule out other possible problems, one plausible explanation for the client's abdominal symptoms could be her inability to digest lactose (altered GI function) related to her recent increase in milk consumption.

DECISION MAKING Making a decision involves making a choice. The activities involved in decision making include the following:

- Identify and define a problem or situation.
- Assess all options for solving the problem.
- Weigh each option against a set of criteria.
- Test possible options.
- Consider the consequences of the decision (examine the positive and negative aspects of each option).
- Make a final decision.

The activities do not necessarily take place in a particular sequence. The clinician is usually moving back and forth and considering things simultaneously. The outcome of this process is a decision that is informed and supported by evidence and reasoning. Continuing the previous example, if it is determined that a patient has lactose intolerance, options for solving this problem could include: discontinue the use of milk but take a calcium supplement; drink smaller quantities of milk more frequently; or continue to consume milk with every meal but use products containing lactase.

DIAGNOSTIC REASONING Diagnostic reasoning is defined as a series of clinical judgments that result in an informal judgment or a formal diagnosis.[19] Physicians are responsible for making a patient's medical diagnosis. However, dietitians continually use the process of diagnostic reasoning to determine the nutrition diagnosis and monitor a patient's progress and/or response to nutrition therapy.

For example, a patient with protein-energy malnutrition (PEM) will manifest specific signs and symptoms associated with this condition. Once nutrition intervention has begun, the dietitian continues to observe anthropometric and laboratory values and compare them with those common to PEM. This diagnostic reasoning process allows the dietitian to monitor the patient's progress.

Attitudes The fourth component of critical thinking is related to attitudes. Attitudes reflect the dietitian's values and should ensure that clinical judgment is made fairly and responsibly. Table 1.7 summarizes the attitudes necessary for effective critical thinking. In the previous example concerning standards for postoperative feeding, the dietitian's attitude includes integrity, thinking independently, and risk taking. His or her questioning of a standard not based on the latest scientific evidence (integrity) may change a traditional practice of postoperative feeding (thinking independently); it takes courage to take action even when change is based on solid research that improves client care (risk taking).

Table 1.7 Critical Thinking Attitudes

Attitude	Application
Confidence	When you are confident, the patient is more trusting of your competence.
Thinking independently	Consider all viewpoints, and base your decision on your own conclusions about the issue.
Fairness	Listen to both sides of a discussion; weigh all facts.
Responsibility and authority	Ask for help when you need it. Follow established Standards of Professional Performance.
Risk taking	If you have reason to question the judgment of others, do so.
Discipline	Be thorough at all times. Follow established procedures.
Perseverance	Be determined to find the most effective solution. Don't settle for quick solutions.
Creativity	Look for different options when outcomes are not as expected.
Curiosity	Always ask "Why?" Find out as much as you can before making a judgment.
Integrity	Question and test your personal knowledge and beliefs. Be willing to admit inconsistencies in your beliefs.
Humility	Admit your limitations. Be willing to rethink a situation and seek additional knowledge.

Source: R. Paul (1993). The art of redesigning instruction. In J. Willsen & A. Blinker (Eds.), *Critical Thinking: how to prepare students for a rapidly changing world.* Santa Rosa, CA: Foundation for Critical Thinking.

Table 1.8 Intellectual Standards That Universally Apply to Critical Thinking

• Accurate	• Fair
• Adequate	• Logical
• Broad	• Plausible
• Clear	• Precise
• Complete	• Relevant
• Consistent	• Significant
• Deep	• Specific

Source: R. Paul (1993). The art of redesigning instruction. In J. Willsen & A. Blinker (Eds.), *Critical Thinking: how to prepare students for a rapidly changing world.* Santa Rosa, CA: Foundation for Critical Thinking.

Standards The final component of critical thinking is standards, both intellectual and professional (see Table 1.8). Application of intellectual standards involves a rigorous approach to critical thinking and ensures that clinical decisions are sound.

In the client with diarrhea presented earlier, these standards should be used so a nutritional diagnosis can be determined and an intervention plan developed. The RD should seek to ensure that the dietary information obtained is adequate, that any confusing statements made by the client are clarified, and that the nutritional diagnosis is plausible and consistent with the assessment data collected.

Professional standards for critical thinking include ethical standards, criteria-based evaluation of outcomes, and Standards for Professional Performance. The ADA has established a Code of Ethics and Standards of Professional Performance, and both include the need for measurable, evidence-based evaluation of outcomes. For example, one principle in the Code of Ethics is, "The dietetic practitioner practices dietetics based on scientific principles and current information."[20] While the Standards of Professional Performance[11] are broader statements to help guide the practice of dietetics, they do require that dietitians continuously improve their knowledge and skills, and that they regularly evaluate the quality of their practice, revising it if necessary. An important link between the nutrition care process, the Standards of Professional Performance, and the Code of Ethics is that **outcomes research**

may be a consequence of regular evaluation of practice quality. If abdominal postoperative diets are changed from clear liquid diets to regular house diets, data should be collected on patient outcomes. Did this change improve, worsen, or have no effect on surgical outcome? Published research on the outcomes would be added to the evidenced-based research library, possibly resulting in the release of updated clinical nutrition care intervention guidelines for postoperative feeding. In the classroom, critical thinking typically is used for exams and assignments, but for the practitioner, critical thinking leads to higher levels of clinical reasoning that could influence the practice of dietetics.

Levels of Clinical Reasoning

When a dietetic student completes the required supervised practice and then passes the Dietetic Registration Examination, he or she is considered to have entry-level competence. As the new dietitian develops professionally and moves beyond entry level, he or she becomes more proficient and develops expertise in his or her area of practice.

For an entry-level dietitian, critical thinking may be at a basic level. The dietitian has only limited experience and relies on the facts and sets of rules or principles to make decisions. These facts and principles are perceived as appropriate because they are established by the authorities. There may be little adaptation of the principles to the patient's own unique needs. As the dietitian becomes more experienced, she or he begins to examine alternatives independently and systematically, disconnecting from the authorities. The dietitian is better prepared to anticipate possible outcomes and identify a broader range of options. It is evident that there is more than one solution to a problem and that the patient's own unique needs will influence which solutions are viable.

The highest level of critical thinking involves analysis of the entire situation: the person, the illness, the meaning of the illness to that person, the person's lifestyle, the family's needs, the social influences, and the physical environment in which the person lives.[12] At this level, the dietitian is acting in support of the patient, the principles of the nutrition care process, and the professional standards that underlie the discipline of dietetics. A specific characteristic of a dietitian at the highest level of critical thinking is accountability for decisions and continuous quality-of-care assessment.

Conclusion

During the nutrition therapy course, you will be required to complete assignments that will help you apply much of the information presented in this text. You may find some assignments so overwhelming that you do not know where to start. Do not despair; after you have completed the first such assignment, the subsequent ones will become easier, and you will gain confidence just as the inexperienced driver does with daily practice driving a car. As with driving, after you start your supervised practice, your fears will decrease and your competence will grow with each new patient.

WEB LINKS

The American Dietetic Association (ADA)   Information and resources for professionals, dietetic students, and the public. Members can access the *Journal of the American Dietetic Association*, Evidence Analysis Library, *International Dietetics & Nutrition Terminology (IDNT)*, and the *Nutrition Care Manual*.
www.eatright.org

National Certification Board for Diabetes Education (NCBDE)   Information on eligibility requirements and the written examination for diabetes educators.
www.ncbde.org

National Board of Nutrition Support Certification (NBNSC)   Information on eligibility requirements and the written examination for nutrition support dietitians.
www.ptcny.com/clients/NBNSC

The International Board of Lactation Consultant Examiners (IBLCE)   Information on eligibility requirements and the written examination for lactation consultants.
www.iblce.org/index.php

Occupational Outlook Handbook   Handbook on career information published by the U.S. Department of Labor, Bureau of Labor Statistics.
www.bls.gov/oco/home.htm

The Critical Thinking Community   The Foundation and Center for Critical Thinking provides resources on critical thinking for instruction in primary and secondary schools, colleges, and universities.
www.criticalthinking.org

END-OF-CHAPTER QUESTIONS

1. Identify members of the clinical nutrition care team. What are the major tasks performed by the clinical dietitian and the chief clinical manager?

2. What are the five components needed for critical thinking skills? Why is supervised practice a requirement for becoming a registered dietitian?

3. Why is continuing education necessary for the practice of dietetics?

4. What are the components of medical problem solving? How does evidence-based dietetics practice contribute to critical thinking skills?

5. A new friend finds out you are a nutrition major and asks your advice about overeating late in the day. She tells you that she has no time to eat lunch and wants to save money, but then she eats too much when she gets home. Suggest three possible solutions. What are the possible consequences (both positive and negative) of each solution? How could your attitude affect each solution?

6. Why is outcomes research necessary for the advancement of dietetics practice?

REFERENCES

1. Cassell JA. Carry the flame. The history of ADA, Chicago (IL): American Dietetic Association; 1990.

2. Butterworth E. The skeleton in the hospital closet. Nutrition Today. 1974;9:4–8.

3. Winterfeldt E, Bogle M, Ebro L. Dietetics: practice and future trends. 2nd ed. Sudbury (MA): Jones and Bartlett Publishers; 2005.

4. Lacey K, Pritchett E. Nutrition care process and model: ADA adopts road map to quality care and outcomes management. J Amer Diet Assoc. 2003;103:1061–72.

5. Wagner EH. The role of patient care teams in chronic disease management. BMJ. 2000;320:569–72.

6. Occupational Outlook Handbook (OOH), 2008-09 Ed. [Internet] Washington, DC. U.S. Bureau of Labor Statistics. Available at http://www.bls.gov/oco/

7. DeNavas-Walt C, Proctor BD, Smith JC, Poverty, and Health Insurance Coverage in the United States: 2007 [Internet]. Washington DC, U.S. Department of Commerce, Economics and Statistics Administration, US Census Bureau. Available at http://www.census.gov/prod/2008pubs/p60-235.pdf

8. New RD-Provided Nutrition Counseling Coverage for Obese/Overweight Children [Internet]. Chicago (IL): American Dietetic Association. Available at http://www.eatright.org/alliance

9. CMS Program Memorandum, MNT Services for Beneficiaries with Diabetes or Renal Disease. Published August 7, 2001 [Internet]. Baltimore, MD, Center for Medicare and Medicaid Services. Available at http://www.cms.hhs.gov/transmittals/downloads/B0148.pdf

10. Delphi Report. Critical thinking: a statement of expert consensus for purposes of educational assessment and instruction (Vol. ERIC ED 315-423). Millbrae (CA): California Academic Press; 1990.

11. The American Dietetic Association Quality Management Committee. American Dietetic Association Revised 2008 Standards of Practice for Registered Dietitians in Nutrition Care; Standards of Professional Performance for Registered Dietitians; Standards of Practice for Dietetic Technicians, Registered in Nutrition Care; Standards of Professional Performance for Dietetic Technicians, Registered. J Am Diet Assoc. 2008;108:1538–42.e9.

12. Kataoka-Yahiro M, Saylor A. Critical thinking model for nursing judgment. J Nurs Educ. 1994;33:351.

13. Commission on Accreditation for Dietetics Education. Eligibility Requirements and Accreditation Standards for Didactic Programs in Dietetics DPD, DI & CP. Chicago (IL): American Dietetic Assn; 2008.

14. Elstein AS, Shulman LS, Sprafka SA. Medical problem solving: an analysis of clinical reasoning. Cambridge (MA): Harvard University Press; 1978.

15. Evidence Analysis Library [Internet]. Chicago (IL): American Dietetic Association. Available at http://www.adaevidencelibrary.com

16. Jeffery, KM, Harkins, B, Cresci GA, Martindale RG. The clear liquid diet is no longer a necessity in the routine postoperative management of surgical patients. American Surgery. 1996;63:167–70.

17. Pearl ML, Frandina M, Mahler L, Valea FA, DiSilvestro PA, Chalas EA. Randomized controlled trial of a regular diet as the first meal in gynecologic oncology patients undergoing intraabdominal surgery. Obstet Gynecol. 2002;100 Suppl:230–34.

18. Lewis SJ, Andersen HK, Thomas S. Early enteral nutrition within 24 h of intestinal surgery versus later commencement of feeding: a systematic review and meta-analysis. J Gastrointest Surg. 2009;13(3):569–75.

19. Carnevali D, Thomas MD. Diagnostic reasoning and treatment decision making in nursing. Philadelphia (PA): Lippincott; 1993.

20. American Dietetic Association. American Dietetic Association/Commission on Dietetic Registration Code of Ethics for the Profession of Dietetics and Process for Consideration of Ethics Issues. J Am Diet Assoc. 2009;109:1461–67.

Part 2

The Nutrition Care Process

2

The Nutrition Care Process

Karen Lacey, MS, RD, CD
University of Wisconsin–Green Bay

Improving Health and Nutritional Status through Nutrition Care

Health Status

A person's state of health is a continuum that can change from (1) being totally healthy and resistant to disease, to (2) having an acute illness, to (3) living with a chronic disease or condition that significantly alters one's capacity to function well, and finally to (4) having a terminal illness. Regardless of the state of health, the role of nutrition is very important. Registered dietitians are uniquely qualified to provide nutrition care to persons in different states of health to improve their nutritional status. Table 2.1 illustrates how the focus of providing nutrition care is different for various states of health. An example of a primary prevention strategy for a healthy adult may be promoting appropriate caloric balance and physical activity to prevent undesirable weight gain, and thus maintain health. For a person who has identified health risk factors such as a family history of cardiovascular disease, nutrition education on the relationship between saturated fat and LDL cholesterol may be appropriate, whereas the goal of **nutrition intervention** in patients who have already developed chronic diseases is to help reduce complications.

Nutrition Status

In addition to understanding the goal of nutrition care as it relates to a person's health status, it is also important to evaluate and determine a person's nutritional status. Assessing a person's nutritional status involves not only comparing the amounts and types of nutrients that a person consumes to nutrient requirements at various stages throughout the continuum of growth, health, and illness but also examining the wide variety of factors that influence both nutrient intake and nutrient requirements. These factors are listed in Table 2.2. If one consumes adequate amounts and types of nutrients to support and optimize a given health state, the balance between nutrient intake and nutrient requirements is considered to be "good." However, if there is an inadequate or excessive intake of nutrients, or the form of nutrients is not well utilized by the body, a nutrient "imbalance" is present. Nutrient imbalance can result in significant health consequences. An excess of kilocalories (kcal) and undesirable eating patterns are associated with the progression of a number of chronic diseases such as obesity, diabetes mellitus, coronary artery disease, and hypertension. Inadequate intake of kcal and certain nutrients such as protein, on the other hand, can contribute to a compromised immune system and poor wound healing.

ADA's Evidence-Based Guides for Practice—specific nutrition practice guidelines available on ADA's Evidence Analysis Library

environmental factors—social and economic factors (wages, transportation, etc.) that impact both lifestyle choices and the consumption of food and nutrients; other external factors such as food safety and sanitation determine the quality of food that is consumed, and food availability/access contributes to the amount and type of food consumed

expected outcomes—the desired change(s) to be achieved over time as a result of nutrition intervention

evidence-based practice guides—a series of guiding statements and treatment algorithms that are developed using a systematic review process of identifying, analyzing, and synthesizing scientific evidence

food and nutrient factors—the amount and type of foods and nutrients that are consumed and therefore made available to the body

human biological factors—conditions that determine a person's nutrient requirements; one's age, gender, and stage of growth and development are used to estimate kcal and nutrient needs; illnesses that alter organ function or metabolism influence not only the amount of nutrients required but also the form of nutrients that the body needs and can tolerate

ideal goals—science-based values intended to control or improve specific health conditions

lifestyle factors—a person's knowledge, attitudes/beliefs, and behavior patterns directly impact the choices that are made regarding food and physical activity; assessment of these factors provides information about a person's ability and/or readiness to make behavior changes

nutrition assessment—a systematic process of obtaining, verifying, and interpreting data in order to make decisions about the nature and cause of nutrition-related problems

nutrition care process (NCP)—a systematic problem-solving method developed by the ADA that dietetics professionals use to think critically, make decisions addressing nutrition-related problems, and provide safe, effective, high-quality nutrition care[1]

nutrition diagnosis—the identification and descriptive labeling of an actual occurrence of a nutrition problem that dietetics practitioners are responsible for treating independently[1]

nutrition intervention—a specific set of activities and associated materials used to address a (nutrition-related) problem[1]

nutrition monitoring and evaluation—an active commitment to measuring and recording the appropriate outcome indicators relevant to a nutrition diagnosis in order to determine the degree to which progress is being made and whether or not the client's goals are being met[1]

outcome measures—data used to evaluate the success of interventions; includes direct nutrition outcomes, clinical and health status outcomes, patient/client-centered outcomes, and health care utilization and cost outcomes

outcomes management system—a system that evaluates the effectiveness and efficiency of the entire NCP: assessment, diagnosis, implementation, cost, and other factors; it links care processes and resource utilization with outcomes[1]

patient/client-centered—care that considers patients' cultural traditions, personal preferences and values, family situations, and lifestyles

PES—problem, etiology, and signs and symptoms; the format used in the NCP to write a nutrition diagnosis; it clarifies a specific nutrition problem and logically links the nutrition diagnosis to nutrition intervention and to monitoring and evaluation

screening and referral system—a supportive system to the Nutrition Care Process and Model that helps identify those persons who would benefit from nutrition care[1]

standardized language—a uniform terminology that is used to describe practice

system factors—external factors (health care, education, and food supply systems) that influence the type of services that are available to individuals and how these services are delivered

---

Comparing nutrient intake to nutrient requirements alone, however, does not describe the broader picture of nutritional status. Even though nutrient balance implies that one is consuming all of the necessary nutrients in their appropriate amounts, assessing nutritional status is not merely a simple equation of intake compared to needs. A person's nutritional status implies that a number of internal and external factors are also present that support optimum nutritional health. **Human biological factors** such as age, sex, physiological stages, illness, and physical and functional abilities determine nutrient requirements. For example, a mother who is breastfeeding needs to consume more kcal and protein compared to a non-breastfeeding mother. Kcal

Table 2.1 Health State and Focus of Nutrition Interventions

Health State	Focus of Nutrition Intervention
Resistant and Resilient	Primary prevention strategies to maintain health
	Example: Promotion of benefits of eating adequate fruits and vegetables that provide food sources of antioxidants
Stage of Susceptibility (at risk)	Primary prevention strategies to promote health and reduce risk
	Example: Importance of decreasing intake of saturated fats in individuals with family history of cardiovascular disease
Presymptomatic Disease (subclinical)	Early identification and intervention to prevent or delay progression
	Example: Promotion of DASH diet in patients with prehypertensive blood pressure levels
Clinical Condition (physical or pathological change)	Diagnosis and treatment to reduce severity and duration, and restore or improve health
	Example: Diabetes self-management counseling for carbohydrate counting and blood glucose control in patient with type 2 diabetes mellitus
Chronic Condition, Disease or Disability (permanently diminished or altered capacity)	Disease management to reduce complications, accommodate limitations, and enable optimal functioning
	Example: Nutrition counseling to maintain balance of serum phosphorus and blood urea nitrogen (BUN) provided as part of dialysis center treatment and care
Terminal Illness	Palliative care/comfort care to relieve discomfort and maintain dignity
	Example: Provide adequate fluid and patient-preferred food choices as part of hospice care

Source: Adapted from Conceptual Framework for a Standardized Nutrition Language by P Splett and members of ADA's Standardized Language Task Force, 2004.

Table 2.2 Factors Affecting Nutritional Status

1. Human Biology Factors (determine nutrient requirements—normal, increased, decreased, change in form, etc.)
   a. Biological factors (age, sex, genetics)
   b. Physiological phases (growth, pregnancy, lactation, aging)
   c. Pathological factors (disease, trauma, altered organ function or metabolism)

2. Lifestyle Factors (determine food, physical activity, and related choices)
   a. Attitudes/beliefs
   b. Knowledge
   c. Behaviors

3. Food and Nutrient Factors (determine the type and amount of nutrients available for use by the body)
   a. Intake/composition
   b. Quantity
   c. Quality

4. Environmental Factors (external influences that impact consumption and lifestyle)
   a. Social (cultural food practices and beliefs, parenting, peer influences)
   b. Economic (household finances, economy of the community/country)
   c. Food safety and sanitation (contaminated or unwholesome food, unsafe food handling)
   d. Food availability/access

5. System Factors (external influences that impact on delivery and services)
   a. Health care system
   b. Educational system
   c. Food supply system (industry, agriculture, institutions)

Source: Adapted from Conceptual Framework for a Standardized Nutrition Language by P Splett and members of ADA's Standardized Language Task Force, 2004.

and protein needs are also increased following major surgery. Furthermore, the form of nutrient may need to be altered depending on the degree of organ function. A person who has had a large portion of the small intestine removed may not be able to digest large molecular nutrients such as triglycerides and would benefit from specialized nutrient forms such as short- or medium-chain triglycerides. **Lifestyle factors** including attitudes, knowledge, and behaviors influence the type of choices that one makes about food and physical activity. For instance, understanding which foods contain saturated fats and cholesterol can influence what type and amount of meats and spreads one consumes. **Food and nutrient factors** describe the nutrients that are available for use by the body. Obtaining accurate information about a person's dietary intake is essential to evaluating nutritional status. **Environmental factors** such as social and cultural food preferences and practices are external influences that impact both food consumption and lifestyle choices. For example, people frequently consume more food than usual at a social event where food is served. It is also common that adults prefer the types of foods that were typically consumed in the household where they grew up as a child. Finally, **system factors** such as the health system, educational system, and food supply system impact the delivery of food, nutrition, and health services. A family whose income is near or at the poverty level and that has limited access to health care will likely purchase fewer fresh foods and may use the services of urgent care more frequently.

It is important to determine both a person's health status and nutritional status, because these guide the type of nutrition care provided. This chapter describes how dietetics practitioners use ADA's nutrition care process to provide quality nutrition care to individuals and groups to improve both health and nutrition status.

Purpose of Providing Nutrition Care

The purpose of providing nutrition care is to restore a state of nutritional balance by influencing whatever factors are contributing to the imbalance or altered state of nutritional status. Because of the wide variety of and interaction among the many variables discussed previously and listed in Table 2.2, identifying the underlying causes of a nutritional status imbalance can be a complex process. If a person's caloric intake is less than desired, it is important to determine which, if any, of the following are contributing to the cause of this problem: a disease condition that is increasing the nutrient needs, a lack of knowledge as to how many kcal are in certain foods, a lack of resources (money, food preparation skills, transportation), or a cultural belief about limiting the intake of certain foods. Accurately determining the underlying, or "root," cause of the problem will permit the selection of the most appropriate nutrition intervention. For example, if lack of financial resources is the main reason why a person is not consuming adequate kcal, providing only a list of expensive oral supplements will not be very effective. It might also be necessary to coordinate nutrition care and refer the client for support services and food aid. Thus, knowing the factors that influence a problem is very important to providing nutrition interventions that are most likely to alleviate the problem.

> **Key Concepts: Health Status and Nutritional Status**
>
> · Nutrition is important to promote health and prevent and treat disease states.
>
> · Adequacy of nutrient intake is important but does not completely describe nutritional status.
>
> · Determination of a person's nutritional status is also dependant on a wide variety of factors (biological, pathological, behavioral, cognitive, environmental, and systems).

> **Key Concept: Nutrition Care**
>
> · Providing nutrition care can influence and change the factors that contribute to an imbalance in nutritional status and thus restore an improved state of nutritional health.

ADA's Standardized Nutrition Care Process (NCP)

A Brief History of ADA's NCP

In early 2002, dietetics practitioners identified the need to create a standardized method of providing nutrition care in order to improve both the quality of care and the likelihood of producing positive outcomes. A Nutrition Care Model Workgroup was appointed by ADA's Quality Management Committee to address this important professional issue, and in March 2003, the House of Delegates adopted the Nutrition Care Process and Model for "implementation and dissemination to the dietetic profession and the Association for the enhancement of the practice of dietetics."[1] ADA's **nutrition care process** is defined as "a systematic problem solving method that dietetics practitioners use to critically think and make decisions to address nutrition related problems and provide safe, effective, high quality nutrition care."[1] This NCP consists of four distinct but interrelated and connected steps: (1) **nutrition assessment**, (2) **nutrition diagnosis**, (3) nutrition intervention, and (4) **nutrition monitoring and evaluation**.[1] The second step, nutrition diagnosis, was the newest addition to the nutrition care process. Subsequent to the development of the NCP steps and model, workgroups were appointed to create a system of standardized language for each of the four steps of the NCP.

Standardized Nutrition Language

Standardized language refers to a uniform terminology that is used to describe practice. Many health professionals including physicians, nurses, and physical therapists had developed standardized terminology; however, none existed for nutrition care prior to the development of the NCP. The lack of a standardized nutrition language and common terminology made it very difficult for dietetics practitioners to communicate with each other and other health professionals. Different practitioners in the past would not use common language to describe nutrition problems. They might use similar words to mean different things or different words to mean the same thing. For example, "poor nutritional status" might be noted in the chart as a red flag for nutrition care; however, this could reflect a number of conditions such as a change in weight, poor intake, or difficulty with chewing or swallowing. Furthermore, terms such as "at risk for malnutrition" or "compromised nutritional status" or "nutritional imbalance" were frequently used by practitioners to denote potential or current nutrition problems. Care plans might have also described nutrition interventions using vague terms such as "complete nutrition assessment," "monitor weights or dietary intake," and/or "provide nutrition education." These inconsistent terms made it difficult to establish specific and clear goals.

Furthermore, there was no easy way to classify, measure, and report on the outcomes of nutrition interventions in various patient populations in order to demonstrate the effectiveness of nutrition care. The lack of specific uniform terminology in dietetics practice made it impossible to gather data needed for research, education, and reimbursement justification via outcomes analysis. Most notably missing was language that described specific nutrition problems. Therefore, a Standardized Language Task Force was formed in May of 2003, immediately following the adoption of the NCP, to develop standardized language for nutrition. This standardized language, especially the nutrition diagnostic terminology, now allows dietetics practitioners to make explicit that which had been implicit in the past.

Nutrition diagnostic terminology was the first to be approved by the American Dietetic Association in May 2005. This diagnostic terminology is basically a list of phrases ("terms") that are organized into groups ("domains") and assigned unique alphanumeric code numbers. These terms clearly describe specific types of nutrition problems that contribute to a person's nutritional imbalance. As of 2009, two additional sets of standardized language and terminology have been developed: (1) nutrition assessment, including also monitoring and evaluation terms; and (2) nutrition intervention. Like the diagnostic terminology, these contain standardized terms with identification codes that are organized and grouped into domains. The use of all three sets of terms together is a very logical and systematic process that connects each of the steps of the NCP to the others. For instance, the assessment data of the food/nutrition-related history domain are likely to provide signs and symptoms that will confirm a nutrition diagnosis in the intake domain, whereas if a behavioral-environmental nutrition problem is suspected, then further data involving knowledge/beliefs and attitudes should be obtained to rule in or rule out the presence of that diagnosis. The standardized terms, their classifications, and their use within each step of the NCP are outlined in Figure 2.1 and discussed in more detail in the following "Steps of the NCP." Refer to Appendix C for a complete list of each of the standardized terminologies.

Use of the NCP to Improve Quality of Care

The NCP is a standardized process—not standardized care. A *standardized process* refers to a consistent structure and framework used to provide nutrition care, whereas *standardized care* implies that all clients receive the same care. When professionals use a systematic process with standardized language, there is less variation of practice and a higher degree of predictability in terms of outcomes. The Institute of Medicine defines quality as "the degree to which health services for individuals and populations increase the likelihood of desired health outcomes and are consistent with current professional knowledge."[2] Quality performance can be assessed by measuring clients' outcomes (end results of intervention and treatment) *or* the degree to which providers adhere to an accepted care process. Clients and patients want service that results in positive outcomes. Health care administrators, payers, and the government require cost-effective, high-quality service based on current evidence-based practice. Use of the ADA's NCP is the means by which dietetics

---

Key Concepts: ADA's Standardized Nutrition Care Process

The four steps of the nutrition care process are:

· Nutrition Assessment

· Nutrition Diagnosis

· Nutrition Intervention

· Nutrition Monitoring and Evaluation

By using the nutrition care process, dietetics practitioners can demonstrate that nutrition care improves outcomes because it:

· Is a systematic method used to make decisions to provide safe and effective care.

· Provides a common language for documenting and communicating the impact of nutrition care.

· Relies on an evidence-based approach.

· Uses specific critical thinking skills for each step.

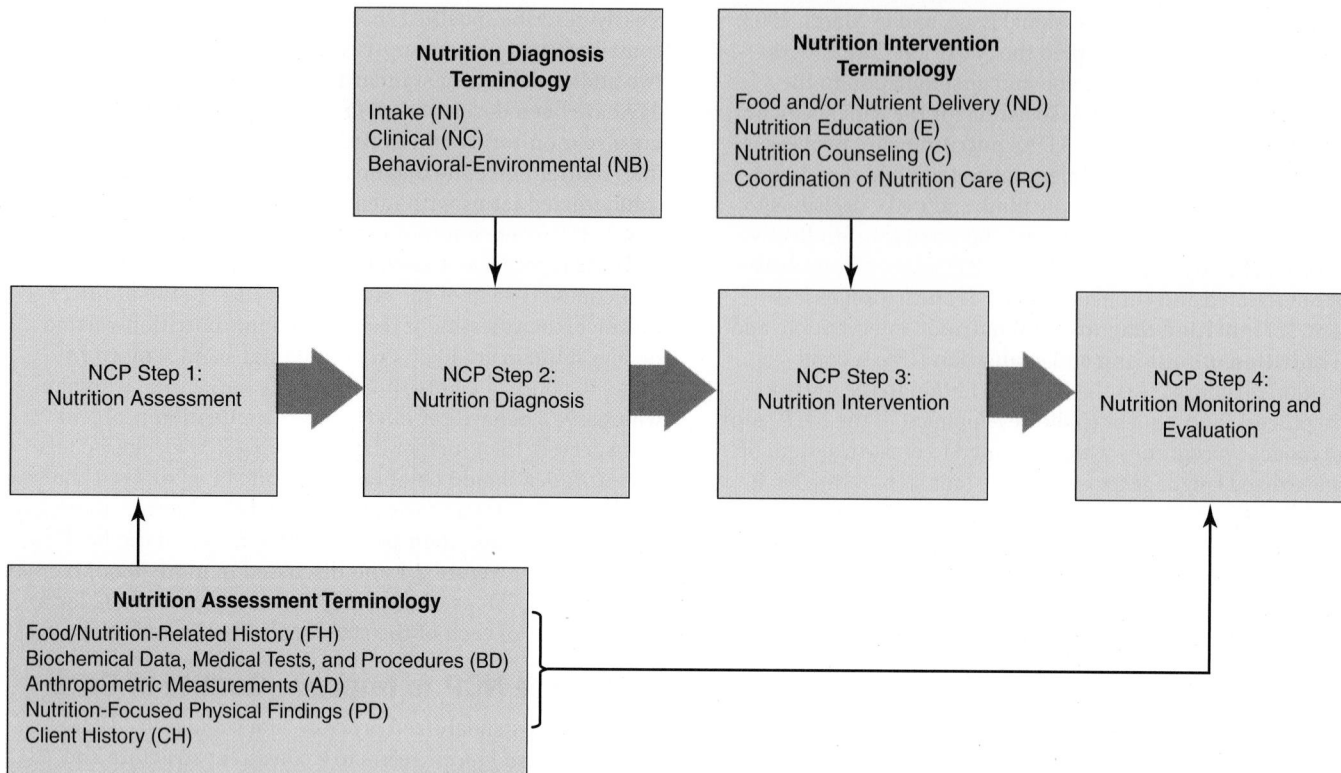

practitioners greatly increase their potential to provide high-quality nutrition care to individuals and groups. It combines both the *process of care* (the steps of the nutrition care process in a systematic and consistent manner) and the *content of care* (incorporation of **evidence-based practice guides**) to produce improved quality of care and improved nutritional status. The other significant benefit of using the NCP is the ability to clearly state patient goals and evaluate outcomes. Figure 2.2 illustrates how combining the process of care with the content of care improves quality.

Critical Thinking

The NCP also enhances the quality of care provided by dietetics practitioners through the use of specific critical thinking skills. Critical thinking integrates facts, informed opinions, active listening, and observations; it is creative and rational, and it requires the ability to conceptualize. Each step of the NCP identifies unique and specific types of critical thinking that, when applied, improve the likelihood that the process is being implemented in an effective manner. These specific critical thinking skills are described in Table 2.3.

Figure 2.2 Demonstrating Quality

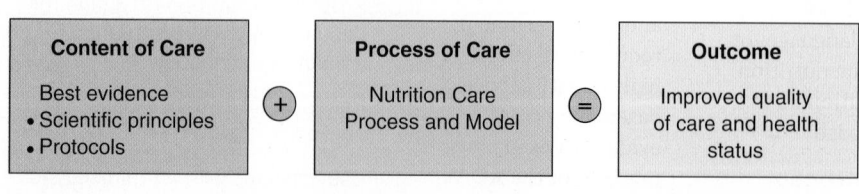

Source: Adapted from Slide #8 of ADA's Nutrition Care Process and Model: Providing Quality Nutrition Care in a Variety of Settings Power Point Prepared by ADA's Nutrition Care Process Task Force, 2004.

Big Picture of Nutrition Care: The Model

The provision of nutrition care does not occur in a vacuum. The Nutrition Care Model in Figure 2.3 is a visual representation that reflects key concepts of each step of the NCP and illustrates the greater context within which nutrition care is provided. The model also identifies other systems that influence and impact the quality of care. It depicts the overlapping relationships of these components and how they interact to result in the best quality nutrition care possible.

Central Core

Central to providing nutrition care is the relationship between the client and the dietetics practitioner or team of dietetics practitioners. The client's previous experiences and readiness for change as well as the ability of the dietetics practitioner to establish trust, demonstrate empathy, and communicate effectively with the client influence this relationship. If a person believes that changing the intake of saturated fat will decrease his or her risk of cardiovascular disease and has had previous nutrition counseling that was helpful, that person is more likely to want to meet again with a dietitian; in contrast, an individual who believes that lifestyle behavior changes will have little to no impact on the risk of disease will probably be less receptive. It is important for the dietetics practitioner to establish trust and be able to communicate effectively with others, and it is essential that the client be actively involved in the care

Table 2.3 Critical Thinking Used in the Nutrition Care Process

Nutrition Assessment

- Observe for nonverbal and verbal cues to prompt effective interviewing methods.
- Determine appropriate data to collect.
- Select assessment tools and procedures.
- Apply assessment tools in valid and reliable ways.
- Distinguish relevant from irrelevant data.
- Distinguish important from unimportant data.
- Validate the data.
- Organize and categorize the data in a meaningful framework that relates to nutrition problems.
- Determine when a problem requires consultation with or referral to another provider.

Nutrition Diagnosis

- Find patterns and relationships among the data and possible causes.
- Make inferences ("If this continues to occur, then this is likely to happen").
- State the problem clearly and singularly.
- Suspend judgment (be objective and factual).
- Make interdisciplinary connections.
- Rule in/rule out specific diagnoses.
- Prioritize the relative importance of problems.

Nutrition Intervention

- Set and prioritize goals.
- Define the nutrition prescription or basic plan.
- Make interdisciplinary connections.
- Initiate behavioral and other interventions.
- Match intervention strategies with client needs, diagnoses, and values.
- Choose from among alternatives to determine a course of action.
- Specify the time and frequency of care.

Nutrition Monitoring and Evaluation

- Select appropriate indicators/measures.
- Use appropriate reference standards for comparison.
- Define where patient/client is now in terms of expected outcomes.
- Explain variance from expected outcomes.
- Determine factors that help or hinder progress.
- Decide between discharge or continuation of nutrition care.

Source: Adapted from Lacey K & Pritchett E. Nutrition Care Process and Model: ADA adopts road map to quality care and outcomes management. *J Amer Diet Assoc.* 2003;103:1061–72.

whenever possible and if culturally acceptable. This means that the client is aware of the purpose of care and participates in the decision-making process of goal setting and intervention selection. This central core reinforces the importance of providing care that is individualized and **patient/client-centered**.

Two Outer Rings

The *outermost ring* of the model identifies environmental factors, including practice settings, health care systems, social systems, and economics, all of which can have an impact on the ability of the client to receive and benefit from the interventions of nutrition care. Dietetics practitioners need to assess these factors and be able to evaluate the degree to which they may either be positive or negative influences on the outcomes of care. A

health care plan that allows for up to three outpatient nutrition counseling sessions per year, where the client pays only a small portion of the cost of the nutrition counseling, is a much more positive external influence than a health care plan in which the client has to pay for the entire visit. The *inner adjoining ring* recognizes the strengths that dietetics practitioners bring to the nutrition care process. These include professional knowledge/ skills and competencies, code of ethics, evidence-based practice, and skills of critical thinking, collaboration, and communication. These are the knowledge and skills that registered dietitians and dietetic technicians obtain through accredited didactic and supervised practice programs. Providing nutrition care that is based on sound scientific evidence increases the likelihood that there will be a positive outcome for the client. Nutrition care also requires a great deal of collaboration with other health care professionals and community services.

Supportive Systems: Screening and Referral System and Outcomes Management System

Although the two supportive systems—the screening and referral system and the outcomes management system—are essential to providing effective and efficient nutrition care, they are not considered steps of the nutrition care process itself, primarily because they may not be accomplished solely by dietetics practitioners. A **screening and referral system** identifies those individuals or groups who would benefit from nutrition care provided by dietetics practitioners. Screening may be completed by nurses, by clients themselves, or through physician referral. Regardless of whether dietetics practitioners are actively involved in conducting the screening process, they are still accountable for providing input into the development of appropriate screening parameters to ensure that the screening process asks the right questions. They should also evaluate how effective the screening process is in terms of correctly identifying the clients who require nutrition care. Screening parameters need to be tailored to the population and to the nutrition care services provided. A referral process may also ensure that the client has an identifiable method of being linked to dietetics practitioners who will ultimately provide the nutrition care or medical nutrition therapy that is necessary. For example, using the DETERMINE checklist (see Chapter 3) at elderly congregate meal sites can identify those clients who are at risk and could then be seen by the dietitian employed at a senior center.

The other system supporting the NCP is the **outcomes management system**. An outcomes management system is used to evaluate

Key Concepts: Nutrition Care Process and Model

- Nutrition care is provided within the context of a larger model that includes a central core focused on individual care and positive relationships.

- Both external (environmental) and internal (resources of dietetics practitioner) factors influence the type of nutrition care provided.

- The steps of the nutrition care process are supported by two other systems, the screening and referral system and the outcomes management system. Dietetics practitioners participate in both of these systems, but may not have sole responsibility for accomplishing the tasks they perform.

Figure 2.3 The Four Distinct but Interrelated and Connected Steps of the Nutrition Care Process and Model

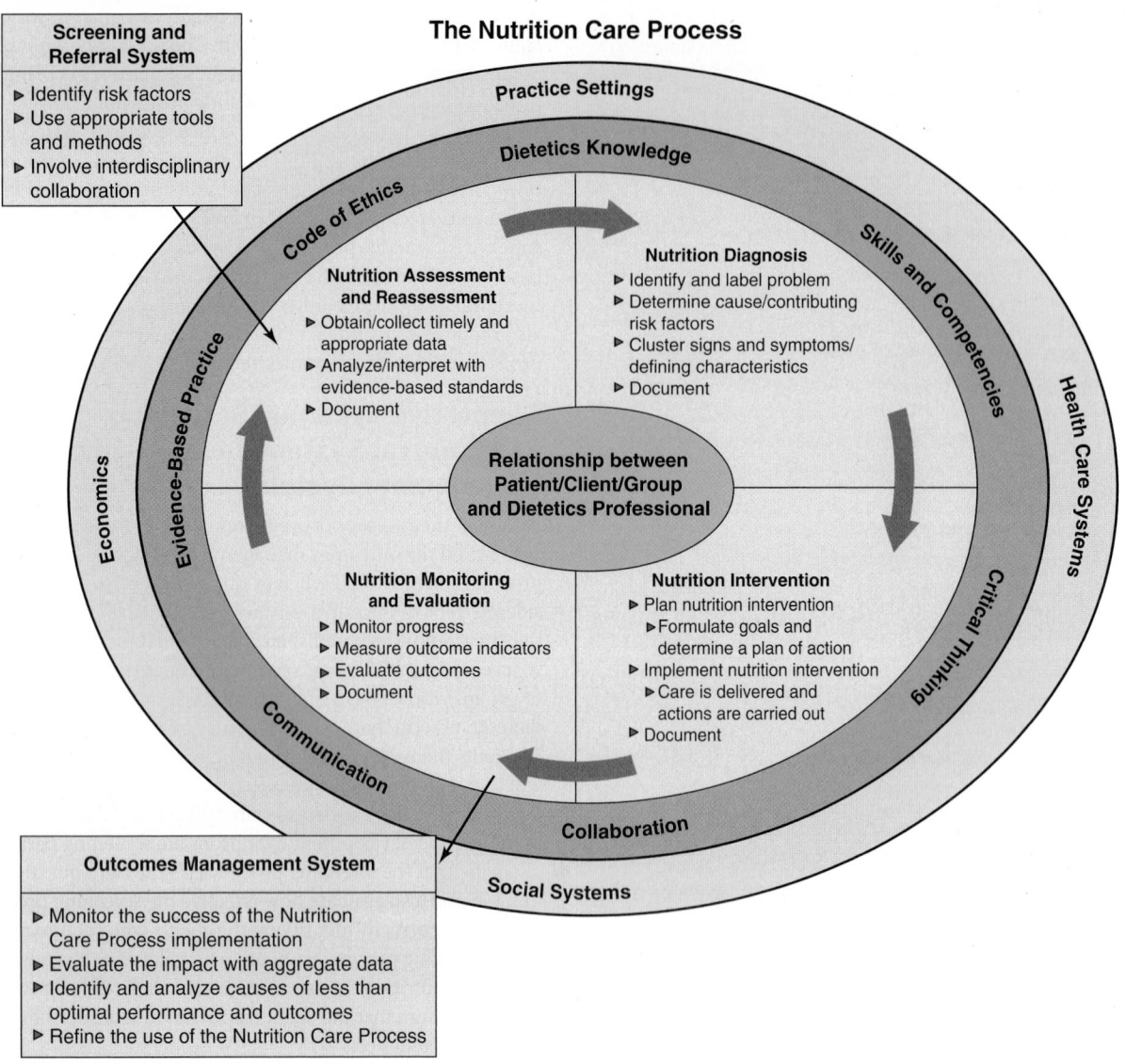

The Nutrition Care Process

Screening and Referral System
- Identify risk factors
- Use appropriate tools and methods
- Involve interdisciplinary collaboration

Practice Settings
Dietetics Knowledge
Code of Ethics
Skills and Competencies
Health Care Systems
Evidence-Based Practice
Economics
Critical Thinking
Communication
Collaboration
Social Systems

Nutrition Assessment and Reassessment
- Obtain/collect timely and appropriate data
- Analyze/interpret with evidence-based standards
- Document

Nutrition Diagnosis
- Identify and label problem
- Determine cause/contributing risk factors
- Cluster signs and symptoms/defining characteristics
- Document

Relationship between Patient/Client/Group and Dietetics Professional

Nutrition Monitoring and Evaluation
- Monitor progress
- Measure outcome indicators
- Evaluate outcomes
- Document

Nutrition Intervention
- Plan nutrition intervention
  - Formulate goals and determine a plan of action
- Implement nutrition intervention
  - Care is delivered and actions are carried out
- Document

Outcomes Management System
- Monitor the success of the Nutrition Care Process implementation
- Evaluate the impact with aggregate data
- Identify and analyze causes of less than optimal performance and outcomes
- Refine the use of the Nutrition Care Process

Source: Writing Group of the Nutrition Care Process/Standardized Language Committee. Nutrition care process and model part I: the 2008 update. *J Am Diet Assoc.* 2008;108:1113–17.

the effectiveness and efficiency of the entire NCP process (assessment, diagnosis, interventions, outcomes, costs, and other factors) when nutrition care is provided to a number of patients. Outcomes management is different from the fourth step, nutrition monitoring and evaluation, which refers to the evaluation of a single patient's/client's progress in achieving goals and desired outcomes. Outcomes management links care processes and resource utilization with outcomes. Health care organizations use complex information management systems to manage resources and track performance. Selected information documented throughout the nutrition care process is entered into these central information management systems and structured databases. Relevant aggregate data (data from a number of individual sources that have been summed together to create a larger whole) can then be collected and analyzed in a timely manner. Performance can be adjusted based on this analysis in order to improve outcomes. For example, data collected over time might reveal that fewer

than 50% of clients seen in an outpatient setting received follow-up appointments, and that of those 50%, fewer than half met desired outcome goals. These data would then be used to more closely examine the system of access and record keeping as well as the type of interventions used to provide care. Such an analysis can assist in the creation of policies for increasing the number of patients who receive follow-up care and in better achieving expected outcomes. When nutrition services have systems in place that can measure and evaluate data from many clients (aggregate data), these data can then be combined with data from other nutrition care providers and be part of evidence-based research that demonstrates the benefit and effectiveness of nutrition care. For example, registered dietitians, all of whom used the NCP, who have collected outcome data on the benefits of nutrition counseling on blood glucose control in patients with type 2 diabetes mellitus could then tabulate the data to summarize and report on a larger number of patients.

Steps of the NCP

Step 1: Nutrition Assessment

The first step of the NCP (see Table 2.4) provides important information that helps determine a person's health and nutritional status. It is initiated by the referral and/or screening of individuals or groups for nutritional risk factors. A nutrition assessment is a very systematic process of obtaining, verifying, and interpreting data in order to make decisions about the nature and cause of nutrition-related problems (see Chapter 3). It is an ongoing, dynamic process that involves not only initial data collection, but also continual reassessment and analysis of a client's needs and condition. A nutrition assessment is not in and of itself a measure of a dietetics practitioner's level of productivity. A nutrition assessment provides data to accurately describe nutrition problems and facilitate the formulation of a nutrition diagnosis at the next step of the NCP. Assessment data also provides a means to reevaluate the nutrition problem as part of nutrition monitoring and evaluation, the fourth step in the NCP.

Nutrition assessment focuses on understanding the wide variety of factors (human biological, lifestyle, food and nutrient, environment, and system) that influence a person's nutritional status, as noted previously in Table 2.2. These data provide information about the types of nutrition problems that exist as well as their likely causes. Data gathered during the assessment step are also used to describe the severity of the problem. For example, if the nutrition problem is NB-1.7, "undesirable food choices" (refer to Appendix C), further clarification of the specific type of food that is undesirable, such as "32 ounces of high-fructose fruit drinks a day," would be used to describe and quantify that problem. These data will then determine what types of outcomes are desired. In this case, the amount and type of beverages consumed would be tracked over time. Once a problem has been accurately defined and quantified, client goals can be established. If a client is consuming too much high-fructose fruit drink, the goal might be to consume 4 ounces of 100% fruit juice in place of the fruit drinks and substitute water for the remainder of the fluid needs in the day. Each piece of nutrition assessment data is collected for a specific purpose. It helps answer the following types of questions:

1. What are these data telling us about a person's nutritional status and all of the possible factors that contribute to nutritional balance?

2. What possible nutrition diagnosis/es might these data provide evidence for?

3. What additional data might be necessary to validate the presence of the suspected nutrition diagnoses?

As dietetics practitioners collect data, they should simultaneously be thinking about the "why" (factors that contribute to or cause imbalance in nutritional status) and the "what" (possible nutrition diagnoses).

Obtain and Verify Appropriate Data The specific type of data gathered in the assessment can vary depending on a number of factors such as practice settings or the individual's/group's present health status. Dietetics practitioners who serve clients at a Women, Infants, and Children (WIC) clinic will obtain anthropometric data on head circumference and height and weight plotted on growth charts in order to assess the development of children. Dietetics practitioners in outpatient clinics will obtain height and weight measurements for adults and may also gather information about body fatness using skinfold thickness or bioelectrical impedance analysis. Recommended practices, as indicated in **ADA's Evidence-Based Guides for Practice**, may also influence the type of data collected in a nutrition assessment. Lipid profiles for patients with type 2 diabetes and cardiovascular diseases would be appropriate, whereas BUN, creatinine, and serum phosphorus should be evaluated when providing nutrition care to patients with renal disease.

The type of data collected depends on whether an initial assessment or a reassessment is being conducted (or done). For example, a thorough, detailed diet history is valuable during an initial assessment, but a brief investigation of a specific type of nutrient such as fiber might be more valuable in a follow-up visit three weeks later, especially if inadequate intake of fiber was one of the nutrition problems identified during the initial appointment. The dietitian needs to know what type of data is most appropriate to collect and to be able to determine whether those data are valid and accurate. For example, a stated weight may or may not represent the current weight of a client. Accurate and valid diet history information is dependent on the ability of the dietitian to establish a trusting and nonthreatening relationship with the client. In all cases, the data that are reviewed should be related to the types of nutrition problems likely to be encountered.

Cluster and Organize Assessment Data As shown in Box 2.1, the nutrition assessment standardized terms are now grouped into five domains: (1) Food/Nutrition–Related History; (2) Anthropometric Measurements; (3) Biochemical

Table 2.4 Overview of the Steps of the Nutrition Care Process

Step One: Nutrition Assessment
1. Obtain and verify appropriate data.
2. Cluster and organize assessment data according to assessment domains and possible nutrition diagnoses.
3. Evaluate the data using reliable standards.
4. Calculate estimated nutrient needs; e.g., nutrition prescription as appropriate.

Step Two: Nutrition Diagnosis
1. Identify possible diagnostic labels.
2. Complete nutrition diagnostic statements using the PES format.
3. Evaluate the quality of PES statements.

Step Three: Nutrition Intervention
1. Prioritize the nutrition diagnoses.
2. Identify ideal goals and expected outcomes.
3. Plan the nutrition interventions using the standardized intervention language.
4. Implement the nutrition interventions.

Step Four: Nutrition Monitoring and Evaluation
1. Monitor progress.
2. Measure outcomes.
3. Evaluate outcomes.

BOX 2.1   **CLINICAL APPLICATIONS**

Nutrition Assessment Standardized Language: Domains and Examples

- **Food/Nutrition-Related History (FH):** Food and nutrient intake, medication/herbal supplement intake, knowledge/beliefs/attitudes and behaviors, food and supply availability, physical activity, and nutrition quality of life
  Examples: Total energy intake (FH-1.2.1.1), fat and cholesterol intake (FH-1.6.1), meal/snack pattern (FH-1.3.2.3), area and level of knowledge (FH-3.1.1), eligibility for community programs (FH-5.2.1), type of physical activity (FH-6.3.1)

- **Anthropometric Measurements (AD):** Height, weight, body mass index (BMI), growth pattern indices/percentile ranks, and weight history
  Examples: Weight (AD-1.1.2), weight change (AD-1.1.4), body mass index (AD-1.1.5)

- **Biochemical Data, Medical Tests, and Procedures (BD):** Lab data (e.g., electrolytes, glucose) and tests (e.g., gastric employing time, resting metabolic rate)
  Examples: BUN : creatinine ratio (BD-1.2.3), fasting glucose (BD-1.5.1), cholesterol, HDL (BD-1.7.2)

- **Nutrition-Focused Physical Findings (PD):** Physical appearance, muscle and fat wasting, swallow function, appetite, and affect
  Examples: Body language (PD-1.1.2), digestive system (PD-1.1.5)

- **Client History (CH):** Personal history, medical/health/family history, treatments and complementary/alternative medicine use, and social history
  Examples: Education (CH-1.1.6), medical treatment/therapy (CH-2.2.1), socioeconomic factors (CH-3.1.1)

Data, Medical Tests, and Procedures; (4) Nutrition-Focused Physical Findings; and (5) Client History. Clustering and organizing the data according to the five domains can reveal possible nutrition problem domains and/or classifications from which a specific nutrition diagnostic statement can then be more accurately formulated.

The dietetics practitioner will examine anthropometric data (refer to Chapter 3) with the intended purpose of ruling in or ruling out the possibility of weight classification problems: underweight, involuntary weight loss or gain, or overweight/obesity. Specific data from a dietary intake assessment reveal important information about the extent of possible intake nutrition diagnoses found in the Intake domain. Information gathered in an interview that reveals a person's knowledge and beliefs about health and nutrition allows the dietetics professional to rule in or rule out possible problems in the Knowledge and Behavior classification of the Behavioral-Environmental domain. This structure, when used to guide nutrition assessment, provides focus and purpose to the gathering of data. Each piece of assessment data helps answer a question regarding the possible impact it may have on the presence, severity, and cause of a specific nutrition problem. Using an organized structure that focuses on nutrition problem areas assists dietetics practitioners to think critically about the meaning of the data and logically move into the next steps of the NCP.

Evaluate the Data Using Reliable Standards It is not only important that data be linked to specific types of problems; it is equally important that the information obtained in a nutrition assessment be compared to reliable standards or ideal goals. It is essential to use current and scientifically valid standards in the determination of whether a nutrition problem actually exists, and if so, to what degree. A nutrition diagnosis is made only after the data gathered are evaluated and compared to an established reference standard or recommendation (taking into account individual physiological needs). For example, calculating accurate estimated energy needs requires the use of appropriate physical activity factors. A reliable standard to use when evaluating the proportions of kcal from fat, carbohydrates, and protein would be the Institute of Medicine's 2002 Acceptable Macronutrient Distribution Ranges (AMDR) for healthful diets, whereas the ideal goals established as part of Adult Treatment Panel III (ATP III) would be used to evaluate and compare intake of saturated fats and monounsaturated fats of clients with cardiovascular risk or disease.[3]

The terminology used for nutrition assessment also includes a sixth domain, Comparative Standards (CS). This domain identifies appropriate reference standards that are used to determine the nutrient needs of an individual. They include standards for energy, macronutrient, fluid, and micronutrient needs as well as recommendations for body weight and growth. The estimated needs, once appropriately determined, provide the basis for nutrition goals and the standards by which to compare current dietary intake and weight to recommendations.[4]

Step 2: Nutrition Diagnosis

Nutrition diagnosis, the second step of the NCP, consists of the identification and descriptive labeling of an actual occurrence of a nutrition problem that dietetics practitioners are responsible for treating independently. Nutrition diagnosis is the heretofore missing link between nutrition assessment and nutrition intervention. A nutrition diagnosis should not be confused with a medical diagnosis; the distinction between the two is extremely important. A medical diagnosis is the art of distin-

> **Key Concepts: NCP Step 1, Nutrition Assessment**
>
> - Nutrition assessment should ensure that appropriate and reliable data are collected for use in determining the existence of specific nutrition problems.
>
> - Organizing and categorizing data utilizing the five domains of the assessment standardized terms improves the efficiency and effectiveness of nutrition assessment and nutrition diagnosis.

guishing one disease from another and describes the nature of that disease. A disease is further defined as any deviation from or interruption of the normal structure or function of a body part, organ, or body system. Treatment of a disease involves the management and care of a patient for the purpose of combating the disorder.[5] Many diseases have profound effects on a person's nutritional balance. The alteration of normal structure and function of organs can result in changes in nutrient intake, losses, requirements, and/or utilization. In some cases, nutrition therapy may be one of the most important ways of treating and managing the disease.

A nutrition diagnosis, in contrast to a medical diagnosis, is written in terms of a client problem for which nutrition-related activities provide the primary intervention. The goal of nutrition care is to improve the health and nutritional status of a client/patient by impacting the underlying cause of the nutritional problem. Nutrition diagnoses and care focus on nutrition issues that may be consequences of or contribute to diseases. Nutrition diagnoses also address behaviors that impact food choices.

Nutrition diagnostic terms are grouped into three domains: (1) Intake, (2) Clinical, and (3) Behavioral-Environmental (see Box 2.2). The **Intake domain** contains nutrition problems that are related to the intake of energy, nutrients, fluids, and bioactive substances through oral diet or nutrition support. Labels such as *inadequate*, *excessive*, or *inappropriate* are used to describe the specific nutrient or substance that is altered. The **Clinical domain** contains nutrition problems that are related to medical or physical conditions. These include problems in swallowing, chewing, digestion, absorption, and maintaining appropriate weight. The **Behavioral-Environmental domain** includes problems that are related to knowledge, attitudes/beliefs, physical environment or access to food, and food safety. Within each of these domains, nutrition problems are further grouped according to classifications and subclassifications (refer to Appendix C).

Each nutrition diagnostic term has a term number and a standard definition. For example, "inadequate protein intake" (NI-52.1) is defined as "lower intake of protein-containing

foods or substances compared to established reference standards or recommendations based upon physiological needs."[5] The use of standard definitions helps dietetics practitioners use the language consistently within the profession. In addition to the numerical coding and standard definition, the American Dietetic Association has published a reference sheet for each diagnostic term that also identifies possible etiologies and signs and symptoms commonly associated with that nutrition problem. These reference sheets provide tools that the practitioner may use to examine the appropriate data and ask key questions when determining whether a nutrition diagnosis is present or not. For instance, the sheet for the diagnosis "inadequate protein intake" names "Lack of access to food" as a potential etiology, and "Report or observation of . . . economic constraints that limit food availability" as a corresponding sign/symptom that might point to this diagnosis.

As the dietetics practitioner gathers nutrition assessment data in order to determine whether or not a patient actually has a nutrition diagnosis of "inadequate protein intake," he or she should attempt to obtain information from the diet and client history that will provide evidence of the problem. A problem should not be identified unless there is adequate evidence to support its presence. In this case, data describing the amount of protein that is consumed would be essential for determining how far below the recommendation the patient's protein intake actually is. Other data from the assessment might provide clues about the cause of the problem, such as physiological reasons for increased need, lack of access to food, knowledge deficit, or psychological causes. By using these reference sheets, dietetics practitioners can be assured that the diagnostic terminology is used consistently and accurately.[6]

PES Statements Nutrition diagnoses are written in a **PES** (problem, etiology, signs/symptoms) format that lists the problem, its cause, and appropriate defining characteristics. The problem (P) is also referred to as the *diagnostic label*. It describes in a general way an alteration in the client's nutritional status. Words like *excessive*, *inadequate*, and *inappropriate* are frequently found in these labels. The etiology (E) or related

---

Nutrition Diagnosis Standardized Language: Domains and Examples

- **Intake (NI):** Domain that contains standardized nutrition diagnostic terms that describe actual problems related to the intake of energy, nutrients, fluids, and bioactive substances (such as plant sterol and stanol esters or soy protein) through oral diet or nutrition support (enteral or parenteral nutrition)[1]
  Examples: Excessive oral food/beverage intake (NI-2.3); evident protein-energy malnutrition

(NI-5.2); inadequate vitamin intake (folate) (NI-54.2)

- **Clinical (NC):** Domain that contains standardized nutrition diagnostic terms that describe nutritional problems that relate to medical or physical conditions[1]
  Examples: Swallowing difficulty (NC-1.1); impaired nutrient utilization (NC-2.1); involuntary weight loss (NC-3.2)

- **Behavioral-Environmental (NB):** Domain that contains standardized

nutrition diagnostic terms that describe nutrition problems related to knowledge, attitudes/beliefs, physical environment, access to food, and food safety[1]
  Examples: Not ready for diet/lifestyle change (NB-1.3); self-feeding difficulty (NB-2.6); intake of unsafe food (NB-3.1)

References
1. American Dietetic Association. Nutrition diagnosis: a critical step in the nutrition care process. Chicago: American Dietetic Association; 2006.

factors are those factors that contribute to the cause or existence of a particular problem. Finally, the signs and symptoms (S) are the defining characteristics obtained from the subjective and objective nutrition assessment data. These data provide evidence that a problem exists and describe the severity of the problem. When these three parts are used to form the nutrition diagnostic statement, it is generally stated in the following way: the problem (P) *related to* the etiology (E) *as evidenced by* the signs and symptoms (S).

For example, consider these nutrition diagnoses:

- "Inadequate energy intake (P) *related to* changes in taste and appetite (E) *as evidenced by* average daily kcal intake 50% less than estimated recommendations (S)"
- "Involuntary weight loss (P) *related to* inadequate energy intake (E) *as evidenced by* eight pounds weight loss within four weeks (S)"

Let's examine how these diagnoses were made. A comprehensive nutrition assessment reveals the following data:

- Client is undergoing chemotherapy for cancer treatment (client's medical history).
- Client complains of meats tasting bitter and most beverages too sweet (food/nutrient-related history).
- Client states, "I have no appetite and no desire to eat" (food/nutrient intake history).
- Three-day food records reveal average kcal intake of approximately 50% of estimated needs (dietary intake data compared to estimated needs).
- Client has experienced weight loss of eight pounds since last outpatient visit one month ago (anthropometric measurements).

In order to evaluate the above nutrition assessment data, the dietetics practitioner applies a number of the critical thinking skills listed in Table 2.3. These include finding patterns and relationships between the data and possible causes, stating each problem clearly and singularly, ruling in/ruling out specific diagnoses, and prioritizing the importance of the diagnoses.

From the relationships that exist among the assessment data just noted, "inadequate energy intake" and "involuntary weight loss" are selected as relevant nutrition problems. It is essential to focus on problems for which nutrition interventions will be the primary treatment. Once the appropriate problems have been selected, the next step is to describe accurately the signs and symptoms. The signs and symptoms are used to validate and confirm the existence of problems. They also indicate the severity of the problems, answering the question "How much?" or "How do I know?"

Finally, after validating the problem by identifying the appropriate signs and symptoms, the etiology or cause of the problem is explored. To determine the etiology, related factors and additional data from the assessment are reviewed. It is important to seek the answer to the question "Why does this problem exist?" and explore all possibilities. It may even be necessary to frequently ask the question "Why?" to uncover the underlying root cause of the nutrition problem. To summarize:

- The problem is the "What?"
- The etiology is the "Why?"

- The signs/symptoms are the "How do I know?" or "How severe is the problem?"

In the present example, two important points about etiology are illustrated. First, even though medical diagnoses (cancer) and/or medical treatment (chemotherapy) contribute to nutrition problems, they should not be used as the primary etiology. Instead, it is best that a nutrition-related cause be part of the etiology. This is consistent with the guiding principle that distinguishes a nutrition diagnosis from other diagnoses. First, *a nutrition diagnosis is written in terms of a client problem for which nutrition-related activities provide the primary intervention.* Second, nutrition diagnostic terminology is always used to identify the nutrition problem (P). This language can also be used as etiology language. Behaviors and patterns of food and nutrient intakes that are undesirable (problems in and of themselves) can produce other problems such as changes in anthropometric, biochemical, or clinical findings. In the present example, inadequate caloric intake is the primary cause of unintentional weight loss.

Traditionally, nutrition care has been driven by diet orders associated with certain disease conditions, such as diet orders for a "renal diet," a "diabetic diet," or a "weight-loss diet." With the advent of the standardized nutrition language and nutrition diagnoses, nutrition care can and should be driven by the extent of a nutrition problem rather than solely by a diet order or medical condition. Medical conditions affect a person's ability to consume, digest, metabolize, and utilize nutrients. They also affect nutrient needs and requirements. However, the specific type of nutrition intervention can now be determined by the nutrition diagnosis. For example, instead of providing nutrition care/education as a result of a diet order for a diabetic diet or renal diet, the dietitian will now carefully assess the nutritional status of each patient to specifically identify what, if any, nutrition problems (diagnoses) exist. For example, a patient with type 2 diabetes could conceivably have inappropriate carbohydrate intake, undesirable food choices, or a self-monitoring deficit. A complete assessment may reveal, however, the absence of any nutritional problems at all. In the case of a patient with chronic renal disease, there could be problems such as excessive potassium intake or excessive fluid intake; but again, a complete assessment may show there are no problems. Another scenario might be two patients with different medical diagnoses who present with a similar nutrition diagnosis, such as involuntary weight loss. In summary, using the nutrition diagnoses to clarify and identify specific nutrition problems may reveal (a) no nutrition problems at the present time, (b) different nutrition problems for patients with similar medical diagnoses, or (c) similar nutrition problems for patients with different medical conditions. Note that in cases where a nutrition assessment reveals no nutrition problems, it is still necessary to record these findings in the patient's record.

Criteria for Evaluating PES Statements Since the intent of nutrition diagnoses is to describe those problems for which nutrition intervention is the primary treatment, it is important to develop PES statements that accurately reflect that intent. The following questions and criteria were developed to ensure that nutrition diagnoses are well written and accurately represent the nutrition problems:

Problem (P)

- Can the dietetics practitioner impact, improve, or resolve the nutrition problem?
- When all things are equal and there is a choice between stating the PES statement using two nutrition diagnoses from different domains, consider the Intake domain as more specific to the role of the RD.

Etiology (E)

- Is the etiology truly the root cause?
- Is there an intervention that will address the root cause, thus increasing the likelihood that a positive change will result?
- If it is not clear that the problem will be resolved by addressing the etiology, can an intervention at least reduce or lessen the significance of the signs and symptoms?

Signs and Symptoms (S)

- Are the signs and symptoms that are used to describe the problem specific enough to be measured?
- Will measuring the signs and symptoms indicate if the problem is resolved or improved?

PES Overall

- Are the problems clearly and singularly stated?
- Does the assessment data used to identify the nutrition diagnosis support and link to the diagnostic statement, etiology, and signs and symptoms?

Box 2.3 demonstrates how these criteria are used to evaluate and refine PES statements for a client with diabetes.

Relationship of Nutrition Diagnosis to the Other Steps of the NCP

Figure 2.4 illustrates the relationship of the nutrition diagnosis to the other steps of the NCP. As stated previously, a nutrition diagnosis is the missing link between nutrition assessment and nutrition intervention. An accurate nutrition diagnosis is generated from a focused nutrition assessment and sets the stage for the next two steps of the NCP: Step 3, nutrition intervention, and Step 4, nutrition monitoring and evaluation.

The signs and symptoms or defining characteristics represent subjective and objective data obtained from the nutrition assessment in Step 1. These data appropriately describe the particular problem by quantifying (generally use of objective data) and qualifying (generally use of subjective data) how that specific problem is present at that point in time. If the problem is an energy intake imbalance (either Inadequate energy intake NI-1.4 or Excessive energy intake NI-1.5), then a measurement of kcal (% of estimated caloric needs, average intake over time, an amount less than or more than desired, etc.) best describes the energy problem; if the problem is one of weight, then an appropriate anthropometric measurement (BMI, relative weight, weight change over time, etc.) should be used to describe the weight problem. Data from the assessment also provide information used to determine the etiology.

These signs and symptoms then become the basis for setting ideal and measurable goals as part of Step 3, nutrition intervention. They are also the **outcome measures** that are used to monitor and evaluate progress toward goals as part of Step 4, nutrition monitoring and evaluation. If kcal are inadequate by 50%, as in the previous example of a nutrition diagnosis, a desired goal might be to meet 75% of estimated caloric needs within two days. A kcal count or food record could be used to track and evaluate that outcome.

Finally, nutrition interventions as part of Step 3 should be logically linked to the cause of problems. If the root cause of inadequate

Figure 2.4 Relationship of the Nutrition Diagnosis to the Other Steps of the NCP

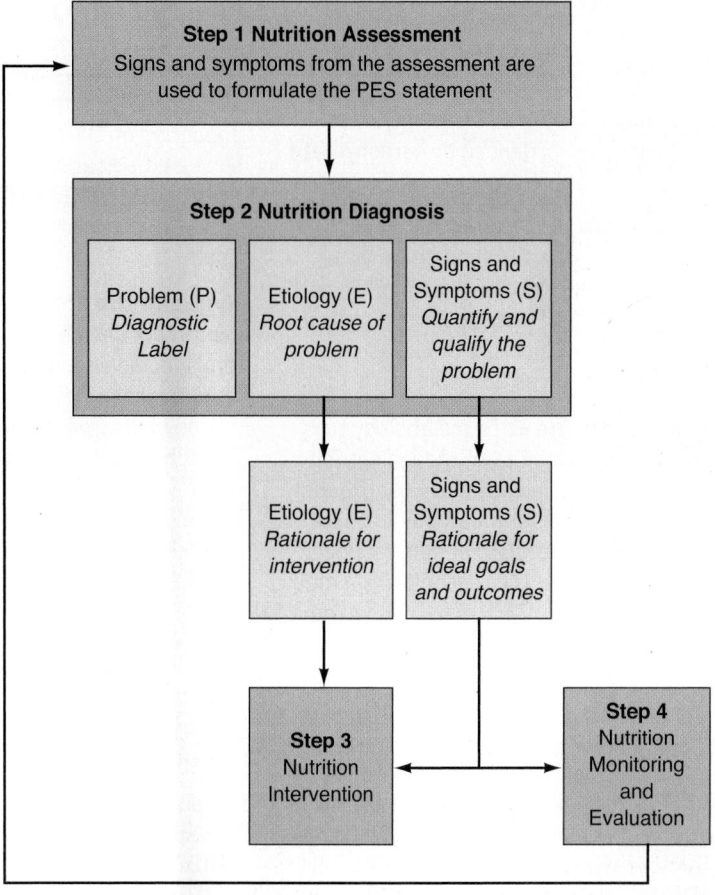

Source: Lacey K and Pritchett E, Nutrition Care Process and Model: ADA adopts road map to quality care and outcomes management. J Amer Diet Assoc. 2003;103:1061–72. © Copyright 2003 with permission from Elsevier.

BOX 2.3 **CLINICAL APPLICATIONS**

Evaluating a Nutrition Diagnosis

When data are obtained from a nutrition assessment, there will be a number of findings that can provide clues to the presence of a particular nutrition diagnosis. The dietetics professional needs to distinguish among (1) data that are associated with a nutrition problem and/or may be a consequence of that problem, (2) data that will be used to document the specific signs and symptoms that describe and quantify that problem, and (3) data that will provide insight into the root cause of the problem.

Which of the following nutrition diagnoses is preferred and why?

A. Inconsistent carbohydrate intake related to not following a diabetic diet as evidenced by elevated blood glucose of 250 mg/dL

B. Inconsistent carbohydrate intake related to inability to read labels correctly for carbohydrate content and lack of knowledge about amount of grams/carbohydrate units as evidenced by carbohydrate units in three meals of 1, 6, and 3

Evaluate the P: Can the dietetics professional impact, improve, or resolve the nutrition problem? In both examples, the nutrition problem "inconsistent carbohydrate intake" is one that can be improved or resolved.

Evaluate the E: Is the etiology truly the *root* cause? Even though not following a diet plan is likely contributing to the inconsistent carbohydrate intake, there needs to be further understanding as to the reasons why a meal

plan is not being followed. In other words, asking "why" uncovers the real reason for not following the plan and is more clearly stated in example B.

Is there an intervention that will address the root cause, thus increasing the likelihood that a positive change will result? Developing an intervention using example A might lead prematurely to a more traditional diet education of a diabetic diet, whereas addressing the real reason for not being able to follow a meal plan gives both the provider and the client a more realistic and specific plan for education. Focusing on the two topics in B should increase the likelihood that a positive change will occur compared with an education plan that is more general.

Can an intervention reduce the significance of the signs and symptoms? In the case of example A, it is not clear that a change in carbohydrate intake alone will improve the blood glucose. There could be other factors that are impacting on the blood glucose levels such as medication, illness, and so on, whereas the signs and symptoms in B are more descriptive of the problem itself.

Evaluate the S: Are the signs and symptoms used to describe the problem specific enough to be measured? In both cases, the signs and symptoms can be measured; however, improvement in carbohydrate units can be evaluated within a shorter time frame than can improvements in the blood glucose. Furthermore, changes in meal patterns can be expected in direct response to

the intervention, whereas changes in blood glucose are influenced by more variables and may not occur directly in response to the nutrition education provided.

Evaluate the PES Overall: Are the problems clearly and singularly stated? In the case of example A, there are really two different types of problems embedded in this diagnosis: inconsistent carbohydrate intake and altered nutrition-related laboratory values. Example B describes a single problem in a straightforward manner, allowing the dietitian to deal with one problem at a time.

Does the assessment data used to identify the nutrition diagnosis support and link to the diagnostic statement, etiology, and signs and symptoms? Even though an elevated blood glucose level provides a clue that there may be a problem with carbohydrate intake, it does not specifically describe the nutrition problem itself. In this case the blood glucose may be a clinical consequence of the intake problem but does not describe and/or quantify the intake problem.

Therefore, after applying the criteria to each of the examples noted previously, example B is the preferred nutrition diagnosis. It states the problem clearly and singularly, and it provides a quantifiable description of the signs and symptoms from which specific goals can be established and measured. It also provides clear direction for a specific intervention targeted at the root cause of the problem.

---

intake is taste alteration and decreased appetite, interventions need to be linked to ways to enhance appetite and improve the taste of foods before a change in caloric intake will occur.

When nutrition diagnoses are written as separate and distinct problems, even though one problem may actually cause another, the dietetics practitioner is able to prioritize which problem should be addressed first as part of Step 3, nutrition intervention. For example, when there is both an intake problem and a weight problem, caloric intake needs to change before a change in weight can be expected.

Step 3: Nutrition Intervention

The third step of the NCP, nutrition intervention, involves both planning and implementing. It is a specific set of activities and associated materials used to address the problem. Nutrition interventions are purposefully planned actions designed with the intent of changing a nutrition-related behavior, risk factor, environmental condition, or aspect of health status for an individual, target group, or the community at large.[1] Dietetics practitioners work collaboratively not only with other health care professionals, but more importantly with the client, family, or caregiver to create a realistic plan that has a good probability of positively influencing the diagnosed problem. This client-driven process is key to successful nutrition intervention, distinguishing it from previous planning steps that may or may not have involved the client to this degree.

Prioritize the Nutrition Diagnoses A nutrition assessment will likely result in the identification and labeling of multiple nutrition diagnoses; therefore, before any action can be taken, it is essential to *prioritize the diagnoses* that are identified in Step 2, nutrition diagnosis. The ranking of nutrition diagnoses permits dietetics practitioners to arrange the problems in the order of their importance and urgency for the client. Once they have been sorted for safety, then another prioritization can be done based on such things as anticipated early response to an intervention, client preference of a behavior change, or the impact of one problem on another. Using the earlier example, it makes sense to first address the primary problem of energy intake before one expects to intervene on the secondary problem of weight:

- "Inadequate energy intake (P) *related to* changes in taste and appetite (E) *as evidenced by* average daily kcal intake 50% less than estimated recommendations (S)"
- "Involuntary weight loss (P) *related to* inadequate energy intake (E) *as evidenced by* eight pounds weight loss within four weeks (S)"

Identify Goals After prioritizing the nutrition diagnoses, it is necessary to *identify ideal goals and patient-focused expected outcomes.* **Ideal goals** are science-based values intended to control or improve specific health conditions. ADA's Evidence-Based Guides for Practice and other practice guides provide resources to assist dietetics practitioners in selecting the appropriate goals for patients.[7] Consuming less than 7% of kcal from saturated fat is an example of an evidence-based ideal goal for the nutrition treatment of hyperlipidemia.[3] This is the desired level of saturated fat that is associated with the least amount of cardiac risk. **Expected outcomes** are the desired change(s)

to be achieved over time as a result of nutrition intervention. Expected outcomes are based on the nutrition diagnosis; for example, decreasing the intake of saturated fat by a specific amount or percentage of kcal is an expected outcome. Expected outcomes can also be defined in terms of a specific behavior that will result in the change in amount of saturated fat consumed; for example, the patient will substitute olive oil for solid margarine as the preferred spread on most breads. Expected outcomes should be written in observable and measurable terms that are clear and concise. They should be client-centered and realistically tailored to the client's circumstances and expectations for treatment. Interventions are then planned that will help a patient to meet these goals and outcomes.

Plan the Nutrition Intervention Finally, as part of the planning, appropriate interventions need to be selected. Like the standardized terms used for nutrition assessment and nutrition diagnosis, intervention terminology is organized into domains. There are four domain categories that identify the various types of nutrition interventions: (1) Food and/or Nutrient Delivery, (2) Nutrition Education, (3) Nutrition Counseling, and (4) Coordination of Nutrition Care. These domains and the major types of interventions in each are summarized in Box 2.4 and detailed in Appendix C.

All interventions must be based on scientific principles and rationales and must be grounded in quality

> **Key Concepts: NCP Step 3, Nutrition Intervention**
>
> - First and foremost is the need to prioritize the nutrition diagnoses.
> - Ideal goals and expected outcomes need to be identified prior to implementing nutrition interventions.
> - Interventions are derived from accurate diagnoses and largely driven by client involvement.
> - ADA's Evidence-Based Guides for Practice provide dietetics practitioners with tools that promote quality service and demonstrate effectiveness of care.

BOX 2.4   CLINICAL APPLICATIONS

Nutrition Intervention Standardized Language: Domains and Examples

- **Food and/or Nutrient Delivery (ND):** An individualized approach for food/nutrient provision including meals and snacks, enteral/parenteral feeding, supplements, assistance, environment, and medication management
  Examples: Modify distribution, type, or amounts of food and nutrients within meals or at specified times (ND-1.1); initiate EN or PN (ND-2.1); table service/set-up (ND-5.5)
- **Nutrition Education (E):** A formal process to instruct or train a

patient/client in a skill or to impart knowledge to help patients/clients to voluntarily manage or modify food choices and eating behavior to maintain or improve health
Examples: Priority modification (E-1.2), skill development (E-2.5)
- **Nutrition Counseling (C):** A supportive, collaborative process; positive relationships that foster responsibility for self-care to treat an existing condition and promote health

Examples: Health belief model (C-1.3), motivational interviewing (C-2.1), problem solving (C-2.4)
- **Coordination of Nutrition Care (RC):** Consultation with, referral to, or coordination of nutrition care with other health care providers that assists in treating nutrition-related problems
Examples: Referral to RD with different expertise (RC-1.2), referral to community agencies/programs (RC-2.2)

research and evidence-based interventions when available. Once again, ADA's Evidence-Based Guides for Practice and other practice guides are invaluable resources for both identifying science-based ideal goals and selecting appropriate interventions. These guides link external scientific evidence regarding nutrition care to a specific health problem, thus giving dietetic practitioners the confidence that they are making the best decisions when providing nutrition care. The evidence-based guides alone do not replace the expertise and judgment of dietetics practitioners, but these tools enhance the value of dietetics practitioners and increase the likelihood that a desired outcome will occur.

Implement the Nutrition Intervention

Implementation is the action phase of the nutrition care process. It is during this phase that dietetics practitioners communicate the plan of action to the client and other professionals. Dietetics practitioners may directly carry out the intervention or may delegate or coordinate care provided by others. Once again, the central core of the Nutrition Care Model (relationship between patient/client/group and dietetics practitioner) recognizes that the client needs to be involved in this decision-making and action step of nutrition care.

Step 4: Nutrition Monitoring and Evaluation

The purpose of monitoring and evaluation is to determine the degree to which progress is being made and whether or not the client's goals or desired outcomes of nutrition care are being met. It is much more than merely "watching" what is happening. It requires an active commitment to measuring and recording the appropriate outcome indicators relevant to the nutrition diagnosis' signs and symptoms. Progress should be

(1) monitored, (2) measured, and (3) evaluated on a planned schedule. Systematic use of each of these components provides consistency in practice, adds value, and demonstrates effectiveness of care.

Monitor Progress

Monitoring refers specifically to determining that the goals and outcomes anticipated by the client and the dietetics practitioner are indeed being achieved. Specific activities that are associated with this level of monitoring include the following:

- Determining whether the intervention is being implemented as planned
- Checking the client's understanding and attainment of goals
- Determining if changes in the client's condition are occurring[1]

Monitoring in this way may require gathering additional information about possible reasons for any lack of progress. Revision of a nutrition diagnosis and/or a change in plan may occur as a result of obtaining additional information. Using the NCP, therefore, may involve performing its steps more than once during the course of nutrition treatment.

Measure Outcomes

Measuring outcomes means that data are collected over time. This is a critical component of the NCP. Interventions have too often been planned and acted upon with little regard for what has really happened as a result of the action taken.

The key to measuring outcomes is knowing what needs to be measured. The NCP provides clear examples of the types of outcomes to be measured.[1] These include the following (among others):

BOX 2.5 CLINICAL APPLICATIONS

Using Standardized Language to Describe Nutrition Care

If a nutrition assessment revealed fiber intake of 10 grams a day due to eating less than one serving of fruit or vegetable daily, the following standardized terms could be used in the NCP:

Step 1: Nutrition Assessment

- Food intake (FH-1.3.1 & 2): Amount and types of food = less than one serving fruit and vegetables daily
- Fiber intake (FH-1.6.4): Total fiber FH-1.6.4.1 = < 10 grams/day
- Factors Affecting Access to Food (FH-5): Limited storage of fresh foods (FH-5.2.5)

Step 2: Nutrition Diagnosis

Inadequate fiber intake (NI-5.8.5) related to infrequent consumption of

fruits and vegetables as evidenced by 10 grams daily compared with recommended daily intake of 25 grams of total fiber

Step 3: Nutrition Intervention

- **Ideal goal:** Average daily fiber intake of 25 grams or more
- **Expected outcome:** Increase servings of fruit and vegetable daily to minimum of 5
- **Intervention selected:** Nutrition Education (E)—Comprehensive Nutrition Education (E-2.0) for the purpose of training leading to in-depth knowledge or skills to change intake of specific food groups, i.e., fruits and vegetables (ND-1.3)

- Purpose of education (E-2.1), recommended modification (E-2.2) and skill development (E-2.5); provision of fruits and vegetables that have longer shelf life, e.g., dried fruits, whole canned or frozen fruits with peels

Step 4: Nutrition Monitoring and Evaluation (refer to Assessment terminology)

Monitor, measure, and evaluate using the same terminology as used in the assessment step, i.e., FH-1.3.1 & 2, FH-1.6.4.1, and FH-5.2.5

- Direct nutrition outcomes such as knowledge gained, behavior changes, food or nutrient intake changes, and improved nutritional status
- Clinical and health status outcomes such as laboratory values, anthropometry and body composition, blood pressure, and risk factor profile
- Patient/client-centered outcomes such as quality of life, satisfaction, self-efficacy, and self-management
- Health care utilization and cost outcomes such as medication changes, special procedures, and planned/unplanned health care visits

The specific outcomes that are measured are determined by the nutrition diagnosis, its etiology, and the signs and symptoms—i.e., data that were obtained directly from the nutrition assessment. Therefore, the standardized terms from monitoring and evaluation are combined with nutrition assessment terms. There are only four domains that reflect the types of data to be monitored: (1) Food/Nutrition-Related History; (2) Anthropometric Measurements; (3) Biochemical Data, Medical Tests, and Procedures; and (4) Nutrition-

Focused Physical Findings. Box 2.5 demonstrates how standardized terms are used to describe each step of the NCP; notice how the same parameters assessed in Step 1 are reassessed as part of Step 4.

Establishing a nutrition diagnosis that clearly describes the signs and symptoms establishes the type of outcome to be measured and thus provides baseline data from which to begin measuring. A variety

BOX 2.6 CLINICAL APPLICATIONS

The NCP in Practice: Nutrition Therapy for Cardiovascular Disease

AJ is a 55-year-old male who works in sales. He has just returned from an annual health physical with his primary care physician. Following the usual exam, the physician initiated a referral to the health care system's outpatient dietitian. The following are pertinent assessment data that the dietitian obtained from both the patient chart and interview.

Step 1: Nutrition Assessment

Client history: Past medical and family history: Positive family history for premature heart disease; recent BP reading of 140/80

Biochemical data: LDL 130 mg/dL, TG 200 mg/dL, TC 200 mg/dL

Anthropometric data: 5'8", 185#, BMI 28.3

Physical activity: Little to no exercise, states he is too busy at work, travels weekly by car and occasionally on plane for business, always takes the elevator to his fourth-floor office and parks in the closest parking lot

Food and nutrient intake summary (typical intake and usual amounts of key nutrients):

Average daily kcal = 3200 kcal (estimated needs based on adjusted body weight of 161# and Mifflin-St. Jeor formula and physical activity factor of 1.4 = 2155 kcal)

Saturated fat (SFA) = 10% of total kcal (approximately 36 g)

- 7–8 oz portions of beef or chicken = 13 g SFA
- Chocolate cake/frosting and ice cream = 8 g SFA
- 3 oz bologna sandwich on white bread or fast-food cheeseburger = 6–8 g SFA
- 3 T butter = 15 g SFA
- 1 c whole milk = 4 g SFA

Sodium intake = 3500–4500 mg

- Cheeseburger = 600 mg
- 3 oz bologna = 226 mg
- Large french fries = 300 mg
- 2 c canned soup = 1600 mg
- 1–2 tsp salt added to foods = 1200–2400 mg

Key Concepts:

1. Assessment data is clustered and organized according to possible nutrition diagnoses.
2. Wherever possible, amounts of key nutrients are estimated.
3. Appropriate standards are applied to evaluate the data.

Step 2: Nutrition Diagnosis

PES #1: Excessive energy intake related to daily intake of whole-fat dairy products, desserts, and large meat portions as evidenced by average caloric intake of 3200 kcal @ 150% in excess of estimated needs of 2100–2200 kcal.

PES #2: Excessive saturated fat intake related to daily intake of whole-fat dairy products and large meat portions as evidenced by saturated fat intake of 36 @ 10% of total kcal compared to recommended 7% of kcal.

PES #3: Excessive sodium intake related to daily consumption of convenience and fast foods as evidenced by typical daily sodium intake of 3500–4500 mg compared to recommended 2400 mg

(continued)

PES #4: Physical inactivity related to client perception of being too busy and frequent business travel as evidenced by little to no regular exercise and very sedentary lifestyle

PES #5: Overweight related to excessive caloric intake and physical inactivity as evidenced by BMI 28.3

Key Concepts:

1. Even though there are five distinct PES statements, two different problems have similar causes; therefore, intervention can address more than one problem simultaneously.

2. One or two problems can actually be the cause of another problem. Therefore, it is necessary to address both the intake problem and physical inactivity before a change in weight will occur.

3. Prioritization results in addressing the intake and physical activity problems first. Addressing the weight problem then becomes a longer-term goal.

4. When there are choices among PES statements from different domains, frequently the intake problem becomes the "default" or priority problem.

Step 3: Nutrition Intervention

Establish Goals:

These are determined by consulting evidence-based practice guides and discussing expectations with the client. They are measurable and realistic and establish the type of outcome to be tracked over time.

1. Average daily caloric intake will be no more than 110% of estimated needs (approximately 2200 kcal; reduction of 1000 kcal/day).

2. Saturated fat intake will be 7% or less of total kcal.

3. Average daily sodium intake will be at or under 2400 mg.

4. Daily physical activity will increase by 2000 steps weekly to goal of 10,000/day.

5. Weight loss over time will average 1–2 pounds per week.

Implementation:

1. Comprehensive nutrition education focusing on:
   a. Recommended modifications to lower fat, saturated fat, and sodium intakes
   b. Skill development in making alternative food choices to lower total fat, saturated fat, and sodium intakes:
      · Choose smaller and leaner meat portions (chicken and fish).
      · Select lower-fat dairy products (low-fat ice cream, 1% milk, etc.).
      · Use plant stanols and/or olive oil in place of butter.
      · Use a variety of seasonings at table in place of salt (provide patient with examples of recipes).
      · Order fast-food hamburger and fresh vegetable or fruit salad in place of cheeseburger and french fries.
      · Decrease frequency of use of canned soups.

2. Nutrition counseling to assist client in increasing physical activity daily.
   · Evaluate and select appropriate theoretical basis (cognitive-behavioral theory, stages of change, etc.) and strategies (goal setting, self-monitoring, rewards, etc.).
   · Arrange for client to obtain a step meter and instruct on its use and record keeping.

Key Concepts:

1. Client is actively involved in establishing realistic behavioral goals.

2. Ideal goals are based on the specific data obtained from the nutrition assessment and the use of evidence-based practice guidelines; for example, current SFA intake is 10% of kcal; the ideal goal based on ATPIII is 7%.

3. The selected interventions are intended to address and alter the etiology in the PES statements.

Step 4: Nutrition Monitoring and Evaluation

Monitor Progress:

The dietetics professional may contact the client to provide support and clarify any questions regarding the plan. This is done in order to determine if the plan is being implemented and whether or not the client fully understands the information provided.

Measure Outcomes:

Direct nutrition outcomes:

· Behavior changes related to decreasing portion sizes, use of low-fat dairy foods and plant stanols, use of alternative seasonings in place of salt, food choices when eating at fast-food establishments, and physical activity

· Intake changes for total kcal, total fat, SFA, and sodium

Clinical and health status outcomes:

· Blood pressure, LDL, and BMI

Patient/client-centered outcomes:

· Satisfaction

· Self-management (food records, physical activity records)

Evaluate Outcomes:

Baseline data will be compared with changes in the preceding outcome data that are tracked over time. Progress will be discussed with the client and any problems or barriers that are identified will be used to revise PES statements, modify interventions, and/or establish new goals.

BOX 2.7 CLINICAL APPLICATIONS

The NCP in Practice: Enteral Nutrition Following Motor Vehicle Accident (MVA)

AT is a 25-year-old female in previously good health who was involved in an MVA. She was admitted to the hospital for extensive oral surgery resulting in a wired jaw. Despite numerous attempts to consume oral supplements, she was able to meet only 20% of her estimated kcal and protein needs. The physician writes the following order to begin tube feeding:

Use of hospital's high-protein, high-kcal tube feeding at goal of 80 mL/hr.

This formula provides 0.0616 g protein/mL, 1.5 kcal/mL, and 75.8% free water. The standard hospital protocol for water flushes of tube feeding is 55 mL water q 8 hours.

According to the hospital's nutrition care protocol, the dietitian was contacted to complete a nutrition assessment.

Step 1: Nutrition Assessment

Anthropometric data: 5'5", 125# (56.8 kg); BMI 20.8

Biochemical data: all WNL

Estimated kcal and protein needs: Based on actual body weight and Mifflin-St. Jeor formula for REE, physical activity factor of 1.3 and injury factor of 1.1. Protein needs based on 1.0–1.2 g/kg. Estimated needs = 1900 kcal and 57–68 grams protein.

Estimated fluid needs based on 1 mL/kcal = 1900 mL

24 hours of high-protein, high-kcal tube feeding at goal rate of 80 mL/hr will provide 2880 kcal, 118 g protein, and 1455 mL free water. Total TF water flushes provide an additional 150 mL water.

Step 2: Nutrition Diagnoses

PES: Inadequate protein-energy intake related to inability to take nutrition orally secondary to oral surgery and wired jaw as evidenced by client meeting only 20% of estimated needs

PES: Excessive intake from enteral/parenteral nutrition related to use of high-protein/high-kcal tube feeding at rate of 80 mL/hr as evidenced by TF exceeding estimated needs for kcal and protein by 50% (2880 kcal and 118 g protein per order compared to estimated needs of 1900 kcal and 57–68 g protein)

PES: Inadequate fluid intake related to use of concentrated tube feeding as evidenced by 24-hour fluid intake 84% of estimated needs (1600 mL compared with estimated needs of 1900 mL)

Step 3: Nutrition Intervention

Goals:

1. Increase fluid to 1900 mL/day within 24 hours.

2. Decrease kcal and protein to 95%–105% of estimated needs within 24 hours.

Intervention:

Recommend change tube feeding to following:

- Standard tube feeding providing 1 kcal/mL, 0.0366 g protein/mL, 83.3% free water.

- Begin tube feeding at 25 mL/hr for 8 hours, then increase to 50 mL next 8 hours, and if tolerated, increase to goal of 80 mL within 24 hours.

- Increase water flushes to 100 cc q 8 hours.

- 24 hours of this tube feeding will provide 1920 kcal, 70 g protein, and 1900 cc free water.

Step 4: Nutrition Monitoring and Evaluation

- Verify that the tube feeding is being provided at desired rate.

- Record nutrients that are being provided.

- Compare nutrition provided to estimated needs and make changes and recommendation as appropriate.

---

of documents and tools can be used to measure and track data, including electronic charting, coding systems, and spreadsheets.

Evaluate Outcomes It is not enough to just measure outcomes, however. Outcomes need to also be evaluated to determine what, if any, changes have occurred as a result of the nutrition intervention. Such an evaluation requires comparing the current findings with the previous signs and symptoms. Outcome evaluation may also involve additional data collection in order to explore why a change has *not* occurred as expected. Additional or revised nutrition diagnoses may also be needed. New goals and new interventions thus may further modify the nutrition care process.

Documentation

Documentation (see Chapter 6) is an ongoing process that supports all of the steps of the NCP. The standardized language that is now part of the NCP improves both the written and oral communication among members of the health care team as well as communication with the patient. It allows dietetics practitioners to more clearly name and document clinical judgments concerning nutrition problems.[8] Documentation should be relevant, accurate, and timely. A variety of charting formats have been used by dietetics practitioners to communicate nutrition care. These formats include Subjective, Objective, Assessment, and Plan (SOAP); focus notes; and Problem, Intervention, and Evaluation (PIE). More recently, a newer form of charting based on the steps of the NCP has been introduced. This new form is the ADIM format (Assessment, Diagnosis, Intervention, and Monitoring). Electronic medical records are also becoming more widely used and many are organized around the nutrition care process and nutrition diagnosis.

As dietetics practitioners implement ADA's NCP and standardized language, more efficient and effective methods of documentation will surely evolve. Regardless of the specific format used, "when the systematic steps of the nutrition care process or the nutrition practitioner's clinical judgments are consistently defined and documented with standardized terms, this information can be collected, compared, and aggregated and therefore used to identify the most effective treatment."[9]

Conclusion

In Step 1, nutrition assessment, adequate and appropriate information is obtained in order to identify the nutrition problem and formulate a complete nutrition diagnosis in Step 2, nutrition diagnosis. A complete nutrition diagnosis includes both an etiology and accurate signs and symptoms. Step 3, nutrition intervention, establishes ideal goals and desired outcomes for which interventions likely to provide positive results are planned. In Step 4, nutrition monitoring and evaluation, appropriate indicators are measured over time to track progress toward desired goals. This completes the cycle of the NCP.

As dietetics practitioners incorporate the NCP into their practices, they recognize that the use of the standardized process along with the standardized nutrition diagnostic terminology changes both the way they think and the way they chart. Early adaptors remark, "You're changing the way you're thinking, you're changing the way you're charting—it's a huge change. There are no shortcuts you can take. . . . It's certainly worthwhile."[10]

WEB LINKS

American Dietetic Association—Quality Management   To find the latest information on the nutrition care process published by the ADA, click on the "Practice" link; on the page that appears, click on "Quality Management" (member only access).
www.eatright.org

American Dietetic Association Evidence Analysis Library   This website, available to ADA members only, summarizes the latest research and evidence-based guidelines related to nutrition and dietetics practice.
www.adaevidencelibrary.com

END-OF-CHAPTER QUESTIONS

1. List the internal and external factors that influence a person's ability to maintain optimal nutritional health.

2. What is the purpose of providing nutrition care?

3. List the four steps of ADA's Standardized Nutritional Care Process (NCP). Briefly describe each step.

4. Why is it important to have standardized nutrition diagnostic terminology in the practice of dietetics?

5. List and briefly describe the domains that are part of the assessment terminology.

6. List and briefly describe the three domains of nutrition diagnostic terms.

7. Explain what P, E, and S are in the nutrition diagnosis.

8. Write an example of a PES nutrition diagnosis.

9. How does the nutrition diagnosis relate to the other steps in the nutrition care process?

10. How does a nutrition diagnosis differ from a medical diagnosis?

11. Describe the criteria used to evaluate the quality of PES statements.

12. What are the domains in the standardized language for nutrition intervention?

13. List and briefly describe the four types of outcome measures that can be monitored and evaluated in the NCP.

14. What is meant by "outcomes management system"? Why is it important in dietetic practice?

REFERENCES

1. Lacey K and Pritchett E. Nutrition Care Process and Model: ADA adopts road map to quality care and outcomes management. J Amer Diet Assoc. 2003; 103:1061–72.

2. Institute of Medicine. Crossing the quality chasm: a health care system for the 21st century. Committee on Quality Health Care in America. Rona Briere, ed. Washington (DC): National Academy Press; 2001.

3. National Cholesterol Education Program. Third Report of the National Cholesterol Education Program Expert Panel on Detection, Evaluation, and Treatment of High Blood Cholesterol in Adults (Adult Treatment Panel III). Bethesda (MD): National Institutes of Health, National Heart, Lung, and Blood Institute; 2002.

4. International Dietetics and Nutrition Terminology (IDNT) Reference Manual: Standardized Language for the Nutrition Care Process. 2nd ed. American Dietetic Association; 2009

5. Dorland WAN. Dorland's illustrated medical dictionary. 30th ed. Elsevier; 2003.

6. American Dietetic Association. Nutrition diagnosis: a critical step in the nutrition care process. Chicago (IL): American Dietetic Association; 2006.

7. ADA Evidence Analysis Library [homepage on the Internet]. Chicago (IL): American Dietetic Association; 2009. Available at http://www.adaevidencelibrary.com/default.cfm?library=EBG

8. Hakel-Smith N and Lewis NM. A Standardized Nutrition Care Process and Language are Essential Components of a Conceptual Model to Guide and Document Nutrition Care and Patient Outcomes. J Amer Diet Assoc. 2004;104:1878–84.

9. Hakel-Smith N, Lewis NM and Eskridge KM. Orientation to nutrition care process standards improves nutrition care documentation by nutrition practitioners. J Amer Diet Assoc. 2005;105:1582–89.

10. Mathieu J, Foust M and Ouellette P. Implementing nutrition diagnosis, step two in the nutrition care process and model: challenges and lessons learned in two health care facilities. J Amer Diet Assoc. 2005;105:1636–40.

3

Nutrition Assessment: Foundation of the Nutrition Care Process

Marcia Nelms, PhD, RD, LD
Diane Habash, PhD, RD, LD
The Ohio State University

CHAPTER OUTLINE

Introduction

Just as the physical examination is the cornerstone of medical assessment, **nutrition assessment** provides the foundation for the nutrition care process. It is from information gathered in the nutrition assessment that nutrition problems are identified and the nutrition diagnosis can be determined. After interventions have been put into place, nutrition assessment data serve as benchmarks with which to measure the effectiveness of treatment.

Methods for nutrition assessment will change with the population, the nutrition diagnosis, and the desired outcomes for the nutrition therapy. The type of assessment required for the healthy individual will correlate with goals for a healthy population. For example, assessment used in screening a healthy population for disease risk will focus on those factors necessary for prevention of disease, such as body mass index or waist circumference. In contrast, if assessment is planned for an at-risk population, data gathered may focus on those factors that confirm a diagnosis of malnutrition, such as specific biochemical tests or physical assessment.

There is no one test that measures nutritional status. That is why assessment draws from many indices to provide a complete picture of nutritional health. It is through experience that a clinician can weigh results of multiple measures to critically evaluate the nutritional status of an individual or a population.

24-hour recall—dietary assessment method in which the clinician interviews the client to obtain a list of all foods/beverages consumed in the previous 24 hours

cirrhosis—end-stage liver disease characterized by damage to hepatic parenchymal cells with modular regeneration and fibrosis, associated with failure of hepatic cell function, interference with hepatic blood flow, frequently jaundice, portal hypertension, ascites, and ultimately hepatic failure

fibroblasts—connective tissue cells capable of forming collagen fibers

food frequency—dietary assessment method in which the client describes the frequency and quantity of his or her consumption of certain foods/food groups

indirect calorimetry—measurement of oxygen consumed and carbon dioxide expired with a subsequent calculation of energy requirements from these data

macrophage—a mononuclear, actively phago-cytic cell arising from monocytic stem cells in the bone marrow

multiple myeloma—a tumor composed of cells derived from hemopoietic tissues of the bone marrow; a plasma cell tumor

nephrotic syndrome—a clinical state characterized by edema, albuminuria, decreased plasma albumin, usually increased blood cholesterol, and increased permeability of the glomerular capillary basement membranes, often caused by diabetes-induced glomerulosclerosis, systemic lupus erythematosus, renal vein thrombosis, or hypersensitivity to toxic agents

nutrition assessment—a systematic method for obtaining, verifying, and interpreting data needed to identify nutrition-related problems, their causes, and significance

nutrition screening—process of identifying patients, clients, or groups who may have a nutrition diagnosis and benefit from nutrition assessment and intervention by a registered dietitian

pathophysiology—alterations from normal anatomy and physiology that occur as a result of disease or injury

percent weight for height—percentage used to evaluate a child's growth pattern relative to population standards

phase angle—calculates a mathematical relationship between resistance and reactance; for use with bioelectrical impedance to calculate body composition; higher values for phase angle appear to be consistent with greater body muscle mass and lower risk for morbidity and mortality; values range from 3 to 12

Prognostic Inflammatory Nutrition Index (PINI)—a combination of serum C-reactive protein (CRP), alpha 1-acid glycoprotein (AAG), prealbumin (PA), and albumin measurements that is scored as a measurement of nutritional risk

prospectively—refers to collecting data as it occurs or happens

protein-losing enteropathy—increased fecal loss of serum protein, especially albumin, causing hypoproteinemia

retrospectively—refers to collecting data from events that have already happened

sensitivity—the likelihood that an individual with a disease or condition will be correctly identified when administered a test designed to detect that particular disease or condition

serum—fluid that is obtained after whole blood is separated into solid and liquid components; this should be differentiated from plasma, which contains serum, proteins, and clotting factors

specificity—the likelihood that an individual who does not have a particular condition or disease will be correctly excluded when administered a test designed to detect that condition or disease

stadiometer—a calibrated device used to measure stature

validity—the quality of producing desired results

Nutrition assessment is defined as "a systematic method for obtaining, verifying, and interpreting data needed to identify nutrition-related problems, their causes, and significance."[1] This chapter provides essential information the clinician will need in order to identify and use the appropriate nutrition assessment techniques and interpret the results for evaluation of nutritional status.

Nutritional Status

Determination of *nutritional status* involves evaluating indices that reflect the body's nutrient stores. Nutritional status is altered when stores of energy, protein, water, vitamins, or minerals fluctuate as a result of increased need, increased utilization, altered intake, or altered utilization. Historically, vitamin, mineral, and other nutrient deficiencies had the largest impact on nutritional status. Today, in most developed countries, concern for nutritional status focuses on the effect of excessive intake of nutrients and energy. In the acute care setting, nutrition problems arise as complications of medical problems that interfere with intake or utilization of nutrients.

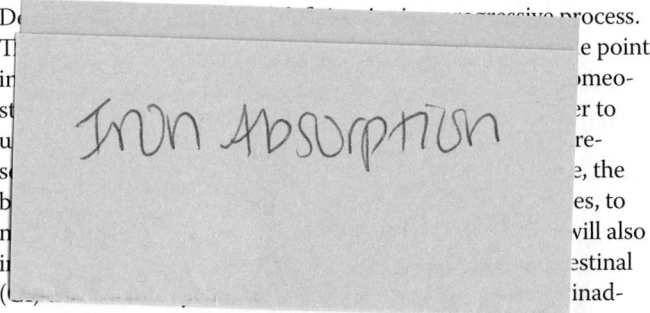

equate iron available to maintain hemoglobin, hematocrit, and mean corpuscular values. At this point, physiological changes may be noted. These may include shortness of breath, paleness of the skin, increased heart rate, and fatigue. Changes in iron status are assessed and confirmed with numerous measures of dietary intake, biochemical levels, and clinical signs.

Determination of *nutritional risk* involves the attempt to predict potential nutritional problems based on the client's current health status.[2, 3] Certain factors increase or decrease a client's nutritional risk; for example, a diagnosis of pancreatic cancer places an individual at a higher nutritional risk than admission for cholestectomy. It is understood that because of such a diagnosis and the likely treatment, nutritional changes and/or problems are probable. Many nutrition problems seen in the hospitalized population are the result of disease or its treatment. The patient will most likely have an increased requirement for certain nutrients and/or an inability to consume enough nutrients or metabolize the ones that can be digested and absorbed. Therefore, knowing the **pathophysiology**, treatment, and clinical course of a disease or diagnosis allows one to identify nutrition problems for an individual and ultimately determine the nutritional diagnosis.

An Overview: Nutrition Assessment and Screening

As a component of the nutritional care process, the nutrition assessment consists of gathering data in the following areas or domains: "food and nutrition related history; biochemical data, medical tests and procedures; anthropometric measurements; nutrition-focused physical findings; and client history."[1] Assessment data from these areas may be both subjective and objective in nature. The assessment process then moves to analysis of data so that current and potential nutritional problems can be identified.

While it is not possible, or necessary, to complete a full nutritional assessment of every patient admitted to a clinic or hospital, it is essential to have a system in place that can quickly identify those patients who may have nutritional problems. The World Health Organization defines screening as "the use of simple tests across a healthy population in order to identify the individuals who have disease, but do not yet have symptoms."[4] The Agency for Healthcare Research and Quality defines screening as "those preventive services in which a test or standardized examination procedure is used to identify patients requiring special intervention."[5] Charney (2008) points out that the American Dietetic Association uses this as their basis for defining **nutrition screening** as the "process of identifying patients, clients, or groups who may have a nutrition diagnosis and benefit from nutrition assessment and intervention by a registered dietitian (RD)."[6, 7] Nutrition screening can be performed by dietetic technicians or other trained personnel, which allows for a more efficient and cost-effective collection and identification of at-risk patients. A dietitian can then perform a full nutrition assessment for those identified as being at nutritional risk. The Joint Commission requires that all patients receive nutrition screening within 48 hours of their admission to a hospital.[8]

Subjective and Objective Data Collection

Types of data include both subjective and objective information. Subjective data include information, usually obtained during interviews, coming directly from the patient, family members, or significant others. Thus, subjective data would include the client's perception of his or her medical condition, dietary intake, lifestyle conditions, current medications or supplement

intake, and family medical history. Subjective data also include the interviewer's observations.

Objective data include information obtained from a verifiable source such as the current medical record and previous medical histories. These data could include anthropometric measurements, biochemical data, and medical tests and procedures. The organization and content of the medical record will vary from institution to institution. See Tables 3.1 and 3.2 for examples of subjective and objective data, respectively.

A crucial skill necessary for conducting a nutrition assessment is the development and use of appropriate interviewing skills. The environment where the interview occurs, the rapport between interviewer and client, and the types of questions and the manner in which they are posed directly affect the quality and accuracy of the information that is obtained. Table 3.3 lists some basic suggestions for conducting an effective interview in order to successfully gather nutrition data.

During the patient interview, information regarding appetite and GI function should be obtained. This includes evaluation of the ability to chew, use and fit of dentures, swallowing ability, nausea, vomiting, constipation, diarrhea, and heartburn, or any other symptoms that might interfere with ability to maintain adequate nutritional intake.

Table 3.1 Examples of Subjective Food-/Nutrition-Related History Assessment

Category	Specific Examples
Food and nutrient intake	Mother reporting child's food intake for previous 24 hours.
Medication and herbal intake	Patient reporting current medications.
Knowledge/beliefs/attitudes	Patient stating that he avoids "sweets and desserts" as his diet modification for diabetes.
Behavior	Patient decides that she will eat only while sitting at kitchen table with family members.
Factors affecting access to food and food-/nutrition-related supplies	Patient reports that he only has access to microwave and does not have a stove or oven.
Physical activity and function	Patient is unable to perform more than 10 minutes of walking physical activity.
Nutrition-related patient-centered measures	Patient voices that she sees no benefit of improving blood glucose levels.

Table 3.2 Objective Nutrition Assessment Information with Examples

Anthropometric Data	
Height	Weight
BMI	Weight change in 1 month, 6 months
% Usual Body Weight	

Biochemical Lab Results That Are of Nutritional Relevance	
Visceral protein assessment	
Albumin	Transferrin
Prealbumin	Total protein
Hematological assessment	*Lipid assessment*
Hemoglobin	Total cholesterol
Hematocrit	HDL-C
Mean corpuscular volume	LDL-C
Mean corpuscular hemoglobin concentration	Triglycerides
Mean corpuscular hemoglobin	
Total iron binding capacity	
Electrolytes	*Other*
Sodium	Blood glucose
Potassium	Glycated hemoglobin
Chloride	Blood urea nitrogen
Calcium	Serum creatinine
Phosphorus	Urinary protein

Nutrition-Focused Physical Findings	
Oral health	Temporal wasting

Client History	
Age, gender, education level	Previous medical history that involves nutrition

Table 3.3 Successful Steps for Interviewing a Patient for Data Collection

- Maintain an environment that is private and assures confidentiality.
- Establish good patient rapport.
- Respect religious, cultural, and familial values and needs.
- Provide for attentive listening skills.
- Structure questions that are both open and neutral.
- Avoid closed and leading questions.

A thorough diet history can identify the patient's usual pattern of intake; food preferences including ethnic, cultural, and religious influences; and the patient's use of alcohol, complementary and alternative medicine, and vitamin, mineral, herbal, or other types of supplements. Any previous nutrition education or nutrition therapy should be evaluated. Questions should also address food allergies or other food intolerances. In order to obtain accurate information, it is crucial to take into consideration environment, culture, religion, and other socioeconomic and psychosocial factors when any component of the nutrition assessment is performed. Interviewing skills, which are an important factor in collecting data, will be addressed later in this chapter (see Table 3.3).

During this interview process, it is important to determine the availability of resources to both purchase and prepare food. This could include identification of kitchen facilities, food preparation skills, access to grocery stores, and any financial or social assistance (such as Meals on Wheels) that the client utilizes.

Client History Many social factors—including socioeconomic status, social support systems, interactions with other people, and lifestyle—impact nutritional status. It is vital to identify these factors during the interview, because they will impact planning and execution of nutrition education and intervention.

Economic situations directly impact nutritional status. Obviously, nutrition education has little meaning for clients who do not have access to adequate food for themselves or their families. (Box 3.1 explores assessment of food insecurity.) Low food security is defined as "reports of reduced quality, variety, or desirability of diet with little or no indication of reduced food intake." Very low food security occurs when there are "reports of multiple indications of disrupted eating patterns and reduced food intake."[9] It is estimated that 11.1% of households in the United States were food insecure at least some of the time during the year 2007, meaning they did not always have access to enough food for active, healthy lives for all household members because they lacked sufficient money or other resources for food.[10]

Support systems and interaction with family members or caretakers should be considered when designing nutrition education and interventions. For example, if the client eats alone most of the time, appetite may be adversely affected. Lifestyle habits such as smoking, exercise, occupation (if still working), and ability to perform activities of daily living are additional pieces of information that should be ascertained.

Information Regarding Education, Learning, and Motivation During the interview, the ability to communicate should be established. The client's primary language,

---

BOX 3.1　Clinical Applications

Limited Access to Food—Assessing Food Access of Clients and Patients
David H. Holben, PhD, RD, LD; *Professor of Nutrition and Director, Didactic Program in Dietetics; Ohio University, Athens, Ohio*

According to the American Dietetic Association, "negative nutritional and non-nutritional outcomes have been associated with food insecurity in adults, adolescents, and children, including poor dietary intake and nutritional status, poor health, increased risk for the development of chronic diseases, poor psychological and cognitive functioning, and substandard academic achievement."[1] Therefore, as part of the nutrition care process,[2] a comprehensive nutrition assessment includes systematically obtaining timely and appropriate information to identify nutrition-related problems. Three domains have been utilized to cluster nutrition diagnoses and problems—food and/or nutrient intake,

clinical, and behavioral/environmental.[3] The behavioral-environmental cluster includes diagnosis NB-3.2, "Limited access to food."[4] This occurs when clients and patients cannot acquire a sufficient quantity and variety of healthful food.[4]

Limited access to food can stem from various etiologies and be evidenced by a variety of signs and symptoms. Regardless, it may hinder purchasing of food and prevent compliance to a prescribed diet.[1] To provide appropriate nutrition care, food access and availability-related information should be obtained during the assessment. As part of a food and nutrition history, factors such as food and beverage intake (amount/variety), food planning, purchasing,

preparation abilities and limitations, food safety practices, and food/nutrition program utilization, as well as information related to building and utilizing social networks, should be gathered. Anthropometric measurements, including poor growth pattern and weight changes, should also be obtained.

Knowing and understanding the culture of the community will assist the RD in asking appropriate questions related to food access. The interview may include questions related to the following:[1, 5, 6]

- Money for dietary prescriptions and/or medications
- Availability of a refrigerator/freezer, utilities, and transportation

(continued)

- Participation in food and nutrition assistance programs
- Gardening practices
- Other means of acquiring foods (hunting/fishing; begging, borrowing, or stealing food)
- Unintentional weight loss
- Quality of the diet
- Nutrition education need regarding meal planning and purchasing, label reading, and food safety

It has also been suggested that registered dietitians and dietetic technicians screen clients for lack of access to food due to resource constraints by using a single food sufficiency question: "Which of the following statements best describes the food eaten in your household?: (1) Enough of the kinds of food we want to eat, (2) Enough but not always the kinds of food we want to eat, (3) Sometimes not enough to eat; or (4) Often not enough to eat."[7] Incorporating questions from the United States Department of Agriculture food security survey module[8] may also be helpful.

References
1. Holben DH. Position of the American Dietetic Association on Food Insecurity and Hunger in the United States. *J Am Diet Assoc.* 2006;106:446–58.
2. American Dietetic Association Writing Group of the Nutrition Care Process/Standardized Language Committee. Nutrition Care Process and Model Part I: The 2008 Update. *J Am Diet Assoc.* 2008;108:1113–17.
3. American Dietetic Association Writing Group of the Nutrition Care Process/Standardized Language Committee. Nutrition Care Process Part II: Using the International Dietetics and Nutrition Terminology to Document the Nutrition Care Process. *J Am Diet Assoc.* 2008;108:1287–93.
4. American Dietetic Association. *International Dietetics & Nutrition Terminology (IDNT) Reference Manual—Standardized Language for the Nutrition Care Process. Second Edition.* Chicago, IL: American Dietetic Association. 2009.
5. Boeing KL, Holben DH. Self-identified food security knowledge and practices of licensed dietitians in Ohio: Implications for dietetics and clinical nutrition practice. *Top Clin Nutr.* 2003;18:185–91.
6. Holben DH, Myles W. Food Insecurity in the United States: How It Affects Our Patients. *American Family Physician.* 2004;69;1058–63.
7. Kaiser LL, Townsend MS. Food insecurity among US children: Implications for Nutrition and Health. *Top Clin Nutr.* 2005;20:313–20.
8. Bickel G, Nord M, Price C, Hamilton W, Cook J. *Guide to measuring household food security: Revised 2000.* Alexandria (VA): U.S. Department of Agriculture, Food and Nutrition Service. 2000.

as well as the ability to speak, read, write, and comprehend both that primary language and English, can be determined. Educational level, attention span, long-term and recent memory, and readiness to learn can all be established during either the initial interview or subsequent sessions.

The client's comments about previous prescribed diets, medical treatment, and any issues regarding compliance can and will impact acceptance of newly established goals and interventions. Perceptions of health status and the client's desire to improve health or be involved with health care are crucial determinants of successful nutrition education and interventions.

Tools for Data Collection

Client interview and subsequent data collection can be organized using a number of tools and instruments. Many facilities design their own tools and instruments so that they can collect and organize the health information for easy use. Additionally, there are standardized forms such as the DETERMINE checklist[11] Subjective Global Assessment,[12, 13] and numerous other instruments that are used for specific populations (see Appendix D). ADA's evidence-based analysis for nutrition screening identified two such tools from the literature as having the highest **sensitivity** and **specificity**. This means that those who have nutrition problems will be correctly identified and those who do not will fall into the category of low risk.[6] These include the Malnutrition Screening Tool (MST) and the Malnutrition Universal Screening Tool (MUST).[6, 7, 14, 15] Components of the MUST are included in Figure 3.1.

Food- and Nutrition-Related History

In general, food and nutrition information is assessed either by collecting data **retrospectively** or by summarizing data gathered **prospectively**. All methods have imperfections. The accuracy (or **validity**) of the information and the reliability of the data depends on the experience and skill of the clinician as well as the cooperation and accurate reporting of the client. Information gathered will include both dietary information and physical activity patterns. The ultimate goal of collecting dietary information is to determine the nutrient content of food that is consumed and to then assess the appropriateness of the nutritional intake for that particular individual.

Nutrition Care Indicator: Twenty-Four-Hour Recall

When using a **24-hour recall** as the dietary assessment method, the clinician guides the client through a recall of all food and drink that has been consumed in the previous 24-hour period (see Figure 3.2). The clinician asks what food or beverage was consumed most recently prior to the interview and then works backward through the previous 24 hours. The clinician questions the client about activities during the period in order to stimulate the client's memory. At the end of the recall, the clinician reviews the information to verify serving sizes and preparation methods, and to clarify any other uncertainties. The USDA multiple-pass approach, a variation of this method, is a widely accepted and validated method that includes five reviews of information.[16, 17, 18]

Advantages of this method include short administration time, very little cost, and negligible risk for the client. One disadvantage is that a 24-hour recall does not always show typical eating patterns, since day-to-day dietary intake may vary considerably. A second disadvantage is that clients may report information they feel the clinician wants to hear. Research indicates clients may over- or underreport their intake. Additionally, the information obtained may be inaccurate, since this method requires dependence on the client's memory. Accuracy of the method can be strengthened by use of food models, cups, and spoons to improve recall of portion sizes.[16, 17, 19–21]

Figure 3.1 Malnutrition Universal Screening Tool

Step 1 + Step 2 + Step 3

Step 1
BMI score

BMI kg/m²	Score
>20(>30 Obese)	= 0
18.5–20	= 1
<18.5	= 2

Step 2
Weight loss score

Unplanned weight loss in past 3–6 months

%	Score
<5	= 0
5–10	= 1
>10	= 2

Step 3
Acute disease effect score

If patient is acutely ill **and** there has been or is likely to be non-nutritional intake for >5 days

Score 2

If weight and height cannot be obtained, see "MUST" Explanatory Booklet for alternative measurements and use of subjective criteria

Step 4
Overall risk of malnutrition

Add Scores together to calculate overall risk of malnutrition
Score 0 Low Risk   Score 1 Medium Risk   Score 2 or more High Risk

Step 5
Management guidelines

0
Low Risk
Routine clinical care

- Repeat screening
  Hospital–weekly Care
  Homes–monthly
  Community–annually for
  special groups
  e.g. those >75 yrs

1
Medium Risk
Observe

- Document dietary intake
  for 3 days if subject in
  hospital or care home
- If improved or adequate
  intake–little clinical
  concern; if no
  improvement–clinical
  concern-follow local policy
- Repeat screening
  Hospital–weekly Care
  Home–at least monthly
  Community–at least every
  2-3 months

2 or more
High Risk
Treat

- Refer to dietitian, Nutritional
  Support Team or implement
  local policy
- Improve and increase overall
  nutritional intake
- Monitor and review care plan
  Hospital–weekly
  Care Home–monthly
  Community–monthly
- * Unless detrimental or no benefit
  is expected from nutritional
  support e.g.imminent death.

All risk categories:
- Treat underlying condition and provide help and
  advice on food choices, eating and drinking when
  necessary.
- Record malnutrition risk category.
- Record need for special diets and follow local policy.

Obesity:
- Record presence of obesity. For those with
  underlying conditions, these are generally
  controlled before the treatment of obesity.

Re-assess subjects identified at risk as they move through care settings
See *The 'MUST' Explanatory Booklet* for further details and *The 'MUST' Report* for supporting evidence.

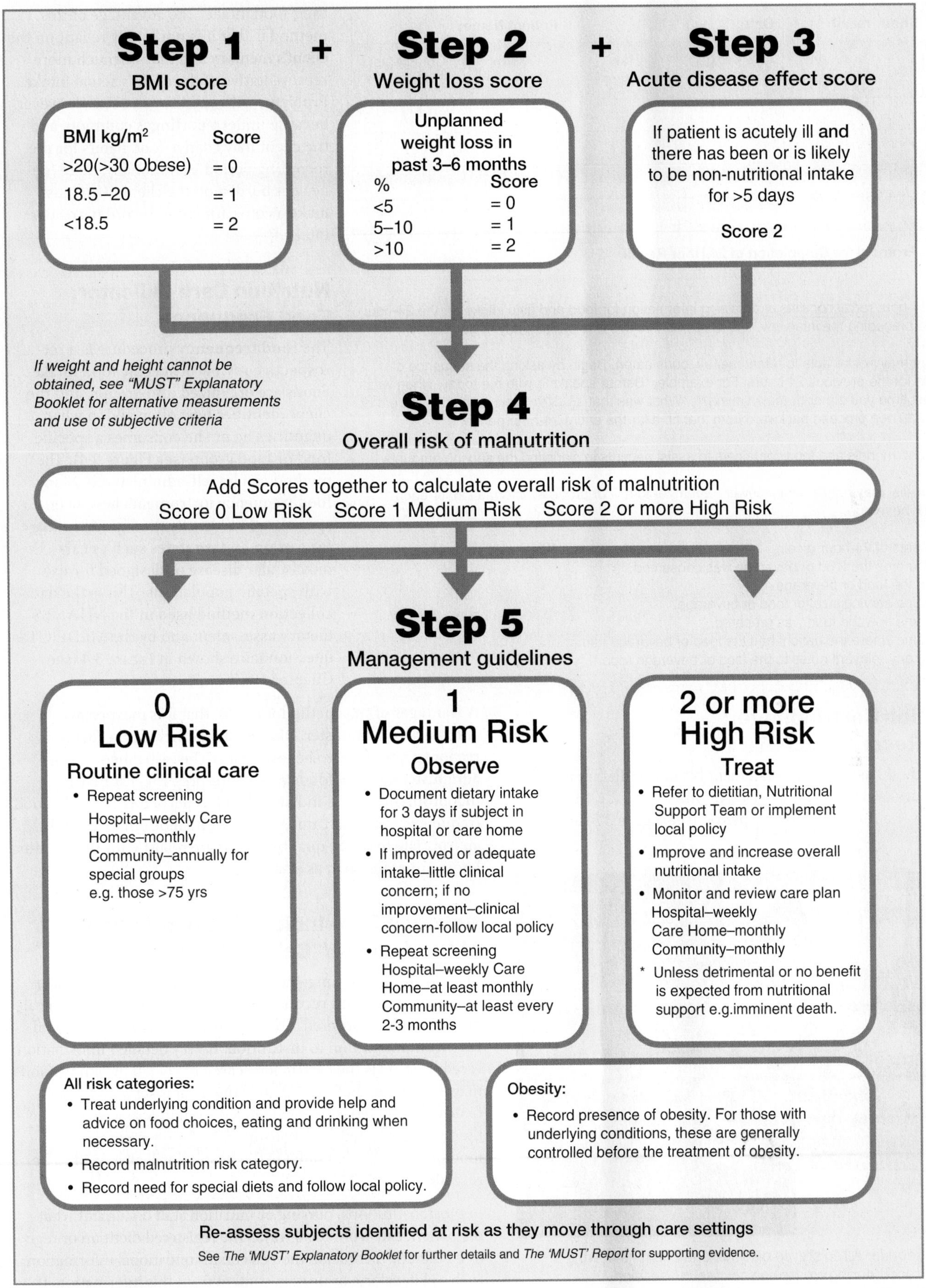

Source: The "Malnutrition Universal Screening Tool' ('MUST') is reproduced here with the kind permission of BAPEN (British Association for Parenteral and Enteral Nutrition).

Figure 3.2 24-Hour Recall Form

24-hour recall		Date:		Patient Name:	
Time	Foods and Beverages	Serving Size	How prepared	Where	Comments:

Sample Protocol for Completion of 24-Hour Recall

1. The 24-hour recall consists of obtaining information for food and fluid intake for the 24-hour period preceding the interview. It is assumed that this is a "typical" day. If not, clarify.

2. Patient may not be able to remember all foods eaten. Begin by asking the sequence of events for the previous 24 hours. For example, "Before speaking with me today, when was the last time you ate or drank anything?"; "What was that ____?"; "How much did you eat of ____?" Then proceed backward from that time for the entire 24-hour period.

3. Use food models and food containers to assist patients in clarifying the serving amounts.

4. A checklist may help the interviewer remember to ask or probe all information for each food or beverage.

Components of 24-hour recall:
• Note the time the food or beverage was consumed.
• Record the food or beverage.
• Determine serving size for food or beverage.
• Determine how the food was prepared.
• Determine where the patient had the food or beverage item.
• Include any relevant notes to the food or beverage report.

Nutrition Care Indicator: Food Record/Food Diary

This method has the client document his or her dietary intake as it occurs over a specified period of time. Typically, the record is kept over a three- or five-day period (see Figure 3.3) and

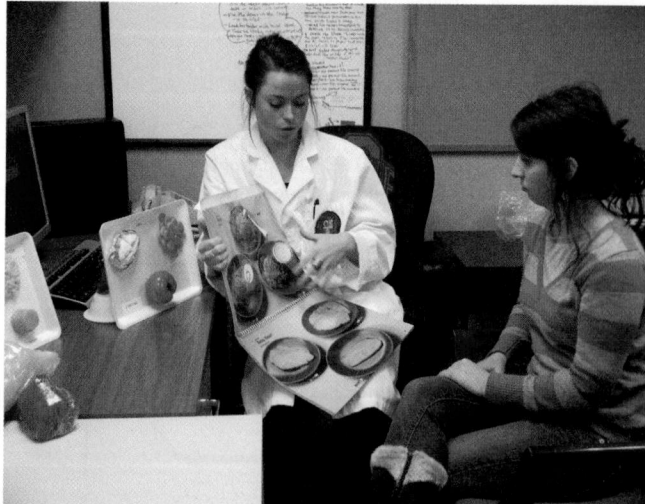

Data Collection: A registered dietitian and client reviewing a 24-hour recall during a nutrition counseling session.
Source: Courtesy of Marcia Nelms.

should include a sampling of both weekdays and weekends. Clients estimate or weigh their food intake. The advantage of this method is that it is not totally reliant on the client's memory and may be much more representative of the client's actual intake. Problems with validity can occur, however, because underreporting is common, and the client may change food habits for the recording period. Additionally, there is a heavier burden on the client, who must make a commitment to record his or her intake.[19]

Nutrition Care Indicator: Food Frequency

The **food frequency** procedure is a retrospective review of specific food intake. Foods are organized into groups, and the client identifies how often and in what quantities he or she consumes a specific food or food group (see Figure 3.4). The method can be self-administered. Many food frequency instruments have been specialized to identify food group intake for certain disease states such as cardiovascular disease or designed for use with specific populations. This is the data collection method used in the NHANES dietary assessment and by the MEDFICTS questionnaire shown in Figure 3.4 (see Chapter 13).[22]

Advantages of this methodology are that it is inexpensive and quick to administer. Disadvantages include the fact that response rates tend to be lower since the instrument is self-administered. Also, foods on the pre-prepared list may be inappropriate for the individual who is participating in the food frequency.[19] The instrument may not include ethnic or child-appropriate foods or quantities that are realistic for those eating larger amounts, such as athletes.

Nutrition Care Indicator: Observation of Food Intake/"Calorie Count"

In an acute care or long-term care setting, actual food intake can be observed and recorded when a kilocalorie (kcal) or kcal-protein count is ordered. Specific procedures for this method vary from institution to institution. If very detailed information is required, as is the case in a research or metabolic study, food may be weighed before and after the meal is served. The patient's food intake is then calculated from differences between the two. Additional food consumed by family members or food brought in from outside the hospital will also need to be recorded.

In most institutions, nursing or nutrition staff document what the patient eats from meal trays. The registered dietitian or registered dietetic technician then calculates nutritional information such as kilocalorie or protein content from this information. If the RD or DTR collects this information, it provides an excellent

Figure 3.3 Food Diary

Date/Time	List all foods and drinks	Amount/serving size	Preparation/ cooking method	Seasonings/ Condiments	Where did you eat?	Who were you with?

Directions for Use of Food Diary

READ THE FOLLOWING INSTRUCTIONS CAREFULLY
Record amounts and descriptions of ALL food and drink (including water) for three consecutive days. These days should be "typical" to the way you eat on a normal basis. Please do not try to change your eating habits on the days you are recording. **Please pick two weekdays and one weekend day that are most like your usual daily intake.**

Helpful Hints:
- Record your intake immediately after you have eaten and NOT at the end of the day. This makes it much easier to remember and to record accurately.
- Include all meals and snacks, granola bars, sandwiches, chocolate, sweets, ice cream, fruits—whatever you eat.
- Include all drinks (e.g., water, tea, coffee, beer, sports drinks, and fruit juice).
- Record any additions to food such as mustard, ketchup, mayonnaise, cream or sugar, steak sauce, salsa, dressings, gravy, pickles, honey, or butter.

Describe foods accurately:
- Record cooking methods (e.g., fried, baked, broiled, grilled, frozen, canned, added water, low sodium, and the amount of fat or oil used for cooking).
- Record brand names and the descriptions (e.g., KRAFT, General Mills, Breyers, Campbell's, Del Monte, and whether regular, 2% reduced fat, light, fat free, low carb, or sweetened).
- Name the types of cheese, fish, or meat (e.g., cheddar, American, cod, tilapia, ground, sirloin, shredded).

Describe the amounts as accurately as possible:
- To help with measuring portion size, try to avoid terms such as "one bowl" or "a handful."
- Visualize the following comparisons when figuring portion size:
  - 3 ounces of meat is about the size of a deck of cards or audiotape cassette.
  - A medium-size piece of fruit is about the size of a tennis ball.
  - 1 ounce of cheese is about the size of 4 stacked dice.
  - 1/2 cup of ice cream is about the size of a tennis ball.
  - 1 cup of mashed potatoes or broccoli is about the size of your fist.
  - 1 teaspoon of butter is about the size of the tip of your thumb.
- Use weights marked on packages (e.g., half of a 425-gram can of corn; half of a 16-ounce can; half of a 6-ounce bag of frozen corn).
- Use cups, teaspoons, and tablespoons to record amounts.

 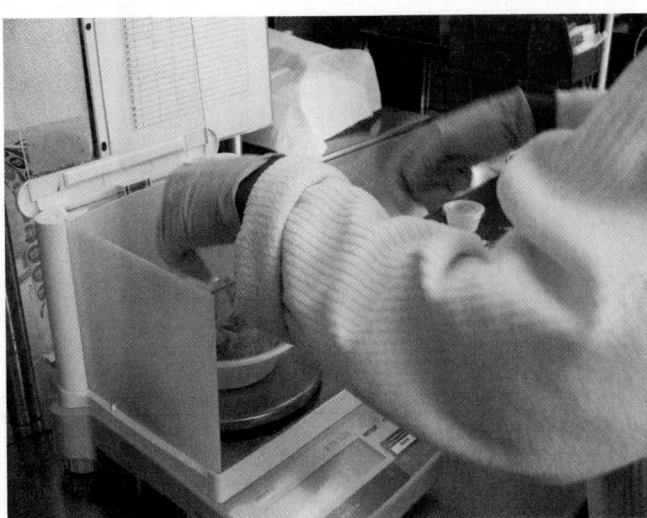

In the clinical setting, visual observation of food consumed is used to estimate nutritional intake. In the research setting, actual weighing of the food prior to and after the meal provides a precise measurement.

Source: Courtesy of Marcia Nelms.

Figure 3.4 Example of a Food Frequency: MEDFICTS

In each food category for both Group 1 and Group 2 foods check one box from the "Weekly Consumption" column (number of servings eaten per week) and then check one box from the "Serving Size" column. If you check Rarely/Never, do not check a serving size box. See next page for score.

	Weekly Consumption			Serving Size			Score
	Rarely/ never	3 or less	4 or more	Small <5 oz/d 1 pt	Average 5 oz/d 2 pts	Large >5 oz/d 3 pts	

Food Category

Meats

- Recommended amount per day: 5 oz (equal in size to 2 decks of playing cards).
- Base your estimate on the food you consume most often.
- Beef and lamb selections are trimmed to 1/8" fat.

Group 1. 10 g or more total fat in 3 oz cooked portion
Beef—Ground beef, Ribs, Steak (T-bone, Flank, Porterhouse, Tenderloin), Chuck blade roast, Brisket, Meatloaf (w/ground beef), Corned beef
Processed meats—1/4 lb burger or lg. sandwich, Bacon, Lunch meat, Sausage/knockwurst, Hot dogs, Ham (bone-end), Ground turkey
Other meats, Poultry, Seafood—Pork chops (center loin), Pork roast (Blade, Boston, Sirloin), Pork spareribs, Ground pork, Lamb chops, Lamb (ribs), Organ meats†, Chicken w/skin, Eel, Mackerel, Pompano

		3 pts	7 pts	✗ 1 pt	2 pts	3 pts	_____

Group 2. Less than 10 g total fat in 3 oz cooked portion
Lean beef—Round steak (Eye of round, Top round), Sirloin‡, Tip & bottom round‡, Chuck arm pot roast‡, Top Loin‡
Low-fat processed meats—Low-fat lunch meat, Canadian bacon, "Lean" fast food sandwich, Boneless ham
Other meats, Poultry, Seafood—Chicken, Turkey (w/o skin)§, most Seafood†, Lamb leg shank, Pork tenderloin, Sirloin top loin, Veal cutlets, Sirloin, Shoulder, Ground veal, Venison, Veal chops and ribs‡, Lamb (whole leg, fore-shank, sirloin)‡

				✗		6 pts	_____

Eggs – Weekly consumption is the number of times you eat eggs each week
Check the number of eggs eaten each time

				1	2	≥3	
Group 1. Whole eggs, Yolks		3 pts	7 pts ✗	1 pt	2 pts	3 pts	_____
Group 2. Egg whites, Egg substitutes (1/2 cups)							_____

Dairy

Milk—Average serving 1 cup

Group 1. Whole milk, 2% milk, 2% buttermilk, Yogurt (whole milk)		3 pts	7 pts ✗	1 pt	2 pts	3 pts	_____
Group 2. Fat-free milk, 1% milk, Fat-free buttermilk, Yogurt (Fat-free, 1% low fat)							_____

Cheese—Average serving 1 oz

Group 1. Cream cheese, Cheddar, Monterey Jack, Colby, Swiss, American processed, Blue cheese, Regular cottage cheese (1/2 cup), and Ricotta (1/4 cup)		3 pts	7 pts ✗	1 pt	2 pts	3 pts	_____
Group 2. Low-fat & fat-free cheeses, Fat-free milk mozzarella, String cheese, Low-fat, Fat-free milk & Fat-free cottage cheese (1/2 cup) and Ricotta (1/4 cup)							_____

Frozen Desserts—Average serving 1/2 cup

Group 1. Ice cream, Milk shakes		3 pts	7 pts ✗	1 pt	2 pts	3 pts	_____
Group 2. Low-fat ice cream, Frozen yogurt							_____

Source: NCEP, National Heart, Lung and Blood Institute, NIH Reference: NIH Publication no. 02-5215, Diet Appendix A; available from URL: http://www.nhlbi.nih.gov/guidelines/cholesterol/atp3_rpt.htm

Figure 3.4 Example of a Food Frequency: MEDFICTS (*Continued*)

	Weekly Consumption			Serving Size			Score
	Rarely/ never	3 or less	4 or more	Small <5 oz/d 1 pt	Average 5 oz/d 2 pts	Large >5 oz/d 3 pts	

Food Category

Frying Foods – Average servings: see below. This section refers to method of preparation for vegetables and meat.

Group 1. French fries, Fried vegetables (1/2 cup), Fried chicken, fish, meat (3 oz)	☐	☐ 3 pts	☐ 7 pts	**×** ☐ 1 pt	☐ 2 pts	☐ 3 pts	_____
Group 2. Vegetables, not deep fried (1/2 cup), Meat, poultry, or fish—prepared by baking, broiling, grilling, poaching, roasting, stewing: (3 oz)	☐	☐	☐	☐	☐	☐	_____

Baked Goods – 1 Average serving

Group 1. Doughnuts, Biscuits, Butter rolls, Muffins, Croissants, Sweet rolls, Danish, Cakes, Pies, Coffee cakes, Cookies	☐	☐ 3 pts	☐ 7 pts	**×** ☐ 1 pt	☐ 2 pts	☐ 3 pts	_____
Group 2. Fruit bars, Low-fat cookies/cakes/pastries, Angel food cake, Homemade baked goods with vegetable oils, breads, bagels	☐	☐	☐	☐	☐	☐	_____

Convenience Foods

Group 1. Canned, Packaged, or Frozen dinners: e.g., Pizza (1 slice), Macaroni & cheese (1 cup), Pot pie (1), Cream soups (1 cup), Potato, rice & pasta dishes with cream/cheese sauces (1/2 cup)	☐	☐ 3 pts	☐ 7 pts	**×** ☐ 1 pt	☐ 2 pts	☐ 3 pts	_____
Group 2. Diet/Reduced calorie or reduced fat dinners (1), Potato, rice & pasta dishes without cream/cheese sauces (1/2 cup)	☐	☐	☐	☐	☐	☐	_____
Table Fats—Average serving: 1 Tbsp **Group 1.** Butter, Stick margarine, Regular salad dressing, Mayonaisse, Sour cream (2 Tbsp)	☐	☐ 3 pts	☐ 7 pts	**×** ☐ 1 pt	☐ 2 pts	☐ 3 pts	_____
Group 2. Diet and tub margarine, Low-fat & fat-free salad dressing, Low-fat & fat-free mayonnaise	☐	☐	☐	☐	☐	☐	

Snacks

Group 1. Chips (potato, corn, taco), Cheese puffs, Snack mix, Nuts (1 oz), Regular crackers (1/2 oz), Candy (milk chocolate, caramel, coconut) (about 1 1/2 oz), Regular popcorn (3 cups)	☐	☐ 3 pts	☐ 7 pts	**×** ☐ 1 pt	☐ 2 pts	☐ 3 pts	_____
Group 2. Pretzels, Fat-free chips (1 oz), Low-fat crackers (1/2 oz), Fruit, Fruit rolls, Licorice, Hard candy (1 med piece), Bread sticks (1–2 pcs), Air-popped of low-fat popcorn (3 cups)	☐	☐	☐	☐	☐	☐	_____

Total from page 1 _____

† Organ meats, shrimp, abalone, and squid are low in fat, but high in cholesterol.

‡ Only lean cuts with all visible fat trimmed. If not trimmed of all visible fat, score as if in Group 1.

¥ Score 6 pts if this box is checked.

§ All parts not listed in Group 1 have <10 g total fat.

Total from page 2 _____

Final Score _____

To Score: For each food category, multiply points in weekly consumption box by points in serving size box and record total in score column. If Group 2 foods checked, no points are scored (except for Group 2 meats, large serving = 6 pts).

Example:

☐	☐ 3 pts	☑ 7 pts	**×** ☐ 1 pt	☐ 2 pts	☑ 3 pts	21 pts

Add score on page 1 and page 2 to get final score.

Key:
≥70   Need to make some dietary changes
40–70 Heart-Healthy Diet
<40   TLC Diet

opportunity to assess the patient's understanding of any dietary interventions and to teach specific nutrition information such as portion control strategies or nutrient content of the meal. This method also allows the RD or DTR to establish rapport and determine food preferences and tolerances.

Evaluation and Interpretation of Dietary Analysis Information

After data are collected and analyzed, it is the clinician's job to compare the information to established scientific criteria. These criteria may include the individual patient's needs and may be as general as the U.S. Dietary Guidelines or as specific as milligrams of vitamin C that should be consumed. Limitations of each assessment method make the assessment of intake an estimation rather than an exact measurement, but in general, the appropriate criteria are determined by how the information will be used. For example, in order to determine whether an intervention to change the patient's food choices has improved intake, a direct observation of food intake may be made and then analyzed by simply estimating the energy value and protein content of the food recorded, using an established method such as the Exchange Lists for Diabetics. This is the data that can be used to support the documentation of a nutrition problem, shown in the Sample PES Statement box.

Sample PES Statement for Intake Domain: Inadequate protein intake related to aversion to meat as evidenced by reported intake of 45% of estimated protein requirements of 70–75 g/day.

Nutrition Care Criteria: Evaluation and Interpretation Using the U.S. Dietary Guidelines

The U.S. Dietary Guidelines, published jointly by the USDA and the U.S. Department of Health and Human Services (USDHHS), provide general recommendations for dietary intake that promote health and prevent disease.[23] These guidelines are based on decades of nutrition research and reflect the most up-to-date understanding of nutritional requirements. These guidelines have been revised five times with the newest release planned for 2010. Although the U.S. Dietary Guidelines are an important tool in nutrition education and planning, they are not the most efficient tool available for evaluating an individual's diet and really only provide a very broad overview. These guidelines are much more useful in setting goals for nutrition and health education or in translating nutrition research into simpler terms for consumers.

Nutrition Care Criteria: Evaluation and Interpretation Using USDA's MyPyramid

Analysis using MyPyramid will quantify food consumed into quantities from each of the groups on the food pyramid.[24] These data give the clinician an overview of adequacy, variety, moderation, and balance. A MyPyramid analysis does not quantify macro- or micronutrients; on the other hand, when one simply looks at total energy and protein intakes, there is no way to determine the source of these nutrients. Using MyPyramid in conjunction with macronutrient data allows the overall quality of the diet to be assessed (see Box 3.2).

BOX 3.2   Clinical Applications

Comparison Assessment of Dietary Intake

Diet 1: 1 egg, 1 slice of toast, coffee with 2 Tbsp half-and-half
2 oz ham sandwich on 2 slices of white bread with 1 Tbsp mayonnaise
1 oz potato chips
4 oz chicken breast fried, 1 roll, iced tea with 2 Tbsp sugar

Diet 2: 1 cup whole-grain cereal, 1 banana, 1 c. skim milk
2 oz ham sandwich, ½ cup chopped fresh vegetables on 2 slices of whole-grain bread with 1 Tbsp mustard, 1 oz pretzels, 1 medium apple
4 oz chicken breast baked, 1 cup fresh broccoli, asparagus, carrots stir-fried with 1 cup brown rice

Sample 24-hour recall	Analysis for energy and protein (using the dietary exchanges)		Analysis for components of food intake pattern for 1600 kcal (using MyPyramid)	
Diet 1	Nutrient	Total	Grains (5 oz with ≥3 oz whole grain)	4 – 0 whole grain
	Kcalories	1099.3	Vegetables (2 cups)	0
	Pro (g)	53.08	Fruits (1.5 cups)	0
	Fat (g)	51.15	Milk (3 cups)	0
	Carb (g)	106.2	Meat and Beans (5 oz)	6
			Solid Fats/ Added Sugars	3 +
Diet 2	Nutrient	Total	Grains (6 oz with 3 oz whole grain)	8.2 – 5 whole grain
	Kcalories	1087.86	Vegetables (2.5 cups)	1.3
	Pro (g)	70	Fruits (2 cups)	2.4
	Fat (g)	24.02	Milk (3 cups)	3
	Carb (g)	184.18	Meat and Beans (5 oz)	5.4

Nutrition Care Criteria: Evaluation and Interpretation Using the Diabetic Exchanges/Carbohydrate Counting

This method of analysis uses the dietary exchanges established jointly by the American Diabetes Association and the American Dietetic Association.[25] Use of the exchanges provides a quick, rough estimate of kcal, protein, carbohydrate, and fat in the diet. Carbohydrate counting concentrates on estimation of carbohydrate only and is used primarily by individuals with diabetes who are balancing their insulin dosages with dietary intake of carbohydrate. (See Chapter 17.)

Nutrition Care Criteria: Evaluation and Interpretation Using Individual Nutrient Analysis

The United States Department of Agriculture (USDA) first published food composition values in 1896. The Nutrient Data Laboratory of the USDA maintains and updates these data banks. Historically, this information has been published in a series of *Agriculture Handbooks*, but now this information is available only online (http://www.ars.usda.gov). Even though this database contains information for over 6000 foods and 100 nutrients, it is difficult for its administrators to keep up with new name-brand foods that are constantly being produced.

Other sources of data can be found on nutrition labels, from food manufacturers, and from some restaurants and fast-food establishments. Data on food labels may not be 100% reliable, especially for products from small companies or imported foods.

Computerized Dietary Analysis Nutrition professionals have access to many sources of computerized data analysis programs. These programs vary significantly in terms of number of food items, nutrients included in the program, sources for the nutrient database, how often the database is updated, cost of the program, and ease of use.

Nutrition Care Criteria: Evaluation and Interpretation Using Daily Values/ Dietary Reference Intakes

One method of evaluating dietary analysis for macro- and micronutrient amounts is use of the Daily Values (DV) and the Dietary Reference Intakes (DRI). DRI are standards established by the National Academy of Sciences. These standard reference values allow evaluation of energy, protein, vitamin, and mineral intake for healthy people. There are four different sets of standards within the DRI: Adequate Intakes (AI), Recommended Dietary Allowances (RDA), Tolerable Upper Intake Levels (UL), and Estimated Average Requirements (EAR). The RDA, AI, and UL can be used to assess diets of individuals.[26, 27]

It is important when using these standards to understand the context in which the references are established. Values for RDA are determined at approximately two standard deviations above the average (mean) requirement within the healthy population. This margin of safety allows the value of the RDA to meet the needs of most healthy people. Therefore, if the evaluated diet falls below the RDA or AI for a specific nutrient, it does not necessarily mean the client is deficient in this nutrient. Diagnosis of specific nutrient deficiencies would require additional confirmation using other components of nutrition assessment. Still, the DRI serve as important benchmarks for evaluating the patient's dietary intake, not only from food, but also from dietary supplements. The UL values can assist in assessing a patient's use of supplements and whether their current dosage poses any health risk.

In the clinical setting, many patients have specific diseases or medical conditions that may have unique nutrient requirements. For example, an individual with a burn injury may require significantly higher doses of vitamin C and zinc to ensure appropriate wound healing. Additionally, medications and treatments may alter absorption, utilization, excretion, or storage of specific nutrients. In these situations, patients may need higher or lower levels of these nutrients. The DRI are established for the healthy population and hence may not be appropriate in all clinical situations. Nonetheless, they can always be used as a starting point in dietary evaluation, and as the medical condition and subsequent nutrition therapy are established, adjustments can be made for specific nutrient requirements.

The DV were established by the Food and Drug Administration to assist consumers in interpreting nutrition labeling information. These standards use the RDA and U.S. Dietary Guidelines as their theoretical base and set target goals for fat, saturated fat, cholesterol, total carbohydrate, fiber, sodium, potassium, and protein. These reference standards are expressed with a 2000- and 2500-kcal reference diet. In general, though, the DV are much more useful to the consumer purchasing groceries than the dietitian performing a nutrition assessment. The dietitian will use much more specific, individualized reference data.

Anthropometric/Body Composition Measurements

"Anthropometry is the measurement of body size, weight, and proportions."[19] *Body composition* refers to the distribution of body compartments (e.g., muscle mass and body fat) as part of the total body weight. Evaluating both anthropometric and body composition data allows the clinician to fully assess these compartments.

Because nutrition is a crucial component of normal growth and development, it is accepted practice to measure body compartments in order to evaluate infants and children for appropriate growth. In a normal, healthy individual, the relationships among body storage compartments are relatively stable. But when disease or stress is present, changes in the storage compartments are an important component of determining nutritional status and risk. Results of anthropometric assessment can be used to both identify goals for nutrition intervention and monitor changes that occur as a result of either those interventions or continued effects of disease and stress. NHANES, the U.S. nationwide survey used to obtain health and nutrition information, collects a variety of anthropometric measurements and provides standardized procedures for practitioners and researchers to apply as they use these techniques in health and disease assessment.[28]

Anthropometrics

Nutrition Care Indicator: Height/Stature

Measurement of supine or standing height is necessary for monitoring growth of infants and children and for interpretation of weight in adults. For children under the age of two years, length is measured recumbently using a length board. This device has a stationary headboard and a movable footboard (Figure 3.5). This measurement requires two clinicians, one of whom holds the child's head touching the headboard while the other extends the leg and bottom of the heel to the footboard. Length is recorded to the nearest 0.1 cm. Over the age of two years, standing height is measured using a tape measure or **stadiometer** (see Figure 3.6). Procedures for measuring height include having the client stand barefoot and look forward with shoulders, buttocks, and heels touching the vertical surface of either a wall or the stadiometer with the Frankfort plane. This ensures the head is not tilted incorrectly.[19]

When a client cannot stand for the measurement of height, there are several estimation methods that may be used. Arm span is one method of height estimation for adults. The client extends the arms from the body at a 90-degree angle and distance is measured between the tips of the two middle fingers. The length of the dominant arm can be measured in the same fashion and multiplied by 2 to estimate height. A limitation of this method is that it is an estimation of maximum adult height and not actual, current height.

Knee height is another method of height estimation (Figure 3.7). Measurement of knee height, using a knee-height caliper, can be taken when the client is sitting or in a supine position. (Measuring supinely is considered to be more accurate.) The client lies supine with right knee and ankle flexed to 90 degrees. The clinician should place the fixed portion of the caliper under the heel and position the other blade over the anterior portion of the thigh above the knee. The shaft of the caliper is parallel to the tibia. The measurement (repeated two to three times) is recorded to the nearest 0.1 cm. Height is then estimated using the following equations:[29]

Figure 3.5 Measuring Infant Length

Children under the age of 2 are measured using a stationary headboard and movable footboard.

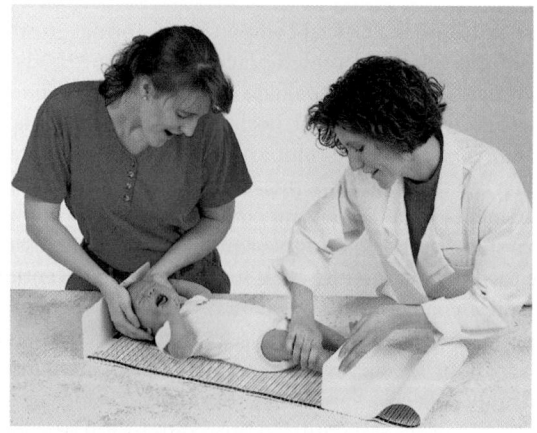

Source: E. Whitney and S. Rolfes, *Understanding Nutrition*, 10e, Copyright © 2005, p. 591.

Figure 3.6 Stadiometer

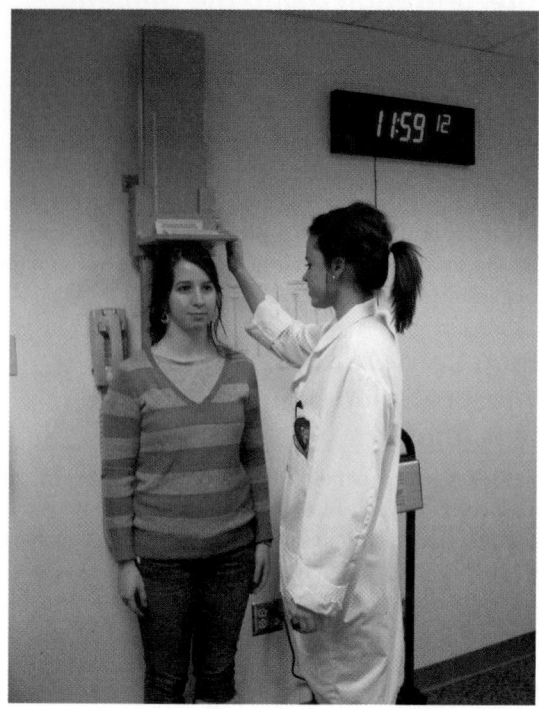

Source: Courtesy of Marcia Nelms.

Age 18–60 years:

- White male = 71.85 + (1.88 × knee height)
- Black male = 73.42 + (1.79 × knee height)
- White female = 70.25 + (1.87 × knee height) − (0.06 × age)
- Black female = 68.10 + (1.86 × knee height) − (0.06 × age)

Age 60–80 years:

- White male = 59.01 + (2.08 × knee height)
- Black male = 95.79 + (1.37 × knee height)
- White female = 75.00 + (1.91 × knee height) − (0.17 × age)
- Black female = 58.72 + (1.96 × knee height)

Height or length has been noted to be one of the most inaccurate measures. In clinical settings, it is often either estimated or recorded from the patient's memory.[30] Accurate measurement is crucial, because height is used to interpret weight, measure growth for children, calculate energy and protein requirements, and calculate creatinine-height index.

Nutrition Care Indicator: Weight

Weight can be measured using a variety of scales, including balance beam and electronic scales. Bathroom scales and those that are moved frequently are not recommended due to problems with calibration. Wheelchair and bed scales are available for nonambulatory patients. Ideally, the client should be weighed with minimal clothing and without shoes, at the same time daily, and after urination.

For those patients with amputation, weight has historically been adjusted using the following factors:[31]

Figure 3.7 Knee-Height

Knee-height calipers are used when height must be measured for an individual who cannot stand.

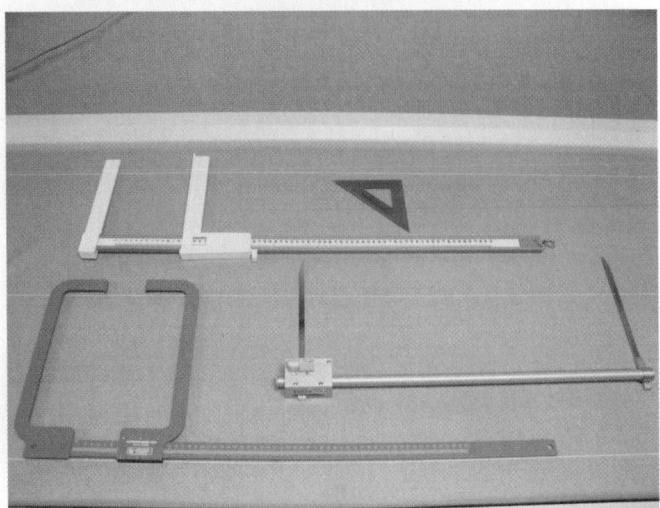

Source: Courtesy of Marcia Nelms.

- Hand: 0.8%
- Forearm and hand: 3.1%
- Entire arm: 6.5%
- Foot: 1.8%
- Lower leg (below knee) and foot: 7.1%
- Entire leg: 18.6%

For example, for an individual who has had an entire leg amputated and currently weighs 165 lbs, weight would be adjusted by using the following equation:

Adjusted body weight = actual measured weight divided by 100 − % amputation. The whole equation then is multiplied by 100.

$$\frac{165 \text{ lbs}}{(100 - 18.6)} \times 100 = 202 \text{ lbs is the estimated body weight.}$$

This calculation allows for weight to be compared to a criterion standard such as body mass index.

Weight is the most common measure of anthropometrics. Unfortunately, it is a gross measurement of all body compartments and does not distinguish body composition or fluid shifts. Nevertheless, due to its common availability and its relationship to growth, development, and health, it is a vital component of nutrition assessment.

Nutrition Care Criteria: Evaluation and Interpretation of Height and Weight in Infants and Children

Growth Charts Weight and height for infants and children are evaluated using growth charts developed by the Centers for Disease Control and Prevention (CDC) and the National Center for Health Statistics[32] (see Appendix). Determination of height for age and weight for age allows comparison of an infant or child to a reference population. Data for the CDC

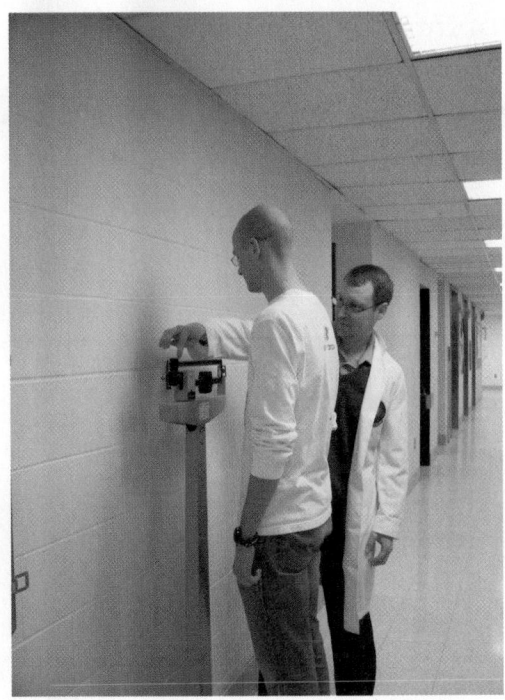

Weight is the most common anthropometric measure.
Source: Courtesy of Marcia Nelms.

growth charts are based on the National Health and Nutrition Examination Survey, and were most recently updated in 2000. When infants and children are either <3rd percentile or >97th percentile, further assessment should be made to confirm any health problems. There are specific clinical diagnoses, such as genetic and endocrine disorders, that negate use of these standard growth charts. Alternative growth charts have been developed for children with specific health care needs.

Weight for height and **percent weight for height** can also be evaluated using CDC growth charts. These measurements allow evaluation to be independent of age and can be used to monitor acute malnutrition (<5th percentile) or the incidence of obesity (>95th percentile).

Body Mass Index Revision of the CDC growth charts in 2000 added the measurement of body mass index (BMI). BMI is weight (kg)/[height (m)]^2.[32] Calculation and interpretation of BMI in children and adolescents has increased for this population in the past several years. Assessment does not use adult standards, however; instead, overweight is defined as 85th to <95th percentile of BMI-for-age, obese is defined as ≥95th percentile of BMI-for-age, and underweight is defined as <5th percentile.

Nutrition Care Criteria: Evaluation and Interpretation of Height and Weight in Adults

Usual Body Weight In the clinical setting, variations from usual body weight have been strongly linked to nutritional risk and health complications. Such variation may be more clinically useful than comparison to ideal body weight standards. In general, an adult is considered at nutritional risk if there is a >5% unexplained weight change in less than one month or >10% in a six-month period (see Table 3.4). The absolute number of lbs or kg change is also used to describe rapid weight gain as well.

Table 3.4 Interpretation of Unintentional Weight Change

Time Frame	Significant Weight Loss	Severe Weight Loss
1 week	1–2% UBW	>2% UBW
1 month	5% UBW	5% UBW
3 months	7.5% UBW	>7.5% UBW
6 months	10% UBW	>10% UBW

Percent Usual Body Weight and Percent Weight Change Percent usual body weight is calculated as:

$$\frac{\text{current weight}}{\text{usual body weight}} \times 100$$

Percent weight change is calculated as:

$$\frac{\text{usual body weight} - \text{present weight}}{\text{usual body weight}} \times 100$$

or:

$$100 - \% \text{ usual body weight} = \% \text{ change}$$

Reference Weights Creation of height-weight tables and estimation of recommended body weight from these data were initially conducted by the life insurance industry. The data have been criticized and their use is discouraged because the original population does not correlate with the current population in the United States. Other criticisms of the development of these standards include the assumption of the weight of an individual's clothes, use of a standard heel height on shoes, and estimation of frame size. The only situation where these standards are routinely used is in the spinal cord-injured population (see Chapter 20).

In some clinical settings, reference body weight is calculated using the Hamwi equation even though it does not adjust for differences in age, race, or frame size.[33] Calculation of body mass index (see the next two sections) is considered to be the only validated method for estimating desirable or ideal body weight. Though the Hamwi method (shown here) has been used historically, the validity is questionable.

Men: 106 lbs for 5 foot + 6 lbs per inch over 5 foot

or 6 lbs per inch under 5 foot

Women: 100 lbs for 5 foot + 5 lbs per inch over 5 foot

or 5 lbs per inch under 5 foot

Body Mass Index (BMI) BMI or Quetelet's Index, as stated previously, is calculated as:

- weight (kg) / [height (m)]^2

or

- weight (lbs) / [height (inches)]^2 $\times$ 704.5.

The use of BMI has been correlated with overall mortality and nutritional risk. It still does not estimate body composition, but it is better at indicating obesity than mere height and weight alone. A client with a BMI $\geq$25 is considered to be overweight, and client with a BMI = 18.5 is considered to be underweight. The Expert Panel on Healthy Weight established target healthy weight to be a BMI between 18 and 25 kg/m^2 for adults.[34] See

Table 3.5 Interpretation of BMI in Adults

For adults, classification of weight based on body mass index by the National Institutes of Health is as follows:

BMI	Weight	Status Health Risk
Below 18.5	Underweight	With <16 suggesting possible eating disorder and other disease risk*
18.5–24.9	Normal	Healthy, low health risk
25.0–29.9	Overweight	Associated with increased risk of disease**
30.0 and above	Obese	Associated with further increased risk of disease

*Diseases associated with underweight include chronic obstructive pulmonary disease, cancer, and congestive heart failure.
**Diseases associated with overweight and obesity include diabetes mellitus, cardiovascular disease, and hypertension.

Tables 3.5 and 3.6 for complete interpretation of BMI in adults and children, respectively.

Waist Circumference Waist circumference of $\geq$40 inches (102 cm) for men or $\geq$35 inches (88 cm) for women is considered to be predictive of obesity and chronic disease risk in Caucasian, African-American, Hispanic, and Native American populations. Within Asian populations, risk is defined at $\geq$90 cm in men and $\geq$80 cm in women. Fat accumulation, primarily in the abdominal region, has been linked to an increased risk of type 2 diabetes mellitus and other obesity-related diseases. The current recommendations from the Endocrine Society Practice Guideline (2008) and NHLBI/AHA are that waist circumference should be used as a part of routine physical examination.[35, 36] See Figure 3.8 for guidelines for measurement of waist circumference.

Body Composition

Height and weight, though crucial components of nutrition assessment, are only gross measurements and cannot distinguish among body compartments. It is not unusual, then, for very muscular individuals to fall into an overweight or obese categorization when only height and weight are used in the nutrition assessment. Measurements of skinfolds, waist circumference, and other body composition techniques allow the clinician to make a more thorough and complete nutrition assessment. However, these are not routinely done in most clinical settings where acute care is the primary goal. In settings where clients are followed for a length of time, these assessments are clinically useful.

Table 3.6 Interpretation of BMI in Children

BMI-for-Age Percentile	Interpretation
<5th percentile	At risk for underweight
5–85th percentile	Normal
$\geq$85th percentile	Overweight
$\geq$95th percentile	Obese

Sources: Barlow SE and the Expert Committee. (2007) Expert Committee recommendations regarding the prevention, assessment, and treatment of child and adolescent overweight and obesity: Summary Report. Pediatrics. 120(4):S164-S192. Centers for Disease Control. National Health and Nutrition Examination Survey. Available at http://www.cdc.gov/nchs/nhanes.htm.

Figure 3.8 Measuring Waist Circumference

The line shows the appropriate position for the tape measure.

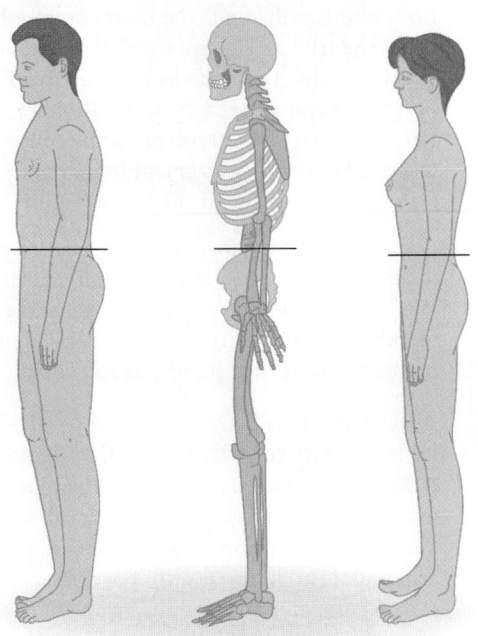

High-Risk Waist Circumference

Men	>40 in (102 cm)
Women	>35 in (88 cm)

Source: National Heart, Lung, and Blood Institute Guidelines on Overweight and Obesity: Electronic Textbook. http://www.nhlbi.nih.gov/guidelines/obesity/e_txtbk/txgd/4142.htm

Body composition refers to distribution and size of all components contributing to total body weight. In most clinical settings, body composition refers to two major components of total body weight: fat mass and fat-free mass. A more thorough definition would include fat mass, total body water, osseous mineral (bone and teeth), and protein. Nutrition professionals are most concerned about metabolically active tissue and fluid

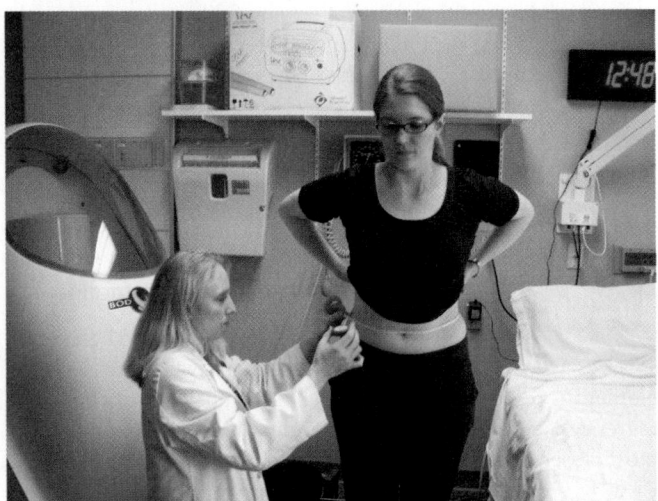

Waist circumference should be measured around the abdomen at the level of the iliac crest.
Source: Courtesy of Marcia Nelms.

Figure 3.9 Abdominal Skinfold

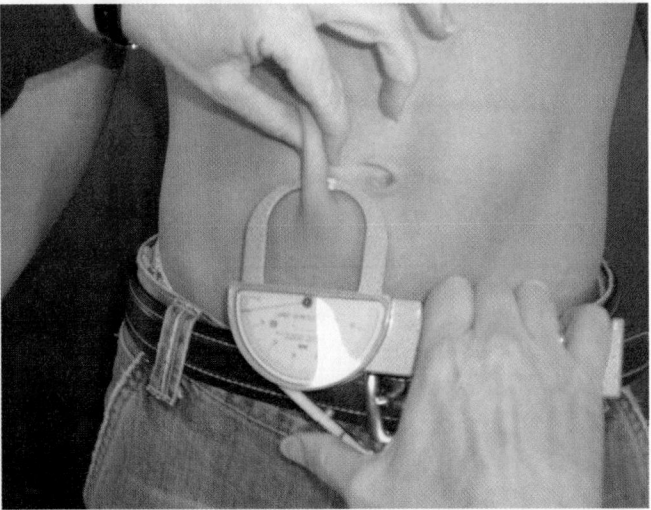

Source: © 2007 Cade Fields-Gardner

status. Techniques that allow assessment of these compartments are necessary for the design of appropriate nutrition interventions.

Measurement techniques such as skinfolds and bioelectrical impedance use portable equipment and can be easily accomplished in any setting. Other measurements using hydrodensitometry, air displacement plethysmography, or dual energy X-ray absorptiometry (DXA) are used in outpatient and research settings but are less practical for use in the daily acute care setting. These assessment methods are covered in this text because the basic understanding of these techniques allows one to compare various assessment methods in a variety of populations.[37]

Nutrition Care Indicator: Skinfold Measurements Skinfold measurement is used to estimate energy reserves—both fat and somatic protein—in subcutaneous tissue. As shown in Figure 3.9, skinfold measurement involves measuring a double fold of skin and subcutaneous adipose

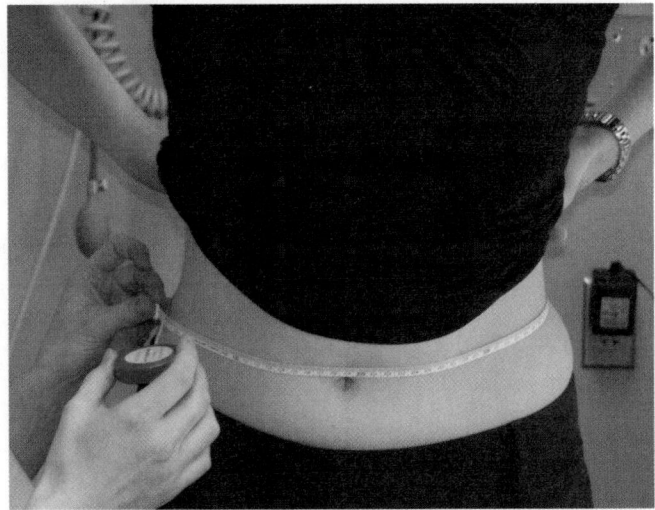

Figure 3.10 Mid Upper Arm Muscle Area in Adults

Triceps skinfold is measured halfway between the olecranon and acromial processes on the nondominant arm, as shown here. Midarm circumference is measured with a tape measure, and these measurements are inserted into the formulas below to calculate arm muscle area.

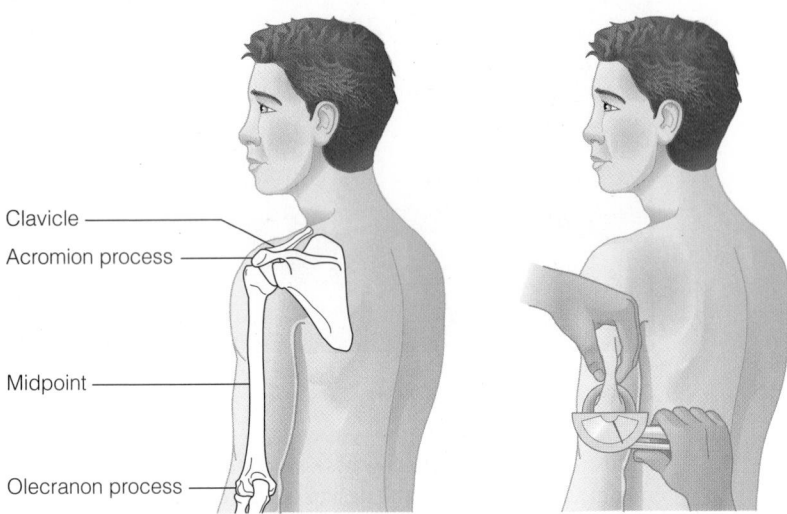

Clavicle

Acromion process

Midpoint

Olecranon process

Source: SR Rolfes, K Pinna & E Whitney, *Understanding Normal and Clinical Nutrition* 7e, Fig E7, p E8.

$$\text{AMA for females} = \frac{[\text{MAC} - (\pi \times \text{TSF})]^2}{4 \times \pi} - 6.5$$

AMA = arm muscle area in mm²; MAC = midarm circumference in cm; and TSF = triceps skinfold in cm.

$$\text{AMA for males} = \frac{[\text{MAC} - (\pi \times \text{TSF})]^2}{4 \times \pi} - 10$$

AMA = arm muscle area in mm²; MAC = midarm circumference in cm; and TSF = triceps skinfold in cm.

Source: Frisancho AR. 1990. Anthropometric standards for assessment of growth and nutritional status. Ann Arbor: University of Michigan Press.

Mid Upper Arm Muscle Area in Adults

Percentage of Standard (%)	Men (cm²)	Women (cm²)	Muscle Mass
100 ± 20*	54 ± 11	30 ± 7	Adequate
75	40	22	Marginal
60	32	18	Depleted
50	27	15	Wasted

*Mean mid upper arm muscle mass ± 1 standard deviation.

From National Health and Nutrition Examination Surveys I and II. Available from: Morley J. Undernutrition.

Source: From the Merck Manual of Diagnosis and Therapy, edited by Robert S. Porter. Copyright 2007 by Merck & Co., Inc., Whitehouse Station, NJ. Available at http://www.merck.com/mmpe. Assessed December 3, 2009.

Percentiles for interpretation of Arm Muscle Area found in Appendix A, pp. A-82 to A-85.

tissue while excluding muscle tissue.[19] Skinfolds are minimally invasive for the client, and require only calipers and a tape measure. Even though more accurate methods for body composition measurement exist, these advantages have led to the common use of skinfolds. It is important to recognize, however, that it takes a significant amount of repetition and experience to obtain consistent, reliable results.

Sites for skinfold measurement include the chest, triceps, subscapular, midaxillary, suprailiac, abdomen, thigh, and calf. Using more than one site for measurement may provide a more accurate "picture" of the individual. The most commonly used site is the triceps. Triceps skinfolds are taken on the nondominant arm, halfway between the olecranon and acromial processes. Additionally, midarm circumference measurement can be combined with triceps skinfold measurement to indirectly estimate arm muscle area and arm fat area (Figure 3.10).

Equipment needed for skinfold measurement includes a tape measure and a skinfold caliper. The Harpendon or Lange calipers are generally recommended because these brands were used in development of reference standards and equations.

Nutrition Care Criteria: Interpretation and Evaluation of Skinfold Measurements

Skinfold and midarm circumference measurements, as well as the index calculations for mid-upper arm muscle circumference, mid-upper arm muscle area, and mid-upper arm fat area, can be compared to references that are based on NHANES summary data. Comparison data are age, gender, and race specific. They may also be specific to frame size. A recent study indicated that using midarm circumference alone was not a sensitive measure for malnutrition when compared to other measures.[38] An individual who is below the 5th percentile or greater than the 95th percentile may be at nutritional risk (Table 3.7). See Appendix A, pp. A-82 to A-85 for these reference standards.

When interpreting these measurements in nutrition assessment, it is important to recognize that these reference standards were developed using a healthy population and cannot be used for a patient in a disease state. Furthermore, a one-time skinfold calculation may not really contribute any clinically useful information in the acute care setting. Hydration state, fluid shifts, and skin elasticity (especially in the elderly) can affect skinfold measurement. When the individual can serve as his or her own control with repeated measures over time, long-term changes in energy stores can be assessed and provide clinically useful information. Accuracy of measurement may be difficult to achieve if more than one clinician is performing the assessment, but error can be minimized when a single well-trained clinician uses the same equipment and method each time the assessment is performed.

Multiple-site skinfold measurements can be used to estimate body density and body fat percentage. Numerous regression equations have been developed. The regression equation originally designed for a population that most closely matches the client should be used.

Nutrition Care Indicator: Bioelectrical Impedance Analysis (BIA)

BIA is an increasingly common procedure used to estimate body composition. It is considered

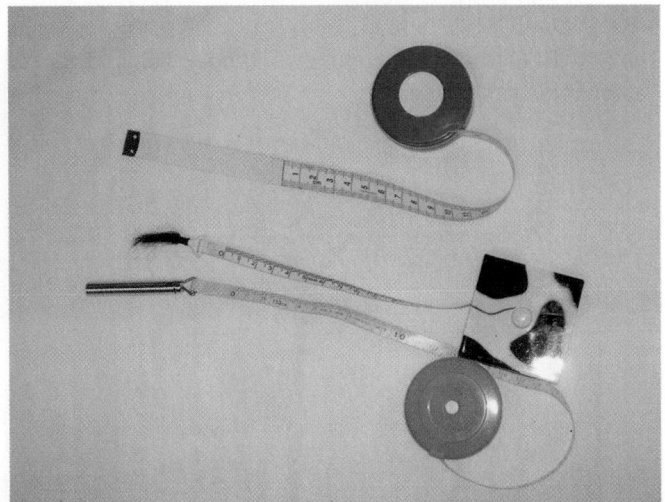

Using appropriate standardized equipment allows for consistent results when taking anthropometric measurements.

Source: Courtesy of Marcia Nelms.

Figure 3.11 Bioelectrical Impedence Analysis (BIA)

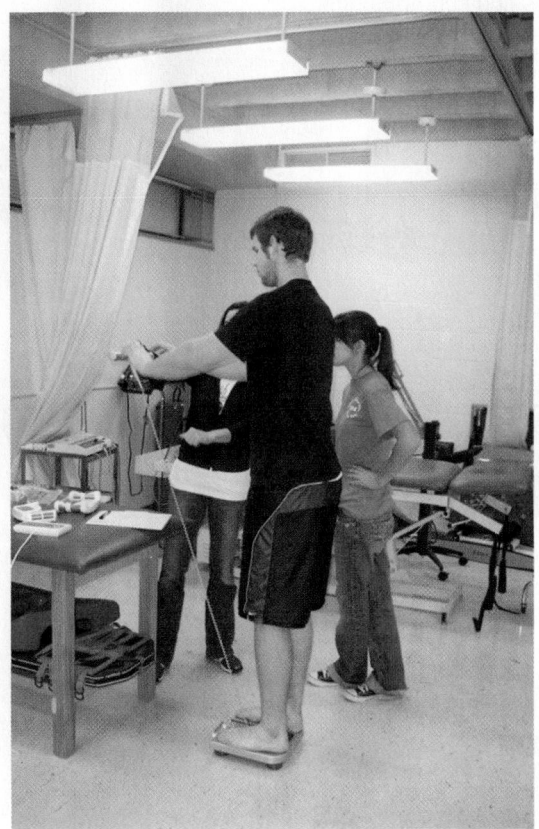

Source: Courtesy of Marcia Nelms.

to be a precise, rapid, safe, portable, and noninvasive method of assessment. There are numerous types of equipment available to measure BIA, but they are all based on the same scientific principle. A small, low-frequency, alternating electrical current is administered at one extremity of the body (see Figure 3.11). Measurement of impedance of the electrical current can then be used to estimate components of body composition, including body cell mass, fat-free mass, fat mass, and total body water. Tissues that contain low amounts of water such as fat and bone are poor conductors of electricity and therefore have a greater resistance or impedance to flow of the current. Other tissues that have greater water content, such as blood, muscle, and vital organs, are good conductors and therefore have lower impedance. More recent advances in technology provide for segmental measurement, which allows for estimation of regional body

Table 3.7 Interpretation of Triceps Skinfold Measurements

Comparing the percentile ranking of a specific individual on the various anthropometric measurements with a classification scheme is the basis for interpreting these values.

Reference data in percentiles for triceps skinfold (TSF), arm muscle area (AMA), and arm fat area (AFA) appear in the appendices. Because reference data are those compiled by Frisancho from the NHANES I and NHANES II data, it is appropriate to use classification categories derived statistically from these data. The table below displays percentile categories and their interpretation for arm muscle and arm fat areas as well as total body weight.

Percentile Rank	AMA	AFA	Total Body Weight
<5	Muscle deficit	Fat deficit	Total body wasting
5.1–15	Below average	Below average	Below average
15.1–85	Average	Average	Average
>85	Above average Musculature	Excess fat	Excess total body weight

Source: Reprinted from Contemporary Nutrition Support Practice, Matarese L and Gottschlich M; Philadelphia: WB Saunders Company; 2003, Box 3-1, p. 36, with permission from Elsevier. Ref: Frisancho AF: Anthropometric standards for the assessment of growth and nutritional status, Ann Arbor, Michigan, 1990. The University of Michigan Press, pp. 28, 41, 51, 54, 60–63.

fat. Additionally, multiple-frequency BIA may provide a more accurate assessment than a single frequency.

Nutrition Care Criteria: Interpretation and Evaluation of BIA Measurements Regression equations used to estimate body composition with BIA measurements include age, gender, weight, height, resistance, and reactance. From this information, components of body composition can be calculated. Additionally, equations have been validated specific to race and physical activity level.[39]

Even though the use of BIA has expanded in recent years, BIA should not be used to assess body composition in patients who have experienced major shifts in water balance and distribution. **Phase angle**, as a measure of body composition, has been used as an additional measure of prognosis in many chronic conditions. Newer research is using BIA as the means to monitor fluid accumulation in clinical conditions. Published studies have utilized BIA in subjects with HIV, chronic renal failure, cancer, chronic obstructive pulmonary disease, and cirrhosis.[40–42] There has previously been a lack of reference data for BIA, but recent NHANES III data include new references for it.

Nutrition Care Indicator: Hydrostatic (Underwater) Weighing Underwater weighing is generally accepted as the most accurate method of measuring body composition. Other body composition methods have historically been validated against underwater weighing results. Hydrostatic weight (hydrodensitometry) does evaluate body composition in

Figure 3.12 DXA

DXA is increasingly recognized as a reference method to assess body composition.

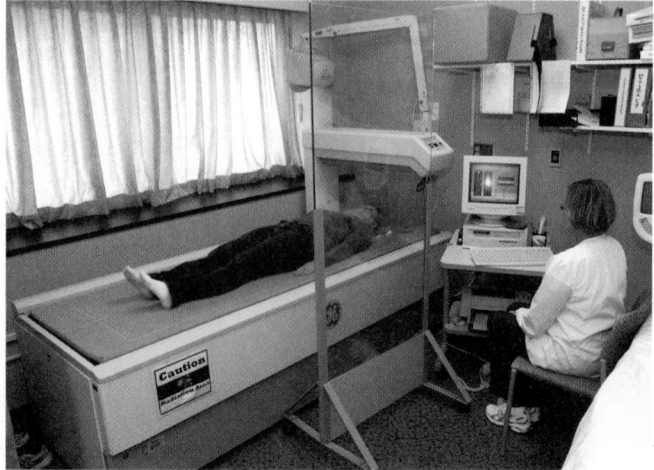

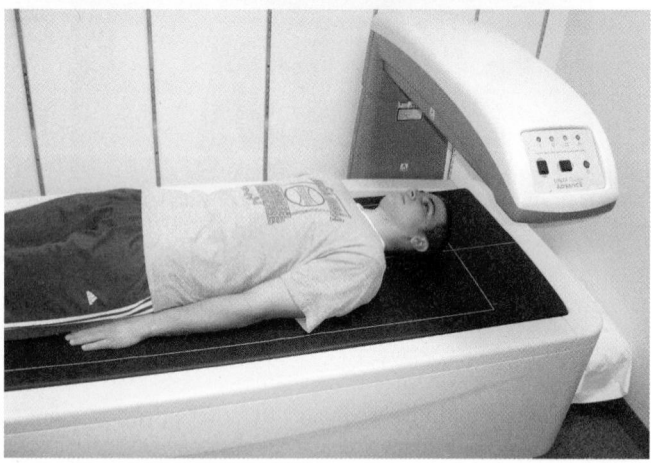

Source: Courtesy of Marcia Nelms.

Figure 3.13 BOD POD

The BOD POD is a highly popular method of body composition assessment.

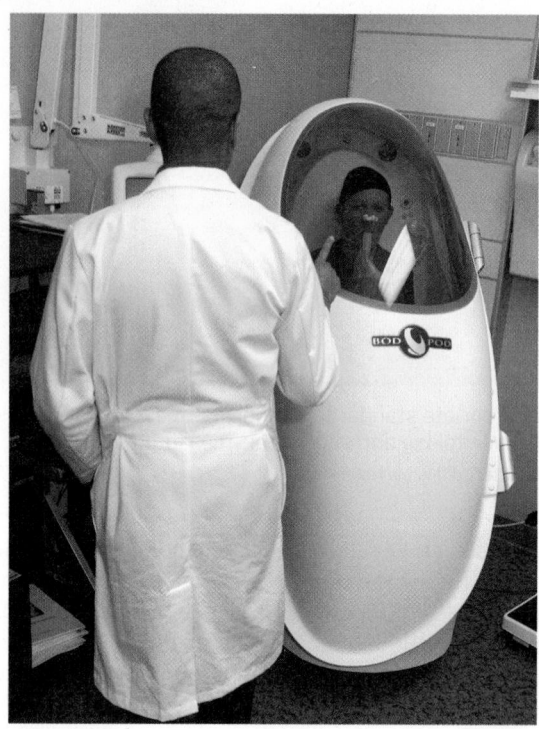

Source: Courtesy of Marcia Nelms.

the more traditional view, that of the two-compartment model. This method measures body volume (density) and relies on the assumption that the density of fat mass and components of fat-free mass are constant.

Obviously, access to underwater weighing facilities is not readily available, and thus this method has limited clinical use. It is also a difficult procedure for the subject to complete. Additionally, the reference equations were based on only Caucasians and may need changes for use with other ethnicities. Other limitations of this methodology include the possible overestimation of fat mass due to the assumption that all components of fat-free mass are constant. A correction for residual lung volume must also be made, and this volume is difficult to estimate.

Nutrition Care Indicator: Dual Energy X-Ray Absorptiometry (DXA) DXA was first developed to measure bone mineral content and density. DXA measures three compartments—mineral mass, mineral-free mass, and fat mass. In this method, the body is scanned with radiation photons at two different energy levels. Absorption of the photons by body tissues is then measured at each of these two levels. The absorption rate of different body tissues allows for calculation of

the three body compartments (see Figure 3.12). Despite some discrepancies in the literature, DXA is increasingly recognized as a reference method to assess body composition.[1, 37, 43, 44] The precision of DXA in assessing body composition and measuring total body fat is considered to be acceptable for measurement of both short- and long-term changes for a variety of conditions.[37] Unlike underwater weighing and air displacement plethysmography, but similar to skinfolds, DXA provides information on fat distribution by providing information on right and left arm, leg, and trunk masses.

Nutrition Care Indicator: Air Displacement Plethysmography In air displacement plethysmography, the client's total volume is measured indirectly by estimating the amount of air that is displaced within a sealed chamber. Since the early 1990s, the use of air displacement plethysmography has increased due to the availability of the equipment known as BOD POD, shown in Figure 3.13. This method is recognized as comparable to hydrostatic weighing and DXA.[19, 37]

Biochemical Assessment and Medical Tests and Procedures

In acute care, assessment of data from numerous sources supports the appropriate nutrition intervention and plan of care. These data include information gathered during medical procedures such as a barium swallowing test, as well as biochemical nutritional markers and indicators of organ function, which are found in blood, urine, feces, and tissue samples. Interpretation of biochemical measurements must be made in the context of diagnosis and medical treatment. Disease states, hydration status, and subsequent treatments can have considerable effect on the levels of the biochemical indices. Since reference values may also vary from lab to lab, data should be interpreted using those values provided by the laboratory that conducted the tests.

Protein Assessment

Protein's unique function in supporting cellular growth and development elevates its significance in nutrition assessment. Even though there is a large amount of protein stored in muscle and viscera, the body strives to protect it from being used as an energy source. Under healthy conditions or normal energy deficits, additional energy is drawn from fat and glycogen stores. But when the body is under metabolic stress, it draws protein from the muscles to meet its needs (see Chapters 9 and 22). No specific laboratory value can determine the precise protein status of an individual. Each test has its own limitations. Therefore, a comprehensive nutrition assessment must use a battery of tests in order to delineate the changes that mark the development of a nutrient deficiency.

Historically, protein assessment has focused on evaluation of two compartments: somatic and visceral. Somatic protein refers to skeletal muscle. Visceral protein refers to nonmuscular protein making up the organs, structural components, erythrocytes, granulocytes, and lymphocytes, as well as other proteins found in the blood.

Somatic Protein Assessment In addition to using anthropometric measures such as midarm muscle area, midarm circumference, and overall body weight and subjective global assessment, biochemical tests such as creatinine height index and nitrogen balance can be used to more specifically analyze somatic protein status.

NUTRITION CARE INDICATOR: CREATININE HEIGHT INDEX Creatinine is formed at a constant daily rate from muscle creatine phosphate. Creatine phosphate, which is stored in muscle, provides the phosphate group needed to regenerate ATP during high-intensity exercise. Creatinine is not stored in muscle but is cleared and excreted by the kidney. Daily urine output of creatinine can be correlated with total muscle mass.

NUTRITION CARE CRITERIA: INTERPRETATION AND EVALUATION OF CREATININE HEIGHT INDEX For this test, a 24-hour urine collection is performed and the total amount of creatinine excreted in that 24-hour period is compared with either a standard based on height or (as a percentage) to a standard excretion for a particular reference individual of a specific height, gender, and age.

In Table 3.8, expected creatinine excretion is shown for various heights. Creatinine height index (CHI) is usually expressed as a percentage of a standard value.

Equation to Calculate Creatinine Height Index:

$$\text{CHI} = \frac{24\text{-hour urine creatinine (mg)}}{\text{expected 24-hour urine creatinine (cm)}} \times 100$$

A value calculated to be 60%–80% of the standard suggests mild skeletal muscle depletion, 40%–59% suggests moderate skeletal

Table 3.8 Expected 24-Hour Creatinine Excretion in Men and Women of Ideal Weight

Adult Males*		Adult Females†	
Height (cm)	Creatinine (mg)	Height (cm)	Creatinine (mg)
157.5	1288	147.3	830
160.0	1325	149.9	851
162.6	1359	152.4	875
165.1	1386	154.9	900
167.6	1426	157.5	925
170.2	1467	160.0	949
172.7	1513	162.6	977
175.3	1555	165.1	1006
177.8	1596	167.6	1044
180.3	1642	170.2	1076
182.9	1691	172.7	1109
185.4	1739	175.3	1141
188.0	1785	177.8	1174
190.5	1831	180.3	1206
193.0	1891	182.9	1240

*Creatinine coefficient for males = 23 mg/kg of "ideal" body weight
†Creatinine coefficient for females = 18 mg/kg of "ideal" body weight
Source: JOURNAL OF PARENTERAL AND ENTERAL NUTRITION by Blackburn, Bistian, Maini, Schlamm, Smith. Copyright 1977 by SAGE PUBLICATIONS INC. JOURNALS. Reproduced with permission of SAGE PUBLICATIONS INC. JOURNALS in the format Textbook via Copyright Clearance Center.

muscle depletion, and <40% is considered to be a severe loss of skeletal muscle.

One limitation of this measurement is that its accuracy depends on a complete collection of urine for 24 hours, which is a common source of error and makes the test more difficult to conduct. Interpretation of this test must also take into account the fact that creatinine excretion can be either higher or lower than the standard depending on certain clinical conditions. Creatinine excretion is increased by meat consumption, sepsis, trauma, fever, and strenuous exercise, and during the second half of the menstrual cycle. Creatinine excretion is decreased with compromised renal function, low urine output, aging, and muscle atrophy unrelated to malnutrition. Furthermore, standards that are used do not account for creatinine excretion changes with age, disease, physical training, frame size, or weight status.[19]

NUTRITION CARE INDICATOR: NITROGEN BALANCE In healthy individuals, nitrogen excretion should be equal to nitrogen intake—thus indicating a state of nitrogen balance. A negative nitrogen balance would occur when nitrogen excretion is greater than nitrogen intake, indicating catabolism or inadequate nitrogen intake, whereas a positive nitrogen balance would occur when nitrogen intake is greater than excretion. Measuring nitrogen balance assesses overall protein status. Additionally, it can serve as a method of assessing the effectiveness of a nutrition intervention. Nitrogen balance is not routinely used for all acute-care patients but is more commonly seen in critical care, with nutrition support, and in research settings.

NUTRITION CARE CRITERIA: MONITORING AND EVALUATION OF NITROGEN BALANCE In order to measure nitrogen balance, the dietary intake of protein for a 24-hour period is estimated while a 24-hour urine collection is gathered to measure total excretion of nitrogen. Nitrogen loss through other routes such as fecal excretion, normal skin breakdown, wound drainage, and nonurea nitrogen is estimated using a constant value of either 3 or 4. The following equation is used for calculation:

$$N_2 \text{ balance} = \frac{\text{dietary protein intake}}{6.25} - \text{urine urea nitrogen}$$
$$- 4 \; (6.25 \text{ g protein} = 1 \text{ g nitrogen})$$

Example: JM consumed approximately 55 grams of protein in the past 24 hours. His 24-hour urine collection indicated a urine urea nitrogen (UUN) of 13 grams. His N_2 balance would be calculated as:

$$\frac{55}{6.25} - 13 - 4 = -8.2 \text{ g}$$

This is interpreted as JM currently being in negative nitrogen balance.

Limitations of measuring nitrogen balance include the inherent error of 24-hour urine collection, failure to account for renal impairment, and inability to measure nitrogen losses from some wounds, burns, diarrhea, and vomiting. Nitrogen intake may also pose difficulties. Oral protein intake may be difficult to measure except when the patient is exclusively on enteral or parenteral nutrition support.

Visceral Protein Assessment As stated previously, visceral protein refers to nonskeletal protein making up the organs, structural components, erythrocytes, granulocytes, and lymphocytes, as well as other proteins found in the blood. Thus, visceral protein assessment indirectly measures these protein stores by assessing proteins made by these organs (primarily the liver) present in blood or lymph fluid (see Figure 3.14). Theoretically, **serum** protein measurement is affected by a change in the amount of amino acids needed for protein synthesis by the liver. Thus, a change in serum protein levels would be consistent with changes in visceral protein status. However, the synthesis rate of these transport proteins can be affected by factors other than protein intake or protein requirements. In general, transport protein synthesis is inhibited when the acute-phase protein synthesis rate is increased in response to inflammation, stress, or trauma. In the acute care setting where most patients are affected by these conditions, using transport proteins to measure protein status becomes difficult. The sensitivity and specificity of these assessment measures are often described by the term "half-life" of the protein. Half-life, in this clinical situation, means the amount of time it takes before half of the protein is either eliminated or broken down by the body. Therefore, a shorter half-life means that actual changes in these levels will be reflected more quickly than in those proteins with a longer half-life.

Acute-phase proteins are defined as "those whose plasma concentration increases (positive acute-phase proteins) or decreases (negative acute-phase proteins) by at least 25%" during inflammation, illness, and/or metabolic stress.[45] C-reactive protein (CRP) is a common indicator for inflammation and stress that has also been correlated with visceral protein markers. Increasing levels of CRP, indicating acute inflammation, are consistent with lower visceral protein markers as well as poor outcome measures for the individual.[46–49]

An overview of visceral protein assessment is presented in Table 3.9. The specificity, sensitivity, and reliability of each visceral protein measurement are different. An accurate interpretation for nutrition assessment should take these differences into consideration.

NUTRITION CARE INDICATOR: ALBUMIN Albumin is probably the most well-known measure of visceral protein status even though it is not the best. It is also the most abundant serum protein. Approximately 60% of the albumin in the body is found in extravascular space—in skin, muscle, and organs. The remaining albumin is found within the vascular space, which allows for its measurement. Normal serum levels of albumin are ≥3.5 g/dL. Synthesized by the liver (120–170 mg/kg/day), albumin serves many significant functions within the body, most commonly as a transport protein and as a component of vascular fluid and electrolyte balance. Decreases in serum albumin occur due to a decreased synthesis rate, an increased degradation rate, or a change in fluid distribution (either total volume or between compartments). Using albumin as a nutritional assessment marker in acute care is complicated by the fact that most patients are experiencing at least one of these factors and thus, albumin changes often reflect illness and not necessarily nutritional status.[50]

NUTRITION CARE CRITERIA: EVALUATION AND INTERPRETATION OF ALBUMIN Albumin has been the subject of significant nutrition research and thus serves as a good prognostic screening tool, though it is

Figure 3.14 Serum Proteins in 100 mL of Blood

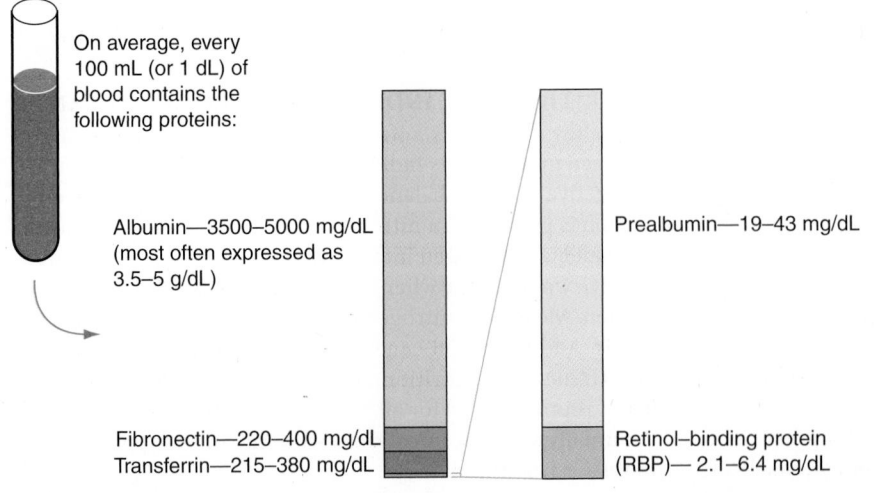

On average, every 100 mL (or 1 dL) of blood contains the following proteins:

Albumin—3500–5000 mg/dL (most often expressed as 3.5–5 g/dL)

Fibronectin—220–400 mg/dL
Transferrin—215–380 mg/dL

Prealbumin—19–43 mg/dL

Retinol–binding protein (RBP)— 2.1–6.4 mg/dL

Table 3.9 Visceral Protein Assessment Overview

Serum Protein	Normal Range	Half-Life	Primary Function	Comments
Albumin	3.5–5.0 g/dL	17–21 days	Blood transport protein; component of vascular fluid and electrolyte balance	Trauma, surgery, and metabolic stress affect levels; affected by hydration status—decreases with overhydration, increases with dehydration
Transferrin	215–380 mg/dL	8–10 days	Iron transport	Negative acute-phase respondent; affected by iron status
Prealbumin/transthyretin	19–43 mg/dL	2–3 days	Transport of thyroxine	Negative acute-phase protein—decreases with illness, infection, trauma, surgery, and metabolic stress; decreases with diagnoses of liver disease such as hepatitis or cirrhosis, malabsorption, and hyperthyroidism
Retinol binding protein	2.1–6.4 mg/dL	10–12 hours	Transport molecule for vitamin A	Negative acute-phase respondent; elevated with renal failure; decreased with hyperthyroidism, cystic fibrosis, liver failure, vitamin A deficiency, zinc deficiency, and metabolic stress
Fibronectin	220–400 mg/dL	15 hours	Wound healing and vascular integrity; cell and differentiation	Affected by coagulation, inflammation, and injury process

not as reliable in the acute care setting as an overall indicator of protein and nutritional status due to the effects of disease, hydration, and numerous other factors. Still, because it is easily measured and has an abundant body pool, albumin is readily available in the lab reports at the acute care setting. Decreased albumin levels have been correlated with increased morbidity, mortality, and length of hospital stay.

Some of the same factors contribute to its limitations as well. Albumin has a long half-life (approximately 20 days), which decreases its sensitivity to short-term changes in protein status or to short-term interventions to improve protein status. Albumin synthesis is also affected by acute stress and the inflammatory response. Albumin loss occurs with burn injuries, **nephrotic syndrome**, **protein-losing enteropathy**, and **cirrhosis**. Other medical conditions that may result in hypoalbuminemia include infection, **multiple myeloma**, acute or chronic inflammation, and rheumatoid arthritis. Levels also decrease with aging. On the other hand, albumin levels will be higher with dehydration and when individuals are prescribed anabolic hormones and corticosteroids. Albumin levels must be interpreted carefully—the levels are a better indicator of stress and inflammation than of overall protein nutriture, even though they historically have been widely used for that purpose.

NUTRITION CARE INDICATOR: TRANSFERRIN Synthesized by the liver, transferrin serves as a transporter for iron throughout the body. Due to its shorter half-life (8 to 10 days), it can also serve as an indicator of protein status, because it is sensitive to acute changes in protein intake or requirements. Normal serum levels are 215–380 mg/dL. Transferrin can be calculated from total iron binding capacity (TIBC) with the use of the following equation: Transferrin saturation (%) = (Serum iron level × 100%)/TIBC.

NUTRITION CRITERIA: EVALUATION AND INTERPRETATION OF TRANSFERRIN Transferrin's primary limitation is that its concentration is directly affected by iron status. When iron stores are decreased, transferrin levels will increase to accommodate the need for increased levels of transport. Other disease states such as hepatic and renal disease, inflam-

mation, and congestive heart failure can also affect transferrin levels. Transferrin can be measured directly or calculated from total iron binding capacity (TIBC).

NUTRITION CARE INDICATOR: PREALBUMIN (THYROXINE-BINDING PREALBUMIN OR TRANSTHYRETIN) Prealbumin is another example of an acute-phase transport protein synthesized by the liver. Prealbumin is responsible for transport of thyroxine and is associated with retinol-binding protein.

NUTRITION CRITERIA: EVALUATION AND INTERPRETATION OF PREALBUMIN Research has shown that because of its very short half-life (two days), prealbumin levels respond to short-term modifications in nutritional intake and interventions. Prealbumin appears to be a consistent indicator of risk for malnutrition.[19, 51] Normal serum levels range from 19 to 43 mg/dL.

Prealbumin is a more expensive test than albumin, but research has indicated that if it were used routinely during admission screening, approximately 44% more hospitalized patients would be identified as being at nutritional risk. Mears confirmed the usefulness of prealbumin in the clinical setting as a means for identifying nutritional risk.[52] Raguso, Dupertuis, and Pichard (2003)[53] found that measured prealbumin was indicative of adequate nutrition support during critical illness when other markers for inflammation were stable. The clinician evaluating prealbumin should recognize that levels are increased with renal disease and Hodgkin's disease (a form of lymphoma), and decreased with liver disorders such as hepatitis or cirrhosis, malabsorption, and hyperthyroidism. Furthermore, like albumin, prealbumin levels may decrease as a result of illness and stress and not necessarily malnutrition.[50, 54]

NUTRITION CARE INDICATOR: RETINOL BINDING PROTEIN (RBP) Retinol-binding protein (RBP), synthesized by the liver, is an acute-phase respondent, and

Sample PES Statement for Clinical Domain: Altered prealbumin related to increased protein requirements and inadequate protein consumption (NPO) status as evidenced by current level of 13 mg/dL.

serves as the transport molecule for vitamin A (retinol). RBP has the smallest body pool and shortest half-life (12 hours) of the serum proteins.

NUTRITION CRITERIA: EVALUATION AND INTERPRETATION OF RBP RBP is considered to be one of the more sensitive indicators of protein status.[19] It will reflect short-term changes and responses to nutrition support interventions. Note that RBP levels are elevated with renal failure, and decreased with hyperthyroidism, liver failure, vitamin A deficiency, zinc deficiency, and metabolic stress.

NUTRITION CARE INDICATOR: FIBRONECTIN (FN)
Fibronectin (FN) is a glycoprotein composed of protein strands and simple carbohydrate molecules that is an important component of many cell types, including endothelial cells, **macrophages**, hepatocytes, and **fibroblasts**. This protein appears to have many functions, with its primary function being the regulation of cell growth and differentiation during cell development. FN plays a crucial role in wound healing and vascular integrity. FN can be identified from lymph, amniotic fluid, cerebrospinal fluid, and plasma. Fetal fibronectin (fFN) is a protein produced during pregnancy that functions to attach the fetal sac to the uterine lining. Recent research indicates that fFN may serve as a predictor for preterm labor during pregnancy.[55]

NUTRITION CRITERIA: EVALUATION AND INTERPRETATION OF FN Because FN is less affected by acute stress than other blood proteins, many believe FN will be increasingly used as a marker of nutritional status as well as a good indicator for the efficacy of nutrition support. FN may be especially pertinent in assessment of patients under acute stress because of its short, 15-hour half-life.[56] At the same time, due to FN's role in wound healing, serum levels may be affected by conditions such as burns, where there is increased deposition at the site of injury and inflammation. Because FN plays a role in thrombosis, a patient's coagulation status may also affect FN levels. Normal serum levels range from 220 to 400 mg/dL. At this time, FN is not routinely used in the acute care setting.

NUTRITION CARE INDICATOR: INSULIN-LIKE GROWTH HORMONE-1 (IGF-1) Insulin-like growth hormone-1 (IGF-1) is a hepatic protein synthesized when the liver is stimulated by growth hormone. IGF-1 alterations have been observed in conjunction with several diagnoses, including cancer and Crohn's disease.[57, 58] IGF-1 is often a component of protein status measurements in research settings.[59, 60] It is thought that IGF-1 may be more effective than other markers in measuring nutritional status during the acute phase of stress, but at this time it is not routinely used in the clinical setting.

NUTRITION CARE INDICATOR: C-REACTIVE PROTEIN (CRP) C-reactive protein is a positive acute-phase protein that is released during periods of inflammation and infection. The levels of acute-phase proteins generally increase as transport protein levels (such as prealbumin or albumin) decrease. Higher CRP levels have been associated with increased nutritional risk during stress, illness, and trauma.[46–49] CRP is a component of the **Prognostic Inflammatory Nutrition Index (PINI)**, which combines several assessment indicators to monitor nutritional status.[54, 61] This will be discussed in greater detail within Chapter 22, "Metabolic Stress Conditions."

Immunocompetence

Historically, evaluation of immunocompetence has been included as a part of any discussion of protein and nutrition assessment. This is logical, since adequate and appropriate immune function is dependent in part on adequate protein status. Protein deficiency routinely results in increased risk of infection as well as altered immune and inflammatory response. In clinical practice, the use of this type of nutrition assessment is complicated by the presence of disease and infection, which of course also affect all components of the immune system. The most valuable use of these measures may be in predicting nutritional risk as well as outcome.[54]

Nutrition Care Indicator: Total Lymphocyte Count (TLC) When evaluating a complete blood count (CBC) and differential count, calculation for TLC can be completed as follows:

$$TLC = \frac{WBC \times \% \text{ lymphocytes}}{100}$$

Total lymphocyte count will be affected by presence of infection, trauma, stress, and diseases such as cancer and HIV, as well as medications that influence the immune system (e.g., chemotherapy and corticosteroids). Normal values of total lymphocyte counts range between 1000 and 3500 mm³ or are assessed as 10%–45% of total white blood cell count.

Hematological Assessment

Evaluation of erythrocytes (red blood cells, or RBC) can be an important component of nutrition assessment, and is key to diagnosis of all anemia types. A complete blood count includes measurement of the total number of blood cells in the volume of blood. Many types of anemias exist, including those caused by deficiencies of iron, folate, or vitamin B_{12} and those arising from chronic diseases such as renal failure and congestive heart failure. Anemias are diagnosed by evaluation of the complete blood count and by the microscopic evaluation of the size, shape, and color of erythrocytes. (See Chapter 19 for detailed information on hematological disorders.)

Nutrition Care Indicators for Hematological Assessment

Hemoglobin (Hgb) Hemoglobin is a protein found in erythrocytes that functions to deliver oxygen to cells and to pick up carbon dioxide for expiration by the lungs. Measurement of hemoglobin is common in diagnosis of anemias, particularly iron-deficiency anemia. Additionally, hemoglobin is decreased in some chronic diseases and protein-energy malnutrition. Even though it is commonly measured, it is not the most sensitive or the most specific of hematological assessments of nutritional status. For example, in iron deficiency, iron stores may be depleted before serum hemoglobin levels will be affected.

Hematocrit (Hct) Hematocrit is defined as the percentage of blood that is actually composed of red blood cells. Hematocrit, like hemoglobin, will be decreased only in the final stages of iron deficiency. Hematocrit is affected by other nutrient deficiencies as well as by hydration status.

Mean Corpuscular Volume (MCV) Mean corpuscular volume is a measure of the size of red blood cells. Because it reflects the average size of the red blood cell, the value for MCV will be changed in a variety of anemias. MCV is reduced in iron and copper deficiencies and elevated in folic acid and vitamin B_{12} deficiencies.

Mean Corpuscular Hemoglobin (MCH) Mean corpuscular hemoglobin estimates the amount of hemoglobin in each cell. These values can reflect total serum hemoglobin levels. In some situations, MCH can be normal while the number of red blood cells is low, resulting in low Hgb. Abnormalities are generally specific to iron deficiency and other nutritional anemias.

Mean Corpuscular Hemoglobin Concentration (MCHC) Mean corpuscular hemoglobin concentration also estimates the amount of hemoglobin in each red blood cell, but it expresses the value as a percentage.

Ferritin Ferritin is a protein that serves as a storage form of iron; therefore, serum ferritin is an estimate of iron stores. Ferritin is a sensitive and specific measure of iron status and will be one of the first indices to change in iron deficiency.

Transferrin Saturation As discussed earlier under "Protein Assessment," transferrin is a serum protein responsible for transport of iron systemically. Each molecule of transferrin can transport two molecules of iron. Under normal conditions, approximately 30% of iron binding sites on the transferrin molecule are saturated. The body's requirement for iron and overall iron status will be reflected by changes in transferrin saturation. When iron status is low, transferrin is less saturated. Transferrin is calculated by using the ratio of serum iron levels to total iron binding capacity (TIBC). TIBC is the test used to measure the saturation ability for transferrin. TIBC is higher during iron deficiency and lower after repletion. There are numerous equations to calculate transferrin from TIBC but as mentioned earlier, transferrin is not the most reliable indicator of protein status due to the effect of iron status.

Protoporphyrin When there is inadequate iron available for hemoglobin synthesis, zinc is substituted for iron within hemoglobin. Consequently, zinc protoporphyrin (the protein transporter for zinc) levels rise during iron deficiency and are considered a sensitive measure of iron-deficiency anemia.

Serum Folate Coenzymes associated with folate are necessary factors for amino acid metabolism, including many one-carbon transfer reactions such as the conversion of histidine to glutamate. Folate coenzymes also play a crucial role in the synthesis of purine needed for DNA. Folate deficiency can be diagnosed when megaloblastic, macrocytic red blood cells are present and serum folate and red cell folate are decreased, while serum B_{12} remains within normal limits. If folate levels are inadequate for conversion of histidine to glutamate, an intermediate product, formiminoglutamate, is formed. Urinary levels of formiminoglutamate (FIGlu) are thus elevated in folate deficiency and serve as a diagnostic tool for the condition.

Serum B_{12} Anemia associated with B_{12} deficiency can be diagnosed in several ways. Clinically, it will be similar to folate deficiency but can be distinguished by measuring serum B_{12}

levels or by performing the Schillings test. In this test, B_{12} is given as an injection and the amount excreted in urine is measured. This allows distinction between different steps of B_{12} absorption (see Chapter 16). In recent studies, the most sensitive indicators for B_{12} deficiency were homocysteine and methylmalonic acid levels.[62]

Vitamin and Mineral Assessment

Laboratory tests are available for the assessment of most vitamins and minerals. These tests vary from high-performance liquid chromatography to microbiological, radioisotopic, and chemiluminescence assays. There is concern that using different types of assays does not always provide comparable results. Clinicians should be aware that serum levels and tissue concentrations vary considerably, and take this into consideration when interpreting data for each specific vitamin or mineral. Common examples of situations where vitamins are routinely measured include serum vitamin D assessment for osteoporosis and thiamin levels for individuals who are on long-term diuretics or who have a history of alcoholism.

Other Labs with Clinical Significance

Many other biochemical labs are routinely assessed and monitored. These may include measures of lipid status such as LDL cholesterol, HDL-C, or triglycerides. Total cholesterol is often <100 mg/dL in protein-energy malnutrition or in conditions causing malabsorption. Electrolytes and measures of blood urea nitrogen (BUN), creatinine (Cr), and serum glucose are components of routine admission labs. Depending on the patient's diagnosis, hydration status, and medical care, specific labs will be monitored by members of the health care team. See Table 3.10 for a summary of routine admission laboratory measurements.

Table 3.10 Routine Admission Laboratory Measurements

Chem – 7 Panel	
BUN (blood urea nitrogen)	Glucose
Serum chloride	Serum potassium
CO_2 (carbon dioxide)	Serum sodium
Creatinine	
Chem – 20 Panel	
Albumin	Glucose
Alkaline phosphatase	LDH (lactate dehydrogenase)
ALT (alanine transaminase)	Phosphorus serum
AST (aspartate aminotransferase)	Potassium test
BUN (blood urea nitrogen)	Serum sodium
Calcium serum	Total bilirubin
Serum chloride	Total cholesterol
CO_2 (carbon dioxide)	Total protein
Creatinine	Uric acid
Direct bilirubin	
Gamma-GT (gamma-glutamyl transpeptidase)	

Nutrition-Focused Physical Findings

In addition to the physical examination performed by medical and nursing staff, the dietitian should perform a nutrition-focused physical examination.[63] The purpose of this physical exam is to assess the patient for signs and symptoms consistent with malnutrition or specific nutrient deficiencies. Techniques of inspection, palpation, percussion, and auscultation are used to examine the body for these signs and symptoms. Inspection is a visual assessment of the body conducted in a systematic manner by the clinician in order to note any variations from normal features. Palpation is examination of the body using the sense of touch. Percussion involves the use of sound to identify deviations from standard sounds produced when the body surface is tapped by the fingers of the practitioner. The presence of body organs and cavities will change the resonance and quality of sounds. In the nutrition-focused physical assessment, percussion may be used to assess status of the gastrointestinal tract when assessing feeding route or to identify fluid in the lungs, which may necessitate the need for medical intervention and possibly fluid restriction.[64] Auscultation also uses the sense of hearing to identify deviations from standard sounds. In this technique, a stethoscope is used to evaluate sounds produced by the heart, lungs, and gastrointestinal tract. An example is the identification of bowel sounds.

The methodology and interpretation for the Nutrition-Focused Physical Assessment are included in Appendices G–H.

Functional Assessment

Functional assessment focuses on measurements that assess skeletal muscle function or strength. Additionally, functional assessment could be expanded to include those activities that require adequate strength. For example, in the Subjective Glob-

Figure 3.15 Dynamometer
Dynamometry is a common measurement of muscle strength.

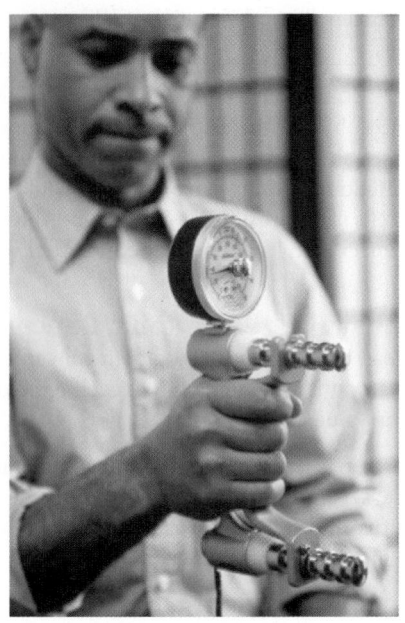

Source: Courtesy of Keith Brofsky/Jupiter Images.

al Assessment (SGA) (see Appendix B2), questions that focus on activities and function are included. The SGA identifies the patient's perception of his or her ability to accomplish self-care and the environment where the patient spends the majority of his or her time (e.g., bedridden, in chair, or normal activity). Another method of functional assessment is the identification of specific activities of daily living (ADL) and instrumental activities of daily living (IADL) related to nutritional status. Different scales have been developed to measure an individual's ability to perform normal daily activities and are scored according to the specific assessment tool. Table 3.11 serves as an example checklist for these activities.

A common assessment of muscle strength is handgrip dynamometry. This is a standardized, simple, and quick means of assessing nutrition in relation to skeletal muscle function. In this assessment, the patient is asked to grip the dynamometer (see Figure 3.15) device as tightly as possible. Handgrip standards are ≥35 kg for males and ≥23 kg for females. Handgrip measure has long been a part of fitness assessment but is now more common in nutrition assessment.[65, 66] This evaluation is especially useful for long-term follow-up in outpatient or rehabilitation settings.[67] The test is not valid in the presence of neuromuscular junction, muscle, or joint disease. Motivation to perform the test must be considered as well. Physical assessment can also assist in determining muscle strength and functional status.

Nutrition Care Criteria: Energy and Protein Requirements

The final component of a nutrition assessment is determination of energy and protein requirements for the patient. This is nec-

Table 3.11 Activities of Daily Living (ADL) and Instrumental Activities of Daily Living (IADL)

Activities of Daily Living	
Bathing, showering	Personal device care
Bowel and bladder management	Personal hygiene and grooming
Dressing	Personal mobility
Eating	Sexual activity
Feeding	Sleep/rest
Functional mobility	Toilet hygiene

Instrumental Activities of Daily Living	
Care of others	Health management and maintenance
Care of pets	
Child rearing	Home establishment and management
Communication device use	
Community mobility	Meal preparation and cleanup
Financial management	Safety procedures and emergency responses
	Shopping

Source: Reprinted from AJOT (1978) by American Occupational Therapy Assoc. Copyright 2002 by Am Occupational Therapy Assn. Reproduced with permission of Am Occupational Therapy Assn in the format Textbook via Copyright Clearance Center.

essary in order to compare intake with needs and to establish nutrition goals. In most clinical settings today, protein and energy requirements are estimated using a variety of established equations. In some situations, it may be possible to measure energy needs by using **indirect calorimetry** and protein needs by performing a nitrogen balance study.

The total amount of energy required by an individual consists of three basic components: basal energy expenditure (BEE) or basal metabolic rate (BMR) + energy for physical activity or exercise (PA) + thermic effect of food (TEF) = total energy expenditure (TEE). Basal energy expenditure is defined as energy used for physiological functions to maintain life, such as respiration and heartbeat. Basal energy expenditure accounts for approximately 60% of an individual's energy requirement. When the term *basal energy expenditure* is used, it refers to a measurement of oxygen consumed by a patient who has gone without food for at least 12 hours, and has been lying down with little movement in a constant temperature environment overnight.[68] Additionally, during the actual measurement, the patient should not be moving, talking, sleeping, or using muscles other than those for breathing (i.e. completely still and relaxed). Due to these strict measurement requirements, actual basal expenditure is in a practical sense theoretical and thus difficult to measure. Therefore, in many discussions regarding energy requirements, the term REE (resting energy expenditure) or RMR (resting metabolic rate) is used. The term "resting" refers to measurement conditions where the individual is resting in a comfortable position without any other restrictions. RMR is usually estimated to be approximately 10% higher than BMR/BEE.[68]

Physical activity (PA) is the most variable portion of an individual's energy needs and fluctuates depending on the type, time, and intensity of physical activity. In most individuals, PA accounts for approximately 15%–20% of energy requirements. TEF is estimated to be approximately 10% of an individual's caloric intake and represents the energy needed for absorption, transport, and metabolism of nutrient intake.

In hospitalized and other diseased populations, many individuals are hypermetabolic and have additional energy and protein requirements. These will be discussed in general here, but will be covered in more detail for specific disease states in the appropriate chapters within this text.

Nutrition Care Criteria: Measurement of Energy Requirements

The most accurate method of measuring REE/RMR in a clinical setting is to use indirect calorimetry. The equipment (metabolic carts) used for this process has steadily become more sophisticated over the last several decades, but the basic principles for measuring REE/RMR remain the same. The amounts of oxygen and carbon dioxide in both inspired and expired air (VO_2 and VCO_2) are measured, and the volume (V) of gas exchanged is equated to known energy constants (specific numbers of kcal per mL of oxygen consumed). These values then are converted to REE/RMR using computer software within the equipment. Calculations are based on the Weir equation REE (kcal/day) = 1.44 (3.9 VO_2 + 1.1 VCO_2).[69] Cost and availability of equipment limit the use of indirect calorimetry in many acute care hospitals.[70]

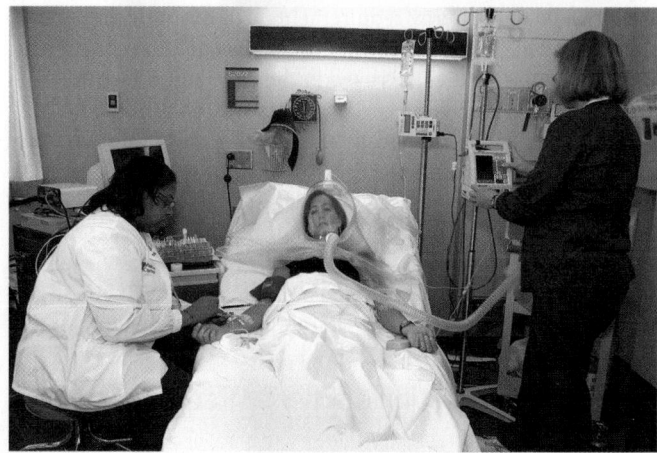

The most accurate method of assessing resting energy requirements is to use indirect calorimetry.
Source: Courtesy of Marcia Nelms.

Estimation of Energy Requirements

As stated previously, it is more common in clinical situations to calculate an estimation of energy requirements and, therefore, clinicians must rely on prediction equations to determine an individual's energy requirement in healthy populations, in the critically ill, and in the mechanically ventilated. Several reviews published in the last five years have attempted to determine the equation that is most consistent with measured energy requirements.[70–73] Walker and Heuberger examined the literature supporting seven different equations that are frequently used in the acute care population. Because each equation was established in a particular population, the accuracy changes significantly when it is applied to a different subset of patients.[70] Factors that affect the accuracy include presence of obesity, mechanical ventilation, stress, and trauma. Another retrospective chart analysis found that in the 395 patients evaluated, none of the equations accurately predicted the measured energy expenditure.[74] The ADA Evidence Analysis work group could not support the use of the Harris-Benedict, Ireton-Jones, or Fick equations in hospitalized, critically ill populations.[72] Their review also stated that there was not enough data to either support or reject the use of the Mifflin-St. Jeor equation in this population. In the most recent analysis of 202 mechanically ventilated patients, the Penn State equation had the highest accuracy when compared to measured energy expenditure.[73]

Historically, the most widely used standard for estimation of energy requirements is the Harris-Benedict equation, first published in 1919.[75] The Food and Agriculture Organization (FAO) of the United Nations and the World Health Organization (WHO) have also established equations to estimate basal energy requirements, which are also gender and age specific. The Mifflin-St. Jeor equation was established in 1990 and has been validated in over 10 different studies in the past decade.[76] Other equations include the American College of Chest Physicians, Ireton-Jones, Penn State, and Swinamer. All equations are outlined in Table 3.12.

Various institutions have differing recommendations on the body weight that should be used with each of these equations.[77, 78] Either actual body weight, ideal body weight, or adjusted body weight for obese patients could potentially be used to calculate energy requirements. For example, in the American College of Chest Physicians equation, BMI determines the weight that is used in the calculation (see Table 3.12). As discussed earlier in this section, lean body mass is the most metabolically active tissue and represents the largest proportion of the REE. Historically, the rationale for use of an adjusted body weight was based on the assumption that 20%–25% of fat mass is metabolically inactive. As Walker and Heuberger (2009) point out in their recent review, there is no research to support this calculation. Clinicians, including the MD and RD, feel that the risk of initially overfeeding in the critically ill justifies the use of ideal body weight in overweight or obese patients. Otherwise, in normal-weight individuals, actual body weight is used in the calculations for REE.

Energy Requirements Based on DRI The Dietary Reference Intakes for macronutrients are standards of intake that are age and gender specific and are designed to meet the nutrient requirements of about 98% of the healthy population. The DRI also include Estimated Energy Requirements (EER) that provide guidelines to meet the energy needs of approximately 50% of the healthy population. Because energy requirements vary considerably from individual to individual, the EER values are not meant to be goals of nutrient intake for individuals and hence are not recommended for estimating patients' energy requirements in a clinical setting. See the front inside cover of the book for these values.

Activity Factor After resting energy expenditure has been determined, energy used in activity also must be estimated in order to estimate total energy requirements. In hospitalized patients, activity is generally estimated using a figure of 1.2 for bed rest and 1.3 for all other patients, which is then multiplied by the REE. Specifically, REE is increased by 20% to account for PA. For example, if REE were 1350, an activity factor of 1.2 would determine the energy requirements to be 1620 kcal (see Table 3.13).

There are many methods used to estimate the amount of energy needed for physical activity, especially in the nonhospitalized population. One total energy requirement formula that was recently developed by the Food and Nutrition Board[79] incorporates a physical activity coefficient. Both the CDC and the American College of Sports Medicine (ACSM) use the exercise metabolic rate, or MET, to estimate the amount of energy used in various physical activities.[80] One MET is estimated to be the energy expenditure for sitting quietly, which for the average adult approximates 3.5 mL of oxygen uptake per kilogram of body weight per minute (1.2 kcal/min for a 70-kg individual).[81, 82] Table 3.14 presents an overview of physical activities and the equivalent energy expended.

Stress Factors Disease, infection, and trauma can affect an individual's energy requirements. Hospitalized patients can be hypermetabolic; estimation of their energy needs should take this fact into account. When indirect calorimetry is not available, it is a common practice to estimate energy needs using a stress factor in addition to an activity factor if the

Table 3.12 Estimation of Energy Requirements

Harris-Benedict Equation

REE for females = 655.1 + 9.6 W + 1.9 H − 4.7 A

REE for males = 66.5 + 13.8 W + 5.0 H − 6.8 A

[W = weight in kg; H = height in cm; A = age in years]

Mifflin-St. Jeor Equation

REE for females: 10 W + 6.25 H − 5 Age − 161

REE for males: 10 W + 6.25 H − 5 Age + 5

[W = weight in kg; H = height in cm; A = age in years]

American College of Chest Physicians Equation

REE = 25 × weight (kg)

[If BMI = 16–25, use usual body weight; If BMI >25, use ideal body weight; and if BMI <16, use existing body weight for first 7–10 days, then use ideal body weight.]

Penn State 2003 Equation

REE = (0.85 × value from Harris-Benedict equation) + (175 × T_{max}) + (33 × V_e) − 6443

[V_e = minute volume (in L/min); T_{max} = maximum body temp in previous 24 hours]

Ireton-Jones 1997 Equation

REE = (5 × weight) − (11 × age) + (244 if male) + (239 if trauma present) + (840 if burns present) + 1784

[W = weight in kg; H = height in cm; A = age in years]

Swinamer 1990 Equation

REE = (945 × body surface are) − (6.4 × age) + (108 × temperature) + (24.2 × respiratory rate) + (817 × V_T) − 4349

[V_T = tidal volume in liters]

FAO/WHO Basal Energy Estimation Equations

	Age	Equation*
Men	18–30 years	Kcal/day = (15.3 × weight) + 679
	30–60 years	Kcal/day = (11.6 × weight) + 879
	>60 years	Kcal/day = (8.8 × weight) + (1128 × height) − 1071
Women	18–30 years	Kcal/day = (14.7 × weight) + 496
	30–60 years	Kcal/day = (8.7 × weight) + 829
	>60 years	Kcal/day = (9.2 × weight) + (637 × height) − 302

*Height in m; weight in kg

Sources: Adapted from: Frankenfield D, Roth-Yousey L, Compher C. Comparison of predictive equations for resting metabolic rate in healthy nonobese and obese adults: a systematic review. J Am Diet Assoc. 2005;105(5):775–89. Frankenfield D, Hise M, Malone A, Russell M, Gradwell, Compher C, Evidence Analysis Working Group. Prediction of resting metabolic rate in critically ill adult patients: Results of a systematic review of the evidence. J Am Diet Assoc. 2007;107:1552–61. Frankenfield DC, Coleman A, Alam S, Cooney RN. Analysis of estimation methods for resting metabolic rate in critically ill adults. J Parenter Enteral Nutr. 2009;33(1):27–36.

Walker RN, Heuberger RA. Predictive equations for energy needs for the critically ill. Respr Care. 2009;54: 509–21.

Table 3.13 Calculation of Total Energy Requirements for the Hospitalized Patient

To calculate total energy requirements for the hospitalized patient: REE (Resting Energy Expenditure) × Activity Factor × Injury Factor

Activity Factors	Average Injury Factors
Out of bed 1.3	Surgery 1.0–1.3
Confined to bed 1.2	Infection 1.0–1.4
	Skeletal trauma 1.2–1.4
	Head injury 1.5

Source: Adapted from: Kudsk K, Sacks G. Ch.91. Nutrition in the care of the patient with surgery, trauma, and sepsis. In Shils M, ed. Modern Nutrition in Health and Disease. Tenth Ed. Philadelphia: Lippincott Williams & Wilkins, 2005. Long, CL. The energy and protein requirements of the critically ill patient. In Wright RA, Heymsfield SB, eds. Nutritional Assessment. Boston: Blackwell Scientific, 1984.

Table 3.14 Energy Requirements of Common Daily Activities*

Leisure Activities	METs†
Mild	
Playing the piano	2.3
Canoeing (leisurely)	2.5
Golf (with cart)	2.5
Walking (2 mph)	2.5
Dancing (ballroom)	2.9
Moderate	
Walking (3 mph)	3.3
Cycling (leisurely)	3.5
Calisthenics (no weight)	4.0
Golf (no cart)	4.4
Swimming (slowly)	4.5
Walking (4 mph)	4.5
Vigorous	
Chopping wood	4.9
Tennis (doubles)	5.0
Ballroom (fast) or square dancing	5.5
Cycling (moderately)	5.7
Skiing (water or downhill)	6.8
Climbing hills (no load)	6.9
Swimming	7.0
Walking (5 mph)	8.0
Jogging (10 min mile)	10.2
Rope skipping	12.0
Squash	12.1
Activities of Daily Living	
Lying quietly	1.0
Sitting; light activity	1.5
Walking from house to car or bus	2.5
Loading/unloading car	3.0
Taking out trash	3.0
Walking the dog	3.0
Household tasks, moderate effort	3.5
Vacuuming	3.5
Lifting items continuously	4.0
Raking lawn	4.0
Gardening (no lifting)	4.4
Mowing lawn (power mower)	4.5

*These activities can often be done at variable intensities, assuming that the intensity is not excessive and that the courses are flat (no hills) unless so specified. Categories are based on experience or tolerance; if an activity is perceived to be more than indicated, it should be judged accordingly.

† MET indicates metabolic equivalent. One MET is the amount of energy used by a 70-kg person when sitting quietly.

Source: Fletcher GF, Balady GJ, Amsterdam EA, et al. Exercise standards for testing and training: a statement for healthcare professionals from the American Heart Association. Circulation 2001;104:1694–740. Available at http://www.hsph .harvard.edu/nutritionsource/staying-active/mets-activity-table/index.html

has deepened. Over-feeding may be much more detrimental than underestimation of energy needs. Table 3.13 estimates stress factors for hypermetabolic conditions. This topic is covered in greater detail in Chapter 22.

> **Sample PES Statement for Energy Balance:** Involuntary weight loss related to inadequate oral intake as evidenced by documentation of PO intake meeting 54% of estimated energy requirements with subsequent weight loss of 3.2 kg.

Nutrition Care Criteria: Measurement of Protein Requirements

A nitrogen balance assessment measures urine urea nitrogen and compares it to protein intake. This allows for some degree of measurement of protein requirements. A negative nitrogen balance reflects catabolism and can sometimes be due to inadequate protein intake. A positive nitrogen balance is consistent with anabolism and generally will indicate that the patient is receiving adequate amounts of protein to support current requirements. For those patients who have no renal function, measurement of nitrogen balance is accomplished through a urea kinetic modeling equation. See Chapter 18 for further details.

Estimation of Protein Requirements

RDA for Protein The Recommended Dietary Allowances provide the best reference for protein requirements in the nonstressed population. For adults, this level is set at 0.8 g protein/ kg of body weight. Additional levels are set for infants and children. These can be found on the inside front cover of this book.

Protein Requirements in Metabolic Stress, Trauma, and Disease Protein requirements, like energy requirements, are affected by metabolic stress, trauma, and disease. The type of protein required and the need for specific amino acids may additionally be altered within certain diseases. This will be discussed within the context of those specific diagnoses throughout the text. In general, though, if patients are receiving adequate kcal, protein requirements can be met by providing 1.0–1.5 g protein/kg/day.

Protein-Kilocalorie Ratio Another traditional approach to estimating protein requirements is based on the concept that energy should be provided by lipids and carbohydrates, and protein intake should be reserved for synthesis requirements. Historically, many clinicians have calculated energy requirements to include only carbohydrates and lipid. Protein estimations were considered separately. For healthy individuals, the ratio of 1:200 (protein: kcal) is recommended. In individuals who have higher protein requirements, ratios range from 1:150 to 1:100.

This subject is controversial and has fueled many professional debates. No standardization of this process currently exists. It is important to realize that biochemistry does not support this practice, since it is known that certain cells use protein for fuel (e.g., enterocyte use of glutamine). Furthermore, overfeeding carbohydrates and lipids can jeopardize medical care by interfering with the function of the lungs, liver, and immune system. This approach is not superior to other systems of estimating protein requirements and does not serve any additional advantage in nutrition assessment.

patient is ambulatory. Unfortunately, these stress factors have not been validated consistently, and thus estimations of energy requirements during stress have been modified considerably over the past decade as understanding of the stress response

Interpretation of Assessment Data: Nutrition Diagnosis

In the nutrition care process, after all components of nutrition assessment have been compiled, the clinician determines the nutritional status of the patient and identifies the specific nutrition-related problems. An essential guideline to use for diagnosing the level and type of malnutrition is the International Classification of Disease[83] which provides criteria for diagnosing specific types of malnutrition, including kwashiorkor, marasmus, other severe PEM (protein-energy malnutrition), malnutrition of a mild degree, and malnutrition of a moderate degree.

Nutrition diagnoses arise from the nutritional problems identified during the nutrition assessment.[1] Using the nutrition assessment information, the clinician identifies the nutrition-related problems, determines the probable cause for each of these problems, and substantiates the problem through specific signs and symptoms. The PES (problem, etiology, symptoms) format, as discussed in Chapter 2, is the recommended manner in which to document the nutrition diagnosis.

Conclusion

Health is strongly influenced by nutritional status. It is no surprise that the Joint Commission for Accreditation of Hospital Organizations requires that all patients receive nutrition screening within 48 hours of their hospital admission. No one test exists to assess nutritional status. In this chapter, numerous measures of dietary intake, physical health, and biochemical status have been presented; in practice, these measures allow the clinician to determine not only current nutritional state but also future nutritional risk. Accurate measurement and interpretation of nutritional status can allow a nutritional diagnosis to be made and interventions that improve patient outcomes to be put in place, ultimately decreasing morbidity and mortality.

WEB LINKS

Agricultural Research Service   Nutrient database of 13,000 commonly consumed foods in the United States.
www.ars.usda.gov/foodsearch

Aim for a Healthy Weight   A link from the National Heart, Lung, and Blood Institute provides excellent information for assessment of weight.
http://www.nhlbi.nih.gov/health/public/ heart/obesity/lose_wt/index.htm

Anthropometric Standardized Procedures
This link is a component of the NHANES data website. It provides specific instructions through video demonstrations on how to conduct various anthropometric procedures.
http://www.cdc.gov/nchs/products/ elec_prods/subject/video.htm

Dietary Guidelines for Americans   This link is also part of the mypyramid.gov website. Provides all of the background and supporting materials for the 2005 U.S. Dietary Guidelines.
http://www.mypyramid.gov/guidelines/ index.html

International Network of Food Data Systems   This website is part of the international Food and Agriculture Organization (FAO). Numerous links for international sources of food composition and nutrition data are provided.
http://www.fao.org/infoods/software_ en.stm

Lab Tests Online   This website is sponsored by the American Association for Clinical Chemistry. This interactive tool demonstrates the components of a typical lab report and their definitions.
http://www.labtestsonline.org/inc/ reports/SampleReport.html

MyPyramid.gov   United States Department of Agriculture (USDA) website outlining all materials pertaining to the customizable food pyramid. Provides individual client information as well as information for the professional to use in teaching and counseling.
http://www.mypyramid.gov/ professionals/index.html

National Heart, Lung, and Blood Institute
This government website provides patient education material as well as clinical practice guidelines.
http://www.nhlbi.nih.gov/guidelines/ index.htm

Nutrient Data Laboratory   New database that allows assessment of foods, beverages, and dietary supplements.
http://www.ars.usda.gov/ba/bhnrc/ndl

The Food Safety Risk Analysis Clearinghouse   The Food Safety Risk Analysis Clearinghouse is the responsibility of the Joint Institute for Food Safety and Applied Nutrition (JIFSAN), a collaboration of the University of Maryland (UM) and the Food and Drug Administration (FDA). This site provides a wealth of nutrition assessment forms and tools as well as links for nutrient analysis.
http://www.foodrisk.org/resource_types/ tools/nutrition_assessment.cfm

The National Guideline Clearinghouse™
The National Guideline Clearinghouse (NGC) is a comprehensive database of evidence-based clinical practice guidelines and related documents. NGC is an initiative of the Agency for Healthcare Research and Quality (AHRQ), U.S. Department of Health and Human Services. NGC was originally created by AHRQ in partnership with the American Medical Association and the American Association of Health Plans (now America's Health Insurance Plans [AHIP]). The NGC mission is (1) to provide physicians, nurses, and other health professionals, health care providers, health plans, integrated delivery systems, purchasers and others an accessible mechanism for obtaining objective, detailed information on clinical practice guidelines and (2) to further their dissemination, implementation, and use.
www.guideline.gov

END-OF-CHAPTER QUESTIONS

1. What is the difference between nutritional status and nutritional risk?

2. How is nutritional screening different from nutritional assessment?

3. Describe the difference between subjective data and objective data that are collected for a nutritional assessment. List three pieces of objective information and three pieces of subjective information that could be collected for nutritional assessment.

4. Name and briefly describe four methods used to collect dietary assessment data. List the advantages and disadvantages of each method.

5. Describe two methods that are used to analyze dietary intake.

6. Which anthropometric measurements are collected for nutritional assessment? Briefly describe each measurement and explain the accuracy of each in determination of body composition and/or health status.

7. List four blood proteins used in nutrition assessment. Describe the effectiveness of each as markers in measuring nutritional status.

8. Describe how energy requirements can be determined or estimated. How is the energy requirement affected by stress?

9. List the hematological measurements collected for nutritional assessment.

REFERENCES

1. American Dietetic Association. Nutrition Care Process Step 1: Nutrition Assessment. In: International Dietetics & Nutrition Terminology. Chicago (IL): American Dietetic Association; 2009.

2. U.S. Department of Health and Human Services. Agency for Healthcare Research and Quality. U.S. Preventive Services Task Force. Available at http://www.ahrq.gov/clinic/ajpmsupp/harris1.htm

3. Valentini L, Schaper L, Buning C, Hengstermann S, Koernicke T, Tillinger W, Guglielmi FW, Norman K, Buhner S, Ockenga J, Pirlich M, Lochs H. Malnutrition and impaired muscle strength in patients with Crohn's disease and ulcerative colitis in remission. Nutrition. 2008;24:694–702.

4. World Health Organization. Screening and early detection of cancer. 2008. Available at http://www.who.int/cancer/detection/en/

5. U.S. Department of Health and Human Services. Agency for Healthcare Research and Quality. U.S. Preventive Services Task Force. Available at http://www.ahrq.gov/clinic/ajpmsupp/harris1.htm

6. Charney P. Nutrition screening vs nutrition assessment? How do they differ? Nutr Clin Prac. 2008;23:366–72.

7. American Dietetic Association. Nutrition Screening Evidence Analysis Project 2007. Available at https://www.adaevidencelibrary.com/topic.cfm?cat=3583

8. Joint Commission on Accreditation of Healthcare Organizations. 2009 comprehensive accreditation manual for hospitals: the official handbook (CAMH). Chicago (IL) and Oak Brook (IL): Joint Commission on Accreditation of Healthcare Organizations; 2009.

9. Nord M, Andrews M, Carlson S. Household Food Security in the United States, 2007. Definitions of Hunger and Food Security. Available at http://www.ers.usda.gov/Briefing/FoodSecurity/labels.htm#labels

10. Holben DH. Position of the American Dietetic Association on food insecurity and hunger in the United States. J Am Diet Assoc. 2006;106:446–58.

11. American Dietetic Association. Nutrition Screening Initiative, a project of the American Academy of Family Physicians, American Dietetic Association and the National Council on the Aging, Inc.; 1991.

12. Ottery FD, Kasenic S, DeBolt S, Roger K. Volunteer network accrues >1900 patients in 6 months to validate standardized nutritional triage. Abstract 282. Meeting of the American Society of Clinical Oncology, 1987.

13. Baker JP, Detsky AS, Wesson DE, Wolman SL, Stewart S, Whitewell J, et al. A comparison of clinical judgment and objective measurements. N Engl J Med. 1982;306:969–72.

14. Stratton RJ, King CL, Stroud MA, Jackson AA, Elia M. Malnutrition Universal Screening Tool predicts mortality and length of hospital stay in acutely ill elderly. Br J Nutr. 2006;95:325–30.

15. Ferguson ML, Capra S, Bauer J, Banks M. Development of a valid and reliable malnutrition screening tool for adult acute hospital patients. Nutrition. 1999;15:458–64.

16. Conway, JM, Ingwersen, LA, Moshfegh, AJ. Effectiveness of the USDA 5-step Multiple-Pass Method to assess food intake in obese and non-obese women. American Journal of Clinical Nutrition. 2003;77:1171–78.

17. Conway, JM, Ingwersen, LA, Moshfegh, AJ. Accuracy of dietary recall using the USDA five-step multiple-pass method in men: an observational validation study. J Am Diet Assoc. 2004;104(4):595–603.

18. Blanton CA, Moshfegh AJ, Baer DJ, Kretsch MJ. The USDA Automated Multiple-Pass Method accurately estimates group total energy and nutrient intake J Nutr. 2006 Oct;136(10):2594–99.

19. Lee RD, Nieman DC. Ch. 3 Measuring Diet. In: Nutritional Assessment. 4th ed. New York (NY): The McGraw-Hill Companies, Inc.; 2007.

20. Novotny JA, Rumpler WV, Riddick H, Herbert JR, Rhodes D, Judd JT, Baer DJ, McDowell M, Briefel R. Personality characteristics as predictors of underreporting of energy intake on 24-hour dietary recall interviews. J Am Diet Assoc. 2003;103(9):1146–51.

21. Tapsell LC, Brenninger V, Barnard J. Applying conversation analysis

to foster accurate reporting in the diet history interview. J Am Diet Assoc. 2000;100(7):818–24.

22. Dwyer Jl. Dietary Assessment in Modern Nutrition in Health and Disease 12th ed. Philadephia (PA): Lippincott Williams & Wilkins; 2003.

23. U.S. Department of Agriculture/U.S. Department of Health and Human Services. Nutrition and Your Health: Dietary Guidelines for Americans. 2nd ed. Hyattsville (MD): U.S. Department of Agriculture. Center for Nutrition Policy and Promotion; 2005.

24. U.S. Department of Agriculture. MyPyramid.gov. Available at: www .mypyramid.gov., 2010.

25. American Diabetes Association (Alexandria VA) & American Dietetic Association Chicago (IL): Exchange List for Meal Planning; 2007.

26. Dietary Reference Intakes for Energy, Carbohydrate, Fiber, Fat, Fatty Acids, Cholesterol, Protein, and Amino Acids (Macronutrients). A Report of the Panel on Macronutrients, Subcommittees on Upper Reference Levels of Nutrients and Interpretation and Uses of Dietary Reference Intakes, and the Standing Committee on the Scientific Evaluation of Dietary Reference Intakes, Food and Nutrition Board, Institute of Medicine, 2002.

27. Dietary Reference Intakes for Vitamin A, Vitamin K, Arsenic, Boron, Chromium, Copper, Iodine, Iron, Manganese, Molybdenum, Nickel, Silicon, Vanadium, and Zinc Panel on Micronutrients, Subcommittees on Upper Reference Levels of Nutrients and of Interpretation and Use of Dietary Reference Intakes, and the Standing Committee on the Scientific Evaluation of Dietary Reference Intakes, Food and Nutrition Board, Institute of Medicine, 2000.

28. Centers for Disease Control. National Health and Nutrition Examination Survey. Available at http://www.cdc.gov/nchs/nhanes.htm

29. Chumlea WCC, Guo SS, Steinbaugh ML. Prediction of stature from knee height for black and white adults and children with application to mobility-impaired or handicapped persons J Am Diet Assoc. 1994;94(12):1385–91.

30. Kuczmarski MF, Kuczmarski RJ, Najjar M. Effects of age on validity of self-reported height, weight, and body mass index: findings from the Third

National Health and Nutrition Examination Survey, 1988–1994. J Am Diet Assoc. 2001;101(1):28–34.

31. Smith LK, Weiss EL, Lehmkuhl LD. Brunnstrom's Clinical Kinesiology. 5th ed. Philadelphia (PA): FA Davis Company; 1996.

32. National Center for Health Statistics in collaboration with the National Center for Chronic Disease Prevention and Health Promotion (2000). Available at www.cdc.gov/growthcharts

33. Hamwi GJ. Changing dietary concepts. In Donowski TS, editor Diabetes Mellitus: Diagnosis and Treatment. New York: American Diabetes Association; 1964.

34. National Institutes of Health. National Heart, Lung, and Blood Institute. NHLBI Obesity Education Initiative North American The Practical Guide: Identification, Evaluation, and Treatment of Overweight and Obesity in Adults. NIH Publication Number 00-4084, October 2000. Available at: http://www .nhlbi.nih.gov/guidelines/obesity/prctgd_b.pdf.

35. Grundy SM, Cleeman JI, Daniels SR, Donato KA, Eckel RH, Franklin BA, Gordon DJ, Krauss RM, Savage PJ, Smith SC Jr, Spertus JA, Costa F; American Heart Association; National Heart, Lung, and Blood Institute. Diagnosis and management of the metabolic syndrome: an American Heart Association/National Heart, Lung, and Blood Institute Scientific Statement. Circulation. 2005;112:2735–52.

36. Rosenzweig JL, Ferrannini E, Grundy SM, Haffner SM, Heine RJ, Horton ES, Kawamori R; Endocrine Society. Primary prevention of cardiovascular disease and type 2 diabetes in patients at metabolic risk: an endocrine society clinical practice guideline. J Clin Endocrinol Metab. 2008;93:3671–89

37. Lee RD, Nieman DC. Ch. 3 Measuring Diet. In: Nutritional Assessment. 4th ed. New York (NY): The McGraw-Hill Companies, Inc., 2007.

38. Burden ST, Stoppard E, Shaffer J, Makin A, Todd C. Can we use mid upper arm anthropometry to detect malnutrition in medical inpatients. A validation study. J Hum Nutr Diet. 2005;18(4):287–94.

39. Kotler DP, Burastero S, Wang J, Pierson RN, Prediction of body cell mass, fat-free mass, and total body water with bioelectrical impedance analysis: effects

of race, sex, and disease. Am. J. Clinical Nutrition, Sep 1996;64:489S–497S.

40. Schiesser M, Kirchhoff P, Müller MK, Schäfer M, Clavien PA. The correlation of nutrition risk index, nutrition risk score, and bioimpedance analysis with postoperative complications in patients undergoing gastrointestinal surgery. Surgery. 2009;145:519–26.

41. Gupta D, Lammersfeld CA, Vashi PG, King J, Dahlk SL, Grutsch JF, Lis CG. Bioelectrical impedance phase angle as a prognostic indicator in breast cancer. BMC Cancer. 2008;8:24.

42. Tang WH, Tong W. Measuring impedance in congestive heart failure: current options and clinical applications. Am Heart J. 2009;157:402–11.

43. Andreoli A, Scalzo G, Masala S, Tarantino U, Guglielmi G. Body composition assessment by dual-energy X-ray absorptiometry (DXA). Radiol Med. 2009;114:286–300

44. Williams JE, Wells JC, Wilson CM, Haroun D, Lucas A, Fewtrell MS. Evaluation of Lunar Prodigy dual-energy X-ray absorptiometry for assessing body composition in healthy persons and patients by comparison with the criterion 4-component model. Am J Clin Nutr. 2006;83:1047–54.

45. Gabay C, Kushner I. Acute-phase proteins and other systemic responses to inflammation. NEJM. 1999;448–54.

46. Sherwood L. Body Defenses. In: Human Physiology from Cells to Systems. 7th ed. Belmont (CA): Brooks-Cole/Cengage; 2010.

47. Gariballa S, Forster S. Effects of acute-phase response on nutritional status and clinical outcome of hospitalized patients. Nutrition. 2006;22:750–57.

48. Winkelmayer WC, Lorenz M, Kramar R, Födinger M, Hörl WH, Sunder-Plassmann G. C-reactive protein and body mass index independently predict mortality in kidney transplant recipients. Am J Transplant. 2004;4:1148–54.

49. Walsh D, Mahmoud F, Barna B. Assessment of nutritional status and prognosis in advanced cancer: interleukin-6, C-reactive protein, and the prognostic and inflammatory nutritional index. Support Care Cancer. 2003;11:60–62

50. Marshall WJ. Nutritional assessment: its role in the provision of nutrition support. J Clin Pathol. 2008;61:1083–88.

51. Devoto G, Gallo F, Marchello C, Racchi O, Garbarini R, Bonassi S, Albalustri G, Haupt E. Prealbumin serum concentrations as a useful tool in the assessment of malnutrition in hospitalized patients. Clin Chem. 2006;52:2281–85.

52. Mears E. Outcomes of continuous process improvement of a nutritional care program incorporating serum prealbumin measurements. Nutrition. 1996 Jul-Aug;12(7–8):479–84.

53. Raguso CA, Dupertuis YM, Pichard C. The role of visceral proteins in the nutritional assessment of intensive care unit patients. Curr Opin Clin Nutr & Met Care. 2003;6(2):211–16.

54. Elamin EM, Camporesi E. Evidence-based nutritional support in the intensive care unit. International Anesthesiology Clinics. 2009;47:121–38.

55. Andersen HF. Use of fetal fibronectin in women at risk for preterm delivery. Clinical Obstetrics and Gynecology. 2000;43(4):746–48.

56. Changjiang G, Qishou X, Jingyu W, Chinfan G. Responses of plasma fibronectin to the changes of dietary protein levels in rats. Nutrition. 1997;13(4):327–29.

57. Huang Q, NaiYJ, Jiang ZW, Li JS. Change of the growth hormone-insulin-like growth factor-I axis in patients with gastrointestinal cancer: related to tumour type and nutritional status. Am J Clin Nutr. 2005 May;81(5):1163–67.

58. Reimund JM, Arondel Y, Escalin G, Finck G, Baumann R, Duclos B. Immune activation and nutritional status in adult Crohn's disease patients. Dig Liver Dis. 2005 Jun;37(6):424–31. Epub 2005 Mar 17.

59. Ballard TL, Clapper JA, Specker BL, Binkley TL, Vukovich MD. Effect of protein supplementation during a 6-mo strength and conditioning program on insulin-like growth factor I and markers of bone turnover in young adults. Am J Clin Nutr. 2005 Jun;81(6):1442–48.

60. Shiraishi T, Kawahara K, Yamamoto S, Maekawa T, Shirakusa T. Postoperative mangagement after esophagectomy: is TPN the standard of nutritional care? Int Surg. 2005 Jan-Mar;90(1):30–35.

61. McMillan DC. An inflammation based prognostic score and its role in the nutrition based management of patients with cancer. Proc Nutr Soc. 2008 Aug;67(3):257–62

62. Oh RC, Brown DL. Vitamin B_{12} deficiency. Am Fam Phys. 2003;67(5):979–93.

63. Mackle TJ, Touger-Decker R, Maillet JO, Holland BK. Registered dietitians' use of physical assessment parameters in professional practice. J Am Diet Assoc. 2003;103(12):1632–38.

64. Shopbell JM, Hopkins B, Shronts EP. Nutrition Screening and Assessment. In: Gottschlich MM,Ed. The Science and Practice of Nutrition Support. Dubuque (IA): Kendall-Hunt; 2001.

65. Kenjle K, Limaye S, Ghurgre PS, Udipi SA. Grip strength as an index for assessment of nutritional status of children aged 6–10 years. J Nutr Sci Vitaminol. 2005;51(2):87–92.

66. Bin CM, Flores C, Alvares-da-Silva MR, Francesconi CF. Comparison between handgrip strength, subjective global assessment, anthropometry, and biochemical markers in assessing nutritional status of patients with Crohn's disease in clinical remission. Dig Dis Sci. 2010;55;137–44.

67. Theander K, Jakobsson P, Jörgensen N, Unosson M. Effects of pulmonary rehabilitation on fatigue, functional status and health perceptions in patients with chronic obstructive pulmonary disease: a randomized controlled trial. Clin Rehabil. 2009;23(2):125–36.

68. Gropper SS, Smith JL, Groff JL. Body composition and energy expenditure. In: Human Nutrition and Metabolism. 5th ed. Belmont (CA): Wadsworth/Thomson Learning; 2009.

69. Weir JB de V. New methods for calculating metabolic rate with special reference to protein metabolism. J Physiol. 1949;109:1–9.

70. Walker RN, Heuberger RA. Predictive equations for energy needs for the critically ill. Resp Care. 2009;54:509–21.

71. Frankenfield D, Roth-Yousey L, Compher C. Comparison of predictive equations for resting metabolic rate in healthy nonobese and obese adults: a systematic review. J Am Diet Assoc. 2005 May;105(5):775–89. Review.

72. Frankenfield D, Hise M, Malone A, Russell M, Gradwell, Compher C. Evidence Analysis Working Group. Prediction of resting metabolic rate in critically ill adult patients: Results of a systematic review of the evidence. J Am Diet Assoc. 2007;107:1552–61.

73. Frankenfield DC, Coleman A, Alam S, Cooney RN. Analysis of estimation methods for resting metabolic rate in critically ill adults. J Parenter Enteral Nutr. 2009;33(1):27–36.

74. Boullata J, Williams J, Cottrell F, Hudson L, Compher C. Accurate determination of energy needs in hospitalized patients. J Am Diet Assoc. 2007;107:393–401.

75. Harris JA, Benedict FG. A Biometric Study of Basal Metabolism in Man. Publication No. 279. Washington (DC): Carnegie Institute; 1919.

76. Mifflin MD, St Jeor ST, Hill LA, Scott BJ, Daugherty SA, Koh YO. A new predictive equation for resting energy expenditure in healthy individuals. Am J Clin Nutr. 1990;51:241–47.

77. KrenitskyJ. Adjusted body weight: pro: evidence to support use of adjusted body weight in calculating calorie requirements. Nutr Clin Pract. 2005;20:468–73.

78. Ireton Jones CS, Turner WW. Adjusted body weight, con: why adjust body weight in energy-expenditure calculations. Nutr Clin Pract. 2005;20:474–79.

79. Walker RN, Heuberger RA. Predictive equations for energy needs for the critically ill. Respr Care. 2009;54:509–21.

80. Food and Nutrition Board, Dietary Reference Intakes for Energy Carbohydrates, Fiber, Fat, Protein, and Amino Acids. Washington (DC): National Academy Press; 2002.

81. U.S. Department of Health and Human Services, Public Health Service, Centers for Disease Control and Prevention, National Center for Chronic Disease Prevention and Health Promotion, Division of Nutrition and Physical Activity. Promoting physical activity: a guide for community action. Champaign (IL): Human Kinetics; 1999.

82. Ainsworth BE, Haskell WL, Leon AS, et al. Compendium of physical activities: classification of energy costs of human physical activities. Medicine and Science in Sports and Exercise. 1993;25(1):71–80.

83. World Health Organization. International Statistical Classification of Diseases and Related Health Problems, 1989 Revision, Geneva, World Health Organization, 1992.

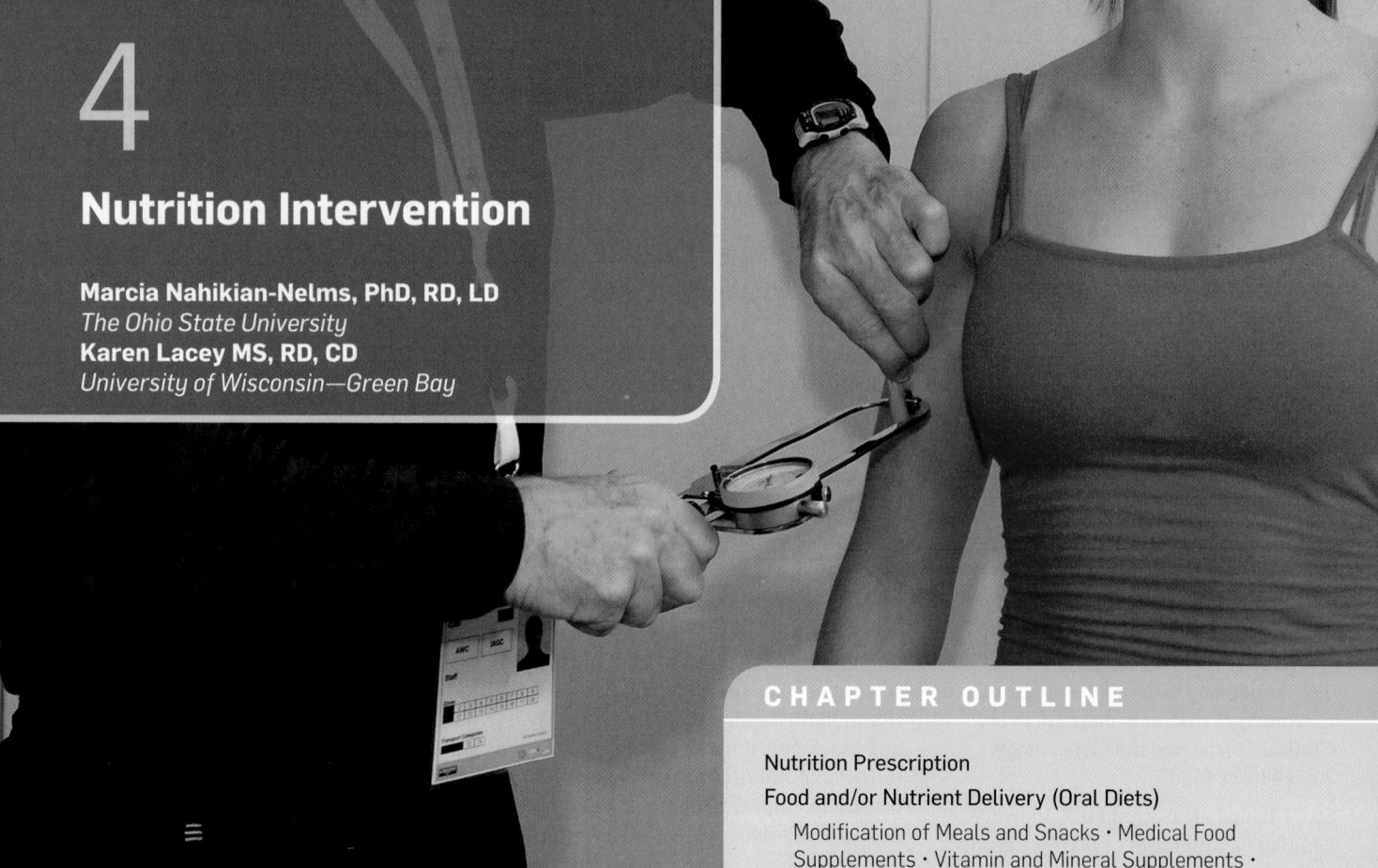

4

Nutrition Intervention

Marcia Nahikian-Nelms, PhD, RD, LD
The Ohio State University
Karen Lacey MS, RD, CD
University of Wisconsin—Green Bay

Introduction

Medical nutrition therapy, as discussed in Chapter 1, is defined as "nutritional diagnostic therapy, and counseling services for the purpose of disease management, which are furnished by a registered dietitian or nutrition professional."[1] The responsibility and scope of practice for the registered dietitian center on the provision of nutrition care utilizing the nutrition care process (refer to Chapter 2). The specific purpose of nutrition intervention, the third step in this process, is to resolve or improve the identified nutrition problems. This is accomplished by planning and implementing appropriate nutrition interventions that are tailored to the patient/client needs.[2] Intervention strategies are selected based on the nutrition diagnoses and their etiologies (refer to Chapter 2) and are intended to change (a) nutritional intake, (b) nutrition-related knowledge or behavior, (c) environmental conditions, or (d) access to supportive care and services. The ultimate goal of the nutrition intervention will, of course, be to improve the individual's overall nutritional status and to support the medical care of that individual. This chapter discusses the types of nutrition interventions that registered dietitians provide within each of the four categories listed previously. Specific interventions for patients requiring enteral and parenteral nutrition are covered in Chapter 5.

Nutrition Prescription

The nutrition prescription outlines the steps to best meet the nutritional needs for the patient and support the medical care that is prescribed by the health care team. "The nutrition prescription concisely states the patient/client's individualized recommended dietary intake of energy and/or selected foods or nutrients based on current reference standards and dietary guidelines and the patient/client's health condition and nutrition diagnosis" (p. 343).[2] The role of the registered dietitian in nutrition interventions is outlined in Table 4.1. Again, as discussed in Chapter 1, the American Dietetic Association defines **evidence-based dietetics practice** as "the incorporation of systematically reviewed scientific evidence into food and nutrition practice decisions. It integrates professional expertise and judgment with client, customer and community values and evaluates outcomes."[3] Throughout this text, as pathophysiology and nutrition therapy are discussed, evidence is provided to support the clinician's ability to develop an appropriate nutrition prescription.

Food and/or Nutrient Delivery (Oral Diets)

The first step in prevention and treatment for malnutrition is an adequate supply of acceptable food composing a diet that has

GLOSSARY

clear liquid diet—diet consisting of liquids that contribute minimal residue to the gastrointestinal tract; includes fruit juices without pulp, carbonated sodas, broth, tea, coffee, water, popsicles, fruit ice, Jell-O (gelatin), and liquid nutritional supplements (e.g., Resource Breeze®)

dysphagia—difficulty swallowing

evidence-based dietetics practice—dietetics practice in which systematically reviewed scientific evidence is used to make food and nutrition practice decisions

full liquid diet—diet consisting of all beverages allowed on clear liquid diets with addition of milk, ice cream, yogurt, and liquid nutritional supplements (e.g., Ensure®, Boost®)

hyperosmolar—having a higher osmolality than body fluids (>300 mOsm/kg)

isotonic—having the same osmolality as body fluids (approximately 300 mOsm/kg)

medical foods—foods administered under the supervision of a physician and intended for the specific dietary management of a disease for

which distinctive nutritional requirements are established

NPO—nil per os, which is Latin meaning "nothing per mouth"

osmolality—number of water-attracting particles per weight of water in kilograms (expressed as mOsm/kg)

structured lipid—chemically modified triglyceride that allows change to the fatty acid composition

---

Table 4.1 Standards of Practice: The Registered Dietitian's Role in Nutrition Intervention

Each RD Plans the Nutrition Intervention
- Prioritizes the nutrition diagnoses based on problem severity, safety, patient/client needs, likelihood that nutrition intervention will influence problem, and patient/client perception of importance
- Bases intervention plan on best available evidence (e.g., national guidelines, published research, evidence-based libraries, and databases)
- Refers to policies and program standards
- Confers with patient/client and caregivers
- Determines patient/client-focused goals and expected outcomes
- Details the nutrition prescription
- Defines time and frequency of care
- Utilizes standardized language for describing interventions
- Identifies resources and/or referrals needed

Each RD Implements the Nutrition Intervention
- Collaborates with colleagues
- Communicates the plan of care
- Initiates the plan of care
- Continues data collection
- Individualizes nutrition intervention
- Follows up and verifies that nutrition intervention is occurring
- Adjusts intervention strategies, if needed, as response occurs
- Documents: Date and time; specific treatment goals and expected outcomes; recommended interventions; adjustments to the plan and justification; client/community receptivity; referrals made and resources used; other information relevant to providing care and monitoring progress over time; plans for follow up and frequency of care; rationale for discharge if applicable and actions are carried out

Source: The American Dietetic Association Quality Management Committee. American Dietetic Association Revised 2008 Standards of Practice for Registered Dietitians in Nutrition Care; Standards of Professional Performance for Registered Dietitians; Standards of Practice for Dietetic Technicians, Registered, in Nutrition Care; and Standards of Professional Performance for Dietetic Technicians, Registered. J Am Diet Assoc. 2008;108:1541.

Collaboration with colleagues allows for appropriate nutrition interventions.

Source: Courtesy of Marcia Nelms.

The general-purpose, healthful diet—commonly referred to as "regular" or "house" diet—served in hospitals and long-term care is supplied in a minimum of three meals each day. Regulations govern the timing, frequency, and nutrient content of meals.[4] Menus are written and approved by a Registered Dietitian and are designed to provide the Dietary Reference Intake for all nutrients. When patients have the option to select the food items they prefer, as they do in most institutions, one of the simplest yet most helpful interventions may be to assist a patient with menu selection. Offering suggestions and appropriate substitutions can be an efficient method of ensuring that the patient's diet remains adequate and acceptable. Restaurant-style menus, room service ordering systems, a la carte food carts, and individual unit kitchens and galleys are all examples of methods to assure patient satisfaction and ability to individualize menu selection.[5, 6]

Modifications of a general diet are often recommended as part of the nutrition prescription for patients under the care of a registered dietitian. These changes result in "therapeutic diets" that have several important functions. They may be used to maintain or restore health and nutritional status. They may be modified to accommodate changes in digestion, absorption, or organ function. Texture and consistency can be adjusted to alleviate mechanical problems. Therapeutic diets can provide the appropriate nutrition therapy to support weight loss or gain, or

been individualized to age, height, weight, activity level, and medical condition. Malnutrition in acute care is often the result of chronic illness experienced by patients before admission, but it is exacerbated by pain, anxiety, depression, and unfamiliar foods or meal schedules associated with admission to a health care setting. Insufficient food intake may be related to the factors depicted in Figure 4.1.

Figure 4.1 Factors Affecting Nutritional Intake during Illness

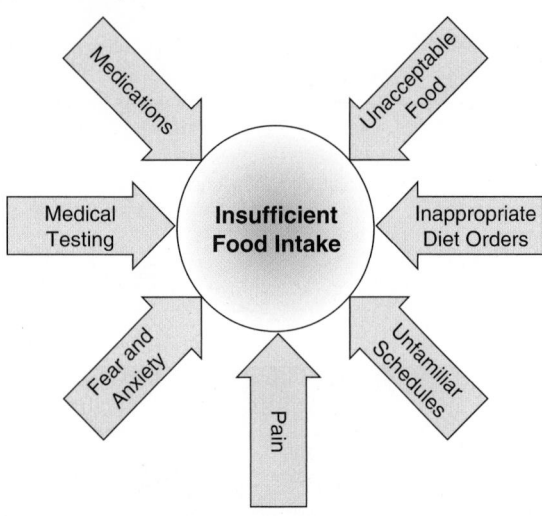

to assist with treatment of a particular diagnosis through nutrient content changes. Further changes to the "house" or "regular" diet can include calorie level; consistency; single nutrient manipulation; method of preparation; specific food restriction; number, size, or frequency of meals; and/or the addition of supplements.

Modification of Meals and Snacks

The regular diet may be modified for patients with impaired chewing ability so that softer foods are served. Soft diets contain foods that are easy to chew and usually omit raw fruits and vegetables. Individuals with **dysphagia** (difficulty swallowing) may require more specific modifications of texture and consistency. Diets for these individuals are discussed more thoroughly in Chapter 14.

For very short periods (two or three meals), liquid diets consisting of broth, juice, cream soups, and milk may be served to patients who are beginning to eat after a long period without food (*nil per os*, or **NPO**). These diets are often referred to as **clear** or **full liquid diets**. A clear liquid diet is intended to provide fluid and energy in a form that requires minimal digestion

and contributes to limited residue in the gastrointestinal tract. It may be used during acute gastrointestinal distress, during gastrointestinal medical testing (such as a colonoscopy), or prior to surgery. Clear liquid diets are inadequate in kilocalories (kcal), protein, vitamins, and minerals, so they should be used only when medically necessary. Historically, the clear liquid diet has been used as a progression toward solid food after a surgical procedure or when the GI tract required minimal stimulation, but more recent opinions question this necessity.[7-11] A full liquid diet also has been used as a transitional diet between liquids and solid foods. Because this diet includes milk and milk products, it may cause intolerance due to the large amounts of lactose. Table 4.2 outlines the basic principles of these liquid diets.

An important component of ensuring tolerance to oral diets, especially clear and full liquid diets, is the consideration of the **osmolality** of the particular liquids that are provided. **Hyperosmolar** liquids may not be tolerated during these transitional periods or when the gastrointestinal tract has not been stimulated. Table 4.3 provides the osmolality of common liquids that are used in these diets. Choosing those with a lower osmolality may assist in ensuring a successful tolerance for the transition to oral feeding.

Oral diets may also be modified to prepare patients for a specific medical test. For example, a high-fat diet (100 g/day) may be administered for two to three days prior to a test for fat malabsorption. Details of the types of diets served are recorded in an institution's diet manual, such as the American Dietetic Association's Nutrition Care Manual.

Given adequate appetite along with sufficient resources to purchase and prepare food, most malnourished individuals can be rehabilitated with oral diet alone. But maximizing oral intake within the hospital setting is often challenging, because this environment is not always conducive to eating. Add to this environment the stress, fear, pain, and isolation of illness and it is a wonder that anyone who is hospitalized can eat adequately. For these individuals, a number of alternatives exist. A primary function of nutrition services in health care institutions is to be the patient's nutrition advocate. When patients present with a suboptimal intake, nutrition services staff members work with the patient and health care team to provide nutritional options.

Table 4.2 Principles of Clear and Full Liquid Diets

Diet	Purpose	Foods Acceptable	Limitations
Clear liquids	Intended to supply fluid and energy in a form that requires minimal digestion and stimulation of the GI tract	• Clear fluids or foods that are liquid at body temperature and leave minimal residue • Clear fruit juices • Bouillon, consommé, clear broth • Gelatin, fruit ice, plain hard candy, sugar, honey • Commercially prepared low-residue, lactose-free nutritional supplements	• Not nutritionally adequate. • Should be limited to 24 to 48 hours unless supplements are added.
Full liquids	Transition between clear liquids and solid food	• Consists of foods or fluids that are or become liquid at body temperature • All clear liquids • Cream soups • Milk, ice cream, pudding, yogurt	• May present problem with large amounts of lactose.

Table 4.3 Osmolality of Selected Liquids

Note: Beverages with an asterisk (*) are considered **isotonic**.

Beverage	(mOsm/kg)	Beverage	(mOsm/kg)
Milk*	275	Prune juice	1265
Malted milk	940	Grape juice	863
Ice cream	1905	Apple juice	683
Eggnog	695	Orange juice	614
Fruit yogurt	871	Tomato juice	595
Sherbet*	125	Punch with sugar	448
Ensure/Boost	590/640	Sugar-free punch*	29
Ensure Plus/ Boost Plus	680/720	Mineral water*	74
Boost Breeze	920	Broth	445
Enlive!	840	Polycose	900
Resource fruit beverage	750	Flavored gelatin	735
Enteral formulas	250–710	Popsicles	720

Source: Rees Parrish C. The clinician's guide to short bowel syndrome. Nutrition issues in gastroenterology, series #31. Practical Gastroenterology. Sept. 2005, pp. 88–89.

Medical Food Supplements

Modified Foods and Beverages Oral intake may be increased by supplementing a balanced diet with between-meal or evening snacks of nutrient-dense foods acceptable to the individual patient. For these snacks, traditional foods such as fruit, crackers, sandwiches, milk shakes, custard, or pudding

Table 4.4 Nutrition Interventions to Increase Nutrient Density

Increasing Energy Content

- Add butter or margarine to cooked cereals, soups, vegetables, or casseroles.
- Add jam, jelly, or honey to toast or other breads and crackers.
- Use whole milk or cream with soups, casseroles, creamed vegetables, or shakes and smoothies.
- Add sour cream or yogurt to soups, casseroles, creamed vegetables, or shakes and smoothies.
- Add nut butters or cream cheese to raw vegetables, bread, or crackers.

Increasing Protein Content

- Add powdered milk to any beverage, soup, or casserole.
- Add liquid egg substitutes to shakes, soups, vegetables, or casseroles.
- Wherever possible, add nuts, nut butters, chopped meats, cooked eggs, cheese, or yogurt to prepared foods.
- Add tofu or soy crumbles to any prepared vegetable, soup, or casserole.

may be served. Increasing nutrient density without actually increasing volume can be an effective tool for the individual who is suffering from decreased appetite. For example, instead of using skim milk, the patient could receive whole milk with 2 T. of dry milk powder to boost both kcal and protein. Adding peanut butter to toast for breakfast is an excellent method to increase both kcal and protein. Table 4.4 provides examples of methods to increase nutrient density using readily available foods, and Figure 4.2 presents a specific breakfast meal that was modified in this manner.

Figure 4.2 Increasing the Nutrient Density of a Breakfast

Increasing nutrient density is an important tool to maximize the nutritional intake of patients.

Breakfast #1

Breakfast #2

Breakfast #1 uses oatmeal with added skim milk powder, raisins, almonds, and brown sugar; bran muffin; apple; and fruit yogurt smoothie to provide approximately 874 kcal and 25 grams of protein.

Source: Courtesy of Marcia Nelms.

A typical breakfast of ¾ c. dry cereal, ½ c. skim milk, whole-grain bread, 4 oz. juice, and apple will provide approximately 340 kcal and 8 grams of protein.

A Review of the MCT Modular Supplement

Medium-chain triglycerides (MCT) are eight- and ten-carbon-chain fatty acids, liberated from coconut oil and then re-esterified to glycerol. Because MCT do not depend on pancreatic lipase or emulsification for digestion, they are used clinically to supply kcal to patients with a variety of pancreatic and gastrointestinal disorders.

MCT are hydrolyzed more readily than LCT (long-chain triglycerides) by lipase, even in the absence of emulsification by bile salts. After transport into the enterocyte, they are not packaged into chylomicrons but instead are absorbed into the portal bloodstream and transported bound to albumin where they can be metabolized in the liver to energy.

Medium-chain fatty acids (MCFA) and long-chain fatty acids (LCFA) also differ in their metabolism. In the liver, LCFA must be transformed into acyl carnitine derivatives before they can enter the mitochondria for subsequent beta-oxidation. Carnitine acyl transferase I (CAT I) and carnitine acyl transferase II (CAT II) on the inner mitochondrial membrane are both necessary for the entry of LCFA into the mitochondria. MCFA do not need CAT I or CAT II for entry and their subsequent beta-oxidation.

The oxidation of MCFA in the fed or fasted state will result in increased production of acetoacetate, 3-hydroxybutyrate, and acetone, three molecules known as ketone bodies. LCFA only produce ketones in the fasted state since malonyl CoA, an intermediate of carbohydrate metabolism, inhibits CAT I, thus decreasing the entry of LCFA into the mitochondria in the fed state. Although ketones have a "bad" reputation as a result of high blood concentrations seen during diabetic ketoacidosis, they can be efficiently utilized as oxidative fuels and converted to fatty acids.

Examples of supplements that contain MCT include MCT Oil (Mead-Johnson), MCT Fuel (Twin Labs), and MCT Power (Universal Nutrition). One tablespoon provides 6 grams of MCT and 50 kcal. Most recently Abbot Nutrition has developed a "**structured lipid**" which chemically produces a combination of MCT and LCT on one triglyceride. This allows the delivery of MCT to the peripheral tissues where it can be used directly as substrate within that cell.

Another means to modify foods is to add single nutrients such as protein or fiber through the use of "modular" products. Appendix J provides information for currently available modular products. A carbohydrate modular supplement, such as Polycose®, provides glucose polymers and generally does not alter the taste of the food to which it is added. Protein modulars, such as ProMod® or Beneprotein®, can be added to both foods and beverages but will need to be mixed well. There is a slight change in taste and consistency, which is also the case for a lipid modular such as medium-chain triglyceride (MCT) oil. In general, modulars are not as cost-efficient as other types of supplements, and they also increase the labor costs. Box 4.1 provides more information about the MCT supplement.

Commercial Beverages Liquid meal replacement formulas may provide a convenient alternative to between-meal snacks. These products typically come in single-portion containers providing 250 to 350 kcal with 7 to 15 grams of protein in 250 mL, and may be available in a variety of flavors. These products are lactose free. Some contain fiber, and others are more calorically dense or higher in protein. Examples of these products include Ensure®, Boost®, MightyShakes®, Resource® Health Shake™, and Carnation Instant Breakfast®.

Manufacturers have introduced many variations of these products for specific medical conditions, such as wound healing or diabetes. Some of these products may be available in liquid form, as puddings, or as cereal-type bars. Because unopened supplement packages do not require refrigeration, these products may be served at a time convenient to the patient. In nursing homes, they may be administered in place of water with medications as a means of increasing nutrient intake.

Commercial supplements are popular because of their convenience and also because patients and caregivers may be familiar with them due to marketing through the media. However, acceptability and intake are highly individual. Patients receiving oral supplements frequently develop "taste fatigue" a few days after

A commercial oral supplement beverage.

Source: Courtesy of © Nestlé Healthcare Nutrition, Inc.

supplements are initiated, and supplement intake then decreases. It is also important to remember that merely providing supplemental feedings will not increase appetite; in fact, many patients complain that extra portions, frequent meals, and supplemental snacks are overwhelming and reduce appetite. This is why it is essential to include the patient in the decision-making process for changes in the meal plan. The patient needs to understand why oral supplements are being offered and how they could improve his or her current medical status. Providing supplement taste tests could be one way to help patients decide which supplement they would prefer to add to their diet. Developing a rotation for snacks or supplements and setting portion goals with the patient may also improve acceptance. Regular follow-up and monitoring are necessary in order to coordinate successful interventions and minimize waste associated with unused products.

Vitamin and Mineral Supplements

If during the nutrition assessment, the patient's diet is determined to be inadequate to meet the standard recommended amounts of essential vitamins and minerals, supplements of these nutrients should be prescribed. Additionally, many medical conditions interfere with digestion, absorption, or utilization of these micronutrients. Thus, making appropriate, evidenced-based recommendations for supplementation of vitamins and/or minerals is an expected step in nutrition intervention. For example, if the patient is diagnosed with osteopenia, the registered dietitian would recommend supplementation with the Adequate Intake (AI) for calcium that is consistent with the patient's age and gender. Furthermore, the RD would support this recommendation with instructions on appropriate sources of supplements to maximize absorption and utilization as well as any specific guidelines on potential drug-nutrient interactions. Resources to assist with this decision-making process are included in the web links at the end of this chapter.

Bioactive Substance Supplements

Bioactive substances are defined as food substances added to a food product or taken as a supplement that have a specific intended health purpose. For example, a patient with hyper-

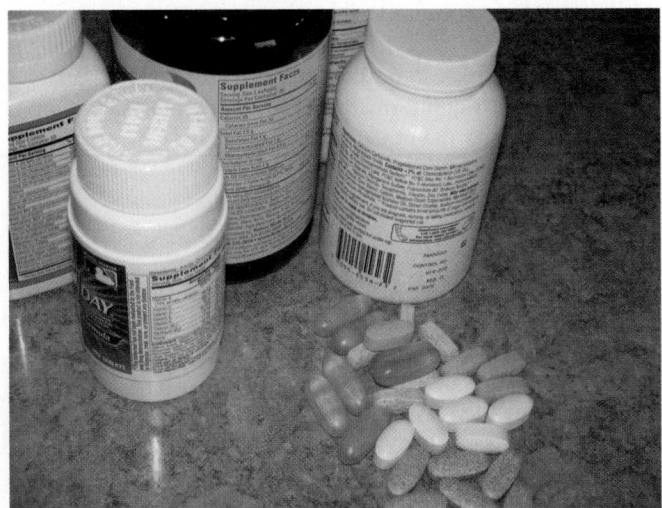

The registered dietitian should make recommendations for supplementation using the latest evidence-based research.
Source: Courtesy of Marcia Nelms.

Benecol is an example of a bioactive substance supplement prescribed as part of the nutrition therapy for hyperlipidemia.
Source: © mediablitzimages (uk) Limited/Alamy.

lipidemia may be instructed to consume stanol esters as a supplement. The registered dietitian, using the American Heart Association's guideline for the National Cholesterol Education Program, would instruct the patient on the amount of stanol esters recommended for the intended reduction of lipid levels. Next, the RD would describe the current products available for purchase and assist the patient with the incorporation of the products into their dietary plan. Throughout this text, many bioactive substances are discussed as a component of nutrition therapy and medical care. Other examples include pro-/ prebiotics or specific types of fiber.

Nutrition Intervention: Feeding Assistance and Feeding Environment

During nutrition assessment and the subsequent identification of nutrition problems, it is not uncommon to discover that even though the appropriate and adequate diet is available to the patient, he or she is unable to consume adequate amounts. Changing the environment to allow for food choice— for instance, organizing family-style meals in a rehabilitation facility—can significantly improve the patient's ability to eat. Preparing the patient to eat may include helping her/him to sit at the appropriate height and distance from the tray or scheduling appropriate mouth care for the patient prior to the meal. Other nutrition interventions may include recommendations for adaptive equipment and providing assistance with eating. Ensuring an appropriate and conducive environment is a team effort requiring the expertise of occupational and physical therapists, speech-language pathologists, and all levels of nursing care. Communication is key to ensure that all components of the nutrition intervention will be successful.

Nutrition-Related Medication Management

During nutrition assessment, each patient's medications are evaluated for possible drug-nutrient interactions. Chapter 12, "Pharmacology," covers this topic in depth. Interventions can address how dietary intake affects drug dissolution, absorption, metabolism, and excretion. Additionally, interventions may target the effects of prescribed drugs on nutrient ingestion, absorption, metabolism, and excretion.

The registered dietitian may also be involved in the coordination of medications with meal planning. An important example would be the development of insulin-to-carbohydrate

Evidence-Based Guidelines

Ethan A. Bergman, PhD, RD, CD, FADA *Central Washington University*

Susan N. Hawk, PhD, RD *Central Washington University*

It is important that the dietitian use the most up-to-date information when providing care for patients. This information must be based on evidence supported by well-controlled research studies and clinical practice. Evidence-based recommendations or guidelines are scientifically developed to assist health care professionals in making appropriate decisions about patient care.

The following are sources of evidence-based research designed to help health care professionals choose the best clinical approach to patient care:

1. American Dietetic Association (ADA). Evidence Analysis Library. Available at www.adaevidencelibrary.com. The Library has been created to summarize the best available research in dietetics and nutrition. Access to the ADA Evidence Analysis Library is free to ADA members but requires a subscription for nonmembers.

2. National Guideline Clearinghouse (NSG). Available at www.guideline.gov. The NSG is a resource for

evidence-based clinical practice guidelines for physicians, nurses, and other health care professionals, including dietitians.

3. Cochrane Library. Available at www.cochrane.org. The Cochrane Library publishes Cochrane Reviews, which are based on the best available information about health care interventions. The reviews explore the evidence for and against the effectiveness and the appropriateness of various types of medical treatment.

4. Agency for Health Care Research and Quality (AHRQ). Available at www.ahrq.gov. AHRQ is the research arm of the U.S. Department of Health and Human Services (HHS). It examines how people access health care, its cost, and the results of this care. The main goals of AHRQ are to identify the most effective ways to organize, manage, finance, and deliver high-quality health care.

ratios as the RD assists the patient with the more complex components of medical nutrition therapy for diabetes (see Chapter 17). Another example would be the use of pancreatic enzyme dosages with meals. The RD provides specific instructions for dosing of enzymes for individuals with cystic fibrosis or other conditions of pancreatic insufficiency. Chapter 14 discusses the role of pro-/prebiotics as a complement to other nutrition therapies and medical care for a number of gastrointestinal conditions—just one more example of the crucial role the RD plays in coordination of patient care.

For patients who are unable to eat amounts of food sufficient to maintain their weight, it is necessary to identify other factors that impair intake. Nonfood causes of poor intake range from poorly fitting dentures to lack of interest in unfamiliar foods to depression. If these factors cannot be resolved and poor intake persists, drugs that stimulate appetite are sometimes ordered. These drugs, including prednisone, megestrol acetate, and dronabinol, are available by prescription.[12] Like all drugs, appetite stimulants can produce significant side effects in some patients.

Prednisone is an inexpensive steroid that is effective over a short period of time and induces an increased sense of well-being and short-term increases in appetite. No studies have shown this drug to be beneficial in increasing body weight over the long term.[13] Patients receiving prednisone may experience muscle weakness, disordered body fat distribution, hyperglycemia (elevated blood glucose), and impaired immunity.

Megestrol acetate will increase appetite and body weight when administered with exercise and nutrition support, but it can

take several weeks to improve intake.[14–16] The drug is also expensive—it can cost several hundred dollars per month—and its side effects can include impotence, vaginal bleeding, and deep vein thrombosis.

Dronabinol is a derivative of marijuana that may improve appetite, but it has not been associated with weight gain. Dronabinol is expensive, and users have experienced nausea, vomiting, and mental status changes, including euphoria and somnolence. As new information becomes available, drug doses may change. Thus, recommendations to use appetite stimulants should be preceded by a thorough review of updated dosing and complication information from reliable sources such as Drug Facts and Comparisons[17] or the American Hospital Formulary Service's AHFS Drug Information.[18] For more information on appetite stimulants and other interventions, see Chapters 23 and 24.

Nutrition Education

In addition to the many types of nutrition interventions discussed previously that focus on the delivery of foods and nutrients, the registered dietitian also provides nutrition-related information to patients and clients in order to change or reinforce eating behaviors. This can either be provided as nutrition education or nutrition counseling; however, it is important to recognize that these are very different processes. Nutrition education is defined as: "a formal process to instruct or train a patient/client in a skill or to impart knowledge to help patients/clients voluntarily manage or modify food choices and eating behavior to maintain or improve health" (p. 345);[2] whereas nutrition counseling is defined as "a supportive process, characterized by a collaborative counselor-patient/client relationship, to set priorities, establish goals, and create individualized action plans that acknowledge and foster responsibility for self-care to treat an existing condition and promote health" (p. 354).[2] Nutrition counseling is discussed in the next section of this chapter.

There are potential opportunities for providing education to patients and clients in nearly every encounter. Nutrition education may occur in a variety of environments or through various mediums including a group class, individual instruction, written instructions, or via telephone or electronic communication. An initial or brief nutrition education is defined as: "Instruction or training to build or reinforce basic nutrition-related knowledge or to provide essential nutrition-related information until patient/client returns" (p. 354).[2]

The acute care setting is certainly not as conducive to education as the outpatient setting can be. Unfortunately, for many patients their hospitalization is the only time that they will have contact with the RD. Furthermore, illness, pain, and an uncomfortable environment can hinder the educational process. We all understand that it is difficult to adjust to illness and even more difficult to understand numerous pieces of information from a variety of individuals. If adequate education or follow-up is needed, referral to an outpatient RD is optimal. While in the hospital, the RD may initially provide basic education that will allow the patient to develop "survival skills" until further education is available.

Basic guidelines for providing nutrition education include the following:

- Clearly communicate the purpose of the education.
- Prioritize the nutrition issues or problems so that education is not complex.
- Tailor the education to fit the individual patient by understanding the level of baseline knowledge, skills, and learning style of the client.

Comprehensive nutrition education is defined as: "Instruction or training intended to lead to in-depth nutrition-related knowledge and/or skills in given topics" (p. 354).[2] These more complex topics or skills require the use of numerous resources. Certainly, this level of education will most likely require follow-up with more time for the patient. Examples of topics for this type of education include carbohydrate counting; developing carbohydrate-to-insulin ratios; or balancing activity levels with diet, insulin, and blood glucose records (see Chapter 17).

The goals of both the brief and comprehensive nutrition education encounter are to develop knowledge and/or skills to make appropriate dietary and nutrition-related changes that promote positive health and nutrition outcomes. The skills and resources needed by the RD include effective communication, use of terms that can be understood by patient, reading materials appropriate to client, visual aids to support the verbal information provided, and sensitive listening skills. Box 4.3 provides an overview of writing skills required for developing nutrition education materials.

BOX 4.3   CLINICAL APPLICATIONS

Writing for Nonmedical Audiences: Instructional Materials for Patients, Their Families, and the Public

Ralph G. Nelms, PhD *The Ohio State University*

Writing to the community of health professionals within your institution or agency will become easier with practice. But writing for other health professionals is not the only kind of writing you may be called on to do. You may also find yourself writing for various nonprofessional audiences, including the public in general. This section describes some of the most common of these forms of writing.

The purpose of these materials is to inform, sometimes with the goal of persuading the reader to take action, if necessary, after reading the material. Examples include informational material outlining nutrition therapies and lifestyle changes. The audience for

such materials obviously differs greatly from the professional audience that charting addresses. You should avoid most medical abbreviations, because your readers simply will not be familiar with them. You cannot use professional or even academic jargon. Words like "data" will need to be replaced with more generally understood words like "information." In other words, use commonsense language. Remember, the goal here is to instruct. Your reader cannot be instructed if she or he cannot understand what is being said. Establish the appropriate reading level and even have a member of your target audience evaluate your instructional material.

Tips for Writing Instructional Materials

Ask nonprofessionals you know to give you feedback on the instructional materials you write before you prepare them for distribution so that you can revise the text if it is unclear to your test audience. You can also establish a focus group that is representative of your intended audience for the purpose of evaluating and responding to your writing. Their insight can help ensure that you will meet the audience's needs.

Use numbering and bullets to create easy-to-read lists. As a model, consider

(continued)

the way bullets and numbers are used in this chapter.

Spend time planning the document you want to produce before beginning to actually write it. Put together a rough organizational plan of the information. This prewriting planning will make writing a lot easier.

Leave yourself plenty of time to revise the document once it is drafted. Read through it first to make sure you included all the information (the content) that you planned to convey. Read

through it a second time to make sure that the organization makes sense; that information introduced early in the document, for example, is explained immediately rather than much later in the document. Read through the document a third time to check the language and make sure it is understandable to your audience. Make sure you check your spelling, too.

A warning: There is one important similarity between writing to professional peers and writing to nonmedical pro-

fessionals: the need for clarity. Sometimes, when shifting from technical writing intended for the professional community to writing for nonprofessionals, writers also shift from an ideal of clarity in their writing to an ideal of eloquence. Eloquence is fine for novelists, but it is irrelevant here. Be clear, be concise, and use language appropriate for your audience.

Nutrition Counseling

Nutrition education and nutrition counseling share the common goal of assisting the patient to make appropriate diet and lifestyle changes to improve his/her health and nutrition status. Nonetheless, there are significant differences in how education and counseling are provided. As discussed previously, nutrition education primarily involves the transfer of knowledge and/or skill building. That does not imply that nutrition education isn't important or relevant; however, having information and/or knowing how to complete a specific task such as label reading or recipe modification alone does not necessarily translate into a behavior that is sustainable. In other words, "knowing is not always doing." The role of the RD as counselor has evolved from that of a clinician who mainly provides information on what and how to eat to that of one who is able to evaluate and take into consideration the complex social and physiological factors that influence food and lifestyle choices. It is important that the counselor develop a collaborative relationship with the patient/client that enables careful examination of nutrition problems to establish goals and plans. The ultimate goal of counseling is for the patient/client to take responsibility for behaviors that improve his/her nutritional status in order to treat an existing condition and promote health.[2]

Counseling Skills

Effective nutrition counseling is greatly influenced by the relationship between the patient/client and the dietetic practitioner, as illustrated in the central core of the Nutrition Care Model (see Figure 2.3 in Chapter 2). The definition of nutrition counseling that is now incorporated into the standardized nutrition intervention terminology supports the importance of this relationship by noting the following important assumptions and characteristics of the role of the RD as counselor:

- *Supportive:* The counselor's role is to encourage and positively guide the patient as changes are made. Being a champion of change and advocating for the benefits of a healthy lifestyle can be especially useful to patients who are undergoing dietary changes.
- *Process:* In other words, counseling is not a one-time encounter in which everything important about nutrition can be explained. Counseling is a process that involves

important follow-up and continued contact in order to be most effective. ADA's Evidence-Based Guidelines for Medical Nutrition Therapy for Diabetes Mellitus (DM) conclude that there is strong support for the improved effectiveness of an initial series with a registered dietitian of three to four encounters each lasting 45–90 minutes. These encounters should be completed within three to six months of diagnosis of DM. Furthermore, at least one follow-up encounter is recommended annually to reinforce lifestyle changes and to evaluate and monitor progress.[3]

- *Collaborative:* Because effective counseling requires working together with the patient/client as a partner to solve problems and create new ideas, the role of the counselor is more subordinate than authoritarian. Changes are more likely to be made when the patient values the benefits of the change and can take personal ownership rather than only being told what to do.
- *Relationship:* Developing a professional relationship with the patient/client that is built on trust and honesty is extremely valuable. Sharing information about one's dietary and lifestyle behaviors is very personal. Many persons are already aware that some of their practices are not necessarily in the best interest of their health, and they may feel ashamed or fear being judged when answering diet history questions. Therefore, characteristics and communication skills that demonstrate good listening and acknowledgement of the client are essential to building trust and promoting openness. Tables 4.5 and 4.6 illustrate many of these important characteristics and communication skills.
- *Individualized:* Unlike nutrition information that describes healthy eating guidelines for a group or population, such as the dietary guidelines, nutrition counseling can take on many different shapes and forms, based on well-known and researched counseling theories and strategies that are tailored to an individual's needs and environment.
- *Self-care:* Even though the nutrition counselor is a champion of change and a partner in this process, the long-term expectation is that the client him- or herself will be able to maintain appropriate changes and solve problems in order to make the diet and lifestyle choices long lasting. That does not limit the opportunity for on-going support and follow-up by the counselor in order to evaluate and monitor progress.

As important as building a positive relationship and demonstrating active listening are, it is equally important to obtain adequate and accurate information from patients and clients. Therefore, developing communication skills of inquiry and appropriate questioning techniques are also essential for the counselor. Examples of these types of questions and skills are summarized in Box 4.4.

Table 4.5 Characteristics of Counselors that Promote a Positive Relationship

- Behave naturally: Appear authentic and sincere; encourages spontaneity and openness on the part of the client
- Have a sense of humor: Helps the client to not take problems too seriously; helps break down barriers between the counselor and the client.
- Be flexible: Do not have unrealistic expectations.
- Be optimistic and hopeful: Clients respond well and appreciate the support.
- Encourage clients to talk: They may not have had opportunities to in the past; provide verbal and nonverbal responses such as nods and occasional "hmms."
- Maintain appropriate eye contact: Look at clients but do not stare.
- Develop attentive body language: Use relaxed gestures and sit or stand with an open and welcoming, calm posture.
- Listen with an open mind and spirit of inquiry: Be sure that responses and nonverbal language are not judgmental.

Source: Adapted from: Bauer K & Sokolik C. Basic Nutrition Counseling Skill Development. Belmont CA: Wadsworth Thomson/Cengage. 2002.

Table 4.6 Effective Communication Skills That Demonstrate Active Listening and Undivided Attention to Clients

Clarifying (Probing): Confirm the accuracy of a client's statement by asking a question or prompting the client to continue talking.
- "Can you explain further. . .?"
- "Tell me more about. . . ."

Paraphrasing (Summarizing): Lets clients know that the counselor is listening; allows the client to clarify any misunderstanding.
- "Let me summarize what I think you just said. . . ."

Responding with empathy: Demonstrates that a counselor understands what a person feels from their frame of reference; clients can feel they are no longer alone. It is not effective to simply state that you know how another person feels.
- "It sounds like. . . ."
- "It seems like. . . ."

Conveying respect: Show consideration and appreciation for the time and information that the client shares.
- "I am impressed with how you. . . ."
- "You have done a great job of. . . ."

Source: Adapted from: Bauer K & Sokolik C. Basic Nutrition Counseling Skill Development. Belmont CA: Wadsworth Thomson/ Cengage. 2002.

BOX 4.4	CLINICAL APPLICATIONS

Effective Communication Skills for Obtaining Accurate Information from Clients

Questioning: When used appropriately, questioning can be very effective; however, it is important to know when to ask a question and what type of question best meets the need of inquiry at a particular point in time during the interview or counseling session. As a general rule, it is advisable to avoid questions that begin with "why" as they are often perceived as judgmental and accusatory. Clients can become defensive and less willing to provide accurate and truthful information if they feel that they need to defend their answer. Following are examples of common types of questions used by nutrition counselors:

- **Closed-ended questions:** Questions designed to obtain either a yes or no response or very brief answer. It is generally recommended that use of this type of question be limited, as a closed-ended question

tends to prompt a response thought be correct or preferred.
  - *Examples of how these questions may begin:* "is," "are," "did," "how many"
- **Open-ended questions:** Although these types of questions are valuable in that they allow the responder to provide a great deal of information and do not limit a response to a short answer, they do require that the counselor listen very carefully to what is being said. Answers may become lengthy and it may be necessary to redirect the client. The major advantage to using open-ended questions is that clients are less likely to feel threatened and more likely to provide honest responses. They may also provide information that guides appropriate secondary and probing questions.

  - *Examples of how these questions may begin:* "how," "what"
- **Secondary Questions:** These are questions that stem from information that has been provided by the client, about which further detail is needed. These types of questions are very similar to the clarifying or probing questions described in Table 4.6.
  - *Examples of how these questions may begin:* "tell me more," "can you explain further"
- **Funneling Questions:** These are questions that are logically arranged so that a broad topic is first introduced and then subsequent questions narrow the subject or topic into more specific and detailed information. Unlike secondary questions that can be asked at any point in the interview, funneling questions

(continued)

assume a more logical and sequential order of questions.

- *Examples of how these questions may be structured:*
  - "What is your usual meal and snack pattern throughout the day?"
  - "Given that you generally eat lunch at work, what would you typically have?"
  - "What kind of salads and dressings are you likely to purchase at work?"

Giving Feedback or Noting Discrepancy: Occasionally it is necessary to confront a client who may be providing information that is either inconsistent or contradictory or excuses for not making changes. This form of questioning is especially valuable if a client is in denial or is expressing resistance to change. Bauer and Sokolik note a variety of ways that the counselor can bring these discrepancies to the attention of the client.[20]

- State observation without adding "but"; use "and" as the connector. Following the "and" is the observation of the discrepancy.
- Begin the statement with "on the one hand . . . on the other hand . . ."
- Directly say "I see/ hear an inconsistency . . ."

Adapted from: Bauer K & Sokolik C. Basic Nutrition Counseling Skill Development. Belmont CA: Wadsworth Thomson/Cengage; 2002.

Theoretical Framework for Nutrition Counseling

The foundation and supporting principles used to facilitate behavior change in nutrition counseling draw from numerous areas of research within the realms of education and psychology. "Behavior change theories and models provide a research-based rationale for designing and tailoring nutrition interventions to achieve the desired effect. A theoretical framework for curriculum and treatment protocols, it guides determination of (1) what information patients/clients need at different points in the behavior change process, (2) what tools and strategies may be best applied to facilitate behavior change, and (3) outcome measures to assess effectiveness or components of interventions" (p. 345).[2] Table 4.7 provides a summary of the major theories that support nutrition education and counseling methods.

Strategies to Accomplish Nutrition Interventions

Registered dietitians can and should use a number of strategies to assist the client in achieving healthful behavior change. Some of these include motivational interviewing, self-monitoring, and cognitive restructuring. For example, Snetselaar outlines the major motivational strategies as: giving advice, identifying and removing barriers, providing choices, decreasing desirability of a present behavior, practicing empathy, providing feedback, clarifying goals, and active helping.[19] Self-monitoring, as discussed in Chapter 3, may include using a food diary to increase the client's awareness of actual food intake. Cognitive restructuring teaches the client appropriate steps in addressing failures in behavior change. Providing positive approaches to particular dietary changes facilitates the client's efforts to deal with problem behaviors.[19]

Table 4.7 Examples of the Theoretical Foundations for Nutrition Counseling and Education

Theory	Role in Nutrition Counseling and Education
Cognitive-Behavioral	Uses the concept that behavior is learned and is directly related to both internal and external factors. Nutrition counseling and education should focus on interventions that will change environment and learning in order to facilitate behavior change.
Transtheoretical	This description of behavior change uses specific steps to guide a client through development of health behaviors.
Social Learning	Albert Bandura is the primary source of this theoretical model that allows the practitioner to understand the factors that predict and change behavior.
Behavior Modification	These learning concepts from developmental psychology include operant conditioning, imitation, and modeling to evaluate current behaviors and determine steps to initiate behavior change.
Planned Behavior/ Reasoned Action	These theories focus on a client's attitudes and beliefs that support or prevent behavior change.
Person Centered	Developed by Carl Rogers; uses the interventions that will allow the patient to develop individual skills to change his or her own behavior.
Health Belief	Developed originally by Rosenstock in 1960s; identifies the client's perceived ability to accomplish a behavior change.
Precede/Proceed	This model identifies those factors that precede a behavior and those factors that follow with the understanding that their identification will facilitate the behavior change.

Source: Adapted from: Snetselaar L. Nutrition Counseling Skills for the Nutrition Care Process. 4th ed. Boston MA: Jones and Bartlett. 2009; ADA Evidence Analysis Library. Behavior Change Theories. Available from: http://www.adaevidencelibrary.com/topic.cfm?cat=1397; Holli B, Calabrese RJ, and O'Sullivan-Maillet J. Ch. 10 Principles and Theories of Learning. IN: Communication and Education Skills for Dietetics Professionals. Philadelphia (PA): Lippincott Williams and Wilkins. 2003; Bauer K & Sokolik C. Basic Nutrition Counseling Skill Development. Belmont (CA): Wadsworth Thomson/Cengage. 2002.

Coordination of Care

Providing health care is a team effort. Registered dietitians work with numerous other health care professionals with the ultimate goal of improving the lives and health of their patients. The coordination of care encompasses those steps necessary to ensure a successful transition from the health care facility to the patient's home, rehabilitation, or a long-term care facility. Depending on the specific nutrition intervention outlined in the nutrition care plan, the coordination of care can include referrals to community agencies for access to food or nutrition support, or to another RD who specializes in a particular nutritional intervention. Many health care facilities regularly coordinate the care of their patients through team meetings where all of the health goals and interventions can be addressed.

Conclusion

Nutrition and health curriculums for medical practitioners will provide extensive training for both education and counseling. This chapter has simply described the major components of nutrition intervention with the understanding that practicing dietitians must cultivate a much deeper understanding of them. Throughout the study of nutrition therapy, you will need to draw from these principles and apply them in order to execute the appropriate nutrition therapy.

WEB LINKS

Annual Bibliography of Significant Advances in Dietary Supplement Research: 1999–2006 Office of Dietary Supplements, National Institutes of Health. Bibliography of 25 annotated top research articles from peer-reviewed journals published annually.
http://ods.od.nih.gov/Research/Annual_Bibliographies.aspx

Complementary/Integrative Medicine Education Resources M.D. Anderson Cancer Center, University of Texas. Provides evidence-based reviews of complementary or alternative therapies for cancer treatment.
www.mdanderson.org/departments/CIMER/

Computer Access to Research on Dietary Supplements (CARDS) Database Office of Dietary Supplements, National Institutes of Health. Database of federally funded research projects pertaining to dietary supplements.
http://dietary-supplements.info.nih.gov/Research/CARDS_Database.aspx

ConsumerLab.com, LLC Provides independent testing and information on nutrition products (subscription required).
333 Mamaroneck Avenue
White Plains, NY 10605
Phone: 914-722-9149 or
(toll free) 888-502-5100
Email: info@consumerlab.com
www.consumerlab.com/index.asp

Dietary Supplements: An Advertising Guide for Industry Federal Trade Commission, 1998. Reviews the rules for advertising dietary supplements.
www.ftc.gov/bcp/conline/pubs/buspubs/dietsupp.htm

Dietary Supplements Labels Database National Library of Medicine, U.S. Department of Health and Human Services. Offers information about ingredients in more than 2000 selected brands of dietary supplements.
http://dietarysupplements.nlm.nih.gov/dietary/

Dietary Supplements—Warnings and Safety Information Center for Food Safety and Applied Nutrition, Food and Drug Administration (FDA). Lists recent alerts and safety information from FDA on dietary supplements.
www.cfsan.fda.gov/~dms/ds-warn.html

Drugs and Dietary Supplements Federal Trade Commission in cooperation with the Food and Drug Administration. Provides information about fraudulently marketed health products and tips for how to spot false claims.
www.ftc.gov/bcp/menus/consumer/health.shtm

EPC Evidence Report/Technology Assessments The AHRQ produced evidenced based reviews on variety of dietary supplements.

Agency for Healthcare Research and Quality, Department of Health & Human Services
www.ahrq.gov/clinic/epcindex.htm#dietsup

International Bibliographic Information on Dietary Supplements (IBIDS) Database Office of Dietary Supplements, National Institutes of Health and the Food and Nutrition Information Center, National Agricultural Library. A comprehensive bibliographic database that helps health care providers, researchers, and consumers find credible, scientific literature on dietary supplements.
http://dietary-supplements.info.nih.gov/databases/ibids.html

International Food Information Council (IFIC) Newsletters, fact sheets (antioxidants and stanol esters) and web site information on functional foods and supplements.
1100 Connecticut Avenue, NW, Suite 430
Washington, DC 20036
Phone: 202-296-6540 Fax: 202-296-6547
Email: foodinfo@ific.org
http://ific.org

Institute of Food Technologists The Institute of Food Technologists is a nonprofit scientific society advancing the science

and technology of food through the exchange of knowledge.
525 W. Van Buren, Suite 1000
Chicago, IL 60607
Phone: 312-782-8424 or
(toll free) 800-438-3663
Fax: 312-782-8348
Email: info@ift.org
www.ift.org/cms/

The Johns Hopkins Center for Complementary and Alternative Medicine   Research center focusing on complementary and alternative medicine as it relates to cancer and cancer treatments. Provides links to ongoing research studies.
Johns Hopkins CAM Center
Room 7400, 1830 Building
1830 E. Monument Street
Baltimore, MD 21287
Phone: 410-614-5678
Email: jhcam@jhmi.edu
www.hopkinsmedicine.org/cam

MedlinePlus for Dietary Supplements
MedlinePlus Health Information, National Library of Medicine, National Institutes of Health. Links, including overviews, organizations, and latest research, on dietary supplements.
www.nlm.nih.gov/medlineplus/dietarysupplements.html

Micronutrient Information Center   Provides scientific information on the role of vitamins, minerals, and phytochemicals in preventing disease and promoting health.
The Linus Pauling Institute, Oregon State University
571 Weniger Hall
Corvallis, OR 97331-6512
Phone: 541-737-5075 Fax: 541-737-5077
Email: lpi@oregonstate.edu
http://lpi.orst.edu/infocenter/

National Center for Complementary and Alternative Medicine, National Institutes of Health   A subset of PubMed which offers free access to over 270,000 citations of journal articles related to complementary and alternative medicine research from the National Library of Medicine's

MEDLINE database and other life science journals.
www.nlm.nih.gov/nccam/camonpubmed.html

Position of the American Dietetic Association: Fortification and Nutritional Supplements (2005)   Outlines the position of the American Dietetic Association related to fortification of foods and nutritional supplements.
http://eatright.org/ada/files/fortnp.pdf

USDA National Nutrient Database for Standard Reference—Release 20   A database of the nutrient content of over 7500 foods.
Agricultural Research Service, U.S. Department of Agriculture,
www.nal.usda.gov/fnic/foodcomp/search/

Reprinted from: Lindsey AT. Food and Nutrition Information Center. Available at www.nal.usda.gov/fnic/pubs/bibs/gen/**dietarysupplements**consumers.pdf

END-OF-CHAPTER QUESTIONS

1. Describe two ways that the house or regular diet can be modified to accommodate patient needs.

2. What is the difference between clear and full liquid diets? When are they used, and what are their limitations?

3. Identify at least two steps to increase energy and protein in a regular diet.

4. Nutrition interventions can include changes to the environment that could improve nutritional intake. Give examples of interventions that can assist with improving nutritional intake within the environment.

5. What is the major difference between nutrition education and nutrition counseling?

6. Give an example of a closed-ended question and an open-ended question that might be used during a patient encounter. What situation would be best in which to use an open-ended question?

REFERENCES

1. Final MNT regulations. American Dietetic Association website. http://www.eatright.org/ada/files/MNTLegislativeLanguage.doc

2. International Dietetics and Nutrition Terminology (IDNT) Reference Manual: Standardized Language for the Nutrition Care Process. Chicago (IL): American Dietetic Association; 2009.

3. Evidence Analysis Library [Internet]. Chicago (IL): American Dietetic Association. Available at http://www.adaevidencelibrary.com

4. Joint Commission on Accreditation of Healthcare Organizations. 2009 Comprehensive Accreditation Manual for Hospitals: The Official Handbook (CAMH). Chicago (IL): 2009 Comprehensive Accreditation Manual for Hospitals. Oak Brook (IL): Joint Commission on Accreditation of Healthcare Organizations; 2009.

5. Burns J, Gregory S. Changing food-service systems: a balancing act between patient satisfaction and cost. J Foodservice Business Research. 2007;20:63–78.

6. Keller M. That's progress. Advancements in hospital foodservice. Today's Dietitian. 2009;8:28–31.

7. Sathiaraj E, Murthy S, Mansard MJ, Rao GV, Mahukar S, Reddy DN. Clinical trial: oral feeding with a soft diet compared with clear liquid diet as initial meal in mild acute pancreatitis. Aliment Pharmacol Ther. 2008;28:777–81.

8. Schulman AS, Sawyer RG. Have you passed gas yet? Time for a new approach to feeding patients postoperatively. Practical Gastroenterology. 2005;10:82–88.

9. Lassen K, Kjaeve J, Fetveit T, Tranø G, Sigurdsson HK, Horn A, Revhaug A. Allowing normal food at will after major upper gastrointestinal surgery does not increase morbidity: a randomized multicenter trial. Ann Surg. 2008 May;247(5):721–29.

10. Charoenkwan K, Phillipson G, Vutyavanich T. Early versus delayed (traditional) oral fluids and food for reducing complications after major abdominal gynaecologic surgery. Cochrane Database Syst Rev. 2007 Oct 17;(4):CD004508.

11. Lewis SJ, Andersen HK, Thomas S. Early enteral nutrition within 24 h of intestinal surgery versus later commencement of feeding: a systematic review and meta-analysis. J Gastrointest Surg. 2009;13:569–75.

12. Yeh SS, Lovitt S, Schuster MW. Pharmacological treatment of geriatric cachexia: evidence and safety in perspective. J Am Med Dir Assoc. 2007;8:363–77.

13. Mantovani G, Maccio A, Massa E, Madeddu C. Managing cancer related anorexia/cachexia. Drugs. 2001;61: 499–514.

14. Berenstein EG, Ortiz Z. Megestrol acetate for the treatment of anorexia-cachexia syndrome. Cochrane Database Syst Rev. 2005;CD004310.

15. Reuben D, Hirsch S, Zhou K, Greendale G. The effects of megestrol acetate suspension for elderly patients with reduced appetite after hospitalization: a phase II randomized clinical trial. J Am Ger Soc. 2005;53:970–75.

16. Simmons SF, Walker KA, Osterweil D. The effect of megestrol acetate on oral food and fluid intake in nursing home residents: A pilot study. J Am Med Dir Assoc. 2005;6(3 Suppl):S5–S11.

17. Drug facts and comparisons. St. Louis (MO): Walters Klewer; 2009.

18. AHFS Drug Information. Bethesda (MD): American Hospital Formulary Service; 2009.

19. Snetselaar L. Nutrition counseling skills for the nutrition care process. 4th ed. Boston (MA): Jones and Bartlett; 2009.

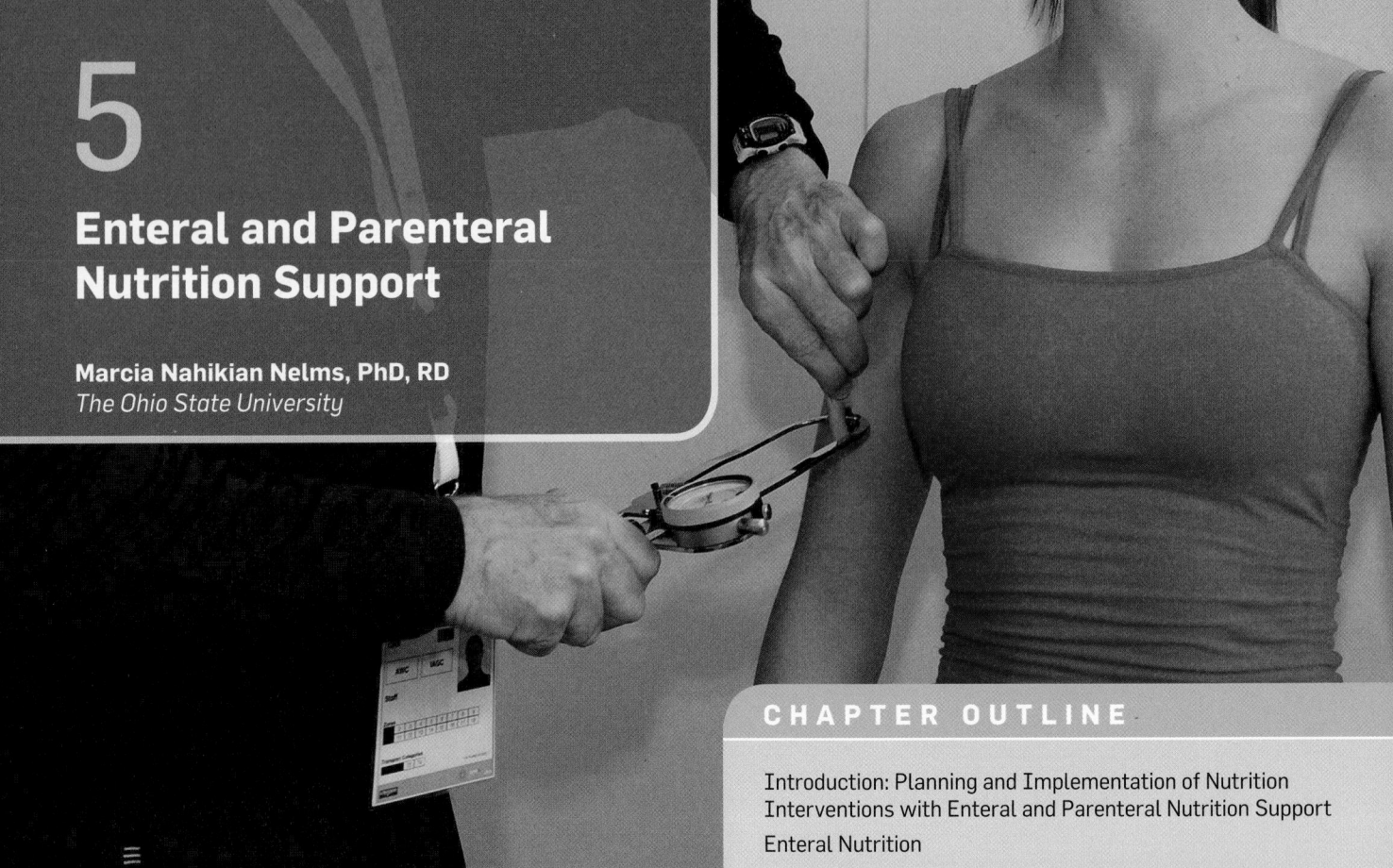

5

Enteral and Parenteral Nutrition Support

Marcia Nahikian Nelms, PhD, RD
The Ohio State University

Introduction: Planning and Implementation of Nutrition Interventions with Enteral and Parenteral Nutrition Support

With increased focus on the issues of overweight, obesity, and chronic disease in health care, it is important not to lose sight of the fact that insufficient food intake and the resulting protein-energy malnutrition remain prevalent among certain populations. As many as 30% to 50% of hospitalized patients and up to 95% of nursing home patients exhibit some signs of malnutrition.[1-4] Malnutrition is not restricted to the elderly, either: Estimations for malnutrition in children admitted to acute care hospitals range from 6% to 14% in Germany, France, the UK, and the United States.[5]

The consequences of malnutrition include increased risk of infection, delayed wound healing, and delayed return to home, work, or baseline activities of daily living following hospitalization.[6-8] All of these consequences contribute to increased health care costs and diminished quality of life. Health care institutions—and in particular, registered dietitians—prevent or treat malnutrition and other diseases by providing the patient with appropriate nutritional support. As was covered in Chapter 4, this support can be provided through an oral diet designed to meet the patient's specific nutritional requirements and modified to meet his or her specific physical needs. This might include modification of type or amount of food and nutrients within meals or at specified times between meals, or the addition of medical food supplements and/or vitamin and mineral supplements. In situations where modification of the oral diet is not enough, enteral nutrition (or tube feeding), and/or parenteral nutrition can be administered. Building on the nutrition intervention principles discussed in Chapter 4, this chapter will focus on food and nutrient delivery for the primary purpose of preventing malnutrition when nutrient needs cannot be met with an oral diet.

For patients who are unable to maintain their nutritional status using oral diets or supplements, nutrition support via enteral and parenteral nutrition is the next alternative. This form of nutrition support became an important nutrition intervention in the 1970s, as methods were developed to provide adequate feedings by vein. Since then, significant technical advancements have occurred as a result of both research and economic necessity. Nutrition support has developed rapidly, and practice is challenging as clinicians strive to keep up with changes in the field. See Box 5.1 for a brief history of nutrition support.

A comprehensive nutrition assessment is needed to determine not only the extent of inadequate nutrition but also the under-

aspiration—inspiration of foreign matter into the lung

bolus feedings—rapid administration of 250 to 500 mL of formula several times daily

central venous catheter (CVC)—intravenous access device inserted into large veins such as the subclavian, jugular, or femoral veins in the center of the body

colonocyte—epithelial cell of the large intestine or colon

continuous feedings—administration of formula for 10 to 24 hours daily, using a pump to control the feeding rate

enteral nutrition (EN)—feeding through the gastrointestinal tract using a tube, catheter, or stoma that delivers nutrients distal to (or beyond) the oral cavity

gastrostomy—an opening into the stomach

hydrophilic—water loving, or attracting water

implantable port—intravenous access device that is completely under the skin, is placed in the vein on the upper chest wall, and exits the body near the xyphoid process, axilla, or abdominal wall

intermittent feedings—administration of formula several times daily, over 20 to 30 minutes

intravenously (IV)—by vein, in reference to administration of drugs or nutrients

jejunostomy—an opening into the jejunum

nasogastric feeding tube—a tube that is inserted nasally (through the nose) into the stomach

nasointestinal feeding tube—a tube that is inserted nasally (through the nose) past the stomach into the intestine

NPO—nil per os, which is Latin meaning "nothing per mouth"

orogastric feeding tube—a tube that is inserted orally (through the mouth) into the stomach

osmolality—number of water-attracting particles per weight of water in kilograms (expressed as mOsm/kg)

osmolarity—number of millimoles of liquid or solid in a liter of solution

ostomy—an artificial opening created by surgical procedure

parenteral nutrition (PN)—administration of nutrition directly into the circulatory system (also known as total parenteral nutrition [PN], central venous nutrition [CVN], or intravenous hyperalimentation [IVH])

percutaneous endoscopic gastrostomy (PEG)—a procedure used by a physician to insert a feeding tube through the skin and into the stomach using an endoscope

peripheral parenteral nutrition (PPN)—administration of nutrition into a vein in the arm or back of the hand (also known as peripheral venous nutrition [PVN])

peripherally inserted central catheter (PICC)—intravenous access device inserted into the arm and threaded into the subclavian vein to the vena cava

propofol—lipid based drug that is used to maintain sedation during mechanical ventilation

refeeding syndrome—metabolic alterations that may occur during nutritional repletion of starved patients

stoma—an opening

stylet—wire guide within the enteral tube that assists with insertion

surgical gastrostomy—an opening into the stomach that requires a surgical procedure

tunneled catheter—intravenous access device that is placed in the vein on the upper chest wall and exits the body near the xyphoid process, axilla, or abdominal wall

viscosity—thickness of a liquid

---

lying cause of poor oral intake (if present). Common nutrition diagnoses that may necessitate alternate routes for nutrition interventions include the following: Inadequate energy intake (NI-1.5), inadequate oral food/beverage intake (NI-2.1), inadequate intake from enteral/parenteral nutrition (NI-2.3), inadequate fluid intake (NI-3.1), inadequate bioactive substance intake (NI-4.1), increased nutrient needs (NI-5.1), malnutrition (NI-5.2), inadequate protein-energy intake (NI-5.3), swallowing difficulty (NC-1.1), altered GI function (NC-1.4), impaired nutrient utilization NC-2.1), underweight or involuntary weight loss (NC-3.1, 3.2), and self-feeding difficulty (NB-2.6). Nutritional support is a significant and extremely important component of nutrition interventions within the nutrition care process, and the nutrition practitioner will use its principles over and over again. Though interventions will be modified depending on the disease course and the patient's individual needs, each intervention begins with the foundation principles presented here.

BOX 5.1   HISTORICAL PERSPECTIVES

The History of Parenteral Nutrition

Providing solutions intravenously is now standard medical practice. Yet safe infusions of saline and dextrose solution were not available until after World War I, and the technology to adequately nourish patients intravenously was not developed until the 1960s. Successful parenteral nutrition began in 1937, when it was reported that you could "feed" a patient peripherally using IV dextrose and protein hydrolysate solutions.[1] Yet adequate protein and kcal were difficult to provide. As much as 5 liters of the dilute nutritional solutions had to be administered through the peripheral hand and arm veins. It was Dr. Stanley Dudrick, a resident in general surgery, who, building on the work of Drs. Rhoads and Vars, perfected a technique in beagle puppies that allowed hypertonic dextrose and protein solutions to be administered through a central vein.[2]

The adoption of central vein parenteral nutrition and its increasing use in the 1970s led to development of new parenteral solutions, such as crystalline amino acid solutions and IV fat emulsions, which provided essential fatty acids.[2] Parenteral nutrition also led to greater recognition for nutrition support of hospitalized patients, resulting in the increased use of enteral nutrition as well. Today, parenteral nutrition is an accepted mode of treatment for the patient who cannot be adequately nourished via the gastrointestinal tract.

References
1. Elman R., Weiner D. Intravenous alimentation with special reference to protein (amino acid) metabolism. JAMA. 1939;112:796–802.
2. Rhoads J, Dudrick S. History of intravenous nutrition. In: Rombeau J, Caldwell M, editors. Clinical nutrition: parenteral nutrition. 2nd ed. Philadelphia: WB Saunders; 1986.

Ethical considerations impact decisions related to specialized nutrition support. Patients and their families should be involved in making these decisions. Furthermore, options for nutrition support therapy should be consistent with the level of medical care that the patient is receiving.[9]

Enteral Nutrition

Enteral nutrition (from the Greek *enteron* or intestine) refers to feeding through the gastrointestinal tract via a tube, catheter, or **stoma** that delivers nutrients distal to (or beyond) the oral cavity.[10] The terms "enteral feeding" and "tube feeding" are used interchangeably in the clinical setting. Medical and nutritional research has increased understanding of the need for gastrointestinal (GI) tract stimulation and the overall health benefits of continuing to provide nutrition via the GI tract, especially in the critically ill. "Nutrition support therapy in the form of enteral nutrition (EN) should be initiated in the critically ill patient who is unable to maintain volitional intake" (p. 279).[10]

Indications

Enteral feeding is indicated for patients who have a functioning gastrointestinal tract but cannot feed themselves adequately. Specifically, enteral nutrition may be recommended for patients with altered mental status, swallowing dysfunction, or disorders of the upper gastrointestinal tract that can be bypassed by inserting a feeding tube below the dysfunction. Examples of nutrition diagnoses that would lead the practitioner to recommend enteral nutrition as an intervention include malnutrition, increased energy expenditure, involuntary weight loss, inadequate oral food/beverage intake, inadequate fluid intake, increased nutrient needs, biting/chewing difficulty, involuntary weight loss, impaired swallowing, and impaired nutrient utilization. Enteral feeding is contraindicated if patients have serious medical conditions that affect the gastrointestinal tract, including diffuse peritonitis (inflammation and infection of the peritoneal lining of the abdominal cavity), GI bleeding, obstruction or ileus that prevents intestinal contents from passing through the intestine, or intractable vomiting or diarrhea not responsive to medical treatment.[11–13]

Research supports the use of enteral nutrition support because of the following advantages:[9–18]

- Cost-effectiveness
- Reduced hospital length of stay

Figure 5.1 Selecting a Feeding Route

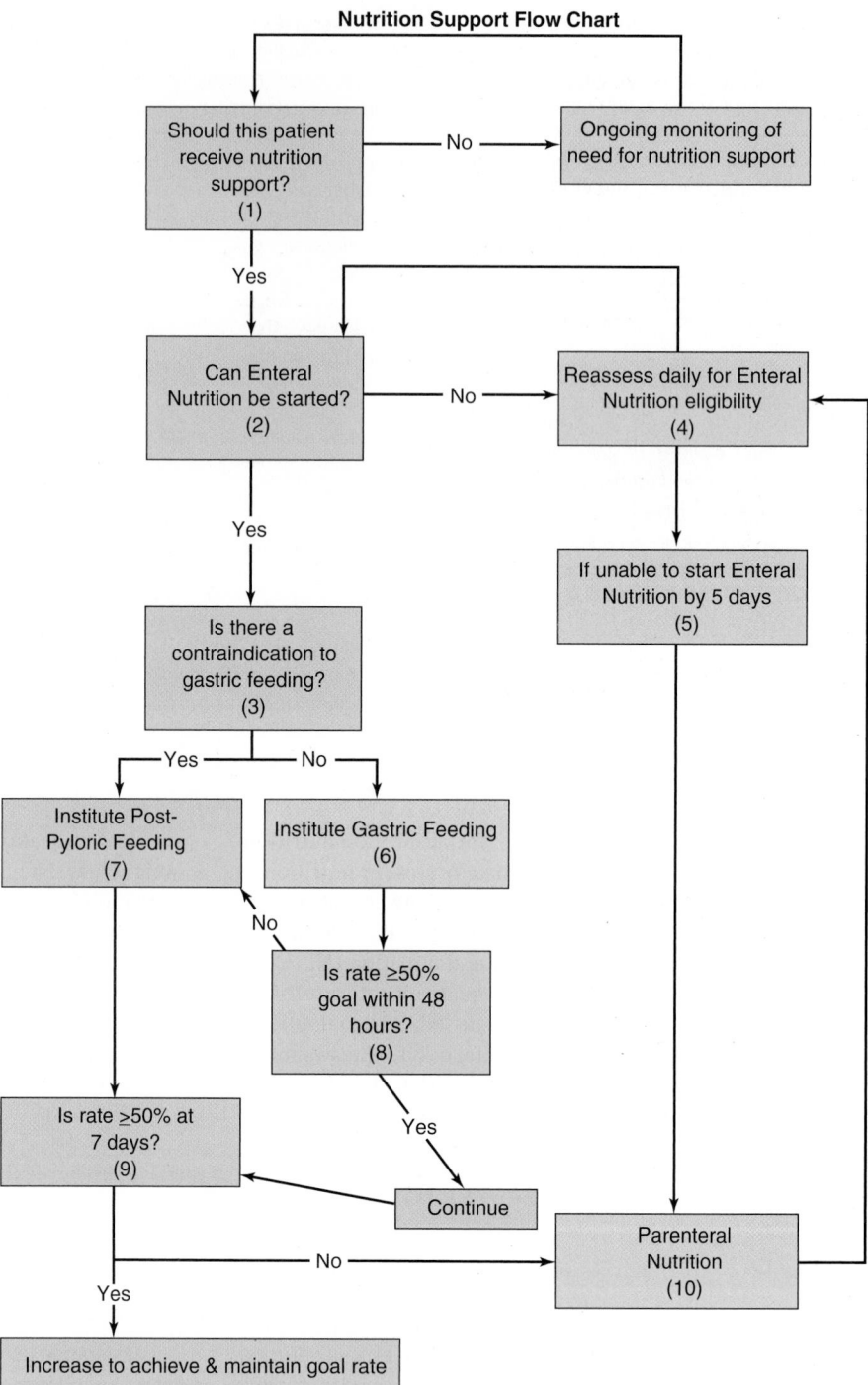

Source: Adapted with permission from: O'Keefe GE, Shelton M, Cuschieri J, Moore EE, Lowry SF, Harbrecht, Maier RV and the Inflammation and the host response to injury collaborative research program. Inflammation and the Host Response to Injury, a Large-Scale Collaborative Project: Patient-Oriented Research Core-Standard Operating Procedures for Clinical Care VIII-Nutritional Support of the Trauma Patient. J Trauma, 2008;65(6):1520–28.

- Reduced surgical interventions
- Reduced rate of infectious complications in critically ill patients
- Improved wound healing
- Maintenance of gastrointestinal function

Even when a patient is unable to meet all nutritional requirements completely with enteral nutrition, the use of trophic or "trickle" enteral nutrition may be prescribed along with parenteral nutrition. This means that a very small rate is prescribed with the goal to minimize villous atrophy and prevent bacterial translocation.[10, 11]

Disadvantages of enteral feeding include the potential difficulty of administration, poor tolerance, and difficulty meeting nutritional requirements of some patients.[11] These disadvantages may be minimized by careful patient selection, thorough nutrition-focused physical assessment, and use of standardized protocols.[11, 14]

Gastrointestinal Access

After a thorough and complete nutrition assessment is completed and it has been determined that enteral nutrition support is both feasible and therapeutic, several important decisions must be made in order to design the enteral nutrition prescription. The first is how to establish access to the gastrointestinal tract. The access route is often determined by the primary physician according to the patient's diagnosis and the anticipated amount of time the patient will require support (see Figure 5.1). Access is achieved when a feeding tube is placed into the stomach or intestine. Figure 5.2 demonstrates the sites for access.

The type of feeding access is described according to (1) where it enters the body, and (2) where the tip is located. Thus, the tube may extend from the nose (naso-) or mouth (oro-) into the stomach, becoming a **nasogastric** or **orogastric feeding tube**.

Nasogastric (nose to stomach) feeding is the most common, the easiest to achieve, and the easiest to maintain. It is also the least expensive and is acceptable in most clinical situations. Small bowel or **nasointestinal feeding tubes** enter the gastrointestinal tract through the nose and reside in the duodenum or jejunum. Nasointestinal access or postpyloric feeding is preferred in some circumstances. For example, postpyloric access is used to bypass the stomach in cases of gastroparesis (delayed gastric emptying), gastric outlet obstruction, or when previous gastric surgery precludes feeding into the stomach. Postpyloric feeding may also minimize accidental aspiration (inhalation) of formula into the lung, but data supporting this proposed benefit are inconclusive.[10, 19]

A disadvantage of nasogastric and nasointestinal feeding tubes is discomfort for the patient. Smaller tubes made of pliable material have been developed to improve patient comfort. Also, nasal tubes are easily dislodged and may have to be replaced frequently.[11, 20] Typically, tubes entering the body through the nose or mouth are used for short-term therapy (usually defined as less than six weeks).

Registered dietitians may be responsible for insertion of nasogastric or nasointestinal feeding tubes in some institutions. Box 5.2 summarizes the protocol for the insertion of tubes. (See the "Equipment" section further in this chapter for more information about the tubes and other supplies involved.)

A more permanent feeding tube can be inserted through the skin, and is usually referred to by the location of the tip of the tube followed by the suffix **ostomy**, from the word *stoma*, meaning opening. Thus, a tube delivering feedings to the

Figure 5.2 Sites for Enteral Access

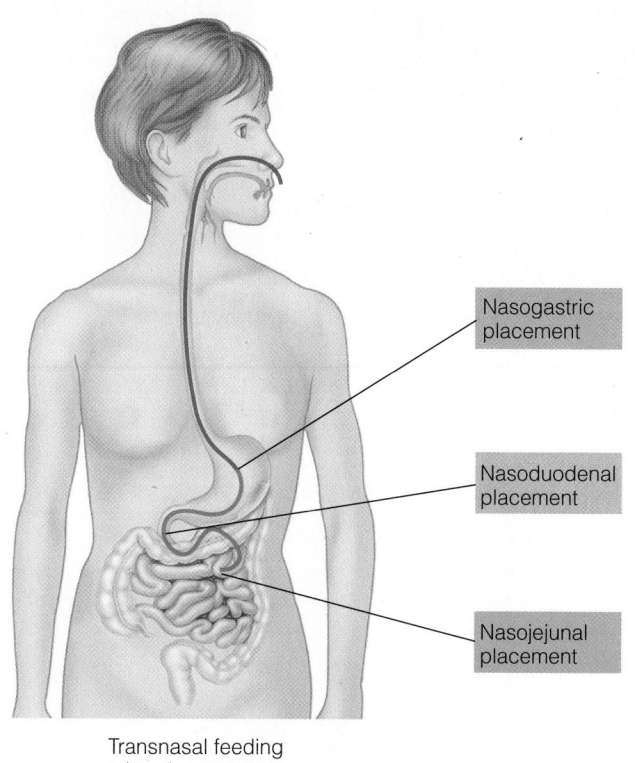

Nasogastric placement

Nasoduodenal placement

Nasojejunal placement

Transnasal feeding tube placements

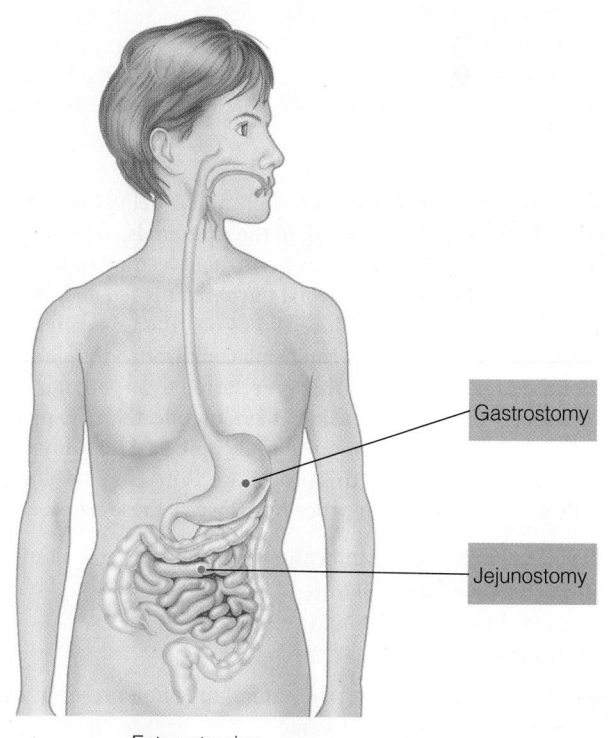

Gastrostomy

Jejunostomy

Enterostomies

Source: S. Rolfes, K. Pinna and E. Whitney, *Understanding Normal and Clinical Nutrition*, 7e, copyright © 2006, p. 655.

BOX 5.2   **CLINICAL APPLICATIONS**

Protocol for Bedside Placement of Feeding Tubes

Supplies:

- 10 Fr feeding tube, with stylet, 43 inch, nonweighted
- 10 mg IV metoclopramide
- Water-soluble lubricant
- 60 cc syringe with Luer Lock tip
- Silk tape
- pH paper (optional)
- Cup with warm water
- Gloves
- Stethoscope

Procedure:

1. Position patient on back or in sitting position, if tolerated, with head of bed at least 30 degrees.

2. Administer 10 mg IV metoclopramide (Reglan)—takes about 10 minutes to begin action.

3. Drape towel over patient's chest.

4. Secure stylet into the feeding tube, and then instill 20 to 30 cc warm water through the tube to activate the internal lubricant.

5. Using the tube, measure for gastric placement on the patient (usually 50 to 65 cm).

6. Insert the feeding tube into the patient's nose, adding water-soluble lubricant as needed for easy advancement. Advance the tube to predetermined gastric placement. If patient is awake/alert, encourage relaxing and swallowing as tube is advanced.

7. Check gastric placement by instilling 15 to 20 cc air utilizing the syringe and simultaneously auscultating over the epigastric area with a stethoscope. Repeat procedure over lungs to assure gastric rather than pulmonary placement.

8. Once gastric placement is confirmed, flush feeding tube with 1 to 15 cc warm water.

9. Loosely tape tube in place.

10. Begin advancing the feeding tube approximately 5 cm every two to three minutes in a corkscrew, clockwise fashion. After every advancement, tape the tube in place.

11. After each advancement, instill 15 to 20 cc air in quick short bursts and auscultate to determine direction of tube movement. Pull back on syringe to obtain aspirate, if available, and then check the pH. (Note if patient is on H_2 blocker to aid with interpreting results.) The pH should be acidic (pH 4 to 5.5) until the tube passes into the duodenum where it will become neutral (pH 6–7.5). Flush feeding tube with 10 to 15 mL warm water with each advancement.

12. Advance the tube to the 100 cm marking; flush tube with 30 mL water.

13. Tape the feeding tube securely in place.

14. Total procedure takes approximately 15 to 30 minutes.

15. Obtain abdominal radiograph to confirm small bowel placement.

16. Upon radiographic placement confirmation, reinsert nasogastric tube if needed.

17. Remove stylet wire from feeding tube and follow standard feeding administration and tube care procedures.

Source: Cresci, G. In: Charney P, Malone A. ADA Pocket Guide to Enteral Nutrition. Chicago, IL: American Dietetic Association, 2006, pp. 39–40.
© 2006 American Dietetic Association. Reprinted with permission.

---

stomach is called a **gastrostomy**, while a tube delivering feedings through the abdominal wall to the jejunum is called a **jejunostomy**. A physician places permanent feeding tubes (like the one shown in Figure 5.3) while the patient is sedated. If a surgeon performs the procedure, it may be called a **surgical gastrostomy**. Feeding tubes can be placed through the skin without a surgical incision, which is referred to as percutaneous gastrostomy. If this is done using an endoscope, then the procedure is called a **percutaneous endoscopic gastrostomy** or **PEG**. Figure 5.4 illustrates this procedure. A gastroenterologist, usually in an outpatient GI procedure, inserts these tubes.

Table 5.1 summarizes the indications, advantages, and disadvantages for each access site.

Formulas

The next step in establishing the enteral nutrition prescription is to consider the choice of an enteral formula (see Figure 5.5). Historically, enteral feedings were composed of liquid mixtures, blenderized foods from a regular diet, or combinations of baby food thinned with milk or juice. Concerns about labor costs,

Figure 5.3 Gastrostomy Tube

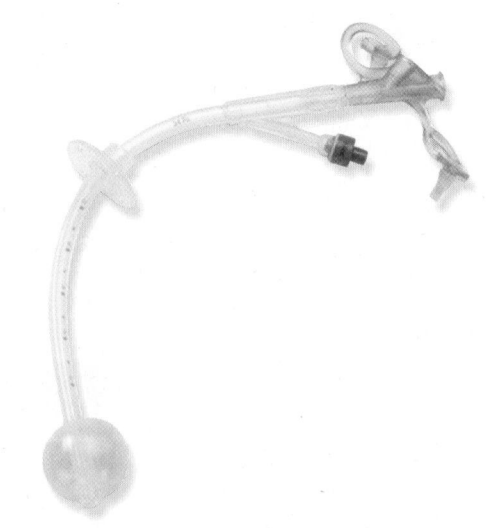

Source: Photo provided courtesy of © Nestlé HealthCare Nutrition, Inc.

Figure 5.4 PEG Tube Placement

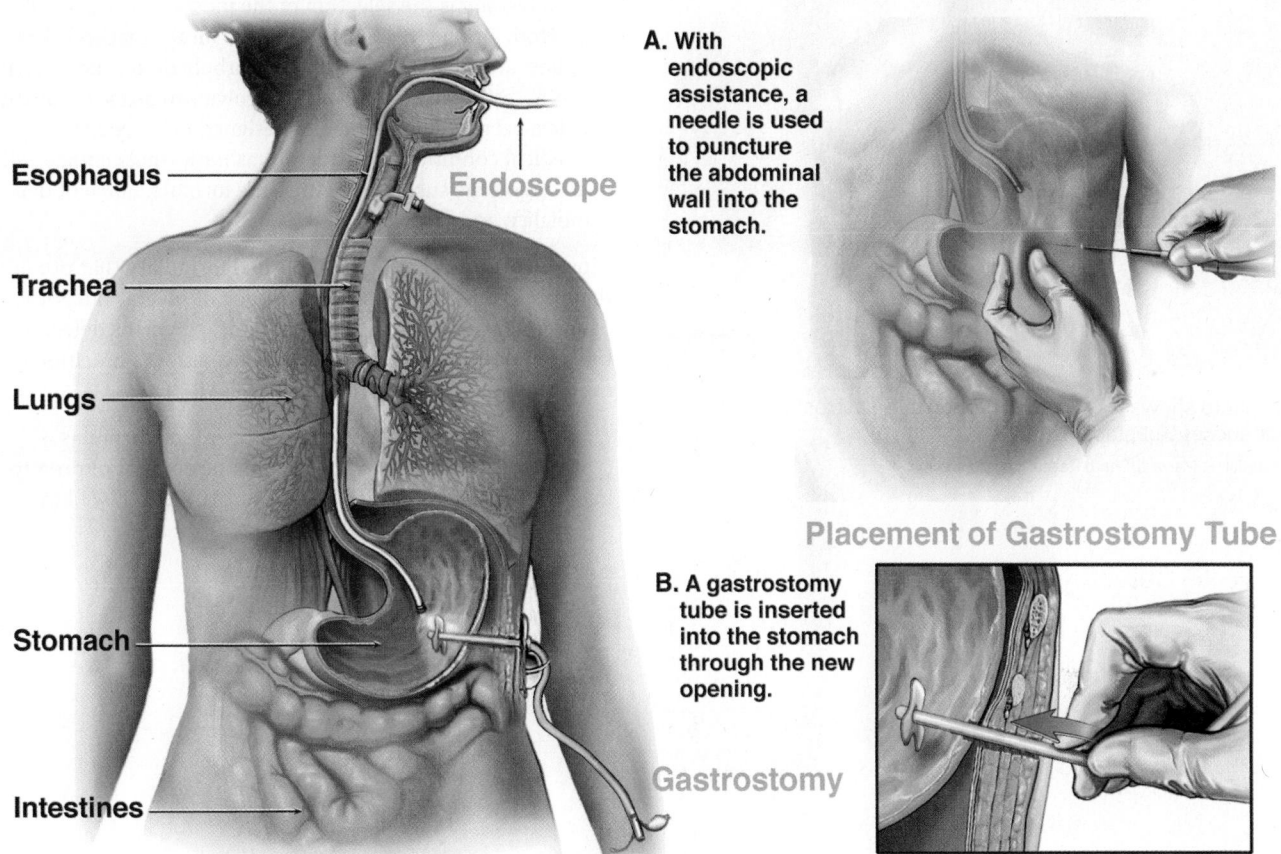

Esophagus

Trachea

Lungs

Stomach

Intestines

Endoscope

A. With endoscopic assistance, a needle is used to puncture the abdominal wall into the stomach.

Placement of Gastrostomy Tube

B. A gastrostomy tube is inserted into the stomach through the new opening.

Gastrostomy

Table 5.1 Summary of Enteral Access Sites

Enteral Access Site	Indications	Advantages	Disadvantages
Nasogastric	Normal GI function	Uses and stimulates normal digestive function; flexibility in administration; medications can be placed in this tube; tube insertion at bedside	Aspiration; discomfort for patient; nasal irritation; tube displacement
Nasoduodenal	Normal small intestine function; need to bypass stomach as primary site of feeding	Tube insertion at bedside	Discomfort for patient; tube displacement
Nasojejunal	Normal small intestine function; need to bypass stomach as primary site of feeding	Tube insertion at bedside	Discomfort for patient; tube displacement
Gastrostomy	Normal GI function but need to bypass upper GI tract; longer-term feeding access	Longer-term feeding access; reduced risk of tube displacement; allows for bolus feedings	Surgical procedure; risk of irritation and infection for insertion site
PEG (percutaneous endoscopic gastrostomy)	Normal GI function but need to bypass upper GI tract; longer-term feeding access	Outpatient procedure without risk of anesthesia; longer-term feeding access; less expensive than surgical insertion; reduced risk of tube displacement; allows for bolus feedings	Risk of irritation and infection for insertion site
Jejunostomy	Normal GI function but need to bypass components of GI tract; longer-term feeding access	Increased tolerance for early initiation of enteral feeding	Surgical procedure; risk of irritation and infection for insertion site; with smaller lumen of tube, the risk of clogging may be greater

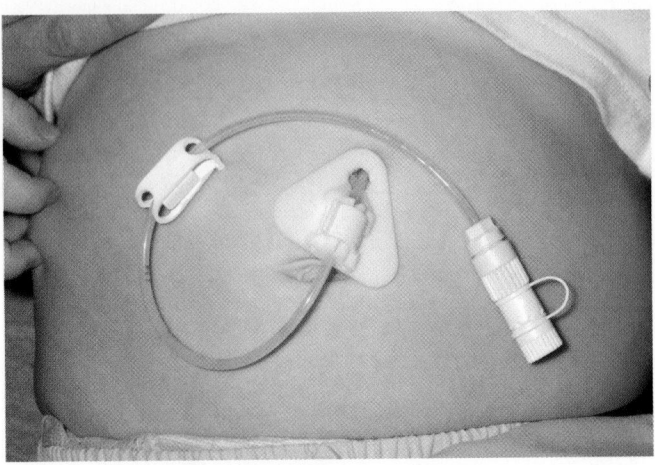

This photo shows the external portion of a gastrostomy tube after successful placement.

Source: Dr. P. Marazzi/Photo Researchers, Inc.

quality control, uncertain formula composition, and sanitation gave rise to commercial formulas. A major consideration in enteral feeding is the selection of the most appropriate formula from among the dozens available on the market. These products are composed of protein, carbohydrate, and fat, with vitamins, minerals, water, and electrolytes in proportions that mimic a balanced diet, or a diet designed for a specific disease or medical condition. Considerations for formula choice will be based on the substrates within the formula, nutrient density, osmolality, and **viscosity**.

Most institutions have an established formulary that provides the most cost-effective choice for each of several categories of enteral formulas. In other situations, clinicians determine the formula choice based solely on the patient's nutritional requirements.

Protein The protein component of enteral formulas is typically derived from soy or casein. The amount of protein in enteral formulas ranges from about 10% to 15% of kcal

Figure 5.5 Selecting a Formula

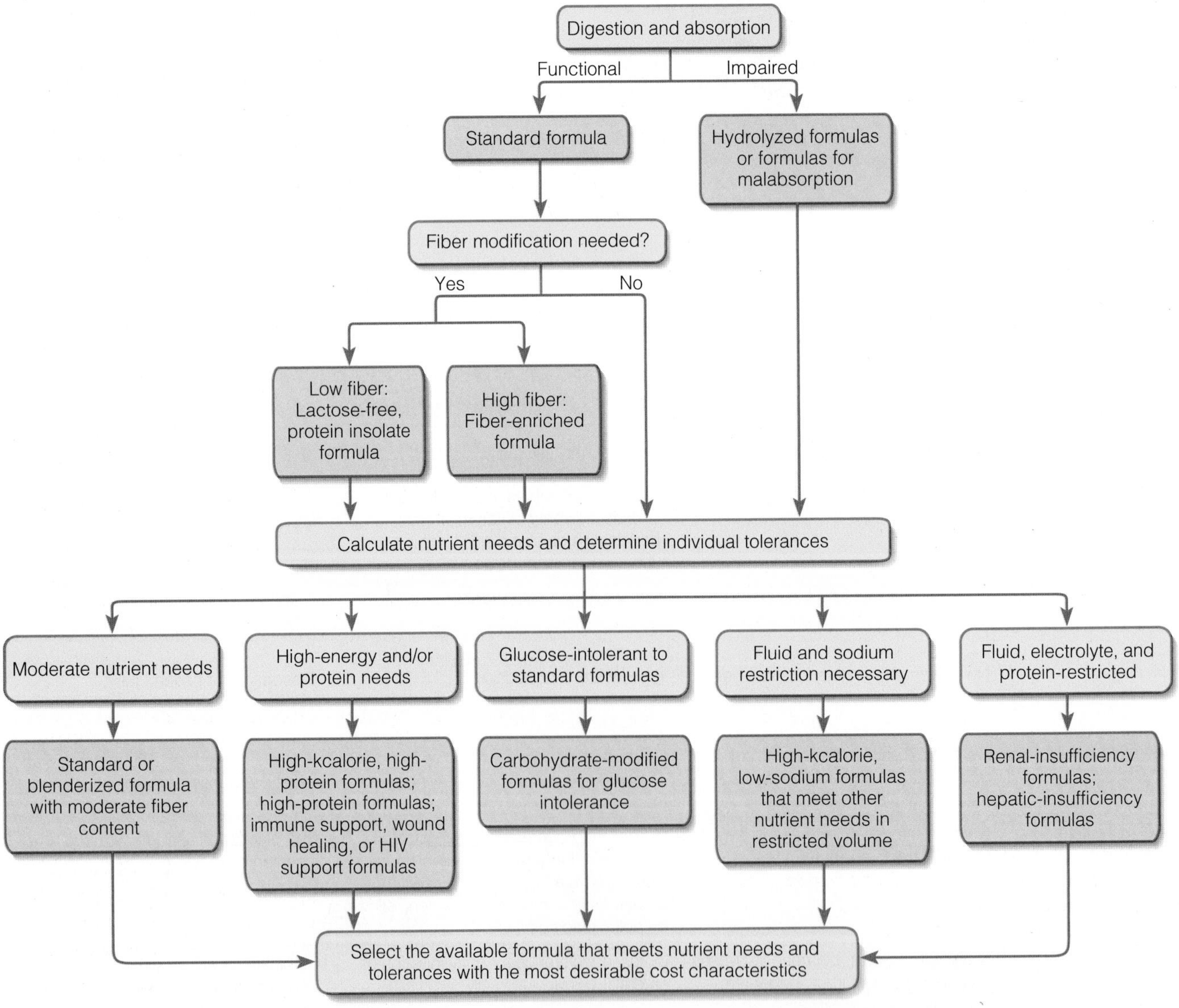

Digestion and absorption

Functional → Standard formula

Impaired → Hydrolyzed formulas or formulas for malabsorption

Fiber modification needed?

Yes → Low fiber: Lactose-free, protein insolate formula

No → High fiber: Fiber-enriched formula

Calculate nutrient needs and determine individual tolerances

Moderate nutrient needs → Standard or blenderized formula with moderate fiber content

High-energy and/or protein needs → High-kcalorie, high-protein formulas; high-protein formulas; immune support, wound healing, or HIV support formulas

Glucose-intolerant to standard formulas → Carbohydrate-modified formulas for glucose intolerance

Fluid and sodium restriction necessary → High-kcalorie, low-sodium formulas that meet other nutrient needs in restricted volume

Fluid, electrolyte, and protein-restricted → Renal-insufficiency formulas; hepatic-insufficiency formulas

Select the available formula that meets nutrient needs and tolerances with the most desirable cost characteristics

Source: S. Rolfes, K. Pinna and E. Whitney, *Understanding Normal and Clinical Nutrition*, 7e, copyright © 2006, p. 658.

for standard formulas to up to 25% of kcal from protein for high-protein formulas. The majority of formulas provide "intact" protein that requires enzymes to split the nutrient into peptides before absorption across the gastrointestinal tract. Formulas containing protein from peptides (also called "hydrolyzed" or "chemically defined" formulas) are used for patients with enzyme deficiency, malabsorption, or other conditions resulting in maldigestion. Formulas with specialized amino acid profiles for renal failure, hepatic failure, stress, and inborn errors of metabolism have been developed from crystalline amino acids (these are also called "elemental" or "chemically defined" diets). Some "elemental" formulas are supplemented with additional amounts of specific amino acids such as glutamine or arginine. The disadvantages of peptide and crystalline amino acid products include increased osmolality and higher cost.

Carbohydrate The carbohydrate sources for enteral formulas are large molecules such as monosaccharides, oligosaccharides, dextrins, and maltodextrins. Formulas are lactose free, and sucrose is rarely used. The carbohydrate composition of some enteral products includes added fructo-oligosaccharides (FOS), which, like all oligosaccharides, are fermented into short-chain fatty acids. Short-chain fatty acids are used by the **colonocytes** (intestinal cells) as fuel and play a role in maintaining gastrointestinal integrity.[21]

Originally, formulas were fiber free and low in residue, but today many products are also available with fiber added. Benefits attributed to fiber, particularly improved bowel function, have more often been associated within soluble fiber. Typically, enteral products contain only small amounts of soluble fiber because it is **hydrophilic** (attracts water). This hydrophilic property causes enteral formulas to thicken and form a gel when fiber is added. Insoluble fibers such as soy polysaccharides are most often found in enteral feedings because they are less hydrophilic.[22]

Lipid The fat or lipid sources for enteral formulas include corn and soy oil, which are long- and medium-chain fatty acids. Concern about the immunosuppressive properties of long-chain fatty acids has increased interest in fat blends containing omega-3 fatty acids, which may play a role in maintaining immune function.[23] Newer products contain structured lipids and omega-3-fatty acids from fish and plant sources.[24, 25] Structured lipids are triglycerides that have had specific fatty acids added or changed. These modified triglycerides have specific physical, chemical, or nutritional characteristics that affect the nutritional or health benefit of the product.[26, 27]

Vitamins/Minerals Enteral formulas meet the Dietary Reference Intakes for vitamins and minerals for adult males and females within a specified volume (usually 1500 mL within 24 hours). Some special formulas contain supplemental amounts for stress and wound healing. It is often necessary to compare the amounts of vitamins and minerals required by individual patients with the amounts provided in the formula and to adjust vitamin and mineral supplements as needed to meet the patient's needs. The discussions of disease-induced variations in nutrient needs within the individual nutrition therapy chapters of this book will provide guidance for planning nutrition support.

Fluid and Nutrient Density Many patients, particularly those with impaired renal, cardiac, or pulmonary function, are unable to tolerate large volumes of fluid. Therefore, nutrient density is of concern in product selection. The nutrient density of an enteral formula is measured in kcal per mL, and usually ranges between 1.0 and 2.0 kcal per mL. Standard feedings contain 1 kcal per mL of fluid, which is consistent with the World Health Organization's recommendations for fluid intake, whereas nutrient-dense formulas are manufactured with less water so that they contain 1.5 or 2.0 kcal per mL of fluid.

Enteral formulas are often the sole source of water for patients receiving them; thus, it is important to ensure adequate fluid intake. Patients who receive these products require careful monitoring of their fluid status to ensure that they remain adequately hydrated. The precise water content of formulas may be obtained from the product literature, but in practice the free water content of formulas may be estimated as about 80% water for 1 kcal per mL formulas and about 65% for 2 kcal per mL formulas.

An additional characteristic to note when choosing a formula is its osmolality and/or **osmolarity**. Osmolality refers to the number of water-attracting particles per weight of water in kilograms (expressed as mOsm/kg).

Osmolality can be an important consideration in selecting enteral feedings. Generally, those formulas that are partially hydrolyzed or considered to be chemically defined have a higher osmolality. The osmolality of body fluids is 300 mOsm/kg. Iso-osmolar (or isotonic—the same osmolality as body fluids) enteral feedings were developed to minimize "dumping syndrome," or diarrhea resulting from rapid movement of fluids into the gastrointestinal tract to dilute hyperosmolar or concentrated fluids. A series of studies conducted decades ago demonstrated tolerance of high-osmolality enteral formulas.[28–31] Consequently, concern about osmolality and diarrhea from enteral feedings has been reduced. Presently, most commercial formulas are of moderate osmolality (300 to 600 mOsm/kg) and similar in osmolality to the popular beverages previously listed in Chapter 4, Table 4.3.

Regulation of Enteral Formula Manufacture
According to the Food and Drug Administration (FDA), enteral formulas are **medical foods** rather than drugs, and hence they are subject to FDA regulations for the accuracy of label claims and standards of manufacturing. There is no requirement that enteral formulas be tested for efficacy or benefit for a particular disease or condition prior to marketing them to professionals or the public. Thus, it is important to evaluate the research conducted to support use of many specialized enteral formulas on the market.[22]

Cost Cost is an important consideration in selecting enteral products. Traditionally, enteral products have been inexpensive, but newer products and those for specialized indications are increasingly expensive. The cost of products varies a great deal according to volume of items purchased. As mentioned previously, many institutions implement a formulary system that limits the total number of products, resulting in an overall reduced cost.

Patients who purchase a few cans of formula from a retail grocery store or pharmacy typically pay a much higher price than large institutions that purchase thousands of cases per year.

Standard products, purchased in bulk by health care institutions, cost as little as a few dollars a case (usually 24 250-mL cans). However, products with specialized amino acid or lipid profiles cost several dollars per 250-mL can.

Feeding Techniques

After the access and formula have been determined, delivery of the enteral feeding should be considered. Several methods to administer enteral feedings have been used successfully.[11, 32] **Bolus feedings** consist of the rapid administration of 250 to 500 mL of formula several times daily. A syringe may be used to inject feedings through the tube (see Figure 5.6). **Intermittent feedings** are also administered several times daily, over 20 to 30 minutes. A pump is typically used to control the flow rate. If pumps are unavailable, formula may flow slowly by gravity into the feeding tube from a container suspended above the patient.

Continuous feedings are administered over 8 to 24 hours daily, using a pump to control the feeding rate. Continuous feedings are typically preferred in the acute care setting as GI tolerance is maximized with this delivery. Additionally, they are easier and less time consuming for staff to administer than bolus or intermittent feedings. Using a pump to deliver feedings slowly at a continuous rate may improve feeding tolerance, but intermittent feedings are also well tolerated by individual patients. A disadvantage of continuous feeding is the expense of the pump and the disposable equipment that is required. Another disadvantage of continuous feeding is restricted mobility if the pump and other equipment are difficult to move. To overcome this disadvantage, the feeding schedule can be adjusted so that feedings are cycled over 8 to 12 hours rather than 24 continuous hours.

Equipment

Sophisticated feeding tubes, feeding administration sets, and pumps have been developed so that enteral feedings are more comfortable for the patient and easier for patients or caregivers to administer. Feeding tubes have been improved so that they are soft and pliable, resist clogging, and have flexible, weighted tips. Most are made of polyvinyl, silicone, or polyurethane. The outer lumen diameter is described using a measurement called *French size* (1 Fr = 0.33 mm). Most tubes range from 10 to 14 Fr.

Enteral formulas can be provided in cans or in pre-packaged sealed containers. In years past, pouring formula into bags (like the one in the photograph) was a standard procedure for formula delivery.

Enteral feeding is most often delivered using a small pump similar to those used to control intravenous fluids (see photo on page 89). Pumps are available that are small enough for

Figure 5.6 Bolus Feeding

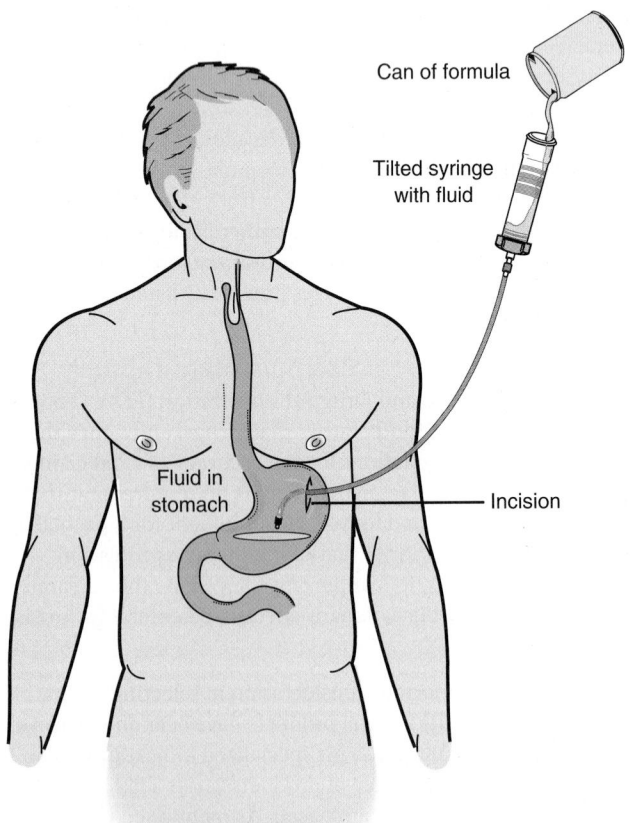

Can of formula

Tilted syringe with fluid

Fluid in stomach

Incision

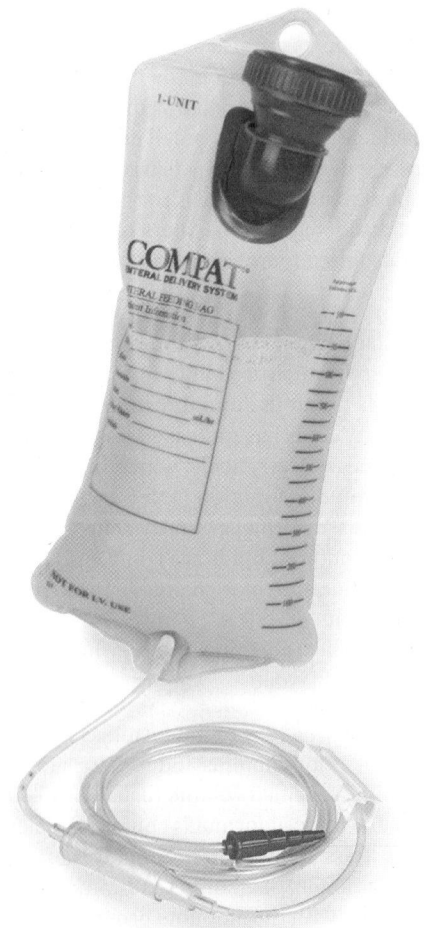

Source: Timby, Website for Fundamental Nursing Skills and Concepts 8e, Copyright © 2005 Lippencott Williams & Wilkins. Instructor's Resource CD-ROM to Accompany Timby's Fundamental Nursing Skills and Concepts 8e, Diana L. Rupert and Geralyn Frandsen.

Formula can be poured into a bag that is then connected to the feeding tube for delivery.
Source: Photo provided courtesy of © Nestlé HealthCare Nutrition, Inc.

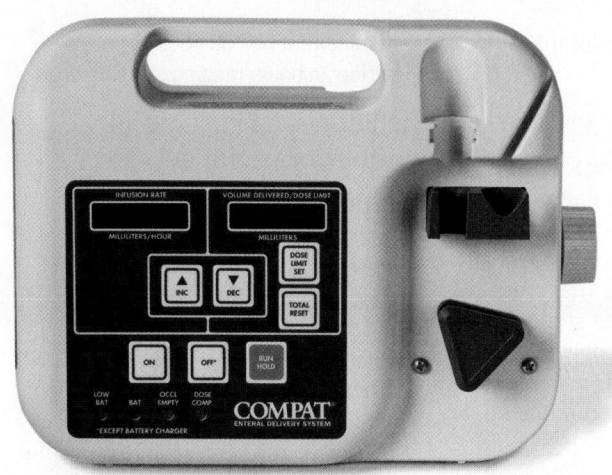

Enteral nutrition delivery is usually regulated by a small, programmable feeding pump.

Source: Photo provided courtesy of © Nestlé HealthCare Nutrition, Inc.

use in ambulatory home care situations and that can be programmed to automatically flush the tube with water. Formula and equipment manufacturers are likely to continue product innovation in order to improve the ease of use and cost-effectiveness of pumps.

Nutrition Assessment and Intervention: Determination of the Enteral Nutrition Prescription

The enteral nutrition prescription will be based on the dietitian's nutrition assessment and recommendations, as described in Box 5.3. The steps in determining the nutrition prescription are the following:

1. Establish dosing weight (the patient's stable weight from which calculations should be made) and protein, energy, and fluid requirements.

2. Identify an appropriate enteral formula that is consistent with the patient's nutritional needs, digestive and absorptive capability, and diagnosis.

3. Calculate the total energy needs and divide by the caloric density of the formula (kcal/mL).

4. Divide the total volume by the total number of hours over which formula will be administered.

Many institutions have specific protocols for initiation, advancement, and transition of feedings. Table 5.2 provides examples of these protocols. Polymeric, isotonic formulas can be initiated at 10–50 mL/hr. The rate is advanced in increments of 10–25 mL/hr every six to eight hours until goal rate is established.[11, 32, 33]

BOX 5.3   CLINICAL APPLICATIONS

Applying the Nutrition Care Process to Develop an Enteral Nutrition Prescription

Nutrition Assessment:

You have been consulted concerning a 76-year-old man who has just had a percutaneous endoscopic gastrostomy (PEG) tube placed and will be discharged to a nursing home in a few days. The patient is recovering from a stroke, which left him with severe swallowing difficulty and persistent aspiration. There is a small possibility that he will recover some swallowing function, but for now, he is NPO. The patient weighed 180 pounds on hospital admission, but his weight has declined to 168 pounds at the present time. He is 5'10" tall. He has been maintained on tube feeding since shortly after hospital admission. Your physical exam reveals that he is appropriately hydrated. His weight has not changed since his medical condition stabilized, and he was transferred from the Neurological Intensive Care Unit several days ago.

Nutrition Diagnosis:

Inadequate oral food/beverage intake related to swallowing difficulty and persistent aspiration secondary to CVA as evidenced by current NPO status and recent unintended weight loss of 12#.

Determine the Enteral Nurition Prescription:

Step 1: Determine a "dosing" weight

A. **Critical Thinking:** The patient has lost weight but is at his ideal weight for height. In this case, the ideal weight is a good choice for the "dosing" weight.

B. **Calculations for the Nutrient Prescription:** Convert the weight to kilograms by dividing weight in pounds by 2.2.

168 pounds/2.2 = 76 kg

Step 2: Determine a kcal goal

A. **Critical Thinking:** Despite the recent weight loss, it is probably desirable to maintain this patient's current weight, as it is also his desirable weight. Weight gain is probably not desirable as it might further limit his mobility and impede his recovery.

To maintain weight, 25 to 30 kcal/kg is often used. As his activity level changes, adjustments in the nutrient prescription may be needed to maintain the desired weight. You could also calculate energy needs using a variety of predictive equations including the Mifflin-St. Jeor.

B. **Calculations for the Nutrient Prescription:** Multiply the weight by the number of kcal/g selected.

76 kg × 25 kcal = 1900 kcal

(continued)

Or, use the Mifflin-St. Jeor equation:

$$10 (76) + 6.25 (177.5) - 5 (76) + 5 = 1494 \times 1.2 \text{ (activity factor)} = 1800 \text{ to } 1900 \text{ kcal}$$

C. **More Critical Thinking:** Calculations of kcal goal using two different methods are within 100 kcal of each other. This verifies that you are on the right track.

Step 3: Adjust for activity and injury

A. **Critical Thinking:** This patient has limited activity due to his disability, but he may also expend considerable energy during physical therapy. You should read the physical therapy consultation note or ask the therapist directly about activities performed. In this case, the patient will be discharged from physical therapy when he is transferred to the nursing home. There is no reason to add to the maintenance weight based on his current activity level. This patient's medical condition is stable, and there is no reason to add an injury factor at this time.

Step 4: Calculate a protein goal

A. **Critical Thinking:** This patient is stable from a nutritional standpoint, with no excessive protein losses. The RDA for protein is 0.8 g/kg of body weight, which would equate to about 10% of kcal from protein. The protein recommendation for older adults is 1 g protein/kg.[1]

B. **Calculations for the Nutrient Prescription:** Multiply weight by 0.8 g/kg to obtain the minimum grams of protein needed. Then multiply the grams of protein by 4 to obtain the kcal provided by protein.

$$76 \text{ kg} \times 0.8 \text{ g of protein} = 60 \text{ g of protein per day}$$

$$60 \times 4 = 240 \text{ protein kcal}$$

Alternatively, if you decide to use the higher range of protein requirements:

$$76 \text{ kg} \times 1.0 \text{ g of protein} = 76 \text{ g of protein per day}$$

$$76 \times 4 = 304 \text{ protein kcal}$$

Divide protein kcal into total kcal requirements:

$$304/1900 = 16\% \text{ of kcal from protein}$$

C. **More Critical Thinking:** You will need to identify a formula that derives about 16% of total kcal from protein. This patient has no medical conditions or diagnoses that will affect ability to tolerate standard sources of protein. His needs are not elevated, so there is no need to look for a high-protein formula. A standard polymeric formula should be tolerated and is formulated to provide approximately 16% of total kcal from protein.

Step 5: Consider electrolyte needs

A. **Critical Thinking:** This patient has no abnormal electrolyte losses, and at present renal function is normal, so his electrolyte needs should be similar to the DRI. Standard polymeric formulas provide the recommended amounts of electrolytes.

Step 6: Consider vitamin and mineral needs

A. **Critical Thinking:** Consult the DRI tables to identify appropriate vitamin and mineral requirements. Remember to check the vitamin levels in the enteral product so that you will know how much the patient is getting. Standard polymeric formulas provide the recommended amounts of vitamins and minerals within a specified volume.

Step 7: Determine fluid needs

A. **Critical Thinking:** Patient is not dehydrated. For this patient, 30 mL/kg/body weight or 1 mL/kcal provided could be used to calculate fluid requirements.

B. **Calculation for Nutrient Prescription:** Fluid requirements are approximately the same as the energy requirements—1900 to 2000 mL.

Step 8: Establish administration and delivery methods

A. **Critical Thinking:** This patient has a PEG tube and will be transferred to a nursing home. It will be best to start on a continuous feeding while keeping

in mind that he will need to be prescribed a bolus feeding for discharge.

Step 9: Write final enteral nutrition prescription

A. **Critical Thinking:** You have determined that this patient needs a standard polymeric formula. From the choices available (see Appendix C2), note that Osmolite® provides 1 kcal/mL and 0.034 g protein/mL. All other nutrient levels are within the calculations. Since the patient needs 1900 kcal, the total volume of formula will be 1900 mL. Multiply the total volume 1900 mL × 0.034 g to verify the amount of protein provided by the feeding. 1900 mL will provide 64.6 g protein. Verify that this is within the prescribed range of protein.

B. **Calculation for Nutrient Prescription:** For a continuous feeding, take the patient's formula volume and divide by 24 (hours/day) to establish goal rate.

$$1900/24 = 80 \text{ mL/hr}$$

For the bolus feeding, divide the volume of feeding by 4 feedings/day:

$$1900/4 = 475 \text{ mL}$$

C. **More Critical Thinking:** Determine initial start rate and recommended progression. As you learned earlier in this text, protocols vary, but a standard isotonic polymeric formula can be initiated at full strength at 10–50 mL/hr.

D. **Calculation for Nutrient Prescription:** Begin Osmolite at 50 mL/hr and increase in 10–20 mL increments to goal rate of 80 mL/hr every 8 hours. Prior to discharge, switch patient to bolus regimen of 475 mL/feeding four times daily.

A note on rounding numbers: Clinicians typically round the results of whole numbers, such as kilocalories, milligrams of sodium, or milliequivalents of potassium to the nearest five or ten. This practice varies widely, however, and attention to local customs is advised.

References

1. Helphingstine C, Bistrian B. New Food and Drug Administration requirements for inclusion of vitamin K in adult parenteral multivitamins. JPEN J Parenter Enteral Nutr. 2003;27:220–24.

Table 5.2 Sample Enteral Protocol

Continuous/Nocturnal Feeding

Initiation: Full strength (all products except 2 kcal/mL) at 50 mL/hour and increase by 25 mL every eight hours to goal rate. A 2.0 cal/mL product is started at 25 mL/hour (as few patients need >50 mL/hour to meet estimated needs). The final goal rate is dependent on the patient's caloric requirements and GI comfort.

Bolus/Intermittent Feeding

Initiation: 125 mL, full strength (regardless of product) every three hours for two feedings; increase by 125 mL every two feedings to final goal volume per feeding during waking hours.

Source: University of Virginia Health System Nutrition Support, Traineeship Syllabus.

Monitoring and Evaluation: Complications

Enteral feeding is not a simple procedure. Patients who receive enteral feeding may experience a variety of complications, and some of these—such as aspiration or tube misplacement—may be serious. Complications of enteral feedings may develop at any point during a course of therapy. High-risk patients who have concurrent illnesses require an experienced dietitian to successfully manage enteral feedings.

In order to ensure that complications do not develop, it is important to determine whether the patient is medically stable and how long it has been since the tube feeding started. Patients for whom tube feeding is newly initiated should have more intense monitoring during the time the feeding is being progressed to goal. The frequency of monitoring can be diminished as the patient becomes stable. Table 5.3 outlines recommendations for monitoring the individual patient receiving enteral nutrition.

Tube-Related Complications The enteral tube should be placed by a well-trained physician, nurse, or dietitian in order to minimize the risk of tube misplacement. Clogged, twisted, or kinked feeding tubes can occur and may result in reduced or delayed feeding. Clogged tubes most often result from administration of medications through the tube or from inadequate flushing.

To prevent clogged tubes, the feeding tube is flushed with a syringe containing at least 25 mL or more of tap water several times daily. (Note: Some institutions use sterile, distilled, or bottled water for this purpose.) A number of home remedies, usually involving various types of soda or juice, are sometimes recommended to unclog tubes. However, none of these have been shown to be superior to warm water in unclogging tubes.[10, 11] In some institutions, a combination of bicarbonate and pancreatic enzymes is used, and in others, commercial products for unclogging tubes are available. In no circumstances should the **stylet** used to place the tube be reinserted into the tube, because this may perforate the feeding tube. Using a small-volume syringe to force liquid into the tube can also rupture the tube, and a large-volume syringe is recommended in order to decrease pressure on the feeding tube wall.

Gastrointestinal Complications Diarrhea in enterally fed patients may result as a complication of antibiotics or as a consequence of administration of other hyperosmolar medications via the feeding tube. It is important to rule out infectious organisms such as *Clostridium difficile*, another common cause of diarrhea. Determining whether the diarrhea is infectious or osmotic, completing stool cultures, and performing exams are all steps in evaluating this complication (see Chapter 15). Changes of the bacterial environment of the colon may also contribute to the onset of diarrhea. Whelan documented reduced amounts of bifidobacteria in a small sample of patients who developed diarrhea while on enteral feeding.[34] In other clinical situations such as traveler's diarrhea (mild diarrhea related to microorganisms in food and water), probiotics are recommended to assist in treatment. Larger clinical trials will need to be conducted to determine whether additional probiotic supplementation can assist in the treatment and/or prevention of diarrhea in enteral nutrition.[35] Manipulation of formula rate, strength, or type is often recommended as a means of reducing diarrhea in tube-fed patients, but data demonstrating the effectiveness of these practices are limited. As mentioned earlier in this section, changing to a formula with added fiber can normalize stool output. Antidiarrheal medications are also prescribed. These medications may include octreotide, diphenoxylate-atropine (Lomotil©), loperamide (Immodium©), deodorized tincture of opium, and paregoric.

Aspiration **Aspiration** occurs when fluid is inspired into the lungs. Patients who are sedated, who have an endotracheal tube (a tube that allows oxygen into the lungs of patients receiving mechanical ventilation), or who have difficulty swallowing are at risk for aspiration. It is a potentially serious condition that may result in pneumonia or even death. To avoid aspiration, it is important that the patient's head be elevated higher than her or his stomach, or at an angle of >30 degrees, during feeding. Common practices to assess for aspiration such as blue dye or glucose oxidase are no longer recommended due to lack of sensitivity and potential harm to the patient.[11, 36] Residual volumes of liquid in the

Table 5.3 Suggested Monitoring for Enteral or Parenteral Feedings in Acute Care

Parameter	Medically or Nutritionally Unstable	Medically or Nutritionally Stable
Sufficiency of nutrient intake: intake/output	Daily	Weekly
Electrolytes, BUN, creatinine	Daily, then 3 × week	3 × week
Magnesium, phosphorus, calcium	Daily, then 3 × week	3 × week
Liver function tests	Weekly	As needed
Triglycerides	Weekly	Every 1–2 weeks
Weight	Daily	Weekly
Hydration/fluid status: physical assessment of skin turgor, presence of edema, temperature; oral cavity for color, texture, moisture/dryness	Daily	3 × week
Vital signs: blood pressure, respirations, pulse	Daily	3 × week
Bowel function	Daily	As needed
Blood glucose	3 × daily until stable	Every 1–2 weeks
Nitrogen balance	PRN (as necessary)	PRN

stomach have also been used to determine whether a feeding was emptied from the gastrointestinal tract, but guidelines for this practice are not well established.[10, 11, 37] In the Consensus Statement presented by the North American Summit on Aspiration in the Critically Ill Patient, it was recommended that enteral feeding be stopped only if there is definite regurgitation or aspiration of gastric contents or if a residual greater than 500 mL is measured. In the absence of such a circumstance, careful monitoring and clinical assessment, combined with the residual volume, should be used to make a decision about tolerance of enteral feeding.[19]

Dehydration Patients with insufficient fluid intake who are receiving enteral feedings may develop hyperosmolar, non-ketotic dehydration over a short two- to four-day period. This condition may be prevented by providing sufficient fluid (about 1 mL/kcal) with the feeding. Patients receiving less fluid should be monitored with a daily fluid status assessment (see Chapter 7 for more on fluid status and its assessment). Fluid status may be assessed by physical exam (e.g. oral mucosa, skin turgor), and by estimating the adequacy of fluid intake and output. If the results of the assessment indicate that the patient is taking insufficient fluid, and there is no reason for a fluid restriction, then the amount of fluid administered is increased. Laboratory values such as specific gravity and osmolality as well as BUN and creatinine assist in confirmation of hydration status.

Electrolyte Imbalances In stable patients receiving enteral feedings, the DRI for sodium, potassium, calcium, magnesium, and phosphorus are often used as a guide for electrolyte intake (see Chapter 7 for more on electrolytes). The electrolyte content of the formula in use may be compared to requirements, and supplemental electrolytes may be provided as needed. Magnesium and potassium administered via the feeding tube may produce a cathartic effect, and gastrointestinal calcium absorption may be poor. Thus, intravenous electrolyte supplementation is sometimes preferred.

It is imperative for the clinician to understand that enteral and parenteral electrolyte requirements differ because of the variable of absorption. Another key distinction is that parenteral electrolytes are measured in mEq or mmol, while oral requirements are stated in milligrams. Details of parenteral and enteral requirements may be found in Table 5.4.

Underfeeding or Overfeeding Enteral feeding is based on a "dosing" weight established by the dietitian. Every attempt is made to feed the patient an appropriate amount of nutrients based on this weight. Both underfeeding and overfeeding can be detrimental to the patient. Underfeeding may delay nutritional repletion and wound healing. ASPEN guidelines state: "efforts to provide >50%–65% of goal calories should be made in order to achieve the clinical benefit of EN over the first week of hospitalization" (p. 291).[9] It is noted that in critically ill patients, however, "permissive underfeeding" may assist with preventing acute metabolic and respiratory complications.[9, 38, 39] This will be discussed in greater detail in Chapter 22. Overfeeding, on the other hand, may result in hyperglycemia, hypertriglyceridemia, and hepatic steatosis (fatty liver).[9, 40]

While many clinicians prefer to feed hospitalized patients 25 to 30 kcal per kg, there are others who prefer to use smaller amounts (18 to 20 kcal/kg) for overweight patients.[41] Current ASPEN guidelines state that: "For all classes of obesity where BMI is >30,

Table 5.4 Electrolyte Requirements

	Dietary Reference Intake for Oral/ Enteral Feeings	Recommendations for Parenteral Intake
Potassium		
Adults over the age of 14	4700 mg	1 to 2 mEq/kg
Sodium		
14–50 years	1500 mg	1 to 2 mEq/kg
51–70 years	1300 mg	
> 70 years	1200 mg	
Chloride		
14–50 years	2300 mg	To maintain acid-base balance
51–70 years	2000 mg	
> 70 years	1800 mg	
Bicarbonate	- - -	To maintain acid-base balance
Calcium		
14–18 years	1300 mg	10 to 15 mEq
19–50 years	1000 mg	
> 51 years	1200 mg	
Magnesium		
Males 14–18 years	410 mg	8 to 20 mEq
19–30 years	400 mg	
> 31 years	420 mg	
Females 14–18 years	360 mg	
19–30 years	310 mg	
> 31 years	320 mg	
Phosphorus		
14–18 years	1250 mg	20 to 40 mmol
> 18 years	700 mg	

These are standard intake ranges for generally healthy people with essentially normal organ function who do not have abnormal needs or losses.

the goal of the EN regimen should not exceed 60%–70% of target energy requirements or 11–14 kcal/kg actual body weight per day (or 22–24 kcal/kg ideal body weight per day)" (p. 291).[9]

Hyperglycemia During periods of physiological stress, such as those caused by severe illness or severe infection (sepsis), hyperglycemia can appear even in patients with no previous history of diabetes. Recently, the use of intensive insulin therapy to maintain normal blood glucose levels has resulted in a reduction of morbidity and mortality for critically ill patients. Insulin therapy not only controls hyperglycemia seen in metabolic stress but may affect the catabolic state, reduce inflammation, and improve the immune response.[42–44] (See Chapter 22.) The hyperglycemia associated with stress usually resolves as the stress response subsides, and nondiabetic patients do not experience long-term complications.

Refeeding Syndrome **Refeeding syndrome** is a term used to describe several common metabolic alterations that may occur during nutritional repletion of starved patients.[9–11] This syndrome has been observed in the surviving victims of famine since the beginning of medical history. With the advent of parenteral nutrition, refeeding syndrome gained attention because of its often dramatic and sometimes fatal presentation.

With starvation lasting more than a few days, liver gluconeo-genesis slows, free fatty acids are used to produce energy in the form of ketones, and basal metabolic rate declines. The reintroduction of carbohydrate, whether in oral, enteral, or parenteral form, results in a shift from ketones to glucose as the primary energy source. Glucose metabolism requires large quantities of phosphorus. Magnesium, potassium, and thiamin requirements may also increase to meet anabolic needs. The result is a drop in serum levels of phosphorus, which, if severe, may result in hemolysis, impaired cardiac function, impaired respiratory function, and even death. Hypomagnesemia (low serum magnesium) may result in tremor, muscle twitching, cardiac arrhythmias, and even paralysis (see Chapter 7). Hypokalemia (low serum potassium) is also associated with cardiac abnormalities. Thiamin deficiency has been documented infrequently, but may result in Wernicke's encephalopathy (see Chapter 16).

Patients at risk for refeeding syndrome include those who present with malnutrition, those who have a history of long-term inadequate oral intake, and those who have had minimal intake for several days as a result of **NPO** status or poor appetite. It is critical to monitor serum levels of phosphorus, magnesium, and potassium, and to provide supplementation as needed until the patient is receiving goal feedings. Clinicians have used the strategy of beginning feedings slowly and avoiding overfeeding in order to prevent refeeding syndrome.

Parenteral Nutrition

The word "parenteral" means "alongside" or "outside" the gastro-intestinal tract, and is now used to describe the administration of drugs or nutrients by vein (**intravenously, or IV**). **Parenteral nutrition (PN)**, developed in the 1960s to sustain the lives of individuals with severe gastrointestinal impairment, may also be called total parenteral nutrition (TPN), central venous nutrition (CVN), or intravenous hyperalimentation (IVH). Generally, *parenteral nutrition* is the preferred term. The term "hyperalimentation" originally described the practice of "hyperalimenting" or overfeeding patients. Although deliberately overfeeding or "hyperalimenting" patients is no longer common clinical practice, the term persists in many institutions. The distinguishing feature of PN is administration of concentrated macronutrients, vitamins, minerals, and electrolytes into a large central vein so that the volume of blood flow is sufficient to immediately dilute the concentrated parenteral solutions.

The term **peripheral parenteral nutrition (PPN)** refers to the administration of large-volume, dilute solutions of nutrients into a vein in the arm or back of the hand and is used infrequently. PPN requires large volumes to meet nutritional needs and this makes this route unacceptable for any fluid-restricted patient. The high osmolality of PPN may cause small veins to collapse, and peripheral access is difficult to maintain for more than a few days.

Indications

Parenteral nutrition is indicated in those clinical situations where the patient is unable to meet nutritional needs either by an oral diet or through the use of enteral nutrition. The clinical conditions that may require parenteral nutrition include an inability to digest and absorb nutrients, such as in massive bowel resection or short bowel syndrome; intractable vomiting, as in hyperemesis gravidarum; GI tract obstruction; impaired GI motility; and abdominal trauma, injury, or infection. Nutrition diagnoses associated with candidates for PN are the same as those for EN.

Decisions related to parenteral nutrition, like those for enteral nutrition, are based on the nutrition care process. The patient's nutrition assessment, the length of time the patient will require nutrition support, and the patient's diagnosis and current medical condition will assist the clinician in making the decisions that are required to build the parenteral nutrition prescription. Certification of medical necessity for PN must be established in order to ensure that the patient's care is financially feasible.

Venous Access

The primary difference between enteral and parenteral feedings is that nutrients are provided via the veins rather than the gastrointestinal tract in PN. Feeding directly into the venous system and large-diameter veins allows for use of high-osmolality solutions of >900 mOsm/L. The illustration in Figure 5.7 may be of assistance in visualizing the types and locations of vascular access used for parenteral nutrition.

Figure 5.7 Sites for Parenteral Access

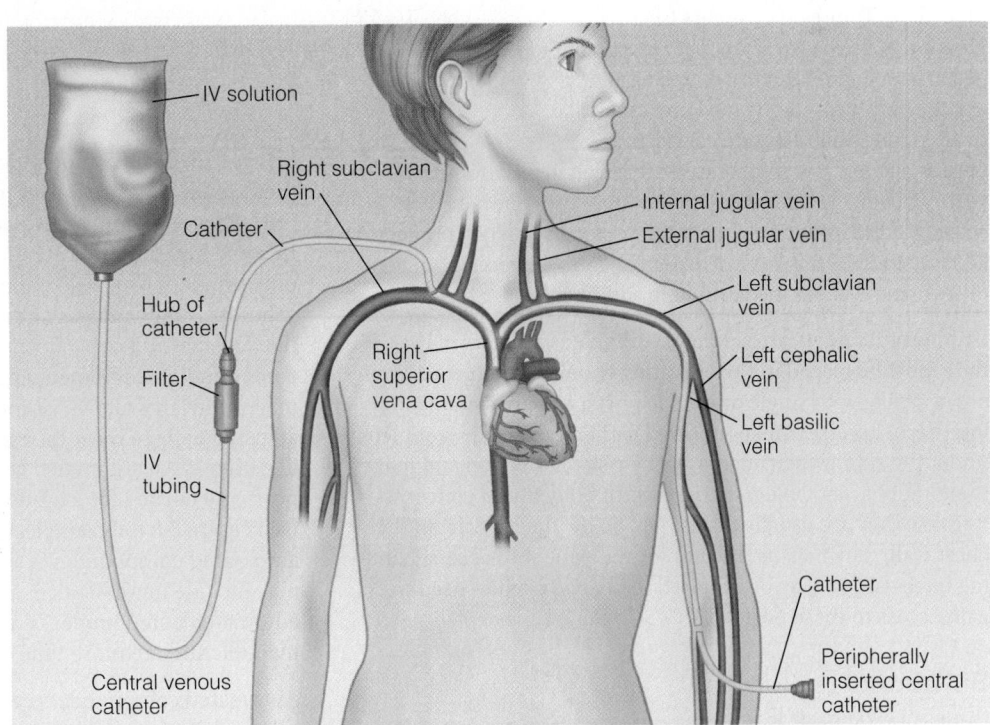

Source: S. Rolfes, K. Pinna and E. Whitney, *Understanding Normal and Clinical Nutrition*, 7e, copyright © 2006, p. 677.

Short-Term Venous Access The most common parenteral access is a **central venous catheter (CVC)** or central line inserted percutaneously (through the skin) at the bedside while the patient is under local anesthesia. Central catheters are inserted into large veins such as the subclavian, jugular, or femoral veins in the center of the body. Ultimately, these catheters reside in the superior vena cava, or in the inferior vena cava, in the case of femoral placement. These catheters are available in single-, double- or triple-lumen models. The lumen of the catheter refers to the interior of the tube through which the PN solution passes. If a catheter with sufficient lumens is available, a patient may receive medications, fluids, and nutrients at the same time. Catheters are usually changed every few days to help decrease the risk of infection inherent with an opening from the skin into a large, central vein.

The **peripherally inserted central catheter (PICC)** is also frequently used. Unlike the central catheter, which requires a bedside surgical procedure by an MD for insertion, the PICC can be inserted by specially trained nurses; this increases the availability of the procedure and decreases costs. PICC lines are inserted into the arm and threaded into the subclavian vein to the vena cava.

Long-Term Venous Access For long-term use, or for home PN, a catheter is tunneled under the skin during a surgical procedure. **Tunneled catheters** (such as Hickman®, Broviac®, Groshong®) most often enter the vein on the upper chest wall and exit the body near the xyphoid process, axilla, or abdominal wall. They are considered permanent, and with proper care can be left in place for several years. If a tunneled catheter contains more than one lumen, it can accommodate infusions of medications, fluids, or blood products in addition to PN. Thus, it is useful for patients who receive frequent doses of intravenous medications in addition to PN.

Implantable ports are similar to tunneled catheters in that they must be placed in the operating room by a surgeon. They are available with single or double ports, and are suitable for long-term access. Unlike tunneled catheters, they lie completely under the skin, which decreases the risk of infection and makes them more acceptable to patients with body image concerns. Because they are usually placed just below the clavicle on the chest wall, they may be difficult for the patient to access. Nursing intervention may be required to change needles used to gain access to these ports.

Solutions

Unlike enteral formulas, which are most often purchased in a form appropriate for patient administration, parenteral formu-

Figure 5.8 Standard PN Label Templates

(A) Standard PN label template for neonate or pediatric patient

Institution/Pharmacy Name, Address and Pharmacy Phone Number		
Name	Dosing Weight	Location
Administration Date/ Time	Do Not Use After: Date/time	
Base Formula	Amount/kg/day	Amount/day
Dextrose	g/kg	g
Amino acids[a]	g/kg	g
Electrolytes		
Sodium chloride[b]	mEq/kg	mEq
Sodium acetate[b]	mEq/kg	mEq
Potassium chloride[b]	mEq/kg	mEq
Potassium acetate[b]	mEq/kg	mEq
Potassium phosphate[b]	mmol of P/kg (mEq of K)/kg	mmol of P (mEq of K)
Sodium phosphate[b]	mmol of P/kg (mEq of Na)/kg	mmol of P (mEq of Na)
Calcium gluconate	mEq/kg	mEq
Magnesium sulfate	mEq/kg	mEq
Vitamins, trace elements and medications		
Multiple vitamins[a]	mL/kg	mL
Multiple trace elements[a]	mL/kg	mL
L-crysteine	mg/kg	mg
H$_2$ antagonists[a]	mg/kg	mg
L-Carnitine	mg/kg	mg
Rate ——— mL/hour	Volume ——— mL	Infuse over 24 hours
Admixture contains ____ mL plus ____ mL overfill		
Central Line Use Only*		

[a]Specify product name.
[b]Since the admixture usually contains multiple sources of sodium, potassium, chloride, acetate, and phosphorus, the amount of each electrolyte/kg provided by the PN admixture is determined by adding the amount of electrolyte provided by each salt.

Source: JOURNAL OF PARENTERAL AND ENTERAL NUTRITION by Mirtallo, Canada, Johnson, Jumpf, Petersen, Sacks, Seres, Guenter. Copyright 2004 by SAGE PUBLICATIONS INC. JOURNALS. Reproduced with permission of SAGE PUBLICATIONS INC. JOURNALS in the format Textbook via Copyright Clearance Center.

las are mixed or "compounded" in the hospital pharmacy. The method the pharmacy uses to compound PN preparations is critical to development of parenteral nutrient prescriptions. In some institutions, a standardized approach includes not only standard ordering and monitoring but also standardized solutions.[45] Advantages of this approach include improvements in efficiency and cost-effectiveness.[45] The use of automated compounding equipment and bulk packaging of concentrated macronutrients allows for individualized formulas that are adjusted daily to meet the rapidly changing needs of critically ill patients. An automated compounder is used to combine all nutrients needed for a 24-hour infusion into a single container (see Figure 5.8 for examples of PN solution labels). When an automated compounder is available, the parenteral prescription may include ingredients in as small as 1 mL increments. If an automated compounder is not available, formula changes and manufacture are more time consuming for the pharmacist.

Automated compounders can be used to manufacture nutrient solutions that combine dextrose and amino acids (two-in-one formulas) or dextrose, amino acids, and lipids (three-in-one

Figure 5.8 Standard PN Label Templates (*Continued*)

(B) Standard PN label template for adult patient

Institution/Pharmacy Name, Address and Pharmacy Phone Number		
Name	Dosing Weight	Location
Administration Date/ Time	Do Not Use After: Date/time	
Base Formula	Amount/kg/day	Amount/day
Dextrose	g	(g/L)
Amino acids[a]	g	(g/L)
IVFE[a]	g	(g/L)
Electrolytes		
Sodium chloride	mEq	(mEq/L)
Sodium acetate	mEq	(mEq/L)
Sodium phosphate	mmol of P	(mmol/L)
	(mEq of Na)	(mEq/L)
Potassium chloride	mEq	(mEq/L)
Potassium acetate	mEq	(mEq/L)
Potassium phosphate	mmol of P	(mmol/L)
	(mEq of K)	(mEq/L)
Calcium gluconate	mEq	(mEq/L)
Magnesium sulfate	MEq	(mEq/L)
Vitamins, trace elements and medications		
Multiple vitamins[a]	mL	
Multiple trace elements[a]	mL	
Insulin	Units	(Units/L)
H_2 antagonists[a]	mg	
Rate ——— mL/hour	Volume ——— mL	Infuse over——hours

Formulation contains ____ mL plus ____ mL overfill

Discard any unused volume after 24 hours

Central Line Use Only*

[a]Specify product name.

Source: JOURNAL OF PARENTERAL AND ENTERAL NUTRITION by Mirtallo, Canada, Johnson, Jumpf, Petersen, Sacks, Seres, Guenter. Copyright 2004 by SAGE PUBLICATIONS INC. JOURNALS. Reproduced with permission of SAGE PUBLICATIONS INC. JOURNALS in the format Textbook via Copyright Clearance Center.

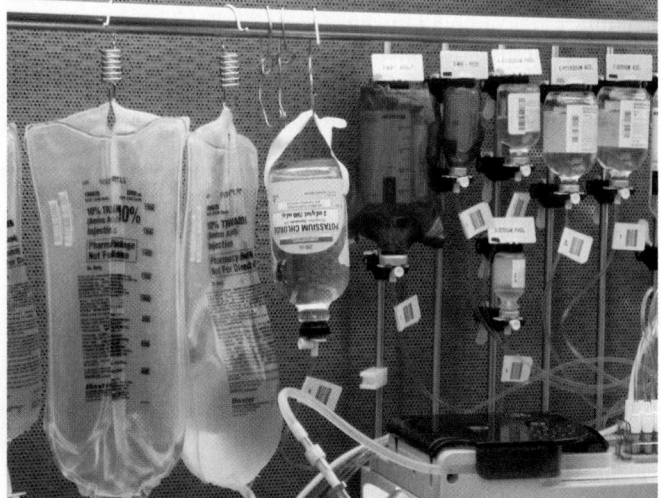

Parenteral nutrition is compounded within the pharmacy under aseptic conditions. An automated compounder like the one shown here minimizes risks and errors in parenteral nutrition.

Source: Courtesy of Marcia Nelms.

formulas). PN can be provided in either a two-in-one or three-in-one system, and each system has both advantages and disadvantages. In the two-in-one, lipids are added separately based on the available container sizes (100 mL, 250 mL, 500 mL). This system provides a greater degree of flexibility in the amounts of dextrose and amino acids that can be given. Another advantage of the two-in-one system is that formulas containing only dextrose and amino acids are clear, and any precipitate can be observed. A disadvantage of the two-in-one system is the need for an additional administration set (intravenous tubing and other devices required for the delivery of parenteral nutrition) for the lipids.

The three-in-one system requires a single administration set, which saves nursing time and reduces costs. On the other hand, the addition of lipids with the three-in-one system results in an opaque solution, which obscures precipitate and increases the risk of particulate being infused into the patient. Addition of lipid into the three-in-one solution limits the electrolytes and final concentration of amino acids in solution.

Parenteral solutions are compounded from as many as 40 different items under the supervision of a licensed pharmacist. In order to maintain sterility, compounding is completed in a "clean room" under a laminar flow hood. Because PN is compounded from amino acid, dextrose, and lipid solutions, solubility is an important consideration affecting both the maximum amount of nutrients and the minimum amount of fluid that can be incorporated. Precipitates may form in PN solutions if greater than recommended maximum amounts of electrolytes and minerals are added, especially when PN is subjected to changes in temperature or pH. Likewise, minimum volumes are impacted by the concentration of amino acids, dextrose, and lipid that are available for compounding.

Parenteral Nutrition Substrates

Protein Protein is included in parenteral nutrition in the form of individual amino acids in amounts consistent with the recommendations of the Food and Agriculture Organization and the World Health Organization. Modified products have been developed and marketed for renal failure, hepatic failure, and stress. Commercial amino acids are available from various manufacturers in concentrations of 3.5% (35 g/L) to 20% (200 g/L). Lower concentrations (3.5%–5.5%) are used for peripheral administration, while higher ones (8.5%, 10%, 11%, 15%, 20%) are used for central administration. Details for these products are available on websites of the major manufacturers and summarized in Table 5.5. Parenteral nutrition is typically designed to provide individualized protein requirements, which range from 0.8 g/kg for adults to 1.5 to 2.0 g/kg for patients with burns, trauma, or healing wounds.[11]

Table 5.5 Amino Acid Solutions Used in Parenteral Nutrition

Brand Name	Type/Indication	Stock Concentrations
Aminosyn II™	Standard	3.5%, 4.25%, 5%, 7%, 8.5%, 10%
Travasol™	Standard	3.5%, 4.25%, 5.5%, 8.5%, 10%
Aminosyn II™	Standard/fluid restriction	15%
Clinisol™	Standard/fluid restriction	15%
Novamine™	Standard/fluid restriction	15%
Prosol™	Standard/fluid restrictions	20%
Hepatamine™	Hepatic failure	8%
Hepatasol™	Hepatic failure	8%
Aminosyn HBC™	Metabolic stress	7%
Freamine HBC™	Metabolic stress	6.9%
Branchamin™ (contains only leucine, isoleucine, and valine; use to supplement standard amino acid base)	Metabolic stress	4%
Amino PF™	Pediatric	7%, 10%
Trophamine™	Pediatric	6%, 10%
Aminess™ (essential amino acids plus histidine)	Renal	5.2%
Aminosyn RF™ (essential amino acids plus arginine)	Renal	5.2%
Nephramine™ (essential amino acids plus histidine)	Renal	5.4%
Renamin™ (essential and some nonessential amino acids)	Renal	6.5%

Source: Madsen H, Frankel EH. Practical Gastroeterology. 2006;7:53. Table 9, "Commercially Available Crystalline Amino Acid Solutions."

Source line that appears in above article for table:
Adapted from Barbar JR, Miller SJ, Sacks GS. Parenteral feeding formulations. In: Gottschlich MM. Ed. *The Science and Practice of Nutrition Support: A Case-Based Core Curriculum.* Dubuque, IA: Kendall/Hunt Publishing Co.: 2001;251–68 with permission from the American Society for Parenteral and Enteral Nutrition (A.S.P.E.N.). A.S.P.E.N. does not endorse the use of this material in any form other than its entirety.

Factors that increase protein requirements above the DRI include diagnoses such as trauma, burns, sepsis, wounds, and bone marrow transplant. The lower range of protein (0.8 g pro/kg) may be needed for patients with acute renal failure who are not receiving dialysis. It is important to remember, however, that without adequate energy from carbohydrate and lipid, the subsequent catabolism of lean body mass defeats the effort to control uremia in this manner.[46]

Carbohydrate The primary function of parenteral carbohydrate is to serve as an energy source. In the United States, dextrose monohydrate is used as the carbohydrate source for parenteral nutrition. The kcal content of this particular form of carbohydrate is 3.4 kcal/g. The minimum carbohydrate intake was recently specified in the DRI as 130 g/day, and it is known that approximately 100 g of carbohydrate is required daily to allow for protein sparing. The amount of 1 mg/kg/min is often used as the reference for the minimal amount of carbohydrate needed to spare protein. The maximum for glucose oxidation was originally studied in burn patients and found to be 7 g/kg/day (5 mg/kg/minute). In practice, lower figures of 3–4 mg/kg/min have been recommended.[47] Dextrose is commercially available in 5%, 10%, 50%, and 70% concentrations, but other concentrations may also be available.

Excessive carbohydrate may contribute to hyperglycemia, hepatic steatosis, and excessive carbon dioxide production. The standard 100 g of dextrose (appropriate for the reference 70-kg male) would be equivalent to an initial dextrose infusion rate as high as 2.2 mg/kg/min for a small patient—increasing the risk for refeeding syndrome. Elevated carbon dioxide also occurs with overfeeding, and it may jeopardize respiratory status and result in difficulty weaning from mechanical ventilation (see Chapter 21).

Lipid The lipid in parenteral solutions is an emulsion of soybean or safflower oil. The lipid emulsion provides essential fatty acids and vitamin K, as well as a concentrated source of energy—an avenue to meet energy needs if the patient is unable to tolerate a higher carbohydrate load. Lipid formulations are 10% (1.1 kcal/mL); 20% (2 kcal/mL) and 30% (3 kcal/mL). Caloric values per gram depend on the energy provided by glycerol so that 10% solutions provide 11 kcal/g and the others provide 10 kcal/g. See Table 5.6 for nutrient profiles of representative lipid emulsions. An estimated 2%–4% of energy from linoleic acid is recommended to prevent essential fatty acid deficiency. Complications of lipid infusions, including hyperlipidemia and impaired immune response, support a more conservative prescription of lipids within the PN prescription. In the critically ill, the medication **propofol** is often used and, since it is lipid based, must be considered as a source of energy.[9, 46]

Electrolytes Using the standards established by the DRI as the beginning benchmark, electrolyte requirements in PN are based on body weight, existing electrolyte deficiencies, ongoing electrolyte losses, and changes in organ function. Because electrolyte requirements are also inextricably linked with the amount of macronutrients provided in the PN, it is impossible to manage PN without a thorough understanding of these complex relationships (see Chapter 8). Recommendations for standard electrolyte intake are found in Table 5.4. Note, however, that in practice electrolytes are individualized according to patient needs and are often considered to be the most difficult component for new registered dietitians working with nutrition support.

Vitamins and Minerals In 1979, the AMA released recommendations for vitamin and mineral additives to PN. These vitamin recommendations were used until 2003, when they

Table 5.6 Nutrient Content of Lipid Emulsions Used in Parenteral Nutrition

Lipid Emulsion	kcal/mL*	Soybean Oil (g/L)	Safflower Oil (g/L)	Vitamin K (mcg/dL)
Intralipid 10%	1.1	100	0	30.8
Intralipid 20%	2	200	0	67.5
Intralipid 30%	3	300	0	93
Liposyn II 10%	1.1	50	50	13.2
Liposyn II 20%	2	100	100	26
Liposyn III 10%	1.1	100	0	31
Liposyn III 20%	2	200	0	62
Liposyn III 30%	2.9	300	0	93

*kcal/mL differ according to lipid and glycerol content of lipid emulsion.

Source: Reprinted from: Madsen H, Frankel EH. Practical Gastroeterology. 2006; 7:56. Table 10, "Energy and Vitamin K Content of Commonly Used IVFE (32–35)."

Table 5.8 Daily Trace Element Additions to Adult PN Formulations*

Trace Element	Standard Intake
Chromium	10–15 mcg
Copper	0.3–0.5 mg
Iron	Not routinely added
Manganese	60–100 mcg†
Selenium	20–60 mcg
Zinc	2.5–5 mg

*Standard intake ranges based on generally healthy people with normal losses.

†The contamination level in various components of the PN formulation can significantly contribute to total intake. Serum concentrations should be monitored with long-term use.

Source: JOURNAL OF PARENTERAL AND ENTERAL NUTRITION by Mirtallo, Canada, Johnson, Kumpf, Petersen, Sacks, Seres, Guenter. Copyright 2004 by SAGE PUBLICATIONS INC. JOURNALS. Reproduced with permission of SAGE PUBLICATIONS INC. JOURNALS in the format Textbook via Copyright Clearance Center.

were revised to include vitamin K.[48] Rather than add individual amounts of vitamins to PN, most pharmacies purchase commercial multiple vitamin infusion products that meet the recommendations. Because the vitamins are administered intravenously, there is no issue with absorption, and the amounts administered may differ from what is recommended for oral intake. The amounts of vitamins in commercial products have been increased over those for well persons based on the assumption that patients receiving PN will have wounds or critical illness. This presents a monitoring challenge for clinicians following patients on long-term PN. Vitamins may be given every other day in situations where excess vitamin intake is of concern. Table 5.7 lists daily adult parenteral vitamin requirements.

TRACE MINERALS Originally, zinc, copper, chromium, and manganese were added to PN.[49] Based on reports of deficien-

cies, newer products have been introduced that contain the original trace minerals plus selenium, iodide, and molybdenum.[50] Trace element preparations are purchased commercially and contain four, five, six, or seven trace elements. In situations where reduced excretion or potential toxicities exist, trace elements are removed from the PN, and individual trace minerals are added according to need. Trace element additions for adult PN are listed in Table 5.8.

Medications PN may be used to deliver medications. It is possible that albumin, aminophylline, cimetidine, famotidine, ranitidine, heparin, or regular insulin may be included in PN. Prior to recommending medications be added to PN, the clinician should gain a thorough understanding of the practice at his or her institution through observation and consultation with pharmacists.

Table 5.7 Adult Daily Requirements for Parenteral Vitamins

Vitamin	Requirement
Thiamin	6 mg
Riboflavin	3.6 mg
Niacin	40 mg
Folic acid	600 mcg
Pantothenic acid	15 mg
Pyridoxine (B$_6$)	6 mg
Cyanocobalamin (B$_{12}$)	5 mcg
Biotin	60 mcg
Ascorbic Acid	200 mg
Vitamin A	3300 International Units
Vitamin D	200 International Units
Vitamin E	10 International Units
Vitamin K	150 mcg

Source: JOURNAL OF PARENTERAL AND ENTERAL NUTRITION by Mirtallo, Canada, Johnson, Kumpf, Petersen, Sacks, Seres, Guenter. Copyright 2004 by SAGE PUBLICATIONS INC. JOURNALS. Reproduced with permission of SAGE PUBLICATIONS INC. JOURNALS in the format Textbook via Copyright Clearance Center.

Nutrition Assessment and Intervention: Determination of the Parenteral Nutrition Prescription

The parenteral nutrition prescription will be based on the dietitian's nutrition assessment and the physician and pharmacist's recommendations, and often using a form (or electronic version) like that shown in Figure 5.9. Box 5.4 details the process of developing the PN prescription. The basic steps include the following:

1. Establish dosing weight and energy requirements.

2. Calculate a protein goal.

3. Distribute remaining kcal between carbohydrate and lipid.

4. Consider the electrolyte needs for this patient.

5. Consider vitamin and mineral requirements.

6. Establish fluid requirements.

7. Calculate the final parenteral prescription.

Many institutions have specific protocols for initiation, advancement, and transition of feedings. Box 5.5 provides examples of these protocols.

Figure 5.9 Sample Adult PN Order Form

Physician Orders
PARENTERAL NUTRITION (PN) – ADULT

Primary Diagnosis: _____ Ht: _____ cm **Dosing Wt:** _____ kg

PN Indication: _____ **Allergies** _____

Instructions: This form must be completed for a new order or continuation of PN and faxed to the Pharmacy by [Insert Time] to receive same day preparation. PN administration begins at [Insert Time]. Contact the Nutrition Support Service at (XXX) XXX-XXXX for additional information.

Administration Route: CVC or PICC   *Note: Proper tip placement of the CVC or PICC must be confirmed prior to PN infusion*

Peripheral IV (PIV) *(Final PN Osmolarity ≤ _____ mOsm/L)*

Monitoring: Daily weights, Strict input & output, Bedside glucose monitoring every _____ hours

Na, K, Cl, CO_2, Glucose, BUN, Scr, Mg, PO_4 every _____

T, Bili, Alk Phos, AST, ALT, Albumin, Triglycerides, Calcium every _____

Base Solution: *Parenteral nutrition MUST be administered through a dedicated infusion port and filtered with a 1.2-micron in-line*
Select one   *filter at all times. Discard any unused volume after 24 hours.*

PERIPHERAL 2-in-1	**CENTRAL 2-in-1**	**CENTRAL 3-in-1**
Dextrose _____ g	Dextrose _____ g	Dextrose _____ g
Amino Acids (*Brand _____*) _____ g	Amino Acids (*Brand _____*) _____ g	Amino Acids (*Brand _____*) _____ g
		Fat Emulsion (*Brand _____*) _____ g
For patients with PIV and established glucose tolerance; Provides _____ kcal; Maximum Rate not to exceed _____ mL/hour	*For patients with CVC or PICC and established glucose tolerance; Provides _____ kcal; Maximum Rate not to exceed _____ mL/hour*	*For patients with CVC or PICC and established glucose/fat emulsion tolerance; Provides _____ kcal; Maximum Rate not to exceed _____ mL/hour*

RATE & VOLUME: _____ mL/hour for _____ hours = _____ mL/day
Must specify

Use of additional fat emulsion not required with 3-in-1 base solution

or **CYCLIC INFUSION:** _____ mL/hour for _____ hours, then _____ mL/hour for _____ hours = _____ mL/day

---

Fat Emulsion (*Brand _____*) – via PIV or CVC with 2-in-1 base solutions   *(Select caloric density & volume)*

10%	250 mL	Infuse at _____ mL/hour over _____ hours	Frequency _____
20%	500 mL	*(Note: infusions < 4 or > 12 hours not recommended)*	*Discard any unused volume after 12 hours.*

Additives: *(per day)* | **Normal Dosages** | **Additives:** *(per day)*

Sodium Chloride _____ mEq	*1-2 mEq Sodium/kg/day*	**Regular Insulin** _____ units
as Acetate _____ mEq	*pH or CO_2 dependent*	*Recommend if hyperglycemic, start*
as Phosphate _____ mmol of PO_4	*Consider if hyperkalemic*	*with 1 unit for every 10 g of dextrose*
Potassium Chloride _____ mEq	*1-2 mEq Potassium/kg/day*	
as Acetate _____ mEq	*pH or CO_2 dependent*	**Pharmacy Use Only:** Ca/PO_4
as Phosphate _____ mmol of PO_4	*20-40 mmol/day (1 mmol Phos = 1.5 mEq K)*	**Limit Checked** _____
Calcium **Gluconate** _____ mEq	*5-15 mEq/day*	*(Note: Some brands of amino acids contain phosphate)*
Magnesium Sulfate _____ mEq	*8-24 mEq/day*	
Adult **Multivitamins** _____ mL/day	*Contains Vitamin K 150 mcg*	
Adult **Trace Elements** _____ mL/day	*Zn ___ mg, Cu ___ mg, Mn ___ mg, Cr ___ mcg, Se ___ mcg (with normal hepatic function)*	
H_2 **Antagonist** _____ _____ mg	*____ mg/day with normal renal function*	
Other:		

Physician's Signature: _____ Pager Number: _____ Date/time: _____

Orders transcribed by: _____ Date/time: _____ Orders verified by: _____ Date/time: _____

SEND COMPLETED ORDERS TO PHARMACY

Source: JOURNAL OF PARENTERAL AND ENTERAL NUTRITION by Mirtallo, Canada, Johnson, Jumpf, Petersen, Sacks, Seres, Guenter. Copyright 2004 by SAGE PUBLICATIONS INC. JOURNALS. Reproduced with permission of SAGE PUBLICATIONS INC. JOURNALS in the format Textbook via Copyright Clearance Center.

BOX 5.4 **CLINICAL APPLICATIONS**

Applying the Nutrition Care Process to Develop a Parenteral Nutrition Prescription

Nutrition Assessment:

You have been consulted to provide nutrition support recommendations for a 35-year-old male who recently underwent small bowel resection. He was admitted with a small bowel obstruction secondary to severe fulminant Crohn's disease. His postoperative course was complicated by infection. The surgeon notes that the patient will not resume a trial of enteral feeding or oral diet for 7–10 days. Ht. 5'9", Wt. 155#, UBW 165# (6 months ago).

Nutrition Diagnosis:

Inadequate protein-energy intake related to altered GI function secondary to small bowel resection as evidenced by patient unable to take food or beverages enterally.

Determine the Parenteral Nutrition Prescription:

Step 1: Determine a "dosing" weight

A. **Critical Thinking:** The hospital bed has a built-in scale, so you can easily weigh the patient. He has lost some weight over the previous six months (7%) and is currently at 97% of his IBW (160 ± 10#). Due to his weight loss and consistency with IBW, his actual body weight is recommended for dosing weight in calculating energy requirements.

B. **Calculations for the Nutrient Prescription:** Convert the weight to kilograms by dividing weight in pounds by 2.2.

$$155 \text{ pounds}/2.2 = 70.4 \text{ kg}$$

Step 2: Determine a kcal goal

A. **Critical Thinking:** You could use several different methods to determine energy requirements. Research tells us that indirect calorimetry is the most accurate in acute care but this patient unit does not have access to this equipment. Many acute care settings and intensive care units use the American College of Chest Physicians recommendations for 25 × weight (kg). (If BMI is 16–25, use usual body weight; if BMI > 25, use ideal body weight; and if BMI < 16, use existing body weight for first 7–10 days, then use ideal body weight.) The patient is within the range of ideal body weight so 25 kcal/kg is appropriate. One could additionally use a prediction equation such as the Mifflin-St. Jeor formula.

B. **Calculations for the Nutrient Prescription:** Multiply the weight by the number of kcal/g selected.

$$70 \text{ kg} \times 25 \text{ kcal} = 1750 \text{ kcal}$$

Or use the Mifflin-St. Jeor equation:

$$10 (70) + 6.25 (167.4) - 5 (48) + 5 = 1511 \text{ kcal}$$

C. **More Critical Thinking:** Calculations using two different methods vary by approximately 200 kcal.

Step 3: Adjust for activity and injury

A. **Critical Thinking:** The 25 kcal per kilogram factor is recommended for critically ill patients in intensive care units. For the Mifflin-St. Jeor equation, one also needs to use activity and injury factors to calculate total energy requirements. This patient is currently confined to bed, is postoperative, and has an infection.

B. **Calculations for the Nutrient Prescription:** Multiply the resting energy needs by appropriate activity and injury factors to calculate total energy needs.

$$1511 \times 1.1 \times 1.3 = 2160 \text{ kcal}$$

C. **More Critical Thinking:** Since the two are approximately 400 kcal different, setting the calorie goal between the two is reasonable, if indirect calorimetry is not available.

$$EER = 1900\text{–}2000 \text{ kcal/day}$$

Step 4: Calculate a protein goal

A. **Critical Thinking:** There is minimal drainage from the surgical wound (100 mL over the last two shifts), which would not result in significant protein loss. To support postoperative wound healing, 1.5 g/kg of protein is appropriate.

B. **Calculations for the Nutrient Prescription:** Multiply the protein requirement by the patient's weight.

$$70 \times 1.5 = 105 \text{ g protein/day}$$

Step 5: Determine fluid requirements (this provides a working volume for the parenteral solution)

A. **Calculations for the Nutrient Prescription:**

$$1 \text{ mL/kcal} = 2000 \text{ mL}$$

Step 6: Determine lipid concentration

A. **Critical Thinking:** This patient has no special requirements and laboratory values are WNL, so starting with 30% of kcal from lipid is reasonable. Remember that the minimum lipid emulsion concentration is approximately 3%, and the maximum lipid that should be administered to the patient is 1.2 g/kg (70 × 1.2 = 84 g).

B. **Calculations for the Nutrient Prescription:** Determine kcal to be supplied from 10% lipid.

$$2000 \times 0.30 = 600 \text{ kcal}$$

Determine grams of lipid by dividing kcal from lipid by 11.

Using 30% calculation: $600/11 \text{ kcal/g} = 54.5 \text{ g}$ (round to 55 g)

Divide lipid grams by total daily volume (= fluid needs or final rate × 24) in mL and multiply by 100 to determine % lipid.

$$55/2000 \times 100 = 2.75\% \text{ lipid or } 27 \text{ g/L (round to 3\%)}$$

(continued)

Step 7: Determine protein concentration

A. **Calculations for the Nutrient Prescription:** Divide protein needs (in grams) by total daily volume (in mL) and multiply by 100.

$$105/2000 \times 100 = $$
5% amino acid solution

Step 8: Determine grams of dextrose to meet remaining energy requirements

A. **Calculations for the Nutrient Prescription:** Subtract kcal from lipid and protein from total kcal to determine remaining kcal needs.

105 g protein $\times$ 4 kcal/g = 420 kcal

55 g lipid $\times$ 11 kcal/g = 605 kcal

605 + 420 = 1025 kcal from lipid and protein

2000 − 1025 = 975 kcal from dextrose

Divide "remaining kcal" by 3.4 kcal/g to determine grams of dextrose.

$$975/3.4 = 288 \text{ g}$$

Determine dextrose concentration by dividing dextrose grams by total daily volume and multiply by 100.

287/2000 $\times$ 100 = 14% dextrose

Final solution for macronutrients: 14% dextrose, 3% lipid, and 5% amino acids within total volume of 2000 mL.

Step 9: Consider electrolyte needs

A. **Calculation for Nutrient Prescription:** Consult Table 5.4 to identify suggestions for electrolyte concentration in the PN. Based on his dosing weight, this patient would need about 70 mEq of sodium (1 to 2 mEq/kg), 70 mEq of potassium (1 mEq/kg), 16 mEq of magnesium, 15 mMol of phosphorus, and 10 mEq of calcium with bicarbonate and acetate to balance the solution.

Step 10: Consider vitamin and mineral needs

A. **Critical Thinking:** Consult Tables 5.7 and 5.8 to identify appropriate vitamin and mineral requirements. Remember to check the package insert for the vitamin and mineral preparations so that you will know how much the patient is getting.

Step 11: Write final parenteral nutrition prescription

Parenteral nutrition to provide 105 grams of protein, 288 grams of dextrose, and 66 grams of 3% lipid with 70 mEq of sodium, 70 mEq of potassium, 15 mMol of phosphorus, 16 mEq of magnesium, and 10 mEq of calcium with bicarbonate and acetate to balance, with 1 vial of multiple vitamin infusion, and 1 vial of multiple trace element infusion in a volume of 2 L daily to run at 83 mL/hour over 24 hours each day. Provide 1 L of parenteral formula to run over 24 hours with advancement to goal rate on day 2.

BOX 5.5   CLINICAL APPLICATIONS

Sample Parenteral Nutrition Protocol:

A. Definitions: IV Nutrition Support using a formulation of amino acids, carbohydrates, lipids, electrolytes, MVI, minerals, and supplemental medications (insulin or H_2 blockers).

B. Patient Selection: Inability to use the gut at goal feeds within 5 days.

C. Patient Exclusion: Ability to use the gut at goal feeds or oral intake within 5 days.

D. IV Access: Central Access (TLC, PICC, Hickman, Port-A-Cath).

E. Formula Selection: Based on patient's requirements, critical illness, organ failure, and comorbid disease.

F. Estimating Nutritional Needs:
   1. Ideal Body Weight (IBW) will be used for nutritional estimates for the majority of patients. Patients greater than 120% of IBW/Ht,

registered dietitian will calculate best weight estimate.

- To calculate IBW/height: Range plus or minus 10%
   i. Males: 2.3 $\times$ (inches over 5') + 50 kg
   ii. Females: 2.3 $\times$ (inches over 5') + 45 kg
- Nutritional Calculations: to formulate PN prescription
   i. Energy: 25 to 30 kcal/kg (aim for 25 kcal/kg)
   ii. Protein: 1.0 to 1.8 grams protein/kg (aim for 1.5 gram protein/kg)
   iii. Lipids: 30 to 70 grams/day (5 to 13 mL/hr), 20 to 30% total
- Monitoring/management of patient care
   i. Labs: Basic Metabolic Panel, C-Reactive Protein, Mg and Phos, Day 1, 2 & PRN (LFTs, Pre-albumin, Triglyceride levels check q Thursdays as routine)
   ii. Metabolic Carts in patients on PN greater than two weeks

PN Protocol:

A. The Adult Nutrition Support Service should be consulted to assist with prescribing parenteral nutrition (ASPEN guidelines).

B. All PN is to be ordered or reordered daily, according to the age-appropriate order form. Orders must be received by the appropriate time:

   1. Adult Medicine and Surgical units by 5 p.m.

C. Monitoring

   2. Blood glucose: See intense glucose control protocol for all ICU patients.
      a. For nonunit patients: Blood glucose testing, adult every 6 hr $\times$ 72 hours. Thereafter, renewal is required.

Source: Used with permission: Critical Care Nutrition Vanderbilt University Medical Center. Nashville, TN, 2004.

Administration

A.S.P.E.N. (American Society for Parenteral and Enteral Nutrition) recommends a standardized protocol for delivery of PN.[45] It is recommended that a standardized order form be used to ensure basic nutritional needs for patients are met and that errors are minimized. Box 5.5 includes an example of a standardized order.

Monitoring and Evaluation: Complications

Patients receiving PN can suffer serious, life-threatening consequences including death if the PN is not appropriately monitored and managed. Thus, standard monitoring protocols are in place in many institutions. Monitoring is intense during the first few days, but it decreases as the patient reaches goal feedings and becomes stable.

Intake and output monitoring is usually initiated. Laboratory monitoring includes testing for hyperglycemia three to four times per day and daily measurements of serum electrolytes, BUN and creatinine, magnesium, and phosphorus. At baseline, serum triglycerides are drawn to assess lipid tolerance, and if abnormal, they may be drawn weekly thereafter. The sample protocol for monitoring PN found in Box 5.5 may serve as a guide, although many institutions have protocols in place.

Complications of parenteral feeding may be severe, and they are best prevented through patient monitoring by nutrition support experts. Many of the complications experienced with enteral nutrition occur with parenteral nutrition as well. Patients receiving parenteral feeding may experience electrolyte imbalance, underfeeding and/or overfeeding, hyperglycemia, and refeeding syndrome, just as patients receiving enteral feedings do. These conditions were described in detail earlier in the "Monitoring and Evaluation: Complications" section within the "Enteral Nutrition" section.

Gastrointestinal (GI) complications of parenteral feedings have been reported, primarily in those patients whose GI tract is at complete rest. These complications include cholestasis (a condition in which bile accumulates in the gallbladder because it contracts infrequently without enteral stimulation). Increased permeability to bacteria has been noted when atrophic intestinal cells result from lack of enteral stimulation. For this reason, many patients who require PN may receive trophic or "trickle" amounts of enteral feedings.

If PN is administered continuously for several weeks, transient elevations in liver enzymes may be noted. These usually disappear after PN is discontinued, but may respond to intermittent or cyclic feedings, adjustments in the lipid-to-dextrose ratio, and kcal reduction if overfeeding is operative.

Patients receiving PN can develop serious infections, and they may have a higher infection rate overall than patients receiving oral or enteral nutrition. Infections may be caused by improperly prepared PN solutions, and therefore most pharmacies institute rigorous monitoring to minimize this risk. Infection may be introduced into a patient's bloodstream while the vascular access device is placed or while a dressing around the line is being changed. Another route for infection is the GI tract, which, according to the indications for PN, should be nonfunctioning. With disuse, a nonfunctioning GI tract may become permeable to intestinal bacteria, and infection may result. Finally, it has been noted that an increase in infection rate may occur with higher amounts of energy provided by parenteral nutrition.[51]

Conclusion

Many health care providers regard improved nutrition, including the development of parenteral and enteral nutrition, as among the most important medical advances of the twentieth century. Both enteral and parenteral nutrition provide lifesaving therapy to those who cannot eat. Yet enteral and parenteral nutrition can be difficult to manage and may require the specialized expertise of dietitians credentialed in this area of practice. Oral diet is still the preferred method of nutrition, and it is the goal of nutrition intervention.

Application of the Nutrition Care Process: Parenteral Nutrition Support

Introduction:

(Case Study data provided courtesy of Kathy Fitzpatrick, MS, RD, CNSD, Special Design, Cape Girardeau, Missouri)

Fifty-two-year-old female with intractable N & V—unable to control with medications. Referred for home start PN. Diagnosis: Pancreatic cancer currently being aggressively treated with chemotherapy; recent hospitalizations for dehydration / N & V.

Nutrition Assessment:

Food/Nutrition-Related History:

Taking only clear liquids (<200 mL), small amounts one week prior to initiation of PN (very poor prior to that)

Anthropometric Measurements:

Ht: 165 cm, wt: 66 kg, usual wt (previous six months): 81 kg

(continued)

Biochemical Data:

DATE TIME LOCATION	NORMAL	1/8 ADMIT	1/12	1/19	UNITS
Albumin	3.6–5	2.6			g/dL
Prealbumin	19–43	17			mg/dL
Sodium	135–155	132	130	131	mmol/L
Potassium	3.5–5.5	4.6	4.7	3.9	mmol/L
Chloride	98–108	90	101	98	mmol/l
PO_4	2.5–4.5	3.9	3.4	3.4	mmol/L
Magnesium	1.6–2.6	2.20	2.10	1.70	mmol/L
Total CO_2	24–30	20.8	22.6	23.8	mmol/L
Glucose	70–120	144	122	104	mg/dL
BUN	8–26	16.9	18.9	17.9	mg/dL
Creatinine	0.6–1.3	0.5	0.5	0.5	mg/dL
Calcium	8.7–10.2	9.2	9.2	8.7	mg/dL
Hb A_{1c}	4.8–7.8	7.2			%

Questions:

Nutrition Diagnosis:

1. Identify at least two nutrition problems based on the nutrition assessment and medical history. Determine the diagnostic term for each nutrition problem. Next, identify the etiology of each nutrition problem. Finally, identify the signs and symptoms that support the evidence for these nutrition problems. An example for this case has been provided.

Diagnosis	Related to	Etiology	As Evidenced by	Sign/Symptoms
Inadequate food and beverage intake	R/T	N/V	AEB	Consumption of only sips of clear liquids providing less than 200 kcal/day.

On 1/8 the following was ordered:

PN formula: 1392 mL, 58 mL/hour continuous 24-hour infusion (21 mL/kg)

Macronutrients:

200 mL 50% dextrose, 800 mL 10% amino acids, 300 mL 10% lipids

Electrolytes:

NaCl—68 mEq, Na Acetate 32 mEq, K Acetate 36 mEq, KPO_4 30 mmol, $MgSO_4$ 16 mEq, Ca Gluconate 9.4 mEq, MVI (multivitamin injection) 10 mL, MTE 5 (multiple trace elements) 1 mL

Added medications:

40 mg Pepcid/day

2. Determine the amount of energy (kcal) and protein provided by the initial PN solution.

3. Calculate the grams of carbohydrate, protein, and lipid provided by this prescription. How many kcal/kg and grams of protein/kg does it provide? Calculate the patient's nutritional needs. Compare the two.

4. Is this patient at risk for refeeding syndrome? Why? What can be done to prevent it?

5. What clue in the patient's admission history gives support to the patient's low chloride level at the initiation of PN?

6. On 1/10, the RD recommended that NaCl be increased by 30 meq, and then by 20 mEq of Na Acetate on 1/12. Why?

Nutrition Intervention:

7. On 1/10, the RD recommended that NaCl be increased by 30 mEq, and then by 20 mEq of Na Acetate on 1/12. Why?

Nutrition Monitoring and Evaluation:

8. Determine nutrition criteria for monitoring and evaluation for each nutrition diagnosis that you identified.

EXAMPLE: Nutrition Care Outcome

Involuntary weight loss	Measure weight weekly.	Patient's weight will stabilize at 66 kg.

American Society for Parenteral and Enteral Nutrition (ASPEN) Information on the American Society for Parenteral and Enteral Nutrition may be obtained from this site.
www.nutritioncare.org

Dietitians in Nutrition Support Information on Dietitians in Nutrition Support—an American Dietetic Association practice

group—may be obtained from this site.
www.dnsdpg.org

National Board of Nutrition Support Certification Inc. Information on the Certified Nutrition Support Dietitian exam.
www.nutritioncare.org/nbnsc/

Information on enteral products can be obtained from manufacturer websites:

Nestle Nutrition
www.nestle-nutrition.com

Abbott Nutrition
http://abbottnutrition.com/

END-OF-CHAPTER QUESTIONS

1. Describe two ways that the house or regular diet can be modified to accommodate patient needs.

2. What is the difference between clear and full liquid diets? When are they used, and what are their limitations?

3. What are the advantages and disadvantages of enteral and parenteral nutrition support?

4. Describe three ways enteral and two ways parenteral nutrition can be delivered to the patient.

5. List five factors that might influence selection of an enteral formula (e.g., viscosity). Explain why each factor is important when choosing a formula.

6. What are medium-chain triglycerides (MCT), and why are they added to some enteral products? What is the most common source that is currently used?

7. List four complications that might occur when feeding a patient enterally. Describe and provide the rationale for three factors that should be monitored.

8. Calculate the caloric content and protein amount in one liter of parenteral solution composed of 25% dextrose and 4.25% amino acids. If 250 milliliters of a 20% fat emulsion were added, how many more kcal would be provided?

REFERENCES

1. Silver, H.J. Oral strategies to supplement older adults' dietary intakes: comparing the evidence. Nutr Rev. 2009;67:21–31.

2. Kruizenga HM, Seidell JC, de Vet HC, Wierdsma NJ, van Bokhorst-de van der Schueren MA. Development and validation of a hospital screening tool for malnutrition: the short nutritional assessment questionnaire (SNAQ). Clin Nutr. 2005;24:75–82.

3. Thomas DR, Zdrowski CD, Wilson M-M, Conright KC, Lewis C, Tariq S, Morley JE. Malnutrition in subacute care. Am J Clin Nutr. 2002;75:308–13.

4. Wendland BE, Greenwood CE, Weinberg I, Young KW. Malnutrition in institutionalized seniors: the iatrogenic component. J Am Ger Soc. 2003;51:85–90.

5. Koen FM, Hulst JM. Prevalence of malnutrition in pediatric hospital patients. Current Opinion in Pediatrics. 2008;20:590–96.

6. Norman K, Pichard C, Lochs H, Pirlich M. Prognostic impact of disease-related malnutrition. Clin Nutr. 2008;27:5–15.

7. Stratton RJ, King CL, Stroud MA, Jackson AA, Elia M. Malnutrition Universal Screening Tool predicts mortality and length of hospital stay in acutely ill elderly. Br J Nutr. 2006;95:325–30.

8. Pirlich M, Schütz T, Norman K, Gastell S, Lübke HJ, Bischoff SC, Bolder U, Frieling T, Güldenzoph H, Hahn K, Jauch KW, Schindler K, Stein J, Volkert D, Weimann A, Werner H, Wolf C, Zürcher G, Bauer P, Lochs H. The German hospital malnutrition study. Clin Nutr. 2006;25:563–72.

9. McClave SA, Martindale RG, Vanek VW, McCarthy M, Roberts P, Taylor B, Ochoa JB, Napolitano L, Cresci G; A.S.P.E.N. Board of Directors; American College of Critical Care Medicine; Society of Critical Care Medicine. Guidelines for the Provision and Assessment of Nutrition Support Therapy in the Adult Critically Ill Patient: Society of Critical Care Medicine (SCCM) and American Society for Parenteral and Enteral Nutrition (A.S.P.E.N.). J Parenter Enteral Nutr. 2009;33:277–316.

10. Bankhead R, Boullata J, Brantley S, Corkins M, Guenter P, Krenitsky J, Lyman B, Metheny NA, Mueller C, Robbins S, Wessel J; A.S.P.E.N. Board of Directors. Enteral nutrition practice recommendations. J Parenter Enteral Nutr. 2009;33:122–67.

11. Marian M, Charney P. Patient selection and indications for enteral feedings. In: Charney P, Malone A. ADA Pocket Guide to Enteral Nutrition. Chicago (IL): American Dietetic Association; 2006.

12. American Dietetic Association Evidence Library. Effects of Enteral Versus Parenteral Nutrition. Available at http://www.adaevidencelibrary.com/topic.cfm?cat=1032

13. Charney P, Malone A. ADA Pocket Guide to Enteral Nutrition. Chicago (IL): American Dietetic Association, 2006.

14. Byrne TA, Wilmore DW, Iyer K, Dibaise J, Clancy K, Robinson MK, Chang P, Gertner JM, Lautz D. Growth hormone, glutamine, and an optimal diet reduces parenteral nutrition in patients with short bowel syndrome: a prospective, randomized short bowel syndrome: a prospective, randomized, placebo-controlled, double-blind clinical trial. Ann Surg. 2005;242:655–61.

15. Jeejeebhoy KN. Enteral feeding. Curr Opin Gastroenterol. 2005;21:187–91.

16. McClave SA, Heyland DK. The physiologic response and associated clinical benefits from provision of early enteral nutrition. Nutr Clin Pract. 2009;24:305–15.

17. Braunschweig CL, Levy P, Sheean PM, Wang X. Enteral compared with parenteral nutrition: a meta-analysis. Am J Clin Nutr. 2001;74:534–42.

18. Marik PE, Zaloga GP. Early enteral nutrition in acutely ill patients: a systematic review. Crit Care Med. 2001; 29:2264–70.

19. McClave SA, DeMeo MT, DeLegge MH, DiSario JA, Heyland DK, Maloney JP, Metheny NA, et al. North American Summit on Aspiration in the Critically Ill Patient: consensus statement. JPEN. 2002;26:S80–85.

20. Cresci G. In: Charney P, Malone A. ADA Pocket Guide to Enteral Nutrition. Chicago (IL): American Dietetic Association; 2006, pp. 39–40.

21. Bengmark S. Bio-ecological control of acute pancreatitis: the role of enteral nutrition, pro and synbiotics. Curr Opin Clin Nutr Metab Care. 2005;8:557–61.

22. Malone A. Enteral formula selection: a review of selected product categories. Practical Gastroenterology. 2005;44–74.

23. Wirtitsch M, Wessner B, Spittler A, Roth E, Volk T, Bachmann L, Hiesmayr M. Effect of different lipid emulsions on the immunological function in humans: a systematic review with meta-analysis. Clin Nutr. 2007;26:302–13.

24. Mayer K, Seeger W. Fish oil in critical illness. Curr Opin Clin Nutr Metab Care. 2008;11:121–27.

25. Lasztity N, Hamvas J, Biro L, Nemeth E, Marosvolgyi T, Decsi T, Pap A, Antal M. Effect of enterally administered n-3 polyunsaturated fatty acids in acute pancreatitis—a prospective randomized clinical trial. Clin Nutr. 2005;24(2):198–205.

26. Marik PE, Zaloga GP. Immunonutrition in critically ill patients: a systematic review and analysis of the literature. Intensive Care Med. 2008;34:1980–90.

27. Ryan AM, Reynolds JV, Healy L, Byrne M, Moore J, Brannelly N, McHugh A, McCormack D, Flood P. Enteral nutrition enriched with eicosapentaenoic acid (EPA) preserves lean body mass following esophageal cancer surgery: results of a double-blinded randomized controlled trial. Ann Surg. 2009;249:355–63.

28. Keohane P, Attrill H, Love M, Frost P, DB S. Relation between osmolality of diet and gastrointestinal side effects in enteral nutrition. Br Med J. 1983;288:678–80.

29. Rees R, Keohane P, Grimble G, Frost P, Attrill H, Silk D. Tolerance of elemental diet administered without starter regimen. BMJ. 1985;290:1869–70.

30. Rees R, Keohane P, Grimble G, Frost P, Attrill H. Elemental diet administered nasogastrically without starter regimens to patients with inflammatory bowel disease. JPEN. 1986;10:258–62.

31. Zarling E, Parmar J, Mobarhan S, Clapper M. Effect of enteral formula infusion rate, osmolality, and chemical composition upon clinical tolerance and carbohydrate absorption in normal subjects. JPEN.1986;10:588–90.

32. Thompson C. Initiation, advancement and transition of enteral feedings. In: Charney P, Malone A, editors. ADA pocket guide to entereal nutrition. Chicago (IL): American Dietetic Association; 2006.

33. Gottschlich MM, ed. The ASPEN Nutrition Support Core Curriculum. A case-based approach—the adult patient. Silver Spring MD: American Society for Parenteral and Enteral Nutrition, 2007.

34. Whelan K, Judd PA, Tuohy KM, Gibson GR, Preedy VR, Taylor MA. Fecal microbiota in patients receiving enteral feeding are highly variable and may be altered in those who develop diarrhea. Am J Clin Nutr. 2009;89:240–47.

35. DeLegge MH. Enteral feeding. Curr Opin Gastroenterol. 2008;24:184–89.

36. Klein L. Is blue dye safe as a method of detection for pulmonary aspiration? J Am Diet Assoc. 2004;104:1651–52.

37. McClave SSA, Lukan JK, Stefater JA, Lowen CC, Looney SW, Matheson PJ, et al. Poor validity of residual volumes as a marker for risk of aspiration in critically ill patients. Crit Care Med. 2005;3:324–30.

38. Malone AM. Permissive underfeeding: its appropriateness in patients with obesity, patients on parenteral nutrition, and non-obese patients receiving enteral nutrition. Curr Gastroenterol Rep. 2007;9:317–22.

39. Kudsk KA, Sacks GS. Nutrition in the care of the patient with surgery, trauma and sepsis. In: Modern nutrition in health and disease. 10th ed. Philadelphia (PA): Lippincott Williams & Wilkins; 2004.

40. Kraft MD, Btaiche IF, Sachs GS. Review of refeeding syndrome. Nutr Clin Prac. 2005;20:625–33.

41. Reeds DN. Nutrition support in the obese, diabetic patient: the role of hypocaloric feeding. Curr Opin Gastroenterol. 2009;25:151–54.

42. Fahy BG, Sheehy AM, Coursin DB. Glucose control in the intensive care unit. Crit Care Med. 2009;37:1769–76.

43. NICE-SUGAR Study Investigators, Finfer S, Chittock DR, Su SY, Blair D, Foster D, Dhingra V, Bellomo R, Cook D, Dodek P, Henderson WR, Hébert PC, Heritier S, Heyland DK, McArthur C, McDonald E, Mitchell I, Myburgh JA, Norton R, Potter J, Robinson BG, Ronco JJ. Intensive versus conventional glucose control in critically ill patients. N Engl J Medjou. 2009;360:1283–97.

44. Griesdale DE, de Souza RJ, van Dam RM, Heyland DK, Cook DJ, Malhotra A, Dhaliwal R, Henderson WR, Chittock DR, Finfer S, Talmor D. Intensive insulin therapy and mortality among critically ill patients: a meta-analysis including NICE-SUGAR study data. CMAJ. 2009;180:821–27.

45. Kochevar M, Guenter P, Holcombe B, Malone A, Mirtallo J. ASPEN Board of Directors and Task Force on Parenteral Nutrition Standardization, ASPEN Statement on Parenteral Nutrition Standardization. J Parenter Enteral Nutr. 2007;31:441–48.

46. Madsen H, Frankel EH. The hitchhiker's guide to parenteral nutrition management for adult patients. Prac Gastroenteology. 2006;7:46–68.

47. Mirtallo J, Canada T, Johnson D, Kumpf V, Petersen C, Sacks G, et al. Task Force for the Revision of Safe Practices for Parenteral Nutrition. Safe practices for parenteral nutrition. JPEN J Parenter Enteral Nutr. 2004;28:S39–S70.

48. Helphingstine C, Bistrian B. New Food and Drug Administration requirements for inclusion of vitamin K in adult parenteral multivitamins. JPEN J Parenter Enteral Nutr. 2003;27:220–24.

49. American Medical Association Department of Foods and Nutrition. Multivitamin preparations for parenteral use; a statement by the Nutrition Advisory Group. JPEN. 1979;3:258–62.

50. Skipper A. Parenteral Nutrition. In: Matarese L, Gottschlich M, eds. Contemporary nutrition support practice. Philadelphia (PA):W.B. Saunders, 2002; 714.

51. Dissanaike S, Shelton M, Warner K, O'Keefe GE. The risk for bloodstream infections is associated with increased parenteral caloric intake in patients receiving parenteral nutrition. Crit Care. 2007;11:R114.

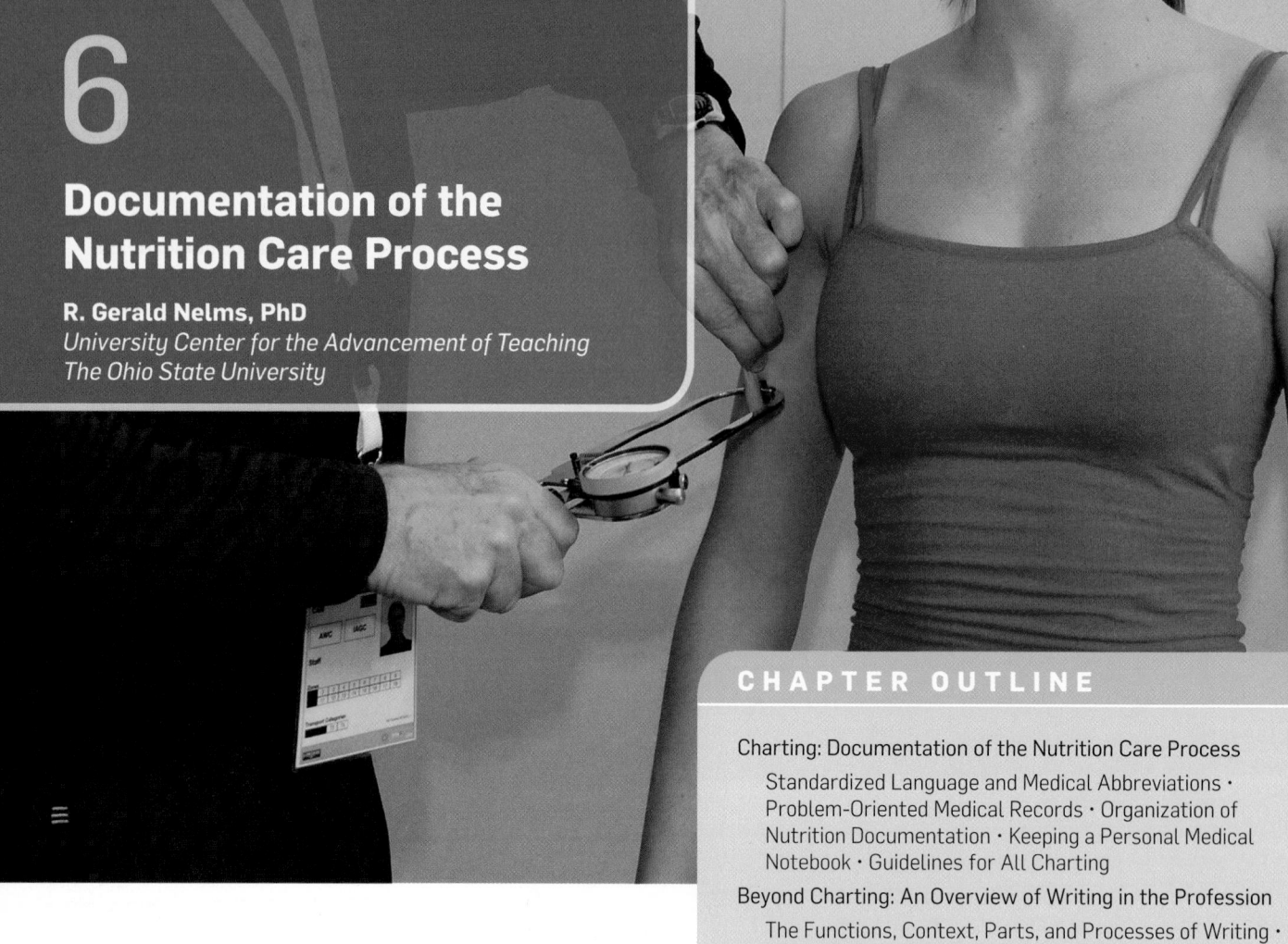

6

Documentation of the Nutrition Care Process

R. Gerald Nelms, PhD
University Center for the Advancement of Teaching
The Ohio State University

Introduction

Students often are surprised to learn how much writing they have to do in clinical settings. As explained in Chapter 2, the nutrition care process (NCP) consists of four interrelated and connected steps: (1) nutrition assessment, (2) nutrition diagnosis, (3) nutrition intervention, and (4) nutrition monitoring and evaluation. Each of these steps includes writing as a vital part of the process.[1,2] The amount of writing and "paperwork" can be extraordinary. In fact, health professionals probably ought to do even more than they do, because keeping personal notes in addition to required documentation of nutrition care provides greater opportunities for insights—and thus better diagnoses and more effective treatment plans.

Charting: Documentation of the Nutrition Care Process

Every medical institution or agency creates a medical record or chart for each individual patient served by that institution or agency. A medical record is a systematic recording of a patient's care, a location in which all data relating to the problem that brought the patient in for medical care are collected. But a medical chart also represents ongoing conversations among the different members of the medical team working on an individual patient's care. The aim of this ongoing conversation is the creation of consensus about the appropriate care and treatment for the patient. The chart is the basis for determining patient care, for documenting communication among health profes-

sionals dedicated to that individual patient's care, and for keeping a clear and comprehensive record of all that is done for the patient for legal reasons (see Table 6.1). Several conventional forms for charting have been developed. Different medical institutions and agencies will prefer different forms. When entering a particular workplace, the novice dietitian will be oriented to the form of charting in that institution. The information that is documented will be consistent, but the way that information is organized may differ.

The driving forces that impact medical record keeping also include accrediting agencies for health care facilities,[3] continuous quality improvement programs, and insurance reimbursement for medical care. Decisions regarding how to organize information and what language to use are guided by the need to meet the specific standards of these agencies.

In order for a health care provider to receive payment, the patient's medical record at discharge must contain documentation of the **Current Procedural Terminology (CPT) codes** and indicate that the patient received the appropriate care. Clear, concise wording in the medical record, using terminology consistent with the **prospective payment system**, will facilitate reimbursement for services. Trained medical coders are responsible for assigning codes for diagnosis and treatment so that reimbursement occurs.

Current Procedural Terminology (CPT) codes—numeric codes used to describe a medical service; these codes were developed by American Medical Association with the Health Care Financing Administration

prospective payment system—a system developed by U.S. government to reimburse health care providers for inpatient health services at a predetermined rate for a particular diagnosis and level of care

Table 6.1 Purposes of Medical Record Charting

- Legal documentation of medical care that the client has received
- Communication between members of the health care team
- Evaluation of medical care for that client
- Funding and resource management
- Continuous quality improvement
- Third-party reimbursement
- Accreditation
- Research

Medical record charts are used to audit and monitor the health care provided by an institution or a specific group of health care providers. Each state licensing agency (as well as The Joint Commission) requires that all health care facilities monitor, evaluate, and seek ways to improve the quality of care for their patients.

Finally, it is important to remember that the medical record is a legal document. The record serves as a description of exactly what happened during the medical care. Clients frequently request copies of their medical records, and they have the right to read those records. Each institution has policies for controlling the manner in which records are shared.

The electronic medical record (EMR) is defined as: "An application environment composed of the clinical data repository, clinical decision support, controlled medical vocabulary, order entry, computerized provider order entry, pharmacy, and clinical documentation applications."[4] This environment supports the patient's electronic medical record across inpatient and outpatient environments, and is used by healthcare practitioners to document, monitor, and manage health care delivery within a care delivery organization (CDO). The data in the EMR is the legal record of what happened to the patient during their encounter at the CDO and is owned by the CDO.[3] A majority of institutions and agencies maintain at least a portion of the medical record in an electronic format. Research has demonstrated significant advantages for use of EMR, including an expedition for transcription and delivery of physician orders.[4–6]

The use of electronic medical records often means that the format for charting is preset, and your style of charting is determined by the established chart guidelines (see Figure 6.1 for a sample template). If this is the case, you will be trained in how to enter charting notes into the electronic system. Familiarity with the various charting formats discussed in this chapter will help you understand the format and conventions of the charting system you are expected to use, whether it is paper-based or electronic.

Standardized Language and Medical Abbreviations

Steps to ensure accuracy of the medical record include the use of standard language and medical abbreviations. Typically, each health care facility designates a list of acceptable abbreviations. The Joint Commission recently recommended that certain abbreviations not be used because they are more likely to contribute to patient care errors. See Table 6.2 for their guidelines.

As discussed in Chapter 2, the American Dietetic Association has also developed standardized language—the Nutrition Diagnostic Terminology.[2, 7] Most institutions are in the process of developing individual protocols for integrating this terminology into their documentation system.[8–10] A previous study indicated that consistent use of the nutrition care process and standardized language within two midwestern hospitals resulted in improved documentation of nutritional care. The study authors predicted that the adoption of standardized language terms will improve accountability and reimbursement and enhance patient care overall.[11]

Problem-Oriented Medical Records

One common type of medical record is the problem-oriented medical record (POMR). The POMR is divided into five parts: data, problem list, care plan, progress notes, and discharge summary. The data make up a collection of subjective and objective information about the patient and is the basis of the problem list. As the problem list is established, a care plan is constructed for each problem. The plan should include expected outcomes, plans for further data collection, and, if needed, a patient teaching plan. Each health care team member composes progress notes in a narrative format. These notes are used in monitoring a client's care. The frequency of the entry of progress notes is determined by the facility's policies and procedures as well as the individual care plan. Finally, the discharge summary addresses each problem on the problem list and notes whether it was resolved or not. If it was not resolved, a plan is developed to treat the problem after discharge. Such a plan may provide for communication with other facilities, home health agencies, and the patient. The discharge summary often is the only part of the hospital record that other facilities (such as long-term care) receive. Providing accurate nutrition documentation will help ensure that consistent care is given to patients with nutrition problems.

Organization of Nutrition Documentation

Nutrition information within the medical record may be organized using any of several different styles. No matter what style is preferred by the practitioner or the institution, the same data are used to document the nutrition care process. The

Figure 6.1 Example of an Electronic Medical Record

This EMR template, used at Fairfield Medical Center in Lancaster, OH, incorporates the nutrition care process by prompting the user to enter nutrition problems, diagnoses, and interventions as well as nutrition assessment data.

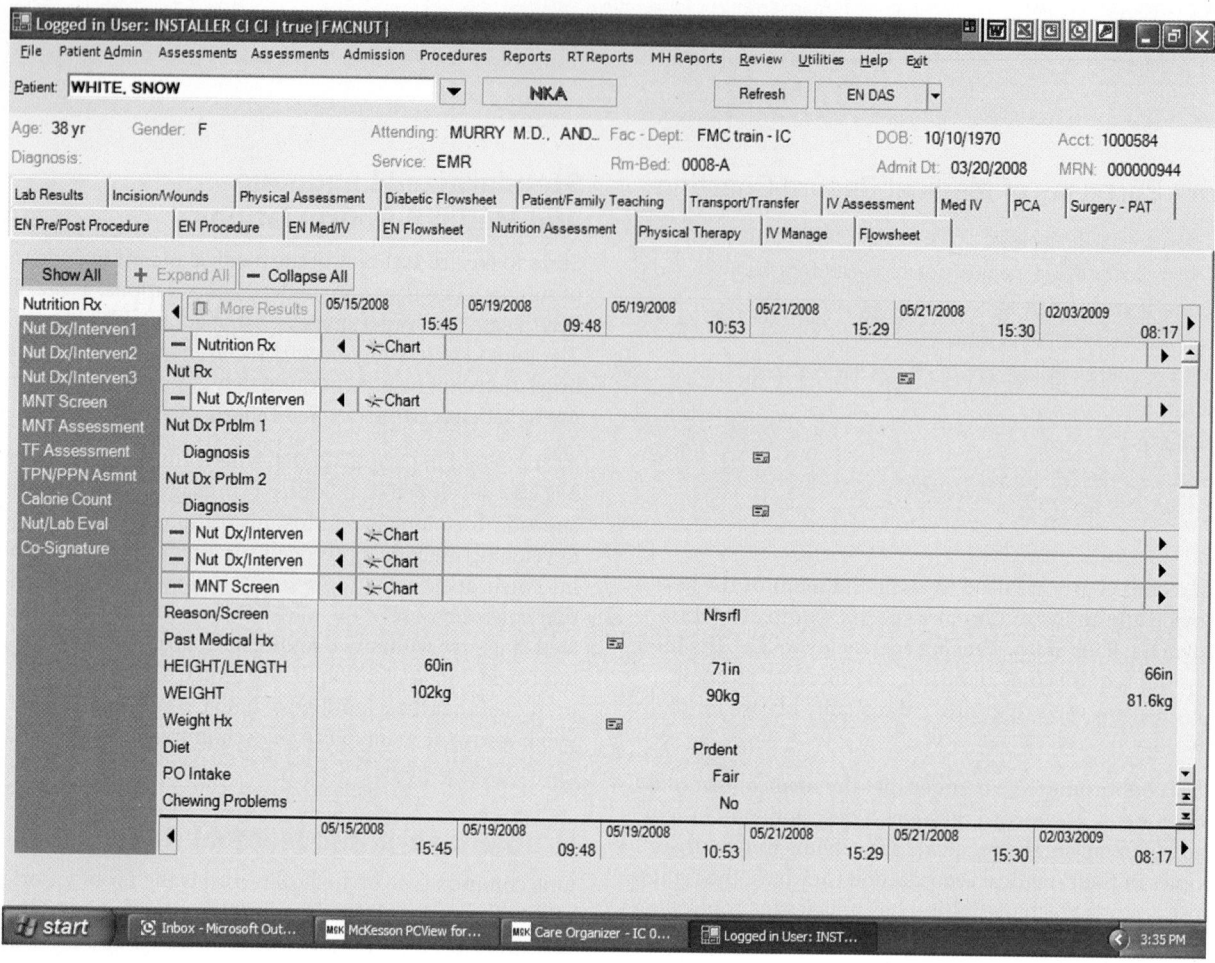

following sections describe the numerous organization styles used in nutrition documentation.

SOAP SOAP is the oldest and most well-known form of medical documentation. It has historically been the format that dietitians have used for daily progress notes. The label "SOAP" refers to the four sections of each entry in the medical chart: subjective data, objective data, assessment, and plan.

Subjective data (see Table 6.3) includes patient information or data collected from the patient or caregiver. This information can be placed into four major categories: diet-related information, lifestyle/psychosocial or emotional information, medical history information, and learning/motivation information. This section may include symptoms expressed by the patient; descriptions by the patient of her or his pain, discomfort, and/or dysfunction; dietary history; or the presence of symptoms that interfere with the ability to eat.

Objective data (see Table 6.4) includes the empirical information—that is, information drawn from physical tests and medical staff observations that are of consequence to the patient's nutritional status. This information can come from physical examinations, X-ray examinations, other imaging techniques,

or biochemical tests that are of nutritional relevance. Examples of objective data would be the patient's age, gender, and anthropometric information. It could include information from the physical examination such as temperature, pulse, blood pressure, or respiratory rate. Medical diagnosis and current medical care are noted here. Objective data can also include sensory information noted by the medical staff member such as smells, how an organ feels during a physical exam, or the visual recording of patient skin coloration.

Assessment (see Table 6.5) is the nutrition diagnosis or interpretation of the patient's nutrition problems. Assessment should include the nutrition problems with their supporting data, stated as nutrition diagnoses in the PES format. In the assessment section of a SOAP note, conclusions are drawn from the subjective and objective data in order to support the nutrition diagnosis.

Plan (see Table 6.6) will include an outline of interventions necessary to treat each nutrition problem. The plan may include requests for additional information needed to address the patient's nutrition problems. Specific nutrition therapy recommendations are stated here. Finally, goals and objectives may be included with a specific measure and timeline for evaluation of the intervention.

Table 6.2 The Joint Commission Official "Do Not Use" List[3]

This list applies to all orders and all medication-related documentation that is handwritten (including free-text computer entry) or on preprinted forms.

Do Not Use	Potential Problem	Use Instead
U (unit)	Mistaken for "0" (zero), the number "4" (four) or "cc"	Write "unit"
IU (International Unit)	Mistaken for IV (intravenous) or the number 10 (ten)	Write "International Unit"
Q.D., QD, q.d., qd (daily)	Mistaken for each other	Write "daily"
Q.O.D., QOD, q.o.d., qod (every other day)	Period after the Q mistaken for "I" and the "O" mistaken for "I"	Write "every other day"
Trailing zero (X.0 mg)*	Decimal point is missed	Write X mg
Lack of leading zero (.X mg)		Write 0.X mg
MS	Can mean morphine sulfate or magnesium sulfate	Write "morphine sulfate"
MSO$_4$ and MgSO$_4$	Confused for one another	Write "magnesium sulfate"
Additional Abbreviations, Acronyms and Symbols **(For *possible* future inclusion in the Official "Do Not Use" List)**		
Do Not Use	Potential Problem	Use Instead
> (greater than)	Misinterpreted as the number "7" (seven) or the letter "L"	Write "greater than"
< (less than)	Confused for one another	Write "less than"
Abbreviations for drug names	Misinterpreted due to similar abbreviations for multiple drugs	Write drug names in full
Apothecary units	Unfamiliar to many practitioners	Use metric units
	Confused with metric units	
@	Mistaken for the number "2" (two)	Write "at"
cc	Mistaken for U (units) when poorly written	Write "mL" or "ml" or "milliliters" ("mL" is preferred)
μg	Mistaken for mg (milligrams) resulting in one thousand-fold overdose	Write "mcg" or "micrograms"

*__Exception:__ A "trailing zero" may be used only where required to demonstrate the level of precision of the value being reported, such as for laboratory results, imaging studies that report size of lesions, or catheter/tube sizes. It may not be used in medication orders or other medication-related documentation.

Source: © The Joint Commission, 2010. Reprinted with permission.

Table 6.3 Subjective Section of SOAP Note

Diet Related

- Eating habits and feeding abilities
- Use and fit of dentures
- Appetite and digestion problems
- Nausea, vomiting, constipation, diarrhea, heartburn, physical problems interfering with adequate oral intake
- Recent weight change
- Diet history/previous diet modification or Rx
- Usual pattern of food intake
- Food allergies/aversions
- Vitamin, mineral, and nutrient supplement intake
- Complementary/alternative nutrition therapy
- Nutritional history and family nutritional history
- Adequacy of prior dietary intake
- Method of obtaining foods/nutrients (e.g., Meals on Wheels)
- Previous nutrition education/counseling

Lifestyle/Psychosocial/Emotional

- Economic situation/income
- Ability to purchase/prepare/store food
- Living or eating alone
- Health promotion and exercise practices
- Exercise
- Smoking
- Interaction with/between other family members or caretakers
- Support systems
- Coping mechanisms
- Occupation

Medically Related

- Personal and family medical history
- Especially, diseases with nutritional implications (e.g., type 2 diabetes)
- Use of complementary/alternative medical therapies (CAM)
- Medications, previous to admission or current PE
- Prescriptions, OTC (antacids, laxatives, etc.), and any CAM medications
- Other physically related problems

Learning and Motivation Related

- Ability to communicate in English (speaking, comprehending, reading, and writing)
- Patient's comments about previous prescribed diets/medical treatment and compliance issues
- Psychosocial problems, including addiction
- Perception of health status/reasons for seeking health care
- Desire to improve health or be involved in their own treatment or treatment decisions
- Learning style/problem-solving abilities
- Intellectual performance
- Educational level
- Communication patterns
- Attention span
- Long-term and recent memory
- Readiness to learn
- Barriers to change
- Growth and maturation

Table 6.4 Objective Section of SOAP Note

A. Age, ethnicity, gender, height/weight, BMI; any anthropometric measurement in addition to height and weight
B. Biochemical lab results that are of nutritional relevance
C. Clinical diagnosis, medication, treatment orders (including diet orders), any additional clinical findings of nutritional relevance
D. Dietary information, including current intake that has been observed (not subjective) or analysis of diet quality; protein/kcalorie requirements

Table 6.5 Assessment Section of SOAP Note

1. Current nutrition **P**roblems, **E**tiology, and **S**igns/symptoms (PES)
2. Potential nutritional problems (due to prognosis or clinical course of the disease, noncompliance and/or drug nutrient interactions)
3. Prioritization of the nutrition diagnoses

Table 6.6 Plan Section of SOAP Note

Gather	Additional information you need or would like (for current and potential nutritional problems; for instance, whether the patient is lactose intolerant)
Referral	Referral to other health or social professional (examples: psychologist for an eating disorder or depression; social worker if patient is homeless)
Nutrition	Specific nutritional recommendations for the client/patient to address *current* nutritional problem(s) (these may be different than those stated under Assessment; for example, fewer kcalories to help achieve weight loss)
Goals/Education	1. What is/are your short-term goal(s) for this client/patient? 2. For each goal, state the expected outcome(s) of dietary compliance as behavioral objectives for change, or expected outcome of the nutrition support. (Remember that outcomes should be measurable; specify a time frame and criterion [by how much], and encourage client participation, if possible.) • Nutrition support to be recommended (when it is a medical procedure). • Visuals, models, printed material to be given or used, if appropriate. ***Example:*** (Goal) Patient will increase dietary fiber consumption. • Patient will eat whole-grain bread instead of white bread and increase consumption by one additional fruit and vegetable every day. • Patient will be given a handout on whole-grain products.
Evaluation	Timeline and measures for nutrition monitoring and evaluation (When and how will you evaluate the outcome of your nutrition plan goals?)

The other health care team members involved in determining and carrying out the individual patient's treatment constitute the audience for SOAP notes. Their purpose is to help create a continuity of appropriate treatment for the patient. The ethos for SOAP notes must be authoritative, knowledgeable, and

Table 6.7 Sample SOAP Note

Nutrition	
11/10/09 11:30 a.m.	
S:	Patient's mother relates that Denise's mouth hurts so badly that she can hardly talk. She has had limited oral intake. Patient's mother also describes an "anti-cancer" diet that Denise's aunt and uncle introduced them to. Denise states that she doesn't want to make anyone mad, but those foods on the anti-cancer diet make her mouth hurt worse, and she doesn't know what to believe.
O:	21 yo ♀ Dx: Stage II diffuse large B-cell lymphoma. Admitted with immunosuppression, fungal infection, dehydration R/O pneumonia. s/p first round of chemotherapy/CHOP Ht. 5'6" Wt. 108# Last adm wt: 120# Preillness wt: 130# Labs: Alb 3.0 WBC 1100 mm³ EER: 1700–1800 kcal EPR: 75–80 g protein
A:	Unintentional weight loss r/t inadequate energy and protein intake AEB 22 lb. weight loss over previous three months. Inadequate oral food/beverage intake r/t oral mucositis AEB intake of <300 kcal/day. Harmful beliefs/attitudes r/t "anti-cancer" diet AEB inadequacy of food choices suggested by this diet.
P:	1. Modify PO diet to accommodate soft, easily chewed foods and liquids. Increase nutrient density with use of modular supplements and high-kcalorie/high-protein liquid supplements as patient tolerates. 6–8 small feedings/day. 2. Initiate calorie count to monitor adequacy of PO intake. 3. Recommend low bacterial/neutropenic precautions. 4. Consult with nursing/MD to ensure adequate pain coverage before attempts at oral intake. 5. Provide evidence-based information about all of the components of the diet recommended by her family members for Denise. 6. Monitor daily through patient visitation, calorie counts, daily weights to assess adequacy of current interventions.

Signature:

M. Nahikian-Nelms, PhD, RD, LD

professional. All other forms of charting are either extensions of SOAP notes or reductions of SOAP notes (see Table 6.7 for a sample SOAP note).

PES, or Problem, Etiology, Signs/Symptoms Statements PES is the organizational structure or format in which the nutrition diagnosis is written: "[Problem] related to [etiology] as evidenced by [signs/symptoms]." Chapters 2, 3, 4, and 5 explain the process by which a nutrition diagnosis is determined; the PES statement is simply the manner in which the nutrition diagnosis is documented.[1, 2, 7]

Assessment, Diagnosis, Intervention, Monitoring/ Evaluation (ADIME) The ADIME format is organized to reflect the nutrition care process.[1, 2, 7] Relevant data about the patient's condition are recorded in the A (*assessment*) section. These might include, but would not be limited to, referral medical diagnosis; pertinent social, family, and medical history; summary of pertinent data collected; and comparison with standards and/or food-related behaviors. The D (*diagnosis*) section is where the actual PES statements are listed and prioritized. The I (*intervention*) section provides documentation of the specific treatment goals and expected outcomes, interventions, and response of the client. Finally, the ME (*monitoring and evaluation*) section records documentation of progress toward goals, factors that are facilitating or hampering progress, any changes in the client's level of understanding or behavior, and future plans for care. (See Table 6.8.)

IER Notes IER is a simplified version of SOAP:

- *Intervention* refers to what has been done for the patient and the patient's response to that treatment.
- *Evaluation* refers to the assessment part of SOAP, the diagnosis and evaluation based on the data gathered. This section often includes a brief summary of the plan and an evaluation of the treatment's effectiveness.
- *Revision of care* refers to any changes recommended or ordered in the patient's treatment.[12] Table 6.9 provides an example of IER charting.

Table 6.8 Sample ADIME Note

Assessment: Patient's mother relates that patient's mouth hurts so badly that she can hardly talk. She has had limited oral intake. Patient's mother also describes an "anti-cancer" diet that patient's aunt and uncle introduced them to.

21 yo ♀ Dx: Stage II diffuse large B-cell lymphoma

Admitted with immunosuppression, fungal infection, dehydration R/O pneumonia. s/p first round of chemotherapy/CHOP

Ht. 5'6" Wt. 108# Last adm wt: 120# Preillness wt: 130#

Labs: Alb 3.0 WBC 1100 mm³

EER: 1700–1800 kcal EPR: 75–80 g protein

Diagnosis: Inadequate oral food/beverage intake R/T mucositis AEB admitting dehydration and recent 12 lb. weight loss in previous two weeks.**

Swallowing difficulty R/T mucositis AEB patient interview and physical assessment.

Involuntary weight loss R/T inappropriate food choices and inadequate energy intake AEB 12 lb. weight loss over last two weeks.

**Highest priority

Intervention:

1. Modify PO diet to accommodate soft, easily chewed foods and liquids. Increase nutrient density with use of modular supplements and high-kcalorie/high-protein liquid supplements as patient tolerates. 6–8 small feedings/day.
2. Recommend low bacterial/neutropenic precautions during periods of immunosuppression.
3. Consult with nursing/MD to ensure adequate pain coverage before attempts at oral intake.
4. Provide evidence-based information regarding nutritional needs during treatment for lymphoma.

Monitoring/Evaluation:

1. Patient weight will stabilize as measured by daily weights.
2. Caloric intake will increase to a minimum of 65% of current recommendations as measured by daily calorie count.
3. Patient will express tolerance to current food choices during daily patient visitations.
4. Patient will state understanding of current nutritional needs during chemotherapy and during periods of immunosuppression.

Signature:

M. Nahikian-Nelms, PhD, RD, LD

Table 6.9 Sample IER Note

11/10/09 11:30 a.m. Nutrition Progress Note

Intervention: Modify oral intake to 6–8 small feedings; increased nutrient density; addition of high-calorie, high-protein supplement; modification of texture; pain medication prior to meals and supplements.

Evaluation: Per calorie counts and patient visitation, oral intake improved, currently meeting 65% of estimated energy and protein requirements. No further weight loss documented since admission.

Revision: Check prealbumin to monitor visceral protein status.

Signature:

M. Nahikian-Nelms, PhD, RD, LD

Focus Notes Focus notes are a blending and reduction of the SOAP and IER formats:

- *Data* simply collapses SOAP's subjective and objective data sections.
- *Action* refers to the SOAP assessment and IER evaluation sections—that is, the diagnosis and evaluation, based on the data gathered, and the treatment(s) applied.
- *Response* represents the SOAP plan and/or any changes in treatment, the same thing as the IER revision of care.[12] See Table 6.10.

PIE Notes PIE (problem, intervention, evaluation) notes are also a blending of other types of charting:

- *Problem* identifies the specific nutrition problems for the client.
- *Intervention* refers to the nutrition treatments and steps designed to resolve the problems.
- *Evaluation* outlines the progress toward resolution of the nutrition problems. See Table 6.11.

Table 6.10 Sample Focus Note

11/10/09 Nutrition Progress Note

Time	Focus	Data
11:30 a.m.	Inadequate oral intake	**Data:** 22 lb. weight loss over previous 3 months. Albumin 3.0; 24-hour recall indicates <25% of kcal and protein requirements met.
		Action: 6–8 small feedings; increased nutrient density; addition of high-calorie, high-protein supplement; modification of texture.
		Response: Caloric intake has increased by 45%. No further weight loss documented.
	Swallowing difficulty	**Data:** Mucositis; dehydration and inadequate caloric intake.
		Action: Pain medication prior to meals and supplements; modification of texture of meals and food choices to minimize pain.
		Response: Mucositis resolving; oral intake improved.

Signature:

M. Nahikian-Nelms, PhD, RD, LD

Table 6.11 Sample PIE Note

11/10/09 11:30 a.m. Nutrition Progress Note

Problem: Involuntary weight loss

Intervention: Modify oral intake to 6–8 small feedings; increased nutrient density; addition of high-calorie, high-protein supplement; modification of texture; pain medication prior to meals and supplements.

Evaluation: Per calorie counts and patient visitation, oral intake improved currently, meeting 65% of estimated energy and protein requirements. No further weight loss documented since admission.

Signature:

M. Nahikian-Nelms, PhD, RD, LD

Charting by Exception Charting by exception (CBE) is an even more abbreviated approach to medical charting that involves recording *only* unusual or out-of-the-ordinary events. The CBE format includes a standardized nutritional care plan. After the initial charting of the care plan, only significant data and/or unanticipated responses to the proposed plan should be included in the record. In most CBE formats, a flowchart is used to document assessments and interventions. An asterisk (*) indicates an abnormal finding on an assessment or an abnormal response to an intervention. The findings are explained in the comments section of the form. The progress notes are used to document revisions in the plan of care and specific interventions.

As you can see, the movement in charting over the last decade or so has tended toward reducing the size of the medical record. All of the formats share similar pieces of information. Regardless of the format or style, there should be a method to document all stages of the nutrition care process.

Keeping a Personal Medical Notebook

Novice health professionals may not have enough experience with patient care to know what might be important and what is probably not going to be important. Therefore, it is better to include more information than is needed instead of less information. Medical charting is inherently paradoxical, however. It must be comprehensive, yet it needs to be easy to read; it must be clear, but it also must be concise and to the point. Doctors, nurses, and other health professionals who care for a patient do not have time to read through long chart entries that may seem irrelevant. The goal is to create complete, concise documentation—for legal reasons, and, more importantly, for medical reasons.

With this last point in mind, health professionals, especially novice health professionals, should consider keeping a personal medical notebook where they can chart everything involved with each patient's treatment, where they can brainstorm problems, and, perhaps most importantly, where they can include their subjective experiences and express their own emotional and intellectual responses to their experiences.

The institutional administration and staff typically determine which information and how much of that information goes into a medical record. As indicated earlier, the current trend is toward streamlining the chart; but as records are simplified, potentially important information can be neglected. Also, health professionals face human anxiety, pain, and mortality daily, and they need a way of addressing the emotional and intellectual stress that comes with the job. Keeping a personal medical notebook can help tremendously.

It might mean simply writing in a small notepad or, if charting is done electronically, writing your chart notes into a portable storage device and then cutting and pasting what you feel is most relevant into the official chart. Later, you might take time to review your notes on a patient and write your own personal notes. You might want to follow the SOAP note format, adding in a section at the end for your personal thoughts and musings about the patient's condition and experience.

Another possibly useful format for your personal medical notebook is what is called "the double-entry notebook."[13] Devised for journal writing, the double-entry format simply involves dividing your notes into "objective" ones and "subjective" ones. One double-entry format involves drawing a vertical line down the middle of the page, but you also could just write your objective notes first and then respond to them in the next subjective section. "Subjective" here refers to your own subjectivity, not the patient's, although you could certainly include the patient's feelings and statements. You need not be tied to a particular formula, though; organize your personal notes in whatever way works for you.

A crucial point to keep in mind *at all times* is that any notes you take on a patient's condition, care, or treatment must be kept confidential. The official medical chart is protected institutionally, but *you* are responsible for keeping your personal medical notebook safe and confidential.

Guidelines for All Charting

- Make sure to chart whatever you see as significant, even if others may not perceive these changes or events as significant.

- The Joint Commission standards require that "all entries in medical records be dated and authenticated, and a method is established to identify the authors of entries."[3] Any medical record entry must include the dietitian's full name and status (e.g., RD, CDE). Students must enter their full names and their status as a student or intern. A student's note will need to be cosigned by the preceptor.

- Always be timely in your charting. Chart just after or shortly after you meet with the patient, receive lab work you ordered, or receive any new data that you find significant.

- Never ask someone else to do your charting, and never do someone else's charting. Legally, an error in the chart could be costly. Medically, an error in the chart could be deadly.

- Never chart a procedure until it is done, and then chart it as soon as possible.

- Remember that medical charts are legal documents as well as medical documents.

- When you chart in handwriting, use only black ink and write legibly. Make sure that your notes will photocopy easily.

- Write clearly. You do not need to use complete sentences, but the meaning of your notes must be obvious and unambiguous to others. Read what you have written and ask yourself, "Will my readers understand what I'm saying?"

- Avoid abbreviations unless you are absolutely certain that anyone needing to read the chart will understand immediately what the abbreviations mean and that they are approved for use within your institution.

- Don't leave "white space." Always begin your notes right after the last note recorded. The chart should follow chronological order, and you should not give someone who handles the chart later the opportunity to add notes that might look like your notes.

- Include too much detail rather than too little. While it is undesirable to write overly long notes, you must make sure that everything significant is noted. Note lengths can differ.

- When writing objective notes, record only what you see, hear, smell, or feel, along with what has been measured. Do not assume or infer anything. Save assumptions, inferences, conclusions, and opinions for the assessment or evaluation and plan sections or for your own subjective notes.

- Bracket your biases. Bracketing simply means you set them aside temporarily. We all meet new experiences with old biases—that is, with our personal preferences, leanings, values, and beliefs—but these biases have no place in objective notes. This does not mean that you have to rid yourself of your values. You will always face patients and situations that run counter to your personal feelings: patients who are drunk, obnoxious, obstinate, and/or abusive; you will always observe doctors and nurses whose treatment of patients is not what you would have ordered. Nonetheless, you can bracket your biases when you are charting (and performing other professional duties) by following these steps:

  1. Identify your biases—either prior to patient care or as you become aware of them during patient care. Make yourself aware of your feelings, beliefs, and preferences.

  2. Consciously imagine a box in your mind into which you place your particular bias.

  3. Imagine how the unbiased health professional would act, and act that way.

  4. If you have a model of unbiased professionalism, imagine how that person would act in the particular situation you find yourself in, and act that way.

- Use neutral language—that is, avoid emotionally charged words. For example, stating that a patient is receiving 65% of nutritional needs with current nutrition support would be preferable to stating that current nutrition support is inappropriate.

- When referring to a patient, use the term "patient" or "client" instead of using his or her name.

- Always keep the medical record intact. Never discard a page from that record for any reason. Such an action could result in legal issues at a later date. Moreover, you can never tell what might prove to be important; diagnoses that are rejected early on may be reconsidered and adopted later. If you spill coffee on the chart and blur some entries, do not discard the page. Simply copy it over and leave both pages in the medical chart. Cross-reference them by writing something like "Recopied from page 2" at the top of the new page.

- If you make a mistake, simply cross the note out with a single horizontal line, write "error" above it, and initial it. Do not scribble through the mistake. It should remain readable.

For legal reasons, there should be no question about what was written there.

- Always sign notes after printing your name.

- Also, be sure to include the date and time for each note. Military time is often used in order to prevent confusion between a.m. and p.m.

- Make sure you always follow your institution's or agency's policies and procedures for charting.

Confidentiality All medical record information is confidential. A confidential communication is given by one person to another with the trust and confidence that such information will not be disclosed. The federal U.S. law that assures patients of the confidentiality of their medical information is the Health Insurance Portability and Accountability Act of 1996, or HIPAA (see Box 6.1).[14] HIPAA protects information about clients that is gathered by examination, observation, conversation, or treatment. A dietitian cannot discuss a client's status with other clients or staff who are uninvolved in the client's care. A legal suit can be brought against a dietitian who has disclosed information about clients without their consent.

Dietitians and other health care professionals use records not only to provide care for individual patients, but also as an information source for continuous quality improvement, data gathering, research, or continuing education. If this is the case, assurance of patient anonymity is required. This is not a breach of confidentiality as long as the records are used as specified and permission is granted from the institution's internal review board.[15]

Beyond Charting: An Overview of Writing in the Profession

Writing to the community of health professionals within your institution or agency through chart notes will become easier with practice. But charting is not the only kind of writing you may be called on to do. You may also find yourself writing educational materials for various nonprofessional audiences, including the public in general, or even an article for publication in a journal. This section outlines general principles that are applicable to these writing tasks.

The Functions, Context, Parts, and Processes of Writing

Writing *functions* in both our personal and professional lives in a variety of different ways:

- To record information, such as agreements among people, documentation of health care in a medical facility, or documentation for legal purposes
- To inform, either as a report of an experience or research findings, or as dissemination of information developed by others
- To persuade
- To entertain

Medicine, including nutrition and dietetics, uses its own language, jargon, and contexts in which practitioners function every day. Sometimes, health professionals may lose sight of the fact that the rest of the "nonmedical" world doesn't participate in, or perhaps even understand, these conversations. Communication is necessary to keep everything moving forward, and thus, writing is a crucial skill that all newcomers to professions must learn.

Three important areas of writing need to be understood: the rhetorical norms—that is, the universal contextual framework within which all communication exists; the levels of discourse—that is, the different levels of writing that one can focus attention on; and writing processes—that is, the cognitive processes and stages of writing.

Rhetorical Norms
Each form of writing within the workplace is produced within a context, and every context involves at least four elements:

- *Subject Matter:* What the text is about
- *Purpose:* A reason for writing the text
- *Audience:* A set of readers to whom the text is directed
- *Ethos:* The personality or voice that comes through the text and characterizes the writer for the reader—that is, the person the reader assumes the writer to be, based on the ideas, organization, and style of the text.

For example, in medical record documentation, the subject matter is the nutrition care process for a patient. The purposes may include not only the legal documentation of care, but also communication with other health care providers or collection of research data. These purposes will influence the writing style, because the presentation of this information will need to be appropriate for the audience. Your professionalism, clinical knowledge, and expertise will characterize the ethos of the writing. Reflecting on these contextual elements—subject matter, purpose, audience, and ethos—before and during the production of a text can improve its effectiveness.

Levels of Discourse
Just as every text exists within a context of rhetorical norms, it also consists of an overall organization of words, sentences, punctuation, paragraphs, and larger passages, all of which communicate the writer's ideas, the content of the writing. These items—rhetorical norms, ideas, organization, and grammar (the sentence structures, punctuation, word choice, and spelling)—are generally referred to as the levels of discourse. Decisions regarding the rhetorical norms at the highest level—that is, decisions about the purpose, audience, subject matter, and writer's ethos—will determine decisions at these other, lower levels of discourse. During the process of writing, successful writers will move—sometimes effortlessly, it seems—between and through these levels of discourse. As they write, successful writers focus on different levels of discourse at different times, as needed. They may be generating ideas at one moment, then reorganizing sections at another. They may consider their choice of particular words and then move on to generating and then revising sentences, and determine the correct punctuation all the way through.

Steps in the Writing Process
While texts appear linear, moving from a beginning to a middle to an end, the actual process of writing is virtually never that linear. Successful writers are always stopping and returning to already produced sentences and ideas in order to reorient themselves so that they can go forward again. They may pause to make sure that what they have written will lead the reader logically from one thought to the next. They may pause to reestablish their own connection to the logical flow of their writing, to give themselves direction. A better metaphor for writing than the straight line is the spiral, because writing tends to be "recursive"—that is, writing moves backward in order to then move forward again. Composition scholars have identified the following major steps in the process of writing:

Prewriting or invention, whatever it is that the writer does before actually writing. For longer pieces of writing, successful writers tend to spend time planning a text before beginning to write it. They develop their ideas about the subject matter. For example, in planning a new patient education resource, RDs may read the most current literature supporting nutrition therapy for a particular diagnosis. They identify their purpose and audience. They produce a rough plan (sometimes in writing, sometimes just in their heads) for how the text is to be organized. From that point, an outline may be developed that organizes the most important issues for the patient.

Drafting, or actual sentence generation. As mentioned above, this is not a linear process. It involves setting goals and subgoals and recalibrating one's thinking as more and more text is produced. Successful writers tend to do some revising during drafting, too.

Revision. In addition to revising while drafting, successful writers virtually always make revisions to their text after it has been completely drafted.

Editing. After the text has been drafted and revised, successful writers also spend time making minor corrections at the lowest, sentence and subsentence levels of discourse. This is when proofreading takes place, and these writers make their changes based on errors they identify during that proofing of their manuscripts. Typically, successful writers wait to make editing changes to a text until revision is complete, to avoid wasting time proofing sentences that may later be revised or even eliminated from the text.

As indicated, all writing involves subject matter, a purpose, an audience, and the writer's ethos. But as writers move from being outsiders to being insiders within their communities, they adopt the discourse conventions of their communities—that is, the purposes, audiences, ethos, subject matter, writing processes, textual organizations, and writing styles specific to their communities. These differences in the writing of different communities—different disciplines, different workplaces—coalesce into what are referred to as "genres." As you have seen throughout this chapter, registered dietitians typically write into *charts*; but they also write memos, brochures, handouts, and other health information texts as well as research reports. These are all different genres adopted by the dietetics profession.

Reporting Your Own Research

When you become a clinical dietitian, you automatically become a member of the professional community of clinical dietitians and other health professionals. You are expected to keep up with the research in your field and to go to conferences and participate in other forms of continuing education. You may also want to contribute to the ongoing deliberations in your field. You may want to do original research and report the findings of that research either in professional journals and books or at professional conferences. Discussion of how to do such research and how to write professional articles,

book chapters, and conference papers is beyond the scope of this text. Nevertheless, you need to be aware that this research and professional writing is possible for you. It is considerably more formal than writing instructions for patients, in the sense that there are certain organizational and stylistic conventions that you must adopt when reporting your research.[16] The Web links at the end of the chapter provide information about doing research in dietetics and reporting that research professionally.

Conclusions: Your Ethos— Establishing Expertise

When you write professionally, you establish your professional ethos by making wise recommendations and orders and, most importantly, by establishing your expertise both in your actions and in your writing. The 2009 Code of Ethics outlines your professional obligation to convey accurate information for all audiences.[17, 18] Fulfilling this obligation means continuing your professional education throughout your career, and drawing on that ongoing knowledge when you chart and when you create and update informational and educational material. This is a crucial component that helps to define a profession and is the foundation for maintaining competence.

Documentation of the Nutrition Care Process

Introduction:

Mr. J, a 52-year-old man, is referred by his physician to the nutrition outpatient clinic for counseling on a weight-reduction diet. You discuss his goals with him and find out that he does want to lose weight because of his family history of diabetes.

Nutrition Assessment:

Food/Nutrition-Related History:

While talking with Mr. J, you obtain a quick diet history, which you feel is reasonably accurate. You calculate that his diet contains approximately 2800 kcal per day. Mr. J tells you that he dislikes sweets and fats, rarely eats vegetables, and drinks about 8 cups of fruit juice daily in addition to coffee and tea (which have added sugar) throughout the day, and that he is fairly inactive, eats two large meals a day, and never eats breakfast.

Anthropometrics:

You measure Mr. J and find that he is 5'7" tall and weighs 195 pounds.

Biochemical Data:

None available.

Nutrition Diagnosis:

Diagnosis	Related to	Etiology	As Evidenced by	Signs/Symptoms
EXAMPLE: Excessive energy intake	R/T	Excessive fruit juice consumption	AEB	Consumption of approximately 720 kcal from juice daily documented in 24-hour recall.

Nutrition Monitoring/Evaluation:

One goal that is determined is to reduce fruit juice to 1 cup per day and change to artificial sweetener for coffee and tea. He will keep a food record for one week and then return for a follow-up appointment.

Questions:

1. Outline the subjective information that should be included in a chart note for this session.

2. List all the objective information.

3. Write a PES statement for one other nutrition problem.

4. Write your assessment.

5. Determine appropriate interventions for the nutrition problem you identified.

6. Design your evaluation/outcome measures for this problem.

END-OF-CHAPTER QUESTIONS

1. What is the primary purpose of the medical record in the clinical setting? What other functions does it serve?

2. Why should standardized language and abbreviations be used in the medical record?

3. Describe each of the four sections that constitute a SOAP note.

4. How does medical record documentation fit within the nutrition care process?

5. How does charting by exception differ from the other charting methods?

6. Why might a dietitian keep a personal notebook in the clinical setting?

7. Why must information in a chart note be kept confidential?

REFERENCES

1. Lacey K, Pritchett E. Nutrition care process and model: ADA adopts road map to quality care and outcomes management. J Am Diet Assoc. 2003;103(8):1061–72.

2. International Dietetics and Nutrition Terminology (IDNT) Reference Manual: Standardized Language for the Nutrition Care Process. Chicago (IL): American Dietetic Association; 2009.

3. Joint Commission on Accreditation of Healthcare Organizations. 2009 Comprehensive Accreditation Manual for Hospitals: The Official Handbook (CAMH). Chicago (IL): 2009. Comprehensive Accreditation Manual for Hospitals. Oak Brook (IL): Joint Commission on Accreditation of Healthcare Organizations, 2009.

4. HIMSS Electronic Health Record Committee. HIMSS Electronic Health Record Definitional Model, 2003. Version 1.1. Healthcare Information and Management Systems Society (HIMSS). Available at http://www.himss.org

5. Peregrin T. Personal and electronic health records: sharing nutrition information across the health care community. J Am Diet Assoc. 2009;109:1988–91.

6. Shamliyan TA, Duval S, Du J, Kane RL. Just what the doctor ordered. Review of the evidence of the impact of computerized physician order entry system on medication errors. Health Serv Res. 2008;43:32–53.

7. Writing Group of the Nutrition Care Process/Standardized Language Committee. Nutrition Care Process Part II: Using the International Dietetics and Nutrition Terminology to Document the Nutrition Care Process. J Am Diet Assoc. 2008;108:1287–93.

8. Suen LF. Implementing Electronic health records with standardized language for the nutrition care process in acute care hospital. J Am Diet Assoc. 2009;109:A18.

9. Staten KD, Harden HC, Drew DA, Goolsby SL. The development of standardized forms for the implementation of nutrition care process in a pediatric hospital. J Am Diet Assoc. 2009;109:A31.

10. Mueller DC, Hancock CR, Ewalt GE, Hoskins CL, Simper AL, Gentry BM, Kliewer KL, Progressive implementation of the nutrition care process and standardized language into medical nutrition therapy documentation. J Am Diet Assoc. 2008;108:A43.

11. Hakel-Smith N, Lewis N, Ethridge K. Orientation to nutrition care process standards improves nutrition care documentation by nutrition practitioners. J Am Diet Assoc. 2005;105:1582–89.

12. Klein CJ, Bosworth JB, Wiles CE. Physicians prefer goal-oriented note format more than three to one over other outcome-focused documentation. J Am Diet Assoc. 1997;97(11):1306–10.

13. Nahikian-Nelms ML, Nelms RG. Use of double-entry journals in a required dietetic course: encouraging critical thinking skills. J Nutr Ed. 1994;26:93–96.

14. HIPPA. Available at www.cns.hhs .gov/HIPAAGenInfo/

15. Willison DJ, Keshavjee K, Nair L, Goldsmith C, Holbrook AM. Patients' consent preferences for research uses of information in electronic medical records: interview and survey data. Br Med J. 2003;326:373–78.

16. Boushey CJ, Harris J, Bruemmer B, Archer SL. Publishing nutrition research: a review of sampling, sample size, statistical analysis, and other key elements of manuscript preparation, Part 2. J Am Diet Assoc. 2008;108:679–88.

17. Berning JR, Karmally W. Ethics Opinion: The RD and DTR are obligated to follow ethical standards when writing for the popular press. J Am Diet Assoc. 2007;107:2052–54.

18. American Dietetic Association/ Commission on Dietetic Registration. Code of ethics for the profession of dietetics and process for consideration of ethics issues. J Am Diet Assoc. 2009;109:1461–67.

Part 3

Introduction to Pathophysiology

7

Fluid and Electrolyte Balance

Marcia Nahikian Nelms, PhD, RD
The Ohio State University

Introduction

Humans have long known the importance of water for survival. The geographical distribution of population groups throughout the world has been shaped by the availability of water. In medical care, restoration of normal fluid status is often the first priority in reestablishing homeostasis.

The functions of water in the body include transporting nutrients, transporting and excreting metabolic waste, supporting cell shape and structure, lubricating friction-generating surfaces, and sustaining normal body temperature. A variety of solutes are found in solution with water throughout the body. An important group of solutes found in body fluids is the **electrolytes**. Maintenance of fluid balance is significantly integrated with maintenance of electrolyte balance.

Daily losses of fluid from the body are normal and require replacement. The presence of injury or illness can result in increased fluid losses and increased fluid needs. Detection of fluid and electrolyte imbalance is an essential component of nutrition assessment.

Normal Anatomy and Physiology of Fluids and Electrolytes

Total Body Water

Total body water accounts for approximately 60% of total body weight in the adult male, and for somewhat less, an average of 50%, in the adult female. At birth, total body water accounts for approximately 75% of the infant's weight. Body water content declines throughout the life span and often falls below 50% in the elderly. In general, this is because the proportion of lean body mass to body fat influences the amount of water as a percentage of body weight. Fat tissue has the lowest percentage of water in comparison with all other tissues in the body. As body fat increases, the percentage of body water decreases (see Table 7.1).

Fluid Compartments

Membranes separate body fluids into compartments. Approximately two-thirds of body water is found within cells (**intracellular fluid**). The remaining body water is found outside of cells (extracellular fluid). **Extracellular fluid (ECF)** is divided into three compartments: interstitial, intravascular, and transcellular (or transitional). Interstitial fluids surround the cells. Intravascular fluid is found within blood. Transcellular fluids are those fluids found in secretions within organs. These include gastrointestinal secretions, cerebrospinal fluid, and intraocular fluid.

GLOSSARY

arginine vasopressin (AVP)—previously known as antidiuretic hormone; a hormone that acts on the renal tubules to reduce urine output in response to dehydration and hyperosmolality

ascites—abnormal accumulation of fluid in the abdominal cavity

baroreceptor—in general, any sensor of pressure changes

colloid osmotic pressure (oncotic pressure)—the osmotic pressure attributed to proteins and other macromolecules

cyclosporine—an immunosuppressant drug

dehydration—a deficit of water in the body

diabetes insipidus—chronic excretion of very large amounts of pale urine of low specific gravity

diuresis—the production of excessive amounts of urine

edema—the accumulation of excess fluid in cells, tissue, or a cavity, resulting in swelling

electrolytes—those substances that bear an electrical charge (ions)

extracellular fluid (ECF)—the interstitial fluid and the plasma, constituting about 20% of the weight of the body; sometimes used to mean all fluid outside of cells, usually excluding transcellular fluid

facultative urine—excess water that is excreted through urination

hypercalcemia—high serum calcium

hyperkalemia—high serum potassium

hypermagnesemia—high serum magnesium levels

hypernatremia—abnormally high levels of serum sodium

hyperosmolar hyperglycemic nonketotic syndrome—a complication of type 2 diabetes mellitus that usually develops after a period of hyperglycemia combined with inadequate fluid intake

hyperphosphatemia—high serum phosphorus

hypervolemia—increased blood volume

hypocalcemia—low serum calcium

hypokalemia—low serum potassium

hypomagnesemia—low serum magnesium levels

hyponatremia—abnormally low concentrations of sodium ions in the circulating blood

hypophosphatemia—low serum phosphorus

hypovolemia—decreased blood volume

insensible losses—fluid loss that cannot be easily measured (usually refers to fluid lost via sweat and respirations)

intracellular fluid (ICF)—the fluid within the tissue cells

Kayexalate—a medication used to reduce high serum potassium; exchanges sodium for potassium in the intestine

leukocytosis—high white blood cell count

metabolic water—water that is produced through nutrient metabolism

obligatory urine—the amount of fluid necessary for the body to excrete waste products and solutes (approximately 500 mL)

osmolality—the number of osmols (standard unit of osmotic pressure) per kilogram of solvent (water) (mOsm/Kg)

osmolarity—the number of osmols (standard unit of osmotic pressure) per liter of solution (mOsm/L)

osmotic pressure—the pressure that must be applied to a solution to prevent the passage of solvent into it when solution and pure solvent are separated by a membrane permeable only to the solvent

paresthesias—symptoms of tingling in fingers and toes; often consistent with electrolyte imbalance

preeclampsia—development of hypertension, with symptoms of proteinuria and edema, during pregnancy

rales—abnormal respiratory sounds made when air flows through liquid present in the airways

sensible losses—fluid loss that can be measured (usually refers to fluid lost via urine excretion)

specific gravity—the weight of a solution (e.g., urine) in comparison to an equal amount of distilled water; this is used to measure the concentrating ability of the kidney

"third space" fluid—shift of fluid from the intravascular space to a nonfunctional space

thrombocytosis—low number of platelets

water intoxication—uncontrolled, excessive water consumption resulting in dilutional complications

Fluids can accumulate within body cavities in spaces between organs. These are often called the "**third spaces**" and include peritoneal, pericardial, and thoracic cavities as well as the joints and bursae. For the normal healthy individual, these spaces hold insignificant amounts of fluid. However, in illness or injury, fluid accumulation in these spaces may become significant. For example, fluid may accumulate in the peritoneal cavity with liver disease, causing the condition known as **ascites**.

Movement of Fluid between Blood and Interstitial Spaces
Fluid status in the body is in a state of dynamic equilibrium. Water is constantly moving but total volume and concentration remain the same. Fluids move freely between fluid compartments by the processes of osmosis and filtration. In osmosis, only water moves between compartments. But with filtration, water and solutes (except plasma proteins and red blood cells) move. Two types of pressure influence the movement of water and solutes: osmotic and hydrostatic.

Osmosis is the movement of fluid across a semipermeable membrane from an area of low concentration (of solute) to an area of high concentration (of solute). The force that pulls water across the membrane is **osmotic pressure**, which is determined

Table 7.1 Body Fluids

Compartment	Volume of Fluid (in Liters)	Percentage of Body Fluid	Percentage of Body Weight
Total body fluid	42	100	60
Intracellular fluid (ICF)	28	67	40
Extracellular fluid (ECF)	14	33	20
Plasma	2.8	6.6 (20% of ECF)	4
Interstitial fluid	11.2	26.4 (80% of ECF)	16
Lymph	Negligible	Negligible	Negligible
Transcellular fluid	Negligible	Negligible	Negligible

Source: Reprinted from: Sherwood L. Table 15.1 Classification of Body Fluid. In: Human Physiology from Cells to Systems. 7e. Belmont (CA): Brooks-Cole/ Cengage, 2010.

by the number of solute particles in solution. Solutes that do not form a true solution, such as large protein molecules, are called colloids. They also contribute to the osmotic pressure (colloid osmotic pressure). Serum albumin is the protein that exerts the greatest effect on the **colloid osmotic pressure (oncotic pressure)**. The purpose of the movement of fluid is to equalize the concentration of solute, and thus osmotic pressure, on both sides of the membrane.

Hydrostatic pressure is pressure exerted by the fluid on the membrane. For intravascular fluid, hydrostatic pressure (pressure of blood on the arterial walls) is more commonly known as blood pressure. When hydrostatic pressure differs on the two sides of the membrane, fluid is pushed from the area of high pressure to the area of low pressure. The goal of this fluid movement (filtration) is to equalize pressure exerted by the fluids on both sides of the membrane.

Osmotic and hydrostatic pressure work together in favor of moving fluid out of the blood into interstitial areas at the arterial end of the capillary and restoring fluid back into blood at the venous end of the capillary. This phenomenon is called Starling's Law of capillaries. Anatomical differences between capillaries of different organs also affect permeability and therefore affect both osmotic and hydrostatic pressure.

Movement between Extracellular Fluid and Intracellular Fluid

Fluid movement between extracellular fluid (ECF) and intracellular fluid (ICF) is directed by osmotic pressure in order to establish osmotic equilibrium. Osmotic pressure can be expressed as either **osmolarity** or **osmolality**. Though technically they have different meanings, they are similar enough that the terms are used interchangeably. Osmolality is the more precise term since the amount of solvent does not vary.

Osmolality of the blood is used as the normal physiologic range for body fluids. Normal osmolality of the blood is 280 to 320 mOsm/kg H_2O. Estimation of blood osmolality uses the serum concentrations of sodium, potassium, glucose, and urea using the following formula:

$$\text{mOsm/kg blood} = 2\,(Na^+\ mEq/L + K^+\ mEq/L)$$
$$+ \frac{\text{glucose mg/dL}}{18} + \frac{\text{BUN mg/dL}}{2.8}$$

In this equation used to calculate osmolality, sodium and potassium concentrations are expressed in mEq/L, and glucose and BUN are expressed in mg/dL. Na^+ and K^+ (cations) are multiplied by 2 to account for the accompanying anions that are needed for electroneutrality (see Box 7.1). The ionic composition of the major body-fluid compartments is discussed later in this chapter (and shown in Figure 7.1).

Fluids that have an osmolality equal to blood are called isotonic. Solutions with an osmolality greater than that of blood

Figure 7.1 Ionic Composition of the Major Body-Fluid Compartments

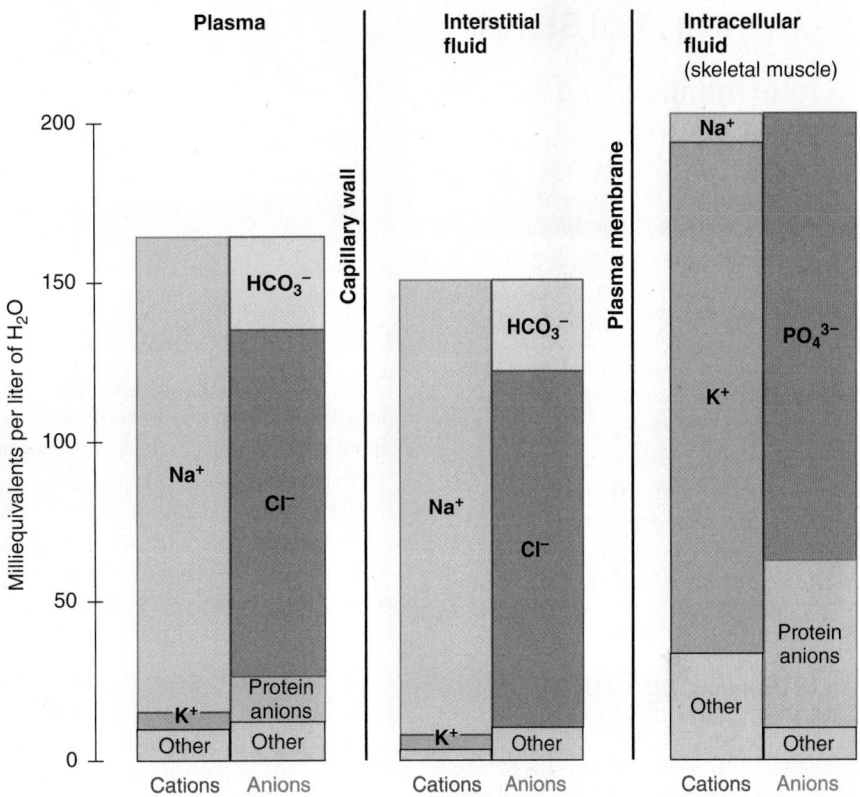

Source: Sherwood L. Figure 15-2 In: Human Physiology From Cells to Systems. 7e. Belmont CA: Brooks-Cole/Cengage, 2010, p. 560.

are called hypertonic, and those with osmolality less than blood are called hypotonic. In the normal state, osmolalities of the ECF and ICF are assumed to be equal. When cells are exposed to hypertonic solutions, fluid moves out of the cell in an attempt to establish osmotic equilibrium. The result is cellular **dehydration**. Conversely, when a cell is exposed to hypotonic solutions, fluid moves into the cell, resulting in cell swelling.

BOX 7.1   CLINICAL APPLICATIONS

Calculation of Osmolality Using Potassium Chloride (KCl)

Milliosmolality (mOsm)

$$mOsm = \frac{\text{atomic wt in mg/}}{\text{particles exerting osmotic pressure}}$$

Example using potassium chloride (KCl):

Step 1: Atomic weight of KCl = 74.5
Step 2: 2 particles in solution: K^+, Cl^-
Step 3: mOsm = 74.5/2 = 32.75 mg
Step 4: 1 mOsm of KCl = 32.75 mg

BOX 7.2   CLINICAL APPLICATIONS

Assessing Fluid Status

Determining Fluid Intake for 24 Hours:

Accuracy in estimating fluid intake is dependent upon recording of all oral intake on the I & O (intake and output) sheet. In the hospital setting, the I & O sheet is usually kept at the bedside and recording is done by nursing staff.

Fluid content of solid foods and liquids can be determined using commercial nutrient database programs. If I & O sheets do not include oral intake provided by the nutrition services, a separate record may need to be established to record all solid food and liquid intakes not recorded on the I & O sheet.

Determining Fluid Output for 24 Hours:

Fluid lost by feces, lungs, and skin is relatively constant in the normal adult

in the absence of extremely hot ambient temperatures and/or extreme exercise. If diarrhea or vomiting is present, fluid losses are higher than normal. Urine output should be collected and measured for the 24-hour period being evaluated.

Determining Obligatory and Facultative Urine:

For patients with decreased renal, hepatic, pulmonary or cardiac function, fluid restrictions are often implemented to decrease fluid retention. The goal of a fluid restriction is to eliminate facultative urine production since this represents fluid in excess of requirements and is most likely to be retained.

Fluid losses from feces, skin, and lungs, as well as that required for excretion of solutes (obligatory urine), must be replaced. To determine obligatory

urine, the renal solute load (RSL) must be known. RSL can be determined in the laboratory using the 24-hour urine collection. The following equation can then be used:

Obligatory urine = RSL (mOsm) ÷ 1200 to 1400 mOsm/L

Facultative urine is determined by subtracting the obligatory urine volume from the total urine volume. The following equation can be used:

Facultative urine = Total urine − Obligatory urine

Example: RSL = 950 mOsm; total 24-hour urine volume = 1800 mL

Obligatory urine = 950 mOsm ÷ 1300 mOsm/L

= 0.731 L (731 mL)

Facultative urine = 1800 mL − 731 mL

= 1069 mL

Total Body Water Balance

The physiological roles of water in excretion of metabolic waste and maintenance of body temperature result in its continual loss from the body. Losses must be replaced in order to maintain equilibrium of fluid in the body. In the clinical setting, this balance is referred to as *intake* and *output*, as discussed in Box 7.2.

Fluid Intake Water is taken into the body as part of food and beverages consumed by an individual. Beverages range between 84% and 100% water, with fruit juices being at the lower end of the range and water being 100%. Solid foods range from 0% to 96%, with oils being 0% and cucumbers 96%. In a clinical setting, anything fluid at room temperature will be considered part of fluid intake. For example, ice cream would be calculated as fluid consumption.

Metabolic reactions often produce water but do not contribute to actual fluid intake in a practical sense. Anabolic reactions such as synthesis of glycogen, triglycerides, or protein (condensation reactions) release water. Some catabolic reactions release water; for example, during aerobic respiration, water is synthesized as a by-product of energy production. In contrast, catabolic reactions that reduce large molecules to smaller molecules (e.g., proteins → amino acids) are hydrolytic reactions and therefore require the addition of water. Precise determination of **metabolic water** is not possible; it is thus usually estimated by using intakes of carbohydrate, protein, and fat as variables.

The role of metabolic water is usually insignificant except in cases of decreased organ function.

Fluid Output Fluid losses from the body are categorized as sensible and insensible. **Sensible losses** are those that are visible and measurable, such as in urine and feces. **Insensible losses** are usually not seen or measured. These include losses through respiration or through the skin by evaporation. Average daily loss from feces is 200 mL. Insensible losses will range from 700 to 900 mL/day.

Water loss from urine is the most variable and will accommodate changes in dietary fluid intake. Total urine output is the sum of **obligatory urine** and the **facultative urine**. Obligatory urine is the amount that must be excreted in order to remove waste products; it is dependent on the concentrating ability of the kidney. Waste products are referred to as renal solute load (RSL) and include primarily sodium, potassium, chloride, and urea. Obligatory urine is formed even when patients are fasting. The kidneys can concentrate urine to approximately 1200 mOsm. Therefore, minimal RSL of 600 to 700 mOsm/day would require a minimum of 500 mL of obligatory urine even in fasting or starvation. The solute concentration in the urine is determined by measuring its **specific gravity**. Water has a specific gravity of 1.00 that will rise with each additional solute.

It is assumed the normal adult will have fluid output equal to fluid intake, thus maintaining equilibrium. This equilibrium is demonstrated by the typical daily intake and output of fluid for a healthy adult shown in Box 7.3. Facultative urine can vary from negligible amounts when fluid intake is low, to large volumes when fluid intake is high (see Box 7.2).

Fluid Requirements Several methods have been developed for estimating fluid requirements. Four methods are listed in Table 7.2, but it should be noted that these methods only estimate fluid needs. Box 7.4 shows how fluid needs for an adult male would be calculated using each method. Clinical assessment should be used to evaluate whether recommended fluid intake is appropriate. Assessment of hydration status includes evaluation of daily weights and intake and output records; physical evaluation of skin, eyes, lips and oral cavity; evaluation of the respiratory rate and lung sounds, and blood pressure and capillary fill; and assessment for peripheral edema. Specific

biochemical assessments of blood and urine are discussed later in this chapter.

Body Solutes

Types of Solutes

Solutes in body fluids include both electrolytes and other molecules. Electrolytes (ions) dissociate in fluid to form one or more charged particles. Other molecules, such as glucose, protein, urea, lactate, and other organic acids, remain stable in solution. The major electrolytes in the body are sodium, potassium, calcium, magnesium, chloride, bicarbonate, phosphate, and sulfate. Ions with a positive charge are referred to as *cations*, and negatively charged ions are called *anions*.

Distribution of Electrolytes

Some electrolytes are found only in the ECF or the ICF, while others are found in both. Concentrations of electrolytes inside and outside of the cell also differ (see Figure 7.1). The key to maintaining normal conditions, however, is ensuring that the amounts of cations and anions in the ECF are equal. Likewise, equal amounts of cations and anions must be present in the ICF. The law of thermodynamics has been used to determine that the sum of the cations must be equal to the sum of anions within a given compartment in order to maintain electroneutrality.

In the ECF, the major cation is sodium, and major anions are chloride and bicarbonate. These are found in only small amounts in the ICF. The major cation in the ICF is potassium,

Table 7.2 Calculating Fluid Requirements

Method 1 (based on energy intake): 1 mL of fluid per kcal	
Method 2 (based on body weight):	

Age/Gender	mL/kg
Infants and Children	
1–10 kg	100–150
11–20 kg	Add 50 mL/kg over 10 kg
≥ 21 kg	Add 25 mL/kg over 20 kg
Adolescents	40–60
Young adult 16–30 yrs	35–40
Average adult	30–35
Adult, 55–65 yrs	30
Adult > 65 yrs	25

Method 3 (based on nitrogen and energy intake):
1 mL/kcal + 100 mL/g N

Method 4 (based on body surface area–BSA): 1500 mL/m²

and the major anion is phosphate. Potassium and phosphate concentrations are low in the ECF. The distribution of ions in the ECF and ICF are shown in Figure 7.1.

Movement of Solutes

While fluids generally move freely through the semi-permeable membranes of the body, cellular membranes can obstruct movement of solutes. Factors influencing movement of solutes include molecular size (smaller molecules move more easily than larger molecules), electrical charge of the molecule, hydrostatic pressure, and method of solute transport. Solutes transported across the membrane by active transport move more easily than those transported by facilitated diffusion or simple diffusion.

Electrolyte Requirements

Adequate Intake (AI) levels for sodium, potassium, and chloride were established in 2005.[1] Table 7.3 summarizes these levels. Electrolyte requirements are generally met by normal dietary intake without difficulty. For example, the AI for sodium is 1500 mg per day and only one teaspoon of table salt contains 2300 mg sodium.

Normal serum levels of these electrolytes are listed in Table 7.4. When electrolyte imbalances occur, serum levels are altered

Table 7.3 Adequate Intakes (AI) for Sodium, Chloride, and Potassium

Age (yr)	Sodium AI (mg/day)	Chloride AI (mg/day)	Potassium AI (mg/day)
Infants			
0–0.5	120	180	400
0.5–1	370	570	700
Children			
1–3	1000	1500	3000
4–8	1200	1900	3800
Adults			
9–13	1500	2300	4500
14–50	1500	2300	4700
51–70	1300	2000	4700
>70	1200	1800	4700
Pregnancy	1500	2300	4700
Lactation	1500	2300	5100

Source: Dietary Reference Intakes for Water, Potassium, Sodium, Chloride, and Sulfate. Reprinted with permission from the National Academies Press Copyright © 2005 National Academy of Sciences.

Table 7.4 Normal Serum Values for Sodium, Potassium, and Chloride

Electrolyte	Normal Serum Level
Sodium	136 to 146 mEq/L
Potassium	3.5 to 5.0 mEq/L
Chloride	98–106 mEq/L

Pagana KD, Pagana TJ. Mosby's Diagnostic and Laboratory Test Reference. 9th ed. 2009.

to maintain electroneutrality. The kidneys accomplish primary regulation of sodium, potassium, and chloride.

Physiological Regulation of Fluid and Electrolytes

Regulation of fluid and electrolytes is complex and utilizes several integrated mechanisms. The influence of osmotic and hydrostatic pressures has already been discussed. Additional factors necessary for fluid regulation include the hypothalamic thirst mechanism, renal function, and hormonal control.

Thirst Mechanism

Sensors within the interstitial fluid are affected by changes in the fluid around them. They trigger the hypothalamus to interpret these signals as thirst and as a result, the individual will be stimulated to increase his or her fluid intake. This thirst mechanism cannot always be relied on, however. In the elderly, thirst sensation decreases. In the trained, elite athlete, thirst sensation may not be a valid indication of need for additional fluid.

Renal Function

As mentioned earlier, hydrostatic pressure is pressure exerted by fluid on a membrane. When blood volume increases, hydrostatic pressure also increases. This increase in pressure results in larger amounts of fluid moving from the capillaries into the renal tubules. This fluid is then excreted as urine by the kidney.

Hormonal Influence: Renin-Angiotensin-Aldosterone System (RAAS) and Vasopressin

Decreasing hydrostatic pressure is the impetus for the RAAS (renin-angiotensin-aldosterone system) regulation of fluids and electrolytes illustrated in Figure 7.2. **Baroreceptors** within blood vessels are stimulated by low hydrostatic pressure, which is indicative of a decrease in blood volume. The hormone renin is released from the kidney and stimulates conversion of angiotensinogen to angiotensin I. A second activation converts angiotensin I to angiotensin II. Increasing amounts of angiotensin II stimulate release of the hormone aldosterone from the adrenal cortex. Aldosterone directly influences the kidney to retain Na^+. When Na^+ levels increase, increased osmotic pressure will pull fluid back into the blood; blood volume will thus increase back to its normal range. For individuals with hypertension (high blood pressure), the heart has to work harder to handle a higher blood volume. Control of the dietary intake of sodium is an important component of nutrition therapy for management of high blood pressure.

As shown in Figure 7.3, two major factors stimulate the pituitary to release the hormone **arginine vasopressin**, formerly known as antidiuretic hormone (ADH). The first and most important factor is an increasing osmolality of the ECF. The second factor that stimulates release of vasopressin is detection of a decrease in hydrostatic pressure by baroreceptors in blood vessels. Vasopressin causes fluid to be reabsorbed in the tubules of the kidney, which increases blood volume and lowers blood osmolality.

Figure 7.2 Renin-Angiotensin-Aldosterone System (RAAS)

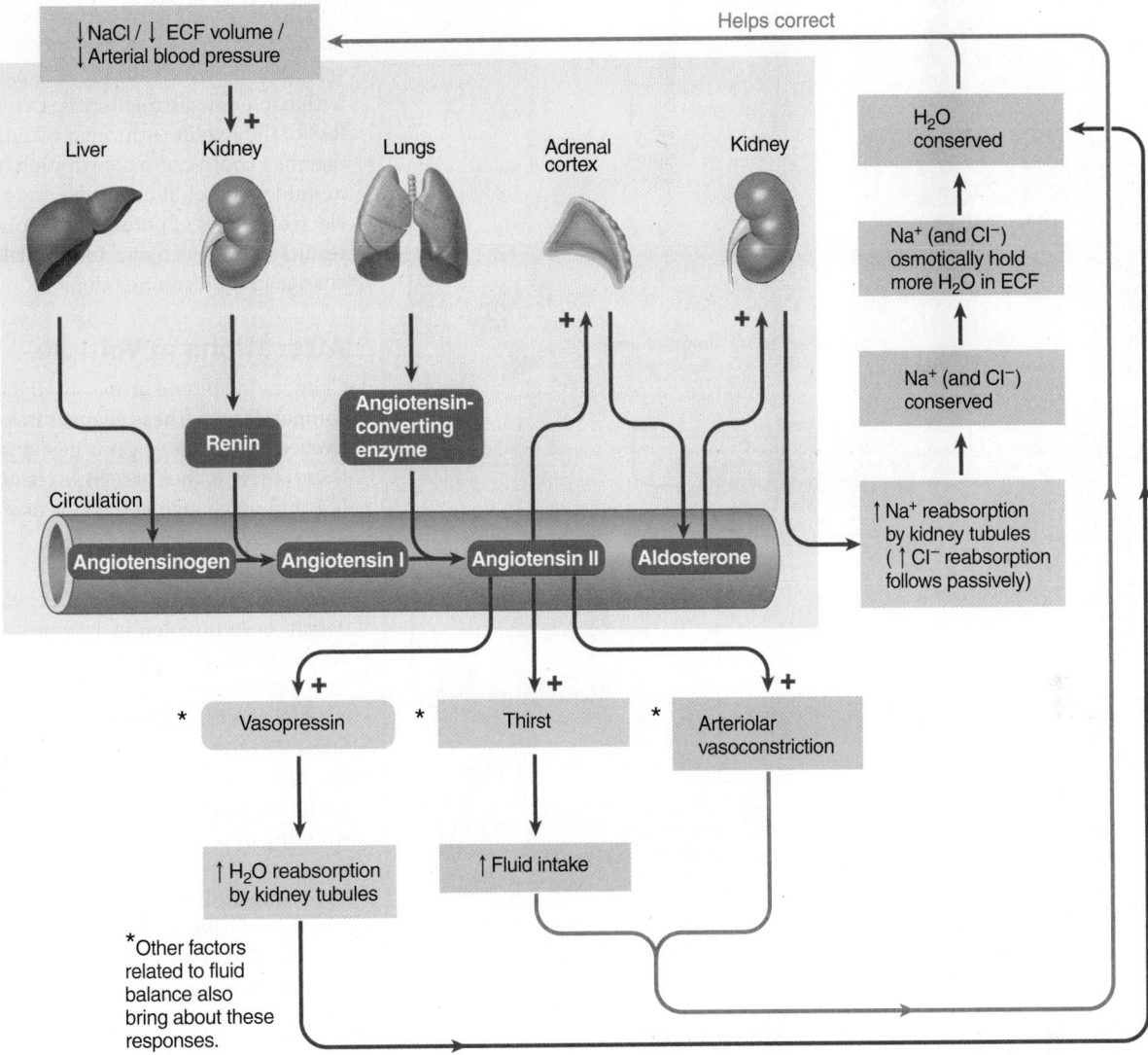

Source: Sherwood L. Figure 14-16 In: Human Physiology From Cells to Systems. 7e. Belmont CA: Brooks-Cole/Cengage, 2010, p. 528.

Electrolyte Regulation

Sodium Controls of electrolytes are interdependent, and are linked to controls of fluid balance, as sodium and water balance are closely related. When vasopressin and aldosterone act to regulate osmolality and blood volume, sodium is regulated as well. Additionally, atrial natriuretic peptide (ANP) assists in the control of sodium. ANP is released when arterial vessels stretch (as occurs when blood volume increases). ANP is an agonist to the RAAS. This effect results in increased urinary output of sodium and fluid and a decrease in blood volume, and indirectly, an increase in osmolality.

Potassium Aldosterone has an independent effect on potassium levels. High levels of potassium cause the adrenal glands to release aldosterone. Aldosterone secretion results in increased excretion of K^+ by the kidney.

Two of the most important components of acid-base balance involve both hydrogen ions and bicarbonate (see Chapter 8 for a detailed discussion of acid-base balance). Since they are both electrolytes, acid-base changes will affect concentrations of other electrolytes, including potassium, in both ECF and ICF. For example, when the body attempts to decrease the concentration of H^+ to restore acid-base balance, K^+ is often exchanged in order to maintain electroneutrality.

Calcium and Phosphorus Serum concentrations of calcium and phosphorus are dependent on intestinal absorption, exchange between extracellular fluid and bone, and renal excretion of these minerals. These routes for maintenance of serum levels are primarily controlled by hormonal influence. Calcium and phosphorus exist in a reciprocal relationship; this means that when serum calcium levels are high, serum phosphorus levels will be low.

Parathyroid hormone (PTH) is secreted from the parathyroid glands when serum calcium levels are low. PTH works to raise serum calcium levels by pulling calcium from the bone and decreasing excretion of calcium in urine, as shown in Figure 7.4. PTH also stimulates activation of vitamin D. Vitamin D works to maintain serum calcium levels by increasing absorption of calcium in the

Figure 7.3 Control of Increased Vasopressin and Thirst during a Water Deficit

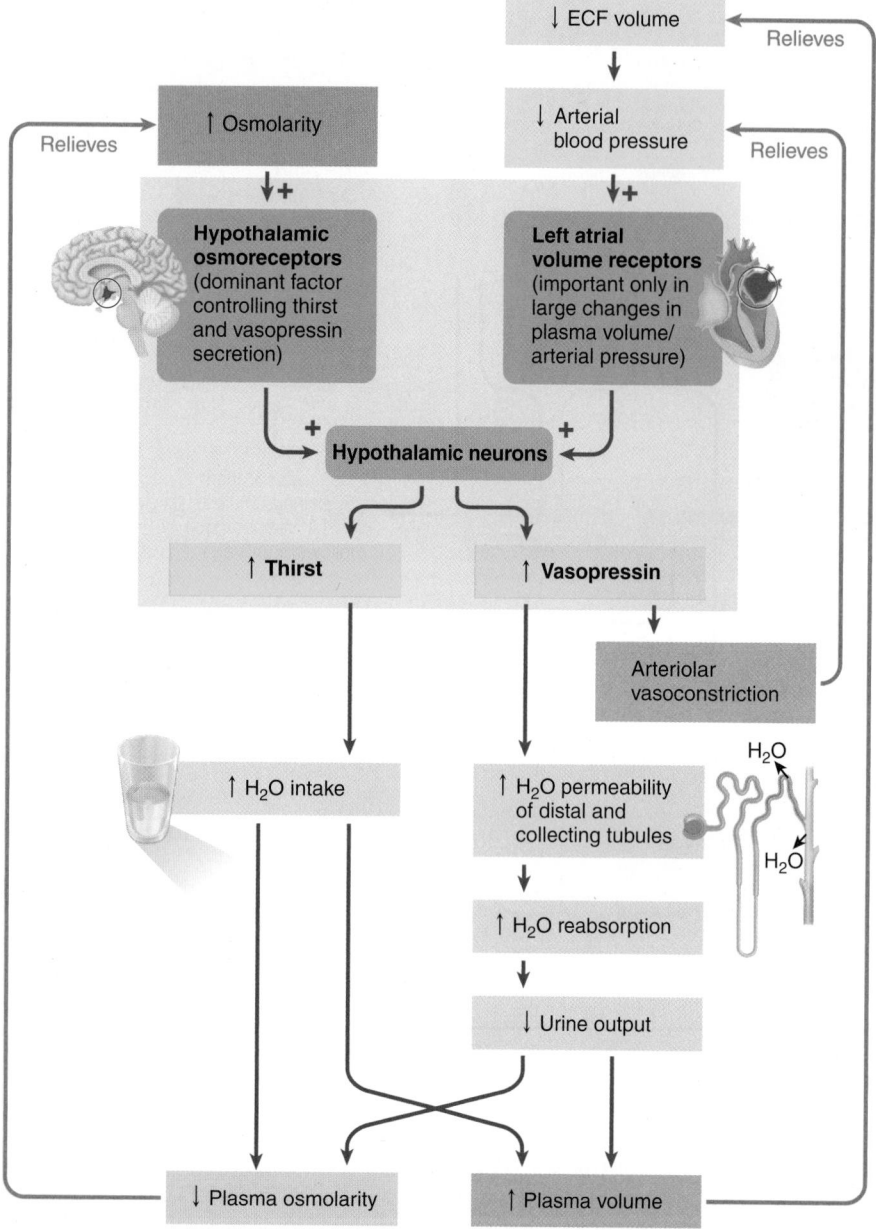

Source: Sherwood L. Figure 15-4 In: Human Physiology From Cells to Systems. 7e. Belmont CA: Brooks-Cole/Cengage, 2010, p. 567.

small intestine. PTH also acts to increase phosphorus excretion when necessary. Calcitonin, another hormone, originates from the thyroid gland. It acts in opposition to PTH by *inhibiting* osteoclasts (cells within the bone that function to break down and resorb bone tissue) and therefore lowering serum calcium levels.

Disorders of Fluid Balance

The following three general categories of alterations occur in fluid balance:

1. Changes in fluid volume

2. Changes in fluid concentration or osmolality

3. Changes in fluid composition

It is uncommon for these alterations to occur in isolation. Changes in fluid, electrolyte, and acid-base balance more commonly occur together, as the following typical scenario suggests. Consider the patient with bacterial gastroenteritis. Excessive loss of fluid volume through vomiting and diarrhea could cause dehydration (change in fluid volume). If untreated, the resulting electrolyte loss of potassium (hypokalemia) results in both a change in osmolality and a change in fluid composition.

Alterations in Volume

Changes in volume primarily affect the ECF compartments. These changes involve relatively equal losses or gains in sodium and water; there is thus very little change in the ICF and no change in the ECF osmolality.

Hypovolemia

PATHOPHYSIOLOGY Extracellular fluid deficit, or **hypovolemia**, is almost always related to renal or extrarenal loss of fluids. This will occur more rapidly when the loss is coupled with decreased oral intake of fluids. Extrarenal losses include any excess loss of fluid outside of renal excretion, including gastrointestinal losses, such as in vomiting or diarrhea. Losses through the skin occur during exposure to heat such as increasing body temperature (fever) or increased environmental heat. An endurance athlete who is involved in physical activity for more than an hour and a half can produce up to 3 liters of sweat per hour.

Excess loss through the skin can occur through burns or draining wounds. A fistula (abnormal opening between the gastrointestinal tract and other organs or peritoneal cavity) can also contribute to ECF losses. These extrarenal losses can be extreme in some clinical situations—as high as 5 to 6 liters in one day.

Other extrarenal losses occur when fluids are trapped in body spaces such as in development of ascites, in congestive heart failure, in pulmonary edema, or in burns. This "third spacing" of fluid results in a net loss of ECF.

Renal losses occur in conditions that increase urinary excretion above what the individual has consumed orally. Excessive renal losses may occur as a component of a disease process; for instance, they may occur secondary to **diuresis**, as seen in the recovery phase of acute renal failure. In uncontrolled type 2 diabetes mellitus or **hyperosmolar hyperglycemic nonketotic syndrome**, the body attempts to correct acid-base imbalances and hyperosmolality by increasing urine excretion.

Medications such as diuretics are often prescribed to purposefully decrease ECF. For example, in the treatment of hyperten-

Figure 7.4 Interactions Between PTH and Vitamin D in Controlling Plasma Calcium

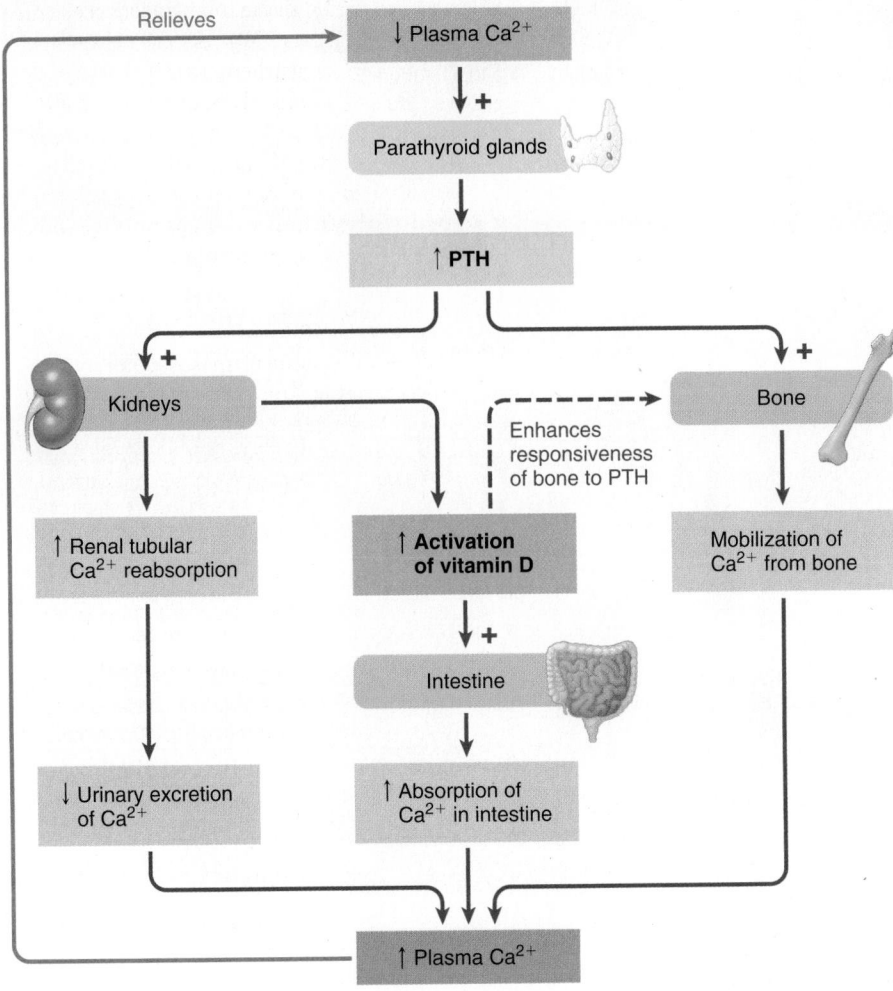

Source: Sherwood L. Figure 19-27 In: Human Physiology From Cells to Systems. 7e. Belmont CA: Brooks-Cole/Cengage, 2010, p. 736.

sion, diuretics are prescribed to decrease blood volume and therefore reduce blood pressure. Dietary composition can also affect urinary excretion. High-protein diets result in an increase in urine excretion due to the increased renal solute load.

CLINICAL MANIFESTATIONS The severity of the signs and symptoms correspond to the severity of the volume deficit. As the ECF volume decreases, the corresponding blood volume will be reduced. The decrease in blood volume will lower blood pressure and decrease cardiac output. In mild cases of volume deficit, compensatory mechanisms will maintain homeostasis and symptoms may not be noticed by the patient. In more severe cases, blood pressure will be low, especially upon changing body position. This is referred to as *orthostatic hypotension.* The change in cardiac output can also result in tachycardia, weak pulse, and dizziness.[2]

Other physical findings can include poor skin turgor and dry skin and mucous membranes. Rapid weight loss can also be monitored to substantiate the diagnosis of hypovolemia. Actually, any rapid changes in weight should initially be considered to be an indication of fluid changes. See Table 7.5 for summary of clinical evaluation of fluid and electrolyte disorders.

LABORATORY FINDINGS No single laboratory finding will confirm hypovolemia. Measurements of blood and urine will correspond to the underlying cause of the hypovolemia. Tables 7.6 and 7.7 provide a summary of laboratory assessment.

Hemoconcentration occurs in hypovolemia unless there is also a loss of blood. Hemoconcentration results from the kidneys' compensation by decreasing urinary output. For example, serum sodium and chloride, blood urea nitrogen, hemoglobin, hematocrit, and albumin may be abnormally elevated in dehydration. As urinary output decreases, urinalysis results will reveal concentrated urine with elevated specific gravity, a darker color, and a cloudy appearance.

TREATMENT Treatment is prescribed according to the underlying cause for the fluid deficit. In mild cases, increasing intake of both sodium and water will allow gradual correction of hypovolemia. In more severe cases, fluid and electrolyte replacement through intravenous fluids will need to be prescribed (see Table 7.8).

Hypervolemia

PATHOPHYSIOLOGY In normal, healthy individuals the kidney will excrete excess water, but ECF excess can be a common occurrence in clinical situations. The most common cause of **hypervolemia** is a decrease in urinary output such as that seen in acute renal failure. Excess intravenous fluids or the failure of the kidney to accommodate a rapid ingestion of fluids quickly enough may also cause hypervolemia. Excessive secretion of vasopressin can also result in excessive volume retention.

When there is ECF excess, fluid shifts into interstitial spaces so a balance between ECF and ICF is maintained. Accumulation of fluid in interstitial spaces, as previously discussed, is called **edema**.[1] Overall blood volume will be increased in hypervolemia. This increases blood pressure and overall work of the heart.

CLINICAL MANIFESTATIONS Increased blood volume results in elevated blood pressure and jugular venous distention (see Table 7.5). The presence of edema can be noted in peripheral regions such as ankles and feet, or in the face and scrotal areas. When the clinician presses on areas of edema, an indentation will occur. The depth of the "pitting" corresponds to the amount of edema (see Figure 7.5).

Respiratory symptoms include difficulty breathing (dyspnea) and the presence of **rales**. Rapid weight changes also correlate with hypervolemia. A weight gain of 1 kilogram is equivalent to the retention of approximately 1 liter of fluid.

Table 7.5 Clinical Changes in Fluid and Electrolyte Disorders

Assessment	Evaluation
Daily weights: 2% ↑↓: mild fluid volume deficit or excess	Rapid changes reflect fluid changes
5% ↑↓: moderate deficit or excess	Body weight does not change when fluid shifts to third spaces
8% ↑↓: severe deficit or excess	
Eyes: dry conjunctiva, decreased tearing	Fluid volume deficit
Periorbital edema	Fluid volume excess
Lips and oral cavity: dry, cracked lips; small multifurrowed tongue	Fluid volume deficit
Decreased skin turgor	Fluid volume deficit
Tachycardia	Fluid volume deficit
Slowed pulse, increased BP	Fluid volume excess
Orthostatic BP	Fluid volume deficit
Hand veins	Prolonged filling: volume deficit Prolonged emptying: volume excess
Central venous pressure (CVP)	↓ CVP: volume deficit ↑ CVP: volume excess
Jugular vein distention (JVD)	Flat neck veins when supine: volume deficit Extended JVD: volume excess
Cardiac dysrhythmias	May indicate deficit or excess of K, Mg, Ca, PO₄
Lungs: pulmonary congestion; ↑ respiratory rate, moist rales, rhonchi	Fluid volume excess
Oliguria	Severe fluid volume deficit
Extremities; localized swelling; sacrum: edema present	Fluid volume excess

Table 7.6 Biochemical Evaluation of Fluid and Electrolyte Status

Blood Tests	Normal Value	Discussion: Additional Factors That May Affect Levels
Potassium	3.5 to 5.0 mEq/L	↑ in acidosis ↓ in alkalosis
Sodium	135 to 145 mEq/L	Consistent with current osmolality. ↓ Na = hyperosmolar body fluids. Rare that it would indicate high Na levels. ↑ Na = dilutional body fluids in relationship to solute.
Chloride	98 to 106 mEq/L	↑ Cl may indicate metabolic acidosis. ↓ Cl often with metabolic alkalosis and hypokalemia.
Calcium	8.7 to 9.2 mg/dL	Evaluate with serum albumin levels—total Ca ↓ when albumin is low, but ionized calcium does not change.
Phosphate	2.5 to 4.5 mg/dL	Elevated in chronic renal failure.
Hematocrit	37% to 47% (women) 40% to 54% (men)	↑ in fluid volume deficit ↓ in fluid volume excess
Glucose	70 to 110 mg/dL	Hyperglycemia causes osmotic diuresis and ↓ blood volume.
BUN	8 to 26 mg/dL	↑ in fluid volume deficit ↓ in fluid volume excess
Osmolality	275 to 295 mOsm/kg	↑ in fluid volume deficit ↓ in fluid volume excess

Hemodilution can also affect laboratory values. Sodium, chloride, hemoglobin, hematocrit, blood urea nitrogen (BUN), and albumin would be abnormally low in this clinical scenario.

TREATMENT Treatment for hypervolemia will target the underlying cause. Control of other symptoms generally involves restriction of both fluid and sodium.[3]

Alterations in Osmolality

In general, alterations in osmolality occur when there is a shift of water without a corresponding shift in solute (electrolytes). This most commonly involves changes in the concentration of sodium in the ECF or hyperglycemia associated with uncontrolled diabetes mellitus.

A change in osmolality within the ECF directly influences osmolality of the ICF. This would also be consistent in the opposite situation where change in the osmolality in the ICF directly influences the osmolality of the ECF.

Table 7.7 Evaluation of Fluid and Electrolyte Status: Urine Tests

Urine Tests	Normal Value	Comments
Sodium	100 to 260 mEq/24 hr >40 mEq/L in random sample	<10 mEq/24 hr = hyponatremia/edema/volume depletion
Potassium	25 to 100 mEq/24 hr	↑ hyperaldosteronism ↓ adrenal insufficiency
Chloride	110 to 250 mEq/24 hr	<10 mEq/L in metabolic alkalosis secondary to volume deficit >20 mEq/L in metabolic alkalosis caused by hyperaldosteronism or ↓ K⁺
Color	Pale yellow	Dark, amber, hazy in dehydration/fluid deficit
Urine osmolality	50 to 1400 mOsm	Reflects concentrating or diluting ability
Specific gravity	1.003 to 1.030	↑ in fluid deficit/dehydration and hyperosmolar urine

Table 7.8 Commonly Prescribed Intravenous Solutions*

Intravenous Solutions	Content	Osmolality mOsm/L	Use
5% Dextrose	No electrolytes—5 g dextrose/dL; 170 kcal/L	252	Free water; correction of fluid balance and hypernatremia; provides some energy
10% Dextrose	No electrolytes—10 g dextrose/dL; 340 kcal/L	505	
0.45% Saline ("Half Normal Saline")	77 mEq Na^+/L, 77 mEq Cl^+/L	154	No energy provided; correction of fluid balance but doesn't necessarily correct electrolyte imbalances
0.9% Saline ("Normal Saline")	154 mEq Na^+/L, 154 mEq Cl^+/L	308	Na^+ and Cl^+ are greater than in plasma levels; can be administered with blood products
Ringer's Solution	130 mEq Na^+, 4 mEq K^+, 2 mEq Ca^{2+}, 109 mEq Cl^+	309	Similar to plasma composition; does not provide free water or energy
Lactated Ringer's (Hartmann's Solution)	130 mEq Na^+, 4 mEq K^+, 2 mEq Ca^{2+}, 109 mEq Cl^+, 29 g lactate	273	Similar to plasma composition; does not provide free water or energy
Dextrose in Saline: 5% in 0.225%	170 kcal/L, 38.5 mEq Na^+/L, 38.5 mEq Cl^+/L	355	Provides energy, free water, sodium, and chloride
Dextrose in Saline: 5% in 0.455%	170 kcal/L, 77 mEq Na^+/L, 77 mEq Cl^+/L	406	
Dextrose in Saline: 5% in 0.9%	340 kcal/L, 154 mEq Na^+/L, 154 mEq Cl^+/L	560	

*Reference values only. Content may vary slightly between formulations.

Source: Modified from: Whitmire SJ . Fluid and electrolytes. IN: The Science and Practice of Nutrition Support. Dubuque IA: Kendall/Hunt Publishing Company. Heitz UE, Home MM (2001).

Heitz UE & Horne MM. Pocket Guide to Fluid, Electrolyte, and Acid-Base Balance. 4th ed. St. Louis MO: Mosby/Elsevier Science (2004).

Figure 7.5 Pitting Edema

An indentation of the skin known as pitting will result when pressure is applied to an area of edema. The degree of pitting is scaled from +1 to +4 as a subjective evaluation.

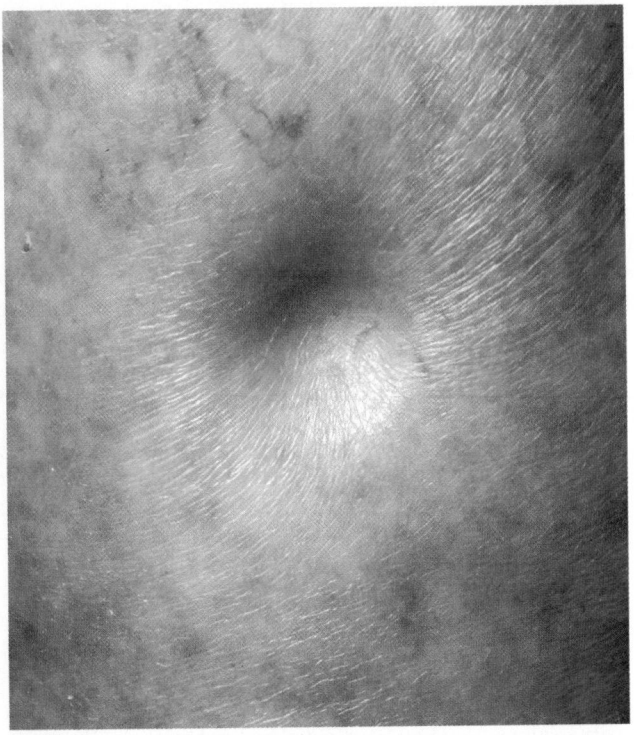

Source: Dr. P. Marazzi/Photo Researchers, Inc.

Sodium Imbalances

Maintaining sodium levels involves an intricate balance between water and sodium. The renin-angiotensin-aldosterone system (RAAS), ECF volume, and overall renal function are most important in maintenance of sodium homeostasis.

Hyponatremia

PATHOPHYSIOLOGY Either decreasing amounts of sodium, increasing amounts of water in the ECF, or a combination of both can cause **hyponatremia**. In most cases of hyponatremia, serum sodium reflects the ratio of water to sodium. It is rare for a deficit in sodium to occur simply from lack of nutritional intake. It could occur through a combination of a sodium restriction used for nutrition therapy and use of diuretics (such as in congestive heart failure or in liver disease) as part of the medical regimen.

An increase in fluid without a corresponding increase in sodium can also result in hyponatremia. This might occur, for instance, in a situation in which intravenous fluids are administered without electrolytes. This might also occur in the condition of **water intoxication**, though it is rare. In this situation, usually seen in psychiatric conditions, the patient is unable to control excessive water intake.

The syndrome of inappropriate antidiuretic hormone (SIADH) may also result in hyponatremia. In this condition, total body water increases without a subsequent increase in sodium, producing a dilutional effect on serum sodium. It is characterized by high levels of vasopressin without the normal stimuli for vasopressin release.

The only situation associated with hyponatremia that does not involve either decreased amounts of sodium or increased amounts of water in the ECF is the increase of solute in the plasma. Hyperglycemia would be the most likely situation that would precipitate this etiology of hyponatremia. The increase in glucose causes a shift of water from the ICF to the ECF in attempts to normalize the osmolality. When this occurs, sodium is diluted and hyponatremia results[4]

CLINICAL MANIFESTATIONS Often, clinical manifestations of hyponatremia do not appear until levels of sodium fall below 120 mEq/L. Signs and symptoms that do occur are consistent with changes in osmolality. As serum osmolality falls, water enters the brain cells. Nausea, vomiting, lethargy, confusion, and even seizures can result. All signs and symptoms are more pronounced if hyponatremia occurs rapidly. Laboratory measurements will confirm a serum sodium of <135 mEq/L. Plasma osmolality will be <287 mOsm/kg.

TREATMENT Treatment focuses on elevating serum sodium and treating underlying causes. This can be accomplished by restricting water intake to less than urinary output. Administration of sodium is reserved for severe cases of hyponatremia. In the case of hyperglycemia, a reduction in serum glucose is accomplished through the administration of insulin.

Hypernatremia

PATHOPHYSIOLOGY **Hypernatremia** can theoretically be caused by either an increase in sodium or decrease in water. But generally, in a normal, healthy person, an increase in sodium results in compensatory increase in renal excretion of sodium, thus maintaining homeostasis. Hence, situations that change water balance are much more likely to cause hypernatremia than are increases in sodium intake.

Insufficient water intake can result in hypernatremia. For example, the elderly have a decreased thirst sensation and may not take in adequate amounts of fluid. Hypernatremia may also result from excessive water losses. Nonrenal losses might be seen in patients who are febrile, who hyperventilate, or who have open wounds. Conditions such as **diabetes insipidus** or SIADH (discussed earlier) are also potential causes.

Hyperosmolality can also result in hypernatremia. This is seen in conditions with hyperglycemia (such as uncontrolled diabetes mellitus) or in increased urea production (high-protein diets). Although administration of excessive sodium in intravenous solutions would be the most likely culprit in the situation of hypernatremia due to an absolute excess of sodium, this situation is not common.

CLINICAL MANIFESTATIONS As in hyponatremia, the central nervous system is most sensitive to changes in fluid and electrolyte balance. Cellular dehydration results in an increasing severity of neurological symptoms as the condition becomes more severe. Possible symptoms range from lethargy and agitation to seizures and coma. Body temperature can be elevated, skin is flushed, and mucous membranes are dry. In cases of hypernatremia, laboratory measurements confirm a serum sodium of >145 mEq/L and a plasma osmolality of >295 mOsm/kg.

TREATMENT The goal of treatment is to gradually lower serum sodium to a normal concentration. If possible, the patient should simply increase his or her water intake, which may return the serum Na+ levels to normal. When levels fall

too quickly, cerebral edema can occur, endangering the patient. Intravenous treatment usually involves slow administration of fluids (such as dextrose) without sodium. Diuretics and dialysis may be required in more severe situations.

Potassium Imbalances

The body strives to maintain potassium levels within a very narrow range. Abnormalities in potassium are often life threatening, due primarily to the role of potassium in neurotransmission and muscle contraction.

Small changes in the ECF concentration of potassium may reflect significant abnormalities in the ICF concentration of potassium. Though only a small amount of potassium is found in the ECF, it is essential for neuromuscular function. Distribution of potassium in both ECF and ICF is influenced by acid-base balance and hormonal secretion.

Hypokalemia

PATHOPHYSIOLOGY **Hypokalemia** can result from inadequate nutritional intake of potassium, increased renal loss of potassium, or increased loss from the gastrointestinal tract. Hypokalemia also results from a shift of potassium from the ECF to the ICF.

Hypokalemia Secondary to Gastrointestinal Losses Vomiting, nasogastric suction, and diarrhea are common gastrointestinal origins of hypokalemia. Compensatory actions in acid-base balance contribute to hypokalemia. For example, loss of gastric acids through vomiting can lead to metabolic alkalosis. Metabolic alkalosis in turn increases bicarbonate excretion in the distal tubules. Increased amounts of potassium are excreted along with bicarbonate; thus, loss of gastric acids can lead to hypokalemia (see Chapter 8). As a result of gastrointestinal losses during vomiting, there is a decrease in ECF volume.[5] This decrease in volume stimulates the release of aldosterone. While aldosterone causes the retention of sodium (and the subsequent increase in volume), potassium excretion is increased as a secondary consequence.

Hypokalemia Secondary to Renal Losses One of the most common causes of hypokalemia is use of loop diuretics. Drugs such as furosemide (e.g., Lasix™) cause increased urine excretion with accompanying loss of potassium. Nutrition therapy for those individuals receiving loop diuretics will include education for a higher dietary intake of potassium. Notably, natural licorice and chewing tobacco, if swallowed, contain an aldosterone compound that will also stimulate increases in urinary potassium excretion.

Refeeding syndrome (see Chapter 5) may result in hypokalemia as levels of potassium shift from the ECF to ICF to accommodate increased metabolism. When a malnourished individual begins nutrition support, there is potential for inadequate amounts of the intracellular electrolytes required to support anabolism. Carbohydrate metabolism also results in shifts of these electrolytes.

CLINICAL MANIFESTATIONS Any body system involving muscular action can be affected by hypokalemia. There will generally be muscle weakness and diminished deep tendon reflexes. In advanced hypokalemia, respirations become shallow due to poor lung muscle action. Within the cardiovascular system, rhythm changes can be noted; untreated, these dysrhythmias can lead to cardiac arrest.

Normal serum potassium is 3.5 to 5.0 mEq/L, and hypokalemia is diagnosed when serum levels fall below 3.5 mEq/L. However, it is important to remember that if hypokalemia results from ECF-ICF shifts, serum levels may not reflect the true level of electrolyte abnormality. Acid-base imbalance can be confirmed by a serum pH <7.45.

TREATMENT Potassium levels in the body can be corrected through dietary sources of potassium, oral supplementation, or intravenous administration of potassium. The underlying cause of potassium loss should be addressed as well.

Hyperkalemia

PATHOPHYSIOLOGY The most common cause of **hyperkalemia** is inadequate excretion of potassium, which is commonly observed in renal failure. Excessive use of potassium-sparing diuretics, used to treat hypertension, may also result in inadequate excretion of potassium.

Shifts in potassium from the ICF to the ECF can result in hyperkalemia. Numerous clinical situations can lead to this shift. Elevated concentrations of potassium in the ECF can occur when there is an increased hemolysis of red blood cells, during **leukocytosis** or **thrombocytosis**. Catabolism and strenuous exercise increase serum potassium. Acidosis also results in hyperkalemia: When hydrogen ions are excreted to correct acidosis, potassium ions are retained (see Chapter 8).

Although it is rare for hyperkalemia to result from excessive ingestion, this can occur with consumption of potassium-containing salt substitutes, especially if the patient does not have normal renal function. (Salt substitutes are composed of potassium chloride [KCl], while salt is composed of sodium chloride [NaCl].) Hyperkalemia can also occur from blood transfusions or when excessive amounts of potassium are given in intravenous solutions.

CLINICAL MANIFESTATIONS Signs and symptoms are a result of the neuromuscular effects of altered potassium levels. A gradual rise in potassium levels, as in chronic renal failure, is better tolerated than the rapid rise in potassium levels that leads to potentially fatal symptoms.

Muscle weakness, paralysis, **paresthesias**, and cardiac dysrhythmias are a consequence of inactivation of electrical transmissions across the cell membrane and the interruption of normal polarization. Severe hyperkalemia, like hypokalemia, can lead to cardiac arrest. Toxic effects of hyperkalemia are enhanced by the presence of other electrolyte imbalances such as hypocalcemia or hyponatremia and acid-base imbalance.

Serum potassium levels in hyperkalemia are measured at >5.5 mEq/L. Further diagnosis can be made by electrocardiogram (ECG) changes.

TREATMENT As with other electrolyte abnormalities, treatment of the underlying cause is crucial in correcting hyperkalemia. In a short-term emergency situation, calcium gluconate can be given intravenously to decrease the abnormalities in cardiac cells that could lead to cardiac arrest. Additionally, both glucose and insulin can be used to shift potassium in the ECF to the ICF. Correction of any acid-base imbalance also results in

potassium movement into the ICF. Cation exchange resins such as **Kayexalate** can be given to allow the exchange of sodium for potassium in the large intestine.

For long-term treatment, dialysis and a potassium-restricted diet are primary interventions to control hyperkalemia. It is also crucial to prevent malnutrition through adequate nutrition support. Catabolism, as noted previously, contributes to the increasing levels of potassium.

Calcium Imbalances

Most calcium found in the body is located within bones and teeth. The remaining 1% found in body fluids is highly regulated due to its function in muscle contraction. Normal serum calcium levels range from 9 to 10.5 mg/dL (4.5 to 5.5 mEq/L). Serum calcium is either protein (albumin) bound (40%), complexed (13%), or ionized (47%). Ionized calcium is essential for cellular processes. Misinterpretation of serum calcium levels is common. Serum calcium is not an indication of bone calcium. Furthermore, total serum calcium must be assessed in relation to serum albumin levels. A decrease in serum albumin by 1 g/dL will decrease serum calcium by 0.8 mg/dL. Thus, a patient with a low albumin level may falsely appear to have low blood calcium.

$$\text{Corrected serum Ca}^+ \text{ (mg/dL)} = (0.8 \times [\text{Normal Albumin} - \text{Pt's Albumin}]) + \text{Serum Ca}$$
(Formula is not valid when serum pH is altered.)

As discussed previously, plasma homeostasis is maintained through the interaction of three hormones: parathyroid hormone (PTH), calcitonin, and 1,25-dihydroxycholecalciferol (activated vitamin D). When serum calcium levels fall, PTH is released. PTH acts to decrease bone resorption of calcium, decrease calcium excretion, and increase calcium absorption in the gastrointestinal tract (see Figure 7.4). Activated vitamin D acts to increase absorption of calcium by increasing production of the protein necessary for calcium absorption in the GI tract. Vitamin D also affects uptake of calcium by the bone. Calcitonin, produced by the thyroid gland, inhibits osteoclast activity, and therefore reduces bone resorption.

Hypocalcemia

PATHOPHYSIOLOGY **Hypocalcemia** most commonly results from a deficit of PTH or from abnormal vitamin D metabolism, such as that seen in patients with renal or liver failure. Activated vitamin D requires normal function of both liver and kidneys, and thus abnormalities of organ function are common causes of hypocalcemia. Alkalosis can cause a decrease in ionized calcium in the ECF, and can result in symptoms consistent with hypocalcemia. Finally, because calcium and phosphorus levels are tightly linked, changes in serum phosphorus levels can result in subsequent changes in serum calcium levels.

CLINICAL MANIFESTATIONS Hypocalcemia can result in symptoms that will be seen in all body systems. Symptoms are a result of altered nerve transmission and electrical activity of the cell. Neuromuscular symptoms include muscle spasms, tetany, and cardiac dysrhythmias. Untreated hypocalcemia is life threatening. Serum Ca$^+$ <9 mg/dL or ionized Ca$^+$ <4.5 mg/dL is consistent with hypocalcemia.

TREATMENT Treatment of the underlying cause is crucial for long-term control. Intravenous administration of calcium is only given in extreme cases due to the potential side effects.

Hypercalcemia

PATHOPHYSIOLOGY The majority of cases of **hypercalcemia** stem from either hyperparathyroidism or from a malignancy. Malignant tumors of the breast, prostate, and cervix commonly metastasize to the bone. Resulting bone resorption can lead to hypercalcemia. Other malignancies produce factors that act in a similar fashion to PTH and cause an increase in bone resorption.

CLINICAL MANIFESTATIONS Signs and symptoms will vary according to rapidity of onset and degree of hypercalcemia. Early symptoms may be vague and include fatigue and weakness. Bone pain, confusion, and cardiac dysrhythmias may also be present. This is often seen when a malignancy has spread to bone and causes abnormal release of calcium. Serum calcium levels >10.5 mg/dL are diagnostic for hypercalcemia.

TREATMENT The underlying cause of hypercalcemia is the focal point for treatment. Hyperparathyroidism can be treated surgically by removal of the parathyroid gland. Agents can be given to bind calcium in the serum when the situation is a clinical emergency. Intravenous fluids, diuretics, and dialysis can also be used to dilute serum calcium and increase its excretion.

Phosphorus Imbalances

Phosphate is a crucial anion essential for metabolism of all substrates. It is a component of the cellular energy reservoir, ATP, and an integral part of DNA and RNA. Phosphate also participates in maintenance of acid-base balance and is a structural component of bones, teeth, and phospholipids.

The average diet contains 1 to 1.6 g of phosphorus, which is easily absorbed. Phosphorus imbalances will rarely originate solely from nutritional intake. Serum phosphate exists as inorganic phosphate ions and only about 10% is bound to protein. Calcium and phosphate interact in a reciprocal fashion. Urinary excretion of phosphate increases or decreases in inverse proportions to serum calcium levels.

Hypophosphatemia

PATHOPHYSIOLOGY **Hypophosphatemia** can result from vitamin D deficiency or from decreased activation of vitamin D. Hyperparathyroidism can also lead to low serum levels of phosphate. Consumption of aluminum-containing antacids may also bind phosphate.

In respiratory alkalosis, hyperventilation causes a shift of phosphate from the ECF to the ICF that results in hypophosphatemia. In refeeding syndrome, a rapid shift of phosphate from the ECF to ICF occurs in response to increased metabolism. In hospitalized alcoholic patients, hypophosphatemia can result from withdrawal of alcohol.

CLINICAL MANIFESTATIONS When phosphate is unavailable to support ATP and 2,3-diphosphoglycerate in glycolysis,

changes are seen in every body system. Respiratory insufficiency and central nervous system abnormalities will eventually lead to encephalopathy and coma.

Low levels of phosphate lead to the mobilization of calcium and phosphorus in the bone, causing osteomalacia and rickets. Metabolic acidosis is often a result of hypophosphatemia, due to decreased hydrogen ion secretion. Hypophosphatemia is defined as a serum level <2.5mg/dL.

TREATMENT Treatment should be focused on the underlying cause of the phosphate abnormality. Oral phosphate as food or supplements is the primary route to increase phosphorus levels. Intravenous phosphate is given only in emergency situations due to the risk of precipitation of calcium—the deposition of calcium into soft tissues where it could cause organ damage.

Hyperphosphatemia

PATHOPHYSIOLOGY Acute or chronic renal failure is the most common clinical condition associated with **hyperphosphatemia**. As glomerular filtration rate decreases, the ability to excrete phosphorus decreases proportionally. Other causes of hyperphosphatemia involve low levels of PTH and other endocrine disorders.

Phosphate is released when cells break down, resulting in a significant shift of phosphate from the ICF to the ECF. Drugs or medications that contain phosphorus or a high intake of vitamin D may also result in hyperphosphatemia.

CLINICAL MANIFESTATIONS Most signs and symptoms associated with hyperphosphatemia are a result of concurrent hypocalcemia. These symptoms would originate from altered nerve transmission and muscle contraction. Hyperphosphatemia is defined as a serum level >4.5 mg/dL.

TREATMENT For treatment of chronic hyperphosphatemia, dietary restriction of phosphorus is necessary. Foods highest in phosphorous include milk, dairy products, and animal protein sources. (See Chapter 18 for complete listing.) Medications that will bind phosphate are also used. In renal failure, where high serum levels of phosphorous are common, calcium supplements are used as a phosphate binder to help control these levels.

Magnesium Imbalances

Magnesium is an abundant mineral in the ICF and is necessary for cellular energy metabolism. Its function is closely related to both calcium and potassium, because it assists in maintaining calcium and phosphorus homeostasis and, like calcium and phosphorous, is partly controlled by vitamin D.

Approximately 50%–60% of magnesium in the body is located in the bone, with the rest found primarily in the ICF. Approximately 50% of dietary magnesium is absorbed in the small intestine and partially controlled by vitamin D. Because serum magnesium levels do not accurately reflect total body magnesium, the most accurate method of assessing magnesium status is to use the magnesium challenge test. In this test, urinary excretion is measured after a specific dose of supplemental magnesium is administered. If there is a reduced excretion of magnesium in the urine, this is indicative of magnesium absorption due to deficient levels.[6]

Hypomagnesemia

PATHOPHYSIOLOGY The most common cause of magnesium imbalance originates from chronic alcoholism and the withdrawal of alcohol. Some medications, such as **cyclosporine**, cause excessive urinary losses of magnesium. In some transplant patient populations, routine magnesium supplementation is necessary due to the excessive excretion in the urine. Magnesium losses in these situations may be as high as 25 mg/day. Malabsorption syndromes can result in both calcium and magnesium malabsorption, which are common in steatorrhea. Other causes of hypomagnesemia can include volume overload, diuretics, low levels of phosphorous, and high levels of ionized calcium.[6]

CLINICAL MANIFESTATIONS Signs and symptoms of **hypomagnesemia** are difficult to distinguish from those of hypokalemia and hypocalcemia. In practice, the clinician will first rule out other electrolyte deficiencies before attempting to replace magnesium. General symptoms include personality changes, depression, anorexia, nausea and vomiting, and ileus, as well as neuromuscular irritability. Serum magnesium levels <1.5 mEq/L or <1.8 mg/dL are considered to be low.

TREATMENT Imbalances can be treated with diet and oral supplements. The underlying cause of hypomagnesemia should be addressed, and other coexisting electrolyte imbalances should be corrected. Oral magnesium supplements can cause diarrhea and other gastrointestinal symptoms. Magnesium should be given cautiously via intramuscular or intravenous routes, because high levels of magnesium will cause cardiac arrest.

Hypermagnesemia

PATHOPHYSIOLOGY High serum levels of magnesium are uncommon, and are generally due to declining renal function or excessive supplementation. Supplementation of magnesium-containing antacids or laxatives (such as Milk of Magnesia) is a common pathway for excessive ingestion. Magnesium is used to treat **preeclampsia**, which is a high-risk condition for both the fetus and the mother.

CLINICAL MANIFESTATIONS Signs and symptoms include nausea, vomiting, facial flushing, and hypotension.[6] Neuromuscular action is impaired, possibly leading to decreased reflexes, muscle weakness, and paralysis. Cardiac function is also impaired, and if uncorrected, may lead to cardiac arrest. Serum levels >2.5 mEq/L or 3.0 mg/dL are diagnostic for **hypermagnesemia**.

TREATMENT In mild hypermagnesemia, discontinuation of medications containing magnesium is warranted. Calcium gluconate, given intravenously, can reverse effects of magnesium in more severe cases. Dialysis can also be used to decrease high levels of magnesium.

Conclusion

Many clinical conditions involve impairment of fluid and electrolyte balance. Nutrition therapy must be coordinated with medical care in the correction of imbalances. Metheny outlines six important points to address before assessing fluid and electrolyte balance.[7] These include the following:

- Does the patient have a disease or injury that could affect fluid/electrolyte balance?
- Is there a medication or treatment that could affect fluid/electrolyte balance?
- Is there fluid loss?
- Has nutrition therapy restricted any nutrient that could affect fluid/electrolyte balance?
- Has oral intake for both water and nutrients been adequate?
- Do the intake and output records balance?

The clinician will use this information to rule out any possible fluid and electrolyte disorders. This information will be confirmed by clinical assessment and laboratory values. Tables 7.5, 7.6, and 7.7 provide a summary of the assessment guidelines. Fluid and electrolyte assessment is not as easy as it may seem—many disorders produce clinical manifestations that are vague and that often overlap each other. With experience, however, the clinician will become proficient in assessment of fluid and electrolyte balance.

Lynne Schonder, RD, MS, CNSC, *Valley Care Medical Center, Pleasanton, CA*
La Paula Sakai, RD, MS, CNSD, *Retired*

Do you assess fluids and electrolytes in your practice?

Yes, it is an integral component of a nutrition assessment, especially in the patients requiring nutrition support. I use the assessment of fluid status and electrolyte balance to help understand the pathophysiology of the disease statues, which helps me in making a decision on what type of parenteral or enteral nutrition I should recommend. Without knowing the patient's fluid and electrolyte status, a dietitian cannot make an appropriate recommendation. Over time, I reviewed and researched fluids and electrolytes in order to know what electrolyte and fluid balance looks like in terms of laboratory values and fluid input and outputs (I & O). For example, a patient I was working with had an ileostomy with a high output (>1500/24 hours). I knew this could result in magnesium and calcium losses, so I recommended that those electrolytes be administered with his IV fluids. This helped minimize the patient's fluctuations in serum magnesium. However, even for a stable post-op patient, I have had to inform nurses and physicians that the patient, now eating, was receiving excessive IV fluids. Dietetic students need to know that fluids and electrolyte status must be assessed even for the lowest complexity patient.

Has the importance of fluid and electrolyte assessment changed over time?

The importance of fluid and electrolyte assessment has not changed over time but I think the *value* of dietitians assessing them has. If a dietitian wants to be a credible and valuable clinician, he/she must understand the importance of fluids and electrolytes, their part in the pathology of the diseases, and their relationship to nutritional assessment and support.

What indicators do you use for assessing fluid and electrolyte status?

First I look at the I & O totals from the previous day. Then I assess all the current sources of fluids in—the volume and type of IV fluids, oral intake, nasogastric feedings, and IV medications—and total fluid losses—urine, stool, emesis, and outputs from chest tube, nasogastric tube, surgical drains, and/or wound vacs. Finally, I compare the current I & O with the previous days. For electrolytes, I look at sodium, potassium, chloride, and bicarbonate but I also look at blood glucose values, since serum sodium can be impacted by hyperglycemia and acid/base balance can be impacted by severe hyperglycemia. I also pay attention

to what the physician is documenting in terms of acidosis and blood gas values. Dietetic interns often think that they can look in just one place in the chart to determine the fluid and electrolyte status. For me it's a combination of eyeballing the patient, reading the chart notes, and talking to the appropriate person(s) for more information.

What are the most common situations (diagnoses) that present problems with fluid and electrolyte?

Common situations are syndrome of inappropriate ADH (SIADH); failure of certain organs: heart, liver, and kidney; critically ill patient in the ICU d/t fluid shifts; patients receiving large amounts of steroids; chronic diarrhea or fistula losses; emesis losses; excessive wound losses; dehydration of older clients often d/t hot days and illness and of the very young (infants); and complications of eating disorders.

How do you interface with the health care team regarding fluid and electrolyte assessment?

I recommend establishing a good rapport and working relationship with all members of the health care team, especially the pharmacist. I try to communicate directly with the team, either in person or by phone. Of course I also document in the medical record, but if I see a tube feeding running at 65 mL/hour and IV fluids running at 100 mL/hour, I call the physician. It is very important to communicate to everyone involved in the patient's care what the nutrition goals/recommendations are. I find that many health care personnel don't understand how diet or nutrition support can interface with fluid and electrolyte balance. I believe there is no such thing as over-communication.

What would be most important for practitioners just starting to develop their skills in fluid and electrolyte assessment?

You need to start with basic textbook knowledge. If you covered the material in your Medical Nutrition Therapy (MNT) course, then you need to read it again before seeing patients for the first time. I even hear doctors remind medical residents to review fluids and electrolytes because they have forgotten what they learned in medical school. As you gain clinical knowledge and later become an RD, you still need to review and research this topic. The interrelationship between nutrition support and fluid and electrolyte needs is so complex, I find myself always going back and reviewing the material again and again.

Application of the Nutrition Care Process: Fluid and Electrolyte Balance

Introduction:

Max Williams is an 18-year-old male who has recently returned from a two-week trip to Quito, Ecuador. He has felt ill for a period of several days, which began the day before he left Ecuador. He describes 5 to 10 episodes of diarrhea each day that have not resolved after taking Kaopectate. Fecal smear indicates gross blood with leukocytes. Admitting diagnosis is moderate dehydration with R/O bacterial versus viral gastroenteritis.

Nutrition Assessment:

Food/Nutrition-Related History:

The registered dietitian's interview indicates that prior to this illness, Max had a good appetite with consumption of a wide variety of foods. The patient did remark about consumption of a "lot of new foods" while in Ecuador, some of which were purchased from street vendors.

Anthropometric Measurements:

Ht. 6'2" Wt. 178 lbs UBW 185 lbs

Physical Assessment:

Physical Exam: General appearance: Lethargic 18-year-old Caucasian male.

Vitals: Temperature 101.5°F BP:80/65 HR: 89 BPM

Respiratory Rate: 22 BPM. Heart: Moderately elevated pulse.

HEENT: Eyes: Sunken; Sclera clear without evidence of tears. Ears: Clear. Nose: Dry mucous membranes. Throat: Dry mucous membranes; no inflammation. Genitalia: Unremarkable. **Neurologic:** Alert, oriented × 3. Irritable.

Extremities: No joint deformity or muscle tenderness. No edema.

Skin: Warm, dry. Reduced capillary refill (approximately 2 seconds).

Chest/lungs: Clear to auscultation and percussion. Abdomen: Tender, nondistended, minimal bowel sounds.

1. What signs and symptoms in the physical assessment provide evidence for the diagnosis of dehydration?

Biochemical Data:

Total Protein 7.2 g/dL, Albumin 4.9 g/dL, Na 154 mmol/L, K^+ 3.2 mmol/L, Cl^- 107, PO_4 4.0 mmol/L, BUN 21 mg/dL, Cr 1.4 mg/dL, Hgb 15.5 g/dL, Hct 41%, WBC $17 \times 10^3/mm^3$

2. How might Max's laboratory values be affected by his hydration status?

3. What factors should be identified in a urinalysis that may also be consistent with dehydration?

Nutrition Diagnosis:

4. Identify at least two nutrition problems based on the nutrition assessment and medical history. Determine the diagnostic term for each nutrition problem. Next, identify the etiology of each nutrition problem. Finally, identify the signs and symptoms that support the evidence for these nutrition problems.

Nutrition Intervention:

5. Calculate Max's fluid needs.

6. Outline the nutrition therapy recommendations you would make for Max as he begins to try oral intake. Address both liquids and progression of his diet.

Nutrition Monitoring and Evaluation:

7. Determine nutrition criteria for monitoring and evaluation for each nutrition diagnosis that you identified.

1. What are electrolytes? What are anions and cations?

2. List the electrolytes that are primarily found in extracellular fluid and intracellular fluid. What is the normal range of concentration for these electrolytes in the serum?

3. What is the difference between osmolality and osmolarity?

4. Describe three factors that influence the movement of solutes through semipermeable membranes.

5. List three mechanisms by which the body regulates the movement of fluid and solutes to ensure homeostasis.

6. Explain how the renin-angiotensin-aldosterone system can affect blood volume.

7. Explain how aldosterone and vasopressin can affect urine volume.

8. Discuss how calcium and phosphate balance are maintained in the body.

9. Physiologically, what does hyper- or hypovolemia describe? What are the common causes of hyper- and hypovolemia?

10. Is there a difference between hypervolemia and hyponatremia? Explain your answer.

11. Do laboratory values of serum Na^+ > 145 mEq/L; serum osmolality > 295; and urine osmolality > 800 mOsm/kg indicate hypernatremia or hyponatremia? List three signs/symptoms that accompany this condition.

12. Describe three common conditions that can result in hypokalemia. What are common signs and symptoms of hypokalemia? Hyperkalemia?

13. How do changes in blood pH affect blood potassium levels?

REFERENCES

1. Dietary Reference Intakes for Water, Potassium, Sodium, Chloride, and Sulfate, Copyright © 2004 by the National Academies of Sciences, courtesy of the National Academies Press, Washington (DC).

2. Armstrong LE. Assessing hydration status: the elusive gold standard. J Am Coll Nutr. 2007;26:575S–584S.

3. Beich KR, Yancy C. The heart failure and sodium restriction controversy: challenging conventional practice Nutr Clin Pract.2008;23:477–86.

4. Schrier RW, Bansal S. Diagnosis and management of hyponatremia in acute illness. Curr Opin Crit Care. 2008;Dec;14(6):627–34.

5. Gennari FJ, Weise WJ. Acid-base disturbances in gastrointestinal disease. Clin J Am Soc Nephrol. 2008;3:1861–68.

6. Musso C. Magnesium metabolism in health and disease. Int Urol Nephrol. 2009;41:357–62.

7. Metheny NM. Fluid and electrolyte balance: nursing considerations. Philadelphia (PA):Lippincott Williams & Wilkins; 2000.

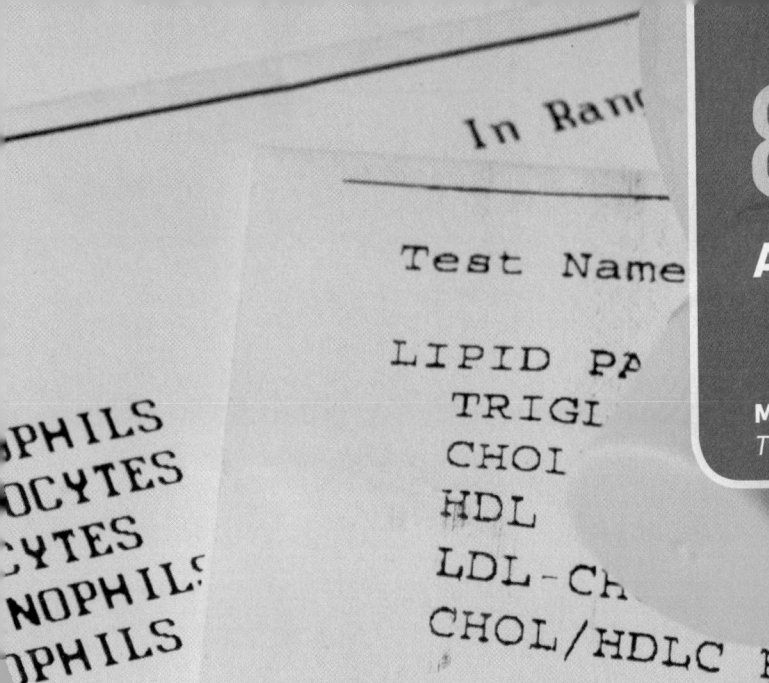

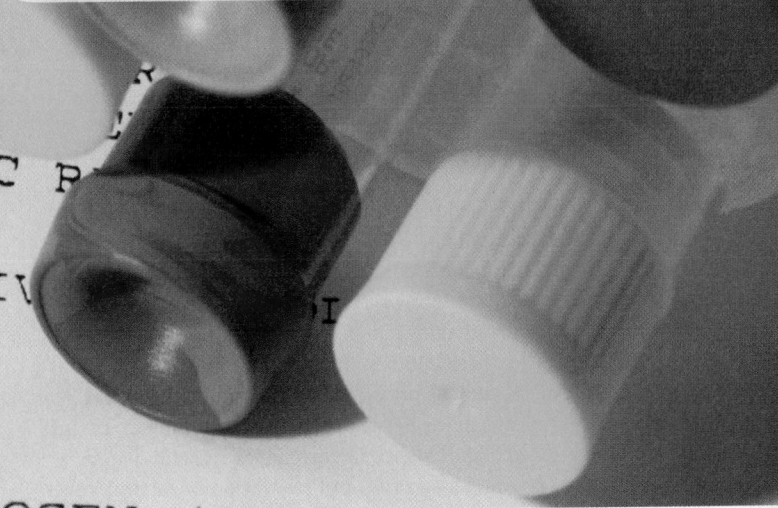

8

Acid-Base Balance

Marcia Nahikian Nelms, PhD, RD
The Ohio State University

Introduction

Even minor changes in pH can have significant effects on physiological function. Maintenance of a normal pH, and thus the body's homeostasis, allows for normal cellular function, enzyme activity, and membrane stability. Alterations of pH at the cellular level are manifested in often-dramatic systemic signs and symptoms.

Acid-base imbalances can occur throughout the life span. During infancy, immature kidneys coupled with a high metabolic rate increase risk of acidosis. Changes in respiration result in rapid changes in CO_2 levels in infants due to their small lung capacities. Diseases affecting the kidney and lungs, which are common in the elderly, reduce their ability to maintain homeostasis.

It is crucial for the registered dietitian to have a strong understanding of **acid-base balance**, since many nutrition therapies, such as parenteral nutrition, are used to address these metabolic changes. When these concepts are understood, the RD can assist in appropriate clinical interventions to return the body to its normal state. This chapter defines acids, bases, and pH, and describes conditions of acid-base balance.

Basic Concepts: Acids, Bases, and Buffers

Acids

Substances that can donate or give up hydrogen ions (H^1) are considered to be acids. In human physiology, two groups of acids are important: volatile and nonvolatile. Volatile acids are those that can be converted to a gaseous form and eliminated by the lungs. Nonvolatile acids include those inorganic acids that are formed during metabolism of carbohydrate, protein, and lipid. The lungs cannot eliminate nonvolatile acids.

Carbonic acid (H_2CO_3) is the most important volatile acid because it is produced in the largest amount and provides the major source of H^1. The body produces an average of 20,000 mmol of carbonic acid daily. This acid readily dissolves in solution as follows:

$$H_2CO_3 \Leftrightarrow CO_2 + H_2O$$

Because H_2CO_3 does dissolve so readily, it is not possible to measure its exact concentration. Instead, the concentration of CO_2 is used as an indirect measure of acidity. The concentration of CO_2 is expressed as $PaCO_2$, which is a measurement of *partial pressure* exerted by carbon dioxide in blood (see chapter endnote 1). It is considered to be partial because additional gases present in the blood such as nitrogen and oxygen also exert their own pressure.[1]

Nonvolatile or fixed acids are produced as end products of carbohydrate, protein, and lipid metabolism.[2] Of course, the amount of these nutrients consumed will affect the amount of fixed acids produced, but on average 50 to 100 mmol are produced each day. Fixed acids can be either organic or inorganic. Protein metabolism contributes the most with its addition of inorganic acids such as phosphoric acid and sulfuric acid. Examples of organic acids produced during metabolism include lactic acid and ketoacids such as hydroxybutyric acid. Fixed acids are also produced as a result of starvation or fasting, fever, exercise, and some disease states.

Bases

Bases are substances that can accept or receive a hydrogen ion. The most predominant base involved in human acid-base balance is bicarbonate (HCO_3^-). Other alkaline (basic) substances are added to the body through ingestion of fruits and vegetables. The kidneys provide primary regulation of HCO_3^- concentration by controlling the amount of free hydrogen ions and the amount of bicarbonate that is removed or retained in the body.

Buffers

A buffer is a substance or a group of substances that reacts with either acid or base in order to decrease the effect of acid or base on the pH of a solution. The most important buffer systems will be discussed in detail later in this chapter.

pH

The unit for measuring relative acidity or alkalinity of a fluid is called pH. Simply stated, pH is the ratio of acids to bases. Hydrogen ion concentration (H^+) is the negative logarithm of hydrogen ions in solution:

$$pH = \log 1/[H^+] = -\log [H^+]$$

Because the scale of (H^+) is logarithmic, in order for the pH to change by one unit (e.g., changing from 3 to 4), there must be a tenfold change in (H^+).

The pH of a substance is measured in a range from 1 to 14. A 1 on the pH scale indicates the most acidic, and 14 indicates the

Figure 8.1 pH of Body Fluids

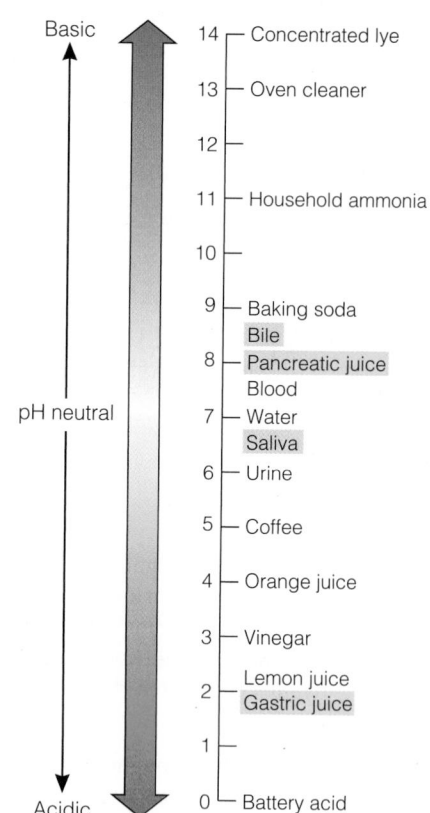

pH's of common substances:

Source: Whitney and S Rolfes, *Understanding Nutrition*, 10e, Copyright © 2005, p. 81.

most alkaline. Water is considered neutral at 7.0. For humans, a normal serum pH is within the range of 7.35–7.45. As shown in Figure 8.1, the pH of other body fluids varies, with gastric juice being the most acidic.

Terms Describing pH

Acidosis is the process (or processes) that leads to accumulation of acid or loss of base. **Acidemia** is the actual decrease in pH within the body to <7.35. Likewise, **alkalosis** is the process (or processes) that leads to accumulation of base or loss of acid, whereas **alkalemia** is the condition where an actual increase of pH >7.45 is observed.

Regulation of Acid-Base Balance

As illustrated in Figure 8.2, the body has to have several lines of defense to accommodate all the hydrogen ions constantly being produced in the body. These include (1) chemical buffers, (2) the respiratory regulation of pH, and (3) the kidney regulation of pH.

Chemical Buffering

As stated earlier, a buffer reacts with free (H$^+$) in order to maintain acid-base equilibrium. Effectiveness or power of the particular buffer is determined by its association with a cellular salt (**pK**) and by its overall concentration in the fluid compartment. Buffers are present in all body fluids—both extracellularly and intracellularly. Table 8.1 summarizes the chemical buffers.

The Bicarbonate-Carbonic Acid Buffer System

The primary buffer in extracellular fluid (ECF) is the bicarbonate-carbonic acid buffer system. This buffer system accommodates more than 80% of the required buffering in the ECF. This buffer system is outlined as follows:

$$H^+ + HCO_3^- \Leftrightarrow H_2CO_3 \Leftrightarrow CO_2 + H_2O$$

As the buffer system reacts with fixed acids, H_2CO_3 is produced. As discussed previously, H_2CO_3 readily dissolves to CO_2 and H_2O. Therefore, the lungs will accommodate the increased load of acids by increasing rate and depth of breathing and by expiring the CO_2 (see Figure 8.2). The kidney helps with this buffer system by either reabsorbing HCO_3^- or regenerating additional HCO_3^- from CO_2 and H_2O.[1, 2]

The **Henderson-Hasselbach equation** helps explain the interrelationships between H_2CO_3, HCO_3^-, and pH. In humans, the pH, or ratio of acids to bases, is 1 part H_2CO_3 to 20 parts HCO_3^-. In order for pH to remain within the normal range, this ratio has to be maintained. Any change in H_2CO_3 must be accompanied by a proportional change in HCO_3^-. If one part of the equation changes without the other and the ratio is not maintained, pH will move out of the normal range.

Other Chemical Buffer Systems
The body has additional buffer systems in place to compensate for changes that

TABLE 8.1 Chemical Buffers and Their Primary Roles

Buffer System	Major Functions
Bicarbonate-carbonic acid buffer system	Primary ECF buffer against non-carbonic-acid changes
Protein buffer system	Primary ICF buffer; also buffers ECF
Hemoglobin buffer system	Primary buffer against carbonic acid changes
Phosphate buffer system	Important urinary buffer; also buffers ICF

Note: ECF = extracellular fluid, ICF = intracellular fluid

Source: Sherwood L. Human physiology: from cells to systems. 7th ed. Belmont, CA: Brooks/Cole, 2010. Table 15-6, p. 573.

Figure 8.2 Overall Schema for Maintenance of Acid-Base Balance

On the usual mixed diet, pH is threatened by production of strong acids (e.g., sulfuric, hydrochloric, and phosphoric), which result mainly from protein metabolism. These strong acids are buffered by chemical buffers in the body. Removal of extra H$^+$s and the accompanying anions from the body is accomplished by renal excretion. When the kidneys excrete H$^+$s, they add new bicarbonate to the blood, thereby restoring depleted body buffer bases. The respiratory system eliminates CO_2 produced by metabolism. CO_2 is not a threat to acid-base balance, provided its partial pressure in arterial blood is kept at a normal value.

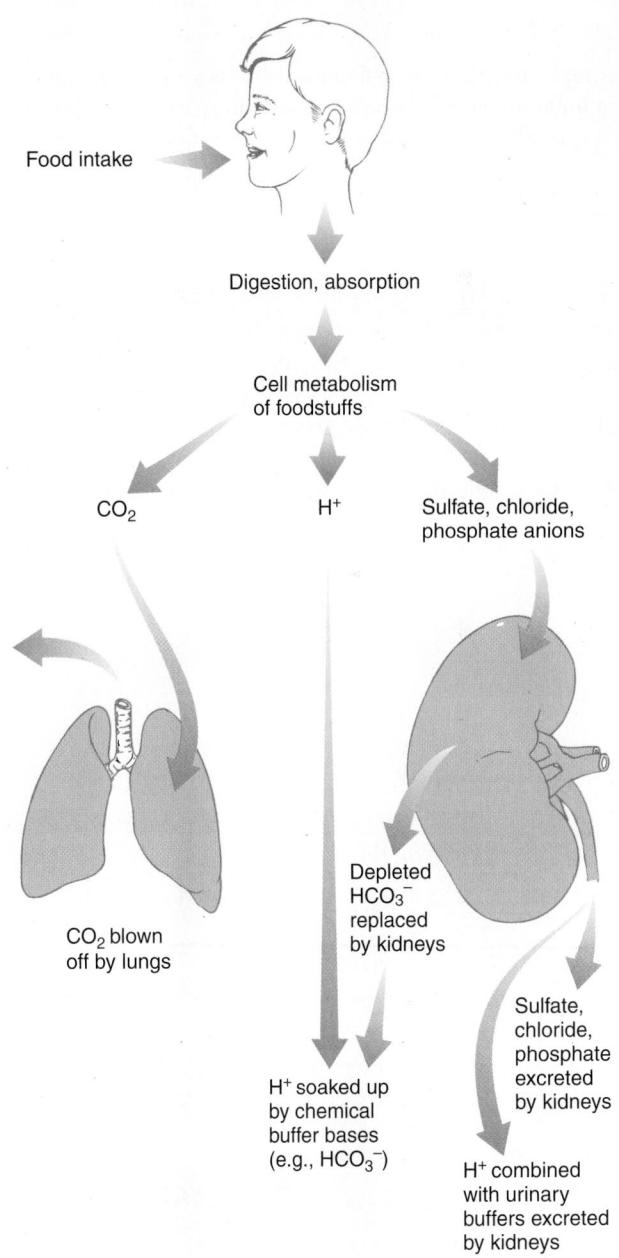

Food intake

Digestion, absorption

Cell metabolism of foodstuffs

CO_2　　H$^+$　　Sulfate, chloride, phosphate anions

CO_2 blown off by lungs

H$^+$ soaked up by chemical buffer bases (e.g., HCO$_3^-$)

Depleted HCO$_3^-$ replaced by kidneys

Sulfate, chloride, phosphate excreted by kidneys

H$^+$ combined with urinary buffers excreted by kidneys

Source: Lauralee Sherwood, Human Physiology: From Cells to Systems, 5e, copyright © 2004, p. 793.

could occur from other sources of acid. An important buffer system within red blood cells and tubules of the kidney is the disodium/monosodium phosphate (Na_2HPO_4) buffer. Excretion of H^+ could potentially make urine so acidic that excretion would be physically damaging to the kidney. Fortunately, phosphate accepts the H^+ and a weaker acid is formed. This is much less harmful to the kidney. This buffer system is outlined as follows:

$$Na_2HPO_4 + H^+ \Leftrightarrow NaH_2PO_4 + Na^+$$

Proteins present in the plasma can act as buffers. Their contribution as buffers is most important intracellularly. This buffer system acts in the same fashion as the bicarbonate-carbonic acid buffer system in that protein accepts the H^+. Many proteins can also release the H^+ if alkalinity increases. The proteins' ability to act in both situations increases the effectiveness of this buffer system.

Hemoglobin within the red blood cell acts as the most important buffer in blood. Carbon dioxide diffuses into the blood as it is produced throughout the body. Most of the CO_2 will combine with water, forming carbonic acid. As stated earlier, carbonic acid will dissociate to bicarbonate and free H^+. Hemoglobin binds the H^+.

The reaction is reversed as blood passes through the lungs and becomes oxygenated. Oxygenated hemoglobin gives up the H^+ to HCO_3^2 and thus carbonic acid (H_2CO_3) is generated. As stated earlier, H_2CO_3 dissolves to CO_2 and H_2O. CO_2 is then expired via the lungs. (See Figure 8.3.)

Respiratory Regulatory Control

The next line of defense in maintaining acid-base balance is respiratory control. The lungs have the ability to change respiratory rate and depth of breathing to control either release or retention of CO_2, and hence to assist in management of acid-base balance. This control system is very sensitive and is able to respond spontaneously.

The level of CO_2 in the blood controls the pH of the cerebrospinal fluid, since H^+ and HCO_3^2 do not cross the blood-brain barrier. Changes in pH—specifically the level of CO_2—are

Figure 8.3 Transport and Exchange of Carbon Dioxide and Oxygen

Carbon dioxide (CO_2) picked up at the tissue level is transported in the blood in three ways: (1) physically dissolved, (2) bound to hemoglobin (Hb), and (3) as bicarbonate ion (HCO_3^-). Hemoglobin is present only in the red blood cells, as is carbonic anhydrase, the enzyme that catalyzes the production of HCO_3^-. The H^+ generated during the production of HCO_3^- also binds to Hb. Bicarbonate moves by facilitated diffusion down its concentration gradient out of the red blood cell into the plasma, and chloride (Cl^-) moves by means of the same passive carrier into the red blood cell down the electrical gradient created by the outward diffusion of HCO_3^-.

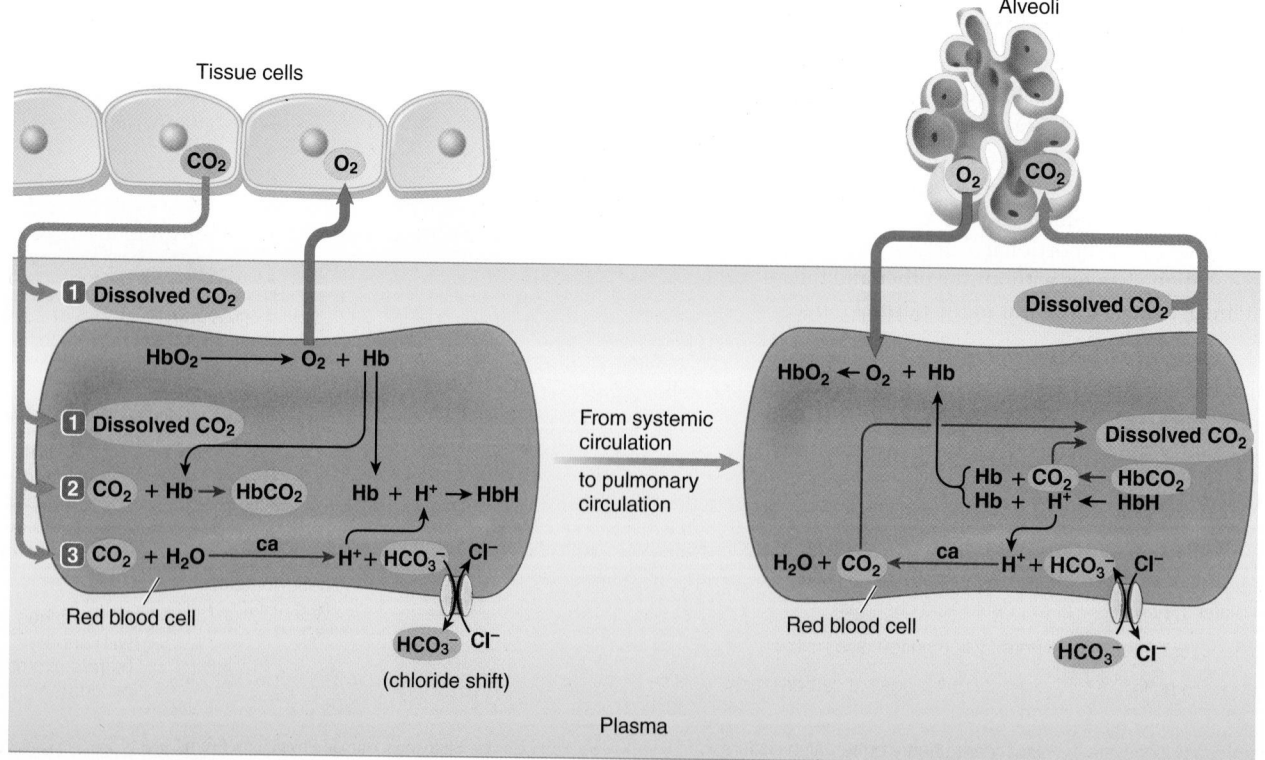

ca = Carbonic anhydrase

Source: L. Sherwood, *Human Physiology: From Cells to Systems*, 7e, Figure 13-31, p. 495

detected in cerebrospinal fluid by respiratory centers in the brain. In response, respiratory rate changes, which will normalize pH.

For example, when acidosis occurs, respiratory rate and depth will increase (hyperventilation). This allows larger amounts of CO_2 to be expired. In a situation where $PaCO_2$ (see chapter endnote 1) has decreased (alkalosis), respirations will slow, CO_2 concentrations will increase, and pH will normalize. Any change in anatomy or physiology of the respiratory system, nervous system control of respiration, or muscles that assist in breathing will affect the ability of the respiratory system to respond to changes in pH.

Renal Regulatory Control

Control of Hydrogen and Bicarbonate Ions The kidney's role in controlling both H^+ and HCO_3^- is a critical component for the maintenance of pH homeostasis. Respiratory control is ineffectual in dealing with the large amount of nonvolatile (fixed) acids constantly produced since these cannot be expired as gases. To maintain pH, a healthy, normally functioning kidney will reabsorb the majority of all HCO_3^- that is needed (see Figure 8.4). This function requires the kidney to secrete H^+ (see Figure 8.5), which combines with HCO_3^-, forming H_2CO_3. H_2CO_3 dissolves to form CO_2 and H_2O, which then forms HCO_3^- and free (H^+). This allows for constant regeneration of bicarbonate, which is needed to buffer the fixed acids being continuously released.

In the situation where alkalosis occurs, the kidney will respond by reducing the amount of HCO_3^- reabsorbed. On the other hand, if acidosis occurs, the kidney will increase secretion of (H^+) and increase the amount of HCO_3^- reabsorbed. Renal regulatory control is much slower than respiratory regulation and may take as much as 24 hours to respond to imbalances.[1,2]

Secretion of (H^+) is a vital component of the renal regulation of acid-base balance. The minimum pH of urine in humans is 4.5. If pH drops below this, the urine's acidity becomes harmful. The kidney cannot use bicarbonate as a buffer since it cannot be excreted at the same time as the hydrogen ions. Thus, the kidney uses two other buffers (dibasic phosphate and ammonium) to prevent damage from acidic urine, as described in the next section.

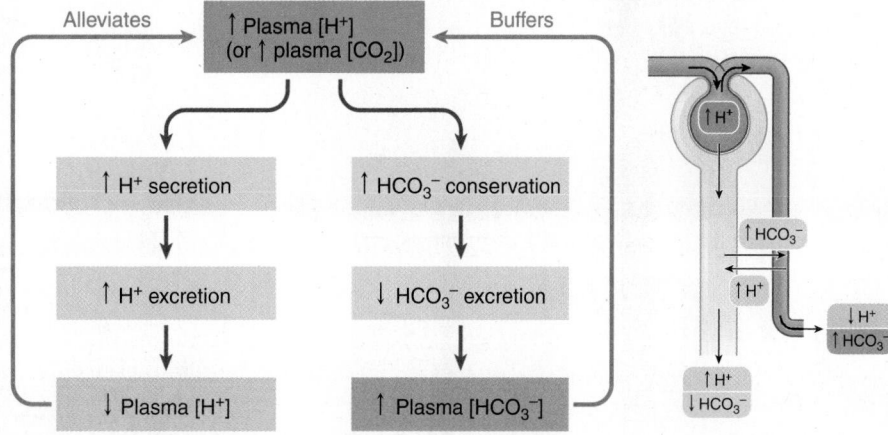

Figure 8.4 Control of the Rate of Tubular H^+ Secretion and HCO_3^- Reabsorption

Source: L. Sherwood, *Human Physiology: From Cells to Systems*, 7e, Figure 15-12, p. 579.

Other Renal Regulatory Controls The base NH_3 (ammonia) is formed in renal tubular cells from the amino acid glutamine. Free H^+ combines with NH_3 to form ammonium (NH_4^+). Ammonium cannot cross back across the cell membrane, so H^+ is trapped and is thus excreted in the urine.

Dibasic phosphate and sulfur both function to accept H^+ in order to control acid-base balance. In a situation where a large load of fixed acids is produced, the kidney will respond by increasing formation of acids within this buffer system.[3,4] Approximately one-third of the free H^+ is excreted as phosphoric acid (H_2PO_4) and sulfuric acid (H_2SO_4). Table 8.2 summarizes renal responses to changes in acid-base balance.

Effect of Acid and Base Shifts on Electrolyte Balance

Hydrogen ions and bicarbonate are both electrolytes. Acid-base changes will therefore affect concentrations of other electrolytes in both ECF and intracellular fluid (ICF). For example, movement of HCO_3^- to the plasma requires the exchange of another negatively charged ion so that **electroneutrality** is maintained. Chloride (Cl^-) is the ion that moves in the opposite direction of the HCO_3^- (as shown in Figure 8.5). Changes in potassium (K^+), chloride (Cl^-), and sodium (Na^+) may accompany acid-base disorders.[2–6]

Table 8.2 Summary of Renal Responses to Acidosis and Alkalosis

Acid-Base Abnormality	H^+ Secretion	H^+ Excretion	HCO_3^- Reabsorption and Addition of New HCO_3^- to Plasma	HCO_3^- Excretion	pH of Urine	Compensatory Change in Plasma pH
Acidosis	↑	↑	↑	Normal (zero; all filtered is reabsorbed)	Acidic	Alkalinization toward normal
Alkalosis	↓	↓	↓	↑	Alkaline	Acidification toward normal

Source: Reprinted from Sherwood L. Human physiology: from cells to systems, 7th ed. Belmont, CA: Brooks/Cole, 2010. Table 15-8, p. 580.

Figure 8.5 Hydrogen Ion Secretion Coupled with Bicarbonate Reabsorption in a Kidney Tubular Cell

Because the disappearance of a filtered HCO_3^- from the tubular fluid is coupled with the appearance of another HCO_3^- in the plasma, HCO_3^- is considered to have been "reabsorbed."

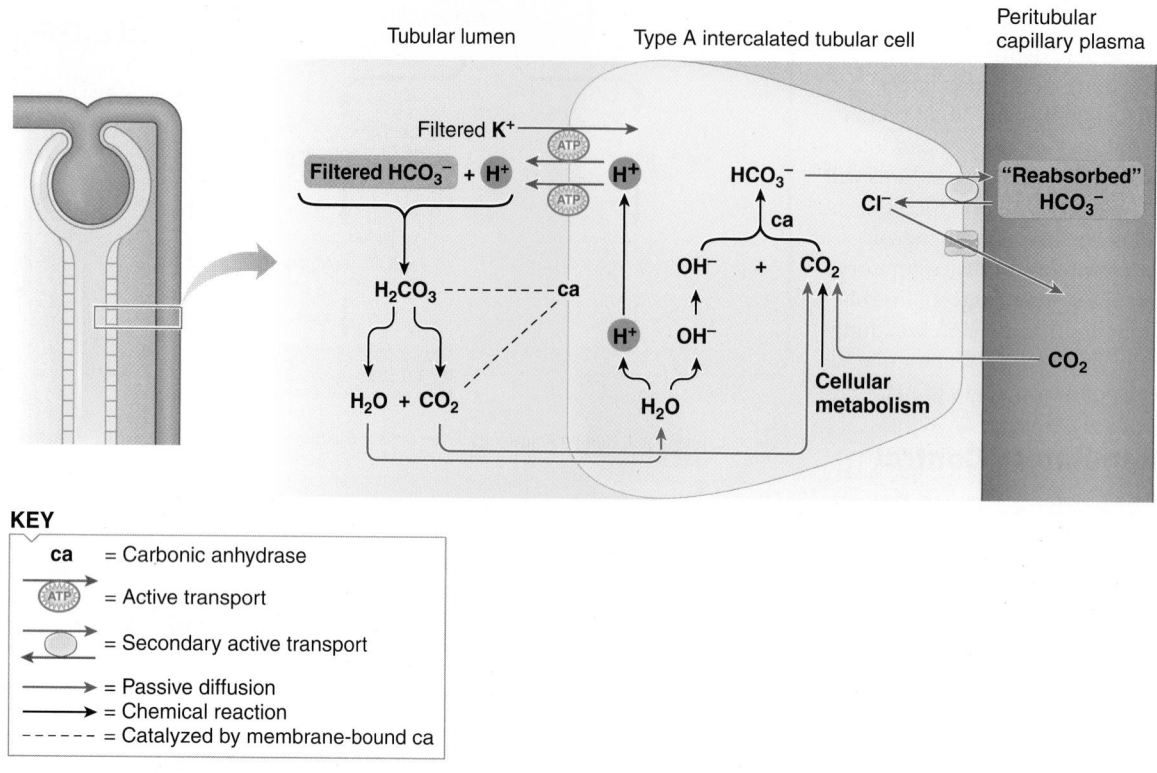

KEY

ca	= Carbonic anhydrase
(ATP)	= Active transport
	= Secondary active transport
	= Passive diffusion
	= Chemical reaction
- - - - -	= Catalyzed by membrane-bound ca

Source: L. Sherwood, *Human Physiology: From Cells to Systems*, 7e, Figure 15-9, p. 577.

Assessment of Acid-Base Balance

It is an understatement to say that assessment of acid-base disturbances is difficult. This difficulty arises because of the body's attempt to self-correct changes in pH. These compensatory responses confuse the clinical situation, making origin of the disturbance difficult to assess. Many times, assessment of acid-base balance requires more than simply examining laboratory values—it needs to be put into the context of the patient's current medical condition. In truth, examining laboratory values elicits only the current state of blood pH. This, of course, is difficult for novice clinicians, but with time and experience, one begins to be able to piece together the puzzle of acid-base disturbances.

Common laboratory measurements of arterial blood gases (ABGs) (see Table 8.3) and serum chemistries will provide values needed to initially assess acid-base balance.[6] These include arterial measures of both CO_2 and O_2 ($PaCO_2$ and PaO_2). Additionally, pH, CO_2, HCO_3^-, base excess, and **anion gap** are also measured. Even though both base excess and HCO_3^- are measured, they directly correlate, so it is not necessary to evaluate both values. See Table 8.3 for an outline of normal values of arterial blood gas parameters and analysis of arterial blood gases.

When evaluating pH, remember that in humans this measures the ratio of acids to bases. If both acid and base increase (or decrease) within the same proportion, pH will remain steady. It

Table 8.3 Normal Arterial Blood Gas (ABG) Values for Assessment of Acid-Base Balance

pH	7.35–7.45
pO$_2$, mmHg	80–100
pCO$_2$, mmHg	35–45
HCO$_3^-$, mEq/L	21–28
Base excess	−2.4–+2.3
Anion gap, mEq/L	8–16
O$_2$ saturation, %	>95

Source: This table was published in Pagana K, Pagana T. Mosby's Diagnostic and Laboratory Test Reference. 7e, Copyright Elsevier.

is only when one changes out of proportion to the other that a change in pH will be measurable. In other words, just because pH is within a normal range, this does not exclude the possibility of an acid-base disturbance.[5–7] A pH <7.35 or >7.45 is considered to indicate acidosis or alkalosis, respectively.

The anion gap represents the difference between unmeasured anions and cations. This calculation is important in distinguishing types of acid-base disorders. To calculate anion gap, the following equation is used: $Na^+ - (Cl^- + HCO_3^-)$ = anion gap in mEq/L. Most laboratories do not use K^+ in the calculation due its variability in acid-base imbalances. Application of these values will be discussed throughout the following sections.

Table 8.4 Summary of CO_2, HCO_3^- and pH in Uncompensated and Compensated Acid-Base Abnormalities

Acid-Base Status	pH	[CO_2] (Compared to Normal)	[HCO_3^-] (Compared to Normal)	[HCO_3^-]/[CO_2]
Normal	Normal	Normal	Normal	20/1
Uncompensated respiratory acidosis	Decreased	Increased	Normal	20/2 (10/1)
Compensated respiratory acidosis	Normal	Increased	Increased	40/2 (20/1)
Uncompensated respiratory alkalosis	Increased	Decreased	Normal	20/0.5 (40/1)
Compensated respiratory alkalosis	Normal	Decreased	Decreased	10/0.5 (20/1)
Uncompensated metabolic acidosis	Decreased	Normal	Decreased	10/1
Compensated metabolic acidosis	Normal	Decreased	Decreased	15/0.75 (20/1)
Uncompensated metabolic alkalosis	Increased	Normal	Increased	40/1
Compensated metabolic alkalosis	Normal	Increased	Increased	25/1.25 (20/1)

Source: Reprinted from Sherwood L. Human physiology: from cells to systems, 7th ed. Belmont, CA: Brooks/Cole, 2010. Table 15-9, p. 585.

Acid-Base Disorders

There are four major types of acid-base disorders. These include (see Table 8.4) respiratory acidosis, respiratory alkalosis, metabolic acidosis, and metabolic alkalosis. Combinations of each of these—indicating a mixed disorder—can also occur. The only combination of imbalances that is not a possibility is simultaneous respiratory acidosis and respiratory alkalosis, since obviously hypoventilation and hyperventilation cannot happen together.[2, 5, 8]

Respiratory Acidosis

Respiratory acidosis occurs when there is an excess of acid in relationship to base caused by retention of carbon dioxide. This generally occurs when there is an inability of the lungs to expire CO_2. As the level of CO_2 rises, **hypercapnia** occurs, more carbonic acid (H_2CO_3) is formed, and pH is shifted toward acidosis.[2, 5, 6, 8, 9]

Etiology Any factor that inhibits the medullary respiratory center can affect ventilation and thus the ability to breathe off CO_2. Medications such as opiates or sedatives can inhibit respiration. Chronic conditions such as sleep apnea or acute events such as cardiac arrest can also affect normal ventilation.

Diseases that affect musculature of the respiratory system and chest wall can result in poor ventilation. These may include neurological conditions such as myasthenia gravis or extreme obesity such as seen in Pickwickian syndrome. Additionally, any injury or trauma to the chest wall can potentially result in an inability to expire adequate amounts of carbon dioxide.

Respiratory diseases, such as chronic obstructive pulmonary disease, result in inability to maintain adequate oxygenation or release of carbon dioxide. Other conditions such as pneumonia, acute pulmonary edema, or pneumothorax can result in respiratory acidosis.[5, 10] Common causes are outlined in Table 8.5.

Pathophysiology In respiratory acidosis, the major cellular buffering defense available is the ability of the lungs to expire CO_2. But since the major reason for respiratory acidosis is respiratory dysfunction, this buffering system is not as efficient. Body stores of HCO_3^- are released in order to maintain the appropriate ratio of CO_2 to HCO_3^-, allowing pH to stay within

a normal range. During acute respiratory acidosis, the kidney regulatory systems do not have time to compensate, since these only begin to react within 12 to 24 hours. Chronic respiratory acidosis is less critical because the kidneys have more time to provide for ongoing compensation. Renal compensatory mechanisms work over a longer period and include increased excretion of H^+ and resorption of HCO_3^-. Other renal buffer systems such as the use of ammonium (NH_4) will also provide compensation.

Clinical Manifestations Laboratory values in acute respiratory failure will indicate a decreased pH and an elevated pCO_2. Bicarbonate levels will be slightly elevated if renal compensation has begun. Compensatory mechanisms allow the pH to remain normal but serum bicarbonate and arterial pCO_2 are elevated. Serum electrolytes will show an increase in serum Ca^+, K^+, and possibly Cl^- due to changes in renal controls.[5, 11]

In both acute and chronic respiratory acidosis, hypoxemia is present. This reduced level of oxygen is responsible for most

Table 8.5 Common Causes of Respiratory Acidosis

- Hypoventilation
- Chronic obstructive pulmonary disease
- Severe pneumonia or asthma
- Acute pulmonary edema
- Pneumothorax
- Drugs: opiate, sedative, anesthetic overdose (acute)
- Excessive oxygen treating chronic hypercapnia
- Sleep apnea
- Neuromuscular disease such as amyotrophic lateral sclerosis (ALS), Guillain-Barré syndrome, spinal cord injury
- Morbid obesity; Pickwickian syndrome
- Chest wall injury or skeletal deformity
- Aspiration of foreign body or vomitus
- Laryngospasm, laryngeal edema, severe bronchospasm
- Excessive production of CO_2
- Overfeeding, especially with high-carbohydrate components of nutrition support

symptoms associated with the acidosis. In general, the more acute onset will result in an increase in severity of symptoms. Alterations in respiration will include increased respiratory rate (hyperventilation) and an increase in depth of respirations. Other symptoms are a result of the change in oxygenation in the brain and/or a decrease in neurotransmission. These include restlessness, apprehension, lethargy, muscle twitching, tremors, convulsions, and finally, coma.[5]

Treatment Treatment will focus on correcting the underlying condition causing respiratory changes. Presence of hypoxemia would focus treatment on increasing oxygenation through administration of oxygen or provision of mechanical ventilation. In those patients with chronic hypoxemia, it is crucial to realize that the hypoxemia may be providing the stimulus for ventilation. If oxygen therapy reduces hypoxemia, ventilation may worsen without this stimulus.

Respiratory Alkalosis

Respiratory alkalosis (see Table 8.6) is characterized by a relative excess amount of base (HCO_3^-) as a result of a reduction of CO_2. This acid-base disturbance is generally a result of conditions causing hyperventilation. Rapid breathing results in a decreased $PaCO_2$.

Etiology Hyperventilation is commonly a result of a reduction in serum oxygen levels (hypoxemia). Hypoxemia can be a result of respiratory diseases such as pneumonia, asthma, pulmonary embolism, or pulmonary edema, or of exposure to high altitudes.

Direct stimulation of the respiratory center in the brain can also cause hyperventilation and resulting loss of CO_2. For example, disorders of the CNS such as a malignancy or stroke can affect respiratory centers and result in hyperventilation. Hypermetabolic states such as in fever and sepsis can directly stimulate hyperventilation. Drugs—including theophylline, salicylates, progesterone, doxapram, and catchecholamines—and even anxiety or other types of emotional distress can also result in hyperventilation. Hyperventilation can also occur as an adaptive response to high oxygen demands during strenuous physical activity.

Pathophysiology The acute response to respiratory alkalosis (within the first 24 hours) is a shift of acid from the ICF to ECF with an accompanying movement of bicarbonate into cells in exchange for chloride. Additional H^+ is synthesized by an increase in lactic acid derived from pyruvate within cells. Shifts in H^+ are generally not adequate to handle a continued decrease

in $PaCO_2$. For chronic respiratory alkalosis (lasting longer than 24 hours), renal compensation occurs. The kidneys reduce their secretion of H^+ (which also reduces regeneration of HCO_3^-) and increase their excretion of bicarbonate (HCO_3^-).

Clinical Manifestations In acute respiratory alkalosis, pH is >7.45 and $PaCO_2$ is decreased. In chronic respiratory alkalosis, pH is >7.45 and plasma HCO_3^- is low. In both situations, alkalosis may be accompanied by electrolyte imbalances. There may be low serum levels of K^+ and Ca^+ as well as high levels of Cl^-.

Other symptoms of respiratory alkalosis are seen in the cardiovascular, central nervous, and respiratory systems. Cardiac arrhythmias may be noted. Symptoms of the respiratory system vary but may include frequent yawning and deeper breaths. Symptoms of the central nervous system are most obvious, and may include "lightheadedness," mental confusion, anxiety, and seizures. Patients also relate parathesias and cold and clammy extremities.

Treatment Correction of the underlying cause of respiratory alkalosis is the only significant treatment. Correction of hypoxia by providing oxygen therapy would be a common first step. If the cause is psychological hyperventilation, rebreathing (see chapter endnote 2) of CO_2 can correct the symptoms.

Metabolic Acidosis

Metabolic acidosis refers to all types of acidosis that are not caused by excessive CO_2. It can result from either excessive loss of base (HCO_3^-) or an excessive gain of fixed (nonvolatile) acids (see Table 8.7). Metabolic acidosis can be seen in both acute and chronic conditions, but due to respiratory compensation, it is most often a chronic condition (see Table 8.8). Metabolic acidosis is sometimes characterized by using the anion gap calculation to determine origin of the disorder.[3, 4]

Etiology Conditions that result in excessive loss of bicarbonate from the gastrointestinal system or from renal excretion of bicarbonate can result in metabolic acidosis. Diarrhea is the most common cause. Additionally, losses from an ileostomy or from pancreatic, biliary, or intestinal fistulas can be another source of excessive HCO_3^- loss.[5]

Carbonic anhydrase inhibitors such as the drug Acetazolamide (used as an antiseizure medication and diuretic) can result in excessive loss of base while inhibiting production of carbonic acid in the kidney. In the condition of renal tubular acidosis,

Table 8.6 Common Causes of Respiratory Alkalosis

- Hyperventilation
- Respiratory infection; pneumonia
- Asthma
- Change in altitude environment—i.e., high altitude
- Drugs that stimulate respirations: theophylline, catecholamines
- Anxiety
- Cerebrovascular accident
- Fever and sepsis

TABLE 8.7 Common Causes of Metabolic Acidosis

- Kidney loss of HCO_3^-
- Chronic kidney disease
- Systemic loss of HCO_3^-
- Diarrhea
- Fistula drainage
- Excessive production of acid
- Ketoacidosis secondary to conditions such as diabetes mellitus, alcoholism, or starvation
- Lactic acidosis secondary to conditions such as diabetes mellitus, salicylate overdose

Table 8.8 Respiratory Adjustments to Acidosis and Alkalosis Induced by Nonrespiratory Causes

Respiratory Compensation	Normal (pH 7.4)	Nonrespiratory (Metabolic) Acidosis (pH 7.1)	Nonrespiratory (Metabolic) Alkalosis (pH 7.7)
Respiratory rate	Normal	↑	↓
Tidal volume*	Normal	↑	↓
Ventilation	Normal	↑	↓
Rate of CO_2 removal	Normal	↑	↓
Rate of carbonic acid formation	Normal	↓	↑
Rate of H^+ generation from CO_2	Normal	↓	↑

*Volume of oxygen inhaled and exhaled during a normal breath

Source: Adapted from Sherwood L. Human physiology: from cells to systems, 7th ed. Belmont, CA: Brooks/Cole, 2010. Table 15-7, p. 576.

ability to reabsorb bicarbonate is decreased. In chronic kidney failure, ability to restore bicarbonate may fail as well. Other mechanisms to correct acid-base disturbances, such as production of NH_4^+, may also fail as renal function declines.

Metabolic acidosis may also result from situations that increase the amount of acid, such as administration of ammonium chloride and rapid administration of IV saline. Accidental poisoning with substances such as salicylate (aspirin), ethylene glycol (antifreeze), or formaldehyde can result in metabolic acidosis.

Figure 8.6 Lactic Acid Production

When NAD concentration is decreased compared to $NADH + H^+$ the scale leans in the direction of lactic acid production and not pyruvic acid. Conditions that result in decreased concentrations of NAD include: decreased oxygenation of the tissues, excessive ketone body production (as in diabetes), and metabolism of ethanol.

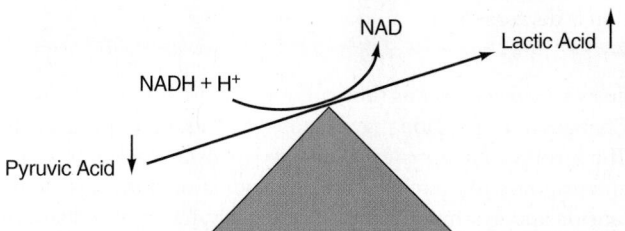

Lactic acidosis (see Figure 8.6) occurs as a result of increased production of lactate or ketoacids. High levels of lactate may result when the kidney or liver fails to convert lactate to pyruvate or bicarbonate. Diabetic ketoacidosis, one of the most common causes, results in metabolic acidosis due to both increased production and inability to metabolize ketones (see Box 8.1).

In both starvation and chronic alcoholism, synthesis of ketoacids is increased. Starvation, with its reliance on fat stores, can also increase synthesis of ketoacids and result in acidosis.[12]

Pathophysiology When H^+ levels increase, the bicarbonate-carbonic acid buffer system is stimulated. This shift of

---

BOX 8.1 CLINICAL APPLICATIONS

Diabetic Ketoacidosis (DKA)

One out of every four emergency room visits for patients with type 1 diabetes mellitus is for diabetic ketoacidosis (DKA). It has been estimated that over one billion dollars is spent each year treating this condition and its complications.[1]

What is DKA, and how is this condition related to the acid-base imbalances you have learned about in this chapter?

Diabetes mellitus is the disease caused by either the absence or inefficient use of the hormone insulin (see Chapter 17). Ketoacidosis is one of the most serious acute complications of type 1 diabetes mellitus. Diabetic ketoacidosis typically develops as a result of infections, or because the patient does not take adequate amounts of insulin.

Without adequate insulin, there is an increased dependence on lipids as the primary fuel source. The increased rate of lipolysis results in the production of ketones: aceto-acetic acid and hydroxybutyric acid. These ketone bodies are acids that lower serum pH. The kidney reacts by excreting the ketone bodies in the urine (ketonuria). The increased levels of hydrogen ions are buffered by plasma bicarbonate. This combination of events results in metabolic acidosis. High levels of ketoacids lead to an increase in the plasma anion gap.[2]

As explained in this chapter, both the respiratory and renal systems serve as compensatory mechanisms in acid-base imbalance. In DKA, as pH lowers, respiratory ventilation changes to accommodate the need to reduce pCO_2. Respirations are deep and labored—

Kussmaul's respirations. Secondly, the renal system compensates by conserving bicarbonate (HCO_3^-). Furthermore, there is increased urinary excretion of positively charged cations (Na^+, K^+, NH_4^+).

DKA must be treated quickly and accurately to prevent coma and death. Providing adequate insulin, fluids, and electrolytes allows the correction of the metabolic acidosis and prevents these complications.[2]

References
1. Kitabchi AE, Umpierrez GE, Fisher JN, Murphy MB, Stentz FB. Thirty years of personal experience in hyperglycemic crises: diabetic ketoacidosis and hyperglycemic hyperosmolar state. J Clin Endocrinol Metab. 2008;93:1541–52.
2. De Beer K, Michael S, Thacker M, Wynne E, Pattni C, Gomm M, Ball C, Walsh D, Thomlinson A, Ullah K. Diabetic ketoacidosis and hyperglycemic hyperosmolar syndrome – clinical guidelines. Nurs Crit Care. 2008;13:5–11.

H^+ into ECF reduces serum K^+ at the same time in order to maintain equilibrium between the ECF and ICF.

High levels of H^+ in the blood stimulate respiratory centers in the brain. The lungs respond by increasing rate and depth of breathing. Finally, the kidneys begin their compensatory response by increasing their excretion of H^+ and retaining HCO_3^-. Renal compensation is much slower than respiratory, and if kidney disease is present, effectiveness of the compensation is decreased.

Clinical Manifestations Symptoms of metabolic acidosis are not as clear as those of other acid-base disturbances. Changes in respiration may be noted as **Kussmaul breathing**. The cardiovascular system is affected by decreased contractility and response to catacholamines. Vasodilation may cause hypotension and dysrhythmias. Neurologically, lethargy and stupor with eventual coma are observed as pH falls in cerebrospinal fluid. In chronic renal failure, ongoing metabolic acidosis relies on carbonate from bone to handle the acid load. This results in growth failure in children and renal osteodystrophy in adults.[2-5]

Treatment Treatment is focused on the underlying cause of the acidosis. Correcting pH too quickly can cause additional complications. The goal is to raise systemic pH to a safe level.

Metabolic Alkalosis

Presence of an excessive amount of base (HCO_3^-) results in **metabolic alkalosis** (see Table 8.9). Generally, this acid-base disturbance is caused by a loss of nonvolatile acids, the over-administration of bicarbonate transfusion of whole blood.

Etiology Clinical situations resulting in alkalosis can be categorized as either those conditions involving fluid imbalance (alkalosis with volume decrease) or those without fluid imbalances (alkalosis without volume contraction). Conditions that involve fluid imbalance include prolonged vomiting and/or nasogastric suction or use of diuretics.

Conditions leading to alkalosis that do not involve fluid imbalance would include hyperaldosteronism, excessive use of corticosteroids, blood transfusions, chronic use of antacids, and excessive administration of sodium bicarbonate.[4]

Pathophysiology Initiation of metabolic alkalosis begins with the underlying event that causes either an excessive loss of acid or accumulation of base. This could be, for example, prolonged vomiting that results in decreased concentration of HCl^-. Normally, the kidney will compensate for the decrease in nonvolatile acid by generating (H^+) and decreasing reabsorp-

Table 8.9 Common Causes of Metabolic Alkalosis

- Loss of acid
- Vomiting
- Nasogastric suctioning
- Hypokalemia
- Excessive base
- Intravenous therapy
- Blood transfusion
- Excessive or chronic use of antacids

tion of bicarbonate. In order for metabolic alkalosis to progress, other events need to occur that prevent adequate compensation by the kidney.

In situations where there is also volume depletion, such as in use of diuretics, there is both a fluid loss and a subsequent decrease in K^+ levels. To maintain serum K^+ levels within a safe range, the kidney will excrete H^+ in exchange for K^+. Even though K^+ may increase, the loss of H^+ results in generation of more bicarbonate. This further contributes to alkalosis.

Stored blood contains citrate as a preservative. If an individual receives a large amount of transfused blood, it is possible the body will convert the citrate to bicarbonate. This potentially could lead to metabolic alkalosis.

Another example of non-volume-related alkalosis is in the condition of primary or secondary hyperaldosteronism. Increased secretion of aldosterone causes the kidney to increase reabsorption of sodium. This is accompanied by secretion of H^+, which increases regeneration of HCO_3^-.

Clinical Manifestations Arterial blood gases in metabolic alkalosis will indicate a pH of >7.45 and elevated levels of HCO_3^- / >26 mEq/L. Accompanying electrolyte imbalances may indicate a $K^+ < 3.5$ mEq/L and $Cl^- < 98$ mEq/L. If compensation by the respiratory system is in place (see Table 8.8), $PaCO_2$ will remain within a normal range or be slightly elevated.

There are no specific signs and symptoms for metabolic alkalosis. Signs and symptoms are determined by accompanying conditions of volume deficit or electrolyte abnormalities. For example, in the situation where there is hypokalemia and a decrease in ECF, the patient may experience muscle cramping, weakness, and cardiac arrythmias.

Treatment In chloride-responsive metabolic alkalosis, correcting volume imbalance with isotonic saline with additional KCl^- will correct alkalosis. Metabolic alkalosis that does not involve a fluid deficit requires treatment of underlying causes before alkalosis can be corrected. In severe conditions, use of a carbonic anhydrase inhibitor will enhance HCO_3^- excretion.

Mixed Acid-Base Disorders

Several acid-base disturbances (see Table 8.10) can coexist in complex medical problems. For instance, this may occur in a patient who is respiratory-compromised and is unable to respond to a situation producing a metabolic disorder. This could also occur in situations of drug overdose where different medications cause both respiratory and metabolic responses.[5] When examining ABGs, a mixed disorder should be suspected when $PaCO_2$ and HCO_3^- are not consistent with the measured pH. A mixed disorder may also be present when the compensatory response is exaggerated. For example, metabolic alkalosis and respiratory acidosis may occur when a patient with chronic obstructive pulmonary disease receives diuretics.

Assessment of Acid-Base Disorders

To place all of this into perspective, Table 8.4 summarizes major components needed to assess all acid-base disorders. As was discussed earlier in this chapter, arterial blood gas measurements will provide all data needed to evaluate acid-base status and begin steps toward intervention.

Table 8.10 Common Mixed Acid-Base Disorders

Dual Mixed Disorder	Examples
Additive Effect on pH Change	
Metabolic acidosis + respiratory acidosis • $PaCO_2$ too high • HCO_3^- too low • pH very low	• Cardiopulmonary arrest • Patient with COPD goes into shock • Chronic renal failure with fluid volume excess and pulmonary edema • Patient with DKA receives potent opiate or barbiturate
Metabolic alkalosis + respiratory alkalosis • $PaCO_2$ too low • HCO_3^- too high • pH very high	• Patient with previously compensated respiratory acidosis caused by COPD overventilated on mechanical respirator • Hyperventilating patient with CHF or hepatic cirrhosis who is vomiting or is treated with potent diuretics or nasogastric suction • Head trauma patient with hyperventilation treated with diuretics
Offsetting Effect on pH Change	
Metabolic acidosis + respiratory alkalosis • $PaCO_2$ too low • HCO_3^- too low • pH near normal	• Lactic acidosis complicating septic shock • Hepatorenal syndrome • Salicylate intoxication
Metabolic alkalosis + respiratory acidosis • $PaCO_2$ too high • HCO_3^- too high • pH near normal	• Patient with COPD who is vomiting or who is treated with NG suction or potent diuretics • Adult respiratory distress syndrome

CHF, congestive heart failure; COPD, chronic obstructive pulmonary disease; DKA, diabetic ketoacidosis; NG, nasogastric

Source: Reprinted from *Pathophysiology: Clinical Concepts of Disease Processes*, 6th ed. by S.A. Price and L.M. Wilson, Table 22-5, p. 306, © 2003, with permission from Elsevier.

Conclusion

Remembering basic concepts for acid-base balance will keep you on track in evaluating these complex clinical situations.

- The scale for measuring acidity or alkalinity of a fluid is the measurement of pH. Simply stated, pH is the ratio of bases to acids. Normal pH in humans is 7.35–7.45. The pH is maintained in a ratio of 20:1 base to acid within body fluids. There will be no change in pH if the ratio remains stable, but changes in either portion of the ratio will result in changes in pH.
- The largest source of acid within the body is carbonic acid. We measure concentration of CO_2 as an indirect measure of acidity. Concentration of CO_2 is expressed as $PaCO_2$.
- The lungs primarily regulate CO_2 levels.
- The largest source of base is HCO_3^-, which is regulated primarily by the kidneys.
- Respiratory acidosis is a result of retention of CO_2, whereas respiratory alkalosis is a result of hyperventilation and a subsequent decrease in CO_2 levels.
- Metabolic acidosis occurs when there is retention of fixed acids or excessive loss of bases. Metabolic alkalosis is a result of excessive loss of fixed acids or retention of bases.

Application of the Nutrition Care Process: Acid-Base Imbalance

Introduction:

Mr. N has presented to the physician with complaints of the following: pain, dizziness, and difficulty breathing. Mr. N's daughter also indicates he has become more and more confused over the previous 24 hours. As the physician proceeds to his physical exam, he notes blood pressure is out of the normal range, and respiratory rate and heart rate are high. Further tests indicate Mr. N's hemoglobin is low while his white count is elevated.

1. Outline items from Mr. N's case that fall into each of the following categories:
   a. Signs
   b. Symptoms
   c. Laboratory abnormalities

2. Upon examination of his medical record, you find Mr. N has the following diagnoses: renal insufficiency, chronic obstructive pulmonary disease, and a history of coronary heart disease. Which of these might interfere with his ability to maintain a normal acid-base balance?

Nutrition Assessment: Biochemical Data:

3. The physician has ordered arterial blood gases. Values you note as abnormal are as follows: pH 7.47; pCO_2 46 mmHg; pO_2 83 mmHg; HCO_3^- 32 mEq/L.
   a. Classify the pH.
   b. Assess pCO_2.
   c. Assess HCO_3^-.
   d. Do you see any indication of compensation? Why or why not?
   e. Identify the primary acid-base disorder.
   f. How do his medical diagnoses relate to this acid-base imbalance?

WEB LINKS

A. Grogano. Acid-Base Tutorial—Tulane Department of Anesthesiology, Tulane School of Medicine This tutorial is an excellent method of review for basic concepts of acid-base balance for the health professional.
http://www.acid-base.com/index.php

Merck Manual of Medical Information: Acid-Base Imbalance The Merck Manual historically has provided information for basic medical concepts. The manual describes the components of acid-base balance and how disease may affect this balance.
http://www.merck.com/mmhe/sec12/ch159/ch159a.html

END-OF-CHAPTER QUESTIONS

1. Define the following terms: pH, volatile acid, nonvolatile (fixed) acid, buffer.

2. What organ controls the level of pCO_2 in the blood? What organ controls HCO_3^- in the blood?

3. What is the basic problem in respiratory acidosis? Respiratory alkalosis?

4. What is the most important difference between metabolic acid-base disorders and those of a respiratory origin?

5. Name some conditions that might result in respiratory acidosis.

6. Name some conditions that might result in metabolic acidosis.

7. What is an anion gap?

8. How can respiratory mechanisms compensate for a metabolic alkalosis? Are there any major limitations to this compensation?

ENDNOTES

1. Reported blood gas abbreviations. The letter "P" before CO_2 or O_2 is the partial pressure of the gas but it could be of arterial, venous, or mixed blood. When the letters "Pa" are in front of the gas, it is the partial pressure of the gas in the arterial blood.

2. Rebreathing into a paper bag: Have you heard of doing this for someone when he or she is nervous and breathing too fast (e.g., for hyperventilation)?

The rationale is that rebreathing into a paper bag will allow the person to replace the carbon dioxide "blown off" while hyperventilating.

REFERENCES

1. Rhoades R, Pflanzer R. Regulation of acid-base physiology. IN: Human Physiology. Pacific Grove (CA):Thomson Learning Inc.; 2003.

2. Sherwood L. Fluid and acid-base balance. In: Human Physiology. From Cells to Systems. Belmont (CA):Brooks-Cole/ Thomson Learning; 2010.

3. Morris CG, Low J. Metabolic acidosis in the critically ill: part 2. Causes and treatment. Anaesthesia. 2008;63:396–411.

4. Morris CG, Low J. Metabolic acidosis in the critically ill: part 1. Classification and pathophysiology. Anaesthesia. 2008;63:294–301.

5. DuBose, Jr. Thomas D, "Chapter 48. Acidosis and Alkalosis" (Chapter). Fauci AS, Braunwald E, Kasper DL, Hauser SL, Longo DL, Jameson JL, Loscalzo J. Harrison's principles of internal medicine. 17th ed. Available at http://www .accessmedicine.com/content .aspx?aID=2866352

6. Ayers P, Warrington L. Diagnosis and treatment of simple acid-base disorders. Nutr Clin Prac. 2008;23:122–27.

7. Horne C, Derrico D. Mastering ABGs. The art of arterial blood gas measurement. Am J Nurs. 1999;99:26–32; quiz 33.

8. Fall PJ. A stepwise approach to acid-base disorders. Practical patient evaluation for metabolic acidosis and other conditions. Postgraduate medicine [Postgrad Med] 2000;107:249–50, 253–54, 257–58.

9. Kraut JA, Madias NE. Approach to patients with acid-base disorders. Respir Care. 2001;46:392–403.

10. Madias NE, Adrogué HJ. Cross-talk between two organs: how the kidney responds to disruption of acid-base balance by the lung. Nephron Physiol. 2003;93:61–66.

11. Martinu T, Menzies D, Dial S. Re-evaluation of acid-base prediction rules in patients with chronic respiratory acidosis. Can Respir J. 2003;10:311–15.

12. Reddy ST, Wang CY, Sakhaee K, Brinkley L, Pak CY. Effect of low-carbohydrate high-protein diets on acid-base balance, stone-forming propensity, and calcium metabolism. Am J Kidney Dis. 2002;40:265–74.

9

Cellular and Physiological Response to Injury: The Role of the Immune System

Marcia Nahikian Nelms, PhD, RD
The Ohio State University
Christina Frazier, PhD
Southeast Missouri State University

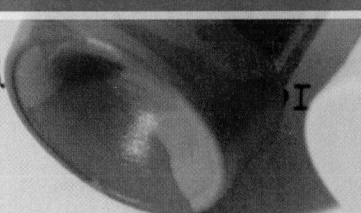

CHAPTER OUTLINE

Introduction

Epidemiology, etiology, pathogenesis, clinical manifestations, and disease outcome: this chapter provides the framework and foundation for understanding the disease process. The processes that result in cellular injury and the characteristic response of the cell to injury are similar among various disease states. The basic concepts outlined in this chapter—cellular responses to injury including inflammation and the overall immune response—provide the foundation for understanding the complexities of each individual diagnosis and disease process discussed later in this text.

Humans are exposed to numerous pathogens as we eat, breathe, and come into contact with environmental objects and other humans. We are protected from disease-causing organisms by natural resistance and the immune system. Natural resistance involves anatomical structures and physiological mechanisms that have other functions in the body. These work predominantly by keeping organisms from entering the body and becoming established in the tissues. The immune system includes organs, cells, and other secretions that respond to pathogens and injured cells that have not been stopped by our natural resistance. Immunity is defined as all those physiological mechanisms that endow the body with the ability to recognize material as foreign and to neutralize, eliminate, and/or metabolize it with or without damage to the body's tissues.

acquired immunity—immune response that results from exposure to an antigen or immunoglobulin

active immunity—immunity produced due to exposure to an antigen (e.g., infection or vaccination)

adhesion—scar tissue that forms between two body surfaces, usually as a result of surgery or injury

allergen—an antigen that triggers an allergic response

allergy—an inappropriate and harmful immune reaction to a harmless nonpathogenic substance; also called *hypersensitivity*

allogeneic—having a different genetic composition; in bone marrow transplant, refers to receipt of bone marrow from a donor of different genetic composition.

allograft—a tissue/organ graft between two genetically different individuals from the same species

amyloid—a starch-like substance present in diseased tissues

anaphylactic shock—a life-threatening IgE-mediated allergic reaction; in humans, symptoms include swelling (especially of the lips and face), vomiting, diarrhea, difficulty in breathing, and a sudden drop in blood pressure; also called *anaphylaxis*

anergy—antigen-specific nonresponsiveness by a T or B cell in which the cell is present but cannot respond

antibody—a protein molecule found in serum and tissues that is secreted by B cells in response to a specific antigen that can bind to that antigen and neutralize or help destroy it; also called *immunoglobulin*

antigen—a substance that is specifically bound by an antibody or lymphocytes; used by the immune system to recognize pathogens and altered cells; see *immunogen*

antigen-presenting cell (APC)—a cell capable of displaying fragments of antigens from pathogens and altered cells joined to major histocompatability molecules on its surface in a manner that can be recognized by T cells

antiseptics—agents that kill microbes within living tissue

antitoxin—an antibody to an exotoxin

apoptosis—genetically programmed cell death

asthma—a chronic inflammatory lung disease that is triggered by either an IgE allergic reaction or nonallergic factors and results in inflammation of the airway and reversible airway obstruction

attenuated—refers to an antigen rendered less virulent but still capable of eliciting an immune response

atrophy—reduction in size of muscle cells

autoantibody—an antibody to self-antigens

autograft—a tissue graft from one area to another on the same individual

autoimmunity—an immune response to one's own tissues

autologous—transplant of one's own body tissue; in bone marrow transplant, refers to treatment through receipt of one's own bone marrow

BALT (bronchial-associated lymphatic tissue)—secondary lymphoid organs of the bronchial tree

basophils—polymorphonuclear leukocytes containing granules that stain with basic dyes; they have much in common with mast cells, including the release of histamine and leukotrienes, which contribute to allergic responses and inflammation

B cell—a lymphocyte derived from the bone marrow, which differentiates into a plasma cell that makes an antibody

betadine—a povidone-iodine containing solution that is used topically to destroy microorganisms

bone marrow—soft tissue in the cavities of bones where stem cells become red and white blood cells

CD—"cluster designation"; an international nomenclature system of leukocyte cell surface molecules (CD number)

CD4—a marker found predominantly on helper T cells that interacts with MHC class II molecules on antigen-presenting cells

CD8—a marker found predominantly on cytotoxic T cells that interacts with MHC class I molecules on target cells

cellular immunity—immune protection provided by the action of immune cells, especially T cells, polymorphonuclear leukocytes, and macrophages

clinical manifestations—unique signs and symptoms

collagen—a fibrous protein found in connective tissue

complement—a group of serum proteins activated in a cascade that produces compounds that lyse cells and mediate immune reactions

contracture—shortening of muscle tissue resulting in immobility

Cushing's syndrome—a disorder resulting from prolonged exposure to high levels of glucocorticoid hormones; symptoms include: muscle weakness, thinning of the skin, moon-shaped face, weight gain, and diabetes mellitus

cytokines—soluble substances secreted by one cell that cause it or other cells to proliferate, differentiate, migrate, or become activated

cytotoxic T cells (Tc)—T lymphocytes that kill cells infected by viruses or transformed by cancer

dehiscence—separation of wound edges

delayed hypersensitivity—a cell-mediated inflammatory allergic reaction in the skin (e.g., poison ivy) that takes 24 to 48 hours to appear

dendritic cells—antigen-trapping and antigen-presenting white blood cells with nerve-like processes (e.g., Langerhans cells and interdigitating cells)

disinfectants—agents that kill microbes on inanimate objects or surfaces

dysplasia—abnormal cell growth

endotoxins—toxins found in bacteria, often as part of the cell wall, that stimulate an immune response

eosinophil—a polymorphonuclear leukocyte containing granules that produce substances that damage parasites and decrease inflammation; these granules stain with acid dyes

eosinophilic esophagitis—abnormal infiltration of eosinophils into the esophagus; may be associated with food allergy as well as other conditions such as scleroderma, gastroesophageal reflux disease, and infection

epidemiology—the study of the rates of disease within a given population

epinephrine—a chemical made by the adrenal gland that relaxes smooth muscles and constricts blood vessels; when it is used to treat severe allergic reactions, it is sometimes referred to as adrenaline

etiology—the cause of disease

exercise-induced allergic syndrome—an allergic reaction that occurs when a food allergen is consumed in combination with physical activity

exotoxins—toxins produced by bacteria

exudate—fluid produced by and released from cells that are inflamed and/or injured

fibrin—a filamentous protein; for blood clotting to occur, fibrinogen must be converted to fibrin

first-set rejection—rejection of a foreign tissue graft due to antibodies and activated cells formed in response to the graft; usually occurs one to two weeks after the tissue is transplanted

food-induced eczema—atopic dermatitis; a chronic skin disorder that usually begins in infancy and causes itching; many children with this diagnosis also suffer from other allergic conditions such as asthma and rhinitis

- *atopic*—refers to a milder IgE-mediated allergic response

GALT (gut-associated lymphatic tissue)—lymphoid tissue including Peyer's patches, the appendix, and solitary lymph nodes in the submucosa

gamma globulins—a group of serum proteins, including most antibody molecules, that migrate fastest toward the cathode during electrophoresis

graft-versus-host (GVH) rejection—a life-threatening reaction in which transplanted immunocompetent cells, usually T cells, attack the tissues of the immunocompromised recipient

hapten—a nonimmunogenic, low-molecular weight molecule that can be recognized by an antibody; it can initiate an immune response if it is conjugated to a "carrier" molecule

helper T cells (TH)—a subset of T cells that triggers B cells to make antibodies, activates macrophages, and promotes the differentiation of other T cells

hematopoietic stem cell—an undifferentiated bone marrow cell that is a precursor for multiple cell types; also called pluripotential stem cells

histamine—a vasoactive amine that contributes to inflammation and IgE-mediated allergic reaction by causing the dilation of local blood vessels and smooth muscle contraction; histamine release produces some of the symptoms of immediate hypersensitivity reactions

hives—an itchy skin condition with raised red lumps, often due to an allergic reaction; also called *urticaria*

humoral immunity—immunity due to soluble factors such as antibodies circulating in the body's fluids, mainly serum and lymph; "humors" is an old term for body fluids

hyaline—a histological term used to describe tissue injury that has a glassy, pink appearance

hyperemia—increased blood flow to a body tissue

hyperplasia—increased number of cells

hypersensitivity—an inappropriate and harmful immune reaction to a harmless, nonpathogenic substance; also called *allergy*

hypertrophy—increase in cell size

IgA (immunoglobulin A)—the predominant immunoglobulin in secretions

IgD (immunoglobulin D)—an immunoglobulin present on the surfaces of B cells

IgE (immunoglobulin E)—the immunoglobulin class that is the predominant mediator of immediate hypersensitivity reactions (allergies)

IgG (immunoglobulin G)—the predominant immunoglobulin class produced during secondary immune responses; the most prevalent immunoglobulin in the blood

IgM (immunoglobulin M)—the predominant immunoglobulin class expressed by virgin B lymphocytes and secreted during primary immune responses

IL-2—interleukin-2; a lymphokine required by activated T cells for growth

immediate hypersensitivity—a hypersensitivity reaction that appears within minutes after the exposure to the allergen

immune complex—a cluster of antibodies bound to antigens

immunodeficiency—decrease in or lack of an immune response due to absence or defect of one or more components of the immune system

immunogen—an antigen capable of inducing an immune response because it is foreign to the host

innate immunity—immune response resulting from natural barriers and resistance that are present at birth

interferon (INF)—a group of cytokines that regulate the immune system and protect cells from viruses

interleukin—now used primarily as a naming convention for cytokines/lymphokines/chemokines/growth factors (IL-number)

ischemia—inadequate supply of oxygen

isograft—tissue transplanted between two genetically identical individuals; also called *syngraft*

jaundice—a symptom that occurs when excessive bilirubin accumulates in the bloodstream, causing body tissues to become tinted yellow

Langerhans cell—dendritic cell that traps and processes antigens in the epidermal layer of the skin and then migrates through lymphatics to lymph nodes where it presents the antigen to T cells

leukotrienes—metabolic products of arachidonic acid that promote inflammation

lymph—extracellular fluid containing white blood cells (mostly lymphocytes) and antibodies that bathe tissues

lymphatic system—a system of vessels through which lymph travels, consisting of lymphatic vessels and lymph nodes at the intersection of vessels

lymph nodes—small organs of the immune system where mature B and T lymphocytes respond to an antigen; they are distributed widely throughout the body and linked by lymphatic vessels that bring in antigens from surrounding tissue

lymphocyte—a small mononuclear cell with a thin rim of cytoplasm that has antigen-specific receptors

lymphokine—a soluble molecule used for communication between lymphocytes and other cells

lysosomes—cytoplasmic granules that contain hydrolytic enzymes and are involved in the digestion of phagocytosed material

macrophage—a large phagocytic antigen-presenting cell derived from the blood monocyte and found in tissues

major histocompatibility complex (MHC)—a cluster of genes encoding polymorphic cell-surface molecules (MHC class I and class II) that help the organism identify pathogens as foreign; they are important in antigen presentation to T cells, play a role in transplantation rejection, and influence the susceptibility to certain autoimmune diseases; MHC antigens are also called *HLA antigens*

  · *Class I MHC antigen*—glycoproteins found on nucleated cells and encoded by the A, B, and C locus of the major histocompatibility complex; they present antigens to cytotoxic (CD8 +) T cells
  · *Class II MHC antigen*—glycoproteins found on nucleated cells and encoded by the Dr, Dq, or DP locus of the major histocompatibility complex; they present antigens to helper (CD4 +) T cells

MALT (mucosa-associated lymphatic tissue)—lymphoid tissue found in the surface mucosa of the respiratory, gastrointestinal, and genitourinary tracts

mast cell—a tissue cell found primarily in mucosal and connective tissue that is similar to the basophil (which is found in blood)

membrane attack complex—the final product of the complement cascade that forms a pore on the surface of the target cell, which results in lysis of the cell

memory cells—lymphocytes produced on the first encounter with an antigen that produce a rapid, more vigorous response upon subsequent exposures, which often prevents reinfection

metaplasia—replacement of one cell type with another

minor histocompatibility antigens—cell surface-processed peptides not encoded by the MHC that can contribute to graft rejection

monoclonal antibody—an antibody produced by an immortal B cell line that reacts with a single antigenic determinant (a specific part of an immunogen that stimulates a specific immune response)

monocyte—a large, mononuclear, phagocytic white blood cell that develops into a macrophage when it enters tissue

morbidity—the state of being diseased

mortality—the incidence of death in a population

natural killer cells (NK cells or K cells)—large granular lymphocyte cells that attack tumors and virally infected cells but do not exhibit antigenic specificity; also called *killer cells (K cells)* and *null cells*

necrosis—general term referring to cell death

neutrophil—the most numerous type of polymorphonuclear leukocyte, with granules that stain with acid and basic dyes; it is phagocytic and enters tissues early in inflammation

nonspecific immune system—all aspects of immunity not directly mediated by antigen-specific lymphocytes

oral allergy syndrome—food allergy symptoms of the mouth and pharynx, which usually occur within minutes of contact between the allergen and the oral mucosa

outcome—the measurable consequence of disease

passive immunity—immunity due to the transfer of antibodies or activated T cells produced by another individual

pathogenesis—the clinical course of disease

pathophysiology—the study of disease

Peyer's patches—distinct lymphoid nodules in the intestine that are part of the gut-associated lymphoid tissue (GALT)

phagocytosis—the engulfment of a particle or a microorganism by leukocytes such as macrophages and neutrophils, normally followed by destruction of the particle

plasma cells—large antibody-producing cells that develop from activated B cells; also called *antigen-forming cells (AFC)*

polymorphonuclear leukocytes (PMN)—leukocytes with a multilobed nucleus and cytoplasmic granules that take up acid and basic dyes; also known as *granulocytes, PMNs,* and *polys*

primary immune response—the immune response that occurs when the naive lymphocyte first encounters its antigen

privileged sites—nonvascularized locations in the body where foreign grafts are not rejected

prognosis—expected outcome; expected response to treatment

secondary immune response—rapid, more vigorous immunologic response by memory lymphocytes after the first encounter with an antigen; produced upon subsequent exposures to the antigen; often prevents reinfection

second-set rejection—accelerated rejection of an allograft due to previous exposure to some of the antigens on the graft

serum sickness—a Type III hypersensitivity response following the administration of a passive antibody in foreign serum

severe combined immune deficiency (SCID)—disease due to several mechanisms that produce an early block in differentiation pathways of both B and T lymphocytes, resulting in infants who are born lacking all major immune defenses

signs—observable phenomena such as heart or respiratory rate

specific immune system—body system responsible for immunity mediated by antigen-specific lymphocytes

spleen—a lymphoid organ in the abdominal cavity that filters blood

sterilization—a process that destroys all living organisms

symptoms—complaints experienced/verbalized by a patient

T cells—lymphocytes that differentiate in the thymus
- *suppressor T cell*—a T lymphocyte that suppresses (turns off) specific immune responses; this may or may not be a separate subclass of T cells

Th1—a subset of the T helper cells that secretes cytokines, which trigger cell-mediated immune responses that promote inflammation and antiviral responses

Th2—helper T cells that predominate in the response to allergens and parasites and that make cytokines that promote antibody responses

thymus—a primary lymphoid organ located in the chest, where T lymphocytes differentiate, proliferate, and are positively and negatively selected
- *positive selection*—the rescue from apoptosis of T cells in the thymus that can recognize self-MHC molecules
- *negative selection*—the process in which B and T cells that react to self molecules are deleted or functionally inactivated during their development

transplantation—grafting an organ (e.g., kidney or heart) or cells (e.g., bone marrow) from one individual to another

tumor necrosis factor (TNF)—a cytokine that induces programmed cell death, primarily in tumor cells but for any cell with a receptor; also involved in immunoregulation

vaccine—a substance made from the whole organism or parts that contain critical antigenic components or genes for those components; it stimulates a primary immune response that produces antibodies and memory cells that protect against subsequent infection by that organism

vasomotor—referring to nerves that innervate smooth muscles in the walls of arteries and veins and can cause their constriction or dilation

xenograft—tissue transplantation between individuals from different species

As you will learn in this chapter, the immune system is not always beneficial. Processes involved in countering organisms that cause infectious disease can damage tissues either as part of the response to a pathogen or when directed at a harmless target, as in allergic reactions or autoimmune disease. Symptoms associated with an infectious disease are often partially or totally caused by the immune response.

Defining Disease and Pathophysiology

Disease is defined as a process that interferes with or disrupts the body's normal function. The human body strives to maintain a delicate balance among body systems and processes, and it is quite efficient at doing so. But disease or injury leads to a state where that balance is interrupted and homeostasis cannot be maintained. **Pathophysiology** is the study of the disruption of normal physiologic processes. Pathophysiology includes understanding structural changes that occur as a result of disease or injury as well as the clinical course that follows regulatory, metabolic, and structural changes. The clinical course includes impact and duration of the disease, and is monitored throughout diagnoses and conditions. **Pathogenesis** is the clinical course of the disease. Understanding disease mechanisms provides the basis for developing treatment.

Disease Process: Epidemiology, Etiology, Pathogenesis, Clinical Manifestations, Outcome

It is common to organize the study of disease process by analyzing patterns of disease occurrence through epidemiology. **Epidemiology** is the study of the distribution of disease within populations. Epidemiology provides data for outcome measures such as **morbidity** and **mortality**, and identifies risk factors associated with disease. Epidemiology commonly provides the first hypotheses for determining disease etiology.

Etiology is the description and identification of the cause of disease. Etiology can be narrowed to a specific causative agent or may be a combination of factors that influence or change disease development. Influential factors could be age, nutrition-

al status, or other coexisting diseases. Etiology of disease can be categorized as *genetic, acquired, multifactorial, idiopathic,* or *iatrogenic.*

Diseases of genetic origin are those that develop from abnormalities in the genetic control of cellular development. These disorders may either be congenital (present at birth) or noncongenital (becoming evident later in life). Examples of genetic diseases include cystic fibrosis, sickle cell anemia, and hemophilia. Acquired disease is one that originates from exposure to environmental agents. Infectious disease may be classified as an acquired disease.

Many disease processes involving nutrition therapy are those considered to be multifactorial. These etiologies include a combination of factors including genetic, environmental, and infectious. Examples include atherosclerosis, osteoporosis, and diabetes mellitus.

An etiology is considered to be idiopathic if the origin is unknown. An iatrogenic disease or complication is any illness or symptom resulting from a medical intervention, treatment, procedure, or error.

Clinical manifestations of disease are evident as cellular injury moves toward systemic changes. You are probably most familiar with this level of discussion of disease as it involves signs, symptoms, and the measurement of laboratory abnormalities. **Signs** are observable phenomena that can be verified. Signs are measurable and include factors such as blood pressure, respiratory rate, weight, or body temperature. Many signs will be noted in the physical examination. **Symptoms** are those factors verbalized by the patient (and/or caregiver) and are frequently noted by the chief complaint in the physician's history and physical. Symptoms are subjective and dependent on the ability of the patient and caregiver to report this information accurately. Laboratory measurements of body fluids and tissue reflect changes occurring from the disease process. Laboratory values falling outside the norm may include blood chemistries, urinalysis, or tissue biopsy. See Box 9.1 for more examples of clinical manifestations.

Outcome of disease is sometimes referred to as the *sequelae* of disease. **Prognosis** is considered to be the expected or usual outcome of the disease. Outcomes include cure, remission, development of chronic disease, or death. Generally, the disease is

Clinical Manifestations of Disease

Consider the following patient: A 35-year-old woman was admitted through the emergency room. She was febrile to 105°F and complained of chest pain and dyspnea. Further evaluation of her condition indicated abnormally elevated electrolytes, a chest x-ray that showed areas of infiltrate in the lower left lobe of the lung, and a positive blood culture for legionella pneumonia.

Can you determine which of these descriptions would be signs, symptoms, or laboratory measurements? Symptoms include the patient complaints of difficulty breathing (dyspnea) and chest pains. She most probably would describe the symptom of fever. Signs include the measurement of fever and the infiltrate viewed on chest X-ray. The physical examination would also find signs of shortness of breath with changes in respiratory rate or the signs of fever including warm skin, dry mucosal membranes, or changes in heart rate. Laboratory measurements confirmed the etiology of disease with the blood culture and identified the abnormalities in electrolytes that might be consistent with dehydration.

resolved as a result of treatment or the patient's own defense—the outcome is a return to the patient's pre-illness state. Complications could potentially occur as a component of disease outcome. And, as mentioned before, an iatrogenic complication could arise from the treatment of the disease. For example, when a prescribed antibiotic causes diarrhea, nausea, or vomiting, these complications would be considered to be iatrogenic.

Cellular Injury

Practitioners may be much more comfortable discussing clinical manifestations of disease than they are describing the underlying events at the cellular level. But to truly understand pathophysiology, practitioners need to examine causes of cellular injury and cellular response to this injury. The more that is understood about this process, the more efficient and effective treatment can become. Cellular injury may result from physical injury to the cell (trauma), a deficiency of a necessary substance required for cellular function, or interruption of normal cellular processes after exposure to a toxin.[1] The cell's response to injury can include changes in cell growth, inflammation and healing, and/or cell death.

Mechanisms of Cellular Injury

Individual cells can be harmed as a result of hypoxia (a deficiency of oxygen), nutritional imbalances, or microbiological agents. Processes that interfere with cell function include damage by free radicals, physical agents, immunologic reactions, nutritional imbalances, and genetic defects. Mechanisms that damage the structure of the cell can include physical and microbiological agents, immunologic reactions, genetic defects, and nutritional imbalances.

Hypoxia is a common reason for cell injury. An insufficient oxygen supply (commonly called **ischemia**) will interfere with cellular metabolism. This is easily demonstrated when you consider the process of a myocardial infarction (heart attack), where a reduction in the oxygen supply to a portion of the heart causes death of those cardiac muscle cells.

A free radical is an atom or group of atoms with a single unpaired electron. Free radicals are chemically unstable and are searching for additional electrons. In their search, free radicals can damage cell membranes or alter DNA, resulting in cellular injuries that interrupt normal function of the cell. An example of a free radical is the reactive oxygen species found in air pollution or produced during the inflammation process.

Physical agents such as ethanol, poisons, or lead and other heavy metals can cause cell damage through several different mechanisms such as oxygen deprivation (hypoxia) or destruction of cell membranes. Some are fast acting, like poisons, while others exert more gradual effects over many years, as seen in neurological changes after lead exposure. Other categories of physical injury include effects of burns, radioactive radiation, or actual trauma to the cell such as that seen in wounds and tissue destruction. Intoxication, where toxic by-products accumulate, may inadvertently arise from genetic abnormalities and lead to systemic changes. This process occurs in the metabolic disorder phenylketonuria (PKU), which prevents the normal metabolism of the amino acid phenylalanine.

Cellular Response to Injury

The cell's response to injury depends on its ability to react, adapt, and repair after being injured. The response may be temporary and completely reversible. But in some situations, the cell is permanently damaged and is no longer functional. Cell responses may include inappropriate accumulation of substances within the cell; changes in cell size, number, or shape; and the inflammatory response.

Cellular Accumulations Injury and disease can result in excessive accumulation of water, lipids, proteins, pigments, and minerals within the cell, which disrupts normal metabolism. For example, when a cell is unable to produce adequate ATP to maintain transport of sodium and potassium, fluid will shift, causing a disruption in fluid balance. In some diseases, triglycerides can accumulate in cells of the liver, heart, and pancreas. Alcoholic liver disease, hepatitis, or carbon tetrachloride poisoning can result in abnormal triglyceride deposits (fatty liver). The abnormal fat and protein deposits within arterial walls of the circulatory system characterize the common chronic disease atherosclerosis.

A common cellular response to injury and disease is deposition of a substance called **hyaline** within and between the cells. Most hyaline deposits are a mixture of different types of proteins such as **collagen**, **fibrin**, and **amyloid**. Hyaline deposits can be found in many different cell types including neurons, hepatocytes, and cells in damaged arteries. The predominant type of protein within the hyaline accumulation can vary depending on the individual disease process. Fibrin masses are found in inflammatory conditions, and collagen predominates in scar tissue. Amyloidosis is a condition where amyloid is deposited in soft tissues, eventually resulting in cell death and organ dysfunction. Amyloid deposits have also been identified in the brains of patients who died with Alzheimer's disease.

Pigment accumulation after cell injury can include melanin and derivatives of hemoproteins. Hemosiderin is a yellow-brown pigment produced when hemoglobin is broken down—for example, when you experience a bruise. Bruised skin first appears red-blue, and then breakdown of the red blood cells occurs, causing hemoglobin to be transformed to hemosiderin. Accumulation of this pigment is what causes a bruise to turn the common yellowish green. Bilirubin, another yellow-green pigment, is found in bile. In some disease conditions, excess bilirubin accumulates in the body, causing the symptom of **jaundice**.

Cellular Alterations in Size and Number

Changes in cell size are both common and a central component of many disease states (see Figure 9.1). Cells respond to variations of hormonal or neurological stimulation by alteration in their size. The cell can increase in number, increase in size, shrink, or have additional functional changes.

Atrophy results from a decrease in cell size. Decreased workload, loss of innervation, diminished blood supply, inadequate nutrition, loss of hormonal stimulation, and aging all contribute to atrophy. A common example of this change is immobility of skeletal muscle. Prolonged bed rest or disuse due to fracture will result in loss of skeletal muscle mass. This can also occur from the lack of neurological stimulation that could be seen in a spinal cord injury.

Hypertrophy is defined as an increase in cell size. This cellular response is prominent in cells that are unable to undergo cell division. Hypertrophy may involve increased synthesis of structural proteins or increased size of organelles. Normal physiologic cellular hypertrophy is demonstrated by increase in skeletal muscle after resistance exercise. Pathologic hypertrophy is seen in valvular heart disease, interstitial lung disease, or even in tonsillitis associated with infectious mononucleosis.

In **hyperplasia** (see Figure 9.2a), there is an increase in overall cell number. Various types of anemia, for example, result in an increase of the total number of erythrocytes. In **Cushing's syndrome**, there is hyperplasia of the adrenal glands due to excess stimulation of cortisol. After partial resection of the liver (hepatectomy), there is regeneration of hepatocytes to accommodate resulting changes from surgery.

Metaplasia occurs when disease or injury results in displacement of one cell type for another that may be less mature. This is classically seen in vitamin A deficiency, since vitamin A is necessary for cellular differentiation. In deficiency, more mature, functioning cells are replaced with less mature cells. Other examples of metaplasia occur in *Helicobacter pylori* infection or in Barrett's esophagus, a premalignant condition of the esophagus (see Figure 9.2b).

Dysplasia is deranged cellular growth and can result in abnormalities of size, shape, or function. Most often, dysplastic changes are considered to be precancerous. For example, dysplastic cells are often noted in an abnormal pap smear (a routine screening test for cervical cancer; see Figure 9.2c).

Cellular Injury from Infection

Cellular injury can also occur as a result of invasion from microorganisms. This

Figure 9.1 Alterations in Cell Size

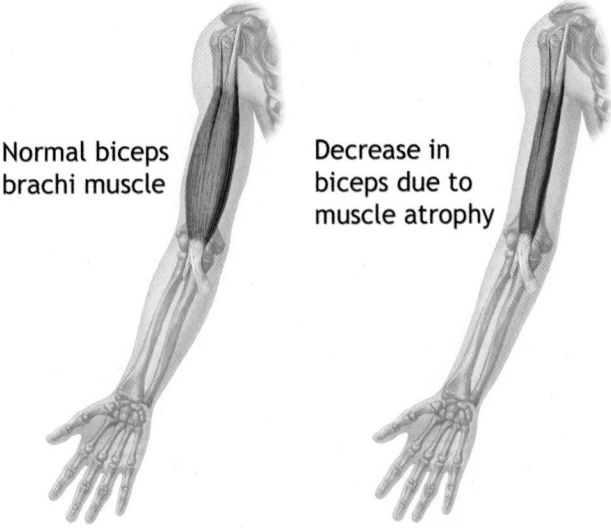

Normal biceps brachi muscle

Decrease in biceps due to muscle atrophy

a. Atrophy: cell wasting

b. Hypertrophy: an increase in size of a cell; can be induced by a number of stimuli

Source: (a) A.D.A.M., Inc., (b) copyright © Fabio Cardoso/Corbis.

presence of microorganisms is most commonly referred to as infection. In order for an infection to occur, three contributing factors must be in place: (1) the pathogen must be present, (2) the host must be susceptible to the infection, and (3) the environment must be conducive for the pathogen to thrive and proliferate.

Figure 9.2 Alterations in Cell Number, Type, or Growth

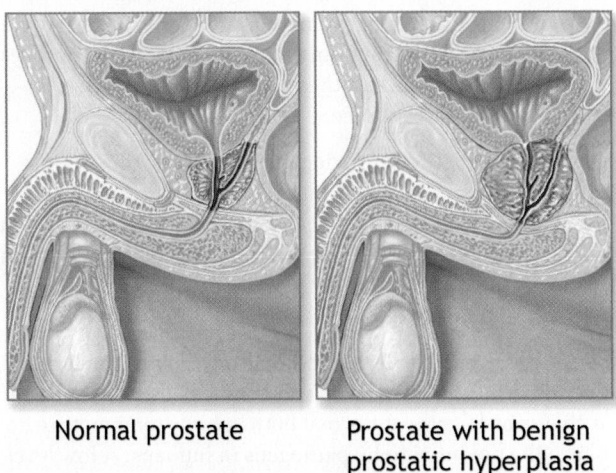

Normal prostate   Prostate with benign prostatic hyperplasia

a. Hyperplasia: increase cell production in normal tissue or an organ

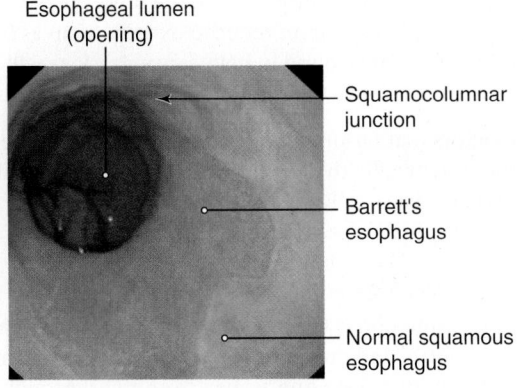

Esophageal lumen (opening)

Squamocolumnar junction

Barrett's esophagus

Normal squamous esophagus

b. Metaplasia: transformation of one mature cell type into another mature cell type as an adaptive response to insult or injury

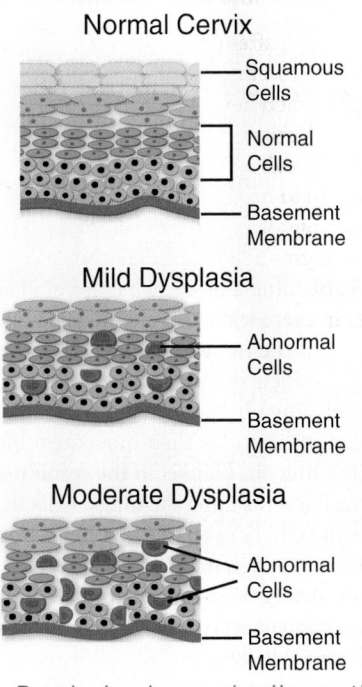

Normal Cervix

Squamous Cells

Normal Cells

Basement Membrane

Mild Dysplasia

Abnormal Cells

Basement Membrane

Moderate Dysplasia

Abnormal Cells

Basement Membrane

c. Dysplasia: abnormal cell growth

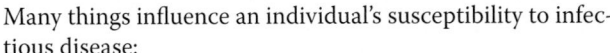

Source: (a) A.D.A.M., Inc., (b) copyright © www.gastrolab.com, (c) MJ Bovo, M.D. Newport Media Concepts, Inc.

Many things influence an individual's susceptibility to infectious disease:

- Gender. Although gender sometimes plays a role, in many cases the underlying mechanism is differential exposure due to occupational and recreational activities.

- Age. The immune system takes time to develop; therefore, the young do not have the full spectrum of immunological defenses available to the adult. As humans grow older, several immune mechanisms decrease, including secretion of mucous and sebaceous glands and the production of cytokines, including an interferon. However, natural killer cells that attack infected cells and tumor cells increase.

- Nutritional status. Malnutrition is a major cause of **immunodeficiency**. Research has suggested that certain nutrients may play a role in supporting the immune response.[2, 3]

- Physical Activity. Intensive physical training may lead to periods of depressed immune response and lead to increased risk of upper respiratory tract infections.[4]

- Hormones. Levels of various hormones play a role in an individual's susceptibility to infectious disease. Individuals with diabetes have an increased risk of fungal and staphylococcal infections, while women with low estrogen have a higher vaginal pH and thus are more susceptible to vaginal infections.

- Stress. Stress activates the fight-or-flight response, resulting in several physiological changes that impact the immune response.[5]

TYPES OF MICROORGANISMS Microorganisms involved in human disease include bacteria, fungi, helminth, protozoa, prions, and viruses. Microbes are more likely to result in infection if they produce **exotoxins** and **endotoxins**, produce destructive enzymes, produce spores, or develop a bacterial capsule. For example, *Clostridium difficile* produces exotoxins, *E. coli* produces endotoxins, and *Streptococcus pneumoniae* exerts its effect by producing destructive enzymes. Any of these characteristics increase pathogenicity of the microorganism.

HOST RESISTANCE In order for infection to occur, the host has to be susceptible to the infection. Susceptibility occurs when there is a break in the host's defense mechanism or in host resistance. The first line of defense against infection is the host's intact skin and mucous membranes. These provide not only a physical barrier against infection and injury but the first chemical response (see Figure 9.3). Intact epithelial surfaces such as skin and the lining of the body's tubular structures such as the gastrointestinal, respiratory, and genitourinary tracts are excellent barriers to most pathogens.

In addition to providing a physical barrier, skin and mucous membrane components produce chemical barriers, such as the lysozyme produced by sweat glands. This enzyme damages peptidoglycan, a critical component of bacterial cell walls. The surface of mucous membranes can glue or trap microorganisms so they cannot continue their movement into the body. For example, the mucous blanket in the respiratory tract can keep organisms from reaching the lungs. Cilia in respiratory mucosa create the ciliary escalator that helps bring the organisms, which are trapped in mucus, to the surface so they can be coughed out. An increased number of respiratory infections are noted in individuals who damage their respiratory cilia by smoking or abusing alcohol.[6]

Body secretions such as saliva, tears, and gastric juices also protect the host. Tears, urine, and saliva serve to wash organisms out of the eyes, the genitourinary tract, and the mouth. Urine is also acidic, and tears and saliva contain lysozyme and a number of other protective enzymes. Saliva produced in the mouth is one of the first protection mechanisms against microorganisms entering the gastrointestinal tract. The presence of lysozomes in saliva provides a nonspecific form of immunity against invad-

ing microorganisms. The pH of the stomach protects humans from many of the organisms taken in by mouth, since very few can survive in the acid environment. Digestive enzymes in the upper GI tract also destroy some microorganisms. Any process that reduces the presence of these body secretions would decrease normal host resistance.

The skin's epithelial cells provide physical protection against injury and prevent easy entry into the body. If there is a break in the skin barrier, as might be caused by a splinter in a person's finger, there is a direct route for microorganisms to enter. Skin additionally has a slightly acidic pH that provides a chemical barrier against microorganisms.

For a microorganism to grow and multiply, environmental temperature must be within its viable range. Therefore, organisms that cannot grow at normal human body temperature do not have the potential to be pathogens in humans. A low level of fever is beneficial, since it enhances the action of the immune system. However, if an individual's temperature rises too high, brain damage can result.

Anaerobic pathogens cannot grow in the presence of oxygen, and microaerophiles require reduced oxygen, so areas in the body where oxygen is found in high concentrations will not provide a good growth environment for these organisms.

Other factors that ensure host resistance include an effective immune system, effective inflammatory response, and absence of underlying disease. These factors will be discussed in greater detail later in this chapter.

PREVENTING TRANSMISSION OF INFECTION Preventing transmission is the basis of the clinical approach for infection control and prevention of disease. Microorganisms can be transmitted from one human to another through contact with blood and body fluids, as seen in the transmission of hepatitis B (HBV). Contact with respiratory droplets is a common mode of transmission, especially for upper respiratory viruses and tuberculosis. Many food-borne illnesses—such as hepatitis A (HAV) or *E. coli*—are transmitted via fecal-oral spread of microorganisms. Finally, microorganisms can be transmitted across the placenta from mother to fetus. This is what occurs in transmission from an HIV-positive mother to her child.

Public health and medical systems attempt to protect both the individual and the health care provider from infection by identifying sources and contacts when infection does occur. "Standard precautions" are the set of guidelines or procedures developed by the Centers for Disease Control and Prevention that include all personal and environmental procedures that should be followed to prevent any transmission of infection.[7] Table 9.1 summarizes the basic guidelines for standard precautions.

Eliminating a favorable environment includes not only preventing transmission but also interrupting conditions that would allow microorganisms to thrive. This is frequently accomplished through use of disinfectants, antiseptics, and sterilization. **Disinfectants** are chemical and physical agents applied to inanimate objects to kill any microbes. Applying a bleach solution to clean kitchen counters is a typical use of a disinfectant. **Antiseptics** are agents that kill microbes within living tissue. When drawing blood from a patient, the technologist often

Figure 9.3 Components of Natural Resistance

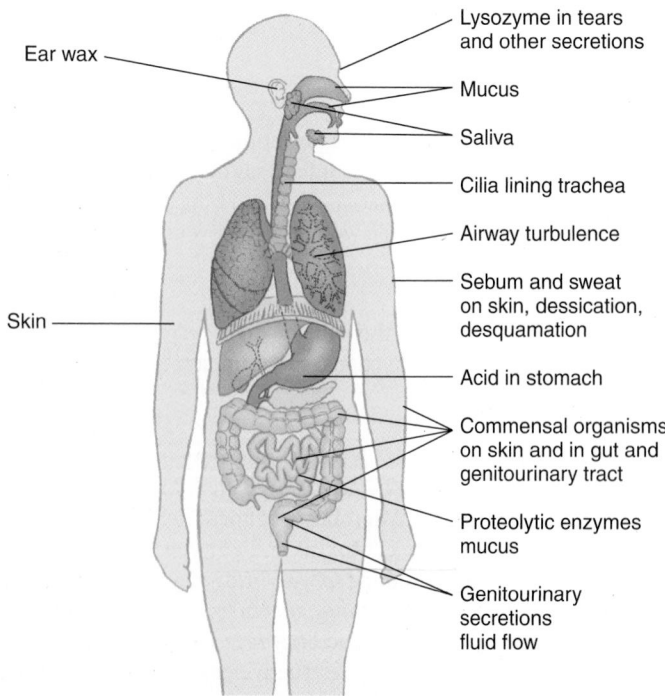

- Lysozyme in tears and other secretions
- Ear wax
- Mucus
- Saliva
- Cilia lining trachea
- Airway turbulence
- Sebum and sweat on skin, dessication, desquamation
- Skin
- Acid in stomach
- Commensal organisms on skin and in gut and genitourinary tract
- Proteolytic enzymes mucus
- Genitourinary secretions fluid flow

Source: Rhoades/Pflanzer, *Human Physiology*, 4e, copyright © 2003, p. 857.

Table 9.1 Standard Precautions

Personal Protective Equipment

Should be used at all times to prevent skin and mucous membrane contamination with blood, body fluids containing visible blood, or other body fluids (cerebrospinal, synovial, pleural, peritoneal, pericardial, and amniotic fluids, semen, and vaginal secretions). The type of barrier protection used should be appropriate for the type of procedures being performed and the type of exposure anticipated. Examples of barrier protection include disposable lab coats, gloves, and eye and face protection.

Gowns

Wear a gown that is appropriate to the task to protect skin and prevent soiling or contamination of clothing during procedures and patient-care activities when contact with blood, body fluids, secretions, or excretions is anticipated. Remove gown and perform hand hygiene before leaving the patient's environment. Do not reuse gowns, even for repeated contacts with the same patient.

Gloves

Are to be worn when there is potential for hand or skin contact with blood, other potentially infectious material, or items and surfaces contaminated with these materials. Remove gloves after contact with a patient and/or the surrounding environment (including medical equipment), using proper technique to prevent hand contamination. Do not wear the same pair of gloves for the care of more than one patient. Do not wash gloves for the purpose of reuse since this practice has been associated with transmission of pathogens.

Face Protection

Use to protect the mucous membranes of the eyes, nose, and mouth during procedures and patient-care activities that are likely to generate splashes or sprays of blood, body fluids, secretions, and excretions. Select masks, goggles, face shields, and combinations of each according to the need anticipated by the task to be performed.

Wash Hands or Other Skin Surfaces

Wash thoroughly and immediately if contaminated with blood, body fluids containing visible blood, or other body fluids to which standard precautions apply. Wash hands immediately after gloves are removed.

Avoid Accidental Injuries

Avoid injuries that can be caused by needles, scalpel blades, laboratory instruments, etc., when performing procedures, cleaning instruments, handling sharp instruments, and disposing of used needles, pipettes, etc.

Sharp Items

Place the following in puncture-resistant containers for disposal: used needles, disposable syringes, scalpel blades, pipettes, and other items marked with a biohazard symbol.

Source: Adapted from: USDHHS. Centers for Disease Control. Guideline for Isolation Precautions: Preventing Transmission of Infectious Agents in Healthcare Settings 2007.

BOX 9.2 CLINICAL APPLICATIONS

Infection Control: Care of the Environment

- Clean and disinfect surfaces that are likely to be contaminated with pathogens, including those that are in close proximity to the patient (e.g., bed rails, over-bed tables) and frequently-touched surfaces in the patient care environment (e.g., door knobs, surfaces in and surrounding toilets in patients' rooms) on a more frequent schedule compared to that for other surfaces (e.g., horizontal surfaces in waiting rooms).

- Use EPA-registered disinfectants that have killing activity against the pathogens most likely to contaminate the patient-care environment. Use in accordance with manufacturer's instructions.

- Review the efficacy of in-use disinfectants when evidence of continuing transmission of an infectious agent (e.g., rotavirus, *C. difficile*) may indicate resistance to the in-use product and change to a more effective disinfectant as indicated.

- In facilities that provide health care to pediatric patients or have waiting areas with child play toys (e.g., obstetric/gynecology offices and clinics), establish policies and procedures for cleaning and disinfecting toys at regular intervals. Use the following principles in developing these policies and procedures:

  - Select play toys that can be easily cleaned and disinfected
  - Do not permit use of stuffed furry toys if they will be shared
  - Clean and disinfect large stationary toys (e.g., climbing equipment) at least weekly and whenever visibly soiled
  - If toys are likely to be mouthed, rinse with water after disinfection; alternatively, wash in a dishwasher
  - When a toy requires cleaning and disinfection, do so immediately or store in a designated labeled container separate from toys that are clean and ready for use

USDHHS. Centers for Disease Control and Prevention. Standard Precautions. IN: Guideline for Isolation Precautions: Preventing Transmission of Infectious Agents in Healthcare Settings 2007. Available at http://www.cdc.gov/ncidod/dhqp/gl_isolation_standard.html

---

swabs the skin with either alcohol or **betadine**. This is a common example of antiseptic use. Finally, **sterilization** destroys all microbes. Sterilization is accomplished by use of heat, chemicals, or radiation. For example, microorganisms within foods may be destroyed by exposure to ionizing radiation or the high temperatures used during the canning process. The current recommendations for controlling environmental infection risks are summarized in Box 9.2.

COURSE OF INFECTION When an infected patient is evaluated, particular symptoms and signs will correlate with the stage of infection. Stages of infection include incubation, prodromal, acute, and recovery/convalescence. The incubation period is the time between entry of the microorganism and appearance of any clinical signs or symptoms. The prodromal period is when the individual begins to experience the first, even vague, symptoms of infection. During the acute period of infection, the disease will fully develop. Clinical manifestations of the disease will be noted. Finally, as symptoms begin to resolve and signs subside, recovery and convalescence will proceed. Box 9.3 describes this clinical scenario. Unfortunately, in some situations, acute infection progresses to a chronic infection.

Stages of Infection

In the following example, see if you can identify the correlating stages of infection: Missy has spent the last several days traveling with the golf team on the team bus. Her roommate, who is also on the golf team, has been sick with "tonsillitis." As Missy prepares for the last leg of this tournament, she notices she is more tired than usual and is not playing as well as she did in earlier tournaments. That night she wakes with a fever and severe sore throat. On the second day, her coach sends her to the health center. She tests positive for strep throat. She immediately starts on antibiotics, and by the next morning, she feels much better, has no fever, and is able to return to class. By the next weekend, she feels "back to normal" and shoots an 84 in the next golf tournament.

Missy's travel with her sick roommate represents the incubation period of illness. Her fatigue and decreased performance are consistent with the prodromal stage. The presence of fever and sore throat occur during the acute stages of the illness. Finally, after treatment, her fever subsides and within the week she is playing well: recovery and convalescence.

Cellular Death If the injurious agent or disease process is severe enough or continued for long enough, the cell will reach a point where adaptation can no longer occur. Compensation is not possible and cellular metabolism ceases. There are noticeable structural changes in an injured cell, but as a cell approaches death, distortions become more significant. Increased membrane permeability allows contents of the cell to spill out, cell structures such as mitochondria and the endoplasmic reticulum will be grossly altered, cell enzymes (**lysosomes**) are released to begin cellular destruction, and the nucleus is permanently damaged.

Necrosis refers to the cellular changes that occur during cell death. Different types of necrosis will occur in different organs or tissues and often will indicate the cause of cell death. Coagulation necrosis, caseous necrosis, gangrenous necrosis, and liquefaction necrosis are all specific terms describing the process of cell death.

Apoptosis is a different pattern of cell death. This form of cell death appears to be genetically programmed, which allows for removal of cells in a systematic, orderly fashion. For example, this is the process that organizes removal of cells after inflammation. Likewise, when a woman stops breastfeeding, apoptosis directs actions that clear cells that are no longer needed from the breasts.

Foundations of the Immune System

As stated previously, the first line of defense against disease and injury is our system of natural defense or host resistance. But if these barrier forms of host resistance are insufficient to protect the body, our next defense is dependent on the actions of our immune system. The immune system not only defends the body from pathogens but is also very active in homeostasis by helping the body to remove damaged and dead cells. It also functions in surveillance by recognizing abnormal cells, such as those infected by viruses and other pathogens. The surveillance function has been adapted to help the immune system identify and attack tumor cells, and it is also the underlying mechanism by which the immune system recognizes transplants and mounts a rejection response.

There are four basic requirements of an effective immune system: (1) specificity, the ability to react with one and only one **antigen**, which lowers the chance that a reaction to a pathogen will also harm the person; (2) diversity, so it can respond to many pathogens; (3) adaptivity, the ability to pick the best response to counter the pathogen; and (4) the ability to respond to stimuli not encountered previously. The current definition of an antigen is a structure that can combine with a cell of the immune system or an antibody but does not necessarily induce activation of the cell or formation of an antibody. An **immunogen** is an antigen that can induce an immune response. The key difference between an antigen and an immunogen is that the immunogen is foreign to the host whose immune system interacts with it.

The immune system is often described as falling into two major categories: **innate immunity** and **acquired immunity**. Innate or nonspecific immunity is the first line of defense against infectious agents, chemical irritants, and tissue injury. The cellular response to this type of injury is inflammation, discussed later within this chapter.

Acquired immunity relies on specific responses selectively targeted against a particular foreign material to which the body has already been exposed. The response consists of two phases: first, the immune system must recognize the antigen; secondly, it must mount a reaction to it. An antigen is generally a large protein or polysaccharide attached to foreign substances. Membranes of neoplastic cells can be antigens and the body's own cell membranes can also serve as self-antigens. The term **hapten** refers to an incomplete antigen which when combined with a larger molecule can elicit an immune response. An example of this occurs in an allergic response.

Organs of the Immune System

Organs of the immune system (see Figure 9.4) are classified as either central or peripheral. Central organs, where immune cells are produced, include bone marrow and the thymus. Peripheral organs, where adaptive immune responses are initiated, include the lymphatic system, spleen, **mucosa-associated lymphatic tissue (MALT)**, **bronchial-associated lymphatic tissue (BALT),** and **gut-associated lymphatic tissue (GALT)**.

The major organ of the immune system is the bone marrow. Bone marrow is located in the central core of long bones and in other bones such as the cranium. Before birth, the fetal liver acts like bone marrow. All cells in the immune system are formed in the **bone marrow**, where they mature to varied degrees and are released. Like all body cells, cells of the immune system originate from **stem cells**. Hemopoietic-inducing factors such as erythropoietin act on the stem cells to cause them

Figure 9.4 Organs of the Immune System

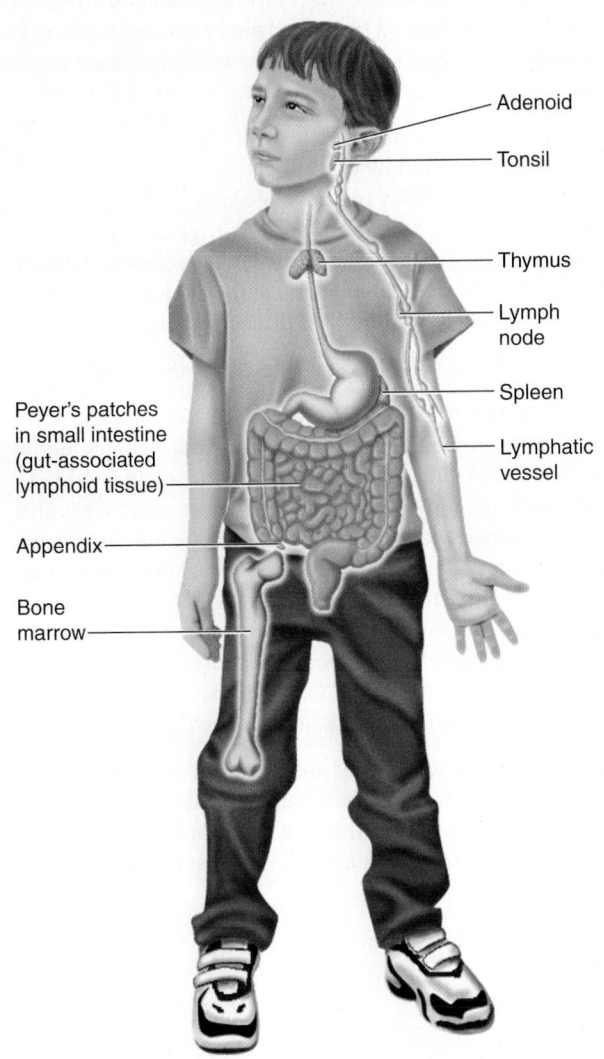

- Adenoid
- Tonsil
- Thymus
- Lymph node
- Spleen
- Lymphatic vessel
- Peyer's patches in small intestine (gut-associated lymphoid tissue)
- Appendix
- Bone marrow

Source: Lauralee Sherwood, *Human Physiology: From Cells to Systems*, 7e, copyright © 2010, Figure 12.1, p. 418.

to differentiate, proliferate, and eventually become red blood cells (RBC) or one of the cells of the immune system called leukocytes or white blood cells (WBC).

The **thymus**, a pouch of epithelial cells filled with lymphocytes, is located below the thyroid in the neck, above the heart. It weighs about 0.5 ounces at birth, grows to about 1 to 1.5 ounces at puberty, and atrophies to about 0.5 ounces by age 40. Although the thymus is very important early in life, removal of the thymus has minimal impact on the ability to respond to infections once humans reach their late teens. Removal of the thymus can be part of the treatment for the autoimmune disease myasthenia gravis.

The **lymphatic system** (see Figure 9.5) is an extensively branched network of walled vessels with one-way valves that lead to lymph nodes. Interstitial fluid, plasma that leaks out of blood vessels and carries nutrients and WBC, enters lymph vessels to become **lymph**. Antigens in intracellular spaces are swept into the lymphatic system by lymph and transported to lymphoid organs, where they encounter cells of the immune system.

Figure 9.5 The Lymphatic System

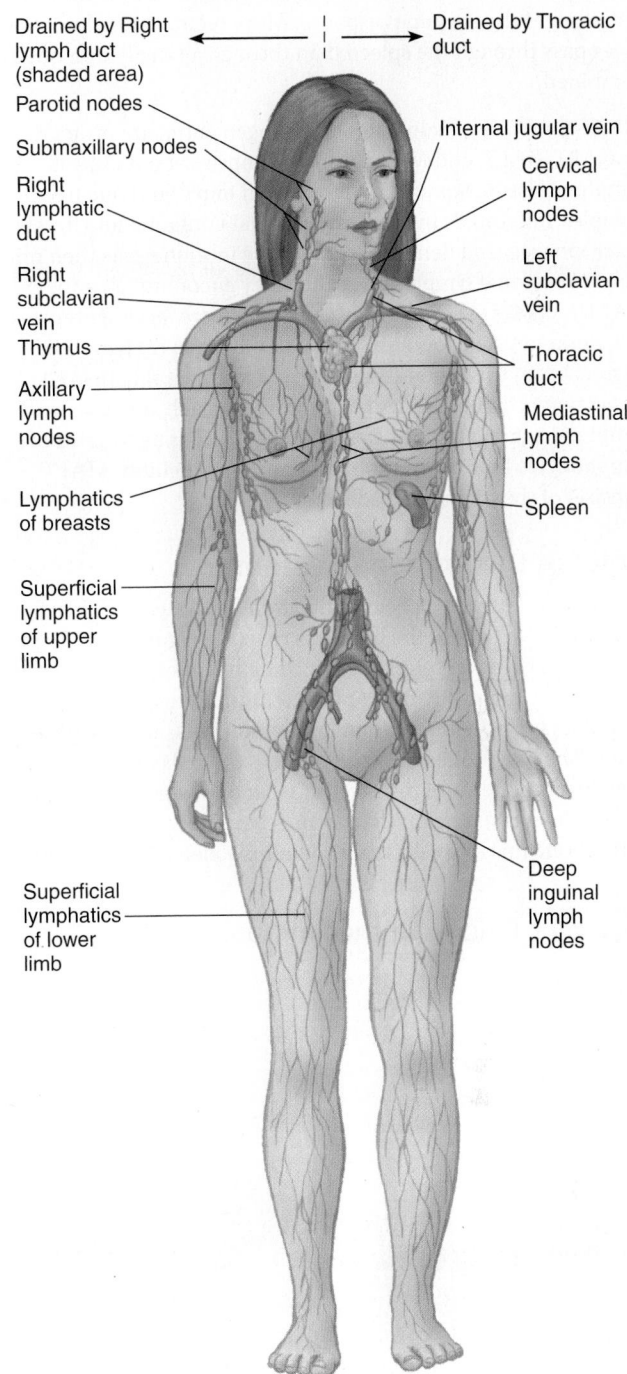

- Drained by Right lymph duct (shaded area)
- Drained by Thoracic duct
- Parotid nodes
- Submaxillary nodes
- Right lymphatic duct
- Right subclavian vein
- Thymus
- Axillary lymph nodes
- Lymphatics of breasts
- Superficial lymphatics of upper limb
- Superficial lymphatics of lower limb
- Internal jugular vein
- Cervical lymph nodes
- Left subclavian vein
- Thoracic duct
- Mediastinal lymph nodes
- Spleen
- Deep inguinal lymph nodes

Source: Rhoades/Pflanzer, *Human Physiology*, 4e, copyright © 2003, p. 852.

The major functions of the lymphatic system are to concentrate antigens from all over the body into a few lymphoid organs, to circulate lymphocytes through lymphoid organs in order to allow antigens to interact with antigen-specific cells, and to carry antibodies and other immune cells to the bloodstream and tissues. Lymph nodes are examined after cancer surgery for cancer cells in order to determine if any of them have spread (metastasized), because some cells leaving the tumor enter the lymph. Lymph nodes become swollen when an immune reaction occurs (some refer to them as swollen glands), but long-term swelling, lymphadenopathy, can be a sign that the immune system is not functioning properly.

The **spleen**, a fist-sized organ behind the stomach, is responsible for destruction of old RBC and is a major site for the initiation of the immune response. More recirculating lymphocytes pass through the spleen than through all the lymph nodes combined.

Mucosal surfaces, major sites of pathogen entry, are protected by GALT, BALT, and MALT, which comprise about 50% of lymphoid tissue. Some epithelial cells in mucosal tissue have complex microfolds in their surfaces that contain B and T cells, macrophages, and dendritic cells. These immune cells then migrate to regional lymph nodes after they encounter an antigen. GALT includes (1) **Peyer's patches**, which are large aggregates of lymphoid tissue found in the small intestine; (2) lymphoid aggregates in the appendix and large intestine; (3) lymphoid tissue that accumulates with age in the stomach; and (4) small lymphoid aggregates in the esophagus. BALT is aggregates of lymphocytes that protect the respiratory epithelium. MALT consists of the tonsils and adenoids.

Cells of the Immune System

As discussed earlier, all cells of the immune system originate from the stem cells of the bone marrow (see Figure 9.6). These immune cells, often referred to as white blood cells, are divided into three groups, the **macrophages/monocytes**, icrophages/granulocytes/**polymorphonuclear leukocytes**, and **lymphocytes**. The lymphoid stem cell produces the cells of the specific immune system—**T cells** (T lymphocytes) and **B cells** (B lymphocytes). The myeloid stem cell produces megakaryocytes (bone marrow cells that produce platelets) and cells of the

innate immune system, including macrophages/monocytes and polymorphonuclear leukocytes. **Natural killer cells (NK cells or K cells)** are produced from the lymphoid stem cell; however, they work as a component of the nonspecific or innate immune response.

White blood cells are not actually white, but they are colorless when compared to red blood cells and are found in a number of other tissues in the body besides the blood. Although there appear to be many more RBC than WBC in a blood smear, WBC outnumber RBC three to one in the body. A complete blood count (CBC) like that depicted in Figure 9.7 determines the percentage of each type of WBC discussed below.

Cells Derived from the Myeloid Stem Cell

MONOCYTES AND MACROPHAGES As shown in Figure 9.6 the mononuclear phagocyte system includes monocytes, which can differentiate into macrophages. In a stained blood smear, monocytes are larger than most other WBC and have approximately equal amounts of nucleus and cytoplasm. The nucleus may either be roughly circular or horseshoe-shaped. Cytoplasm is grayish in most common stains and appears to contain many little holes due to the presence of the vacuoles. Monocytes circulate in the blood, where they make up approximately 1% to 3% of the total WBC; then they migrate into tissues where they divide and differentiate into macrophages.

Monocytes and macrophages are highly specialized; they ingest and destroy bacteria, aged cells, and neoplastic cells in a process called **phagocytosis**. They are major **antigen-presenting cells (APC)** that can break down antigens into small pieces that

Figure 9.6 Origins of the Immune System

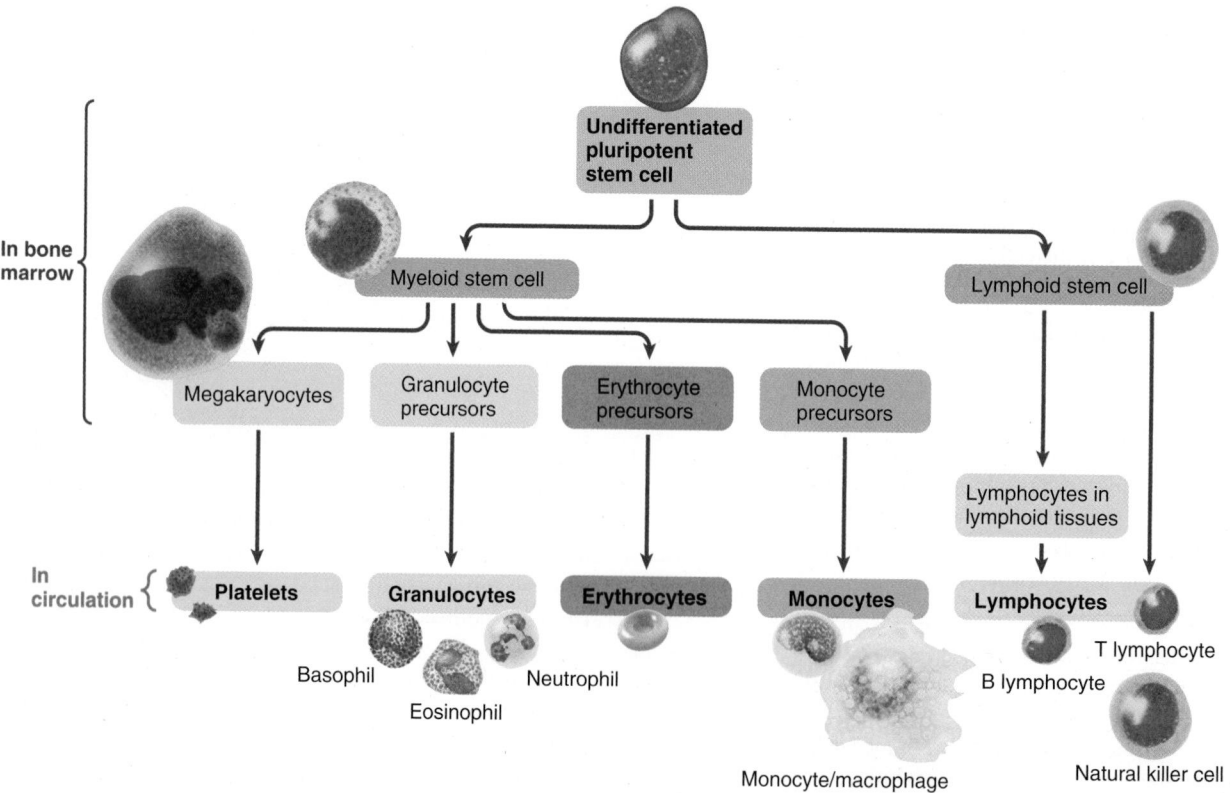

Source: Lauralee Sherwood, *Human Physiology: From Cells to Systems*, 7e, copyright © 2010, Figure 11.9, p. 404.

Figure 9.7 Complete Blood Count

The medical record for a patient may include a lab report for a complete blood count (CBC) listing levels of each type of WBC in the patient's blood.

Hematology, CBC
WBC (4.5-11.0) K/uL
RBC (4.3-5.7) M/uL
Hgb (13.2-17.3) g/dL
Hct (39.0-49.0) %
Mean Cell Volume (80.0-99.0) fL
Mean Cell Hgb Concentration (32-36) g/dL
RBC Distribution (11.6-14.8)
Platelet Count (150-400) K/uL
Mean Platelet Volume (7.5-11.2) fL

Hematology, Electronic Diff
Grans, Electronic (40-70) %
Lymphs, Electronic (22.0-44.0) %
Monocytes, Electronic (0-7.0) %
Eosinophils, Electronic (0-5.0) %
Basophils, Electronic (0-2) %
Grans, Absolute (1.8-7.7) K/uL
Monos, Absolute (0-0.8) K/uL
Lymphs, Absolute (1.0-4.8) K/uL
Eos, Absolute (0-0.5) K/uL
Baso, Absolute (0-0.2) K/uL

Source: Provided by Marcia Nelms.

they "present" on their cell surface. This function is very important for initiating an immune response and will be discussed later in the chapter.

POLYMORPHONUCLEAR LEUKOCYTES Polymorphonuclear leukocytes (PMN), also called granulocytes, are a second group of cells involved in the nonspecific immune response. The term *granulocyte* refers to cytoplasmic granules that are visible in commonly used stains. The term *polymorphonuclear leukocyte* refers to nuclei that look very different from cell to cell. In mature cells, the nucleus is segmented; however, in immature cells (also called band cells) the nucleus is in one segment. PMN move easily between blood and tissues, so their number in the blood increases or deceases in infections depending on the type of organism involved. A large pool of PMN is available in bone marrow for rapid response to an infection.

Three types of PMN can be distinguished by the shape of the nucleus and the stains taken up by the granules (see Figure 9.8). The **neutrophil** is the first type and the most common; it comprises about 60% of WBC in blood and about 90% of PMN. Neutrophils are identified by a nucleus that has three or more connected lobes that may appear unattached in stained cells. Their granules have affinity for both

acidic and basic dyes, so they stain purple. The granules contain lysozyme, which these cells use to destroy organisms that they ingest by phagocytosis.

The second type, **eosinophils**, has a bilobed nucleus and are named for their granules, which are bright red when stained with eosin. In the typical adult, they constitute between 1% and 5% of circulating WBC, but the number often increases in a person experiencing an allergic reaction or worm infection. Eosinophils remain in the blood for a short time and then migrate to tissues. Although they are phagocytic, this is not their major role in the immune system since they are adapted to attack larger pathogens, especially worms. Their granules do not contain lysozyme, but eosinophils do produce other chemicals that are secreted and damage pathogens.

The third type of PMN is the **basophil**, so named because the granules, which appear blue-black, take up the basic dye. Basophils are rare, usually representing less than 1% of the WBC in a typical adult. The nucleus is not always well segmented and is obscured by the granules in some cells. Phagocytic function of these cells is uncertain. Basophils produce a number of chemicals, most notably **histamine** and serotonin, which are associated with allergic responses in humans.

OTHER CELLS Megakaryocytes divide into platelets, which are pieces that aggregate to help form a blood clot. In addition to their role in blood clotting, platelets are involved in inflammation and are a component of certain types of allergic responses.

Mast cells, which are important in some forms of allergy, share many anatomical and physiological characteristics with basophils, and the relationship between the two is controversial. Mucosal mast cells are found in mucosal surfaces, while connective mast cells are found in connective tissues. The myeloid precursor also gives rise to other antigen-presenting cells, including **dendritic cells**, named for their long, nerve-like membranous extensions, and found in **lymph nodes**, the spleen, and blood; **Langerhans cells** in skin; and interdigitating cells in the lymph nodes and thymus.

Figure 9.8 Cell Identification Diagram

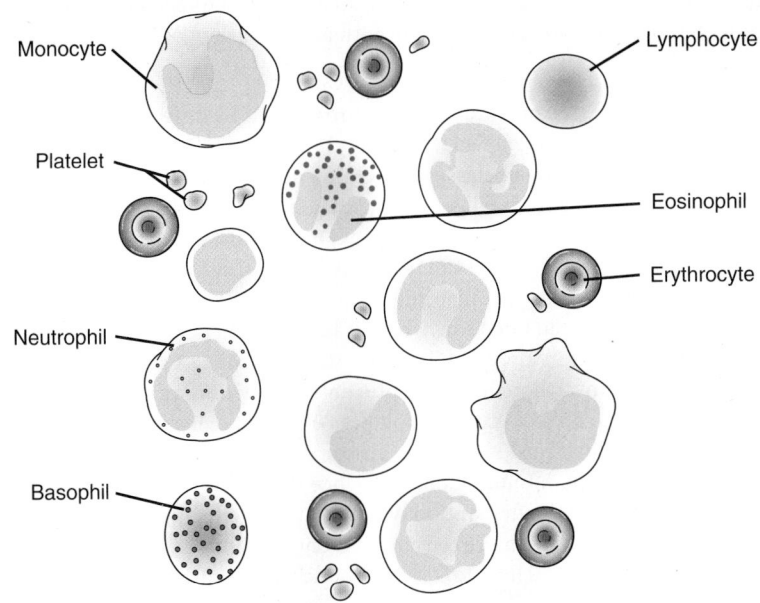

Cells Derived from the Lymphoid Stem Cell The lymphoid precursor differentiates into cells of the specific immune system, including B and T cells or lymphocytes (see Figure 9.6). B and T cells share 98% of expressed genes and cannot be distinguished from each other in a stained blood smear, but numerous antigenic markers and biological characteristics can be used to separate them. They are smaller than monocytes and PMN (6–10 μm), they are agranular, and their cytoplasm stains blue. Resting cells are mostly nucleus with a thin ring of cytoplasm, but activated cells have more cytoplasm.

T CELLS T cells were named for the thymus gland, the site where they mature and differentiate. Most of the lymphocytes in blood, lymph nodes, and lymph are T cells. T cells are divided into subcategories based on their role in the immune response and include T-helper cells and cytotoxic T cells. **Helper T cells (TH)**, also known as T4 cells because of one of the characteristic molecules on their surface, **CD4**, are very important in directing the immune response. They determine how the immune system will respond to various antigens. They stimulate cell and cytokine action, and they interact with other cells of the immune system, causing them to become more immunologically active and to proliferate. There are two subsets of helper T cells called **Th1** and **Th2**. Th1 helper cells activate the cellular immune system, while Th2 cells increase the production of antibodies. **Cytotoxic T cells (Tc)**, also called T8 or **CD8** cells, are capable of killing targeted infected, tumor, or transplant cells directly.

B CELLS B cells differentiate in bone marrow. When stimulated by antigens and T cells, B cells divide and differentiate into plasma cells and memory B cells. Plasma cells are full of endoplasmic reticulum that facilitates efficient production of protein antibodies. Memory B cells produce a rapid **antibody** response the next time the person is exposed to the antigen, and they normally block infection and prevent symptoms.

Antibodies, proteins made in response to an antigen that assist in the destruction or neutralization of that antigen, are found primarily in the **gamma globulin** portion of serum when they are in blood. Thus, they are also called immunoglobulins, which is abbreviated Ig. An antibody molecule must bind specifically to antigens from the pathogen, so part of the molecule must be variable to react specifically with a large variety of antigens.

There are five classes of antibodies that are defined by differences in their structure: (1) IgG, (2) IgA, (3) IgM, (4) IgD, and (5) IgE. There are four subclasses of IgG that differ in amino acid sequence and two subclasses of IgA. Each class or subclass differs in structure and function. **IgG** is the most abundant (75%). Its size allows it to move easily between serum and tissues and to cross the placenta. It is the second antibody made during an initial infection, where it functions to bring the ongoing infection under control, and the first made in subsequent infections, where it usually prevents reinfection. **IgA** constitutes 5% to 15% of immunoglobulins in serum, where it is a monomer. Most **IgM** is found in serum, where it accounts for 5% to 10% of the immunoglobulins, since its large size limits its mobility. It is the first antibody found in new infections, where its 10 binding sites help ensure that once it binds to the antigen it will stay attached. **IgD** is also found on the surface

of B cells. **IgE** is involved in allergy to food and respiratory **allergens**, such as animal dander and pollens, and in countering worm infections.

Antigen Recognition Molecules: Major Histocompatibility Complex (MHC)/Human Leukocyte Antigens (HLA)

The initiation of the immune response depends on our body's recognition of the "foreign" substance—the antigen. But how is this accomplished? In other words, how does our body distinguish self from non-self? Recognition of antigens is possible because our body's cells are different from the antigen, and this difference can be detected and communicated to other cells by a type of molecule referred to as the histocompatibility complex molecule. Genetic variations in these molecules among individuals can result in differences in susceptibility to disease as well as allergic or autoimmune conditions.

In the early 1900s, scientists noted that acceptance or rejection of tumors transplanted between mice was influenced by several dominant genes. This was followed by the discovery that skin transplants between identical twins were not rejected. The genes responsible for transplant rejection or acceptance mapped to an area on chromosome 6 called the **major histocompatibility complex (MHC)**, which contains genes for MHC antigens (also called human leukocyte antigens or HLA) and other proteins involved in the immune response. Activation of T cells requires that an antigen be "presented" to the T cells—meaning the antigen is attached to the MHC molecules on the surface of the antigen-presenting cell (APC). When a pathogen or foreign material invades and is engulfed by a cell, the MHC molecule selects fragments of it, transports them to the cell surface, and displays them for the T cell. The cell that picked up the invader has now become an antigen-presenting cell (APC). The term APC refers to the cell, such as a macrophage, that presents the antigen alerting the B and T cells to the presence of a "non-self" substance such as a bacterium. The T cell is not activated unless its receptor recognizes both the antigen and the MHC molecule it is attached to.

An individual's MHC haplotype is the combination of closely linked genes that code for his/her MHC antigens. For example, if A refers to a gene from the father and B refers to a gene from the mother, then an individual would have genes from both parents which are expressed as: two A, two B, and two C antigens (one inherited from each parent). These genes are found on nucleated cells of the body in varying amounts. Thus, few individuals, other than identical twins, have the exact same haplotype. MHC antigens play an important role in transplant rejection, since the presence of an MHC antigen on the transplanted organ or tissue that is different from the MHC antigens on the recipient's tissues signals the presence of the transplanted tissue and initiates an immune response. The immune system attacks the transplanted cells that display MHC antigens that are different from those found on the recipient's cells.

MHC antigens play a role in susceptibility to some diseases, including many autoimmune and some infectious diseases, since they select the antigenic fragments from the pathogen that will be presented to the T cells. If an MHC selects a fragment from a pathogen that is similar to an antigen on a human tissue, the resulting immune response might target the human tissue.

Soluble Mediators: Communication between Immune Cells

Two important soluble mediators (complement and cytokines) assist the immune system's cells in their function. The complement proteins are activated by other elements of the immune response and are involved in destroying infected cells and some pathogens. Cytokines mediate communication among the cells of the immune system as well as between the cells of the immune system and other body systems, including the nervous system.

Complement **Complement** is a system of plasma proteins that react in tightly regulated cascades activated by antibodies or specific molecules found on pathogens. These cascades result in a final product, the **membrane attack complex**, that lyses cells and produces by-products that trigger inflammation, attract phagocytes to the area, assist in phagocytosis, contribute to activation of naive B lymphocytes, and remove **immune complexes**.

Complement cascades function in both specific and nonspecific modes. Complement is an important part of the body's defenses against infected and tumor cells, but it can also kill beneficial neighboring cells and participate in transplant rejection.

Cytokines In order for cells in the immune system to work together, they must have mechanisms of communication and regulation. **Cytokines** are proteins produced in cells that, in small amounts, affect behavior of other cells. An inducing stimulus causes a cell to make a cytokine, which then acts on a target cell that has a receptor for it. The result is altered biological activity in the target cell. The target cell may be activated, stimulated to proliferate, or stimulated to differentiate into another cell. If the cytokine acts on the cell that produced it, it is called an autocrine; if it acts on a nearby cell, it is called a paracrine; and if acts at a distance, it is called an endocrine. Cytokines may be pleiotropic—meaning the same cytokine will have different effects on different cells—or redundant—meaning two or more cytokines have the same function. They may work alone or as part of the cascade, where production of one cytokine stimulates production of another cytokine. When more than one cytokine is present, they may be synergetic (work together) or antagonistic (counteract one another).

Initially, cytokines were called **lymphokines** because the first ones discovered were produced by lymphocytes. Later, many of the newly discovered cytokines were called **interleukins**, because it was thought they communicated only between WBC. Thus, many cytokines are identified by IL and a number (e.g., IL-2). One of the most widely studied is **IL-2**, which is very active in causing cells of the immune system to proliferate and become more immunologically active. Several cytokines are used in medical treatment (see Box 9.4), including IL-2, which is given to individuals with compromised immune systems. **Interferon (INF)**, a major component of the body's defense against viruses, is also a cytokine, as is **tumor necrosis factor (TNF)**. Three classes of interferons have been identified: (1) alfa, (2) beta, and (3) gamma.

The Immune Response

Humoral and Cellular Immunity

The immune system is divided into two arms: **humoral** and **cellular immunity**. The humoral arm of the immune system refers to the antibodies that appear in serum and B cells that become **plasma cells**, which produce antibodies. The cellular part of the immune system consists of the T cells, macrophages, monocytes, and polymorphonuclear leukocytes (also known as PMNs, macrophages, and granulocytes) that interact with pathogens at the cellular level.

Specific and Nonspecific Immunity

The immune system is also divided into two branches, both of which contain elements of both the humoral and cellular immune systems. One is referred to as the innate, nonadaptive, or **nonspecific immune system**. The other is called either the acquired, adaptive, or **specific immune system**.

These two immune systems, although often described separately, are interdependent and are both required for a strong immune response. Cells of the nonspecific immune system (macrophages, monocytes, natural killer cells, and polymorphonuclear leukocytes) react with any antigen; they can thus react immediately. This initial reaction is quite often sufficient to eliminate the pathogen or reduce its numbers significantly enough to prevent initiation of the disease process. This response does not increase or improve with repeated exposures.

Nonspecific Immune Response: Inflammation

Inflammation is defined as an innate, nonspecific response to cellular injury. As discussed earlier in this chapter, cellular injury may result from foreign invasion—such as infection, chemical exposure, and allergens—physical damage, and/or trauma. Inflammatory response is localized, but the process of inflammation can result in further systemic response (see Table 9.2). The inflammatory response occurs within seconds to minutes of injury, whereas the immune response occurs much more slowly. In general, the inflammatory response is classified as a protective mechanism, even though symptoms may result in systemic signs and symptoms.

STAGES OF INFLAMMATION Inflammation begins with the onset of cellular injury, as demonstrated in Figure 9.9. The body reacts to injury with both a **vasomotor** and a cellular response. Vasomotor response begins with a brief period of vasoconstriction that serves to limit any bleeding to the area of

Table 9.2 Inflammation versus Inflammatory Systemic Response

Concept	Inflammation	Inflammatory Response
Location	Localized	Systemic
Response time	Seconds to minutes	Hours to days
Responses	Redness, heat, swelling, pain	Fever, neutropenia, anorexia, fatigue

injury. Immediately following this period, capillaries and other blood vessels within the area of injury experience vasodilation. Increased blood flow to the injury, referred to as **hyperemia**, results in several classic symptoms of inflammation: redness,

Figure 9.9 Stages of Inflammation

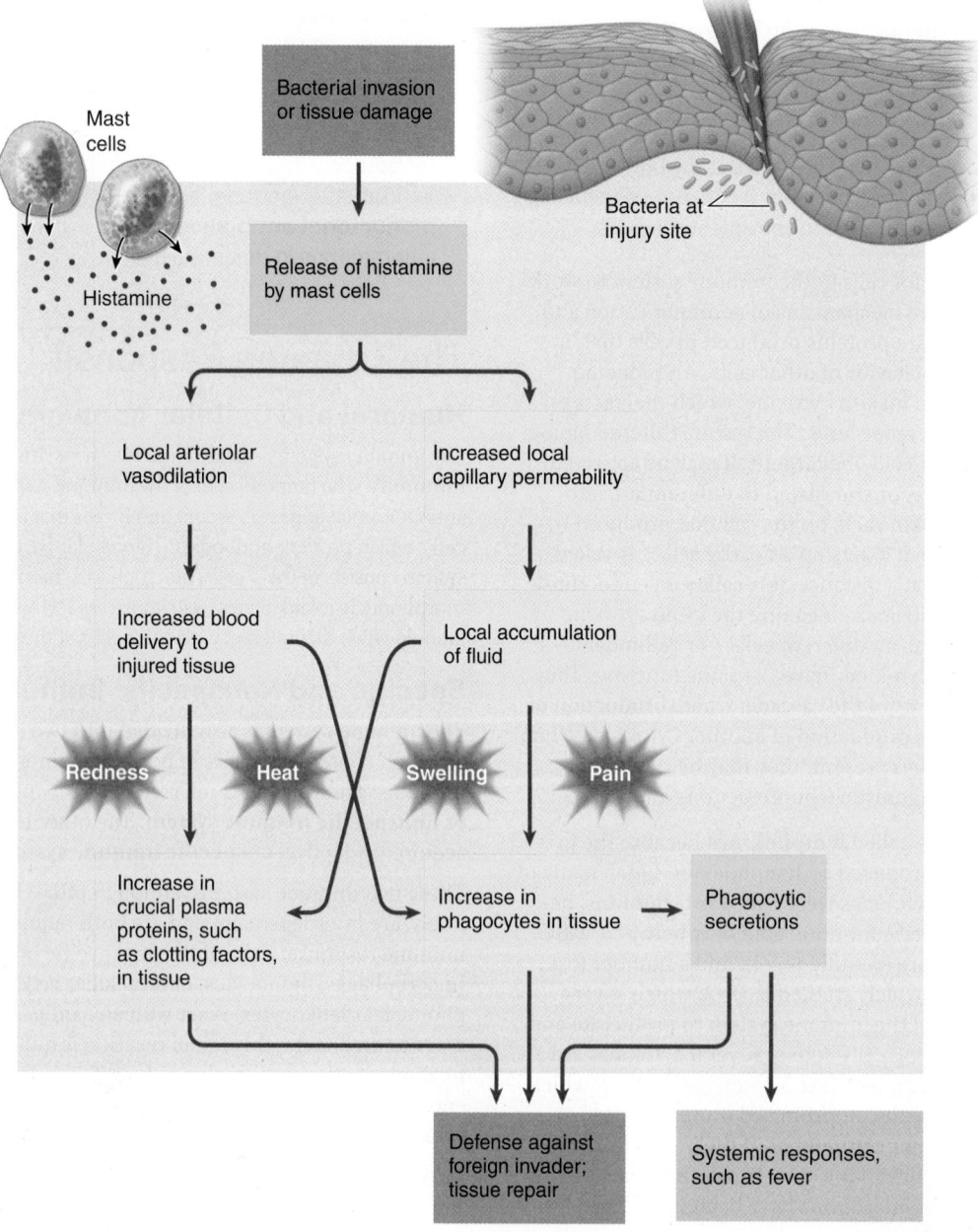

Source: Lauralee Sherwood, *Human Physiology: From Cells to Systems,* 7e, copyright © 2010, Figure 12.3, p. 422.

heat, swelling, pain, and altered function (sometimes referred to by the Latin terms *rubor, calor, tumor, dolor,* and *functio laesa*; see Figure 9.9). Increased blood flow to the injury will supply additional oxygen and nutrients needed for the healing process. It will serve to increase transport of cells needed for repair, healing, and prevention of infection, and will dilute any toxins present. Finally, the pain caused by increased blood flow will limit movement, which serves as additional protection against further injury.

During vasodilation, the increased vascular permeability allows proteins and immune cells to pass from blood to tissue spaces where the injury has occurred. Material that accumulates in tissue spaces is called **exudate**. The type of exudate will be consistent with the type and extent of injury. Exudate can contain proteins and general immune cells (serous exudate), blood (hemorrhagic exudate), or microorganisms (purulent exudate).

Chemical messengers or mediators direct vasodilation and changes in vascular permeability. Examples (see Table 9.3) include histamine, nitric oxide, serotonin, prostaglandins, thromboxanes, **leukotrienes**, complement, and cytokines such as interleukins (IL-1; IL-8). These chemical messengers/mediators have particular interest not only for medical and pharmaceutical research but also for nutrition practice. The precursor for many of these lipid mediators is arachidonic acid (see Figure 9.10). Since many signs and symptoms are caused by release of these messengers, many current treatments for inflammation center on interrupting the cytokine communication. Furthermore, these acute-phase proteins are also used as markers for assessment of the inflammatory process and as an indirect measure of nutritional status. Cytokines are also used therapeutically as the basis of numerous medications used for treatment of various conditions, as discussed in Box 9.4.

Cellular response to inflammation involves the action of particular immunocompetent cells that accomplish both phagocytosis (engulfment of particles or microorganisms) and initiation of the next stages of the immune response, if that is necessary. These cells (neutrophils and macrophages) move into the area and begin destruction of any microorganisms and foreign debris. They also work to dispose of any dead cells that result from the injury.

Table 9.3 Mediators of Inflammation

Mediator	Action
Chemokines	Chemotaxis: Movement of phagocytes toward the site of injury
Histamine	Increases blood flow as well as seepage of fluid and proteins from blood
Reactive oxygen species (ROS)	Toxic for microorganisms but also damage tissue
Interleukin 1 (IL-1)	Triggers blood clotting, T cell activation, decrease in blood pressure, fever, and release of prostaglandins
Prostaglandins	Increase vascular permeability and influence platelet aggregation
Leukotrienes	Prolong the response; have vasoactive properties

Source: Table created by Dr. Christina Lee Frazier of Southeast Missouri State University.

Figure 9.10 Arachidonic Acid

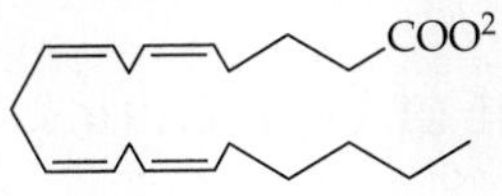

Arachidonic acid

Finally, proteins necessary for healing are produced. As noted previously, during the period of vascular permeability, proteins escape from the circulatory system into tissue spaces. Fibrinogen (which is activated to fibrin) serves as a major component in wound healing. Fibrin not only works to provide the framework for healing tissue, but also seals off the area of injury and contributes to blood clotting.

SIGNS AND SYMPTOMS OF INFLAMMATION The cardinal or classic signs of inflammation, as mentioned previously, are redness, warmth, swelling, and pain resulting from hyperemia and vascular permeability. Table 9.2 compares the clinical features of a localized inflammation to a systemic inflammatory response.

Systemic effects of inflammation may include fatigue, malaise, and fever, an elevation in body temperature. When macrophages begin phagocytosis, they also cause release of prostaglandins, which in turn stimulates the hypothalamus to increase body temperature. Increased body temperature is a protective factor, because most microorganisms do not thrive at higher temperatures. Heat also promotes phagocytosis, which is a natural step in the inflammatory response. These benefits of higher temperature make the common practice of taking medications to interrupt the fever process problematic. Suppressing a mild fever may do more harm than good.

Laboratory markers for inflammation can include increased white cell count (*leukocytosis*). Other components of white cell count, as determined by measuring the proportions of cells within the total white cell count (differential white cell count), may give an indication of the source of infection. Elevated levels of acute phase proteins, such as C-reactive protein (CRP) and fibrinogen, can also be indicative of inflammation (see also Chapter 22). Increased levels of fibrinogen affect another marker of inflammation, erythrocyte sedimentation rate (ESR). Technically, ESR measures the distance erythrocytes have fallen ("sedimentation" or "sed rate") after one hour. When there is an increase in fibrinogen, there is an increase in clotting for erythrocytes that in turn affects the rate of sedimentation. In fact, CRP, ESR, and fibrinogen are often used as markers for exacerbations of chronic inflammatory conditions such as rheumatoid arthritis. Box 9.5 describes clinical assessment of cellular injury.

CLINICAL MANAGEMENT OF INFLAMMATION Even though inflammation is a natural and protective response to cellular injury, symptoms of inflammation can be painful and interrupt activities of daily life. Therefore, it is common to treat inflammation with both physical and pharmaceutical interventions. Acutely, applying cold to localized inflammation will result in vasoconstriction to that area. This will limit blood flow and reduce swelling and pain associated with vasodilation and increased vascular permeability. Additionally, pressure and elevation will assist in treatment of these same symptoms.

Assessment of Cell Injury

Once fully understood, basic concepts of the disease process can be applied to many different diagnoses and conditions. Furthermore, the same basic methods of measuring cellular injury provide data to direct appropriate clinical interventions in many conditions. When a specific diagnosis or disease process is studied, characteristic responses to cellular injury are apparent.

When cells are injured—whether it is due to the lack of oxygen, nutritional imbalances, or physical injury—cell components may be released into the body, as noted in the section on inflammation. Part of clinical diagnosis and practice is measuring release of these cellular components. For example, many diagnoses depend on measurement of cellular enzymes. In liver disease, enzymes such as alkaline phosphatase (ALT) and aspartate amino transferase (AST) are present in abnormally high levels in the blood. When a myocardial infarction occurs, the contractile protein, troponin, is released from the damaged cells and can be detected in the blood.

Abnormal electrical activity of the cell can also be monitored. Standard medical practice uses the electrocardiogram (ECG, EKG), electroencephalogram (EEG), and electromyogram (EMG) to identify abnormalities in the function of the heart, brain, or muscles. For example, when cardiac tissue is damaged, changes in heart rate or rhythm will be noted on the electrocardiogram.

Finally, actual tissue cells can be examined. This is referred to as a biopsy. Under microscopic examination, abnormal cells can be detected and clinically evaluated for specific diagnosis and treatment.

Elevating the injury decreases blood flow as well and will assist in reducing swelling. Constricting the area with pressure may reduce accumulation of exudate.

Several different classes of medications can assist in treating inflammation. Nonsteroidal anti-inflammatory medications (NSAIDs), selective COX-2 inhibitors, and glucocorticoids are all used to treat inflammation.[8] All of these medications block this process in one or more steps. Salicylates (aspirin), an NSAID, inhibit prostaglandin production. Other nonsteroidal anti-inflammatory medications, such as ibuprofen, further prevent synthesis of prostaglandins by blocking cyclooxygenase enzyme 2 (COX-2; see Figure 9.11). Unfortunately, NSAIDs also block COX-1, which has a protective effect on the gastric mucosa. When this enzyme (COX-1) is reduced, side effects of gastritis and upper GI bleeding are possible.[9] Selective COX-2 inhibitors only block COX-2 and do not have GI side effects. These medications are generally prescribed for musculoskeletal inflammation.[10]

Glucocorticoids or steroids can be given not only for musculoskeletal inflammation, but also for systemic inflammatory conditions. Depending on the source of inflammation, glucocorticoids can be given intravenously, orally, topically, or by injection directly to the site of inflammation. Glucocorticoids are synthetic derivatives of endogenously produced cortisone. These medications act on several different aspects of the inflammatory response, affecting vascular permeability or blocking one of the enzymes required for production of prostaglandins.[11] Long-term use of glucocorticoids is associated with many possible side effects, including hyperglycemia, osteoporosis, and protein loss/muscle wasting.

Figure 9.11 Action of NSAIDs and Selective COX-2 Inhibitors

Both NSAIDs and selective COX-2 inhibitors block COX-2 enzymes. Traditional NSAIDs block COX-1 enzymes as well.

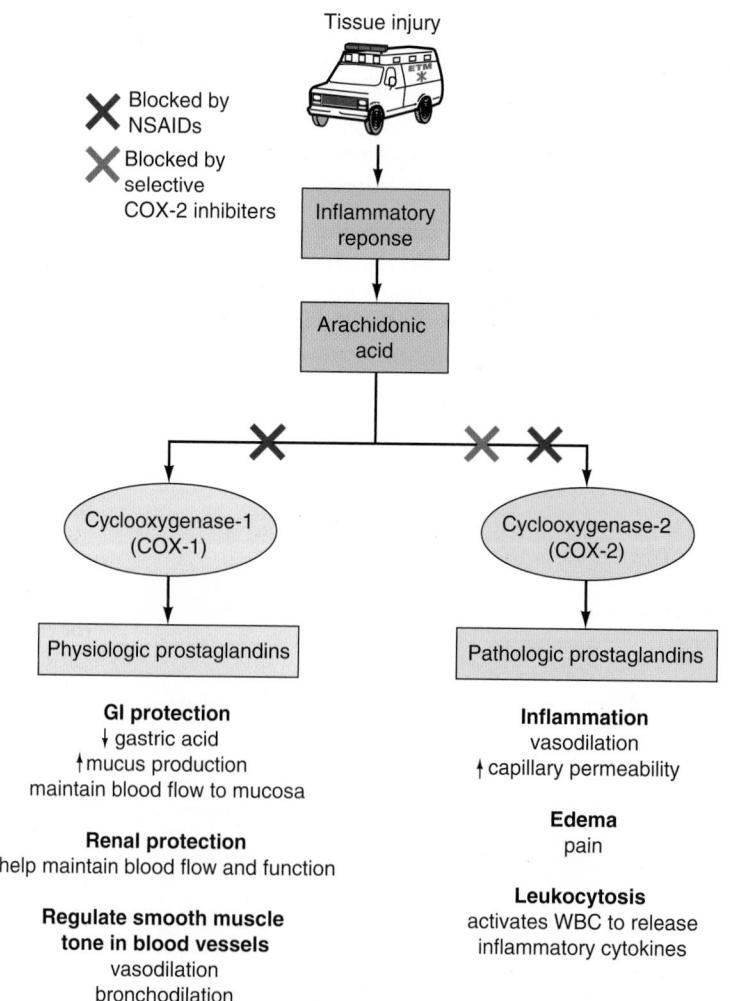

Source: Dr. David Gotlieb doc on-line, http://www.arthritis.co.za/nsaids2.htm

Figure 9.12 Healing by First and Second Intention

(a) Healing by first intention takes place in an injury that has even and closely opposed edges. Cuts or incisions typically heal in this manner. (b) Healing by secondary intention results when tissue loss leads to a gaping lesion or when purulent infections prevent direct association of the wound edges. Lacerations commonly heal by secondary intention.

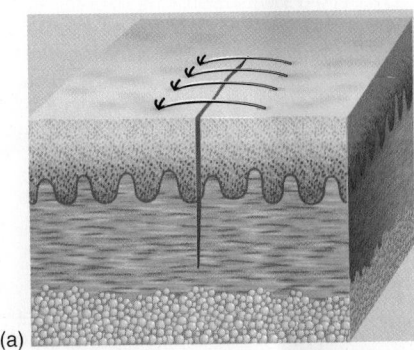
(a)

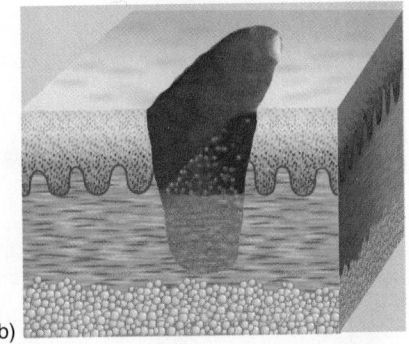
(b)

CELLULAR RESPONSE WITH HEALING Inflammation is the body's natural response to cellular injury. An important component of inflammation is the body's ability to begin the steps toward healing. Healing is defined as repair and restoration of damaged tissue and cells, and is the process by which structure and function are restored after an injury.

Wounds and injuries heal by restoring cells, repairing cells, or replacement of cells. The type of healing will depend on the original injury. If the injured cell is a type able to undergo mitosis (cellular replication), those cells can usually be restored if optimal conditions are maintained. Other cells (specifically cardiac, skeletal, and neurological) are unable to be fully restored after injury. Healing of tissues composed of these cells incurs replacement of original tissue with scar tissue.

Healing by first intention (see Figure 9.12) applies to most wounds that are smaller, where cell loss is minimal and the edges of the wound are close together. In this process, epithelial cells are replaced. Below the surface, granulation tissue is formed with the support of collagen. For deep or large wounds, healing will proceed from the bottom of the wound upward. This is referred to as healing by second intention. The normal progression of wound healing is outlined in Table 9.4.

Nutrition is an important component of successful wound healing, even though clinical research has not provided adequate evidence regarding absolute amounts for supplement recommendations.[12–14] A review of the literature indicates adequate energy, protein, and fluid are the foundation of healing support. Requirements for arginine and glutamine are increased, but specific levels for these amino acids have not been verified. Specific vitamins and minerals that are involved with wound healing, and that may need supplementation, include vitamin C, vitamin A, vitamin K, selenium, and zinc (see Table 9.5). Further research will clarify the role of supplementation in wound healing. Ensuring an adequate blood supply and keeping the wound clean and undisturbed during healing are crucial components for success.

Healing will be delayed if a foreign material is present or if infection develops. Wound healing will be impaired if the wound is exposed to radiation. The elderly and those individuals who are bedridden or immobile are especially at risk for poor wound healing due to changes in their skin and increased risks of poor circulation and poor nutritional status. These wounds may include a pressure ulcer, defined as: "localized injury to the skin and/or underlying tissue usually over a bony prominence, as a

Table 9.4 Progression of Normal Wound Healing

Phase	Typical Duration for Acute Wound Healing	Primary Events
Inflammatory	Begins at the time of injury and continues for about four to six days	• Coagulation cascade and fibrin clot formation control bleeding. • Vasodilation and increased capillary permeability occur. • Neutrophils phagocytize bacteria. • Macrophages remove debris and necrotic tissue and secrete growth factors.
Proliferative	Begins about the third to fifth day and continues for two to three weeks	• Epithelial cells form a protective covering and framework over the wound. • Angiogenesis enables development of granulation tissue. • Fibroblasts produce collagen and matrix protein, forming granulation tissue. • Collagen deposition and cross-linking begin to strengthen the wound. • Myofibroblasts induce wound contraction and the wound begins to close.
Remodeling	Begins about two to three weeks after injury; can continue for up to two years	• Collagen maturation and stabilization occur. • Fibrous scar tissue matures (decreases in fibroblasts and vascularization) but skin and fascia never regain full strength.

Source: NUTRITION IN CLINICAL PRACTICE by Thomson, Fuhrman. Copyright 2005 by SAGE PUBLICATIONS INC. JOURNALS. Reproduced with permission of SAGE PUBLICATIONS INC. JOURNALS in the format Textbook via Copyright Clearance Center.

Table 9.5 Nutrition Intervention for Wound Healing for Adults

Nutrient	Function in Wound Healing	Dietary Reference Intake: RDA, AI, UL	Recommended Intake for Wound Healing	Additional Notes
Kcalories	Energy for hypermetabolic response.	—	25 to 30 kcal/kg to 30 to 35 kcal/kg depending on type of wound	Monitoring actual kcalories received is essential. Kcalorie goals should be adjusted for ebb and flow phases of the hypermetabolic response, severity of wound, response to healing, comorbidities, and other factors that affect metabolic rate.
Protein	Immune response, nitrogen losses from wounds.	RDA: 0.8 g protein/kg for adults	1.2 to 1.5 g/kg	Unless severe catabolic states or exogenous protein losses are present, >1.5 to 2.0 g/kg may indicate overfeeding.
Fat n-3/n-6 fatty acids	May play a role in collagen formation.	AI Men, 19–70+ years: 1.6 g n-3 fatty acids/day AI Women, 19–70+ years: 1.1 g n-3 fatty acids/day n-3 to n-6 ratio for adults: 10.6:1 (<51 y); 10:1 (>51 y)	Not yet determined	One small study found topical use of oils rich in essential fatty acids may prevent skin breakdown (Cardoso 2004).
Fluid	Maintains skin turgor and blood flow to wounded tissue to prevent breakdown of skin.	AI Men, 19–70 years: 3.7 L AI Women, 19–70 years: 2.7 L	>18–55 yo: 35 mL/kg >55 yo: >30 mL/kg or minimum of 1.5 L/d (unless contraindicated)	Provide additional fluid to maintain adequate hydration and compensate for additional losses from fever, draining wounds, etc. Additional fluid needed if fever present.
Vitamin A	Supplementation may assist wound healing that has been retarded by vitamin A deficiency, radiation, chemotherapy, or DM.	RDA Men: 900 RAE/~ 3000 IU RDA Women: 700 RAE/~ 2300 IU UL: 3000 RAE or 10,000 IU	20,000–25,000 IU orally for 10 days if deficiency suspected or confirmed (low serum retinol and/or functional tests such as dark adaptation)	Supplementation rarely warranted in chronic kidney disease. Patients with fat malabsorption should receive water-soluble form.
Vitamin C	Deficiency can delay wound healing.	RDA Women: 75 mg RDA Men: 90 mg Additional 35 mg/d for smokers UL: 2000 mg due to GI side effects (nausea, abdominal cramps, or diarrhea)	Up to 1 to 2 mg/day for deficiency; during acute illness and injury for up to three months	Evaluation of serum ascorbic acid levels reflects dietary intake and not tissue levels. White blood cell ascorbic acid and ascorbic acid tissue saturation tests are better indices.
Zinc	Low serum zinc levels associated with impaired healing. Supplementation may enhance wound healing, but only in those who are zinc deficient.	RDA Women: 8 mg RDA Men: 11 mg (Requirements may be up to 50% higher for vegans.) UL: 40 mg	15 to 25 mg elemental zinc/day	Zinc intakes below RDA are commonly seen in the elderly. Increased losses can occur from large skin wounds or urine (after trauma, closed head injury, etc.). Impaired absorption from GI tract can occur with high volume diarrhea or fistulae output.
Iron	Routine supplementation not recommended for wound healing.	RDA premenopausal women: 18 mg RDA postmenopausal women: 8 mg RDA men: 8 mg UL: 45 mg	Not applicable	Low hemoglobin concentration does not seem to impair wound healing, provided adequate tissue perfusion is maintained.
Vitamin E	No clear role for vitamin E supplementation. Use may adversely affect healing of some types of wounds.	RDA: 15 mg alpha-tocopherol equivalents	Ensure 100% of DRI for all nutrients.	Large doses of vitamin E should be avoided before surgery unless deficiency is present. Topical vitamin E is often recommended to reduce scar formation and improve cosmetic appearance of scar tissue, but this benefit has not been documented by research.
Other vitamins, minerals, and trace elements	Thiamin, riboflavin, and pantothenic acid play a role in collagen production. Copper and manganese are required for tissue regeneration. Vitamin K is required for prothrombin and synthesis of other clotting factors.	Varies	Ensure 100% of DRI for all nutrients.	No studies verifying deficiency of these nutrients as having an impact on wound healing could be found.

Nutrient	Function in Wound Healing	Dietary Reference Intake: RDA, AI, UL	Recommended Intake for Wound Healing	Additional Notes
Arginine	Supplementation may benefit wound healing by increasing collagen deposition, improving nitrogen balance, and enhancing several parameters of immune function. Most, but not all, research demonstrated enhanced nitrogen retention and immune function.	Not applicable	Oral dose of 17.0–18.7 g/day free arginine was used in previous human studies. The recommended dosage, and populations who may benefit most, are unknown at this time.	Supplementation may be contraindicated in patients with severe renal or hepatic failure.
Glutamine	Preferred fuel for enterocytes, lymphocytes, and macrophages. Precursor for nucleotides and glutathione.	Not applicable	Oral doses up to 0.57 mg/kg/d in adults appear safe.	Supplementation may be contraindicated in patients with severe renal or hepatic failure.

Adapted from Thompson, C.W., Nutrition and adult wound healing. Nutrition Week January 18, 2003. Available at http://www.nutritioncare.org/listserv/wound%20healing.pdf.

result of pressure, or pressure in combination with shear and/or friction."[15] It is common for nursing staff to assess risk for pressure ulcers during a hospitalization or in a long-term care or rehabilitation facility. One tool to assess risk is the Braden scale, which consists of six components (sensory perception, mobility, activity, moisture, nutrition, and friction and shear) with scores from 6 to 23. The lower the score, the greater the person's risk of developing pressure ulcers.[16, 17]

Describing pressure ulcers includes staging the depth of the wound. The National Pressure Ulcer Advisory Panel staging criteria are outlined in Box. 9.6.

COMPLICATIONS OF INFLAMMATION AND WOUND HEALING Complications of inflammation that may occur include development of chronic inflammatory conditions. When the injurious agent is not quite strong enough to cause a systemic response but continues to be present, chronic low-level inflammation can occur. This underlying constant inflammation draws cells of the immune system (particularly macrophages) to the site of injury. These cells accumulate and result in chronic inflammation.

Complications can also occur from wound healing. Ineffective wound healing is called **dehiscence**. This simply means that a

---

BOX 9.6   CLINICAL APPLICATIONS

National Pressure Ulcer Advisory Panel Staging of Pressure Ulcers

Suspected Deep Tissue Injury:

Purple or maroon localized area of discolored intact skin or blood-filled blister due to damage of underlying soft tissue from pressure and/or shear. The area may be preceded by tissue that is painful, firm, mushy, boggy, warmer or cooler as compared to adjacent tissue.

Further description: Deep tissue injury may be difficult to detect in individuals with dark skin tones. Evolution may include a thin blister over a dark wound bed. The wound may further evolve and become covered by thin eschar. Evolution may be rapid, exposing additional layers of tissue even with optimal treatment.

Stage I

Intact skin with non-blanchable redness of a localized area usually over a bony prominence. Darkly pigmented skin may not have visible blanching; its color may differ from the surrounding area.

Further description: The area may be painful, firm, soft, warmer or cooler as compared with adjacent tissue. Stage I may be difficult to detect in individuals with dark skin tones. May indicate "at risk" persons (a heralding sign of risk).

Stage II

Partial thickness loss of dermis presenting as a shallow open ulcer with a red pink wound bed, without slough. May also present as an intact or open/ruptured serum-filled blister.

Further description: Presents as a shiny or dry shallow ulcer without slough or bruising.* This stage should not be used to describe skin tears, tape burns, perineal dermatitis, maceration or excoriation.

Stage III

Full thickness tissue loss. Subcutaneous fat may be visible but bone, tendon or muscle are not exposed. Slough may be present but does not obscure the depth of tissue loss. May include undermining and tunneling.

*Bruising indicates suspected deep tissue injury

Source: Reprinted from: National Pressure Ulcer Advisory Panel. Pressure Ulcer Stages. 2007. Available at http://www.npuap.org/pr2.htm

wound reopens. Dehiscence may result from any of the situations described earlier, such as poor circulation or malnutrition. Obese patients are at higher risk for dehiscence.[14] Other complications include contractures and adhesions. During the healing process, scar tissue can result in a shrinking of the connective tissue. This **contracture** can distort the area around the injury and limit the ability to use that part of the body. Healing of burns often results in development of contractures. **Adhesions** result when two previously unconnected tissues are abnormally joined together. This most often occurs after a surgical procedure, particularly after abdominal surgery.

Specific Immune Response   In the specific immune response, each B and T cell is programmed to attack one specific antigen, but these cells can interact with other antigens that are closely related or very similar. In rheumatic fever, for example, an immune response stimulated by antigens on a Group A *Streptococcus* can attack similar antigens on the valves of the heart. The specific immune system takes time to respond initially, but it improves with additional exposures and responds more rapidly on subsequent encounters with the organism. Thus, it normally protects the human from reinfection.

BOX 9.7   CLINICAL APPLICATIONS

Types of Immunity

Specific immunity can be described as either **active immunity**, where individuals synthesize their own antibodies or activate immune cells, or **passive immunity**, where they receive antibodies or activated cells produced by another individual. Both active and passive immunity can be described as either natural (occurring without human intervention) or artificial (resulting from human intervention). Thus, the four types of immunity are as follows:

- *Active natural immunity:* Mounting an immune response to an infectious organism.
- *Active artificial immunity:* Mounting an immune response to vaccination.
- *Natural passive immunity:* An antibody from the mother goes to the fetus across the placenta. Both regular breast milk and colostrum contain antibodies, but the concentration is higher in colostrum.
- *Passive artificial immunity:* Transferring antibodies or immune cells produced in one organism to another organism to prevent the action of a virus or toxin before it does damage. Examples of clinically used antibodies include antirabies or hepatitis globulin, antivenom for snake bites, or antitoxin for tetanus or botulism. Commercially available intravenous immune globulins (Gamimune N, Gammagard, Gammar, Iveegam, Polygam, Sandoglobulin) contain gamma globulins from a number of individuals and are used to boost the body's natural defense system against infection in persons with a weakened immune system.

The response to a pathogen normally involves an initial contact with the nonspecific immune system, which often is capable of eliminating the organism by itself. The nonspecific immune system then stimulates the specific immune system to seek out and target any remaining pathogens. In some cases, the specific immune system can eliminate the pathogen; in others, it merely tags it or alters it in such a way that it becomes more susceptible to the cells of the nonspecific immune system. The two systems thus work together and are interdependent.

THE PRIMARY RESPONSE   The **primary immune response** (see Figure 9.13) normally begins with phagocytosis, which can be sufficient to block infection if virulence of the pathogens is low and/or the exposure is small. This process can remove about 90% of an antigen by the time the initial exposure has circulated through the body once. As discussed previously (under "Antigen Recognition Molecules"), once a macrophage has engulfed a pathogen, its MHC proteins select and present antigens from that pathogen to the T cells. The phagocytic cells thus become APC for the helper T cells (TH).

TH activation requires recognition of the MHC II and antigen presented by the APC and interactions between molecules on both the T cell and the APC. In addition to providing the MHC-antigen signal, the macrophages secrete cytokines that contribute to activation of TH, B cells, PMN, and NK.

The activated TH cell begins to divide and secretes cytokines that will activate other cells of the immune system. IL-2 made by T cells is the key cytokine involved in inducing proliferation and activating other T cells, B cells, monocytes, and NK. Activated TH cells secrete cytokines that induce proliferation of B cells that can react with antigens on the pathogen and trigger the B cells' differentiation into a plasma cell or antibody-forming cell (AFC). Plasma cells are larger than B cells and packed full of endoplasmic reticulum to produce antibodies at a rate of 30,000 Ig/sec.

After about five days, the plasma cell will produce IgM for about three days before switching to the production of IgG. Some cells will switch to either IgE or IgA instead of IgG due to the influence of cytokines. Since it takes time to produce sufficient antibodies and activated T cells, the response does not prevent disease but terminates the ongoing infection. The longevity of the antibody, which will degrade over time, depends on a number of characteristics of the antigen and host.

THE SECONDARY RESPONSE   During the primary response to an antigen, some of the T and B cells that can react with the antigen are partially activated but do not participate in the immune response. These cells are called **memory cells** and are responsible for the **secondary (or memory) response**. The secondary response is more rapid than the primary response since it does not require activation of naive TH cells by APC. IgM is not produced, and significantly more IgG is made. Memory cells are long-lived and protect from reinfection for at least twenty years and sometimes for life. Thus, humans rarely become ill from the same organism twice unless the organism has mechanisms to evade the memory response, such as changing its antigens.

Figure 9.13 Interactions among Macrophages, B Cells, and Helper T Cells

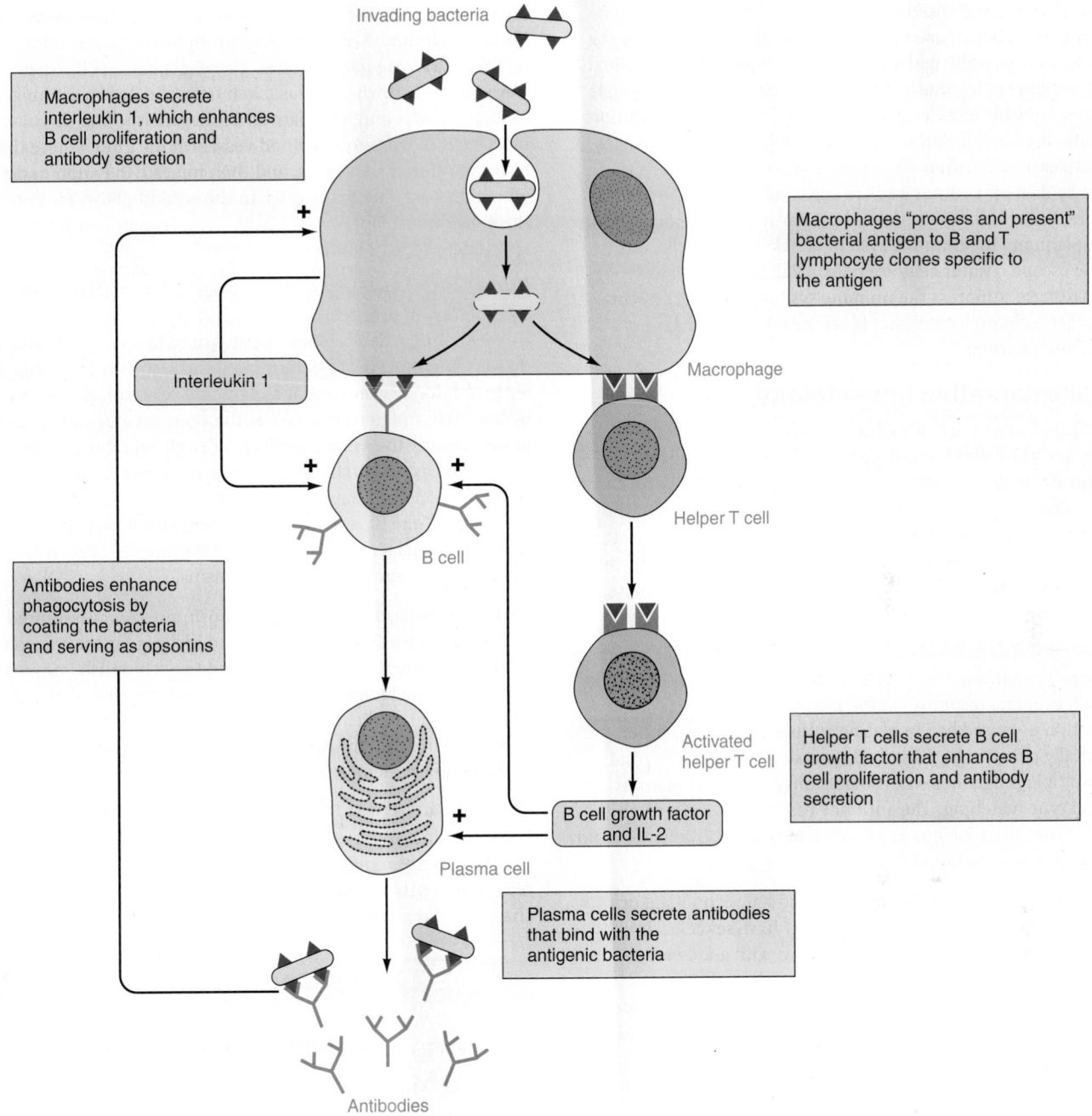

Invading bacteria

Macrophages secrete interleukin 1, which enhances B cell proliferation and antibody secretion

Macrophages "process and present" bacterial antigen to B and T lymphocyte clones specific to the antigen

Interleukin 1

Macrophage

Antibodies enhance phagocytosis by coating the bacteria and serving as opsonins

B cell

Helper T cell

Helper T cells secrete B cell growth factor that enhances B cell proliferation and antibody secretion

Activated helper T cell

B cell growth factor and IL-2

Plasma cell

Plasma cells secrete antibodies that bind with the antigenic bacteria

Antibodies

Source: Lauralee Sherwood, *Human Physiology: From Cells to Systems,* 5e, copyright © 2004, p. 434.

Attacking Altered and Foreign Cells: Tumors and Transplants

Research and development for the medical treatments used in cancer as well as for **transplantation** involve investigation of how our immune system reacts to different types of antigens. A tumor is a mass of cells that contains antigens normally found on the person's cells as well as some new antigens that provoke an immune response like viral or bacterial antigens. Similarly, a transplant is a mass of cells whose antigens are matched to some degree with those of the recipient. Thus, tumors and transplants appear the same to the immune system, and the mechanisms used by the immune system to attack tumors also contribute to transplant rejection.

Tumor Immunology

Cancer is caused by progressive, uncontrolled growth of a single transformed cell (see Chapter 23). There is ample evidence that tumors are attacked by the immune system. Immune cells, including lymphocytes, macrophages, PMN, and dendritic cells, are found in tumors.

Tumor cells have new antigens in addition to those on similar noncancerous cells. When a virus causes a tumor, some of the antigens are specific to the virus, while other antigens, found on tumors of both viral and chemical origin, are unique to the tumor. Cancer cells shed antigens that are picked up by APC and recognized by multiple elements of the immune response as new antigens or an excess of existing antigens. Solid tumors are attacked by TC, NK cells, complement, and macrophages that release anti-tumor chemicals. Metastatic cells (cells that have broken off a tumor and are traveling to a new site) and single cancer cells are destroyed primarily by antibodies and complement. The immune system attacks cancer cells, but some escape. Tumor cells are poor APC. Some, such as Hodgkin's disease, suppress the immune system, while others cover themselves in molecules that block lymphocyte attachment (antigen masking).

Transplantation Immunology

Transplants can normally occur between one part of a person's body and another (**autograft**) or from an identical twin (**isograft**) without a problematic immune response. Most transplants, however, are between non-twin members of the same species (**allografts**), and immunological rejection is an important factor. The vast majority of **xenografts**, which come from a different species, are not vascularized, so immunological rejection is not a concern.

Transplant Rejection

Host-versus-graft (HVG) rejection occurs when immunocompetent cells (those capable of mounting an immune response) in the host reject the graft. Acute rejection is caused by circulating antibodies that are either naturally occurring IgM or IgG against xenografts, the result of multiple pregnancies or transfusions, or a result of improper blood type matching. The antibody reacts with antigens on the surface of cells, blocking establishment of good circulation, and the graft does not succeed.

HVG rejection may be classified as first-set, second-set, or chronic depending on the onset. When **first-set rejection** occurs, the graft at first appears healthy, with good vasculature and blood supply, and if it is an organ, starts to function. However, rejection occurs in 11 to 17 days and tissue is infiltrated with macrophages, lymphocytes, and plasma cells. **Second-set rejection** occurs in cases involving a graft with the same MHC antigens as a previous graft. A memory response is mounted so the graft is sloughed off in three to four days. In contrast, it takes months to years for chronic rejection to occur. In properly matched grafts, chronic rejection is due to antibodies to **minor histocompatibility antigens**, immune complexes, or viruses that stimulate immune responses by placing new antigens on infected cells in the graft. The life span of a kidney transplant is about seven to eight years.

In **graft-versus-host (GVH) rejection**, which occurs in bone marrow but not heart or kidney transplants, immunocompetent cells are found within the graft, and the host is incompetent due to age, disease, or immunosuppression. Lymphocytes in the graft are sensitized to the recipient's antigens and mount an immunological attack on multiple tissues and organs of the recipient.

Acceptance or rejection is immunological, and rejection is heavily dependant on T cells. T cells recognize donor-derived peptides in association with the donor's MHC. In the first phase of rejection, the immune system must detect the presence of the graft. The graft releases soluble MHC antigens that initiate an immune response when they are presented by APC to TH in draining lymph nodes. Passenger cells, WBC that were in the graft when it was taken from the donor, can migrate to lymph nodes and stimulate an immune response. Circulating T cells move through blood vessels in the graft, where they encounter foreign antigens, and then migrate to lymph nodes where they activate many cells. In the second phase, part of the entire transplant is attacked by complement and antibodies specific for MHC I antigens, Tc cells, and NK cells.

Matching

Matching of MHC antigens is critical in most transplants. In practice, only A, B, and DR are typed (see Box 9.8). The patient's blood is also tested for reactivity with the donor's lymphocytes to determine if the recipient has antibodies to the donor's cells and if hyperacute rejection is likely to occur. MHC antigens that test as the same serologically are not always exactly the same genetically. Family members are often considered as potential donors. Since the genes for MHC are inherited as a unit (haplotype), each individual will share at least one A, one B, and one C with each parent; hypothetically, with three-quarters of his or her siblings; and possibly with grandparents, aunts, uncles, and cousins.

Immunosuppression

Immunosuppression is required both at the time of the transplant and lifelong for the transplant recipient. Examples of the drugs used to accomplish immunosuppression include corticosteroids, cyclosporine, and tacrolimus.[18] These medications have numerous side effects, some of which can affect nutritional status. Table 9.6 summarizes the most common medications.

Transplantation of Specific Organs and Tissues

For some sites and tissues, known as **privileged sites**, MHC matching is not required. Allograft tissue is protected from rejection in privileged sites such as the brain, the anterior chamber of the eye, and the cornea, because they are not vas-

---

BOX 9.8   CLINICAL APPLICATIONS

Finding the Right Donor: MHC Matching

In a cytotoxic assay, lymphocytes are mixed with antisera to each MHC antigen in the presence of complement, and cell lysis due to antigen-antibody reaction is detected by dyes such as trypan blue or Cr51 release. Unless a person inherited the same alleles from both parents, their lymphocytes will react with antisera to two of the A antigens, two of the B antigens, two of the C antigens, and at least two alpha and two beta chains each for DR, DP, and DQ.

In mixed lymphocyte culture, lymphocytes from the donor and recipient are mixed in culture with the donor's lymphocytes inactivated. If they are incompatible, proliferation (measured by increased DNA synthesis) is initiated. Absence of proliferation is a strong predictor of graft survival.

Table 9.6 Nutrition Side Effects of Immunosuppressive Medications

Immunosuppressive Drug	Nutrition Side Effects
Azathioprine	Nausea, vomiting, sore throat, altered taste acuity
Corticosteroids	Hyperglycemia, hypertension, sodium retention, electrolyte disturbances, impaired wound healing, hyperphagia, calciuria
Calcineurin inhibitors (cyclosporine and tacrolimus)	Hyperlipidemia, hyperglycemia, hypomagnesemia, hyperkalemia, hypertension
Sirolimus	Hyperlipidemia, gastrointestinal symptoms
Mycophenolate mofetil; mycophenolic acid	Diarrhea
OKT-3	Nausea, vomiting, diarrhea, loss of appetite

Source: NUTRITION IN CLINICAL PRACTICE by McPartland KJ, Pomposelli JJ. Copyright 2007 by SAGE PUBLICATIONS INC. JOURNALS. Reproduced with permission of SAGE PUBLICATIONS INC. JOURNALS in the format Textbook via Copyright Clearance Center.

cularized. However, trauma at the site can lead to inflammation and then rejection. Privileged tissues such as bone, cartilage, heart valves, and blood vessels are usually not rejected no matter where they are transplanted, since they are more structural than cellular. Most xenografts, such as porcine heart valves in humans, are privileged tissues.

The degree of MHC matching varies significantly according to the tissue, since tissues differ immunologically (e.g., expressing different levels of MHC II antigens). Matching is very important in hemopoietic stem cell transplants (i.e., bone marrow), significant in kidney transplants, and desirable with heart transplants, but it appears to have no net beneficial effect in liver transplants. Outcomes for individuals receiving liver transplants are not improved by MHC matching. Corneal transplants are usually not matched, but if the recipient's cornea has become vascularized, or if previous grafts have been rejected, MHC I matching may be warranted. Skin transplants are usually the patient's own skin, so incompatibility and rejection are not an immunological issue.

Kidneys for transplantation can come from either brain-dead individuals or living donors. MHC matching currently emphasizes three loci—HLA-A, HLA-B, and HLA-DR—but the closer the overall match, the better the success rate. HLA-DR antigens have a powerful impact in heart transplantation, but heart size and availability often take precedence.

Peripheral blood, bone marrow, and cord blood can all provide cells for hemopoietic stem cell transplants. Such transplants of healthy bone marrow are used to replace bone marrow that is nonfunctioning, has been damaged by high levels of chemotherapy or radiation, or has genetic defects. With **autologous** bone marrow transplants, the recipient is also the donor. Stem cells are harvested, stored, and returned to the individual after radiation or chemotherapy, and there is no risk of rejection. Both GVH and HVG rejection can occur in **allogeneic** bone marrow transplants, but GVH is normally the larger concern because the immunocompromised status of bone marrow recipients limits HVG rejection. Unrelated donor transplants using genetically matched marrow or stem cells from donors on national bone marrow registries can be used. The donor and recipient must share some MHC I and II antigens.

Immunization

Passive Immunization

In passive immunization, antibodies are administered to either prevent the disease or decrease the severity of the symptoms. Viruses, including rabies, measles, hepatitis A and B, and chicken pox; bacterial toxins such as tetanus, diphtheria, and botulism; and bites from spiders and snakes are common targets of vaccinations. Immunity is fast acting but the antibodies are short lived, and no immunological memory is induced. The major risk is **serum sickness** (see "Type III Allergic Responses").

Active Immunization

In active immunization, the individual is exposed to an antigen in a harmless form, and produces antibodies as well as activated T and B cells and memory cells. The memory immune response, initiated upon exposure to the pathogen or a booster, prevents infection and provides additional antibodies and memory cells. In some cases, such as influenza, the residual antibodies are more important than the memory response due to the short incubation period of the disease, and more frequent vaccinations are often required to maintain the antibody level.

Types of Vaccines

The ability to accomplish acquired immunity without the person contracting a disease has relied on the scientific progress in producing vaccines. Most **vaccines** are whole-organism vaccines (see Table 9.7). Live natural vaccines immunize with an organism similar to the organism that is being vaccinated against. Killed vaccines are inactivated using heat and/or chemicals that may alter the antigens. They are safe but not as effective as **attenuated** vaccines, since they produce primarily IgG and often require more frequent boosters.

Attenuated vaccines are live mutants that have lost their pathogenicity while retaining immunogenicity, and provide more natural protection. Attenuated vaccines are made by passage (serial infections) of the organism in a different species, either using cell culture or living animals, and selecting for mutants. Yellow fever vaccine 17D was made by passing a human isolate in chicken cells. As genes associated with virulence are identified, attenuation may be produced by causing mutations in the virulence genes while retaining the genes needed for immunization. Attenuated vaccines prolong the immune system's exposure to antigens and often can be given by the portal of entry used by the pathogen, resulting in mucosal immunity from IgA in addition to IgG in the blood and tissue. The increased immunogenicity can result in activation of TC and a stronger memory response, which results in a need for fewer boosters. The chief drawback of attenuated vaccines is reversion to the pathogenic form. The Sabin vaccine (attenuated) has been a powerful weapon in the control of polio, but the Salk vaccine (inactivated) is now used in the United States because wild type polio has been eradicated, and all reported cases were caused by vaccine reversions. Sabin vaccine is used where the virus is still endemic and to control epidemics.

Table 9.7 Common Vaccines

Disease	Preparation	Disease	Preparation
Chickenpox (varicella)	Attenuated virus	Anthrax	Extract of attenuated bacteria
Measles	Attenuated virus	*Hemophilus influenzae*, type b (HIB)	Capsular polysaccharide conjugated to protein
Mumps	Attenuated virus	Hepatitis B	HBsAg surface protein
Polio Sabin	Attenuated virus	Influenza	Hemagglutinins
Rubella	Attenuated virus	Meningococcal disease	Polysaccharides
Smallpox	Attenuated virus	Pertussis	Purified components (acellular pertussis = "aP")
Tuberculosis	Attenuated bacteria (BCG)	Pneumococcus	Capsular polysaccharides
Yellow fever	Attenuated virus	Staphylococcus	2 capsular polysaccharides conjugated to protein
Hepatitis A	Inactivated virus	Diphtheria	Toxoid
Polio Salk	Inactivated virus	Tetanus	Toxoid
Rabies	Inactivated virus		

Immunodeficiency

In developing countries, most immunodeficiency is caused by malnutrition, whereas in developed countries most immunodeficiency results from genetic disorders, infectious disease (HIV), or immunosuppressive therapy for other medical conditions. Babies experience transient immunodeficiency of the newborn since the antibodies provided via the placenta and in colostrum wane before the infant is able to achieve normal levels. Low levels of "natural" IgM antibodies, derived from neonatal lymphocytes and formed without direct immunization with foreign Ag, are found circulating in the umbilical cord and the neonate. Adult levels of IgM are found at 2 years of age, but mucosal secretory IgA antibodies do not reach adult levels until age 6 to 8 years. The subclasses of IgG attain adult levels from 1 to 5 years of age.

Malnutrition and Immunodeficiency

Malnutrition can result in immunodeficiency. A combination of factors, including insufficient protein, energy, and micronutrients, and not just an insufficient amount of food, are involved. Thus, undernutrition can result from personal dietary choices such as fad diets with limited food choices as well as socioeconomic factors. An individual's need for proper nutrition for optimal immune function begins in utero with maternal nutrition, and continues throughout life. Maternal nutritional deficiencies—both large-scale deficiencies due to lack of access to sufficient food and specific nutrient deficiencies due to dietary choice—impair fetal development. Maternal nutrition can impact immune functioning throughout life, not just in the fetal and neonatal stages. For example, adolescents who were prenatally and are currently undernourished produce a significantly lower antibody response to vaccination.[19] In a series of vicious circles, infants with weakened immune function may benefit less from vaccines and are more susceptible to infections such as diarrhea, which can in turn result in worsened nutrition status. Nutritional deficiencies in the elderly can exacerbate the decline in immune responses associated with aging.

Protein-energy malnutrition in infancy and early childhood has adverse effects on the thymus, including a significant reduction in thymic weight, lowered thymic hormone levels, and fewer maturing T cells. It can also cause alterations in the thymic microenvironment and peripheral T-cell function. Lowered helper T cell function will have a negative impact on both the cellular and humoral branches of the immune system. Short-term provision of a high-protein, high-kcalorie diet later can increase levels of serum IgG and IgM and improve the functioning of the cellular immune system, but cell-mediated immune responses diminish within a year of such treatment.

Nutrients critical to the development and effective functioning of the immune system include vitamins A, C, B_6, and E, essential fatty acids, beta-carotene, and the minerals manganese, selenium, zinc, copper, iron, sulfur, magnesium, and germanium.[20] Zinc deficiency promotes apoptosis in B and T lymphocytes (especially helper T cells), hinders the function of the macrophage, alters the production and potency of several cytokines, and is linked to poor thymic development in infants. Low maternal selenium is associated with lowered numbers of cytotoxic and helper T cells, B cells, and NK cells in neonates, while neutrophils and helper T cells are affected by selenium deficits later in life.[21] Vitamin B_6 is a cofactor for many enzymes involved in protein metabolism and is important for cellular growth and maintenance of the thymus, spleen, and lymph nodes. Vitamin A deficiency hinders normal regeneration of mucosal barriers; decreases the function of neutrophils, macrophages, and NK cells; negatively impacts the development of helper T cells and B cells; and diminishes antibody-mediated responses. While the essential omega-3 and omega-6 fatty acids are needed for the production and maintenance of immune cells, reduction in total fat intake enhances the immune response by increasing the numbers of monocytes and T and B lymphocytes in the blood. Vitamin C supports phagocyte oxidative burst activity as well as B cell and T cell function.[22] While nutritional supplements may be necessary to provide the desirable levels of nutrients, excessive intake of some required nutrients can create adverse reactions.

Inherited Immunodeficiencies

Most congenital/inherited immunodeficiencies are detected in young children because they experience recurrent and/or overwhelming infections, often from opportunists. Males are more apt to have immunodeficiencies, because many such deficiencies involve recessive genes, often on the X chromosome.

Some immunodeficiencies involve just one part of the immune response. In X-linked agammaglobulinemia, individuals have few or no B cells and produce no IgA, IgM, or IgE and small amounts of IgG. They suffer from numerous staphylococcal and streptococcal infections but can be treated with passive antibodies. Some individuals are deficient in a single antibody class, with IgA being the most common. Individuals with deficiencies in phagocytic cells have difficulties in killing intracellular and ingested extracellular bacteria.

Other disorders impact more than one part of the immune response. In DiGeorge syndrome, the thymus epithelium fails to develop, so T cells cannot mature, which affects the production of cell-mediated immunity and T-dependant antibodies. In Wiskott-Aldrich syndrome, a defect in a gene on the X chromosome coding for Wiskott-Aldrich syndrome protein affects B and T lymphocytes and platelets, which results in overwhelming pyogenic and opportunistic infections. Bare lymphocyte syndrome, an autosomal recessive condition, is due to a defect in genes that regulate MHC expression; in this condition, there are no MHC II antigens on cells, and thus APC can't stimulate the TH. There are several types of **severe combined immune deficiencies (SCID)**, which are characterized by extreme susceptibility to infection due to the absence of T and B lymphocyte function and often NK cells. Some forms can be treated with bone marrow transplants, and gene therapy has been used in others.

Acquired Immunodeficiencies

Some immunodeficiencies are acquired in later life. The immune system can be suppressed by many cancer drugs; by infectious agents, including HIV (see Chapter 24); or in clinical situations such as burns where there is a severe loss of immunoglobulins through wound output. **Anergy** is the term used to refer to a situation when a lymphocyte fails to respond when stimulated by its antigen-specific receptor.

Hypersensitivity

Overview of Hypersensitivity

Definition **Hypersensitivity (allergy)** occurs when extrinsic antigens (allergens) are recognized by presensitized individuals. Allergic responses are identical to responses to pathogens, but the antigen is usually harmless, so all pathology is due to immune response. Since the allergic reaction requires presen-

sitization, a response is not seen on first exposure, and reactions can become worse with subsequent exposures due to the memory response.

Epidemiology Over 50 million Americans have allergic diseases, making allergies the sixth leading cause of chronic disease.[23]

Etiology: Classifications of Allergic Reactions

Gell and Coombs (see Table 9.8) grouped allergic reactions into four classes. Types I, II, and III, which involve antibodies, are considered **immediate hypersensitivities** since initial signs and symptoms can occur within minutes to a few hours after exposure. Type IV, which is driven by T cells, is considered delayed, because it takes one to three days before a reaction is noticed.

IgE (Type I) allergies have a genetic component, and most humans with allergies are allergic to more than one substance. Some MHC types have been linked with specific allergies due to the role of MHC in antigen presentation. Some individuals have genes for high levels of IgE production and often react to numerous antigens. Several other factors influence IgE allergic reactions, including nutritional status, level of exposure to the allergen, chronic infections, acute viral infections, and exposure to environmental factors. Environmental factors include sulfur dioxide, nitrous oxide, and diesel fuel, which may increase mucosal permeability and enhance allergen entry.

Pathophysiology Type I or IgE allergies involve reactions to food and respiratory allergens, including pollens, spores, animal dander, and dust that diffuse across the mucous membrane of nasal passages and activate mucosal mast cells. In the case of respiratory allergies, these cells release mediators that produce sneezing, watery red eyes, runny noses, and respiratory distress. If food containing an allergen is ingested, activation of mucosal mast cells can cause oral inflammation, canker sores, cramps, nausea, diarrhea, gas, **hives** (urticaria), and sometimes respiratory distress. Hives occur when histamine, released from skin cells due to an allergic reaction, causes blood vessels to dilate, leak fluid, and produce swelling, which in turn irritates nerve endings and results in itching.

The most severe reaction, **anaphylactic shock** or anaphylaxis, is potentially fatal. Risk of anaphylaxis is greatest when the allergen is injected directly into circulation so that it activates cells all over the body, as occurs with insect stings and IV drugs.

Antigens involved in Type I respiratory allergies are small, highly soluble molecules presented at very low doses that are often inhaled in desiccated particles that diffuse into the

Table 9.8 Characteristics of Type I, II, III, and IV Allergic Responses

	I	II	III	IV
Time to onset of symptoms	20 to 30 min	5 to 8 hrs	2 to 8 hrs	24 to 72 hrs
Name	Anaphylactic	Cytotoxic	Immune complex	Delayed
Immunoglobulin	IgE	IgG, IgM	IgG, IgM, etc.	None (T cells)
Antigens involved	Heterologous	Autologous or hapten modified	Autologous or heterologous	Autologous or heterologou...
Cellular involvement	Mast cells and basophils	RBC, WBC, platelets, etc.	Host tissue cells	Host tissue cells

mucosa. Cells in mucosa bind allergens and transport them to lymph nodes. Transmucosal presentation causes Th2 cells to release IL-4, which causes more B cells to make IgE. IgE attaches to basophils and mast cells found throughout the body, and is highly concentrated in connective tissue, the lungs, the uterus, and around blood vessels.

The histamine concentration in mast cells is the cause of many of the symptoms experienced during the allergic reaction. In the early phase of the allergic response, granules release preformed chemicals, including histamine, the major mediator of IgE allergy in humans. The result is smooth muscle contraction and an increase in vasopermeability. In the late phase, which starts within four to six hours and lasts for one to two days, leukotrienes are released. This release increases vascular permeability; at the same time, mucus secretions contract smooth muscle in the airway, and attract and activate inflammatory cells. Even though food allergens have been studied less than respiratory ones, it does appear that the same sequence of events occurs in the gastrointestinal tract, though the reaction may be delayed due to the travel time within the tract.[24] Continued exposure to the allergen can initiate a chronic phase of inflammatory response.

Medical Diagnosis and Treatment Most treatments for IgE allergic reactions contain antihistamines, which act as competitive inhibitors and block histamine from combining with receptors on nerve endings. This treatment will block further early phase reactions, but corticosteroids are required to block the late phase. In severe reactions, **epinephrine** is used to counteract the actions of histamine.

ALLERGY TESTING Skin scratch testing is relatively inexpensive, safe, and easy to perform, but can produce discomfort in the individual being tested. Commercial inhalant allergens are available for respiratory allergies, but the stability of extracts of food allergens makes testing for food allergies a greater challenge (see "Adverse Reactions to Food" for more on diagnosis of food allergies). A small amount of suspected allergens is placed on the skin, pricked into it, or injected under the surface. A reaction, usually swelling and redness, occurs in about 20 minutes at the site of the substance(s) to which the person is allergic. Tests employing radioisotopes are used to look for levels of IgE to a specific allergen (RAST) or to any antigen (RIST) in blood.[25]

ALLERGY SHOTS Allergy shots contain a regulated dose of the compound(s) to which the person is allergic. Since the antigen does not enter transmucosally, IgG production is stimulated, and little, if any, IgE is produced. The shot series builds and maintains a high level of IgG so that allergens encounter IgG when they enter the body and do not reach the IgE on the mast cells. Since the shot contains the substance(s) to which the person is allergic, there is always a risk that the shot will initiate an allergic reaction. Therefore, recipients are often observed for a period after the shot with epinephrine ʼily available.

ʼtions to Food: Food Allergy

individuals assume that all adverse reac-
from food allergy. Though there may be
, they are quite different in their origin.

Food intolerance is defined as an abnormal reaction to food that is not immune mediated—for example, lactose intolerance or a reaction to monosodium glutamate (MSG). A food allergy, in contrast, is an immunologically based abnormal response to a food.[24, 26] Food allergies can be further classified as those that are IgE mediated, those that are non-IgE, and those that are considered to be mixed. Other allergic disorders include **food-induced eczema**, **oral allergy syndrome**, **eosinophilic esophagitis** and **exercise-induced allergic syndromes**.[27–29]

Epidemiology The overall prevalence of food allergy is 6%–8% among children and 2%–4% among adults in the United States. Interestingly, the incidence of food allergy in children under the age of 18 has increased by approximately 18% over the past ten years.[30]

Etiology

FOOD INTOLERANCE There are numerous examples of food intolerance reactions. Histamine, the major mediator in IgE-based allergic reactions, is found in cheese, some wines, and some fish, including tuna and mackerel, and may cause intolerance reactions. Lactase deficiency, which affects about 1 out of 10 people, is the most common intolerance. Deficiency of the lactase enzyme leads to gas formation, bloating, abdominal pain, and diarrhea when dairy products are consumed, because bacteria degrade the lactose that the person cannot (see Chapter 15). Food additives underlie some intolerance reactions: yellow dye number 5 can cause hives; monosodium glutamate has been linked to flushing, sensations of warmth, headache, facial pressure, chest pain, or feelings of detachment; and sulfites can irritate the lungs and lead to severe bronchospasm in people with **asthma**.[31, 32]

FOOD ALLERGY The American Academy of Allergy, Asthma & Immunology states that approximately 3 million adults in the United States report allergy to peanuts and tree nuts (almonds, Brazil nuts, hazelnuts, and walnuts). Another 6 million report an allergy to fish and shellfish. The most common allergen sources in infants and children include cow's milk, eggs, peanuts, tree nuts, wheat, soy, fish, and shellfish.[33]

Researchers have sought to understand whether the timing of food introduction for infants affects the development of food allergies. The results of a prospective cohort study recently published in the *Journal of Pediatrics* indicated that there was not a significant difference in incidence of allergic conditions (asthma, allergic rhinitis, or food allergy) when solid foods were introduced to the infant.[34] The results were not as clear for the relationship between introduction of solid foods and eczema. Nonetheless, introduction of solids after 4 months of age continues to be the standard recommendation due to the physical development of the infant.

Pathophysiology In most allergic reactions to food, a person with an inherited predisposition produces IgE in response to proteins that cross the gastrointestinal lining (see Figure 9.14). These proteins enter the bloodstream because they are not broken down by cooking, stomach acids, or enzymes. The IgE attaches to mast cells, and subsequent exposures to the food result in the reaction of the allergen with the attached IgE and the release of chemical mediators, especially histamine, by the mast cells.

Figure 9.14 Pathophysiology of the Type I Allergic Reaction

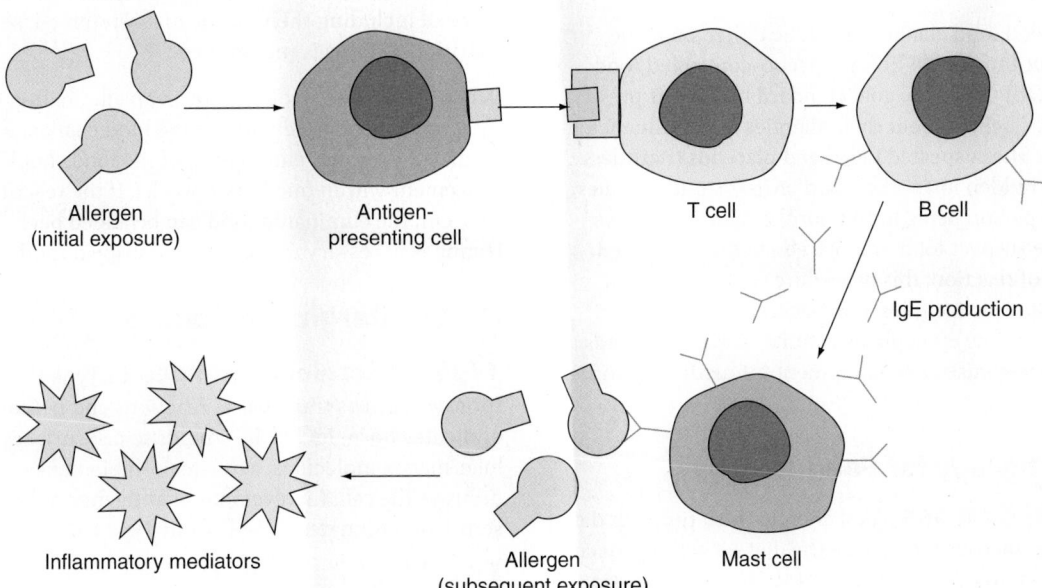

Allergen (initial exposure) → Antigen-presenting cell → T cell → B cell

IgE production

Inflammatory mediators ← Allergen (subsequent exposure) / Mast cell

Symptoms, which appear within minutes to hours, are influenced by the location of the histamine release. Reactions in the ears, nose, and throat may result in itching in the mouth or trouble breathing or swallowing, whereas interactions in the gastrointestinal tract can lead to abdominal pain, vomiting, or diarrhea. Hives can be a product of histamine release by skin mast cells. Food-initiated anaphylaxis is a severe allergic reaction involving the whole body, with the lungs the major target in humans. Histamine causes constriction of the airways, which causes difficulty in breathing; blood vessel dilation, which lowers blood pressure; and fluid leakage from the bloodstream to tissues, which results in shock, hives, and gastrointestinal symptoms such as abdominal pain, cramps, vomiting, and diarrhea.

Medical Diagnosis The first step in diagnosing a food allergy is to obtain a detailed history and perform a complete physical examination to rule out other causes of symptoms. Timing of symptoms and reproducibility are important characteristics to note. A thorough diet history and use of a food diary will provide invaluable information for diagnosis.

Several tests can be used to diagnose IgE-mediated food allergy: radioallergosorbent tests (RAST) and the CAP System fluorescent-enzyme immunoassay (FEIA), which use serum and skin prick–puncture tests; elimination diet tests; and food challenge (single- and double-blind) tests that expose the individual to the potential allergen.[35–37]

Some commercial laboratories offer RAST and FEIA testing with food allergy panels. Unlike the other methods, these do not require that the individual be exposed to potential allergens and thus pose no risk of an adverse allergic reaction during the test. They use a blood sample and provide a convenient method for both patient and physician. Nonetheless, due to lack of consistent quality control from laboratory to laboratory, there are questions about the reliability of the RAST panel. The CAP-FEIA results are as effective as the skin prick tests in predicting food allergy.[35–37]

In the skin prick test, a drop of food extract is put on the skin and then the top layer of skin is pricked with a small needle, or a pricking device is presoaked in the food extract. The skin is then observed for signs of a reaction such as swelling, as shown in Figure 9.15. Skin prick tests can be used as a preliminary test to narrow the list of potential problem foods, since the negative predictive value is greater than 95%. However, the positive predictive value is about 50%, so the test may not be sufficient to establish an absolute diagnosis.[35–37]

Another diagnostic approach, the RAST test, measures for a specific IgE antibody. If the test is positive, then that food is eliminated from the diet for up to six weeks; if symptoms have not resolved, then that food can be reintroduced. In elimination/challenge diet testing, a single food or a combination of suspect foods is not consumed for two weeks. If the symptoms disappear, suspect foods are added back, one at a time, in increasing amounts until normal levels are reached or symptoms occur. Elimination/challenge tests require a

Figure 9.15 Results of a Skin Prick Test

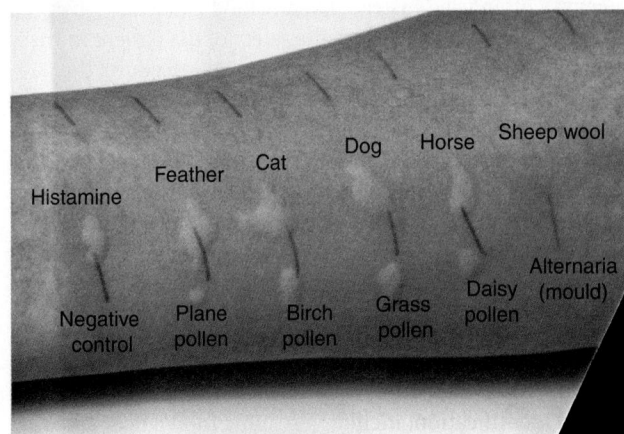

Histamine, Feather, Cat, Dog, Horse, Sheep wool, Negative control, Plane pollen, Birch pollen, Grass pollen, Daisy pollen, Alternaria (mould)

Source: Southern Illinois University/Photo Researchers, Inc.

high degree of patient motivation and compliance during the elimination phase.

Placebo-controlled food challenges can be either single or double blind, but the double-blind, placebo-controlled food challenge test (DBPCF) is the gold standard because it prevents individuals' beliefs about their allergies from influencing their responses. The suspected foods and placebos (harmless substances) are hidden in another food or in opaque capsules, and neither the person being tested nor the provider knows whether it is the suspect food or a placebo being consumed. Due to the risk of reaction, this procedure requires trained personnel and specific facilities: rapid access to emergency medications, including epinephrine, antihistamines, steroids, and inhaled beta agonists, and equipment for cardiopulmonary resuscitation.[36]

Nutrition Therapy for Food Allergy

Nutrition Implications Food allergies may result in the elimination of numerous foods from the diet. When this occurs without proper guidance for nutrition adequacy, delayed or stunted growth in children as well as additional nutrient deficiencies in both children and adults may develop.

Nutrition Assessment Standard nutrition assessment procedures (see Chapter 3) should be used for all individuals with food allergies. It is especially crucial to obtain an extensive dietary history, both retrospectively and prospectively. Numerous data collection tools may be utilized, including food frequency questionnaires and food diaries, some of which are designed to elicit a dietary history that is detailed enough to correlate foods with allergy symptoms. This is the most important method to connect specific nutritional intake with allergic reactions.

Nutrition Diagnosis Common nutrition diagnoses associated with food allergy include: inadequate food/oral beverage intake; undesirable food choices; food and nutrition-related knowledge deficit; disordered eating pattern; and poor nutrition quality of life.

Nutrition Intervention Both the diagnosis and the treatment of food allergies involve nutrition interventions. The double-blind, placebo-controlled food challenge has long been the standard procedure in diagnosing food allergies. Though it reduces bias, this method is cumbersome, and open food challenges are often used instead. For any food challenge, physician support is necessary to ensure the safety of the patient in case of anaphylaxis. Prior to the food challenge procedure, the patient is instructed to discontinue any use of antihistamines and asthma medications for at least two weeks prior to the tests. The suspect food is then given to the patient in such a way that food is not known by either the patient or the clinician (in CF). If the patient does not react to the initial dose, are increased in 10- to 15-minute intervals, with a as the final challenge after 90 minutes.[37]

food allergy has been established, of avoidance of all sources of that ill include development of skills uding reading food labels; steps turers; and methods to avoid cross-

contamination. Finally, nutritional equivalents for the food that is being eliminated need to be identified, and the importance of including them in the diet reinforced, so that adequate nutritional status is maintained.

When the suspect food is known, an elimination diet that omits that food can be tried without the food challenge. For IgE-mediated allergies, elimination of the food should result in improvement within one to two weeks. If the reaction continues to occur, the eliminated food can be added back, but further testing is necessary to confirm the suspect food.[35]

Other Allergic Responses

Type II Allergic Responses In Type II allergic responses, such as transfusion reactions and hemolytic disease of the newborn, IgG or IgM binds to cell surfaces or extracellular matrix molecules, activates complement and ultimately destroys the cell. Involvement of antigens on the cell and not in serum differentiates Type II from Type III.

Type III Allergic Responses: Immune Complexes Immune complexes (clusters of antibodies bound to antigens) are usually removed by macrophages in the liver and spleen. In Type III allergies, however, these complexes are deposited in blood vessel walls and tissues, especially the synovial membrane of joints and glomerular basement membrane of the kidney. This reaction contributes to the pathology of infectious diseases such as leprosy, malaria, dengue, and viral hepatitis; the mechanism of some autoimmune diseases; allergic pneumonitis; and serum sickness.

Immune complexes form due to the continued presence of antigens in blood from low-grade, persistent infections in infectious disease, continual inhalation of an antigen in allergic pneumonitis, and passive antibodies in serum sickness. In serum sickness, immune complexes deposit in blood vessels and tissues, resulting in hives; edema in the face, neck, and joints; joint pain; malaise; and fever that lasts 7 to 10 days. Long-lasting sequelae and fatalities are very rare. Serum sickness develops in 50% of individuals who receive a foreign antibody during passive immune therapy (e.g., tetanus **antitoxin**), due to the presence of the antigen (passive antibody) and the antibody (anti-antibody) in blood at same time.

Type IV Allergic Responses: Delayed Hypersensitivity Type IV allergic reactions—also called **delayed hypersensitivity**, T cell-mediated allergy, or contact dermatitis—include allergies to nickel, latex, plant chemicals (poison ivy or poison oak), and the tuberculin used in TB testing. Small, nonprotein antigens or haptens complex with skin proteins when they penetrate the skin, sometimes due to scratching. APC (Langerhans cells) present complexed proteins to TH cells, which become activated and produce memory cells that reside in the skin. Upon subsequent exposure, activated memory cells produce cytokines, including IL-17 and IFNg, that cause skin keratinocytes to secrete IL-1, IL-6, TNF GM-CSF, and chemokines. Chemokines attract monocytes and activate macrophages. These start a generalized attack that results in the characteristic skin irritation. The treatment for delayed hypersensitivity usually involves immunosuppression by medications such as steroids.

Autoimmunity

Autoimmune disease occurs when a specific adaptive immune response is mounted against self, and is a consequence of the open repertoires of B and T cells that allow them to recognize any pathogen. Since many antigens on human cells and pathogens are similar, immune cells targeted at pathogens can cross-react with human cells. This potential for cross-reaction cannot be eliminated, or there would be a limited response to pathogens. When an **autoantibody** is found in association with disease, the autoimmune response usually produces lesions, but in rare cases, such as the anticardiac antibody found after myocardial infarction, tissue damage simulates an autoantibody.

Autoimmune disease affects 5% to 7% of adults in Europe and North America, with autoantibodies more common in older people. Many clinically normal individuals have low titers of antibodies against some of their own tissues (e.g., against erythrocytes), and these increase with age. Babies can have autoimmune responses for a short time due to maternal antibodies. Most autoimmune diseases are more common in females. Estrogen and testosterone are thought to play a role, because

they activate cells to express different genes. In many autoimmune diseases, including rheumatoid arthritis (RA), disease severity decreases during pregnancy but rapidly rebounds after pregnancy termination.

Autoimmune diseases may be organ specific, with the thyroid, adrenals, stomach, and pancreas common targets, or non-organ specific (see Table 9.9). Systemic lupus erythematosus (SLE) involves all or almost all tissues in the body. Many autoimmune disorders have spontaneous exacerbations and remissions due to fluctuations between positive and negative regulatory factors. An affected individual can have more than one autoimmune disorder (e.g., the RA cluster), and approximately 15% of all autoimmune patients have two.

Autoimmune diseases have a strong tendency to run in families, with a 40% chance that a family with one affected adult will have another. Genetic factors are clearly involved in autoimmune disease, with combinations of alleles rather than a single predisposing allele the norm. Identical twins both develop a common autoimmune disease only 20% to 40% of the time, so it is highly likely that environmental factors such as diet are also important.

Table 9.9 Major Autoimmune Diseases

Disease	Organ	Mechanism
Hashimoto's thyroiditis	Thyroid	Inflammation is linked to antibodies against thyroglobulin (TG) and thyroid peroxidase (TPO); autoreactive cytotoxic T cells and natural killer cells destroy the thyroid gland.
Graves' disease	Thyroid	The antibody to the thyroid-stimulation hormone receptor on thyroid cells reacts with the receptor and has the same effect as thyroid-stimulating hormone, but it is not subject to feedback control, which results in overproduction of thyroid hormone.
Pernicious anemia	Red blood cells	An autoantibody reacts with intrinsic factor produced by parietal cells, resulting in decreased B_{12} absorption in the small intestine
Addison's disease	Adrenal	Antibodies attack and destroy the adrenal cortex cells that make cortisol and aldosterone.
Premature onset menopause	Ovary	Destruction of ovarian function that is linked to autoimmune responses.
Male infertility	Sperm	Antisperm antibodies bind to the sperm and impair motility, cause them to clump together, and interfere with fertilization of the egg.
Type 1 diabetes mellitus	Pancreas	Insulin-producing ß cells are destroyed by cytotoxic T cells or antibodies.
Myasthenia gravis	Muscle	Cells from the immune system cause inflammation in the bowel wall; may also involve antibodies generated in response to an infection that cross-react with cellular antigens.
Goodpasture's syndrome	Kidney, lung	Autoantibodies are deposited in the membranes of the lung and kidneys, causing both inflammation in the kidney and bleeding in the lungs.
AI hemolytic anemia	Red blood cells, platelets	Antibodies bind to cell membrane antigens, causing cell lysis.
Ulcerative colitis/Crohn's disease	Gastrointestinal tract	Cells from the immune system cause inflammation; the condition may also involve antibodies generated in response to an infection that cross-react with cellular antigens.
Sjögren's syndrome	Secretory glands	The exact trigger and target are unknown, but WBC invade and destroy glands that produce moisture, resulting in dry mouth and dry eyes; also negatively affects joints, lungs, muscles, kidneys, nerves, thyroid gland, liver, pancreas, stomach, and brain.
Rheumatoid arthritis (RA)	Skin, kidney, joints	The etiology of RA is not fully understood; the presence of rheumatoid factor (an autoantibody, usually IgM, that reacts with IgG), cytokines, and cells of the immune system are indications of an autoimmune link to an acute or chronic inflammation of synovial joints that causes pain, damage, and loss of function.
Systemic lupus erythematosus	Joints, etc.	Immune complexes containing antibodies to DNA, RNA, and nucleoproteins are deposited in the walls of small blood vessels in the kidney and joints.
Rheumatic fever	Heart	Antibodies generated in response to Group A *Streptococcus* cell wall antigens cross-react with cardiac muscle and heart valves, causing damage to the heart.

Several mechanisms can cause damage associated with autoimmune disease. Antibodies can bind to cell membrane antigens, causing cell lysis (autoimmune hemolytic anemia), or bind to receptors, stimulating them (Graves' disease). Auto-antibodies can also bind to receptors and either block or damage the receptor (myasthenia gravis). Immune complex deposition in walls of small blood vessels in the kidney and joints is a key characteristic of systemic lupus erythematosus. In Sjögren's syndrome, WBC invade and destroy glands that produce moisture, resulting in dry mouth and dry eyes. Rheumatoid arthritis is characterized by acute and chronic inflammation of synovial joints causing pain, damage, and loss of function. Etiology of RA is not fully understood, and several etiological factors may cause rheumatoid arthritis even in the same individual. In type 1 diabetes mellitus, insulin-producing beta cells are selectively destroyed by cytotoxic T cells or antibodies. The role of autoimmunity in multiple sclerosis (MS) is a subject of intense study. In MS, immune cells attack and destroy the myelin sheath of neurons in the brain and spinal cord, resulting in a decrease in speed and efficiency when nerve messages are sent.

Conclusion

The study of both health and disease is dependent on the successful comprehension of how the body responds to cellular injury and the role of the human immune system in this response. Because malnutrition is the leading cause of immunodeficiency, it is especially important for one to understand the interdependent relationship between nutrition and immunity. Equally crucial, however, is the concept that disease response, as the practitioner may see first-hand in many hospitalized patients, is linked to both nutritional status and immunocompetence. Understanding this complex relationship will only further improve the clinician's ability to enhance both.

WEB LINKS

Genes and Disease This website sponsored by the National Center for Biotechnology Information provides an excellent foundation on the etiology of disease, emphasizing the genetic contribution.
http://www.ncbi.nlm.nih.gov/bookshelf/br.fcgi?book=gnd

General Immunology This site provides a helpful review of key immunology concepts, including the many types of cells and organs involved.
http://www.cehs.siu.edu/fix/medmicro/genimm.htm

Microbiology and Immunology On-line, University of South Carolina School of

Medicine An online textbook is available at this site.
http://pathmicro.med.sc.edu/book/immunol-sta.htm

The American Academy of Allergy, Asthma & Immunology The largest professional medical specialty organization in the United States.
http://www.aaaai.org/

The Food Allergy & Anaphylaxis Network (FAAN) FAAN serves as the communication link between the patient and health care professionals.
http://www.foodallergy.org/

American College of Allergy, Asthma & Immunology You can locate an allergist and find patient education materials at this site.
http://www.acaai.org

Medline Plus: Food Allergy This site includes descriptions of various conditions and their treatments as well as general overview information on food allergies.
http://www.nlm.nih.gov/medlineplus/foodallergy.html

TransWeb Links to sites with information on organ transplantation.
http://www.transweb.org

END-OF-CHAPTER QUESTIONS

1. When researchers study the prevalence of atherosclerosis in developing countries and compare this to the prevalence in industrialized nations, this is an example of:
   A. etiology.
   B. epidemiology.
   C. disease incidence.

2. Infectious disease is an example of which etiological category of disease?
   A. multifactorial
   B. acquired
   C. genetic
   For the other two answers that you did not choose, give an example of that category of disease.

3. Mrs. J is meeting with her physician to discuss her recent diagnosis of breast cancer. As her physician outlines the probable response to therapy and how she expects Mrs. J to respond, the physician is actually discussing:
   A. remission.
   B. prognosis.
   C. cure.

4. Mrs. J's physician also states that the initial goal of her treatment is to find no indication of disease after five years post-chemotherapy. In this discussion, the MD is outlining what we could call the _____ of her disease.

   A. remission
   B. prognosis
   C. cure

5. When Mark twists his ankle in practice, the trainer immediately places cold packs and elevates the ankle. What symptoms do these two actions prevent?

6. How do nonsteroidal anti-inflammatory drugs (NSAIDs) treat the acute inflammatory process? Give an example of an NSAID.

7. List and describe factors that can influence an individual's susceptibility to infectious disease.

8. Describe an example of natural resistance.

9. What are the differences between antigens, haptens, and immunogens?

10. Describe humoral and cellular immunity, specific and nonspecific immunity, and active and passive immunity.

11. Briefly describe the function of each of the three groups of white blood cells: macrophages/monocytes, microphages/granulocytes/polymorphonuclear leukocytes, and lymphocytes and natural killer cells.

12. How are mast cells involved in the symptoms of allergies?

13. Briefly describe the functions of T helper cells, Th1 and Th2 cells, and cytotoxic T cells. What are CD4 and CD8 cells?

14. Briefly describe the functions of B cells—plasma cells, memory B cells, and antibody-producing cells. What are the immune functions for each of the five antibodies produced by B cells?

15. What is meant by "antigen-presenting cell," and which cells in the body can serve this function?

16. What is the immune function of the lymphatic system?

17. There are several soluble mediators of the immune system. List and briefly describe their function.

18. How do major histocompatibility complexes I and II aid the immune system in distinguishing between self and non-self?

19. How can T helper cells be activated? After activation, what is their response?

20. Why is it critical to match MHC antigens for tissues used in transplantation? What might happen if they are not matched?

21. What is the difference between active and passive immunization?

22. Describe one way that malnutrition can compromise immunity.

23. List common food allergies. What are the roles of nutrition intervention in the diagnosis and treatment of food allergies?

REFERENCES

1. Nowak TJ, Handford AG. Chapter 1. Cell Injury. In: Pathophysiology: concepts and applications for health care professionals. New York:McGraw-Hill; 2004.

2. McClave SA, Martindale RG, Vanek VW, McCarthy M, Roberts P, Taylor B, Ochoa JB, Napolitano L, Cresci G; A.S.P.E.N. Board of Directors; American College of Critical Care Medicine; Society of Critical Care Medicine. Guidelines for the Provision and Assessment of Nutrition Support Therapy in the Adult Critically Ill Patient: Society of Critical Care Medicine (SCCM) and American Society for Parenteral and Enteral Nutrition (A.S.P.E.N.). J Parenter Enteral Nutr. 2009;33:277–316.

3. Marik PE, Zaloga GP. Immunonutrition in critically ill patients: a systematic review and analysis of the literature. Intensive Care Med. 2008;34:1980–90.

4. Nieman DC. Immunonutrition support for athletes. Nutr Rev. 2008;Jun;66(6):310–20. Review.

5. Kiecolt-Glaser JK, McGuire L, Robles TF, Glaser R. Psychoneuroimmunology and psychosom med: back to the future. Psychom Med. 2002;64:15–28.

6. Vander Top EA, Wyatt TA, Gentry-Nielsen MJ. Smoke exposure exacerbates an ethanol-induced defect in mucociliary clearance of streptococcus pneumoniae. Alcohol Clin Exp Res. 2005;29:882–87.

7. USDHHS. Centers for Disease Control and Prevention. Standard Precautions. IN: Guideline for Isolation Precautions: Preventing Transmission of Infectious Agents in Healthcare Settings 2007. Available at http://www.cdc.gov/ncidod/dhqp/gl_isolation_standard.html

8. Sherwood L. Human physiology: from cells to systems. 7th ed. Belmont (CA):Brooks/Cole/Cengage Learning; 2010.

9. Lehman FS, Beglinger C. Impact of COX-2 inhibitors in common clinical practice a gastroenterologist's perspective. Curr Top Med Chem. 2005;5:449–64.

10. Brown TJ, Hooper L, Elliott RA, Payne K, Webb R, Roberts C, Rostom A, Symmons D. A comparison of the cost-effectiveness of five strategies for the prevention of non-steroidal anti-inflammatory drug-induced gastrointestinal toxicity: a systematic review with economic modelling. Health Technol Assess. 2006 Oct;10(38):iii–iv, xi–xiii, 1–183.

11. Barnes PJ, Adcock IM. Glucocorticoid resistance in inflammatory diseases. Lancet. 2009;373:1905–17.

12. Sriram K, Lonchyna VA. Micronutrient supplementation in adult nutrition therapy: practical considerations. J Parenter Enteral Nutr. 2009;33:548–62.

13. Campos AC, Groth AK, Branco AB. Assessment and nutritional aspects of wound healing. Curr Opin Clin Nutr Metab Care. 2008;11:281–88.

14. Thompson C, Fuhrman MP. Nutrients and wound healing: still searching for the magic bullet. Nutr Clin Pract. 2005;20:333.

15. National Pressure Ulcer Advisory Panel. Available at http://www.npuap.org/pr2.htm

16. Comfort EH. Reducing pressure ulcer incidence through Braden Scale risk assessment and support surface use. Adv Skin Wound Care. 2008;21:330–34.

17. Stechmiller JK, Cowan L, Whitney JD, Phillips L, Aslam R, Barbul A, Gottrup F, Gould L, Robson MC, Rodeheaver G, Thomas D, Stotts N. Guidelines for the prevention of pressure ulcers. Wound Repair Regen. 2008 Mar–Apr;16(2):151–68.

18. McPartland KJ, Pomposelli JJ. Update on immunosuppressive drugs used in solid-organ transplantation and their nutrition implications. Nutr Clin Pract. 2007;22:467–73.

19. McDade TW, Beck MA, Kuzawa C, Adair LS. Prenatal undernutriiton, postnatal environments, and antibody response to vaccination in adolescence. Am J Clin Nutr. 2001; 74:543–48.

20. Lomax AR, Calder PC. Prebiotics, immune function, infection and inflammation: a review of the evidence. Br J Nutr. 2009;101:631–32.

21. Dylewski ML, Mastro AM, Picciano MF. Maternal selenium nutrition and neonatal immune system development. Biol Neonate. 2002;2:122–27.

22. Bowers JM. Nutrition and immunity: you are what you eat [monograph on the Internet]. The Body; 2002. Available at http://www.thebody.com/cria/spring02/nutrition_immunity.html

23. National Institute of Allergy and Infectious Disease. Food Allergy. Available at http://www3.niaid.nih.gov/topics/foodAllergy/

24. Bischoff SC, Sellge G. Immune mechanisms in food-induced disease. In: Food allergy: adverse reactions to foods and food additives. 3rd ed. Malden (MA):Blackwell Publishing; 2003.

25. Du Toit G, Santos A, Roberts G, Fox AT, Smith P, Lack G. The diagnosis of IgE-mediated food allergy in childhood. Pediatr Allergy Immunol. 2009 Jun;20(4):309–19.

26. Sampson HA, Sicherer SH, Birnbaum AH. AGA technical review on the evaluation of food allergy in gastrointestinal disorders. American Gastroenterological Association. Gastroenterology. 2001;120:1026–40.

27. Rothenberg ME. Biology and treatment of eosinophilic esophagitis. Gastroenterology. 2009;137:1238–49.

28. Pastorello EA, Ortolani C. Oral allergy syndrome. In: Food allergy: adverse reactions to foods and food additives. 3rd ed. Malden (MA):Blackwell Publishing; 2003.

29. O'Connor ME, Schocket AL. Exercise and pressure induced syndromes. In: Food allergy: adverse reactions to foods and food additives. 3rd ed. Malden (MA):Blackwell Publishing; 2003.

30. Branum AM, Lukacs SL. Food allergy among U.S. children: trends in prevalence and hospitalizations. NCHS data brief, no 10. Hyattsville (MD):National Center for Health Statistics; 2008.

31. American Academy of Allergy Asthma and Immunology. Food Allergy. Available at http://www.aaaai.org/patients/virtual_allergist/food.asp

32. Randhawa S, Bahna SL. Hypersensitivity reactions to food additives. Curr Opin Allergy Clin Immunol. 2009;9:278–83.

33. American Academy of Allergy Asthma and Immunology. Allergy Statistics. Available at http://www.aaaai.org/media/statistics/allergy-statistics.asp

34. Zutavern A, Brockow I, Schaaf B, von Berg A, Diez U, Borte M, et al. LISA Study Group. Timing of solid food introduction in relation to eczema, asthma, allergic rhinitis, and food and inhalant sensitization at the age of 6 years: results from the prospective birth cohort study LISA. Pediatrics. 2008;121:e44–e52.

35. American College of Allergy, Asthma, & Immunology. Food allergy: a practice parameter. Ann Allergy Asthma Immunol. 2006;96(3 Suppl 2):S1–68.

36. Nowak-Wegrzyn A, Assa'ad AH, Bahna SL, Bock SA, Sicherer SH, Teuber SS; Adverse Reactions to Food Committee of American Academy of Allergy, Asthma & Immunology. Work Group report: oral food challenge testing. J Allergy Clin Immunol. 2009;123:S365–83.

37. Sicherer SH. In vivo diagnosis: skin testing and challenge procedures. In: Food allergy: adverse reactions to foods and food additives. 3rd ed. Malden (MA):Blackwell Publishing; 2003.

10

Nutritional Genomics

Melissa Hansen-Petrik, PhD, RD, LDN
Department of Nutrition, The University of Tennessee—Knoxville

Introduction

In 2003, the International Human Genome Sequencing Consortium published the finished version of the human **genome** sequence, thereby marking a historic milestone in science with great implications for the future of health care (see Figure 10.1 on page 186).[1] The human genome is the blueprint for approximately 20,000–25,000 different proteins.[2] In many respects, the human body is a system of proteins. Proteins serve as structural components, hormones, neurotransmitters, and cell-signaling agents that ensure the body is operating smoothly. Production and degradation of each of these proteins is tightly regulated but is also influenced by environmental factors such as nutrition. This interaction between nutrients and other bioactive dietary components and the genome is known as **nutritional genomics**.[3,4] The promise of nutritional genomics, ultimately, is translation of knowledge of such interactions into healthcare applications that improve health for both individuals and populations.[4] Nutritional genomics is further subdivided into the areas of **nutrigenetics** and **nutrigenomics**.[3] Nutrigenetics studies the role of interactions between

individual gene sequence variations (**genotype**) and dietary components in determining health.[5] The field of nutrigenomics focuses on the influence of dietary components on gene expression (whether genes are turned on or off) and, ultimately, production of proteins and metabolites. In the evolution of "omics," other terms including **proteomics** and **metabolomics** have also been introduced to describe the complement of proteins and of metabolites, respectively, produced by an organism.[6,7] They, like the genome, are also responsive to dietary components and all are integrated to establish the role of diet in determining **phenotype**.[8,9]

Nutritional Genomics: Nutrigenetics and Nutrigenomics

Over the last few decades, various dietary guidelines have been developed for the purpose of optimizing overall health, preventing or treating cardiovascular disease, preventing cancer, treating hypertension, and treating diabetes.[10–15] While based upon the best available knowledge of the relationship between diet and disease, these guidelines do not yet take into account the genetic and epigenetic variation within the population and how that variation can determine individual response to dietary factors and, hence, the propensity to develop disease. Such information has heretofore been unavailable, but completion of the Human Genome Project and a heavy research emphasis on identifying the specific and complex interactions between diet and the genome are now yielding results.[4]

GLOSSARY

acetylation—in genomics, modification of histones by attachment of acetyl groups

allele—a copy of a specific gene situated in a given locus on a chromosome

anticodons—tRNA coding sequences; these sequences are complementary to the codons in mRNA and thus serve as anticodons

antisense strand—the noncoding strand of DNA

autosomal dominant—an inheritance pattern of a dominant allele on an autosome

autosomal recessive—an inheritance pattern of a recessive allele on an autosome

autosomes—non-sex-determining chromosomes; a human has 22 autosomes

chromatin—the entire complement of DNA plus the histone proteins with which it is associated

chromosomes—units of the genome, each consisting of a long molecule of DNA that encodes numerous genes plus histone proteins; there are 22 autosomes and 2 sex chromosomes located within the nucleus of a human cell

codon—a series of three nucleotides in mRNA that encodes a specific amino acid

dinucleotides—paired nucleotide sequences

epigenetics—inheritance of information based on gene expression levels rather than gene sequence; regulated by genomic modifications such as DNA methylation and histone methylation or acetylation

exons—expressed sequences in mRNA; sequences that are translated into the final protein product

gene expression—the level of activity of a specific gene in producing mRNA and, subsequently, protein; expression can be regulated by many variables, including diet

genome—the entire set of genes of a given organism

genomic imprinting—expression of specific genes, which depends on the parent of origin; some genes are expressed only from the maternal allele and others are expressed only from the paternal allele

genotype—the specific variants of a gene present in the two alleles in an individual that can result in specific traits or disorders

haplotype—a group of gene variants that occur together

heterozygous—having two different alleles or variants of a given gene

histone—a protein around which DNA is wrapped

homozygous—having two identical alleles or variants of a given gene

introns—intervening sequences in mRNA that are enzymatically excised during posttranscriptional processing prior to translation into a protein

karyotype—a chart that displays chromosome pairs in order according to size

meiosis—cell division to produce gametes (sperm and ova) that results in the production of cells with half the complement of chromosomes

metabolomics—study of the collection of all metabolites present in a living organism

methylation—the addition of methyl (-CH$_3$) groups; DNA methylation patterns can be inherited and impact patterns of gene expression

microarray—technology used to measure expression of thousands of genes simultaneously

mitosis—cell division that produces two cells that are genetically identical to the progenitor cell

monogenic—arising from a single gene

non-coding DNA—sequences of DNA that lie within or between expressed genes and whose function is largely unknown; over 95% of DNA in humans is made up of non-coding DNA; sometimes referred to as "junk DNA"

nucleotide—the building block of a nucleic acid, consisting of a ribose sugar, a phosphate group, and a nitrogenous base

nutrigenetics—the interaction between an individual's genetic profile, i.e., genotypes of specific genes and the function of proteins encoded by those genes, with nutrients and other bioactive food components

nutrigenomics—the study of mechanisms by which nutrients and other food-derived bioactive substances interact with the genome to influence gene expression

nutritional genomics—a field of study that describes the application of genetic technology to food and nutrition and includes nutrigenetics and nutrigenomics

pharmacogenomics—the interaction between drugs and an individual's genome that can impact drug efficacy and toxicity

phenotype—the expressed or physical properties of an organism

polygenic—arising from multiple genes interacting with each other

polymorphisms—DNA sequences of specific genes that vary among individuals

posttranslational modification—modification of a newly synthesized protein to its active form through changes such as phosphorylation or cleavage of specific sections

posttranscriptional processing—the processing of newly transcribed RNA to excise introns, thus creating the final mRNA product prior to translation of mRNA into a protein

promoter region—regulatory sequence in a gene to which molecules, such as fatty acids, can bind in order to induce expression of that specific gene; molecules can also bind to the promoter region to suppress transcription of a specific gene

proteomics—study of the complement of proteins produced by a living organism

sense strand—the coding strand of DNA that is transcribed into RNA

single nucleotide polymorphisms (SNPs)—situations in which one nucleotide is replaced by another in a gene, potentially leading to altered function

stop codon (nonsense codon)—the codon in mRNA that signals completion of translation

transcription—the manufacture of RNA from DNA

transcription factor—a protein that activates transcription of a gene or genes by interacting with RNA polymerase in a gene promoter region

transcriptome—the complement of transcripts (mRNA) produced during gene expression

translation—the assembly of a polypeptide chain based on the sequence of mRNA

xenobiotics—chemicals that are found in an organism but are not produced by it or expected to be there, such as drugs or pollutants

X-linked dominant—an inheritance pattern of a dominant allele on the X chromosome; such disorders are relatively rare

X-linked recessive—an inheritance pattern of a recessive allele on the X chromosome; related disorders are more common in males, who carry only one X chromosome

Y-linked—inheritance based on the Y chromosome; disorders are extremely rare and occur only in males

Attention to nutritional genomics and **pharmacogenomics** has been escalating and this trend is likely to continue in the coming years. For example, the American Dietetic Association has identified nutritional genomics as a "Mega Issue" of strategic importance to dietetics practice. In the 2008 HOD backgrounder on nutritional genomics, it is stated:

> Though still in its infancy, nutritional genomics has revealed much about the complex interactions between diet and genes. But it is in its potential applications that nutritional genomics promises to revolutionize the ways to manage human health and combat

disease in the years ahead. Great progress has already been made in modeling "personalized" nutrition for optimal health and longevity as well as in genotype-based dietary interventions for the prevention, mitigation, or possible cure of a variety of chronic diseases and some types of cancer.[16]

While genomics holds potential for revolutionizing health care, there is much that is not yet known and the advent of genomics—including nutritional genomics—is not without its ethical challenges (see Box 10.1).

BOX 10.1   RESEARCH TO PRACTICE

Ethics and the ELSI Research Program

The National Human Genome Research Institute (NHGRI) established the Ethical, Legal and Social Implications (ELSI) Research Program in 1990 as a part of the Human Genome Project.[1] Its purpose is to support research on the ethical, legal, and social implications of genetics and genomics research. Although there is great public interest in the application of personal genetic knowledge to improved health, there is also concern regarding misuse of personal genetic information.[2] In May of 2008, the Genetic Information Nondiscrimination Act (GINA) was signed into federal law by President George W. Bush. This legislation prohibits insurers from requesting or requiring genetic testing of an individual or family. It also prohibits insurers from using genetic information to establish eligibility or premiums. Furthermore, it prohibits employers from requesting or requiring genetic testing and from using genetic testing for hiring or promotional decisions.[3]

While the passage of GINA is a step in the right direction, the application of genomics itself can still be an ethical minefield. For example, examining the complex relationship of genomics with race, ethnicity, and behavioral characteristics goes beyond studying the relationship of the genome to disease propensity. Linking specific genotypes to intelligence or sexual orientation, for example, has the potential to overstate the role of genetics and confer stigmatization by suggesting alleles associated with perceived negative traits are more common in some populations than in others. Thus, the implications for individuals and society in uncovering the genomic contribution to specific behaviors or traits are immense. These implications must be considered, along with input from a diverse group of individuals and organizations, before such research is undertaken.[2] To date, "grand challenges" for the future of genomics research have been identified by NHGRI. The ELSI Research Program funds and manages studies and supports workshops, research consortia, and policy conferences related to these topics:

- **Intellectual Property Issues Surrounding Access to and Use of Genetic Information.** Projects in this area examine the impact that laws, regulations, and practices in the area of intellectual property have on (1) the development and commercialization of genomic technologies and derived products and (2) the access to and use of such technologies and information by both researchers and the public.

- **Ethical, Legal, and Social Factors That Influence the Translation of Genetic Information to Improved Human Health.** Projects in this area address issues of access to and use of new genetic information and technologies to improve human health.

- **Issues Surrounding the Conduct of Genetic Research.** Projects in this area explore ethical ways to conduct cutting-edge genetic and genomic research that involves human participants.

- **Issues Surrounding the Use of Genetic Information and Technologies in Non-Health Care Settings.** Projects in this area examine the ethical, legal, and social implications of using genetic information and technologies in non-health care settings, such as in the arenas of employment, insurance, education, adoption, criminal justice, or civil litigation.

- **The Impact of Genomics on Concepts of Race, Ethnicity, Kinship, and Individual and Group Identity.** These projects examine the complex historical, social, and psychological contexts of genomics-derived data as they relate to concepts of race, ethnicity, kinship, and identity.

- **The Implications, for Both Individuals and Society, of Uncovering Genomic Contributions to Human Traits and Behaviors.** Research in this area explores the individual and societal implications of the discovery of genetic contributions related to diseases, nondisease attributes, and various behavioral traits such as cognition, mental illness, diurnal rhythms, and aging.

- **How Different Individuals, Cultures, and Religious Traditions View the Ethical Boundaries for the Uses of Genomics.** Research in this area explores how different individuals, cultures, and religious traditions view the use of genomics.

One of the hottest areas of debate regarding the interplay between ethics and genetics has to do with direct-to-consumer (DTC) marketing of genetic tests.[4] Direct marketing companies are increasingly offering genetic testing over the Internet or via in-store sales, thereby bypassing health care providers. Health care practice, particularly in clinical genetics, is very focused on informed consent, confidentiality, and appropriate counseling and guidance regarding test results.[5] These elements may be missing in the DTC environment, potentially misleading consumers and resulting in inappropriate healthcare choices. Consumers may also be subject to dubious interpretation of scant research regarding the strength of relationships between specific genes and disease. Thirty-nine companies currently offer DTC genetic testing.[6] Notably, neither DTC genetic testing nor the information provided to consumers with testing results is regulated by the U.S. Food and Drug Administration; state regulation varies widely. According to a 2009 update of a survey conducted by the Genetics and Public Policy Center at John Hopkins University, ten states restrict surreptitious DNA collection, analysis, and/or disclosure for both health- and non-health-related purposes, fifteen states restrict surreptitious testing for health-related purposes only, six states have restrictions in the context of court-ordered parentage proceedings, and two states have employment-related restrictions only. No laws relevant to surreptitious DNA testing were identified in twenty-one states or the District of Columbia.[7,8] Where DTC is prohibited, it is generally due to a legal requirement that diagnostic tests be ordered by and results received by a licensed physician.[7] For further exploration of ethics specifically relating to nutritional genomics, see the section entitled Nutritional Genomics and the Practice of Dietetics near the end of this chapter.

References

1. National Human Genome Research Institute (NHGRI). The ethical, legal and social implications (ELSI) research program [monograph on the Internet]. Bethesda (MD): NHGRI; 2009. Available at http://www.genome.gov/10001618
2. Collins FS, Green ED, Guttmacher AE, Guyer MS. A vision for the future of genomics research: A blueprint for the genomic era. *Nature.* 2003;422:835–47.
3. Human Genome Project Information. Breaking news: GINA becomes law May 2008. Available at http://www.ornl.gov/sci/techresources/Human_Genome/elsi/legislat.shtml
4. Hogarth S, Javitt G, Melzer D. The current landscape for direct-to-consumer genetic testing: Legal, ethical, and policy issues. Ann Rev Genomics Hum Genet. 2008;9:161–82.
5. American College of Medical Genetics. ACMG Statement on Direct-to-Consumer Genetic Testing [monograph on the Intenet]. Bethesda (MD): American College of Medical Genetics; 2009. Available at http://www.acmg.org
6. Genetics & Public Policy Center. Summary: Direct-to-consumer genetic testing companies [monograph on the Internet]. Washington (DC): Genetics and Public Policy Center; 2009. Available at http://www.dnapolicy.org/resources/DTCcompanieslist.pdf
7. Genetics & Public Policy Center. Survey of direct-to-consumer testing statutes and regulations [monograph on the Internet]. Washington (DC): Genetics and Public Policy Center; 2007. Available at www.DNApolicy.org
8. Genetics & Public Policy Center. Summary: Analysis of state laws on surreptitious testing [monograph on the Internet]. Washington (DC): Genetics and Public Policy Center; 2009. Available at http://www.dnapolicy.org/resources/SurreptitiousDNAtestingsummary.pdf

Figure 10.1 Timeline of Genetics and Genomics from Discovery by Mendel of the Laws of Genetics in 1865 to Completion of the Human Genome Project in 2003

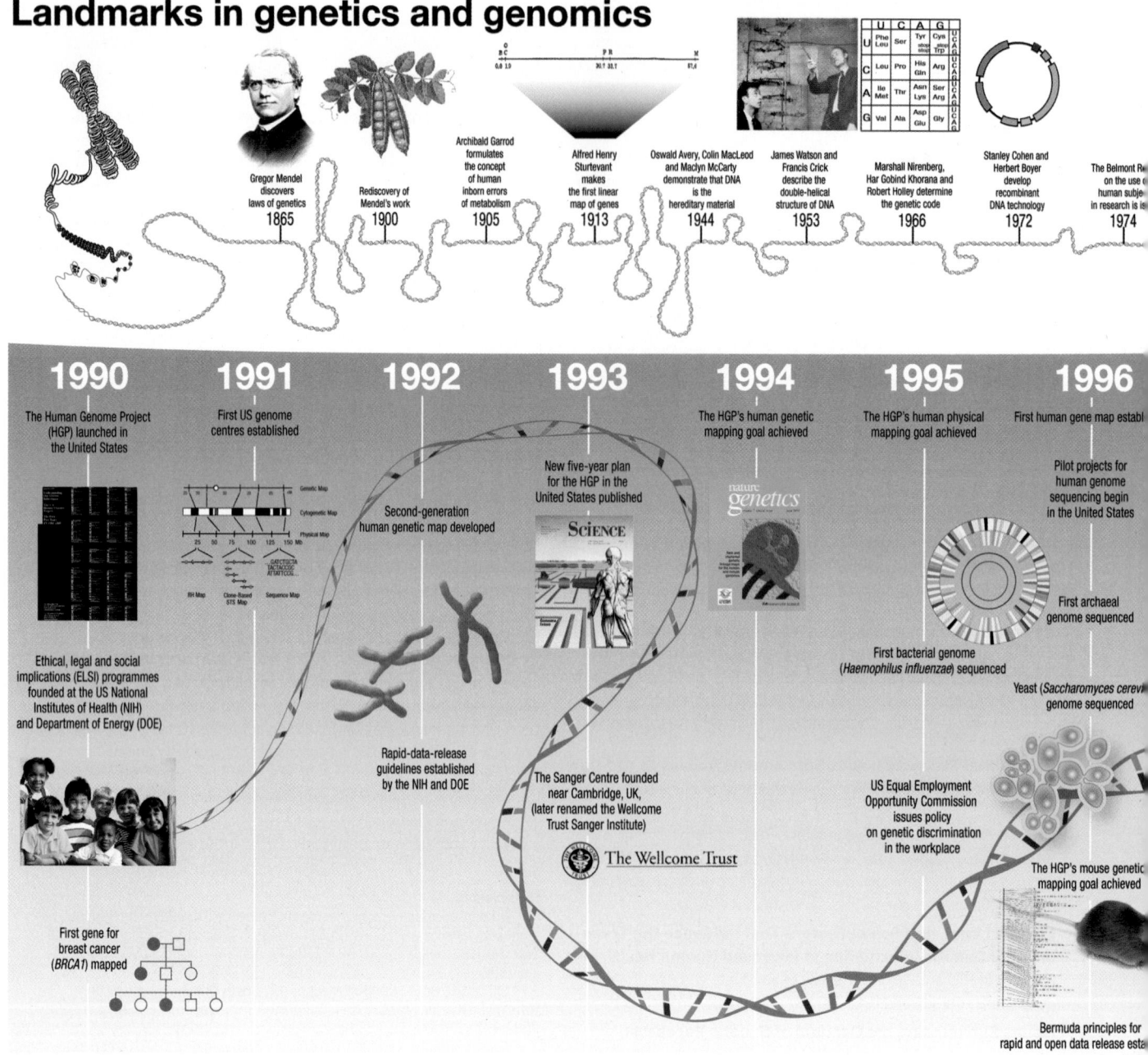

Landmarks in genetics and genomics

Gregor Mendel discovers laws of genetics
1865

Rediscovery of Mendel's work
1900

Archibald Garrod formulates the concept of human inborn errors of metabolism
1905

Alfred Henry Sturtevant makes the first linear map of genes
1913

Oswald Avery, Colin MacLeod and Maclyn McCarty demonstrate that DNA is the hereditary material
1944

James Watson and Francis Crick describe the double-helical structure of DNA
1953

Marshall Nirenberg, Har Gobind Khorana and Robert Holley determine the genetic code
1966

Stanley Cohen and Herbert Boyer develop recombinant DNA technology
1972

The Belmont Re[port] on the use [of] human subje[cts] in research is is[sued]
1974

1990
The Human Genome Project (HGP) launched in the United States

Ethical, legal and social implications (ELSI) programmes founded at the US National Institutes of Health (NIH) and Department of Energy (DOE)

First gene for breast cancer (*BRCA1*) mapped

1991
First US genome centres established

1992
Second-generation human genetic map developed

Rapid-data-release guidelines established by the NIH and DOE

1993
New five-year plan for the HGP in the United States published

The Sanger Centre founded near Cambridge, UK, (later renamed the Wellcome Trust Sanger Institute)

The Wellcome Trust

1994
The HGP's human genetic mapping goal achieved

1995
The HGP's human physical mapping goal achieved

First bacterial genome (*Haemophilus influenzae*) sequenced

US Equal Employment Opportunity Commission issues policy on genetic discrimination in the workplace

1996
First human gene map establ[ished]

Pilot projects for human genome sequencing begin in the United States

First archaeal genome sequenced

Yeast (*Saccharomyces cerevi[siae]*) genome sequenced

The HGP's mouse genetic mapping goal achieved

Bermuda principles for rapid and open data release esta[blished]

Source: U.S. Department of Energy Human Genome Program, Oak Ridge National Laboratories, Oak Ridge, Tennessee, http://genomics.energy.gov.

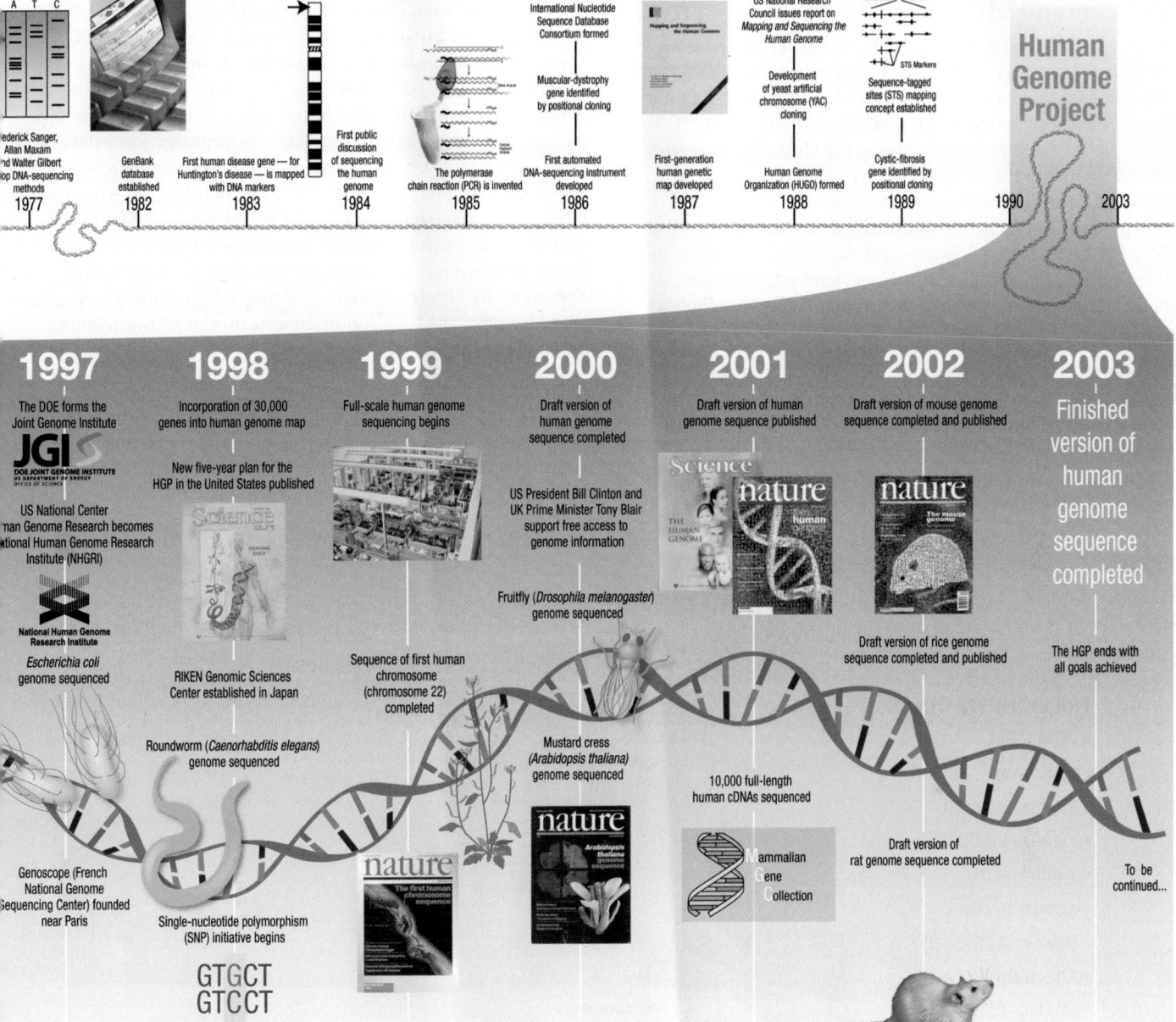

A T C

Frederick Sanger, Allan Maxam and Walter Gilbert develop DNA-sequencing methods
1977

GenBank database established
1982

First human disease gene — for Huntington's disease — is mapped with DNA markers
1983

First public discussion of sequencing the human genome
1984

The polymerase chain reaction (PCR) is invented
1985

International Nucleotide Sequence Database Consortium formed

Muscular-dystrophy gene identified by positional cloning

First automated DNA-sequencing instrument developed
1986

First-generation human genetic map developed
1987

Mapping and Sequencing the Human Genome

US National Research Council issues report on *Mapping and Sequencing the Human Genome*

Development of yeast artificial chromosome (YAC) cloning

Human Genome Organization (HUGO) formed
1988

STS Markers

Sequence-tagged sites (STS) mapping concept established

Cystic-fibrosis gene identified by positional cloning
1989

1990

Human Genome Project

2003

1997

The DOE forms the Joint Genome Institute

JGI
DOE JOINT GENOME INSTITUTE
US DEPARTMENT OF ENERGY
OFFICE OF SCIENCE

US National Center Human Genome Research becomes National Human Genome Research Institute (NHGRI)

National Human Genome Research Institute

Escherichia coli genome sequenced

Genoscope (French National Genome Sequencing Center) founded near Paris

1998

Incorporation of 30,000 genes into human genome map

New five-year plan for the HGP in the United States published

RIKEN Genomic Sciences Center established in Japan

Roundworm (*Caenorhabditis elegans*) genome sequenced

Single-nucleotide polymorphism (SNP) initiative begins

**GTGCT
GTCCT**

Chinese National Human Genome Centers established in Beijing and Shanghai

1999

Full-scale human genome sequencing begins

Sequence of first human chromosome (chromosome 22) completed

Mustard cress (*Arabidopsis thaliana*) genome sequenced

2000

Draft version of human genome sequence completed

US President Bill Clinton and UK Prime Minister Tony Blair support free access to genome information

Fruitfly (*Drosophila melanogaster*) genome sequenced

10,000 full-length human cDNAs sequenced

Executive order bans genetic discrimination in US federal workplace

2001

Draft version of human genome sequence published

2002

Draft version of mouse genome sequence completed and published

Draft version of rice genome sequence completed and published

Draft version of rat genome sequence completed

2003

Finished version of human genome sequence completed

The HGP ends with all goals achieved

To be continued...

An Overview of the Structure and Function of Genetic Material

Deoxyribonucleic Acid (DNA) and Genome Structure

While Mendel, the father of modern-day genetics, discovered the laws of genetics in 1865, deoxyribonucleic acid (DNA) was not itself identified as the blueprint of life until 1944, and its double-helical structure was not discovered until 1953.[17] DNA makes up the genome, which does not itself build an organism, but provides the instructions that tell *how* to build an organism. From a human perspective, the genes contained in each person's DNA encode essentially the same proteins, but this code varies from person to person, thus yielding inherited differences in physical characteristics, intellect, and behavioral characteristics as well as the propensity for developing disease.[17]

The genetic material or genome lies within each nucleus of each cell in the body (except mature red blood cells, which do not contain nuclei) (see Figure 10.2).[17] In humans, the genome is comprised of 23 pairs of **chromosomes**. During **mitosis** (cell division) within an individual, all 23 pairs of chromosomes are copied during the creation of a new daughter cell. These chromosomal pairs can be visualized in a **karyotype** (see Figure 10.3). During **meiosis** (reproduction), only one member of each pair of chromosomes is passed on to each ovum or sperm cell; the result is offspring that contain chromosomal pairs created by the donation of one copy of each chromosome from each parent. Of the 23 pairs, 22 are **autosomes** and one pair is comprised of the sex chromosomes. Males have one X and one Y chromosome, while females have two X chromosomes, so the gender of offspring is determined by the sex chromosome passed on by the male parent. Each chromosome consists of DNA containing a linear sequence of genes, each encoding a specific protein. The copy of each gene inherited from the father is the paternal **allele**, and the one inherited from the mother is the maternal allele. Each gene inhabits a particular location on a particular chromosome called its "locus." For example, Figure 10.4 shows a map of chromosome 4 that identifies the location of defects on this chromosome associated with specific disease states.

Each gene is itself a linear sequence of **nucleotides** that are actually responsible for encoding proteins (see Figure 10.5). Nucleotides have three primary components: a purine or pyrimidine nitrogenous base, a ribose (a pentose sugar), and a phosphate group. The backbone of the chain is an alternating

Figure 10.2 DNA—the Molecule of Life

Genetic material is located in the nucleus of each cell in the body except mature red blood cells, which do not contain nuclei. DNA is arranged in chromosomes, and specific sequences of DNA are divided into genes, each of which encodes a protein.

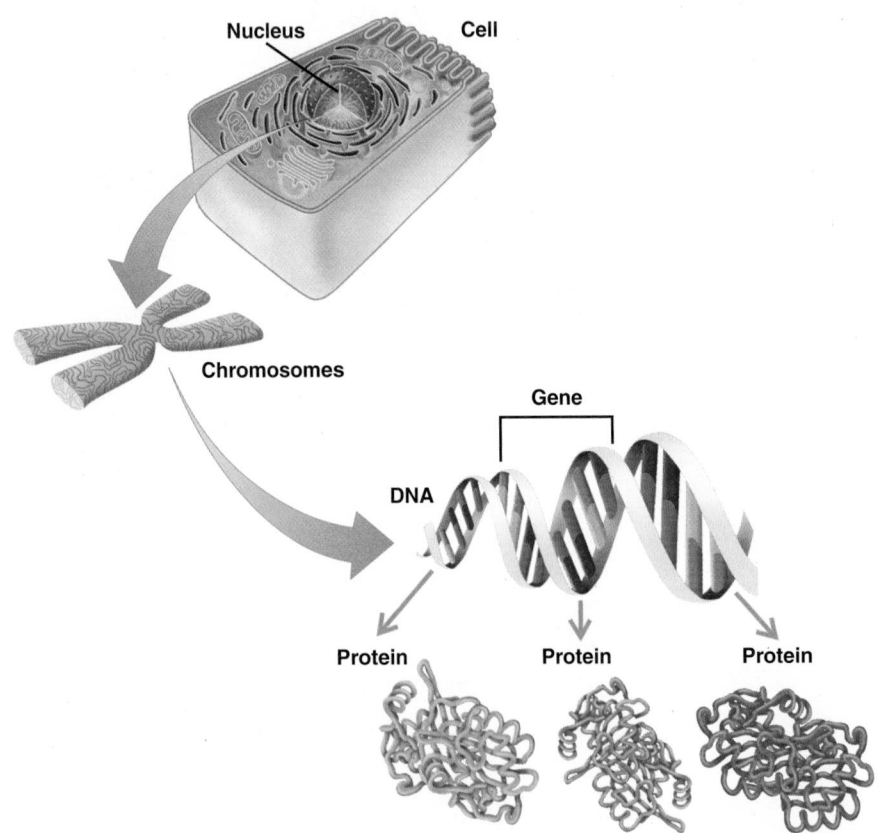

TRILLIONS OF CELLS

EACH CELL:

- **46 human chromosomes**
- **2 meters of DNA**
- **3 billion DNA subunits (the bases: A, T, C, G)**
- **Approximately 20,000–25,000 genes code for proteins that perform most life functions**

Source: F. Sizer and E. Whitney, *Nutrition: Concepts and Controversies*, 10e, copyright © 2006.

Figure 10.3 Karyotype: Down Syndrome

Microscopic examination of chromosome size and banding patterns allows medical laboratories to identify and arrange each of the 23 different chromosomes (22 pairs of autosomes and one pair of sex chromosomes) into a karyotype, which then serves as a tool in the diagnosis of genetic diseases. The presence of one X and one Y chromosome indicates this person is male. The extra copy (trisomy) of chromosome 21 in this karyotype identifies this individual as having Down syndrome. The presence of a third chromosome is often referred to as trisomy.

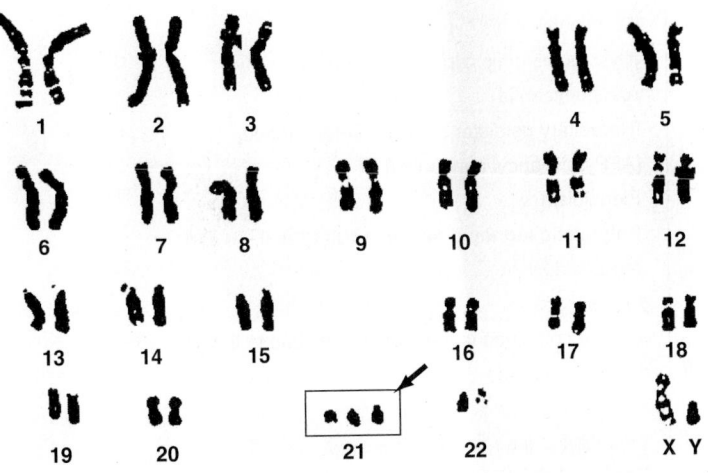

Source: U.S. Department of Energy Human Genome Program, http://genomics.energy.gov

strand of the ribose and phosphate residues. The nitrogenous bases project from this backbone and include adenine (A) and guanine (G) (both purines) as well as thymine (T) and cytosine (C), which are both pyrimidines. As DNA, this chain is paired with a complementary strand in which As always pair with Ts and Gs always pair with Cs to form a double-stranded molecule. The DNA is tightly twisted into a double-helical form, which makes each chromosome extremely compact.[18]

Translating the Message from DNA to Protein

The Genetic Code The code responsible for translation of DNA into proteins was identified as a triplet code in 1961. In other words, a series of three nucleotide bases, called a **codon**, encodes a specific amino acid. Thus, a specific sequence of nucleotides (the genetic code) translates into a specific chain of amino acids.[17] This specific chain of amino acids is a protein. Proteins have various functions, including serving as hormones, enzymes, receptors, transporters, cell-signaling agents (transcription factors, etc.), and antibodies. DNA also contains noncoding regulatory sequences called **promoter regions** to which molecules can bind in order to signal unwinding of a specific region of DNA for creation of a needed protein. Furthermore, over 95% of DNA in humans is made up of **noncoding DNA**, which lies within or in between expressed genes and whose function is largely unknown.[17] Due to the apparent lack of purpose, non-coding DNA was initially classified as "junk DNA."[17,18] This view has been challenged as research-

ers continue to identify functions of these non-coding regions. For example, numerous sections of non-coding DNA have been found to be transcribed into RNA but this does not result in production of proteins. Rather, these RNA transcripts bind to mRNA to block translation of the mRNA into protein.[19] Thus, non-coding DNA may play a largely regulatory role. Table 10.1 shows how each of the 64 codons translates into its respective amino acid. Note that in place of the "T" base there is a "U" for *uracil*, which takes the place of thymine in RNA during the process of creating a protein.[18]

Transcription and Translation Progression from code to protein involves two major steps: **transcription** and **translation**.[18] In transcription, DNA unwinds in the area encoding the gene of interest. The code is then transcribed (copied) by means of complementary base pairing (see Figure 10.6) into messenger RNA (mRNA), a single-stranded molecule consisting of the bases U, C, A, and G. mRNA is the medium by which the code for a needed protein is carried from the DNA to the cytosol, where the new protein is created. Transcription is accomplished by the enzyme RNA polymerase, which first complexes with **transcription factors** in a gene's promoter region before facilitating production of mRNA. Binding of transcription factors can either prevent RNA polymerase from binding, thus repressing transcription of a specific gene, or enhance RNA polymerase binding, thereby increasing transcription of that specific gene. Transcription is very tightly regulated and is dependent in part upon environmental variables such as dietary factors. For example, intracellular cholesterol levels (derived from diet as well as endogenous synthesis) regulate the expression of genes that regulate cholesterol synthesis and uptake from the circulation.[20] The DNA strand that serves as the template for mRNA synthesis is known as the **sense strand**, whereas the noncoding strand is the **antisense strand**. Once transcription is complete, mRNA undergoes **posttranscriptional processing**. Enzymes in the nucleus excise segments of the mRNA known as **introns** (intervening sequences), while leaving the segments known as **exons** (expressed sequences). Thus, only exons are ultimately translated into the final protein product. While DNA remains in the nucleus, the messenger RNA carries the code out of the nucleus into the cytosol, where ribosomes on the rough endoplasmic reticulum (RER) are prepared for protein assembly.[18] The complement of RNA transcripts produced when genes are "expressed" is referred to as the **transcriptome**.[21]

During translation, the triplet codons come into play. As shown in Table 10.1, most amino acids have multiple codons, but each codon only encodes one specific amino acid.[18] Small molecules of another form of RNA, transfer RNA (tRNA), serve as **anticodons**. The tRNA molecules each consist of a three-base sequence that is complementary to the codons found in mRNA. After the corresponding amino acids are transferred to the appropriate tRNA, the tRNAs carry each amino acid to the ribosomes, which serve as the protein-making machinery in the cell, and attach to the mRNA via complementary base pairing (A with U and G with C) (see Figure 10.6). After the amino acids are positioned in sequence, peptide bonds are formed between adjacent amino acids and the new protein elongates until a **stop (nonsense) codon** is reached and the newly

Figure 10.4 Chromosome 4

Sequencing and analysis of human chromosomes have enabled researchers to characterize in detail a number of genes associated with diseases. Identifying the genes on all human chromosomes offers scientists worldwide an invaluable resource for improving human health and combating disease. Knowledge about genes will increase understanding of how genetics influences the development of disease, help researchers find genes associated with particular diseases, and aid in the identification of appropriate dietary interventions and development of new pharmaceuticals. Chromosome 4 (pictured below) contains 203 million bases and is one of the larger human chromosomes. Among the many disease genes it contains is the gene for Huntington's disease, a rare single-gene disorder.

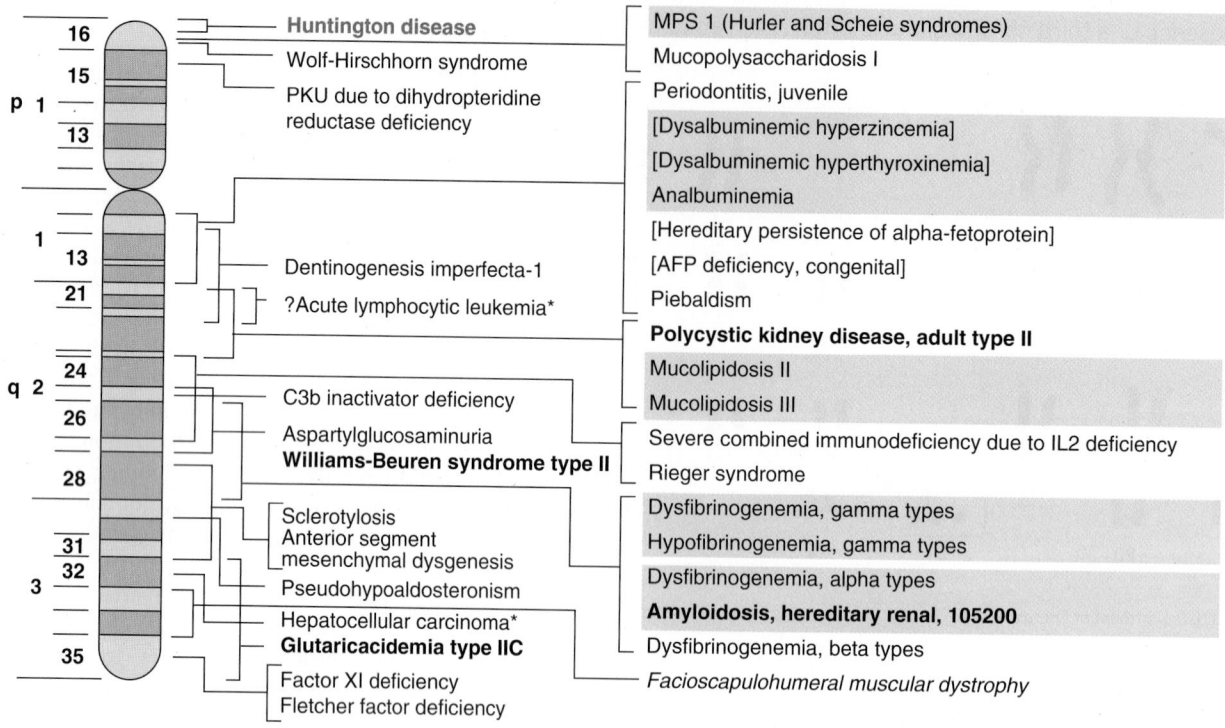

Source: U.S. Department of Energy Human Genome Program, http://genomics.energy.gov

created protein is released. Additional processing of new proteins is called **posttranslational modification**.[18] For example, the insulin polypeptide folds and forms two disulfide bonds, after which it is cut twice in the middle to remove a center section. What remains is two polypeptide chains connected by two disulfide bonds—the active form of insulin.

Genetic Variation

Polymorphisms (variations) exist within genes throughout the population.[17] Most of these variations are not a cause for concern. The outcome of a given variation depends on its nature and location within a given gene. In other words, a specific variation may have no appreciable effect on the production and function of the protein product. However, it is also possible that a single nucleotide change or a more complex alteration in a single gene can have profound effects.

Inheritance Inheritance of specific genes can be classified as **autosomal dominant**, **autosomal recessive**, **X-linked dominant**, **X-linked recessive**, or **Y-linked**.[18] Because individuals inherit one copy of each gene from each of their parents, the actual expression of an inherited gene can vary and gene expression is what determines phenotype. For example, brown eyes are autosomal dominant whereas blue eyes are autosomal

recessive. If an individual inherits the gene for brown eyes from one parent and the gene for blue eyes from the other, his or her eyes will be brown because that is the dominant gene. While the genotype includes genes for both blue and brown eyes, the eye color phenotype is brown. Thus, whether a trait is recessive or dominant determines whether that trait is phenotypically expressed. When the alleles from each parent differ from each other, as in this case, an individual is **heterozygous** for that gene (*hetero* = different). If the alleles from both parents are a match, then that individual is **homozygous** for that gene (*homo* = same).

Autosomal recessive or dominant traits can be inherited by both males and females. One common example of an autosomal recessive trait with nutritional implications is phenylketonuria, in which affected individuals must inherit one mutated copy of the phenylalanine hydroxylase gene from each parent (homozygous). The resulting inability to convert phenylalanine to tyrosine requires lifelong phenylalanine restriction to prevent mental retardation (see Chapter 26). Cystic fibrosis is another common autosomal recessive disease (see Chapter 16 and Chapter 21). Familial hypercholesterolemia is an autosomal dominant disorder characterized by absence or mutation of LDL receptors leading to severely elevated LDL-cholesterol levels and risk of early myocardial infarction and

Figure 10.5 Nucleotides within the DNA Molecule

The four nitrogenous bases of DNA are arranged along the sugar-phosphate backbone in a particular order, encoding all genetic instructions for an organism. Adenine (A) pairs with thymine (T), while cytosine (C) pairs with guanine (G). The two DNA strands are held together by weak bonds between the bases.

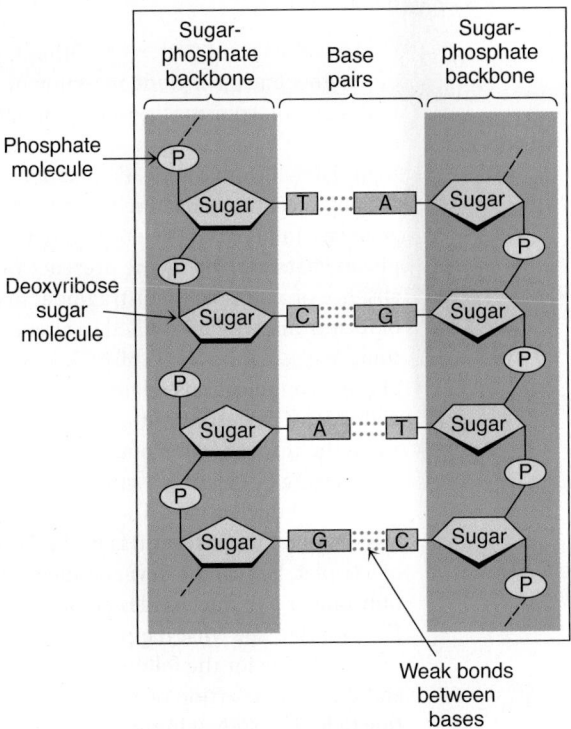

Source: R. Rhoades and R. Pflanzer, Human Physiology, 4e, copyright © 2003 p. 57.

TABLE 10.1 The Triplet Code

	U		C		A		G	
U	UUU	Phe	UCU	Ser	UAU	Tyr	UGU	Cys
	UUC	Phe	UCC	Ser	UAC	Tyr	UGC	Cys
	UUA	Leu	UCA	Ser	UAA	STOP	UGA	STOP
	UUG	Leu	UCG	Ser	UAG	STOP	UGG	Try
C	CUU	Leu	CCU	Pro	CAU	His	CGU	Arg
	CUC	Leu	CCC	Pro	CAC	His	CGC	Arg
	CUA	Leu	CCA	Pro	CAA	Gln	CGA	Arg
	CUG	Leu	CCG	Pro	CAG	Gln	CGG	Arg
A	AUU	Ile	ACU	Thr	AAU	Asn	AGU	Ser
	AUC	Ile	ACC	Thr	AAC	Asn	AGC	Ser
	AUA	Ile	ACA	Thr	AAA	Lys	AGA	Arg
	AUG	Met	ACG	Thr	AAG	Lys	AGG	Arg
G	GUU	Val	GCU	Ala	GAU	Asp	GGU	Gly
	GUC	Val	GCC	Ala	GAC	Asp	GGC	Gly
	GUA	Val	GCA	Ala	GAA	Glu	GGA	Gly
	GUG	Val	GCG	Ala	GAG	Glu	GGG	Gly

The left-hand column represents the first nucleotide base for each codon in the row, while the row across the top of the table represents the second nucleotide base in each codon. All 64 triplet codons have meaning, with 61 of them encoding amino acids and 3 serving as STOP codons to signal the end of a coding sequence. Methionine (Met) is always the first amino acid in a protein, and its codon, therefore, serves as a START codon. Standard amino acid abbreviations are as follows:

- Ala, alanine
- Asn, asparagine
- Asp, aspartic acid
- Arg, arginine
- Cys, cysteine
- Gln, glutamine
- Glu, glutamic acid
- Gly, glycine
- His, histidine
- Ile, isoleucine
- Leu, leucine
- Lys, lysine
- Met, methionine
- Phe, phenylalanine
- Pro, proline
- Ser, serine
- Thr, threonine
- Try, tryptophan
- Tyr, tyrosine
- Val, valine

death (see Chapter 13). Familial hypercholesterolemia homozygotes are rare and have a much more severe manifestation of the disorder than do heterozygotes.

The sex chromosomes also contain genes that can result in recessive or dominant disorders.[18] Two examples of X-linked recessive disorders are red-green colorblindness and hemophilia. In red-green colorblindness, individuals are unable to distinguish shades of red and green in the color spectrum, whereas in hemophilia, individuals most commonly lack clotting factor VIII, so their blood does not clot normally. Hemophilia requires transfusions to supply the clotting factor and for replacement of blood losses (see Chapter 19). Because these X-linked disorders are recessive disorders, individuals require only one normal copy of the gene for normal function. However, because males have only one X chromosome and, thus, only one copy of this gene, they are much more susceptible to inheriting these disorders. Occurrence in females is rare because they would need to inherit a defective copy of the gene from both the mother *and* father, who would himself have the disorder. Male offspring of affected fathers will not inherit the disorder because only a Y chromosome is inherited from the father. However, female offspring of affected fathers are carriers (heterozygotes), and male children born to them have a 50% chance of having the disorder, depending wholly on which copy of the maternal X chromosome is passed on. X-linked dominant traits are relatively rare. Y-linked disorders are extremely rare, occurring only in males as a result of the inheritance of mutations in the Y chromosome from the father. Y-linked disorders are not considered dominant or recessive because only one copy of the affected chromosome can exist in an individual.

Single Nucleotide Polymorphisms Understanding **monogenic** disorders such as those described above helps lay the groundwork for comprehending the complexities of **polygenic** diseases such as obesity, diabetes, cancer, and cardiovascular disease.[22] The study of gene-nutrient interactions that are dependent upon gene variance (nutrigenetics) is focused primarily on **single nucleotide polymorphisms** (SNPs, pronounced "snips").[23] SNPs are defined as those genetic variants or polymorphisms in which a single nucleotide has been exchanged for another. For example, the codon UGU is mutated to UGC. Because both encode the amino acid cytosine, this particular SNP results in no alteration in function. However, if UGA is present in place of UGU, that is a potential problem, because UGA is a nonsense or *stop* codon (see example of a SNP in Figure 10.7). Depending on the location of the mutation within a gene, such a mutation could have deleterious effects. If the affected codon is near the end of a coding sequence, it is possible that the final protein product will not be functionally altered. However, if UGU is mutated to UGG, then altering that one amino acid from cytosine to tryptophan has the potential for altering the shape and function of the protein product.

Figure 10.6 Transcription and Translation

When genes are expressed, the genetic information (base sequence) on DNA is first transcribed (copied) to a molecule of messenger RNA in a process similar to DNA replication. The mRNA molecules then leave the cell nucleus and enter the cytoplasm, where triplets of mRNA bases (codons) forming the genetic code specify the particular amino acids that make up an individual protein. This process, called translation, is accomplished by ribosomes (cellular components composed of proteins and another class of RNA) that read the genetic code from the mRNA, and by transfer RNAs (tRNAs) that transport the corresponding amino acids to the ribosomes for attachment to the growing protein.

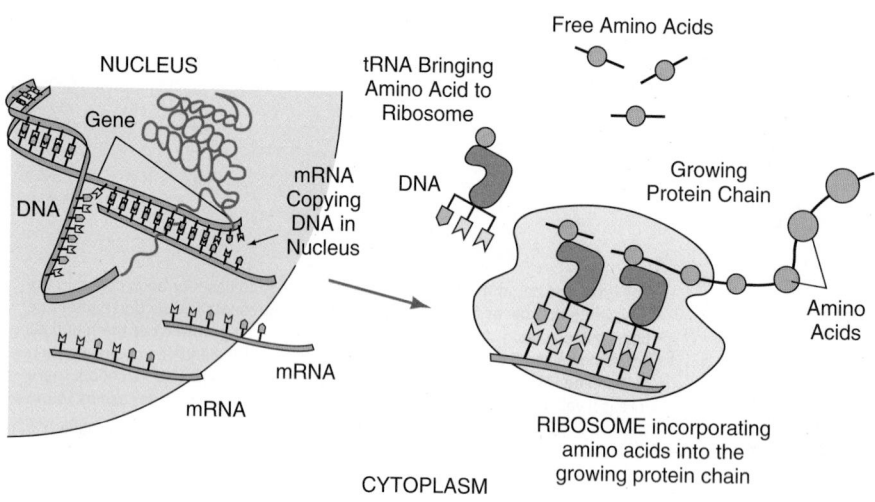

Source: U.S. Department of Energy Human Genome Program, http://genomics.energy.gov

SNPs are generally identified by the gene name, the location of the affected nucleotide within the gene sequence, the common nucleotide in that position, and an arrow indicating that a less common nucleotide is present. For example, *MTHFR* 667→T (ala→val) indicates that there is a SNP at nucleotide number 667 in the methylene tetrahydrofolate reductase gene characterized by a thymine in place of the more common cytosine. This may also be signified by *MTHFR* 667C>T. As shown in parentheses, this SNP results in an amino acid change in that position from the typical alanine to the less typical valine. This particular SNP has implications for folate metabolism and cancer risk, as discussed later in this chapter. SNPs may also be defined by the amino acid change; for instance, *PPARA*-L162V indicates the 162nd amino acid in the protein sequence for peroxisome proliferator activated receptor-α is a valine (V) when the typical amino acid in this position is a leucine (L). The nomenclature has not been standardized so other similar variations may be seen in the literature as well.

Identification of SNPs has been a primary focus of genomics research since completion of the Human Genome Project in 2003. In late 2005, it was reported that the human genome has approximately 10 million polymorphisms, meaning that any two unrelated humans have millions of genetic differences.[24] These polymorphisms are not all independent of each other. Rather, when a specific gene variant is present on a chromosome, it is associated with other particular gene variants on that same chromosome. This group of gene variants

that associate together is referred to as a **haplotype**, and these variants may work in concert to produce a specific phenotype. The focus in genomics research now is to determine which of these millions of polymorphisms is likely to be functionally important, and to continue to identify how each might relate to environment and health.[25]

Other Polymorphisms Other types of polymorphisms include insertion or deletion polymorphisms, in which a number of nucleotide base pairs are either added to or deleted from a gene. For example, the angiotensin-converting enzyme (ACE) gene has an insertion/deletion polymorphism characterized by the presence or absence of a 287-base pair fragment in one of its introns, which is linked to alterations in circulating levels of ACE and risk of renal complications related to type 2 diabetes.[26] Frameshift mutations can occur when the reading frame of a gene is altered by inserting or deleting a single nucleotide or series of nucleotides. These tend to have less impact if the insertion is in the form of a triplet, but can be devastating when only one or two nucleotides are inserted. For example, see what happens when the reading frame for the following sequence is shifted by an insertion of a single nucleotide (adenine, shown in red):

... CUU	AUG	UUA	CGU	AAG ...
Leu	Met	Leu	Arg	Lys

... CUU	AAU	GUU	ACG	UAA	G...
Leu	Asn	Val	Thr	STOP	

Other syndromes can occur as a result of inheriting extra copies of chromosomes, as in Down syndrome, or deletions of sections of chromosomes, which is one cause of the neurological disorder Angelman syndrome.

Epigenetic Regulation

Epigenetics relates not to the genome sequence itself, but to the pattern of gene expression regulated by modifications to DNA.[27, 28] Gene expression is regulated in many ways, including DNA methylation, histone methylation, acetylation, or phosphorylation, and transcription factors.[27] All of these regulatory mechanisms can be influenced by early programming in response to nutrition and other environmental factors in fetal life or infancy as well as throughout the life span. For example, monozygotic (identical) twins have an identical genotype, but in a recent study were found to have differing epigenetic patterns as they got older.[29] This was especially true of those who spent more of their lifetime apart and had the greatest lifestyle differences. This could explain differences in disease risk in twin pairs and indicates that modifying the epigenome to activate or suppress genes through diet has potential to modify risk of chronic disease and cancer. Epigenetic patterns may also be passed from one generation to the next;[30, 31] thus, individual patterns may reflect environmental exposures of previous generations.

DNA Methylation Although humans have the full complement of genetic material in all nucleated cells, not all genes are expressed in all cells, and the actual level of expression varies based on DNA **methylation** patterns. Each tissue type in the body has a distinctive methylation pattern, which results in the tissue-specific gene expression,[31, 32] such as insulin being expressed only in the beta cells of the pancreas. Approximately 2% to 5% of cytosines in mammalian DNA are methylated, primarily in CpG **dinucleotides** present in the promoter regions of genes, and the pattern of methylation is inherited.[32] This methylation (a CH_3 group is donated by *S*-adenosylmethionine) provides tight control over genes by keeping **chromatin** (DNA plus the histone proteins with which it is associated) condensed and thereby suppressing gene expression or keeping the genes "silenced."[28, 33]

For most genes, both maternal and paternal alleles contribute to production of the protein product, but for others, **genomic imprinting** takes place. In other words, for specific genes, *only* the maternal or paternal allele is expressed. For example, the gene encoding insulin-like growth factor II is expressed only from the paternal allele, while the maternal allele in mammals is silenced.[32] In humans it is predicted there are 156 imprinted genes.[34] Imprinting errors can result in devastating outcomes in offspring, including the neurological disorders Angelman syndrome and Prader-Willi syndrome.[32] It is proposed that imprinted genes (as opposed to gene sequence) hold the most promise for rapid evolutionary adaptation to changes in the nutritional environment and may therefore be implicated in the current epidemic of obesity, metabolic syndrome, and type 2 diabetes mellitus.[35]

Methyl groups are derived in the diet from sources including folate, choline, methionine, and vitamin B_{12}.[33] As shown in Figure 10.8, MTHFR catalyzes conversion of 5,10-methylene-tetrahydrofolate to 5-methyl-tetrahydrofolate, which then donates its methyl group to vitamin B_{12}. Vitamin B_{12}, thus activated, then methylates homocysteine in order to form methionine. Alternatively, choline can be converted to betaine, which can also methylate homocysteine to form methionine. Methionine adenosyl transferase then unites methionine with adenosine to form *S*-adenosylmethionine (SAM), which methylates DNA via the action of DNA methyltransferases.[33] Thus, dietary adequacy plays a role in maintaining appropriate DNA methylation.[28, 33] A deficiency of methyl groups related to lack of the previously listed nutrients means that as cells

Figure 10.7 DNA Sequence Variation in a Gene

Specific codons direct the cell's protein-synthesizing machinery to add specific amino acids. For example, the base sequence ATG codes for the amino acid methionine. Since 3 bases code for 1 amino acid, the protein coded by an average-sized gene (3000 bp) will contain 1000 amino acids. The DNA code is thus a series of codons that specify which amino acids are required to make up specific proteins. Some variations in a person's genetic code will have no effect on the protein that is produced; others can lead to disease or an increased susceptibility to disease.

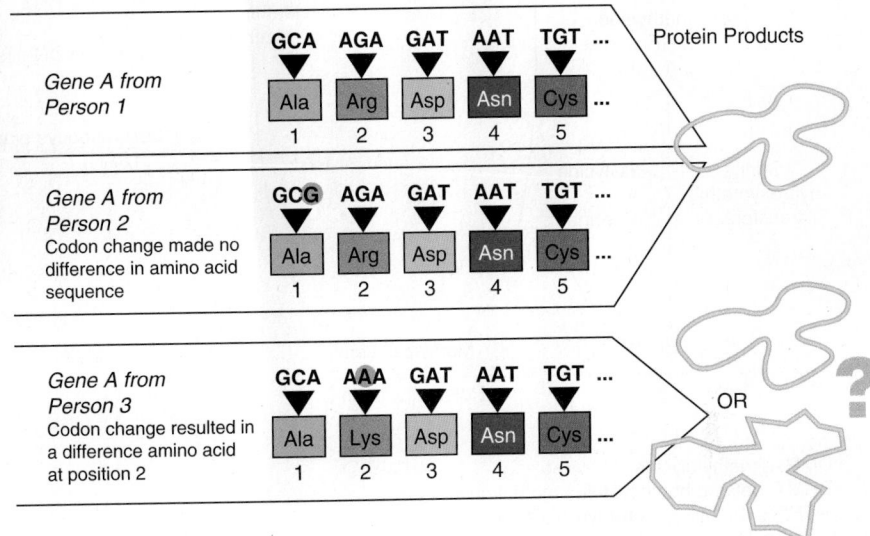

DNA Sequence Variation in a Gene Can Change the Protein Produced by the Genetic Code

Source: U.S. Department of Energy Human Genome Program, http://genomics.energy.gov

divide, methylation may be reduced, and some of that transcriptional regulation is lost. Impaired methylation of DNA is related strongly to impaired fetal development and cancer.[28] For example, hypomethylation of DNA is related to chromosomal instability, including gain or loss of entire chromosomes or increased gene mutation rates during mitosis, both of which can contribute to cancer.[33]

Histone Modification In addition to DNA methylation, **histone** modification is another form of epigenetic regulation.[36] Histones are small proteins around which DNA is wrapped. The histone tail can be modified by methylation, acetylation, phosphorylation, ubiquitination, biotinylation, and so forth, which helps to regulate transcription, DNA repair, apoptosis (programmed cell death), mitosis, and meiosis. This pattern is often referred to as the histone code. Histone modifications work in concert with DNA methylation to determine shape and accessibility of chromatin for transcription. For example, enzymes called histone acetyltransferases attach acetyl groups to histones, and this **acetylation** is associated with unfolding and accessibility of chromatin for transcription, whereas histone deacetylases, which remove acetyl groups, promote folding of chromatin and block gene transcription. Several dietary factors—sulforaphane in cruciferous vegetables, diallyl disulfide in garlic, and butyric acid derived from fermentable fibers—have been identified as inhibitors of histone deacetylase.[36]

Figure 10.8 The Resynthesis of Methionine from Homocysteine, Showing the Roles of Folate and Vitamin B$_{12}$

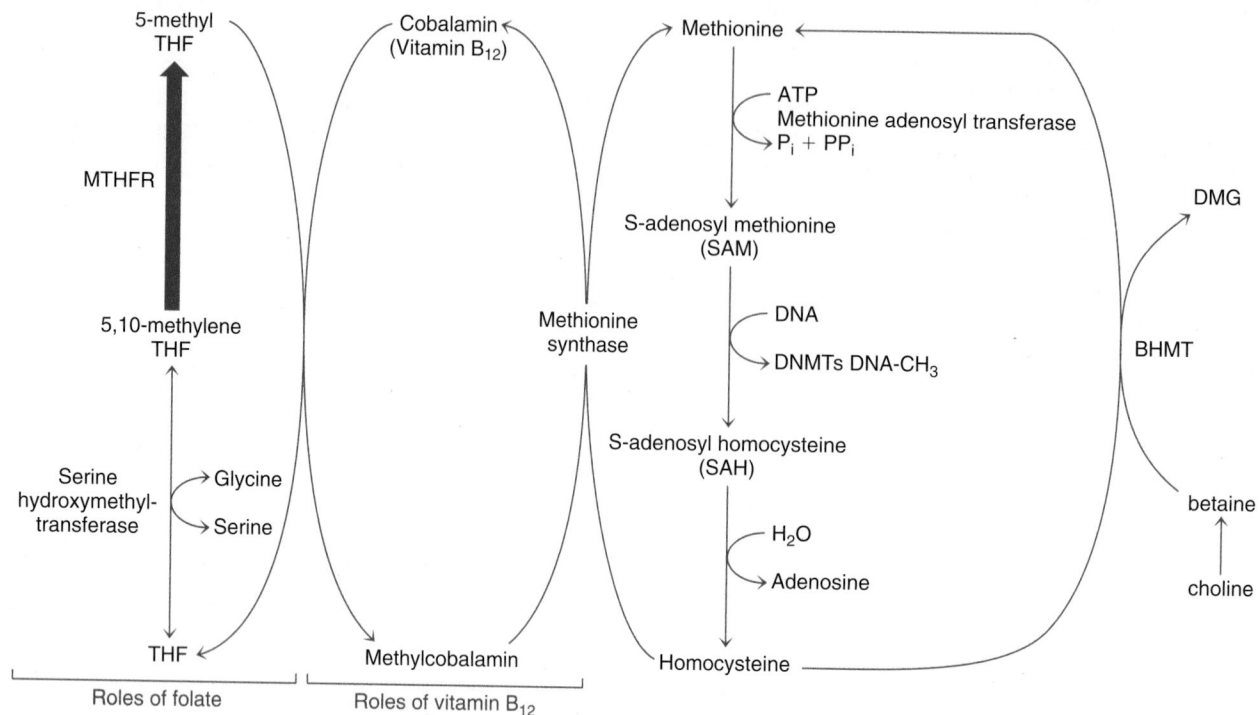

DMG: dimethylglycine
BHMT: betaine homocysteine methyltransferase
MTHFR: methylene tetrahydrofolate reductase
DNMTs: DNA methyltransferases

Source: J. Smith, J. Groff, and S. Gropper, *Advanced Nutrition and Human Metabolism*, 4e, copyright © 2005, p. 305.

The Epigenotype Because the epigenotype (an individual's unique pattern of DNA methylation and histone modification) displays greater variability than the genotype, it may be more responsive to environmental influences.[31] Several reviews on the topic have been published.[35, 37, 38] The roles of dietary folic acid, vitamin B$_{12}$, choline, and methionine are of particular interest in this regard, since these are primary sources of methyl groups, and dietary adequacy may influence DNA methylation patterns and thus genomic stability and gene expression. Methionine deficiency is unlikely, but folic acid, vitamin B$_{12}$, and choline adequacy is of concern.[33, 39] Research is only at the early stages of elucidating how dietary modifications may manipulate the epigenome. To date, an optimal epigenotype and nutrient intake levels necessary to achieve it have not been defined.

Developmental Origins of Adult Disease Epigenetic variation has been implicated in obesity and related comorbidities. This stems back to the "developmental (or fetal) origins of adult disease" or "thrifty phenotype" paradigm, which relates metabolic status and fetal adaptation in the womb to disease risk in later life.[36, 40, 41] Fetal adaptation likely involves innumerable histone modifications and DNA methylation patterns determined by the maternal diet. Beyond maternal and fetal genome sequence and interaction with the immediate environment, nutrient availability during fetal life is predictive of future growth trajectory and disease. For example, nutrient deprivation *in utero* has the outcome of low birth weight but also leads to fetal adaptation to a deprived environment by an increased efficiency in use of nutrients. This has been termed a "predictive adaptive response" in which the fetus predicts the postnatal environment based on fetal nutritional conditions and adapts in order to maximize ability to survive postnatal life.[40] This adaptation is epigenetically regulated[27] and is referred to as metabolic imprinting or metabolic programming.[42] It may manifest itself through alterations in appetite regulation, decreased physical activity, altered adipocyte metabolism, and altered mitochondrial function.[35]

As a consequence of the previously listed adaptations, nutritional deprivation in both the fetal environment and early childhood has been linked to a predisposition for metabolic syndrome, obesity, diabetes, and cardiovascular disease in later life.[27, 40, 41, 43, 44] This has been particularly true of type 2 diabetes mellitus, in which disease susceptibility is determined by an as-yet-undetermined number of genes and a cumulative effect of the environment over a lifetime.[45] For example, dietary protein restriction in pregnant rats has been shown to alter methylation and expression of genes involved in glucose and lipid metabolism and to alter glucose metabolism in offspring. The same parameters were even altered in grand-offspring who experienced nutritional adequacy in their own fetal environment, indicating the environment of the grandmother during pregnancy may influence disease risk in grandchildren related to epigenetic programming.[46, 47] This has been termed the "transgenerational effect." In another recent study examining the role of epigenetic programming and obesity it was shown that Agouti mice whose mothers were fed a diet high in methyl groups during pregnancy were less likely to be obese than mice whose mothers were fed a low-methyl diet.[48] Adaptive responses such as this provide excellent examples of phenotypic changes that can occur without regard to genotype (specific gene sequence) and profoundly impact health risk. In fact, it has been hypothesized that the fetal environment is the most critical determining factor in the development of type 2 diabetes.[41]

Genomics and Technology

Individual SNPs may not be the best measure of genotype. Rather, looking at the totality of a gene, including all SNPs in coding, non-coding, and regulatory regions, may be more appropriate as a measure of gene function in combination with epigenetic modifications.[32, 49] Additionally, regardless of individual genotypes, environmental factors play a large role in regulating epigenetics and, therefore, **gene expression**. In other words, diet, activity, smoking, and so forth can turn specific genes on or off and thus determine the quantity of specific protein products produced as well as the activity of related metabolic pathways. Thus, it is important to understand when and where a gene is expressed as well as the circumstances that influence its expression level.

Until relatively recently, it was a laborious process to conduct experiments to determine the effects of diet on alterations in expression of a single gene. However, the recent advent of **microarray** technology permits large-scale exploration of the effects of diet on the expression of thousands of genes simultaneously.[21, 50] While the entire complement of DNA is present in all nucleated cells, and genotyping can be done on any sample containing such cells, the sample used must come from the tissue of interest, because not all genes are expressed in all tissues. For example, determining the effects of diet on expression of lipogenic genes (those involved in synthesizing fat in the body, such as fatty acid synthase) would be best accomplished by analyzing liver tissue, because the liver is where lipogenesis occurs. Because tissue biopsy is not realistic in human research, much of the gene expression information is derived from animal and cell research following exposure to various experimental diets. Figure 10.9 shows an example of a microarray gene chip. mRNA is isolated from tissue samples, and the quantity of mRNA for a specific gene provides information about how highly that gene is being expressed at the time the sample was collected. Each spot on the microarray gene chip represents a single gene, and the color intensity of each spot corresponds to the level of expression of that particular gene. Animals on different experimental diets often show differing expression levels of multiple genes.[50] Further analysis is required to follow up and confirm altered expression of specific genes of interest

Figure 10.9 Microarray Gene Chip

A cDNA microarray can be used to determine how the expression of specific genes changes in response to diet. Each spot on the grid represents a specific gene. The ones with the greatest color intensity are being expressed at the highest levels.

Source: Courtesy of Julia Stair Gouffon, Affymetrix Core Facility, The University of Tennessee-Knoxville.

once identified via microarray.[51] This technology is invaluable in determining specific impacts of dietary components or dietary patterns on large-scale gene activity. SNP microarrays and a number of other high-throughput DNA sequencing methods that can be used to conduct genome-wide association studies to pinpoint gene variants linked with specific diseases have been developed.[52] These technologies allow simultaneous screening of thousands of genes and analysis of entire genomes. A more recent technological advance allows study of DNA methylation patterns and is being used to study the potential role of locus-specific DNA methylation in obesity.[53, 54]

Nutritional Genomics in Disease

Completion of the Human Genome Project and recent advances in exploration of the epigenome have thrust the interactions between the genome and environment into the limelight. Efforts to identify genes and gene variants linked to disease have escalated, as have efforts to identify environmental factors that can modulate the effects of genotype for optimization of health. While in most cases it is premature to base medical nutrition therapy on specific gene variants or epigenomic differences, discussion in this section will provide a glimpse into the future by examining a few of the gene-environment interactions and their potential roles in chronic disease. See Chapters 12, 13, 17, and 23 for current medical nutrition therapy guidelines for obesity, cardiovascular disease, diabetes, and cancer.

Cancer

From Single Gene Inherited Cancers to Gene-Nutrient Interactions
Several well-defined and relatively rare cancers have a clearly established genetic inheritance based on mutations in a single gene. An inherited mutation in the adenomatous polyposis coli (APC) tumor suppressor gene, for example, carries a 100% risk of developing the disease familial adenomatous polyposis (FAP).[55] The *APC* mutation causes FAP because it encodes a truncated, or shortened, and therefore dysfunctional, protein product that is unable to act as a tumor suppressor. FAP is characterized by development of thousands of tumors, primarily in the gastrointestinal tract, and requires intensive treatment.[55] Remarkably, this well-defined path to intestinal tumorigenesis can be thwarted to some extent by dietary means. For example, mice bearing an inherited *APC* mutation develop 50% fewer tumors when consuming a diet supplemented with the long-chain omega-3 polyunsaturated fats stearidonic acid (SDA, 18:3 n-3) or eicosapentaenoic acid (EPA, 20:5 n-3).[56] Fortunately, strictly inherited mutations are rare, although they are devastating to those affected. Heretofore, all other "noninherited" cancers have been attributed primarily to environmental exposures including diet, physical activity, alcohol intake, and tobacco use.[57] These links between cancer and environmental exposure have formed the basis for public health initiatives and education (e.g., World Cancer Research Fund/American Institute for Cancer Research, American Cancer Society), although it has not been completely clear who will benefit the most from the broad recommendations presented in these initiatives and educational efforts.

Predicting benefit is difficult because people do not all respond in the same way to environmental exposures.[57] Individuals have consequently been classified as responders or nonresponders to a specific treatment. For example, not all who smoke develop lung cancer. Not all who eat red meat, which has long been associated with higher rates of colon cancer in epidemiological studies, develop cancer. One important variant has been the lack of knowledge as to each individual's genetic background and how that may interact with nutrients or non-nutritive substances in food.[58] Nutrients have the potential to alter carcinogen metabolism, hormonal status, cell signaling, apoptosis, cell-cycle control, angiogenesis, or a combination thereof. Therefore, current research strives to identify less penetrating polymorphisms or groups of polymorphisms in a single metabolic pathway that may interact with environmental variables such as diet to increase or decrease risk of various disease states, including cancer.[58] While genotyping itself is a straightforward undertaking, linking individual foods, nutrients, or other bioactive components in foods to an interaction with each common genetic variant remains a daunting task that will take time and perseverance to accomplish. Additionally, growing evidence of the importance of epigenetics in determining cancer risk has stimulated much research in the area of chemoprevention.

Variations in Xenobiotic Metabolism Influence Risk
A study published by Le Marchand et al. illustrates the complexities of gene-environment interactions.[59] N-acetyl transferase 2 (NAT2) and cytochrome P450 1a2 (CYP1A2) enzymes in the liver are both involved in biotransformation of incoming **xenobiotics** into harmless substances for excretion. Individuals exhibiting different phenotypes of these enzymes metabolize xenobiotics at different rates and are thus often classified as slow, intermediate, or rapid acetylators.[58, 59] This phenotypic variation has potential implications for cancer risk, because such enzymes transform some xenobiotics into genotoxic substances. For example, both NAT2 and CYP1A2 are integral to the biotransformation of heterocyclic amines from cooked meat into genotoxic substances, which by definition have the potential to cause cancer. Furthermore, smoking is known to induce CYP1A2—that is, it increases production of the CYP1A2 enzyme. Epidemiological research has long linked cooked meats to increased risk of colon cancer, and the authors in this study examined how that risk is modified by phenotypic variation in these two enzymes. The only group experiencing a statistically higher risk were those with a rapid NAT2 phenotype combined with an above-average CYP1A2 phenotype who were also smokers and consumed their red meat well done.

A more recent study examined the potential role of red meats in modulating colorectal cancer risk according to genotypes of different sets of genes—those encoding various DNA repair enzymes. Findings indicate that components in red meats may interact with DNA repair mechanisms to result in higher colorectal cancer risk in individuals with specific genotypes.[60] These findings clearly illustrate the oversimplification associated with stating that "eating red meat increases colon cancer risk" when that appears to be true only for a small, well-defined subset of the population. It also illustrates the growing necessity of individualizing dietary recommendations based on specific genomic and environmental variables as research findings begin to more clearly establish such relationships.

MTHFR and ADH Polymorphisms Interact with Dietary Folate and Alcohol Cancer is characterized by altered DNA methylation patterns. Regulatory genes, i.e., those that would normally suppress tumor formation, are frequently hypermethylated or "silenced." Alternatively, genome-wide hypomethylation can result in inappropriate gene activation, chromosomal instability, and cancer. Histone methylation patterns are thought to play a role as well.[36] Low intakes of folate have a long association with cancer risk, including cancer of the colon, and this risk appears to escalate in the presence of high alcohol intake. However, results are not always consistent, suggesting individual effects may vary based on genomic characteristics. Polymorphisms in the methylenetetrahydrofolate reductase gene, primarily [MTHFR 667C>T (ala>val)], can reduce MTHFR activity.[32] As mentioned previously, MTHFR plays a critical role in metabolizing 5,10-methylenetetrahydrofolate (5,10-methylene THF) to 5-methyl-tetrahydrofolate (5-methyl THF), which is necessary for remethylation of homocysteine to methionine, formation of S-adenosylmethionine, and, therefore, DNA methylation. Because 5,10-methylene THF itself is also necessary for production of thymine, both forms of folate are needed in adequate quantities for genome health. Folate deficiency and reduced activity of MTHFR can thus contribute to both compromised genome integrity and the risk of acquiring genetic damage and cancer.[32, 33] But it is possible high folate intake and consequent hypermethylation of tumor suppressors could increase cancer risk as well. Examples of both are present in the research literature.

The interaction between folate status and MTHFR polymorphisms in carcinogenesis is illustrated by findings from the Health Professional Follow-Up Study.[61] That study showed that individuals homozygous for the TT mutation at nucleotide 667 (thus encoding valine from both copies of the gene) tend to accumulate 5,10-methyl THF intracellularly. They appear to be hyperresponders to folate status, meaning they are at low risk for colon cancer if following a low-risk diet (high in folate, low in alcohol), presumably due to accumulation of 5,10-methyl THF and optimal chromosomal stability, but may be at higher risk for developing colon cancer if consuming a low-folate, high-alcohol diet. Folate intakes were relatively high among the study population, and it is possible that more dramatic effects would have been observed with respect to the relationship of folate intake to risk had there been a wider spread in consumption levels. That will likely be difficult to see in U.S.-based studies due to the folate fortification of the food supply in place since 1998.

The interaction of folate status and MTHFR genotype with alcohol appears to be a critical point. The same researchers also examined the *alcohol dehydrogenase* (*ADH*) genotype of this cohort and found those with a slow metabolizing genotype had significantly higher risk of developing colon cancer with an alcohol intake ≥20 g/day combined with folate intakes <338 mcg/day.[61] The low *ADH* activity could result in slower alcohol metabolism and magnification of alcohol's effects. These effects may include inducing malabsorption of folate, blocking folate release from hepatocytes, and blocking remethylation of 5-methyl THF back to 5,10-dimethyl THF, thereby depleting the latter. In addition, acetylaldehyde, an alcohol metabolite, can cleave and destroy folate. Thus, a slow *ADH* genotype could contribute to colon cancer risk, dependent upon levels of alcohol and folate intake.

Current dietary intake recommendations call for alcohol in moderation.[10] However, even within moderate intake levels, individuals with the slow *ADH₃* genotype, encoding one of several known ADH isozymes, appear to be at risk.[61] Another study shows, though, that those with the intermediate *ADH₃* genotype benefit from a reduced risk of myocardial infarction with moderate alcohol intake.[62]

In contrast to published research showing the benefits of folate intake, other findings have pointed to risks associated with high folate intake in some subgroups. Ericson et al. compared dietary folate intake to breast cancer incidence and MTHFR genotype among women participating in the population-based Malmö Diet and Cancer cohort in Sweden.[63] Among women with the MTHFR 677CT or TT variants in combination with the MTHFR 1298AA genotype (a combination of two distinct SNPs in MTHFR), high dietary folate intake was associated with greater likelihood of being diagnosed with breast cancer. Among women who were heterozygous at both positions (677CT-1298AC), higher dietary folate intake was associated with a lower likelihood of breast cancer diagnosis. Thus, folate may be protective against breast cancer or may increase risk depending on genotype. The subject is reviewed in a 2006 article by Kim who concludes that routine dietary folate supplementation as a chemopreventive measure is not advisable for women and may, in fact, be harmful.[64] Again, this is a clear illustration of the critical relationship between the individual genome and nutrition as well as the complications associated with basing dietary recommendations on a single genotype. Broad dietary recommendations are still likely to be generally accepted, but in the future, as research more clearly elucidates the multitude of gene-gene and gene-nutrient relationships, individualized recommendations will be the key to optimizing health.

Fruits and Vegetables Dietary intake of fruits and vegetables has long been associated with a lower risk of colon cancer, although research results have been contradictory and specific mechanisms have remained somewhat elusive.[65, 66] One complicating factor is that each individual fruit or vegetable is made up of a wide array of nutrients and other bioactive compounds. Many of these have been studied individually, but little research has focused on the potential synergistic effects or interactions of this mix of compounds within a whole vegetable. Furthermore, genomic variation means that individual responses are likely to vary. Van Breda et al. took the interesting approach of feeding four different whole vegetables (cauliflower, peas, carrots, or onions) to mice and determining how each of these vegetables impacted gene expression in colonic tissue as a clue to their cancer-preventive mechanisms.[67] Expression of several genes known to have either promoting or protective effects with respect to colorectal cancer was altered. For example, mice fed cauliflower or carrots expressed lower levels of the ornithine decarboxylase (ODC) enzyme, which is the rate-limiting step in synthesis of polyamines from the amino acid ornithine (i.e., low ODC levels result in limited polyamine production). Polyamines have a long association with increased risk of colon cancer. ODC is also known to be regulated by the protein product of the *APC* tumor suppressor gene, whose function is lost in many cases of noninherited colon cancer as well as in FAP. Similar effects were observed with high vegetable intake in a parallel human study, along with suppression of genes encoding several cytochrome P450 isozymes.[68] Thus, the effects of these vegetables may be protective against colon cancer via interaction with this gene and/or a host of others.

Other evidence of the interaction between vegetables and genes includes the observation that colon cancer risk reduction in humans via intake of cruciferous vegetables is linked to glutathione *S*-transferase genotype.[69] Isothiocyanates derived from *cruciferae* are known to induce phase II detoxification enzymes, which, like the phase I cytochrome P450 enzymes, are involved in metabolism and removal of potential carcinogens. The glutathione *S*-transferase (GST) family of enzymes is among the most important in this regard. In the Singapore Chinese Health Study, it was demonstrated that individuals with GST genotypes leading to absence of activity among some of the GST subtypes are the only ones who benefited (lesser risk of colon cancer) from a higher intake of cruciferous vegetables.[69] This is biologically plausible because less active GST would presumably lead to slower clearance of carcinogens and greater cancer risk. A higher intake of cruciferous vegetables would lead to higher isothiocyanate levels and increased activity of remaining GST subtypes to compensate for the loss of others. Isothiocyanates are also known inhibitors of histone deacetylase and may exert anti-cancer properties via that mechanism.[36] See section "Histone Modification" earlier in this chapter for more information.

Similarly, a study of gene-nutrient interactions in breast cancer causation explored the possible role of antioxidants derived from fruits and vegetables.[70] The human body produces endogenous antioxidants, including manganese superoxide dismutase (MnSOD), and can also acquire antioxidants exogenously via dietary intake of fruits and vegetables. Since oxidative DNA damage is thought to play a role in carcinogenesis, adequacy of antioxidants is suggested to be anticarcinogenic. Results of a large case-control study found that premenopausal women homozygous for a valine to alanine change at the 9 position (resulting in loss of function) of MnSOD were at increased risk for breast cancer, which was attenuated by high consumption of fruits and vegetables. Presumably, exogenous antioxidant consumption compensated, in part, for the deficiency in endogenous antioxidant function. Although researchers are in the early stages of defining the relationships between genotype, diet, and health, and these are but a few examples, it seems likely that effects will vary among individuals depending on numerous specific genotypes and numerous environmental variables. However, the fact that nutrients and other bioactive compounds in food interact with the genome and thereby impact health and disease, including cancer risk, cannot be disputed.

Obesity

Obesity has a clear link to genetics demonstrated by numerous studies showing that obesity does persist in families even where food intake and physical activity patterns differ. However, the escalating obesity epidemic in recent decades supports the idea that, while *susceptibility* to obesity is genetically and epigenetically determined, the development of obesity itself is the result of a susceptible genome in the presence of a conducive environment. In other words, placing susceptible individuals in an obesigenic environment—one characterized by plentiful energy-dense, high-fat foods and technological advances requiring little in the way of physical activity—results in obesity. The questions then become: How do we identify those individuals who are genetically susceptible? Beyond that, what is the role of epigenetics? How do we intervene to prevent and/or treat obesity?

The answers are not simple, however. Unlike single-gene inherited traits, obesity susceptibility most often involves multiple genes and is, therefore, a complex (polygenic) genetic trait. Each gene itself may make only a small contribution to obesity risk, but a multitude of gene variants working together within the context of epigenetic programming may have a profound effect. These may include genes involved in energy and appetite regulation, metabolism, and storage. Additionally, interactions may involve multiple environmental factors; thus, the issue is complex. At this time, information is inadequate to utilize genes as markers for screening people or to develop an intervention other than current interventions for obesity treatment and prevention. In order for genetic screening to be useful, the relationship between the gene and obesity must be clearly established and there must be a useful intervention available for those deemed at risk. Research has not yet progressed to that point. However, exploring some of the most well-studied obesity-susceptibility genes sheds light on what the big picture might look like.

Eleven genes to date have been identified in monogenic obesity, primarily in severely obese individuals with onset at an early age.[71] In all, over 127 gene candidates have been linked to human obesity.[71] One gene that appears particularly promising as a marker of obesity risk is the fat mass and obesity-associated gene (FTO). This was first identified in a genomewide association study looking for type 2 diabetes susceptibility genes.[72] In a cohort of 38,759 participants the 16% who were homozygous for a specific gene variant (in this case an obesity risk allele) weighed 3 kg more and had a 1.67-fold higher likelihood of being obese. This association was observed from the age of 7 years onward. In another study of nearly 3,000 children ages 4–10 years, the same risk allele was not associated with differences in energy expenditure.[73] However, children with the risk allele had significantly greater energy intakes, suggesting that this *FTO* variant may play a role in control of food intake and possibly a preference for energy-dense foods.

Another example of a gene linked to obesity is that encoding perilipin. Perilipins are localized on the surface of fat droplets inside adipocytes and play a regulatory role, primarily by blocking release of stored triglycerides and thereby helping to preserve stored fat.[74] Mice lacking perilipin are resistant to diet-induced obesity.[75] Furthermore, obese humans have higher perilipin expression,[76] and variations in the perilipin gene are predictive of obesity risk, particularly among women.[77] A recent study found that perilipin polymorphism PLIN4 11482G>A was associated with a lower baseline body weight (234 versus 251 pounds among obese subjects enrolled in the study) and the researchers' findings further suggest that this SNP confers resistance to weight loss while following a reduced-kilocalorie diet for one year.[78] A more recent study of 234 obese children participating in a 20-week weight loss intervention examined the relationship between PLIN genotype and weight loss success.[79] Those with the PLIN4 allele were more likely to have metabolic syndrome at baseline, whereas those with the PLIN6 14995A>T variant were the most successful weight losers. Such findings require further confirmation with larger samples, but have important implications if specific genotypes are definitively proven predictive of weight loss success. Moreover, it is likely that there are numerous variables that predict weight loss success, and one must consider the ethics of advising a client to refrain from attempting weight reduction due to a prediction of failure.

Another interesting polymorphism relating to obesity is one occurring in the serotonin (5-hydroxytryptamine or 5-HT) receptor gene promoter. The neurotransmitter serotonin is a key regulator of food intake whose function has been related to obesity and anorexia. A study of 370 children and adolescents ages 10 to 20 suggests that the 1438G>A polymorphism in the 5-HT$_{2A}$ gene promoter does indeed impact food intake.[80] While there was no difference among study subjects in way of age, height, weight, or BMI, there was a significantly higher intake of energy and fat in children with two G alleles compared to those with two A alleles, and an intermediate effect for those with one G and one A allele. While the differences in fat and energy intake were not linked to overweight in these children, another study of the same polymorphism in middle-aged men observed significantly higher BMIs and abdominal fat associated with the GG genotype.[81] It is possible that these gene polymorphisms simply coexist with polymorphisms in other genes that predict eating behavior and obesity. It is also possible that 5-HT$_{2A}$ receptor promoter polymorphisms result in mood or personality characteristics that impact food intake.[80] In all, the polygenic nature of obesity paired with its complex environmental interactions will require large studies with thousands of subjects to accurately identify the contribution of various gene polymorphisms.

Diabetes

Hundreds of genes have been examined for potential roles in the development of type 2 diabetes, including those that may play a role in pancreatic beta cell function, insulin signaling, and so forth.[45] The closely related metabolic syndrome, characterized by insulin resistance, dyslipidemia, abdominal obesity, and hypertension, has garnered much recent attention due to its association with an increased risk of both type 2 diabetes and cardiovascular disease.[82] Growing evidence suggests a proinflammatory state underlies the abnormalities associated with metabolic syndrome and predisposition to diabetes and cardiovascular disease. The Genetics of Lipid Lowering Drugs and Diet Network (GOLDN) study has been undertaken to identify the genetic variables associated with response of triglycerides to diet intervention. The researchers found that participants with a polymorphism in the interleukin-1ß gene, which encodes a proinflammatory cytokine, had a greater likelihood of developing metabolic syndrome. However, when stratified by intakes of dietary polyunsaturated fat, it was found that the increased risk was only present if their polyunsaturated

fatty acid (PUFA) intake was also low. Other inflammatory markers involved in metabolic syndrome are also under investigation for potential gene-diet interaction.[83] Many other findings (described in this section) beyond the thrifty phenotype theory have focused on modulation of insulin resistance or prevention of type 2 diabetes in susceptible individuals.[84–87]

The Finnish Diabetes Prevention Study, which aimed to measure the effect of lifestyle intervention in preventing conversion from impaired glucose tolerance (prediabetes) in obese subjects to type 2 diabetes, has also examined the roles of genes.[86] The researchers observed that polymorphisms in the gene encoding GLUT2, or glucose transporter 2, which helps the pancreatic beta cells detect glucose and secrete insulin accordingly, are related to type 2 diabetes risk. Those in the intervention group had an equally low risk of developing type 2 diabetes regardless of genotype, but those in the control group (continuing their usual lifestyle) were significantly more likely to develop type 2 diabetes if they had a polymorphism in the gene for GLUT2 versus the common allele. Because this subset of the population is at high risk for conversion to type 2 diabetes and benefits from lifestyle intervention, these polymorphisms may serve as a trigger for early and intensive nutritional and physical activity intervention.

One other gene heavily investigated in type 2 diabetes is that for peroxisome proliferator-activated receptor gamma (PPARγ), which is a receptor on the cell nucleus that plays a central role in adipocyte development and function. PUFAs are natural ligands for this receptor—meaning that they bind to and activate it—but thiazolidinedione drugs (i.e., rosiglitazone) treat diabetes also by interacting with PPARγ to enhance insulin sensitization. Thus, PPARγ activation is associated with greater insulin sensitivity. Because fatty acids with longer chain length and greater desaturation have a higher affinity for PPARγ, diets high in saturated fat are likely to have little effect on PPARγ and have been associated with insulin resistance.[85] Studies of humans have shown that a relatively common P12A variant, in which alanine (Ala) is substituted for the amino acid proline at the 12 position in one allele, is associated with a protective effect, resulting in a 25% lower risk of type 2 diabetes.[84] It has been suggested that this polymorphism may interact with dietary fatty acid composition and physical activity level to determine diabetes risk by influencing fasting insulin levels. For example, high levels of physical activity and a high dietary polyunsaturated fat-to-saturated fat (P:S) ratio independently contributed to lowering fasting insulin levels among proline allele homozygotes. In contrast, Ala allele carriers did not benefit at all unless the high P:S ratio and high physical activity level were present simultaneously.[85] Thus, although the Ala allele carriers may be at lower risk of diabetes, disease development seems to be subject to critical environmental determinants.

While a diet high in saturated fat has been linked to insulin resistance, and diets higher in monounsaturated fats or carbohydrate are linked to improved insulin sensitivity,[88] this does not necessarily hold true in all people. For example, the apolipoprotein E, or APOE, genotype has also been linked to insulin resistance in response to dietary fat. Apolipoprotein E plays an important role in lipoprotein metabolism, and specific genotypes have been linked to insulin resistance. In a study of healthy subjects, those with a specific variant in the APOE gene promoter did not experience a lowering of glucose and insulin levels when switching from a diet high in saturated fats to diets high in monounsaturated fatty acids (MUFA) or carbohydrates.[87] Identification of such polymorphisms can assist in determining which patients will benefit or fail to benefit from specific diet prescriptions to improve insulin sensitivity. However, it must be kept in mind when determining appropriate nutritional intervention that lowering intake of saturated fat in these subjects may have other benefits related to cardiovascular disease risk, even if there is not a direct impact on insulin resistance.

Beyond the gene polymorphisms discussed here, many other genes involved in glucose regulation and insulin secretion are also under investigation for potential roles in influencing risk of type 2 diabetes and diabetes complications.[45]

Cardiovascular Disease

Examples of interactions of dietary components with the genome in modulation of lipid metabolism are many. One of the most illustrative and well-studied examples involves polyunsaturated fatty acids (PUFAs) and their derivatives, which are ligands for (i.e., they bind to and activate) the peroxisome proliferator-activated receptors (PPARs). PPARγ was introduced earlier as an example in the section on diabetes. PPARs (alpha, gamma, and delta) are nuclear receptors, meaning they reside in the nucleus. When activated by binding of a ligand such as a PUFA or PUFA-derived eicosanoid, they respond by altering expression levels of genes involved in lipid metabolism. Effects include adipocyte differentiation, increased fatty acid catabolism and ß-oxidation, lower serum triglyceride levels, and improved insulin sensitivity. Recent evidence suggests that soy isoflavones may also mediate their effects on improved lipid metabolism—that is, lower total cholesterol, LDL cholesterol, and triglycerides—by serving as PPAR ligands.[89] This role of diet in activating PPARs is without regard to individual variations in PPAR gene sequence and, thus, is classified as a nutrigenomic effect rather than a nutrigenetic effect. However, genetic factors contributing to hyperlipidemia have long been known to have an interplay with environmental factors—diet, tobacco use, physical activity—and these environmental influences can impact occurrence, age of onset, and the severity of cardiovascular disease.[90]

Individual Variation in Response to Environmental Influences While dietary guidelines aimed at the public have long been in place, an individual's genomic sequence itself plays a primary role in determining which of these modulations, such as a decrease in saturated fat intake, actually have beneficial effects. It has been known for many years that some individuals are more responsive to dietary intervention than others with respect to hyperlipidemia.[91, 92] What has not been possible in the past, though, is to determine specifically who will or will not respond to specific dietary measures. In fact, while a low-fat diet is beneficial for many, for others it increases atherogenesis. Like obesity and diabetes, cardiovascular disease is also a complex area of study with numerous gene-gene and gene-diet interactions that have not been completely elucidated. However, more is known about the dyslipidemias, offering perhaps some early opportunities for individualized dietary intervention that are not yet possible for obesity or diabetes. The fact that there is substantial interindividual variation in

response to diet provides a clearer basis for individualized rather than generalized population intervention.[90] Nonetheless, research designs to date have varied widely, and relationships between various genotypes and diet with respect to CVD require much additional study.[93]

Dietary Modification Is Effective in Monogenic Disease Monogenic disorders of lipid metabolism are fairly well understood and provide a basis for examining the more complex polygenic dyslipidemias.[90] Familial hypercholesterolemia, which results from mutations in the LDL-receptor gene, is perhaps the most readily recognizable example. Over 800 different mutations have been identified as causative, and the variance in mutations results in equally varying phenotypes. Null alleles are mutations resulting in no LDL receptor protein being produced. Other mutations can impair the ability of LDL to bind to the receptor, or impair post-translational processing, and thus function, of the LDL receptor protein. Most people affected by this disorder are heterozygotes, so their functional LDL receptor allele continues to work normally despite being unable to compensate for loss in function of the other. However, clinical presentation continues to vary even among individuals with the same mutation. This is explained by two modulating factors: other genes involved in lipid metabolism can affect the phenotype, and dietary factors can likewise affect the phenotype. In other words, variations in other genes and variations in diet both influence the course of atherosclerosis and life expectancy related to the LDL receptor genotype in familial hypercholesterolemia. This evidence illustrating the efficacy of diet in monogenic disease establishes the likelihood that other gene variations related to lipid metabolism will also be responsive to dietary intervention.[90] A recent large-scale study identified 30 loci on the human genome that contribute to variations in lipoprotein concentration in humans.[94] This opens the door for further research to determine how a number of genes interact with each other and the environment to produce polygenic dyslipidemia.

Dietary Fats Interact with Various Genotypes to Influence Outcomes In population studies, PUFA intake has been shown to have a differential effect on HDL-cholesterol concentrations depending on whether the nucleotide base located at the 75 position of the APOAI gene promoter is an A or a G.[95] Women in the Framingham Offspring Study with a G/G genotype were observed to have a decrease in HDL-cholesterol levels as PUFA intake increased, whereas HDL-cholesterol levels increased in those with an A/A or G/A genotype. In another example, a variant in the APOC3 gene promoter region was observed to determine the effectiveness of omega-3 polyunsaturated fatty acids in lowering triglyceride levels.[96] APOC3 encodes apolipoprotein C-III, which is associated with triglyceride-rich lipoproteins and is a known marker of cardiovascular disease risk. Long-chain omega-3 PUFAs are known to reduce apolipoprotein C-III levels, but specific polymorphisms in the APOC3 promoter result in a lack of response.[96] As described previously, dietary PUFAs primarily alter lipid metabolism by interacting with transcription factors that regulate genes involved in lipid metabolism. Peroxisome proliferator-activated receptor α (PPARα) is a well-studied nuclear transcription factor. PUFAs are natural ligands for PPARα as well as PPARγ, and polymorphisms in PPARα have also been shown to determine the effect of dietary PUFAs on triglyceride levels and apolipoprotein C-III levels.[97] Specifically, there was no significant difference in triglyceride or apolipoprotein C-III levels based on level of PUFA in the diet among subjects with the common allele of PPARα. However, a mutation resulting in valine being substituted for leucine at position 162 (PPARA-L162V) was associated with significantly lower triglycerides and apolipoprotein C-III levels in response to a high-PUFA diet.[97]

When it comes to discussing the role of individual genotypes in cardiovascular disease, no gene has been more studied than the one coding for apolipoprotein E. APOE circulates in the blood associated with chylomicrons, very low-density lipoproteins (VLDLs), and high-density lipoproteins (HDLs). In the human population, three common APOE alleles exist: APOE ε2, APOE ε3, and APOE ε4, with frequencies in the white population of approximately 15%, 77%, and 8%, respectively.[98] Cholesterol levels, particularly low-density lipoprotein-cholesterol (LDL-C) levels, are highest among individuals with the APOE ε4 variant and lowest among those with the APOE ε2 variant.[98] It was proposed that this difference occurred only in the context of an atherogenic diet and that widely varying responsiveness of LDL-C to dietary intervention may be related to APOE genotype. Further study of this potential association has yielded inconsistent results.[99] Most consistent are the observations that an interaction of diet with APOE genotype appears to occur only in men, positive findings occurred only among subjects who were hyperlipidemic at baseline, and interactions were more likely to be found when both total dietary fat and cholesterol were modified. Thus it is possible that dietary cholesterol may be particularly important. Some studies have found cholesterol absorption to vary by APOE genotype.[99] Additionally, LDL-C levels in the Framingham Offspring Study were found to vary only among drinkers of alcohol.[100] Whereas LDL-C levels were lower in male drinkers versus nondrinkers with the APOE ε2 variant, LDL-C levels were higher in male drinkers versus nondrinkers with the APOE ε4 variant. Among women, the effect of alcohol on LDL-C did not differ by APOE genotype. Similarly, cardiovascular disease risk in the Framingham Offspring Study did not differ by APOE genotype in nonsmokers. However, male smokers who were also APOE ε4 carriers had a significantly greater risk of cardiovascular events compared with male smokers who were not carriers of the APOE ε4 allele. There was no difference in risk by APOE genotype among women.[101] Physical activity as a protective factor has also been examined with regard to APOE status and found to be most protective among male APOE ε4 carriers.[102] It is also likely that obesity is a modulator of phenotype with regard to APOE genotype as well as other genes. For example, obese men who were APOE ε4 carriers had significantly higher insulin and glucose levels than obese men with the other APOE variants in the Framingham Offspring Study. There was no difference in insulin and glucose levels among obese women according to APOE genotype.[103] Based on available APOE research to date, it appears that male APOE ε4 carriers are most likely to be responsive to dietary and other lifestyle interventions to minimize cardiovascular disease risk. It is possible that this information could be used to implement disease prevention efforts early in life. However, there are other implications to consider. For example, a considerable body of research has established that the APOE ε4 variant is also associated with greater risk of Alzheimer's disease whereas the APOE ε2 variant has been linked to prenatal risk to both the fetus and mother.[104] This example illustrates how variance in genotype of genes involved in lipid metabolism has the potential to play a role in dictating the most appropriate lifestyle interventions to prevent and treat cardiovascular disease on an individual basis.

Nutritional Genomics and the Practice of Dietetics

Grasping the intricate interactions between the genome and innumerable dietary factors is critical to the future of nutrition and dietetics practice. As evidenced by the examples in this chapter and the many others that might have been mentioned (but were beyond its scope), this is a complex issue. Appropriate intervention is not simply a matter of genotyping an individual and matching each polymorphism to a specific dietary change as is often the case in DTC marketing. Depending on the outcome sought—decreased risk of colon cancer, breast cancer, or pancreatic cancer; lower triglyceride levels; increased HDL-cholesterol—the interventions may end up contradicting each other. The fact is that much remains unknown about the interactions of genes with each other. Even less is known of nutrient interactions with genotype to define an individualized diet that achieves the best outcome based on the genotypes of 20,000+ genes.[90] The recommendation of one diet intervention in response to one polymorphism is too simplistic, and as research evolves, practice will in time evolve as well.[90] In the future, diet counseling is likely to include sequencing of the entire genome to determine disease risk profiling for an individual and planning appropriate lifestyle interventions in accordance with the results of genomic sequencing. In addition to the genome sequence itself, the role of epigenetics and modulation of gene expression by dietary factors must also be considered.

Individual Testing in the Marketplace

Despite the fact that nutritional genomics research is not deemed by many to be ready for general clinical application, some practitioners already offer genetic testing and an individualized diet.[105, 106] For example, Berkeley HeartLab offers the proprietary 4myheart genetic testing and counseling service to consumers.[107] The genetic testing is designed to assess risk of cardiovascular disease and must be ordered by a physician. Berkeley HeartLab clinical educators, many of whom are registered dietitians, then help patients develop a personalized plan for reduction of CVD risk.[107] Direct-to-consumer genetic testing also took hold as the Human Genome Project unfolded, but DTC purveyors have been under fire by consumer watchdog groups (see Box 10.1).[108] For example, Carolyn Katzin, a Certified Nutrition Specialist in California, has trademarked the "DNA Diet."[109] Clients can mail in a buccal (inner cheek) swab for genotyping of 19 disease-related genes and a personalized diet via telephone or in person. While the advice offered is unlikely to cause harm, it is also unlikely to differ from general dietary recommendations.[108] Based on genetic profile, advice may include such recommendations as eating more cruciferous vegetables, legumes, whole grains, and fish—the same advice provided free by the federal government and nonprofit agencies.[108] In contrast, the DNA Diet baseline testing costs clients $625.[109] While it is not yet clear how various genotypes interact with each other or with diet to determine risk for polygenic diseases such as cancer, obesity, diabetes, and cardiovascular disease, this is the reality of the marketplace.[108]

A 2006 report from the United States Government Accountability Office (GAO) explored the practices of DTC companies.[110] They found that DTC nutrigenetic testing may be misleading to consumers because information provided from the four nutrigenetics websites they investigated was scientifically unproven or too ambiguous to be meaningful. Most advice provided entailed only general health recommendations, rather than the promised gene-specific recommendations. Additionally, some companies advised consumers to purchase expensive personalized products to repair their DNA or spend $1,200 annually on personalized supplements. To address this, the Evaluation of Genomic Applications in Practice and Prevention (EGAPP) Project was developed by the Office of Genomics and Disease Prevention (OGDP) of the Centers for Disease Control and Prevention (CDC).[111] The EGAPP Project brings together experts in health care, epidemiology, genomics, public health, laboratory practice, and evidence-based medicine. The goal is to establish a coordinated process for evaluating genetic tests and translating genomic applications that are in transition, such as those predictive for common diseases, from research to clinical practice and health policy. Information regarding the efficacy and cost-effectiveness of testing will ensure that available tests are safe, effective, and used appropriately.

Other ethical dilemmas are possible in the practice of nutritional genomics beyond what a registered dietitian would be expected to encounter in complying with Health Insurance Portability and Accountability Act (HIPAA) regulations. For example, in managing a child with an autosomal recessive disease, the dietitian may discover that the husband does not carry the expected mutation. What is the role of the registered dietitian when he/she unexpectedly discovers nonpaternity? In another example, an individual may be identified with a gene variant placing him or her at high risk for cardiovascular disease. What is the responsibility of the registered dietitian to siblings or children of that individual who may likewise be at risk? In another example, the APOE gene has been the target of nutritional genomics intervention to modulate cardiovascular disease risk. However, the same gene variant placing an individual at risk for CVD is also associated with high risk of Alzheimer's disease. Is the registered dietitian as counselor prepared adequately to discuss the risk of this devastating illness as well as diet modification? Additionally, dietitians practicing in the area of nutritional genomics must consider from an ethical perspective whether research on the link of the gene to disease and diet is adequately robust to warrant intervention and whether the testing lab is certified and has a sufficient privacy policy. The dietitian should also have a clear written policy in place detailing how sensitive private information will be managed. Furthermore, he or she must be familiar with state and federal laws governing his or her actions regarding handling of genetic information.[105, 112–114]

Evolving Knowledge and Practice Requirements for Dietitians

Clearly, registered dietitians must be knowledgeable in general genetics and genomics concepts. They must also be able to understand the role of diet in interactions with the genome and knowledgably read and interpret the current research literature. As the research base in this area continues to expand, as practice evolves, and as patients more routinely undergo gene sequencing to determine disease risk, registered dietitians with a solid grasp of genomics and diet will perhaps be among the health professionals best equipped to provide genetic counseling to optimize health. Such growth in the field of dietetics will require a substantial expansion of the knowledge base to include pathophysiology of disease at the genomic level as well as at the biochemical, metabolic, and dietary manipulation levels. Practitioners will also need to effectively communicate with consumers not only about diet, but also about the intricacies of genomics and, specifically, how the dietary interventions exert their beneficial effects. It has been proposed that a graduate degree, substantial clinical experience, and perhaps a certification exam would be desirable in order to effectively practice in this arena.[4]

With regard to implementation of genomics into the practice of dietetics, it is important for practitioners to remember that nutritional genomics "is not an end in itself." Rather, integrated as part of the nutrition care process, it provides an additional component to the nutritional assessment for arrival at an appropriate nutrition diagnosis and implementation of an appropriate intervention.[4] Nutritional genomics research is the key to closing gaps in the evidence base and strengthening evidence-based nutrition practice—but practitioners must be up to the task.[4]

Conclusion

Beyond evolution in clinical practice, there will be other changes as well.[3, 5] Research will continue to identify bioactive food components and examine how they interact with specific genes and specific genotypes, and how they influence epigenotype and gene expression to yield changes in health risk.[115] Dietary guidelines will be informed by genomics research and focus on disease prevention. Food scientists will measure bioactive components in foods and develop new functional foods to meet the demand.[116] Clinical trials will examine how functional foods and dietary supplements prevent or slow progression of disease. Medical nutrition therapy will be more individualized and public health interventions will be better targeted to meet the needs of a genetically diverse population. Dietitians have the opportunity to be involved every step along the way, from development of functional food products to serving as clinical trial coordinators or principal investigators. The new knowledge base will be immense, and dietitians will be called upon to translate new research findings into something consumers can understand and apply. It is also important that dietitians be involved in developing nutrition policies that reflect this new knowledge and find effective ways to communicate to the public dietary recommendations that may contain individualized guidance based on gene polymorphisms and epigenetic profiles. The merit of dietitian involvement is further underscored by the American Dietetic Association's identification of nutritional genomics as a "Mega Issue."[16] Many opportunities unique to the intersection of nutrition with genomics will arise in the near future and, if the dietetics profession is prepared, dietetics practice will undergo an exciting metamorphosis that will shape the future of health care.

American College of Medical Genetics
The ACMG provides education and resources for medical genetics professionals. A 2009 ACMG webcast entitled "Is Direct-to-Consumer Genetic Testing Ready for Prime-Time?" is accessible from the site and includes presenters from the Centers for Disease Control, a direct-to-consumer genetic testing company, ACMG, and others.
http://www.acmg.net

Centers for Disease Control Office of Genomics and Disease Prevention (OGDP)
The CDC focus is on the relationship of genetics and genomics to public health, including family history and genetic testing.
http://www.cdc.gov/genomics

Duke Institute for Genome Sciences and Policy   The IGSP at Duke University has developed a multidisciplinary approach to study of the genome.
http://www.genome.duke.edu

Genetics and Public Policy Center   The GPPC was founded in 2002 at Johns Hopkins University and aims to influence policymakers and inform the public in response to advances in genetics and their application to human health.
http://www.dnapolicy.org

GeneWatch UK   GeneWatch UK is a not-for-profit policy research and public interest group that focuses on how genetic science and technologies will impact our food, health, agriculture, environment, and society.
http://www.genewatch.org

Genomic Imprinting Website resource for students and researchers   This site was established in 1995 to provide information on genomic imprinting for researchers, students, and others and is maintained by the Jirtle Lab at Duke University.
http://www.geneimprint.com

Human Epigenome Project (HEP)   HEP aims to identify DNA methylation patterns in the human genome. The website provides basic information about the project.
http://www.epigenome.org

NCMHD Center of Excellence for Nutritional Genomics at the University of California—Davis   This organization is sponsored by the National Center for Minority Health and Health Disparities at the National Institutes of Health. It is dedicated to developing culturally competent methods and technologies to examine the interactions between the environment and the genome related to health disparities.
http://nutrigenomics.ucdavis.edu/

National Coalition for Health Professional Education in Genetics   NCHPEG is an organization made up of several health professional organizations dedicated to inclusion of genetics and genomics education in the training of health professionals. The American Dietetic Association is a member of NCHPEG. The site includes their publication of core competencies in genetics for health care professionals.
http://www.nchpeg.org

National Human Genome Research Institute (NHGRI), National Institutes of Health   The NHGRI home page provides links to a wide array of resources and information for professionals and the public relating to the Human Genome Project. Subjects include grants and research, genomics and health, policy and ethics, educational resources, and careers and training information.
http://www.genome.gov

National Society of Genetic Counselors (NSGC)   The official website of NSGC details the profession of genetic counseling.
http://www.nsgc.org

END-OF-CHAPTER QUESTIONS

1. What is a genome? How does knowledge of its content possibly affect dietary recommendations for individuals?

2. What are the differences between genotype, haplotype, epigenotype, and phenotype?

3. Define the following terms: autosomal dominant; autosomal recessive; X-linked dominant; X-linked recessive; Y-linked; heterozygous alleles; and homozygous alleles. Name one autosomal recessive disorder, one autosomal dominant disorder, and an X-linked recessive disorder.

4. What is the difference between a monogenic disorder and a polygenic disorder?

5. Define single nucleotide polymorphisms. How are they identified? Give an example of one and explain what it means.

6. What is meant by epigenetic regulation? How could the nutrients folate, choline, methionine, and vitamin B_{12} affect gene expression?

7. For each of the following disorders, list at least one gene that is linked to its occurrence: obesity, type 2 diabetes, and colon cancer. For each gene listed, describe its possible role in the development of the disorder.

8. Describe an example of "developmental origins of adult disease."

1. Collins FS, Green ED, Guttmacher AE, Guyer MS. A vision for the future of genomics research: A blueprint for the genomic era. *Nature.* 2003;422:835–47.

2. International Human Genome Sequencing Consortium. Finishing the euchromatic sequence of the human genome. Nature. 2004;431:931–45.

3. Stover PJ, Caudill MA. Genetic and epigenetic contributions to human nutrition and health: managing genome-diet interactions. J Am Diet Assoc. 2008;108:1480–87.

4. DeBusk R. Diet-related disease, nutritional genomics, and food and nutrition professionals. J Am Diet Assoc. 2009;109:410–13.

5. DeBusk RM, Fogarty CP, Ordovas JM, Kornman KS. Nutritional genomics in practice: Where do we begin? J Am Diet Assoc. 2005;105:589–98.

6. Anderson NL, Anderson NG. Proteome and proteomics: new technologies, new concepts, and new words. Electrophoresis. 1998;19:1853–61.

7. Gieger C, Geistlinger L, Altmaier E, Hrabé de Angelis M, Kronenberg F, Meitinger T, Mewes HW, Wichmann HE, Weinberger KM, Adamski J, Illig T, Suhre K. Genetics meets metabolomics: a genome-wide association study of metabolite profiles in human serum. PLoS Genet. 2008;4:e1000282.

8. Trujillo E, Davis C, Milner J. Nutrigenomics, proteomics, metabolomics, and the practice of dietetics. J Am Diet Assoc. 2006;106:403–13.

9. Milner JA. Nutrition in the 'omics' era. Forum Nutr. 2007;60:1–24.

10. U.S. Department of Health and Human Services. Dietary Guidelines for Americans 2005 [monograph on the Internet]. Washington (DC): U.S. Department of Health and Human Services; 2005. Available at http://www.healthierus .gov/dietaryguidelines/

11. National Heart, Lung, and Blood Institute. Third Report of the Expert Panel on Detection, Evaluation, and Treatment of High Blood Cholesterol in Adults (ATP III Final Report) [monograph on the Internet]. Bethesda (MD): National Heart, Lung, and Blood Institute; 2002. Available at http://www.nhlbi.nih.gov/guidelines/ cholesterol/atp3_rpt.htm

12. National Heart, Lung, and Blood Institute. In Brief: Your Guide to Lowering your Blood Pressure with DASH [monograph on the Internet]. Bethesda (MD): National Heart, Lung, and Blood Institute; 2006. Available at http://www .nhlbi.nih.gov/health/public/heart/hbp/ dash/dash_inbrief.htm

13. World Cancer Research Fund, American Institute for Cancer Research. Food, Nutrition, Physical Activity, and the Prevention of Cancer: A Global Perspective [monograph on the Internet]. 2007. Available at http://www .dietandcancerreport.org/

14. Kushi LH, Byers T, Doyle C, Bandera EV, McCullough M, Gansler T, Andrews KS, Thun MJ, and The American Cancer Society 2006 Nutrition and Physical Activity Guidelines Advisory Committee. American Cancer Society Guidelines on Nutrition and Physical Activity for Cancer Prevention: Reducing the Risk of Cancer With Healthy Food Choices and Physical Activity. CA Cancer J Clin. 2006;56:254–81.

15. American Diabetes Association. Clinical Practice Recommendations [monograph on the Intenet]. Alexandria (VA): American Diabetes Association; 2009. Available at http://professional .diabetes.org/CPR_search.aspx

16. American Dietetic Association. Nutritional Genomics—Backgrounder [monograph on the Internet]. Chicago (IL): American Dietetic Association; 2008. Available at http://www.eatright .org/cps/rde/xchg/ada/hs.xsl/ governance_17801_ENU_HTML.htm

17. DeSalle R, Yudell M. Welcome to the genome: a user's guide to the genetic past, present, and future. New York: Wiley-Liss; 2005.

18. Lewin B. *Genes IX.* Sudbury (MA): Jones & Bartlett Publishers; 2007.

19. Lin S-L, Miller JD, Ying S-Y. Intronic micro-RNA (miRNA). J Biomed Biotech. 2006;Article ID 26818:1–13.

20. Horton JD, Goldstein JL, Brown MS. SREBPs: Activators of the complete program of cholesterol and fatty acid synthesis in the liver. J Clin Invest. 2002;109:1125–31.

21. Barnes S. Nutritional genomics, polyphenols, diets, and their impact on dietetics. J Am Diet Assoc. 2008;108:1888–95.

22. Ordovas JM, Corella D. Nutritional genomics. Annu Rev Genomics Hum Genet. 2004;5:71–118.

23. Dennis C, Gallagher R, editors. The human genome. Nature Publishing Group; 2001.

24. International HapMap Consortium. A haplotype map of the human genome. Nature. 2005;437:1299–1320.

25. Goldstein DB, Cavalleri GL. Understanding human diversity. *Nature.* 2005;437:1241–42.

26. Ng DP, Tai BC, Lim XL. Is the presence of retinopathy of practical value in defining cases of diabetic nephropathy in genetic association studies? The experience with the ACE insertion/deletion polymorphism in 53 studies comprising 17,791 subjects. Diabetes. 2008;57:2541–46.

27. Gallou-Kabani C, Junien C. Nutritional epigenomics of metabolic syndrome. Diabetes. 2005;54:1899–1906.

28. Oommen AM, Griffin JB, Sarath G, Zempleni J. Roles for nutrients in epigenetic events. J Biol Chem. 2005;16:74–77.

29. Fraga MF, Ballestar E, Paz MF, Ropero S, Setien F, Ballestar ML, Heine-Suñer D, Cigudosa JC, Urioste M, Benitez J, Boix-Chornet M, Sanchez-Aguilera A, Ling C, Carlsson E, Poulsen P, Vaag A, Stephan Z, Spector TD, Wu YZ, Plass C, Esteller M. Epigenetic differences arise during the lifetime of monozygotic twins. Proc Natl Acad Sci USA. 2005;102:10604–09.

30. Cooney CA. Germ cells carry the epigenetic benefits of grandmother's diet. Proc Natl Acad Sci USA. 2006;103:17071–72.

31. Jiang Y, Bressler J, Beaudet AL. Epigenetics and human disease. Annu Rev Genomics Hum Genet. 2004;5:479–510.

32. Beck S, Olek A (eds). The epigenome: molecular hide and seek. Weinheim, Germany: Wiley-VCH Verlag GmbH & Co. KGaA; 2003.

33. McCabe DC, Caudill MA. DNA methylation, genomic silencing, and links to nutrition and cancer. Nutr Rev. 2005;63:183–95.

34. Luedi PP, Dietrich FS, Weidman JR, Bosko JM, Jirtle RL, Hartemink AJ. Computational and experimental identification of novel human imprinted genes. Genome Res. 2007;17:1723–30.

35. Junien C, Nathanielsz P. Report on the IASO Stock Conference 2006: early and lifelong environmental epigenomic programming of metabolic syndrome, obesity and type II diabetes. Obes Rev. 2007;8:487–502.

36. Delage B, Dashwood RH. Dietary manipulation of histone structure and function. Annu Rev Nutr. 2008;28:347–66.

37. Feinberg AP. Phenotypic plasticity and the epigenetics of human disease. Nature. 2007;447:433–40.

38. Gallou-Kabani C, Vigé A, Gross MS, Junien C. Nutri-epigenomics: lifelong remodelling of our epigenomes by nutritional and metabolic factors and beyond. Clin Chem Lab Med. 2007;45:321–27.

39. Zeisel SH. Genetic polymorphisms in methyl-group metabolism and epigenetics: Lessons from humans and mouse models. Brain Res. 2008;1237:5–11.

40. Gluckman PD, Cutfield W, Hofman P, Hanson MA. The fetal, neonatal, and infant environments—the long-term consequences for disease risk. Early Human Dev. 2005;81:51–59.

41. Hales CN, Barker DJP. The thrifty phenotype hypothesis. Br Med Bull. 2001;60:5–20.

42. Waterland RA, Garza C. Potential mechanisms of metabolic imprinting that lead to chronic disease. Am J Clin Nutr. 1999;69:179–97.

43. Barker DJP, Osmond C, Forsén TJ, Kajantie E, Eriksson JG. Trajectories of growth among children who have coronary events as adults. N Engl J Med. 2005;353:1802–09.

44. Syddall HE, Sayer AA, Simmonds SJ, Osmond C, Cox V, Dennison EM, Barker DJP, Cooper C. Birth weight, infant weight gain, and cause-specific mortality: The Hertfordshire Cohort Study. Am J Epidemiol. 2005;161:1074–80.

45. McCarthy MI. Progress in defining the molecular basis of type 2 diabetes mellitus through susceptibility-gene identification. Hum Mol Genet. 2004;13:R33–R41.

46. Benyshek DC, Johnston CS, Martin JF. Glucose metabolism is altered in the adequately-nourished grand-offspring (F3 generation) of rats malnourished during gestation and perinatal life. Diabetologia. 2006;49:1117–19.

47. Burdge GC, Slater-Jefferies J, Torrens C, Phillips ES, Hanson MA, Lillycrop KA. Dietary protein restriction of pregnant rats in the F0 generation induces altered methylation of hepatic gene promoters in the adult male offspring in the F1 and F2 generations. Br J Nutr. 2007;97:435–39.

48. Jirtle, RL, Skinner MK. Environmental epigenomics and disease susceptibility. Nat Rev Genet. 2007;8:253–62.

49. Syvanen AC. Toward genome-wide SNP genotyping. Nat Genet. 2005;37:S5-S10.

50. Davis CD, Milner J. Frontiers in nutrigenomics, proteomics, metabolomics and cancer prevention. Mutat Res. 2004;551:51–64.

51. Chuaqui RF, Bonner RF, Best CJM, Gillespie JW, Flaig MJ, Hewitt SM, Phillips JL, Krizman DB, Tangrea MA, Ahram M, Linehan WM, Knezevic V, Emmert-Buck MR. Post-analysis follow-up and validation of microarray experiments. Nat Genet. 2002;32:509–14.

52. Walker EJ, Siminovitch KA. Primer: Genomic and proteomic tools for the molecular dissection of disease. Nat Clin Pract Rheumatol. 2007;3:580–89.

53. Shen L, Kondo Y, Guo Y, Zhang J, Zhang L, Ahmed S, Shu J, Chen X, Waterland RA, Issa JP. Genome-wide profiling of DNA methylation reveals a class of normally methylated CpG island promoters. PLoS Genet. 2007;3:2023–36.

54. Waterland RA. Epigenetic epidemiology of obesity: Application of epigenomic technology. Nutr Rev. 2008;66:521–23.

55. Ficari F, Cama A, Valanzano R, Curia MC, Palmirotta R, Aceto G, Esposito DL, Crognale S, Lombardi A, Messerini L, Mariani-Costantini R, Tonelli F, Battista P. APC gene mutations and colorectal adenomatosis in familial adenomatous polyposis. Br J Cancer. 2000;82:348–53.

56. Hansen Petrik, MB, McEntee MF, Johnson BT, Obukowicz MG, Whelan J. Highly unsaturated (n-3) fatty acids, but not α-linolenic, conjugated linoleic or γ-linolenic acids, reduce tumorigenesis in Apc$^{Min/+}$ mice. J Nutr. 2000;130:2434–43.

57. Le Marchand L. The predominance of the environment over genes in cancer causation: Implications for genetic epidemiology. Cancer Epidemiol Biomarkers Prev. 2005;14:1037–39.

58. Nowell SA, Ahn J, Ambrosone CB. Gene-nutrient interactions in cancer etiology. Nutr Rev. 2004;62:427–38.

59. Le Marchand L, Hankin JH, Wilkens LR, Pierce LM, Frank A, Kolonel LN, Seifried A, Custer LJ, Chang W, Lum-Jones A, Donlon T. Combined effects of well-done red meat, smoking and rapid N-acetyltransferase 2 and CYP1A2 phenotypes in increasing colorectal cancer risk. Cancer Epidemiol Biomarkers Prev. 2001;10:1259–66.

60. Joshi AD, Corral R, Siegmund KD, Haile RW, Le Marchand L, Martínez ME, Ahnen DJ, Sandler RS, Lance P, Stern MC. Red meat and poultry intake, polymorphisms in the nucleotide excision repair and mismatch repair pathways and colorectal cancer risk. Carcinogenesis. 2009;30:472–79.

61. Giovannucci E, Chen J, Smith-Warner SA, Rimm EB, Fuchs CS, Palomeque C, Willett WC, Hunter DJ. Methylenetetrahydrofolate reductase, alcohol dehydrogenase, diet, and risk of colorectal adenomas. Cancer Epidemiol Biomarkers Prev. 2003;12:970–79.

62. Hines LM, Stampfer M, Ma J, Gaziano JM, Ridker PM, Hankinson SE, Sacks F, Rimm EB, Hunger DJ. Genetic variation in alcohol dehydrogenase and the beneficial effect of moderate alcohol consumption on myocardial infarction. N Engl J Med. 2001;344:549–55.

63. Ericson U, Sonestedt E, Ivarsson MI, Gullberg B, Carlson J, Olsson H, Wirfält E. Folate intake, methylenetetrahydrofolate reductase polymorphisms, and breast cancer risk in women from the Malmö Diet and Cancer cohort. Cancer Epidemiol Biomarkers Prev. 2009;18:1101–10.

64. Kim YI. Does a high folate intake increase the risk of breast cancer? Nutr Rev. 2006;64:468–75.

65. McCullough ML, Robertson AS, Chao A, Jacobs EJ, Stampfer MJ, Jacobs DR, Diver WR, Calle EE, Thun MJ. A prospective study of whole grains, fruits, vegetables

and colon cancer risk. Cancer Causes Control. 2003;14:959–70.

66. Fung T, Hu FB, Fuchs C, Giovannucci E, Hunter DJ, Stampfer MJ, Colditz GA, Willett WC. Major dietary patterns and the risk of colorectal cancer in women. Arch Int Med. 2003;163:309–14.

67. van Breda SGJ, van Agen E, van Sanden S, Burzykowski T, Kienhuis AS, Kleinjans JCS, van Delft JHM. Vegetables affect the expression of genes involved in anticarcinogenic processes in the colonic mucosa of C57Bl/6 female mice. J Nutr. 2005;135:1879–88.

68. van Breda SGJ, van Agen E, Engels LGJB, Moonen EJC, Kleinjans JCS, van Delft JHM. Altered vegetable intake affects pivotal carcinogenesis pathways in colon mucosa from adenoma patients and controls. Carcinogenesis. 2004;25:2207–16.

69. Seow A, Yuan J-M, Sun C-L, Van Den Berg D, Lee H-P, Yu MC. Dietary isothiocyanates, glutathione S-transferase polymorphisms and colorectal cancer risk in the Singapore Chinese Health Study. Carcinogenesis. 2002;23:2055–61.

70. Ambrosone CB, Freudenheim JL, Thompson PA, Bowman E, Vena JE, Marshall JR, Graham S, Laughlin R, Nemoto T, Shields PG. Manganese superoxide dismutase (MnSOD) genetic polymorphisms, dietary antioxidants, and risk of breast cancer. Cancer Res. 1999;59:602–06.

71. Bouchard C. Gene-environment interactions in the etiology of obesity: Defining the fundamentals. Obesity. 2008;16:S5–S10.

72. Frayling TM, Timpson NJ, Weedon MN, Zeggini E, Freathy RM, Lindgren CM, Perry JR, Elliott KS, Lango H, Rayner NW, Shields B, Harries LW, Barrett JC, Ellard S, Groves CJ, Knight B, Patch AM, Ness AR, Ebrahim S, Lawlor DA, Ring SM, Ben-Shlomo Y, Jarvelin MR, Sovio U, Bennett AJ, Melzer D, Ferrucci L, Loos RJ, Barroso I, Wareham NJ, Karpe F, Owen KR, Cardon LR, Walker M, Hitman GA, Palmer CN, Doney AS, Morris AD, Smith GD, Hattersley AT, McCarthy MI. A common variant in the FTO gene is associated with body mass index and predisposes to childhood and adult obesity. Science. 2007;316:889–94.

73. Cecil JE, Tavendale R, Watt P, Hetherington MM, Palmer CAN. An obesity-associated FTO gene variant and increased energy intake in children. New Engl J Med. 2008;359: 2558–66.

74. Mottagui-Tabar S, Rydén M, Löfgren P, Faulds G, Hoffstedt J, Brookes AJ, Andersson I, Arner P. Evidence for an important role of perilipin in the regulation of human adipocyte lipolysis. Diabetologia. 2003;46:789–97.

75. Tansey JT, Sztalryd C, Gruia-Gray J, Roush DL, Zee JV, Gavrilova O, Reitman ML, Deng CX, Li C, Kimmel AR, Londos C. Perilipin ablation results in a lean mouse with aberrant adipocyte lipolysis, enhanced leptin production, and resistance to diet-induced obesity. Proc Natl Acad Sci USA. 2001;98:6494–99.

76. Kern PA, Di Gregorio G, Lu T, Rassouli N, Ranganathan G. Perilipin expression in human adipose tissue is elevated with obesity. J Clin Endocrinol Metab. 2004;89:1352–58.

77. Qi L, Shen H, Larson I, Schaefer EJ, Greenberg AS, Tregouet DA, Corella D, Ordovas JM. Gender-specific association of a perilipin gene haplotype with obesity risk in a white population. Obesity Res. 2004;12:1758–65.

78. Corella D, Qi L, Sorlí JV, Godoy D, Portolés O, Coltell O, Greenberg AS, Ordovas JM. Obese subjects carrying the 11482G>A polymorphism at the perilipin locus are resistant to weight loss after dietary energy restriction. J Clin Endocrinol Metab. 2005;90:5121–26.

79. Deram S, Nicolau CY, Perez-Martinez P, Guazzelli I, Halpern A, Wajchenberg BL, Ordovas JM, Villares SM. Effects of perilipin (PLIN) gene variation on metabolic syndrome risk and weight loss in obese children and adolescents. J Clin Endocrinol Metab. 2008;93:4933–40.

80. Herbeth B, Aubry E, Fumeron F, Aubert R, Cailotto F, Siest G, Visvikis-Siest S. Polymorphism of the 5-HT$_{2A}$ receptor gene and food intakes in children and adolescents: the Stanislas Family Study. Am J Clin Nutr. 2005;82:467–70.

81. Rosmond R, Bouchard C, Bjorntorp P. Increased abdominal obesity in subjects with a mutation in the 5-HT(2A) receptor gene promoter. Ann N Y Acad Sci. 2002;967:571–75.

82. Roche HM, Phillips C, Gibney MJ. The metabolic syndrome: the crossroads of diet and genetics. Proc Nutr Soc. 2005;64:371–77.

83. Ordovas JM, Shen J. Gene-environment interactions and susceptibility to metabolic syndrome and other chronic diseases. J Periodontol. 2008;79:1508–13.

84. Altshuler D, Hirschhorn JN, Klannemark M, Lindgren CM, Vohl MC, Nemesh J, Lane CR, Schaffner SF, Bolk S, Brewer C, Tuomi T, Gaudet D, Hudson TJ, Daly M, Groop L, Lander ES. The common PPARγ Pro12Ala polymorphism is associated with decreased risk of type 2 diabetes. Nat Genet. 2000;26:76–80.

85. Franks PW, Luan J, Browne PO, Harding AH, O'Rahilly S, Chatterjee VK, Wareham NJ. Does peroxisome proliferator-activated receptor γ genotype (Pro-12ala) modify the association of physical activity and dietary fat with fasting insulin level? Metabolism. 2004;53:11–16.

86. Laukkanen O, Lindström J, Eriksson J, Valle TT, Hämäläinen H, Ilanne-Parikka P, Keinänen-Kiukaanniemi S, Tuomilehto J, Uusitupa M, Laakso M. Polymorphisms in the SLC2A2 (GLUT2) gene are associated with the conversion from impaired glucose tolerance to type 2 diabetes. Diabetes. 2005;54:2256–60.

87. Moreno JA, Pérez-Jiménez F, Marín C, Pérez-Martínez P, Moreno R, Gómez P, Jiménez-Gómez Y, Paniagua JA, Lairon D, López-Miranda J. The apolipoprotein E gene promoter (-219G/T) polymorphism determines insulin sensitivity in response to dietary fat in healthy young adults. J Nutr. 2005;135:2535–40.

88. Riccardi G, Giacco R, Rivellese AA. Dietary fat, insulin sensitivity and the metabolic syndrome. Clin Nutr. 2004;23:447–56.

89. Ricketts M-L, Moore DD, Banz WJ, Mezei O, Shay NF. Molecular mechanisms of action of the soy isoflavones includes activation of promiscuous nuclear receptors. A review. J Nutr Biochem. 2005;16:321–30.

90. Corella D, Ordovas JM. Single nucleotide polymorphisms that influence lipid metabolism: Interaction with dietary factors. Annu Rev Nutr. 2005;25:341–90.

91. Jacobs DR, Anderson JT, Hannan P, Keys A, Blackburn H. Variability in individual serum cholesterol response to change in diet. Arteriosclerosis. 1983;3:349–56.

92. Katan MB, Beynen AC, De Vries JH, Nobels A. Existence of consistent hypo- and hyperresponders to dietary cholesterol in man. Am J Epidemiol. 1986;123:221–34.

93. Masson LF, McNeill G. The effect of genetic variation on the lipid response to dietary change: Recent findings. Curr Opin Lipidol. 2005;16:61–67.

94. Kathiresan S, Willer CJ, Peloso GM, Demissie S, Musunuru K, Schadt EE, Kaplan L, Bennett D, Li Y, Tanaka T, Voight BF, Bonnycastle LL, Jackson AU, Crawford G, Surti A, Guiducci C, Burtt NP, Parish S, Clarke R, Zelenika D, Kubalanza KA, Morken MA, Scott LJ, Stringham HM, Galan P, Swift AJ, Kuusisto J, Bergman RN, Sundvall J, Laakso M, Ferrucci L, Scheet P, Sanna S, Uda M, Yang Q, Lunetta KL, Dupuis J, de Bakker PI, O'Donnell CJ, Chambers JC, Kooner JS, Hercberg S, Meneton P, Lakatta EG, Scuteri A, Schlessinger D, Tuomilehto J, Collins FS, Groop L, Altshuler D, Collins R, Lathrop GM, Melander O, Salomaa V, Peltonen L, Orho-Melander M, Ordovas JM, Boehnke M, Abecasis GR, Mohlke KL, Cupples LA. Common variants at 30 loci contribute to polygenic dyslipidemia. Nat Genet. 2009;41:56–65.

95. Ordovas JM, Corella D, Cupples LA, Demissie S, Kelleher A, Coltell O, Wilson PWF, Schaefer EJ, Tucker K. Polyunsaturated fatty acids modulate the effects of the APOA1 G-A polymorphism on HDL-cholesterol concentrations in a sex-specific manner: the Framingham Study. Am J Clin Nutr. 2002;75:38–46.

96. Olivieri O, Marinelli N, Sandri M, Bassi A, Guarini P, Trabetti E, Pizzolo F, Girelli D, Friso S, Pignatti PF, Corrocher R. Apolipoprotein C-III, n-3 polyunsaturated fatty acids, and "insulin-resistant" T-455C APOC3 gene polymorphism in heart disease patients: Example of gene-diet interaction. Clin Chem. 2005;51:360-367.

97. Tai ES, Corella D, Demissie S, Cupples LA, Coltell O, Schaefer EJ, Tucker KL, Ordovas JM. Polyunsaturated fatty acids interact with the PPARA-L162V polymorphism to affect plasma triglyceride and apolipoprotein C-III concentrations in the Framingham Heart Study. J Nutr. 2005;135:397–403.

98. Ordovas JM, Litwack-Klein L, Wilson PW, Schaefer MM, Schaefer EJ. Apolipoprotein E isoform phenotyping methodology and population frequency with identification of apoE1 and apoE5 isoforms. J Lipid Res. 1987;28:371–80.

99. Ordovas JM. Genotype-phenotype associations: Modulation by diet and obesity. Obesity. 2008;16:S40–S46.

100. Corella D, Tucker K, Lahoz C, Coltell O, Cupples LA, Wilson PW, Schaefer EJ, Ordovas JM. Alcohol drinking determines the effect of the APOE locus on LDL-cholesterol concentrations in men: the Framingham Offspring Study. Am J Clin Nutr. 2001;73:736–45.

101. Talmud PJ, Stephens JW, Hawe E, Demissie S, Cupples LA, Hurel SJ, Humphries SE, Ordovas JM. The significant increase in cardiovascular disease risk in APOEepsilon4 carriers is evident only in men who smoke: potential relationship between reduced antioxidant status and ApoE4. Ann Hum Genet. 2005;69:613–22.

102. Bernstein MS, Costanza MC, James RW, Morris MA, Cambien F, Raoux S, Morabia A. Physical activity may modulate effects of ApoE genotype on lipid profile. Arterioscler Thromb Vasc Biol. 2002;22:133–40.

103. Elosua R, Demissie S, Cupples LA, Meigs JB, Wilson PW, Schaefer EJ, Corella D, Ordovas JM. Obesity modulates the association among APOE genotype, insulin, and glucose in men. Obes Res. 2003;11:1502–08.

104. Ordovas JM. Genetic influences on blood lipids and cardiovascular disease risk: tools for primary prevention. Am J Clin Nutr. 2009;89:1509S–1517S.

105. Bergmann MM, Gorman U, Mathers JC. Bioethical considerations for human nutrigenomics. Annu Rev Nutr. 2008;28:447–67.

106. Kraft P, Hunter DJ. Genetic risk prediction – are we there yet? N Engl J Med. 2009;360:1701–03.

107. Berkeley HeartLab, Inc. Home page. Available at http://www.bhlinc.com/

108. Sinha, G. News feature: Designer diets. Nat Med. 2005;11:701–02.

109. Katzin C. The DNA Diet [homepage on the Internet]. Mirage (CA): Carolyn Katzin; 2009. Available at http://www.thednadiet.com

110. United States Government Accountability Office. Nutrigenetic Testing: Tests Purchased from Four Web Sites Mislead Consumers [monograph on the Internet]. Washington (DC): U.S. Government Accountability Office 2006. Available at http://www.gao.gov

111. Centers for Disease Control and Prevention, Office of Genomics and Disease Prevention. Genomics and Population Health 2005. Atlanta (GA): CDC; 2005.

112. Reilly PR, DeBusk RM. Ethical and legal issues in nutritional genomics. J Am Diet Assoc. 2008;108:36–40.

113. Castle D, DeBusk R. The electronic health record, genetic information, and patient privacy. J Am Diet Assoc. 2008;108:1372–74.

114. Castle D, Ries NM. Ethical, legal and social issues in nutrigenomics: The challenges of regulating service delivery and building health professional capacity. Mutat Res. 2007;622:138–43.

115. Kauwell GPA. Epigenetics: What it is and how it can affect dietetics practice. J Am Diet Assoc. 2008;108:1056–59.

116. Ferguson LR. Nutrigenomics approaches to functional foods. J Am Diet Assoc. 2009;109:452–58.

11

Pharmacology

Marcia Nahikian Nelms, PhD, RD
The Ohio State University
Kathryn Sucher, ScD, RD
San Jose State University

Introduction to Pharmacology

The use of drugs has been a significant component of medical care since ancient times. Historically, drugs were available without a prescription, and alcohol, cocaine, marijuana, and opium were common components of drugs. The Pure Food and Drug Act of 1906, along with the subsequent Food, Drug, and Cosmetic (FD&C) Act, enacted in 1938, began government regulation for drugs in the United States through the Food and Drug Administration (FDA).[1] As medical care has advanced, so has the development of **pharmacotherapy**.

Today, more than two-thirds of all physician visits include a written prescription.[2] In 2006, over 39% of noninstitutionalized individuals in the United States purchased one or more prescribed medications, with the number of unique prescriptions increasing with age (5–84 years old). The magnitude of medication use is reflected in its contribution to health care costs: Spending on prescription drugs increased fivefold between the years 1990 and 2006 (40.3 to 216.7 billion dollars). Prescription medications represent one of the fastest growing segments of health care expenditure, due in part to the implementation of Medicare Part D, which provides prescription drug coverage.[3, 4]

Pharmacotherapy is defined as the use of drugs for treatment of disease and health maintenance. A medical drug (or medicine) is defined as a chemical used for the diagnosis, prevention, treatment of symptoms, or cure of diseases. Drugs can be classified by structure or pharmacological action. Many drugs require a physician's prescription, while others are classified as over-the-counter (OTC) medications (not requiring a prescription). **Complementary and alternative medicine (CAM)** includes the use of herbs, botanicals, and megadoses of nutrients, which may have pharmacological properties as well (Box 11.1). In 2007, approximately 17% of adults and 5% of children under 18 years old reported using biologically based therapies.[5] Those most commonly used are listed in Table 11.1.

Pharmacology is the study of drugs, their properties and their effects; **pharmacokinetics** is the study of drug absorption, distribution, metabolism, and excretion. This chapter focuses on the basic principles of pharmacology, with an emphasis on the interaction of medications with nutrition.

An understanding of the basic principles of pharmacology is especially valuable for registered dietitians (RDs) as they work toward coordination and integration of nutrition therapy with pharmacotherapy. Nutrition therapy (NT) is a "specific nutrition service or procedure used to treat an illness, injury or condition."[6] Lifestyle, behavior changes, and complementary

biotransformation—modification of a drug through metabolism

buccal—refers to placement of a drug in the cheek

complementary and alternative medicine (CAM)—a collective term for:
- **complementary medicine**—unconventional modalities used by clients in addition to conventional biomedicine; may involve practitioner, but often self-prescribed
- **alternative medicine**—unconventional therapeutic systems used by clients in place of or parallel to conventional biomedicine; typically administered by trained practitioner

creatinine clearance—rate at which creatinine is filtered through the kidney; often used as a measure of kidney function

cyclosporine—immunosuppressant medication that is often prescribed after organ transplant

CYP 3A4—a specific cytochrome enzyme involved in drug metabolism

cytochrome P-450 isoenzymes (CP450)—family of enzyme systems responsible for drug metabolism

digoxin—cardiac glycoside that is prescribed to alter the contractions of the heart

dissolution—dissolving of a medication

epidural—refers to placement of a drug into the spinal fluid

excipients—those substances added to formulations of medications, such as color or coating agents

gentamycin—an antibiotic

H₂ blockers—medications that interrupt the production of acid in the stomach

inhalation—refers to placement of a drug so that it is breathed into the respiratory system

intradermal (ID)—refers to injection under the outermost layer of skin

intramuscular (IM)—refers to injection into the muscle

intraperitoneal (IP)—refers to injection into the body's peritoneal cavity

intrathecal—refers to injection of a drug into the membrane surrounding the central nervous system

intravenous (IV)—refers to injection directly into a vein

ionization—process of producing negatively or positively charged ions

monoamine oxidase (MAO) inhibitors—group of medications that block the enzyme system that inactivates some neurotransmitters

omeprazole—a type of proton pump inhibitor used to treat GERD and peptic ulcer disease

ophthalmic—refers to placement of a drug into the eye

otic—refers to placement of a drug into the ear

parenteral—refers to injection into the body's circulatory system through a blood vessel

pharmacokinetics—study of drug absorption, distribution, metabolism, and excretion

pharmacology—study of drugs, their properties, and their effects

pharmacotherapy—use of drugs for treatment of disease and health maintenance

pressor agents—substances that cause blood pressure to increase

prokinetics—medications that increase peristalsis

protease inhibitor—a medication that prevents protein replication; a common class of drug that is used to prevent human immunodeficiency virus replication

proton pump inhibitors—drugs that reduce acid secretion in the stomach

statin—a type of medication that is used to treat hyperlipidemias

subcutaneous (SC)—refers to injection into the body under the skin

sublingual—refers to placement of a drug under the tongue

topically—refers to placement of a drug on the skin

and alternative medicines, which include nutrition therapy, are important elements in treatment for many conditions, but use of medications remains a cornerstone of most disease treatment. An understanding of all aspects of medical care, including pharmacotherapy, among practitioners should result in improved patient outcomes, maximized nutritional status, and decreased complications or risks of the prescribed medical care.

The Joint Commission on Accreditation of Hospitals (JCAHO), the organization that accredits medical facilities, requires monitoring, documentation, and patient education for food-drug interactions. Ensuring that this requirement is met necessitates the coordinated efforts of all health care practitioners.

Role of Nutrition Therapy in Pharmacotherapy

Consider the situation of a 52-year-old male currently being treated for hypertension and hyperlipidemia. His physician has prescribed 40 mg Inderal twice daily (BID) to control his blood pressure; 20 mg of Zocor each day; and Niacor 500 mg three times per day (TID) to treat his hyperlipidemia. What is the role of nutritional care within this typical patient situation? Though nutrition's role in pharmacotherapy can be approached from several perspectives, it has traditionally been discussed within the context of the effect of nutrition on the action of the prescribed medication or the effect of the medication on an aspect of nutrition. Drug-nutrient interactions are defined as "undesirable/harmful interaction(s) between food and over-the-counter (OTC) medications, prescribed medications,

Table 11.1 Most Common Non-vitamin, Non-mineral Biological Products Used for Health Reasons in the United States, 2007*

Adults (18 years and older)		Children (Under 18 years old)	
Product	Percentage	Product	Percentage
Fish oil/omega-3 fatty acids/DHA	37.4	Echinacea	37.2
Glucosamine	19.9	Fish oil/omega-3 fatty acids/DHA	30.5
Echinacea	19.8	Combination herb pill	17.9
Flaxseed oil/pills	15.9	Flaxseed oil/pills	16.7
Ginseng	14.1	Prebiotic/Probiotic	13.6
Combination herb pill	13.0	Goldenseal	8.6
Ginkgo biloba	11.3	Garlic supplements	5.9
Chondroitin	11.2	Melatonin	5.8
Garlic supplements	11.0		
Coenzyme Q-10	8.7		

*Barnes PM, Bloom B, Nahin RL. Complementary and alternative medicine use among adults and children: United States, 2007. National health statistics reports; no 12. Hyattsville, MD, National Center for Health Statistics, 2008.

BOX 11.1 **CLINICAL APPLICATIONS**

Complementary and Alternative Medicine

In the last two decades, complementary and alternative medicine (CAM)—including the use of natural products such as nonvitamin, nonmineral supplements (primarily botanicals and herbals) and megavitamin therapy—has become increasingly popular. Consumer spending on CAM more than tripled during the last two decades, rising from $11 billion annually to approximately $40 billion.[1, 2] The market for dietary supplements alone now exceeds $24 billion each year.[3] The 2007 National Health Interview Surveys (NHIS) CAM study reported that nearly 18% of the general population had used natural products in the last 12 months.[4]

Natural products are used to maintain health, prevent disease, treat pain, lose weight, reduce stress, induce sleep, and improve strength, stamina, speed, and mental acuity. They are taken as whole foods, teas, cold beverages and beverage supplements, nutritional bars, injections, tablets, capsules, powders, and suppositories. According to data from the NHIS report, fish oil/omega 3 fatty acids or DHA, glucosamine, and echinacea were consumed most often, followed by flaxseed oils/pills, ginseng, and gingko biloba. Bilberries, evening primrose oil, milk thistle, goldenseal, grape seed extract, peppermint, saw palmetto, St. John's wort, and valerian are other examples of commonly used botanicals. In addition to single remedies, blends of botanicals (and sometimes other substances) are promoted for a variety of ailments. (See Tables 11.8, 11.9, and 11.10 located at the end of this chapter, for information about specific products.) Although many natural products have proved effective for certain conditions, research has questioned or refuted some of the claims made for these products.[5–10]

Beyond questions of efficacy, there are several potential problems with natural product consumption. Quality and content can vary from the label information.[11] Mild toxicity to some substances has been reported by regional poison centers.[12, 13] More serious reactions can occur from interactions with prescription drugs or with other natural products (see "Cautions" in Tables 11.8, 11.9, and 11.10). For example, therapeutic doses of garlic may act synergistically with fish oil to inhibit blood platelet aggregation; valerian may enhance depressants, such as barbiturates; ginkgo biloba reduces the effectiveness of some drugs, such as certain antacids and anti-anxiety medications, while potentiating others, including anticoagulants, antidepressants, and antipsychotics; and black cohosh may alter the response of cells to chemotherapy in patients under treatment for breast cancer.[14] St John's wort has been shown to cause multiple drug interactions, primarily through induction of the cytochrome P450 system and enzyme CYP 3A4—resulting in increased metabolism of those drugs, with decreased concentration and clinical effect.[15] Further, natural products can be adulterated with pesticides, heavy metals (such as mercury), or prescription drugs (such as warfarin or alprazolam).[16]

Natural products are regulated by the 1994 Dietary Supplement and Education Act (DSHEA). The act defines dietary supplements as neither foods nor drugs but in a separate category, and thus not subject to federal monitoring by the Food and Drug Administration (FDA). Safety evaluation, efficacy testing, and quality control are left up to manufacturers. Many in the industry have adopted uniform manufacturing standards; however, variation in potency is still common. The American Herbal Products Association has developed a numerical rating system for botanical safety: (1) safe when consumed appropriately, (2) restricted for certain uses, (3) use only under supervision of an expert qualified in the appropriate use of this product, and (4) insufficient data to make a safety claim. (These ratings are available from Blumenthal M, *Herbal Medicine: The Expanded Commission E Monographs 2000*, American Botanical Council.) The FDA has the authority to protect the public from harmful natural products, but the government has the burden of proving a product is unsafe. Manufacturers may make statements regarding the structure and function of a product, but no claims regarding its use to prevent or cure specific illnesses and conditions can be stated. Ultimately, it is up to the consumer to make informed choices regarding natural product selection and use.

Vitamin/Mineral Supplements and Megavitamin Therapy

There is no consistent definition for the term "megavitamin therapy," but it usually encompasses mineral as well as vitamin intake, often in amounts of over 10 times the RDA. It is sometimes also called *orthomolecular medicine*, a system that uses vitamin, mineral, and enzyme supplements to address the individual biochemical differences and needs of each client. A comparison of data from the 1987, 1992, and 2000 National Health Interview Surveys found that the daily intake of multivitamin/mineral supplements has increased dramatically over the period, from 17% to 27% of the total population.[17] The 2002 NHIS CAM report found that approximately 3% of respondents used megavitamin therapy.[18]

Most practitioners of orthomolecular medicine are medical doctors (MDs) who typically prescribe a regimen of injections followed by tablets taken several times daily. However, many megavitamin consumers self-diagnose or rely on the advice of supplement salespeople in health food and other stores. Over-the-counter megavitamin therapy is used frequently for minor complaints. For instance, vitamin C or zinc is taken for colds and chromium picolinate for carbohydrate cravings or to increase metabolism. Combinations of multiple vitamins and minerals are suggested for more serious conditions. Megadoses of vitamins A and C, copper, selenium, and zinc are suggested for osteoarthritis, for instance. Diabetes is sometimes self-treated with high amounts of vitamins B_{12}, C, and E, biotin, chromium picolinate, and zinc. Patent mixtures of vitamins and minerals are marketed for specific health problems; for example, thyroid-stimulating compounds are marketed for hypothyroidism. Megavitamin therapy is especially associated with psychiatric conditions; for example, B_6, magnesium, and zinc are used for autism, and B_3, B_6, B_{12}, C, folic acid, chromium, selenium, and zinc for depression. Megavitamin therapy is also used for children with behavioral disorders or developmental delays.

Proponents argue that megavitamins can remedy nutrient deficiencies that damage DNA, improve enzyme-to-coenzyme binding in numerous genetic disorders and in certain diseases, and reduce oxidant leakage from decaying mitochondria, thus slowing aging.[19] Advocates believe that megavitamin therapy is relatively inexpensive and safe. Little research on long-term use of megavitamin therapy has been reported.[20, 21] Excess intake of zinc and deficiencies of copper have been reported.[22] Hepatotoxicity and carcinogenicity from beta-carotene intake in smokers and drinkers suggest that even supplements presumed safe in the majority population may be dangerous for some clients.[23] Further, megadose side effects are not uncommon. Headaches, insomnia, nausea, constipation or diarrhea, anorexia, mood changes, kidney stones, and allergic reactions (from dermatitis to anaphylactic shock) are possible (see Table 11.10).

References

1. Eisenberg DM, Kessler RC, Foster C, Norlock FE, Calkins DR, Delbanco TL. Unconventional medicine in the United States. N Engl J Med. 1993;328:246–52.
2. Sizing Up the Market for Alternative Medicine, Highly Educated, High-Income Americans Most Likely to Try Everything from Herbal Supplements to Reiki. Thomson Medstat Research Brief. December 2006. Available at http://pharma.thomsonhealthcare.com/uploadedFiles/docs/Research_Brief—Alternative_Medicine—2006.pdf
3. Charles Thurston Dietary Supplements: The Latest Trends & Issues, Nutraceutical World. Available at http://www.nutraceuticalsworld.com/articles/2008/04/dietary-supplements-the-latest-trends-issues
4. Barnes PM, Bloom B, Nahin RL. Complementary and alternative medicine use among adults and children: United States, 2007. National health statistics reports; no 12. Hyattsville (MD): National Center for Health Statistics. 2008. Available at http://nccam.nih.gov/news/2008/nhsr12.pdf
5. Sleivert G, Burke V, Palmer C, Walmsley A, Gerrard D, Haines S, et al. The effects of deer antler velvet extract or powder supplementation on aerobic power, erythropoiesis, and muscular strength and endurance characteristics. Int J Sport Nutr Exerc Metab. 2003;13:251–65.
6. Turner RB, Bauer R, Woelkart K, Hulsey TC, Gangemi JD. An evaluation of Echinacea angustifolia in experimental rhinovirus infections. N Engl J Med. 2005;353:341–48.
7. Van Hasselt P, Gashe BA, Ahmad J. Colloidal silver as an antimicrobial agent: fact or fiction? J Wound Care. 2004;13:154–55.
8. Cefalu WT, Ye J, Zuberi A, Ribnicky DM, IRaskin I, et. al. Botanicals and the metabolic syndrome. Am J Clin Nutr. 2008;87:481S—487S.
9. Birt DR, Widrlechner MP, LaLone CA, Wu L, Bae J, Solco ADS, et al. Echinacea in infection. Am J Clin Nutr. 2008;87:488S-492S.
10. Chilton FH, Rudel LR, Parks JS, Arm JP, Seeds MC. Mechanisms by which botanical lipids affect inflammatory disorders. Am J Clin Nutr. 2008;87:498S–503S.
11. Krochmal R, Hardy M, Bowerman S, Lu QY, Wang HJ, Elashoff R, Heber D. Phytochemical Assays of Commercial Botanical Dietary Supplements. Evid Based Complement Alternat Med. 2004;Dec;1(3):305–13.
12. Robinson RF, Griffith JR, Nahata MC, Mahan JD, Casavant MJ. Herbal weight-loss supplement misadventures per a regional poison center. Ann Pharmacother. 2004;38:895–97.
13. Yang S, Dennehy CE, Tsourounis C. Characterizing adverse events reported to the California poison control system on herbal rememdies and dietary supplements: a pilot study. Journal of Herbal Pharmacotherapy. 2003;2:1–11.
14. NCCAM, National Institutes of Health, Bethesda, Maryland. Available at http://nccam.nih.gov/ updated February 17, 2009.
15. Rommel G Tirona and David G Bailey, Herbal product–drug interactions mediated by induction. Br J Clin Pharmacol. 2006 June;61(6):677–81.
16. Ernst E. Toxic heavy metals and undeclared drugs in Asian herbal medicines. Trends Pharmacol Sci. 2002 Mar;23(3):136–39
17. Gunther S, Patterson RE, Kristal AR, Stratton KL, White E. Demographic and health-related correlates of herbal and specialty supplement use. J Am Diet Assoc. 2004;104:27–34.
18. Barnes PM, Powell-Griner E, McFann K, Nathin RL. Complementary and alternative medicine use among adults: United States, 2002. Advance Data from Vital and Health Statistics, 343. Hyattsville (MD): National Center for Health Statistics; 2004.
19. Bjelakovic G, Nikolova D, Gluud LL, Simonetti RG, Gluud C. Mortality in randomized trials of antioxidant supplements for primary and secondary prevention: systematic review and meta-analysis. JAMA. 2007;297(8):842–57.
20. Ames BN. The metabolic tune-up: metabolic harmony and disease pervention. J Nutr. 2003;133:1544S–1548S.
21. Neuhouser ML, Wassertheil-Smoller S, Thomson C, et al. Multivitamin use and risk of cancer and cardiovascular disease in the Women's Health Initiative cohorts. Arch Intern Med. 2009;169(3):294–304.
22. Igic PG, Lee E, Harper W, Foach KW. Toxic effects associated with consumption of zinc. Mayo Clinic Proceedings. 2002;77:713–16.
23. Leo MA, Lieber CS. Alcohol, vitamin A, and beta-carotene: adverse interactions, including hepatotoxicity and carcinogenicity. Am J Clin Nutrition. 1999;69:1071–85.

herbals, botanicals, and/or dietary supplements that diminishes, enhances, or alters effect of nutrients and/or medications."[7] *The Position of the American Dietetic Association: Integration of Medical Nutrition Therapy and Pharmacotherapy* emphasizes a collaborative model of health care that allows maximum benefit from the use of both pharmacotherapy and nutrition therapy.[8]

In the patient scenario just presented, the well-trained RD would first recognize that a therapeutically important drug-nutrient interaction could occur between Zocor and grapefruit.[9] Grapefruit interferes with absorption of Zocor and could significantly change availability of this medication. Next, it is important to note that first-pass hepatic metabolism of propranolol (Inderal) may be decreased when this medication is taken with food. Drug levels in the body may be increased due to this interaction; to avoid such an event, the patient would be counseled to take this medication on an empty stomach.

Though important, preventing interactions is only one goal of understanding nutrition's contribution to pharmacotherapy. Current recommendations for treatment of hyperlipidemia include the Therapeutic Lifestyle Changes (TLC) from the National Cholesterol Education Program. These recommendations incorporate nutrition therapy as a major component of treatment for cardiac disease, hyperlipidemia, and hypertension.[10] Weight loss, if the patient were overweight, could lower his blood pressure, reduce the required dosage of Inderal, and improve his lipid profile so that his dosage of these medications could be reduced or eliminated. Incorporating principles of the DASH (Dietary Approaches to Stop Hypertension) diet with his weight-loss program may result in a further decrease in his blood pressure (see Chapter 13). Complete counseling for this patient would encompass the following recommendations: avoid grapefruits and grapefruit juice, take Inderal on an empty stomach, and follow a low-kcal, low-saturated fat, and low-sodium diet that is rich in fruits and vegetables and fat-free or low-fat dairy products. Over time, successful nutrition therapy could decrease his yearly prescription costs by $4800. Hence, intervention by the RD provides clinical and economic benefits while also meeting all legal responsibilities.[11] This chapter addresses the complex relationship between nutrition therapy and pharmacotherapy in detail, focusing on information the RD will need to successfully integrate the two.

Drug Mechanisms

The most common mechanisms for drug action involve binding of the drug to specific receptors on the cell membrane, which initiates changes in specific enzyme reactions. Drugs react with a cellular receptor site due to their design and shape—as a lock and key might fit together. When this occurs, physiological functions are altered. Most drugs can interact with more than one cell receptor, which may account for various side effects of medication use (see Figure 11.1).

Alterations in enzyme systems by medications are caused either by stimulating (induction) or inhibiting an enzyme system (see Pharmacokinetics: Metabolism of Drugs). An example of enzyme system inhibition is the action of the class of medications called ACE (angiotensin-converting enzyme) inhibitors, illustrated in Figure 11.2. In normal control of blood pressure, angiotensin-converting enzyme stimulates conversion of angiotensin I to angiotensin II. The function of angiotensin II is to constrict blood vessels and cause an increase in blood pressure (see Chapters 7 and 13 for more detail). If this enzyme system is inhibited, blood vessels will vasodilate, causing a decrease in blood pressure. Of course, other physiological functions can

Figure 11.1 Cellular Receptor Site

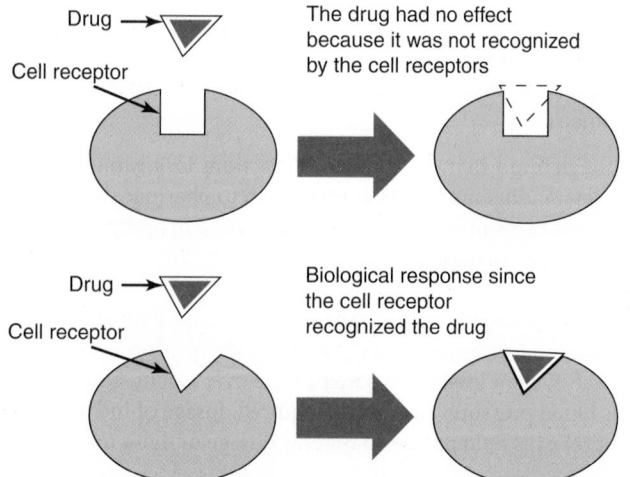

Figure 11.2 Inhibition of Enzyme System

(a) Normal action of ACE; (b) Inhibition of ACE through medication causes blood pressure to drop.

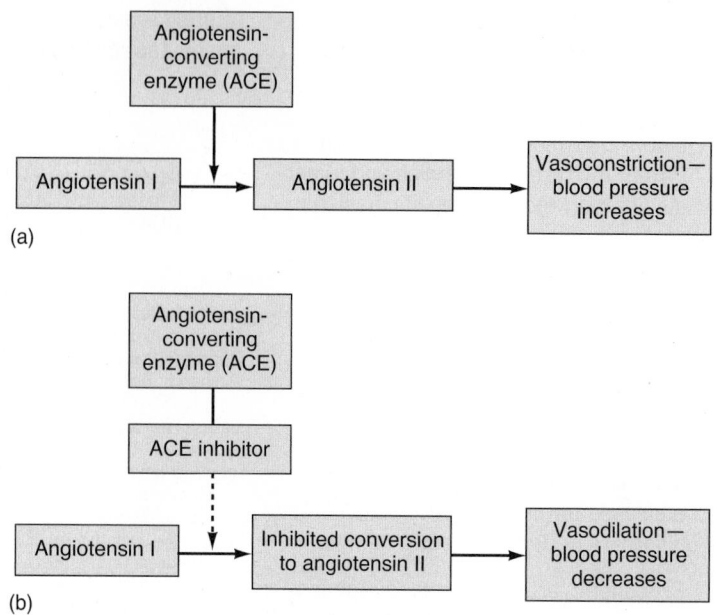

change as a result of drug action. These nonspecific responses can either be therapeutic or fall into the role of side effects and drug interactions.

Administration of Drugs

Drugs can be administered in multiple ways. The administrative route depends on the chemical properties of the drug, the type of effect desired, and, of course, patient characteristics that affect how the medication could be administered. The oral route of administration requires that the patient be able to swallow medication and that the slower rate of absorption of this administration method is acceptable. **Sublingual** or **buccal** administration means the drug is placed under the tongue or in the cheek, respectively. It dissolves there, so it is quickly absorbed across mucous membranes into the circulatory system. When an individual takes nitroglycerin for angina, it is usually via a sublingual route.

Routes of administration can also be **parenteral**, **topical** via skin and mucous membranes, or through **inhalation**. Parenteral administration requires an injection into the body through routes that are either **subcutaneous (SC)**, **intradermal (ID)**, **intramuscular (IM)**, **intraperitoneal (IP)**, or **intravenous (IV)**. Topical medications are applied to skin for a direct effect, but can also be absorbed via skin or mucous membranes; for example, Estraderm® is an estrogen patch worn to increase circulating levels of estrogen. Drugs that are inhaled have the opportunity to act locally within the respiratory system or to have a systemic effect. When anesthesia is inhaled, systemic effects occur, whereas the medication Combivent® uses two different types of bronchodilators, which act locally to treat asthma and other respiratory conditions. Medications can also be placed directly into target tissues such as the eye (**ophthalmic**), ear (**otic**), or spinal canal (**epidural** or **intrathecal**).

Pharmacokinetics

Absorption of Drugs

Absorption of the drug/medication involves several steps as the substance is transferred from the administrative site (e.g., oral, sublingual, intravenous) to the circulatory or lymphatic system. Absorptive mechanisms for drugs encompass the same basic processes as those for nutrients (see Chapters 14 and 15). Collectively, these processes include passive diffusion, facilitated diffusion, and active transport. The rate and effectiveness of absorption for drugs is dependent on several key factors. First, solubility of the medication determines where in the gastrointestinal tract the medication will dissolve and thus be absorbed. **Dissolution** or dissolving of the medication has to occur before absorption is successful. **Excipients** are those substances added to formulations of medications that affect dissolution. Binders, lubricants, and coating agents decrease dissolution, whereas disintegrants (ingredients that dissolve readily in water) increase dissolution. Coloring and flavoring agents have varying effects on dissolution. Tablet formulation is also a factor; hard, round, and large tablets dissolve more slowly. Dissolution rates of generic equivalents to the original medication may also vary.[12, 13]

The amount of time a medication is present in a specific portion of the gastrointestinal (GI) tract, the pH of that portion of the GI tract, and the surface area of the GI tract also affect absorption capability. The largest surface areas for drug absorption are located in the small intestine and lungs. Other factors that affect absorption include the chemical properties of the drug, the integrity of the gastrointestinal tract and other tissues, and the circulation and blood supply.[14] Anatomical regions with the highest blood flow, including the small intestine, lungs, muscle, and buccal and nasal cavities, have efficient rates of absorption and distribution.

The most important chemical properties of medications related to drug absorption include the solubility of the drug in lipid or water and the **ionization** of the medication. Lipid-based drugs will be absorbed across cell membranes quickly, since cell membranes are primarily lipid based. Drugs that are not ionized will also be absorbed much more readily. If the drug is ionized, absorption will be dependent on the pH of the solution where it will be absorbed. For example, if a medication is mildly acidic, absorption will be enhanced in solutions that are also acidic, such as gastric juices.[14] Aspirin is a good example of a medication that is absorbed in the stomach but can also damage the gastric mucosa.

Distribution of Drugs

After absorption, distribution of the drug occurs. Distribution is defined as the movement of the drug throughout the body to the target sites where it can act. Distribution is variable and is affected by the circulation, the binding of the drug to proteins within the circulation (e.g., albumin, 1-acid glycoprotein), capillary permeability, the drug's solubility in water, and the binding of the drug to other tissues within the body. Overall, the greater the amount of the drug that binds to another substance, the smaller the amount of active or free drug within circulatory or storage tissues. Physiological or anatomical features also affect distribution of the drug. For example, some drugs cannot cross the placenta or cell membrane into the central nervous system, while others, mostly lipid soluble or neutral pH, are readily distributed to those sites. The blood-brain barrier of juxtaposed cells (cells that form a tight junction) usually inhibits the passage of polar or ionized drugs.

Metabolism of Drugs

The metabolism of drugs involves **biotransformation** (changing the physical form), which renders the drug inactive so it may be excreted via urine or bile. The liver is the major site for biotransformation, but metabolism occurs within other organs as well. Enzyme systems, including the family of **cytochrome P-450 isoenzymes (CP450)**, are often responsible for metabolizing drugs. Approximately 30 enzymes are responsible for the numerous reactions that oxidize drugs within the liver.[15]

A substance may interact with the CP450 enzymes as either an inhibitor or inducer. An inhibitor reacts with the specific enzyme by competition for the receptor site. An inducer works to stimulate synthesis of the enzymes, increasing action potential. Inhibitors decrease metabolism and generally lead to increased drug effect, whereas inducers will increase metabolism and generally lead to decreased drug effect. Phenobarbital and theophylline are examples of inducers of the CP450 enzymes. Examples of drugs known to be inhibitors include chloramphenicol, cimetidine, valproic acid, allopurinol, and erythromycin. Drug dosages must be adjusted to accommodate metabolism of each medication. Therapeutic levels are determined by measuring blood levels in order to establish the correct effective dose for each person (see Figure 11.3). The dosage range with therapeutic efficacy is referred to as the "therapeutic window." Levels below this window may not be effective, and those above may result in toxicity.

Excretion of Drugs

Generally, after drugs have been metabolized, the remaining compounds are eliminated from the body. There are exceptions; some drugs can be excreted before they are metabolized. Most drugs are removed by either urinary or biliary excretion, but some can be excreted via the lungs or bowel, depending on the chemical structure of the metabolite. It

Figure 11.3 Therapeutic Levels of Drugs

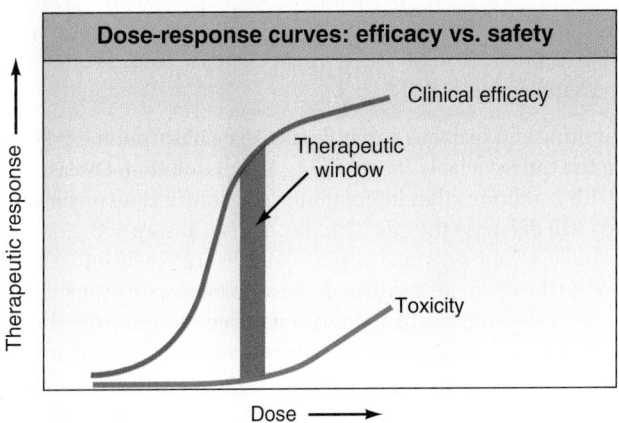

Source: Bottorll MD. Evans WE. Drug concentration monitoring. In: Progress in Clinical Biochemistry and Medicine. 1988.

is important to be aware that some drugs can be excreted in breast milk as well, which means that the nursing infant would be exposed to that drug.

Urinary excretion of drugs can occur in all three stages of urinary filtration and concentration within the nephron, the functional unit of the kidney (see Chapter 18). Each of the over one million nephrons consists of a glomerulus and tubule. Each tubule is divided into several sections, depending on the type of epithelial cells it contains. Sections are referred to as the proximal tubule, Loop of Henle, distal tubule, and the collecting duct. (See Chapter 18, Figure 18.2.) All collecting ducts drain into the ureter and ultimately into the bladder. Most drugs of low molecular weight are filtered out of the blood in the glomerulus unless they are bound to large molecules such as proteins or to erythrocytes. Drugs can be reabsorbed within the tubules. Reabsorption depends on the pH of the urine and the solubility of the drug. Lipid-soluble drugs are more readily reabsorbed. Since the acidity of the urine is quite variable, there is a significant variation in drug reabsorption.

Alterations in Drug Pharmacokinetics

No two people will react in the same way to any given medication. Age; gender; cardiovascular, hepatic, and renal function; presence of disease or infection; diet; and even genetic differences will affect how an individual will respond to a drug dosage. The following sections describe potential alterations in each pharmacokinetic phase.

Altered GI Absorption   GI absorption will be altered as health conditions, disease, and treatment modalities interrupt normal absorption processes. Simultaneous consumption of food with medication is one of the more common factors that may change the effectiveness of absorption.[16, 17] The presence of food stimulates normal digestion and absorption mechanisms, such as changes in rate of gastric emptying and the release of enzymes and hydrochloric acid. All of these normal mechanisms may alter the GI environment so that it is not suitable for absorption of the medication. The presence of food also increases the chance for adhesion of the drug to a food component. Directions for a medication should indicate whether the drug should be taken with or without food.

A classic example of this situation is the effect of different foodstuffs on iron absorption. The absorption rate for iron supplements can vary tremendously depending on the type of food consumed with them. Vitamin C enhances absorption, whereas milk, tea, and coffee would decrease absorption of the iron supplement.[18]

Vomiting and diarrhea can influence drug absorption by reducing the time available for solubility and dissolution. Diseases or health conditions that interrupt normal transit time or surface area will decrease the effectiveness of drug absorption. For example, Crohn's disease or other malabsorptive diagnoses will change the ability of the drug to be absorbed across the membrane of the enterocyte. Circulation deficits to and from the GI tract could also reduce the effectiveness of absorption from the small intestine to the rest of the body.

Drugs, nutrients, and other substances may compete for the carriers needed for active transport. For example, Levodopa, a standard medication for treatment of Parkinson's disease, is transported using the same pathways as neutral amino acids such as leucine and isoleucine. This medication should be taken on an empty stomach so that adequate absorption can be ensured.[19] As mentioned previously, the pH at the absorption site can alter ionization of the drug, which may change the speed and effectiveness of absorption. For example, Ketoconazole, an antifungal agent, must be in an acidic environment for appropriate dissolution and absorption.

Altered Distribution   Major factors that change distribution of a drug include variations in circulation, body size and composition, and protein binding of the medication. Factors that could alter circulation include age and disease. The drugs propranolol and dextropropoxyphene increase blood flow to the liver, and thus increase circulation or distribution of other medications. Any factor that causes vasodilation would theoretically increase distribution of the drug; for example, physical activity and increased body temperature increase vasodilation and thus drug distribution. Body size and body composition can alter drug distribution. The elderly individual may have decreased muscle mass requiring a downward adjustment for drug dosing.[20] Large amounts of body fat may slow distribution of a medication. Many medications are bound to a protein carrier—most often, albumin. Any situation that could alter albumin concentrations, such as liver or kidney disease or malnutrition, would increase the amount of unbound medication, multiplying the amount of active drug within the body.

Altered Metabolism   Age is also a major factor in how drugs are metabolized. Neonates, infants, and young children have vastly different levels of liver function and enzyme systems than adults do, which affects their reactions to different medications.[21] On the other end of the spectrum, the elderly may also have a decreased ability to metabolize drugs because of the normal physiological changes of aging. For instance, circulation within the liver decreases by approximately 35% by age 70 years with concurrent decreases in liver mass.[22] Drug metabolism alterations may appear as decreased effectiveness of some medications or may surface as toxicity symptoms.[20, 22, 23]

Appropriate metabolism of drugs requires adequate function of organs—especially the liver. When disease and injury interrupt organ functioning, drug metabolism may change as well. The types of drugs or alternative regimens will need to be considered when concurrent drug treatment interferes with metabolism.

Genetic factors may also play a major role. Phenotypic differences are often attributed to differences in genetic coding of metabolic enzyme systems.[24] For example, differences in metabolism for **proton pump inhibitors** (such as omeprazole) can affect treatment effectiveness for *H. pylori* infections.[25] In 2007, the FDA announced the addition of genetic information to the label for warfarin, an anticoagulant drug. Individuals who have the phenotype(s) for vitamin K epoxide reductase may not be able to metabolize warfarin, or may metabolize it more slowly, resulting in a greater tendency to bleed and a need to lower the dosage.[26] Five common enzyme variants associated with warfarin sensitivity have been described.[27] Gender differences are also apparent in metabolism for some drugs.[28]

One of the most common mechanisms for alteration of drug metabolism is concurrent use of other medications, which may interrupt enzyme systems and prevent clearance of metabolites. Numerous drug-drug interactions have been identified that, unless monitored closely, can cause significant adverse symptoms.[29]

Altered Urinary Excretion Urinary excretion of drugs can change as a result of numerous mechanisms. As stated earlier, the pH of the urine has a direct effect on the type of drugs easily excreted. Nutritionally, different foods can affect the pH of the urine, though these effects are difficult to predict due to variations in digestion and metabolism.[30] Excretion can also be changed by the presence of a competitor for active transport across the renal tubule. Finally, urinary excretion can be altered by changes in urinary flow rates or kidney function. This may occur as a result of another medication, as a result of disease or injury, or as a consequence of aging. Changes in **creatinine clearance** significantly alter the effectiveness of medications. If an individual has renal insufficiency from any etiology, drug levels must be adjusted to ensure therapeutic levels. **Digoxin**, **cyclosporine**, and **gentamycin** are examples of medications affected by changes in kidney function. Other medications such as ampicillin or cephalosporins are nephrotoxic and could themselves change kidney function.[31]

How Do Food and Drugs Interact?

As stated earlier, drug-nutrient interactions can be organized by examining the effect of nutrition on the action of the prescribed medication, the effect of the medication on nutritional status, or the role of nutrition therapy in maximizing the prescribed effect of pharmacotherapy and/or minimizing the side effects.[7]

Effect of Nutrition on Drug Action

This section will discuss the effects of food and nutrition on dissolution, absorption, metabolism, and excretion of medications. Since it is virtually impossible to have a working knowledge of all potential reactions, health professionals in specialty areas become very familiar with medications of their typical patient population. This textbook highlights specific drug-nutrient interactions for each diagnosis. Heightened awareness of potential interactions makes the integration of nutrition and pharmacotherapy a routine component of patient care.

Effect of Nutrition on Drug Dissolution In order for oral drugs to be absorbed, dissolution of the medication is necessary. The pH of the stomach and the gastric emptying rate are two of the most important nutrition-related factors impacting drug dissolution. Medications may require an acidic environment for dissolution. Achlorhydria, which is decreased production of hydrochloric acid, occurs in aging as well as some medical conditions such as HIV and AIDS. Medications that could affect gastric acidity include use of **H₂ blockers** (cimetidine, famotidine), proton-pump inhibitors (omeprazole, lansoprazole), and antacids (TUMS, Rolaids) (see Chapter 14). Gastric emptying rate influences the amount of time in which dissolution can occur; medications that affect gastric emptying time include **prokinetics** such as metoclopramide.

Any disease, injury, or surgery that affects oral intake or gastric function can affect dissolution of medications. For example, vomiting and diarrhea would certainly decrease dissolution.

Gastric surgical resections can dramatically change the rate of gastric emptying as well as the amount of gastric secretions (see Chapter 14 for a discussion of these surgical procedures). Any client who presents with this medical history will need adjustments in the form of the medication to ensure appropriate dissolution. Medications in liquid form are more easily dissolved than those in capsule or tablet form.

Effect of Nutrition on Drug Absorption The presence of food, alcohol, or dietary supplements can interact with drugs in several important ways that interfere with drug action. Interactions may increase absorption of medications, hence increasing the amount of available drug. In contrast, if absorption of a medication is decreased by food, therapeutic levels may not be achieved. The presence of food in the stomach will increase gastric emptying time (i.e., slow emptying rate), especially when a high-fat meal is consumed, and this could potentially affect absorption. For example, the presence of food dramatically reduces absorption of Fosamax, a medication used to treat osteoporosis. Saquinavir, a **protease inhibitor** used to treat HIV, is another dramatic example of a drug for which food affects absorption. Taking these medications at the same time as food can reduce absorption considerably. On the other hand, it is recommended that some medications be taken with food in order to decrease the gastric distress associated with them. Examples include amoxicillin, ketoconazole, and erythromycin. Chelation is another mechanism that affects absorption. Chelation, the binding of a nutrient or food component with a drug, makes the drug unabsorbable. For example, consumption of calcium with the antibiotic tetracycline causes chelation of the drug, which decreases absorption. Patient education should include specific guidelines for consuming a medication with or without food, if applicable.

Effect of Nutrition on Drug Metabolism Some of the most important food-drug interactions fall into the category of metabolism changes. Research has identified several mechanisms; a summary of these findings is that, in general, some nutrients act either as an inducer or as an inhibitor for metabolic enzyme systems. These actions can change drug effectiveness as well as produce toxic side effects, which increase the potential for morbidity and mortality.[16, 32, 33] Nutrients can also compete for carrier systems involved in normal drug metabolism.

For example, a recent study found that St. John's wort, an herbal supplement used to treat depression, significantly induced activity of **CYP 3A4**. Long-term use of St. John's wort may result in diminished clinical effectiveness or increased dosage requirements for at least 50% of all marketed medications.[34] These types of interactions appear to pose a much more common and serious risk than was previously recognized. The use of herbal therapies prior to surgical anesthesia, for instance, could prolong the effect of the anesthesia.[35, 36]

The potential for nutrient-drug interactions with anticoagulation therapy, a standard component of clinical care in prevention of stroke and heart attack, provides an important illustration of how foods interrupt drug metabolism. Vitamin K improves blood clotting. When foods high in vitamin K or vitamin K supplements are taken during the same time period as warfarin (Coumadin), a vitamin K antagonist, the amount of warfarin needed is increased. Vitamin K intake should therefore be consistent in order to maintain the levels of warfarin within a therapeutic level. Additionally, the dietary supplements feverfew, garlic, gingko biloba, ginger, cayenne, and omega-3 fatty acids can also affect blood coagulation. A change in the dosage of anticoagulation drugs may be required in order to compensate for a patient's dietary intake of these foods and supplements.

A classic example of a drug-nutrient interaction resulting in harmful side effects is the interaction between **pressor agents** in foods (tyramine, dopamine, histamine, phenylethylamine) and **monoamine oxidase (MAO) inhibitors** (e.g., Nardil). This interaction can result in sudden increases in blood pressure with resulting complications. Box 11.2 outlines the specifics for this drug-nutrient interaction.

BOX 11.2 CLINICAL APPLICATIONS

Monoamine Oxidase Inhibitors (MAOIs) and Nutrient Interactions

Monoamine oxidase (MAO) is an intricate enzyme system distributed predominantly in nervous tissue, liver, and lungs. This enzyme system is responsible for inactivating the neurotransmitters dopamine, norepinephrine, and serotonin once they have played their part in sending messages to the brain. Monoamine oxidase inhibitors (MAOIs) are drugs that block this activity. When the excess neurotransmitters are not destroyed, they accumulate in the brain.

In addition to inactivating these neurotransmitters, MAO breaks down another amine called tyramine. When MAO is blocked by an MAOI, levels of tyramine also rise. Excess tyramine can cause sudden, sometimes fatal increases in blood pressure. To avoid this life-threatening side effect, those taking MAOIs must avoid or limit foods that contain high levels of tyramine.

Tyramine occurs naturally in foods, but it is difficult to quantify the exact amount of tyramine in foods. Tyramine can also vary among different brands of certain foods based on processing, storage, and preparation methods. It is also formed from bacterial breakdown of protein in foods as they age.

MAOIs are most often prescribed for depression, bacterial and protozoal infections, and Hodgkin's disease.

Class	Generic Name	Trade Names
Antidepressants	Isocarboxazid	Marplan
	Phenelzine	Nardil
	Tranylcypromine	Parnate
	Selegiline	Ernsam
Antimicrobials	Furazolidone	Furoxone
Antineoplastic	Procarbazine hydrochloride	Matulane

Foods high in tyramine that should be avoided include:

- Aged foods
- Alcoholic beverages (especially chianti, sherry, liqueurs, beer)
- Alcohol-free or reduced-alcohol beer or wine
- Anchovies
- Bologna, pepperoni, salami, pastrami, mortadella, summer sausage, any fermented sausage
- Caviar

- Cheeses (especially strong or aged varieties), except cottage cheese, cream cheese, ricotta, part-skim mozzarella, American
- Chicken and beef livers, smoked or pickled fish, herring
- Fermented foods
- Figs (canned)
- Fruit: raisins, bananas (or any overripe fruit)
- Broad-beans (fava beans), lima beans, bean curd (tofu), eggplant, Chinese pea pods, tomatoes, tomato sauce including ketchup, chili sauce, marmite (vegetable extract)
- Meat prepared with tenderizers; unfresh meat extracts
- Smoked or pickled meat, poultry, or fish
- Soy sauce, teriyaki sauce, soybean paste, fermented bean curd (fermented tofu), miso soup, tamari, natto, shoyu, tempeh

Foods that can be eaten in moderation are:

- Avocados
- Caffeine (including chocolate, coffee, tea, cola)
- Chocolate
- Raspberries
- Sauerkraut
- Soup (canned or powdered)
- Sour cream
- Yogurt
- Buttermilk

All foods should be very fresh or properly frozen. Meat products should not be refrigerated more than three to four days. Refrigerated cheeses should be eaten within two to three weeks. Combination foods like cheese crackers, submarine sandwiches, and stir-fried dishes containing soy sauce should be avoided. Pizza, lasagna, and other cheese-containing dishes may be eaten only if made with "allowed" cheeses and toppings.

Sources: Monoamine oxidase inhibitors (MAOIs) Updated Dec 10, 2008. Available at http://www.mayoclinic.com

University of Pittsburgh Medical Center. (2003). *MAOI Diet Facts.* Available at http://patienteducation.upmc.com/Pdf/MaoiDiet.pdf

National Institutes of Health Drug Nutrient Interactions Task Force. Warren Grant Magnuson Clinical Center. Drug-Nutrient Interactions: Monoamine oxidase inhibitor (MAOI) medications. Available at http://www.cc.nih.gov

Table 11.2 Grapefruit-Drug Interactions

Confirmed Grapefruit-Drug Interactions		
Drug Class	Generic	Trade Name
Antiarrhythmic	amiodarone ranolazine	Cordarone® Ranexa®
Antihistamines	fexofenadine terfenadine astemizole	Allegra® Seldane® Hismanal®
Anti-infective agents	halofantrine (antimarlarial) primaquine (antimarlarial) indinavir	Halfan® No trade name Crixivan®
Benzodiazepines	diazepam	Valium®
Calcium Channel Blockers	felodipine nimodipine nisoldipine nitrendipine	Plendil® Nimotop® Sular® Bayotensin
Cholesterol-lowering (HMG-CoA reductase inhibitors) ("-statins")	lovastatin simvastatin atorvastatin cerivastatin ezetimibe	Mevacor® Zocor® Lipitor® Baycol® Zetia®
Immunosuppressants	sirolimus cyclosporine tacrolimus	Rapamune® Neoral® Prograf®
Psychiatric	buspirone pimozide ziprasidone carvedilol diazepam sertraline	BuSpar® Orap® Geodon® Coreg® Valium® Zoloft®
Miscellaneous	cisapride sildenafil cilostazol budesonide colchicine eletriptan etoposide mifepristone eplerenone telithromycin ixabepilone nilotinib nifedipine temsirolimus	Prefulside®, Propulsid® Viagra® Pletal Entocort® None Relpax® Vepesid® Mifeprex® Inspra® Ketek® Ixempra® Tasigna® Procardia® Torisel®

Sources: Elbe D. Grapefruit-Drug Interactions. Available at http://www .powernetdesign.com. Last updated Jan 29 2009; Elbe D. Grapefruit-Drug Interactions. In Pronsky ZM. Food-Medications Interactions, 15th ed. Birchrunville (PA): Food-Medication Interactions; 2008. ADA, Nutrition Care Manual.

The interaction of drugs with grapefruit and grapefruit juices has been the subject of recent clinical investigations. Numerous drugs subject to such metabolic interactions, including **statin** medications used to treat hyperlipidemia, several medications used in cardiac care (talinol, nifedipine), and cyclosporines (which are immunosuppressants), have been identified and are a targeted patient education issue for clinicians.[37–40] See Table 11.2 for a summary of these interactions.

Effect of Nutrition on Drug Excretion
The pH of the urine can vary widely and is one of the most important concerns related to maintenance of consistent drug excretion. Variable urine pH can alter reabsorption of the drug, resulting in fluctuating therapeutic levels. Dietary intake, kidney and respiratory function, acid-base balance, hydration status, and the presence of disease or infection can alter urinary pH and necessitate evaluation of drug dosage. Modification of dietary intake to control urine pH has been applied in the treatment of urolithiasis (kidney stones; see Chapter 18).[41, 42] These interventions include increased water intake, limited protein, and overall reduced dietary oxalate.

Nutritional Complications Secondary to Pharmacotherapy

The previous sections of this chapter have focused on the effect of diet and nutrition on drug pharmacokinetics. The other side of drug-nutrient interactions is the effect of drug action on nutritional status. Drugs affect nutrient ingestion, digestion, absorption, and metabolism. The clinical expertise of the registered dietitian is a critical component in the identification, prevention, and correction of these interactions.

Drug Consequence: Effect on Nutrient Ingestion
One only has to evaluate the possible side effects of any medication to understand their potential effect on nutrient ingestion. Nausea, vomiting, diarrhea, constipation, increased appetite, and decreased appetite are all common side effects that dramatically affect dietary intake. Further complicating this situation is the fact that many individuals are prescribed numerous medications and that most senior citizens take multiple medications each day.[43] Next, consider the additive effect of over-the-counter medications as well as herbal supplements.[35] Recently, an evaluation of 100 patients with renal disease indicated an average of one to five dietary supplements were used daily.[44]

Appetite and subsequent food ingestion can be affected by taste, smell, and saliva production. Many medications alter saliva production by either increasing or decreasing saliva, or even by changing its consistency. For example, amitriptyline, a common antidepressant, may cause a decrease in saliva production. Since adequate solution is necessary for taste, many clients on these medications will report difficulty eating, decreased appetite, or anorexia, ultimately due to dry mouth.

Other medications may actually result in a perceived abnormal taste. Patients have reported experiencing metallic, salty, sweet, and simply foul tastes after taking some medications. Chemotherapy agents, analgesics (pain relievers), cardiac drugs, antibiotics, central nervous system drugs, and antifungal agents are common groups of medications that result in these patient complaints. For example, captopril and cisplatin ingestion may result in a metallic or altered sense of taste.[45]

Increased appetite secondary to medications can result in unplanned weight gain. A common example is treatment with prednisone or other corticosteroids, antiseizure medications, or antidepressants. Zyprexa (olanzapine) and Clozaril (clozapine), used to treat schizophrenia, almost always result in weight gain. These medications appear to block the serotonin receptor associated with satiety, inhibit histamine and dopamine, and increase the hormone prolactin.[46] Other antidepressant medications, such as Prozac, can result in the opposite effect—decreased appetite and weight loss. See Table 11.3 for common medications that can result in weight gain and Table 11.4 for those that can result in weight loss.

Table 11.3 Medications That May Cause Weight Gain

Drug Class	Generic Name	Trade Name
Antiarthritic	celecoxib	Celebrex®
Anti-anxiety	alprazolam	Xanax®
	prochlorperazine	Compazine®
	venlafaxine	Effexor XR®
Anticonvulsants	valproic acid (sodium valproate)	Depakote®
	chlordiazepoxide	Librium®
	gabapentin	Neurontin®
	topiramate	Topamax®
Bipolar Disorders	lithium carbonate	Eskalith, Eskalith CR, Lithobid®
Antidepressants	nefazodone	Serzone®
	trazodone	Desyrel®
	phenelzine	Nardil®
	tranylcypromine	Parnate®
	fluoxetine	Prozac®, Sarafem®
	sertraline	Zoloft®
	paroxetine	Paxil®
	fluvoxamine	Luvox®
	amitriptyline	Elavil®, Vanatrip®
	doxepin	Sinequan®
	clomipramine HCl	Anafranil®
	imipramine	Tofranil®
	nortriptyline	Aventyl®, Pamelor®
	trimipramine	Surmontil®
	isocarboxazide	Marplan®
	mirtazapine	Remeron®
Antihistamines	diphenhydramine	Nytol®, Benadryl®
	loratadine	Claritin®, Claritin RediTabs®
Antihypertensives	prazosin	Minipress®
	doxazosin	Cardura
	terazosin	Hytrin®
	propranolol	Inderal®, Inderal LA®
	metoprolol	Lopressor®, Toprol XL®
	atenolol	Tenormin®
Anti-osteoporosis	raloxifene	Evista®
Antipsychotics	haloperidol	Haldol®
	loxapine	Loxitane®
	clozapine	Clozaril®
	perphenazine	Trilafon®
	olanzapine	Zyprexa®
	risperidone	Risperdal®
	olanzapine	Zyprexa®
	chlorpromazine HCl	Thorazine®
	quetiapine fumarate	Seroquel®
	thioridazine	Mellaril®
	thiothixene	Navane®
	ziprasidone	Geodon®
Appetite Stimulant	dronabinol	Marinol®
	megestrol acetate	Megace®
Bronchodilator	albuterol sulfate	Proventil®, Proventil®, Repetabs (SR)®, Ventolin®, Ventolin Repetabs (SR)®
Corticosteroids	methylprednisolone	Medrol®
	prednisolone	Prelone®
	dexamethasone	Decadron®
	prednisone	Deltasone®, Orasone®, Prednicen-M®, Liquid Pred®

Drug Class	Generic Name	Trade Name
Hormone	danazol	Danocrine®
	medroxyprogesterone acetate	Cycrin®
	estrogen	Cenestin®, Estrace®, Estradiol oral®, Ogen®, Premarin®, Climara®, Estraderm®
	estrogen/progesterone	Activella®, Femhrt®, Prempro®, Premarin®, CombiPatch®
	testosterone	Androgel, Striant, Androderm
	oxandrolone	Oxandrin
Insulin	none	Humalog®, Novolog®
Oral Hypoglycemic	glipizide	Glucotrol®, Glucotrol XL®
	glyburide	Diabeta®, Micronase®, Glynase®
	glimepiride	Amaryl®
	chlorpropamide	Diabinese®
	tolbutamide	Orinase®
	repaglinide	Prandin®
	rosiglitazone	Avandia®
	pioglitazone	Actose®

Source: Ness-Abramof R, Aprovian CM. Drug-induced weight gain. Drugs of Today. 2005 Aug;41(8):547–55. Pronsky ZM. Food Medication Interactions, 15th ed. Birchrunville (PA): Food-Medication Interactions; 2008; Leslie WS, Hankey CR, Lean MEJ. Weight gain as an adverse effect of some commonly prescribed drugs: a systematic review. QJM. 2007;100(7):395–404.

Drug Consequence: Effect on Nutrient Absorption

Any drug that affects gastrointestinal function has the potential to interrupt nutrient absorption. This includes medications that cause side effects such as nausea, vomiting, diarrhea, and constipation. Adequate and efficient nutrient absorption requires exposure to enzymes in the appropriate metabolic environment, adequate transit time, sufficient GI tract surface area, and any transporters necessary for absorption. Any medication that speeds gastric emptying or affects the pH of gastric juices could therefore interfere with nutrient absorption. For example, since calcium supplements are absorbed best in an acidic environment, the chronic use of proton pump inhibitors may affect calcium absorption by decreasing stomach acidity.[47] Other examples include **omeprazole** and H_2 blockers, both of which can impair the absorption of vitamin B_{12}. Medications that interfere with lipid metabolism or absorption can interfere with fat-soluble vitamin absorption.

Chronic use of corticosteroids, which are anti-inflammatory and immune-suppressing medications, is a mainstay of treatment for several medical conditions, including rheumatoid arthritis, COPD, and others. This class of medications results in decreased absorption of calcium from the GI tract as well as increased urinary loss of calcium. This significant drug consequence places the patient at high risk for bone fracture and osteoporosis.[48]

Drug Consequence: Effect on Nutrient Metabolism

Drugs can interfere with macronutrient, vitamin, and mineral metabolism. For example, corticosteroids increase

Table 11.4 Medications That May Cause Weight Loss

Drug Class	Generic Name	Trade Name
Anti-Alzheimer's	donepezil rivastigmine galantamine	Aricept® Exelon® Reminyl®
Antiarrhythmia	digitalis digoxin hydroxychloroquine sulfate	Digitoxin® Digoxin, Lanoxin® Plaquenil®
Antianxiety	venlafaxine alprazolam	Effexor®, EffexorXR® Xanax®
Antibiotic	clindamycin gentamicin sulfate	Cleocin® Garamycin®
Anticonvulsant	ethosuximide	Zarontin®
Anticonvulsant/ antiglaucoma	acetazolamide	Diamox®
Anticonvulsant/ antipanic	clonazepam	Klonopin®
Antidepressant	bupropion fluoxetine fluvoxamine maleate sertraline	Wellbutrin®, Wellbutrin SR® Prozac®, Prozac Weekly®, Sarafem® Luvox® Zoloft®
Anti-ADHD	amphetamines methylphenidate	Adderall®, Dexedrine® Ritalin®
Antigout	colchicines	none
Antifungal	amphotericin B	Abelcet®, AmBisome®, Amphotec®, Fungizone®
Antihypertensive	captopril indapamide hydralazine	Capoten® Lozol® Apresoline®
Antihyperlipidemia	cholestyramine	Questran®
Anti-inflammatory	mesalamine	Asacol®, Entasa®, Canasa®, Rowasa®
Antineoplastic	cytarabine aldeleukin/ interleukin 2 capecitabine carboplatin fluorouracil (5-FU) tamoxifen citrate anastrozole cisplatin cyclophosphamide bicalutamide bleomycin sulfate dacabazine mitomycin alpha 2a alpha 2b hydroxyurea imatinib mesylate irinotecan HCL methotrexate vinblastine sulfate vinorelbine tartrate	Cytosar-U® Proleukin® Xeloda® Paraplatin® Adrucil,® Nolvadex®, Nolvadex-D® Arimidex® Platinol-AQ® Cytoxan®, Cytoxan lyophilized® Casodex® Blenoxane® DTIC-Dome® Mutamycin® Roferon-A® Intron-A® Hydrea® Gleevec® Camptosar® Methotrexate®, Rheumatrex® Velban,® Oncovin® Navelbine®

Drug Class	Generic Name	Trade Name
Anti-Parkinson's	levodopa pramipexole	Depar®, Larodopa® Mirapex®
Antipsychotic	loxapine	Loxitane®
Antiviral	ganciclovir sodium	Cytovene®
Calcium regulator	calcitriol	Rocaltrol®, Calcijex®
Laxative	bisacodyl mineral oil	Dulcolax® Agoral plain®
Oral hypoglycemic	metformin	Glucophage®
Thyroid preparations	levothyroxine sodium	Synthroid®, Levoxyl®, Unithroid®
Weight control agent	orlistat phentermine phentermine resin sibutramine	Xenical® Alli® (OTC) Adipex-P®, Fastin® Lonamin® Meridian®

Source: Pronsky ZM. Food Medication Interactions, 15th ed. Birchrunville (PA): Food-Medication Interactions; 2008.

the rate of gluconeogenesis, resulting in hyperglycemia and increased nitrogen loss. Numerous medications interfere with vitamin and mineral metabolism. Phenytoin (Dilantin), used for treatment of seizures, inhibits both vitamin D and folate metabolism. Long-term use may result in megaloblastic anemia secondary to folate deficiency. See Table 11.5 for examples of common interactions for nutrient metabolism.

Drug Consequence: Effect on Nutrient Excretion

Since most drugs are excreted in urine, any drug that increases urinary output places the patient at risk for accelerated nutrient excretion as well. A classic example is the use of diuretics that are potassium wasting. Use of the diuretic Lasix (furosemide) or any other medications in this class can result in hypokalemia (low serum potassium). Any medication that affects renal function in a significant way—reducing reabsorption of nutrients, for instance—can also cause excessive loss of a nutrient in the urine. An example of a tubular reabsorption deficit involves the use of immunosuppressant medications called cyclosporins (e.g., Neoral, Sandimmune, SangCya). These medications have been associated with large amounts of magnesium loss in the urine. See Table 11.6 for examples of common medications affecting nutrient excretion.

At-Risk Populations

As stated previously, it would be a daunting task to acquire a working knowledge of all potential drug-nutrient interactions. However, a study of the basic principles of pharmacology and categories of interactions reveals that certain situations place individuals at risk. These may include disease state, organ function, or treatment modality. Furthermore, certain groups of individuals are more likely than others, not only to take more medications, but to also have an increased risk of improper or inadequate pharmacokinetics. Knowing these populations are at risk allows the practitioner to target them for monitoring and education.

Table 11.5 Medications That Interfere with Nutrient Metabolism

Nutrient	Drug(s)	Effect on Nutrient(s)
Minerals	Diuretics (thiazides), corticosteroids, purgatives; amphotericin B (antifungal), digoxin	Potassium depletion
	Cortisol, desoxycorticosterone, aldosterone, estrogen-progestogen oral contraceptives, phenylbutazone	Sodium and water retention
	Sulfonylureas, phenylbutazone, cobalt, lithium	Impair uptake or release of iodine
	Oral contraceptives; ethambutol (antitubular agent); amphotericin B (antifungal)	Lower plasma zinc, elevate copper
	Corticosteroids, biphophonates	Calcium depletion
	H_2 receptors antagonists, proton pump inhibitors	Decreased absorption of iron
	Laxatives	Malabsorption of electrolytes and calcium
Vitamins	Phenobarbital and phenytoin (Dilantin), carbamazepine	Increase metabolism of folic acid, vitamins D and K
	Isoniazid (INH), hydralazine	Pyridoxine and niacin antagonists
	Laxatives	General malabsorption of fat-soluble vitamins
	Pyrimethamine, sulfadoxine, methotrexate, metformin, oral contraceptives, H_2 receptor antagonists, proton pump inhibitors	Vitamin B_{12} &/or folate antagonists
Amino acids	Oral contraceptives, selective serotonin reuptake inhibitors, trazodone	Altered tryptophan metabolism
Blood glucose	Metprolol, cholorpromazine	Increases blood glucose
	Niacin	Decreases blood glucose

Source: Pronsky ZM. Food Medication Interactions, 15th ed. Birchrunville (PA): Food-Medication Interactions; 2008

Table 11.6 Medications That Affect Nutrient Excretion

Nutrient	Drug(s)	Effect on Nutrient
Minerals	Loop diuretics	Increase excretion of sodium, potassium, chloride, magnesium, calcium
	Thiazide diuretics	Increase excretion of most electrolytes
	Antifungals	Increase excretion of potassium
	NSAIDs	Increase excretion of potassium
	Caffeine	Increases sodium excretion
	Calcitonin	Increases excretion of phosphorus, magnesium, potassium, chloride, and sodium; may increase or decrease excretion of calcium
	Antihyperlipidemic	Increases excretion of calcium and magnesium
	Antineoplastics	Increase excretion of magnesium, calcium potassium, zinc, copper
	Clonidine	Decreases excretion of sodium and chloride
	Corticosteroids	Decrease excretion of sodium; increase excretion of potassium, calcium, nitrogen, zinc
	Cyclosporine	Increases excretion of magnesium; decreases excretion of potassium
	Digitalis	Increases urinary excretion of magnesium
Vitamins	NSAIDs	Increase excretion of vitamin C
	Corticosteroids	Increase excretion of vitamin C
	Tetracycline	Increases urinary excretion of riboflavin, folacin
Macronutrients	NSAIDs	Increase excretion of protein
	Calcitriol	Increases excretion of albumin
	Antineoplastics	Increase excretion of amino acids

Source: Pronsky ZM. *Food Medication Interactions,* 15th ed. Birchrunville (PA): Food-Medication Interactions; 2008.

Drug-Nutrient Interactions in the Elderly The elderly population represents one group with an exceptionally high risk for drug-nutrient interactions.[32, 49, 50] This risk exists for several reasons. Older individuals generally have the highest rate of chronic disease and are therefore prescribed the largest number of medications; this sheer volume increases risk. Furthermore, the use of over-the-counter and complementary medications compounds the incidence of interactions.[49, 51] In addition, drug pharmacokinetics is affected by physiological changes that occur with aging. Decreased muscle mass and impaired cardiac, liver, and renal function all are common in the elderly and can change how a drug is absorbed, metabolized, and excreted. For example, the elderly may experience an exacerbation of drug-related confusion if other neurological diseases are present. Finally, compliance with drug regimens can be an important issue for this population. Financial burdens, complex regimens, or lack of proper drug education can lead to inappropriate drug dosing.

Polypharmacy, a term that is often associated with the elder population, is defined as administration of excessive drugs at one time or concurrent use of a large number of drugs, which increases the risk of interactions. Other features of polypharmacy may include the use of medications without a reason; the use of multiple medications for the same condition; the use of medications that interact with one another; the use of inappropriate dosages; the use of additional drugs to treat side effects of medications; and overall improvement when medications are discontinued.

Protocols and clinical guidelines have been developed to prevent adverse drug effects in this population. Beer's Criteria (see Table 11.7) have identified the medications most likely to result

Table 11.7 Beer's Criteria: Potentially Inappropriate Medications Used with Older Adults with and without Concomitant Diagnoses or Conditions

Key: ↔ Neutral risk; ↑ High Risk; ↓ Low Risk

Medications to Avoid (or Use within Specified Dose/Duration Ranges)

Medications	Problem(s)	Risk
Anti-infectives		
Oral antibiotics	Therapy > four weeks should be avoided except when treating osteomyelitis prostatitis, tuberculosis, or endocarditis.	↔
Cardiac Agents		
Digoxin (Lanoxin)	Decreased renal clearance, which may increase toxic effects. Doses >0.125 mg should be avoided except for treatment of atrial arrhythmias.	↑
Disopyramide (Norpace)	May induce heart failure.	↑
EENT Agents		
Antihistamines (alone or in combination, including chlorpheniramine [Clor-Trimeton], diphenhydramine [Benadryl], hydroxyzine [Vistaril and Atarx], cyproheptadine [Periactin], promethazine [Phenergan], and dexchlorpheniramine [Polaramine])	Strong anticholinergic activity.	↓
Decongestants (oxymetazoline [Afrin], phenylephrine [Neo-Synephrine, Vicks Sinex], pseudoephedrine [Sudafed, Suphedrin, Triaminic, Dimetapp])	Avoid daily use for longer than two weeks.	↔
Diphenhydramine (Benadryl)	Should not be used as sleep aid. May cause confusion. Use lowest possible dose for allergies.	↔
Endocrine Agents		
Chlorpropamide (Diabinese)	May cause prolonged and serious hypoglycemia.	↑
Gastrointestinal Agents		
Bisacodyl (Dulcolax), cascara sagrada, and Neoloid except in presence of opiate analgesic use	Long-term use of stimulant laxatives may exacerbate bowel dysfunction.	↑
Cimetidine (Tagamet)	Avoid doses >900 mg/day; do not use >12 weeks.	↔
Dicyclomine (Bentyl), hyoscyamine (Levsin & Levsinex), propantheline (Pro-Banthine), belladonna alkaloids (Donnatal and others), clidinium-chloridiazepoxide (Librax)	Strong anticholinergic activity. Questionable efficacy as antispasmodic agents. Should be avoided, especially long-term use.	↑
Mineral oil	Potential for aspiration.	↑
Ranitidine (Zantac)	Avoid doses >300 mg/day; do not use >12 weeks.	↔
Trimethobenzamide (Tigan)	Produces extrapyramidal side effects; one of the least effective antiemetic agents.	↓

(continued)

Table 11.7 Beer's Criteria: Potentially Inappropriate Medications Used with Older Adults with and without Concomitant Diagnoses or Conditions (Continued)

Key: ↔ Neutral risk; ↑ High Risk; ↓ Low Risk

Medications	Problem(s)	Risk
Hematopoietic Agents		
Ferrous sulfate iron supplements >325 mg	Cause constipation; higher doses not effective.	↓
Musculoskeletal Agents		
Indomethacin (Indocin and Indocin SR)	Has more CNS side effects than any other NSAID.	↓
Methocarbamol (Robaxin), carisoprodol (Soma), oxybutynin (Ditropan), chlorzoxazone (Paraflex), metaxalone (Skelaxin), cyclobenzaprine (Flexeril)	Anticholinergic side effects, sedation, weakness. Effectiveness of tolerated doses questionable.	↓
Naproxen (Naprosyn, Avaprox, Aleve), oxaprozin (Daypro), piroxicam (Feldane)	Potential to produce GI bleeding, renal failure, high blood pressure and heart failure.	↑
Phenylbutazone (Butazolidin; not on U.S. market)	Serious hematologic side effects.	↓
Psychotropic Agents		
Amitriptyline (Elavil), alone or in combination products (Limbitrol, Triavil)	Anticholinergic and sedating properties.	↑
Barbiturates (other than phenobarbital)	Side effects and addictive properties.	↑
Chlordiazepoxide (Librium), alone or in combination; or diazepam (Valium)	Risk of sedation and increased falls.	↑
Doxepin (Sinequan)	Sedating properties and powerful anticholinergic.	↑
Ergot mesylates (Hydergine), cyclandelate isoxsuprine (Cyclospasmol)	Not proven effective.	↓
Flurazepam (Dalmane)	Risk of sedation and increased falls.	↑
Haloperidol (Haldol)	Avoid doses >3 mg/day.	↔
Lorazepam (Ativan [3 mg]), oxazepam (Serax [60 mg]), alprazolam (Xanax [2 mg]), temazepam (Restoril [15 mg]), zolpidem (Ambien [5 mg]), triazolam (Halcion [0.25 mg])	Avoid higher doses. Avoid single oxazepam dose >30 mg or >0.25 mg triazolam.	↓
Meperidine (Demerol)	Not effective orally. More disadvantageous than other narcotic analgesics.	↑
Meprobamate (Miltown and Equanil)	Highly addictive and sedating.	
Pentazocine (Talwin)	Many CNS side effects, including confusion & hallucinations.	↑
Propoxyphene (Darvon)	Few advantages over acetaminophen, produces adverse effects of other narcotic drugs.	↓
Thioridazine (Mellaril)	Avoid doses >30 mg/day.	↔
Vascular Agents		
Dipyridamole (Persantine)	Causes orthostatic hypotension. Useful only in patients with artificial heart valves.	↓
Hydrochlorothiazide (Hydrodiuril)	Avoid doses >50 mg/day.	↔
Methyldopa (Aldomet, alone or in combination [Aldoril])	Bradycardia and exacerbates depression.	↑
Propranolol (Inderal)	Better beta-receptor selectivity and less CNS penetration in other beta-blockers.	↔
Reserpine (Harmonyl), alone or in combination	Depression, impotence, sedation, and orthostatic hypotension.	↓

Table 11.7 Beer's Criteria: Potentially Inappropriate Medications Used with Older Adults with and without Concomitant Diagnoses or Conditions (*Continued*)

Medications to Avoid with Specific Concomitant Diseases

Disease	Medication	Problem	Risk
Anorexia			
Malnutrition	CNS stimulants: dextroamphetamine (Adderall), methylphenidate (Ritalin), methamphetamine (Desoxyn), pemoline, and fluoxetine (Prozac)	Appetite-suppressing effects.	↑
Bleeding Disorders			
Blood clotting disorders or receiving anticoagulant therapy	Aspirin, NSAIDs, dipyridamole (Persantin), ticlopidine (Ticlid), and clopidogrel (Plavix)	May prolong clotting time.	↑
Cardiac Disorders			
Arrhythmias	Tricyclic antidepressants (imipramine hydrochloride [Tofranil, Tofranil PM, Janimine], doxepin hydrochloride [Sinequan], amitriptyline hydrochloride [Adepril, Endep, Enovil, Trepiline])	May induce arrhythmias.	↑ if started recently
Heart failure	Disopyramide (Norpace)	May exacerbate symptoms of heart failure.	↑
	Drugs with high sodium content (sodium and sodium salts [alginate bicarbonate (Di-Gel, Maalox, Mylanta), biphosphate (Fleet enema), citrate (Bicitra), phosphate (K-Phos), salicylate (Alka-Seltzer, Pepto-Bismol), and sulfate (colyte)])	May lead to fluid retention and worsen heart failure.	↓
Endocrine Disorders			
Diabetes	Beta-blockers (Inderal, Lopressor)	May exacerbate symptoms in patients treated with insulin or oral hypoglycemic agents.	↓
	Corticosteroids (started recently)	May decrease glycemic control.	↓
Gastrointestinal Disorders			
Constipation	Anticholinergics (Levbid, Anaspaz)	Exacerbate constipation.	↓
	Calcium channel blockers (Procardia, Cardizem)	Exacerbate constipation.	↓
	Narcotics	Exacerbate constipation.	↓
	Tricyclic antidepressants (imipramine hydrochloride [Tofranil], doxepin hydrochloride [Sinequan], and amitriptyline hydrochloride [Limbitrol])	Exacerbate constipation.	↓
Ulcers	NSAIDs	May exacerbate ulcer disease, gastritis, GERD.	↑
	Aspirin	May exacerbate ulcer disease, gastritis, GERD.	↓
	Potassium supplements	May exacerbate ulcer disease, gastritis, GERD.	↓
Neurologic Disorders			
Cognitive impairment	Barbiturates, anticholinergics/antispasmodics (Levbid, Symax), and muscle relaxants (Paraflex, Remular, Skelaxin), CNS stimulants: dextroamphetamine (Adderall), methylphenidate (Ritalin), methamphetamine (Desoxyn)	CNS-altering effects.	↑
Epilepsy	Clozapine (Clozaril), chlorpromazine (Thorazine), thioridazine (Mellaril), thiothixene (Navane)	Lower seizure threshold.	↓
	Metoclopramide (Reglan)	Lowers seizure threshold.	↑
Parkinson's disease	Metoclopramide (Reglan), conventional antipsychotics, and tacrine (Cognex)	Antidopaminergic/cholinergic effects.	↑
Seizure disorder	Bupropion (Wellbutrin)	May lower seizure threshold.	↑

(continued)

Table 11.7 Beer's Criteria: Potentially Inappropriate Medications Used with Older Adults with and without Concomitant Diagnoses or Conditions (*Continued*)

Disease	Medication	Problem	Risk
Psychiatric Disorders			
Depression	Long-term benzodiazepine use	May exacerbate depression.	↑
	Methyldopa (Aldomet), reserpin, & guanethidine (Ismelin)	May exacerbate depression.	↑
Insomnia	Decongestants	May cause or worsen insomnia.	↓
	Theophylline (Theodur)	May cause or worsen insomnia.	↓
	Methylphenidate (Ritalin)	May cause or worsen insomnia.	↓
	Desipramine, SSRIs, MAOIs	May cause or worsen insomnia.	↓
Respiratory Disorders			
Asthma	Beta-blockers	May worsen respiratory function.	↑
COPD	Beta-blockers	May worsen respiratory function.	↑
	Sedative-hypnotics	May slow respirations and increase CO_2 retention.	↑
Urologic Disorders			
Benign prostatic hypertrophy	Anticholinergic antihistamines	May cause obstruction.	↑
	Gastrointestinal antispasmodics	May cause obstruction.	↑
	Muscle relaxants	May cause obstruction.	↓
	Narcotic drugs (including propoxyphene)	May cause obstruction.	↓
	Flavoxate, oxybutynin	May cause obstruction.	↓
	Bethanechol	May cause obstruction.	↓
	Anticholinergic antidepressants	May cause obstruction.	↑
Incontinence	Alpha blockers (Doxazosin, Prazosin, Terazosin)	May produce polyuria.	↑
	Anticholinergics	May produce polyuria.	↑
	Tricyclic antidepressants (Imipramine hydrochloride, doxepin hydrochloric, amitriptyline hydrochloride)	May produce polyuria.	↑
	Long-acting benzodiazepines: (chlordiazepoxide [Librium], alone or in combination; or diazepam [Valium])	May produce polyuria.	↑
Vascular Disorders			
Clotting disorders treated with anticoagulants	Aspirin	May cause bleeding.	↑
Hypertension	Amphetamines & other weight control agents	May increase blood pressure.	↑
Peripheral vascular disease	Beta-blockers	Negative chronotropic and inotropic activity.	↓
Syncope	Beta-blockers	Negative chronotropic and inotropic activity.	↓
	Long-acting benzodiazepines	May contribute to falls.	↑
Weight Disorders			
Obesity	Olanzapine (Zyprexa)	May stimulate appetite and increase weight gain.	↓

Adapted from: Beers MH, Ouslander JG, Rollingher I, Rueben DB, Brooks J, Beck JC, Explicit criteria for determining inappropriate medication use in nursing home residents. Arch Intern Med. 1991;151:1825–32.

Beers MH. Explicit criteria for determining potentially inappropriate medication use by the elderly: an update. Arch Intern Med. 1997;157:1531–36.

Fick DM, Cooper JW, Wade WF, Waller JL, Maclean R, Beers MH. Updating the Beers criteria for potentially inappropriate medication use in older adults. Results of a US consensus panel of expert. Arch Intern Med. 2003;163:2716–24.

BOX 11.3 CLINICAL APPLICATIONS

Prevention of Adverse Drug Reactions in the Elderly

Adverse drug reactions (ADRs) are any harmful, unintentional drug reactions that take place at customarily prescribed doses. These reactions contribute to hospitalizations, disability, morbidity, and mortality, consequently adding billions of dollars to health care expenditures. Elderly patients are considered vulnerable to ADRs as a consequence of adverse physiologic changes that take place as a result of the aging process, a high frequency of comorbid conditions, and the large numbers of medications prescribed to them.

Many ADRs are the result of inescapable patient eccentricities, but many others are believed to be preventable. One way to prevent ADRs is to avoid prescribing inappropriate medications. The Beers criteria (see Table 11.7) are some of the most commonly used methods for assessing appropriateness of prescribing medications for elderly patients, though they are not evidence-based. The most common reasons for ADRs are:

- Decline in physiological functions that naturally occur with aging (May influence disposition of drugs)
- Impaired organ function from prior disease or aging (Alters drug kinetics, organ responses, and homeostatic counter-regulatory drug effects)
- Number of medications prescribed (Probability of toxicity increases with number of medications prescribed)

Noncompliance with medication regimens is another cause for ADRs in elderly patients.

Noncompliance may be a result of:

- Inadequate instructions for taking medications
- Switching to alternative medical practices
- Illiteracy
- Poverty
- Misconceptions
- Inability to recall complicated medical regimens

A list of 10 drug interactions frequently identified in long-term care facilities has been developed by the Multidisciplinary Medication Management Project:

- Warfarin and NSAIDs[1]
- Warfarin and sulfa drugs
- Warfarin and macrolides
- Warfarin and quinolones[2]
- Warfarin and phenytoin
- ACE inhibitors and potassium supplements
- ACE inhibitors and spironolactone
- Digoxin and amiodarone
- Digoxin and verapamil
- Theophylline and quinolones

Sources: Beard K. Adverse reactions as a cause of hospital admissions for the aged. Drug Aging. 1992;2:356–63.

Brown KE. Top ten dangerous drug interactions in long-term care. Multidisciplinary Medication Management Project. Available at http://www.scoup.net/M3Project/topten

Chang C, Liu PY, Yang YK, Yang Y, Wu C, Lu F. Use of the Beers criteria to predict adverse drug reactions among first-visit elderly outpatients. Pharmacotherapy. 2005;25(6):831–38.

Malhotra S, Karan RS, Pandhi P, Jain S. Drug related medical emergencies in the elderly: role of adverse drug reactions and non-compliance. Postgrad Med. J. 2001;77:703–07.

Montamat SC, Cusack BJ, Verstal RE. Management of drug therapy in the elderly. N Engl J Med. 1989;321:303–09.

World Health Organization. International drug monitoring: the role of national centres. WHO technical report series no. 498. Geneva, Switzerland: World Health Organization; 1972.

Endnotes

[1]NSAID class does not include COX-2 inhibitors.

[2]Quinolones does not include ciprofloxacin, enoxacin, norfloxacin, and ofloxacin.

---

in adverse effects.[52] General components of these criteria state that if a patient uses more than five drugs, is noncompliant with medication regimens, and has a history of adverse effects, the risk of continued interactions is high.[53] Box 11.3 provides guidance for prevention of adverse drug reactions in the elderly.

Drug-Nutrient Interactions in HIV and AIDS

Antiretroviral therapy requires concomitant use of multiple medications (see Chapter 24). These medications represent a unique situation that places this population at high risk for drug-nutrient interactions.[54] Many of these medications have specific guidelines for consumption with or without food due to the effect of food on absorption and utilization, and many of them cause significant nutritional side effects such as nausea, vomiting, and diarrhea.

Drug-Nutrient Interactions in Nutrition Support

The use of specialized nutrition support (SNS) is another clinical measure that poses a high risk for drug-nutrient interactions (see Chapter 7). Tube feedings have been documented to decrease absorption of some medications (e.g., warfarin, phenytoin, and tetracycline).[16, 55] Macronutrients present in the tube feeding may cause chelation of some medications. The following are the 2009 guidelines from the American Society for Enteral and Parenteral Nutrition for medication and tube feedings:[56]

- Do not add medication directly to an enteral feeding.
- Avoid mixing together medications that are intended to be administered through an enteral feeding tube. Additional medications can be given sequentially but not mixed.
- Liquid medication formulations should be used, when available, for administration via enteral feeding tubes.
- When medications are administered via an enteral feeding tube, the tube should be flushed before and after each medication is administered.
- Restart the feeding after administering the medication to avoid compromising nutrition status.
- Consult with pharmacist for patients receiving medication co-administered with enteral nutrition.

Nutrition Therapy

Nutritional Implications

Any use of prescribed drugs, over-the-counter medications, or complementary treatments has the potential to affect nutritional status, interfere with drug pharmacokinetics, and/or alter nutrient metabolism. Additionally, regulatory agencies for health care, including The Joint Commission on Accreditation of Healthcare Organizations (JCAHO) and Centers for Medicare and Medicaid Services, require an established protocol for identification of, intervention for, and patient education for drug-nutrient interactions.

Nutrition Assessment

Nutrition assessment will focus on factors that could affect absorption, distribution, metabolism, or excretion of drugs (see Chapter 3). First, the clinician should evaluate past and current medical history for any diagnosis affecting kidney, liver, or cardiac function. Baseline laboratory measurements for kidney function (blood urea nitrogen, creatinine), liver function (ALT, AST, bilirubin, alkaline phosphatase, prothrombin time), and glucose should then be evaluated. The medical history should identify any treatment regimens (for example, enteral nutrition or dialysis) that may potentiate drug-nutrient interactions or adverse effects. Overall, nutritional status will need to be quantified to ensure that consistent physiological response to medications is possible. If the patient is malnourished, for example, the amount of protein-bound drug can be reduced due to hypoalbuminemia, increasing the effect of the medication.

Next, all drugs, over-the-counter medications, dietary supplements, and other complementary medical regimens should be identified. Patient interviews and social history should identify any potential barriers to compliance with, understanding of, or access to medical or nutrition therapies. For each prescription drug, over-the-counter medication, and dietary supplement, drug-drug interactions should be identified, along with any nutrition implications.[57, 58] Figure 11.4 provides an overview of the assessment process.

Consider the nutrition assessment data presented in Box 11.4. Figure 11.5 outlines the factors considered during

---

BOX 11.4 CLINICAL APPLICATIONS

Assessment of a Patient at Risk for Drug-Nutrient Interactions

Client History:
Age and medical history

65-year-old adult male; Hx of hypertension; myocardial infarction; 4 vessel coronary artery bypass graft; type 2 diabetes mellitus; prostate cancer s/p TURP; long-term use of alcohol

Medical diagnoses

Cardiac function: hypertension, previous MI, and cardiac surgery

Liver function: probable alcohol abuse

Renal function: related to type 2 diabetes mellitus; hypertension

Biochemical Data, Medical Tests and Procedures:

Treatment regimens that may potentiate drug-nutrient interactions: none at this time

Glucose 180 mg/dL; BUN 21 Cr 1.2

Interpretation of data: Poor glycemic control; possible renal insufficiency

Food/Nutrition-Related History:

Medication and herbal supplement use to determine the nutrition implications of the current or ordered medications. This includes determining the dosage and timing of the medications, the medical effect of the medication(s), and possible drug-drug interactions and drug-nutrient interactions.

Medication regimen

Once daily: Toprol 50 mg; Plavix 5 mg; Aspirin 325 mg; Altace 5 mg;

Amaryl 2 mg twice daily.

Define current drugs

- *Toprol (metoprolol)*—beta-blocker used to reduce the overall workload of the heart
- *Plavix (clopidogrel)*—inhibits platelet aggregation; used to prevent stroke and myocardial infarction in patients with cardiac history or history of previous stroke
- *Aspirin*—inhibits platelet aggregation; used to prevent stroke and myocardial infarction in patients with cardiac history or history of previous stroke
- *Altace (ramipril)*—ACE inhibitor used to treat hypertension
- *Amaryl (glimepiride)*—oral agent that stimulates insulin release from the beta cells of the pancreas and improves insulin resistance in peripheral tissues in patients with type 2 diabetes mellitus

Drug-drug interactions

- *Toprol and Amaryl*—Beta-blockers may increase the risk of hypoglycemia in patients taking Amaryl.
- *Altace and Amaryl*—ACE inhibitors may increase the risk of hypoglycemia in patients taking Amaryl.
- *Plavix and Aspirin*—These drugs together may increase chance of bleeding. Patients should avoid other over-the-counter medications that contain aspirin.

Drug-nutrient interactions

- *Altace* may cause hyperkalemia (high serum potassium). Patients should be instructed to avoid foods high in potassium, especially salt substitutes.
- *Toprol* absorption increases with food intake. Daily dosages should be consistently taken with meals so that therapeutic levels can be reached.
- The patient should avoid all alcohol, because there is an interaction between alcohol, *Altace*, and *Amaryl*.

Figure 11.4 Nutrient Assessment of Drug-Nutrient Interactions

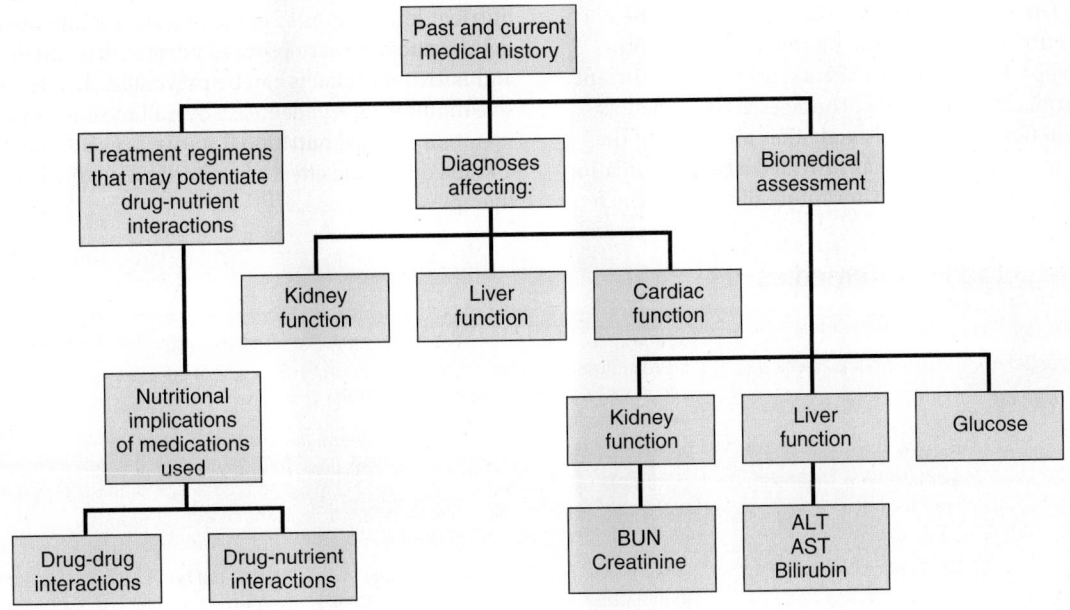

Figure 11.5 Nutrition Assessment of Drug-Nutrient Interactions: A Clinical Example

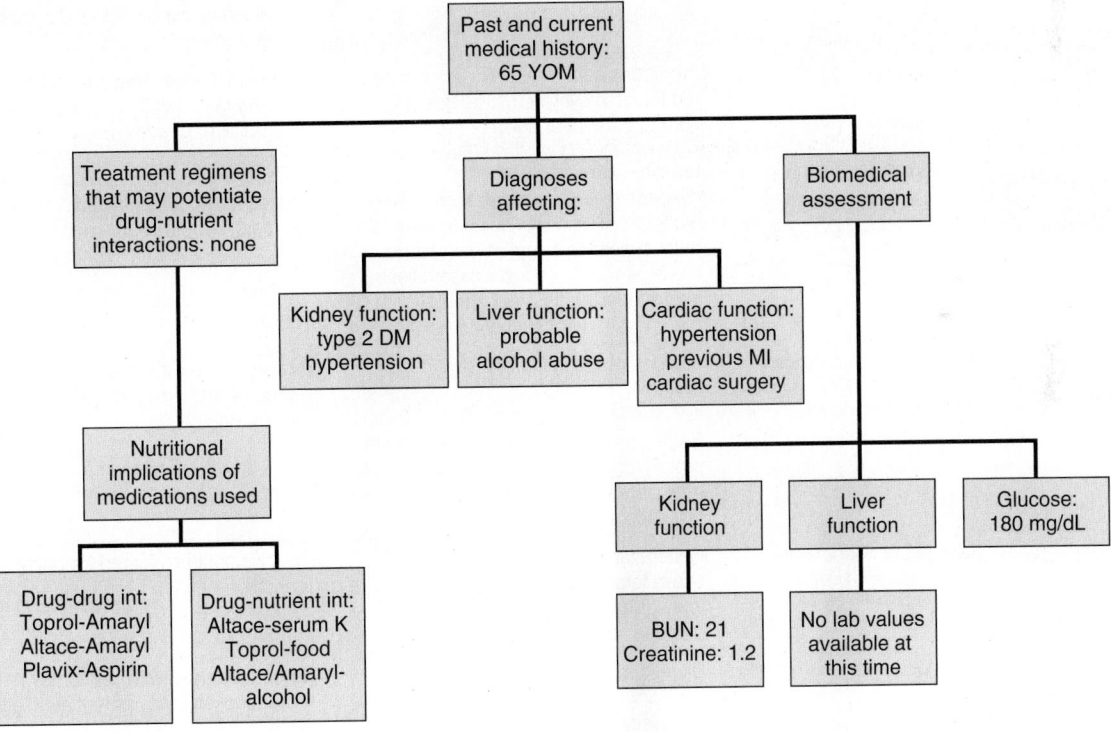

assessment of this patient's risk for drug-nutrient interactions. Based on this assessment, the nutrition diagnosis will be *food-medication interaction*[6] and the nutrition interventions will center around modification of diet and/or medication to reduce the risk of compromising the patient's nutritional or health status. Interventions to optimize both drug efficacy and nutritional status include recommending appropriate timing of medication (Toprol should be taken with meals), avoiding alcohol, and avoiding foods high in potassium. The patient should be supplied with educational materials on foods high in potassium.

The last step in the nutrition care process will be to determine when and what to monitor and evaluate based on the nutrition diagnosis and intervention. If the patient was hospitalized, the evaluation might be obtained within 48 hours:

1. "Calorie count" to evaluate consumption of potassium and alcohol and determine whether a meal was consumed when the Toprol was administered.

2. Review of laboratory values for serum potassium.

At discharge, the patient's knowledge of foods high in potassium may also be determined.

Conclusion

The American Dietetic Association position regarding the integration of nutrition therapy and pharmacotherapy "promotes a team approach to care for clients receiving concurrent MNT and pharmacotherapy and encourages active collaboration among dietetics professionals and other members of the health care team."[7] This chapter has provided a basic foundation in the principles of pharmacology for future dietitians, which is the first step in ensuring this collaboration. If practitioners use their understanding of the common categories for drug-nutrient interactions to guide assessment and intervention during the nutrition care process, adverse drug, herbal/botanical, and nutritional effects can be prevented. This level of aggressive monitoring will decrease overall morbidity and mortality, maintain optimal nutritional status, prevent polypharmacy, and maximize the effect of prescribed medical and nutrition therapies.

Table 11.8 Selected Folk Remedies

Remedy	Preparation	Common CAM Use	Cautions*
Bearberry (Manzanita) *Arctostaphylos uva-ursi*	Leaf, stem tea, infusion or extract	Diuretic; treat diabetes; treat urinary tract infections, kidney problems (esp. stones); treat bronchitis	Should not be used by pregnant/lactating women; prolonged use or excessive doses toxic
Bitter Root (Dogbane) *Apocynum spp.*	Root, fruit decoction or extract	Purgative; contraceptive/abortive; treat cardiovascular problems; treat kidney problems; treat liver ailments, gallstones; treat gout; treat edema; headache	May cause nausea; may increase heart rate and arterial blood pressure; excessive doses toxic
Black Cohosh *Cimicifuga racemosa*	Root decoction or extract	Treat menstrual problems; ameliorate menopause symptoms; ease labor; treat hypertension; treat kidney problems; treat arthritis; treat diarrhea; treat cough	Should not be used by pregnant/lactating women; should be used with caution by children, adolescents, and women with history of breast cancer or undergoing chemotherapy for breast cancer; prolonged use or excessive doses may cause mild nausea, vomiting, headaches, hypotension; dizziness; mastalgia, weight gain
Black Nightshade (Zhoa ia) *Solanum nigrum*	Leaf juice	Promote sleep; treat pain; ameliorate menopause symptoms; treat toothache, sore throats; treat colds, cough	Excessive doses may cause nausea, vomiting, disorientation; cardiac arrhythmia; respiratory depression; death
Bloodroot, Red Root, Red Puccoon *Sanguinaria canadensis*	Root juice	Emetic; stomach "cleansing"; treat dyspepsia, peptic ulcers; treat "weak" blood; treat kidney problems; stimulate mucus-clearing cough, treat asthma, croup, whooping cough, tuberculosis; treat liver ailments; treat arthritis, rheumatism; treat skin cancers	Topical applications can be caustic; excess doses may cause nausea, vomiting, dizziness, tremors, hypotension, shock, coma, death
Burdock *Arctium lappa*	Root juice, tea or extract	Prevent/treat cancer; improve immune system in HIV/AIDS: treat diabetes; reduce serum cholesterol levels; cleanse blood toxins; treat prostate cancer; treat kidney problems (esp. stones); treat hemorrhoids; treat back pain; treat gout; treat venereal diseases; treat asthma; treat acne	May interfere with absorption of some medications; may reduce need for insulin in type 1 diabetes
Guava Leaf *Psidium guajava*	Leaf tea or decoctions	General antioxidant; reduce serum cholesterol levels; reduce blood glucose levels; treat digestive tract disorders, diarrhea, dysentery	May be contraindicated for persons with heart conditions
Hawthorn *Crategus oxyacantha*	Root decoction; berry tea or extract	Diuretic; prevent/treat cardiovascular conditions (e.g., angina, arrhythmia, congestive heart failure); hypertension; reduce serum cholesterol levels; treat blood disorders, insomnia, sore throat	Should be used with caution by persons using beta-blockers; may potentiate digitalis; may cause nausea, headache
Kava Kava *Piper methysticum*	Root, rhizome tea, decoction, extract, powder additive for beverages	Sedative; euphoric; reduce stress, anxiety; treat urinary tract infections; treat bronchitis, asthma; treat venereal diseases; treat headache, backache; treat obsessive compulsive disorder	Should not be used by pregnant/lactating women or persons being treated for depression, hypertension, or Parkinsonism; may be hepatotoxic (esp. when consumed with alcohol, Echinacea, or aspirin); may potentiate antiepileptic drugs; may cause central nervous system depression when taken with valerian or chamomile; may cause rash, drowsiness

*Adverse side effects and/or interactions may occur even if not indicated.

Table 11.8 Selected Folk Remedies (*Continued*)

Remedy	Preparation	Common CAM Use	Cautions*
Licorice Root *Glycyrrhiza glabra*	Root juice, tea, extract	Purgative; treat dyspepsia, gastric ulcers; stimulate endocrine system (esp. in HIV/AIDS); treat sore throat; treat tuberculosis, cough; liquefy mucus in cystic fibrosis; treat liver ailments; treat kidney tumors; treat arthritis, rheumatism; treat lupus erythematosis; menstrual problems	Should be used with caution by persons with heart or renal disease; may increase sensitivity to digitalis; may interfere with hypertension drugs; may potentiate insulin, corticosteroids, MAO inhibitors; prolonged use or excessive doses may cause hypokalemia, hypertension, headache, dizziness, edema
Mandrake (Mayapple) *Podophyllum peltatum*	Root decoction, extract, resin	General tonic; purgative; emetic; treat treat anorexia; stomach problems, dyspepsia, constipation; treat urinary tract infections, incontinence; treat liver ailments (esp. hepatitis); lung conditions; treat rheumatism, arthritis	Should not be used by pregnant/lactating women (even topically—may cause fetal abnormalities or miscarriage); should not be used by children; excessive doses may cause rash, irritation, nausea, vomiting, renal failure, hepatotoxicity, cerebrotoxicity
Mango *Mangifera indica*	Leaf tea, decoction; fresh fruit; bark extract (Vimang)	Antioxidant; improve immune system, treat flu; treat diabetes; treat hypertension; treat liver ailments	Persons allergic to poison ivy or oak may also be allergic to mango sap
Mistletoe *Phoradendron leucarpum*	Leaf tea or extract	Treat cancer (esp. breast, ovarian, prostate); treat hypertension; treat cardiovascular disease; treat blood conditions, hemorrhaging, stomach disorders, diarrhea; ease anxiety, panic disorder	Should not be used by pregnant/lactating women or by children; may suppress immune system; may potentiate hypertension drugs and sedatives; berries highly toxic
Morning Glory *Ipomoea spp.*	Root tea or decoction	Purgative; treat diabetes; treat diarrhea; treat kidney problems; treat urinary tract infections; treat menstrual cramps; treat epilepsy, hysteria	Should not be used by pregnant/lactating women; may cause nausea, diarrhea
Noni (Indian Mulberry) *Morinda citirolia*	Fruit juice; leaf tea; bark extract	Improve immune system; prevent/treat cancer; treat diabetes; treat hypertension; treat anorexia, stomach problems, parasites; treat liver ailments; treat urinary tract infections; treat tuberculosis; treat edema; treat sore throat; treat eye problems	Should be used with caution by persons limiting potassium intake; may cause constipation; may turn urine pink
Prickly Pear Cactus *Opuntia spp.*	Fruit juice; pad extract; root decoction	Diuretic; treat diabetes; treat hypertension; reduce serum cholesterol levels; treat kidney problems (esp. stones); treat urinary tract infections; treat hangovers	Should be used with caution by pregnant/lactating women; some persons may experience allergic reactions, e.g., rash, hives, shortness of breath, and chest pain
Poke (Fitolaca) *Phytolacca Americana*	Root, shoots decoction	Purgative; improve immune system; treat cancer; treat inflammation, fungal infections (esp. in HIV/AIDS); treat stomach problems; treat liver ailments; treat kidney problems; improve "weak" blood; treat prostate cancer; treat lung conditions; croup; treat arthritis; bursitis; rheumatism; treat toothache	Should not be used by pregnant/lactating women or by children or by persons using antidepressants or oral contraceptives; may be hepatotoxic; improper preparation or excessive doses may cause nausea, vomiting; berries toxic
Raspberry *Rubus spp.*	Leaf tea	Treat anorexia, diarrhea, stomach problems; treat hemorrhaging, anemia; menstrual problems; induce vomiting; induce labor	Should not be used by pregnant women during the first trimester
Willow *Salix spp.*	Bark decoction or extract	Treat inflammation, fever; treat diarrhea; treat osteoarthritis, rheumatism; aphrodisiac; treat premature ejaculation; treat headache, chronic pain; treat bedwetting	Should not be used by pregnant/lactating women or persons with sensitivity to salicylates; should not be given to children under age 16 with flu-like symptoms (to prevent Reyes syndrome); should be used with caution by persons being treated for diabetes, hemophilia, asthma, peptic ulcers, gout; may potentiate anticoagulant and antiplatelet therapies; excessive doses may cause rash, nausea, vomiting, kidney inflammation, tinnitus
Yellowroot *Xanthorhiza simplicissma*	Root tea	Improve the immune system; treat diabetes; treat hypertension; treat liver ailments (e.g., jaundice); treat stomach problems, dysentery	Should not be used by pregnant/lactating women or by infants; may interfere with B-vitamin metabolism

*Adverse side effects and/or interactions may occur even if not indicated.

Table 11.9 Selected Natural Products

Product	CAM Dose	Common CAM Use	Cautions*
Bilberry *Vaccinium myrtillus*	80–160 mg	Prevent diabetic retinopathy; treat cataracts, macular degeneration; improve night vision; treat infections and inflammation; treat diarrhea, dyspepsia; treat mouth, throat problems; treat varicose veins	May cause allergic reaction or diarrhea; may interfere with iron absorption; prolonged use or excessive doses may be toxic
L-Carnitine	1000–6000 mg	Improve immune system; facilitate metabolism, protect heart in HIV/AIDS; prevent cardiovascular disease; treat angina; treat arrythmia; treat congestive heart failure; improve physical endurance; improve athletic performance; treat chronic fatigue syndrome; reduce memory loss; treat sports injuries; increase fat metabolism, weight loss	High doses may cause nausea, vomiting, and diarrhea
Chondroitin	400–600 mg	Treat osteoarthritis	May potentiate anticoagulant drugs (e.g., aspirin); excessive doses may cause nausea, diarrhea
Coenzyme Q_{10}	100–400 mg	Treat angina; treat arrhythmia; treat congestive heart failure; prevent heart disease; reduce hypertension; reduce serum cholesterol levels; reduce cancer risk; alleviate fibromylagia symptoms; increase energy levels, improve stamina in HIV/AIDS; treat chronic fatigue syndrome; treat fibrocystic disease; treat Parkinsonism; weight loss	Should be avoided by pregnant/lactating women; excessive doses may cause anorexia, diarrhea, fatigue, twitching
DHEA (dehydroepian-drosterone)	5–200 mg	Improve immune system; slow aging; treat chronic fatigue syndrome; alleviate fibromylagia symptoms; prevent muscle wasting in HIV/AIDS; treat lupus erythematosis; promote weight loss	Should not be taken by pregnant/lactating women or persons with ovarian, adrenal or thyroid tumors; may increase risk of breast, ovarian, liver cancer; may decrease serum HDL cholesterol levels; may precipitate mania in mood disorders; large doses may cause acne, facial hair on women, lowering of the voice—should be taken under supervision of health provider
Echinacea *Echinacea purpurea*, *E. angustifolia*	200–600 mg	Prevent/treat colds, flu; improve immune system; treat HIV/AIDS; treat chronic respiratory infections; treat urinary tract infections; maintain prostate health; treat chronic fatigue syndrome; treat fungal infections; prevent cancer	Should not be used by pregnant/lactating women, persons allergic to sunflower-family plants, or persons with certain systemic diseases, such as AIDS, tuberculosis, diabetes, multiple sclerosis, leukemia, and lupus erythematosis; may cause rash; may be hepatotoxic or nephrotoxic when combined with kava, salicylate (in herbs or aspirin), or hepatotoxic drugs such as anabolic steroids; prolonged use may be toxic
Evening Primrose Oil *Oenothera biennis*	2500–3000 mg	Treat diabetic neuropathy; treat impotency and female infertility; alleviate premenstrual syndrome and menopause symptoms; treat osteoarthritis; treat attention deficit hyperactivity disorder; treat memory loss; improve athletic performance; treat symptoms of alcohol withdrawal; treat skin disorders	Should not be used by persons being treated for epilepsy or schizophrenia (may cause seizures); may cause headaches, nausea; prolonged use may suppress immune system
Garlic *Allium sativum*	1200–2000 mg (or 4 g fresh)	Prevent/treat colds, flu, sore throat; reduce hypertension; lower serum cholesterol levels; inhibit platelet aggregation; reduce risk of colon, esophageal, lung, stomach cancers; stimulate immune system in HIV/AIDS; stimulate mucus-clearing cough, treat coughs, bronchitis; treat fungal infections; improve nails	May potentiate antihypertensive, hypoglycemic, and anticoagulant drugs (e.g., aspirin); may interfere with certain protease inhibitors; large doses may cause nausea, heartburn, flatulence, diarrhea, body odor
Ginkgo Biloba *Ginkgo biloba*	120–240 mg (extract)	Treat diabetic neuropathy; inhibit platelet aggregation; improve circulation; reduce macular degeneration, cataracts; improve hearing loss; treat involuntary ejaculation, impotence; treat depression, anxiety; treat headache, dizziness; treat Alzheimer's disease, reduce memory loss; optimize brain function	Should not be used by persons being treated for epilepsy or schizophrenia (may cause seizures); may increase blood pressure when taken with certain diuretics; may potentiate certain antidepressant, antipsychotic, and anticoagulant therapies (e.g., aspirin, warfarin); may interfere with hypoglycemic drugs, antianxiety drugs, and some antacids; excessive doses may cause nausea, headache, or rash

*Adverse side effects and/or interactions may occur even if not indicated.

Table 11.9 Selected Natural Products (*Continued*)

Product	CAM Dose	Common CAM Use	Cautions*
Ginseng *Panax ginseng*, *P. quinquefolius*	300–2000 mg	Improve immune system; reduce stress and fatigue; improve physical, mental performance; reduce memory loss; regulate sugar metabolism, reduce blood glucose levels in diabetes, treat hypoglycemia; treat impotence and male infertility	Should not be used by persons being treated for acute infections or heart arrhythmia; should be used with caution by pregnant/lactating women and persons receiving immunosuppressive therapies; may potentiate anticoagulant, corticosteroid, hypoglycemic, and estrogen drugs, also MAO inhibitors, NSAIDS, and stimulants, such as caffeine and Ritalin; may interfere with calcium channel blockers, opiates, and antipsychotic drugs; excessive doses may cause nausea, headache, insomnia, rash, breast tenderness
Glucosamine	900 mg/100 lb. of body weight	Treat osteoarthritis; alleviate back and joint pain	Should not be taken by persons allergic to shellfish; may increase insulin resistance; may interact with diuretics; excessive doses may cause nausea, diarrhea
Goldenseal *Hydrastis canadensis*	125–650 mg (extract)	Improve immune system; prevent/treat colds, flu; treat HIV/AIDS; inhibit lung cancer growth; treat urinary tract infections; treat chronic fatigue syndrome; treat fungal infections; treat diarrhea	Should not be used by pregnant/lactating women; should not be used by persons with high blood pressure, heart disease, or glaucoma; may potentiate other natural products; prolonged use or excessive doses may cause mouth irritation, nausea, vomiting, diarrhea, nosebleed, and lethargy
Grape Seed Extract *Vitis vinifera*, *V. coignetiae*	50–300 mg	General antioxidant; prevent/treat cancer; treat HIV/AIDS; lower serum cholesterol levels; reduce LDL oxidation; ameliorate fibromylagia symptoms; treat allergies; treat eczema, psoriasis; treat macular degeneration, cataracts, other vision problems	May interfere with hypocholesterolemic drugs; may potentiate anticoagulant drugs; may act synergistically with vitamin C
Melatonin	1–3 mg	Improve immune system, slow aging; treat cancer, adjunct to chemotherapy, radiation; treat seasonal affective disorder; treat insomnia	May cause excessive drowsiness when combined with sedatives, antihistamines, narcotic pain relievers; may interfere with corticosteroid drugs; may stimulate autoimmunity conditions
Peppermint *Mentha piperita*	450–2500 mg 3–4 cups (tea)	Relieve symptoms of irritable bowl syndrome, diverticulits, morning sickness; improve digestion; treat nausea, vomiting, diarrhea; dissolve gallstones; treat allergies, asthma; treat headache, chronic pain	Should be used with caution by pregnant women in the last trimester and persons with hiatal hernia; large doses may cause heartburn, muscle tremor, or rash
St. John's Wort *Hypericum perforatum*	900–4000 mg	Treat depression, anxiety, stress; prevent/treat infections, inhibit growth of HIV/AIDS; reduce serum cholesterol levels; treat breast cancer (prevent infiltration of chest wall); treat premenstrual syndrome; treat chronic fatigue syndrome; alleviate fibromylagia symptoms; reduce memory loss; weight loss	Should be used with caution by pregnant/lactating women; adverse interactions with numerous over-the-counter and prescription drugs, e.g., cold and flu medications, antibiotics, MAO inhibiters, oral contraceptives, and protease inhibitors; may interfere with chemotherapy; may precipitate mania in mood disorders; may increase sunburn damage
Saw Palmetto *Serenoa repens*, *Sabal serrulata*	160–320 mg	Improve immune system; treat prostate problems, e.g., cancer; treat impotence; slow aging	Should not be used by pregnant/lactating women
Valerian *Valeriana spp.*	400–3000 mg	Treat flu; treat stress; treat insomnia; ease anxiety, panic disorder, obsessive compulsive disorder; treat headache; treat alcoholism	Should not be used by pregnant/lactating women; may potentiate other sedatives, e.g., alcohol, barbiturates; may impair driving and operation of machinery; prolonged use may cause headaches, irritability, insomnia, arrythmia

*Adverse side effects and/or interactions may occur even if not indicated.

Table 11.10 Selected Individual Vitamin and Mineral Supplements

Supplement	CAM Dose	Common CAM Use	Cautions*
Fat-soluble Vitamins			
Vitamin A	2000–25,000 IU	Improve immune system; prevent infection; prevent/treat cancer, cardiovascular disease; treat osteoarthritis; treat colds, flu	Toxic at high doses: total vitamin A intake may exceed 100,000 IU when combined with other sources of A (food and supplement) for acute infections; high intake may be associated with osteoporosis; may interfere with anticoagulants and anticonvulsants
Beta-carotene	15–100 mg	General antioxidant; reduce oxidation of LDL cholesterol; reduce cardiovascular disease risk; reduce cancer risk (esp. cervical cancer); protect lungs in cystic fibrosis	Hepatotoxic when combined with alcohol intake; may promote pulmonary cancer when used by smokers who drink; may interfere with prescription drugs
Vitamin E	800–1200 IU	General antioxidant; improve immune system; prevent infection; improve glucose tolerance; reduce oxidation of LDL and increase HDL cholesterol; improve circulation; reduce colon cancer risk; protect lungs in cystic fibrosis; reduce pain in osteoarthritis; treat depression, Alzheimer's, memory loss	May interfere with anticoagulants; high doses may suppress immune system
Water-soluble Vitamins			
Ascorbic acid/vitamin C	1000–20,000 mg	Improve immune system, prevent infection; shorten duration of colds, flu; prevent/treat cancer; improve iron absorption; reduce hypertension; reduce serum cholesterol levels and increase glutathione levels; reduce oxidation of LDL cholesterol; reduce pain of angina; improve glucose tolerance, reduce diabetic vascular damage; treat fungal infections, treat HIV/AIDS; protect lungs in cystic fibrosis; treat osteoarthritis; treat depression	May enhance iron absorption (increasing oxidative cellular stress) and decrease copper absorption; may increase risk of kidney stones; may increase in vitro conversion of amygdalin (natural laetrile) to cyanide; high doses can cause diarrhea
Biotin	300–16,000 mcg	Improve glucose metabolism, reduce blood glucose; prevent cracking, peeling nails	No adverse effects reported for oral intake
Cobalamin/vitamin B_{12}	100–2000 mcg	Reduce plasma homocysteine levels and treat cardiovascular disease; treat anemia; treat diabetic neuropathy; improve brain function in HIV/AIDS, treat psychosis, depression, Alzheimer's, memory loss	No adverse effects for oral intake: B-vitamins are interdependent, excess of one may cause deficiency of others
Folic acid	800–50,000 mcg	Reduce plasma homocysteine levels; reduce risk of cervical, colon cancer; treat anemia; treat depression, insomnia, irritability, dementia	Megadoses may inhibit cobalamine absorption; should not be used by persons with epilepsy
Niacin/vitamin B_3	25–1000 mg	Reduce serum cholesterol, LDL, triglyceride levels, increase HDL levels; treat depression, mania, anxiety, dementia, memory loss	Megadoses may impair glucose tolerance; may increase plasma homocysteine; can induce hyperuricemia; should not be used by persons taking high-dose aspirin or uricosuric drugs, or those with liver dysfunction, diabetes, or who abuse alcohol; can cause transient flushing, cramps, nausea, diarrhea
Pantothenic acid	50–1000 mg	Prevent infections; treat hypertension; treat depression, irritability; increase longevity	May reduce thiamin absorption and produce deficiency symptoms; can cause diarrhea
Pyridoxine/vitamin B_6	25–1800 mg	Improve immune system; improve glucose tolerance; reduce homocysteine levels; treat artherosclerosis; reduce cervical cancer risk; alleviate premenstrual syndrome; treat morning sickness; treat depression	High doses may reduce folate levels; can interfere with medications for Parkinsonism; may cause neuropathy; may cause rash
Riboflavin/vitamin B_2	2–400 mg	Reduce frequency, severity of migraines; treat depression; improve immune system	No adverse effects for oral intake: B-vitamins are interdependent, excess of one may cause deficiency of others
Thiamin/vitamin B_1	9–100 mg	Treat psychosis, anxiety, depression, irritability, Alzheimer's disease, memory loss	No adverse effects for oral intake: B-vitamins are interdependent, excess of one may cause deficiency of others

*Adverse side effects and/or interactions may occur even if not indicated.

Table 11.10 Selected Individual Vitamin and Mineral Supplements (*Continued*)

Supplement	CAM Dose	Common CAM Use	Cautions*
Minerals			
Boron	1–9 mg	Treat arthritis; improve bone density	Total intake may include additional amounts found in other supplements or foods; toxic in large doses: diarrhea, vomiting, death
Chromium	300–1000 mcg	Lower blood glucose levels; improve glucose tolerance; reduce LDL and increase HDL cholesterol levels; increase metabolism/ weight loss; increase lean muscle, maintain muscle mass in HIV/AIDS	May cause rash; may cause renal or liver damage in large doses; may accumulate in body tissues causing oxidative damage; may be mutagenic; alters serotonin, dopamine, and norepinephrine metabolism in brain, can contribute to mood changes; may potentiate antidepressants
Copper	2–6 mg	Improve immune system; prevent infections; reduce risk of cardiovascular disease; prevent/ treat cancer; treat osteoarthritis; treat anemia	Large doses may impair memory, cause depression, insomnia, depress immune system, cause oxidative tissue damage
Magnesium	400 mg	Improve pancreatic function, glucose tolerance; reduce birth defects, spontaneous abortion in pregnant women with diabetes; reduce diabetic retinopathy; relieve angina; treat hypertension; reduce cancer risk; treat autism, attention deficit hyperactivity disorder	Excessive doses may cause diarrhea; extremely large doses (usually due to antacid or Epsom salt abuse) may cause shock, coma, or cardiopulmonary arrest; should not be used by persons with poor kidney function
Potassium	200–500 mg	Improve glucose tolerance; prevent/treat hypertension; prevent muscle cramps	May interact with certain prescription and over-the-counter drugs (e.g., nonsteroidal anti-inflammatory drugs—NSAIDS); can cause hyperkalemia resulting in heart arrythmias, death
Selenium	50–400 mcg	Improve immune system; prevent infection; reduce risk of cardiovascular, cerebrovascular disease; reduce oxidation of LDL cholesterol; prevent/treat cancer (esp. prostate cancer); treat HIV/AIDS; treat osteoarthritis; protect lungs in cystic fibrosis; detoxify body of heavy metals; improve hair, nails, skin	Large doses may cause rash, irritability, gastrointestinal problems, impair immune system; may be hepatotoxic; may interact with lipid-lowering drugs
Zinc	10–75 mg	Improve immune system; prevent/treat colds, flu; prevent diabetes, reduce blood glucose levels; prevent/treat cancer; treat HIV/AIDS; treat osteoarthritis; treat prostate problems; treat fungal infections; treat acne	May cause stomach upset; large doses inhibit copper and iron absorption; may suppress immunity; may affect absorption levels of prescription drugs

*Adverse side effects and/or interactions may occur even if not indicated.

PRACTITIONER INTERVEW

Gretchen Vannice, MS, RD *Managing Director, Omega-3 RD™ Nutrition Consulting*
www.omega3RD.com

What motivated you to follow your professional career path in nutrition counseling?

After getting a BS in Nutrition and Dietetics and working a few years, I earned an MS in Nutrition Science and became an RD. I have been a health educator in the medical setting, and have worked in research and development, education, and clinical research in the dietary supplement industry. I'm currently an industry consultant to academia and health care professionals, bridging my expertise. Being active with dietetic practice groups at the state and national level has been a useful and rewarding way to stay in touch with colleagues, continue to learn, and stay abreast of growth and change in our profession.

I've been interested in Complementary and Alternative Nutrition/Medicine (CAN/M) for more than two decades. My initial work in nutrition demonstrated that food, supplements, mind and body work, and therapies from other countries (e.g., ayurvedic) could benefit human health and that quality research was needed to discern benefits from hype and guide responsible use. That motivated me to earn my MS, RD.

(continued)

If you would counsel individuals, is there anything unique about their assessment that students should know about?

We start with our standard assessment tools and build on them. A new breed of "functional tests" is available. It's important to discern which tests use validated methodology.

Are certain medical problems more common among your clients or do they just want to be more healthy and eat better?

People turn to CAN/M for several reasons. Some want to eat better and live healthier and don't know where else to look. Some turn to CAN/M when diagnosed with a serious health condition to 1) avoid surgical and drug therapy, 2) use complementary approaches before allopathic medicine, or 3) work with their physician and dietitian to integrate complementary and allopathic approaches. For others, practicing CAN/M is influenced by family culture or religious beliefs. Unfortunately, there is also the group who self-diagnose and treat; this can be the most irresponsible and potentially life-threatening.

In general, the greater public is looking to CAN/M for help with health conditions where allopathic medicine is perceived as lacking. Examples include chronic immune conditions, irritable bowel syndrome, or diet and supplement therapy for diabetes. Disease prevention is the new frontier in CAN/M.

What changes have you seen in a CAN/M acceptance among RDs and the public in the last several years?

Big changes. CAN/M is becoming accepted by the public. When I began working in CAN/M, the idea that eating salmon could prevent a heart attack or fresh ginger could sooth nausea was considered alternative. There is a wide range of people promoting themselves as CAN/M practitioners, some of whom are qualified, many of whom are not. Interest has increased, more evidence is being published (some with positive findings, some not), and Internet-based information is free. Discerning reliable information and responsible therapies is where RDs can take the lead.

How should a student/intern keep current on this area of practice?

Choosing to work in CAN/M will require continual self-education and learning; join dietetic practice groups and national health organizations; attend continuing education seminars; be strategic and selective about where you learn.

Any other advice you would give to dietetic students/interns/new RDs?

- Registered dietitians are well-trained to be leaders in the complementary and alternative arena; we are skilled at applying nutrition in health and medicine.
- Individuals referring to themselves as CAN/M practitioners may or may not have valid credentials. Check and see if their diploma is issued by a credible accrediting organization.
- Internet websites can appear professional and be funded by sales and marketing companies that are intent on making profit, not improving health.
- Dietary supplement companies are not all the same. Use products from reputable companies. Call the company; talk with someone in the product development or quality department, if they have one. They may market science but not employ individuals trained in science. Don't assume research being promoted has been completed on the products they are selling. Ask to see publications.
- Cultivate relationships with "alternative practitioners" in your community. Learn from each other as professionals.
- Keep an open mind to ancient practices in nutrition therapy from other countries. Countries such as India (ayurvedic) and China (Traditional Chinese Medicine) use traditional practices that have come to bear evidence.
- CAN/M is still an emerging field. It is exciting to be part of something new and easy to be seduced by the passion, stories, and powerful beliefs. We, RDs, are health care professionals with a Standards of Practice and Ethics Code to adhere to. Patients rely on our training and professionalism.

WEB LINKS

Food and Drug Administration   The home site of the federal agency responsible for the regulation for all drugs in the United States.
http://www.fda.gov

Healthfinder   A Service of the National Health Information Center, U.S. Department of Health & Human Services. Excellent sources for general information about treatments including common medications.
http://www.healthfinder.gov

Medline Plus—Drugs Information   Site provided by U.S. National Library of Medicine. Drugs, herbs, and supplements are alphabetically linked. Other links from this site are easily followed to product recalls, clinical trials, and other medical information.
http://medlineplus.gov/

National Center for Complementary and Alternative Medicine (NCCAM)   The National Center for Complementary and Alternative Medicine (NCCAM) is the Federal Government's lead agency for scientific research on complementary and alternative medicine.
http://nccam.nih.gov/

1. Match the following examples to their routes of administration. (Choose from oral, sublingual, buccal, rectal, intramuscular, intravenous, and inhalation.)
   a. Tablet dissolved under the tongue =
   b. Insulin given into the muscle =
   c. Dextrose given into a peripheral vein =
   d. Asthma medication that is delivered by puffs through a breathing device =

2. What factors could affect the dissolution of a medicine?

3. Distribution of the drug is defined as ___. What is the major physiological factor that can affect this?
   a. Body temperature
   b. Blood flow
   c. Presence of food

4. What organ is primarily involved in the metabolism of a drug?

5. Name three factors that can affect metabolism of a drug.

6. How are drugs excreted? Give an example of how disease (affecting an organ function) affects drug excretion.

7. Polypharmacy means: _____.
   Who is most at risk?

8. Determine how each of the following could be considered a drug-nutrient interaction:
   a. When the use of methotrexate causes a change in taste
   b. When antacids bind phosphorus
   c. When Lasix increases the renal excretion of potassium
   d. When phenobarbital decreases folate metabolism

REFERENCES

1. History of the FDA. FDA History Office. Available at http://www.fda.gov/AboutFDA/WhatWeDo/History/default.htm, updated 06/18/2009.

2. Stagnitti, MN. Average Number of Total (Including Refills) and Unique Prescriptions by Select Person Characteristics, 2006. Statistical Brief #245. Agency for Healthcare Research and Quality, Rockville (MD): May 2009. Available at http://www.meps.ahrq.gov/mepsweb/data_files/publications/st245/stat245.pdf

3. National Health Expenditure Accounts, Historical. Centers for Medicare & Medicaid Services. Available at http://www.cms.hhs.gov/NationalHealthExpendData/

4. Stagnitti, MN Average Number of Total (Including Refills) and Unique Prescriptions by Select Person Characteristics, 2006. Statistical Brief #245. Agency for Healthcare Research and Quality, Rockville, MD. May 2009. Available from http://www.meps.ahrq.gov/mepsweb/data_files/publications/st245/stat245.pdf

5. Barnes PM, Bloom B, Nahin RL. Complementary and alternative medicine use among adults and children: United States, 2007. National health statistics reports; no 12. Hyattsville (MD): National Center for Health Statistics; 2008

6. Lacey K, Pritchett E. Nutrition care process and model: ADA adopts road map to quality care and outcomes management. J Amer Diet Assoc. 2003;103:1061–72.

7. American Dietetic Association. International Dietetics & Nutrition Terminology (IDNT) Reference Manual. 2nd ed. American Dietetic Assn. Chicago (IL); 2009.

8. American Dietetic Association. Position of the American Dietetic Association: integration of medical nutrition therapy and pharmacotherapy. J Am Diet Assoc. 2003;103:1363–70.

9. Walsky RL, Gaman EA, Obach RS. Examination of 209 drugs for inhibition of cytochrome P450 2C8. J Clin Pharmacol. 2005;45:68–78.

10. Grundy SM, Cleeman JI, Bairey Merz CN, Brewer HB, Clark, LT, Hunninghake DB, et al. Implications of Recent Clinical Trials for the National Cholesterol Education Program Adult Treatment Panel III Guidelines. Circulation. 2004;110:227–39.

11. Pronsky ZM & Crowe JP. Ch 16. In: Mahan K, Escott-Stump S, editors. Krause's Food and Nutrition Therapy. Philadelphia (PA): Saunders; 2008.

12. Epstein S, Cryer B, Ragi S, Zanchetta JR, Walliser J, Chow J, et al. Disintegration/dissolution profiles of copies of Fosamax (alendronate). Curr Med Res Opin. 2003;19:781–89.

13. Perkins AC, Blackshaw PE, Hay PD, Lawes SC, Atherton CT, et al. Esophageal transit and in vivo disintegration of branded risedronate sodium tablets and two generic formulations of alendronic acid tablets: a single-center, single-blind, six-period crossover study in healthy female subjects. Clin Ther. 2008;30(5):834–44.

14. Porter RS, Kaplan JL and Homeier BH, editors. Merck Manuel of Diagnosis and Therapy 18e, Merck Research Laboratories, Whitehouse Station (NJ): 2006–2008.

15. Wilkinson GR. Drug metabolism and variability among patients in drug response. New Engl J Med. 2005;352:2211–21.

16. Chan LN. Drug-nutrient interaction in clinical nutrition. Curr Opin Clin Nutr Metab Care. 2002;5(3):327–32.

17. Genser D. Food and drug interaction: consequences for the nutrition/health status. Ann Nutr Metab. 2008;52 Suppl 1:29–32.

18. Gropper SS, Smith JL. Advanced nutrition and human metabolism. 5th ed. Belmont (CA):Cengage; 2009

19. Harvey FA, Champe PC, Finkel R, Cubeddu L, Clarke MA. Lippincott's illustrated reviews: pharmagolog. 4th ed. Philadelphia (PA): Lippincott Williams & Wilkins; 2008.

20. Fulton MM, Allen ER. Polypharmacy in the elderly: a literature review. J Am Acad Nurse Pract. 2005;17:123–32.

21. de Wildt SN, Johnson TN, Choonara I. The effect of age on drug metabolism. Paediatric and Perinatal Drug Therapy. 2002;5:101–06.

22. Wynne H. Drug metabolism and ageing. J Br Menopause. Soc. 2005;11:51–56.

23. Bowie MW, Slattum PW. Pharmacodynamics in older adults: a review. Am J Geriatr Pharmacother. 2007 Sep;5(3):263–303.

24. Okey AB, Boutros PC, Harper PA. Polymorphisms of human nuclear receptors that control expression of drug-metabolizing enzymes. Pharmacogenet Genomics. 2005;5:371–79.

25. Padol S, Yuan Y, Thabane M, Padol IT, Hunt RH. Clinical impact of CYP2C19 polymorphism on the action of proton pump inhibitors: a review of a special problem. Int J Clin Pharmacol Ther. 2007 Mar;45(3):188; author reply 189–90.

26. Riley, KM (2007, August 16). Transcript of FDA press conference on warfarin. Available at www.fda.gov/bbs/transcripts/transcript081607.pdf

27. Wadelius M, Chen LY, Downes K, Ghori J, Hunt S, Eriksson N, et al. Common VKORC1 and GGCX polymorphisms associated with warfarin dose. Pharmacogenomics J. 2005;5(4):262–70.

28. Cotreau MM, von Moltke LL, Greenblatt DJ. The influence of age and sex on the clearance of cytochrome P450 3A substrates. Clin Pharmacokinet. 2005;44(1):33–60.

29. Roden DR. Principles of clinical pharmacology. In: Fauci AS, Braunwald E, Kasper DL, Hauser SL, Longo DL, Jameson JL, Loscalzo J, Eds. Harrison's principles of internal medicine. 17th ed. New York: McGraw-Hill; 2008. Available at http://www.accessmedicine.com/

30. Remer T, Manz F. Potential renal acid load of foods and its influence on urine pH. J Am Diet Assoc. 1995;95:791–97.

31. Loboz KK, Shenfield GM. Drug combinations and impaired renal function—the 'triple whammy'. Br J Clin Pharmacol. 2005;59:239–43.

32. Sood A, Sood R, Brinker FJ, Mann R, Loehrer LL, Wahner-Roedler DL. Potential for interactions between dietary supplements and prescription medications. Am J Med. 2008;121(3):207–11.

33. Pal D, Mitra AK. MDR- and CYP3A4-mediated drug-herbal interactions. Life Sci. 2006 Mar 27;78(18):2131–45.

34. Markowitz JS, Donovan JL, DeVane CL, Taylor RM, Ying, Ruan Y, et al. Effect of St. John's Wort on drug metabolism by induction of Cytochrome P450 3A4 Enzyme. JAMA. 2003;290:1500–04.

35. Peng CC, Glassman PA, Trilli LE, et al. Incidence and severity of potential drug-dietary supplement interactions in primary care patients. Arch Intern Med. 2004;164:630–36.

36. Norred CL. Complementary and alternative medicine use by surgical patients. AORN J. 2002;76:1013–21.

37. Lilja JJ, Neuvonen M, Neuvonen PJ. Effects of regular consumption of grapefruit juice on the pharmacokinetics of simvastatin. Br J Clin Pharmacol. 2004;58:56–60.

38. Odou P, Ferrari N, Barthelemy C, Brique S, Lhermitte M, Vincent A, et al. Grapefruit juice-nifedipine interaction: possible involvement of several mechanisms. J Clin Pharmacol Ther. 2005;30:153–58.

39. Paine MG, Criss AB, Watkins PB. Two major grapefruit components differ in time to onset of intestinal CYP3A4 inhibition. J Pharmacol Exp Ther. 2005;312:1151–60.

40. Schwarz UI, Seemann D, Oertel R, Miehlke S, Kuhlisch E, Fromm MF, et al. Grapefruit ingestion significantly reduces talinolol bioavailablity. J Clin Pharmacol Ther. 2005;77:291–301.

41. Asplin JR, Coe FL, Favus MJ. Nephrolithiasis. In: Fauci AS, Braunwald E, Kasper DL, Hauser SL, Longo DL, Jameson JL, Loscalzo J, Eds.. Harrison's principles of internal medicine. 17th ed. New York: McGraw-Hill; 2008. Available at http://www.accessmedicine.com/

42. Remer T, Manz F. Potential renal acid load of foods and its influence on urine pH. J Am Diet Assoc. 1995;95:791–97.

43. Hajjar ER, Cafiero AC, Hanlon JT. Polypharmacy in elderly patients. Am J Geriatr Pharmacother. 2007 Dec;5(4):345–51.

44. Spanner ED, Duncan AM. Prevalence of dietary supplement use in adults with chronic renal insufficiency. J Ren Nutr. 2005;15:204–10.

45. Pronsky ZM. Food-Medications Interactions. 15th ed. Birchrunville (PA): Food-Medication Interactions; 2008.

46. Hellings HA, Zarcone JR, Crandall K, Wallace D, Schroder SR. Weight gain in a controlled study of Risperidone in children, adolescents and adults with mental retardation and autism. J Child Adolesc Psychopharmacol. 2001;11:229–38.

47. O'Connell MB, Madden DM, Murray AM, Heanery RP, Kerzner LJ. Effects of proton pump inhibitors on calcium carbonate absorption in women: a randomized crossover trial. Am J Med. 2005;118:778–81.

48. Lindsay R, Cosman F. Chapter 348, Osteoporosis. In: Fauci AS, Braunwald E, Kasper DL, Hauser SL, Longo DL, Jameson JL, Loscalzo J, Eds. Harrison's principles of internal medicine. 17th ed. New York: McGraw-Hill; 2008. Available at http://www.accessmedicine.com/

49. Bergman-Evans B. Evidence-based guideline. Improving medication management for older adult clients. J Gerontol Nurs. 2006;32(7):6–14.

50. Lindblad CI, Artz MB, Pieper CF, Sloane RJ, Hajjar ER, Ruby CM, et al. Potential drug–disease interactions in frail, hospitalized elderly veterans. Ann Pharmacother. 2005;39:412–17.

51. Bruno JJ, Ellis JJ. Herbal use among US elderly: 2002 National Health Interview Survey. Ann Pharmacother. 2005;39:643–48.

52. MacLaughlin EJ, Raehl CL, Treadway AK, Sterling TL, Zoller DP, Bond CA. Assessing medication adherence in the elderly: which tools to use in clinical practice? Drugs Aging. 2005;22:231–55.

53. Chang CM, Liu PY, Yang YH, Yang YC, Wu CF, Lu FH. Use of the Beers criteria to predict adverse drug reactions among first-visit elderly outpatients. Pharmacotherapy. 2005;25(6):831–38.

54. Panel on Antiretroviral Guidelines for Adults and Adolescents. Guidelines for the use of antiretroviral agents in HIV-1-infected adults and adolescents. Department of Health and Human Services. November 3, 2008; 1–139. Available at http://www.aidsinfo.nih.gov/ContentFiles/AdultandAdolescentGL.pdf

55. AuYeung SC, Ensom, MH. Phenytoin and enteral feedings: does evidence support an interaction? Ann Pharmacother. 2000;34:896–905.

56. Bankhead R, Boullata J, Brantley S, Corkins M, Guenter P, Krenitsky J, Lyman B, Metheny NA, Mueller C, Robbins S, Wessel J; A.S.P.E.N. Board of Directors. Enteral nutrition practice recommendations. J Parenter Enteral Nutr. 2009;33(2):122–67.

57. Sanford MG, Ryan C, Cummings AD, Hunt A, Hackes B. Protocols for identifying drug-nutrient interactions in patients: the role of the dietitian. J Am Diet Assoc. 2002;102:729–30; discussion 730–31.

58. Santos CA, Boullata JI. An approach to evaluating drug-nutrient interactions. Pharmacotherapy. 2005;25(12):1789–1800.

Part 4
Nutrition Therapy

12

Energy Balance and Body Weight

Robert D. Lee, DrPH, RD
Central Michigan University

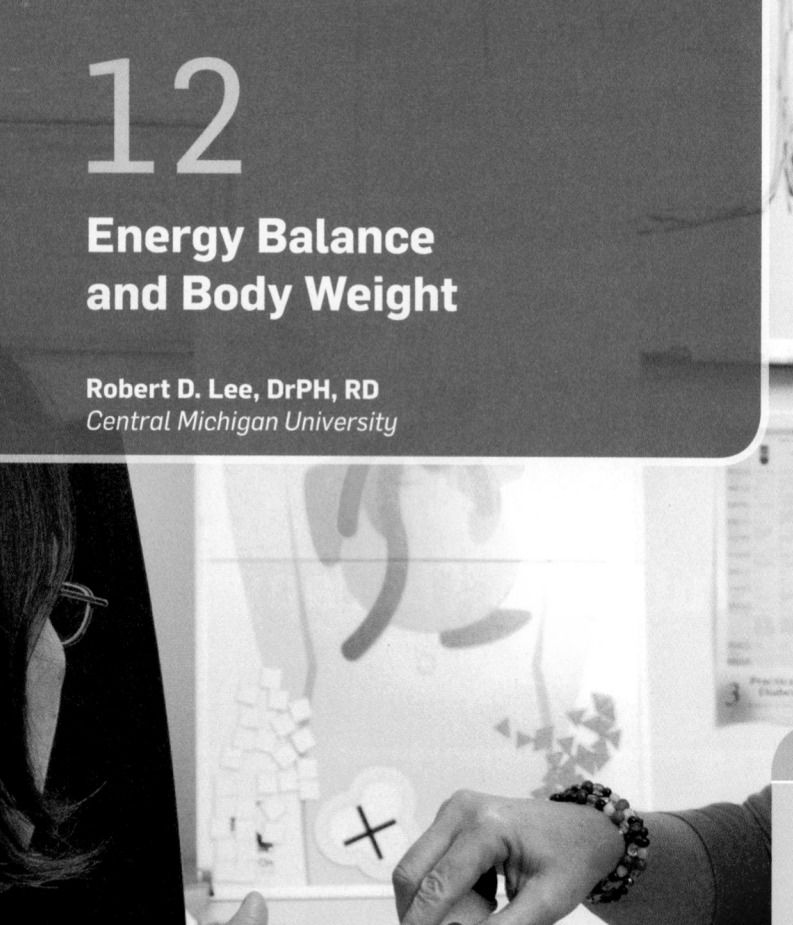

CHAPTER OUTLINE

Introduction

Throughout most of recorded history, humans have spent a large proportion of their time and energy obtaining an adequate amount of calories and essential nutrients. Historically, scarcity of food has been a more common condition than one in which there has been a surplus of food. Hunger, nutrient-deficiency diseases, and starvation have been constant threats for most population groups. Tragically, in many **developing nations**, hunger, malnutrition, and starvation continue, resulting in untold suffering, misery, and death. Only within the past century has the mechanization of agriculture produced the abundant harvests that have become commonplace in **developed nations**. Food in developed nations is so readily available and inexpensive that some nutritionists refer to the food situation in these countries as a "toxic food" environment. This has contributed to a marked increase in the prevalence of **overweight** and **obesity** in these countries, and to diseases associated with these conditions, such as type 2 diabetes, hypertension, stroke, coronary heart disease, sleep apnea, gallbladder disease, osteoarthritis, and cancer of the endometrium, breast, prostate, and colon.[1, 2, 3]

Energy Balance

Energy Intake

Humans obtain the energy and nutrients that their bodies need from the foods and beverages they consume. The human body derives energy from the oxidation of the macronutrients

GLOSSARY

24-hour energy expenditure—the total amount of energy expended by a human in a 24-hour period, made up of three main components: resting energy expenditure, thermic effect of food, and physical activity-related energy expenditure

adaptive thermogenesis—energy expenditure above and beyond the thermic effect of food and resting energy expenditure that is seen in response to overfeeding, traumatic injury, changes in hormonal status, and exposure to a cold environment

algorithm—a finite set of well-defined instructions for accomplishing a task; given an initial state, an algorithm will terminate in a corresponding recognizable end-point

anorexigenic—appetite inhibiting

basal energy expenditure—the minimum level of energy expended by the body to sustain life; it is measured in the morning when a subject is in a postabsorptive state, comfortably lying motionless in a supine position, and in a thermally neutral environment

body mass index (BMI)—weight in kilograms divided by height in meters squared (BMI = kg ÷ m²); although technically not a body composition assessment technique, it correlates well with estimates of body composition derived from skinfold measurements and underwater weighing (hydrodensitometry), and can easily be calculated from weight and height; it is also known as Quetelet's index, named after its developer, Adolphe Quetelet (1796–1874), a Belgian statistician, astronomer, mathematician, and sociologist; the formula for calculating body mass index is:

$$\text{body mass index} = \frac{\text{weight (kg)}}{\text{height (m)}^2}$$

developed nation—a nation that is generally regarded as one with a high standard of living, a high per capita income, a well-developed infrastructure (e.g., public utilities and systems for transport, public health, and public education), high literacy, long life-expectancy, and so on, when compared with the global average

developing nation—a nation that is generally regarded as one with a low standard of living, a low per capita income, a relatively poorly developed infrastructure (e.g., public utilities and systems for transport, public health, and

public education), low literacy rates, low life expectancy, and so on, when compared with the global norm

direct calorimetry—a technique to determine energy expenditure using a highly sophisticated chamber capable of measuring the amount of heat released by a subject's body through evaporation, convection, and radiation

doubly labeled water—a technique to determine energy expenditure in which subjects drink a known amount of water containing two different stable isotopic forms of water: $H_2^{18}O$ and ^2H_2O; the rate that this water disappears from the subject's body is used to calculate the subject's energy expenditure

estimated energy requirement (EER)—the average dietary energy intake that is predicted to maintain energy balance in a healthy adult of a defined age, gender, weight, height, and level of physical activity, consistent with good health; in children and pregnant and lactating women, the EER includes the needs associated with the deposition of tissues or the secretion of milk at rates consistent with good health

iatrogenic—an adverse condition in a patient resulting from treatment, usually by a physician; iatrogenic literally means "brought forth by a physician"

indirect calorimetry—an approach to determine energy expenditure by measuring a subject's oxygen consumption, carbon dioxide production, and minute ventilation (the amount of air a subject breathes in one minute)

kilocalorie (kcalorie or kcal)—the amount of heat required to raise 1000 mL (1 liter) of water 1° Celsius

kilojoule (kjoule or kJ)—the SI (Système International d'Unités or International System of Units) unit of measurement for energy; the amount of work required to move 1 kilogram for 1 meter with the force of 1 newton; 1 kcal = 4.2 kJ (to convert kcal to kJ, multiply kcal by 4.2)

lipogenesis—the synthesis of triglyceride from carbohydrates and proteins

nonexercise activity thermogenesis (NEAT)—the energy expended through physical activity involved in performing the ordinary activities of daily life; it excludes energy expended in

activities to obtain physical exercise or involving sports-like activity

obesigenic—promoting or encouraging the development of obesity; an obesigenic environment is one that promotes weight gain and the development of obesity by encouraging consumption of energy and discouraging physical activity

obesity—an excess of body fat or adipose tissue; obesity can be defined as a proportion of body weight that is adipose tissue (percent body fat) that is greater than some standard; because it is often impractical in the clinical setting to measure the percent of body fat using body composition analysis, obesity is generally defined as a BMI ≥30.0 kg/m² for adults; for children and adolescents, obesity is defined as a BMI-for-age at or above the 95th percentile using the CDC growth charts; the term *obesity* comes from the Latin *obesus*, meaning, "one who has become plump through eating"

orexigenic—appetite stimulating

overweight—an excess of body weight in relationship to height; for adults, overweight is generally defined as a body mass index or BMI of 25.0 kg/m² to 29.9 kg/m²; for children and adolescents, overweight can be defined as a BMI-for-age-and-sex at or above the 85th percentile but less than the 95th percentile using the CDC growth charts

physical activity-related energy expenditure—energy expended in voluntary body movement resulting from the daily activities of life, physical exercise, sports, and play, and nonvoluntary behaviors such as spontaneous muscle contractions, maintenance of posture, and fidgeting; it is the most variable component of 24-hour energy expenditure, depending on how physically active a person is

resting energy expenditure—energy expended by the body at rest to keep vital organ systems functioning, including the heart, kidneys, brain, liver, and lungs; it accounts for approximately 60% to 75% of 24-hour energy expenditure and is roughly 1 kcal/kg body weight/hour

thermic effect of food—energy expended by the body to digest, absorb, and metabolize food; it accounts for about 10% of 24-hour energy expenditure

carbohydrate, protein, and fat, and from alcohol. Internationally, the most commonly used unit of measurement of food energy is the **kilojoule (kJ)**, whereas the **kilocalorie (kcal)** is the unit of measurement of food energy most familiar to those living in the United States. The amounts of energy released by the oxidation of carbohydrate, protein, fat, and alcohol are shown in Box 12.1. These values, rounded for the sake of convenience, initially were derived from experiments in which a small amount of each macronutrient was incinerated in a device known as a bomb calorimeter, which allowed scientists to accurately measure the amount of heat released from the macronutrient when it was burned. Subsequent experiments have shown that the amounts of energy released by the oxidation of these macronutrients within the human body

are similar to the release of energy when burned in the bomb calorimeter.[4] Today, the energy content of a food or beverage is generally determined by first measuring the amount of carbohydrate, protein, fat, and alcohol it contains using relatively simple laboratory techniques, and then multiplying the number of grams of carbohydrate, protein, fat, and alcohol in the food or beverage by the energy values for each of the macronutrients shown in Box 12.1. Information on the energy and nutrient content of foods is widely available to consumers and health professionals from a variety of sources, including the Nutrition Facts labels on commercially available food containers, brochures provided by fast-food restaurants, food composition tables and databases, and dietary analysis software.

Energy Content of Food Components

The energy contents in kilocalories (kcal) and kilojoules (kJ) per gram for each of the macronutrients and alcohol are listed below (1 kilocalorie = 4.2 kilojoule).

Food Component	kcal/g	kJ/g
Carbohydrate	4	17
Protein	4	17
Fat	9	38
Alcohol	7	29

Energy Expenditure

The total amount of energy expended in one day is referred to as **24-hour energy expenditure** or total energy expenditure, and can be divided into three major components: **resting energy expenditure**, the **thermic effect of food**, and **physical activity-related energy expenditure**. Figure 12.1 illustrates the relative proportions of each of these three components for the majority of people living in developed countries, where much of the work is done by labor-saving devices.

Resting Energy Expenditure Resting energy expenditure (REE) is the energy necessary to sustain life and to keep such vital organs as the heart, lungs, brain, liver, and kidneys functioning. For the average North American, REE accounts for approximately 60% to 75% of 24-hour energy expenditure and is roughly 1 kcal/kg body weight/hour. Factors affecting REE are shown in Box 12.2. Of these factors, the most important determinant is lean body mass (or fat-free mass), with REE being greater in persons having a higher lean body mass. **Basal energy expenditure** (BEE) is defined as the lowest rate of energy expenditure of an individual. It is measured in the

Figure 12.1 Components of Energy Expenditure

For most North Americans who are sedentary and rely on labor-saving devices to accomplish most of their work, energy expended in physical activity accounts for less than one quarter of the energy expended in a typical 24-hour period. Surprisingly, resting energy expenditure accounts for about 67% of 24-hour energy expenditure and the thermic effect of food accounts for the remaining 10%

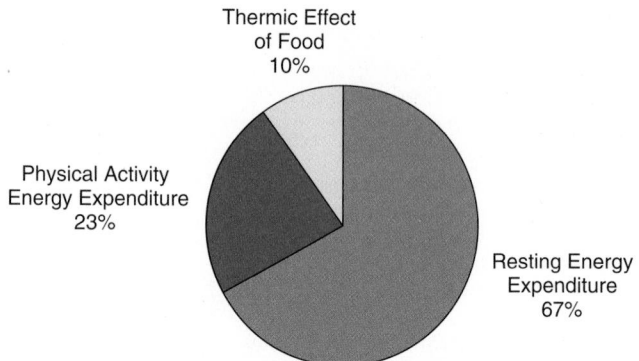

morning when a subject is in a postabsorptive state (no food consumed during the previous 12 to 14 hours) and is comfortably lying motionless in a supine position (lying on one's back) in a thermally neutral environment (a room temperature that is perceived as neither hot nor cold). These strict conditions often make obtaining a true BEE impractical in the clinical setting. REE, on the other hand, can be measured at any time of day after a subject has quietly rested for the previous 30 minutes. Basal energy expenditure is generally 10% to 20% less than resting energy expenditure.[4, 5]

Thermic Effect of Food The thermic effect of food (TEF) is a measurable increase in energy expenditure over and above resting energy expenditure that can be measured for several hours following a meal. The thermic effect of food is

Factors Affecting Resting Energy Expenditure (REE)

Lean Body Mass
Because muscle and other lean tissues are generally more metabolically active than adipose tissue, the greater the lean body mass (also known as the fat-free mass), the greater the REE. This is the primary determinant of REE.

Male Sex
Because males tend to have greater percentage of lean body mass than females, males tend to have a greater REE.

Body Temperature
REE increases in persons who have a fever or an elevated body temperature.

Age
REE decreases about 2% for every decade after age 30 years, even after adjusting for changes in lean body mass.

Energy Restriction
After several weeks of energy restriction, for example to lose weight, resting energy expenditure declines.

This is at least part of the reason that, after several weeks of dieting, some people experience a decline in the rate of weight loss or a phenomenon some refer to as a "plateau."

Genetics and the Endocrine System
Depending on genetic influences, some people inherit a predisposition to a higher REE while others are predisposed to a lower REE. Hypothyroidism and hyperthyroidism can dramatically decrease or increase REE, respectively.

the energy required to digest, absorb, metabolize, and store the nutrients contained in foods that are consumed and to eliminate the resulting by-products and wastes. Originally referred to as the specific dynamic action of food, it accounts for about 10% of the 24-hour energy expenditure for a person consuming a typical mixed meal.[4] The TEF of a meal is influenced primarily by the amount and macronutrient composition of the food consumed. Large meals have a greater TEF than small meals. Fat has the lowest TEF, while protein has the highest TEF due to the relatively high energy cost of processing the amino acids released from the proteins in food, including the synthesis of urea. TEF peaks at about 60 to 120 minutes following a meal and can last up to four to six hours, depending on the size and composition of the meal. TEF is generally not factored into calculations of BEE due to the high variability of its contribution to energy requirements.

Physical Activity Energy Expenditure The most highly variable component of 24-hour energy expenditure is that expended in physical activity. For most people in developed nations it accounts for about 20% to 25% of 24-hour energy expenditure. However, in very active individuals, such as heavy laborers and some athletes, the amount of energy expended in physical activity can exceed REE by twofold or more.[4, 5] Physical activity energy expenditure is influenced by the person's body weight, the number of muscle groups used in the activity, and the intensity, duration, and frequency of the activity. For any given activity, heavy people expend more energy than lighter-weight people because heavier people have a greater body mass to move. Activities requiring multiple groups of large muscles (e.g., cross-country skiing or handball) expend more energy than those requiring fewer groups of muscles (e.g., walking or golfing).

Estimating and Measuring Energy Requirements

An individual's energy requirements can either be estimated using a predictive equation or, if a more accurate determination is necessary, using such methods as indirect calorimetry, doubly labeled water, or direct calorimetry. For most patients in the clinical setting, it is usually adequate to estimate energy requirements by means of a predictive equation that uses variables such as sex, age, weight, stature (height), and physical activity level. However, in critically ill patients, it may be necessary to more accurately determine energy requirements using indirect calorimetry. Doubly labeled water is commonly used in human metabolic research, while use of direct calorimetry is limited by the small number of research facilities having the necessary technology.

Equations In most instances, an individual's energy requirements are estimated using one of several empirically derived equations. Two examples are the Harris-Benedict equations developed in the early 1900s by the researchers J. A. Harris and F. G. Benedict, and those developed in the 1980s by the World Health Organization (WHO), both of which are shown in Table 12.1. Harris and Benedict measured the resting energy expenditures (REE) of 239 healthy young adult males and females using indirect calorimetry (discussed later in this chapter) and then developed a set of regression equations that best predicted REE using the variables sex, weight, stature (height), and age.[6] Although originally published in 1919, their equations remain in use today for healthy individuals. (See Chapters 5

Table 12.1 Examples of Equations for Estimating Resting Energy Expenditure in Healthy Persons[1]

Harris-Benedict		
Females	REE = 655.096 + 9.563 W + 1.850 S − 4.676 A	
Males	REE = 66.473 + 13.752 W + 5.003 S − 6.755 A	
Harris-Benedict (Values Rounded for Simplicity)		
Females	REE = 655.1 + 9.6 W + 1.9 S − 4.7 A	
Males	REE = 66.5 + 13.8 W + 5.0 S − 6.8 A	
World Health Organization (WHO)		**SD**[2]
Females	3–9 years old	22.5 W + 499	± 63
	10–17 years old	12.2 W + 746	± 117
	18–29 years old	14.7 W + 496	± 121
	30–60 years old	8.7 W + 829	± 108
	>60 years old	10.5 W + 596	± 108
Males	3–9 years old	22.7 W + 495	± 62
	10–17 years old	17.5 W + 651	± 100
	18–29 years old	15.3 W + 679	± 151
	30–60 years old	11.6 W + 879	± 164
	>60 years old	13.5 W + 487	± 148

[1] W = weight in kilograms; A = age in years; S = stature in cm.

[2] SD = standard deviation of the differences between actual and computed values—68% of the time actual REE will be within ± 1 standard deviation of the predicted REE.

Source: From Harris JA, Benedict FG. 1919. A biometric study of basal metabolism in man. Publication 279. Washington, DC: Carnegie Institution of Washington; World Health Organization. Energy and protein requirements. Report of a joint FAO/WHO/UNU expert consultation. Technical Report Series 724. Geneva, Switzerland: World Health Organization; 1985.

and 22 for discussion of estimation of energy requirements in acute care and the critically ill.) The World Health Organization equations were developed by a group of experts using an approach similar to that used by Harris and Benedict. A key difference between the two sets of equations is that the WHO equations do not include stature as a variable, because it was not found to improve their predictive ability.[5, 7]

A more recent set of prediction equations are those established by the Institute of Medicine (as part of its development of the Dietary Reference Intakes) to calculate the **estimated energy requirement (EER)**. The formulas for calculating EER are shown in Table 12.2. The EER is defined as the average dietary energy intake that is predicted to maintain energy balance in a healthy person of a defined age, gender, weight, height, and level of physical activity consistent with good health.[4] For infants, children, and adolescents, the EER includes the energy needed for a desirable level of physical activity, as well as the energy needed for optimal growth and development at an age- and gender-appropriate rate that is consistent with good health, including maintenance of a healthy body weight and appropriate body composition. For females who are pregnant or lactating, the EER includes the energy needed for physical activity, maternal and fetal development, and for secretion of milk at a rate consistent with good health.

The EER equations are based on the measurement of 24-hour total energy expenditure using the doubly labeled water technique (discussed later in this chapter) from more than 1,200 healthy-weight subjects of all ages.[4] The Dietary Reference Intake (DRI) Committee used these measurements of 24-hour

Table 12.2 Equations for Calculating Estimated Energy Requirement (EER) in Kilocalories per Day[1]

EER for Infants and Young Children

EER = TEE + Tissue Deposition[2]

0–3 months	(89 × weight − 100) + 175
4–6 months	(89 × weight − 100) + 56
7–12 months	(89 × weight − 100) + 22
13–35 months	(89 × weight − 100) + 20

EER for Males 3 through 8 Years

EER = TEE + Tissue Deposition

EER = 88.5 − 61.9 × age + PA × (26.7 × weight + 903 × height) + 20

Where PA is the physical activity coefficient:

PA = 1.00 for sedentary

PA = 1.13 for low active

PA = 1.26 for active

PA = 1.42 for very active

EER for Females 3 through 8 Years

EER = TEE + Tissue Deposition

EER = 135.3 − 30.8 × age + PA × (10.0 × weight + 934 × height) + 20

Where PA is the physical activity coefficient:

PA = 1.00 for sedentary

PA = 1.16 for low active

PA = 1.31 for active

PA = 1.56 for very active

EER for Males 9 through 18 Years

EER = TEE + Tissue Deposition

EER = 88.5 − 61.9 × age + PA × (26.7 × weight + 903 × height) + 25

Where PA is the physical activity coefficient:

PA = 1.00 for sedentary

PA = 1.13 for low active

PA = 1.26 for active

PA = 1.42 for very active

EER for Females 9 through 18 Years

EER = TEE + Tissue Deposition

EER = 135.3 − 30.8 × age + PA × (10.0 × weight + 934 × height) + 25

Where PA is the physical activity coefficient:

PA = 1.00 for sedentary

PA = 1.16 for low active

PA = 1.31 for active

PA = 1.56 for very active

EER for Males 19 Years of Age and Older

EER = TEE

EER = 662 − 9.53 × age + PA × (15.91 × weight + 539.6 × height)

Where PA is the physical activity coefficient:

PA = 1.00 for sedentary

PA = 1.11 for low active

PA = 1.25 for active

PA = 1.48 for very active

EER for Females 19 Years of Age and Older

EER = TEE

EER = 354 − 6.91 × age + PA × (9.36 × weight + 726 × height)

Where PA is the physical activity coefficient:

PA = 1.00 for sedentary

PA = 1.12 for low active

PA = 1.27 for active

PA = 1.45 for very active

EER for Pregnancy

EER = EER for age + Pregnancy Energy Needs[3] + Tissue Deposition

1st trimester = EER for age + 0

2nd trimester = EER for age + 160 + 180

3rd trimester = EER for age + 272 + 180

EER for Lactation

EER = EER for age + Milk Energy Output[4] − Weight Loss[5]

1st six months = EER for age + 500 − 170

2nd six months = EER for age + 400 − 0

[1]EER = Estimated Energy Requirement; TEE = Total Energy Expenditure; PA = Physical Activity Coefficient; age is in years; height is in meters; weight is in kilograms.

[2]Tissue Deposition represents the energy cost of growth during infancy, childhood, adolescence, and pregnancy as measured in kilocalories.

[3]Pregnancy Energy Needs represents the additional energy required to support the metabolic demands of pregnancy.

[4]Milk Energy Output represents the energy needed to produce the milk during lactation. Milk output is somewhat greater in the first six months than in the second six months of breastfeeding.

[5]Weight Loss represents an average decline in EER of 170 kcal/day that well-nourished lactating women experience during the first six months postpartum, resulting in an average weight loss of 0.8 kg/month.

Sources: Adapted from Panel on Macronutrients, Panel on the Definition of Dietary Fiber, Subcommittee on Upper Reference Levels of Nutrients, Subcommittee on Interpretation and Uses of Dietary Reference Intakes, Standing Committee on the Scientific Evaluation of Dietary Reference Intakes. 2002. Dietary reference intakes for energy, carbohydrate, fiber, fat, fatty acids, cholesterol, protein, and amino acids. Washington, DC: National Academy Press.

total energy expenditure to develop a series of regression equations that best predicted the energy requirements of healthy-weight individuals using such variables as age, sex, life stage (pregnant or lactating), body weight, stature, and physical activity level. As shown in Table 12.2, EER equations have been developed for infants and young children of both sexes age 0 to 35 months, males and females age 3 to 8 years, males and females age 9 to 18 years, males and females age 19 years and older, and females who are pregnant or lactating. Additionally, the DRI for both energy and protein can be used when estimating nutritional needs for infants and children. See Table 12.3.

Except in the case of infants and young children ages 0 to 35 months, a physical activity coefficient (PA) is used in the equations. The PA represents one of four different categories of physical activity level: sedentary, low active, active, and very active. Energy expenditure at each of these levels is as follows:

- *Sedentary:* Includes basal energy expenditure, the thermic effect of food, and physical activities required for independent living.

- *Low active:* Roughly equivalent to the energy expended by a 70 kg (154 lb) adult walking 2.2 miles per day at a rate of 3 to 4 miles per hour (or an equivalent amount of energy

Table 12.3 Estimated Energy Requirements (EER) for Infants and Children

Age (years)	EER (kcal/day) for Males	EER (kcal/day) for Females
0–0.5	570	520
0.5–1	743	676
1–2	1046	992
3–8	1742	1642

Sources: Data taken from Table S-1 (Criteria and Dietary Reference Intake Values for Energy by Active Individuals by Life Stage Group), page 5, from *Dietary Reference Intakes for Energy, Carbohydrate, Fiber, Fat, Fatty Acids, Cholesterol, Protein, and Amino Acids (Macronutrients)*. 2005: The National Academies Press. Available at http://books.nap.edu/openbook.php?record_id=10490&page=5

expended in other activities) in addition to the activities necessary for independent living.

- *Active:* Roughly equivalent to the energy expended by a 70 kg (154 lb) adult walking 7 miles per day at a rate of 3 to 4 miles per hour in addition to the activities related to independent living.

- *Very active:* Equivalent to walking 17 miles per day in addition to the activities a normal person would ordinarily engage in.

The extra energy needed for growth during infancy, childhood, adolescence, and pregnancy is included in an allowance referred to as "tissue deposition." During pregnancy, the metabolic rate is also increased due to the energy requirements of the uterus and fetus and the increased work of the maternal cardiovascular system. During lactation, extra energy is needed to support milk production, which is somewhat greater in the first six months of breastfeeding than in the second six months. Because most women lose an average of 0.8 kg per month in the first six months postpartum (i.e., after delivery), EER is, on average, 170 kcal per day less.[4]

It is important to note that the EER equations apply only to persons having a healthy weight and that EER values have not been established for persons who are overweight or obese.[4] Instead, the DRI Committee has developed a separate set of equations for calculating total energy expenditure (TEE) for the maintenance of weight for adults age 19 years and older who are overweight (i.e., have a **body mass index** or BMI between 25.0 kg/m^2 and 29.9 kg/m^2) and/or obese (BMI ≥30.0 kg/m^2), and an additional set was developed for children and adolescents age 3 to 18 years who are obese (a BMI for age and sex ≥95th percentile).[4] These are shown in Table 12.4. The DRI Committee adopted the definition of healthy weight for adults (age 19 years and older) used by the *Dietary Guidelines for Americans*, which is a BMI ≥18.5 kg/m^2 but ≤24.9 kg/m^2. Healthy weight for persons age 2 to 18 years is defined as a BMI that is >5th percentile but <85th percentile of BMI for age and sex, as discussed in Chapter 3. The equations developed for overweight or obese persons shown in Table 12.4 allow calculation of the TEE necessary for weight maintenance using the variables gender, age, weight, height, and physical activity level. For an adult desiring to lose weight, a recommended approach is reducing energy intake to 500 to 1000 kcal per day less than that needed for maintenance,

Table 12.4 Equations for Calculating Total Energy Expenditure (TEE) for Weight Maintenance in Kilocalories per Day for Overweight and Obese Adults and for Overweight Children and Adolescents[1]

TEE for Overweight and Obese Males Aged 19 Years and Older

TEE = 1086 − 10.1 × age + PA × (13.7 × weight + 416 × height)

Where PA is the physical activity coefficient:

PA = 1.00 for sedentary

PA = 1.12 for low active

PA = 1.29 for active

PA = 1.59 for very active

TEE for Overweight and Obese Females Aged 19 Years and Older

TEE = 448 − 7.95 × age + PA × (11.4 × weight + 619 × height)

Where PA is the physical activity coefficient:

PA = 1.00 for sedentary

PA = 1.16 for low active

PA = 1.27 for active

PA = 1.44 for very active

TEE for Overweight Males Aged 3 through 18 Years

TEE = 114 − 50.9 × age + PA × (19.5 × weight + 1161.4 × height)

Where PA is the physical activity coefficient:

PA = 1.00 for sedentary

PA = 1.12 for low active

PA = 1.24 for active

PA = 1.45 for very active

TEE for Overweight Females Aged 3 through 18 Years

TEE = 389 − 41.2 × age + PA × 15.0 × weight + 701.6 × height

Where PA is the physical activity coefficient:

PA = 1.00 for sedentary

PA = 1.18 for low active

PA = 1.35 for active

PA = 1.60 for very active

[1]TEE = Total Energy Expenditure; PA = Physical Activity Coefficient; age is in years; height is in meters; weight is in kilograms. In persons age 19 years and older overweight is defined as a BMI between 25.0 kg/m^2 and 29.9 kg/m^2 and obese is defined as a BMI ≥ 30.0 kg/m^2. In persons age 3 to 18 years, obesity is defined as a BMI for age and sex ≥ 95th percentile.

Source: Adapted from Panel on Macronutrients, Panel on the Definition of Dietary Fiber, Subcommittee on Upper Reference Levels of Nutrients, Subcommittee on Interpretation and Uses of Dietary Reference Intakes, Standing Committee on the Scientific Evaluation of Dietary Reference Intakes. Dietary reference intakes for energy, carbohydrate, fiber, fat, fatty acids, cholesterol, protein, and amino acids. Washington, DC: National Academy Press; 2002.

and increasing energy expenditure by gradually increasing the level of physical activity to 300 minutes per week of moderate-intensity aerobic activity or 75 minutes per week of vigorous-intensity aerobic activity.[2, 8, 9]

Indirect Calorimetry The most commonly used approach for measuring energy requirements in critically ill patients and in human metabolic research is **indirect calorimetry**. It is based on the fact that energy expenditure is proportional to the body's oxygen consumption and carbon dioxide production. Expired air contains less oxygen and more carbon dioxide than inspired air. When the differences in oxygen and carbon dioxide in inspired and expired air are known and the volume of air moving through a subject's lungs is measured, the body's energy expenditure can be calculated.

In the laboratory or clinical settings, indirect calorimetry is accomplished using a portable computerized metabolic monitor that can be brought to the bedside or positioned next to a subject exercising on a treadmill or cycle ergometer. A mask or hood is placed over the subject's face and the amount of air flow through the lungs per minute (known as minute ventilation) is measured by various types of instruments that are built into the mask. Gas analyzers in the monitor measure the oxygen and carbon dioxide content of both inspired and expired air. Technological advances have resulted in the development of lightweight, portable indirect calorimetry units, which can be worn by subjects while working or participating in sports. A typical unit weighs 2.1 lbs (950 g) and allows accurate testing while the subject is engaged in practically any activity at any time or location, without being confined to an artificial laboratory environment. In the clinical setting, indirect calorimetry is recommended as a means to accurately determine the energy requirements of critically ill and/or mechanically ventilated patients, and to monitor the adequacy and appropriateness of nutritional support.

Doubly Labeled Water The **doubly labeled water** (DLW) technique is a relatively new approach for measuring total energy expenditure in subjects who are engaged in their normal daily routines (i.e., "free-living individuals") over a one- to two-week period, without the use of the instrumentation used in indirect calorimetry.[4, 5, 10] It has been found to be accurate within 1% to 2% when compared to indirect calorimetry, and is considered the "gold standard" for measuring energy expenditure and physical activity in free-living subjects over a one- to two-week period.[10, 11]

DLW involves subjects drinking a known amount of water containing two different stable isotopic forms of water: $H_2^{18}O$ and ^2H_2O. Ordinary water is a molecule composed of two atoms of hydrogen, each having an atomic mass of one (^1H), and one atom of oxygen having an atomic mass of 16 (^{16}O). Hydrogen atoms with an atomic mass of two (^2H or deuterium) and oxygen atoms with an atomic mass of 18 (^{18}O) are only naturally present in the environment in extremely minute quantities. Consequently, essentially all of the ^2H and ^{18}O present in the body of test subjects comes from the doubly labeled water. After the subject drinks the two different isotopic forms of water, they mix with the body's water and are gradually eliminated from the body.[4, 5, 10, 11] Over the next one to two weeks, the subject provides several urine samples that are used to measure the rate at which the two isotopes disappear from the body. The rate of disappearance is then used to calculate energy expenditure. The method is noninvasive, provides an accurate measurement of energy expenditure over a period of one to two weeks, and, because the two isotopes are stable (nonradioactive), the procedure is considered safe to use even on infants and females who are pregnant or lactating.[4]

Direct Calorimetry **Direct calorimetry** involves using a highly sophisticated chamber that is capable of determining a subject's total energy expenditure by measuring the amount of heat given off by the subject's body through evaporation, convection, and radiation.[5, 12, 13] The size of direct calorimeters varies from a chamber just large enough to accommodate a subject lying down to those the size of a small bedroom. Once inside the calorimeter, the subject's activity is monitored, and the subject's response to clinically prepared meals can be studied. If necessary, samples of urine and feces can be collected for analysis. In some instances, subjects may remain inside the calorimeter for up to 24 hours or longer, making it an impractical approach for use with critically ill patients or those afraid of being in an enclosed space for several hours. Because direct calorimetry requires equipment that is bulky, very expensive, and technologically sophisticated, it is rarely used.

Regulation of Energy Balance

The regulation of energy balance and body weight is dependent upon the complex interaction of the nervous system and various hormones.[5, 14] A decrease in energy intake and loss of body fat mass typically result in **orexigenic** neural and hormonal stimuli that lead to increased appetite and decreased resting energy expenditure. Modest increases in energy intake and increased body fat mass typically result in **anorexigenic** stimuli that lead to decreased appetite and an increase in energy expenditure known as **adaptive thermogenesis**.

Appetite

Appetite is influenced by a number of signals to the brain that are primarily orchestrated by the hypothalamus region, including neural signals from mouth, stomach, and small intestine during and following eating and the secretion of pancreatic and gastrointestinal hormones such as insulin, glucagon, amylin, cholecystokinin, glucagon-like peptide-1, peptide YY, and ghrelin.[14–16] Pleasurable taste sensations within the mouth stimulate appetite and encourage eating. As the stomach fills, it becomes distended, stimulating stretch receptors in the stomach wall that provide neural signals to the hypothalamus that inhibit appetite. Proteins, monosaccharides, and fatty acids in the chyme (semiliquid mass of partially digested food) leaving the stomach stimulate neural and endocrine receptors in the mucosa of the small intestine, resulting in neural signals to the brain that decrease appetite and food intake, and the release of the hormones cholecystokinin, glucagon-like peptide-1, and peptide YY, which also decrease appetite and food intake.[15, 16]

As plasma glucose level rises following a meal, the beta-cells of the pancreas release insulin and amylin, which decrease appetite and food intake. During fasting, the beta-cells of the pancreas release glucagon, which also decreases appetite and food intake.[14–16] Ghrelin is a peptide hormone that is mainly produced by the stomach and stimulates appetite. Ghrelin levels are normally increased during fasting, but immediately following food intake, ghrelin levels decline. This appears to decrease appetite and food intake.[15, 17] However, in patients with Prader-Willi syndrome, a genetic disorder characterized by voracious appetite and massive obesity, ghrelin levels are increased by as much as threefold or fourfold compared to individuals of similar age, sex, and BMI.[14, 18, 19] These numerous and diverse neural and hormonal signals influence the release of various peptides from the hypothalamus, resulting in the final expression of appetite and eating behavior.[14]

The Adipocyte and Adipose Tissue

The adipocyte (fat cell) is a large, rounded cell primarily filled with a droplet of triglyceride. The cytoplasm containing the nucleus, mitochondria, and other cell organelles is forced to occupy a thin layer immediately beneath the plasma membrane (see Figure 12.2).[20] Throughout the body, adipocytes occur individually or in small groups joined by connective tissue.[21] When found in large aggregations in conjunction with fibrous connective tissue, they form adipose tissue, which serves as the storage site for more than 90% of the body's energy reserves.[14, 21] In addition, adipose tissue fills body crevices, provides thermal insulation to the body, surrounds and shields internal organs, gives shape and form to the body, and cushions such body areas as the feet, hands, shoulders, and buttocks.[21]

There are two types of adipose tissue: white adipose tissue (WAT) and brown adipose tissue (BAT). The predominant type is WAT, which in reality is a light-yellow in color due to the presence of carotenoids. The cells of WAT store triglycerides derived from dietary fats or those synthesized from carbohydrates and proteins through the process of **lipogenesis**. BAT derives its color from the large number of mitochondria in the adipocytes and from its abundance of blood vessels. BAT is primarily found in fetuses, infants, and young children, and accounts for up to 6% of an infant's body weight. As humans age, the amount of BAT diminishes. In WAT, triglyceride is stored within a single large droplet, whereas in BAT, there are multiple smaller droplets. It appears that the primary function of BAT is maintaining body temperature in human neonates and in hibernating animals by generating heat through a process known as diet-induced thermogenesis or nonshivering thermogenesis. However, because of the small amount of BAT in adult humans, it has a minimal effect on energy expenditure.[20, 21]

Figure 12.2 An Adipocyte or Fat Cell

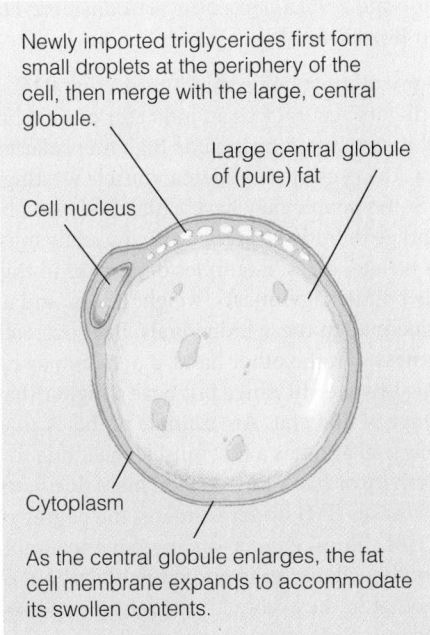

Newly imported triglycerides first form small droplets at the periphery of the cell, then merge with the large, central globule.

Large central globule of (pure) fat

Cell nucleus

Cytoplasm

As the central globule enlarges, the fat cell membrane expands to accommodate its swollen contents.

Source: S. Rolfes, K. Pinna, and E. Whitney, Understanding Normal and Clinical Nutrition, 7e, copyright © 2006, p. 157.

Although once thought to be a relatively inert storage site for energy consumed in excess of the body's needs, the adipocyte is metabolically active in the uptake, synthesis, storage, and mobilization of triglycerides. There is a constant turnover of triglycerides in the adipocyte as new triglycerides are synthesized and stored and older triglycerides are hydrolyzed and released from the adipocyte into the circulation.[20] Recent research has shown the adipocyte to be a metabolically active endocrine cell releasing numerous hormones involved in regulating appetite, energy balance, body fat content, and reproduction. In addition, adipocytes release numerous molecules that influence blood pressure, blood coagulation, insulin sensitivity, lipid oxidation, and serum lipid levels. In obesity, the increased release by adipocytes of several potent proinflammatory cytokines such as tumor necrosis factor-α, interleukin-6, and C-reactive protein results in obesity being associated with a state of low-grade chronic inflammation which appears to be linked to many of the adverse health risks discussed later in this chapter.[14, 17, 22–24]

Two hormones produced by adipose tissue that are involved in energy balance and fat storage are adiponectin and leptin. Research suggests that adiponectin signals that the body has the capacity to store fat, while leptin appears to signal that ample fat has been stored.[17] Adiponectin levels increase as the body fat content decreases and decrease as body fat mass increases.[17] Increased levels of adiponectin improve the body's sensitivity to insulin which, in turn, enhances the body's capacity to store fat. Higher levels of adiponectin are associated with decreased risk of coronary artery disease, whereas low levels accompany obesity and its associated health complications such as type 2 diabetes and coronary heart disease. In contrast, leptin levels increase as body fat mass increases. Although understanding of this hormone is limited, it is known that leptin regulates body mass by inhibiting food intake through its action on the hypothalamus. In addition, it appears to regulate reproduction by promoting fertility and initiating puberty when the body's energy stores are adequate for the demands of reproduction, and inhibiting reproduction when energy stores are inadequate.[14, 17]

Adipose tissue mass increases in two ways. First, fully mature adipocytes can increase in size (undergo hypertrophy) as they accumulate more triglyceride during periods when energy intake exceeds energy expenditure. Second, adipocytes can increase in number (undergo hyperplasia) as immature adipocytes divide to produce more cells.[21] Overweight (BMI 25.0 to 29.9 kg/m^2) and moderate obesity (BMI 30.0 to 34.9 kg/m^2) are characterized by hypertrophy (enlargement) of adipocytes, and with weight loss, these enlarged adipocytes become smaller. However, as BMI approaches extreme obesity (BMI ≥40.0 kg/m^2), adipocytes reach their maximum size and then experience hyperplasia (an increase in number). As persons with extreme obesity lose weight, the adipocytes become smaller in size but the number of adipocytes does not decrease. The clinical implication of this fact is that overweight and moderately obese people who have experienced only fat cell hypertrophy are more successful at maintaining their weight loss than are extremely obese people who have undergone fat cell hypertrophy and hyperplasia. Achieving and maintaining a healthy body weight is more likely if an increase in fat cell number can be avoided.

During the first year of life, the proportion of fat in the human body typically increases from approximately 15% at birth to about 30% at one year as adipocytes undergo hypertrophy and hyperplasia.[25] A higher percentage of body fat is seen in infants who have a high birth weight and infants of diabetic mothers, and these infants are at increased risk of being overweight in later childhood and adolescence.[26] Between the ages of 1 and 6 years, the percentage of body fat generally decreases, and then begins to increase at 6 to 8 years of age in a process known as "adiposity rebound" or "BMI rebound."[25, 26] Children who experience their adiposity rebound or BMI rebound before age 4 to 6 years are at increased risk of increased BMI in later life.[26] It is estimated that 25% to 80% of overweight children remain overweight as adults. This is particularly likely for adolescent females, who have three times the risk of being overweight as adults compared to overweight adolescent males.[26]

Definitions of Overweight and Obesity

Body Composition

Technically, obesity is an excess of adipose tissue or body fat. It can be defined as a proportion of body weight composed of adipose tissue (percent body fat) that exceeds a range that is considered healthy. Adult males are generally considered obese when their percent body fat is ≥25% and adult females are considered obese when their percent body fat is ≥33%.[27] The problem with this definition is that it requires that the body's composition be assessed in order to determine the relative proportions of fat and lean tissue.

The human body is composed of different types of tissues—adipose tissue or body fat, muscle, bone, blood, cartilage, ligaments, tendons, the brain and nervous tissue, and the viscera located within the thoracic and abdominal cavities. The most common approach to body composition analysis views the body as consisting of two different compartments: fat and fat-free. This is referred to as the "two-compartment model." Using this model, body composition is expressed as a ratio of fat to fat-free mass, or as a percentage of the body composed of adipose tissue or lean tissue.

As discussed in Chapter 3, a variety of methods are available to clinicians for assessing body composition, and the most common of these body composition assessment methods are based on the two-compartment model. They include the following:

- Skinfold measurements
- Underwater weighing, or hydrodensitometry
- Bioelectrical impedance analysis
- Air-displacement plethysmography
- Dual-energy X-ray absorptiometry

Most clinicians do not have the time, expertise, or equipment to accurately assess body composition using the techniques just mentioned. In contrast, accurately measuring weight and height is relatively easy and quick, and the necessary equipment is inexpensive and readily available. Consequently, weight and height measurements are often used in place of body composition analysis to determine whether a person is obese. The problem with this approach is that body composition cannot be determined by merely evaluating body weight and height, because measurements of body weight and height alone are in-

capable of differentiating between the weight of the body's lean tissue and the weight of the body's adipose tissue.

Body Mass Index (BMI)

Because it is often impractical to determine body composition in the clinical setting, and because accurate measurements of height and weight can be easily obtained, obesity in adults is often defined as a BMI ≥30.0 kg/m². Also known as Quetelet's index, BMI is not a direct measure of body fatness. BMI can be considered a proxy for measures of body fatness and is regarded as a convenient and reliable indicator of obesity. It is reasonable to assume that for most people a high BMI (i.e., ≥30 kg/m²) represents an increased amount of adipose tissue in the body rather than unusually well-developed musculature or a large, dense skeleton. Because changes in BMI parallel changes in body composition obtained by direct measures of body fat such as underwater weighing and dual-energy X-ray absorptiometry, it is a convenient and useful approach for tracking improvements in body composition. Throughout North America and Europe, BMI is regarded as the best and most convenient clinical approach to use in evaluating the body weights of patients.[2, 9, 28–31]

Overweight is a body weight in excess of some standard weight, and usually includes a consideration of height. In adults, overweight is generally defined as a BMI of 25.0 kg/m² to 29.9 kg/m²,[2, 32] "healthy weight" is defined as a BMI of 18.5 kg/m² to 24.9 kg/m², and underweight is defined as a BMI <18.5 kg/m².[32] Box 12.3 discusses the calculation of BMI and shows these BMI classifications for adults.

For children and adolescents, overweight is defined using the U.S. Centers for Disease Control and Prevention (CDC) growth charts that provide the body mass index-for-age percentiles for males and females 2 to 20 years of age (see Chapter 3 and Appendix B1). Using these charts children and adolescents having a BMI-for-age ≥ 85th percentile but <95th percentile are considered overweight, while those having a BMI-for-age ≥ 95th percentile are considered obese, as described in Box 12.4.[33–35]

Clinical judgment must be used in interpreting BMI in situations affecting its accuracy as an indicator of total body fat. Examples of these situations include high muscularity, large skeletal mass, the presence of edema, muscle wasting, and osteoporosis. Some people may have a high body weight but have a low percentage of body fat if they are unusually muscular or have a large skeletal mass. Examples of persons in this category include body builders, gymnasts, weight lifters, and other highly muscular people. In these individuals, BMI overestimates the degree of fatness. On the other hand, a person may have a BMI within the healthy weight range but have a higher than desirable percentage of body fat. An example of this is an elderly person who is frail and has a low muscle mass due to prolonged physical inactivity or has a low skeletal mass due to osteoporosis; in these cases BMI underestimates the degree of fatness. The presence of clinical edema (an excessive accumulation of fluid in the body) will increase body weight without changing the amount of fat in the body. Consequently, in persons with significant edema, BMI overestimates body fatness. In addition, females generally have more body fat for a given BMI than males. Despite these circumstances, in most instances BMI

BOX 12.3 CLINICAL APPLICATIONS

Calculating BMI and Using BMI to Classify Adults

Formulas for calculating body mass index or BMI are as follows:

$$BMI = \text{weight in kilograms} \div (\text{height in meters})^2$$

To convert weight in pounds to weight in kilograms:

$$\text{pounds} \div 2.2 = \text{kilograms}$$

To convert height in inches to height in meters:

$$\text{inches} \times 0.0254 = \text{meters}$$

For those who have difficulty using the SI units of measurement, the following formula can also be used to calculate BMI using weight in pounds and height in inches:

$$BMI = (\text{weight in pounds} \times 703) \div (\text{height in inches})^2$$

The BMI classifications for adults shown here are recommended by the National Institutes of Health in the publication *Clinical Guidelines on the Identification, Evaluation, and Treatment of Overweight* and *Obesity in Adults*[1] and the World Health Organization in the publication *Obesity: Preventing and Managing the Global Epidemic. Report of a World Health Organization Consultation*.[1] Similar classification values are used in the latest edition of the *Dietary Guidelines for Americans*[2] and by a number of highly respected national and international scientific groups.[3, 4–7]

Classification	BMI
Underweight	<18.5 kg/m²
Healthy weight	18.5 to 24.9 kg/m²
Overweight	25.0 to 29.9 kg/m²
Obesity (Class 1)	30.0 to 34.9 kg/m²
Obesity (Class 2)	35.0 to 39.9 kg/m²
Extreme obesity (Class 3)	≥40.0 kg/m²

References

1. World Health Organization. 2000. Obesity: Preventing and managing the global epidemic. Report on a WHO consultation. Technical Report Series 894. Geneva, Switzerland: World Health Organization.
2. U.S. Department of Health and Human Services and U.S. Department of Agriculture. Dietary Guidelines for Americans, 2005. 6th ed. Washington (DC): U.S. Government Printing Office; 2005.
3. Kushner RF. Evaluation and treatment of obesity. In Fauci AS, Braunwald E, Kasper DL, Hauser SL, Longo DL, Jameson JL, Loscalzo J (eds.), Harrison's principles of internal medicine. 17th ed. New York: McGraw-Hill, 2008, 468–73.
4. International Obesity Task Force, European Association for the Study of Obesity. Obesity in Europe. London: International Obesity Task Force; 2002. http://www.iotf.org/media/euobesity.pdf
5. International Obesity Task Force, European Association for the Study of Obesity. Obesity in Europe 2: Waiting for a Green Light for Health. London: International Obesity Task Force; 2003. http://www.iotf.org/media/euobesity2.pdf
6. Shields M. Overweight Canadian children and adolescents. Ottawa: Statistics Canada; 2005.
7. Tjepkema M. Adult Obesity in Canada, Measured Height and Weight. Ottawa: Statistics Canada; 2005.

BOX 12.4 Life Cycle Perspectives

Classifying Pediatric BMI

When the CDC's BMI-for-age charts were originally published in 2000, experts recommended that a BMI-for-age ≥85th percentile but <95th percentile be classified as *at risk of overweight* and a BMI-for-age ≥95th percentile be classified as *overweight*.[1, 2] The experts who developed the clinical guidelines on overweight in children and adolescents made a deliberate effort to avoid using the term *obese* to describe those whose BMI-for-age was 95th percentile because of the stigma associated with the term *obesity* and because weight and height data, including BMI, are incapable of accurately determining body composition.[1, 2] Since their publication in 2000, experts have recommended a change in how the CDC growth charts are used to classify BMI in children and adolescents age 2 years and older, as shown below.[3–5] Experts now recommend that a BMI-for-age ≥85th percentile but <95th percentile be classified as overweight and that children and adolescents be classified as obese when their BMI-for-age is ≥95th percentile or they have a BMI ≥30 kg/m², whichever represents the lower weight. For example, it is possible for an adolescent to have a BMI ≥30 kg/m² but have a BMI-for-age that is <95th percentile, in which case the adolescent would still be classified as obese. The BMI of an adolescent is classified as obese when it is ≥30 kg/m² even if it is <95th percentile curve.[3–5]

Percentile Cut-Off Value	Classification of BMI
<5th percentile	underweight
≥5th and <85th percentile	healthy weight
≥85th and <95th percentile	overweight
≥95th percentile *or* ≥30 kg/m² (whichever represents the lower weight)	obese

References

1. Kuczmarski RJ, Ogden CL, Grummer-Strawn LM, et al. CDC growth charts: United States. Advance data from vital and health statistics; no. 314. Hyattsville, Maryland (MA): National Center for Health Statistics; 2000.
2. Kuczmarski RJ, Ogden CL, Guo SS, et al. 2000 CDC growth charts for the United States: Methods and development. National Center for Health Statistics. Vital Health Stat. 11(246); 2002.
3. Barlow SE. Expert committee recommendations regarding the prevention, assessment, and treatment of child and adolescent overweight and obesity: summary report. Pediatrics. 2007;120(suppl 4):S164–S192.
4. Daniels SR, Jacobson MS, McCrindle BW, Eckel RH, Sanner BM. American Heart Association Childhood obesity research summit report. Circulation. 2009;119:e489–e517.
5. Krebs NS, Himes JH, Jacobson D, Nicklas TA, Guilday P. Assessment of child and adolescent overweight and obesity. Pediatrics. 2007; 120(suppl 4):S193–S228.

remains a valuable tool for classifying individuals into broad categories of overweight and obesity in order to monitor the weight status of individuals in clinical settings.[2, 9]

Body Fat Distribution

The location or distribution of adipose tissue within the body is an important concept when considering the health implications of overweight and obesity.[2, 36, 37] Body fat distribution can be divided into two clinically significant categories: (1) abdominal or central body fat distribution, and (2) lower body fat distribution. Abdominal or central fat placement refers to fat located primarily within the abdominal region of the body, both surrounding the organs of the abdomen (intra-abdominal or visceral fat) and located just under the skin around the waist (subcutaneous fat). Abdominal fat distribution is more often seen in males, tends to give the body a shape resembling that of an apple, and is sometimes referred to as *android*, which literally means "manlike." Lower body fat placement refers to fat located primarily in the lower region of the body, particularly within the hips and thighs, and tends to give the body a shape resembling a pear. Lower body fat distribution is more often seen in females and is sometimes referred to as *gynoid* ("womanlike").[27]

While it is possible to accurately quantify adipose tissue within the regions of the abdomen, hips, and thighs using magnetic resonance imaging (MRI) or computed tomography (CT), routine use of these imaging techniques is not practical in the clinical setting because the instruments are not readily available to most clinicians; they are expensive to acquire, maintain, and operate; and their use is time consuming.

Waist Circumference A much more practical approach to quantifying abdominal fat is to measure waist circumference (see Chapter 3). Estimates of abdominal adiposity based on waist circumference measurements compare favorably to abdominal fat measurements using MRI and CT scans.[1, 2, 9, 27] A practical approach for estimating the amount of adipose tissue in the hips and thighs is to measure the circumference of the hips or buttocks. This measurement is taken at the point yielding the maximum circumference around the hips or buttocks.

Excessive adipose tissue located deep within the abdomen and surrounding the intestines and liver (abdominal obesity) is associated with increased risk of type 2 diabetes, hypertension, dyslipidemia, coronary heart disease, and metabolic syndrome, even when BMI is within the healthy weight range.[2, 36, 38] In contrast, an increased amount of adipose tissue located within the hips and thighs is not associated with these increased risks; in fact, there is research suggesting that adipose tissue located within the lower body is inversely related to (i.e., lowers) risk of these conditions.[36, 39, 40] Epidemiologic research shows that in adults with a BMI between 25.0 kg/m^2 and 34.9 kg/m^2, risk of type 2 diabetes, hypertension, dyslipidemia, coronary heart disease, and metabolic syndrome increases when the waist circumference exceeds 40 inches in males and 35 inches in females.[1, 2, 32] A waist circumference measurement in excess of these values is regarded as "high-risk," as shown in Table 12.5.[1, 2, 32] Waist circumference is particularly useful in assessing the disease risk of patients who are categorized as having a healthy body weight (BMI of 18.5 kg/m^2 to 24.9 kg/m^2), who are considered overweight (BMI of 25.0 kg/m^2 to

Table 12.5 High-Risk Waist Circumference in Adult Males and Females

Males	>40 in (>102 cm)
Females	>35 in (>88 cm)

Source: National Heart, Lung, and Blood Institute. *The practical guide: identification, evaluation, and treatment of overweight and obesity in adults.* Bethesda (MD): U.S. Department of Health and Human Services, National Institutes of Health; 2000.

29.9 kg/m^2), or who are mildly obese with a BMI of 30.0 kg/m^2 to 34.9 kg/m^2. Waist circumference is a better indicator of disease risk than is BMI for individuals of Asian descent, and it assumes greater value for estimating risk for obesity-related diseases in older persons who often experience a loss of muscle mass and an increase in abdominal fat without marked changes in BMI.[1, 2, 36, 37] In persons with a BMI >35.0 kg/m^2, waist circumference is of little value in improving disease risk assessment; therefore, it is not recommended that waist circumference be measured in persons having a BMI >35.0 kg/m^2.[1, 2]

Waist-to-Hip Ratio (WHR) An alternative approach to evaluating the impact of body fat distribution on disease risk is the waist-to-hip ratio (WHR), which is calculated by dividing the waist circumference measurement by the hip circumference measurement. Some clinicians prefer using the WHR instead of the waist circumference, citing research suggesting that the WHR is somewhat better at predicting risk of coronary heart disease than is waist circumference alone.[36, 39, 40] Disease risk increases when the WHR is >0.95 in males and >0.8 in females.[36] A WHR >1.0 results when waist circumference is greater than hip circumference and suggests that the amount of abdominal fat is unhealthful. One plausible explanation for the potential superiority of the WHR is that it takes into account the protective effects of larger hip circumferences which may result from increased adipose tissue or from increased muscle mass in the hips and thighs, both of which are associated with a lower risk of type 2 diabetes, hypertension, dyslipidemia, coronary heart disease, and metabolic syndrome.[36, 39, 40] However, both waist circumference and WHR have been shown useful in assessing body fat distribution and evaluating disease risk. The key concept is that fat deep within the abdomen and around the intestines and liver increases disease risk; the technique used to measure it is less critical. Because BMI does not distinguish between lean tissue and adipose tissue or indicate how fat is distributed, it cannot predict disease risk when used alone. This is particularly the case for older persons who, as they age, tend to lose muscle mass and gain fat mass. When evaluating a patient's disease risk in relation to their weight, height, and body fat distribution, BMI and circumferences of the waist and hip should be used. Table 12.6 illustrates how BMI and waist circumference can be used to classify overweight and obesity and to provide an indication of relative disease risk.

Epidemiology of Overweight and Obesity

In nearly every country of the world, the average body weight of children and adults is increasing to such an extent that the World Health Organization (WHO) has coined the term "globesity" to describe what it calls a "global epidemic of

Table 12.6 Classification of Overweight and Obesity by BMI, Waist Circumference, and Associated Disease Risk[1]

| | BMI (kg/m²) | Obesity Class | Disease Risk[1] (Relative to Normal Weight and Waist Circumference) | |
			Men ≤ 40 in (≤ 102 cm) Women ≤ 35 in (≤ 88 cm)	Men > 40 in (> 102 cm) Women > 35 in (> 88 cm)
Underweight	< 18.5		—	—
Normal[2]	18.5–24.9		—	—
Overweight	25.0–29.9		Increased	High
Obesity	30.0–34.9	I	High	Very high
	35.0–39.9	II	Very high	Very high
Extreme Obesity	≤ 40.0	III	Extremely high	Extremely high

[1]Disease risk for type 2 diabetes, hypertension, and cardiovascular disease.

[2]Increased waist circumference can also be a marker for increased risk even in persons of normal weight.

Source: National Heart, Lung, and Blood Institute. *The practical guide: identification, evaluation, and treatment of overweight and obesity in adults.* Bethesda (MD): U.S. Department of Health and Human Services, National Institutes of Health; 2000.

obesity."[41] While the term "epidemic" is generally used in the context of infectious disease, it can appropriately be applied to any condition or situation having an adverse effect on health, including overweight and obesity.[42] While primarily considered a problem affecting developed nations, overweight and obesity are common in urban areas of many developing nations, where, paradoxically, they coexists with undernutrition occurring in the rural areas of the same country. According to the WHO, the prevalence of obesity ranges from less than 5% of the population of China, Japan, and certain African nations to more than 75% in urban Samoa. Even in a relatively low-prevalence country like China, obesity rates can be as high as 20% in some cities.[41]

Because of the ease of accurately measuring weight and height, estimates of the prevalence of overweight and obesity are typically based on BMI. Attempting to measure body composition on large numbers of people using such methods as skinfold measurements or underwater weighing is impractical. There are some instances when it is not possible to obtain measured weights and heights on subjects, in which case researchers must rely on self-reported weights and heights. However, estimates based on measured weight and height are more accurate than those based on self-reported weight and height, which tend to underestimate the true prevalence of overweight and obesity.

Overweight and Obesity in the United States

In the early 1960s, the National Center for Health Statistics launched a series of surveys examining the health and nutritional status of the U.S. population, in which participants completed questionnaires evaluating diet and lifestyle habits, and underwent diagnostic and laboratory testing as well as extensive anthropometric assessment. In these surveys (initially called the National Health Examination Survey and then renamed the National Health and Nutrition Examination Survey), all anthropometric measurements, including height and weight, were taken by trained health technicians using standardized measuring procedures and equipment yielding highly accurate data

for calculating BMI. A key finding from these surveys is that the percentage of people in the United States who are overweight or obese has increased since the early 1960s (see Figures 12.3 and 12.4). When looking at Figure 12.3, note that the percentage of U.S. adults who are *either* overweight *or* obese increased to a much lesser extent than did the percentage of U.S. adults who are obese, as shown in Figure 12.4. While the body weights of Americans have increased in recent decades, most of this increase has been due to increases in the obesity category, whereas only minor increases occurred in the prevalence of persons who are overweight but not obese.

Two of the national health objectives for the year 2010 were to increase the proportion of U.S. adults who are at a healthy weight to 60% and to decrease the proportion of U.S. adults who are obese to 15%.[43] Figure 12.5 shows how the percentage of the U.S. adult population categorized as having a healthy

Figure 12.3 Prevalence of U.S. Adults Who Are Either Overweight or Obese, 1960 to 2006

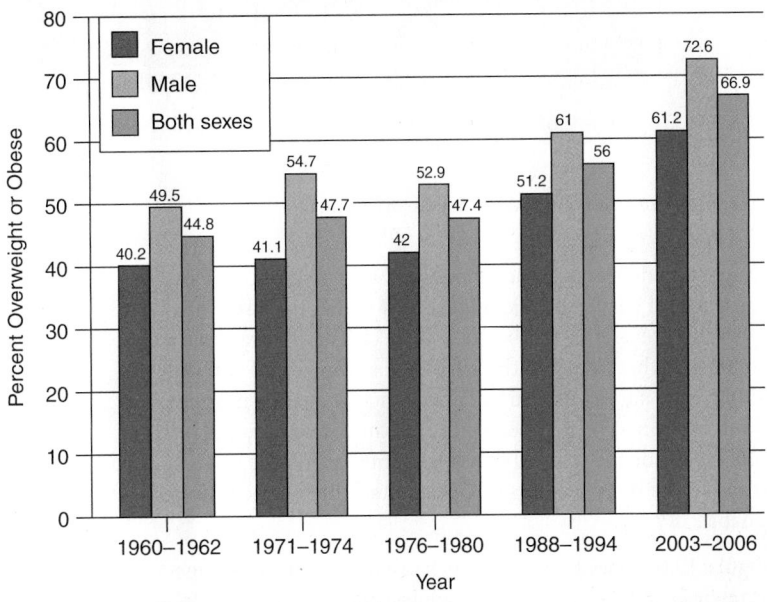

Source: Data from the National Center for Health Statistics.

Figure 12.4 Prevalence of Obesity among U.S. Adults, 1960 to 2006

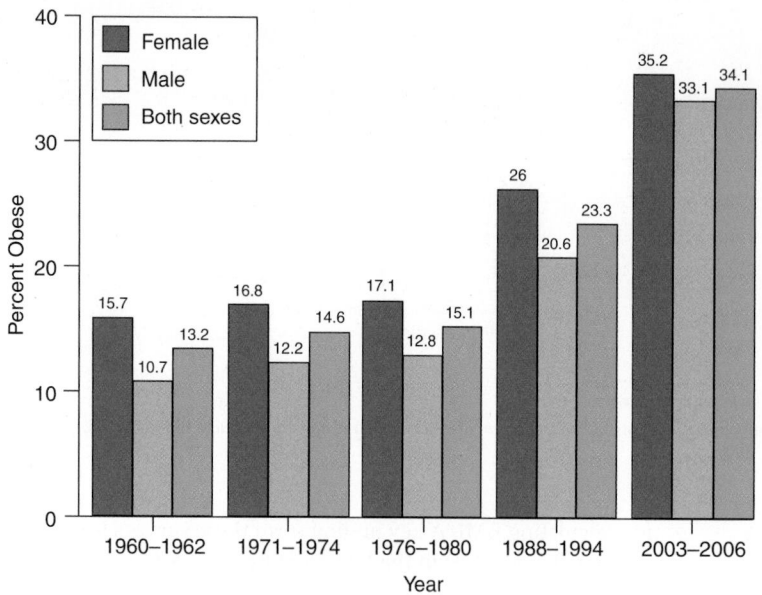

Source: Data from the National Center for Health Statistics.

Figure 12.5 Prevalence of Healthy Weight among U.S. Adults, 1960–2006

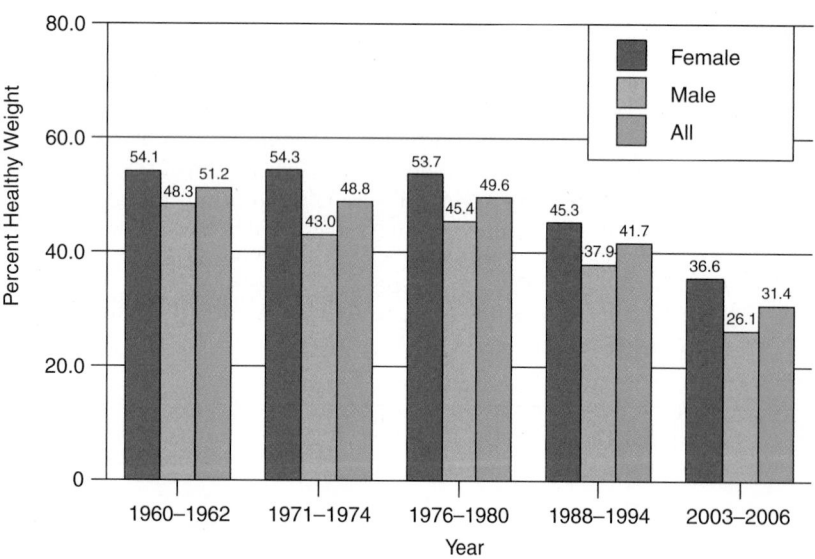

Source: Data from the National Center for Health Statistics.

weight has decreased since the early 1960s to about 31% in the most recent survey period. Figure 12.4 shows that from 1960 to 1980 the prevalence of obesity among U.S. adults was relatively stable, but that between the 1976–80 survey period and the 2003–2006 survey period the prevalence of obesity more than doubled from 15% to 34%. Neither of these two national health objectives for the year 2010 has been met, given the upward trend in the body weights of U.S. adults, which has continued unabated.[44]

Figure 12.6 shows how the prevalence of obesity in two age categories of U.S. children and adolescents (6–11 years old and 12–19 years old) has changed since the 1960s. As discussed earlier, in children and adolescents ages 2 to 19 years, obesity is now defined as a BMI at or above the 95th percentile for sex and age using the 2000 Centers for Disease Control and Prevention (CDC) BMI-for-age growth charts for the United States.[33–35] Figure 12.6 illustrates that from the 1960s to 1980 the prevalence of obesity among U.S. children and adolescents was relatively stable, and that between the 1976–1980 survey period and the 2003–2006 survey period the prevalence of obesity nearly tripled from approximately 6% to roughly 17%. The prevalence of obesity among children and adolescents in the United States is increasing at a faster rate than among U.S. adults.[37] One of the national health objectives for the year 2010 was to reduce the proportion of obese children and adolescents to 5%.[43] But instead of decreasing or even leveling off, the prevalence of obesity in these two groups is increasing to even higher levels. The data on adolescents are of particular concern in light of the fact that obese adolescents are at increased risk of becoming obese adults and experiencing the health risks associated with obesity.[33, 34, 44]

Overweight and Obesity in Canada

Changes in the prevalence of healthy weight, overweight, and obesity among adult Canadians between 1978–79 and 2004 are shown in Figures 12.7 to 12.10. The data used in these figures come from two different surveys conducted by Statistics Canada: the 1978–79 Canada Health Survey and the 2004 Canadian Community Health Survey, in which the weight and height of subjects were measured by trained health technicians as part of a more comprehensive assessment of nutritional and health status.[31] In these figures, the definitions of healthy weight, overweight, and obesity in adults are the same as those used in the United States. Between 1978–79 and 2004, the percentage of Canadian adults who were either overweight or obese increased, as shown in Figure 12.7, and most of this increase was due to a marked rise in the prevalence in obesity between the two survey periods, as shown in Figure 12.8. The prevalence of overweight among Canadian adults remained relatively static between the two survey periods, as shown in Figure 12.9. As of 2004, 36% of Canadian adults of both sexes were overweight and 23% were considered obese.[31] Figure 12.10 shows the decline in the percentage of Canadian adults who had a healthy weight between the two surveys.

The prevalence of overweight and obesity among Canadian children and adolescents is summarized in Figure 12.11 and Figure 12.12. Statistics Canada defines overweight and obesity using criteria developed by the International Obesity Task Force (IOTF), which are similar to those used in the United States by the National Center for Health Statistics but are considered better suited when evaluating multi-national populations.[45] The IOTF definitions of overweight and obesity

Figure 12.6 Prevalence of Obesity among U.S. Children and Adolescents, 1963–2006

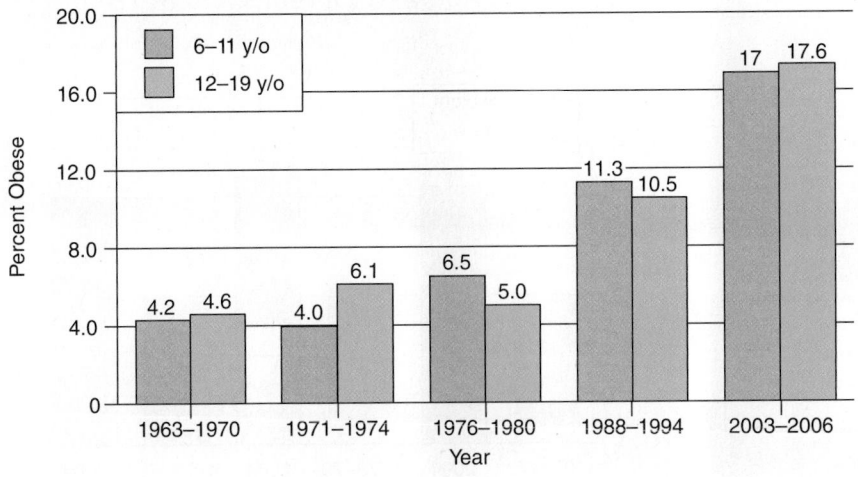

Source: Data from the National Center for Health Statistics.

obesity more than doubled from 14% to 29%, and the obesity rate tripled from 3% to 9%.

Overweight and Obesity in Europe

The most reliable comparative data on the prevalence of overweight and obesity in Europe come from the World Health Organization's MONICA (Multinational MONItoring of trends and determinants in CArdiovascular disease) Project, an international survey conducted in the 1980s and 1990s to monitor global trends in cardiovascular disease in persons 35 to 64 years of age.[46, 47] Included among the various types of data collected from participants in the MONICA Project were measured weight and height, from which BMI was calculated. Figures 12.13 and 12.14 show the prevalence of overweight and obesity for males and females, respectively, from selected European countries.

In recent decades, the prevalence of overweight and obesity among children and adults in most western European countries has increased. In many of those countries, more than half of adults are overweight, and as many as 30% are clinically obese. However, in some central and eastern European countries, such as the Czech Republic, Lithuania, Serbia and Montenegro, and Russia, the average BMI of adults is declining.[28, 29, 47, 48]

Comparing the prevalence of overweight and obesity among children and adolescents in different European countries is difficult, because the various data sets do not uniformly define overweight and obesity, sometimes rely on self-reported data, and are not representative of the demographic, cultural, and socioeconomic composition of the European population.[49, 50] Despite these shortcomings, the data indicate that the prevalence of overweight and obesity is increasing throughout most European countries, that the prevalence of obesity in young children is relatively low compared to that of adolescents, and that the highest rates of obesity are observed in eastern and

are based on BMI calculated from weight and height measurements obtained from nearly 200,000 children and adolescents ages 2 to 18 years old from Brazil, Great Britain, Hong Kong, the Netherlands, Singapore, and the United States.[45] In the 25-year interval between the 1978–79 Canada Health Survey and the 2004 Canadian Community Health Survey, the prevalence of overweight and obesity among Canadian children and adolescents increased by about 70%, and the obesity rate increased 250%, as shown in Figure 12.11.[30] There are some notable differences among different age groups, as shown in Figure 12.12. Among children ages 2 to 5 years, the prevalence of obesity increased, although when overweight and obesity rates were combined, there was no change. Among children 6 to 11 years of age, the prevalence of overweight and obesity combined doubled from 13% to 26%, and there was a marked increase in the prevalence of obesity. Among 12 to 17 year olds, the prevalence of overweight and

Figure 12.7 Prevalence of Canadian Adults Who Are Either Overweight or Obese, 1978–79 to 2004

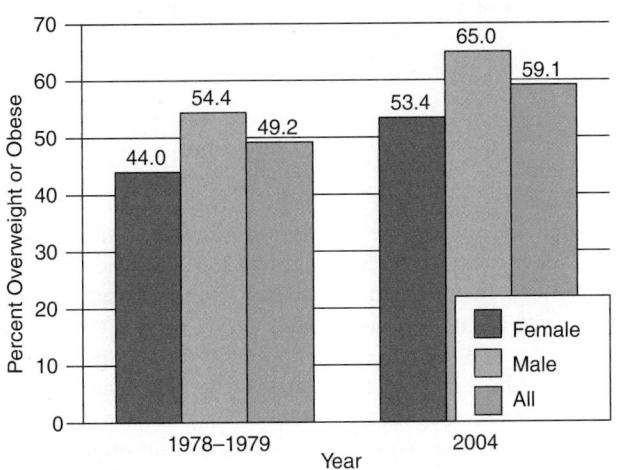

Source: Data from Statistics Canada.

Figure 12.8 Prevalence of Obesity among Adults in Canada, 1978–79 to 2004

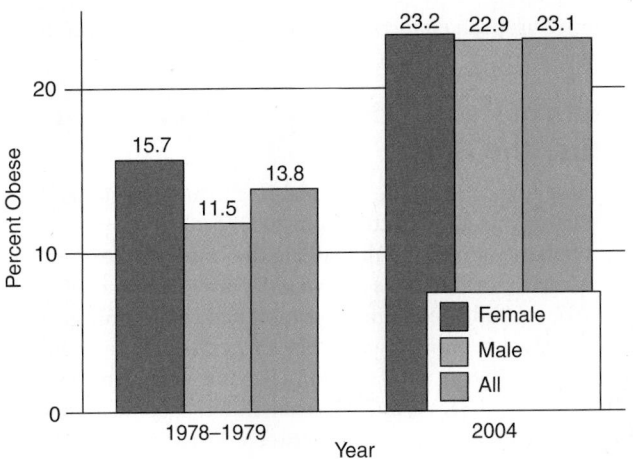

Source: Data from Statistics Canada.

Figure 12.9 Prevalence of Overweight among Adults in Canada, 1978–79 to 2004

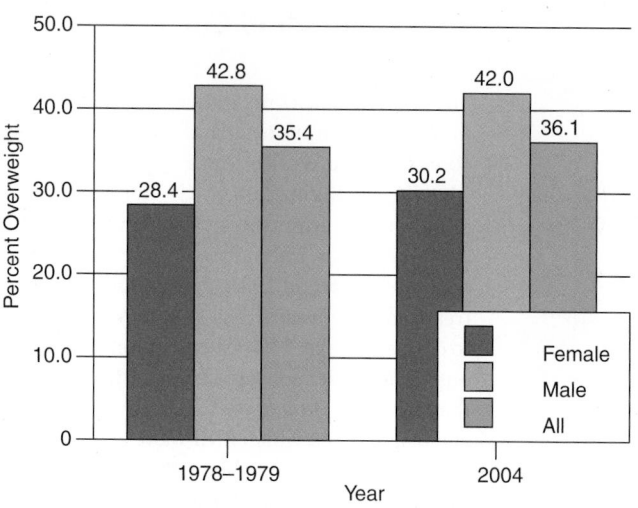

Source: Data from Statistics Canada.

Figure 12.10 Prevalence of Healthy Weight among Canadian Adults, 1978–79 to 2004

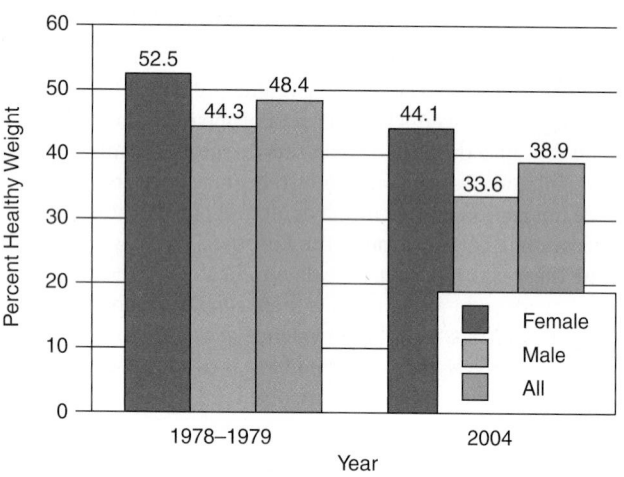

Source: Data from Statistics Canada.

Figure 12.11 Prevalence or Overweight and Obesity among Canadian Children and Adolescents Age 2 to 17 years, 1978–79 and 2004

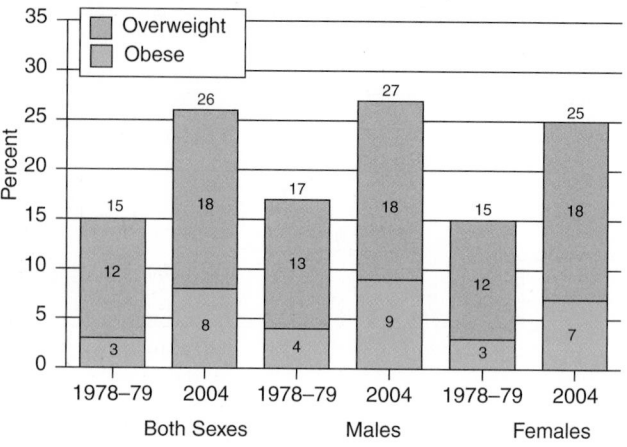

Source: Data from Statistics Canada.

Figure 12.12 Prevalence of Overweight and Obesity among Canadian Children and Adolescents 2 to 5 Years of Age, 6 to 11 Years of Age, and 12 to 17 Years of Age, 1978–79 and 2004

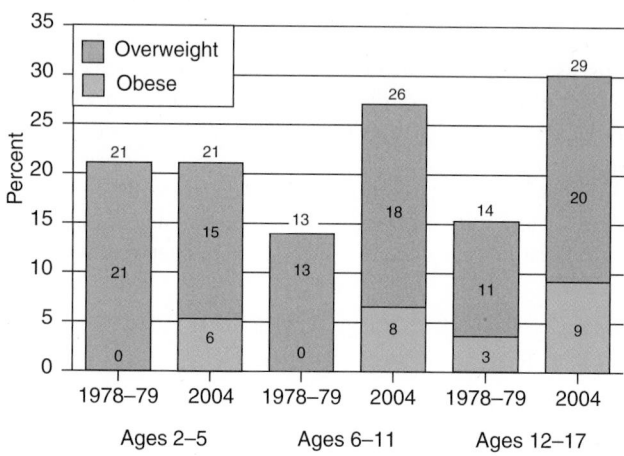

Source: Data from Statistics Canada.

southern European countries, particularly Italy, Greece, and Portugal.[49, 50]

Effects of Race, Ethnicity, Socioeconomic Status, and Age

According to data from the National Health and Nutrition Examination Survey collected from 2003 to 2006, the prevalence of obesity among adult males in the United States varied little by race or ethnicity, as shown in Figure 12.15. In contrast, there were considerable differences among racial/ethnic groups for adult females. Non-Hispanic black females had the highest prevalence of obesity at 53.4%, non-Hispanic white females had the lowest rate at 31.6%, and Mexican-American females had an intermediate prevalence rate between the two groups. Data from the same survey indicate that higher socioeconomic status is associated with a lower prevalence of obesity.

Between 2003 and 2006, the prevalence of obesity among U.S. adults whose income was below the poverty threshold was 34.7%, while the obesity rate of those whose income was 200% or more above the poverty threshold was 28.7%. As shown in Figure 12.16, the prevalence of obesity among U.S. adults increases with age until approximately age 64 years, after which the prevalence of obesity declines. Some researchers suggest that because obese individuals die at a younger age than those who are not obese, the average BMI of older Americans appears to decline. In addition, many of the chronic conditions commonly seen in the elderly are associated with diminished food intake and weight loss.[37]

Figure 12.13 Prevalence of Overweight (BMI 25.0 kg/m² to 29.9 kg/m²) and Obesity (BMI ≥30.0 kg/m²) among Adult Males in Select European Countries

BMI calculated from measured weight and height.

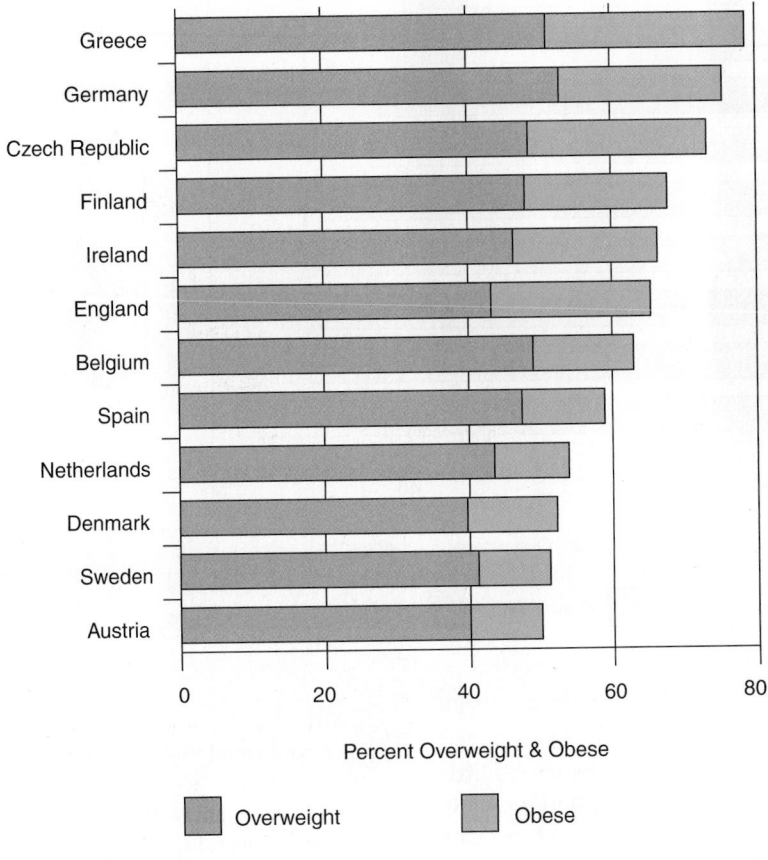

Percent Overweight & Obese

■ Overweight   ■ Obese

Source: Data from the International Obesity Task Force.

Adverse Health Consequences of Overweight and Obesity

Psychosocial and Emotional Consequences

In North America, the combination of a thin standard of beauty with fat ways of living has resulted in the current era being referred to by some as "the age of caloric anxiety." The media are relentless in promoting the consumption of foods and beverages having a high caloric density while simultaneously advancing an "ideal" body shape that is impossible to attain for practically all females and males. Because of the strong pressures from society to be thin, overweight and obese people often suffer feelings of guilt, depression, anxiety, and low self-worth.[51, 52]

Physiological Consequences

Type 2 Diabetes Type 2 diabetes is three times as prevalent among the obese as compared with normal-weight persons. Excess body fat, especially when located within the abdominal region, elevates fasting and postprandial levels of plasma free fatty acids, and is a major risk factor for metabolic syndrome.[22, 24] Elevated plasma free fatty acids can stimulate

secretion of insulin from the beta-cells of the pancreas, cause insulin resistance in peripheral tissues, inhibit cellular uptake of glucose from the blood, reduce glycogen storage, and increase hepatic glucose production, all of which lead to hyperglycemia, hyperinsulinemia, impaired glucose tolerance, and eventual development of type 2 diabetes.[53] Even modest weight loss in persons with type 2 diabetes can result in dramatic improvements in blood glucose control and a reduced need for medications to control blood glucose levels (see Chapter 17).[14, 54, 55] For example, the Diabetes Prevention Program showed that in middle-aged, obese subjects who had impaired glucose tolerance (see Chapter 17), a 7% weight loss and at least 150 minutes of exercise per week reduced the chance of developing type 2 diabetes by 58%.[56]

High Blood Pressure High blood pressure is three times more common in the obese than in normal-weight persons. Even among schoolchildren, increases in obesity are associated with corresponding increases in blood pressure, and weight loss may be an effective treatment for high blood pressure, as it is in adults.[14] It is thought that the hyperglycemia and hyperinsulinemia associated with obesity increase blood pressure through several mechanisms that are not well understood.[57]

Lipid Abnormalities Obese adults are more likely than normal-weight adults to have elevated serum levels of total and low-density lipoprotein (LDL) cholesterol and triglycerides, as well as lower serum levels of high-density lipoprotein (HDL) cholesterol. Elevated serum LDL-cholesterol and low serum HDL-cholesterol are major risk factors for coronary heart disease. Consequently, obesity places individuals at greater risk of coronary heart disease. Obesity results in the overproduction of very-low-density lipoprotein (VLDL) by the liver. Because the body eventually converts VLDL to LDL, increased serum levels of VLDL result in elevations of serum LDL.[58] The prevention of the onset of obesity in early life may be important for reducing the risk of coronary heart disease in later life (see Chapter 13).[59]

Hepatobiliary Disorders There is a six-fold increased risk of symptomatic gallstones (cholelithiasis) in persons whose body weight is 50% greater than recommended.[14] Central or abdominal obesity is a major risk factor for nonalcoholic fatty liver disease (NAFLD), considered the most common chronic liver condition in the Western world.[60, 61] Although the histological damage to the liver is similar to that seen in alcoholic liver disease, NAFLD arises from causes other than the consumption of alcoholic beverages.[61, 62] Although generally asymptomatic, NAFLD can range from simple steatosis (increased accumulation of fat in the hepatocytes) to nonalcoholic steatohepatitis (steatosis with inflammation and necrosis of the hepatocytes), and eventual liver failure.[62, 63] NAFLD is a fairly recent health concern, being first described in the 1950s. The prevalence of NAFLD is currently estimated to range from 14% to 20% in the United States and Europe.[63] Risk factors associated with NAFLD include central obesity, type 2 diabetes, hyperlipidemia, and metabolic syndrome.[62, 63]

Figure 12.14 Prevalence of Overweight (BMI 25.0 kg/m² to 29.9 kg/m²) and Obesity (BMI ≥30.0 kg/m²) among Adult Females in Select European Countries

BMI calculated from measured weight and height.

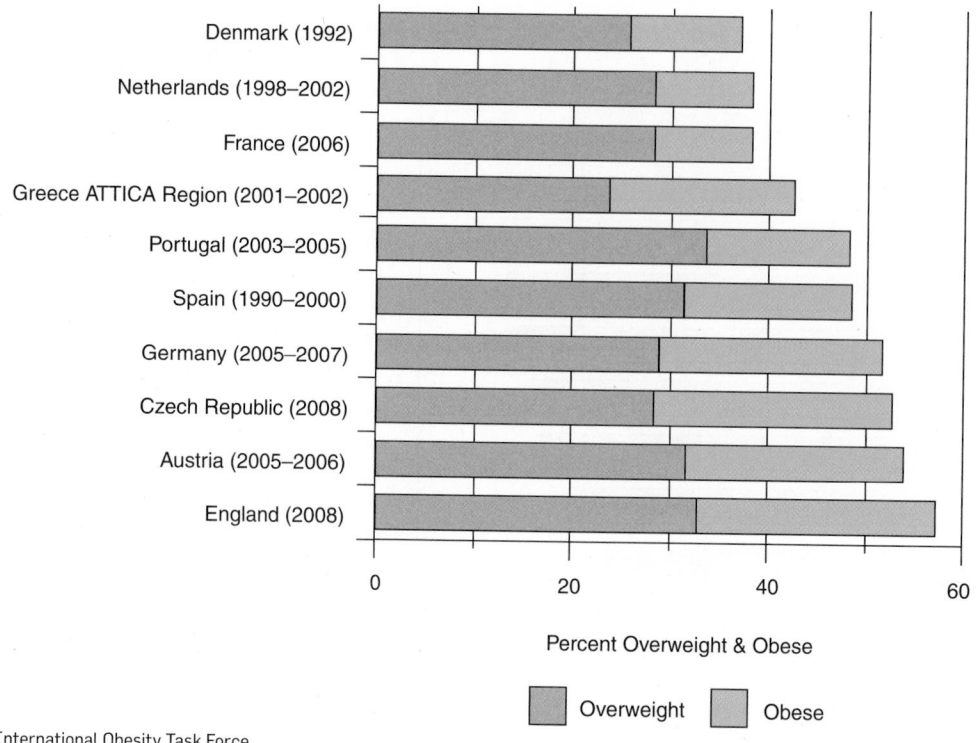

Percent Overweight & Obese

Overweight · Obese

Source: Data from the International Obesity Task Force.

Figure 12.15 Prevalance of Obesity among U.S. Adults by Race/Ethnicity, 2003 to 2006

Between 2003 and 2006 the prevalence of obesity among adult males varied little by racial or ethnic group. Among adult females, non-Hispanic blacks had the highest prevalence of obesity and non-Hispanic whites had the lowest. Mexican-American females had a prevalence that was intermediate between the other two groups. Obesity is defined as a BMI ≥30.0 kg/m².

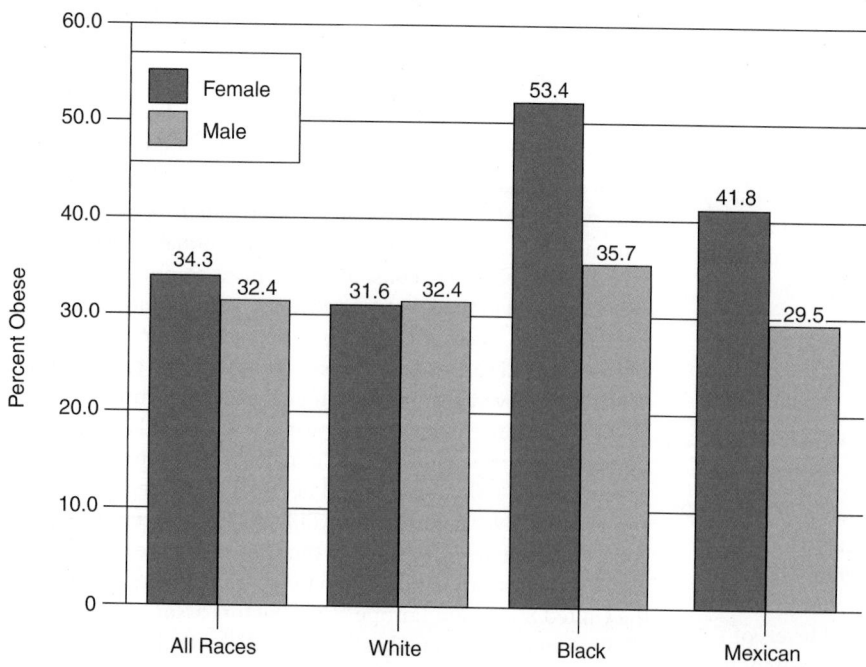

Data from the National Center for Health Statistics.

Cancers A number of studies have confirmed that obesity is a significant risk factor for death from cancer generally and from cancer in several specific sites. Obesity in males is associated with increased death from cancer of the esophagus, colon, rectum, pancreas, liver, and prostate. In females, obesity increases risk of death from cancer of the gallbladder, bile duct, breast, endometrium, cervix, and ovaries.[14, 64] Obesity accounts for 14% and 20% of cancer deaths in U.S. males and females, respectively.[14]

Reproductive Disorders Obesity is associated with reproductive disorders in both sexes. In males, obesity is associated with gynecomastia (enlarged mammary glands in the male), hypogonadism, reduced testosterone levels, and elevated estrogen levels. In females, obesity is associated with menstrual abnormalities (particularly in those with abdominal or central body fat distribution) and polycystic ovarian syndrome (PCOS), a common endocrine

Figure 12.16 Prevalence of Obesity among U.S. Adults by Age, 2003 to 2006

As adult Americans age, the prevalence of obesity increases until about age 65 years, at which point the obesity rate declines. Data are from the National Health and Nutrition Examination Survey collected between 2003 and 2006. Obesity is defined as a BMI ≥30.0 kg/m².

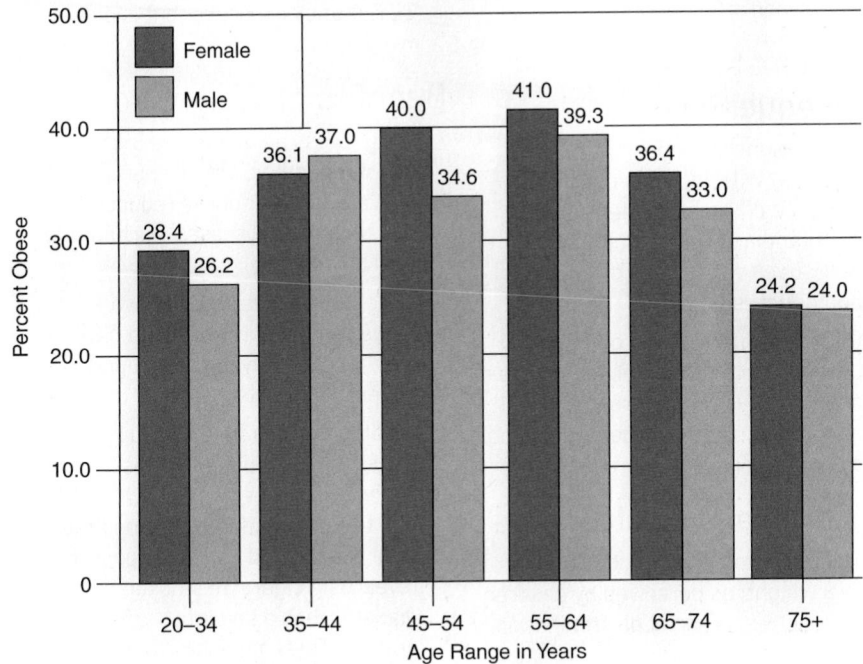

Data from the National Center for Health Statistics.

disorder affecting 5% to 10% of females of reproductive age. Approximately 50% of women diagnosed with PCOS are obese. The disorder is characterized by menstrual irregularities, acne, excess body hair, male pattern hair loss, and a chronic failure to ovulate leading to infertility.[14, 65]

Premature Death

Numerous studies by a variety of researchers have consistently shown that, on average, the obese experience earlier death than lean persons. The lowest mortality for men and women occurs among those whose body mass index is somewhat less than average. Mortality from heart disease, cancer, and diabetes increases with increasing fatness. A person whose body weight is more than double his or her healthy body weight has a risk of death that is 12 times greater than it would be if he or she were at a healthy weight.[14, 66] Additional information on the health consequences of overweight and obesity is shown in Box 12.5.

Etiology of Obesity

Obesity develops when the body's chronic energy intake exceeds its energy expenditure. At first glance this may seem simple

---

Health Consequences of Overweight and Obesity

Premature Death

- An estimated 300,000 deaths per year in the United States may be attributable to obesity.
- The risk of death rises with increasing weight.
- Even moderate weight excess (10 to 20 pounds for a person of average height) increases the risk of death, particularly among adults aged 30 to 64 years.
- Individuals who are obese (BMI >30 kg/m²) have a 50% to 100% increased risk of premature death from all causes, compared to individuals in the healthy weight range (BMI 18.5 kg/m² to 24.9 kg/m²).

Heart Disease

- The incidence of heart disease (myocardial infarction, conges-

tive heart failure, sudden cardiac death, angina, and abnormal heart rhythm) is increased in persons who are overweight or obese (BMI >25 kg/m²).
- High blood pressure is twice as common in adults who are obese as in those who are at a healthy weight.
- Obesity is associated with elevated serum triglycerides and decreased serum HDL-cholesterol.

Diabetes

- A weight gain of 11 to 18 pounds increases a person's risk of developing type 2 diabetes to twice that of individuals who have not gained weight.
- Over 80% of people with T2 diabetes are overweight or obese.

Cancer

- Overweight and obesity are associated with an increased risk of some types of cancer including endometrial, colon, gallbladder, prostate, kidney, and postmenopausal breast cancer.
- Women gaining more than 20 pounds from age 18 to midlife double their risk of postmenopausal breast cancer, compared to women whose weight remains stable.

Breathing Problems

- Sleep apnea is more common in obese persons.
- Obesity is associated with a higher prevalence of asthma.

(continued)

Arthritis

- For every 2-pound increase in weight, the risk of developing arthritis is increased by 9% to 13%.
- Symptoms of arthritis can improve with weight loss.

Reproductive Complications

- Obesity is associated with increased risk of menstrual abnormalities and polycystic ovary syndrome (PCOS) in females and in reduced levels of testosterone, increased levels of estrogen, and gynecomastia (enlarged mammary glands) in males.
- Obesity during pregnancy is associated with an increased risk of fetal and maternal death and increases the risk of maternal high blood pressure tenfold.
- In addition to many other complications, women who are obese during pregnancy are more likely to have gestational diabetes and problems with labor and delivery.
- Infants born to women who are obese during pregnancy are more likely to have high birthweights and, therefore, are more likely to be delivered by Cesarean section delivery and experience hypoglycemia, which can be associated with brain damage and seizures.

- Obesity during pregnancy is associated with an increased risk of birth defects, particularly neural tube defects such as spina bifida.
- Obesity in premenopausal women is associated with irregular menstrual cycles and infertility.

Additional Health Consequences

- Overweight and obesity are associated with increased surgical risk as well as increased risks of gallbladder disease, nonalcoholic fatty liver disease, incontinence, and depression.
- Obesity can affect the quality of life through limited mobility and decreased physical endurance as well as through social, academic, and job discrimination.

Children and Adolescents

- The most immediate consequence of overweight, as perceived by children themselves, is social discrimination.
- Risk factors for heart disease, such as hyperlipidemia and hypertension, occur more frequently in overweight and obese individuals than those in the healthy weight range.
- The prevalence of type 2 diabetes, often considered a disease primarily affecting adults, has increased

dramatically in children and adolescents. Overweight and obesity increase the risk of type 2 diabetes.
- Overweight adolescents have a 70% chance of becoming overweight or obese as adults. This increases to 80% if one or more parent is overweight or obese.

Benefits of Weight Loss

- Weight loss as modest as 5% to 15% of total body weight in a person who is overweight or obese reduces the risk of certain diseases, particularly heart disease.
- Weight loss can result in lower blood pressure, lower blood glucose, and improved serum lipid and lipoprotein levels.
- A person with a BMI > 25.0 kg/m² may benefit from weight loss, especially if he or she has other health risk factors such as high blood pressure or elevated lipid and lipoprotein levels, is a cigarette smoker, has diabetes, has a sedentary lifestyle, or has a personal and/or family history of heart disease.

Adapted from: U.S. Department of Health and Human Services. The surgeon general's call to action to prevent and decrease overweight and obesity. Rockville (MD): U.S. Department of Health and Human Services, Public Health Service, Office of the Surgeon General; 2001.

---

and straightforward. However, because of the multiple and complex neuroendocrine and metabolic systems influencing energy intake and energy expenditure, obesity is actually a heterogeneous group of disorders. Its etiology remains elusive, and its successful, long-term treatment is difficult.[14, 37] Among the key factors contributing to obesity are specific medical and psychiatric disorders or their treatment, genetics, and an **obesigenic** environment that promotes a high energy intake and discourages physical activity.

Medical Disorders, Pharmacological Treatments, and Smoking Cessation

Obesity can result from a specific medical disorder such as Cushing's syndrome, hypothyroidism, or Prader-Willi syndrome, but these are relatively rare. Certain pharmacologic agents (see Table 12.7) are also associated with weight gain.[37] (When an adverse health condition results from some treatment administered by a physician or other health-care provider, the condition is said to be **iatrogenic** or literally "brought forth by a physician.") Because nicotine in tobacco smoke increases 24-hour energy expenditure by about 10%,

smoking cessation results in a normalization of 24-hour energy expenditure and a tendency for people to gain weight when they quit smoking. In addition, some people tend to snack more and drink more alcoholic beverages after they quit smoking. Compared to the weight gain of males and females who continue to smoke, males who quit smoking gain 9.7 lb (4.4 kg) over 10 years and females who quit smoking gain 11 lb (5.0 kg) over a 10-year period.[67]

Disordered Eating Patterns

Two non-normative eating patterns or forms of disordered eating known to contribute to weight gain are night eating syndrome and binge eating.[68, 69] Night eating syndrome, a common practice among the obese, is defined as consumption of at least 25% of total energy intake between the evening meal and the next morning. However, some patients with night eating syndrome consume as much as 50% of their total energy intake at night after their evening meal. It is estimated that 10% to 25% of obese persons experience at least occasional episodes of eating large quantities of food in relatively short periods of time, usually in the evening.[37, 68, 69]

Table 12.7 Medical Conditions and Pharmacologic Agents Known to Cause Obesity

Congenital Causes
- Prader-Willi syndrome
- Down syndrome
- Bardet-Biedel syndrome
- Alstrom syndrome
- Cohen syndrome
- Carpenter syndrome

Neuroendocrine Disorders
- Cushing's syndrome
- Hypothalamic disorders
- Hypothyroidism
- Polycystic ovary syndrome
- Growth hormone deficiency

Pharmacologic Agents

Psychiatric Medications
- Olanzapine, clozapine
- Selective serotonin reuptake inhibitors
- Monoamine oxidase inhibitors
- Gabapentin
- Valproate
- Carbamazepine

Steroid Hormones
- Hormonal contraceptives
- Corticosteroids
- Progestational agents

Antidiabetic Agents
- Insulin
- Sulfonylureas
- Thiazolidinediones

Miscellaneous
- Antihistamines
- α-adrenergic inhibitors
- ß-adrenergic inhibitors
- Protease inhibitors

Source: Hill JO, Catenacci VA, Wyatt HR. Obesity: Etiology. In: Shils ME, Shike M, Ross AC, Cabellero B, Cousins RJ editors. *Modern nutrition in health and disease.* 10th ed. Philadelphia: Lippincott Williams & Wilkins; 2006. 1013–28.

Genetics

Genetics affects body weight and body composition by influencing such factors as appetite, taste preferences, energy intake, resting energy expenditure, the thermic effect of food, **nonexercise activity thermogenesis (NEAT)**, and the body's efficiency in storing energy. For example, it has been observed that despite some daily variation in energy intake and energy expenditure, most people maintain their body weight within a fairly narrow range. One explanation for this is the idea that each person's body has a genetically determined metabolic "set-point" that maintains a preferred body weight. While this appears to hold true if the environment remains fairly consistent, significant changes in the past several decades in eating habits and activity levels throughout most of the world have led to a gradual increase in average body weights.[37]

Understanding the etiologic role of genetics in obesity is complicated by the fact that obesity is not inherited in families in a predictable manner as are other diseases such as sickle cell anemia, cystic fibrosis, or Huntington's disease. This lack of predictability indicates that multiple genes are involved, with each making a small contribution to body weight and how a person responds to environmental factors like diet, physical activity, and culture.[70]

Furthermore, separating the influence of genetics from the impact of environmental and cultural factors on body weight is difficult. To explore the question of genetics versus environment, investigators have studied individuals within the family unit, pairs of twins, and body weights of adoptees in relation to their biologic and adoptive parents. Having obese family members increases one's risk of obesity, even if the family members do not live together or have similar dietary or physical activity patterns. Studies comparing the body weights of parents and their offspring show that 80% of the offspring of two obese parents eventually become obese, that 40% of offspring of one obese parent eventually become obese, and that when neither parent is obese, the likelihood of obesity in a child is 14%.[71]

Studies comparing the BMI and percent body fat/total body fat (as determined by hydrostatic weighing) of identical or monozygotic twins (MZ) and fraternal or dizygotic twins (DZ) have shown that body weights and adiposity of MZ twins tend to be much closer than those of DZ twins. This suggests that genetics plays a role in determining body weight and adiposity. In one study, researchers took 12 pairs of MZ twins who were fed 1000 kcal/day more than that necessary to maintain their body weight while kept in a sedentary mode of life.[72] This was done for six days a week during a period of 100 days. There was considerable variation in weight gain and change in fat and lean body mass between the 24 individuals. Weight gain ranged from 4 to 13 kg, with mean weight gain being 8.1 kg. However, the variation was not random—there was significant within-pair similarity in weight gain and change in fat and lean body mass in response to the overfeeding. Results of the study suggest that genetics influences the amount of weight gained and the change in fat and lean body mass in response to over-feeding.[73, 74] A study of 540 Danish adoptees gave evidence for a smaller yet still substantial genetic contribution. In this study, the BMI of the adoptees correlated strongly with that of their biologic parents, but not at all with that of their adoptive parents. This finding suggests that in this Danish population, early family environment had apparently little influence in determining the degree of fatness.[73]

The weight of scientific evidence indicates that some people are more prone to obesity than others due to genetic factors, and that 40% to 50% of the variation in BMI is explained by genetic factors.[37, 70] However, environmental factors probably play a greater etiologic role for most people, particularly in light of the fact that famine prevents obesity even in the most obesity-prone individuals. For persons who are genetically predisposed to obesity, it appears that the severity of the disease is largely determined by lifestyle and environmental factors. When the environment changes from one where access to high-energy foods is limited and regular physical activity is required (a "restrictive environment") to one where high-energy foods are easily accessible and the humans are largely sedentary (an "obesigenic environment"), most humans will gain weight. However, those who are genetically predisposed to obesity will gain the most weight while those who are not genetically predisposed to obesity will gain little if any weight.[75] As important as genetic influences are, persons born with a genetic predisposition to obesity are not necessarily destined to a life of obesity.

Obesigenic Environment

The term "toxic food environment" aptly describes the convenient availability of low-cost, tasty, energy-dense foods, in large portion sizes, in North America and the developed world. The toxic food environment, a key component of our obesigenic environment, encourages a high energy intake and has been a major contributing factor in the epidemic of overweight

Environmental Influences on Eating Habits

Environmental changes occurring within the past several decades that have had an impact on the eating habits of North Americans include:

- Growth of the fast-food industry—more food is now eaten out of the home and much of this is relatively high in fat.

- Average portion sizes of food consumed in and out of the home have increased.

- Increased availability of foods and beverages, especially those with a high energy density.

- Increased numbers of snack and convenience foods.

- Aggressive marketing of foods, particularly to children.

- Decrease in the proportion of disposable income spent on food—average income has increased faster than the increase in the price of food.

and obesity. This is in sharp contrast to what was the norm throughout most of human history, when considerable energy and time were spent in obtaining food, obesity was rare, and hunger, malnutrition, and starvation were common. In the past, genes favoring the efficient use and storage of energy allowed our ancestors to survive periods of food shortages. Now, these same genes work against maintaining a healthy weight in the present environment where food is plentiful, inexpensive, accessible, and energy-dense.[76]

Over the past several decades, important changes in the eating habits of North Americans have contributed to the increased prevalence of overweight and obesity. These are outlined in Box 12.6. As shown in Table 12.8, U.S. government surveys attempting to estimate the energy and nutrient intake of Americans suggest that between two surveys conducted from 1971 to 1974 and between 2005 and 2006, average energy intake of U.S. females and males increased 243 kcal and 188 kcal, respectively. However, it should be noted that between the two surveys, there were changes in the methodology used to quantify dietary intake that could account for some of the differences in reported energy intake between the two surveys.

Food portion size affects energy intake. When offered larger food portion sizes, subjects tend to consume more energy during mealtimes and while snacking than when eating smaller portion sizes.[77, 78] Also, when consuming larger portion sizes, subjects generally do not report an increased or earlier sense of fullness.[77] Energy density, the energy content of a food relative to its weight (kcal/g), is another important factor. High-energy-dense foods—such as many fast and processed foods—tend to contain less water and more fat and added sugars than those of lower energy density. When eating a high-energy-dense diet containing foods high in fat, refined carbohydrates, and added sugars, subjects tend to consume a greater number of kcal than when eating a low-energy-dense diet containing more vegetables, fruits, and broth-based soups.[37, 77, 78]

Two aspects of the obesigenic environment promoting a high-energy-dense diet are cost and convenience. There is an inverse relationship between energy density and cost.[37] Refined grains, added sugars, and added fats are among the lowest-cost sources of dietary energy.[79] On a per kcal basis, high-energy-dense foods such as hamburgers and french fries cost considerably less than low-energy dense foods such as fresh fruits and vegetables.[37] Heavily processed foods high in added fats and sugars tend to have a longer shelf-life and are generally more convenient to prepare than low-energy-dense foods that require refrigeration, tend to spoil faster, and require time and effort to cook or prepare. Fast-food restaurants are ubiquitous throughout North America and not only offer the consumer convenience, but may be an effective way for families to save money. The high density of fast-food restaurants, convenience stores, bars, and vending machines and the aggressive mass marketing of energy-dense foods promote consumption of an energy-dense diet. Many areas lack convenient access to supermarkets offering high-quality, low-energy-dense foods such as whole grains, vegetables, and fruits at competitive prices. Compared to wealthier neighborhoods, poorer neighborhoods have one-third as many supermarkets but more convenience stores, fast-food restaurants, and bars.[80] Is it "elitist" for nutritionists to encourage low- and middle-income families to consume healthier but more costly low-energy-dense foods that they may not be able to afford, have easy access to, or have time to prepare? Some experts in the field are examining whether the increasing disparities in income and wealth and the declining value of the minimum wage in North America contribute significantly to an obesigenic diet.[79]

Changes in Physical Activity

Of the three major components of 24-hour energy expenditure illustrated in Figure 12.1, energy expended through physical activity is the most highly variable and the one humans can most easily control. Physical activity energy expenditure includes movement from the performance of the routine activities of daily life and purposeful exercise as well as energy expended by maintaining posture, fidgeting, and spontaneous muscle contraction. Most studies indicate that obese children and adults are less physically active than their leaner counterparts. However, when obese persons engage in physical activity, they expend more energy than leaner persons performing the same activity. Overall, daily energy expenditure from physical activity by obese persons appears to be no different than that of leaner persons.[37] Because obese persons have a greater amount of weight to carry than do lean persons, their lean body mass is greater. Consequently, the obese have a greater resting energy expenditure compared to leaner persons.[37]

Table 12.8 Changes in Reported Energy Intakes in the U.S. Adult Population, 1971–1974 to 2005–2006

	1971–1974	2005–2006	Change
U.S. Adult Females	1542 kcal	1785 kcal	+243 kcal (16%)
U.S. Adult Males	2450 kcal	2638 kcal	+188 kcal (8%)

Source: Data from the National Center for Health Statistics.

Environmental Influences on Physical Activity Habits

Environmental changes occurring within the past several decades that have had an impact on the physical activity habits of North Americans include:

- The variety of electronic media has increased (television, Internet, video games, DVD, wireless communication devices, etc.), which has increased time spent in sedentary activities.
- Physical education programs have been markedly reduced in public schools.
- Many neighborhoods lack sidewalks for safe walking.
- The workplace has become increasingly automated.
- Household chores are assisted by labor-saving machinery.
- Walking or bicycling has been replaced by automobile travel for all but the shortest distances.

In the past several decades, a number of environmental changes (outlined in Box 12.7) have impacted the physical activity habits of North Americans by providing inducements to be sedentary and discouraging physical activity. Considerable attention is being focused on how body weight is affected by the amount of time spent viewing television, playing computer games, or using computers (collectively referred to as "screen time"). Research has consistently shown that in children and adults, BMI increases as the amount of screen time increases.[81-83] As screen time increases, children tend to eat fewer fruits and vegetables and consume more energy-dense snacks.[81] Screen time occupies children for hours in a sedentary activity and exposes them to aggressive marketing of energy-dense foods and beverages, often resulting in increased energy intake. Research supports the recommendation to limit screen time as an approach to preventing the development of obesity and promoting weight loss in children who are overweight.[81] The majority of scientific evidence suggests that the high prevalence of overweight in developed countries is more a function of excessive energy intake than of low activity level. However, increased physical activity is important for the long-term prevention of weight gain and management of healthy body weight.[81]

The choices an individual makes about energy expenditure have an important impact on his or her body weight. However, an individual's environment is an important factor influencing that person's behavior, either by facilitating or impeding regular physical activity.[76, 80] In addition to encouraging energy consumption, the current obesigenic environment in North America and throughout the developed world discourages expenditure of energy, and is widely regarded as a causal factor in the increased prevalence of obesity in North America in the past several decades.[76, 80, 84] For example, most areas in the United States and Canada have been designed to be accessed by motor vehicles with little thought, if any, given to the needs of pedestrians or bicyclists. For many people, physical activity is impeded by an environment lacking convenient, pleasant, and safe areas for walking, bicycling, or other forms of recreation. Urban sprawl and lack of public transportation force most to resort to the automobile for commuting to work and school, and for shopping for food and other items. There is a growing awareness among researchers and public health experts that successfully addressing the problem of overweight and obesity will require identifying feasible ways to cope with and to change the current environment.[76, 80] A first step in this process would be to give people strategies to better manage within the current environment and to better resist the many factors promoting weight gain. A second, long-term approach would be to build an environment that is more conducive to the adoption and maintenance of healthy dietary and exercise habits.[76]

Treatment of Overweight and Obesity

The medical treatment of overweight and obesity is a two-step process: assessment and management.[2] Assessment (see the following "Nutrition Assessment" section) includes determining the degree of overweight and obesity by calculating BMI, measuring waist circumference, checking for the presence of life-threatening conditions often accompanying obesity, evaluating dietary and exercise habits, and determining the patient's readiness to lose weight. Management includes applying therapies to lose weight and maintain weight loss, and applying measures to control other disease risk factors.[1, 2] An **algorithm** for the treatment of overweight and obesity developed by the National Institutes of Health is shown in Figure 12.17.

General Guidelines for Medical Management

Management of overweight and obesity involves the appropriate use of the recommended therapies for initial and long-term successful weight loss, and control of the factors known to increase risk of morbidity and mortality in overweight and obese persons.[2] Recommended therapies for overweight and obesity include diet, physical activity, and behavioral therapy. For some patients, pharmacologic treatment and bariatric surgery are indicated, as shown in Table 12.9.

Pharmacologic Treatment

Drug therapy can be useful as an adjunct to diet, physical activity, and behavior therapy in patients whose BMI is ≥30 kg/m² or in patients whose BMI is ≥27 kg/m² and who have obesity-related risk factors or diseases.[2] The modest benefits of drug treatment are offset by its cost and side effects, and rebound weight gain following cessation of drug use.[14] The U.S. Food and Drug Administration has approved two drugs for long-term use for weight loss and the maintenance of weight loss: sibutramine and orlistat. Several have been approved for short-term treatment (6–12 weeks), including mazindol, diethylpropion, benzphetamine, phendimetrazine, and phentermine. Phentermine, the drug most commonly used for short-term treatment, is an amphetamine-like drug with a low addictive potential that acts on the hypothalamus to suppress appetite.[14]

Sibutramine, only available by prescription and marketed under the trade name Meridia, is a serotonin-norepinephrine reuptake

Figure 12.17 An Algorithm for the Treatment of Overweight and Obesity Developed by the National Institutes of Health

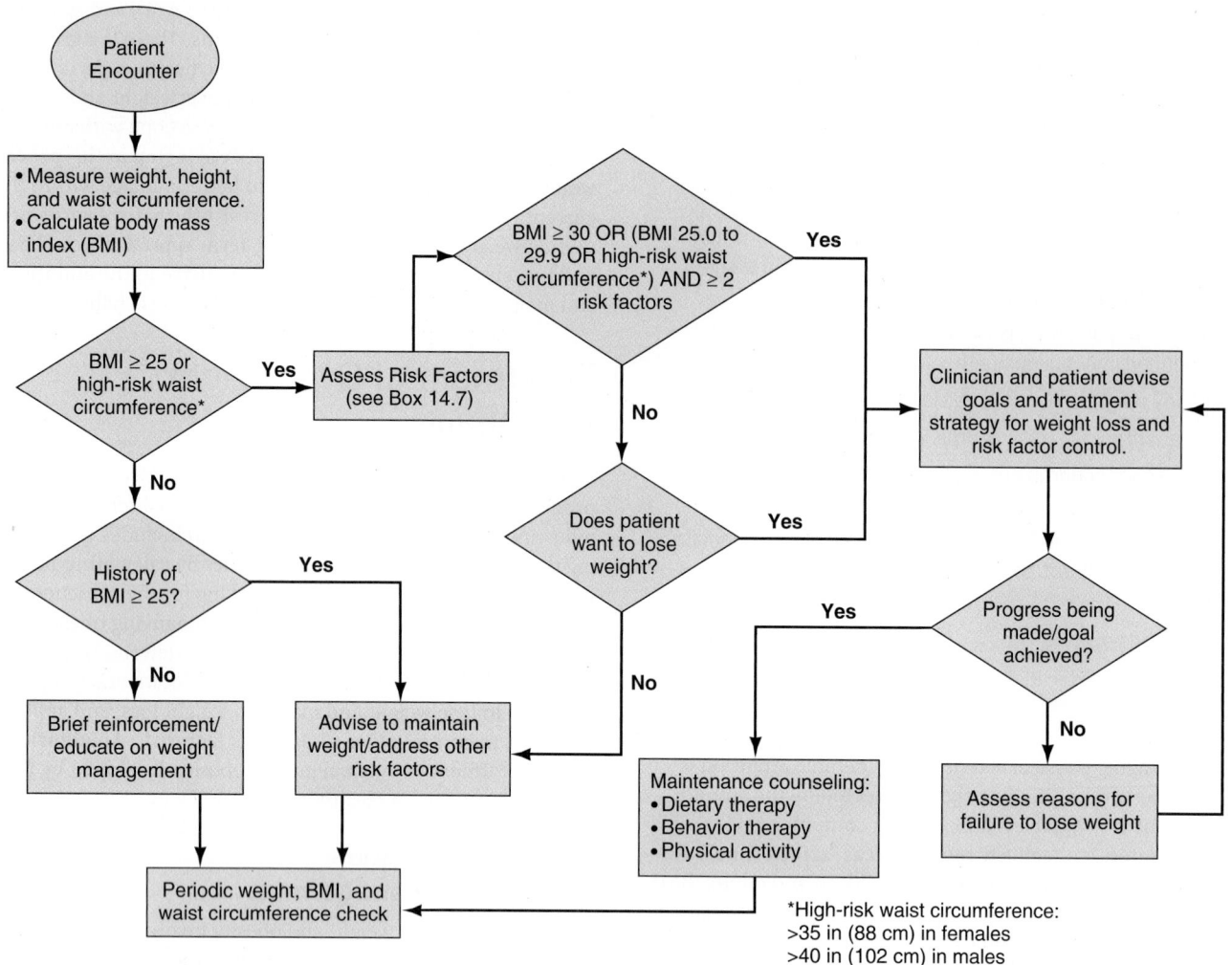

Source: Adapted from National Heart, Lung, and Blood Institute. *The practical guide: identification, evaluation, and treatment of overweight and obesity in adults.* Bethesda (MD): U.S. Department of Health and Human Services, National Institutes of Health; 2000.

Table 12.9 A Guide to Selecting Treatment of Overweight and Obesity

Treatment	BMI Category (kg/m²)				
	25.0–26.7	27.0–29.9	30.0–34.9	35.0–39.9	≥ 40.0
Diet, physical activity, and behavioral therapy	with comorbidities	with comorbidities	+	+	+
Pharmacotherapy		with comorbidities	+	+	+
Surgery			with comorbidities	with comorbidities	with comorbidities

- Prevention of weight gain with lifestyle therapy is indicated in any patient with a BMI ≥ 25 kg/m², even without comorbidities, while weight loss is not necessarily recommended for those with a BMI of 25–29.9 kg/m² or a high waist circumference, unless they have two or more comorbidities.
- Combined therapy with a low-kcalorie diet (LCD), increased physical activity, and behavior therapy provides the most successful intervention for weight loss and weight maintenance.
- Consider pharmacotherapy only if a patient has not lost 1 pound per week after 6 months of combined lifestyle therapy. The + represents the use of indicated treatment regardless of comorbidities.

Source: National Heart, Lung, and Blood Institute. *The practical guide: identification, evaluation, and treatment of overweight and obesity in adults.* Bethesda (MD): U.S. Department of Health and Human Services, National Institutes of Health; 2000.

inhibitor that acts on receptors in the hypothalamus to suppress appetite. Because it increases heart rate (4–5 beats/minute) and blood pressure (1–2 mm Hg), it should not be used by patients with CVD or uncontrolled hypertension.[9, 85] It can result in a 7% reduction in body weight after one year, but when used in conjunction with an intensive program of diet, exercise, and behavior therapy, losses can increase to 10% to 15% of body weight.[85] Orlistat, marketed under the trade names Xenical (only available by prescription) and Alli (available over-the-counter beginning in 2007 at half the dosage of the prescription version), inhibits the action of gastrointestinal lipase and reduces the digestion of triglyceride.[9] Undigested triglyceride is not absorbed, does not provide energy to the body, and is eliminated in the feces. Each capsule of Xenical contains 120 mg of orlistat and each capsule of Alli contains 60 mg. Both are intended to be taken three times a day with reduced-energy, low-fat meals (providing less than 30% of energy from fat). When taken as directed, Xenical results in a 25% to 30% reduction in fat digestion and the loss of 150 to 180 kcal/day. If orlistat is taken with high-fat meals, there is an increased risk of side effects that include fecal staining of underwear, flatus with fecal discharge, fecal urgency, fatty/oily stools, increased urgency and frequency of defecation, and fecal incontinence. These side effects are generally less severe

among those taking the lower-dose, over-the-counter version of orlistat than among those taking the higher-dose prescription version. After discontinuation of the drug, fecal fat levels return to normal within 48 to 72 hours.[9] Because orlistat inhibits the absorption of fat-soluble vitamins, patients are advised to take a multivitamin supplement to offset losses of vitamins A, D, E, and K.[9, 14, 85] Clinical trials of diet plus orlistat resulted in a 10% weight loss after 1 year compared to a 6% loss from diet alone. A 4-year trial resulted in a 6.4% weight loss with a significant reduction in the development of type 2 diabetes in persons who had impaired glucose tolerance.[85]

Surgery

Weight loss or bariatric surgery is reserved for patients who have failed to lose weight by other, less invasive means and who have clinically severe obesity: BMI ≥40 kg/m² or BMI ≥35 kg/m² in a patient with a high-risk condition such as sleep apnea, cardiovascular disease, or diabetes mellitus.[9, 85–87]

There are four different types of bariatric surgical procedures in common use today, as illustrated in Figure 12.18: adjustable gastric banding, vertical sleeve gastrectomy, Roux-en-Y gastric bypass, and biliopancreatic diversion with duodenal switch.

Figure 12.18 The Four Most Commonly Used Surgical Procedures for Weight Loss

(a) Adjustable gastric banding; (b) vertical sleeve gastrectomy; (c) Roux-en-Y gastric bypass; (d) biliopancreatic diversion with duodenal switch

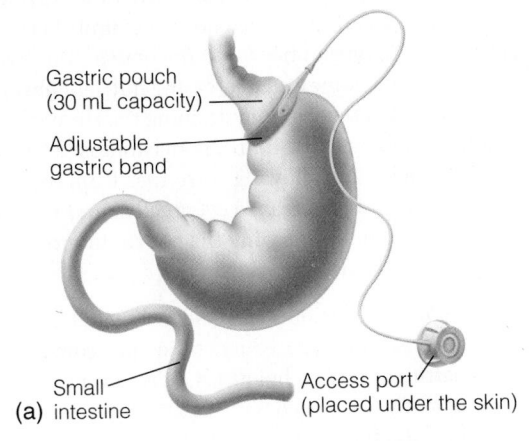

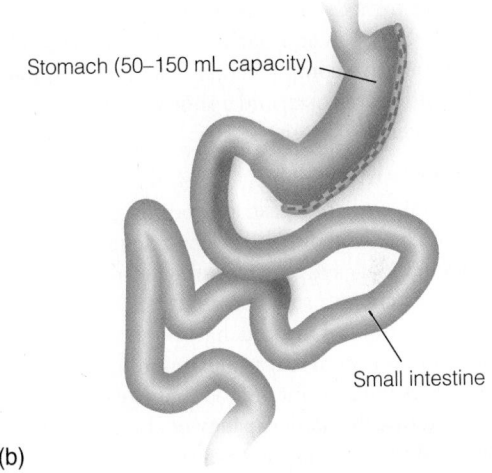

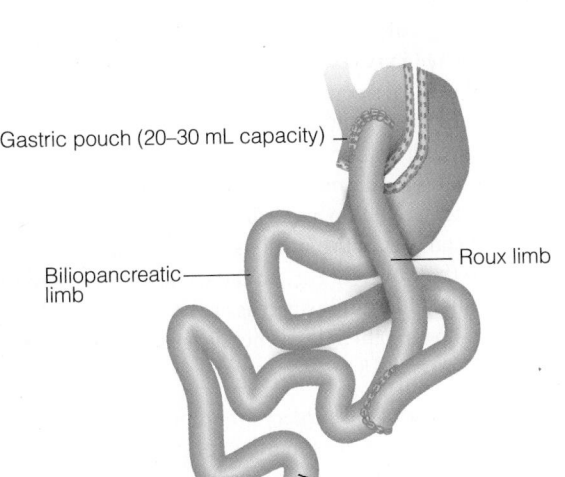

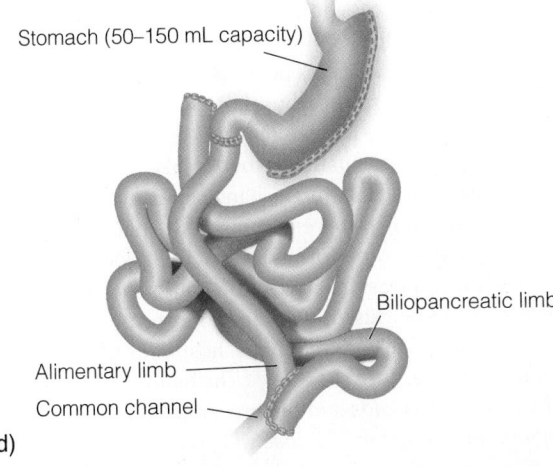

These procedures fall into one of two categories: restrictive and restrictive-malabsorptive.[9, 87] The restrictive procedures are adjustable gastric banding and vertical sleeve gastrectomy. These restrictive procedures create a small gastric reservoir (or "pouch"), markedly limiting the volume of food and beverage that the stomach can hold and narrowing the outlet of the reservoir, which delays gastric emptying. Consequently, the patient feels "full" sooner and the appetite is decreased. The restrictive-malabsorptive procedures—Roux-en-Y gastric bypass and biliopancreatic diversion with duodenal switch—reduce the volume of food and beverage that the stomach can hold in addition to interfering with food digestion and reducing nutrient absorption by bypassing part of the stomach and/or small intestine.[14, 85–87] Because the restrictive-malabsorptive procedures result in greater weight loss than the restrictive ones, they are regarded as better suited for very obese patients. The two most common bariatric procedures are Roux-en-Y gastric bypass and adjustable gastric banding, which account for approximately 90% of all bariatric surgeries performed worldwide.[88] Adjustable gastric banding has replaced laparoscopic vertical banded gastroplasty (not shown in Figure 12.18), which once was commonly performed but is rarely used today.[9]

Bariatric surgery can be performed using the "open" approach requiring relatively large incisions into the abdominal cavity or using the less invasive laparoscopic approach in which sophisticated surgical instruments (including a light source and video camera) are inserted into the abdominal cavity though much smaller (0.5 to 1.5 cm) incisions. Most bariatric surgery is performed using the laparoscopic approach which, compared to the open approach, results in less tissue damage, fewer postoperative complications, a shorter hospitalization, reduced health-care costs, and a shorter period of postoperative convalescence.[9, 87]

In adjustable gastric banding (AGB) (also known as laparoscopic adjustable silicone gastric banding or the "lap band"), an inflatable silicone ring or band is laparoscopically introduced into the abdominal cavity and secured around the upper part of the stomach to create a small pouch with a narrow opening or stoma at the bottom of the pouch through which food passes into the rest of the stomach, as illustrated in Figure 12.18. The pouch restricts the amount of food that can be consumed at one time and initially has a capacity of about 30 mL, with the potential to stretch somewhat to eventually hold up to 90 mL. The band is connected by a tube to an access port that is placed under the skin of the abdomen and is inflated when saline is injected into the access port using a syringe. The inflation of the band can be adjusted by the amount of saline injected into or removed from the access port. As the band is inflated, the stoma becomes narrowed, thus delaying the emptying of the pouch and giving the patient a greater sense of fullness and further restricting the patient's food intake. If it is necessary to permit greater food intake, the band can be deflated to enlarge the stoma, allowing faster emptying of the pouch.[14, 89, 90] AGB is the most commonly performed restrictive procedure.[9] It is also the simplest and least invasive of the four procedures illustrated in Figure 12.18 because it does not require any stapling or cutting of the stomach. Consequently, hospitalization and postoperative recovery are shorter.[9, 86] The band can be adjusted to suit the patient's needs and the procedure is fully reversible.[89–91] Patients receiving the AGB lose weight at a slower rate com-

pared to patients receiving the Roux-en-Y gastric bypass, but over time the total amount of weight lost is comparable.[90]

In the vertical sleeve gastrectomy (VSG), up to 85% of the stomach is surgically removed, leaving a narrow, tubular, banana-shaped portion of the stomach (a "gastric sleeve") between the esophagus and the duodenum, restricting the remaining stomach's holding capacity to 50 to 150 mL while preserving its functions, leaving the pylorus (the stomach's outlet) intact, and resulting in minimal, if any, nutrient malabsorption.[9, 86, 92] The surgery is typically performed laparoscopically using an instrument that inserts two adjacent rows of staples through both walls of the stomach and then cuts through both walls of the stomach between the lines of staples, eventually separating the stomach into two sections. Once the two sections are separated, the section detached from the esophagus and duodenum is removed through a small abdominal incision. Because the detached section of the stomach is primarily responsible for the production of the hormone ghrelin, weight loss is further enhanced by the markedly lower serum levels of ghrelin seen in patients who have had the VSG.[92]

The Roux-en-Y gastric bypass (RYGB) is the best accepted and most commonly performed restrictive-malabsorptive procedure in North America.[9] The procedure creates a small (20–30 mL) pouch at the top of the stomach, restricting food intake and quickly inducing satiety, and bypasses the remainder of the stomach, the duodenum, and the first part of the jejunum, reducing food digestion and nutrient absorption. The jejunum is cut and the distal end ("roux limb") is surgically connected (or anastomosed) to the pouch, thus bypassing part of the small intestine and decreasing nutrient absorption. The proximal end of the jejunum draining the stomach and duodenum ("biliopancreatic limb") is anastomosed to a lower segment of the jejunum, allowing secretions from the stomach, liver, gallbladder, and pancreas to eventually enter the jejunum and mix with the contents leaving the pouch. In most cases the procedure is done laparoscopically.

The biliopancreatic diversion with duodenal switch (BPDDS), a restrictive-malabsorptive procedure, is the most complicated of the four bariatric surgical techniques. Of the four procedures, it is the least frequently performed but results in the greatest amount of weight lost.[93] Because of its complexity it is typically reserved for very obese patients (e.g., those whose BMI ≥50.0 kg/m^2) who are at high risk for surgical complications due to severe obesity and/or other medical conditions linked to obesity. The procedure is accomplished in two stages: the first is a laparoscopically-performed vertical sleeve gastrectomy, resulting in the partial attainment of the weight loss goal and reducing the risk of complications in preparation for the second stage of the procedure, which re-routes food from the stomach past most of the small intestine (the duodenum and jejunum), resulting in further weight loss. The distal part of the small intestine (the ileum) is surgically attached to the stomach. Also in the second stage, secretions from the liver, gallbladder, and pancreas are rerouted (via a biliopancreatic limb), allowing them to eventually enter the small intestine, as shown in Figure 12.18.[9, 86, 87, 93]

The results of bariatric surgery are impressive, making it the most effective weight-loss therapy for patients with clinically severe obesity.[9] Most patients lose 30%–35% of total body weight in 1 to 2 years postoperatively with successful maintenance of weight loss in about 60% of patients after a decade or more.[9, 85] Other benefits of the surgery include dramatic improvements in diabetes, sleep apnea, hypertension, and CVD risk factors.[14, 85, 88] For example, in a study of nearly 8,000 patients who underwent RYGB surgery, there was a 40% reduction in mortality from such causes as coronary heart disease, diabetes, and cancer compared to the control group.[94]

Postoperative complications include pulmonary embolism, anastomotic leaks, injury to the spleen, wound infection, and, rarely, death.[9, 85, 95] However, a recent study of nearly 5,000 patients who underwent bariatric surgery showed that when performed by highly-skilled surgeons in high-volume centers, the risks of RYGB and the AGB were no greater than gallbladder or hip-replacement surgery.[88, 95] Among these patients, the overall risk of death was 0.3% and 4.3% experienced at least one major complication; however, these rates were greater in patients with very high BMI and who had a history of obesity-related complications.[95] Careful preoperative screening and education will ensure that candidates for surgery meet the BMI requirements, are free of major psychopathology, have the physical and emotional stamina to tolerate the procedure, are willing to make the necessary dietary and lifestyle changes, and are committed to long-term follow-up. Also important are long-term follow-up and the availability of individual or group counseling, as needed.[14, 85]

In general, complications are more likely to result from the restrictive-malabsorptive procedures than from the restrictive procedures, and are least often seen in patients undergoing AGB. Some patients may experience nausea and vomiting when too much food is consumed at one time and others may experience "dumping syndrome," which is characterized by nausea, flushing, bloating, and diarrhea after eating a food or beverage that is high in refined carbohydrates. For patients undergoing AGB and VSG, there are no intestinal malabsorption problems, only a reduction in gastric volume and rate of gastric emptying. In these two procedures, nutrient deficiencies are uncommon unless the diet is unbalanced and has a low nutrient density.[9] Nutrient deficiencies are more common following the restrictive-malabsorptive procedures (RYGB and BPDDS), and common deficiencies include the fat-soluble vitamins (A, D, E, K), vitamin B_{12}, folate, iron, and calcium. Counseling, monitoring, and life-long supplementation with these nutrients are required in these patients.[9, 85]

These same surgical procedures are used for other gastric diagnoses. These procedures and nutrition therapy to prevent dumping are discussed in Chapter 14.

Nutrition Therapy for Overweight and Obesity

Nutrition Assessment

Anthropometric Measurements
Determining the degree of overweight or obesity is based on BMI calculated from an accurate measurement of the patient's weight and height. Clinical judgment must be used in interpreting the BMI of persons who are very muscular, have lost significant amounts of lean body mass, are short, or have edema or ascites.[2] Waist circumference is used as an index of abdominal adiposity, and is interpreted using the classifications shown in Table 12.5. In

patients with a BMI ≥35 kg/m², measuring waist circumference is not necessary, because it does not materially contribute to disease risk classification. Table 12.6 incorporates BMI and waist circumference to arrive at a disease risk relative to normal weight and low-risk waist circumference for patients having a BMI <35 kg/m². BMI and waist circumference are used to initially assess the degree of obesity and for monitoring the patient's response to treatment.[2, 9, 85]

Client History
Patients should be evaluated for the presence of diseases that place them at high risk of morbidity and mortality and that require aggressive treatment. These include established coronary heart disease, the presence of other atherosclerotic diseases (peripheral arterial diseases, abdominal aortic aneurysm, and symptomatic carotid artery disease), type 2 diabetes, impaired glucose tolerance, and sleep apnea.[2] In overweight and obese persons, excess weight is associated with elevated blood pressure, blood glucose, serum levels of total cholesterol, LDL-cholesterol, and triglycerides, and reduced levels of HDL-cholesterol. Patients who have three or more of the cardiovascular disease risk factors listed in Box 12.8 are at high risk for obesity-related disorders and will likely require intensive treatment of dyslipidemia and/or management of hypertension, as discussed in Chapter 13. Obese patients should be evaluated for the presence of metabolic syndrome as discussed in Chapter 13.

Food/Nutrition-Related History
The dietary and physical activity habits of patients are important considerations. Approaches for evaluating diet are outlined in Chapter 3. Assessment of physical activity involves asking patients how much time they spend sleeping, sitting, watching television, driving or riding in motorized vehicles, walking, and standing, and how often they climb stairs, engage in work in and around the home, et cetera.[85] Patients should also be queried about their formal exercise habits such as walking, jogging, cycling, swimming, and participating in other types of aerobic activities and weight training. Pedometers can be used to count the number of steps walked daily and accelerometers can be used to record the intensity and duration of body movement.[85]

Assessing a patient's readiness to lose weight and identifying and addressing potential barriers to that patient's ability to maintain long-term behavior change are important for understanding the patient's needs and achieving successful weight loss.[2, 85] Factors associated with successful long-term weight management include a high initial BMI and resting metabolic rate, positive coping skills, and self-efficacy (a patient's belief that he or she can perform the behaviors necessary for weight management). Depression, anxiety, and binge eating tend to be associated with poor success at weight management.[2] A brief behavioral assessment is shown in Box 12.9. Table 12.10 summarizes nutrition assessment of overweight and obese patients. Note that patients who are pregnant, lactating, or have anorexia nervosa, bulimia nervosa, a serious uncontrolled psychiatric illness such as major depression, or active substance abuse should be excluded from weight loss therapy.[2]

Nutrition Diagnosis

The following are commonly used nutrition diagnoses among patients who are overweight or obese: excessive fat intake, food

BOX 12.8 **CLINICAL APPLICATIONS**

Cardiovascular Risk Factors Placing Patients at a High Priority for Treatment of Overweight and Obesity*

- Cigarette smoking.
- Hypertension (systolic blood pressure of ≥140 mm Hg or diastolic blood pressure >90 mm Hg) or current use of antihypertensive agents.
- Increased low-density lipoprotein (LDL) cholesterol (serum concentration >160 mg/dL). A borderline high-risk LDL-cholesterol (130 to 159 mg/dL) plus two or more other risk factors also confers high risk.
- Decreased high-density lipoprotein (HDL) cholesterol (serum concentration <35 mg/dL).

- Impaired fasting glucose (IFG) (fasting plasma glucose between 110 and 125 mg/dL). IFG is considered by many authorities to be an independent risk factor for cardiovascular (macrovascular) disease, thus justifying its inclusion among risk factors contributing to high absolute risk. IFG is well established as a risk factor for type 2 diabetes.
- Family history of premature CHD (myocardial infarction or sudden death experienced by the father or other male first-degree relative

≤55 years of age, or experienced by the mother or other female first-degree relative ≤65 years of age).
- Age ≥45 years for men or age ≥55 years for women (or postmenopausal).

*Patients with three or more of these risk factors are at high risk for obesity-related disorders and may be candidates for intensive treatment of dyslipidemia and/or management of hypertension.

Source: National Heart, Lung, and Blood Institute. The practical guide: identification, evaluation, and treatment of overweight and obesity in adults. Bethesda (MD): U.S. Department of Health and Human Services, National Institutes of Health; 2000.

---

and nutrition-related knowledge deficit, disordered eating pattern, undesirable food choices, overweight/obesity, involuntary weight gain, and physical inactivity.

Nutrition Intervention

A minimum goal is to avoid additional weight gain with age once a person reaches his or her healthy, adult weight. Those who are at their normal or healthy weight (BMI 18.5 to 24.9 kg/m²) should be counseled about effective dietary and physical activity habits that can prevent further weight gain. Those who are overweight or obese should set as their goal an initial weight loss of about 10% of their body weight over a six-month period at a rate of about 1 to 2 pounds lost per week.[2, 85]

Successful weight maintenance is defined as a regain of weight that is less than 6.6 pounds (3 kg) in 2 years and a sustained reduction in waist circumference of at least 1.6 inches (4 cm).[2] Success in weight maintenance is dependent on permanent adoption of a low-energy-dense diet and regular physical activity, and will be enhanced by long-term practitioner monitoring and encouragement through regular clinic visits, group meetings, postal mailings, telephone calls, and e-mails.

The recommended approach is for patients to reduce their energy intake by 500 to 1000 kcal/day. Theoretically, this should result in a 26- to 52-pound weight loss after 6 months, but a more typical loss is between 20 and 25 pounds after six months. Continued weight loss after six months is difficult for most patients, in large part because 24-hour energy expenditure declines in response to restricted energy intake and in response to the losses of metabolically active lean body mass that invariably accompany weight loss.[2] Resting energy expenditure begins to decline within days of restricting energy intake, and by 3 to 4 weeks will fall by as much as 25% to 35% below normal in response to total fasting.[96] With loss of body weight, there is loss of both fat and fat-free tissue. A 10% decrease in body weight results in a 15% reduction in 24-hour energy expenditure.[97] These compensatory reductions in energy expenditure

make it difficult to maintain the weight loss. After six months of weight loss, achieving additional weight loss beyond the initial 10% requires further energy restriction and increased energy expenditure, which many patients find difficult to maintain over a long period of time.[2]

The cornerstone of weight reduction therapy is an individually planned low-kcal diet (LCD) that reduces energy intake by 500 to 1000 kcal/day and achieves a slow but progressive weight loss of 1 to 2 pounds per week.[2, 14, 85] The key features of this approach, as recommended by the National Institutes of Health, are shown in Table 12.11. Although greater energy deficits may be useful during the period of active weight loss to provide needed motivation to some patients, a very-low-kcal diet (VLCD) providing less than 800 kcal per day should not be used for routine weight loss. VLCDs require special monitoring and nutritional supplementation and should be used only in very limited circumstances by specialized practitioners experienced in their use.[2] Clinical trials indicate that VLCDs are no more effective in achieving weight loss after 1 year than are LCDs.[2, 85] In addition to reducing energy intake, the diet should be modified to minimize CVD risk factors by following the National Cholesterol Education Program's Therapeutic Lifestyle Change diet, which is discussed in Chapter 13.[98] A meal plan providing 1000 to 1200 kcal/day is generally recommended for most women. A meal plan providing 1200 to 1600 kcal/day is generally recommended for most men and may be suited for women who exercise more or weigh 165 lb or more. A greater reduction in energy intake may be necessary for patients failing to respond to these energy levels, whereas patients complaining of hunger or having difficulty adhering to these recommendations may need a somewhat more liberal intake.[2] Sample, one-day meal plans providing 1200 kcal and 1600 kcal are provided in Table 12.12 and examples of common chores and sporting activities that provide moderate amounts of physical activity are given in Table 12.13.

A growing body of scientific research is demonstrating the value of consuming low-energy-dense foods (i.e., foods low in kcal

BOX 12.9 CLINICAL APPLICATIONS

A Brief Behavioral Assessment

Clinical experience suggests that health care practitioners briefly consider the following questions when assessing an obese individual's readiness for weight loss.

"Has the individual sought weight loss on his or her own initiative?" Weight loss efforts are unlikely to be successful if patients feel that they have been forced into treatment by family members, their employer, or their physician. Before initiating treatment, health care practitioners should determine whether patients recognize the need for and benefits of weight reduction and want to lose weight.

"What events have led the patient to seek weight loss now?" Responses to this question will provide information about the patient's weight loss motivation and goals. In most cases, individuals have been obese for many years. Something has happened to make them seek weight loss. The motivator differs from person to person.

"What are the patient's stress level and mood?" There may not be a perfect time to lose weight, but some times are better than others. Individuals who report higher than usual stress levels with work, family life, or financial problems may not be able to focus on weight control. In such cases, treatment may be delayed until the stressor passes, thus increasing the chances of success. Briefly assess the patient's mood to rule out major depression or other complications. Reports of poor sleep, a low mood, or lack of pleasure in daily activities can be followed up to determine whether intervention is needed; it is usually best to treat the mood disorder before undertaking weight reduction.

"Does the individual have an eating disorder, in addition to obesity?" Approximately 20% to 30% of obese individuals who seek weight reduction at university clinics suffer from binge eating. This involves eating an unusually large amount of food and experiencing loss of control while overeating. Binge eaters are distressed by their overeating, which differentiates them from persons who report that they "just enjoy eating and eat too much." Ask patients which meals they typically eat and the times of consumption. Binge eaters usually do not have a regular meal plan; instead, they snack throughout the day. Although some of these individuals respond well to weight reduction therapy, the greater the patient's distress or depression, or the more chaotic the eating pattern, the more likely the need for psychological or nutritional counseling.

"Does the individual understand the requirements of treatment and believe that he or she can fulfill them?" Practitioner and patient together should select a course of treatment and identify the changes in eating and activity habits that the patient wishes to make. It is important to select activities that patients believe they can perform successfully. Patients should feel that they have the time, desire, and skills to adhere to a program that you have planned together.

"How much weight does the patient expect to lose? What other benefits does he or she anticipate?" Obese individuals typically want to lose 2 to 3 times the 8% to 15% often observed and are disappointed when they do not. Practitioners must help patients understand that modest weight losses frequently improve health complications of obesity. Progress should then be evaluated by achievement of these goals, which may include sleeping better, having more energy, reducing pain, and pursuing new hobbies or rediscovering old ones, particularly when weight loss slows and eventually stops.

Source: National Heart, Lung, and Blood Institute. The practical guide: identification, evaluation, and treatment of overweight and obesity in adults. Bethesda (MD): U.S. Department of Health and Human Services, National Institutes of Health; 2000.

relative to weight) as an approach to maintaining satiety while controlling energy intake and promoting healthier weights.[77, 78] Low-energy-dense foods are relatively low in energy while having a relatively high weight, and include such foods as high-water vegetables and fruits, cooked whole grains, and broth-based soups. A successful strategy for reducing energy intake while maintaining satiety is providing as a first course of a meal satisfying portions of low-energy-dense foods such as vegetable salads or broth-based soups. Larger portion sizes of low-energy-dense foods have the advantage of maintaining satiety with a lower total energy intake than when high-energy-dense foods are consumed. Greater use of cooked vegetables as side dishes can be an effective way of decreasing the energy density of a meal. An additional strategy to reduce energy density is to prepare the main course of a meal using ingredients that reduce its fat content and increase its water content. Fat can be reduced by using smaller amounts of high-fat ingredients such as meats, dairy products, and oils, or by using leaner cuts of meat and/or reduced-fat dairy products. By using more vegetables in the preparation of a dish such as a pasta salad or casserole, one can increase the water content of that dish while decreasing the energy density.[77, 78]

Among the most controversial issues related to dietary therapy for body weight management is whether overweight people are more successful at long-term weight management when following a diet that emphasizes a specific macronutrient composition. For example, is a low-carbohydrate, high-protein diet superior to a more balanced hypocaloric diet such as the one outlined in Table 12.11? While several short-term clinical trials (three to six months in length) have shown that low-carbohydrate, high-protein diets result in greater weight loss than diets providing carbohydrate in the range of 50% to 60% of kcal,[85, 99, 100] other clinical trials lasting one to two years have not shown that low-carbohydrate, high-protein diets are superior to those that are higher in carbohydrate.[101–103] A one-year randomized trial of four popular weight-loss programs (the Atkins, Ornish, Weight Watchers, and Zone diets) having widely different

Table 12.10 Nutrition Assessment for Overweight/Obese Patients

Client History	• Education—primary language
	• Ethnic, cultural, and religious influences
	Patient/client/family medical/health history
	• Comorbidities that may indicate risk of metabolic syndrome: hypertension, diabetes mellitus/impaired glucose tolerance, dyslipidemia
	Treatments/therapy/alternative medicine
	• Medications (especially medications that might cause weight gain: antidepressants, lithium, beta-blockers, corticosteroids)
	Social history
	• Socioeconomic status/food security
	• Support systems
Food-/Nutrition-Related History	Food and nutrient intake
	• 24-hour recall, diet history, food frequency; focus on portion sizes, meals eaten away from home, food preparation methods, sources of high energy density (fat, concentrated sugar content)
	• Use of alcohol, vitamin and mineral supplements
	• Previous methods used for weight loss if applicable
	• Food allergies, preferences, or intolerances
	Medication and herbal supplement use
	• Herbal or other type of supplements
	Knowledge/beliefs/attitudes
	• Previous nutrition education or nutrition therapy
	Mealtime behaviors
	• Previous food restrictions
	• Eating pattern
	Factors affecting access to food and food/nutrition-related supplies
	• Ability to consistently purchase adequate amounts of food on a daily basis
	• Ability to feed self
	• Ability to cook and prepare meals
Anthropometrics	• Height
	• Current weight
	• Weight history: highest adult weight; usual body weight
	• Reference weight/BMI
	• Waist and hip circumferences
Biochemical Data, Medical Tests and Procedures	• Laboratory measures for comorbidities/assessment of metabolic syndrome: serum glucose, HgBA₁C, total cholesterol, LDL, HDL, triglycerides
	• Blood pressure
	• Visceral protein assessment: standard
	• Hematological assessment: standard

Table 12.11 Low-Calorie Diet (LCD) Recommended by the National Institutes of Health

Nutrient	Recommended Intake
Calories[1]	Approximately 500 to 1000 kcal/day reduction from usual intake
Total fat[2]	30% or less of total calories
Saturated fatty acids[3]	8%–10% of total calories
Monounsaturated fatty acids	Up to 15% of total calories
Polyunsaturated fatty acids	Up to 10% of total calories
Cholesterol[3]	<300 mg/day
Protein[4]	Approximately 15% of total calories
Carbohydrate[5]	55% or more of total calories
Sodium chloride	No more than 100 mmol/day (approximately 2.4 g of sodium or approximately 6 g of sodium chloride)
Calcium[6]	1000 to 1500 mg/day
Fiber[5]	20 to 30 g/day

[1]A reduction in calories of 500 to 1000 kcal/day will help achieve a weight loss of 1 to 2 pounds/week. Alcohol provides unneeded calories and displaces more nutritious foods. Alcohol consumption not only increases the number of calories in a diet but has been associated with obesity in epidemiologic studies as well as in experimental studies. The impact of alcohol calories on a person's overall caloric intake needs to be assessed and appropriately controlled.

[2]Fat-modified foods may provide a helpful strategy for lowering total fat intake but will only be effective if they are also low in calories and if there is no compensation by calories from other foods.

[3]Patients with high blood cholesterol levels may need to use the National Cholesterol Education Program's Therapeutic Lifestyle Changes (TLC) diet to achieve further reductions in LDL-cholesterol levels; in the TLC diet, saturated fats are reduced to less than 7% of total calories, and cholesterol levels to less than 200 mg/day.

[4]Protein should be derived from plant sources and lean sources of animal protein.

[5]Complex carbohydrates from different vegetables, fruits, and whole grains are good sources of vitamins, minerals, and fiber. A diet rich in soluble fiber, including oat bran, legumes, barley, and most fruits and vegetables, may be effective in reducing blood cholesterol levels. A diet high in all types of fiber may also aid in weight management by promoting satiety at lower levels of calorie and fat intake. Some authorities recommend 20 to 30 grams of fiber daily, with an upper limit of 35 grams.

[6]During weight loss, attention should be given to maintaining an adequate intake of vitamins and minerals. Maintenance of the recommended calcium intake of 1000 to 1500 mg/day is especially important for women who may be at risk of osteoporosis.

Source: Adapted from National Heart, Lung, and Blood Institute. The practical guide: identification, evaluation, and treatment of overweight and obesity in adults. Bethesda (MD): U.S. Department of Health and Human Services, National Institutes of Health; 2000.

macronutrient compositions showed that each of the four diets modestly reduced body weight, waist circumference, and several CVD risk factors regardless of the diet that was followed.[103] A two-year randomized trial of four different diets varying in macronutrient composition resulted in modest yet clinically significant improvements in body weight, waist circumference, lipid-related risk factors, and fasting insulin levels regardless of which of the four different diets the subjects followed.[101] In these and other studies, the key determinant of successful weight loss was personal commitment and adherence to the diet, not what particular diet was followed.[104]

A growing body of scientific evidence suggests that long-term improvements in body weight, waist circumference, and CVD risk factors appear to be determined more by the total number of kcal consumed and expended than by the proportion of macronutrients in the diet.[85, 99, 103–105] However, kcal-restricted diets with a modest increase in the proportion of kcal from mono-

Table 12.12 Sample Reduced-Energy Meal Plan for One Day Providing 1200 and 1600 kilocalories

Food Item	Amount for 1,200 kcal Plan	Amount for 1,600 kcal Plan
Breakfast		
Whole-wheat bread	1 medium slice	1 medium slice
Jelly, regular	2 tsp	2 tsp
Cereal, shredded wheat	½ cup	1 cup
Milk, 1%	1 cup	1 cup
Orange juice, unsweetened	¾ cup	¾ cup
Coffee, regular	1 cup	1 cup with 1 fl oz of 1% milk
Lunch		
Roast beef sandwich:		
Whole-wheat bread	2 medium slices	2 medium slices
Lean roast beef, unseasoned	2 oz	2 oz
American cheese, low fat and low sodium	—	1 slice, ¾ oz
Lettuce	1 leaf	1 leaf
Tomato	3 medium slices	3 medium slices
Mayonnaise, low calorie	1 tsp	2 tsp
Apple	1 medium	1 medium
Water	1 cup	1 cup
Dinner		
Salmon	2 oz edible	3 oz edible
Vegetable oil	1½ tsp	1½ tsp
Baked potato	¾ medium	¾ medium
Margarine	1 tsp	1 tsp
Green beans, seasoned with margarine	½ cup	½ cup
Carrots	½ cup	—
Carrots, seasoned, with margarine	—	½ cup
Dinner roll	1 small	1 medium
Ice milk	—	½ cup
Iced tea, unsweetened	1 cup	1 cup
Water	2 cups	2 cups
Snack		
Popcorn	2½ cup	2½ cup
Margarine	½ tsp	¾ tsp
Nutritional analysis of diets:		
Kilocalories	1247	1613
Total carbohydrate, percent of kilocalories	58	55
Total fat, percent of kilocalories	26	29
Sodium*, mg	1043	1341
Saturated fat, percent of kilocalories	7	8
Cholesterol, mg	96	142
Protein, percent of kilocalories	19	19

*No salt added in recipe preparation or as seasoning

Source: National Heart, Lung, and Blood Institute. Aim for a Healthy Weight. Bethesda (MD): U.S. Department of Health and Human Services, National Institutes of Health, 2005. http://www.nhlbi.nih.gov/health/public/heart/obesity/aim_hwt.pdf

Table 12.13 Examples of Common Chores and Sporting Activities That Provide Moderate Amounts of Physical Activity

	Common Chores	Sporting Activities
Less vigorous, more time ↑ ↓ More vigorous, less time	• Washing and waxing a car for 45 to 60 minutes • Washing windows or floors for 45 to 60 minutes • Gardening for 30 to 45 minutes • Wheeling self in wheelchair 30 to 40 minutes • Pushing a stroller 1½ miles in 30 minutes • Raking leaves for 30 minutes • Walking 2 miles in 30 minutes (15 minutes/mile) • Shoveling snow for 15 minutes • Stair-walking for 15 minutes	• Playing volleyball for 45 to 60 minutes • Playing touch football for 45 minutes • Walking 1¾ miles in 35 minutes (20 minutes/mile) • Basketball (shooting baskets) for 30 minutes • Bicycling 5 miles in 30 minutes • Dancing fast (social) for 30 minutes • Water aerobics for 30 minutes • Swimming laps for 20 minutes • Basketball (playing game) for 15 to 20 minutes • Bicycling 4 miles in 15 minutes • Jumping rope for 15 minutes • Running 1½ miles in 15 minutes (10 minutes/mile)

Source: National Heart, Lung, and Blood Institute. Aim for a Healthy Weight. Bethesda (MD): U.S. Department of Health and Human Services, National Institutes of Health, 2005. http://www.nhlbi.nih.gov/health/public/heart/obesity/aim_hwt.pdf

unsaturated fats and protein from plant products, poultry, and fish may increase satiety, facilitate weight loss, and improve CVD risk factors in some individuals.[85, 98, 99, 106] Any reduction in the proportion of kcal coming from carbohydrates should be accomplished by reducing intake of foods and beverages containing refined sugars and milled grains, not by sacrificing consumption of whole grains, legumes (dried beans and peas), vegetables, and fruits, which have a low energy density and are associated with reduced CVD risk.[107] Adherence to a diet is improved when the patient's food preferences and lifestyle are carefully assessed and modified incrementally. The practitioner and patient must collaboratively establish goals for modifying dietary and physical activity patterns, and the patient must see these modifications as desirable and achievable. In helping the patient be a better informed consumer, particular attention should be given to the topics listed in Box 12.10.[2]

Collaboration/Referral to Other Providers: Physical Activity Although physical activity is less important than an energy-restricted diet in promoting initial weight loss, it is nevertheless considered an important component of weight loss therapy. Moreover, it appears to be crucial for maintaining weight loss.[2, 14, 85] Physical activity has the added benefit of minimizing loss of lean body mass, reducing LDL-cholesterol levels, increasing HDL-cholesterol levels, improving insulin sensitivity, and improving fitness.[85] A study of 22,000 men showed that fitness level was a stronger predictor of cardiovascular disease and all-cause mortality than was fatness. Fat but fit men had a significantly lower risk of health complications than did lean men who were unfit.[108]

BOX 12.10   CLINICAL APPLICATIONS

Suggested Educational Topics for the Overweight and Obese Patient

- Energy value of different foods.
- Food composition—fats, carbohydrates (including dietary fiber), and proteins.
- Evaluation of nutrition labels to determine caloric content and food composition.
- New habits of purchasing—give preference to low-kcalorie foods.
- Food preparation—avoid adding high-kcalorie ingredients during cooking (e.g., fats and oils).
- Avoiding overconsumption of high-kcalorie foods (both high-fat and high-carbohydrate foods).
- Adequate water intake.
- Reducing portion sizes.
- Limiting alcohol consumption.

Source: National Heart, Lung, and Blood Institute. The practical guide: identification, evaluation, and treatment of overweight and obesity in adults. Bethesda (MD): U.S. Department of Health and Human Services, National Institutes of Health; 2000.

A minimum initial physical activity goal for adults is 150 minutes (2 hours and 30 minutes) per week of moderate-intensity physical activity or 75 minutes (1 hour and 15 minutes) of vigorous physical activity. The goal for children and adolescents is 60 minutes or more of physical activity per day.[8] For the sedentary and obese, physical activity should be initiated slowly and then gradually increased in duration and intensity. Physical activity can involve either programmed or lifestyle activities. Programmed or formal activities include regularly scheduled periods of swimming, brisk walking, running, jumping rope, or other aerobic activities. Lifestyle activity involves moving the body more throughout the day in the discharge of the activities of daily life. Examples include walking or bicycling instead of riding in a motor vehicle, climbing stairs instead of using an elevator or escalator, decreasing time spent in sedentary behaviors such as watching television, and increasing time spent performing common chores such as house cleaning and yard work.[8]

Nutrition Counseling Strategies for Behavior Therapy Behavior therapy provides patients with a set of techniques (self-monitoring, stimulus control, rewards, etc.) with which to identify and overcome barriers to positive dietary, exercise, and other lifestyle habits.[109] The practitioner collaborates with the patient to establish specific, achievable, and measurable goals related to food intake, physical activity, and weight loss. Patients are taught to observe and record their food intake, physical activity, and body weight. Self-monitoring of behavior generally changes behavior in the desired direction and is associated with long-term weight loss.[2, 85] Self-monitoring also helps the patient identify social or environmental stimuli that lead to undesirable behaviors or that block the adoption of desirable behaviors. Once these stimuli

are identified, steps can be taken to prevent them from occurring or to change one's reaction to them. This is referred to as stimulus control. Rewards are used to encourage attainment of the established goals. Behavioral therapy is a valuable adjunct to diet and physical activity, resulting in marked improvements in weight loss and weight maintenance.[85]

Eating Disorders

Eating disorders are psychiatric conditions characterized by severe disturbances in eating behavior, resulting in significant physiologic impairment and, in some instances, death.[110–112] The American Psychiatric Association recognizes three categories of eating disorders: anorexia nervosa, bulimia nervosa, and eating disorders not otherwise specified.[110] The diagnostic criteria for these are listed in Box 12.11.

The most prominent clinical characteristic of anorexia nervosa (AN) is a refusal or inability to maintain a minimally normal body weight, leading to a body weight that is less than 85% of what is expected for age and height.[110, 113, 114] Bulimia nervosa (BN) is characterized by repeated episodes of binge eating followed by abnormal compensatory weight-loss behaviors such as self-induced vomiting, fasting, excessive exercise, and misuse of laxatives and diuretics.[110, 114] Unlike AN, patients with BN generally maintain a body weight within normal limits. The category *eating disorder not otherwise specified (EDNOS)* encompasses behaviors that fail to meet the specific diagnostic criteria for either AN or BN. EDNOS also includes binge-eating disorder (BED), in which the patient engages in binging behavior without the compensatory weight-loss behaviors characteristic of BN.

While in theory these three conditions exist as distinct categories, in practice they share common features: all patients with eating disorders have a disturbed body image that leads them to overestimate their body shape and weight, perceive themselves as being obese even though they may have a very low body weight, have an intense fear of weight gain and obesity, and have a relentless drive to lose weight.[110–112, 114, 115] This body image disturbance drives a set of abnormal behaviors. While most people evaluate themselves on the basis of their perceived performance in such areas as relationships, work, parenting, athletics, and accumulation of possessions and wealth, persons with eating disorders base their self-worth primarily on the ability to control their shape and body weight.[115] Further blurring the distinctions among the three categories is the fact that, over time, patients with eating disorders tend to migrate back and forth between the diagnostic categories of AN, BN, and EDNOS, as illustrated in Figure 12.19.[111, 115] Many BN patients have a past history of AN, and many AN patients have engaged in binge eating and compensatory weight-loss behaviors characteristic of BN. Body weight is a critical distinction between the two disorders: patients with AN are significantly underweight while those with BN generally have a body weight within normal limits.[114, 115]

Common characteristics of AN and BN are shown in Table 12.14. Although males are diagnosed with eating disorders, approximately 90% of cases of AN and BN are seen in white females living in Western societies where food is plentiful and where being thin is associated with attractiveness.[114, 115] AN is more often seen

BOX 12.11   CLINICAL APPLICATIONS

American Psychiatric Association's Diagnostic Criteria for Anorexia Nervosa, Bulimia Nervosa, and Eating Disorders Not Otherwise Specified, and Research Criteria for Binge-Eating Disorder

Diagnostic Criteria for Anorexia Nervosa:

A. Refusal to maintain body weight at or above a minimally normal weight for age and height (e.g., weight loss leading to maintenance of body weight less than 85% of that expected).

B. Intense fear of gaining weight or becoming fat, even though underweight.

C. Disturbance in the way in which one's body weight or shape is experienced, undue influence of body weight or shape on self-evaluation, or denial of the seriousness of the current low body weight.

D. In postmenarchal females, amenorrhea, i.e., the absence of at least three consecutive menstrual cycles. (A woman is considered to have amenorrhea if her periods occur only following hormone, e.g., estrogen, administration.)

Subtypes of Anorexia Nervosa:

Restricting Type

During the current episode of anorexia nervosa, the person has not regularly engaged in binge-eating or purging behavior (i.e., self-induced vomiting or the misuse of laxatives, diuretics, or enemas)

Binge-Eating/Purging Type

During the current episode of anorexia nervosa, the person has regularly engaged in binge-eating or purging behavior (i.e., self-induced vomiting or the misuse of laxatives, diuretics, or enemas)

Diagnostic Criteria for Bulimia Nervosa:

A. Recurrent episodes of binge eating. An episode of binge eating is characterized by both of the following.

1. Eating, in a discrete period of time (e.g., within any two-hour period), an amount of food that is definitely larger than most people would eat during a similar period of time and under similar circumstances

2. A sense of lack of control over eating during the episode (e.g., a feeling that one cannot stop eating or control what or how much one is eating)

B. Recurrent inappropriate compensatory behavior in order to prevent weight gain, such as self-induced vomiting; misuse of laxatives, diuretics, enemas, or other medications; fasting; or excessive exercise.

C. The binge eating and inappropriate compensatory behaviors both occur, on average, at least twice a week for three months.

D. Self-evaluation is unduly influenced by body shape and weight.

E. The disturbance does not occur exclusively during episodes of anorexia nervosa.

Subtypes of Bulimia Nervosa:

Purging Type

During the current episode of bulimia nervosa, the person has regularly engaged in self-induced vomiting or the misuse of laxatives, diuretics, or enemas

Nonpurging Type

During the current episode of bulimia nervosa, the person has used other inappropriate compensatory behaviors, such as fasting or excessive exercise, but has not regularly engaged in self-induced vomiting or the misuse of laxatives, diuretics, or enemas

Eating Disorder Not Otherwise Specified:

The eating disorder not otherwise specified category is for disorders of eating that do not meet the criteria for any specific eating disorder. Examples include:

1. For females, all of the criteria for anorexia nervosa are met except that the individual has regular menses.

2. All of the criteria for anorexia nervosa are met except that, despite significant weight loss, the individual's current weight is in the normal range.

3. All of the criteria for bulimia nervosa are met except that the binge-eating inappropriate compensatory mechanisms occur at a frequency of less than twice a week or for a duration of less than three months.

4. The regular use of inappropriate compensatory behavior by an individual of normal body weight after eating small amounts of food (e.g., self-induced vomiting after the consumption of two cookies).

5. Repeatedly chewing and spitting out, but not swallowing, large amounts of food.

6. Binge-eating disorder; recurrent episodes of binge eating in the absence of the regular use of inappropriate compensatory behaviors characteristic of bulimia nervosa.

Research Criteria for Binge-Eating Disorder:

A. Recurrent episodes of binge eating. An episode of binge eating is characterized by both of the following:

1. Eating, in a discrete period of time (e.g., within any two-hour period), an amount of food that is definitely larger than most

(continued)

among adolescents, with the peak age of onset being 14 to 19 years, although it is diagnosed in prepubertal children and, much less frequently, in middle-aged and older adults.[110, 116] On the other hand, BN is generally diagnosed in persons in their mid-to-late 20s with a history of binge eating and purging for as long as 5 to 10 years.[114, 115] Although there is disagreement among experts regarding whether eating disorders are now being diagnosed with greater frequency than in past decades, it appears that BN is more prevalent than in past decades, while the observed increase in AN is more likely due to a greater number of AN patients seeking help and better detection of cases by clinicians.[115–117]

Mortality from AN is one of the highest of any psychiatric disorder, with 5% of AN patients dying every decade. In contrast, mortality from BN is much less likely, but is greater than in similar women in the general population who do not have an eating disorder.[118] Patients with AN often deny that they have a problem and seldom seek medical treatment unless urged by concerned family or friends. Compared with patients with AN, those with BN are painfully aware that they have a problem and are more likely to seek treatment because of their inability to control their chaotic eating behavior. When interviewed in a supportive, nonjudgmental environment, patients with BN are able to provide details about eating behaviors and discuss their condition.[114]

Etiology of Eating Disorders

Eating disorders often begin when an individual starts dieting, initially in a manner similar to that followed by many adolescent and young adult females.[112, 114, 115] As weight loss progresses, persons predisposed to eating disorders experience an intense fear of gaining weight, diet more strictly, and begin developing the characteristic psychological, behavioral, and medical problems associated with eating disorders. For those developing AN, there is a sustained and obsessive pursuit of self-starvation. These patients view weight loss as a desired accomplishment rather than an affliction, and they have little motivation to change their behavior. In patients developing BN, similar attempts to lose weight and control body size and shape are thwarted by regular episodes of uncontrolled overeating followed by abnormal compensatory behaviors to lose weight. Persons with bulimia sometimes describe themselves as "failed anorexics."[115]

Although the etiology of eating disorders is unknown, it is clear that multiple factors are associated with an increased risk of being diagnosed with an eating disorder, including environmental factors, certain character traits, and genetics.[112, 114, 115] Environmental risk factors include a family history of mood disturbances, childhood physical or sexual abuse, the perception that there is a low degree of social support from family members, and societal pressures on females to attain a degree of thinness that is not only unhealthy but impossible for most females to attain. Of particular relevance to BN are childhood and parental obesity, early menarche, and parental alcoholism.[114, 115] Character traits associated with both AN and BN include low self-esteem and elevated harm avoidance, while traits more often seen in patients with AN include perfectionism, conscientiousness, persistence, and obsessiveness. Traits more often associated with BN include impulsiveness, novelty seeking, negative emotionality, and stress reactivity.[112] Although no specific gene for AN, BN, or EDNOS has yet been conclusively identified, the observation that an

Figure 12.19 Patient Migration among Eating Disorder Categories

Over time patients with eating disorders tend to migrate back and forth between the various categories of eating disorders. An arrow's size represents the likelihood of patients migrating from one category to another in the indicated direction. Recovery is represented by an arrow pointing outside the circle.

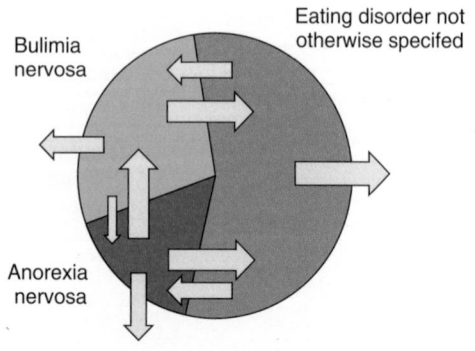

Source: Fairburn CG, Harrison PJ. Eating disorders. Lancet. 2003;361:407–416.

Table 12.14 Common Characteristics of Anorexia Nervosa (AN) and Bulimia Nervosa (BN)

	Anorexia Nervosa	Bulimia Nervosa
Global and ethnic distribution	Predominantly seen in Western societies among white people	Predominantly seen in Western societies among white people
Sex	90% female 10% male	Female to male proportion ranges from 10:1 to 20:1
Age of onset	Mid-adolescence	Late adolescence, early adulthood
Socioeconomic status	Appears evenly distributed across all social classes	Appears evenly distributed across all social classes
Prevalence in females	0.5% to 0.7% of adolescent females	1% to 3% in females 16 to 35 years of age
Incidence (per 100,000 per year)	19 in females, 2 in males	29 in females, 1 in males
Weight status	Markedly decreased	Typically within normal limits
Menstruation	Absent	Usually normal
Mortality	Approximately 5% per decade	Lower than in AN but higher than similar women in the general population who do not have an eating disorder
Prognosis	Less favorable than BN; full recovery seen in 25%–50% of patients with treatment	More favorable than AN; full recovery occurs in 50% of patients within 10 years

Source: Adapted from Fairburn CG, Harrison PJ. Eating disorders. Lancet. 2003;361:407–16; Bulik CM, Reba L, Siega-Riz AM, Reichborn-Kjennerud T. Anorexia nervosa: definition, epidemiology, and cycle of risk. Int J Eat Disord. 2005;37:S2–S9; Mehler PS. Bulimia nervosa. N Eng J Med. 2003;349:875–81; Walsh BT. Eating disorders. In Kasper DL, Braunwald E, Fauci AS, Hauser SL, Longo DL, Jameson JL, eds. Harrison's principles of internal medicine. 16th ed. New York: McGraw-Hill; 2005, 430–33.

individual's risk of developing an eating disorder is greater when another first-degree family member is affected suggests that genetics plays an etiologic role. The role of genetics in AN has been more clearly elucidated than its role in BN.[110, 114, 115] A useful approach for examining the etiologic role of genetics is to study twins. For example, if one identical twin has AN, the other identical twin has an approximately 55% chance of having AN, whereas if one fraternal twin has AN, the other fraternal twin has an approximately 5% chance of having AN.[114, 119]

Anorexia Nervosa

As outlined in Box 12.11, the diagnostic criteria for AN are a refusal to maintain a minimally normal body weight for age and height, an intense fear of gaining weight, a disturbed perception of body shape and/or size, and amenorrhea in postmenarchal females.[110] In the *Diagnostic and Statistical Manual of Mental Disorders*, 4th ed. (DSM-IV), the American Psychiatric Association recognizes two mutually exclusive subtypes of AN: restricting subtype and binge eating/purging subtype.[110] Those with the restricting subtype accomplish their weight loss through dieting, fasting, or excessive exercise and do not regularly engage in binge eating or any of the compensatory weight loss behaviors such as purging. Those with the binge eating/purging subtype engage in regular binge eating and/or purging.

Patients with AN experience a severe and selective restriction of food intake, particularly of foods perceived as fattening, resulting in a marked weight loss. The term "anorexia," which literally means "without appetite," is a misnomer considering the fact that most AN patients do not lose their desire for food and experience profound hunger.[110, 115] AN patients are generally obsessed with exercising and dieting, and preoccupied with thoughts of food to the extent that some AN patients collect cookbooks and recipes and are drawn to food-related occupations. As weight loss progresses, AN patients typically become irritable, moody, socially withdrawn, and isolated, and they experience loss of libido.[114, 115] Although amenorrhea in postmenarcheal females is a diagnostic criterion for AN, some females continue to menstruate while meeting every other criterion for the disease.[114]

In applying the criterion that AN patients refuse to maintain body weight at or above a minimally normal body weight for age and height, the DSM-IV suggests that a body weight less than 85% of that expected for age and height be used as a guideline. Clinical judgment is necessary in applying this criterion, and consideration must be given to the patient's body build, weight history, and developmental stage, if the patient is an adolescent.[111, 113] For patients 20 years of age and older, a BMI ≤18.5 kg/m² has been suggested by some as meeting this guideline,[114] while others suggest using a BMI ≤17.5 kg/m² as the cutpoint.[111, 120] Another approach for evaluating the body weight of an adult is the Hamwi equation (see Chapter 3). For patients less than 20 years of age, underweight is defined as a BMI for age and sex that is ≤5th percentile using the CDC growth charts (see Chapter 3).[37, 38, 111] Because body weight can be manipulated by such means as excessive water or fluid intake or severe restriction of water or fluid intake, a body weight that has unexpectedly changed or is inconsistent with other clinical findings should be interpreted with caution.[111]

For AN patients with the binge eating/purging subtype, compensatory weight-loss behaviors include self-induced vomiting, fasting, excessive exercise, and misuse of laxatives, diuretics, and enemas.[110, 114] The most common compensatory weight-loss behavior is purging through self-induced vomiting.

Health Complications of Anorexia Nervosa Numerous health complications and abnormalities, some of which can be life threatening, are seen in patients with AN, as outlined in Table 12.15 and illustrated in Figure 12.20. Most of these are the direct result of malnutrition due to the self-imposed state of starvation and are reversed as healthy eating habits are restored, body weight returns to a more normal level, and nutritional status improves, especially if AN is diagnosed early in the course of the illness and is treated by a skilled interdisciplinary team.[114, 121] One notable exception to this pattern of recovery is bone mineral density, which may not reach the level expected for the patient's sex and age, particularly if AN occurs during adolescence, a critical time period for bone development because the rate of bone mineralization is at its peak (see Chapter 25).[114, 121, 122]

Table 12.15 Physical and Diagnostic Findings in Patients with Anorexia Nervosa

Skin & Extremities
- Cold hands & feet
- Dry skin
- Lanugo
- Alopecia
- Acrocyanosis
- Dependent edema

Cardiovascular
- Bradycardia
- Hypotension
- Orthostatic hypotension
- Cardiac arrhythmias
- Electrocardiographic abnormalities

Gastrointestinal
- Salivary gland enlargement
- Delayed gastric emptying
- Constipation

Plasma/Serum Values
- Elevated BUN & creatinine
- Hyponatremia
- Hypokalemia
- Hypercholesterolemia
- Hypoglycemia
- Low T_3
- Low-normal T_4
- Low luteinizing hormone
- Low follicle-stimulating hormone
- Hypophosphatemia (during refeeding)

Bone
- Decreased bone mineral density

Sources: American Psychiatric Association. Practice guideline for the treatment of patients with eating disorders. *Am J Psychiatry.* 2000;157(Suppl):1–39. Fairburn CG, Harrison PJ. Eating disorders. *Lancet.* 2003;361:407–416. Walsh BT. Eating disorders. In Kasper DL, Braunwald E, Fauci AS, Hauser SL, Longo DL, Jameson JL, editors. *Harrison's Principles of Internal Medicine.* 16th ed. New York: McGraw-Hill; 2005, 430–33.

Common physical findings of AN include cold intolerance, reduced gastric emptying, constipation, and the presence on the skin of fine, downy-like hair known as lanugo. In some instances patients experience alopecia (hair loss). Despite profound weight loss, the face may have a fullness due to salivary gland enlargement resulting from both starvation and frequent vomiting. The fingers and toes may take on a bluish tint (known as acrocyanosis), while patients consuming large quantities of vegetables rich in carotenoids may experience hypercarotenemia, which can result in the skin having a slightly yellow or orange color to it. There may be abnormalities in vital signs, including bradycardia (heart rate <60 beats/minute), hypotension (systolic blood pressure <90 mm Hg), orthostatic hypotension (low blood pressure upon standing), and hypothermia.[110, 112, 114]

Abnormal laboratory test values can include anemia, leukopenia (abnormally low white blood cell count), low plasma glucose, elevated serum total cholesterol, and low-normal serum values for the thyroid hormones triiodothyronine (T_3) and thyroxine (T_4). Dehydration can result in slight increases in blood urea nitrogen and serum creatinine. Self-induced vomiting can result in hypokalemia (low serum potassium), hypochloremia (low serum chloride), and metabolic alkalosis (elevated serum bicarbonate). Excessive water consumption, a common tactic to increase body weight, can result in hyponatremia (low serum sodium). AN has a marked impact on the endocrine system, particularly on the female reproductive system, resulting in amenorrhea, which is the absence of menses when they would normally be expected to occur. AN results in decreased secretion of luteinizing hormone and follicle-stimulating hormone from the anterior pituitary; this results in decreased estrogen, which causes the amenorrhea.[20, 110, 112, 114]

Reduced bone mineral density, a common feature of AN, results from multiple nutritional deficiencies and amenorrhea.[112, 114, 121] Depending on the severity and length of the disease, AN during adolescence can result in premature cessation of linear bone growth, failure to achieve expected adult height, and reduced bone mineral density. Even after several years of recovery from AN, these patients may never achieve their peak bone mineral density, and as they enter middle and late adulthood are at increased risk of painful fractures, disfiguring kyphosis, loss of height, and increased risk of death.[122] On average, more than 50% of females with AN eventually develop osteoporosis and more than 50% of males with AN eventually develop a marked reduction in the mineral density of the femoral neck and the lumbar vertebrae. The importance of early diagnosis and treatment of AN is underscored by the observation that clinically significant bone mineral loss does not usually occur in the first 12 months of the illness.[122]

Bulimia Nervosa

As outlined in Box 12.11, the American Psychiatric Association's diagnostic criteria for BN are recurrent episodes of binge eating and use of inappropriate compensatory behaviors to prevent weight gain that occur, on average, at least twice a week for 3 months and that do not occur exclusively during episodes of AN. In addition, BN patients are preoccupied with their body shape and weight and their self-esteem is primarily based on their perceptions of body shape and weight, which are often distorted.[110, 114, 118] The diagnostic criteria specify that the episode of binge eating must be characterized by eating an amount of food during a relatively short period of time (e.g., within a two-hour time period) that is

Figure 12.20 Physical Changes and Health Complications Seen in Patients with Anorexia Nervosa

Hypotension
Orthostatic hypotension

Dry skin
Lanugo

Decreased bone mineral density

Delayed gastric emptying
Constipation

Amenorrhea

Dizziness
Alopecia

Bradycardia
Cardiac arrhythmias
Electrocardiographic abnormalities

Marked weight loss
Muscle wasting

Cold hands & feet
Acrocyanosis

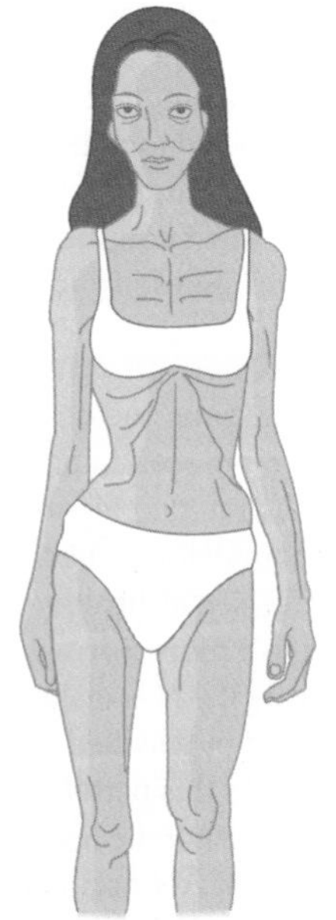

Source: http://www.pdrhealth.com/content/nutrition_health/chapters/fgnt09.shtml

definitely larger than what most people would ordinarily eat in a similar situation. In addition, the episode of binge eating must be characterized by the subject sensing a lack of control (e.g., the subject feels he or she cannot stop or control what or how much he or she is eating).[110] When defining a binge, it is important to consider the context in which the food is consumed. For example, normal food consumption during a holiday or celebration might be considered excessive during a typical meal. Foods consumed during a binge are often those that are sweet and kcal-rich, such as ice cream or cake.

The American Psychiatric Association recognizes two mutually exclusive subtypes of BN: purging and nonpurging. In the purging subtype, the patient regularly engages in some type of purging behavior such as self-induced vomiting or the misuse of laxatives, diuretics, or enemas. The most common form of purging is self-induced vomiting, which is used by 80% to 90% of persons treated for BN. Initially, these subjects may use their fingers or some object to stimulate the gag reflex. Eventually, however, many patients develop the ability to initiate vomiting at will. Although rarely used to induce vomiting, syrup of ipecac can be toxic to the myocardium (the heart muscle) if regularly used.[118] The next most common purging method is laxative abuse, which is used by about one-third of patients with BN.[110] In the nonpurging subtype, the patient does not resort to purging, but does use some other abnormal compensatory weight loss behavior such as fasting or excessive exercise.[110] Exercise is considered excessive if it significantly interferes with important activities, it occurs at inappropriate times or in inappropriate settings, or if the subject continues to exercise despite an injury or medical complication. Subjects with type 1 diabetes may resort to weight loss by omitting or reducing insulin doses.[110]

Patients with BN are ashamed of their chaotic eating habits, binge eat in secret, and keep their disorder hidden from friends and family members.[110, 114, 115] Although the diagnostic criteria for BN focus on the binge/purge cycles, much of the time patients with BN are restricting their eating in an attempt to control body shape and weight. Triggers for binge eating include hunger resulting from restricting food intake, anxiety, depression, and low self-esteem. The amount of food consumed in a binge is variable, with energy intake per binge ranging between 1000 and 2000 kcal.[115] The binge typically results in shame and an unpleasant feeling of fullness which then triggers the purging or other compensatory weight loss behavior; this behavior initially provides some degree of relief, but this is often soon followed by guilt and shame.[111] Although the diagnostic criteria in Box 12.11 indicate a minimum frequency of twice a week for three months, the frequency of purging can vary considerably, with some patients reporting purging as often as 5 to 10 times per day.[118] Despite purging and other abnormal compensatory behaviors, body weights of BN patients tend to be within normal limits, indicating that these attempts to control body weight are not as effective as some patients believe.

Table 12.16 Physical and Diagnostic Findings in Patients with Bulimia Nervosa

Skin & Extremities
- Callus on back of hand from stimulating gag reflex to induce vomiting (Russell sign)

Cardiovascular
- Cardiomyopathy from syrup of ipecac
- Cardiac arrythmias from syrup of ipecac
- Electrocardiographic changes from syrup of ipecac

Gastrointestinal
- Loss of dental enamel
- Esophageal tearing (Mallory-Weiss tears)
- Dental caries
- Constipation & laxative dependence
- Salivary gland enlargement
- Esophagitis
- Gastroesophageal reflux disease

Plasma/Serum Values
- Alkalosis
- Hypokalemia
- Hypochloremia
- Hyponatremia

Sources: American Psychiatric Association. Practice guideline for the treatment of patients with eating disorders. *Am J Psychiatry.* 2000;157(Suppl):1–39. Fairburn CG, Harrison PJ. Eating disorders. *Lancet.* 2003;361:407–416. Walsh BT. Eating disorders. In Kasper DL, Braunwald E, Fauci AS, Hauser SL, Longo DL, Jameson JL, editors. *Harrison's Principles of Internal Medicine.* 16th ed. New York: McGraw-Hill; 2005, 430–33.

Health Complications of Bulimia Nervosa The health complications of BN are usually the result of regular purging or the other compensatory weight loss behaviors and are generally not life threatening.[110, 114, 115, 118] These are outlined in Table 12.16 and illustrated in Figure 12.21. Frequent self-induced vomiting using the fingers to stimulate the gag reflex can result in the teeth leaving a scar or callus on the back of the hand, referred to as the Russell sign.[112] Painless, bilateral enlargement of the salivary glands may give the face a full or puffy

Figure 12.21 Physical Changes and Health Complications Seen in Patients with Bulimia Nervosa

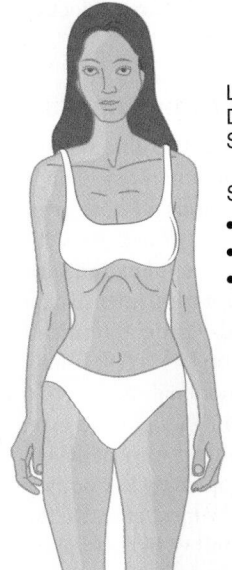

Esophagitis
Gastroesophageal reflux disease
Esophageal tearing

Constipation
Laxative dependence

Callus on back of hand from using fingers to stimulate gag reflex to induce vomiting

Loss of dental enamel
Dental caries
Salivary gland enlargement

Syrup of ipecac can cause
- Cardiomyopathy
- Cardiac arrhythmias
- Electrocardiographic abnormalities

Source: http://www.pdrhealth.com/content/nutrition_health/chapters/fgnt09.shtml

appearance. Frequent vomiting can permanently erode the enamel of the teeth, especially the lingual surface of the front teeth, giving them a ragged or "moth-eaten" appearance. Frequent vomiting can increase the risk of dental caries and cause esophagitis, gastroesophageal reflux disease (GERD), and, less frequently, tearing of the esophagus (Mallory-Weiss tears) and rupture of the stomach. Both vomiting and laxative abuse can cause fluid and electrolyte imbalances including hypokalemia, hyponatremia, hypochloremia, and alkalosis. Emetine, the active ingredient in syrup of ipecac, is cardiotoxic, has a long half-life, and tends to accumulate in cardiac muscle.[112] Frequent use of syrup of ipecac to induce vomiting can cause cardiomyopathy, electrocardiographic changes, and cardiac arrhythmias, potentially leading to congestive heart failure and death.[110, 111, 114, 115, 118] Laxative abuse can result in laxative dependence, constipation, dehydration, and potentially renal damage.[112] Patients with BN generally have normal bone density.[118]

Eating Disorders Not Otherwise Specified

Patients who have atypical eating disorders or with disordered eating behaviors that fail to meet all the criteria for either AN or BN are categorized as having an eating disorder not otherwise specified (EDNOS), as outlined in Box 12.11.[110, 112] A common example of EDNOS is a female who meets all the criteria for AN but who still has regular menses or whose body weight has not fallen below 85% of that expected for age and height. Another example of EDNOS is the patient meeting all the diagnostic criteria for BN except that the binge eating and purging or other compensatory weight loss behaviors occur at a frequency of less than twice a week or for a duration of less than three months. Some individuals may engage in repeatedly chewing and then spitting out food rather than swallowing it.[110, 111]

The diagnostic criteria for binge-eating disorder (BED), which is classified as an EDNOS, are outlined in Box 12.11.[110] BED is characterized by a person binge eating at least twice a week for six or more months and sensing a lack of control over eating during the binge. Patients with BED do not engage in purging or the other compensatory weight loss behaviors seen in patients with BN.[110] The estimated prevalence of BED is 1% to 2% of the population.[111] Most BED patients are obese, seek treatment because of their obesity, and are subject to the same medical problems as other obese patients.[111]

Nutrition Therapy for Eating Disorders

Because eating disorders are psychiatric illnesses with major medical complications, their treatment requires an interdisciplinary team of health care professionals whose primary focus is on psychiatric management.[111, 114] For patients with AN, there are a range of psychiatric treatment options and various treatment settings, including outpatient care, intensive outpatient care, partial hospitalization as a day patient, residential treatment, and inpatient hospitalization.[123–125] Very briefly, patients with AN require considerable emotional support to overcome their strenuous resistance to gaining weight and extensive counseling to help them find healthier ways of developing self-esteem than achieving an inappropriately low body weight.[114] The psychiatric model shown to be most effective for treating patients with BN is known as cognitive behavioral therapy, and is an approach focusing on modifying the specific behaviors and ways of thinking that maintain the patient's disordered

eating.[114, 118, 125] Pharmacologic agents are infrequently used in the treatment of AN but have been shown effective in treating patients with BN.[114]

Because of the numerous dietary and nutritional implications of eating disorders and because nutrition therapy (in addition to psychotherapy) is an integral part of treating eating disorders, the dietitian is a key member of the interdisciplinary team. It is important to note that each team member has specific professional responsibilities related to that member's discipline, and that it is essential for each member to understand his or her personal and professional boundaries and to not overstep them.[111] The role of the dietitian is to initially assess the patient's nutritional status, to address the patient's food and nutrition issues and the behaviors associated with those issues, to periodically monitor the patient's response to treatment, and to communicate with and support the efforts of the other interdisciplinary team members as is appropriate. The dietitian develops the nutrition component of the treatment plan, takes a leading role in implementing the nutritional component of the treatment plan, and provides ongoing support to the patient in accomplishing the goals set out in the treatment plan.[111]

Nutrition Therapy for Anorexia Nervosa In theory, the primary care physician's decision as to where and how to treat the patient with AN should be primarily determined by the subject's current weight, the rapidity of recent weight loss, the severity of medical and psychological complications, and the necessity of removing the patient from an unhealthful environment.[123, 124] In practice, however, the treatment setting is more often determined by the availability of care and its cost. For example, a suitable residential or inpatient treatment program may be several hours away from where the patient lives, or the cost of intensive treatment may be prohibitive, depending on the patient's insurance coverage or whether the patient is insured.[124] Medical conditions warranting residential or inpatient treatment include severe electrolyte imbalances or a body weight <75% of expected regardless of the patient's electrolytes or other laboratory values. In less severe cases of AN, the day patient or outpatient treatment can be less expensive and somewhat more convenient to the patient and family members. The primary treatment goal in AN is restoring the patient's weight to at least 90% of the expected weight, and in most cases this degree of nutritional repletion can be accomplished by normal oral feedings without resorting to administering nutrients enterally (via nasogastric tube) or very rarely, parenterally.[111, 114] Enteral feedings are generally used as a last resort—a life-saving intervention for those with AN who are hospitalized as in-patients. Feeding severely malnourished patients requires careful management because of the risk of the patient developing "refeeding syndrome," which is characterized by hypophosphatemia (low blood phosphorus), hypomagnesemia (low blood magnesium), hypokalemia (low blood potassium), and potential cardiac failure.[114, 126] Additional treatment goals are cessation of weight loss behaviors, improvement in eating behaviors, and improvement in emotional and psychological health. Regardless of the treatment setting, the goals remain the same; the only difference is the intensity of treatment. In the outpatient setting, the treatment is somewhat less intense than in the inpatient setting.[111]

Key to the success of residential treatment or inpatient hospitalization is a duration of treatment sufficient to allow adequate weight gain and weight stabilization, and sufficient to provide the therapy necessary to allow the patient to adjust emotionally to the healthier weight. A low body weight at the beginning of treatment and inadequate weight gain during inpatient treatment are both associated with poor treatment outcome and inpatient readmission.[124] The recommended weight gain is 2 to 3 lb/week for inpatient treatment and 0.5 to 1 lb per week for outpatient treatment.[123] Initially, the energy intake should be 30 to 40 kcal/kg of body weight per day, which can then be advanced as tolerated by the patient. During the phase of active weight gain, the energy intake may need to be as high as 70 to 100 kcal/kg of body weight while an intake of 40 to 60 kcal/kg of body weight may be sufficient for weight maintenance and to support adequate growth and development in children and adolescents.[123] For some patients, vitamin and mineral supplements may be helpful, and it is important to ensure adequate intake of vitamin D (400 IU/d) and calcium (1500 mg/d) to minimize bone losses.[114] Meals should be supervised by staff members who firmly stress the importance of adequate food consumption, are empathetic about the patient's challenges, and provide encouragement and reassurance about the patient's eventual recovery.[114]

The patient's response to nutrition therapy can be assessed by monitoring the patient's vital signs, food intake, fluid intake and output, and changes in weight, height, body mass index, body composition, and laboratory test values. An unexpected increase in body weight may indicate fluid retention during refeeding or excessive water or fluid intake by the patient to artificially increase body weight. Treatment programs generally have specific protocols for weighing patients that address when patients are weighed, who weighs the patients, and whether patients are informed of their weights. In some instances, a patient may be weighed with his or her back to the scale and not immediately informed of the value.[111] Children's and adolescents' height and BMI can be assessed using the CDC growth charts. Changes in body composition can be monitored by measuring skinfold thicknesses (see Chapter 3). Patients should also be observed for signs of congestive heart failure and gastrointestinal problems such as constipation and bloating. Cardiac monitoring may be warranted for severely malnourished patients whose weight is <70% of expected.[111, 123]

During nutritional repletion, serum electrolytes should be closely monitored for signs of refeeding syndrome (see Chapter 5), which is characterized by serum electrolyte depletion, fluid shifts, cardiac arrhythmias, and glucose derangements occurring in severely malnourished patients when they receive nutritional repletion either orally, enterally, or parenterally.[111, 123, 127] Common electrolyte disturbances seen in refeeding syndrome include hypokalemia, hypomagnesemia, and, most notably, hypophosphatemia. Adverse effects of hypophosphatemia include cardiac failure, muscle weakness, immune dysfunction, and possibly death.[127]

Nutrition Therapy for Bulimia Nervosa BN can often be treated on an outpatient basis, and compared to AN there is a greater likelihood of recovery. The primary treatment goal in BN is to reduce the chaotic cycle of binging and purging and to normalize the patient's eating habits.[111, 123] Because the weight of most patients with BN is within the normal range, weight restoration is not the focus of therapy as it is with patients who have AN. Some BN patients may initially be under-

weight and could benefit emotionally and physiologically from weight gain. Others, because of their disturbed body image, overestimation of their body shape and weight, and intense fear of obesity and weight gain, may insist on losing weight. Any efforts to change the patient's body weight should be postponed until after eating habits are normalized.[111, 123]

The role of the dietitian is to work with the other members of the interdisciplinary team to develop an eating plan to normalize the patient's eating habits. A common recommendation for patients with BN is to consume three meals per day with one to three snacks per day in a structured manner, providing order to eating and breaking the chaotic eating pattern of binging and purging. Initially, food intake should be sufficient to prevent hunger, which is a common trigger of binge eating. Nutrition counseling should focus on helping the patient expand the diet to include the patient's self-imposed "forbidden" or "feared" foods. As eating habits normalize, patients may experience fluid retention and will need education and support to deal with this temporary and disturbing phenomenon. The laxative-dependent patient will benefit from information on prevention of bowel obstruction resulting from laxative withdrawal through consumption of foods rich in dietary fiber and adequate amounts of fluid.[111, 123]

Nutrition Therapy for Eating Disorders Not Otherwise Specified Medical nutrition therapy for EDNOS will depend on the patient's specific abnormal eating behaviors. For example, if the EDNOS patient presents with many, but not all, of the criteria for AN, the nutrition therapy will more closely follow that recommended for AN. Likewise, if the EDNOS patient presents with binge eating but at a frequency less than that required for a diagnosis of BN, or if the patient regularly purges or resorts to some compensatory weight loss behavior after eating a very small amount of food, the treatment will more closely resemble that used in treating patients with BN.

Conclusion

Throughout human history, energy balance has been an issue of concern for most people. Prior to the mechanization of agriculture in the middle of the twentieth century, most humans faced an uncertain food supply and hunger, malnutrition, and starvation were common. Tragically, these conditions remain a threat to many people living in developing nations; yet a much more common condition, particularly in developed nations, is an obesigenic environment that promotes the consumption of high-energy dense foods and tends to discourage regular physical activity. The dramatic rise in the prevalence of overweight and obesity in recent decades has been aptly described as an epidemic, and the term "globesity" has been coined to represent the global nature of the obesity epidemic. The complex etiology of obesity requires a multifactorial approach to its prevention and treatment. This will involve better understanding of the biological basis of body weight regulation and the control of hunger and appetite, more informed personal choices about food intake and physical activity, and modifying the environment to promote the adoption and maintenance of healthy dietary and physical activity behaviors. At the same time, eating disorders remain a threat to health and demand attention, particularly considering the high mortality associated with anorexia nervosa.

Betty Kovacs, MS, RD *Co-Director & Director of Nutrition, New York Obesity Research Center Weight Loss Program—New York Obesity Research Center St. Luke's-Roosevelt Hospital, New York, NY*

How long have you been an RD? How long have you worked in weight management?

I have been an RD for thirteen years and worked in weight management for twelve years.

Describe a typical client that you see at the weight-loss program. The clients in our program have a minimum of 50 lbs to lose. The average age is mid-40s and 80% are women. By the time they come to us they have tried most diets and options available for weight loss.

In your practice, what do you find is key information for you to obtain in order to assess a client? Are there common nutritional problems associated with these clients? If so, what are they?

Before beginning the program, each client has a 30-minute screening appointment. The point of this appointment is to assess if the client is appropriate for the program and if the program is appropriate for him or her. Just because someone comes to us to lose weight does not mean that it's the best way for them to do it. It's important to discuss what they have tried to do to lose weight and what did and did not work for them. It can give you insight into what to do with them next. Some people will do better with individual appointments and others will benefit from being a part of a group. Most people don't know what will work best so it's our job to help them figure that out.

Over the years I have changed how they are counseled about their diet. I used to have an individual appointment with each person before they began the groups. I realized that they were starting the group at different points because of this. Some of them did really well with what I told them and others didn't. Now, they are all taught together in group. They learn the information at the same time and they can support each other while learning it. It's so easy to want to teach them everything, but it's too much. I start with the basics and once they see that they can lose weight with that they are ready for more information.

Their nutritional problems tend to be related to their health conditions and medications that they are taking. The most common health conditions are diabetes, hypertension, and high cholesterol. Along with the medications for these conditions, psychotropic drugs are very common and may be contributing to their weight problem. It's a very difficult situation when the medication is helping them psychologically but I know that it's affecting their weight. I try not to say it to the client; instead I speak with the doctor or recommend that they speak with him or her. The decision to change the medication has to be made by the prescribing physician, not me. One last nutritional concern is deficiencies. A lot of people assume that if you are overweight you could never be deficient in any nutrients. Once I discuss what they are eating and evaluate their labs, it's clear what vitamins and minerals they are not getting enough of. There are some cases when their intake of a macronutrient is too low as well. You can't use their weight as an indication of their nutritional status.

Do you have a specific goal or approach for counseling clients?

My first goal is to make them feel comfortable opening up to me, so I try to make light of what it's like to see a dietitian. During their orientation to find out about the program, I ask what they don't want to "have" to do or give up. They tend to be very honest and start to relax when they hear that this isn't about deprivation or punishment. I use that time to assure them that they are not coming to hear how I think that they should eat. My goal is to help them figure out how to take what they like to eat and adjust it so that they can lose weight. I tell them that I believe that they have never failed a diet—that the diets out there have failed them. There is no one "right" way to eat for weight loss and the diets out there try to say that there is. They need to know that this isn't easy and there is nothing wrong with them for struggling with this. They developed their habits over their lifetime and it will take time to change those. If I can find a way to keep their favorite foods in their plan and help them lose weight, they will be much more likely to stick with this for the long term.

What is the biggest challenge in working with patients who are overweight?

This would be undoing the misconceptions about nutrition and weight loss. I spend more time telling people about why other diets don't work and explaining what the latest study in the news is really about. I am fortunate that I get a year or more with my clients, so there is time for them to learn to trust what I say. If I only had a couple of appointments, it would be difficult to get them to believe me over the media. The public doesn't really understand how a registered dietitian is the nutrition expert. They hear physicians and that is who they listen to.

Has the treatment of obesity changed much in the last ten years? How do you stay abreast of changes in this field?

There have been a lot of changes in our treatment options and understanding of the causes and consequences of being overweight and obese. The National Heart, Lung, and Blood Institute's Obesity Education Initiative convened a panel of experts to develop The Clinical Guidelines on the Identification, Evaluation, and Treatment of Overweight and Obesity in Adults.

Objectives of the Guidelines:

- To identify, evaluate, and summarize published information about the assessment and treatment of overweight and obesity
- To provide evidence-based guidelines for physicians, other health care practitioners, and health care organizations for the evaluation and treatment of overweight and obesity in adults
- To identify areas for future research

In order to keep abreast of the latest research, it's imperative that I read the appropriate journals, am a member of the professional associations related to this field, attend meetings, and read any news and books that my clients may be exposed to. The most important membership to have in this field is The Obesity Society. Their journal, *Obesity Research*, and their annual meeting keep us updated on all of the advances happening in our field.

Any advice for dietetic students about counseling clients with weight issues?

My advice is to help the clients improve on what they are doing; don't try to get them to eat a "perfect" diet. This is a chronic problem and the progress is slow and difficult at times. You need to be patient and understand that there is a lifetime of habits behind the way that they eat and think about food. When I was a recent graduate, I can remember thinking that I could teach people about the Food Guide Pyramid and serving sizes and that would be enough to help them lose weight. I didn't agree with using meal replacements, medication, or surgery. I now know that all of these tools exist because they are needed and that my job is to know how and when to use them. I have been doing this for twelve years, and I am still learning. The more that I have learned, the more I realize how much there is to still learn. My training in school was the foundation, and it's my responsibility to become an expert in this and to learn all that I can to be able to do whatever possible to help them. As long as my client is willing to continue trying, I have to be willing to keep looking for ways to help. Obesity research clearly indicates that weight loss requires three interventions: behavioral, nutrition, and exercise. Ideally, you want to work with experts in the other two areas. When that's not possible, it's my responsibility to know when to refer my clients to experts in the other fields. Clients can tell that I genuinely want to do what is best for them and that I believe that they can be successful. Society is constantly judging them for their weight, so they need to be able to seek help from people who will not do the same.

Application of the Nutrition Care Process: Disordered Eating

Introduction:

NL is a 16-year-old premenarchal junior in high school who competes on high school cross country and track teams. She is an endurance athlete competing at regional and state events. After a training run today, she complained of dizziness. Mother states that patient has lost a lot of weight over the past six months. Her mother does not know exact amount but seven months ago patient began training for her first marathon. Patient states she has not lost enough weight and her recent performance has been hampered by weight. Mother states she trains constantly, even going out for additional runs after homework in the evening.

Nutrition Assessment:

Food/Nutrition-Related History:

AM: 8 oz skim milk, bagel with 1 tsp peanut butter; lunch: 1 c ice milk, 1 banana, pretzels—about 20, water; PM: meat (usually chicken)—skinless, baked 3 oz, 1 c vegetable, 2 c salad without dressing, skim milk—8 oz Mother says that patient rarely eats at the hours of the rest of the family and often will not eat evening meal until after 8 p.m. Meals, particularly breakfast and lunch, are missed at least twice per week. Patient states she eats no red meat and no cheese and never adds sugar or salt to food.

Trains 2.5 hours/day seven days per week. Includes a minimum of 2 hours of running at 8 miles/hour with weight lifting at least 4 times per week for 30–45 minutes.

1. Using her diet history, estimate her daily energy intake. Compare this with her estimated energy and protein requirements.

Anthropometrics/Physical Assessment:

Ht. 5'4" Wt. 91 lbs. UBW 105 lbs. (nine months previous)

% body fat—10.6%, triceps skinfold—11.75 mm, upper arm circumference 8.75 inches.

2. Calculate NL's BMI. Calculate NL's percent usual body weight and the % weight loss.

3. Compare her body fat percentage, triceps skinfold, and upper arm circumference to population standards.

Biochemical Data:

Alb 3.8 mg/dL, Prealbumin 22 mg/dL; Na$^+$ 137 mmol/L, K$^+$ 3.2 mmol/L, Cl 104 mmol/L, Hgb 14 mg/dL, Hct 37%, Ferritin 159 pg/mL, Glu 48 mg/dL

4. Evaluate her biochemical data. Which are not within a normal range?

(continued)

Nutrition Diagnosis:

5. Identify at least two nutrition problems that can be identified as a result of the nutrition assessment and medical history. Determine the diagnostic term for each nutrition problem. Next, identify the etiology of each nutrition problem. Finally, identify the signs and symptoms that support the evidence for these nutrition problems.

6. Why is this patient at age 16 still premenarchal? Does this place her at any medical risk?

7. Using the diagnostic criteria for eating disorders and your nutritional assessment data, identify her risk for an eating disorder. What are your conclusions?

END-OF-CHAPTER QUESTIONS

1. How is the energy content of food determined today? What is a kilojoule? How does it differ from a kilocalorie?

2. Describe the three main components of energy expenditure. What is the difference between basal energy expenditure and resting energy expenditure?

3. Describe three methods that can be used to estimate or determine a person's energy requirement.

4. List five substances produced in the body that can affect appetite. Pick two of them and describe them in more detail (source, function, and effect).

5. What is the clinical implication of excess adipose tissue in specific body locations? How should it be measured—BMI, waist circumference, or waist-to-hip ratio (WHR)? Why?

6. List at least five factors that have contributed to growing problem of obesity. Pick one factor and explain how you would address this concern with a client who needs to lose weight.

7. What are the key elements of nutrition therapy for weight loss? What additional benefits are associated with increasing physical activity?

8. What medications are currently approved for weight loss? Pick one, describe its mechanism of action, and list possible side effects. Describe one surgery that is performed for weight reduction. List possible complications associated with the surgery.

1. National Heart, Lung, and Blood Institute. Clinical guidelines on the identification, evaluation, and treatment of overweight and obesity in adults. Bethesda (MD): U.S. Department of Health and Human Services, National Institutes of Health; 1998.

2. National Heart, Lung, and Blood Institute. The practical guide: identification, evaluation, and treatment of overweight and obesity in adults. Bethesda (MD): U.S. Department of Health and Human Services, National Institutes of Health; 2000.

3. Saltzman E. Obesity as a health issue. In Bowman BA, Russell RM (eds.), Present knowledge in nutrition. 9th ed. Washington (DC): International Life Science Institute; 2006, 637–48.

4. Panel on Macronutrients, Panel on the Definition of Dietary Fiber, Subcommittee on Upper Reference Levels of Nutrients, Subcommittee on Interpretation and Uses of Dietary Reference Intakes, Standing Committee on the Scientific Evaluation of Dietary Reference Intakes. Dietary Reference Intakes for Energy, Carbohydrate, Fiber, Fat, Fatty Acids, Cholesterol, Protein, and Amino Acids. Washington (DC): National Academy Press, 2002.

5. Das SK, Roberts SB. Energy Metabolism. In Bowman BA, Russell RM (eds.), Present knowledge in nutrition. 9th ed. Washington (DC): International Life Science Institute; 2006, 45–55.

6. Roza AM, Shizgal HM. The Harris Benedict equation reevaluated: resting energy requirements and the body cell mass. Am J Clin Nutr. 1984;40:168–82.

7. World Health Organization. 1985. Energy and protein requirements. Report of a joint FAO/WHO/UNU expert consultation. Technical Report Series 724. Geneva, Switzerland:World Health Organization.

8. U.S. Department of Health and Human Services. Physical Activity Guidelines for Americans, 2008. Washington (DC): U.S. Government Printing Office, 2008. www.health.gov/paguidelines

9. Kushner RF. Evaluation and treatment of obesity. In Fauci AS, Braunwald E, Kasper DL, Hauser SL, Longo DL, Jameson JL, Loscalzo J (eds.), Harrison's principles of internal medicine. 17th ed. New York: McGraw-Hill, 2008, 468–73.

10. Trabulsi J, Troiano RP, Subar AF, Sharbaugh C, Kipnis V, Schatzkin A, Schoeeler DA. Precision of the doubly labeled water method in a large-scale application: evaluation of a streamlining-dosing protocol in the Observing Protein and Energy Nutrition (OPEN) study. Eur J Clin Nutr. 2003;57:1370–77.

11. Hoos MB, Plasqui G, Gerver WJM, Westerterp KR. Physical activity level measured by doubly labeled water and accelerometry in children. Eur J Appl Physiol. 2003;89:624–26.

12. Seale JL, Rumpler WV, Moe PW. Description of a direct-indirect room-sized calorimeter. Am J Physiol. 1991;260:E306–E320.

13. Committee on Metabolic Monitoring for Military Field Applications, Standing Committee on Military Nutrition Research, Food and Nutrition Board. Monitoring metabolic status: predicting decrements in physiological and cognitive performance. Washington (DC): National Academy Press, 2004.

14. Flier JS, Maratos-Flier E. Biology of obesity. In Fauci AS, Braunwald E, Kasper DL, Hauser SL, Longo DL, Jameson JL, Loscalzo J (eds.), Harrison's principles of internal medicine. 17th ed. New York: McGraw-Hill; 2008, 462–68.

15. Anderson JW. Diabetes mellitus: medical nutrition therapy. In Shils ME, Shike M, Ross AC, Cabellero B, Cousins RJ (eds.), Modern nutrition in health and disease. 10th ed. Philadelphia (PA): Lippincott Williams & Wilkins; 2006, 1043–66.

16. Smith GP. Controls of food intake. In Shils ME, Shike M, Ross AC, Cabellero B, Cousins RJ (eds.), Modern nutrition in health and disease. 10th ed. Philadelphia (PA): Lippincott Williams & Wilkins; 2006, 707–19.

17. Brodsky IG. Hormones and growth factors. In Shils ME, Shike M, Ross AC, Cabellero B, Cousins RJ (eds.), Modern nutrition in health and disease. 10th ed. Philadelphia (PA): Lippincott Williams & Wilkins; 2006, 636–54.

18. Paik KH, Jin DK, Song SY, Lee JE, Ko SH, Song SM, et al. Correlation between fasting plasma ghrelin levels and age, body mass index (BMI), BMI percentiles, and 24-hour plasma ghrelin profiles in Prader-Willi syndrome. J Clin Endocrinol Metab. 2004;89:3885–89.

19. Chanoine JP. Ghrelin in growth and development. Horm Res. 2005;63:129–38.

20. Saladin KS. Anatomy and physiology: the unity of form and function. 4th ed. Boston (MA): McGraw-Hill; 2007.

21. Pleuss J, Matfin G. Alterations in Nutritional Status. In Porth CM, Matfin G. (eds.) Pathophysiology: concepts of altered health states. 8th ed. Philadelphia (PA): Lippincott Williams & Wilkins; 2009, 982–1005.

22. Hsing AW, Sakoda LC, Chua SC. Obesity, metabolic syndrome, and prostate cancer. Am J Clin Nutr. 2007;86(suppl):843S–857S.

23. Forsythe LK, Wallace JM, Livingstone MB. Obesity and inflammation: the effects of weight loss. Nutr Res Rev. 2008;21:117–23.

24. Pradhan A. Obesity, metabolic syndrome, and type 2 diabetes: inflammatory basis of glucose metabolic disorders. Nutr Rev. 2007;65:S152–S156.

25. Norgan NG. Body composition. In Ulijaszek SJ, Johnston FE, Preece MA (eds.), *The Cambridge Encyclopedia of Human Growth and Development.* London: Cambridge University Press, 1998; 212–15.

26. Dietz WH. Childhood Obesity. In Shils ME, Shike M, Ross AC, Cabellero B, Cousins RJ (eds.), Modern nutrition in health and disease. 10th ed. Philadelphia (PA): Lippincott Williams & Wilkins; 2006, 979–90.

27. Lee RD, Nieman DC. Nutritional assessment. 5th ed. Boston (MA): McGraw-Hill; 2010.

28. International Obesity Task Force, European Association for the Study of Obesity. Obesity in Europe. London: International Obesity Task Force; 2002. http://www.iotf.org/media/euobesity.pdf

29. International Obesity Task Force, European Association for the Study of Obesity. Obesity in Europe 2: Waiting for a Green Light for Health. London: International Obesity Task Force; 2003. http://www.iotf.org/media/euobesity2.pdf

30. Shields M. Overweight Canadian children and adolescents. Ottawa: Statistics Canada; 2005.

31. Tjepkema M. Adult Obesity in Canada, Measured Height and Weight. Ottawa: Statistics Canada; 2005.

32. U.S. Department of Health and Human Services and U.S. Department of Agriculture. Dietary Guidelines for Americans, 2005. 6th ed. Washington (DC): U.S. Government Printing Office; 2005.

33. Barlow SE. Expert committee recommendations regarding the prevention, assessment, and treatment of child and adolescent overweight and obesity: summary report. Pediatrics. 2007; 120(suppl 4):S164–S192.

34. Daniels SR, Jacobson MS, McCrindle BW, Eckel RH, Sanner BM. American Heart Association Childhood obesity research summit report. Circulation. 2009;119:e489–e517.

35. Krebs NS, Himes JH, Jacobson D, Nicklas TA, Guilday P. Assessment of child and adolscent overweight and obesity. Pediatrics. 2007;120(suppl 4):S193–S228.

36. Yusuf S, Hawken S, Ounpuu S, Bautista L, Franzosi MG, et al. Obesity and the risk of myocardial infarction in 27,000 participants from 52 countries: a case-control study. Lancet. 2005;366:1640–49.

37. Hill JO, Catenacci VA, Wyatt HR. Obesity: etiology. In Shils ME, Shike M, Ross AC, Cabellero B, Cousins RJ (eds.), Modern nutrition in health and disease. 10th ed. Philadelphia (PA): Lippincott Williams & Wilkins; 2006, 1013–28.

38. Pi-Sunyer FX. The epidemiology of central fat distribution in relation to disease. Nutr Rev. 2004;62:S120–S26.

39. Snijder MB, Dekker JM, Visser M, Yudkin JS, Stehouwer CD, et al. Larger thigh and hip circumferences are associated with better glucose tolerance: the Hoorn Study. Obes Res. 2003;11:104–111.

40. Snijder MB, Zimmet PZ, Visser M, Dekker JM, Seidell JC, et al. Independent and opposite associations of waist and hip circumferences with diabetes, hypertension and dyslipidemia: the AusDiab Study. Int J Obes Relat Metab Disord. 2004;28:402–09.

41. World Health Organization. Obesity and Overweight Facts [monograph on the Internet]. Geneva (Switzerland):World Health Organization; 2005. Available at http://www.who.int/dietphysicalactivity/publications/facts/obesity/en/index.html

42. U.S. Department of Health and Human Services. The Surgeon General's Call to Action to Prevent and Decrease Overweight and Obesity. Rockville (MD): U.S. Department of Health and Human Services, Public Health Service, Office of the Surgeon General; 2001.

43. U.S. Department of Health and Human Services. Healthy People 2010. 2nd ed. Washington (DC): U.S. Government Printing Office; 2000.

44. Hedley AA, Ogden CL, Johnson CL, Carroll MD, Curtin LR, Flegal KM. Overweight and obesity among US children, adolescents, and adults, 1999–2002. JAMA. 2004;291:2847–50.

45. Cole TJ, Bellizzi MC, Flegal KM, Dietz WH. Establishing a standard definition for child overweight and obesity worldwide: international survey. Brit Med J. 2000;320:1240–45.

46. Petersen S, Peto V, Rayner M, Leal J, Luengo-Fernandez R, Gray A. European cardiovascular disease statistics. London:British Heart Foundation; 2005.

47. Silventoinen K, Sans S, Tolonen H, Monterde D, Kuulasmaa K, Kesteloot K, Tuomilehto J. Trends in obesity and energy supply in the WHO MONICA Project. Int J Obes. 2004;28:710–18.

48. Fry J, Finley W. The prevalence and costs of obesity in the EU. Proc Nutr Soc. 2005;64:359–62.

49. Livingstone B. Epidemiology of childhood obesity in Europe. Eur J Pediatr. 2000;159(Suppl 1):S14–S34.

50. Lobstein T, Baur L, Uauy R. Obesity in children and young people: a crisis in public health. Obes Rev. 2004;5:4–85.

51. Garner DM, Garfinkel PE, Schwartz D, Thompson M. Cultural expectations of thinness in women. Psychol Rep. 1980;47:483–91.

52. Katzmarzyk PT, Davis C. Thinness and body shape of Playboy centerfolds from 1978 to 1998. Int J Obes Relat Disord. 2001;25:590–92.

53. Guven S, Kuenzi JA, Matfin G. Diabetes mellitus and the metabolic syndrome. In: Porth CM (ed.). Pathophysiology: concepts of altered health status. 7th ed. Philadelphia (PA): Lippincott Williams & Wilkins; 2005, 987–1015.

54. Mokdad AH, Ford ES, Bowman BA, Dietz WH, Vinicor F, Bales VS, Marks JS. Prevalence of obesity, diabetes, and obesity-related health risk factors, 2001. JAMA. 2003;289:76–79.

55. Manson JE, Skerrett PJ, Greenland P, VanItallie TB. The escalating pandemics of obesity and sedentary lifestyle: a call to action for clinicians. Arch Intern Med. 2004;164:249–58.

56. Diabetes Prevention Program Research Group. Reduction in the incidence of type 2 diabetes with lifestyle intervention or metformin. N Engl J Med. 2002;346:393–403.

57. Fisher NDL, Williams GH. Hypertensive vascular disease. In: Kasper DL, Braunwald E, Fauci AS, Hauser SL, Longo DL, Jameson JL, editors. Harrison's principles of internal medicine. 16th ed. New York: McGraw-Hill; 2005, 1463–81.

58. Grundy SM. Nutrition in the management of disorders of serum lipids and lipoproteins. In: Shils ME, Shike M, Ross AC, Cabellero B, Cousins RJ, editors. Modern nutrition in health and disease. 10th ed. Philadelphia (PA): Lippincott Williams & Wilkins; 2006, 1076–94.

59. Wessell TR, Arant CB, Olson MB, Johnson BD, Reis SE, Sharaf BL, et al. Relationship of physical fitness vs. body mass index with coronary artery disease and cardiovascular events in women. JAMA. 2004;292:1179–87.

60. Adams LA, Lindor KD. Nonalcoholic fatty liver disease. Ann Epidemiol. 2007;17:863–69.

61. Angulo P. Obesity and Nonalcoholic Fatty Liver Disease. Nutr Rev. 2007;65:S57–S63.

62. Porth CM. Disorders of Hepatobiliary and Exocrine Pancreas Function. In Porth CM, Matfin G. (eds.), Pathophysiology: concepts of altered health states. 8th ed. Philadelphia (PA): Lippincott Williams & Wilkins; 2009, 949–81.

63. Bacon BR. Genetic, metabolic, and infiltrative diseases affecting the liver. In Fauci AS, Braunwald E, Kasper DL, Hauser SL, Longo DL, Jameson JL, Loscalzo J (eds.), Harrison's principles of internal medicine. 17th ed. New York: McGraw-Hill; 2008, 1980–83.

64. Strom SS, Wang X, Pettaway CA, Logothetis CJ, Yamamura Y, Do KA, et al. Obesity, weight gain, and risk of biochemical failure among prostate cancer

patients following prostatectomy. Clin Cancer Res. 2005;11:6889–94.

65. Mehring PM. Disorders of the Female Reproductive System. In Porth CM, Matfin G. (eds.) Pathophysiology: concepts of altered health states. 8th ed. Philadelphia (PA): Lippincott Williams & Wilkins; 2009, 1129–65.

66. Fontaine KR, Redden DT, Wang C, Westfall AO, Allison DB. Years of life lost due to obesity. JAMA. 2003;289:187–93.

67. Flegal KM, Troiano RP, Pamuk ER, Kuczmarski RJ, Campbell SM. The influence of smoking cesssation on the prevalence of overweight in the United States. N Engl J Med. 1995;333:1165–70.

68. Stunkard AJ, Allison KC. Two forms of disordered eating in obesity: binge eating and night eating. Int J Obes Relat Metab Disord. 2003;27:1–12.

69. Tanofsky-Kraff M, Yanovski SZ. Eating disorder or disordered eating? Non-normative eating patterns in obese individuals. Obes Res. 2004;12:1361–66.

70. Lyon HN, Hirschhorn JN. Genetics of common forms of obesity: a brief overwiew. Am J Clin Nutr. 2005;82(suppl):215S–217S.

71. Mayer J. Genetic factors in human obesity. Ann NY Acad Sci. 1965;131: 412–21.

72. Bouchard C, Tremblay A, Despres JP, Nadeau A, Lupien PJ, et al. The response to long-term overfeeding in identical twins. N Engl J Med. 1990;322:1477–82.

73. Stunkard AJ. Genetic contributions to human obesity. Res Publ Assoc Res Nerv Ment Dis. 1991;69:205–18.

74. Sorensen TI. The genetics of obesity. Metabolism. 1995;44(Suppl. 3):4–6.

75. Loos RJF, Rankinin T. Gene-diet interactions on body weight changes. J Am Diet Assoc. 2005;105:S29–S34.

76. Hill JO, Wyatt HR, Reed GW, Peters JC. Obesity and the environment: where do we go from here? Science. 2003;299:853–55.

77. Ello-Martin JA, Ledikwe JH, Rolls BJ. The influence of food portion size and energy density on energy intake: implications for weight management. Am J Clin Nutr. 2005;82(suppl):236S–241S.

78. Rolls BJ, Drewnowski A, Ledikwe JH. Changing the energy density of the diet as a strategy for weight management. J Amer Diet Assoc. 2005;105:S98–S103.

79. Drewnowski A, Darmon N. The economics of obesity: dietary energy density and energy cost. Am J Clin Nutr. 2005;82(suppl):265S–273S.

80. Booth KM, Pinkston MM, Poston WSC. Obesity and the built environment. J Am Diet Assoc. 2005;105:S110–S117.

81. Jackson DM, Djafarian K, Stewart J, Speakman JR. Increased television viewing is associated with elevated body fatness but not with lower total energy expenditure in children. Am J Clin Nutr. 2009;891031–36.

82. Danner FW. A national longitudinal study of the association between hours of TV viewing and the trajectory of BMI growth among US children. J Pediatr Psychol. 2008;33:1100–07.

83. Anderson SE, Economos CD, Must A. Active play and screen time in US children aged 4 to 11 years in relation to sociodemographic and weight status characteristics: a nationally representative cross-sectional analysis. BMC Public Health. 2008, 8:366. doi:10.1186/1471-2458-8-366. http://www.biomedcentral.com/1471-2458/8/366

84. Jeffery RW, Utter J. The changing environment and population obesity in the United States. Obes Res. 2003;11(suppl):12S–22S.

85. Wadden TA, Byrne KJ, Krauthamer-Ewing S. Obesity: Management. In Shils ME, Shike M, Ross AC, Cabellero B, Cousins RJ (eds.), Modern Nutrition in Health and Disease. 10th ed. Philadelphia (PA): Lippincott Williams & Wilkins; 2006, 1029–42.

86. Kral JG. ABC of obesity management: part III—surgery. Brit Med J. 2006;333:900–03.

87. DeMaria EJ. Bariatric surgery for morbid obesity. N Engl J Med. 2007;356:2176–83.

88. Robinson MK. Surgical treatment of obesity—weighing the facts. N Engl J Med. 2009;361:520–21.

89. Vella M, Galloway DJ. Laparoscopic adjustable gastric banding for severe obesity. Obes Surg. 2003;13:642–48.

90. Provost DA. Laparoscopic adjustable gastric banding: an attractive option. Surg Clin North Am. 2005;85:789–805.

91. Al-Momen A, El-Mogy I, Ibrahim A. Initial experience with Swedish adjustable gastric band at Saad Specialist Hospital, Al-Khobar, Saudi Arabia. Obes Surg. 2005;15:506–09.

92. Frezza EE. Laparoscopic vertical sleeve gastrectomy for morbid obesity. The future procedure of choice? Surg Today. 2007;37:275–81.

93. Hess DS, Hess DW, Oakley RS. The biliopancreatic diversion with the duodenal switch: results beyond 10 years. Obes Surg. 2005;15:408–16.

94. Adams TD, Gress RE, Smith SC, Halverson RC, Simper SC, et al. Long-term mortality after gastric bypass surgery. N Engl J Med. 2007;357:751–61.

95. Longitudinal Assessment of Bariatric Surgery (LABS) Consortium. Perioperative safety in the longitudinal assessment of bariatric surgery. N Engl J Med. 2009;361:445–54.

96. Hoffer LJ. Metabolic Consequences of Starvation. In Shils ME, Shike M, Ross AC, Cabellero B, Cousins RJ (eds.), Modern nutrition in health and disease. 10th ed. Philadelphia (PA): Lippincott Williams & Wilkins; 2006, 730–48.

97. Leibel RL, Rosenbaum M, Hirsch J. Changes in energy expenditure resulting from altered body weight. N Engl J Med. 1995;332:621–28.

98. National Cholesterol Education Program. Third Report of the National Cholesterol Education Program Expert Panel on Detection, Evaluation, and Treatment of High Blood Cholesterol in Adults (Adult Treatment Panel III). Bethesda (MD): National Institutes of Health, National Heart, Lung, and Blood Institute; 2002.

99. Eckel RH. The dietary approach to obesity: is it diet or the disorder. Am J Clin Nutr. 2005;293:96–97.

100. Noakes M, Keogh JB, Foster PR, Clifton PM. Effect of an energy-restricted, high-protein, low-fat diet relative to a conventional high-carbohydrate, low-fat diet on weight loss, body composition, nutritional status, and markers of cardiovascular health in obese women. Am J Clin Nutr. 2005;81:1298–1306.

101. Sacks FM, Bray GA, Carey VJ, Smith SR, Ryan DH, et al. Comparison of weight-loss diets with different composition of fat, protein, and carbohydrates. N Engl J Med. 2009;360:859–73.

102. Brinkworth GD, Noakes M, Buckley JD, Keogh JB, Clifton PM. Long-term

effects of a very-low carbohydrate weight loss diet compared with an isocaloric low-fat diet after 12 mo. Am J Clin Nutr. 2009;90:23–32.

103. Dansinger ML, Gleason JA, Griffith JL, Selker HP, Schaefer EJ. Comparison of the Atkins, Ornish, Weight Watchers, and Zone diets for weight loss and heart disease risk reduction. JAMA. 2005;293:43–53.

104. Katan MB. Weight-loss diets for the prevention and treatment of obesity. N Engl J Med. 2009;360:923–24.

105. Melanson K, Dwyer J. Popular Diets for Treatment of Overweight and Obesity. In Wadden TA, Mitchell JE, Cook-Myers T, Wonderlich SA. Diagnostic criteria for anorexia nervosa: looking ahead to DSM-V. Int J Eat Disord. 2005;37:S95–S97.

106. Hu FB, LI TY, Colditz GA, Willett WC, Manson JE. Television watching and other sedentary behaviors in relation to risk of obesity and type 2 diabetes mellitus in women. JAMA. 2003;289:1785–91.

107. Schaefer EJ, Gleason JA, Dansinger ML. The effects of low-fat, high-carbohydrate diets on plasma lipoproteins, weight loss, and heart disease risk reduction. Curr Atheroscler Rep. 2005;7:421–27.

108. Lee CD, Blair SN, Jackson AS. Cardiorespiratory fitness, body composition, and all-cause and cardiovascular disease mortality in men. Am J Clin Nutr. 1999;69:373–80.

109. Berkel LA, Poston WSC, Reeves RS, Foreyt JP. Behavioral interventions for obesity. J Am Diet Assoc. 2005;105: S35–S43.

110. American Psychiatric Association. Diagnostic and Statistical Manual of Mental Disorders. Fourth Edition, Text Revision (DSM-IV-TR). Washington (DC): American Psychiatric Association; 2000. (This is APA 2000 DSM)

111. American Dietetic Association. Position of the American Dietetic Association: nutrition intervention in the treatment of anorexia nervosa, bulimia nervosa, and eating disorders not otherwise specified (EDNOS). J Am Diet Assoc. 2001;101:810–19.

112. Coughlin JW, Guarda A. Behavioral Disorders Affecting Food Intake: Eating Disorders and Other Psychiatric Conditions. In Shils ME, Shike M, Ross AC, Cabellero B, Cousins RJ (eds.), Modern Nutrition in Health and Disease. 10th ed. Philadelphia (PA): Lippincott Williams & Wilkins; 2006, 1353–61.

113. Mitchell JE, Cook-Myers T, Wonderlich SA. Diagnostic criteria for anorexia nervosa: looking ahead to DSM-V. Int J Eat Disord. 2005;37:S95–S97.

114. Walsh BT. Eating disorders. In Fauci AS, Braunwald E, Kasper DL, Hauser SL, Longo DL, Jameson JL, Loscalzo J (eds.), Harrison's principles of internal medicine. 17th ed. New York: McGraw-Hill; 2008, 473–77.

115. Fairburn CG, Harrison PJ. Eating disorders. Lancet. 2003;361:407–16.

116. Bulik CM, Reba L, Siega-Riz AM, Reichborn-Kjennerud T. Anorexia nervosa: definition, epidemiology, and cycle of risk. Int J Eat Disord. 2005;37:S2–S9.

117. Hoek HW, van Hoeken D. Review of the prevalence and incidence of eating disorders. Int J Eat Disord. 2003;34:383–96.

118. Mehler PS. Bulimia nervosa. N Engl J Med. 2003;349:875–81.

119. Bulik CM, Sullivan PF, Tozzi F, Furberg H, Lichtenstein P, Pedersen NL. Prevalence, heritability, and prospective risk factors for anorexia nervosa. Arch Gen Psychiatry. 2006;63:305–12.

120. Becker AE, Grinspoon SK, Klibanski A, Herzog DB. Eating disorders. N Engl J Med. 1999;340:1092–98.

121. Mehler PS. Diagnosis and care of patients with anorexia nervosa in primary care settings. Ann Intern Med. 2001;134:1048–59.

122. Mehler PS. Osteoporosis in anorexia nervosa: prevention and treatment. Int J Eat Disorder. 2003;33:113–26.

123. American Psychiatric Association. Practice guideline for the treatment of patients with eating disorders. Am J Psychiatry. 2000;157(suppl):1–39. (This is APA 2000)

124. Vandereycken W. The place of inpatient care in the treatment of anorexia nervosa: questions to be answered. Int J Eat Disord. 2003;34:409–22.

125. Gowers S, Bryant-Waugh R. Management of child and adolescent eating disorders: the current evidence base and future directions. J Child Psychol Psychiatry. 2004;45:63–83.

126. Cockfield A, Philpot U. Feeding size 0: the challenges of anorexia nervosa. Managing anorexia from a dietitian's perspective. Proc Nutr Soc. 2009;68:281–88.

127. Marinella MA. The refeeding syndrome and hypophosphatemia. Nutr Rev. 2003;61:320–23.

13

Diseases of the Cardiovascular System

Thomas J. Pujol, EdD, FACSM
Southeast Missouri State University
Joshua E. Tucker, MS
Michigan State University College of Osteopathic Medicine
Jeremy T. Barnes, Ph.D., CSCS, HFI, CHES
Southeast Missouri State University

CHAPTER OUTLINE

Introduction

Diseases of the cardiovascular system are the leading causes of death in the United States, accounting for 34.2% of all deaths in 2006, or 1 in every 2.98 deaths (see Figure 13.1). Cardiovascular disease (CVD) was an underlying or contributing cause in over 1.37 million deaths (or about 56% of all deaths) in 2005. Approximately 80 million American adults live with one or more cardiovascular diseases, and 38 million of these individuals are over the age of 60. These diseases include high blood pressure, coronary heart disease, cerebrovascular disease, peripheral vascular disease, and congenital heart defects. Projections for 2009 indicated that the direct and indirect costs of cardiovascular disease would exceed $475 billion.[1] In 2006, over 7 million inpatient cardiovascular operations and procedures were performed in the United States.[1] In the past thirty years, however, the age-adjusted cardiovascular disease death rate has dropped by almost 40% as a result of developments in treatment and prevention of these diseases.[2] CVD deaths decreased by 27% between 1994 and 2004.[1] A study comparing coronary heart disease (CHD) deaths between and 1980 and 2000 estimated that about 47% of the decrease in CHD deaths was attributable to treatments and about 44% was attributable to changes in risk factors.[3]

In this chapter, all major forms of cardiovascular disease—hypertension, atherosclerosis, ischemic heart disease, peripheral vascular disease, and heart failure—will be discussed. These conditions are interrelated and more often than not will coexist. For example, atherosclerosis is often synonymous with ischemic heart disease and peripheral vascular disease, while

acute coronary syndrome—condition characterized by an episode of acute unstable angina

aneurysm—a weakened portion of the blood vessel wall

angina—chest pain caused by oxygen deficit to the heart; two forms are stable and unstable
- **stable angina**—chest pain associated with increased oxygen demand such as occurs with physical exertion
- **unstable angina**—chest pain that occurs at rest

ankle brachial index (ABI)—ratio of Doppler-recorded systolic blood pressures between upper and lower extremities; a measure of peripheral vascular disease

apolipoprotein—protein portion of the lipoprotein; provides cellular stability and allows for cellular recognition and binding

arteriosclerosis—a general term for thickening of the walls of the blood vessels with a resulting loss of vascular elasticity and narrowed lumen

atherosclerosis (AS)—thickening of the blood vessel walls specifically caused by the presence of plaque

cardiac cachexia—CVD-associated malnutrition/wasting syndrome characterized by extreme skeletal muscle wasting, fatigue, and anorexia

cardiac output—the volume of blood ejected from the left ventricle each minute; mathematically defined as heart rate × stroke volume

claudication—pain in arms and legs due to inadequate blood flow to those muscles

congestive heart failure (CHF) or heart failure (HF)— impairment of the ventricles' capacity to eject blood from the heart or to fill with blood

coronary artery disease (CAD)—general term for all causes of heart disease characterized by narrowing of vessels supplying blood to the heart

diastole—relaxation phase of the cardiac cycle; during this phase, ventricles empty and blood fills the atria

diastolic blood pressure—pressure that occurs as ventricles relax (diastole phase of the cardiac cycle)

ejection fraction—the percentage of the LVEDV that is ejected in the systolic phase; in normal, apparently healthy adults, the typical ejection fraction is 50% to 60%; defined mathematically as stroke volume ÷ LVEDV

embolus—blood clot that breaks from the cellular surface and freely moves through the circulation

foam cells—macrophage cells containing lipid; found within the fatty streaks in the development of atherosclerosis

hepatomegaly—enlargement of the liver

hydrogenation—the addition of hydrogen atoms; in the food industry, the addition of hydrogens to unsaturated fatty acids in order to increase their degree of saturation; results in the formation of *trans* fatty acids (with hydrogens on opposite sides of the C-C double bond)

hypertension—condition of chronically elevated blood pressure

hypertrophic cardiomyopathy—a genetic disorder causing abnormal thickening of the left ventricular wall

infarct—cellular necrosis as a result of lack of oxygen

ischemic heart disease (IHD)—heart disease characterized by inadequate blood supply to the heart

left ventricular end diastolic volume (LVEDV)—the amount of blood in the left ventricle at the end of the diastolic phase and immediately prior to systolic ejection of blood

left ventricular end systolic volume (LVESV)—the amount of blood that remains in the left ventricle at the conclusion of the systolic phase

left ventricular hypertrophy (LV hypertrophy)—enlargement of the left ventricle; most commonly related to hypertension and/or congestive heart failure

monounsaturated fats—sources of fat that have a predominant amount of fatty acids with one carbon-carbon double bond within their chemical structures

myocardial cells—cells found in the myocardium

myocardial infarction (MI)—necrosis of the myocardial cells as a result of oxygen deprivation.

orthopnea—shortness of breath associated with lying in the supine position

peripheral arterial disease (PAD)—atherosclerotic heart disease of all vessels except specific coronary vessels; term used interchangeably with peripheral vascular disease

peripheral vascular disease (PVD)—atherosclerotic heart disease of all vessels except specific coronary vessels

polyunsaturated fats—sources of fat that have a predominant amount of fatty acids that contain more than one double bond in their chemical structures

salt resistant—describes an individual whose body presents resistance to change in blood pressure as a result of salt intake

salt sensitive—describes an individual who experiences an increase in blood pressure as a result of salt intake

saturated fats—sources of fat that have a predominant amount of fatty acids that contain all single bonds within their chemical structures

splenomegaly—enlargement of the spleen

stearic acid—an 18-carbon saturated fatty acid

stroke volume—the volume of blood that is ejected from the left ventricle with each systolic phase; defined mathematically as LVEDV −LVESV

systole—contraction phase of the cardiac cycle; during this phase blood is ejected from the ventricles into the aorta and pulmonary artery

systolic blood pressure—pressure exerted when ejected from the ventricles (systole phase of the cardiac cycle)

thrombus—blood clot

ulceration—break in skin or tissue surface

ventricular fibrillation—uncontrolled contractions of the ventricle; often associated with myocardial infarction

ventricular tachycardia—rapid heartbeat originating from the ventricle

---

hypertension is a risk factor for all other cardiovascular diseases. As with all chronic disease, the risk of developing a cardiovascular disease is determined by a combination of hereditary, environmental, and lifestyle factors. The lifestyle modifications described in this chapter aid in the prevention and treatment of these conditions.

Anatomy and Physiology of the Cardiovascular System

The role of the cardiovascular system is to regulate blood flow to the tissues in order to deliver oxygenated blood and nutrients as well as to retrieve waste products from cellular metabolism. Other major functions include thermoregulation, hormone transport, maintenance of fluid volume, regulation of pH (to control acidosis and alkalosis), and gas exchange. The cardiovascular system forms a closed loop of blood vessels for which the heart acts as two pumps.

The Heart

The heart is a hollow, muscular organ whose walls are composed of three layers. The outer layer is the epicardium, and the inner layer is the endocardium. The middle layer, the myocardium, is responsible for the muscle contraction that pumps the blood from the heart. The heart is divided into four chambers (see Figure 13.2). The upper two chambers are the left and right atria, and the lower two chambers are the left and right ventricles. The right atrium and right ventricle make up the right heart and pump blood through the pulmonary circulation for oxygenation. The left atrium and left ventricle make up the left heart and pump oxygenated blood through the systemic circulation.

Figure 13.1 Age-Adjusted Death Rates for Cardiovascular Diseases, 2001–2003*

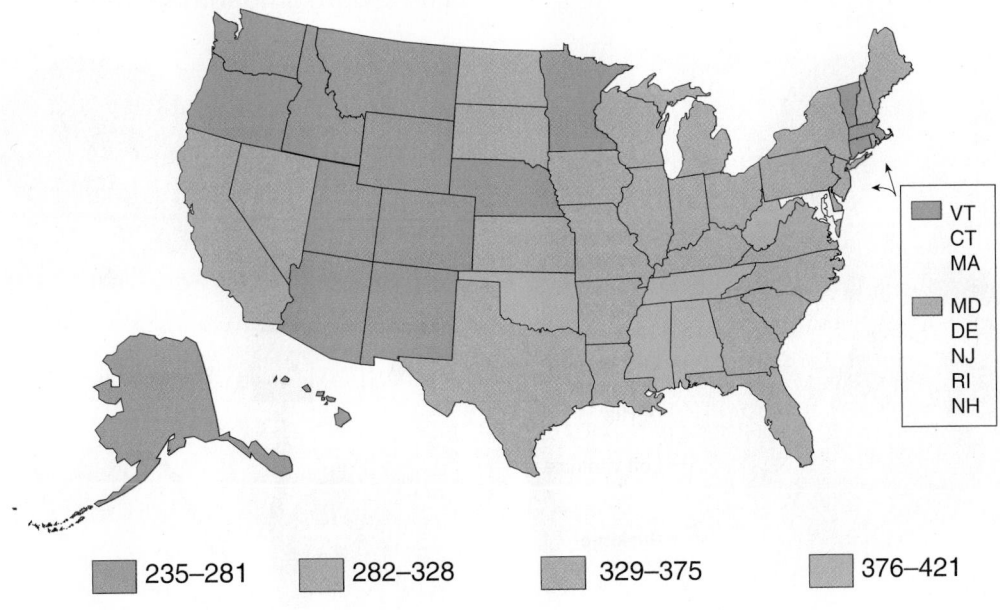

	VT
	CT
	MA
	MD
	DE
	NJ
	RI
	NH

■ 235–281   ■ 282–328   ■ 329–375   ■ 376–421

*Deaths per 100,000 population.

Source: National Heart, Lung, and Blood Institute. Morbidity & Mortality: 2007 Chart Book on Cardiovascular, Lung, and Blood Diseases, Chart 3-11, p. 27.

Figure 13.2 Blood Flow through the Heart

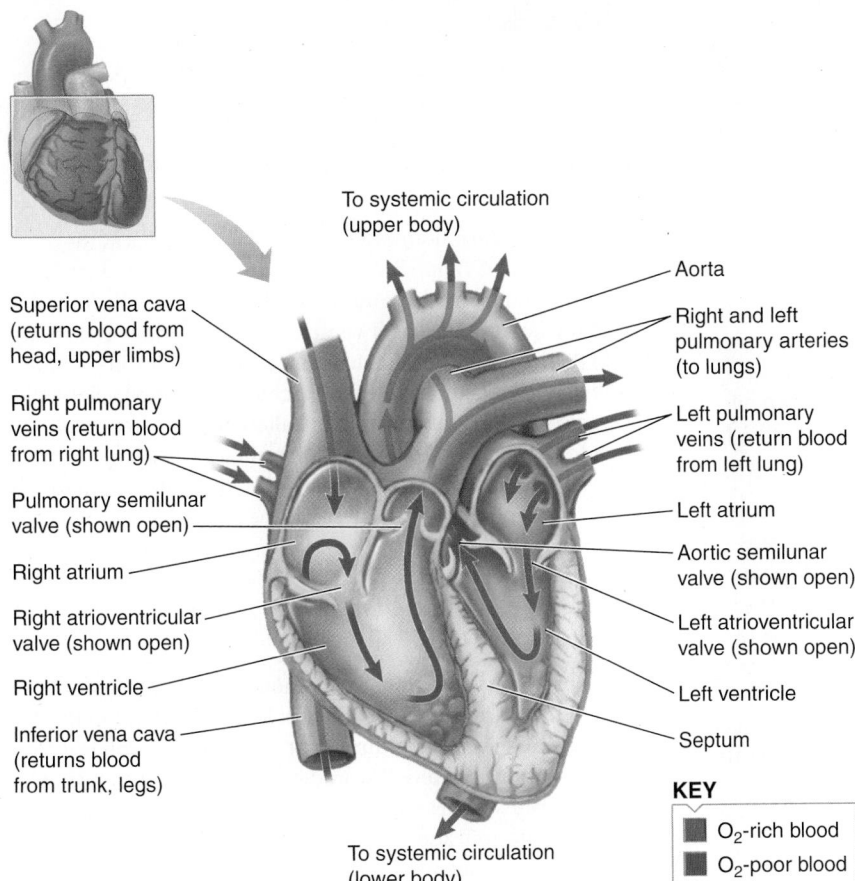

To systemic circulation (upper body)

Superior vena cava (returns blood from head, upper limbs)

Right pulmonary veins (return blood from right lung)

Pulmonary semilunar valve (shown open)

Right atrium

Right atrioventricular valve (shown open)

Right ventricle

Inferior vena cava (returns blood from trunk, legs)

Aorta

Right and left pulmonary arteries (to lungs)

Left pulmonary veins (return blood from left lung)

Left atrium

Aortic semilunar valve (shown open)

Left atrioventricular valve (shown open)

Left ventricle

Septum

To systemic circulation (lower body)

KEY
■ O₂-rich blood
■ O₂-poor blood

Source: Sherwood's *Human Physiology: From Cells to Systems* 7e, Figure 9-2 (a), p. 305.

Blood moves from the right atrium into the right ventricle, where it is pumped through the pulmonary trunk into the pulmonary circulation. As blood flows through the pulmonary circulation, it enters the lungs, where carbon dioxide is removed and oxygen is added. The blood returns to the heart via the pulmonary veins and enters the left atrium. Then it moves from the left atrium to the left ventricle (LV), from which it is pumped into the systemic circulation through the aorta, the largest artery. Adequate function of the LV is crucial for maintaining adequate cardiovascular function; this becomes obvious during the conditions of hypertension, congestive heart failure, and **left ventricular hypertrophy (LVH)**. The aorta branches into smaller arteries, which direct blood to the organs of the body. As an artery enters into an organ, it branches into progressively smaller arteries and eventually into arterioles. The arterioles then branch into the body's smallest vessels, the capillaries, where gas exchange occurs.

The blood is then returned to the heart through progressively larger vessels. The capillaries unite to form venules, and then the venules unite to form veins. The veins from the upper part of the body drain into and form the superior vena cava, and those from the lower part of the body drain into and form the inferior vena cava. The superior and inferior vena cava unite to return the blood to the right atrium.

Figure 13.3 Specialized Conduction System of the Heart

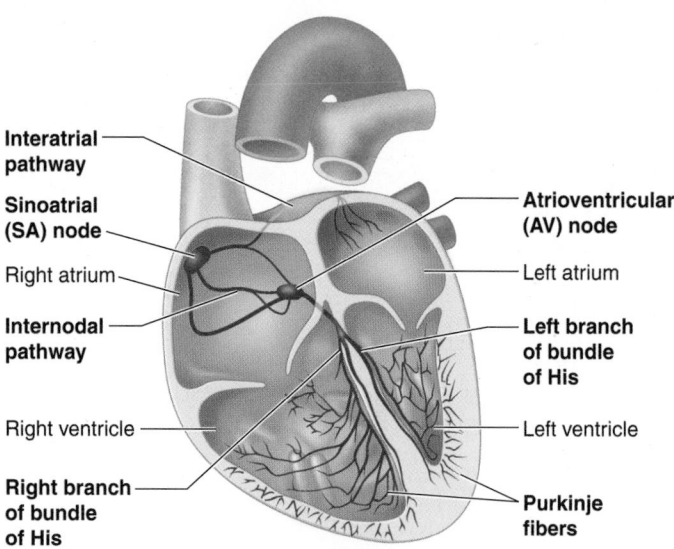

- Interatrial pathway
- Sinoatrial (SA) node
- Right atrium
- Internodal pathway
- Right ventricle
- Right branch of bundle of His
- Atrioventricular (AV) node
- Left atrium
- Left branch of bundle of His
- Left ventricle
- Purkinje fibers

Source: Sherwood's *Human Physiology: From Cells to System*, 7e, Figure 9-8 (a), p. 311.

Electrical Activity of the Heart Heart muscle cells are connected to one another by membranes called intercalated discs, which allow electrical impulses to pass from one cell to the next. Though many **myocardial cells** have the capability of generating spontaneous electrical activity, under normal conditions electrical activity is initiated in the heart at the sinoatrial (SA) node (see Figure 13.3). The change in electrical membrane potential (depolarization) in the SA node causes the contraction of the atria. The depolarization is carried from the atria to the ventricles by way of the atrioventricular (AV) node located in the base of the right atrium. The depolarization of the AV node is carried into the ventricles by the atrioventricular bundle (bundle of His), which splits into the right and left bundle branches. The depolarization is carried down the bundle branches, and then spread throughout the ventricles by the Purkinje fibers. As resting membrane potential decreases, the SA node will reach its depolarization threshold, and the process will repeat. The electrical activity of the heart can be measured through an electrocardiogram (ECG) (see Box 13.12 later in the chapter).

Cardiac Cycle The repeating contraction and relaxation of the heart is termed the cardiac cycle. The heart alternates between the phase of contraction (the **systole**) and the phase of relaxation (the **diastole**). Since the atria and ventricles depolarize and thus contract separately, there is an atrial systole and diastole as well as a ventricular systole and diastole. The force exerted by the blood on the walls of blood vessels during the contraction of the ventricles is termed **systolic blood pressure;** the force exerted during relaxation of the ventricles is **diastolic blood pressure** (see Figure 13.4).

Cardiac Function

The volume of blood ejected with each contraction of the left ventricle is termed **stroke volume**. It is regulated by **end diastolic volume (EDV)**, mean aortic blood pressure

Figure 13.4 Measurement of Blood Pressure Using a Sphygmonomanometer

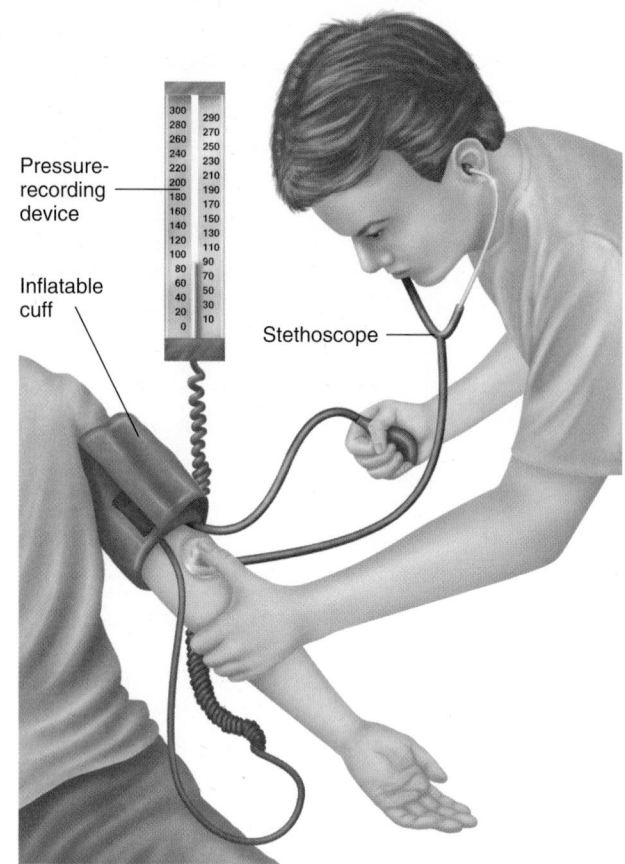

- Pressure-recording device
- Inflatable cuff
- Stethoscope

Source: L. Sherwood *Human Physiology: From Cells to System*, 5e, copyright 2004, p. 350.

(mean arterial pressure/MAP), and strength of ventricular contraction.[4]

EDV refers to the amount of blood in the ventricles at the end of diastole. Often EDV is referred to as preload, since it is indicative of the possible amount of blood that can be forced out of the heart on the next ventricular contraction. The greater the EDV, the more the ventricles are stretched. Starling's Law of the heart indicates that this stretching of the ventricles will increase their force of contraction, allowing a greater amount of blood to be ejected. (According to Starling's Law, osmotic and hydrostatic pressure work together in favor of moving fluid out of the blood into interstitial areas at the arterial end of the capillary and restoring fluid back into blood at the venous end of the capillary; see Chapter 7.) The fraction of EDV that is ejected from the heart by contraction of the left ventricle is termed the **ejection fraction (EF)**.

EDV is primarily determined by venous return, the amount of blood that is returned to the heart by the veins. Several factors increase venous return. First, venoconstriction reduces the amount of blood stored in the veins by decreasing their capacity. Venous return is also increased by respirations. During inspiration, abdominal pressure increases and pressure within the thorax decreases, promoting blood flow back to the heart. Rhythmic skeletal muscle contractions, termed a skeletal muscle pump, also affect venous return. As the muscles contract, blood is pushed toward the heart because the muscles compress the veins. Valves within the veins prevent blood from flowing away from the heart.[4]

The mean arterial pressure (MAP) also affects stroke volume. MAP (also termed afterload) is the average force exerted by blood against the walls of the arteries over a cardiac cycle. MAP represents the resistance against which the ventricles must contract in order to eject blood into systemic circulation. MAP can be calculated using the values of systolic and diastolic blood pressure: MAP = [(2 × diastolic) + systolic]/3. The normal range for MAP is 70–110, which is adequate to perfuse all tissues and organs. If there is increased resistance such as that seen in atherosclerosis, the ventricles will eject less blood, and this could eventually lead to the complications of heart failure.

The strength of ventricular contraction is affected by circulating epinephrine and norepinephrine, as well as the sympathetic stimulation of the heart by cardiac accelerator nerves. These mechanisms increase the amount of calcium available to the myocardial cells, thus increasing contractility.[4]

Regulation of Blood Pressure

MAP is determined through a combination of **cardiac output** and total peripheral resistance. The MAP must be regulated so that it is high enough to force blood through systemic circulation without being so high as to cause vascular damage.

The regulation of MAP (see Figure 13.5) involves the sympathetic nervous system, the renin-angiotensin-aldosterone system, and renal function. All three affect cardiac output and thus blood pressure (BP). Cardiac output is equal to heart rate multiplied by stroke volume. Heart rate is dependent upon the balance between parasympathetic activity, which decreases heart rate, and sympathetic activity, which increases heart rate. The parasympathetic nervous system acts to decrease heart rate through part of the tenth cranial nerve (the vagus nerve), which stimulates both the SA and AV nodes. When stimulated, the fibers release acetylcholine, which causes a decrease in heart rate. The sympathetic fibers, which are part of the cardiac accelerator nerves, stimulate the SA node and ventricles. When stimulated, these fibers release norepinephrine, which causes an increase in heart rate.[4]

Blood flow is directly proportional to the change in pressure and inversely proportional to resistance to flow.

$$\text{Flow} = \Delta \text{ pressure/resistance}$$

$$\text{Resistance} = \frac{(\text{length of vessel} \times \text{viscosity of the blood})}{(\text{radius})^4}$$

Figure 13.5 Factors Influencing Arterial Blood Pressure

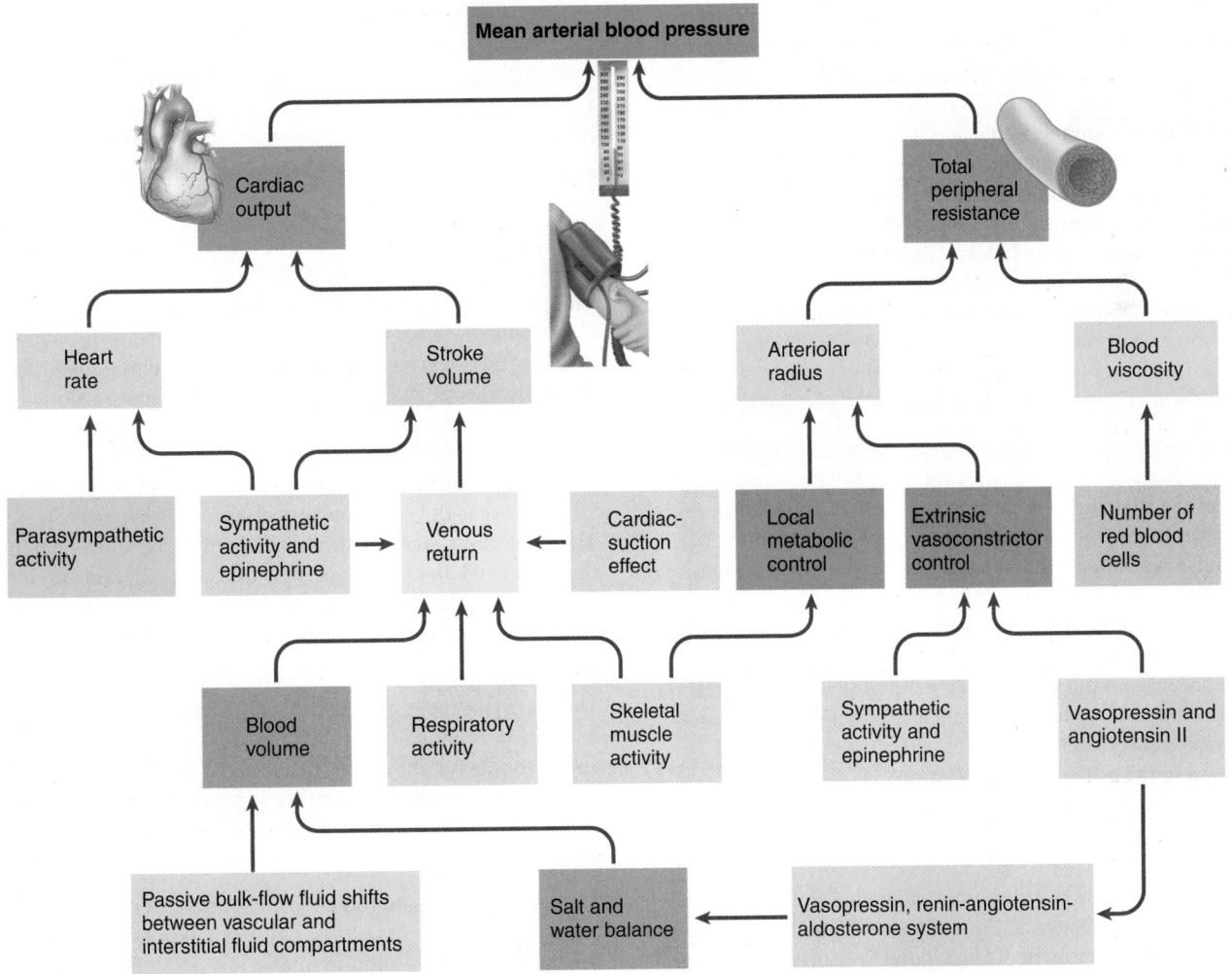

Source: Sherwood's *Human Physiology: From Cells to System*, 7e, Figure 10-34, p. 377.

Resistance is dependent upon the radius of all arterioles, length of the vessel, and the blood viscosity. Arteriolar radius is the most important factor in determining peripheral resistance. Resistance to flow is inversely proportional to the fourth power of the radius of the vessel. Thus, a small reduction in the radius of a vessel would cause a great increase in resistance.[5] Blood viscosity is determined by the number of formed elements (such as hematocrit). If the hematocrit is higher than normal, this will result in greater viscosity and greater resistance to flow. For example, an athlete who is supplementing with erythropoietin could potentially have a greater blood viscosity than normal.

The radius is controlled by several factors. Local metabolic controls in skeletal muscles within a particular region of the body may cause vasodilation and increase blood flow to those muscles in order to match metabolic needs. The vasodilation would decrease resistance by increasing the radius of the vessel. Vasoconstriction will decrease vessel radius and increase resistance. This is caused by sympathetic activity and epinephrine. The hormones vasopressin and angiotensin II also control blood vessel radius by causing vasoconstriction (see Chapter 7).

Vasopressin and angiotensin II also affect BP in other ways. Vasopressin, also known as antidiuretic hormone, is stored in the posterior pituitary gland, and its release is controlled by the hypothalamus. When there is a water deficit, vasopressin is released, which causes an increase in the reabsorption of water. This will increase blood volume, thus increasing BP. Angiotensin II is part of the renin-angiotensin-aldosterone system, depicted in Figure 7.2 (in Chapter 7). When there is a decrease in sodium, plasma volume, and arterial BP, the hormone renin is secreted by the granular cells of the juxtaglomerular apparatus within the kidney. Renin acts as an enzyme and activates the plasma protein angiotensinogen into angiotensin I. Since angiotensin-converting enzyme (ACE) concentrations are high in the lungs, angiotensin I is converted to angiotensin II by ACE via pulmonary circulation. Angiotensin II stimulates the adrenal cortex to secrete aldosterone, which causes an increase in sodium and chloride reabsorption. This promotion of salt retention causes water to be retained and BP to be increased. As discussed later in this chapter, inhibition of ACE is a major pharmaceutical pathway for treatment of hypertension.

The circulatory system contains pressure sensors, called baroreceptors, which constantly monitor BP. An increase or decrease in BP triggers a baroreceptor reflex. The carotid sinus baroreceptor and aortic arch baroreceptor both monitor MAP and pulse pressure, which is the difference between systolic and diastolic BP. Baroreceptors make short-term adjustments to BP by using the autonomic nervous system to alter cardiac output and total peripheral resistance. For long-term adjustments, urine output and thirst are regulated in order to restore normal sodium, chloride, and water balance. If either cardiac output or peripheral resistance increases without a compensatory decrease in the other, BP will increase.[4]

Hypertension

Hypertension refers to a chronic elevation in BP. The Seventh Report of the Joint National Committee on Prevention, Detection, Evaluation, and Treatment of High Blood Pressure (JNC 7) classifies hypertension according to the criteria shown in Box 13.1.

A measurement of blood pressure is expressed using the reading for systolic pressure as the first (higher) number and the reading for diastolic pressure as the second (lower) number. A reading greater than or equal to 140/90 mmHg is considered to be hypertensive. However, it is not necessary for both systolic and diastolic blood pressure to be elevated for an individual to be considered hypertensive; thus, readings of 140/80 mmHg or 120/90 mmHg are both high—i.e., they represent elevations in either systolic BP or diastolic BP. An individual who is currently taking antihypertensive medication is considered to have hypertension even if the blood pressure normalizes as a result of the medication.[6]

Hypertension is important, not only because it affects so many Americans, but because it often also goes undiagnosed in its early stages. It is frequently referred to as the "silent killer," because there are typically no symptoms. Hypertension can cause congestive heart failure, kidney failure, myocardial infarction, stroke, and aneurysms if left untreated. Vision problems may occur due to blood vessels bursting or bleeding within the eyes.[6] Hypertension may also cause decreased left ventricular ejection fraction, ventricular arrhythmias, and sudden cardiac death.[7] According to statistics compiled by the American Heart Association, 77% of individuals who have a first stroke, 69% who have a first myocardial infarction (MI), and 74% who have congestive heart failure have hypertension.[1] Thus, hypertension is a strong risk factor for subsequent CVD morbidity/mortality.

Epidemiology

Approximately 74 million American adults, or 1 in 3 adults, have hypertension. In 2005, more than 57,356 people in the United States died as a direct result of hypertension. In addition, hypertension was listed as a primary or contributing cause of more than 319,000 deaths in the U.S. that year. The age-adjusted death rate related to high blood pressure increased 25.2% from 1995 to 2005.[1]

The rates of hypertension vary by gender and ethnic group. Hispanics/Latinos have the lowest prevalence among ethnic groups, at 19.0%. Mexican Americans have higher rates, at 27.8% and 28.7% for males and females, respectively. White males, at 30.6%, have a lower prevalence of hypertension than white females, at 31.0%. The prevalence of hypertension is highest among blacks, at 41.8% for males and 45.4% for females.[8] Prevalence rates for hypertension rise dramatically with age. For example, only 7% of males and 2.7% of females in the 20- to 34-year-old age group are classified as hypertensive. These numbers increase dramatically to 47.5% and 54.5% for those 55 to 64 and 67.1% and 82% for those 75 years and older for men and women, respectively.[9]

Etiology

There are two types of hypertension. Primary or essential hypertension is idiopathic, which means there is no known cause, and accounts for about 90% of all cases.[2] Secondary hypertension occurs as a result of another primary problem, such as renal disease, other cardiovascular disease, endocrine disorders, or neurogenic disorders.[4]

Though its cause is unknown, primary hypertension may be a result of a variety of factors. Lifestyle factors such as diet (including excessive sodium intake, low potassium intake, excessive alcohol intake), lack of exercise, smoking, stress, and

JNC 7 Guidelines for Evaluation of Hypertension

Classification of Blood Pressure (BP)

Category	Systolic BP (mm Hg)		Diastolic BP (mm Hg)
Normal	<120	and	<80
Prehypertension	120–139	or	80–89
Hypertension, Stage 1	140–159	or	90–99
Hypertension, Stage 2	≥160	or	≥100

Diagnostic Workup of Hypertension

- Assess risk factors and comorbidities.
- Reveal identifiable causes of hypertension.
- Assess presence of target organ damage.
- Conduct history and physical examination.
- Obtain laboratory tests: urinalysis, blood glucose, hematocrit and lipid panel, serum potassium, creatinine, and calcium. Optional: urinary albumin/creatinine ratio.
- Obtain electrocardiogram.

Assess for Major Cardiovascular Disease (CVD) Risk Factors

- Hypertension
- Obesity (body mass index ≥ 30 kg/m^2)
- Dyslipidemia
- Diabetes mellitus
- Cigarette smoking
- Physical inactivity
- Microalbuminuria, estimated glomerular filtration rate <60 mL/min
- Age (>55 for men, >65 for women)
- Family history of premature CVD (men age <55, women age <65)

Assess for Identifiable Causes of Hypertension

- Sleep apnea
- Drug induced/related
- Chronic kidney disease
- Primary aldosteronism
- Renovascular disease
- Cushing's syndrome or steroid therapy
- Pheochromocytoma
- Coarctation of aorta
- Thyroid/parathyroid disease

Source: Reference Card from the Seventh Report of the Joint National Committee on Prevention, Detection, Evaluation, and Treatment of High Blood Pressure (JNC 7). National Heart, Lung, and Blood Institute, NIH Publication No. 03-5231, May 2003.

---

obesity all contribute to the development of primary hypertension. Poor lifestyle choices may exacerbate the problem, since it appears to have a strong genetic component. Numerous genes that contribute to the management of sodium balance and most probably play a role in the development of hypertension have been identified.[7, 10–13] The development and progression of hypertension may also be due to inflammatory responses and individual differences within the renin-angiotensin-aldosterone control of blood pressure.[14] As already mentioned, dietary factors also play a role in the development of hypertension.[15–23] This is discussed in detail under "Nutrition Therapy for Hypertension" later in this chapter.

Pathophysiology

Vasopressin and angiotensin II, as previously mentioned, cause vasoconstriction and fluid retention. Both will increase BP. Often hypertensive individuals have excessive secretion of vasopressin from the hypothalamus. Hypertensive individuals may also have a variation in the gene that produces angiotensinogen. An increased production of angiotensinogen may increase production of angiotensin II, thus increasing BP.

Though the mechanisms are not fully understood, smoking is known to be a risk factor for the development of hypertension. Smoking causes acute and chronic elevations in blood pressure. The relationship may be partially explained by the fact that cigarette smoking interferes with the action of nitrous oxide, thus impairing endothelial relaxation and vasodilation.[24]

In renal disease, blood flow is reduced through the kidney, because of either atherosclerosis within the lumen of a renal artery or compression of a vessel by a tumor. Other factors include gene mutations and renal endothelial cell dysfunction. In order to improve blood flow, angiotensin II is released. This causes vasoconstriction and promotes sodium, chloride, and water retention, which increase blood volume. The increase in blood volume and vasoconstriction both act to increase arterial pressure.[23] In addition, renal endothelial cell dysfunction may cause a decrease in mediators such as nitric oxide and prostacyclin that cause vasodilation. These changes may occur with or without increases in mediators of vasoconstriction such as endothelin and thromboxane A_2.[25]

Hypertension related to the endocrine system may occur with adrenal disorders that cause excessive secretion of epinephrine and norepinephrine. As previously discussed, this will increase cardiac output and peripheral resistance by vasoconstriction, both of which increase BP. Hyperinsulinemia may also play a role in the development of hypertension in some individuals, though the relationships remain unclear and controversial.[26, 27]

Neurological disease impacting the medulla oblongata can cause changes in blood pressure control. This is because the cardiovascular control center, located in the medulla oblongata, helps to maintain the balance between the sympathetic and parasympathetic nervous system. If the balance is disrupted, appropriate BP will not be maintained.[4] For example, in patients with untreated obstructive sleep apnea, intermittent hypoxia causes increased sympathetic activity throughout the day.[28] This increased sympathetic activity increases heart rate, sodium reabsorption, cardiac output, and peripheral resistance, thus increasing blood pressure.[29]

Treatment

Figure 13.6 summarizes medical treatment of hypertension. The goals of treatment for hypertension are (1) reduction in the risk of cardiovascular and renal disease, and (2) reduction of BP to <140/80 mmHg (or to <130/80 mmHg in those individuals with diabetes or chronic renal disease). This is achieved through a comprehensive plan involving weight reduction,

Figure 13.6 JNC 7 Guidelines for Treatment of Hypertension

Principles of Hypertension Treatment
- Treat BP <140/90 mm Hg or BP <130/80 mm Hg in patients with diabetes or chronic kidney disease.
- Majority of patients will require two medications to reach goal.

Algorithm for Treatment of Hypertension

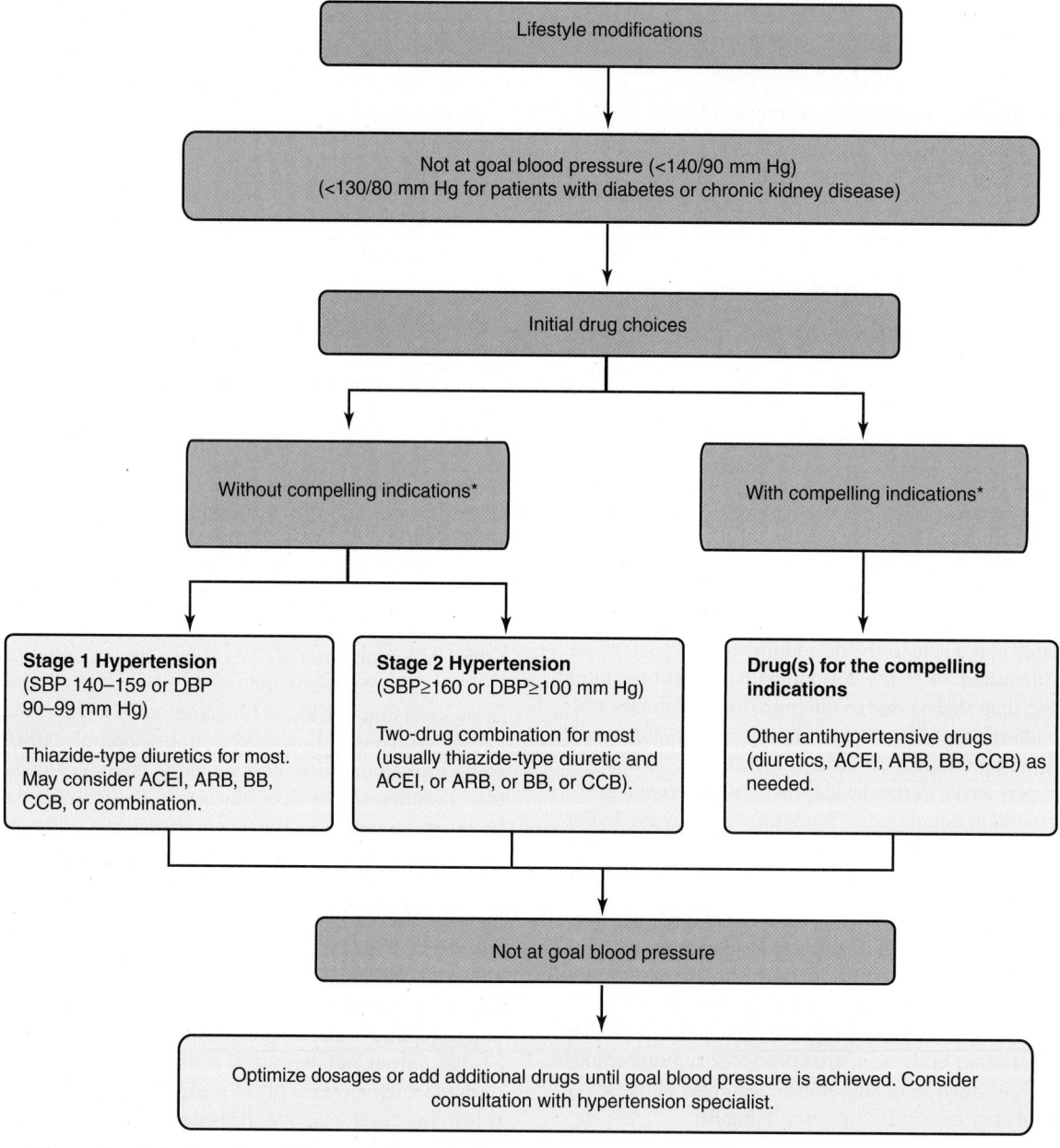

*Compelling indications: heart failure, post myocardial infarction, high CVD risk, diabetes, chronic kidney disease, or recurrent stroke prevention.

Source: Reference Card from the Seventh Report of the Joint National Committee on Prevention, Detection, Evaluation, and Treatment of High Blood Pressure (JNC 7). National Heart, Lung, and Blood Institute, NIH Publication No. 03-5231, May 2003.

physical activity, nutrition therapy, and pharmacological interventions.[19, 20] Lifestyle modifications, nutrition therapy, and physical activity will be discussed later in this section and within this chapter.

In order to change BP, either cardiac output or peripheral resistance must be altered. Pharmacological interventions include several major classes of medications that use one or both of these mechanisms (see Table 13.1). Major groups of diuretics include "loop" diuretics, thiazides, carbonic anhydrase inhibitors, and potassium-sparing diuretics. Loop diuretics

(furosemide, bumetanide, torsemide) act by inhibiting sodium, chloride, and potassium reabsorption in the loop of Henle of the kidney (see Chapter 18 for the anatomy/physiology of the kidneys). Loop diuretics also increase prostaglandins, resulting in vasodilation. Thiazides (hydrochlorothiazide) also inhibit the resorption of sodium, chloride, and potassium, but primarily act in the distal tubule and the ascending loop of Henle. Carbonic anhydrase inhibitors (acetazolamide, methazolamide) prevent the exchange of hydrogen ions with sodium and water by blocking the enzyme carbonic anhydrase. Potassium-sparing

Table 13.1 Selected Drugs Used in Cardiac Care

Classification	Mechanism	Generic	Brand Name	Possible Food–Drug Interactions/Side Effects
Diuretics	Decrease blood volume by increasing urinary output; inhibit renal sodium and water reabsorption	Furosemide Hydrochlorothiazide	Lasix HydroDIURIL	Hypokalemia, hyperlipidemia, hypertriglyceridemia hypercholesterolemia, glucose intolerance; N/V, anorexia, dry mouth, diarrhea, constipation; potassium supplements may be necessary; contraindications: effect antagonized by NSAIDS, avoid natural licorice
ACE Inhibitors	Vasodilators that reduce BP by decreasing peripheral vascular resistance by interfering with the production of angiotensin II from angiotensin I and inhibiting degradation of bradykinin	Captopril Benazepril Enalapril Lisinopril Ramipril	Capoten Lotensin Vasotec Prinivil/Zestril Altace	Hypotension, esp. in elderly patients; can worsen renal function, hyperkalemia, dysgeusia; causes dry, nonproductive cough, hyperkalemia; contraindications: pregnancy, avoid natural licorice, avoid salt substitutes
Beta-1-Blocker	Blocks ß-receptors in heart to decrease heart rate and cardiac output	Metoprolol Atenolol Acebutolol	Lopressor Tenormin Sectral	Nausea, diarrhea; calcium may interfere with absorption; upset stomach, dry mouth, stomach pain, gas or bloating, heartburn
Alpha Adrenergic Blockers	Blocks the vascular muscle response to sympathetic stimulation; reduces stroke volume	Alfuzosin Terazosin Tamsulosin Prazosin		Avoid natural licorice; nausea/vomiting, diarrhea, mouth dryness
Calcium Channel Blockers	Affect the movement of calcium, cause blood vessels to relax; therefore, reduce vasoconstriction	Nisoldipine Nifedipine Nicardipine Bepridil Diltiazem Verapamil	Sular Adalat/Procardia Cardene Vascor Cardizem Calan/Isoptin	Edema, nausea, heartburn; contraindications: heart failure or greater than first degree heart block, avoid natural licorice, limit caffeine, avoid or limit alcohol
Aldosterone Antagonists	Interrupt aldosterone, which increases sodium and water excretion	Spironolactone	Aldactone	May increase serum potassium; avoid salt substitutes; dysgeusia, upset stomach, vomiting, diarrhea, stomach pain
Angiotensin II Receptor Blockers	Interferes with renin-angiotensin system without inhibiting degradation of bradykinin	Candesartan Eprosartan Irbesartan Telmisartan Valsartan	Atacand Teveten Avapro Micardis Cozaar	May increase serum potassium; avoid salt substitutes; nausea, dysgeusia
Nitrate	Vasodilation	Nitroglycerin	Nitro Bid Nitro Dur Nitrostat Transderm-Nitro Minitran Deponit Nitrol	Nausea, vomiting, abdominal pain, dryness of mouth
Digitalis	Increases strength of heart contractions; slows the electrical conduction between the atria and ventricles	Digoxin	Lanoxin	Diarrhea, loss of appetite, lower stomach pain, nausea, and/or vomiting
Fibrinolytic Therapy	Interrupts prothrombin, which reduces ability of blood to clot	Heparin Alteplase Reteplase Streptokinase	Heparin Activase Retavase Streptase	Abdominal pain, nausea, constipation
Positive Inotropic Drugs	Stimulate heart rate; increase heart contractions	Dopamine HCL Milrinone	Dopamine Primacor	May decrease serum potassium; proteinuria; nausea, vomiting

diuretics (spironolactone, amiloride) act within the collecting and convoluted tubules, where they prevent sodium-potassium exchange and reduce aldosterone stimulation.

Medications called ACE inhibitors (Captopril) competitively block the enzyme that converts angiotensin I into angiotensin II. This results in vasodilation, a decrease in vasopressin release, and a resulting decrease in BP.

Beta-adrenergic blocking agents (propanolol, atenolol, acebutolol) block ß-receptors in the heart to decrease rate and cardiac output. Alpha-receptor antagonists (Cardura, Minipress, Hytrin) block the vascular muscle action that normally responds to sympathetic stimulation. This reduces stroke volume and thus BP. Calcium channel blocking agents (Verapamil) affect the movement of calcium, which causes the blood vessels to relax and therefore reduces vasoconstriction. The final class of medications that can be used in treatment for hypertension are the aldosterone antagonists (spironolactone and eplerenone), which suppress the actions of aldosterone. The type of medication regimen is determined by the classification of hypertension and other risk factors (see Figure 13.6).

Nutrition Therapy for Hypertension

Nutritional treatment of hypertension includes both lifestyle modifications and nutrition therapy. Increased physical activity, smoking cessation, and weight loss, as well as reduction of sodium and alcohol intake, are primary strategies.[19, 20, 22, 23, 30] In the past decade, several clinical trials—including the landmark Dietary Approaches to Stop Hypertension (DASH) and the PREMIER trials—have revealed that nutrition interventions that include decreasing sodium, saturated fat, and alcohol while increasing calcium, potassium, and fiber have demonstrated significant effects for lowering blood pressure (see Table 13.2).[19, 20, 22, 23, 31, 32, 33–39]

Nutrition Assessment

The dietetic practitioner should obtain nutrition assessment data that will assist in evaluating factors that are associated with an increased risk or incidence of hypertension. Focusing on the dietary factors and patterns that have been associated with the presence of hypertension (identified in Table 13.2) can guide the assessment process. Data from the following domains of the standardized terminology for nutrition assessment need to be obtained to accurately determine specific nutrition problems and nutrition diagnoses: Food and Nutrient Intake, Knowledge/beliefs/attitudes, Behavior, Physical Activity and Function, and pertinent Biochemical Data.

It is important to evaluate the need for weight reduction in order to move toward the goal of reaching a BMI of 18.5–24.9.[30] Since the DASH diet (discussed next) is the foundation of nutrition therapy for hypertension, the clinician should also assess dietary intake for alcohol, sodium, potassium, calcium, and fiber. The practitioner and client should then work together to prioritize the methods to meet the DASH dietary goals. It is just as important to assess the client's physical activity levels and then tailor exercise goals to the individual, in order to meet the current recommendations for 30–60 min of aerobic exercise on a minimum of four days per week.[19, 23, 40] Table 13.3 summarizes nutrition assessment for individuals with cardiovascular diseases.

Table 13.2 Effects of Dietary Factors and Dietary Patterns on BP: A Summary of the Evidence

	Hypothesized Effect	Evidence
Weight	Direct	+ +
Sodium Chloride (salt)	Direct	+ +
Potassium	Inverse	+ +
Magnesium	Inverse	+/−
Calcium	Inverse	+/−
Alcohol	Direct	+ +
Fat: Saturated	Direct	+/−
Omega-3 Polyunsaturated Fat	Inverse	+ +
Omega-6 Polyunsaturated Fat	Inverse	+/−
Monounsaturated Fat	Inverse	+
Protein: Total	Uncertain	+
Protein: Vegetable	Inverse	+
Protein: Animal	Uncertain	+/−
Fiber	Inverse	+
Cholesterol	Direct	+/−
Dietary Patterns: Vegetarian Diets	Inverse	+ +
DASH Type Dietary Patterns	Inverse	+ +

Key: +/− indicates limited or equivocal evidence; + suggestive evidence, typically from observational studies and some clinical trials; + + persuasive evidence, typically from clinical trials.
Source: Table 2, p. 305: Appel LJ, Brands MW, Daniels SR, Karania N, Elmer PJ, Sacks FM; American Heart Association. Dietary approaches to prevent and treat hypertension: a scientific statement from the American Heart Association. *Hypertension.* 2006;Feb;47(2):296–308.

Nutrition Diagnosis

Common nutrition diagnoses for individuals with hypertension include excessive energy intake; excessive or inappropriate intake of fats; excessive sodium intake; inadequate calcium, fiber, potassium, or magnesium intake; overweight/obesity; food and nutrition-related knowledge deficit; and physical inactivity.

Nutrition Intervention

A comprehensive approach that addresses multiple lifestyle factors has the most significant effect on blood pressure control for hypertensive individuals (see Tables 13.2 and 13.4).[16, 19, 21, 23, 32, 41, 42] Evidence-based guidelines should be utilized to provide nutrition therapy to patients/clients. Education is a key component of providing nutrition therapy. A recent Cochrane data analysis of twenty-three clinical trials confirmed that nutrition education increased fiber, fruit, and vegetable intake; lowered total dietary fat intake; and reduced blood pressure, LDL cholesterol, and total serum cholesterol.[35] Nutrition therapy is guided by the patient's hypertension history, other medical risk factors, current medical treatment, and readiness for behavior change (see Box 13.2). The sections that follow address each component of the current recommendations for blood pressure control.

DASH—Dietary Approaches to Stop Hypertension As discussed earlier in this section, the concept of approaching nutrition therapy for hypertension with a compre-

Table 13.3 Nutrition Assessment for the Cardiovascular System

Medical/Social History	• Diagnoses/date of diagnosis • Comorbidities • Medications • Previous medical conditions or surgeries • Socioeconomic status/food security • Support systems • Education level—primary language
Food-/Nutrition-Related History (adapted from U. of Wisconsin School of Medicine)	• Meals/snacks—patterns, frequency • Portion sizes • Saturated and *trans* fat from dairy products and fatty meats, commercial snack foods and pastries, fried foods, and added fats and oils • Refined carbohydrates from baked products, desserts, cookies, and other sweets • Sweetened beverages (juice drinks, soda) and alcohol • Major sources of sodium from processed foods, eating out, and added salt • Frequency of restaurant meals, fast food, take-out food
Physical Assessment	• Blood pressure
Anthropometric Measurements	• Height/length • Current weight • Weight history if adult: highest adult weight; usual body weight • Body mass index • Waist circumference
Biochemical Data, Medical Tests and Procedures	• Erythrocyte thiamin pyrophosphate effect • Glucose, BUN, Cr • Electrolytes
Lipid Assessment:	• Triglyceride, total cholesterol, HDL, LDL, LDL:HDL ratio
Visceral Protein Assessment:	• Albumin • Prealbumin

Table 13.4 Effects of Lifestyle Modification to Manage Hypertension

Reduction	Average Systolic BP Recommendation
Weight reduction to maintain a BMI 18.5–24.9	5–20 mmHg/10 kg
Diet rich in fruits, vegetables, & low-fat dairy products with reduced saturated & total fat–the Dietary Approaches to Stop Hypertension (DASH) eating plan	8–14 mmHg
Potassium—Levels below the DRI may increase BP	N/A
Sodium intake of not >100 mEq/day (2.3 g sodium or 6 g sodium chloride)	2–8 mmHg
Aerobic activity, such as brisk walking for 30 min/day, most days of the week	4–9 mmHg
Most men: Not >2 drinks/day Women & lighter weight men: Not >1 drink/day	2–4 mmHg

Sources: Appel LJ, et al. A clinical trial of the effects of dietary patterns on blood pressure. (*N Engl J Med* 1997;336:1117–24). Sacks FM, et al. Effects on blood pressure of reduced dietary sodium and the dietary approaches to stop hypertension (DASH) diet. (*N Engl J Med* 2001;344:3–10). JNC 7 USDHHS, 2003. ADA Evidence Analysis Hypertension Guideline 2009. Funk KL, Elmer PJ, Stevens VJ, Harsha DW, Craddick SR, Lin P-H, Young DR, Champagne CM, Brantley PJ, McCarron PB, Simons-Morton DG, and Appel LJ. PREMIER—A Trial of Lifestyle Interventions for Blood Pressure Control: Intervention Design and Rationale *Health Promot Pract*, July 2008;9:271–80.

hensive dietary method was brought to the forefront with the Dietary Approaches to Stop Hypertension (DASH) in the late 1990s (see Box 13.3).[43] These clinical trials focused on the use of a variety of foods that not only reduced sodium intake but increased potassium, magnesium, calcium, and fiber intakes within a moderate energy intake. At 2000 kcal a day, the DASH Sodium Diet provides approximately 4700 mg (120 mEq) potassium, 500 mg magnesium, 1240 mg calcium, 90 g protein, 30 g fiber, and 2400 mg (100 mEq) sodium.[38] The Canadian Hypertension Education Program Evidence Based Recommendations Task Force has further supported this multi-faceted plan for

---

BOX 13.2   CLINICAL APPLICATIONS

Brief Nutrition Counseling for Hypertension:
Steps in Behavioral Counseling ("5 A's")

Assess (Food and Nutrient Intake, Knowledge/beliefs/attitudes, Behavior, Physical Activity and Function, and Biochemical Data)

- Food intake and diet habits in the context of health risks
- Current physical activity
- Readiness to change behavior

- From WAVE Nutrition Counseling Tool (http://bms. brown.edu/nutrition/acrobat/wave.pdf)
  - W = Weight: Review BMI, blood pressure, lipids, blood sugar to screen for metabolic syndrome.
  - A = Activity: Conduct physical activity assessment. Ask about:
    - Moderate physical activity? Goal—30 minutes/day or more

- V = Variety and E = Excess: Based on DASH Sodium Diet:
  - Number of low-fat dairy foods? Goal—2 to 3 servings/day
  - Number of fruits and vegetables? Goal—8 to 10 servings/day
  - Salt added to food? Goal—no salt added at the table, only half the usual amount in cooking
  - Use of frozen, canned, or dried processed foods (soup, spaghetti

(continued)

sauce, frozen dinners, helper mixes)? Goal—reduced sodium from processed foods
- Alcohol intake? Goal—no more than one drink/day for women and two drinks/day for men

Advise

- Give clear, specific, and personalized behavior change advice. You might say:
  - "Diet changes, exercise, and weight loss can reduce your blood pressure as much as medicine."
  - For patients taking medication for diabetes, lipids, or hypertension: "Diet choices are important even if you are taking medication, since eating carefully helps the medicine do a better job. You may be able to save money by cutting down on the amount of medicine you take."

- For patients NOT ready to change behavior, add: "I'd like to help you when you are ready to make changes in your diet and be more active."

Agree

- Collaborate with patient to select treatment goals and methods.
- Base goals on readiness to change behavior.
- For patient NOT ready to change behavior: "Is it okay if I ask you again at our next visit?"
- Possible goals for patient ready to change:
  - Return for further discussion in 2–4 weeks.
- Keep food and exercise records to increase awareness, if patient is willing.
  - Refer for additional registered dietitian visit.

Assist

- Help patient acquire knowledge, skills, and support for behavior.
- Provide hand-outs and Web resources, based on patient interest and need.
- Provide lists and recommendations for community resources (exercise and diet programs, health clubs, etc.).

Arrange

- Schedule follow-up appointments.

Source: Adapted from: Medical Nutrition Handbook, Department of Medicine, University of Wisconsin School of Medicine and Public Health (http://www.medicine.wisc.edu/mainweb/includes/viewfile.php?fileid=899&viewtype=inline&section=naa)

---

nutrition therapy and lifestyle change.[44] It has been proposed that the blood pressure reductions seen in both the DASH and OMNI HEART trials are most likely a synergistic effect of increasing potassium, magnesium, calcium, and fiber while reducing sodium and saturated fat.[23, 36, 45] Initiating these broad dietary changes and then continuing them lifelong is a realistic approach to both the treatment of hypertension and the prevention other diseases.[19, 35, 46] Box 13.4 provides an overview of the DASH eating plan, and following sections discuss aspects of its nutrient composition, as well as other lifestyle recommendations, in reference to hypertension treatment.

Weight Loss Weight reduction is a standard component of nutrition therapy for treatment of hypertension.[19, 20, 21, 41, 42] Evidence analysis of the literature continues to support this recommendation. Neter (2003) evaluated 26 studies and found that weight loss of greater than 5 kg reduced both diastolic and systolic BP. An approximate 20-lb weight loss will result in lowered systolic BP, and even less than 10% weight loss has a sustained effect on BP.[19, 23, 41, 46] See Box 13.5 for practical suggestions for reducing kcal intake for weight loss while following the DASH diet.

Though waist circumference (see Figure 3.8 in Chapter 3) is related to body weight, it is an independent predictor of hypertension risk. For those patients who fall within a normal or overweight BMI, waist circumference should be measured. It is not necessary to measure waist circumference for those patients with a BMI >35, because it adds no additional predictive power.[47]

Sodium Although the use of sodium restriction to manage BP has been a controversial issue in the past, recent consistent evidence has supported the efficacy of a reduction of sodium for controlling BP.[15–21, 46] Large population studies, such as the INTERSALT Study, have confirmed that urinary sodium excretion has a significant and direct relationship with systolic blood pressure.[48–50] It has been estimated that sodium modifications may reduce incidence of hypertension by as much as 17%.[50] The DASH trials have further supported the role of sodium reduction in treatment for hypertension.[22, 51] BP control through sodium restriction could reduce the incidence of cardiovascular disease, renal disease, and stroke.

Individual response to sodium restriction can vary.[52] Nonetheless, Appel et al. argue that, despite descriptions of people as **"salt sensitive"** or **"salt resistant,"** changes in BP as a result of sodium restriction are seen across all ages and all ethnicities.[23] The DASH Sodium trial tested the response to three different sodium levels and provided further evidence of a reduction in blood pressure with sodium restriction. The most significant reductions were seen in blacks, older individuals, and individuals with comorbidities such as diabetes mellitus.[15, 23, 32, 46, 53]

Americans consume high amounts of sodium, in part because of the amounts used in processed foods.[15, 23] Average sodium intake for Americans ranges between 3000 and 4500 mg/day (130–195 mEq Na or 8–10 g of sodium chloride). The *Dietary Guidelines for Americans* recommend an intake of less than 2300 mg of sodium, the equivalent of 6 g of sodium chloride (table salt), each day.[54] This goal is supported by the most recent statement from the American Heart Association.[23]

Because only small amounts of sodium occur naturally in food, effective reduction of sodium intake requires limiting the intake of highly processed foods, avoiding those foods that are cured using salt, and omitting salt during the cooking and preparation

The History of DASH

Scientists from Brigham and Women's Hospital (Boston, MA), Duke University Medical Center (Durham, NC), Johns Hopkins University (Baltimore, MD), and Pennington Biomedical Research Center, Louisiana State University (Baton Rouge, LA), who were supported by the National Heart, Lung, and Blood Institute (NHLBI), conducted two key studies in the 1990s. The first was called "DASH," and it tested the effects of nutrients, as they occur together in food, on BP. This study found that blood pressures were reduced with an eating plan that is low in saturated fat, cholesterol, and total fat, and that emphasizes 8–9 servings of fruits and vegetables, and three servings of low-fat dairy foods. This eating plan—known as the DASH eating plan—also includes whole-grain products, fish, poultry, and nuts. It is reduced in red meat, sweets, and sugar-containing beverages. It is rich in magnesium, potassium, and calcium, as well as protein and fiber.

The DASH study involved 459 adults with systolic blood pressures of less than 160 mmHg and diastolic pressures of 80–95 mmHg. About 27% of the participants had hypertension. About 50% were women and 60% were African-Americans. DASH compared three eating plans: a plan similar in nutrients to what many Americans consume; a plan similar to what Americans consume but higher in fruits and vegetables; and the DASH eating plan. All three plans included about 3000 milligrams of sodium daily. None of the plans was vegetarian or used specialty foods. Results were dramatic: both the fruits and vegetables plan and the DASH eating plan reduced BP. But the DASH eating plan had the greatest effect, especially for those with high BP. Furthermore, the BP reductions came fast—within two weeks of starting the plan.

The second study was called "DASH Sodium," and it looked at the effect on blood pressure of a reduced dietary sodium intake as participants followed either the DASH eating plan or an eating plan typical of what many Americans consume. DASH Sodium involved 412 participants. Their systolic blood pressures were 120–159 mmHg and their diastolic blood pressures were 80–95 mmHg. About 41% of them had high blood pressure.

About 57% were women and about 57% were African-Americans. Participants were randomly assigned to one of the two eating plans and then followed for a month at each of three sodium levels. The three sodium levels were as follows: a higher intake of about 3300 milligrams per day (the level consumed by many Americans); an intermediate intake of about 2400 milligrams per day; and a lower intake of about 1500 milligrams per day. Results showed that reducing dietary sodium lowered BP for both eating plans. At each sodium level, BP was lower on the DASH eating plan than on the other eating plan. The biggest BP reductions were for the DASH eating plan at the sodium intake of 1500 milligrams per day. Those with hypertension saw the biggest reductions, but those without it also had large decreases. These reductions occurred even when body weight remained stable. The magnitude of BP reduction with this dietary pattern was similar to the reduction noted with BP lowering medications.

Source: Modified from: U.S. Dept. of Health and Human Services, National Institutes of Health, National Heart, Lung, and Blood Institute. Facts about the DASH eating plan. NIH Publication No. 03–4082. Bethesda (MD): NHLBI; 2003, pp. 3–4. Available at http://www.nhlbi.nih.gov/health/public/heart/hbp/dash/new_dash.pdf

of foods. The practitioner should teach the client strategies for limiting intake to 2300 mg/day (100 mEq) and provide information on the sodium content of foods (see Table 13.5). The DASH diet (Box 13.4) gives specific guidelines for comprehensive nutrition therapy. Boxes 13.6 and 13.7 list practical steps for controlling sodium intake.

Alcohol As alcohol intake increases above two drinks per day for men (and one drink/day for women), the risk of hypertension increases accordingly, in a dose-dependent relationship.[23, 55] The U.S. *Dietary Guidelines* and JNC 7 recommend limiting alcohol intake to ≤2 drinks per day for men and ≤1 drink per day for women. One drink is defined as 12 oz. of beer, 5 oz. of wine, or 1.5 oz. of 80 proof distilled spirits.[30, 54]

Potassium, Calcium, and Magnesium Potassium, calcium, and magnesium have all been positively correlated with reduction of BP and treatment of hypertension. The role of these minerals as part of the nutrition therapy for hypertension is highlighted by the results of the DASH studies. All three minerals appear to have an inverse relationship to

hypertension—suggesting that as dietary intakes increase, BP decreases.[22, 23, 32, 51, 56]

The relationship between potassium and BP is a strong inverse relationship.[57] The diet used in the DASH trials provided an average of 4–6 g of potassium/day from fruits and vegetables (see Chapter 18 for food sources of potassium). These intakes were associated with reduced blood pressures. As the ADA Evidence-Based Guideline states: "Dietitians should advise individuals to consume adequate food sources of potassium as part of Medical Nutrition Therapy to reduce blood pressure. Research suggests that potassium intake lower than recommended levels (DRI) is associated with increased blood pressure." Appel et al. emphasize that an increased potassium intake does not pose a health risk in healthy individuals.[23] In those clients who may have an impaired urinary excretion of potassium, however, these recommendations may need to be modified.

The relationship between calcium and hypertension has been studied for over 25 years. The most dramatic relationship

BOX 13.4 CLINICAL APPLICATIONS

Dash Eating Plan

The dash eating plan shown below is based on 2000 calories a day. The number of daily servings in a food group may vary from those listed, depending on your caloric needs. Use this chart to help you plan your menus or take it with you when you go to the store.

Food Group	Daily Servings (except as noted)	Serving Sizes	Examples and Notes	Significance of Each Food Group to the DASH Eating Plan
Grains and grain products	7–8	1 slice bread 1 oz dry cereal* ½ cup cooked rice, pasta, or cereal	Whole-wheat bread, English muffin, pita bread, bagels, cereals, grits, oatmeal, crackers, unsalted pretzels and popcorn	Major sources of energy and fiber
Vegetables	4–5	1 cup raw leafy vegetable ½ cup cooked vegetable 6 oz vegetable juice	Tomatoes, potatoes, carrots, green peas, squash, broccoli, turnip greens, collards, kale, spinach, artichokes, green beans, lima beans, sweet potatoes	Rich sources of potassium, magnesium, and fiber
Fruits	4–5	6 oz fruit juice 1 medium fruit ¼ cup dried fruit ½ cup fresh, frozen, or canned fruit	Apricots, bananas, dates, grapes, oranges, orange juice, grapefruit, grapefruit juice, mangoes, melons, peaches, pineapples, prunes, raisins, strawberries, tangerines	Important sources of potassium, magnesium, and fiber
Low-fat or fat-free dairy foods	2–3	8 oz milk 1 cup yogurt 1½ oz cheese	Fat-free (skim) or low-fat (1%) milk, fat-free or low-fat buttermilk, fat-free or low-fat regular or frozen yogurt, low-fat and fat-free cheese	Major sources of calcium and protein
Meats, poultry, and fish	2 or less	3 oz cooked meats, poultry, or fish	Select only lean; trim away visible fats; broil, roast, or boil, instead of frying; remove skin from poultry	Rich sources of protein and magnesium
Nuts, seeds, and dry beans	4–5 per week	⅓ cup or 1½ oz nuts 2 Tbsp or ½ oz seeds ½ cup cooked dry beans/peas	Almonds, filberts, mixed nuts, peanuts, walnuts, sunflower seeds, kidney beans, lentils	Rich sources of energy, magnesium, potassium, protein, and fiber
Fats and oils	2–3†	1 tsp soft margarine 1 Tbsp low-fat mayonnaise 2 Tbsp light salad dressing 1 tsp vegetable oil	Soft margarine, low-fat mayonnaise, light salad dressing, vegetable oil (such as olive, corn, canola, or safflower)	DASH has 27% of calories as fat, including fat in or added to foods
Sweets	5 per week	1 Tbsp sugar 1 Tbsp jelly or jam ½ oz jelly beans 8 oz lemonade	Maple syrup, sugar, jelly, jam, fruit-flavored gelatin, jelly beans, hard candy, fruit punch, sorbet, ices	Sweets should be low in fat

*Equals ½–1¼ cups, depending on cereal type. Check the product's Nutrition Facts Label.

†Fat content changes serving counts for fats and oils: For example, 1 Tbsp of regular salad dressing equals 1 serving; 1 Tbsp of a low-fat dressing equals ½ serving; 1 Tbsp of a fat-free dressing equals 0 servings.

between calcium and blood pressure reduction was seen in the DASH trials. The DASH diets provided the equivalent of 3 cups of dairy products as their major source of calcium. In a recent trial, intakes of lower-fat milk and milk products were correlated with lower rates of hypertension, but this relationship was not sustained for whole milk products.[58] The ADA Evidence Analysis states that there is not adequate research to make a definitive claim regarding the role of calcium and its effect on blood pressure.[42] At present, more specific recommendations for calcium intake in hypertension have not been established beyond the recognized DRI levels.[23]

It is important to remember that the nutritional effects demonstrated by the DASH study—and in particular, the relationship between K, Ca, and Mg and blood pressure reduction—were a result of a dietary *pattern* rich in these nutrients rather than mineral intake from *supplements*.

Physical Activity According to the JNC 7, physical activity of 30 minutes per day does decrease blood pressure by 4–9 mm Hg. Moreover, increasing physical activity decreases the relative workload on the heart for all forms of activity, a benefit important for all forms of cardiovascular disease. For

Table 13.5 Sodium Content of Foods

Sodium content varies by processing. The AI for Sodium is 1500 mg

Food Groups	Sodium (mg)
Grains and grain products	
Cooked cereal, rice, pasta, unsalted, ½ cup	0–5
Ready to eat cereal, 1 cup	100–360
Bread, 1 slice	110–175
Vegetables	
Fresh or frozen, cooked without salt, ½ cup	1–70
Canned or frozen with sauce, ½ cup	140–460
Tomato juice, canned ¾ cup	820
Fruit	
Fresh, frozen, canned, ½ cup	0–5
Low-fat or fat-free dairy foods	
Milk, 1 cup	120
Yogurt, 8 oz	160
Natural cheeses, 1½ oz	110–450
Processed cheeses, 1½ oz	600
Nuts, seeds, and dry beans	
Peanuts, salted, ⅓ cup	120
Peanuts, unsalted, ⅓ cup	0–5
Beans, cooked from dried, or frozen, without salt, ½ cup	0–5
Beans, canned, ½ cup	400
Meats, fish, and poultry	
Fresh meat, fish, poultry, 3 oz	30–90
Tuna canned, water pack, no salt added, 3 oz	35–45
Tuna canned, water pack, 3 oz	250–350
Ham, lean, roasted, 3 oz	1020

Source: Reprinted from: U.S. Dept. of Health and Human Services, National Institutes of Health, National Heart, Lung, and Blood Institute. Facts about the DASH eating plan. NIH Publication No. 03–4082. Bethesda (MD): NHLBI; 2003, Box 5, p. 7. Available at http://www.nhlbi.nih.gov/health/public/heart/hbp/dash/new_dash.pdf

instance, mowing the lawn requires a certain percentage of one's maximal functional capacity. If a person starts a walking program and improves his or her cardiorespiratory fitness, then mowing the lawn will require a lower percentage of his or her functional capacity. Since the relative strain on the cardiovascular system will be reduced, the BP response to the activity will be reduced as well. Furthermore, increasing physical activity will facilitate weight management.

Smoking Cessation When an individual quits smoking, he or she realizes health benefits almost immediately. Smoking cessation may be the most important change any individual can make to reduce the risk of hypertension and all forms of cardiovascular disease.[2] All smoking cessation plans are not equal, and each individual should seek out a program that suits his or her needs. In order to achieve success, the smoker should also be able to identify his or her reasons for quitting. The American Lung Association has developed a "Freedom from Smoking" program that can be accessed online at http://www.lungusa.org[59]

BOX 13.5 CLINICAL APPLICATIONS

Recommendations to Promote Weight Loss on the DASH Eating Plan

To increase fruits—

- Eat a medium apple instead of four shortbread cookies. *You'll save 80 calories.*
- Eat ¼ cup of dried apricots instead of a 2-ounce bag of pork rinds. *You'll save 230 calories.*

To increase vegetables—

- Have a hamburger that's 3 ounces of meat instead of 6 ounces. Add ½ cup serving of carrots and ½ cup serving of spinach. *You'll save more than 200 calories.*
- Instead of 5 ounces of chicken, have a stir-fry with 2 ounces of chicken and 1½ cups of raw vegetables. Use a small amount of vegetable oil. *You'll save 70 calories.*

To increase low-fat or fat-free dairy products—

- Have a ½ cup serving of low-fat frozen yogurt instead of a 1½-ounce milk chocolate bar. *You'll save about 110 calories.*

And don't forget these calorie-saving tips—

- Use low-fat or fat-free condiments.
- Use half as much vegetable oil, soft or liquid margarine, or salad dressing, or choose fat-free versions.
- Eat smaller portions—cut back gradually.
- Choose low-fat or fat-free dairy products to reduce total fat intake.
- Check the food labels to compare fat content in packaged foods—items marked "low fat" or "fat free" are not always lower in calories than their regular versions.
- Limit foods with lots of added sugar, such as pies, flavored yogurts, candy bars, ice cream, sherbet, regular soft drinks, and fruit drinks.
- Eat fruits canned in their own juice.
- Add fruit to plain yogurt.
- Snack on fruit, vegetable sticks, unbuttered and unsalted popcorn, or bread sticks.
- Drink water or club soda.

Atherosclerosis

Definition

The term **atherosclerosis (AS)** comes from the Greek *athero,* meaning gruel, and *sclerosis,* meaning hardening. The terms AS and arteriosclerosis are often used interchangeably. **Arteriosclerosis** is a general term defined as a thickening of the walls of the vessels and a loss of vascular elasticity.

AS is the development of atherosclerotic plaque in the vascular wall that will occlude the lumen of the vessel and create ischemic conditions (see Figure 13.7). The plaque begins as a fatty and fibrous growth, and over time may calcify. The devel-

opment of an atherosclerotic plaque can result in a restriction of blood flow severe enough to cause an **infarct** resulting in a **myocardial infarction (MI)** or in a cerebrovascular accident (stroke). Therefore, AS is the root cause of two of the three leading causes of death in the United States, **coronary artery disease (CAD)** and stroke. In addition, an atherosclerotic plaque in the leg can result in **peripheral vascular disease (PVD)** that may result in the tissue death associated with gangrene and loss of a limb. Severe CAD may impair cardiac function to the point that **congestive heart failure (CHF)** results.

Epidemiology

CVD is the leading cause of death in the United States and throughout the world, and greater than 50% of all diagnoses related to CVD result from atherosclerosis.[8] Most of what we know about the epidemiology of AS and heart disease has come from large epidemiological studies that have provided

Figure 13.7 Stages of Plaque Progression

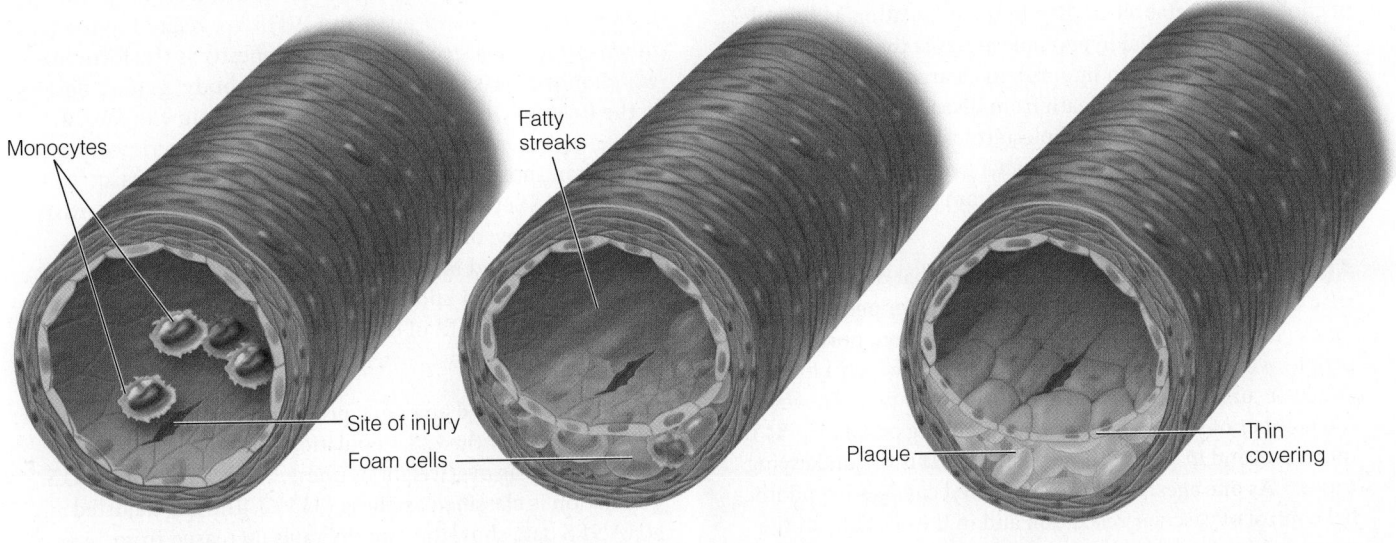

Monocytes—phagocytic white blood cells—circulate in the bloodstream and respond to injury on the artery wall.

Monocytes slip under blood vessel cells and engulf LDL cholesterol, becoming foam cells. The thin layers of foam cells that develop on artery walls are known as *fatty streaks*.

A fatty streak thickens and forms plaque as it accumulates additional lipids, smooth muscle cells, connective tissue, and cellular debris.

The artery may expand to accommodate plaque. When this occurs, the plaque that develops often contains a large lipid core with a thin fibrous covering and is vulnerable to rupture and thrombosis.

Source: S. Rolfes, K. Pinna, and E. Whitney, *Understanding Normal and Clinical Nutrition*, 7e, copyright 2006, p. 821.

an endless flow of data for over three decades. These studies—particularly the Framingham Study, National Health and Nutrition Examination Survey (NHANES), National Cholesterol Education Program (NCEP), and Heritage Study—continue to provide us with valuable information about factors related to the development of heart disease. The American Heart Association statistics indicate that over 12 million individuals are affected by AS, and that it results in more than one half million deaths each year.[8]

Etiology

These long-term studies have allowed researchers to identify risk factors for CAD, PVD, and stroke. These risk factors include family history, age, sex, obesity, dyslipidemia, hypertension, diabetes, physical inactivity, and cigarette smoking. The risk factors are additive in their predictive power; thus, the more risk factors one demonstrates, the greater the risk of development of AS, and hence of CVD. Risk factors are often said to work together in a synergistic manner. There is an extensive body of literature which shows that as an individual's number of CVD risk factors increases, so do CVD morbidity and mortality.

A study utilizing 366,000 men and women participants from the Multiple Risk Factor Intervention Trial (MRFIT) compared the mortality rates for those individuals defined as low risk based on the absence of certain risk factors—i.e., serum cholesterol level <200 mg/dL, untreated blood pressure ≤120/80 mm Hg, non-cigarette smokers—to those of individuals with those risk factors. Those participants defined as

low risk had between a 73% and an 85% lower risk for CVD mortality, 40% to 60% lower total mortality rates, and 6 to 10 years' greater life expectancy.[60]

A study of 84,129 women enrolled in the Nurses' Health Study identified five healthy lifestyle factors. These were the absence of cigarette smoking, drinking half a glass or more of wine per day (or equivalent alcohol consumption), 30 minutes or more per day of moderate or vigorous physical activity, BMI <25 kg/m², and dietary score in the top 40% (including diets with lower amounts of *trans* fats, lower glycemic load, higher cereal fiber, higher omega-3 fatty acids, higher folate, and higher polyunsaturated-to-saturated fat ratio). When three of the five healthy lifestyle factors were present, risk for CHD over a fourteen-year period was reduced by 57%; when four were present, risk was reduced by 66%; and when all five factors were present, risk was reduced by 83%.[61]

Typically, the AS risk factors are divided into categories based on whether they are alterable or unalterable. Family history, ethnicity, age, and sex are considered unalterable risk factors. Alterable risk factors include obesity, dyslipidemia, hypertension, physical inactivity, atherogenic diet, and cigarette smoking. Many sets of risk factors include a category termed *impaired fasting glucose*, which may be alterable; diabetes itself is an unalterable risk factor because, although one can be in metabolic control, this does not change that individual's risk status (see Chapter 17).

Family History There is certainly a genetic component to the development of AS. This genetic component may be related to endothelial function or cholesterol metabolism. CAD death rates are higher in individuals with these disorders of cholesterol metabolism than in the population as a whole. In some cases, familial hypercholesteremia that is caused by genetic abnormalities in lipoprotein clearance and lipid metabolism results in early death from disease (see Chapter 10). In many other cases, hypercholesterolemia is the result of environmental influences rather than a genetic trait. Dietary and physical activity habits are learned from the social network in which one is reared.

Age and Sex Because AS is a disease that develops over a span of years, a greater age allows for a greater period of time for the disease to develop. Heart disease is more prevalent in people over the age of 65. The average age at which a first heart attack occurs is 64.5 for men and 70.3 for women.[1] However, the number of sudden cardiac deaths among people 15–34 has increased, and in 70% of these deaths CHD is the underlying cause.[62] As one ages, there are associated changes in endothelial control of vascular relaxation and in the elasticity of the arteries. Males tend to develop AS at a faster rate than females, though the differences between the sexes tend to decline after the woman reaches menopause. The purported reason for the differences has to do with a protective effect of estrogen. Thus, as women age past the point of menopause, their risk of development of clinically significant AS increases greatly in the absence of estrogen therapy. The lifetime risk of development of ischemic heart disease (discussed later in this chapter) after the age of 40 is reported to be one in two for men and one in three for women.[63]

Obesity Only in the last decade has obesity been listed as a separate and independent risk factor for CAD. Obesity can be defined in several different ways when used for risk assessment, but is most commonly identified as a body mass index (BMI, measured in kg/m^2) of 30 or greater (see Chapter 12). A BMI of 30 or greater is associated with a proportionally higher all-cause mortality rate than is a BMI of 25–29.9 (see Table 13.6). Alternatively, obesity may be defined by waist girth or waist-to-hip ratio (WHR). The NCEP Adult Treatment Panel III (ATP III) report identified waist girth alone as a suitable predictor of risk.[64] A waist girth of >102 cm for men or >88 cm for women is used as the criterion for increased risk.[65] Waist girth and WHR are related to the way in which humans store fat. Android obesity is the form in which more mass is stored in the upper body; gynoid obesity is the form in which more mass is stored below the waist. Those who store more fat below the waist are at lower risk for CVD than those who store more fat above their waist. This relationship between weight distribution and AS risk exists for two reasons: one is a purported relationship between abdominal fat and insulin resistance, and the second is that men store more fat above their waist. Men, who do not enjoy the protective benefit of estrogen, are more likely to develop AS at an earlier age.

Obesity has reached epidemic proportions in the United States. In 2006, 66.7% of the U.S. population aged 20 years and over was classified as overweight or obese. Almost 34% of the U.S. population is classified as obese (33.9%). The age-adjusted NHANES data show that obesity rates increased from 15% (from 1976–1980 data) to 23% (1988–1994 data) and then to 34.3% in 2003–2004.[1, 66] Data from 2005–2006 indicate that the prevalence of obesity has since remained stable;[67] nonetheless, of grave concern is the prevalence of extreme obesity (BMI ≥40 kg/m^2), which is now 5.9%.[68]

Obesity is positively associated with dyslipidemia, hypertension, physical inactivity, and diabetes, all of which are also associated with AS.[69] These associations make it impossible to determine just how many deaths are associated with obesity alone, but obesity is estimated to be the cause of over 300,000 deaths annually (see Chapter 12).

Hypothyroidism (see Chapter 17) leading to obesity has also been identified as a factor leading to increased risk of coronary AS. This relationship is due at least in part to the altered lipid metabolism in this population. There is a decrease in the activity of the lipogenic enzyme that downregulates LDL receptors. Studies of patients with poorly managed hypothyroidism have revealed evidence of greater progression of coronary atherosclerotic lesions than in patients who were well treated.[70]

Table 13.6 Classification of Overweight and Obesity by BMI, Waist Circumference, and Associated Disease Risk*

| | BMI (kg/m²) | Obesity Class | Disease Risk* Relative to Normal Weight and Waist Circumference | |
			Men ≤102 cm (≤40 in.) Women ≤88 (≤35 in.)	Men >102 cm (>40 in.) Women >88 cm (>35 in.)
Underweight	<18.5		—	—
Normal⁺	18.5–24.9		—	—
Overweight	25.0–29.9		Increases	High
Obesity	30.0–34.9	I	High	Very high
	35.0–39.9	II	Very high	Very high
Extreme Obesity	≥40	III	Extremely high	Extremely high

*Disease risk for type 2 diabetes, hypertension, and CVD.

⁺Increased waist circumference can also be a marker for increased risk even in persons of normal weight.

National Heart, Lung, and Blood Institute. Determination of relative risk status based on overweight and obesity parameters [monograph on the Internet]. Table IV 2. Bethesda (MD): National Heart, Lung, and Blood Institute; 1998. Available at http://www.nhlbi.nih.gov/guidelines/obesity/e_txtbk/txgd/4121.htm

Table 13.7 Chemical and Physical Properties of Plasma Lipoproteins in Humans

Property	Chylomicrons	VLDL	IDL	LDL	HDL
Density (g/mL)	<1.006	<1.006	1.006–1.019	1.019–1.063	1.063–1.21
Diameter (nm)	80–500	40–80	24.5	20	7.5–12
Lipids (% by wt)	98	92	85	79	50
Cholesterol	9	22	35	47	19
Triglyceride	82	52	20	9	3
Phospholipid	7	18	20	23	28
Apolipoproteins (%)	2	8	15	21	50
Major	A-1, A-2; B-48; C-1,2,3; E	B-100; C-1,2,3; E	B-100	B-100	A-1, A-2; C-1,2,3; D; E

Adapted from Table 5.4 Gropper SS, Smith JL, Groff JL. Advanced Nutrition and Human Metabolism. 2009, p. 146.

Dyslipidemia

Lipids are transported via lipoproteins comprised of a lipid interior and protein shell. There are several different types of lipoproteins. These lipoproteins vary in their protein makeup, in their lipid-to-protein ratio, and in the proportion of lipid components they contain. The protein portion of the lipoprotein is called the **apolipoprotein**. The apolipoprotein provides structural integrity and allows for receptors to recognize the lipoprotein particle. The lipid-to-protein ratio and lipid composition of a lipoprotein affect the density of the structure and allow for classification. Table 13.7 identifies the lipid composition of different classifications of lipoproteins and the apolipoproteins associated with each type.

Chylomicrons transport dietary lipids after intestinal absorption. The other lipoproteins transport endogenous lipid from the liver to the rest of the body. Note from Table 13.7 that very-low-density lipoproteins (VLDL) are similar to chylomicrons in density but are smaller in size. The triglyceride content is much higher in VLDL than in low-density lipoproteins (LDL), which are much higher in cholesterol content. VLDL is produced in the liver, and as it travels through the body, triglycerides are removed until the particle density and size are reduced. It becomes a transient intermediate-density lipoprotein (IDL), then finally an LDL. Thus, the cholesterol-rich LDL may be viewed as the end point of the forward transport of lipids to tissues. Ultimately, LDL particles are removed from circulation by tissues in need of cholesterol for structural purposes. Cells have LDL receptors that are activated when the need arises. The LDL particle is bound to the receptor and then engulfed by the cell. LDL receptors are found on liver cells and cells in a number of other tissues.

Dyslipidemia refers to a lipid profile that increases the risk of atherosclerotic development. Typically, dyslipidemia is a condition in which LDL levels are elevated and high-density lipoprotein (HDL) levels are low. (A variety of other dyslipidemic conditions can also exist, such as the combination of normal LDL and high triglyceride levels.) HDL particles are involved in reverse cholesterol transport, in that they transport cholesterol from tissues and other lipoproteins to the liver. ATP III indicated that serum LDL levels are the single strongest indicator of CVD risk. Among the lipoproteins, LDL are most heavily involved in the atherosclerotic process. Oxidation of the LDL causes this lipoprotein to be altered and can initiate damage,

starting the atherosclerotic process. Additionally, oxidized LDL is more likely to be taken up into the atherosclerotic plaque (see the Pathophysiology section). Thus, the higher the serum LDL levels, the greater the risk of the initiation of an atherosclerotic plaque and the greater the risk of continual development of an existing atherosclerotic plaque. In the ATP III, recommendations included maintaining an LDL level below 100 mg/dL (see Table 13.8).[65]

Table 13.8 Interpretation of Laboratory Values

Total Cholesterol Levels	
Less than 200 mg/dL	"Desirable" level that puts you at lower risk for heart disease; a cholesterol level of 200 mg/dL or greater increases your risk
200 to 239 mg/dL	Borderline High
240 mg/dL and above	"High" blood cholesterol; a person with this level has more than twice the risk of heart disease compared to someone whose cholesterol is below 200 mg/dL
HDL Cholesterol Levels	
Less than 40 mg/dL	A major risk factor for heart disease
40 to 59 mg/dL	Normal
60 mg/dL and above	An HDL of 60 mg/dL and above is considered protective against heart disease
LDL Cholesterol Levels	
Less than 100 mg/dL	Optimal
100 to 129 mg/dL	Near optimal/Above optimal
130 to 159 mg/dL	Borderline high
160 to 189 mg/dL	High
190 mg/dL and above	Very high
Triglyceride Levels	
Less than 150 mg/dL	Normal
150 to 199 mg/dL	Borderline high
200–499 mg/dL	High
500 mg/dL or above	Very high

Source: Grundy SM, Cleeman JI, Bairey Merz CN, Brewer HB, Clark LT, Hunninghake DB, Pasternak RC, Smith SC, Stone NJ, for the Coordinating Committee of the National Cholesterol Education Program. Implications of Recent Clinical Trials for the National Cholesterol Education Program Adult Treatment Panel III Guidelines Circulation. Adult Treatment Panel III Guidelines 2004; 110:227–39.

HDL offers a protective effect against AS. Since HDL removes cholesterol from tissues and returns that cholesterol to the liver, it reduces cholesterol in plaques. ATP III recommends that an HDL <40 mg/dL be considered low, indicating a greater level of risk. An HDL of >60 mg/dL is considered high and will in turn reduce risk for AS.[65]

In assessing risk for AS, one would use LDL as the primary marker; if LDL is unavailable, then HDL should be used as the indicator of risk. Total cholesterol should be used as an indicator of risk only if it is the only measure available. Total cholesterol levels are recommended to be below 200 mg/dL.

Approximately 45% of adults in the United States have a total cholesterol level of 200 mg/dL or above, and about 16% have a total cholesterol level above 240 mg/dL. While these numbers may seem shocking, the mean total serum cholesterol of adults in the United States decreased from 204 mg/dL to 199 mg/dL between 1999 and 2006.[1] However, the only significant declines were observed for men >40 years of age and for women >60 years of age. Approximately 33% of adults have a high LDL cholesterol level (≥130 mg/dL), while only 16% have a low HDL cholesterol level (<40 mg/dL).[1] HDL levels average about 45 mg/dL in males and about 55 mg/dL in females. In 2005–2006 approximately 65% of women and 70% of men reported that they had had their cholesterol level checked in the past five years.[1]

Hypertension
Hypertension is both a cardiovascular condition and a risk factor for other forms of cardiovascular disease. From 40 to 70 years of age, an increase of systolic BP by 20 mmHg or diastolic BP by 10 mmHg increases risk of cardiovascular disease twofold. One estimate is that a reduction of 12 mmHg in systolic blood pressure of hypertensives will prevent one death for every 11 patients treated.[2] An increase in BP increases the forces applied to the endothelium and can cause the initiation of an atherosclerotic lesion (see the "Pathophysiology" section). Changes in pressure may also cause established plaques to rupture, which not only can initiate an event such as an infarct but can also cause a proliferation of existing plaques. The sheer forces of blood against the arterial wall alone can cause endothelial damage. AS that occurs at bifurcations and trifurcations (branch points) of blood vessels, where these forces are amplified, demonstrates the importance of BP in the initiation of AS. Obstructive AS in the epicardial coronary arteries occurs most frequently in the first 5 cm where the forces associated with the branching of the arteries are greatest.

Physical Inactivity
Many professionals think of AS and heart disease in general as hypercaloric diseases. In other words, they think that the increase in the patient's mass due to chronic positive caloric balance, along with the associated dyslipidemia, hypertension, and insulin resistance, results in AS. The exact mechanism by which physical inactivity increases AS risk has not been identified. However, increasing physical activity is known to impact several factors related to AS by lowering blood pressure and triglycerides, increasing HDL, improving endothelial function, and decreasing platelet aggregation.[71] Increases in physical activity also aid in weight maintenance and reduce the relative workload of any activity of daily living on the cardiovascular system. For example, if vacuuming a room required a person to work at 70% of his or her functional capacity (or maximal workload) prior to beginning an exercise program, then after beginning the exercise program, the same activity may only require 65% of that individual's functional capacity. This would mean that the relative strain on the cardiovascular system would be reduced.

The relative risk of CHD associated with physical inactivity ranges from 1.5 to 2.4. That makes physical inactivity a risk factor with an impact similar to dyslipidemia, hypertension, or cigarette smoking.[1]

Atherogenic Diet
Naturally, diet plays a role in obesity, which is directly and indirectly associated with AS. This is discussed in much more detail under "Nutrition Therapy for Artherosclerosis" later in this chapter.

Research over the years has used the term "Westernized Diet" to describe a diet high in saturated fat, high in sodium, and low in fiber. When compared with populations that eat diets high in saturated fat, populations that consume diets higher in fruits, vegetables, whole grains, and unsaturated fats have lower rates of atherosclerotic disease than can be explained by other risk factors.[2, 21] The National Cholesterol Education Program (NCEP) Adult Treatment Plan (ATP III) includes an LDL-lowering diet (Therapeutic Lifestyle Changes diet) that limits saturated fat intake to <7% of total kcal or less than 16 g for an individual on a 2000 kcal/day diet.[72]

Diabetes Mellitus
CAD is the most common cause of death among patients with diabetes (see Chapter 17). The risk of death from cardiovascular disease in patients with type 1 and type 2 diabetes is 2–4 times greater than in those without diabetes.[1] In 2003, 3.5 million patients with diabetes age 35 or older self-reported having CAD.[67] Age-adjusted prevalence of CHD in patients with diabetics remained steady between 1997 and 2003 at around 22.3%.[73]

Impaired Fasting Glucose and Metabolic Syndrome
Impaired fasting glucose and diabetes are closely associated with the risk of cardiac death. An impaired fasting glucose is defined as plasma glucose between 110 and 125 mg/dL (fasting plasma glucose of >126 mg/dL is diagnostic for diabetes). Although impaired fasting glucose is listed by many organizations as a separate and independent risk factor, it does have a close association with other factors of the *metabolic syndrome*. Though there is not a universally accepted definition, the metabolic syndrome is a constellation of metabolic risk factors, including abdominal obesity, insulin resistance, dyslipidemia, hypertension, and prothrombotic state (a state in which the formation of blood clots is facilitated).[74]

According to the NCEP guidelines, displaying three of the five risk factors classifies an individual as having metabolic syndrome (see Table 13.9). The NCEP criteria predict both all-cause and cardiovascular mortality, while the World Health Organization criteria predict cardiovascular mortality but not all-cause mortality.[75] The age-adjusted prevalence of metabolic syndrome is 34.6%. Approximately 76 million Americans ages 20 years and older, as well as 44% of overweight and obese adolescents, have metabolic syndrome.[1]

Given that the metabolic syndrome consists of multiple CAD risk factors, it is a potent predictor of risk for AS. Risk is increased with the number of metabolic syndrome factors present in an individual. Persons with metabolic syndrome are

Table 13.9 Diagnosis of Metabolic Syndrome

Risk Factor	NCEP ATP III Criteria	IDF Criteria	WHO Criteria†
Abdominal Obesity*	Men: waist circumference** >102 cm (>40 in) Women: waist circumference >88 cm (>35 in)	Europoid, Sub-Saharan, Eastern Mediterranean, and Middle Eastern (Arab) men: waist circumference ≥94 cm for men Europoid, Sub-Saharan, Eastern Mediterranean, and Middle Eastern (Arab) women: waist circumference ≥80 cm South Asian, Chinese, Ethnic South and Central American men: waist circumference ≥90 cm South Asian, Chinese, Ethnic South and Central American women: waist circumference ≥80 cm Japanese men: waist circumference ≥85 cm Japanese women: waist circumference ≥90 cm	Men: waist-hip ratio >0.9 Women: waist-hip ratio >0.85
Triglycerides	≤150 mg/dL (1.7 mmol/L)	>150 mg/dL (1.7 mmol/L) or treatment for this lipid abnormality	≥150 mg/dL
HDL Cholesterol	Men: <40 mg/dL (0.9 mmol/L) Women: <50 mg/dL (1.1 mmol/L)	Men: <40 mg/dL (0.9 mmol/L) Women: <50 mg/dL (1.1 mmol/L) or specific treatment for this lipid abnormality	Men: <35 mg/dL Women: <39 mg/dL
Blood Pressure	≥130/≥85 mmHg	≥130/≥85 mmHg or treatment of previously diagnosed hypertension	≥140 (systolic) or ≥90 (diastolic) mmHg or treatment of previously diagnosed hypertension
Insulin Resistance	Fasting plasma glucose ≥100 mg/dL	Fasting plasma glucose ≥100 mg/dL (5.6 mmol/L) or previously diagnosed T2DM	Impaired fasting glucose or impaired glucose tolerance or glucose uptake below the lowest quartile for background population under investigation under hyperinsulinemic, euglycemic conditions (in patients with normal fasting glucose levels <110 mg/dL) or previously diagnosed T2DM
Other	—	—	BMI >30 Urinary albumin excretion rate ≥20 mcg/min or albumin/creatinine ratio ≥30 mg/g

*Overweight and obesity are associated with insulin resistance and the metabolic syndrome. However, the presence of abdominal obesity is more highly correlated with the metabolic risk factors than is an elevated body mass index (BMI). Therefore, the simple measure of waist circumference is recommended to identify the body weight component of the metabolic syndrome.

**Some male patients can develop multiple metabolic risk factors when the waist circumference is only marginally increased, e.g., 94–102 cm (37–39 in). Such patients may have a strong genetic contribution to insulin resistance. They should benefit from changes in life habits, similarly to men with categorical increases in waist circumference.

†WHO diagnostic criteria include insulin resistance plus any two of the other factors.

References: National Cholesterol Education Program (NCEP) Expert Panel on Detection, Evaluation, and Treatment of High Blood Cholesterol in Adults (Adult Treatment Panel III). Third Report of the National Cholesterol Education Program (NCEP) Expert Panel on Detection, Evaluation, and Treatment of High Blood Cholesterol in Adults (Adult Treatment Panel III) final report. *Circulation.* 2002;106:3143–3421. Grundy SM, Cleeman JJ, Daniels SR, Donato KA, Eckel RH, Franklin BA, Gordon DJ, Krauss RM, Savage PJ, Smith SC, Spertus JA, Costa F. Diagnosis and Management of the Metabolic Syndrome. Circulation. 2005;112:2735–2752. International Diabetes Federation: The IDF consensus worldwide definition of the metabolic syndrome. Available at http://www.idf.org/webdata/docs/Metabolic_syndrome_definition.pdf

more likely to exhibit what may be termed a fibrinolytic profile (i.e., an impaired ability to dissolve blood clots), which in turn would translate to atherosclerotic development and arterial stiffness.[2, 76] The evidence of this lies in the C-reactive protein (CRP) levels and plasminogen activator inhibitor type 1 (PAI 1) levels, which are typically higher in persons with metabolic syndrome. CRP and PAI 1 are likely related to the excess of adipose tissue. CRP is a powerful predictor of coronary events associated with AS. The full importance of CRP to the atherosclerotic process, hypertension, and acute coronary events is just beginning to be recognized.[14]

Cigarette Smoke Despite the fact that cigarette smoking declined over 50% among adults from 1965 to 2006, cigarette smoking remains a major health problem. Over 44 million adult Americans smoke, comprising about 22% of males and 17.5%

of females. Prevalence of cigarette smoking is highest among American Indians/Alaska Natives (36.4%), followed by African-Americans (19.8%), whites (21.4%), Hispanics (13.3%), and Asians (9.6%).[77] Between 1990 and 2000, the largest increases in smokers were in the group aged 18–24 years and in Hispanic females. At least in part because of smoking and obesity, younger white males and females aged 18–44 years are at higher risk for noncommunicable diseases than in previous years.

Over 443,000 Americans die each year as the result of smoking-related illnesses,[77] and about 35% of these deaths are cardiovascular disease related.[1] In fact, cigarette smokers are 2–4 times more likely to develop heart disease than nonsmokers.[77] Use of low-tar cigarettes likewise increases risk of cardiovascular disease and MI compared to that of nonsmokers. Even passive smoke exposure is associated with an increase in cardiovascular disease.[78] Compared to nonsmokers, smokers have significantly higher levels of serum cholesterol, triglycerides, and LDL cholesterol, as well as lower HDL cholesterol levels. Exercise may attenuate the effect of smoking on lipid profile.

A number of studies support cigarette smoking as a strong risk factor for AS and causative factor in CAD mortality. Environmental exposure to cigarette smoke has also been associated with a significant increase in CAD risk. In addition to the relationship noted for coronary AS, aortic and peripheral AS are associated with smoking as well.[78]

Endothelial dysfunction, inflammation, and modification of lipids that initiate and progress atherosclerotic development are affected by cigarette smoke. Endothelial relaxation is impaired by cigarette smoking. Nitric oxide (NO), which is primarily responsible for vasodilation of the endothelium, is decreased in endothelial cells exposed to components of cigarette smoke such as nicotine. Cells exposed to blood from cigarette smokers demonstrate a decrease in the activity of the endothelial NO synthase enzyme.[78]

Inflammatory markers such as CRP are increased in response to cigarette smoke. The increased leukocyte count and proinflammatory cytokines that occur in response to cigarette smoking increase endothelium leukocyte interaction.[78] Cigarette smokers also have significantly higher total cholesterol and LDL and lower HDL than nonsmokers. Moreover, cigarette smoking increases oxidative modification of LDL, which is a major step in the development of atherosclerosis. LDL exposed to cigarette smoke extract in culture was taken up by macrophages after modification of the lipoprotein. The clearance of LDL by macrophages is an integral part of atherosclerotic plaque formation. Increased LDL oxidation is likely the result of a decrease in the activity of protective enzymes in the plasma.[78]

Obstructive Sleep Apnea Obstructive sleep apnea (OSA) is not normally included in a list of risk factors for AS, and no organization has recognized it as such; however, emerging evidence suggests a link between OSA and CVD. OSA is repeated interruption of sleep by ventilator disruptions that result from mechanical collapse of the airway. The prevalence of OSA in CVD patients is two to three times greater than that in the non-CVD population. One study showed that in an eleven-year follow-up, risk of CVD death was 35% higher in the OSA patients.[79]

While it should be noted that OSA patients typically have existing comorbidities such as obesity and hypertension, OSA and CVD share common risk factors such as male sex, age, overweight, cigarette smoking, android fat deposition pattern, and physical inactivity. Though the close relationship between risk factors might suggest that the risk associated with OSA is simply a product of these associations, recent studies indicate that OSA may be associated with other mechanisms that can be considered atherogenic. Many of these mechanisms are associated with the cycle of hypoxemia and reoxygenation which occurs in OSA, while some are tied to sleep deprivation.[79]

Pathophysiology

The prevailing theory for the pathophysiology of AS is that the onset of disease begins as a response to injury that results in an inflammatory process. The injury is damage to the endothelial

lining of the arterial wall. This injury may be caused by pressure on the wall exerted by the blood (as a result of hypertension) or by vasospasm.[80] Some arterial sites are more susceptible to endothelial damage due to their hemodynamic characteristics. Because curved and branching vessel geometries simply have more abnormalities in blood flow, the artery is more likely to suffer endothelial damage at these sites.[81, 82] Chemical irritants from tobacco, oxidized LDL, glycated substances resulting from diabetic metabolism, and homocysteine are also implicated in the pathogenesis.

Historically the renin-angiotensin system has drawn attention of researchers in this field due to the role this system plays in blood pressure regulation. More recent research indicates that angiotensin II is a proinflammatory mediator which plays a key role in AS plaque development.[83]

Nitric oxide (NO) is a substance naturally produced by endothelial cells and a number of other cells. NO produced in the endothelial cells controls the normal relaxation of smooth muscle in the arteries and arterioles. NO also helps regulate other mechanisms important to the atherosclerotic process, including leukocyte adhesion, platelet adhesion, and thrombosis. Research has indicated that increased oxidative stress observed in atherosclerosis results in an imbalance of NO.[14, 69, 78, 84, 85] For example, insulin resistance decreases NO availability due to changes in blood flow for endothelial cells.[85] NO also appears to be inactivated in the presence of angiotensin II.[83] Low levels of NO exacerbate the inflammatory state that contributes to the atherosclerotic process.

At the onset of lesion formation, the damage to the endothelial layer attracts platelets to the area (see Chapter 19). These platelets attach to the endothelium and form a small clot termed a mural thrombus. As platelets adhere to the subendothelial surfaces at the site of injury, they begin to secrete both adenosine diphosphate (ADP), which promotes platelet aggregation, and platelet-derived growth factor (PDGF), which attracts monocytes and promotes smooth muscle cell mitosis from the media. Growth factors such as PDGF have a chemotactic or attracting effect, drawing smooth muscle cells, fibroblasts, and other cells to the injured area. The net result is an increase in collagen and a harder, more fibrous growth. Platelets also secrete thromboxane A2, which causes vasoconstriction and promotes additional vascular injury, as well as a number of proinflammatory substances and growth factors that promote plaque formation.[86] Greater amounts of CRP and PAI 1 promote a fibrinolytic state, impairing the ability to dissolve thrombi.

Monocytes adhere to the damaged section, and migrate between endothelial cells and into subendothelial spaces where they convert to macrophages. The macrophages will express receptors for oxidized LDL, and PDGF stimulates an increase in LDL receptors. Oxidized LDL enters the cell at a much more rapid rate than does non-oxidized LDL. Smooth muscle cells, which migrate to the intima and macrophages, take up LDL until they are transformed into **foam cells**. Foam cells are filled with cholesterol and are released into the extracellular spaces where they form fatty streaks.[80, 87] The fatty streak, which may occur as early as age 5, is the earliest visible sign of AS.[69] Since clot formation is integral to the AS process, it is understood that AS thrives on a prothrombotic state—one in which the formation of clots is facilitated. In addition to the factors already mentioned, catecholamine secretion as the result of stress can also introduce a prothrombotic state.[80]

The atherosclerotic lesion progresses with continued migration of cells into the area, proliferation of the plaque, and growth of tissue. Over time, this plaque develops into a fibromuscular complex with a fibrous cap. Smooth muscle cells migrate into the intima from the media by mitosis, and platelets secrete collagen, adding to the fibrous makeup of the plaque's exterior surface. Inside the complex is a mix of connective tissue, lipids, macrophages, smooth muscle cells, **thrombus**, and calcium. As this plaque grows, the artery compensates by expanding outward and leaving the lumen size basically unchanged. As the plaque continues to grow, it eventually decreases the size of the lumen.[80, 87]

Plaques are replete with the products of oxidative damage. The inflammatory cells that move into the plaque secrete a number of oxidants, which cause cell necrosis. Cells such as the macrophage can produce superoxide radicals, which dismutate into hydrogen peroxide, which in turn can damage cells and oxidize LDL. When hydrogen peroxide is combined with other substances, such as metal ions like iron, it can cause cell death. As a result of a number of reactions involving cellular-derived antioxidants and metal ion oxidation, the cells within the plaque, including the macrophage, can be destroyed. This oxidative stress is thought to only enhance the prothrombotic state and promote progression of the plaque.[88]

The rate at which a plaque progresses varies. Some plaques may grow rapidly and then stabilize; some may slowly and steadily progress, while still others may grow at a rapid rate. The structure and composition affect the likelihood for rupture; cigarette smoke increases the risk of plaque rupture.[78] Rupture may mean a portion of the fibrous cap is lost as the plaque ruptures, exposing the underlying tissue to the blood. The process begins anew at that location; thus, a new thrombus is formed. This scenario may occur repeatedly, with layers of thrombi being continually incorporated into the lesion.[80, 87] Stability of the plaque varies considerably; those plaques that contain more collagen fibers are more stable. Some plaque components such as macrophages and smooth muscle cells secrete substances that inhibit collagen production or degrade existing collagen.[89] Instability of the plaque makes it more vulnerable to rupture and thus more likely to result in myocardial infarction or stroke.

Clinical Manifestations

AS is asymptomatic except when the patient begins to experience the symptoms of ischemic heart disease (see the Ischemic Heart Disease section later in this chapter). Signs of AS are assessed by a lipid profile and determination of other specific risk factors. These are summarized in steps 1 through 4 of the ATP III Treatment Guidelines (see Table 13.10).

Medical Treatment

Current treatment for and prevention of AS are guided by the Adult Treatment Panel III Guidelines developed by the National Cholesterol Education Program (NCEP) and approved by the National Heart, Lung, and Blood Institute, the American College of Cardiology, and the American Heart Association (see steps 5–9 in Table 13.10).[72, 90] These practice guidelines are a synthesis of existing research and are the currently accepted practice. The most recent change in treatment occurred with the release of the ATP III guidelines, when the focus of treatment shifted from total serum cholesterol levels to LDL and triglyceride levels.

Table 13.11 outlines current classifications of medications used to reduce the risk of AS. Given recent evidence regarding the proinflammatory role of angiotensin II and the role of this hormone in plaque development, treatment with ACE inhibitors and angiotensin receptor blockers shows promise. ACE inhibitors improve vasodilatory function and NO production. Angiotensin receptor blockers also appear to provide benefits.[83]

Table 13.10 Summary of ATP III Guidelines

STEP 1: Determine lipoprotein levels—obtain complete lipoprotein profile after 9- to 12-hour fast.	STEP 4: If 2+ risk factors (other than LDL) are present without CHD or CHD risk equivalent, assess 10-year (short-term) CHD risk.
STEP 2: Identify presence of clinical atherosclerotic disease that confers high risk for coronary heart disease (CHD) events (CHD risk equivalent):	Three levels of 10-year risk:
• Clinical CHD	• >20%—CHD risk equivalent
• Symptomatic carotid artery disease	• 10%–20%
• Peripheral arterial disease	• <10%
• Abdominal aortic aneurysm	STEP 5: Determine risk category:
STEP 3: Determine presence of major risk factors (other than LDL):	• Establish LDL goal of therapy
Major Risk Factors (Exclusive of LDL Cholesterol) that Modify LDL Goals	• Determine need for therapeutic lifestyle changes (TLC)
• Cigarette smoking	• Determine level for drug consideration
• Hypertension (BP ≥140/90 mmHg or on antihypertensive medication)	
• Low HDL cholesterol (<40 mg/dL)*	
• Family history of premature CHD (CHD in male first degree relative <55 years; CHD in female first degree relative <65 years)	
• Age (men ≥45 years; women ≥55 years)	

*HDL cholesterol ≥60 mg/dL counts as a "negative" risk factor; its presence removes one risk factor from the total count.

Note: in ATP III, diabetes is regarded as a CHD risk equivalent.

(continued)

Table 13.10 Summary of ATP III Guidelines (*Continued*)

LDL Cholesterol Goals and Cutpoints for Therapeutic Lifestyle Changes (TLC) and Drug Therapy in Different Risk Categories

Risk Categories	LDL Goal	LDL Level at Which to Initiate Therapeutic Lifestyle Changes Risk Category (TLC)	LDL Level at Which to Consider Drug Therapy
CHD or CHD Risk Equivalents (10-year risk >20%)	<100 mg/dL	≥100 mg/dL	≥130 mg/dL (100–129 mg/dL: drug optional)*
2+ Risk Factors (10-year risk ≥20%)	<130 mg/dL	≥130 mg/dL	10-year risk 10%–20%: ≥130 mg/dL 10-year risk <10%: ≥160 mg/dL
0–1 Risk Factor**	<160 mg/dL	≥160 mg/dL	≥190 mg/dL (160–189 mg/dL: LDL-lowering drug optional)

*Some authorities recommend use of LDL-lowering drugs in this category if an LDL cholesterol <100 mg/dL cannot be achieved by therapeutic lifestyle changes. Others prefer use of drugs that primarily modify triglycerides and HDL, e.g., nicotinic acid or fibrate. Clinical judgment also may call for deferring drug therapy in this subcategory.

**Almost all people with 0–1 risk factor have a 10-year risk <10%; thus, 10-year risk assessment in people with 0–1 risk factor is not necessary.

STEP 6: Initiate therapeutic lifestyle changes (TLC) if LDL is above goal.

TLC Features

• TLC Diet:

• Saturated fat <7% of kcal, cholesterol <200 mg/day

• Consider increased viscous (soluble) fiber (10–25 g/day) and plant stanols/sterols (2 g/day) as therapeutic options to enhance LDL lowering

• Weight management

• Increased physical activity

STEP 7: Consider adding drug therapy if LDL exceeds levels shown in Step 5 table:

• Consider drug simultaneously with TLC for CHD and CHD equivalents.

• Consider adding drug to TLC after 3 months for other risk categories.

STEP 8: Identify metabolic syndrome and treat, if present, after 3 months of TLC.

Clinical Identification of the Metabolic Syndrome—Any 3 of the risk factors defined in Table 13.9.

Treatment of the metabolic syndrome

• Treat underlying causes (overweight/obesity and physical inactivity):

• Intensify weight management

• Increase physical activity

• Treat lipid and non-lipid risk factors if they persist despite these lifestyle therapies:

• Treat hypertension

• Use aspirin for CHD patients to reduce prothrombotic state

• Treat elevated triglycerides and/or low HDL (as shown in Step 9 below)

STEP 9: Treat elevated triglycerides.

ATP III Classification of Serum Triglycerides (mg/dL)

<150	Normal
150–199	Borderline high
200–499	High
≥500	Very high

Treatment of elevated triglycerides (≥150 mg/dL)

• Primary aim of therapy is to reach LDL goal.

• Intensify weight management.

• Increase physical activity.

• If triglycerides are ≥200 mg/dL after LDL goal is reached, set secondary goal for non-HDL cholesterol (total HDL) 30 mg/dL higher than LDL goal.

Comparison of LDL Cholesterol and Non HDL Cholesterol Goals for Three Risk Categories

Risk Category	LDL Goal (mg/dL)	Non HDL Goal (mg/dL)
CHD and CHD Risk Equivalent (10-year risk for CHD >20%)	<100	<130
Multiple (2+) Risk Factors and 10-year risk ≥20%	<130	<160
0–1 Risk Factor	<160	<190

If triglycerides 200–499 mg/dL after LDL goal is reached, consider adding drug if needed to reach non-HDL goal:

• intensify therapy with LDL-lowering drug, or

• add nicotinic acid or fibrate to further lower VLDL.

If triglycerides ≥500 mg/dL, first lower triglycerides to prevent pancreatitis:

• very-low-fat diet (≤15% of calories from fat)

• weight management and physical activity

• fibrate or nicotinic acid

• when triglycerides <500 mg/dL, turn to LDL-lowering therapy

Treatment of low HDL cholesterol (<40 mg/dL)

• First reach LDL goal, then:

• Intensify weight management and increase physical activity

• If triglycerides 200–499 mg/dL, achieve non-HDL goal

• If triglycerides <200 mg/dL (isolated low HDL) in CHD or CHD equivalent, consider nicotinic acid or fibrate

Source: U.S. Department of Health and Human Services, Public Health Service, National Institutes of Health, National Heart, Lung, and Blood Institute, NIH Publication No. 01–3305 May 2001.

Table 13.11 Drug Therapy for Primary Prevention

Drug Class	Agents and Daily Doses	Lipid/Lipoprotein Effects	Side Effects	Contraindications
HMG CoA Reductase Inhibitors (statins)	Lovastatin (20–80 mg), Pravastatin (10–40 mg), Simvastatin (20–80 mg), Fluvastatin (20–80 mg), Atorvastatin (10–80 mg), Pitavastatin (1–4 mg), Rosuvastatin (10–40 mg)	LDL-C ↓ 18%–55% HDL-C ↑ 5%–15% TG ↓ 7%–30%	Myopathy Increased liver enzymes	Absolute: Active or chronic liver disease Relative: Concomitant use of certain drugs*
Bile Acid Sequestrants	Cholestyramine (4–16 g), Colestipol (5–20 g), Colesevelam (2.6–3.8 g)	LDL-C ↓ 15%–30% HDL-C ↑ 3%–5% TG No change or increase	Gastrointestinal distress Constipation Decreased absorption of other drugs	Absolute: dysbeta-lipoproteinemia TG > 400 mg/dL Relative: TG > 200 mg/dL
Nicotinic Acid	Immediate release (crystalline) nicotinic acid (1.5–3 mg), extended release nicotinic acid (Niaspan(r)®) (1–2 g), sustained release nicotinic acid (1–2 g)	LDL-C ↓ 5%–25% HDL-C ↑ 15%–35% TG ↓ 20%–50%	Flushing Hyperglycemia Hyperuricemia (or gout) Upper GI distress Hepatotoxicity	Absolute: Chronic liver disease Severe gout Relative: Diabetes Hyperuricemia Peptic ulcer disease
Fibric Acids	Gemfibrozil (600 mg BID), Fenofibrate (200 mg), Clofibrate (1000 mg BID)	LDL-C ↓ 5%–20% (may be increased in patients with high TG) HDL-C ↑ 10%–20% TG ↓ 20%–50%	Dyspepsia Gallstones Myopathy	Absolute: Severe renal disease Severe hepatic disease

*Cyclosporine, macrolide antibiotics, various anti-fungal agents, and cytochrome P 450 inhibitors (fibrates and niacin should be used with appropriate caution).

Sources: Goldenberg N, Glueck C. Efficacy, effectiveness and real life goal attainment of statins in managing cardiovascular risk. *Vasc Health Risk Manag.* 2009;5: 369–376. U.S. Department of Health and Human Services, Public Health Service, National Institutes of Health, National Heart, Lung, and Blood Institute, NIH Publication No. 01–3305 May 2001.

Severe AS involving obstruction of blood flow often requires surgical intervention. Current procedures include the percutaneous coronary intervention (PCI), laser angioplasty, or coronary artery bypass graft (CABG). Each of these procedures is described in Box 13.8.

Nutrition Therapy for Atherosclerosis

Poor nutrition has historically been considered a risk factor for the development of AS.[91–93] In general, it is believed that nutrition therapy affects AS by interfering with plaque formation and/or by inhibiting the inflammatory response that causes the physiological changes within the blood vessels.[21, 94, 95]

But specific nutritional risk factors and recommendations have changed as scientific understanding of the disease process has deepened. The use of a single nutrition intervention to reduce cardiovascular risk reflects a rather simplistic view of the cardiovascular disease process. Certainly, diet is a modifiable risk factor, but to impact the disease process with interventions based on one dietary component is improbable. The clinician should focus on the cumulative effect of the entire diet as well as other lifestyle factors when planning dietary changes.[96] The currently accepted nutrition therapy for the prevention of atherosclerosis is the Therapeutic Lifestyle Changes (TLC) plan developed as a component of the ATP III guidelines.[65, 72] Box 13.9 outlines the TLC diet.

Nutrition Assessment

The first step in the nutrition care process is completion of a thorough nutrition assessment. As noted above and in Box 13.9, the TLC Dietary plan is the nutrient standard by

---

BOX 13.8 CLINICAL APPLICATIONS

Surgical Treatment Procedures for Atherosclerosis and Ischemic Heart Disease

- **Percutaneous Coronary Intervention (PCI):** Slender balloon-tipped tube—a catheter—from an artery in the groin to a trouble spot in an artery of the heart. The balloon is then inflated, compressing the plaque and dilating (widening) the narrowed coronary artery so that blood can flow more easily. In about 90% of all cases, this is accompanied by insertion of an expandable metal stent. Stents are wire mesh tubes used to prop open arteries after PCI.

- **Atherectomy/Endardectomy:** Procedure that removes plaque from arterial wall.

- **Laser Angioplasty:** Procedure similar to PCI, but the catheter uses a laser that is able to remove enough plaque to permit a balloon to be inflated in order to dilate the stenosis.

- **Percutaneous Transmyocardial Revascularization (PTMR):** Catheter and laser are directed through an artery in the leg to the heart. Small holes within the blocked vessel that are created by the laser to increase blood flow to heart.

- **Coronary Artery Bypass Graft (CABG):** Surgical procedure that typically uses the saphenous vein or internal mammary artery to "bypass" the blocked vessel.

Table 13.12 Dietary CAGE Questions for Assessment of Intakes of Saturated Fat and Cholesterol

C: Cheese	and other sources of dairy fats—whole milk, 2% milk, ice cream, cream, whole-fat yogurt
A: Animal fats	hamburger, ground meat, frankfurters, bologna, salami, sausage, fried foods, fatty cuts of meat
G: Got it away from home	high-fat meals either purchased and brought home or eaten in restaurant
E: Eat (extra) high-fat commercial products	candy, pastries, pies, doughnuts, cookies

Source: Table V 2–4 Adopting Healthful Habits to Lower LDL Cholesterol and Reduce CHD Risk. ATP Guidelines III, 2001, p. V-5.

which a comprehensive nutrition assessment should be evaluated. Specifically, dietary assessment should focus on those components, such as dietary fat intake and saturated fat intake, that will assist the clinician to identify specific nutrition problems and develop individualized nutrition therapy. Assessment tools that help target these specific nutrients include the MEDFICTS assessment tool (see Chapter 3), the Dietary CAGE questions (see Table 13.12),[64] and the REAP (see Table 13.13).[97] During the nutrition assessment, the patient's target weight is also calculated. Any of the established methods for estimating energy requirements can be used to determine weight maintenance energy requirements and/or energy requirements to facilitate weight loss. Optimally, the modifications that are priorities for the TLC (reducing fat, increasing physical activity, and increasing fruits, vegetables, and fiber) result in subsequent weight loss.

Nutrition Diagnosis

Next, using the assessment of the patient's dietary history, nutrition problems are prioritized in order to determine nutrition diagnoses. For example, suppose the dietary assessment indicates that the patient's overall fat intake is >45% of total kcal and that most of the dietary fat comes from large servings of animal protein several times a day as well as high-fat dairy products. An appropriate PES using nutrition diagnosis NI 51.1 would be "Excessive fat intake related to large servings of animal protein several times a day and high-fat dairy products as evidenced by fat intake >45% of total calories."

Other nutrition diagnoses likely to be present in patients with CAD are the following: excessive energy intake, excessive fat intake, inappropriate intake of fats (saturated fat, *trans* fats, cholesterol), inadequate intake of fat types (monounsaturated fat, omega-3 fatty acids), inadequate bioactive substance intake (plant sterols), inadequate fiber intake, inadequate vitamin intake (folate), overweight/obesity, food and nutrition-related knowledge deficit, undesirable food choices, and physical inactivity.

Nutrition Intervention

The nutrition recommendations in the TLC diet (presented in Box 13.9) can be summarized using the current approved health claim for food labeling: "Diets low in saturated fat and cholesterol and rich in fruits, vegetables, and grain products that contain some types of dietary fiber, particularly soluble fiber, may reduce the risk of heart disease."[54] Long-term clinical trials also support the following major components for dietary intervention in reducing cardiac risk: modification of dietary fat, saturated fat, and cholesterol, with increased amounts of fruits, vegetables, and fiber.[21, 45, 94, 95, 98]

Weight Loss Achieving a BMI within the normal range of 18.5–24.9 kg/m^2 is thought to decrease cardiovascular risk and complications associated with cardiovascular disease. Obesity negatively affects many of the known risk factors for atherosclerosis including dyslipidemia, high BP, and insulin resistance.[69]

Achieving weight reduction along with reduced waist circumference and visceral/abdominal obesity should be a priority goal in development of nutrition therapy interventions for treatment and prevention of AS and cardiovascular disease. Chapter 12 provides an in-depth discussion of the risks of overweight as well as treatment options.

Physical Activity Patients who experience angina or who are post MI and/or post operative should participate in medically supervised exercise programs, commonly referred to as cardiac rehabilitation. Cardiac rehabilitation is designed to improve functional capacity while controlling risk factors in order to improve outcomes. Data indicate that cardiac rehabilitation is underused after an MI, particularly in women and older

---

BOX 13.9 CLINICAL APPLICATIONS

Guide to Therapeutic Lifestyle Changes (TLC)

Food Items to Choose More Often

- **Breads and Cereals:**
  - ≥6 servings per day, adjusted to caloric needs
  - Breads, cereals, especially whole grain; pasta; rice; potatoes; dry beans and peas; low-fat crackers and cookies
- **Vegetables:** 3–5 servings per day fresh, frozen, or canned, without added fat, sauce, or salt

- **Fruits:** 2–4 servings per day fresh, frozen, canned, dried
- **Dairy Products:**
  - 2–3 servings per day
  - Fat-free, ½%, 1% milk, buttermilk, yogurt, cottage cheese; fat-free & low-fat cheese
- **Eggs:**
  - ≤2 egg yolks per week
  - Egg whites or egg substitute

- **Meat, Poultry, Fish:**
  - ≤5 oz per day
  - Lean cuts: loin, leg, round; extra-lean hamburger; cold cuts made with lean meat or soy protein; skinless poultry; fish
- **Fats and Oils:** Amount adjusted to caloric level: Unsaturated oils; soft or liquid margarines and vegetable oil spreads, salad dressings, seeds, and nuts
- **TLC Diet Options:** Stanol/sterol-containing margarines; viscous fiber food sources: barley, oats, psyllium, apples, bananas, berries, citrus fruits, nectarines, peaches, pears, plums, prunes, broccoli, brussels sprouts, carrots, dry beans, peas, soy products (tofu, miso)

Food Items to Choose Less Often:

- **Breads and Cereals:**
  - Many bakery products, including doughnuts, biscuits, butter rolls, muffins, croissants, sweet rolls, Danish, cakes, pies, coffee cakes, cookies
  - Many grain-based snacks, including chips, cheese puffs, snack mix, regular crackers, buttered popcorn
- **Vegetables:** Vegetables fried or prepared with butter, cheese, or cream sauce
- **Fruits:** Fruits fried or served with butter or cream
- **Dairy Products:** Whole milk/2% milk, whole-milk yogurt, ice cream, cream, cheese
- **Eggs:** Egg yolks, whole eggs
- **Meat, Poultry, Fish:** Higher-fat meat cuts: ribs, t-bone steak, regular hamburger, bacon, sausage; cold cuts: salami, bologna, hot dogs; organ meats: liver, brains, sweetbreads; poultry with skin; fried meat; fried poultry; fried fish
- **Fats and Oils:** Butter, shortening, stick margarine, chocolate, coconut

Recommendations for Weight Reduction:

- **Weigh Regularly:** Record weight, BMI, & waist circumference
- **Lose Weight Gradually:** Goal: lose 10% of body weight in 6 months. Lose ½ to 1 lb per week.
- **Develop Healthy Eating Patterns:**
  - Choose healthy foods (see Food Items to Choose More Often)
  - Reduce intake of unhealthy foods (see Food Items to Choose Less Often)
  - Limit number of eating occasions
  - Select sensible portion sizes
  - Avoid second helpings
  - Identify and reduce hidden fat by reading food labels to choose products lower in saturated fat and calories, and ask about ingredients in ready-to-eat foods prepared away from home
  - Identify and reduce sources of excess carbohydrates such as fat-free and regular crackers; cookies and other desserts; snacks; and sugar-containing beverages

Recommendations for Increased Physical Activity:

- **Make Physical Activity Part of Daily Routines:**
  - Reduce sedentary time
  - Walk, wheel, or bike-ride more, drive less
  - Take the stairs instead of an elevator
  - Get off the bus a few stops early and walk the remaining distance
  - Mow the lawn with a push mower
  - Rake leaves
  - Garden
  - Push a stroller
  - Clean the house
  - Do exercises or pedal a stationary bike while watching television
  - Play actively with children
  - Take a brisk 10-minute walk or wheel before work, during your work break, and after dinner
- **Make Physical Activity Part of Exercise or Recreational Activities:**
  - Walk, wheel, or jog
  - Bicycle or use an arm pedal bicycle
  - Swim or do water aerobics
  - Play basketball
  - Join a sports team
  - Play wheelchair sports
  - Golf (pull cart or carry clubs)
  - Canoe
  - Cross-country ski
  - Dance
  - Take part in an exercise program at work, home, school, or gym

Nutrient Composition of the TLC Diet:

Nutrient	Recommended Intake
Saturated fat	Less than 7% of total kcal
Polyunsaturated fat	Up to 10% of total kcal
Monounsaturated fat	Up to 20% of total kcal
Total fat	25%–35% of total kcal
Cholesterol	<200 mg/day
Carbohydrate	50%–60% of total kcal
Fiber	20–30 g/day
Protein	Approximately 15% of total kcal
Sodium	<2400 mg/day
Stanol esters	3–4 g/day

Source: U.S. Department of Health and Human Services, Public Health Service, National Institutes of Health, National Heart, Lung, and Blood Institute, NIH Publication No. 01–3305 May 2001.

Table 13.13 Rapid Eating Assessment for Patients (REAP)

Client is asked to respond to each of the following questions with "usually/often," "sometimes," "rarely/never," or "does not apply to me."

Topic	In an average week, how often do you:
Meals	1. Skip breakfast?
	2. Eat 4 or more meals from sit-down or take-out restaurants?
Grains	3. Eat less than 3 servings of whole-grain products a day? *Note:* Serving = 1 slice of 100% whole-grain bread; 1 cup whole-grain cereal like Shredded Wheat, Wheaties, Grape Nuts, high-fiber cereals, oatmeal, 3–4 whole-grain crackers, ½ cup brown rice or whole-wheat pasta
Fruits and Vegetables	4. Eat less than 2–3 servings of fruit a day? *Note:* Serving = ½ cup or 1 med. fruit or 4 oz. 100% fruit juice
	5. Eat less than 3–4 servings of vegetables/potatoes a day? *Note:* Serving = ½ cup vegetables/potatoes, or 1 cup leafy raw vegetables
Dairy	6. Eat or drink less than 2–3 servings of milk, yogurt, or cheese a day? *Note:* Serving = 1 cup milk or yogurt; 1½–2 ounces cheese
	7. Use 2% (reduced-fat) or whole milk instead of skim (non-fat) or 1% (low-fat) milk?
	8. Use regular cheese (like American, cheddar, Swiss, Monterey jack) instead of low-fat or part-skim cheeses as a snack, on sandwiches, pizza, etc.?
Meats/Chicken/Turkey	9. Eat beef, pork, or dark-meat chicken more than 2 times a week?
	10. Eat more than 6 ounces (see sizes below) of meat, chicken, turkey, or fish per day? *Note:* 3 ounces of meat or chicken is the size of a deck of cards or ONE of the following: 1 regular hamburger, 1 chicken breast or leg (thigh & drumstick), or 1 pork chop.
	11. Choose higher-fat red meats like prime rib, T-bone steak, hamburger, ribs, etc. instead of lean red meats?
	12. Eat the skin on chicken and turkey or the fat on meat?
	13. Use regular processed meats (like bologna, salami, corned beef, hot dogs, sausage, or bacon) instead of low-fat processed meats (like roast beef, turkey, lean ham; low-fat cold cuts/hot dogs)?
Fried Foods	14. Eat fried foods such as fried chicken, fried fish, or french fries?
Snacks	15. Eat regular potato chips, nacho chips, corn chips, crackers, regular popcorn, nuts instead of pretzels, low-fat chips or low-fat crackers, air-popped popcorn?
Fats and Oils	16. Use regular salad dressing and mayonnaise instead of low-fat or fat-free salad dressing and mayonnaise?
	17. Add butter, margarine, or oil to bread, potatoes, rice, or vegetables at the table?
	18. Cook with oil, butter, or margarine instead of using non-stick sprays like Pam or cooking without fat?
Sweets	19. Eat regular sweets like cake, cookies, pastries, donuts, muffins, and chocolate instead of low-fat or fat-free sweets?
	20. Eat regular ice cream instead of sherbet, sorbet, low-fat or fat-free ice cream, frozen yogurt, etc.?
	21. Eat sweets like cake, cookies, pastries, donuts, muffins, chocolate, and candies more than 2 times per day?
Soft Drinks	22. Drink 16 ounces or more of non-diet soda, fruit drink/punch or Kool-Aid a day? *Note:* 1 can of soda = 12 ounces
Sodium	23. Eat high-sodium processed foods like canned soup or pasta, frozen/packaged meals (TV dinners, etc.), chips?
	24. Add salt to foods during cooking or at the table?
Alcohol	25. Drink more than 1–2 alcoholic drinks a day? *Note:* One drink = 12 oz. beer, 5 oz. wine, one shot of hard liquor or mixed drink with 1 shot.
Activity	26. Do less than 30 total minutes of physical activity 3 days a week or more? *Examples:* Walking briskly, gardening, golf, jogging, swimming, biking, dancing, etc.
	27. Watch more than 2 hours of television or videos a day?

Client is asked to respond to each of the following questions with "yes" or "no."

Do you . . .

28. Usually shop and prepare your own food?

29. Ever have trouble being able to shop or cook?

30. Follow a special diet, eat or limit certain foods for health or other reasons?

Client is asked to circle the number that best describes how he/she feels on a scale of 5 to 1, with 5 = "Very willing" and 1 = "Not at all willing."

31. How willing are you to make changes in what, how, or how much you eat in order to eat healthier?

Source: © 2005 Institute for Community Health Promotion, Brown University, Providence, RI, All rights reserved. Available at http://bms.brown.edu/nutrition/acrobat/REAP%206.pdf

patients. Of those 70 years of age and older, only 32% participate in cardiac rehabilitation compared to 66% of 60–69-year-olds and 81% of the patients under 60 years.[8]

Total Dietary Fat Current recommendations by the NCEP ATP III and the ADA evidence-based guidelines include maintenance of dietary fat intake within 20%–35% of total caloric intake. Fat that is consumed as a part of the normal diet is a mixture of **saturated, polyunsaturated**, and **monounsaturated** fatty acids. For example, olive oil is composed primarily of monounsaturated fat but also contains lesser amounts of saturated and polyunsaturated fat. The specific type of fat that is consumed has been the emphasis for research in nutrition and cardiovascular disease for the previous five decades.[21, 72, 99, 100] Current guidelines emphasize reducing amounts of saturated and *trans* fatty acids rather than a strict adherence to a reduced-fat diet.

The benefit from restricting *total* dietary fat remains controversial. A recent controlled trial with over 48,000 women in the Women's Health Initiative study indicated that a lower-fat diet that was higher in fruits and vegetables affected cardiovascular risk only slightly.[101] Critics of this study, however, point out that the reduction of total dietary fat was only minimal, and suggest that a larger reduction is needed to achieve significant results. Adherence plays a significant role in any dietary intervention. A comparison of popular diets for cardiovascular disease indicated that patients who lost weight were able to reduce cardiovascular risk as measured by LDL-to-HDL ratios.[102] Very-low-fat diets combined with other lifestyle modifications such as increased physical activity and smoking cessation appear to have the most dramatic results in reducing cardiovascular risk factors.[103–105]

Saturated Fat A saturated fatty acid is a fatty acid that has only single bonds between carbons in its chemical structure. Saturated fats are primarily found in animal sources, though there are highly saturated plant sources, such as palm and coconut oil, as well. Research over the past five decades has led to the recommendation that no more than 7% of total kcal should be from saturated fat sources.[8, 21, 64] But controversy remains, because not all saturated fatty acids appear to affect serum lipids in the same manner.[45, 106, 107] For example, **stearic acid**, found in beef, has neither a positive nor a negative effect on cholesterol and LDL levels. While it is difficult to identify genetic differences in responses to dietary fat modification, it is these differences that may ultimately, in the future, allow more individualized nutrition therapy based on genetic profile to evolve. In the interim, current practice recommends moderation in saturated fat intake.

Trans Fatty Acids In the **hydrogenation** of fatty acids, monounsaturated and polyunsaturated fats are made into solid fats so that they can be used as margarine or shortening. When hydrogen is introduced, a double bond is formed. If the carbons are on the opposite side of the carbon-carbon double bond (which occurs during hydrogenation), this fat is designated a

Figure 13.8 Fatty Acids

Saturated Fatty Acid

$$-\overset{\overset{\displaystyle H}{|}}{C} - \overset{\overset{\displaystyle H}{|}}{\underset{\underset{\displaystyle H}{|}}{C}} -$$

Unsaturated Fatty Acid (*cis* fatty acid)

$$-\overset{\overset{\displaystyle H}{|}}{C} = \overset{\overset{\displaystyle H}{|}}{C} -$$

***Trans* Fatty Acid**

$$-\overset{\overset{\displaystyle H}{|}}{C} = \overset{\displaystyle C}{\underset{\underset{\displaystyle H}{|}}{}} -$$

trans fatty acid (see Figure 13.8). In nature, carbons are generally found on the same side of the carbon-carbon double bond, in what is designated as a *cis* configuration.[108]

A *trans* fat is used in food products to increase the shelf life. Unsaturated products have a lower melting point and can reach rancidity faster than saturated fatty acids. In the diet, *trans* fatty acids appear to behave similarly to saturated fatty acids in that they increase total cholesterol and LDL levels and perhaps lower HDL levels.[109] Additional research indicates that an increased amount of *trans* fatty acids is associated with an increased risk of myocardial infarction.[110, 111] Furthermore, *trans* fatty acids appear to be associated with an increased inflammatory response that may contribute to the atherogenic process.[112] This has led to the change in food labeling (which took effect in January 2006) that requires the listing of *trans* fatty acids on the Nutrition Facts panel.[113] Current recommendations are to consume less than 1% of energy from *trans* fat (see Box 13.10).

Monounsaturated Fat A monounsaturated fatty acid is a fatty acid with one carbon-carbon double bond within its chemical structure. Intakes of monounsaturated fats are related to cardiovascular disease, since they affect serum lipid levels positively.[21, 92, 93, 114] Specifically, monounsaturated fatty acid intake appears to lower LDL while having no affect on HDL levels. Evidence indicates that monounsaturated fatty acids tend to lower both LDL and apolipoprotein AII lipoprotein levels.[100, 115]

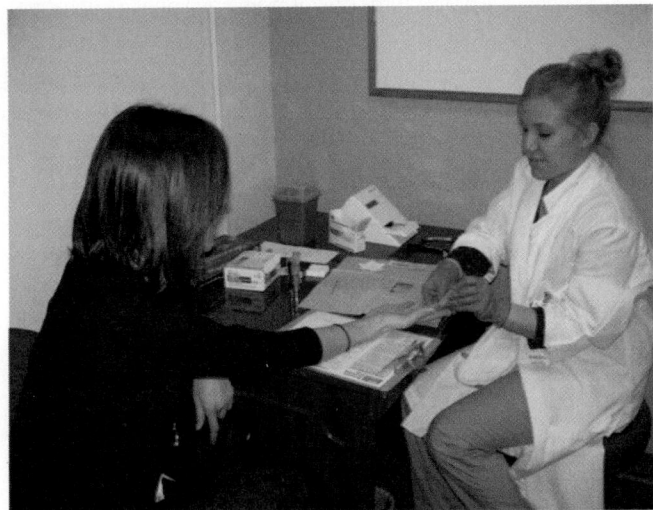

Point of care testing allows for rapid screening for blood lipid levels and determination of CVD risk.

Source: Courtesy of Marcia Nelms.

BOX 13.10 CLINICAL APPLICATIONS

Practical Tips for Consumers (Use with sample food package label)

- Check the Nutrition Facts panel to compare foods, because the serving sizes are generally consistent in similar types of foods. Choose foods lower in saturated fat, *trans* fat, and cholesterol. For saturated fat and cholesterol, use the Quick Guide to %DV: 5%DV or less is low and 20%DV or more is high. (Remember, there is no %DV for *trans* fat.)
- *Choose Alternative Fats*. Replace saturated and *trans* fats in your diet with monounsaturated and polyunsaturated fats.
  - Sources of monounsaturated fats include olive and canola oils.
  - Sources of polyunsaturated fats include soybean oil, corn oil, sunflower oil and foods like nuts and fish.
  - Choose vegetable oils (except coconut and palm kernel oils) and soft margarines (liquid, tub, or spray) more often because the amounts of saturated fat, *trans* fat, and cholesterol are lower than the amounts of these substances in solid shortenings, hard margarines, and animal fats, including butter.
- *Consider Fish*. Most fish are lower in saturated fat than meat is. Some fish, such as mackerel, sardines, and salmon, contain omega-3 fatty acids that are being studied to determine if they offer protection against heart disease.
- *Choose Lean Meats*. These include poultry (without skin and not fried), lean beef, and pork (visible fat trimmed and not fried).
- *Ask When Eating Out*. A good tip to remember is to ask which fats are being used in the preparation of your food when eating or ordering out.
- *Watch Calories*. Don't be fooled! Fats are high in calories. All sources of fat contain 9 kcalories per gram, making fat the most concentrated source of kcalories. By comparison, carbohydrates and protein have only 4 kcalories per gram.
- Here are two actions consumers can take to keep their intake of saturated fat, *trans* fat, and cholesterol "low":

- Look at the Nutrition Facts panel when comparing products. Choose foods low in the combined amount of saturated fat and *trans* fat and low in cholesterol as part of a nutritionally adequate diet.
- When possible, substitute alternative fats that are higher in monounsaturated and polyunsaturated fats like olive oil, canola oil, soybean oil, sunflower oil, and corn oil.

Figure 13.9 Sample Label for Macaroni and Cheese

Sample Label for Macaroni and Cheese

Nutrition Facts		
Serving Size 1 cup (228g)		
Servings Per Container 2		
Amount Per Serving		
Calories 250	Calories from Fat 110	
		% Daily Value*
Total Fat 12g		18%
Saturated Fat 3g		15%
Trans Fat 1.5g		
Cholesterol 30mg		10%
Sodium 470mg		20%
Total Carbohydrate 31g		10%
Dietary Fiber 0g		0%
Sugars 5g		
Protein 5g		
Vitamin A		4%
Vitamin C		2%
Calcium		20%
Iron		4%

* Percent Daily Values are based on a 2,000 calorie diet. Your Daily Values may be higher or lower depending on your calorie needs.

	Calories	2,000	2,500
Total Fat	Less than	65g	50g
Sat Fat	Less than	20g	25g
Cholesterol	Less than	300mg	300 mg
Sodium	Less than	2,400mg	2,400mg
Total Carbohydrate		300g	375g
Dietary Fiber		25g	30g

Start Here

Limit these Nutrients

Get Enough of these Nutrients

Footnote

Quick Guide to % DV

5% or less is low
20% or more is high

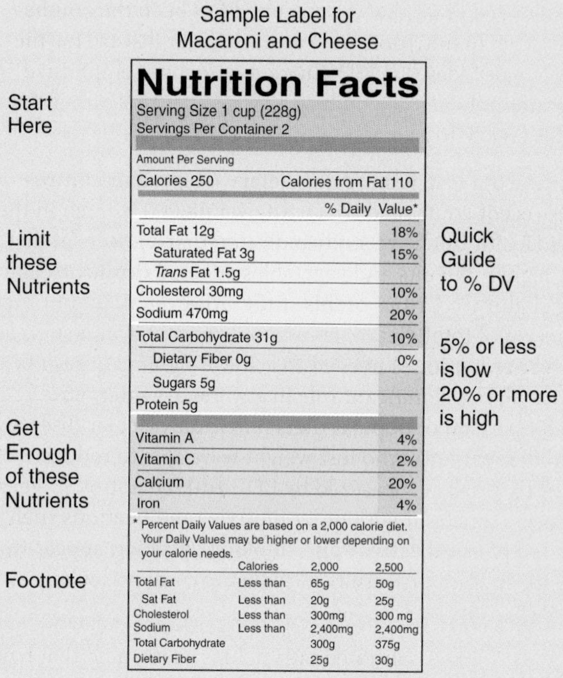

Source: Center for Disease Control and Prevention.

Reprinted from: Food and Drug Administration, Center for Food Safety and Applied Nutrition. *Trans* Fat Now Listed With Saturated Fat and Cholesterol on the Nutrition Facts Label [monograph on the Internet]. College Park (MD): FDA/CFSAN; 2006. Available at http://www.cfsan.fda.gov/□dms/transfat.html

Though olive and canola oils are rich sources of monounsaturated fatty acids, a recent review of longitudinal data indicates that the majority of monounsaturated fat intake in the United States is from oleic acid. In the United States, unfortunately, the most common food sources of oleic acid include french fries, whole milk, peanut butter, and pizza.[116]

The benefits of monounsaturated fat intake have been linked to the dietary habits of populations living within the Mediterranean region of the world. This connection is based on the epidemiological evidence that, despite higher dietary fat intakes, individuals in this area have lower cardiovascular disease rates than people from other regions.[31, 117] It is suggested that this dietary pattern may be protective for other chronic diseases such as the metabolic syndrome and type 2 diabetes mellitus

as well.[31] The Mediterranean diets that have been examined do indeed have greater amounts of monounsaturated fat, but additionally include more seafood—which provides increased amounts of omega-3 fatty acids—as well as more whole grains, fruits, and vegetables than the typical U.S. diet. Finally, overall amounts of saturated fat from animal sources are much lower than in typical diets in the United States and other western countries. It is obvious that more than one dietary factor may be involved in the positive effects seen in these population studies.

Ω-3 (Omega-3) Fatty Acids: Linolenic Acid
Linolenic acid, an omega-3 fatty acid, is considered an essential fatty acid because humans lack the enzymes Δ^{12} and Δ^{15} desaturases that are necessary to add double bonds to this 18-carbon fatty

acid.[108] Two additional fatty acids—eicosapentaenoic acid (EPA) and docosahexenoic acid (DHA)—are considered conditionally essential fatty acids because their synthesis is dependent on adequate amounts of linolenic acid. EPA is a 20-carbon fatty acid that is a precursor of the important eicosanoids. Eicosanoids include families of substances called thromboxanes, prostaglandins, and leukotrienes, as were discussed in Chapter 9. EPA, therefore, is important in cellular processes, vasoconstriction, vasodilation, platelet function, immune system response, and inflammatory response, and has been implicated as a mediator in asthma and allergic reactions.

Cold-water fishes and fish oils are particularly rich sources of linolenic acid. Flaxseed and flaxseed oil are also significant sources of alpha-linolenic acid (ALA), which the body converts to EPA to a limited extent. Several large clinical trials have examined these sources of linolenic acid and cardiovascular disease outcomes. This research has demonstrated reduced mortality with increased intakes of these specific types of lipids, though other studies have had mixed results.[21, 95, 118, 119] The AHA recommends that patients with CHD consume 1 g of EPA and DHA daily.[8]

Polyunsaturated Fatty Acids Polyunsaturated fat sources have the predominant amount of fatty acids that contain more than one double bond in their chemical structures. In the typical American diet, these fatty acids are primarily n-6 linoleic acid, which is an essential fatty acid. Polyunsaturated fat sources are oils from vegetables such as corn, cottonseed, soybean, safflower, and sunflower oils. When substituted for saturated fatty acids, polyunsaturated fatty acids have been linked to a reduction of LDL and are associated with decreased cardiovascular disease risk.[65] Table 13.14 provides guidance for making food choices to replace saturated with polyunsaturated fats.

Cholesterol Humans have a consistent requirement for cholesterol, as it is a precursor for hormones such as estrogen and testosterone and for the vitamin D provitamin (dehydrocholesterol). Cholesterol is also a major component of cell membranes and other cellular structures. Total daily cholesterol absorption and synthesis is approximately 1000 mg/day. The multistep process for cholesterol synthesis (see Figure 13.10), occurring primarily in the liver, is a negative feedback reaction: as the body pool of cholesterol increases, the synthesis rate will decrease.[108] The rate-limiting step in synthesis involves the enzyme HMG CoA reductase, which has been a target for pharmaceutical intervention in treating hypercholesterolemia through the use of statin medications.

Historically, dietary cholesterol intake was a major point of nutrition therapy for treatment of heart disease. It is now understood that the primary concern for dietary cholesterol intake centers around its effect in raising LDL levels, not on its effect on serum cholesterol levels.[21, 98, 120] In the United States, dietary intake of cholesterol has steadily decreased over the past two decades. The National Cholesterol Education Program currently recommends an intake of less than 200 mg/day.[65]

Fiber Current recommendations for dietary intake of soluble fiber in cardiovascular disease are based on its ability to reduce LDL and total serum cholesterol levels. Meta-analyses of clinical trials have supported the hypothesis that diets high in soluble fiber decrease total and LDL cholesterol.[21, 121, 122]

Figure 13.10 An Overview of the Pathway of Cholesterol Biosynthesis in the Hepatocyte

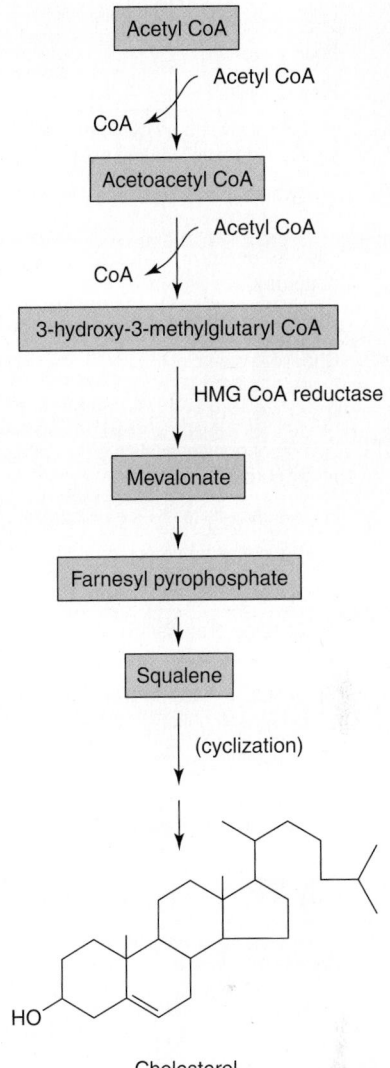

Source: From Advanced Nutrition and Human Metabolism 5e, Figure 5.33, pg. 167.

Fiber is a type of polysaccharide—the classification of complex carbohydrates that includes starches, dietary fiber, and functional fibers.[108] Dietary fiber is defined as "nondigestible carbohydrates and lignin that are intact and intrinsic in plants" (p. 107).[108] Functional fiber is defined as "Nondigestible carbohydrates that have been isolated, extracted or manufactured and have beneficial physiological effects in humans" (p. 107).[108] Though fiber has many characteristics that impact human health, fiber's ability to bind molecules is the most important for its impact on cardiovascular disease. Soluble, viscous fiber may reduce serum cholesterol and LDL levels in several ways. First, soluble fiber may decrease overall absorption of lipids. Secondly, soluble fiber is thought to bind bile acids and increase their excretion rather than allow bile to enter enterohepatic circulation. This excretion decreases the overall body pool of cholesterol. Finally, in response to the decreasing amounts of cholesterol, the body transfers low-density lipoproteins (LDL) and the cholesterol they contain to the liver to support the increased synthesis of bile.[21, 123]

Table 13.14 Dietary Modification to Decrease Saturated Fat and Increase Polyunsaturated Fat Intake

Food Group	Choose	Decrease
Lean meat, poultry, and fish	Beef, pork, lamb—lean cuts well trimmed before cooking	Regular hamburger, fatty cuts of beef, spare ribs, organ meats
	Poultry w/o skin	Poultry with skin, fried chicken
	Fish, shellfish	Fried fish, fried shellfish
	Processed meats prepared from lean meats, e.g., lean ham, lean frankfurters, lean meat with soy protein or carrageen	Regular luncheon meat (bologna, salami, sausage, frankfurters)
Eggs	Egg whites, cholesterol-free egg whites	Egg yolks (if more than the recommended); includes eggs used in baking and cooking
Low-fat dairy products	Milk—skim ½% or 1% fat (fluid, powdered, evaporated, buttermilk)	Whole milk, regular yogurt (fluid, evaporated, condensed), 2% milk, imitation milk
	Yogurt—non-fat or low-fat yogurt or yogurt beverages	Whole milk yogurt
Dairy products	Cheese—low-fat natural or processed cheese	Regular cheeses (American, blue, Brie, cheddar, Colby, Edam, Monterey Jack, whole-milk mozzarella, Parmesan, Swiss), cream cheese, Neufchatel cheese
	Low-fat or nonfat varieties, e.g., cottage cheese-low-fat, nonfat, or dry curd (0% to 2%)	Cottage cheese (4% milk fat)
	Frozen dairy dessert—ice milk, frozen yogurt (low-fat or nonfat)	Ice cream
	Low-fat coffee creamer	Cream, half & half, whipping cream
	Low-fat or nonfat sour cream	Non-dairy creamer, whipped topping, sour cream
Fats and oils	Unsaturated oils—safflower, sunflower, corn, soybean, cottonseed, canola, olive, peanut	Coconut oil, palm kernel oil, palm oil
	Margarines—made from unsaturated oils listed above, especially soft or liquid forms, low in *trans* fatty acids	Butter, lard, shortening, bacon fat, hard margarine high in *trans* fatty acids
	Salad dressings—made with unsaturated oils, low-fat or fat-free	Dressings—made with egg yolk, cheese, sour cream, whole milk
	Seeds and nuts—peanut butter, other nut butters	Coconut
	Cocoa powder	Milk chocolate
Breads and cereals	Breads—whole-grain bread, English muffins, bagels, buns, corn or flour tortillas	Bread in which eggs, fat, and/or butter are a major ingredient; croissants
	Cereal—oat, wheat, corn, multi-grain	Most granolas
	Pasta	
	Rice	
	Dry beans and peas	
	Crackers, low-fat—animal type, graham, soda crackers, breadsticks, melba toast	High-fat crackers
	Homemade baked goods using unsaturated oil, skim or 1% milk, and egg substitute—quick breads, biscuits, cornbread muffins, bran muffins, pancakes, waffles	Commercial baked pastries, muffins, biscuits
Soups	Reduced- or low-fat and reduced-sodium varieties, e.g., chicken or beef noodle, minestrone, tomato, vegetable, potato, reduced-fat soups made with skim milk	Soup containing whole milk, cream, meat fat, poultry fat, or poultry skin
Vegetables	Fresh, frozen, or canned, without added fat or sauce	Vegetables fried or prepared with butter, cheese, or cheese sauce
Fruits	Fruit—fresh, frozen, canned or dried	Fried fruit or fruit served with butter or cream sauce
	Fruit juice—fresh, frozen or canned	Beverages—fruit—fruit-flavored drinks, lemonade, fruit punch
Sweets and modified fat desserts	Sweets—sugar, syrup, honey, jam, preserves, candy made without added fat (candy corn, gumdrops, hard candy), fruit-flavored gelatin	Candy made with milk chocolate, coconut oil, palm kernel oil, palm oil
	Frozen dessert—low-fat and nonfat yogurt, ice milk, sherbet, sorbet, fruit ice, popsicles	Ice cream and frozen treats made with ice cream
	Cookies, cake, pie, pudding—prepared with egg whites, egg substitutes, skim milk or 1% milk, and unsaturated oil; or fruit bar cookies, fat-free cookies, angel food cake	Commercial baked pies, cakes, doughnuts, high-fat cookies, cream pies

Reprinted from: National Institutes of Health. From: National Cholesterol Education Program. *Second Report of the National Polyesterol Education Program Expert Panel on Detection, Evaluation, and Treatment of High Blood Cholesterol in Adults (Adult Treatment Panel II)*. Bethesda, MD: National Institutes of Health, National Heart, Lung, and Blood Institute; 1993. NIH Publication 93-3095. Available at http://www.nhlbi.nih.gov/guidelines/choleterol/atp3full.pdf

Since people eat a mixture of fibers, distinguishing the physical effects from individual types of fiber is difficult. The U.S. *Dietary Guidelines* recommend consuming 14 g of fiber for every 1000 kcal, while the Food and Nutrition Board recommends that men under 50 years of age consume 38 g of fiber/day and women in the same age group consume 25 g of fiber/day.[54] The TLC dietary recommendations additionally support the use of 20–30 grams of fiber/day. The best sources of soluble fiber with the binding ability that can lower serum cholesterol and LDL levels include gums, beta-glucans, psyllium, resistant starches, and pectin. Food sources that are recommended include fruits, vegetables, and oat and soy products.

Nuts The consumption of nuts has been correlated with a reduced cardiac risk. This may be due to their specific lipid composition (high-ALA, low-saturated fat, high-unsaturated fat). When substituted for animal protein, they may favorably improve the overall nutrient content of the diet. Van Horn summarizes the recommendation to consume up to 1 oz of unsalted nuts as a substitution for equivalent energy from higher-saturated fat food sources in the diet.[21]

Plant/Sterols (Phytosterols) Plants do not contain cholesterol but they do have similar sterol components. There are over 60 different types of plant sterols, but the most common is sitosterol. Humans do not synthesize these sterols as they do cholesterol, nor are they well absorbed. Research has demonstrated that when these plant sterols are esterified to a common fatty acid (stanol ester or sterol ester), they can assist in lowering serum cholesterol and LDL levels.[21, 99, 124] Plant sterols inhibit lipid absorption at the micelle, which reduces dietary and biliary cholesterol absorption. No safe levels have been established for pregnant women or children. The FDA approved the following health claim for plant stanol/sterol esters and reduced risk of heart disease: "Diets low in saturated fat and cholesterol that include at least 1.3 grams of plant sterol esters or 3.4 grams of plant stanol esters, consumed in 2 meals with other foods, may reduce the risk of heart disease."[125, 126] Even though typical daily diets include small amounts of plant stanols, supplementation with products such as Benecol® or Promise activ® (formerly Take Control®) is necessary in order to consume the amounts demonstrated to have a therapeutic effect. Exceeding this amount does not appear to have additional benefit.

Folate Folate and vitamin B_{12} are required coenzymes for the conversion of the amino acid homocysteine to methionine. It has been consistently observed that high levels of serum homocysteine are correlated with negative cardiac risk, and thus there is interest in modifying nutritional intake to reduce this risk. Dietary intake of folate does lower serum homocysteine levels but controlled trials have not been able to consistently demonstrate that this actually improves cardiac outcomes.[21]

Nutrition Education and/or Nutrition Counseling Identifying ways to assist the patient to reduce serving sizes and choose substitutions for each of the high-fat foods is the first step toward accomplishing several of the target TLC goals (see Box 13.11) and is incorporated into the third step of the nutrition care process—nutrition intervention. For many individuals, it is overwhelming to make these dietary changes all at one time. Dietary intake and physical activity plans should be used for a minimum of six weeks, and if results are not achieved, it is recommended that pharmaceutical intervention be considered by the primary physician as a means to assist with reducing LDL and total cholesterol levels.[65]

Monitoring and Evaluation

The American Dietetic Association has made the following recommendations: "Referral to a registered dietitian for Medical Nutrition Therapy (MNT) is recommended whenever an individual has an abnormal lipid profile, based on ATP III Risk category and LDL-C goals, or has CHD. A planned initial visit lasting from 45–90 minutes and at least two to six planned follow-up visits (30–60 minutes each, with an RD) can lead to improved dietary pattern; improved lipid profile; reduced plasma total cholesterol, LDL-C, and triglycerides; and improved weight status. The number and duration of visits in the course of Medical Nutrition Therapy will need to be greater if the client is in a higher risk category, if there is a large number of Therapeutic Lifestyle Changes (TLC) that need to be made, and if the individual is not motivated to make TLC changes. Increasing the number of visits and length of time spent with a dietitian can improve serum lipid levels and CVD risk."[94]

Ischemic Heart Disease

Definition

A sedentary individual may develop an atherosclerotic plaque that occludes up to 50% of the lumen of a coronary artery, and remain completely asymptomatic. However, if the individual becomes active they may experience a symptom, called **angina**, which is directly associated with reduced blood flow to parts of the heart. When the coronary arteries are occluded to the point that the blood flow to portions distal to the blockage is compromised, the individual is said to have myocardial ischemia. The term **ischemic heart disease (IHD)** is often used interchangeably with the term CAD.

Severe and prolonged myocardial ischemia can precipitate a myocardial infarction (MI), during which necrosis of heart tissue occurs due to the lack of oxygen. Depending upon the site of the infarct, the result may be necrosis of a small area of myocardium, cardiac rhythm abnormalities due to damage to neural pathways such as the AV node or bundle branches, or sudden cardiac death.

Epidemiology

IHD is the single largest killer of Americans, accounting for 20% of all deaths in 2003.[8] It has also been estimated that by the year 2020, IHD will be the leading cause of death and disability worldwide.[127] About 37% of individuals who suffer a coronary event in a given year will die as a result. The estimated average number of years of life lost as a result of an MI is 15. Among individuals who have had an MI the sudden cardiac death rate is four to six times greater than in the general population.[1]

As large as some of these numbers are, it is important also to note that deaths from IHD have declined dramatically over the last sixty years. In the last half of the twentieth century, IHD death rates declined by 59%, and from 1999 to 2003 the death rate decreased by 30.2%. This suggests that progress is being

BOX 13.11 | CLINICAL APPLICATIONS

Brief Nutrition Counseling for Hyperlipidemia:
Steps in Behavioral Counseling ("5 A's")

Due to limited time, your approach will be more directive, with less opportunity for patient input.

Assess (Food and Nutrient Intake, Knowledge/Beliefs/Attitudes, Behavior, Physical Activity and Function, and Biochemical Data)

- Food intake and diet habits in the context of health risks
- Current physical activity
- Readiness to change behavior
- From WAVE Nutrition Counseling Tool (http://bms. brown.edu/nutrition/acrobat/wave.pdf)
  - W = Weight: Review BMI, blood pressure, blood sugar, lipids to screen for metabolic syndrome.
  - A = Activity: Conduct physical activity assessment. Ask about:
    - Moderate physical activity? Goal—30 minutes/day or more
  - V = Variety and E = Excess: Conduct brief diet assessment. Ask about:
    - High-saturated fat foods like cheese, ice cream, butter, fatty meats? Goal—low-fat dairy, lean meat, vegetable oils
    - High-fiber foods like oats, barley, or legumes? Goal—daily or several times/week

- Number of fruits and vegetables? Goal—at least five/day
- Number of meals and snacks? Goal—at least three meals/day
- Use of sweetened beverages? Goal—if triglycerides high, eliminate or reduce significantly

Advise

- Give clear, specific, and personalized behavior change advice. You might say:
  - "Changes in your diet and exercise habits can lead to significant improvement in your blood fats and reduce your risk of heart disease."
- For patients taking medication for lipids, blood pressure, diabetes: "Diet choices are important even if you are taking medication, since eating carefully helps the medicine do a better job. You may be able to save money by cutting down on the amount of medicine you take."
- For patients NOT ready to change behavior, add: "I'd like to help you when you are ready to make changes in your diet and be more active."

Agree

- Collaborate with patient to select treatment goals and methods.

- Base goals on readiness to change behavior.
- For patient NOT ready to change behavior: "Is it okay if I ask you again at our next visit?"
- Possible goals for patient ready to change:
  - Return for further discussion in 2–4 weeks.
  - Keep food and exercise records to increase awareness, if patient willing.
  - Refer for registered dietitian visit.

Assist

- Help patient acquire knowledge, skills, and support for behavior change.
- Provide hand-outs and Web resources, based on patient interest and need.
- Provide lists and recommendations for community resources (exercise and diet programs, health clubs, etc.).

Arrange

- Schedule follow-up appointments.

References
Medical Nutrition Handbook, Department of Medicine, University of Wisconsin School of Medicine and Public Health (http://www.medicine.wisc .edu/mainweb/includes/viewfile.php?fileid=893& viewtype=inline&section=naa)

---

made. However, this condition still impacts the duration and the quality of life of millions of Americans.

Ischemia and MI can lead to heart failure and rhythmic abnormalities, which result in death. In individuals over the age of 35, 80% of all sudden cardiac death is related to IHD; in contrast, before the age of 35, only 10% of these deaths result from IHD and almost 50% result from **hypertrophic cardiomyopathy**.[128] Even though there have been declines in IHD and deaths from MI, there have been increases in the prevalence of risk factors. From 1991 to 2001, the prevalence of hypertension, dyslipidemia, diabetes, and obesity increased, while the prevalence of smoking remained stable. The result is a decreased prevalence of individuals who have no risk factors for IHD. From 1991 to 2001, the prevalence of persons with one or more risk factors increased from 58.2% to 64%, and the prevalence of persons with no known risk factors decreased from 41.8% to 36%.[62]

Given that risk factors account for 90% of the risk of an initial MI, this leads one to predict that IHD and cardiovascular diseases as a whole will increase.[8, 62]

Etiology

Acute coronary syndrome is a term used to describe the condition of persons who present with either an acute MI or **unstable angina**. An estimated 879,000 persons were discharged from hospitals in 2003 with acute coronary syndrome.[8] The causes of an acute MI or unstable angina are plaque erosion, rupture of a plaque resulting in formation of a thrombus, and vasoconstriction. Over half (55%–60%) of sudden cardiac death cases are the result of a plaque rupture, while 30%–35% are due to plaque erosion.[89] The type of acute coronary syndrome depends on the duration of the occlusion.[80] Unstable angina is likely to be caused by a transient occlusion of the artery due to vasoconstriction or a thrombus that dissolves rapidly. Longer-term occlusion would result in MI.

Traditional risk factors for AS apply to IHD. A prospective study of lipid and non-lipid risk factors among healthy middle-aged females indicated that the addition of C-reactive protein (CRP) improved prediction of increased risk for MI.[129]

Individuals who survive an MI have a chance of illness and death that is 1.5 to 15 times that of the general population. Within five years of suffering a first MI:[1]

- At age 40 or older, 18% of men and 23% of women will die.
- At ages 40–69, 15% of Caucasian men, 22% of Caucasian women, 27% of African-American men, and 32% of African-American women will die.
- At age 70 and older, 50% of Caucasian men, 56% of Caucasian women, 56% of African-American men, and 62% of African-American women will die.

Pathophysiology

Any of the following four mechanisms can initiate an MI or angina in an individual with IHD:

- Sudden blockage of a coronary artery
- Hemorrhage into an atherosclerotic plaque
- Arterial spasm
- Increase in myocardial oxygen demand

All of the aforementioned mechanisms have one factor in common: an atherosclerotic plaque that is contributing to the occlusion of the lumen of the artery. Recall that resistance to flow is inversely proportional to diameter of the vessel. A very minor occlusion results in a great increase in resistance; thus, after the lesion has caused remodeling of the artery and has progressed beyond that point, resistance to flow will increase with every small increase in plaque size. While all plaques have similar features, they are not all the same and some plaques are more likely to rupture and form thrombi.

Large, hardened plaques contain more smooth muscle cells that have migrated into the plaque than smaller, softer ones. These types of plaques are less likely to rupture, but over time will gradually create occlusion of more than 70% of the lumen. These plaques may cause angina, but because of collateral circulation that can develop over time, will occlude the artery without infarct. Thus, the large, hardened plaques are not typically the cause of acute events.[87]

Soft, lipid-rich plaques are more likely to cause acute MIs. A lipid-rich plaque is one with more than 50% lipid by volume, and while only approximately 15% of plaques fall into this category, they account for an estimated 80% of all acute MI and episodes of unstable angina.[87] These plaques are less likely to occlude large areas within the lumen, but are more prone to rupture. These plaques have a fibrous cap covering the lipid-rich core, which contains a great number of macrophages. Inflammation from within the plaque weakens the bond between the cap and the interior. Macrophages within the core secrete matrix metalloproteinases and other substances that break down the cap. As the cap's collagen is dismantled, the bonding plaque becomes more unstable.[88, 130] This weakening, in combination with the physical forces of blood flow and sometimes augmented by vasoconstriction, along with nicotine and immune complexes, causes a rupture. The rupture or tearing of the cap exposes the plaque to the flowing blood. Blood pushes into the fissure, where there is hemorrhaging into the plaque, and clotting ensues. The result may be abrupt thrombotic occlusion of the artery.[80, 87] In some cases, occlusion will occur because blood seeping into the plaque causes it to enlarge and eventually obstruct the coronary artery (p. 300).[131]

The rupture may be predicated by a change in blood pressure or flow dynamics in the area of the plaque. If a vasospasm occurs in the area of the plaque, it can certainly cause angina because of temporarily obstructed blood flow. Spasms do occur in the area of atherosclerotic plaques, possibly the result of endothelial dysfunction and impaired NO production. The spasm would change flow dynamics and could initiate a plaque rupture that would lead to occlusion.

It is not unusual for individuals to experience episodes of angina in response to an increase in cardiac workload. An increase in physical activity will cause the heart rate to increase with a concomitant increase in BP. As these variables increase the workload on the heart, the oxygen requirements of the myocardium are also increased. As the oxygen demand increases, the ischemic artery becomes unable to supply adequate flow to satisfy this demand. In such a case, the individual will experience anginal pain. If the activity is stopped, the pain will likely fade because myocardial oxygen needs will be met when the workload placed on the heart is reduced. However, it is also possible that the increasing pressure in the coronary arteries will cause rupture of a plaque, occlusion, and a myocardial infarction. Severe ischemia in the coronary arteries can even cause an abnormal rhythm that results in sudden cardiac death.

These rupture-prone lesions are usually fairly small, occluding less than 50% of the lumen, and therefore most individuals who experience an acute MI have had no symptoms previously. Occlusion typically has to be much more significant to cause angina. Even though the smaller lesions are more likely to rupture, angiographically significant blockage does provide an indication of the extent of the disease.[80] The plaques that are more likely to rupture are those with a larger lipid core relative to total plaque area. In men, cigarette smoke increases the risk of lipid-rich plaque rupture and resultant sudden cardiac death.[78]

Though efforts to lower lipids tend to have only a minimal effect on the harder, more fibrous plaques, they do result in regression of the lipid-rich plaques. Lipid lowering efforts retard further plaque formation and reduce the chance for plaque rupture.[87]

The same type of atherothrombotic occlusion that was described earlier can initiate **ventricular tachycardia** or **ventricular fibrillation** and sudden cardiac death. In sudden cardiac death, there is an abrupt loss of heart function and death occurs immediately or within an hour of the appearance of symptoms.[80] There are other potential causes of sudden cardiac death, but not all are triggered by ischemia and/or infarct. Hypertrophic cardiomyopathy, left ventricular hypertrophy, and valvular disease are all causes of sudden cardiac death. In adults 35 years of age and under, hypertrophic cardiomyopathy is the cause of almost half of all sudden cardiac death, particularly in competitive athletes. In combination with left ventricular hypertrophy and congenital coronary abnormalities, cardiomyopathy accounts for over 75% of all sudden cardiac deaths in this population.[128]

When an MI occurs, necrosis of myocardial cells will occur because of the ischemia caused by prolonged arterial occlusion. If the ischemia is severe enough to cause an infarct, then the damage to the myocardium will be irreversible. This means that the membrane of the myocardial cell is disrupted and the contents escape the cell. These include certain biological markers that are used to diagnose an MI: creatine kinase and cardiac troponin T (see Table 13.15). The necrotic cardiac cells do not regenerate and are replaced instead by scar tissue. How long scar tissue formation takes is dependent upon the size of the area impacted by the infarct. The size of the tissue damage is determined by the location of the occlusion and by the amount of collateral circulation to the area affected. As blood flow to tissues may be compromised for a number of years, a secondary pathway for blood flow may be developed from smaller vessels; this is collateral circulation. An infarct may only affect part of the cardiac wall; this is known as a subendocardial infarct. If the entire width of the cardiac wall is damaged by the infarct, it is termed a transmural infarct.

In some cases the infarct may result in adverse changes to the left ventricle. These changes include an expansion of the left ventricular chamber and a thinning of the left ventricular wall, both of which severely impact ventricular contractility and result in heart failure.[80]

The left ventricle and the septum of the heart contain most of the muscle mass of the heart and therefore most of the blood flow. The oxygen requirements of the left ventricle are much greater than those of the right ventricle. To visualize this, consider that the left ventricle must move blood against an average systolic BP of around 120 mmHg, whereas the right ventricle moves against a force that is around ⅙ of that. The workloads of the right ventricle are low enough that collateral circulation can often meet the needs of the tissue if flow is disturbed. Since most of the blood flow and oxygen demands are made by the

myocardium of the left ventricle and the septum, these areas are most susceptible to disturbances in oxygen supply. Infarcts involve these areas of the heart almost exclusively.[131]

Persons who suffer an MI are subject to complications, which may be classified as:

- Disturbances of cardiac rhythm
- Heart failure
- Intracardial thrombi
- Pericarditis
- Cardiac rupture
- Papillary muscle dysfunction
- Ventricular aneurysm

Cardiac muscle tissue adjacent to an infarcted area will become irritable and can cause arrhythmias. The most serious of these is ventricular fibrillation (mentioned previously). Ventricular fibrillation causes circulation to come to a halt. In some cases the nervous tissue that transmits impulses from the atria to the ventricles will become impaired, causing what is termed heart block. In heart block the atria and ventricles may be contracting on separate rhythms, depending on the severity of the heart block. Over time the rhythmic disturbances resulting from the infarct may subside.[131]

In cases where the infarct affects the endocardium, a thrombus may form on the interior surface of the ventricular wall. This thrombus will cover the damaged area, forming what is known as a mural thrombus. Some parts of this thrombus may break loose and travel as **emboli**. These are particularly dangerous and may cause an infarct in other tissues, such as the brain, with catastrophic results.

Infarcts involving the epicardium may cause fluid accumulation in the pericardial sac. The inflammation associated with the damage to the myocardial cells will be the source of the fluid. The fluid accumulation in the pericardial sac is known as pericarditis.

Much more serious is cardiac rupture. In the case of a transmural infarct, it is possible that a leak may develop in the cardiac

Table 13.15 Cardiac Biomarkers

Biomarker	Normal Levels (Standard reference range dependent on individual patient factors and laboratory diagnostic methods)	Time to Initial Elevation	Time to Peak Elevation	Time to Return to Normal
CK-MB	Total CPK Male 55–170 U/L Female 30–135 U/L CPK MB – 0%	4–8 hours	12–24 hours	72–96 hours
CK-MB Isoforms	0%	2–6 hours	18 hours	<24 hours
Myoglobin	<90 m/L	2–4 hours	8–10 hours	24 hours
LD-I	313–618 U/L	10–12 hours	48–72 hours	7–10 days
cTnI	<0.5 ng/dL	4–6 hours	12 hours	3–10 days
cTnT	<0.5 ng/dL	4–6 hours	12–48 hours	7–10 days

CK-MB, MB isoenzyme of creatine kinase; LD-I, lactate dehydrogenase isoenzyme; cTnI, cardiac troponin I; cTnT, cardiac troponin T.

Source: Abbott Laboratories. Troponin—Physician's Brochure [monograph on the Internet]. Table 2. Abbott Park (IL): Abbott Laboratories; 2006. Available at http://www.abbottdiagnostics.com/Your_Health/Heart_Disease/troponin-physicians-brochure.cfm

wall through the necrotic tissue. If this occurs, the pericardial sac will fill with blood, eventually placing enough pressure on the heart so that it cannot expand to accept blood during diastole, a condition known as cardiac tamponade. Circulation will cease because the heart can no longer move blood. In transmural infarcts involving the septum, this same scenario can occur between the right and left ventricles. The perforation of the septum causes blood to move from the left to the right ventricle during systole. This will compromise cardiac output, resulting in heart failure.

Papillary muscles contract during systole to hold the leaflets of the valves closed, restricting blood from moving back into the atria. Infarcts damage papillary muscles and can result in their inability to keep valve leaflets in place. Mitral valve insufficiency causes blood to leak back into the left atrium during ventricular systole, compromising the stroke volume.

A ventricular **aneurysm** is an outward bulging of the healing infarct during ventricular systole. The aneurysm fills with blood during systole and therefore impacts the total distribution of blood. This can result in heart failure. The damaged tissue does not contract and overall cardiac efficiency is reduced.

Clinical Manifestations

Practitioners should be aware that the majority of men and women who die from heart disease have reported no previous symptom. In fact, less than 20% of coronary attacks are preceded by angina, the primary symptom of ischemic heart disease.[1] Angina occurs as a result of an oxygen supply that does not match oxygen demand of the tissues. If blood flow to an area of the heart is compromised, the oxygen demands of the section of heart tissue distal to the blockage will not be met due to restricted blood flow. **Stable angina** is the substernal pain experienced when the workload on the heart is increased due to physical or emotional stress. Unstable angina is angina pain that is not associated with increases in workload, and may even occur in a person at rest. While angina is typically referred to as a dull ache in the substernal area radiating to the arm and neck, it should be noted that individuals experience angina pain differently. Other symptoms of angina may include indigestion, nausea, vomiting, sweating, shortness of breath, weakness, and fatigue. Ischemic heart disease is often misdiagnosed or undiagnosed in females because they tend to experience angina pain differently than men, reporting more intense levels of pain.[132] Diabetics may present differently as well.[133]

There are other symptoms associated with heart disease that may precede a significant coronary event. These include a number of signs and symptoms that are most commonly associated with rhythmic abnormalities and may not necessarily be caused by coronary ischemia. This is particularly true in individuals who have no history of previous ischemic disease. These may include sudden-onset bradycardia, palpitations, and syncope and/or dizziness.

Medical Diagnosis

Diagnostic procedures for ischemic heart disease may include a variety of noninvasive tests, as previously discussed in the section for AS (see Box 13.12). These include chest x-ray, electrocardiogram, or exercise stress tests. Imaging tests include radionucleotide imaging, PET/CT scan, echocardiogram, and cardiac catheterization.

The World Health Organization criteria for diagnosis of myocardial infarction state that an individual must meet two out of the following three criteria: clinical history of ischemic-type chest pain, changes in serial ECG readings, and a rise and fall of serum cardiac enzymes.[134, 135] These diagnostic criteria focus on measurement of a series of enzymes or proteins that are released from damaged or dying cells. During necrosis, cellular contents, such as enzymes and proteins, are released (see Chapter 9). If the patient's blood contains enzymes/proteins normally found in large amounts within myocardial cells, this would be indicative of myocardial injury. These include MB isoenzyme of creatine kinase (CK MB), lactate dehydrogenase isoenzyme (LD I), cardiac troponin I (cTnI,), cardiac troponin T (cTnT), and myoglobin (see Table 13.15).

Treatment

Advances in medical and surgical treatments have greatly reduced the mortality rates attributable to IHD. One study found that approximately 47% of the decline in IHD deaths was attributable to treatments. Lipid-lowering and antihypertensive therapies along with other primary prevention therapies account for 12% of the decline. Secondary prevention after MI or revascularization accounts for 11%, with initial treatment of MI or angina accounting for 10% of the decline.[3]

The goals of immediate medical treatment after myocardial infarction are to reduce pain, reduce the work of the heart, stabilize cardiac function, and prevent or limit complications. Medical interventions include the use of oxygen, aspirin (for antithrombolytic effect), and morphine (for pain). A variety of medications may be used to provide fibrinolytic therapy, reduce the overall work of the heart, and treat other cardiac dysfunctions. See Table 13.1 for a description of these classes of medications.

Myocardial infarction treatment protocols provide a structure for initiation of activity. Usually bed rest is only recommended for the initial 24–48 hours after the cardiac event. Stages of physical activity slowly increase and the patient is usually discharged within 5–7 days.[135]

Nutrition Therapy for Ischemic Heart Disease

Immediate medical care after myocardial infarction strives to reduce pain, stabilize cardiac function, and, when appropriate, begin the rehabilitation post MI. Nutrition therapy after MI will be consistent with these medical goals.

Nutrition Intervention

During the immediate post-MI period, oral intake may be decreased due to pain, anxiety, fatigue, and shortness of breath. Many institutions' treatment protocols limit initial oral intake to clear liquids without caffeine in order to prevent arrythmias and to decrease risk of vomiting or aspiration.[136] Oral diets usually progress from liquids to soft, easily chewed foods with smaller, more frequent meals. As the patient stabilizes, the goals of nutrition therapy will be individualized according to the patient's risk factors and should follow the Therapeutic Lifestyle Changes Dietary Recommendations.[65, 137]

BOX 13.12   CLINICAL APPLICATIONS

Cardiac Diagnostic Procedures

Noninvasive

- *Chest X-ray:* Assesses anatomy of the heart; pulmonary congestion.
- *Electrocardiogram (ECG/EKG):* Graphic recording of the heart's electrical activity.
- *Holter Monitor:* Portable ECG.
- *Exercise Stress Test:* Heart rate, blood pressure, and ECG are measured after a session of prescribed exercise on a treadmill or bicycle.

Imaging Tests

- *Radionucleotide Imaging:* Involves the intravenous injection of small quantities of radioactive isotopes into a peripheral vein. The distribution of the radioactive tracers can be detected by gamma cameras from the radiation emitted as the radionuclide decays. Depending on the radioactive isotope used, RI is used to measure myocardial perfusion and detect ischemia, perform infarct imaging, evaluate ventricular function, and to detect and evaluate coronary artery disease.
- *MUGA (Multiple Gated Acquisition scan):* Radionucleotides attached to red blood cells allow for visualization of the heart while beating. Es-

pecially useful to measure ejection fraction of the ventricles.

- *Thallium Stress Test:* After performing an exercise stress test, the patient is injected with thallium, which allows for visualization of blood flow to the heart after stress. As the patient rests, continued visualization allows for documentation of blood flow during rest.
- *PET/CT (Positron Emission Tomography/Computer Tomography):* The combined use of PET and CT allows for anatomical visualization as well as metabolic assessment of tissues and organs.
- *SPECT (Single photon emission computed tomography):* Allows for 3 D visualization of the heart as well as blood flow.
- *Echocardiogram:* Uses sound waves to document anatomy and function of the heart.
- *Coronary angiography/cardiac catherization:* Catheter is inserted from appropriate artery or vein into the ventricle. Allows visualization of heart anatomy, function of the valves. When contrast dye is injected, blood flow can be documented in order to identify any obstructions in circulation.

Physical Assessment and Laboratory Tests

- *Pulse:* Measurement of heart rate.
- *Blood Pressure:* Measurement of cardiac output and peripheral resistance.
- *Doppler Studies:* Measurement of audible blood flow within the peripheral vessels.
- *Auscultation:* Listening to heart sounds with stethoscope.
- *Cardiac Enzymes:* Enzymes and isoenzymes are released as cells die from oxygen deprivation. These are not necessarily specific to cardiac tissue. These include lactic dehydrogenase (LDH 1), aspartate aminotransferase (AST), creatinine phosphokinase (CK-MB or CPK-2).
- *Cardiac Troponin I, T:* Protein released from myocardial cells—elevated after cardiac injury.
- *Myoglobin:* Equivalent to hemoglobin that is present in skeletal muscle. Will be released when tissue is damaged.
- *Lipid Profile:* Serum cholesterol, HDL-C, LDL-C and TC:HDL-C ratio; serum triglyceride. Others may include apolipoprotein B.

Peripheral Arterial Disease

Definition

Peripheral arterial disease (PAD) is a term used to describe occlusion of blood flow in non-coronary arteries, and for the purpose of this section of the text is limited to the lower extremities. Most PAD occurs in the pelvis and legs. *Peripheral vascular disease* is a term that includes diseases of the veins.[138] PAD has been used by some to describe all non-coronary atherosclerotic disease, including involvement in the carotid arteries.[127]

Epidemiology

PAD affects about 8 million Americans.[1] The prevalence of the condition increases with age and disproportionately affects blacks.[139] Hispanics have a slightly higher risk than whites.[139] One study found a prevalence of PAD of 2% to 3% by age 50 and

approximately 20% by age 75. This study found that 10% of the individuals with PAD had **claudication**, 50% had atypical leg pain, and 40% had no leg pain associated with physical activity.[127] In the Framingham Heart Study, the incidence of PAD was based on symptoms of intermittent claudication in subjects 29 to 62 years of age. Annual incidence of intermittent claudication per 10,000 subjects at risk rose from 6 in men and 3 in women between the ages of 30 and 44 years to 61 in men and 54 in women between the ages of 65 and 74 years. The incidence of intermittent claudication has declined since 1950, but survival among persons with intermittent claudication has remained low.[140]

There is a strong association between vascular disease and damage to the coronary, cerebral, and carotid arteries; thus, the 5-year mortality rate is relatively high. The long-term prognosis for symptomatic PAD patients is poorer than that of asymptomatic patients. Likewise, those with severe symptoms have a much worse prognosis compared to those who have mild symptoms.[127]

Risk factors for the development of AS are important to the development of PAD and are targets for treatment. However, the impacts of the individual risk factors on atherosclerotic disease development and progression in the periphery are not the same as in the coronary arteries. The most influential independent risk factors for PAD development and progression are cigarette smoking and diabetes mellitus.[127, 130, 141] In smokers, PAD development is increased 2 to 5 times. Smokers also are 8–10 times more likely to develop intermittent claudication. Cessation of smoking is associated with decreased amputation rates and increased longevity.[130]

The single strongest risk factor for PAD is diabetes. Data from the Framingham study indicate that 20% of symptomatic PAD patients were diabetics.[127, 142] Diabetic men have a higher rate of claudication than any other group. Diabetes in women eliminates the protective effect of estrogen so that their risk is elevated to that of men with similar risk profiles. Diabetes is the leading cause of non-traumatic amputation in the United States (see Chapter 17).[130]

While dyslipidemia and hypertension are important and do increase risk of PAD, they do not appear to be as influential as diabetes and cigarette smoking. Elevated LDL, low HDL, and elevated serum triglycerides do impact risk of PAD. It has been estimated that each 10 mg/dL rise in total cholesterol increases the relative risk of PAD by approximately 1.1 times. Thus, the risk is increased by 10% over adults with the same risk profiles and optimal total cholesterol levels.[130] Studies of lipid and non-lipid risk factors for PAD have shown that the total cholesterol-to-HDL ratio is the strongest lipid predictor. When CRP was added to lipid screening, the predictive value was greatly enhanced.[129]

The effect of hypertension on PAD appears to be much more subdued than the effect in the cerebral or coronary arteries. Because the data from several studies appear to provide mixed results, the degree of influence hypertension exerts on atherosclerotic development in the peripheral arteries is unclear.[130]

Pathophysiology

In PAD, the occlusion of an artery, typically in the pelvis or lower leg, restricts blood flow to tissues. The pathophysiology of PAD is quite similar to that of AS and IHD. In these conditions, an inflammatory response precedes the plaque rupture with subsequent embolus formation. The presence of PAD is also an indicator of ischemic disease in other vascular beds; thus, the risk of MI, stroke, unstable angina, and sudden cardiac death is increased in PAD patients. There are some small differences between the pathophysiology of PAD and that of IHD. The focus in this section will be on those differences.

While thrombosis is known to play a critical role in all acute ischemic incidents, it has been postulated that it has an even more important role in acute ischemic events in those with PAD. Fibrinolytic therapy in PAD patients is more effective in reducing cardiovascular events than the same therapy in those with IHD. Reducing platelet adhesion lowers the risk of emboli and thrombosis, the formation of which is key in PAD and other ischemic events, such as stroke and transient ischemic attack.[130]

In previous sections, the pathophysiology of thrombus formation and plaque rupture was described as occurring in a pro-

thrombotic state. That prothrombotic state included high levels of PAI 1 and CRP. The inflammatory process is a part of both the development of the plaque and its rupture. An elevated level of CRP is a serum marker of inflammation and is commonly found not only in persons suffering acute coronary events such as MI and unstable angina, but also in persons with PAD.[130, 143] The importance of CRP to the progression of PAD is unknown at this time, but it is suspected that the severity of acute events and symptoms are associated with increased CRP levels.[130]

There is a poor correlation between variations in BP across the leg and ischemic symptoms. The variables that contribute to claudication and their function in eliciting this symptom are somewhat complex. What is known is that in response to ischemia, collateral vessels will be developed to allow some blood flow. The extent to which these vessels are formed and their contribution to function seem to be determined to a great degree by the region in which the ischemia occurs. In some cases the new vasculature may be sufficient to cause regression of symptoms.[130]

Patients with PAD will eventually suffer from denervation of affected muscle tissue. The ischemia reperfusion of the tissue over time will cause this damage, which is thought to be mediated by oxidative damage. The damage of denervation and alterations in muscle fiber type will reduce muscle function. There is additional evidence of abnormal muscle cell metabolism in affected tissues.[130, 144]

There are several changes in muscle tissue as a result of ischemia in PAD. Mitochondrial expression increases in patients with PAD. While this is not uncommon to conditions that impair mitochondrial function, the implication is that activities inside the muscle lead to its dysfunction. Accumulation of metabolic intermediates and products of incomplete metabolism lend credence to the idea that there is a state of metabolic disorder inside the muscle. Lactate levels at rest and during light workloads are elevated beyond that which might be attributable to low blood/oxygen supply. Thus, the muscle may have made an adaptation to shunt pyruvate away from complete oxidation. This may also be a function of the relative activity of pyruvate dehydrogense and lactate dehydrogenase. Pyruvate dehydrogenase has been found to be altered in patients with PAD.[144] Acylcarnitines, a class of intermediates of substrate oxidation, accumulate in skeletal muscle and blood of PAD patients. The degree of accumulation of acylcarnitines at rest is strongly correlated with functional impairment of the individual during exercise.[130, 144]

The electron transport chain is not only a major source of reactive oxygen species formation, but also suffers extensive oxidative damage as a result of PAD. Muscle tissue from PAD patients shows specific defects in the enzymes of the electron transport chain. This damage can lead to increased reactive oxygen species formation and may further metabolic injury. In addition, the increased free radical production can contribute to further endothelial injury.[144]

PAD can cause ulcerations of the lower leg. **Ulceration** is defined as a nonhealing break in the skin. Inadequate perfusion to the tissues (i.e., lack of oxygen) is the primary etiology for skin breakdown. The ischemic conditions cause a breakdown of tissue, forming the ulcer. The arterial ulcer associated with PAD commonly occurs on the foot or toes. This type of ulcer is said to be more painful than other types and has a pitted or punched out appearance.[138] This ulceration occurs more often in those with an **Ankle Brachial Index (ABI)** < 0.4.[127]

Clinical Manifestations and Medical Diagnosis

The symptom associated with the ischemic conditions of PAD is intermittent claudication.[138, 144] Intermittent claudication is a cramp-like pain that is associated with activity and then subsides with rest. This symptom is most common in the calf but may occur in the thigh, buttocks, or feet. Intermittent claudication has an earlier onset and is more intense if the activity is more strenuous, such as walking up a slope or stairs. Therefore, individuals with PAD may unconsciously alter their physical activity patterns.

Intermittent claudication may be used as a means of diagnosing PAD.[145] An individual who walks a certain distance and senses claudication pain, then rests until symptoms subside, then walks the same distance and experiences the same level of pain is more likely to have PAD. This is particularly true if the pulses in the foot and ankle are not present.[145]

Though the presence of PAD may be identified by claudication with activity and absence of a peripheral pulse, a more accurate indicator is the Ankle Brachial Index (ABI). The ABI is also more reproducible and is a good indicator of future ischemic cardiac and cerebral events. This test is more sensitive and specific for PAD than observation of symptoms.[146] The ABI is the ratio of the systolic blood pressures of the upper and lower extremities measured with Doppler recordings, which use low-intensity ultrasound to detect blood flow in arteries or veins. The ABI possesses one flaw: poorly compressible vessels, particularly in the elderly and diabetics, yield falsely high readings (greater than 1.3). Poorly compressible vessels have a great degree of calcification compared to normal vessels. The use of ABI is less reliable in these cases.[142]

In some cases, a treadmill test may be needed for diagnostic purposes. Individuals who experience claudication will typically have a decrease in ankle BP of 20 mmHg after exercise.[142] In cases of mild obstruction the postexercise ABI will drop to 0.5, and in moderate obstruction, to 0.2; below 0.2 indicates severe obstruction.[80]

Heart Failure

Definition

Heart failure (HF) is an impairment of the ventricles' capacity to eject blood from the heart or to fill with blood. The underlying cause of this disorder can be either structural or functional in nature. Heart failure represents the end stage of all forms of cardiovascular disease. Many heart failure patients will have well-preserved left ventricular function, while others may display signs of significant left ventricular impairment.[63, 147]

Epidemiology

Data for 2005 indicate that 5 million Americans were diagnosed with heart failure and that it was a contributing cause of death in 292,215 individuals. HF was listed as the "underlying cause"

in 58,933 of those deaths. Between 1999 and 2003, the overall death rate declined 2%, while the death rate from heart failure increased 20.5%. At least part of this change in death rate can be explained by improvements in treatment and management of MI and other conditions, which extend life expectancy. Death rates from heart failure are higher for black males and females compared to their white counterparts.[1]

Data from the Framingham Heart Study indicate that:[1]

- After age 65, the incidence of heart failure is slightly below 10 per 1,000 in the population.
- 75% of persons diagnosed with heart failure have hypertension; lifetime risk of heart failure doubles for those with a resting blood pressure of 160/90 mmHg compared to those with a blood pressure below 140/90 mmHg.
- Among MI survivors, 22% of males and 46% of females will be disabled with heart failure within six years.
- At age 40, lifetime risk of developing heart failure is one in five; for individuals who have not suffered an MI, the risk drops to one in nine for males and one in six for women.

Heart failure is closely associated with aging. Women have a higher relative risk of heart failure primarily because women comprise over 60% of the population over 65 years of age and 75% of the population over the age of 85 years.

The prevalence, incidence, and mortality for heart failure in diabetics are very high. For every 100 individuals with diabetes who are free of heart failure at the start of a year, twelve will develop heart failure and six will die within that year. The prevalence and incidence of heart failure in diabetics is associated with age and comorbidities such as nephropathy, IHD, and PAD.[148, 149]

Etiology

Heart failure may result from disorders of the pericardium, myocardium, endocardium, or vessels, but the majority is due to impaired left ventricular myocardial function. Heart failure is a broad term, and may be used to describe conditions in which left ventricular size and ejection fraction are maintained as well as those in which the left ventricle is dilated and ejection fraction is severely reduced.[147] Classifications for heart failure are outlined in Table 13.16. There are common differences between men and women: women tend to have preserved left ventricular systolic function, whereas men tend to have greater impairment in systolic function.[63]

The primary causes of heart failure are IHD, hypertension, and dilated cardiomyopathy. In women, hypertension is the most common cause, while in men, IHD is the most common cause. Of those with dilated cardiomyopathy, approximately 30% may have a genetic cause. Valvular disease is an additional common cause of heart failure.[63, 147]

Pathophysiology

Heart failure is now being referred to as a "final common pathway of many risk factors and cardiovascular illnesses."[150] Heart failure is a process beginning with an injury to the heart (as described earlier in this chapter) or with left ventricle hypertrophy that impairs overall function of the heart. To compensate for

Table 13.16 Stages of Heart Failure

Stage	Definition
A	Patients who are at high risk for developing heart failure but have no structural abnormalities
B	Patients who have structural heart disease but demonstrate no symptoms of heart failure
C	Patients with past or current symptoms of heart failure who have underlying structural heart disease
D	Patients with end-stage disease requiring specialized treatment, such as mechanical circulatory support, procedures to facilitate fluid removal

Source: Hunt AS, Abraham WT, Chin MH, Feldman AM, Francis GS, Ganiats TG, et al. ACC/AHA 2005 guideline update for the diagnosis and management of chronic heart failure in the adult—summary article: a report of the American College of Cardiology/American Heart Association Task Force on Practice Guidelines (Writing Committee to Update 2001 Guidelines for he Evaluation and Management of Heart Failure). *Circ.* 2005;112:1825–52.

the impairment in function, the renin-angiotensin-aldosterone system initiates changes in BP that exacerbate the dysfunction. As a result, the heart becomes weakened and dilated, myocardial fibrosis limits the ability of the walls to respond to stresses, oxidative damage further impairs contractility, and the overall structure of the heart is changed in such a way that it cannot function properly (see Figure 13.11).

Often the pathophysiology of heart failure can be traced to damage resulting from IHD. The most common scenario is that damaged sections of tissue resulting from an MI impair contractile function of the left ventricle and reduce ejection fraction. Heart failure is described as a progressive disorder because even if there is no additional injury to the myocardium, there will be a continued deterioration in function. Post MI, the left ventricle will undergo a change in structure and geometry. The change, referred to as cardiac remodeling, will result in a hypertrophied and/or dilated left ventricular chamber. The change in structure impairs performance of the heart and may even cause some regurgitation through the mitral valve back into the left atria.[147]

Cardiac remodeling precedes symptoms and will continue after symptoms appear. The progression may result in a worsening of symptoms even when treatment is ongoing. Progression of heart failure may be accelerated by progression of IHD, diabetes, hypertension, or the onset of atrial fibrillation.[147, 151]

The progression of heart failure and cardiac remodeling is mediated to some extent by neurohormonal systems. Heart failure patients typically have elevated blood and tissue levels of norepinephrine, angiotensin II, aldosterone, endothelin, vasopressin, and cytokines. All of these substances, either alone or in tandem, can have adverse effects on cardiac structure. Sodium retention and peripheral vasoconstriction result in increases in arterial BP, thereby increasing myocardial workload. Other substances mediate oxidative stress, causing myocardial cell damage and myocardial fibrosis that further alter the structure of the heart and reduce its function.[7, 63, 147]

Left ventricular hypertrophy as a result of extended periods of hypertension can initiate heart failure by reducing extensibility of the left ventricular wall and contractility. A characteristic of hypertension-related cardiac hypertrophy is

a phenotype change of cardiac fibroblasts. When stimulated, myofibroblasts proliferate and increase production of fibrous substances such as fibronectin and collagens. Under normal circumstances, matrix metalloproteinases increase the degradation of fibrillar collagen, but an inhibitor will prevent their activation in the presence of soluble collagen. The balance between the matrix metalloproteinases and their inhibitors is disrupted in heart failure. The metalloproteinases destroy normal collagens and leave poorly crosslinked collagens intact. This weakens the structure of the cardiac wall and results in dilatation.[7]

The dilatation of the left ventricle impairs contractility, resulting in a decreased cardiac output and ejection fraction. A decrease in cardiac output will also cause decreases in renal blood flow and glomerular filtration (see Chapter 18). The kidneys respond by activating the renin-angiotensin-aldosterone system to raise blood pressure and restore blood flow to the filtration units. This attempt to maintain homeostasis increases circulating levels of angiotensin II and aldosterone. Subsequently, afterload increases, edema develops, and heart failure progresses. The typical heart failure patient has serum levels of aldosterone 20 times higher than what is considered normal. These increased aldosterone levels are also associated with myocardial fibrosis. Administration of aldosterone antagonists and angiotensin-converting enzyme inhibitors to MI patients after revascularization surgery decreases fibrosis and increases left ventricular ejection fraction.[152] Increased aldosterone concentrations have been implicated in a number of other mechanisms associated with heart failure. These include endothelial dysfunction, reduced variability in heart rate, reduced cardiac norepinephrine uptake, and increased risk of cardiac arrhythmias.[152]

Reductions in renal blood flow and glomerular filtration will reduce the rate of solute and water delivery to the distal diluting segment of the nephron. The end results are an inability to excrete a dilute urine and hyponatremia. Hyponatremia is the most common electrolyte abnormality in heart failure patients and is more common in severe heart failure as compared to mild to moderate heart failure. Only about 5% of heart failure patients suffer from hyponatremia; however, the hormonal abnormalities that cause this imbalance are present in most heart failure patients.[153]

Conditions such as aortic regurgitation and mitral regurgitation can result in impaired left ventricular function. In these conditions, structural abnormalities, infection, and/or damage to tissues can result in hypertrophy of the left ventricle as a result of volume overload. While mitral regurgitation may occur as a result of structural changes to the heart in cases of severe heart failure, it often occurs as a genetic disorder or secondary to damage to the chordae tendinea or valve leaflets.[80, 154]

Clinical Manifestations

Signs and symptoms of heart failure are a result of the basic pathophysiology of the disease. They will vary depending on the predominance of the disorder—either left- or right-sided failure. In general, decreased blood flow and oxygen supply lead to dyspnea, fatigue, weakness, exercise intolerance, and poor adaptation to cold temperatures. Dyspnea is an unusual shortness

Figure 13.11 Effects of Congestive Heart Failure

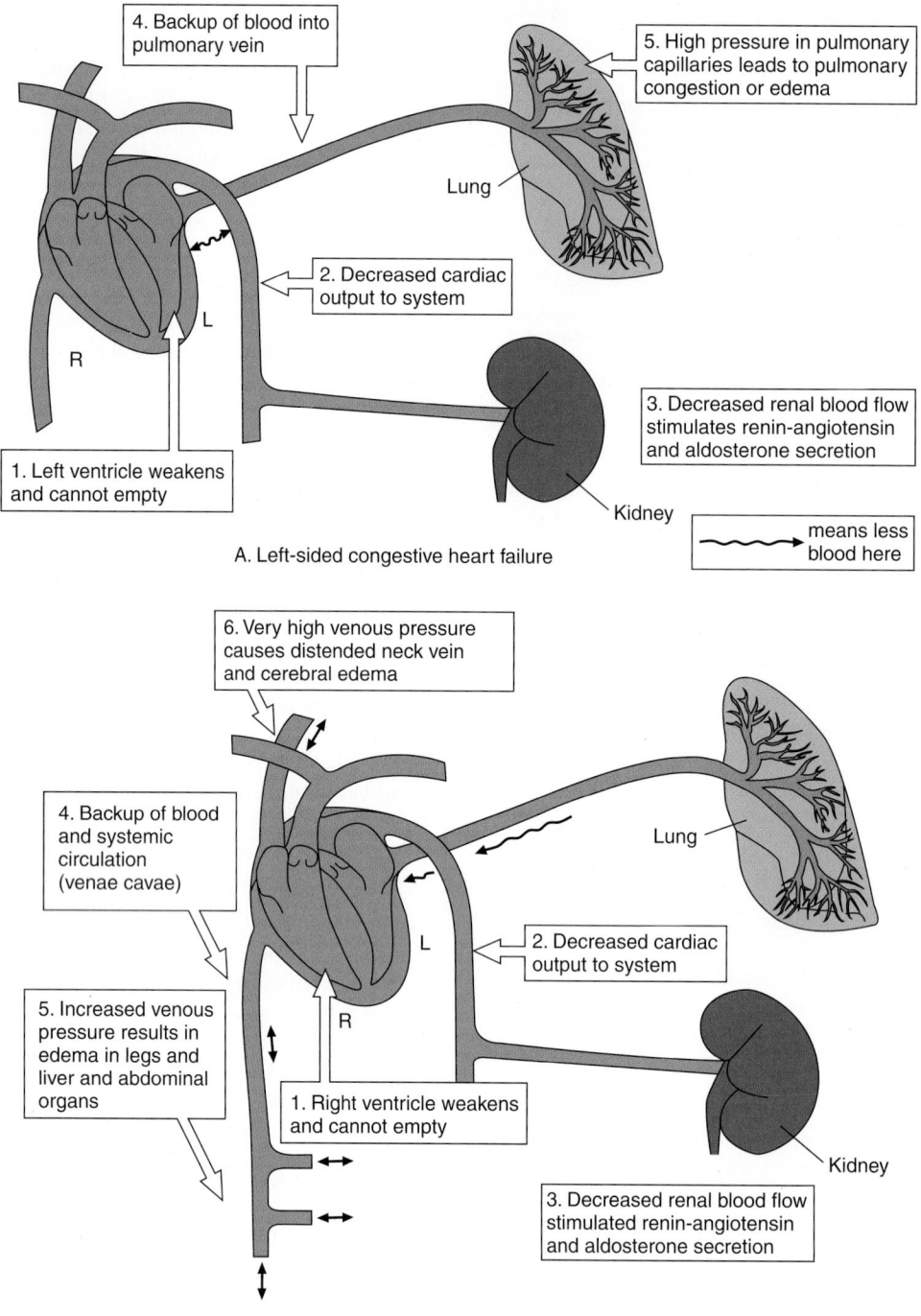

4. Backup of blood into pulmonary vein

5. High pressure in pulmonary capillaries leads to pulmonary congestion or edema

Lung

2. Decreased cardiac output to system

3. Decreased renal blood flow stimulates renin-angiotensin and aldosterone secretion

R

L

1. Left ventricle weakens and cannot empty

Kidney

→ means less blood here

A. Left-sided congestive heart failure

6. Very high venous pressure causes distended neck vein and cerebral edema

4. Backup of blood and systemic circulation (venae cavae)

Lung

2. Decreased cardiac output to system

5. Increased venous pressure results in edema in legs and liver and abdominal organs

L

R

1. Right ventricle weakens and cannot empty

Kidney

3. Decreased renal blood flow stimulated renin-angiotensin and aldosterone secretion

B. Right-sided congestive heart failure

Source: Pathophysiology for the Health Professions Third Edition, Philadelphia, PA: Saunders Elsevier. Figure 18-22, p. 333.

of breath not appropriate to the workload, and this symptom in conjunction with fatigue may limit the individual's functional capacity. When left-sided failure is predominant, dyspnea is more predominant and also includes **orthopnea**.

Right-sided failure is characterized by the signs and symptoms caused by systemic backup of the circulatory system. This fluid retention can cause pulmonary congestion and edema in the periphery. Eventually, edema affects the gastrointestinal tract and results in **hepatomegaly**, **splenomegaly**, and ascites (fluid retention within the abdominal cavity). These may further im-

pair respirations by pressing up on the diaphragm and limiting pulmonary function. Other signs of right-sided failure include distended neck veins, headache, and a flushed face. Signs of compensation include tachycardia, pallor, polycythemia, and oliguria. Dyspnea, fatigue, and edema affect quality of life and limit functional capacity.[147, 151]

Treatment

The goals of medical treatment for heart failure are to treat the underlying cause of the cardiac disorder, control the associated symptoms, and prevent continued damage to the heart.[147, 155] Many of the same medications used to treat other cardiovascular disorders are also used to treat heart failure (see Table 13.1). Diuretics such as furosemide, bumetanide, and torsemide are crucial for control of edema and fluid retention. Some diuretics, such as Spironolactone, are not as effective in this stage of heart disease but may be used in conjunction with other diuretics. Control of BP is essential and clinical trials indicate an improved outcome for those individuals treated with ACE inhibitors. Other medications associated with improved outcome include beta-adrenergic blockers.[147] Medications used to improve heart function by increasing myocardial contraction include digitalis, dopamine, and dobutamine.[156] The newest classes of medications for treatment—levosimendan, nesiritide, and L NAME—may prove to be alternatives that can improve long-term outcomes.[157] Other components of care include prevention of respiratory infections, exercise, and nutrition therapy.

Nutrition Therapy for Heart Failure

It has been estimated that as many as 50% of patients with heart failure are malnourished.[158, 159] Nutrition therapy that restricts both sodium and fluid is crucial to control acute symptoms and may assist with reducing the overall work of the heart. But at the same time, individuals with heart failure have difficulty eating and many experience a syndrome of malnutrition called **cardiac cachexia**. Cardiac cachexia is a form of malnutrition similar to the wasting syndrome seen in AIDS and cancer and characterized by extreme skeletal muscle wasting, fatigue, and anorexia.[158, 160, 161] The etiology is not completely understood, but it is assumed that it is multifactorial and involves both metabolic and hormonal abnormalities.[161, 162] Additional contributing mechanisms for cachexia in heart failure include myocardial nutrient deficiencies of L-carnitine, coenzyme Q10, creatine, thiamin, and taurine.[158, 163–166]

Further complications from heart failure that contribute to nutrition problems include (1) decreased blood flow to the gastrointestinal tract causing slowed peristalsis and early satiety, (2) possible impairment of nutrient absorption due to this decreased blood flow, and (3) side effects from drugs such as nausea, vomiting, and anorexia, which are common with the use of ACE inhibitors, beta blockers, cardiac glycosides, and digoxin.[167–171] Nutrient deficiencies are also a common side effect from the use of diuretics and other medications. Table 13.17 summarizes the factors that promote malnutrition in heart failure patients.

Nutrition Assessment and Diagnosis

Since the intake of sodium and fluid can directly impact the progression and management of HF, accurate assessment of both of these nutrients is essential. Furthermore, problems with early satiety and possible drug/nutrient interactions as noted above need to be carefully examined before determining the presence of common nutrition diagnoses such as excessive sodium and/or fluid intake, inadequate oral food/beverage intake, food-medication interactions, or impaired ability to prepare foods/meals. Undesirable food choices and limited adherence to nutrition-related recommendations might also be observed in patients with HF.

Table 13.17 Physiologic Contributors to Malnutrition and Cachexia in Heart Failure

Intestinal ischemia	• Decreased splanchnic circulation from increased sympathetic nervous system activity: 1. impaired epithelial function and increased mucosal permeability; 2. exposure to endotoxin and inflammatory cytokine release
Hypercatabolism	• Increased levels of stress hormones and neurohormones: 1. epinephrine; 2. norepinephrine; 3. cortisol • Decreased levels of anabolic hormones: 1. dehydroepiandrosterone sulfate; 2. insulin-like growth factor-1 • Increased circulating levels of cytokines
Malabsorption	• Reduced gut circulation • Gut edema: 1. decreased fat absorption; 2. protein loss
Anorexia	• Nausea and early satiety from gut/hepatic edema • Multiple medication use • Clinical depression

Reprinted from: Dunn SP, Bleske B, Dorsch M, Macaulay T, Van Tassell B, Vardeny O. Nutrition and heart failure: impact of drug therapies and management strategies. *Nutr Clin Pract*. 2009;24:60–75. Table 1.

Nutrition Intervention

Nutrition counseling for individuals with HF is a priority. In a study by Kuehneman, Saulsbury, Splett, and Chapman, readmissions and cost of hospital stays were found to have a direct relationship to excessive sodium intake.[172] When a registered dietitian provides specific nutrition education for a patient, it can lead to fewer readmissions and an overall improved response to medical treatment.[173, 174]

Nutrition therapy for HF focuses on the control of signs and symptoms associated with the diagnosis and on the promotion of overall nutritional rehabilitation. Components of nutrition therapy include sodium and fluid restriction, correction of nutrient deficiencies, and nutrition education for increasing nutrient density and making food choices that enhance oral intake.[169, 171, 172, 175] Education and awareness for drug-nutrient interactions should be a priority for nutritional care.[158] Table 13.18 provides guidance for prevention and treatment of these potential interactions.

Sodium A 2000 mg sodium diet is a standard initial recommendation for individuals with HF.[148, 172] Adjustments to levels of 2000 mg, 1000 mg, or 500 mg may be prescribed depending on the patient's individual medical condition—specifically, fluid and volume states as well as overall oral intake. Guidelines for low sodium diets are outlined in Table 13.19. Because it is a challenge to manage this level of restriction outside of a hospitalized setting, it is crucial to critically evaluate the patient's actual oral food and beverage intake to determine the level of sodium the patient is consuming prior to putting any further

Table 13.18 Recommendations for the Treatment and Prevention of Detrimental Drug-Induced Nutrition Effects in Patients with Heart Failure

Effect	Drug-Induced Cause	Recommendations*
Hypokalemia	Loop diuretics	• Monitor serum potassium level every 3–6 months if stable; monitor every 2–4 weeks until stable if initiation or change in dosage of any of the following: (1) ACEI/ARB; (2) loop or thiazide diuretic; (3) potassium supplement; (4) aldosterone antagonist • Maximize ACEI/ARB therapy • Consider the addition of an aldosterone inhibitor • Consider dietary or pharmacologic potassium supplementation • Anticipate use of diuretic therapy
Hyperkalemia	Potassium supplements; ACEIs; aldosterone antagonists; beta-blockers; aggressive diuretic use	• Monitor serum potassium level every 3–6 months if stable; monitor every 2–4 weeks until stable if initiation or change in dosage of any of the following: (1) ACEI/ARB; (2) loop or thiazide diuretic; (3) potassium supplement • Monitor serum potassium within a week of initiation of aldosterone antagonist, then every 2–4 weeks until stable or following dose adjustment; monitor every 3–6 months if stable • Monitor serum potassium closely when combination inhibitors of RAAS are used; guidelines recommend clinical trial protocols of 48 hours after initiation, then at 1, 4, and 5 weeks • Adjust use of potassium supplementation with diuretic use or when combination inhibitors of RAAS are used
Hypomagnesemia	Loop diuretics	• Monitor serum magnesium level weekly after initiation or change in diuretic dose and every 3–6 months afterwards if stable • Consider dietary and pharmacologic magnesium supplementation • Anticipate use of diuretic therapy
Thiamin deficiency	Loop diuretics	• Consider supplementation with 100–200 mg/d if receiving chronic, high-dose, loop diuretic therapy and deficient in TPPE
Calcium deficiency	Loop diuretics	• Aggressively supplement if hypocalcemia is the suspected cause of heart failure • Monitor serum calcium levels and supplement if receiving chronic loop diuretic therapy
Zinc deficiency	ACEIs or ARBs; thiazide diuretics	• Routine monitoring and supplementation not recommended • Consider supplementation if dysgeusia symptoms present in patient receiving ACEI or ARB therapy

*ACEI = angiotensin-converting enzyme inhibitor; ARB = angiotensin receptor blocker; RAAS = renin-angiotensin-aldosterone system; TPPE = thiamin pyrophosphate effect.

Reprinted from: Dunn SP, Bleske B, Dorsch M, Macaulay T, Van Tassell B, Vardeny O. Nutrition and heart failure: impact of drug therapies and management strategies. *Nutr Clin Pract*. 2009;24:60–75. Table 2.

modifications into place. Anorexia, fatigue, and shortness of breath lead to such poor oral intake that many patients consume much less than 2000 mg.

Fluid Fluid requirements are typically calculated at 1 mL/kcal or 35 mL/kg (see Chapter 5). To treat fluid overload in HF, a fluid limitation of 1500 mL/day is the standard recommendation, with an upper level of 2000 mL. Again, adjustments will need to be made based on renal and cardiac status in order to prevent volume overload. Weighing the patient daily will allow the practitioner to monitor fluid status.

Fluid restriction is one of the most difficult diet modifications for patients to tolerate. When providing nutrition education on fluid restriction, the clinician should make sure the patient understands the specific volume that is allowed, what items are considered to be fluids, and the suggestions to aid with controlling thirst. Visually demonstrating the amount of fluid the patient is allowed may support the patient's understanding and compliance. All beverages and foods such as soups, Popsicles®, sherbet, ice cream, yogurt, custard, and gelatin should be counted within the fluid allowance. Finally, good mouth care, rinsing the mouth frequently, and using cold or frozen foods

can help control thirst. (See Table 18.10 in Chapter 18 for additional tips for controlling fluid intake.)

Drug-Nutrient Interactions The use of multiple diuretics in the medical treatment of HF may lead to losses of multiple water-soluble nutrients, including potassium, magnesium, thiamin, riboflavin, and pyridoxine.[176–178] Nutrition education for increasing these nutrients within the diet is the first level of intervention. However, since meeting the patient's increased needs may be difficult because of the patient's overall decreased oral intake, supplementation may be warranted, as noted in Table 13.18. A multivitamin should be recommended daily. Additional levels of supplementation will start with providing the DRI for each and then adjusting dosages after monitoring biochemical indices. Serum levels of magnesium (normal = 1.6–2.6 mmol/L) and potassium (normal = 3.5–5.5 mg/dL) would be used as benchmark levels for comparison. Thiamin levels are assessed by measuring the activity of erythrocyte thiamin pyrophosphate (thiamin pyrophosphate effect). Adequate or normal levels are evaluated at 0% to 15%; mild deficiency is indicated at >15% to 25%; and severe deficiency is indicated at greater than 25% stimulation.

Table 13.19 Guidelines for a 2-Gram Sodium Diet

Sodium in Table Salt

¼ teaspoon salt	= 575 mg sodium
½ teaspoon salt	= 1150 mg sodium
¾ teaspoon salt	= 1725 mg sodium
1 teaspoon salt	= 2300 mg sodium
1 teaspoon baking soda	= 1000 mg sodium

Food Group	High-Sodium Foods	Low-Sodium Alternative
Meats, Poultry, Fish, Legumes, Eggs and Nuts	Smoked, cured, salted or canned meat, fish or poultry including bacon, cold cuts, ham, frankfurters, sausage, sardines, caviar, and anchovies Frozen breaded meats and dinners, such as burritos and pizza Canned entrees, such as ravioli, spam, and chili Salted nuts	Fresh meats and fish (check labels for frozen products) Tuna packed in water Dried beans and peas Edamame (fresh soybeans) Eggs, especially egg whites
Dairy Products	Buttermilk, chocolate milk Regular and processed cheese, cheese spreads, and sauces Cottage cheese	Milk, eggnog, yogurt, ice cream, and ice milk Low-sodium cheeses, cream cheese, ricotta cheese, and mozzarella
Breads, Grains and Cereals	Bread and rolls with salted tops Quick breads, self-rising flour, biscuits, pancake and waffle mixes Pizza, croutons, and salted crackers, commercial bread stuffing or bread crumbs Prepackaged, processed mixes for potatoes, rice, pasta, and stuffing	Enriched white, wheat, rye, and pumpernickel bread, hard rolls and dinner rolls; muffins; waffles and most ready-to-eat cereals All rice and pasta, but do not to add salt when cooking Corn and flour tortillas and noodles Low-sodium crackers and breadsticks Unsalted popcorn, chips, and pretzels
Fruits and Vegetables	Regular canned vegetables and vegetable juices Olives, pickles, sauerkraut, and other pickled vegetables Vegetables made with ham, bacon or salted pork Packaged mixes, such as scalloped or au gratin potatoes, frozen hash browns and tater tots Commercially prepared pasta and tomato sauces	Fresh and frozen vegetables without sauces Low-sodium canned vegetables, sauces, and juices Fresh potatoes, frozen French fries, and instant mashed potatoes Low-salt tomato or V-8 juice. Most fresh, frozen, and canned fruit Dried fruit if processed without sodium All fruit juices
Soups	Regular canned and dehydrated soup, broth, and bouillon	Low-sodium canned and dehydrated soups, broth, and bouillon Homemade soups without added salt;
Fats, Desserts and Sweets	Bottled salad dressings, regular salad dressing with bacon bits Salted butter or margarine Instant pudding and cake	Unsalted butter or margarine Vegetable oils and sodium-free salad dressings All desserts made without salt
Seasoning and Condiments	Seasoning made with salt; sea salt, rock salt, kosher salt; meat tenderizers; monosodium glutamate; regular soy sauce, barbecue sauce, teriyaki sauce, steak sauce, Worcestershire sauce, and most flavored vinegars; canned gravy and mixes; regular condiments; salted snack foods; olives	Salt substitute with physician's approval; pepper, herbs, spices; vinegar, lemon, or lime juice; hot pepper sauce; low-sodium soy sauce (1 tsp or 5 mL); low-sodium condiments (catsup, chili sauce, mustard); fresh ground horseradish; salsa (2 tbsp or 30 mL)

Some treatment protocols for thiamin supplementation include thiamin prescribed at 200 mg/day orally for six weeks, while others recommend an initial parenteral dose of 100 mg followed by daily supplementation.[177, 179–181]

Other Nutrients of Concern Additional conditionally essential nutrients that have been examined to determine their possible role in treatment for heart failure include arginine, carnitine, and taurine (as mentioned earlier, they have been linked to cardiac cachexia as well). Hawthorn, an herbal supplement, has also been studied as a complementary treatment for heart failure but is without significant demonstrated benefit (see the CAM feature at the end of the chapter).[182] The rationale for use of arginine supplementation in heart failure is to increase the production of nitric oxide. As discussed earlier, nitric oxide plays a significant role in intiating vasodilation in the vascular endothelium. Initial studies have indicated a possible role for L-arginine supplementation in heart failure, although more research is necessary to confirm this benefit.[94, 183, 184] Carnitine is responsible for carrying fatty acids intracellularly into the mitochondria for oxidation. Patients with heart failure have been shown to have lower levels of carnitine, and when supplemented with carnitine have demonstrated positive outcomes, though most have been in small clinical studies.[94, 185] It will be important to monitor future research in larger clinical trials for substantiated evidence to support supplementation.

Conclusion

This chapter has illustrated the impact of nutrition as a controllable risk factor, as a means to prevent disease, and as a critical component of medical treatment. Management of hypertension, dyslipidemia, and diabetes, weight management, physical activity, and smoking cessation are common targets. Lifestyle modifications begin with management of these controllable risk factors.

The initial strategy should be weight management and increased physical activity. A well-established relationship exists between increased body weight, hypertension, dyslipidemia, and a prothrombotic state. In addition to its relationship with high triglycerides and low HDL cholesterol, visceral adiposity is associated with thrombotic and inflammatory markers such as CRP, homocysteine, and fibrinogen, and type 2 diabetes mellitus.[186] Since visceral obesity is associated with dyslipidemia, prothrombotic state, glucose intolerance, and hypertension, it is strongly implicated as the root cause of metabolic syndrome. Over 47 million (age-adjusted prevalence, 23.7%) Americans have metabolic syndrome.[8]

Close to 100 million American adults (49.8% of the adult population) have a total serum cholesterol level greater than 200 mg/dL, and 17.3% have a total serum cholesterol greater than 240 mg/dL; 39.5% of adults have an LDL cholesterol level greater than 130 mg/dL, and 22.6% have an HDL cholesterol level lower than 40 mg/dL.[8] Using the research and principles of the DASH diet, the Therapeutic Lifestyle Changes diet, and extensive skills for individualization of patient education and behavior modification, the registered dietitian is uniquely situated to impact the extent of cardiovascular disease within the population and to impact an individual's quality of life with overall improvement of health and well-being.

PRACTITIONER INTERVIEW

Eileen MacKusick, MS, RD *Lead Clinical Dietitian, Watsonville Community Hospital, Watsonville California*

I have been a clinical dietitian for over 15 years. In my facility, roughly 25% of the patients I see have congestive heart failure (CHF). Only one of the attending physicians routinely classifies the CHF patient, but the most common etiology is cardiomyopathy or DCM (dilated cardiomyopathy).

How do you nutritionally assess these patients, and is there anything unique that would be useful for students to be aware of?

Physical: Obtaining a dry weight can be challenging because of edema, especially if the patient is obese. Whenever a dry weight can be established, it is helpful. It is also important to do a visual assessment, because patients don't appear to be underweight based on their weight but they are actually cachectic with edema. It seems that patients with CHF are either morbidly obese with poor somatic protein stores or cachectic. I recently had a patient that weighed 484 pounds. We removed 80 lbs of water off of him over a 1½-month period. In cases such as this, it is very difficult to determine the water weight vs. the actual weight loss. Our goals for him were weight loss while increasing his visceral protein, which were achieved.

Diet Assessment: This generally focuses on salt/sodium and fluid intake. However, overall nutritional intake in terms of adequacy or deficiency must also be considered, especially with underweight individuals. Most of our physicians are using a no-added-salt diet vs. a low-sodium diet these days. Trying to maintain a 2 g sodium diet was resulting in undernutrition and exacerbating fluid shifts and edema. The fluid limits are individualized by the physicians. We monitor the patients for dehydration due to overdiuresis.

Biochemical: I always look at albumin/prealbumin with C-reactive protein (CRP) to distinguish acute-phase reaction. Visceral protein determination is difficult in this population because of the fluid shifts and dilution of several of your biochemical markers. B-type natriuretic peptide (BNP) provides the degree or severity of the CHF. When the BNP is extremely high, you may see a more severe fluid/sodium restriction. You must also monitor kidney function with BUN, creatinine, and GFR in patients that are being aggressively diuresed.

What are the common nutritional problems associated with CHF?

There are several nutrition problems associated with CHF. Patients find it difficult complying with dietary fluid and sodium restrictions while consuming adequate nutrition to increase or maintain adequate blood protein levels to prevent third spacing of extracellular fluid. Adequate nutrition intake is also compromised by shortness of breath (SOB), decreased appetite, and lethargy; many patients have additional complications as the result of polypharmacy with drug-nutrient interactions.

What changes have you seen with CHF nutrition therapy in the last ten years?

The use of the BNP is relatively new, and sodium restrictions have been liberalized over the last 10 to 15 years. When I first started practicing, you would see 500–1000 mg sodium diets ordered. A 2 gram sodium was the standard diet. Now we barely use 2 grams and never use 500–1000 mg diets. Journal articles and CEU programs are primarily how I have stayed abreast of changes for this disorder.

Application of the Nutrition Care Process: Hypertension

Introduction:

GG is a 49-year-old African-American male diagnosed with Stage 2 essential HTN six months ago. He states: "I really want to control this blood pressure—my parents both died of complications from blood pressure and heart disease."

Current treatment includes lower-salt diet, smoking cessation, and a sporadic walking program.

His labs include the following: BP 160/100; TChol 300 mg/dL; HDL-C 35 mg/dL; LDL-C 135 mg/dL; TG 250 mg/dL. GG's physician has discussed starting both a thiazide diuretic and ACE inhibitor with him.

1. What are the criteria for diagnosis with Stage 2 essential HTN? What factors allow for that diagnosis for this patient?

2. What would be the complications of untreated hypertension?

3. What are the mechanisms through which thiazide diuretics and ACE inhibitors treat high blood pressure?

Nutrition Assessment:

Food/Nutrition-Related History:

The registered dietitian obtained the following description of GG's usual intake:

AM: oatmeal or cold cereal, 2% milk, coffee. Snack: coffee, sweet roll, or doughnut. Lunch: sandwich (ham or salami with cheese) or soup from home; chips; diet Coke. Dinner: 6–8 oz meat, baked, broiled, or grilled; salad; potato, pasta, or rice; roll or bread; diet Coke. Bedtime snack: peanut butter and crackers or popcorn.

4. Assess the client's usual intake using appropriate nutrition criteria.

Anthropometric Measurements:

Ht. 6'2"; Wt. 245#; UBW/Highest adult weight 245–250#

5. Evaluate the patient's weight history. Is this a risk factor? What other possible risk factors does this patient present with?

Nutrition Diagnosis:

6. Identify at least two nutrition problems that can be identified as a result of the nutrition assessment and medical history. Determine the diagnostic term for each nutrition problem. Next, identify the etiology of each nutrition problem. Finally, identify the signs and symptoms that support the evidence for these nutrition problems.

Nutrition Intervention:

7. Identify the nutrition prescription for GG by recommending the appropriate nutrition therapy for his diagnosis.

8. Recommend energy and protein requirements for GG.

Nutrition Monitoring and Evaluation:

9. List the data that should be collected for monitoring and evaluation for each nutrition diagnosis that you identified.

National Heart Lung Blood Institute Provides all guidelines and patient teaching materials for the DASH diet.
http://www.nhlbi.nih.gov/health/public/heart/hbp/dash/index.htm

National Heart Lung Blood Institute/ National Cholesterol Education Program Provides clinical guidelines for treatment of dyslipidemia.
http://www.nhlbi.nih.gov/about/ncep

American Heart Association Provides education materials and clinical information for all aspects of cardiovascular disease.
http://www.americanheart.org

Start! Physical activity program Information regarding exercise and nutrition designed by the American Heart Association.
http://www.americanheart.org/presenter.jhtml?identifier=3053031

American Heart Association Journal Provides access to all professional journals published by American Heart Association.
http://www.ahajournals.org

National Institutes of Health Compilation of other web sources for information about all aspects of diagnosis and treatment for cardiovascular disease.
http://www.nlm.nih.gov/medlineplus/heartdiseases.html

University of Wisconsin—Department of Medicine Nutrition Academic Award Program. Grant-developed educational tools and guidelines for hypertension, lipid disorders, obesity, and diabetes.
http://www2.medicine.wisc.edu/home/naa/medicalnutritionhandbook/

The Framingham Heart Study This website chronicles this important longitudinal study that began in 1948.
http://www.framingham.com/heart

END-OF-CHAPTER QUESTIONS

1. Define the following terms: systolic, diastolic, stroke volume, and cardiac output.

2. Describe the factors that will influence stroke volume and mean arterial pressure (MAP).

3. What is the definition of *hypertension*? Explain the pathophysiology for the lifestyle factors known to contribute to the development of hypertension.

4. List the major classifications of medications used to treat hypertension.

   Describe their mechanism of effect. Describe the DASH diet.

5. List the risk factors associated with development of atherosclerosis. Which risk factors are alterable? List the dietary changes in the ATP III guidelines.

6. What are the four mechanisms that can initiate a myocardial infarction (MI)? How does the rupture of an atheromatous plaque result in an MI?

7. Compare peripheral arterial disease (PAD) to atherosclerosis—how are

   they similar and how do they differ? Describe several complications associated with PAD.

8. What are the primary causes of heart failure? Describe the clinical progression of heart failure. What is meant by "cardiac remodeling" and what is its etiology?

9. What are the nutritional implications associated with heart failure and cardiac cachexia?

Complementary and Alternative Medicine Remedies Used in Diseases of the Cardiovascular System

Remedy	Scientific Name	CAM Use	Side Effects and/or Risks
Amalaka, amla; Indian gooseberry	*Emblica officinalis*	Heart tonic	Adverse reactions rare.
Ascorbic acid/ vitamin C		Treat hypertension; reduce oxidation of LDL cholesterol; lower serum cholesterol levels; increase glutathione levels; treat angina	High intakes may cause GI upset or mild diarrhea. May increase activity of anticoagulants. May interfere with B_{12}, copper, chromium absorption and metabolism. Aspirin, corticosteroids, & indomethacin increase urinary vitamin C losses. High doses may reduce efficacy of select chemotherapeutic medications. Supplementation above DRI (or >70–90 mg/d) not recommended. May increase plasma estrogen levels in patients taking HRT or oral contraceptives. Large doses may interfere with anticoagulant medications.
Astragalus, milk vetch; huang qi	*Astragalus membranaceus*	Treat cardiovascular disease	May antagonize effects of immunosuppresants such as tacrolimus and cyclosporine.

Complementary and Alternative Medicine Remedies Used in Diseases of the Cardiovascular System (Continued)

Remedy	Scientific Name	CAM Use	Side Effects and/or Risks
Beta-carotene		Reduce oxidation of LDL cholesterol; reduce risk of cardiovascular disease	No adverse reactions reported.
Black cohosh	Cimicifuga racemosa	Treat hypertension	Gastrointestinal upset and rashes are most common, followed by dizziness, headaches, nausea, and vomiting when higher than normal doses are taken.
Burdock; niu bang zi	Arctium lappa	Reduce serum cholesterol levels	May increase hypoglycemic effects of insulin and oral hypoglycemic agents. Burdock products may be significantly contaminated with atropine at high enough levels to cause toxicity.
L-carnitine		Prevent cardiovascular disease; treat angina, arrhythmia, congestive heart failure	Appears to be safe. Might have antithyroid actions; possibly contraindicated in individuals with hypothyroidism.
Cat's claw; gambir; gao teng	Uncaria rhynchophylla	Treat hypertension	May cause diarrhea and lower blood pressure.
Chinese, Baikal skullcap; huang qin	Scutellaria baicalensis	Treat hypertension	Hepatotoxicity, pneumonitis. Stupor, confusion, seizures. Products have been found to be contaminated with a similar-looking plant known as germander (**Teucrium chamaedrys**) that can cause hepatitis.
Chromium		Reduce LDL cholesterol levels, increase HDL cholesterol levels	Long-term effects of high doses and increased cellular concentrations not known.
Cobalamin/ vitamin B_{12}		Reduce homocysteine levels; treat cardiovascular disease	Folic acid supplementation may mask vitamin B_{12} deficiency. Supplemental vitamin C or iron may interfere with bioavailability of vitamin B_{12}. The following drugs may interfere with B_{12} absorption: zidovudine, antacids, clofibrate, colchicines, cycloserine, erythromycin, isoniazid, metformin, aldomet, neomycin, nitrous oxide, oral contraceptives, sufanamides, tetracycline.
Coenzyme Q10		Prevent cardiovascular disease; treat angina, arrhythmia, congestive heart failure; treat hypertension; reduce serum cholesterol levels	Anorexia, diarrhea, epigastric discomfort, mild nausea.
Copper		Reduce risk of cardiovascular disease	Nausea may occur with doses of 10 mg, and 60 mg may cause vomiting.
Dong quai; angelica	Angelica sinensis	Heart tonic	Should not be combined with blood-thinning medications (warfarin, heparin, aspirine) or supplements (garlic, ginkgo, vitamin E, fish oil).
Folic acid		Reduce homocysteine levels	Supplementation ≥400 μg can mask pernicious anemia caused by vitamin B_{12} deficiency; dosages ≥5 mg may cause abdominal cramps, diarrhea, and rash; doses ≥15 mg can alter sleep patterns or cause vivid dreaming; large doses exacerbate neuropathy in those with B_{12} deficiency; doses >1 mg/d interfere with anticonvulsant medications. The following medications can affect absorption or activity of folate: oral contraceptives, aspirin, indomethacin, methotrexate, famotidine, tetracycline, erythromycin, folfonamides, & cholestyramine.
Garlic; ajo; lasuna	Allium satvium	Reduce hypertension; lower serum cholesterol levels; inhibit platelet aggregation	Headache, fatigue, altered platelet function with potential for bleeding, offensive odor, GI upset, diarrhea, sweating, changes in the intestinal flora, hypoglycemia. Discontinue use of garlic at least 7 days prior to surgery.
Gentian; bitter root; long dan cao	Gentiana spp.	Treat hypertension	Stomach irritation, nausea, vomiting, changes in menstrual cycle.

(continued)

Complementary and Alternative Medicine Remedies Used in Diseases of the Cardiovascular System (*Continued*)

Remedy	Scientific Name	CAM Use	Side Effects and/or Risks
Gingko biloba; ying xing	*Gingko biloba*	Inhibit platelet aggregation; improve circulation; treat coronary thrombosis	No agreed-upon standards exist; different formulations will vary in actual content. Affects platelet activating factor; use with caution in patients taking atnicoagulant or antiplatelet therapy. May cause elevated blood glucose levels in those taking insulin or oral hypoglycemic agents. Increases blood pressure when combined with thiazide diuretics. Do not use with trazodone.
Gotu kola; brahmi	*Centella asiatica*	Treat hypertension	Well tolerated. High doses may interfere with actions of antidiabetic medications & cholesterol-lowering drugs. Contraindicated in pregnancy and breast-feeding.
Grape seed extract	*Vitis vinifera, V. coignetiae*	Lower serum cholesterol levels; reduce LDL cholesterol oxidation	Essentially nontoxic. Might potentiate anticoagulant and antiplatelet agents.
Guggul; bedellium	*Commiphora mukul*	Reduce serum cholesterol levels	Headache, mild nausea, eructation, hiccough, loose stools, hypersensitivity rash. Theoretically may potentiate effects of aspirin, NSAIDs, and warfarin. May have thyroid-stimulating activities. Can induce CYP3A4 activity and may interact with substances metabolized by the same enzyme.
Hare's ear; chai hu	*Bupleurum chinense*	Reduce serum cholesterol, triglyceride levels; treat hypertension	Large doses can cause nausea and vomiting, facial and extremity edema, gastrointestinal distention, and constipation.
Hawthorn	*Crategus oxyacantha*	Prevent/treat cardiovascular diseases, esp. angina, arrhythmia, congestive heart failure	All plant parts of mistletoe are toxic. May increase hypotensive effects of antihypertensive medications. Increases sedative effects of CNS depressants. Causes cytotoxic & immunostimulant effects.
Magnesium		Treat angina	Wide margin of safety in individuals with healthy kidneys. Can mutually interfere with absorption of antibiotics in the tretracycline and fluoroquinolone families, as well as nitrofurantoin, penicilamine, ACE inhibitors, phenytoin, and H2 blockers. Might increase effectiveness of sulfonylureas, potentially creating risk of hypoglycemia.
Mistletoe	*Phorandendron leucarpum*	Treat cardiovascular disease; treat hypertension	All plant parts of mistletoe are toxic. May increase hypotensive effects of antihypertensive medications. Increases sedative effects of CNS depressants. Causes cytotoxic & immunostimulant effects.
Niacin/vitamin B$_3$		Reduce serum cholesterol, triglyceride levels; reduce LDL cholesterol, increase HDL cholesterol	Dosages >100 mg daily frequently cause skin flushing, especially of the face, which may be accompanied by stomach distress, itching, and headache. Slow-release niacin may induce hepatic inflammation. High doses may present risks in individuals with history of ulcer disease, gout, and excess alcohol consumption. May increase serum levels of anticonvulsant medications.
Noni, Indian mulberry; ashyulka	*Morinda citirolia*	Treat hypertension	No known side effects, but no safety studies have been conducted. Use by patients with severe liver disease not recommended. May interact with potassium-sparing diuretics due to high potassium content.
Potassium		Prevent/treat hypertension	May produce hyperkalemia when taken in combination with ACE inhibitors (aptopril, enalapril). Do not use with potassium-sparing diuretics.
Pyroxidine/ vitamin B$_6$		Reduce homocysteine levels; treat artherosclerosis	Excessive intake (>2000 mg daily) can cause sensory neuropathy. May cause or worsen acne symptoms. Doses >5 mg may interfere with effects of levodopa when drug is taken alone.
Reishi; ling zhi	*Ganoderma lucidum*	Lower serum cholesterol, triglyceride levels; treat hypertension	No known drug interactions. Widely believed to be safe.

Complementary and Alternative Medicine Remedies Used in Diseases of the Cardiovascular System (*Continued*)

Remedy	Scientific Name	CAM Use	Side Effects and/or Risks
St. John's wort	*Hypericum perforatum*	Reduce serum cholesterol levels	Avoid administration with alcohol, MAOIs, narcotics, OTC cold & flu medications, digoxin, drugs metabolized by CYP3A, indinavir, amphetamines, SSRIs, tricyclic antidepressants. May result in phototoxicity. May influence hepatic microsomal enzymes.
Schizandra; magnolia vine; gomeishi; wu wei zi	*Schizandra chinensis*	Treat arrhythmia	CNS depression, heartburn, may induce cytochrome P450 and other hepatic metabolic pathways.
Selenium		Reduce risk of cardiovascular, cerebrovascular disease; reduce oxidation of LDL cholesterol	May reduce efficacy of statin medications (HMG-CoA-reductase inhibitors).
Selfheal; all heal; xi ku cao	*Prunella vulgaris*	Treat hypertension	None reported.
Schizandra; magnolia vine; gomeishi; wu wei zi	*Schizandra chinensis*	Treat arrhythmia	Found as an ingredient in herbal formulas; considered generally safe; in some it may cause heartburn, acid indigestion, decreased appetite, stomach pain, or allergic skin rashes.
Vitamin A		Treat cardiovascular disease	Toxic in doses 10 times RDA (900 mcg/d for men; 700 mcg/d for women). HMG-CoA-reductase inhibitors and oral contraceptives are associated with increased blood levels of vitamin A. Bile acid sequestrants, mineral oil, orlistat, and neomycin may reduce vitamin A absorption. Antacids may interfere with vitamin A and beta-carotene absorption.
Vitamin E		Reduce oxidation of LDL cholesterol; increase HDL cholesterol; improve circulation	Relatively safe, although high doses (>50 IU/day) may cause mild antiplatelet effects. May potentiate anticoagulant or antiplatelet medications. Could interact with herbs and supplements that possess anticoagulant or antiplatelet effects (garlic, policosnol ginkgo).

Sources: Fetrow CW, Avila JR. *Professional's Handbook of Complementary and Alternative Medicines*, 3rd ed. Philadelphia: Lippincott Williams & Wilkins, 2004.
Bratman S, Girman AM. *Mosby's Handbook of Herbs and Supplements and Their Therapeutic Uses*. St. Louis: Mosby/Elsevier, 2003.
Fragakis AS, Thomson C. *The Health Professional's Guide to Popular Dietary Supplements*, 3rd ed. Chicago: American Dietetic Association, 2007.
Memorial Sloan-Kettering Cancer Center. About Herbs, Botanicals & Other Products. Available at http://www.mskcc.org/mskcc/html/11570.cfm

REFERENCES

1. American Heart Association. Heart Disease and Stroke Statistics_2009 Update: A Report From the American Heart Association Statistics Committee and Stroke Statistics Subcommittee. Circulation. 2009;119:e21–e181.

2. Gordon NF, Leighton RF, Mooss A. Factors associated with increased risk of coronary heart disease. In: Kaminsky LA, editor. American College of Sports Medicine's Resource Manual for Guidelines for Exercise Testing and Prescription. 5th ed. Baltimore (MD): Lippincott Williams & Wilkins; 2006. 95–114.

3. Ford ES, Ajani UA, Croft JB, Critchley, JA, LaBarthe DR, Kottke TE, Giles WH & Capewell S. Explaining the decrease in U.S. deaths from coronary disease, 1980–2000. N Engl J Med. 2007;356:2388–98.

4. Sherwood L. Human physiology. 7th ed. Belmont (CA): Thomson Brooks/Cole; 2010.

5. Rhoades R, Flanzer R. Human physiology. 4th ed. Pacific Grove (CA): Thomson Learning; 2003.

6. National Institutes of Health. Bethesda (MD): U.S. Department of Health and Human Services, National Institutes of Health; c2004 [updates 2004 August]. What is high blood pressure?; [about 2 screens]. Available at http://www.nhlbi.nih.gov/health/dci/Diseases/Hbp/HBP_WhatIs.html

7. Diamond JA, Phillips RA. Hypertensive heart disease. Hypertens Res. 2005;28:191–202.

8. American Heart Association. Heart Disease and Stroke Statistics—2006 Update. Dallas: American Heart Association; 2006.

9. Ostchega Y, Yoon SS, Hughes J, Louis T. Hypertension awareness, treatment, and control—continued disparities in adults: United States, 2005–2006. NCHS Data Brief No. 3. Hyattsville (MD): National Center for Health Statistics; 2008.

10. Patel TV, Williams GH, Fisher ND. Angiotensinogen genotype predicts abnormal renal hemodynamics in young hypertensive patients. J Hypertens. 2008;26:1353–59.

11. Hasenkamp S, Telgmann R, Staessen JA, Hagedorn C, Dördelmann C, Bek M, Brand-Herrmann SM, Brand E. Characterization and functional analyses of the human G protein-coupled receptor kinase 4 gene promoter. Hypertension. 2008;52:737–46.

12. Cambien F. Coronary heart disease and polymorphisms in genes affecting lipid metabolism and inflammation. Curr Athero Rep. 2005;7:188–95.

13. Meneton, P, Jeunemaitre, X, DeWardener, HE, MacGregor, GA. Diseases handling, blood pressure, and cardiovascular links between salt intake, and renal salt. Physiology Reviews. 2005;85:679–715.

14. Savoia C and Schiffrin EL. Inflammation in hypertension. Cur Opin Nephrol Hypertens. 2006;15:152–58.

15. He FJ, MacGregor GA. A comprehensive review on salt and health and current experience of worldwide salt reduction programmes. J Hum Hypertens. 2009 Jun;23(6):363–84. Epub 2008 Dec 25. Review.

16. Cook NR. Salt intake, blood pressure and clinical outcomes. Curr Opin Nephrol Hypertens. 2008 May;17(3): 310–14. Review.

17. Titze J, Ritz E. Salt and its effect on blood pressure and target organ damage: new pieces in an old puzzle. J Nephrol. 2009 Mar–Apr;22(2):177–89. Review.

18. Dumler F. Dietary sodium intake and arterial blood pressure. J Ren Nutr. 2009 Jan;19(1):57–60. Review.

19. Campbell NR, Khan NA, Hill MD, Tremblay G, Lebel M, Kaczorowski J, McAlister FA, Lewanczuk RZ, Tobe S; Canadian Hypertension Education Program. 2009 Canadian Hypertension Education Program recommendations: the scientific summary—an annual update. Can J Cardiol. 2009 May;25(5):271–77. Review.

20. Farsang C, Naditch-Brule L, Avogaro A, Ostergren J, Verdecchia P, Maggioni A, van de Borne P, Lins R, Roca-Cusachs A. Where are we with the management of hypertension? From science to clinical practice. J Clin Hypertens (Greenwich). 2009 Feb;11(2):66–73. Review.

21. Van Horn L, McCoin M, Kris-Etherton PM, Burke F, Carson JA, Champagne CM, Karmally W, Sikand G. The evidence for dietary prevention and treatment of cardiovascular disease. J Am Diet Assoc. 2008 Feb;108(2):287–331. Review.

22. Svetkey, RL, Simons-Morton, DG, Proschan, MA, Sacks, FM, Conlin, PR, Harsha, et al. Effect of the dietary approaches to stop hypertension diet and reduced sodium intake on blood pressure control. J Clin Hypertens. 2004;6:373–81.

23. Appel, LJ, Brands, MW, Daniels, SR, Karania, N, Elmer, PJ, Sacks, FM. American Heart Association. Dietary approaches to prevent and treat hypertension: a scientific statement from the American Heart Association. Hypertension. 2006;47:296–308.

24. Marks B. Tobacco Exposure and Chronic Illness. In: Roitman JL, editor. American College of Sports Medicine's Resource Manual for Guidelines for Exercise Testing and Prescription. 4th ed. Baltimore (MD): Lippincott Williams & Wilkins; 2001, 41–46.

25. Ponnuchamy B, Khalil R. Cellular mediators of renal vascular dysfunction in hypertension. Am J Physiol Regul Integr Comp Physiol. 2009;296:R1001–18.

26. Davy KP, Hall JE. Obesity and hypertension: two epidemics or one? Am J Physiol Regul Integr Comp Physiol. 2004;286:R803–13.

27. Krenkel J, St. Joer S, Kulick D. Relationship of Nutrition to Chronic Diseases. In: Kaminsky LA, editor. American College of Sports Medicine's Resource Manual for Guidelines for Exercise Testing and Prescription. 5th ed. Baltimore (MD): Lippincott Williams & Wilkins; 2006, 146–64.

28. Khayat R, Patt B, Hayes Jr. D. Obstructive sleep apnea: the new cardiovascular disease. Part I: obstructive sleep apnea and the pathogenesis of vascular disease. Heart Failure Reviews. 2009;14:143–53.

29. Parish JM, Somers VK. Obstructive sleep apnea and cardiovascular disease. Mayo Clin Proc. 2004;79:1036–46.

30. U.S. Department of Health and Human Services, National Institutes of Health, National Heart, Lung, & Blood Institute. Seventh Report of the Joint National Committee on Prevention, Detection, and Treatment of High Blood Pressure. (JNC 7.) NIH Publication 03-5231; 2003. Available at http://www.nhlbi.nih.gov/guidelines/hypertension/jncintro.htm

31. Giugliano D, Esposito K. Mediterranean diet and metabolic diseases. Curr Op in Lipidology. 2008;19:63–68.

32. Houston MC, Harper KJ. Potassium, magnesium, and calcium: their role in both the cause and treatment of hypertension. J Clin Hypertens (Greenwich). 2008 Jul;10(7 Suppl 2):3–11. Review.

33. Svetkey LP, Harsha DW, Vollmer WM, Stevens VJ, Obarzanke E, Elmer PJ, et al. Premier: a clinical trial of comprehensive lifestyle modification for blood pressure control: rationale, design and baseline characteristics. Ann Epidemiol. 2003;13:462–71.

34. McGuire HL, Svetkey LP, Harsha DW, Elmer PJ, Appel LJ, Ard JD. Comprehensive lifestyle modification and blood pressure control: a review of the PREMIER trial. J Clin Hypertens. 2004;6: 383–90.

35. Brunner EJ, Thorogood M, Rees K, Hewitt G. Dietary advice for reducing cardiovascular risk. Cochrane Database Syst Rev. 2005 Oct 19;(4):CD002128.

36. Carey VJ, Bishop L, Charleston J, Conlin P, Erlinger T, Laranjo N, et al. Rationale and design of the Optimal Macro-Nutrient Intake Heart trial to prevent heart disease (OMNI-Heart). Clin Trials. 2005;2:529–37.

37. Geleijnse JM, Grobbee DE, Kok FJ. Impact of dietary and lifestyle factors on the prevalence of hypertension in Western populations. J Hum Hypertens. 2005;19 Suppl 3:S1.

38. National Heart, Lung, and Blood Institute. The DASH eating plan. 2005. Available at http://www.nhlbi.nih.gov/health/public/heart/hbp/dash/

39. Steffen LM, Kroenke CH, Yu X, Periera MA, Slattery ML, Van Horn L, et al. Associations of plant food, dairy product, and meat inakes with 15-y incidence of elevated blood pressure in young black and white adults: the Coronary Artery Risk Development in Young

Adults (CAR-DIA) Study. Am J Clin Nutr. 2005;82:1169–77.

40. Koutroumpi M, Pitsavos C, Stefanadis C. The role of exercise in cardiovascular rehabilitation: a review. Acta Cardiol. 2008 Feb;63(1):73–79. Review.

41. Reisin E, Jack AV. Obesity and hypertension: mechanisms, cardiorenal consequences, and therapeutic approaches. Med Clin North Am. 2009 May;93(3):733–51. Review.

42. ADA Hypertension (HTN) Evidence-Based Nutrition Practice Guideline. 2008. Available at www.adaevidencelibrary.com

43. Appel LJ, Moore TJ, Obarzanek E, Vollmer WM, Svetkey LP, Sacks FM, et al. A clinical trial of the effects of dietary patterns on blood pressure. DASH Collaborative Research Group. N Engl J Med. 1997;336:1117–24.

44. Khan NA, McAlister FA, Lewanczuk RZ, Touyz RM, Padwal R, Rabkin SW, et al. Canadian Hypertension Education Program. The 2005 Canadian Hypertension Education Program recommendations for the management of hypertension: part II—therapy. Can J Cardiol. 2005;21:657–72.

45. Appel LJ, Sacks FM, Carey VJ, Obarzanek E, Swain JF, Miller Er, McCarron P, Bishop LM. Effects of protein, monounsaturated fat and carbohydrate intake on blood pressure and serum lipids. Results of the OmniHeart randomized trial. JAMA. 2005;294:2455–64.

46. Funk KL, Elmer PJ, Stevens VJ, Harsha DW, Craddick SR, Lin PH, Young DR, Champagne CM, Brantley PJ, McCarron PB, Simons-Morton DG, Appel LJ. PREMIER—A Trial of Lifestyle Interventions for Blood Pressure Control: Intervention Design and Rationale Health Promot Pract. 2008 July;9:271–80.

47. National Heart, Lung, and Blood Institute: The practical guide: identification, evaluation and treatment of overweight and obesity in adults. Bethesda (MD): U.S. Department of Health and Human Services, National Institutes of Health; 2000.

48. Stamler J. The INTERSALT Study: background, methods, findings, and implications. Am J Clin Nutr. 1997;65: 626S–642S.

49. Freedman DA, Petitti DB, Salt and blood pressure. Conventional wisdom reconsidered. Eval Rev. 2001;25:267–87.

50. Beevers DG. The epidemiology of salt and hypertension. Clin Auton Res. 2002;12:353–57.

51. Vollmer WM, Sacks Fm, Ard J, Appel LJ, Bray GA, Simons-Morton DG, et al. DASH-Sodium Trial Collaborative Research Group. Effects of diet and sodium intake on blood pressure: subgroup analysis of the DASH-Sodium trial. Ann Intern Med. 2001;135:1019–28.

52. Obarzanek E, Proschan MA, Vollmer WM, Moore TJ, Sacks FM, Appel LJ, et al. Individual blood pressure responses to changes in salt intake: results from the DASH-Sodium trial. Hypertension. 2003;42:459–67.

53. Sacks FM, Svetkey LP, Vollmer WM, Appel LJ, Bray GA, Harsha D, et al. DASH-Sodium Collaborative Research Group. Effects on blood pressure of reduced dietary sodium and the Dietary Approaches to Stop Hypertension (DASH) diet. DASH-Sodium Collaborative Research Group. N Engl J Med. 2001;344:3–10.

54. U.S. Food and Drug Administration. United States Department of Agriculture and U.S. Department of Health and Human Services, Dietary Guidelines for Americans 2005. 6th ed. Home and Garden Bulletin no. 232. Washington (DC): 2005. Available at http://www.cnpp.usda.gov/DGAs2005Guidelines.htm

55. Xin X, He J, Frontini MG, Ogden LG, Motsamai OI, Whelton PK. Effects of alcohol reduction on blood pressure: a meta-analysis of randomized controlled trials. Hypertension. 2001;38:1112–17.

56. Jee SH, Miller ER 3rd, Guallar E, Singh VK, Appel LJ, Klag MJ. The effect of magnesium supplementation on blood pressure: a meta-analysis of randomized clinical trials. Am J Hypertens. 2002;15:691–96.

57. Geleijnse JM, Kok FJ, Grobbee DE. Blood pressure response to changes in sodium and potassium intake: a metaregression analysis of randomised trials. J Hum Hypertens. 2005;19 Suppl 3:S1–S4.

58. Alonso A, Beunza JJ, Delgado-Rodriguez M, Martinez JA, Martinez-Gonzalez MA. Low-fat dairy consumption and reduced risk of hypertension: the Seguimiento Universidad de Navarra (SUN) cohort. Am J Clin Nutr. 2005;82:972–79.

59. American Lung Association. New York: American Lung Association [updated 2009; cited 2009 29 Oct]. Freedom from Smoking. Available at http://www.lungusa.org/stop-smoking/quit-smoking/how-to-quit/getting-help/

60. Stamler J, Stamler R, Neaton JD, Wentworth D, Daviglus ML, Garside D, Dyer AR, Liu K, Greenland P. Low risk-factor profile and long-term cardiovascular and noncardiovascular mortality and life expectancy: findings for 5 large cohorts of young adult and middle-aged men and women. JAMA. 1999;282: 2012–18.

61. Stampfer MJ, Hu FB, Manson JE, Rimm EB, Willett WC. Primary prevention of coronary heart disease in women through diet and lifestyle. N Engl J Med. 2000;343:16–22.

62. Centers for Disease Control and Prevention. Declining prevalence of no known major risk factors for heart disease and stroke among adults—United States, 1991–2001. Morbidity and Mortality Weekly Reports. 2004;53:4–7.

63. Barnard DD. Heart failure in women. Curr Cardiol Rep. 2005;7:159–65.

64. Expert Panel on Detection, Evaluation, and Treatment of High Blood Cholesterol in Adults. Executive summary of the Third Report of the National Cholesterol Education Program (NCEP) Expert Panel on Detection, Evaluation, and Treatment of High Blood Cholesterol in Adults (Adult Treatment Panel III). JAMA. 2001;285:2486–97.

65. U.S. Department of Health and Human Services. Public Health Service National Institutes of Health National Heart Lung and Blood Institute Third Report on the National Cholesterol Education Program Adult Treatment Panel III Guidelines NIH Publication No. 01-3305; 2001.

66. National Center for Health Statistics. Health, United States, 2004, with Chartbook on Trends in the Health of Americans. 2004.

67. Ogden CL, Carroll MD, McDowell MA, Flegal KM. Obesity among adults in the United States— no change since 2003–2004. NCHS data brief no 1. Hyattsville, MD: National Center for Health Statistics. 2007.

68. Ogden CL, Carroll MD, Flegal KM. High body mass index for age among U.S.

children and adolescents, 2003–2006. JAMA. 2008;299:2401–05.

69. Poirier P, Giles TD, Bray GA, Hong Y, Stern JS, Pi-Sunyer, X, et al. Obesity and cardiovascular disease: Pathophysiology, evaluation, and effect of weight loss. Circ. 2006;113:898–918.

70. Nichols N. Hypothyroidism and cardiovascular disease. Canadian J Cardiovasc Nurs. 2005;15:68–73.

71. Altena TS, Michaelson JL, Ball SD, Guilford BL, Thomas TR. Lipoprotein subfraction changes after continuous or intermittent exercise training. Med Sci Sports Exerc. 2006;38:367–72.

72. Grundy SM, Cleeman JI, Bairey Merz CN, Brewer HB, Clark LT, Hunninghake DB, et al. for the Coordinating Committee of the National Cholesterol Education Program Circulation. Adult Treatment Panel III Guidelines 2004;110:227–39.

73. Centers for Disease Control and Prevention. National diabetes fact sheet: general information and national estimates on diabetes in the United States, 2007. Atlanta (GA): U.S. Department of Health and Human Services, Centers for Disease Control and Prevention; 2008.

74. Huang PL. A comprehensive definition for metabolic syndrome. Dis Model Mech. 2009 May–Jun;2(5–6):231–37.

75. Vitarius JA. The metabolic syndrome and cardiovascular disease. Mount Sinai J Med. 2005;72:257–62.

76. Bodary PF, Yasuda N, Watson DD, Brown AS, Davis JM, Pate RR. Effects of short-term exercise training on Plasminogen Activator Inhibitor (PAI-1). Med Sci Sport Exer. 2003;35:1853–58.

77. Centers for Disease Control and Prevention. Smoking and tobacco use fast facts. Available at http://www.cdc.gov/tobacco/data_statistics/fact_sheets/index.htm

78. Ambrose JA, Barua RS. The pathophysiology of cigarette smoking and cardiovascular disease. J Am Coll Cardiol. 2004;43:1731–37.

79. Somers, VK, White, DP, Amin, R, Abraham, WT, Costa, F, et al. Sleep apnea and cardiovascular disease: an American Heart Association/American College of Cardiology Foundation scientific statement. Circulation. 2008;118:1080–1111.

80. Squires RW. Pathophysiology and clinical features of cardiovascular diseases. In: Kaminsky LA, editor. American College of Sports Medicine's Resource Manual for Guidelines for Exercise Testing and Prescription. 5th ed. Baltimore (MD): Lippincott Williams & Wilkins; 2006, 411–26.

81. Van der Giessen, AG, Wentzel, JJ, Meijboom, WB, Mollet, NR, van der Steen, AFW, et al. Plaque and shear stress distribution in human coronary bifurcations: a multislice computed tomography study. Eurointervention. 2009;4:654–61.

82. Davies, PF. Endothelial mechanisms of flow mediated athero-protection and susceptibility. Circulation Research. 2007;101:10–12.

83. Montecucco, F, Pende, A, Mach, F. The rennin-angiotensin system modulates inflammatory processes in atherosclerosis: evidence from basic research and clinical studies. Mediators of Inflammation. 2009:1–13.

84. Pepine CJ. The impact of nitric oxide in cardiovascular medicine: untapped potential utility. Am J Med. 2009 May;122(5 Suppl):S10–S5.

85. Kearney, MT, Duncan, ER, Kahn, M, Wheatcroft, SB. Insulin resistance and endothelial cell dysfunction: studies in mammalian models. Experimental Physiology. 2008;93:158–63.

86. Jennings, LK, Role of platelets in atherothrombosis. American Journal of Cardiology. 2009;103:4A–10A.

87. Pradka LR. Lipids and their role in coronary heart disease: What they do and how to manage them. Nurs Clin North Am. 2000;35:981–91.

88. Gieseg, SP, Leake, DS, Flavall, EM, Amit, Z, Reid, L, Yang, Y. Macrophage antioxidant protection within atherosclerotic plaques. Frontiers in Bioscience. 2009;14:1230–46.

89. Burnier, L, Fontana, P, Angelillo-Scherrer, A, Kwak, BR. Intercellular communication in atherosclerosis. Physiology. 2009;24:36–44.

90. Fletcher B, Berra K, Ades P, Braun LT, Burke LE, Durstine JL, et al. Managing abnormal blood lipids: a collaborative approach. Circulation. 2005;112:3184–209.

91. Pollak OJ. Reduction of blood cholesterol in man. Circulation. 1953;2:702–06.

92. Keys A, Arvanis C, Blackburn H. Seven countries: a multivariate analysis of death and coronary heart disease. Cambridge (MA): Harvard University Press; 1980, 381.

93. Keys A, Menotti A, Aravanis C, Blackburn H, Djordjevic BS, Buzina R, et al. The seven countries study: 2,289 deaths in 15 years. Prev Med. 1984;13:141–54.

94. ADA Disorders of Lipid Metabolism Guideline. 2007. Available at www.adaevidencelibrary.com

95. De Caterina R, Zampolli A, Del Turco S, Madonna R, Massaro M. Nutritional mechanisms that influence cardiovascular disease. Am J Clin Nutr. 2006;83:421S–426S.

96. Koertge J, Weidner G, Elliott-Eller M, Scherwitz L, Merritt-Worden TA, Marlin R, et al. Improvement in medical risk factors and quality of life in women and men with coronary artery disease in the Multicenter Lifestyle Demonstration Project. Am J Cardiol. 2003;91:1316–22.

97. Institute for Community Health Promotion, Brown University. Rapid eating assessment for patients. 2005. Available at http://bms.brown.edu/nutrition/acrobat/REAP%206.pdf

98. Jenkins DJ, Kendall CW, Marchie A, Faulkner DA, Wong JM, de Souza R, et al. Aldosterone receptor antagonism in heart failure. Pharmacotherapy. 2005;25:1126–33.

99. American Dietetic Association. Position of the American Dietetic Association and Dietitians of Canada: Dietary Fatty Acids. J Am Diet Assoc. 2007;107:1599–1611

100. U.S. Department of Health and Human Services. USDA dietary guidelines for Americans. 6th ed. 2005. Available at http://www.health.gov/dietaryguidelines

101. Howard BV, Van Horn L, Hsia J, Manson JE, Stefanick ML, Wassertheil-Smoller S, et al. Low-fat dietary pattern and risk of cardiovascular disease: the Women's Health Initiative Randomized Controlled Dietary Modification Trial. JAMA. 2006;295:655–66.

102. Dansinger ML, Gleason JA, Griffith JL, Selker HP, Schaefer EJ. Comparison of the Atkins, Ornish, Weight Watchers, and Zone diets for weight loss and heart disease risk reduction. JAMA. 2005;293:43–53.

103. Aldana SG, Greenlaw R, Thomas D, Salberg A, DeMordaunt T, Fellingham GW, Avins AL. The influence of an intense cardiovascular disease risk factor modification program. Prev Cardiol. 2004;7:19–25.

104. Marshall DA, Vernalis MN, Remaley AT, Walizer EM, Scally JP, Taylor AJ. The role of exercise in modulating the impact of an ultralow-fat diet on serum lipids and apolipoproteins in patients with or at risk for coronary artery disease. Am Heart J. 2006;151:484–91.

105. Mohanka M, Irwin M, Heckbert SR, Yasui Y, Sorensen B, Chubak J, et al. Serum lipo-proteins in overweight/obese postmenopausal women: a one-year exercise trial. Med Sci Sports Exerc. 2006;38:231–39.

106. German JB, Dillard CJ. Saturated fats: what dietary intake? Am J Clin Nutr. 2004;80:550–59.

107. Knopp RH, Retzlaff BM. Saturated fat prevents coronary artery disease? An American paradox. Am J Clin Nutr. 2004;80:1102–03.

108. Gropper SS, Smith JL. Advanced nutrition and human metabolism. 5th ed. Belmont (CA):Wadsworth-Cengage Learning; 2009.

109. Eckel RH, Borra S, Lichenstein AH, Yin-Piazza SY. Understanding the complexity of trans fatty acid reduction in the American Diet: American Heart Association Trans Fat Conference 2006: Report of the Trans Fat Conference Planning Group. Circulation. 2007;115:2231–46.

110. Baylin A, Kabagambe EK, Ascherio A, Spiegelman D, Campos H. High 18:2 trans-fatty acids in adipose tissue are associated with increased risk of nonfatal acute myocardial infarction in Costa Rican adults. J Nutr. 2003;133:1186–91.

111. American Dietetic Association Evidence Analysis Library. Disorders of lipid metabolism. Available at http://www.ebg.adavidencelibrary.com

112. Mozaffarian D, Pischon T, Hankinson SE, Rifai N, Joshipura K, Willett WC, et al. Dietary intake of trans fatty acids and systemic inflammation in women. Am J Clin Nutr. 2004;79:606–12.

113. U.S. Food and Drug Administration. Trans fatty acids in nutrition labeling; consumer research to consider nutrient content and health claims and possible footnote or disclosure statements; final rule and proposed rule. July 11, 2003. Available at http://www.cfsan.fda.gov/~lrd/fr03711a.html

114. Keys A. Food items, specific nutrients, and "dietary" risk. Am J Clin Nutr. 1986;43:477–79.

115. Rodenas S, Rodriguez-Gil S, Merinero MC, Sanchez-Muniz FJ. Dietary exchange of an olive oil and sunflower oil blend for extra virgin olive oil decreases the estimate cardiovascular risk and LDL and apolipoprotein AII concentrations in postmenopausal women. J Am Coll Nutr. 2005;24:361–69.

116. Nicklas TA, Hampl JS, Taylor CA, Thompson VJ, Heird WC. Monounsaturated fatty acid intake by children and adults: temporal trends and demographic differences. Nutr Rev. 2004;62:132–41.

117. Trichopoulou A, Costacou T, Bamia C, Trichopoulos D. Adherence to a Mediterranean diet and survival in a Greek population. New England Journal of Medicine. 2003;348:2599–2608.

118. De Caterina R, Massaro M. Omega-3 fatty acids and the regulation of expression of endothelial pro-atherogenic and pro-inflammatory genes. J Membr Biol. 2005;206:103–16.

119. Harper CR, Jacobson TA. Usefulness of omega-3 fatty acids and the prevention of coronary heart disease. Am J Cardiol. 2005;96:1521–29.

120. Connelly PW. Direct comparison of a dietary portfolio of cholesterol-lowering foods with a statin in hypercholesterolemic participants. Am J Clin Nutr. 2005;81(2):380–87

121. Brown L, Rosner B, Willett WW, Sacks FM. Cholesterol-lowering effects of dietary fiber: a meta-analysis. Am J Clin Nutr. 1999;69:30–42.

122. Pereira MA, O'Reilly E, Augustsson K, Fraser GE, Goldbourt U, Heitmann BL, et al. Dietary fiber and risk of coronary heart disease: a pooled analysis of cohort studies. Arch Intern Med. 2004;164:370–76.

123. Slavin JL. Dietary fiber and body weight. Nutrition. 2005;21:411–18.

124. Pinedo S, Vissers MN, von Bergmann K, Elharchaoui K, Lutjohann D, Luben R, Wareham NJ, Kastelein JJ, Khaw KT, Boekholdt SM. Plasma levels of plant sterols and the risk of coronary artery disease: the prospective EPIC-Norfolk Population Study. J Lipid Res. 2007;48:139–44.

125. U.S. Food and Drug Administration. 2000. Food labeling: health claims; plant sterol/stanol esters and coronary heart disease. Interim final rule. Fed. Register 65:54686–739.

126. U.S. Food and Drug Administration. FDA authorizes new coronary heart disease health claim for plant sterol and plant stanol esters. FDA Talk Paper. September 5, 2000. Available at http://www.cfsan.fda.gov/~lrd/tpsterol.html

127. Pasternak RC, Criqui MH, Benjamin EJ, Fowkes FGR, Isselbacher EM, McCullough PA, et al. Atherosclerotic vascular disease conference: Writing group I: Epidemiology. Circ. 2004;109:2605–12.

128. Saffitz JE. The pathology of sudden cardiac death in patients with ischemic heart disease—arrhythmology for anatomic pathologists. Cardiovasc Pathol. 2005;14:195–203.

129. Smith SC, Milani RV, Arnett DK, Crouse JR, McDermott MM, Ridker PM, et al. Atherosclerotic vascular disease conference: Writing group II: Risk factors. Circ. 2004;109:2613–16.

130. Faxon DP, Fuster V, Libby P, Beckman, JA, Hiatt WR, Thompson RW, et al. Atherosclerotic vascular disease conference: writing group III: pathophysiology. Circulation. 2004;109:2617–25.

131. Crowley LV. An Introduction to Human Disease: Pathology and Pathophysiology Correlations. 5th ed. Boston: Jones and Bartlett; 2001.

132. Barsky, AJ, Peekna, HM, Borus, JF. Somatic symptom reporting in women and men. J Gen Intern Med. 2001;16:266–75.

133. Smitherman TC, Reis, SE. Heart disease in women with diabetes. Diabetes Spectrum. 1997;10:207–15.

134. Ryan TJ, Antman EM, Brooks NH, et al. 1999 update: ACC/AHA guidelines for the management of patients with acute myocardial infarction. A report of the American College of Cardiology/American Heart Association Task Force on Practice Guidelines (Committee on Management of Acute Myocardial Infarction). J Am Coll Cardiol. 1999;34:890–911 and Circulation. 1999;100:1016–30.

135. Beers MH, Berkow R, editors. Merck manual of diagnosis and therapy 17th ed. Ch. 202 Coronary Heart Disease. Available at http://www.merck.com/mrkshared/mmanual/sections.jsp

136. Escott-Stump S. Myocardial Infarction. In: Nutrition and Diagnosis Related Care. Philadelphia (PA): Lippincott Williams & Wilkins; 2008.

137. Grundy SM. Nutrition in the management of disorders of serum lipids and lipoproteins. In: Shils ME, Shike M, Ross AC, Cabellero B, Cousins RJ, editors. Modern nutrition in health and disease. 10th ed. Philadeplphia (PA): Lippincott Williams & Wilkins; 2006, 1076 –94.

138. Chant T. Peripheral vascular disease. Primary Health Care. 2004 Oct;14(8):29–34.

139. Allison MA, Ho E, Denenberg J, Langer R, Newman A, Fabsitz R, Criqui M. Ethnic-specific prevalence of peripheral arterial disease in the United States. Am J Prev Med. 2007;32:328–33.

140. Murabito JM, Evans JC, D'Agostino RB Sr, Wilson PW, Kannel WB. Temporal trends in the incidence of intermittent claudication from 1950 to 1999. Am J Epidemiol. 2005;162:430–37

141. Hirsch AT, Haskal ZJ, Hertzer NR, Bakal CW, Creager MA, Halperin JL, Hiratzka LF, Murphy WR, Olin JW, Puschett JB, Rosenfield KA, Sacks D, Stanley JC, Taylor LM Jr, White CJ, White J, White RA, Antman EM, Smith SC Jr, Adams CD, Anderson JL, Faxon DP, Fuster V, Gibbons RJ, Hunt SA, Jacobs AK, Nishimura R, Ornato JP, Page RL, Riegel B; American Association for Vascular Surgery; Society for Vascular Surgery; Society for Cardiovascular Angiography and Interventions; Society for Vascular Medicine and Biology; Society of Interventional Radiology; ACC/AHA Task Force on Practice Guidelines Writing Committee to Develop Guidelines for the Management of Patients With Peripheral Arterial Disease; American Association of Cardiovascular and Pulmonary Rehabilitation; National Heart, Lung, and Blood Institute; Society for Vascular Nursing; TransAtlantic Inter-Society Consensus; Vascular Disease Foundation. ACC/AHA 2005 Practice Guidelines for the management of patients with peripheral arterial disease (lower extremity, renal, mesenteric, and abdominal aortic):

a collaborative report from the American Association for Vascular Surgery/Society for Vascular Surgery, Society for Cardiovascular Angiography and Interventions, Society for Vascular Medicine and Biology, Society of Interventional Radiology, and the ACC/AHA Task Force on Practice Guidelines (Writing Committee to Develop Guidelines for the Management of Patients With Peripheral Arterial Disease): endorsed by the American Association of Cardiovascular and Pulmonary Rehabilitation; National Heart, Lung, and Blood Institute; Society for Vascular Nursing; TransAtlantic Inter-Society Consensus; and Vascular Disease Foundation. Circulation. 2006;113:e463–e654

142. American Diabetes Association/American College of Cardiology. Peripheral arterial disease in diabetes. Diab Cardiovasc Dis Rev. 2004;(6):1–8.

143. Goldenberg-Cohen N, Cohen Y, Monselise Y, Eldar I, Axer-Siegel I, Weinberger D, Kramer M. C-reactive protein levels do not correlate with retinal artery occlusion but with atherosclerosis. Eye. 2009;23:785–90.

144. Brass EP, Hiatt WR, Green S. Skeletal muscle metabolic changes in peripheral arterial disease contribute to exercise intolerance: a point-counterpoint discussion. Vasc Med. 2004;9:293–301.

145. Jude E and Gibbons J. Identifying and treating intermittent claudication in people with diabetes. The Diab Foot. 2005;8:84–92.

146. Craeger MA, White CJ, Hiatt WR, Criqui MH, Josephs SC, Alberts MJ, et al. Atherosclerotic peripheral vascular disease symposium II: Executive summary. Circulation. 2008;118:2811–25.

147. Hunt AS, Abraham WT, Chin MH, Feldman AM, Francis GS, Ganiats TG, et al. ACC/AHA 2005 guideline update for the diagnosis and management of chronic heart failure in the adult—summary article: a report of the American College of Cardiology/American Heart Association Task Force on Practice Guidelines (Writing Committee to Update 2001 Guidelines for he Evaluation and Managmenet of Heart Failure). Circ. 2005;112:1825–52.

148. Bertoni AG, Hundley WG, Massing MW, Bonds DE, Burke GL, Goff DC. Heart failure prevalence, incidence, and mortality in the elderly with diabetes. Diab Care. 2004;27:699–703.

149. Centers for Disease Control and Prevention. Self-reported heart disease and stroke among adults with and without diabetes—United States, 1999–2001. Morbidity and Mortality Weekly Reports. 2003;52:1065–70.

150. Schocken DD, Benjamin EJ, Fonarow GC, Krumholz HM, Levy D, et al. Prevention of heart failure: a scientific statement from the American Heart Assocation Councils on Epidemiology and Prevention, Clinical Cardiology, Cardiovascualr Nursing, and High Blood Pressure Research; Quality of Care and Outcomes Research Interdisciplinary Working Group; and Functional Genomics and Translational Biology Interdisciplinary Working Group. Circ. 2008;117:2544–65.

151. Vader JM, Drazner MH. Clinical assessment of heart failure: utility of symptoms, signs, and daily weights. Heart Fail Clin. 2009 Apr;5(2):149–60.

152. Jennings DL, Kalus JS, O'Dell KM. Aldosterone receptor antagonism in heart failure. Pharmacotherapy. 2005;25:1126–33.

153. Farmakis D, Filippatos G, Parissis J, Kremastinos DT, Gheorghiade M. Hyponatremia in heart failure. Heart Failure Reviews. 2009;14:59–63.

154. Rogers, JH, Bolling, SF. The tricuspid valve: current perspective and evolving management of tricuspid regurgitation. Circulation. 2009;119:2718–25.

155. Little WC, Brucks S. Therapy for diastolic heart failure. Prog Cardiovasc Dis. 2005;47:380–88.

156. Veterans Health Administration, Department of Veterans Affairs. The pharmacologic management of chronic heart failure. Washington (DC): Veterans Health Administration, Department of Veterans Affairs; 2003.

157. Rauch H, Motsch J, Bottiger BW. Newer approaches to the pharmacological management of heart failure. Curr Opin Anaesthesiol. 2006;19:75–81.

158. Dunn SP, Bleske B, Dorsch M, Macaulay T, Van Tassell B, Vardeny O. Nutrition and heart failure: impact of drug therapies and management strategies. Nutr Clin Pract. 2009;24:60–75.

159. Anker SD, Steinborn W, Strassburg S. Cardiac cachexia. Ann Med. 2004;36:518–29.

failure. J Am Diet Assoc. 1995;95:541–45.

164. Sole MJ, Jeejeebhoy KN. Conditioned nutritional requirements and the pathogenesis and treatment of myocardial failure. Curr Opin Clin Nutr Metab Care. 2000;3:417–24.

165. Sole MJ, Jeejeebhoy KN. Conditioned nutritional requirements: therapeutic relevance to heart failure. Herz. 2002;27:174–78.

166. Witte KK, Clark AL, Cleland JG. Chronic heart failure and micronutrients. J Am Coll Cardiol. 2001;37:1765–74.

167. Sharma R, Anker SD. From tissue wasting to cachexia: changes in peripheral blood flow and skeletal musculature. Eur Heart J Supp. 2002;4:D12–D17.

168. Tangalos EG. Congestive heart failure. Nutrition management for older adults. Washington (DC): Nutrition Screening Initiative (NSI); 2002.

169. Berger MM, Mustafa I. Metabolic and nutritional support in acute cardiac failure. Curr Opin Clin Nutr Metab Care. 2003;6:195–201.

170. Anker SD, Steinborn W, Strassburg S. Cardiac cachexia. Ann Med. 2004;36:518–29.

171. Carson JAS, Grundy SM, Van Horn L, Stone N, Binkoski A, s-Etherton P. Medical nutrition therapy for the prevention and management coronary heart disease. In: Carson S, Burke F, Hark L. eds. Cardiovascular nutrition: Disease Prevention and Management. Chicago (IL): American Dietetic Association; 2004:109–48

2. Kuehneman T, Saulsbury D, Splett Chapman DB. Demonstrating the pact of nutrition intervention in a art failure program. J Am Diet Assoc. 02;102:1790–94.

3. Colin Ramirez E, Castillo Martinez Orea Tejeda A, Rebollar Gonzalez V, Narvaez David R, Asensio Lafuente E. Effects of a nutritional intervention on body composition, clinical status, and quality of life in patients with heart failure. Nutr. 2004;20(10):890–95.

174. Arcand JA, Brazel S, Joliffe C, Choleva M, Berkoff F, Allard JP, Newton GE. Education by a dietitian in patients with heart failure results in improved adherence with a sodium-restricted diet: a randomized trial. Am Heart J. 2005;150:716.

175. Neily JB, Toto KH, Gardner EB, Rame JE, Yancy CW, Sheffield MA, et al. Potential contributing factors to noncompliance with dietary sodium restriction in patients with heart failure. Am Heart J. 2002;143:29–33.

176. Keith ME, Walsh NA, Darling PB, Hanninen SA, Thirugnanam S, Leong-Poi H, Barr A, Sole MJ. B-vitamin deficiency in hospitalized patients with heart failure. J Am Diet Assoc. 2009;109:1406–10.

177. Wooley JA. Characteristics of thiamin and its relevance to the management of heart failure. Nutr Clin Pract. 2008 Oct–Nov;23(5):487–93.

178. Hanninen SA, Darling PB, Sole MJ, Barr A, Keith ME. The prevalence of thiamin deficiency in hospitalized patients with congestive heart failure. J Am Coll Cardiol. 2006;47:354–61.

179. Seligmann H, Hallkin H, Rauchfleisch S, Kaufmann N, Motro M, Vered Z, et al. Thiamine deficiency in patients with congestive heart failure receiving long-term furosemide therapy: a pilot study. Am J Med. 1991;91:151–55.

180. Shimon I, Almog S, Vered Z, Seligmann H, Shefi M, Peleg E, et al. Improved left ventricular function after thiamine supplementation in patients with congestive heart failure receiving long-term furosemide therapy. Am J Med. 1995;98:485–90.

181. McCabe-Sellers BJ, Sharkey JR, Browne BA. Diuretic medication therapy use and low thiamin intake in homebound older adults. J Nutr Elder. 2005;24:57–71.

182. Pittler MH, Schmidt K, Ernst E. Hawthorn extract for treating chronic heart failure: meta-analysis of randomized trials. Am J Med. 2003;114:665–74.

183. Bednarz B, Jaxa-Chamiec T, Gebalska J, Herbaczynska—Cedro K, Ceremuzynski L. L-arginine supplementation prolongs exercise capacity in congestive heart failure. Kardiol Pol. 2004;60:348–53.

184. Tousoulis D, Charakida M, Stefanadis C. Inflammation and endothelial dysfunction as therapeutic targets in patients with heart failure. Int J Cardiol. 2005;100:347–53.

185. Ferrari R, Merli E, Cicchitelli G, Mele D, Fucili A, Ceconi C. Therapeutic effects of L-carnitine and propionyl-L-carnitine on cardiovascular diseases: a review. Ann N Y Acad Sci. 2004;1033: 79–91.

186. Mora A, Lee IM, Buring JE, Ridker PM. Association of physical activity and body mass index with novel and traditional cardiovascular biomarkers in women. JAMA. 2006;295:1412–19.

14

Diseases of the Upper Gastrointestinal Tract

Marcia Nahikian Nelms, PhD, RD
The Ohio State University

Introduction

No other system of the human body is so intimately involved in preservation of an optimal nutritional status than the gastrointestinal (GI) tract. Any pathology involving the GI tract can have a significant effect on nutritional status. A wide array of diagnoses affect normal digestion and absorption. Nausea, vomiting, diarrhea, constipation, and malabsorption, all common symptoms in GI disease, can potentially jeopardize the individual's nutritional status. In some diseases, such as celiac disease, nutrition therapy is the only treatment.

GI disease is estimated to affect more than 70 million people in the United States each year. More than 35.9 million physician office visits are related to GI disease. With advances in diagnosis and treatment, hospitalizations for GI disease have decreased over the past ten years.[1] Still, health care costs for these individuals are staggering, at over $40 billion dollars per year. Leading medications commonly prescribed include **proton pump inhibitors (PPI)**.[2, 3]

Normal Anatomy and Physiology of the Upper Gastrointestinal Tract

A good working knowledge of normal anatomy and physiology of the GI tract is essential to understanding pathophysiology, and subsequent medical and nutritional care of GI disease. The GI tract is often described as a long tube approximately 15 feet in length. This description, though technically accurate, minimizes the complexity of this body system. Cell and organ function and intricacy of control factors throughout the GI tract are highly differentiated. The upper GI tract is composed of the mouth, pharynx, esophagus, and stomach, while the lower GI tract includes the small and large intestine (see Figure 14.1). These continuous organs differ from one another in both anatomical structure and specialized function. Accessory or ancillary organs that contribute to the function of the GI tract include the liver, biliary system, and pancreas.

GLOSSARY

acarbose—also known as Precose™; a medication—an alpha glucoside inhbitor—that slows the digestion of starch; used in diabetes treatment and to prevent dumping syndrome

acetylcholine—excitatory neurotransmitter involved in stimulation of parietal cells

achalasia—motility disorder characterized by an absence or weakened peristalsis within the esophagus

achlorhydria—lack of gastric hydrochloric acid secretions

ageusia—inability to taste

anastomosis—the surgical connection of body parts, especially hollow tubular parts like those of the GI tract

anticariogenic—describes foods or conditions that assist in prevention of dental caries

aspiration—the accidental inhalation of food particles or fluids into the lungs

autocrine—a type of communication between hormones and other chemical messengers that is released from a cell at a distance from the target cell

bariatric—referring to medical treatment of morbid obesity

Barrett's esophagus—a complication of severe chronic GERD involving changes in the cells of the tissue that line the bottom of the esophagus

borborygmas—stomach "growling"

calculus—calcified deposits that have formed around the teeth

cariogenic—describes foods or conditions that contribute to dental caries

cariostatic—describes foods or conditions that neither contribute to nor prevent dental caries

Chagas disease—a parasitic disease caused by *Trypanosoma cruzi*

cheilosis—fissures that develop at the edges of the mouth

cholinergic—resembling acetylcholine; stimulated by or releasing acetylcholine or a related compound

chyme—partially digested food in a semifluid state

dental caries—decay of the teeth that begins when acid dissolves the enamel that covers the tooth

dentin—the hard tissue of the tooth surrounding the central core of nerves and blood vessels

dumping syndrome—a group of symptoms that occurs with rapid passage of large amounts of food into the small intestine; symptoms include dizziness, sweating, decreased blood pressure, and diarrhea

dysgeusia—abnormalities or reduced ability to taste

dyspepsia—vague upper abdominal symptoms that may include upper abdominal pain, bloating, early satiety, nausea, or belching

dysphagia—difficulty swallowing

edentulous—without any teeth

enamel—hard outer layer of teeth consisting of hydroxyapatite; this mineral is composed of calcium, phosphorous, fluoride, chloride, sodium, and magnesium

endoscopy—examination of the interior of a canal by means of an endoscope

epigastric—referring to the upper abdominal region

eructation—belch or burp

esophageal phase of swallowing—esophageal peristalsis carries the bolus through the esophagus and LES and into the stomach

fundoplication—a surgical technique used to suture the fundus of the stomach around the esophagus to prevent reflux

gastrectomy—surgery to resect a portion of or the entire stomach

gastrin—primary hormone released to stimulate digestion and production of hydrocholric acid in the stomach

gastritis—inflammation of the gastric mucosa

gastroesophageal reflux disease (GERD)—chronic or recurrent gastric pain due to reflux of gastric secretions into the lower esophagus

gingiva—the gums

glossitis—inflammation of the tongue

hematemesis—the vomiting of blood

hemorrhage—bleeding

hiatal hernia—protrusion of part of the stomach through the diaphragm into the space normally occupied by the esophagus, heart, and lungs

histamine—paracrine released from parietal cells involved in production of hydrochloric acid; also released from mast cells and basophils as a component of inflammatory and immune responses

hyperosmolar—having a higher osmolality than body fluids (>300 mOsm/kg)

hypoglycemia—a low serum glucose; generally considered to be <70 mg/dL

laparoscopically—describes the process of using a laparoscopic procedure through which an instrument is used to see structures within the abdomen and pelvis; in this way, a number of surgical procedures can be performed without the need for a large surgical incision

learned food aversion—avoidance of certain foods due to association with unpleasant GI symptoms

lower esophageal sphincter (LES)—the junction between the esophagus and the stomach

obstruction—blockage

octreotide—medication that mimics the action of somatostatin

oral preparatory phase—tongue, teeth, and mandible involved in chewing of food and preparation of bolus; food is mixed with saliva, pressed against the hard palate, and formed into a bolus

oral transit phase of swallowing—tongue moves bolus to back of throat

osmolality—number of water-attracting particles per kilogram of water (expressed as mOsm/kg)

paracrine—a name for a neurotransmitter that is released from a cell that is close to the target cell

parietal cell—one of the gastric gland cells that lies on the basement membrane covered by chief cells, and secretes hydrochloric acid

peptic ulcer disease—ulceration or perforation in the lining of the stomach, duodenum, or esophagus

perforation—a break in the integrity of the tissue

periodontal disease—a bacterial infection that destroys the attachment fibers and supporting bone that hold the teeth in the mouth

pharyngeal phase of swallowing—the involuntary swallowing reflux begins, and the bolus is carried through the pharynx to the top of the esophagus; the entrance to the trachea (larynx) closes, and the soft palate lifts and closes off entrance to the nose

plaque—the noncalcified accumulation of oral microorganisms and their by-products that adhere to the teeth

proton pump inhibitors—class of medications that block the H^+, K^+-ATPase enzyme, a component in HCL production

pyloroplasty—enlarging the pyloric sphincter

reduction and fixation of fracture—a method to surgically repair a bone fracture

Sjögren's syndrome—a chronic systematic inflammatory disorder, etiology unknown, characterized by dryness of mucous membranes

somatostatin—a hormone and neurotransmitter that inhibits release of peptide hormones in several tissues

sphincter—a circular muscle that prevents movement or passage through the circle when contracted; sphincter muscles are located throughout the GI tract and are crucial control factors for peristalsis

steatorrhea—excessive fat in the feces

stomatitis—inflammation of the membrane in the mouth

syncope—temporary loss of consciousness; fainting

vagotomy—severing of the vagus nerve; often a component of gastric surgery

vagus nerve—tenth cranial nerve; one of its major functions is to coordinate the autonomic nervous system communication between organs of digestion

xerostomia—decreased saliva production and dry mouth

Figure 14.1 Anatomy of the Digestive System

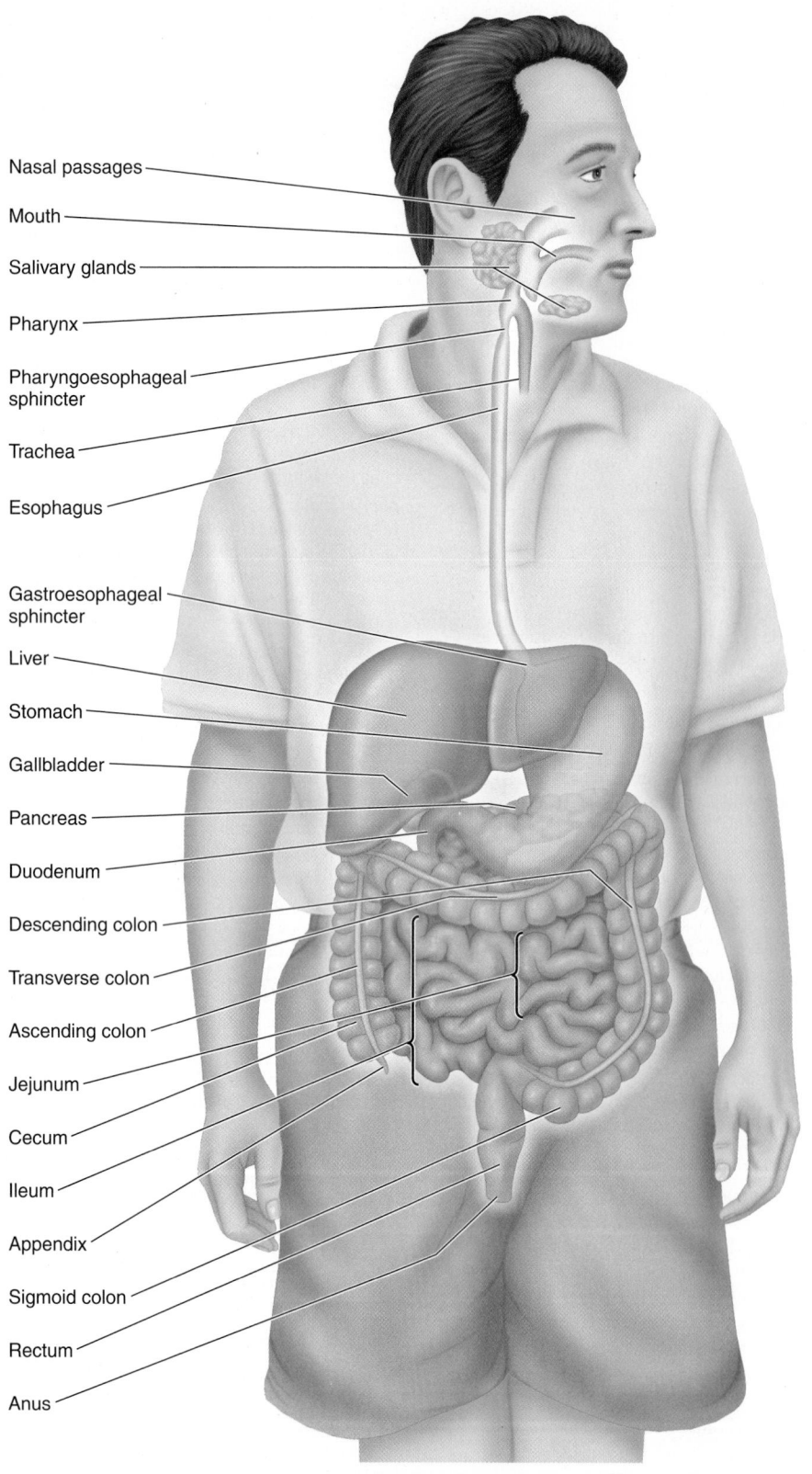

- Nasal passages
- Mouth
- Salivary glands
- Pharynx
- Pharyngoesophageal sphincter
- Trachea
- Esophagus
- Gastroesophageal sphincter
- Liver
- Stomach
- Gallbladder
- Pancreas
- Duodenum
- Descending colon
- Transverse colon
- Ascending colon
- Jejunum
- Cecum
- Ileum
- Appendix
- Sigmoid colon
- Rectum
- Anus

Source: Table 16.1, p. 592, of Sherwood's *Human Physiology: From Cells to Systems*, 7e.

Motility, Secretion, Digestion, and Absorption

Motility is the movement of the food consumed along the GI tract. Both propulsive contractions and mixing movements serve not only to move foodstuffs toward sites of digestion and absorption, but to mix foods with digestive secretions and maximize potential absorption. Secretions of the GI tract include water, electrolytes, enzymes, bile salts, and mucus. Through the process of digestion, complex molecules are converted to their simplest form. Carbohydrates are digested from their most complex form of polysaccharides to the monosaccharides: glucose, fructose, and galactose. Proteins are converted from polypeptides to single amino acids, di-, and tripeptides. Lipids are digested to their simplest forms: free fatty acids, monoglycerides, glycerol, phospholipids, and cholesterol. After digestion, these basic molecules are absorbed along with water, electrolytes, vitamins, and minerals to provide essential nutrients to every cell.

Contraction and motility are regulated through a complex network of communication paths involving **autocrine**, **paracrine**, and neuronal control systems. In particular, the GI tract's own specialized pacemaker cells called interstitial cells of Cajal control smooth muscle activity. The autonomic nervous system relays input for the smooth muscle of the GI tract through a specialized enteric nervous system. Two types of neurons regulate contraction of smooth muscle, motility, and secretory functions of the GI tract. Communication from one of these neurons includes a variety of substances including neuropeptides, hormones, and neurotransmitters. A neuropeptide is a protein substance originating from the nerve which can alter functional activity of an organ. As you may remember, a hormone is defined as a chemical substance formed in one organ and carried via the circulatory system to another target organ or cell. Hormones involved with control of the GI tract include gastrin, cholecystokinin, and secretin. These will be discussed in more detail later in this section.

Primary neuropeptides involved in transmission of impulses include acetylcholine, somatostatin, vasoactive intestinal polypeptide (VIP), Substance P, histamine, neurotensin, serotonin, and nitric oxide. Other neurotransmitters and modulators involved in neuroregulation for the GI tract include norepinephrine, neuropeptide Y, and g-aminobutyric acid (GABA). (See Table 14.1.)

This section will focus on each organ of the upper GI tract and discuss its normal function in the context of the four basic functions of the GI tract: motility, secretion, digestion, and absorption. GI disease can affect any one or all of these functions.

Table 14.1 Neurotransmitters of the Enteric Nervous System and Their Effects on Gastrointestinal Motility

Neurotransmitter	Effect on Motor Activity
Nonpeptides	
Acetylcholine	Typically excitatory
Serotonin	Excitatory
ATP	Inhibitory
Dopamine	Inhibitory
Nitric oxide	Inhibitory
Peptides	
Enkephalins	Excitatory
Gastrin-releasing peptide	Excitatory
Neuropeptide Y	Excitatory
Substance P	Excitatory
Gastric inhibitory peptide	Inhibitory
Somatostatin	Inhibitory
Vasoactive intestinal peptide	Inhibitory

Source: Rhoades R, Pflanzer, R, *Human Physiology*, 4th ed. Belmont, CA: Brooks/Cole; 2003. Table 22-1, p. 690.

Major innervation for the enteric nervous system is supplied by parasympathetic and sympathetic fibers of the autonomic nervous system. Most sympathetic impulses are carried along the splanchnic nerves. Parasympathetic impulses are carried by the **vagus nerve**.

Anatomy and Physiology of the Oral Cavity

The oral cavity or mouth serves as entry into the digestive tract. The mouth consists of the lips, teeth, tongue, and palate. The lips assist in directing food into the oral cavity. By separating the mouth and nasal cavity, the palate (roof of the mouth) allows for chewing, swallowing, and breathing to occur all at the same time. The tongue serves its primary role in moving food from the front of the mouth to the pharynx in preparation for swallowing. Major taste buds are also located on the tongue. The tongue and lips also play a primary role in speech.

Oral Cavity Motility After food is voluntarily placed in the oral cavity, teeth begin their work of mastication (chewing). The purposes of mastication are (1) to break food into smaller pieces, (2) to mix food with saliva, and (3) to stimulate taste buds. The tongue assists by moving food in place for chewing. When the jaw is closed, upper and lower teeth fit together to mash, grind, and tear food.

Oral Cavity Secretions The primary secretion in the oral cavity is saliva. Saliva is produced in the mouth by three pairs of salivary glands. These include the parotid, submandibular, and sublingual. Saliva is made of water (99.5%), electrolytes, and protein. Electrolytes include sodium chloride, bicarbonate, and potassium. Proteins include enzymes, mucus, and lysozyme. Lysozyme functions as part of the first level of defense for the immune system and is capable of destroying bacteria in the mouth.

One to two liters of saliva are produced each day. Both autonomic and acquired reflexes can stimulate salivation. An example of an autonomic reflex is production of saliva in response to the pressure exerted by the presence of food in the mouth. An acquired reflex is learned; for example, the "mouth waters" at the sight or smell of food (saliva is produced without oral stimulation).

Functions of saliva include (1) moistening and lubricating food to facilitate swallowing; (2) initiating digestion of carbohydrate; (3) providing antibacterial protection with lysozyme and by rinsing away food from the oral cavity; (4) enhancing taste by providing a solution that can interact with taste buds; (5) serving as a buffer—the pH of saliva is approximately 6.8, which neutralizes acids and protects the teeth from dental caries; (6) promoting oral hygiene by dissolving food, dead cells, and foreign substances; and (7) assisting speech by allowing free movement of the lips and tongue.[4]

After food is chewed and mixed with saliva, it is shaped into a sticky ball called a bolus. The bolus is moved from the front of the oral cavity to the back where swallowing is then initiated. Pressure of the bolus on the pharynx stimulates nerve impulses to the swallowing center. The swallowing center, located in the medulla, then stimulates the sequence of muscle actions that coordinate the swallow. Thus, the initiation of swallowing is voluntary, but thereafter swallowing is under autonomic control.

Normal Anatomy and Physiology of the Esophagus

The esophagus is a straight, hollow tube approximately 25 cm long and 2 cm in diameter. The esophagus has **sphincter** muscles at either end.

Walls of the esophagus consist of four layers of tissue (see Figure 14.2). The inner layer is the mucosa, which is made of stratified squamous epithelial cells. The next layer is the submucosa, which contains secretory cells that produce mucus to facilitate movement of the bolus during swallowing. The muscle layer consists of both longitudinal and circular muscles that coordinate movement of the food bolus by alternately contracting. This "squeezing" contraction easily moves the bolus down the esophagus. The outer layer (or adventitia) of tissue for the esophagus is connective tissue and has no additional outer covering.

The chief function of the esophagus is motility. Transporting the bolus of food from the oral cavity to the stomach is its primary task. Though it sounds simplistic, several disorders of the upper GI tract actually involve derangements in this task.

An individual unconsciously swallows over 600 times each day.[5] Each swallow is composed of four phases. The first phase, described previously, is the **oral preparatory phase** where food is chewed and mixed with saliva. The second phase or **oral transit phase** was also described previously and consists of voluntary movement of the bolus of food from the front of the oral cavity to the back. The third phase of swallowing is known as the **pharyngeal phase**. The most important part of this phase is to ensure the bolus is directed into the esophagus and is prevented from entering the trachea. This is initially accomplished when the uvula seals off the nasal passage so food does not leak into the nose. Next, laryngeal muscles contract and seal off the glottis (entrance to the larynx). The epiglottis also tilts upward to assist in preventing food from entering the larynx.

Figure 14.2 Layers of the Digestive Tract Wall

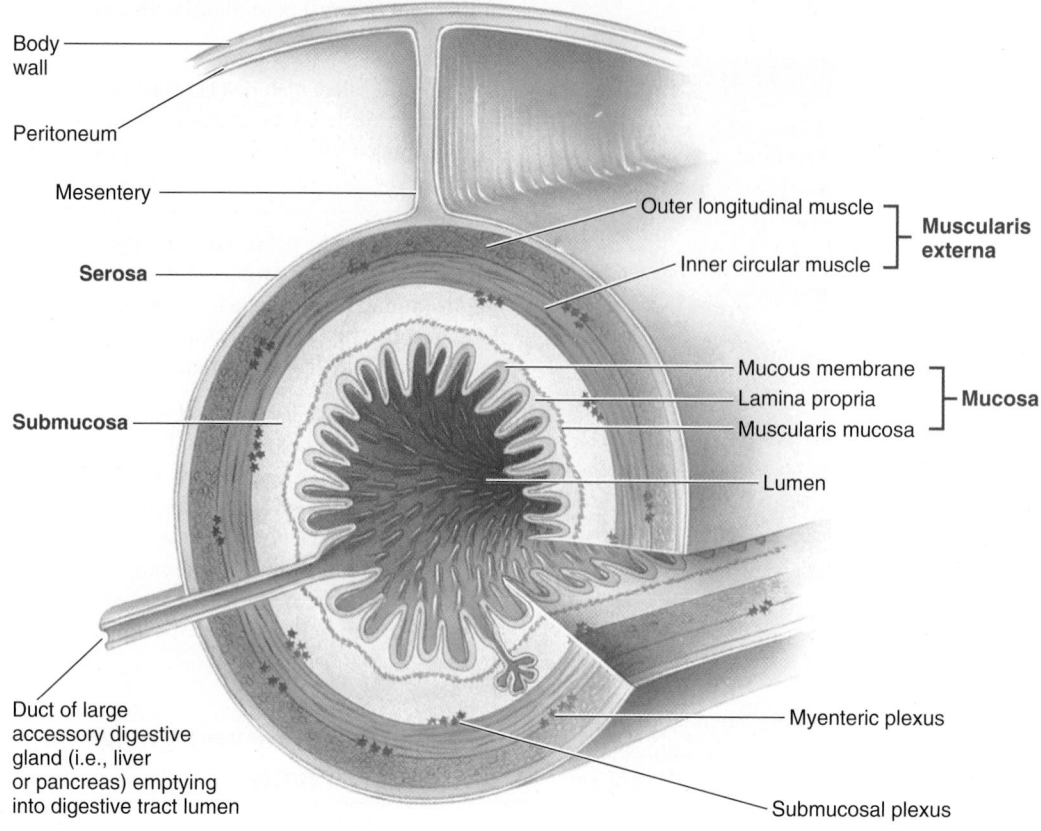

Source: Figure 16.2, p. 594, of Sherwood's *Human Physiology: From Cells to Systems* 7e.

The final phase of swallowing is the **esophageal phase**. The upper esophageal sphincter (UES) or pharyngoesophageal sphincter is located at the top of the esophagus. This sphincter, when open, allows the bolus to enter the esophagus. When the sphincter is closed, it prevents air from entering the GI tract during breathing. After the bolus of food moves through the UES into the esophagus, the sphincter closes and normal breathing will resume.

When the esophageal phase of swallowing begins, autonomic control initiates the peristaltic wave that moves the bolus of food down the esophagus into the stomach (see Figure 14.3). Mucus secreted by the esophagus lubricates the food bolus and aids in its successful passage. At the end of the esophagus, another sphincter muscle—the **lower esophageal sphincter (LES)**—controls release of the bolus from the esophagus into the stomach. This sphincter is closed except during swallowing. It serves as a barrier to protect the esophageal mucosa from stomach contents. Atmospheric pressure is greater in the esophagus than in the stomach under normal conditions. This positive pressure of approximately 30 cm H_2O assists in preventing stomach contents from refluxing back into the esophagus and preventing large amounts of air from entering the stomach. (Gastric pressure is approximately 10 cm H_2O.) Two major neurotransmitters are responsible for allowing the LES to relax and oppose the stimulatory action of acetylcholine. Nitric oxide and VIP inhibit closure of the LES, allowing it to relax so that the bolus slides from the esophagus into the stomach. The swallow is complete when the bolus

of food moves through the LES. Pharyngeal and esophageal phases of swallowing take only 6 to 10 seconds under normal conditions. If for some reason all food is not cleared from the esophagus, secondary peristaltic waves are initiated. This might occur when a sticky substance is eaten that does not move as readily down the esophagus. The individual, in this case, would usually be unaware of these secondary peristaltic waves.[6]

Figure 14.3 Peristalsis in the Esophagus

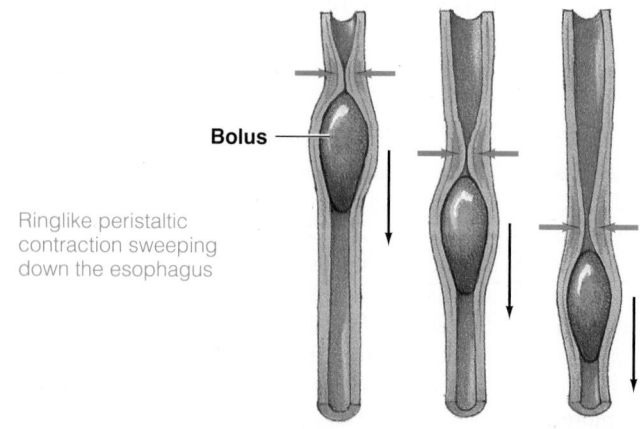

Bolus

Ringlike peristaltic contraction sweeping down the esophagus

Source: L. Sherwood, *Human Physiology: From Cells to Systems*, 5e, copyright © 2004, p. 603.

Normal Anatomy and Physiology of the Stomach

The final portion of the upper GI tract is the stomach (see Figure 14.4). This organ lies from left to right across the upper abdomen directly under the diaphragm. Portions of the stomach (fundus, corpus, antrum, and pylorus) differ by anatomy and function. Sphincters at both ends of the stomach regulate flow of foodstuffs from the esophagus to the small intestine. Major functions of the stomach include all four digestive processes: motility, secretion, digestion, and absorption.

Gastric Motility Gastric motility includes filling of the stomach, storage of foodstuffs, mixing with gastric juices, and finally emptying into the small intestine. When empty, the stomach's volume is only about 50 mL, but it can stretch to hold more than 1000 mL. Storage occurs primarily in the body (corpus) of the stomach. Mixing occurs in the antrum where muscle is much thicker and can accommodate the strong peristaltic waves. Each peristaltic wave moves foodstuffs toward the pyloric sphincter at the bottom of the stomach. Rate of gastric emptying through the pyloric sphincter into the upper portion of the small intestine is controlled by the anatomical structure of the stomach, nutrient content of the foodstuffs, the nervous system, and the influence of specific hormones.

Gastric Secretions The stomach secretes approximately 1 to 3 liters of gastric juices each day. Gastric juice is composed of water, mucus, hydrochloric acid, enzymes, and electrolytes.

The mucosa, which lines the fundus and the body of the stomach, contains gastric glands. Several different types of cells are located within the gastric glands. The mucous cells secrete mucus, which protects the lining of the stomach from mechanical or acid insult (see chapter endnote 1). Chief cells secrete the zymogen pepsinogen and the enzyme gastric lipase. Remember that a zymogen is an inactive enzyme. Pepsinogen, when activated, will begin protein digestion. Gastric lipase provides some preliminary digestion for lipids. **Parietal cells** secrete hydrochloric acid and intrinsic factor. Hydrochloric acid serves to activate pepsinogen, kill microorganisms, and denature proteins. Intrinsic factor is a protein necessary for the absorption of vitamin B_{12}.

Figure 14.4 Anatomy of the Stomach

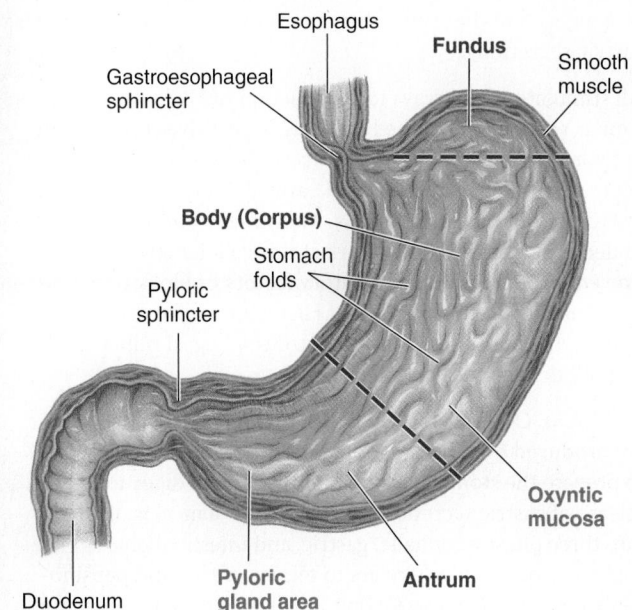

Source: Figure 16.7, p. 601, of Sherwood's *Human Physiology: From Cells to Systems* 7e (0495391840).

In the pylorus, enterochromaffin (ECL) cells secrete histamine, G cells secrete gastrin, and D cells secrete somatostatin. All three of these substances assist in overall control and production of gastric juices.

CONTROL OF GASTRIC SECRETIONS Control of gastric secretions is accomplished through complementary actions of nervous and endocrine systems, and involves four major chemical messengers: acetylcholine, histamine, and gastrin, which stimulate gastric secretions, and somatostatin, which inhibits gastric secretions (see Table 14.2). **Acetylcholine** is a neurotransmitter that stimulates parietal, chief, and ECL cells. **Histamine**, a paracrine, acts on parietal cells to increase hydrochloric acid (HCl) release. **Gastrin**, a hormone, stimulates chief and parietal cells as well as ECL cells to release histamine. **Somatostatin** works as an inhibitory paracrine

Table 14.2 Control of Gastric Secretions

The Stomach Mucosa and the Gastric Glands			
Type of Secretory Cell	**Product Secreted**	**Stimuli for Secretion**	**Function(s) of Secretory Product**
Exocrine cells			
Mucous cells	Alkaline mucus	Mechanical stimulation by contents	Protects mucosa against mechanical, pepsin, and acid injury
Chief cells	Pepsinogen	Acetyl choline (ACh), gastrin	When activated, begins protein digestion
Parietal cells	Hydrochloric acid	ACh, gastrin, histamine	Activates pepsinogen, breaks down connective tissue, denatures proteins, kills micro-organisms
	Intrinsic factor		Facilitates absorption of vitamin B_{12}
Endocrine/paracrine cells			
Enterochromaffi-like (ECL) cells	Histamine	ACh, gastrin	Stimulates parietal cells
G cells	Gastrin	Protein products, ACh	Stimulates parietal, chief, and ECL cells
D cells	Somatostatin	Acid	Inhibits parietal, G, and ECL cells

Source: L. Sherwood, *Human Physiology: From Cells to Systems,* 7th ed., copyright © 2010, p. 606.

by providing negative feedback to the stimulatory pathways. When gastric pH falls (becomes more acidic), somatostatin acts on each of the stimulatory mechanisms to slowly decrease gastric secretions.

All stimulatory pathways for production of HCl work in similar ways. The enzyme H^+, K^+-ATPase drives production of hydrogen ions (H^+). Gastrin, acetylcholine, and histamine act to increase the amount of H^+ available for the formation of HCl within the parietal cells. Many medications designed to decrease gastric acidity work at this cellular level to either prevent production of stimulatory factors or block transport of H^+ needed for production of HCl. This will be discussed in more detail in the sections on gastroesophageal reflux disease and peptic ulcer disease.

RELEASE OF GASTRIC SECRETIONS Gastric secretions are produced even before food enters the stomach and serve to prepare the stomach for its eventual role in digestion. Thus, release of gastric secretions in response to a meal is divided into three phases: cephalic, gastric, and intestinal phases. The cephalic ("head") phase refers to release of HCl and pepsinogen when stimulated by tasting, smelling, or even seeing food. The gastric phase begins when food enters the stomach. Both cephalic and gastric phases are stimulatory phases, meaning they result in production of gastric secretions. Several factors contribute to stimulation of gastric juices, including presence of protein, distention of the stomach, caffeine, and alcohol. Stimulation of HCl by alcohol and caffeine may be damaging if food is not present in the stomach. Thus, alcohol and caffeine may be restricted for persons with peptic ulcer disease (PUD) or gastroesophageal reflux disease (GERD). (See "Nutrition Therapy—Gastroesophageal Reflux Disease" for further discussion.)

In contrast to cephalic and gastric phases that stimulate gastric juices, the intestinal phase is inhibitory in that it slows gastric secretions and prepares the small intestine for receipt of the acidic **chyme**. Distention of the stomach, accumulation of acid, presence of fat, and increasing **osmolality** of partially digested food result in release of two major hormones: cholecystokinin and secretin. These hormones act on the smooth muscle of the antrum to slow gastric motility. Additionally, somatostatin is released from the antrum of the stomach. Together these chemical messengers inhibit action of the parietal and chief cells, which reduces the amount of gastric juices.

Gastric Digestion and Absorption

DIGESTION IN THE STOMACH Digestion in the stomach is both mechanical and chemical in nature. Mechanical digestion occurs as contractions of the stomach shear the foodstuffs and mix the bolus of food with gastric juices. HCl causes proteins to unravel or denature, allowing for more efficient digestion. Additionally, HCl converts pepsinogen to the active enzyme, pepsin. Pepsin works to cleave amino acids, di- and tripeptides from inner portions of the protein, and results in shorter chains of amino acids called peptones. Of the three macronutrients, proteins are subjected to the most active chemical digestion in the stomach; carbohydrate and lipid digestion are fairly limited. Gastric lipase preferentially hydrolyzes fatty acids at the 3rd carbon position, releasing fatty acids and diglycerides. Short- and medium-chain fatty acids are more easily hydrolyzed than long-chain fatty acids in the stomach. Carbohydrate digestion that began in the oral cavity is reduced because salivary amylase is inactivated by HCl. A small amount of carbohydrate digestion may continue in the inner portions of the food bolus, which are not exposed to the acid.

ABSORPTION IN THE STOMACH Absorption in the stomach is limited; no food and only a small amount of water are absorbed. Exceptions include alcohol and some medications. Alcohol can be absorbed in the gastric mucosa and enters the bloodstream through capillaries of the stomach. Because the presence of food in the stomach slows this process considerably, consuming food prior to or with alcohol will slow systemic effects of alcohol ingestion. Most medications are absorbed in the small intestine, but acetylsalicylic acid (aspirin) is readily moved across epithelial cells of the gastric mucosa.

The gastric mucosa is well protected against injury. Anatomical design provides one of the strongest sources of protection. Cell membranes of the gastric mucosal cells are impermeable to hydrogen ions, so the increasing acidity cannot harm them. Secretion of mucus by the cells also provides a layer of protection. Furthermore, gastric mucosal cells have a high turnover rate and are sloughed off approximately every 72 hours—too short a lifespan for these cells to experience any significant damage from acidic conditions of the stomach. Still, if for some reason injury does occur, the high acidity will cause tissue to erode. This tissue erosion occurs in peptic ulcer disease and will be described later in this chapter.

Pathophysiology of the Oral Cavity

In this section, general problems associated with mouth, teeth, and gums will be discussed. These are of nutritional concern because most of these conditions involve problems that interfere with adequate oral intake and can lead to nutritional deficits.

Conditions affecting oral health include the most common dental diseases: dental caries, gingivitis, and **periodontal disease**. Over the past several decades, the U.S. population has seen significant improvement in prevalence of most dental diseases. According to the report *Oral Health in America*, both the incidence and prevalence of dental caries has decreased considerably.[7] Even with these improvements, dental disease remains one of the most common health problems in our country. Approximately all children in third grade have had at least one cavity or filling in either a primary or a permanent tooth.[8] This number is approximately 80% to 85% for adults. Differences in the prevalence of untreated dental caries exist in association with race/ethnicity and poverty level. For all age groups, non-Hispanic blacks and Mexican Americans have higher numbers of untreated decayed teeth than non-Hispanic Caucasian individuals.[8] According to the National Oral Health Surveillance System, approximately 46% of all Americans over the age of 65 have lost six or more teeth.[9]

Dental Caries

Dental caries or tooth decay is destruction of tooth structure through development of small pits or cavities. The cause of dental caries is multifactorial. Individual factors may include structure of the tooth, immunologic response to bacteria, and composition of saliva in the mouth. It is currently understood that several different types of bacteria commonly found in the mouth, including *Streptococcus mutans*, are involved in development of dental caries. Steps in caries development can be described as follows: after dietary carbohydrates are hydrolyzed by salivary amylase, bacteria preferentially will use those simple sugars as fuel and produce acids as a waste product. As a result, pH falls. The lowered pH causes demineralization of the tooth and ultimately results in decay of tooth **enamel** and **dentin**.[10] Foods that contribute to this lowered pH are referred to as **cariogenic**, those that raise the oral pH are referred to as **anticariogenic**, and those that have no effect are **cariostatic**.[11] Table 14.3 outlines foods that are consistent with each of these definitions. Foods that are sticky adhere to the surface of the tooth and also are more likely to cause caries. Frequent snacking increases the time acids are in contact with the tooth and additionally increases risk of caries. The use of liquids with eating helps clear the mouth of food debris and can reduce the potential for caries.

The combination of acids, bacteria, food, and saliva make up biofilm, a sticky substance commonly called **plaque**. Plaque adheres to the tooth and, if it is not removed, will mineralize into tartar or **calculus**. Factors that contribute to development of caries and plaque include frequency of eating, stickiness of the food, composition of saliva, presence of buffers, and overall oral hygiene.[12]

Cavities are usually painless until they are large enough to destroy internal structures of the tooth. Dental caries can also lead to death of the nerve and blood vessels in the tooth.

Individuals at Risk for Dental Disease

As stated earlier, in 2004, it was estimated that 46% of all Americans over the age of 65 have lost at least six permanent teeth. Dental caries, tooth loss, and ill-fitting dentures may contribute to inadequate or improper intake of nutrients. At the same time, malnutrition and weight loss may contribute to the poor fit for dentures and loss of teeth. This may be a significant issue for the elderly, as it is estimated that one-third of those persons over 65 years are **edentulous**. Individuals who have difficulty

Table 14.3 Foods Associated with Dental Health (ADA 2007)

Description	Examples
Cariogenic foods/drinks containing fermentable carbohydrates that can cause a decrease in salivary pH to <5.5	Foods high in simple sugars such as candy, cake, pies, cookies, soft drinks, fruit juices, sweetened beverages
	Honey, syrup, jelly, white bread, crackers, pretzels, chips, sweetened cereals
	Fruits—fresh, canned, frozen, and dried
Cariostatic foods that are not metabolized by microorganisms in plaque; do not subsequently cause a drop in salivary pH to <5.5 within 30 minutes	Protein foods such as eggs, fish, beef, chicken, pork, nuts, and seeds
	Vegetables
	Popcorn
Anticariogenic foods or components of foods that can raise salivary pH to an alkaline level and protect the enamel	Cheese, milk, nuts
	Xylitol-containing chewing gum and mints

chewing may rely on soft foods with limited variety, resulting in an inadequate nutrient intake.[13]

Other medical conditions may be associated with increased dental caries. Eating disorders are frequently linked with dental disease. In bulimia nervosa, for example, the exposure to gastric juices during repetitive vomiting contributes to the demineralization of the teeth.[14]

Infants and children are considered to be at high risk for dental disease. "Baby bottle tooth decay" can occur when an infant falls asleep with a bottle containing a high-sugar beverage such as fruit juice or infant formula. During sleep, the beverage may pool in the mouth, contributing to widespread caries. Cavities usually form on upper front teeth and back molars. It is recommended that bottles be removed from a child's mouth during sleep and that infants be weaned from a bottle starting at 12 months of age. Other preventions include brushing a child's teeth as soon as they erupt, or wiping them with a wet cloth. A child's first dental examination should occur by the end of his or her first year.[11]

Prevention of Dental Disease

Prevention of dental caries focuses on both nutrition and public health interventions. Fluoridation of water supplies, use of topical fluoride treatments, and use of dental sealants are primary methods of preventing dental caries. Fluoride ingested when teeth are developing is incorporated into the structure of the enamel and protects it against the action of acids. Topical fluoride and sealants also provide protection for the outer covering of the tooth.[15] Nutritional choices can play a role in the prevention of dental disease. As described earlier, some foods have been identified as anticariogenic because they serve as a buffer that prevents decreases in pH and resulting demineralization.[16, 17] Sugar alcohols, such as xylitol, are used to sweeten products such as chewing gum, toothpaste, and mouthwashes. These carbohydrates are not metabolized by bacteria in the mouth, and current research indicates that xylitol may have an additional antimicrobial effect against dental caries.[18–20] The following recommendations can assist in the prevention of dental caries:

- Choose fiber-rich vegetables and whole grains often.
- Choose and prepare foods and beverages with little added sugars or caloric sweeteners, such as amounts suggested by the USDA Food Guide and the DASH Eating Plan.
- Rinse mouth after meals and snacks.
- Chew sugarless gum with xylitol after meals.
- Reduce the incidence of dental caries by practicing good oral hygiene and consuming sugar- and starch-containing foods and beverages less frequently.[11, 20, 21]

Inflammatory Conditions of the Oral Cavity

It is not uncommon for patients to suffer from inflammatory conditions of the mouth. When the mouth is inflamed or infected, maintaining oral intake is very difficult. Inflammatory conditions may result from poor dental hygiene and lack of dental care, and can also be seen in persons who are immunosuppressed or who have undergone chemotherapy or radiation therapy. Gingivitis is an inflammation of the **gingiva** (gums). Gums appear red and swollen and often bleed. The tissue is very tender and painful. Other symptoms include fever, loss of appetite, foul breath, and a bad taste in the mouth.

Table 14.4 Nutrition Intervention: Stomatitis

- Prevention is key—good oral hygiene with frequent mouth rinses is important (avoid alcohol-based products).
- Treat oral lesions pharmacologically as appropriate (antifungal medications if needed).
- Consider using oral topical agents and anesthetics, such as viscous lidocaine and institution-specific mouth rinses, which are combinations of nystatin, Maalox® (Novartis, Parsippany, NJ), diphenhydramine, hydrocortisone, and viscous lidocaine.
- Adjust texture and temperature as tolerated. Extremes of very hot or cold are often not tolerated.
- Avoid carbonated beverages.
- Avoid caffeine, alcohol, and tobacco products.
- Avoid other irritants (e.g., acidic, spicy foods).
- Try oral glutamine supplementation—optimal dose is 10 g three times daily.
- Consider enteral nutrition support if unable to maintain nutritional status orally.

Source: Adapted from "Appendix C. Nutrition Impact Symptoms and Interventions" (pp. 348–49), in V.J. Kogut and S.K. Luthringer (Eds.), *Nutritional Issues in Cancer Care*, 2005, Pittsburgh, PA: Oncology Nursing Society. Copyright 2005 by the Oncology Nursing Society. Reprinted with permission.

Stomatitis or mucositis is inflammation of the oral mucosa and is often associated with fungal infections such as *Candida albicans* or with herpes-like viruses. It is common in stomatitis to have open ulcerations on the oral mucosa, gingiva, and palate. Nutritional strategies for individuals with stomatitis are presented in Table 14.4.[22]

Glossitis and **cheilosis** are inflammatory symptoms of the oral cavity classically associated with vitamin deficiencies. Glossitis involves increased redness, swelling, and pain of the tongue and lips. Cheilosis is fissuring and scaling at the corners or angles of the mouth. Both cheilosis and glossitis may be a result of riboflavin, niacin, or pyridoxine deficiency.

Conditions Resulting in Altered Salivary Gland Function

The importance of saliva production in ensuring normal chewing and swallowing is emphasized above (see Oral Cavity Secretions). In many clinical conditions and with the use of numer-

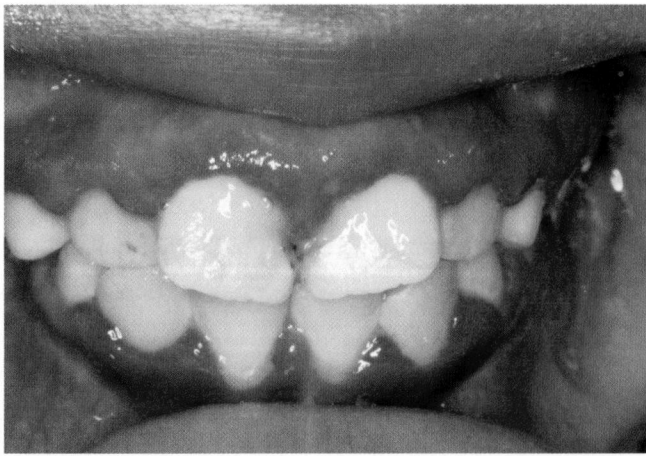

Stomatitis

Copyright © Gill/Custom Medical Stock Photo

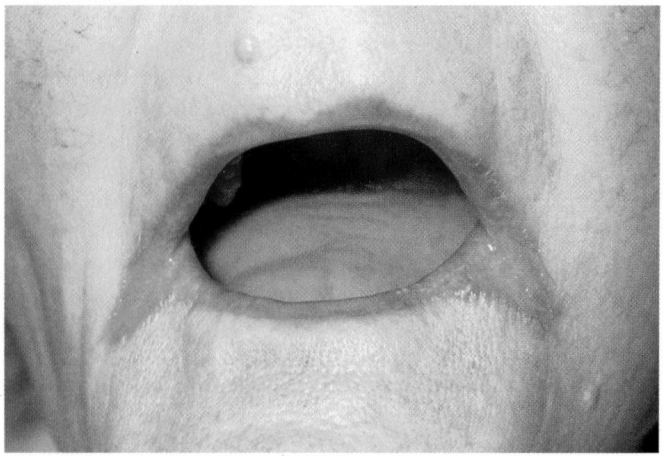

Cheilosis

Copyright © Gill/Custom Medical Stock Photo

ous medications, saliva production may be altered, resulting in decreased oral intake and ultimately placing the patient at nutritional risk.

Xerostomia may occur as a result of a disease process, but can also be a result of medical treatment (see Table 14.5). Infection or damage to salivary glands through surgical resection or radiation therapy can interrupt normal saliva production. Blockage of the salivary ducts may also occur as a result of a tumor or other medical condition. Systemic changes such as those seen in **Sjögren's syndrome** or with dehydration can also change saliva production. Many medications cause a reduction of saliva by affecting the parasympathetic nervous system. For instance, medications called anticholinergics, which act to block the effect of acetylcholine, reduce the amount of saliva as a major side effect. Groups of medications that have anticholinergic effects include antihistamines and antidepressants. Table 14.6 outlines interventions to assist with xerostomia.

Excessive saliva production may also be a concern, but in general does not pose the nutritional problems that xerostomia can. Nervous system diseases such as Parkinson's disease may interfere with autonomic control of saliva production. Other situations that increase saliva production, such as seeing, tasting, or smelling food or the consumption of sour-tasting food, involve the autonomic reflex.

Table 14.5 Possible Causes of Xerostomia

Side effects of some medicines	More than 400 medicines can cause the salivary glands to make less saliva. Medicines for high blood pressure and depression often cause dry mouth.
Disease	Some diseases affect the salivary glands, for example, Sjögren's syndrome, HIV/AIDS, diabetes, and Parkinson's disease.
Radiation therapy	The salivary glands can be damaged if they are exposed to radiation during cancer treatment. Consistency of saliva may also change.
Chemotherapy	Drugs used to treat cancer can make saliva thicker, causing the mouth to feel dry.
Nerve damage	Injury to the head or neck can damage the nerves to the salivary gland.

Table 14.6 Nutrition Interventions: Xerostomia

- Try tart foods to stimulate saliva.
- Sip on liquids or suck on ice chips throughout the day.
- Avoid caffeine, alcohol, and tobacco products.
- Try using a cool mist humidifier at bedtime.
- Try drinking through a straw.
- Rinse mouth frequently with mild saline solution.
- Add extra sauces and gravies to foods.

Source: From "Appendix C. Nutrition Impact Symptoms and Interventions" (pp. 348–49), in V.J. Kogut and S.K. Luthringer (Eds.), *Nutritional Issues in Cancer Care*, 2005, Pittsburgh, PA: Oncology Nursing Society. Copyright 2005 by the Oncology Nursing Society. Reprinted with permission.

Surgical Procedures for the Oral Cavity

Surgical resections of the tongue, palate, or pharynx occur with head/neck malignancies. Fractures of the mandible (lower jaw) are a common consequence of accidents, trauma, and certain disease states. Repair of these fractures requires surgery and immobilization to ensure adequate healing. Jaw fractures are described as follows: open (exposed to open area of mouth or skin); closed (within tissue); complete (in two separate pieces); incomplete (splintered); and comminuted (multiple pieces). The type of fracture, as well as other medical and psychosocial factors, will determine the type of surgery that the patient will have and, ultimately, the type of nutrition therapy that is required. Surgical repair includes **reduction and fixation of the fracture**. Fixation will include either an external or internal use of the upper jaw to hold the lower jaw in place while healing occurs. This is what is commonly referred to as "wiring the jaw," or maxillomandibular fixation.[23]

These procedures require significant nutrition interventions to ensure maintenance of nutritional status. Nutritional intake is limited to those liquids and blenderized foods that can be put through a straw or syringe (see Table 14.7—Nutrition Therapy for Jaw Fracture).[24]

Nutrition Assessment and Diagnosis Since this procedure severely limits the type of foods the patient can consume, a past diet history may not be that useful other than as a means to obtain a sense of food likes and dislikes. Depending on the age of the patient, interviewing a parent or caregiver helps the clinician to evaluate behaviors associated with food and meal preparation and social network and support. Common nutritional diagnoses associated with a jaw fracture may include inadequate oral food/beverage intake; inadequate fluid intake; increased nutrient needs; biting/chewing difficulty; involuntary weight loss; and food and nutrition knowledge-related deficit.

Nutrition Intervention Nutrition interventions will center around modification of texture and consistency, increasing the nutrient density of nutritional intake, and nutrition education to ensure adequacy and safety of food preparation. All foods must be prepared so that each easily moves through a straw or syringe. Table 14.7 summarizes the blenderized diet recommended for a jaw fracture. If the patient is unable to maintain their nutritional intake through these interventions or if the nutritional status of the patient warrants, enteral feeding may be recommended.[25–27]

Impaired Taste: Dysgeusia/Ageusia

Dysgeusia is the condition of altered or impaired sense of taste. **Ageusia** is the inability to taste or "mouth blindness." Many clinical conditions affect the ability to taste. For example, patients undergoing treatment for cancer relate changes in both taste and smell.[28] Epithelial cells of the oral mucosa have a high turnover rate and can be affected by either chemotherapy or radiation therapy (see Chapter 23). Diseases of the tongue and palate can interfere with normal function of taste buds. Nervous system diseases can also affect transmission of sensory information. Many patients relate changes in taste that are associated with certain medications.[29] For example, the use of methotrexate (a common medication used in treatment of cancer and autoimmune disease) can cause a strong metallic taste.

BOX 14.1 CLINICAL APPLICATIONS

The Nutrition Care Process for Jaw Fracture

Nutrition Diagnosis: Chewing Difficulty

PES: Chewing difficulty related to recent jaw fracture and postoperative status for maxillomandibular fixation as evidenced by inability to consume food of normal texture and consistency.

Sample Intervention

1. Modify texture, viscosity, and consistency of all foods to ensure passage of blenderized diet through opening between teeth after surgical repair.

2. Provide nutrition education regarding selection, preparation, and storage of foods and supplements, and selection and use of feeding equipment.

3. Recommend liquid multivitamin daily.

4. Maintain adequate hydration by consuming a minimum of 3000 mL of fluid daily.

Goals

1. Client will maintain weight within 3 lbs. of pre-surgical weight during consumption of blenderized diet.

2. Client's 24-hour recall will demonstrate consumption of target kilocalories and protein.

Monitoring and Evaluation

1. Evaluate patient for weight status

2. Evaluate tolerance and acceptance of blendarized diet

3. Calorie counts and/or food diary to evaluate adequate intake

Table 14.7 Nutrition Therapy for Jaw Fracture and Blenderized Diets

General recommendations:

- Prepare all foods to a very smooth and thin consistency.
- Cut foods into small pieces and place in a strong blender or food processor.
- Commercially prepared baby food is available to use (do not use "junior" foods), but you will need to experiment with herbs and spices to add flavor.

Adding extra protein:

- Dry milk powder—this can be added to thin cooked cereal, cocoa, eggnog, blenderized soups and casseroles, milk shakes, pudding, and gravy.
- Double strength milk—add 1 cup dry milk powder to 1 cup whole milk. Use this as a beverage or in place of milk in recipes.
- Liquid egg substitutes—Add EggBeaters or other pasteurized egg substitutes.
- Blenderized meats.
- Cottage cheese.
- Yogurt, smooth blended.
- Commercial liquid nutritional supplements such as Ensure, Boost, or Carnation Instant Breakfast.

To add extra kcalories:

- Use whole milk, half and half, or evaporated milk in place of water as the fluid when blenderizing.
- Add extra butter or margarine to blenderized vegetables, soups, or thin cereals. Add gravy to blenderized meats.
- Add sugar or honey to blenderized fruits, squash, pumpkin, or carrots.

Food Group	Recommended Foods	Food Group	Recommended Foods
Milk and Milk Products	All	Fruits	Any cooked or canned fruits without seeds and skins
			Fresh, peeled soft fruits (like peaches and bananas) that can be blended until smooth
Meat and Other Protein Foods	Tender, well-cooked meat, poultry, or fish prepared without bones, skin, or added fat	Fats and Oils	Any oils
	Well-cooked eggs prepared without added fat		Melted butter or margarine
	Soft soy foods (like tofu)		
	Smooth nut butters		
Grains	Rice	Beverages	Any
	Pasta		Look for liquid supplements that provide both calories and protein, such as Carnation Instant Breakfast, Boost, or Ensure
	Couscous without seeds and nuts		
	Cooked cereals, such as oatmeal and cream of wheat		
Vegetables	Any cooked or canned vegetables without seeds and skins	Other	Ground spices, seeds, and nuts
			Smooth condiments, such as mustard

Sample 1-Day Menu			
Breakfast	Mix well in blender: 1 cup cooked oatmeal 1 cup 2% milk 1 teaspoon cinnamon 1 teaspoon brown sugar	**Midafternoon Snack**	8–12 ounces liquid supplement, such as Carnation Instant Breakfast, Boost, or Ensure
Midmorning Snack	Eggnog—Mix well in blender: 1 cup liquid pasteurized eggs (such as Egg Beaters) 1 very ripe banana 1 cup 2% milk 1 tablespoon vanilla 1 teaspoon nutmeg	**Evening Meal**	Mix well in blender: 1 cup pasta sauce 2 ounces tender, cooked ground beef 1 cup cooked pasta Enough vegetable juice to make the blended food drinkable
Lunch	Mix well in blender: 2 cups cream of chicken soup ¼ cup tender-cooked chicken ½ cup mashed potatoes ½ cup cooked vegetables 1 cup orange mango juice	**Bedtime Snack**	2 cups ice cream blended with 1 cup milk

Source: Nahikian-Nelms M. Nutrition Therapy for Fractured Jaw. In: Gastrointestinal Disease. Nutrition Care Manual. Chicago, IL: American Dietetic Association, 2009. Used with permission.

Nutrition Therapy for Pathophysiology of the Oral Cavity

Nutritional Implications The primary nutrition problem for diseases involving the oral cavity is the inability to maintain adequate oral intake. Foods may be difficult to swallow due to lack of saliva, thickened saliva, or the pain of an inflammatory condition or dental disease. When saliva production or quality is impaired, there is an additional risk of infection and dental caries. Dysgeusia can reduce overall oral intake due to changes in appetite. If appropriate medical and nutrition interventions do not occur, and without subsequent improved dietary intake, the patient can become malnourished.

Nutrition Diagnosis Common nutrition diagnoses associated with pathology of the oral cavity include inadequate oral food/beverage intake; inadequate fluid intake; inadequate protein-energy intake; biting/chewing difficulty; swallowing difficulty; and involuntary weight loss.

Nutrition Intervention Nutrition intervention strategies will primarily be focused on modification of the distribution, type or amount of foods provided. Texture modification of the current diet will be necessary for most of these clinical conditions. Changing the diet to include soft, moist foods, liquids, or blenderized foods will allow for increased intake. Use of gravies, sauces, and soft casseroles should be encouraged, while foods that are dry, crunchy, or have sharp edges should be avoided. Liquids should be consumed with meals.

Medical food supplements may also be appropriate. Kcalorie and protein density can be increased by using modular components (such as dry skim milk powder) and high-kcalorie, high-protein liquid supplements (such as Ensure®, Boost®, or milk shakes), and by increasing fat intake (serving cream soups or adding margarine to mashed potatoes). Increasing frequency of meals may also allow for overall improved intake. Many patients may need to be prescribed a general multivitamin supplement. If specific deficiencies are confirmed, then appropriate supplementation can proceed.

Extreme temperatures and spices in foods may increase pain and intolerance to oral intake. Foods at room temperature and foods that are bland in flavor such as custards, yogurt, or pudding are generally well tolerated.

Feeding assistance in terms of mouth care may also be beneficial. Liberally using fluids will increase moisture in the mouth and assist with increasing solid food intake. Ensure adequate

---

BOX 14.2 CLINICAL APPLICATIONS

Nutrition Assessment of the Upper GI Tract

Client History

Personal history: Socioeconomic status/food security; support systems; education level—primary language

Medical and health history for client and family: Diagnoses; medications; previous medical conditions or surgeries

Anthropometric Measurements

Height; current weight; weight history: highest adult weight, usual body weight, reference weight (BMI)

If other areas of the nutrition assessment indicate that the patient may be at nutritional risk, other anthropometric measures would be assessed to substantiate the presence of malnutrition.

Biochemical Data

Laboratory measures of protein status are evaluated to establish any visceral protein deficit. Other laboratory indices would be assessed depending on the underlying medical condition, such as those that might be seen in dehydration and/or anemia: Albumin; prealbumin; hemoglobin; hematocrit.

Nutrition-Focused Physical Findings

Physical assessment of the head, neck, and oral cavity is a standard component of physical assessment. It is a crucial component of nutrition assessment for conditions of the oral cavity. Review results of swallowing evaluations. Assess for clinical symptoms of malnutrition; identify signs and symptoms that may be associated with dehydration. These may include increased pulse and orthostatic hypotension. Physical exam may reveal: weight loss, lethargy, sunken eyes, absence of tears, dry mucous membranes, dry skin, decreased capillary refill, and decreased skin turgor. Patient interview can confirm actual fluid intake. (See Chapters 3 and 7.)

Food/Nutrition-Related History

Dietary assessment will focus on determining the changes in oral intake that have occurred due to the disease condition. Methods to obtain this information can be a 24-hour recall, diet history, or direct observation of the patient's intake. Evaluate tolerance to different textures and consistencies. Evaluate for supplement intake.

- Ability to chew; use and fit of dentures; problems swallowing
- Nausea, vomiting; constipation, diarrhea; heartburn
- Any other symptoms interfering with ability to ingest normal diet
- Ability to feed self; ability to cook and prepare meals
- Food allergies, preferences, or intolerances: spicy foods, high-fat foods, pepper, caffeine, coffee, tea, alcohol, spearmint, peppermint, chocolate
- Previous food restrictions
- Ethnic, cultural, and religious influences
- Use of alcohol, vitamin, mineral, herbal, or other type of supplements
- Previous nutrition education or nutrition therapy
- Eating pattern: 24-hour recall, food history, food frequency

hydration at all times. Spraying the mouth with water or sucking on ice chips may be helpful. Sugar-free beverages containing citric acid (lemonade, etc.) may stimulate saliva production. Though usually a temporary intervention, using sugar-free gum or mints can also help. Cold and frozen foods are sometimes preferred. In extreme conditions, artificial saliva can be used.

Oral hygiene is an important part of nutritional care. Frequently rinsing the mouth to remove food particles can help prevent a bad taste in the mouth and an increase in dental caries. Alcohol-containing mouthwashes tend to dry the mouth, so other choices such as lemon-glycerine solutions or warm water with baking soda can be used. Using ¼ tsp of baking soda in 1 cup of water is recommended.

If pain is severe, coordinating pain medication with oral intake is important. Oral agents that provide localized numbing can be used, but in severe inflammatory conditions, systemic pain medications are frequently prescribed.

Monitoring and Evaluation Oral intake can be monitored by observation, a kcalorie count, or a food diary. Adequacy of intake is compared to estimated kcalorie and protein requirements. Every effort should be made to meet the patient's food preferences and tolerances. Success of nutrition interventions will be measured by weight gain or weight maintenance and the evaluation of biochemical parameters.

Pathophysiology of the Esophagus

This section will discuss the most common diseases or conditions involving the esophagus, including gastroesophageal reflux disease (GERD), Barrett's esophagus, achalasia, hiatal hernia, and dysphagia.

Gastroesophageal Reflux Disease (GERD)

Each year, more than 20 million Americans suffer daily symptoms of gastroesophageal reflux, while more than 100 million suffer occasional symptoms.[30] **Gastroesophageal reflux disease (GERD)** occurs as a result of reflux of gastric contents into the esophagus. As discussed earlier in this chapter, the lower esophageal sphincter (LES) normally serves as a barrier between the esophagus and stomach. Under normal conditions, atmospheric pressure is greater in the esophagus than in the stomach, and this pressure differential prevents reflux of gastric contents. The signs and symptoms experienced with GERD are attributed to a reflux of gastric acid and pepsin and occur during transient LES relaxations. These periods occur outside of the swallowing period and may be stimulated by the presence of food in the stomach after a meal.[31–33] The etiology of the reflux is multifactorial and can include both physical and lifestyle factors. Factors that can lower LES pressure and thus contribute to LES incompetence include (1) increased secretion of the hormones gastrin, estrogen, and progesterone; (2) presence of other medical conditions, such as hiatal hernia, scleroderma, or obesity; (3) cigarette smoking; (4) use of medications, including dopamine, morphine, and theophylline; and (5) specific foods. Foods high in fat, chocolate, spearmint, peppermint, alcohol, and caffeine all may decrease LES pressure.[29, 34–36]

Symptoms of GERD may include **dysphagia** (difficulty swallowing), heartburn, increased salivation, and belching. In some situations, pain is severe and may radiate to the back, neck, or jaw. In fact, pain from GERD can be confused with pain that is cardiac in origin because of the diffuse spread of pain into these other areas. For many patients, pain is worse at night when they are lying down. Complications of untreated or unresponsive GERD may include impaired swallowing, aspiration of gastric contents into the lungs, ulceration, and perforation or stricture of the esophagus. Barrett's esophagus is also considered to be a complication of GERD (see "Barrett's Esophagus—A Complication of GERD").

Treatment for GERD includes three major goals: (1) increasing LES competence; (2) decreasing gastric acidity, and thus decreasing the severity of symptoms; and (3) improving clearance of contents from the esophagus. Surgical intervention may be warranted if the disease is unresponsive to medical management or if the patient experiences complications.[31–33, 37–40]

Lifestyle factors that compromise LES competence such as cigarette smoking, use of medications, obesity, and nutritional history should be addressed.[41] Patient education can focus on ways to improve clearance in the esophagus; for example, patients may be instructed to remain upright after eating, to lose weight, to wear loose-fitting clothing, and to raise the head of their bed for sleeping.[42–44]

Decreasing gastric acidity involves use of medications and nutrition therapy. Medications fall into five major categories: (1) antacids or buffering agents, (2) histamine blocking agents, (3) prokinetic agents, (4) proton pump inhibitors, and (5) mucosal protectants. Gastric secretions are controlled in several different ways (see "Control of Gastric Secretions") and medications used to treat GERD interfere with control of gastric secretions by blocking several of those control pathways.[45] See Table 14.8 for a summary of these medications.

The surgical procedure used for GERD is **fundoplication** (see Figure 14.5). This procedure takes the fundus of the stomach and wraps it around the lower esophagus. This provides additional strength to the LES and assists in preventing the reflux. This procedure can be done **laparoscopically**, which avoids an abdominal incision and considerably reduces the recov-

Figure 14.5 Fundoplication

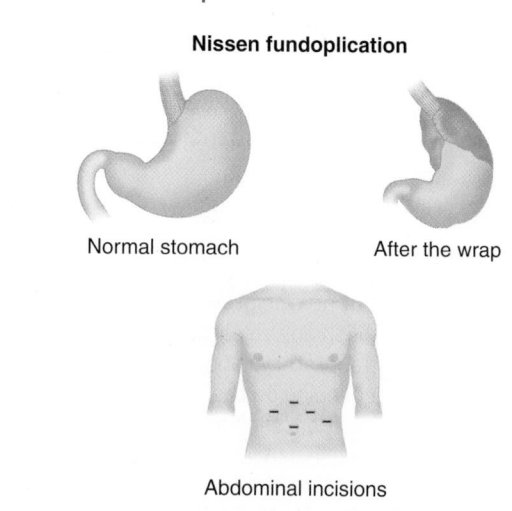

Nissen fundoplication

Normal stomach   After the wrap

Abdominal incisions

Source: www.mayoclinic.org/gerd/refluxsurgery.html

Table 14.8 Medications for Treatment of GERD

Classification of Medication	Generic and/or Trade Names and Description	Precautions
Antacids	These medications use different combinations of three basic salts—magnesium, calcium, and aluminum—with hydroxide or bicarbonate ions to neutralize HCL. Examples include Alka-Seltzer, Maalox, Mylanta, Pepto-Bismol, Rolaids, and Riopan.	Antacids may have side effects. Magnesium salt can lead to diarrhea, and aluminum salts can cause constipation. Aluminum and magnesium salts are often combined in a single product to balance these effects. Calcium carbonate antacids, such as Tums, Titralac, and Alka-2, can also be a supplemental source of calcium. They can cause constipation as well.
Foaming agents	These are also combinations of aluminum, magnesium, and sodium bicarbonate. Reduce the symptoms associated with reflux. Examples include Gaviscon, Foamicon.	
H$_2$ Antagonists	These medications block histamine receptors that are a component of one of the stimulatory paths for acid secretion. Medications available include: cimetidine (Tagamet HB), famotidine (Pepcid AC), nizatidine (Axid AR), and ranitidine (Zantac 75). They are available in prescription strength and over-the-counter.	These drugs provide short-term relief, but over-the-counter H$_2$ antagonists should not be used for more than a few weeks at a time. They are effective for about half of those who have GERD symptoms.
Proton pump inhibitors	Proton pump inhibitors block the H$^+$, K$^+$-ATPase enzyme, a component in HCL production. They include omeprazole (Prilosec, Zegerid), lansoprazole (Prevacid), pantoprazole (Protonix), rabeprazole (Aciphex), and esomeprazole (Nexium), which are all available by prescription. Over-the-counter strength Prilosec is also available.	
Prokinetics	Help strengthen the pyloric sphincter and increase speed of gastric emptying. This group includes bethanechol (Urecholine) and metoclopramide (Reglan).	These drugs may have frequent side effects that can limit their usefulness in treatment of GERD.

Sources: National Digestive Diseases Information Clearinghouse [homepage on the Internet]. Bethesda: National Digestive Diseases Information Clearinghouse; June 2007. Heartburn, Hiatal Hernia, and Gastroesophageal Reflux Disease (GERD); NIH Publication No. 03-0882. Available at http://digestive.niddk.nih.gov/ddiseases/pubs/gerd/#5

ery time.[31, 38, 40, 46] Other treatment options include the Stretta procedure, in which radiofrequency energy is delivered to the lower esophageal sphincter and gastric cardia.

Barrett's Esophagus—A Complication of GERD

Barrett's esophagus, or Barrett's metaplasia, involves a change in the epithelial cells of the esophageal mucosa and is usually considered a complication of GERD. Barrett's esophagus is detected in approximately 10% of patients undergoing **endoscopy** for GERD. Furthermore, it has been established that patients with refractory GERD (unresponsive to treatment) are much more likely to develop Barrett's esophagus. For example, patients with Barrett's esophagus have been shown to have persistent abnormal pH monitoring results even on maximum pharmacological treatment.[47, 48]

In this condition, the normal squamous cell epithelium of the esophagus changes to metaplastic columnar cell epithelium. This dysplastic cellular change is considered to be a precursor to a malignancy. Those patients with Barrett's esophagus are at higher risk for adenocarcinoma of the esophagus.[47] Current research is focusing on determining better ways to identify high-risk patients. Such research might establish biomarkers and would allow for early detection and treatment.[39, 49]

Patients do not experience any specific symptoms with this condition outside of those experienced with GERD. It is generally undetected unless the patient has a biopsy done as part of an upper GI diagnostic work-up, usually for the treatment of GERD. There are no specific nutritional concerns unless the patient is diagnosed with esophageal cancer. In the case of a malignancy, nutritional issues are addressed as the patient begins treatment for that diagnosis.

Nutrition Therapy for Gastroesophageal Reflux Disease

Nutritional Implications Most patients identify foods they feel make their symptoms worse and thus decrease intake of those foods. In these situations, restriction of food groups may result in weight loss or nutritional deficiency. Nutritional therapy may assist not only by addressing these nutritional problems but also by decreasing the symptoms that the patient is experiencing. Long-term use of medications for GERD may impair calcium absorption and iron and B$_{12}$ status.[50–54]

Nutrition Assessment For the patient with GERD, a 24-hour recall, diet history, or food diary should be used to focus on consumption of foods that lower LES pressure, increase gastric acidity, or cannot be tolerated by the patient. Additionally, lifestyle factors such as smoking and physical activity patterns are important for their contribution to LES incompetence.

Nutrition Diagnosis Common nutrition diagnoses that may be associated with GERD include inadequate food/oral beverage intake; excessive fat intake; swallowing difficulty; food-medication interaction; overweight/obesity; inadequate

Gastroesophageal Reflux in Infants

Obviously, many of the problems covered in this chapter can occur in infants and children as well as adults. Gastroesophageal reflux (GER) commonly occurs in approximately 2%–7% of young infants but generally resolves before 1 year of age in children without other medical problems.[1]

The signs and symptoms associated with GER include regurgitation, vomiting, coughing, irritability, and difficulty feeding. Many of these same signs and symptoms can be associated with more serious medical problems such as obstruction, hiatal hernia, pyloric stenosis, or infection. The type of regurgitation will yield clues to its etiology. For example, projectile vomiting is not normal in the healthy infant. In general it is recommended that immediate medical advice be sought if there is poor growth, refusal to feed due to pain, presence of blood in vomitus and/or stool, or breathing problems.[2,3] The specific etiology for GER may be related to overfeeding, poor positioning, food allergy, eosinophilic esophagitis, decreased LES pressure, delayed gastric emptying, or excessive acid production.[4–7]

Nutritional Implications   Nutrition concerns for GER in infants are related to inadequate nutritional intake with the resulting effects on growth and development.

Nutrition Diagnosis   Nutrition diagnoses associated with GER include malnutrition, inadequate protein-energy intake; underweight; impaired GI function; and swallowing difficulty.

Nutrition Intervention   Treatment of any underlying medical cause will be the first step for treating GER. Infants suspected of GER will undergo many of the same diagnostic tests that adults do. These include X-ray, ultrasound imaging, and pH monitoring.[3] H_2-receptor agonists and proton pump inhibitors are also prescribed for infants with GER.[1] Those infants diagnosed with a specific food allergy will be treated with removal of the antigen, such as cow's milk protein.[5,8]

Nutrition interventions for GER include caregiver education. Education will address positioning the baby for feeding, modification of formula choice, adding rice cereal for a

thickened feeding of formula or breast milk, and elevation of the head of the crib or bassinet. The effectiveness of thickened feedings has recently been evaluated by a meta-analysis of the literature. Horvath found that thickened formulas may reduce regurgitation and vomiting but did not affect incidence of reflux.[9] Thickened formulas also promoted weight gain when compared to standard formulas. It is recommended that babies be held upright for at least 30 minutes after each feeding. Babies should also be put to sleep on their backs.[10]

Monitoring and Evaluation   Interventions are evaluated by measurement of specific outcomes. These could include tolerance to feedings, amount of formula consumed, and weight gain.

References

1. Malcolm WF, Gantz M, Martin RJ, Goldstein RF, Goldberg RN, Cotten CM. National Institute of Child Health and Human Development Neonatal Research Network. Use of medications for gastroesophageal reflux at discharge among extremely low birth weight infants. Pediatrics. 2008 Jan;121(1):22–27.
2. NIDDK. Gastroesophageal Reflux in Infants. Available at http://digestive.niddk.nih.gov/ddiseases/pubs/gerdinfant/
3. Cohen HL, Strain JD, Fordham L, Gelfand MJ, Gunderman R, McAlister WH, Slovis TL, Smith WL. Expert Panel on Pediatric Imaging. Vomiting in infants up to 3 months of age. [online publication]. Reston (VA): American College of Radiology (ACR); 2005, 7.
4. Sretenović A, Perisić V, Simić A, Zivanović D, Vujović D, Kostić M, Pesko P, Krstić Z. Sretenović A, Perisić V, Simić A, Zivanović D, Vujović D, Kostić M, Pesko P, Krstić Z. Gastroesophageal reflux in infants and children Acta Chir Iugosl. 2008;55(1):47–53.
5. Heine RG. Allergic gastrointestinal motility disorders in infancy and early childhood. Pediatr Allergy Immunol. 2008 Aug;19(5):383–91.
6. Gilger MA, El-Serag HB, Gold BD, Dietrich CL, Tsou V, McDuffie A, Shub MD. Prevalence of endoscopic findings of erosive esophagitis in children: a population-based study. J Pediatr Gastroenterol Nutr. 2008 Aug;47(2):141–46.
7. Ozdemir O, Mete E, Catal F, Ozol D. Food intolerances and eosinophilic esophagitis in childhood. Dig Dis Sci. 2009 Jan;54(1):8–14. Epub 2008 Jul 2.
8. Martorell A, Plaza AM, Boné J, Nevot S, García Ara MC, Echeverria L, Alonso E, Garde J, Vila B, Alvaro M, Tauler E, Hernando V, Fernández M. Allergol Immunopathol (Madr). Cow's milk protein allergy. A multi-centre study: clinical and epidemiological aspects 2006, Mar–Apr;34(2):46–53.
9. Horvath A, Dziechciarz P, Szajewska H. The effect of thickened-feed interventions on gastroesophageal reflux in infants: systematic review and meta-analysis of randomized, controlled trials. Pediatrics. 2008;122:1268–77.
10. Corvaglia L, Rotatori R, Ferlini M, Aceti A, Ancora G, Faldella G. The effect of body positioning on gastroesophageal reflux in premature infants: evaluation by combined impedance and pH monitoring. J Pediatr. 2007 Dec;151(6):591–96, 596.e1. Epub 2007 Oct 24.

iron and calcium intake; impaired nutrient utilization; food and nutrition-related knowledge deficit; and undesirable food choices.

Nutrition Intervention   The goals of nutrition therapy are consistent with the goals of medical care discussed earlier. These goals include reducing gastric acidity and a trial of food restriction excluding those foods that may lower LES pressure. To reduce gastric acidity, black and red pepper, coffee (both caffeinated and decaffeinated), and alcohol can initially be avoided, because all have been identified as stimulants for gastric acid production. Likewise, meals of larger quantity tend to produce more acid, delay gastric emptying, and increase the risk of reflux. Thus, smaller, more frequent meals may be

indicated. Foods that have been identified as potentially lowering LES pressure may also be restricted. An initial trial to avoid chocolate, mint, and foods with a high fat content is recommended. Furthermore, any food the client identifies as irritating should be avoided. If the patient is obese, weight reduction is a critical component of the plan for nutrition therapy. Additional supplementation for calcium, iron, and other micronutrients may be warranted.[34,42–44,55] Nutrition interventions for GERD are outlined in Table 14.9.

Dysphagia

Dysphagia, or difficulty swallowing, is not generally considered a diagnosis but a symptom caused by a variety of disorders.

Table 14.9 Nutrition Interventions for GERD

Foods to Avoid

I. Foods that may relax the lower esophageal sphincter

Peppermint or spearmint

Chocolate

Fried foods or those with high amounts of added fat

Alcohol

Coffee (decaffeinated and caffeinated)

II. Foods that may increase gastric acid secretion

Coffee (decaffeinated and caffeinated)

Alcohol

Pepper

Food Group	Foods to Avoid (if symptomatic)
Beverages	Carbonated beverages, caffeinated and decaffeinated coffee and tea, cocoa, alcohol
Milk and milk products	2% milk, whole milk, cream, high-fat yogurts, chocolate milk
Eggs	Fried or scrambled using high-fat cooking methods
Cereals/grains	High-fat choices such as pastries
Meat and protein sources	Fried meats, bacon, sausage, pepperoni, salami, bologna, frankfurters/hot dogs
Vegetables	Only those that aggravate individual symptoms
Fruits	Only those that aggravate individual symptoms
Fat	As tolerated within current recommendations of U.S. Dietary Guidelines—Less than 8 teaspoons daily
Desserts	Those considered high fat or those that are fried
Miscellaneous	Pepper

Source: Adapted from: Nahikian-Nelms M. Gastrointestinal Disease. Nutrition Care Manual. Chicago, IL: American Dietetic Association. © 2009 American Dietetic Association. Adapted with permission.

Since many conditions of the esophagus involve dysphagia, it is important to understand this "symptom" and the importance of nutrition therapy in its treatment. There are numerous medical conditions and treatments that can ultimately affect one or more of the four phases of swallowing: oral preparatory, oral, pharyngeal, and esophageal. See Table 14.10 for an outline of potential causes of dysphagia.

Symptoms of dysphagia that a patient will experience depend on the phase of impaired swallowing. For example, if the problem originates in the oral preparation phase, food may be pocketed in the buccal mucosa (cheek area) because the patient cannot propel the bolus of food effectively from the front of the oral cavity to the pharyngeal area. Other general symptoms may include drooling, coughing, and choking. Many patients will experience weight loss and generalized malnutrition due to inadequate nutritional intake.[56–58] **Aspiration** or inhalation of oropharyngeal contents is a primary complication. This can lead to aspiration pneumonia with the accompanying infections and is the primary reason enteral nutrition support is recommended.[59]

Diagnosis and treatment of dysphagia involve many different members of the health care team. Many institutions have

Table 14.10 Diseases and Conditions Associated with Dysphagia

Acute Neurological Diseases	Stroke, closed head injury
Chronic Neurological Diseases	Amyotrophic lateral sclerosis, Parkinson's, multiple sclerosis, myasthenia gravis, Alzheimer's, Huntington's, neuropathy associated with type 1 diabetes mellitus
Muscle Disorders	Myositis, myopathies; scleroderma
Gastrointestinal Disease	GERD, hiatal hernia, achalasia, gastroparesis
Malignancy	Head and neck cancers; stomatitis, mucositis, esophagitis associated with chemotherapy and radiation therapy
Other	Inflammatory secondary to infection; post intubation; injury secondary to thermal or caustic exposure; aspiration; esophageal varices; drug side effects; aging

Sources: Goyal Raj K, "Chapter 38. Dysphagia" (Chapter). Fauci AS, Braunwald E, Kasper DL, Hauser SL, Longo DL, Jameson JL, Loscalzo J: Harrison's Principles of Internal Medicine, 17th ed. Available at http://www.accessmedicine.com/content.aspx?aID=2888607. Logemann JA. Swallowing disorders. Best Pract Res Clin Gastroenterol. 2007;21(4):563–73. Review.

a dysphagia team consisting of physicians, nurses, a speech-language pathologist, a dietitian, a physical therapist, and an occupational therapist. Diagnosis of dysphagia begins with a clinical bedside evaluation and bedside swallowing assessment usually performed by the speech-language pathologist.[60, 61] Conclusive evaluation uses a videofluoroscopy swallowing study or fiberoptic endoscopic evaluation of swallowing (see Figure 14.6 and Box 14.4). In these diagnostic procedures, barium is added to a variety of textures of foods and liquids. The patient is then monitored to determine his or her ability to swallow each of these foods. From these evaluations, a specific site of the dysphagia can be determined and a care plan can be developed.

Nutrition Therapy for Dysphagia

Nutritional Implications and Assessment The primary nutrition implications are weight loss and subsequent development of nutritional deficiencies that can occur due to an inadequate dietary intake. Therefore it is very important to accurately evaluate anthropometric measures and obtain as much information as possible regarding dietary intake and food patterns. After reviewing results of swallowing diagnostic tests, the health care team will be able to determine how the patient handles various textures of foods and liquids.

Nutrition Diagnosis Common nutrition diagnoses associated with dysphagia may include inadequate oral food/beverage intake; inadequate fluid intake; malnutrition; inadequate protein-energy intake; and swallowing difficulty.

Nutrition Intervention The registered dietitian can then use acceptable textures for the development of an adequate menu. The standard definitions for foods, liquids, and levels of nutrition intervention for dysphagia diets were developed by a United States task force in 2002. These levels of diet intervention are the *National Dysphagia Diets 1, 2, and 3*. Table 14.11 outlines the foods allowed on each level of the National Dysphagia Diet.[62, 63]

Figure 14.6 Endoscopic Evaluation

(a) Endoscopy; (b) barium swallow

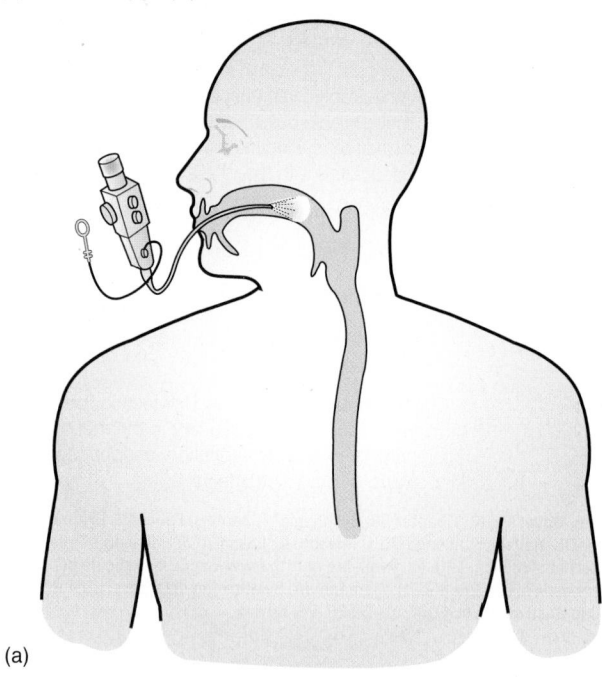

(a)

Source: http://www.artwiredmedia.com/elements/endob.htm

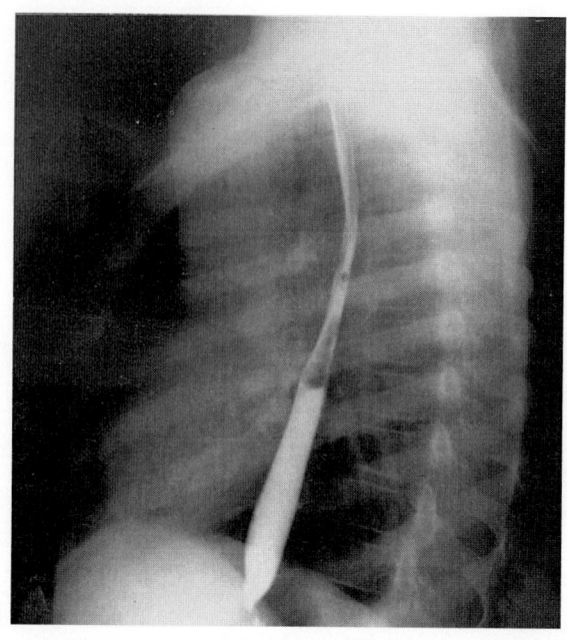

(b)

Source: From the University of Alabama at Birmingham Department of Pathology
PEIR Digital Library © (http://peir.net)

BOX 14.4 **CLINICAL APPLICATIONS**

Diagnostic Procedures: Upper GI Pathology

A variety of standard procedures are available to assist in the diagnosis of pathology involving the upper GI system, and most specifically for the esophagus and stomach. These procedures can be divided into those tests that involve visualization of the GI tract, and those that involve the measurement of secretions by these organs.

Endoscopy

During esophagogastroduodenoscopy (EGD), a fiberoptic endoscope (see Figure 14.6) is introduced through the oral pharynx and is moved through the esophagus and stomach and into the duodenum. This is also used for the fiberoptic endoscopic evaluation of swallowing (FEES). The endoscope is an optical instrument that includes a lens viewer, a long flexible tube, and a light source. The lens viewer allows the physician to visually inspect the mucosa of the organs and determine any abnormalities. The patient may have clear liquids from midnight the night before and then should be NPO

for the last 6 hours prior to the procedure. The patient will also receive a variety of premedications to assist in the procedure. These may include Versed, a medication that is given as an agent to induce sedation and/or amnesia prior to and during the procedure. Other medications include pain medications and antibiotic prophylaxis. A needle biopsy can be passed through the endoscope that allows for determination of the presence of abnormalities in the tissue or the presence of infection.

Barium Radiology Studies

In these procedures, the patient is given a contrast medium to drink. The most common medium is barium sulfate, which is a chalky, white radiopaque substance that the patient drinks like a milk shake. The barium can be visualized by fluoroscopy or by x-ray. This is the procedure used for a videofluorographic swallowing study (VFSS). This study allows the physician to monitor swallowing and the movement through the stomach into the duodenum. It also can distinguish

many other abnormalities such as ulcers, tumors, or inflammation.

Esophageal Manometry

Motor function of the esophagus and the lower esophageal sphincter is evaluated in this procedure. A tube is passed through the oral pharynx into the stomach. An instrument called a transducer is attached to the outer end of the tube, which records the barometric pressure as the tube is pulled back from the stomach through the LES. The patient should be NPO for 8 hours prior to the procedure, and is premedicated with both antianxiety and sedative drugs. A normal manometry will indicate that the LES pressure is normal and relaxes during a swallow. It will also indicate that the pattern of muscle contractions is systematic and coordinated during a swallow.

24-Hour pH Monitoring

This test previously involved placing a pH probe into the distal esophagus for a 12- to 24-hour period in order to generate a graph depicting continuous pH readings. New technology allows for the placement of a capsule which contains an acid sensing probe, a battery, and a transmitter. The probe monitors the acid in the esophagus and transmits the information to a recorder that is worn by the patient on a belt. The capsule transmits for two days, and then the battery dies. Within a week, the capsule is passed in the stool. Information is obtained regarding quantity and pattern of gastroesophageal reflux events, the correlation with symptoms, and the efficiency of esophageal acid clearance.

Bernstein Test (Acid Perfusion Test)

The Bernstein test may be used to differentiate between chest pain that is cardiac in origin and pain caused by acid reflux. A nasogastric tube is inserted through the nasopharynx into the esophagus. The physician attempts to replicate the symptoms of acid reflux by alternately injecting either a mild hydrochloric acid solution or a saline solution through the tube. The patient reports the differences in symptoms with each solution.

Urea Breath Test

This test is both specific and sensitive for diagnosis of *Helicobacter pylori*. The client ingests urea that is labeled with radioactive carbon-14 or nonradioactive carbon-13. The *Helicobacter pylori*, if present, will digest the urea. The by-product after digestion will be expired along with carbon dioxide and will indicate the presence of *H. pylori*.

Electrogastrography (EGG)

This procedure is used to diagnose and study stomach rhythm. Disturbances in the stomach's pacemaker can produce nausea and vomiting. This procedure can assist in determining the etiology of nausea and vomiting.

Antroduodenal Manometry

The purpose of 24-hour antroduodenal manometry is to measure the pressure in the small intestine before and after a meal and during sleep and waking hours. The measurements provide documentation of gastric emptying and peristalsis of the small intestine.

Gastric Analysis (Basal Acid Output)

This procedure measures the amount of hydrochloric acid produced by the stomach under baseline conditions. This test is often used to diagnose pernicious anemia, achlorhydria, and Zollinger-Ellison syndrome or to evaluate the effectiveness of surgical or pharmacological interventions.

Table 14.11 National Dysphagia Diet 1, 2, and 3

National Dysphagia Diet 1 (NDD-1) "Dysphagia Pureed"

Food Allowed	Food Not Allowed	Sample Menu
Includes foods of "pudding-like" consistency that are smooth or pureed with no lumps.	Gelatin desserts, fruited yogurt, peanut butter, unblenderized cottage cheese, scrambled, fried, or hard-cooked eggs.	Pureed chicken, mashed potatoes with gravy, pureed carrots, applesauce, and chocolate pudding.

National Dysphagia Diet 2 (NDD-2) "Dysphagia Mechanically Altered"

Food Allowed	Food Not Allowed	Sample Menu
Foods that are moist and soft textured such as tender ground or finely diced meats, soft cooked vegetables, soft ripe or canned fruit, and some moistened cereals.	Bread, dry cake, rice, cheese cubes, corn, and peas.	Scrambled egg, pancake with syrup, flaked cold cereal with milk, banana, orange juice (beverages thickened as appropriate).

Dysphagia Mixed is a term used by some institutions to designate a customized puree (NDD-1) diet that also allows one mechanically altered (NDD-2) item. Sample meal: orange juice, vanilla yogurt, cream of wheat cereal, scrambled egg.

Mechanical Soft is another alternative diet that allows bread, cake, and rice in addition to the NDD-2 mechanically altered diet. Sample menu: diced chicken with gravy, steamed rice, Harvard beets, pound cake, and fresh strawberries.

National Dysphagia Diet 3 (NDD-3) "Dysphagia Advanced"

Food Allowed	Food Not Allowed	Sample Menu
Includes most regular foods except very hard, sticky, or crunchy items. Bread, rice, cake, shredded lettuce, and tender, moist meats are allowed.	Not allowed are hard fruit and vegetables, corn skins, nuts, and seeds.	Vegetable soup, shredded lettuce salad with dressing, turkey sandwich with mayonnaise, fresh ripe melon, and chocolate chip cookie with no nuts.

Liquids

The following terminology is recommended by the National Dysphagia Diet Task Force to describe the viscosity of beverages and other liquids on the dysphagia diet:

- Spoon-thick
- Honey-like
- Nectar-like
- Thin liquids: Allows all liquids, including water, ice, milk, milk shakes, juice, coffee, tea, carbonated beverages, frozen desserts, and gelatin.

Sources: Reprinted from *Journal of the American Dietetic Association*, 103(3), SL McCallum, The National Dysphagia Diet: implementation at a regional rehabilitation center and hospital system. pages 381–84. Copyright 2003, with permission from American Dietetic Association.

Table 14.12 Thickening Agents and Specialty Food Products Used to Treat Dysphagia

Product	Manufacturer
Thickening Agents for Liquids	
Thicken up	Nestle Nutrition
Frutex	Crescent Foods
Thick & Easy	Hormel Health Labs
NutraThik™ Fortified Instant Food Thickener	Hormel Health Labs
Thick-it	Precision Foods, Inc.
Thixx, Ultra Thixx	Bernard Fine Foods, Inc.
Specialty Food Products	
Shape&Serve™	Hormel Health labs
Magic Cup™	Hormel Health labs
Cliffdale Farms™	Hormel Health labs
Thick & Easy Thickened Beverages	Hormel Health labs
TrePuree®	Campbell Soups—Hormel Health Labs
Puree Appeal	Novartis
NutraBalance	Ross Labs
Pureed foods	Tavis Meats, Inc.

Figure 14.7 Dysphagia Products

Can you tell that the foods in this photo are pureed foods shaped with commercial thickeners?

Source: S. Rolfes, K. Pinna and E. Whitney, *Understanding Normal and Clinical Nutrition,* 7e, copyright © 2006 p. 719.

There are many specialty products available to assist in development of foods for dysphagia diets. They include a variety of thickening agents and specialized products (see Figure 14.7) pre-prepared to meet a specific consistency for the diets. See Table 14.12 for thickening agents and specialty food products. Thickening agents are primarily made from carbohydrates, and many foods found within the dysphagia diets are also high in simple sugars. Successfully managing blood glucose for individuals with diabetes is challenging in this context.

For those individuals who are unable to safely swallow or who are unable to consume adequate nutrition orally, enteral feeding should be considered (see Chapter 5). Enteral feeding may be transitional until the patient successfully rehabilitates or may be a long-term support. Long-term nutritional support often poses ethical challenges for those individuals with chronic, terminal illnesses. (See Chapter 5, pp. 82–93.)

Monitoring and Evaluation Depending on the origin of the patient's dysphagia, tolerance of an oral diet may improve with treatment. The registered dietitian, with the other health care team members, will reevaluate the ability of the patient to progress in use of the prescribed diet. If problems arise, the diet may also need to be further restricted or changed in texture or consistency. Patients' weight, nutritional parameters, and hydration should be monitored closely to ensure adequacy of nutritional intake.

Achalasia

Achalasia is a motility disorder in which there is an absence of peristalsis or a weakened peristalsis within the esophagus. Additionally, there is often elevated LES pressure and impaired relaxation of the LES. This condition is relatively uncommon. It is estimated that there is approximately one case per 100,000 individuals in the United States.[64]

The etiology of achalasia is unknown at this time, but current research has focused on the role of infectious disease and on autoimmune origins.[3, 31] There are two distinct types of achalasia, primary and secondary. Primary achalasia is the most common type and results from loss of ganglion cells in the myenteric plexus of the lower portion of the esophagus. Secondary achalasia is due to other disease states such as diabetes mellitus, **Chagas disease**, and certain malignancies.

In primary achalasia, damaged ganglion cells result in a subsequent loss of the inhibitory neurotransmitters nitric oxide and VIP. The autonomic swallow is coordinated by autonomic nervous system control with nitric oxide and VIP acting as the primary inhibitory neurotransmitters. Without these neurotransmitters, the LES will not relax and appropriate swallowing cannot occur.[64, 65]

Treatments for achalasia include medications and invasive procedures. Primary medications focus on relaxation of smooth muscle. Calcium channel blockers such as nifedipine or nitrates can provide temporary relief. *Botulinum* toxin has been used to block acetylcholine, resulting in prolonged release of nitric oxide and relaxation of the LES.[61, 65] Unfortunately, most patients experience a recurrence within one year of treatment.

Pneumatic dilatation involves mechanical dilation of the LES. Esophageal myotomy is a surgical procedure that divides the muscle fiber of the LES. This procedure has been done via laparoscopy, which allows for reduced complications and shorter hospital stays. The most common complication is gastroesophageal reflux, which occurs in approximately 25% of patients after the procedure.[65]

Nutrition Therapy for Achalasia

Nutritional Implications In achalasia, patients experience dysphagia, vomiting, and substernal pain upon swallowing. Foods and fluids accumulate in the lower esophagus, causing the body of the esophagus to lose its muscle tone and become dilated or stretched. These symptoms result in poor oral intake with subsequent weight loss. Nutrition assessment would be similar to that for dysphagia.

Nutrition Diagnosis Nutrition diagnoses common for achalasia would also be similar to dysphagia and may include difficulty swallowing; inadequate oral food/beverage intake; and involuntary weight loss.

Nutrition Intervention Prior to treatment, patients with achalasia will need to have texture-modified diets with increased caloric and protein density. Foods that are extreme in temperature or very spicy should also be avoided in order to prevent damage to esophageal mucosa. Smaller, more frequent feedings will be tolerated best. After myotomy or dilatation, patients should receive a texture-modified diet such as the NDD-3 outlined in Table 14.11. A regular diet can be resumed within 5 to 7 days of the procedure.

Hiatal Hernia

Hiatal hernia is a condition where the upper portion of the stomach protrudes through the esophageal hiatus into the thoracic cavity. Most cases of hiatal hernia are designated as type 1 (sliding) where both the LES and some portion of the upper stomach protrude through the esophageal hiatus or diaphragm into the chest (see Figure 14.8). In type 2 (rolling hiatal hernia) the LES remains below the diaphragm. Incidence of hiatal hernia increases with age. Any factor that increases intra-abdominal pressure, such as obesity or pregnancy, will also increase the risk of hiatal hernia.

Symptoms of hiatal hernia are consistent with those of GERD. First-line interventions, both medically and nutritionally, are the same as those previously discussed for GERD. Some patients do require surgical repair of the hernia. In this procedure, which may be performed conventionally or laparoscopically, the surgeon retracts the hernia and repairs the hole in the diaphragm. Fundoplication (previously described in the section on GERD) can also be done at this time, if needed. The combination of surgical repair with fundoplication provides additional support of the LES, which prevents the stomach from sliding back through the diaphragm.[6, 31]

Pathophysiology of the Stomach

Disease and clinical disorders that affect the stomach can certainly influence normal nutritional status. Some disorders, such as indigestion, are mild and temporary conditions that resolve easily. Others, such as peptic ulcer disease, are chronic and require aggressive medical intervention.

Indigestion

Indigestion, or **dyspepsia**, is not considered to be a specific condition. Most people use the term "indigestion" to refer to a wide range of symptoms that may include abdominal pain, abdominal fullness, gas, bloating, belching, nausea, or even gastroesophageal reflux.

Nausea and Vomiting

Nausea is the unpleasant sensation that there is a need to vomit; vomiting is the expulsion of gastric contents. Even though nausea does not always lead to vomiting, they are often

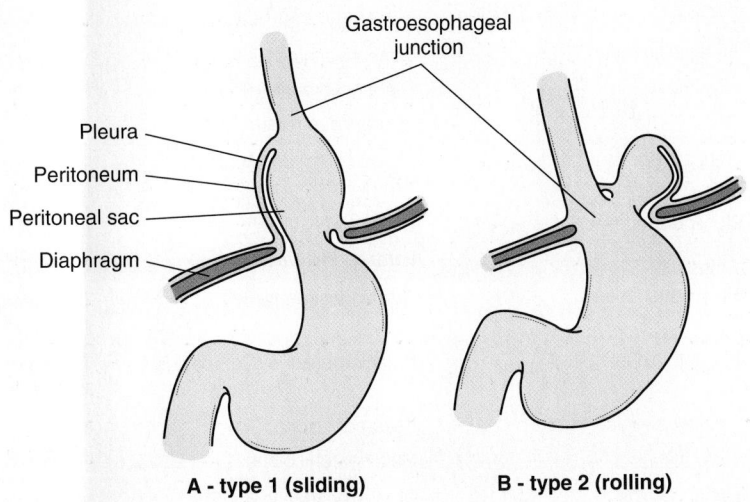

Figure 14.8 Hiatal Hernia

Gastroesophageal junction

Pleura
Peritoneum
Peritoneal sac
Diaphragm

A - type 1 (sliding)   **B - type 2 (rolling)**

Reprinted from Price and Wilson: *Pathophysiology: Clinical Concepts of Disease Processes*, 6e © 2006 Mosby with permission from Elsevier.

considered together because they are controlled through the same neural pathways. Neural signals are sent to the vomiting center located in the medulla. As a result of these stimuli, the steps of vomiting or emesis occur. In this sequence of events, gastric contents are pushed upward by the constriction of the respiratory muscles, the esophageal sphincter opens, the glottis closes (to prevent aspiration), and gastric contents are expelled through the mouth. Additionally, chemoreceptor zones in the medullary nucleus can also trigger the vomiting center. Drugs, toxins, metabolic conditions (such as renal failure or acid-base imbalances), and motion affect chemoreceptor zones, which can lead to nausea and emesis. Vomiting may also occur as a result of stress or extreme emotions.

Nausea and vomiting occur with many different medical conditions. These may include infection, pain, pregnancy, **syncope**, headache, metabolic disorders, motion sickness, kidney failure, myocardial infarction, and a host of other possibilities. Therefore, treatment of the underlying cause is the most important step in treating nausea and vomiting. The patient's history and physical examination will assist in determining the cause of nausea and vomiting. The etiology may be further clarified after assessment of the symptoms the patient experiences prior to and after vomiting. For example, if vomiting occurs within a very short time after eating, it may be indicative of an obstruction. Abdominal pain is symptomatic of an inflammatory process. Simple regurgitation of food occurs when gastric contents move easily from stomach to the mouth and is not a forceful expulsion like that seen in vomiting.

After determining the etiology of nausea and vomiting, the next step for treatment is use of medications or antiemetics. Table 14.13 provides a summary of antiemetics used to treat nausea and vomiting. Medication action may decrease the sensitivity of the chemoreceptor trigger zones. In many situations, such as in the use of chemotherapy, antiemetics are prescribed at the onset of treatment to prevent nausea. Controlling nausea from the very beginning of treatment prevents anticipatory nausea and/or vomiting, which can occur when there is a direct association between the nausea and

Table 14.13 Antiemetic Agents Used in the Treatment of Nausea and/or Vomiting

Classification of Medication	Generic and/or Trade Name	Uses/Mechanism
H₁ antihistamines	Dimenhydrinate (Dramamine®) Diphenhydramine (Benadryl®) Hydroxyzine (Atarax®)	Work by blocking histamine—may treat mild nausea such as motion sickness; also provide benefit of mild relaxation. Not effective for severe nausea and vomiting.
Benzamides	Metoclopramide (Reglan®)	This medication blocks dopamine and therefore affects the vomiting center in the brain. Side benefit of increasing gastric emptying.
Benzodiazepines	Diazepam (Valium®) Lorazepam (Ativan®)	Primarily used as tranquilizers, but can also increase the effectiveness of other antiemetics.
Butyrophenones	Droperidol (Inapsine®)	This medication blocks dopamine and therefore affects the vomiting center in the brain.
Phenothiazines	Prochlorperazine (Compazine®)	This medication blocks dopamine and therefore affects the vomiting center in the brain.
Corticosteroids	Dexamethasone Methylprednisolone	Reduces the effect of prostaglandins and can also improve the effectiveness of other antiemetics.
Cannabinoids	Dronabinol (Marinol®)	Mechanism unclear but produces feelings of euphoria and antiemetic effect.
NK1-receptor antagonists	Aprepitant (Emend®)	Blocks Substance P in the brain, which appears to have direct affect on vomiting center.
5-HT₃ receptor antagonists/ serotonin antagonists	Ondansetron (Zofran®) Tropisetron (Navoban®) Granisetron (Kytril®) Dolasetron (Anzemet®)	Blocks serotonin—usually given in combination with dexamethasone.

vomiting and a specific event, food, or smell. For instance, an individual who has gotten sick after eating a specific food may experience similar symptoms when faced with that food again, because it continues to remind the person of how sick he/she was previously. Complementary and alternative medicine (CAM) may provide some additional avenues for controlling and treating symptoms experienced with nausea as well as other symptoms experienced with diagnoses involving the upper gastrointestinal tract (see Appendix F.1). These may include the use of ginger and peppermint oil in treatment of nausea. Acupuncture, yoga, meditation, and guided imagery have also been successfully used to assist with control of nausea.[66–68]

Prolonged nausea and vomiting can have significant clinical consequences. Forceful vomiting can either rupture the esophagus (Boerhaave's syndrome) or tear the lower esophageal sphincter (Mallory-Weiss tear). Bleeding or **hematemesis** is a serious outcome of these injuries. Continued vomiting also can result in dehydration and acid-base imbalances. Malnutrition can be a long-term consequence for the patient if he or she is not able to ingest an adequate diet for a prolonged amount of time. If gastric contents are aspirated into the lungs, aspiration pneumonia is a likely result.

Sample PES statement: Inadequate oral food/beverage intake related to nausea and vomiting as evidenced by <300 kcal ingestion for previous 48 hours reported by patient and nursing records.

Nutrition Therapy for Nausea and Vomiting

Nutritional Implications Nausea and vomiting can result in inadequate nutrient intake, dehydration, and acid-base imbalances, and over time can lead to **learned food aversions**. This is similar to anticipatory nausea and vomiting. When a negative consequence is linked to a particular food, most people choose to avoid eating that food.

Nutrition Diagnosis Nutrition diagnoses secondary to nausea and vomiting may include altered GI function; involuntary weight loss; inadequate fluid intake; and inadequate food/beverage oral intake.

Nutrition Intervention Nutrition therapy does not necessarily treat nausea and vomiting but can minimize symptoms and discomfort. Appropriate nutrition therapy can assist in maintaining nutritional status during periods of nausea and vomiting. If patients can manage oral intake, foods that are cold and have minimal smell usually are best tolerated. Table 14.14 outlines suggestions for foods and food-related activities that may reduce nausea and vomiting. Close monitoring of hydration status and length of time that the patient is without adequate oral intake will be crucial in preventing long-term nutritional consequences. Nutritional support will be necessary for those individuals who are unable to meet their nutritional needs orally.

Gastritis

Gastritis is inflammation of the gastric mucosa. This condition is not a single disorder and may be a result of numerous conditions. Under normal conditions, the gastric mucosa is protected against injury. The production of mucus provides a barrier that prevents damage to the cells, and their high turnover rate allows for efficient recovery from injury. Prostaglandins also assist in support of the mucosal defense by stimulating mucus production, inhibiting acid production and release, and regulating blood flow to the mucosal cells. Acute gastritis is due to local irritation of the gastric mucosa. This irritation can result from infections, such as with *Helicobacter pylori* (*H. pylori*), food poisoning, alcohol ingestion, or medications such as nonsteroidal anti-inflammatory drugs (NSAIDs). By

Table 14.14 Nutrition Interventions to Reduce Nausea and Vomiting

1. Liquids to try after vomiting has stopped
- Water
- Apple juice
- Sports drink
- Warm or cold tea
- Lemonade

Instructions

First suck on ice chips if over 3 years of age. If tolerated, start with 1 teaspoon (5 g) every 10 minutes. Increase to 1 tablespoon (15 g) every 20 minutes. Double the amount of fluid every hour. Progress to the other liquids as tolerated. If diarrhea is present, use only rehydration beverage.

2. Solid Food Introduction

When there has been no vomiting for at least eight hours, start oral intake slowly by adding one solid food at a time in very small increments. Avoid food that high in fat or fiber as well as food that has a strong odor or is gas producing. The use of ginger to treat nausea and vomiting may help. Take medications after eating.

Foods Recommended for Initial Introduction of Solid Food

Grains	Dry toast, crackers, pretzels, rice or rice cereal, potato
Milk and Dairy Products	Yogurt, sherbet
Meat, Poultry, and Fish	Clear broths, baked chicken, eggs

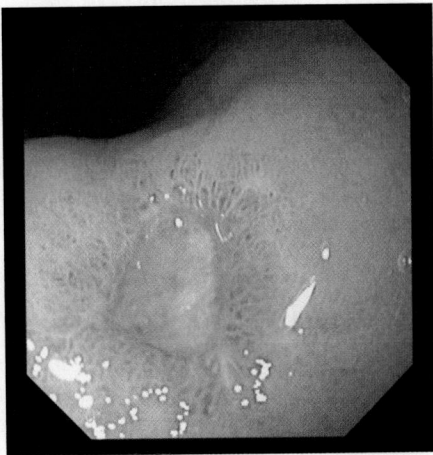

Peptic Ulcer Disease

Source: Atlas of Gastrointestinal Endoscopy www.EndoAtlas.com

blocking prostaglandin release, NSAIDs assist in control of inflammation but may also inhibit their protective function for the gastric mucosa. Generally, gastritis caused by NSAIDs is short lived and causes no long-term problems. Symptoms of gastritis can include belching, anorexia, abdominal pain, vomiting, and, in the more severe cases, bleeding and hematemesis.

Chronic gastritis is usually classified by either the etiology or the region of the stomach involved. Type A chronic gastritis involves the fundus and is associated with an autoimmune process, which results in the formation of antibodies against the parietal cells. Type A chronic gastritis also occurs with pernicious anemia. Type B chronic gastritis results in atrophy of the gastric mucosa and is most frequently associated with infection from *H. pylori*.[31] Incidence of chronic gastritis increases with age and is often seen with **achlorhydria.** Achlorhydria is associated with nutritional implications including B[12], iron, and calcium malabsorption.[54, 55, 69, 70] Treatment for gastritis includes identifying and treating the cause of the gastritis; for example, an antibiotic and medication regimen may be used to treat infections caused by *H. pylori* that are responsible for gastritis.

Peptic Ulcer Disease

Peptic ulcer disease (PUD) involves ulcerations of the gastric or duodenal mucosa that penetrate the submucosa. They usually occur in the antrum of the stomach or in the first few centimeters of the duodenum. Erosion may proceed to other levels of tissue and can eventually perforate. Breakdown in the tissue allows for continued insult by the highly acidic environment of the stomach as well as damage from other secretions of the stomach, such as pepsin.

Peptic ulcer disease has been redefined over the last decade because *H. pylori* is now recognized as a pivotal factor in development of gastric and duodenal ulcers. It is estimated that 92% of duodenal ulcers and 70% of gastric ulcers are caused by *H. pylori*. Recent research indicates that there is also an increased risk of gastric cancer associated with *H. pylori* infection.[71, 72] Nonetheless, even with this progress in research on PUD, there are still quite a large number of individuals who suffer from ulcer disease and are not infected with *H. pylori*. This section will describe the role of *H. pylori* and other factors that have been correlated to the development of PUD (see Box 14.5.)

Helicobacter pylori is a spiral-shaped, flagellated, Gram-negative rod that lives under the mucous layer of the stomach and attaches to mucus-secreting cells lining the stomach. These organisms break down urea to produce ammonia, which helps neutralize acid in the immediate vicinity of these bacteria and enhances their survival. The *H. pylori* organisms subsequently produce various proteins that damage mucosal cells, attracting lymphocytes and causing persistent inflammation.[71–73] By-products released by the organism result in damage to the epithelium and impair the mucous barrier within the stomach.

The etiology of PUD also involves factors that may decrease mucosal integrity, such as a reduction of protective prostaglandins through the use of NSAIDs (e.g., ibuprofen) or alcohol, excessive glucocorticoid secretion or steroid medication, and factors that decrease the blood supply, such as smoking, stress, or shock. Factors that increase acid secretions, including certain foods, rapid gastric emptying, or increased gastrin secretions, also contribute to the development of PUD. The genetic link to PUD has also been explored; ulcers are approximately three times more common in first-degree relatives than in the general population. This may be related to an increased susceptibility to infection from *H. pylori*.[74]

The most common symptom related to PUD is **epigastric** pain, but this pattern is not consistent. In general, patients will complain of abdominal pain and a burning sensation, which may be precipitated by certain types of foods or accentuated by food intake. For others, epigastric pain may be relieved by food intake due to its ability to dilute any irritants. For a duodenal ulcer, pain characteristically occurs from 90 minutes to 3 hours after eating, and is usually relieved within minutes either by eating or by use of antacids. Unfortunately, partial neutralization of gastric acid is followed by a rebound of gastrin release, causing additional stimulation of HCl and probably more pain.[75, 76]

History of Ulcer Diagnosis and Treatment

The road to a cure for ulcers has been a long and bumpy one. Research over the previous decade that indicates that ulcers are caused by a bacterium and can be cured with antibiotics has changed traditional thinking.

Early Twentieth Century

Ulcers are believed to be caused by stress and dietary factors. Treatment focuses on hospitalization, bed rest, and prescription of special bland foods. Later, gastric acid is blamed for ulcer disease. Antacids and medications that block acid production become the standard of therapy. Despite this treatment, there is a high recurrence of ulcers.

1982

Australian physicians Robin Warren and Barry Marshall first identify the link between *Helicobacter pylori* (*H. pylori*) and ulcers, concluding that the bacterium, not stress or diet, causes ulcers. The medical community is slow to accept their findings.

1994

A National Institutes of Health Consensus Development Conference concludes that there is a strong association between *H. pylori* and ulcer disease and recommends that ulcer patients with *H. pylori* infection be treated with antibiotics.

1995

Data show that about 75% of ulcer patients are still treated primarily with antisecretory medications, and only 5% receive antibiotic therapy. Consumer research by the American Digestive Health Foundation finds that nearly 90% of ulcer sufferers are unaware that *H. pylori* causes ulcers. In fact, nearly 90% of those with ulcers blame their ulcers on stress or worry, and 60% point to diet.

1996

The Food and Drug Administration approves the first antibiotic for treatment of ulcer disease.

1997

The Centers for Disease Control and Prevention (CDC), with other government agencies, academic institutions, and industry, launches a national education campaign to inform health care providers and consumers about the link between *H. pylori* and ulcers. This campaign reinforces the news that ulcers are a curable infection and the fact that health can be greatly improved and money saved by disseminating information about *H. pylori*. Medical researchers sequence the *H. pylori* genome. This discovery can help scientists better understand the bacterium and design more effective drugs to fight it.

2005

Nobel Prize for physiology or medicine awarded to Drs. Barry J. Marshall and Robin Warren for proving that bacteria and not stress was the main cause of painful ulcers of the stomach and intestine.

Sources:
Centers for Disease Control and Prevention [homepage on the Internet]. Atlanta: Centers for Disease Control and Prevention; 2001 Feb 2. History of Ulcer Disease and Treatment. Available at http://www.cdc.gov/ulcer/history.htm
Helicobacter pylori in Peptic Ulcer Disease, National Institutes of Health Consensus Development Panel on Helicobacter pylori in Peptic Ulcer Disease, Journal of the American Medical Association. 1994 July 6;272(1):65–69. Source: Centers for Disease Control and Prevention [homepage on the Internet]. Atlanta: Centers for Disease Control and Prevention; 2001 Feb 2. History of Ulcer Disease and Treatment. Available at http://www.cdc.gov/ulcer/history.htm
Marshall BJ, ed. Helicobacter pioneers: firsthand accounts from the scientists who discovered helicobacters, 1892–1982. Victoria, Australia: Blackwell; 2002.
Munnangi S., Sonnenberg A. Time trends of physician visits and treatment patterns of peptic ulcer disease in the United States. Arch Intern Med. 1997 July 14;175:1489–94.
Parsonnet J Clinician-discoverers—Marshall, Warren, and H. pylori. N Engl J Med. 2005 Dec 8;353(23):2421–23.

The presence of blood in stool or vomit may be indicative of active bleeding from the ulcer. Changes in hematological indices such as hemoglobin or hematocrit will also be indicative of active bleeding. If there is an active infection, changes in white blood cell count will be consistent with the inflammatory process.

The same diagnostic procedures that have been discussed earlier in this chapter will allow for a definitive diagnosis. Endoscopy coupled with a tissue biopsy will allow for visualization of the ulcer and confirmation of *H. pylori* infection. Less invasive testing for *H. pylori* includes the urea breath test, stool antigen test, and serum testing for antibodies. At present, the most accurate and reliable test for both diagnosis and follow-up is the urea breath test.[71, 73]

Treatment of peptic ulcer disease associated with *H. pylori* infections includes regimens of three to four medications (triple/quadruple therapy). The recommended therapy involves a 7- to 14-day course of two antibiotics with bismuth and one of the proton pump inhibitors.[71, 73, 77, 78] Eradication rates associated with triple/quadruple therapy range from 86% to 98% if patients comply with triple/quadruple therapy treatment regimens (see Table 14.15). However, frequently occurring adverse effects such as nausea, vomiting, and abdominal pain associated with these regimens significantly hinder patient compliance and most often the 7-day treatment is recommended.[71, 73, 77–79]

Other treatment for PUD focuses on use of medications to suppress acid secretion, which will ultimately promote healing of the ulceration (see Table 14.16). These medications, as discussed in the section on treatment for GERD, include antacids, proton pump inhibitors, histamine blocking agents, prokinetic agents, and mucosal protectants. Because salicylates (aspirin) and NSAIDs are linked to increased gastric

Table 14.15 FDA-Approved Treatment Options for Eradication of *H. pylori* Infection

Omeprazole 20 mg bid + clarithromycin 500 mg bid + amoxicillin 1 g bid × 7–14 days

—OR—

Omeprazole 20 mg bid + clarithromycin 500 mg bid + Metronidazole (500 mg bid) × 7–14 days

—OR—

Omeprazole 20 mg bid + bismuth subsalicylate (2 tabs qid) + tetracycline HCL (500 mg qid) + Metronidazole (500 mg bid) × 7–14 days

Sources: Malfertheiner P, Megraud F, O'Morain C, et al. Current concepts in the management of Helicobacter pylori infection: the Maastricht III Consensus Report. Gut 2007;56:772–81.

Atherton John C, Blaser Martin J, "Chapter 144. Helicobacter pylori Infections" (Chapter). Fauci AS, Braunwald E, Kasper DL, Hauser SL, Longo DL, Jameson JL, Loscalzo J: Harrison's Principles of Internal Medicine, 17th ed. Available at http://www.accessmedicine.com/content.aspx?aID=2894613

Table 14.16 Drugs Used in Treatment of Peptic Ulcer Disease

Antibiotics	Metronidazole; tetracycline; clarithromycin; amoxicillin
Antacids	Mylanta, Maalox; Tums; Gaviscon
H$_2$ blockers	Cimetidine; ranitidine; famotidine; nizatidine
Proton pump inhibitors	Lansoprazole; rabeprazole; esomeprazole; omeprazole; pantoprozole
Cytoprotective agents	Sucralfate; prostaglandin analogue (Misoprostol); bismuth subsalicylate

irritation, these medications should not be taken by someone with PUD.

For those patients who are refractory to treatment or who suffer from complications such as hemorrhage, perforation, or gastric outlet obstruction, surgical resection may be necessary (this will be discussed later in this chapter).

Nutrition Therapy for Peptic Ulcer Disease

Nutritional Implications and Assessment For patients with PUD, symptomatic abdominal pain can impair oral intake and result in weight loss and/or nutrient imbalances. Therefore it is important to obtain as much information as possible regarding weight and dietary intake changes and to evaluate this data in the context of the medical history of abdominal pain and symptoms.

Nutrition Diagnosis Nutrition diagnoses associated with peptic ulcer disease include inadequate oral/food beverage intake; altered GI function; involuntary weight loss; and food and nutrition-related knowledge deficit.

Nutrition Intervention For several decades, dietary factors have gained and lost favor as a significant component in both the cause and treatment of peptic ulcers. Currently, goals for nutrition therapy include supporting medical treatment, maintaining or improving nutritional status, and providing a diet that minimizes symptoms of PUD. There are no data indicating that diet is a causative factor for PUD.[73]

Current nutrition therapy for PUD (see Table 14.17) recommends a trial of restricting foods that may increase acid secretion or cause direct irritation to gastric mucosa. These foods include black and red pepper, caffeine, coffee (including decaffeinated), and alcohol. Additionally, it is recommended that patients avoid any foods they do not individually tolerate. Historically, milk and cream were used to treat PUD, but it is now known that their consumption increases both gastrin and pepsin secretion. Furthermore, pH of a food prior to its consumption has little effect after it is consumed. Restricting acidic juices or other foods is not consistently warranted unless the patient identifies intolerance to them.[34, 42–44, 55]

Other components of NT will include timing and size of meals. Patients should not lie down after eating and avoid eating large meals close to bedtime.[80] Smaller, more frequent meals may be better tolerated.

Micronutrients of concern may include iron, calcium, and B$_{12}$, as previously discussed in this chapter. The need for

Table 14.17 Nutrition Therapy for Peptic Ulcer

Food Groups	Foods Recommended	Food Not Recommended if Symptomatic
Beverages	Non-cola carbonated beverages, Postum®, herbal teas	Cola, coffee, tea, cocoa, alcohol
Milk and milk products	Skim, 1%, buttermilk, low-fat yogurt	2% or whole milk, cream, high-fat yogurt, chocolate milk
Eggs	Poached, hard-cooked, or scrambled using low-fat cooking methods	Fried or scrambled using high-fat cooking methods
Cereals	All ready-to-eat or cooked	None
Meats and Protein Sources	Baked, roasted, broiled, grilled, stewed; trimmed of visible fat; beef, veal, lamb, pork, poultry, fish, low-fat cottage cheese, low-fat cheese, peanut butter	Fried meats, bacon, sausage, pepperoni, salami, bologna, frankfurters/hot dogs
Potatoes/Rice/Pasta	All except fried	None except fried
Vegetables	All	Only individual intolerance
Fruits	All	Only individual intolerance
Fat	As tolerated consistent with current dietary guidelines	As tolerated consistent with current dietary guidelines
Dessert	All except those considered high fat or those fried such as pastries, doughnuts	Those considered high fat or those fried such as pastries, doughnuts
Miscellaneous	All except pepper	Pepper

additional supplementation should be evaluated for each client individually.

Monitoring and Evaluation Follow-up for the patient with PUD will focus on adequacy of the patient's nutritional intake and tolerance to the oral diet. Normal nutrition assessment indices will be monitored to ensure maintenance of nutritional status.

Gastric Surgery

When peptic ulcer disease does not respond adequately to medical treatment, or when the patient experiences a complication of PUD, surgery is often the next step. Complications from PUD may include **hemorrhage**, **perforation**, or **obstruction** of the pyloric sphincter. These surgical procedures are based on the patient's current medical status and prior surgical history. The same procedures are used for surgical resections required for other diagnoses such as gastric malignancy.

Vagotomy The purpose of the **vagotomy** is to eliminate the **cholinergic** stimulation of the stomach. Selective vagotomy eliminates innervations from the vagus nerve to parietal cells, resulting in decreased acid production and a decreased response to gastrin. Other functions of the vagus nerve remain intact, and the normal pathway for gastric emptying and peristalsis continues. In many patients, total vagotomy with **pyloroplasty** is chosen. In this procedure, innervations to parietal cells are severed, and the portion of the vagus nerve controlling gastric emptying is also eliminated. Pyloroplasty enlarges the pyloric sphincter. Gastric resection is also an option depending on location of the ulcer and extent of the stomach that requires removal. Reconstruction after pyloroplasty or gastric resection will generally use one of three procedures: a gastroduodenostomy (Billroth I), gastrojejunostomy (Billroth II), or Roux-en-Y procedure (see Figure 14.9).

Gastroduodenostomy (Billroth I); Gastrojejunostomy (Billroth II); Roux-en-Y Procedure Figure 14.9 illustrates these types of surgery. In the procedure gastroduodenostomy, or Billroth I, a partial **gastrectomy** or pyloroplasty is performed with a reconstruction that consists of an **anastomosis** of the proximal end of the duodenum to the distal end of the stomach. A gastrojejunostomy, or Billroth II, is a partial gastrectomy with a reconstruction that consists of an anastomosis of the proximal end of the jejunum to the distal end

Figure 14.9 Gastric Surgeries

(a) Roux-en-Y gastric bypass; (b) Billroth I; (c) Billroth II

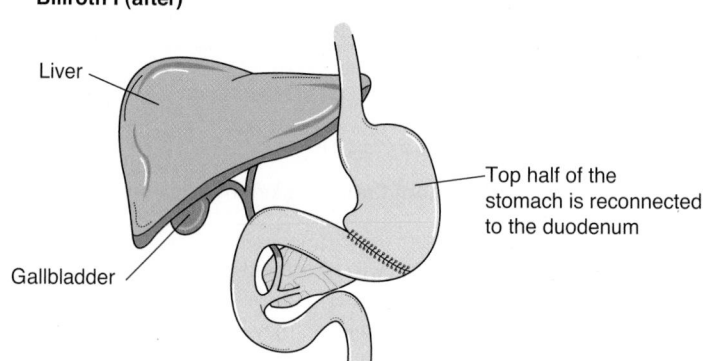

Billroth I (after)

Liver

Gallbladder

Top half of the stomach is reconnected to the duodenum

(b) This operation removes part of the stomach

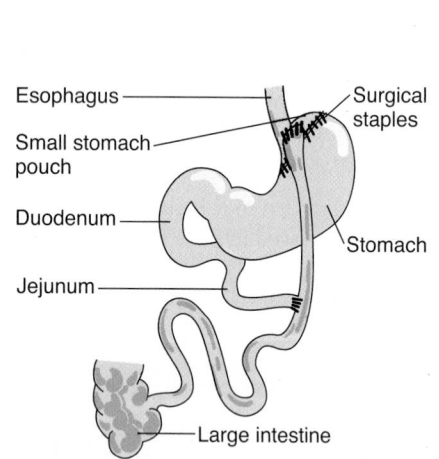

Esophagus

Small stomach pouch

Duodenum

Jejunum

Surgical staples

Stomach

Large intestine

In gastric bypass, the surgeon constructs a small stomach pouch and creates an outlet directly to the jejunum.

(a)

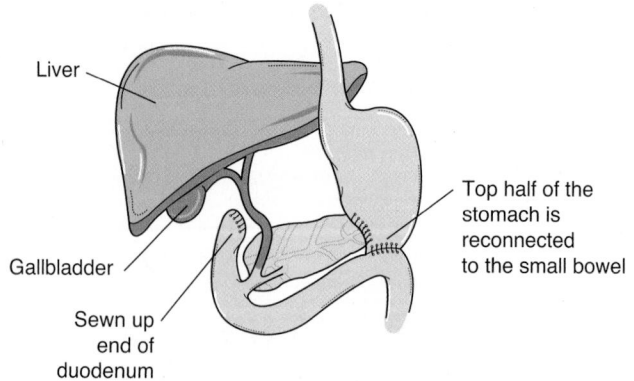

Billroth II (after)

Liver

Gallbladder

Sewn up end of duodenum

Top half of the stomach is reconnected to the small bowel

(c) This operation removes part of the stomach

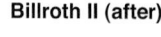

Sources: (a) John E. Pandolfino, Brintha Krishnamoorthy, Thomas J. Lee. Gastrointestinal Complications of Obesity Surgery, Medscape General Medicine 6(2), 2004. http://www.medscape.com/viewarticle/471952; (b, c) Diagrams reproduced with permission from CancerHelp UK, a free information service about cancer and cancer care for people with cancer and their families. It is brought to you by Cancer Research UK. www.cancerhelp.org.uk

of the stomach. In this surgical procedure, a blind loop of the duodenum is created. The Roux-en-Y procedure accomplishes the same thing as the Billroth II but creates a very small pouch after the gastric resection and connects the jejunum to the upper portion of the stomach. Although the Roux-en-Y procedure (or gastric bypass) has most recently featured prominently as a treatment for morbid obesity, it originated as a treatment for PUD and other gastric diseases.[81]

Nutrition Therapy for Gastric Surgery

Nutritional Implications Nutritional risk is due to reduced capacity of the stomach and potential changes in gastric emptying and transit time when the normal pathway for digestion and absorption is interrupted. Additionally, when portions of the stomach are resected, valuable components of digestion may be altered or lost. These issues combine to place the patient at significant nutritional risk due to decreased oral intake, maldigestion, and/or malabsorption.

Additional nutrition concerns include the potential for vitamin and mineral deficiencies. With changes in gastric anatomy, there may be a reduction in or absence of intrinsic factor secretion. This would prevent normal B_{12} absorption and lead to a subsequent deficiency. Research has confirmed that patients who have had gastric surgery have a high prevalence of vitamin B_{12} deficiency. Treatment of the deficiency can prevent cardiovascular, hematologic, and neurologic abnormalities seen with B_{12} deficiency or pernicious anemia.[82] Levels of methylmalonic acid and homocysteine are measured in addition to serum B_{12} to determine deficient levels, but it is also standard practice to prescribe prophylactic B_{12} injections for these patients. Iron deficiency is also common. The cause is multifactorial, including a decrease in HCl, decreased dietary intake, and possible malabsorption. Risk of osteoporosis is also increased due to decreased absorption of calcium. It is recommended that both calcium and vitamin D supplements be prescribed for these patients.[82]

Nutrition Diagnosis Nutrition diagnoses that may occur with or as a consequence of gastric surgery include increased energy expenditure; inadequate oral food/beverage intake; inadequate protein-energy intake; inappropriate intake of carbohydrates (simple or lactose); altered GI function; impaired nutrient utilization; involuntary weight loss; and food- and nutrition-related knowledge deficit.

Dumping Syndrome One of the most common complications after gastric surgery is **dumping syndrome**. Dumping syndrome occurs when an increased osmolar load enters the small intestine too quickly from the stomach. Severity of the symptoms varies depending on the extent of gastric surgery and the overall change in gastric emptying. When the stomach is removed or partially resected, important steps in digestion are missed. As discussed earlier, in a healthy person food may remain in the stomach anywhere from 1 to 3 hours as it becomes liquefied and partially digested. It then enters the duodenum via the pyloric sphincter slowly, so the acidic chyme is neutralized by pancreatic bicarbonate. When the pyloric portion of the stomach is removed, bypassed, or destroyed, the rate

of gastric emptying is increased. Additionally, when the duodenum is bypassed, feedback inhibition is lost. Furthermore, surgery will affect the release of hormones, enzymes, and other secretions. If this process is altered (as it is in gastric resections), food "dumps" into the small intestine (see Figure 14.10). Because the chyme is **hyperosmolar**, fluid is drawn into the small intestine from the vascular compartment in an attempt to dilute intestinal contents. These processes result in cramping, abdominal pain, hypermotility, and diarrhea. Furthermore, fluid changes in the vascular compartment result in dizziness, weakness, and tachycardia. These symptoms constitute what is generally referred to as "early" dumping syndrome, which actually occurs within 10 to 20 minutes after eating. "Intermediate" dumping syndrome occurs approximately 20 to 30 minutes after eating. As foodstuffs enter the colon, fermentation and action of microflora cause the production of gas, abdominal pain, cramping, and diarrhea. "Late" dumping syndrome can occur anywhere from 1 to 3 hours after eating, and is especially common after consuming simple carbohydrates. In this situation, rapid absorption in the small intestine stimulates insulin release. After quick movement and absorption of food through the small intestine, there is no longer any substrate for the insulin to act upon. This results in **hypoglycemia** and its symptoms of shakiness, sweating, confusion, and weakness. The Sigstad diagnostic scoring system (see Table 14.18) assesses these symptoms.[83, 84]

Medications that have been used for the prevention of dumping symptoms include **acarbose** and **octreotide**.[83]

> **Sample PES statement** Inappropriate intake of carbohydrates (simple vs. complex) related to food and nutrition knowledge deficit as evidenced by frequent ingestion of simple carbohydrates such as regular sweetened sodas with meals.

Figure 14.10 Dumping Syndrome

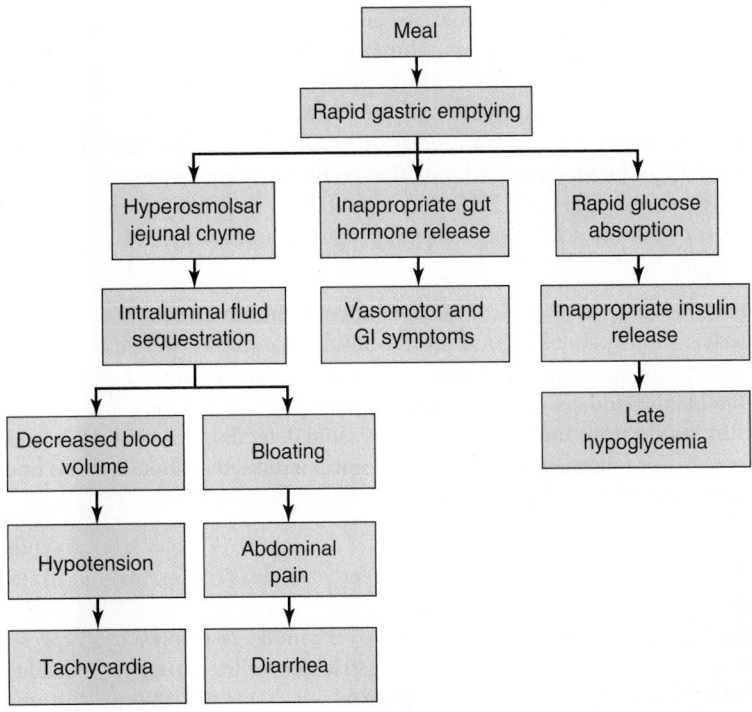

Pathophysiology of Dumping Syndrome

Table 14.18 Dumping Symptoms According to the Sigstad's Scoring System

Shock +5

Fainting, syncope, unconsciousness +4

Desire to lie or sit down +4

Breathlessness, dyspnea +3

Weakness, exhaustion +3

Sleepiness, drowsiness, apathy, falling asleep +3

Palpitation +3

Restlessness +2

Dizziness +2

Headaches +1

Feeling of warmth, sweating, pallor, clammy skin +1

Nausea +1

Abdominal fullness, meteorism +1

Borborygmus +1

Eructation −1

Vomiting −4

A score of greater than 7 is consistent with a diagnosis of dumping syndrome.

Source: Sigstad H. A clinical diagnostic index in the diagnosis of the dumping syndrome. Changes in plasma volume and blood sugar after a test meal. *Acta Med Scand.* 1970;188:479–86.

Table 14.19 Nutrition Interventions after Gastric Surgery

- Prescribe adequate energy and protein intake to ensure appropriate healing and recovery postoperatively.
  - Recommend appropriate nutrition support if progression to solid food does not proceed successfully.
- Initiate slow progression of solid food to prevent onset of early and late dumping syndromes.
  - Initially avoid all simple sugars. Do not start clear liquids with simple sugars as first oral feeding. Broth is acceptable.
  - The first meals should consist of protein, fat, and complex carbohydrate, but with only one to two food items at a time.
- Patients may be initially lactose intolerant and dairy products should be avoided.
- Slowly progress to five to six small meals each day.
- Consume liquids 30 minutes to 1 hour after solid food.
- Lie down after eating.
- Consider addition of functional fibers to delay gastric emptying and assist with treatment of diarrhea.
- Prevent development of nutrient deficiencies.
  - Liquid multivitamin and mineral supplements should be initiated to meet minimally the Dietary Reference Intakes for all established nutrients.
  - Vitamin B_{12} injections are initiated if extent of gastric surgery warrants this intervention.
  - Monitor serum levels of other micronutrients and supplement as necessary.
- Provide nutrition education that will promote optimal nutritional intake and minimize symptoms of malabsorption and/or maldigestion.

Source: Nahikian-Nelms, M. Gastrointestinal Disease. Nutrition Care Manual. Chicago, IL: American Dietetic Association, 2009. © 2009 American Dietetic Association. Adapted with permission.

Nutrition Intervention Nutrition therapy can prevent and treat many complications of gastric surgery (see Table 14.19 for an overview). The post-gastrectomy or "anti-dumping" diet encourages a well-balanced diet slightly higher in protein and fat than what is recommended by the U.S. Dietary Guidelines. Simple sugars (including clear liquids) are avoided in order to prevent hyperosmolality and hypoglycemia associated with the dumping syndrome. Lactose is often not tolerated. If the patient is lactose intolerant, commercial products that provide lactase or are lactose-free can be recommended. This is an additional reason to recommend calcium and vitamin D supplements. Liquids should be consumed between meals to prevent their contribution to dumping syndrome, because they facilitate quick movement through the small intestine. The patient is encouraged to consume five to six small meals throughout the day and, if necessary, lie down after meals. See Table 14.20 for an outline of nutrition recommendations for gastric surgery patients. Nutrition support is recommended if advancement to solid food does not progress successfully. See Chapter 5.[85–87]

Monitoring and Evaluation Patients who have had gastric surgery should be monitored closely to assess for weight loss and for symptoms of malabsorption and **steatorrhea**. Biochemical indices monitoring hemoglobin, hematocrit, ferritin, serum iron, and serum B_{12} will assist in detecting deficient iron, B_{12}, or folate levels. Other biochemical indices that should be monitored include those assessing visceral protein status: albumin and prealbumin (see Chapter 3).

Bariatric Surgery As discussed earlier in this section, the treatment of morbid obesity may include gastric surgery. Restricted access procedures include vertical banded gastroplasty, adjustable banded gastroplasty, and gastric sleeve. The most common surgical procedures include Roux-en-Y gastric bypass

and biliopancreatic diversion (BPD). These are discussed fully in Chapter 12.

Other Conditions of Gastric Pathophysiology

Stress Ulcers Acute illness and trauma can result in multiple ulcerations in the gastric mucosa. Conditions of shock, sepsis, burns, closed head injuries or trauma to the head (also known as Cushing's ulcers) have been linked to stress ulcerations. Decreased blood supply to the gastric mucosa causes breakdown of normal protective barriers in the stomach. Loss of the protective barrier allows for continued exposure to the highly acidic gastric juices. Ulcers generally develop very early in the trauma or shock period.

The most common symptoms are acute hemorrhage or perforation. It is standard practice in the clinical treatment of trauma and sepsis to provide preventive care with the use of a continuous infusion of H_2 blockers or liquid antacids every 2 to 3 hours.

Zollinger-Ellison Syndrome Zollinger-Ellison (ZE) syndrome is a condition of gastric acid hypersecretion. Symptomatically, ZE is initially similar to PUD, but is typically unresponsive to standard therapy. Hypersecretion is caused by presence of a non-B-cell endocrine tumor, or what is commonly called a gastrinoma. Zollinger-Ellison syndrome can be distinguished by measuring serum gastrin levels. In ZE patients,

Table 14.20 Nutrition Therapy after Gastric Surgery

Foods Recommended for the Recovery Period

Food Group	Recommended Foods	Notes
Milk and Milk Foods	Buttermilk Evaporated, skim, and 1%-fat milk Soy milk with no added sugar Yogurt with no added sugar Powdered milk Cheese Low-fat, low-sugar ice cream	• Choose lactose-free products if you have lactose intolerance after surgery. (If you have this condition, you will have symptoms after drinking regular milk or eating foods made from milk. Symptoms include diarrhea, nausea, stomach pain, and bloating.) • If you eat yogurt, choose ones that include live, active cultures. (The food label will list this information.) • Do not drink milk or other beverages with meals or snacks. After eating solid foods, wait 30 to 60 minutes before having a beverage.
Meat and Other Protein Foods	Tender, well-cooked meats, poultry, fish, eggs, or soy foods prepared without added fat Smooth nut butters	• Make sure to include a protein food in every meal and snack.
Grains	White flour Bread, bagels, rolls, crackers, and pasta made from white or refined flour Cold or hot cereals made from white or refined flour	• Choose grain foods with less than 2 grams of fiber per serving. (The grams of dietary fiber in one serving are listed on the Nutrition Facts label of packaged foods.) • Choose cereals that have no added sugar.
Vegetables	Most well-cooked vegetables without seeds or skins Potatoes without skin Lettuce Strained vegetable juice	• See the Foods Not Recommended chart for specific vegetables to avoid.
Fruits	Canned, soft fruits without added sugar Bananas, melon	
Fats	Oils, butter, margarine Cream, cream cheese Mayonnaise	
Beverages	Decaffeinated coffee Caffeine-free tea Sugar-free soft drinks without caffeine	• After eating solid foods, wait 30 to 60 minutes before having a beverage. Do not have beverages with meals. • Sweeten coffee or tea with artificial sweeteners only.
Other	Any allowed foods made with artificial sweeteners	• Allowed artificial sweeteners include saccharin (Sweet 'N Low), aspartame (Equal, NutraSweet), sucralose (Splenda), and acesulfame potassium (Sunette, SweetOne).

Foods Not Recommended

Food Group	Foods Not Recommended	
Milk and Milk Products	Chocolate milk Other milk foods made with added sugar If you have lactose intolerance, avoid regular milk and foods made with regular milk. Choose lactose-free products or soy milk instead.	Do not drink milk or other beverages with meals or snacks. After eating solid foods, wait 30 to 60 minutes before having a beverage.
Meat and Other Protein Foods	Fried meat, poultry, or fish Luncheon meats, such as bologna and salami Sausage, hot dogs, bacon	Tough or chewy meats Dried beans and peas, such as pinto or kidney beans Nuts, chunky nut butters
Vegetables	All raw vegetables except lettuce Any cooked vegetables served with skins or seeds Beets Broccoli, brussels sprouts, cabbage	Cauliflower Collards, mustard and turnip greens Corn Potato skins
Fruits	All raw fruits except banana and melons Dried fruits including prunes and raisins Fruit juice Canned fruit in sugar or syrup	
Beverages	Caffeinated coffee, tea with caffeine Alcoholic beverages Beverages made with sugar, corn syrup, or honey Fruit juices and fruit drinks	Do not drink beverages with meals or snacks. After eating solid foods, wait 30 to 60 minutes before having a beverage.

(continued)

Table 14.20 Nutrition Therapy after Gastric Surgery (*Continued*)

Foods Not Recommended

Food Group	Foods Not Recommended	
Other	Sugar	Foods that list sugar, honey, syrup, xylitol, or sorbitol as one of the first three ingredients on the food label
	Honey, syrup	
	Sorbitol, xylitol	

Sample 1-Day Menu

Note: This menu is suitable when you have recovered enough to eat six meals and snacks per day.

Breakfast	Scrambled egg
	1 slice white toast with 2 teaspoons margarine
30–60 minutes after breakfast	8 ounces decaffeinated coffee with half-and-half
Snack	2 ounces cheddar cheese and 6 saltines; ½ cup canned pears and peaches without added sugar
30–60 minutes after snack	1 cup calcium-fortified soy milk without added sugar
Lunch	½ cup tuna salad with 6 saltines
	1 ounce potato chips
	Sugar-free applesauce
30–60 minutes after lunch	12 ounces sugar-free soda
Snack	1 cup yogurt without added sugar
Evening Meal	5 ounces roast beef
	1 cup mashed potatoes
	1 cup green beans
30–60 minutes after meal	Caffeine-free herbal tea
Snack	½ plain bagel with 1 ounce cream cheese
30–60 minutes after snack	1 cup calcium-fortified soy milk without added sugar

Source: Nahikian-Nelms, M. Gastric Surgery Nutrition Therapy. In: Gastrointestinal Disease. Nutrition Care Manual. Chicago, IL: American Dietetic Association, 2009. © 2009 American Dietetic Association. Adapted with permission.

gastrin levels are >150 to 200 pg/mL compared to the normal levels of <150 pg/mL. Treatment consists of the use of proton pump inhibitors or surgical resection for those unresponsive to medical intervention.[88]

Conclusion

A thorough understanding of the pathophysiology involving the upper gastrointestinal tract is a fundamental component of nutrition therapy. This chapter has covered numerous physical signs and symptoms—such as dysphagia, nausea, and vomiting—that are a concern for the hospitalized patient with any number of conditions. Additionally, this chapter has covered multiple diagnoses, conditions, and symptoms that contribute to significant nutritional risk and cause frequent nutrition problems. For many, nutrition therapy is the major form of medical treatment. The role of the registered dietitian in planning, implementing, and evaluating nutrition therapy for the upper gastrointestinal tract should be of primary importance.

PRACTITIONER INTERVIEW: PRACTITIONER'S PERSPECTIVE ON BARIATRIC SURGERY

Valerie Simler, MS, RD, *Bariatric Surgery Program Manager*
Valley Care Medical Center, Pleasanton, California

Our facility performs predominantly Roux-en-Y (RNY) gastric bypass surgeries, but also offers laparoscopic adjustable gastric banding.

Nutrition after Bariatric Surgery. In the early post-op period, I recommend clear, then full, liquids for the first two weeks; pureed diet for weeks three and four; soft diet for the second month; and finally a return to solid foods in the third month. The most common problem immediately after RNY is an inability to consume adequate fluids.

It takes considerable effort on the part of the patient to drink enough to stay well hydrated, when all liquids must be sipped very slowly to prevent nausea and vomiting. It is also challenging to consume enough protein over the first several months. Patients are instructed to use liquid protein supplements between meals until they are able to consume enough protein from foods eaten at meal times. For many, they will always need to include a protein supplement as a snack, since the meal capacity remains too small to eat adequate protein as well as a variety of foods from other food groups. Some patients experience some early post-op nausea, which, if persistent, can be treated with medication and invariably improves within a few months. Vomiting is always a sign that something is wrong, either medically or from eating/drinking too fast, too much, or foods of the wrong texture. Constipation may be a problem until the body adapts to the new physiology and fiber is reintroduced. Dumping syndrome may occur early on after surgery or long term. Fifty to seventy percent of patients will experience the various gastrointestinal or vasomotor symptoms associated with dumping syndrome after eating simple sugars, which tends to be a strong deterrent against these nutrient-poor food choices.

After about three months, most patients have really good food tolerance and are able to consistently meet their protein and fluid needs. Some may have varying intolerances to certain foods (beef, lactose, and so on), but this is not usually a significant problem. It is recommended that patients take chewable multivitamins as vitamin pills tend to be too large to swallow and are often enteric-coated, which may be poorly absorbed. Calcium, vitamin D, and B_{12} supplements are also needed. Compliance is extremely important to prevent micronutrient deficiencies and possible osteoporosis.

The dietary recommendations once solid food tolerance has been achieved are:

1. Eat three meals each day. Avoid snacking except for protein supplements. Meal size will start at 2 to 3 oz, but will enlarge over the first 6 to 12 months to a very comfortable size of 1–1.5 cups of food per meal.

2. Eat slowly, chew food well, and stop eating at the first sign of fullness. Extra bites are likely to cause nausea and/or vomiting.

3. Choose healthful, mostly solid-texture, foods. Strive to include a lean protein food and a fruit or vegetable in each meal.

4. Don't drink fluids with meals. Wait 60 minutes or longer after finishing the meal before drinking. Choose only noncarbonated, low-sugar beverages, mostly decaffeinated. Avoid alcoholic beverages.

5. Read labels for protein, sugar, and fat. Choose foods and supplements that will contribute substantially to your protein goal. Limit saturated fat as is appropriate for everyone. Strictly limit sugar intake to avoid dumping syndrome.

6. Take multivitamin(s) with iron, calcium + vitamin D, and B_{12} supplements daily. Have labs checked regularly to assess for nutritional adequacy.

In addition to following this structured, nutrient-dense diet, exercise daily for 30–60 minutes for preservation of lean mass and for long-term weight control.

Suggested bariatric nutrition practices, compiled from published research and from the successful experiences of dietitians throughout the country, were recently summarized in the journal, *Surgery for Obesity and Related Disorders* 4 (2008) S73–S108. While the published article is not a nutrition mandate, it does serve as a guide for bariatric dietitians as they create the nutrition protocols for their various hospital programs.

Would I recommend the surgery?

Of course there are surgical risks, which should be investigated thoroughly and not taken lightly. For many obese individuals, it seems like the surgical risks are worth taking, given their medical risks living with obesity. Even for healthy obese people, quality of life is an important consideration. If your health is good and you are not unhappy being obese, why take the risk? But if you are dissatisfied, self-conscious, unable to do what you would really like to be doing in your life because of your weight, then I think it is worth it. Almost every single patient I have worked with—even those who have some complication, or find that surgery did not cure all their problems—still says it has been worth all the hard work and sacrifices. Even though there are significant challenges to keeping the weight off permanently, most people say that is better than what they were dealing with before the surgery. Struggling to keep off 20 to 30 lbs puts them on the same playing field with most other Americans, and that is someplace my patients say they want to be.

This is an amazing, satisfying career choice. People who come to me for help in achieving success with weight-loss surgery are usually very motivated individuals. Not only that, they are also very hopeful and optimistic about the expected outcome. This gives me a unique opportunity to provide education that is likely to be well received. I try to give patients a realistic perspective of their choice of surgery to achieve weight loss: what the role of the surgery is *and* what their personal role is in maximizing their weight loss and keeping the weight off permanently. Seeing the improvements in their health and quality of life is extremely rewarding.

In addition to being a bariatric dietitian for nine years, I am also the manager of the weight-loss surgery program at my hospital. This job has traditionally been held by an RN, which is still the case at most institutions. I feel that a dietitian can do this job equally well and there are many of us throughout the country proving this is true. Our multidisciplinary team consists of surgeons, nurses, dietitians, psychotherapists, pharmacists, physical therapists, and more. Managing the program allows me to put protocols in place to ensure patients have a positive experience from first phone call through years of support group meetings. One of my most rewarding career experiences has been earning Valley Care the distinction as a Bariatric Surgery Center of Excellence. This designation by the American Society for Metabolic and Bariatric Surgery recognizes programs with a demonstrated track record of favorable outcomes. It indicates that we have excellent performance in patient safety and in the provision of quality care. At the end of the day, making a positive impact on the lives of patients is what it's all about.

Introduction:

Sara Flores is a 45-year-old Hispanic female who has a five-year history of gastroesophageal reflux disease. Mrs. Flores presented to her physician with complaints of chronic indigestion and increased abdominal pain. She is now s/p endoscopy, which revealed a 2 cm duodenal ulcer and generalized gastritis. Biopsy positive for *Helicobacter pylori*. She was prescribed a 14-day course of Omeprazole 20 mg bid + clarithromycin 500 mg bid + Metronidazole 500 mg bid.

1. Mrs. Flores's endoscopy indicated that her biopsy was positive for *Helicobacter pylori*. What is this and how is it related to her duodenal ulcer?

2. This patient was prescribed three different medications. How do each of these work? What are the current recommendations for treatment of *Helicobacter pylori* infection? Are there any drug-nutrient interactions that need to be addressed?

Nutrition Assessment:

Food/Nutrition-Related History:

The registered dietitian's interview indicates that the patient describes appetite as poor. States that she is afraid to eat because it makes the pain worse. Specific food intolerances include anything fried or "spicy," coffee, and chocolate. Patient relates her usual weight to be about 145 pounds. The last time she weighed was 6 weeks ago. Her admission weight is 110 pounds.

Usual dietary intake (prior to current illness):

AM: coffee, dry toast. On weekends, cooks large breakfasts for family, which include omelets, rice or grits, or pancakes, waffles, fruit. Lunch: sandwich from home, fruit, cookies. Dinner: rice, some type of meat, fresh vegetables, coffee. Has previously consumed 8 to 10 cups coffee daily. Drinks one to two sodas each day.

Anthropometrics:

Ht 5'2" Wt 110# UBW 145#

Biochemical Data:

Labs: Total Protein 5.9 g/dL, Albumin 3.4 g/dL, Prealbumin 22 mg/dL, Hgb 11.5 g/dL, Hct 36%.

3. What admission laboratory values are abnormal? Interpret their significance in relationship to both her diagnosis and nutritional status.

Nutrition Diagnosis:

4. Identify at least two nutrition problems suggested by the nutrition assessment and medical history. Determine the diagnostic term for each nutrition problem. Next, identify the etiology of each nutrition diagnosis. Finally, identify the signs and symptoms that support the evidence for the diagnoses.

Diagnosis	Related to	Etiology	As Evidenced by	Sign/Symptoms
EXAMPLE:				
Involuntary weight loss	R/T	poor dietary intake secondary to fear of eating and pain	AEB	25% unplanned weight loss in 6 weeks

WEB LINKS

National Digestive Diseases Information Clearinghouse A service of the National Institute of Diabetes and Digestive Diseases. This government resource provides basic information and statistics on diagnoses of all digestive tract diseases. *http://digestive.niddk.nih.gov*

Directory of Digestive Disease Organizations for Patients Contact information for patient advocate, nonprofit organizations serving those individuals with diagnoses of digestive disease. *http://digestive.niddk.nih.gov/resources/patient.htm*

Digestive Health Center of Excellence, University of Virginia Health System This site provides excellent summaries of current research and patient education materials. *http://www.healthsystem.virginia.edu/internet/digestive-health*

International Foundation for Functional Gastrointestinal Disorders (IFFGD) Nonprofit education and research organization publishing several quarterly newsletters and patient education pamphlets. *http://www.iffgd.org*

Pediatric/Adolescent Gastroesophageal Reflux Association (PAGER) Provides information on pediatric gastroesophageal reflux and related disorders. *http://www.reflux.org*

1. Define and describe the four basic functions of the GI tract as discussed in this chapter.

2. Considering the basic functions of saliva, what are the possible consequences of xerostomia?

3. An imbalance of pressure at the lower esophageal sphincter (LES) may result in the symptoms associated with gastroesophageal reflux disease. What factors may affect LES pressure?

4. Explain the potential nutritional and metabolic consequences of prolonged vomiting.

5. Peptic ulcer disease (PUD), in many cases, is linked to an infection. What is the origin of this infection, and how is it treated?

6. Identify three major goals for nutrition interventions to assist in the control of symptoms associated with PUD.

7. Complications of PUD may result in surgical resection. What are the physiological consequences of gastric resection?

8. What are the potential nutritional complications of gastric resection?

9. Explain the symptomatic and etiological differences between early and late dumping syndrome.

Complementary and Alternative Medicine Remedies Used in Diseases of the Upper Gastrointestinal System

Remedy	Scientific Name	CAM Use	Side Effects and/or Risks
Barley	*Hordeum* spp.	Treat gastritis, diarrhea, and inflammatory bowel conditions	No known issues.
Birch	*Betula* spp.	Treat GI disorders	Keep out of reach of children; contraindicated in pregnancy & lactation.
Bitter orange	*Citrus aurantium*	Treat GI disorders	Loss of appetite, indigestion; avoid use in pregnancy & lactation.
Black catechu	*Acacia catechu*	Topical agent for sore gums & mouth ulcers; treat diarrhea & other GI problems	Increases risk of constipation if taken with anticholinergics. Increases risk of hypotension if taken with antihypertensives. Increases additive hypotensive effect with Captopril. Increases risk of fungal infections if taken with immunosuppressants. May bind iron products, creating an insoluble complex.
Blessed thistle	*Cnicus benedictus*	Treat dyspepsia and loss of appetite	Nausea, vomiting. May cause cross-sensitivity if used with other herbal drugs based on *Asteraceae* family.
Brewer's yeast		Treat flatulence, diarrhea, infectious diarrhea, & loss of appetite	May increase blood pressure if used with MAOIs.
Buchu	*Barosma* spp.	Treat stomachaches	Diarrhea, nausea, vomiting; hepatoxicity. Not recommended for use.
Butterbur	*Petasites* spp.	Treat GI disorders and GI-related pain	Abdominal pain or pressure, sustained constipation, discoloration of stool, dysphagia, severe nausea, vomiting. Use with anticholinergics not advisable.
Boldo	*Peumus boldus*	Treat digestive disorders	Exaggerated reflexes, lack of coordination, seizures. Cannot be recommended for human use.
Caraway	*Carum carvi*	Treat colic, constipation, flatulence, hiatal hernia, indigestion, mild spastic conditions of the GI tract, stomach ulcers	Diarrhea, hepatic dysfunction; pregnant women should avoid.
Chamomile	*Matricaria chamomilla*	Treat abdominal cramps & inflammatory GI conditions, hemorrhoids	May enhance effects of anticoagulants; may decrease absorption of other drugs taken concurrently. Should be avoided during pregnancy or breast-feeding.
Chinese rhubarb	*Rheum palmatum*	Treat GI bleeding, indigestion, antidiarreal, laxative	Contraindicated in pregnant & breast-feeding women, children <2 years of age, persons with ulcers or colitis.
Curcumin	*Curcuma longa*	Dyspepsia	Stimulates contraction of the gallbladder; provides full or partial relief of symptoms. Could present risks in individuals with gallbladder disease; maximum safe doses in individuals with severe hepatic or renal disease not known.

(continued)

Complementary and Alternative Medicine Remedies Used in Diseases of the Upper Gastrointestinal System (*Continued*)

Remedy	Scientific Name	CAM Use	Side Effects and/or Risks
Evening primrose	*Oenothera biennis*	Treat GI disorders	Nausea. Increases risk of seizures if taken with phenothiazines; avoid administration with EPO.
Ginger	*Zingiber officinale*	Nausea	As effective for motion sickness as standard pharmaceutical agents; also effective in reducing nausea in pregnancy. No drug interactions are known; ginger should be used with care in patients using anticoagulants or antiplatelet agents.
Goldenseal	*Hydrastis canadensis*	Appetite stimulant, gastritis, GI disorders, mouth ulcers, peptic ulcer disease, laxative	CNS depression, abdominal cramps & pain, diarrhea, nausea, vomiting. Might inhibit CYP3A4.
Grapefruit seed extract	*C. paradisi*	GI upset, sore throat	May increase bioavailability of anticonvulsants, benzodiazepines, some calcium channel blockers, certain antibiotics & antivirals, some co-A reductase inhibitors, cyclosporine, hormonal contraceptives, intestinal P-450,3A4 system, nonsedating antihistamines, quinidine.
Iceland moss	*Cetraria islandica*	Gastritis	Hepatotoxicity, indigestion, nausea with large doses or prolonged use.
Nutmeg	*Myristica fagrans*	Chronic diarrhea, indigestion, nausea	Dry mouth, constipation, nausea, vomiting, tachycardia, seizures, delusions (with excessive doses), euphoria, hallucinations. May interfere with neuroleptic dugs. May be carcinogenic. Imported nutmeg has been found to be contaminated with aflatoxins.
Peppermint	*Mentha piperita*	Treat nausea, vomiting, morning sickness, irritable bowel syndrome	Enteric-coated peppermint oil believed to be reasonably safe; nonenteric-coated peppermint oil can cause heartburn.
Probiotics	*Lactobacillus plantarum*	Peptic ulcer disease	Evidence of efficacy is mixed; may reduce intestinal gas and pain. No known safety issues.
Propolis	Resin collected by *Apis mellifera* bees	Gum disorders, mouth ulcers, sore throats, toothaches	Should not be used in children <1 year of age. Contraindicated in patients with asthma or history or anaphylaxis.
Raspberry	*Rubus* spp.	Induce vomiting	Increases effectiveness of hypoglycemic actions of antidiabetic medications.
Rue	*Ruta graveolens*	Digestive disorders	Hypotension, increased risk of spontaneous abortion.
Self-heal	*Prunella vulgaris*	Sore throat, colic, diarrhea, flatulence	None reported.
Slippery elm	*Ulmus rubra* Muhl	Soothe GI discomfort	Spontaneous abortion with whole bark preparations. Should be avoided by pregnant or breast-feeding women.
Turmeric	*Curcuma longa*	Cholelithiasis, GI bacterial overgrowth, flatulence, gastritis, ulcers, hepatic disorders	GI ulceration with high doses or prolonged use. Avoid administration with anticoagulants. Decreases immunosuppressive effects. Avoid administration with NSAIDs; may inhibit platelet function and increase risk of bleeding.
Wild yam	*Dioscarea villosa*	Abdominal cramps, dysentery, gallstones	Headache, menstrual irregularities, acne, hair loss, hirsutism, oily skin.
Yarrow	*Achillea millefolium*	Stimulate GI tract	Avoid use in pregnancy or breast-feeding as effects are unknown. Should be used cautiously by men due to potential inhibitory effect on spermatogenesis.

Sources: Fetrow CW, Avila JR. *Professional's Handbook of Complementary and Alternative Medicines*, 3rd ed. Philadelphia: Lippincott Williams & Wilkins, 2004.

Bratman S, Girman AM. *Mosby's Handbook of Herbs and Supplements and their Therapeutic Uses*. St. Louis: Mosby/Elsevier, 2003.

Fragakis AS, Thomson C. *The Health Professional's Guide to Popular Dietary Supplements*, 3rd ed., Chicago: American Dietetic Association, 2007.

ENDNOTE

1. "Mucus" is a noun; "mucous" is an adjective.

REFERENCES

1. Zhao Y, Encinosa W. Hospitalizations for Gastrointestinal Bleeding in 1998 and 2006. HCUP Statistical Brief #65. December, 2008. Rockville (MD):Agency for Healthcare Research and Quality.

2. Cherry DK, Hing E, Woodwell DA. National Ambulatory Medical Care Survey: 2006 Summary Number 3. August 6, 2008.

3. Talley NT, Locke RG, Saito YA. GI epidemiology. Blackwell; 2007.

4. Rhoades R, Pflanzer R. Human physiology. 4th ed. Belmont (CA): Thomson/Brooks/Cole. 2003, p. 686.

5. Baker DM. Assessment and management of impairments in swallowing. Nursing Clinics of North America. 1993;28(4)793–805.

6. Goyal Raj K. "Chapter 38. Dysphagia" (Chapter). Fauci AS, Braunwald E, Kasper DL, Hauser SL, Longo DL, Jameson JL, Loscalzo J. Harrison's principles of internal medicine. 17th ed. Available at http://www.accessmedicine.com/content.aspx?aID=2888607.

7. U.S. Department of Health and Human Services. Surgeon General's conference on children and oral health. June 12–13, 2000. Available at http://www.nidcr.nih.gov/AboutNIDCR/SurgeonGeneral/Children.htm

8. Centers for Disease Control. National Oral Health Surveillance System Caries Experience. Available at http://apps.nccd.cdc.gov/nohss/IndicatorV.asp?Indicator=2

9. Centers for Disease Control. National Oral Health Surveillance System Lost 6 or More Teeth. Available at http://apps.nccd.cdc.gov/nohss/ListV.asp?qkey=7&DataSet=2

10. Touger-Decker R, van Loveran C. Sugars and dental caries. Am J Clin Nutr. 2003;78(4):S881.

11. American Dietetic Association. Position of the American Dietetic Association: Oral Health and Nutrition J Am Diet Assoc. 2007;107:1418–28.

12. Kerr AR, Touger-Decker R. Nutritional consequences of oral conditions and diseases. In: Touger-Decker R, Sirois D, Mobley C, eds. Nutrition and Oral Medicine. Totowa (NJ): Humana Press; 2005, 143–66.

13. Bailey RL, Ledikwe JH, Smiciklas-Wright H, Mitchell DC, Gordon L, Jensen GL. Persistent oral health problems associated with comorbidity and impaired diet quality in older older adults. J Am Diet Assoc. 2004;104:1273–76.

14. Aranha AC, Eduardo Cde P, Cordás TA. Eating disorders part II: clinical strategies for dental treatment. J Contemp Dent Pract. 2008 Nov 1;9(7):89–96.

15. Marinho VCC, Higgins J PT, Logan S, Sheiham A. Topical fluoride (toothpastes, mouthrinses, gels or varnishes) for preventing dental caries in children and adolescents Cochrane Database of Systematic Reviews, Issue 1, 2009.

16. Kashket S, DePaola DP. Cheese consumption and the development and progression of dental caries. Nutrition Reviews. 2002;60(4):97–104.

17. Stephenson J. Combating cavities. JAMA. 2002;287(22):2937.

18. Soderling E. Nutrition, diet, and oral health in the 21st century. Int Dent J. 2001; 51(supp):389–91.

19. Haukioja A, Söderling E, Tenovuo J. Acid production from sugars and sugar alcohols by probiotic lactobacilli and bifidobacteria in vitro. Caries Res. 2008;42(6):449–53. Epub 2008 Oct 16.

20. Holgerson PL, Sjöström I, Stecksén-Blicks C, Twetman S. Dental plaque formation and salivary mutans streptococci in schoolchildren after use of xylitol-containing chewing gum. Int J Paediatr Dent. 2007 Mar;17(2):79–85.

21. United States Department of Agriculture. Dietary Guidelines for Americans 2005: Chapter 7 carbohydrates [monograph on the Internet]. United States Department of Agriculture; 2005. Available at http://www.health.gov/dietaryguidelines/dga2005/document/html/chapter7.htm

22. Kogut VJ, Luthringer SK. (eds), Appendix C. "Nutrition Impact Symptoms and Interventions" (pp. 348–49), in Nutritional Issues in Cancer Care. Pittsburgh (PA): Oncology Nursing Society; 2005.

23. Stacey DH, Doyle JF, Mount DL, Snyder MC, Gutowski KA. Management of mandible fractures. Plast Reconstr Surg. 2006 Mar;117(3):48e–60e. Review.

24. Nahikian-Nelms M. Nutrition Therapy for Fractured Jaw. In: Gastrointestinal Disease. Nutrition Care Manual. Chicago (IL): American Dietetic Association; 2009.

25. Manus RC, Dodson TB, Miller EJ, Perciaccante VJ. Nutritional status of substance abusers with mandible fractures. J Oral Maxillofac Surg. 2000;58(2):15.

26. Ziccardi VB, Ochs MW, Braun TW. Indications for enteric tube feedings in oral and maxillofacial surgery. J Oral Maxillofac Surg. 1993;51(11):1250–54.

27. Stacey DH, Doyle JF, Mount DL, Snyder MC, Gutowski KA. Management of mandible fractures. Plast Reconstr Surg. 2006 Mar;117(3):48e–60e. Review.

28. Wismer WV. Assessing alterations in taste and their impact on cancer care Curr Opin Support Palliat Care. 2008 Dec;2(4):282–87.

29. Doty RL, Shah M, Bromley SM. Drug-induced taste disorders. Drug Saf. 208;31(3):199–215. Review.

30. National Center for Health Statistics [homepage on the Internet]. Centers for Disease Control and Prevention. Hyattsville (MD): Centers for Disease Control and Prevention, United States Department of Health and Human Services. Available at http://www.cdc.gov/nchs/

31. Kahrilas PJ, Shaheen NJ, Vaezi MF. American Gastroenterological Association Institute technical review on the management of gastroesophageal reflux disease. Gastroenterology. 2008;135:1392–413.

32. Richter JE. Gastroesophageal reflux disease. Best Pract Res Clin Gastroenterol. 2007;21(4):609–31.

33. Valle John D. "Chapter 287. Peptic Ulcer Disease and Related Disorders" (Chapter). Fauci AS, Braunwald E, Kasper DL, Hauser SL, Longo DL, Jameson JL, Loscalzo J. Harrison's principles of internal medicine. 17th ed. Available at http://www.accessmedicine.com/content.aspx?aID=2899580

34. Dore MP, Maragkoudakis E, Fraley K, Pderoni A, Tadeu V, Realdi G, Graham DY, Delitala G, Malaty HM. Diet, lifestyle and gender in gastro-esophageal reflux disease. Dig Dis Sci. 2008;53(8):2027–32.

35. Fox M, Barr C, Nolan S, Lomer M, Anggiansah A, Wong T. The effects of dietary fat and calorie density on esophageal acid exposure and reflux symptoms. Clin Gastroenterol Hepatol. 2007;5(4):439–44.

36. Nocon M, Labenz J, Jaspersen D, Meyer-Sabellek W, Stolte M, Lind T, Malertheiner P, Willich SN. Association of body mass index with heartburn, regurgitation and esophagitis: results of the progresion of gastroesophageal reflux disease. J Gastroenterol Hepatol. 2007;22(11):1728–31.

37. Nocon M, Labenz J, Jaspersen D, Meyer-Sabellek W, Stolte M, Lind T, Malertheiner P, Willich SN. Long-term treatment of patients with gastro-oesophageal reflux disease in routine care—results from the ProGERD study. Aliment Pharmacol Ther. 2007 Mar 15;25(6):715–22.

38. Richter JE. Update on the management of achalasia: balloons, surgery and drugs. Expert Rev Gastroenterol Hepatol. 2008 Jun;2(3):435–45. Review.

39. Vakil N, Vaira D. Sequential therapy for Helicobacter pylori: time to consider making the switch? JAMA. 2008 Sep 17;300(11):1346–47.

40. DeVault KR, Castell DO, The Practice Parameters Committee of the American College of Gastroenterology. Updated Guidelines for the Diagnosis and Treatment of Gastroesophageal Reflux Disease. Am J Gastroenterol. 2005;100:190–200.

41. Jacobson BC, Somers SC, Fuchs CS, Kelly CP, Camargo CA Jr. Body-mass index and symptoms of gastroesophageal reflux in women. N Engl J Med 2006;354:2340–48.

42. Kahrilas PJ. Gastroesophageal reflux disease. N Engl J Med 2008;359:1700–07.

43. Shapiro M, Green C, Bautista JM, Dekel R, Risner-Adler S, Whitacre R, Graver E, Fass R. Assessment of dietary nutrients that influence perception of intra-oesophageal acid reflux events in patients with gastro-oesophageal reflux disease. Aliment Pharmacol Ther. 2007 Jan 1;25(1):93–100.

44. Kaltenbach T, Crockett S, Gerson L. Are lifestyle measures effective in patients with gastroesophageal reflux disease? Arch Intern Med. 2006;166:965–71.

45. van Marrewijk CJ, Mujakovic S, Fransen GA, Numans ME, de Wit NJ, Muris JW, van Oijen MG, Jansen JB, Grobbee DE, Knottnerus JA, Laheij RJ. Effect and cost-effectiveness of step-up versus step-down treatment with antacids, H_2-receptor antagonists, and proton pump inhibitors in patients with new onset dyspepsia (DIAMOND study): a primary-care-based randomised controlled trial. Lancet. 2009 Jan 17;373(9659):215–25.

46. Society for Surgery of the Alimentary Tract. SSAT patient care guidelines. Surgical treatment of reflux esophagitis. J Gastrointest Surg. 2007;11(9):1207–09.

47. Zhang HY, Spechler SJ, Souza RF. Esophageal adenocarcinoma arising in Barrett esophagus. Cancer Lett. 2009 Mar 18;275(2):170–77. Epub 2008 Aug 13. Review.

48. Souza RF, Krishnan K, Spechler SJ. Acid, bile, and CDX: the ABCs of making Barrett's metaplasia. Am J Physiol Gastrointest Liver Physiol. 2008 Aug;295(2):G211–18. Epub 2008 Jun 12. Review.

49. Spechler SJ. Screening and surveillance for Barrett's esophagus—an unresolved dilemma. Nat Clin Pract Gastroenterol Hepatol. 2007 Sep;4(9):470–71. Epub 2007 Jul 3.

50. Yang YX, Lewis JD, Epstein S, Metz DC. Long-term proton pump inhibitor therapy and risk of hip fracture. JAMA. 2006;296:2947–53.

51. O'Connell MB, Madden DM, Murray AM, Heaney RP, Kerzner LJ. Effects of proton pump inhibitors on calcium carbonate absorption in women; a randomized control trial. Am J Med. 2005;117(7):778–81

52. Carmel R. Current concepts in cobalamin deficiency. Annu Rev Med 2000;51:357–75.

53. Andrès E, Noel E, Abdelghani MB. Vitamin B(12) deficiency associated with chronic acid suppression therapy. Ann Pharmacother 2003;37:1730

54. Annibale B, Capurso G, Delle Fave G. The stomach and iron-deficiency anaemia: a forgotten link. Dig Liver Dis. 2003;35:288–95.

55. Nahikian-Nelms M. Nutrition Therapy for Gastroesophageal Reflux Disease. In: Gastrointestinal Disease. Nutrition Care Manual. Chicago (IL):American Dietetic Association; 2009.

56. Logemann JA Oropharyngeal dysphagia and nutritional management. Curr Opin Clin Nutr Metab Care. 2007 Sep;10(5):611–14. Review.

57. Finestone HM, Greene-Finestone LS. Rehabilitation medicine: 2. Diagnosis of dysphagia and its nutritional management for stroke patients. Canadian Medical Association Journal. 2003;169(10).

58. Preshaw R. Management of dysphagia. Can. Med. Assoc. J. 2004;170(7):1079.

59. Logemann JA, Rademaker A, Pauloski BR, Antinoja J, Bacon M, Bernstein M, Gaziano J, Grande B, Kelchner L, Kelly A, Klaben B, Lundy D, Newman L, Santa D, Stachowiak L, Stangl-McBreen C, Atkinson C, Bassani H, Czapla M, Farquharson J, Larsen K, Lewis V, Logan H, Nitschke T, Veis S. What information do clinicians use in recommending oral versus nonoral feeding in oropharyngeal dysphagic patients? Dysphagia. 2008 Dec;23(4):378–84. Epub 2008 Aug 1.

60. Manual of Clinical Dietetics. 6th ed. American Dietetic Association; 2000, 672–74. © 2005 American Dietetic Association.

61. The Joint Commission. Stroke: Disease Specific Care Performance Measurement Implementation Guide. 2nd ed. The Joint Commission; 2007.

62. McCallum SL. The national dysphagia diet: implementation at a regional rehabilitation center and hospital system. J Am Diet Assoc. 2003;103:381–84.

63. National Dysphagia Diet Task Force. National Dysphagia Diet: Standardization for Optimal Care. Chicago (IL):American Dietetic Association; 2002.

64. Liang, Chi-Yen, Lin, Ming-Shian Achalasia N Engl J Med. 2009;360:801.

65. Fisichella PM, Patti FG. Achalasia. Available at http://emedicine.medscape.com/article/169974-overview

66. National Center for Complementary and Alternative Medicine. Ginger. NCCAM Publication No. D320 Created May 2006 Updated May 2008 http://nccam.nih.gov/health/ginger/

67. National Center for Complementary and Alternative Medicine. Peppermint oil NCCAM Publication No. D320 Created May 2006. Updated May 2008. Available at http://nccam.nih.gov/health/peppermintoil/

68. Streitberger K, Ezzo J, Schneider A. Acupuncture for nausea and vomiting: an update of clinical and experimental studies. Auton Neurosci. 2006 Oct 30;129(1–2):107–17. Epub 2006 Sep 1. Review.

69. Hershko C, Ronson A, Souroujon M, Maschler I, Heyd J, Patz J. Variable hematologic presentation of autoimmune gastritis: age-related progression from iron deficiency to cobalamin depletion. Blood. 2006 Feb 15;107(4):1673–79. Epub 2005 Oct 20.

70. Hershko C, Hoffbrand AV, Keret D, et al. Role of autoimmune gastritis, Helicobacter pylori and celiac disease in refractory or unexplained iron deficiency anemia. Haematologica. 2005;90:585–95.

71. Malfertheiner P, Megraud F, O'Morain C, et al. Current concepts in the management of Helicobacter pylori infection: the Maastricht III Consensus Report. Gut 2007;56:772–81.

72. Take S, Mizuno M, Ishiki K, Nagahara Y, Yoshida T, Yokota K, Oguma K, Okada H, Shiratori Y. The effect of eradicating helicobacter pylori on the development of gastric cancer in patients with peptic ulcer disease. Am J Gastroenterol. 2005 May;100(5):1037–42.

73. Atherton John C, Blaser Martin J, "Chapter 144. Helicobacter pylori Infections" (Chapter). Fauci AS, Braunwald E, Kasper DL, Hauser SL, Longo DL, Jameson JL, Loscalzo J. Harrison's principles of internal medicine. 17th ed. Available at http://www.accessmedicine.com/content.aspx?aID=2894613

74. Valle John D. "Chapter 287. Peptic Ulcer Disease and Related Disorders" (Chapter). Fauci AS, Braunwald E, Kasper DL, Hauser SL, Longo DL, Jameson JL, Loscalzo J. Harrison's principles of internal medicine. 17th ed. Available at http://www.accessmedicine.com/content.aspx?aID=2899580

75. Chan FL, Leung WK. Peptic ulcer disease. The Lancet. 2002;360:933–42.

76. Harbison SP, Dempsey DT. Peptic ulcer disease. Curr Prob Surg. 2005;42(6):346–454.

77. Makola D, Peura DA, Crowe SE. Helicobacter pylori infection and related gastrointestinal diseases [review]. J Clin Gastroenterol. 2007;41(6):548–58.

78. Vaira D, Zullo A, Vakil N, et al. Sequential therapy versus standard triple-drug therapy for Helicobacter pylori eradication: A randomized trial. Ann Intern Med. 2007 Apr 17;146(8):556–63.

79. Ford AC, Delaney BC, Forman D, Moayyedi P. Eradication therapy for peptic ulcer disease in Helicobacter pylori positive patients. Cochrane Database Syst Rev. 2006;(2):CD003840. Review.

80. Fujiwara Y, Machida A, Watanabe Y, Shiba M, Tominaga K, Watanabe T, Oshitani N, Higuchi K, Arakawa T. Association between dinner-to-bed time and gastro-esophageal reflux disease. Am J Gastroenterol. 2005;100(12):2633–36.

81. Society for Surgery of the Alimentary Tract. SSAT patient care guidelines. Surgical treatment of reflux esophagitis. J Gastrointest Surg. 2007;11(9):1207–09.

82. O'Donnell K. Small but mighty: Selected micronutrient issues in gastric bypass patients. Pract Gastroenterol. 2008;XXXII(5):37–48.

83. Ukleja A. Dumping syndrome. Pract Gastroenterol. 2006:32–46.

84. Sigstad H. A clinical diagnostic index in the diagnosis of the dumping syndrome. Changes in plasma volume and blood sugar after a test meal. Acta Med Scand. 1970;188:479–86.

85. Chin KF, Townsend S, Wong W, Miller GV. A prospective cohort study of feeding needle catheter jejunostomy in an upper gastrointestinal surgical unit. Clin Nutr. 2004;23(4):691–96.

86. Farreras N, Artigas V, Cardona D, Rius X, Trias M, Gonzalez JA. Effect of early postoperative enteral immuno-nutrition on wound healing in patients undergoing surgery for gastric cancer. Clin Nutr. 2005;24(1):55–65.

87. Malhotra A, Mathur AK, Gupta S. Early enteral nutrition after surgical treatment of gut perforations: a prospective randomised study. J Postgrad Med. 2004;50(2):102–06.

88. Ellison EC, Johnson JA. The Zollinger-Ellison syndrome: a comprehensive review of historical, scientific, and clinical considerations. Curr Probl Surg. 2009 Jan;46(1):13–106. Review.

15

Diseases of the Lower Gastrointestinal Tract

Marcia Nahikian Nelms, PhD, RD
The Ohio State University

Introduction

An exploration of pathophysiology affecting the small and large intestine quickly reveals that nutrition therapy is the foundation of treatment for many of these diagnoses. For some of these, such as celiac disease, nutrition therapy is the only treatment. Additionally, many gastrointestinal symptoms such as diarrhea, constipation, or malabsorption place an individual at significant nutritional risk by impairing adequate or appropriate utilization of nutrients.

The discussion of the upper gastrointestinal tract (mouth, esophagus, and stomach) in Chapter 14 focused on four basic functions: motility, secretion, digestion, and absorption. These four functions are also of primary importance in both the small and large intestine, because more than 98% of all digestion and absorption occurs in the lower GI tract.

Normal Anatomy and Physiology of the Lower Gastrointestinal Tract

The small intestine is composed of three distinct parts: the duodenum, jejunum, and ileum. These are not separate compartments, but each differs in anatomy, motility, secretion, digestion, and absorption.

Small Intestine Anatomy

The anatomy of the small intestine is both unique and highly functional. This anatomy is organized to provide maximum

surface area and allow for complete digestion and absorption of most foodstuffs. First, the tissue of the small intestine is circularly folded into what are referred to as the folds of Kerckring. Rising from the mucosal surface are fingerlike projections called villi. On the surface of villi are fine hairlike projections called microvilli. This area is often referred to as the "brush border." The combined features of these anatomical structures increase the surface area of the small intestine to such an extent that it is 600 times greater than it would be if the intestine were a straight, flat tube. The spaces between villi are called crypts. These crypts are the location of **stem cells** from which specialized epithelial cells (enterocytes) for the small intestine develop.

bacterial overgrowth syndrome—malabsorption and malnutrition that result from cross-contamination of bacteria from the colon to the small intestine

bilirubin—the breakdown product of hemoglobin molecules; it is normally excreted from the body via bile secretions

celiac disease (CD)—inflammation of the small intestine caused by gluten found in various grains, including wheat

colostomy—a procedure in which the rectum only is surgically removed, and the end of the colon is attached to the stoma

compassionate use—the use of an investigational drug outside of a clinical trial to treat a patient with a serious or immediately life-threatening disease or condition who has no comparable or satisfactory alternative treatment options; also called *expanded access* (Food and Drug Administration, http://www.fda.gov/ForConsumers/ ByAudience/ForPatientAdvocates/Accessto InvestigationalDrugs/ucm176098.htm)

constipation—a decrease in frequency of bowel movements with straining with defecation and/ or hard stools

Crohn's disease—a chronic inflammatory bowel disease (IBD) that can affect the entire gastrointestinal tract but most commonly affects the ileum and colon

diarrhea—frequent or unusually liquid bowel movements

dietary fiber— the component of plant matrix that serves as the plant cell wall or intercellular structure and is not digestible

diverticulitis—an acute inflammation of the diverticula

diverticulosis—an abnormal presence of outpockets or pouches (diverticula) on the surface of the small intestine or colon

exudate—fluid and cellular debris that seeps from blood vessels, usually as a result of inflammation

fibromyalgia—a condition characterized by chronic pain (in muscle and soft tissues surrounding joints) and fatigue

fistula—an abnormal opening or passage between two internal organs or from an internal organ to the surface of the body

flatulence—perceived excess gas in the intestinal tract

fructose—a disaccharide absorbed by a facilitated transport mechanism but not against a concentration gradient; when the concentration

of fructose in the small intestine is greater than that of glucose, its rate of absorption slows and the unabsorbed fructose is fermented in the colon, causing diarrhea; osmotic diarrhea has been reported in persons who have overconsumed sodas sweetened with high-fructose corn syrup or fruit juices

functional fiber—undigested plant components that have an established physiological function for humans

high-fiber diet—a diet high in fiber (6 to 10 g above the usual recommendation of 20 to 35 g/day)

ileostomy—a procedure in which the colon and rectum are surgically removed, and the end of the ileum is attached to the stoma

ileus—decreased or absent motility of the bowel and forward movement of bowel contents

inflammatory bowel disease (IBD)—an autoimmune, chronic inflammatory condition of the gastrointestinal tract; IBD is actually the term designating a syndrome consisting of two diagnoses: **ulcerative colitis** and **Crohn's disease**

inulin—a fructooligosaccharide derived from chicory; intravenous inulin is used as a diagnostic test for kidney function since it is not utilized by the body and is excreted in the urine

irritable bowel syndrome (IBS)—a bowel disorder characterized by abdominal pain with diarrhea and/or constipation

lactoferrin—a protein in plasma and secretions (milk, mucus, bile), secreted by leukocytes, that can bind iron; it helps prevent infection by depriving bacteria of the iron necessary for their growth

low-residue diet—a diet low in fiber and other food constituents that may contribute to bulk in the large intestine

medium-chain triglycerides (MCTs)—triglycerides composed of fatty acids with 8 carbons (octanoic and decanoic fatty acids)

migrating motility complex (MMC)—weak contractions of the gastrointestinal tract that serve to assist in clearing waste

Na^+/K^+ pump—the enzyme-based mechanism that moves potassium ions into and sodium ions out of a cell by active transport

oncotic pressure—pressure exerted by large protein molecules in blood plasma, which usually do not cross the capillaries; these molecules decrease the fluid that can leak out of the capillaries into the tissue

oxidative stress—a disturbance in the pro-oxidant–antioxidant balance in favor of the former, leading to potential damage; indicators

include damaged DNA bases, protein oxidation products, and lipid peroxidation products

pelvic floor—refers to the pelvic diaphragm, the sphincter mechanism of the lower urinary tract, the upper and lower vaginal supports, and the internal and external anal sphincters; it is a network of muscles, ligaments, and other tissues that hold up the pelvic organs (vagina, rectum, uterus and bladder)

pelvic floor dysfunction—weakening of the pelvic floor that can cause the organs to shift, bulge, and push outward against each other, resulting in urinary or fecal incontinence or obstruction, vaginal prolapse or pain, sexual dysfunction, and other problems.

prebiotics—substances in food that stimulate the beneficial flora of the large intestine

probiotics—products containing microorganisms manufactured and sold as food products and supplements

refractory celiac disease—initial or subsequent failure of a strict gluten-free diet to restore normal intestinal architecture and function in patients who have celiac-like enteropathy

resistant starch—indigestible starch that can be found naturally in foods such as beans and peas; produced during food processing or from chemical modification

short bowel syndrome (SBS)—decreased digestion and absorption that result from a large resection of the small intestine

sorbitol—a sugar alcohol; it is used as a sugar substitute

steatorrhea—excess fat in the stool (>6 g/24 hrs)

stem cells—nondifferentiated, primitive cells that have the ability both to multiply and to differentiate into more specialized cells that display unique functions

stoma—a surgically created artificial opening into the abdomen

synbiotics—products that contain both prebiotics and probiotics

temporomandibular joint (TMJ) syndrome—a condition of facial pain in the joints of the lower jaw

toxic megacolon—a very inflated colon with abdominal distention, and sometimes fever, abdominal pain, or shock

ulcerative colitis (UC)—a chronic inflammatory bowel disease (IBD) primarily located in the colon and rectum

volvulus—the twisting of the bowel causing obstruction

They migrate up the villi where, after serving their particular physiological function, they are sloughed off and replaced with newly generated enterocytes.

As you may remember from previous study of the small intestine, this rapid turnover of enterocytes makes the nutritional needs of the small intestine particularly crucial. Enterocytes, therefore, have high nutrient needs, and the health and function of the GI tract are strongly influenced by nutritional status

and disease. Malnutrition and disease affect the ability of these cells to regenerate and result in decreased villous height. Ultimately, they reduce the ability of the small intestine to perform digestion and absorption. Maintaining nutritional health of the small intestine has been the focus of much nutrition research over the past decade.[1–4]

The small intestine is very adaptive and can adjust its function rather efficiently. More than 50% of the small intestine has to be removed before any significant reduction in its capability is observed. The duodenum and jejunum can perform each other's role in both digestion and absorption. The ileum can also adapt in this way—up to a certain point. The ileocecal sphincter (valve) protects the small intestine from bacteria translocation from the large intestine by remaining closed except during the digestive process (see Figure 15.1). This sphincter also maintains an appropriate transit time to ensure adequacy of both digestion and absorption. This sphincter relaxes when stimulated by either increasing pressure, such as seen in the presence of fluid, or by chemical irritation. At that time, relaxation allows slow movement of remaining digestive contents into the upper portion of the large intestine. As discussed later in this chapter, preservation of the ileocecal valve is often a strong prognostic indicator for intestinal adaptation after surgery.

Small Intestine Motility

Motility of the small intestine is controlled by the enteric nervous system and influenced by a variety of hormones, peptides,

Figure 15.1 Sphincters of the Gastrointestinal Tract

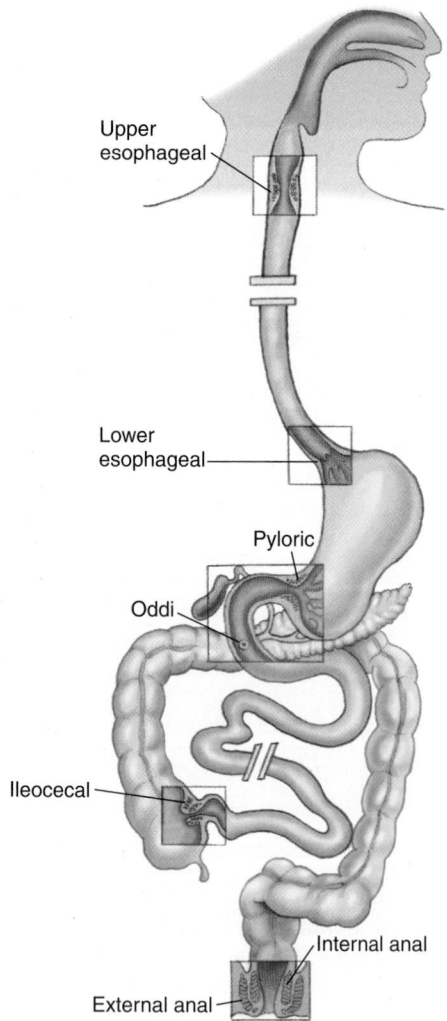

Upper esophageal

Lower esophageal

Pyloric

Oddi

Ileocecal

Internal anal

External anal

Source: R. Rhoades and R. Pflanzer, *Human Physiology*, 4e, copyright © 2003 p. 693.

and neurotransmitters. An understanding of small intestine peristalsis has led not only to an understanding of digestion and absorption but also to researchers' deeper understanding of conditions where motility may be disturbed, such as irritable bowel syndrome (IBS).[5, 6]

As the chyme enters the duodenum from the stomach, the hormone gastrin and the intestinal pacemaker cells (the interstitial cells of Cajal) stimulate the onset of segmental contractions.[5] (See Figure 15.2.) This segmentation motility allows mixing of chyme with digestive secretions. In the duodenum, contractions occur approximately every 9 minutes, with a slower rate further down the small intestine. It may take as long as 3 to 5 hours to complete the process.[7]

Additionally, when the small intestine is empty, motility continues with the action of the **migrating motility complex (MMC)**. The MMC, first described in 1969, consists of much weaker contractions that occur approximately every 100 to 150 minutes and serves the purpose of cleaning out the small intestine of any leftover bacteria or waste.[8] **Motilin**, a hormone secreted by the small intestine, assists in the control of the MMC. Additional hormones and neuropeptides assist in the regulation of small intestine motility and release of intestinal secretions. These include somatostatin, vasoactive inhibitory peptide, neurotensin, serotonin, cholecystokinin, orexin, leptin, and ghrelin.[7–9]

Motility of the small and large intestine is of much interest due to its possible role in several diseases and its importance in enteral nutrition support.[6, 10] Understanding motility and identifying target receptors for control of motility may assist in the treatment of disease. For example, irritable bowel syndrome,

Figure 15.2 Segmentation

Segmentation consists of ringlike contractions along the length of the small intestine. Within a matter of seconds, the contracted segments relax and the previously relaxed areas contract. The oscillating contractions thoroughly mix chyme within the small intestine lumen.

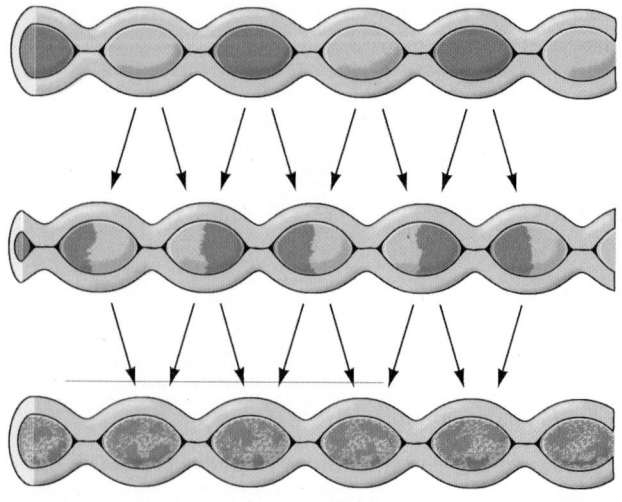

Source: L. Sherwood, Human Physiology: From Cells to Systems, 5e, copyright © 2004, p. 623.

chronic constipation, and diarrhea, which are discussed later in this chapter, are common disorders and may originate from abnormal motility.

The anatomy of the small intestine also contributes to motility and rate of transit. In particular, the ileocecal valve controls the rate of movement of food stuffs from the ileum (shown in Figure 15.3) to the ascending colon. Later sections of this chapter discuss how various disease states interrupt this normal function. This contributes to a faster transit time with subsequent risk for maldigestion and malabsorption.

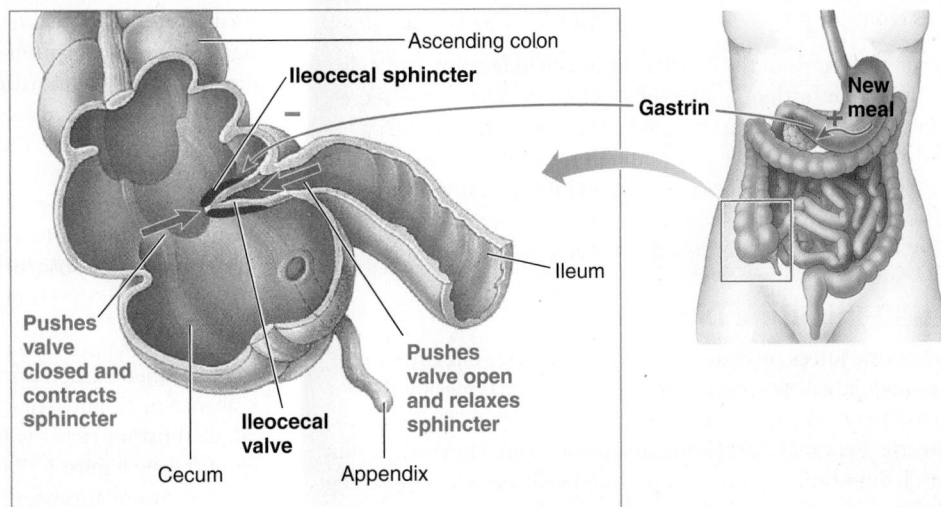

Figure 15.3 Ileocecal Valve

Small Intestine Secretions

The small intestine produces its own secretions and also receives secretions from ancillary organs of digestion, including the pancreas and gallbladder. These secretions include hormones, digestive enzymes, bicarbonate, and bile.

As chyme moves from the stomach into the duodenum, the hormones cholecystokinin, gastrin, and secretin stimulate release of pancreatic and gallbladder secretions (see Table 15.1). Bicarbonate from the pancreas allows neutralization of the very acidic chyme as it enters from the stomach. Neutralization protects the

Table 15.1 Gastrointestinal Hormones and Neuropeptides

Hormone	Source	Stimuli for Release	Function/Action
Gastrin	G cells in gastric atrium and proximal duodenum	Gastrin-releasing peptide Ingestion of protein, amino acids, peptides, coffee, alcohol, calcium; gastric distension, vagal stimulation, HCl in contact with gastric mucosa	Stimulates: acid secretion, pancreatic HCO_3 secretion, pancreatic enzyme secretion, gallbladder contraction, gastric motility, intestinal motility, insulin release, gastric oxyntic gland mucosa growth, pancreatic growth Relaxes ileocecal sphincter Inhibits gastric emptying
Secretin	S cells in duodenal mucosa	Acid in duodenal lumen	Stimulates: pancreatic HCO_3 secretion, pancreatic enzyme secretion, gallbladder contraction, insulin release, pancreatic growth Inhibits: acid secretion, gastric emptying and gastric motility, intestinal motility, mucosal growth
Cholecystokinin (CCK)	I cells of proximal duodenal mucosa	Nutrients in duodenal lumen, especially fat and to a lesser extent protein	Stimulates: acid secretion, pancreatic HCO_3 secretion, pancreatic enzyme secretion, gallbladder contraction, intestinal motility, insulin release, mucosal growth, pancreatic growth Inhibits: gastric emptying, gastric motility
Glucose-dependent insulinotropic peptide [gastric inhibitory peptide (GIP)]	K cells of duodenum and jejunum mucosa	Glucose, amino acids, fatty acids Interdigestive state	Stimulates: insulin release Inhibits: gastric emptying, gastric motility
Glucagon-like peptides (GLP)	Duodenum	Presence of food; glucose concentration	Stimulates insulin release Inhibits intestinal motility
Vasoactive intestinal polypeptide (VIP)	Inhibitory neuro-transmitter; enteric nervous system	Presence of food in digestive tract	Relaxes intestinal contractions Stimulates release of secretions in the colon
Motilin	M cells of duodenum and jejunum mucosa	Interdigestive state	Stimulates gastric motility between meals

Sources: Gropper SS, Smith JL, Groff JL: *Advanced Nutrition and Human Metabolism,* 5th ed., Belmont, CA: Cengage Learning, 2009; Sherwood L: *Human Physiology: From Cells to Systems,* 7th ed., Belmont, CA: Cengage Learning, Inc, 2010.

duodenum from acidity and supports a favorable environment for both digestion and absorption. Bile from the gallbladder supplies emulsification needed for adequate lipid digestion.

Other secretions of the small intestine include approximately 1.5 liters of intestinal "juices" or *succus entericus*. These secretions, which are primarily water and mucus, provide the appropriate water-soluble environment for digestion and provide protection to the mucosa of the small intestine. Important digestive enzymes are secreted at the brush border of the small intestine and will be discussed in the next section.

Small Intestine Digestion

Pancreatic juices provide the primary digestive enzymes in the small intestine. These protein digestive enzymes include trypsinogen, chymotrypsinogen, procarboxypeptidases, and elastase. Pancreatic amylase is the primary enzyme involved in starch digestion. Pancreatic lipase and colipase accomplish the largest proportion of lipid digestion.

Brush border enzymes in the small intestine include lactase, alpha-dextrinase, sucrase, maltase, and glucosidase, which provide final digestion of all carbohydrates (see Figure 15.4). Other brush border enzymes include enterokinase, which activates the pancreatic enzyme trypsinogen. Trypsin then activates other trypsinogen molecules and pancreatic proenzymes. Together, these enzymes degrade protein into smaller units (oligopeptides of two to six amino acids and free amino acids). Peptidases, located in the brush border, are responsible for digestion of oligopeptides into free amino acids, dipeptides, and tripeptides that can then be absorbed (see Figure 15.5).

Small Intestine Absorption

The anatomy of the small intestine, as discussed earlier, is uniquely constructed to accomplish maximal digestion and absorption. Each villus contains access to the circulatory and lymphatic systems via capillaries and lymphatic vessels, and villi thus provide necessary routes for absorbed nutrients.

Absorption for end products of digestion occurs primarily through active transport and may utilize a **Na$^+$/K$^+$ pump** system at the brush border. Glucose, galactose, and amino acids utilize this type of absorption mechanism. Fructose uses facilitated/carrier-mediated transport as its absorptive mechanism and is affected by the presence of glucose, amino acids, and the total amount of fructose.[11, 12] Fructose malabsorption has been of recent interest for its potential contribution to gastrointestinal symptoms and will be discussed later in this chapter.[12, 13] Additional research indicates some nutrients, including glucose, may be absorbed in part through a paracellular route. This means that small amounts of nutrients may leak between epithelial cells. Other recent research has indicated the amount of glucose absorbed may be directly related to motility of the duodenum.[14]

Lipid absorption is much more difficult due to its insolubility in water. For lipids to be successfully absorbed, they must undergo several steps and utilize several protein carriers. Fatty acids and other lipid components must be first incorporated into micelles in the gut lumen before absorption into the enterocytes. Then, they must be incorporated into chylomicrons (a type of lipoprotein) that can then enter the lymphatic system via passive absorption (see Figure 15.6). Due to this complex absorption mechanism, many diseases of the small intestine can interrupt normal fat digestion and absorption. Understanding steps of fat digestion and absorption can assist in differentiation of diseases affecting the small intestine or ancillary organs of digestion. For example, pancreatic disease may reduce the amount of pancreatic lipase available for adequate digestion. Crohn's disease may decrease transit time and reduce the ability of the small intestine to accomplish all the steps required for digestion and absorption. **Steatorrhea** is the condition that exists when lipid is not digested or absorbed correctly. The result is an abnormal amount of fat in the stool (see "Fat Malabsorption" for further discussion).

As mentioned earlier, portions of the small intestine can adapt to absorb most nutrients. However, in a normal, healthy individual, most nutrients will be absorbed in the duodenum and jejunum (see Figure 15.7 for sites of absorption). The ileum can also accommodate absorption of many nutrients if foodstuffs remain there long enough.[15] One exception to this is the absorption of B$_{12}$, which can occur only at specific receptor sites in the ileum.

The ileum is also the primary site for reabsorption of bile acids. This process is referred to as the enterohepatic circulation of bile acids. Some bile acids may also be reabsorbed in the jejunum and colon. Since bile acids are exclusively produced in the body, these substances need to be recirculated back to the liver in order to maintain an adequate body pool of approximately 4 grams. Bile acids recirculate from the small and large intestine back to the liver approximately six to eight times per day. When disease interrupts enterohepatic circulation, fat malabsorption can occur. Furthermore, in hypercholesterolemia, the medication cholestyramine is used to bind bile acids in order to decrease the body pool and reduce serum levels of cholesterol.

Large Intestine Anatomy

The anatomy of the large intestine has both significant differences from and important similarities to the anatomy of the small intestine. First, mucosa of the large intestine form three nearly straight portions rather than the circular folds found in the small intestine. The colon's major portions are referred to as the ascending colon, the transverse colon, and the descending colon (see Figure 15.8). The final section of the colon is referred to as the sigmoid colon due to its "S" shape. The sigmoid colon ends in the rectum where another sphincter (the anal sphincter) controls voluntary release of intestinal contents.

Secondly, the large intestine does not have villi or microvilli. But there are large pits or crypts (crypts of Lieberkuhn) that are similar to the crypts between villi in the small intestine. Again, similar to the small intestine, cells are generated within these crypts and, after migration, differentiate into specialized epithelial cells such as goblet cells, which produce mucus.

Large Intestine Motility

The musculature of the large intestine provides the basic structure that supports its motility. The large intestine has repeating bands of longitudinal skeletal muscle (called taeniae coli) that follow the length of the colon and circular smooth muscle that covers the entire organ.

Figure 15.4 Carbohydrate Digestion and Absorption

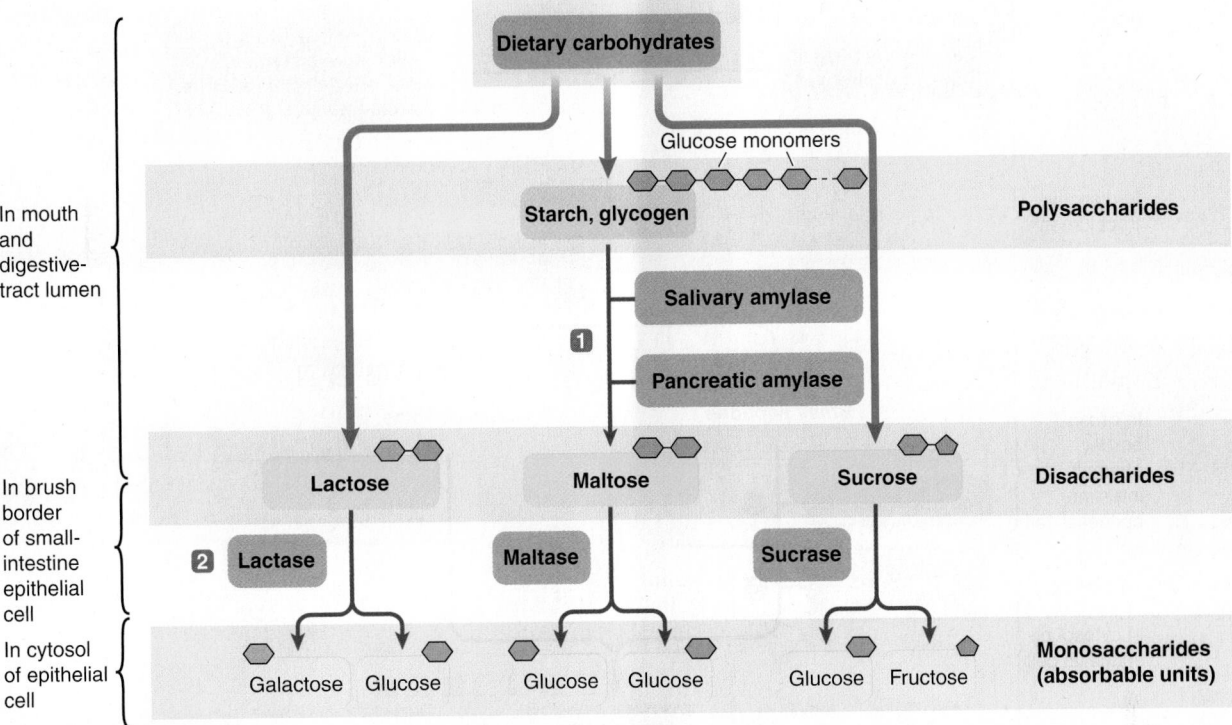

(a) Carbohydrate digestion

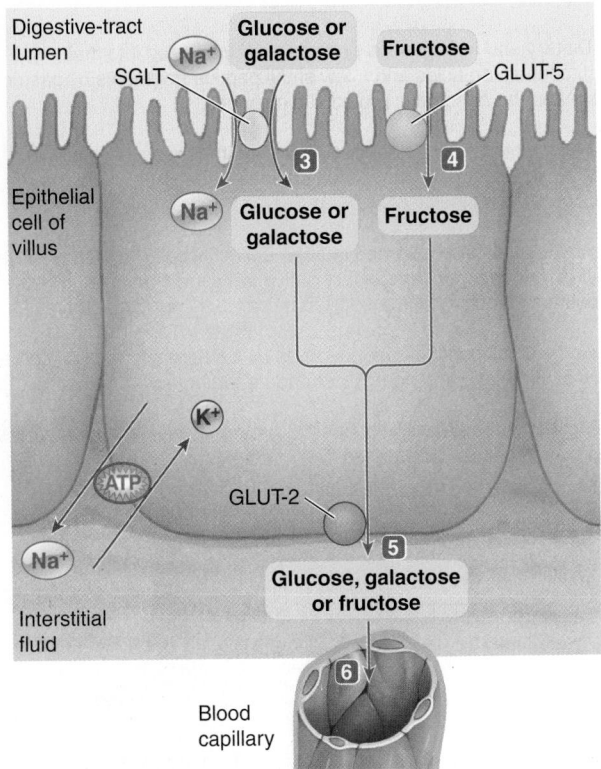

1 The dietary polysaccharides starch and glycogen are converted into the disaccharide maltose through the action of salivary and pancreatic amylase.

2 Maltose and the dietary disaccharides lactose and sucrose are converted to their respective monosaccharides by the disaccharidases (maltase, lactase, and sucrase) located in the brush borders of the small-intestine epithelial cells.

3 The monosaccharides glucose and galactose are absorbed into the epithelial cells by Na^+- and energy-dependent secondary active transport (via the symporter SGLT) located at the luminal membrane.

4 The monosaccharide fructose enters the cell by passive facilitated diffusion via GLUT-5.

5 Glucose, galactose, and fructose exit the cell at the basal membrane by passive facilitated diffusion via GLUT-2.

6 These monosaccharides enter the blood by simple diffusion.

KEY

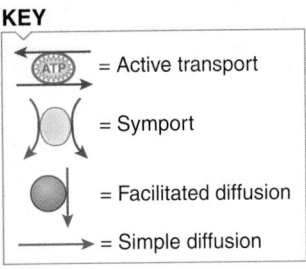

= Active transport	
= Symport	
= Facilitated diffusion	
= Simple diffusion	

(b) Carbohydrate absorption

Source: L. Sherwood, Human Physiology: From Cells to Systems, 7e, copyright © 2010, p. 627, F 16-24.

Figure 15.5 Protein Digestion and Absorption

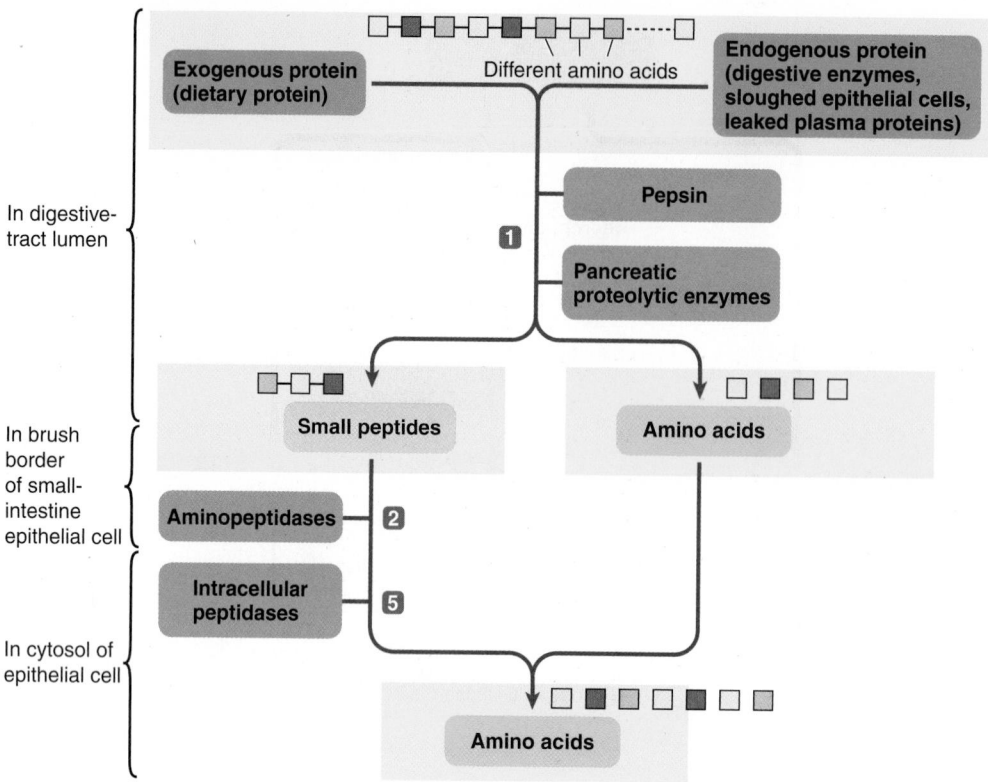

Exogenous protein (dietary protein)

Different amino acids

Endogenous protein (digestive enzymes, sloughed epithelial cells, leaked plasma proteins)

In digestive-tract lumen

1 — **Pepsin** — **Pancreatic proteolytic enzymes**

In brush border of small-intestine epithelial cell

Small peptides

Amino acids

Aminopeptidases — **2**

In cytosol of epithelial cell

Intracellular peptidases — **5**

Amino acids

(a) Protein digestion

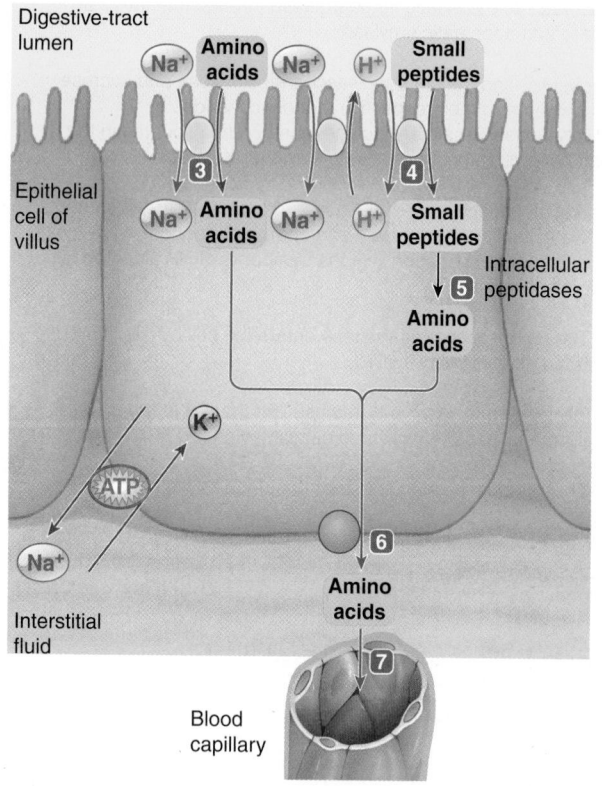

(b) Protein absorption

1 Dietary and endogenous proteins are hydrolyzed into their constituent amino acids and a few small peptide fragments by gastric pepsin and the pancreatic proteolytic enzymes.

2 Many small peptides are converted into their respective amino acids by the aminopeptidases located in the brush borders of the small-intestine epithelial cells.

3 Amino acids are absorbed into the epithelial cells by means of Na^+- and energy-dependent secondary active transport via a symporter. Various amino acids are transported by carriers specific for them.

4 Some small peptides are absorbed by a different type of symporter driven by H^+, Na^+-, and energy-dependent tertiary active transport.

5 Most absorbed small peptides are broken down into their amino acids by intracellular peptidases.

6 Amino acids exit the cell at the basal membrane via various passive carriers.

7 Amino acids enter the blood by simple diffusion. (A small percentage of di- and tripeptides also enter the blood intact.)

KEY

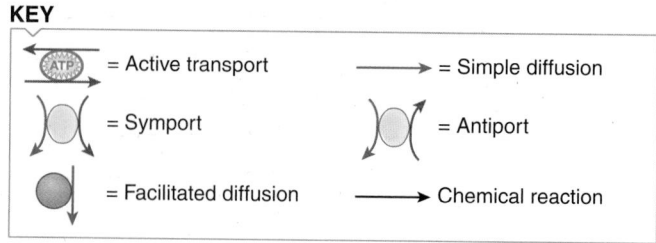

= Active transport

= Simple diffusion

= Symport

= Antiport

= Facilitated diffusion

= Chemical reaction

Source: L. Sherwood, Human Physiology: From Cells to Systems, 7e, copyright © 2010, p. 629, F 16-25.

Figure 15.6 Lipid Digestion and Absorption

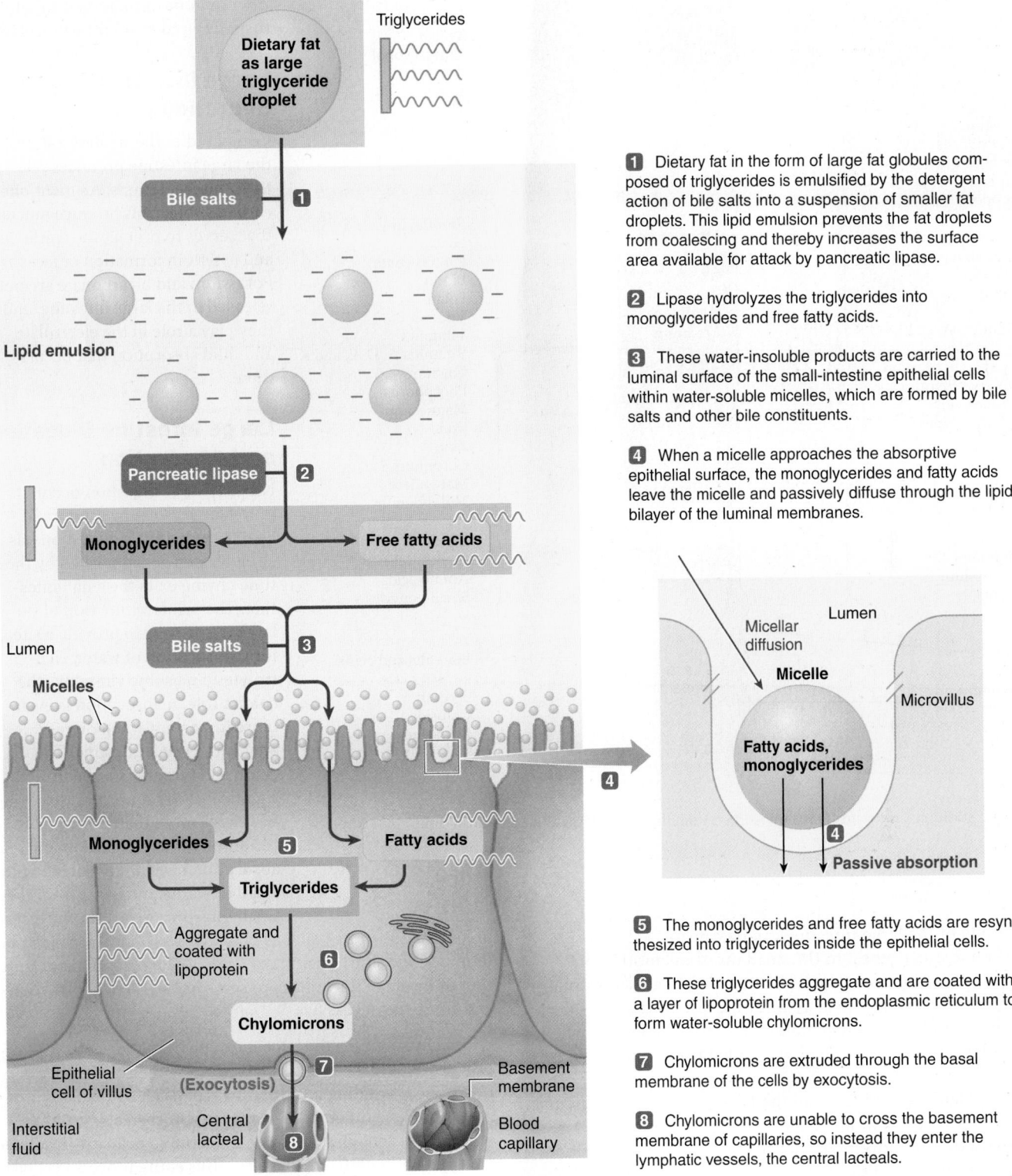

1 Dietary fat in the form of large fat globules composed of triglycerides is emulsified by the detergent action of bile salts into a suspension of smaller fat droplets. This lipid emulsion prevents the fat droplets from coalescing and thereby increases the surface area available for attack by pancreatic lipase.

2 Lipase hydrolyzes the triglycerides into monoglycerides and free fatty acids.

3 These water-insoluble products are carried to the luminal surface of the small-intestine epithelial cells within water-soluble micelles, which are formed by bile salts and other bile constituents.

4 When a micelle approaches the absorptive epithelial surface, the monoglycerides and fatty acids leave the micelle and passively diffuse through the lipid bilayer of the luminal membranes.

5 The monoglycerides and free fatty acids are resynthesized into triglycerides inside the epithelial cells.

6 These triglycerides aggregate and are coated with a layer of lipoprotein from the endoplasmic reticulum to form water-soluble chylomicrons.

7 Chylomicrons are extruded through the basal membrane of the cells by exocytosis.

8 Chylomicrons are unable to cross the basement membrane of capillaries, so instead they enter the lymphatic vessels, the central lacteals.

Source: L. Sherwood, Human Physiology: From Cells to Systems, 7e, copyright © 2010, p. 630, F 16-26.

Figure 15.7 Sites of Nutrient Absorption

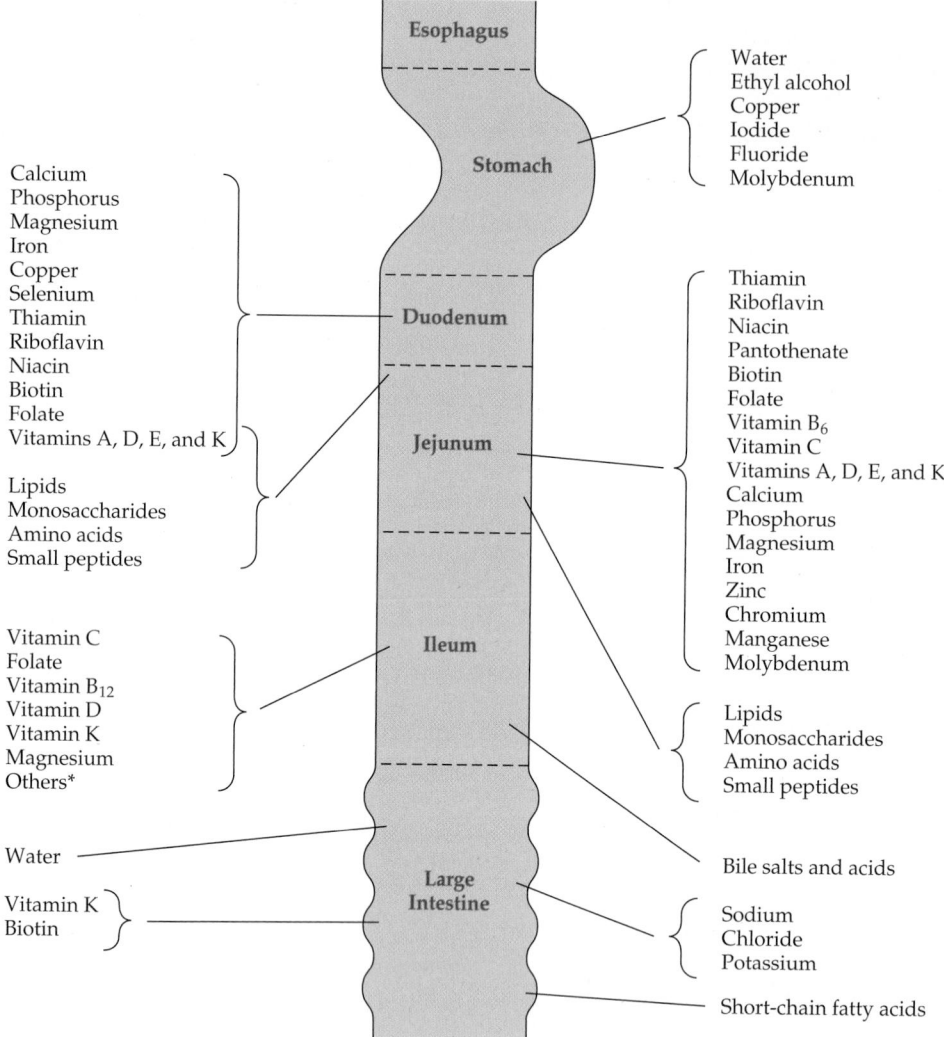

Esophagus

Stomach
- Water
- Ethyl alcohol
- Copper
- Iodide
- Fluoride
- Molybdenum

Calcium
Phosphorus
Magnesium
Iron
Copper
Selenium
Thiamin
Riboflavin
Niacin
Biotin
Folate
Vitamins A, D, E, and K

Lipids
Monosaccharides
Amino acids
Small peptides

Duodenum

Jejunum
- Thiamin
- Riboflavin
- Niacin
- Pantothenate
- Biotin
- Folate
- Vitamin B_6
- Vitamin C
- Vitamins A, D, E, and K
- Calcium
- Phosphorus
- Magnesium
- Iron
- Zinc
- Chromium
- Manganese
- Molybdenum

Vitamin C
Folate
Vitamin B_{12}
Vitamin D
Vitamin K
Magnesium
Others*

Ileum
- Lipids
- Monosaccharides
- Amino acids
- Small peptides

Water

Vitamin K
Biotin

Large Intestine
- Bile salts and acids
- Sodium
- Chloride
- Potassium
- Short-chain fatty acids

*Many additional nutrients may be absorbed from the ileum depending on transit time.

Source: J. Smith, J. Groff, and S. Gropper, *Advanced Nutrition and Human Metabolism*, 4e, copyright © 2005, p. 47.

tion occurs when distention of the rectum relaxes the anal sphincter. This final movement is ultimately (usually) under voluntary control.

Large Intestine Secretions

Compared to the small intestine, the large intestine produces relatively few secretions. As mentioned earlier, goblet cells produce mucus that serves to protect the epithelium and assists in formation of feces. Potassium and bicarbonate are both released in the large intestine, and they play a role in the electrolyte and fluid absorption that occurs there.

Large Intestine Digestion and Absorption

No enzymatic digestion occurs in the large intestine. In normal, healthy individuals, digestion has already been accomplished by the time chyme exits the small intestine. The primary function of the large intestine is to provide a site for reabsorption of water, electrolytes, and some vitamins. The colon's role in absorption is even more important when disease affects the small intestine. The colon can increase its absorption significantly—as much as three to five times more than normal.[15] In conditions where digestion and absorption have not occurred in the small intestine, nutrients from the small intestine are lost in the feces unless the substrate (such as fiber and resistant starch) can be fermented to short-chain fatty acids.[16] The second major function of the large intestine is to serve as the site for formation and storage of feces.

When chyme enters the large intestine from the ileum, it is primarily liquid. During movement along the colon, water is reabsorbed, resulting in a drier mass of fecal matter. Sodium, potassium, and other electrolytes are absorbed along with water. Feces contain undigested foodstuffs—primarily insoluble fiber, only **bilirubin**, and bacteria. This entire process may take anywhere from 12 to 72 hours.[17]

As many as 400 different species of bacteria—including bifidobacteria, coliforms, bacteroides, peptococci, clostridia, lactobacteria, and methanogens—live within the colon. They provide fermentation of fiber, resistant starch, and sugar alcohols. As these substrates undergo fermentation, short-chain fatty acids (SCFA) (acetate, propionate, butyrate) and lactate are produced. Some of the energy produced during fermentation is

Motility of the large intestine can be categorized into several distinct types. In the small intestine, motility includes segmentation, which allows for mixing of intestinal contents. Likewise, the large intestine also uses a type of segmentation, called haustration. Haustration occurs when circular muscle forms small sacs called haustra. Haustra hold amounts of chyme as it is mixed with secretions of the colon. Haustra can form and then disappear when intestinal contents are moved through the colon (see Figure 15.8). Other types of movement that accomplish motility include propulsion, mass movements, and defecation. Propulsion is accomplished by alternating waves of relaxation and contraction of smooth muscle lasting for several minutes. Intestinal contents can move in both directions within the colon, allowing for adequate absorption of fluid and electrolytes. Mass movements occur when there is a significant contraction of a large portion of the colon. This generally occurs several times a day and will accomplish moving a large portion of intestinal contents along the colon. Finally, defeca-

Figure 15.8 Large Intestine Anatomy

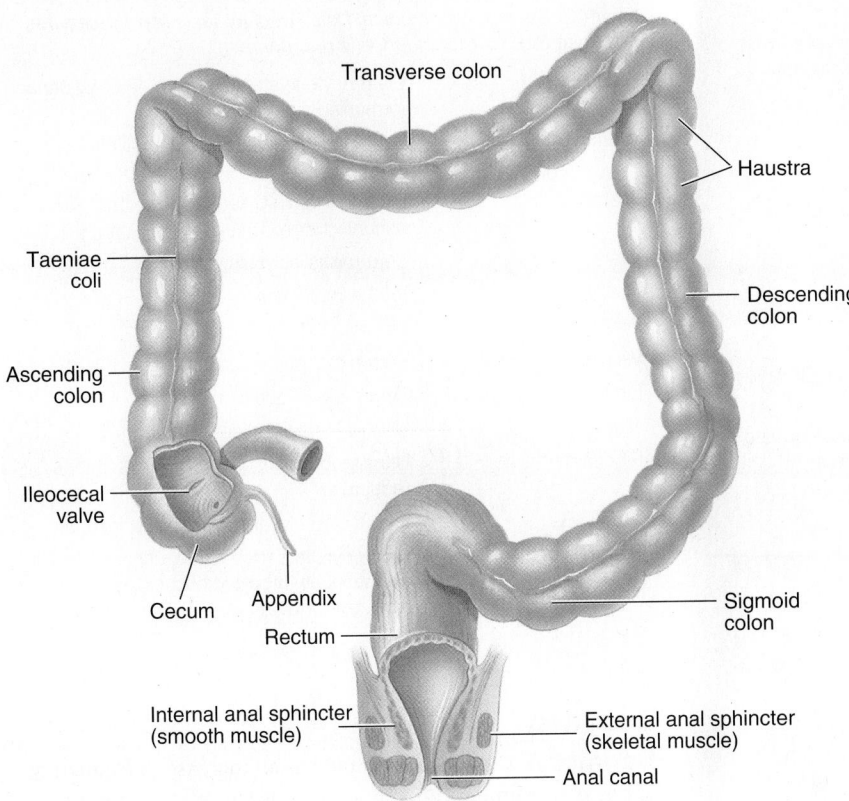

Labels: Transverse colon, Haustra, Taeniae coli, Descending colon, Ascending colon, Ileocecal valve, Cecum, Appendix, Rectum, Sigmoid colon, Internal anal sphincter (smooth muscle), External anal sphincter (skeletal muscle), Anal canal

Source: L. Sherwood, Human Physiology: From Cells to Systems, 7e, copyright © 2010, p. 635, F 16-28.

starch and soluble fiber, for example, benefit the large intestine not only by their physical presence but by fermentation that results with the interaction of these undigested foodstuffs and the colonic flora.[23] Lactate and short-chain fatty acids that result from the fermentation can then be absorbed from the colon and utilized elsewhere in the body. Probiotics, by altering the flora of the large intestine, may improve the microbial balance, normalize transit time, and enhance the immune defenses provided in the gut mucosa.[23] Other by-products from metabolism of these bacteria include gas and ammonia. As described in Chapter 16, in order to assist in control of abnormal ammonia levels in persons with liver disease, medications are given to induce diarrhea, which decreases the ability of the colon to reabsorb this ammonia.

Vitamin K and biotin are two endogenously produced vitamins. Biotin is produced by normal intestinal flora in the colon and is absorbed via passive diffusion.[28] *E. coli* and *Bacteroides fragilis* in the colon synthesize vitamin K. The absorption route for this endogenous vitamin K is not clear at this time. Truly, for both of these vitamins, it is difficult to estimate the contribution of endogenous synthesis.[29] Diseases of the lower GI tract, the use of antibiotics, and the presence of prebiotics and probiotics may all potentially interfere with and/or promote endogenous synthesis.

used directly by bacteria for their own support, but the SCFA they release can provide 500 to 1200 kcalories per day for the human host.[15, 18–20] These SCFA are either utilized by the colon for support of its own tissue growth or absorbed for utilization by the body elsewhere. When there is excessive substrate in the colon—such as undigested carbohydrates—gas and **flatulence** may result. This will be discussed in greater detail later in this chapter.

Many factors affect not only the amount but also the types of bacteria in the colon. These factors include age and health status, composition of the diet, transit time, stress, and alcohol intake. Maintaining an optimal balance of intestinal flora is of significant interest in human health and nutrition. The use of **resistant starch**, **prebiotics**, **probiotics**, and **synbiotics** is currently being studied in an effort to determine their role in the promotion of the health of the colon and in the prevention and treatment of disease.[21–26]

Resistant starch is defined as starch components that enter the large intestine undigested.[22, 27] Examples of resistant starch include potato, banana, and some legumes. Prebiotics are substances in food (such as **inulin** and oligosaccharides) that stimulate the beneficial flora of the large intestine. Probiotics are products containing live microorganisms that are sold as foods and supplements. Synbiotics are those products that contain both prebiotics and probiotics.[22]

Intestinal flora use undigested carbohydrate and small amounts of protein to support their own growth. Resistant

Nutrition Assessment for Lower Gastrointestinal Tract Conditions

As covered in Chapter 3, nutrition assessment involves gathering the appropriate data in order to identify nutrition problems and establish nutritional status. In the following sections, multiple disorders and conditions will be covered. Though these conditions have distinct etiologies and medical care, they share common factors that guide the nutrition assessment. Table 15.2 summarizes nutrition assessment for lower gastrointestinal conditions.

Pathophysiology of the Lower Gastrointestinal Tract

This section will begin with a focus on several common conditions that may occur by themselves but may also be symptoms associated with specific diseases. These conditions, which include diarrhea, constipation, and malabsorption, often require medical and nutritional intervention. It is important to remember that their etiology is multifactorial and may be a symptom of an underlying disorder. In the latter sections of this chapter, the discussion of pathophysiology will focus on specific disorders and diagnoses.

Table 15.2 Nutrition Assessment for the Lower Gastrointestinal Tract

Client History (personal, patient, family, medical and social)	• Previous medical conditions or surgeries • Medications: both prescription and over-the-counter (laxatives, fiber supplements) • Socioeconomic status/food security • Support systems • Education—primary language
Food-/Nutrition-Related History	• Ability to chew; use and fit of dentures • Problems swallowing • Nausea, vomiting • Constipation, diarrhea • Heartburn • Any other symptoms interfering with ability to ingest normal diet • Ability to consistently purchase adequate amounts of food on a daily basis • Ability to feed self • Ability to cook and prepare meals • Food allergies, preferences, or intolerances • Previous food restrictions • Ethnic, cultural and religious influences • Use of alcohol, vitamin, mineral, herbal or other type of supplements • Previous nutrition education or medical nutrition therapy • Eating pattern: 24-hour recall, diet history, food frequency
Anthropometric Measurements	• Height • Current weight • Weight history: highest adult weight; usual body weight • Reference weight (BMI)
Biochemical Data, Medical Tests and Procedures	
Visceral Protein Assessment:	• Albumin • Prealbumin • Transferrrin • Retinol Binding Protein • C Reactive Protein
Immunocompetence:	• Total Lymphocyte Count
Hematological Assessment:	• Hemoglobin • Hematocrit • MCV • MCHC • MCH • TIBC
Lipid Assessment:	• Total cholesterol • HDL • LDL • Triglyceride

Lower GI-Specific Biochemical Data, Medical Tests and Procedures (Also see Box 15.4 for diagnostic procedures):

Folate	Erythrocyte folate, free folate, (FIGLU) urinary formiminoglutamic acid
Vitamin B_{12}	Schilling test, erythrocyte B_{12}, DUMP test, serum B_{12}
Vitamin C	Plasma vitamin C, leukocyte vitamin C, urinary vitamin C
Vitamin D	Serum alkaline phosphatase
Vitamin K	Prothrombin time
Vitamin A	Serum carotene, retinal binding protein
Vitamin E	Serum tocopherol, erythrocyte hemolysis
Biotin	Serum biotin, urinary biotin
Niacin	Urinary N-methyl nicotinamide
Riboflavin	Urinary riboflavin, erythrocyte glutathione reductase
Vitamin B_6	Whole blood level of pyridoxal phosphate
Thiamin	Blood pyruvate and lactate, urinary thiamin excretion, erythrocyte transketolase, apoenzyme levels

Diarrhea

Definition **Diarrhea** is defined as an increase in frequency of bowel movements and/or an increase in water content of stools that affects either the consistency or the volume of fecal output. Other definitions describe abnormality in stool production as >200 g/day for adults and >20 g/kg for children.[30]

Epidemiology Incidence of infectious diarrhea is estimated to be approximately 99 million new cases each year, resulting in 3100 deaths each year in the United States. Food-borne illnesses are a major cause of these cases. According to the Foodborne Diseases Active Surveillance Network there were more than 45 cases of bacterial and parasitic infections per 100,000 persons in the United States during the year 2008.[31] The highest incidence was for *Salmonella* and *Campylobacter* infections. More than 500,000 deaths worldwide have been associated with rotavirus.[32] Because over 2 million children will die from dehydration secondary to diarrhea each year, it is a primary focus of public health intervention throughout the world.[33]

Etiology Diarrhea can be described as either acute or chronic in origin. Diarrhea can also be classified as either osmotic or secretory. The etiology of the diarrhea can also serve as a framework for discussion and understanding of diarrhea-related conditions. Acute diarrhea is short-term (less than two weeks), whereas diarrhea lasting longer than four weeks is considered chronic.[30] Diarrhea can be associated with a number of health concerns, such as electrolyte imbalances, malabsorption, dehydration, and malnutrition.

Osmolality is a measurement of concentration of particles in solution. Normal osmolality of the gastrointestinal tract is approximately 300 mOsm/L. When there is an increase in osmotically active particles in the intestine, the body responds by pulling water into the lumen in an attempt to normalize osmolality. When this occurs, increased water efflux results in what we refer to as osmotic diarrhea. Osmotic diarrhea can be caused by maldigestion of nutrients, excessive **sorbitol** or **fructose** intake, enteral feeding, and some laxatives. In general, when the causative agent is removed, osmotic diarrhea will cease. An example would be a patient whose diarrhea resolves when he is made NPO. This is one of the major differentiations between osmotic and secretory diarrhea.[34]

Secretory diarrhea also results from excessive fluid and electrolyte secretions into the intestine. The difference here is that the underlying disease is what causes excessive secretions, not the hyperosmolality. Furthermore, secretory diarrhea does not resolve when the patient is made NPO. Bacterial infections often produce enterotoxins that result in this type of diarrhea. Protozoa, viruses, and other infections can also cause secretory diarrhea. Traveler's diarrhea is a common health problem affecting those who travel to other countries. The major infectious agents resulting in traveler's diarrhea are enterotoxigenic *Escherichia coli*, enteroaggregative *E. coli*, and *Shigella* spp, *Salmonella*, *Campylobacter*, *Yersinia*, *Aeromonas*, and *Plesiomonas* spp.[35] Other factors that could potentially cause secretory diarrhea include medications, hormone-producing tumors, prostaglandins, and excessive amounts of bile acids or unabsorbed fatty acids in the colon.[30, 34]

Antibiotics and other medications may cause diarrhea as a side effect. These medications generally cause diarrhea either by increasing GI motility or by altering the normal flora of the colon.[26] See Table 15.3 for a list of frequently used antibiotics and other medications that can cause diarrhea.

Table 15.3 Medications That May Cause Diarrhea

Classification	Medication	
	Generic Name	**Trade Name**
Antacids		
Magnesium-containing antacids		Milk of Magnesia
H₂ receptor antagonists	Ranitidine	Zantac
	Cimetidine	Tagamet
	Famotidine	Pepcid
	Nizatidine	Axid
Proton pump inhibitors (PPI therapy)	Omeprazole	Prilosec
	Lansoprazole	Prevacid
	Pantoprazole	Protonix
	Rabeprazole	AcipHex
	Esomeprazole	Nexium
Antibiotics	Clindamycin	Cleocin
	Ampicillin	Penicillin, Unasyn
	Cephalosporins	Ancef, Cefizox, Cefobid, Cefotan, Ceptaz, Claforan, Fortaz, Kefzol, Mandol, Maxipime, Mefoxin, Monocid, Rocephin, Tazicef, Tazidime, Zefazone

Table 15.3 Medications That May Cause Diarrhea *(Continued)*

Classification	Medication	
	Generic Name	**Trade Name**
Antibiotics	Erythromycin	E-Mycin
	Etracycline	Achromycin-V, Sumycin, Tetracycline
	Sulfonamides	Gantanol, Gantrisin, Novo-Soxazole
	Any broad-spectrum antibiotic	
Anti-Inflammatory Medications		
NSAIDs	Acetylsalicylic acid	Aspirin, Ascriptin, Bufferin, Ecotrin
	Ibuprofen	Advil, Motrin
	Naproxen	Aleve, Anaprox
Antigout	Colchicine	Colchicine
Cardiac Drugs		
Sodium channel blockers	Quinidine	Quinaglute
	Procainamide	Pronestyl
	Disopyramide	Norpace
	Phenytoin	Dilantin
	Bretylium	Bretylol
Beta blockers	Esmolol hydrochloride	Brevibloc
	Carvedilol	Coreg
	Timolol	Blocadren
	Propranolol	Inderal
	Metoprolol	Lopressor
	Atenolol	Tenormin
	Nadolol	Corgard
ACE inhibitors	Captopril	Capoten
	Enalapril	Vasotec
	Lisinopril	Prinivil, Zestril
	Fosinopril	Monopril
	Benazepril	Lotensin
	Moexipril	Univasc
	Perindopril	Aceon
	Quinapril	Accupril
	Ramipril	Altace
	Trandopril	Mavik
Antiarrhythmic	Digitalis	Digitek, Digitoxin, Digoxin, Lamoxin
Antihypertensives	Reserpine	Serpalan, Serpasi
	Guanethidine	Ismelin
	Methyldopa	Aldomet, Aldoril
	Guanabenz	Wytensin
	Guanadrel	Hylorel
	Hydralazine	Apresoline

(continued)

Table 15.3 Medications That May Cause Diarrhea (Continued)

Classification	Medication	
	Generic Name	Trade Name
Cholinergics	Bethanechol	Duvoid, Reglan, Urabeth,
	Metoclopramide	Urecholine
	Neostigmine	
Lipid Lowering Medications	Clofibrate	Abitrate, Atromid-S, Novofibrate
	Gemfibrozil	Lopid
	HMG-CoA reductase Inhibitors	Lovastatin, fluvastatin, pravastatin
Laxatives		Castor oil, Citrucel, Colace, Correctol, Dulcolax, Ex-Lax, Fiberall, Fleet laxative, herbal laxatives (senna, etc.), Metamucil, Milk of Magnesia, Purge, Senokot
Neuropsychiatric drugs	Lithium	Eskalith CR (SR), Lithobid (SR), Lithotabs
	Fluoxetine	Prozac, Prozac Weekly, Sarafem
	Alprazolam	Xanax
	Valproic acid	
	Ethosuximide	Zarontin
Miscellaneous agents	Theophylline (bronchodilator)	Elixophyllin, Slo-Phyllin, Theo-24, Theobid, Theo-Dur, Theolair, Uniphyl
	Thyroid hormones	Synthroid, Levoxyl, Unithroid
	Misoprostol	Cytotec, PGEI gel
	Some chemo-therapeutic agents	Methotrexate

Sources: Family Practice Notebook.com. Diarrhea secondary to medications: drug-induced diarrhea. Available at http://www.fpnotebook.com/GI/Pharm/DrhScndryTMdctns.htm. Medline Plus. Available at http://www.nlm.nih.gov/medlineplus/druginformation.html

Clostridium difficile is a Gram-positive anaerobic bacterium that is the major cause of antibiotic-related diarrhea and is generally found in hospital and long-term care environments. When antibiotics are prescribed, their use can disturb the balance of normal flora of the colon. When this occurs, *C. difficile* has the potential to proliferate. Its endotoxins result in injury and inflammation to the colon. The most common antibiotics associated with *C. difficile* infection are ampicillin, amoxicillin, cephalosporins, and clindamycin. Symptoms can range from mild diarrhea to severe colitis. In mild cases, stopping the prescribed antibiotic will be enough to stop the infection. In more severe cases, antibiotics, such as metronidazole or vancomycin, are generally used to treat *C. difficile* infections.[36]

Enteral feedings are often implicated as a cause of diarrhea. Even though changing types of formulas or reducing the rate of feeding is often proposed, the evidence for the effectiveness of this strategy is limited.[37] Antibiotics and other prescribed medications are much more likely to be the cause.[38] As will be discussed later in this section, the use of fiber-enriched formulas provides an additional route of treatment. Traditional sources of fiber added to enteral formulas are soy polysaccharides (primarily insoluble), guar, and pectin (primarily soluble). Newer sources include fructooligosaccharides (FOS) and galacto-oligosaccharides (GOS). Most fiber-enriched enteral formulas contain a blend of both FOS, GOS, and insoluble fiber sources.[39, 40] Delegge and Berry state that for the patient with diarrhea, it is appropriate to continue to feed enterally unless there is abdominal pain, fever, or distention; or fluid output is greater than input.[37] If the stool output looks the same as the feeding, this indicates rapid transit and malabsorption.

Many gastrointestinal diseases have diarrhea as a common symptom. If the underlying disease disrupts normal digestion and absorptive capabilities, diarrhea will most often result. Examples of these conditions that will be covered in this chapter include Crohn's disease, ulcerative colitis, and celiac disease. These diagnoses can also result in malabsorption of lipids and other nutrients, which further contributes to the diarrhea.

Other diseases that do not originate in the gastrointestinal tract can also present with symptoms of diarrhea. These may include, but are not limited to, AIDS enteropathy with HIV infection, thyroid dysfunction, and some malignancies.

Clinical Manifestations Diarrhea presents as a change from the normal bowel function. This is generally a watery stool that is increased in frequency. Other characteristics of stool output will vary depending on the etiology of the diarrhea. For example, foul-smelling, frothy stools are associated with steatorrhea.

Composition and volume of stool is also consistent with the etiology of the diarrhea. Blood may be present in the stool, and is characterized as "frank" blood, occult blood, or melena. Frank blood is bright red blood on the surface of the stool, and represents contamination with blood from the rectum or anus. Occult blood is detected by testing the stool and usually results from bleeding in the lower gastrointestinal tract. Melena is a dark stool and is caused by contamination with blood within the upper GI tract. Hemoglobin from blood contributes to the dark color. Mucus in the stool may also be indicative of secretory diarrhea and may be infectious in origin. High amounts of electrolytes are also consistent with secretory diarrhea.[30] The presence of leukocytes in the stool indicates an inflammatory process such as inflammatory bowel disease, which will be discussed later in this chapter.

Other clinical manifestations that may occur with diarrhea are abdominal pain and cramping. When defecation relieves cramping, diarrhea is generally from the distal colon. If abdominal pain and cramping continue after defecation, the origin is generally from the small bowel. Other symptoms such as dehydration, weight loss, and electrolyte and acid-base imbalances are dependent on volumes of stool lost and represent one of the most serious consequences of diarrhea.

Medical Diagnosis Diagnosis of the underlying etiology is the most important step in determining treatment of diarrhea. Considerations that will direct diagnostic procedures are the age of the patient, hydration status, the presence of blood in the stool, and whether the patient is immunocompromised. Other important symptoms to note are any recurring characteristics of diarrheal episodes, including time of day or any relationship to food intake.

Typical diagnostic work-up will begin with stool cultures that will be examined for microorganisms, ova and parasites, leukocytes, **lactoferrin**, and the presence of blood. Further invasive procedures such as upper endoscopy, flexible sigmoidoscopy, or colonoscopy may assist in diagnoses not determined with initial stool cultures.

Osmolality and electrolyte content of the stool can also be determined, and will assist in differentiation between osmotic and secretory diarrhea, but is not as often used in clinical practice.[41] Other clinical tests will measure complete blood count, electrolytes, albumin, and thyroid-stimulating hormone. Specific diagnostic tests for *C. difficile* include stool sample culture and tests using the cytotoxicity assay or the enzyme immunoassay for Toxins A and B.[36]

Treatment Treatment of the underlying disorder is the most important component of therapy. If the diarrhea is infectious in nature, antibiotics will be the first line of treatment. Restoring normal fluid, electrolyte, and acid-base balance is crucial. This is accomplished through either intravenous therapy or the use of rehydration solutions (discussed in the Nutrition Therapy section).[30, 42]

Other medications can be used to treat the symptoms of diarrhea. These agents work either to decrease motility or to thicken the consistency of the stool. These include medications such as LoMotil®, Immodium, Tincture of Opium, paregoric, Kaopectate®, or bismuth subsalicylate. It is important to note any medication side effects for these drugs. See Table 15.4 for a list of antidiarrheal medications and possible side effects/drug-nutrient interactions.

Table 15.4 Common Medications Used in Lower Gastrointestinal Disorders

Selected Medications for Lower Gastrointestinal Diseases

Classification	Mechanism	Generic	Brand Names	Possible Food-Drug Interactions
Antidiarrheal: Adsorbents	Provide protective coating for intestinal walls; absorb toxins, viruses, or bacteria	Kaolin, pectin, methylcellulose, activated attpulgite, magnesium aluminum silicate	Kaopectate Advanced Formula, Donnagel, Diasorb, Rheaban Maximum Strength, Equalactin	Constipation, bloating
		Polycarbophil	FiberCon, Fiberall, Mitrolan, Equalactin	Constipation, bloating
Antidiarrheal: Anticholinergics	Decrease intestinal muscle tone and peristalsis of GI tract to slow movement of fecal material through GI tract	Belladonna		Dry mouth/constipation
		Alkaloids		Dry mouth/constipation
		Atropine		Dry mouth/constipation
		Hyoscyamine	Anaspaz, Cystospaz, Cystospaz-M, Levsin, Neoquess	Dry mouth/constipation
Aminosalicylates: Anti-inflammatory	Works as an anti-inflammatory agent in the colon and may also act as an immune suppressant	Sulfasalazine	Azulfadine	Take with 8 oz water after meals or with food, take Fe or folate separately, folate supplement (1 mg/d), ↑ dietary folate, adequate hydration, anorexia, N/V, GI distress, diarrhea
		Olsazine	Dipentum	
		Mesalamine	Asacol, Pentasa	
Antibiotics	Enters bacteria and destroys DNA	Metronidazole	Flagyl	Anorexia, dry mouth, metallic taste, N/V, epigastric distress, diarrhea, constipation; avoid alcohol during use and 3 days after
Antidiarrheal: Bismuth compounds	Decrease fluid secretions; reduce stool output	Bismuth subsalicylate	Pepto Bismol	Dry mouth/constipation
Antidiarrheal: Intestinal flora modifiers	Cultures of *lactobacillus* organisms supply missing bacteria in GI tract and suppress growth of diarrhea causing bacteria	*L. acidophilus*	Lactinex	None

(continued)

Selected Medications for Lower Gastrointestinal Diseases

Classification	Mechanism	Generic	Brand Names	Possible Food-Drug Interactions
Antidiarrheal: Opiates	Inhibit acetylcholine and decrease peristalsis	Loperamide	Immodium	Dry mouth, N/V, abdominal pain, bloating, constipation
		Diphenoxylate	Lomotil, Logen	
		Paregoric	Camphorated Tincture of Opium	
Glucocorticoids Anti-inflammatory	Mimics the action of cortisol; redistribution of white blood cells—reduction of lymphocytes; increased neutrophils; decreased production of prostaglandins	Prednisone Hydrocortisone Prednisolone	Deltasone Depo-Medrol, Solumedrol, Medrol	Caution with ↑ DM; glucose, highly protein bound, may need ↑ K, PO$_4$, Ca, and ↑ vits A, C, D, ↑ protein, and ↓ dietary Na, avoid alcohol
		Dexamethasone	Decodron, Hexadrol, Dexameth, Dexone	
		Budesonide	Entocort EC	Headache, nausea
Biologic therapies	Inhibits TNF-α	Infliximab	Remicade	Abdominal pain, nausea, fatigue, vomiting
Immunosuppressive	Antagonizes purine metabolism and may inhibit synthesis of DNA, RNA, and proteins; it may also interfere with cellular metabolism and inhibit mitosis	Azathioprine	Imuran	Anorexia, N/V, diarrhea, steatorrhea
		6-mercaptopurine	6-MP, Purinethol	
		Cyclosporine	Sandimmune, Neoral	
Laxative: Bulking agent	Soluble fiber forms gel in the colon; retains water and increases peristalsis	Psyllium	Fiberall (powder and wafers), Metamucil (powder, wafers, sugar-free formula) Perdiem Fiber	Diarrhea, cramping, malabsorption of nutrients
		Calcium polycarbophil	FiberCon (tablets)	Diarrhea, cramping, malabsorption of nutrients
		Methylcellulose	Citrucel (powder and sugar-free formula)	Diarrhea, cramping, malabsorption of nutrients
		Guar gum	Benefiber	Diarrhea, cramping, malabsorption of nutrients
Laxative: Stimulants	Irritate intestinal mucosa or directly stimulate submucosal and myenteric plexus	Anthraquinone (cascara extract [Casanthranol]) and senna extract, diphenylmethane (bisacodyl and phenolphthalein)	Senokot Dulcolax Correctol Ex-Lax	Diarrhea, cramping, malabsorption of nutrients
Laxative: Enemas and suppositories	Evacuation induced by distended colon, mechanical lavage	Tap water (500 mL rectally); soapsuds (1500 mL rectally); glycerine suppository	Fleets enema	Diarrhea, cramping, malabsorption of nutrients
Laxative: Stool softener	Cyclic AMP stimulates secretion of water, sodium, and chloride into GI lumen	Glycerin Mineral oil Docusate sodium		Diarrhea, cramping, malabsorption of nutrients
Laxative: Osmotic	Poorly absorbed polyvalent ions (e.g., Mg, phosphate, sulfate) or disaccharides (lactulose, sorbitol) remain in colon, increasing intraluminal osmosis	Lactulose, sorbitol, magnesium laxatives, sodium salts, polyethylene glycol lavage solution	Milk of Magnesia GoLytely	Cramping, flatulence, diarrhea, malabsorption of nutrients

Sources: Lichtenstein GR, Hanauer SB, and Sandborn WJ. Management of Crohn's disease in adults. *The American Journal Of Gastroenterology* [serial online]. February 06, 2009;104(2):465. Available at MEDLINE with Full Text, Ipswich, MA.

Pronsky, Zaneta. *Food Medication Interactions*, 15th ed. Birchrunville, PA: Food-Medication Interactions, 2008.

Emedicine. Available at http://www.emedicine.com

U.S. National Library and National Institutes of Health. MedlinePlus. Available at http://www.nlm.nih.gov/medlineplus/druginformation.html

Chapter 27 Diarrhea and Constipation. The Merck Manual. Available at http://www.merck.com

Prevention of diarrhea should be a major focus of any discussion regarding this condition. Recommendations for the prevention of diarrhea worldwide include strategies such as:[43]

- Improving access to clean water and safe sanitation
- Promoting hygiene education
- Exclusive breast-feeding
- Improving weaning practices
- Immunizing all children, especially against measles
- Using latrines
- Keeping food and water clean
- Washing hands with soap (the baby's as well) before touching food
- Sanitary disposal of stools

Nutrition Therapy for Diarrhea

Nutritional consequences of diarrhea are initially dependent on the volume of gastrointestinal losses and then on the length of the disease course. Large-volume losses can quickly lead to dehydration, and electrolyte and acid-base imbalances. Hyponatremia and hypokalemia are both common with diarrhea. Metabolic acidosis may occur due to excessive loss of bicarbonate ions in stool output (see Chapters 8 and 9). Infants and elderly are at particular risk because their systems are much more sensitive to rapid shifts in both fluids and electrolytes. Maintaining homeostasis is much more difficult for both of these populations, in part due to the inability of their renal systems to act quickly enough for adequate compensation. Chronic diarrhea can cause fluid and electrolyte complications and can result in malnutrition and specific nutrient deficiencies. Diarrhea can affect appetite and thus impair adequate ingestion. Diarrhea also results in decreased transit time, which interferes with the ability of the gastrointestinal tract to perform adequate digestion and absorption.

Nutrition Assessment When completing a nutrition assessment for a patient with diarrhea, it is especially important to review the following types of assessment data: Fluid and beverage intake, energy and mineral intake, medication and herbal supplement use, weight change, biochemical data (reflective of hydration status), nutrition-focused physical findings (especially those related to the digestive system and skin), and past surgical history.

Nutrition Diagnosis Common nutrition diagnoses associated with diarrhea and its physical impact on nutritional status include inadequate energy intake; inadequate oral food/beverage intake; inadequate fluid intake; altered GI function; and involuntary weight loss. Other potential diagnoses may be a result of inadequate intake or excessive losses. A common example in diarrhea is excessive potassium losses.

Nutrition Interventions Nutrition therapy for diarrhea is designed to restore normal fluid, electrolyte, and acid-base balance; decrease gastrointestinal motility; thicken the consistency of the stool; repopulate the gastrointestinal tract with normal flora; and gradually introduce foods that allow the patient to return to a normal diet without aggravation of symptoms.[44–46] Historically, nutrition therapy for diarrhea has included making the patient NPO or prescribing clear liquids.

Current practice recognizes the importance of stimulating the gastrointestinal tract by feeding the patient. This speeds recovery of damaged cells. In addition, clear liquids are typically high in simple carbohydrates, which increase osmolality of the gastrointestinal tract. This actually can make diarrhea worse due to hyperosmolality.

Oral rehydration solutions are designed to both restore fluid and electrolyte balance and enhance absorption in the intestinal tract. There are several commercially prepared rehydration solutions such as Pedialyte® and Rehydralyte®. The World Health Organization has a standard recipe for an oral rehydration solution: ⅓ to ⅔ tsp table salt, ¾ tsp sodium bicarbonate, ⅓ tsp potassium chloride (salt substitute), 1⅓ Tbsp sugar or rice powder, and 1 L bottled or sterile water. (See Table 15.5.) The use of rice powder yields a solution with reduced osmolality that appears to be just as effective as those with sucrose as a carbohydrate source.[47]

Nutrition therapy to decrease motility should focus on avoiding high-sugar beverages and foods high in simple carbohydrates (lactose, sucrose or fructose); sugar alcohols (sorbitol, xylitol, mannitol); caffeine; and alcoholic beverages. Gas-producing foods should be avoided.

As mentioned previously, adding sources of soluble fiber and resistant starch has been the most typical route to thicken the consistency of the stool. For infants this may include the use of banana flakes, apple powder, or other pectin sources which can be added to infant formula and to other foods for children and adults. If the infant has begun solid foods, use of strained bananas, applesauce, and rice cereal are the best initial food choices; historically, the BRAT (bananas, rice, applesauce, and toast) eating pattern has been used to guide the initial food choices for acute diarrhea, but it does not provide a sufficient variety of nutrients for long periods of use.[45]

Another step in treating diarrhea is the use of probiotics and prebiotics.[48–52] These foods and supplements support growth

Table 15.5 WHO Rehydration Solution

Reduced-Osmolarity Oral Rehydration Solution (ORS)	Grams/Liter
Sodium chloride	2.6
Glucose, anhydrous	13.5
Potassium chloride	1.5
Trisodium citrate, dehydrate	2.9
Total weight	20.5

Reduced-Osmolarity Oral Rehydration Solution (ORS)	mmol/liter
Sodium	75
Chloride	65
Glucose, anhydrous	75
Potassium	20
Citrate	10
Total weight	245

Source: WHO Dept. of Child and Adolescent Health and Development. The treatment of diarrhoea. A manual for physicians and other senior health workers. 2005. ISBN 52-4-159318.0.

of healthy flora and/or repopulate the intestinal tract with healthy bacteria (see Box 15.1 for suggestions of food sources of probiotics). As mentioned previously, undigested substrates are fermented to short-chain fatty acids. Probiotics and prebiotics will increase the amount of short-chain fatty acids (SCFA) produced. Recent research indicates SCFA promote water and electrolyte absorption in the colon, which reduces the incidence

of diarrhea.[53] Other research has suggested that probiotics and prebiotics may improve the mucosal defense in the GI tract and may also reduce the growth of harmful bacterial.[48, 54] This may play a significant role in reducing diarrhea associated with enteral feeding.[39] Other research has studied the effect of using probiotics and prebiotics as part of the treatment for radiation-induced diarrhea, diarrhea secondary to rotavirus, and traveler's diarrhea.[49, 51, 54] Dosages for probiotics and prebiotics are still being established and current research has not provided consistent enough data to make dosage recommendations.[48, 55] Another challenge in using probiotics as nutrition therapy is the difficulty in maintaining the viability of the organisms. Heat, light, and aging are just some of the issues that can affect the actual dosage provided in the food product or supplement. Currently the FDA does not allow for any health claims regarding probiotics, but has established manufacturing guidelines.[56]

In adults and older children, introducing solid foods should begin with a **low-residue diet**. Even though the term *low residue* is used in practice, there can be confusion as to what it means. The typical scenario for a low-residue diet begins with adding refined starches (for example, rice) and then slowly adding other foods to the diet. See Table 15.6: Nutrition Therapy for Diarrhea.[57]

Constipation

Definition There are many subjective definitions for **constipation**. Individuals commonly describe constipation as a decrease in frequency of bowel movements, but it is difficult to interpret an individual's complaints of gastrointestinal distress. In order to provide some general framework for this diagnosis, The Rome Consensus Criteria were established. The Rome III Consensus Criteria define constipation as a condition where at least two of the following symptoms have occurred in the previous year for at least 12 nonconsecutive weeks:[58–60]

· Must include 2 or more of the following:
  · Straining during at least 25% of defecations
  · Lumpy or hard stools in at least 25% of defecations
  · Sensation of incomplete evacuation for at least 25% of defecations
  · Sensation of anorectal obstruction/blockage for at least 25% of defecations
  · Manual maneuvers to facilitate at least 25% of defecations (e.g., digital evacuation, support of the **pelvic floor**)
  · Fewer than three defecations per week
· Loose stools are rarely present without the use of laxatives
· There are insufficient criteria for irritable bowel syndrome

Epidemiology As previously stated, constipation is one of the most common gastrointestinal complaints. Studies have estimated that as much as 17% of the U.S. population experiences symptoms of constipation. These symptoms result in more than 7 million physician visits each year.[61]

Etiology Constipation can be a result of several different distinct causes.[62, 63] Slowed colonic transit can result in constipation. Constipation can be due to rectal outlet obstruction or other sources of obstruction such as fecal impaction, adhesions, or even the presence of a tumor. **Pelvic floor dysfunction** can cause not only slowed colonic transit but also storage of fecal

BOX 15.2 CLINICAL APPLICATIONS

Probiotic Products Guide for Practitioners

Minimum criteria to be considered a probiotic:

- Purified strain of the candidate microbe (usually bacterium or yeast)
- Identified to the strain level using biochemical and genetic techniques
- Shown in human studies to improve some parameter of human health
- Safe for target consumers

Adapted from: Douglas LC, Sanders ME. Probiotics and prebiotics in dietetics practice. J Am Diet Assoc. 2008;108:512.

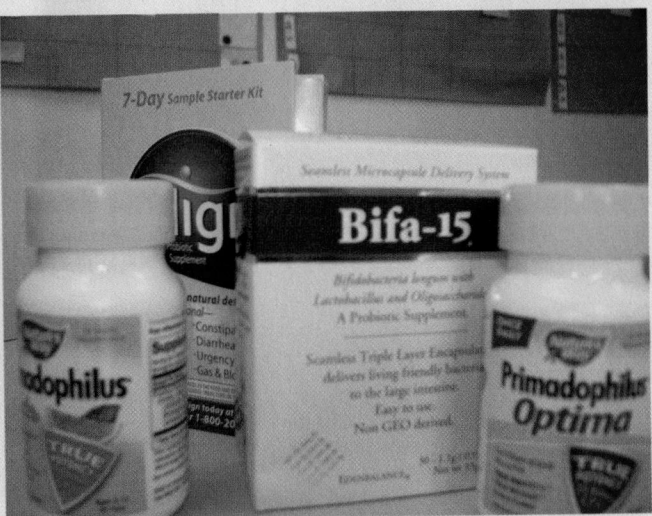

Probiotics are available in various forms, including dietary supplements and foods.

Courtesy of Marcia Nelms.

Examples of Probiotic Products Available in the United States

Product (Manufacturer)	Active Organism(s)	Colony Count	Package Directions
Activia yogurt (Dannon/Danone)	*Bifidobacterium animalis* DN 173-010 (*Bifidus regularis*™), *Lactobacillus bulgaricus*, *Streptococcus thermophilus*	Manufactured to contain >100 million active cultures per g; guaranteed 10 million active cultures per g at end of shelf life	One to three 4-oz servings daily for at least 10–14 days
Align (Proctor & Gamble)	*Bifidobacterium infantis* 35624 (Bifantis)	Manufactured to contain 1 billion live cells per 4 mg capsule	One capsule daily
Culturelle (Amerifit Brands)	*Lactobacillus GG* (*L. rhamnosus*)	Manufactured to contain >30 billion live cells; guaranteed 10 billion live cells when consumed	One capsule daily
DanActive (Dannon)	*Lactobacillus casei* strain DN-114-001 (*L. Casei Immunitas*)	Manufactured to contain 10 billion cells per bottle	At least one bottle per day (100 mL)
FLORA PROBIOTIC PLUS (Golden Health Products)	15 different species of *Lactobacillus*, *Bifidobacterium*, *Lactococcus*, and *Streptococcus*	Manufactured to contain 22 billion viable cells per capsule	Capsules: Two capsules 30 minutes before breakfast with water, juice, or milk for two weeks, then one capsule daily Powder: ¼ tsp mixed with liquid 30 min prior to breakfast
Flora Q 2 (Kenwood Therapeutics)	*Lactobacillus acidophilus*, *Bifidobacterium*, *Lactobacillus paracasei*, and *Streptococcus thermophilus*	375 mg	One capsule daily
Florastor Kids (Biocodex)	*Saccharomyces boulardii*	Manufactured to contain 5 billion live freeze-dried cells per 250 mg packet	Take packet contents and mix in drink or semi-solid food
Florastor (Biocodex)	*Saccharomyces boulardii*	Manufactured to contain 5 billion live freeze-dried cells per 250 mg capsule	One capsule daily; capsule contents can be mixed in drink or semi-solid food
Lactinex (Becton, Dickenson and Company)	*Lactobacillus acidophilus*, *L. bulgaricus*	1 million cells per tablet or 100 million per packet	Four tablets 3–4 times daily

(continued)

Product (Manufacturer)	Active Organism(s)	Colony Count	Package Directions
Primadophilus for Kids (Nature's Way)	*Lactobacillus rhamnosus, Bifidobacterium longum, L. acidophilus*	Manufactured to contain 3 billion per tablet	For children 2–12 years; 1 tablet daily between meals (3/day)
VSL#3 (VSL Pharmaceuticals, Inc.)	*Lactobacillus acidophilus, L. plantarum, L. casei, L. bulgaricus, Bifidobacterium breve, B. infantis, B. longum, Streptococcus thermophilus*	Packets: 450 billion bacteria per packet Capsules: 250 billion bacteria per 2 capsules	Bloating associated with diarrhea-predominant irritable bowel syndrome: one packet twice daily Pouchitis and ulcerative colitis: one to four packets daily, depending on number of bowel movements

Reprinted from: Ulbrich T, Plogsted S, Geraghty, Reber K, Valentine CJ. Probiotics and prebiotics: Why are they "bugging" us in the pharmacy? J Pediatr Pharmacol Ther 2009;14:17–24

Cost Comparisons of Selected Products

Product	Quantity	Consumer Price*	Package Directions for Use	Estimated Daily Price
Activia	4 pk–4 oz (113 g)	2.44	1 yogurt daily (minumum)	$0.61
Align	28	29.99	1 capsule daily	$1.07
Culturelle	30	14.99	1 capsule daily	$0.50
Culturelle kids	10	5.99	1 capsule daily	$0.60
DanActive	4 pk–3.3 fl oz	2.58	1 bottle daily	$0.64
FLORA SOURCE	240	104 (6 m supply)	2 daily for two weeks, 1 daily thereafter	$0.58
FLORA SOURCE	120	52 (3 mo supply)	3 daily for two weeks, 1 daily thereafter	$0.58
FLORA SOURCE	1.25 oz	29.95 (6 wk supply)	¼ tsp of powder daily	$0.68
Florastor	10	8.93	1 capsule twice a day	$1.79
Florastor	50	32.85	1 capsule twice a day	$1.31
Florastor kids	10	9.90	1 capsule twice a day	$1.98
Lactinex	50	9.86	4 tablets 3–4 times daily	$2.37
Lactinex	12	11.44	1 packet 3–4 times daily	$2.86
Primadophilus Kids	30	6.99	1 tablet daily, between meals	$0.70
VSL #3	10	29.68	½–8 packets per day (depends on use)	$1.48–$23.74
VSL #3	30	79.50	½–8 packets per day (depends on use)	$1.33–$21.20
VSL #3	60	45.00	4 capsules daily	$3.00

*Prices estimated in USD from currently available online sources and in stores December 2007.

Reprinted from: Ulbrich T, Plogsted S, Geraghty, Reber K, Valentine CJ. Probiotics and prebiotics: Why are they "bugging" us in the pharmacy? J Pediatr Pharmacol Ther 2009;14:17–24

contents in the rectum for long periods of time. Constipation can be a major component of irritable bowel syndrome, which will be discussed later in this chapter. Constipation can be secondary to other medical conditions, including scleroderma, amyloidosis, and neurological diseases such as multiple sclerosis (MS) or Parkinson's disease. Finally, constipation can be a side effect of many different classes of medications. These include very common prescription drugs such as calcium channel blockers, antidepressants such as amitriptyline, pain medications such as morphine, diuretics, and antihistamines. Other over-the-counter medications that often cause constipation include iron, calcium, and other vitamin supplements, and, for some individuals, even nonsteroidal anti-inflammatory drugs can result in constipation.

Clinical Manifestations Symptoms of constipation include decreased frequency of bowel movements. Bowel movements are often hard and pellet-like. Abdominal pain, bloating, and gas are common accompanying symptoms.

Medical Diagnosis Diagnosis will include a complete history and physical identifying specifically the Rome Consensus III criteria. Laboratory screening tests will include a complete blood count, thyroid-stimulating hormone, and serum chemistries. Additional tests can include a colonoscopy or flexible sigmoidoscopy if further evaluation is needed.

Treatment Treatment of the underlying etiology will direct medical care for constipation. Common interventions include bowel retraining and use of enemas or cathartic and laxative

Table 15.6 Nutrition Therapy for Diarrhea

Recommended Foods

Food Group	Recommended Foods	Notes
Milk and Milk Products	• Buttermilk • Evaporated, skim, and low-fat milk • Soy milk • Yogurt with live active cultures • Powdered milk • Cheese • Low-fat ice cream • Sherbet	If you have lactose intolerance, drinking milk products may aggravate diarrhea. Try lactose-free products. Avoid yogurts with nuts or dried fruit.
Meat and Other Protein Foods	• Tender, well-cooked meat, poultry, fish, eggs, or soy foods made without added fat • Smooth nut butters	
Grains	• White flour • Bread, bagels, rolls, crackers, and pasta made from white or refined flour • Cold or hot cereals made from white or refined flour	Choose grain foods with less than 2 grams (g) dietary fiber per serving. (To find out how much fiber is in a serving of a packaged food, look on its Nutrition Facts label.)
Vegetables	• Most well-cooked vegetables without seeds or skins • Potatoes without skin • Lettuce • Strained vegetable juice	See the Foods Not Recommended chart for vegetables to avoid.
Fruits	• Fruit juice without pulp, except prune juice • Ripe bananas • Melons • Canned soft fruits	See the Foods Not Recommended chart for fruits to avoid.
Fats	• Fats include oil, butter, cream, cream cheese, margarine and mayonnaise	Limit fats to less than 8 teaspoons a day
Beverages	• Decaffeinated coffee • Caffeine-free teas • Soft drinks without caffeine • Rehydration beverages	Healthy people need 8 to 10 cups of fluid each day. You may need to drink more to replace fluids lost to diarrhea.

Foods Not Recommended

Food Group	Foods Not Recommended	
Milk and Milk Products	• Whole milk • Half-and-half • Cream	• Sour cream • Regular (whole milk) ice cream • Yogurt with berries, dried fruit, or nuts
Meat and Other Protein Foods	• Fried meat, poultry, or fish • Luncheon meats, such as bologna or salami • Sausage and bacon • Hot dogs	• Fatty meats • Nuts • Chunky nut butters
Grains	• Whole-wheat or whole-grain breads, rolls, crackers, or pasta • Brown or wild rice • Barley, oats, and other whole grains	• Cereals made from whole grain or bran • Breads or cereals made with seeds or nuts • Popcorn
Vegetables	• Raw vegetables (except for lettuce) • Fried vegetables • Beets • Broccoli • Brussels sprouts	• Cabbage • Cauliflower • Collard, mustard, and turnip greens • Corn • Potato skins
Fruits	• All raw fruits except banana and melons • Dried fruits, including prunes and raisins • Fruit juice with pulp	• Canned fruit in heavy syrup • Any fruits sweetened with sorbitol • Prune juice

(continued)

Table 15.6 Nutrition Therapy for Diarrhea (*Continued*)

Foods Not Recommended

Food Group	Foods Not Recommended
Fats	• Limit fats to less than 8 teaspoons per day
Beverages	• Beverages containing caffeine, including regular coffee, regular tea, colas, and energy drinks • Limit beverages containing high-fructose corn syrup to 12 oz. per day • Avoid beverages sweetened with sorbitol • Alcoholic beverages
Other foods to avoid	• Sugar alcohols such as xylitol and sorbitol; honey

Sample 1-Day Menu

Breakfast	1 cup Rice Krispies 1 cup vanilla soy milk ½ ripe banana
Snack	1 cup decaffeinated tea 6 ounces yogurt or 2 graham cracker rectangles
Lunch	2 cups chicken rice soup with 2 ounces of added chicken ¼ cup cooked carrots 1 slice white toast with a thin spread of jelly ½ cup applesauce
Snack	2 or 3 saltine crackers 1 cup fruit juice without pulp
Evening Meal	4 to 6 ounces baked fish topped with breadcrumbs, a squeeze of lemon, and 1 teaspoon butter or margarine ½ cup mashed potato without skins ½ cup green beans, cooked well 1½ cups water or other caffeine-free beverage
Snack	½ cup sorbet 1 cup sugar-free hot chocolate made with water or soy milk

Meal Planning Tips

• Limit foods and beverages that contain sugar, lactose, fructose, high-fructose corn syrup, and sorbitol.

• Avoid beverages with caffeine.

• Eat a small meal or snack every 3 or 4 hours.

• If any food makes diarrhea worse, stop eating that food. You can try it again when the diarrhea stops.

• Avoid spicy foods if they make diarrhea worse.

Cooking Tips

Some types of diarrhea can spread through food or from person to person. To avoid this:

• Clean: Always wash hands, silverware and dishes, cooking tools, and cooking surfaces thoroughly. Use soap and hot water. Also wash fresh fruits and vegetables before cooking or eating them.

• Separate: Keep raw and cooked foods separate from one another. Don't eat or serve with forks, spoons, knives, or other cooking tools that were used to prepare raw meat, eggs, or fish (unless the tools were washed after use).

• Cook: Foods should be cooked completely. Hot foods should be kept and served at a temperature above 140 degrees Fahrenheit. Avoid rare and raw meat.

• Chill: Store meats, fish, poultry, eggs, milk, dairy foods, fresh fruits and vegetables, and all cooked foods in the refrigerator or freezer. Keep these foods at 40 degrees Fahrenheit or colder until you are ready to cook or eat them.

Source: Reprinted with permission: American Dietetic Association. Nutrition Care Manual. 2009; Diarrhea Nutrition Therapy. Available at www.nutritioncaremanual.org

medications. Other medications involve bulking agents and stool softeners.[61] See Table 15.4 for a listing of medications used in the treatment of constipation.

Nutrition Therapy for Constipation

Historically, nutritional treatment of constipation has concentrated on the role of adequate fiber and fluid intake. This is despite the overall lack of outcomes research for the role of fiber in gastrointestinal disease.[63, 64] Survey data have indicated most Americans consume an average of 15 grams of fiber each day. A more recent report indicates that Americans aged 1 year and older are estimated to consume approximately 4.9 g resistant starch per day (range 2.8 to 7.9 g/day).[27] Adequate intake of whole grains, fruits, and vegetables has been one of the primary focuses of both the *Dietary Guidelines for Americans* and the *Nutrition Recommendations for Canadians*.[65, 66]

Nutrition Assessment Relevant nutrition assessment data in cases of constipation would include fluid and beverage intake, dietary fiber intake, bioactive substance intake, beliefs and attitudes, and misuse of medication, especially laxatives.

Nutrition Diagnosis Nutrition diagnoses commonly associated with constipation may include inadequate fluid intake; inadequate fiber intake; altered GI function; not ready for diet/ lifestyle change; or undesirable food choices.

Nutrition Interventions Twenty-five to thirty-five grams of dietary fiber are recommended for adults each day. Based on caloric intake, this would be approximately 10 to 13 g of dietary fiber per 1000 kcal. For children over the age of 2 years, fiber intake is recommended to be the amount equal to their age plus 5 grams/day.[64]

Ensuring adequate fiber and fluid intake has been the foundation for nutrition therapy in constipation treatment. Fiber results in increased stool weight that assists in providing for consistent bowel movements and a colonic transit time of 2–4 days.[64] Despite the lack of current outcome evidence, the relationship between fiber and normal laxation provides the basis for the clinician to make recommendations for a gradual increase in fiber to the daily goal of 25–35 grams. This should begin slowly with adding one to two high-fiber foods each day. Total fiber in foods includes a mixture of **dietary fiber** and **functional fiber**—including resistant starch. See Table 15.7 for ways to increase fiber to meet the goal of 25 to 35 grams each day, and a sample menu. Even though food is recommended as the optimal route for fiber sources, due to individual tolerance, some individuals—especially the elderly—may not be able to achieve levels in the highest ranges without using fiber supplementation. Table 15.7 outlines high-fiber nutrition therapy to be used for treatment of constipation. Table 15.4 lists suggestions for fiber supplements that can be used in treatment (under Laxatives: Bulking Agents).

Table 15.7 High Fiber Nutrition Therapy

Tips for Adding Fiber to Your Eating Plan

- Eat whole-grain breads and cereals. Look for choices with 100% whole wheat, rye, oats, or bran as the first or second ingredient.
- Have brown or wild rice instead of white rice or potatoes.
- Enjoy a variety of grains. Good choices include barley, oats, farro, kamut, and quinoa.
- Bake with whole-wheat flour. You can use it to replace some white or all-purpose flour in recipes.
- Enjoy baked beans more often! Add dried beans and peas to casseroles or soups.
- Choose fresh fruit and vegetables instead of juices.
- Eat fruits and vegetables with peels or skins on.
- Compare food labels of similar foods to find higher-fiber choices. Packaged foods have the amount of fiber per serving listed on the Nutrition Facts label.
- Drink plenty of fluids. Set a goal of at least 8 cups per day. You may need even more with higher amounts of fiber. Fluid helps your body process fiber without discomfort.
- If you are taking calcium or iron supplements check with your doctor or dietitian. You may be able to take smaller amounts several times a day.

Recommended Foods

Foods with at Least 4 g Fiber per Serving

Food Group	Choose
Grains	⅓–½ cup high-fiber cereals; check Nutrition Facts labels and choose products with 4 or more grams (g) dietary fiber per serving
Dried beans and peas	½ cup cooked red beans, kidney beans, large lima beans, navy beans, pinto beans, white beans, lentils, or black-eyed peas

Table 15.7 High Fiber Nutrition Therapy *(Continued)*

Food Group	Choose
Vegetables	1 artichoke (cooked)
Fruits	½ cup blackberries or raspberries
	4 prunes (dried)

Foods with 1 to 3 g Fiber per Serving

Food Group	Choose
Grains	• 1 bagel (3.5-inch diameter) • 1 slice whole-wheat, cracked wheat, pumpernickel, or rye bread • 2-inch square corn bread • 4 whole-wheat crackers* • 1 bran, blueberry, cornmeal, or English muffin • ½ cup cereal with 1–3 grams fiber per serving (check dietary fiber on the product's Nutrition Facts label) • 2 tablespoons bran, rice, or wheat cereal • 2 tablespoons wheat germ or whole-wheat flour
Fruits	• 1 apple (3-inch diameter) or ½ cup applesauce • ½ cup apricots (canned) • 1 banana • ½ cup cherries (canned or fresh) • ½ cup cranberries (fresh) • 3 dates (whole) • 2 medium figs (fresh) • ½ cup fruit cocktail (canned) • ½ grapefruit • 1 kiwi fruit • 1 orange (2½-inch diameter) • 1 peach (fresh) or ½ cup peaches (canned) • 1 pear (fresh) or ½ cup pears (canned) • 1 plum (2-inch diameter) • ¼ cup raisins • ½ cup strawberries (fresh) • 1 tangerine
Vegetables	• ½ cup bean sprouts (raw) • ½ cup beets (diced, canned) • ½ cup broccoli, brussels sprouts, or cabbage (cooked) • ½ cup carrots • ½ cup cauliflower • ½ cup corn • ½ cup eggplant • ½ cup okra (boiled) • ½ cup potatoes (baked or mashed) • ½ cup spinach, kale, or turnip greens (cooked) • ½ cup squash—winter, summer, or zucchini (cooked) • ½ cup sweet potatoes or yams • ½ cup tomatoes (canned)
Other	• 2 tablespoons almonds or peanuts • 1 cup popcorn (popped)

(continued)

Table 15.7 High Fiber Nutrition Therapy (Continued)

Sample 1-Day Menu
(Approximately 25 to 30 grams of fiber)

Meal	Food Choices	Dietary Fiber
Breakfast	½ cup Total Raisin Bran with 1 cup skim milk	2.5 grams
	½ cup orange juice with pulp	0.25 gram
	1 cup coffee	
Lunch	1½ cups chili made with ½ cup kidney beans and ¼ cup soy crumbles per serving and topped with 2 tablespoons shredded cheese	11.2 grams
	8 wheat crackers	0.7 gram
	1 fresh apple (with skin)	2.5 grams
	2 cups water or sugar-free lemonade	
Snack	8 oz yogurt	
	2 cups water	
Evening Meal	2 cups mixed fresh vegetables, with 2 ounces sliced chicken and 1 ounce firm tofu	6 grams
	1 cup jasmine rice	1.5 grams
	½ cup fresh raspberries, blueberries, and sliced bananas	3.5 grams
	1 cup hot tea	
Snack	2 tablespoons almonds	3.3 grams
	1 cup hot chocolate	

Reprinted with permission: American Dietetic Association. Nutrition Care Manual. 2009; Diarrhea Nutrition Therapy. Available at www.nutritioncaremanual.org

At the same time that fiber intake is increased, the clinician should also emphasize adequate fluid intake. This should be at a minimum of 2000 mL/day (approximately 8–10 cups/ day).

Probiotics and prebiotics have also been recommended for treatment of constipation. For example, consumption of fructans or fructooligosaccharides (FOS) has been shown to soften feces and to assist in relieving constipation.[23, 67] Fructans are defined as chains of fructose molecules with glucose on the terminal end of the chain. They are also commonly referred to as inulin and levan. They naturally occur in wheat-containing grain products such as breads, cereals, and pasta.

Malabsorption

Definition Malabsorption is a general term referring to malabsorption of fat, carbohydrate, or protein as a result of maldigestion or from damage to the anatomy and physiology of the small intestine.

Etiology Damage to the anatomy and physiology of the small intestine due to disease is the most common cause of malabsorption. Conditions such as celiac disease, Crohn's disease, and even protein-calorie malnutrition result in decreased villous height, decreased enzyme production, and subsequent malabsorption and/or maldigestion. Dysfunction of the accessory organs of digestion (liver, pancreas, and gallbladder) may also serve as the origin of the maldigestion.

Decreased transit time, as seen in diarrhea or from surgical changes in the anatomy, also can result in either maldigestion or malabsorption. For example, after a gastrectomy, dumping syndrome causes a rapid transit through the small intestine, which prevents adequate exposure to enzymes and adequate time for the absorptive mechanisms. See Table 15.8 for a list of potential causes of malabsorption.

Pathophysiology Nutrient digestion and absorption are dependent on normal anatomy; normal physiology with adequate production of enzymes, hormones, and other secretions such as bile; and appropriate motility. Malabsorption will be discussed here in the context of each of the main macronutrients: lipid, carbohydrate, and protein.

FAT MALABSORPTION The digestion and absorption process for lipid or fat is the most complex, and therefore the easiest to disrupt. Fat malabsorption is called steatorrhea—literally meaning fat in the stool. Digestion and absorption of fat requires adequate colipase and pancreatic lipase, adequate emulsifier—bile—from the liver and gallbladder, and adequate secretion through the common bile duct and pancreatic ducts. Motility needs to be normal due to the lengthy process lipid has to undergo from micelle to chylomicron for absorption. When any of these processes is disrupted, fat remains in the stool and travels undigested and unabsorbed to the large intestine. Fat-soluble vitamins are malabsorbed as well. An additional concern for fat malabsorption is the potential presence of excess oxalate. Under normal conditions, calcium within the GI tract binds with oxalate and allows for its excretion or metabolism. In fat malabsorption, calcium often binds with the malabsorbed fat. This allows oxalate to be absorbed and then excreted through the kidney. Excessive amounts of oxalate have been linked to development of urothiasis or kidney stones. Hyperoxaluria (excessive oxalate in the urine) is responsible for about 30% of kidney stones and is considered to be the most common cause.

Individuals with fat malabsorption will experience abdominal pain, cramping, and diarrhea. Stools produced will be frothy, foul-smelling, and greasy in appearance.

Specific laboratory tests can diagnose steatorrhea and then assist in determination of the etiology of the malabsorption. Diagnostic tests include the 72-hour quantitative fecal fat test that involves collection of stool output for three days after ingesting 100 grams fat/day. If more than 6 grams of fat are present in the stool after 24 hours, the diagnosis of steatorrhea can be made. This establishes malabsorption but does not determine the specific cause. This test is costly and unpleasant for the patient. It often takes up to 10 days to receive results.[15]

The D-xylose absorption test assists in distinguishing between pancreatic dysfunction and small bowel malabsorption. D-xylose is easily absorbed in the small intestine and is not metabolized. Its absorption does not require pancreatic or biliary function. In this test, after the patient drinks 25 grams of D-xylose, urine and blood samples are collected. Normal findings would show that blood levels of D-xylose are 25 to 40 mg/dL after 2 hours. Excretion in the urine should be 80% to 95% after 5 hours. Normal val-

Table 15.8 Possible Causes of Malabsorption

Cause	Examples
Inadequate Digestion	
Postgastrectomy*	
Deficiency or inactivation of pancreatic lipase	
Exocrine pancreatic insufficiency	Chronic pancreatitis
	Pancreatic carcinoma
	Cystic fibrosis
	Pancreatic insufficiency, congenital or acquired
Gastrinoma-acid inactivation of lipase*	
Drugs	Orlistat
Reduced Intraduodenal Bile Acid Concentration/Impaired Micelle Formation	
Liver disease	Parenchymal liver disease
	Cholestatic liver disease
Bacterial Overgrowth in Small Intestine	
Anatomic stasis	Afferent loop stasis/blind loop/strictures/fistulae
Functional Stasis	
Diabetes*	
Scleroderma*	
Intestinal pseudoobstruction	
Interrupted Enterohepatic Circulation of Bile Salts	
Ileal resection	
Crohn's disease*	
Drugs (bind or precipitate bile salts)	Neomycin, cholestyramine, calcium carbonate
Impaired Mucosal Absorption/Mucosal Loss or Defect	
Intestinal resection or bypass	
Inflammation, infiltration, or infection	Crohn's disease*
	Amyloidosis
	Scleroderma*
	Lymphoma*
	Eosinophilic enteritis
	Mastocytosis
	Tropical sprue
	Celiac disease
	Collagenous sprue
	Whipple's disease
	Radiation enteritis
	Folate and vitamin B_{12} deficiency
	Infections: salmonellosis, giardiasis
	Graft-vs.-host disease
Genetic disorders	Disaccharidase deficiency
	Agammaglobulinemia
	Abetalipoproteinemia
	Hartnup disease
	Cystinuria

Cause	Examples
Impaired Nutrient Delivery to and/or from Intestine	
Lymphatic obstruction	Lymphoma*
	Lymphangiectasia
Circulatory disorders	Congestive heart failure
	Constrictive pericarditis
	Mesenteric artery atherosclerosis
	Vasculitis
Endocrine and Metabolic Disorders	
Diabetes*	
Hypoparathyroidism	
Adrenal insufficiency	
Hyperthyroidism	
Carcinoid syndrome	

*Malabsorption caused by more than one mechanism.

ues point to malabsorption originating from pancreatic or biliary dysfunction. In patients with intestinal malabsorption, blood and urine levels would be diminished.

The Schilling test is used to determine vitamin B_{12} absorption. B_{12} absorption requires not only intrinsic factor produced in the stomach but also adequate ileal receptors. The test can be performed in two phases—one with intrinsic factor and one without. The patient ingests radioactive B_{12} and then urinary amounts of B_{12} are measured. With normal absorption, B_{12} is absorbed and the excess is excreted in the urine. Without absorption, no urinary B_{12} will be measured.

Finally, a small bowel x-ray with contrast dye can indicate delays of motility such as an obstruction or **ileus**. A faster motility can also support the diagnosis of malabsorption.

CARBOHYDRATE MALABSORPTION The most common example of carbohydrate malabsorption is lactose malabsorption, commonly referred to as lactose intolerance. When there is inadequate lactase available for digestion, or if anatomy or motility does not allow adequate exposure to lactase, lactose will travel to the large intestine undigested and unabsorbed. Bacteria in the large intestine will cause the lactose to undergo fermentation, which creates increased gas and abdominal cramping. Undigested lactose also pulls additional water into the large intestine, contributing to abdominal cramping and resulting diarrhea. Lactose intolerance is commonly associated with other gastrointestinal diagnoses including irritable bowel syndrome, celiac disease, and inflammatory bowel disease.[68, 69]

Lactose malabsorption can be diagnosed using either a lactose tolerance test or lactose hydrogen breath test. After measuring the baseline breath hydrogen concentration, the patient consumes 25 to 50 grams of lactose. Breath hydrogen concentration is re-measured in 3 to 8 hours. An increase >20 ppm suggests lactose malabsorption. This test is preferred over the lactose tolerance test and has approximately 90% sensitivity.

PROTEIN MALABSORPTION Protein malabsorption is most commonly referred to as protein-losing enteropathy. This is not a specific disease, but it occurs, as do most all malabsorption disorders, as a result of other diseases. Excessive protein loss in the gastrointestinal tract has been noted in over 65 different diagnoses.[70] Excessive protein is lost in the stool, and the patient will experience reduced serum levels of proteins and an increasing amount of peripheral edema due to the reduced **oncotic pressure**.

Treatment Appropriate treatment for malabsorption will depend on the nutrient that is malabsorbed and the underlying disease causing malabsorption. These will be discussed in much more depth as specific diseases are covered in this chapter.

Nutrition Therapy for Malabsorption

Nutrition Assessment Conditions of malabsorption can result in weight loss, vitamin and mineral deficiencies, and chronic protein-calorie malnutrition. Therefore, it is important to accurately assess these possible conditions. Anthropometric data, especially weight change; dietary intake of macro- and micronutrients; and biochemical data reflective of fluid status and malnutrition are especially relevant.

Nutrition Diagnosis Nutrition diagnoses that may be associated with malabsorption include malnutrition; altered GI function; impaired nutrient utilization; underweight; and involuntary weight loss.

Nutrition Intervention The purpose of nutrition therapy is to provide structure for elimination of the malabsorbed nutrient from the diet, but at the same time provide appropriate substitutions in order to ensure maintenance of nutritional status.

NUTRITION THERAPY FOR FAT MALABSORPTION Restriction of fat to 25 to 50 grams per day is a standard first step in reducing the symptoms of fat malabsorption. Table 15.9 outlines the dietary interventions for a fat-restricted diet. Additionally, **medium-chain triglyceride (MCT)** supplements can be used to increase caloric intake. Triglycerides that contain fatty acids considered medium chain are those with 6 to 12 carbons. Most MCT oil products contain primarily caprylic (C8) and capric (C10) fatty acids. One half-ounce (15 mL) is 115 kcal or 8.3 kcal/g. MCT is absorbed directly into the circulatory system from the small intestine and does not require the normal lipid digestion and absorption processes that long-chain fatty acids require. MCT oil can be used as a supplement to add kcal for individuals who malabsorb lipid. This supplement can be added to soups and other hot foods. It has a neutral flavor but, because it is an oil, cannot be mixed adequately into cold foods.

If the etiology for steatorrhea originates from pancreatic dysfunction, use of pancreatic enzymes is a primary mode of treatment. An example is the product Pancrease® that consists of pancreatic lipase, amylase, and proteases. Dosages are individualized and taken with each meal or snack to ensure adequate digestion (see Chapter 16 for more information).

Table 15.9 Nutrition Therapy for Steatorrhea: Fat-Restricted Diet

Food Group	Foods Recommended	Foods Not Recommended
Grains	Whole-wheat and enriched breads Tortillas Low-fat crackers Brown and wild rice Pasta and couscous Ready-to-eat and hot breakfast cereals • Choices should be prepared without added fat. • Choose whole grains (such as whole wheat, brown or wild rice, oats) for at least half of your daily grains.	Products made with added fat (such as biscuits, waffles, and regular crackers)
Vegetables	All • Prepare without added fat. • Avoid fried or breaded vegetables.	Breaded or fried vegetables Vegetables with cheese, cream, butter, or oil-based sauces
Fruits	All fruits except avocado All juices	Avocado Fruit dishes prepared with added fat
Meat and Beans	Very lean cuts of meat Skinless poultry (except duck); eggs and beans • Prepare without added fat. Avoid fried foods. • Foods may be baked, grilled, roasted, broiled, or steamed. • Trim all visible fat from meat. Remove skin from poultry.	Fatty cuts of beef, pork, and lamb Regular (75% to 85% lean) ground beef Regular sausages, hot dogs, and bacon; high-fat luncheon meats High-fat types of poultry, such as duck; poultry with skin Nuts and seeds
Milk	Skim or reduced-fat (1%) milk Skim and reduced-fat cheeses Nonfat yogurt	Whole or low-fat (2%) milk Cream; half-and-half Regular (not reduced-fat) cheese Dairy products made from whole milk or cream
Fats and Oils	Limit fats and oils to less than 8 teaspoons per day	

Source: Pancreatitis Nutrition Therapy. Nutrition Care Manual 2009. © American Dietetic Association. Reprinted with permission.

NUTRITION THERAPY FOR LACTOSE MALABSORPTION Lactose is the simple carbohydrate found in milk and dairy products. Lactose is also found as an ingredient in many other food products in which it is often used as a filler or browning agent. Milk provides approximately 11 grams of lactose per cup. Other dairy products have varying amounts, with ice cream having approximately 9 grams per cup and cheese having 1 to 2 grams per ounce. Restriction of all milk and dairy products is the major step to treat lactose malabsorption, though individuals do vary in the amounts of lactose they can tolerate. If the lactose intolerance is secondary to another condition or due to acute illness, lactose exclusion is recommended for approximately four weeks. Then small amounts of lactose can be reintroduced—this is important to ensure adequacy of the diet for nutrients such as calcium and vitamin D. Dairy products such as cheese and yogurt have smaller amounts of lactose and may be adequately tolerated. Up to 1 cup of milk can be introduced in small increments throughout a 24-hour period.[69] Careful monitoring will guide the practitioner and patient on the amounts of lactose that can be tolerated. Products such as Lact-Aid® provide the lactase enzyme and can be used when milk and dairy foods are ingested. See Table 15.10 for an overview of nutrition therapy for lactose malabsorption/intolerance.

Table 15.10 Nutrition Therapy for Lactose Malabsorption and Intolerance

Recommended Foods

Food Group	Recommended Foods	Notes
Lactose-free milk and nondairy foods	Lactose-free milk Nondairy creamers* Nondairy whipped topping* Almond, rice, or soy milk Soy yogurt or soy cheese Almond milk cheese Soy-based sour cream	Foods marked with a star (*) may contain lactose. • Read ingredients lists and avoid products that list butter, cream, milk, milk solids, or whey. • Also avoid products when the ingredients list states, "May contain milk."
Low-lactose dairy foods	Some people with lactose intolerance can safely eat dairy foods that contain a little lactose (less than 1 gram lactose per serving).	
	You may wish to try: • 1–2 ounces aged cheese, such as Swiss, cheddar, or parmesan • 2 tablespoons cream cheese • ⅓ cup cottage cheese • ½ cup ricotta cheese	• Try these foods one at a time, in small amounts. • Stop eating them if symptoms return or get worse.

Table 15.10 Nutrition Therapy for Lactose Malabsorption and Intolerance (*Continued*)

Food Group	Recommended Foods	Notes
Meat, poultry, fish, dry beans, eggs, and nuts	All, unless prepared with ingredients that contain lactose	To know whether a food is made with ingredients that contain lactose: • Check ingredients lists. Avoid foods made with butter, cream, milk, milk solids, or whey. • Also avoid products when the ingredients list states, "May contain milk."
Grains	All, unless prepared with ingredients that contain lactose	
Vegetables	All, unless prepared with ingredients that contain lactose	
Fruit	All, unless prepared with ingredients that contain lactose	
Fats and Oils	Vegetable oils Oils from nuts or seeds	
Desserts	Fruit ices Sorbet Gelatin Soy yogurt Soy ice cream Rice milk ice cream	
Beverages	Coffee Tea Vitamite nondairy beverage	
Other	All spices and herbs	

Note: Milk and dairy foods are a primary source of calcium, a mineral that is important to good health. When you recommend that a client avoid milk and dairy foods, you will need to recommend other calcium sources:

- Sardines
- Canned salmon
- Tofu (calcium-fortified)
- Shellfish
- Turnip greens
- Collards
- Kale
- Dried beans
- Broccoli
- Calcium-fortified orange juice
- Calcium-fortified soy milk
- Blackstrap molasses
- Almonds

Foods Not Recommended

Avoid all foods made with ingredients that contain lactose. To find out whether a food has lactose:

- Check ingredients. Avoid foods made with butter, cream, milk, milk solids, or whey.
- Also avoid products when the ingredients list states, "May contain milk."

Food Group	Foods Not Recommended
Milk and dairy foods	Avoid *all* milk and dairy foods except those listed on the Recommended Foods chart.
Meat, poultry, fish, dry beans, eggs, and nuts	Any prepared with ingredients that contain lactose
Grains	Any prepared with ingredients that contain lactose
Vegetables	Any prepared with ingredients that contain lactose
Fruits	Any prepared with ingredients that contain lactose

(continued)

Food Group	Foods Not Recommended
Fats and oils	Butter
	Margarine
	Cream cheese
Desserts and beverages	Any prepared with ingredients that contain lactose
Sample 1-Day Menu	
Breakfast	1 cup oatmeal with slivered almonds, dried apricots, and brown sugar
	½ cup lactose-free milk
	¾ cup calcium-fortified orange juice
Snack	4 ounces soy yogurt
	Small banana
Lunch	Tossed salad with mixed vegetables
	2 tablespoons vinegar and oil salad dressing
	3 ounces salmon
	1 whole-grain roll with a thin spread of hummus
	1 pear
Snack	1 cup trail mix with toasted oats cereal, nuts, and raisins
Evening meal	2 chicken fajitas with onions and peppers, sliced chicken breast, and salsa in a flour or corn tortilla
	½ cup rice mixed with ½ cup pinto beans
Snack	1 slice whole-grain bread
	1 tablespoon peanut butter
	1 cup soy milk

Source: Nutrition Therapy for Lactose Intolerance. Nutrition Care Manual 2009. © American Dietetic Association. Reprinted with permission.

Celiac Disease

Definition Celiac disease (CD) is a complex disease whose etiology originates from genetic, environmental, and autoimmune factors triggered by the body's abnormal reaction to gluten.[71] In this disease, exposure to the gliadin component of gluten results in an inflammatory response damaging the intestinal mucosa.

Epidemiology Epidemiological studies indicate CD is much more common than previously thought, at approximately 1% worldwide.[72, 73] In 2003, the prevalence of CD in the United States was estimated to be 1:22 in first-degree relatives, 1:39 in second-degree relatives, and 1:56 in symptomatic patients. Overall, prevalence of CD in not-at-risk groups was 1:133.[74] More recently, CD was diagnosed in 30 out of a group of 976 at-risk patients (see "Etiology").[73] Other countries, including Argentina, Italy, Germany, Denmark, and Finland, have published similar studies indicating a much higher prevalence in these countries.

Etiology It is well understood that the damage to the intestinal mucosa observed in CD occurs when the small intestine is exposed to the prolamin fraction—α-gliadin and other protein components of gluten. Gluten is found in wheat, rye, malt, barley, and, in smaller amounts, in oats.

Damage to the intestinal mucosa is accompanied by an infiltration of white blood cells into the mucosa. This inflammatory response is also reflected in production of IgA anti-tissue transglutaminase (anti-tTG) and antiendomysial (EMA) and antigliadin (AGA) antibodies. These antibodies, which can serve as components of the diagnostic procedures for CD, reflect the autoimmune nature of this disease.

The major genes for CD that have been isolated include HLA-DQ2 and HLA-DQ8. These are present in 95% of all patients with CD.[73] It appears that other non-HLA genes may also influence the development of the disease even though this is not completely clear.[71]

Influential environmental factors include a younger age when gluten is introduced, a shorter length of breastfeeding, and the presence of viral infections during infancy, all of which appear to increase risk.[71]

Pathophysiology When the small intestine is exposed to specific sequences of amino acids found in gluten of wheat (gliadin), rye (secalin), and barley (hordein), there is a toxic and inflammatory response.[73] The immune response to gluten signals T-lymphocytes to produce various cytokines, which in turn direct the inflammatory response. Innate immunity is also a component of the pathophysiology and results in activated cytoxic T-cells that ultimately damage the enterocytes. Both inflammatory and innate immune responses damage the villi; height is reduced, and they are flattened in appearance (see Figure 15.9). The damage to the enterocytes results in a reduced absorptive surface area and loss of digestive enzymes.[71, 75] In Chapter 9, you will find more detail about the immune response with in-depth coverage of both innate and adaptive immunity. Celiac disease is often accompanied by other systemic autoimmune disorders, including type 1 diabetes mellitus, thyroid disease, systemic lupus erythematous, primary biliary cirrhosis, rheumatoidarthritis, Sjögren's, Down, Turner, William syndromes, and IgA deficiency. Persons with CD are considered to be at higher risk for lymphoma and osteoporosis as well as the complications of nutrient deficiencies and malnutrition.

Clinical Manifestations Classic clinical symptoms of CD include diarrhea, abdominal pain and cramping, bloating, and gas production. Other symptoms that can occur in the absence of GI problems include bone and joint pain, muscle cramping, fatigue, peripheral neuropathy, seizures, skin rash, and mouth ulcerations. However, many individuals are diagnosed without these classic GI signs and symptoms. Many individuals present with iron-deficiency anemia and the screening for other conditions such as thyroid disease, chronic fatigue, constipation, and irritable bowel syndrome has resulted in an increased rate of diagnosis. Catassi et al. used the following as criteria for screening for CD: positive family history, chronic fatigue, unexplained anemia, abnormal liver function, autoimmune disorders, Down's syndrome, Turner's syndrome, infertility, and epilepsy/ataxia.[16] Using this criteria, the authors revealed a 43-fold increase in the rate of diagnosis.

Medical Diagnosis Previously, diagnosis for CD was confirmed by biopsy of the small intestine mucosa and subsequent indication of villous atrophy, crypt hyperplasia, and

Figure 15.9 Small Intestine Villi in Celiac Disease

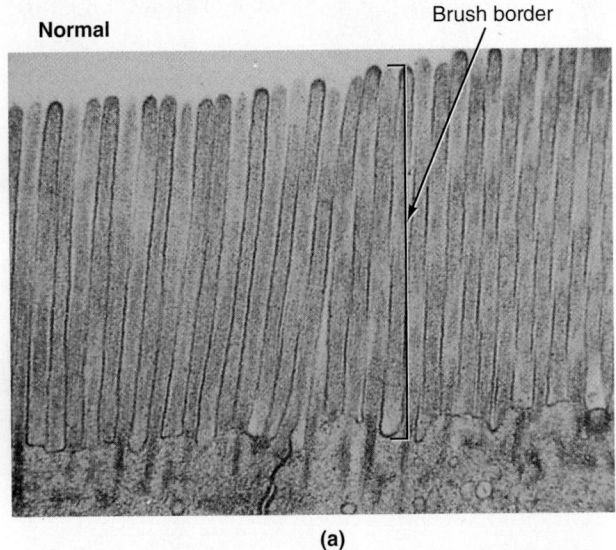

Normal

Brush border

(a)

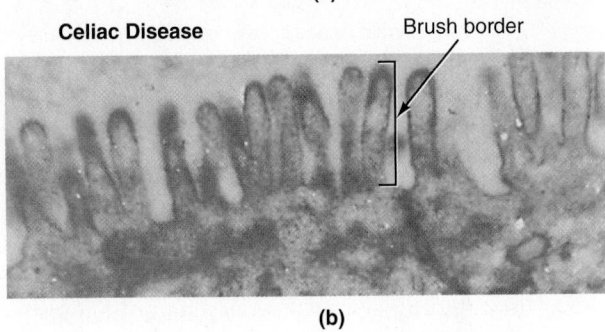

Celiac Disease

Brush border

(b)

Source: Thomas W. Sheehy, MD; Robert L. Slaughter, MD: The Malabsorption Syndrome by Medcom, Inc. Reproduced by permission of Medcom, Inc.

(a) Copyright © 2009 Medcom Inc. (b) Michael C. Webb/Visuals Unlimited

lymphocytic and plasma cell infiltrate in the lamina propria. Reversal of symptoms after restriction of gluten provided the final evidence. Even though biopsy remains the gold standard for diagnosis, it is common now to diagnose CD after identifying antibodies to gluten including anti-tissue transglutaminase (anti-tTG), endomysial IgA (EMA), and anti-gliadin (AGA). Anti-tTg is the most commonly used and EMA is considered to have greater than 90% sensitivity.[71] Additional tests are currently being developed that allow for rapid and inexpensive screening for anti-tTG.[76]

Prognosis and Treatment The only current treatment for CD is nutrition therapy consisting of a gluten-free diet. After avoidance of all gluten, villous height generally returns to normal. As the anatomy returns to normal, maldigestion and malabsorption (if present) resolve, and there is a resolution of physical signs and symptoms. Unfortunately, there is a percentage of patients who continue to experience intestinal damage despite a gluten-free diet.[77, 78] Abdulkarim and colleagues found the most common reasons for continued villous damage (or **refractory CD**) were unknown gluten contamination and the presence of coexisting diseases such as pancreatic insufficiency, irritable bowel syndrome, and malignancy.[79] Gluten cross-contamination is common and the amount of gluten that is

considered "safe" is not consistently defined. In some countries, 100 ppm of gluten is allowed for certain "gluten free" foods. In the United States, 20 ppm has been set as the tolerated level. Catassi demonstrated that as little as 50 mg of gluten per day for three months resulted in further intestinal damage.[73]

Nutrition Therapy for Celiac Disease

The acute nutritional consequences of CD are dependent on the extent of malabsorption present, which correlates directly with the extent of intestinal damage.

Nutrition Assessment Severe malabsorption will result in significant weight loss, vitamin and mineral deficiencies, and, ultimately, protein-energy malnutrition. Many patients do not present with extensive weight and protein deficiencies but may have micronutrient deficiencies. Nutrition interventions are lifelong and require significant education and adherence to allow for successful treatment. Nutrition assessment not only should evaluate that extent of anthropometric changes and nutrient imbalances that may be present, but also explore social networks, knowledge and beliefs, and adherence behaviors.

Nutrition Diagnosis Nutrition diagnoses that may result from CD and its nutritional consequences can include impaired nutrient utilization and altered GI function.

Nutrition Intervention Nutrition therapy will be consistent with the level of damage to the intestinal mucosa and the degree of malabsorption. Most often, the individual diagnosed with CD will initially be prescribed a lactose-free, gluten-free diet. Secondary lactase deficiency can be common in this disorder due to damaged villi and enzyme secretion. As villi are regenerated and absorptive capability returns, lactose can be added back to the diet slowly and usually does not require a lifelong restriction.

On the other hand, gluten does require a lifelong restriction. The patient will need to avoid all foods and other products that contain wheat, rye, barley, and malt. Restriction of oats is still controversial, but most recent research has indicated individuals may tolerate oats, as long as the oats are from a pure, uncontaminated source.[71, 80, 81] The major controversy regarding use of oats is contamination within oat products by wheat, barley, or rye. Recent nutrition practice guidelines for celiac disease indicate that up to 50 grams of oats per day is safe and is generally tolerated.[80] But the patient should be well-established on the gluten-free diet before adding oats to the diet. Measures to avoid contamination would include, in part, contacting manufacturers regarding the methods of production and avoiding products sold in bulk bins.[82]

Specific modifications that allow for avoidance of all gluten are outlined in Table 15.11. It is important that patients understand that fillers used in prescription and over-the-counter medications may contain gluten. The manufacturers of such products should be contacted to confirm any gluten content. Current proposed labeling for gluten-free products in the United States states that a product does not contain any more than 20 ppm of prolamin.[83] The standard for European countries (CODEX) has recommended this level but these have not been made standard. Current labeling does not require sources of ingredients be

Table 15.11 Nutrition Therapy for Celiac Disease: Gluten-Free Diet

Gluten-Free Diet by Food Groups[(1)]

Foods Allowed[(a)]	Foods to Question[(b)]	Foods to Avoid[(c)]
Milk & Dairy		
Milk, cream, most ice cream, buttermilk, plain yogurt, cheese, cream cheese, processed cheese, processed cheese foods, cottage cheese	Flavored yogurt, frozen yogurt, cheese sauces, cheese spreads, seasoned (flavored) shredded cheese or cheese blends	Malted milk, ice cream made with ingredients not allowed
Grains & Starches		
Breads, Baked Products and Other Items: Made with amaranth, arrowroot, buckwheat, corn bran, corn flour, cornmeal, cornstarch, flax, legume flours (bean, garbanzo or chickpea, Garfava™, lentil, pea), mesquite flour, millet, Montina™ flour (Indian ricegrass), nut flours (almond, chestnut, hazelnut), potato flour, potato starch, pure uncontaminated oat products (oat flour, oat groat, oatmeal), quinoa, rice bran, rice flours (brown, glutinous, sweet, white), rice polish, sago, sorghum flour, soy flour, sweet potato flour, tapioca (cassava, manioc), taro, teff	Items made with buckwheat flour	Items made with wheat bran, wheat farina, wheat flour, wheat germ, wheat-based semolina, wheat starch*, durum flour, gluten flour, graham flour, atta, bulgur, einkorn, emmer, farro, kamut, spelt, barley, rye, triticale, commercial oat products (oat bran, oat flour, oat groats, oatmeal) *Imported foods labeled "gluten-free" made with wheat starch
Cereals: Hot Puffed amaranth, cornmeal, cream of buckwheat, cream of rice (brown, white), hominy grits, pure, uncontaminated oatmeal, quinoa, rice flakes, soy flakes, soy grits	Rice and soy pablum	Cereals made from wheat, rye, triticale, barley and commercial oats
Cold: Puffed (amaranth, buckwheat, corn, millet, rice), rice crisps or corn flakes (with no barley malt extract or barley malt flavoring), rice flakes, soy cereals	Rice and corn cereals	Cereals made with added barley malt extract or barley malt flavoring
Pastas: Macaroni, spaghetti and noodles made from beans, corn, lentils, peas, potato, quinoa, rice, soy, wild rice	Buckwheat pasta	Pastas made from wheat, wheat starch and other ingredients not allowed (e.g., orzo)
Rice: Plain (e.g., basmati, black, brown, jasmine, red, white, wild)	Seasoned or flavored rice mixes	
Miscellaneous: Corn tacos, corn tortillas, rice tortillas		Wheat flour tacos and tortillas Matzoh, matzoh meal, matzoh balls, couscous, tabouli
Plain rice crackers, rice cakes & corn cakes	Multi-grain or flavored rice crackers, rice cakes & corn cakes	
Gluten-free communion wafers	Low-gluten communion wafers	Regular communion wafers
Meats & Alternatives		
Meat, Fish, Poultry: Plain (fresh or frozen)	Deli or luncheon meats (e.g., bologna, salami), wieners, frankfurters, sausages, pâte, meat and sandwich spread, frozen burgers (meat, fish, chicken), meatloaf, ham (ready to cook), dried meats (e.g., beef jerky), seasoned/flavored fish in pouches, imitation fish products (e.g., surimi), meat substitutes, meat product extenders	Canned fish in vegetable broth containing hydrolyzed wheat protein Frozen turkey basted or injected with hydrolyzed wheat protein; frozen or fresh turkey with bread stuffing Frozen chicken breasts containing chicken broth (made with ingredients not allowed) Meat, poultry or fish breaded in ingredients not allowed
Eggs: Fresh, liquid, dried or powdered	Flavored egg products (liquid or frozen)	

Foods Allowed[a]	Foods to Question[b]	Foods to Avoid[c]
Others: Dried beans (e.g., black, garbanzo [also known as chickpea, besan, channa, gram], kidney, navy, pinto, soy, white), dried peas, lentils	Baked beans	
Plain nuts and seeds (chia, flax, sesame, pumpkin, sunflower)	Seasoned or dry roasted nuts, seasoned pumpkin or sunflower seeds Nut butters (e.g., almond, peanut)	
Plain tofu	Flavored tofu Tempeh, miso	Fu, Seitan
Fruits & Vegetables		
Fruits: Fresh, frozen and canned fruits & juices	Dates, fruits with sauces	
Vegetables: Fresh, frozen, and canned vegetables & juices	Vegetables with sauces, French-fried potatoes cooked in oil also used for gluten-containing products, French fries (various shapes)	Scalloped potatoes (containing wheat flour), battered deep-fried vegetables
Soups		
Homemade broth, gluten-free bouillon cubes, cream soups and stocks made from ingredients allowed	Canned soups, dried soup mixes, soup bases, and bouillon cubes	Soups made with ingredients not allowed, bouillon cubes containing hydrolyzed wheat protein
Fats		
Butter, margarine, lard, shortening, vegetable oils, salad dressings with allowed ingredients	Salad dressings, suet, baking cooking spray	Salad dressing made with ingredients not allowed
Desserts		
Ice cream, sherbet, whipped toppings, whipping cream, milk puddings, custard, gelatin desserts, cakes, cookies, pies, and pastries made with allowed ingredients	Cake icings and frostings	Bread pudding, ice cream made with ingredients not allowed (e.g., cookie crumbs), cakes, cookies, muffins, pies, and pastries made with ingredients not allowed
Gluten-free ice cream cones, wafers, and waffles		Ice cream cones, wafers, and waffles made with ingredients not allowed
Others		
Sweets: Honey, jam, jelly, marmalade, corn syrup, maple syrup, molasses, sugar (brown and white), icing sugar (confectioner's)	Honey powder	
Gluten-free licorice, marshmallows	Hard candies, Smarties®, chocolates, chocolate bars	Licorice and other candies made with ingredients not allowed
Snack Foods: Plain popcorn, nuts, soy nuts, potato chips, taco (corn) chips	Seasoned (flavored) potato chips, taco (corn) chips, nuts, soy nuts	Potato chips with ingredients not allowed
Gluten-free pizza		Pizza made with ingredients not allowed
Beverages: Tea, instant or ground coffee (regular or decaffeinated), cocoa, soft drinks	Flavored and herbal teas, flavored coffees, coffee substitutes, hot chocolate mixes	Cereal and malt-based beverages (e.g., Ovaltine [chocolate malt and malt flavor], Postum)
Distilled alcoholic beverages (e.g., bourbon, gin, rum, rye whiskey, scotch whiskey, vodka, and liqueurs), wine	Flavored alcoholic beverages (e.g., coolers, ciders, Caesar vodka beverage)	
Gluten-free beer, ale, and lager		Beer, ale, and lager derived from barley
Most non-dairy beverages made from nut, potato, rice, and soy		Non-dairy beverages (nut, potato, rice, soy) made with barley malt extract, barley malt flavoring, or oats

(continued)

Table 15.11 Nutrition Therapy for Celiac Disease: Gluten-Free Diet (*Continued*)

Foods Allowed[a]	Foods to Question[b]	Foods to Avoid[c]
Condiments/Sauces: Ketchup, relish, plain prepared mustard, pure mustard flour, herbs, spices, salt, pepper, olives, plain pickles, tomato paste, vinegars (apple or cider, balsamic, distilled white, grape or wine, rice, spirit), gluten-free soy sauce, gluten-free teriyaki sauce, other sauces and gravies made with allowed ingredients	Specialty prepared mustards, prepared mustard flour, mustard pickles, worcestershire sauce, salsa, curry paste, seasoning mixes	Malt vinegar, soy sauce (made from wheat), teriyaki sauce (made with soy sauce containing wheat), other sauces and gravies made with wheat flour and/or hydrolyzed wheat protein
Miscellaneous: Plain cocoa, pure baking chocolate, carob chips and powder, chocolate chips, baking soda, cream of tartar, coconut, monosodium glutamate (MSG), vanilla, pure vanilla extract, artificial (synthetic, imitation) vanilla extract, vanillin, yeast (active dry, autolyzed, baker's, nutritional, torula), xanthan gum, guar gum	Baking powder, wasabi peas	Brewer's yeast

Excerpted from: Gluten-Free Diet: A Comprehensive Resource Guide—Revised and Expanded Edition, January 2010. Shelley Case, RD, Case Nutrition Consulting Inc., Regina, SK, Canada. www.glutenfreediet.ca

(a), (b), (c) See following tables for further background information on foods allowed, foods to question and foods to avoid.

Notes on Foods Allowed	
Food Products	**Notes**
Grains	
Garfava™ Flour	A specialty flour from garbanzo beans (chickpeas) and fava beans developed by Authentic Foods.
Mesquite Flour	Made from the ground pods of the mesquite tree.
Montina™ Flour	Made from Indian ricegrass.
Quinoa	A small seed of a South American plant that can be cooked and eaten whole or ground into flour or flakes.
Glutinous Rice Flour	Also known as sweet, sticky, or sushi rice flour. Made from a sticky short-grain rice that is higher in starch than brown or white rice. Does not contain any gluten.
Sago	An edible starch derived from the pith of the stems of a certain variety of palm trees. Usually ground into a powder and used as a thickener or dense flour.
Tapioca (Cassava, Manioc, Yuca)	A tropical plant that produces a starchy edible root that is peeled and can be boiled, baked, or fried. The peeled root can also be dried and washed with water to extract the starch (known as tapioca starch), which can be used to make baked products and tapioca pearls.
Taro (Dasheen, Eddo)	A tropical plant harvested for its large, starchy tubers, which are consumed as a cooked vegetable or made into breads, puddings, or Poi (a Polynesian dish).
Teff	A tiny seed of a grass native to Ethiopia that can be cooked and eaten whole or ground into flour.
Hominy Grits (Corn Grits)	Corn kernels that are coarsely or finely ground that are cooked and eaten as a hot breakfast cereal or side dish.
GF Communion Wafers	No-gluten host made from soy and rice flour by Ener-G Foods. These hosts are allowed by most major denominations except the Catholic Church.
Meats & Alternatives	
Chia	An oilseed of the ancient plant species (*Salvia hispanica L.*) belonging to the mint family, which is grown in Central and South America. Available in a natural brown and white seed, sold as "Chia" and a pure white variety sold under the trademark name "Salba." It is high in omega-3 fatty acids and fiber. The seed should be ground in order to get the maximum benefit of all the nutritional components.
Other	
Distilled Alcoholic Beverages	Rye whiskey, scotch whiskey, gin, vodka, and bourbon are distilled from a mash of fermented grains. Even though they are derived from a gluten-containing grain, the distillation process removes the gluten from the purified final product. Rum (distilled from sugar cane) and brandy (distilled from wine) are also gluten free. Liqueurs (also known as cordials) are made from an infusion of a distilled alcoholic beverage and flavoring agents such as nuts, fruits, seeds, or cream.
Gluten-Free Beer, Ale and Lager	Can be made from fermented rice, buckwheat, millet, and/or sorghum.
Plain Prepared Mustard	Made from distilled vinegar, water, mustard seed, salt, spices, and flavors.
Pure Mustard Flour	A powder made from pure ground mustard seed.

Table 15.11 Nutrition Therapy for Celiac Disease: Gluten-Free Diet (*Continued*)

Food Products	Notes
Vinegars	Produced from various ingredients: Balsamic (grapes), cider (apples), rice (rice wine), white distilled (corn, wheat, or both), wine (red wine). All these vinegars are gluten-free (including distilled white derived from wheat as the distillation process removes the gluten from the final purified product) except for malt vinegar.
Vanilla	Pure vanilla and pure vanilla extract are derived from the vanilla bean pods of a climbing orchid grown in tropical locations. The vanilla beans are chopped and soaked in alcohol and water; aged, and then filtered. It must contain at least 35% ethyl alcohol by volume. The pure vanilla is bottled or the pure extract can be mixed with sugar and a stabilizer and then bottled.
Natural Vanilla Flavor	Derived from vanilla beans but contains less than 35% ethyl alcohol. May also contain sugar and a stabilizer.
Artificial (Imitation, Synthetic) Vanilla-Vanillin Extract/Flavoring	Made from a by-product of the pulp and paper industry or a coal-tar derivative that is chemically treated to mimic the flavor of vanilla. Also contains alcohol, water, color, and a stabilizer.
Baker's Yeast	A type of yeast grown on sugar beet molasses. It is available as active dry yeast granules (sold in packets or jars) or compressed yeast (also known as wet yeast, cake yeast, or fresh yeast), which must be refrigerated.
Autolyzed Yeast/Autolyzed Yeast Extract	A special process that causes yeast to be broken down by its own enzymes, resulting in the production of various compounds that can be used as flavoring agents. Autolyzed yeast is almost always derived from baker's yeast.
Torula Yeast	A yeast grown on wood sugars (a by-product of waste products from the pulp and paper industry). Used as a flavoring agent that has a hickory smoke characteristic.
Nutritional Yeast	A specific strain of an inactive form of baker's yeast that is grown on a mixture of sugar beet molasses, which is fermented, washed, pasteurized, and dried at high temperatures. Used as a dietary supplement as it contains protein, fiber, vitamins, and minerals. Available in pills, flakes, or powder.
Xanthan Gum	It is produced from the fermentation of corn sugar. This powder is used to thicken sauces and salad dressings. Also used in gluten-free baked products to improve the structure and texture.
Guar Gum	A gum extracted from the seed of an East Indian plant. Available as a powder that is used as a thickener and stabilizer. Can be substituted for xanthan gum in gluten-free baked products. It is high in fiber and may have a laxative effect if consumed in large amounts.

Notes on Foods to Question

Food Products	Notes
Milk & Dairy	
Cheese Spreads, Cheese Sauces (e.g., Nacho), Seasoned (flavored) Shredded Cheese or Cheese Blends	May be thickened with wheat flour or wheat starch. Seasonings may contain hydrolyzed wheat protein, wheat flour, or wheat starch.
Flavored Yogurt, Frozen Yogurt	May contain granola, cookie crumbs, or wheat bran.
Grains & Starches	
Buckwheat Flour	Pure buckwheat flour is gluten free; however, some buckwheat flour may be mixed with wheat flour.
Rice & Corn Cereals	May contain barley malt, barley malt extract, barley malt flavoring.
Buckwheat Pasta	Also called Japanese Soba noodles. Some Soba pasta contains pure buckwheat flour, which is gluten free, but others may also contain wheat flour.
Seasoned or Flavored Rice Mixes	Seasonings may contain hydrolyzed wheat protein, wheat flour, or wheat starch or have added soy sauce that contains wheat.
Multi-grain or Flavored Rice Crackers, Rice Cakes & Corn Cakes	Multi-grain products may contain barley and/or oats. Some contain soy sauce (made from wheat), seasonings containing hydrolyzed wheat protein, wheat flour, or wheat starch.
Low-Gluten Communion Wafers	The Catholic Canon Law, code 924.2, requires the presence of some wheat in communion wafers and will not accept the gluten-free hosts made with other grains. A very low-gluten host made with a small amount of specially processed wheat starch is available from the Benedictine Sisters of Perpetual Hope. The level of gluten in these hosts is extremely small (less than 37 micrograms or 0.037 milligrams per wafer). The Italian Celiac Association's scientific committee approved the use of the low-gluten host. Many health professionals allow the use of this host. Some recommend consuming only ¼ of a wafer per week. The decision of whether to use this host should be discussed with your health professional. The hosts can be purchased by contacting 1-800-223-2772 or email: altarbreads@benedictinesisters.org or write to Benedictine Sisters Altar Bread Department, 31970 State Highway P, Cyde, MO, 64432, USA. More information for Catholics with celiac disease can be found at www.catholicceliacs.org

(continued)

Table 15.11 Nutrition Therapy for Celiac Disease: Gluten-Free Diet (*Continued*)

Food Products	Notes
Meats & Alternatives	
Deli/Luncheon Meats, Hot Dogs & Sausages, Dried Meats	May contain fillers made from wheat. Seasonings may contain hydrolyzed wheat protein, wheat flour, or wheat starch.
Meat & Sandwich Spreads	Products such as pâte may contain wheat flour or seasonings containing hydrolyzed wheat protein, wheat flour, or wheat starch.
Frozen Burgers (Meat, Poultry and Fish) and Meatloaf	May contain fillers (wheat flour, wheat starch, bread crumbs). Seasonings may contain hydrolyzed wheat protein, wheat flour, or wheat starch.
Ham (ready to cook)	Glaze may contain hydrolyzed wheat protein, wheat flour, or wheat starch.
Seasoned/Flavored Fish in Pouches	May contain wheat or barley.
Imitation Fish Products (e.g., Surimi)	Imitation crab/seafood sticks may contain fillers such as wheat starch.
Meat Substitutes (e.g., vegetarian burgers, sausages, roasts, nuggets, textured vegetable protein)	Often contain hydrolyzed wheat protein, wheat gluten, wheat starch, or barley malt.
Flavored Egg Products (frozen or liquid)	May contain hydrolyzed wheat protein.
Baked Beans	Some are thickened with wheat flour.
Seasoned or Dry Roasted Nuts, Pumpkin, or Sunflower Seeds	May contain hydrolyzed wheat protein, wheat flour, or wheat starch.
Nut Butters (e.g., almond, peanut)	Most brands are gluten free; however, some specialty brands may contain wheat germ.
Flavored Tofu	May contain soy sauce (made from wheat) or other seasonings that contain hydrolyzed wheat protein, wheat flour, or wheat starch.
Tempeh	A meat substitute made from fermented soybeans and millet or rice. Often seasoned with soy sauce (made from wheat).
Miso	A condiment used in Oriental cooking made from fermented soybeans and/or barley, wheat, or rice. Wheat or barley are the most common grains used.
Fruits & Vegetables	
Dates	Chopped, diced, or extruded dates are packaged with oat flour, dextrose, or rice flour. Oat flour or dextrose are the most common sources used.
French-Fried Potatoes	Often cooked in the same oil as gluten-containing foods (e.g., breaded fish and chicken fingers), resulting in cross-contamination. Some French fries in various shapes may also contain wheat or barley flour.
Soups	
Canned Soups, Dried Soup Mixes, Soup Bases & Bouillon Cubes	May contain noodles or barley. Cream soups are often thickened with wheat flour. Seasonings may contain hydrolyzed wheat protein, wheat flour, or wheat starch.
Fats	
Salad Dressings	May contain wheat flour, malt vinegar, or soy sauce (made from wheat). Seasonings may contain hydrolyzed wheat protein, wheat flour or wheat starch.
Suet	The hard fat around the loins and kidneys of beef and sheep. Flour may be added to packaged suet. Suet can be used to make mincemeat, steamed Christmas pudding, and Haggis (a traditional Scottish dish).
Cooking sprays	Baking cooking spray may contain wheat flour or wheat starch.
Desserts	
Cake Icing & Frostings	May contain wheat flour or wheat starch.
Sweets	
Honey Powder	This commercial powder is used in glazes, seasoning mixes, dry mixes, and sauces. May contain wheat flour or wheat starch.
Hard Candies & Chocolates	May contain barley malt flavoring and/or wheat flour.
Smarties®	Canadian product contains wheat flour.
Chocolate Bars	May contain wheat flour or barley malt flavoring.

Table 15.11 Nutrition Therapy for Celiac Disease: Gluten-Free Diet (*Continued*)

Food Products	Notes
Snack Foods	
Seasoned Potato Chips, Taco (corn) Chips, Nuts, Soy Nuts	Some potato chips contain wheat starch. Seasoning mixes may contain hydrolyzed wheat protein, wheat flour, or wheat starch.
Beverages	
Flavored or Herbal Teas, Flavored Coffees	May contain barley malt flavoring. Some specialty coffees may be prepared with a chocolate chip-like product that contains cookie crumbs.
Coffee Substitutes	Roasted chicory is the most common coffee substitute and is gluten-free. Other coffee substitutes are derived from wheat, rye, barley, and/or malted barley.
Hot Chocolate Mixes	May contain barley malt or wheat starch.
Flavored Alcoholic Cooler Beverages	May contain barley malt.
Caesar Vodka Beverage Mix	May contain hydrolyzed wheat protein.
Condiments/ Sauces	
Specialty Prepared Mustards	Some brands contain wheat flour.
Prepared Mustard Flour	Made from ground mustard seed, sugar, salt, and spices, which are gluten free. However, some brands also contain wheat flour.
Mustard Pickles	May contain wheat flour and/or malt vinegar.
Worcestershire Sauce	May contain malt vinegar.
Salsa	Some brands contain wheat flour, wheat starch, hydrolyzed wheat protein, or malt extract.
Curry Paste	Made from the pulp of the tamarind pod and a variety of spices. Some curry pastes may also contain wheat flour or wheat starch.
Baking Powder	Most brands contain cornstarch, which is gluten free. However, some brands contain wheat starch.
Seasoning Mixes	Some brands contain wheat flour, wheat starch, or hydrolyzed wheat protein as the carrier agent.
Wasabi Peas	Roasted green peas coated in wasabi. Some brands contain wheat flour or wheat starch.

Notes on Foods to Avoid

Food Products	Notes
Milk Products	
Malted Milk	Contains malt powder derived from malted barley.
Grains	
Semolina	A coarsely ground grain (usually made from the refined portion of durum wheat) that can be used to make porridge or pasta.
Atta	A fine whole-meal flour made from low-gluten, soft texturized wheat used to make Indian flat bread. Also known as chapatti flour.
Bulgur (Burghul)	Quick-cooking form of whole wheat. Wheat kernels that are parboiled (partially cooked), dried, and then cracked. Used in soups, pilafs, stuffing, or salad (e.g., Tabouli).
Einkorn, Emmer, Farro, Kamut, Spelt	Types of wheat. Many "wheat-free" foods are made from these varieties of wheat, especially kamut and spelt. Remember that "wheat-free" does not always mean "gluten-free."
Triticale	A cereal grain that is a cross between wheat and rye.
Orzo	A type of pasta that is the size and shape of rice. Used in soups and as a substitute for rice.
Matzoh	Unleavened bread made with wheat flour and water that comes in thin sheets. Used primarily during Passover.
Matzoh Meal	Ground matzoh.
Matzoh Balls	Dumplings made of matzoh meal, which is not gluten free. However, can be made with potato flour which is gluten free.
Couscous	Granules of semolina (made from durum wheat) that are precooked and dried. Cooked couscous is served hot or cold as a dish or salad.
Tabouli	A salad usually made with bulgur wheat, or couscous which are not gluten free. Can also be made with quinoa, which is gluten free.

(continued)

Table 15.11 Nutrition Therapy for Celiac Disease: Gluten-Free Diet (*Continued*)

Food Products	Notes
Meats & Alternatives	
Fu	A dried gluten product derived from wheat that is sold as thin sheets or thick round cakes. Used as a protein supplement in Asian dishes such as soups and vegetables.
Seitan	A meat-like food derived from wheat gluten used in many vegetarian dishes. Sometimes called "wheat meat."
Other	
Licorice	Regular licorice contains wheat flour.
Cereal & Malted Beverages	Contain malted barley or other grains such as wheat or rye (e.g., Postum, Ovaltine).
Beer, Ale & Lager	Basic ingredients include malted barley, hops (a type of flower), yeast, and water. As this mixture is only fermented and not distilled, it contains varying levels of gluten.
Potato Chips	Some brands of plain potato chips contain added wheat flour and/or wheat starch.
Soy Sauce	Many brands are a combination of soy and wheat.
Malt Vinegar	Made from malted barley. As this vinegar is only fermented and not distilled, it contains varying levels of gluten.
Brewer's Yeast	A dried inactive yeast that is a bitter by-product of the brewing industry. It is not commonly used as a flavoring agent in foods. ELISA tests are unable to accurately confirm the amount of residual gluten in this type of yeast.

Gluten-Free Additives and Ingredients

Additives
- Acetic Acid
- Adipic Acid
- Benzoic Acid
- BHA
- BHT
- Calcium Disodium EDTA
- Fumaric Acid
- Glucono-delta-lactone
- Lactic Acid
- Lecithin
- Malic Acid
- Mono and diglycerides
- Polysorbate 60; 80
- Propionic Acid
- Propylene Glycol
- Rennet
- Silicon Dioxide
- Sodium Benzoate
- Sodium Metabisulphite
- Sodium Nitrate
- Sodium Nitrite
- Sodium Sulphite
- Sorbate
- Sorbic Acid
- Stearic Acid
- Tartaric Acid
- Titanium Dioxide

Coloring Agents
- Natural Colors [e.g. annatto, caramel color: carotene, beta carotene, paprika]
- Artificial Colors [e.g., tartrazine*, sunset yellow FCF, erythrosine, citrus red No. 2, brilliant blue FCF, fast green FCF, titanium dioxide]

Flavoring Agents
- Ethyl Maltol
- Maltol
- Monosodium Glutamate (MSG)
- Vanilla
- Vanilla Extract
- Vanilla Flavoring
- Vanillin

Sugars/Sweeteners
- Acesulfame-potassium
- Agave
- Aspartame
- Brown Sugar
- Corn Syrup/Solids
- Dextrose
- Fructose
- Glucose
- Glucose Syrup
- Honey
- Invert Sugar
- Isomalt
- Lactose
- Maltitol
- Maltitol Syrup
- Maltose
- Mannitol
- Molasses
- Saccharin
- Stevia
- Sucralose
- Sucrose
- White Sugar
- Xylitol

Vegetable Gums
- Acacia Gum (Gum Arabic)
- Agar (Agar-Agar)
- Algin (Alginic Acid)
- Carageenan
- Carboxymethylcellulose (Cellulose Gum)
- Carob Bean (Locust Bean)
- Guaiac Gum
- Guar Gum
- Karaya Gum
- Methylcellulose
- Tragacanth Gum
- Xanthan Gum

Miscellaneous
- Ascorbic Acid
- Autolyzed Yeast
- Baker's Yeast
- Beta Carotene
- Cream of Tartar
- Gelatin
- Lecithin
- Maltodextrin
- Modified Food Starches (except wheat starch)
- Nutritional Yeast
- Papain
- Pectin
- Psyllium
- Starches (except wheat starch)
- Torula Yeast

*A very small number of individuals may experience an allergic-type reaction to the yellow food color tartrazine; however, this is unrelated to gluten.

Note: This is not an all-inclusive listing. For a more comprehensive listing of ingredients see the Canadian Celiac Association's *Pocket Dictionary: Acceptability of Foods and Food Ingredients for the Gluten-Free Diet*.

designed on the label; this would be an important improvement to ensure adherence to a gluten-free diet. Many specialty products that provide alternatives for wheat, rye, and barley are now available. Food products using rice, corn, amaranth, millet, and soy allow for greater variety today than was previously available. National organizations, including the Celiac Disease Foundation, Canadian Celiac Association, and the Gluten Intolerance Group of North America, for example, provide excellent resources for persons diagnosed with CD.

Irritable Bowel Syndrome

Definition As early as 1849, symptoms of this disorder were described.[84] Throughout the last century, varying names such as spastic colon, irritable colon syndrome, and neurogenic mucous colitis have been given to this syndrome. Currently, **irritable bowel syndrome (IBS)** is defined "by abdominal pain or discomfort that occurs in association with altered bowel habits over a period of at least 3 months" (p. S2).[85] Three subtypes of IBS include IBS-D (diarrhea predominant), IBS-C (constipation predominant), and IBS-M (mixed diarrhea and constipation).[85]

Epidemiology Irritable bowel syndrome is the most common gastrointestinal complaint in the United States and Canada. Worldwide prevalence ranges from 1%–20%, and in North America, is estimated to be approximately 7%. This condition affects women more than men and occurs frequently before age 50. Both the direct and indirect health care costs for IBS are estimated to be as much as $20 billion dollars.[85, 86]

Etiology IBS historically has been designated as a "functional" disorder. This means a diagnosis is made after ruling out all other possible organic causes of the patient's symptoms. A variety of criteria have been used to expedite the diagnosis of IBS, including the Manning and the Rome Criteria III discussed earlier in this chapter. The specific cause of IBS is unknown, but current research is focused on multiple factors including genetic predisposition; altered immune response stimulated by food sensitivity and altered microbial environment; an elevated inflammatory response to gastroenteritis; small intestinal bacterial overgrowth (SIBO); abnormal release, transport, or recognition of serotonin; and an increased sensitivity of the enteric nervous system that causes abnormal motility and pain.[85, 87–90] Figure 15.10 presents an overview of the factors thought to contribute to IBS symptoms. Other conditions that are commonly associated with IBS include celiac disease and lactose maldigestion. The American College of Gastroenterology recommends that all patients be screened for celiac disease as a component of the differential diagnosis for IBS.[85]

Pathophysiology The pathophysiology of IBS is complex and, as previously stated, not completely understood. Proposed etiological factors may be examined in context of what we know about the normal physiology of the gastrointestinal tract. In IBS, abnormal motility is considered to be one of the major factors involved in symptoms of abdominal pain and altered bowel habits.[91] As discussed in Chapter 14, contraction and motility are regulated through the gastrointestinal tract's own specialized pacemaker cells called the interstitial cells of Cajal. The autonomic nervous system relays input for the smooth muscle of the GI tract through a specialized enteric nervous system. Two types of neurons regulate the contraction of smooth muscle, motility, and secretory functions of the GI tract. Primary neurotransmitters involved in transmission of impulses include acetylcholine, substance P, vasoactive intestinal polypeptide (VIP), and nitric oxide. Other neurotransmitters and modulators involved in neuroregulation for the GI tract include norepinephrine, serotonin, neuropeptide Y, melatonin, and G-aminobutyric acid (GABA).

Serotonin (also known as 5-HT$_4$ [5-hydroxytryptamine]) is synthesized from the amino acid tryptophan. More than 95% of serotonin is found within the GI tract, with the remaining 5% active within the brain. Serotonin can activate both excitatory and inhibitory neurons in the gastrointestinal tract. Serotonin stimulates both the release of acetylcholine, causing smooth muscle contraction, and the inhibitory neurons that release nitric oxide, which results in relaxation of smooth muscle. Altered serotonin levels have been documented in all types of IBS and may lead to abnormal motor and secretory function.[85, 92–94] The mechanism for these effects is not clear, even though reduced levels have been documented in IBS-C and increased levels have been measured in IBS-D. Research is focused on determining whether the abnormal levels result from altered release, altered synthesis, or altered transport.[92] Genes for IBS may be located among candidate genes that control serotonin transporters. The migrating motility complex (MMC), discussed earlier, is an additional type of motility whose weak contractions serve to constantly sweep through the small intestine and remove leftover waste. Patients with IBS appear to have abnormal periods of MMC contractions when compared to controls.[93]

Individuals with IBS have an increased sensitivity to stimulation of the gastrointestinal tract. This means that, when exposed to the same stimuli, most individuals do not develop the symptoms that patients with IBS experience: abdominal pain, urgency, diarrhea, or constipation. For example, when IBS patients were evaluated using balloon-distention and duodenal lipid infusion, they experienced significantly more abdominal pain and gastrointestinal symptoms than controls.[85, 95, 96] There is also interest in the role of bacterial overgrowth (SIBO) as a component of IBS. Testing for SIBO is neither consistently sensitive nor specific and, therefore, it has been difficult to fully establish this contribution.[5, 97] The use of antibiotics for treatment of IBS has resulted in a decrease of symptoms such as bloating or gas production but is not FDA approved as a component of the treatment protocol.[85, 97]

Other studies have observed development of IBS after infectious enteritis. Specific organisms that have been documented include *Blastocystis hominis*, *Campylobacter*, *Salmonella*, and parasites such as *Trichinella spiralis*.[98] Abnormal cellular immune responses to certain nutrients have also been documented.[90, 98]

At least two-thirds of IBS patients associate the signs and symptoms they experience with characteristics of the foods that they eat, including the size of the meal; the fat content; and the inclusion of lactose, fructose, sorbitol and other sugar alcohols, caffeine, and carbonated drinks. The "food triggers" reported by an individual patient may become the foundation for nutrition therapy in IBS. These will be discussed in more detail under the nutrition therapy section.

Figure 15.10 Potential Etiological Factors in IBS

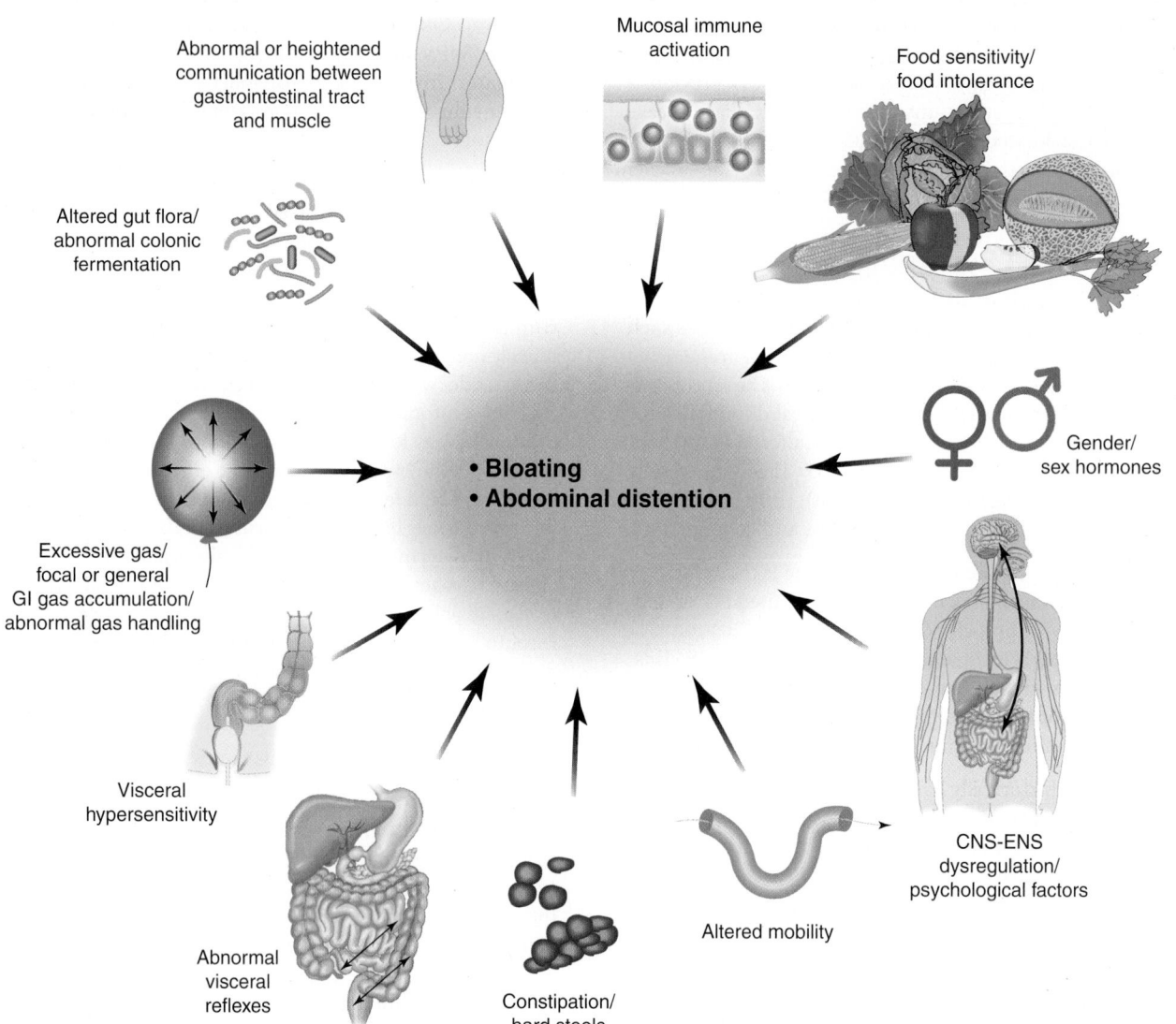

Abnormal or heightened communication between gastrointestinal tract and muscle

Mucosal immune activation

Food sensitivity/ food intolerance

Altered gut flora/ abnormal colonic fermentation

Excessive gas/ focal or general GI gas accumulation/ abnormal gas handling

- Bloating
- Abdominal distention

Gender/ sex hormones

Visceral hypersensitivity

CNS-ENS dysregulation/ psychological factors

Altered mobility

Abnormal visceral reflexes

Constipation/ hard stools

Source: Reprinted from Gastroenterology 136, Simren M., Bloating and abdominal distention: not so poorly understood anymore!, 1487–90, © Copyright 2009, with permission from Elsevier.

Other concurrent diagnoses that occur with IBS include **fibromyalgia**, chronic fatigue syndrome, **temporomandibular joint (TMJ) syndrome**, and food allergies. Though correlations with certain psychiatric conditions have been documented frequently in patients with IBS, no specific connections have been made. Some researchers feel previous physical and sexual abuse may increase sensitivity to GI stimulation and thus may contribute to symptoms. Stress is known to exacerbate symptoms of IBS in some individuals.

Clinical Manifestations Abdominal pain, alteration in bowel habits or motility, gas, and flatulence as well as some upper GI symptoms (reflux and noncardiac chest pain) are major symptoms for IBS. Abdominal pain can be acute and relieved by defecation. At the same time, some patients with IBS experience constant, chronic abdominal pain.

Alterations in bowel habits are seen in both major types of IBS—constipation (IBS-C) and diarrhea (IBS-D). In some patients, both constipation and diarrhea are experienced (IBS-M).

Increased levels of gas and flatulence are also experienced. Gas is produced when food passes into the large intestine and is only partially digested. Intestinal bacteria act on these foodstuffs, and by-products of the fermentation result in gas production. The registered dietitian can assist by providing nutrition therapy that may alter the signs and symptoms that occur as a result of specific food ingestion.

Medical Treatment For those patients with IBS-D, antidiarrheal agents can be used. These medications assist by decreasing motility and increasing consistency of the stool. (See previous discussion regarding diarrhea.) These medications include diphenoxylate (Lomotil®), loperamide, atropine, and cholestyramine. Loperamide is the only medication that has been used in controlled studies of IBS, and its effectiveness appears limited to treatment of diarrhea without affecting any other IBS symptoms.[85] Antispasmodics include peppermint oil, which has been noted to improve abdominal pain and cramping.[99]

Specific medications for IBS include those that work as either agonists or antagonists for the serotonin receptors. The class of medications that work as 5-HT$_4$ agonists include renzapride, cisapride, and tegaserod. These medications are not currently available in the United States except for **compassionate use** request. Alosetron®, which is a 5-HT$_3$ receptor antagonist, is used for treatment of IBS-D. This medication is prescribed in the United States under a strict, limited marketing program. Severe constipation and ischemic colitis have been reported as potential side effects of the drug.

Another class of medications developed for use in IBS includes selective C-2 chloride channel activators. These medications affect chloride transport, which in turn increases fluid in the intestine and improves GI motility.[85] Lubiprostone (Amitiza) is the medication in this class that is currently available. New drugs in all of these classes are in various phases of development and testing.

Tricyclic antidepressants and selective serotonin reuptake inhibitors (SSRIs) are also used to treat IBS. These medications include amitriptyline, desipramine, doxepin, or trazodone and fluoxetine, paroxetine, or sertraline, respectively. A summary of the research indicates that these medications can be used effectively in treating the pain of IBS. Antispasmodics, including dicyclomine and hyoscyamine, are also commonly prescribed, but the American College of Gastroenterology Functional Gastrointestinal Disorders Task Force suggests these are no more effective than a placebo in controlling symptoms of altered motility and abdominal pain.

Other medications used to treat IBS-C include bulking agents and osmotic laxatives. Bulking agents are supplements or medications that add psyllium, bran, or other sources of fiber to the diet. Research to date has indicated that bulking agents are an effective component of constipation treatment, but only psyllium is effective in treatment of IBS-C. Osmotic laxatives such as Milk of Magnesia®, polyethylene glycol, and sorbitol have increased stool frequency of IBS-C patients but do not treat the other symptoms of IBS.

Other treatments for IBS include cognitive behavioral therapies (hypnosis, relaxation techniques, guided imagery), probiotics, and nutrition therapy.[85, 90, 99, 100]

Nutrition Therapy for Irritable Bowel Syndrome

Nutrition Assessment Symptoms of IBS can lead to changes in oral intake that in turn lead to nutrient deficiencies, potential underweight, and malnutrition. Since IBS may present with either diarrhea, constipation, or both, the specific assessment data needed to provide appropriate nutrition care will vary. Refer to Table 15.2 for guidance with relevant nutrition assessment data.

Nutrition Diagnosis Common nutrition diagnoses associated with IBS include inadequate food/oral beverage intake; altered GI function; undesirable food choices; food and nutrition-related knowledge deficit; disordered eating pattern; and poor nutrition quality of life.

Nutrition Intervention Nutrition therapy goals for IBS will focus on interventions for the patient's specific symptoms and their response to specific food triggers. These may include exclusion of individual foods that are not tolerated or that are suspected to cause an allergic reaction.[90, 100] Nutrition education should focus on normalizing overall dietary patterns, ensuring adequate nutritional intake, eliminating those foods that trigger symptoms, and taking the necessary steps to reduce gas production and abnormal motility.

One intervention for treating IBS is the use of a traditional exclusion diet. This approach eliminates all possible foods related to the patient's symptoms. Then each food is added back within a specific time period. If the patient does not experience any symptoms when the food is added back to the diet, that food is considered to be tolerated and allowed to be included in the regular diet. The ACG states that the research studies on exclusion diets have demonstrated variable success and, in general, were not randomized controlled trials.[85] Methodology to identify food allergies (skin prick, IgA, IgE) is also problematic (see Chapter 9 for more information on food allergies and exclusion diets).

The presence of lactose indigestion does not necessarily mean that it is a cause of IBS, but certainly the two can co-exist. Breath hydrogen tests can confirm this. Additionally, taking a careful diet history relating signs and symptoms with lactose ingestion or presenting a milk challenge would be other steps to determine if lactose should be avoided. Generally, elimination of lactose for 1–2 weeks is adequate to determine the success of this intervention.[100]

An additional proposed nutrition therapy for IBS focuses on restriction of those foods contributing fermentable oligosaccharides, disaccharides, monosaccharides, and polyols (sugar alcohols), collectively referred to as FODMAPs.[101–103] Food sources of FODMAPs are listed in Table 15.12. These foods are not well digested and contribute to fermentation—potentially leading to those specific signs and symptoms associated with IBS, as demonstrated in Figure 15.11. To date, results of only one controlled trial using FODMAP exclusion as nutrition therapy for IBS have been published, but it did demonstrate a significant reduction in the signs and symptoms of IBS within the participants.[101, 103] As discussed in Table 15.2, nutrition assessment includes food diaries or a diet history that should focus on dietary components that the patient has associated with any increase in gastrointestinal symptoms. It is hoped that future research will be able to clearly identify the steps in nutrition therapy for IBS, but for now, a reasonable approach is to begin with a careful examination of symptoms related to specific food triggers, including those foods with higher amounts of fructose, sugar alcohols, and lactose.[68, 90, 100]

Once a baseline nutritional history has been established and nutrition problems are identified, the RD and patient can begin to identify any needed changes in diet. Overall nutritional adequacy should be addressed first. Many patients with IBS tend to eat erratically due to their gastrointestinal symptoms, and often eating is associated with a high level of anxiety and stress. Establishing a regular eating pattern that does not exacerbate symptoms is a crucial initial step.

Historically, the most common nutrition intervention for IBS has been increasing fiber intake, but research does not support the effectiveness of this intervention. In fact, insoluble

Table 15.12 Fermentable Oligo-, Di-, and Monosaccharides, and Polyols (FODMAP)* Checklist

The following food list may assist in identifying those foods that serve as food triggers for patients with IBS.

- **Fruit:** apple, pear, guava, honeydew melon, mango, nashi fruit (Asian pear), pawpaw/papaya, quince, star fruit (carambola), watermelon
- **Stone fruits:** apricots, peaches, cherries, plums, nectarines
- **Fruits with high sugar content:** grapes, persimmon, lychee
- **Dried fruit**
- **Fruit juice, canned packing juice**
- **Dried fruit bars**
- **Fruit pastes and sauces:** tomato paste, chutney, relish, plum sauce, sweet and sour sauce, barbecue sauce
- **Fruit juice concentrate**
- **Fructose** as an added sweetener
- **High-fructose corn syrup or corn syrup solids including:** Fruit drinks, carbonated drinks, pancake syrups, catsup, jams, jellies, pickle, relish, etc. and/or liquid cough remedies and liquid pain relievers, etc.

- **Honey**
- **Coconut:** milk, cream
- **Fortified wines:** sherry, port, etc.
- **Vegetables:** Onion, leek, asparagus, artichokes, cabbage, brussels sprouts, beans
- **Legumes:** Baked beans, kidney beans, lentils, black eye peas, chickpeas, butter beans
- **Wheat or white bread**
- **Wheat pasta, noodles**
- **Wheat-based breakfast cereal**
- **Wheat-based cakes, cookies, crackers**
- **Chicory-based coffee-substitute beverages**
- **Artificial sweeteners:** sorbitol, mannitol, isomalt, xylitol
- **If lactose malabsorption:** milk, ice-cream, yogurt

*FODMAPs consumed on a regular basis are potential triggers for functional gut symptoms and a trial of limiting them should be undertaken. Please note, this checklist has not undergone evaluation for its usefulness in clinical practice.

Source: Reprinted from: Table 1 in: Barrett JS, Gibson PR. Clinical Ramifications of Malabsorption of Fructose and other short-chain carbohydrates. Practical Gastroenterology. 2007;8:51–65.

---

fiber sources may exacerbate symptoms.[85, 100] The only current recommendation for fiber supplementation in IBS patients is the use of ispaghula husk, which is more commonly known as psyllium. For those patients with IBS-C, psyllium supplementation may assist in normalizing bowel movements. Of course, increased fluid is also necessary as fiber intake is increased.

Probiotics have received attention for their potential use in IBS. Translating the research to practice poses some difficulty, though. Many of the studies have used a variety of strains with a variety of dosages. In the largest studies, probiotics have demonstrated improvement for the symptoms of bloating and

gas production.[94, 104] As discussed earlier in this chapter, adding these foods and supplements may be beneficial within the overall nutrition therapy plan.[104]

Due to problems with gas and flatulence, providing recommendations to relieve these symptoms will also be beneficial. Simple carbohydrates that cause gas are raffinose, lactose, fructose, and sorbitol[101] (see Table 15.13 for examples of gas-producing foods). Avoiding foods that produce gas and taking steps to decrease swallowed air will decrease gas production. Products such as Beano® or Bean-zyme® provide alpha-galactosidase, which potentially decreases the volume of undigested carbo-

Figure 15.11 Effects of Fermentable Carbohydrate Consumption

Malabsorption and fermentation of these carbohydrates within the GI tract can result in several symptoms characteristic of IBS.

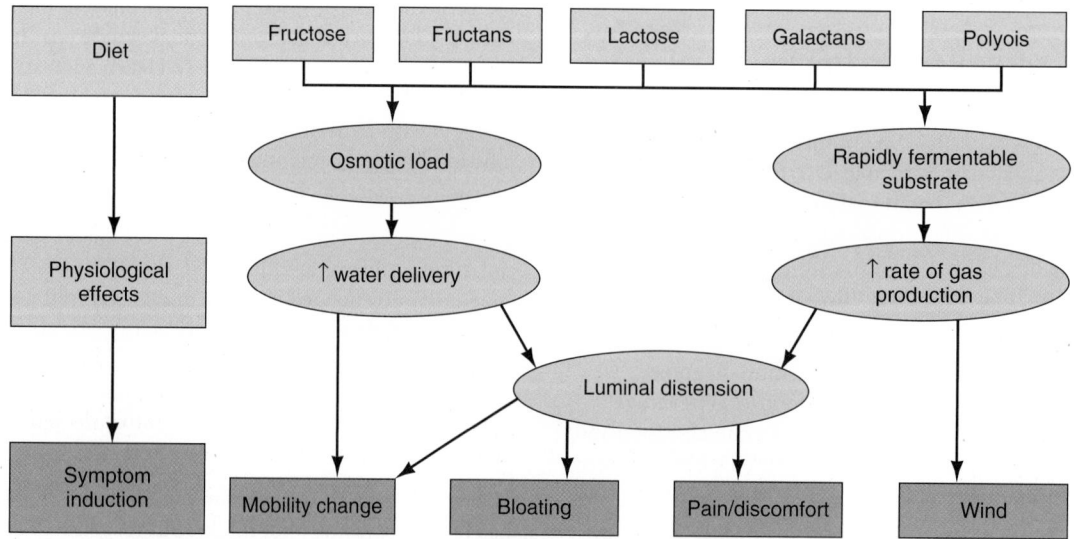

Source: Barrett JS, Gibson PR. Clinical Ramifications of Malabsorption of Fructose and other short-chain carbohydrates. Practical Gastroenterology. 2007;8:51–65.

Table 15.13 Gas-Producing Foods

Vegetables	Legumes
Beets	Black-eyed peas
Broccoli	Bog beans
Brussels sprouts	Broad beans
Cabbage	Chickpeas
Carrots	Field beans
Cauliflower	Lentils
Corn	Lima beans
Cucumbers	Mung beans
Leeks	Peanuts
Lettuce	Peas
Onions	Pinto beans
Parsley	Red kidney beans
Peppers, sweet	Soybeans

Grains/Cereals/Seeds/Nuts	Others
Barley	Bagels
Breakfast cereals	Baked beans
Granola	Bean salads
Oat bran	Chili
Oat flour	Lentil soup
Pistachios	Pasta
Rice bran	Peanut butter
Rye	Soy milk
Sesame flour	Split-pea soup
Sorghum, grain	Stir-fried vegetables
Sunflower flour	Stuffed cabbage
Wheat bran	Tofu
Whole-wheat flour	Whole-grain breads

Source: http://www.beanogas.com/Smart_Problem.aspx

Table 15.14 Steps to Decrease Gas Production

Avoid swallowing air	• Foods should be chewed slowly before swallowing. • Avoid use of straws. • Chewing gum or hard candy should not be used. • Don't smoke or use tobacco products. • If dentures are worn, they should be checked by a dentist for proper fit.
Avoid gas-forming foods	• See Table 15.13. Identify and avoid problem foods. • Assess potential for lactose intolerance. • Soak dried beans in water overnight; then discard water and cook soaked beans in new water.
Eating tips	• Cut back on fried and higher-fat foods. • Eat smaller meals throughout the day. • Avoid eating while anxious, upset, or on the run. • Avoid carbonated beverages.
Personal habits	• Increase physical activity for a minimum of thirty minutes daily.
Medications	• Nonprescription antigas products (e.g., Beano™) help break down sugars found in vegetables and grains. • Lactase enzyme supplements (such as Dairy Ease™ and Lact-Aid™) can be taken with dairy products to help break down lactose in foods. • Peppermint tea contains menthol which may ease cramping and abdominal pain. (Caution: peppermint tea may contribute to heartburn or acid reflux.)

Sources: American Gastroenterological Association. *Gas in the Digestive Tract.* Available at http://www.gastro.org

Mayo Clinic. *Diseases and Conditions: Gas and gas pains.* MayoClinic.com. Available at http://www.mayoclinic.com

WebMDHealth. *Gas, Bloating, and Burping Prevention.* WebMD.com. Available at http://www.webmd.com

hydrate entering the large intestine and thus decreases the amount of gas produced. Table 15.14 outlines steps to decrease gas production. Table 15.15 summarizes nutrition interventions for IBS.

Inflammatory Bowel Disease

Definition **Inflammatory bowel disease (IBD)** is characterized as an autoimmune, chronic inflammatory condition of the gastrointestinal tract. IBD is actually the general term for either of two diagnoses: **ulcerative colitis (UC)** and **Crohn's disease**. These diagnoses are very similar but have very distinct differences. See Box 15.3 for a comparison of the characteristics of UC and Crohn's disease.

Epidemiology Prevalence of IBD is higher in countries within the Northern Hemisphere—North America and Northern European countries. Respectively, prevalence is much lower in the Southern Hemisphere, in countries of Southern Europe, and in Australia. There is very low prevalence in Asia and South America and in African-Americans.[105, 106] Incidence of IBD in the United States ranges from 5 to 15 per 100,000 persons. Prevalence is estimated to be approximately 388–444 per 100,000. Prevalence is fairly equal in both males and females, but it is higher in those populations with Caucasian and Ashkenazi Jewish ancestry. The estimated cost of medical care

Table 15.15 Nutrition Therapy for IBS

- Avoid large meals
- Reduce lactose (eliminate milk, ice cream, and yogurt)
- Reduce fat to no more than 40 to 50 g/day
- Reduce sorbitol, mannitol, xylitol (mainly "sugarless" gum; read labels)
- Reduce fructose in all forms, including high-fructose corn syrup, honey, and high-fructose fruits (e.g., dates, oranges, cherries, apples, and pears)
- Reduce gas-producing foods (e.g., beans, peas, broccoli, cabbage, and bran)

Source: Reprinted from American Dietetic Association 109, Heizer WD, Southern S, McGovern S. J., 1204–1214, © Copyright 2009, with permission from Elsevier.

for these patients is approximately $1.6 billion per year in the United States.

Etiology The complete etiology for both Crohn's disease and UC is unknown at this time, though it is understood that multiple factors play a role in these conditions.[107, 108] These may include environmental factors such as smoking, infectious agents, intestinal flora, and physiological changes in the small intestine from which an abnormal inflammatory response

BOX 15.3   CLINICAL APPLICATIONS

Ulcerative Colitis versus Crohn's Disease

Ulcerative Colitis	Crohn's Disease
Etiology:	
Abnormal immune response resulting in inflammatory damage of gastrointestinal mucosa; genetic susceptibility; association with ex-smokers	Abnormal immune response resulting in inflammatory damage of gastrointestinal mucosa; genetic susceptibility; association with cigarette smoking
Epidemiology:	
Both sexes affected equally; higher prevalence in North America, northern Europe, and UK populations; increased prevalence for Ashkenazi Jews; approximately 10% of those with UC have a first-degree relative with disease; peak onset 20 to 30 years, secondary peak in middle age	Both sexes affected equally; higher prevalence in North America, northern Europe, and UK populations; Western European Jews (Ashkenazi Jews) have highest incidence; peak age of onset is teens to twenties.
Pathology:	
GI tract unable to distinguish foreign from self-antigens; characterized by chronic inflammation of colonic mucosa and submucosa, atrophy and possible dysplasia limited to colon; extent of disease varies and may involve only the rectum (ulcerative proctitis) left side of colon to splenic flexure, or entire colon (pancolitis)	Localized inflammation in bowel mucosa progressing through bowel wall; tends to be localized in terminal ileum and colon but can involve any portion of the GI tract.
Signs and Symptoms:	
Bloody diarrhea with mucus	Chronic diarrhea
Abdominal and/or rectal pain	Abdominal pain and cramping
Fever	Blood and/or mucus in stool
Weight loss	Anorexia
Possibly constipation and rectal spasm	Weight loss
Arthritis	Malnutrition
Dermatological changes	Fever
Ocular manifestations	Delayed growth in adolescents
Complications:	
Severe bleeding	Malabsorption
Toxic colitis	Malnutrition
Toxic megacolon	Abdominal fistulas and abscesses
Strictures	Intestinal obstruction
Perforation	Bacterial overgrowth (blind loop syndrome)
Intolerance to immunosuppression	Gallstones
Colonic strictures	Kidney stones
Dysplasia	Urinary tract infections
Carcinoma	Thromboembolic complications
	Perianal disease
	Neoplasia
Diagnosis:	
Abdominal ultrasound	Clinical presentation—CDAI score
MRI	Abdominal ultrasound
CT	MRI
Antiglycan antibodies (ASCA/ANCA)	CT
Calprotectin (Cal), lactoferrin (Lf), and polymorphonuclear neutrophil elastase (PMN-e)	Antiglycan antibodies (ASCA/ANCA)
	Calprotectin (Cal), lactoferrin (Lf), and polymorphonuclear neutrophil elastase (PMN-e)
Prognosis:	
Chronic with repeated exacerbations and remissions; nearly 30% of those with extensive ulcerative colitis require surgery; patients with localized UC have best prognosis, surgery rarely required, and life expectancy normal.	Rarely cured, but characterized by intermittent exacerbations. Approx 70% require surgery.

Ulcerative Colitis	Crohn's Disease
Treatment:	
Reduce acute and chronic inflammation eventually resulting in remission	Based on severity of disease
Drugs:	Drugs:
• Immunosuppressants (6-mercaptopurine, azathioprine cyclosporine)	• Immunosuppressants (6-mercaptopurine, azathioprine cyclosporine)
• Adrenocorticosteroids	• Antibiotics
• Antiinflammatory—5-aminosalicylic acid (5-ASA)	• Steroids
• Biologic therapies—Infliximab	• Methotrexate
• Antidiarrheals	• Biologic therapies—Infliximab
• Steroids	Surgery: Surgical removal of affected areas—may include ileocolic resections, segmental resections, total proctocolectomy and ileostomy.
• Antibiotics	
Surgery:	
• Colectomy	
• Subtotal colectomy	
• Total proctocolectomy with Brooke ileostomy	
• Restorative proctocolectomy with ileal pouch–anal anastomosis	

Sources:
Ali S, Tamboli CP. Advances in epidemiology and diagnosis of inflammatory bowel diseases. *Current Gastroenterology Reports*. December 2008;10(6): 576–84.
Chaudhri S, Rooney P. Surgical management of inflammatory bowel disease. Intestinal Surgery. 2008;26:352–56.
Cho JH. The genetics and immunopathogenesis of inflammatory bowel disease. *Nature Reviews. Immunology* [serial online]. 2008;8(6):458–66. Available from: MEDLINE with Full Text, Ipswich, MA.
Harris M, Kaufman H, Talamini M, Dassapoulus T, Norwitz L, Kalloo AN. Ulcerative colitis. The Johns Hopkins Medical Institutions Gastroenterology and Hepatology Resource Center. Available at: http://hopkins-gi.nts.jhu.edu
Lashner BA. Inflammatory bowel disease. The Cleveland Clinic Disease Management Available at: http://www.clevelandclinicmeded.com/medicalpubs/diseasemanagement/gastroenterology/inflammatory-bowel-disease/
Lichtenstein GR, Hanauer SB, and Sandborn WJ. Management of Crohn's disease in adults. *The American Journal Of Gastroenterology* [serial online]. February 06, 2009;104(2):465. Available from: MEDLINE with Full Text, Ipswich, MA.

is triggered[109, 110] (see Figure 15.12). There is a strong genetic association for IBD as evidenced by positive family history in approximately 5% to 15% of patients with IBD. In identical twins, the incidence of IBD is 44% versus only 3.8% in fraternal twins. Genetic associations have been identified within both the innate and acquired immune response.[109, 110]

Pathophysiology It is theorized that those individuals who are genetically susceptible and who are exposed to certain triggers experience an abnormal immune response. This immune response results in release of cytokines that direct an excessive and abnormal inflammatory reaction that ultimately destroys the intestinal mucosa. UC and Crohn's are also characterized by exacerbations of the disease process interspersed with periods of remission.

Approximately 50% of patients with UC have disease only involving the rectum. Damage to intestinal mucosa in UC usually only involves the first two layers of tissue (mucosa and superficial submucosa) within the colon and rectum. But with chronic disease, the intestinal wall can become so thin that the mucosa is ulcerated. This is referred to as **toxic megacolon**. UC disease usually affects one section of the gastrointestinal tract at a time, whereas Crohn's disease often presents with a "skipping" pattern affecting multiple portions of the gastrointestinal tract.

Crohn's disease can affect any portion of the gastrointestinal tract from mouth to anus, but it most commonly affects the ileum and colon. Crohn's disease can damage all layers of gastro-intestinal mucosa. This inflammatory process is characterized by the development of **fistulas** that when healed are replaced by fibrotic tissue. The fibrosis can result in recurrent strictures and bowel obstructions.

Clinical Manifestations Patients with UC present with signs and symptoms including abdominal pain, bloody diarrhea, and tenesmus (urgency for defecation). Patients with severe disease often are febrile, are tachycardic, and have diarrhea that contains pus and mucus. Disease activity is rated using the Truelove and Witts Criteria shown in Table 15.16.[111] Radiological testing often will show severe inflammation of the large bowel with thickened walls and superficial ulcerations, and over time, the haustra become edematous and thickened. Serological testing for markers of IBD include serum levels of cytokines (interleukins and tumor necrosis factors) and anti-glycan antibodies (ASCA and ANCA). The use of these markers has increased, and is proposed not only for diagnosis but for evaluating response to treatment. Other acute-phase reactants such as C-reactive protein and the erythrocyte sedimentation rate (ESR) are elevated but are not necessarily specific to intestinal inflammation. Most recently, elevated calprotectin (Cal), lactoferrin (Lf), and polymorphonuclear neutrophil elastase (PMN-e) levels in stool have been found to be indicative of exacerbations for IBD.[112, 113] White blood cell (WBC) count can also be elevated, and leukocytes in stool confirm the inflammatory process. Biochemical indices for anemia are generally depressed, and in severe disease often confirm significant anemia.

Figure 15.12 Disease Pathogenesis of IBD

Comparison of colonic mucosa in normal, Crohn's, and ulcerative colitis patients; gross (top); histological (center); endoscopic appearance (bottom).

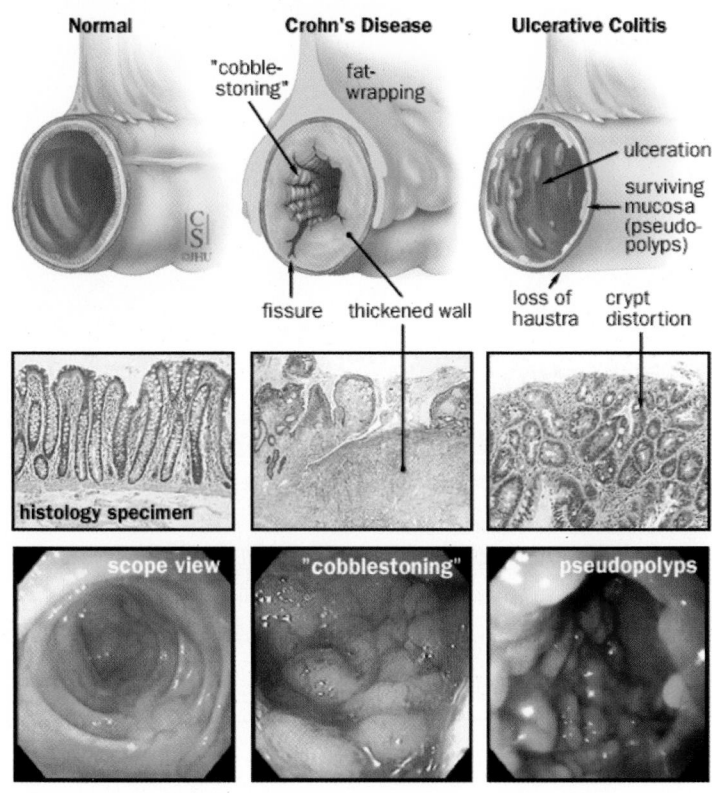

Source: Artwork is reproduced, with permission, from the Johns Hopkins Gastroenterology and Hepatology Resource Center, www.hopkins-gi.org, copyright 2006, Johns Hopkins University, all rights reserved.

Patients with Crohn's disease experience abdominal pain, diarrhea, and tenesmus. They are much less likely to have blood in their stool but usually experience more abdominal pain and cramping than patients with UC. Disease involving the small intestine places the individual at higher nutritional risk than disease involving only the colon. Weight loss is very common

in Crohn's due to both increased requirements and decreased oral intake. Crohn's disease can also be insidious, presenting with only mild symptoms or only those that are extraintestinal in nature. In research and clinical trials, Crohn's disease is described using the CDAI (Crohn's Disease Activity Index), which suggests that patients who score over 150 are experiencing a flare-up of the disease, and those who score over 300 are experiencing severe exacerbation of the disease.[114] Factors such as diarrhea, abdominal pain, abdominal mass, decreased sense of well-being, extraintestinal manifestations, weight loss, and laboratory features are evaluated in this index. As discussed previously, calprotectin (Cal), lactoferrin (Lf), and polymorphonuclear neutrophil elastase (PMN-e) levels in stool have been found to be indicative of exacerbations of Crohn's disease and may be as good a predictor as the CDAI.[112, 113] In severe exacerbations, low albumin levels and elevated WBC are common. Imaging tests may demonstrate deep ulcerations that often skip over portions of the GI tract. Fistulas and tracts between ulcerations can be observed in severe disease (see Box 15.4). Current AGA practice guidelines describe Crohn's using the four definitions shown in Table 15.17: mild-moderate disease, moderate-severe disease, severe-fulminant disease, and remission.[115]

As discussed earlier in this section, serological markers include antibody testing. It has been used to distinguish between UC and Crohn's disease as well as to distinguish IBD from other GI diseases. Anti-saccharomyces cerevisiae antibodies (ASCA) are more specific for Crohn's and perinuclear antineutrophil cytoplasmic autoantibodies (pANCA) are more specific for ulcerative colitis. Due to the presence of these antibodies with other GI diseases, these markers may not correctly identify IBD. Future research is focused on development of more sensitive markers.[112]

Patients with IBD can experience disease manifestations (referred to as extraintestinal) outside the GI tract. These include osteopenia and osteoporosis, dermatitis, rheumatological conditions such as ankylosing spondylitis, ocular symptoms, and hepatobiliary complications.[107]

Table 15.16 Truelove and Witts Criteria for Assessing Disease Activity in Ulcerative Colitis

	Mild Activity	Severe Activity
Daily bowel movements (no.)	≤5	>5
Hematochezia	Small amounts	Large amounts
Temperature	<37.5°C	≥37.5°C
Pulse	<90/min	≥90/min
Erythrocyte sedimentation rate	<30 mm/h	≥30 mm/h
Hemoglobin	>10 g/dL	≤10 g/dL

Patients with fewer than all 6 of the above criteria for severe activity have moderately active disease.

Source: Reproduced with permission from BMJ Publishing Group: Truelove SC, Witts LJ. Cortisone in ulcerative colitis: final report on a therapeutic trial. *Br Med J.* 1955;2:1041–48.

Treatment Treatments for both UC and Crohn's disease include antibiotics, immunosuppressive medications, immunomodulators, and biologic therapies as well as surgical intervention.[107] Medical treatment for ulcerative colitis historically has used combinations of both antibacterial coverage with sulfapyridine and anti-inflammatory 5-aminosalicylic acid (5-ASA) therapy (see Table 15.4 for information about this and other medications used for diseases of the lower GI tract). The most commonly used medications in this category today include olsalazine, balsalazide, Asacol, Claversal, and Pentasa. Immunomodulators work to inhibit inflammatory cell proliferation by interrupting cellular RNA and by inhibiting the overall immune response. These medications include azathioprine (AZA) and 6-mercaptopurine (6-MP). Corticosteroids work to inhibit the overall inflammatory response and are commonly used to treat UC. Antibiotics are used in UC only when there is an acute infection.

Diagnostic Testing for Lower Gastrointestinal Tract Disorders

In addition to a comprehensive medical history and physical examination, diagnostic evaluation including laboratory tests, imaging tests, and/or endoscopic procedures may be used to diagnose lower gastrointestinal disorders.

Category	Procedure	Can Be Used to Diagnose/Monitor
Laboratory tests	Fecal occult blood	Irritable bowel syndrome
	Stool culture	Antibiotic-associated diarrhea and colitis, irritable bowel syndrome
	Serologic markers: ASCA, ANCA, ALCA, ACCA	IBD, celiac disease
	Fecal markers of inflammation: calprotectin, lactoferrin	IBD: Crohn's disease, UC; IBS
	Enzyme-linked immunoassay	Antibiotic-associated diarrhea and colitis
	CBC, electrolytes, ESR, urinalysis	Irritable bowel syndrome
Imaging tests	CT scan MRI	Diverticulitis, Crohn's disease
	Ultrasound	Diverticulitis, ulcerative colitis
	Barium enema with air contrast	Crohn's disease
	Small bowel follow through	Diverticulitis, Crohn's disease
Endoscopic procedures	Wireless capsule endoscopy Colonoscopy	Diverticulitis, ulcerative colitis, Crohn's disease
	Sigmoidoscopy	Ulcerative colitis, antibiotic-associated diarrhea and colitis
Other procedures	Mucosal biopsies	Ulcerative colitis, Crohn's disease, Celiac disease

Sources: Ali S, and Tamboli CP. Advances in epidemiology and diagnosis of inflammatory bowel diseases. *Current Gastroenterology Reports.* December 2008;10(6):576–84.
Lichtenstein GR, Hanauer SB, and Sandborn WJ. Management of Crohn's disease in adults. *The American Journal of Gastroenterology* [serial online]. February 06, 2009;104(2):465.

Table 15.17 Definitions of Stages of Crohn's Disease

Stage	Definitions
Mild-Moderate Disease (CDAI 150–220)	Ambulatory individuals able to tolerate oral alimentation without development of dehydration, toxicity (high fevers, rigors, prostration), abdominal tenderness, painful mass, obstruction, or >10% weight loss
Moderate-Severe Disease (CDAI 220–450)	Individuals who have failed to respond to treatment for mild-moderate disease or those with more major symptoms of fevers, significant weight loss, abdominal pain or tenderness, intermittent nausea or vomiting (without obstructive findings), or significant anemia
Severe-Fulminant Disease (CDAI >450)	Individuals with persisting symptoms in spite of introduction of steroids or biologic agents as outpatients, or those presenting with high fever, persistent vomiting, evidence of intestinal obstruction, rebound tenderness, cachexia, or evidence of an abscess
Remission	Asymptomatic individuals or those without inflammatory sequelae and includes those who have responded to acute medical intervention or have undergone surgical resection without gross evidence of residual disease

Note: Individuals requiring steroids to maintain well-being are considered to be "steroid-dependent" and are usually not considered to be "in remission."
Source: Reprinted by permission from Macmillan Publishers Ltd: *The American Journal of Gastroenterology* [serial online]. February 06, 2009;104(2):465.

as prednisone or budenoside, are often used in acute exacerbations, especially in severe-fulminant disease, but patients are at risk for becoming steroid dependent. Antibiotics used include metronidazole and ciprofloxacin.[107] Biologic therapy for Crohn's disease includes infliximab, adalimunab, and certolizumab pegol. These medications work to interrupt tumor necrosis factor-alpha (TNF-alpha) and thus the cytokine-directed inflammatory activity. Other classifications of biologic therapies include selective anti-adhesion molecules and anti-interleukin antibodies.[108]

Surgical intervention is required in both UC and Crohn's disease in over 60% of patients. The most common procedure in UC is a total colectomy, and in Crohn's disease, the ileostomy. Surgery is performed due to nonresponsive disease and due to acute complications such as perforation, obstruction, or abscess. These surgical procedures will be described in greater detail later in this chapter in the section Common Surgical Interventions for the Lower GI Tract.

Nutrition Therapy for Inflammatory Bowel Disease

A significant portion of individuals diagnosed with active Crohn's disease experience weight loss and nutritional deficits (see Box 15.5).[107, 116, 117] Nutritional support is routinely required during periods of exacerbation and recovery from surgery. Crohn's and IBD affect normal digestion and absorption; may increase caloric, protein, and micronutrient requirements;

Treatment for Crohn's disease can utilize all categories of medical treatments listed previously. Aminosalicylate medications are typically used in Crohn's disease that has ileal and colon involvement. These include mesalamine and sulfasalazine. As in UC, the immunomodulators azathioprine (AZA) and 6-mercaptopurine (6-MP) are used. Corticosteroids, such

BOX 15.5 CLINICAL APPLICATIONS

Common Nutrient Deficiencies Seen with Crohn's Disease

Nutrient Deficiency	Probable Cause
Calories	Insufficient intake
	Anorexia
	Increased energy requirements
	Fear of abdominal pain and diarrhea after eating
Protein	Increased protein needs (losses from GI tract caused by inflammation)
	Catabolism (steroid-induced or when infection or abscesses present)
	Healing from surgery
Fluid and electrolytes	Short bowel syndrome
	High-volume diarrhea
Iron	Blood loss
	Malabsorption
Magnesium, zinc	Intestinal losses, especially from short bowel syndrome or high-volume diarrhea
Calcium and vitamin D	Long-term steroid use
	Decreased intake of dairy foods as result of lactose-restricted diets
B₁₂	Surgical resections of stomach (loss of intrinsic factor) and/or terminal ileum (site of absorption)
Folate	Medications used to treat IBD
Water-soluble vitamins	Surgical resections—loss of terminal ileum
Fat-soluble vitamins	Steatorrhea

Sources:
Smith PA. Nutritional therapy for active Crohn's disease. World J Gastroenterol. 2008;14:4420–23.
Lochs H, Dejong C, Hammarqvist F, Hebuterne X, Leon-Sanz M, Schütz T, van Gemert W, et al. 2006. ESPEN Guidelines on Enteral Nutrition: Gastroenterology. *Clinical Nutrition (Edinburgh, Scotland)* [serial online]. April 15, 2006;25(2):260–74.
Eiden KA. Nutritional considerations in inflammatory bowel disease. *Practical Gastroenterology.* 2003 May:33–54.

can result in protein-energy malnutrition; and additionally may require nutrition therapy to minimize symptoms. Medications that are used to treat IBD, especially corticosteroids, may impact nutritional status by either increasing nutrient requirements or exacerbating nutrient losses. The use of bowel rest with parenteral nutrition and the use of enteral nutrition as major components of the medical care for IBD have been debated.[116–119] The exact mechanism for the role of nutrition therapy in Crohn's pathophysiology is not clearly established but is assumed to be related to improvement in intestinal permeability.

Nutrition Assessment Due to the significant nutritional complications and problems associated with IBD, a comprehensive nutrition assessment is frequently required.

Nutrition Diagnosis Nutrition diagnoses related to IBD include malnutrition; inadequate energy intake; inadequate oral food/beverage intake; increased nutrient needs; inadequate vitamin/mineral intake; impaired nutrient utilization; food-medication interaction; and altered nutrition-related laboratory values.

Malnutrition can be present even when Crohn's disease is in remission. Protein-calorie malnutrition and other nutrient deficiencies can be caused by decreased nutrient intake, malabsorption, drug-nutrient interactions, anorexia, and protein-losing enteropathy. These, in turn, can lead to growth retardation (in children), anemia, osteoporosis, poor wound healing, and a compromised immune system.

Nutrition therapy is a crucial component of treatment for active disease. Nutritional needs and deficits are significantly different during periods of remission. Acute symptoms of both UC and Crohn's involve diarrhea and abdominal pain. Because increased motility decreases the success of digestion and absorption, severe diarrhea can result in malabsorption of all nutrients. These symptoms can also result in reduced oral intake. Many patients, during acute exacerbations of the disease, electively restrict eating in order to minimize symptoms. Pain can cause generalized anorexia, which further decreases dietary intake.

When infection and inflammation are present or when the patient is febrile, energy needs are higher than normal. Protein needs are increased, in some cases up to 150% of normal requirements. A negative nitrogen balance has been demonstrated to occur in approximately 50% of individuals with Crohn's. This is due, in part, to increased protein losses in inflammatory **exudate**. Crohn's patients are at risk for deficiency of micronutrients, especially iron, zinc, magnesium, and electrolytes, due to their loss in blood and diarrhea, and malabsorption. Because IBD commonly presents in children and young adults, meeting nutritional needs of the growing child or adolescent poses its own challenge. It is crucial for nutrition therapy to be designed to ensure adequate nutrients to support growth and development.

Since the mainstay of treatment for IBD involves multiple medications and often surgery, these nutritional risks compound those of the disease process. For example, use of corticosteroids can result in hyperglycemia, nitrogen wasting, and increased risk of osteoporosis. Another example is the use of sulfasalazine, which interferes with folate metabolism. Surgery increases calorie and protein requirements and additional nutrients are needed to support wound healing. Depending on the extent and type of surgery, normal absorption and digestive pathways may be interrupted. Specific nutrition therapy may be required if the patient has either an ileostomy or colostomy, as discussed in the section Common Surgical Interventions for the Lower GI Tract.

Nutrition Intervention

NUTRITION THERAPY DURING EXACERBATION OF DISEASE During acute exacerbations of both UC and Crohn's disease, the extent of intestinal involvement, previous surgeries, diarrheal output, and bleeding direct the amount and route for nutrition intervention. The literature supports

Unique Nutritional Considerations for Pediatric Crohn's Disease

Nutritional problems are quite common for individuals with Crohn's disease. But because 25% of Crohn's cases are diagnosed during childhood and the adolescent years, an additional concern arises when the nutritional requirements for growth must also be met. Growth failure is characterized by a delay in skeletal formation with subsequent reduction in height and a delay in onset of puberty. It has been estimated that up to 40% of children diagnosed with Crohn's disease during childhood will fail to reach their optimal adult height.[1] Other nutritional complications include osteoporosis, anemia, and other vitamin and mineral deficiencies.[2]

Nutritional deficits in children can occur for the same reasons as those in adults—malabsorption, increased energy expenditure, and inability to consume adequate nutrients. When steroids are used in treatment for children, there is additional risk due to the effect on bone growth. Stunted stature and delayed growth are significant concerns in

pediatric Crohn's. To avoid these complications, those medications used for adults—Infliximab and other biological therapies—are being used earlier in treatment. Another route is to use enteral feeding as the primary treatment. Enteral feeding has resulted in the reduction of clinical signs and symptoms along with an overall reduction in Crohn's Disease Activity Index scores, and does not have any negative effect on growth for children.[3,4] But despite this evidence, enteral feeding is not consistently used as a first-line therapy. Registered dietitians can and should play a significant role in research and clinical application for these therapies.

References

1. Kappelman MD, Bousvaros A. Nutritional concerns in pediatric inflammatory bowel disease patients. Mol Nutr Food Res. 2008;52:867–874.
2. Day AS, Whitten KE, Lemberg DA, Clarkson C, Vitug-Sales M, Jackson R, Bohane TD. Exclusive enteral feeding as primary therapy for Crohn's disease in Australian children and adolescents: a feasible and effective approach. J Gastro Hepatol. 2006;21:1609–14.
3. Heuschkel R, Salvestrini C, Beattie RM, Hildebrand H, Walters T, Griffiths A. Guidelines for the management of growth failure in childhood inflammatory bowel disease. Inflamm Bowel Dis. 2008;14:839–49.
4. Heuschkel R. Inflammatory bowel disease in children. Clin Med. 2008;8:297–98.

that enteral nutrition plays a crucial role in treatment of active CD in children (see Box 15.6). In children, enteral nutrition is preferred over the use of corticosteroids in treatment, as it allows for support of growth and development without the side-effects associated with medications.[115–117, 120, 121] For adults, enteral nutrition is recommended when use of medications is not feasible and when additional nutrition is needed to improve or maintain nutritional status. Most research indicates parenteral nutrition (PN) with bowel rest is not necessary and the limited amount of clinical trials indicate there are not improved outcomes of remission.[117, 118] For those individuals with extensive intestinal involvement or short bowel syndrome, PN may be required immediately postoperatively or until there is adequate intestinal adaptation (see section Short Bowel Syndrome). The type of enteral formula (chemically defined or polymeric) depends on the individual patient's medical and nutritional status. Enteral nutrition research has demonstrated that supplementation may play a role in modifying inflammatory response in the disease process. The use of enteral nutrition has been associated with improvement in clinical markers of inflammation (such as CRP), but has not consistently shown a positive effect on remission state or other clinical outcomes.[116, 118, 119]

Energy needs for adults can be estimated using the Harris-Benedict or Mifflin-St. Jeor equation with appropriate stress factor (1.3–1.5). The amount of prior weight loss and the presence of infection will support the need for higher energy provision. To meet growth needs of infants, children, and adolescents, specific attention to their unique requirements is important. As much as 120 kcal/kg for infants and 80 kcal/kg for adolescents may be required (see Chapter 5). If available,

indirect calorimetry provides the most reliable indicator of energy needs in the hospitalized patient.

Estimation of protein requirements will be based on the presence of any lean body mass wasting and biochemical parameters measuring protein status such as prealbumin and albumin. Protein needs may be as high as 1.5 to 1.75 g protein/kg for adults and 2.0 to 2.5 g/kg for infants, children, and adolescents.

If oral intake can be initiated, a low-residue, lactose-free diet with small, frequent meals is best tolerated. If steatorrhea is present, then fat should be reduced with added MCT or an MCT-containing supplement to assist with meeting energy requirements. As the patient responds to medical therapy, adding small amounts of fiber and then lactose as the patient can tolerate will advance the diet. Restricting fiber is generally only necessary during the acute exacerbation or if a stricture is present.[122] Other foods that may need to be initially restricted may be gas-producing foods, spicy or fried foods, caffeinated beverages, or any other food the individual patient identifies as problematic. The addition and advancement of an oral diet will need to be highly individualized.

All patients should receive a multivitamin that meets the RDA or AI for all nutrients. Patients with IBD are at higher risk for deficiencies of vitamin B_{12} and iron. In a normal small intestine, the ileum has specific receptor sites that allow for B_{12} absorption. Therefore, disease affecting the ileum specifically can potentially result in B_{12} deficiency. Supplementation of B_{12} to prevent pernicious anemia can be accomplished using nasal gel or oral tablets, or by intramuscular injection.[122, 123] Micronutrient requirements are additionally increased during exacerbations of disease. It is recommended that additional supplementation should include zinc (12 to 15 mg/liter of stool output); calcium (10 to 25 mEq/day); magnesium (15 to 30 mEq/day); and copper (0.5 to 1.5 mg/day).[122, 124]

Research has shown patients with Crohn's disease have lower serum levels of antioxidants (vitamin E, vitamin C, and beta-carotene).[107, 116] It is thought that this might contribute to higher levels of **oxidative stress** in this disease. Higher levels of antioxidants may be warranted, but specific levels have not been established at this time; nor is it clear that supplement forms produce the same effect as foods. The most convincing evidence for the relationship between antioxidants and disease prevention has been in epidemiologic studies where strong associations have been demonstrated between dietary sources of fruits and vegetables and disease risk.[125] Patients with IBD may avoid fruits and vegetables due to their disease symptoms and perceived intolerance to these foods.

NUTRITION THERAPY FOR REHABILITATION DURING PERIODS OF REMISSION Maximizing energy and protein intake to facilitate rehabilitation should be the primary goal. Weight gain within a normal healthy range combined with physical activity will ensure rebuilding of protein stores and muscle mass. Depending on the extent of disease and the response to treatment, specific dietary modifications will need to be individualized. It is always a goal to normalize dietary patterns and encourage a variety of all foods as the patient is able to tolerate them.

Consumption of foods high in antioxidants (for example, carotenoids, vitamin E, vitamin C, and selenium) and omega-3 fatty acids has been associated with protection against inflammation. These would include fruits, vegetables, vegetable oils, nuts, and fishes such as tuna and salmon. Although some reports have indicated that glutamine, short-chain fatty acids, antioxidants, and immunonutrition with omega-3 fatty acids are an important therapeutic alternative in the management of inflammatory bowel diseases, the reported beneficial effects have yet to be translated into clinical practice. The real efficacy of these nutrients still needs further evaluation through prospective and randomized trials.[120]

Foods high in oxalate may increase risk for urolithiasis or kidney stones, which can occur in IBD. These foods include, for example, cocoa, tea, wheat germ, strawberries, nuts, spinach, beets and baked beans, peanut butter, tofu, and high doses of vitamin C supplements (>2 g/day).

As has been previously discussed in this chapter, use of probiotics and prebiotics enhances the normal flora of the GI tract. In several recent studies, consumption of foods and supplements with probiotics and prebiotics has been associated with decreased symptoms for patients with IBD and a positive change in anti-inflammatory markers.[126–131] (See the previous discussion in this chapter regarding prebiotics and probiotics.)

Diverticulosis/Diverticulitis

Definition **Diverticulosis** is defined as the abnormal presence of outpockets or pouches on the surface of the small intestine or colon. Diverticulosis is most common in the adult but Meckel's diverticulum is a type of diverticulosis present at birth. Meckel's diverticula are usually found near the ileocecal valve and may cause gastrointestinal bleeding or obstruction for the newborn.

Epidemiology Estimations of prevalence or incidence are difficult because diverticulosis, in most people, is asymptomatic. Best estimates, however, indicate diverticulosis is most common in Western and industrialized countries where it is thought that approximately 5% to 10% of the population will have diverticula by age 50. Incidence will increase with age, with some estimates as high as 65% to 70% in persons over the age of 85.[131]

Etiology Though the etiology has not specifically been determined, development of diverticulosis is related to low fiber intake, history of constipation, and the resulting long-term increased colonic pressure.[132] Factors that may increase risk for development of diverticulosis include obesity, decreased physical activity, steroids, alcohol and caffeine intake, and cigarette smoking.[133, 134]

Pathophysiology Diverticula do occur in the small intestine but are most common in the colon. Factors that affect integrity of the mucosa of the colon appear to contribute to development of the diverticula. The aging process and differences within parts of the colon may account for the pattern of development that has been observed. It is thought that within the colon, two or more of the muscular bands (taeniae coli) contract at the same time. This hinders motility of the colon, and thus its ability to move waste products. Fecal matter becomes trapped and exerts excessive pressure against the wall of the colon. This pressure causes development of small pouches on the wall of the colon, which are referred to as diverticula.[132, 135, 136] (See Figure 15.13.) Constipation increases colonic pressure through the excessive straining involved in bowel movements. This further increases the probability for development of diverticula.[137]

Diverticulitis is acute inflammation of the diverticula. Foodstuffs and bacteria can collect in diverticula and the mucosa can become infected. Further complications can include development of bleeding, abscess, obstruction, fistula, or perforation.[131, 132]

Clinical Manifestations Diverticulosis is asymptomatic for most individuals. Diverticula are usually only diagnosed when other tests such as a colonoscopy identify them. In approximately 20% of individuals, complications from diverticula, including diverticulitis, may develop. Signs and symptoms of

Figure 15.13 Diverticula

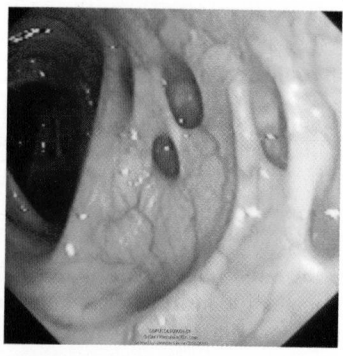

Courtesy of © ISM/Phototake.

diverticulitis can include fever, abdominal pain, gastrointestinal bleeding, and elevated white blood cell count. Radiology testing (ultrasound and CT scan) may be used to diagnose diverticulitis. Test results can demonstrate thickened walls of the colon, abscess, or inflammation.[131, 132]

Treatment Treatment for diverticulosis involves only nutrition therapy, with a specific focus on fiber intake, and use of probiotic and prebiotic supplementation. Treatment for acute diverticulitis begins with making the patient NPO with complete bowel rest until symptoms (bleeding and/or diarrhea) subside. Antibiotics are used to treat any infection. The most common antibiotic regimens involve treatment for Gram-negative rods and anaerobes.[132] For those patients with complications (such as abscess or sepsis), surgical resections may be necessary.

Nutrition Therapy for Diverticulosis/Diverticulitis

Research indicates that dietary habits may be strongly linked to the etiology of diverticulosis. Nutrition therapy should then focus on those nutrition interventions that could impact disease course. The patient with diverticulosis is not at any more risk for malnutrition than any other individual. The presence of diverticulitis with infection and inflammation does impact nutritional requirements if this condition is prolonged or if other complications, such as sepsis, occur.

Nutrition Assessment Evaluation of dietary fiber intake is very important. Since some patients who have had diverticular disease for an extended period of time may have been advised to decrease fiber, it is also important to assess a patient's prior education and knowledge of the disease.

Nutrition Diagnosis Nutrition diagnoses commonly associated with diverticulosis include altered GI function and inadequate fiber intake.

Nutrition Intervention As mentioned earlier in the discussion of constipation, many Americans consume only limited amounts of fiber.[138] Nutrition therapy to treat and prevent diverticulosis will include a **high-fiber diet** of 6 to 10 grams above and beyond the recommendations of 25 to 35 grams/day.[131] Historically, it has been common practice to avoid nuts, seeds, and hulls with the theory that these may aggravate the condition or result in inflammation. There is not adequate evidence to support this advice and the restriction is no longer recommended.[139, 140] Many patients, especially the elderly, will need to use a fiber supplement if they are unable to consume adequate fiber from foods. Products that may be used to normalize GI function include Fiberall® and Metamucil®, bulk-forming agents made from psyllium, a source of insoluble fiber, and Benefiber®, soluble dietary fiber extracted from guar gum. Other sources of fiber supplementation such as methylcellulose have also been used successfully. Certainly the preferred method to increase fiber intake is through foods, but, if needed, these supplements are available.

The patient with acute diverticulitis will be progressed from bowel rest to clear liquids. The patient can then move toward a low-residue diet until inflammation and bleeding are no longer a risk.

Common Surgical Interventions for the Lower GI Tract

Surgical resections may be warranted for many diagnoses discussed in this chapter, including Crohn's disease, UC, and diverticulitis. Specific details of each individual's disease course will determine need for surgical intervention. These may include disease refractory to current medical treatment, abscess not responding to antibiotic therapy and bowel rest, or acute emergencies such as peritonitis or gastrointestinal bleeding. The extent of the surgical resection and procedure used again depend on the individual patient's disease course. The most common procedures will be discussed here with a review of the nutritional implications.

Ileostomy and Colostomy Surgical resection of the colon and rectum requires development of a new path for feces to be excreted from the body. Any of these surgeries creates a **stoma**—a surgically created artificial opening into the abdomen—from which waste products can be excreted. An **ileostomy** is removal of the colon and rectum. The end of the ileum is surgically attached to the stoma. The individual then uses an appliance (pouch) where feces and other waste products are collected. A **colostomy** exists when the rectum only is removed and the end of the colon is surgically attached to the stoma. Again, the individual utilizes a pouch appliance to collect waste products.

Alternatives for these procedures where the outside surgical appliance can be avoided are available. These procedures, which create internal pouches where waste products can collect, include ileoanal reservoir surgery, ileal pouch-anal anastomosis, or continent ileostomy (see Figure 15.14).

Nutrition Therapy for Ileostomy and Colostomy

Implications of intestinal resection may be analyzed in the context of anatomy and physiology of the gastrointestinal tract. When a certain part of the intestinal tract is removed, normal physiology and function of that portion is lost to the individual. This loss of function will produce changes in motility, in absorption, and in how waste products are handled—all of which potentially can impact nutritional status. Larger resection will result in the most nutritional complications. Resections of the terminal ileum and loss of the ileocecal valve tend to result in significant fluid, electrolyte, vitamin, and mineral deficiencies. The ileocecal valve controls the rate of movement from the small intestine to the large; hence, when it is absent, motility is much faster, interrupting normal absorption.

The location of the stoma on the GI tract will also determine the type of fecal matter produced. As explained earlier, the function of the colon is to reabsorb water and electrolytes. Fecal matter further along the colon will produce firmer, less watery stool. Output with an ileostomy, then, will be much more liquid, while output with a colostomy, depending upon where it is located along the colon, will result in firmer, more normal stool.

Nutrition Intervention The goals for nutrition therapy are to decrease risk of obstruction, maintain normal fluid and electrolyte balance, reduce excessive fecal output and/

Figure 15.14 Surgical Options for Treatment of Ulcerative Colitis

(a) Proctocolectomy; (b) Brooke ileostomy; (c) Koch pouch ileostomy; (d) restorative proctocolectomy.

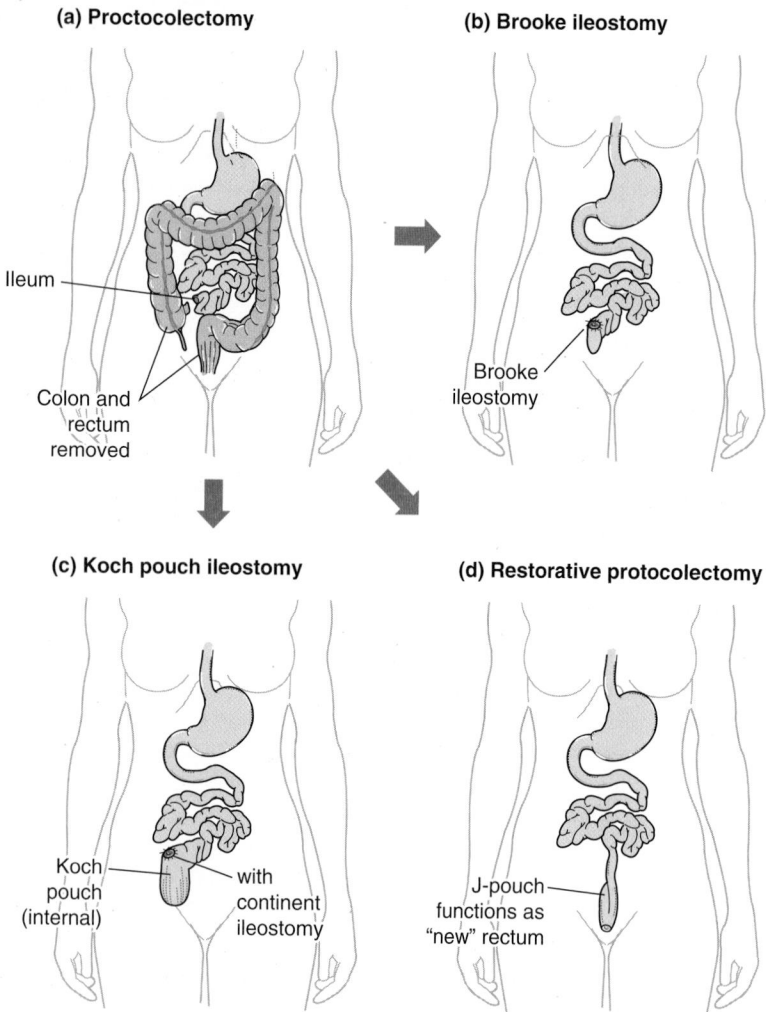

(a) Proctocolectomy

Ileum

Colon and rectum removed

(b) Brooke ileostomy

Brooke ileostomy

(c) Koch pouch ileostomy

Koch pouch (internal)

with continent ileostomy

(d) Restorative protocolectomy

J-pouch functions as "new" rectum

Source: Artwork is reproduced, with permission, from the Johns Hopkins Gastroenterology and Hepatology Resource Center, www.hopkins-gi.org, copyright 2009, Johns Hopkins University, all rights reserved.

or change consistency of output, and minimize gas and flatulence (to reduce odor and inflation of the appliance).[141] After surgery, the patient will be transitioned to an oral diet. Evidence regarding the optimal amount of time between postoperative status and initiation of oral feeding has recently been evaluated. A review of the literature indicates that early oral feeding (postoperative day 1) after intestinal surgery is safe and tolerated in the majority of patients.[143] This begins with liquids and progresses as tolerated to a low-residue diet with four to six small feedings each day. Foods that may not be completely digested and that can cause stoma obstruction should be avoided for the first 6 to 8 weeks after surgery. These include tough fibrous meats; vegetables such as spinach, corn, and peas; dried fruits such as raisins; fruit skins and seeds; and popcorn. The patient will need to be instructed to eat slowly, chew thoroughly, and drink adequate fluids. (See Table 15.18—Nutrition Therapy for Ostomies.) Generally, oral intake should resemble the regular diet, meeting all nutritional needs by the eighth week postoperatively.

If the patient experiences excessive or watery fecal output, the amount of insoluble fiber should be reduced while the amount of soluble fiber is increased. Applesauce, bananas, tapioca, potatoes, oatmeal, oat bran, rice, and pasta may help decrease diarrhea. Foods that cause gas and flatulence should be avoided (see Table 15.13), as these same foods can cause difficulty for the patient with an ostomy. Use of yogurt, parsley, and buttermilk may decrease gas and odor.

Vitamin and mineral requirements for the patient after bowel surgery will be affected by the length and location of the remaining portions. Earlier in this chapter, absorption sites for vitamins and minerals were discussed (see Figure 15.7). More common micronutrient issues will involve vitamins K and B_{12}. Vitamin B_{12} absorption is of specific concern because it requires adequate ileal receptors and a normal transit time. Research focused on absorption of dietary vitamins and minerals in patients with ileostomy has determined that most vitamins, minerals, and phytochemicals appear to be adequately absorbed in this population.[144, 145] Livny and colleagues did find that beta-carotene was best absorbed from cooked carrots rather than raw carrots in these patients.[146] However, since most patients' intake and tolerance vary widely, a general multivitamin is recommended.

Typical daily fluid output from a colostomy ranges from 200 to 600 mL after adaptation has occurred.[147] Individuals with an ileostomy are at higher risk for nutrient complications. Outputs average 1200 mL per day initially, though they will decrease to close to 600 mL per day. Nutrients of concern after an ileostomy include primarily fluid and electrolytes, especially if the ileocecal valve is not intact. All nutrients may be at risk if transit time is sufficiently altered. It is estimated that individuals will lose approximately 90 mmol/L sodium within fecal output if the stoma is at the duodenojejunal flexure and up to 140 mmol/L sodium if the stoma is at the terminal ileum.[147, 148] Immediately postoperatively and during times of increased output, fluid intake should be monitored closely to avoid the complications of dehydration. (See Chapter 8 for methods to estimate fluid requirements.)

Short Bowel Syndrome

Definition **Short bowel syndrome (SBS)** results from a large resection of the small intestine. Each patient presents with a unique situation where the underlying diagnosis, the condition of the ileocecal valve, and the amount of colon that is preserved are important factors affecting the long-term prognosis.[149] O'Keefe et al. have proposed the following definition: "Short bowel syndrome intestinal failure results from surgical resection, congenital defect or disease-associated loss of absorption and is characterized by the inability to maintain protein, energy, fluid, electrolyte, or micronutrient balances when on a conventionally accepted, normal diet" (pp. 9–10).[146]

Epidemiology Incidence of SBS is estimated to be approximately two to three cases per million individuals per year. Prevalence is approximately four cases per million individuals

Table 15.18 Nutrition Therapy for Ostomy

- Avoiding practices that may contribute to swallowed air and gas formation, such as the following:
  - Chewing gum
  - Use of drinking straws
  - Carbonated beverages
  - Smoking
  - Chewing tobacco
  - Eating quickly
- Taking small bites of foods and chewing thoroughly
- Adding foods that may decrease odor, such as the following:
  - Buttermilk
  - Parsley
  - Yogurt
  - Kefir
  - Cranberry juice
- Adding foods that may thicken stool, such as the following:
  - Banana flakes
  - Applesauce
  - Pectin
  - Pasta
  - Potatoes
  - Cheese

Source: Reprinted with permission. American Dietetic Association. Bowel Surgery: Colostomy Nutrition Care. In: Nutrition Care Manual. 2008. http://nutritioncaremanual.org/

per year.[150] The most common causes of SBS are malignancy, damage from radiation therapy (radiation enteritis), Crohn's disease with resulting multiple resections, vascular accident, trauma, or **volvulus**.[142, 151]

Etiology Surgical resections of the small intestine and colon due to disease and trauma can result in extensive loss of surface area of the small intestine and colon. Without normal anatomy and physiology, malabsorption of nutrients, fluids, and electrolytes will result (see Tables 15.19 and 15.20).

Pathophysiology Several factors will determine the prognosis of this condition: extent of remaining small intestine, presence of the colon, presence of the ileocecal valve, health of

Table 15.19 Etiology of Short Bowel Syndrome in Children and Adults

Children	Adults
• Necrotizing enterocolitis	• Massive surgical resection
• Intestinal atresia (volvulus, hernia, intussusception)	• Crohn's disease
• Congenital short bowel syndrome	• Malignancy
• Trauma	• Radiation enteritis
• Gastroschisis	• Trauma
• Apple peel anomaly	• Vascular catastrophes (embolus/thrombus)
• Crohn's disease	• Volvulus
• Abdominal tumors	• Strangulated hernias
• Radiation enteritis	• SB fistulas
• Hirschsprung's disease	• Surgical bypass
	• Obesity treatment
	• Chronic intestinal pseudo-obstruction

Source: Ree Parrish, C. The Clinician's Guide to Short Bowel Syndrome. Practical Gastroenterology. September 2005, p. 70.

Table 15.20 Clinical Consequences of SBS

- Jejunal resection of 50–60% is usually well tolerated.
- Greater than 30% ileal resections are poorly tolerated.
- Severe malabsorption occurs with residual small bowel <60 mm.
- Deficiencies include fluid and electrolytes (mild and moderate cases)/plus nutrient absorption (severe cases)
- Severe fluid and electrolyte loss is associated with end jejunostomy.
- Magnesium, calcium, and zinc deficiencies are common.

Reprinted from: Clinical Gastroenterology and Hepatology, 4:7, O'Keefe, S. J. D., Buchman, A. L., Fisbein, T. M., Jeejeebhoy, K. N., Jeppesen, P. B., Shaffer, J. "Short Bowel Syndrome and Intestinal Failure: Consensus Definitions and Overview." © Copyright 2006, with permission from Elsevier.

the remaining gastrointestinal tract, and any comorbid conditions the individual may have. Though each case is highly individualized, most research agrees that a resection of more than 70% of the GI tract will result in severe nutritional and metabolic complications.[149, 151]

The postoperative period for SBS generally follows three distinct phases. The first period ranges anywhere from 7 to 10 days and is characterized by extensive fluid and electrolyte losses within large volumes of diarrhea. During this phase, patients are dependent on parenteral nutrition, which not only provides required nutrients but manages fluid and electrolyte balance.

The second postoperative phase may last for several months and is characterized by reduction in diarrhea volumes with the initial stages of adaptation of the remaining bowel. It is during this phase that enteral nutrition can be introduced with a gradual transition to an oral diet.[147]

During the third phase, there is continued adaptation of the remaining bowel. This includes increased blood flow, secretions, and mucosal cell growth. The inner lumen of the remaining small intestine increases in both length and diameter with additional increase in villous height. This time frame varies, but may range from 1 to 2 years.[149, 151] Enteral feeding, especially early exposure to enteral nutrition, supports the successful adaptation.[150] Other determinants of adaptation include the health of remaining bowel and whether the colon is present.

As mentioned earlier, the amount of remaining bowel determines the extent of this condition. Loss of the ileum prevents B_{12} absorption and reabsorption of bile salts. Reduction in bile salts further contributes to fat malabsorption. No other part of the intestinal tract can compensate for these losses. The ileocecal valve not only controls intestinal motility but also prevents translocation of bacteria from the colon to the small intestine. When this control is lost, nutritional and metabolic complications are much more prominent.

Vitamin and mineral losses are major issues in SBS. When there is fat malabsorption, there is an inability to absorb adequate amounts of vitamins A, D, E, and K. These will need to be supplemented appropriately, and levels within the body will need to be evaluated. Other nutrients often deficient include sodium, magnesium, iron, zinc, selenium, and calcium, because they are often lost in the large volumes of diarrhea.[148]

Treatment Initially, medical treatment will focus on managing fluid and electrolyte balance. This is generally managed by parenteral nutrition and intravenous support initially, and then as the patient is able, by oral rehydration solutions. Motility is controlled by medications used to treat symptoms of diarrhea. These agents work to either decrease motility or to thicken consistency of the stool, and include medications such as LoMotil® (diphenoxylate), Immodium (loperamide), paregoric, codeine, Tincture of Opium, Kaopectate, or bismuth subsalicylate. Octreotide (LAR® Depot), which is given intravenously, is a somatostatin analog that reduces the levels of growth hormone and has been used to treat diarrhea in short bowel syndrome.[148, 152] These medications, while decreasing motility, may also improve digestion by increasing the amount of time nutrients and enzymes are exposed within the bowel. Gastric hypersecretion, common after extensive bowel resection, is treated by use of proton pump inhibitors or H_2 antagonists. In general, though, oral medications are not consistently absorbed and will need to be monitored closely.

The newest class of drugs for SBS have been used to enhance cell proliferation in the remaining portions of the intestinal tract. These include growth hormone and GLP-2 (glucagon-like peptides). Clinical trials with Teduglutide (GATTEX, ALX-0600; NPS Allelix Corp) indicate an overall improvement for patients with SBS.[153]

Surgical treatment for short bowel syndrome includes intestinal lengthening procedures and intestinal transplant.[154] These procedures are indicated for those individuals who meet the criteria for intestinal failure—meaning that "intestinal failure results from obstruction, dysmotility, surgical resection, congenital defect, or disease-associated loss of absorption and is characterized by the inability to maintain protein-energy, fluid, electrolyte, or micronutrient balance" (p. 10).[146]

Nutrition Therapy for Short Bowel Syndrome

Nutrition Assessment   Maintenance of nutritional and hydration status is critical for individuals with SBS; therefore assessment of fluid and beverage intake is especially important. Aggressive nutrition support and safe progression to an oral diet require careful attention by the entire health care team. Accurate assessment of bowel function and estimation of needs are essential in the care of patients receiving nutrition support and transitional feedings.

Nutrition Diagnosis   Nutrition diagnoses associated with short bowel syndrome may include inadequate energy intake; inadequate protein intake; malnutrition; inadequate vitamin/mineral intake; involuntary weight loss; altered GI function; and impaired nutrient utilization.

Nutrition Intervention   Immediately postoperatively, patients will receive parenteral nutrition. Prescription for this therapy will be based on energy, protein, and micronutrient requirements (see Chapters 5 and 7). As diarrhea begins to decrease (anywhere from 2 to 6 weeks postoperatively), oral diets can begin.[155] Many patients require combinations of parenteral, enteral, and oral nutrition support in order to accommodate degrees of malabsorption and patient requirements for nutrition and fluids.

Sugar-free, isotonic clear liquids should be the first items offered. Table 5.2 (in Chapter 5) outlines examples of isotonic beverage choices. Diet is then progressed slowly to a low-residue, low-fat, lactose-free, low-oxalate diet. Caffeine and alcohol should not be initially consumed. Alcohol sugars such as xylitol, mannitol, and sorbitol are usually not tolerated. Insoluble fiber is generally not tolerated initially, but sources of soluble fiber may actually assist in promoting mucosal health. Soluble fiber, like other sources of prebiotics, assists in production of short-chain fatty acids that are a primary fuel for the colon. Soluble fiber may also enhance water absorption in the colon and reduce amount of diarrhea.[150] Overall, it is crucial to remember that the diet truly needs to be designed for the individual patient and that significant differences between patients will be likely. As Rees-Parrish states, "Fat is an important calorie source. Maximize medication delivery before imposing strict dietary guidelines—no diet is a good diet if not eaten."[148] Table 15.21 summarizes nutrition therapy for short bowel syndrome.

One item at a time is added to the diet to ensure tolerance. If GI symptoms are exacerbated, the food added should be removed from the diet. It may be added again at a later date, depending on the patient's adaptation after surgery. Lykins and Stockwell suggest it may be best to retry categories of restricted foods even as long as 6 months postoperatively, since bowel adaptation can take as long as 1 to 2 years.[156]

BOX 15.7   CLINICAL APPLICATIONS

Sample Nutrition Diagnosis and Intervention for Short Bowel Syndrome

PES statement: Altered GI function related to decreased functional length of GI tract as evidenced by postoperative status >65% of small bowel resected, diarrhea, and dehydration.

Nutrition Intervention:

1. Initiate central parenteral nutrition 200 mL 50% dextrose, 600 mL 10% amino acids, 300 mL 10% lipids in total volume of 1392 mL, 58 mL/hour continuous 24-hour infusion (21 mL/kg).

2. Increase rate 80 mL/hr as patient tolerates to meet EER 1500–1600 kcal/ EPR of 80–85 grams.

3. Adjust electrolytes per daily laboratory values.

Nutrition Monitoring and Evaluation:

1. Patient will meet goal rate of 80 mL/hr to meet 95% of estimated energy and protein needs within 48 hours.

2. Evaluate laboratory indices per PN protocol.

Table 15.21 Nutrition Therapy for Short Bowel Syndrome

General Guidelines

- Patients with jejunostomies/ileostomies (higher fat): approximately 20%–30% CHO, 20%–30% protein, 50%–60% fat.
- Patients with intact colon (higher CHO): approximately 50%–60% CHO, 20%–30% protein, 20%–30% fat.
- Avoid concentrated sweets and fluids.
- Chew foods well.
- Add salty meals and snacks if no colon.
- Eat smaller meals, more often.
- Decrease total nutrient load over the day and space out over time.
- Trial of oral rehydration solutions.
- Limit fluids with meals; drink isotonic beverages.
- Separate solids and liquids at meals as much as possible (solids before liquids).
- Solids slow emptying.
- Too much liquid creates a column effect (imagine the swelling of a stream when it rains and the increased flow generated).
- Use MCT-containing beverages if necessary vs. MCT oil (45).
- Lactose restriction if necessary (may try Lactaid).
- Avoid high-oxalate foods in those patients with kidney stones.
- Liquid or chewable vitamin/mineral supplements if necessary.
- Limit or avoid enteral stimulants such as alcohol and caffeine.

Good Choices	Avoid
Starches/breads	
Breads, pita bread, rolls	Donuts, sweet rolls, pastries, Pop-Tarts
Bagels, English muffins	
Plain waffles or pancakes	
Corn bread, plain muffins	
Banana or zucchini bread	
Tortillas—whole wheat or white flour, corn—toasted	
Pasta, macaroni, noodles	
Rice, brown rice, wild rice	
Cereals	
Unsweetened cereals (wet or eaten dry as a snack)	Sugary cereals, high-fiber cereals (>1–2 grams fiber/serving), bran cereals
Cheerios, cornflakes, Rice Krispies, Rice Chex, Spoonfuls, Special K, Kix, puffed rice or wheat	Flavored hot cereals
Hot cereals: cream of rice or wheat, grits, oatmeal	
Vegetables	
Canned or cooked vegetables	Creamed vegetables, legumes such as lima, kidney, pinto beans, etc.
Potatoes, sweet potatoes, yams	
Small amounts of lettuce (½ cup)	

Good Choices	Avoid
Fruits	
Bananas, melons, unsweetened canned fruits (applesauce, pears, peaches, mandarin oranges, apricots, cherries, plums, etc.)	Dried fruits, fruit canned in syrup, fruit juice, fruit drinks, watch out for high-fructose corn syrup in drinks (e.g., Capri Sun)
Meats/fish/poultry	
Meats, fish, shellfish, poultry, tuna fish, ham	Heavily fried meats, fish, poultry
Dairy/Soy	
Cheese, cottage cheese, plain yogurt or yogurt sweetened with artificial sweeteners, cream cheese	Highly sweetened yogurts or kefir, chocolate or other flavored milks, cream, half and half, Go-GURT, flavored soy milks
Plain soy milk	
Eggs	
Poached, hard or soft cooked, omelet, scrambled	Eggs prepared with ingredients not allowed
Nut butters	
Peanut, almond, cashew	Nutella, peanut butter with jam/jelly mixed in it
Beverages	
Oral rehydration solution	>4 oz coffee, tea, ice tea, flavored coffees or teas, hot cocoa, Ovaltine, Quick, fruit juices or fruit drinks (watch out for high-fructose corn syrup in drinks), Kool-Aid, Tang, regular sodas (all kinds), alcohol, water, sugar-free beverages, supplements such as Boost or Ensure
Soups, broth—4 oz per day	
Lactaid milk	
Snacks	
Crackers—saltines, soda, and so on	
Pretzels, matzo	
Corn or potato chips	
Bagel snack crackers	
Desserts	
Animal crackers, graham crackers, angel food cake, vanilla wafers, shortbread, plain pound cake, cake donuts—no icing, marshmallows	Iced cakes, cookies, Little Debbie Cakes, pie, ice cream, sherbet, candies, donuts, sweetened gelatin
Miscellaneous	
Salt, pepper, herbs, spices, dill pickles, Splenda, Equal, Sweet 'n Low	Sugar, sorbitol-containing sweets, maple or other syrups, jams, jellies, chocolate syrup, honey, molasses

Source: Rees Parrish C. The clinician's guide to short bowel syndrome. Nutrition issues in gastroenterology, series #31. Practical Gastroenterology. Sept. 2005, pp. 88–89.

Many patients with SBS are discharged on home parenteral nutrition (PN) or home enteral nutrition support (EN) in addition to the limited oral diet. PN or EN is usually cycled over 10 to 12 hours at home, which will allow a patient to resume normal activity.

Bacterial Overgrowth

Definition **Bacterial overgrowth syndrome** results from cross-contamination of bacteria from the colon to the small intestine. This may be a result of surgery, disease, or trauma to the GI tract.

Pathophysiology In this condition, motility of the gastrointestinal tract is delayed due to disease, surgery, or trauma, and stasis develops. There is a high risk for development of small bowel bacterial overgrowth for those individuals with short bowel syndrome. Bacteria numbers increase and begin to compete with the host for nutrients. Malabsorption, maldigestion, and malnutrition can result.[157]

Clinical Manifestations Signs and symptoms are similar to all conditions of malabsorption. Diarrhea, steatorrhea, anemia, and weight loss all may be present in this condition. Hydrogen breath tests (see previous section "Carbohydrate Malabsorption") can assist in diagnosis, but according to Rees-Parrish, most clinicians initiate treatment with antibiotics due to the cost and complications of administering the hydrogen breath test.[148]

Treatment Bacterial overgrowth syndrome is treated by both correcting the underlying cause and administering broad-spectrum antibiotics.[157]

Nutrition Therapy for Bacterial Overgrowth

Nutrition therapy will be consistent with the level of malabsorption that is present. Nutrients most commonly malabsorbed (fat and lactose) should be eliminated from the diet initially until the underlying condition is treated. Methods to increase nutrient density will accompany any restrictions, and steps should be made to maximize caloric and protein intake to replenish nutrient stores.

Conclusion

Nutrition is intimately involved in treatment for all diseases of the gastrointestinal tract. This chapter has discussed digestion, absorption, and transport of nutrients, which are crucial when evaluating the effect of disease on the gastrointestinal tract. Any diagnosis involving maldigestion and malabsorption has the potential to alter the nutritional status of the patient. The genetics of disease and the use of nutrition supplementation to treat disease (such as the use of fructooligosaccharides, glutamine supplementation, and probiotics) are current topics that have been addressed and will be of utmost interest to any student. Finally, nutrition therapy must be based on a thorough understanding of the effect of malnutrition on the integrity and function of the intestinal tract. This foundational knowledge allows the clinician not only to understand the disease process but also to plan nutritional rehabilitation for the patient. This chapter has discussed nutrition interventions for the lower GI tract, including specific diet modifications to assist in the treatment of disease.

PRACTITIONER INTERVIEW

Shelly Case, BSc, R.D. *Nutrition Counselor and Author of the Book* Gluten-Free Diet

Working with patients with celiac disease is very rewarding. Nutrition therapy is the only treatment for this disorder. I see many patients or family members who don't feel well, and by educating them about what they can and can't eat, they start to feel better, often within two weeks. They think of me as a miracle worker. What could be better than that?

We now know that celiac disease is more common than previously thought, and more individuals are being diagnosed. As a result, I am seeing more clients. When I first meet with the client I try to focus on the positive; for example: "This is a 'good' autoimmune disorder with a known treatment; the diet will make you feel better and improve your quality of life; and the diet is healthful and may help prevent other chronic diseases." But at the same time I am up front with them since the diet is complex and challenging, and explain that I will give them the tools they need so they know which food to eat or not eat.

First, I work with them so they can identify the foods they should avoid and then plan meal menus. I use the grocery store as an education tool. I give them a list of the cereals that are allowed and those to be avoided, and have them walk the perimeter of the store to identify foods they can eat. Less-processed food is located in these aisles. I then have them move into the inside aisles and again identify foods that they can eat that are more processed. Obviously, learning to read the ingredient label is very important. Lastly, I go over the special gluten-free products. The breakfast and lunch menus are the hardest, since they are often wheat based. Kids are more willing to try nontraditional items, like fruit smoothies for breakfast. I encourage dinner leftovers for lunch—meat, rice, cheese, fruit, and so on. For dietitians, food is our profession; but for patients newly diagnosed with celiac disease, we forget that they now have to think about every bite they take, and it's overwhelming. I try to hook them up with a support group, so they can network with others, and it helps them cope with the diet and the disease.

Although working with celiac disease clients is very rewarding, I have frustrating moments as well. Some clients who have been diagnosed with the disease have no symptoms, and they are often noncompliant, even though one of the long-term complications of this disorder, if untreated, is an increased incidence of several types of cancer. I also have to convince doctors to not tell patients to try the diet before they have the diagnostic serological tests and biopsy as these tests may come back negative if the patient is avoiding gluten.

When I started out, I wasn't an expert in celiac disease, but I had clients who needed help, so I did my homework. Remember—you are not taught everything you need to know in school or your internship, but you are taught how to look for information. My search for information resulted in a career just by developing resources for clients. I networked with celiac patients, which led to being the local advisor of a celiac chapter, which led to being on the national board. I was asked to be a speaker, which led to more referrals and then a book. I am now an expert RD on celiac disease and have been on the *Today Show* and a National Institutes of Health panel. My advice to young dietitians is to be on the lookout for a niche that you are interested in, do your homework, and be a bit of a risk taker.

Application of the Nutrition Care Process: Celiac Disease

Introduction:

KM is a 36-year-old female whose small bowel biopsy indicates flat mucosa with villous atrophy and hyperplastic crypts—inflammatory infiltrate in lamina propria. Additional labs include positive EMA antibodies and t-TG. Her gastroenterologist has informed her of a positive diagnosis of celiac disease with secondary malabsorption and anemia.

1. What is the etiology of celiac disease? How do t-TG, and EMA antibodies assist in this diagnosis?

2. What do the results of her small bowel biopsy indicate about changes to the anatomy of the small intestine? How is celiac disease related to this change?

Nutrition Assessment:

Food/Nutrition-Related History:

The registered dietitian's interview with the patient reveals that the patient feels hungry all the time. "I do eat, but it seems that every time I eat in any large amount I almost immediately have diarrhea. I do not have nausea or vomiting." Foods that are fried and meat—especially beef—tend to make the diarrhea worse. KM tells the RD that, for the last several days, she has only eaten chicken noodle soup, crackers, and Sprite. Her estimated energy intake is 275 kcal with 3 g protein per day. But before this past week, I really have eaten pretty normally, I think.

Anthropometric Measurements:

Ht. 5'3" Wt. 92 lbs UBW 112 lbs

Biochemical Data:

Hgb 10.5 g/dL, Hct 35%, Ferritin 12 g/dL + EGA, t-TG

Nutrition Diagnosis:

3. Identify at least two nutrition problems that can be found as a result of the nutrition assessment and medical history. Next, identify the etiology of each nutrition problem. Finally, identify the signs and symptoms that support the evidence for these nutrition problems. An example has been provided.

Diagnosis	Related to	Etiology	As Evidenced by	Sign/Symptoms
EXAMPLE: Involuntary weight loss	R/T	diarrhea and malabsorption, especially following consumption of fried foods and meats	AEB	18% unexplained weight loss and diet hx meeting < 5% of estimated energy needs

Nutrition Intervention:

4. Identify the nutrition prescription for this patient by recommending the appropriate dietary modification for her diagnosis.

5. Recommend energy and protein requirements for KM.

Nutrition Monitoring and Evaluation:

6. Determine nutrition criteria for monitoring and evaluation for each nutrition diagnosis that you identified.

EXAMPLE: Nutrition Care Outcome

Involuntary weight loss	Measure weight weekly.	Patient's weight will stabilize at 66 kg.

National Institutes of Health—National Institute of Diabetes and Digestive and Kidney Diseases This site provides information about diagnoses and treatment for diseases of the gastrointestinal tract, including excellent information about anatomy and physiology. There are links to current research and clinical trials.
http://www.niddk.nih.gov

National Guideline Clearinghouse Resource for evidence-based clinical practice guidelines which is a collaboration between the Agency for Healthcare Research and Quality, U.S. Department of Health and Human Services, American Medical Association, and American Association of Health Plans.
http://www.guideline.gov

Natural Medicine Comprehensive Database This evidence-based website provides safety and efficacy information for complementary and alternative medicine. *www.naturaldatabase.com*

Crohn's Colitis Foundation of America This organization provides information about these diseases and current treatments for patients. Additionally, this organization funds research, provides and sponsors educational workshops and symposia, and publishes the journal *Inflammatory Bowel Diseases*.
http://www.ccfa.org

Celiac Foundation This organization provides support, information, and assistance to people affected by celiac disease/

dermatitis herpetiformis (CD/DH). Links to current research and gluten-free products are prominently included at this site. *http://www.celiac.org*

U.S. Probiotics Organization Background research about probiotics along with consumer information about products. *http://www.usprobiotics.org*

Canadian Celiac Association—CCA Resource Guide for Health Professionals This association provides excellent resources for assessment, education, and counseling for use by health professionals in their practice and care for patients with celiac.
http://www.celiacguide.org

1. What are pre- and probiotics? How do they affect the health of the GI tract?

2. Describe the types of diarrhea and compare/contrast their possible etiologies. Are there nutritional consequences of diarrhea? Describe dietary measures that are commonly recommended for diarrhea.

3. Describe the pathophysiology of irritable bowel syndrome (IBS) and its recommended medical treatment. What is the role of NT in the treatment of IBS?

4. Describe the pathophysiology of inflammatory bowel disease (IBD) by comparing Crohn's disease and ulcerative colitis. Medically, what is recommended for the treatment of IBD? What are the potential nutritional consequences of IBD? Describe common NT recommendations for IBD.

5. How can diet help prevent and treat diverticulosis? Describe the pathophysiology of diverticulitis.

6. What is short bowel syndrome, and what factors increase its incidence after surgery? Describe the role of NT in the treatment of short bowel syndrome.

7. Describe the primary nutrition-related concerns of people who have undergone colostomies and ileostomies.

Complementary and Alternative Medicine Remedies Used in Diseases of the Lower Gastrointestinal System

Remedy	Scientific Name	CAM Use	Side Effects and/or Risks
Aloe vera; aloe latex	*Aloe* spp.	Stimulates colonic motility, propulsion, & transit time	Causes active secretion of fluids & electrolytes (hypokalemia) in lumen, inhibits reabsorption of fluids from colon.
Avens	*Geum* spp.	Diarrhea, dysentery, gastric irritation	Little scientific information exists concerning pharmacologic actions; keep away from children and pets. May antagonize antihypertensive medications
Barley	*Hordeum* spp.	Treat gastritis, diarrhea, and inflammatory bowel conditions	No known issues.
Betony	*Stachys officinalis*	Treat diarrhea	GI irritation, hepatic dysfunction. May increase hypotensive effect of antihypertensive medications.

Complementary and Alternative Medicine Remedies Used in Diseases of the Lower Gastrointestinal System (*Continued*)

Remedy	Scientific Name	CAM Use	Side Effects and/or Risks
Black catechu	*Acacia catechu*	Topical agent for sore gums & mouth ulcers; treat diarrhea & other GI problems	Increases risk of constipation if taken with anticholinergics. Increases risk of hypotension if taken with antihypertensives; increases additive hypotensive effect with Captopril. Increases risk of fungal infections if taken with immunosuppressants. May bind iron products, creating an insoluble complex.
Brewer's yeast		Treat flatulence, diarrhea, infectious diarrhea, & loss of appetite	May increase blood pressure if used with MAOIs.
Caraway	*Carum carvi*	Treat colic, constipation, flatulence, hiatal hernia, indigestion, mild spastic conditions of the GI tract, stomach ulcers	Diarrhea, hepatic dysfunction; pregnant women should avoid.
Cat's claw	*Uncaria* spp.	Crohn's disease, diverticulitis, dysentery, gastritis, ulcerations	May enhance effects of antihypertensive medications.
Chamomile	*Matricaria chamomilla*	Treat abdominal cramps & inflammatory GI conditions, hemorrhoids	Should be avoided during pregnancy or breast-feeding.
Chinese rhubarb	*Rheum palmatum*	Treat GI bleeding, indigestion, antidiarreal, laxative	Contraindicated in pregnant & breast-feeding women, children < 2 years of age, persons with ulcers or colitis. May enhance effects of anticoagulants; may decrease absorption of other drugs taken concurrently.
Chlorophyll		Reduce odor of feces, colostomy appliances; constipation	None reported.
Cumin	*Cuminum cyminum*	Used as antispasmotic and to treat GI problems	No human trials have been conducted to substantiate claims.
Daisy	*Bellis perennis*	Antidiarrheal, antispasmodic	Avoid in pregnancy & breast-feeding.
Essential fatty acids	DHA, EPA, omega-3 fatty acids, omega-3 oils	Ulcerative colitis	Might be helpful for reducing symptoms of UC; regular use does not appear to help prevent disease flare-ups.
Fish oil	Docosahexaenoic acid (DHA), eicosapentaenoic acid (EPA), omega-3 fatty acids, omega-3 oils	Crohn's disease	Possible risk of bleeding complications.
Flaxseed	*Linum usitatissimum*	Irritable bowel syndrome (IBS)	Helps relieve constipation, abdominal pain, and bloating. Not associated with any significant adverse effects.
Ground ivy	*Glechoma hederacea*	Diarrhea	None reported
Khella	*Ammi visnaga*	Spastic colon	Phototoxic in those predisposed to skin cancer. Headache, insomnia, vertigo, anorexia, constipation, elevated liver function test results, nausea, vomiting.
Mugwort	*Artemisia vulgaris*	Abdominal cramps, colic, constipation, diarrhea, "weak digestion"	May increase anticoagulant effects. Contraindicated in pregnant or breast-feeding women, and those with bleeding abnormalities. Use cautiously with GERD.
Myrtle	*Myrtus communis*	Digestive disorders	May increase hypoglycemic effects in those using insulin or sulfonylureas.
Nutmeg	*Myristica fagrans*	Chronic diarrhea, indigestion, nausea	Dry mouth, constipation, nausea, vomiting, tachycardia, seizures, delusions (with excessive doses), euphoria, hallucinations. May interfere with neuroleptic dugs. May be carcinogenic. Imported nutmeg has been found to be contaminated with aflatoxins.
Papaya	*Cymbogogon papaya*	Digestive disorders	Severe gastritis with prolonged use. Pregnant and breast-feeding women should avoid.

(continued)

Complementary and Alternative Medicine Remedies Used in Diseases of the Lower Gastrointestinal System (*Continued*)

Remedy	Scientific Name	CAM Use	Side Effects and/or Risks
Parsley	*Petroselinum crispum*	Antispasmodic	Fatty liver, GI bleeding. May increase hypotensive effects of antihypertensives. May promote or produce serotonin syndrome in those taking MAOIs.
Peppermint oil	*Menthe piperita*	Irritable bowel syndrome (IBS)	Provides antispasmodic properties; *may* provide some relief from crampy abdominal pain. Enteric-coated peppermint believed to be reasonably safe in healthy adults; non-enteric-coated peppermint oil can cause heartburn; maximum dosages in individuals with severe hepatic or renal disease not known.
Probiotics	*Lactobacillus plantarum*	Irritable bowel syndrome, ulcerative colitis (UC)	Evidence of efficacy is mixed; may reduce intestinal gas and pain. No known safety issues.
Rue	*Ruta graveolens*	Digestive disorders	Hypotension, increased risk of spontaneous abortion.
Self-heal	*Prunella vulgaris*	Sore throat, colic, diarrhea, flatulence	None reported.
Soapwort	*Saponaria officinalis*	Laxative	Use of high doses longer than 2 weeks should be avoided due to potential gastrotoxicity.
Turmeric	*Curcuma longa*	Cholelithiasis, GI bacterial overgrowth, flatulence, gastritis, ulcers, hepatic disorders	GI ulceration with high doses or prolonged use. Avoid administration with anticoagulants. Decreases immunosuppressive effects. Avoid administration with NSAIDs; may inhibit platelet function and increase risk of bleeding.

Sources: Bratman S, Girman AM. *Mosby's Handbook of Herbs and Supplements and their Therapeutic Uses.* St. Louis: Mosby/Elsevier, 2003.

Fetrow CW, Avila JR. *Professional's Handbook of Complementary and Alternative Medicines*, 3rd ed. Philadelphia: Lippincott Williams & Wilkins, 2004.

Fragakis AS, Thomson C. *The Health Professional's Guide to Popular Dietary Supplements*, 3rd ed, Chicago: American Dietetic Association, 2007.

REFERENCES

1. Reeds PJ, Burrin DG. Glutamine and the bowel. J Nutr. 2001;131:2505S–2508S.

2. Coëffier M, Claeyssens S, Lecleire S, Leblond J, Coquard A, Bôle-Feysot C, Lavoinne A, Ducrotté P, Déchelotte P. Am J Clin Nutr. 2008 Nov;88(5):1284–90. Combined enteral infusion of glutamine, carbohydrates, and antioxidants modulates gut protein metabolism in humans.

3. Peters JH, Mulder CJ, van Bodegraven AA. Fasting plasma citrulline concentrations do not reflect intestinal absorption capacity. J Parenter Enteral Nutr. 2008;32(3):288.

4. Moinard C, Cynober L. Nutritional consequences of weight-loss surgery. J Nutr. 2007 Jun;137(6 Suppl 2):1621S–1625S. Review.

5. Jones MP, Bratten JR. Small intestinal motility. Curr Opin Gastroenerol. 2008;24:164–72.

6. Furness JB. The enteric nervous system: normal functions and enteric neuropathies. Neurogastroenterol Motil. 2008 May; 20 Suppl 1:32–38. Review.

7. Sherwood L. Human physiology: from cells to systems. 7th ed. Belmont (CA): Brooks/Cole/Cengage Learning; 2010.

8. Ehrstom M, Naslund E, Ma J, et al. Physiological regulation and NO-dependent inhibition of migrating myoelectric complex in the rat small bowel by orexin A. Am J Physio Gastrointest Liver Physiol. 2003;285:688–95.

9. Wu CL, Hung CR, Change FY, et al. Involvement of cholecystokinin receptors in the inhibition of gastrointestinal motility by estradiol in ovariectomized rats. Scand J Gastroenterol. 2002;37:1133–39.

10. Smout AJP. Small intestinal motility. Curr Opin Gastroenterol. 2004;20:77–81.

11. Riby JE, Takuji F, Kretchmer N. Fructose absorption. Am J Clin Nutr. 1993;58(suppl):S748–S753.

12. Shepherd SJ, Gibson PR. Fructose malabsorption and symptoms of irritable bowel syndrome: guidelines for effective dietary management. J Am Diet Assoc. 2006;106:1631–39.

13. Beyer PL, Caviar EM, McCallum RW. Fructose intake at current levels in the United States may cause gastrointestinal distress in normal adults. J Am Diet Assoc. 2005;105:1559–66.

14. Schwartz MP, Samsom M, Renooij W, et al. Small bowel motility affects glucose absorption in a healthy man. Diabetes Care. 2002;25:1857–61.

15. Rees-Parrish C. The clinician's guide to short bowel syndrome. Practical Gastroenterology. 2005 Sept:67–106.

16. Grabitske HA, Slavin JL. Gastrointestinal effects of low-digestible carbohydrates. Crit Rev Food Sci Nutr. 2009 Apr;49(4):327–60.

17. Gropper S, Smith J, Groff JL. Advanced nutrition and human metabolism.

5th ed. Belmont (CA): Brooks/Cole/ Cengage Learning; 2009.

18. Hamer HM, Jonkers D, Venema K, Vanhoutvin S, Troost FJ, Brummer RJ. Review article: the role of butyrate on colonic function. Aliment Pharmacol Ther. 2008;27(2):104–19.

19. Grabitske HA, Slavin JL. Low-digestible carbohydrates in practice. J Am Diet Assoc. 2008;108:1677–81.

20. O'Keefe SJ. Gastric versus postpyloric feeding. Curr Opin Gastroenterol. 2008;24(1):51–58. Review.

21. Ulbrich T, Plogsted S, Geraghty ME, Reber KM, Valentine CJ. Probiotics and prebiotics: Why are they "bugging" us in the pharmacy? J Pediatr Pharmacol Ther. 2009;14:17–24.

22. Douglas LC, Sanders ME. Probiotics and prebiotics in dietetics practice. J Am Diet Assoc. 2008;108:510–21.

23. Spiller R. Review article: probiotics and prebiotics in irritable bowel syndrome. Aliment Pharmacol Ther. 2008;28(4):385–96.

24. Borowiec AM, Fedorak RN. The role of probiotics in management of irritable bowel syndrome. Curr Gastroenterol Rep. 2007;9(5):393–400.

25. Jonkers D, Stockbrugger R. Review article: probiotics in gastrointestinal and liver diseases. Aliment Pharmacol Ther. 2007;26(Suppl 2):133–48.

26. Johnston BC, Supina AL, Ospina M, Vohra S. Probiotics for the prevention of pediatric antibiotic-associated diarrhea. Cochrane Database of Systematic Reviews 2007, Issue 2. Art. No.: CD004827. DOI10.1002/14651858.CD004827.pub2.

27. Murphy MM, Douglass JS, Birkett A. Resistant starch intakes in the United States. J Am Diet Assoc. 2008 Jan;108(1):67–78. Erratum in: J Am Diet Assoc. 2008 May;108(5):890.

28. Said H. Cell and Molecular Aspects of Human Intestinal Biotin Absorption. J. Nutr. 2009;139:158–62.

29. Suttie JW. Vitamin K. In: Shils ME, Shike M, Ross CA, Caballero BC, Cousins RJ, eds. Modern nutrition in health and disease. 10th ed. Philadelphia: Walters Kluwer, Lippincott, Williams & Wilkins; 2005, 412–25.

30. Camilleri M, Murray JA. "Chapter 40. Diarrhea and Constipation" (Chapter). Fauci AS, Braunwald E, Kasper

DL, Hauser SL, Longo DL, Jameson JL, Loscalzo J. Harrison's principles of internal medicine. 17th Ed. http://www .accessmedicine.com/content .aspx?aID=2888656

31. United States Department of Health and Human Services. Centers for Disease Control and Prevention. Foodborne Diseases Active Surveillance Network. Available at http://www.cdc.gov/foodnet

32. United States Department of Health and Human Services. Centers for Disease Control and Prevention. Rotavirus. Available at http://www.cdc.gov/rotavirus

33. Green ST, Small MJ, Casman EA. Determinants of national diarrheal disease burden. Environ Sci Technol. 2009;43(4):993–99.

34. Binder HJ. Causes of chronic diarrhea. N Engl J Med. 355;3:236–39.

35. Shah N, DuPont HL, Ramsey DJ. Global etiology of travelers' diarrhea: systematic review from 1973 to the present. Am J Trop Med Hyg. 2009;80:609–14.

36. Nelson R. Antibiotic treatment for Clostridium difficile-associated diarrhea in adults. Cochrane Database Syst Rev. 2007;18(3):CD004610.

37. Delegge MH, Berry A. Enteral feeding: should it be continued in the patient with Clostridium difficile enterocolitis? Practical Gastroenterology. 2009;XXXIII(3):40–49.

38. McErlean A, Kelly O, Bergin S, Patchett SE, Murray FE. The importance of microbiological investigations, medications and artificial feeding in diarrhea evaluation. Ir J Med Sci. 2005;174:21–25.

39. Whelan K. Enteral-tube-feeding diarrhoea: manipulating the colonic microbiota with probiotics and prebiotics. Proc Nutr Soc. 2007;66(3):299–306.

40. Yang G, Wu XT, Zhou Y, Wang YL. Application of dietary fiber in clinical enteral nutrition: A meta-analysis of randomized controlled trials. World J Gastroenterol. 2005;11(25):3935–38.

41. Binder HJ. Chapter 286: Disorders of absorption. In: Eugene Braunwald, Anthony S. Fauci, Kurt J. Isselbacher, Dennis L. Kasper, Stephen L. Hauser, Dan L. Longo, J. Larry Jameson, editors. Harrison's principles of internal medicine. 15th ed. New York: McGraw-Hill, Inc; 2005. Online edition available at http:// harrisons.accessmedicine.com

42. Murphy C, Hahn S, Volmink J. Reduced osmolarity oral rehydration solution for treating cholera. Cochrane Database Syst Rev. 2004;18(4):CD003754.

43. Bateman M, McGahey C. A framework for action: child diarrhea prevention. Global Health Link; 2001. Available at http://www.ehproject.org/Pubs/ GlobalHealth/GH-CArticle.htm

44. Steffen R, Gyr K. Diet in the treatment of diarreha: from tradition to evidence. Clin Infect Dis. 2004;39:472–73.

45. Duro D, Duggan C. The BRAT Diet for acute diarrhea. Pract Gastroenterol. 2007;60–68.

46. American Dietetic Association. Diarrhea. In: Nutrition Care Manual. Available at www.nutritioncaremanual.org

47. Khan AM, Sarker SA, Alam NH, Hossain MS, Fuchs GJ, Salam MA. Low osmolar oral rehydration salts solution in the treatment of acute watery diarrhea in neonates and young infants: a randomized, controlled clinical trial. J Health Popul Nutr. 2005;23:52.

48. de Vrese M, Marteau PR. Probiotics and prebiotics: effects on diarrhea. J Nutr. 2007;137:803S–811S.

49. Takahashi O, Noguchi Y, Omata F, Tokuda Y, Fukui T. Probiotics in the prevention of traveler's diarrhea: meta-analysis. J Clin Gastroenterol. 2007;41(3):336–37.

50. Johnston BC, Supina AL, Ospina M, Vohra S. Probiotics for the prevention of pediatric antibiotic-associated diarrhea. Cochrane Database Syst Rev. 2007;(2):CD004827.

51. Jonkers D, Stockbrugger R. Review article: probiotics in gastrointestinal and liver diseases. Aliment Pharmacol Ther. 2007;26(Suppl 2):133–48.

52. Canani RB, Cirillo P, Terrin G, Cesarano L, Spagnuolo MI, De Vincenzo A, Albano F, Passariello A, De Marco G, Manguso F, Guarino A. Probiotics for treatment of acute diarrhea in children: randomized clinical trial of five different preparations. BMJ. 2007;335:340–42.

53. O'Keefe SJ. Nutrition and colonic health: the critical role of the microbiota. Curr Opin Gastroenterol. 2008 Jan;24(1):51–58. Review.

54. Hamer HM, Jonkers D, Venema K, Vanhoutvin S, Troost FJ, Brummer RJ. Aliment Pharmacol Ther. 2008 Jan 15;

27(2):104–19. Epub 2007 Oct Review article: the role of butyrate on colonic function.

55. Douglas LC, Sanders ME. Probiotics and prebiotics in dietetics practice. J Am Diet Assoc. 2008;108:510–21.

56. United States Department of Health and Human Services. US Food and Drug Administration. Complementary and Alternative Medicine Products and their Regulation by the Food and Drug Administration. Available at http://www.fda.gov/RegulatoryInformation/Guidances/ucm144657.htm

57. American Dietetic Association. Nutrition Therapy for Diarrhea. In: Nutrition Care Manual. Available at www.nutritioncaremanual.org

58. Drossman DA. The functional gastrointestinal disorders and the Rome III. Gastroenterology. 2006;130:1377–90.

59. Longstreth GF, Thompson WG, Chey WD, Houghton LA, Mearin F, Spiller RC. Functional bowel disorders. Gastroenterology. 2006;130:1480–91.

60. Ternent CA, Bastawrous AL, Morin NA, Ellis CN, Hyman NH, Buie WD. Standards Practice Task Force of The American Society of Colon and Rectal Surgeons. Practice Parameters for the Evaluation and Management of Constipation. Dis Colon Rec. 2007;50:2013–22.

61. Choung RK, Locke GR, Schleck CD, Zinsmeister AR, Talley NJ. Cumulative incidence of chronic constipation: a population-based study 1988–2003. Gastroenterology. 2006;130:A-508.

62. Müller-Lissner SA, Kamm MA, Scarpignato C, Wald A. Myths and misconceptions about chronic constipation. Am J Gastroenterol. 2005;100:124–29.

63. Leung FW. Etiologic factors of chronic constipation: review of the scientific evidence. Dig Dis Sci. 2007;52(2):313–16.

64. American Dietetic Association. Position Paper: Health Implications of Dietary Fiber. J Am Diet Assoc. 2008;108:1716–31.

65. United States Department of Health and Human Services and U.S. Department of Agriculture. Dietary Guidelines for Americans 2. 6th ed. Washington (DC): U.S. Government Printing Office; 2005.

66. Health and Welfare Canada. Canada's food guide to healthy eating. Minister of Supply and Services Canada; 2005.

67. Bu LN, Chang MH, Ni YH, Chen HL, Cheng CC. Lactobacillus casei rhamnosus Lcr35 in children with chronic constipation. Pediatr Int. 2007;49(4):485–90.

68. Monsbakken KW, Vandvik PO, Farup PG. Perceived food intolerance in subjects with irritable bowel syndrome—etiology, prevalence and consequences. Eur J Clin Nutr. 2006;60(5):667–72.

69. Lomer MCE, Parkes GC, Sanderson JD. Review article: lactose intolerance in clinical practice— myths and realities. Aliment Pharmacol Ther. 2008;27:93–103.

70. Binder HJ. Chapter 286: Disorders of absorption. In: Harrison's principles of internal medicine. 15th ed. Eugene Braunwald, Anthony S. Fauci, Kurt J. Isselbacher, Dennis L. Kasper, Stephen L. Hauser, Dan L. Longo, J. Larry Jameson, Eds. New York: McGraw Hill, Inc; 2009. Online edition available at http://harrisons.accessmedicine.com

71. Green PHR and Cellier C. Celiac Disease. N Engl J Med. 2007;357:1731–43.

72. Dubé C, Rostom A, Sy R, Cranney A, Saloojee N, Garritty C, Sampson M, Zhang L, Yazdi F, Mamaladze V, Pan I, Macneil J, Mack D, Patel D, Moher D. The prevalence of celiac disease in average-risk and at-risk Western European populations: a systematic review. Gastroenterology. 2005 Apr;128(4 Suppl 1):S57–67. Review.

73. Catassi C, Fabiani E, Iacono G, D'Agate C, Francavilla R, Biagi F, Volta U, Accomando S, Picarelli A, De Vitis I, Pianelli G, Gesuita R, Carle F, Mandolesi A, Bearzi I, Fasano A. A prospective, double-blind, placebo-controlled trial to establish a safe gluten threshold for patients with celiac disease. Am J Clin Nutr. 2007 Jan;85(1):160–66.

74. Fasano A, Berti I, Gerarduzzi T, Not T, Colletti RB, Drago S, Elitsur Y, Green PH, Guandalini S, Hill ID, Pietzak M, Ventura A, Thorpe M, Kryszak D, Fornaroli F, Wasserman SS, Murray JA, Horvath K. Prevalence of celiac disease in at-risk and not-at-risk groups in the United States: a large multicenter study. Arch Intern Med. 2003;163:286–92.

75. Setty M, Hormaza L, Guandalini S. Celiac disease: risk assessment, diagnosis, and monitoring. Mol Diagn Ther. 2008;12(5):289–98.

76. Nemec G, Ventura A, Stefano M, Di Leo G, Baldas V, Tommasini A, Ferrara F, Taddio A, Citta A, Sblattero D, Mazari R, Not T. Looking for celiac disease: diagnostic accuracy of 2 commercial assays. Available at www.medscape.com/viewarticle/540961

77. Catassi C, Fasano A. Celiac Disease. Current Opinion in Gastroenterology. 2008;24(6):687–91.

78. Bardella MT, Velio P, Cesana BM, Prampolini L, Casella G, Di Bella C, Lanzini A, Gambarotti M, Bassotti G, Villanacci V. Coeliac disease: a histological follow-up study. Histopathology. 2007;50:465–71.

79. Abdulkarim AS, Burgart LJ, See J, Murray JA. Etiology of nonresponsive celiac disease: results of a systematic approach. Am J Gastroenterology. 2002;97:2016–18.

80. American Dietetic Association. Evidence-based Nutrition Practice Guideline on Celiac Disease published January 2009 at http://www.adaevidencelibrary.com/ and copyrighted by the American Dietetic Association.

81. Peräaho M, Collin P, Kaukinen K, Kekkonen L, Miettinen S, Maki M. Oats can diversify a gluten-free diet in celiac disease and dermatitis herpiformes. J Am Diet Assoc. 2004;104:1148–50.

82. Thompson T. Gluten contamination of commercial oat products in the United States. N Engl J Med. 2004;351:2021–22.

83. Food and Drug Administration. Federal Register Proposed Rule—72 FR 2795 January 23, 2007: Food Labeling; Gluten-Free Labeling of Foods. Available at http://www.fda.gov/Food/LabelingNutrition/FoodAllergensLabeling/GuidanceComplianceRegulatoryInformation/ucm077926.htm

84. Cummings W. Electro-galvanism in a peculiar affection of the mucous membrane of the bowels. London Med Gazette. 1849;NS9:969–73.

85. Brandt LJ, Chey WD, Foxx-Orenstein AE, Schiller LR, Spiegel BM, Talley N, Quigley E, Moayyedi P. An evidenced-based systematic review on the management of irritable bowel syndrome. Am J Gastroenterology. 2009;104:S1–S35.

86. Sharara AI, Aoun E, Abdul-Baki H, Mounzer R, Sidani S, Elhajj I. Am J Gastroenterol. 2006;101:326–33.

87. Saito Y. Genes and Irritable Bowel Syndrome: Is There a Link? Current Gastroenterology Reports. 2008;10:355–62.

88. Furness JB. Neurogastroenterol Motil. 2008 May;20 Suppl 1:32–38. Review

89. Akbar A, Walters JRF, Ghosh S. Visceral hypersensitivity in irritable bowel syndrome: molecular mechanisms and therapeutic agents. Aliment Pharmacol Ther. 2009 Sep 1;30(5):423–35.

90. Drisko J, Bischoff B, Hall M, McCallum R. Treatiang irritable bowel syndrome with a food elimination diet followed by food challenge and probiotics. J Am Coll Nutr. 2006;25(6):514–22.

91. Bharucha AE. The enteric nervous system: normal functions and enteric neuropathies. Neurogastroenterol Motil. 2008 May;20 Suppl 1:103–13. Review.

92. Sikander A, Rana SV, Prasad KK. Role of serotonin in gastrointestinal motility and irritable bowel syndrome. Clin Chim Acta. 2009;403:47–55.

93. Thor PJ, Krolczyk G, Kil K, Zurowski D, Nowak L. Melatonin and serotonin effects on gastrointestinal motility. J Physiolo Pharmacol. 2007;58 Suppl 6:97–103.

94. Spiller R. Review article: probiotics and prebiotics in irritable bowel syndrome. Aliment Pharmacol Ther. 2008 Aug 15;28(4):385–96.

95. Simren M. Bloating and abdominal distention: not so poorly understood anymore! Gastroenterology. 2009;136: 1487–90.

96. Akbar A, Walters JRF, Ghosh S. Visceral hypersensitivity in irritable bowel syndrome: molecular mechanisms and therapeutic agents. Aliment Pharmacol Ther. 2009 Sep 1;30(5):423–35.

97. Saad R, Chey WD. Recent developments in the therapy of irritable bowel syndrom. Expert Opin Investig Drugs. 2008;17:117–30.

98. Gomez-Escudero O, Schmulson-Wasserman MJ, Valdovinos-Diaz MA. Post-infectious irritable bowel syndrome. A review based on current evidence. Rev Gastroenterol Mex. 2003;68(1):55–61.

99. Merat S, Khalili S, Mostajabi P, Ghorbani A, Ansari R, Malekzadeh R. The effect of enteric-coated, delayed-release peppermint oil in irritable bowel syndrome. Dig Dis Sci. 2009 Jun 9. [Epub ahead of print]

100. Heizer WD, Southern S, McGovern S. The role of diet in symptoms of irritable bowel syndrome: a narrative review. J Am Diet Assoc. 2009;109:1204–14.

101. Shepherd SI, Parker FC, Muir JG, Gibson PR. Dietary triggers of abdominal symptoms in patients with irritable bowel syndrome: randomized placebo controlled evidence. Clin Gastroenterol Hepatol. 2008;6:765–71.

102. Barrett JS, Gibson PR. Clinical ramifications of malabsorption of fructose and other short-chain carbohydrates. Practical Gastroenterology. 2007;8:51–65.

103. Austin G, Dalton CB, Hu Y, Morris CB, Hankins J, Weinland SR, Westman EC, Yancy WS, Drossman DA. A very low-carbohydrate diet improves symptoms and quality of life in diarrhea-predominant irritable bowel syndrome. Clin Gastro Hep. 2009;7:706–08

104. Hoveyda N, Heneghan C, Mahtani KR, Perera R, Roberts N, Glasziou P. A systematic review and meta-analysis: probiotics in the treatment of irritable bowel syndrome. BMC Gastroenterology. 2009;9:15–26.

105. Kappelman MD, Rifas-Shiman SL, Kleinman K, et al. The prevalence and geographic distribution of Crohn's disease and ulcerative colitis in the United States. Clin Gastroenterol Hepatol 2007, 5:1424–29.

106. Loftus CG, Loftus EV Jr, Harmsen WS, et al. Update on the incidence and prevalence of Crohn's disease and ulcerative colitis in Olmsted County, Minnesota, 1940–2000. InflammBowel Dis. 2007;13:254–61.

107. Lichtenstein GR, Hanauer SB, and Sandborn WJ. Management of Crohn's disease in adults. The American Journal of Gastroenterology [serial online]. February 06, 2009;104(2):465.

108. Rutgeerts P, Vermeire S, Van Assche G. Biological therapies for inflammatory bowel disease. Gastroenterology. 2009;136:1182–97.

109. Cho JH. The genetics and immunopathogenesis of inflammatory bowel disease. Nature Reviews. Immunology [serial online]. 2008;8(6):458–66. Available from: MEDLINE with Full Text, Ipswich, MA.

110. Macfarlane S, Steed H, Macfarlane GT. Intestinal bacteria and inflammatory bowel disease. Crit Rev Clin Lab Sci. 2009;46(1):25–54.

111. Tuelove SC, Witts LJ. Cortisone in ulcerative colitis: final report on a therapeutic trial. Br Med J. 1955;2:1041–48.

112. Ali S, and Tamboli CP. Advances in epidemiology and diagnosis of inflammatory bowel diseases. Current Gastroenterology Reports. December 2008;10(6):576–84.

113. Langhorst J, Elsenbruch S, Koelzer J, Rueffer A, Michalsen A, Dobos GJ. Noninvasive markers in the assessment of intestinal inflammation in inflammatory bowel diseases: performance of fecal lactoferrin, calprotectin, and PMN-elastase, CRP, and clinical indices. Am J Gastroenterol. 2008;103:162–69.

114. Best WR, Becktel JM, Singleton JW and Kern F. Development of a Crohn's disease activity index: National Cooperative Crohn's Disease Study. Gastroenterology. 1976;70(3):438–44.

115. Kappelman MD, Bousvaros A. Nutritional concerns in pediatric inflammatory bowel disease patients. Mol Nutr Food Res. 2008;52:867–74.

116. Smith PA. Nutritional therapy for active Crohn's disease. World J Gastroenterol. 2008;14:4420–23.

117. Lochs H, Dejong C, Hammarqvist F, Hebuterne X, Leon-Sanz M, Schütz T, van Gemert W, et al. 2006. ESPEN Guidelines on Enteral Nutrition: Gastroenterology. 2006;25:260–74.

118. Wiese DM, Rivera R, and Seidner DL. Is there a role for bowel rest in nutrition management of Crohn's disease? Nutrition In Clinical Practice. 2008;23:309–17.

119. Zachos M, Tondeur M, Griffiths AM. Enteral nutritional therapy for induction of remission in Crohn's disease. Cochrane Database Syst Rev. 2007;Jan 24;(1):CD000542.

120. El-Matary W. Enteral nutrition as a primary therapy of Crohn's Disease: the pediatric perspective. Nutrition in Clinical Practice. 2009;24:91–97.

121. Day AS, Whitten K, Lembery DA, Clarkson C, Vitug-Sales M, Jackson R, Bohane TD. Exclusive enteral feeding

as primary therapy for Crohn's disease in Australian children and adolescents: a feasible and effective approach. J of Gastroenterol and Hepatology. 2006; 21:1609–14.

122. Eiden KA. Nutritional considerations in inflammatory bowel disease. Practical Gastroenterology. 2003;May:33–54.

123. Little DR. Ambulatory management of common forms of anemia. Am Fam Physician. 1999;59:1598–604.

124. Jeejeehboy KN. Clinical Nutrition 6. Management of nutritional problems of patients with Crohn's Disease. CMAJ. 2002;166:913–18.

125. McDermott JH. Antioxidant nutrients: current dietary recommendations and research update. J Am Pharm. 2000. Assoc. 40:785–99.

126. Douglas LC, Sanders ME. Probiotics and prebiotics in dietetics practice. J Am Diet Assoc. 2008;108:510–21.

127. Baroja ML, Kirjavainen PV, Hekmat S, Reid G. Anti-inflammatory effects of probiotic yogurt in inflammatory bowel disease patients. Clinical and Experimental Immunology. 2007;149:470–79.

128. Galvez J, Rodriguez-Cabezas ME, Zarzuelo A. Effects of dietary fiber on inflammatory bowel disease. Mol Nutr Food Res. 2005;49(6):601–08. Review.

129. Gassull, MA. Macronutrients and bioactive molecules: is there a specific role in the management of inflammatory bowel disease? JPEN. 2005 Jul–Aug; 29(4) Suppl:S179–83.

130. Guarner F. Inulin and oligofructose: impact on intestinal diseases and disorders. Br J Nutr . 2005 Apr; 93 Suppl 1: S61–65

131. Shanahan F. Probiotics in inflammatory bowel disease—therapeutic rationale and role. Adv Drug Deliv Rev. 2004 Apr 19;56(6):809–18.

132. Ternent CA, Bastawrous AL, Morin NA, Ellis CN, Hyman NH, Buie WD. Standards Practice Task Force of The American Society of Colon and Rectal Surgeons. Practice Parameters for the Evaluation and Management of Constipation. Dis Colon Rec. 2007;50:2013–22.

133. Sheth AA, Longo W, Floch MH. Diverticular disease and diverticulitis. Am J Gastroenterol. 2008;103(6):1550–56.

134. Aldoori WH, Giovannucci EL, Rimm EB, et al. Prospective study of physical activity and the risk of symptomatic diverticular disease in men. Gut. 1995;36:276–82.

135. Aldoori WH, Giovannucci EL, Rockett HR, Sampson L, Rimm EB, Willett WC. A prospective study of dietary fiber types and symptomatic diverticular disease in men. J. Nutr. 1998;128:714–19.

136. Beitz JM. Diverticulosis and diverticulitis spectrum of a modern malady. J Wound Ostomy Continence Nurs. 2004 Mar–Apr 31;(2):75–82.

137. West AB, Losada M. The pathology of diverticulosis coli. J Clin Gastroenterol. 2004 May–Jun;38(5 Suppl):S11–6.

138. Stollman NH, Raskin JB. Diagnosis and management of diverticular disease of the colon in adults. Ad Hoc Practice Parameters Committee of the American College of Gastroenterology. Am J Gastroenterol. 1999;94:3110–21.

139. Cordain L, Eaton SB, Sebastian A, Mann N, Lindeberg S, Watkins BA, O'Keefe JH, Brand-Miller J. Origins and evolution of the Western diet: health implications for the 21st century. Am J Clin Nutr (The American journal of clinical nutrition.) 2005 Feb;81(2):341–54.

140. Makola D. Diverticular disease: Evidence for dietary intervention? Prac Gastenterol. 2007;Feb:38–68.

141. Strate LL, Liu YL, Syngal S, Aldoori WH, Giovannucci EL. Nut, Corn, and Popcorn Consumption and the Incidence of Diverticular Disease. JAMA. 2008;300(8):907–14.

142. American Dietetic Association. Position of the American Dietetic Association: Health implications of dietary fiber. J Am Diet Assoc. 2008;108:1716–31.

143. Bedi MS, Ramesh H. Annals of Surgery. 2008;248(6)1108–09; author reply 1109–10.

144. Chen J, Lindmark-Mansson H, Drevelius M, Tidehag P, Hallmans G, Hertervig E, Nilsson A, Akesson B. Bioavailability of selenium from bovine milk as assessed in subjects with ileostomy. European Journal of Clinical Nutrition. 2004 Feb;58(2):350–55.

145. Faulks RM, Hart DJ, Brett GM, Dainty JR, Southon S. Kinetics of gastrointestinal transit and carotenoid absorption and disposal in ileostomy volunteers

fed spinach meals. European Journal of Nutrition. 2004 Feb;43(1):15–22.

146. Livny O, Reifen R, Levy I, Madar Z, Faulks R, Southon S, Schwartz B. Beta-carotene bioavailability from differently processed carrot meals in human ileostomy volunteers. European Journal of Nutrition. 2003 Dec; 42(6):338–45.

147. Willcutts K, Scarano K, Eddins CW. Ostomies and fistulas: a collaborative approach. Pract Gastroenterol. 2005;63–79.

148. Rees-Parrish C. The clinician's guide to short bowel syndrome. Practical Gastroenterology. 2005 Sept:67–106.

149. O'Keefe SJD, Buchman AL, Fisbein TM, Jeejeebhoy KN, Jeppesen PB, Shaffer J. Short bowel syndrome and intestinal failure: consensus definitions and overview. Clin Gastroenterol Hepatol. 2006;4:7.

150. Tilg H. Short bowel syndrome: searching for the proper diet. Eur J of Gastroenterology & Hepatology. 2008;20:1061–63.

151. Buchman AL. The Medical and Surgical Management of Short Bowel Syndrome. Available at http://www .medscape.com/viewarticle/474629_1 .Medscape.com

152. Gomez-Herrera E, Farias-Llamas OA, Gutierrez-de la Rosa JL, Hermosillo-Sandoval JM. The role of long-acting release (LAR) depot octreotide as adjuvant management of short bowel disease. Cir Cir. 2004 Sep–Oct;72(5):379–86.

153. Mardini HE, de Villiers WJ. Teduglutide in intestinal adaptation and repair: light at the end of the tunnel. Expert Opin Investig Drugs. 2008;17:945–51.

154. Duro D, Kamin D, Duggan C. Overview of pediatric short bowel syndrome.J Pediatr Gastroenterol Nutr. 2008;47 Suppl 1:S33–6.

155. Sundaram A, Koutkia P, Apovian CM. Nutritional management of short bowel syndrome in adults. J Clin Gastroenterol. 2002;34:207–20.

156. Lykins TC and Stockwell J. Comprehensive modified diet simplifies nutrition management of adults with short-bowel syndrome. J Am Diet Assoc. 1998;98:309–15.

157. Abu-Shanab A, Quigley EM. Diagnosis of small intestinal bacterial overgrowth: the challenges persist! Expert Rev Gastroenterol Hepatol. 2009;3:77–87.

16

Diseases of the Liver, Gallbladder, and Exocrine Pancreas

Kathryn Sucher, ScD, RD
Department of Nutrition, Food Science & Packaging, San Jose State University
Mildred Mattfeldt-Beman, PhD, RD LD
Department of Nutrition and Dietetics, Saint Louis University

Introduction

The liver, pancreas, and gallbladder are accessory organs to the gastrointestinal (GI) tract (see Figure 16.1). All three organs are important in digestion, absorption, and/or metabolism of nutrients from food, which constitute the hepatobiliary processes. (The Greek word for liver is "hepatos," hence "hepatic.") The liver is one of the largest organs in the body, weighing about three pounds in an adult. It influences nutritional status through the synthesis of **bile** salts and metabolism of protein, carbohydrate, fat, and vitamins. It is the first stop for nutrient-rich blood from the intestines and protects the body through modification of toxic substances. It is central to hemopoiesis and blood clotting, both synthesizing and storing compounds needed for these processes. An indication of the importance of the liver is its ability to regenerate itself: Even if up to 70% of a healthy liver is removed, it will regenerate to its original size.[1]

Anatomy and Physiology of the Hepatobiliary System

Anatomy of the Liver

The liver is located in the upper right quadrant of the abdomen, following the curve of the diaphragm. The healthy human liver is brownish-red in color; when highly infiltrated with fat, it becomes a dark, yellowish-brown mustard color. There are four anatomical lobes of the liver: the right lobe (largest), quadrate lobe, caudate lobe, and left lobe. Based on blood supply and biliary drainage, there are two functional lobes: right and left. The right lobe receives blood from the right hepatic artery, and the left lobe (including the caudate and quadrate lobes) receives blood from the left hepatic and middle hepatic arteries. The right and left hepatic ducts drain each lobe, respectively.

The basic functional unit of the liver is the lobule, which is a cylindrical structure several millimeters in length and 0.8 to 2 mm in diameter (see Figure 16.2). Each lobe is made up of thousands of lobules (50,000–100,000 individual lobules).[2] The liver lobule is constructed around a central vein that empties into the hepatic vein and then into the vena cava. The lobule itself is composed of many hepatic cellular plates that radiate from the central vein like spokes of a wheel. Each plate is about

GLOSSARY

acute (fulminant) hepatic/liver failure—the severe impairment of hepatic functions in the absence of pre-existing liver disease

alcoholic liver disease (ALD)—liver disease associated with alcoholism

aromatic amino acids (AAA)—amino acids containing an aromatic side chain (phenylalanine, tyrosine, and tryptophan)

ascites—accumulation or retention of free fluid within the peritoneal cavity

asterixis—abnormal involuntary movements that primarily affect the extremities

bile—an emulsifying agent produced in the liver and eventually secreted into the duodenum

biliary cirrhosis—liver cirrhosis in which there is interference with intrahepatic bile flow

biliary sludge—a mixture of particulate matter and mucus that forms in bile

biliary stasis—intrahepatic impairment of bile flow

biliary tract (tree)—the common anatomical term for the path by which bile that has been secreted by the liver travels on its way to the small intestine

bilirubin—the yellow breakdown product of normal heme catabolism

branched-chain amino acids (BCAA)—the amino acids that have a branched side chain (leucine, isoleucine, and valine).

cholecystectomy—surgical removal of the gallbladder

cholecystitis—inflammation of the gallbladder

choledocholithiasis—gallstones that are present in the common bile duct but are usually formed in the gallbladder

cholelithiasis—the presence or formation of gallstones

cholestasis—condition in which the flow of bile from the liver is blocked

cirrhosis—any pathological condition where fibrous connective tissue replaces healthy tissue in an organ, usually as a consequence of inflammation or other injury

fatty liver—yellow discoloration of the liver due to fatty degeneration of liver parenchymal cells

hepatic encephalopathy—a syndrome characterized by central nervous system dysfunction in association with liver failure

hepatitis—inflammation of the liver and liver disease involving degenerative or necrotic alterations of hepatocytes

hepatosteatosis—accumulation of fat in the interstitial tissue of the liver

jaundice—a clinical manifestation of hyperbilirubinemia, consisting of deposition of bile pigments in the skin, resulting in a yellowish staining of the skin and mucous membranes

Korsakoff's psychosis—condition characterized by amnesia, confabulation (false memories), and hallucinations

Kupffer cells—specialized phagocytic (macrophage) cells of the reticuloendothelial system found on the luminal surface of the hepatic sinusoids; they filter bacteria and small foreign proteins out of the blood and dispose of worn-out red blood cells

nonalcoholic fatty liver disease (NAFLD)—a wide spectrum of non-alcohol-related liver diseases ranging from fatty liver (steatosis), to nonalcoholic steatohepatitis (NASH), to cirrhosis

nonalcoholic steatohepatitis (NASH)—non-alcohol-related liver inflammation caused by a buildup of fat in the liver

pancreatitis—inflammation of the pancreas

paracentesis—a procedure in which fluid is withdrawn from a body cavity via a trocar and cannula, needle, or other hollow instrument

portal hypertension—abnormally increased pressure in the portal venous system; frequently seen in cirrhosis of the liver and in other conditions that cause obstruction of the portal vein

Wernicke-Korsakoff syndrome—manifestation of thiamin deficiency usually seen in individuals suffering from alcoholism

Wernicke's encephalopathy—condition characterized by confusion, nystagmus (involuntary eye movement), anisocoria (unequal size of the pupils), ataxia, and sluggishness

---

two cells thick and between the adjacent cells are bile canaliculi, which collect bile produced in the cells and carry it to the hepatic terminal bile duct.

The liver's structure allows for the many functions of this organ. Understanding the vascular supply, both serum and lymph, is essential. The main blood and lymph vessels, bile ducts, and nerves enter and leave the liver through the hilus. The liver receives blood from two sources. As you look at the schematic diagram of the lobule (Figure 16.2b), you will see one branch of the portal vein (venule). The portal vein drains the intestine, spleen, and pancreas, bringing nutrients to the liver. The branches of the hepatic portal vein flow into the flat, branched sinusoids that lie between the hepatic cell plates and into the central hepatic vein. Three-quarters of the blood flow to the liver comes from the central hepatic vein. The hepatic arteries carry oxygenated blood from the abdominal aorta to the hepatic cellular plates (see Figure 16.3). In the liver sinusoids, there is a mixing of the oxygenated blood coming from the hepatic artery with the venous blood (rich in nutrients) coming from the portal vein. The size of the venous sinusoids is a function of the amount of the blood within them. Smaller blood vessels within the sinusoids (capillaries) differ from other capillaries in that they have a greater permeability to macromolecules, especially protein.

The venous sinusoids are lined with at least four types of cells: typical endothelial cells, large **Kupffer cells**, specialized macrophage white blood cells (see chapter 9 for information on macropages), perisinusoidal fat (and vitamin A) storing cells, and pit cells (least common cells with natural killer cell functions).[3] The endothelial lining of the venous sinusoids has very large pores, allowing hepatocytes to have ready access to nutrients in the plasma. A narrow space is present between the surface of the hepatocyte and the surface of the endothelial cell. This is called the space of Disse; it is filled with numerous microvilli from the hepatocytes, allowing an increase in the surface area coming in contact with the blood and facilitating the exchange of molecules between hepatocytes and the blood. Substances of the plasma move freely into the space. There are also many terminal lymphatics to remove excess fluid from these spaces.

In summary, the blood flow through the liver can be traced as follows:

- The portal vein from the gastrointestinal tract divides into very fine branches that discharge portal blood into the venous sinusoids.

- Each branch of the portal vein is accompanied by networking branches of the hepatic artery.

- The blood eventually flows into the sinusoids. Kupffer cells lining the sinusoids system remove **bilirubin**, dyes, bacteria, damaged red blood cells, and other debris from the plasma through phagocytosis.

- The content of the sinusoids enters the central vein and finally reaches the vena cava.

- A network of lymphatic vessels in the portal canals drain the space of Disse.

Figure 16.1 Gross Anatomy of the Hepatobiliary System

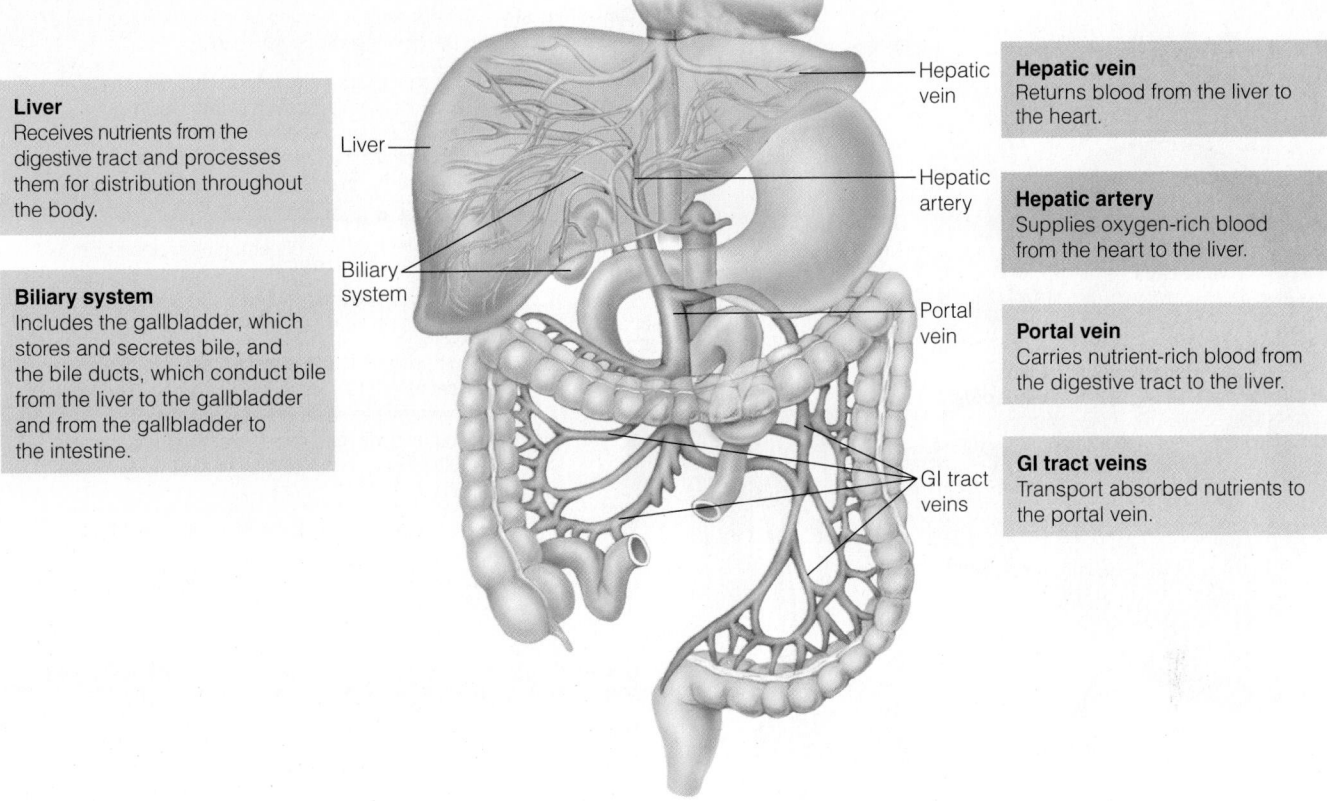

Liver
Receives nutrients from the digestive tract and processes them for distribution throughout the body.

Biliary system
Includes the gallbladder, which stores and secretes bile, and the bile ducts, which conduct bile from the liver to the gallbladder and from the gallbladder to the intestine.

Hepatic vein
Returns blood from the liver to the heart.

Hepatic artery
Supplies oxygen-rich blood from the heart to the liver.

Portal vein
Carries nutrient-rich blood from the digestive tract to the liver.

GI tract veins
Transport absorbed nutrients to the portal vein.

Liver — Hepatic vein — Hepatic artery — Portal vein — GI tract veins — Biliary system

Source: S. Rolfes, K. Pinna, and E. Whitney, *Understanding Normal and Clinical Nutrition*, 8e, copyright © 2009, p. 788.

Figure 16.2 Anatomy of the Lobule

(a) Hepatic lobule; (b) wedge of a hepatic lobule.

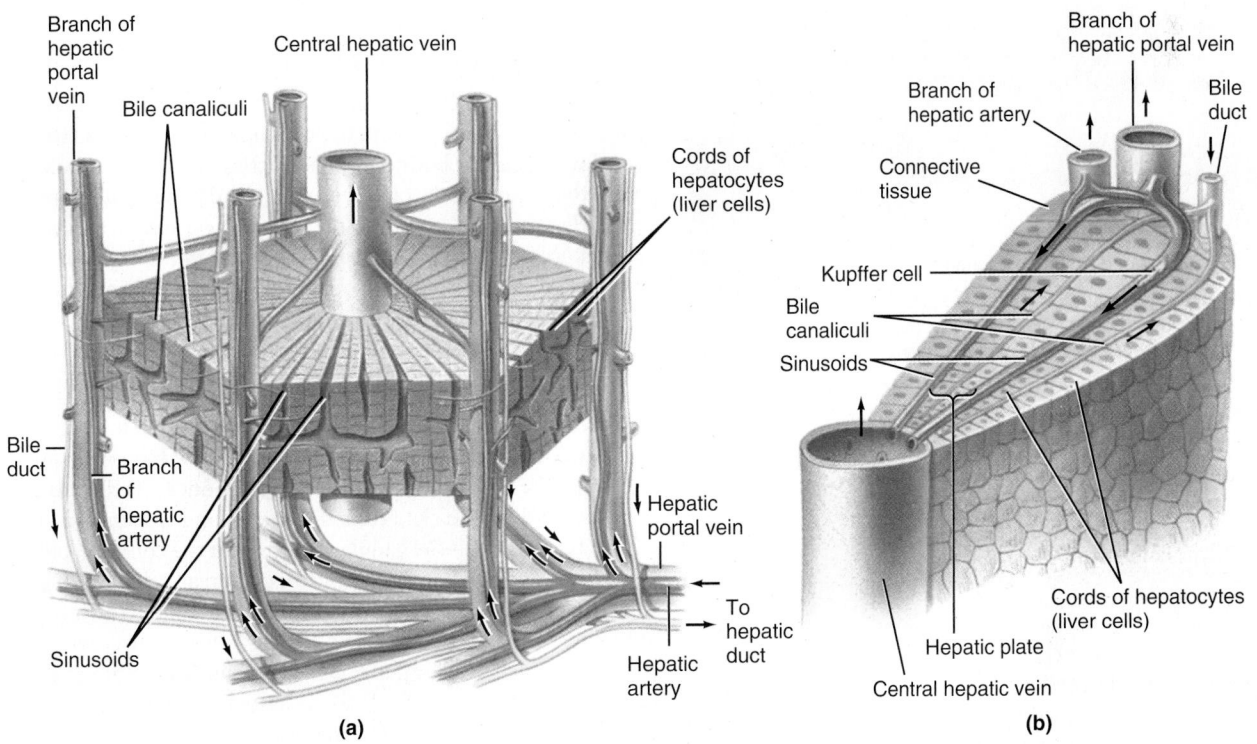

(a)

Branch of hepatic portal vein — Bile canaliculi — Central hepatic vein — Cords of hepatocytes (liver cells) — Bile duct — Branch of hepatic artery — Sinusoids — Hepatic portal vein — To hepatic duct — Hepatic artery

(b)

Branch of hepatic artery — Connective tissue — Kupffer cell — Bile canaliculi — Sinusoids — Branch of hepatic portal vein — Bile duct — Cords of hepatocytes (liver cells) — Hepatic plate — Central hepatic vein

Source: L. Sherwood, *Human Physiology: From Cells to Systems,* 5e, copyright © 2004, page 619.

Figure 16.3 Circulation of Blood to and from the Liver

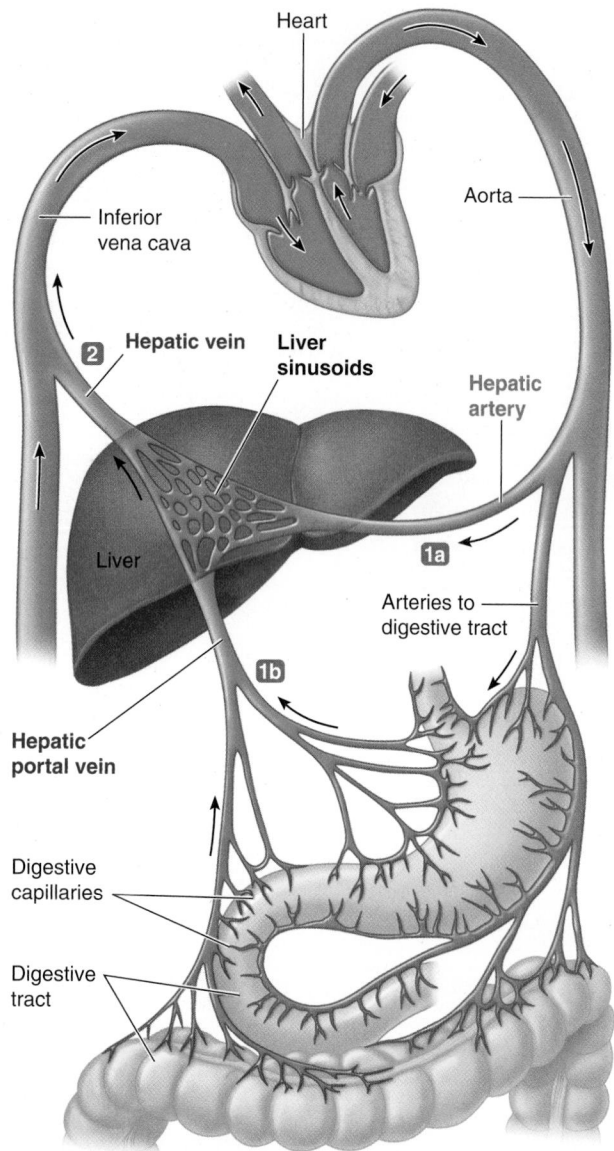

The liver receives blood from two sources:

1a Arterial blood, which provides the liver's O_2 supply and carries blood-borne metabolites for hepatic processing, is delivered by the **hepatic artery**.

1b Venous blood draining the digestive tract is carried by the **hepatic portal vein** to the liver for processing and storage of newly absorbed nutrients.

2 Blood leaves the liver via the **hepatic vein**.

Source: Figure 16-14, p. 616 from: Sherwood L., Human Physiology From Cells to Systems. 7e. ISBN 0495391840. Belmont CA: Brooks-Cole/Cengage, 2010.

The liver is one of the primary blood reservoirs in the body, and can store 200–400 mL of blood.[4] If an individual hemorrhages, and large amounts of blood are lost from the circulatory system, much of the blood normally in the liver sinusoids drains into the general circulation to help replace blood loss.

Table 16.1 Summary of Liver Functions

Metabolic Functions	
Carbohydrate Metabolism	Glycogenesis, gluconeogenesis, oxidation via TCA cycle, glycogenolysis, glycolysis
Lipid Metabolism	Lipogenesis, lipolysis, saturation/desaturation, ketogenesis, esterification of fatty acids, fatty acid oxidation, uptake/formation/breakdown of phosphotides, synthesis/degradation/esterification/excretion of cholesterol, formation of lipoproteins
Protein Metabolism	Synthesis of serum proteins, synthesis of prothrombin, globin of hemoglobin, apoferritin, nucleoproteins and serum mucoprotein, degradation of some proteins to peptides and amino acids, synthesis of urea
Enzyme Metabolism	Synthesis of alkaline phosphatase, mono-amine oxidases (MAOs), acetylcholine esterase, oxidases, cholesterol esterase, dehydrogenases, beta glucuronidase, glutamic oxalacetic transaminase (SGOT-AST), and glutamic pyruvic transaminase (SPGT-ALT)
Vitamin Metabolism	Formation of acetyl CoA from pantothenic acid, hydroxylation of vitamin D to 25-OH D_3, formation of 5-methyl tetrahydrofolic acid (THFA), methylation of niacinamide, phosphorylation of pyridoxine, dephosphorylation of thiamin, formation of coenzyme B_{12}
Bile Acid Metabolism	Transformation of cholesterol to 7-hydroxycholesterol to cholic acid and chenodeoxycholic acid
Heme Metabolism	Heme is oxidized to biliverdin, which is then reduced to bilirubin; bilirubin is transported to the liver where it is converted to bilirubin diglucuronide to be excreted with the bile pigments
Storage	Storage of glycogen, fats, fatty acids, and fat-soluble vitamins
Other Functions	Conjugation, detoxification and degradation, reticuloendothelial system (RES) activity, water movement regulation, fetal hematopoiesis, excretion

Functions of the Liver

The liver has over 500 known functions (see Table 16.1). Most of the liver's activities depend upon its unique structure as well as its location in the body. The construction of the liver lobule meets the requirements of a "secreting gland." The lobule contains cells that form secretions (i.e., bile), blood vessels to provide raw materials, and a system of ducts to carry the secretions away. The versatile liver cells also play an important role in monitoring all the nutrient-rich blood delivered by the portal vein.

Bile Secretion Bile is a complex aqueous solution secreted by the liver (see Box 16.1—Bile and Its Composition). Ultimately, all bile drains into one large duct from each lobe of the liver. Two main trunks, one from the right lobe and one from the left, unite to form the common hepatic duct. The hepatic duct descends to the right for a few inches and is then joined by the cystic duct from the gallbladder to form the common bile duct. The common bile duct joins the pancreatic duct, forming a single tube called the ampulla of Vater. There is a strong sphincter of Boyden in the bile duct just before it fuses with the pancreatic duct. The ampulla opens into the duodenum at the duodenal papilla. The muscle tissue associated with the ampulla forms a weak sphincter called the sphincter of Oddi.

Bile is secreted continually by the liver cells and enters the canaliculi to be drained into the bile ducts and finally taken to

BOX 16.1 **CLINICAL APPLICATIONS**

Bile and Its Composition

Bile (or gall) is a bitter yellowish, blue and green fluid secreted by the liver and stored in the gallbladder. It is composed of cholesterol, bile pigments, bile acids (glycocholic and taurocholic acid), phospholipids (mainly lecithin), bicarbonate, and waste products that are metabolized by the liver. Bilirubin, the yellow breakdown product of heme, is excreted in bile. Bilirubin is delivered to the liver in an unconjugated form that is not soluble in water. In the liver it is conjugated to glucuronic acid, making it soluble, and then excreted with bile. Interestingly, the Greek term for yellow bile is the origin of the prefix "choler" and is the source of the medical words "choleric" and "cholesterol."

the gallbladder. The components of bile must remain in their normal ratio to prevent cholesterol from precipitating and forming gallstones. The liver cells synthesize and secrete 600–1000 mL of bile/day, although the maximum volume of the gallbladder is only 30–60 mL. Nevertheless, as much as 12 hours of bile secretion (usually about 450 mL) can be stored in the gallbladder because water, sodium, and most other electrolytes are continually absorbed by the gallbladder mucosa, concentrating the remaining bile constituents, including the bile salts, cholesterol, lecithin, and bilirubin. Bile is normally concentrated 5-fold, but can be concentrated up to a maximum of 20-fold.[5]

Bile is an emulsifying agent, decreasing the surface tension and allowing intestinal agitation to break up fat globules. Bile salts help in the absorption of fatty acids, monoglycerides, cholesterol, and other lipids by forming micelles that are soluble in the chyme. Without bile salts, a large percentage of ingested fats would be lost in the feces.

Enterohepatic Circulation of Bile

About 95% of the bile salts are reabsorbed into the blood from the small intestine, about half by diffusion through the mucosa in the upper small intestine and half by active transport though the intestinal mucosa in the distal ileum. This recirculation of bile salts is called the enterohepatic circulation (see Figure 16.4). Once absorbed, the bile salts enter the portal blood and are returned to the liver, where the venous sinusoids absorb them almost entirely into the hepatic cells and then resecrete them into the bile. About 95% of all the bile salts are recirculated into the bile; this happens, on average, at least two to three times per meal, and 16 times before the salts are excreted in the feces.[6] The small quantities of bile salts lost into the feces are replaced by new amounts formed continually by the liver cells. Should bile be lost, the total circulation quantity of bile salts is depressed, as is the output by the liver. After

several days to several weeks, the liver will increase production of bile salts as much as 6- to 10-fold, increasing the rate of bile secretion to nearly normal levels.[7] This demonstrates that the daily rate of bile salt secretion is actively controlled in the enterohepatic circulation.

Anatomy and Physiology of the Gallbladder

The gallbladder is on the underside of the liver and the right side of the abdomen (see Figure 16.1). The **biliary tract** (or **biliary tree**) is the anatomical path that bile takes after it is secreted by the liver and concentrated in the gallbladder, and before it arrives in the duodenum. Bile leaves the liver via the common hepatic duct, which merges with the cystic duct from the gallbladder to form the common bile duct. The common bile duct merges with the pancreatic duct, forming the ampulla of Vater, which enters into the duodenum. Bile is secreted into the intestine in response to the presence of food (particularly fat). The functions of the gallbladder are as follows:

- Removal of water and some inorganic electrolytes from the bile, thus increasing the concentration of larger organic solutes.

Figure 16.4 Enterohepatic Circulation

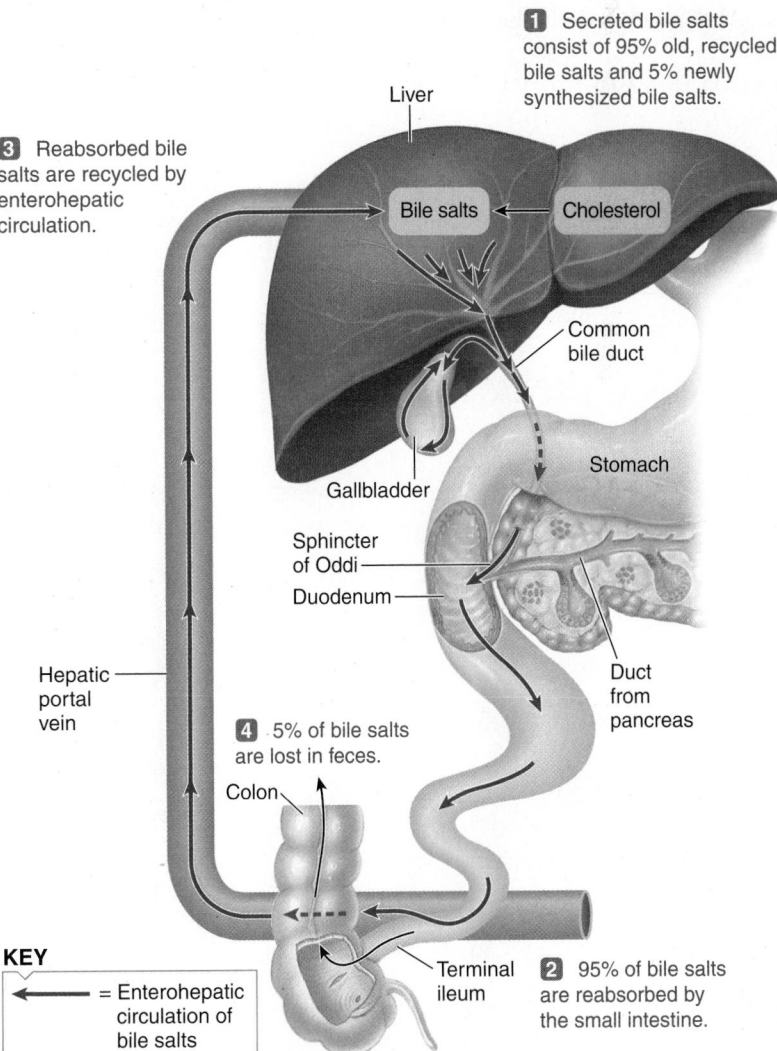

1 Secreted bile salts consist of 95% old, recycled bile salts and 5% newly synthesized bile salts.

3 Reabsorbed bile salts are recycled by enterohepatic circulation.

Liver

Bile salts

Cholesterol

Common bile duct

Stomach

Gallbladder

Sphincter of Oddi

Duodenum

Hepatic portal vein

4 5% of bile salts are lost in feces.

Colon

Duct from pancreas

Terminal ileum

2 95% of bile salts are reabsorbed by the small intestine.

KEY

⟵ = Enterohepatic circulation of bile salts

Source: Figure 16-16, p. 618 from: Sherwood L., Human Physiology From Cells to Systems. 7e. ISBN 0495391840. Belmont CA: Brooks-Cole/Cengage, 2010.

- Storage of bile.
- Control of the delivery of bile salts to the duodenum.

For the gallbladder to empty, it must contract with enough force to expel the bile and move it along the common bile duct, and the sphincter of Oddi must relax and allow bile to flow into the duodenum. Following the eating of a meal, the enzyme cholecystokinin (CCK) is released from the mucosa of the small intestine, stimulated by the hydrolytic products of digestion. CCK stimulates the gallbladder to contract and deliver bile into the proximal small intestine. Bile salts are essential for the emulsification of fat that must occur before lipids can be digested and absorbed.

Anatomy and Physiology of the Pancreas

The pancreas (see Figure 16.5) is a large accessory digestive gland located behind the stomach and opposite both the duodenum and the spleen. The pancreas is divided into different portions: the head, which is the wider portion of the pancreas, followed by the neck, the body, and the tail.

The pancreas has two major functions:

- Exocrine function: Produces enzymes necessary for digestion (see Table 16.2).
- Endocrine function: Produces hormones to regulate the use of body fuels, mainly glucose (see Chapter 17).

The pancreas is the only organ of the body that has both exocrine and endocrine functions. The pancreas is composed of two major types of tissues: the acini, ducted exocrine tissues that are responsible for secreting digestive juices into the duodenum; and the islets of Langerhans, ductless endocrine tissues (meaning they have no means of secreting externally) that secrete the hormones insulin and glucagon directly into the blood. The pancreas has almost a million islets of Langerhans, organized around small capillaries (making it very well vascularized) into which its cells secrete their hormones. There are three major types of endocrine cells: alpha—which compose 25% of the cells and secrete glucagons; beta—which make up 60% of the cells and are responsible for the synthesis and secretion of insulin; and delta—which make up the remaining 10% of cells and secrete somatostatin. Exocrine secretions are carried to the digestive tract via the pancreatic duct, which then joins with the common bile duct and enters the duodenum.

Pertinent Laboratory Values

Liver Laboratory tests, often referred to as liver function tests (LFTs), are useful in the evaluation and management of patients with hepatic dysfunction. Table 16.3 lists the laboratory methods commonly used in the clinical management of hepatic disease. In general they can be grouped based on three types of liver functions: (1) detoxification and excretory; (2) biosynthesis; and

Figure 16.5 Anatomy of the Pancreas

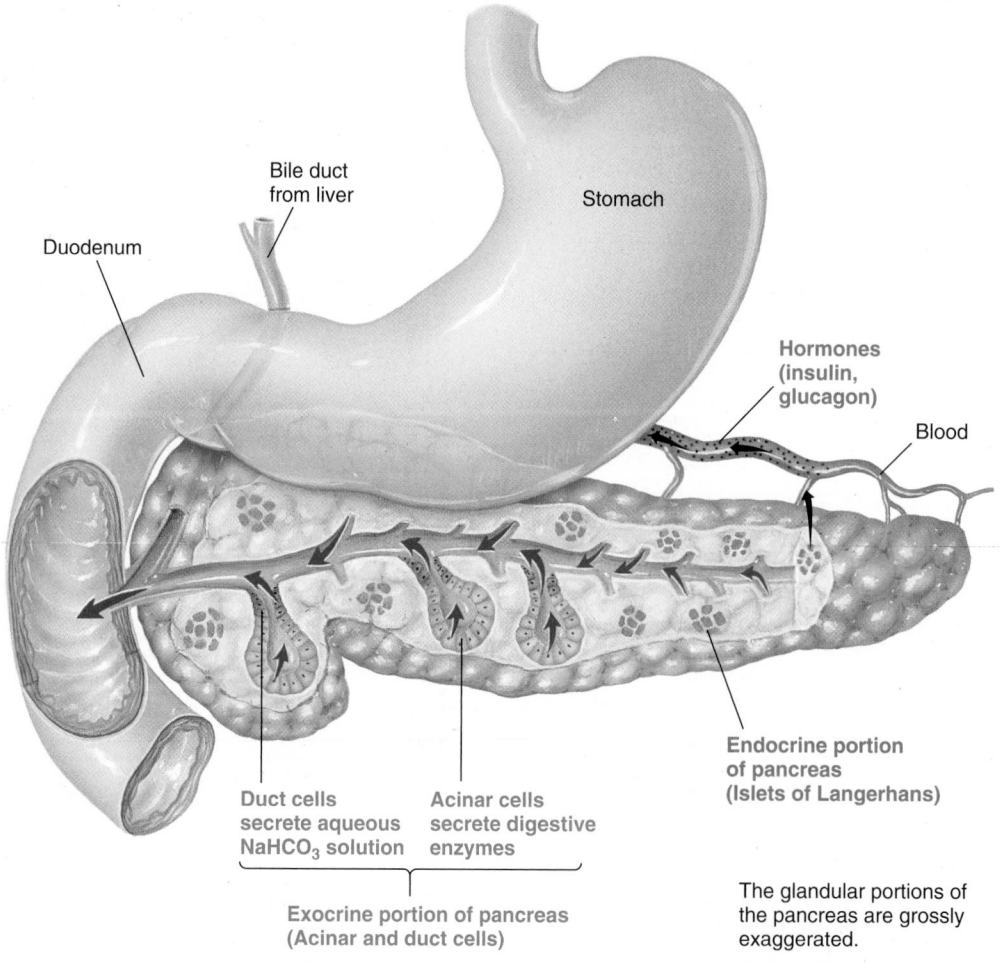

Bile duct from liver

Stomach

Duodenum

Hormones (insulin, glucagon)

Blood

Duct cells secrete aqueous NaHCO₃ solution

Acinar cells secrete digestive enzymes

Endocrine portion of pancreas (Islets of Langerhans)

Exocrine portion of pancreas (Acinar and duct cells)

The glandular portions of the pancreas are grossly exaggerated.

Source: L. Sherwood, *Human Physiology: From Cells to Systems, 5e,* copyright © 2004, p. 616.

Table 16.2 Pancreatic Exocrine Secretions

Enzyme	Substrate	Action & Product of Action
Trypsin	Protein and polypeptides	Hydrolysis of interior peptide bonds to form polypeptides
Chymotrypsin	Protein and polypeptides	Hydrolysis of interior peptide bonds to form polypeptides
Carboxypolypeptidase	Polypeptides	Hydrolysis of terminal peptide bonds to form amino acids
Ribonuclease Deoxyribonuclease	Ribonucleic acids Deoxyribonucleic acids	Hydrolysis to form mononucleotides
Elastrase	Fibrous protein	Hydrolysis to form peptides and amino acids
Lipase	Fat	Hydrolysis to form simple monoglycerides, fatty acids, and glycerol
Cholesterol Esterase	Cholesterol	Hydrolysis to form cholesterol and fatty acids
α-Amylase	Starch and dextrin	Hydrolysis to form dextrins and maltose

Table 16.3 Laboratory Methods Used in Clinical Management of Hepatic Disease

Test	Normal Value	Clinical Implications
Blood		
Ammonia	19–60 μg/dL	Increased in cirrhosis, liver failure, and with portacaval shunting of the blood
Cholesterol Ester	60%–75% of cholesterol	Elevated in biliary obstruction; decreased in parenchymal liver disease
Dye Clearances		
BSP (sulfobromophthalein Na)	4% retention	Normal clearance depends on hepatic blood flow, functioning liver cell mass, & lack of obstruction; retention associated with hepatic damage
Indocyanine Green	0% retention	Similar sensitivity in detecting liver dysfunction compared with BSP
Protein Studies		
Albumin	3.5–4.5 g/dL	Decreased value associated with hepatic disease; generally parallels the functional status of parenchymal cells, but normal with considerable cellular damage
Globulin	1.5–3.8 g/dL	Often increased as it reflects inflammation
Total Protein	6.5–8.3 g/dL	Decreased value associated with hepatic disease
Prothrombin Time	9–11 seconds, 100% return	Prolonged with hepatic disease
Urine		
Bilirubin	0	Presence in the urine is indicative of biliary obstruction or RBC hemolysis
Urobilinogen	0–4 mg/24 hrs	Same as bilirubin
Stool		
Color	Brown	Alterations in color occur due to a decrease or absence of urobilinogen.
Urobilinogen	75–400 mg/24 hrs	Stools are clay colored with biliary obstruction and light brown with hepatocellular damage
Enzymes		
Alkaline Phosphatase	30–95 U/L	Increased activity occurs in hepatic disease & malignancy & in chronic obstruction of the biliary duct, but is non-specific; also increased in bone diseases, bone trauma, & bone growth
GGT	30 U/L or less	Elevation highly indicative of hepatocellular injury secondary to ethanol abuse
AST or SGOT	30–40 U/L or less	Less specific enzyme to detect hepatic disease secondary to cellular necrosis; also elevated with severe cardiac and muscle damage
ALT or SGPT	7–40 U/L or less	Most sensitive test to detect hepatocellular injury secondary to exacerbation of infectious hepatitis; high levels of 300 observed in acute hepatocellular damage
SGOT/SGPT Ratio	1	Ratio is useful in differential diagnosis; SGOT/SGPT >2—most pts have alcoholic liver disease; increased ratio is due primarily to the low activity of SGPT in the liver of alcoholic pts
LDH	280 IU/L or less	Iso-enzyme used to differentiate hepatitis from mononucleosis
Pigment Studies		
Serum Bilirubin (total)	0.2–0.9 mg	Reflects the ability of the liver to conjugate and excrete bilirubin; increased in liver and biliary disease causing jaundice clinically
Direct	0.1–0.3 mg	Indicates biliary tree obstruction
Indirect (unconjugated)	0.1–0.5 mg	Indicates RBC hemolysis or liver damage

Sources: Pratt DS, Kaplan MM. *Evaluation of the liver: laboratory tests.* In: Schiff, ER, Sorrell, MF, Maddrey WC, editors. *Schiff's diseases of the liver.* 9th ed. Baltimore: Lippincott Williams and Wilkins; 2003. Friedman LS, Martin P, Muñoz SJ. *Laboratory Evaluation of the Patient with Liver Disease.* In: Zakim D, Boyer TD, editors. *Hepatology: a textbook of liver disease.* 4th ed. Philadelphia: Saunders; 2003. Bishop ML, Fody EP, Schoeff L, editors. *Clinical chemistry: principles, procedures, correlations.* Baltimore: Lippincott Williams & Wilkins; 2005.

(3) coagulation. The most commonly used tests are bilirubin, aminotransferases, alkaline phosphatase, albumin, and pro-thrombin time. The liver carries out thousands of biochemical functions, and laboratory tests measure only a limited number of these functions. Thus, no one test accurately assesses the liver's total functional capacity.

Pancreas Autodigestion of the pancreas is prevented by storing the digestive enzymes in their inactive forms, as zymogens. In addition, low calcium concentrations within the pancreas decrease trypsin activity. Loss of any protective mechanisms leads to zymogen activation and the presence of enzymes in the blood stream. The laboratory test that is most commonly used measures two of these enzymes, lactase and amylase.[8]

Pathophysiology Common to the HepatobiliaryTract

Jaundice

Jaundice, also known as icterus, causes a yellowish tint to the body tissues and is usually the result of elevated bilirubin concentration in the extracellular fluids, either unconjugated (also called indirect) or conjugated (direct) bilirubin (see Box 16.1). The normal plasma concentration of bilirubin, including both the unconjugated and conjugated forms, averages less than 1.1 mg per dL of plasma.[9] The skin begins to appear jaundiced when serum bilirubin exceeds 2.4 to 3.0 mg/dL.[10]

Jaundice is a symptom rather than its own condition, and generally results from one of three causes:

- Increased destruction of red blood cells with rapid release of bilirubin into the blood—hemolytic jaundice (pre-hepatic);
- Decreased uptake of bilirubin and/or decreased liver function (hepatic); or
- Obstruction of the bile ducts, which prevents excretion of bilirubin into the gastrointestinal tract—obstructive jaundice (post-hepatic).[11]

Portal Hypertension/Ascites

Portal hypertension is elevated blood pressure in the portal vein (the vein in the abdominal cavity that drains blood primarily from the gastrointestinal tract and spleen). It is usually defined as a portal pressure gradient (the difference in pressure between the portal vein and the hepatic veins) of 5 mm Hg or greater.[12] The primary symptoms and complications of portal hypertension include **ascites** (accumulation of fluid within the peritoneal cavity; see Figure 16.6), along with gastrointestinal bleeding from varices (extremely dilated veins), and encephalopathy (reduced mental capacity and consciousness).[13]

Other factors that contribute to fluid retention with ascites are reduced osmotic pressure of plasma due to failure of the liver to synthesize adequate amounts of serum proteins, particularly albumin, and increased retention of sodium. The excessive renal absorption of sodium is a result of increased aldosterone production and decreased inactivation of aldosterone. A reduction in renal blood flow and the subsequent decrease in renal filtration rate contribute to the sodium retention.

Figure 16.6 Ascites

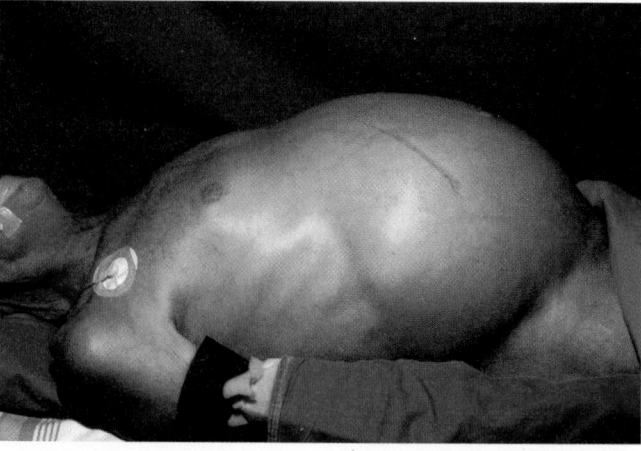

Courtesy of © M. English, MD/Custom Medical Stock Photo.

The medical treatment for many cases of ascites is sodium restriction and/or diuretics. The fluid in the abdominal cavity can amount to as much as 15.5 liters and contains protein concentrations of 1–2 g/100 mL.[14] This fluid can be removed by **paracentesis** (withdrawing fluid from the abdomen via a catheter), but this results in protein loss as well. For instance, if 4 L of ascitic fluid are removed from a patient, he or she loses 40–80 g of protein via this route. The transjugular intrahepatic portosystemic shunt (TIPS) is a surgical procedure (see Figure 16.7) for ascites that reroutes blood flow to the liver and reduces pressure in all auxiliary veins, not only in the stomach and esophagus, but also in the bowel and the liver.[15]

Encephalopathy

Hepatic encephalopathy is a syndrome of impaired mental status and abnormal neuromuscular function that results from major failure liver. Important contributing factors are the degree of liver failure, the diversion of the portal blood through the venous systemic circulation, bleeding from varices, and exogenous factors such as sepsis.[16]

The symptoms and signs of hepatic encephalopathy vary, and include: (1) changes in mental status and personality, and (2) neuromuscular changes. These neurological changes are generally "graded." The most common grading system uses the four-staged West Haven scale, but recently, the CHESS scale has also been used (Table 16.4). If the patient is in a coma, the complete assessment includes the Glasgow Coma Scale score (see Chapter 20). An example of a neuromuscular change that is included in the grading of hepatic encephalopathy is **asterixis**, which is often called "flaps." Asterixis presents as small, brief, intermittent movements of individual fingers, either in flexion or laterally in an ulnar direction, with a rapid return of the fingers to the original position. To exhibit this, non-comatose patients are asked to raise both arms horizontally (palms downward), to bend back (dorsiflex) the wrists and spread the fingers wide apart, and to hold this posture for about 15 seconds. With more severe asterixis, the flap spreads proximally, and movements involve the wrist and even the shoulders, and in extreme cases, the head.[17] Patients may also have trouble writing, drawing simple geometric figures, or completing the number connection test in which a patient is asked to follow the numbers (1 to 25) randomly distributed on a sheet of paper.

Figure 16.7 Diagrams of Portal Hypertension Blood Flow and TIPS Procedure

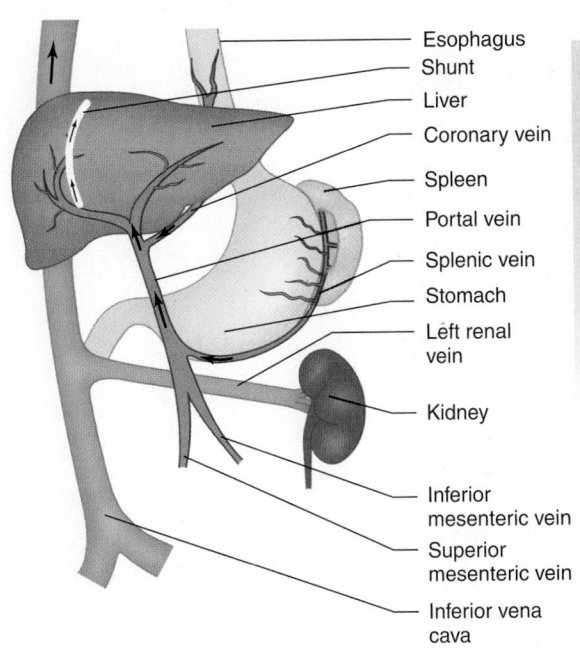

Esophagus
Liver
Coronary vein
Spleen
Portal vein
Splenic vein
Stomach
Left renal vein
Kidney
Inferior mesenteric vein
Superior mesenteric vein
Inferior vena cava

Portal Hypertension before the TIPS procedure is performed.

Portal hypertension causes blood flow to be forced backward, causing veins to enlarge and varices to develop across the esophagus and stomach from the pressure in the portal vein. The backup of pressure also causes the spleen to become enlarged.

Esophagus
Shunt
Liver
Coronary vein
Spleen
Portal vein
Splenic vein
Stomach
Left renal vein
Kidney
Inferior mesenteric vein
Superior mesenteric vein
Inferior vena cava

After the TIPS procedure is performed.

A radiologist makes a tunnel through the liver with a needle, connecting the portal vein to one of the hepatic veins. A metal stent is placed in this tunnel to keep the track open.

The shunt allows the blood to flow normally through the liver to the hepatic vein. This reduces portal hypertension, and allows the veins to shrink to normal size, helping to stop variceal bleeding.

The pathogenesis of hepatic encephalopathy is unknown, though research points to the inability of the liver to metabolize products that are toxic to the brain. There are three major hypotheses that try to explain the impairment of the neurotransmission: (1) the ammonia hypothesis, (2) the synergistic neurotoxin hypothesis, and (3) the false neurotransmitter.

Ammonia is thought to be a direct toxin to the brain. It is generated from the catabolism of proteins, amino acids, purines, and pyrimidines. The normal liver disposes of nitrogen from amino acids by transamination and the formation of glutamic acid. The nitrogen is then released as ammonia and enters the urea cycle, resulting in the eventual formation of urea. Ammonia synthesized in the gut is absorbed and transported in the intestinal venous blood to the liver, where it is also metabolized to urea. Figure 16.8 illustrates this cycle.

Liver disease interferes with this detoxification process and shifts ammonia metabolism to skeletal muscle, where it is used in the conversion of glutamate to glutamine. Muscle wasting reduces the capacity of patients with liver failure to detoxify ammonia. Ammonia therefore accumulates in arterial blood, which leads to elevated levels in the brain. Because the brain lacks significant activity of urea cycle enzymes and cannot synthesize urea, the ammonia is detoxified via the conversion of glutamate to glutamine. According to the ammonia

Table 16.4 Staging Scales for Hepatic Encephalopathy

West Haven Scale

Stage	West Haven Criteria	Adapted-West Haven Criteria
0	• No abnormality detected	Alert and attentive (oriented in time and space) without signs of encephalopathy (neither dysarthria, ataxia, flapping tremor or obvious decrease in the speed of mental processing)
1	• Trivial lack of awareness • Euphoria or anxiety • Shortened attention span • Impairment in performance of addition	Alert and attentive, but with at least one of the following signs: dysarthria, ataxia, flapping tremor, or obvious decrease in the speed of mental processing.
2	• Lethargy or apathy • Minimal disorientation for time or place • Subtle personality change • Inappropriate behavior • Impaired performance of subtraction	Awake but inattentive: disoriented, somnolent, easy to distract, unable to perform easy mental tests (addition, subtraction, remember a list of numbers). Patient's speech is easy to understand.
3	• Somnolence to semi-stupor but responsive to verbal stimuli • Confusion • Gross disorientation	Marked somnolence or psychomotor agitation. Speech is difficult to understand.
4	• Coma (unresponsive to verbal or noxious stimuli)	Coma—The patient does not speak and does not follow simple commands (such as raising an arm or opening the mouth).

Clinical Hepatic Encephalopathy Staging Scale (CHESS)

The CHESS is a linear scale that scores hepatic encephalopathy from 0 (normal mental state) to 9 (deep coma) according to presence or absence of nine items.

1. Does the patient know which month he/she is in (e.g., January, February)?	0. Yes. 1. No, or he/she does not talk.
2. Does the patient know which day of the week he/she is in (e.g., Thursday, Friday, Sunday…)?	0. Yes. 1. No, or he/she does not talk.
3. Can he/she count backwards from 10 to 1 without making mistakes or stopping?	0. Yes. 1. No, or he/she does not talk.
4. If asked to do so, does he/she raise his/her arms?	0. Yes. 1. No.
5. Does he/she understand what you are saying to him/her? (based on the answers to questions 1 to 4)	0. Yes. 1. No, or he/she does not talk.
6. Is the patient awake and alert?	0 Yes. 1. No, he/she is sleepy or fast asleep.
7. Is the patient fast asleep, and is it difficult to wake him/her up?	1. Yes. 0. No.
8. Can he/she talk?	0. Yes. 1. He/she does not talk.
9. Can he/she talk correctly? In other words, can you understand everything he/she says, and he/she doesn't stammer?	0. Yes. 1. No, he/she does not talk or does not talk correctly.

Source: Reprinted from: The West-Haven adapted criteria, the list of manifestations of hepatic encephalopathy and the Clinical Hepatic Encephalopathy Staging Scale (CHESS). M. Ortiz, J. Córdoba, E. Doval, C. Jacas, F. Pujadas, R. Esteba and J. Guardia Alimentary Pharmacology & Therapeutics. 2007;26(6):Appendix S1.

hypothesis, the depletion of glutamate and accumulation of glutamine contribute to hepatic encephalopathy.

The following evidence indicates that altered ammonia metabolism may be responsible for the symptoms of hepatic encephalopathy:

• There is a positive correlation between blood ammonia levels and the abnormalities in mental state, though 10% of patients with hepatic encephalopathy have normal blood ammonia levels.

• Hepatic encephalopathy can be induced in animals by giving them toxic doses of ammonium salts.

• Hepatic encephalopathy can be precipitated in the cirrhotic patient by feeding a high-protein diet or by increased endogenous production of ammonia.

• Cirrhotic patients often have abnormal ammonium tolerance tests.

• The symptoms of hepatic coma are often corrected by decreasing endogenous ammonia production.[18]

Figure 16.8 Normal Utilization of Ammonia and Formation of Urea

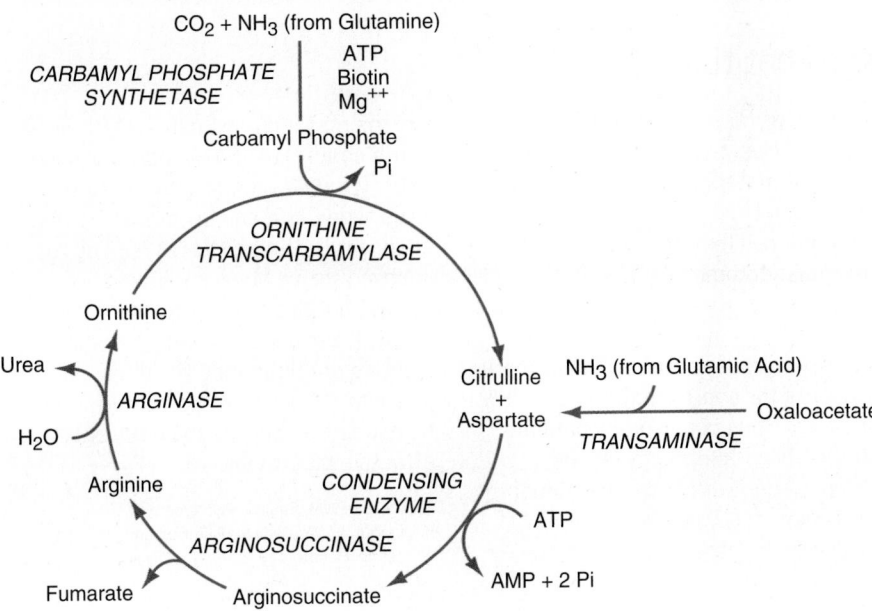

Source: Robert E. Hodges (ed.), *Nutrition-Metabolic and Clinical Applications,* New York: Plenum Press; 1979, p. 147.

Mercaptans, ammonia, tyramine, octopamine, beta-phenylethanolamines, manganese, and gamma-aminobutyric acid also accumulate in liver failure. The synergistic neurotoxin hypothesis proposes that these neurotoxins, many of which are produced by intestinal bacteria, are involved in hepatic encephalopathy. The neuronal gamma aminobutyric acid (GABA) receptor complex contains a binding site for neurosteroids. Today, some investigators contend that neurosteroids play a key role in hepatic encephalopathy. In experimental models, neurotoxins, like ammonia and manganese, increase neurosteroid production by indirectly stimulating the conversion of cholesterol to pregnenolone to neurosteroids. These neurosteroids are capable of binding to their receptor within the neuronal GABA receptor complex and can increase inhibitory neurotransmission.[19]

Hyperammonemia is thought to stimulate glucagon secretion and enhance gluconeogenesis from amino acids. The resulting hyperglycemia stimulates hyperinsulinemia, enhancing the uptake of **branched-chain amino acids (BCAA)** by muscle and lowering the plasma levels of BCAA relative to **aromatic amino acids (AAA)**. The two types of amino acids compete for the same carrier transport system across the blood-brain barrier. This leads to an accumulation of AAA in the brain. The subsequently raised tryptophan level leads to formation of the inhibitory neurotransmitter serotonin, and the raised phenylalanine level leads to the inhibition of dihydroxyphenylalanine (DOPA) production and formation of so called "false neurotransmitters." The false neurotransmitter hypothesis suggests that these false transmitters may displace catecholamines from their receptors.

Aromatic amino acids, such as tryptophan, tyrosine, and phenylalanine, are elevated in the brains of patients with **acute hepatic/liver failure** (discussed further in Box 16.2) and hepatic encephalopathy. A pronounced decrease in

BCAA to AAA ratio has been described in chronic liver disease but correlates poorly with the presence or absence of hepatic encephalopathy. In theory, administration of BCAA metabolized by skeletal muscle rather than the liver would restore this imbalance between BCAA and AAA, and would decrease production of the toxic false neurotransmitters. This hypothesis has provided the rationale for the use of BCAA in the treatment of hepatic encephalopathy.[20]

Medical treatment for hepatoencephalopy is primarily directed at deceasing blood ammonia levels. Medications include oral administration of lactulose, a nonabsorbable disaccharide, and antibiotics (e.g., Neomycin). Lactulose stimulates the passage of ammonia from body tissues into the gut lumen and inhibits intestinal ammonia production. Antibiotics decrease the colonic concentration of bacteria that can produce ammonia.[21]

Pathophysiology of the Liver

Hepatitis

Hepatitis is an inflammation of the liver caused by a virus, bacteria, toxins, obstruction, parasites, or chemicals (chloroform, carbon tetrachloride). Viral hepatitis is caused by five identified viruses (types A, B, C, D, E), with at least 10 other viruses under study. The clinical manifestations are similar for all types, and common symptoms include jaundice, dark urine, anorexia, fatigue, headache, nausea, vomiting, and fever. The liver becomes enlarged (hepatomegaly) and, in some cases, the spleen enlarges (splenomegaly). Bilirubin, alkaline phosphatase, and serum AST are generally elevated.[22]

Hepatitis A Virus (HAV) In 2007 the rate of hepatitis A, formerly called infectious hepatitis, in the United States was the lowest it had been in 40 years. Hepatitis A vaccination is the mostly likely reason for the decline. NHANES III found that slightly more than a third of the U.S. population had serological evidence of having had HAV infection; however, worldwide, about half of reported viral hepatitis cases are still HAV.[23]

HAV is generally transmitted via the oral-fecal route. Sources of contamination include drinking water, food (many times seafood), and sewage. Symptoms usually last less than two months, although some people can be ill for as long as six months, and symptoms are more likely to occur in adults than children. There are no special treatments for hepatitis A. Rest, adequate nutrition, and fluids are usually recommended, although a few people will need to be hospitalized.[24] Rarely, hepatitis A may result in acute liver failure (previously referred to as fulminant hepatitis). See Box 16. 2 for more details regarding this syndrome.

Serum Hepatitis, or Hepatitis B (HBV) HBV is transmitted through transfusions of blood or blood-derived fluids, or through improperly sterilized medical instruments, dental

BOX 16.2 CLINICAL APPLICATIONS

Acute Liver Failure (Fulminant Hepatitis)

Acute liver failure (ALF), or fulminant hepatitis, is a rare but potentially fatal syndrome marked by the sudden loss of hepatic function in a person with no prior history of liver disease. Currently in the United States, the mortality rate is 30% in adults without transplantations, while the survival rate with liver transplantation is 25%. Higher rates of spontaneous recovery (56%) and transplantation (31%) with lower rates of death (13%) occur in children. ALF in adults is characterized by the development of hepatic necrosis, coagulopathy, and encephalopathy.

Acute liver failure is usually secondary to a virus, toxin, metabolic (inborn errors of metabolism), and/or immune-mediated attack. Viral hepatitis and drug-induced hepatotoxicity are the two most common causes of ALF. In the United States, acute viral hepatitis accounts for approximately 50% of cases, whereas acetaminophen toxicity accounts for approximately 20%–35% of cases. However, in many patients, no specific viral etiology can be found. Hepatitis A and B are thought to account for 7.3% and 3% of ALF, respectively. Other ALF-associated

viruses may include Epstein-Barr virus; cytomegalovirus (CMV); paramyxovirus; varicella-zoster virus; herpes virus types 1, 2, and 6; parvovirus; and adenovirus. The processes leading to such extensive hepatic damage are unknown, but are thought to be multifactorial. Treatment is directed towards treating or eliminating the underlying cause of the liver failure but liver transplant is thought to be the only effective treatment.

Source: Lee WM, Squires RH, Nyberg SL, Do E, Hoofnagle JH. Acute Liver Failure: Summary of a Workshop. Hepatology. 2008;47:1401–15.

---

drills, tattooing needles, or other skin-puncturing instruments that have come in contact with contaminated blood. It can also be transmitted by other than the parenteral route, which makes its specific cause more obscure. Some people may be carriers of the hepatitis B antigen (HBsAG) while remaining asymptomatic. The prevalence of chronic HBV in the United States is low (5%) and about 30% of these individuals have no signs or symptoms. Hepatitis B vaccinations are now available and usually required for attending public educational institutions.[25]

Hepatitis B can be either an acute or chronic infection. Acute hepatitis B virus infection is a short-term illness, much like HAV infection, and occurs within the first 6 months after someone is exposed to the virus. About 15%–25% of the 1.4 million people with chronic HBV will develop serious liver conditions, such as **cirrhosis** or liver cancer. The risk for chronic infection varies according to the age of exposure and is higher among young children. Approximately 90% of infants and

25%–50% of children aged 1–5 years will remain chronically infected with HBV.[26] In contrast, approximately 95% of adults recover completely from HBV infection and do not become chronically infected. Asian Americans and Native Hawaiian and other Pacific Islanders have the highest rates of chronic hepatitis B among all racial/ethnic groups in the United States and also a disproportionately high risk for liver cancer. Most Asian Americans are infected with HBV as infants or children. The HBV infection-related death rate among Asian Americans and Native Hawaiians and Other Pacific Islanders is seven times greater than the rate among whites.[27]

There are several different blood tests available to diagnose hepatitis B (see Box 16.3) and monitor treatment. For acute infection, no medication is available; treatment is supportive, much like that for HAV. For chronic infection, several antiviral drugs (adefovir dipivoxil, interferon alfa-2b, pegylated interferon alfa-2a, lamivudine, entecavir, and telbivudine) are available. (Table 16.5 provides

---

BOX 16.3 CLINICAL APPLICATIONS

HBV Antigen/Antibody Tests

- **Hepatitis B Surface Antigen (HBsAg)** is a protein on the surface of the hepatitis B virus. It can be detected in the blood during acute or chronic hepatitis B virus infection. The body normally produces antibodies to HBsAg as part of the normal immune response to infection.

- **Hepatitis B Surface Antibody (anti-HBs)** is an antibody that is produced by the body in response to the hepatitis B surface antigen.

- **Total Hepatitis B Core Antibody (anti-HBc)** is an antibody that is produced by the body in response to a part of the hepatitis B virus called the "core antigen." The meaning of this test often depends on the results of two other tests, anti-HBs and HBsAg.

- **IgM Antibody to Hepatitis B Core Antigen (IgM anti-HBc)** is used to detect an acute infection.

- **Hepatitis B "e" Antigen (HBeAg)** is a protein found in the blood when the hepatitis B virus is present during an active hepatitis B virus infection.

- **Hepatitis B e Antibody (HBeAb or anti-HBe)** is an antibody that is produced by the body in response to the hepatitis B "e" antigen.

- **Hepatitis B Viral DNA** refers to a test to detect the presence of hepatitis B virus DNA in a person's blood.

Table 16.5 Drug–Nutrient Interactions for Hepatobiliary Medications

Classification	Mechanism	Generic	Brand Names	Possible Food–Drug Interactions
K-sparing diuretic	Aldosterone receptor antagonist	Spironolactone	Aldactone, Aldactazide	Avoid excess K intake & K supplementation, avoid salt substitutes; ↓ kcal and Na may be recommended; avoid natural licorice; anorexia, ↑ thirst, N/V, diarrhea, gastritis
K-sparing diuretic	Aldosterone receptor antagonist	Amiloride	Midamor	N/V, dry mouth, diarrhea; avoid salt substitutes.
Antibiotic	Interferes with bacterial protein synthesis	Neomycin	Mycifradin, Neo-Fradin, Neo-Tab	Neomycin impairs absorption (and may also increase excretion) of a broad variety of nutrients including carbohydrates, fats, calcium, iron, magnesium, nitrogen, potassium, sodium, folic acid, and vitamins A, B_{12}, D, and K.
Laxative, anti-hyperammonemic		Lactulose, lactitol	Cephulac, Chronulac, Duphalac, Kristalose	High fiber with 1500–2000 mL fluid/d to prevent constipation, N/V, belching, cramps, borborygmi, diarrhea, flatulence
Benzodiazepine receptor antagonist	Benzodiazepine receptor antagonist	Flumazenil	Romazicon	N/V
Glucocorticoids Anti-inflammatory, immunosuppressant, hormone	Mimics the action of cortisol	Prednisone	Deltasone	Caution with DM—↑ glucose, highly protein bound, may need ↑ K, PO_4, Ca, and ↑ vitamins A, C, D, ↑ protein, and ↓ dietary Na; avoid alcohol.
Immunosuppressant	Attacks the white blood cells	Cyclosporine, tacrolimus	Neoral, Sandimmune	No K supplement or salt sub, caution with grapefruit; anorexia, N/V, diarrhea, ↑ glucose, esophagitis, steatorrhea
Immunosuppressant	Purine synthesis inhibitor	Azathioprine	Imuran; Azasan	Diarrhea, steatorrhea, negative nitrogen balance
Immunosuppressant	Inhibits the response to interleukin-2	rapamune	Sirolimus	↑ cholesterol, hypertriglyceridemia
Immunosuppressant	Purine synthesis inhibitor	mycophenolate mofetil	CellCept	Take on an empty stomach; anorexia, stomatitis, dyspepsia, N/V, abdominal pain, colitis, diarrhea, constipation
Antivirals	Inhibits reverse transcriptase	adefovir dipivoxil	Hepsera	Dyspepsia
Antivirals	Immunoregulatory cytokine	interferon alfa-2b	Intron A	Nausea, anorexia, stomach pain, diarrhea, taste alterations, dry mouth, hyperglycemia
		peg interferon alfa-2a	Pegasys	N/V, heartburn, anorexia, change in the way things taste, dry mouth, weight loss, diarrhea, colitis
Antivirals	Inhibits reverse transcription	lamivudine	Epivir	Diarrhea, fatigue, N/V, anorexia, abdominal pain, cramps, dyspepsia
Antivirals	DNA analogue (guanine) that inhibits reverse transcription	entecavir	Baraclude	Diarrhea, N/V, dyspepsia
Antivirals	Synthetic thymidine nucleoside analogue	telbivudine	Tyzeka	Diarrhea, fatigue, N/V, anorexia, abdominal pain, cramps, dyspepsia, pharyngolaryngeal pain
Antacid, antiflatulent			Mylanta	Take 1 hr after meals, take Fe or folate supplement separately by 2 hr, take separately from citrus fruit/juice or Ca citrate by 3 hr (juice ↑ Al abs); ↓ absorption folate, PO_4, Fe; diarrhea
Antacid		Aluminum hydroxide, magnesium hydroxide	Maalox	Take 1 hr after meals, take Fe or folate supplement separately by 2 hr, take separately from citrus fruit/juice or Ca citrate by 3 hr (juice ↑ Al abs); ↓ absorption folate, PO_4, Fe; diarrhea
Antiulcer, anti-GERD, antisecretory	H_2 receptor antagonist	Famotidine	Pepcide AC, Pepcid, Pepcid IV	Bland diet may be recommended; take >1 hour after Fe supplement, take Mg supplement or Al/Mg antacids separately by >2 hr; limit caffeine; may ↓ absorption of Fe and B_{12}.
Antiulcer, anti-GERD, antisecretory	H_2 receptor antagonists	Gaviscon, cimetidine	Tagamet HB, Zantac	Bland diet may be recommended; take at least 2 hr after Fe supplement, take Mg supplement or Al/Mg antacids separate by at least 2 hr; limit caffeine; ↓ Fe and vitamin B_{12} absorption; Mg or Al/Mg antacids ↓ drug abs; liquid cimetidine precipitates tube feeding.

(continued)

Table 16.5 Drug–Nutrient Interactions for Hepatobiliary Medications (*Continued*)

Classification	Mechanism	Generic	Brand Names	Possible Food–Drug Interactions
Antiulcer, anti-GERD, antisecretory	H_2 receptor antagonists	Nizatidine	Axid AR, Axid	Bland diet may be rec.; take at least 2 hr after Fe supplement, take Mg supplement or Al/Mg antacids sep by at least 2 hr; limit caffeine; ↓ Fe and vitamin B_{12} abs; Mg or Al/Mg antacids ↓ drug abs; liquid cimetidine precipitates tube feeding.
Antiulcer, anti-GERD, antisecretory	H_2 receptor antagonists	Ranitidine	Zantac 75	Bland diet may be rec.; take at least 2 hr after Fe supplement, take Mg supplement or Al/Mg antacids separate by at least 2 hr; limit caffeine; ↓ Fe and vitamin B_{12} abs; Mg or Al/Mg antacids ↓ drug absorption; liquid cimetidine precipitates tube feeding.
Antiulcer, anti-GERD, antisecretory	Proton pump inhibitors; block production of acid by the stomach	Esomeprazole	Nexium	May ↓ absorption of Fe and B_{12}.
Antiulcer, anti-GERD, antisecretory	Proton pump inhibitor	Lansoprazole	Prevacid	Take 30–60 min before meals; may ↓ Fe and vitamin B_{12} absorption; diarrhea
Antiulcer, anti-GERD, antisecretory	Proton pump inhibitor	Omeprazole	Prilosec	Take 30–60 min before meals; may ↓ Fe and vitamin B_{12} absorption; diarrhea
Anti-GERD	Proton pump inhibitor	Pantoprozole	Protonix	Take 30–60 min before meals; may ↓ Fe and vitamin B_{12} absorption; diarrhea
Anti-GERD	Proton pump inhibitor	Rabeprazole	Aciphex	Take 30–60 min before meals; may ↓ Fe and vitamin B_{12} absorption; diarrhea

more information about these and other medications.) Persons with chronic HBV infection require medical evaluation and regular monitoring to determine whether disease is progressing and to identify liver damage or hepatocellular carcinoma.[25]

Hepatitis C (HCV) Hepatitis C (HCV) occurs when an individual is exposed to blood or body fluids from an infected person—by sharing needles (most are from this cause) or having been a recipient of blood clotting factors before 1987. Hemodialysis patients and infants born to infected mothers are also at increased risk. Unlike HAV and HBV, HCV cannot be prevented by vaccination. There are several blood laboratory tests that are used to diagnose HCV infections, with the most common one being measurement of antibodies to hepatitis C virus (anti-HCV); negative tests for antibodies to HAV and HBV help confirm the diagnosis.[28]

While the number of new infections per year has declined, about 75%–85% with acute infections will become chronically infected, 60%–70% will go on to develop chronic liver disease, and 5%–20% will go on to develop cirrhosis over a period of 20–30 years. HCV is now the most common reason for needing a liver transplant. Individuals with HCV who develop cirrhosis are also at an increased risk of developing hepatocellular carcinoma. Chronic HCV infection accounts for an estimated 8,000–10,000 deaths each year in the United States.[29] It is unknown how to predict which patients will have a benign course and which will develop cirrhosis or carcinoma.[30]

There are six known genotypes and more than 50 subtypes of hepatitis C, and knowing the genotype or serotype (genotype-specific antibodies) of HCV is helpful in counseling and making recommendations regarding therapy. The therapy for chronic hepatitis C has become more effective over the 10 years since alpha interferon was first approved for use in this disease. At the present time, the recommended current therapy is a 24- or 48-week course of the combination of pegylated alpha interferon and ribavirin, an oral antiviral agent[31] (see Table 16.5).

Hepatitis D and E (HDV; HEV) HDV is uncommon in the United States and requires HBV to replicate. In order to be infected with HDV, someone must already be infected with HBV. Transmission of the virus occurs via contact with infected blood. Hepatitis E usually causes an acute infection and does not lead to a chronic infection. While rare in the United States, hepatitis E is common in many parts of the world and is transmitted in a manner similar to hepatitis A virus.[32]

Nutrition Therapy for Viral Hepatitis

Nutritional Implications The primary nutritional implications are weight loss and nutritional deficiencies due to inadequate dietary intake, which may affect recovery. Drug-nutrient interactions may also occur as a result of HCV treatment (see Table 16.5).

Nutrition Assessment For the patient with acute hepatitis, weight and weight history should be assessed and a 24-hour recall, diet history, or food diary should be obtained to determine consumption of foods/liquids and those that may not be tolerated by the patient. Additionally, lifestyle factors such as alcohol intake should be assessed.

Nutrition Diagnosis Common nutrition diagnoses that may be associated with acute hepatitis include inadequate food/oral beverage intake; inadequate protein and calorie intake; food-medication interaction; and impaired nutrient utilization.

Nutrition Intervention The primary objective is to spare the liver and provide it with the nutrients needed for regeneration. Treatment includes adequate rest, fluids, good nutrition, and avoidance of further damage to the liver—in particular, avoidance of alcohol to prevent further liver cell damage. Patients suffering from anorexia are frequently unable to consume an adequate diet, and the first effort should be to increase dietary intake through the use of foods. Frequent small feedings are generally better tolerated. Patients with acute hepatitis should be encouraged to consume a meal plan adequate in energy, protein, and micronutrients with a regular meal schedule. Food restrictions, other than alcohol, are usually not required.

Monitoring and Evaluation Interventions are evaluated by measurement of specific outcomes. These could include tolerance to feedings, amount of nutrients consumed, weight (dry) changes, laboratory values (see diagnostic indicators of liver disease), and evaluation of the patient's sense of well being.[33]

Alcoholic Liver Disease

Alcoholism/alcohol abuse represents one of the largest health problems in the United States. It is a costly disease with serious physical, psychosocial, and nutritional implications. "Alcoholism or alcohol dependence is a diagnosable disease characterized by several factors, including a strong craving for alcohol, continued use despite harm or personal injury, the inability to limit drinking, physical illness when drinking stops, and the need to increase the amount drunk to feel the effects."[34]

From 2001 through 2005 there were approximately 79,000 deaths attributable to excessive alcohol use each year in the United States, making this the 3rd leading lifestyle-related cause of death for the nation.[35] Alcohol abuse is also associated with an annual economic loss of $184.6 billion in the United States.[36]

A large part of the morbidity and mortality associated with alcohol abuse is attributable to its toxic effects on all the major organ systems, with **alcoholic liver disease (ALD)** the primary cause of chronic medical illness and death. It is the twelfth leading cause of death in the United States.[37] For Hispanics, deaths from alcoholic liver disease and cirrhosis were 1.5 times higher than for whites, and for American Indian/Alaska Natives, they were 2.5 times higher.[37] The hepatotoxic alcohol threshold at which ALD may develop is 40 g (4 drinks) and 20 g (2 drinks) daily for men and women, respectively.[38, 39] The pathophysiology of ALD comprises a spectrum of three disorders: fatty liver (hepatic steatosis), alcoholic hepatitis, and cirrhosis.

Fatty Liver **Fatty liver** or steatosis occurs in about 25% of the general U.S. population and 90% of chronic alcohol abusers.[40] As shown in Table 16.6, which lists the potential causes of fatty liver, this disorder can also develop in the absence of alcohol; in this case it is known as **nonalcoholic steatohepatitis (NASH)**. A broader term, **nonalcoholic fatty liver disease (NAFLD)**, refers to a wide spectrum of liver disease ranging from simple fatty liver (steatosis), to nonalcoholic steatohepatitis (NASH), to cirrhosis. Individuals who are diagnosed with NASH tend to be middle-aged and overweight or obese (see Chapter 12). Over 50% percent of bariatric surgery patients have been found to have NASH.[41] Individuals with cystic fibrosis, an inherited disorder, have an increased risk of developing NASH as well (see Box 16.4).

Most patients with fatty liver are virtually asymptomatic. Hepatomegaly is the most common clinical sign. Severe forms of fatty liver may present a clinical picture mimicking extrahepatic obstructive jaundice with dark urine and light stools. Typical abnormalities in laboratory tests are slightly or moderately elevated gamma glutamyltransferase (GGT) and serum transaminases (AST and ALT). In alcoholic fatty liver, in contrast to more advanced liver injury, all the abnormalities in laboratory

Table 16. 6 Conditions Associated with Steatohepatitis

1. Alcoholism
2. Insulin resistance
3. Disorders of lipid metabolism
   a. Abetalipoproteinemia
   b. Hypobetalipoproteinemia
   c. Andersen's disease
   d. Weber-Christian syndrome
4. Cystic fibrosis
5. Total parenteral nutrition
6. Severe weight loss
   a. Jejunoileal bypass
   b. Gastric bypass[a]
   c. Severe starvation
7. Iatrogenic
   a. Amiodarone
   b. Diltiazem
   c. Tamoxifen
   d. Steroids
   e. Highly active antiretroviral therapy
8. Refeeding syndrome

[a]Much less common than after jejunoileal bypass.

Adapted from *Gastroenterology*, Vol. 123, American Gastroenterological Association, American Gastroenterological Association medical position statement: Nonalcoholic fatty liver disease, pp. 1702–04, Copyright 2002, with permission from American Gastroenterological Association.

tests tend to return to normal rapidly within the first days of hospitalization.[42] Though fatty liver is a benign and reversible condition with the cessation of alcohol use, its progression to alcoholic hepatitis and cirrhosis is life-threatening. The metabolism of alcohol and its role in the pathogenesis of fatty liver are described in Box 16.5. Although obesity is a risk factor for progression of ALD, aggressive weight reduction may not help in the treatment of alcoholic fatty liver and can even be associated with a worsening of injury.[43] It is estimated that 10% of patients with NASH will ultimately develop cirrhosis.[44]

Alcoholic Hepatitis Alcoholic hepatitis is a form of toxic liver injury associated with chronic ethanol consumption. Classically, alcoholic hepatitis presents with fever, jaundice, hepatomegaly, and, occasionally, signs of more advanced liver disease, such as ascites, portal hypertensive bleeding, and hepatic encephalopathy.[45] Patients with alcoholic hepatitis frequently have increased susceptibility to infections—pneumonia, spontaneous bacterial peritonitis, cellulitis, and even septicemia. Antibiotics are often required, and occasionally corticosteroids are recommended in severe alcoholic hepatitis with **cholestasis**. The catabolic effect of these medications adds to the complexity of the nutritional management. Diagnostic tests include blood labs. A serum aspartate aminotransferase/alanine aminotransferase (AST/ALT) ratio greater than 2.0 helps differentiate ALD from other liver diseases with ratios of less than 2.0, such as nonalcoholic steatohepatitis and chronic viral hepatitis.[46] Ultrasonography or computed tomography imaging (see Box 16.6) may be used to determine evidence of cirrhosis and portal hypertension or to rule out an alternative diagnosis. If needed, a liver biopsy will confirm the diagnosis of ALD. Abstinence is

BOX 16.4   CLINICAL APPLICATIONS

Cystic Fibrosis

Cystic fibrosis (CF) is an inherited disorder that affects epithelial transport in exocrine tissues. It is one of the most common life-limiting inherited disorders among Caucasians, affecting 1 in every 3,200–3,300 live births.[1] In CF, the chloride ion is prevented from leaving the cell and water cannot exit. Without water, mucus thickens, cilia (hairlike structures on cells) cannot function properly, and bacteria can collect on the cells, which can lead to infections. Manifestations of CF occur in the lungs, gastrointestinal tract, hepatobiliary tract, and exocrine pancreas. Specific disorders associated with CF and their occurrence are listed in the table below.[1, 2]

	Approximate Frequency
Liver	
Hepatic steatosis and steatohepatitis	20%–60%
Cirrhosis	2%
Focal biliary cirrhosis	25%
Multilobular cirrhosis	5%
Gallbladder	
Cholelithiasis and cholecystitis	10%–12%
Pancreas	
Pancreatic insufficiency	85%–90%
Pancreatitis	<2%

Most CF patients with cirrhosis are asymptomatic, with the most common clinical presentation being hepatomegaly or splenomegaly. The usual clinical signs of liver disease, such as jaundice, ascites, and encephalopathy, are rarely present or occur very late. The most deleterious complication of cystic fibrosis with cirrhosis is portal hypertension. Complications of portal hypertension need to be monitored, because variceal bleeding occurs in one-third of CF patients with associated liver cirrhosis. Fortunately, only a small proportion progress to advanced liver disease. Liver transplantation is an option, and has been successfully performed in CF patients with end-stage liver disease.[3]

Ursodeoxycholic acid (UDCA) may be used to prevent bile-related disorders. The mechanisms of action are not well understood; however, it most likely works by reducing bile viscosity and preventing plugging of the bile ducts. Early studies have indicated that functional improvement is substantially lower in CF patients with advanced liver disease, suggesting that treatment should be started as early as possible in the course of liver disease. The long-term risks or benefits of taking this medication are not known, though currently under evaluation.

Pancreatic insufficiency is common with CF and enterically coated synthetic pancreatic enzymes are taken to help improve digestion and absorption of nutrients. More detailed information on CF and the associated nutrition therapy is located in Chapter 21.

References
1. Giannouli E, Sharma S, Maycher B. Cystic Fibrosis, Thoracic. Available at http://emedicine.medscape.com/article/354931-overview. Updated: Apr 13, 2009.
2. Beers MH, Berkow R, editors. Merck manual of diagnosis and therapy. 18th ed. Whitehouse Station (NJ): Merck & Co.; 2006–08. Available at http://www.merck.com/mmpe/sec19/ch278/ch278a.html?qt=cystic fibrosis&alt=sh. Content last reviewed and modified August 2008.
3. Curry MP, Hegarty JE. The gallbladder and biliary tract in cystic fibrosis. Curr Gastroenterol Rep. 2005;7:147–53.

BOX 16.5   CLINICAL APPLICATIONS

Alcohol Metabolism

The liver contains three pathways for ethanol oxidative metabolism, each located in a different subcellular compartment of the hepatocyte: the ADH pathway in the cytosol, the microsomal ethanol oxidizing system (MEOS) in the endoplasmic reticulum, and catalase in peroxisomes (responsible for metabolizing less than 10% of ethanol). Each of these pathways produces acetaldehyde, a highly toxic metabolite that is converted to acetate.

In the ADH pathway, ADH catalyzes the conversion of ethanol to acetaldehyde, coupled with the reduction of NAD^+ to NADH. This acetaldehyde, in turn, is converted to acetate and reducing equivalents by ALDH. The acetate may be oxidized further to carbon dioxide and water or enter the TCA cycle. Metabolism of ethanol via the ADH/ALDH pathway produces a marked increase in the $NADH/NAD^+$ ratio, which produces an alteration in the cell's redox potential.[1] Damage to the mitochondria occurs, inhibiting reoxidation of NADPH and causing lactic acidosis, hypoglycemia, and hyperuricemia. Another important consequence of this redox change is decreased fatty acid oxidation related to decreased TCA cycle activity. The increased fatty acids form triglyceride by combining with alpha-glycerophosphate, causing an increase in plasma triglycerides.

With chronic alcoholism, an additional pathway contributes to the oxidation of ethanol, namely, the microsomal ethanol oxidizing system (MEOS).[2] The MEOS can be used for both ethanol and drug metabolism. The MEOS system differs from the ADH/ALDH system in that it is triggered by exposure to ethanol, utilizes oxygen and NADPH (using up energy rather than generating it), and has a lower optimum pH.

The cause of fatty liver after alcohol ingestion is the increased availability of fatty acids in the liver. The sources of fatty acids are adipose tissue, lipids synthesized by the liver itself, and dietary lipids. The source of fatty acids depends on the fat content of the diet and whether alcohol is ingested acutely or chronically, since fatty acids originate from adipose tissue after the acute ingestion of a large dose of alcohol. Fatty liver develops in most people who abuse alcohol for a period of days.[3]

Increased synthesis and decreased degradation of fatty acids in the liver occur during the chronic ingestion of alcohol. The latter effects of alcohol are principally related to the increase in the NADH/NAD ratio that occurs during the metabolism of alcohol. Synthesis of fatty acids is stimulated by increases in NADPH produced when reducing equivalents from NADH are transferred to $NADP^+$, while the oxidation of fatty acids is reduced by the depressant effect of the increased $NADH/NAD^+$ ratio on the TCA cycle.

References
1. Svensson S, Some M, Lundsjo A, Helander A, Cronholm T, Hoog JO. Activities of human alcohol dehydrogenases. In: The metabolic pathways of ethanol and serotonin. Eur J Biochem. 1999;262:324–29.
2. Lieber CS. The discovery of the microsomal ethanol oxidizing system and its physiologic and pathologic role. Drug Metab Rev. 2004;36:511–29.
3. Menon KV, Gores GJ, Shah VH. Pathogenesis, diagnosis, and treatment of alcoholic liver disease. Mayo Clinic Proceedings. 2001;76:1021–29.

BOX 16.6 CLINICAL APPLICATIONS

Hepatobiliary and Pancreatic Diagnostic and Surgical Procedures

- **Computerized tomography (CT) scan.** The CT scan is a noninvasive X-ray that produces three-dimensional pictures of parts of the body. The person lies on a table that slides into a donut-shaped machine.

- **Abdominal ultrasound.** A handheld device, which a technician glides over the abdomen, sends sound waves toward the abdomen. The sound waves bounce off the liver, gallbladder, and other organs, and their echoes create a picture of the liver and biliary system on a video monitor.

- **Magnetic resonance imaging (MRI).** MRI machines use radio waves and magnets to scan internal organs and tissues to create a picture.

- **Liver scan.** The liver is visualized by inserting a laparoscope into the abdomen. A laparoscope is an instrument with a camera that relays pictures to a computer screen.

- **Cholescintigraphy (HIDA scan).** The patient is injected with a small amount of nonharmful radioactive material that is absorbed by the gallbladder, which is then stimulated to contract. The test is used to diagnose abnormal contraction of the gallbladder or obstruction of the bile ducts.

- **Endoscopic retrograde cholangiopancreatography (ERCP).** ERCP is used to locate and remove stones in the bile ducts. After the patient is lightly sedated, the doctor inserts an endoscope—a long, flexible, lighted tube with a camera—down the throat and through the stomach and into the small intestine. The endoscope is connected to a computer and video monitor. The doctor guides the endoscope and injects a special dye that helps the bile ducts appear better on the monitor.

The following procedures can be performed using ERCP:

- **Sphincterotomy.** Using a small wire on the endoscope, the doctor finds the muscle that surrounds the pancreatic duct or bile ducts and makes a tiny cut to enlarge the duct opening. When a pseudocyst is present, the duct is drained.
- **Gallstone removal.** The endoscope is used to remove pancreatic or bile duct stones with a tiny basket. Gallstone removal is sometimes performed along with a sphincterotomy.
- **Stent placement.** Using the endoscope, the doctor places a tiny piece of plastic or metal that looks like a straw in a narrowed pancreatic or bile duct to keep it open.
- **Balloon dilatation.** Some endoscopes have a small balloon that the doctor uses to dilate, or stretch, a narrowed pancreatic or bile duct. A temporary stent may be placed for a few months to keep the duct open.
- **Percutaneous transhepatic cholangiography.** This procedure involves inserting a needle through the skin and placing a thin tube into a duct in the liver. Dye is injected through the tube and x-rays are taken.

- **Endoscopic ultrasound (EUS).** After spraying a solution to numb the patient's throat, the doctor inserts an endoscope—a thin, flexible, lighted tube—down the throat, through the stomach, and into the small intestine. The doctor turns on an ultrasound attachment to the scope that produces sound waves to create visual images of the pancreas and bile ducts.

- **Magnetic resonance cholangiopancreatography (MRCP).** MRCP uses magnetic resonance imaging, a noninvasive test that produces cross-section images of parts of the body. After being lightly sedated, the patient lies in a cylinder-like tube for the test. The technician injects dye into the patient's veins, which helps show the pancreas, gallbladder, and pancreatic and bile ducts. This safe and painless test is increasingly being used for diagnosis.

- **Hepatobiliary scintigraphy.** A radioactive tracer substance is injected into an arm vein. The liver clears the tracer from the bloodstream and excretes it through bile. A special (gamma) camera takes pictures of the radioactive compound as the bile moves through the liver, bile ducts, gallbladder, and small intestine.

the foundation of therapy for alcoholic steathepatitis. Corticosteroids improve the short-term survival of patients with severe alcoholic hepatitis by approximately 20%.[47]

Nutrition Therapy for Individuals with Alcohol Dependency/Alcoholic Hepatitis

Nutritional Implications The chronic alcoholic with liver disease is usually also malnourished, which can aggravate the liver disease. Halsted lists the following reasons for malnutrition in the alcoholic:[48]

- Imbalanced diet and/or anorexia. Alcohol often replaces food in the diet. Though it is possible to obtain maintenance energy needs from the energy in the alcohol consumed, malnourishment results because of an inadequate intake of nutrients; though high in calories, alcohol contains no protein, vitamins, or minerals. Alcohol is also preferentially metabolized over whatever nutrients are obtained through food. Release of cytokines, such as tumor necrosis factor, and leptin from the cells also contributes to malnutrition.

- Intestinal maldigestion and malabsorption. Alcohol causes inflammation of the stomach, pancreas, and intestine, interfering with the normal processes of digestion and absorption and resulting in secondary malnutrition. The subsequent decrease in pancreatic enzyme secretion, intestinal transporters, and micelle formation (due to inadequate bile salt secretion) results in altered absorption of fat-soluble vitamins, thiamin, and folic acid. In addition, alcohol and acetaldehyde have a hepatotoxic effect that interferes with metabolism and activation of vitamins by liver cells.

- Increased excretion of selected vitamins. The metaboism of alcohol increases the need for certain nutrients, particularly the B-vitamins and magnesium. Alcohol causes increased urinary excretion of folate and pyridoxine (B_6), as well as increased retinoid metabolism. (See Box 16.7 for more information.)

BOX 16.7 CLINICAL APPLICATIONS

Effect of Alcoholism on Vitamins and Minerals

Excessive chronic alcohol intake is generally associated with vitamin deficiency due to malnutrition, malabsorption, and ethanol toxicity, as well as lower vitamin intake.[1] Alcoholism can also result in changes in nutritional requirements since alcohol can promote nutrient degradation or impaired activation.[2]

Folic Acid

Folic acid deficiency is the most frequent nutritional deficiency in alcoholics, and is related to several interacting factors. Alcoholics ingest folate-deficient diets. Additionally, alcohol interferes, directly and indirectly, with absorption of folate, its transport to tissues, and its storage and release by the liver. A major effect of the acute ingestion of alcohol is the reversible sequestration of folate in the hepatocyte and a failure to release folate into the bile, thus disrupting the enterohepatic circulation of folate.[3] Chronic alcohol exposure impairs folate absorption by inhibiting expression of the reduced folate carrier and decreasing the hepatic uptake and renal conservation of circulating folate. At the same time, folate deficiency decreases alcohol-induced changes in hepatic methionine metabolism, promoting enhanced oxidative

liver injury and the histopathology of alcoholic liver disease.[4] Folate deficiency can cause morphological changes in the intestine, including villus shortening, decreased mitosis, macrocytosis, and enlargement of epithelial cell nuclei. The most common sign of folate deficiency is megaloblastic anemia.[5]

Thiamin

The most common cause of thiamin deficiency in the United States is alcoholism. Alcohol affects thiamin uptake and other aspects of thiamin utilization, and these effects may contribute to the prevalence of thiamin deficiency in alcoholics. The major manifestations of thiamin deficiency in humans involve the cardiovascular (wet beriberi) and nervous (dry beriberi, and **Wernicke-Korsakoff syndrome**) systems. A number of mechanisms may be involved in the pathogenesis of thiamin deficiency in the alcoholic population, though this subject remains controversial. Among the mechanisms proposed are an inadequate thiamin intake, impairment of thiamin absorption, and decreased conversion to the biologically active form of thiamin (thiamin pyrophosphate). The Wernicke-Korsakoff syndrome is characterized by weakness of eye movements, gait

disturbance (ataxia), and confusion. **Korsakoff's psychosis** is characterized initially by anterograde amnesia (inability to create memory after an event occurs), retrograde amnesia (loss of memories prior to an event) to a lesser extent, a disordered time sense, and often confabulation of the acute senses. Cognitive deficits have also been observed. Ophthalmoplegia (loss of eye movement control) in **Wernicke's encephalopathy** responds rapidly to thiamin administration, while the ataxia and confusion respond more slowly. The rapidity of response depends upon the conversion of thiamin to its active form in the liver; patients with advanced liver disease such as cirrhosis may therefore have a delayed response.

Vitamin B_6

Low plasma levels of pyridoxine (vitamin B_6) have been reported in over 50% of alcoholics without hematologic findings or abnormal liver function tests.[6] Inadequate intake may explain some of it, but increased destruction and reduced formation may play a role. Clinical management involves the provision of pyridoxine in the usual multivitamin dosage. Large doses (as little as 200 mg) should be avoided, because there is a danger of

B_6 toxicity causing ataxia due to sensory neuropathy.

Niacin

Evidence of niacin deficiency is difficult to find in alcoholics. Nevertheless, the concurrent administration of vitamin B_6 and niacin returns the tryptophan metabolites excreted in the urine of alcoholics to normal. There have been case reports of pellagra associated with excessive intake of alcohol. In one reported case, supplementation with niacin had a prompt effect on correcting the skin changes associated with pellagra.[7]

Vitamin C

Vitamin C is deficient in alcoholics with and without liver disease, and levels correlate with dietary intake. Low leukocyte concentrations of ascorbic acid (a measure of tissue stores) are found in patients with alcoholic cirrhosis. Daily supplementation with 175–500 mg of ascorbic acid may be necessary for weeks or months to restore plasma and urinary ascorbate to normal.

Vitamin D

In chronic alcoholics, the deficiency of active metabolites of vitamin D is often observed.[8] Alcoholics have been found to have low, normal, and increased levels of 25-hydroxy vitamin D. In patients with alcoholic liver disease, vitamin D deficiency probably derives from too little vitamin D substrate, which is a result of poor intake, malabsorption due to cholestasis or pancreatic insufficiency, and insufficient sunlight. Insufficient intake of calcium and phosphorus or decreased calcium absorption in the presence of normal 25-hydroxy vitamin D might accelerate bone loss in alcoholics. In addition, ethanol impairs osteoblastic activity, which results in reduced bone formation and mineralization. By affecting the osteoblasts' function, alcohol is also able to induce bone resorption. In addition, ethanol has an indirect influence on the metabolism of hormones participating in bone homeostasis. Bone disease in those with liver disease should be treated through increased intake of vitamin D,

ultraviolet light therapy, and correction of fat malabsorption to keep plasma calcium and phosphorus normal, along with abstinence from alcohol.[2]

Vitamin K

Vitamin K deficiency may be present in alcoholics experiencing fat malabsorption due to pancreatic insufficiency, biliary obstruction, or intestinal mucosal abnormality secondary to folate deficiency. It is unlikely to be due to low intake, since the microflora of the gut are a reliable source of the vitamin. Any prolongation of the prothrombin time is probably related to the associated hepatic disease causing failure of prothrombin synthesis. Vitamin K may be given intramuscularly to clinically test whether hepatocellular dysfunction or lack of availability of vitamin K to the liver is responsible for low levels of vitamin K-dependent clotting factors in the blood.[2]

Vitamin A and Zinc

Alcoholics may suffer from vitamin A deficiency. Vitamin A malnutrition in cirrhotics may be caused by poor diet, malabsorption, decreased hepatic vitamin A uptake, and decreased hepatic storage capacity for vitamin A. An important clinical consequence of low tissue vitamin A is night blindness. Abnormal dark adaptation occurs in alcoholics with and without cirrhosis and is more common among those with cirrhosis, those with elevated bilirubin, and those at an older age.[9] Several factors make vitamin A therapy complicated when alcoholism is present: assessment of tissue stores of vitamin A is difficult; vitamin A in high doses is toxic, and even the usual doses of vitamin A are potentially toxic with continued intake of alcohol; and monitoring vitamin A hepatotoxicity is difficult in the presence of continued alcohol intake. Replacement of vitamin A via supplementation should only be considered for patients who are confirmed as deficient and who assuredly practice abstinence from alcohol. Low serum levels and night-blindness can be considered evidence of a deficiency.

There are clear interactions between zinc and vitamin A, with zinc participating in the absorption, mobilization, transport, and metabolism of vitamin A. Similarly, vitamin A affects zinc absorption and use. Therefore, changes in the status of either could affect the metabolism of the other. Vitamin A and zinc metabolism are affected both by ethanol and by hepatic cirrhosis. Ethanol causes abnormal dark adaptation by acting as a competitive inhibitor with retinol alcohol dehydrogenase in the eye. In some cirrhotic patients, zinc deficiency and/or protein deficiency may limit the ability to respond to vitamin A. In animals, oral ethanol intake results in increased losses of zinc by the urinary and fecal routes. Combined vitamin A and zinc deficiencies are common in cirrhotics and either may result in abnormal dark adaptation or impaired taste or smell. A low serum zinc level can be treated with doses of 600 µg $ZnSO_4$ per day; considering the interrelationship of vitamin A and zinc metabolism, zinc therapy might be tried when vitamin A therapy fails. However, recent literature suggests that patients with ALD who have hypozincemia responded poorly to oral zinc supplementation.[10] Some clinical trials suggest that vitamin A should be parenterally replaced when there is documented fat malabsorption.

Iron

Chronic alcohol abuse is associated with both an altered response to infection and deranged iron homeostasis.[11] Iron metabolism is important in alcoholism because there may be a deficiency or an excess of iron in the body. The question of the metabolism of iron is particularly relevant because of the association of hepatic injury with excess iron. It is not clear whether an increase in hepatic iron results from alcohol increasing intestinal absorption or hepatic uptake of iron from serum in established alcoholic liver disease. Alcoholics may receive excessive dietary iron from the beverages they drink, such as certain wines, or through inadvertent treatment with iron-containing vitamin preparations.

(continued)

In addition, anemias unrelated to iron deficiency may be incorrectly treated with iron. Pancreatic insufficiency, folate deficiency, portosystemic shunting, and cirrhosis may increase iron absorption. Of greatest potential significance is the contribution hepatic iron may make to liver damage via its role in lipid peroxidation and its possible role in promoting fibrogenesis.[2] Alcoholics may be iron deficient as a result of GI lesions that may bleed (esophagitis, esophageal varices, gastritis, duodenitis). The usual laboratory tests (serum iron, serum binding capacity) are helpful. Iron supplements should be restricted to clearly diagnosed cases of deficiency. Plasma ferritin concentration has been positively associated with alcohol use among men.[12]

Calcium

Acute and chronic exposure to ethanol influences intracellular calcium homeostasis.[13] Illnesses associated with abnormalities of calcium homeostasis in alcoholics include decreased bone density and bone mass, increased susceptibility to fractures, increased bone cell death (osteonecrosis), and an earlier occurrence of osteoporosis.

Potassium and Magnesium

Hypokalemia (low serum potassium) is common in patients with alcoholic liver disease. Even if the serum potassium level is normal, body stores of potassium are likely to be low due to poor dietary intake, vomiting, and particularly diarrhea. In patients with ascites, secondary aldosteronism and the administration of diuretics contribute to hypokalemia.[5] Decreased blood levels of magnesium are also common in alcoholics (30%–80%). This is thought to be due to increased secretion of magnesium as a result of the diuretic effect of alcohol.[14]

Supplementation

Vitamin supplementation is commonly provided in the treatment of alcoholism, whether deficiencies are present or not. It is generally recommended that a multiple water-soluble vitamin be given at twice the RDA. During "drying out" or recovery, the alcoholic also requires increased amounts of vitamins.

References

1. Van Den Berg H, Van Der Gaag M, Hendricks H. Influence of lifestyle on vitamin bioavailability. Int J Vitamin Nutrition Resource. 2001;72:53–59.
2. Lieber CS. Alcohol: its metabolism and interaction with nutrients. Annu Rev Nutr. 2000;20:395–430.
3. Watson, Watzl. Nutrition and Alcohol. Florida: CRC Press, Inc.; 1992.
4. Halsted CH, Villanueva JA, Devlin AM, Chandler CJ. Metabolic interactions of alcohol and folate. J Nutr. 2002;132:85.
5. Halsted CH. Nutrition and alcoholic liver disease. Semin Liver Dis. 2004;24:289–304.
6. Gloria L, Cravo M, Camilo ME, Resende M, Cardoso JN, Oliveira AG, et al. Nutritional deficiencies in chronic alcoholics: relation to dietary intake and alcohol consumption. Am J Gastroenterol. 1997;92:485–89.
7. Lorentzen HF, Fugleholm AM, Weismann K. Zinc deficiency and pellagra in alcohol abuse. Ugeskr Laeger. 2000;162:6854–56.
8. Medras M, Jankowska EA. The effect of alcohol on bone density in men. Przeglad Lekarski. 2000;57:743–46.
9. Hussaini SH, Henderson T, Morrell AJ, Losowsky MS. Dark adaptation in early primary biliary cirrhosis. Eye. 1998;12(Pt 3a):419–26.
10. McClain CJ, Hill DB, Song Z, Deaciuc I, Barve S. Monocyte activation in ALD. Alcohol. 2002;27:53–61.
11. Potter BJ, Wang F. Molecular regulation of iron homeostasis and resistance to infection in alcoholics. Front Bioscience. 2002;7:1396–1409.
12. Peach HG, Bath NE. Post test probability that men in the community with raised plasma ferritin concentrations are hazardous drinkers. J Clin Pathol. 1999;52:853–55.
13. Bondy B, Engel RR, de Jonge S, Schutz CG, Soyka M. Phytohemagglutinin-stimulated calcium signed in lymphocytes of alcoholics before, during, and after detoxification. Psychiatry Res. 1998;81:157–62.
14. Novello NP, Blumstein HA. Hypomagnesemia: Treatment & Medication. Available at http://emedicine.medscape.com/article/767546-treatment. Updated: Aug 18, 2009.

Nutrition Assessment Weight and weight history should be obtained. A physical assessment may reveal skeletal muscle myopathy, as shown by reduced muscle strength and reduced musculature.[49] A 24-hour recall, diet history, or food diary should be used to assess consumption of foods/liquids and identify those that may not be tolerated by the patient. Additionally, data about lifestyle factors, including alcohol intake, diagnostic indicators of liver disease, vitamin and mineral deficiencies (see Table 16.8), and use of complementary or alternative medical/dietary remedies (see the CAM feature at the end of this chapter) should be obtained, along with information about current medical diagnoses and treatments plus previous medical history.

Nutrition Diagnosis Common nutrition diagnoses that may be associated with alcoholic liver disease include increased energy expenditure; inadequate energy intake; inadequate oral food/beverage intake; inadequate protein-energy intake; malnutrition; inappropriate intake of AAA/BCAA; inadequate vitamin/mineral intake (e.g. thiamin, etc.); altered GI function; impaired nutrient utilization; underweight; altered nutrition-related laboratory values; food-medication interactions; and involuntary weight loss.

Nutrition Intervention The primary objective is to spare the liver and provide it with the nutrients needed for regeneration. Treatment includes adequate rest, fluids, good nutrition, and in particular, to prevent or avoid further damage to the liver, avoidance of alcohol. Patients suffering from anorexia are frequently unable to consume an adequate diet, and the first effort should be to increase dietary intake through the use of foods. Frequent small feedings are generally better tolerated. Patients with alcoholic dependency should be encouraged to consume a meal plan adequate in energy (30–35 kcal/kg body weight) and protein (1.5–2 g/kg body wt); high in carbohydrate (6–8 g/kg body wt); moderate in or without sugar intake; and moderate in fat, with a regular meal schedule. Table 16.7 provides guidance in food choices for a fat-restricted diet. Use of a multivitamin with additional nutrients as indicated by nutrition assessment and blood chemistry information is recommended. See Table 16.8 for more details on supplements.[50]

Monitoring and Evaluation Interventions are evaluated by measurement of specific outcomes. These could include tolerance to feedings, amount of nutrients consumed, (dry) weight changes, laboratory values (see diagnostic indicators of liver disease), and evaluation of the patient's sense of well being.[51]

Cirrhosis

Cirrhosis represents the end of the pathophysiology spectrum for a wide variety of chronic liver diseases in which healthy

Table 16.7 Fat-Restricted Diet Foods

Allowed	Excluded or Limited Use
Fats	
	• Butter, margarine, oils, lard, mayonnaise, salad dressings
	• Bacon
	• Gravy or cream sauce
	• Nondairy creamers
	• Cocoa butter, found in chocolate
Protein Foods	
• Skim (fat-free)/low-fat (1 percent) milk	• Full-fat milk/reduced-fat (2%) and other dairy products
• Fat-free or low-fat dairy products, such as yogurt and cheese	• Organ meats, such as liver
• Egg whites or egg substitutes	• Egg yolks
• Fish	• Fatty and marbled meats
• Skinless poultry	• Spareribs
• Legumes	• Cold cuts, frankfurters, hot dogs and sausages, bacon
• Soybeans and soy products; for example, soy burgers	• Fried, breaded, or canned meats or fish
• Lean ground meats	• Fish packed in oil
Fruits and Vegetables, Nuts	
• Almost all fruits and vegetables	• Coconut
	• Fried, au gratin, or creamed vegetables
	• Olives
	• Nuts
	• Peanut butter
	• Avocado
Grains	
• Whole-wheat flour	• Muffins, waffles, pancakes, fritters
• Whole-grain bread, preferably 100% whole-wheat or 100% whole-grain bread	• Corn bread
	• Doughnuts, cakes
• High-fiber cereal with 5 or more grams of fiber per serving	• Biscuits
	• Cakes
• Brown rice	• High-fat snack crackers
• Whole-grain pasta	• Potato chips
• Oatmeal (steel-cut or regular)	• Popcorn with butter
• Ground flaxseed	• Pie crust
Other	
• Desserts made with skim milk	• Ice cream
	• Chocolate

Table 16.8 Oral Nutritional Supplementation Often Given to Chronic Alcoholics

Thiamin	50–100 mg for 7 to 14 days
Folic acid	1 mg (1000 mcg) daily
Riboflavin	Amount in a typical daily multivitamin
Vitamin B$_6$	1–3 mg as part of a multivitamin. Use of large amounts should be avoided because of potential for toxicity.
Vitamin B$_{12}$	6–12 mg as part of a multivitamin
Vitamin C	175–500 mg/day
Vitamin A	Amount in typical multivitamin. More may be given only for clear cases of deficiency because of potential for hepatotoxicity.
Vitamin D	200–500 IU (International Units)
Vitamin E	10–50 IU as part of a multivitamin
Iron	Standard amount in multivitamin for premenopausal women. For men and postmenopausal women, a multivitamin with no iron may be the safest route. There is a potential for overload, so additional iron therapy should be restricted to clear cases of deficiency.
Magnesium	100–400 mg/day if magnesium deficiency is suspected or confirmed. With severe magnesium deficiency, parenteral replacement may be necessary.
Selenium	5–50 mcg daily
Zinc	Amount available in multivitamin. Serum zinc levels may not respond to supplementation, and zinc toxicity may occur with large-dose supplementation. Zinc replacement should only be given for night blindness.

Source: Adapted from: Markowitz JS, McRae AL, Sonne SC. Oral nutritional supplementation for the alcoholic patient: a brief overview. Ann Clin Psychiatry. 2000;12(3):153–58.

tissue is replaced by scar tissue, blocking the flow of blood through the organ and resulting in the loss of liver function. The most common causes of cirrhosis are chronic HCV and alcoholism, but not all heavy drinkers develop the disease. (See Table 16.9 for other common causes of cirrhosis.) Genetic factors can increase susceptibility.[52]

Even when irreversible liver complications are present, therapeutic intervention, primarily nutritional treatment, can alleviate major complications of cirrhosis, such as encephalopathy, and manifestations of portal hypertension, such as ascites (which responds favorably to salt restriction). At present, a major task is to avoid the development of these serious complications at an early stage and to arrest the disease process prior to the medical or social disintegration of the individual. For patients with end-stage liver disease, average 1- and 5-year survival rates are approximately 80% and 50%, respectively.[53] The clinical tools most widely used to determine prognosis in patients with cirrhosis are the Child-Turcotte-Pugh (CTP) classification (shown in Table 16.10) and the prognostic model for end-stage liver disease (MELD) developed by investigators at the Mayo Clinic.[54] The MELD score is based on three blood tests: international normalized ratio (INR)—tests the clotting tendency of blood; bilirubin—tests the amount of bile pigment in the blood; and creatinine—tests kidney function.

Clinical Manifestations The cirrhotic patient has an enlarged liver as a result of fat accumulation and necrosis of the liver cells. Symptoms of cirrhosis often include fatigue, weakness, nausea, poor appetite, and malaise and the more liver-specific symptoms of jaundice, dark urine, light stools, steatorrhea, itching, abdominal pain, and bloating. As with **hepatosteatosis**, malnutrition is common. Vitamin and mineral deficiencies may cause or contribute to depressed hematocrit and hemoglobin values. Bruising and bleeding (coagulopathy) are related to decreased vitamin K absorption and ability of the liver to synthesize protein clotting factors. (Refer to Tables 16.8 and 16.11 for additional information.)

The major clinical complications associated with a cirrhosis are portal hypertension, hepatic encephalopathy, ascites (discussed earlier in the chapter), hepatorenal syndrome, and esophageal

Table 16.9 Causes of Cirrhosis in the United States

Most common causes of cirrhosis in the United States

- Hepatitis C (26%)
- Alcoholic liver disease (21%)
- Hepatitis C plus alcoholic liver disease (15%)
- Cryptogenic causes (18%)
- Hepatitis B, which may be coincident with hepatitis D (15%)
- Miscellaneous (5%)

Miscellaneous causes of chronic liver disease and cirrhosis

- Nonalcoholic steatohepatitis (NASH)
- Autoimmune hepatitis
- Primary biliary cirrhosis
- Secondary biliary cirrhosis (associated with chronic extrahepatic bile duct obstruction)
- Primary sclerosing cholangitis
- Hemochromatosis
- Cystic fibrosis
- Wilson's disease
- Alpha-1 antitrypsin deficiency
- Granulomatous disease (e.g., sarcoidosis)
- Type IV glycogen storage disease
- Drug-induced liver disease (e.g., methotrexate, alpha methyldopa, amiodarone)
- Venous outflow obstruction (e.g., Budd-Chiari syndrome, veno-occlusive disease)
- Chronic right-sided heart failure
- Tricuspid regurgitation

Source: Reprinted from Wolf DC. Cirrhosis. Emedicine. Available at http://emedicine.medscape.com/article/185856-overview. Updated: Aug 11, 2008.

Table 16.10 Child-Turcotte-Pugh Scoring System for Cirrhosis

Clinical Variable	1 Point	2 Points	3 Points
Encephalopathy	None	Stages 1–2	Stages 3–4
Ascites	Absent	Slight	Moderate
Bilirubin (mg/dL)	<2	2–3	>3
Bilirubin in PBC or PSC (mg/dL)	<4	4–10	10
Albumin (g/dL)	>3.5	2.8–3.5	<2.8
Prothrombin time (seconds prolonged or INR)	<4 s or INR <1.7	4–6 s or INR 1.7–2.3	>6 s or INR >2.3

Interpretation of the Child-Turcotte-Pugh Scoring System

Points	Risk (Grade)	Survival Rate (%)	
		1-yr	2-yr
5–6	Low (A)	100	85
7–9	Moderate (B)	80	60
10–15	High (C)	45	35

Source: Reprinted from Pugh RN, Murray-Lyon IM, Dawson JL, et al. Transection of the oesophagus for bleeding oesophageal varices. *Br J Surg.* Aug 1973;60(8):646–49.

vessels in the esophagus, called esophageal varices. If they burst, serious bleeding can occur in the esophagus, requiring immediate medical attention. All these complications may be due to liver failure and portal hypertension or to factors such as tumor necrosis factor (TNF), which is thought to play an etiologic role in liver disease.[56]

Treatment The primary medical treatments for cirrhosis are abstention from alcohol, treatment of the complications mentioned previously, and nutrition therapy.

Nutrition Therapy for Cirrhosis

Interest in nutrition therapy for cirrhosis was stimulated when several studies showed that a nutritious diet or nutrition sup-

varices. Hepatorenal syndrome (HRS) is the development of renal failure in patients with advanced chronic liver disease. Treatment, which includes albumin infusions combined with vasoconstrictor medication, significantly decreases mortality.[55] When portal hypertension occurs, it may cause enlarged blood

Table 16.11 Anemias of Liver Disease

Type of Anemia	RBC MCV	Serum Fe	TIBC	Transferritin Percentage of SAT	Ferritin	Folate	B$_{12}$	Type of Liver Disease
Megaloblastic normochromic	>110	N	N	N	N	↓	N	Alcoholism with injury
		N	N	N	N/↑	↓	↑	Acute injury
		↓	↓	↓	N	N	N	Chronic injury
Microcytic normochromic	<80	↓	↑/N	↓	↓	N	N	Chronic disease with blood loss
Hemolytic	<95	N	N	N	N/↑	N	↑/N	Autoimmune
		N	N/↓	N	N	N	↑/N	Zieve's syndrome Wilson's disease
Normochromic normocytic	80–95	↓	↓	↓	N	N	N	Chronic disease
Pancytopenis	80–95	N	N	N	N	N	↑	Acute viral hepatitis, drug toxicity, hypersplenism
Megaloblastic hyperchomic	>95	↑	↑	↑	↑	N	N	Hemohypochromic
		↑	↑	↑	N	↓	N	Thalassemia

port improved the five-year outcome of patients with alcoholic cirrhosis compared to patients consuming an inadequate diet.[54] The goal of nutrition therapy is to provide adequate nutrition to maintain or replete nutrition stores without triggering or interfering with treatment of clinical complications.

Nutritional Implications In addition to the implications previously listed under Nutrition Therapy for Individuals with Alcohol Dependency/Alcoholic Hepatitis, patients with cirrhosis may be unable to consume adequate amounts of food due to early satiety caused by the effect of ascites on the stomach's capacity to expand. They may also experience impaired nutrient digestion and absorption due to portal hypertension, decreased pancreatic enzyme production and/or secretion, and villus atrophy. Cirrhosis increases energy expenditure because of vasodilation and expanded blood volume.[57] Blood sugar levels can be erratic. When cirrhosis reaches levels at which 80% of hepatocytes are dysfunctional, hypoglycemia is a frequent occurrence due to hyperinsulinemia. The lack of adequate amounts of glycogen reserves as a result of the liver damage may cause hypoglycemia after an overnight fast and also prompt mobilization of amino acids from the skeletal muscles for gluconeogenesis.[55] However, 40%–50% of all patients with end-stage liver disease suffer from insulin-resistant diabetes mellitus (see Chapter 17).[58] Therefore, hyperglycemia may also be evident.

Nutrition Assessment As with alcoholic hepatitis, assessment should include attention to food/liquid intake and intolerances; lifestyle factors including alcohol intake; lab values related to liver function and vitamin/mineral status; and current and past medical diagnoses and treatments. Many of the methods often used for anthropometric and biochemical assessment of nutritional status are not reliable for individuals with cirrhosis. Due to the edema and ascites, dry weight may not be obtainable, and electrical impedance analysis will not be accurate. Many of the liver protein lab values used to determine visceral protein status will be abnormal because of hepatic dysfunction. In 2006, the European Society for Parenteral and Enteral Nutrition (ESPEN) recommended that the Subjective Global Assessment (SGA) (see Appendix D) and measurement of mid-arm muscle circumference (MAMC) or mid-arm circumference (MAC) and triceps skin fold thickness (TST) be used to nutritionally assess the patient[59] (see Chapter 3, Figure 3.10). The 2009 ASPEN guidelines for critical care recommend that, in addition, calorie requirements be assessed using indirect calorimetry.[60]

Nutrition Diagnosis Nutrition diagnoses associated with cirrhosis include those listed for alcoholic hepatitis above (see "Nutrition Therapy for Individuals with Alcohol Dependency/ Alcoholic Hepatitis"). Nutrition diagnoses for cirrhosis may relate to particular complications of the disorder; for example:

- *Ascites*—excessive sodium intake; excessive fluid intake
- *Esophageal varices*—swallowing difficulty; inadequate oral/ food beverage intake; disordered eating patterns
- *Encephalopathy*—excessive protein intake; inappropriate intake of AAA/BCAA
- *Hepatorenal syndrome*—excessive sodium intake

Nutrition Intervention
ENERGY NUTRIENTS Calorie and protein recommendations for patients with liver cirrhosis are 35–40 kcal/kg per day

with a protein intake up to 1.6 g/kg per day depending on the degree of malnutrition and other medical complications.[61, 62] Protein should only be restricted with severe forms of encephalopathy, and in the case of a coma that does not respond to standard treatment, an enteral product high in branched-chain amino acids should be administered.[58] Greater amounts of protein from vegetable and dairy sources have been recommended for patients with mild encephalopathy because these sources are lower in aromatic amino acids.[63] Fat restriction to less than 30% of calories (with or without medium-chain triglyceride supplements) is commonly recommended for patients with steatorrhea. Because diabetes is common in patients with cirrhosis, carbohydrate intake should be spread out by offering multiple meals throughout the day in order to minimize hypo- and hyperglycemia.

VITAMINS AND MINERALS Management of ascites usually requires a sodium restriction of 2 grams per day; fluids, however, are not usually restricted. Use of a multivitamin/mineral supplement as indicated by nutrition assessment and blood chemistry information is appropriate.[48] See Table 16.8 for more details on supplementation.

OTHER CONSIDERATIONS Adequate nutrition should be provided through meals and supplements whenever possible, but enteral and parenteral support should be considered in situations that result in inadequate oral intake and certainly are indicated for those patients who are comatose. For esophageal varices, a mechanically soft diet may be recommended as a preventative measure to reduce risk of bleeding, even if there are no evident nutrition diagnoses related to this complication.

Monitoring and Evaluation In general, the patient should be monitored for tolerance to feedings, amount of nutrients consumed, (dry) weight changes, laboratory values (see diagnostic indicators of liver disease in Table 16.3), and cognitive status.

Liver Transplant

Transplantation is considered in cases where the effects of liver disease have a greater potential to cause mortality than the short- and long-term complications of liver transplant. The most common medical conditions requiring a transplant in adults are: chronic active hepatitis, cirrhosis, and biliary-related disorders. A patient's eligibility to be placed on the waiting list for a liver transplant is determined based on the Model for End-Stage Liver Disease (MELD) scoring system. After transplant, all patients require immunosuppressive drugs to prevent rejection of the new liver (see Table 16.5 and Chapter 9). Overall patient survival rates at one and five years after a liver transplant are 86.4% and 72.9%, respectively.[64]

Nutrition Therapy for Liver Transplant

Nutritional Implications Dietary management before and after transplantation is individualized. Patients with end-stage liver disease (ESLD) on the transplant waiting list frequently display a gradual decline of their nutritional status. A general nutrition goal before transplant is to lessen the effects of malnutrition and complications of liver disease, such as ascites, discussed previously. After transplant, most nutritional deficiencies and metabolic disturbances common in patients with ESLD improve. However, the risks for preoperative malnutrition, surgical stress, post-interventional complications, postoperative protein catabolism, fasting periods, and side effects of immunosuppressant medication (Table 16.6) suggest the need for early nutritional support after the transplant. After recovery, patients are more susceptible to food-borne infections as a result of the immune-suppressing medication.

Nutrition Assessment Initially, SGA, anthropometric measurements (e.g., MAC), and dietary intake should be monitored as well as GI symptoms. As complications of ESLD lessen, biochemical lab values should be monitored, and if the patient no longer is retaining fluid, weight should be tracked. Side effects of post-transplant medications (see Table 16.5), such as hyperglycemia, hyperlipidemia, sodium retention, nausea, and vomiting, should be assessed.

Nutrition Diagnosis Possible diagnosis for liver transplant include increased energy expenditure; inadequate energy intake; inadequate oral food/beverage intake; inadequate protein-energy intake; malnutrition; inadequate vitamin/mineral intake (e.g. thiamin); altered GI function; impaired nutrient utilization; underweight; altered nutrition-related laboratory values; food-medication interactions; food and nutrition-related knowledge deficit; and involuntary weight loss.

Nutrition Intervention After transplant, the nutritional goal is to meet the needs for healing. Preferred nutrition support should be either oral or enteral since they are both associated with lower postoperative infection rates compared to parenteral nutrition. Calorie and protein recommendations are 15%–30% above basal needs and 1.5–2.0 g/kg, respectively.[65] Other nutrients are individualized, based on the immunosuppressant drug regimen used to prevent rejection and the functioning of the new liver. Hyperglycemia can be managed by decreasing simple sugars, with carbohydrates providing 50% to 60% of total kcal; and sodium retention, by a 2–4 g sodium restriction. Calcium supplements along with a multivitamin may be recommended post surgery to help maintain bone health and ensure overall nutritional needs are being met. Patients should be educated on food safety because of their increased susceptibility to food-borne illnesses (see Chapter 24).

Monitoring and Evaluation Post-transplant patients should be monitored regularly for recovery and healing and evaluated for dietary intake, tolerance to the diet, weight changes, incidence of food-borne illness, and nutrient-related complications associated with the medications.

Pathophysiology of the Biliary System

Biliary disease includes a wide variety of disorders of bile composition or biliary anatomy or function. The liver is the source of bile, but bile's composition is further modified by the gallbladder. Gallstones (**cholelithiasis**) and biliary tract diseases are common problems in the United States and are thought to affect 20 million Americans each year.[66] Gallstone related diseases are responsible for about 10,000 deaths per year in the United States, and 7,000 deaths are attributable to acute gallstone complications, such as acute pancreatitis. About 2,000–3,000 deaths are caused by gallbladder cancers, of which

80% occur in individuals who have had previous gallstones, chronic infections, and/or inflammation of the gallbladder (**cholecystitis**) or other disorders of the gallbladder.[67]

Cholelithiasis (Gallstones)

Cholelithiasis is the formation of stones (calculi) within the gallbladder or biliary duct system and is one of the most common medical problems leading to surgery. There are basically three types of stones: cholesterol (80%, with only ~10% being purely cholesterol), pigment, and mixed stones.[68] Cholesterol, a major component of bile, is normally kept in solution by bile acids, lecithin, and phospholipids. However, when bile is supersaturated with cholesterol, it crystallizes and gallstones are formed. **Biliary sludge**, a mixture of microscopic particulate—usually cholesterol crystals and calcium salts—is associated with many of the same risk factors as gallstones. Usually the presence of sludge has few symptoms, and it may appear and disappear over time. Biliary sludge, however, is thought to grow in size and result in larger gallstones.[69]

There are approximately 500,000 surgeries to remove gallstones (cholecystectomies) performed each year. Nearly two-thirds of patients with gallstones are asymptomatic.[70] The exact cause of this disorder is not known. However, the presence of obesity, diabetes, inflammatory bowel disease, or cystic fibrosis increases the risk for developing gallstones, as do rapid weight loss, fat-restricted diets, bariatric surgery, and cholesterol-lowering medications. Prolonged parenteral nutrition and short bowel syndrome can also cause **biliary stasis** and therefore increase risk of stones. Other risk factors include: multiple pregnancy in women, use of estrogen medication, and a genetic propensity. Caucasians, Mexican Americans, and Native Americans have a relatively high prevalence of gallstones. Approximately 50% of Caucasian women and 30% of men will develop gallstones during their lifetime.[71] Gallstone disease is less common in Asians and Africans.

The etiology of the formation of gallstones has been under investigation for years. In the process of bile salts secretion, approximately one-tenth as much free cholesterol is also secreted into the bile. Cholesterol is insoluble in water, but the bile salts and lecithin keep it in solution in the form of micelles. Because bile salts and lecithin are concentrated with cholesterol in the gallbladder, cholesterol remains in solution. In abnormal conditions, however, cholesterol precipitates as gallstones.

Inflammation of the gallbladder epithelium often results from low-grade chronic infection; this changes the absorptive characteristics of the gallbladder mucosa, sometimes allowing excessive absorption of water, bile salts, or other substances that are necessary to keep the cholesterol in solution. As a result, cholesterol begins to precipitate, usually forming many small crystals of cholesterol on the surface of the inflamed mucosa. These, in turn, act as nidi (points of origin) for further precipitation of cholesterol, and the crystals grow larger and larger. Occasionally, tremendous numbers of sand-like stones develop, but much more frequently they coalesce to form a few large gallstones, or even a single stone that fills the entire gallbladder. Four conditions that may cause the precipitation of gallstones are illustrated in Figure 16.9.

The most common treatment for symptomatic cholelithiasis is surgical removal of the gallbladder (**cholecystectomy**) laparoscopically (see Box 16.6). Medication (ursodeoxycholic acid/

Figure 16.9 Formation of Gallstones

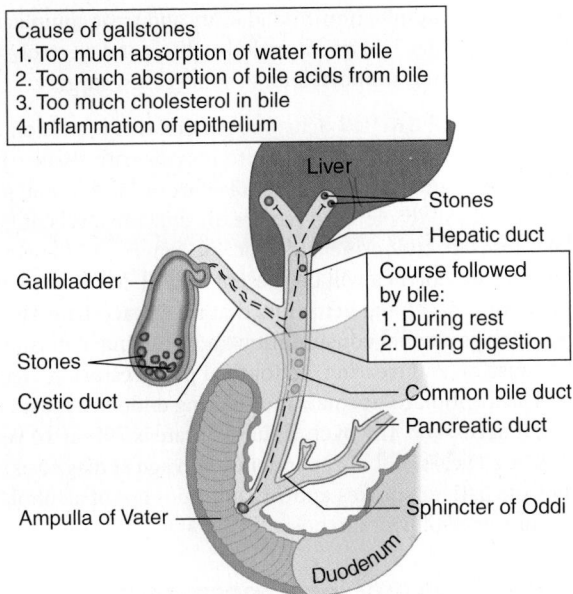

Cause of gallstones
1. Too much absorption of water from bile
2. Too much absorption of bile acids from bile
3. Too much cholesterol in bile
4. Inflammation of epithelium

Liver
Stones
Hepatic duct
Gallbladder
Course followed by bile:
1. During rest
2. During digestion
Stones
Cystic duct
Common bile duct
Pancreatic duct
Ampulla of Vater
Sphincter of Oddi
Duodenum

Source: AA Guyton, *A Textbook of Medical Physiology*, W.B. Saunders Co.; 2000. Fig. 64-12.

ursodeoxycholic acid) can be administered to help dissolve stones but is only used for patients who may not survive surgery or in clients who are prone to developing stones. To treat any accompanying infections, antibiotics are provided, along with analgesics and antiemetics for nausea and vomiting.

Biliary Obstruction When a gallstone passes from the gallbladder through the cystic duct and lodges in the common bile duct, or in the head of the pancreas, this condition is called **choledocholithiasis**. The bile is no longer carried to the duodenum and the excretion of bile pigments into the urine gives the urine a dark color. The feces are no longer colored by bile pigments and hence become grayish (clay colored). Additionally, maldigestion and malabsorption of fat develop. Patients usually experience severe right upper quadrant pain (biliary colic). If uncorrected, backed-up of bile can result in jaundice and liver damage (secondary **biliary cirrhosis**). A stone blocking the ampulla of Vater may transiently obstruct the pancreatic duct, triggering an attack of acute pancreatitis.[72] Biliary obstruction is treated with endoscopic retrograde sphincterotomy, a medical procedure used to remove gallstones from the common bile duct (see Box 16.6). In this procedure, the surgeon enters the bile duct via the ampulla of Vater from the duodenum. For infections, antibiotics are provided, along with analgesics and antiemetics for nausea and vomiting.

Cholecystitis Inflammation of the gallbladder, cholecystitis, generally develops secondary to obstruction, infection, and ischemia of the gallbladder. This condition can be chronic or acute. Obstruction of the cystic duct by gallstones is the most common cause of gallbladder inflammation, which can lead to infection and necrosis.

Cholangitis Cholangitis is an inflammation of the biliary ducts, usually secondary to obstruction of the common bile duct leading to infection. The infection can ascend into the hepatic ducts, then into the biliary canaliculi, hepatic veins,

and perihepatic lymphatics, leading to sepsis.[73] It is a life-threatening complication of biliary obstruction, particularly in the elderly. Initial therapy generally consists of antibiotics, fluid resuscitation, and correction of blood clotting.[74]

Nutrition Therapy for Cholelithiasis

Nutritional Implications Symptomatic gallstones and acute attacks require a diet low in fat to relieve symptoms before surgery. Patients experiencing an acute attack of cholangitis should take nothing by mouth (NPO) for twelve hours before surgery. Indigestion, decreased ability to digest fat, and increased abdominal gas can contribute to decreased food intake and altered nutritional status. After surgery, patients adjust rapidly to a regular diet since bile from the liver continues to enter the duodenum via the common bile duct. Diarrhea may occur after surgery, probably due to bile in the large intestines. Other causes for the diarrhea should be excluded before nutrition intervention is suggested.[75]

Nutrition Assessment For the patient with gallbladder disorders, weight and weight history should be obtained, and a 24-hour recall, diet history, or food diary should be used to determine intake and tolerance of specific foods or nutrients (e.g., fatty foods). Albumin and prealbumin levels should be assessed if the patient is NPO and/or has a concurrent infection. Medication for treatment of gallstones, infection, and pains should be noted.

Nutrition Diagnosis Possible nutrition diagnoses for biliary disorders include: inadequate oral food/beverage intake; altered GI function; food-medication interaction; and—if the patient has an infection—increased nutrient needs.

Nutrition Intervention A low-fat nutrition prescription (<30% energy as fat) with a modest protein content may assist in controlling symptoms until surgery is performed to remove the gallstones. Small, frequent feedings may also help improve the total nutrient intake to meet patients' needs. An acute attack almost always occurs in connection with an obstruction. When it does occur, the gallbladder should be kept as inactive as possible, which is achieved through an NPO order and complete bowel rest until symptoms lessen, with nutrition administered parenterally as needed.[76] The diet is advanced as tolerated to liquids, though only low-fat liquids are typically used. Due to poor absorption of fat, a water-soluble form of vitamins A, D, E, and K may be necessary. Postsurgery diarrhea may be managed through increased fiber intake to increase fecal bulk, and patient avoidance of foods that are known to cause diarrhea.[77]

Monitoring and Evaluation Adherence to, as well as, tolerance of the diet should be determined. If the patient is NPO and/or has an infection, weight and biochemical labs should be monitored as needed, based on nutritional status.

Pathophysiology of the Exocrine Pancreas

Pancreatitis

Pancreatitis, an inflammation of the pancreas, can be either acute or chronic. The disease is characterized by edema, autodigestion, fat necrosis, and hemorrhage of pancreatic

Table 16.12 Common Causes of Acute Pancreatitis

- Gallstones (including microlithiasis)
- Alcohol (acute and chronic alcoholism)
- Hypertriglyceridemia
- Endoscopic retrograde cholangiopancreatography (ERCP), especially after biliary manometry
- Trauma (especially blunt abdominal trauma)
- Postoperative (abdominal and nonabdominal operations)
- Drugs (azathioprine, 6-mercaptopurine, sulfonamides, estrogens, tetracycline, valproic acid, anti-HIV medications)
- Sphincter of Oddi dysfunction

tissue. There are approximately two hundred thousand new cases of acute pancreatitis each year in the United States.[78] Thirty to sixty percent of acute pancreatitis cases are due to a gallstone passing into the bile duct and temporarily lodging at the sphincter of Oddi. Alcohol accounts for 15%–30% of acute pancreatitis cases, and hypertriglyceridemia (serum triglyceride levels often higher than 1000 mg/dL) is the cause in 1.3%–3.8% of cases. Other common causes are listed in Table 16.12.

Chronic pancreatitis is the result of irreversible damage to the pancreas that is caused by repeated inflammation and results in destruction of both the exocrine and eventually endocrine tissue. In the United States, alcoholism is the most common cause of chronic pancreatitis in adults, and cystic fibrosis is the most frequent cause in children (see Box 16.4 and Chapter 21 for more information on cystic fibrosis). Twenty-five percent of chronic pancreatitis is of unknown origin, but it is now thought that up to 15% of cases are due to genetic defects.[79]

Acute Pancreatitis   While some cases of acute pancreatitis are asymptomatic, presenting symptoms may include upper abdominal pain radiating to the back, generally worsening with ingestion of food. Clinical presentation may also include nausea, vomiting, abdominal distention, and steatorrhea. Severe cases are complicated by hypotension and dehydration. The exact mechanisms that lead to pancreatic injury are not fully understood. However, a common characteristic seems to be premature activation of trypsin within the pancreas, resulting in autodigestion of the pancreatic cells. The enzymes released by destroyed pancreatic cells eventually reach the bloodstream, causing elevated serum amylase and lipase levels. For acute pancreatitis, a diagnosis is based on the clinical symptoms and the abnormal serum amylase and/or lipase.[78] Various criteria (e.g., Ranson and APACHE II scores) have been used to help predict the severity and outcome of acute pancreatitis, but none are currently recognized as a diagnostic standard. Recently, research has indicated that blood urea nitrogen (BUN) levels within 24 hours of admission may be a predictor of severity.[80] In patients with severe disease (organ failure), the mortality rate is approximately 30%.[81]

Medical treatment for mild acute pancreatitis includes administration of intravenous fluids for hydration and analgesics for pain relief. Recovery is generally uneventful. In contrast, patients with severe acute pancreatitis require intensive care since a number of complications (e.g., shock, pulmonary failure, renal failure, gastrointestinal bleeding, multiorgan system failure) may develop within hours or days of admission. The goals of medical management are to provide aggressive supportive care, decrease inflammation, limit infection, and identify and treat complications as appropriate. Treatment includes fluids and electrolytes, antibiotics, nutrition support, and, in some cases, surgery.[80]

Chronic Pancreatitis   Chronic pancreatitis is a chronic, irreversible inflammation that leads to fibrosis with tissue calcification. Signs and symptoms include chronic abdominal pain and normal or mildly elevated pancreatic enzyme levels. If the pancreas loses its endocrine and exocrine functions, diabetes mellitus and steatorrhea will be present as well. Medical treatment includes modifying behaviors that may exacerbate the disease, such as alcohol consumption and smoking; determining the cause of, and treating, abdominal pain; restoring digestion and absorption to normal; and treating endocrine insufficiency, if necessary. The overall survival rate is 70% at 10 years and 45% at 20 years, but is dependent upon age at diagnosis, whether the patient smokes and/or continues use of alcohol, and the presence of liver cirrhosis.[82]

Nutrition Therapy for Pancreatitis

Nutritional Implications   For the mild form of acute pancreatitis, research now supports feeding patients without delay, as tolerated. Nutrition support during this period is not usually necessary unless the patient is malnourished or cannot tolerate oral feeding for longer than a week. The diet is generally modified to accommodate digestive abilities, but little research has been conducted on nutrient composition of the diet in mild acute pancreatitis. For the severe form of acute pancreatitis, early initiation of enteral nutritional supplementation is recommended in order to maintain nutritional status or minimize losses and to prevent intestinal bacterial translocation.[83] If additional organs are compromised, the amount and composition of the nutritional support may need to be modified. For chronic pancreatitis, the nutritional status of the patient will depend on the underlying etiology of the disease and level of endocrine and exocrine function. In general, chronic pancreatitis often leads to steatorrhea; thus, fat intake is limited and enteric-coated pancreatic digestive enzymes are provided (see Tables 16.7 and 16.13). Additionally, there is a decrease in vitamin B_{12} absorption.[81] Absorption of fat-soluble vitamins may also be impaired.

Nutrition Assessment   For patients with pancreatitis, height, weight, and weight history should be obtained. Diet history should be used to determine intake and tolerance of specific foods or nutrients (e.g., fatty foods). Lifestyle factors, such as frequency and amount of alcohol intake, should be determined. Because pancreatitis is an inflammatory state causing a reprioritization of hepatic proteins, the following laboratory values may not reflect the nutritional status of the patient: albumin, prealbumin, and transferrin. Labs that should be monitored are blood glucose, triglycerides, hematocrit, hemoglobin, and white blood cells. Medication should be assessed for drug-nutrient interactions.

Nutrition Diagnosis   Possible nutritional problems include: altered GI function, malnutrition, inadequate oral food/beverage intake, impaired nutrient utilization, altered nutrition-related laboratory values, and increased nutrient (protein/energy) needs.

Table 16.13 Pancreatic Enzymes

Enteric-coated pancreatic enzyme microspheres assist in digestion of protein, starch, and fat and contain various amounts of lipase, protease, and amylase.

Name	Composition
Creon	
5	Delayed-release 5,000 units lipase/18,750 units protease/16,600 units amylase
10	Delayed-release 10,000 units lipase/37,500 units protease/33,200 units amylase
20	Delayed-release 20,000 units lipase/75,000 units protease/66,400 units amylase
Kutrase	2,400 units lipase/30,000 units protease/30,000 units amylase
Kuzyme	1,200 units lipase/15,000 units protease/15,000 units amylase
Ku-Zyme HP	8,000 units lipase/30,000 units protease/30,000 units amylase/lactose
Lipram	
4500	Delayed release: 4,500 units lipase/25,000 units protease/20,000 units amylase
PN10	Delayed release: 10,000 units lipase/30,000 units protease/30,000 units amylase
PN16	Delayed release: 16,000 units lipase/48,000 units protease/48,000 units amylase
PN20	Delayed release: 20,000 units lipase/44,000 units protease/56,000 units amylase
UL12	Delayed release: 12,000 units lipase/39,000 units protease/39,000 units amylase
UL18	Delayed release: 18,000 units lipase/58,500 units protease/58,500 units amylase
UL20	Delayed release: 20,000 units lipase/65,000 units protease/65,000 units amylase
Pancrease	
MT 4	4,000 units lipase/12,000 units protease/12,000 units amylase
MT 10	10,000 units lipase/30,000 units protease/30,000 units amylase
MT 16	16,000 units lipase/48,000 units protease/48,000 units amylase
MT 20	20,000 units lipase/44,000 units protease/56,000 units amylase
Pancrecarb	
MS-4	Delayed release: 4,000 units lipase/25,000 units protease/25,000 units amylase
MS-8	Delayed release: 8,000 units lipase/45,000 units protease/40,000 units amylase
Pancrelipase	
	Capsules: 4,500 units lipase/25,000 units protease/20,000 units amylase
	Tablets: 8,000 units lipase/30,000 units protease/30,000 units amylase
	Capsules: 16,000 units lipase/48,000 units protease/48,000 units amylase
	Tablets: 16,000 units lipase/60,000 units protease/60,000 units amylase
Panokase	8,000 units lipase/30,000 units protease/30,000 units amylase
Plaretase8000	8,000 units lipase/30,000 units protease/30,000 units amylase
Ultrase	
—	4,500 units lipase/25,000 units protease/20,000 units amylase
MT12	12,000 units lipase/39,000 units protease/39,000 units amylase
MT18	18,000 units lipase/58,500 units protease/58,500 units protease
MT20	20,000 units lipase/65,000 units protease/65,000 units protease
Viokase	
—	Powder: 16,000 units lipase/60,000 units protease/60,000 units amylase
8	Tablet: 8,000 units lipase/30,000 units protease/30,000 units amylase
16	Tablet: 16,000 units lipase/60,000 units protease/60,000 units amylase

Source: Drugs.com. Available at http://www.drugs.com/ppa/pancrelipase.html

Nutrition Intervention

ACUTE PANCREATITIS During mild acute attacks and in patients with adequate nutrition stores, oral feeding has traditionally been withheld until pain and nausea/vomiting have abated and bowel sounds returned. However, current research has found that immediate refeeding shortens hospital length of stay and that there are no statistical differences between fed and fasted patients as regards pain, lab values, or GI symptoms.[84] Traditionally, patients were started on clear liquids; however, current research now supports oral intake as tolerated. One study suggests that an unrestricted diet is appropriate, while another suggests a diet that provides at least 1000 kcal/day, in the form of easily digested foods with a lower fat content.[85] Foods may be better tolerated if they are divided into six small meals to compensate for the diminished exocrine function of the pancreas.[86] Supplement the patient's diet with a multivitamin and minerals until nutritional intake is adequate to meet all nutrient needs.[6]

Early, aggressive nutrition support is suggested in severe, acute pancreatitis. Protein catabolism increases by 80% and caloric needs increase by 20%; nutrition support in these patients may help prevent nutrient deficits while preserving or minimizing loss of lean body mass.[87] The use of nutrition support to resolve negative nitrogen balance in acute pancreatitis is associated with improved outcomes.[88] Goals, therefore, are to provide adequate kcal and protein, minimize nitrogen losses, and manage imbalances. These include hypo-/hyperglycemia, hypertriglyceridemia, and imbalances of micronutrients such as calcium, magnesium, and vitamins. At the same time, pancreatic stimulation should be minimized.

While in the past parenteral feedings were the norm, enteral feedings have now become the preferred method. Recent studies suggest that, compared to parenteral nutrition support, early enteral feedings are associated with a reduction in infection, mortality, and need for surgery and are less costly to administer.[87, 89] In addition, they maintain gut integrity. A prospective study of pancreatitis patients comparing the effects of TPN and gut disuse to those of enteral feeding found that enteral nutrition maintained near-normal villi, while TPN and gut disuse resulted in significant villous atrophy.[90] Even hypocaloric jejunal feedings have been shown to be superior to TPN.[91] Gastric feeding should be tried first.[92] If duodenal ileus or obstruction from the inflammatory mass occurs, jejunal feeding should be initiated. However, paralytic ileus is common in severe pancreatitis, and in this case parenteral nutrition is recommended if enteral feeding has failed after five to seven days.[89, 93]

Recommended initiation of enteral feeding is 25 mL/hour with advancement to a goal kcal level of 25 kcal/kg over the first 24–48 hours.[94] If the tube is low enough, any formula may be used, but nearly fat-free elemental formulas result in the least stimulation of the pancreas; small-peptide formulas with 70% of the fat as medium-chain triglycerides stimulate the pancreas slightly more, but are better absorbed. Some small trials have indicated that this population may benefit from immune-enhanced formulas. Advancement to an oral diet may occur when amylase and lipase begin decreasing towards normal levels and the patient has been pain free for at least 24 hours.[91] For parenteral nutrition, the volume should be increased slowly to a goal kcal level of 25 kcal/kg. Fat emulsion should provide less than 15% to 30% of kcal, and protein should be individualized to meet patient needs.

CHRONIC PANCREATITIS The goals of medical and nutrition therapy are to prevent further damage to the pancreas, forestall further attacks of acute inflammation, alleviate pain, treat steatorrhea, and correct malnutrition, preventing weight loss and promoting weight gain as appropriate. The frequency of attacks may be reduced by frequent small meals of a moderate- to low-fat diet (Table 16.7). Pancreatic enzymes are prescribed to be taken at each meal or snack to improve digestion and absorption (Table 16.13). The amounts of these enzymes necessary with each meal may vary depending on the fat content of the food consumed. Because pancreatic bicarbonate secretion is frequently defective, medical management may also include maintenance of an optimal intestinal pH to facilitate enzyme activation. Antacids, H2-receptor antagonists, or proton pump inhibitors that reduce gastric acid secretion may be used to achieve this effect.[87]

To promote weight gain, the level of fat in the diet should be the maximum a patient can tolerate without increased steatorrhea or pain. Medium-chain triglycerides may be added to the diet, since they do not require lipase for digestion. Supplementation of vitamins and minerals is recommended. Water-soluble forms of the fat-soluble vitamins and parenteral administration of vitamin B_{12} may be necessary.

Dietary recommendations should be adjusted for other medical conditions associated with chronic pancreatitis, such as diabetes, obesity, and alcoholism.[6] For those patients with a history of alcoholism, thiamin (100 mg by mouth once a day) and folate (1 mg by mouth once a day) should be provided.[6]

Monitoring and Evaluation Generally, weight measurements should be obtained at least weekly and adjustments in the nutrition regimen made accordingly to prevent rapid weight loss as well as excessive gain. Overall improvement in symptoms with progression toward resuming oral feedings is the primary outcome goal in nutrition treatment of pancreatitis. Electrolytes, especially calcium, should be monitored closely, as should triglyceride and blood sugar levels, which should be maintained as near to normal as possible.[94]

Conclusion

Diseases of the hepatobiliary system have a significant impact on the nutritional status of the patient. The clinical manifestations—jaundice, anorexia, fatigue, abdominal pain, steatorrhea, and malabsorption—all impact nutritional status. Furthermore, the disease processes have the potential to interrupt normal metabolism, placing the patient at significant nutritional risk. Hence, nutrition therapy is a vital component of medical treatment.

Mary Ellen Beindorff, RD, LD *Abdominal Organ Transplant Nutrition Specialist,*
Barnes-Jewish Hospital, St. Louis, MO

Background

I have been a dietitian for twenty-five years, and currently my primary responsibility is patients with abdominal organ transplant (liver, kidney, pancreas) and hepatobiliary surgery.

How many liver patients do you see in a day, and what are the primary disorders?

I see four to ten patients with liver disease every day. This is significantly higher than the number a dietitian would see in a community hospital, which would be more likely to be a handful a month. I see liver patients with a wide variety of diagnoses, but the most common liver diseases are hepatitis C (HCV) and alcoholic liver disease. However, I also see patients with primary biliary cirrhosis (PBC), primary sclerosing cholangitis (PSC), non-alcoholic steatohepatitis (NASH), hepatitis B (HBV), autoimmune hepatitis (AIH), and other genetic liver disorders (Wilson's disease, hemochromatosis, biliary atresia, etc.). I see less hepatitis B (HBV) patients because the vaccine has eliminated the virus in patients under the age of 40. Now most of our HBV patients are foreign-born.

How has MNT practice for liver disease changed over the years?

When I first started practice, all liver patients were put on low-protein diets. Now we know that only in cases of refractory encephalopathy (in 5%–10% of patients) should protein be restricted. We also used to severely restrict fluid in liver patients; now we restrict only when hyponatremia occurs (usually with serum Na levels less than 125 mEq/L). So both protein and fluid restrictions may occur, but only temporarily. Unfortunately, I find that many MDs who do not see liver patients on a regular basis still recommend these restrictions and many RDs (not knowing any better) educate the patient on these restrictions.

How do you nutritionally assess a liver patient?

The key information needed for nutritional assessment is a detailed weight history, thorough diet history, functional capacity, physical assessment, past medical history, current medications (including herbals), and labs. By functional capacity I mean, are they still working? Are they independent with sufficient activities of daily living (ADL)? Do they need help for some things? Do they mostly stay in bed or on the couch all day? Most labs are not helpful in assessing liver patients, but I do look at electrolytes, to correct if necessary. I also look at bilirubin and international normalized ratio (INR), which measures the speed of prothrombin time for coagulation, to determine the extent of liver disease.

What is a common nutrition diagnosis?

The "typical" diagnosis for liver disease patients is some degree of malnutrition. In the past years, I have seen many more severely obese liver patients that are malnourished. To assess this, I rely on a thorough diet history and physical assessment. For diet history, I look to see if adequate protein and other nutrients are ingested, and on the physical exam, I look for muscle wasting and to what degree in addition to noting volume of ascites and/or edema. Physicians often assume that because patients are obese they are well nourished, but this is not necessarily the case.

What are the challenges working with this population?

I see the greatest challenges in working with the moderately to severely malnourished patients who, because of symptoms of their liver disease, have a difficult time getting adequate dietary intake. They usually have ascites, which often causes early satiety and reflux. If they're getting tapped often, they're losing protein there too. They often are also taking lactulose to control encephalopathy, which causes malabsorption and diarrhea.

What resources to you use to stay current?

I currently rely on many resources for information on liver disease and nutrition support. Both the Dietitian in Nutrition Support (DNS) practice group, ADA, and ASPEN have numerous resources available on liver disease and nutrition support. *Nutrition in Clinical Practice*, an ASPEN publication, and DNS's Support Line have articles related to liver disease and nutrition often. In addition, I scan the medical journals available each month in the hepatologists' and liver surgeons' offices. Lastly, I find it invaluable to "network" with my colleagues at national meetings.

Any advice for the nutrition students?

In my opinion, many interns are not confident of their clinical skills at first. Jump right in and try to do it. (We all learn from our mistakes.) Be self-confident; we are the nutrition experts, and most of the time, we know much more than our audience.

Introduction:

Anthony Cowan is a 45-year-old Caucasian male who is employed as a sales representative for a midwestern trucking firm. He is married with two children. Mr. Cowan presented to his physician with acute abdominal pain extending into the lower back; severe nausea and vomiting; and a fever of 101°F. His blood labs revealed: TP 3.6 g/dL; Alb 3.5 mg/dL; Lipase 521 U/L; Amylase 925 U/L. He was admitted to the hospital. Dx: mild form of acute pancreatitis; R/O obstructive biliary stone. Tx:, analgesic for pain, and IV fluids for hydration.

1. Describe the pathophysiology of acute pancreatitis. What signs and symptoms support the diagnosis?

2. What are the common causes of acute pancreatitis?

3. Why is Mr. Cowan not NPO?

Nutrition Assessment:

Food/Nutrition-Related History:

Normally consumes 3 meals/day with breakfast and evening meal at home with family. Rarely eats out. Has no food intolerances. Describes alcohol intake as 5 beers/week. Denies regular "hard liquor" intake but does consume occasionally. The weekend before his hospitalization, he went to Las Vegas for a bachelor party for his best friend and admits to drinking too much. Denies any drug use except prescribed medicines. *Past Med Hx*: Hypertension 1 years; *Rx* Captopril 25 mg; Hydrochlorothiazide 50 mg.

Anthropometrics:

Ht: 6' Wt: 192#; UBW: 200–210#

Biochemical Data:

TP 3.6 g/dL; Alb 3.5 mg/dL; Lipase 521 U/L; Amylase 925 U/L.

4. What admission laboratory values are abnormal? Interpret their significance in relationship to both his diagnosis and his nutritional status.

5. What factor in this client's history is consistent with risk for acute pancreatitis?

Nutrition Diagnosis:

6. Identify at least two nutrition problems suggested by the nutrition assessment and medical history. Determine the diagnostic term for each nutrition problem. Next, identify the etiology of each nutrition diagnosis. Finally, identify the signs and symptoms that support the diagnoses.

Diagnosis	Related to	Etiology	As Evidenced by	Sign/Symptoms
EXAMPLE: Involuntary weight loss	R/T	poor dietary intake secondary to abdominal pain	AEB	25% unplanned weight loss in 6 weeks

END-OF-CHAPTER QUESTIONS

1. List the major functions of the liver, pancreas, and gallbladder.

2. What is jaundice? What is the difference between conjugated and unconjugated bilirubin? What disorders could cause elevated unconjugated jaundice, and what disorders could cause elevated conjugated jaundice?

3. What is portal hypertension (include causes, signs, and symptoms)? Why would portal hypertension cause ascites and esophageal varices? What is the medical and nutrition treatment for portal hypertension?

4. List the types of viral hepatitis and their modes of transmission. How is alcoholic hepatitis different from viral hepatitis?

5. What is cirrhosis? List some of the causes of this liver disorder. What are the common complications of cirrhosis? How does cirrhosis cause hypoglycemia and hyperglycemia? What parameters can you use to nutritionally assess a patient with cirrhosis?

6. What are the possible biochemical causes of hepatic encephalopathy?

What is the amount and type of protein used in the medical nutrition therapy for hepatic encephalopathy?

7. What is recommended for a post-operative cholecystectomy diet?

8. Describe the difference between acute and chronic pancreatitis. List the pertinent labs for each type of pancreatitis. What are common nutritional problems associated with chronic pancreatitis?

Complementary and Alternative Medicine Remedies Used in the Diseases of the Hepatobiliary System

Remedy	Scientific Name	CAM Use	Side Effects and/or Risks
Colloidal silver	Silver	Hepatitis C	Silver has no known function in the human body, and is not an essential mineral supplement. Claims of silver "deficiency" in the body are unfounded. Silver builds up in body tissues: argyria—a bluish gray discoloration of the body, especially skin, other organs, deep tissues, nails, and gums. Argyria is not treatable or reversible. Other possible problems include neurologic effects (such as seizures), kidney damage, stomach distress, headaches, fatigue, and skin irritation. May interfere with absorption of the following drugs: penicillamine, quinolones, tetracyclines, and thyroxine.
Licorice root (plant)	Glycyrrhiza glabra	Hepatitis C; peptic ulcer disease	Licorice intake over a long period of time can lead to hypertension, salt and water retention, swelling, depletion of potassium, headache, and/or sluggishness; can worsen ascites; can interact with certain drugs (diuretics, digitalis, antiarrhythmic agents, and corticosteroids).
Milk thistle (plant)	Silybum marianum	Cirrhosis	Generally well tolerated; can cause laxative effect, allergic reactions in people allergic to ragweed, chrysanthemum, marigold, and daisy.
Neem; nimb	Azadirachta india	Cleanse liver	None known; no full scientific evaluation of toxicity and side effects has been conducted.
Noni, Indian mulberry; ashyulka	Morinda citirolia	Treat liver ailments	No known side effects, but no safety studies have been conducted. Use by patients with severe liver disease not recommended. May interact with potassium-sparing diuretics due to high potassium content.
Peppermint	Mentha piperita	Dissolve gallstones	No evidence of effectiveness.
SAMe (S-adeno-sylmethionine)		Treat cirrhosis	Generally well tolerated. Long-term use may decrease efficacy of L-dopa.
Schisandra; gomishi; wu wei zi	Schisandra chinensis	Treat hepatitis	Side effects rare, but in large doses it may cause GI distress or appetite depression. May enhance elimination of hepatically metabolized drugs. Avoid use in pregnant or breast-feeding women.
Reishi; ling zhi	Ganoderma lucidum	Treat hepatitis	Thought to be safe, but safety studies have not been performed. Occasional side effects include mild digestive upset, dry mouth, and rash. No known drug interactions.
Schisandra (plant)	Schisandra chinensis, Schisandra sphenanthera	Hepatitis C	Found as an ingredient in herbal formulas; considered generally safe; in some it may cause heartburn, acid indigestion, decreased appetite, stomach pain, or allergic skin rashes.
Thymus extract (gland)	(Should not be confused with the prescription drug thymosin alpha-1)	Hepatitis C	Thrombocytopenia (drop in number of platelet cells in blood); concern about possible contamination from diseased animal parts (people on immunosuppressive drugs should use caution).
Turmeric	Curcuma longa	Cholelithiasis, GI bacterial overgrowth, flatulence, gastritis, ulcers, hepatic disorders	Avoid administration with alcohol, MAOIs, narcotics, OTC cold and flu medications, digoxin, drugs metabolized by CYP3A, indinavir, amphetamines, SSRIs, tricyclic antidepressants. May result in phototoxicity. May influence hepatic microsomal enzymes.

Sources: Fetrow CW, Avila JR. *Professional's Handbook of Complementary and Alternative Medicines*, 3rd ed. Philadelphia: Lippincott Williams & Wilkins, 2004.

Bratman S, Girman AM. *Mosby's Handbook of Herbs and Supplements and their Therapeutic Uses*. St. Louis: Mosby/Elsevier, 2003.

Fragakis AS, Thomson C. *The Health Professional's Guide to Popular Dietary Supplements*, 3rd ed. Chicago: American Dietetic Association, 2007.

1. Fausto N. Liver regeneration. J Hepatol. 2000;32(1 Suppl):19–31.

2. Sherwood L. Human Physiology. 7th ed. Belmont (CA): Brooks/Cole, Cengage Learning; 2010.

3. LJ Worobetz, RJ Hilsden, EA Shaffer et al. The Liver, Chapter 14 in First principles of gastroenterology: the basis of disease and an approach to management, A. B.; Shaffer, ed. Canadian Association Pf Gastroenterology/Astra Pharma. Available at http://www.gastroresource.com/gitextbook/en/Default.htm

4. Shier D, Butler J, Lewis R. Hole's human anatomy and physiology. 9th ed. St. Louis: McGraw Hill; 2002.

5. Way LW, Doherty GM. Current surgical diagnosis and treatment. 11th ed. St. Louis: McGraw Hill; 2003.

6. Pauli-Magnus C, Stieger B, Meier Y, Kullak-Ublick GA, Meier PJ. Enterohepatic transport of bile salts and genetics of cholestasis. J Hepatol. 2005;43:342–57.

7. Elsas LJ, Gilat T. Cholecystocolonic fistula with malabsorption. Ann Intern Med. 1965;63:481–86.

8. Pancreatitis. Nutrition Care Manual. Chicago (IL): American Dietetic Association, © 2009 American Dietetic Association. Available at www.nutritioncaremanual.com

9. Tintinalli JE, Kelen GD, Stapczynski JS, Ma OJ, Cline DM. Tintinalli's emergency medicine: a comprehensive study guide. 6th ed. St. Louis: McGraw Hill; 2004.

10. LaBrecque DR, Moody FG. Diseases of the liver and biliary tract. St. Louis: Mosby Year Book; 1991.

11. Jaundice. Medline Plus U.S. National Library of Medicine, 8600 Rockville Pike, Bethesda (MD): 20894. National Institutes of Health, Department of Health & Human Services. Date last updated: 24 August 2009. Available at http://www.nlm.nih.gov/medlineplus/jaundice.html#cat58

12. Bacon BR. Ch 302. Cirrhosis and Its Complications In. Harrison's principles of internal medicine. 17th ed. Edited by Fauci AS, Braunwald E, Kasper DL, et al. 2008 The McGraw-Hill Companies, Inc.

13. The Cleveland Clinic Foundation 1995–2009. Available at http://my.clevelandclinic.org/disorders/portal_hypertension/hic_portal_hypertension.aspx

14. Wang SS, Chen CC, Chao Y, Wu SL, Lee FY, Lin HC, et al. Sequential hemodynamic changes for large volume paracentesis in post-hepatitic cirrhotic patients with massive ascites. Proc Natl Sci Counc Repub China B. 1996;20:117–22.

15. Wolf DC. Cirrhosis. eMedicine. Available at http://emedicine.medscape.com/article/185856-overview

16. Gerber T, Schomerus H. Hepatic encephalopathy in liver cirrhosis: pathogenesis, diagnosis, and management. Drugs. 2000;60:1353–70.

17. Charles PD, Esper GJ, Davis TL, Maciunas RJ, Robertson D. (1999). Classification of tremor and update on treatment. Am Fam Physician. 59(6):1565–72.

18. Cordoba J, and Blei AT. (1997). Treatment of hepatic encephalopathy. American Journal of Gastroenterology. 92:1429–39.

19. Butterworth RF. The astrocytic ("peripheral-type") benzodiazepine receptor: role in the pathogenesis of portal-systemic encephalopathy. Neurochem Int. Apr 2000;36(4–5):411–16.

20. Hazell AS, Butterworth RF. Hepatic encephalopathy: an update of pathophysiologic mechanisms. Proceedings of the Society for Experimental Biology and Medicine. 1999;222:99–112.

21. Wolf DC. Cirrhosis. eMedicine. Available at http://emedicine.medscape.com/article/185856-overview

22. The ABCs of Hepatitis. Division of Viral Hepatitis, National Center for HIV/AIDS, Viral Hepatitis, STD, and TB Prevention. Available at http://www.cdc.gov/hepatitis/Resources/Professionals/PDFs/ABCTable_BW.pdf. Page last modified: March 26, 2009.

23. Bell BP, Kruszon-Moran D, Shapiro CN, et al. Hepatitis A virus infection in the United States: serologic results from the Third National Health and Nutrition Examination Survey. Vaccine. 2005;23:5798–5806

24. Hepatitis A. Division of Viral Hepatitis, National Center for HIV/AIDS, Viral Hepatitis, STD, and TB Prevention. Available at http://www.cdc.gov/hepatitis/ Page last modified: July 24, 2009.

25. Center for Disease Control and Profession. Surveillance for Acute Viral Hepatitis—United States, 2007. Daniels D, Grytdal S, Wasley A. Division of Viral Hepatitis, National Center for HIV/AIDS, Viral Hepatitis, STD, and TB Prevention, MMWR 2009;58 (Vol. 58/SS-3). Available at http://www.cdc.gov/hepatitis/statistics.htm

26. Hepatitis B. Division of Viral Hepatitis National Center for HIV/AIDS, Viral Hepatitis, STD, and TB Prevention. Available at http://www.cdc.gov/hepatitis/HBV. Page last reviewed: July 8, 2008. Page last modified: June 9, 2007.

27. National Hepatitis B Initiative for Asian Americans and Pacific Islanders. MMWR May 15, 2009;58(18);503. Available at mmwrq@cdc.gov. Date last reviewed: 5/14/2009.

28. Hepatitis C virus infection, acute. Laboratory criteria for diagnosis. Available at http://www.cdc.gov/ncphi/disss/nndss/casedef/hepatitiscacutecurrent.htm. Updated January 9, 2008.

29. CDC Division of Viral Hepatitis. Available at http://www.cdc.gov/hepatitis/HCV.htm. Page last modified: March 26, 2009.

30. Changing Trends in Hepatitis C–Related Mortality in the United States, 1995–2004 Matthew Wise, Stephanie Bialek, Lyn Finelli, Beth P. Bell, and Frank Sorvillo. Hepatology 2008;47:1128–35. Available at http://www.cdc.gov/hepatitis/statistics.htm 1995–2004.

31. Chronic Hepatitis C: Current Disease Management. NIH Publication No. 07–4230 November 2006. National Institute of Diabetes and Digestive and Kidney Diseases, National Institutes of Health. Available at http://digestive.niddk.nih.gov/ddiseases/pubs/chronichepc/#g

32. Viral Hepatitis. Division of Viral Hepatitis National Center for HIV/AIDS, Viral Hepatitis, STD, and TB Prevention Cited August 9, 2009. Available at http://www.cdc.gov/hepatitis/. Last modified: July 24, 2009.

33. Hepatitis. [cited August 9, 2009]. Nutrition Care Manual. Chicago (IL): American Dietetic Association, © 2009 American Dietetic Association. Available at www.nutritioncaremanual.com

34. Diagnostic and Statistical Manual of Mental Disorders (DSM-IV), 4th ed., Text Revision. Washington (DC): American Psychiatric Association; 2000.

35. Alcohol-Related Disease Impact. Available at http://www.cdc.gov/alcohol/quickstats/general_info.htm Division of Adult and Community Health, National Center for Chronic Disease Prevention and Health Promotion. Page last modified: August 6, 2008.

36. Harwood H. Updating Estimates of the Economic Costs of Alcohol Abuse in the United States: Estimates, Update Methods, and Data. Report prepared by The Lewin Group for the National Institute on Alcohol Abuse and Alcoholism, 2000. Based on estimates, analyses, and data reported in Harwood H, Fountain D, and Livermore, G. The Economic Costs of Alcohol and Drug Abuse in the United States 1992. Report prepared for the National Institute on Drug Abuse and the National Institute on Alcohol Abuse and Alcoholism, National Institutes of Health, Department of Health and Human Services. NIH Publication No. 98-4327. Rockville (MD): National Institutes of Health, 1998. Available at http://pubs.niaaa.nih.gov/publications/economic-2000/printing.htm

37. Heron MP, Hoyert DL, Murphy SL, Xu JQ, Kochanek KD, Tejada-Vera B. Deaths: Final data for 2006. National vital statistics reports; Vol 57 no 14. Hyattsville (MD): National Center for Health Statistics. 2009. Available at http://www.cdc.gov/NCHS/data/nvsr/nvsr57/nvsr57_14.pdf.

38. McCullough AJ, O'Connor JF. Alcoholic liver disease: proposed recommendations for the American College of Gastroenterology. Am J Gastroenterol. 1998;93:2022–36.

39. Becker U, Deis A, Sorensen TI, et al. Prediction of risk of liver disease by alcohol intake, sex, and age: a prospective population study. Hepatology. 1996;23:1025–29.

40. Sherman DIN, Williams R. Liver damage: mechanisms and management. Br Med Bull. 1994;50:124–38.

41. Sears D. Fatty liver [monograph on the Internet]. Omaha:eMedicine/WebMD; 2009. Available from http://emedicine.medscape.com/article/175472-overview. Updated: Jan 6, 2009.

42. Lieber CS. Alcoholic liver injury: pathogenesis and therapy in 2001. Pathol Biol. 49:738–52.

43. Capron JP, Delamarre J, Dupas JL, et al. Fasting in obesity: another cause of liver injury with alcoholic hyaline? Dig Dis Sci. 1982;27:265–68.

44. Wolf DC. Cirrhosis. Emedicine from Medscape.com. Available at http://emedicine.medscape.com/article/185856-overview. Updated Aug 11, 2008.

45. Mendenhall CL. Alcoholic hepatitis. Clin Gastroenterol. 1981;10:417–41.

46. Josh Levitsky, MD; Mark E. Mailliard, MD. Diagnosis and therapy of alcoholic liver disease. Semin Liver Dis. 2004;24(3).

47. Mathurin P, Mendenhall CL, Carithers RL, et al. Corticosteroids improve short-term survival in patients with severe alcoholic hepatitis (AH): individual data analysis of the last three randomized placebo controlled double blind trials of corticosteroids in severe AH. J Hepatol. 2002;36:480–87.

48. Halsted CH. Nutrition and alcoholic liver disease. Semin Liver Dis. 2004;24:289–304.

49. Nicolas JM, Garcia G, Fatjo F, Sacanella E, Tobias E, Badia E, Estruch R, Fernandez-Sola J. Influence of nutritional status on alcoholic myopathy. Am J Clin Nutr. 2003;78(2):326–33.

50. Alcoholism. Nutrition Care Manual. Chicago (IL): American Dietetic Association, © 2009 American Dietetic Association. Available at www.nutritioncaremanual.com

51. Hepatitis. Nutrition Care Manual. Chicago (IL): American Dietetic Association, © 2009 American Dietetic Association. Available at www. www.nutritioncaremanual.com

52. Bohinjec M. Clinical immunogenetics and cell therapy. Transpl Immunol. 2005;14:171–74.

53. Mendenhall CL. Alcoholic hepatitis. Clin Gastroenterol. 1981;10:417–41.

54. Malinchoc M, Kamath PS, Gordon FD, Peine CJ, Rank J, ter Borg PC. A model to predict poor survival in patients undergoing transjugular intrahepatic portosystemic shunts. Hepatology. 2000;31:864–71.

55. Ortega R, et al. Terlipressin therapy with and without albumin for patients with hepatorenal syndrome: results of a prospective, nonrandomized study. Hepatology. 2002; 36:941–48.

56. Bergheim I, McClai CJ, Arteel GE. Treatment of Alcoholic Liver Disease Dig Dis. 2005;23:275–84.

57. Tsiaousi ET, Hatzitolios AI, Trygonis SK, Savopoulos CG. Malnutrition in End Stage Liver Disease: Recommendations and Nutritional Support J Gastroenterol Hepatol. 2008;23(4):527–33.

58. Matos C, Porayko MK, Francisco-Ziller N, DiCecco S. Nutrition and chronic liver disease. J. Clin. Gastroenterol. 2002;35:391–97.

59. Plauth M, Cabre E, Riggio O, Assis-Camilo M, Pirlich M, Kondrup J. ESPEN guidelines on enteral nutrition: liver disease. Clin. Nutr. 2006;25:285–94.

60. McClave SA, Martindale RG, Vanek VW, McCarthy M, Roberts P, Taylor B, et al; Guidelines for the Provision and Assessment of Nutrition Support Therapy in the Adult Critically Ill Patient: Society of Critical Care Medicine (SCCM) and American, Society for Parenteral and Enteral Nutrition (A.S.P.E.N.) Journal of Parenteral and Enteral Nutrition. 2009;33(3):277–316.

61. Kondrup J. Nutrition in end stage liver disease. Best Pract. Res. Clin. Gastroenterol. 2006;20:547–60.

62. Plauth M, Cabre E, Riggio O, Assis-Camilo M, Pirlich M, Kondrup J. ESPEN guidelines on enteral nutrition: liver disease. Clin. Nutr. 2006;25:285–94.

63. Amodio P, Caregaro L, Patteno E, Marcon M, Del Piccolo F, Gatta A. Vegetarian diets in hepatic encephalopathy: facts or fantasies? Dig Liver Dis. 2001;33:492–500.

64. Manzarbeitia C. Liver Transplantation. eMedicine. Available at http://emedicine.medscape.com/article/431783-overview. Updated: November 1, 2007.

65. Stickel F, Inderbitzin D, Candina D. Role of nutrition in liver transplantation for end-stage chronic liver disease. Nutrition in Clinical Care. 2007;66(1):47–54.

66. Everhart JE, Khare M, Hill M, Maurer KR. Prevalence and ethnic differences in gallbladder disease in the United States. Gastroenterology. 1999;117:632–39.

67. Heuman DM, Anastasios A, Mihas, A, Allen, J, Cuschieri JA. Cholelithiasis. Emedicine. Available at http://emedicine.medscape.com/article/175667-overview. Updated: Aug 2, 2006.

68. Ahmed A, Cheung RC, Keeffe EB. Management of gallstones and their complications. Am Fam Physician. 2000;61:1673–80, 1687–88.

69. Heuman DM, Anastasios A, Mihas, A, Allen, J, Cuschieri JA. Cholelithiasis. Emedicine. Available at http://emedicine.medscape.com/article/175667-overview. Updated: Aug 2, 2006.

70. Gupta SK, Shukla VK. Silent gallstones: a therapeutic dilemma. Trop Gastroenterol. 2004;25:65–68.

71. Heuman DM, Mihas A, Allen J, Cuschieri JA. Cholelithiasis [cited August 16, 2009]. Emedicine. Available at http://emedicine.medscape.com/article/175667-overview. Updated: Aug 2, 2006.

72. Sugiyama M, Atomi Y. Risk factors for acute biliary pancreatitis. Gastrointest Endosc. 2004;60:210–12.

73. Santen S. Cholangitis. Available at http://www.emedicine.com/emerg/topic96.htm

74. Bornman PC, van Beljon JI, Krige JE. Management of cholangitis. J Hepatobiliary Pancreat Surg. 2003;10:406–14.

75. Gallbladder. Nutrition Care Manual. Chicago (IL): American Dietetic Association, © 2009 American Dietetic Association. Available at www.nutritioncaremanual.com

76. Kalloo AN, Kantsevoy SV. Gallstones and biliary disease. Prim Care. 2001;28:591–606.

77. Mayo Clinic. Diarrhea: a concern after gallbladder removal? Available at http://www.mayoclinic.com/health/gallbladder-removal/AN00067. Updated: March 31, 2009.

78. Greenberger NJ, Toskes PP. Chapter 307. Acute and Chronic Pancreatitis In: Harrison's principles of internal medicine. 17th ed. Edited by Fauci AS, Braunwald E, Kasper DL, et al. 2008. The McGraw-Hill Companies, Inc.

79. Hartmann D, Felix K, Ehmann M, et al. Protein expression profiling reveals distinctive changes in serum proteins associated with chronic pancreatitis. Pancreas. 2007;35:334–42.

80. Wu BU, Johannes RS, Sun X, Conwell DL, Peter A, Banks PA. Early changes in blood urea nitrogen predict mortality in acute pancreatitis. Gastroenterology. 2009:137(1):129–35.

81. Gardner TB, Berk, BS, Yakshe, P. Pancreatitis, Acute: Treatment & Medication. eMedcine. Available at http://emedicine.medscape.com/article/181364-treatment. Updated: Jun 10, 2008.

82. Obideen K, Yakshe P, Wehbi M. Pancreatitis, Chronic. eMedicine. Available at http://emedicine.medscape.com/article/181554-overview. Updated: Jun 16, 2008.

83. Gardner TB, Berk BS, Yakshe P. Pancreatitis, Acute: Treatment & Medication. eMedcine. Available at http://emedicine.medscape.com/article/181364-treatment. Updated: Jun 10, 2008.

84. Eckerwall GE, Tingstedt BB, Bergenzaun PE, Andersson RG. Immediate oral feeding in patients with mild acute pancreatitis is safe and may accelerate recovery—a randomized clinical study. Clin Nutr. 2007;26:758e63.

85. Sathiaraj E, Murthy S, Mansard MJ, Rao GV, Mahukar S, Reddy DN. Clinical trial: oral feeding with a soft diet compared with clear liquid diet as initial meal in mild acute pancreatitis. Aliment Pharmacol Ther. 2008 Sep 15;28(6):777–81.

86. Schneider A, Singer MV. Conservative treatment of chronic pancreatitis. Schweiz Rundsch Med Prax. 2005;18:94:831–38.

87. Abou-Assi S, O'Keefe SJ. Nutrition support during acute pancreatitis. Nutrition. 2002;18:938–43.

88. Forsmark CE, Baillie J. AGA Institute Clinical Practice and Economics Committee, AGA Institute Governing Board. AGA institute technical review on acute pancreatitis. Gastroenterology. 2007;132:2022e44.

89. Petrov MS, Pylypchuk RD, Emelyanov NV. Systematic review: nutritional support in acute pancreatitis. Aliment Pharmacol Ther. 2008 Sep 15;28(6):704–12. Review.

90. Groos S, Hunefeld G, Luciano L. Parenteral versus enteral nutrition: morphological changes in human adult intestinal mucosa. J Submicrosc Cytol Pathol. 1996;28:61–74.

91. Abou-Assi S, Craig K, O'Keefe SJ. Hypocaloric jejunal feeding is better than total parenteral nutrition in acute pancreatitis: results of a randomized comparative study. Am J Gastroenterol. 2002;97:2255–62.

92. Eatock FC, Chong P, Menezes N, et al. A randomized study of early nasogastric versus nasojejunal feeding in severe acute pancreatitis. Am J Gastroenterol. (2005)100:432–39.

93. Nathens AB, Curtis JR, Beale RJ, et al. Management of the critically ill patient with severe acute pancreatitis. Crit Care Med. 2004;32:2524–36.

94. McClave SA. Nutrition support in acute pancreatitis. Missouri Dietetic Association Annual Meeting Program; 2005.

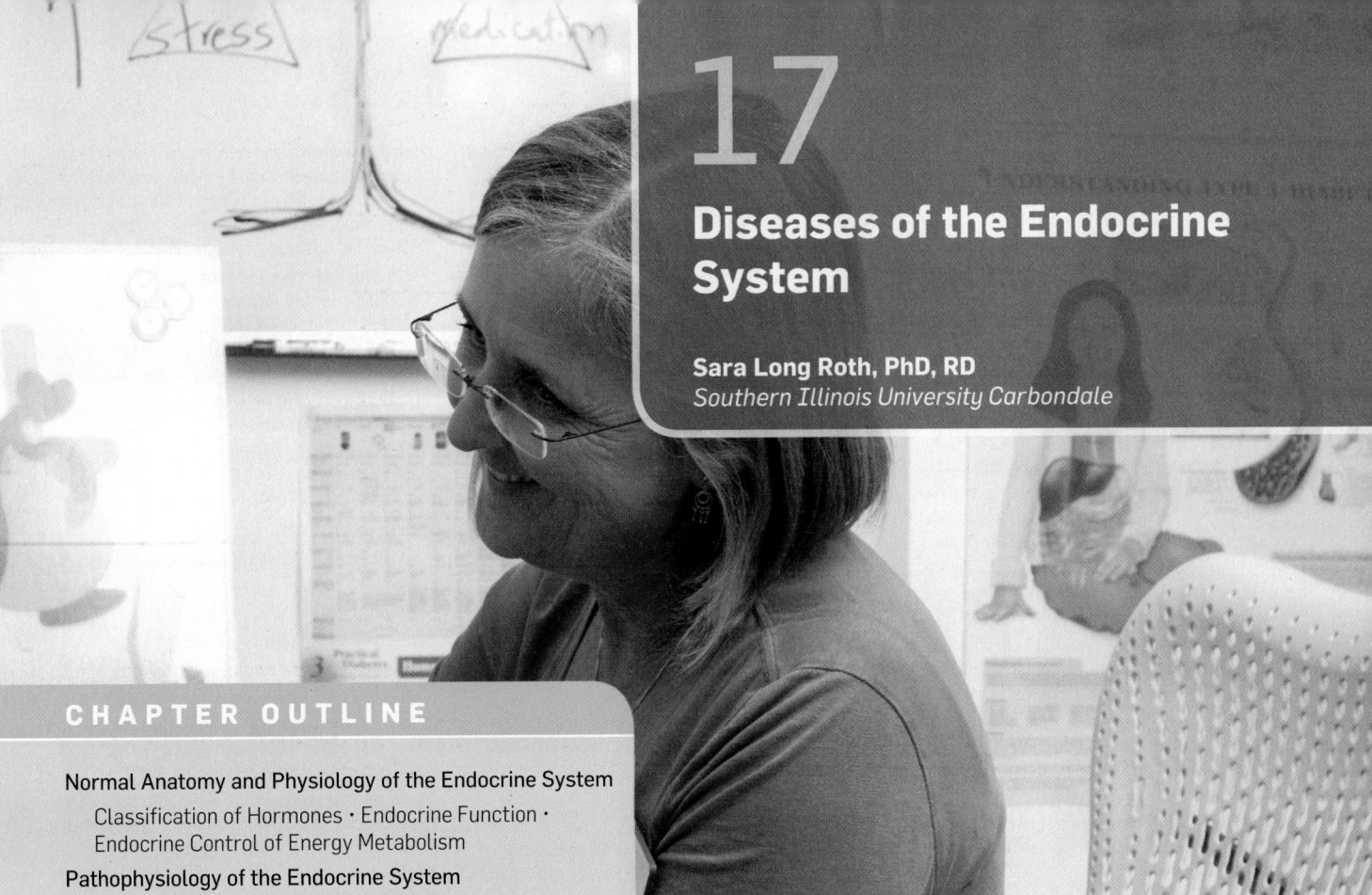

17
Diseases of the Endocrine System

Sara Long Roth, PhD, RD
Southern Illinois University Carbondale

Introduction

The endocrine system is more of a complex functional system
than an anatomical system. Endocrine glands that make up the
endocrine system are not attached anatomically, but scattered
all through the body. All the same, these glands make up a
system in a functional sense, since their regulatory activities are
often interdependent and must be closely coordinated.

Endocrine glands carry out their functions by secreting
hormones (chemical messengers) into the blood, and numerous interactions occur between the various glands (see
chapter endnote 1). A single endocrine gland may secrete
several hormones: the pituitary gland secretes six hormones
that have distinct functions and are under different control
mechanisms. A single hormone may be secreted by more than

one endocrine gland; for example, somatostatin is secreted by
both the hypothalamus and the pancreas. Some endocrine organs, like the anterior pituitary, exclusively secrete hormones,
while other endocrine organs perform additional functions.
The testes, for example, both produce sperm and secrete
testosterone.[1]

Once released into the blood, hormones travel to target
organs—the intended recipients of the chemical "message." A
single hormone can have more than one type of target organ
and thus can generate more than one type of effect. Vasopressin acts as a vasoconstrictor throughout the body in addition to
promoting H_2O reabsorption by kidney tubules. Some chemical
messengers may function as a hormone or neurotransmitter
depending on the sources and mode of delivery to the target
cell. Norepinephrine is secreted as a hormone by the adrenal
medulla and released as a neurotransmitter from sympathetic
postganglionic nerve fibers. Some single hormones have as-
sorted target-cell types and are capable of coordinating and
integrating activities of various tissues toward a common end.
Such is the effect of insulin on muscle, liver, and fat in the storage of nutrients after absorption.[1]

Secretion rates of specific hormones fluctuate in a cyclic pattern over the course of time, providing chronological syn-
chronization of function. Additionally, some hormones have a
diurnal variation meaning that there are fluctuations that occur
during the waking hours. Reproductive cycles, such as the menstrual cycle, are managed by endocrine hormones. Single target

GLOSSARY

acanthosis nigricans—diffuse, velvety-thickening hyperpigmentation of the skin; it may be present at the nape of the neck, axillae, area beneath the breasts, intertriginous areas (where skin touches or rubs together), and exposed areas (elbows, knuckles); thought to be the result of insulin resistance

achlorhydria—absence of hydrochloric acid from the gastric juice

Addison's disease—also known as chronic adrenal insufficiency; a rare endocrine disorder in which the adrenal glands do not produce enough steroid hormones; some common symptoms include fatigue, dizziness, muscle weakness, weight loss, difficulty in standing up, vomiting, anxiety, diarrhea, headache, sweating, changes in mood and personality, and joint and muscle pains

adrenocortical hormones—steroids derived from the precursor cholesterol

alpha-glucosidase—a digestive enzyme found in the brush border cells of the small intestine that cleaves more complex carbohydrates into sugars

amylin—a hormone synthesized by pancreatic ß-cells that contributes to glucose control during the postprandial period

anabolic—refers to building up or synthesis of larger organic molecules from smaller organic molecular subunits

anovulation—lack of ovulation during the menstrual cycle

autoantibodies—self-antibodies; in the case of autoimmunity affecting the pancreas, these include islet cell autoantibodies, autoantibodies to insulin, and autoantibodies to glutamic acid decarboxylase (GAD$_{65}$)

catecholamines—the chemical classification of adrenomedullary hormones

dawn phenomenon—an increase in blood glucose in the early morning, most likely due to increased glucose production in the liver after an overnight fast

diabetes mellitus—a diverse group of disorders that share the primary symptom of hyperglycemia resulting from defective insulin production, insulin action, or both

epinephrine—a hormone that is secreted from the adrenal medulla; regulates arterial blood pressure and prepares body for "fight or flight" responses; formerly referred to as adrenaline

exocrine pancreas—part of the pancreas that secretes digestive enzymes and bicarbonate into the duodenal lumen

fasting hypoglycemia—also called postabsorptive hypoglycemia; low blood glucose that is often related to an underlying disease and can be diagnosed from a blood glucose level below 50 mg/dL after an overnight fast, between meals, or after physical activity; symptoms are similar to those of diabetes-related hypoglycemia and may include hunger, sweating, shakiness, dizziness,

light-headedness, sleepiness, confusion, difficulty speaking, anxiety, and weakness

glucagon—a hormone produced by the α cells of the Islets of Langerhans in the pancreas that works in concert with insulin to maintain blood glucose levels; it promotes glycogenolysis, gluconeogenesis, and lipolysis, while inhibiting glycogenesis and triglyceride synthesis

GLUT-4—glucose transporter that transports glucose between blood and cells; it is the only glucose transporter responsive to insulin

glycemic control—control of blood glucose

glycosuria—the presence of glucose in the urine

hormones—blood-borne chemical messengers that act on target cells located in a different part of the body from the endocrine gland that produces them

hyperthyroidism—excess thyroid secretion

hyporesponsiveness—hormone resistance on the part of target cells/tissues

hypothyroidism—deficient thyroid secretion

insulin—a hormone produced by the ß cells of the Islets of Langerhans in the pancreas to regulate blood glucose; it promotes uptake, utilization, and storage of nutrients

insulin resistance—resistance of body cells to the action of insulin

ketoacidosis—an acid-base imbalance caused by an increase in concentration of ketones in the blood

Kussmaul respirations—rapid, deep, and labored breathing commonly seen in people who have ketoacidosis or who are in a diabetic coma; named for Adolph Kussmaul, the 19th century German doctor who first noted this sign

latent autoimmune diabetes of adulthood (LADA)—sometimes called T1.5DM, a slowly progressive form of T1DM; individuals are often diagnosed as T2DM, but have positive pancreatic islet antibodies, especially to glutamic acid decarboxylase (GADA)

macrosomia—refers to the condition of abnormally large infants whose mothers have diabetes

myxedematous—refers to nonpitting edema; noun form is *myxedema*

negative feedback—a regulatory mechanism in which a change in a controlled variable triggers a response that opposes the change, thus maintaining a relatively steady state for the regulated factor

negative nitrogen balance—net loss of protein in the body

nephropathy—renal disease that results from damage to blood vessels from hyperglycemia

neuroglycopenia—inadequate glucose supply to the brain

neuropathy—disorder of the nerves; symptoms depend on the type of nerves affected

norepinephrine—a neurotransmitter released from sympathetic postganglionic fibers; formerly referred to as noradrenaline; also a stress hormone that affects parts of the brain where attention and responding actions are controlled

oral glucose tolerance test (OGTT)—timed glucose challenge to examine efficiency of the body in metabolism of glucose

oxytocin—a hormone that stimulates contraction of the uterus during childbirth, and promotes ejection of milk from mammary glands during breast-feeding

pancreatic polypeptide—a polypeptide produced by the F cells of the Islets of Langerhans in the pancreas; its function is not yet known

polydipsia—excessive thirst

polyhydramnios—excessive accumulation of amniotic fluid

polyphagia—excessive hunger

polyuria—frequent urination

positive nitrogen balance—net accumulation of protein in the body

pre-diabetes mellitus—blood glucose levels that are higher than normal but not yet high enough to be diagnosed as diabetes

reactive hypoglycemia—low blood glucose levels that occur within 4 hours after a meal; also called postprandial hypoglycemia; symptoms are similar to those of diabetes-related hypoglycemia and may include hunger, sweating, shakiness, dizziness, light-headedness, sleepiness, confusion, difficulty speaking, anxiety, and weakness; a blood glucose level below 70 mg/dL at the time of symptoms and relief after eating will confirm the diagnosis

renal threshold—a concentration level of glucose in the blood above which the kidneys pass it through into the urine

retinopathy—disorder of the retina, the tissue layer within the back of the eye that senses light and transmits sensory information to the brain

secretagogues—medications that increase secretion of insulin

serum osmolality—a measure of the concentration of solute molecules in the blood

somatostatin—a hormone produced by the Δ cells of the Islets of Langerhans in the pancreas to control secretion of growth hormone from the anterior pituitary gland

tropic hormone—a hormone that regulates secretion of another hormone

U-100—refers to the units of insulin in each mL; U-100 is equivalent to 100 units per mL

vasopressin—the primary endocrine factor that regulates urinary H_2O loss and overall H_2O balance; regulates blood pressure via this hormone's pressor effects on blood vessels; also known as antidiuretic hormone (ADH)

cells may be influenced by more than one hormone. Some cells contain an assortment of receptors for interacting in different ways with different hormones. Insulin promotes conversion of glucose into glycogen in liver cells by stimulating one particular hepatic enzyme, whereas glucagon activates another hepatic enzyme to enhance degradation of glycogen into glucose in liver cells. Specific functions of major hormones are listed in Table 17.1.

Table 17.1 Summary of the Major Hormones

Endocrine Gland	Hormones	Target Cells	Major Functions of Hormones
Hypothalamus	Releasing and inhibiting hormones (TRH, CRH, GnRH, GHRH, GHIH, PRH, PIH)	Anterior pituitary	Controls release of anterior pituitary hormones
Posterior pituitary (hormones stored in)	Vasopressin (antidiuretic hormone)	Kidney tubules	Increases H_2O reabsorption
		Arterioles	Produces vasoconstriction
	Oxytocin	Uterus	Increases contractility
		Mammary glands (breasts)	Causes milk ejection
Anterior pituitary	Thyroid-stimulating hormone (TSH)	Thyroid follicular cells	Stimulates T_3 and T_4 secretion
	Adrenocorticotropic hormone (ACTH)	Zona fasciculata and zona reticularis of adrenal cortex	Stimulates cortisol secretion
	Growth hormone	Bone; soft tissues	Essential but not solely responsible for growth; stimulates growth of bones and soft tissues; metabolic effects include protein anabolism, fat mobilization, and glucose conservation
		Liver	Stimulates somatomedin secretion
	Follicle-stimulating hormone (FSH)	*Females:* ovarian follicles	Promotes follicular growth and development; stimulates estrogen secretion
		Males: seminiferous tubules in testes	Stimulates sperm production
	Luteinizing hormone (LH) (interstitial cell stimulating hormone—ICSH)	*Females:* ovarian follicle and corpus luteum	Stimulates ovulation, corpus luteum development, and estrogen and progesterone secretion
		Males: interstitial cells of Leydig in testes	Stimulates testosterone secretion
	Prolactin	*Females:* mammary glands	Promotes breast development; stimulates milk secretion
		Males	Uncertain
Thyroid gland ·follicular cells	Tetraiodothyronine (T_4 or thyroxine); triiodothyronine (T_3)	Most cells	Increases the metabolic rate; essential for normal growth and nerve development
Thyroid gland C cells	Calcitonin	Bone	Decreases plasma calcium concentration
Adrenal cortex *zona glomerulosa*	Aldosterone (mineralocorticoid)	Kidney tubules	Increases Na^+ reabsorption and K^+ secretion
Zona fasciculata and zona reticularis	Cortisol (glucocorticoid)	Most cells	Increases blood glucose at the expense of protein and fat stores; contributes to stress adaption
	Androgens (dehydroepiandrosterone)	*Females:* bone and brain	Responsible for the pubertal growth spurt and sex drive in females
Adrenal medulla	Epinephrine and norepinephrine	Sympathetic receptor sites throughout the body	Reinforces the sympathetic nervous system; contributes to stress adaption and blood pressure regulation
Endocrine pancreas (islets of Langerhans)	Insulin (β cells)	Most cells	Promotes cellular uptake, use, and storage of absorbed nutrients
	Glucagon (α cells)	Most cells	Important for maintaining nutrient levels in blood during postabsorptive state
	Somatostatin (Δ cells)	Digestive system	Inhibits digestion and absorption of nutrients
		Pancreatic islet cells	Inhibits secretion of all pancreatic hormones
		Anterior pituitary gland	Controls secretion of growth hormone
	Pancreatic polypeptide (F or PP cells)	Pancreas	Plays possible role in reducing appetite and food intake by inhibiting postprandial pancreatic exocrine secretion

(continued)

Table 17.1 Summary of the Major Hormones (*Continued*)

Endocrine Gland	Hormones	Target Cells	Major Functions of Hormones
Parathyroid Gland	Parathyroid hormone (PTH)	Bone, kidneys, intestine	Increases plasma calcium concentration; decreases plasma phosphate concentration; stimulates vitamin D activation
Gonads *Female: ovaries*	Estrogen (estradiol)	Female sex organs; body as a whole	Promotes follicular development; governs development of secondary sexual characteristics: stimulates uterine and breast growth
		Bone	Promotes closure of the epiphyseal plate
	Progesterone	Uterus	Prepares for pregnancy
Male: testes	Testosterone	Male sex organs; body as a whole	Stimulates sperm production; governs development of secondary sexual characteristics; promotes sex drive
		Bone	Enhances pubertal growth spurt; promotes closure of the epiphyseal plate
Testes and ovaries	Inhibin	Anterior pituitary	Inhibits secretion of follicle-stimulating hormone
Pineal gland	Melatonin	Brain; anterior pituitary; reproductive organs; immune system; possibly others	Entrains body's biological rhythm with external cues; believed to inhibit gonadotropins; initiation of puberty possibly caused by a reduction in melatonin secretion; acts as an antioxidant; enhances immunity
Placenta	Estrogen (estradiol); progesterone	Female sex organs	Help maintain pregnancy; prepare breasts for lactation
	Chorionic gonadotropin	Ovarian corpus luteum	Maintains corpus luteum of pregnancy
Kidneys	Renin (→ angiotensin)	Zona glomerulosa of adrenal cortex (acted on by angiotensin, which is activated by renin)	Stimulates aldosterone secretion
	Erythropoietin	Bone marrow	Stimulates erythrocyte production
Stomach	Gastrin	Digestive-tract exocrine glands and smooth muscles; pancreas; liver; gallbladder	Control of motility and secretion to facilitate digestive and absorptive processes
Duodenum	Secretin; cholecystokinin	Same as for gastrin	Same as for gastrin
	Glucose-dependent insulinotropic peptide	Endocrine pancreas	Stimulates insulin secretion
Liver	Somatomedins	Bone; soft tissues	Promotes growth
	Thrombopoietin	Bone marrow	Stimulates platelet production
Skin	Vitamin D	Intestine	Increases absorption of ingested calcium and phosphate
Thymus	Thymosin	T lymphocytes	Enhances T lymphocyte proliferation and function
Heart	Atrial natriuretic peptide	Kidney tubules	Inhibits Na+ reabsorption

Adapted from: Sherwood L. *Human physiology: from cells to systems.* 7th ed. Belmont (CA): Thomson-Brooks/Cole; 2010.
Brashers Vl, Jeuter SE. Mechanisms of hormonal regulation. In Understanding Pathophysiology. 4th ed. Philadelphia, PA: Mosby/Saunders, 2008.

Normal Anatomy and Physiology of the Endocrine System

Many functional interactions take place among the various ductless glands (see Figure 17.1) that make up the endocrine system.[1] This chapter will focus only on endocrine glands and disorders related to nutrition and nutritional status.

Classification of Hormones

Hormones released from endocrine glands regulate activities throughout the body. In a healthy state, hormones are released when their actions are required and inhibited when effects are achieved. Endocrine diseases manifest through either hyperfunction (exceptionally high blood concentrations of a hormone), hypofunction (depressed levels of hormones in the blood), or abnormal target-cell responsiveness.[1, 2]

The functions of hormones may be grouped into four categories:[3]

- Reproduction and sexual differentiation
- Growth and development
- Homeostasis
- Regulation of metabolism and nutrient supply

There are three chemical classes of hormones: (1) peptides and proteins, (2) amines, and (3) steroids.[1] The majority of hormones fall into the category of peptides and proteins, which are amino acid derivatives. Peptide hormones are secreted by the following glands: hypothalamus, anterior and posterior

Figure 17.1 The Endocrine System

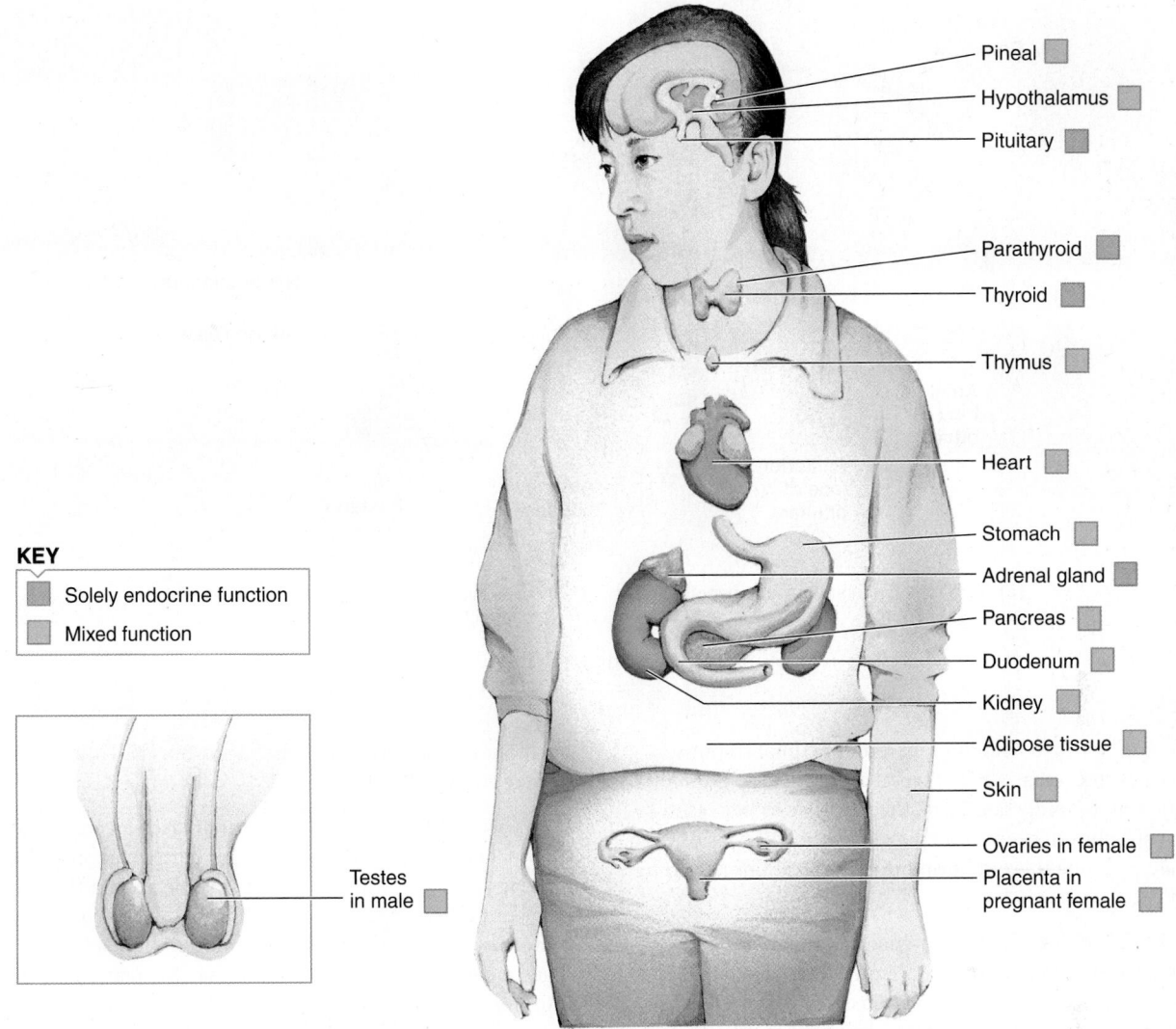

KEY

▣	Solely endocrine function
▣	Mixed function

Pineal ▣
Hypothalamus ▣
Pituitary ▣
Parathyroid ▣
Thyroid ▣
Thymus ▣
Heart ▣
Stomach ▣
Adrenal gland ▣
Pancreas ▣
Duodenum ▣
Kidney ▣
Adipose tissue ▣
Skin ▣
Ovaries in female ▣
Placenta in pregnant female ▣

Testes in male ▣

Source: L. Sherwood, *Human Physiology: From Cells to Systems*, 7e, copyright © 2010, Figure 18-1, p. 662.

pituitary glands, pineal gland, pancreas, parathyroid gland, gastrointestinal tract, kidney, liver, thyroid C cells, heart, and thymus. Amines are derivatives of the amino acid tyrosine. They are secreted by the thyroid gland and adrenal medulla. Steroid hormones, derived from cholesterol, are secreted by the adrenal cortex, gonads, and placenta.[1]

Endocrine Function

Pituitary Gland The pituitary gland is located in the bony cavity at the base of the brain just below the hypothalamus (see Figure 17.2). It is connected to the hypothalamus by a thin connecting stalk called the pituitary stalk.[1, 4]

The pituitary gland actually consists of two anatomically and functionally distinct glands: the anterior pituitary and the posterior pituitary. Location is the only thing they have in common. The anterior pituitary secretes six hormones (see Figure 17.3) that control secretion of various other hormones. None of the hormones is secreted at a constant rate; rather, secretion is regulated by hypothalamic hormones and

feedback from target gland hormones. The posterior pituitary releases **vasopressin** and **oxytocin**, hormones synthesized by the hypothalamus.[1]

Thyroid Gland The thyroid gland, which is responsible for controlling metabolic rate, lies over the trachea just below the larynx, and consists of two lobes connected by a thin strip called the isthmus (see Figure 17.4).[1]

The thyroid hormones are two iodine-containing hormones derived from the amino acid tyrosine: thyroxine (T_4 or tetraiodothyronine) and triiodothyrone (T_3). The prefixes *tetra-* and *tri-* and subscripts 4 and 3 denote the number of iodine atoms incorporated into each of these hormones. T_4 is the major hormone secreted by the thyroid, but T_3 is more active. The conversion of T_4 to T_3 within the anterior pituitary, liver, and kidney accounts for approximately two-thirds of T_3 production.[1, 3]

Response from increased secretion of thyroid hormone (which may affect several different organs and processes in the body)

Figure 17.2 Anatomy of the Pituitary Gland

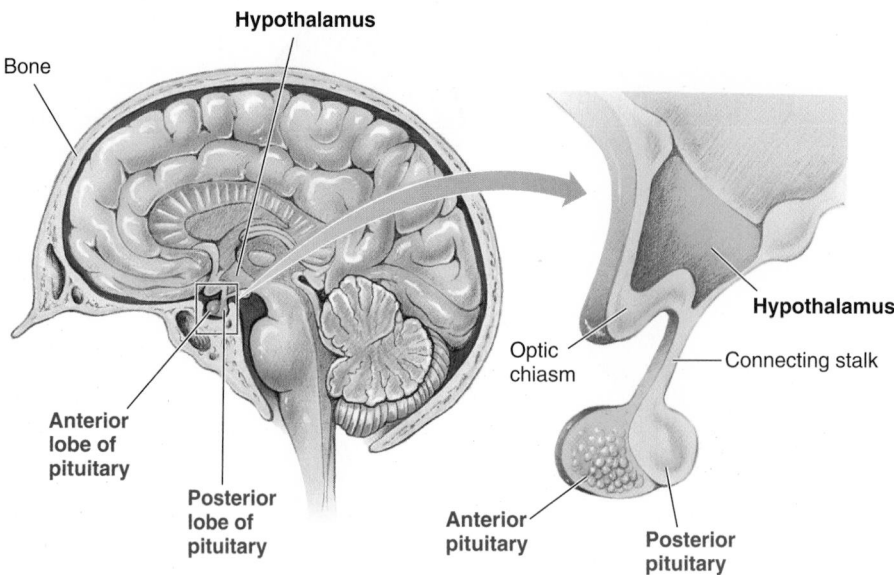

(a) Relation of pituitary gland to hypothalamus and rest of brain

(b) Enlargement of pituitary gland and its connection to hypothalamus

Source: L. Sherwood, *Human Physiology: From Cells to Systems*, 7e, copyright © 2010, Figure 18-4, p. 670.

takes several hours to become apparent. Maximal response does not become apparent for several days. Because thyroid hormone is not rapidly degraded, the response to increased secretion continues to be expressed over a period of days or even weeks after plasma thyroid hormone concentrations return to normal.[1]

Adrenal Glands The two adrenal glands are embedded above each kidney (see Figure 17.1) and encapsulated in fat. Each adrenal gland is composed of two endocrine organs.[1] The inner portion, the adrenal medulla, forms part of the sympathetic nervous system and secretes **epinephrine**, **norepinephrine**, and **catecholamines**.[1, 3]

The outer layers, known as the adrenal cortex, compose 80% to 90% of the adrenal gland and produce over 50 known **adrenocortical hormones**. Structural variations in these hormones confer different functional capabilities and allow them to perform different primary actions.[1, 3] Aldosterone directly influences the kidney to retain sodium, and thus has an integral role in regulating fluid balance and blood volume. Cortisol, a glucocorticoid, is secreted in response to physical (trauma, surgery, intense heat or cold), chemical (reduced O_2 supply), physiologic (heavy exercise, hemorrhagic shock, pain), psychological or emotional (anxiety, fear sorrow), and social (personal conflict, change in lifestyle) stresses. Cortisol stimulates hepatic gluconeogenesis; inhibits glucose uptake and use by tissues (except brain); stimulates protein degradation (especially in muscle) for use in gluconeogenesis or protein synthesis; and facilitates lipolysis. Administration of glucocorticoids inhibits almost every step of the inflammatory process, making them effective anti-inflammatory drugs. Lastly, cortisol permits catecholamines to induce vasoconstriction. Dehydroepiandrosterone (a male "sex" hormone) produced in the adrenal cortex is identical or similar to that produced by the gonads.[1]

Endocrine Pancreas The pancreas is located in the abdominal cavity adjacent to the upper part of the small intestine (see Figure 17.1). Different groups of cells within the pancreas carry out different functions. Cells making up the **exocrine pancreas** are responsible for secretion of fluid and various digestive enzymes that are secreted via the pancreatic duct into the duodenum (see Chapter 16). The endocrine cells of the pancreas (see Figure 17.5) include the alpha cells, which secrete glucagon and GLP-1 (glucagon-like-peptide); beta cells, which secrete insulin; delta cells, which secrete **somatostatin**; and the F cells, which secrete **pancreatic polypeptide**. These cells make up an anatomically small portion of the pancreas, but the hormones they secrete play a vital role in energy regulation and fuel homeostasis.

Endocrine Control of Energy Metabolism

Energy use in the body is constant, but ingestion of energy-yielding nutrients is sporadic. This means that excess energy taken in meals must be stored for later use between meals. About 1500 kcalories (less than a day's worth of energy) are stored as carbohydrate in the form of glucose (circulating in blood) and as glycogen (liver and muscle cells). Carbohydrate is the body's primary energy source and the preferred source of energy for brain cells. Excess carbohydrate (beyond what is used for glucose and glycogen) is converted to and stored as fat (triglycerides). Fat, the body's primary energy reservoir, can provide about two months' worth of energy during a prolonged fasting period. Fat is stored in adipose tissue in the form of triglycerides, and also circulates in the blood as

Figure 17.3 Functions of the Anterior Pituitary Hormones

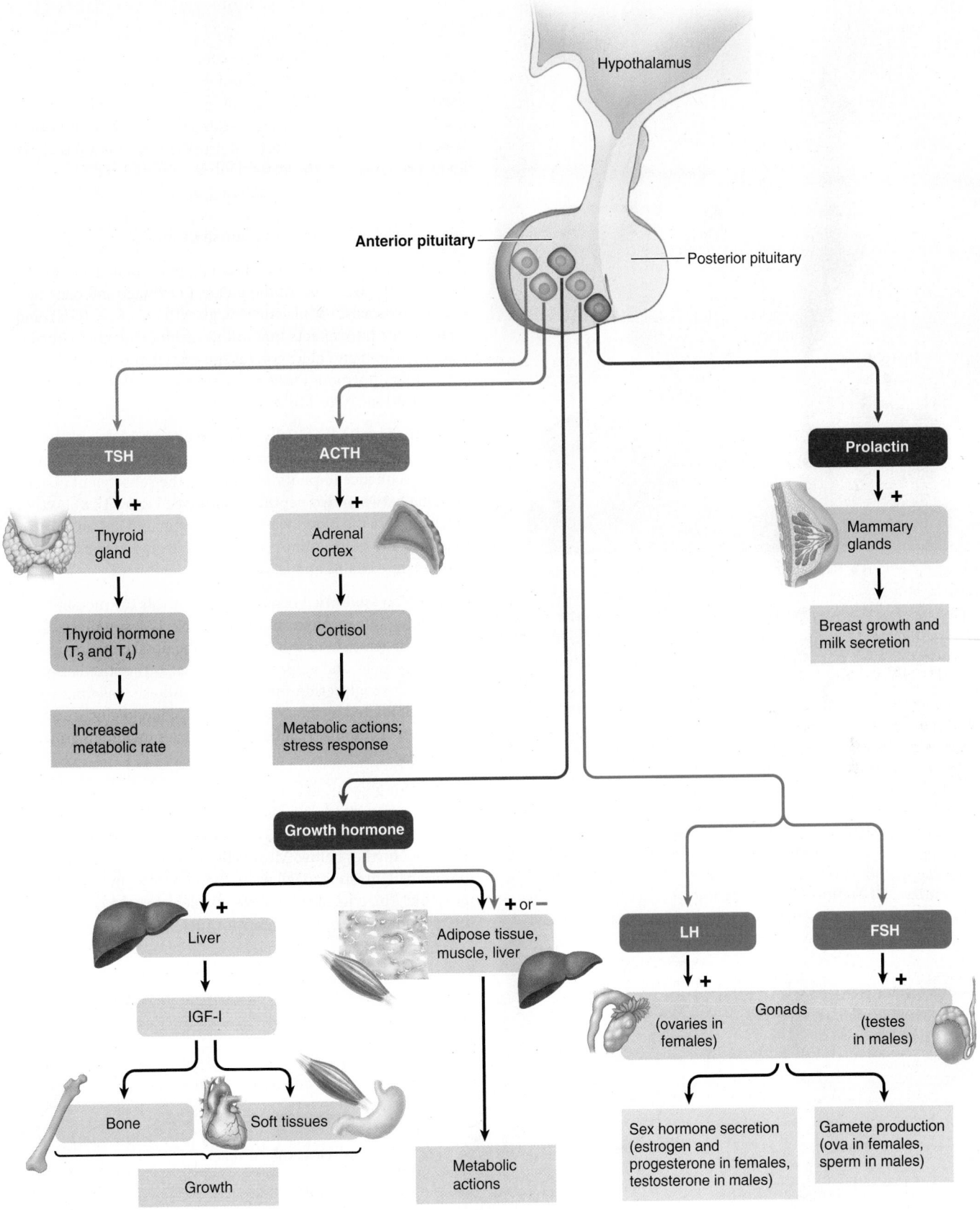

Source: L. Sherwood, *Human Physiology: From Cells to Systems,* 7e, copyright © 2010, Figure 18-6, p. 673.

Figure 17.4 Anatomy of the Thyroid Gland

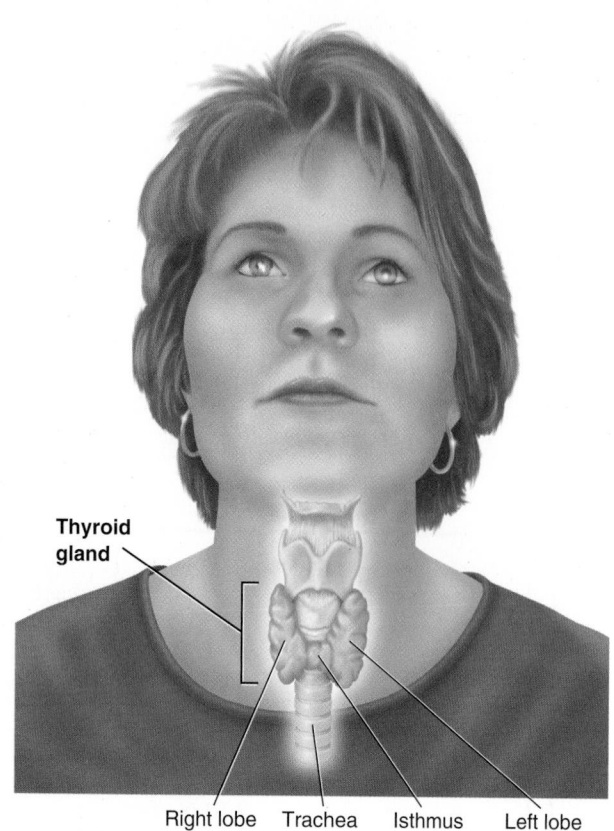

Thyroid gland

Right lobe Trachea Isthmus Left lobe

Source: L. Sherwood, Human Physiology: From Cells to Systems, 5e, copyright © 2004, p. 702.

free fatty acids. Protein is not stored as an energy source in the same manner as carbohydrate and fat, but can be used for energy as a "last resort." Protein can be converted to glucose (gluconeogenesis) to provide energy for the brain during a prolonged fast.[1]

Following meals, ingested nutrients are absorbed and enter the bloodstream; this period is termed the fed state, and its major metabolic pathways are represented by the blue anabolism arrows in Figure 17.6. During this period of time, glucose functions as the main energy source, because most cells have

a preference to use glucose. Additional amounts of glucose or fat not immediately used for energy or structural repairs are converted into their storage forms: glycogen or triglycerides, respectively.[1]

It takes approximately four hours for a typical meal to be absorbed. During the time period when no nutrients are in the gastrointestinal tract (fasting state), endogenous energy stores are mobilized for energy, as indicated by the red catabolism arrows in Figure 17.6. Synthesis of protein and fat is abbreviated, and stored forms of these nutrients are catabolized for glucose formation and energy production, respectively. Through mechanisms of gluconeogenesis and glucose sparing, the blood glucose level is sustained to nourish the brain.[1]

As outlined in Table 17.2, hormones—in particular the pancreatic hormones—afford the means to manage and control fuel homeostasis.[4] While cortisol, growth hormone (GH), and epinephrine have effects that influence blood glucose concentration, insulin and glucagon (as illustrated in Figure 17.7) are the primary hormones that maintain normal blood glucose concentration (70 to 110 mg/100 mL).

Insulin **Insulin** is initially secreted as a prohormone, proinsulin. The activation of proinsulin involves the removal of a peptide sequence (c-peptide) and then the subsequent disulfide bonding between two peptide chains (see Figure 17.8). Active insulin enters the blood via the portal vein and has an approximate half-life of five minutes. It is then degraded back into the two separate chains and is inactive.

Insulin is an **anabolic** hormone as it controls the metabolic fate of carbohydrate, protein, and lipid (Table 17.2 summarizes its effects). In general, insulin promotes the uptake of glucose into hepatic, muscle, and adipose cells as well as the stimulation of glycogen, triglyceride, and protein synthesis. Insulin secretion is stimulated by an increased level of blood glucose and by the action of counter-regulatory hormones including growth hormone.

In order for glucose, fructose, or galactose to be absorbed into the cell, transport molecules—GLUT-1 through -5, listed in Table 17.3—are necessary. **GLUT-4** is insulin dependent, as described further below. Most cells, such as the enterocytes in the small intestine, need more than one type of transport molecule. The type of transporter available on the cell reflects the individual cell requirements for fuel.

Figure 17.5 Release of Hormones from the Pancreas

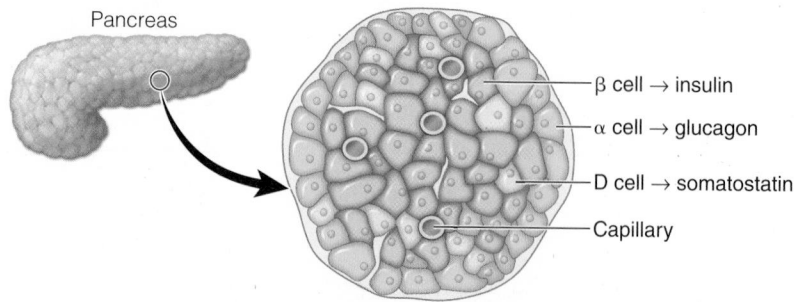

Pancreas

β cell → insulin
α cell → glucagon
D cell → somatostatin
Capillary

Source: L. Sherwood, *Human Physiology: From Cells to Systems*, 7e, copyright © 2010, p. 715. Figure 19-15, part (b) (without photo).

Figure 17.6 Summary of Major Pathways Involving Nutrient Absorption and Metabolism

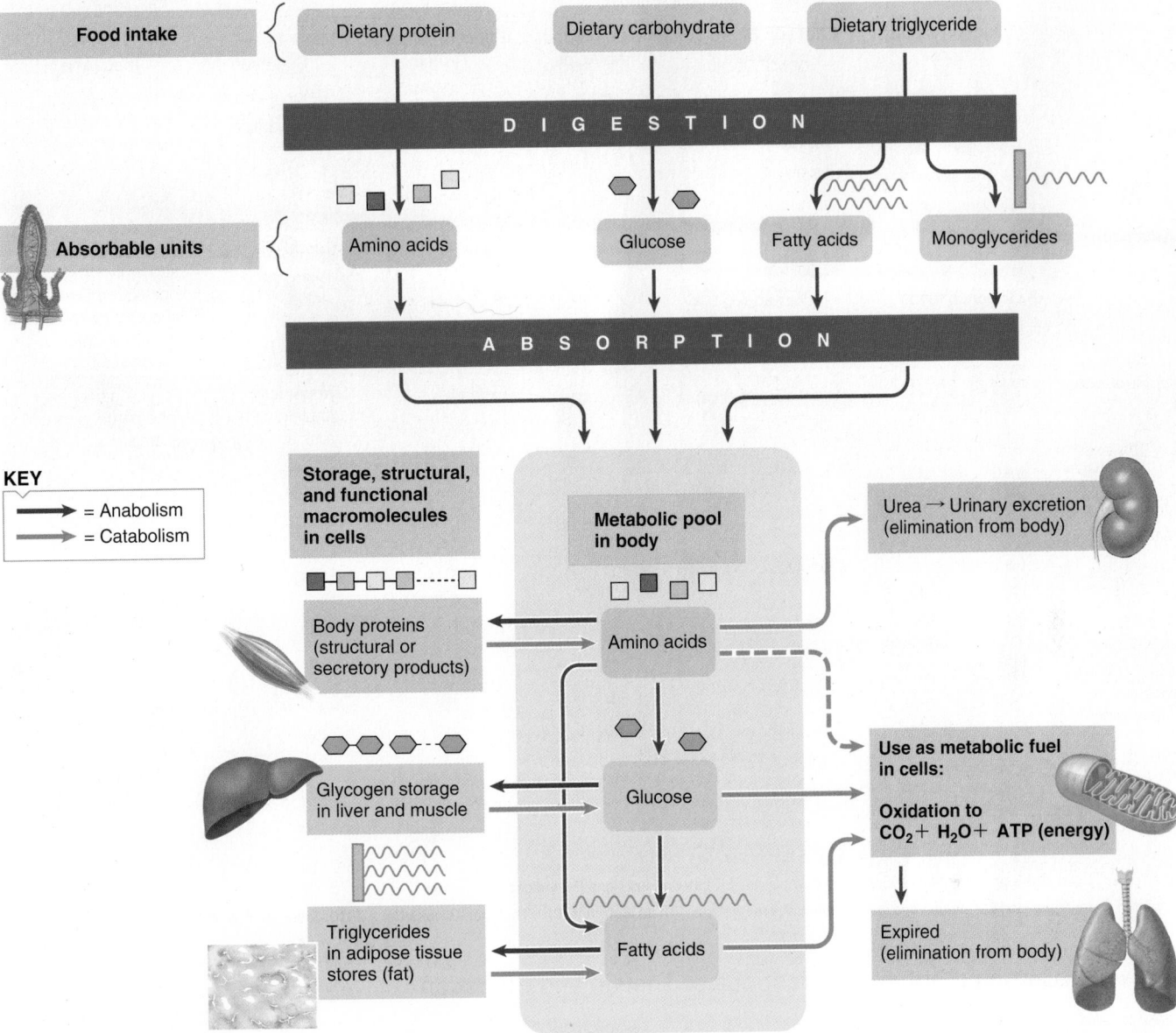

Source: L. Sherwood, *Human Physiology: From Cells to Systems*, 7e, copyright © 2010, Figure 19-14, p. 711.

Most tissues in the body depend on insulin for transportation of glucose from the bloodstream into cells to be used for energy. There are three exceptions: cells of the brain, liver, and working muscles are readily permeable to glucose even in the absence of insulin.[1,4] The GLUT-4 transporter (which is insulin-dependent) is present in skeletal and cardiac muscle and in the adipocytes. Insulin allows the translocation of the GLUT-4 from the interior of the cell to the cell membrane, where it transports glucose into the cell (see Figure 17.9). The net effect of insulin is to promote glucose oxidation, glycogen storage, and triglyceride storage.[1,4]

The most pronounced effect of insulin on protein metabolism is seen in skeletal muscle and the liver. It promotes active transport of amino acids from the blood into muscle and other tissues, thus promoting protein synthesis within cells.

This anabolic effect of insulin on protein metabolism produces a **positive nitrogen balance**. When insulin is deficient, there is net loss of protein, or **negative nitrogen balance**. These effects demonstrate the importance of insulin in tissue growth.[1,4]

Glucagon **Glucagon** is the hormone released from alpha cells of the pancreas when blood glucose levels fall below the normal range, necessitating a source of energy to maintain homeostasis. Glucagon stimulates the breakdown of stored glycogen (glycogenolysis) and the production of new glucose from amino acids (gluconeogenesis), and thus raises blood glucose. Glucagon also stimulates lipolysis, which provides additional substrate to meet energy requirements.

The glucagon-like-peptides 1 and 2 (GLP-1 and GLP-2) are derived from the precursor for glucagon and act to stimulate

Table 17.2 Summary of Hormonal Control of Energy Metabolism

| Hormone | Major Metabolic Effects | | | | Control of Secretion | |
	Effect on Blood Glucose	Effect on Blood Fatty Acids	Effect on Blood Amino Acids	Effect on Muscle Protein	Major Stimuli for Secretion	Primary Role in Metabolism
Insulin	↓ + Glucose uptake + Glycogenesis − Glycogenolysis − Gluconeogenesis	↓ + Triglyceride synthesis − Lipolysis	↓ + Amino acid uptake	↑ + Protein synthesis − Protein degradation	↑ Blood glucose ↑ Blood amino acids	Primary regulator of absorptive and postabsorptive cycles
Glucagon	↑ + Glycogenolysis + Gluconeogenesis − Glycogenesis	↑ + Lipolysis − Triglyceride synthesis	No effect	No effect	↓ Blood glucose ↑ Blood amino acids	Regulation of absorptive and postabsorptive cycles in concert with insulin; protection against hypoglycemia
Epinephrine	↑ + Glycogenolysis + Gluconeogenesis − Insulin secretion + Glucagon secretion	↑ + Lipolysis	No effect	No effect	Sympathetic stimulation during stress and exercise	Provision of energy for emergencies and exercise
Cortisol	↑ + Gluconeogenesis − Glucose uptake by tissues other than brain; glucose sparing	↑ + Lipolysis	↑ + Protein degradation	↓ + Protein degradation	Stress	Mobilization of metabolic fuels and building blocks during adaptation to stress
Growth Hormone	↑ − Glucose uptake by muscles; glucose sparing	↑ + Lipolysis	↓ + Amino acid uptake	↓ + Protein synthesis − Protein degradation + Synthesis of DNA and RNA	Deep sleep Stress Exercise Hypoglycemia	Promotion of growth; normally little role in metabolism; mobilization of fuels plus glucose sparing in extenuating circumstances

Up arrows (↑) and plus signs (+) indicate increases; down arrows (↓) and minus signs (−) indicate decreases.

Reprinted from: Sherwood L. *Human Physiology: From Cells to Systems.* 7th ed. Belmont (CA): Thomson-Brooks/Cole; 2010. Table 19–6, p. 726.

Figure 17.7 Complementary Interactions of Insulin and Glucagon

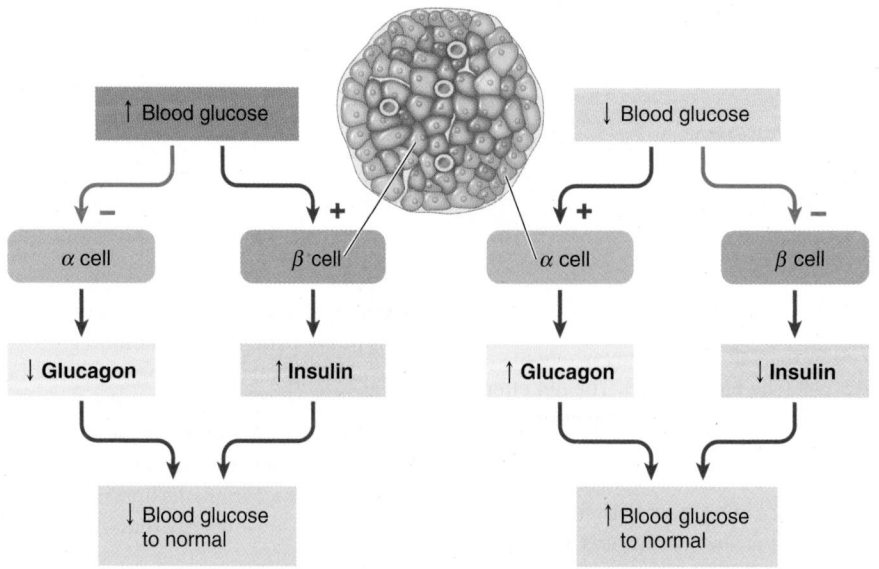

Source: L. Sherwood, *Human Physiology: From Cells to Systems*, 7e, copyright © 2010, Figure 19-20, p. 724.

Table 17.3 Glucose Transportation

Glucose Transporter (GLUT)[1]	Substrate	Tissue Distribution	Function
GLUT-1	Glucose	Most cells	Transports glucose across blood-brain barrier
GLUT-2	Glucose, galactose, fructose	Liver, β cells of pancreas, hypothalamus, basolateral membrane of small intestine, kidney tubules	Transports glucose from kidney and intestinal cells into adjacent bloodstream by means of co-transport carriers
GLUT-3	Glucose	Brain (neurons), kidney, placenta, testes	Transports glucose into neurons
GLUT-4	Glucose	Skeletal and cardiac muscle, white and brown adipose tissue	Transports the majority of glucose used by most cells of the body via the influence of insulin
GLUT-5	Fructose	Mucosal surface of small intestine, adipose tissue, skeletal muscle, sperm	Fructose transport

[1]All glucose is transported across plasma membrane by passive facilitated diffusion.

Source: Sherwood L. *Human Physiology: From Cells to Systems.* Belmont (CA): Thomson; 2010.

Figure 17.8 The Structure of Insulin

In its active form, insulin consists of two polypeptide chains (a total of 51 amino acids) linked by disulfide bridges between cysteine molecules. When the chains separate, it becomes inactive.

Source: Understanding Nutrition 12e by Whitney/Rolfes, Figure 6-4, page 175.

the action of insulin and inhibit that of glucagon. Because GLP-1 and GLP-2 have a very short half life, their action potential is limited. Nonetheless, maximizing their function allows for an additional path to control abnormal blood glucose levels, and thus has recently been a focus in the development of new medications such as Byetta, discussed later in this chapter.

Pathophysiology of the Endocrine System

As mentioned previously, endocrine disorders are the result of hyposecretion or hypersecretion of hormones, or of **hyporesponsiveness** of target organs.[1,2] Table 17.4 outlines the most common causes of endocrine dysfunction.

Primary hyposecretion occurs when an endocrine organ releases an inadequate amount of hormone to meet physiological needs. Secondary hyposecretion occurs when secretion of a **tropic hormone** is inadequate to cause an endocrine organ to secrete adequate amounts of a hormone. For instance, if the thyroid gland produces inadequate amounts of thyroid hormone, this would be considered primary hyposecretion, whereas inadequate production of thyroid hormone that is

caused by insufficient secretion of a tropic hormone such as thyroid-stimulating hormone (TSH) is secondary hyposecretion. The interrelationship of the various hormones makes diagnosis of hormone deficiency quite complex. Evaluation of multiple lab values may assist in differentiating between primary and secondary deficiencies. In a primary thyroid hormone deficiency, for example, thyroid hormone would be low, but TSH levels would be high; in secondary thyroid hormone deficiency, both thyroid hormone and TSH levels would be abnormally low.[2]

Hypersecretion disorders can also be primary or secondary. When an endocrine gland is secreting abnormally high amounts of a hormone due to a primary disorder, the tropic hormone will be at unusually low levels. When hypersecretion is secondary (to elevated tropic hormone levels), plasma concentrations of both hormones will be elevated.[2]

Hyporesponsiveness of the target organ will cause the same symptoms as hyposecretion, but hormone levels will be normal or high instead of low. Most cases of hyporesponsiveness are caused by a lack or deficiency of hormone receptors on the target cells.[2] An example of this type of disorder is type 2 diabetes mellitus.

Table 17.4 Most Common Causes of Endocrine Dysfunction

Effect of Hyposecretion	Effect of Hypersecretion

• Too little hormone secreted by the endocrine gland (hyposecretion)[1]
 • Increased removal of the hormone from the blood
 • Abnormal tissue responsiveness to the hormone
   • Lack of target cell receptors
   • Lack of an enzyme essential to the target cell response
	• Too much hormone secreted by the endocrine gland (hypersecretion)
	 • Reduced plasma-protein binding of the hormone (too much free, biologically active hormone)
	 • Decreased removal of the hormone from the blood
	   • Decreased inactivation
	   • Decreased excretion

[1]Most common causes of endocrine dysfunction.

Reprinted from: Sherwood L. *Human Physiology: From Cells to Systems.* 7th ed. Belmont (CA): Thomson-Brooks/Cole; 2010. Table 18–1, p. 665.

Figure 17.9 The Role of Insulin in Cellular Uptake of Glucose

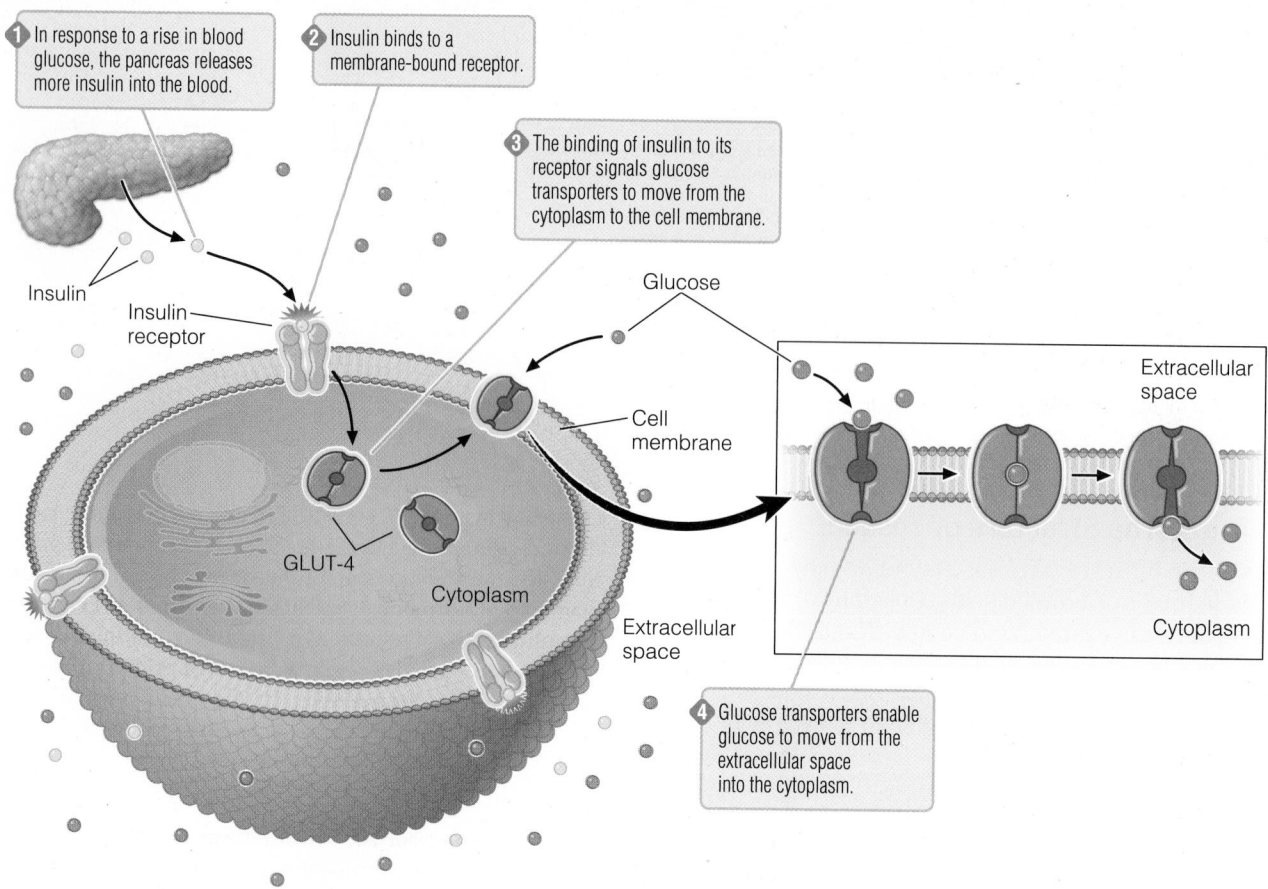

1 In response to a rise in blood glucose, the pancreas releases more insulin into the blood.

2 Insulin binds to a membrane-bound receptor.

3 The binding of insulin to its receptor signals glucose transporters to move from the cytoplasm to the cell membrane.

4 Glucose transporters enable glucose to move from the extracellular space into the cytoplasm.

Insulin

Insulin receptor

Glucose

GLUT-4

Cytoplasm

Cell membrane

Extracellular space

Extracellular space

Cytoplasm

Source: Nutritional Sciences: From Fundamentals to Food 2e by McGuire/Beerman, Figure 4-19, page 147.

Diabetes Mellitus

Diabetes mellitus, by far the most common of all endocrine disorders, is one of the foremost public health concerns confronting the world today.[5] Over 23 million individuals in the United States, or 8% of the population, have diabetes. An estimated 17.5 million have been diagnosed, but 5.5 million (nearly one fourth) are unaware they have the disease.[6] The average medical costs for a patient with diabetes are more than twice that of others receiving care in the health care system and most of these costs are associated with the chronic complications that accompany diabetes.[5] In 2007, 2.35 million new cases of diabetes were diagnosed in people aged 20 years and older. It is the seventh leading cause of death in the United States and is likely to be underreported. Risk for death among people with diabetes is about twice that of those without diabetes.[7]

Diabetes mellitus is not a single disease but a diverse group of disorders that differ in origin and severity.[1, 3, 4] Yet all forms of diabetes mellitus share one common characteristic: hyperglycemia resulting from defects in insulin production, insulin action, or both.[3, 4, 6, 8] Chronic hyperglycemia is correlated with organ dysfunction and damage, progressing to failure of numerous organs, particularly the eyes, kidneys, nerves, heart, and blood vessels.[6]

Insulin deficiency is generally due to either insufficient insulin secretion by beta (β) cells or comparative deficient response by target tissue cells to insulin.[4] Whatever the cause of insulin deficiency, it results in glucose intolerance. Various conditions associated with glucose intolerance were used by the Expert Committee on the Diagnosis and Classification of Diabetes Mellitus in 2009 to diagnose and classify diabetes (see Box 17.1).[8]

Type 1 Diabetes Mellitus

Over the years, several terms have been used to classify the different types of diabetes. In order to prevent individuals from being classified by treatment modality rather than disease characteristics, the terms insulin-dependent diabetes mellitus (IDDM), juvenile diabetes, brittle diabetes, non-insulin-dependent diabetes mellitus (NIDDM), adult-onset diabetes, or borderline diabetes should not be used.[3]

Epidemiology Type 1 diabetes mellitus (T1DM—see chapter endnote 2) accounts for 5% to 10% of all diagnosed cases of diabetes.[6, 9] While this form of diabetes develops most frequently in children and adolescents, it is being increasingly noted later in life, even in individuals in their 80s and 90s.[6] Gender distribution of T1DM is equal.[3] T1DM can occur at any age; about half acquire the illness before age 20, and the other half develop the disease as adults.[10]

Etiology Immune-mediated type 1 diabetes mellitus results from a cellular-mediated autoimmune destruction of β-cells of the pancreas. One or more **autoantibodies** are present in 85%

Etiologic Classifications of Diabetes Mellitus

Type 1 Diabetes Mellitus

· Immune mediated
· Idiopathic

Type 2 Diabetes Mellitus

Gestational Diabetes Mellitus (GDM)

Statistical Risk of Diabetes Mellitus

· Impaired glucose tolerance (IGT)
· Impaired fasting glucose tolerance (IFG)

Specific Types of Diabetes Due to Other Causes

· Drug or chemical induced
· Cystic fibrosis or other diseases of the pancreas

Source: American Diabetes Association. Diagnosis and classification of diabetes mellitus. Diabetes Care, 2009; 32:S13.

to 89% of individuals diagnosed with T1DM.[11] Rate of β-cell destruction is variable, fast in certain individuals (primarily infants and children) and slow in others (primarily adults). The first sign of T1DM in children and adolescents can be **ketoacidosis**, but the disease can also present with moderate fasting hyperglycemia that can quickly transform into severe hyperglycemia and/or ketoacidosis in the presence of physiological stress such as infection. Residual β-cell function that is sufficient to prevent ketoacidosis may be preserved in adults diagnosed with T1DM. Nonetheless, these adults may ultimately develop dependence upon exogenous insulin for survival and be at risk for ketoacidosis.[11] Causes of the autoimmune destruction of β-cells are not clearly understood, but multiple genetic predispositions and unidentified environmental factors appear to contribute to T1DM.[4, 8] Determining the environmental agent has been difficult because of the time lapse between exposure and the development of DM, but research has identified the coxsackie virus, cow's milk proteins, and rubella as potential triggers.[4]

Some forms of T1DM have no known cause and are referred to as idiopathic diabetes. Individuals with idiopathic diabetes produce no insulin and are prone to ketoacidosis, but have no evidence of autoimmunity. Individuals with T1DM who fall into this category represent a very small minority, and most are of African or Asian ancestry.

Pathophysiology and Clinical Manifestations

T1DM is characterized by an absolute deficiency of insulin due to destruction of pancreatic β-cells, resulting in the inability of cells to use glucose for energy.[2, 10] By the time clinical symptoms occur, 60% to 80% of β-cells have been destroyed. Cells that produce glucagon, somatostatin, and pancreatic polypeptide are typically conserved but may be redistributed within the islets.[10]

As shown in Figure 17.10, the acute consequences of an insulin deficit are numerous and potentially fatal. When glucose cannot enter cells, two things happen: plasma glucose levels rise (hyperglycemia) and cells starve. To compensate for the hyperglycemia, excess glucose is lost in the urine because the kidneys can filter only so much glucose from the blood. As a result, **glycosuria** and frequent urination (**polyuria**) occur. Loss of fluid stimulates the thirst mechanism and leads to **polydipsia**. Cells dependent on glucose for energy have none available. In turn, the body responds to this emergency by promoting hunger (**polyphagia**).[2]

As the insulin deficiency persists, production of additional hormones (catecholamines, cortisol, glucagon, and growth hormone) increases, leading to lipolysis. As the body breaks down fat stored in adipose tissue, the resulting fatty acids are transformed into keto acids in the liver. In the nondiabetes state, keto acids can be used for energy by muscle and brain cells. As increased production of keto acids occurs, pH falls (7.3 to 6.8), and ketone bodies are secreted in the urine. Metabolic acidosis develops as bicarbonate concentration is reduced, and ketoacidosis results.[2] The body tries to offset metabolic acidosis through deep, labored respirations (**Kussmaul respirations**).[2]

As total body water decreases, potassium, sodium, magnesium, and phosphorus are also lost. Serum levels of these ions may be normal or elevated due to decreased fluid volume in the body (hypovolemia). Hypovolemia also accounts for increased hematocrit, hemoglobin, protein, white blood cell count, creatinine, and **serum osmolality**. Hypovolemia and muscle catabolism are the cause for considerable, imminent weight loss in persons with ketoacidosis, and often present at diagnosis of T1DM. Hypovolemic shock can lead to death if left untreated (see Figure 17.10).

Diagnosis General criteria for diagnosis of diabetes are outlined in Box 17.2. There are three ways to diagnose diabetes. If diagnosis based on one of these three methods is made in the absence of hyperglycemia, then that diagnosis must be confirmed on a subsequent day by any one of the three methods. Use of glycosylated hemoglobin (A1C) is not recommended for diagnosis of diabetes.[8]

Diagnosis of T1DM can be made on the basis of a casual plasma glucose ≥200 mg/dL (≥11.1 mmol/L) in addition to certain symptoms (unexplained weight loss, polydipsia, polyuria), or fasting plasma glucose ≥126 mg/dL (≥7.0 mmol/L). Other tests and laboratory measurements used in DM diagnosis and risk assessment are described below.

ORAL GLUCOSE TOLERANCE TEST (OGTT) Oral glucose tolerance tests (OGTTs) are rarely needed to diagnose T1DM (19) due to the sudden onset of symptoms accompanied by hyperglycemia. In fact, OGTT is contraindicated in infants and young children, but it is commonly used to diagnose gestational diabetes, impaired glucose tolerance (IGT), and impaired fasting glucose (IFG).[8, 12] An OGTT is administered after at least 3 days of an unrestricted diet providing at least 150 grams of carbohydrate daily and normal physical activity. The test is preceded by an overnight fast of 8 to 14 hours, during which water may be drunk. Smoking is not permitted during the test. After collection of a fasting blood glucose sample, a drink containing 75 grams of anhydrous glucose (100 grams for pregnant

Figure 17.10 Acute Effects of Insulin Deficiency

Acute consequences of insulin deficiency can be grouped according to effects on carbohydrate, protein, and fat metabolism. These effects ultimately cause death through a variety of pathways.

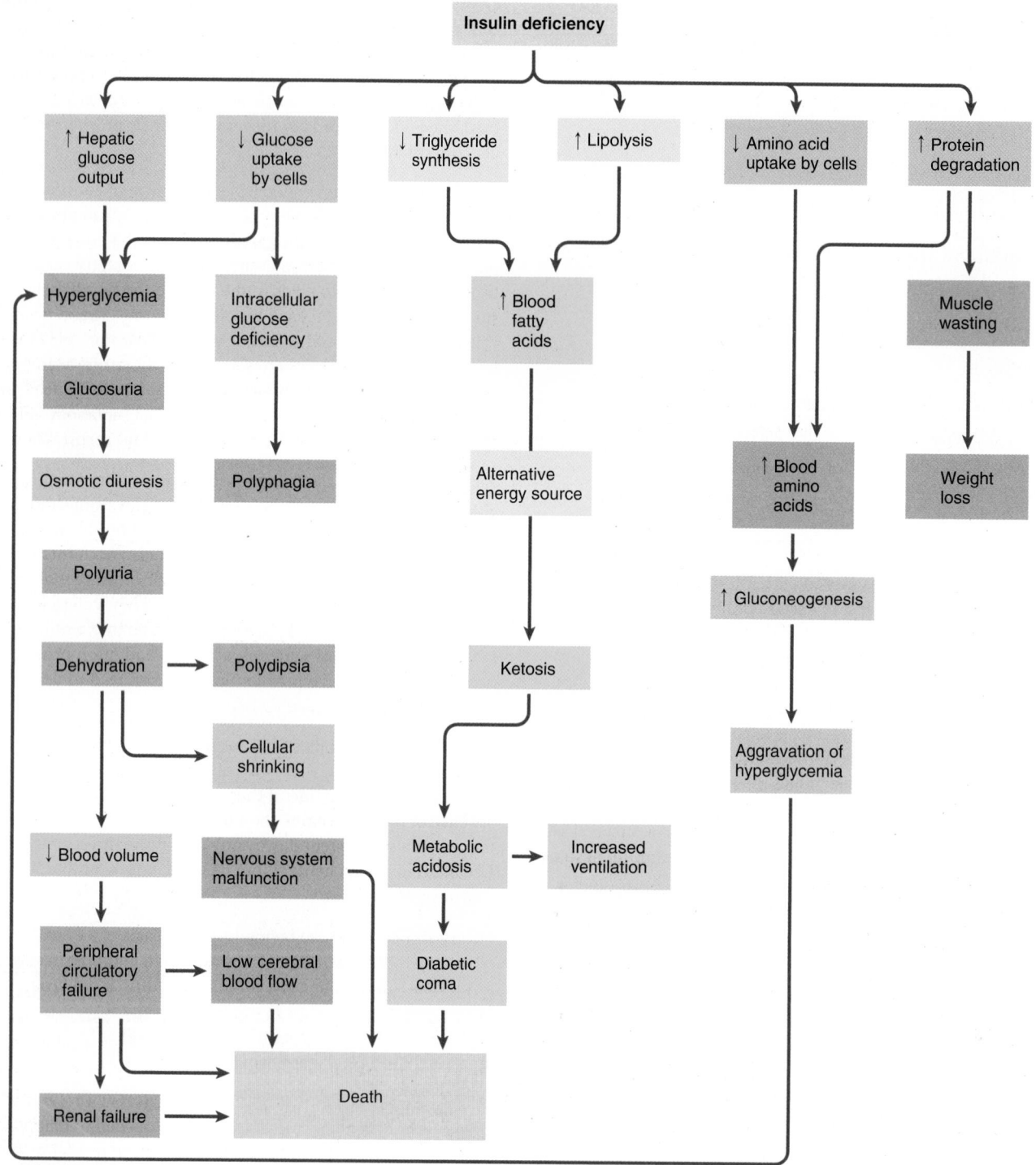

Source: L. Sherwood, *Human Physiology: From Cells to Systems*, 7e, copyright © 2010, Figure 19-19, p. 722.

women) in 250 to 300 mL of water should be consumed over a 5-minute period. Timing of the test begins at the beginning of the drink. Blood samples are collected 2 hours after the test load. In a person without diabetes, blood glucose levels rise, then fall quickly to normal. In individuals with diabetes, blood glucose levels rise higher than normal, then fall slowly back to normal. IFG is diagnosed when fasting plasma glucose is

found to be 100 to 125 mg/dL; IGT is diagnosed when 2-hour postprandial load is found to be between 140 and 199 mg/dL. Details of diagnostic criteria for diabetes are found in Box 17.2.

DIABETES-RELATED AUTOANTIBODIES Autoantibody testing is not compulsory to diagnose T1DM, but it can be valuable in screening individuals at high risk for developing

Criteria for Diagnosis of Diabetes Mellitus[1]

Symptoms of diabetes[2] plus casual[3] plasma glucose concentration ≥200 mg/dL (11.1 mmol/L)

OR

Fasting plasma glucose[4] ≥126 mg/dL (7.0 mmol/L)

OR

2-hour post-prandial glucose ≥200 mg/dL (11.1 mmol/L) during an oral glucose tolerance test (OGTT)[5]

1. In absence of unequivocal hyperglycemia, these criteria should be confirmed by repeat testing on a different day.
2. Polyuria, polydipsia, and unexplained weight loss.
3. Casual is defined as any time of day without regard to time since last meal.
4. Defined as no caloric intake for at least 8 hours.
5. OGTT should be performed as described by WHO, using a glucose load containing the equivalent of 75 g anhydrous glucose dissolved in water. OGTT is not recommended for routine clinical use.

Source: Adapted from American Diabetes Association. Diagnosis and classification of diabetes mellitus. *Diabetes Care.* January 2009 vol. 32 no. Supplement 1 S62-S67.

American Diabetes Association. Standards of medical care in diabetes. *Diabetes Care.* 2008;31(Suppl 1):S12.

diabetes (e.g., siblings of individuals with T1DM and offspring of parents with T1DM) up to 7 years before clinical onset.[13] Diabetes-related autoantibody testing is largely performed to differentiate between autoimmune T1DM and diabetes resulting from obesity and/or **insulin resistance**. These autoantibodies serve as indicators of the body's destructive immune response against its own β-cells.[13] Tests used to measure diabetes-related autoantibodies include islet cell cytoplasmic autoantibodies (ICA), insulin autoantibodies (IAA), glutamic acid decarboxylase autoantibodies (GADA), and insulinoma-associated-2 autoantibodies (IA-2A).

GLUTAMIC ACID DECARBOXYLASE AUTOANTIBOD-IES (GADA)
Tests for glutamic acid decarboxylase autoantibodies (GADA) measure specific islet cell antigens. GADA have been found in 70% to 90% of individuals with T1DM, and have been shown to be the most sensitive marker for identifying persons at risk for developing T1DM. They are generally more prevalent in older children and individuals with **latent autoimmune diabetes of adulthood (LADA)**.

ISLET CELL AUTOANTIBODIES (ICA)
The ICA test measures a group of islet cell autoantibodies. ICA have been found in 70% to 80% of individuals younger than 30 years of age with newly diagnosed diabetes. Among individuals with T1DM, prevalence of ICA decreases the longer an individual has diabetes. Existence of ICA in relatives without diabetes has been shown to be a sign of increased risk for the disease.[11] There is an increased risk for T1DM in individuals without diabetes who test positive for one or more islet autoantibodies. The more islet autoantibodies present, the greater the individual's risk for developing T1DM.[13]

INSULIN AUTOANTIBODIES (IAA) The presence of IAA is evidence of ongoing destruction of β-cells.[11] IAA testing must be performed before insulin therapy is initiated since the test does not determine whether the body's immune system is making autoantibodies against endogenous or exogenous insulin.[13] IAA are found primarily, though not exclusively, in young children developing T1DM as an early predictive marker. They are rarely established in adults with T1DM.[14]

C-PEPTIDE As discussed earlier in this chapter, insulin is secreted as two polypeptide chains joined by disulfide bonds. C-peptide is released when the two chains separate. C-peptide levels then can be used to measure the insulin production in the body. Autoantibodies, as previously outlined, are not always useful in determining the beta cell function of the pancreas. C-peptide levels allow that determination in both T1 and T2DM.

Medical Treatment To survive, individuals with T1DM must depend on daily administration of exogenous insulin in conjunction with nutrition therapy and physical activity to mimic the insulin secretion in an individual without diabetes.[8, 14] Treatment goals include avoiding hyperglycemia and retarding development of complications within an acceptable level of treatment side effects. The closer to the normal range blood glucose can be maintained over the long term, the lower the risk of complications.[15] Box 17.3 describes risk factors and treatments for short- and long-term complications of diabetes.

Medical care of diabetes mellitus should be the coordinated effort of a team with expertise and special interest in diabetes. The team should be comprised of (but not limited to) the individual with diabetes and the following care providers:[11]

- Physicians
- Registered dietitians (R.D.) or dietetic technicians, registered (D.T.R.)
- Certified Diabetes Educators (CDE)
- Nurse practitioners
- Nurses
- Physician's assistants
- Pharmacists
- Mental health professionals

The management plan should be individualized for the patient and family. Diabetes self-management education (DSME) is a vital element of care. Development of the management strategy should integrate the patient's age, school/work schedule and circumstances, eating patterns, physical activity, social environment and personality, cultural issues, other medical conditions, and presence of complications. It is important that each aspect of the management plan is understood and agreed upon by the patient and care providers, and that the goals and treatment plan are reasonable.[11]

Currently there are a variety of insulin options that permit design of suitable insulin regimens corresponding to an individual's preferential meal routine, food choices, and lifestyle.[16] A clear understanding of insulin pharmacokinetics and the ability to recognize trends in blood glucose data are essential in order to monitor patients' meal and insulin plans and determine appropriate insulin-to-carbohydrate ratios.[14] The best possible insulin management can only be achieved by evaluating blood

BOX 17.3 CLINICAL APPLICATIONS

Risk Factors for and Treatment of Complications of Diabetes Mellitus

Short-Term Complications

Complication/Symptoms[1]	Causes	Treatment
Hyperglycemia *Symptoms:* Polyuria, polydipsia, blurred vision, polyphagia, weight loss, fatigue, low energy, delayed healing, irritability	Excess food and/or CHO; large meals or excess snacking	Eat less food and/or CHO; distribute food/CHO appropriately
	Physical inactivity	Gradually increase activity
	Lack of blood glucose monitoring	Regular self-monitoring of blood glucose
	Inadequate diabetes medication	Add, adjust, or change medication(s)
	Inappropriate timing of medications	Coordinate timing of medication(s) and food
	Over-treatment of hypoglycemia	Use appropriate amounts of CHO sources
	Adverse effect of nondiabetes medications	Seek information about effect of medication on glucose
	Illness	Know how to manage diabetes during illness
	Variability in insulin absorption	Proper site rotation; proper insulin storage; check expiration date and discard 30 days after opening
	Variability in rates of digestion/absorption of food	Address issues of gastroparesis (delayed stomach emptying)
	Stress	Practice relaxation techniques
Ketoacidosis *Symptoms:* Nausea and/or vomiting; stomach pain, fruity (acetone) breath, heavy (or Kussmal) breathing, mental status change	Lack of blood glucose self-monitoring	Regular self-monitoring; test for ketones if glucose >250 mg/dL
	Severe illness or infection	Closely monitor effects of illness on blood glucose; increase frequency of glucose measurement; treat illness if indicated; take DM medications even when eating less; maintain hydration; plan for sick-day management
	Insulin omitted	Investigate rationale
	Increased insulin needs with growth spurts	Frequent blood glucose monitoring
	Inappropriately stored insulin	Discard expired insulin; protect insulin from excessive heat or cold
Hyperglycemic Hyperosmolar Syndrome *Symptoms:* Polyuria, polydipsia, polyphagia, weight loss; symptoms persist and worsen over several days or hydration status worsens	Dehydration	Monitor fluid intake; establish plan to take fluids regularly
	Excessive fluid losses	Monitor fluid status; address causal factors, replace fluids
	Prolonged hyperglycemia	Monitor blood glucose regularly; treat mild hyperglycemia
Mild Hypoglycemia *Symptoms:* Trembling, nervousness, trouble concentrating, anxiety, blurred vision, sweating, irritability, rapid heart rate, inability to think clearly, tingling in extremities, dizziness, hunger, nausea, fatigue, weakness, headache	Excess medication, or inappropriate timing of medications	Adjust amount and/or type of medication; coordinate timing of medications with food and activity
	Overcorrection of hyperglycemia with insulin	Use appropriate amount of insulin for correction
	Too little food and/or CHO	Consume appropriate amount of food and/or CHO
	Missed or delayed meal	Eat meals on time, or eat snack if meal will be late
	Increased activity	Increase food intake or reduce insulin
	Side effects from non-diabetes medication	Seek information about effect of medication on glucose
	Variability in insulin absorption	Proper site rotation; proper insulin storage; check insulin expiration date
	Variability in rates of digestion/absorption of food	Address issues of gastroparesis
	Alcohol consumption	Consume food when drinking alcohol; limit amount of alcohol consumed
Severe Hypoglycemia *Symptoms:* Mental confusion, argumentative, combative, lethargy, seizures, unconsciousness	Glucose level 51–70 mg/dL	Consume 15 g CHO[2], repeat if glucose does not return to normal range after 15 minutes
	Glucose ≤50 mg/dL	Consume 20–30 g CHO, repeat if glucose does not return to normal range after 15 minutes

Long-Term Complications

Complication	Risk Factors	Treatment
Macrovascular Complications		
Cardiovascular disease	T2DM	• Reduce LDL-cholesterol to 110 mg/dL (↓ foods high in saturated or *trans* fats) • Lower triglycerides to <150 mg/dL (wt loss, ↑ consumption of fish and n-3 vegetable sources, fish oil supplements) • Increase HDL-cholesterol to >40–50 mg/dL (weight loss, increased physical activity, smoking cessation)
Microvascular Complications		
Nephropathy	• Hypertension • Hyperglycemia • Native American, Hispanic American, or African American descent	• Aggressive BP control (<130/80 mm Hg) • Optimize glycemic control (preprandial plasma glucose 90–130 mg/dL; postprandial plasma glucose >180 mg/dL)
Retinopathy	• Duration of DM • Hyperglycemia • Hypertension	• Optimal glycemic control • Optimal BP control
Nervous System Disease		
Peripheral neuropathy	• DM ≥10 years • Poor glucose control • Other DM-related complications	• Optimal glycemic control • Daily foot care, walking, gentle stretching, relaxation exercises • Medication for pain relief (topical capsaicin, antidepressants, and anticonvulsants)
Autonomic neuropathy • Cardiovascular (postural hypotension and "silent" heart disease) • Genitourinary (sexual dysfunction, bladder emptying problems) • Gastroparesis (delayed emptying of stomach)	Long-standing DM	• Optimized glycemic control • Dietary modifications • Therapeutic lifestyle changes (small, frequent meals, reduce fat intake, reduce fiber intake, use foods with soft consistency, exercise after meals, adjust insulin doses and timing) • Pharmacologic options (cholinergic drugs, dopamine antagonists, motilin-receptor agonists) • Surgical treatment (jejunostomy, gastrectomy)

1. All symptoms may not be experienced by all individuals
2. ½ cup fruit juice or regular (nondiet) soft drink, 3–4 glucose tablets, 3–5 hard candies.

Sources: Adapted from: Arnold MS. Hypoglycemia and hyperglycemia. In: Ross TA, Boucher JL, O'Connell BS, editors. *American Dietetic Association guide to diabetes: medical nutrition therapy and education.* Chicago: American Dietetic Association, 2005; Wheeler ML. Long-term complications. In: Ross TA, Boucher JL, O'Connell BS, editors. *American Dietetic Association guide to diabetes: medical nutrition therapy and education.* Chicago: American Dietetic Association, 2005; American Diabetes Association. Standards of medical care in diabetes. *Diabetes Care.* 2008;31(Suppl 1):S12.

glucose monitoring records, adjusting food and exercise activities, and proposing insulin adjustments.[17]

TYPES OF INSULIN Types of insulin that have different patterns and rates of activity are frequently combined in an effort to mimic normal physiological action of insulin. To accomplish this, insulin is classified based on expected onset of action, peak time of action, and duration of action. It is important for health care professionals to thoroughly understand these features before developing a care plan and meal plan.[14] Table 17.5 summarizes the features of available insulin types.

DETERMINING INSULIN DOSES The initial insulin dosage is often determined by using algorithms based on body weight. For example, 0.3–0.5 units/kg is a common starting dosage for insulin in an individual who is within 120% of their IBW. These algorithms (see Box 17.4) can be used as a general guideline, but insulin dosages must then be adjusted based on blood glucose patterns over time.[14]

INSULIN REGIMENS A single dose of insulin is rarely capable of providing optimal glycemic control in T1DM.[18] There are three basic types of insulin administration regimens: fixed (conventional or standard therapy), flexible (intensive insulin therapy), and continuous subcutaneous insulin infusion (CSII) (see Box 17.5).

Conventional or standard insulin therapy consists of a constant dose of basal (or background) insulin combined with short- or rapid-acting (or bolus) insulin. This is referred to as a mixed dose and the individual may mix the insulins or use premixed insulins (for example, 30 units of 70/30 insulin). If more than one injection is used, such as 15 units lispro plus 25 units NPH before breakfast and 10 units lispro and 20 units NPH before the evening meal, this is referred to as a split (mixed) dose. Individuals using conventional therapy must synchronize administration of their insulin and food intake to avoid hypoglycemia. A good understanding of onset, peak, and duration of their insulin dose

Table 17.5 Types of Insulin and Pramlintide

Insulin Type	Onset of Action	Peak of Action (hours)	Duration of Action (hours)	Comments
Rapid-Acting Insulin Analogs				
Lispro Aspart Glulisine	5–15 min	30–90	3–5	Can be used in pump therapy
Short-Acting				
Regular	30–60 min	2–4	5–8	Can be mixed with longer-acting insulin
Intermediate-Acting				
NPH	2–4 hrs	4–10	10–16	Usually given in 2 daily doses
Extended Long-Acting Analog				
Insulin glargine	2–4 hrs	Peakless	20–24	Cannot be mixed with other insulins
Insulin detemir	2–4 hrs	6–14	16–20	
Premixed				
70/30	30–60 min	Dual	10–16	70% NPH, 30% regular
75/25 lispro analog mix	5–15 min	Dual	12–20	75% intermediate, 25% lispro
70/30 aspart analog mix	5–15 min	Dual	12–20	70% intermediate, 30% aspart
50/50 human mix	30–60 min	Dual	10–16	50% NPH, 50% regular
50/50 lispro analog mix	5–15 min	Dual	12–20	50% intermediate, 50% lispro
Antihyperglycemic Drug				
Pramlintide (synthetic analog amylin)	Slows transit of digesting food through intestine; given at mealtimes to increase efficacy of insulin; should not be mixed with insulin			

Source: Skyler JS. Insulin treatment. In American Diabetes Association, *Therapy for Diabetes Mellitus and Related Disorders*, 5th ed. Alexandria, VA: American Diabetes Association, 2009.

in relation to their meals and snacks in addition to consistency of food intake is also important.[14, 18] Nutrition goals are based on overall diabetes management goals: target glycemic goals and nutrition-related behaviors that affect these goals.

Flexible or intensive insulin therapy requires multiple daily injections (MDIs) of bolus insulin before meals in addition to basal insulin once or twice daily. Insulin can be adjusted to correspond to food intake, therefore replicating endogenous insulin secretion in a person without diabetes. This also allows for adjustment of insulin dose in response to hyperglycemia, variable carbohydrate intake, or alteration in usual physical activity.[14] Intensive insulin therapy (when compared to conventional therapy) delays onset and slows the progression of complications such as retinopathy, nephropathy, and neuropathy in patients with T1DM.[15] These medical issues will be discussed later in this chapter.

It is still important to integrate the insulin regimen with the patient's lifestyle.[19] Individuals employing intensive therapy are required to know their basic doses for background and bolus (meal time) insulins. This permits them to fine-tune the short- and rapid-acting insulin dose when they digress from customary meal plans and/or exercise programs.[20]

BOX 17.4 CLINICAL APPLICATIONS

Algorithms Used to Determine Insulin Dose

Diabetes Type	Daily Insulin Dose
T1DM	0.6 units/kg actual body weight
T1DM with trace to small amounts of ketones or T2DM with BMI ≤27	0.3–0.5 units/kg actual body weight
T1DM with moderate to large amounts of ketones or T2DM with BMI ≥27	0.5–0.7 units/kg actual body weight
T2DM with oral hypoglycemic medications	0.1–0.3 units/kg ideal body weight

For example, Susan has recently been diagnosed with T1DM. She is 15 years old and weighs 100 lbs. Once Susan's blood glucose levels were under control, her doctor prescribed 27 units insulin. How did the doctor arrive at this dosage?

0.6 units/kg actual body wt

100 lb = 45.5 kg body wt

$45.5 \times 0.6 = 27.3$ units

Sources: Adapted from Rystrom JK. Insulin therapy. In: Ross TA, Boucher JL, O'Connell BS, editors. *American Dietetic Association guide to diabetes: medical nutrition therapy and education.* Chicago: American Dietetic Association; 2005 and Nelms, MN, Long S, Lacey K. Medical Nutrition Therapy: A Case Study Approach, 3rd ed., Belmont (CA): Wadsworth/Cengage Learning; 2009.

Insulin Regimens

Continuous subcutaneous insulin infusion (CSII) Provides basal rapid- or short-acting insulin pumped continuously in micro-amounts through a subcutaneous catheter and is monitored 24 hours a day. Boluses of rapid- or short-acting insulins are given before meals.

Intensive insulin therapy (multiple daily injections [MDIs]) Intermediate insulin given once or twice daily and rapid- or short-acting insulin is given prior to meals. Allows more flexibility in type and timing of meals. Amount of rapid- or short-acting insulin can be adjusted based on meal composition and/or its carbohydrate content.

Conventional therapies ("split" or "mixed" dose) There are two basic conventional therapy regimens. Option #1: Short- or rapid-acting insulin mixed with intermediate-acting insulins[1] given before breakfast and before evening meal. Option #2: Combination of short- and intermediate-acting insulins before breakfast, short-acting insulin before evening meals, and intermediate-acting insulin at bedtime. Used to control **dawn phenomenon**.

Source: Adapted from: Tamborlane WV, Sikes KA, Swan K, Weignzimer SA. Type 1 diabetes in children. In American Diabetes Association. *Therapy for Diabetes Mellitus and Related Disorders*, 5th ed. Alexandria, VA: American Diabetes Association, 2009.

1. See Table 17.5 for descriptions of insulin types.

This type of therapy may not be appropriate for everyone. Patients should be reminded that out-of-target blood glucose results can be brought about by circumstances (stress, illness, unpredictable insulin absorption, changes in exercise) other than food intake.

Continuous subcutaneous insulin infusion (CSII) (or pump therapy) is a form of intensive therapy. Basal rapid- or short-acting insulin is pumped continuously in micro-amounts through a subcutaneous catheter and is received 24 hours a day. Boluses of rapid- or short-acting insulins are given before meals.

SYRINGES AND PENS The two conventional methods of insulin administration are by means of syringes and pens (see Figure 17.11). Insulin syringes are disposable (should only be used once), have short, fine beveled needles, and are designed for U-100 insulin. To make the injection easier and reduce tissue damage, needles are lubricated.[14]

Insulin pens look a lot like large marking pens. Reusable (with prefilled cartridges) or disposable pens are available and come filled with either 150 units or 300 units of insulin. Cartridges and prefilled pens are available for rapid-acting, regular, and extended long-acting insulins, some premixed insulins, and glargine. Needles in pens are used once, discarded, and then replaced for each injection.[14]

INSULIN PUMPS Insulin pumps (see Figure 17.11) are approximately the size of pagers, and are powered by batteries. Regular or rapid-acting (aspart, lispro, glulisine, and apridra) insulin is delivered through flexible tubing and is attached to the individual via an infusion set. Continuous subcutaneous insulin infusion (CSII) allows creation of variable and adjustable insulin dosing to meet specific, individual insulin needs.[14] Pump therapy duplicates endogenous insulin secretion more closely than other methods of insulin delivery. Detailed instructions and training are necessary, and mastering this method requires time and effort.

Nutrition Therapy for Type 1 Diabetes Mellitus

Nutritional Implications The impact of dietary modification on overall health, metabolic control, and treatment for acute and chronic complications is substantial.[11] Individualized nutrition therapy is required to achieve treatment goals.[11]

Nutrition Assessment A comprehensive nutrition assessment, a self-care treatment plan, and the client's health status, learning ability, readiness to change, and current lifestyle should be the basis for nutrition therapy and DSME. Table 17.6 outlines the components of nutrition assessment in diabetes.

Nutrition Diagnosis Common nutrition diagnoses associated with diabetes include inadequate energy intake, inappropriate intake of types of carbohydrates, inconsistent carbohydrate intake, inadequate fiber intake, altered GI function (gastroparesis), altered nutrition-related laboratory values, food-medication interaction, underweight, food- and nutrition-related knowledge deficit, not ready for diet/lifestyle change, self-monitoring deficit, undesirable food choices, physical inactivity, or inability or lack of desire to manage self-care.

Nutrition Intervention It is important to address individual nutritional needs with regard to personal and cultural preferences and lifestyles while respecting the individual's wishes and willingness to change. It will be important to build or reinforce the patient's basic or essential nutrition-related knowledge, and that of his/her family and friends. It is also crucial to continue instruction and/or training to promote acquisition of in-depth nutrition-related knowledge or skills. Other intervention strategies may include, but are not limited to, motivational interviewing, goal-setting, self-monitoring, problem-solving, stress management, and relapse prevention.

GOALS OF NUTRITION THERAPY Four main goals of nutrition therapy are as follows:[11, 21, 22]

- Attain and maintain optimal metabolic outcomes, including:
  - Glucose level in normal range, or as close to normal range as is safely possible, to prevent or reduce risk of complications;
  - Lipid or lipoprotein profile that reduces risk for macrovascular disease;
  - Blood pressure levels that reduce risk for vascular disease.
- Prevent and treat chronic complications. Modify nutrient intake and lifestyle as appropriate for prevention and

Figure 17.11 Insulin Syringes, Insulin Pens, Insulin Pumps, and Continuous Glucose Sensors

(a) Insulin syringe

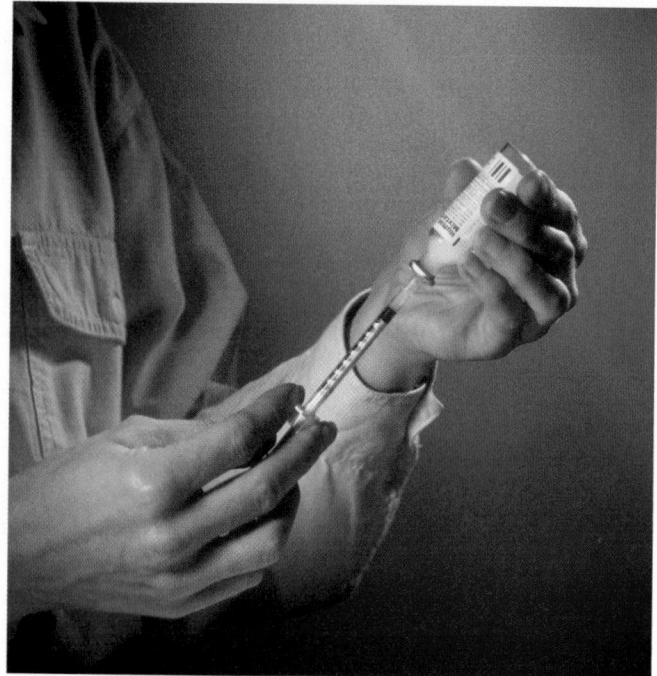

Source: Saturn Stills/Photo Researchers, Inc.

(b) Insulin pen (disposable or refillable)

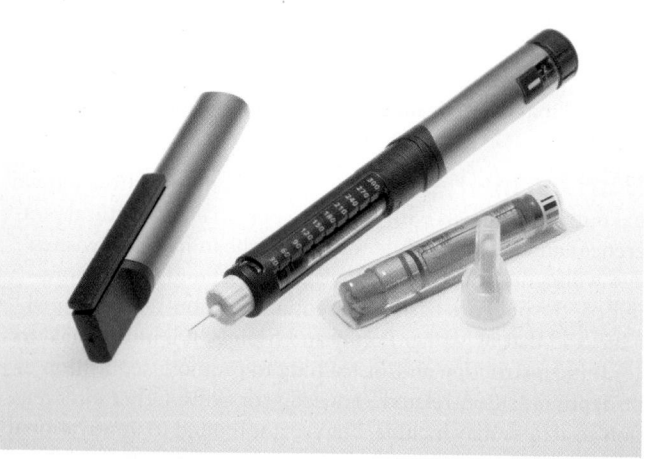

Copyright © Creatas/Fotosearch.

(c) Insulin pump

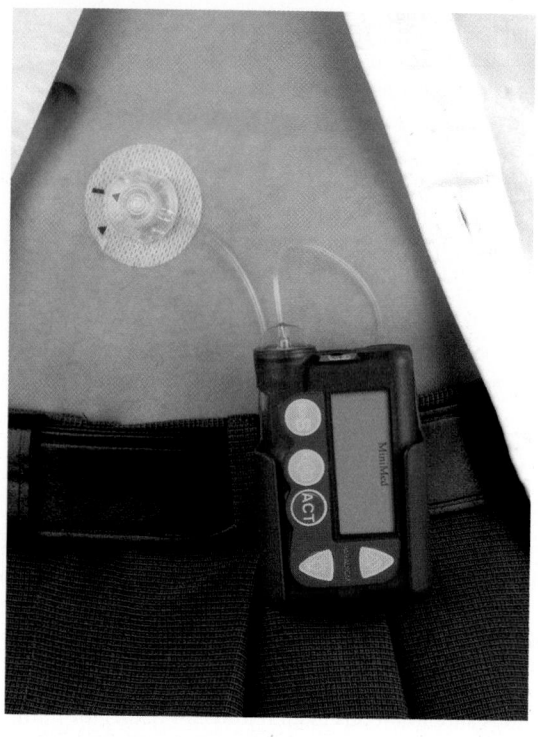

Source: Photo courtesy of the Diabetes division of Medtronic, Inc.

(d) Continuous glucose sensor

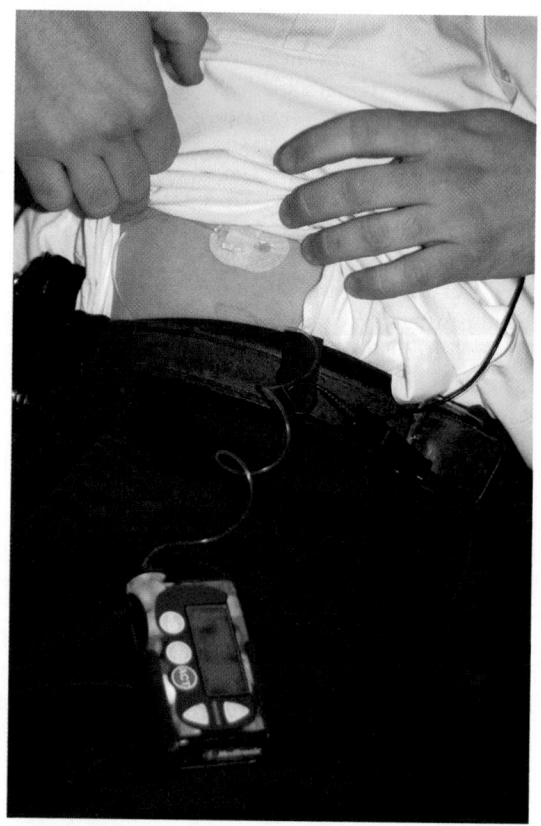

Source: Fotosearch.

Table 17.6 Nutrition Assessment for Diabetes Mellitus

Client History (personal, patient, family, medical and social)	• Previous medical conditions or surgeries • Medications: both prescription and over-the-counter (laxatives, fiber supplements) • Socioeconomic status/food security • Support systems • Education—primary language	**Biochemical Data, Medical Tests and Procedures** *Visceral Protein Assessment:* *Hematological Assessment:*	• Albumin • Prealbumin • Transferrrin • Retinol Binding Protein • C Reactive Protein • Hemoglobin • Hematocrit • MCV • MCHC • MCH • TIBC
Food-/Nutrition-Related History	• Ability to chew; use and fit of dentures • Problems swallowing • Nausea, vomiting • Constipation, diarrhea • Heartburn • Any other symptoms interfering with ability to ingest normal diet • Ability to consistently purchase adequate amounts of food on a daily basis • Ability to feed self • Ability to cook and prepare meals • Food allergies, preferences, or intolerances • Previous food restrictions • Ethnic, cultural and religious influences • Use of alcohol, vitamin, mineral, herbal, or other type of supplements • Previous nutrition education or medical nutrition therapy • Eating pattern: 24-hour recall, diet history, food frequency	*Lipid Assessment:* *Renal Assessment:* *Endocrine-Specific Biochemical Data, Medical Tests and Procedures:*	• Total cholesterol • HDL • LDL • Triglyceride • BUN • Creatinine • Creatinine clearance • Spot urinalysis for albumin: creatinine ratio • Glomerular filtration rate (GFR) • Fasting plasma glucose • Oral glucose tolerance test • Hemoglobin A1c • C Peptide • Screening for celiac disease: transglutaminase or antiendomysial antibodies • T_3 (triiodothyronine) • T_4 (thyroxine) • TSH (thyroid-stimulating hormone) • TRH (thyrotropin-releasing hormone)
Anthropometric Measurements	• Height • Current weight • Weight history: highest adult weight; usual body weight • Reference weight (BMI)		

treatment of obesity, dyslipidemia, cardiovascular disease, hypertension, and **nephropathy**.

- Enhance health through food choices and physical activity.
- Address individual nutritional needs with regard to personal and cultural preferences and lifestyles while respecting the individual's wishes and willingness to change.

NUTRITION PRESCRIPTIONS Nutrition recommendations for total fat, saturated fat, cholesterol, fiber, vitamins, and minerals are the same for individuals with diabetes as for the general population. Protein intake can range from 15% to 20% of daily kcalories from animal and vegetable protein sources. If the patient has nephropathy, lower intakes of protein (about 10% of daily energy intake) may be warranted. Carbohydrate recommendations are individualized based on the individual's eating habits, blood glucose goals, and lipid goals, but at least 130 grams/day are recommended. Blood glucose control is not impaired by use of sucrose within the meal plan, but sucrose-containing foods should not necessarily be substituted for other carbohydrates and foods, and should not be eaten in addition to a meal plan. Blood glucose levels are not affected by moderate alcohol use if diabetes is well controlled. Alcohol kcal should be considered an addition to regular food or meals, and alcohol should not be substituted for food.[23]

There is no one "diabetic diet" or "ADA diet." Even though the term "ADA diet" has never been clearly defined, in the past it usually meant a physician-determined kcal level with explicit percentages of carbohydrate, protein, and fat based on the exchange lists. The American Diabetes Association (ADA) recommends the term "ADA diet" not be used since the ADA no longer sanctions any single meal plan or specified percentages of nutrients.[24]

MEAL PLANNING Every meal planning method has advantages and drawbacks. Consequently, tailoring the meal planning approach to each individual's needs is key.[25] Individuals receiving conventional insulin therapy must be consistent with timing of their meals and amounts of food consumed. Those receiving intensive insulin therapy have more flexibility in when and what they eat.[21] Regardless of which meal planning method is used, it should be individualized based on customary food intake and the patient's preferences, which will permit diet modification to be put into practice with less difficulty. Meals and snacks should be distributed in a manner that is consistent with the individual's way of life, activity patterns, and diabetes medications.[25] Meal planning approaches range from simple guidelines to more complex counting methods.[22] The approach selected should be the one that best fits the individual patient's situation. Carbohydrate counting and the exchange lists are described in the next section; other meal planning approaches are outlined in Appendix E4.

CARBOHYDRATE COUNTING Of the four meal planning approaches used in the Diabetes Control and Complications Trial (DCCT), the most successful was carbohydrate counting.[26] The basic concept of carbohydrate counting is that the carbohydrate found in foods is the major macronutrient influencing postprandial glucose variations, and that it influences pre-meal insulin requirements more than the protein and fat content of the meal.[22, 27] The total amount of daily carbohydrate intake, not its source, is the focus of this meal planning approach.

Emphasis on eating consistent amounts of carbohydrate at meals and snacks can make carbohydrate counting a simpler method of meal planning.[22] Food carbohydrate sources are starches, fruits, milk/yogurt, and sweets. (Nonstarchy vegetables do not need to be counted unless eaten in servings containing >15 g of carbohydrates.) Carbohydrates can be counted in one of the following two ways:[22]

- The amount of food containing 15 g carbohydrate counts as one carbohydrate choice.

- Total grams of carbohydrate in a meal or snack can be counted by use of food label information or other sources of nutrient analysis information.

This is not to say that meats and fats can be totally ignored. The kcal and fat content of meats and fats can contribute to weight gain and/or lipid abnormalities. By allowing individuals to make appropriate adjustments in their diabetes management, carbohydrate counting empowers them to learn relationships between food, insulin, physical activity, and blood glucose levels.[21] Box 17.6 outlines carbohydrate counting (including matching carbohydrate content to insulin doses).

EXCHANGE SYSTEM Since 1950, a widely used method for planning food intake has been the exchange list.[28] It provides uniformity in meal planning and allows a wide variety of foods to be included in the diet. This method uses the concept of "exchange" or substitution of different foods within each of three groups: carbohydrate (starch, fruit, milk, non-starchy vegetables, and other carbohydrates), meat and meat substitutes (very lean meats, lean meats, medium-fat meats, high-fat meats), and fats (monounsaturated, polyunsaturated, saturated).[25] Each food portion on a particular list can be exchanged with any other food portion on the same list.[22]

The following guidelines can be used to calculate a meal plan using exchanges:[16, 21, 22, 29]

1. Assess current food intake and eating pattern using diet history or food records.

2. Categorize usual food intake into exchange amounts based on portions and foods consumed at each meal and snack. Calculate total grams of carbohydrate, protein, and fat and translate into energy. Refer to Chapter 3 for guidelines for calculation and Appendix L1 for the exchange lists.

3. Determine appropriate energy prescription. Subtract energy if weight loss is desired; add energy if weight gain is desired. Generally 250 to 500 kcal per day can be subtracted/added for a ½ to 1 pound per week weight loss/gain.

4. Translate energy prescription into exchanges, staying as close to current pattern of intake (from diet history) as pos-

sible. Calculate grams of carbohydrate, protein, and fat from exchanges and determine percentages of energy contributed by each macronutrient:

- % carbohydrate = grams carbohydrate × 4/total kcal
- % protein = grams protein × 4/total kcal
- % fat = grams fat × 9/total kcal

5. Adjust exchanges as needed to reach goal percentages for each macronutrient.

6. Compare usual intake to energy prescription and mutually determine how to distribute exchange groups among meals and snacks.

While there are many advantages to using the exchange lists, they are often not the most appropriate meal planning system for individuals with diabetes. The exchange lists are written at a ninth to tenth grade reading level; therefore, individuals using them must be able to read at this level and understand the concept of "exchanging" foods. It may require several educational sessions and practice for them to be used effectively.[22]

Monitoring and Evaluation Whatever type of intervention is necessary can be monitored with the use of self-monitoring records review and laboratory tests. Evaluation may reveal a need for further education or adjustment of insulin dosages based on self-monitoring. Follow-up sessions will provide an opportunity to evaluate how well the patient is able to manage his/her diabetes.

Glycemic control is fundamental to the management of diabetes and provides the basis for monitoring and evaluation of the condition. Improved glycemic control is correlated with sustained reduced rates of retinopathy, nephropathy, and neuropathy.[15, 30] Efficacy of glycemic control and the management plan can be evaluated by health providers and the individual with diabetes through several methods (see Table 17.7 for an overview).[15, 30] Recommended glycemic goals are outlined in Table 17.8. The combination of self-monitoring of blood glucose (SMBG) and A1C is the best indicator of glycemic control.

GLYCATED HEMOGLOBIN ASSAYS (A1C) Glycated hemoglobin assays (hemoglobin A1C, or A1C) measure the amount of glucose bound to hemoglobin protein. The higher the glucose concentration in the blood, the more hemoglobin is glycated (via addition of a glucose molecule to amino acid side-chains), thus making it a valid test to measure degree of hyperglycemia. Because red blood cells have a lifespan of 120 days, A1C can measure the average glucose concentration for the previous two to three months. Table 17.9 shows the correlation between A1C levels and mean plasma glucose levels. Because A1C values cannot be significantly affected by manipulating diet or treatment in the week before a clinic appointment, they are valuable tools to assess glycemic control for a period of time. A1C is a monitoring tool for individuals with known diabetes and is not specific or sensitive enough to be used for diagnosis of diabetes.[31] In addition, since it measures hemoglobin-bound glucose, A1C is inappropriate as a gauge of glycemic control for individuals with blood disorders such as anemia (see Chapter 19).

A1C should be tested at least twice yearly in individuals who are meeting treatment goals and who have stable glyce-

BOX 17.6 CLINICAL APPLICATIONS

Carbohydrate Counting

Carbohydrate counting is a meal-planning approach that concentrates on the total amount of carbohydrate eaten at meals and in snacks. It is based on research demonstrating that consistent intake of a wide variety of carbohydrates results in similar post-prandial glucose responses. Awareness of total carbohydrate intake and distribution has also been shown to improve metabolic control. The American Diabetes Association (ADA) states that monitoring carbohydrate remains a significant tactic in realizing glycemic control. The ADA also states that sucrose-containing foods can be used as other carbohydrates in the meal plan or, if added to the meal plan, balanced with insulin or other glucose-lowering medications.

Carbohydrate Choices

There is no fixed quantity of carbohydrate that is recommended for everyone. The amount of carbohydrate is established in consultation with the individual with diabetes. Amounts can then be adjusted based on blood glucose monitoring results and what is determined to be reasonable in context of the individual's lifestyle.

Example for establishing carbohydrate choices:

Mike, a 24-year-old male, has T1DM. He is 5'10" and weighs 165 lbs. Using the recommended standards for percentage of kcal from carbohydrate and his diet history, you determine that he should consume approximately 55% of his kcal from CHO.

Estimated energy requirement (EER): 2200–2400 kcal/day

Step One: Take EER and multiply by 55%. $2200 \times 0.55 = 1210$ kcal from CHO

Step Two: 1210 kcal/4 = 302 grams of CHO

Step Three: Determine the number of CHO choices for the 24 hour period— 302 g/15 g per choice = 20. This is the number of CHO choices for Mike to consume in the 24-hour period. This amount will be adjusted according to his weight, hunger, and physical activity levels.

Step Four: Divide the CHO choices between meals and snacks so that Mike will know the amount of rapid-acting insulin he should use at each meal. After he has stabilized and has records from his self-monitoring of blood glucose, the amount of insulin required for each carbohydrate choice can be fine-tuned. This is discussed below.

Insulin-to-Carbohydrate Ratio

The insulin-to-carbohydrate ratio is a mechanism for determining insulin dosage based on carbohydrate intake. Generally, 1 unit of rapid-acting insulin is taken for every 10–15 grams of carbohydrate. This is used as a starting point, then adjusted based on SMBG records. The "500 rule" is used to calculate the initial insulin-to-carbohydrate ratio. "500" is divided by an individual's total daily dose of rapid-acting insulin. ("450" is used for regular insulin.) For example:

If total daily insulin dose is 50 units of rapid-acting insulin, 500 is divided by 50 to equal 10 grams of carbohydrate covered by each unit of rapid-acting insulin.

The insulin-to-carbohydrate ratio is 1 unit of insulin for every 10 grams of carbohydrate.

Correction Factor

If necessary, a correction factor can be used with the insulin-to-carbohydrate ratio to assist in returning blood glucose levels to the target range. Blood glucose values will indicate whether this is necessary. One unit of rapid-acting insulin is given for every 50 mg/dL that blood glucose rises above 150 mg/dL. For example, if blood glucose is 151–200 mg/dL, 1 unit of insulin is added; if blood glucose is 201–250, 2 units are added; and blood glucose levels 251–300 would require addition of 3 units of insulin. Blood glucose levels of 300 mg/dL and above require addition of 4 units of insulin.

Meal-Time Insulin

In summary, meal-time insulin dosage equals insulin-to-carbohydrate ratio plus correction factor.

Sources: Nutrition Care Manual [database on the Internet]. Chicago (IL): American Dietetic Association; 2008. Available at http://nutritioncaremanual.org (by subscription only); Franz MJ, Bantle JP, Beebe CA, Brunzell JD, Chiasson J-L, Garg A, Holzmeister LA, Hoogwerf BJ, Mayer-Davis E, Mooradian AD, Purnell JQ, Wheeler M. Evidence-based nutrition principles and recommendations for the treatment and prevention of diabetes and related complications (Technical Review). *Diabetes Care.* 2002;25:148–98; American Diabetes Association. Nutrition recommendations and interventions for diabetes. *Diabetes Care.* 2008;31(suppl 1):S61–S78.

---

mic control. For individuals not meeting glycemic goals, or whose therapy has changed, A1C testing should be performed quarterly.[11] The values of A1C are used to calculate the estimated average glucose (EAG). This new calculation is now a routine component of biochemical assessment.

SELF-MONITORING OF BLOOD GLUCOSE (SMBG)
Daily home glucose monitoring indicates what an individual's glucose level is at the very moment the measurement is taken. Information provided by SMBG can assist in adjusting daily eating patterns and medications as necessary to maintain glycemic control. SMBG is also useful in identifying patterns and the ways in which food, exercise, or other factors affect glycemic control. Adjustments to an individual's treatment program can be made immediately in order to prevent hyperglycemia, hypoglycemia, and long-term complications of diabetes.[11]

A typical SMBG test includes a drop of blood obtained via a finger prick that is applied to a chemically treated reagent strip. Home monitors are generally used to determine results. Frequency and timing of SMBG should be determined by the specific needs and goals of the individual with diabetes and the health care team. As a means to monitor for asymptomatic hypoglycemia and hyperglycemia,

Table 17.7 Techniques Used to Assess Glycemic Control

Technique	Benefit	Recommendations for T1DM	Recommendations for T2DM	Comments
Self-Monitoring of Blood Glucose (SMBG)	• Allows patients to individualize response to therapy • Useful in preventing hypoglycemia • Useful in adjusting medications, nutrition therapy, and physical activity	3+ times daily	• Nutrition therapy and exercise alone: 1–2 times daily at alternating times throughout day (e.g., Monday before dinner and bedtime, Tuesday before breakfast and lunch) • Oral glucose-lowering medication: 1–2 times daily, rotating test times each day • Insulin: ≥3 times	• Accuracy instrument- and user-dependent • Evaluate patients' monitoring techniques initially and subsequently at regular intervals • Regularly evaluate patients' ability to use SMBG data to adjust food intake, exercise, or pharmacological therapy to achieve specific glycemic goals
A1C	• Allows measurement of average glycemia over preceding 2–3 months	Every 3 months	• Patients whose therapy has changed or who are not meeting glycemic goals: test every 3 months • Patients with stable glycemic control test (at least) twice yearly	• Regular A1C testing allows detection of departures from target management goals

Source: American Diabetes Association. Standards of medical care in diabetes. *Diabetes Care.* 2008;31(Suppl 1):S12.

Table 17.8 Recommendations for Glycemic Control

Glycemic Indicator	Normal	Goal[1]	Goal in Pregnancy
Preprandial glucose	<100 mg/dL (<6.7 mmol/L)	70–130 mg/dL (5.0–7.2 mmol/L)	60–99 mg/dL (3.3–5.0 mmol/L)
Postprandial glucose[4]	<140 mg/dL (<7.2 mmol/L)	<180 mg/dL (<10.0 mmol/L)	100–129 mg/dL (<6.7 mmol/L)
A1C[2,3]	4–6%	<7%	≤6%

[1]For nonpregnant adults. Different treatment goals may be warranted by individuals with comorbid diseases, the very young, older adults, and others with unusual conditions or circumstances.

[2]Primary target for glycemic goals.

[3]Goals of <6% may further reduce complications at increased risk or hypoglycemia (particularly in T1DM).

[4]Measured 1–2 hours after the beginning of a meal.

Adapted from: American Diabetes Association. Standards of medical care in diabetes. *Diabetes Care.* 2009;32(Suppl 1):S13. Garg SK, Ramachandra GN. Monitoring diabetes. American Diabetes Association. *Therapy for Diabetes Mellitus and Related Disorders,* 5th ed. Alexandria, VA: American Diabetes Association, 2009.

Table 17.9 Correlations between A1C Level and Mean Plasma Glucose Levels[1]

A1C (%)	Mean Plasma Glucose mg/dL
6	135
7	170
8	205
9	240
10	275
11	310
12	345

[1]Based on a normal A1C of 6.

Source: American Diabetes Association. Standards of medical care in diabetes. *Diabetes Care.* 2009;32(Suppl 1):S13.

daily SMBG is particularly valuable for all individuals with diabetes. SMBG is recommended three or more times daily for most individuals with T1DM and diabetic pregnant women. Individuals with T2DM taking insulin usually need to perform SMBG more often than those not using insulin. Individuals with T1DM or T2DM ought to test more often than usual when therapy is modified.[11]

Accuracy of SMBG is instrument- and user-dependent. For this reason, the patient's monitoring techniques should be assessed at the onset and at regular intervals thereafter. Patients should also be taught how to use SMBG data to modify food intake, exercise, and pharmacological therapy in order to realize individual glycemic goals.[11]

CONTINUOUS GLUCOSE MONITORING Continuous glucose monitoring uses a device that places a sensor right under the skin (see Figure 17.11 d). The sensor transmits the blood glucose reading to a receiver device worn around the waist—similar to a pager. This allows for constant reading of blood glucose levels every five minutes. It is not meant to replace SMBG but does provide a more detailed picture of blood glucose fluctuations and can help direct more detailed insulin prescriptions.

FRUCTOSAMINE TEST Fructosamine assays have nothing to do with fructose; instead, the name refers to the chemical structure formed when glucose attaches to a molecule of protein, which resembles a fructose molecule. The fructosamine test is another index of time-averaged plasma glucose. This assay is used to monitor glycemic control over a 1- to 3-week period and is neither influenced by variant hemoglobin nor attached to the lifespan of a red blood cell. This assay is not reliable in individuals with renal failure or liver disease and is not frequently used in the clinical setting.[32]

URINE TESTING FOR GLUCOSE Glycosuria occurs when the **renal threshold** for glucose is exceeded. Renal threshold varies between individuals, but usually occurs when blood glucose levels are >250 mg/dL. Hypoglycemia cannot be detected by urine testing. SMBG is a much more accurate method of monitoring glycemic control.

URINE TESTING FOR KETONES The test for presence of ketones in urine should be performed regularly during periods of illness or stressful situations when glucose levels are likely to be elevated. In individuals with T1DM, urine ketones should be tested when blood glucose is consistently over 300 mg/dL.[11]

OTHER TESTING In addition to monitoring glycemic control, other parameters that should be monitored include lipids and blood pressure. Total cholesterol, low-density lipoproteins (LDL-cholesterol), high-density lipoprotein (HDL-cholesterol), and triglycerides should be monitored annually or more frequently as needed. The accuracy of these tests is dependent upon an overnight fast. Goal values are as follows:[11]

- Total cholesterol: <200 mg/dL
- LDL-cholesterol: <70mg/dL
- HDL-cholesterol: >40 mg/dL (men); >50 mg/dL (women)
- Triglycerides: <150 mg/dL

Physical Activity

For most individuals with diabetes, the following benefits of physical activity far exceed the risks:[33, 34]

- Improved glycemic control (A1C)
- Improved blood lipids and blood pressure, with subsequent lower cardiovascular risks and overall mortality
- Positive impact on metabolic abnormalities characteristic of T2DM
- Prevention or delay of onset of T2DM for individuals at high risk for developing diabetes or with **pre-diabetes mellitus**
- Reduced risk of development of cardiovascular disease, since physical inactivity and diabetes are independent risk factors for it
- Improved coping and stress management and reduced feelings of depression
- Improved physical fitness and functional capacity
- Enhanced quality of life

Nutritional Implications Both hypoglycemia and hyperglycemia are acute risks of exercise. Hypoglycemia can occur during exercise that lasts longer than 1 hour, and for up to 24 hours after unusually strenuous, prolonged, and/or sporadic exercise. Blood glucose levels should be monitored, and carbohydrates should be increased and/or insulin adjustments should be made.[35] In those individuals whose diabetes is poorly controlled (underinsulinized), exercise can cause hyperglycemia. When insulin is deficient, the rise in counter-regulatory hormones that takes place during exercise causes an increase in hepatic glucose production and free fatty acids. Cellular uptake of glucose is minimal, resulting in both hyperglycemia and increased production of ketones.[11]

Nutrition Assessment Self-management records should be evaluated to determine how blood glucose levels are affected by carbohydrate intake before and after exercise as well as insulin administration.[35]

Nutrition Intervention Blood glucose levels should be monitored both before and after exercise to understand how diabetes affects glycemic control and to determine appropriate insulin and carbohydrate adjustments. For moderate exercise lasting less than 30 minutes, additional carbohydrate or insulin adjustment is rarely necessary. On the other hand, if blood glucose levels are within normal limits and exercise will last longer than 30 minutes, a small snack is needed. As a rule, an additional 15 grams of carbohydrate should be adequate for one hour of moderate physical activity. For more strenuous exercise, 30 grams of carbohydrate per hour may be required. For exercise before breakfast or later in the afternoon, extra carbohydrate should be consumed before the exercise. For exercise after meals, the additional carbohydrate can be taken after exercise.[34]

It may be necessary to adjust insulin dosage before exercise to avoid hypoglycemia, especially if exercise lasts for 45 to 60 minutes. For most, a modest decrease (~20%) in the insulin component corresponding to the period of exercise is a good place to start. More prolonged or vigorous exercise may necessitate a larger reduction in initial insulin dosage (by as much as one-third to one-half) to avoid hypoglycemia. Regular exercise (at least every other day) usually does not require adjustments to insulin dosage. Since the bodies of patients who exercise regularly will have adjusted to this activity level, the total insulin doses prescribed will already be lower.[34]

Acute Complications Associated with T1DM

Side Effects and Complications of Insulin Therapy Although individuals with T1DM would not survive without exogenous insulin treatment, it is not without risks. The appropriate insulin dose and regimen necessary to avoid hyperglycemia (and its complications) may produce side effects. This is especially true of intensive insulin therapy.[15]

The most universal side effect of insulin is hypoglycemia (blood glucose level <70 mg/dL). Thus, it is imperative that individuals taking insulin be educated about signs, symptoms, and remedies for hypoglycemia. Each individual may experience hypoglycemia at varying blood glucose levels, and while symptoms are individual, the most common are weakness, shakiness, perspiration, hunger, and rapid heartbeat.

Mild hypoglycemia can be self-treated by following several steps:[35]

1. If blood glucose level is <70 mg/dL, the individual should consume 10 to 15 grams of any carbohydrate that contains fast-acting carbohydrate (for example: ½ cup fruit juice, ½ cup regular soft drink, 3–4 glucose tablets). If blood glucose is <50 mg/dL, 20 to 30 grams of carbohydrates should be consumed.

2. Fifteen minutes after treatment, blood glucose levels should be rechecked to ascertain whether blood glucose levels have been restored to the normal range. If not, the process outlined in #1 should be repeated.

3. The individual should determine whether additional snacks are required. If blood glucose normalizes, but the individual will not eat in less than an hour, has recently exercised, or is going to bed, an additional snack may be needed. Blood glucose should be tested and treated as appropriate.

Severe hypoglycemia is defined as hypoglycemia that cannot be self-treated.[20] With severe hypoglycemia, an individual can lose consciousness or become so confused that he or she is unable to think clearly. If conscious, an individual may seem lethargic or belligerent. Conscious or unconscious, the individual may not be grateful for help, but he or she still needs it. A trained family member, friend, or emergency medical professional must inject glucagon or glucose to restore blood glucose levels to normal. All persons at risk for severe hypoglycemia should have a glucagon emergency kit available at all times.[35]

Weight gain from improved glycemic control is another common side effect. Instead of losing glucose through urine, the body actually utilizes glucose or stores it as glycogen or fat.

Diabetic Ketoacidosis (DKA)

Diabetic ketoacidosis (DKA), a severe form of hyperglycemia, is a life-threatening situation that commands prompt medical attention.[36] DKA occurs more often in T1DM, but is also a risk for individuals with type 2 diabetes mellitus (T2DM) during acute illness and/or when they have become insulin deficient.[37] Risk for DKA intensifies during illness, infection, and emotional stress. Omission of insulin is also a frequent cause. Individuals may not take their insulin when they feel too sick to eat, or because they are afraid of developing hypoglycemia.[36, 37] Symptoms of DKA include nausea and/or vomiting; stomach pain; fruity or acetone breath; Kussmaul respirations; and mental status changes.[37]

PATHOPHYSIOLOGY When adequate insulin is not available, glucose production via gluconeogenesis and lipolysis is stimulated by counter-regulatory hormones in an effort to avoid starvation. One of the by-products of lipolysis is the generation of ketones. As glucose and ketones accumulate in the bloodstream, osmotic diuresis occurs, resulting in dehydration and electrolyte imbalances (see Figure 17.8). As fluid is lost, the blood becomes concentrated, bringing about hyperglycemia.[36, 37]

MEDICAL TREATMENT Treatment typically involves hospitalization for assessment and/or administration of IV fluids, insulin, and electrolytes. Supplemental doses of insulin are administered until metabolic stability returns.[37] While the vast majority of DKA cases are resolved, 2% to 5% of cases are fatal.[36]

Short-Term Illness

Everyday maladies like colds, fever, nausea, vomiting, and diarrhea can cause havoc with glycemic control for individuals with diabetes. If hyperglycemia is left untreated, diabetic ketoacidosis (DKA) or hyperglycemic hyperosmolar nonketotic syndrome (described later in the chapter) can develop. Treatment includes supplemental insulin; replacement fluids, electrolytes, and glucose; blood glucose monitoring; and urine testing for ketones. Sometimes, medical intervention is necessary.[22]

NUTRITION ASSESSMENT Standard nutrition assessment should focus on fluid, food, and carbohydrate intakes; insulin dosages and delivery methods; SMBG records; urine ketone levels; and knowledge regarding self-care.

NUTRITION DIAGNOSIS Nutrition diagnoses associated with short-term illness may include inadequate fluid intake, altered GI function, and inability or lack of desire to manage self-care.

NUTRITION INTERVENTION Individuals with diabetes should be provided with a list of carbohydrate-containing foods that are tolerated during acute illness and are easy to digest.

Furthermore, for illnesses lasting less than 24 hours, the following guidelines are recommended:[22]

- Take usual insulin doses during acute illness. Insulin is still necessary, and insulin needs may even increase, because fever, infection, or stress can trigger release of counter-regulatory hormones.

- Monitor blood glucose and test for ketones at least 4 times a day: before each meal and at bedtime. Additional insulin is needed if blood glucose >240 mg/dL and moderate to large amounts of ketones are present.

- Drink a large glass of liquid (e.g., water, tea, broth) every hour. Small sips of 1 to 2 tablespoons every 15 to 30 minutes should be consumed if nausea or vomiting is present. A primary care provider should be notified if vomiting persists longer than 6 hours.

- If regular foods are not tolerated, replace meals with small amounts of liquid or soft carbohydrate-containing foods eaten every 3 to 4 hours. Consume 3 to 4 carbohydrate servings (45–60 grams of carbohydrate) of foods such as: regular soft drinks (do not use sugar-free soft drinks); soup; juices; Jell-O®; and ice cream.

- Contact a primary care provider if illness continues >24 hours or if unable to eat regular foods for >24 hours.

- Call a physician if any of the following DKA symptoms develop, especially in children: moderate to large amount of ketones in urine with elevated blood glucose levels; fruity-smelling breath; severe nausea; vomiting; diarrhea; abdominal pain; and rapid breathing.

Long-Term Complications of Hyperglycemia

Diabetes is a complicated, chronic metabolic disorder that requires attention to issues beyond glycemic control.[11] Long-term hyperglycemia from either type 1 or type 2 DM results in microvascular (see Figure 17.12) and macrovascular complications that substantially increase morbidity and mortality associated with the disorder and reduce quality of life.[19] These complications typically occur 15 to 20 years after onset of diabetes.[1] Occurrence and rate of development of chronic complications of diabetes can be reduced, however, as demonstrated by two groundbreaking studies.[38] Intensive treatment (see the section Insulin Regimens) of individuals with T1DM results in significantly lower A1C (glycated hemoglobin, a measure of long-term glucose control) and reductions in incidence and rate of progression of **retinopathy**, **nephropathy**, and **neuropathy**.[15] Likewise, individuals with newly diagnosed T2DM who received intensive treatment (insulin, sulfonylureas, or metformin) experienced relatively improved A1C results, significant reductions in all microvascular complications, and reductions in cardiovascular disease outcomes.[30]

Macrovascular Complications: Cardiovascular Disease (CVD)

Cardiovascular lesions are the most common cause of premature death in individuals with diabetes.[1] About 65% of deaths among individuals with diabetes are due to heart disease or stroke, and adults with diabetes have heart disease-related death rates about 2 to 4 times higher than those of adults without diabetes.[6, 7]

Hyperglycemia makes all blood vessels prone to endothelial damage, leading to thickening and changes in composition of

Figure 17.12 Biology of Microvascular Complications

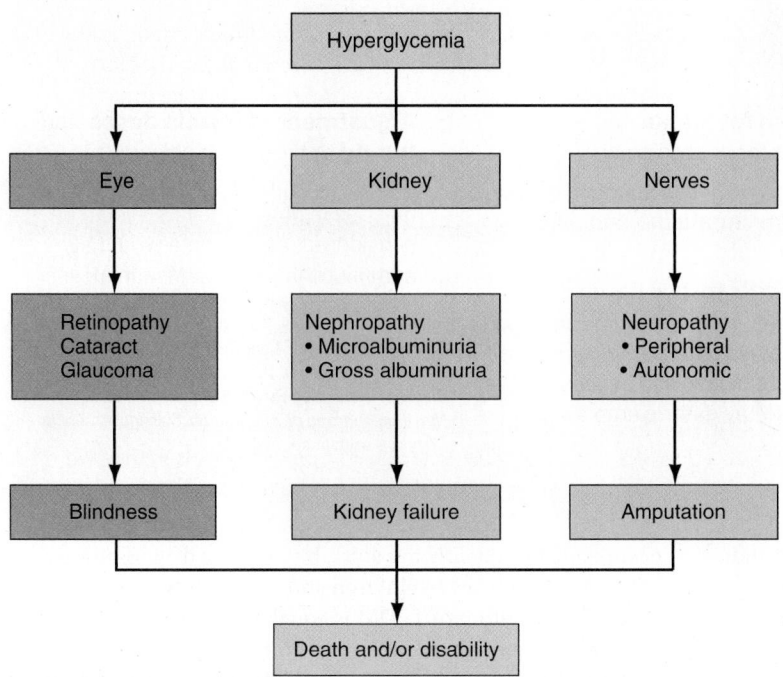

Source: CADRE—Council for the Advancement of Diabetes Research and Education. Section 3—Diabetes Slides: Microvascular and Macrovascular Complications: Epidemiologic Studies, Accessed September 27, 2009 from http://www.cadre-diabetes.org/r_slide_show.asp?id=4.

the subendothelial (intimal) layer. Hyperglycemia also directly affects the structure of the basement membrane of the vessels. This thickening and decreased flexibility of the vessel increase blood pressure and contribute to the acceleration of atherosclerosis seen in diabetes. In larger vessels, the intima can enlarge significantly and amass both intracellular lipid and extracellular lipid, necrotic debris, and calcium. This results in an advanced, complex atherosclerotic lesion. Ulceration of the lesion (plaque) into the lumen can also occur, leading to embolization, thrombus formation, or both.[39]

Diabetes is an independent risk factor for macrovascular disease in addition to the common coexisting risk factors of hypertension and dyslipidemia (see Chapter 13). Hypertension is not only a major risk factor for CVD, but also a complication for microvascular complications of DM such as retinopathy and nephropathy. Hypertension is often the consequence of underlying nephropathy in individuals with T1DM. In T2DM, hypertension may manifest as part of the metabolic syndrome (see Chapter 13), which is accompanied by high rates of CVD. Hypertension can be improved by increasing physical activity and consumption of fruits, vegetables, and low-fat dairy products; by decreasing sodium intake and body weight (when indicated); and by avoiding excessive alcohol intake. Dyslipidemia can be improved by a reduction of saturated fat and cholesterol intake, weight loss if indicated, and increased physical activity.[11]

Nephropathy—A Microvascular Complication

Nephropathy occurs in 20% to 40% of individuals with diabetes, and is the single leading cause of chronic kidney disease (CKD)—kidney failure that must be treated with dialysis or

transplantation (see Chapter 18).[11] Diabetes is the cause of 44% of new cases of CKD.[6,7] Hyperglycemia results in changes in the structure of the blood vessels of the glomerulus, the functioning unit of the kidney, which is comprised of a tuft of capillaries. Changes in the capillary structure result in increased permeability and decreased filtering ability. The earliest stage of nephropathy in T1DM, which is also a marker for development of nephropathy in T2DM, is persistent albuminuria in the range of 30–299 mg/24 hours (microalbuminuria). Onset of microalbuminuria and progression to macroalbuminuria in individuals with diabetes can be delayed by intensive diabetes management (defined as achieving near normoglycemia). Protein restriction and use of ACE inhibitors may also slow progression of albuminuria, glomerular filtration rate decline, and occurrence of CKD.[11]

Retinopathy—A Microvascular Complication

Retinopathy is the most frequent cause of new cases of blindness in adults, and prevalence of retinopathy is strongly associated with duration of diabetes.[6,7,11] Though the mechanisms are not completely understood, the damage to the eye appears to be directly related to hyperglycemic damage to its blood vessels. The eye is highly vascularized and has a significant oxygen demand. Changes in the blood vessels and the accumulation of sorbitol appear to be the major factors associated with retinopathy. In addition, other eye ailments, including glaucoma and cataracts, occur earlier in individuals with diabetes.[11] Hypertension, an established risk factor for development of macular edema, is also associated with the development of retinopathy.[11] Progression of retinopathy can be decreased by glycemic control and lowering of blood pressure.[30]

Nervous System Diseases

Approximately 60% to 70% of individuals with diabetes have some form of nervous system damage that causes impaired sensation or pain in the feet or hands, slowed digestion of food in the stomach, carpal tunnel syndrome, and other nerve problems.[6,7] It is once again the continued presence of hyperglycemia that leads to these complications. The accumulation of abnormal substances such as sorbitol and glycated proteins results in cellular damage, disrupting the normal nervous system pathways.

The significance of autonomic neuropathy, a serious and common complication of diabetes, is often underappreciated.[40] It can involve one or more mechanisms of the autonomic nervous system.[38] Autonomic neuropathy commonly coexists with other peripheral neuropathies and other complications of diabetes. It may affect many organ systems throughout the body, including the gastrointestinal tract, genitourinary tract, and cardiovascular system. Gastrointestinal (GI) disturbances are common and can occur along any section of the GI tract.[40] Gastroparesis, delayed gastric emptying, results from damage to the vagus nerve, which controls peristalsis. It can cause anorexia, nausea, vomiting, early satiety, postprandial bloating, and erratic glycemic control.[38] Box 17.7 describes nutrition therapy for treatment of gastroparesis. Constipation is the most frequent lower GI symptom, but can alternate with episodes of diarrhea. Bladder and/or sexual dysfunction are common genitourinary

BOX 17.7 | CLINICAL APPLICATIONS

Nutrition Therapy for Gastroparesis

No controlled trial of nutrition therapy for management of symptoms of gastroparesis has been reported. The following recommendations have been established using professional judgment, clinical practice, and interpretation of gastric physiology.

- **Small, frequent meals** may decrease the feeling of bloating and early satiety, thus decreasing the possibility of impaired nutritional status.

- **Reduced fat intake** may shorten the time for gastric emptying.
- **Physical activity, such as walking, after meals** may increase gastric emptying rates.
- **Foods with soft or liquid consistency** may be more easily digested, although hypertonic enteral formulas should be avoided because they further delay gastric emptying.

- **Adjustment of insulin doses and timing** to better match delayed nutrient absorption and postprandial rise in glucose levels should be considered. An example would be administering regular insulin after meals instead of before meals.

Source: Wheeler ML. Long-term complications. In: Ross TA, Boucher JL, O'Connel BS. *American Dietetic Association Guide to Diabetes Medical Nutrition Therapy and Education.* Chicago: American Dietetic Association, 2005.

---

tract disturbances associated with autonomic neuropathy and may manifest as recurrent urinary tract infections, pyelonephritis (injury to kidneys caused by bacterial infection), or incontinence.[40] Males and females may suffer sexual dysfunction, including retrograde ejaculation in males.[11, 40] Cardiovascular autonomic neuropathy (CAN) is considered the most clinically important form of autonomic neuropathy. It may manifest through resting tachycardia (>100 bmp), orthostatic hypotension (a fall in systolic blood pressure >20 mmHg upon standing), or increased risk of silent heart disease.[11, 40]

Nutrition Therapy for Long-Term Complications of Hyperglycemia

Nutrition problems related to the medical complications associated with diabetes include inappropriate intake of types of carbohydrates; inconsistent carbohydrate intake; inadequate fiber intake; altered GI function (gastroparesis); altered nutrition-related laboratory values; food-medication interaction; underweight; food- and nutrition-related knowledge deficit; harmful beliefs/attitudes about food or nutrition-related topics; not ready for diet/lifestyle change; self-monitoring deficit; undesirable food choices; physical inactivity; and inability or lack of desire to manage self-care. Many of these problems are common to diabetes patients with or without complications, and should be addressed using the nutrition intervention and monitoring strategies for hyperglycemia discussed throughout the "Diabetes Mellitus" section of the chapter. Additional strategies for addressing nutrition problems that many diabetic patients experience are covered throughout this text and can be incorporated into the nutrition intervention. For example, Chapter 13 provides details on lipid management, which is an important component of preventing heart disease for individuals with diabetes. Chapter 14 provides suggestions for nausea, vomiting, and delayed gastric emptying.

Type 2 Diabetes Mellitus

Type 2 diabetes was once called non-insulin diabetes mellitus (NIDDM) or adult-onset diabetes, but these terms perform a disservice to individuals with diabetes, because they classify them by treatment modality rather than disease characteristics.[31]

Epidemiology

In the United States and worldwide, about 90% to 95% of all diagnosed cases of diabetes are T2DM. T2DM occurs most frequently in adults, but is being diagnosed with increasing frequency in children and adolescents as well.[6, 7, 8] Gender distribution of T2DM is equal, but prevalence increases with age.[3] Other risk characteristics for T2DM include obesity, family history, history of gestational DM, impaired glucose metabolism, and physical inactivity (CDC diabetes fact sheet ref).

T2DM is not an equal-opportunity disease. The elderly and persons of color are disproportionately affected, as shown in Figure 17.13.[41] Prevalence of diabetes in non-Hispanic whites 20 years or older is 6.6% of this population. The prevalence of diabetes for other groups of individuals of similar age is as follows: 29% among American Indians in the southern United States; 6% of Alaska Natives; 7.5% of Asian Americans; 10.4% of Hispanics; and 11.8% of non-Hispanic blacks: 8.2% of Cubans, 11.9% of Mexican Americans, and 12.6% of Puerto Ricans.[7] Worldwide, diabetes prevalence is predicted to increase from approximately 170 billion currently to over 330 billion in the year 2030.[42] Areas of the world with significant prevalence of diabetes include India, Latin America and the Caribbean, the Middle East, and China.

Etiology

In some, heredity may be a factor in development of T2DM. A clear autosomal dominant pattern of inheritance (maturity-onset diabetes of the young [MODY]) has been found in those who develop T2DM before the age of 25 years. (Box 17.8 outlines criteria for testing for T2DM in children.) Identifiable gene defects have been found in most families with this pattern of diabetes. Obesity—body fat distribution in particular—also appears to play a role in development of T2DM. Central body adiposity seems to increase the degree of insulin resistance, but the mechanism is unclear.[3]

Physical inactivity increases risk of T2DM unrelated to body weight. Exercise seems to reduce risk of T2DM by enhancing whole-body insulin sensitivity. High birth weight also appears to increase the risk for T2DM during adulthood, which may be an indication of the genetic connection for T2DM. Low-birth-weight infants tend to develop T2DM later in life when compared to higher-birth-weight babies. Poor placental growth (or food insecurity for the mother during pregnancy) can bring about poor fetal nutrition, producing defective pancreatic organogenesis in utero. Later in life, obesity-related insulin

Figure 17.13 Prevalence of Diagnosed DM by Race, Sex, and Age Group U.S., 2007

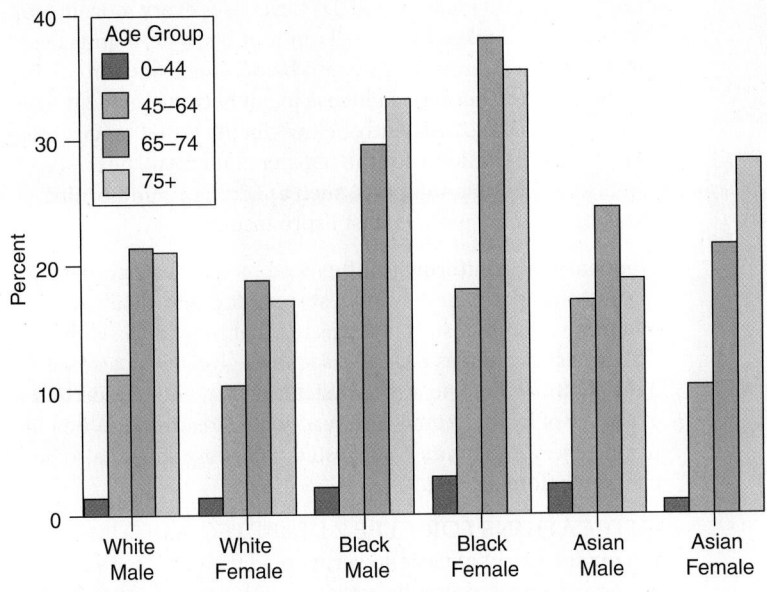

Source: Centers for Disease Control and Prevention, National Center for Chronic Disease Prevention and Health Promotion, Division of Diabetes Translation. http://www.cdc.gov/diabetes/statistics/prev/national/fig2004.htm

resistance results in increased stress on the pancreas, causing essential insulin deficiency, which ultimately leads to diabetes.[3]

Pathophysiology Whereas T1DM results from lack of insulin caused by destruction of β-cells, individuals with T2DM produce insulin, but their tissues are insulin resistant. This causes increased need for insulin, so the pancreas increases production. Eventually the pancreas loses its ability to produce

---

BOX 17.8 CLINICAL APPLICATIONS

Testing for T2DM in Children

Criteria:	Obesity (BMI >85th percentile for age and gender, or weight for height >85th percentile, or weight >120% of ideal for height)
	Plus any 2 of the following risk factors:
	• Family history of T2DM in first- or second-degree relative
	• Race/ethnicity (Native American, African-American, Latino, Asian American, Pacific Islander)
	• Signs of insulin resistance or conditions associated with insulin resistance (**acanthosis nigricans**, hypertension, dyslipidemia, or polycystic ovarian syndrome
Age of initiation:	10 years or at onset of puberty, if puberty occurs at a younger age
Frequency:	Every 2 years
Test preferred:	Fasting plasma glucose

Source: Adapted from: American Diabetes Association. Standards of medical care in diabetes. *Diabetes Care.* 2008;31(Suppl 1):S12.

---

insulin.[2, 3, 6, 7, 8] Consequently, two metabolic defects are observed in individuals with T2DM: insulin resistance and relative insulin deficiency. Although insulin resistance develops many years before onset of diabetes in individuals with predisposition to T2DM, clinical onset is correlated with the diminishing pancreatic release of insulin.[3]

T2DM is typified by peripheral insulin resistance with an insulin secretory defect that varies in severity. Insulin resistance is caused by a cell-receptor defect resulting in the body's inability to use insulin. When cells cannot respond to insulin by translocating glucose transporters to their outer membrane, they are unable to take up glucose from the blood for fuel. Since insulin normally serves to inhibit glycogenolysis and gluconeogenesis when blood glucose is high, defective insulin secretory response results in excess production of glucose from the liver.[12] For T2DM to manifest, both defects must be present. At first, postprandial glucose levels rise due to the inability of the cells to utilize glucose; subsequently, hepatic gluconeogenesis steps up to compensate for this lack of glucose, resulting in fasting hyperglycemia.[43]

METABOLIC SYNDROME Another condition related to insulin resistance is metabolic syndrome, which shares some characteristics of T2DM.[44] Central obesity and insulin resistance are significant contributing features, along with atherosclerotic risk factors including dyslipidemia and hypertension.[45, 46] Diagnostic criteria are outlined in Chapter 13. Metabolic syndrome places individuals at increased risk for coronary artery disease. Treatment of metabolic syndrome is multifaceted and includes diet, exercise, and pharmacologic therapy including statins, fibrates, angiotensin-converting enzyme (ACE) inhibitors, and thiazolidinediones.[46] Over one-third of the U.S. population over age 20 meet the criteria for metabolic syndrome.[47]

Clinical Manifestations While onset of T1DM is sudden, onset of T2DM is insidious. Many individuals will be asymptomatic for as long as 6–10 years but present with complications associated with diabetes.[3] For example, an optician may detect retinopathy during a visit provoked by blurred vision.[3] The estimate that as many as one-third of all individuals with T2DM are undiagnosed reinforces the need for screening of individuals at high risk. Criteria for testing and screening for diabetes in asymptomatic, undiagnosed adults are listed in Box 17.9, and one of these criteria—polycystic ovarian syndrome—is discussed further in Box 17.10.

HYPERGLYCEMIC HYPEROSMOLAR SYNDROME (HHS) HHS, which occurs most often in individuals with T2DM, is characterized by blood glucose levels >600 mg/dL, serum osmolality >320 mOsm/kg of water, and absence of significant ketoacidosis. Infection and dehydration are precipitating factors of HHS.[35, 36, 37]

Symptoms of HHS are comparable to moderate hyperglycemia: polyuria, polydipsia, polyphagia, and weight loss. These symptoms develop gradually, and for that reason are less conspicuous and more easily overlooked than DKA. Without judicious monitoring, elderly individuals with T2DM who are unable or

Criteria for Testing for Diabetes in Asymptomatic Adults

	BMI	Other Criteria
Individuals ≤45 years of age	≥25[1] kg/m²	• If normal, test every 3 years
All individuals	≥25[2] kg/m²	• Habitually physically inactive • Have first-degree relative w/diabetes • Are members of high-risk ethnic populations • Delivered a baby weighing >9 lbs or been diagnosed with GDM[3] • Hypertensive (≥140/90 mmHg) • Have HDL-cholesterol, <35 mg/dL (0.90 mmol/L), and/or triglyceride >250 mg/dL (2.82 mmol/L) • Have polycystic ovarian syndrome (see Box 17.10) • Had IGT or IFG on previous testing (see section on impaired glucose tolerance and impaired fasting glucose) • Have other clinical conditions associated with insulin resistance (acanthrosis nigricans) • History of vascular disease

1. May not be correct for all ethnic groups.
2. May not be correct for all ethnic groups.
3. Gestation diabetes mellitus.

Source: Adapted from: American Diabetes Association. Standards of medical care in diabetes. *Diabetes Care.* 2008;31(Suppl 1):S12.

unwilling to self-hydrate may slide into a spiraling process of gradual but steady fluid losses and rising blood glucose levels leading to severe dehydration.[35, 36, 37]

Treatment entails hospitalization for slow rehydration as well as treatment for complications and underlying medical problems (e.g., infection). Insulin may or may not be required to adequately reduce hyperglycemia.[2, 35] Mortality rate for HHS is approximately 15%, much higher than for DKA.[35, 37] Box 17.11 compares and contrasts HHS with DKA.

Medical Treatment Three factors contribute to glycemic control: hepatic glucose production, glucose uptake by the periphery, and absorption of glucose from food. Medical management of T2DM entails a combination of nutrition therapy, physical activity, and medication when required to counteract abnormalities of glycemic control. Successful management requires awareness of the potential of each therapy, the synergistic connection between therapies, and maximal use of each therapy.

The optimal method for achieving glycemic control is targeted blood glucose control, whereby individuals know and attempt to achieve their blood glucose goals for various times of the day. This is accomplished using feedback from daily SMBG and

routine laboratory evaluations (A1C).[12] This has been found to be effective in individuals with T1DM, those with T2DM using insulin, and those with T2DM who are not using insulin.[48] For example, an individual with T2DM exercises every morning for 30 minutes and takes 18 units glargine at bedtime, 6 units lispro before breakfast, and 5 units lispro before lunch and dinner. In reviewing his or her blood glucose log, it becomes evident that glucose levels are consistently below glucose target range before lunch. In consultation with the registered dietitian, he or she can discuss possible solutions, such as adding a midmorning snack or reducing prebreakfast lispro insulin.[29]

Frequency of monitoring is influenced by selected tightness of control, capacity to execute tests unaided, affordability, and motivation to test. It is suggested that individuals with T2DM test as often as needed to achieve glycemic goals (see Table 17.8), before and after physical activity, and to ascertain existence of hypoglycemia and reaction to treatment. When ill, testing every 4–6 hours is suggested.[12] Glucose should also be checked before driving.

MEDICATIONS FOR TYPE 2 DIABETES As T2DM progresses, use of glucose-lowering medications is indicated if glycemic control cannot be achieved with nutrition therapy and regular physical activity alone. There are seven classes of medications used to treat T2DM, as shown in Table 17.10:[43, 49]

- **Alpha-glucosidase** inhibitors (AGIs)
- **Amylin** analogs
- Biguanides
- Incretin mimetics
- Meglitinides
- Sulfonylurea agents
- Thiazolidinediones

Drug-nutrient interactions for these medications are listed in Table 17.11.

PHYSICAL ACTIVITY Because the benefits of physical activity for glycemic control are well documented, exercise (along with nutrition therapy) is generally prescribed for all individuals with T2DM. Physical activity improves blood glucose levels by enhancing muscle blood glucose uptake during or shortly after activity and by improving insulin sensitivity. Furthermore, it enhances weight loss efforts, which in turn improve insulin sensitivity and glycemic control.

Thirty to forty-five minutes of moderate-intensity physical activity three to five days a week is recommended to improve glycemic control, assist with weight maintenance, and reduce risk of CVD. No more than two consecutive days should go by without physical activity. As long as there are no contraindications, resistance exercise targeting major muscle groups should be performed three times a week. If the client begins an exercise program more vigorous than brisk walking, however, conditions that could be related to increased likelihood of CVD or cause injury to the individual should be assessed. Age and previous physical activity should also be considered.[11]

In those taking insulin and/or insulin **secretagogues** (sulfonylurea or meglitinides), physical activity can cause hypoglycemia if medication dose or carbohydrate consumption is not changed. Hypoglycemia rarely occurs in individuals with

BOX 17.10   CLINICAL APPLICATIONS

Polycystic Ovary Syndrome: More than Infertility

What is polycystic ovary syndrome (PCOS)?

PCOS is a health problem that can affect a woman's menstrual cycle, fertility, hormones, insulin production, heart, blood vessels, and appearance. Women with PCOS have these characteristics:

- High levels of male hormones, also called androgens.
- An irregular or no menstrual cycle.
- May or may not have many small cysts (fluid-filled sacs) in their ovaries.

PCOS is the most common hormonal reproductive problem in women of childbearing age.

How many women have PCOS?

An estimated 5% to 10% of women of childbearing age have PCOS.

What causes PCOS?

No one knows the exact cause of PCOS. Women with PCOS frequently have a mother or sister with PCOS. But there is not yet enough evidence to say there is a genetic link for this disorder. Many women with PCOS have a weight problem, so researchers are looking at the relationship between PCOS and the body's ability to make insulin. Since some women with PCOS make too much insulin, it's possible the ovaries react by making too many male hormones, called androgens. This can lead to acne, excessive hair growth, weight gain, and ovulation problems.

Why do women with PCOS have trouble with their menstrual cycle?

In women with PCOS, the ovary doesn't make all of the hormones it needs for any of the eggs to fully mature. They may start to grow and accumulate fluid. But no one egg becomes large enough. Instead, some may remain as cysts (see Figure 17.14). Since no egg matures or is released, ovulation does not occur and the hormone progesterone is not made. Without progesterone, a woman's menstrual cycle is irregular or absent. Also, the cysts produce male hormones, which continue to prevent ovulation.

What are the symptoms of PCOS?

These are some of the symptoms of PCOS:

- Infrequent menstrual periods, no menstrual periods, and/or irregular bleeding
- Infertility or inability to get pregnant because of not ovulating
- Increased growth of hair on the face, chest, stomach, back, thumbs, or toes
- Acne, oily skin, or dandruff
- Pelvic pain
- Weight gain or obesity, usually carrying extra weight around the waist
- Type 2 diabetes
- High cholesterol
- High blood pressure
- Male-pattern baldness or thinning hair

Figure 17.14 Polycystic Ovarian Syndrome

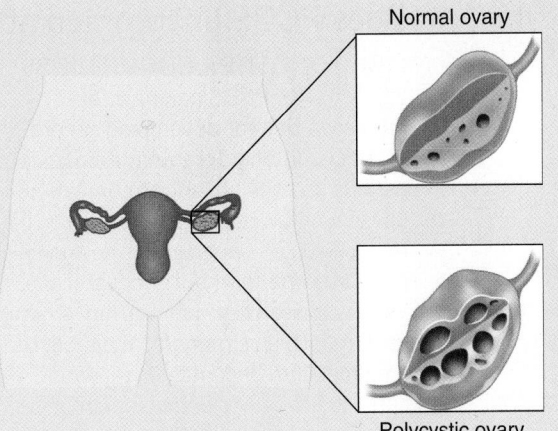

Normal ovary

Polycystic ovary

Source: http://www.4woman.gov/faq/pcos.html

- Patches of thickened and dark brown or black skin on the neck, arms, breasts, or thighs
- Skin tags, or tiny excess flaps of skin in the armpits or neck area
- Sleep apnea—excessive snoring and cessation of breathing at times while asleep

What tests are used to diagnose PCOS?

There is no single test to diagnose PCOS. Physicians will take a medical history, perform a physical exam—possibly including an ultrasound—check hormone levels, and measure glucose in the blood. At the physical exam, the doctor will want to evaluate the areas of increased hair growth. During a pelvic exam, the ovaries may be enlarged or swollen by the increased number of small cysts. This can be seen more easily by vaginal ultrasound, or screening, to examine the ovaries for cysts and the endometrium (the lining of the uterus). The uterine lining may become thicker if there has not been a regular period.

How is PCOS treated?

Because there is no cure for PCOS, it needs to be managed to prevent problems. Treatments are based on the symptoms each patient is having and whether she wants to conceive or needs contraception. Below are descriptions of treatments used for PCOS.

Birth control pills. For women who don't want to become pregnant, birth control pills can regulate menstrual cycles, reduce male hormone levels, and help to clear acne. However, the birth control pill does not cure PCOS. The menstrual cycle will become abnormal again if the pill is stopped. Women may also think about taking a pill that only has progesterone, like Provera, to regulate

(continued)

the menstrual cycle and prevent endometrial problems. But progesterone alone does not help reduce acne and hair growth.

Diabetes medications. The medicine Metformin, also called Glucophage, which is used to treat T2DM, also helps with PCOS symptoms. Metformin affects the way insulin regulates glucose and decreases the testosterone production. Abnormal hair growth will slow down and ovulation may return after a few months of use.

Fertility medications. The main fertility problem for women with PCOS is the lack of ovulation. Even so, her husband's sperm count should be checked and her tubes checked to make sure they are open before fertility medications are used. Clomiphene (pills) and Gonadotropins (shots) can be used to stimulate the ovary to ovulate. PCOS patients are at increased risk for multiple births when using these medications. In vitro fertilization (IVF) is sometimes recommended to control the chance of having triplets or more. Metformin can be taken with fertility medications and helps to make PCOS women ovulate on lower doses of medication.

Medicine for increased hair growth or extra male hormones. If a woman is not trying to get pregnant there are some other medicines that may reduce hair growth. Spironolactone is a blood pressure medicine that has been shown to decrease the male hormone's effect on hair. Propecia, a medicine taken by men for hair loss, is another medication that blocks this effect. Both of these medicines can affect the development of a male fetus and should not be taken if pregnancy is possible. Other nonmedical treatments such as electrolysis or laser hair removal are effective at getting rid of hair. A woman with PCOS can also take hormonal treatment to keep new hair from growing.

Surgery. Although it is not recommended as the first course of treatment, surgery called ovarian drilling is available to induce ovulation. The physician makes a very small incision above or below the navel, and inserts a laparoscope. The ovary is then punctured with a small needle carrying an electric current to destroy a small portion of the ovary. This procedure carries a risk of developing scar tissue on the ovary. This surgery can lower male hormone levels and help with ovulation. But these effects may only last a few months. This treatment does not help with increased hair growth and loss of scalp hair.

Healthy weight. Maintaining a healthy weight is another way women can help manage PCOS. Since obesity is common with PCOS, a healthy diet and physical activity help maintain a healthy weight, which will help the body lower glucose levels and use insulin more efficiently, and may help restore a normal period. Even loss of 10% of her body weight can help make a woman's cycle more regular.

How does PCOS affect a woman while pregnant?

There appears to be a higher rate of miscarriage, gestational diabetes, pregnancy-induced high blood pressure, and premature delivery in women with PCOS. Researchers are studying how the medicine Metformin prevents or reduces the chances of having these problems while pregnant, in addition to looking at how the drug lowers male hormone levels and limits weight gain in women who are obese when they get pregnant. No one yet knows if Metformin is safe for pregnant women. Because the drug crosses the placenta, doctors are concerned that the baby could be affected by the drug. Research is ongoing.

Does PCOS put women at risk for other conditions?

Women with PCOS can be at an increased risk for developing several other conditions. Irregular menstrual periods and the absence of ovulation cause women to produce the hormone estrogen, but not the hormone progesterone. Without progesterone, which causes the endometrium to shed each month as a menstrual period, the endometrium becomes thick, which can cause heavy bleeding or irregular bleeding. Eventually, this can lead to endometrial hyperplasia or cancer. Women with PCOS are also at higher risk for diabetes, high cholesterol, high blood pressure, and heart disease. Getting the symptoms under control at an earlier age may help to reduce this risk.

Does PCOS change at menopause?

Researchers are looking at how male hormone levels change as women with PCOS grow older. They think that as women reach menopause, ovarian function changes and the menstrual cycle may become more normal. But even with falling male hormone levels, excessive hair growth continues, and male pattern baldness or thinning hair gets worse after menopause.

For More Information . . .

You can find out more about PCOS by contacting the National Women's Health Information Center (NWHIC) at (800) 994–WOMAN (9662) or the following organizations:

National Institute of Child Health and Human Development (NICHD), NIH, HHS
Phone: (800) 370–2943
Internet Address: http://www.nichd.nih.gov/womenshealth
American Association of Clinical Endocrinologists (AACE)
Phone: (904) 353–7878
Internet Address: http://www.aace.com
American Society for Reproductive Medicine (ASRM)
Phone: (205) 978-5000
Internet Address: http://www.asrm.org
Center for Applied Reproductive Science (CARS)
Phone: (423) 461-8880
Internet Address: http://www.ivf-et.com
InterNational Council on Infertility Information Dissemination, Inc. (INCIID)
Phone: (703) 379-9178
Internet Address: http://www.inciid.org
Polycystic Ovarian Syndrome Association, Inc. (PCOSA)
Phone: (877) 775-7267
Internet Address: http://www.pcosupport.org
The Hormone Foundation
Phone: (800) 467-6663
Internet Address: http://www.hormone.org

Source: Adapted from: The National Women's Health Information Center. Polycystic Ovarian Syndrome (PCOS) [monograph on the Internet]. Washington (DC): U.S. Department of Health and Human Services Office on Women's Health; 2007. Available at http://www.womenshealth.gov/faq/polycystic-ovary-syndrome.cfm.

BOX 17.11 CLINICAL APPLICATIONS

Hyperglycemic Hyperosmolar Syndrome and Diabetic Ketoacidosis

	Hyperglycemic Hyperosmolar Syndrome (HHS)	Diabetic Ketoacidosis (DKA)
Characteristics	• Adequate insulin to prevent lipolysis and ketogenesis but inadequate to maintain normoglycemia • Occurs most often in T2DM • Occurs in those between 55 and 70 years old • Frequently occurs in residents of long-term care facilities	• Hyperglycemia • Metabolic acidosis • Ketogenesis • Occurs most often in T1DM & those with new-onset T2DM who are obese & have impaired insulin secretion & insulin action
Causes	• Dehydration from inadequate fluid intake or excess fluid losses • Prolonged hyperglycemia	• Infections • Acute illnesses (CVA, alcohol/drug abuse, pancreatitis, pulmonary embolism, MI, trauma) • Psychological stress • Lack of SMBG • Insulin omitted • Increased insulin needs with growth spurts • Pump malfunction • Drug abuse
Symptoms	• Undiagnosed diabetes • Progresses slowly (over days and weeks) • Polyuria • Polydipsia • Progressive decline in level of consciousness • Fever (due to underlying infection) • Volume depletion	• Develops rapidly • Polyuria • Polydipsia • Weight loss • Vomiting • Abdominal pain • Dehydration (loss of skin turgor, dry mucous membranes, tachycardia, hypotension) • Acetone breath • Kussmaul respirations

Laboratory Findings

Plasma glucose	>600 mg/dL	>250 mg/dL
Arterial pH	>7.3	<7.0 to 7.30
Serum bicarbonate	>15 mEq/L	<10 to 18 mEq/L
Urine ketones	Small	Positive
Serum ketones	Small	Positive
Serum osmolality	>320 mOsm/kg	Variable
Treatment	• Hospitalization for slow rehydration • Treatment for underlying medical problems • Insulin may or may not be required	• Hospitalization for administration of IV fluids • Insulin • Assessment of serum electrolytes
Prevention	• Routine hydration • Adequate monitoring	Identify cause(s) to determine approach to prevention. Can include: • Regular self-monitoring • Test for ketones if BG >250 mg/dL • Monitor effects of illness on BG closely • Take medications even when eating less • Sick day management plan • Probe rationale for omitting insulin

Sources: Adapted from: Arnold MS. Hypoglycemia and hyperglycemia. In: Ross TA, Boucher JL, O'Connell BS, editors. *American Dietetic Association guide to diabetes: medical nutrition therapy and education.* Chicago: American Dietetic Association; 2005, and Kitabchi AE, Gosmanov AR. Diabetetic ketoacidosis and Hyperosmolar hyperglycemic syndrome in adults. In American Diabetes Association. *Therapy for Diabetes Mellitus and Related Disorders,* 5th ed. Alexandria, VA: American Diabetes Association, 2009.

Table 17.10 Types of Diabetes Medications

Class	Generic	Trade Name	Action	Susceptibility to Hypoglycemia	Disadvantages	Advantages
α-Glucosidase inhibitors (AGIs)	Acarbose Miglitol Voglibose	Precose Glyset Volix	Delays intestinal absorption of glucose	No	Flatulence, diarrhea, less efficacy frequent dosing Contraindicated in individuals with intestinal diseases, must take with meals 3 times/day	Safety, postprandial effect
Amylin analogs (injectable medication)	Pramlintide acetate	Symlin	Delays gastric emptying, decreases postprandial glucagon release, suppresses appetite	Increases risk of insulin-induced hypoglycemia (can be used to treat T1DM or T2DM taking insulin)	GI complaints, must be used in syringe separate from insulin, hypersensitivity to pramlintide	Improves long-term control (A1c) compared to insulin alone; lowers insulin use and body weight
Biguanides	Metformin	Glucophage	Decreases hepatic glucose production, increases insulin uptake in muscles	No	Transient diarrhea, nausea, bloating, anorexia, flatulence, lactic acidosis (rare); contraindicated in individuals with renal insufficiency, liver failure, or treated CHF	Weight control, no hypoglycemia with monotherapy, may be CV benefits
Incretin Mimetics (injectable medication)	Exenatide	Byetta	Mimics glucose-dependent insulin secretion, suppresses elevated glucagon secretion, delays gastric emptying	Can cause hypoglycemia when used with sulfonylureas	May decrease absorption of orally administered drugs (drugs requiring rapid absorption such as oral contraceptives, antibiotics)	Better glycemic control
Meglitinides	Repaglinide Nateglinide	Prandin Starlix	Stimulates insulin secretion in presence of glucose, short-acting	Yes	Hypoglycemia, frequent dosing, expensive	Short action with less hypoglycemia at night or with missed meal, glucose-dependent effect on insulin, postprandial effect
Sulfonylurea agents First Generation Second Generation	Acetohexamide Chlorpropamide Tolazamide Tolbutamide Glipizide Glipizide-GITS Glyburide Glimepiride	Dymelor Diabinese Tolinase Orinase Glucotrol Glucotrol XL DiaBeta Micronase PresTab Glynase Amaryl	Stimulates insulin secretion	Yes	Hypoglycemia (more with glyburide); contraindicated in individuals with renal insufficiency, weight gain	Inexpensive, long history of effectiveness, only needed once daily for most patients
Thiazolidi-nediones	Pioglitazone Rosiglitazone	Actos Avandia	Decreases insulin resistance	No	Weight gain, edema, worsened CHF, most expensive, slow onset of action; contraindicated in individuals with CHF	Very effective in highly insulin-resistant individuals, okay with renal insufficiency, potential CV benefit, usually needed only once daily
Insulin (not an oral medication, but often used to treat T2DM)	Rapid, short, intermediate, long	See Table 17.5	Replaces endogenous insulin	Yes	See Table 17.5	See Table 17.5

Sources: Lebovitz HE. Insulin secretagogues: sulfonylureas, repaglinide, and nateglinide. In American Diabetes Association, *Therapy for Diabetes Mellitus and Related Disorders*, 5th ed. Alexandria, VA: American Diabetes Association, 2009; Bailey CJ. Metformin. In American Diabetes Association, *Therapy for Diabetes Mellitus and Related Disorders*, 5th ed. Alexandria, VA: American Diabetes Association, 2009; Rabasa-Lhoret R, Chiasson J. α-glucosidase inhibitors in the treatment of hyperglycemia. In American Diabetes Association, *Therapy for Diabetes Mellitus and Related Disorders*, 5th ed. Alexandria, VA: American Diabetes Association, 2009; Lebovitz H. Thiazolidinediones. In American Diabetes Association, *Therapy for Diabetes Mellitus and Related Disorders*, 5th ed. Alexandria, VA: American Diabetes Association, 2009; Buse JB, Weyer C, Maggs DG. Amylin replacement with pramlintide in type 1 and type 2 diabetes mellitus: a physiological approach to overcome barriers with insulin therapy. *Clinical Diabetes*. 2002;20(3):137.

Table 17.11 Drug-Nutrient Interactions for Medications Used to Treat Diabetes Mellitus

Classification	Mechanism	Brand Names	Possible Food-Drug Interactions
Insulin	Replaces endogenous insulin.	See Table 17.5	Increased weight; use alcohol with caution.
α-Glucosidase inhibitors (AGIs)	Delays intestinal absorption of glucose.	Precose, Glyset	Take with first bite of meal; limit alcohol.
Amylin analogs	Delays gastric emptying, decreases postprandial glucagon release, suppresses appetite.	Symlin	Caution with alcohol.
Biguanides	Decreases hepatic glucose production, increases insulin uptake in muscles.	Glucophage	Decreases folate and vitamin B_{12} absorption; avoid alcohol; take with meals to decrease GI distress.
Incretin mimetics	Mimics glucose-dependent insulin secretion, suppresses elevated glucagon secretion, delays gastric emptying.	Byetta	Caution with alcohol; may cause GI disturbances.
Meglitinides	Stimulates insulin secretion in presence of glucose.	Prandin, Starlix	Limit alcohol.
Sulfonylurea agents (first generation)	Stimulates insulin secretion.	Dymelor, Diabinese, Tolinase, Orinase	Avoid alcohol.
Sulfonylurea agents (second generation)	Stimulates insulin secretion.	Glucotrol, Glucotrol XL, DiaBeta, Micronase, PresTab, Glynase	Avoid alcohol.
Thiazolidinediones	Decreases insulin resistance.	Actos, Avandia	None.

Sources: Pronsky ZM. *Food medication interactions*, 15th ed. Birchrunville (PA): Food-Medication Interactions; 2008. Ahmann AJ, Riddle MC. Current oral agents for type 2 diabetes; many options, but which to choose when? *Postgrad Med.* 2002;111:32. Beebe CA. Nutrition therapy of type 2 diabetes. In: Franz MJ, Bantle JP, editors. *American Diabetes Association guide to medical nutrition therapy for diabetes*. Alexandria (VA): American Diabetes Association; 1999. Inzucchi SE. Oral antihyperglycemic therapy for type 2 diabetes. *JAMA.* 2002;287(3):360–72. Luna B, Feinglos MN. Oral agents in the management of type 2 diabetes mellitus. *American Family Physician.* 2001;63(9):1747–56.

T2DM not being treated with insulin or insulin secretagogues. If pre-exercise glucose levels are <100 mg/dL, additional carbohydrate should be ingested. Supplementary carbohydrate is generally not necessary for individuals treated with diet alone, metformin, α-glucosidase inhibitors, and/or TZDs without insulin or a secretagogue.[11, 34]

Nutrition Therapy for Type 2 Diabetes Mellitus

T2DM is a complex disorder that is heterogeneous in nature, in that it affects individuals of different ages, lifestyles, and cultural backgrounds. Therefore, nutrition therapy should be implemented by prioritizing metabolic problems and instituting a plan based on those priorities.

Nutritional Implications Treatment goals and lifestyle changes that the patient is willing and able to make, rather than predetermined energy levels and percentages of carbohydrate, protein, and fat, should be paramount in determining the nutrition prescription. The aim of nutrition intervention is to support and facilitate lifestyle and behavior modifications that will result in improved metabolic control.[23]

Nutrition Diagnosis Common nutrition diagnoses associated with T2DM include excessive energy intake; inadequate fluid intake (especially for HHS); excessive fat intake; inappropriate intake of fats or carbohydrates; inconsistent carbohydrate intake; inadequate fiber intake; altered GI function (gastroparesis); altered nutrition-related laboratory values; food-medication interaction; overweight/obesity (T2DM); food and nutrition-related knowledge deficit; harmful beliefs/attitudes about food or nutrition-related topics; not ready for diet/lifestyle change; self-monitoring deficit;

undesirable food choices; and inability or lack of desire to manage self-care.

Nutrition Intervention WEIGHT MANAGEMENT Overweight and obesity are strongly associated with development of T2DM. Moreover, obesity is an independent risk factor for hypertension, dyslipidemia, and CVD, the major cause of death in those with diabetes. Moderate weight loss improves glycemic control and reduces CVD risk; therefore, weight loss is recommended for individuals with BMIs >25.0 kg/m². Therapeutic lifestyle changes (see Chapters 12 and 13) that include a reduction in energy intake and an increase in physical activity are recommended.[11]

CARBOHYDRATES Because the total amount of dietary carbohydrate is a strong predictor of glycemic response, monitoring total grams of carbohydrate by either the use of exchanges or carbohydrate counting is strategic in achieving glycemic control. Low-carbohydrate diets are not suggested, because carbohydrates are significant sources of energy, water-soluble vitamins and minerals, and fiber. Less than 130 grams of carbohydrate per day is not recommended because the brain and central nervous system have an absolute requirement for glucose as an energy source.[11]

PROTEIN Intake of dietary protein exceeding 20% of energy intake may be a risk factor for development of nephropathy. Protein intake for individuals with diabetes who have nephropathy should not exceed 0.8 g/kg or ~10% of total kcal.[11]

FAT The goal for dietary fat intake (amount and type) for individuals with diabetes is the same as for those without diabetes who have a history of CVD. Total fat intake should not exceed 25% to 35% of total kcal, and saturated fat intake should not exceed 7%. Intake of *trans* fat should be minimal.[11]

FIBER A variety of fiber-containing foods such as legumes and fiber-rich cereals (≥5 g fiber/serving), as well as fruits, vegetables, and whole-grain products, are recommended.[11] Foods contain a mixture of fibers, but those foods that have high amounts of gums, beta-glucans, psyllium, resistant starches, and pectin appear to have the biggest positive effect on serum glucose levels by slowing the absorption of glucose from the small intestine. The foods highest in these types of fiber include legumes, fruits, vegetables, and oats. The U.S. Dietary Guidelines recommend consuming 14 grams of fiber for every 1000 kcal, while the Food and Nutrition Board recommends that men under 50 consume 38 grams of fiber/day and women consume 25 grams of fiber/day.[50, 51]

Monitoring and Evaluation As with T1DM, necessary interventions can be monitored with use of self-monitoring records review and education. Follow-up sessions will provide an opportunity to evaluate how well the patient is able to manage his/her diabetes.

Gestational Diabetes Mellitus (GDM)

Definition Gestational diabetes (GDM) is a form of glucose intolerance first diagnosed during pregnancy.[6, 7, 52]

Epidemiology Approximately 7% of all pregnancies are complicated by GDM, and women who have had GDM have a 20% to 50% chance of developing diabetes in the next 5 to 10 years. Women at risk for GDM have the following characteristics:[6, 7, 53, 54]

- Obesity (BMI >30.0)
- Personal history of GDM
- Glycosuria
- Strong family history of diabetes (1st degree relative)
- Prior poor obstetrical outcome (stillbirth, birth defects, or baby >9 lbs)
- Member of a high-risk ethnic group (Hispanic, African American, Native American, South or East Asian, Pacific Islander)

Etiology During the second or third trimesters of pregnancy, metabolic alterations occur to meet maternal and fetal demands for energy and nutrients. In addition to alterations in insulin secretion, these alterations affect glucose, amino acid, and lipid metabolism.[54] Although most women with GDM revert to normal glucose tolerance postpartum, there is increased likelihood of developing GDM in subsequent pregnancies and T2DM later in life. Increasing physical activity and reducing postpartum weight gain can reduce risk of subsequent diabetes.[54, 55, 56]

Pathophysiology GDM is pathophysiologically similar to T2DM. Islet cell function abnormalities or peripheral insulin resistance are thought to decrease insulin secretory response and insulin sensitivity. Inability of the β-cells to meet increased insulin needs during pregnancy results in higher levels of circulating glucose.[54]

GDM also affects the fetus. When maternal blood glucose levels are elevated, the fetus is constantly exposed to these levels as well, and fetal insulin production is increased. It appears maternal hyperglycemia induces fetal hyperglycemia, leading to fetal hyperinsulinemia and **macrosomia**.[54]

Table 17.12 Diagnosis of Gestational Diabetes Mellitus

Time	50 g Glucose Challenge Test (GCT)[1]	100 g Oral Glucose Tolerance Test[2]
Fasting		95 mg/dL (5.3 mmol/L)
1 hour	≥140 mg/dL (7.8 mmol/L)	180 mg/dL (10.0 mmol/L)
2 hour		155 mg/dL (8.6 mmol/L)
3 hour		140 mg/dL (7.8 mmol/L)

[1]Administered regardless of time of day or if food was consumed. If results exceed 140 mg/dL threshold, the 3-hour 100 g OGTT is administered.

[2]Two or more venous plasma concentrations must be met or exceeded for a positive diagnosis. Test should be done in the morning after an overnight fast of 8 to 14 hours and after at least 3 days of unrestricted diet (≥150 g carbohydrate per day) and unlimited physical activity.

Sources: Adapted from: American Diabetes Association. Gestational diabetes mellitus. *Diabetes Care.* 2004;27(Suppl 1):S88, and Thomas AM. Classification, screening, and diagnosis. In: Thomas AM, Gutierrez YM, editors. *American Dietetic Association guide to gestational diabetes mellitus.* Chicago: American Dietetic Association; 2005.

Clinical Manifestations Maternal complications associated with GDM include hypertension (preeclampsia), **polyhydramnios**, difficult birth, preterm delivery (before 38 weeks gestation), and a higher rate of cesarean sections. Fetal and neonatal complications include macrosomia, hypoglycemia, respiratory distress syndrome, hypocalcemia, hyperbilirubinemia, and polycythemia.[53, 57]

Diagnosis In the U.S., a two-step approach is generally used to diagnose GDM.[53] Table 17.12 outlines the diagnosis of GDM using the glucose challenge test (GCT; 50 g) and 100 g OGTT. Use of this two-step approach identifies approximately 80% of women with GDM.[53, 58]

Medical Treatment The American Diabetes Association recommends all women with GDM receive nutrition counseling from a registered dietitian. Nutrition therapy should be individualized and based on maternal weight and height. Energy and nutrients adequate to meet the needs of pregnancy should be incorporated with established maternal blood glucose goals.[58] When nutrition therapy alone fails to maintain SMBG at the following levels, insulin therapy is added to reduce fetal mortality:[53]

Fasting plasma glucose ≤105 mg/dL (5.8 mmol/L)

or

1 hour postprandial plasma glucose ≤155 mg/dL (8.6 mmol/L)

or

2 hour postprandial plasma glucose ≤130 mg/dL (7.2 mmol/L)

Use of oral diabetes medications is generally not recommended during pregnancy.[53]

Monitoring Maternal hyperglycemia increases fetal risk. SMBG, not A1C, is considered to be the best method to detect maternal hyperglycemia. Postprandial monitoring is superior to preprandial monitoring when insulin therapy is used. Postprandial blood glucose levels are directly related to rates of macrosomia, neonatal hypoglycemia, and cesarean delivery. Additionally, maternal blood pressure and urine protein should be monitored to detect hypertensive disorders.[53, 59]

Nutrition Therapy for Gestational Diabetes Mellitus

Nutritional Implications Nutrition therapy goals for GDM include a goal shared by all pregnant women: to promote nutrition for maternal and fetal health while providing adequate energy for appropriate gestational weight gain. Furthermore, achievement and maintenance of normoglycemia and absence of ketones are goals specific to treatment of GDM.[53, 59]

Nutrition Diagnosis Nutrition diagnoses related to GDM include excessive energy intake, excessive fat intake, inappropriate intake of fats, inappropriate intake of types of carbohydrates. inconsistent carbohydrate intake, inadequate fiber intake, altered nutrition-related laboratory values, food-medication interaction, food- and nutrition-related knowledge deficit, harmful beliefs/attitudes about food or nutrition-related topics, not ready for diet/lifestyle change, self-monitoring deficit, undesirable food choices, physical inactivity, and inability or lack of desire to manage self-care.

Nutrition Intervention Adequate energy is necessary for desirable weight gain during pregnancy.[23, 53] Energy needs are evaluated indirectly by monitoring the woman's physical activity, appetite, food intake, blood glucose levels, ketone records, and weight change.[53, 59] It is not necessary to calculate energy needs unless problems with excessive weight loss or gain are experienced.[59] Formulas useful for estimating energy requirements for women with GDM are outlined in Box 17.12.

Protein requirements increase during the second and third trimesters of pregnancy to 25 grams per day or 1.1 g protein per kg desirable body weight.[51] Two factors should be considered when making recommendations concerning dietary fat intake: impact on the woman's body weight and plasma lipoprotein profiles. Reduced fat intake may be necessary if total energy intake should be decreased, and saturated fat, *trans* fat, and cholesterol intake should be curtailed.[59]

Consequences of folate deficiency in pregnancy (i.e., neural tube defects) have been well documented. All women of reproductive age capable of becoming pregnant should take 400 mcg additional folate daily from food or supplements.[51] Whereas about 10% of the iron is absorbed from food in the nonpregnant state, iron absorption increases to 25% at the beginning of the second trimester.[59] Supplementation of 30 mg ferrous iron in the second and third trimester is recommended.[51] Nutrition recommendations for GDM are outlined in Table 17.13.

Monitoring and Evaluation Follow-up appointments are crucial for the well-being of the infant as well as the mother. Reclassification of the mother's glycemic status should be performed by the physician at least six weeks after delivery. If glucose levels are normal, glycemic status should be reassessed at a minimum of three-year intervals.[53]

Hypoglycemia

Definition Hypoglycemia is an abnormally low blood glucose level. It occurs when glucose is utilized too rapidly, glucose release rate falls behind tissue demands, or excess insulin enters the bloodstream. Spontaneous (see chapter endnote 3) hypoglycemia in adults is either fasting or postprandial.[60, 61]

BOX 17.12   CLINICAL APPLICATIONS

Estimating Energy Requirements in GDM

To calculate a woman's (19 years and older) energy needs during pregnancy, estimated energy requirements (EER) must first be calculated:

$$EER = 354 - (6.9 \times A) + PA \times (9.36 \times W + 726 \times H)$$

Where A = age in years; PA = physical activity coefficient [1.0 (sedentary), 1.12 (low active), 1.27 (active), 1.45 (very active)]; W = weight in kg; H = height in meters.

To estimate energy requirements for pregnant women who have normal weight:

    1st trimester = Adult EER + 0
    2nd trimester = Adult EER + 160 kcal[1] + 180 kcal
    3rd trimester = Adult EER + 272 kcal[2]

There is no formula supported by research to determine energy needs in overweight or obese pregnant women. For these women, weight gain should be monitored regularly to maintain an approximate weight gain of 0.5 lb/week.

1. 8 kcal/wk × 20 wk.
2. 8 kcal/wk × 34 wk.

Sources: Gutierrez YM, Reader DM. Medical nutrition therapy. In: Thomas AM, Gutierrez YM, editors. *American Dietetic Association guide to gestational diabetes mellitus.* Chicago: American Dietetic Association; 2005.

Etiology **Reactive hypoglycemia** may occur in individuals with diabetes due to administration of too much insulin or oral diabetes medications. In those without diabetes, reactive hypoglycemia may occur due to a sharp increase in insulin release after a meal. It usually disappears when the individual eats something.[60]

Fasting hypoglycemia usually results from excess insulin or insulin-like substance from external factors such as alcohol or drug ingestion.[60]

Pathophysiology When blood glucose levels fall too low, glucagon releases stored hepatic glucose to raise blood glucose levels. Epinephrine is also released, causing the symptoms of weakness, fatigue, sweating, and tachycardia.

FASTING HYPOGLYCEMIA Fasting hypoglycemia can be a primary or secondary manifestation. Primary causes of fasting hypoglycemia include hyperinsulinism due to pancreatic β-cell tumors or surreptitious administration of insulin or oral diabetes medications, and non-insulin-producing extrapancreatic tumors. Secondary fasting hypoglycemia may be caused by certain endocrine disorders such as hypopituitarism, **Addison's disease**, or myxedema; liver disorders such as acute alcoholism or liver failure; and renal failure, especially in individuals undergoing dialysis.[61]

POSTPRANDIAL (REACTIVE) HYPOGLYCEMIA Postprandial hypoglycemia is classified as either early (within two

Table 17.13 Nutrition Recommendations for GDM

Nutrient or Food Type	Recommendation	Meal-Planning Tips
Energy	Intake should be sufficient to promote adequate, but not excessive, weight gain and to avoid ketonuria.	Include 3 small- to moderate-sized meals and 2–4 snacks. Space snacks and meals at least 2 h apart. A bedtime snack (or even a snack in the middle of the night) is recommended, to diminish the number of hours fasting.
Carbohydrate	Recommendations are based on effect of intake on blood glucose levels. Intake should be distributed throughout the day. Frequent feedings, smaller portions, with intake sufficient to avoid ketonuria.	Common carbohydrate guidelines: 2 carbohydrate choices (15–30 g) at breakfast, 3–4 choices (45–60 g) for lunch and evening meal, and 1–2 choices (15 to 30 g) for snacks. Recommendations should be modified based on individual assessment and blood glucose self-monitoring test results.
High-Sucrose/High-Energy Foods	Inclusion should be based on individual's ability to maintain blood glucose goals, nutritional adequacy of diet, and contribution of these foods to total meal plan.	Eliminate foods containing large amounts of carbohydrates, such as sweets and sweetened drinks.
Protein	RDA for adult women (0.8 g/kg DBW) + 25 g/day, or 1.1 g/kg DBW.	Protein foods do not raise post-meal blood glucose levels. Add protein to meals and snacks, to help provide enough calories and to satisfy appetite.
Fat	Limit saturated fat.	Fat intake may be increased because of increased protein intake; focus on leaner protein choices.
Sodium	Not routinely restricted.	
Fiber	For relief of constipation, gradually increase intake and increase fluids.	Use whole grains and raw fruits and vegetables. Activity and fluids help relieve constipation.
Nonnutritive Sweeteners	Generally safe in pregnancy. Use in moderation.	Saccharin crosses the placenta but has not been shown to be harmful.
Vitamins and Minerals	Preconception folate. Assess for specific individual needs: multivitamin throughout pregnancy, iron at 12 weeks, and calcium especially in last trimester and while lactating.	Take prenatal vitamin. If it causes nausea, try taking at bedtime.
Caffeine	Limit to <300 mg/day.	
Alcohol	Avoid.	

DBW = desired body weight; RDA = Recommended Dietary Allowance.

Source: © 2005 American Dietetic Association. Reprinted with permission. Ross TA, Boucher JL, O'Connell BS, editors. *American Dietetic Association guide to diabetes: medical nutrition therapy and education.* Chicago: American Dietetic Association, 2005, Chapter 17, Table 17.1, p. 191.

to three hours after eating) or late (three to five hours after eating). Early hypoglycemia occurs after rapid discharge of ingested foods from the stomach into the small intestine. This is followed by rapid glucose absorption and hyperinsulinemia. This is often observed after gastrointestinal surgery, particularly with dumping syndrome after gastrectomy.[61]

Clinical Manifestations Symptoms of hypoglycemia occur when plasma glucose levels reach 70 mg/dL. Impairment of brain function occurs at approximately 50 mg/dL. Fasting hypoglycemia usually presents with **neuroglycopenia**, while reactive hypoglycemia manifests with symptoms of sweating, palpitations, anxiety, and tremulousness.[61]

Regardless of cause, characteristics of hypoglycemia consist of:[61]

- History of hypoglycemic symptoms
- Fasting blood glucose ≤40 mg/dL
- Immediate recovery upon administration of glucose

Hypoglycemic symptoms often develop in the early morning or after missing a meal, and may occasionally occur after exercise. Symptoms can include blurred vision, headache, feelings of detachment, slurred speech, and weakness. Personality and mental changes vary from anxiety to psychotic behavior. Sweating and palpitations may not occur.[22, 61]

Medical Treatment Anticholinergic agents may be used in treatment of reactive hypoglycemia in order to slow gastric emptying and intestinal motility and inhibit vagal stimulation of insulin release. Surgery and drug therapy are generally necessary for fasting hypoglycemia. Tumor removal is the treatment of choice for patients with insulinoma (insulin-producing tumor). Medications may include nondiuretic thiazides such as diazoxide to inhibit insulin secretions; streptozocin; and hormones, such as glucagon or glucocorticoids.[60]

Nutrition Therapy for Reactive Hypoglycemia

Nutritional Implications Effective treatment of reactive hypoglycemia requires dietary modification to help delay glucose absorption and gastric emptying.

Nutrition Diagnosis Nutrition diagnoses related to reactive hypoglycemia include inappropriate intake of carbohydrates (simple), excessive carbohydrate intake, inconsistent protein intake, food- and nutrition-related knowledge deficit, and undesirable food choices.

Nutrition Intervention Small, frequent meals of complex carbohydrates, fiber, and a protein source are used to treat reactive hypoglycemia.[60] Simple carbohydrates (such as candy, sugar, jam, jelly, syrup, honey, and soft drinks) and alcohol should be avoided. It may be beneficial to restrict caffeine, which may reduce cerebral blood flow and consequently glucose supply to the brain. Use of carbohydrate counting may be helpful in regulating total carbohydrate intake.[22]

Other Endocrine Disorders

Other pathologies of the endocrine system include thyroid, pituitary, and adrenal cortex disorders. Thyroid disorders are common endocrine disorders and fall into two major categories: **hypothyroidism** and **hyperthyroidism**. Specific causes of these conditions are outlined in Table 17.14.

Hypothyroidism

Definition Hypothyroidism is the clinical state resulting from decreased production and secretion of thyroid hormones and is the most common pathologic hormone deficiency.[3, 62] Many medical conditions may directly or indirectly affect the thyroid gland. Thyroid hormone influences growth, development, and many cellular processes; therefore, insufficient thyroid hormone has extensive consequences throughout the body.

Cretinism refers to congenital hypothyroidism, which affects 1 in 4,000 newborns.[7] Because adequate levels of thyroid hormone are essential for normal growth and central nervous system (CNS) development, cretinism is characterized by dwarfism and mental retardation in addition to other general symptoms of thyroid deficiency (discussed later).[1] This discussion will focus specifically on hypothyroidism in adults.

Epidemiology In the United States, hypothyroidism is found in 4.6% of the population and is most prevalent in the elderly. Generally, thyroid disease is more common in females, increases with age, and is more common in whites and Mexican-Americans than in blacks.[63] Death from hypothyroidism is uncommon.[62]

Etiology Worldwide, the most frequent cause of hypothyroidism is iodine deficiency. In the United States and other areas where iodine intake is adequate, autoimmune thyroid disease and previous treatment for hyperthyroidism are the most common causes.[62] Table 17.15 outlines common causes of hypothyroidism.

Table 17.14 Types of Thyroid Dysfunctions

Thyroid Dysfunction	Cause	Plasma Concentrations of Relevant Hormones	Goiter Present?
Hypothyroidism	Primary failure of the thyroid gland	$\downarrow$ T$_3$ and T$_4$, $\uparrow$ TSH	Yes
	Secondary to hypothalamic or anterior pituitary failure	$\downarrow$ T$_3$ and T$_4$, $\downarrow$ TRH and/or $\downarrow$ TSH	No
	Lack of dietary iodine	$\downarrow$ T$_3$ and T$_4$, $\uparrow$ TSH	Yes
Hyperthyroidism	Abnormal presence of thyroid-stimulating immunoglobulin (TSI) (Graves' disease)	$\uparrow$ T$_3$ and T$_4$, $\downarrow$ TSH	Yes
	Secondary to excess hypothalamic or anterior pituitary secretion	$\uparrow$ T$_3$ and T$_4$, $\uparrow$ TRH and/or $\uparrow$ TSH	Yes
	Hypersecreting thyroid tumor	$\uparrow$ T$_3$ and T$_4$, $\uparrow$ TSH	No

TSH = thyroid-stimulating hormone; TRH = thyrotropin-releasing hormone.
Reprinted from: Sherwood L. *Human Physiology: From Cells to Systems.* 7th ed. Belmont (CA): Thomson-Brooks/Cole; 2010. Table 19–1, p. 696.

Table 17.15 Common Causes of Hypothyroidism

Condition	Pathophysiology	Signs/Symptoms
Hashimoto's thyroiditis	Inherited autoimmune disease; 5–10 times more common in women	Presence of anti-TPO (antithyroid peroxidase) antibodies in blood
Lymphocytic thyroiditis (following hyperthyroidism)	Inflammation of thyroid gland; common after pregnancy, majority affected ultimately regain normal thyroid function (remaining hypothyroid is a possibility)	Usually a hyperthyroid phase (excessive amounts of thyroid hormone leak from inflamed gland) followed by hypothyroid phase
Thyroid destruction	Secondary to radioactive iodine treatment (for Graves' disease, for example) or surgery to remove thyroid gland	Little or no functioning thyroid tissue
Pituitary or hypothalamic	Pituitary gland or hypothalamus unable to signal thyroid gland to produce thyroid hormones; labeled *secondary hypothyroidism* if caused by pituitary tumors or disease, and *tertiary hypothyroidism* if caused by hypothalamic tumors or disease	Decreased levels of T$_4$ and T$_3$ even if thyroid gland is normal
Pituitary injury	May result after brain surgery or decreased supply of blood to area. In cases of pituitary injury, TSH (produced by pituitary gland) deficient and blood levels are low	TSH levels elevated
Medications	Medications used to treat overactive thyroid (methimazole [Tapezole] and propylthiouracil [PTU]); psychiatric medication lithium; drugs containing large amounts of iodine (amiodarone [Cardorone], SSKI, Lugol's solution)	Decreased thyroid function; low blood levels of thyroid hormone

Source: Reprinted with permission from Medscape.com 2009. Available at http://eme.medscape.com

Pathophysiology Decreased thyroid hormone production resulting from localized disease of the thyroid gland is the most common cause of hypothyroidism. The thyroid gland normally releases 100–125 μg of T_4 (thyroxine) and only small amounts of T_3 daily. The half-life of T_4 is approximately 7–10 days. As mentioned previously, T_4 is converted to T_3 in peripheral tissues. Early in the disease process, compensatory mechanisms maintain T_3 levels. Decreased production of T_4 causes increased secretion of TSH by the pituitary gland. TSH stimulates hypertrophy and hyperplasia of the thyroid gland, causing the thyroid to release more T_3.[62]

Thyroid hormone deficiency causes a variety of system-wide effects throughout the body. **Myxedematous** infiltration of heart tissue results in decreased contractility, cardiac enlargement, pericardial effusion, decreased pulse rate, and decreased cardiac output. Infiltration of the GI tract can cause **achlorhydria** and increased intestinal transit time with gastric stasis. Other common occurrences are delayed puberty, **anovulation**, menstrual irregularities, and infertility. Additionally, hypothyroidism can cause increased levels of total cholesterol and LDL-cholesterol, and increased insulin resistance.[62]

Clinical Manifestations Symptoms of hypothyroidism are generally subtle. They are not specific (meaning they can mimic symptoms of many other conditions) and are often attributed to aging. Patients with mild hypothyroidism may have no signs or symptoms, but symptoms usually become more noticeable as the condition worsens and are, for the most part, correlated with reduction in overall metabolic activity. Patients with hypothyroidism may present with the following symptoms:[1, 62]

- Reduced basal metabolic rate
- Intolerance of cold
- Weight gain
- Easily fatigued
- Bradycardia
- Slow reflexes and movement
- Slow mental responsiveness (diminished alertness, slow speech, poor memory)
- Pitting edema of lower extremities
- Periorbital puffiness
- Myxedema
- Goiter (see Figure 17.15)
- Loss of scalp hair, axillary hair, pubic hair, or a combination
- Abdominal distension

Medical Treatment Treatment of hypothyroidism consists of administration of exogenous thyroid hormone to supplement or replace endogenous thyroid hormone production, with one exception. If hypothyroidism is caused by iodine deficiency, it can be remedied through adequate intake of dietary iodine.[1, 62]

Clinical benefits of either treatment begin in three to five days and level off after four to six weeks. Patients should be monitored for signs and symptoms of over treatment (tachycardia, palpitations, nervousness, tiredness, headache, increased excitability, sleeplessness, tremors, possible angina).[62]

Figure 17.15 Woman with Goiter

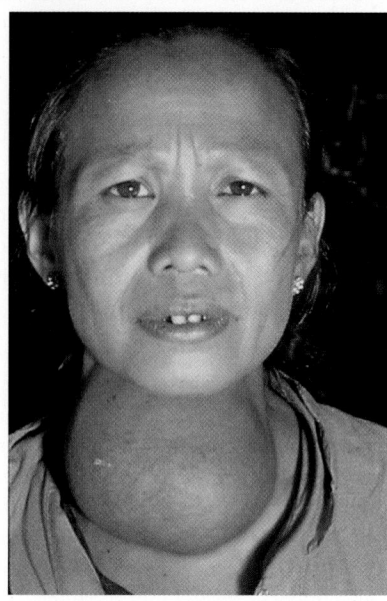

Source: L. Sherwood, Human Physiology: From Cells to Systems, 5e, copyright © 2004, p. 707.

Courtesy of © Lester V. Bergman/Corbis.

Nutrition Therapy for Hypothyroidism

No special nutrition therapy is required for patients with hypothyroidism other than correcting iodine deficiency where it exists.[62] However, there are nutrition implications for patients receiving hypothyroidism medications. Drug-nutrient interactions for these medications (and others used to treat endocrine disorders) are outlined in Table 17.16.

Hyperthyroidism

Definition Hyperthyroidism is characterized by excessive secretion of thyroid hormones.[3]

Epidemiology and Etiology The most common cause of hyperthyroidism is Graves' disease (see Table 17.17), which accounts for approximately 60%–80% of cases.[1, 3, 64, 65] Graves' disease is hereditary and most commonly found in iodine-replete regions.[3] Autoimmune thyroid disease occurs with the same frequency in whites and Asians, and less frequently in blacks. All thyroid diseases occur more frequently in women than men. Peak incidence occurs between the ages of 20 and 40 years.[64]

Pathophysiology Graves' disease is an organ-specific autoimmune disease in which an antibody (thyroid-stimulating immunoglobulin, or TSI) mistakenly targets TSH receptors on the thyroid cells. This antibody stimulates both secretion and growth of the thyroid in a manner similar to TSH, but it is not subject to **negative-feedback** inhibition by thyroid hormone.[1, 64]

Clinical Manifestations The hypermetabolic effects of hyperthyroidism affect every organ in the body. Key symptoms include palpitation, nervousness, sweating, hyperdefecation, heat intolerance, and oligomenorrhea. General signs of hyperthyroidism include weight loss despite increased appetite, drooping eyelids and stare, sinus tachycardia, atrial fibrillation, tremor, and muscle weakness.[64] A well-known characteristic of

Table 17.16 Drug-Nutrient Interactions for Medications Used to Treat Non-Diabetic Endocrine Disorders

Classification	Mechanism	Generic (Brand Name)	Possible Food-Drug Interactions
Thyroid Hormone Replacement	Influences growth and maturation of tissues; involved in normal growth, metabolism and development; produces stable levels of T_3 and T_4.	Levothyroxine (Synthroid, Levoxyl, Levothroid, Unithroid)	• Take iron, calcium, or magnesium supplements separately from drug by ≥4 hr (may decrease absorption). • Decreased absorption also reported with soy, soy milk, soy infant formula, walnuts, cottonseed meal, and high-fiber foods. • Appetite changes may result in weight loss.
Antithyroid Medications	Blocks oxidation of iodine in thyroid gland, inhibiting T_4 to T_3 conversion. Inhibits thyroid hormone by blocking iodine oxidation (not known to inhibit peripheral conversion of thyroid hormone).	Propylthiouracil (Propylthiour) Methimazole (Tapazole)	• Antivitamin K activity. • Inhibits vitamin K activity.
Beta-Adrenergic Receptor Blockers	Used to reduce symptoms of tachycardia, tremor, and anxiety in hyperthyroidism.	Propranolol (Inderal, Betachron E-R)	• Low-sodium, low-kcal diet may be recommended. • Avoid natural licorice. • Avoid alcohol.
Corticosteroids	Used to restore corticosteroid levels.	Cortisone (Cortone)	• Low sodium, high calcium, high vitamin D, high protein. • May need high potassium, vitamin A, vitamin C, phosphorus (or supplements). • Calcium-vitamin D supplement recommended with long-term use. • Caution with grapefruit/ grapefruit juice (with methylprednisolone). • Increased appetite, increased weight. • Avoid alcohol. • Negative N balance due to protein catabolism. • Calcium wasting with long-term use. • Chromium deficiency may increase risk of steroid-induced diabetes.
	Used for partial replacement therapy in adrenocortical insufficiency.	Fludrocortisone (Florinef)	• Decrease dietary sodium unless increased sodium is used to manage hypotension. • Increase calcium, vitamin D. • May need increased potassium (or supplement). • Calcium-vitamin D supplement recommended with long-term use. • Avoid alcohol.

Sources: Lee SL, Anathahrishnan S. Hyperthyroidism. Last updated 6/8/09. eMedicine.com, Inc. Available at http://www.emedicine.com. Liotta EA, Elston DM. Addison disease. Last updated 1/2/09. eMedicine.com, Inc. Available at http://www.emedicine.com. Bharaktiya S, Orlander PR, Woodhouse WR, Davis AB. Hypothyroidism. Last updated 7/23/09. eMedicine.com, Inc. Available at http://www.emedicine.com. Pronsky ZM. *Food Medication Interactions*, 15th ed. Birchrunville, PA; 2008.

Table 17.17 Common Causes of Hyperthyroidism

Graves' Disease	Autoimmune thyroid disease distinguished by overactive thyroid gland
Iodine Induced	Usually seen in those who already have underlying abnormal thyroid gland; a number of medications (such as amiodarone [Cordarone], used to treat heart problems) contain large amounts of iodine and may be associated with thyroid function irregularity
Excessive Intake of Thyroid Hormones	Arises often due to lack of follow-up of patients taking thyroid medications
Toxic Multi-modular Goiter	Thyroid gland becomes lumpier as people age; the lumps do not produce thyroid hormones, but occasionally a nodule may become "autonomous," fail to respond to pituitary regulation via TSH, and produces thyroid hormones independently
Abnormal Secretion of TSH	Pituitary gland tumor may produce abnormally high secretion of TSH, which excessively signals thyroid gland to produce thyroid hormones
Inflammation of Thyroid (thyroiditis)	May come about after viral illness; associated with a fever and sore throat that is often painful upon swallowing; thyroid gland tender to touch

Sources: Axford J, O'Callaghan C (eds.). *Medicine,* 2nd ed. Oxford, UK: Blackwell Science Ltd.; 2004, and Mathur R, Shiel WC. Hyperthyroidism. Last editorial review 7/22/08. MedicineNet.com. Available at http://www.medicinenet.com.

Figure 17.16 Exophthalmos

Abnormal fluid retention behind eyeballs causes them to bulge forward.

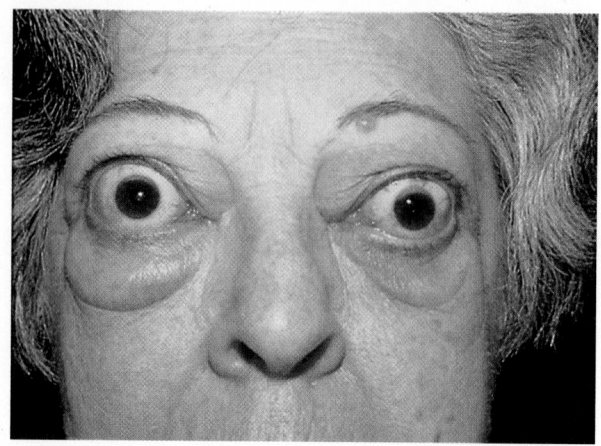

Source: L. Sherwood, Human Physiology: From Cells to Systems, 5e, copyright © 2004, p. 706.

Courtesy of L. Sherwood/Cengage Learning.

Graves' disease, but not other types of hyperthyroidism, is exophthalmos (bulging eyes) (see Figure 17.16). Water-retaining carbohydrates are deposited behind the eyes and the resulting fluid retention pushes the eyeballs forward so they bulge.[1]

Medical Treatment Three general methods of treatment are used to suppress excess thyroid hormone secretion:[1, 64]

- Surgical removal of the over-secreting portion of the thyroid gland
- Administration of radioactive iodine to selectively destroy thyroid glandular tissue
- Antithyroid drugs that specifically interfere with thyroid hormone synthesis

Thyroidectomy is the oldest manner of treatment for hyperthyroidism. Nevertheless, because hyperthyroidism can usually be successfully treated with antithyroid medications and radioactive iodine, thyroidectomy is reserved for cases where these treatments fail. Antithyroid drugs (e.g., methimazole, propylthiouracil) have been used since their introduction in the 1940s. These drugs inhibit synthesis of T_4 and T_3, leading to a gradual decline in thyroid hormone levels over a period of two to eight weeks. Radioactive iodine therapy is the most widespread treatment of hyperthyroidism in the U.S. Administered orally as a single dose (capsule or liquid), radioactive iodine acts less rapidly than antithyroid medications or thyroidectomy, but it is successful and reliable, and does not necessitate hospitalization.[64]

Nutrition Therapy for Hyperthyroidism

While no special diet must be followed by individuals with hyperthyroidism, they should be monitored for drug-nutrient interactions related to any medications they receive (see Table 17.16).[64]

Pituitary Disorders

Pituitary disorders are a result of hypersecretion or hyposecretion. The most common disorders include pituitary tumors, acromegaly, Cushing's disease, and diabetes insipidus.

Pituitary Tumors Pituitary tumors are adenomas found in the pituitary gland. They are slow growing and most are benign. Autopsy data from the National Institute of Diabetes and Digestive and Kidney Diseases indicates that 25% of the U.S. population has some form of small pituitary tumor. Pituitary tumors often present with signs and symptoms related to hypofunction or hyperfunction. Tumors that produce hormones are called functioning tumors, whereas those that do not produce hormones are called nonfunctioning tumors.

Hyperpituitarism Increased secretion of pituitary hormones can result in prolactinoma, acromegaly, Cushing's disease, and thyroid-stimulating-hormone-secreting tumor. Prolactinoma and thyroid-stimulating-hormone-secreting tumor are summarized in Table 17.18.

Acromegaly Acromegaly is a Greek word that means "extremities" and "enlargement." It results from growth hormone (GH) hypersecretion, which causes excessive growth (see Figure 17.17). Affecting mostly middle-aged (35- to 50-year-old) adults, elevated GH levels are associated with changes in appearance (coarsening of facial features as bones grow, enlarged hands and feet, protruding jaw, enlarged lip, nose, tongue; thickened ribs [creating barrel chest], coarsening of body hair as skin thickens and/or darkens); headaches; excessive sweating and oily skin; vision disturbances; sleep apnea (and snoring); joint pain; degenerative arthritis; enlarged heart; fatigue; irregular menstrual cycle and/or breast milk production in women; and impotence in men. Untreated, it can lead to severe illness and death.

Treatment is dependent upon the cause of the disease. Ninety percent of acromegaly cases are caused by benign tumors; therefore, treatment may include surgery to remove the tumor, radiation, and injection of a GH-blocking drug. Untreated acromegaly can lead to diabetes mellitus and hypertension. It also increases risk for cardiovascular disease and colon polyps that may lead to cancer.

Cushing's Syndrome Cushing's syndrome, also called hypercortisolism, is a rare disorder caused by chronic exposure to excessive circulating cortisol. The most common causes of Cushing's syndrome are Cushing's disease (an ACTH-secreting

Figure 17.17 Acromegaly

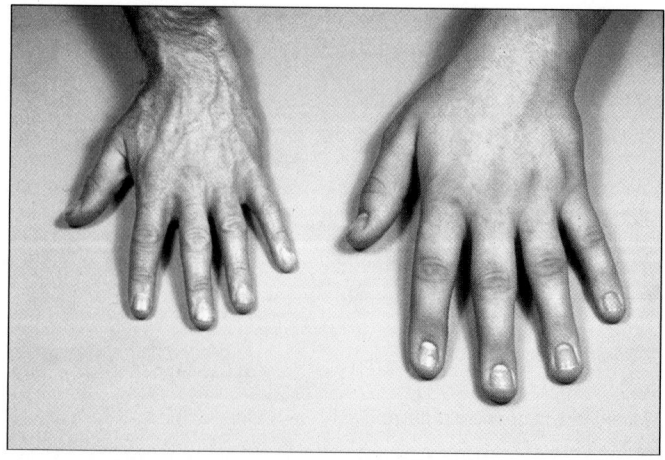

Source: R. Rhoades and R. Pflanzer, Human Physiology, 4e, copyright © 2003, p. 422.

Courtesy of Cengage Learning.

Table 17.18 **Hyperpituitarism and Hypopituitarism**

Disorder	Symptoms	Therapy
Hyperpituitarism		
Prolactinoma		Dopamine agonists, surgery for those intolerant of or refractory to medical therapy; radiation therapy may be considered for those intolerant of dopamine agonists and not likely to be cured by surgery
Women of child-bearing age	Oligomenorrhea, amenorrhea, galactorrhea (spontaneous lactation), or infertility	
Men and postmenopausal women	Headaches and visual field defects	
Thyroid-stimulating-hormone-secreting (TSH) tumor	Hyperthyroidism and goiter	Surgery and possibly radiation
Hypopituitarism		
Growth hormone deficiency	Decreased muscle strength, exercise intolerance, reduced sense of well-being, increased body fat (particularly intra-abdominally), decreased lean body mass	GH replacement
Gonadotropin deficiency		
Women	Infertility, oligomenorrhea or amenorrhea, lack of libido, hot flashes, dyspareunia (painful intercourse)	Estrogen replacement to prevent osteoporosis and treat hot flashes, decreased libido, and vaginal dryness
Men	Decreased libido, impotence	IM (intramuscular) testosterone injections or transdermal testosterone (patch or gel)
Adrenocorticotropic hormone deficiency (ACTH)	Chronic malaise, fatigue, anorexia, hyponatremia	Replacement regimen of hydrocortisone
Thyrotropin (TSH) deficiency	Malaise, leg cramps, fatigue, dry skin, cold intolerance	Levothyroxine replacement therapy

tumor) or use of synthetic steroids to treat other conditions. Overproduction of adrenocorticotropic hormone (ACTH) in Cushing's disease stimulates the adrenal glands to overproduce the steroid hormone cortisol. Cushing's syndrome is found more often in women than men and affects all age groups, but peak incidence is seen in middle age. Common symptoms of Cushing's syndrome (see Figure 17.18) include upper body obesity, moon face, redness of the face, severe fatigue and muscle weakness, infertility, hypertension, backache, hyperglycemia, easy bruising, bluish-red striae on skin, and development of diabetes mellitus. Women may experience increased growth of facial hair and hirsutism (excessive body and facial hair), oligomenorrhea, or amenorrhea; males may experience impotence. Patients may also present with neurological symptoms, including memory difficulties and neuromuscular disorders.

Figure 17.18 Cushing's Syndrome

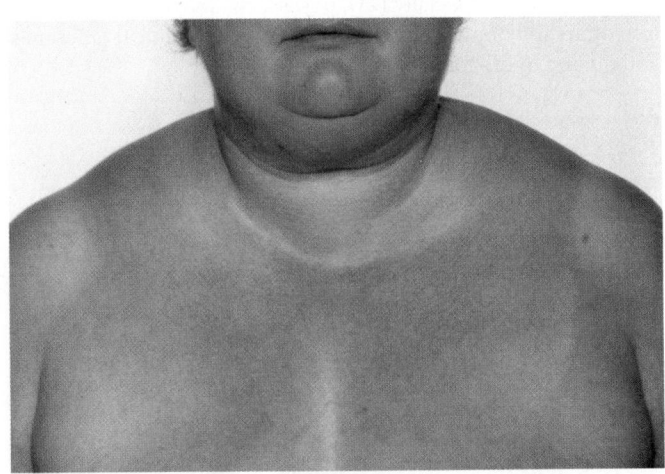

Source: Copyright © NMSB/Custom Medical Stock Photo.

Cushing's progresses slowly and gradually in most cases, and can therefore go unrecognized for some time.

Treatment of Cushing's syndrome is dependent on the cause of the excess cortisol. If the cause is long-term use of synthetic steroid medications, dosage may be reduced until symptoms are controlled. Surgery or radiation may be used to treat pituitary tumors. Surgery, radiation, chemotherapy, immunotherapy, or a combination may be used to treat ectopic ACTH syndrome. Prognosis is also dependent upon cause of the disease. Most cases can be cured, although recovery may be complicated by various aspects of the causative illness.

Hypopituitarism Pituitary tumors are the most common cause of hypopituitarism, which results in deficiencies of GH, gonadotropin, adrenocorticotropic hormone (ACTH), and thyrotropin (TSH). Manifestations of these disorders are shown in Table 17.18.

Diabetes Insipidus Diabetes insipidus (DI), not to be confused with diabetes mellitus, results from insufficient production of antidiuretic hormone (a.k.a. vasopressin) by the hypothalamus (the portion of the brain that stimulates the pituitary gland). Antidiuretic hormone is produced by the hypothalamus, but stored and released into the bloodstream by the pituitary gland. Normally, antidiuretic hormone controls the kidneys' output of urine. DI causes polyuria (>3 L/24 hr) and polydipsia resulting from excessive loss of fluid.

Treatment of DI depends upon its cause. Treatment of the cause usually resolves DI. Common causes of DI include the following:

- Malfunctioning hypothalamus
- Malfunctioning pituitary gland
- Brain injury
- Tumor

- Tuberculosis
- Blockage of cerebral arteries
- Encephalitis
- Meningitis
- Sarcoidosis

Adrenal Cortex Disorders

A number of common deficiencies result from either insufficient or excess secretion of adrenal cortex hormones. And, as with other endocrine disorders, symptoms are the result of either the absence or magnification of effects of the hormones involved.

Excess Secretion of Glucocorticoids
Prolonged exposure to high levels of endogenous or exogenous glucocorticoids results in the condition Cushing's syndrome.

Insufficient Secretion of Adrenal Cortex Steroids
Both adrenal glands must be nonfunctional (or removed) before adrenocortical insufficiency can occur.[1] As a result of either occurrence, both glucocorticoid (cortisol) and mineral corticoid (aldosterone) hormone production is lacking. Death may result from untreated adrenocortical insufficiency.

Primary adrenal insufficiency is uncommon, but iatrogenic (caused by medical treatment) adrenal insufficiency is more frequent, although exact incidence is unknown. Autoimmune Addison's disease is the more common form of adrenal insufficiency. Addison's can occur at any age, but is most common in people 30–50 years of age, and afflicts men and women.

Adrenal insufficiency can be classified as either primary or secondary. Primary adrenal insufficiency (Addison's disease) is caused by a dysfunctional adrenal cortex that impairs both glucocorticoid and mineral corticoid production. Secondary adrenal insufficiency results from inadequate ACTH production by the anterior pituitary, resulting primarily in deficient glucocorticoid secretion. Adrenal insufficiency can further be classified as congenital or acquired.

Primary adrenal insufficiency results from destruction of the adrenal cortex. Aldosterone is produced by the medulla of the adrenal gland; cortisol is produced in the adrenal cortex. Clinical findings manifest after 90% of the adrenal cortex has been destroyed. Causes of this destruction are as follows:

- Autoimmune
- Infectious (e.g., mycobacterial, fungal)

- Neoplastic (e.g., primary, metastatic)
- Traumatic
- Iatrogenic (e.g., surgery, medication)
- Vascular (e.g., hemorrhage, emboli, thrombosis)
- Metabolic (e.g., amyloidosis)

With destruction of the adrenal cortex, feedback inhibition of the hypothalamus and anterior pituitary gland is interrupted. Symptoms associated with aldosterone deficiency in Addison's disease progress slowly and insidiously. Typical symptoms of Addison's disease reflect loss of glucocorticoid and mineral corticoid action. Since aldosterone is essential for life, the condition can be fatal. Aldosterone deficiency causes reduced potassium loss in the urine, resulting in hyperkalemia, which in turn results in disturbed cardiac rhythm. Hyponatremia caused by excessive urinary loss of sodium results in hypotension.

Cortisol deficiency results in poor response to stress, hypoglycemia (resulting from reduced gluconeogenesis), and hyperpigmentation (from excessive secretion of ACTH)[1]. Addison's disease may coexist with other autoimmune disorders, especially thyroid disease, premature ovarian failure, and T1DM.

Treatment of Addison's disease involves replacing or substituting hormones not being produced by the adrenal gland. Medications used to treat Addison's disease, as well as potential drug-nutrient interactions, are outlined in Table 17.16.

Patients with Addison's disease should not restrict salt in their diets. Patients with concurrent primary hypertension may restrict salt intake rather than discontinue mineral corticoid replacement. Patients living in warm climates should increase salt intake due to increased loss of salt through perspiration.

Conclusion

This chapter has described the physiology behind endocrine function, endocrine control of metabolism, thyroid disorders, diabetes mellitus, and hypoglycemia, and their clinical consequences for the patient and practitioner. Nutrition therapy must address not only endocrine disorders but the effect of the medications used to treat these disorders on nutrition and nutrition status. Therapeutic nutrition interventions have been discussed, with particular emphasis on specific modifications to assist in the treatment of disease. Nutrition therapy for endocrine disorders, especially DM, poses a challenge for the RD but also offers the opportunity to greatly improve clinical outcomes and quality of life for clients and patients.

PRACTITIONER INTERVIEW

Jill Weisenberger, MS, RD, CDE, *Research Dietitian, National Clinical Research, Inc.—Norfolk in Norfolk, VA, consultant to the food and healthcare industries, and health & nutrition writer*

How did you become a Certified Diabetes Educator?

The opportunity to become a CDE fell into my lap and I jumped at the chance. As a consulting dietitian, I felt the CDE would make me more marketable by widening my scope of practice, and it did. I was asked to apply for the part-time position of an outpatient diabetes dietitian. I had just your usual diabetes background, but an interest and a willingness to learn more. I worked in that job long enough to earn the right to take the CDE exam. Now the CDE has helped me get other clinical jobs and even public relations and writing jobs.

How has dietetics changed since you have begun your practice?

During my career, I have seen substantial changes in how we treat diabetes. Today the medications are better, the diet is much less restrictive, and we focus on the clients being able to manage themselves. When I started out, I mostly just educated the client on what to eat via the exchange system and to not eat much sugar. Now I counsel people. I help them to use information to solve their own problems and recognize their barriers. This change didn't happen overnight but rather with experience. I always say I'm not in the business of telling people what to eat and what to do. Rather, let's figure out together what works for you and what doesn't. The improvement and options in diabetes medications has helped tremendously by giving patients more options and freedom in planning meals.

What is your typical patient like? Tell us about how you accomplish nutrition therapy with your clients.

Almost all my clients have type 2 diabetes but many of them are on insulin. Some are as young as 20, yet most (75%) are over 45 years old. About half my clients are African Americans. One of my clients was a 45-year-old gentleman who came in with HbA$_1$C of 13.7%, overweight, and on oral medications. We worked on: (1) monitoring blood sugar, (2) eating three meals a day, 60 g of carbohydrate per meal, (3) eating 0–20 g carbohydrate for a snack if hungry, and (4) identifying foods in the diet high in saturated fat. Over three visits (3 months), his HbA$_1$C dropped to 6.5%. He lost only a small amount of weight but was able to go off diabetes medications.

During a client's first visit I have a basic plan—chart the outcomes for specific behavioral changes. For example, the goal may be to normalize blood sugar by spreading the consumption of carbohydrates throughout the day. The behavioral outcome would be to consume three meals a day (60 g of carbohydrate per meal) and if hungry a snack of 0–20 g carbohydrates. On the second visit, I find out what worked and what didn't work for the clients and if they were happy with the plan. Then, I find out if they anticipate any issues coming up (wedding, vacation, work) that might interfere with the plan. Lastly, I look at blood glucose records, but they don't always bring them.

When I work with people I suggest only small changes in their diet each time I see them. But first I always ask, "What is the first thing you'd like to change and why? How confident are you that you can do this, and how motivated are you? Why?" Changing someone's diet is very hard. They're busy, tired, have kids, and often have few cooking skills. But I do insist that they monitor their blood sugars. If they refuse, we negotiate. They learn so much from knowing their blood sugar levels, and it allows them to manage their own disease by seeing first hand the impact of food and exercise. Working on preventing heart disease is as important as the rest, but it's sometimes hard to get patients to focus on this. So often patients want to go low, low carb, but they don't consider the quality of their food choices. Working on the big picture is always a challenge.

My advice to new dietitians is to have compassion for each patient. You will do a better job and will enjoy what you do. Wherever you work, surround yourself with smart, capable people who work hard, because that is how you become successful. Keep up to date on all aspects of nutrition and feel confident to discuss your patients with their physician.

Application of the Nutrition Care Process

Type 2 Diabetes Mellitus

Introduction:

71-year-old female who arrives at the emergency room presents with nonhealing wound on right foot between the 2nd and 3rd digits. Patient has a history of frequent bladder infections, slight tingling and numbness in her feet, and today, serum blood glucose of 325 mg/dL. She is admitted for antibiotics, probable surgical debridement of a wound, and stabilization and treatment for T2DM. Past medical history includes a diagnosis of HTN treated with 50 mg Captopril two times daily.

1. What is the difference between type 1 and type 2 DM? Why is it assumed that this patient has type 2?

2. What risk factors does this patient present with? What symptoms may indicate that she has complications of type 2 DM?

Nutrition Assessment:

Food/Nutrition-Related History:

Lives with sister who has T2DM; prepares own meals—rarely eats at restaurants. Likes all foods but avoids "foods with sugar." 24-hour recall indicates the following: Breakfast: 2–3 slices of toast with margarine (about 1 T); coffee with milk. Lunch: 1 can tomato soup prepared with water; saltine crackers (12) with peanut butter (about 2 T). Dinner: pork chop, 3 oz.; 1 cup of corn; cornbread—1 slice; and applesauce—1 cup.

Anthropometrics:

Ht. 5'0", Wt. 155 lbs., Usual adult body weight—145–165 lbs.

3. Evaluate this patient's weight. Calculate and interpret her BMI.

(continued)

Biochemical Data:

Labs: BUN 26 mg/dL; Cr 1.2 mg/dL; Chol 300 mg/dL; HDL 35 mg/dL; LDL 140 mg/dL; Glucose 325 mg/dL; HbA1C 8.5%

4. What admission laboratory values are abnormal? Interpret their significance in relationship to both her diagnosis and nutritional status.

Nutrition Diagnosis:

5. Identify at least two nutrition problems suggested by the nutrition assessment and medical history. Determine the diagnostic term for each nutrition problem. Next, identify the etiology of each nutrition diagnosis. Finally, identify the signs and symptoms that support the evidence for the diagnoses.

WEB LINKS

National Diabetes Education Program (NDEP) (1-800-860-8747)   NDEP is a federally sponsored initiative that involves public and private partnerships to improve treatment and outcomes for people with diabetes. Single copies of most materials are available free of charge or can be downloaded from the NDEP website. *http://ndep.nih.gov*

American Diabetes Association (ADA) (1-800-DIA-BETES)   The American Diabetes Association publishes many health professional and client materials in addition to its scientific journals. In an annual supplement to *Diabetes Care*, ADA reissues practice guidelines that address a wide array of clinical issues including nutrition. These guidelines can be downloaded from the ADA website free of charge. *http://diabetes.org*

Diabetes Care and Education (DCE) Dietetic Practice Group of the American Dietetic Association   DCE publishes *On the Cutting Edge*, a theme-centered newsletter on timely topics related to nutrition and diabetes. Educational materials developed by the DCEP are available for purchase from the ADA. The DCE website provides information about Medicare MNT benefits for persons with diabetes, Health Care Financing Administration (HFCA) rules for ADA Education Recognition Programs reimbursement, and CPT codes for MNT. *http://eatright.org or http://www.dce.org*

Division of Diabetes Translation of the Centers for Disease Control and Prevention (1–877–CDC–DIAB)   The website has information about the Diabetes Control Program in each state and diabetes-related statistics such as the rise in the prevalence and incidence of diabetes. The *National Diabetes Fact Sheet CDC Information* is available as a downloadable file. *http://www.cdc.gov/diabetes*

National Diabetes Information Clearinghouse (NDIC)   The National Institute of Diabetes and Digestive and Kidney Diseases (NIDDK) website provides information about its diabetes-related clinical trials and other research programs, a directory of diabetes organizations, health education programs, and diabetes-related topics. Downloadable files include *Diabetes Dateline*, an NIDDK newsletter; client education materials; and information for health professionals. *http://diabetes.niddk.nih.gov/*

American Association of Diabetes Educators (AADE)   As a multidisciplinary organization of health professionals who teach about diabetes, the AADE and its website provide information about the scope of practice and standards related to diabetes education. The website also has information and AADE publications. *http://www.diabeteseducator.org/*

END-OF-CHAPTER QUESTIONS

1. List the three chemical classes of endocrine hormones. For each class, pick one hormone, name its production site, and briefly describe its function.

2. Describe the action of insulin on carbohydrate, lipid, and protein metabolism.

3. What is the definition of diabetes mellitus (DM)? List the classifications for DM and briefly explain similarities and differences for their epidemiology, etiology, pathophysiology, and clinical manifestations. Describe three ways diabetes can be diagnosed.

4. What is meant by glycemic control, and why is it important? Describe the physiological consequences of poor glycemic control. Which laboratory measurements are indicators of short- and long-term glycemic control? How often are they checked?

5. List and describe the types of insulin that are now available. How is insulin dosage determined, and how can insulin be administered? What are the differences in nutrition therapy recommendations for a person with diabetes who is using insulin on the conventional plan versus a person using intensive insulin therapy?

6. Briefly describe several meal planning approaches that are used with individuals with diabetes mellitus.

Select a meal plan you would use for a 70-year-old man with an 8th-grade education (type 2 DM), a 13-year-old teenage female athlete (type 2 DM), and a 32-year-old pregnant women (gestational DM), and justify your answers.

7. List the 7 classes of diabetes medications. Briefly describe their effects and mechanisms of action.

8. For individuals with type 2 diabetes, why is weight management often included as a component of nutrition therapy? Why is it important for the treatment of type 2 diabetes?

ENDNOTES

1. Only endocrine disorders with nutritional implications will be discussed.

2. The abbreviations T1DM and T2DM used in this chapter are not standardized abbreviations supported by the ADA, but are used for the sake of brevity.

3. Spontaneous hypoglycemia, as opposed to hypoglycemia related to diabetes, is the subject of this section.

REFERENCES

1. Sherwood, L. Human physiology from cells to systems. 7th ed. Belmont (CA): Thomson; 2010.

2. Copstead LC, Banasik JL. Pathophysiology. 3rd ed. St. Louis: Elsevier Saunders; 2005.

3. Kaplan F, Conway G. Endocrine disease. In: Axford J, O'Callaghan C, editors. Medicine. 2nd ed. Oxford (UK): Blackwell Science Ltd.; 2004.

4. Rhoades R, Pflanzer R. Human physiology. 4th ed. Pacific Grove (CA): Thomson Learning; 2003.

5. Lebovitz HE. Introduction: goals of treatment. In American Dietetic Association. Therapy for Diabetes Mellitus and Related Disorders. 4th ed. Alexandria (VA): American Dietetic Association; 2009.

6. Centers for Disease Control and Prevention. National diabetes fact sheet, general information and national estimates on diabetes in the United States, 2007. Atlanta, GA: United States Department of Health and Human Services, Centers for Disease Control and Prevention, 2008. Available at http://www.cdc.gov/diabetes/pubs/pdf/ndfs_2007.pdf

7. National Diabetes Information Clearinghouse (NDIC). National diabetes statistics, 2007. Washington (DC): NIH Publication # 08–3892, 2008. Available at http://diabetes.niddk.nih.gov

8. American Diabetes Association. Diagnosis and classification of diabetes mellitus. Diabetes Care. 2009;32(Suppl 1):S62–67.

9. American Diabetes Association. Economic costs of diabetes in the U.S. in 2007. Diabetes Care. 2008;31(3):596–615.

10. Barker JM, Eisenbarth GS. Genetic counseling for autoimmune type 1 diabetes. In American Diabetes Association, Therapy for Diabetes Mellitus and Related Disorders. 5th ed. 2009.

11. American Diabetes Association. Standards of medical care in diabetes—2009. Diabetes Care. 2009;32 (Suppl 1):S13–61.

12. Mulcahy K, Lumber T. The diabetes ready reference for health professionals. 2nd ed. Alexandria (VA): American Diabetes Association; 2004.

13. American Association for Clinical Chemistry. Diabetes-related autoantibodies. Washington (DC): Lab Tests Online. Available at http://www.labtestsonline.org. Updated: 2006 Jan 30.

14. Rystrom JK. Insulin therapy. In: Ross TA, Boucher JL, O'Connell BS, editors. American Dietetic Association Guide to Diabetes: Medical Nutrition Therapy and Education. Chicago: American Dietetic Association; 2005.

15. Diabetes Control and Complications Trial Research Group. The effect of intensive treatment of diabetes on the development and progression of long-term complications in insulin-dependent diabetes mellitus. N Engl J Med. 1993;329:977–86.

16. American Diabetes Association. Nutrition principles and recommendations in diabetes. Diabetes Care. 2004;27 (Suppl 1):S36–46.

17. Garg SK, Naik RG. Monitoring diabetes. In American Dietetic Association. Therapy for Diabetes Mellitus and Related Disorders. 4th ed. Alexandria (VA): American Dietetic Association; 2009.

18. Skyler JS. Insulin treatment. In American Dietetic Association. Therapy for Diabetes Mellitus and Related Disorders. 4th ed. Alexandria (VA): American Dietetic Association; 2009.

19. American Diabetes Association. Implications of the United Kingdom prospective diabetes study. Diabetes Care. 2002;25(Suppl 1):S28–32.

20. Cryer PE, Davis SN, Shamoon H. Hypoglycemia in diabetes (American Diabetes Association Technical Review). Diabetes Care. 2003;26:1902–12.

21. Daly A, Powers MA. Medical nutrition therapy. In American Dietetic Association. Therapy for Diabetes Mellitus and Related Disorders. 4th ed. Alexandria (VA): American Dietetic Association; 2009.

22. Nutrition Care Manual [database on the Internet]. Chicago: American Dietetic

Association; 2008. Available at http://nutritioncaremanual.org (by subscription only).

23. Franz MJ, Bantle JP, Beebe CA, Brunzell JD, Chiasson J, Garg A, Holzmeister LA, Hoogwerf B, Mayer E, Mooradian AD, Purnell JQ, Wheeler M. Evidence-based nutrition principles and recommendations for the treatment and prevention of diabetes and related complications. Diabetes Care. 2002;25(1):148–98.

24. American Diabetes Association. Translation of the diabetes nutrition recommendations for health care institutions. Diabetes Care. 2002;35 (Suppl 1):S61–63.

25. Pastors JG, Waslaski J, Gunderson H. Diabetes meal-planning strategies. In: Ross TA, Boucher JL, O'Connell BS, editors. American Dietetic Association Guide to Diabetes: Medical Nutrition Therapy and Education. Chicago (IL): American Dietetic Association; 2005.

26. Anderson EJ, Delahanty L, Richardson M, Castle G, Cercone S, Lyon R, Mueller D, Snetselaar L. Nutrition interventions for intensive therapy in the diabetes control and complications trial. J Am Diet Assoc. 1993;93:768–72.

27. Tamborlane WV, Sikes KA, Swan K, Wienzimer SA. Type 1 diabetes in children. In American Dietetic Association. Therapy for Diabetes Mellitus and Related Disorders. 4th ed. Alexandria (VA): American Dietetic Association; 2009.

28. American Dietetic Association. Choose Your Foods: Exchange Lists for Diabetes. Chicago: American Dietetic Association; 2008.

29. Kulkarni K. Pattern management. In: Ross TA, Boucher JL, O'Connell BS, editors. American Dietetic Association Guide to Diabetes: Medical Nutrition Therapy and Education. Chicago (IL): American Dietetic Association; 2005.

30. UK Prospective Diabetes Study (UKPDS) Group. Intensive blood-glucose control with sulphonylureas or insulin compared with conventional treatment and risk of complications in patients with type 2 diabetes (UKPDS 33). Lancet. 1998;352:937–53.

31. Baynes K, and Betteridge DJ. Diabetes mellitus, lipoprotein disorders and other metabolic diseases. In: Axford J, O'Callaghan C, editors. Medicine. 2nd ed. Oxford (UK): Blackwell Science Ltd.; 2004.

32. American College of Physicians. Diabetic testing and monitoring. Philadelphia (PA): Focus On; 1998, Issue 6. Available at http://www.acponline.org/mle/diabetic_test.htm

33. Hayes C. Physical activity and exercise. In: Ross TA, Boucher JL, O'Connell BS, editors. American Dietetic Association Guide to Diabetes: Medical Nutrition Therapy and Education. Chicago (IL): American Dietetic Association; 2005.

34. American Diabetes Association. Physical activity/exercise and diabetes. Diabetes Care. 2004;27(Suppl 1):S58–62.

35. Arnold MS. Hypoglycemia and hyperglycemia. In: Ross TA, Boucher JL, O'Connell BS, editors. American Dietetic Association Guide to Diabetes: Medical Nutrition Therapy and Education. Chicago (IL): American Dietetic Association; 2005.

36. Umpierrez GE, Murphy MB, Kitabchi AE. Diabetic ketoacidosis and hyperglycemic hyperosmolar syndrome. Diabetes Spectrum. 2002;15(1):28–36.

37. American Diabetes Association. Hyperglycemic crisis in diabetes. Diabetes Care. 2006; 29(12):2739–48.

38. Wheeler ML. Long-term complications. In: Ross TA, Boucher JL, O'Connell BS, editors. American Dietetic Association Guide to Diabetes: Medical Nutrition Therapy and Education. Chicago (IL): American Dietetic Association; 2005.

39. Council for the Advancement of Diabetes Research and Education (CADRE). Section 3—Diabetes Slides: Microvascular and Macrovascular Complications: Epidemiologic Studies. Brooklyn (NY): Available at http://www.cadre-diabetes.org/r_slide_show.asp?id=4

40. Vinik AI, Maser RE, Mitchell BD, Freeman R. Diabetic autonomic neuropathy (technical review). Diabetes Care. 2003;26:1552–79.

41. Centers for Disease Control and Prevention. National diabetes fact sheet, general information and national estimates on diabetes in the United States, 2007. Atlanta (GA): United States Department of Health and Human Services, Centers for Disease Control and Prevention, 2008. Available at http://www.cdc.gov/diabetes/pubs/pdf/ndfs_2007.pdf

42. World Health Organization. Diabetes Programme: Country and regional data: World. Geneva (Switzerland): World Health Organization. Available at http://www.who.int/diabetes/facts/world_figures/en/

43. Votey SR, Peters AL. Diabetes mellitus, type 2—a review. New York: eMedicine.com, Inc. Available at http://www.emedicine.com. Updated 2009 Aug 6.

44. O'Connell BS. Diabetes classification, Pathophysiology, and diagnosis. In: Ross TA, Boucher JL, O'Connell BS, editors. American Dietetic Association Guide to Diabetes: Medical Nutrition Therapy and Education. Chicago: American Dietetic Association; 2005.

45. American Heart Association. Metabolic syndrome. Available at http://www.americanheart.org/presenter.jhtml?identifier=4756

46. Scott CL. Diagnosis, prevention, and intervention for the metabolic syndrome. Am J Cardiol. 2003;82(suppl):35i–42i.

47. Ervin RB. Prevalence of metabolic syndrome among adults 20 years of age and over, by sex, age, race and ethnicity, and body mass index: United States, 2003–2006. U.S. Department of Health and Human Services National Health Statistics Reports, No. 13, 2009.

48. Welschen LMC, Bloemendal E, Nupels G, Dekker JM, Heine RJ, Stalman WAB, Bouter LM. Self-monitoring of blood glucose in patients with type 2 diabetes who are not using insulin. Diabetes Care. 2005;28:1510–17.

49. Freeman J. Oral diabetes medications. In: Ross TA, Boucher JL, O'Connell BS, editors. American Dietetic Association Guide to Diabetes: Medical Nutrition Therapy and Education. Chicago (IL): American Dietetic Association; 2005.

50. U.S. Department of Health and Human Services, U.S. Department of Agriculture. Dietary guidelines for Americans 2005 [monograph on the Internet]. Washington (DC): U.S. Department of Health and Human Services/U.S. Department of Agriculture; 2005. Available at http://www.health.gov/dietaryguidelines/dga2005/document/

51. Institute of Medicine. Dietary Reference Intake for Thiamin, Riboflavin, Niacin, Vitamin B6, Folate, Vitamin B12, Pantothenic Acid, Biotin, and Choline. Washington (DC): National Academy Press; 2000.

52. Buchanan TA. Gestational diabetes mellitus. In American Dietetic Association. Therapy for Diabetes Mellitus and Related Disorders. 4th ed. Alexandria (VA): American Dietetic Association; 2009.

53. Alwan N, Tuffnell DJ, West J. Treatments for gestational diabetes. Cochrane Database Syst Rev. 2009 Jul 8;(3):CD003395. Review.

54. Thomas AM. Classification, screening, and diagnosis. In: Thomas AM, Gutierrez YM, editors. American Dietetic Association Guide to Gestational Diabetes Mellitus. Chicago: American Dietetic Association; 2005.

55. Bentley-Lewis R, Levkoff S, Stuebe A, Seely EW. Gestational diabetes mellitus: postpartum opportunities for the diagnosis and prevention of type 2 diabetes mellitus. Nat Clin Pract Endocrinol Metab. 2008 Oct;4(10):552–58.

56. England LJ, Dietz PM, Njoroge T, Callaghan WM, Bruce C, Buus RM, Williamson DF. Preventing type 2 diabetes: public health implications for women with a history of gestational diabetes mellitus. Am J Obstet Gynecol. 2009 Apr;200(4):365.e1–8

57. Thomas AM, Gutierrez YM. Maternal and fetal complications associated with gestational diabetes mellitus. In: Thomas AM, Gutierrez YM, editors. American Dietetic Association Guide to Gestational Diabetes Mellitus. Chicago (IL): American Dietetic Association; 2005.

58. Thomas AM. Pathophysiology of gestational diabetes mellitus. In: Thomas AM, Gutierrez YM, editors. American Dietetic Association Guide to Gestational Diabetes Mellitus. Chicago (IL): American Dietetic Association; 2005.

59. Gutierrez YM, Reader DM. Medical nutrition therapy. In: Thomas AM, Gutierrez YM, editors. American Dietetic Association Guide to Gestational Diabetes Mellitus. Chicago (IL): American Dietetic Association; 2005.

60. Brabson TA, Chussil JT, Daack-Hirsch S, Dixon D, Falk KM, Ferrelra BF, Gruener RC, et al. Professional guide to diseases. 7th ed. Springhouse (PA): Springhouse Corporation; 2001.

61. Masharani U, Karam JH. Diabetes mellitus and hypoglycemia. In: Tierney LM, McPhee SJ, Papadakis MA, editors. Current Medical Diagnosis and Treatment 2003. New York: Lange Medical Books/McGraw-Hill; 2003.

62. Orlander PR, Woodhouse WR, Davis AB. Hypothyroidism. New York: eMedicine.com, Inc. Available at http://www.emedicine.com. Updated: 2009 Jul 23.

63. Hollowell JG, Staehling NW, Flanders WD, Hannon WH, Gunter EW, Spencer CE, Braverman LE. Serum TSH, T$_4$, and thyroid antibodies in the United States population (1988 to 1994): National Health and Nutrition Examination Survey (NHANES III). J Clin Endocrinol Metab. 2002;87:489–99.

64. Lee SL. Hyperthyroidism. New York: eMedicine.com, Inc. Available at http://www.emedicine.com. Updated: 2009 Jun 8.

65. Lee SL, Ananthakrishnan, S. Hyperthyroidism. New York: eMedicine.com, Inc. Available at http://www.emedicine.com. Updated: 2009 Jun 8.

18

Diseases of the Renal System

Karen Lacey, MS, RD, CD
University of Wisconsin–Green Bay, WI

Marcia Nahikian-Nelms, PhD, RD
The Ohio State University

Introduction

An estimated 13.1% of adults ages 20 or older (26 million adults) have physiological evidence of chronic kidney disease, as determined from data collected through the National Health and Nutrition Examinations Survey between 1999 and 2004.[1] The medical interventions for these diseases are among the most expensive treatments. The annual cost for treatment of end-stage renal disease (ESRD) as reported in 2006 by the National Institute of Diabetes and Digestive and Kidney Disease of the National Institutes of Health (NIH) was $33.61 billion in both public and private monies.[2] The primary causes of kidney disease are directly related to diabetes mellitus and hypertension. As the prevalence of these two diagnoses rises, so will the incidence of kidney disease. In light of these alarming statistics, this chapter will focus on chronic kidney disease (CKD), its effect on nutritional status, and the crucial role that nutrition therapy plays in successful management of CKD.

The Kidneys

Anatomy

The kidneys are two **retroperitoneal** organs the size of a fist. The right kidney is usually found to be slightly lower than the left kidney (see Figure 18.1). Each kidney is 11–12 cm long, 5–7.5 cm wide, and 2.5–3 cm thick. The average weight of a kidney in adults is 125–170 grams in men and 115–155 grams in women.[3] The kidney is made up of a complex capillary

network and an array of **tubules** that perform regulatory and metabolic functions that are vital to life.[3]

The functioning unit of the kidney is called the **nephron**. Each kidney consists of approximately 1.2 million nephrons. Each nephron is made up of a **glomerulus**, which is a capillary tuft located between two arterioles (the **afferent** and the

GLOSSARY

acute renal failure (ARF)—refers to kidney dysfunction of short duration or any sudden, severe impairment of kidney function

acute tubular necrosis (ATN)—a common cause of acute renal failure when cells of the renal tubules die. Often associated with trauma and other serious illnesses.

afferent—carrying blood to the designated site; for example, the afferent arteriole carries blood to the glomerulus

anasarca—generalized edema with accumulation of serum in the connective tissue

arteriovenous fistula (AVF)—a connection of an artery and vein to provide circulatory access for hemodialysis

arteriovenous graft (AVG)—the planting of an artificial vessel connecting an artery and vein; used when patients' blood vessels are fragile and a fistula is not feasible

azotemia—a buildup of nitrogenous waste products such as urea in the blood and body fluids

bone resorption—a process whereby osteoclasts destroy an area of bone as the first step in bone remodeling

chronic kidney disease (CKD)—kidney damage or GFR <60 mL/min/1.73m^2 for >3 months; kidney damage is defined as pathologic abnormalities or markers of damages

continuous renal replacement therapy (CRRT)—type of renal replacement therapy used to treat patients in acute renal failure, particularly those with multiple organ failure; the types of patients treated tend to be hemodynamically unstable, have poor cardiac output, and be unable to tolerate hemodialysis

cystine—a sulfur-containing amino acid, which is produced by the actions of acids on proteins that contain this compound

dialysate—fluid used by the dialysis procedure to assist in removal of metabolic by-products, wastes, and toxins; composition is determined by individual patient requirements

dialysis—renal replacement procedure that removes excessive and toxic by-products of metabolism from the blood, thus replacing the filtering function of healthy kidneys

diffusion—passage of particles through a semi-permeable membrane

efferent—carrying blood away from the designated site; for example, the efferent arteriole carries blood away from the glomerulus

end-stage renal disease (ESRD)—kidney disease in which kidney function declines to 10% to 15% of normal; the term *CKD* is now preferred over this older term; ESRD is equivalent to CKD Stage 5 or when a patient requires renal replacement therapy

extracorporeal shockwave lithotripsy (ESWL)—a common procedure used to treat kidney stones whereby shock waves are used to break down the stones into smaller pieces

focal segmental glomerulosclerosis (FSGS)—describes scarring in scattered regions of the kidney, typically limited to one part of the glomerulus and to a minority of glomeruli in the affected region; FSGS may result from a systemic disorder, or it may develop as an idiopathic kidney disease, without a known cause

glomerular filtration rate (GFR)—the filtration ability of the glomerulus; used as an index of

kidney function; normal value is approximately 125 mL/min

glomerulosclerosis—development of scar tissue within the glomerulus

glomerulonephritis—nephritis marked by inflammation of the capillaries of the renal glomeruli and membrane tissue that serves as a filter

glomerulus—a network of thin-walled capillaries closely surrounded by a pear-shaped epithelial membrane called the Bowman's capsule (within the kidney)

hematuria—the presence of blood in the urine

hemodialysis (HD)—a type of renal replacement therapy whereby wastes or uremic toxins are filtered from the blood by a semipermeable membrane and removed by the dialysis fluid

hydroxyapatite—the apatite form of calcium phosphate present with calcium carbonate

hypercalciuria—an excess of calcium in the urine

hyperoxaluria—an excess of oxalate in the urine

hyperuricosuria—a disorder of uric acid metabolism

IgA nephropathy—a form of glomerular disease that results when immunoglobulin A (IgA) forms deposits in the glomeruli, where it creates inflammation

intravenous pyelogram (IVP)—radiographic imaging of the kidneys, ureter, and bladder using x-ray and contrast dye that is injected intravenously

intrinsic acute renal failure—acute renal failure associated with damage to the renal anatomy

membranous nephropathy—disease diagnosed when a kidney biopsy reveals unusual deposits of immunoglobulin G and complement C3, substances created by the body's immune system; 75% of cases are idiopathic

microalbuminuria—the leaking of small amounts of albumin into the urine by the kidneys

nephritic syndrome—a condition of inflammation of the glomerulus, resulting in hematuria, proteinuria, and oliguria

nephrolithiasis—kidney stones; a common disorder in the United States

nephron—basic functioning unit of the normal kidney; each nephron has two main parts: the glomerulus and the tubule

nephrotic syndrome—a clinical condition consisting of losses of protein in the urine exceeding 3.5 g/day, hyperlipidemia, and low albumin levels (<3.5 g/dL) with edema

normalized protein equivalent of nitrogen appearance (nPNA)—an assessment of protein catabolic rate

obstructive acute renal failure—acute renal failure related to the obstruction of urine flow

oliguria—urine output less than 400 mL, which is the minimum amount of normal urine that can carry away the daily load of metabolic waste products

osmosis—movement of fluid across a semipermeable membrane from a lower concentration of solutes to a higher concentration of solutes

osteitis fibrosa cystica—a form of high-turnover bone disease caused by overproduction of parathyroid hormone (PTH), which increases the rate of bone turnover

oxalate—a salt of oxalic acid produced by the body's metabolism and excreted in the urine

percutaneous nephrolithotomy—a surgical procedure in which a surgeon makes an incision in the back and creates a tunnel to the kidney to remove a kidney stone

peritoneal dialysis (PD)—a type of renal replacement therapy during which the peritoneal cavity serves as the reservoir for the dialysate and the peritoneum acts as the semipermeable membrane across which excess body fluid and solutes are removed

peritonitis—an inflammation of the peritoneum membrane

prerenal azotemia—uremic symptoms associated with acute renal failure

proteinuria—the presence of too much protein in the urine

pyelonephritis—inflammation of both the parenchyma of a kidney and the lining of its renal pelvis, especially due to bacterial infection

renal osteodystrophy—a general term that refers to bone disease related to CKD, caused by over- or underproduction of PTH or by exposure to aluminum

retroperitoneal—lying behind the peritoneum (lining of the abdominal cavity)

rhabdomyolysis—an acute condition of skeletal muscle destruction

secondary hyperparathyroidism—high levels of PTH in the circulation that stimulate bone turnover, which may be accompanied by hyperplasia of the parathyroid glands

Sgöjren's syndrome—autoimmune disease characterized by dry eyes and mouth—often associated with lupus or rheumatoid arthritis

struvite—a form of kidney stones composed of ammonium and magnesium phosphate; they resemble hard crystals

tubules—component of the nephron responsible for reabsorption and secretion (within the kidney); designated as the proximal convoluted tubule, the loop of Henle, and the distal convoluted tubule

ultrafiltrate—referring to the initial filtration of metabolic by-products from the filtered blood within the tubule

ultrafiltration—a form of filtration that provides additional pressure to achieve more concentrated filtration

urea kinetic modeling—a quantitative method by which an individualized hemodialysis treatment prescription can be developed; requires measures of pre and post dialysis BUN, weights and treatment times

uremia (uremic syndrome)—a general term used to encompass a cluster of symptoms resulting from disordered biochemical processes as chronic kidney disease progresses; early symptoms include fatigue, delayed thinking, and pruritis

ureterorenoscopy—a nonsurgical procedure where a surgeon uses a fiberoptic instrument called a ureteroscope to remove a stone lodged in the ureter

uric acid—a crystalline acid occurring as an end product of purine metabolism; a common constituent of renal calculi

Figure 18.1 Anatomical Overview of the Kidney

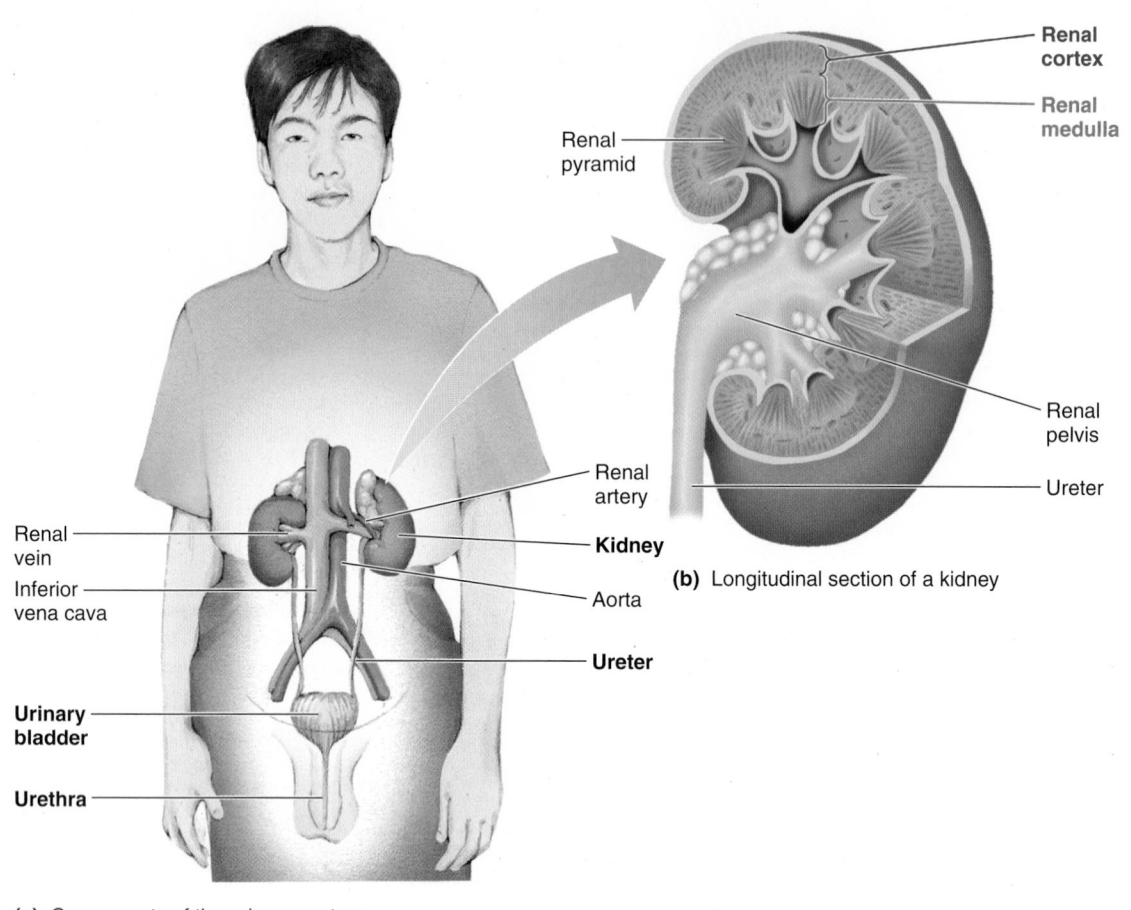

(a) Components of the urinary system

Source: Sherwood, *Human Physiology: From Cells to Systems* 7e, ISBN 0495391840 Figure 14.1 p. 513.

efferent), and a network of tubules lined by epithelial cells (see Figure 18.2).[3] The afferent arteriole carries blood to the glomerulus, and the efferent arteriole carries blood away from it. The nephron extends through three sections of the kidney called the cortex, outer medulla, and inner medulla.[3] Tubules are divided into several sections that differ by the type of epithelial cells that they contain, including the proximal convoluted tubule, the loop of Henle, the distal convoluted tubule and the collecting duct.[3] The cortex contains the glomeruli and the proximal and distal convoluted tubules. The medulla consists of the collecting ducts, loops of Henle, and vasa recta.[3]

The cells located throughout the nephron vary in terms of structure and function. The proximal tubule contains cells with a complex brush border, tight intercellular junctions, and large numbers of mitochondria.[4–5] These cells are suited for active transport, and 65% of filtered sodium and water are reabsorbed here. The cells of the loop of Henle, unlike those in the proximal tubules, are suited for passive diffusion and do not perform active transport.

Physiological Functions

The primary functions of the kidney include maintenance of homeostasis through control of fluid, pH, and electrolyte bal-

ance and blood pressure; excretion of metabolic end-products and foreign substances; and the production of enzymes and hormones. By understanding this normal physiology, you will be able to understand how the direct consequences of the loss of renal function result within various disease states.

Urine formation (see Figure 18.3) is a crucial component in the maintenance of homeostasis. As systemic blood filters through the glomerulus, which filters large proteins and blood cells, the first step in urine formation occurs: the production of approximately 200 L of **ultrafiltrate** each day. The ultrafiltrate, which is similar in composition to the blood, then moves through the tubules where most of its components are reabsorbed during the second phase of urine production. Ultrafiltrate is modified as it passes through the network of tubules, either by reabsorption of amino acids, glucose, selective minerals, and water or by secretion of solutes and water. Under the influence of aldosterone, sodium is reabsorbed in the proximal tubule. Vasopressin controls the final phase of urine production, as it directs concentration of urine and assists in maintaining overall fluid balance (see Chapter 7). Urine osmolality can vary widely from 500 mOsm to 1200 mOsm. Volume varies as well; as little as 500 mL or as much as 12 liters of urine can be produced to maintain normal homeostasis.

Figure 18.2 The Nephron: Functional Unit of the Kidney

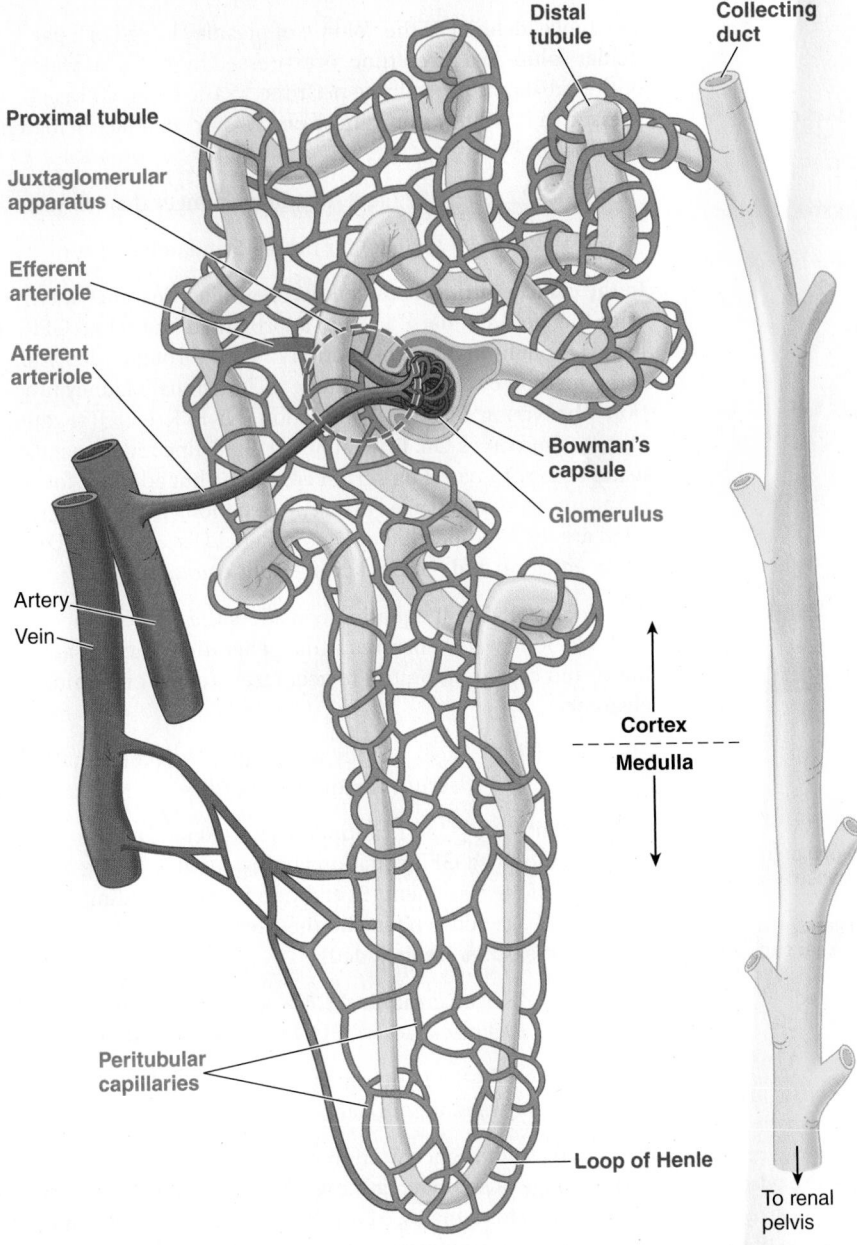

Distal tubule

Collecting duct

Proximal tubule

Juxtaglomerular apparatus

Efferent arteriole

Afferent arteriole

Bowman's capsule

Glomerulus

Artery

Vein

Cortex

Medulla

Peritubular capillaries

Loop of Henle

To renal pelvis

Overview of Functions of Parts of a Nephron

Vascular component

- Afferent arteriole—carries blood to the glomerulus
- Glomerulus—a tuft of capillaries that filters a protein-free plasma into the tubular component
- Efferent arteriole—carries blood from the glomerulus
- Peritubular capillaries—supply the renal tissue; involved in exchanges with the fluid in the tubular lumen

Tubular component

- Bowman's capsule—collects the glomerular filtrate
- Proximal tubule—uncontrolled reabsorption and secretion of selected substances occur here
- Loop of Henle—establishes an osmotic gradient in the renal medulla that is important in the kidney's ability to produce urine of varying concentration
- Distal tubule and collecting duct—variable, controlled reabsorption of Na$^+$ and H$_2$O and secretion of K$^+$ and H$^+$ occur here; fluid leaving the collecting duct is urine, which enters the renal pelvis

Combined vascular/tubular component

- Juxtaglomerular apparatus—produces substances involved in the control of kidney function

Source: Sherwood, Human Physiology: From Cells to Systems 7e, ISBN 0495391840 Figure 14.3 p. 514.

Sodium is regulated by the kidneys, under the control of aldosterone. If serum sodium levels are elevated, sodium is exchanged with potassium so that homeostasis is restored. The kidney additionally plays a significant role in blood pressure control. Cardiac output and blood pressure are dependent on plasma volume. Vasopressin, secreted by the pituitary gland, works at the level of the collecting duct to either increase or decrease absorption of water in order to maintain plasma volume and thus blood pressure (see Chapters 7 and 13).

The formation of urine also serves as the route for excretion of waste products, including the by-products of metabolism such as uric acid, creatinine, and urea. Other wastes excreted are drugs and environmental toxins.

The kidney's role in controlling both hydrogen and bicarbonate ions is a critical component of the maintenance of pH

homeostasis. To maintain pH, a healthy, normally functioning kidney will reabsorb the majority of bicarbonate (HCO$_3^-$) that is needed. This function requires the kidney to secrete hydrogen ions, which combine with the HCO$_3^-$ and form carbonic acid (H$_2$CO$_3$). As discussed in Chapter 8, carbonic acid rapidly dissolves to form CO$_2$ and H$_2$O. The H$_2$O is released and CO$_2$ is reabsorbed, allowing for the constant regeneration of HCO$_3^-$.

The kidney also produces important enzymes and hormones. Renin, previously mentioned in this section, is produced by the kidney and is necessary for the initiation of the renin-angiotensin control of fluid balance. The hormones 1,25-dihydroxycholecalciferol and erythropoietin are also synthesized by the kidney. The active form of vitamin D (1,25-dihydroxycholecalciferol; 1,25-(OH)$_2$D$_3$ is synthesized in the kidney after the inactive direct

Figure 18.3 Filtration Processes along the Renal Tubules

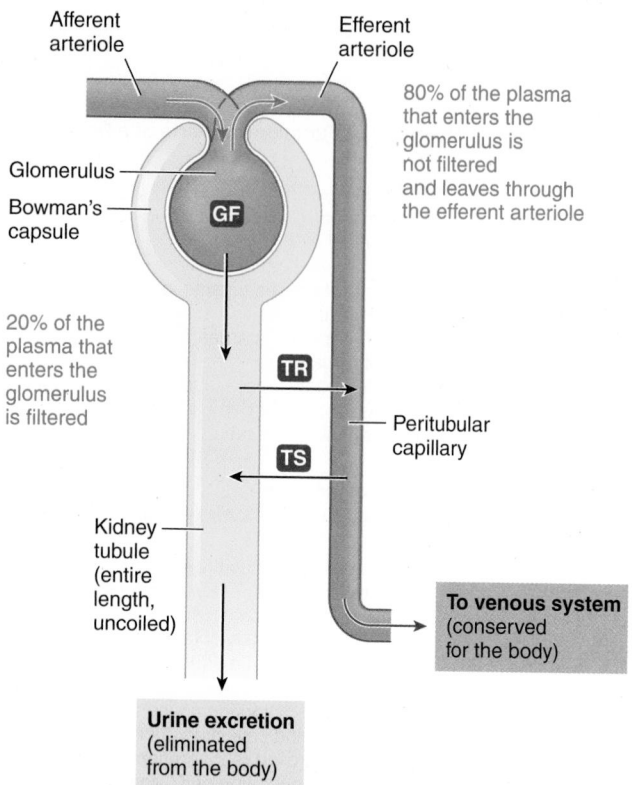

80% of the plasma that enters the glomerulus is not filtered and leaves through the efferent arteriole

Afferent arteriole

Efferent arteriole

Glomerulus

Bowman's capsule

GF

20% of the plasma that enters the glomerulus is filtered

TR

TS

Peritubular capillary

Kidney tubule (entire length, uncoiled)

To venous system (conserved for the body)

Urine excretion (eliminated from the body)

GF = Glomerular filtration—nondiscriminant filtration of a protein-free plasma from the glomerulus into Bowman's capsule

TR = Tubular reabsorption—selective movement of filtered substances from the tubular lumen into the peritubular capillaries

TS = Tubular secretion—selective movement of nonfiltered substances from the peritubular capillaries into the tubular lumen

Source: Sherwood, *Human Physiology: From Cells to Systems* 7e, ISBN 0495391840 Figure 14.6 p. 517.

precursor 25-hydroxycholecalciferol, 25(OH)D₃) is hydroxylated in the liver. Erythropoietin (EPO) is a glycoprotein synthesized in the kidneys that stimulates erythropoiesis (the production of red blood cells) in the bone marrow (see Chapter 19).

Laboratory Evaluation of Kidney Function

Evaluation of kidney function and the diagnosis of kidney disease require biochemical tests and morphological evaluation of the organ's function. Even a simple urinalysis can be evaluated for the appearance of abnormal constituents. For example, the presence of **microalbuminuria** is often one of the first signs of renal insufficiency. Other biochemical tests involve estimation of the kidneys' ability to perform their normal physiological functions. Most commonly, kidney function is measured by the **glomerular filtration rate (GFR)**, which is reflected in clearance tests that measure the rate at which substances are cleared from the plasma by the glomeruli. The normal GFR is 135–200 liters per day. Of this large volume, 98% to 99% of the filtrate is reabsorbed with urine output, usually averaging 1–2 liters per day. The GFR is used to evaluate kidney health,

estimate the severity of diagnosed disease, and monitor kidney disease progression.[6]

Clearance, defined as the volume of plasma cleared of a particular solute in a given time, is expressed in moles, or weight of the substance per volume per time.[7, 8] The mean GFR, expressed in milliliters per minute per 1.73 m², can be calculated as follows:[8–10]

GFR = 122.49 − 0.37 (age) for adults younger than 45 years

GFR = 153.9 − 1.07 (age) for those 45 and older

In the clinical setting, endogenous creatinine clearance was once the "gold standard" used to approximate the actual GFR. It is now thought that approximation of GFR through calculations is the method of choice to evaluate an individual's kidney function. The National Kidney Foundation (NKF) K/DOQI (Kidney Disease Outcomes Quality Initiative) Guideline recommends using equations based on serum creatinine but adjusted for ethnicity, gender, and age.[11] The two equations most frequently cited are the Modification of Diet in Renal Disease (MDRD) Study equation and the Cockcroft-Gault equation.

The Cockcroft-Gault equation considers the effects of age, sex, and body weight on creatinine generation, thereby adjusting serum creatinine values to accurately reflect creatinine clearance:

GFR = [(140 − age) × body weight (kg) × 0.85 if female] ÷ [72 × serum creatinine (mg/dL)]

More recently, the Modification of Diet in Renal Disease (MDRD) modified GFR equation is considered to be the "gold standard" of measurement (in addition to incorporating the influence of age and gender, and the effects of race, three biochemical measures are included):[12]

$$GFR = 170 \times \text{serum creatinine}^{-0.999} \times \text{age}^{-0.176} \times (0.762 \text{ if female}) \times (1.20 \text{ if black race}) \times BUN^{-0.17} \times \text{serum albumin}^{0.320}$$

Web-based GFR calculators, such as the one on the NKF website (http://www.kidney.org), are also available.

Plasma creatinine concentration varies inversely with GFR. The normal range of serum creatinine is 0.8 to 1.2 mg/dL for males and 0.6 to 1.0 for females. Many different laboratory techniques are available for measuring serum creatinine, and the upper limit of the normal range varies significantly. Although the serum creatinine value cannot be used as a measure of GFR, levels start to increase as kidney function decreases. The clinical definition of CKD includes a long-term reduction in GFR, decreased creatinine clearance, and a corresponding increase in serum creatinine concentration.[6] Classifications of CKD correspond with a specific level of GFR and are outlined in the discussion regarding CKD later in this chapter.

Other biochemical assessments may involve tubular function tests. These assessments include evaluation of concentration and dilution, urine acidification, and sodium conservation. Morphological evaluation of the kidney includes microscopic evaluation of the urine (see Table 7.7), radiological evaluation, and biopsy of the organ. Radiological procedures include **intravenous pyelogram (IVP)**, renal ultrasonography, renal radionuclide imaging, computing tomography, MRI, and renal arteriogram.

Pathophysiology Overview

Signs and symptoms associated with inadequate kidney function are directly related to the kidney's inability to perform the normal homeostatic control functions that were discussed in the previous section. Advanced impairment of kidney function results in edema, metabolic acidosis, hyperkalemia, anemia, **uremia**, **azotemia**, hyperphosphatemia, **oliguria**, hypertension, and bone and mineral disorders.

Extracellular fluid volume and plasma volume are largely regulated by sodium and chloride. As kidney function declines below a GFR of 15 mL/min/1.73, the resulting sodium retention and edema lead to hypertension.[13] Metabolic acidosis increases as a result of the kidneys' decreased ability to excrete hydrogen ions.

Under normal conditions, the kidney excretes at least 80% to 90% of the total daily potassium intake, or about 2 to 6 g/day. As GFR begins to decline, the excretion rate of potassium increases; however, as renal function continues to decline, this compensatory mechanism can no longer prevent the accumulation of potassium, which ultimately results in hyperkalemia (elevated blood potassium).

Microcytic anemia and iron deficiency are common in chronic kidney disease. This is due to the kidney's inability to make adequate erythropoietin which stimulates the production of red blood cells. As CKD progresses and kidney function declines, nitrogenous waste excretion declines and blood urea and other nitrogen-containing compounds increase, resulting in azotemia.

In normal individuals, serum calcium is maintained via several regulatory processes: calcium absorption and secretion by the gut, excretion of calcium by the kidney, release of calcium from bone, and calcium deposition in the bone. Parathyroid hormone (PTH) also aids in regulating serum calcium by stimulating **bone resorption** and kidney reabsorption, and by converting the inactive form of vitamin D to the active form $(1,25[OH]_2D_3)$. Deficiency of the active form of vitamin D is associated with impaired intestinal calcium absorption and **secondary hyperparathyroidism**, both of which contribute to the development of bone and mineral disorders (**renal osteodystrophy**).[13] With respect to secondary hyperparathyroidism in CKD patients, it is important to understand the role of the parathyroid gland and PTH in normal individuals. In the healthy individual, the parathyroid gland (PTG) monitors serum calcium and phosphorus levels, identifies conditions of low serum calcium, and responds by synthesizing and secreting PTH into the bloodstream. The circulating PTH targets tissues involved in normalization of blood calcium levels throughout the body. PTH detected by the kidneys will facilitate three processes: (1) calcium reabsorption, (2) phosphorus excretion, and (3) vitamin D activation (1,25-dihydroxycholecalciferol). Activation of vitamin D then increases intestinal absorption of calcium and phosphorus. In addition, PTH targets bone tissue, stimulating osteoclastic activity that results in bone reabsorption and the release of calcium and phosphorus into the bloodstream. All of these processes increase serum calcium levels. Once they are normalized, this same system can relay negative feedback to the PTG and suppress PTH release.

Figure 18.4 CKD Response to Low Serum Calcium and/or High Phosphorus in CKD Stage 5

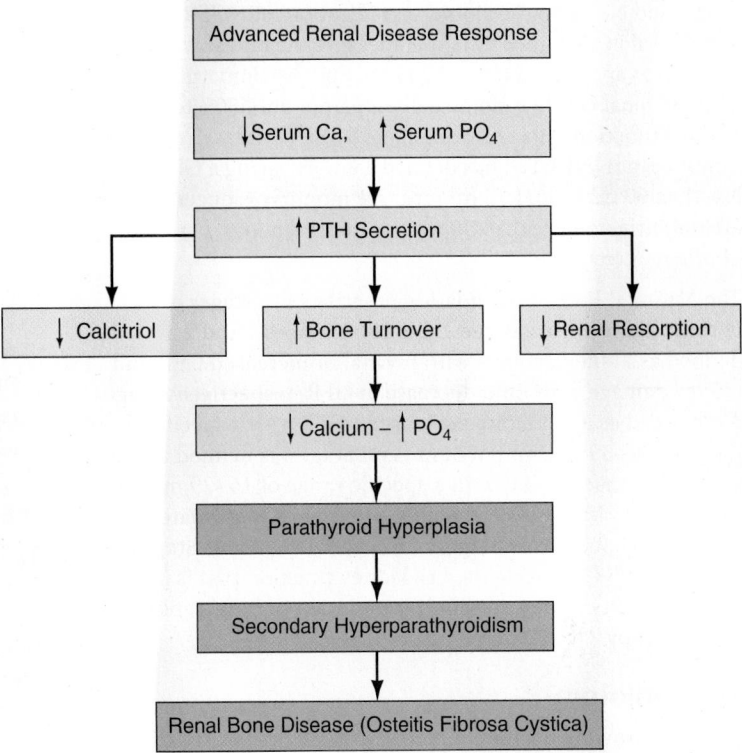

In a person with CKD, the normal PTH feedback loop is disabled, as shown in Figure 18.4. Damaged kidneys no longer have the ability to convert inactive vitamin D into active calcitriol, thus limiting calcium absorption from the intestine. Additionally, damaged kidneys may not be able to reabsorb calcium and excrete phosphorus, leading to an imbalance in the blood. Phosphorus is excreted primarily in the urine (about 500 to 1,400 mg/day), and to a lesser degree in the feces (about 200 to 600 mg/day). Reabsorption of phosphorus in the renal tubules is the main mechanism by which the body maintains a constant level of phosphorus. The kidneys may reabsorb 70% to 100% of the phosphorus they filter, depending on the level of dietary intake. The amount of phosphorus reabsorbed in the kidneys is controlled primarily by PTH. High serum phosphorus levels trigger the release of PTH, which then increases urinary excretion of phosphorus. As a result, the PTH is constantly stimulated with a continued release of PTH. Because no mechanism is operable to prevent PTH release from occurring, serum levels of PTH continue to rise to consistently high levels. As a result, the parathyroid cells increase in number (parathyroid hyperplasia) and lead to secondary hyperparathyroidism and renal bone disease. Additionally, this condition may impact the nervous system, cardiac function, the endocrine system, the immune system, and cutaneous tissues.

Chronic Kidney Disease

Definition and Medical Diagnosis

Chronic kidney disease (CKD) is a syndrome of progressive and irreversible loss of the excretory, endocrine, and metabolic functions of the kidney secondary to kidney damage. CKD progresses slowly over time, and there may be intervals during which kidney functions remain stable. As stated previously (see "Laboratory Evaluation of Kidney Function"), kidney function is assessed based on the glomerular filtration rate (GFR), which is reflected in clearance tests that measure the rate at which substances are cleared from the plasma by the glomeruli. The onset of renal failure is not usually apparent until 50% to 70% of renal function is lost. The National Kidney Disease Education Program (NKDEP) has defined CKD as having a GFR of less than 60 mL/min/1.73 m² for three months or longer and/or albuminuria of more than 30 mg of urinary albumin per gram of urinary creatinine.

The National Kidney Foundation describes five stages of chronic kidney disease (see Table 18.1). Stages 1 and 2 are defined as kidney damage with normal or increased GFR and kidney damage with mild decrease in GFR, respectively. Stage 3 is defined as a moderate decrease in GFR with a specific range of 30–59 mL/min/1.73 m², and Stage 4 is defined as a severe decrease in GFR with a specific range of 15–29 mL/min/1.73 m². In the United States, prevalence is estimated at 7,000,000 people for Stage 3 CKD and 400,000 for Stage 4 CKD. Stage 5 CKD is defined as kidney function that is inadequate to sustain life and requires initiation of renal replacement therapy.[11]

Epidemiology

CKD is a growing health concern. As noted previously, the incidence of CKD is very high among the U.S. adult population; nearly one in every nine adults has CKD and millions more are at risk for developing the condition. An estimated 11.5% of adults ages 20 or older (23 million adults) have evidence of CKD as documented by the National Health and Nutrition Examination Survey.[1] Furthermore, the incidence of CKD is even higher among patients with diabetes mellitus, cardiovascular disease, and hypertension. Between 1990 and 2000, the number of persons with kidney failure requiring dialysis or transplantation more than doubled, and in 2006, more than 500,000 individuals in the United States were receiving dialysis.[2]

Etiology

Diabetes (see Box 18.1), hypertension, and glomerulonephritis are the leading causes of kidney failure; however, there are the following additional causes and risk factors associated with the disease:[2, 14]

- Ethnicity—African-Americans are nearly four times as likely to develop kidney failure as white Americans; Native Americans are nearly two times as likely, and Hispanic Americans have nearly twice the risk of non-Hispanic whites
- Family history—CKD runs in families, so one's risk is greater if a family member has kidney failure
- Hereditary factors such as polycystic kidney disease (PKD)
- A direct and forceful blow to the kidneys
- Prolonged consumption of over-the-counter painkillers that combine aspirin, acetaminophen, and other medicines such as ibuprofen

Treatment

The National Kidney Foundation Kidney Disease Outcomes Quality Initiative (NKF KDOQI) has developed evidence-based clinical practice guidelines for all stages of CKD and related complications since 1997. These guidelines are recognized throughout the world for improving the diagnosis and treatment of kidney disease. To date, there are thirteen evidence-based clinical practice guidelines for CKD care, as follows:

- Diabetes and Chronic Kidney Disease
- Anemia in Chronic Kidney Disease
- Chronic Kidney Disease: Evaluation, Classification and Stratification
- Bone Metabolism and Disease in Chronic Kidney Disease
- Bone Metabolism and Disease in Children with Chronic Kidney Disease
- Hypertension and Antihypertensive Agents in Chronic Kidney Disease
- Managing Dsylipidemia in Chronic Kidney Disease

Table 18.1 Stages of Chronic Kidney Disease

Stage	Description	GFR (mL/min/1.73 m²)	Action*
—	At increased risk	≥60 (with CKD risk factors)	Screening CKD risk reduction
1	Kidney damage with normal or increased GFR	≥90	Diagnosis and treatment
			Treatment of comorbid conditions
			Slowing progression
			CVD risk reduction
2	Kidney damage with mild decrease in GFR	60–89	Estimating progression
3	Moderate decrease in GFR	30–59	Evaluating and treating complications
4	Severe decrease in GFR	15–29	Preparation for kidney replacement therapy
5	Kidney failure	<15 (or dialysis)	Replacement (if uremia present)

Chronic kidney disease is defined as either kidney damage or GFR <60 mL/min/1.73 m² for 3 months or longer. Kidney damage is defined as pathologic abnormalities or markers of damage, including abnormalities in blood or urine tests or imaging studies.

*Includes actions from preceding stages.

Abbreviations: GFR, glomerular filtration rate; CKD, chronic kidney disease; CVD, cardiovascular disease.

Source: K/DOQI Clinical Practice Guidelines for Chronic Kidney Disease: Evaluation, Classification, and Stratification. Guideline 1. Definition and Stages of Chronic Kidney Disease." pp. S7–S10, Copyright 2002, with permission from National Kidney Foundation. Available from: http://www.kidney.org/professionals/kdoqi/guidelines.cfm

BOX 18.1   **CLINICAL APPLICATIONS**

How Does Diabetes Lead to CKD?

Diabetic nephropathy is the most common cause of CKD in the United States. People with either type 1 or type 2 diabetes are at increased risk. The risk is greater if blood sugars are not controlled. The earliest detectable change in the course of diabetic nephropathy is a thickening in the glomerulus, perhaps caused by hyperglycemia and a change in the basement membrane of the tissue. Since the glomerulus is responsible for filtering the blood and the fluid that eventually forms urine, as these glomeruler changes occur, the kidney may start allowing more protein (albumin) than normal to be excreted in the urine. As diabetic nephropathy progresses, increasing numbers of glomeruli are destroyed and increasing amounts of albumin are excreted, which can be detected by a urinalysis. As the number of functioning nephrons declines, each remaining nephron must clear an increasing solute load. Eventually, the limit to the amount of solute that can be cleared is achieved and the concentration in body fluids increases, leading to azotemia and uremia. Because the progression is slow (microalbuminuria can continue up to 5–10 years before other symptoms develop), the body can partially adapt to the changes. At this point, a kidney biopsy clearly shows diabetic nephropathy.

- Nutrition in Children with CKD
- Nutrition in CKD
- Hemodialysis Adequacy
- Peritoneal Dialyisis Adequacy
- Vascular Access
- Cardiovascular Disease in Dialysis Patients (http://www.kidney.org/professionals/kdoqi/guidelines_commentaries.cfm#guidelines)

The goal of medical and nutritional management of kidney disease is to treat the underlying renal pathophysiology in order to delay the progression of the disease. Medical and nutritional care correlates with the level of kidney dysfunction. For example, those individuals with Stage 1 or 2 CKD may initially only require EPO replacement and supplementation of vitamin D. Progression of the disease is highly individualized, and many patients may remain at these initial stages for months to years. However, when CKD progresses to **end-stage renal disease (ESRD** or CKD stage 5) and harmful wastes build up in the blood, blood pressure rises, and excess fluid is retained, more extensive treatment is needed to replace the work of the kidneys. Treatment options include hemodialysis, peritoneal dialysis, and kidney transplantation.

Dialysis   **Dialysis** is a renal replacement procedure that removes excessive and toxic by-products of metabolism from the blood, thus replacing the filtering function of healthy kidneys. It can maintain life once CKD progresses to the end stage, even though endocrine and metabolic functions of the kidney are not totally replaced. The decision to initiate dialysis depends on the severity of symptoms. Unnecessary delay should be avoided in order to prevent medical complications of advanced uremia and subsequent patient debilitation and deterioration. Symptoms considered to be definite indications for dialysis therapy include pericarditis, uncontrollable fluid overload, pulmonary edema, uncontrollable and repeated hyperkalemia, coma, and lethargy. Less severe symptoms such as azotemia, nausea, and vomiting require a subjective determination that takes into consideration the patient's quality of life.

Currently two major types of renal replacement therapy are used for patients with CKD Stage 5: **hemodialysis (HD)** and **peritoneal dialysis (PD)**. The most common method is hemodialysis. Patients and their nephrologist (a physician specializing in kidney disease) choose the type of dialysis based on several factors, including underlying kidney disease and other comorbid factors such as cardiovascular disease, age, family support, and proximity to a dialysis center.

Regardless of the modality, both methods require a selective, semipermeable membrane that allows passage of water and small- to middle-molecular weight molecules and ions but excludes large-molecular weight molecules such as proteins. In hemodialysis, the selective membrane is a man-made dialyzer sometimes referred to as an artificial kidney. The most common types are the hollow fiber and parallel-plate dialyzers. An example of a dialysis system is shown in Figure 18.5. In peritoneal dialysis (see Figure 18.7), the lining of the patient's peritoneal wall serves as the selective membrane. A third group of dialysis methods are termed continuous renal replacement therapies (CRRT). These are used in the acute care setting during acute renal failure or as a temporary treatment mode until the patient is able to start either hemodialysis or peritoneal dialysis.

Waste products and excess fluids are removed from the body by the actions of **diffusion, ultrafiltration,** and **osmosis**. During the removal of unwanted solutes, fluid and electrolyte balance must be maintained. This is accomplished by passing blood across the semipermeable membrane that is exposed to some rinsing fluid (dialysate). **Dialysates** have varying ion and mineral compositions.[15] The dialysate does not come into direct contact with the blood.

HEMODIALYSIS   Since dialysis therapy requires access to the circulatory system, patients receiving hemodialysis first need to undergo a procedure that allows continual access to the bloodstream. The preferred permanent access site is an **arteriovenous fistula (AVF)**, created surgically by fashioning in the forearm a subcutaneous joining of the radial artery and the cephalic vein (see Figure 18.5).[15, 16] If the patient's veins are not adequate for this procedure, an **arteriovenous graft (AVG)** can be created with polytetrafluoroethylene (Teflon) (see Figure 18.6).[16] The AV fistula requires four to six weeks to become fully functional. The subclavian route may be used temporarily if HD is required before the AV fistula is ready for use.

The grafts can be punctured repeatedly by the arterial and venous needlesticks required for each dialysis treatment. Blood travels through a needle placed into the arterial side of the graft. The needle is attached to tubing that leads to the hollow fibers of the dialyzer, or between the sheets of membranes in the parallel plate design.[15] While blood passes through the dialyzer, dialysate simultaneously passes around the artificial

Figure 18.5 Example of a Hemodialysis System

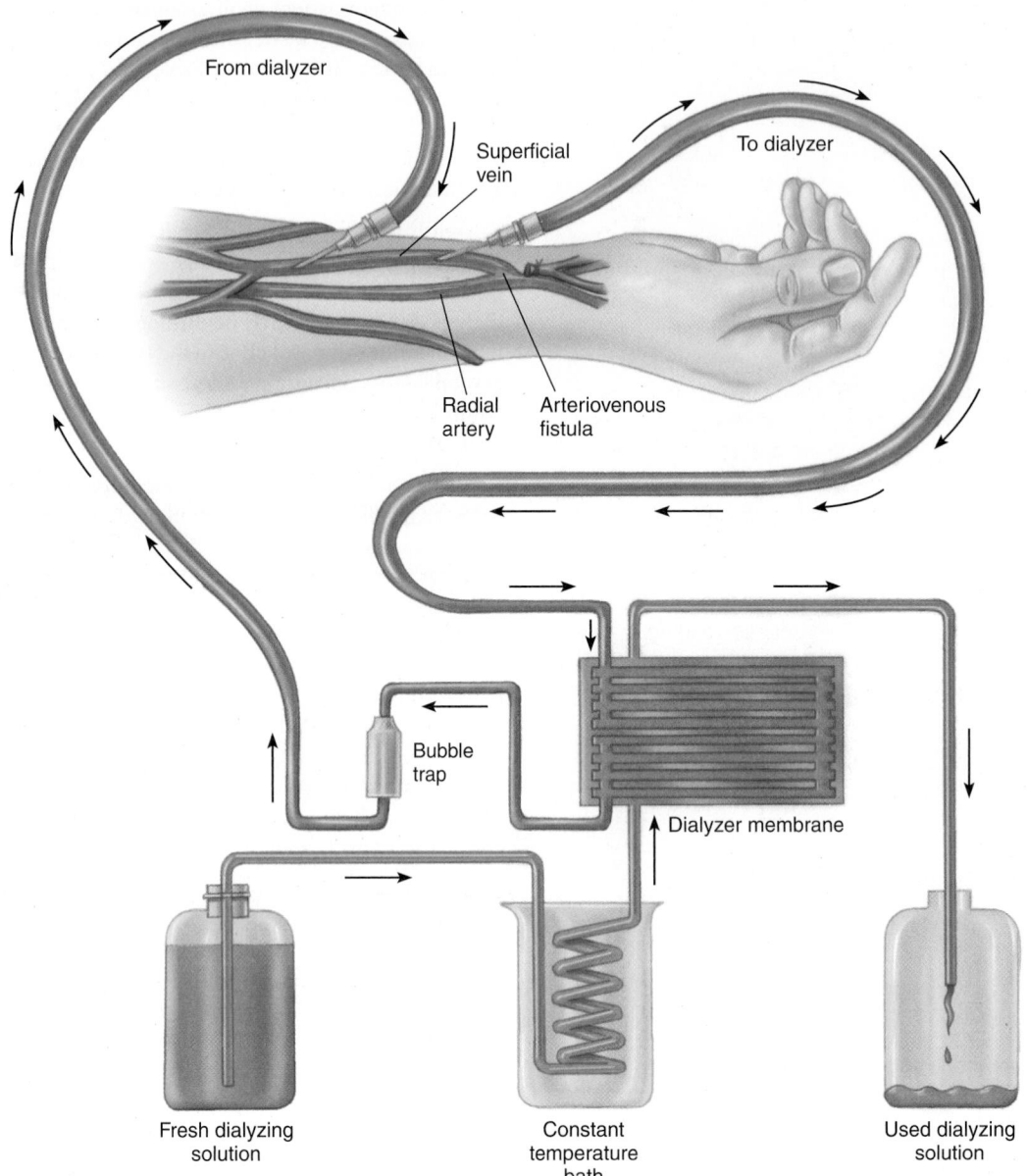

From dialyzer

Superficial vein

To dialyzer

Radial artery

Arteriovenous fistula

Bubble trap

Dialyzer membrane

Fresh dialyzing solution

Constant temperature bath

Used dialyzing solution

Source: R. Rhodes, and R. Pflanzer, *Human Physiology*, 4e, copyright © 2003, p. 782.

membrane. Because the electrolyte content of the dialysate is similar to that of normal plasma, this results in the removal of waste products and excess electrolytes from the blood to the dialysate via diffusion, ultrafiltration, and osmosis. The filtered blood then returns to the patient through the venous side (see Figure 18.5).

Hemodialysis treatments are typically prescribed three times a week for an average of 4 hours per treatment; Medicare covers this regimen. Additional hemodialysis therapies under investigation include short-duration high-efficiency (SDHD) and long-duration, slow, daily nocturnal dialysis (NHD).[17] Multicenter investigations are currently exploring whether these types of dialysis therapies will improve the high morbidity and mortality rates associated with the current hemodialysis treatment. Box 18.2 describes the roles of health care team members in a dialysis unit.

Although most hemodialysis treatments are done at a dialysis center, home treatments can be an option for some patients. Daily home hemodialysis (DHHD) is conducted five to seven days per week for two to three hours at a time, and nocturnal home hemodialysis (NHHD) is performed three to six nights per week during sleep. The major advantage of home dialysis is the ability to set one's own schedule; however, it is necessary to have a trained partner and it may also be stressful to the patient's family.

PERITONEAL DIALYSIS In peritoneal dialysis, access to the patient's blood supply is gained via a catheter of silicone rubber or polyurethane, placed surgically into the peritoneal cavity.[18, 19] In this procedure, which is shown in Figure 18.7, dialysate is introduced into the peritoneum through the peritoneal catheter. Solutes from the plasma circulating in the vessels and capillaries perfusing the peritoneal wall pass across the peritoneal

Figure 18.6 Diagram of an Arteriovenous Graft (AVG)

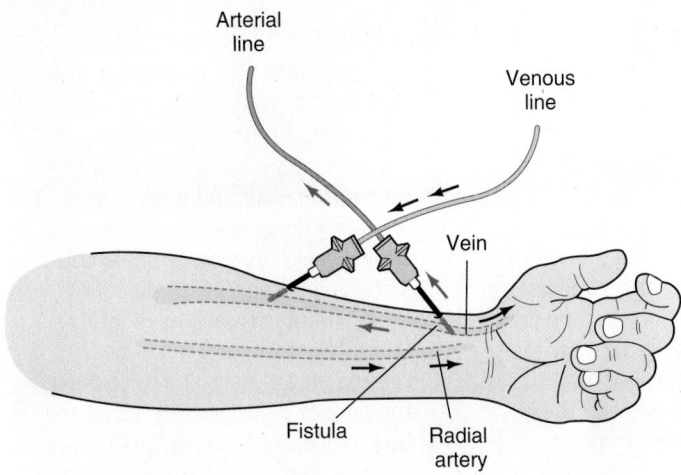

Source: *Handbook of Dialysis*, 3e, Lippincott, Williams & Wilkins.

membrane into the dialysate, which is subsequently removed and discarded.[18, 19]

The dialysate for PD is available with a range of dextrose concentrations that alter its osmolality and assist in fluid removal.[18, 19] In addition, the dwell time (i.e., how long the dialysate remains in the peritoneum) and the number of exchanges (i.e., how many bags of dialysate and the total volume of each used in twenty-four hours) also affect the amount of fluid and solute removal.

There are two main types of peritoneal dialysis (PD): continuous ambulatory peritoneal dialysis (CAPD) and continuous

cycling peritoneal dialysis (CCPD). CAPD requires no machine and can be done in any clean, well-lit location. The usual dwell time is four to six hours, followed by the draining of used dialysate and its replacement with fresh solution requiring an additional thirty to forty minutes. Most patients change the dialysate solution at least four times a day and sleep with the solution in their abdomens at night. CCPD does require a machine called a cycler to fill and empty the abdomen three to five times during the night. One exchange is then done during the day with a dwell time that lasts the entire day. An additional exchange can be added in the afternoon for some patients.[20]

Renal Transplantation Approximately 18,000 transplants are performed annually, while more than 70,000 individuals are waiting for a donor kidney to become available.[2] In order for an organ transplant to occur, the immunological characteristics of the donated organ must be matched with the recipient's medical and immunological characteristics. In Chapter 9, the role of the major histocompatibility complex (MHC) in determining acceptability for a transplanted organ was discussed. The antigens for MHC (often referred to as human leukocyte antigens [HLA]) provide the basis for the MHC haplotype (a combination of closely linked genes on a chromosome inherited as a unit from one parent). MHC antigens play an important role in transplant rejection, since the presence of a MHC antigen on the transplanted organ or tissue that is different from the MHC antigens on the recipient's tissues signals the presence of the transplanted tissue and initiates an immune response. The immune system attacks the transplanted cells presenting MHC antigens that are different from those found on the recipient's tissues. A national database—United Network for Organ Sharing—provides the information and coordination that allow the recipient to be matched with a potential donor.[21]

Figure 18.7 Peritoneal Dialysis

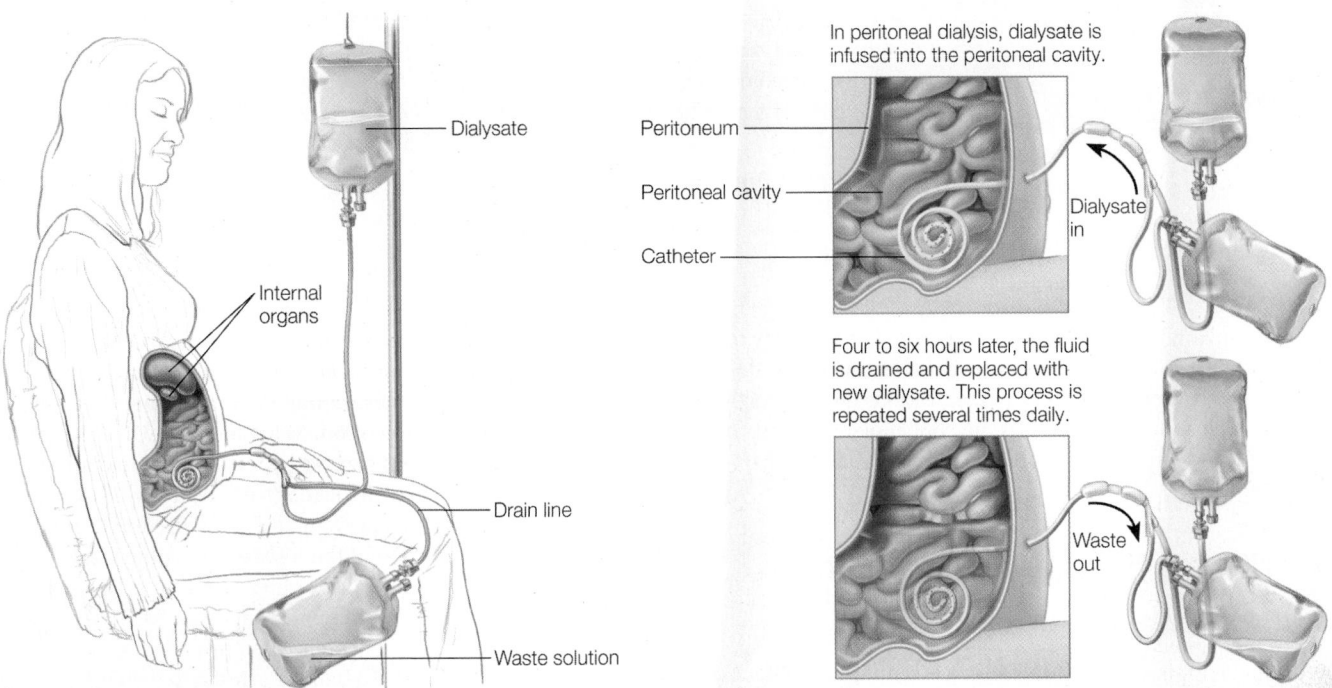

Source: S. Rolfes, K. Pinna, and E. Whitney, *Understanding Normal and Clinical Nutrition*, 7e, copyright © 2006, p. 876.

BOX 18.2 CLINICAL APPLICATIONS

Roles and Responsibilities of the Health Care Team in a Dialysis Unit

- **The Patient:** The CKD patient and/or significant other should be engaged in the overall treatment plan. This means that the patient should seek to understand and follow the instructions provided by the health care team. The patient should also strive to make informed decisions and to assume responsibility for predetermined short-term and long-term goals.

- **Dialysis Nurse:** The nurse is responsible for ongoing assessment of the CKD patient and is usually the individual who initiates multidisciplinary team conferences. Patient and family education, support for self-care, and ongoing reinforcement are critical services provided by the nurse.

- **Dialysis Technician:** Basically, there are two different roles for the technicians: one technician is responsible for assembly and maintenance of dialysis equipment, and the other technician focuses on monitoring the patient during dialysis. Technicians spend the most time with hemodialysis patients and can be a valuable resource for the health care team. Dialysis nurses supervise technicians and delegate the patient care activities to the technicians.

- **Nephrologist:** A nephrologist is an internist with 2–3 years of specialty training in the field of renal disease. The nephrologist determines when a patient has reached end-stage renal disease and thus needs renal replacement therapy (RRT). Typically, a patient is referred early in the continuum of the disease (preferably at CKD Stage 2 or 3) to facilitate disease management and minimize associated comorbidities. With early intervention, appropriate medical and medical nutrition therapy (MNT) can be begun in order to slow the progression of CKD.

  As CKD progresses and the need for RRT becomes imminent, the nephrologist and office nurse will review all of the RRT modalities with the patient so that he or she can make an informed decision regarding the most suitable type of therapy. The nephrologist is then responsible for writing the orders for the dialysis modality chosen and monitoring the patient's progress while on RRT.

- **Nephrology Social Worker:** The primary role of the social worker is to assist the CKD patient and immediate family in their adjustment to the patient's chronic illness. This involves a psychosocial assessment, provision of emotional support, and educational reinforcement. The social worker is a valuable resource for information pertaining to the patient's behavior, history, and functioning that may influence the patient's care and course of treatment.

- **Renal Dietitian:** The dietitian is primarily responsible for nutritional assessment and nutrition therapy. Nutrition therapy includes developing a plan for nutritional management, counseling the patient, monitoring nutrient intake, and providing educational activities and ongoing reinforcement to maximize dietary adherence. One aspect of assessment is to review monthly laboratory tests and nutrition-related parameters and to counsel the patient appropriately. Laboratory indicators relating to bone and mineral disorders, anemia, and adequacy of dialysis are also reviewed, and appropriate recommendations are made to the nephrologist.

In some cases, the role of the RD in a dialysis facility is expanded beyond the traditional role outlined here. Some renal dietitians monitor protocols and make recommendations on anemia and bone and mineral disorder therapies. In some instances, they are instrumental in leading continuous quality improvement efforts.

Although randomized prospective controlled trials have not been done, it has been suggested that a full-time RD should be responsible for the care of approximately 100 hemodialysis patients—not to exceed 150 patients. Administration should strongly consider a higher ratio in the case where the RD is involved in many other nontraditional roles.

Source: National Kidney Foundation Renal Dietitian Standard's for Clinical Practice Monograph 2004.

After transplantation, patients are maintained on a variety of immunosuppressive regimens to prevent rejection of the donated kidney. Immunosuppressive medications include corticosteroids, cyclosporine, tacrolimus, mycophenolate mofetil, and sirolimus. These medications are reviewed later in this chapter (see Table 18.17).

Nutrition Therapy for Chronic Kidney Disease

Medical nutrition therapy (MNT) is an essential component of medical care for early, progressive, and end-stage CKD. Malnutrition, cardiovascular disease, bone and mineral disorders, and anemia are the most common complications that accompany kidney disease. Each of these complications requires both medical and nutritional intervention. Many patients with CKD also have comorbid conditions such as hypertension and diabetes mellitus that also require medical nutrition therapy. Nutrition therapy for CKD can help prevent and manage complications as well as slow progression of the disease and compensate for impaired renal function and/or limitations of treatment modalities. In recognition of their significance, MNT services are a distinct reimbursable benefit when provided by registered dietitians to Medicare part B beneficiaries with diabetes or renal disease.

Nutrition Assessment

Malnutrition is very common in patients with CKD, especially those who are on dialysis. Evidence suggests that the prevalence of protein-energy malnutrition (PEM) ranges from approximately 20% to 70% among adult dialysis patients. Signs and symptoms of malnutrition present themselves when the GFR declines to 30 mL/minute, and may progress to significant malnutrition at a GFR of 10 mL/minute. Those patients who have PEM prior to dialysis have an increased morbidity and mortality, whereas those who do not exhibit PEM are better able to maintain their healthy status. Therefore, it is extremely important to evaluate the many complex factors that contribute to poor nutritional status in patients with CKD as part of a comprehensive nutrition assessment. Table 18.2 highlights the various factors that contribute to the increased incidence of PEM in these patients.

The dietary restrictions associated with CKD add to the potential for inadequate nutrient intake. It is especially important to assess for usual dietary patterns and food intake, intolerance of specific foods, nutrient restrictions of the diet, and fear of eating "wrong" foods. Careful assessment of anthropometric measurements can also assist in the identification of nutrition problems and PEM.

In summary, a nutrition assessment of the CKD patient should include:

- Review of the medical record for comorbid conditions, drug/nutrient interactions, or potential issues with digestion and absorption of nutrients
- Nutrition interview for usual food/nutrient intake as well as changes in appetite, changes in food intake with the onset/progression of CKD, and changes in elimination (urine output and stool)
- Interview to evaluate social barriers to adequate nutritional intake
- Physical assessment, including height, weight, frame size, subjective global assessment (see Chapter 3), and physical signs/symptoms of nutrient deficits
- Review of biochemical/laboratory indices that might be affected by CKD (see Table 18.3)
- Assessment of current food intake—kcalories, protein, fat, sodium, potassium, calcium, phosphorus, fluid, vitamins, and minerals

Nutrition Diagnosis

Given the complexity of nutrition implications associated with CKD, many nutrition diagnoses may be present. The following is a list of possible nutrition diagnoses for patients with CKD:

- Inadequate energy intake
- Inadequate oral/food beverage intake
- Excessive fluid intake
- Malnutrition
- Excessive protein intake
- Excessive mineral intake (potassium, phosphorus, sodium)
- Altered GI function
- Altered nutrition-related laboratory values

Table 18.2 Multiple Factors Leading to Malnutrition in CKD

Inadequate food intake due to physiological factors:
- Nausea and vomiting
- Diabetic retinopathy resulting in impaired vision
- Taste alterations
- Fatigue
- Anorexia caused by uremia
- Emotional distress
- Unpalatable diet
- Concurrent medications resulting in anorexia

Inadequate food intake due to social barriers:
- Limited income
- Inability to prepare foods and meals
- Living and eating alone
- Depression
- Lack of motivation
- Lack of family support
- Missed meals due to travel and/or treatments

Endocrine disorders associated with uremia, such as:
- Hyperparathyroidism
- Hyperglucagonemia
- Resistance to the actions of insulin and IGF-1

Dialysis procedure, which causes:
- Protein loss due to amino acid removal
- Inflammatory response with subsequent increased protein requirements (this is due to use of a bio-incompatible membrane)

Loss of blood due to:
- GI bleeding
- Frequent blood sampling
- Blood lost in the dialyzer and tubing

Metabolic acidosis results in:
- Increased protein catabolism

Source: Adapted from: National Kidney Foundation. Diabetes, Nutrition and Chronic Kidney Disease: New Guidelines for a New Clinical Approach. Available at http://www.kidney.org/professionals/kls/cmeprograms.cfm

- Food-medication interaction
- Involuntary weight loss
- Involuntary weight gain
- Food and nutrition-related knowledge deficit
- Disordered eating pattern
- Limited adherence to nutrition-related recommendations
- Undesirable food choices
- Impaired ability to prepare foods/meals
- Poor nutrition quality of life
- Limited access to food

Nutrition Intervention

CKD is not a static medical condition. As noted in Table 18.1, the disease progresses through a series of stages. The nutrition goals and interventions therefore need to be individualized based on the severity of nutrition diagnoses, progression of

Table 18.3 Biochemical/Laboratory Indices Used in Nutrition Assessment in CKD

Test	Ref. Range	CKD Range	Significance of Abnormal
Albumin	3.5–5.0 g/dL **SI units:** 35–50 g/L	WNL for the laboratory or ideal >4.0	**High:** severe dehydration, albumin infusion **Low:** fluid overload, chronic liver/pancreatic disease, steatorrhea, nephrotic syndrome, protein-energy malnutrition, inflammatory GI disease, infection
Alkaline Phosphatase	30–85 IU/L **SI Units:** 42–128 U/L	WNL	**High:** renal osteodystrophy, healing of fractures, malignancies **Low:** congenital hypophosphatemia, possibly in kwashiorkor, general debility, anemia, nephrotic syndrome
Aluminum	<7 µg/L	<21 µg/L >60 perform DFO test	**High:** Ingestion of aluminum-containing medications Other potential sources: parenteral fluids, injections, antiperspirants, dialysate
Ammonia Levels	15–110 µg/dL **SI Units:** 47–65 µmol/L	WNL	**High:** primary hepatocellular disease, Reye's syndrome, portal HTN, GI bleeding/obstruction w/mild liver disease **Low:** essential or malignant hypertension
B₁₂	100–700 pg/ml **SI Units:** 74–517 pmol/L	WNL	**High:** leukemia, polycythemia vera, severe liver dysfunction **Low:** pernicious anemia, atrophic gastritis, malabsorption syndrome, inflammatory bowel disease, Zollinger-Ellison syndrome, achlorhydria, pregnancy, vitamin C or folic acid deficiency
Blood Urea Nitrogen (BUN)	10–20 mg/dL **SI Units:** 3.6–7.1 mmol/L	60–80 mg/dL (anuric, well dialyzed and eating adequate protein)	**High:** w/excessive protein intake. GI bleeding, dehydration, hypercatabolism, CHF (a ↓ in cardiac output causes a ↓ GFR), transplant rejection, inadequate dialysis **Low:** hepatic failure, over-hydration, acute low protein intake, malabsorption, ↑ secretion of anabolic hormones
Serum Calcium* *May be adjusted for low albumin, but value is questionable	9.0–10.5 mg/dL **SI Units:** 2.25–2.75 mmol/L	WNL (low end)	**High:** excess vit D/calcium, ↑ GI absorption, osteolytic disease, excess vit A, carcinoma, immobilization, primary SHPT, ABD, dehydration, prolonged **Low:** insufficient vit D, during bone building, malabsorption, post-parathyroidectomy, long-term Dilantin therapy, hypoparathyroidism with low albumin (lack of carrier), but ionized is usually WNL
Ceruloplasmin (Cp)	23–43 mg/dL	WNL	**High:** acute inflammatory response, cancer, biliary cirrhosis, pregnancy, copper intoxication **Low:** nephrotic syndrome, infants, kwashiorkor, sprue, hyperalimentation
Carbon Dioxide CO₂	23–30 mEq/L **SI Units:** 23–30 mmol/L	WNL ≥22	**High:** metabolic alkalosis **Low:** metabolic acidosis
Chloride	100–106 mEq/L **SI Units:** 98–106 mmol/L	WNL	**High:** excess salt, dehydration, some forms of metabolic acidosis, excessive use of chloride-containing meds, primary hypoparathyroidism **Low:** diabetic acidosis, K⁺ deficiency, metabolic alkalosis, excessive sweating, starvation, abnormal GI losses, chronic pyelonephritis, dilution (fluid excess), chloride is affected by the same conditions as sodium moves in same direction
Cholesterol	<200 mg/dL **SI Units:** <5.2 mmol/L	WNL	**High:** high chol/saturated fat diet, disorders, of lipid metabolism, nephrotic syndrome, glucocorticoid use **Low:** acute infection, starvation. PEM
CHr Reticulocyte Hemoglobin Content	24.5–31.8 pg/cell	>29 pg/cell	**High:** iron supplementation **Low:** iron deficiency
Creatinine	0.5–1.1 mg/dL ♀ 0.6–1.2 mg/dL ♂) **SI Units:** 44–97 µmol/L ♀ 53–106 µmol/L ♂	2–15 mg/dL (based on muscle mass, GFR and/or dialysis clearance)	**High:** muscle damage, catabolism, MI, muscular dystrophy, ARF/CKD, use of cephalothin/cimetidine, excess protein intake, inadequate dialysis, transplant rejection **Low:** in chronic dialysis <10 may indicate PEM/wasting of muscle
C-Reactive Protein	<0.8 mg/dL **SI Units:** N/A	WNL	**High:** arthritis, Crohn's disease, lupus, tissue infarction or damage, acute MI, kidney or bone marrow transplant rejection, soft tissue trauma, bacterial infection, postoperative wound infection, UTI, TB, malignant disease
Ferritin	12–300 ng/mL ♂ 10–150 ng/mL ♀ **SI Units:** 12–300 µg/L ♂ 10–150 µg/L ♀	**HD** ≥200 ng/mL **PD/CKD** >100 ng/mL >500 unknown benefit/harm	**High:** iron overload, many transfusions, dehydration, inflammatory state, falsely elevated in active liver disease **Low:** iron deficiency

Test	Ref. Range	CKD Range	Significance of Abnormal
Folic Acid	5–20 μg/mL **SI Units:** 14–34 mmol/L	WNL	**High:** pernicious anemia, recent massive blood transfusion, vegetarianism **Low:** folic acid deficiency, hemolytic anemia, malnutrition, malabsorption, malignancy, liver disease, pregnancy, alcoholism, anorexia nervosa
Glucose (Fasting)	70–105 mg/dL **SI Units:** 3.9–5.8 mmol/L Peak post prandial capillary glucose <180 mg/dL (<10 mmol/L)	WNL <200 non-fasting	**High:** DM, chronic hepatic disease, hyperthyroidism, malignancy, acute/emotional stress, burns, diabetic acidosis, pancreatic insufficiency, glucose intolerance **Low:** hyperinsulinemia, ETOH abuse, pancreatic tumors, liver failure, pituitary dysfunction, malnutrition, extreme exercise
Hematocrit	42–52% ♂ 37–47% ♀ **SI Units:** 0.42–0.52 ♂ 0.37–0.47 ♀ volume fraction	33%–36% <39%	**High:** polycythemia, dehydration **Low:** anemias, blood loss (endogenous & dialysis), CKD, insufficient ESA
Hemoglobin	14–18 g/dL ♂ 12–16 g/dL ♀ **SI Units:** mmol/L 8.7–11.2 ♂ 7.4–9.9 ♀	Variable 10–12 g/dL <13g/dL FDA ≤12 g/dL	**High:** dehydration **Low:** over-hydration, prolonged iron deficiency, anemias, blood loss, CKD
Hemoglobin A$_{1c}$ (Glycosolated hemoglobin GHb, GHB)	Adult: 4–8% "Good" control <7% "Fair" control 10% <6 not desirable **SI Units:** N/A	WNL <7%	**High:** newly diagnosed/poorly controlled DM, splenectomy, pregnancy, non-diabetic hyperglycemia **Low:** hemolytic anemia, chronic blood loss, early CKD
Phosphorus	3.0–4.5 mg/dL **SI Units:** 0.97–1.45 mmol/L	WNL 3.5–5.5 mg/dL (KDIGO rec "towards normal")	**High:** CKD, osteodystrophy, vit D intoxication, diurnal rhythm—evening or afternoon as much as 2x the am level, excessive intake, inadequate P binder **Low:** vit D deficiency, low intake, excess P binding, malabsorption/diarrhea/vomiting, alkalosis, diabetic acidosis, diuretic therapy, alcoholism, refeeding syndrome, post parathyroidectomy, osteomalacia
Potassium	3.5–5.0 mEq/L **SI Units:** 3.5–5.0 mmol/L	WNL 3.5–6.0 mEq/L	**High:** CKD, tissue destruction, shock, acidosis, dehydration, hyperglycemia, aldosterone antagonistic overuse, diuretics, false ↑ w/tourniquet, excessive oral intake, inadequate dialysis, inappropriate dialysate K^+, compression/fist clenching prior to sample **Low:** diuretic therapy, ETOH abuse, diarrhea/vomiting/laxative or enema abuse, malabsorption, correction of diabetic acidosis
Prealbumin/ Transthyretin	15–36 mg/dL **SI Units:** 150–360 mg/L	≥30 mg/dL	**High:** administration of corticoids **Low:** neonate, liver disease, malnutrition, inflammation
Protein, Total	6.4–8.3 g/dL **SI Units:** 64–83 g/L	WNL	**High:** dehydration, acute/chronic infectious disease, leukemia/multiple myeloma **Low:** malnutrition, malabsorption, cirrhosis, steatorrhea, edema, nephrotic syndrome
Intact Parathyroid Hormone (iPTH)	Intact: 10–65 pg/mL **SI Units:** N/A	KDOQI: 150–300 pg/mL KDIGO: between 2 and 9× normal limit; avoid extremes	**High:** hyperparathyroidism, non-PTH producing tumors, lung or kidney cancer, hypocalcemia, malabsorption, vit D deficiency, rickets **Low:** hypoparathyroidism, hypercalcemia, metastatic bone tumor, sarcoidosis, vit D intoxication, hypomagnesemia
	**Editorial Note:** Variance between third and second generation iPTH depends on the patient-specific level of PTH fragment (7–84) that is measured by the second generation iPTH. Observational studies suggest PTH levels that may increase relative risk of death: however, there are no RCTs that show correction of PTH to a specific level absolutely translates into improved patient level outcomes. KDIGO stresses that there is insufficient research to establish absolute PTH targets, but suggests avoiding extremes and using trends rather than single measurements to guide therapy.		
Reticulocyte Count	0.5%–2%	Variable in response to EPO	Index of bone marrow activity; reflects early change in RBC production **High:** hemolytic anemia, acute bleed **Low:** certain anemias due to ineffective erythropoiesis (defic. of iron, B_{12}, folic acid, B_6) or anemia of chronic disease

(continued)

Test	Ref. Range	CKD Range	Significance of Abnormal
RBC Count Multiply automatic counter values × 1 million for total #	million/mm³ 4.7–6.1 ♂ 4.2–5.4 ♀ **SI Units:** N/A	WNL	**High:** high altitude, temporarily w/strong emotion, diurnally, cold shower, reduced plasma volume, dehydration **Low:** anemia, hemorrhage, infectious disease, iron deficiency
Sodium	136–145 mEq/L **SI Units:** 136–145 mmol/L	WNL	**High:** dehydration, diabetes insipidus, often masked by water retention **Low:** overhydration, inappropriate ADH diuretic use, burns, starvation, adrenal insufficiency, nephritis, hyperglycemia, diabetic acidosis, hyperproteinemia
TIBC Transferrin = (0.8 × TIBC) − 43	250–420 µg/dL **SI Units:** 45–73 µmol/L	WNL Varies with iron stores	**High:** chronic iron deficiency, acute hepatitis, pregnancy, alcoholism **Low:** cirrhosis, malnutrition, collagen or chronic disease/infection/inflammation
Transferrin	Adult: mg/dL 215–365 ♂ 250–380 ♀ **SI Units:** N/A	WNL	**High:** chronic iron deficiency, acute hepatitis, pregnancy, alcoholism **Low:** cirrhosis, malnutrition, collagen or chronic disease/infection/inflammation
Transferrin Saturation	20%–50% ♂ 15%–50% ♀ **SI Units:** N/A	≥20%	**High:** iron overload, pregnancy, acute hepatitis, alcoholism **Low:** cirrhosis, malnutrition, collagen or chronic disease/infection, iron deficiency
Triglycerides	40–160 mg/dL ♂ 35–135 mg/dL ♀ **SI Units:** 0.45–1.81 mmol/L 0.40–1.52 mmol/L	WNL <200 mg/dL	**High:** liver disease, gout, pancreatitis, ETOH abuse, MI, diabetes, PD, use of steroids, nephrotic syndrome **Low:** malnutrition, malabsorption

Source: Reprinted from: National Kidney Foundation. Pocket Guide to Nutrition Assessment of the Patient with Chronic Kidney Disease (CKD) 2009. Fourth Edition [Revised] Online, National Kidney Foundation, New York, NY. "Laboratory Tests" table.

CKD, and overall health status of the patient. The following section of this chapter discusses the general principles of nutrition therapy for the various stages of CKD as well as more in-depth nutrient recommendations frequently provided for patients receiving renal replacement therapies (dialysis or transplantation).

CKD Stages 1 and 2 Specific nutrition goals for Stages 1 and 2 have not been identified, however nutrition therapy should focus on the comorbid conditions—diabetes, hypertension, and hyperlipidemia—and on slowing the progression of potential cardiovascular disease. Glucose control, blood pressure reduction, and lipid management all have well-documented nutrition guidelines.[11] Refer to Chapters 13 and 17.

Nutrition counseling recommendations include regular assessment (at 1–3 month intervals) of the patient's nutritional status. Interventions that maintain or improve nutritional status during the progression of kidney disease are associated with improved patient survival.

CKD Stages 3 and 4 The panel of experts that comprised the K/DOQI Nutrition Work Group developed three clinical practice guidelines specifically for adults with CKD who are not on dialysis (Stage 4). These guidelines outline recommended nutrition measures, protein intake, energy requirements, and nutrition counseling.[11] Nutrition measures identified for individuals with GFR <20 mL/min include:

- Serum albumin and actual or percent standard body weight and/or subjective global assessment (SGA) every 1 to 3 months
- Dietary interviews and food intake records and/or **normalized protein equivalent of nitrogen appearance (nPNA)** every 3 to 4 months

The American Dietetic Association has also developed guides for the CKD patient (Stages 3 and 4). These practice guidelines outline outcome assessment factors (biochemical, hematological, anthropometric, and clinical signs and symptoms), expected outcome of therapy, ideal goal of each of the clinical assessments, patient/caregiver behavioral outcomes, and NT goals.[22] ADA's evidence-based practice guidelines for CKD are currently being updated to include the most recent recommendations.

Recommendations for nutrition intervention for patients with CKD (GFR <25 mL/min) who are not currently undergoing dialysis primarily come from the Modification of Diet in Renal Disease Study and subsequent meta-analysis papers. The most recent evidence analysis indicates that low-protein diets are associated with a slowed progression of renal disease and a delayed need for renal replacement therapy, but there is not a consistent agreement regarding the level of protein restriction that is effective.[23]

The purpose of medical nutrition therapy (MNT) in Stages 3 and 4 is to provide adequate kcal to prevent malnutrition; provide adequate protein to preserve muscle mass and serum proteins; treat abnormalities of vitamin and mineral absorption, utilization, and excretion present in CKD; and normalize blood lipids.

As kidney function deteriorates, the ability of the kidney to excrete metabolic products of protein, regulate acid/base balance, produce adequate amounts of erythropoietin, activate vitamin D, and regulate calcium, phosphorus, potassium, sodium, and fluid excretion diminishes. Therefore, nutrients that are usually affected by CKD include protein, energy, sodium, potassium, phosphorus, calcium, vitamins, minerals, and fluid. As a result,

BOX 18.3 CLINICAL APPLICATIONS

Unique Challenges in Evaluating Protein-Energy Nutritional Status in Patients on HD and PD

As already noted, many patients with CKD, especially those on HD and PD, have protein-energy malnutrition resulting from a wide variety of factors. Anthropometric measurements, biochemical assessment values, and nutrient intake records provide important nutrition assessment data to verify the cause and extent of nutrition problems, especially malnutrition. Given the complexity of CKD and its nutritional consequences, interpretation and evaluation of laboratory results and protein intake require unique critical thinking skills.

Serum albumin is a measure of both visceral protein status and acute-phase inflammatory response, and there is evidence that it is a predictor of morbidity and mortality in individuals with CKD. Serum albumin can be adversely affected by fluid status, inflammation, urinary protein losses, changes in kidney function, recent surgery, and other factors. These conditions must be taken into consideration before assuming an albumin value (normal or abnormal) is indicative of nutritional status. A recent study investigated the effects of inflammation and nutritional status on the association between serum albumin and mortality in patients on HD and PD. It was concluded that in dialysis patients, a 1-gram/dL decrease in serum albumin was associated with an increased mortality risk of 47% in HD patients and 38% in PD patients. The changes in serum albumin associated with mortality risk were influenced to a greater degree by inflammatory markers than as a consequence of malnutrition. The authors concluded that nutritional status cannot be assessed with precision by the measurement of serum albumin alone in dialysis patients.[1]

Although prealbumin is not identified in the 2000 NKF K/DOQI Nutrition Practice Guides as a nutritional index for the patient with CKD, many practitioners use this marker. A recent study indicates that prealbumin is an important marker of patient outcome.[2] The advantage is the short half-life of prealbumin (transthyretin) as compared to albumin (3 days vs. 21 days). However, prealbumin also has limitations; it is an acute-phase reactant (it decreases in response to inflammation and infection), it may not be well correlated with other nutritional markers, and it is increased with CKD.

Since the protein requirement in CKD patients on HD and PD is generally higher than in healthy individuals, it is important to accurately assess protein status in these patients. During steady-state healthy conditions (where nitrogen balance is neutral) the net amount of protein catabolized is nearly the same as the amount of protein ingested. The sum of all nitrogen losses plus the change in body nonprotein nitrogen (mainly urea nitrogen) is expressed as total nitrogen appearance (TNA). The TNA reflects the net breakdown of protein, which again in the steady state should be nearly equal to the intake of nitrogen, of which protein is the major source. In patients on HD and PD, additional losses of nitrogen in the dialysate must also be considered; therefore, the TNA in CKD patients is equal to the sum of dialysate, urine, and fecal nitrogen losses, and the postdialysis increment in body urea-nitrogen content. Because the nitrogen content of protein is relatively constant at 16%, the protein equivalent to total nitrogen appearance (PNA) can be estimated by multiplying TNA by 6.25 (PNA is mathematically identical to the protein catabolic rate or PCR). PNA is a valid and clinically useful measurement of net protein degradation and protein intake in a stable dialysis patient. PNA is expressed in terms of grams of protein per day (g/day). It should be noted that PNA may be affected by protein intake and anabolic and catabolic factors. If the patient is catabolic, PNA will be overestimated, and in anabolism, PNA will underestimate actual protein intake.

Because protein needs are determined by edema-free and fat-free body mass, PNA is typically normalized (nPNA) to some function of body weight. Normalized protein equivalent of total nitrogen appearance (nPNA) is a measure of protein nitrogen appearance and is expressed in terms of grams of protein per kilogram per day (g/kg/day). nPNA can be used to estimate dietary protein intake and can be calculated using the formula below. nPNA is closely correlated to dietary protein intake in the steady state or when protein and energy intake are relatively constant (<10% variance). PNA is calculated automatically as part of routine **urea kinetic modeling** results. nPNA requires a 24-hour urine collection.[3]

Protein Equivalent of Total Nitrogen Appearance (PNA), Previously Referred to as Protein Catabolic Rate (PCR):

$$nPNA \ (g/day) = (6.25 \times (\text{urinary urea N } (g/day) \times 0.031 \text{ weight } (kg)) \times \text{urinary protein losses } (g/day)$$

References
1. de Mutsert R, Grootendorst DC, Indemans Fleur, Boeschoten EW, Krediet RT, Dekker FW, for the Netherlands Cooperative Study on the Adequacy of Dialysis-II Study Group. J Ren Nutr. 2009;19:127–35.
2. Wiggins KJ, Johnson DW. The influence of obesity on the development and survival outcomes of chronic kidney disease. Adv Chronic Kidney Dis. 2005;12:49–55.
3. National Kidney Foundation. NKD K/DOQI Guidelines Adult guidelines #8 Maintenance Dialysis 2000 and Calculation of the protein equivalent of total nitrogen appearance from urea appearance: which formula should be used. Peritoneal Dialysis International 2000;18:467–73.

Table 18.4 Nutrient Recommendations for CKD

	Acute	CKD	Nephrotic Syndrome	HD	PD
Protein g/kg[1]	0.6–0.8 based on renal function/treatment	0.60–0.75 g/kg ≥50% HBV	0.8–1.0 Replacing urine loss is controversial	≥1.2 ≥50% HBV	≥1.2–1.3 ≥50% HBV
Energy kcal/kg[1]	35–50 based on stress/ nutrition status	30–35 >60 yrs 35 <60 yrs	35 unless obese; complex CHO; ↓ cholesterol; <30% fat	30–35 >60 yrs; 35 <60 yrs	30–35 >60 yrs; 35 <60 yrs including dialysate
Na⁺ g/day[2]	1–2 based on BP, edema; replace losses in diuretic phase	Varies from 1–3 to no added salt	1–2	2	2; monitor fluid balance
K⁺ g/day	Maintain serum <5 mEq/L; replace losses in diuretic phase	Usually unrestricted unless serum level is high	Usually unrestricted	2–3 adjust to serum levels	3–4 adjust to serum levels
Phosphorus mg/g pro or mg/day	Maintain serum P WNL	800–1000 mg/d 10–12 mg/g pro Maintain serum P and PTH WNL	Maintain serum P WNL	800–1000 mg/d adjust to meet pro needs; 10–12 mg/g pro	800–1000 mg/d adjust to meet pro needs; 10–12 mg/g pro
Calcium g/day	Maintain serum WNL	DRI; maintain serum levels WNL	Same as predialysis	<2.0 g including binder load; maintain serum levels WNL	<2.0 g including binder load; maintain serum levels WNL
Fluid cc/d	Output plus 500 cc	Usually unlimited	Maintain balance	Output + 1000 cc Limit ID wt gain	Maintain balance
Vitamins/ Minerals (Daily)	DRI: adjust to degree of catabolism; TPN may require MVT and minerals	DRI: B-complex & C; ensure adequate nutritional vitamin D; give 1,25 vitamin D as needed to control PTH; individualize iron, zinc	Same as CKD	C 60–100 mg; B_6 2 mg; folate 1–5 mg: B_{12} 3 µg/d; RDA others; vit E 15 IU/d; zinc 15 mg/d; individualize iron and vit D; replete nutritional D	Same as HD but may need 1.5 to 2 mg of B_1 due to dialysis loss; replete nutritional vitamin D; individualize 1,25 vitamin D as needed for control of SHPT
Fiber	N/A	20–30 g	N/A	20–25 g	20–25 g

CKD Stage 4 = GFR 15–29 mL/min /1.73 m²

[1]Based on standard or adjusted body weight.

[2]Note: A 1 g sodium restriction is generally difficult to achieve in an outpatient setting.

Source: Kopple ID and Massry SG, eds. *Nutritional Management of Renal Disease.* Philadephia: Lippincott Williams & Wilkins; 2004. Mitch WE and Klahr S, eds. *Handbook of Nutrition and the Kidney.* 5th ed. Philadelphia: Lippincott Williams & Wilkins; 2005. NKF K/DOQI Clinical Practice Guidelines for Nutrition in Chronic Renal Failure. *Am J Kidney Dis.* 2000;35:6(suppl 2):SI-S140. NKF KDOQI Clinical Practice Guidelines for Bone Metabolism and Disease in Chronic Kidney Disease. *Am J Kidney Dis.* 2003;42:4(suppl 3)SI-S201.

Reprinted from: National Kidney Foundation. Pocket Guide to Nutrition Assessment of the Patient with Chronic Kidney Disease (CKD) 2009, Fourth Edition [Revised] Online, National Kidney Foundation, New York, NY. "Daily Nutrient Recommendations for CKD" table.

modifications in these nutrients are frequently necessary. See Table 18.4 for the nutrient recommendations for individuals at various stages of kidney disease.

Meal planning can be a challenge for patients with CKD. Emphasizing the use of usual foods and recipes with modifications to meet nutrient needs allows for an individualized approach (see Box 18.4). Table 18.5 provides an example of a two-day menu for a patient with Stage 3 CKD.

CKD Stage 5 The goals of nutrition therapy for patients with CKD Stage 5 (refer to Table 18.1 for the NKF K/DOQI classification stages of CKD) are to meet nutritional requirements, prevent malnutrition, minimize uremia and associated CKD complications (cardiovascular disease, anemia, secondary hyperparathyroidism), and maintain blood pressure and fluid status. Nutrition therapy for CKD patients receiving different dialysis modalities are reviewed in the sections that follow.

In general, the hemodialysis diet is high in protein and controls intake of potassium, phosphorus, fluids, and sodium. Based on nutritional requirements, additional modifications may be needed for fat, cholesterol, and triglycerides. Patients receiving

peritoneal dialysis have a more liberalized diet than hemodialysis patients; this diet is higher in protein, sodium, potassium (due to increased losses during the dialysis process), and fluid, but it is still limited in phosphorus. Refer to Table 18.4 for the summary of the nutrient recommendations for both HD and PD patients.

PROTEIN Factors relating to higher protein requirements include (1) losses of approximately 10 to 12 grams free amino acids per day[24] and 5 to 15 grams per day of albumin;[25] (2) altered albumin turnover; (3) metabolic acidosis, which increases amino acid degradation; (4) inflammation; and (5) infection.

The NKF K/DOQI guidelines on nutrition have established a general recommendation of 1.2 grams of protein/kg body weight for HD patients to ensure adequate intake of essential amino acids. At least 50% of the protein should be of high biological value. These recommendations are based on the level of protein intake that will maintain neutral or positive nitrogen balance and lead to an improvement or maintenance of visceral protein stores for the majority of HD patients.[23]

The protein requirements for PD are slightly higher than for HD. During episodes of **peritonitis** (inflammation of the

BOX 18.4 | CLINICAL APPLICATIONS

National Renal Diet (NRD)

The National Renal Diet was introduced in 1993 in an attempt to provide a "nationwide, consistent tool for teaching CKD patients how to eat well." A collaborative effort between the ADA Renal Practice Group and the NKF Council on Renal Nutrition, the National Renal Diet consists of the following two patient workbooks:

- *A Healthy Food Guide for People on Dialysis* provides sections on protein, fruit/vegetables, dairy/phosphorus, bread/cereals, fluids, calorie and flavorings, diabetes, and dialysis and vegetarian eating. The most common

food lists are provided, with blank lines for adding other choices. The dietitian works with the patient to fill in the serving limits, serving sizes, and additional food choices.

- *A Healthy Food Guide for People with Chronic Kidney Disease* covers sections or "choices" on protein, fruit, calorie and flavoring, diabetes, and kidney disease and vegetarian eating.

The following tables provide the nutrition composition of the various food lists. These tables should be used for planning meals, serving sizes, and amounts.

NRD Nutrition Composition of Foods for People on Dialysis

Food List	Protein (g/serving)	Calories (kcal/serving)	Sodium (mg/serving)	Potassium (mg/serving)	Phosphorus (mg/serving)
Animal protein	6–8	50–100	20–150	50–150	50–100
Higher-sodium, potassium or phosphorus proteins	6–8	50–100	200–500	250–450	100–300
Fruit/Vegetable					
Low	0–3	10–100	1–50	20–150	0–70
Medium	0–3	10–100	1–50	150–250	0–70
High	0–3	10–100	1–50	250–550	0–70
Dairy/Phosphorus	2–8	100–400	30–300	50–400	100–120
Breads/cereals	2–3	50–200	0–150	10–100	10–70
Calorie	0–1	100–150	0–100	0–100	0–100
Flavorings	0	0–20	250–300	0–100	0–20
Vegetarian protein	6–8	70–150	10–200	60–150	80–150

NRD Nutrition Composition of Foods for People Not on Dialysis

Food List	Protein (g/serving)	Calories (kcal/serving)	Sodium (mg/serving)	Potassium (mg/serving)	Phosphorus (mg/serving)
High-Protein	6–8	50–100	20–150	50–150	50–100
High-Phosphorus Proteins	6–8	50–100	20–150	50–350	100–300
High-Sodium Proteins	6–8	50–100	200–450	50–150	50–100
Vegetable					
Low	2–3	10–100	0–50	20–150	10–70
Medium	2–3	10–100	0–50	150–250	10–70
High	2–3	10–100	0–50	250–550	10–70
Fruit					
Low	0–1	20–100	0–10	20–150	1–20
Medium	0–1	20–100	0–10	150–250	1–20
High	0–1	20–100	0–10	250–550	1–20
Higher-Sodium and/or Phosphorus Foods	2–3	50–200	150–400	10–100	100–200
Breads/ Cereals	2–3	50–200	0–150	10–100	10–70
Calorie	0–1	100–150	0–100	0–100	0–100
Flavorings	0	0–20	250–300	0–100	0–20
Vegetarian Protein	6–8	70–150	10–200	60–150	80–150

Source: ADA National Renal Diet: Professional Guide, 2nd ed., American Dietetic Association, 2002.

Table 18.5 Two-Day CKD Menu with Nutrient Analysis (Based on a 60-kg Person)

Day 1	Day 2
Breakfast	**Breakfast**
½ cup applesauce	½ cup apple juice
4 ounces whole milk	4 ounces whole milk
½ English muffin	1 blueberry muffin
Margarine	Rice Krispies
Jelly/jam	Margarine
Corn Flakes	8 ounces coffee
8 ounces of coffee	Sugar
Sugar	**Lunch**
Pepper	½ cup coleslaw
Lunch	2 ounces baked chicken
½ cup canned peaches	½ cup carrots
Turkey sandwich (2 ounces turkey, 2 slices of white bread, mayonnaise, lettuce)	½ cup white rice
	1 slice white bread
½ cup cranapple juice	Margarine
3 squares of graham crackers	**Dinner**
Dinner	Tossed green salad with sodium-restricted salad dressing
2 ounces of roast pork	½ cup noodles with reduced-sodium
½ cup white rice	
½ cup green beans	gravy
White dinner roll	2 ounces of Swiss steak
Margarine	Margarine
1 cup pineapple chunks	1 cup canned pears
1 slice angel food cake	White dinner roll
	Orange sherbet
Nutrient Analysis	**Nutrient Analysis**
~2100 calories	~2000 calories
44 grams protein	45 grams protein
830 mg phosphorus	840 mg phosphorus
1880 mg potassium	1850 mg potassium
1700 mg sodium	2000 mg sodium

peritoneum), even in mild cases, the dialysate protein losses increase by 50% to 100% to an average of 15 to 36 g/24 hours and have been reported to remain elevated for several weeks.[26] Hence, higher protein levels are necessary in cases of peritonitis.

ENERGY Adequate energy intake is important in order to prevent catabolism and achieve optimal nutritional status. Sufficient kcal from carbohydrates and fat may help to prevent muscle and visceral protein from being utilized as energy. The energy recommendation from the K/DOQI Nutrition Guidelines is based on the finding that energy expenditure is similar to or slightly higher than that of healthy persons.[23] Following these recommendations has been shown to provide neutral nitrogen balance and the maintenance of both serum albumin and anthropometric parameters. A lower caloric requirement has been shown to be adequate for older persons who are more sedentary. Caloric requirements should be adjusted accordingly for those with higher activity levels, those who are underweight, and those who display catabolic stress.[27]

When estimating a PD patient's energy requirements, the kcal absorbed from the glucose in the dialysate should be taken into account. It should be noted, however, that caloric intake in this population has been reported to be below recommended levels.[28] There are several methods/formulas used to determine caloric load from the peritoneal dialysate. One common method uses the peritoneal equilibration test (PET) and is described in Box 18.5.

More than half of all hemodialysis patients report dietary protein intakes of less than 1.0 g/kg/day, and in almost every survey done on energy intakes, most patients ingest about 24 to 27 kcal/kg/day.[23, 27]

ADJUSTED EDEMA-FREE BODY WEIGHT (ABW_EF) FOR OBESE AND UNDERWEIGHT PATIENTS There have been discrepancies in the literature regarding the body weight measurement to be used when calculating protein and energy requirements for underweight and obese patients. According

Table 18.6 Adjusted Body Weight for the Obese or Underweight Patient*

Adjusted body weight (BW) = edema-free BW + [(standard BW − edema-free BW) × 0.25]	
Adjust when patient's weight is <95% or >115% of standard body weight	
Examples of KDOQI adjustment:[1]	
Underweight patient:	Actual BW 60, Standard BW 80 so Adjusted BW = 60 + [(80 − 60) × 0.25]
	Adjusted BW = 60 + [20 × 0.25]
	Adjusted BW = 60 + 5 or 65
Overweight patient:	Actual BW 80, Standard BW 60 so Adjusted BW = 80 + [(60 − 80) × 0.25]
	Adjusted BW = 80 + [−20 × 0.25]
	Adjusted BW = 80 − 5 or 75 (Traditional adjustment = 65)

[1]The KDOQI adjustment is less than the traditional adjustment. Patient adherence might be enhanced with more modest stepwise alterations in nutrient intake, initially, with periodic reassessment of weight and recalculation of needs as the patient's status changes.

Source: Reprinted with permission from the National Kidney Foundation. Pocket Guide to Nutrition Assessment of the Patient with Chronic Kidney Disease (CKD) 2009, Fourth Edition [Revised] Online, National Kidney Foundation, New York, NY. "Adjusted Body Weight for the Obese or Underweight Patient."

Table 18.7 Therapeutic Lifestyle Changes for Patients with CKD

Nutrient	Recommended Intake
Saturated Fat	7% of total kcal
Polyunsaturated Fat	Up to 10% of total kcal
Monounsaturated Fat	Up to 20% of total kcal
Total Fat	25–35% of total kcal
Carbohydrates	50–60% of total kcal
Protein	Approximately 15% of total kcal
Cholesterol	<200 mg/day
Total kcal	Balance energy intake and expenditure to maintain desirable body weight/prevent weight gain
Fiber	20–30 g/day with 5–10 grams soluble fiber

Source: NKF-K/DOQI. Clinical Practice Guidelines for Managing Dyslipidemias in Chronic Kidney Disease (2003). *Am J Kidney Dis.* 41; s1–s91 (suppl 3).

to NKF K/DOQI Nutrition Guidelines, it is recommended that the adjusted edema-free body weight (aBW_{ef}) should be based on the National Health Nutrition Evaluation Survey (NHANES) II data.[23] Table 18.6 summarizes this calculation.

FAT In general, patients on HD and PD are at increased risk for CAD (coronary artery disease) and stroke. Only 20% of HD and 15% of PD patients meet the normal lipid parameters as identified by the Adult Treatment Panel (ATP) III Guidelines. Hemodialysis patients typically display normal low-density lipoprotein (LDL) cholesterol, low high-density lipoprotein (HDL) cholesterol, and elevated triglyceride levels. Unlike HD patients, PD patients exhibit higher total serum cholesterol levels as well as LDL levels. Triglyceride levels are especially increased in PD patients due to the absorption of glucose in the dialysate.

It is recommended that both PD and HD patients adhere to the nutrient composition guidelines of the therapeutic lifestyle changes (TLC) diet (see Table 18.7). These guidelines, which were developed by the National Cholesterol Education Program (NCEP) Expert Panel and modified for CKD patients, can be found in the NKF K/DOQI clinical practice guidelines for managing dyslipidemias.[29]

POTASSIUM The potassium restriction for HD and PD patients varies, depending on the degree of kidney function, serum potassium levels, modality, and drug therapy. For the most part, a diet that allows 50 to 70 mmol/day or approximately 2 to 3 g/day of potassium is commonly prescribed in HD.[30] Those who are oliguric or anuric are at an increased risk for hyperkalemia and should have a more stringent dietary restriction. The target range for serum potassium is 3.5–6.0 mEq/L. Severe hyperkalemia (serum K greater than 7.0 mEq/L) may precipitate fatal arrhythmias.

It is also important to note that there are other nondietary causes of hyperkalemia. These include gastrointestinal bleeding, acidosis, catabolism, hypoaldosteronism, hyperglycemia, and certain types of pica.

Patients on HD should be counseled about limiting high-potassium foods. This requires restricted selection of fruits, vegetables, nuts and seeds, chocolate, and dairy products (see Box 18.6 for examples of high-potassium and low-potassium foods). Salt substitutes and low-sodium baking soda and baking powder are concentrated sources of potassium chloride and thus should not be used.

In PD, patients often experience low serum potassium levels due to daily dialysis removal. Most will not need a restriction, and some may require oral potassium supplementation.[31]

FLUID AND SODIUM Fluid and sodium allowances are highly individualized and based primarily on residual urine output and dialysis modality. Other considerations include blood pressure control, interdialytic weight gains in HD patients, presence of edema, and congestive heart failure.

The interdialytic weight gain goal in HD patients should not exceed 5% of body weight. Higher fluid gains can lead to sudden changes in blood volume and hypotension during the hemodialysis treatment. Since most patients become oliguric or anuric within the first 12 months of hemodialysis, it is prudent to recommend a 2 gram sodium diet with a fluid allowance of not more than 1 L (1000 mL) daily. If urine output is greater than 1 L per day, the sodium and fluid allowance can be liberalized to approximately 2 to 4 g sodium per day and 2 L (2000 mL) of fluid per day.

Fluid and sodium requirements for patients on PD therapy are highly individualized and largely based on ultrafiltration. Ultrafiltration can remove 2 to 2.5 kg of fluid per day. If a patient on PD gains too much fluid, higher dextrose concentrations must be used, which leads to greater fluid removal but can also lead to weight gain, higher triglyceride levels, and insulin resistance. Therefore, the use of these higher-dextrose concentrations should be minimized. Typical fluid and sodium restriction for the PD patient includes a fluid allowance of 2 L per day and a sodium allowance between 2 and 4 g per day. Symptoms of fluid

BOX 18.6 | CLINICAL APPLICATIONS

High- and Low-Potassium Food Guide

What foods are high in potassium (greater than 200 milligrams per portion)?

The following table lists foods that are high in potassium. The portion size is ½ cup unless otherwise stated. **Please be sure to check portion sizes.** While all the foods on this list are high in potassium, some are higher than others.

High-Potassium Foods

Fruits	Vegetables	Other Foods
Apricot, raw (2 medium); dried (5 halves)	Acorn squash	Bran/bran products
Avocado (¼)	Artichoke	Chocolate (1.5–2 ounces)
Banana (½)	Bamboo shoots	Granola
Cantaloupe	Baked beans	Milk, all types (1 cup)
Dates (5 whole)	Butternut squash	Molasses (1 tablespoon)
Dried fruits	Refried beans	Nutritional supplements: Use only under the direction of your doctor or dietitian
Figs, dried	Beets, fresh then boiled	
Grapefruit juice	Black beans	
Honeydew	Broccoli, cooked	Nuts and seeds (1 ounce)
Kiwi (1 medium)	Brussels sprouts	Peanut butter (2 tbs.)
Mango (1 medium)	Chinese cabbage	Salt substitutes/ lite salt
Nectarine (1 medium)	Carrots, raw	Salt-free broth
Orange (1 medium)	Dried beans and peas	Snuff/chewing tobacco
Orange juice	Greens, except kale	Yogurt
Papaya (½ whole)	Hubbard squash	
Pomegranate (1 whole)	Kohlrabi	
Pomegranate juice	Legumes	
Prunes	Lentils	
Prune juice	Mushrooms, canned	
Raisins	Parsnips	
	Potatoes, white and sweet	
	Pumpkin	
	Rutabagas	
	Spinach, cooked	
	Tomatoes/tomato products	
	Vegetable juices	

What foods are low in potassium?

The following table list foods that are low in potassium. **A portion is ½ cup** unless otherwise noted. **Eating more than 1 portion can make a lower-potassium food into a higher-potassium food.**

Low-Potassium Foods

Fruits	Vegetables	Other Foods
Apple (1 medium)	Alfalfa sprouts	Rice
Apple juice	Asparagus (6 spears)	Noodles
Applesauce	Beans, green or wax	Pasta
Apricots, canned in juice	Cabbage, green and red	Bread and bread products (not whole grains)
Blackberries	Carrots, cooked	Cake: angel, yellow
Blueberries	Cauliflower	Coffee (**limit to 8 ounces**)
Cherries	Celery (1 stalk)	
Cranberries	Corn, fresh (½ ear); frozen (½ cup)	Pies without chocolate or high-potassium fruit
Fruit cocktail	Cucumber	
Grapefruit (½)	Eggplant	Cookies without nuts or chocolate
Grape juice	Kale	
Grapes	Lettuce	Tea (**limit to 16 ounces**)
Mandarin oranges	Mixed vegetables	
Peaches, fresh (1 small); canned (½ cup)	Mushrooms, fresh	
Pears, fresh (1 small); canned (½ cup)	Okra	
Pineapple	Onions	
Pineapple juice	Parsley	
Plums (1 whole)	Peas, green	
Raspberries	Peppers	
Strawberries	Radish	
Tangerine (1 whole)	Rhubarb	
Watermelon (**limit to 1 cup**)	Water chestnuts, canned	
	Watercress	
	Yellow squash	
	Zucchini squash	

How do I get some of the potassium out of my favorite high-potassium vegetables?

The process of leaching will help pull potassium out of some high-potassium vegetables. It is important to remember that leaching will not pull all of the potassium out of the vegetable. You must still limit the amount of leached high-potassium vegetables you eat. Ask your dietitian about the amount of leached vegetables that you can safely have in your diet.

How to Leach Vegetables

For potatoes, sweet potatoes, carrots, beets, and rutabagas:

1. Peel and place the vegetables in cold water so they won't darken.

2. Slice vegetable ⅛ inch thick.

3. Rinse in warm water for a few seconds.

4. Soak for a minimum of two hours in warm water. Use ten times the amount of water to the amount of vegetables. If soaking longer, change the water every four hours.

5. Rinse under warm water again for a few seconds.

6. Cook vegetable with five times the amount of water as the amount of vegetable.

For squash, mushrooms, cauliflower, and frozen greens:

1. Allow frozen vegetable to thaw to room temperature and drain.

2. Rinse fresh or frozen vegetables in warm water for a few seconds.

3. Soak for a minimum of two hours in warm water. Use ten times the amount of water to the amount of vegetables. If soaking longer, change the water every four hours.

4. Rinse under warm water again for a few seconds.

5. Cook the usual way, but with five times the amount of water as the amount of vegetable.

Sources: Bowes & Church Food Values of Portions Commonly Used, 17th ed., Pennington JA, Lippincott, 1998. Diet Guide for Patients with Kidney Disease, Renal Interest Group-Kansas City Dietetic Association, 1990.

Reprinted with permission from the National Kidney Foundation. Available from: http://www.kidney.org/atoz/atozItem.cfm?id=103.

---

overload can include shortness of breath, hypertension, congestive heart failure, and edema. Fluid intake is best controlled by limiting dietary sodium intake. Table 18.8 lists tips for controlling fluid intake, and Table 18.9 describes how to choose foods appropriate for a low-sodium diet (see Chapter 13 for additional information).

Table 18.8 Tips to Control Fluid Intake

- Limit high-salt foods so you will have less thirst.
- Take your pills with your mealtime liquids, applesauce, or pureed fruits, as allowed.
- Drink from small glasses and cups.
- Drink only when you are thirsty. Reach for very cold beverages. Beverages that are less sweet will quench your thirst.
- Weigh yourself daily. You should not gain more than the prescribed number of pounds each day.
- Use sour candy or sugar-free gum to moisten your mouth. Try special thirst-quencher gums.
- Add some lemon juice to water or ice. The sour taste will help to quench your thirst.
- Try swishing your mouth with very cold water or low-alcohol mouthwash when you are thirsty. *Do not swallow it!*
- Brush teeth often—good mouth hygiene is essential.
- Keep lips moist with lip balm or moisturized lipstick.
- Use ice cubes instead of liquids. One cup of ice is equal to ½ cup of water/juice and will last longer.
- Freeze grapes and eat throughout the day as one of your fruit servings (serving is ½ cup).
- Try frozen blueberries and pineapple tidbits, fruit cocktail, and other recommended fruits.
- Remember that some foods should also be counted as fluids. These include soups, Popsicles®, sherbet, ice cream, yogurt, custard, and gelatin.

Source: Department of Veterans Affairs, Nutrition and Food Service, PFEM#000641. Available at http://www.va.gov/portland/Education/PatientEd/Documents/Nutrition/Tips_to_Control_Your_Fluid_Intake.pdf

PHOSPHORUS In early CKD, hyperphosphatemia is prevented by an adaptive increase in renal excretion and decreased phosphate reabsorption. Hyperphosphatemia is evident when the GFR falls between 20 and 30 mL/min/1.73 m^2. A dietary phosphorus restriction of 800–1000 mg/day or <17 mg/kg of ideal body weight or standard body weight per day has been recommended for both HD and PD patients.[32] Table 18.10 lists high-phosphorus foods that patients should limit, and Box 18.7 discusses phosphate additives in foods.

In addition to dietary restriction of phosphorus, most patients require phosphate binders to prevent gastrointestinal absorption of dietary phosphorus (see Table 18.11). Calcium salts have replaced aluminum- and magnesium-based binders as the binder of choice. Aluminum-based binders were found to cause aluminum-related dementia and to also accumulate in the bone.

CALCIUM Dietary calcium requirements are higher in those with CKD. Low serum calcium levels often accompany CKD due to alterations in vitamin D metabolism, decreased absorption of calcium from the gut, and elevated phosphorus levels. Foods high in calcium are restricted, because they tend to be high in phosphorus as well. If calcium supplements are indicated, they should be taken on an empty stomach—between meals or at bedtime (calcium-based phosphate binders are sometimes used as calcium supplements). The amount of calcium from the diet plus the amount found in supplements and phosphate binders should not exceed 2,000 mg per day. Other sources of calcium include calcium-based phosphate binders and dialysate, which may contain calcium. The use of active vitamin D sterols increases calcium absorption in the intestine.

For the previous decade there has been a great deal of attention given to the adverse consequences of elevated PTH, phosphorus, and Ca × P product (product of the serum calcium level multiplied by the serum phosphorus level), especially as they relate to morbidity and mortality. Block et al. (1998) clearly

Table 18.9 Low-Sodium Diet

Food Groups & High-Salt Food Items	Foods to Limit Because of Their High Sodium Content		Acceptable Substitutes
Meats and Meat Substitutes	• Processed deli meats (pepperoni, bologna, salami, pastrami, ham, turkey, corned beef) • Sausage • Hot dogs • Breaded meats (chicken nuggets, fish sticks) • Canned meats (spam) • Smoked or cured meats (salt pork, bacon) • Lox and herring		Fresh or frozen un-breaded meats, eggs
Dairy	• Buttermilk • Processed cheese spreads (Cheez Wiz, Easy Cheese) • Processed cheese (Velveeta, American cheese, nacho cheese), • Pimento cheese		Limit dairy to daily allowance
Vegetables	• Regular canned vegetables and vegetable juices • Pickles • Relish • Olives • Pepperoni • Sauerkraut		Fresh or frozen vegetables, canned vegetables with no added salt
Fruits	• No need to limit any fruits or fruit juices		All low-potassium fruits or fruit juices
Starches	• Biscuits • Prepared mixes (pancakes, muffins, cornbread) • Seasoned rice, noodle, or potato mixes (such as Rice-a-Roni, macaroni and cheese) • Coating mixes (seasoned bread crumbs, Shake'n'Bake) • Salted snack foods (potato chips, corn chips, pretzels, pork rinds, crackers, tortilla chips, nuts, popcorn, sunflower seeds)		English muffins, bagels, plain pasta, noodles, rice, cooked hot cereals, unsalted or low-sodium snack foods
Fats	• Bacon • Salt pork • Fat back • Commercial salad dressings		Plant oils (olive, canola), tub or squeeze margarine, low-sodium salad dressings
Salt and High-Salt Seasonings/Condiments	• Table salt • Seasoning salt • Garlic salt • Onion salt • Celery salt • Lemon pepper • Lite salt • Meat tenderizer	• Bouillon cubes • Flavor enhancers High-sodium sauces such as: • Barbecue sauce • Steak sauce • Soy sauce • Teriyaki sauce • Oyster sauce	Fresh garlic, fresh onion, garlic powder, onion powder, black pepper, lemon juice, low-sodium/salt-free seasoning blends, vinegar Homemade or low-sodium sauces and salad dressings, vinegar, dry mustard
Mixed/Processed Foods	Canned: • Soups • Tomato products Convenience foods such as: • TV dinners • Canned ravioli • Chili • Macaroni and cheese • Spaghetti with sauce • Commercial mixes • Frozen prepared foods • Fast foods		Homemade or low-sodium soups, canned food without added salt Homemade casseroles without added salt, made with fresh or raw vegetables, fresh meat, rice, pasta, or unsalted canned vegetables

Sources: University of Virginia, Digestive Health Center of Excellence. Available at http://www.healthsystem.virginia.edu/internet/digestive-health/nutrition/lowsoddiet .cfm. The National Kidney Foundation. Available at www.kidney.org/atoz/atozItem.cfm?id=175

Table 18.10 High-Phosphorus Foods to Limit

Beverages	• Ale	• Beer
	• Chocolate drinks	• Cocoa
	• Drinks made with milk	• Dark colas
	• Canned iced teas	
Dairy Products	• Cheese	• Cottage cheese
	• Custard	• Ice cream
	• Milk	• Pudding
	• Cream soups	• Yogurt
Protein	• Carp	• Crayfish
	• Beef liver	• Chicken liver
	• Fish roe	• Organ meats
	• Oysters	• Sardines
Vegetables	• Dried beans and peas	• Soy beans
	• Baked beans	• Black beans
	• Chick peas	• Garbanzo beans
	• Kidney beans	• Lentils
	• Limas	• Northern beans
	• Pork 'n beans	• Split peas
Other Foods	• Bran cereals	• Brewer's yeast
	• Caramels	• Nuts
	• Seeds	• Wheat germ
	• Whole-grain products	

Source: Reprinted with permission from the National Kidney Foundation. Available at http://www.kidney.org/atoz/atozItem.cfm?id=101

Table 18.11 Phosphate-Binders

- Calcium acetate (Phos-Lo)
- Calcium carbonate (Tums, Calci-Chew, Caltrate, Calci-Mix, Nephro-Calci, Titralac, Chooz Gum or Oscal 500)
- Calcium citrate (Citracal 950)
- Magnesium carbonate (MagneBind 200 or MagneBind 300)
- Lanthanum carbonate (Fosrenol)
- Sevelamer hydrochloride (Renagel)
- Aluminum hydroxide (Amphojel, AlternaGEL, Dialume, Alu-Cap, Alu-Tab)
- Aluminum carbonate (Basaljel)

bid conditions, patients with a serum phosphorus level >6.5 mg/dL had a 27% higher risk of death than those patients with levels of 2.4–6.5 mg/dL; 39% of these patients had a serum phosphorus >6.5 mg/dL, placing them at increased risk of mortality. Additionally, patients with a Ca $\times$ P > 72 mg^2/dL^2 had a 34% higher risk of death relative to those with a Ca $\times$ P between 42 and 52 mg^2/dL^2. PTH levels >975 pg/mL also were associated with an increased mortality rate.

Further analyses using the same database confirm and extend the findings just presented.[33] An elevated serum phosphorus level (>6.5 mg/dL) was associated with sudden death and death due to coronary artery disease. More recently, Block et al. (2004) further confirmed the relationship between mineral imbalances and relative risk of cardiovascular mortality.[34] Serum phosphorus concentrations from 5.0 to 5.5 mg/dL and 5.5 to 6.0 mg/dL were associated with significant increases in relative risk of death (1.10 and 1.25 times, respectively).

Strong relationships exist among elevated serum phosphorus, Ca $\times$ P product, PTH, and cardiac causes of death in HD patients. Since these parameters have been found to be independently associated with increased morbidity and mortality, the National Kidney Foundation has established target goals for

delineated the adverse consequences of elevated PTH, phosphorus, and Ca $\times$ P.[31] They found that in two retrospective analyses of United States Renal Data System (USRDS) data from over 6,400 HD patients, elevated serum phosphorus, Ca $\times$ P, and PTH levels were independently associated with increased morbidity and mortality. After adjustment for comor-

---

BOX 18.7 CLINICAL APPLICATIONS

Sodium and Phosphate Additives in Foods

Phosphate additives in foods have become a mainstay in the U.S. food supply. In 1990, phosphate additives contributed an estimated 470 mg of phosphorus/day to the diet. Increased use of these phosphate additives could now contribute up to 1,000 mg/day of phosphorus depending upon one's food choices. This is a concern for the general population, but especially for CKD patients. Restructured meats (chicken nuggets and hotdogs), processed and spreadable cheeses, "instant" products (puddings and sauces), refrigerated bakery products, and beverages are products commonly enhanced with phosphorus additives.

Enhanced meats can also contribute unwanted phosphorus and sodium to the diet. Enhanced meats are fresh meats that have been injected with a water solution containing ingredients such as sodium and phosphate salts, potassium salts, antioxidants, and/or flavorings in order to increase shelf-life as well as enhance taste and appearance. Enhanced meats are significantly higher in sodium than fresh, unadulterated meat. Enhanced meats contain over 300 mg of sodium per 3-oz. serving, while most fresh meats contain much less, approximately 50–75 mg per 3-oz. serving.

Phosphorus additives are highly absorbable. In a typical mixed diet containing grains, meat, and dairy, only 60% of the dietary phosphorus is absorbed, whereas phosphoric acid and various polyphosphates and pyrophosphates are almost 100% absorbed. Diets higher in these inorganic salts will result in higher phosphorus absorption, which can aggravate hyperphosphatemia in CKD patients.

Sources: Karalis M, Murphy-Gutekunst L. Enhanced Foods: Hidden Phosphorus and Sodium in Foods Commonly Eaten. Journal of Renal Nutrition, Vol 16, No 1 January 2006. pp. 79–81. Murphy-Gutekunst L, Uribarri J. Hidden Phosphorus-Enhanced Meats: Part 3. Journal of Renal Nutrition, Vol 15, No 4 October 2005. pp. E1–E4.

these parameters. The serum target goals are as follows:[32] keep Ca × P product less than 55 mg^2/dL^2, calcium within the normal range of 8.4 to 10.2 mg/dL (preferably within low end of normal), and serum phosphorus within the normal range of 3.5 to 5.5 mg/dL. A summary of the nutrient recommendations is provided in Table 18.4.

VITAMIN SUPPLEMENTATION Supplementation of water-soluble vitamins due to increased losses during dialysis, anorexia, and poor dietary intake is typically indicated for HD and PD patients. Other reasons include a renal diet that is low in fresh fruits/vegetables, whole grains, and dairy products; altered metabolism; impaired synthesis; resistance to the actions of some vitamins; and decreased intestinal absorption. Daily requirements for both HD and PD are listed in Table 18.4.[23, 35]

Specially designed vitamins are available and should be used exclusively if supplementation is required. In general, the "renal" vitamin contains B vitamins, folic acid, and vitamin C. Fat-soluble vitamins and minerals are not included. Preparations containing vitamin A or high doses of vitamin C should be avoided.

Serum levels of B$_1$ (thiamin), B$_2$ (riboflavin), pantothenic acid, and biotin are typically normal. Biotin supplementation of 10 mg/day has been found to be helpful in treating patients with peripheral neuropathy, encephalopathy, and intractable hiccups.[36]

Vitamin C should be limited to no more than 100 mg per day. Higher doses can lead to oxalosis—an accumulation of oxalate stones that can deposit in viscera, soft tissues, joints, and blood vessels.

Because serum vitamin A levels are elevated in HD and PD patients, supplementation is not necessary. Reasons for elevated vitamin A include increased serum retinal-binding protein, decreased kidney catabolism, and the failure of dialysis to remove vitamin A. The requirements for vitamin D will be discussed in greater detail in the Secondary Hyperparathyroidism section. Very little is known about the long-term effects of vitamin E supplementation. Vitamin K supplementation may be needed for those patients receiving antibiotic therapy, since the antibiotics may destroy the bacteria found in the gastrointestinal tract (these bacteria are a primary source of vitamin K). However, since most HD patients receive anti-coagulation therapy, caution must be taken due to vitamin K's role in promoting clot formation.

Controversy exists about whether or not to supplement folic acid. As kidney failure progresses, there is decreased metabolism of homocysteine (Hcy) by the renal parenchyma. Additionally, the cofactors involved in Hcy metabolism, such as folic acid, vitamins B$_6$, and B$_{12}$, are less available due to dialysis losses and overall decreased dietary intake due to dietary restrictions.

In the 1980s it was found that end-stage renal disease patients have higher Hcy levels when compared to the general population.[37] Hyperhomocysteinemia has been linked to cardiovascular disease (CVD) in both the dialysis and general populations. Some studies have suggested that hyperhomocysteinemia is an independent risk factor for CVD. However, a causal relationship between hyperhomocysteinemia and CVD has yet to be clarified in CKD because both inflammation and malnutrition also impact Hcy levels in this population.[38] The National Kidney Foundation concludes that there is no evidence at this time to recommend supplementation of folic acid.[39] Supplementation studies have not normalized plasma levels, and researchers have not determined whether lowering Hcy levels is clinically beneficial. Since hyperhomocysteinemia can occur as a result of folic acid, riboflavin, and pyridoxine deficiencies, the recommendation is to be sure dialysis patients are prescribed and routinely take a vitamin supplement that meets the current recommendations (see Table 18.4).

MINERAL SUPPLEMENTATION Magnesium is excreted by the kidneys. As such, magnesium levels are usually normal to mildly elevated in dialysis patients and not generally supplemented. Magnesium-containing phosphate binders, supplements, and antacids should be avoided.

Iron deficiency is common in both HD and PD patients. This is due to the kidney's inability to make adequate erythropoietin for production of red blood cells (see Chapter 19). Most if not all patients receive rHuEPO (recombinant human erythropoietin) during dialysis. Most patients will require iron supplementation, which is individualized depending on serum markers of ferritin, iron, total iron binding capacity, and transferrin saturation. With the advent of intravenous iron, most patients do not take oral supplements. Further discussion can be found in the "Anemia" section.

The allowance for zinc is the same as the RDA, which is 15 mg per day. Some studies have indicated that impaired taste sensation, loss of appetite, and sexual dysfunction may be ameliorated with zinc supplementation.

Nutrition Therapy for Comorbid Conditions and Complications

CARDIOVASCULAR DISEASE CVD has been associated with chronic kidney disease, largely because people with CKD are more likely to die from cardiovascular complications than to progress to CKD Stage 4. The National Kidney Foundation recommends that CKD patients be considered in the "highest risk group" for subsequent CVD events.[40] It is well established that the prevalence of cardiovascular disease in dialysis patients is high and accounts for approximately 50% of deaths in HD and PD patients.[2] It has been reported that there is a high prevalence of CVD in CKD, and that mortality due to CVD is 10 to 30 times higher in dialysis patients than in the general population.[40]

The prevalence rates of heart failure, left ventricular hypertrophy (LVH), and atherosclerosis are very high, with about 40% of HD patients having ischemic heart disease. In the Canadian Prospective Cohort Study, 74% of patients initiating dialysis were found to have LVH at baseline, 44% had concentric LVH, 30% had hypertrophy with left ventricular dilatation, and 15% had systolic dysfunction.[41]

This high mortality rate from CVD may suggest that these patients are subjected to an accelerated process of atherogenesis that may not be related to traditional risk factors. Dialysis patients display other "nontraditional" risk factors: abnormalities in lipoprotein metabolism, hyperparathyroidism, calcium and phosphate imbalances, vascular calcification, malnutrition, elevated Hcy levels, oxidative stress, and inflammation. Of particular interest are the roles of inflammation and oxidative

Table 18.12 Dialysis-Related Causes of Inflammation in Chronic Kidney Disease

Hemodialysis	Peritoneal Dialysis
• Access infection	• Peritonitis
• Bio-incompatible HD membrane	• Access infection
• Endotoxin exposure from contaminated dialysate	• Bio-incompatible dialysate
• Backfiltration	• Endotoxin exposure from contaminated dialysate

Source: Adapted from: Kopple J. and Massry S., editors. *Nutritional Management of Renal Disease,* 2nd ed. Lippincott Williams & Wilkins; 2004. Table 13.3.

stress in contributing to atherosclerosis and the consequent CV morbidity and mortality in CKD patients.

Many HD and PD patients have elevated C-reactive protein (CRP) levels, which have been linked to increased cardiovascular mortality, possibly suggesting an association between inflammation-driven malnutrition and atherosclerosis. Elevated CRP levels indicate an acute-phase response to inflammation. Evidence indicates that elevated CRP levels are indicative of inflammatory activity and reflect the generation of pro-inflammatory cytokines such as IL-1, IL-6, and tumor necrosis factor-α (TNF-α). High levels of pro-inflammatory cytokines predict low serum albumin levels and are associated with malnutrition in dialysis patients. Potential causes of inflammation related to the dialysis procedure in HD and PD patients can be found in Table 18.12.

Within this patient population, CVD not only exhibits a continued high prevalence, but is the number one cause of mortality; it is therefore worthy of concern. Although there are no conclusive nutrition recommendations for CKD patients, the dietary guidelines outlined in Table 18.4 should be applied until solid evidence becomes available. Additionally, more recently published guidelines recommend that well-nourished, stable dialysis patients incorporate omega-3 fatty acids into their diet at least two times per week.[39] Some acceptable food sources of omega-3 fatty acids for dialysis patients include salmon (Atlantic, farmed), mackerel, herring, and canola oil.

SECONDARY HYPERPARATHYROIDISM (SHPT)

Secondary hyperparathyroidism (SHPT) in dialysis patients (see Figure 18.4) can progress to severe, intractable forms of bone disease. Prolonged exposure to elevations of PTH causes the development of **osteitis fibrosa cystica**, a form of bone disease characterized by rapid bone turnover with an excess of collagen production and inadequate mineralization. Although bone mass itself may appear constant, bone quality is poor and the bones are more prone to fracture. The decrease in bone mineralization can cause vascular and soft tissue calcification. Patients at this stage often are resistant to vitamin D therapy and require a surgical parathyroidectomy. Resistance to erythropoietin therapy, related to marrow fibrosis, may also be noted in patients with SHPT. Once the disease progresses to advanced stages, irreversible changes usually have occurred in the patient's parathyroid glands, thus making this disease state difficult to treat.

Most HD and PD patients will require management of SHPT through the use of dietary phosphorus restriction, phosphate binders, and oral or IV supplementation of vitamin D or vita-

min D analogs. Serum PTH, calcium, phosphorus, and Ca $\times$ P product should be closely monitored for those on vitamin D therapy.

A variety of oral and intravenous vitamin D or vitamin D analogs are commercially available and should be dosed and titrated based on individual patient needs to prevent over or under suppression of PTH.

ANEMIA CKD patients are not able to synthesize adequate amounts of endogenous erythropoietin, a hormone that is made by renal tubular cells, which leads to decreased red cell production in the bone marrow and low hemoglobin (Hgb) levels.[42] Unlike PD patients, HD patients also incur increased blood loss from dialysis procedures. The discovery of rHuEPO was a major breakthrough and has significantly improved the quality of life in CKD patients. Treatment with rHuEPO and iron is common practice today. EPO is given intravenously for HD patients and subcutaneously for PD patients. The K/DOQI anemia guidelines recommend subcutaneous (SC) administration instead of intravenous (IV) administration in hemodialysis patients, because SC administration is less expensive (lower doses of rHuEPO can be used to maintain Hct levels >33%) and the EPO is used more efficiently.[43] Most hemodialysis patients prefer to have their rHuEPO delivered in their dialysis lines.

Recommended target levels are 33% to 36% for Hct and 11 to 12 g/dL for Hgb, with Hgb as the primary measurement.[43] Hgb has been shown to be a more accurate indicator than Hct. EPO dosing should be individualized, based on the physician's orders and the manufacturer's guidelines. EPO doses may vary significantly depending on individual patient response.

The effectiveness of EPO therapy depends on adequate iron management. Since there is an increased demand for iron during the production of red blood cells, most patients will require supplemental iron. Iron deficiency has been found to be the most frequent cause of EPO nonresponsiveness.[44] Other causes of EPO nonresponsiveness include malnutrition, infections, chronic inflammation, secondary hyperparathyroidism, chronic blood loss, and folate and vitamin B_{12} deficiency.[45] For effective EPO therapy, iron stores should be measured monthly and adequate stores maintained according to the NKF K/DOQI Clinical Practice Guidelines for Anemia. The goals for transferrin saturation (TSAT) and serum ferritin are listed in Table 18.13. In some patients, a better response or equal response with lower EPO doses may be achieved with higher iron stores (up to 50% TSAT or 800 ng/mL for serum ferritin).

Ferrous gluconate, ferrous sulfate, ferrous fumarate, polysaccharide-iron complex, and a heme iron polypeptide are common oral sources of iron supplements. In HD patients, iron losses are higher due to blood losses—up to 6 mg/day as a result of blood testing, dialyzer/tubing waste, gastrointestinal bleeding, and leakage of vascular access—and thus exceed the amount of iron that can be absorbed from oral preparations. In HD patients, intravenous iron supplementation is more effective than oral administration in treating iron-deficiency anemia.

PD patients have minimal iron losses, and many are able to meet their increased iron requirements through oral supplementation.[43] Dosages for oral iron will vary depending on the source of the supplement (see Table 18.14). They should

Table 18.13 Therapeutic CKD Anemia Targets for Those on EPO

Lab Test	Testing Frequency	HD	ND-/PD-CKD	Pediatric
Hemoglobin (Mid-week sampling)	CKD—at least annually CKD on ESA—at least monthly	**CPR** 11–12 g/dL **CPG** ≤13 g/dL	**CPR** 11–12 g/dL **CPG** ≤13 g/dL	Same*
Serum ferritin	Monthly (initial ESA therapy), then ≥ quarterly for HD, CKD or those not on ESA	>200 ng/mL (inadequate evidence for benefit of >500 ng/mL)	>100 ng/mL	>100 ng/mL
Transferrin saturation or CHr (HD adults only)	Same as above	>20% >29 pg/cell	>20%	>20% N/A

*There are no studies in children to indicate increased risk at Hgb ≥13 g/dL; thus, an individual child's likely benefit from Hgb >13 must be weighed carefully against potential risk.

Source: Reprinted with permission from the National Kidney Foundation. Pocket Guide to Nutrition Assessment of the Patient with Chronic Kidney Disease (CKD) 2009, Fourth Edition [Revised] Online, National Kidney Foundation, New York, NY. "Therapeutic CKD Anemia Targets for Those on ESA" table.

Table 18.14 Oral Iron Supplements

Iron Source	% Elemental	Examples	Elemental	Dose	Possible Side Effects
Carbonyl iron	Pure	Feosol Caplets Ferralet 90	45 mg 90 mg	2–3/day 1/day	Note: contains 120 mg vit C
Ferrous gluconate	12%	Fergon	27 mg/300 mg tablet	6–7/day	GI symptoms, N & V, constipation
Ferrous fumarate	33%	Nephro-Fer Nephron FA	115 mg 200 mg	2/day 2/day	Same as above
Ferrous sulfate	20% 30% dried	Slow-Fe Feosol Tablets	50 mg 65 mg	3–4/day 3/day	Same as above
Polysaccharide	100%	Niferex 150 Nulron 150	150 mg 150 mg	1–2/day 1–2/day	Same as above
Heme iron polypeptide	N/A	Proferrin	7 mg/capsule	1 up to 3 times/day	Less GI distress (per manufacturer)

Source: Used with permission: National Kidney Foundation. Pocket Guide to Nutrition Assessment of the Patient with Chronic Kidney Disease (CKD) 2009, Fourth Edition [Revised] Online, National Kidney Foundation, New York, NY. "Iron Supplements (Oral)" table.

be consumed between meals and should not be taken at the same time as phosphate binders (iron should be taken ≥2 hours before or ≥1 hour after meals). Absorption of iron is inversely related to body iron stores and may decrease with serum ferritin >200 ng/mL and TSAT >20%. Side effects of oral iron supplementation include GI upset, nausea/vomiting, and constipation. Strategies to avoid these side effects include (1) using smaller, more frequent doses, (2) increasing the dose gradually, (3) changing to another iron compound, or (4) taking the supplement at bedtime.

IV iron preparations available in the United States include iron dextran, sodium ferric gluconate, iron sucrose (see Table 18.15). Supramagnetic iron oxide nanoparticles is awaiting FDA approval for use in the United States.[46] A typical course of IV iron consists of 1000 mg given over 10 dialysis treatments. Subsequently, it is recommended that HD patients be placed on a maintenance dose of IV iron, generally between 25 and 65 mg per week.[43] An additional course of IV therapy may be warranted if iron parameters indicate that the patient remains iron deficient.

Untreated anemia can lead to cardiac and ventricular hypertrophy, angina, congestive heart failure, malnutrition, and impaired immunological response, and has also been associated with increased mortality.

Table 18.15 IV Iron Supplements

Iron Source	Examples	Elemental	Dose	Possible Side Effects
Iron dextran	INFeD DexFerrum	50 mg/mL	Test dose required 100–1000 mg over 4–8 hrs	Altered/metal taste, N/V, ↓ BP, flushing, myalgia, rash, anaphalaxis/tachycardia
Sodium ferric gluconate	Ferrlecit	12.5 mg/mL	1 gram over 8 tx (125 mg/ea)	Rarely: ↓ BP, flushing, N/V, cramps, malaise, fatigue
Iron sucrose	Venofer	20 mg/mL	1g over 10 tx (100 mg/ea)	Hypotension, cramps/leg cramps, nausea
Supramagnetic iron oxide nanoparticles	Fermoxytol	30 mg/mL	510 mg (projected) pending FDA approval	Nausea, dizziness, diarrhea, dysgeusia, vomiting, rash, constipation

The iron supplement lists are not all inclusive.

Sources: http://www.proferrin.com/
Auerbach M. Ferumoxytol as a new, safer, easier-to-administer intravenous iron: yes or no? *Am J Kidney Dis.* 2008;52(5):826–9.
Manufacturer's information (Bayer, Watson, AMAG, American Regent).
Used with permission: Chapter 9 Anemia. From: National Kidney Foundation. Pocket Guide to Nutrition Assessment of the Patient with Chronic Kidney Disease (CKD) 2009, Fourth Edition [Revised] Online, National Kidney Foundation, New York, NY. "Iron Supplements (IV)" table.

Medicare Coverage for Medical Nutrition Therapy In 2000, Congress passed the Medicare, Medicaid and SCHIP Benefits Improvement and Protection Act (BIPA). Section 105 of BIPA permitted coverage of Medical Nutrition Therapy, if a patient had Part B Medicare, for renal disease (nondialysis patients) and diabetes. The Center for Medicare and Medicaid Services (CMS), which sets administrative rules for laws passed by Congress, has defined patients who qualify for coverage with a diagnosis of renal disease as those with chronic renal insufficiency (nondialysis) who have a glomerular filtration rate (GFR) of <50 mL/min/1.73 m^2 or who are receiving posttransplant care after discharge from the hospital. Dialysis patients were excluded from this benefit since nutrition services are part of the composite rate paid to dialysis units for the treatment of these patients. The benefit became effective January 1, 2002.

Nutritional Requirements of the Posttransplant Patient

The goal in the acute posttransplant period is to manage the increased metabolic demands of transplant surgery. In addition to achieving optimal nutritional status, the goals of the transplant diet in the long term include the management of obesity, blood pressure, insulin resistance, diabetes, and hyperlipidemia; maintenance of electrolyte balance; and maximized bone health. Nutrition therapy for kidney transplant patients differs between the acute phase (up to 8 weeks following transplant) and the chronic phase (beginning the ninth week following transplant). A summary of the nutrition guidelines for posttransplant patients in the acute and chronic phases is provided in Table 18.16.

PROTEIN AND ENERGY NEEDS Generally, protein and energy requirements are increased for up to 6 to 8 weeks following transplantation due to postoperative stress and the excessive doses of corticosteroids. In the long term (after 8 weeks following kidney transplantation), these patients require a diet that provides the RDA for protein and is low in saturated fat in an effort to manage dyslipidemia, diabetes, obesity, and cardiovascular disease.

CARBOHYDRATE Glucose intolerance is a frequent occurrence post transplant. About 5% to 10% of patients develop hyperglycemia.[47] This is due to corticosteroids and other transplant medications such as cyclosporine A (CyA).[48, 49] Due to this impaired glucose tolerance and the potential for development of posttransplant diabetes mellitus, the majority of carbohydrate content should come from complex sources such as vegetables, fruits, and grains. Special emphasis should also be placed on increasing dietary fiber intake. In addition to dietary modification and exercise, the use of insulin or OHA (oral hypoglycemic agents) may be required.

The control of glucose impairment should be a priority due to its detrimental effects on cardiovascular health. Moreover, it has been shown that posttransplant patients with glucose intolerance have a higher risk for infection and decreased survival rates.[50]

FAT With cardiovascular disease as the leading cause of mortality in kidney transplant patients, a low-fat diet and exercise regimen form the cornerstone of therapy. Fat should comprise no more than 25–35% of the total energy requirement. Saturated fat should be limited to 7% of kcal, with the remainder from monounsaturated and polyunsaturated fatty acids.

SODIUM Hypertension is common after kidney transplantation, occurring in 50% to 80% of patients.[51] Sodium is usually restricted in the acute postoperative period if hypertension is evident or in the presence of poor allograft function. A 4-gram sodium restriction may be necessary if immunosuppressive agents cause hypertension and fluid retention.

POTASSIUM In the acute period, a potassium restriction is warranted in the presence of hyperkalemia. Hyperkalemia is caused by poor graft function as well as impaired potassium excretion with cyclosporine and potassium-sparing diuretics. In

Table 18.16 Nutrition Guidelines for Adult Kidney Transplant Patients

Nutrient	Acute Phase (up to 8 weeks following transplant and during acute rejection)	Chronic Phase (after 8 weeks)
Protein	1.3–1.5 g/kg; based on standard or adjusted body weight	1.0 g/kg; limit with chronic graft dysfunction
Calories	30%–35% kcal/kg; may increase with postoperative complications	Maintain desirable weight
Carbohydrates	50%–60% of total kcal; limit simple CHO if intolerance is apparent	50%–60% of total kcal; emphasis on complex CHO and 20–30 g dietary fiber (5–10 g per day soluble fiber)
Fats	25%–35% of total kcal	25%–35% of total kcal with saturated fat <7% of total kcal; up to 10% of kcal from PUFA, and up to 20% of kcal from MUFA
Cholesterol	—	<200 mg per day; consider plant stanols/sterols, 2 g per day
Potassium	2000–4000 mg if hyperkalemia exists	No restriction unless hyperkalemia exists
Sodium	2000–4000 mg may be necessary	2000–4000 mg with hypertension
Calcium	1200–1500 mg	1200–1500 mg
Phosphorus	1200–1500 mg (supplements may be needed)	1200–1500 mg (supplements may be needed)
Vitamins/Minerals/Trace Elements	Dietary reference intake	Dietary reference intake; may need additional vitamin D
Fluids	No restriction unless graft not functioning	No restriction unless graft not functioning

Sources: Wiggins, KL. *Guidelines for Nutrition Care of Renal Patients*, 3rd ed. American Dietetic Association; 2002.
NKF-K/DOQI. Clinical Practice Guidelines for Managing Dyslipidemias in Chronic Kidney Disease (2003). *Am J Kidney Dis.* 41; s1–s91 (suppl 3).

the chronic period, potassium is not restricted and hypokalemia has been noted in those patients taking potassium-wasting diuretics.

IMMUNOSUPPRESSANTS Immunosuppressant medications are used to prevent acute rejection and maintain long-term survival of the transplanted kidney. Multiple drugs are used in an attempt to lower levels of certain agents and decrease unwanted side effects. Some of the commonly used agents and their potential side effects are listed in Table 18.17 (drug-nutrient interactions for other renal medications are in Table 18.18). Many of these agents, such as cyclosporin and tacrolimus, can be altered by certain foods. Patients are provided detailed instructions on when to take their medications in order to guarantee adequate drug absorption. Grapefruit and grapefruit juice enhance the bioavailability of tacrolimus and cyclosporine.[52] Consequently, patients are advised to avoid these foods while taking these agents.

CARDIOVASCULAR DISEASE Posttransplant patients are at an increased risk of CVD. The prevalence of dyslipidemias in this patient population is high. Approximately 70% of transplant patients display dyslipidemia, with increases in triglycerides, total cholesterol, and LDL. Increased oxidation of LDL and possibly a decreased cardioprotective effect of HDL may also play a role.[53] The causes of dyslipidemia are multifactorial, and include (1) genetic predisposition, (2) pretransplant lipid levels, (3) certain immunosuppressive agents (prednisone, cyclosporine, sirolimus), (4) other medications (diuretics, beta blockers), (5) insulin resistance, (6) obesity, (7) changes in proteinuria, and (8) impairment in renal function. It is not clear how quickly these immunosuppressive agents alter lipoprotein metabolism. It is recommended that a lipid profile be measured during the first six months posttransplant, at one year after transplantation, and yearly thereafter. Additional testing is recommended when there are changes in immunosuppressive agents, graft function, or CVD risk.[54, 55]

In addition to the therapeutic lifestyle changes outlined in Table 18.7, patients often require lipid-lowering agents to reach target lipid levels. HMG-CoA reductase inhibitors have been found to be most beneficial in this population.[56] The ALERT (Assessment of Lescol in Renal Transplantation) study found that dyslipidemia management with statins is associated with a significant reduction in the incidence of cardiac death and of myocardial infarction.[57, 58]

HYPOMAGNESEMIA Low magnesium levels have been linked to cyclosporine and tacrolimus use due to magnesium wasting. Magnesium deficiency has been linked to coronary heart disease, insulin resistance, and type 2 diabetes in the general population. Supplementation has been shown to lower LDL levels and apolipoprotein B in renal transplant patients.[59]

OBESITY Excessive weight gain is a frequent problem that may complicate hyperlipidemia and glucose intolerance. Weight control may be problematic in this patient population, because many of the anti-rejection medications stimulate appetite. Other reasons for weight gain include lack of physical activity, genetic predisposition, age, gender, and race. Excess weight gain may have adverse effects on heart disease, lipids, blood pressure, and diabetes and possibly increase graft rejection.[60] Emphasis should be placed on diet, behavior modification, and exercise.

CALCIUM, PHOSPHORUS, AND ALTERED BONE MINERAL METABOLISM Osteoporosis and altered vitamin D metabolism are significant problems after kidney transplant. Hyperparathyroidism may remain an issue even after transplant, and contributes to low phosphorus and occasionally high serum calcium levels. In one study, 43% of patients were found to have hyperparathyroidism one year after kidney transplant.[61] Long-term corticosteroid therapy results in altered bone formation and increased bone resorption, and can lead to osteoporosis. Corticosteroids can also lead to reduced intestinal absorption of calcium and **hypercalciuria**. Often calcium

Table 18.17 Commonly Used Immunosuppressants and Potential Adverse Effects

Drug	Nutrition-Related Side Effects
Azathioprinc (Imuran)	N & V, stomatitis, esophagitis, pancreatitis, muscle wasting, loss of appetite
Basiliximab (Simulect)	N & V, diarrhea, constipation, abdominal pain
Corticosteroids/prednisone (Deltasone, liquid prednisone, Meticorten, Orasone)	Diarrhea, nausea, abdominal distension, GI hemorrhage, increased appetite, pancreatitis, osteoporosis, poor wound healing, fluid retention, hyperglycemia
Cyclosporin (Sandimmune, Neoral)	N & V, diarrhea, oral candida, gum hyperplasia, pancreatitis, hepatoxicity, nephrotoxicity, hyperkalemia
Muromonab-CD3 (Orthoclone OKT3)	N & V, stomatitis, muscle wasting, peripheral edema, stomach pain, diarrhea, tremors, body aches, flu symptoms
Mycophenolate (CellCept)	GI bleeding, N & V, abdominal pain, dyspepsia, peripheral edema, hypophosphatemia, hypercholesterolemia, hyperglycemia
Sirolimus (Rapamune)	Hyperlipidemia
Tacrolimus (Prograf)	N & V, diarrhea, constipation, albuminuria, proteinuria, hematuria, hyper-/hypokalemia, hypomagnesemia, hyperglycemia, nephrotoxicity, appetite loss
Lymphocyte immune globulin (Anti-thymocyte globulin)	N & V, diarrhea, hiccups, epigastric pain, abdominal distention, stomatitis, hyperglycemia

Sources: http://www.drugs.com; pharmaceutical company Web sites.
Nursing 2006 Drug Handbook. Philadelphia: Lippincott Williams & Wilkins; 2006.
Kopple JD, Massry S, eds. *Nutritional Management of Renal Disease.* 2nd ed. Philadelphia: Lippincott Williams & Wilkins; 2004.
Used with permission: National Kidney Foundation. Pocket Guide to Nutrition Assessment of the Patient with Chronic Kidney Disease (CKD) 2009, Fourth Edition [Revised] Online, National Kidney Foundation, New York, NY. "Potential Nutrition Effects of Common Immunosuppressant Drugs" table.

Table 18.18 Drug-Nutrient Interactions for Medications Used for Diseases of the Renal System

Multiple drugs may have simultaneous and compounding effects depending on dose and timing of administration.

Drug		Nutrient
Alcohol		↑ excretion of Mg^{++}, K^+, zinc, impaired utilization of folic acid
Antacids		↓ absorption of phosphorus, Fe^{++}; bicarbonate decreases folate and Fe^{++} absorption
Antibiotics	Cycloserine	↓ levels of B_{12} B_6, folic acid
	Neomycin	↓ absorption: fat, lactose, protein, vitamins A, D, K, B_{12}, Ca^{++}, K^+, Fe^{++}
	Isoniazid	Pyridoxine deficiency
	Tetracycline	Ca^{++}, Mg^{++}, Fe^{++}, zinc
	Tobramycin	↑ urinary loss of K^+ and Mg^{++}, hypokalemia
Anticoagulants		↓ vitamin K-dependent coagulation factors
Anticonvulsants		↑ needs for B_{12}, folic acid, Ca^{++}, Mg^{++}, vitamins K & D; pyridoxine may increase drug effect
Anti-gout		↑ excretion of K^+, Na^+, Ca^{++}, Mg^{++}, P, Mg^{++}, amino acids, chloride, vitamin B_2
Anti-inflammatory		Folic acid, Na^+, K^+, fat, vitamin C, nitrogen, B_{12}
Corticosteroids		↑ protein catabolism/↓ synthesis; ↓ absorption Ca^{++}, P, K^+; ↑ need B_6 folate, vitamins C and D, zinc
Diuretics		↑ urinary excretion of Mg^{++}, zinc, K^+, thiamin
Hypocholesterolemics		↓ absorption: fat, carotene, vitamins A, D, K, B_{12}, Fe^{++}
Laxatives		↑ fecal loss of fat, Ca^{++}, K^+, Mg^{++}, fluids, most vitamins, carotene
Mineral oil		↓ absorption of vitamins A, D, E, and K^+, Ca^{++}; may not be clinically significant

Sources: Nursing 2006 Drug Handbook. Philadelphia: Lippincott Williams & Wilkins; 2006.

Alpers DH, Stenson WF, Taylor B, Bier DM. *Manual of Nutritional Therapeutics.* Philadelphia: Lippincott Williams & Wilkins; 2008.

Used with permission: National Kidney Foundation. Pocket Guide to Nutrition Assessment of the Patient with Chronic Kidney Disease (CKD) 2009, Fourth Edition [Revised] Online, National Kidney Foundation, New York, NY. "Drugs That May Impair Absorption or Utilization of Nutrients" table.

supplementation, vitamin D, and anti-resorptive agents are warranted.

Even in the absence of hyperparathyroidism, hypophosphatemia occurs in as many as 50% of posttransplant patients.[62] This is common due to renal tubular phosphate wasting and increased urinary phosphate loss from the effects of immunosuppressive medications. Most posttransplant patients will need to increase their phosphorus. Phosphorus supplementation is often needed in the early posttransplant period. Serum potassium levels should be monitored closely, because some phosphorus supplements (Neutraphos-K) contain large amounts of potassium.

REJECTION Because doses of corticosteroids are increased during periods of acute rejection, protein and kcal require-

Table 18.19 Clinical and Patient Behavioral Outcomes in CKD

Index	Ideal Goal
Biochemical and Hematological Parameters	Maintains or achieves appropriate lab values (see Table 18.3)
Anthropometric	BMI 20–25
Clinical Signs and Symptoms	Adequate muscle/fat stores Optimal functional ability Minimal GI symptoms Food intake >80% of recommended Blood pressure within appropriate limits
Meal Planning/Food Selection	Makes appropriate food choices and takes medications as directed
Nutrient Need	Maintains appropriate protein intake Maintains lab values within acceptable limits Maintains stable glucose if diabetic
Potential Food/Drug Interaction	Aware of food and drug combinations to avoid
Exercise	If no contraindications, exercise program

Source: Adapted from Wiggins, KL. *Guidelines for Nutrition Care of Renal Patients,* 3rd ed. American Dietetic Association; 2002. pp. 7, 21.

ments are increased due to increasing catabolism. The same guidelines used for calculating protein and kcal requirements in the acute posttransplant phase apply. In the presence of chronic graft rejection, several studies have alluded to reduced protein intakes (0.55 g per kg body weight) as providing a beneficial effect in protecting the graft.[54, 56] However, the long-term efficacy of lower protein intakes needs to be further studied.

Monitoring and Evaluation

Outcome measures for MNT in the CKD population can be classified as clinical or patient behavioral outcomes. Clinical outcomes would include biochemical measures, hematological measures (see Table 18.3), anthropometrics, and clinical signs and symptoms. Patient behavioral outcomes would include meal planning, meeting nutrient needs, awareness of potential food/drug interactions, and exercise.[22] Anthropometric data would include both height and weight (current, usual, assessment of fluid status to validate the weight as well as assessment of weight gain between dialysis treatments). Clinical signs and symptoms can be assessed using subjective global assessment (SGA) or a more extensive physical assessment. Other considerations should include functional status, appetite, blood pressure control, and maintenance of adequate body mass.[22] Table 18.19 describes these factors and expected outcomes for those measures not previously addressed.

Acute Renal Failure

Definition

Acute renal failure (ARF) is a disorder characterized by abrupt cessation or reduction in GFR and accumulation of nitrogenous wastes.

Epidemiology

The prevalence of ARF is estimated at 1% for all hospitalized patients, 3% to 5% for general medical-surgical patients, 5% to 25% for those in intensive care units, 5% to 20% for open-heart surgery patients, 10% to 30% for those receiving aminoglycoside therapy (a group of antibiotics used to treat Gram-negative bacteria), 20% to 60% for those with severe burns, 20% to 30% for those with rhabdomyolysis (destruction of muscle tissue accompanied by the release of myoglobin into the bloodstream), and 15% to 25% for those treated with cisplatinum, bleomycin, and vinblastine (chemotherapeutic agents).[63–65]

It has been estimated that death occurs in 40% of nonsurgical patients with severe ARF, in as many as 80% of surgical patients, and in 20% of those with noncatabolic conditions. This high mortality is associated with the degree of hypercatabolism and infection. No method has yet been proven to reduce the catabolism observed in this patient population. ARF is associated with many clinical situations that result in a stress- or injury-induced hypercatabolic state.

Etiology

The three major types of acute renal failure are classified as **prerenal azotemia**, **intrinsic** and **obstructive**.[66] Prerenal azotemia generally refers to conditions that reduce perfusion to the kidney. This would include, for example, severe dehydration or circulatory collapse, or fluid losses from the GI tract or from extensive wounds such as seen in burns. **Acute tubular necrosis (ATN)** is an example of a more severe type of prerenal ARF. In ATN, the ischemia damages the epithelial cells, causing significant necrosis of the kidney and, potentially, irreversible renal failure. Intrinsic ARF refers to damage to the anatomical structure of the kidney. This could occur after exposure to toxins such as antibiotics, chemotherapy, or contrast dyes used in various imaging tests. Other causes of intrinsic ARF include infection such as **glomerulonephritis** or inflammation from conditions such as **Sgöjren's syndrome**. Obstructive failure results from a blockage of the ureter or neck of the bladder. Examples of conditions that could result in obstruction include kidney stones, blood clots, or a tumor.

Pathophysiology

In prerenal ARF, the lack of blood flow to kidneys results in a reduction in filtrate pressure within the glomerulus and results in the subsequent uremic symptoms. If blood flow is not restored, necrosis of the cells will occur. Liu (2010) describes four phases of ARF: initiation (when GFR declines); extension (when ischemia and inflammatory damage continue); maintenance (when GFR is at its lowest level); and finally, recovery (when epithelial cells regenerate).[66]

Clinical Manifestations

Normal urine output is 1 to 1.5 L per day. During the period when GFR declines and reaches its lowest level, ARF patients may produce <500 mL of urine.[66] ARF patients are likely to develop fluid and electrolyte disorders, azotemia, and wasting, particularly if they are both oliguric and hypercatabolic (common complications of ARF).

Electrolytes Serum levels of potassium, magnesium, and phosphorus are generally elevated in patients with ARF because of decreased renal clearance and marked net protein breakdown. On the other hand, decreased levels of serum potassium, magnesium, and phosphorus may also occur as a result of intracellular shifts associated with carbohydrate delivery and anabolism. In addition, serum phosphorus may be decreased secondary to severe respiratory alkalosis as a result of increased clearance across the dialysis membrane with **continuous renal replacement therapy (CRRT)** or because of intracellular shifts. Hypophosphatemia also occurs in the refeeding syndrome, malnutrition, and diuretic therapy. Serum levels of potassium, magnesium, and phosphorus should therefore be monitored frequently to assess the need for additional supplementation. Delivery of potassium, magnesium, and phosphorus should be individualized according to serum levels.[67]

Blood Urea Nitrogen and Creatinine Blood urea nitrogen (BUN) and creatinine are elevated in ARF, although the ratio of BUN to creatinine may be normal (10:1 or higher). Insufficient dietary kcal and protein and altered blood levels of proteases contribute to high levels of protein catabolism. Dialysis may be required to remove metabolic wastes and excess water. When recovery of renal function is expected to take several weeks, or when wasting is severe, aggressive dialysis is often recommended. Medical and nutrition management typically aim to maintain BUN in the range of 80 to 100 mg/dL.[67]

Treatment

Treatment options and nutrition support should be based on the underlying cause of renal failure, the specific metabolic changes within each patient, and other complications associated with the illness.[66] It is imperative to be aware of the medical intervention planned for each patient, particularly the type of dialysis therapy to be used, if any. Generally, continuous renal replacement therapy (CRRT) is the mode of treatment for an individual with ARF. CRRT is the term for several different modes of providing dialysis over a continuous period that allow for a significant amount of fluid (1–2 L/hour) to be removed.[66]

Nutrition Therapy for Acute Renal Failure

It is common for the nutritional status of patients with acute renal failure to decline within a short period of time, owing to nitrogen losses (up to 30 g per day), which lead to loss of lean body mass; toxicity-related symptoms (anorexia, nausea, vomiting, bleeding); loss of essential and nonessential amino acids and plasma proteins during intervention dialysis therapy; and intermediary metabolic disturbances (impaired glucose utilization and protein synthesis) from uremia. Energy and protein malnutrition often result. The ARF patient's nutrition requirements are directly influenced by the type of renal replacement therapy (if any), nutritional and metabolic status, and the degree of hypercatabolism.

Nutritional Implications

Trace Minerals and Vitamins The optimal vitamin and mineral requirements of patients with ARF are not known. Patients with ARF can develop trace mineral toxicity because of reduced renal clearance of these nutrients; thus, daily infusion of trace minerals is not recommended. Vitamin A levels can be increased secondary to enhanced hepatic release of retinol and retinol-binding protein, decreased renal catabolism, and decreased degradation of vitamin A transport protein by the kidneys.[68] Levels of vitamin D may be decreased with ARF because of impaired activation of 1,25-dihydroxycholecalciferol. Vitamin K deficiency has been reported in postoperative non-uremic patients receiving antibiotics and no enteral feedings; however, plasma levels are usually elevated, and supplementation is rarely indicated. Water-soluble vitamin deficiencies can also develop because of dialysis losses, with the exception being vitamin C, which is metabolized to oxalate. In fact, vitamin C supplements greater than 250 mg per day may cause ARF from secondary oxalosis. The majority of multivitamin preparations for parenteral solutions contain the vitamin DRI and can be used without ill effect for ARF patients.[68]

Triglycerides The type and amount of fat delivered to catabolic ARF patients is important. Serum levels of triglycerides should be within the normal range. High doses of omega-6 fatty acids may result in hypoxemia, bacteremia, and suppression of tests of immune function in vitro. Serum triglyceride levels can be useful to monitor the efficiency of hepatic removal of long-chain and medium-chain triglycerides. Elevated serum triglyceride levels suggest reduced hepatic capacity for intravenous lipids. When this is observed, a lipid-free nutrition formula (enteral or parenteral) might be in order, or the amount of total lipid given might initially be reduced.

Nutrition Intervention

Energy and protein intake should meet the patient's requirements, which are challenging to identify. The amount of nutrition administered depends on the patient's nutritional status, catabolic rate, residual GFR, and medical and/or surgical interventions (e.g., type of dialysis therapy). Patients with ARF (particularly those with underlying catabolic illness) frequently have metabolic derangements that promote degradation of protein and amino acids and consumption of fuel substrates. Protein, amino acids, and energy substrates may not be utilized efficiently, so it may be difficult to maintain or improve the nutrition status of these patients by enteral or parenteral support. When feasible, patients with ARF should receive oral nutrition. If a patient cannot eat adequately, other forms of nutrition support should be pursued. Enteral nutrition support with standard formulations should be considered first for patients with ARF. If necessary, specific formulas with lower electrolytes can be considered.[69]

General recommendations for protein are 0.6 (no dialysis) to 1.4 (dialyzed) g/kg/d.[23] Adequate kcal should also be provided (30 to 35 kcal/kg/d). The percentage of increase in protein and kcal needs in catabolic ARF patients is generally related to the precipitating event. Not only must protein and kcal requirements be estimated, but altering intake of vitamins and electrolytes should also be considered in view of the degree of renal dysfunction. Noncatabolic patients may not require dialysis therapy if normal fluid and electrolyte balance can be maintained and BUN does not exceed 80 mg/dL.

Protein sources containing both essential and nonessential amino acids should be provided. The patient's nutrition and metabolic status and the renal diagnosis determine the exact dose. For patients receiving acute hemodialysis, the recommendation is 1.2 to 1.4 g/kg of protein. For those on CRRT, protein requirements should be estimated at 1.5 to 2.0 g/kg.[69]

Despite some question about the relative nephrotoxicity of certain amino acids, current recommendations call for formulations containing mixed amino acids, essential and nonessential. There is currently no conclusive evidence that specialized formulations, such as those additionally enriched with branched-chain amino acids, have a beneficial effect on outcome. Traditional recommendations indicate that when energy requirements cannot be measured by indirect calorimetry, they can be estimated using 25 to 35 kcal/kg. According to ASPEN guidelines: "In all ICU patients receiving PN, mild permissive underfeeding should be considered at least initially. Once energy requirements are determined, 80% of these requirements should serve as the ultimate goal or dose of parenteral feeding."[69]

The total fluid intake for any patient depends on the amount of residual renal function (i.e., whether the patient is oliguric or anuric) and fluid and sodium status. In general, fluid intake can be calculated by adding 500 mL (for insensible losses) to the 24-hour urine output.[70] Fluid balance and mineral balance need to be carefully monitored in ARF to prevent over hydration and electrolyte disorders. Total fluid input is recommended to equal output from urine and all other measured sources (e.g., nasogastric aspirate or fistula drainage) plus 400 to 500 mL per day. This regimen takes into consideration the contributions of endogenous water production from metabolism and insensible water losses in breath and perspiration. If the patient is catabolic, weight can be allowed to drop by 0.2 to 0.5 kg per day to avoid excessive accumulation of fluid.[71] Records of daily intake and output, weight changes, serum electrolyte levels, and blood pressure will be useful for assessment of fluid tolerance and requirements.

Supplementation of minerals, electrolytes, and trace elements, when appropriate, is regulated by monitoring serum and urine levels in order to prevent excess or deficiency states from arising. Standard parenteral vitamin and mineral guidelines should be followed.

Nephrotic Syndrome

Definition

Nephrotic syndrome (NS) is an abnormal condition that is marked by a deficiency of albumin in the blood and its excretion in the urine due to altered permeability of the glomerular basement membranes.

Epidemiology

About two in every 10,000 people experience nephrotic syndrome. The prevalence is difficult to establish in adults because the condition is usually a result of an underlying disease. Nephrotic syndrome may occur at any age but is more prevalent in children than in adults. In children, it is most common between the ages 1½ and 4 years, and affects more males than females.

Etiology

NS can occur with many diseases, such as primary glomerular disease. A number of different diseases can result in glomerular dysfunction, including many conditions with a variety of

BOX 18.8   CLINICAL APPLICATIONS

Renal Organizations

Joining a professional interest group is an easy way to network and connect with other health care professionals in the same area of practice. Two organizations that are noteworthy in renal nutrition include the National Kidney Foundation Council on Renal Nutrition (CRN) and the American Dietetic Association Renal Practice Group (ADA RPG).

Their shared goals are to:

- Develop and promote patient and public education.
- Support the profession of the renal dietitian and promote professional education.
- Stimulate, support, encourage, and disseminate nutrition-related research.
- Impact regulatory and legislative issues.
- Promote and encourage quality nutrition care for renal patients.

Both of these organizations work together to meet these shared goals. NKF/CRN and ADA-RPG have jointly developed invaluable resources for the renal dietitian. These include *A Clinical Guide to Nutrition Care in CKD*, materials for the *National Renal Diet*, and *Suggested Guidelines for Nutrition Care of Renal Patients*.

The ADA RPG membership provides networking opportunities, a publication (quarterly newsletter containing professional education credits), stipends for meetings, and academic scholarships for graduate programs. Of particular interest is the lending library that includes current resource materials for preparing for the CSR (Certified Specialist in Renal Nutrition) exam.

The NKF/CRN functions as a professional membership council within the framework of the National Kidney Foundation (NKF) and networks with other organizations to support the National Kidney Foundation's goal of prevention, eradication, and treatment of kidney and urologic diseases. Benefits of membership include networking opportunities, discount registration at the annual scientific and educational meeting (Clinical Nephrology Meeting), a quarterly scientific journal (*The Journal of Renal Nutrition*), and a reduced rate for the renal dietitian's "Bible," the Pocket Guide for Nutritional Assessment of the Adult Renal Patient. Affiliate or chapter memberships are also available at the state level for a small fee if you are already a member at the national level. Supporting the local chapter allows for greater networking opportunities and educational offerings.

---

genetic and environmental causes. It may be the direct result of a sclerotic disease such as **focal segmental glomerulosclerosis (FSGS)** or diabetic nephropathy, or of an infection/drug toxic to the kidneys. It also may result from autoimmune diseases such as **IgA nephropathy** or systemic lupus erythematosus (SLE). Many different kinds of diseases can cause swelling or scarring of the glomerulus. Diabetic nephropathy is the leading cause of glomerular disease and of chronic kidney disease in the United States, followed by **membranous nephropathy**, the second most common cause of **nephrotic syndrome**.

Pathophysiology

As stated above, nephrotic syndrome is one of several conditions that are associated with damage to the glomerulus. Previously in the review of anatomy and physiology, it was discussed that the role of the glomerulus is to receive systemic blood flow for the initiation of filtration. Nephrotic syndrome is characterized by **proteinuria** (urinary protein levels greater than 3.5 g per 1.73 m^2 of body surface area per day), hyperlipidemia, and hypoalbuminemia <3.5 g/dL with edema.[72] Proteinuria is thought to occur through a functional derangement in the glomerular barrier that allows larger molecular weight proteins and sometimes red blood cells to leak into the urine. Superficially, NS is like kwashiorkor or protein-energy malnutrition. In both NS and kwashiorkor, albumin levels are low, plasma volume is expanded, and albumin pools shift from the extravascular space to the vascular space. Muscle wasting is common in those patients with massive and continual proteinuria and can often be masked by edema.

An early sign of NS is frothy urine; this is primarily due to proteinuria. Other symptoms include anorexia, malaise, puffy eyelids, abdominal pain, and muscle wasting. **Anasarca** with ascites and pleural effusions may occur.[72]

Depending on the degree of angiotensin II production, adults may have low, normal, or elevated blood pressure. Oliguria or acute renal failure may develop because of hypovolemia (a decrease in the volume of the circulating blood) and diminished renal perfusion. Orthostatic hypotension and even shock can be seen in pediatric patients.

Protein losses in adults with NS average about 6 to 8 grams per day.[73] Albumin is the principal protein lost in urine, accounting for between 75% and 90% of urinary protein.[74] Micronutrients lost in the urine include those bound to plasma proteins such as zinc, copper, vitamin D, and iron.

Hypoalbuminemia is common in NS due to an inadequate rate of hepatic albumin synthesis—in other words, the synthesis cannot compensate for the excessive urinary protein losses. In adults, serum albumin levels may be less than 2 g/dL.[72]

Edema is another hallmark of NS. Two mechanisms have been proposed. The first is the classic explanation, also known as the "underfill" model. A decrease in plasma albumin leads to a decrease in difference between interstitial and plasma oncotic pressure and thus plasma volume contraction. Edema occurs when the amount of fluid flowing into the interstitium exceeds maximal lymph flow. The second explanation suggests that renal disease creates primary sodium and water retention, leading to plasma volume expansion and increased capillary hydrostatic pressure. Focal edema may present as difficulty breathing (pleural effusion or laryngeal edema), substernal chest pain (pericardial effusion), scrotal swelling, swollen knees, and ascites. In children, signs of abdominal pain may be common due to edema of the mesentery. Most often, the edema is mobile (e.g., detected in the eyelids in the morning and in the ankles after ambulation).

Nephrotic syndrome has also been associated with increased risk of atherosclerosis. This is most likely due to altered lipid metabolism that is characterized by high levels of serum cholesterol and triglycerides, increased low-density lipoprotein (LDL), and either normal or low levels of high-density lipoprotein (HDL). The reduction of lipoprotein lipase, an enzyme responsible for lipoprotein metabolism, is thought to be the reason for reduced clearance of these lipids. Atherogenic lipoprotein and fibrinogen levels are also increased due to increased hepatic synthesis, which raises cardiovascular risk in NS. Animal studies strongly suggest that hyperlipidemia alone can impair renal function, and that reducing lipid levels can slow down the progression of renal injury. Although clinical studies are less conclusive, evidence suggests that this is also true in humans.[73]

Treatment

Medical treatment of nephrotic syndrome focuses on identifying and treating the underlying cause, if possible, and on reducing high cholesterol, blood pressure, and protein in the urine through a protein-controlled diet, the use of angiotensin-converting enzyme (ACE) inhibitors, or both. In diabetic patients with microalbuminuria, research has found that the use of ACE therapy reduces proteinuria and slows down the progression of chronic kidney disease. As renal disease progresses, nephrons are destroyed, and the remaining healthy nephrons must compensate (hyperfiltration) in order to maintain the body's homeostasis. In addition, the remaining nephrons' GFR increase via dilation of the glomerular afferent arteriole. This results in secondary injury to the glomerulus, increased proteinuria, and ultimately destruction of the nephron. ACE inhibitor and angiotensin receptor blockers (ARBs) therapy, including combination therapy, has been shown to reduce proteinuria in patients with diabetic nephropathy and in those with idiopathic nephritic syndrome. Proteinuria may also be decreased by lowering the patient's mean arterial pressure to levels below 92 mm Hg, independent of the class of antihypertensives used.[75] Potassium levels should be checked in those with moderate to severe renal dysfunction, because ACE inhibitors may exacerbate hyperkalemia.

To minimize accelerated atherosclerosis, hyperlipidemia should be managed with HMG-CoA reductase inhibitors or agents such as gemfibrozil in order to lower triglycerides and LDL levels. These types of drugs should not be used at the same time due to the increased risk of **rhabdomyolysis**.

Nutrition Therapy for Nephrotic Syndrome

The goals for nutrition therapy in persons with nephrotic syndrome (who are not yet on dialysis) include minimizing the effects of edema, proteinuria, and hyperlipidemia; replacing nutrients lost in the urine; and reducing the risks of further

renal progression and atherosclerosis. There are many clinical benefits of minimizing proteinuria, including an increase in serum albumin, decreased lipid levels, slowed progression of kidney disease, and reduced edema.

Nutrition Assessment

Nutrition assessment will evaluate biochemical measures of renal function (GFR, BUN, creatinine), acid-base balance, lipid profile, fluid and electrolyte balance, protein status, and vitamin/mineral status. In evaluating calcium and phosphorous, the clinician will need to consider the abnormalities posed by hypoalbuminemia. The equation most often used to calculate corrected serum calcium is shown in Table 18.20.

Table 18.20 Nutrition Assessment for Nephrotic Syndrome

Body Weight	% Standard Body Weight (SBW): (Current Weight/SBW) × 100
	% Usual Body Weight (UBW): (Current Weight/UBW) × 100
Adjusted Body Weight	Adjusted body weight (ABW) is used to calculate nutrient needs for individuals <95% or >115% of SBW:
	• Adjusted Weight (kg) = [(Actual body weight [ABW] − SBW) × Fat-free mass (FFM) factor] + SBW
	• FFM factor = 0.22–0.33 Women, 0.19–0.38 Men
Fluid Status	Total body water – Watson Method (Weight in kg; Height in cm)
	• Men: V (liters) = 2.447 + (0.3362 × Wt) + (0.1074 × Ht) − (0.09516 × Age)
	• Women: V (liters) = −2.097 + (0.2466 × Wt) + (0.1069 × Ht)
Estimated Dry Weight	• Liters of actual body water = 142 mEq/L × Liters of total body water/serum sodium (mEq/L)
	• Liters of actual body water – Liters of total body water = Excess body fluid (kg)
	• Weight (kg) – Excess body fluid (kg) = Estimated Dry Weight
Body Mass Index (BMI)	BMI = Weight (kg)/Height (m)^2
	Underweight: <20.5
	Normal: 20.5–24.9
	Overweight: 25–29.9
	Obese: >30
	KDOQI: 23.6 for women and 24.0 for men*
Corrected Serum Calcium (when serum albumin is low)	Calcium (mg/dL) + 0.0704 × [34 − Serum albumin (g/dL)]
Serum Calcium × Phosphorus Product	Calcium (mg/dL) × Phosphorus (mg/dL)

*Note: Upper 50th percentile BMI may be best for survival in maintenance dialysis patients. Maintenance dialysis patients who are >120% SBW or BMI >30 may benefit from weight reduction, but the safety and efficacy of nutrient modification for weight loss needs to be studied.

Source: Classifications of Body Mass Index by Health and Welfare. Canada: NKF K/DOQI; 1988.

American Dietetic Association Nutrition Care Manual. Renal Disease—Nephrotic Syndrome. Available at http://nutritioncaremanual.org/content.cfm?ncm_content_id=78436

Because anthropometrics will be skewed due to the presence of edema, energy and protein needs should be calculated using usual or ideal body weight. Table 18.20 outlines the tools necessary for interpretation of weight in this diagnosis.

Dietary assessment will focus on usual dietary intake with a careful evaluation of protein, phosphorous, calcium, potassium, and sodium consumption.[23]

Nutrition Diagnosis

Common nutrition diagnoses associated with nephrotic syndrome include: increased nutrient needs, inadequate protein intake, excessive sodium intake, and food/nutrition knowledge deficit.

Nutrition Intervention

The nutrition therapy prescription components for nephrotic syndrome are summarized in Table 18.4. Efforts to minimize edema include sodium restriction and diuretics. A sodium restriction of 2000 mg or less per day is recommended. Fluid intake is generally not restricted.

Current protein recommendations are 0.8 to 1.0 g/kg/day.[23] This level of intake is believed to decrease proteinuria without reducing serum albumin. Several studies have suggested that soy protein or flaxseed-based proteins may be more beneficial than high-quality proteins in reducing proteinuria, lowering lipid levels, and slowing down the progression of renal disease.[73] Maroni et al. (1997) have found that patients with nephrotic syndrome who consumed 0.8 g/kg/day of protein over 24 days maintained positive nitrogen balance.[76] There are no long-term studies to suggest the safety of low-protein diets that provide less than 0.8 g/kg/day, and thus such diets are not recommended. Additionally, protein *supplementation* is not warranted, because there appears to be no benefit for NS patients, as demonstrated in animal and human studies.[73]

Nephrolithiasis

Definition

Kidney stones (renal lithiasis or nephrolithiasis) form as a result of abnormal crystallization of calcium, **oxalate**, **struvite**, **cystine**, **hydroxyapatite**, or **uric acid** that is unable to be excreted normally in the urine.

Epidemiology

Nephrolithiasis affects almost one million individuals in the United States each year.[77] It generally occurs more frequently in men than in women and the prevalence also varies depending on ethnicity and geographic region. It occurs more frequently in Caucasians than African Americans or Asians and peaks between the ages of 20 and 30 years.[77]

Risk factors for kidney stones include family history; certain medical conditions, such as hypercalciuria, **hyperuricosuria**, and **hyperoxaluria**; and low urine volume. Hypercalciuria, an inherited condition, is the cause of more than 50% of all kidney stones. It is defined as urinary calcium excretion greater than 300 mg per 24 hours in men or 250 mg per 24 hours in women. Other causes of kidney stones include gout, excess intake of vitamin D, urinary tract infections (see Box 18.9), and urinary tract blockages.[78]

> ## BOX 18.9   CLINICAL APPLICATIONS
>
> ## Urinary Tract Infections (UTI)
>
> UTI is the most common bacterial infection in all age groups, occurring more frequently in women than men (1 in 5 women will develop a UTI over a lifetime, with some experiencing more than one infection). UTI generally occur when bacteria enter the urinary tract through the urethra and begin to multiply in the bladder. Risk factors include being female and sexually active, diabetes and other chronic illnesses, medications that lower immunity, and prolonged use of catheters in the bladder. Not all people experience symptoms, but many do get a frequent urge to urinate and a painful, burning feeling in the area of the bladder or urethra during urination. Signs and symptoms of a UTI will differ depending on which part of the urinary tract is infected (urethritis, cystitis, prostatitis, and **pyelonephritis** all have specific signs and symptoms). Uncomplicated UTI are generally treated with antibiotics followed by a urinalysis to confirm that the infection has cleared. Measures to reduce UTI, especially in those who have recurring infections, include drinking plenty of fluids, especially water and cranberry juice (the latter has infection-fighting properties). Cranberry juice should not be taken with warfarin, however, due to increased risk of bleeding. Other suggestions include urinating frequently, especially right after sexual intercourse, and avoiding potentially irritating feminine products.

Pathophysiology

Kidney stones generally consist of calcium salts, cystine, uric acid, or struvite (a combination of magnesium, ammonium, and phosphate). About 90% of kidney stones among patients treated in the U.S. are either calcium oxalate or calcium phosphate stones.[79] Generally, the development of kidney stones is not directly related to any one cause. Multiple factors may contribute, including abnormal urine flow, urine composition, and presence of renal calculi that cause retention of urine.[80] Kidney stones typically do not cause any symptoms until a stone acutely blocks urine flow. The pain will migrate depending on the location of the stone. Other symptoms include **hematuria**, nausea and vomiting, pain with urination, and an urgency to urinate.

Treatment

Diagnosis of kidney stones uses an intravenous pyelogram (IVP) and renal ultrasound, as well as urinalysis and serum levels of factors that may contribute to the stone.

Treatment of kidney stones depends on whether or not the patient can pass the stone on his or her own. Although most patients can pass the stone with plenty of fluids and pain medication, hospitalization may be required in some cases if the pain is severe. If the patient cannot pass the stone, there are several procedures available: **extracorporeal shockwave lithotripsy (ESWL)**, **percutaneous nephrolithotomy**, and **ureterorenoscopy**.[77]

ESWL is the most common procedure for the treatment of kidney stones. Shock waves are created outside of the body (extracorporeal) and travel through the skin and body tissues until they reach the denser stones. The stones break down into gravel-like particles, which are then passed through the urinary tract in the urine. Percutaneous nephrolithotomy is used to remove larger stones or when ESWL is presumed to be ineffective. Ureterorenoscopy is a procedure used when the stone is located in the mid- and lower-ureter area.

Nutrition Therapy for Nephrolithiasis

Nutrition therapy can assist with minimizing the factors that may contribute to kidney stone formation. A complete analysis of the stone composition is necessary for the development of appropriate nutrition interventions.

Nutrition Assessment

Dietary assessment should focus on nutrient intake that may affect the development of a specific stone composition. Assessment of fluid intake is also a critical factor in development of appropriate nutrition prescriptions and interventions.

Nutrition Diagnosis

Common nutrition diagnoses include excessive mineral intake, inadequate fluid intake, or food and nutrition-related knowledge deficit.

Nutrition Intervention

The objectives of nutrition therapy are to minimize the super-saturation of urinary components associated with the formation of stones and to prevent stones from recurring. Regardless of the composition of stones, the most effective preventative treatment is to increase fluid intake by 3 L per day, with at least 50% of the increase to consist of water. Fluids should be taken in divided doses, throughout the day and night. This will ensure a minimum urine output of 2 L/day.[79]

Previously it was thought that a diet low in calcium would help prevent the formation of calcium stones. However, it has now been demonstrated that foods high in calcium, including dairy products, can actually prevent the formation of stones. A possible explanation for this is that a reduction in dietary calcium aids in the intestinal absorption of oxalate, thus increasing urinary super-saturation of calcium oxalate.[81, 82] A recent analysis of >4,500 individuals found that those who consumed a diet most consistent with the DASH diet had the lowest risk for development of kidney stones.[82]

In those prone to forming calcium oxalate stones, it may be prudent to limit oxalate intake to 50–60 mg/day. The goal with minimizing oxalate intake is to decrease the bioavailability of oxalate. Some foods, such as parsley and collards, contain high amounts of oxalate, but its bioavailability is low. The following foods should be avoided due to their ability to increase urinary oxalate: beets, chocolate, cola, coffee/tea, nuts/nut butters, berries, wheat bran, spinach, and rhubarb. Avoidance of greater than 2 grams of vitamin C should also be advised.[83] Table 18.21 outlines oxalate content of selected foods.

Table 18.21 Oxalate Content of Selected Foods

High Content (>10 mg/serving)	Moderately High Content (2–10 mg/serving)	Low Content (<2 mg/serving)
Amaranth	Apple and apple juice	Artichokes
Beets* and beet greens	Apricots	Asparagus stalks
Blackberries	Asparagus	Avocado
Raspberries	Bacon	Bamboo shoots
Blueberries	Bottled beer	Bananas
Buckwheat	Brown rice	Beef (lean)
Carrots	Celery	Bing cherries
Chard	Coffee (8 oz)	Broccoli
Chives	Corn	Brussels sprouts
Chocolate*	Fruit juice with berries	Cabbage
Cocoa*	Green beans	Cauliflower
Collard greens	Lettuce	Chayote squash
Cornmeal Lima beans	Mushrooms	Cheese
Currants, red	Nuts (macadamia/pistachio/walnut)	Chicken
Dandelion greens	Oatmeal	Chicory
Eggplant	Onions	Cola sodas
Escarole	Oranges	Cranberries
Figs	Peaches	Cucumber
Fruit juice with berries	Rice milk	Eggs
Fruit salad (canned)	Sardines	Grapefruit
Green bell peppers	Sponge cake	Grapes, green and red
Grits (white corn)	Tomatoes and tomato juice	Jelly
Kiwi	Turnip	Lamb (lean)
Kale	White bread and many products made with white flour	Lemonade/limeade
Leeks		Melons
Lemon/lime peel		Milk and other dairy products
Marmalade		Nectarine
Nuts (peanuts/almonds*; cashews/hazelnuts/pecans)		Noodles (egg)
Okra		Oatmeal
Parsley		Oils
Parsnips		Orange juice (4 oz)
Peanut butter		Pears
Pepper (>2 tsp/day)		Peas
Pokeweed		Peppers
Potatoes		Pineapple
Poppy seeds		Pork (lean)
Rhubarb*		Poultry
Rutabaga		Prunes and plums
Soybeans & all products		Radishes
Spinach*		Rice
Squash, summer and winter		Seafood
Strawberries*		Spaghetti
Star fruit		Tea, herbal
Sweet potatoes		Wine
Tea*		Yogurt
Textured vegetable protein/meat substitutes*		
Tofu		
Tomato paste		
Watercress		
Wheat germ*/bran		

*Have been documented to raise urinary oxalate excretion.

Source: Oxalosis & Hyperoxaluria Foundation (OHF) (http://www.ohf.org/docs/Oxalate2008.pdf)

Table 18.22 Nutrition Therapy for Prevention and Treatment of Kidney Stones

- Fluids: 12 cups to 16 cups to produce a urine volume >2.5 L
- Protein: Normalize intake 0.8 g/kg to 1.0 g/kg of body weight/day; not to exceed the Dietary Reference Intakes (DRI)
- Calcium: Normalize intake 800 mg/day for men;1,200 mg/day for women; do not restrict and balance intake of calcium throughout the day
- Oxalate: <40 mg/day to 50 mg/day
- Sodium: Lower intake to 2,300 mg/day to 3,450 mg/day
- Vitamin C: Limit to <100 mg/day

Source: Adapted from http://www.kidney.org/atoz/content/diet.cfm

For stones that are composed of uric acid, avoidance of foods high in purine is recommended.[80] Foods high in purine include: animal protein, seafood, meat extracts, consommé, gravies and organ meats. The overall nutrition therapy for prevention and treatment of kidney stones is summarized in Table 18.22.

Conclusion

The kidneys are vital organs responsible for maintenance of many homeostatic functions necessary for life. Deterioration of kidney function presents complex medical and nutritional problems. Comorbidities that develop, such as cardiovascular disease, anemia, and osteodystrophy, contribute to the fragile nature of the patient with CKD. Individuals who develop end-stage disease requiring renal replacement therapy lead very altered lives and face many challenges as they try to survive with kidney disease.

Nutrition therapy, by interacting with metabolism and the mechanisms of kidney disease pathology, has the potential to address risk factors for CKD, delay progression of established disease, and contribute toward better patient outcomes for those receiving renal replacement therapy. The correlation of nutritional status and nutrition indices to patient outcomes highlights the relevance of nutrition therapy for both clinical care and research.

PRACTITIONER INTERVIEW

Marianne Hutton, RD, CSR, CDE, *Renal Dietitian, Advertising Editor for the Renal Nutrition Forum, a publication of the Renal Practice Group of the American Dietetic Association and Program Chair, National Kidney Foundation Council on Renal Nutrition 2010 Spring Clinical Meeting, Orlando Florida, April 13–17*

I am currently working as a renal dialysis dietitian in an outpatient kidney dialysis clinic. My career has taken quite a few twists and turns. I started in the field as a registered dietetic technician and enjoyed it so much that I continued in the Coordinated Undergraduate Program (CUP) in dietetics. I still remember the first time I had to see a renal patient. Because of my inexperience, my self-confidence was not great and I thought to myself, "This patient will see right through me and will know I'm a rookie." I went into the room, did the interview, taking one step at a time, and I got through it. My instructor said I did a great job and that most interns felt the same way I had felt.

After I became an RD in 1979, I worked in acute care hospitals, but when I had to find another job in 1986, I went into outpatient renal dialysis. I now work part-time in one dialysis facility, per diem at two local hospitals, and have a private practice (Finely Fit Nutrition Services, www.finelyfitnutrition.com) providing medical nutrition therapy (MNT) and teaching group classes for people with diabetes. When I started working with dialysis patients, my only experience had been during my internship. I didn't really even comprehend the differences in MNT between inpatient acute care and outpatient hemodialysis. Since that time a new staging system for treating kidney disease has been developed. I took the attitude that the patient was the expert about his circumstances, and it has served me well. The first time I see a client who has just started on dialysis, he or she is often frightened or anxious but too ill and weak to really let it show. At the first meeting, I try to build rapport, trust, and the patient's confidence in me. I listen, in order to individualize my consultation, and let him or her lead the way in integrating what I think will best serve his needs. The patient will often say, "I am overwhelmed with so much that has happened—so scared that I don't know what to eat and not eat." All he or she has heard since being diagnosed with end-stage renal disease is that the dietary limitations will be severe. I try to take some of that burden away by telling the patient not to worry—that *together* we can make managing his condition easier.

Ideally, a patient has seen a dietitian before beginning dialysis. But the reality is that he is often referred to a nephrologist when the disease is advanced, and all too often, by the time I see the patient, he is becoming malnourished and has developed secondary hyperparathyroidism.

Why didn't I see him earlier? There are many barriers that come to mind. Some nephrologists won't accept referrals to see patients until they have very late stage renal disease and need renal replacement therapy. Other reasons are lack of early referral by the primary physician to a nephrologist or to a dietitian who specializes in MNT for chronic kidney disease (CKD), the patient's denial of the loss of kidney function, cost of nutritional counseling, or lack of transportation. Medicare covers MNT for people with stage 3 CKD but this is an underutilized benefit.

At the dialysis clinic I meet with patients at least twice a month, but I check their lab results and charts more often. There are a lot of nutritional issues with dialysis, but I would say the most common problem with my clients is insufficient protein intake. If I could only have made one dietary change, it usually is to advise and facilitate the consumption of adequate, high-biological value protein with adequate calories. I'm sure students reading this are thinking, "So what about all those other nutritional issues?" Well, you have to know and practice them, but if the patient is malnourished, protein and energy needs usually take priority. Once the patient's nutritional status is stable, if I could make only one dietary change, it would be limiting dietary sodium. Following that is limiting phosphorus and potassium for some but not all patients. Finally there is the fluid limit.

Since I started as a renal dietitian, I have seen tremendous changes in the delivery of care to renal patients. A few examples are Medicare reimbursement for MNT; ability to delay the need for dialysis through dietary changes and medication; larger variety of better nutritional products available for patients; the use of human erythropoietic agents instead of blood transfusions; better phosphate binders and use of vitamin D hormone and calcimimetics instead of parathyroidectomy. One way I stay current is by being an active member of the American Dietetic Association's Renal Practice Group and the National Kidney Foundation Council on Renal Nutrition.

As a renal dietitian I am a respected member of the treatment team and feel that I make a difference in the patient's well-being. Interestingly, this attitude has carried over from the dialysis clinic setting to my other jobs. Many years ago while working as a part-time RD in an acute care hospital, I recommended phosphate binders and oral vitamin D for a renal patient. It turned out that my suggestions to the doctor helped him to become aware that calcium acetate was used in this circumstance to bind dietary PO_4 and that oral vitamin D hormone or "active" vitamin D controlled the secondary hyperparathyroidism. This led him to be more confident with my skills and knowledge. He asked me if I would consider working with him as a renal case manager and created the position with RD as one of the qualifications. Prior to this, only nurses were case managers at Kaiser in Northern California.

My advice for students is to be involved with your profession and the people you work with. Join the ADA and its practice groups. Take on leadership responsibilities and embrace change. Change will happen—you can't stop it, so stay in front of it by making contributions and being one of the movers and shakers. Save time to promote your profession, educate other health professionals on the role of a dietitian, and stand firm and take action when others—knowingly or unknowingly—give nutritional advice that is within your scope of practice. Above all, have fun and keep learning every day!

Application of the Nutrition Care Process: CKD

Introduction:

Mrs. J is a 26-year-old Native American who presents with a history of renal insufficiency, hypertension, and type 2 diabetes mellitus. Her current symptoms include anorexia, N/V, 4 kg recent weight gain, edema, shortness of breath, pruritus, and inability to urinate. She is admitted with a diagnosis of Stage 5 CKD with plans to initiate hemodialysis. Medications: Captopril; Calcitriol; erythropoietin; vitamin/mineral supplement; Glucophage.

1. Define each of the patient's current symptoms. Explain how they are related to chronic kidney disease.

2. How are hypertension and type 2 diabetes mellitus related to her kidney disease?

3. Identify each of her medications. What is the rationale for each?

Nutrition Assessment:

Food/Nutrition-Related History:

Pt. states that she has a very poor appetite. Her 24-hour recall is as follows: Breakfast: dry toast, Pepsi 12 oz; Lunch: 1 oz ham sandwich on bun with 1 oz American cheese, iced tea; Dinner: spaghetti sauce on 1 c. noodles, iced tea.

4. Assess her dietary intake and compare it to her recommended nutritional needs.

Anthropometric Measurements:

Ht. 5'0" Wt. 170 lbs. UBW 160–162 lbs.

Biochemical Data:

Alb. 3.4 g/dL; Na 130 mEq/L; K$^+$ 5.6 mEq/L; Cl 91 mEq/L; PO_4 9.5 mEq/L; Mg 2.9 mEq/L; BUN 69 mg/dL; Cr 12 mg/dL; Glucose 200 mg/dL; Hgb A$_1$C 8.9%.

Nutrition Diagnosis:

5. Identify at least two nutrition problems based on the nutrition assessment and medical history. Determine the diagnostic term for each nutrition problem. Next, identify the etiology of each nutrition problem. Finally, identify the signs and symptoms that support the evidence for these nutrition problems.

Nutrition Intervention:

6. Calculate the appropriate energy and protein recommendations for Mrs. J.

7. Write the complete nutrition prescription for this patient.

8. Select additional appropriate nutrition interventions.

Nutrition Monitoring and Evaluation:

9. Determine nutrition criteria for monitoring and evaluation for each nutrition diagnosis that you identified.

Atlas of Diseases of the Kidney Online Atlas of Kidney Diseases; numerous clinical PowerPoint presentations.
http://www.kidneyatlas.org

Hypertension, Dialysis and Clinical Nephrology Clinical resource; provides slides from lectures given at major nephrology meetings—some are free and some require a subscription.
http://www.hdcn.com

International Society for Peritoneal Dialysis (ISPD) Clinical information and resources for peritoneal dialysis modality.
http://www.ispd.org

National Kidney Foundation Council on Renal Nutrition Online format of the most widely used "pocket guide." The online format of the 4th edition enables the reader to view and/or print select chapters of the Pocket Guide and link to specific NKF KDOQI Clinical Practice Guidelines that are referenced in the Pocket Guide; contains a collection of practical assessment tools and guidelines to evaluate the nutritional status and calculate the nutritional needs of the adult and pediatric patient with CKD. Access to the online resource available to CRN members only.
http://www.kidney.org/professionals/crn/pocketGuide/login.cfm

National Institute of Diabetes and Digestive and Kidney Diseases (NIDDK) Part of the NIH dedicated to basic and clinical research; extensive consumer health information related to diabetes and kidney disease; websites for National Education Programs for kidney disease and diabetes.
http://www.niddk.nih.gov

The Nephron Information Center One of the most widely used websites among renal dietitians; nutrition section and comprehensive coverage of nephrology, GFR, and other calculators.
http://www.nephron.com

National Kidney Foundation KDOQI (Kidney Disease and Outcome Quality Initiative) Guidelines Original guidelines and updates to NKF KDOQI guidelines including HD, PD, anemia, vascular access, nutrition, CKD, CVD, and dyslipidemia.
http://www.kidney.org/professionals/KDOQI/

United States Renal Data System (USRDS) Extensive collection of end-stage renal disease-related data; funded by NIDDK working with CMS.
http://www.usrds.org

Kidney School Comprehensive, self-paced patient education program on all aspects of chronic kidney disease; from Life Options Rehabilitation Advisory Council (LORAC).
http://www.kidneyschool.org

Renal Web Known as a "neutral, nonaffiliated web site for the dialysis industry." Nutrition is one of many topics covered. Renal WEB is an independent company.
http://www.renalweb.com

Culinary Kidney Cooks Recipes for renal diets updated weekly; includes renal exchanges and helpful dietary hints.
http://www.culinarykidneycooks.com

American Association of Kidney Patients Online brochures, Na-K-Phos Counter, PD & HD advisory, Protein and Kcal Counter.
http://www.aakp.org

The Nephron Information Center—Food Values Find nutritional information instantly (database: USDA).
http://foodvalues.us

Cook's Thesaurus The Cook's Thesaurus is a cooking encyclopedia that covers thousands of ingredients and kitchen tools. Entries include pictures, descriptions, synonyms, pronunciations, and suggested substitutions.
http://www.foodsubs.com

Renal Dialysis—A Team Effort This video webstream is a virtual tour of the Santa Clara Valley Medical Center's Renal Care Center in San Jose California that was designed to help introduce health care students and clients to hemodialysis from the patient and renal professional perspective.
http://www.nufs.sjsu.edu/renaldial/index.html

Reference: Knotek B., McCarthy M., and Smith B. Bookmark these favorites: a guide to getting started on the internet. *Journal of Renal Nutrition,* Vol 13, Issue 4, Pages E1–E4 (October 2003).

END-OF-CHAPTER QUESTIONS

1. List the top three diseases and four risk factors for developing kidney disease.

2. How does development of diabetes lead to kidney disease?

3. How is kidney function assessed? Explain the formula used for this assessment.

4. Symptomatically, how does acute renal failure (ARF) differ from chronic kidney disease (CKD)? Which conditions most commonly lead to ARF?

5. What is nephrotic syndrome? How are its signs and symptoms like protein-energy malnutrition? What is the cause of the edema? Should you recommend a high level of dietary protein for patients with nephrotic syndrome? Why or why not?

6. Describe the five stages of chronic kidney disease and include GFR in your explanation. For each stage, list the recommended frequency of nutrition assessment and components of assessment.

7. Describe the two renal replacement treatments used in kidney failure. How do they differ with respect to access to the circulatory system, type of dialysate, length of dialysis, frequency of dialysis, and kcal obtained from glucose during dialysis? At what stage of kidney disease is dialysis initiated?

8. What are the protein and energy recommendations for patients undergoing HD or PD? Should body weight be adjusted to determine these

amounts? Why or why not? Which formula is used to determine edema-free body weight?

9. Why must dietary potassium, sodium, phosphates, and fluid be restricted during HD? Which vitamins are supplemented during HD and which ones must not be taken in excess?

10. What are the risk factors for cardiovascular disease in CKD? What are the current dietary recommendations used to help prevent cardiovascular disease in patients with CKD?

11. What is osteitis fibrosa? How are calcium, phosphate, PTH, and vitamin D involved in its pathophysiology? What are the dietary recommendations to help control its development?

12. Anemia is very common in patients with CKD. Why does it occur, and how is it treated?

13. List the long-term nutrition therapy goals for post-renal transplant patients.

14. What are the common causes of kidney stone formation?

REFERENCES

1. Levey AS, Stevens LA, Schmid CH, et al. A new equation to estimate glomerular filtration rate. Annals of Internal Medicine. 2009;150:604–12.

2. National Institute of Diabetes and Digestive and Kidney Disease of the National Institute of Health (NIH). Kidney and urologic disease statistics for the United States. Available at http://kidney.niddk.nih.gov/kudiseases/pubs/kustats/index.htm#1

3. Sherwood L. Urinary System. In: Human physiology: from cells to systems. 7th ed. Belmont (CA): Brooks/Cole/Cengage Learning; 2010.

4. Rose, DR. Clinical assessment of renal function. In: Rose, BD (Ed.). Pathophysiology of renal disease. 2nd ed. New York: McGraw-Hill; 1987, 1–39.

5. Madsen K, Verlander J. Renal structure in relation to function. In: Wilcox C and Tisher C. Craig, editors. Handbook of nephrology & hypertension. 5th ed. Philadelphia (PA): Lippincott Williams & Wilkins; 2005, 3–13.

6. Bargman JM, Skorecki K. "Chapter 274. Chronic Kidney Disease" (Chapter). Fauci AS, Braunwald E, Kasper DL, Hauser SL, Longo DL, Jameson JL, Loscalzo J: Harrison's principles of internal medicine. 17th ed. Available at http://www.accessmedicine.com/content.aspx?aID=2880823

7. Barri Y. Vascular disorders of the kidney. In: Carpenter G, Griggs R, Loscalzo J, editors. Cecil essentials of medicine. Philadelphia (PA): WB Saunders; 2001, 278–82.

8. Zawada, E. Initiation of dialysis. In: Daugirdas J, Blake P, Ing T, editors. Handbook of dialysis. Philadelphia (PA): Lippincott Williams & Wilkins; 2001; 3–11.

9. Kopple JD, Massry SG, eds. Nutrition management of renal disease. 2nd ed. Philadelphia (PA): Lippincott Williams & Wilkins; 2004.

10. Goldstein DJ. Assessment of Nutritional Status in Renal Disease. In: Mitch W, Klahr S, editors. Handbook of nutrition and the kidney. 5th ed. Philadelphia (PA): Lippincott-Raven; 2005.

11. National Kidney Foundation. K/DOQI clinical practice guidelines for chronic kidney disease: evaluation, classification, and stratification. Am J Kidney Dis. 2002 Feb;39(2 Suppl 1):S1–266.

12. National Kidney Foundation: Kidney Disease Outcomes Quality Initiative Clinical Practice Guidelines for Nutrition in Chronic Renal Failure. Adult Guidelines. American Journal of Kidney Diseases. 2000;35 (suppl 2):s17–s104.

13. KDIGO clinical practice guideline for the diagnosis, evaluation, prevention, and treatment of Chronic Kidney Disease-Mineral and Bone Disorder (CKD-MBD). Kidney Disease: Improving Global Outcomes (KDIGO) CKD-MBD Work Group. Kidney Int Suppl. 2009;(113):S1–130.

14. National Institutes of Health. National Kidney Disease Education Program (NKDEP). Available at http://nkdep.nih.gov/

15. Daugirdas J, Van Stone J, Boag J. Hemodialysis apparatus. In: Daugirdas J, Blake P, Ing T, editors. Handbook of dialysis. 3rd ed. Philadelphia (PA): Lippincott Williams & Wilkins; 2001, 46–66.

16. Besarab A, Raja R. Vascular access for hemodialysis. In: Daugirdas J, Blake P, Ing T, editors. Handbook of dialysis.

3rd ed. Philadelphia (PA): Lippincott Williams & Wilkins; 2001, 67–101.

17. Lindsay R, Daily hemodialysis: the time has come. Am J Kidney Dis. 2005; 45:793–97.

18. Blake P, Daugirdas J. Physiology of peritoneal dialysis. In: Daugirdas J, Blake P, Ing T, editors. Handbook of dialysis. 3rd ed. Philadelphia (PA): Lippincott Williams & Wilkins; 2001; 281–96.

19. Sorkin M, Blake P. Apparatus for peritoneal dialysis. In: Daugirdas J, Blake P, Ing T, editors. Handbook of dialysis. 3rd ed. Philadelphia (PA): Lippincott Williams & Wilkins; 2001, 297–308.

20. National Institute of Diabetes and Digestive and Kidney Diseases. National Diabetes Statistics, 2007 fact sheet. Bethesda (MD): U.S. Department of Health and Human Services, National Institutes of Health; 2008.

21. United Network for Organ Sharing. Available at http://www.unos.org/

22. Wiggins, KL. Guidelines for nutrition care of renal patients. 3rd ed. American Dietetic Association; 2002.

23. National Kidney Foundation. Pocket Guide to Nutrition Assessment of the Patient with Chronic Kidney Disease (CKD). 4th ed. [Revised] Online, National New York: Kidney Foundation; 2009.

24. Ikizler TA, Flakoll PJ, Parker RA, Hakim, RM. Amino acid and albumin losses during hemodialysis. Kidney Int. 1994;46:830–37.

25. Blumenkrantz MJ, Gahl GM, Kopple JD, Kamdar AV, Jones MR, Kessel M, Coburn JW. Protein losses during peritoneal dialysis. Kidney Int. 1981;19:593–602.

26. Bannister DK, Acchiardo SR, Moore LW, et al. Nutritional effect of peritonitis in continuous ambulatory peritoneal dialysis. J Am Diet Assoc. 1987;87:53–56.

27. Rocco, MV, Blumenkrantz MJ. Nutrition. In: Daugirdas JD, Blake P, Ing T, eds. Handbook of dialysis. 3rd ed. Philadelphia (PA): Lippincott Williams & Wilkins; 2001, 420–45.

28. Heimburger O, Stevinkel P, Lindholm B. Chronic peritoneal dialysis. In: Kopple J, Massry S, editors. Nutritional management of renal disease. 2nd ed. Philadelphia (PA): Lippincott Williams & Wilkins; 2004, 477–511.

29. National Kidney Foundation. K/DOQI Clinical practice guidelines for managing dyslipidemias in chronic kidney disease. Am J Kidney Dis. 2003;41; s1–s91 (suppl 3).

30. Byham-Gray L, Wiesen, K. A clinical guide to nutrition care in kidney disease. American Dietetic Association; 2004.

31. Block GA, Hulbert-Shearon TE, Levin NW, Port FK. Association of serum phosphorus and calcium $\times$ phosphorus product with mortality risk in chronic hemodialysis patients: a national study. Am J Kidney Dis. 1998;31:607–17.

32. National Kidney Foundation. K/DOQI Clinical practice guidelines for bone metabolism and disease in chronic kidney disease, Am J Kidney Dis. 2003;42;s1–s202 (suppl 3).

33. Ganesh SK, Stack AG, Levin NW, Hulbert-Shearon T, Port FK. Association of elevated serum phosphorus, calcium $\times$ phosphorus product and parathyroid hormone with cardiac mortality risk in chronic hemodialysis patients. J Am Soc Nephrol. 2001;12(10):2131–38.

34. Block GA, Klassen PS, Lazarus JM, Ofsthun N, Lowrie EG, Chertow GM. Mineral metabolism, mortality and morbidity in maintenance HD. JASN. 2004;15:2208–18.

35. Chazot C, Kopple J. Vitamin metabolism and requirements in renal disease and renal failure. In: Kopple J, Massry S, editors. Nutritional management of renal disease. 2nd ed. Philadelphia (PA): Lippincott Williams & Wilkins; 2004;315–56.

36. Jones WO, Nidus BD. Biotin and hiccups in chronic dialysis patients. J Ren Nut. 1991;2:80–83.

37. Wilcken DE, Dudman NPB, Tyrrell PA, Robertson MR. Folic acid lowers elevated plasma homocysteine in chronic renal insufficiency: Possible implications for prevention of vascular disease. Metabolism. 1988;37:697–701.

38. Suliman M, Stenvinkel P, Qureshi AR, Kalantar-Zadeh K, Bárány P, Heimbürger O, Vonesh EF, Lindholm B. The reverse epidemiology of plasma total homocysteine as a mortality risk factor is related to the impact of wasting and inflammation. Nephrol Dial Transplant. 2007;22:209–17.

39. National Kidney Foundation. K/DOQI Clinical practice guidelines for cardiovascular disease in dialysis patients. Am J Kidney Dis. 2005;45:S1–S154 (suppl 3).

40. Foley RN, Parfrey PS, Sarnak MJ. Clinical epidemiology of cardiovascular disease in chronic renal disease. AJKD 1998;32:S112–S119.

41. Yan AT, Yan RT, Tan M, Constance C, Lauzon C, Zaltzman J, et al. Canadian acute coronary syndromes (ACS) Registry investigators. Treatment and one-year outcome of patients with renal dysfunction across the broad spectrum of acute coronary syndromes. Can J Cardiol. 2006;22:115–20.

42. McGonigle RSR, Wallin JD, Shadduck RK, Fisher JW. Erythropoietin deficiency and erythropoiesis in renal insufficiency. Kidney Int. 1984;25:437–44.

43. NKF KDOQI Clinical Practice Guidelines and Clinical Practice Recommendations for Anemia in Chronic Kidney Disease. Am J Kidney Dis. 2007;47(suppl 3):S1–S146.

44. Sunder-Plassmann G, Hurl WH. Erythropoietin and iron. Clin Nephrol. 1997;47:141–57.

45. Tarng D, Huang T, Chen TW, Yang W. Erythropoietin hyporesponsiveness: from iron deficiency to iron overload. Kidney Int. 1999;55;S107–S118.

46. Auerbach M. Ferumoxytol as a new, safer, easier-to-administer intravenous iron: yes or no? Am J Kidney Dis. 2008;52:826–29.

47. Markell MS, Armenti V, Danovitch G, et al. Hyperlipidemia and glucose intolerance in the renal transplant patient. J Am Soc Nephrol. 1994;27:117–23.

48. Nakai I, Omoni Y, Aikawa I, Yasumura T, Suzuki S, Yoshimura N, et al. Effect of cyclosporine on glucose metabolism in kidney transplant recipients. Transplant Proc. 1988;20;969–78.

49. Roth D, Milgrom N, Esquenazi V, Fuller L, Burke G, Miller J. Post Transplant hyperglycemia: Increased incidence in cyclosporine treated allograft recipients. Transplantation. 1989;47:278–81.

50. Bordreaux JP, McHugh L, Canfax DM Ascher N, Sutherland DE, Payne W, Simmons RL, Najarian JS, Fryd DS. The impact of cyclosporine and combination immunosuppression on the incidence of posttransplant diabetes in renal allograft recipients. Transplantation. 1987;44: 371–81.

51. Kasiske B, Magdalena AA. Nutritional management of renal transplantation. In: Kopple J, Massry S, editors. Nutritional Management of Renal Disease 2nd ed. Philadelphia (PA): Lippincott Williams & Wilkins; 2004, 513–25.

52. Hasse JM. Recovery after organ transplantation in adults: the role of postoperative nutrition therapy. Top Clin Nutr. 1998;13:15–26.

53. Arnadottir M, Berg A. Treatment of hyperlipidemia in renal transplant recipients. Transplantation. 1997;63:339–45.

54. Kasiske BL, Bazquez MA, Harmon WE, et al. Recommendations for the outpatient surveillance of renal transplant recipients. J Am Soc Nephrol. 2000:11,s1.

55. National Kidney Foundation. KDOQI Clinical Practice Guidelines for Managing Dyslipidemias in Chronic Kidney Disease Available at: http://www .kidney.org/professionals/KDOQI/ guidelines_lipids/index.htm

56. Martinez-Castelao A, Grinyo JM, Gil-Vernet S, et al. Lipid lowering long term effects of six different statins in hypercholesterolemic renal transplant patients under cyclosporine immunosuppression. Transplant Proc. 2002;34: 398–400.

57. Reilly, R. The patient with kidney stones. In: Schrier, R, editor. Manual of nephrology. Philadelphia (PA): Lippincott Williams & Wilkins; 2005, 78–90.

58. Holdaas H, Fellstrom B, Jardine A, et al. Prevention of cardiac death and non-fatal coronary events with fluvastatin in renal transplant patients: a multicentre randomized placebo controlled trial. Lancet. 2003;361:2024–31.

59. Gupta BK, Glicklick D, Tellis VA. Magnesium repletion therapy improves lipid metabolism in hypomagnesemic

renal transplant recipients. Transplantation. 1999:69:1485–87.

60. Patel MD. The effect of dietary intervention on weight gain after renal transplantation. JRN. 1998;8:137–141.

61. Messa P, Sindici C, Cannella G, et al. Persistant SHPT after renal transplantation. Kidney Int. 1998;54:1704–13.

62. Massari PU. Disorders of bone and mineral metabolism after renal transplantation. Kidney Int. 1997;52:1412–21.

63. Anderson R, Schrier R. Acute tubular necrosis. In: Schrier R, Gottschalk CW, editors. Diseases of the Kidney. Boston (MA): Little, Brown; 1993, 1287–1318.

64. Thadhani R, Pascual M, Bonventre J. Acute renal failure. N Engl J Med. 1996;334:1448–60.

65. Albright RC. Acute renal failure: a practical update. Mayo Clin Proc. 2001;76:67–74.

66. Liu Kathleen D, Chertow Glenn M. "Chapter 273. Acute Renal Failure" (Chapter). Fauci AS, Braunwald E, Kasper DL, Hauser SL, Longo DL, Jameson JL, Loscalzo J: Harrison's principles of internal medicine. 17th ed. Available at http://www.accessmedicine .com/content.aspx?aID=2872603

67. Mehta R. Therapeutic alternatives to renal replacement for critically ill patients in acute renal failure. Semin Nephrol. 1994;14(1):64–82

68. Druml W. Nutritional management of acute renal failure. In: Jacobson H, Striker G, Skahr S, editors. The principles and practice of nephrology. St. Louis: CV Mosby; 1995, 745–53.

69. McClave SA, Martindale RG, Vanek VW, McCarthy M, Roberts P, Taylor B, Ochoa JB, Napolitano L, Cresci G. A.S.P.E.N. Board of Directors; American College of Critical Care Medicine; Society of Critical Care Medicine. Guidelines for the Provision and Assessment of Nutrition Support Therapy in the Adult Critically Ill Patient: Society of Critical Care Medicine (SCCM) and American Society for Parenteral and Enteral Nutrition (A.S.P.E.N.). J Parenter Enteral Nutr. 2009;33:277–316.

70. Weiner Feldman R. Nutrition in acute renal failure. J Renal Nutr. 1994;4(2)97–99.

71. Monson P, Mehta, R. Nutrition in acute renal failure: a reappraisal for the 1990s. J Renal Nutr. 1994;(4)2:5–B77.

72. Lewis Julia B, Neilson Eric G, "Chapter 277. Glomerular Diseases" (Chapter). Fauci AS, Braunwald E, Kasper DL, Hauser SL, Longo DL, Jameson JL, Loscalzo J. Harrison's principles of internal medicine. 17th ed. Available at http://www.accessmedicine .com/content.aspx?aID=2897259

73. Kaysen GA, and Yeun, JY. Nephrotic syndrome: nutritional consequences and dietary management. In: Mitch WE, Klahr S, editors. Nutrition and the kidney. 5th ed. Philadelphia (PA): Lippincott Williams & Wilkins; 2005, 160–75.

74. Don B, Kaysen G. Nutritional and nonnutritional management of the nephrotic syndrome. In: Kopple JD, Massry SG, editors. Nutritional management of renal disease. 2nd ed. Philadelphia (PA): Lippincott Williams & Wilkins; 2004, 415–32.

75. Adler S, Fairley K. The patient with hematuria, proteinuria, or both, and abnormal findings on urinary microscopy. In: Schrier R, editor. Manual of nephrology. 6th ed. Philadelphia (PA): Lippincott Williams & Wilkins; 2005, 116–33.

76. Maroni BJ, Staffeld C, Young VR, et al. Mechanisms permitting nephritic syndrome patients to achieve nitrogen equilibrium with a protein restricted diet. J Clin Invest. 1997;99:2749–87.

77. National Kidney Foundation. Kidney Stones. Available at http://www.kidney .org/atoz/atozTopic_KidneyStones.cfm

78. Worcester EM, Coe FL. Nephrolithiasis. Prim Care. 2008 Jun;35(2):369–91, vii. Review.

79. American Dietetic Association. Nutrition Therapy for Urolithiasis/Urinary Stones. In: Nutrition Care Manual. Available at http://www.nutritioncaremanual .org/topic.cfm?ncm_heading= Nutrition%20Care&ncm_toc_id=22599

80. Grases F, Costa-Bauza A, Prieto RM. Renal lithiasis and nutrition. Nutr J. 2006;5:23.

81. Taylor EN, Curhan GC. Diet and fluid prescription in stone disease. Kidney Int. 2006;70:835–39.

82. Taylor EN, Fung TT, Curhan GC. DASH-style diet associates with reduced risk for kidney stones. J Am Soc Nephrol. 2009 Oct;20(10):2253–39.

83. Massey LK. Food oxalate: factors affecting measurement, biological variation, and bioavailability. J Am Diet Assoc. 2007;107:1191–94.

19

Diseases of the Hematological System

Roschelle A. Heuberger, PhD, RD
Associate Professor, Central Michigan University

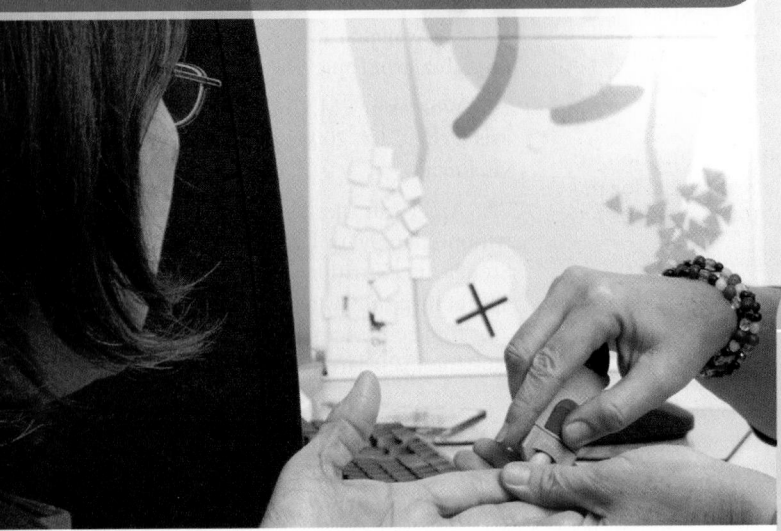

The Hematological System

Early man understood that blood is essential for life, as evidenced by early treatments for the sick, including bloodletting (bleeding of the patient), the use of leeches, and the administration of blood extracts. The hematological system—which includes the blood and blood vessels, bone marrow, spleen, and other tissues—is intertwined with all other major organ systems in the body. Blood requires the pulmonary system for oxygenation, the circulatory system for circulation, the excretory systems for the removal of waste products, and the immune system for the development of white blood cells. The hematological system involves the bone marrow and spleen for **hematopoiesis**, the lymph system for cell maturation and immune function; the liver for protein production, trafficking, and activation; and the endothelial vasculature of the blood vessels for clotting and bleeding. This chapter will address the more common diseases of the hematological system.

Blood Composition

Blood levels of nutrients and other components are kept relatively constant through homeostatic feedback mechanisms. For example, serum levels of calcium are kept in a very tight range, because high or low calcium levels affect so many processes that dysregulation may cause shutdown of all body systems. The composition of blood is used as a diagnostic tool for disease presence, severity, or risk. The ranges of constituents regarded as "normal" are based on population studies of healthy persons. Some constituents in blood have half-lives that are unsuitable for establishing a timeline for the etiology of a disease, while

anemia—abnormal blood constituents resulting from various etiologies; anemia is a symptom and is often a result of the decrement in blood constituents, although some forms of elevated blood components that are nonfunctional may be referred to as an "anemia"

angiogenesis—the formation of new blood vessels and expanded systems for nutrient delivery and waste removal; a result of chemokines and hormonal messages that up-regulate the formation of these processes; cancer cells can produce various messengers that trigger this up-regulation, thus allowing the rapidly dividing abnormal cells to acquire materials for growth and spread

antidiuresis—inhibition of water losses through the kidney's reaction to hormones and abnormal cell signals, which reduce tubular losses

apheresis—removal of harmful blood components or substitution of more desirable constituents in blood

aplastic anemia—idiopathic anemia from abnormal, deficient, or absent red cell production due to bone marrow disorders

atrophic gastritis—atrophy of the lining of the stomach, which contains the parietal cells that produce intrinsic factor, proteases, and hydrochloric acid; this form of inflammation is often accompanied by bacterial overgrowth due to elevation of the gastric pH

bilirubin—the metabolite of heme breakdown excreted with bile; build-up of bilirubin and related compounds causes accumulation in extra-hepatic tissues such as the skin and sclera of the eye

chronic myeloproliferative disease—long-term hyperplasia of hematological tissues, with concomitant overproduction of abnormal cells, growth factors, chemokines, cytokines, and hormones involved in hematopoiesis

cicatrix—scar tissue formation with calcification or hardening of the connective tissue used in repair of tissue damage

cytokines—glycoproteins that act as local messengers or regional hormones; hundreds of different cytokines have been discovered and categorized, all with very distinct functions in the homeostatic and genetic controls over cellular function

dextrans—hydrophilic branched product of carbohydrate metabolism $[(C_6H_{10}O_5)_{11}]$ used to expand plasma volume and chelate iron in solution

DMT-1—transmembrane protein divalent metal transporter that traffics iron

eryptosis—suicidal death of erythrocytes characterized by cell shrinkage, membrane loss, cell disintegration, and engulfing by macrophages

erythroblastosis fetalis—an antigen-induced hemolytic anemia of the newborn or premature infant, as a result of incompatibility of maternal Rh factors with the neonate

erythropoiesis—production of erythrocytes or red blood cells

erythropoietin—the hormone produced in the kidney that regulates marrow production of red blood cells

ferritin—the storage protein for iron

ferroportin—transmembrane iron export regulating protein

hematocrit—packed red blood cell volume expressed as a percentage of whole blood upon centrifugation

hematopoiesis—production of blood cells

heme—iron-containing, nonprotein portion of the hemoglobin molecule that contains iron in the ferrous (+3) state

hemochromatosis—iron overload; elevated levels of iron that can cause tissue damage, especially in the liver

hemoconcentration—the decrease in free water circulating in the blood supply, causing increased levels of proteins, electrolytes, wastes, and nutrients per deciliter of blood; elevations of several laboratory values are present, and dehydration signs and symptoms may be present

hemoglobin—the four-pyrrole ring compound in red blood cells that contains iron centers and is responsible for the transport of oxygen

hemoglobinemia—excess free hemoglobin build-up in circulation

hemoglobinuria—excessive free hemoglobin spillage into the urine

hemojuvelin—protein sensor for iron levels in tissues

hemolytic anemia—an anemia brought on by the rapid, premature destruction of red blood cells in circulation, which may be precipitated by vitamin E deficiency

hemophilia—an inherited disorder of blood clotting, with pronounced bleeding upon tissue injury

hemostasis—normal blood flow and blood clotting

hepcidin—hormone regulating iron homeostasis

hephestin—membrane-bound, copper-containing oxidizing protein that incorporates iron into transferrin

histone—class of simple proteins found in the cell's nucleus

HLA antigens—antigens specific to the individual that cause a rejection reaction in a host receiving transplantation or foreign cells

holotranscobalamin—the fraction of metabolically active B_{12} that is composed of cobalamin linked to transcobalamin in circulation

hypochromic—abnormally pale in color upon visual inspection under a microscope

iatrogenic, iatrogenicity—harm caused by treatment or procedures performed by health care personnel or sustained through hospitalization, medical intervention, or prescription

intrinsic factor (IF)—the protein produced by the parietal cells in the stomach lining that is responsible for the pick-up of cyanocobalamin from protein in foods

kernicterus—infiltration of excessive amounts of bilirubin into the neurons of the spinal cord and brain

leukocytes—white blood cells (WBC); a generic term for several types of WBC that arise from the same parent cell in the bone marrow

leukocytopoiesis—the production of all categories of white blood cells from the pluripotent

(able to differentiate into multiple cell types) stem cells found in the bone marrow

macrocytic—refers to abnormally large cell size

megaloblastic—refers to an immature, large red blood cell that is oval in shape and abnormal

microcytic—refers to abnormally small cell size

osteopetrosis—death of bone cells through excessive calcification

pancytopenia—a reduction in the numbers of all the blood elements—white, red, other cells, and proteins

pernicious anemia—the anemia associated with B_{12} deficiency that is slow, aggressive, and potentially life threatening; it is specific to gastrointestinal dysfunction, namely, to gastric enterocytic atrophy, with diminished availability of intrinsic factor, HCl, and enzymes; neuropathy (especially peripheral) results from prolonged deficiency; the nervous system has a decreased ability to regenerate as well as regain function and feeling in the affected areas

phlebotomy—blood removal through a venous puncture; blood draw

pica—eating of abnormal items, or non-nutritive substances, such as laundry starch, clay, ice, dirt, paint chips, etc.

plasma—the portion of the blood in which blood constituents are dissolved or suspended; it contains water, proteins, electrolytes, gases, nonproteinaceous compounds, wastes, and nutrients

porphyria—a cluster of blood-related disorders characterized by abnormal porphyrin synthesis or metabolism; these disorders are hereditary and vary greatly depending upon which enzyme in the cascade of reactions is affected

prohepcidin—precursor to hepcidin; a marker for hemostasis

prophylaxis—preventative administration of a compound to avoid consequences of a disease state

protoporphyrin—the derivative of hemoglobin containing four pyrrole rings without the iron centers

reticulocytes—immature red blood cells; normal ranges for circulating erythrocytes exist, and levels reflect the ability of the bone marrow to produce precursor cells in normal amounts

serum—the fraction of blood containing water after the removal of cellular components

sickle cell anemia—a hereditary disease of genetically altered red blood cells that have a sickled shape, carry abnormally formed hemoglobin, and have abnormal transport capabilities for oxygen; the disease is thought to confer protection against malaria

sideroblastic anemia—a form of anemia characterized by the appearance of sideroblasts, immature ferritin-containing blast marrow cells in circulation

systemic lupus erythematosus (SLE)—an autoimmune, chronic inflammatory disease that affects the connective tissue; affects skin, joints, kidneys, central nervous system, and mucous membranes and eventually spreads to all tissues, invoking a systemic reaction with pain, fever, sensitivity to light, and skin lesions

others are so variable among individuals that they make poor diagnostic tools. Laboratory values may also differ due to differences in techniques used in the processing of blood samples, or the time of day that the blood was drawn. While laboratory values provide objective measures of nutritional status, it is useful to use several different indicators to rule in or rule out associations of a laboratory value with a pathogenic state or nutritional deficiency. Blood parameters must also be evaluated in terms of clinical signs and symptoms, dietary history, and overall medical status. Blood constituents are often abnormal as a result of another primary pathology, and a thorough evaluation of all facets of the patient's health status, environmental issues, genetics, and lifestyle must be performed.[1, 2]

Blood components include mature and immature red blood cells (RBC), white blood cells (WBC), and platelets; transport proteins such as **transferrin**; nitrogenous wastes; and nutrients, among others. Blood components also include electrolytes (e.g., chloride) and other minerals, some of which are tightly controlled (e.g., calcium) or tightly bound (e.g., copper). These also affect hematological status.[3, 4]

After a venous blood draw, blood samples are put into tubes containing an anticoagulant. When these samples are spun in a centrifuge, the densities of the various cells cause them to separate out of the aqueous portion of the blood, known as **plasma** (see Figure 19.1). Plasma appears at the top of the sample and is approximately 90% water, into which organic acids, proteins, gases, and other small molecular weight compounds are dissolved. This is the portion of the sample from which laboratory values for proteins such as albumin are obtained. A layer of platelets and WBC or **leukocytes** will form beneath the plasma, with erythrocytes (RBC) at the bottom of the tube. A percent measurement of the erythrocytes from this standard volume sampling of whole blood is called a **hematocrit**.[5]

Anatomy and Physiology of the Hematological System

The Cells of the Hematological System

Erythrocytes (red blood cells), containing **hemoglobin**, comprise the largest percentage of the blood volume. Normal RBC production, development, and function are dependent on nutritional and genetic factors as well as environmental influences. Dietary iron, zinc, copper, all the B vitamins, and vitamins A, K, and E are closely tied to normal hematopoiesis.[3] Environmental contaminants such as lead cause abnormal **erythropoiesis**. Erythrocytes are made from undifferentiated cells, called stem cells, in the bone marrow. They live for approximately four months before they are destroyed and their components recycled in a process called **eryptosis.** Eryptosis is suicidal, preprogrammed RBC death characterized by shrinkage, membrane loss, disintegration, and engulfment by WBC. Immature RBC are called **reticulocytes,** and they become functional as they mature.[5, 6]

WBC include granulocytes (eosinophils, basophils, neutrophils), monocytes (macrophages), and lymphocytes (B-cells, T-cells), and their production by the bone marrow is called granulocytopoiesis, monocytopoiesis, and lymphocytopoiesis, respectively. The WBC comprise approximately 1% of the total blood volume, and their normal production, development, and function are dependent upon nutritional status. Vitamins and minerals involved in cellular differentiation, such as vitamin A and zinc, along with iron, B vitamins (especially those involved in DNA synthesis), and protein are essential for **leukocytopoiesis** (WBC production). Granulocytes are phagocytic, and engulf and degrade bacteria and other foreign entities in addition to participating in the inflammatory response. Monocytes are also phagocytic, play pivotal roles in inflammation, and act as cell signals for the recruitment of other immune system cells. Lymphocytes of the T-cell variety are important in the cell-mediated immune response, while B-cell lymphocytes primarily participate in the humoral immune system (see Chapter 9).

B-cells retain a working memory of foreign antigens and produce antibodies. T-cells are found in two forms, helper T-cells

Figure 19.1 Blood Components

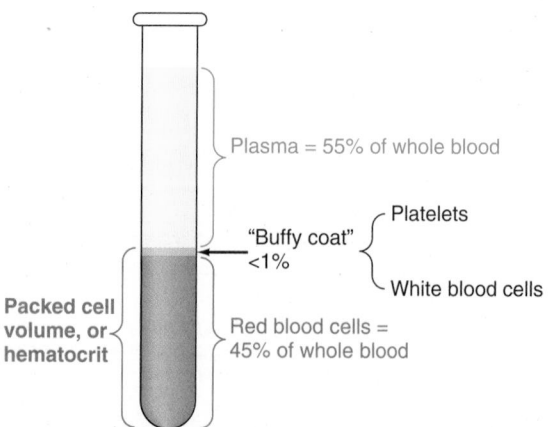

Plasma = 55% of whole blood

Platelets

"Buffy coat"
<1%

White blood cells

Packed cell volume, or hematocrit

Red blood cells = 45% of whole blood

Source: L. Sherwood, *Human Physiology: From Cells to Systems,* 5e, copyright © 2004, p. 392.

(CD3 1 CD4 type cells) and cytotoxic T-cells (CD8 type cells). T-cells also have antigenic memory and work in conjunction with **cytokines** for activation and interactions with other WBC types.[7] T-helper cells are targeted by the human immunodeficiency virus (HIV), and specific indices of T-cell status are important for monitoring AIDS patients.[8]

Platelets or **thrombocytes** are short-lived cell fragments that comprise less than 1% of the total blood volume. Unless they are activated, they are degraded and their components are recycled after approximately one week. If activated by chemical mediators, such as the cytokines, they aggregate (cluster) to form a clot that is attracted to the site of an injury and is very sticky. Individual clotting factors, bleeding time, clotting time, and other measures of platelet activity are available from laboratory panels for standard blood samples. This information is useful for assessing risk of clots or hereditary disorders such as **hemophilia**.

The Development of the Hematological Cells

Almost all forms of hematological disease have some nutritional component associated with either their induction or their treatment.[9] All blood components are generated from the bone marrow stem cells, which have the capability of differentiating into various cell types (see Figure 19.2). These stem cells are located in the red portion of the bone marrow.[5] Red marrow declines as a person ages, and less of the bone's interior remains dedicated to hematopoiesis over time. This has implications for geriatric patients, whose blood cell forming capabilities are decreased. The elderly may exhibit age-related **anemias**, clotting abnormalities, or diminished immune capabilities.[10] Red bone marrow function may be assessed by analyzing a sample obtained through a specialized fine needle inserted into the bone, known as an aspirate or biopsy.[11]

Genetic control over hematopoiesis has been investigated, and many of the genes responsible for the control of these processes have been mapped. It is now possible to genetically engineer blood cell production proteins and produce them in large quantities for use in patients with a variety of hereditary bone marrow disorders or diseases such as the anemia seen in renal failure.[12] One such recombinant protein, EPO (Epoetin, trade name for the biological protein erythropoietin), is being marketed directly to cancer patients as a drug to combat the fatigue and anemia associated with cancer treatments, which tend to destroy a wide variety of normal tissues and blood cells along with the cancerous cells. It is also possible to activate genes

Figure 19.2 An Overview of Hematopoiesis

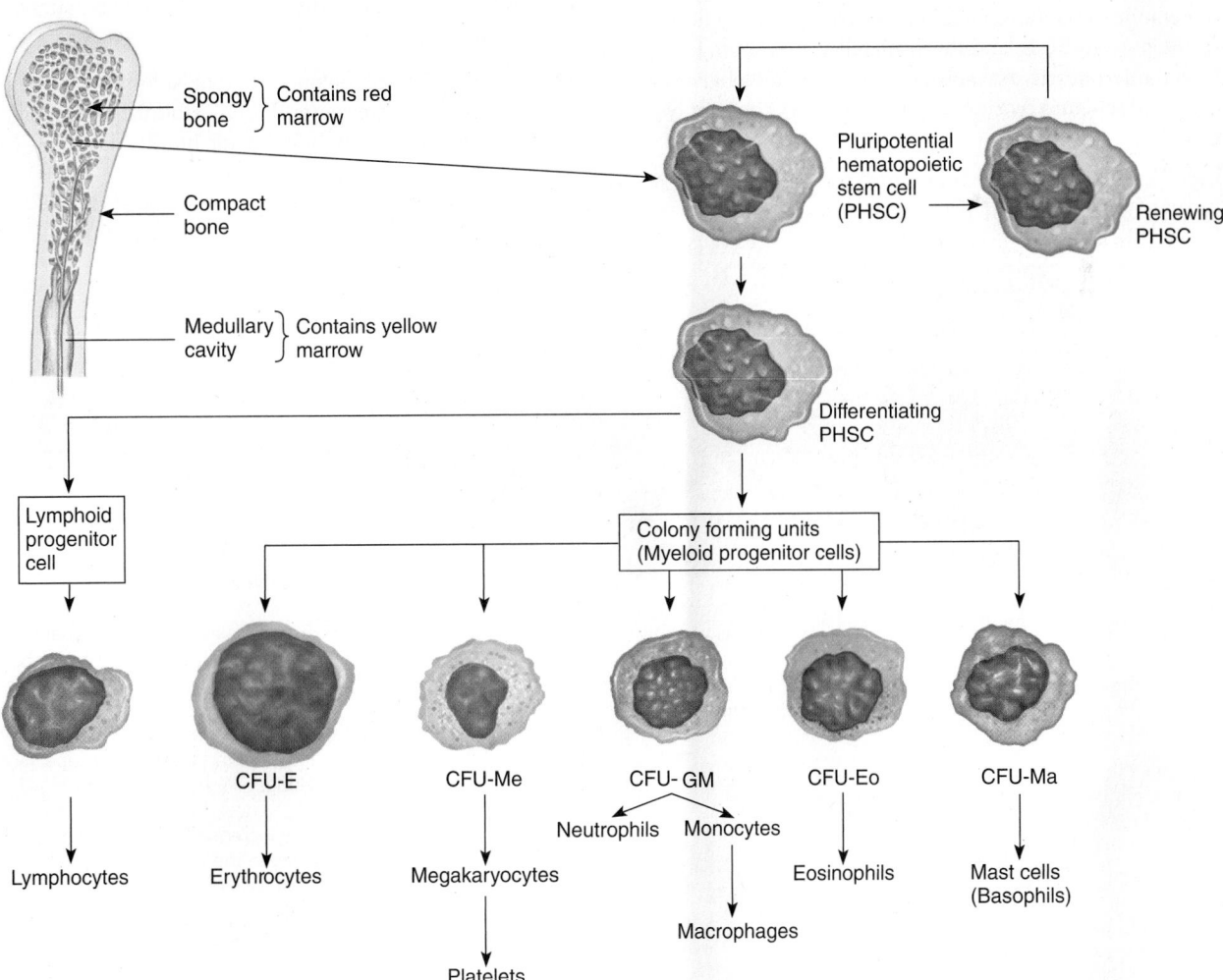

Source: R. Rhoades and R. Pflanzer, *Human Physiology*, 4e, copyright © 2003, p. 529.

that code for specific proteins involved in one or more steps of hematopoiesis through high-dose vitamin administration; this is used for specific forms of hemophilia with mutations in genes responsible for vitamin K-dependent clotting factors.[13]

Fetal hematopoiesis differs from adult cell development in that the stem cells develop both in the red bone marrow and in other tissues such as the spleen and the liver. This has implications for stem cell research, where such rapidly dividing, undifferentiated cells are used to provide a specific cell type to replace abnormal or non-functional cells in persons with a variety of diseases. After birth, the red marrow takes over most of these functions, so that by adulthood little cell proliferation occurs elsewhere. The marrow and spleen contain clusters of cells that migrate and mature into hematopoietic stem cells. The maturation process requires cytokines along with hundreds of other permissive, proliferative, activating, or inhibitory cell signals.[14] Permissive factors allow a cascade of reactions or interactions among cell signals to occur. Genetically regulated proteins that participate in homeostasis act as permissive signals until terminated by inhibitory protein signals. These proteins include **hepcidin**, **hemojuvelin**, **ferroportin**, and **hephestin**, among others.[15]

Erythropoiesis involves the incorporation of hemoglobin into a RBC, which loses its nucleus as it matures. Maturation also alters the shape of the cell, flattening it and thus increasing the surface area for oxygenation. Proper erythropoiesis requires several components: functional red marrow, proper genetic blueprints for synthesis, and the availability of protein, fatty acids, and micronutrients, especially folate, cyanocobalamin (vitamin B_{12}), vitamin A, zinc, copper, and most commonly, iron.[16]

Upon extrusion of the inactive nucleus, the reticulocyte becomes more pliable and able to move through small capillaries and decrease the workload on the heart. The heart must work very hard to move blood components through areas of increased resistance (such as small capillaries, where larger, rigid cells cannot pass through easily); the more pliable the cell, the less resistance there is to overcome. The flexibility and shape of the RBC are determined by fatty-acid composition of the membrane. Polyunsaturated long-chain fatty acids are more flexible, and vitamin E is required to prevent their oxidation and maintain the membrane's stability.[17] Hence, the diet of an individual impacts the flexibility and membrane health of RBC. The membrane also has transferrin receptors and **erythropoietin** receptors, which perform better with a flexible membrane.[18]

Because erythropoietin is produced, activated, and disseminated by the kidney, normal kidney function is essential for erythropoiesis, whereas kidney failure is often associated with anemia (see Figure 19.3). Erythropoietin is a glycoprotein (protein with a carbohydrate moiety) that stimulates the production of red blood cells and has an important role in control over the regulation of blood components.[19] Erythropoietin synthesis is regulated both by the kidney's ability to monitor blood volume and by oxygen status monitors located in the brainstem, the lung, and throughout the blood vessels of the body. Hypoxia (low oxygenation of body tissues), low RBC count, and a low hemoglobin concentration provide feedback signals assisting the kidney and ultimately the red marrow in maintaining homeostasis.

Hemoglobin Hemoglobin is a four-subunit metalloprotein containing iron at the center of each subunit. Its capacity for oxygen transport and carbon dioxide binding is related to the

Figure 19.3 Control of Erythropoiesis

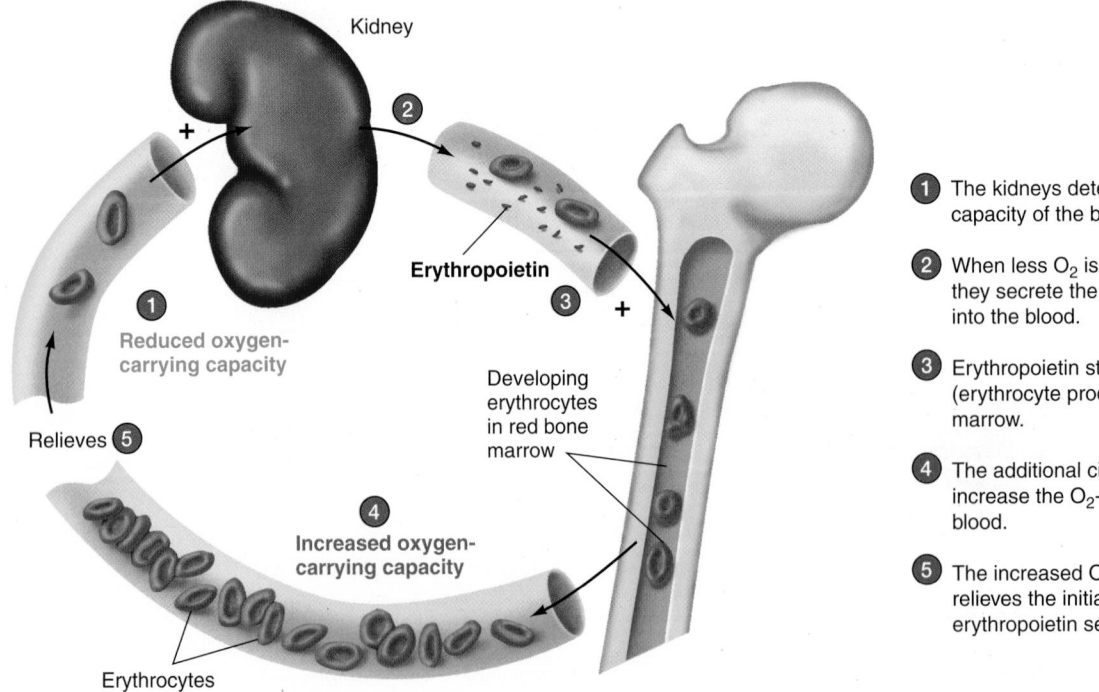

Kidney

+

Erythropoietin

Reduced oxygen-carrying capacity

Relieves 5

Developing erythrocytes in red bone marrow

+

Increased oxygen-carrying capacity

Erythrocytes

1. The kidneys detect reduced O_2-carrying capacity of the blood.

2. When less O_2 is delivered to the kidneys, they secrete the hormone erythropoietin into the blood.

3. Erythropoietin stimulates erythropoiesis (erythrocyte production) by the bone marrow.

4. The additional circulating erythrocytes increase the O_2-carrying capacity of the blood.

5. The increased O_2-carrying capacity relieves the initial stimulus that triggered erythropoietin secretion.

Source: L. Sherwood, *Human Physiology: From Cells to Systems*, 5e, copyright © 2004, p. 395.

Figure 19.4 A Model of the Hemoglobin Molecule

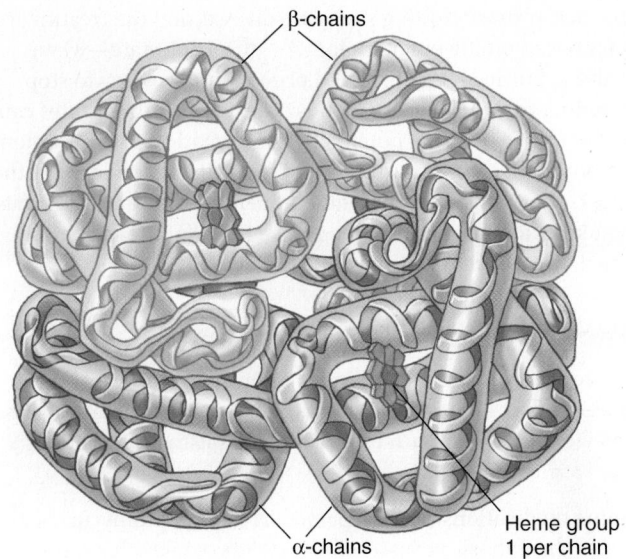

β-chains

Heme group
1 per chain

α-chains

Source: R. Rhoades and R. Pflanzer, *Human Physiology*, 4e, copyright © 2003, p. 530.

redox potential of each subunit. Due to its oxidative potential, iron is not found free in circulation or tissues. Iron is stabilized within the **heme** subunit, thus protecting cellular membranes from oxidation.[20]

The subunits of hemoglobin (see Figure 19.4) are comprised of globin—a protein moiety—and heme, a porphyrin ring that carries the iron. Porphyrin ring structures contain N, C, H, and O. Succinyl Co-A and glycine form delta aminolevulonic acid, which is converted to porphobilinogen; this in turn is converted to protoporphyrin through cyclization, decarboxylation, and desaturation. The chelation of iron yields heme.

There are two sets of globin protein chains, alpha and beta, which differ in their ability to carry oxygen and resist binding to other unwanted substrates. In hereditary diseases of the subunits, such as the thalassemia syndromes, hemoglobin contains nonfunctional or abnormal chains. The type of **thalassemia** is dependent on which abnormal chain predominates; the major beta chain hemoglobinopathies, such as beta thalassemia and sickle cell anemia, are most common.[21]

Heme protein levels in red marrow and the spleen control porphyrin synthesis. Disorders of porphyrin metabolism include the **porphyria** syndromes, where one or more of the intermediates for heme synthesis are abnormal or the enzymes for the steps are inactive. These hereditary disorders have far-reaching, systemic consequences. The erythrocytes are destroyed prematurely, the skin is hypersensitive to light, and the intermediates accumulate in tissues of the body and in the urine. Certain types of porphyrias are thought to be clustered in Eastern Europe and Great Britain, and this condition may have been the origin of the fabled "werewolf," as the skin of patients is tough and mottled, and the skin and teeth reflect red under light. There are often hypertrichosis (excessive body and facial hair) and scarring, which in the hand may produce a claw-like deformity with the loss of fingers.

Sideroblastic anemia, another disorder resulting from abnormal heme synthesis, actually presents clinically as iron overload due to adjustments for poor oxygenation. Iron absorption and RBC production are increased to compensate for ineffective oxygen transport.[22] There are presently eight characterized forms of sideroblastic anemia identified by the genetic mutation that is present, and all have characteristic mitochondrial iron overloading.[23]

The rate of hemoglobin synthesis is also a factor in blood disorders. Iron availability dictates the rate of heme synthesis. If there is low iron intake or absorption, there will be little hemoglobin available. Alternatively, if there is low folate or cyanocobalamin, rapidly dividing and differentiating cells like the RBC are affected, and there will be adequate or excessive hemoglobin inserted into fewer RBC. As the circulating RBC die off without adequate replacement, the oxygen tension falls and erythropoietin is produced in greater quantities by the kidney. Homeostatic mechanisms eventually fail to compensate, resulting in anemia.[5, 24]

As the RBC age and die, they are engulfed by macrophages and the hemoglobin is recycled. The proteins are hydrolyzed into their component amino acids, and those amino acids are picked up from the liver for reutilization. Iron is also recycled and either stored in the liver complexed to **ferritin** or utilized. The heme unit is degraded into **bilirubin**, which is excreted through losses of bile (and gives bile its characteristic color). If excessive bilirubin is produced or accumulates because of ineffective removal, the net effect is jaundice, yellowing of the skin, eyes, and nails. Bilirubin can be measured in laboratory assays in its free or direct form, or as a conjugate (indirect).

Post-hemolysis of the erythrocyte, hemoglobin in circulation is usually bound by haptoglobin, or to a lesser extent by albumin. In some blood disorders, so much red cell destruction occurs that the haptoglobin system becomes saturated and the remaining hemoglobin circulates unbound (**hemoglobinemia**) and is excreted in the urine (**hemoglobinuria**).[5] These disorders are called hemolytic anemias, and they can be due to vitamin E deficiency, hepatitis, sports anemia, transfusion reactions, autoimmune or hereditary disorders.[25, 26] There is also a hemolytic anemia of the newborn that is due to the immaturity of the erythropoietic system and its controls.

Homeostatic Control of the Hematological System

Hemostasis involves the regulation of bleeding, clotting, and blood flow. Under normal circumstances, these processes are tightly regulated, and dietary factors play a central role in that regulation. Vitamin K; dietary fats; antioxidants such as vitamin C, vitamin E, and selenium; pro-oxidants such as iron and copper; and cell signals such as the presence of calcium ions are all integral to normal clotting, bleeding, and blood flow.

Blood Clotting

Upon injury, the lining of the blood vessels (endothelium) produces signals that result in spasms of the blood vessel, causing constriction and a reduction in blood flow at the site of injury.

These signals are mediated by thromboxanes derived from arachidonic acid, an omega-6 fatty acid that is the product of linoleic acid elongation.[5]

As a result of cell signals, platelets in the area of injury aggregate and stick to one another. Another protein, the **von Willebrand factor**, is responsible for the increased stickiness of the platelets. The inflammatory signals also act on red marrow to increase **thrombopoietin**, which regulates platelet production. Activation of protein factors results in conversion of prothrombin to thrombin and fibrinogen to fibrin. The platelets and other blood components are trapped in strands of fibrin and form a clot. Vitamin K (essential for Factors IX, X, VII) and calcium (essential for factor IV) are directly involved in clotting.[27]

The coagulation proteins (Factors IX, X, VII) that are vitamin K-dependent are synthesized in the liver, and the receptors for these proteins are calcium-dependent. Compounds like the drug warfarin, which act as anticoagulants or anticlotting compounds, inhibit the ability of vitamin K and calcium to activate the clotting proteins. Treatment with vitamin K antagonists like warfarin must involve patient counseling on steady-state consumption of vitamin K-containing foods and supplements.[28] Patients must keep their intake of vitamin K from foods constant in order to prevent its interaction with the drug at recommended dosages. Advances in genotyping and the discovery of the genetic underpinnings of vitamin K epoxide reductase and γ glutamyl carboxylase will allow future prescribers to accurately predict how much warfarin and vitamin K can be given to any specific individual and how much flexibility they have in moderating their dosing regimens.[29–31]

Lysis of a clot (a clot is dissolved with special enzymes) and repair of an injured blood vessel require many proteins for signaling and a supply of amino acids and micronutrients for repair.[32] Antithrombin, heparin, Protein C, and thrombomodulin are just a few of the major players involved in lysis and repair following clotting. Hereditary abnormalities in these cascades of reactions can lead to either hyper- or hypo-coagulable states (i.e., more or less propensity to form clots), or varied bleeding and clotting times. Often, these conditions remain undetected in persons until they suffer a serious trauma or injury where it becomes critical that normal clotting and bleeding occur.[33]

There are a number of hereditary diseases in which deficient clotting occurs, such as Glansmann's thrombasthenia, where there are deficiencies in platelets, and von Willebrand's disease, where there are defects of platelet adhesion. Deficiencies in clotting also occur as a secondary outcome in multiple conditions, including thrombocytopenia due to cancer of the bone marrow and spleen or autoimmune diseases like **systemic lupus erythematosus (SLE)**. In addition, drug reactions, artificial valves, and transplant rejection can result in hemostatic disease.[33]

Common thrombotic diseases where hypercoagulability is influenced by heredity include defective inhibition of coagulation factors and impaired clot lysis due to protein deficiency or mutation. Hyper-coagulability diseases that are secondary or acquired include SLE, cancer of the marrow or spleen, advanced diabetes, hyperlipidemia, dehydration, renal disease, and congestive heart failure.

There are also physiological situations that lead to excessive clot formation and possible complications, including obesity, pregnancy, advanced age, immobilization, and postoperative state. The mechanisms between each of these primary, secondary, and tertiary clotting states are diverse, and the treatments vary based on the exact etiology.[27, 33] (Clotting state—when a clot is formed—is considered primary if necessary to stop bleeding from a wound. Secondary/tertiary clotting states can result in unnecessary, potentially problematic clots.) Additional common mechanisms for increased clotting secondary to other disease states or environmental influence include alcohol abuse, smoking, atherosclerosis, congestive heart failure, diabetes, renal disease, dehydration, diet, and heredity.

Factors Affecting Hemostasis

There are many other conditions or primary states that ultimately affect hemostasis, such as trauma, autoimmune disease, malignancy, and inborn errors of metabolism (see Chapters 9, 22, 23, and 26).

Some medications, such as penicillin and streptomycin, can cause thrombosis in susceptible patients, while others, such as aspirin, nonsteroidal anti-inflammatory drugs (NSAIDs) and certain antibiotics, can increase bleeding tendencies. Cancer chemotherapy that targets marrow and spleen tissue also causes hemostatic disorders through cellular destruction. Antiretroviral therapy for HIV and AIDS patients causes abnormal clotting by decreasing the red bone marrow's ability to produce platelets. Autoimmune reactions to some medications (e.g., quinine for malaria) decrease clotting abilities. In addition, several drugs interact with vitamin K by decreasing absorption, transport, activation, or assimilation into or interaction with coagulation protein factors.[5] These include:

- Antibiotics, such as the beta lactam antibiotic series or nitrofurantoin (Macrobid, Nitrofuracot, Furalan).
- Anabolic steroids, such as dehydroepiandrosterone (DHEA, Andro).
- Clofibrate, a lipid-lowering drug (Atromid).

Several other drugs induce adverse reactions featuring increased bruising, bleeding, and hemostatic deregulation in susceptible patients. Adverse event rates vary widely, and it is rarely possible to predict this side effect from drug administration. Drugs that may induce these reactions include the following:

- Statins (lovastatin, pravastatin, simvastatin, etc.) for hypercholesterolemia.
- Antidepressants such as Prozac, and anti-psychotics such as the phenothiazines.
- Antihistamines, such as diphenhydramine.
- Anesthetics, such as cocaine, halothane, procaine.
- Radiographic contrast dyes used for visualization of scans—reaction includes decreased platelet activation.
- Calcium ion flux regulators and calcium channel blockers for hypertension treatment, such as verapamil or nifedipine.

The most familiar compound affecting bleeding through platelet function is salicylic acid or aspirin. Aspirin affects the cyclooxygenase enzyme systems that produce thromboxanes from arachadonic acid. It is common knowledge that daily aspirin dosing for the prevention of heart attacks or for chronic pain can lead to gastrointestinal bleeding and substantial blood losses. Anemia can be seen as a result of prolonged occult (hidden) bleeding.[34]

Supplementation with omega-3 fatty acids to shunt production of eicosanoids away from omega-6 fatty acids will also affect thrombosis and inflammation. Fish oil administration is therefore associated with decreased clotting and slightly increased bleeding. This can also be seen in native peoples, such as the Greenland Eskimos, who consume high quantities of fish oils; for this reason, their cardiovascular health has been studied extensively. Despite a high-fat diet, there is a low incidence of cardiovascular mortality in this population.[35]

Artificial clotting factors can now be made for people with hereditary clotting diseases. Recombinant DNA technology has increased the availability, safety, and quality of these proteins. Previously, clotting factors would have been isolated from pooled blood of several donors, and the risk of contracting diseases such as AIDS or hepatitis was high.[5] Genetic counseling is offered to carriers of these disorders, and in the future, gene therapy for those afflicted may be possible.

Summary

Hemostasis is controlled by many complex cascades that involve macronutrients, micronutrients, and hereditary and environmental factors. Deregulation of the controls for normal clotting and bleeding has far-reaching consequences, from stroke to fatality from blood loss. The primary nutrients involved in normal hemostasis include vitamins K and E, calcium, protein, and long-chain fatty acids. Many diseases, physiological states, and drugs influence hemostasis directly or indirectly. The exact etiology of the underlying cause for clotting or bleeding must be elucidated prior to treatment. Treatment of the disordered hemostasis thus varies depending upon the primary condition. A comprehensive and aggressive approach is usually taken in the advent of a clotting or bleeding disorder.

Nutritional Anemias

The nutritional anemias are commonplace worldwide, and may be categorized as **macrocytic**, **microcytic**, or hemolytic (see Table 19.1). "Macrocytic" (large cell size) and "microcytic" (small cell size) are characteristics of different processes that have been dysregulated. Macrocytic type anemia results from decreased ability to synthesize new cells and DNA, due to deficiencies in cyanocobalamin, folate, thiamin, or pyridoxine. The deficiencies may be dietary or due to genetics.[36]

Microcytic type anemia is caused by impaired heme synthesis, as a result of inability to absorb, transport, store, or utilize iron, or impaired synthetic abilities from deficiencies of protein, iron, ascorbate, vitamin A, pyridoxine, copper, or manganese. Microcytosis can also be due to chronic disease states.[37] The ability to synthesize heme may also be impaired by toxicities of copper, zinc, lead, cadmium, or other heavy metals. **Hemolytic anemia** may be due to deficiencies or excesses of vitamin E. The clinical signs and symptoms of anemia are listed in Table 19.2.

Recycled iron contributes to the body's total iron pool. Levels of stored, dietary, and recycled iron are monitored, and uptake from the endothelium of the small intestine increases

Table 19.1 Nutritional Anemias

Type of Anemia	Related Nutrient Deficiencies/Toxicities	
Macrocytic	Deficiencies in: • cyanocobalamin • folate • thiamin • pyridoxine	
Microcytic	Deficiencies in: • protein • iron • ascorbate • vitamin A • pyridoxine • copper • manganese	Toxicities in: • copper • zinc • lead • cadmium • other heavy metals
Hemolytic	Vitamin E deficiency or toxicity	

Table 19.2 Clinical Signs and Symptoms of Anemia

Fatigue
Lethargy
Cheilosis, glossitis
Pallor
Pale sclera
Spoon-shaped fingernails
Clubbing of joints in the digits
Cold extremities
Muscle aches
Difficulty concentrating
Sleepiness
Irritability
General malaise
Gastrointestinal distress including nausea, vomiting, diarrhea, and cramping
Reproductive dysfunction, including amenorrhea and loss of libido
Cardiovascular sequelae including palpitation, tachycardia, dyspnea, angina
Paresthesias (tingling, numbness of the extremities) as seen in pernicious anemia

when levels fall. This is not the case for folate and vitamin B_{12}, although vitamin B_{12} stores tend to be sufficient for as long as three years despite marginal dietary intake. Folate and B_{12} deficiencies result in specific forms of anemia, where the red cells are **megaloblastic** and macrocytic. Since folate and vitamin B_{12} are interdependent, cells require adequate levels of both for enzymatic activation. Without adequate amounts of vitamin B_{12}, folate is trapped and unavailable in the active form. At the same time, adequate folate intake may mask a vitamin B_{12} deficiency, and resultant neuropathy is common in those individuals.[38]

In iron-deficiency anemia, where iron is deficient in the diet, or competing with another element (as in lead poisoning), or unavailable due to being bound to another compound in food

Figure 19.5 Blood Cells in Iron-Deficiency Anemia

A normal red blood cell smear is shown on the top, using Wright's stain under a microscope. The cell in the center is a neutrophil or white blood cell type. The red blood cell smear shown on the bottom is of severe iron-deficiency anemia. There is a lymphocyte slightly left of center. Notice the cells are microcytic (small sized) and hypochromic (pale colored). There is a good deal of variation in the size of the cells and in their shapes (anisocytosis, poikilocytosis). The red cell is normally rounded and all are about the same size.

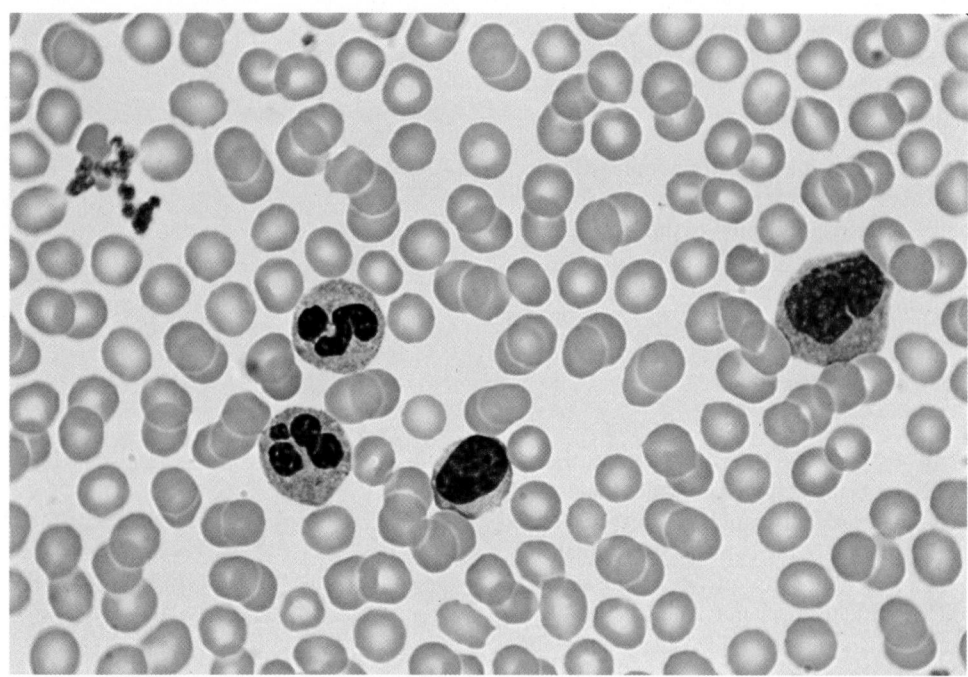

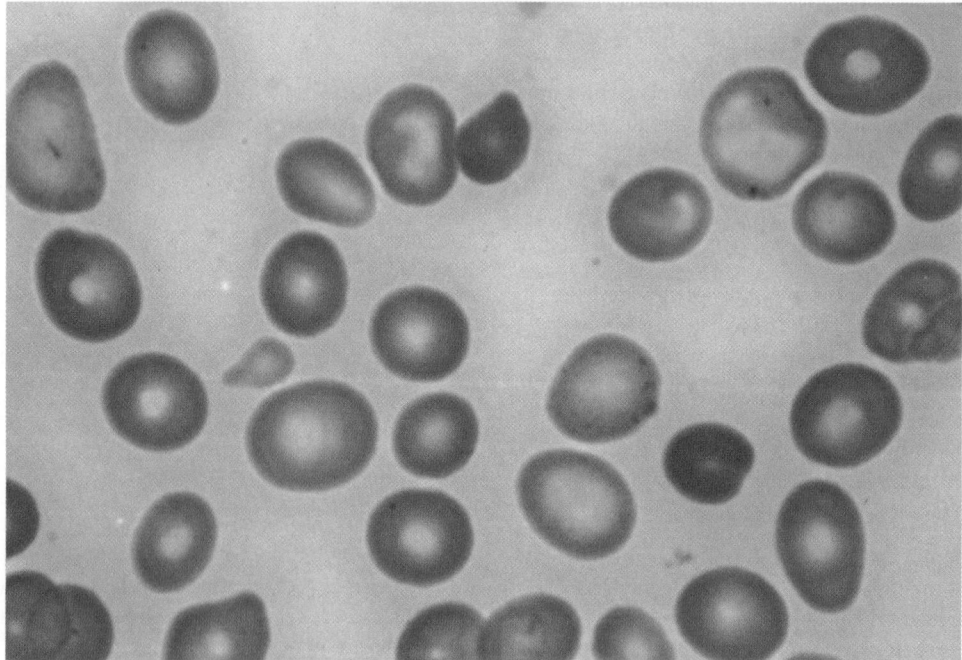

Source: Copyright © Ed Reschke/Peter Arnold, Inc.

(as in millet), the anemia is microcytic and **hypochromic** (see Figure 19.5).[39] The cells are irregularly shaped and vary in size. The rates of iron deficiency are different between genders, races, and age groups.[40] Iron regulatory hormones (such as hepcidin) and their appropriate genetic expression, function, and metabolism are critical to the iron deficiency syndrome's severity, manifestation, and outcome.[41]

Several other nutritional deficiencies or toxicities have been implicated in RBC disorders, such as vitamins A, E, and K, pyridoxine, zinc, copper, manganese, fatty acids, and protein. Severe vitamin C, niacin, riboflavin, and thiamin deficiencies are also associated with abnormal erythropoiesis.[42–44] The normal development of hematological cells is dependent on dietary adequacy, absence of underlying disease, and genetics.

The following sections describe nutritional anemias and effective nutrition therapies used to treat them. In addition, various complementary and alternative therapies are associated with these anemias as well as with other diseases; these therapies are listed in Appendix F4.

Microcytic Anemias: Iron Deficiency and Functional Anemia

Definition Iron-deficiency anemia is a condition where there is either 1) a decrease in the number of normal circulating RBC per cubic millimeter of blood, 2) decreased levels of hemoglobin, or 3) decreased volume of packed RBC per deciliter of blood as a result of greater demand on stored iron than can be supplied. The diagnosis of anemia is based on visual inspection of RBC as well as laboratory indices. Iron-deficiency anemia is a microcytic anemia, and is staged as follows:

- Subclinical with no overt symptoms
- Clinical with laboratory value alterations and some observable signs and symptoms
- Overt clinical iron-deficiency anemia with alterations in laboratory values and observable signs and symptoms upon physical examination of the patient

These stages correspond to negative iron balance, iron-deficient erythropoiesis, and iron-deficiency anemia (see Figure 19.6).

Epidemiology Iron-deficiency anemia is the most common nutritional anemia and affects many different groups. The most vulnerable groups in the United States are children under the age of 2, menstruating females, pregnant women, and frail elderly persons. Anemia in the frail elderly is becoming increasingly worrisome with the rapid rise in the over 85 years of age group and the atypical presentation of the anemia in this population.[45–47]

The epidemiology of iron-deficiency anemia worldwide varies by socioeconomic status, with citizens of poor nations and poor persons within wealthy nations the most susceptible. Because minority groups occupy lower socioeconomic strata, the incidence and prevalence of iron deficiency with overt clinical features is higher in African American and Hispanic women than in white women.[48]

Prevalence of iron deficiency in the United States is estimated to be greatest among females aged 12–49 (12%) and children aged 1–2 (7%). The rates of iron deficiency in the United States increased as a percentage of the population from 1994 to 2000, but then stabilized and now seem to be on the decline. Predictions are that these numbers will increase again with the "graying of America" (the aging of the U.S. population), which will result in greater numbers of older, poorer minority women who are at high risk for iron-deficiency anemia.[49]

Etiology The etiology of iron-deficiency anemia varies greatly. It can result from blood loss, inadequate intake, increased needs (e.g., in pregnancy), poor absorption, mineral excesses, or containments.

BLOOD LOSS Blood loss, as in the event of gastric ulceration or dysmenorrhea (abnormal menses) can deplete the body of significant amounts of iron. Blood losses require homeostatic restoration of blood volume.[50] If volume is restored, but new RBC have not been produced at the same rate, the number of viable RBC in a given volume of blood is decreased; this results in a functional anemia, a situation where oxygen is insufficient due to inability to transport the appropriate amount needed for function. In the event that erythropoiesis cannot keep up with RBC losses, depletion of iron stores in the liver, spleen, and other tissues results. Depletion of iron from ferritin, hemosiderin (another storage form of iron), or transferrin can be measured to ascertain the extent of the depletion. These levels are closely tied to total iron available to the body for use.

INADEQUATE INTAKE AND/OR ABSORPTION Iron deficiency can also be the result of poor intake, particularly in conjunction with greater needs, as with a vegan female who is menstruating but gets insufficient amounts of absorbable iron from plant sources due to inappropriate food choices, or a child

Figure 19.6 Sequential Changes in Iron Status

	Normal	Early Negative Iron Balance	Iron Depletion	Iron-Deficient Erythropoiesis	Iron Deficiency Anemia
Iron stores / Circulating iron / Erythron iron					
Reticuloendothelial marrow iron	2–3+	1+	0–1+	0	0
Transferrin iron-binding capacity (μg/dL)	330±30	330–360	360	390	410
Plasma ferritin (μg/L)	100±60	<25	20	10	<10
Iron absorption (%)	5–10	10–15	10–15	10–20	10–20
Plasma iron (μg/dL)	115±50	<120	115	<60	<40
Transferrin saturation (%)	35±15	30	30	<15	<15
Sideroblasts (%)	40–60	40–60	40–60	<10	<10
Erythrocyte protoporphyrin (μg/dL)	30	30	30	100	200
Erythrocytes	Normal	Normal	Normal	Normal	Microcytic Hypochromic
Serum transferrin receptors	Normal	Normal–high	High	Very high	Very high
Ferritin iron	Normal	Normal–low	Low	Very low	Very low

Source: AMERICAN JOURNAL OF CLINICAL NUTRITION. ONLINE by American Journal of Clinical Nutrition. Copyright 2009 by AMERICAN SOCIETY FOR NUTRITION. Reproduced with permission of AMERICAN SOCIETY FOR NUTRITION in the format Textbook via Copyright Clearance Center.

that is growing rapidly but refuses iron-containing foods or is not provided iron-fortified food products.[51, 52]

Meat and animal products have proteins that increase the bioavailability of iron. These food sources may not be readily available or may not be included in the diet for a variety of reasons. Plant sources of iron are lower in their availability and often contain compounds that bind iron and render it less absorbable.[53] Iron can be bound to phytates in cereal products, and this occurs to some extent in cereals fortified with reduced metallic forms of iron, which are poorly absorbed as a result. Plant sources should be taken with a vitamin C source to enhance absorption. The acidity of vitamin C (ascorbic acid) changes the oxidation state of the iron, making it more absorbable.[54]

The food supply in the United States and abroad is routinely fortified with iron in various forms, some more bioavailable than others.[55] Modifiers for iron absorption include heme versus non-heme availability, as well as phytates and perhaps tea consumption due to the polyphenols, such as tannin, contained in the latter. Tannins prevent iron from being picked up by transporters, making less of it available for absorption. Polyphenols interfere with non-heme iron absorption, and in persons with marginal iron intakes, increased tea consumption with meals may pose a risk for anemia.[56–59]

INFANCY/CHILDHOOD As previously mentioned, young children are vulnerable to iron deficiency related to poor intake and/or absorption. Maternal milk contains lactoferrin, which increases the bioavailability and absorption of iron from milk in infants. Human milk and fortifiers with iron for bottle-fed infants have been shown to decrease anemia in infancy. However, prolonged bottle use is still considered a risk factor for the "milk anemia" seen in infants or children who exclusively consume cow's milk.[60] Poor absorption, poor iron density, and the presence of increased amounts of calcium, another divalent cation that can alter iron uptake from the gastrointestinal tract, all contribute to anemia in infants and children.[61, 62] In some infants, a cell-mediated immune response to cow's milk and cow's milk-based formula can be seen; this causes damage to the jejunum and an enteropathy, with additional blood losses conferring heightened anemia.[63, 64] Children fed soy-based products that are not iron fortified are also at risk for anemia due to iron deficiency.[65]

PREGNANCY Iron needs are greater during pregnancy. Prenatal supplementation with iron and iron-containing multi-nutrient preparations is commonplace in the West, and has been a focus for public health efforts worldwide, especially in regions where vegetarianism and poverty coexist.[66] Iron deficiency complicates the pregnancy and worsens postpartum outcomes for both infant and mother. Premature birth, low birth weight, and other developmental disorders can be seen in the neonate when iron needs of pregnant women are not met.[67, 68]

Fetal needs take precedence over maternal needs for iron delivery, thus depleting maternal stores quickly. Maternal iron needs for increased blood volume over the course of pregnancy increase iron demands. Maternal complications during pregnancy from iron deficiency range from pre-eclampsia to pica.[69–72] Maternal iron-deficiency anemia may also be associated with other adverse outcomes during the postpartum period—such as postpartum depression, cognitive disturbances, and gallstones.[73–76]

MINERAL EXCESSES Excessive intakes of other minerals, such as calcium, zinc, copper or manganese, can interfere with iron absorption.[77–79] Supplementation with both high-dose iron and zinc decreases bioavailability of iron. Absorption of iron depends on its transport by carrier molecules, which are often nonspecific. For example:

- Metallothionein (transports zinc, but may pick up iron)
- Ceruloplasmin (transports copper, but may pick up iron)
- Transferrin and transmanganin (both transport manganese as well as iron)[80–82]

Iron intake may be adequate, but the transport systems for these carriers may overload, leading to a decreased preference for iron uptake and resulting in a deficiency. Heavy supplementation with trace minerals should therefore be avoided. Individualization of treatment strategies are needed, taking into account trace mineral dosing and supplementation scheduling.[83, 84]

CONTAMINANTS Contaminants such as lead also interfere with iron metabolism. Lead paint has been associated with anemia in children, because peeling lead paint chips (which have a sweet taste) in tenement housing are consumed by children. These children are already malnourished, which hastens the onset of both iron-deficiency anemia and lead poisoning.

Both anemia and lead poisoning are associated with impaired cognitive function and neurological damage in children.[85, 86] Children with both anemia and lead poisoning have lowered IQ, difficulty learning, permanent nervous system alterations, and behavioral disturbances. The brain has many receptors for iron, and many proteins that require iron for activation and function.[87]

Iron supplementation in poor, undernourished children has been associated with improvements in mental and motor development.[88] Manufacture of lead paint has been discontinued in the United States, and this type of poisoning is now largely confined to old, urban slum areas. The CDC has a specific branch that deals with the prevention of lead poisoning in children as part of its environmental health program. Monitoring of blood lead levels in children deemed at risk is a public health priority, and the states run public health programs that provide free lead testing for children.

The free erythrocyte **protoporphyrin** levels in blood are a useful laboratory tool for determining the anemia of lead poisoning, because it is a sensitive indicator. Normal ranges are 0.28–0.64 micrograms per deciliter of blood.[40, 89] Free erythrocyte protoporphyrin, zinc protoporphyrin, and **serum** transferrin are all useful indicators of anemia from iron deficiency with or without lead poisoning.[90, 91] Serum-soluble transferrin receptor to ferritin ratio has also emerged as a sensitive marker of deficiency.

Associated Health Conditions

PICA The disorder **pica**, which is the eating of non-nutritive substances like lead paint chips or clay, is thought to result from iron or other mineral deficiencies. Iron deficiency or mineral deficiency increases cravings for these nonfood substances in poor children and pregnant women, respectively. Not much is known about the exact etiology of pica, but it is often overlooked as either a cause or consequence of anemia.[92–94]

CHILDHOOD OBESITY Obesity in children has also been linked to iron-deficiency anemia, despite an increase in overall caloric consumption. Iron-poor food choices that are high in fat but lack micronutrient density are thought to be central to this new trend. Investigations into correlations between levels of the hormone leptin in obese children and alterations in iron status have been inconclusive, but more research needs to be done to rule out intermediate mechanisms by which being overweight might physiologically alter the ability of the child to absorb, transport, store, or assimilate iron from food.[95, 96] Since several other micronutrients are responsible for heme synthesis, it is important to ascertain whether it is dietary iron deficiency or a deficiency in another mineral or vitamin, such as copper or vitamin A, that is responsible for this increase in microcytic anemia.[97, 98] It is of interest to note that data from the National Health and Nutrition Examination Survey (NHANES) found that U.S. children who were overweight were twice as likely to be iron deficient as those who were not overweight.[99] Recent investigations into levels of hepcidin (hormone that regulates iron homeostasis) and its relationship to iron deficiency seen in overweight has shown some correlation. It should also be noted that body fat increases inflammatory cytokines, which contribute to the anemia of chronic disease (ACD), and this may be the mechanism for the anemia seen in this population. Similarly, the iron deficiency seen in obesity with type 2 diabetes and metabolic syndrome is also being studied with regard to hepcidin and/or ACD, as they are comorbid conditions.[100–103]

DECREASED IMMUNITY Iron status and immunity are inter-related. Worldwide, iron-deficiency anemia and infection rates are closely linked. Iron deficiency decreases immune function, but infection and infestation also increase the rate of iron utilization.[104] Organisms and parasites use iron, and if there is a parasitic infestation with blood losses from the gut, the resultant anemia progresses rapidly.[105–108] In addition, low levels of carotenes and ascorbate are seen in diets high in phytate, especially in regions where parasitic diseases such as hookworm are common. This complicates the anemia, through depressing already low levels of iron absorption.[109]

Increased infection rates for antibiotic-resistant tuberculosis and hepatitis C are also associated with iron-deficiency anemia.[110, 111] Zinc deficiency is a confounding factor, since it is involved in both immunity and the onset of hematological abnormalities such as iron-related anemia. Vitamin A deficiency is also a confounding factor, because vitamin A-induced anemia is common in pregnant women and schoolchildren worldwide. Conversely, iron deficiency also alters vitamin A metabolism.[112, 113] The confounding effects of zinc and vitamin A are a result of the increased need for these nutrients for cell division of white and red blood cells. Worldwide, co-morbid conditions typically occur with the iron-deficiency anemia. These include general malnutrition and/or repeated pregnancy with:

- **Low vitamin C levels.** Low vitamin C levels decrease the amount of iron absorbed from plant sources within the gut, especially when the diet lacks animal sources. Vitamin C is also important for immune function.
- **Hypovitaminosis A.** Vitamin A deficiency is linked to both poor immune function and iron-deficiency anemia, through its actions on the liver, where both vitamin A and iron are stored and activated, and through vitamin A's direct involvement in normal differentiation of rapidly dividing stem cells.

- **Protein-energy malnutrition.** Deficiencies in multiple vitamins and minerals as well as protein result in depressed immunity and decreased protein synthesis for iron transport, storage, and utilization.
- **Zinc deficiency.** Zinc deficiency results in decreased immunity and is tied to abnormal erythropoiesis and iron uptake. Zinc-containing foods like shellfish and meats are often expensive and unavailable to impoverished persons.
- **Iodine deficiency.** Globally, iodine deficiency with goiter and impaired thyroid function are seen along with anemia.

PEDIATRIC *H. PYLORI* INFECTION *Helicobacter pylori* (*H. pylori*) infection is associated with iron-deficiency anemia in children. It has long been known to be a cause of anemia in adults, along with esophagitis (inflammation of the esophagus).[114] Gastrointestinal blood losses due to minute ulcerations are causal.[50] *H. pylori* organisms bind to iron under conditions of iron depletion, which further exacerbates the problem. Iron supplementation may or may not show a treatment response for patients when these organisms increase their expression of iron binding receptors for their own growth and differentiation.[115]

Children are susceptible to infections, in general, since their decreased exposure to pathogenic organisms results in lower antibody levels and thus, decreased resistance. In addition, linkages between cells in a child's immature gut may not be as tight as those in an adult, allowing passage of pathogens. Importantly, iron-deficiency anemia from blood loss in children may have greater long-term consequences, such as developmental delay.[116] Children have increased iron needs, and many have marginal iron intakes; the rapidity of onset and severity of anemia with *H. pylori* infection may be greater as a result.[117]

IMPAIRED THYROID FUNCTION Worldwide, thyroid impairment is seen alongside iron-deficiency anemia. There is usually poor intake of both iron and iodine among low-income persons isolated from coastal waters rich in minerals such as iodine and iron.[118] Iron deficiency also disrupts thyroid peroxidase, contributing to adverse effects. Iron is an important cofactor for energy regulation, thermoregulation (control of body temperature), and metabolic rate, and thus exerts depressive systemic effects when deficient. Thermoregulation, metabolism, and thyroid function are intertwined; they are dependent on iodine, iron, and selenium, another important component of de-iodinase enzymes in thyroxine metabolism.[119–121]

CEREBROVASCULAR AND CARDIOVASCULAR DISEASE Cerebrovascular and cardiovascular disease are associated with worsened outcome in the event of concomitant iron-deficiency anemia. Symptoms in heart failure patients with iron-deficiency anemia are worse than in those with normal iron status, and their physical and functional capacity decline. Mortality is increased among persons with co-morbid heart failure and anemia. The presence of anemia is associated with a worse prognostic profile in persons with chronic disease.

The anemia of chronic disease (ACD) is a condition whose severity runs parallel to the level of inflammation and the release of cytokines (interleukin-1 and 6 and tumor necrosis factor). ACD significantly impacts morbidity and mortality. This is seen in both pediatric and adult populations.[122, 123] Iron deficiency develops along with the ACD, paralleling its severity. The reason for anemia is that inflammatory proteins from the chronic disease block iron reutilization, decrease erythropoiesis, and reduce erythropoietin production. More nutrient resources are needed for the body during chronic disease, and fewer are taken in by the person who is ill, leaving little left for hematopoiesis.

CANCER Cancer and its treatment are associated with anemia. Treatments such as radiation or chemotherapy target rapidly dividing cells, such as those in the hematopoietic systems.[124] Chemotherapy-related anemias are often resistant to treatment with EPO or darbepoetin (DA), both potent erythropoietin stimulating agents (ESAs), with or without iron supplementation.[125, 126] Cytokines and the inflammatory process result in cancer cachexia, and iron intake is usually low in these patients. Hypermetabolic states in cancer require an increase in iron.[127] Some cancers themselves cause iron-deficiency anemia and its resultant fatigue through a variety of mechanisms, such as **angiogenesis** for the increased blood supply directed toward the abnormal, rapidly dividing cells and blood losses through tissue destruction.[128] There are also cancers of the hematopoietic system that result in functional anemia.[129, 130] For further information on cancer, see Chapter 23.

TRAUMATIC CONDITIONS, SUCH AS WOUNDS, SEPSIS, AND SURGERY Traumatic conditions are often associated with anemia. When the patient has had significant trauma to the body and anemia is present, the anemia decreases wound healing and increases infection rates. Anemia worsens clinical outcomes and increases morbidity and mortality.[131, 132] Sepsis, or systemic infection, is associated with an increase in inflammatory proteins that decrease iron reutilization, erythropoiesis, and hematological function. Wounds and surgery are generally associated with the same underlying increases in inflammation that cause anemia (see Chapter 9). Septic patients also have significant drains on their nutrient resources with bacterial, viral, and protozoal utilization of available nutrients for their own metabolism and reproduction. Burns are even more draining on those scarce resources.

Bariatric surgery (resectioning of the GI tract) is especially associated with rampant deficiency in vitamin and mineral nutriture, but iron and cyanocobalamin deficiency are the most severe and prevalent. Not only is there trauma, a surgical wound, and co-morbid conditions associated with obesity and chronic disease, but there is a deliberate attempt to reduce nutriture by resecting the intestine or decreasing the stomach volume by procedures such as bypass and vertical banded gastroplasty. These deficiencies are due to general malabsorption, decreased production of hydrochloric acid that results in decreased iron solubility, decreased intake of iron-containing foods, and decreased transport and storage protein synthesis.[133–135] Often these deficiencies are lifelong, respond poorly to treatment, and result in subsequent complications.[136, 137] There is evidence that long-term deficiency in bariatric surgery patients may increase pica or disordered eating.[138]

HIV AND AIDS HIV-related wasting and malabsorptive syndromes are common (see Chapter 24). These are associated with iron-deficiency anemia through the following mechanisms:

- HIV infection-related enterocyte injury with iron malabsorption

- Increased requirements of chronic illness
- Poor oral intake
- Increased secondary opportunistic infection with organisms such as *Cryptosporidium*
- Increased requirements of acute fever and sepsis

Fatigue and anemia are common sequelae affecting quality of life in HIV-infected patients. Physical functioning is impaired, and anemia has been associated with increased morbidity and mortality. Several of the drugs used to treat HIV and concomitant infections, such as hepatitis C, cause anemia and fatigue because they affect the bone marrow's production of RBC. Hematological consequences of both disease and treatment must be carefully monitored.[139] It has also been shown that even with antiretroviral treatment, HIV infection itself alters the metabolism of iron and the ACD which ensues is difficult to correct. There are risks to excessive iron administration in this group, as the increase in circulating iron may decrease immunity, increase oxidative stress, and allow pathogens to replicate at a more rapid rate.[140]

ALCOHOLIC LIVER DISEASE Alcohol abuse and alcoholism are associated with alterations in hepatic function (see Chapter 16), and, indirectly, to sudden onset of iron-deficiency anemia. Cirrhotic patients often present with malnutrition, anorexia, and malabsorption of micronutrients such as iron.[141, 142] The liver is responsible for the storage of iron as ferritin and hemosiderin. Cirrhosis leads to the decreased production of transferrin and ferritin for iron mobilization. Iron is trapped in the liver cells and surrounded by scar tissue. Iron's pro-oxidant nature—its capacity to start a chain reaction in which free radicals are generated—increases tissue injury near the storage vacuole. The etiology of iron-deficiency anemia in liver disease is complex, and high-dose supplemental iron is contraindicated in this population. Increases in hepatic damage with high-dose iron administration lead to further anemia and poorer outcome, such as with the development of systemic or eye-related complications.[143, 144] Alcohol and iron are both hepatotoxic (damaging to liver cells) in high doses.[145]

GASTROINTESTINAL DISEASE Gastrointestinal diseases like celiac disease, Crohn's disease, and ulcerative colitis are associated with iron deficiency through blood losses caused by ulceration and inflammation. Although oral iron supplementation is routine, it has been suggested that this practice may aggravate the condition through free radical generation and increases in bleeding and inflammatory response. Recent trials of injectable ferric carboxymaltose (Ferinject™) for short periods resulted in greater elevations in hemoglobin. However, severe adverse events occurred in ~30% of IBD patients treated.[146]

In celiac disease, where the offending proteins in wheat, barley, and rye cause damage to the mucosa, a gluten-free diet will ameliorate the anemia and further iron supplementation is neither required nor advised.[147–148] The increased incidence of ACD in this population suggests that freely circulating or macrophage sequestered excess iron predisposes patients to greater risk and the usual diagnostic indicators of systemic inflammation are absent.[149]

KIDNEY DISEASE Renal disorders (see Chapter 18) are responsible for many reported cases of anemia in the adult and pediatric populations. The kidney's role in erythropoietin production and erythropoiesis is central. The anemia seen in renal disease is sometimes normocytic (normal cell size) and normochromic (normal cell color), but if it is due to reduction in iron intake along with decreases in erythropoietin and marrow cell function, it results in a microcytic, hypochromic anemia and reduced packed cell volume.[150] Even modest declines in renal function are associated with poor iron status.[151]

Iron-containing foods are often protein-containing foods, and protein restriction in renal patients can decrease iron intake.[152] Anorexia is also reported among patients with renal insufficiency due to the buildup of nitrogenous wastes. Renal disease and anemia both result in fatigue, reduced exercise capacity, and impaired immunity. Anemia-dependent angina results from the increased workload on the heart.[153] Recent work on the relationship between genetic components of hemoglobin synthesis and novel erythropoiesis stimulating agents (ESA) promises to overcome the treatment failure seen in renal patients with current management strategies such as EPO administration.[154]

ANOREXIA NERVOSA Anemia is routinely underreported in patients with eating disorders such as anorexia nervosa (AN), because the iron deficiency is masked by **hemoconcentration** due to dehydration and plasma volume depletion. Associated electrolytic abnormalities impact kidney function and hemostasis. There are multiple deficiencies in both macronutrients and micronutrients that occur in AN, and it is often difficult to isolate and attribute the symptoms.[155, 156]

PHENYLKETONURIA Patients with phenylketonuria (PKU), an inborn error of phenylalanine metabolism, are at an increased risk of iron-deficiency anemia due to both the inability of the red marrow to make RBC and the prescribed low-protein/low-iron diet (see Chapter 26). PKU patients must be routinely monitored for protein and erythropoietic status.[157]

SPECIAL CONDITIONS: SPORTS, SPACE FLIGHT, AND BIO-TERRORISM Environmental conditions may also impact iron status. Anemia is seen in female athletes, as a result of RBC losses with consistent, continuous high-impact landings coupled with menstruation losses, poor oral intake of iron, and increased needs for oxygen carrying capacity with sport (see Box 19.1).[158, 159] Conversely, even marginal tissue iron deficiency decreases the ability of females to train aerobically and may decrease physical capacity overall.[160–162]

Anemia is also seen in space flight, where weightlessness causes changes in the body's adaptive responses, and the system removes newly released RBC from circulation, resulting in a functional anemia.[163]A third environmental factor affecting iron status is exposure to chemical warfare agents such as Arsine gas (AsH_3), or exposure to various infectious agents (e.g., Ebola virus) with potential for use in bioterrorism. The mechanism is rapid hemolysis of circulating RBC, resulting in deficiency.

Clinical Manifestations The overt signs and symptoms of iron-deficiency anemia include cold extremities, pallor, fatigue, malaise, and tachycardia (heart rate >100 beats/min). A complete blood analysis is used to discriminate between iron and other vitamin/mineral deficiency anemias, or to rule out other factors and/or disease states. Hematological values commonly available from routine blood analyses include indices of RBC, WBC, and clotting factor levels or functionality (see Tables 19.3 and 19.4).

BOX 19.1 CLINICAL APPLICATIONS

Anemia in Female Athletes

Female athletes may be at heightened risk for iron-deficiency anemia. There may be multiple reasons for this increased risk. Sports requiring high-impact activities, such as running, gymnastics, dance, cheerleading, and skating, result in small numbers of blood cells being destroyed with each impact. Over time, red blood cell hemolysis requires an increased production of new cells and the increased utilization of iron and iron stores. These sports are also known for image conscious participants, and the risk of disordered eating and nutritionally poor diet are also factors. In addition, there is blood loss from menstruation.

Anemia impedes performance, decreases oxygen delivery to muscles, decreases muscle myoglobin saturation, and dramatically decreases endurance, strength, and overall fitness levels.[1] Subclinical low iron levels in tissues decrease aerobic capacity and impede brain function, affecting concentration as well as physical parameters. Iron is also essential for the neurotransmitters such as dopamine that govern concentration, energy level, and mood.[2] Low iron levels may also contribute to the decrease in ability to perform in sport. Routine screening using serum ferritin, haptoglobin, erythrocyte volume, and hepcidin should be done biannually for these sports. There is some indication for prophylactic treatment of female elite athletes before anemia is detected, using routine administration of lactoferrin (iron binding glycoprotein) to bind sufficient dietary iron for enhanced endurance and performance.[3–7]

References
1. Brownlie T 4th, Utermohlen V, Hinton PS, Haas JD. Tissue iron deficiency without anemia impairs adaptation in endurance capacity after aerobic training in previously untrained women. Am J Clin Nutr. 2004;79:437–43.
2. Erikson KM, Jones BC, Beard JL. Iron deficiency alters dopamine transporter functioning in rat striatum. J Nutr. 2000;130:2831–37.
3. Peeling P, Dawson B, Goodman C, Landers G, Trinder D. Athletic induced iron deficiency: new insight into the role of inflammation, cytokines and hormones. Eur J Appl Physiol. 2008;103(4):381–91.
4. VanHeest JL, Mahoney CE. Female athletes: factors impacting successful performance. Curr Sports Med Rep. 2007;6(3):190–94.
5. Fallon KE. Screening for haematological and iron-related abnormalities in elite athletes-analysis of 576 cases. J Sci Med Sport. 2008;11(3):329–36.
6. Scholl TO. Iron status during pregnancy: Setting the stage for mother and infant. Am J Clin Nutr. 2005;81:1218S–1222S.
7. Koikawa N, Nagaoka I, Yamaguchi M, Hamano H, Yamauchi K, Sawaki J. Preventive effect of lactoferrin intake on anemia in female long distance runners. Biosci Biotechnol Biochem. 2008;72(4):931–35.

Table 19.3 Select Indices of Hematological Function

Blood Values	Normal Range
Ferritin	
Males	30–300 mg/L
Females	10–200 mg/L
Hematocrit	
Males	41%–53%
Females	36%–46%
Hemoglobin	
Males	13.5–17.5 g/dL
Females	12.0–16.0 g/dL
Serum Iron	30–160 mg/dL
Total Iron Binding Capacity	228–428 mg/dL
Iron binding capacity saturation	0.2–0.45 (20%–45%)
Mean Corpuscular Hemoglobin (MCH)	26.0–34.0 pg/cell
Mean Corpuscular Hemoglobin Concentration (MCHC)	31.0–37.0 g/dL of packed erythrocytes
Mean Corpuscular Volume	
Males	78–100 mm³
Females	78–102 mm³
Methemoglobin	Up to 1% total hemoglobin
Protoporphyrin, free erythrocytic	0.28–0.64 mg/dL
Oxygen Percent Saturation (sea level) venous arm	60%–85%
Transferrin Receptor	9.6–29.6 nmol/L
Transferrin	2.3–3.9 g/L
Sickle Cell Test	Negative

Source: Kratz A, Sluss, PM, Januzzi JL, Lewandrowski KB. Appendices: Laboratory Values of Clinical Importance. In: *Harrison's Principles of Internal Medicine,* Kasper Dl, Fauci AS, Longo DL, Braunwald E, Hausner SL, Jameson JL, Eds., 16th ed., New York, McGraw Hill Publishing, 2005. Constructed from data obtained using this source.

Table 19.4 Select Indices of Clotting Function

Blood Values	Normal Range
Activated Clotting Time	70–180 sec
Bleeding Time	2–9.5 minutes
Factor II, Prothrombin	60%–140%
Fibrinogen	150–400 mg/dL
Activated Partial Prothrombin Time (PTT)	22.1–35.1 sec
Prothrombin Time	11.1–13.1 sec
Thrombin Time	16–24 sec
Platelet Count	150–350 × 103 μm³
Platelet Mean Volume	6.4–11.0 μm³
International Normalized Ratio (INR) (Warfarin anti-clot drug administration)	2.0–3.0

Source: Kratz A, Sluss, PM, Januzzi JL, Lewandrowski KB. Appendices: Laboratory Values of Clinical Importance. In: Harrison's Principles of Internal Medicine, Kasper Dl, Fauci AS, Longo DL, Braunwald E, Hausner SL, Jameson JL, Eds., 16th ed., New York, McGraw Hill Publishing, 2005. Constructed from data obtained using this source.

Laboratory indices used to diagnose clinical iron-deficiency anemia are shown in Table 19.5. In uncomplicated iron-deficiency anemia, the remainder of the hematological indices are normal. There has been reanalysis of national data regarding the normal ranges for hematological indices. Trends based on age, gender, and race were noted for many of the tests. It may be useful for clinicians to account for these differences by comparing patient values against normal ranges specific to demographic characteristics.[1]

Worldwide, inexpensive, easy analyses are required for iron-deficiency anemia screening, and capillary blood analysis is often done for hemoglobin alone.[164] Automated techniques

Table 19.5 Laboratory Values for Iron-Deficiency Anemia

Characteristic	Normal Value	Iron-Deficiency Anemia
Hemoglobin		
Males	13.5–17.5 g/dL	Less than normal
Females	12.0–16.0 g/dL	Less than normal
Hematocrit		
Males	41%–53%	Less than normal
Females	36%–46%	Less than normal
Serum iron	30–160 μg/dL	Less than normal
Serum Ferritin		
Males	30–300 μg/L	Less than normal
Females	10–200 μg/L	Less than normal
Serum transferrin	2.3–3.9 g/L	Less than normal
Total Iron Binding Capacity	228–428 μg/dL	Greater than normal
Free erythrocyte protoporphyrin	0.28–0.64 μg/dL	Greater than normal
Blood smear shows pale cells, small cell size with abnormal irregular shapes microscopically	6–8 mm diameter	Smaller than normal
Mean corpuscular hemoglobin and hemoglobin concentration	(MCH) 26.0–34.0 pg/cell (MCHC) 31.0–37.0 g/dL of packed erythrocytes	Less than normal
Mean Corpuscular Volume		
Males	78–100 μm³	Less than normal
Females	78–102 μm³	Less than normal
RBC Distribution Width[a]	0.115–0.145	Greater than normal

[a]Red Blood Cell Distribution Width—a measure of the shape and proximity of the red blood cell in a sample to standard reference values, used in conjunction with other measures of hematological status to rule in iron deficiency or functional anemia.

Sources: Urrechaga E. Discriminant value of microcytic/hypochromic ratio in the differential diagnosis of microcytic anemia. *Clin Chem Lab Med.* 2008;46(12):1752–58.

Aulakh R, Sohi I, Singh T, Kakkar N. Red cell distribution width (RDW) in the diagnosis of iron deficiency with microcytic hypochromic anemia. *Indian J Pediatr.* 2009;76(3):265–68.

Harrington AM, Kroft SH. Pencil cells and prekeratocytes in iron deficiency anemia. *Am J Hematol.* 2008;83(12):927.

for the analysis of RBC volume vary in their accuracy and reliability. Differences among ranges and techniques are amplified in patients with co-morbid conditions. Kidney disease, for example, will alter the **total iron-binding capacity (TIBC)** to the extent that it is no longer a useful diagnostic tool for iron deficiency in those patients. Newer techniques for diagnosing anemia are becoming available. One such technology is noninvasive point-of-care imaging. In an outpatient setting, a patient may be scanned for microvascular density, which is correlated with anemia from nutritional deficiency, without the need for a blood draw and traditional laboratory measurements.[165]

Treatment Iron-deficiency anemia can be corrected through continued iron-dense, nutrient-dense dietary intake (see Table 19.6), supplementation, and correction of any underlying conditions that may be contributing to the deficit. Persons with uncomplicated deficiencies of iron already have up-regulated proteins for enhancing iron absorption from the digestive tract. One such protein (hepcidin) communicates iron status and erythropoietic demands to the small intestine and modulates absorption.[166] This feedback mechanism allows deficient individuals to enhance their absorption of available heme and non-heme iron. Increasing the nutrient density of foods over the long term is the best strategy. Referral to assistance programs may be necessary for clients unable to afford iron-dense foods. Nutrition education regarding micronutrient density and sufficiency is also important.

Nutrition Therapy for Microcytic Anemias

Nutritional Implications Nutrition concerns for iron-deficiency anemia are related to the functions that iron performs in the cell and the subsequent consequences of insufficient levels of iron as a cofactor, transporter, and promoter.

Nutrition Diagnosis Nutrition diagnoses associated with iron deficiency are most often inadequate mineral intake (sp. Iron); imbalance of nutrients; altered nutrition-related laboratory (Hb) values; and increased nutrient (iron) needs.

Nutrition Intervention Iron intake can be increased through dietary changes and/or supplement use, and recent technological developments have provided many options for both. Approaches to improve bioavailability from complemen-

tary foods (i.e., plant foods with complementary amino acids, such as in vegetarian diets) have been undertaken through a variety of methods. In the West, bioengineering has made high-iron cultivars of plant materials available for agricultural use. The pharmaceutical industry has formulated iron supplements and iron-containing formulas for addition to the food supply. Governmental policies ensure that the food industry routinely fortifies the food supply with iron, and the industry has increased its use of iron chelates that are more bioavailable. Education regarding iron-containing foods from complementary sources should be instituted when possible.[167]

It is thought that single supplement administration of iron is more efficacious than a multivitamin preparation for treating iron-deficiency anemia. Iron supplementation is notorious for side effects such as gastrointestinal upset, constipation, and nausea. The amount of iron supplemented is dependent on the individual's age, gender, physiological state, and/or disease co-morbidity. Pregnant women often receive prenatal supplementation in a two-fold higher dose than do non-pregnant women (see Box 19.3). Iron-deficient adult females are prescribed 15 to 60 mg supplemental iron per day.[168] Supplemental iron should be given as ferrous sulfate or ferrous gluconate, which are better absorbed and tolerated than other iron chelates. Weekly doses of high-iron supplements are preferable to daily doses with lower amounts.[169] The upper intake level established for the United States is 45 mg/day, based on gastrointestinal distress symptomatology.[170]

Supplemental use of iron among women has treatment implications. Women who are younger, of poor socioeconomic status, pregnant, or marginally deficient should be advised to take a supplement. Postmenopausal women, who are at risk for increased pro-oxidant damage from excessive iron intake, should be discouraged from taking supplements containing high doses of iron.[171]

Patients on total parenteral nutrition (TPN) should receive iron in small, regular doses, rather than in one large dose, for increased efficacy.[172] TPN patients should be monitored for copper deficiencies, which may present as iron-deficiency anemia.[173] TPN solutions are routinely infused with iron **dextrans**.

Nutrition policy regarding pediatric iron deficiency and its potential for adverse effects on brain development is controversial.[174] Iron is thought to be essential for neuron myelination, growth, differentiation, development, and subsequent brain capacity for memory, cognitive ability, intelligence, psychomotor skill, and attentiveness. Large-scale studies have not been conducted.[175–181] Community-based education and interventions regarding iron and neonatal brain development have been successful. Worldwide, successful implementation of increased iron sufficiency must address culturally, ethnically, and geographically specific issues such as phytates in millet or vegetarianism due to religious observances. Dietary intervention on a community basis in persons with diets low in animal foods and high in phytate requires the introduction of acceptable foods that can be locally grown or obtained (see Box 19.4). Introduction of genetically modified crops has had little success, due to novelty, fear, and lack of cultural substantiation. Routine dietary supplementation is difficult in areas where the majority has no access to health care.[39]

Table 19.6 Food Sources of Iron

Food (Heme Iron Source)	Iron (mg)
Clams—3 ounces	14*
Steak—4 ounces	4*
Poultry—3 ounces	1*
Food (Non-Heme Iron Source)	**Iron (mg)**
Fortified cereal—1 cup	9
Spinach—1 cup	6
Kidney beans—1 cup	5
Tortilla—1 item	2
Baked potato with skin—1 item	2

*Highly bio-available animal source of iron.

Source: Adapted from the USDA Agricultural Handbook Series. Available at http://www.usda.gov

BOX 19.2 CLINICAL APPLICATIONS

Fatigue of Unknown Origin

Nutrition Assessment:

Client History

· **Personal Data:** Mrs. P is a 47-year-old married Caucasian female with a professional occupation.

· **Family Medical/Health History:** She presents with fatigue of unknown origin. Mrs. P has been fatigued, sleeping 12–14 hours per day, and complains of apathy, lethargy, and exhaustion. She has difficulty concentrating, poor performance at work, and general malaise. Parity × 2 (1980, 1984). Past Med Hx: Seasonal allergies. Broken L Clavicle: 4/1979. T and A : 2/1967. Treatments/therapy/medicine: Zyrtec prn for seasonal allergies, Tylenol prn for h/a., MVI qd.

Anthropometrics

Ht: 5'2" Wt: 140 lbs.

Biochemical Data, Medical Tests and Procedures

All WNL except: HgB = 10.2, TSH = 5.0 U/mL, T_4 = 1.0 g/dL

Food-/Nutrition-Related History

NKDA, NKFA

The RD's interview indicates that patient does not like meat. Eats fish, beans, dairy, and eggs. Dislikes okra and beets.

Usual dietary intake:

· Drinks 64 ounces of caffeinated coffee or tea per day. No soda, juice, milk, or water.

· Rarely drinks alcohol, nonsmoker. Eats one time per day in the early evening, usually a large salad and a bowl of fruit. Average kcal—1,600/d, average pro—30g/day, average total fiber—15 g/day. Average micronutrient intake 90% to 260% recommendations.

Evaluation of Assessment Data

A thorough nutritional analysis and examination of repeated laboratory values is essential. The caloric content of the diet is marginal and the protein content is low. The non-heme iron in the patient's diet is poorly absorbed, and there may be additional interference with mineral absorption due to higher fiber intake, as fiber may bind trace elements. Supplementation does not ensure adequacy, as the constituent binders in supplements do not ensure bioavailability or adequate uptake. Ferrous formulations in the supplements must be investigated.

The patient's health status is further clouded by possible hypothyroidism, which often presents as fatigue, depression, weight gain, and slowed metabolic utilization of both macronutrients and micronutrients. Further laboratory testing needs to be done to investigate the nature of the low thyroid values, because hypothyroidism can be primary or secondary to pituitary abnormalities or autoimmune disease. Furthermore, other assessments should include depression testing, hormone panels, and checks for occult blood losses. Chronic caffeinism may result in GI blood losses and over time can present with paradoxical effects (more caffeine is needed due to increased fatigue as the body increases its rate of breakdown of methylxanthines). Overuse of caffeine has diuretic effects, results in increased rates of mineral excretion, and may interfere with absorption for several micronutrients.

Nutrition Diagnosis:

Nutrition Diagnostic Terminology: Altered nutrition-related laboratory values NC 2.2.

PES Statement: Altered nutrition-related laboratory values RT caffeinism, poor heme iron intake, and mineral-mineral interactions AEB HgB laboratory values below normal.

Nutrition Intervention:

Strategies that may be employed to enhance iron availability would be to use ferrous sulfate or gluconate formulations, including iron supplements that are not part of a multivitamin preparation to decrease competition for absorption from other minerals such as copper and zinc; taking iron supplements; and using heme iron if possible. A vitamin C source in conjunction with the iron supplement may be beneficial, dependent upon the type of supplement used, or the ionic state of the iron in the preparation. If the patient refuses to eat meat, then an effort should be made to use cast iron skillets in cooking, eat non-heme sources of iron with high-acid/high-vitamin C foods, and consume many servings of varied plant sources of iron, such as spinach or kale. High-biological value protein sources should be encouraged, and timing of supplementation should be discussed with the patient.

Monitoring and Evaluation Interventions are evaluated by measurement of specific outcomes. These include laboratory values such as: serum ferritin, TIBC, hemoglobin, hematocrit or transferrin saturation. Newer methods for the evaluation of iron status such as levels of the protein hepcidin will become more available to practitioners over time. Repeated evaluation of food and supplement documentation, such as patient food logs or prescribed supplement records, is required.

Nutrition Therapy for Microcytic Anemia with Chronic Kidney Disease Patients with concomitant kidney disease and/or who are undergoing dialysis may be treated with erythropoietin directly, or with novel erythropoiesis-stimulating proteins, which are glycosylated analogues of human recombinant erythropoietin and act as hormones to stimulate erythropoiesis in marrow. Their half lives are three times longer than those of Epogen® or Procrit® (the trade names of the human recombinant protein erythropoietin).[182]

BOX 19.3 **Lifecycle Perspectives**

Preventing Anemia:
Iron Supplementation in Pregnancy

Pregnant women are usually given some form of iron supplementation as a routine part of prenatal care. The hemodilution of pregnancy and the demands of the fetus result in an increased utilization of iron and depletion of liver stores. Fetal iron status regulates maternal iron metabolism.[1] The increased needs are difficult to meet through diet alone for many women. It has been shown that increased tissue stores of iron during pregnancy have a long-term impact on both the mother and the infant post partum: growth, development, and gestation are all positively affected by adequate iron stores.[2]

Iron supplements are generally poorly tolerated by pregnant women. They cause gastrointestinal distress, constipation, cramping, and increased nausea and vomiting, adding to the discomforts of pregnancy. Compliance with supplemental iron varies by the extent to which the client experiences these side effects. It is recommended that iron be taken with a vitamin C source to enhance absorption and not taken with a concentrated mineral supplement containing zinc, calcium, or copper. To alleviate constipation, a high-fiber diet should be encouraged, water intake should be increased, and regular physical activity should be recommended. The nausea and vomiting may be decreased by the use of bland, cold foods, dry foods eaten upon arising (e.g., saltines), and iced herbal teas containing ginseng and chamomile. It has been shown that multivitamin-mineral preparations are not as effective in increasing tissue iron stores as an iron supplement.[3] The ferrous sulfate or ferrous gluconate supplements are generally well absorbed and are available in higher dosages for prenatal use under a physician's prescription.

Nutritional Implications:

Nutritional concerns regarding anemia in pregnancy are related to blood volume expansion with insufficient delivery of oxygen to tissues, decreased availability of iron for maternal needs, and complications of pregnancy with poor fetal outcomes. Long-term complications may include behavioral and cognitive problems, in addition to poor growth and physical development.[4,5]

Nutrition Diagnosis:

Nutrition diagnoses associated with anemia in pregnancy include: inadequate intake of minerals (iron), inadequate intake of protein (containing heme iron), and impaired GI function.[6]

Nutrition Intervention:

Treatment of the anemia associated with blood volume expansion or inadequate maternal intakes of iron requires supplementation with highly bioavailable iron. Education of the pregnant patient should include: increasing heme iron consumption, regular supplementation with prenatal vitamins containing ferrous sulfate/ferrous gluconate, and use of vitamin C-containing foods in conjunction with iron-containing foods.[7–9] Counseling of the patient regarding small, frequent, nutrient-dense meals along with strategies for reducing nausea, GERD, and constipation should also take place. These strategies include eating cold foods, not lying down after eating, and regular exercise with increased fluid consumption. Newer strategies, such as the use of lactoferrin or intravenous administration of iron containing dextrans, may also be utilized for difficult cases.[10, 11]

Monitoring and Evaluation:

Interventions are evaluated by measurements of specific outcomes. These may include laboratory values such as serum iron, transferrin saturation, ferritin, total iron binding capacity (TIBC), or newer methods such as serum hepcidin levels. Food intake assessment via food records, 24 hour recall, or direct observation should be performed on an ongoing basis. Additional measures could include subjective patient report of symptoms such as nausea, fatigue, constipation, and malaise. Compliance in this group is generally good and continued, periodic follow-up visits should include encouragement and praise.

References
1. Gambling L, Czopek A, Anderson HS, Holtrop G, Srai SK, Krejpcio Z, McArdle HJ. Fetal iron status regulates maternal iron metabolism during pregnancy in the rat. Am J Physiol Regul Integr Comp Physiol. 2009;264(4):R1063–70.
2. Ostojic SM, Ahmetovic Z. Weekly training volume and hematological status in female top-level athletes of different sports. J Sports Med Phys Fitness. 2008;48(3):398–403.
3. Moriarty-Craige SE, Ramakrishnan U, Neufeld L, Rivera J, Martorell R. Multivitamin-mineral supplementation is not as efficacious as is iron supplementation in improving hemoglobin concentrations in nonpregnant anemic women living in Mexico. Am J Clin Nutr. 2004;80:1308–11.
4. McCann JC, Ames BN. An overview of the evidence for a causal relation between iron deficiency during development and deficits in cognitive and behavioral function. Am J Clin Nutr. 2007;85(4):931–45.
5. Rao R, Georgieff MK. Iron in fetal and neonatal nutrition. Semin Fetal Neonatal Med. 2007; 12(1):54–63.
6. Baker PN, Wheeler SJ, Sanders TA, Thomas JE, Hutchinson CJ, Clarke K, Berry JL, Jones RL, Seed PT, Poston L. A prospective study of micronutrient status in adolescent pregnancy. Am J Clin Nutr. 2009;89(4): 1114–24.
7. Gautam CS, Saha L, Sekhri K, Saha PK. Iron deficiency during pregnancy and the rationality of iron supplements prescribed during pregnancy. Medscape J Med. 2008;10(12):283.
8. Milman N. Prepartum anaemia: prevention and treatment. Ann Hematol. 2008;87(12):949–59.
9. Reveiz L, Gyte GM, Cuervo LG. Treatments for iron deficiency anaemia in pregnancy. Cochrane Database Syst Rev. 2007(2):CD003094.
10. Paesano R, Torcia F, Berlutti F, Pacifici E, Ebano V, Moscarini M, Valenti P. Oral administration of lactoferrin increases hemoglobin and total serum iron in pregnant women. Biochem Cell Biol. 2006;84(3):377–80.
11. Al RA, Unlubilgin E, Kandemir O, Yalvac S, Cakir L, Haberal A. Intravenous versus oral iron for the treatment of anemia in pregnancy: a randomized trial. Obstet Gynecol. 2005;106(6):1335–40. Comment: Allen R. Obstet Gynecol. 2006;107(3)742.

BOX 19.4 **Research to Practice**

Bioavailable Iron Compounds

Investigations into iron amino acid chelates of ferrous glycinate and ferric glycinate have shown promise. Binding iron to an amino acid results in demonstrable improvements in bioavailability. Further refinement may result in an easily absorbed fortificant (compound used to fortify the food supply) that causes less gastrointestinal distress.[1] Recently, salt fortified with both iodine and microencapsulated iron has been tested, as has a powdered mix of micronutrients with added phytase (enzyme that breaks down phytate, known to decrease the absorption of divalent minerals such as iron), with good results.[2, 3] Ferric pyrophosphate and ascorbate fortification of fruit juice is also a promising strategy for improving bioavailability.[4]

Intrinsically labeled microencapsulated ferrous fumarate sprinkles have been tested in infants. Initial results show that these supplemental granules are bioavailable, well tolerated, and easily administered. Further testing in areas of the world where iron deficiency in infancy is high will reveal whether or not aggressive distribution and supplementation will impact the global nutritional status of neonates. Lactoferrin has also shown considerable promise. Isolated first in breast milk, this complex of iron allows neonates enhanced absorption of iron; the utility in non-neonates and in persons suffering from absorptive issues is great.[5]

References
1. Hertrampf E, Olivares M. Iron amino acid chelates. Int J Vitam Nutr Res. 2004;74:435–43.
2. Yuan JS, Li YO, Ue JW, Wesley AS, Diosady LL. Development of field test kits for determination of microencapsulated iron in double-fortified salt. Food Nutr Bull. 2008;29(4):288–96.
3. Troesch B, Egli I, Zeder C, Hurrell RF, de Pee S, Zimmermann MB. Optimization of a phytase-containing micronutrient powder with low amounts of highly bioavailable iron for in-home fortification of complementary foods. Am J Clin Nutr. 2009;89(2):539–44.
4. Haro-Vicente JF, Perez-Conesa D, Rincon F, Ros G, Martinez-Graci C, Vidal ML. Dose ascorbic acid supplementation affect iron bioavailability in rats fed micronized dispersible ferric pyrophosphate fortified fruit juice? Eur J Nutr. 2008; 47(8):470–78.
5. Paesano R, Torcia F, Berlutti F, Pacifici E, Ebano V, Moscarini M, Valenti P. Oral administration of lactoferrin increases hemoglobin and total serum iron in pregnant women. Biochem Cell Biol. 2006;84(3):377–80.

Sample Nutrition Diagnostic Terminology for Microcytic Anemia: Undesirable food choices (NB-1.7); inadequate bioactive substance intake (NI-4.1)

Sample PES Statements for Microcytic Anemia: Undesirable food choices R/T limited access to food secondary to subsistence farming and use of high-phytate, arid-tolerant crops AEB limited consumption of animal sources of protein and iron

Inadequate bioactive substance intake R/T limited access to a variety of foods/animal sources of protein and iron AEB majority of food intake from low-iron bioavailability foods (high in phytates, tannins, and non-heme iron)

Resistance to recombinant human erythropoietin and inflammatory processes occur in a large percentage of dialysis patients, and malnutrition is often seen in this population.[183] Supplements specifically formulated for renal patients, which contain lower protein with higher levels of iron to ensure absorption, are often administered (see Chapter 18).

NUTRITIONAL IMPLICATIONS The nutritional implications of anemia include fatigue, depression, and difficulty in physical exertion leading to poor oral intake. Tasks of daily living, including shopping and food preparation, may become more taxing. Appetite may be depressed. Iron supplements themselves cause constipation, bloating, and cramping, and depress intake. Administration of iron supplements may lead to nutritional issues if competition with other divalent cations occurs.

NUTRITION DIAGNOSIS Nutrition diagnoses associated with the anemia seen in chronic kidney disease include inadequate intake of minerals (iron), impaired utilization of nutrients, altered nutrition-related laboratory values for iron-specific tests, or impaired psychosocial determinants of intake.

NUTRITION INTERVENTION Dietary iron absorption can be enhanced by consumption of iron-containing foods along with vitamin C sources. Oily fish such as salmon also enhances iron absorption, as do low-fiber fruits and vegetables.[184, 185] The use of cast iron skillets in cooking acidic foods and the increased intake of animal sources of iron are also strategies for enhancing intake and absorption of dietary iron. Refer to Table 19.7 for additional strategies.

MONITORING AND EVALUATION Laboratory values for iron and overall nutriture should be routinely monitored over time, as should indices of kidney function. Compliance with dietary recommendations should be checked on an ongoing basis (food records, surrogate informants); investigation of psychosocial functioning should also be undertaken.

Megaloblastic Anemias

Definition The RBC in a megaloblastic anemia has a decreased capacity for oxygen transfer and is large, irregular, and immature (see Figure 19.7). These cells are found both in marrow and in circulation. Anemias of this type are often observed with deficiencies of folate and cyanocobalamin (vitamin B_{12}).

Epidemiology The megaloblastic anemias are commonly seen in specific population groups. Women who have insufficient stores for pregnancy, infants of deficient mothers, elderly persons, persons with atrophic gastritis, and chronic alcohol consumers may be at risk for megaloblasticity due to decreased intake or absorption of folate and/or vitamin B_{12}.[186, 187]

Pernicious anemia is specific to the gastrointestinal disorder that results from stomach lining atrophy and inflammation. Vitamin B_{12} malabsorption is secondary to gastroenterological dysfunction and may or may not have concomitant megalo-

Table 19.7 Strategies for Increasing Dietary and Supplemental Iron

Strategy	Action
Increase iron through cooking.	Use cast iron skillets for acidic food preparation, such as tomato sauce. Iron will leech out of the iron skillet into the food.
Increase iron through consumption with a source of vitamin C.	Drink orange juice with food sources of iron, such as spinach.
Increase iron through consumption of foods with higher iron bioavailability.	Lean meats contain highly absorbable iron.
Increase the use of high-iron foods in the diet.	Find acceptable ways to prepare spinach, use blackstrap molasses or include dried beans.
Increase iron intake through the use of iron-fortified foods.	Read ingredient labels for iron fortification of cereals and other products.
Increase iron absorption through proper supplementation.	Take an iron supplement at a different time than other mineral supplements or multivitamin-mineral preparations to avoid competitive absorption. Take supplements with a vitamin C source and preferably on an empty stomach. Continue vitamin administration and reduce negative side effects of constipation and cramping through alternative actions.

blastic, microcytic, or other features of the anemia. Regardless of the underlying mechanism, B12 deficiency results in nerve damage that is not reversible.[188, 189]

Pernicious anemia with megaloblastic, microcytic anemia through vitamin B12 and/or folate deficiency has been recognized since the nineteenth century. More recently, deficiencies of folate and B12 have been linked to cardiovascular disease through homocysteine and its metabolism.[190, 191] Homocysteine is a by-product of methionine metabolism and 1 carbon transfer and methylation capacity of cells. Insufficient folate and vitamin B12 result in accumulations of homocysteine, with a decrease in the availability of donor and acceptor molecules.[192, 193] Several genetic defects and B vitamin deficiencies synergistically contribute to the cardiovascular, cerebrovascular, peripheral vascular, cognitive impairment, and Alzheimer's disease risks seen with hyperhomocysteinemia.[194–196]

Pregnant women, older adults, institutionalized patients, and persons of low socioeconomic status with poor dietary choices should be screened for this form of anemia.[74, 197] The prevalence of megaloblastic anemia due to nutrient deficiency increases with age from 3% to 44% among persons over the age of 65. Institutionalization is also associated with higher rates of anemia. Folate and B12 deficiencies have also been observed in conjunction with iron deficiency in 20% of the elderly population.[198]

Etiology Pernicious anemia has been seen among persons over 65 as a consequence of aging, secondary to a disease state, and sometimes due to poor intake of vitamin B12. **Atrophic gastritis** in the elderly, in which the lining of the

Figure 19.7 Megaloblastic Anemia

Late-stage immature red blood cells from the bone marrow are shown. This is usually seen in folate deficiency that impedes DNA synthesis in the marrow. Alternatively, this could be caused by a consecutive cyanocobalamin deficiency or by treatment with a drug (such as methotrexate, which is given to patients with certain forms of cancer and other disease states) that inhibits DNA synthesis.

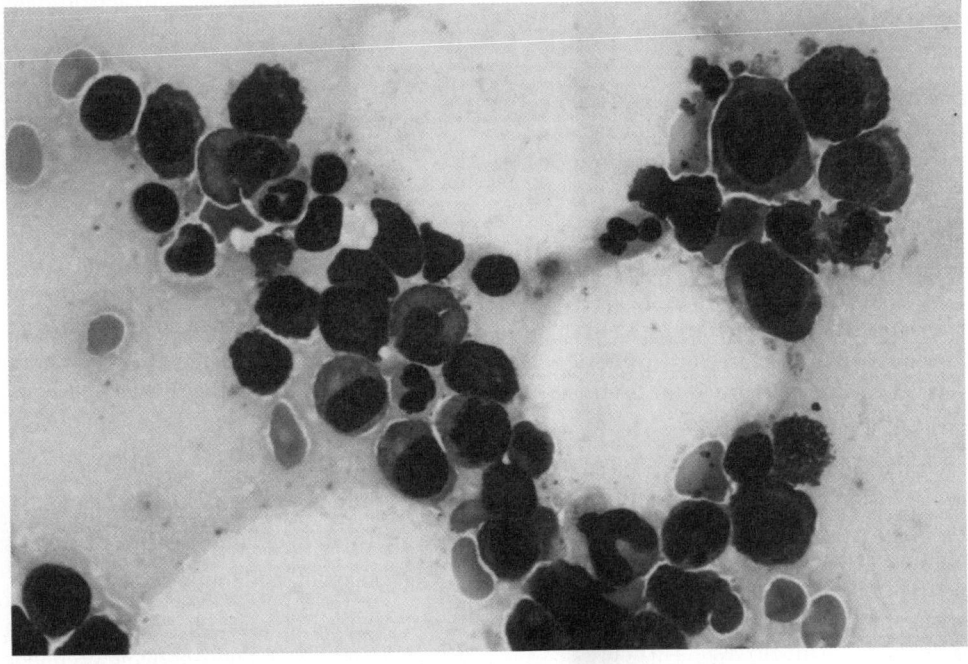

Source: Copyright © Dr. F.C. Skvara/Peter Arnold, Inc.

Figure 19.8 Vitamin B$_{12}$ Absorption

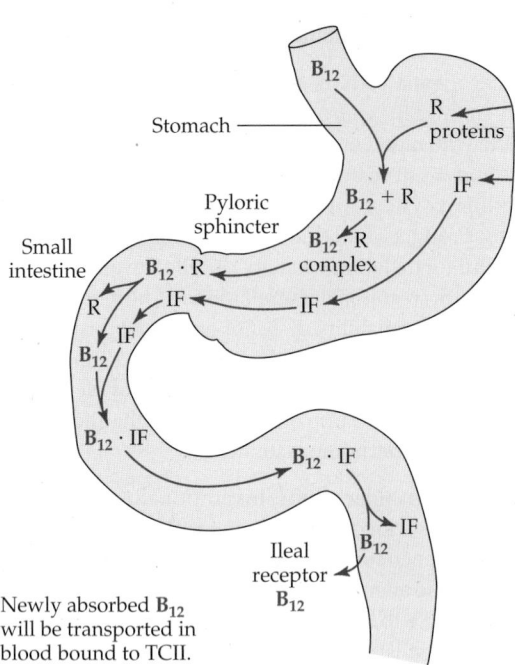

Stomach

Pyloric sphincter

Small intestine

R proteins

B$_{12}$ + R

B$_{12}$·R complex

IF

IF

B$_{12}$·R

R

IF

B$_{12}$

IF

B$_{12}$·IF

B$_{12}$·IF

IF

B$_{12}$

Ileal receptor

B$_{12}$

Newly absorbed **B$_{12}$** will be transported in blood bound to TCII.

Source: J. Smith, J. Groff, and S. Gropper, *Advanced Nutrition and Human Metabolism,* 4e, copyright © 2005, p. 311.

stomach is damaged, is also thought to be age-related. More than half of persons over age 85 have atrophy of the gastric mucosa and hypochlorohydria or achlorohydria (low or absent HCl in the stomach). Parietal cells located in the lining are responsible for secreting **intrinsic factor (IF)** (which picks up B$_{12}$), proteases, and hydrochloric acid, all of which facilitate B$_{12}$ absorption (see Figure 19.8). Vitamin B$_{12}$ is first cleaved from the food source; it is liberated from protein by gastric HCl and the protease breakdown of the foods containing the cobalamin structure. This structure is then solubilized and linked with intrinsic factor, and the two molecules travel into the intestine together. The B$_{12}$ is then transported to the blood side where it is attached to another transport molecule, called **transcobalamin.**[199–201]

Transcobalamin deficiencies are also thought to play a role. **Holotranscobalamin** and the series of transcobalamins (I, II, III) are responsible for the transport of cobalamin, and levels may decline with age as a result of decreased production by the liver or via the disruption in cell signals for homeostasis.

Gastrectomy (stomach resection) and bariatric surgery result in decreased ability to take in, digest, absorb, and transport many vitamins and minerals in food, including folate and B$_{12}$. Bacterial overgrowth, as a result of decreased hydrochloric acid availability, also furthers B$_{12}$ deficiency because the bacteria require and use B$_{12}$.[202]

There has also been substantiation of the megaloblastic anemia of chronic disease (ACD) that often accompanies advanced age and institutionalization.[203] It is often difficult to separate out the numerous contributing factors to anemia in a frail, ill, older person. Persons who are chronically ill also suffer from depression, and the combination of symptoms from pain, inflammation, stress, and dysphoria (sad mood) results in depressed appetite and subsequent malnutrition.

Inflammation in general, whether it is from chronic disease status or acute injury or illness, along with environmental exposure, causes deficits in folate and B$_{12}$ through the increased needs for cellular carbon transfer. Other vitamin and mineral deficits often accompany chronic ailments, thus confounding relationships between the anemia of chronic illness and folate and/or B$_{12}$.[204]

Pathophysiology

CYANOCOBALAMIN Vitamin B$_{12}$ has a corrinoid molecular structure containing cobalt in its rings. The vitamin is a cofactor for methionine synthase and methylmalonyl CoA mutase. These enzymes are responsible for methyl group transfers from methyltetrahydrofolate (mTHF—active folate) to homocysteine and for conversion of succinyl Co-A. An insufficient amount of B$_{12}$ results in megaloblastic anemia, neuropathies, and homocysteinemia.[205] Rapidly dividing cells and cells with increased 1 carbon transfer needs for metabolic intermediate synthesis are adversely affected. Derangements in 1 carbon metabolism affect DNA integrity by altering the methylation and insertion of uracil into the strand.[206] These cause genetic abnormalities that code for abnormal proteins, resulting in a non-functional protein that cannot carry out metabolic reactions.

Active absorption of B$_{12}$ requires parietal cell function, pancreatic function, and a normal terminal ileum. As described earlier, intrinsic factor (IF) attaches to vitamin B$_{12}$ to form a complex that is specific for attachment to receptors on the terminal ileum enterocytic membrane. This complex is phagocytosed and transported by transcobalamin. Absorption can also occur by passive diffusion to some extent, but overall, the absorptive capacity is low. Vitamin B$_{12}$ is recycled and/or stored in the liver. It is secreted into bile and re-circulated enterohepatically, unless there is a deficiency of IF. Without IF, the vitamin is lost in the stool and deficiency progresses rapidly.[207]

There are genetic predispositions to increased numbers of autoantibodies to gastric parietal cells and to IF in white older North American and European populations. Achlorhydria (absence of HCl in the stomach) and atrophic gastritis are often referred to as type A gastritis, whereas the gastritis caused by *H. pylori* infection is type B. Gastric carcinoma is increased threefold in type A gastritis, but the total number of gastric cancer cases induced by gastritis is higher in type B, because *H. pylori* infection is more common. Gastric carcinoma and megaloblastic anemia are usually comorbid.[208]

FOLATE Pteroylmonoglutamic acids (folates) are reduced to tetrahydrofolate or active folate that is poly-glutamated (has glutamic acids attached) and whose activity varies by the number of g-carboxylic acid-linked glutamates available. Methyl tetrahydrofolate (mTHF) participates as a cofactor for many enzymatic reactions, but particularly in 1 carbon transfer for the following functions:

- The synthesis of thymidine for DNA
- The synthesis of purines for RNA
- Remethylation of homocysteine to methionine
- Conversion of methionine to SAMe

These reactions are essential for protein, nucleic acid, **histone**, neurotransmitter, and phospholipid production. If the metabolic reactions for uracil conversion are unable to proceed, due to deficiency in 1 carbon transfer as a result of diminished cycling of folate and B_{12}, uracil accumulates, causing DNA damage. This is how the RBC becomes abnormal.[209] RBC are rapidly dividing cells that require these processes to take place at a rapid rate. If they do not, the RBC cannot be properly blueprinted, and the abnormality is indirectly a result of uracil accumulation.

Clinical Manifestations Many of the clinical signs and symptoms of megaloblastic anemia are comparable to those found in other types of anemias. Anemia is associated with irritability, pallor, pale sclera, and general **pancytopenia**. In chronic deficiencies of B_{12} and folate, megaloblasticity and chromosomal damage are evident . Neuropathies in prolonged cyanocobalamin deficiency and homocysteinemia in milder B_{12} and folate deficiencies are clinical consequences that should be monitored.

Treatment Oral crystalline cyanocobalamin and supplemental folate have been administered with success. Parenteral administration of vitamin B_{12}, alone or in conjunction with other B vitamins, has also been successful in treating the deficiency state. Weekly intramuscular injections (IM) of vitamin B_{12}, alone or in conjunction with other nutrients, are often administered. Blood levels of transcobalamin are routinely monitored, and it is recommended that several consecutive doses be administered before rechecking laboratory values. Patients may be monitored by measuring the numbers of immature red cells (reticulocytes), which are an indication of the patient's response to vitamin therapy.

Holotranscobalamin II levels are thought to be the best markers of B_{12} status, since this value can be used to determine B_{12} absorption as well as adequacy and eliminates the need for the Schilling's test. In practice, the Schilling's test, which is based on gastric functionality, is frequently used to measure B_{12} absorption. Methylmalonic acid (MMA) is also a superior test for the measurement of the patient's ability to utilize vitamin B_{12} and/or folate for homocysteine metabolism.[210] RBC folate is tightly regulated and not directly dependent on intake, and is therefore a poor diagnostic tool for anemia. Both adequacy and functionality tests are required to establish vitamin B_{12} and folate status when other risk factors are present.[211] Multiple laboratory tests, history, and physical exam in addition to dietary assessment procedures are recommended for definitive diagnoses.[212–214]

Nutrition Therapy for Megaloblastic Anemias

Underlying causes for the anemia must be preferentially addressed. Supplementation with vitamin B_{12} and/or folate is recommended. It is often appropriate to provide a multivitamin and mineral preparation in conjunction with single supplement administration in patients with anemia. If the anemia is secondary to impaired absorption, intramuscular or intranasal administration is preferable.

Blood levels of homocysteine, MMA, transcobalamin II, and other measures of hematological status should be routinely monitored. Education of the patient regarding increasing the nutritional density of the diet by incorporation of foods higher in folate and B_{12} is recommended (see Tables 19.8 and 19.9).

Table 19.8 Food Sources of Vitamin B_{12}

Food	Vitamin B_{12} (µg)
Beef liver—1 ounce	32
Clams—1 ounce	16
Oysters—1 ounce	14
Brewer's yeast—2 T	3.0
Lobster—3 ounces	3.0
Pot roast—3 ounces	3.0
Yogurt—1 cup	1

Source: Adapted from: the USDA Agricultural Handbook Series. Available at http://www.usda.gov

Table 19.9 Food Sources of Folate

Food	Folate (µg)
Asparagus—1 cup	263
Spinach (cooked)—1 cup	262
Cooked peas or lentils—½ cup	180
Romaine lettuce—1.5 cups	115
Tortilla—1 item	90

Source: Adapted from: the USDA Agricultural Handbook Series. Available at http://www.usda.gov

Foods rich in B_{12} and/or folate are usually higher in a number of other micronutrients as well.[215]

Nutritional Implications Vitamin B_{12} deficiency increases risk for elevated homocysteine in children as well as in adults.[216] The elderly suffer from heightened risk, especially those receiving supplemental services such as Meals on Wheels. Usually, older people who are homebound and receiving these services are chronically ill and disabled, have poor nutritional status, and are of low socioeconomic status, which means they have less access to health care and caretaking services. Folate status is also poor in elderly patients, especially those who are institutionalized or have functional deterioration, either physical or mental.[217, 218] The nutritional and general health implications of sustained anemia due to folate or vitamin B_{12} deficiency, either clinical or subclinical, are great.

Nutrition Diagnosis Nutrition diagnoses most often associated with these anemias include: Inadequate vitamin intake (specify folate, B_{12}), imbalance of nutrients, and altered nutrition-related laboratory values (specify).

Nutrition Intervention Because of speculation regarding a link between soft drink consumption and folate deficiency,[219] it is recommended that patients consuming excessive amounts of carbonated soft drinks be counseled against this practice. Patients should be educated regarding the increased use of breakfast cereals fortified with folate, pyridoxine, and cyanocobalamin to reduce homocysteinemia.[220] Folate fortification of foods is now standard practice in many countries worldwide, and concerns regarding increased folate masking deficits in cyanocobalamin remain under investigation due to the non-reversible neurological consequences of masked B_{12} deficiency.[221–223]

BOX 19.5 **Research to Practice**

Connections among Nutritional Anemias

Considerations for future research include an investigation into the relationship between folate and/or B_{12} deficiency and iron deficiency, since they often appear together. Megaloblastic anemias increase the absorption of dietary iron from the GI tract, due to feedback mechanisms induced by faulty erythropoiesis, and often increase iron storage, ferritin levels, and saturation levels of transferrin. Despite this phenomenon, spuriously low iron values may be recorded with the use of only hematocrit, hemoglobin, and other non-specific, nonsensitive measures. These spurious findings may in fact be due to rapid mobilization of iron for new erythropoiesis or some other mechanism involving rapid cell division. This may be due to abnormal vitamin A metabolism in iron deficiency,

since hepatic retinol mobilization is adversely affected by iron deficiency and retinol is essential for erythropoiesis.[1] Depending on the circumstances, various abnormalities in the RBC population may co-exist in patients with iron deficiency and megaloblastic anemias of nutritional origin.[2, 3]

References
1. Oliveira JM, Michelazzo FB, Stefanello J, Rondo PH. Influence of iron on vitamin A nutritional status. Nutr Rev. 2008;66(3):141–47.
2. Carmel R. Cobalamin (B12). In: Shils, ME, Shike M, Ross CA, Caballero BC, Cousins RJ, editors. Modern nutrition in health and disease. 10th ed. Philadelphia (PA): Lippincott Williams & Wilkins; 2005, 482–97.
3. Carmel R. Folic Acid. In: Shils, ME, Shike M, Ross CA, Caballero BC, Cousins RJ, editors. Modern nutrition in health and disease. 10th ed. Philadelphia (PA): Lippincott Williams & Wilkins; 2005; 470–81.

Although it has been suggested that folate fortification of grain in this country could lead to a masking of anemia stemming from deficiencies of vitamin B_{12} in susceptible populations, this has not been documented. The U.S. policy of folate, B-vitamin, and iron fortification of the food supply was adopted out of the need to decrease the risk of neural tube defects and deficiency diseases in the population. Fortification does not seem to increase the number of false negatives for the diagnosis of B_{12} anemia and has decreased the prevalence of iron deficiency of nutritional origins in the United States.[224] Other research has investigated the relationships among deficiencies of folate, vitamin B_{12}, and iron, but the phenomenon is complex and requires further study (see Box 19.5).

Worldwide, in geographic areas where strictly megaloblastic anemia from decreased intake of B_{12}-containing foods is commonplace, animal foods are the best way to improve B_{12} status and overall nutritional adequacy. The use of yeast, milk, and meats should be attempted prior to issuing supplements. Supplements may be made from cyanobacterium and algal derivatives, which contain pseudo-vitamin B_{12} which is nonactive in humans.[225] Food sources, such as liver, that contain fat- and water-soluble vitamins and many minerals lacking in the dietary patterns of third-world nations are preferable. If vegetarianism, religious beliefs, or cultural prohibition is not the issue, animal sources should be incorporated rather than supplements.[226]

Sample Nutrition Diagnostic Terminology for Megaloblastic Anemia: Altered nutrition-related laboratory values (NC-2.2)

Sample PES Statement for Megaloblastic Anemia: Altered nutrition-related laboratory values RT inadequate vitamin B_{12} intake secondary to long-standing vegan vegetarianism without incorporation of fortified foodstuffs AEB Holotranscobalamin II laboratory values below normal indicating poor cyanocobalamin status, with no observable absorptive abnormalities, and normal MMA results

Monitoring and Evaluation Laboratory values such as transcobalamin, serum homocysteine, and serum folate levels must be consistently monitored. Food intake should be logged, education regarding fortified products and reduction in carbonated beverages should be continued, and the periodic check of absorptive capacity of the gastrointestinal tract is essential.

Hemochromatosis

Definition

Hemochromatosis (iron overload) is a condition in which a number of regulatory mechanisms for iron are inoperative. The primary result is the buildup of iron and pro-oxidant iron damage of cells. Primary forms of hemochromatosis are hereditary, while secondary hemochromatosis is due to an alternate condition such as liver disease.[227, 228]

Epidemiology

Prevalence of the mutation for hereditary hemochromatosis in the United States is 5/1,000 Caucasians. Approximately 10% of the non-Hispanic white population is heterozygous for the abnormality. There are other forms of hemochromatosis, such as juvenile hemochromatosis, African hemochromatosis, and aceruloplasminemia. The latter is an inherited lack of ceruloplasmin (protein responsible for copper transport and utilization), which results in the inability of the liver to release iron. These diseases are extremely rare and little is known of their etiology or epidemiology.

Etiology

The etiology of the disease varies according to classification as primary hereditary or hemochromatosis secondary to a disease state such as cirrhosis. Genetic predisposition exists for mutations in hepcidin, **prohepcidin,** hemojuvelin, **divalent metal transporter (DMT-1)**, ferroportin, and hephaestin (among other important proteins in iron homeostasis), which result in the following:[229, 230]

- Increased iron absorption from the gut of heme or non-heme iron irrespective of intake.
- Decreased removal of iron from circulation.

- Decreased ability to mobilize iron from the liver, so that excessive amounts build up and cause hepatocytic damage.
- Decreased ability to synthesize hemosiderin, ferritin, transferrin, and iron-containing cytochromes, the specific proteins for iron transport, sensing, storage, and utilization.
- Homozygous and heterozygous mutations of HFE-b2 microglobulin, a protein associated with the affinity for transferrin receptors in the membrane of the enterocytes lining the gastrointestinal tract.

The presence of mutations increase symptomatic hemochromatosis, which is seen frequently in persons of northern European decent. The mutation is thought to be driven by the evolutionary need to prevent iron-deficiency anemia in these populations during times of low animal product intake due to environmental conditions prevalent in those regions.[231]

Other conditions can also lead to iron overload. Nonalcoholic liver disease has been shown to decrease levels of ferroportin and hemojuvelin.[232] Heavy alcohol consumption is considered a risk factor for iron overload (see Box 19.6); cirrhosis and several genetic blood disorder treatments are associated with elevated iron levels and cellular damage. Vitamin C, by enhancing absorption of non-heme iron, may aggravate iron overload with pre-existing conditions. Elderly persons are more likely to exhibit high iron stores versus iron-deficiency anemia in the United States, but this is reversed worldwide.[145, 233, 234]

Pathophysiology

Chronic toxicity from iron overload leads to functional diabetes, hepatomegaly (enlarged liver), **cicatrix** formation, and eventually cirrhosis, through oxidative damage to cellular membranes in nearby locations. Hepatolenticular carcinoma, a cancer of a specific set of liver cells, develops in 30% of cirrhotic patients, due to pro-oxidant changes in the nucleic acids of cells. Cardiomyopathy develops and leads to heart failure in approximately 50% of patients, through a variety of mechanisms.[235]

Clinical Manifestations

The patient may present with non-specific symptoms. Feeling cold, tired, or irritable or being unable to eat are common complaints. Slate grey skin is characteristic, along with abnormal blood glucose levels. Slow onset type 2 diabetes is common. Laboratory assessments show increased levels of free and bound iron (serum iron, serum **transferrin saturation**, bound iron-ferritin), indicators of oxidative stress such as C-reactive protein (CRP) reactive oxygen species (ROS), and alterations in iron regulatory proteins (hepcidin, hemojuvelin, and ferroportin), along with abnormal liver function tests (SGOT and SGTP).[236]

Treatment

Phlebotomy is done routinely for hemochromatosis patients. Removal of 500 milliliters of blood, containing 250 mg of iron,

---

BOX 19.6 **CLINICAL APPLICATIONS**

Alcohol and Iron Overload in Men and Postmenopausal Women

The liver is responsible for the detoxification of alcohol via alcohol dehydrogenase and acetaldehyde dehydrogenase, as well as the microsomal ethanol oxidizing systems and the peroxisomal oxidizing systems. As the liver completes the bulk of ethanol breakdown, increased ingestion of large quantities of ethanol eventually cause scarring and destruction of hepatic cells, leaving less functional capacity for ethanol breakdown.

Mobilization of iron from storage in the liver is dependent on correct cell signaling and liver functionality. An accumulation of circulating ethanol will rapidly decrease the ability of the liver to handle its remaining homeostatic functions, including the following:

- production of transferrin, the transport protein for iron
- maintenance of ferroportin, the iron export regulator
- maintenance of hemojuvelin, the iron level sensor protein

- maintenance of hepcidin, the iron homeostasis hormone
- maintenance of prohepcidin, the precursor to hepcidin and a marker for hemostasis
- function of divalent metal transport-1 (DMT-1), the iron import protein
- function of hephaestin, the protein that incorporates iron into transferrin, and
- ferritin, the storage protein for iron.

Iron, because of its oxidative capacity as a mineral, is thought to cause oxidative damage in surrounding cells if improperly stored. The damaged liver is further decompensated by immobilized, improperly stored iron.[1, 2] Iron overload is exacerbated by alterations in iron absorption resulting from pH changes in the stomach and the duodenum that are secondary to alcohol's effect on the gastrointestinal system. Increased

uptake may also be mediated by the disruption in homeostatic mechanisms for iron absorption and trafficking due to liver and brain abnormalities.[3] There are also additional mechanisms by which heavy alcohol consumption leads to an iron deficiency, especially if dietary iron is low and there is gastrointestinal bleeding as a result of ethanol ingestion.[4]

References
1. Petrak J, Msylivcova D, Man P, Cmejla R, Cmejlova J, Vyoral D, Elleder M, Vulpe CD. Proteomic analysis of hepatic iron overload in mice suggests dysregulation of urea cycle, impairment of fatty acid oxidation, and changes in the methylation cycle. Am J Physiol Gastrointest Liver Physiol. 2007;292(6):G1490–98.
2. Asare GA, Bronz M, Naidoo V, Kew MC. Synergistic interaction between excess hepatic iron and alcohol ingestion in hepatic mutagenesis. Toxicology. 2008;524(1–2):11–18.
3. Halstead CH. Nutrition and alcoholic liver disease. Semin Liver Dis. 2004;24:289–304.
4. Ioannou GN, Dominitz JA, Weiss NS, Heagerty PJ, Kowdley KV. The effect of alcohol consumption on the prevalence of iron overload, iron deficiency, and iron deficiency anemia. Gastroenterology. 2004;126:1293–301.

is performed once a week until laboratory measures are normalized; then the phlebotomy is individualized based on consistent testing. An iron chelator such as deferoxamine may also be administered, but due to poor absorption, it must be infused or injected. Deferiprone, deferasirox, deferitrin (second and third generation iron chelators), and combinations thereof can be orally administered.[237]

Alternate methods include proton pump inhibitors (used for acid reflux disease) to suppress absorption of dietary non-heme iron and black tea tannin isolates to reduce iron absorption from the gut.[238, 239]

Nutrition Therapy

The purpose of nutrition therapy in patients with hemochromatosis is to minimize the risk of excess iron intake.

Nutritional Implications Aggressive phlebotomy in hemochromatosis patients leads to functional acute anemia due to blood losses and RBC destruction. This, however, should not be corrected through dietary supplementation. Patients should still be advised to moderate their iron intake.

Nutrition Diagnosis Nutrition diagnoses most often associated with hemochromatosis are imbalance of nutrients and excessive iron intake.

Nutrition Intervention Patients should be educated on the difference between heme and non-heme iron sources. They should also be knowledgeable of the affect that acidic or vitamin C-rich foods have on the absorption of iron. Extraneous iron sources such as cast-iron skillets or multivitamin preparations replete with iron should be avoided. Patients should be educated regarding minimization of intake of highly absorbable iron obtained from meats and the decreases in iron absorption with diets high in fiber, phytates, and tannins. Patients should be advised to consume little or no alcohol and

---

BOX 19.7 CLINICAL APPLICATIONS

Restless Leg Syndrome—The Iron Connection

Restless leg syndrome (RLS) is a common, poorly understood and often misdiagnosed disorder.[1, 2] It is characterized by uncomfortable sensations in the legs (and sometimes other body parts) that only resolve with movement. These sensations have been described as pain, tightening, crawling, cramping or dripping deep inside the legs. It occurs at rest, is often worse at night, and causes sleep disturbances.[3, 4]

RLS occurs across age groups and racial and ethnic backgrounds, and may have a genetic component. Several genes have been identified as potentially contributory to RLS. Prevalence increases with age, women are disproportionately affected, and the course of the syndrome is chronic.[5–7] There is increasing evidence that the condition may be common in children but largely under-diagnosed because children are unable to describe the symptoms.[8, 9]

The etiology of RLS in many persons may be related to iron deficiency or altered iron metabolism.[10, 11] Primary

RLS occurs in the absence of any other disorder or condition, and patients report similar symptoms in several family members. Secondary RLS may be due to medications, diabetes, kidney disease, rheumatoid arthritis, neurological disease, hemochromatosis, pregnancy, or uncomplicated iron deficiency, among others.[12, 13]

Mechanistically RLS is related to the interplay of brain iron deficiency and impaired dopamine neurotransmission in many regions of the sub-cortex of the brain. Dopamine is a neurotransmitter responsible for a wide variety of functions systemically, including nerve firing for muscular activity.[14–16] Iron is necessary for dopamine synthesis.[17, 18] Iron deficiency alters the ability of the brain to properly metabolize dopamine, and to some extent this alteration changes norepinephrine concentrations as well. This is evidenced by an increase in catabolism of available dopamine by dopamine beta hydroxylase. RLS patients express both higher levels of transferrin receptors and divalent

metal ion (DMT-1) receptors, signifying increased needs for iron, and increased ferroportin, signifying increased export of iron. Greater iron turnover and greater iron needs result in greater iron deficiency in the brain.[19–21] Sufficient peripheral iron may not be a good indicator of brain iron stores or utilization-export patterns. Levels of hepcidin, an iron regulating proteinaceous hormone that signals for iron homeostasis by interacting with ferroportin, is altered in the cells of the brain and cerebrospinal fluid. The type of RLS (primary, secondary, early or late) determines how abnormal iron homeostasis presents in any given RLS patient.[22–24]

Primary treatment options for RLS patients include pharmaceutical administration, with drugs such as opioids, gabapentin, and levodopa given for neurological symptoms.[25, 26] Iron supplementation is now also used for patients that display iron responsive RLS. Iron can be delivered either orally or intravenously, alone or in combination with folate. The iron itself may be administered in a variety of chelations—such as with iron-sucrose, iron-dextrans, or iron-gluconate—and long term; repeated administration is required to maintain reductions in symptoms.[27–30]

Nutritional Implications:
Nutrition concerns in RLS are related to alterations in iron metabolism and iron levels in the brain and cerebrospinal fluid (CSF). The resulting effects include movement disorders, poor sleep quality, unpleasant sensations in the legs (and/or other body parts), depression, and decreased quality of life.[31–33]

Nutrition Diagnosis:
Nutrition diagnoses associated with RLS include: imbalance of nutrients, impaired nutrient utilization, inadequate mineral intake (iron), or altered nutrition-related laboratory values for ferritin, hepcidin or CSF iron carrying compounds.

Nutrition Intervention:
Treatment of iron deficiency in brain and neurological tissues can be implemented through high-dose oral iron intake (>250 mg per dose, multiple doses per day) of a highly soluble, absorbable form of iron. Iron should be administered to maximize absorption, decrease chelation by other minerals/phytates, and decrease GI discomfort. Iron should be taken separately from other mineral supplements, with a vitamin C source, and should not be administered with tea or milk. Heme sources of iron should be encouraged, but will not provide the amounts needed to overcome the decrement in distribution to the brain and CSF. Gastrointestinal concerns may be alleviated with a stool softener, high-fiber diet, plenty of fluids, and timing of supplement administration.

Monitoring and Evaluation:
Laboratory values for iron should be routinely monitored. Serum ferritin and (in the advent of normal serum ferritin values with continued low CSF iron status) other markers should be assessed.[34] Compliance with supplemental iron administration should be documented, intake records evaluated periodically, and level of knowledge regarding the importance of iron in decreasing RLS symptomatology ascertained.

Example: Nutrition Diagnostic Terminology: Inadequate mineral intake (iron) NI 55.1

PES Statement: Inadequate mineral intake (iron) RT insufficient iron availability to brain and CNS with restless legs syndrome AEB abnormal laboratory values indicative of iron deficiency in CNS (low serum ferritin), abnormal CSF iron availability indicators (CSF ferritin), and continued abnormal leg movements, discomfort, and alterations in sleep.

References
1. Agarawal P, Griffith A. Restless Leg Syndrome: A unique case and essentials of diagnosis and treatment. Medscape J Med. 2008;10(12):296–98.
2. Cotter PE, O'Keefe ST. Restless Leg Syndrome: Is it a real problem? Ther Clin Risk Manag. 2006;2(4):465–75.
3. Vergne-Salle P, Coyral D, Dufauret K, Bonnet C, Bertin P, Traves R. Is restless leg syndrome underrecognized? Current management. Joint Bone Spine. 2006;73(4):1297–319.
4. Schapira AH. RLS patients; who are they? Eur J Neurol. 2006;13(S3):2–7.
5. Hening W, Allen RP, Tenzer P, Winkleman JW. Restless legs syndrome: demographics, presentation, and differential diagnosis. Geriatrics. 2007;32(9):26–29.
6. Bayard M, Avonda T, Wadzinski J. Restless leg syndrome. Am Fam Phys. 2008;78(2):235–40.
7. Trotti LM, Bhandriraju S, Rye DB. An update on the pathophysiology and genetics of restless legs syndrome. Curr Neurol Neurosci Rep. 2008;8(4):281–87.
8. Simakajornboon N, Kheirandish L, Gozal D. Diagnosis and management of restless leg syndrome in children. Sleep Med Rev. 2009;13(2):149–56.
9. Patrick LR. Restless legs syndrome: pathophysiology and the role of iron and folate. Altern Med Rev. 2007;12(2):101–12.
10. Agarwal R. Nonhematological benefits of iron. Am J Nephrol. 2007;27(6):565–71.
11. Fitzgerald MA. Restless Leg Syndrome: Iron important. Nurse Pract. 2008;33(4):7–9.
12. Haba-Rubio J, Staner L, Petiau C, Erb G, Schunck T, Macher JP. Restless legs syndrome and low brain iron levels in patients with hemochromatosis. J Neurol Neurosurg Psychiatry. 2005;76(7):1009–10.
13. Winkleman JW. Considering the causes of restless legs syndrome. Eur J Neurol. 2006;13(S3):8–14.
14. Manconi M, Hutchins W, Feroah TR, Zucconi M, Ferini L. On the pathway to an animal model for restless leg syndrome. Neurol Sci. 2007;28:853–60.
15. Godau J, Klose U, Di Santo A, Schweitzer K, Berg D. Multiregional brain iron deficiency in restless legs syndrome. Mov Disord. 2008;23(8):1184–87.
16. Bianco LE, Wiesinger J, Earley CJ, Jones BC, Beard JL. Iron deficiency alters dopamine uptake and response to L-Dopa injection in Sprague Dawley rats. J Neurochem. 2008;106(1):205–15.
17. Oertel WH, Trenkwalder C, Zucconi M, Benes H, Borreguero DG, Bassetti C, Partinen M, Ferinin-Strambi L, Stiasney-Kolster. State of the art in restless legs syndrome therapy: practice recommendations for treating restless legs syndrome. Mov Disord. 2007;22(18):S466–75.
18. Allen RP, Early CJ. The role of iron in restless legs syndrome. Mov Disord. 2007; 22 Suppl 18:S440–48.
19. Connor JR. Pathophysiology of restless legs syndrome; evidence for iron involvement. Curr Neurol Neurosci Rep. 2008;8(2):162–66.
20. Earley CJ, Ponnuru P, Wang X, Patton SM, Connore JR, Beard JL, Taub DD. Altered iron metabolism in lymphocytes from subjects with restless legs syndrome. Sleep. 2008;31(6):847–52.
21. Paulus W, Dowling P, Rijsman R, Stiasny-Kolster K, Trenkwalder C. Update on the pathophysiology of the restless leg syndrome. Mov Disord. 2007;22(18):S431–S439.
22. Clardy SL, Wang X, Boyer PJ, Earley CJ, Allen RP, Connor JR. Is ferroportin-hepcidin signaling altered in restless legs syndrome? J Neurol Sci. 2006;247(2):173–79.
23. Clardy SL, Earley CJ, Allen RP, Beard JL, Connor RJ. Ferritin subunits in CSF are decreased in restless legs syndrome. J Lab Clin Med. 2006;147(2):67.
24. Earley CJ, Barker P, Horska A, Allen RP. MRI determinations regional brain iron concentrations in early and late onset restless legs syndrome. Sleep Med. 2006;7(5):458–61.
25. Hening WA. Current guidelines and standards of practice for restless legs syndrome. Am J Med. 2007;120(1):Suppl.1:S22–S27.
26. Winkleman JW, Allen RP, Tenzer P, Hening W. Restless legs syndrome: non-pharmacological and pharmacological treatments. Geriatrics. 2007;62(10):13–16.
27. Trenkwalder C, Hogl B, Benes H, Kohnen R. Augmentation in restless leg syndrome is associated with low ferritin. Sleep Med. 2008;9(5):572–74.
28. Earley CJ, Horska A, Mohamed MA, Barker PB, Bears JL, Allen RP. A randomized placebo controlled trial of intravenous iron sucrose in restless leg syndrome. Sleep Med. 2009;10(2):206–11.
29. Earley CJ, Heckler D, Allenm RP. Repeated IV doses of iron provides effective supplemental treatment of restless legs syndrome. Sleep Med. 2005;6(4):301–05.
30. Gorder V, Kuntz S, Khosla S. Treatment of restless leg syndrome with iron infusion therapy. JAAPA. 2009;22(3):29–32.
31. Allen RP. Controversies and challenges in defining the etiology and pathophysiology of restless legs syndrome. Am J Med. 2007;120(1):Suppl.1:S13–S21.
32. Trenkwalder C, Hening WA, Montagna P, Gertel WH, Allen RP, Walters AS, Costa J, Stiansny-Kolster K, Sampaio C. Treatment of restless leg syndrome: An evidence based review and implications for clinical practice. Mov Disord. 2008;23(16):2267–302.
33. Shariatpanaahi M, Vahdat Shariatrpanaahi Z, Moshtaaghi M, Shahbaazi SH, Abadi A. The relationship between depression and serum ferritin level. Eur J Clin Nutr. 2007;61(4):532–35.
34. Mizuno S, Mihara T, Miyakota T, Inagaki T, Horiguchi J. CSF iron, ferritin, transferrin levels in restless legs syndrome. J Sleep Res. 2005;14(1):43–47.

no iron-containing supplements or heavily fortified cereal products. Counseling and education should involve consumption of fewer foods of animal origin and vitamin C sources between meals. Iron-fortified meal replacements and energy drinks should not be used unless specified by a physician. If the patient is being treated with binders, the diet should be modified accordingly, and more liberalization of the diet may occur. More iron sources and iron–vitamin C combinations are permitted.

Monitoring and Evaluation Continued monitoring of blood levels should dictate dietary constraints and patient education strategies for dietary adherence. Routine follow-up should consist of inspection of laboratory values, food records, and findings of the history and physical.

Future Research

The genetic mutations for hemochromatosis have been mapped. Isolating the genetic sequence responsible for enhanced iron absorption in hemochromatosis may lead to discoveries of compounds that would enhance iron absorption from nutritionally poor substrates. Such a discovery could decrease the incidence of iron-deficiency anemia worldwide. Iron-deficiency anemia is prevalent and causes many deaths secondary to decreased immunity. This application could improve global health.

Hemoglobinopathies: Non-Nutritional Anemias

The five major classes of hemoglobinopathy are structural (e.g., sickle cell anemia), thalassemias, thalassemic hemoglobin variants, hereditary persistence of fetal hemoglobin, and acquired hemoglobinopathy (e.g., anemia secondary to exposure to toxins or a disease state such as cancer). Table 19.10 summarizes the etiology, manifestations, and medical treatment for selected hemoglobinopathies. Pertinent nutrition interventions, which are typically designed to support medical treatments, are described in the following sections.[240]

Sickle Cell Anemia

Sickle cell anemia is the most common structural hemoglobinopathy, defined as homozygous abnormal hemoglobin polymerization resulting in a sickling of cells in symptomatic individuals.[241] The crescent-shaped cell morphology is obvious when stained and magnified. Combinations of sickle cell and thalassemia occur as a result of inheritance of the traits from each parent and result in variant syndromes not strictly categorized as sickle cell anemia.[5]

Nutrition Therapy Treatment for an acute sickle cell crisis should include increased macronutrients, because there is an increased level of energy expenditure.[242] Oral glutamine may be beneficial for the patient, because glutamine is an important amino acid for rapidly dividing cells. Antioxidant vitamin and mineral status should be evaluated and maintained at recommended levels, as patients have reduced blood levels of antioxidants.[243] Optimal folate, vitamin B_{12}, and pyri-

doxine levels should be maintained through supplementation. Sickle cell patients have documented declines in nutritional adequacy over time.[244–246] Fluids and hydration must be closely monitored in these patients. Fatty acid metabolism is altered and elevations in non-esterified fatty acids are seen postprandially. Fatty acid composition of the diet should be low in saturated fat.[247]

Thalassemia

The thalassemias are inherited disorders of abnormal alpha or beta globin synthesis. Reduction in globin availability results in decreased hemoglobin synthesis. The severity of the resulting anemia is dependent upon the degree to which synthesis is impaired. Inheritance of several abnormal genes worsens the clinical phenotype. The RBC are hypochromic, elliptical, and irregular. The red cell distribution width may be altered even in mild Beta-Thalassemia Minor whose presentation mimics iron-deficiency anemia from poor dietary intake.[248, 249] The bone marrow becomes hyperplastic in severe cases, with increased anemia stimulating excessive erythropoietin production with no concomitant increase in functional hemoglobin synthesis.[250] Thalassemia traits may increase iron damage to the liver, increase co-morbid conditions, and exacerbate adverse consequences due to alterations in hepcidin and failure to distinguish between the disease and severe iron deficiency by practitioners, resulting in improper treatment.[251, 252] A systematic review of degree of microcytosis, hematocrit distribution, RBC width, and hypochromia is necessary to decrease **iatrogenicity** (harm as a result of treatment or procedures performed by healthcare practitioners).[253, 254]

Nutrition Therapy Transfusions are the standard treatment for hemoglobinopathies. Iron overload is often treated with an iron chelator, such as deferoxamine. Patients often undergo periods where they become unresponsive to deferoxamine treatment. Ascorbic acid administration enhances the efficacy of the chelation regimen.[21]

Polycythemia

Defined as an increase in circulating RBC, the syndrome can be spurious or real.[255] Spurious polycythemia relates to decreased plasma volumes showing an increase in RBC above normal within a deciliter of blood; real polycythemia is a result of dysregulated feedback and increased production of red cells by marrow, usually detected by abnormal epoietin levels.[256] Very low erythropoietin levels usually indicate polycythemia vera (**chronic myeloproliferative disease**). Very high erythropoietin levels indicate either polycythemia due to abnormal production of RBC (primary) or polycythemia as a physiological response to hypoxia (secondary).[257]

Nutrition Therapy Routine phlebotomy induces iron-deficiency anemia. Patients should be educated regarding increasing their dietary iron (refer to Table 19.7) and avoiding very high-dose supplementation, which causes complications from oxidative stress and the increased competition among minerals for absorption.[257, 258] Transfusion in polycythemia increases resting energy expenditure, which is normally low in these patients. Treatment modality will alter energy requirements for patients.[259, 260]

Hemolytic Anemia

RBC destruction or hemolysis results in hemolytic anemia. Specific factors that cause hemolytic anemia may include, but are not limited to: sustained blood losses, inherited disease of abnormal reticulocyte formation or autoimmune disease, severe trauma or mild sustained trauma (as in long distance runners), certain dysplastic diseases, and sickle cell phenotype.[261] The losses of RBC either through hemorrhage or premature destruction leads to an increase in RBC production, as evidenced by laboratory testing.[262]

Hemolytic anemia may occur in newborn infants because of either hemorrhage from the umbilicus, internal bleeding, or **erythroblastosis fetalis**, which develops when Rh positive infants are born to Rh negative mothers. The latter occurs in the second child of the mother, after sensitization by the first child's birth. Antibodies from the mother enter the fetal circulation via the placenta and react with RBC antigens in the fetus, resulting in significant RBC destruction and compensatory hyperplasia of blood-forming tissues. In either form, there is usually hyperbilirubinemia from red cell destruction, jaundice, and the possibility of **kernicterus** leading to neurological and brain damage.[263]

Nutrition Therapy If the underlying cause of the hemolysis is being treated (e.g., toxic compound removal, immunosuppressives), the patient should be supplemented with iron, folate, and protein to enhance RBC production and food safety precautions should be closely monitored.[264] If therapeutic **apheresis** (blood compound exchange or removal) is being administered, practitioners should monitor and prevent iron overload that results from repeated RBC exchange.[265, 266]

Anemia of Prematurity

Anemia seen in premature infants is usually related to low levels of erythropoietin due to underdeveloped kidneys and failure of feedback mechanisms for erythropoiesis.[267]

Nutrition Therapy Vitamin E supplementation during erythropoietin treatment of the anemia of prematurity is efficacious for protecting red blood cell membranes and decreasing hemolysis. Iron dextrans in solution should be administered through the parenteral route (if possible), and folate status should be monitored closely.

Aplastic Anemia

True primary **aplastic anemia** is also known as Fanconi's anemia. It is the result of marrow failure and is inherited. Fanconi cells of the marrow have abnormal oxygen metabolism cycles.[268] Secondary aplastic anemias may result from poisoning, trauma, autoimmune disease, or cancer. There are also other inherited syndromes of marrow failure that present as an aplastic anemia, although the true etiology is myeloproliferative.[269]

Nutrition Therapy Nutrition therapy for aplastic and other rare anemias includes maintenance of macronutrient and micronutrient adequacy through transfusion therapy, bone marrow transplant (BMT), and other treatments. Strategies for patients undergoing high-dose corticosteroid treatment should include maintenance of normal fluid and electrolytes status as well as monitoring of calcium and vitamin D.[268, 269]

Other Rare Anemias

Blackfan-Diamond anemia and Schwachmann's syndrome are two rare forms of unilineage (affecting one line of cell types) cytopenia and pancytopenia, respectively. These are inherited bone marrow failure syndromes that result in a severe anemia and hypoxia.[269]

Nutritional Implications of Non-Nutritional Anemias

There are profound nutritional implications associated with immunosuppressive therapy and treatment by BMT (see: Diseases Involving WBC Types).[266] Treatments involve drugs with adverse consequences ranging from anorexia to renal failure. Patients undergoing these modalities exhibit poor intake, absorption, utilization, and disposition for a wide variety of nutrients.[250, 270]

Nutrition Intervention for Non-Nutritional Anemias

Disease progression in many of these anemias is related to oxidative stress. Improvements in cellular antioxidant capabilities may reduce chromosomal instability. Supplementation with compounds that are cofactors for antioxidant enzymes or are antioxidants themselves (selenium, zinc, vitamins A, C, and E) may be beneficial.[271, 272]

Transfusion anemia, corticosteroid-induced Cushinoid syndrome, chemotherapy cachexia, and immunosuppression are all sequelae of treatment.[273] Long-term nutritional interventions should include continued nutrition support, as tolerated, and consistent monitoring.[270, 274]

Clotting and Bleeding Disorders

Hemophilia, hemorrhagic disease of the newborn, and thrombosis are well-known disorders of coagulation. Abnormal coagulation results either in prolonged bleeding times and significant blood loss, as in hemophilia, or in increased clot formation without significant clot lyses, resulting in hypoxia of nearby tissues, as in thrombosis.[275] These disorders are described in detail in Table 19.11, and nutrition therapy for each is discussed in the following sections.[276]

Hemophilia

Hemophilia, an inherited disorder, results in insufficient factors for thrombin generation and clotting. A deficiency in plasma protein factor VIII is classified as hemophilia A; a deficiency in plasma protein factor IX is classified as hemophilia B.

Nutrition Therapy If the patient is being treated with desmopressin, (a drug that is used to control fluid in the body) for hemostatic control, **antidiuresis** occurs with severe hyponatremia and water intoxication. Fluids must be restricted and monitoring of fluid and electrolyte status is essential. Patients should be counseled on healthy eating practices.[277]

Table 19.10 Characteristics of Non-Nutritional Anemias (Hemoglobinopathies)

Anemia	Epidemiology	Etiology	Pathophysiology	Clinical Manifestations	Treatment	Nutritional Therapy
Sickle Cell Anemia	Found in areas where malaria is endemic. 1 in 400 African-Americans in the U.S. are homozygous for sickle cell anemia.[a]	The substitution of valine for glutamic acid in the hemoglobin moiety causes it to turn into a gel. There is occlusion of blood vessels due to increased stickiness of cells.	The crescent-shaped cell and abnormal hemoglobin decrease oxygen carrying capacity, make the cell die off prematurely, and cause cell stickiness.[b]	Sickle cell crisis involves severe pain and fatigue. Blood vessel occlusion can occur anywhere in the body. Infections, jaundice, hyperbilirubinemia are all the result of premature lysis of RBC.[c,d,e]	Transfusions, bone marrow transplant, chronic antibiotic/pain medication.[a]	Increase macronutrients. Optimize mineral intake. Improve antioxidant status. Ensure fluid adequacy and nutrient density.
Thalassemia	Present in 15% of African-Americans and persons of Mediterranean heritage.[f]	Abnormal globin proteins are produced, and severity varies with hetero- or homozygous genetics.	Abnormal proteins cause premature RBC death and inability to carry oxygen.[a]	Stunting, birth defects, organ damage, and severe hypoxia.[f]	Transfusion. Bone marrow transplant.	Ascorbate administration, antioxidant administration, ensure nutritional adequacy.
Polycythemia	Prevalence is estimated at 2 per 100,000 persons.[g]	Hyperproliferation of stem cells in bone marrow, with high levels of circulating blood cells.	Increased cell numbers result in increased blood viscosity.	Impaired circulation, hypertension, brain ischemia, and organ damage.[g]	Phlebotomy, radiation and chemotherapy to reduce cell numbers Continuous airway pressure pumps to deliver oxygen. Anticoagulants.[g]	Increase dietary iron. Ensure nutrient density.
Hemolytic Anemia	Prevalence of autoimmune hemolytic anemia is 1 per 100,000 with boys 2.5x more likely to be affected. Prevalence of Rh alloimmunization is estimated at 11 cases per 100,000 live births.[c]	Anemia results in insufficient oxygen carrying capacity due to fewer numbers of RBC.[h,j]	Immature RBC are produced as a result of negative feedback on hematopoiesis by premature RBC destruction. Hypoxia and tissue damage occur.[c,h,j]	Build-up of the by-products of hemoglobin metabolism result in fatigue, bleeding, infection, jaundice, and presence of breakdown products in the urine (hemoglobinuria).[c,h,j]	Transfusion. Immunosuppressive treatment.[j]	Iron, folate, and protein supplementation.
Anemia of Prematurity	Prematurity occurs in approximately 12% of live births in the U.S. 84% of these births exhibit anemia.[k]	Insufficient or absent production of erythropoietin by the premature kidney.[l]	Bone marrow is understimulated without epoetin; there is decreased iron absorption by the infant and insufficient oxygenation of tissues as a result.	Cyanosis (bluish appearance with lack of oxygen and build-up of carbon dioxide in blood); congestive heart failure; hemorrhage; multiple organ system failure.[l]	Recombinant human erythropoietin administration. Transfusion. Iron supplementation.[k,l]	Vitamin E supplementation. Ensure appropriate infant feeding practices.

Condition	Prevalence	Cause	Consequences	Complications	Treatment	Nutritional Considerations
Aplastic Anemia	Prevalence is estimated at 1 in 100,000 live births (Fanconi's autosomal recessive aplastic anemia).[m]	Bone marrow failure. Marrow fails to produce blood cells. This can be hereditary or secondary to toxic infection or exposure to chemicals.[m]	Marrow failure to make cells reduces clotting ability, immunity, and ability to carry oxygen to tissues.[m]	Sepsis (systemic infection); hemorrhage; failure to thrive in infants.[m]	Transfusion. Bone marrow transplant.	Ensure nutrient density. Maintenance of normal fluid and electrolytic status (particularly sodium) with corticosteroid and immunosuppressive treatment. Maintenance of adequate macro- and micronutrient status. Monitor calcium and vitamin D.
Blackfan-Diamond Anemia	Prevalence is estimated at 7 per 1 million live births.[n]	Inherited bone marrow failure.[n]	Anemia, hypoxia resulting from abnormal stem cells in the marrow failing to produce viable blood constituents.[n]	Tissue and organ damage is widespread.	Bone marrow transplant. Transfusions. Il-3 and corticosteroid administration.[n]	Monitor sodium, calcium and vitamin D. Maintain adequate macro- and micronutrient status.
Schwachmann's Syndrome	Prevalence is estimated at 100 per 1 million live births.[n]	Inherited bone marrow failure.	Anemia, hypoxia due to bone marrow stem cell failure.[n]	Tissue and organ damage is widespread.	Bone marrow transplant. Transfusions. Il-3 and corticosteroid administration.[n]	Monitor sodium, calcium and vitamin D. Maintain adequate macro- and micronutrient status.

[a]Benz, E. Hemoglobinopathies. In Kasper DL, Braunwald E, Fauci AS, Hausner SL, Longo DL, Jameson JL , editors. *Harrison's Principles of Internal Medicine.* 16th ed., New York, McGraw Hill Medical Publishing. 2005. pp. 593–601.

[b]Tiosano D, Hochberg Z. Endocrine complications of thalassemia. *J Endocrinol Invest.* 2001;24:716–23.

[c]Gaspard KJ. The Hematopoietic System. In: Porth CM. *Pathophysiology: Concepts of Altered Health States.* 7th ed. Philadelphia (PA): Lippincott Williams & Wilkins, 2005. pp. 279–340.

[d]Stettler N, Zemel BS, Kawchak DA, Ohene-Frempong K, Stallings VA. Iron status of children with sickle cell disease. *JPEN J Parenter Enteral Nutr.* 2001;25:36–38.

[e]Terlouw DJ, Desai MR, Wannemuehler KA, Kariuki SK, Pfeifer CM, Kager PA, Shi YP, Ter Kuile FO. Relation between the response to iron supplementation and sickle cell hemoglobin phenotype in preschool children in western Kenya. *Am J Clin Nutr.* 2004;79:466–72.

[f]Forget, BG, Cohen AR. Thalassemia Syndromes. In: Hoffman R. *Hematology, Basic Principles and Practice.* 4th ed. Philadelphia (PA). Elsevier Medical Publishing, 2005. pp. 557–86.

[g]Hoffman B, Baker KR, Prchal JT. The Polycythemias. In Hoffman R. *Hematology, Basic Principles and Practice.* 4th ed. Philadelphia (PA): Elsevier Medical Publishing, 2005. pp. 1209–45.

[h]Cunningham MJ, Silberstein LE. Autoimmune Hemolytic Anemia. In Hoffman R. *Hematology, Basic Principles and Practice.* 4th ed. Philadelphia (PA): Elsevier Medical Publishing, 2005. pp. 693–708.

[i]Schrier SL, Reid EG. Extrinsic Non-Immune Hemolytic Anemia. In Hoffman R. *Hematology, Basic Principles and Practice.* 4th ed. Philadelphia (PA): Elsevier Medical Publishing, 2005. pp. 709–18.

[j]Rovelli A, Corti P, Beretta C, Bovo G, Conter V, Mieli-Vergani G. Alemtuzumab for giant cell hepatitis with autoimmune hemolytic anemia. *J Pediatr Gastroenterol Nutr.* 2007;45(5):596–99.

[k]Kramer K, Cohen HJ. Antenatal Diagnosis of Hematological Disorders. In Hoffman R. *Hematology, Basic Principles and Practice.* 4th ed. Philadelphia (PA): Elsevier Medical Publishing, 2005. pp. 2697–712.

[l]Rowe JM, Avivi I. Clinical Use of Hematopoietic Growth Factors. In Hoffman R. *Hematology, Basic Principles and Practice.* 4th ed. Philadelphia (PA): Elsevier Medical Publishing, 2005. pp. 1029–44.

[m]Young NS, Maciejewski JP. Aplastic Anemia. In Hoffman R. *Hematology, Basic Principles and Practice.* 4th ed. Philadelphia (PA): Elsevier Medical Publishing, 2005. pp. 381–418.

[n]Freedman MH. Inherited forms of Bone Marrow Failure. In Hoffman R. *Hematology, Basic Principles and Practice.* 4th ed. Philadelphia (PA): Elsevier Medical Publishing, 2005. pp. 339–80.

Table 19.11 Characteristics of Clotting and Bleeding Disorders

Anemia	Epidemiology	Etiology	Pathophysiology	Clinical Manifestations	Treatment	Nutritional Therapy
Hemophilia	Hemophilia A occurs in 1 per 5,000 live male births. Hemophilia B occurs in 1 in 30,000 male live births.[a]	Hereditary inability to make needed clotting factors.	Hemorrhage and bleeding occurs in the joints.	Bleeding; joint and soft tissue damage; pain and loss of function.[a]	Synthetic clotting factor administration. Gene transfer therapy is under investigation.[a]	Maintain nutritional adequacy. Ensure antioxidant adequacy.
Hemorrhagic Disease of the Newborn	Vitamin K administration at birth results in an extremely low prevalence rate.	Vitamin K deficiency is seen in infants with a sterile GI tract.[b]	Vitamin K is required for clotting factor activation.	Bleeding from the umbilicus after birth; bleeding from the GI tract, skin, or orifices post delivery.[c]	0.5–1.0 mg phylloquinone given by intramuscular injection.[d]	Maintain optimal feeding practices for young infants.
Thrombosis	Prevalence differs according to the exact etiology of the clotting.[e]	Atherosclerosis. Chronic inflammation. Increased blood viscosity. Increased platelet activity. Risk factors such as: high-fat, high-cholesterol diet, smoking, sedentary lifestyle, obesity. Hypertension. Oxidative stress. Hyperhomocysteinemia. Increased coagulation factors.[f,g]	Coagulation processes including: vessel spasms, platelet plug formation, and fibrin formation.[h]	Differs depending on where the clot is formed or where it lodges.[i]	Anticoagulants (Coumarin derivatives, Coumadin, warfarin, Heparin etc.).[j]	Maintain constant vitamin K-rich food intake over time.

[a]Lazier JN, Kessler CM. Clinical Aspects and Therapy of Hemophilia. In Hoffman R. *Hematology, Basic Principles and Practice*. 4th ed. Philadelphia (PA): Elsevier Medical Publishing, 2005. pp. 2047–70.
[b]Suttie, JW. Vitamin K. In: Shils, ME, Shike M, Ross CA, Caballero BC, Cousins RJ, editors. *Modern Nutrition in Health and Disease*. 10th ed. Philadelphia (PA). Lippincott Williams & Wilkins, 2005, pp. 412–25.
[c]Kramer K, Cohen HJ. Antenatal Diagnosis of Hematological Disorders. In Hoffman R. *Hematology, Basic Principles and Practice*. 4th ed. Philadelphia (PA): Elsevier Medical Publishing, 2005. pp. 2697–712.
[d]Brandao L, DiMichele D. Disorders of Coagulation in the Neonate. In Hoffman R. *Hematology, Basic Principles and Practice*. 4th ed. Philadelphia (PA): Elsevier Medical Publishing, 2005. pp. 2189–96.
[e]Handin RI. Disorders of Coagulation and Thrombosis. In: Kasper DL, Braunwald E, Fauci AS, Hausner SL, Longo DI, Jameson JL, editors. *Harrison's Principles of Internal Medicine*. 16th ed. New York (NY). McGraw Hill Medical Publishing, 2005. pp. 680–87.
[f]Crowther MA, Ginsberg JS. Venous Thromboembolism. In Hoffman R. *Hematology, Basic Principles and Practice*. 4th ed. Philadelphia (PA): Elsevier Medical Publishing, 2005. pp. 2228–40.
[g]Crowther MA, Ginsberg JS. Arterial Thromboembolism. In Hoffman R. *Hematology, Basic Principles and Practice*. 4th ed. Philadelphia (PA): Elsevier Medical Publishing, 2005. pp. 2241–48
[h]Gaspard KJ. The Hematopoietic System. In: Porth CM. *Pathophysiology: Concepts of Altered Health States*. 7th ed. Philadelphia (PA): Lippincott Williams & Wilkins, 2005. pp. 279–340.
[i]Handin RI. Disorders of the Platelet and Vessel Wall. In Kasper DL, Braunwald E, Fauci AS, Hausner SL, Longo DI, Jameson JL, editors. *Harrison's Principles of Internal Medicine*. 16th ed. New York (NY). McGraw Hill Medical Publishing, 2005. pp. 673–80.
[j]Deitcher SR. Antiplatelet, Anticoagulant and Fibrinolytic Therapy. In Kasper DL, Braunwald E, Fauci AS, Hausner SL, Longo DI, Jameson JL editors. *Harrison's Principles of Internal Medicine*, 16th ed. New York (NY), McGraw Hill Medical Publishing, 2005. pp. 687–94.

Hemophilia patients are at risk for blood-borne diseases such as hepatitis and HIV. Liver diseases greatly impact nutritional status (see Chapter 16), as does infection with HIV[277] (see Chapter 24).

Hemorrhagic Disease of the Newborn

Hemorrhagic disease of the newborn is characterized by clotting deficiencies due to insufficient amounts of vitamin K available to the neonate.

Nutrition Therapy
Prophylaxis with vitamin K in newborns is commonplace. Breastfeeding mothers should be advised that breast milk is not exceptionally high in vitamin K. Vitamin K status of the neonate should be monitored. In cases of vitamin K deficiency bleeding (VKDB), 5 mg/d oral supplementation is recommended for infants up to 6 months of age.

A 30-day supply of vitamin K can be stored in a normal adult liver, but not in an infant. Infants may need repeated injections or parenteral administration if the VKDB persists and cannot be rectified by oral vitamin K administration.[263]

Thrombosis

Thrombosis, which has multifactorial etiologies and is associated with a wide variety of conditions and environmental factors, is enhanced clotting as a result of platelet aggregation, oxidative influences, inflammatory response, and blood viscosity.[278] There have also been reports of thrombotic complications resulting from iron deficiency and overload. In anemia, the system over-compensates for the lack of iron by hyper-coagulation; in iron overload the increases in oxidative stress, blood viscosity, and vasoconstriction cause clot formation.[279–281]

Nutrition Therapy
NUTRITIONAL IMPLICATIONS Nutrition concerns for patients with thrombotic diagnoses are related to whether or not specific medications are being used to treat the conditions. Excessive bleeding tendencies must be avoided when anti-thrombotic agents are being administered.

NUTRITION DIAGNOSIS Nutrition diagnoses associated with thrombotic or vascular conditions being medicated with anticoagulants include: excessive vitamin intake (sp - steady-state vitamin K, vitamin E), excessive bioactive substance intake (sp - omega fatty acid preparations), food-medication interactions, and impaired nutrient utilization.

NUTRITION INTERVENTION Because warfarin decreases clot formation by interfering with vitamin K's ability to participate in activation of clotting factors, patients on drugs such as warfarin must be counseled regarding steady-state consumption of vitamin K-containing foods such as green leafy vegetables. Patients must also be advised to avoid supplementation with excessive amounts of vitamin E, which may interact with such drugs to prolong bleeding tendencies. Omega-3 fatty acids are thought to be useful in the treatment of hypercoagulability, since they decrease the inflammatory response through alterations in the eicosanoid synthetic pathways. Medical practitioners should advise patients receiving anticoagulant therapy that the use of omega-3 fatty acids, fish oil preparations, and flaxseed supplements may cause drug-nutrient interactions or alter hemostasis. If patients are being medicated, they should seek advice before using omega-3 fatty acid preparations. Hyperhomocysteinemia, a risk factor for thrombosis, should

be addressed and intake of vitamins B_{12} and folate should be increased. Frank megaloblastic anemia does not need to be present for increases in homocysteine to occur.[282–285]

MONITORING AND EVALUATION Interventions are evaluated by measurement of specific outcomes. These include laboratory tests for hematologic indicators of anemia, hepatic function, and nutritional adequacy. Aspirin use for anticoagulation may result in gastrointestinal distress and blood loss; this may cause an iron-deficiency anemia. Heparins, thrombin inhibitors, and fibrinolytic drugs used to treat hypercoagulation are hepatotoxic, and liver damage can significantly impact nutritional status. Liver enzymes should be monitored and nutritional adequacy maintained to decrease the potential for further medical complications that can easily arise in this population. Hyperhomocysteinemia and megaloblasticity should be monitored using blood levels and smears.[286]

Diseases Involving WBC Types and Bone Marrow Failure Requiring Bone Marrow Transplant (BMT)

Definition

Various cancers of the hematopoietic system exist (see Chapter 23). Patients diagnosed with leukemia, lymphoma, and other cancers of white cell types receive 85% of all bone marrow transplant procedures. The remaining patients undergo transplantation for autoimmune disease and inherited marrow failure syndromes that can affect either or both the RBC or WBC. There are also procedures involving marrow transplant for metastatic breast cancer and **osteopetrosis**. Thirty-five percent of patients receiving BMT procedures are under 19 years of age. Only a third of all patients can find a matched donor from family members; the remaining patients must search for unrelated donors through national marrow donor programs.[287]

Epidemiology

Acute lymphoblastic leukemia (ALL) is primarily a childhood onset leukemia and comprises approximately 85% of all cases. Acute myelocytic leukemia (AML) also occurs in younger persons, ages 13–40. Chronic lymphocytic leukemia (CLL) is a disease of the older adult and chronic myelocytic leukemia (CML) is seen primarily in persons of middle age. An estimated 17,000 new cases of myeloid leukemia were diagnosed in the United States (or 4:100,000 people) in the past decade.[288] CLL incidence is estimated at approximately 8,000 per year, and is more common in men and Caucasians. Incidence of non-Hodgkin's lymphoma has risen dramatically, with an estimated 54,000 new cases in the United States. Hodgkin's lymphoma incidence has remained stable at ~8,000 new cases per year, while *all* incidence varies.[289]

Etiology

In hematopoietic cancers, the cells of the marrow are hyperproliferative, and control mechanisms are abnormal. The exact etiology is unknown. There seems to be a genetic component for the leukemias. Increasing amounts of white cells are produced, but the cells are abnormal and ineffective; as a result, widespread tissue injury occurs. Marrow, lymphatics, thymus, spleen, and other organs become increasingly involved and eventually fail. In malignant lymphomas, the cells that are hyper-proliferative are derived from lymphatic tissues. Infectious agents may trigger these diseases in susceptible persons.[289]

Bone marrow failure is a feature of other chronic disease states, excluding inherited or acquired hyper-proliferative diseases such as cancer. For instance, autoimmune diseases compromise marrow function, and marrow failure is common in children with acute liver failure.[290]

Bone marrow transplantation is an example of cell therapy in which marrow progenitor cells are grafted into the host bone.[287] It is effective because the bone marrow has precursor cells that differentiate into so many cell types. Allogenic transplantation is the most common, and involves a donor and a patient that are not identical. The cells are seen as foreign by the host, resulting in graft-versus-host disease if the patient is not adequately immunosuppressed, or if the typing of the **HLA antigens** is not acceptable.

Pathophysiology

The majority of BMT patients have neoplastic disease, resulting in a deficiency of hematopoietic cells, which is treated with BMT. The transplant procedures cause a wide range of toxic reactions when immunosuppressive drug chemotherapy is given at high doses. In addition to decreased immunity and damaged normal ancillary tissues, there are some global pathophysiological outcomes, such as gastrointestinal distress, occlusion of blood vessels, infection, and decompensation, all of which worsen nutritional status.[287, 291]

Clinical Manifestations

Mucositis, ulceration, chronic fatigue, and malaise are only a small fraction of the symptoms exhibited by BMT patients. Poor nutritional status delays healing, compromises host defense mechanisms, and increases lethargy. Low levels of macronutrients and micronutrients are commonplace, and constant monitoring through anthropometric and laboratory assessment is essential (see Table 19.12). Patients exhibit negative nitrogen balance, increased muscle catabolism, and lower levels of important antioxidant nutrients.[291]

Treatment

Several classes of immunosuppressive drugs are used in BMT. Radiation treatment may also be administered, either concurrently or prior to chemotherapy. Postsurgical patients are kept isolated, in "clean room" environments, and personnel and visitors must be gowned, masked, and gloved. Total parenteral or tube feeding may be initiated within 48 hours of surgery, after discontinuation of NPO (nothing by mouth) status, if oral intake is inadequate. Immunosuppressive treatments are continued through immediate postsurgical periods and after discharge.

In the event of graft-versus-host disease (GVHD) or the rejection of foreign cells due to antigenic response (see Chapter 9), the patient's situation is usually treated as an emergency. Antibiotics, steroids, and other medications are added to immuno-

Table 19.12 Laboratory Values for Select Micronutrient Status and Related Factor Assessment

Clinical Parameter	Normal Range for Adults
Serum Vitamin K	0.13–1.19 ng/mL
Serum Vitamin B_{12}	200–800 pg/mL
Red Cell Folate	150–450 ng/mL
Serum Folate	3–16 ng/mL cells
Schilling's Test—Urinary excretion of vitamin B_{12} challenge administration	7%–40%
Gastric pH	1.6–1.8
Serum Lead (adult)	<10–20 mg/dL

Source: Kratz A, Sluss, PM, Januzzi JL, Lewandrowski KB. Appendices: Laboratory Values of Clinical Importance. In: *Harrison's Principles of Internal Medicine*, Kasper Dl, Fauci AS, Longo DL, Braunwald E, Hausner SL, Jameson JL, Eds., 16th ed. New York, McGraw Hill Publishing; 2005. Constructed from data obtained from this source.

suppressive treatment. Multiple organ dysfunction syndrome is a risk in these patients. Methotrexate, cyclosporins, and prednisone all have profound nutritional consequences. Methotrexate interferes with folate metabolism, cyclosporins cause abnormalities in magnesium and potassium homeostasis, and prednisone interferes with glucose, electrolytes, fluid, zinc, calcium, vitamins A, D, and C, and phosphorus. Acute GVHD occurs within three months of transplant; chronic GVHD occurs thereafter. Acute GVHD will develop in approximately 30% of transplant patients, and for those patients surviving six months posttransplant, chronic GVHD will occur in 30% to 50%.[291]

Chronic GVHD is treated prophylactically with antibiotics, and within three years of immunosuppressive therapy, the chronic GVHD begins to resolve. Another consequence, graft failure, where marrow function never returns, is treated with growth factors and high-dose glucocorticoids. Both of these drug classes have extreme nutritional effects. Growth factors and glucocorticoids both cause fluid and electrolytic abnormalities as well as gastrointestinal distress.[287, 291]

Nutrition Therapy

Protein and kcal recommendations for patients undergoing BMT are elevated due to metabolic stressors (see Table 19.13 for suggestions on improving oral intake). Stress factors of 1.5–2.0 may be added to the calculations of protein and energy requirements. Glutamine, with its key role as an energy substrate for gastrointestinal epithelial cells and immune cells, may positively influence nutritional status and decrease susceptibility to infection and mucositis.[292] Glutamine enriched parenteral nutrition may confer some protection from GVHD via immuno-modulation.[293] In addition, several of the micronutrients involved in mitigating stress-related events are required in increased amounts. These include antioxidants, such as vitamin C and the carotenoids, as well as minerals such as zinc. TPN or enteral feeding is often necessary. Enteral nutrition is preferable to TPN because of increased risk of infection with TPN administration through a central line, but if GI toxicity is severe, enteral administration may not be feasible.

Table 19.13 Strategies for Improving Intake in Bone Marrow Transplant Patients

Assess Oral Cavity: Check for glossitis, cheilosis, stomatitis, xerostomia.	Maintain good oral hygiene. Keep mucous membranes moist. Prescription salivary gland stimulants are available. Give ice chips if tolerated. Treat underlying cause if possible.
Assess Sensory Status: Check for new food aversions, changes in smell and taste, or the sensation of pain or burning.	Avoid offensive or strong smells. Offer cold food. Offer bland, minimally salted/spiced food. Include items that are well tolerated.
Assess Dietary Intake: Use standardized formats for assessing food intake, dietary patterns, effects of treatment on appetite, and intake.	Use oral supplementation to decrease the impact of poor oral intake. If macronutrient density is below recommendations, a high-kilocalorie, high-protein supplement should be administered. These are tolerated best when cold and combined with ice cream or similar frozen product. Increase well-tolerated, nutritionally dense foods. Small, frequent meals, with oral supplements or shakes given between meals, are ideal. Present food in an appealing and visually attractive manner. If possible involve a social element during mealtimes. Friends and family at meal times will stimulate appetite.

Daily administration of 20–35 kcal/kg/day and 1.25–2 g protein/kg/day is advised.[291]

Nutritional Implications Posttransplantation proliferative disorders are common in BMT. Cellular proliferation results in increased nutrient requirements for both macronutrients and micronutrients. Oral recurrent herpes infection is common in patients. Pain and sores in the oral cavity decrease oral intake, further compromising marginal nutritional status. Mucositis and stomatitis may be so severe that morphine must be administered for pain. Opiates such as morphine also depress intake and increase constipation and GI distress.[291]

Nutrition Diagnosis Nutrition diagnoses associated with transplant include: malnutrition, inadequate protein-energy intake, underweight, impaired GI function, swallowing difficulty, and increased nutrient needs. If nutrition support is initiated, potential nutrition diagnoses include inadequate intake from enteral/parenteral infusion, inappropriate infusion of enteral/parenteral nutrition, and altered nutrition-related laboratory values.

Nutrition Intervention Transplant patients must be counseled on issues regarding food safety because of their immunocompromised status. Bottled water is recommended for decreasing infection and infestation in persons without access to filtration systems. Information regarding cross-contamination from solid surfaces and directions for thorough sanitation must be given. "No fresh fruit or vegetable" orders are given in institutional settings to avoid potential risk of food-borne infection. Conservative food holding temperatures and internal temperatures/doneness in cooking must be adhered to for BMT patients.

Adequate nutritional status must be maintained in patients despite complaints of anorexia, mucositis, stomatitis, dysguesia, xerostomia, cheilosis, glossitis, oral thrush (yeast infection of the mouth), dysosmia (abnormal smell), fatigue, and general malaise. Mucositis is a cause for morbidity among hematopoietic stem cell transplant patients.[294] Cold, palatable foods that are moist and visually appealing should be offered. If oral intake is insufficient to meet needs, oral supplements should be given. Alternatively, supplemental drinks, several of which have glutamine added, are available for cancer cachexia and metabolic stress. Oral feeding and intravenous fluids should be encouraged.[291]

Data suggest that glutamine should be added to enteral or parenteral formulations to decrease the effects of gastrointestinal toxicity caused by chemotherapy and to decrease infection rates. A recent Cochrane review of glutamine-enriched oral, enteral, and parenteral formulations suggest that infection rate is decreased, but length of hospital stay may not change despite enrichment.[295] Glutamine is not routinely added to formulations in nutrition support due to the compound's inability to stay in solution without undergoing molecular changes. Further research into glutamine's role in mitigating GI complications of chemotoxicity and infection rates is ongoing, as are attempts to better stabilize the compound in solution.[296]

Outpatient parenteral therapy is often given until oral intake can meet needs. Data show that this practice may lead to early satiety and prolong the time before patients consume enough food by mouth. Extended parenteral nutrition has also been shown to increase GVHD. It is recommended that standard PN be tapered off as early as possible.[297]

Hepato veno-occlusive disease is also common posttransplantation. The blood vessels running through the liver and the ancillary internal organs fed by portal circulation become blocked. Insufficient glutathione is thought to play a role in the pathogenesis of this complication. Increased oxidative stress results in endothelial cell injury of the blood vessels surrounding the liver and ancillary organs. A diet high in antioxidants and their precursors along with sufficient energy and protein may modulate pathogenic progression.[291]

Monitoring and Evaluation Interventions are evaluated by measurement of specific outcomes. These include tolerance to enteral/parenteral support; laboratory values indicative of nutritional status, immune function, and organ function (sp. liver function tests); or oral food intake acceptance logs, consumption records, and weight.

Example: Nutrition Diagnostic Terminology: Altered GI function NC 1.4

PES Statement: Altered gastrointestinal function RT chemotherapy damage of mucosal lining AEB physical exam findings of mucositis, stomatitis, cheilosis, glossitis, oral thrush, and esophageal irritation and documentation of anorexia, dysosmia, dysguesia, and malaise.

Table 19.14 Drug-Nutrient Interactions Related to Diseases of the Hematological System

Classification	Mechanism	Generic	Brand Names	Possible Food–Drug Interactions
Recombinant Human Erythropoietin	Stimulates RBC production	Epoetin alfa	Epogen, Procrit	May need Fe, B$_{12}$, and/or folate supplementation, N/V, diarrhea
Alkylating Agents	Destroys cells	Phosphorus-32	Phosphorus-32	Rare
	Alkylating agent, decreases cell division	Chlorambacil	Leukeran	Fatigue, loss of appetite, N/V (nausea and vomiting), diarrhea and mouth ulcers
	Alkylating agent, decreases cell division	Pipobroman	Pipobroman	N/V, diarrhea, abdominal cramps
	Alkylating agent, decreases cell division	Busulfan	Myleran	Abdominal pain; diarrhea; general fatigue, muscle pain; loss of appetite; N/V; weight loss (sudden)
Antiarrhythmic	Inhibits the vagus nerve	Disopyramide phosphate	Norpace, Norpace-CR	Hypoglycemia, anorexia, dry mouth/throat, N/V, GI pain, flatulence, bloating, constipation, diarrhea, avoid alcohol
Anti-Inflammatory Hormone	Diminished eicosanoid production	Corticosteroids	Betamethasone, Budesonide, Cortisone, Dexamethasone, Hydrocortisone, Methylprednisolone, Prednisolone, Prednisone, Triamcinolone	Increased appetite; indigestion; loss of appetite (for triamcinolone only); increased thirst
Antineoplastic Biological Response Modifier	Cytokine, intercellular "hormone"	Interleukin-3	Alferon N	Change in taste or metallic taste; loss of appetite; N/V

Source: Pronsky Z. *Food Medication Interactions*. 14th ed. Birchrunville (PA): Food-Medication Interactions, 2006.

Conclusion

Many types of blood disorders affect bone marrow and red or white cell types, including nutritional deficiency anemias, inherited disorders of RBC synthesis, clotting disorders, and cancers. Nutritional interventions differ and will depend on the etiology of the disease state. Longitudinal assessment of the hematological parameters in a patient with a blood disorder is imperative, with more attention paid to regulatory proteins. Nutrition therapy for patients with nutritional anemias should include adequacy of micronutrient intake, with special attention paid to iron, cyanocobalamin, and folate. The incidence of anemias seems to be declining in the United States, but the long-term consequences of anemia may increase the health care burden in future generations as more people live to older ages.[298–301]

PRACTITIONER INTERVIEW

Mary Ellen Beindorff, RD LD, *Specialty—Abdominal Organ Transplant Barnes-Jewish Hospital, St. Louis, MO*

In your practice, which patients are typically anemic?

From my experience, I expect to see anemia in certain populations—such as chronic kidney disease, cancer, liver disorders—usually resulting from the disease and/or its treatment. Anemia is common in chronic inflammatory diseases and disorders that cause blood loss. It is also not unusual to see anemia in the very young, pregnant women, or seniors. In the very young and during pregnancy, iron deficiency would typically be the issue, but with seniors it is more likely macrocytic pernicious anemia related to decreased intake of vitamin B$_{12}$ and/or impaired absorption resulting from decreased production of intrinsic factor.

How has the treatment of anemia changed?

Since I became a dietitian, the medical treatment of non-nutritional anemias has dramatically improved because of the medication erythropoietin. Yet it too has nutritional implications. The patient must have adequate protein, energy, and iron for the medication to be effective.

Advice to dietetic students?

Nutritional assessment should include looking at the patient's hematological lab values. I find that new dietetic interns frequently overlook theses labs, mainly because they are so overwhelmed with everything else that needs to be taken into consideration for nutritional assessments. My advice to interns is to look at the "big picture." As with all signs and symptoms, the dietitian needs to distinguish between nutritional or non-nutritional causes for altered labs. Understanding the pathophysiology of the disease is essential to determine these distinctions.

Application of the Nutrition Care Process: Anemia

Introduction:

African-American male, 5 years old; DOB: 5/4/2005 birth-weight 5 lbs. 2 oz. at 38 weeks gestation. Chief complaint: Mother relates that child has poor appetite; appears more tired than usual; having some difficulty in kindergarten—when asked, describes behavior issues and "acting out." Past Med Hx: Frequent upper respiratory infections; multiple ear infection

1. Patient was diagnosed with hypochromic microcytic anemia, most likely secondary to iron deficiency. Define the terms *hypochromic* and *microcytic*. How are they related to his medical diagnosis?

2. What are signs and symptoms in the patient's history that are consistent with this diagnosis?

Nutrition Assessment:

Food-/Nutrition-Related History:

Food and Beverage Intake: Consumes 24–48 oz. milk per day; does not like a lot of meat, and mother relates that they really don't have it very often. Favorite foods include peanut butter and jelly sandwiches, french fries, and hot dogs.

Factors Affecting Access to Food and Food-/Nutrition-Related Supplies: Family receives WIC supplemental foods for younger sibling and $200.00 in food assistance each month. Mother is employed, but with four children, appears to have inadequate access to food.

3. What additional dietary history information will be important for the RD to determine?

Anthropometric Measurements:

Ht. 40 inches (5 percentile stature for age) Wt. 32 lbs. (5–10 percentile for weight for age)

BMI: 14 (5 percentile)

4. Evaluate his anthropometric information. Are there any particular concerns?

Biochemical Data:

MCV 65; MCHC 27; Hgb 10.5; Hematocrit 33%

5. Define each of his laboratory values.

Nutrition Diagnosis:

6. Identify at least two nutrition problems based on the nutrition assessment and medical history. Determine the diagnostic term for each nutrition problem. Next, identify the etiology of each nutrition problem. Finally, identify the signs and symptoms that support the evidence for these nutrition problems.

Nutrition Intervention:

7. How will his iron-deficiency anemia be treated?

8. Determine at least one nutrition intervention that you would recommend.

Nutrition Monitoring and Evaluation:

9. Determine nutrition criteria for monitoring and evaluation for each nutrition diagnosis that you identified.

WEB LINKS

National Heart Lung and Blood Institute
This site is the official website of the branch of the National Institute of Health and the Department of Health and Human Services that deals with diseases of the hematopoietic system as well as with other acute and chronic diseases. The site contains scientifically based evidence and links to authoritative sources for further information.
http://www.nhlbi.nih.gov

The National Center for Health Statistics
This site is the official website of the Centers for Disease Control and Prevention and the National Centers for Health Statistics. Data regarding the incidence, prevalence, and mortality from blood-related diseases can be accessed through searchable datasets available from the United States government.
http://cdc.gov/nchs

The National Cancer Institute This site is the official website of the branch of the National Institutes of Health that deals with cancers of the hematopoietic system. The National Cancer Institute funds much of the research into cancer treatment, prevention, and epidemiology. Clinical trial results for promising new drugs can also be accessed.
http://nci.nih.gov

National Anemia Action Council A nonprofit association dedicated to research, public education, and improving health outcomes for persons with all forms of anemia.
http://www.anemia.org

American Sickle Cell Anemia Association
A nonprofit association dedicated to research, public education, and improving health outcomes for persons with sickle cell anemia.
http://www.ascaa.org

Aplastic Anemia and Myelodystrophic Disease International Foundation A nonprofit association dedicated to research, public education, and improving health outcomes for persons with aplastic anemia or myelodystrophic disease.
http://www.aamds.org

Cooley's Anemia Foundation A nonprofit association dedicated to research, public education, and improving health outcomes for persons with thalassemia.
http://www.thalassemia.org

1. Describe the abnormal laboratory values and clinical signs and symptoms indicative of:
   a. Iron-deficiency anemia
   b. Pernicious anemia
   c. Hemolytic anemia
   d. Sickle cell anemia

2. Describe the abnormal laboratory values and clinical signs and symptoms of:
   a. Leukemia
   b. Aplastic anemia
   c. Thalassemia

3. Describe the homeostatic mechanisms involved in hematopoiesis.

4. List dietary strategies that can be used to increase intake of folate, cyanocobalamin, iron, zinc, and vitamin A.
   a. Why are these nutrients especially important in hematopoiesis?

5. The elderly, pregnant women, and children are at increased risk for anemia. Explain the different contributing factors that heighten risk for these groups.

Complementary and Alternative Medicine Remedies Used in Diseases of the Hematological System

Remedy	Scientific Name	CAM Use	Side Effects and/or Risks
Aconite, monkshood; fu zi	*Aconitum carmichaeli*	Improve spleen function	No forms of the plant are recommended for human consumption.
Amalaka, amla; Indian gooseberry	*Emblica officinalis*	Strengthen blood, treat anemia, treat hemorrhaging	Adverse reactions rare.
Ascorbic acid/ vitamin C		Improve iron absorption	High intakes may cause GI upset or mild diarrhea. May increase activity of anticoagulants. May interfere with B_{12}, copper, chromium absorption and metabolism. Aspirin, corticosteroids, & indomethacin increase urinary vitamin C losses. High doses may reduce efficacy of select chemotherapeutic medications. Supplementation above DRI (or >70–90 mg/d) not recommended. May increase plasma estrogen levels in patients taking HRT or oral contraceptives. Large doses may interfere with anticoagulant medications.
Astragalus, milk vetch; huang qi	*Astragalus membranaceus*	Blood, spleen tonic; treat anemia	None reported.
Bloodroot, red root, red puccoon	*Sanguinaria Canadensis*	Treat "weak" blood	Should be used only under strict supervision. Unsafe for use in foods, beverages, or drugs.
Burdock; niu bang zi	*Arctium lappa*	Cleanse blood toxins	May increase hypoglycemic effects of insulin and oral hypoglycemic agents. Burdock products may be significantly contaminated with atropine at high enough levels to cause toxicity.
Castor oil plant; eranda, vatari	*Ricinus communis*	Treat enlarged spleen	Avoid use if pregnant or breast-feeding. Can cause malabsorption of fat-soluble vitamins, fluid, & electrolytes. Prolonged use can result in laxative dependency.
Cat's claw; gambir; gao teng	*Uncaria tomentosa*	Purify blood, esp. during pregnancy, birth	Should not be taken with antihypertensive or anticoagulant medications.
Cobalamin/ vitamin B_{12}		Treat anemia	Folic acid supplementation may mask vitamin B_{12} deficiency. Supplemental vitamin C or iron may interfere with bioavailability of vitamin B_{12}. The following drugs may interfere with B_{12} absorption: zidovudine, antacids, clofibrate, colchicines, cycloserine, erythromycin, isoniazid, metformin, aldomet, neomycin, nitrous oxide, oral contraceptives, sufanamides, tetracycline.
Copper		Treat anemia	Nausea may occur with doses of 10 mg, and 60 mg may cause vomiting.
Dong quai; angelica	*Angelica sinensis*	Purify blood; treat anemia	Should not be combined with blood-thinning medications (warfarin, heparin, aspirine) or supplements (garlic, ginkgo, vitamin E, fish oil).

Complementary and Alternative Medicine Remedies Used in Diseases of the Hematological System (*Continued*)

Remedy	Scientific Name	CAM Use	Side Effects and/or Risks
Folic acid		Treat anemia	Supplementation ≥400 μg can mask pernicious anemia caused by vitamin B_{12} deficiency; dosages ≥5 mg may cause abdominal cramps, diarrhea, and rash; doses ≥15 mg can alter sleep patterns or cause vivid dreaming; large doses exacerbate neuropathy in those with B_{12} deficiency; doses >1 mg/d interfere with anticonvulsant medications. The following medications can affect absorption or activity of folate: oral contraceptives, aspirin, indomethacin, methotrexate, famotidine, tetracycline, erythromycin, folfonamides, & cholestyramine.
Gotu kola; brahmi	*Centella asiatica*	Purify blood	Well tolerated. High doses may interfere with actions of antidiabetic medications & cholesterol-lowering drugs. Contraindicated in pregnancy and breast-feeding.
Hawthorn	*Crategus* spp	Treat blood disorders	Augments the effect of cardiac medications (digoxine) or blood pressure drugs.
Mistletoe	*Phoradendron leucarpum*	Treat blood problems, hemorrhaging	All plant parts of mistletoe are toxic. May increase hypotensive effects of antihypertensive medications. Increases sedative effects of CNS depressants. Causes cytotoxic & immunostimulant effects.
Neem; nimb	*Azadirachta india*	Purify blood	None known; no full scientific evaluation of toxicity and side effects has been conducted.
Poke, pokeweed, inkberry; fitolaca	*Phytolacca americana*	Treat "weak" blood	FDA has classified pokeweed as an herb of undefined safety due to narcotic-like effects. Nearly all parts of pokeweed appear to be toxic.
Turmeric; haridra	*Curcuma longa*	Purify blood	GI ulceration with high doses or prolonged use. Avoid administration with anticoagulants. Decreases immunosuppressive effects. Avoid administration with NSAIDs; may inhibit platelet function and increase risk of bleeding.
Raspberry	*Rubus* spp.	Treat anemia; treat hemorrhaging	Increases effectiveness of hypoglycemic actions of antidiabetic medications.

Sources: Fetrow CW, Avila JR. *Professional's Handbook of Complementary and Alternative Medicines*, 3rd ed. Philadelphia: Lippincott Williams & Wilkins, 2004.
Bratman S, Girman AM. *Mosby's Handbook of Herbs and Supplements and their Therapeutic Uses*. St. Louis: Mosby/Elsevier, 2003.
Fragakis AS, Thomson C. *The Health Professional's Guide to Popular Dietary Supplements*, 3rd ed. Chicago: American Dietetic Association, 2007.

REFERENCES

1. Cheng CK, Chan J, Cembrowski GS, van Assendelft OW. Complete blood count reference interval diagrams derived from NHANES III: Stratification by age, sex, and race. Lab Hematol. 2004;10: 42–53.

2. Greenwood DC, Gilthrope MS, Cade JE. The impact of imprecisely measured covariates on estimating gene-environment interactions. BMC Med Res Methodol. 2006;6:21.

3. Chung J, Prohaska JR, Wessling-Resnick M. Ferroportin-1 is not up-regulated in copper-deficient mice. J Nutr. 2004;134:517–21.

4. Mullally AM, Vogelsang GB, Moliterno AR. Wasted sheep and premature infants: The role of trace metals in hematopoiesis. Blood Rev. 2004;18:227–34.

5. Gaspard KJ. The Hematopoietic System. In: Porth CM. Pathophysiology: concepts of altered health states. 7th ed. Philadelphia (PA): Lippincott Williams & Wilkins; 2005, 279–340.

6. Foller M, Huber SM, Lang F. Erythrocyte programmed cell death. IUMB Life. 2008;60(10):661–68.

7. Monroe JG, Turka LA. Regulation of Activation of B and T Lymphocytes. In: Hoffman R. Hematology, basic principles and practice. 4th ed. Philadelphia (PA): Elsevier Medical Publishing; 2005, 157–77.

8. Davis S. Clinical sequelae affecting quality of life in the HIV-infected patient. J Assoc Nurses AIDS Care. 2004;15(5):28S–33S.

9. Koury MJ, Ponka P. New insights into erythropoiesis: The roles of folate, vitamin B12, and iron. Ann Rev Nutr. 2004;24:105–31.

10. Woodman R, Ferrucci L, Guralnik J. Anemia in older adults. Curr Opin Hematol. 2005;12:123–28.

11. Came NA, Westerman DA. Quality of bone marrow iron assessment. Pathology. 2007;39(6):610.

12. Armstrong SA, Golub TR. Genomic Approaches to the Study of Hematologic Science. In: Hoffman R. Hematology, Basic Principles and Practice. Fourth Edition, Philadelphia (PA): Elsevier Medical Publishing; 2005; 17–27.

13. Ames BN, Elson-Schwab I, Silver EA. High-dose vitamin therapy stimulates

variant enzymes with decreased co-enzyme binding affinity (increased K(m)): Relevance to genetic disease and polymorphisms. Am J Clin Nutr. 2002;75:616–58.

14. Shaheen M, Broxmeyer HE. The Humoral Regulation of Hematopoiesis. In: Hoffman R. Hematology, basic principles and practice. 4th ed. Philadelphia (PA): Elsevier Medical Publishing; 2005, 233–65.

15. Chua AC, Graham RM, Trinder D, Olynyk JK. The regulation of cellular iron metabolism. Crit Rev Clin Lab Sci. 2007;44(5–6):413–59.

16. Wood RJ, Ronnenberg AG. Iron. In: Shils ME, Shike M, Ross CA, Caballero BC, Cousins RJ, editors. Modern nutrition in health and disease. 10th ed. Philadelphia (PA): Lippincott Williams & Wilkins; 2005, 248–70.

17. Carrier J, Aghdassi E, Cullen J, Allard JP. Iron supplementation increases disease activity and vitamin E ameliorates the effect in rats with dextran sulfate sodium-induced colitis. J Nutr. 2002;132:3146–50.

18. Russel RM. Vitamin and Trace Mineral Deficiency and Excess. In: Kasper DL, Braunwald E, Fauci AS, Hausner SL, Longo Dl, Jameson JL, editors. Harrison's principles of internal medicine. 16th ed. New York (NY): McGraw Hill Medical Publishing; 2005, 403–15.

19. Kopple JD. Nutrition, Diet and the Kidney. In: Shils ME, Shike M, Ross CA, Caballero BC, Cousins RJ, editors. Modern nutrition in health and disease. 10th ed. Philadelphia (PA): Lippincott Williams & Wilkins; 2005, 1475–511.

20. Srigiridhar K, Nair KM. Supplementation with alpha-tocopherol or a combination of alpha-tocopherol and ascorbic acid protects the gastrointestinal tract of iron-deficient rats against iron-induced oxidative damage during iron repletion. Br J Nutr. 2000;84:165–73.

21. Forget BG, Cohen AR. Thalassemia Syndromes. In: Hoffman R. Hematology, basic principles and practice. 4th ed. Philadelphia (PA): Elsevier Medical Publishing; 2005, 557–86.

22. Wiley JS, Moore MR. Heme Biosynthesis and its disorders. Porphyrias and sideroblastic anemias. In: Hoffman R. Hematology, basic principles and practice. 4th ed. Philadelphia (PA): Elsevier Medical Publishing; 2005, 499–517.

23. Camaschella C. Recent advances in the understanding of inherited sideroblastic anaemia. Br J Haematol. 2008;143(1):27–38.

24. Metz J. A high prevalence of biochemical evidence of vitamin B12 or folate deficiency does not translate into a comparable prevalence of anemia. Food Nutr Bull. 2008;29(2)Suppl:S74–85.

25. Zamvar V, McClean P, Odeka E, Richards M, Davison S. Hepatitis E virus infection with nonimmune hemolytic anemia. J Pediatr Gastroenterol Nutr. 2005;40:223–25.

26. Miloh T, Manwani D, Morotti R, Sukru E, Shneider B, Kerkar N. Giant cell hepatitis and autoimmune hemolytic anemia successfully treated with rituximab. J Pediatr Gastroenterol Nutr. 2007;44(5):634–36.

27. Handin RI. Bleeding and Thrombosis. In: Kasper DL, Braunwald E, Fauci AS, Hausner SL, Longo Dl, Jameson JL, editors. Harrison's principles of internal medicine, 16th ed. New York (NY): McGraw Hill Medical Publishing; 2005, 337–43.

28. Suttie JW. Vitamin K. In: Shils ME, Shike M, Ross CA, Caballero BC, Cousins RJ, editors. Modern nutrition in health and disease. 10th ed. Philadelphia (PA): Lippincott Williams & Wilkins; 2005, 412–25.

29. Kamali F. Genetic influences on the response to warfarin. Curr Opin Hematol. 2006;13(5):357–61.

30. Kimura R, Kokubo Y, Miyashita K, Otsubo R, Nagatsuka K, Otsuki T, Sakata T, Nagura J, Okayama A, Minematsu K, Naritomi H, Honda S, Sato K, Tomoike H. Polymorphisms in vitamin K-dependent gamma-carboxylation-related genes influence interindividual variability in plasma protein C and protein S activities in the general population. Int J Hematol. 2006;84(5):387–97.

31. Gage BF. Pharmacogenetics-based coumarin therapy. Hematol Am Soc Hematol Educ Prog. 2006:467–73.

32. Brodsky IG. Hormones and growth factors. In: Shils ME, Shike M, Ross CA, Caballero BC, Cousins RJ, editors. Modern nutrition in health and disease. 10th ed. Philadeplphia (PA): Lippincott Williams & Wilkins; 2005, 636–53.

33. Handin RI. Disorders of coagulation and thrombosis. In: Kasper DL, Braun-wald E, Fauci AS, Hausner SL, Longo Dl, Jameson JL, editors. Harrison's principles of internal medicine. 16th ed. New York (NY): McGraw Hill Medical Publishing; 2005, 680–87.

34. Lopez, JA, Thiogarajan P. Acquired Disorders of Platelet Function. In: Hoffman R. Hematology, basic principles and practice. 4th ed. Philadelphia (PA): Elsevier Medical Publishing; 2005, 2347–67.

35. Jones PJ, Kubow S. Lipids, sterols and their metabolites. In: Shils ME, Shike M, Ross CA, Caballero BC, Cousins RJ, editors. Modern nutrition in health and disease. 10th ed. Philadelphia (PA): Lippincott Williams & Wilkins; 2005, 92–122.

36. Moestrup SK. New insights into carrier binding and epithelial uptake of the erythropoietic nutrients cobalamin and folate. Curr Opin Hematol. 2006;13(3):119–23.

37. Koulaozidis A, Saeed AA, Abdallah M, Said EM. Transferrin receptor levels as surrogate peripheral blood marker of iron deficiency states. Scand J Gastroenterol. 2009;44(1):126–27.

38. Mills JL, Von Kohorn I, Conley MR. Low vitamin B-12 concentrations in patients without anemia: The effect of folic acid fortification of grain. Am J Clin Nutr. 2003;77:1474–77.

39. Gibson RS, Yeudall F, Drost N, Mtitimuni BM, Cullinan TR. Experiences of a community-based dietary intervention to enhance micronutrient adequacy of diets low in animal source foods and high in phytate: A case study in rural Malawian children. J Nutr. 2003;133(11) Suppl. 2:3992S–99S.

40. Mei Z, Parvanta I, Cogswell ME, Gunter EW, Grummer-Strawn LM. Erythrocyte protoporphyrin or hemoglobin: Which is a better screening test for iron deficiency in children and women? Am J Clin Nutr. 2003;77:1229–33.

41. Ganz T, Olbina G, Girelli D, Nemeth E, Westerman M. Immunoassay for human serum hepcidin. Blood. 2008;112(10):4292–97.

42. Powers HJ. Riboflavin (vitamin B-2) and health. Am J Clin Nutr. 2003;77:1352–60.

43. Prasad AS. Recognition of zinc-deficiency syndrome. Nutrition 2001;17: 67–69.

44. Finley JW. Manganese absorption and retention by young women is associated with serum ferritin concentration. Am J Clin Nutr. 1999;70:37–43.

45. Artz AS. Anemia and the frail elderly. Semin Hematol. 2008;45(4):261–66.

46. Lee P, Gelbart T, Waalen J, Beutler E. The anemia of ageing is not associated with increased plasma hepcidin levels. Blood Cells Mol Dis. 2008;41(3):252–54.

47. Sanchez C, Lopez-Jurado M, Planells E, Llopis J, Aranda P. Assessment of iron and zinc intake and related biochemical parameters in an adult Mediterranean population from southern Spain: influence of lifestyle factors. J Nutr Biochem. 2009;20(2):125–31.

48. Murray-Kolb LE, Heard JL. Iron deficiency and child and maternal health. Am J Clin Nutr. 2009;89(3):946S–950S.

49. Guralnik JM, Eisenstaedt RS, Ferrucci L, Klein HG, Woodman RC. Prevalence of anemia in persons 65 years and older in the United States: Evidence for a high rate of unexplained anemia. Blood. 2004;104:2263–68.

50. Ashorn M. Acid and iron-disturbances related to helicobacter pylori infection. J Pediatr Gastroenterol Nutr. 2004;38:137–39.

51. Allen LH. Advantages and limitations of iron amino acid chelates as iron fortificants. Nutr Rev. 2002;60(7) Pt.2:S18–21.

52. Waldmann A, Koschizke JW, Leitzmann C, Hahn A. Dietary iron intake and iron status of German female vegans: Results of the German vegan study. Ann Nutr Metab. 2004;48:103–08.

53. Abrams SA. Using stable isotopes to assess mineral absorption and utilization by children. Am J Clin Nutr. 1999;70:955–64.

54. Atanasova B, Mudway IS, Laftah AH. Duodenal ascorbate levels are changed in mice with altered iron metabolism. J Nutr. 2004;134:501–05.

55. Uauy, R, Hertrampf, E, Dangour AD. Food based Dietary Guidelines. In: Shils ME, Shike M, Ross CA, Caballero BC, Cousins RJ, editors. Modern nutrition in health and disease. 10th ed. Philadelphia (PA): Lippincott Williams & Wilkins; 2005, 1701–15.

56. Heath AL, Skeaff CM, Gibson RS. The relative validity of a computer-ized food frequency questionnaire for estimating intake of dietary iron and its absorption modifiers. Eur J Clin Nutr. 2000;54:592–99.

57. Zijp IM, Korver O, Tijburg LB. Effect of tea and other dietary factors on iron absorption. Crit Rev Food Sci Nutr. 2000;40:371–98.

58. Temme EH, Van Hoydonck PG. Tea consumption and iron status. Eur J Clin Nutr. 2002;56:379–86.

59. Nelson M, Poulter J. Impact of tea drinking on iron status in the UK: A review. J Hum Nutr Diet. 2004;17:43–54.

60. Boccio JR, Iyengar V. Iron deficiency: Causes, consequences, and strategies to overcome this nutritional problem. Biol Trace Elem Res. 2003;94:1–32.

61. Bonuck KA, Kahn R. Prolonged bottle use and its association with iron deficiency anemia and overweight: A preliminary study. Clin Pediatr Phila. 2002;41:603–07.

62. Berseth CL, Van Aerde JE, Gross S, Stolz SI, Harris CL, Hansen JW. Growth, efficacy, and safety of feeding an iron-fortified human milk fortifier. Pediatrics. 2004;114:e699–706.

63. Lake AM. Food-induced eosinophilic proctocolitis. J Pediatr Gastroenterol Nutr. 2000;30 Suppl: S58–60.

64. Savilahti E. Food-induced malabsorption syndromes. J Pediatr Gastroenterol Nutr. 2000;30 Suppl: S61–66.

65. Carvalho NF, Kenney RD, Carrington PH, Hall DE. Severe nutritional deficiencies in toddlers resulting from health food milk alternatives. Pediatrics. 2001;107:E46.

66. Penney DS, Miller KG. Nutritional counseling for vegetarians during pregnancy and lactation. J Midwifery Womens Health. 2008;53(1):37–44.

67. Yip R, Brion MJ, Leary SD, Smith GD, McArdle HJ, Ness R. Significance of an abnormally low or high hemoglobin concentration during pregnancy: Special consideration of iron nutrition. Am J Clin Nutr. 2000;72(1):209S–211S.

68. Scholl TO. Iron status during pregnancy: Setting the stage for mother and infant. Am J Clin Nutr. 2005;81:1218S–1222S.

69. Smith TG, Robbins PA. Iron, pre-eclampsia and hypoxia-inducible factor. BJOG. 2007;114(12):1581–82.

70. Lopez LB, Langini Sh, Pita de Portela ML. Maternal iron status and neonatal outcomes in women with pica during pregnancy. Int J Gynaecol Obstet. 2007;98(2):151–52.

71. Kathula SK. Craving lemons: another form of pica in iron deficiency. Am J Med. 2008;121(7):e1.

72. Kawai K, Saathoff E, Antelman G, Msamanga G, Fawzi WW. Geophagy (Soiling-eating) in relation to Anemia and Helminth infection among HIV-infected pregnant women in Tanzania. Am J Trop Med Hyg. 2009;80(1):36–43.

73. Seid MH, Derman RJ, Baker JB, Banach W, Goldberg C, Rogers R. Ferric carboxymaltose injection in the treatment of postpartum iron deficiency anemia: a randomized controlled clinical trial. Am J Obstet Gynecol. 2008;199(4):435.e1–7.

74. Casanueva E, Pfeffer F, Drijanski A, Fernández-Gaxiola AC, Gutiérrez-Valenzuela V, Rothenberg SJ. Iron and folate status before pregnancy and anemia during pregnancy. Ann Nutr Metab. 2003;47:60–63.

75. Corwin EJ, Murray-Kolb LE, Beard JL. Low hemoglobin level is a risk factor for postpartum depression. J Nutr. 2003;133:4139–42.

76. Pamuk GE, Umit H, Harmandar F, Yesil N. Patients with iron deficiency anemia have an increased prevalence of gallstones. Ann Hematol. 2009;88(1):17–20.

77. Greger JL. Nutrition versus toxicology of manganese in humans: Evaluation of potential biomarkers. Neurotoxicology. 1999;20:205–12.

78. Sreedhar B. Conflicting evidence of iron and zinc interactions in humans: Does iron affect zinc absorption? Am J Clin Nutr. 2003;78:1226–27.

79. Wieringa FT, Dijkhuizen MA. Iron and zinc interactions. Am J Clin Nutr. 2004;80:787–88.

80. Olivares M, Pizarro F, Ruz M. New insights about iron bioavailability inhibition by zinc. Nutrition. 2007;23(4):292–95.

81. Jamil KM, Rahman As, Bardhan PK, Khan AI, Chowdhury F, Sarker SA, Khan AM, Ahmed T. Micronutrients and anaemia. J Health Popul Nutr. 2008;26(3):340–55.

82. Hambidge M. Biomarkers of trace mineral intake and status. J Nutr. 2003;133 Suppl 3: 948S–955S.

83. Clark SF. Iron deficiency anemia. Nutr Clin Pract. 2008;23(2):128–41.

84. Alleyne M, Horne MK, Miller JL. Individualized treatment for iron-deficiency anemia in adults. Am J Med. 2008;121(11):493–98.

85. Agarwal KN. Iron and the brain: Neurotransmitter receptors and magnetic resonance spectroscopy. Br J Nutr. 2001;85Suppl 2:S147–50.

86. Beard JL. Iron biology in immune function, muscle metabolism and neuronal functioning. J Nutr. 2001;131:568S–579S; discussion 580S.

87. Beard JL. Why iron deficiency is important in infant development. J Nutr. 2008;138(12):2534–36.

88. Sachdev H, Gera T, Nestel P. Effect of iron supplementation on mental and motor development in children: Systematic review of randomized controlled trials. Public Health Nutr. 2005;8:117–32.

89. Kim H, Lee S, LeeG, Hwangbo Y, Ahn K, Lee B. The protective effect of aminolevulinic acid dehydratase 1–2 and 2–2 isoenzymes against blood lead with higher hematologic parameters. Environ Health Perspect. 2004;112:538–41.

90. Labbé RF, Vreman HJ, Stevenson DK. Zinc Protoporphyrin: A metabolite with a mission. Clin Chem. 1999;45:2060–72.

91. Zimmermann MB, Molinari L, Staubli-Asobayire F, Hess SY, Chaouki N, Adou P, Hurell RF. Serum transferrin receptor and zinc protoporphyrin as indicators of iron status in African children. Am J Clin Nutr. 2005;81:615–23.

92. Rose EA, Porcerelli JH, Neale AV. Pica: Common but commonly missed. J Am Board Fam Pract. 2000;13:353–58.

93. Singhi S, Ravishanker R, Singhi P, Nath R. Low plasma zinc and iron in pica. Indian J Pediatr. 2003;70:139–43.

94. Yilmaz A, Candan F, Turan M. Coffee phagia and iron-deficiency anemia: A possible association with helicobacter pylori. J Health Popul Nutr. 2005;23:102–03.

95. Topaloglu AK, Hallioglu O, Canim A, Duzovali O, Yilgor E. Lack of association between plasma leptin levels and appetite in children with iron deficiency. Nutrition. 2001;17:657–59.

96. Pinhas-Hamiel O, Newfield RS, Koren I, Agmon A, Lilos P, Phillip M. Greater prevalence of iron deficiency in overweight and obese children and adolescents. Int J Obes Relat Metab. 2003;27:416–18.

97. Karp RJ, Kersey M, Cutts, DB. Iron deficiency, obesity, and food insecurity. Arch Pediatr Adolesc Med. 2008;162(12):1194–95.

98. Reeves PG, DeMars LC. Copper deficiency reduces iron absorption and biological half-life in male rats. J Nutr. 2004;134:1953–57.

99. Nead KG, Halterman JS, Kaczorowski JM, Auinger P, Weitzman M. Overweight children and adolescents: A risk group for iron deficiency. Pediatrics. 2004;114:104–08.

100. Tussing-Humphreys LM, Liang H, Nemeth E, Freels S, Braunschweig CA. Excess adiposity, inflammation, and iron-deficiency in female adolescents. J Am Diet Assoc. 2009;109(2):297–302.

101. Zimmerman MB, Zeder C, Muthayya S, Winichagoon P, Chaouki N, Aeberli I, Hurrell RF. Adiposity in women and children from transition countries predicts decreased iron absorption, iron deficiency and a reduced response to iron fortification. Int J Obes. 2008;32(7):1098–104.

102. Sun L, Franco OH, Hu FB, Cai L, Yu Z, li H, Ye X, Qi Q, Wang J, Pan A, Liu Y, Lin X. Ferritin concentrations, metabolic syndrome, and type 2 diabetes in middle-aged and elderly Chinese. J Clin Endocrinol Metab. 2008;93(12):4690–96.

103. Liu Q, Sun L, Tan Y, Wang G, Lin X, Cai L. Role of iron deficiency and overload in the pathogenesis of diabetes and diabetic complications. Curr Med Chem. 2009;16(1):113–29.

104. Malone DL, Genuit T, Tracy JK, Gannon C, Napolitano LM. Surgical site infections: Reanalysis of risk factors. J Surg Res. 2002;103:89–95.

105. Dreyfuss ML, Stoltzfus RJ, Shrestha JB. Hookworms, malaria and vitamin A deficiency contribute to anemia and iron deficiency among pregnant women in the plains of Nepal. J Nutr. 2000;130:2527–36.

106. Oppenheimer SJ. Iron and its relation to immunity and infectious disease. J Nutr. 2001;131:616S–633S; discussion 633S–635S.

107. Bergström S. Infection-related morbidities in the mother, fetus and neonate. J Nutr. 2003;133(5)Suppl. 2:1656S–1660S.

108. Steketee RW. Pregnancy, nutrition and parasitic diseases. J Nutr. 2003;133(5) Suppl. 2:1661S–1667S.

109. García-Casal MN, Leets I, Layrisse M. Beta-carotene and inhibitors of iron absorption modify iron uptake by caco-2 cells. J Nutr. 2000;130:5–9.

110. Das BS, Devi U, Mohan Rao C, Srivastava VK, Rath PK, Das BS. Effect of iron supplementation on mild to moderate anaemia in pulmonary tuberculosis. Br J Nutr. 2003;90:541–50.

111. Streiff MB, Mehta S, Thomas DL. Peripheral blood count abnormalities among patients with hepatitis C in the United States. Hepatology. 2002;35:947–52.

112. Semba RD, Bloem MW. The anemia of vitamin A deficiency: Epidemiology and pathogenesis. Eur J Clin Nutr. 2002;56:271–81.

113. Ikeda R, Uehara M, Takasaki M, Chiba H, Masuyama R, Furosho T, Suzuki K. Dose-responsive alteration in hepatic lipid peroxidation and retinol metabolism with increasing dietary beta-carotene in iron deficient rats. Int J Vitam Nutr Res. 2002;72:321–28.

114. Ruhl CE, Everhart JE. Relationship of iron-deficiency anemia with esophagitis and hiatal hernia: Hospital findings from a prospective, population-based study. Am J Gastroenterol. 2001;96:322–26.

115. Lee JH, Choe YH, Choi YO. The expression of iron-responsible outer membrane proteins in H. pylori and its association with iron deficiency anemia. Heliobacter. 2009;14(1):36–39.

116. Jain S, Kamat D. Evaluation of microcytic anemia. Clin Pediatr. 2009;48(1):7–13.

117. Barabino A, Dufour C, Marino CE, Claudiani F, De Alessandri A. Unexplained refractory iron-deficiency anemia associated with helicobacter pylori gastric infection in children: Further clinical evidence. J Pediatr Gastroenterol Nutr. 1999;28:116–19.

118. Mason J, Bailes A, Beda-Andourou M, Copeland N, Curtis T, Deitchler M, Foster L, Hensley M, Horjus P, Johnson C, Lloren T, Mendez A,

Munoz M, Rivers J, Vance G. Recent trends in malnutrition in developing regions: Vitamin A deficiency, anemia, iodine deficiency, and child underweight. Food Nutr Bull. 2005;26:59–108.

119. Hess SY, Zimmermann MB, Arnold M, Langhans W, Hurrell RF. Iron deficiency anemia reduces thyroid peroxidase activity in rats. J Nutr. 2002;132:1951–55.

120. Zimmermann MB, Köhrle J. The impact of iron and selenium deficiencies on iodine and thyroid metabolism: Biochemistry and relevance to public health. Thyroid. 2002;12:867–78.

121. Zimmermann MB, Wegmueller R, Zeder C, Chaouki N, Biebinger R, Hurrell RF, Windhab E. Triple fortification of salt with microcapsules of iodine, iron, and vitamin A. Am J Clin Nutr. 2004;80:1283–90.

122. Swann IL, Kendra JR. Severe iron deficiency anemia and stroke. Clin Lab Haematol. 2000;22:221–23.

123. Horwich TB, Fonarow GC, Hamilton MA, MacLellan WR, Borenstein J. Anemia is associated with worse symptoms, greater impairment in functional capacity and a significant increase in mortality in patients with advanced heart failure. J Am Coll Cardiol. 2002;39: 1780–86.

124. Al-Hilaly N, Kwiatkowski D, Gould S, Mitchell C, Sullivan PB. Gastric leiomyosarcoma presenting as severe iron-deficiency anaemia. J Pediatr Gastroenterol Nutr. 1999;29:354–57.

125. Schiavetto I, Pedrazzoli P, Basilico V, Siena S. Iron supplementation during treatment with erythropoiesis-stimulating agents for cancer-relation anemia. Chemotherapy. 2008;54(6):417–20.

126. Vadhan-Raj S. Management of chemotherapy-induced thrombocytopenia: current status of thrombopoietic agents. Semin Hematol. 2009;46(1): Suppl 2:S26–32.

127. Argilés JM, Busquets S, López-Soriano FJ. Cytokines in the pathogenesis of cancer cachexia. Curr Opin Clin Nutr Metab Care. 2003;6:401–06.

128. Ioannou GN, Rockey DC, Bryson CL, Weiss NS. Iron deficiency and gastrointestinal malignancy: A population-based cohort study. Am J Med. 2002;113:276–80.

129. Gadducci A, Cosio S, Fanucchi A, Genazzani AR. Malnutrition and cachexia in ovarian cancer patients: Pathophysiology and management. Anticancer Res. 2001;21:2941–47.

130. Mock V, Olsen M. Current management of fatigue and anemia in patients with cancer. Semin Oncol Nurs. 2003;19(4)Suppl. 2:36–41.

131. Frank C. Approach to skin ulcers in older patients. Can Fam Physician. 2004;50:1653–59.

132. Seaman S. Considerations for the global assessment and treatment of patients with recalcitrant wounds. Ostomy Wound Manage. 2000;46(1)Suppl. A: 10S–29S.

133. Brolin RE, LaMarca LB, Kenler HA, Cody RP. Malabsorptive gastric bypass in patients with superobesity. J Gastrointest Surg. 2002;6:195–203; discussion 204–05.

134. Skroubis G, Sakellaropoulos G, Pouggouras K, Mead N, Nikiforidis G, Kalfarentzos F. Comparison of nutritional deficiencies after roux-en-Y gastric bypass and after biliopancreatic diversion with roux-en-Y gastric bypass. Obes Surg. 2002;12:551–58.

135. Alvarez-Leite JI. Nutrient deficiencies secondary to bariatric surgery. Curr Opin Clin Nutr Metab Care. 2004;7: 569–75.

136. Topart P. Iron deficiency and anemia after bariatric surgery. Surg Obes Relat Dis. 2008;4(6):719–20.

137. Vargas-Ruiz AG, Hernandez-Rivera G, Herrera MF. Prevalence of iron, folate, and vitamin B12 deficiency anemia after laparoscopic Roux-en-Y gastric bypass. Obes Surg. 2008;18(3):288–93.

138. Marinella MA. Nocturnal pagophagia complicating gastric bypass. Mayo Clin Proc. 2008;83(8):961.

139. Davis S. Clinical sequelae affecting quality of life in the HIV infected patient. J Assoc Nurse AIDS Care. 2004;15(5) Suppl. 1:28S–33S.

140. Butensky JE, Harmatz P, Lee M, Kennedy C, Petru A, Wara D, Miaskowski C. Altered iron metabolism in children with human immunodeficiency virus disease. Pediatr Hematol Oncol. 2009;26(2):69–84.

141. Cunha DF, Monteiro JP, Ortega LS, Alves LG, Cunha SF. Serum electrolytes in hospitalized pellagra alcoholics. Eur J Clin Nutr. 2000;54:440–42.

142. Halsted CH. Nutrition and alcoholic liver disease. Semin Liver Dis. 2004;24:289–304.

143. Fudenberg HH. The effect of severe iron deficiency anemia on primary photophobia. South Med J. 2009;102(3):335–36.

144. Roncone DP. Xerophthalmia secondary to alcohol-induced malnutrition. Optometry. 2006;77(3):124–33.

145. Ioannou GN, Dominitz JA, Weiss NS, Heagerty PJ, Kowdley KV. The effect of alcohol consumption on the prevalence of iron overload, iron deficiency, and iron deficiency anemia. Gastroenterology. 2004;126:1293–301.

146. Kulnigg S, Stoinov S, Simanenkov V, Dudar LV, Karnafel W, Garcia LC, Sambuelli AM, D'Haens G, Gasche C. A novel intravenous iron formulation for treatment of anemia in inflammatory bowel disease: the ferric carboxymaltose (FERN-IJECT) randomized controlled trial. Am J Gastroenterol. 2008;103(5):1182–92.

147. Treem WR. Emerging concepts in celiac disease. Curr Opin Pediatr. 2004;16:552–59.

148. Guariso G, D'Incà R, Sturniolo GC, Zancan L, Dall'Amico R. Photopheresis treatment in severe Crohn's disease. J Pediatr Gastroenterol Nutr. 2003;37:517–20.

149. Bergamaschi G, Markopoulos K, Albertini R, Di Sabatino A, Biagi F, Ciccocioppo R, Arbustini E, Corazza GR. Anemia of chronic disease and defective erythropoietin production in patients with celiac disease. Haematologica. 2008;93(12):1785–91.

150. Astor BC, Muntner P, Levin A, Eustace JA, Coresh J. Association of kidney function with anemia: The third national health and nutrition examination survey (1988–1994). Arch Intern Med. 2002;162:1401–08.

151. Hsu CY, McCulloch CE, Curhan GC. Epidemiology of anemia associated with chronic renal insufficiency among adults in the United States: Results from the third national health and nutrition examination survey. J Am Soc Nephrol. 2002;13:504–10.

152. Sheashaa H, El-Husseini A, Sabry A. Parenteral iron therapy in treatment of anemia in end-stage renal disease patients:

A comparative study between iron saccharate and gluconate. Nephron Clin Pract. 2005;99:c97–101.

153. Kovesdy CP, Kalantar-Zadeh K. Emerging challenges of anemia management in CKD. Adv Chronic Kidney Dis. 2009;16(2):74–75.

154. Yee J. Anemia of chronic kidney disease: forward to the past. Adv Chronic Kidney Dis. 2009;16(2):71–73.

155. Matsuhashi Y. Thinness: Drives and results. J Adolesc Health. 2000;27:149–50.

156. Caregaro L, Di Pascoli L, Favaro A, Nardi M, Santonastaso P. Sodium depletion and hemoconcentration: Overlooked complications in patients with anorexia nervosa? Nutrition 2005;21:438–45.

157. Arnold GL, Kirby R, Preston C, Blakely E. Iron and protein sufficiency and red cell indices in phenylketonuria. J Am Coll Nutr. 2001;20:65–70.

158. Tsalis G, Nikolaidis MG, Mougios V. Effects of iron intake through food or supplement on iron status and performance of healthy adolescent swimmers during a training season. Int J Sports Med. 2004;25:306–13.

159. Iglesias-Gutiérrez E, García-Rovés PM, Rodríguez C, Braga S, García-Zapico P, Patterson AM. Food habits and nutritional status assessment of adolescent soccer players. A necessary and accurate approach. Can J Appl Physiol. 2005;30:18–32.

160. Mann SK, Kaur S, Bains K. Iron and energy supplementation improves the physical work capacity of female college students. Food Nutr Bull. 2002;23:57–64.

161. Brownlie T 4th, Utermohlen V, Hinton PS, Haas JD. Tissue iron deficiency without anemia impairs adaptation in endurance capacity after aerobic training in previously untrained women. Am J Clin Nutr. 2004;79:437–43.

162. Lukaski HC. Vitamin and mineral status: Effects on physical performance. Nutrition 2004;20:632–44.

163. Smith SM. Red blood cell and iron metabolism during space flight. Nutrition. 2002;18:864–66.

164. Morris SS, Ruel MT, Cohen RJ, Dewey KG, de la Brière B, Hassan MN. Precision, accuracy, and reliability of hemoglobin assessment with use of capillary blood. Am J Clin Nutr. 1999;69:1243–48.

165. Nadeau RG, Groner W. The role of a new noninvasive imaging technology in the diagnosis of anemia. J Nutr. 2001;131(5) Suppl:1610S–1614S.

166. Leong WI, Lönnerdal B. Hepcidin, the recently identified peptide that appears to regulate iron absorption. J Nutr. 2004;134:1–4.

167. Davidsson L. Approaches to improve iron bioavailability from complementary foods. J Nutr. 2003;133(5) Suppl.1:1560S–1562S.

168. Moriarty-Craige SE, Ramakrishnan U, Neufeld L, Rivera J, Martorell R. Multivitamin-mineral supplementation is not as efficacious as is iron supplementation in improving hemoglobin concentrations in nonpregnant anemic women living in Mexico. Am J Clin Nutr. 2004;80:1308–11.

169. Hyder SM, Persson LA, Chowdhury AM, Ekström EC. Do side-effects reduce compliance to iron supplementation? A study of daily- and weekly-dose regimens in pregnancy. J Health Popul Nutr. 2002;20:175–79.

170. Kennedy E, Meyers L. Dietary reference intakes: Development and uses for assessment of micronutrient status of women—a global perspective. Am J Clin Nutr. 2005;81:1194S–1197S.

171. Cogswell ME, Kettel-Khan L, Ramakrishnan U. Iron supplement use among women in the United States: Science, policy and practice. J Nutr. 2003;133(6)Suppl.1:1974S–1977S.

172. Khaodhiar L, Keane-Ellison M, Tawa NE, Thibault A, Burke PA, Bistrian BR. Iron deficiency anemia in patients receiving home total parenteral nutrition. J Parenter Enteral Nutr. 2002;26:114–19.

173. Nagano T, Toyoda T, Tanabe H, Nagato T, Tsuchida T, Kitamura A, Kasai G. Clinical features of hematological disorders caused by copper deficiency during long-term enteral nutrition. Intern Med. 2005;44:554–59

174. Tanner EM, Finn-Stevenson M. Nutrition and brain development: social policy implications. Am J Orthopsychiatry. 2002;72:182–93.

175. Badaracco ME, Ortiz EH, Soto EF, Connor J, Pasquini JM. Effect of transferrin on hypomyelination induced by iron deficiency. J Neurosci Res. 2008; 86(12):2663–73.

176. Lakhal S, Talbot NP, Crosby A, Stoepker C, Townsend AR, Robbins PA, Pugh CW, Ratcliffe PJ, Mole DR. Regulation of growth differentiation factor 15 expression by intracellular iron. Blood. 2009;113(7):1555–63.

177. Carlson ES, Tkac I, Magid R, O'Connor MB, Andrews NC, Schallert T, Gunshin H, Georgieff MK, Petryk A. Iron is essential for neuron development and memory function in mouse hippocampus. J Nutr. 2009;139(4):672–79.

178. Khedr E, Hamed SA, Elbeih E, El-Shereef H, Ahmad Y, Ahmed S. Iron states and cognitive abilities in young adults: neuropsychological and neurophysiological assessment. Eru Arch Psychiatry Clin Neurosci. 2008;258(8):489–96.

179. Buchanan GR. Screening for iron deficiency during early infancy: is it feasible and is it necessary? Am J Clin Nutr. 2009;89(2):473–74.

180. Oner P, Oner O. Relationship of ferritin to symptom ratings children with attention deficit hyperactivity disorder: effect of comorbidity. Child Psychiatry Hum Dev. 2008;39(3):323–30.

181. Mahoney DH. Anemia in at-risk populations—what should be our focus? Am J Clin Nutr. 2008;88(6):1457–58.

182. Avram MM. Improving prognosis for kidney disorders in the 21st century: Hypertension, anemia, nutrition, and lipids: Introduction. Am J Kidney Dis. 2001;38:1334–36.

183. Del Vecchio L, Pozzoni P, Andrulli S, Locatelli F. Inflammation and resistance to treatment with recombinant human erythropoietin. J Ren Nutr. 2005;15:137–41.

184. Navas-Carretero S, Perez-Granados AM, Sarri B, Carbajal A, Pedrosa MM, Roe MA, Fairweather-Tait SJ, Vaquero MP. Oily fish increases iron bioavailability of a phytate rich meal in young iron deficient women. J Am Coll Nutr. 2008;27(1):96–101.

185. Peneau S, Dauchet L, Vergnaud AC, Estaquio C, Kesse-Guyot E, Bertrais S, Latino-Martel, Hercberg S, Galan P. Relationship between iron status and dietary fruit and vegetables based on their vitamin C and fiber content. Am J Clin Nutr. 2008;87(5):1298–305.

186. Zengin E, Sarper N, Caki Kilia S. Clinical manifestations of infants with nutritional vitamin B deficiency due to maternal dietary deficiency. Acta Paediatr. 2009;98(1):98–102.

187. Carmel R. Nutritional anemias and the elderly. Semin Hematol. 2008;45(4):225–34.

188. Khanduri U, Sharma A. Megaloblastic anaemia: prevalence and causative factors. Natl Med J India. 2007;20(4): 172–75.

189. Guerra-Shinohara EM, Morita OE, Pagliusi RA, Blaia-d'Avila VL, Allen RH, Stabler Sp. Elevated serum S-adenosylhomocysteine in cobalamin-deficient megaloblastic anemia. Metabolism. 2007;56(3):339–47.

190. Green R, Miller JW. Folate deficiency beyond megaloblastic anemia: Hyperhomocysteinemia and other manifestations of dysfunctional folate status. Semin Hematol. 1999;36:47–64.

191. Hoffbrand AV, Herbert V. Nutritional anemias. Semin Hematol. 1999;36(4) Suppl. 7:13–23.

192. Chwatko G, Boers GH, Strauss KA, Shih DM, Jakubowski H. Mutations in methylenetetrahydrofolate reductase or cystathionine beta-synthase gene, or a high-methionine diet, increase homocysteine thiolactone levels in humans and mice. FASEB J. 2007;21(8):1707–13.

193. Troen AM, Shea-Budgell M, Shukitt-Hale B, Smith DE, Selhub J, Rosenberg IH. B-vitamin deficiency causes hyperhomocysteinemia and vascular cognitive impairment in mice. Proc Natl Acad Sci. 2008;105(34):12474–79.

194. Rodionov RN, Lentz SR. The homocysteine paradox. Arterioscler Thromb Vasc Biol. 2008;28(6):1031–33.

195. Sauls DL, Arnold EK, Bell CW, Allen JC, Hoffman M. Pro-thrombotic and pro-oxidant effects of diet-induced hyperhomocystinemia. Thromb Res.2008;120(1):117–26.

196. Nenseter MS, Ueland T, Retterstl K, Strom E, Morkid L, Landaas S, Ose L, Aukrust P, Holven KB. Dysregulated RANK ligand/RANK axis in hyperhomocysteinemic subjects: effect of treatment with B-vitamins. Stroke. 2009;40(1): 241–47.

197. U. S. Preventive Services Task Force. Screening for presence of deficiency, toxicity, and disease. Nutr Clin Care. 2003;6:120–22.

198. Arinzon Z, Fidelman Z, Peisakh A, Adunsky A. Folate status and folate related anemia: A comparative cross-sectional study of long-term care and post-acute care psycho-geriatric patients. Arch Gerontol Geriatr. 2004;39:133–42.

199. Baker H. Nutrition in the elderly: hypovitaminosis and its implications. Geriatrics. 2007;62(8):22–26.

200. Andrès E, Noel E, Kaltenbach G. Comment: Treatment of vitamin B(12) deficiency anemia: Oral versus parenteral therapy. Ann Pharmacother. 2002;36:1809–10.

201. Patel KV. Epidemiology of anemia in older adults. Semin Hematol. 2008;45(4):210–17.

202. Schölmerich J. Postgastrectomy syndromes—diagnosis and treatment. Best Pract Res Clin Gastroenterol. 2004;18:917–33.

203. Mitrache C, Passweg JR, Libura J. Anemia: An indicator for malnutrition in the elderly. Ann Hematol. 2001;80: 295–98.

204. Tungtrongchitr R, Pongpaew P, Soonthornruengyot M, Viroonudomphol D, Phnorat A, Pooudong S, Schelp FP. Relationship of tobacco smoking with serum vitamin B12, folic acid and hematological indices in healthy adults. Public Health Nutr. 2003;6: 675–81.

205. Dali-Youcef N, Andres E. An update on cobalamin deficiency in adults. QJM. 2009;102(1):17–28.

206. Choi SW, Friso S, Ghandour H, Bagley PJ, Selhub J, Mason JB. Vitamin B-12 deficiency induces anomalies of base substitution and methylation in the DNA of rat colonic epithelium. J Nutr. 2004;134:750–55.

207. Carmel R. Cobalamin (B12). In: Shils, ME, Shike M, Ross CA, Caballero BC, Cousins RJ, editors. Modern nutrition in health and disease. 10th ed. Philadelphia (PA): Lippincott Williams & Wilkins; 2005, 482–97.

208. Baik HW, Russell RM. Vitamin B12 deficiency in the elderly. Ann Rev Nutr. 1999;19:357–77.

209. Carmel R. Folic Acid. In: Shils, ME, Shike M, Ross CA, Caballero BC, Cousins RJ, editors. Modern nutrition in health and disease. 10th ed. Philadelphia (PA): Lippincott Williams & Wilkins; 2005; 470–81.

210. Chen X, Remacha AF, Sardà MP, Carmel R. Influence of cobalamin deficiency compared with that of cobalamin absorption on serum holo-transcobalamin II. Am J Clin Nutr. 2005;81:110–14.

211. Klee GG. Cobalamin and folate evaluation: Measurement of methylmalonic acid and homocysteine vs vitamin B(12) and folate. Clin Chem. 2000;46:1517–22.

212. Galloway M, Hamilton M. Macrocytosis: pitfalls in testing and summary of guidance. BMJ. 2007;335(7625):884–86.

213. Kaferle J, Strzoda CE. Evaluation of macrocytosis. Am Fam Physician. 2009;79(3):203–08.

214. Aslinia F, Mazza JJ, Yale SH. Megaloblastic anemia and other causes of macrocytosis. Clin Med Res. 2006;4(3):236–41.

215. Ho C, Kauwell GP, Bailey LB. Practitioners' guide to meeting the vitamin B-12 recommended dietary allowance for people aged 51 years and older. J Am Diet Assoc. 1999;99:725–27.

216. Rogers LM, Boy E, Miller JW, Green R, Sabel JC, Allen LH High prevalence of cobalamin deficiency in Guatemalan schoolchildren: Associations with low plasma holotranscobalamin II and elevated serum methylmalonic acid and plasma homocysteine concentrations. Am J Clin Nutr. 2003;77:433–40.

217. Essama-Tjani JC, Guilland JC, Potier de Courcy G, Fuchs F, Richard D. Folate status worsens in recently institutionalized elderly people without evidence of functional deterioration. J Am Coll Nutr. 2000;19:392–404.

218. Johnson MA, Hawthorne NA, Brackett WR, Fischer JG, Gunter EW, Allen RH, Stabler SP. Hyperhomocysteinemia and vitamin B-12 deficiency in elderly using title IIIc nutrition services. Am J Clin Nutr. 2003;77:211–20.

219. Gaudet G, Laplante J. Soft drink abuse, malnutrition, and folic acid deficiency. Am J Hematol. 1999;60:311–12.

220. Tucker KL, Olson B, Bakun P, Dallal GE, Selhub J, Rosenberg IH. Breakfast cereal fortified with folic acid, vitamin B-6, and vitamin B-12 increases vitamin concentrations and reduces

homocysteine concentrations: A randomized trial. Am J Clin Nutr. 2004;79: 805–11.

221. Dary O. Nutritional interpretation of folic acid interventions. Nutr Rev. 2009;67(4):235–44.

222. Chalouhi C, Faesch S, Anthoine-Milhomme MC, Fulla Y, Dulac O, Cheron G. Neurological consequences of vitamin B12 deficiency and its treatment. Pediatr Emerg Care. 2008;24(8):538–41.

223. Wun Chan JC, Yu Liu HS, Sang Kho BC, Yin Sim JP, Hang Lau TK, Luk YW, Chu RW, Fung Cheung FM, Tat Choi FP, Kwan Ma ES. Pernicious anemia In Chinese: a study of 181 patients in a Hong Kong hospital. Medicine (Baltimore). 2006;85(3):129–38.

224. Ganji V, Kafai MR. Hemoglobin and hematocrit values are higher and prevalence of anemia is lower in the post-folic acid fortification period than in the pre-folic acid fortification. Am J Clin Nurt. 2009;89(1):363–71.

225. Wantanabe F. Vitamin B12 sources and bioavailability. Exp Biol Med. 2007;232(10):1266–74.

226. Murphy SP, Allen LH. Nutritional importance of animal source foods. J Nutr. 2003;133(11):3875S–3878S.

227. Beutler E. Iron absorption in carriers of the C282Y hemochromatosis mutation. Am J Clin Nutr. 2004;80:799–800.

228. Hunt JR, Zeng H. Iron absorption by heterozygous carriers of the HFE C282Y mutation associated with hemochromatosis. Am J Clin Nutr. 2004;80:924–31.

229. Johnson EE, Wessling-Resnick M. Flatiron mice and ferroportin disease. Nutr Rev. 2007;65(7):341–45.

230. Roe MA, Spinks C, Heath AL, Harvey LJ, Foxall R, Wimperis J, Wolf C, Fairweather-Tait S. Serum prohepcidin concentration: no association with iron absorption in healthy men; and no relationship with iron status in men carrying HFE mutations, hereditary haemochromatosis patients undergoing phlebotomy treatment, or pregnant women. Br J Nutr. 2007;97(3):544–49.

231. Pietrangelo A. Hereditary hemochromatosis. Annu Rev Nutr. 2006;26:251–70.

232. Aigner E, Theurl I, Theurl M, Lederer D, Haufe H, Dietze O, Strasser M, Datz C, Weiss G. Pathways underlying iron accumulation in human nonalcoholic fatty liver disease. Am J Clin Nurt. 2008;87(5):1374–83.

233. Fleming DJ, Jacques PF, Tucker KL, Massaro JM, D'Agostino RB Sr, Wilson PW, Wood RJ. Iron status of the free-living, elderly Framingham heart study cohort: An iron-replete population with a high prevalence of elevated iron stores. Am J Clin Nutr. 2001;73:638–46.

234. Sharma N, Trope B, Lipman TO. Vitamin supplementation: What the gastroenterologist needs to know. J Clin Gastroenterol. 2004;38:844–54.

235. Heimburger DC, McLaren DS, Shils ME. Clinical manifestations of nutrient deficiencies and toxicities. In: Shils, ME, Shike M, Ross CA, Caballero BC, Cousins RJ, editors. Modern nutrition in health and disease. 10th ed. Philadelphia (PA): Lippincott Williams & Wilkins; 2005, 595–614.

236. Nagababu E, Gulyani S, Early CJ, Cutler RG, Mattson MP, Rifkind JM. Iron-deficiency anaemia enhances red blood cell oxidative stress. Free Radic Res. 2008;42(9):824–29.

237. Mair SM, Weiss G. New pharmacological concepts for the treatment of iron overload disorders. Curr Med Chem. 2009;16(5):576–90.

238. Hutchinson C, Geissler CA, Powell JJ, Bomford A. Proton pump inhibitors suppress absorption of dietary non-haem iron in hereditary haemochromatosis. Gut. 2007;56(9):1291–95.

239. Bring P, Partovi N, Ford JA, Yoshida EM. Iron overload disorders: treatment options for patients refractory to or intolerant of phlebotomy. Pharmacotherapy. 2008;28(3):331–42.

240. Landier W, Bhatia S. Late complications of hematological diseases and their therapies. In Hoffman R. Hematology, basic principles and practice. 4th ed. Philadelphia (PA): Elsevier Medical Publishing; 2005, 1665–82.

241. Papayannopoulou T, D'Andrea AD, Abkowitz JL, Migliaccio AN. Biology of erythropoiesis, erythroid differentiation and maturation. In: Hoffman R. Hematology, basic principles and practice. 4th ed. Philadelphia (PA): Elsevier Medical Publishing; 2005, 267–83.

242. Williams R, Olivi S, Li CS, Storm M, Cremer L, Mackert P. Oral glutamine supplementation decreases resting energy expenditure in children and adolescents with sickle cell anemia. J Pediatr Hematol Oncol. 2004;26:619–25.

243. Ren H, Ghebremeskel K, Okpala I, Lee A, Ibegbulam O, Crawford M. Patients with sickle cell disease have reduced blood antioxidant protection. Int J Vitam Nutr Res. 2008;78(3):139–47.

244. Kawchak DA, Schall JI, Zemel BS, Ohene-Frempong K, Stallings VA. Adequacy of dietary intake declines with age in children with sickle cell. J Am Diet Assoc. 2007;107(5):843–48.

245. Nelson MC, Zemel BS, Kawchak DA, Barden EM, Frongillo EA, Coburn SP, Ohene EA. Vitamin B6 status of children with sickle cell disease. J Pediatr Hematol Oncol. 2002;24:463–69.

246. Van der Dijs FP, Fokkema MR, Dijck-Brouwer DA, Niessink B, Van der Wal T. Optimization of folic acid, vitamin B(12), and vitamin B(6) supplements in pediatric patients with sickle cell disease. Am J Hematol. 2002;69:239–46.

247. Buchowski MS, Swift LL, Akohoue SA, Shankar SM, Flakoll PJ, Abumrad N. Defects in postabsorptive plasma homeostasis of fatty acids in sickle cell disease. JPEN J Parenter Enteral Nutr. 2007;31(4):263–68.

248. Ntaios G, Chatzinikolaou A. Red cell distribution width in iron deficiency anemia and beta-thalassemia minor. Am J Clin Pathol. 2008;130(2):313.

249. Burdick CO. Separating thalassemia trait and iron deficiency by simple inspection. Am J Clin Pathol. 2009;131(3):444.

250. Benz, E. Hemoglobinopathies. In Kasper DL, Braunwald E, Fauci AS, Hausner SL, Longo Dl, Jameson JL, editors. Harrison's principles of internal medicine. 16th ed. New York (NY): McGraw Hill Medical Publishing; 2005, 593–601.

251. Origa R, Galanello R, Ganz T, Giagu N, Maccioni L, Faa G, Nemeth E. Liver iron concentrations and urinary hepcidin in beta-thalassemia. Haematologica. 2007;92(5):583–88.

252. Pamuk GE, Pamuk ON, Set T, Harmandar O, Yesil N. An increased prevalence of fibromyalgia in iron deficiency anemia and thalassemia minor and associated factors. Clin Rheumatol. 2008;27(9):1103–08.

253. Eivazi-Ziaei J, Dastgiri S, Pourebrahim S, Soltanpour R. Usefulness of red blood cell flags in diagnosing and

differentiating thalassemia trait from iron-deficiency anemia. Hematology. 2008;13(4):253–56.

254. Harrington AM, Ward PC, Kroft SH. Iron deficiency anemia, beta-thalassemia minor, and anemia of chronic disease: a morphologic reappraisal. Am J Clin Pathol. 2008;129(3):466–71.

255. Adamson JW, Longo DL. Hematological Alterations. In Kasper DL, Braunwald E, Fauci AS, Hausner SL, Longo Dl, Jameson JL. Editors. Harrison's principles of internal medicine. 16th ed. New York (NY): McGraw Hill Medical Publishing; 2005, 329–37.

256. Spivak Jl. Polycythemia Vera and other Myeloproliferative Diseases. In Kasper DL, Braunwald E, Fauci AS, Hausner SL, Longo Dl, Jameson JL editors. Harrison's principles of internal medicine. 16th ed. New York (NY): McGraw Hill Medical Publishing; 2005, 626–31.

257. Hoffman B, Baker KR, Prchal JT. The Polycythemias. In Hoffman R. Hematology, Basic principles and practice. 4th ed. Philadelphia (PA): Elsevier Medical Publishing; 2005, 1209–45.

258. Adamson, J. Iron deficiency and other hypoproliferative anemias. In Kasper DL, Braunwald E, Fauci AS, Hausner SL, Longo Dl, Jameson JL, editors. Harrison's principles of internal medicine. 16th ed. New York (NY): McGraw Hill Medical Publishing; 2005, 586–93.

259. Dollberg S, Marom R, Mimouni FB, Littner Y. Increased energy expenditure after dilutional exchange transfusion for neonatal polycythemia. J Am Coll Nutr. 2007;26(5):412–15.

260. Dzieczkowski JS, Anderson KC. Transfusion biology and therapy. In Kasper DL, Braunwald E, Fauci AS, Hausner SL, Longo Dl, Jameson JL, editors. Harrison's principles of internal medicine. 16th ed. New York (NY): McGraw Hill Medical Publishing; 2005, 662–68.

261. Abramson N. Traumatic hemolytic anemia. Blood. 2008;111(5):2946–47.

262. Bunn HF, Rosse W. Hemolytic anemias and acute blood loss. In Kasper DL, Braunwald E, Fauci AS, Hausner SL, Longo Dl, Jameson JL, editors. Harrison's principles of internal medicine. 16th ed. New York (NY): McGraw Hill Medical Publishing; 2005;607–17.

263. Kramer K, Cohen HJ. Antenatal diagnosis of hematological disorders. In Hoffman R. Hematology, basic principles and practice. 4th ed. Philadelphia (PA): Elsevier Medical Publishing; 2005, 2697–712.

264. Schrier SL, Reid EG. Extrinsic non-immune hemolytic anemia. In Hoffman R. Hematology, basic principles and practice. 4th ed. Philadelphia (PA): Elsevier Medical Publishing; 2005; 709–18.

265. Rovelli A, Corti P, Beretta C, Bovo G, Conter V, Mieli-Vergani G. Alemtuzumab for giant cell hepatitis with autoimmune hemolytic anemia. J Pediatr Gastroenterol Nutr. 2007;45(5):596–99.

266. McLeod BC. Evidence based therapeutic apheresis in autoimmune and other hemolytic anemias. Curr Opin Hematol. 2007;14(6):647–54.

267. Rowe JM, Avivi I. Clinical use of hematopoietic growth factors. In Hoffman R. Hematology, basic principles and practice. 4th ed. Philadelphia (PA): Elsevier Medical Publishing; 2005;1029–44.

268. Young NS, Maciejewski JP. Aplastic anemia. In Hoffman R. Hematology, basic principles and practice. 4th ed. Philadelphia (PA): Elsevier Medical Publishing; 2005;381–418.

269. Freedman MH. Inherited forms of bone marrow failure. In Hoffman R. Hematology, basic principles and practice. 4th ed. Philadelphia (PA): Elsevier Medical Publishing; 2005;339–80.

270. Cunningham MJ, Silberstein LE. Autoimmune hemolytic anemia. In Hoffman R. Hematology, basic principles and practice. 4th ed. Philadelphia (PA): Elsevier Medical Publishing; 2005;693–708.

271. King LE, Fraker PJ. Zinc deficiency in mice alters myelopoiesis and hematopoiesis. J Nutr. 2002;132:3301–07.

272. Pagano G, Korkina LG. Prospects for the nutritional interventions in the clinical management of Fanconi anemia. Cancer Causes Control. 2000;11:881–89.

273. Wethers DL. Sickle cell disease in childhood: Part I. laboratory diagnosis, pathophysiology and health maintenance. Am Fam Physician. 2000;62:1013–20, 1027–28.

274. Wilson RE, Krishnamurti L, Kamat D. Management of sickle cell disease in primary care. Clin Pediatr Phila. 2003;42:753–61.

275. Handin RI. Disorders of the platelet and vessel wall. In Kasper DL, Braunwald E, Fauci AS, Hausner SL, Longo Dl, Jameson JL, editors. Harrison's principles of internal medicine. 16th ed. New York (NY): McGraw Hill Medical Publishing; 2005;673–80.

276. Deitcher SR. Antiplatelet, anticoagulant and fibrinolytic therapy. In Kasper DL, Braunwald E, Fauci AS, Hausner SL, Longo Dl, Jameson JL editors. Harrison's principles of internal medicine. 16th ed. New York (NY): McGraw Hill Medical Publishing; 2005;687–94.

277. Lazier JN, Kessler CM. Clinical aspects and therapy of hemophilia. In Hoffman R. Hematology, basic principles and practice. 4th ed. Philadelphia (PA): Elsevier Medical Publishing; 2005; 2047–70.

278. Babior BM, Bunn HF. Megaloblastic anemia. In Kasper DL, Braunwald E, Fauci AS, Hausner SL, Longo Dl, Jameson JL , editors. Harrison's principles of internal medicine. 16th ed. New York (NY): McGraw Hill Medical Publishing; 2005;601–07.

279. Franchini M, Targher G, Montagnana M, Lippi G. Iron and thrombosis. Ann Hematol. 2008;87(3):167–73.

280. Naito Y, Tsujino T, Matsumoto M, Sakoda T, Ohyanagi M, Masuyama T. Adaptive response of the heart to long-term anemia induced by iron deficiency. Am J Physiol Heart Circ Physiol. 2009;296(3):H585–H593.

281. Ramel A, Jonsson PV, Bjornsson S, Thorsdottir I. Anemia, nutritional status, and inflammation in hospitalized elderly. Nutrition. 2008;24(11–12):1116–22.

282. Andres E, Federici L, Serraj K, Kaltenbach G. Update of nutrient-deficiency anemia in elderly patients. Eur J Intern Med. 2008;19(7):488–93.

283. Bender DA. Megaloblastic anemia in vitamin B12 deficiency. Br J Nutr. 2003;89:439–41.

284. Jakubowski H. Pathophysiological consequences of homocysteine excess. J Nutr. 2006;136(6):1741S–1749S.

285. Laudicina RJ. Anemia in an aging population. Clin Lab Sci. 2008; 21(4): 232–39.

286. Chan CW, Liu SY, Kho CS, Lau KH, Liang YS, Chu WR, Ma SK. Diagnostic clues to megaloblastic anaemia without macrocytosis. Int J Lab Hematol. 2007;29(3):163–71.

287. Appelbaum, FR. Hematopoietic cell transplantation. In Kasper DL, Braunwald E, Fauci AS, Hausner SL, Longo Dl, Jameson JL , editors. Harrison's principles of internal medicine. 16th ed. New York (NY): McGraw Hill Medical Publishing; 2005;668–73.

288. Wetzler M, Byrd JC, Bloomfield CD. Acute and chronic myeloid leukemia. In Kasper DL, Braunwald E, Fauci AS, Hausner SL, Longo Dl, Jameson JL, editors. Harrison's principles of internal medicine. 16th ed. New York (NY): McGraw Hill Medical Publishing; 2005;631–41.

289. Armitage JO, Longo DL. Malignancies of lymphoid cells. In Kasper DL, Braunwald E, Fauci AS, Hausner SL, Longo Dl, Jameson JL, editors. Harrison's principles of internal medicine. 16th ed. New York (NY): McGraw Hill Medical Publishing; 2005;641–56.

290. Tung J, Hadzic N, Layton M. Bone marrow failure in children with acute liver failure. J Pediatr Gastroenterol Nutr. 2000;31:557–61.

291. Lenssen P, Aker SN. Nutritional support of patients with hematologic malignancies. In Hoffman R. Hematology, basic principles and practice.

4th ed. Philadelphia (PA): Elsevier Medical Publishing; 2005;1591–1610.

292. Ziegler TR. Glutamine supplementation in bone marrow transplant. Br J Nutr. 2002;87:Suppl 1:S9–S15.

293. De Gama Torres HO, Vilela EG, da Cunha As, Goulart EM, Souza MH, Aguirre AC, Azevedo WM, Lodi FM, Silva AA, Bittencourt HN. Efficacy of glutamine-supplemented parenteral nutrition on short-term survival following allo-SCT: a randomized study. Bone Marrow Transplant. 2008;41(12):1021–27.

294. Acquino VM, Harvey AR, Garvin JH, Godder KT, Nieder ML, Adams RH. A double blind randomized placebo controlled study of oral glutamine in the prevention of mucositis in children undergoing hematopoietic stem cell transplantation: A pediatric blood and marrow transplant consortium study. Bone Marrow Transplant. 2005;36:611–16.

295. Murray SM, Pindoria S. Nutrition support for bone marrow transplant patients. Cochrane Database Syst Rev. 2009(1):CD002920.

296. Wernerman J. Clinical use of glutamine supplementation. J Nutr. 2008;138(10):2040S–2044S.

297. Mattson J, Westin S, Edlund S, Remberger M. Poor oral nutrition after allogenic stem cell transplantation correlates significantly with severe graft-versus-host disease. Bone Marrow Transplant. 2006;38(9):629–33.

298. Cusick SE, Mei Z, Freedmand DS, Looker AC, Ogden Cl, Gunter E, Cogswell ME. Unexplained decline in the prevalence of anemia among US children and women between 1988–1994 and 1999–2002. Am J Clin Nutr. 2008;88(6):1611–17.

299. Edison ES, Bajel A, Chandy M. Iron homeostasis: new players, newer insights. Eur J Haematol. 2008;81(6):411–24.

300. Nemeth E, Ganz T. Regulation of iron metabolism by hepcidin. Annu Rev Nutr. 2006;26:323–42.

301. Carlson ES, Magid R, Petryk A, Georgieff MK. Iron deficiency alters expression of genes implicated in Alzheimer disease pathogenesis. Brain Res. 2008;1237:75–83.

20

Diseases and Disorders of the Neurological System

Kathy Jones Irwin MS, RD
Mercy Health Systems, Knoxville, Tennessee
Melissa Hansen-Petrik PhD, RD
The University of Tennessee–Knoxville

CHAPTER OUTLINE

Introduction

Pathology of the neurological system represents a wide array of conditions, many of which have significant nutritional concerns. More than 600 diseases and disorders affecting the nervous system have been identified, and the role of nutrition within the broad spectrum of neurological disease and disorders is far reaching.

Acute traumatic events can result in impairment of nutritional intake and may require alternate paths for nutritional support. Closed head injuries have been shown to represent significant metabolic stress, so nutritional support is critical for a positive outcome. Many chronic neurological disorders are treated with multiple medications that can have serious drug-nutrient interactions. Progressive neurological diseases also involve many possibilities for impairment of nutritional intake and status. Furthermore, progressive neurological disorders can lead to physical changes that directly impact nutritional status. These complicated nutrition problems are often not easily identified or easily solved. Throughout the course of a neurological disease, nutrition therapy plays a crucial role in maintaining optimal quality of life. This chapter describes the pathophysiology, medical treatments, and nutrition therapy for diseases of the neurological system, with a focus on the most common neurological diseases and disorders that involve nutrition therapy as a component of standard care.

Normal Anatomy and Physiology of the Nervous System

The nervous system, certainly the most complex body system, serves to coordinate the action among all components of the body in order to maintain the normal, stable, homeostatic environment required for human functioning. The nervous system has two major divisions: the **central nervous system** (CNS) and the **peripheral nervous system** (PNS). The major

GLOSSARY

acetylcholine—an excitatory neurotransmitter

akinesia—loss of voluntary movement

Alzheimer's disease—the most common form of dementia, characterized by formation of amyloid plaques in the brain and neurofibrillary tangles within neurons

amyloid plaques—cellular deposits found between nerve cells

amyloid precursor protein (APP)—protein from which beta-amyloid is formed

amyotrophic lateral sclerosis—a progressive neurological disease that causes destruction of the motor neurons of the nervous system, resulting in muscle weakness, twitching, and atrophy; also known as Lou Gehrig's disease

aneurysm—weakened area of a wall of a blood vessel

apolipoprotein E—a protein marker found on lipoproteins chylomicron and IDL.

asymmetric muscle weakness—muscle weakness occurring unequally in different parts or sides of the body

atrophy—wasting of body tissue that occurs from disuse, disease, or malnutrition

autonomic division—components of the nervous system that control involuntary functions of the body

autonomic dysreflexia—a complication of spinal cord injury and paralysis; combination of stimuli that result in sudden increase in blood pressure and changes in heart function

axon—part of a neuron that transmits outgoing signals to other neurons

axon terminals—structure at the end of an axon that releases neurotransmitters

beta-amyloid—a part of the amyloid precursor protein found in the insoluble deposits outside neurons, which forms the core of plaques

beta-hydroxybutyrate (BHB)—ketone measured in serum as a marker for patients on the ketogenic diet

bradykinesia—delayed or slowed body movements

brain stem—the part of the brain that connects the brain to the spinal cord and controls autonomic body functions

central nervous system—the brain, spinal cord, and the associated nerves

cerebellum—the part of the brain that is responsible for maintaining the body's balance and coordination

cerebral cortex—the outer layer of nerve tissue surrounding the cerebral hemispheres

cerebral hemisphere—one side of the cerebrum; each side contains four lobes (frontal, parietal, occipital, and temporal)

dendrite—branches extending out from the neuron that assist in transmission of impulses

endorphins—neuropeptides that assist with pain control

epilepsy—a neurological disorder characterized by recurrent seizures, generally more than two unprovoked seizures

excitatory—in the context of the neurological system, refers to a stimulus that results in neural response

fasciculations—involuntary twitching or movement of muscle

frontal lobe—a division of the cerebrum that is responsible for voluntary movement, speech, and complex thought

gastroparesis—delayed emptying of the stomach

Guillain Barré syndrome—an acute peripheral nervous system disease characterized by progressive paralysis

hemorrhagic stroke—stroke caused by rupture of a blood vessel (e.g., aneurysm)

hyperreflexia—overresponse or exaggeration of response to a neural stimulus (e.g., twitching)

inhibitory—in the context of the neurological system, refers to a stimulus that results in a decreased neural response

intractable—resistant to treatment

ischemic stroke—stroke caused by an interruption of blood flow to the tissue

ketogenic diets—nutrition therapy characterized by diets high in fat and restrictive in carbohydrates to produce a therapeutic ketosis (increase in ketones in serum and urine).

limbic lobe—component of the brain involved in control of emotions

lock and key model—description of communication between two cells; action between two substances within the body; in order for action to occur, the two cells must fit together as a lock and key might

multiple sclerosis—a disorder characterized by demyelination of cells within the CNS, inflammation, and development of scar tissue, causing numbness, tingling, incoordination, weakness, and varying degrees of blindness

myasthenia gravis—a neuromuscular disorder that affects the skeletal muscles and causes muscle weakness, particularly of the face, eyes, arms, and legs

myelin—the covering or insulation of the axon that ensures proper communication between neurons

neurofibrillary tangles—collections of twisted tau found in the cell bodies of neurons in Alzheimer's disease

neuromodulator—substance released that will increase or decrease the activity of specific neurotransmitters

neuron—a nerve cell in the central nervous system

neuropeptide—protein messengers within the brain and nervous system that assist in communication between neurons

neurotransmitter—a chemical messenger that communicates between neurons

occipital lobe—portion of the cerebral cortex controlling vision

paraplegia—paralysis involving the lower body below the umbilicus

parasympathetic branch—division of the autonomic nervous system that is involved in control of gastrointestinal, cardiac, and respiratory systems

parietal lobe—portion of the cerebral cortex responsible for the sensations of pain, touch, taste, temperature, and pressure; related to mathematical and logical thinking

Parkinson's disease—a neuromuscular, neurodegenerative disease resulting in the loss of dopamine-producing cells in the substantia nigra portion of the brain, leading to resting tremor, rigidity, slowed movement, stooped posture and postural instability, mask-like facial features, and a shuffling gait

peripheral nervous system—all components of the nervous system except for the brain and spinal cord (central nervous system)

plasmapheresis—treatment that removes blood from the body, separates out certain cells from the plasma, and then returns the blood back to the body

quadriplegia—paralysis involving all arms and legs; also known as tetraplegia

seizure—episode of spontaneous, uncontrolled electrical activity in the brain

shaken baby syndrome—signs and symptoms that occur as a result of brain injury caused by violently shaking or impacting the head of an infant or small child

soma—major body portion of the neuron

somatic division—portion of the peripheral nervous system that carries messages from the body back to the brain

spasticity—involuntary muscle contraction that results in rigidity

sympathetic branch—portion of the peripheral nervous system that prepares the body for action; controls flight or fight response

synapse—space or gap between nerve cells across which neurotransmitters pass

tau—a protein that is a principal component of the paired helical filaments in neurofibrillary tangles; helps to maintain the structure of microtubules in normal nerve cells

temporal lobe—portion of the cerebral cortex responsible for hearing and memory

transient ischemic attacks (TIAs)—"mini-strokes"; an episode of ischemia where blood flow is quickly restored and symptoms resolve within 24 hours

function of the CNS is the processing of sensory information, after which the appropriate responsive motor signal is generated. The PNS, including the paired cranial and spinal nerves, conducts the transmission of impulses between the CNS and peripheral organs. The **somatic division** of the PNS carries messages forward to the CNS from the sensory organs, and outward from the CNS to the muscles for action. The **autonomic division** of the PNS is responsible for the involuntary functions of the body. This autonomic nervous system is divided into the **sympathetic** and the **parasympathetic branches**.[1]

Communication within the Nervous System

At the cellular level, the nervous system is composed of two major classifications of functional cells: the **neurons** and glial cells. Neurons are responsible for transmission of information from one cell to another and are surrounded by glial cells. Many of the neurological disorders that will be discussed in this chapter arise from dysfunction of neurons, glial cells, and their communication with one another and with their target organs.[1]

The components of a neuron are the soma, **dendrites**, axon, and axon terminals (see Figure 20.1). The **soma** is the body of the neuron. The branch-like dendrites, which project out from the soma, are responsible for receiving information from another neuron. The **axon** is a long thin fiber that is responsible for transmission of information from one neuron to another. The **axon terminals** are small nodes that secrete neurotransmitters. A **myelin** sheath covers and insulates the axon, ensuring faster and more consistent transmission of information and preventing inconsistent, random communication.[1]

A neuron passes its message on to another neuron by releasing chemical **neurotransmitters** into the space between each neuron, called the **synapse**. A presynaptic neuron sends the specific information, or impulse, and a postsynaptic neuron receives it. A neurotransmitter can stimulate a postsynaptic neuron only at specific receptor sites on its dendrites and soma. Receptor sites respond to only one type of neurotransmitter. This **lock and key model** ensures that specific neurotransmitters work only at specific kinds of synapses (see Figure 20.2)[1]

Abnormalities in neurotransmitter production and function have been linked to many of the neurological disorders that will be discussed in this chapter. For example, abnormalities in acetylcholine have been linked to both myasthenia gravis and Alzheimer's disease, while deficiency of dopamine is the underlying abnormality in Parkinson's disease (see Table 20.1).[1]

Neurotransmitters may **excite** or **inhibit** the next neuron. Additionally, other substances also act in the same fashion as neurotransmitters. These include **endorphins** and **neuromodulators** (see Table 20.2) as well as gases such as nitric oxide and carbon dioxide. Endorphins are **neuropeptides** that naturally occur in the brain. Endorphins serve multiple functions, including decreasing a person's sensitivity to pain. Neuromodulators do not carry neural messages directly; instead, they can either increase or decrease the activity of specific neurotransmitters. Stimulation of an excitatory synapse makes the neuron more likely to respond, whereas stimulation of an inhibitory synapse makes production of an action potential

Figure 20.1 Anatomy of a Neuron

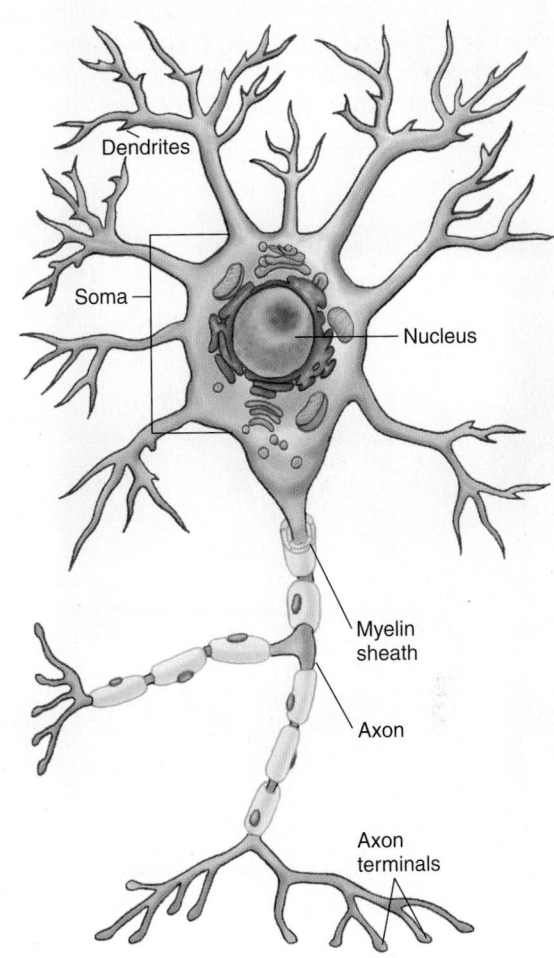

Source: R. Rhoades and R. Pflanzer, Human Physiology, 4e, copyright © 2003, p. 216.

(the mode of electrical impulse transmission from one neuron to the next) less likely.[1]

The ability to conduct the nerve impulse is directed by electrical activity. This electrical activity or "action potential" is generated by differences in electrolyte levels between the cytosol and extracellular fluid of the nervous system. Again, disorders of this electrical conduction are the basis of the disease process for many neurological dysfunctions. For example, epilepsy is linked to abnormalities in the control of membrane potentials that result in excessive stimulation of the excitatory synapse.[1]

The Central Nervous System (CNS)

The brain is the greatly modified and enlarged anterior portion of the CNS. It is surrounded by three protective membranes (meninges) and is enclosed within the cranial cavity of the skull. By means of a prominent groove, called the longitudinal fissure, the brain is divided into two halves called **cerebral hemispheres**. At the base of the brain, a thick bundle of nerve fibers, called the corpus callosum, provides the pathway for communication between the hemispheres. The left hemisphere

Figure 20.2 Communication between Neurons

A neurotransmitter released from the axon terminal of the presynaptic neuron (on the left) travels across the synapse and enters the dendrite of the postsynaptic neuron (on the right) through specific receptor sites.

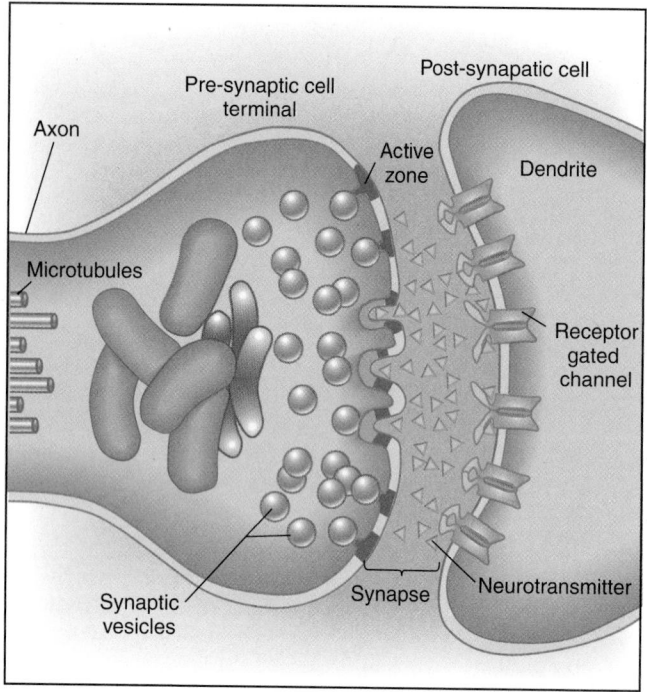

Source: R. Rhoades and R. Pflanzer, *Human Physiology*, 4e, copyright © 2003 p. 232.

Table 20.2 Neurotransmitters and Neuropeptides

Classical Neurotransmitters (small, rapid-acting molecules)	
• Acetylcholine	• Histamine
• Dopamine	• Glycine
• Norepinephrine	• Glutamate
• Epinephrine	• Aspartate
• Serotonin	• Gamma amino butyric acid (GABA)

Neuropeptides (large, slow-acting molecules)	
• β-endorphin	• Substance P
• Adrenocorticotropic hormone (ACTH)	• Motilin
• α-melanocyte-stimulating hormone (MSH)	• Insulin
• Thyrotropin-releasing hormone (TRH)	• Glucagon
• Gonadotropin-releasing hormone (GnRH)	• Angiotensin II
• Somatostatin	• Bradykinin
• Vasoactive intestinal polypeptide (VIP)	• Vasopressin
• Cholecystokinin (CCK)	• Oxytocin
• Gastrin	• Carnosine
	• Bombesin
	• Neurotensin

Reprinted from: Table 4-2 p. 121 Sherwood L. Human Physiology: from cells to systems. 5th ed. Belmont (CA): Brooks/Cole, 2004.

Table 20.1 Neurotransmitters—Their Link to Disease

Neurotransmitter	Function	Disease Link
Acetylcholine (ACh)	Excitatory neurotransmitter	Alzheimer's disease; myasthenia gravis
Dopamine	Inhibitory neurotransmitter that controls posture and movement	Parkinson's disease; schizophrenia
Gamma amino butyric acid (GABA)	Inhibits central nervous system and regulates anxiety and movement	Anxiety disorders; Huntington's disease
Glutamate	Major excitatory neurons in central nervous system; important for learning and memory	Memory loss; Alzheimer's disease
Norepinephrine	Important for psychological arousal, mood changes, sleep, and learning	Bipolar disorder
Serotonin	Regulates sleep, mood, appetite, and pain	Depression

Adapted from: Sherwood L. Human Physiology: From Cells to Systems. 6th ed. Belmont (CA): Brooks/Cole, 2009.

controls the right half of the body while the right hemisphere controls the left half, because of a crossing of the nerve fibers in the medulla. Although the right and left hemispheres seem to be a mirror image of one another, there are important functional differences. As shown in Figure 20.3, distinct areas of the brain, such as the **cerebral cortex**, **cerebellum** and **brain stem**, are responsible for specific functions. Understanding the basis of CNS function allows one to interpret, understand, and predict the possible dysfunction seen in brain injury or neurological disease.[1]

The upper regions of the motor and sensory areas, located at the top of each hemisphere, control the lower parts of the body. The **frontal lobe** is responsible for the elaborate processes of thinking, planning, and emotion. The **parietal lobe** is responsible for the sensations of pain, touch, taste, temperature, and pressure; it is also related to mathematical and logical thinking. The **temporal lobe** controls hearing and auditory function and plays a role in memory and emotions. Damage to this area results in partial or complete deafness. Since the **occipital lobe** processes visual information, damage to this area results in partial or complete blindness. The **limbic lobe** is involved in the emotional and sexual aspects of behavior and in the processing of memory.[1]

The spinal cord—the other essential component of the CNS—is the cylinder of nerve tissue encased within the vertebral column. The spinal cord coordinates neural messages between the brain and the rest of the body. Secondly, the spinal cord integrates information within the PNS.[1] The vertebrae and corresponding nerves are named according to their location along the spinal cord: cervical, thoracic, lumbar, sacral, and coccygeal (see Figure 20.4).[1]

Epilepsy and Seizure Disorders

Definition

Epilepsy is a disorder characterized by periods of spontaneous, disordered electrical discharging of cerebral neurons resulting in a **seizure**. A seizure can occur in either a portion of or the entire brain and typically presents as sudden deviations in the way the person feels or acts.

Epidemiology

More than 3 million Americans have epilepsy.[2] In the United States alone, 200,000 people are diagnosed with epilepsy each year. Worldwide, it is estimated that 1% of the population has epilepsy;[2] 10% of the population can expect to have a seizure sometime in their lifetime.[2]

Etiology

Common known causes of epilepsy involve some type of insult to the brain tissue, such as head trauma, CNS infections, drug and alcohol abuse, CNS tumors, and cerebrovascular disease. However, the majority of patients with epilepsy experience no precipitating event and have no known defect associated with the disorder.[3]

Pathophysiology

During a seizure, periods of electrical discharge can range in duration from a few seconds to several minutes. Extended seizure activity lasting over 5 minutes (often classified as status epilepticus) can require medications resulting in heavy sedation (sometimes to the point of inducing a coma) to stop the seizure activity. Epilepsy can include seizures of several types, including generalized seizures that involve the entire brain and partial seizures affecting only part of the brain. It is important to differentiate a single seizure event from the diagnosis of epilepsy. Single seizures can occur as a result of numerous conditions, including fever (referred to as febrile seizures) and metabolic or electrolyte abnormalities such as hyperglycemia or hypoglycemia.

Clinical Manifestations

Generalized seizures occur when abnormal electrical activity involves the entire brain and can result in a broad array of possible signs such as staring, muscle jerking, twitching, head drops, falls, and/or loss of consciousness. There are several classifications of generalized seizures including tonic-clonic seizures (formerly grand-mal), characterized by a mixture of symptoms; and absence (formerly petit-mal), in which the person appears to be staring without response.

Partial seizures are isolated to one section or lobe of the brain; the area of the brain affected will determine how the seizure presents (what the seizure will look like or feel like). For example, a seizure in the motor cortex of the frontal lobe may result in involuntary movements of a hand or leg.[4] Partial seizures can be further classified as simple partial or complex partial. The patient affected by a simple partial seizure typically remains aware of the event and can recall what took place during the event (consciousness is maintained).[4] Complex partial seizures begin in the temporal or frontal lobe and affect the area of the brain that regulates awareness and alertness; thus, the patient

is unaware of what is happening during the seizure. The patient may have a blank stare and make meaningless movements such as climbing, drooling, or lip smacking.[4] Often, after a complex partial seizure, the patient is confused and tired. The electrical activity of partial or complex partial seizures can remain in one area of the brain and resolve, or can spread throughout the entire brain, thus evolving into a generalized seizure.[4]

Medical Diagnosis

To help confirm a diagnosis of epilepsy, an electroencephalogram (EEG) is used to record the electrical activity of the brain. Video electroencephalograms are sometimes used to help practitioners correlate abnormal electrical activity with clinical features of seizure activity. These diagnostic tools provide information about abnormal electrical activity and seizure classification, and can assist in locating the focal point of the seizure, if identifiable.[4] These tools are also very useful in identifying certain epilepsy syndromes such as Lennox-Gastaut, a difficult-to-treat syndrome with a distinct EEG pattern that involves multiple seizure types.[5] Identifying the seizure type and investigating a focal point can give the practitioner insight into which treatment option would be indicated for a patient.[4] Anti-epileptic drugs (AEDs) are typically FDA approved for particular seizure types, and surgical interventions could be indicated in an individual **intractable** to medications when a clear seizure focal point is found.[4] Nutrition therapies, such as the classic ketogenic diet, are growing in popularity and are now used worldwide for intractable epilepsy.[6]

Treatment

Anti-epileptic Drugs (AEDs) The earliest reported use of AEDs was in the late 1800s, when treatment with bromides was coupled with serious side effects.[7] The turn of the twentieth century brought better options than the bromides for epilepsy, including the barbiturates (such as phenobarbital) and subsequently the introduction of phenytoin (Dilantin®) in the 1930s. In addition to these AEDs, which are still in use today, current pharmacological treatments include valproic acid (Depakote®); carbamazepine (Tegretol®); and the second-generation AEDs introduced in the 1990s and thereafter, such as levetiracetam (Keppra®) and zonisamide (Zonegran®). (See Table 20.3.) Side effect profiles of AEDs appear to have improved with the second-generation AEDs, yet the percentage of those relieved from seizures with AEDs remains around 70%. Therefore, 30% of patients with epilepsy are considered intractable to AEDs.[8–10] For those patients who fail to respond to AEDs, other treatment options include surgical interventions and nutrition therapy.

Surgical Treatments Vagus nerve stimulation has grown in popularity since it was first approved by the FDA in 1997, with over 32,000 stimulation devices implanted since that time.[11] This device is surgically implanted under the clavicle with a wire connected to the left vagus nerve. This wire carries periodic electrical charges to the brain and, in many cases, reduces seizure activity.[11] Currently, other devices that involve non-surgical transcutaneous stimulation are being developed and tested.[12]

Invasive surgical interventions involving the brain directly include temporal lobectomy (removal of a lobe), hemispherectomy (removal of a hemisphere), and corpus callosotomy (disconnection of the 2 hemispheres). During evaluation for candidacy

Figure 20.3 Regions of the Brain

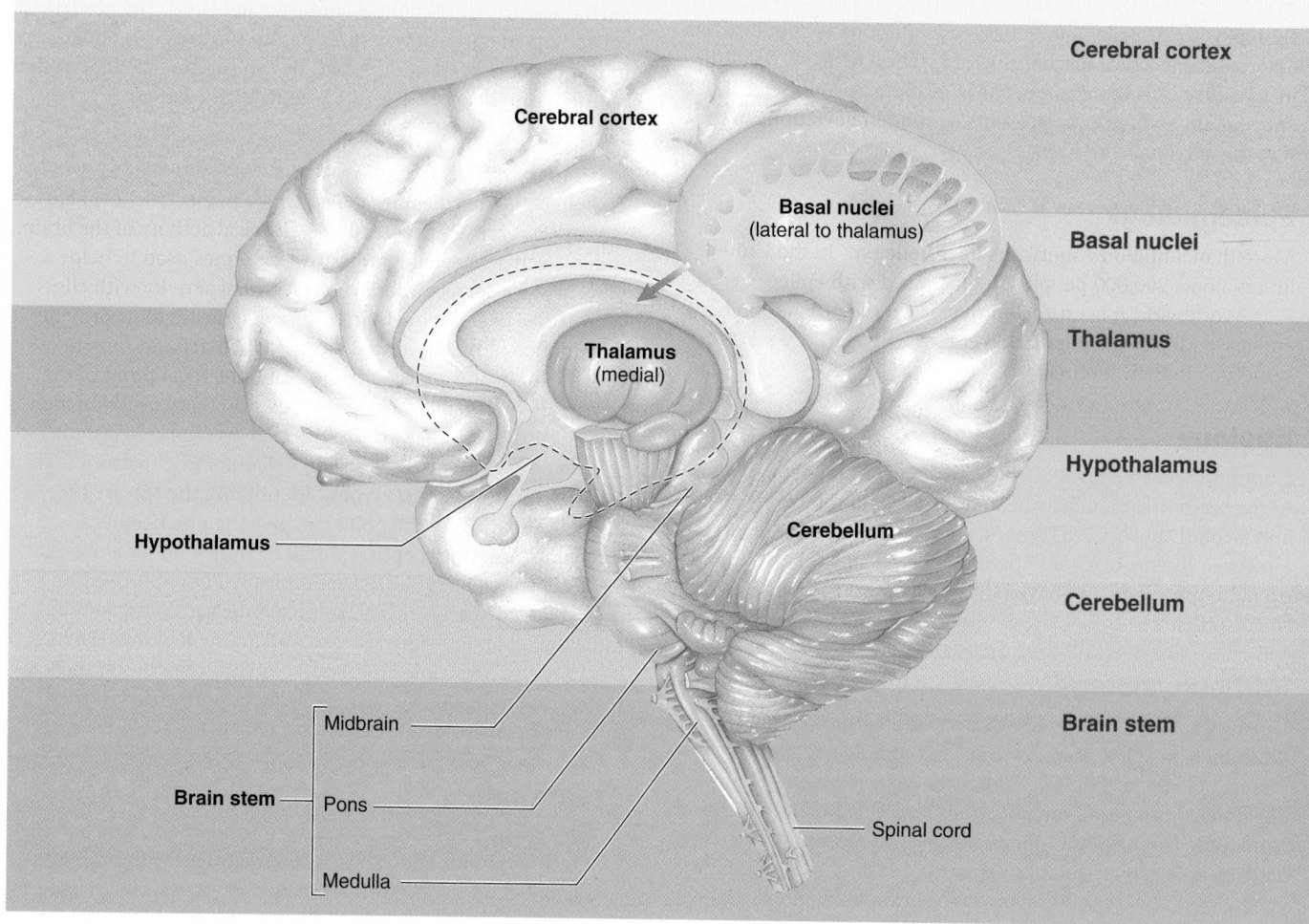

Other: L. Sherwood, *Human Physiology: From Cells to Systems* 7e, ISBN 0495391840, Table 5-2, pp. 144–145.

for lobectomy or hemispherectomy surgeries, patients typically undergo a rigorous screening by a multidisciplinary medical team to determine whether a seizure focus can be identified and safely removed without loss of function.[13] Each of these surgical interventions involves considerable risk and can affect the patient permanently.

Nutrition Therapy for Epilepsy and Seizure Disorders

Nutrition interventions for epilepsy dates back to ancient times when fasting was the mainstay for this disorder.[14] Since starvation is not a practical long-term option, several diets that metabolically mimic starvation through carbohydrate deprivation have been developed over time and continue to be used today. These high-fat, low-carbohydrate diets are considered **ketogenic diets**, as they induce a metabolic shift of energy utilization within some tissues including the brain and produce ketosis (an increase in ketone bodies in serum and urine). Dietary therapy for epilepsy was first recognized by the American Medical Association in the 1920s,[15] but its use diminished with AED development throughout the twentieth century.[14] AEDs became the primary treatment and these diets were often

considered "alternative therapies" and used as a last resort. Yet recent interest generated from parent groups has stimulated newfound research interest with over 500 peer-reviewed publications, many validating the efficacy of this treatment, which is now offered worldwide.[6] A recent meta-analysis of 19 peer-reviewed studies that included 1084 patients reported the ketogenic diet reduced seizures by 90% in a third and by 50% in half of the patients. Recent class I and II data confirmed clear anti-epileptic effects of these diets, with efficacy rates similar to main-line AEDs.[16] With 30% of the epilepsy population intractable to AEDs, ketogenic diets are now considered valid epilepsy treatment options. These treatments were once reserved for children only, but preliminary data indicate they can be effective in adult populations as well.[17, 18]

The major variations of ketogenic diets in use today include the classic ketogenic diet, the medium-chain triglyceride (MCT) diet, the low-glycemic-index treatment (LGIT), and the modified Atkins diet (MAD). The classic ketogenic diet is the most widely used and requires adherence to a strict high-fat dietary regimen. The MCT diet allows for more liberal carbohydrate, yet because the oil is expensive and not reimbursable by insurance companies, this version is not often used in the U.S. The modified Atkins diet (restricted carbohydrate, liberal fat

Figure 20.3 Regions of the Brain (Continued)

1. Sensory perception
2. Voluntary control of movement
3. Language
4. Personality traits
5. Sophisticated mental events, such as thinking, memory, decision making, creativity, and self-consciousness

1. Inhibition of muscle tone
2. Coordination of slow, sustained movements
3. Suppression of useless patterns of movement

1. Relay station for all synaptic input
2. Crude awareness of sensation
3. Some degree of consciousness
4. Role in motor control

1. Regulation of many homeostatic functions, such as temperature control, thirst, urine output, and food intake
2. Important link between nervous and endocrine systems
3. Extensive involvement with emotion and basic behavioral patterns
4. Role in sleep–wake cycle

1. Maintenance of balance
2. Enhancement of muscle tone
3. Coordination and planning of skilled voluntary muscle activity

1. Origin of majority of peripheral cranial nerves
2. Cardiovascular, respiratory, and digestive control centers
3. Regulation of muscle reflexes involved with equilibrium and posture
4. Reception and integration of all synaptic input from spinal cord; arousal and activation of cerebral cortex
5. Role in sleep–wake cycle

and protein) and the low-glycemic-index treatment (selected carbohydrates are restricted, liberal fat and protein) are less restrictive than the classic ketogenic diet and are currently being investigated and used in some circumstances, although clear indications and efficacy for these variations have not yet been established.[19, 20]

Despite the recent resurgence of research interest, the mechanisms of action for this dietary treatment remain poorly understood. It has been proposed that the effect could stem from the altered substrate utilization in brain tissue (the shift from glucose to ketones), possibly the ketones themselves,[21] or simply the lack of free glucose to precipitate a seizure.[22] Brain glucose uptake accelerates during seizures[23] and it is possible that the reduced availability of glucose translates to a lack of readily available energy to support neuronal membrane excitability. It is also possible that ketogenic diets alter neurotransmitter synthesis and action and some investigators suggest that increased synthesis of GABA, the primary inhibitory neurotransmitter, may explain dietary efficacy.[24] There has been some indication that ketogenic diets can be neuroprotective, and thus could be useful in other neurological diseases and disorders. Table 20.3 summarizes diseases or disorders other than epilepsy that are currently being investigated for effectiveness of treatment with ketogenic diets.

Nutritional Implications

Individuals with epilepsy and seizure disorders can be at nutritional risk. For infants, children, and adolescents, impaired ability to consume adequate nutrients, limited food choices (if on a ketogenic diet), and drug-nutrient interactions may interfere with the ability to achieve optimal growth and development.[25] Because the ketogenic diet is low in fiber and calcium and insufficient in other nutrients, supplementation is recommended to avoid deficiencies.[26] Fluid restriction has been commonly used with this treatment in the past, yet most practitioners do not restrict fluid today, as no seizure control benefits have been shown.[27] In fact, due to the diuretic effect of low-carbohydrate diets and risk of kidney stones, fluids are typically encouraged.[27]

Since pharmacotherapy is the primary seizure treatment used today, identifying potential interactions of AEDs as well as nutrients is a critical component of nutritional care. Common issues include weight gain with valproate (Depakote®), carbamazepine (Tegretol®, Carbatrol®), and gabapentin (Neurontin®). Topiramate (Topamax®) and zonisamide (Zonegran®) could cause weight loss as well as kidney stones (especially when used concurrently with a ketogenic diet); adequate fluid intake is essential. Felbamate (Felbatol®) and carbamazepine (Tegretol®, Carbatrol®) should not be taken with grapefruit products.

Phenytoin (Dilantin®) inhibits both vitamin D and folate metabolism. It is estimated that more than 50% of patients on phenytoin have decreased serum levels of folate.[28] Long-term use may result in megaloblastic anemia secondary to folate deficiency (see Chapter 11). When folate is supplemented, the

Table 20.3 Potential Future Uses of Ketogenic Diets

Non-epilepsy Uses of Ketogenic Diets in Published Literature Currently Being Investigated (in alphabetical order):
• Alzheimer's disease
• Amyotrophic lateral sclerosis (ALS)
• Autism
• Bipolar disease and depression
• Brain tumors
• Glycogenesis type V
• Migraine
• Narcolepsy and sleep disorders
• Obesity
• Parkinson disease
• Post hypoxic myoclonus
• Schizophrenia
• Stoke
• Traumatic brain injury

Adapted from: JOURNAL OF CHILD NEUROLOGY by Kossoff, Zupec-Kania. Copyright 2009 by SAGE PUBLICATIONS INC. JOURNALS. Reproduced with permission of SAGE PUBLICATIONS INC. JOURNALS in the format Textbook via Copyright Clearance Center.

Figure 20.4 The Spinal Cord

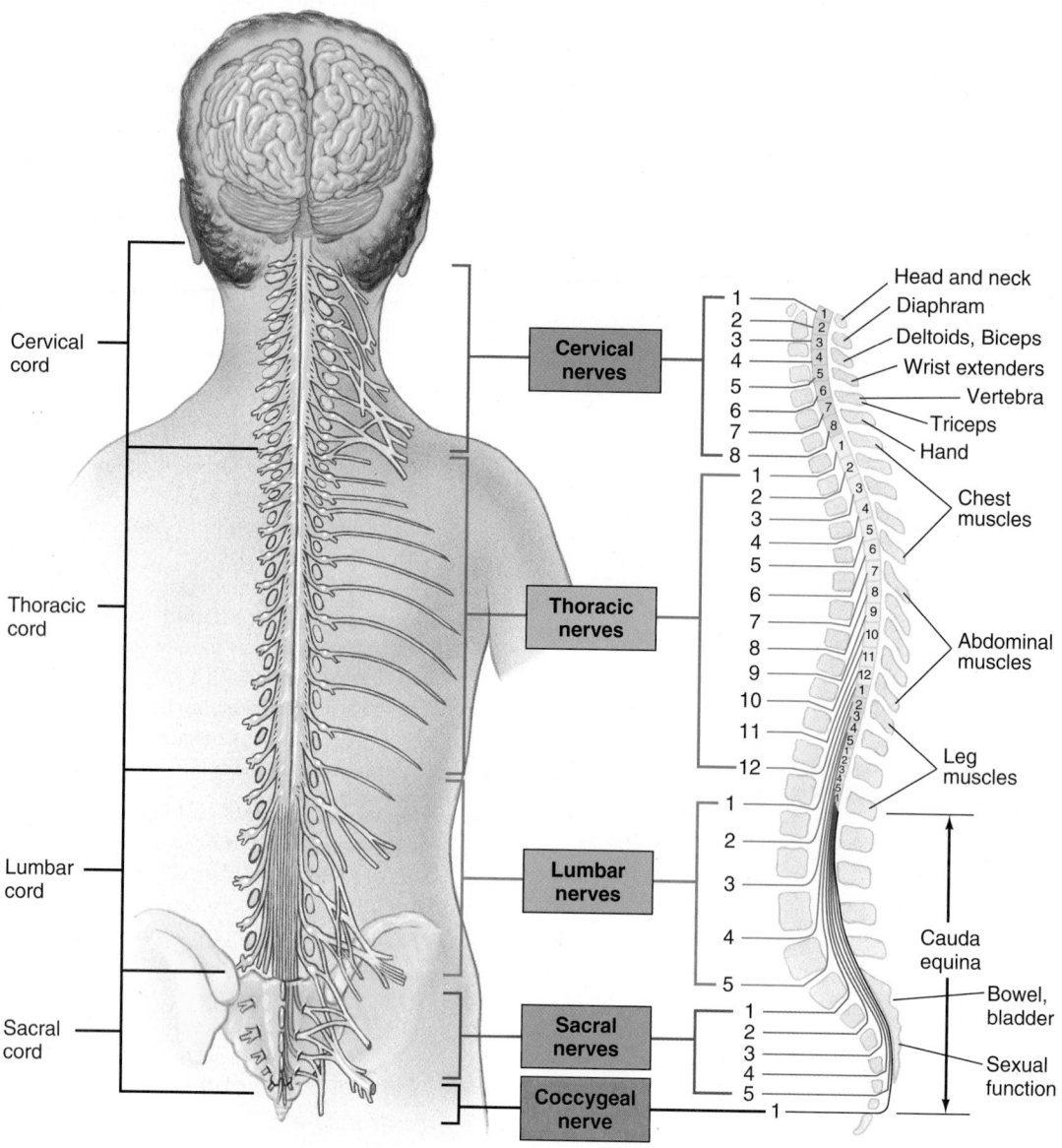

Source: L. Sherwood, *Human Physiology: From Cells to Systems* 5e, copyright © 2004, p. 173.

effectiveness of phenytoin (i.e., seizure control) should be monitored closely so that a steady state is achieved. Table 20.4 provides a listing of commonly used AEDs and possible food-drug interactions.

Nutrition Assessment

Table 20.5 provides guidelines for conducting a thorough nutrition assessment of individuals with epilepsy and other neurological disorders.

Nutrition Diagnosis

Nutrition diagnoses most likely to be identified for individuals with epilepsy include inadequate or excessive energy intake, inadequate oral food/beverage intake, evident protein-energy malnutrition, inadequate protein-energy intake, malnutrition,

inadequate vitamin or mineral intake (if on ketogenic diet or some AEDs), food-medication interaction, altered nutrition-related laboratory values, underweight, involuntary weight loss, overweight/obesity, and involuntary weight gain.

Nutrition Intervention

The classic ketogenic diet is typically initiated with a macronutrient ratio of 4:1 (or 3:1) to provide 4 g of fat to every 1 g of protein + carbohydrate. Kcalories are calculated at 80%–90% of 2004 RDAs for energy and minimal protein to allow for growth are calculated into the meal plans.[29] Recent development of a web-based "ketocalculator" (accessed at www.ketocalculator.com) offered free to registered dietitians has proven a beneficial tool to aid in diet calculations; meal planning; and micronutrient, drug, and fluid management.[29] Table 20.6 provides

Table 20.4 Selected Medications Used in Epilepsy

Brand Names	Generic	Possible Food–Drug Interactions/Side Effects That May Impact Nutritional Status
Tegretol; Carbatrol	Carbamazepine	Do not take with grapefruit juice; consistently take either with food or without food to ensure consistent absorption; contraindicated with alcohol
Felbatol	None available	Do not take with grapefruit juice; aplastic anemia, hepatic failure; upset stomach; decreased appetite; weight loss; liver failure; contraindicated with alcohol
Neurontin	Gabapentin	Do not take with antacids; consistently take either with food or without food to ensure consistent absorption; pedal edema; weight gain; upset stomach; contraindicated with alcohol
Lamictal	Lamotrigine	Consistently take either with food or without food to ensure consistent absorption; upset stomach; contraindicated with alcohol
Lyrica	Pregabalin	Consistently take either with food or without food to ensure consistent absorption; edema; weight gain; contraindicated with alcohol
Keppra	Levetiracetam	Sedation, cognitive effects, dizziness, contraindicated with alcohol
Trileptal	Oxcarbazepine	Do not take with grapefruit juice; upset stomach; hyponatremia; vomiting; contraindicated with alcohol
Luminal	Phenobarbital	Sedation, cognitive effects; upset stomach; vomiting; constipation; contraindicated with alcohol
Dilantin; Phenytek	Phenytoin	Ataxia; cognitive disturbance; swelling or increased growth of the gums
Topamax	Topiramate	Cognitive disturbance; loss of appetite; weight loss; metabolic acidosis; nephrolithiasis; contraindicated with alcohol
Depakote	Valproic acid	Consistently take either with food or without food to ensure consistent absorption; stomach discomfort; nausea; weight gain; liver toxicity; low platelet counts; contraindicated with alcohol
Zonegran	None available	Weight loss; upset stomach; changes in blood glucose and sodium levels; nephrolithiasis; contraindicated with alcohol
Benzal	Rufinamide	Somnolence; vomiting; nausea; decreased appetite
Mysoline	Primidone	Upset stomach; vomiting; loss of appetite; anemia or blood abnormalities; contraindicated with alcohol
Klonopin	Clonazepam	Nausea; loss of appetite; contraindicated with alcohol

Sources: The Epilepsy Therapy Project. Seizure Medicines. Last updated 9/20/2008. http://www.epilepsy.com/EPILEPSY/seizure_medicines
Parke-Davis, Division of Pfizer, Inc. Full Prescribing Information for Lyrica (pregabalin) Capsules, CV. Revised 4/2009. http://www.pfizer.com/files/products/uspi_lyrica.pdf
Eisai, Inc. Banzel Medication Guide. Revised 11/08. http://www.eisai.com/package_inserts/Banzel%20PI%200109.PDF

examples of ketogenic diet meal plans that were developed using the ketocalculator. No/low-carbohydrate forms of all medications and other products (such as toothpaste, soaps, and lotions) must be used. Supplements prescribed include a multivitamin with minerals, a calcium with vitamin D supplement, and others as indicated. Parents receive extensive training on weighing foods on a gram scale, measuring urine ketones, managing sick days, traveling, modifying holiday celebrations (such as birthdays), and minimizing exposure to carbohydrates. Most practitioners prefer in-hospital initiations, although out-patient diet initiation is growing in popularity and has been shown to be safe and effective in some instances with proper medical supervision and parent education.[30, 31]

Ensuring adequate energy, protein, vitamin, and mineral intakes is a major component of nutrition therapy for these populations.[32] Box 20.1 outlines some general principles of the ketogenic diet; nonetheless, it is crucial that each diet be carefully calculated and monitored closely in order to both control seizures and ensure nutrient needs are met.

Monitoring and Evaluation

Periodic monitoring of the child's well-being to include clinical assessment, anthropometric measurements, and laboratory studies is essential for management of the ketogenic diet. In addition to standard laboratory blood studies, **beta-hydroxybutyrate (BHB)**, a blood ketone, is monitored

as a ketosis marker and therefore an indicator of diet compliance (see Box 20.1). Most children maintain the best seizure control when BHB levels are over 4.00 mmol/L.[33] Patients should also be monitored for vitamin and mineral deficiencies, nutritional status, and proper growth. When assessing the effectiveness of the ketogenic diet, seizure frequency data is necessary. Parental records are to be encouraged.

Stroke and Aneurysm

Definition

A stroke (cerebrovascular accident) occurs when a sudden interruption of blood flow to the brain deprives neurons and other brain cells of nutrients and oxygen and thereby alters brain function.[34] When blood vessels supplying blood to the brain are obstructed, as with a clot, the stroke is classified as an **ischemic stroke**. Ischemic strokes are the most common form, accounting for about 80% of all strokes. **Transient ischemic attacks** (TIAs) or "mini-strokes" are defined as an episode of ischemia where blood flow is quickly restored and symptoms resolve within 24 hours. A **hemorrhagic stroke** occurs when a blood vessel in the brain ruptures. This is most likely to occur when blood vessel walls are weakened by hypertension or other conditions, including aneurysms. An **aneurysm** is defined as the dilation and bulging of smooth muscle usually found at the points where cerebral arteries divide or split (bifurcation).

Table 20.5 Nutrition Assessment for Disorders of the Neurological System

Client History (personal, patient, family, medical, and social)	• Diagnoses • Previous medical conditions or surgeries • Neurological impairment of cranial nerves V, VII, IX, X, XI, XII (affect swallowing) • Sensory limitations • Medications • Socioeconomic status/food security • Support systems • Education—primary language, literacy level • Psychological—affective disorders
Food-/Nutrition-Related History	• Ability to chew; use and fit of dentures • Problems swallowing foods and liquids • Nausea, vomiting • Constipation, diarrhea • Heartburn • Any other symptoms interfering with ability to ingest normal diet • Ability to consistently purchase adequate amounts of food on a daily basis • Ability to feed self • Muscular strength and coordination • Ability to cook and prepare meals • Food allergies, preferences, or intolerances • Food-drug interactions • Previous food restrictions • Ethnic, cultural, and religious influences • Use of alcohol, vitamin, mineral, herbal, or other type of supplements • Previous nutrition education or nutrition therapy • Eating Pattern: 24-hour recall, diet history, food frequency
Anthropometric Measurements	• Current weight • Weight history: highest adult weight; usual body weight; recent weight change • Reference weight (use adjustments for quadriplegia or paraplegia, BMI)
Biochemical Data, Medical Tests, and Procedures	• *Visceral Protein Assessment:* Standard • *Renal, Hepatic Function Assessment:* Standard • *Hydration Status Assessment:* Standard • *Immunocompetence:* Standard • *Hematological Assessment:* Standard

Table 20.6 Sample Ketogenic Diet Meals

Example for 5 year old: 1200 kcal and 19 g protein per day (divided among 4 meals per day)

Meal Ingredients	Fat (g)	Protein + CHO (g)
Meal 1: Cream, 56 g Canola oil, 6 g Fresh egg, 16 g Crisp bacon, 5 g Strawberries, 14 g	30	7.5
Meal 2: Cream, 45 g Canola oil, 4 g Ranch dressing, 20 g Iceberg lettuce, 16 g Fresh spinach, 16 g Grilled chicken breast, 9 g	30	7.5
Meal 3: Cream, 46 g Canola oil, 4 g Butter, 6 g Spaghetti squash, 20 g Cheddar cheese, 15 g	30	7.5
Meal 4: Sugar-free mayo, 32 g Shredded chicken breast, 16 g Dill pickles, 11 g Red grapes, 11 g	30	7.5
Total for day	120 g	30 g
Diet Ratio 4:1	1200 kcal per day	

Meals planned using the Ketocalculator (www.ketocalculator.com)

Epidemiology

According to the National Institute of Neurological Diseases and Stroke, stroke is the third leading cause of death in the United States, the fourth leading cause in Canada, and a significant health issue throughout the world.[34]

Etiology

Multiple factors are thought to place an individual at risk for stroke.[34] As with many conditions, risk factors can be categorized as modifiable or unmodifiable. Unmodifiable risk factors include age, gender, and race. Age has the strongest association with stroke risk, with risk doubling for each decade after age 55.[34] Risk by gender varies with age category.[35] In the United States, African Americans are at highest risk of stroke.[35] Genetic risk factors have also been identified.[34]

Modifiable stroke risk factors include the presence of hypertension, cardiovascular disease, diabetes mellitus, dyslipidemia, asymptomatic carotid stenosis, atrial fibrillation, cigarette smoking, physical inactivity, and obesity.[34] Although dietary factors are not specifically identified, diet does play a critical role in prevention and management of hypertension, cardiovascular disease, diabetes, dyslipidemias, and obesity and, therefore, prevention of stroke. Combined findings from the Health Professionals Follow-up Study and Nurses' Health Study showed that people with a healthy lifestyle have an 80% lower risk of stroke compared to those with the least healthy lifestyle.[36] Smoking was the strongest predictor of stroke risk. Overall, healthy lifestyle in this study was defined as not smoking, maintaining a BMI less than 25 kg/m², exercising at least 30 minutes per day at moderate to vigorous intensity, eating a healthy diet, and consuming 5–15 g alcohol daily for women or 5–30 g alcohol daily for men. Alcohol intakes above 30 g daily (>2 drinks per day) were linked to greater risk of stroke,

Outline for Classic Ketogenic Nutrition Therapy Used for Seizure Control

- Strict clinical supervision by a qualified medical team, including a registered dietitian, is required.

- In-hospitalization diet initiation (or out-patient when appropriate) with follow-up visits at least every 3 months for the first year.

- Establish energy requirements and protein requirements for individual patient (adjusted for changes in weight and growth).

- Grams of fat calculated at a 4:1 (or 3:1) ratio (4 grams of fat to each gram of protein plus carbohydrate—approximately 75%–90% of total energy intake).

- Minimal fluid requirements established and fluids encouraged. Fluids are not restricted by most practitioners; no correlation to seizure control has been established.

- Sugar-free multivitamin, calcium + vitamin D supplement to meet AI, other supplements as needed.

- Monitor urine for maintenance of ketosis.

- Laboratory studies typically include baseline and routine monitoring of CBC with platelets, CMP, magnesium, phosphorus, fasting lipid panel, acylcarnitine profile, urinalysis, urine calcium and creatinine, others as indicated.

Adapted from: Optimal clinical management of children receiving the ketogenic diet: Recommendations of the International Ketogenic Diet Study Group, 2008, Kossoff, Zupec-Kania, et al.

---

whereas intake within the healthy lifestyle range appeared to be protective. Healthfulness of the diet was scored based on the Alternative Healthy Eating Index (AHEI), which characterizes a healthy diet as one that is high in vegetables, fruit, nuts, soy, and cereal fiber; has a high ratio of chicken plus fish to red meat; has a high ratio of polyunsaturated fat to saturated fat; is low in *trans* fat; and includes multivitamin use for five or more years.[36] Hyperhomocysteinemia has also been linked with higher risk of stroke, particularly in late midlife.[37] In the Heart Outcomes Prevention and Evaluation 2[38] clinical trial, participants randomized to receive supplements of folic acid, vitamin B_{12}, and vitamin B_6 over five years significantly lowered their homocysteine levels and also had a lower risk of stroke compared to those receiving placebo treatment.[38] However, it remains unclear whether it is the lowering of homocysteine levels that modifies stroke risk or if homocysteine is simply a marker of vascular disease.

Risk factors for aneurysm include cigarette smoking and excessive alcohol ingestion. Other conditions that place an individual at risk for aneurysm include familial intracranial aneurysm syndrome and polycystic kidney disease.

Pathophysiology

In Chapter 9, hypoxia was introduced as a common mechanism for cell injury. An insufficient oxygen supply (or ischemia) will disrupt cellular metabolism. When circulation to a particular region of the brain ceases, as occurs during a stroke, the cells within that region die. Without a blood supply, necrosis (cell death) occurs within 4–10 minutes due to a lack of oxygen and glucose.

Clinical Manifestations

Signs and symptoms of stroke vary depending on the area of the brain that is involved, and include loss of vision or speech, and paralysis or muscle weakness. Other signs and symptoms include a change in mental status. This change can be as dramatic as the onset of coma, but may present more subtly as symptoms of confusion or changes in memory.

Medical Diagnosis

Diagnostic criteria used in acute evaluation for stroke include the National Institutes of Health Stroke Scale (NIHSS), which evaluates stroke severity through an assessment tool including characteristics such as level of consciousness, visual loss, sensory loss, language, and motor skills.[39] Brain imaging techniques, a critical tool in stroke diagnosis, are used to generate three-dimensional images of the brain or visual "slices." This allows physicians to distinguish whether the stroke was ischemic or hemorrhagic, determine the size and location of the stroke, and rule out other potential neurological conditions such as brain tumors.[34] Typical diagnostic testing for stroke includes computed tomography (CT) and magnetic resonance imaging (MRI). CT scans use X-rays from multiple angles to produce brain images and are more readily available and less expensive than MRI. CT images, such as the one in Figure 20.5, show areas of the brain lacking adequate blood flow so the location of a stroke can be pinpointed. MRI uses a magnetic field to generate a detailed image of the brain for diagnosis. Other imaging techniques can be used to specifically examine blood vessels in the brain. This can be useful for detecting the site of blockage or malformations such as aneurysms. Angiography entails injection (via a vein in the arm or leg) of contrast dye, which then travels to the brain, resulting in a map of the brain's blood vessels that shows clearly through imaging. Ultrasound uses sound waves, which are deflected from soft tissues. These echoes allow determination of density and shape of tissues, resulting in generation of an image, including blood cells and blood vessels.[34]

Treatment

Prevention of stroke through lifestyle change and use of medications to lower cholesterol and blood pressure, and thereby minimize atherosclerosis, is an important measure. Aspirin, warfarin, and other anticoagulant or antiplatelet medications can also reduce risk of ischemic stroke in those at risk.[40] However, when prevention fails, other measures must be taken quickly for optimal outcome. The longer the blood supply to the brain is impaired, the more extensive the damage. Thus,

Figure 20.5 CT Scan: Diagnosis of Stroke

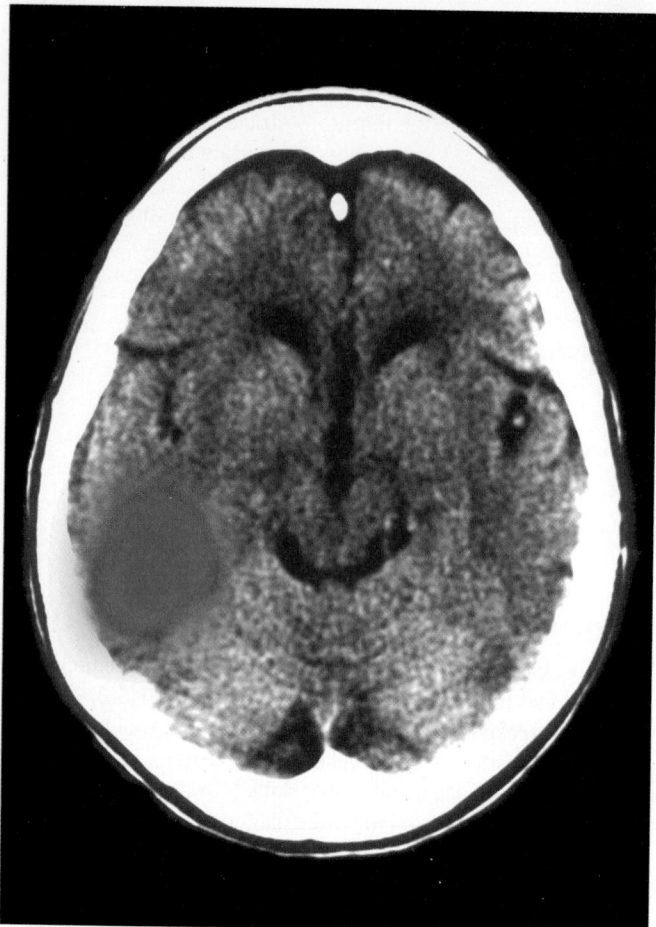

Source: http://www.strokecenter.org/pat/diagnosis/ct.htm
Courtesy of Scott Camazine/Photo Researchers Inc.

the first priority in acute care following ischemic stroke is to eliminate the clot and restore blood flow.[34]

Tissue plasminogen activator (tPA) is a protein produced by the body to dissolve clots. It has been engineered into a thrombolytic drug that has proven very successful in treating both myocardial infarctions and ischemic strokes. Research has demonstrated that patients who receive tPA within three hours of a stroke are more likely to have minimal or no disability three months after treatment as compared to patients receiving placebo.[41] Patients receiving tPA are also less likely to require rehabilitation or nursing home care.[42] It does carry a risk of intracerebral hemorrhage, though, and may not be appropriate for all patients. Research efforts are currently focused on tPA derivatives that may be less likely to cause hemorrhage and on ways to increase efficacy beyond the three-hour mark.[34] Aspirin, an antiplatelet drug, given within 48 hours of stroke onset has been shown to significantly reduce the risk of recurrent ischemic stroke at two weeks.[43] Other methods being tested to eliminate clots include ultrasound, which uses sound waves to break apart a clot, and physical clot removal through new and innovative techniques.[34] Acute treatment in hemorrhagic stroke is focused on lowering blood pressure, and studies are underway to determine the potential value of drugs that promote coagulation.[34]

Neuroprotection is another important component of stroke treatment focused not on restoring blood flow, but on protecting neurons from ongoing damage that may continue even after blood flow has been restored. The most critical damaging reactions include excitotoxicity, reperfusion injury, and damage to the blood-brain barrier. When neuronal ischemia occurs during a stroke, ATP (whose production requires oxygen) is depleted and neurons lose control over their normal signaling. They release large amounts of glutamate, an excitatory neurotransmitter, that excites nearby neurons to an excessive extent. This excitation spreads through the brain and results in degeneration of neurons and production of free radicals, a process termed excitotoxicity.[34] If blood supply is restored following stroke, the surge of oxygen uptake and ATP production in mitochondria results in production of oxygen-containing free radicals. The resulting neuronal injury is termed reperfusion injury.[34] Current research efforts in the area of neuroprotection are aimed toward minimizing these effects. Chronic therapy focuses on prevention of further stroke and rehabilitation for functional recovery. Rehabilitation goals set by the American Heart Association practice guidelines are to prevent complications, minimize impairments, and maximize function.[44] Comprehensive rehabilitation can occur in several different settings, such as an inpatient facility, nursing facility, and outpatient or in-home care setting. The program of care should utilize a multidisciplinary team involving physical therapy, occupational therapy, speech and language pathology, nutrition therapy, kinesiotherapy, and physical medicine plus consideration of psychological treatment.[34, 44]

The goal of treatment for aneurysm depends on the size and symptoms of the aneurysm as well as individual patient characteristics. Options include surgical treatment or conservative intervention intended to prevent further hemorrhage.

Nutrition Therapy for Stroke and Aneurysm

Nutritional Implications

Individuals who have experienced stroke are at varying degrees of nutritional risk, with the severity of nutritional risk and the resulting nutritional interventions depending on the area of the brain that has been affected by the stroke. For example, in the acute period after stroke, an individual may be comatose, cognitively impaired, or unable to swallow successfully. Nutrition intervention will be individualized to maximize nutritional support, whether it is administered orally, enterally, parenterally, or through a combination of routes. Additionally, the nutritional status of patients at the onset of the injury is an important consideration for a successful recovery. The FOOD (Feed or Ordinary Diet) randomized clinical trial for 3,012 stroke patients indicated that poor nutritional status was related to complications and negative outcomes post injury.[45]

Nutrition Diagnosis

Nutrition diagnoses common with stroke include inadequate energy intake, inadequate oral food/beverage intake, inadequate fluid intake, inadequate protein-energy intake, swallowing difficulty, and self-feeding difficulty.

Nutrition Intervention

As stated in the previous section, acute nutritional problems for stroke or aneurysm typically involve impairment in ability to chew, swallow, or self feed that results in inadequate nutritional intake. Enteral nutrition support could be necessary if an oral diet cannot meet nutritional needs. This would depend on the severity of the stroke and the patient's prognosis. Ethical decisions concerning metabolic support (delineated in Box 20.2) should be addressed for patients with poor prognosis. For most patients, evidence supports the early initiation of nutritional support to prevent complications, reduce hospital stay, and promote rehabilitation (see Chapter 5). Ongoing changes in the nutritional plan will certainly be necessary as rehabilitation progresses.

Dysphagia (see Chapter 14) is a common condition that accompanies stroke. The dysphagia symptoms that a patient will experience depend upon the phase of swallowing that is impaired. For example, if the problem originates in the oral preparation phase, food may be pocketed in the buccal mucosa (cheek area) because the patient cannot propel the bolus of food effectively from the front of the oral cavity to the pharyngeal area. Other general symptoms may include drooling, coughing, and choking. Many patients will experience weight loss and generalized malnutrition due to inadequate nutritional intake. Aspiration or inhalation of oropharyngeal contents into the lungs is a primary complication. This can lead to aspiration pneumonia with accompanying infection (see Chapter 21). Table 14.11 (in Chapter 14) provides a review of dysphagia symptoms with nutritional considerations.

Diagnosis and treatment of dysphagia involve different members of the health care team. Many institutions have a dysphagia team consisting of physicians, nurses, a speech-language pathologist, a registered dietitian, a physical therapist, and an occupational therapist, just as they may have a stroke rehabilitation team. Diagnosis of dysphagia begins with a bedside swallowing assessment, usually performed by the speech-language pathologist. Conclusive evaluation uses a videofluoroscopy swallowing study or fiberoptic endoscopic evaluation of swallowing. From these evaluations, the specific site of the swallowing disorder can be determined, a diagnosis can be established, and a care plan can be developed. The nutrition care plan may

Sample PES Statement: Inadequate energy intake related to swallowing difficulty as evidenced by dietary intake meeting <30% of estimated energy requirements.

BOX 20.2 CLINICAL APPLICATIONS

Ethical Decisions in Nutrition Support

1. The patient's expressed desire for extent of medical care is a primary guide for determining the level of nutrition intervention.

2. The decision to forgo hydration or nutrition should be weighed carefully because such a decision may be difficult or impossible to reverse with a period of days or weeks.

3. The expected benefits, in contrast to the potential burdens, of non-oral feeding must be evaluated by the health care team and discussed with the patient. The focus of care should include the patient's physical and psychological comfort.

4. Artificial nutrition and hydration are considered medical interventions.

5. Consider whether or not nutrition, either oral or artificial, will improve the patient's quality of life during the final stages of life.

6. Consider whether or not nutrient support, either oral or artificial, can be expected to provide the patient with emotional comfort, decreased anxiety about disease cachexia, improved self esteem with cosmetic benefits, improved interpersonal relationships, or relief from fear of abandonment.

7. If death is imminent and feeding will not alter condition consider whether or not nutrient support will be burdensome.

8. When oral intake is appropriate:
   a. Oral feeding should be advocated whenever possible. Food and control of food intake may give comfort, pleasure and sense of autonomy and dignity. The most important priority is to provide food according to the individual patient's wishes.
   b. Efforts should be made to enhance the patient's physical and emotional enjoyment of food by encouraging staff and family assistance in feeding the patient.
   c. Nutrition supplements, including commercial products and other alternatives, should be used to encourage intake and ameliorate symptoms associated with hunger, thirst, or malnutrition.
   d. The therapeutic rationale of previous diet prescriptions for an individual patient should be reevaluated. Dietary restrictions can be liberalized. Coordination of medication or medication schedules with the diet should be discussed with the physician, with the objective of maximizing food choice and intake by the patient.
   e. The patient's right to self determination must be considered in determining whether to allow the patient to consume foods that are not generally permitted within the diet prescription.
   f. Suboptimal oral feedings may be more appropriate than burdensome tube or parenteral feeding.

(continued)

9. When tube feeding or parenteral feeding is being considered:
   a. The patient's informed preference for the level of nutrition intervention is primary. The patient or substitute decision maker should be advised on how to accomplish whatever feeding the patient desires.
   b. When palliative care is the agreed goal nutritional support must be part of the palliative plan. A palliative care plan does not automatically preclude aggressive nutrition support. The decision to forgo "heroic" medical treatment does not preclude baseline nutrition support. All options for nutritional support can be considered.
   c. Feeding may not be desirable if death is expected within hours or a few days and the effects of partial dehydration or the withdrawal of nutrition support will not adversely alter patient comfort.
   d. Facilities should provide and distribute written protocols for the provision of and termination of tube feedings and parenteral feedings. The protocols should be reviewed periodically, and revised if necessary, by the health care team. Legal and ethical counsel should be routinely sought during the development and interpretation of the guidelines. The institution's ethics committee, if available, should assist in establishing and implementing defined, written guidelines for nutrition support protocol. The registered dietitian should be a contributing member of or consultant to such a committee.
   e. Conflict within the family or among stakeholders can be resolved by referring to an ethics committee or consultant if available within the institution.

   f. The potential benefits vs. burdens of tube feeding or parenteral feeding should be weighed on the basis of specific facts concerning the patient's medical and mental status, as well as on the facility's options and limitations.
   g. Facility options and limitations—one should consider the following:
      (1) Lack of staffing—no one to manage or monitor feeding
      (2) Too costly without financial help
      (3) If a feeding strategy is started in one site it will have to be stopped when the patient is transferred to another site, which can lead to a sense of abandonment.
10. Either short- or long-term parenteral nutrition should be considered only when other routes are impossible or inadequate to meet the comfort needs of the patient.
11. The physician's written diet order in the medical chart documents the decision to administer or forgo nutrition support.
    a. The registered dietitian should participate in the decision.
    b. If a decision is made that the registered dietitian does not agree with, appeal to the facility's ethics mechanism (committee or consultant) is appropriate.
    c. If the court has ordered feeding or no feeding and there is agreement with the court's decision, appeal to the facility's ethics mechanism is appropriate.

Reprinted from *Journal of the American Dietetic Association*, V108(5), O'Sullivan Maillet J., Position of the American Dietetic Association: Ethical and legal issues in nutrition, hydration, and feeding, pp. 873–882, Copyright 2008, with permission from Elsevier.

include modifying the consistency of food or liquids, positioning of the patient, or swallowing exercises.

Modifiable risk factors for patients who have had or are at risk for a stroke include the secondary medical diagnoses of hypertension, atherosclerosis, and diabetes. Each of these medical diagnoses involves nutrition therapy as an important component of medical care, as discussed in the respective chapters of this text, and should be foremost in the practitioner's plan of care.

Monitoring and Evaluation

Ongoing monitoring must take into consideration any changes in swallowing status and ability to self feed as well as adequacy of intake from an oral diet and/or nutrition support. Additionally, status of underlying disease states such as diabetes or hypertension should be monitored and updated nutrition interventions planned accordingly for prevention of complications, such as another stroke.

Progressive Neurological Disorders

The medical management of progressive neurological (neurodegenerative) disorders can be challenging. Nutritional influences among these disorders are being explored, and although they are likely diverse, some commonalities, such as the positive influence of caloric restriction, are emerging.[46] The physiological changes accompanying progression of the disease can affect both nutritional needs and nutritional status. As symptoms increase in severity, the ability to meet nutritional needs by an oral route can diminish. Nutrition therapy and individualized patient care can benefit patients affected by progressive neurological diseases by addressing complications as the disease progresses or by intervening with nutritional measures. Nutritional concerns in this population include decreased appetite and food intake, chewing and swallowing complications, and altered digestion of food. Metabolic support, such as enteral feeding, may need to be considered as these diseases progress.

Treatment decisions for these conditions often involve complex ethical concerns, such as stem cell transplant or the possibility of long-term nutrition support. Box 20.2 explores ethical decisions related to nutrition support.

The following sections address individual diagnoses including pathophysiology and medical care. Because of the similarities in nutrition diagnoses, nutrition therapy for problems common to all progressive neurological disorders is covered in Table 20.7. More specific interventions appropriate for certain conditions or their medical treatments are discussed below.

Table 20.7 Nutrition Therapy in Progressive Neurological Diseases

The nutrition diagnoses below are common to the following progressive neurological disorders: amyotrophic lateral sclerosis (ALS); Parkinson's disease (PD); myasthenia gravis (MG); Guillain Barré syndrome (GB); multiple sclerosis (MS); and Alzheimer's disease (AD).

Nutrition Diagnosis	Nutrition Interventions	Nutrition Monitoring/Evaluation
Inability to maintain adequate oral intake	• Determine energy and protein requirements. • Establish food preferences. • Increase nutrient density. • Adjust size and timing of meals. • Maximize environmental support: lights, noise, odor. • Allow for extended eating times. • Trial of high-kcalorie, high-protein supplement. • Determine need for other nutrition support routes.	• Monitor body weight and weight trends • Monitor energy and protein intakes • Monitor appetite/satiety • Monitor changes in food preferences • Assess whether food texture changes may be needed • Monitor need for feeding assistance • Monitor labs for visceral protein status • Determine if nutrition support is needed or is appropriate • Assess medication changes • Assess whether adaptive equipment is needed
Involuntary weight loss	• Determine energy and protein requirements. • Establish food preferences. • Increase nutrient density. • Adjust size and timing of meals. • Allow for extended eating times. • Trial of high-kcalorie, high-protein supplement. • Determine need for other nutrition support routes.	• Monitor body weight and weight trends • Monitor energy and protein intakes • Monitor appetite/satiety • Monitor changes in food preferences • Assess whether food texture changes may be needed • Monitor need for feeding assistance • Monitor labs for visceral protein status • Determine if nutrition support is needed or is appropriate • Assess medication changes • Monitor bowel function (diarrhea, constipation) • Monitor hydration status • Assess whether adaptive equipment is needed
Chewing difficulty	• Physical assessment of oral cavity—need for mouth care or dentures. • Adjust texture of foods according to need. • Provide adequate moisture. • Adaptive equipment as needed.	• Monitor body weight and weight trends • Assess whether food texture changes may be needed • Assess oral cavity changes—need for mouth care or dentures • Assess whether moisture is adequate and whether adaptive equipment is needed
Swallowing difficulty	• Seated in armchair, table appropriate height. • Muscular support. • Swallowing evaluation. • Establish level for dysphagia nutrition therapy if required.	• Monitor body weight and weight trends • Monitor energy and protein intakes • Monitor changes in food preferences • Assess whether food texture changes may be needed • Assess whether moisture is adequate and whether adaptive equipment is needed • Determine if nutrition support is needed or is appropriate
Impaired ability to prepare foods/meals	• Assess current dietary intake. • Establish current cooking facilities and access to food. • Referral for other sources of daily meals.	• Monitor body weight and weight trends • Assess medication changes • Assess whether referral for other sources should be made for daily meals, living placement, etc.
Inadequate fluid intake	• Determine fluid requirements and establish method to track intake. • Assess side effects of medications that might interfere with adequate fluid intake. • Schedule small, frequent amounts of fluid. • Determine need for other nutrition support/hydration routes.	• Monitor body weight and weight trends • Monitor appetite/satiety • Determine if nutrition support is needed or is appropriate • Assess medication changes • Monitor bowel function (diarrhea, constipation) • Monitor hydration status • Assess any medication changes
Drug-nutrient interaction	• Complete assessment of current medications. • Adjust timing of medications and meals to allow for absorption and effectiveness of both medications and foods. • Limit specific foods that interfere with drug absorption. • Coordinate drug ingestion to prevent nutrient malabsorption.	• Monitor body weight and weight trends • Monitor appetite/satiety • Assess any medication changes • Assess any changes in times of dosing of medications in relation to meals

Parkinson's Disease

Definition **Parkinson's disease** was first described in 1817 in James Parkinson's publication, "Essay on the Shaking Palsy."[47] Although the exact cause of Parkinson's disease is unclear, this neurodegenerative disease results in the loss of dopamine-producing cells in the substantia nigra portion of the brain.

Epidemiology Parkinson's disease is one of the most common neurological disorders, affecting approximately 1% of the population of the United States and Canada over the age of 70.[48] Although present throughout the world, Parkinson's disease is less common in Asian and African American populations. The incidence of Parkinson's disease increases with age—it is rarely diagnosed before age 40—and it is more commonly diagnosed in men.

Etiology The etiology of Parkinson's disease remains unclear, but proposed contributing factors include genetics, environmental toxins, prions (protein malformations), oxidative stress, imbalance of neurotransmitters leading to excitotoxicity, mitochondrial dysfunction, loss of ability to reproduce dopamine-producing cells (including the loss that occurs with aging), abnormal inflammatory response within certain cells of the brain, improper response to dopamine within the brain, and increased rate of cell death. Twin and family studies indicate that genetics may certainly contribute to early-onset Parkinson's; specifically, several genes associated with Parkinson's have been traced to chromosome 4.[49]

Pathophysiology Parkinson's disease involves abnormalities of cells within the substantia nigra, a portion of the basal ganglia deep within the cerebral hemispheres beneath the cerebral cortex. (See the opening section of this chapter and Figure 20.3 for a review of neurological anatomy.) The neurons within the substantia nigra produce the neurotransmitter dopamine. A balance of excitatory (dopamine) and inhibitory (GABA) neurotransmitters normally maintains slow, coordinated movement, muscle tone, and posture.[1] In Parkinson's disease, however, there is a progressive loss of dopamine, which causes an imbalance between excitatory and inhibitory communication. The loss of dopamine and the resulting imbalance of neurotransmitters appear to cause the myriad of symptoms experienced.

Clinical Manifestations Classic motor symptoms of Parkinson's disease include resting tremor, rigidity, **bradykinesia** (slowed movement) or **akinesia** (loss of movement), stooped posture, postural instability, mask-like facial features, and a shuffling gait. Other symptoms, such as depression, anxiety, sleep disturbances, sensory abnormalities, and pain are sometimes present, often prior to the motor symptoms. Cognitive dysfunction may include problems with memory, inability to complete complex tasks, and inability to retrieve new information, as well as the development of Parkinson's disease dementia.[50]

Medical Diagnosis Diagnosis of Parkinson's disease is difficult. There is no definitive diagnostic test; instead, evaluation of symptoms and response to treatment allows a diagnosis to be made. Diagnosis may include presence of tremor, rigidity, and bradykinesia, loss of postural reflexes, and clinical response to treatment with L-dopa.[50] The TRAP mnemonic is sometimes used as a helpful screening tool: T – Tremors; R – Rigidity; A – Akinesia; P – Postural instability.[51]

Tools for staging of Parkinson's disease has historically included the Hoehn and Yahr Staging Process, but most recently, in-depth staging has included the Unified Parkinson's Disease Rating Scale (UPDRS).[52, 53] The UPDRS assesses more detail for cognition, behavior, mood, activities of daily living, and motor skills.

Treatment Since the symptoms of Parkinson's disease are caused by an inadequate generation of dopamine in the substantia nigra portion of the brain, treatment for Parkinson's disease is centered on medications that will improve the supply of dopamine to the brain. Levodopa, first developed in the 1970s, is the precursor of dopamine. Other medications include dopamine agonists and MAO inhibitors, which may offer neuroprotection. Treatment options include the following:

- **Levodopa (L-dopa).** This medication provides the precursor for dopamine to compensate for its diminished availability in Parkinson's disease.

- **Decarboxylase inhibitors.** These medications include carbidopa and benserazide and they act to reduce the breakdown of L-dopa outside the blood brain barrier, resulting in an improvement in the action of L-dopa.

- **Dopamine agonists.** These medications bind to and activate dopamine receptors in the absence of dopamine; examples include ropinirole (Requip®) and pramipexole (Mirapex®).[54]

- **COMT inhibitors.** Catechol-O-methyltransferase is one of two enzymes necessary for the metabolism of L-dopa. When this enzyme's action is blocked, the overall amount of L-dopa remains higher. COMT inhibitors include entacapone (COMTan®) and tolcapone (Tasmar®). Tolcapone is rarely used due to severe hepatotoxicity.

- **Combination medications.** These medication options include both L-dopa and a decarboxylase inhibitor; Sinemet is the most common combination drug. Another common combination medication includes L-dopa, a decarboxylase inhibitor, and a COMT inhibitor. A combination medication containing L-dopa, carbidopa, and entacapone is marketed as Stalevo®. These combinations allow for reduced dosages of L-dopa with fewer side effects.

- **Anticholinergic medications.** These medications (Trihexyphenidyl and Benztropine) affect dopamine release.[55]

- **Neuroprotective therapies.** These interventions include the medication selegiline (Eldepryl®, Zelapar®, and Anapryl®) and the use of vitamin E. Neuroprotective interventions are designed with the goal of preventing or slowing the progression of disease. This area of treatment is the newest and most controversial.

- **Surgery.** Procedures include deep brain stimulation, pallidotomy, and thalamotomy. Deep brain stimulation involves the implantation of a battery device that provides electrical stimulation to the brain, much like a pacemaker used to treat heart arrhythmias.[56] Pallidotomy destroys the globus pallidus internus to help control tremor symptoms. Likewise, thalamotomy attempts to control tremor symptoms by destroying particular cells in the thalamus.

- **Diet.** Ketogenic diets are now being investigated for multiple neurodegenerative diseases, including Parkinson's disease. These diets are thought to be neuro-protective based on some preliminary data that show relief of Parkinson's disease symptoms;[57, 58] however, more research is needed in this area before these diets are considered for this disease state.

Nutrition Therapy for Parkinson's Disease

Nutritional Implications Parkinson's disease, like any neurodegenerative disease, can significantly impair an individual's nutritional status. Nutritional problems include drug-nutrient interactions, weight loss, inadequate intake (including protein), dehydration, decrease in self-feeding abilities, and consumption and motility dysfunction (chewing, swallowing, and constipation).[59]

Drug-nutrient interactions are important issues for both clients and clinicians' pharmacotherapy management. Chapter 11 indicates that nutritional intake can interfere with drug absorption, metabolism, and excretion. Additionally, drugs may cause a change in overall nutritional intake due to drug side effects, or they may interfere with the normal digestion and absorption of specific nutrients. This is the case for many of the drugs used to treat Parkinson's disease. Table 20.8 summarizes the most important concerns for drug-nutrient interactions.

When evaluating drug-nutrient interactions, it is important to keep in mind that a major goal in Parkinson's disease is to maximize absorption of the medication while minimizing the amount of time without adequate levels. The literature describes this in terms of "on-off fluctuations." "On" refers to the time periods in which medication level, and hence patient functioning, are optimal. Patients will need to be aware of any factor that might slow down or interrupt absorption, such as consuming medications with meals.

Nutrition Diagnosis Nutrition diagnoses most likely to be appropriate for individuals with Parkinson's disease may include inadequate or excessive energy intake, inadequate or excessive oral food/beverage intake, increased nutrient needs, evident protein-energy malnutrition, inadequate protein-energy intake, malnutrition, inadequate vitamin or mineral intake, food-medication interaction, underweight, involuntary weight loss, overweight/obesity, involuntary weight gain, masticatory and swallowing difficulty, inability to manage self care, impaired ability to prepare foods/meals, poor nutrition quality of life, and self-feeding difficulty.

Sample PES Statement: Self-feeding difficulty related to diagnosis of Parkinson's disease with increasing tremor and fatigue as evidenced by observation at mealtime, calorie count, and recent 5% weight loss.

Nutrition Intervention Two of the most important drug-nutrient interactions involve the medication Levodopa (L-dopa). L-dopa is taken orally and transported from the small intestine to the blood brain barrier; at both of these sites, protein carriers transport the medication. Because amino acids also utilize these same carriers, they compete with L-dopa for transport. Therefore, it has been proposed that high protein intake interferes with optimal levels of L-dopa.[60, 61] Manipulation of protein intake may promote increased therapeutic effects of the medication.[62] Unfortunately, evidence indicates that results are not consistent in every patient. The specific method of prescribing and scheduling protein in the diet has varied throughout the literature, but it is generally understood that the benefit of protein manipulation should be apparent within 7–10 days.[63] One strategy is to limit overall protein intake to 0.5–1 g protein/kg with an even distribution of protein throughout all meals. Though this should meet overall protein needs, it is generally less than most patients consume and results in an overall reduction in dietary protein. The second strategy limits protein intake during the waking hours with increased amounts of protein at the evening meal and before bed. This strategy may improve drug effectiveness when the drug is most needed by the patient. Patients receiving enteral feeding should be placed on a bolus feeding schedule as opposed to continuous feeding, as L-dopa dosing can then be administered between the bolus feeds, thus increasing the likelihood of achieving optimal therapeutic levels. If bolus feeds are not possible, the patient's protein intake should be kept to the minimum needed to meet his or her needs.[64]

Pyridoxine (vitamin B_6) is required as a cofactor for dopa decarboxylase (DDC)—one of the enzymes needed for the conversion of L-dopa to dopamine.[65] Vitamin supplements and foods

Table 20.8 Selected Medications Used in Treatment of Parkinson's Disease

Drug to Treat Parkinson's	Generic	Brand Names	Possible Food–Drug Interactions
Dopamine Precursor	Levodopa	Atamet, Lodosyn, Sinemet, Larodopa	Nausea, vomiting, loss of appetite, change in sense of taste, fatigue
Dopamine Agonist	Ropinirole (Requip) and Pramipexole (Mirapex)	Requip, Mirapex	Nausea, heartburn, vomiting, constipation, frequent urination, dry mouth, excessive tiredness
Decarboxylase Inhibitor	Carbidopa	Atamet, Sinemet, Larodopa	Abdominal pain, dryness of mouth, loss of appetite, passing gas, constipation, diarrhea
Catechol-O-Methyltransferase inhibitor	Entacapone (Comtan®) and Tolcapone (Tasmar®)	Comtan®, Tasmar®	Abdominal pain, constipation, diarrhea, fatigue, nausea, GERD, belching
Anticholinergic Medications	Trihexyphenidyl and Benztropine	Akineton, Artane, Artane Sequels, Cogentin, Kemadrin, Parsidol, Trihexane, Trihexy	Constipation; dryness of mouth, nose, or throat; nausea or vomiting
Combination Medications	Sinemet	Atamet, Sinemet, Larodopa	Abdominal pain, dryness of mouth, loss of appetite, passing gas, constipation, diarrhea, hiccups

high in pyridoxine may expedite this conversion before the L-dopa reaches the brain and reduce the amount of dopamine actually transported to the brain.[66] Since most individuals take a combination medication (such as Sinemet), the interaction does not pose as significant a risk as it did when Levodopa was the only available medication. It has been noted that some individuals are more sensitive to B_6 than others, and supplement levels of greater than 15 mg are not recommended.

Decreased intake and unintentional weight loss are common nutrition problems for many individuals with Parkinson's disease.[67] Tremors and fatigue can contribute to decreased intake, as greater effort is required to physically place the food in the mouth. This extra effort can increase fatigue at mealtimes, resulting in less intake. Weight loss can also occur despite maintaining consistent baseline caloric intake. Nutrition interventions should address the possible need for increased energy intake and close monitoring of weight status. Individuals experiencing eating difficulties could possibly benefit from protein modifications to decrease tremors (as previously described). Assistance with eating and extended mealtimes could increase intake. The timing of meals may need to be adjusted to periods in which the individual is well rested.

Nutritional status may be affected by numerous gastrointestinal symptoms that occur with Parkinson's disease.[68] Table 20.7 addresses these issues in more detail. Muscular rigidity, tremor, and bradykinesia can affect the stages of swallowing and increase the risk for aspiration. The muscular dysfunction and nervous system abnormalities could lead to **gastroparesis** (delayed emptying of the stomach). Nutrition intervention for gastroparesis includes encouraging small, frequent, low-fat, low-fiber meals, as fat and fiber delay gastric emptying.[69] Enteral feeding past the stomach to the jejunum could be indicated if the gastroparesis is persistent and intake is inadequate.[70] Further complications such as gastroesophageal reflux disease (GERD), early satiety, nausea, and vomiting are possible. These same abnormalities often cause slowed transit time throughout the small intestine and colon. The goals of nutrition therapy for GERD are discussed in Chapter 14 and focus on reducing gastric acidity with restriction of black and red pepper, coffee (both caffeinated and decaffeinated), and alcohol and encouraging smaller, more frequent meals. Because constipation and difficult defecation are common issues, encouraging adequate fiber and fluid intake is important for this population; yet, increasing fiber could be challenging in the presence of gastroparesis. The changes within the GI tract can also affect absorption of drugs, setting up a cycle of ineffective control of the disease and complication of symptoms.[59] As Parkinson's disease progresses, patients may require assistance with preparing and eating foods. Food modifications may be needed to help with mastication and swallowing concerns.

Monitoring and Evaluation Inadequate oral intake may result from anorexia related to several contributing factors, including those addressed above. Gastroparesis and constipation can cause the sensation of fullness and nausea; these symptoms contribute to early satiety and loss of appetite, thus decreasing nutritional intake. The medications prescribed for treatment may cause anorexia, nausea, and vomiting. Physical and cognitive impairments, fatigue, and depression all potentially contribute to the inability to maintain adequate oral

intake and increase the risk for dehydration as well.[63, 71] As the disease progresses, textures may be modified due to swallowing or chewing difficulties and alternate routes of nutrition support such as enteral feeding may be considered. (See Chapter 14 for a discussion of texture and consistency modification, and Chapter 5 for nutrition support guidelines.)

Amyotrophic Lateral Sclerosis

Definition **Amyotrophic lateral sclerosis** (ALS), commonly known as Lou Gehrig's disease (named for the 1920–30s baseball legend diagnosed with ALS), is a progressive neurological disease that affects the motor neurons of the nervous system. These motor neurons control the voluntary movements in the body, such as the movement of extremities, swallowing, talking, and even breathing (a person has the ability to voluntarily control some movements such as holding their breath). ALS does not affect involuntary types of neurons, such as those involved in memory, reasoning, vision, and hearing, as well as muscular processes such as heartbeat and digestion. This disease has no cure available and a poor prognosis: progressive decrease in function and eventually death within 2–5 years from diagnosis.[72]

Epidemiology The prevalence of ALS is approximately 2–3 cases per 100,000 individuals per year with an estimation of 1 in 400 by the age of 70 years.[72] This disease has been slightly more prevalent in men, yet recent data suggest that the gender gap may be closing.[72]

Etiology The cause of ALS is generally unknown except in familial ALS, with research indicating a genetic link in at least 10% of cases.[73] This condition is likely multifactorial, involving genetic, biological, and environmental aspects. Diagnosis can be delayed by up to one year from onset of the disease due to slow presenting symptoms.[74] For example, it requires destruction of one third of the motor neurons before atrophy begins and symptoms are noticeable. With no cure available, current treatment has focused on controlling symptoms. Abnormalities with the neurotransmitter glutamate have been linked to the pathology of ALS, as anti-glutamatergic agents, such as riluzole, have shown some clinical efficacy.[73]

Pathophysiology This progressive neurological disease is characterized by destruction of neurons involved in controlling stimulation of the muscles. Previously, in Chapter 9, it was stated that cellular response to injury can involve replacement of normal cells with nonfunctional cells. This occurs in ALS, where dead motor neurons are replaced with nonfunctional fibrous cells. As the communication between the CNS and PNS decreases, the components of the peripheral neurons deteriorate. Finally, without the ability for neurons to stimulate muscle action, muscles begin to **atrophy**. Two distinct forms of the disease exist: bulbar and spinal. Initial presentation is due to the differences in where the neuron deterioration begins.

Clinical Manifestations Classic signs and symptoms of ALS include **asymmetric muscle weakness** and atrophy, **hyperreflexia** (hyper-stiffening of the muscles), and **fasciculations** (uncontrolled twitching of the muscles). The manner in which these signs and symptoms manifest in the patient depend upon the area of neuron destruction. There may be difficulty with gross motor actions such as walking, or in fine motor actions such as grasping an object. Swallowing and chewing skills decline as the neurons controlling these skills weaken. As the disease progresses, the extensive muscle atrophy leads to increasing paralysis, requiring mechanical ventilation and nutrition support.

Medical Diagnosis Diagnosing ALS involves a complete neurological examination that allows the exclusion of other diseases, because no single diagnostic test exists. Tests utilized may include an electromyography (EMG), or a nerve conduction velocity test (NCV), imaging studies such as magnetic resonance imaging (MRI), and other testing such as muscle biopsy to rule out other "ALS mimicking" disorders.[72]

Treatment There is only one current medicinal treatment for ALS: Riluzole®, which appears to affect the release of the neurotransmitter glutamate, although the exact mechanism of action is not clear.[75] This treatment has shown some efficacy, extending life on average 2–3 months.[75] Other supportive treatments such as non-invasive respiratory ventilation, early percutaneous endoscopic gastrostomy (PEG) placement, and multidisciplinary care in ALS clinics have shown some promise in prolonging survival and increasing quality of life.[76] Other symptom relief medications are often used to help control excessive saliva or secretions, muscle spasms, depression/anxiety, and pain.

Guillain-Barré Syndrome

Definition Guillain-Barré syndrome (GBS) is defined as an acute peripheral nervous system syndrome characterized by progressive paralysis. The effects of this disease, after a relatively rapid onset, can eventually resolve completely or leave lasting impairment that could result in death. Although this syndrome is not completely understood, an autoimmune response is known to be involved.[77]

Epidemiology Incidence is approximately 1–2 cases per 100,000. Risk increases with age and men are 1–5 times more likely to be affected than women.[78]

Etiology GBS appears to be an autoimmune response to an infectious trigger. Two-thirds of patients report a history of infection in the period prior to the onset of GBS.[79] It is not typically associated with other autoimmune disorders, and occurs in otherwise healthy people.[78] In cases with preceding infections, the predominant types of infection appear to be upper respiratory infections and gastroenteritis including *Campylobacter jejuni* and *Cytomegalovirus*.[79]

Pathophysiology The onset of GBS occurs anywhere from a few hours to 3–4 weeks after the preceding infection.[77] Several variants of the syndrome exist, including acute inflammatory demyelinating polyneuropathy (AIDP), acute motor axonal neuropathy (AMAN), acute motor sensory axonal neuropathy (AMSAN), and Miller Fisher syndrome. In all forms, it appears that autoantibodies attack specific cells of the nervous system after infectious exposure. This results in damage to the myelin sheath, axons, sensory nerves, and nerve roots (beginning of the nerve leaving the central nervous system), depending on the variant of the disease.[80]

Clinical Manifestations Symptoms include a rapidly progressive paralysis that can involve all limbs. This sometimes starts with numbness in the extremities and progresses to the major clinical manifestation of muscular weakness. Most patients will eventually recover without permanent deficits, yet some will develop total motor paralysis that can lead to death. These extreme cases involve the cranial nerves, resulting in severe dysphagia and respiratory failure. Approximately 30% of patients with GBS require mechanical ventilation.[79]

Medical Diagnosis Diagnosing GBS typically begins with a recent medical history (including recent illnesses). A nerve conduction velocity test (NCV) is often performed to determine whether nerve signaling is impaired. Assessing protein content in spinal fluid, via a spinal tap, is often part of the diagnostic process, as patients with GBS typically will have an increased protein level in the cerebrospinal fluid.[77]

Treatment Treatment includes the use of high-dose intravenous immunoglobulin, an antibody produced by the B lymphocytes. **Plasmapheresis**, which is a procedure that removes the antibodies from the plasma, is also used effectively. Eight-five percent of patients with GBS will completely recover within 18 months. The remaining 15% may have longer lasting or permanent neurological deficiencies.[80] Complications could arise from respiratory failure or infection.

Myasthenia Gravis

Definition Myasthenia gravis is an autoimmune neuromuscular disease that affects the skeletal (voluntary) muscles and results in varying levels of muscle weakness and fatigue.

Epidemiology Myasthenia gravis is one of the more common neuromuscular disorders, with an incidence of approximately 1 in 7,500 individuals.[81] It appears to be more prevalent in women prior to age 60, yet among the population aged 60 and older, it is more common in men.[82]

Etiology With the exception of the paraneoplastic form of the disease, which is caused by a tumor (thymoma) and affects only 10%–15% of patients, the exact cause of myasthenia gravis is unknown.[83] In myasthenia gravis, an autoimmune reaction damages or destroys the cellular receptors for **acetylcholine**, a major neurotransmitter of the autonomic nervous system.

Pathophysiology As discussed under anatomy and physiology of the nervous system, acetylcholine is an excitatory neurotransmitter. Normally, acetylcholine receptors allow for stimulation of the muscle for contraction by acetylcholine. When there are reduced numbers of these receptors, the muscle will tire easily and the individual will experience muscle weakness that may improve with rest. The disease may have periods of remission and exacerbation which will vary between patients.[81, 84] Fifty to eighty-five percent of these patients have anti-acetylcholine receptor (anti-AChR) present.[85] Because the thymus gland is abnormal in many individuals with this disorder, it is thought that this gland may be responsible for the autoantibody production.[86] The cause of the autoimmune response remains unknown.

Clinical Manifestations Signs and symptoms include fluctuating levels of skeletal muscle weakness that is exacerbated by physical activity. The most common muscles that are affected include muscles of the face, eyes, arms, and legs. Some individuals have drooping eyelids and double vision.

Medical Diagnosis Diagnostic tests include the Tensilon test, repetitive nerve stimulation test, and assessment for anti-acetylcholine receptor (anti-AChR) antibodies. The Tensilon test requires injection of edrophonium (Tensilon), a drug that inhibits the enzyme anticholinesterase and allows acetylcholine to accumulate in the muscles, resulting in improved muscle functioning. This rapid-acting test can be given at bedside. Repetitive nerve stimulation tests measure the response of a single nerve to a series of electrical stimulations. The muscle's potential action is measured by electrodes and is documented. The third test measures antibodies against acetylcholine receptors in the blood. A positive test for anti-AChR antibodies is almost always diagnostic for myasthenia gravis, but a negative result does not rule out the disease.[87]

Treatment Currently, there is no cure for myasthenia gravis, but effective treatment is available. Medications include anticholinesterase drugs such as Neostigmine and Pyridostigmine as well as immunosuppressive drugs such as glucocorticoids, azathioprine, or intravenous immunoglobulin. Other treatments, such as stem cell transplant, have been used to treat autoimmune diseases.[81] As mentioned previously, the thymus gland is often abnormal in those with myasthenia gravis, and the removal of the gland (thymectomy) has had positive results in these patients.[83] Finally, plasmapheresis can be used to remove antibodies from the plasma for symptomatic relief.[84]

Multiple Sclerosis

Definition Multiple sclerosis (MS) is also thought to be an autoimmune disease and is characterized by destruction of the myelin sheath and the formation of scar tissue.[88] This destruction results in the distortion or interruption of nerve impulses and, consequently, symptoms such as numbness or tingling in the limbs, loss of coordination, weakness, paralysis, or loss of vision. There are four distinct types of MS, which are classified as relapsing-remitting MS, primary-progressive MS, secondary-progressive stage MS, and progressive-relapsing MS. Relapsing-remitting MS is the most common and is characterized by alternating periods of exacerbation and remission of symptoms without progressive worsening of symptoms. Approximately half of individuals diagnosed with relapsing-remitting MS progress to secondary-progressive MS within 10 years, meaning the severity of symptoms begins to progressively worsen.

Epidemiology An estimated 2.5 million people worldwide, including 400,000 people in the United States, have MS.[88, 89] Most are diagnosed between the ages of 20 and 50 years. MS is more common in women than in men;[88, 90] a recent review estimates the lifetime risk for MS to be 2.5% among women compared to a 1.4% lifetime risk among men.[90] While MS is most prevalent in Caucasians, the disease does occur in other ethnic groups as well. Worldwide, MS has reportedly been more common further from the equator, although recent reports suggest this trend may have been attenuated in recent years, with more people being diagnosed in locales nearer to the equator.[88, 90, 91]

Etiology The exact cause of MS has not been established. The major areas of study regarding etiology include both genetics and environmental factors.[88, 92] Family history is the strongest known risk factor for MS, with studies of twins and adopted children indicating that genetics are a key factor,[93] but environment also plays a critical role.[92] Initial work to identify genes associated with MS risk suggests that the human leukocyte antigen (HLA) region on chromosome 6 is highly linked to MS.[94] This region contains a large number of genes involved in immune system function (see Chapter 9). Variations in genes encoding interleukin receptors *IL7R* and *IL2RA,* both of which are involved in the immune response, have also been recently linked to MS risk.[94]

Geographic latitude, as mentioned above, has long been correlated with prevalence of MS. Sunlight exposure is an environmental variable that varies greatly by latitude and several studies have supported a relationship between sunlight exposure and lower MS risk.[93] Endogenous synthesis upon exposure to sunlight is a primary source of vitamin D for many people, and circulating levels of 25-hydroxyvitamin D$_3$ (the main circulating form) correspond closely to latitude. Additionally, dietary intake of vitamin D from food and supplements corresponded to higher circulating levels of 25-hydroxyvitamin D$_3$ and lower risk of MS in the Nurses' Health Study.[95] Thus, some have proposed that lower vitamin D levels could be involved in the development of MS. The vitamin D hypothesis is plausible, as vitamin D plays an important regulatory role in the immune system by dampening the autoimmune response.[92] However, no large prospective randomized trial has yet established whether or not vitamin D supplements are able to reduce risk of MS.[93] Smoking is another environmental factor conferring an estimated 70% higher risk of developing MS in heavy smokers compared to non-smokers. The mechanisms are not yet clear, but research suggests a direct effect on the immune system and a role for elevated levels of nitric oxide. It is estimated that up to 6% of MS cases in the U.S. could be prevented if smoking was eliminated.[93]

The hygiene hypothesis suggests that exposure to multiple infectious agents during early childhood aids in proper development of the immune system and reduction in risk of autoimmune diseases such as type 1 diabetes, systemic lupus erythematosus, and MS.[92, 93, 96, 97] Areas of high MS risk are characterized by delayed exposure to common viruses and intestinal parasites, which is in sharp contrast to areas of low MS risk.[93] One infectious agent that has undergone particular scrutiny is the Epstein-Barr virus (EBV).[96] Almost all children in developing countries are exposed at an early age, whereas in developed countries such as the United States the prevalence among young adults may vary regionally from 50% to 80%. The latitude gradient of EBV infection mirrors that of MS.[93] Although it appears that EBV infection at a later age is an important risk factor for MS (or marker of a more hygienic environment), it likely interacts with other factors to determine overall MS risk. It is also possible that different strains of EBV confer different levels of MS risk.[93] At this time, there is not an effective prevention strategy for MS based on the hygiene hypothesis.[93]

Pathophysiology As explained previously, the myelin sheath covering the nerves ensures rapid, consistent, nonrandom communication. In MS, destruction of the myelin sheath, called demyelination, occurs. Furthermore, it has been found that the axon—the long, thin fiber that transmits information to other neurons—can be damaged as well. If both of these components are damaged or destroyed, the communication between neurons is reduced or lost.

Clinical Manifestations

Any nerve can be affected in MS; therefore, symptoms vary widely. Numbness, tingling (paresthesia), uncoordination (ataxia), and weakness are all common symptoms. Some individuals experience visual problems from optic neuritis such as double vision, blurred vision, or blindness. Other symptoms can include difficulty swallowing, constipation, and bladder dysfunction. As mentioned above, rates of disease progression vary. Some patients experience a rapid progression, while others have periods of remission and relapse. Though most individuals with MS are able to function well and maintain a normal lifestyle, a small percentage progress to a severe, debilitating form of the disease.

There is no single diagnostic test for MS. As the symptoms can be highly variable, definitive diagnosis is not always possible and remains elusive for about 10%–15% of affected individuals.[89] The McDonald Criteria for diagnosis of MS were established in 2001 and most recently revised in 2005. These criteria include clinical evaluation of the patient, MRI, and analysis of cerebrospinal fluid[98] (see Figure 20.6 for MRI results indicating the sclerotic plaques seen in MS).

Treatment There is no cure for MS, but current treatment involving a variety of disease-modifying medications is able to reduce the frequency and severity of relapses, slow lesion development, and overall slow the course of MS.[88] These medications are effective for treatment of relapsing forms of MS, but none are currently approved for treatment of primary progressive MS. They include interferons (Avonex®, Betaseron®, Rebif®), mitoxantrone (Novantrone®), glatiramer acetate (Copaxone®), and natalizumab (Tysabri®). All play a role in modifying immune system function. These medications can be expensive, but a number of patient assistance programs are available to provide financial support for access to the disease-modifying drugs.[99] Other medications used in MS target specific symptoms such as muscle spasms, depression, pain, fatigue, or bladder or bowel dysfunction.[81] Additionally, corticosteroids are often used to minimize inflammation during severe exacerbations.[99] Drug-nutrient interactions can be common. Table 20.9 summarizes the common medications used in the treatment of MS.

Nutrition Therapy for Multiple Sclerosis

Nutritional Implications Nutrition implications are numerous with MS and range from the use of nutrition as an adjunct therapy to the use of nutrition therapy to address individual nutritional problems that may develop and change along with disease progression and relapse. These may include

Figure 20.6 MRI Results: Sclerotic Plaques in MS

The sclerotic plaques of muliple sclerosis as shown by MRI. The pale orange regions indicate areas of plaque.

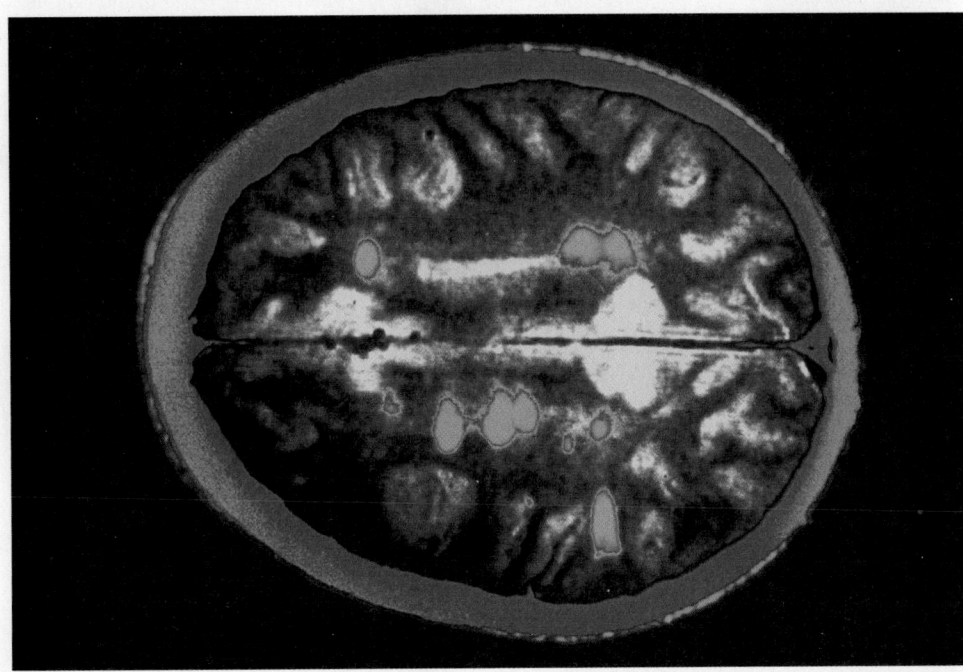

Source: Copyright © Leonard Lessin/Peter Arnold, Inc.

medication side effects such as nausea, vomiting, or dry mouth; self-prescribed fluid restriction due to impaired bladder control; challenges in meal preparation due to fatigue and muscle weakness; need for consistency modification of foods due to dysphagia; weight gain due to physical inactivity; and other issues which vary greatly from individual to individual and by stage of disease. Furthermore, due to the consistent interest in nutrition as an etiological factor, patients often seek complementary and alternative therapies[100] (see the CAM feature at the end of this chapter).

Nutrition Diagnosis Nutrition diagnoses most likely to occur in individuals with MS include inadequate or excessive energy intake, inadequate or excessive oral food/beverage intake, evident protein-energy malnutrition, inadequate protein-energy intake, inadequate vitamin intake (vitamin D if on regular courses of corticosteroids), inadequate mineral intake (calcium, if on regular courses of corticosteroids), swallowing difficulty, chewing (masticatory) difficulty, food-medication interaction, underweight, involuntary weight loss, overweight/obesity, involuntary weight gain, physical inactivity, impaired ability to prepare foods/meals, and self-feeding difficulty.

Nutrition Intervention Nutrition interventions employed as adjunct medical therapies include supplementation with omega-3 fatty acids and restriction of saturated fat.[101, 102] Historically, nutrition therapy for MS began with the Swank diet, which was developed in 1948 and restricted saturated fat while supplementing omega-3 fatty acids in 144 MS patients over a 34-year period. Though anecdotally the symptoms of MS appeared to be reduced by this diet, these studies were not experimentally controlled and have not been replicated.[102, 103] Overall, studies published to date have not definitively

Table 20.9 Medications Used in Treatment of Multiple Sclerosis

Drugs Used in Treatment of Multiple Sclerosis	Rationale	Generic	Brand Names	Possible Side Effects
Interferons IFNβ-1a IFNβ-1b	Prevent autoimmune dysfunction and inflammatory damage.	Interferon beta 1a or 1b	Avonex, Betaseron, Rebif	Immunosuppression, muscle pain, fever, fatigue, headaches, chills, nausea, and vomiting, liver abnormalities, depression, suicidal ideation, and new or worsening other psychiatric disorders; contraindicated during pregnancy.
Mitoxantrone	Antineoplastic; suppresses activity of T cells, B cells, and macrophages. Demonstrated to reduce relapses and slow progression. Not FDA approved for primary-progressive MS.	Mitoxantrone	Novantrone	Immunosuppression, sterility, birth defects, nausea, vomiting, black tarry stools, abdominal pain, decreased appetite, mouth sores, fatigue, headache, painful urination, alopecia. Use contraindicated during breastfeeding.
Natalizumab	Monoclonal antibody; inhibits movement of damaging immune cells across the blood-brain barrier; recommended only after failure or poor tolerance of other MS medications due to risk of PML.	Natalizumab	Tysabri	Carries risk of progressive multifocal leukoencephalopathy (PML), liver abnormalities. Contraindicated during pregnancy and with other immunosuppressing drugs.
Glatiramer Acetate	Protein based on the amino acid sequence of myelin used to prevent autoimmune dysfunction and inflammatory damage.	Glatiramer acetate	Copaxone	Nausea, vomiting, decreased appetite, weight gain; contraindicated during pregnancy.
Corticosteroids	Anti-inflammatory; used during severe exacerbations.	Methylprednisolone	Duralone, Medipred, Medralone, Medrol, Predacorten, Solu-Medrol	Insomnia, muscle weakness, joint pain, nausea, vomiting, immunosuppression; hyperglycemia, increased risk of osteoporosis, muscle wasting, weight gain.
Oxybutynin	Antispasmodic—reduce urinary frequency.	Oxybutynin	Ditropan XL	Decreased secretions of GI tract—dry mouth, throat; nausea, constipation, and abdominal discomfort; taking with food may decrease nausea.
Metoclopramide	Treat gastroparesis.	Metoclopramide	Reglan	Should be taken 30 minutes prior to eating; may cause irreversible Parkinson's-like symptoms.
Baclofen	Muscle relaxer.	Balloted	Lioresal	Drowsiness and dizziness; contraindicated with alcohol.
Tizanidine	Muscle relaxer.	Tizanidine	Zanaflex	Drowsiness and dizziness; contraindicated with alcohol.

Source: Information compiled from: National MS Society 2009 Disease-Modifying Drugs. Available at www.nationalmssociety.org; Payne A. Nutrition and diet in the clinical management of multiple sclerosis. *J Hum Nutr Diet.* 2001;14:349–57. Schwarz S, Leweling H. Multiple sclerosis and nutrition. *Mult Scler.* 2005;11:24–32.

demonstrated clear benefit of omega-3 fatty acids in treatment of MS, but it is possible that an anti-inflammatory effect of omega-3 fatty acids could be beneficial in MS management. Further research in this area is warranted.[101]

Other nutrition interventions that are often proposed include supplementation with antioxidants, including beta-carotene, vitamin C, vitamin E, and selenium. A recent survey suggests that many individuals with MS turn to these alternatives due to lack of satisfaction with conventional treatments.[100] However, clinicians considering this therapy should be aware that these antioxidants do not cross an intact blood-brain barrier, which is necessary for the proposed effect.[104] Supplementation with vitamin D and calcium appears to be warranted in those patients who are at increased risk of osteoporosis because of long-term steroid use.[102]

Overweight and obesity as well as weight loss and malnutrition can all occur with the diagnosis of MS. Due to decreased mobility and the use of antidepressants and steroids, many patients with MS experience overweight and obesity. Anticipating the drug-nutrient interactions associated with both antidepressants and steroids can assist in preventing weight gain. Initiating a prescribed, regular exercise program can also improve weight status and provide the additional benefits of decreased fatigue and improved strength.[105]

Inadequate nutritional intake may occur as a result of decreased mobility, fatigue, difficulty swallowing, pain, drug-nutrient interactions, nausea, vomiting, diarrhea, and/or constipation. Interventions should appropriately address each specific nutrition problem (see Table 20.7). The need for nutrition support should be addressed with the patient and family as the disease progresses so that appropriate decisions can be made (see Chapter 5).

Monitoring and Evaluation For progressive forms of the disease, ongoing monitoring is important for detection of declining ability to acquire and prepare foods, manipulate utensils, and chew and swallow safely. The addition of new drugs to the treatment regimen may likewise result in side effects with nutritional implications. Thus, the registered dietitian should monitor changes in medical status as well as dietary intake, weight, and laboratory values and modify medical nutrition therapy as appropriate.

Dementia

Definition Dementia is the general term for conditions that involve loss of memory and impaired cognition. These conditions may include impairment in language, object recognition, motor skills, abstract thought, and judgment to an extent that interferes with daily life.[106] **Alzheimer's disease** is the most common type of dementia; other types include vascular dementia, dementia with Lewy bodies, and frontotemporal dementia.[106] Mild cognitive impairment (MCI) is characterized by noticeable problems with memory, language, or other cognitive functions that are not significant enough to interfere with daily life. MCI may or may not progress to dementia.[106]

Epidemiology Alzheimer's disease (AD) was first described in the early 1900s, when Dr. Alois Alzheimer described autopsy results for an individual that showed tangles and plaques within the brain tissue.[107] An estimated 5.3 million Americans have Alzheimer's disease, and prevalence of this condition continues to increase with the aging of the population.[106] It is estimated that the number of new cases every year will increase from 411,000 in 2000 to 959,000 by the year 2050.[106] According to the 2009 Alzheimer's disease facts and figures, AD is the sixth leading cause of death in the United States.[107] As prevalence of AD increases, so does the cost of related health care, which was estimated to be $112 billion nationally in Medicare and Medicaid costs alone in 2005.[106]

Etiology The most important risk factor for AD is age, with most cases of AD diagnosed at age 65 or older. AD or dementia affecting individuals younger than 65 years is referred to as early-onset or younger-onset AD or dementia.[106] An estimated less than 5% of AD cases occur due to rare genetic variants in a small number of families.[106]

Recent research identifies a variant of the **apolipoprotein E** (APOE) gene as a risk factor for AD developing at age 65 or older (late-onset). Apolipoprotein E (APOE) is produced in the liver and circulates in the blood associated with very low-density lipoproteins (VLDLs), a subclass of high-density lipoproteins (HDLs), and chylomicrons. It is also produced by cells in the brain, where it delivers cholesterol and lipids to neurons. The APOE ε4 variant is found in approximately 15% of the general population, but in 40% of patients with AD. Individuals with one copy of the ε4 variant have a 3–4 times greater risk of AD than those without ε4.[108] Greater risk for AD due to the ε4 variant may be conferred through numerous pathways under investigation, including effects on **beta-amyloid** (Aß) metabolism, which leads to the pathophysiological changes seen in AD[108] (see the section "Pathophysiology"). Variants of the genes encoding presenelin 1 (PSEN1), presenelin 2 (PSEN2), and amyloid precursor protein (APP) have been linked to early-onset AD.[109] A number of other potential AD susceptibility genes are also under investigation.[109]

Other areas of research have linked risk factors such as cardiovascular disease, diabetes, free radical oxidative damage, Down syndrome, and a history of previous head injury with AD.[110–113] High intakes of fruits and vegetables, fish, and omega-3 rich oils have been associated with lower risk of Alzheimer's disease as well, and research in this area is ongoing.[114, 115]

Pathophysiology In AD, two major changes that are thought to interrupt the normal processes of the brain occur. These include formation of **amyloid plaques** and **neurofibrillary tangles**. Amyloid plaques are composed of beta-amyloid peptide combined with other cells and proteins. The neurofibrillary tangles are made of a protein called **tau**, which is required for normal neuronal function, but clumps together abnormally in AD. These changes are associated with damage and loss of cholinergic neurons in the hippocampus and neocortex, which are areas showing high density of plaques and tangles in AD.[116] This results in abnormalities in communication, function, and repair among the damaged neurons and in shrinkage of the brain as damaged neurons die.[117]

Clinical Manifestations The first stages of AD are characterized by memory loss and confusion. Over time, AD progresses to include impaired decision-making and language skills, disorientation with regard to time and place, behavior and personality changes, and difficulty recognizing friends and family.[106, 117] Affected individuals require assistance in basic functions such as bathing, dressing, eating, and other daily activities. In the final stage, individuals may be bedbound and have completely lost their ability to communicate.[106]

Medical Diagnosis Diagnosing AD is not always straightforward. *DSM-IV-TR* Diagnostic Criteria for Dementia of the Alzheimer's Type are defined as including 1) memory impairment and at least one of the following: 2) aphasia, apraxia, agnosia, and/or disturbance of executive functioning. These deficits must represent significant impairment in either social or occupational functioning and represent gradual onset and progressive decline in cognitive function. Additionally, other potential diagnoses must first be ruled out, making AD a diagnosis of exclusion.[118] Similar diagnostic criteria were published as NINCDS-ADRDA Alzheimer's Criteria in 1984.[119] Using multiple sources of information allow for a probable diagnosis to be made. These sources include patient history from family and friends, psychological testing of the patient with instruments such as the Mini-Mental State Examination (MMSE),[120] thorough neurological examination, and magnetic resonance imaging (MRI) and/or computerized tomography (CT).[110]

Treatment Medications for AD include the second-generation cholinesterase inhibitors (ChEIs)—donepezil (Aricept®), rivastigmine (Exelon®), and galantamine (Razadyne®)—as well as the NMDA receptor antagonist memantine (Namenda®). The ChEIs block degradation of acetylcholine and result in improved cognition and functioning in mild-moderate AD. Gastrointestinal side effects, including nausea, vomiting, and diarrhea, are common. Memantine works by antagonizing

Nutrition Therapy for Depression: Weight Gain as a Side Effect of Medical Treatment

The frequency of diagnosis and subsequent treatment of both acute and chronic depression has steadily risen over the previous two decades. In fact, it is estimated that at any one time, almost 10% of the population will be experiencing a depressive disorder.[1] More than 71 million prescriptions for the leading depression medications have been written.[2] With this high rate of prevalence and the frequency of medication use, understanding potential nutrition concerns should be a priority for the registered dietitian.

"A depressive disorder is an illness that involves the body, mood, and thoughts. It affects the way a person eats and sleeps, the way one feels about oneself, and the way one thinks about things."[2] Three major classes of medications are used to treat depression: selective serotonin reuptake inhibitors (SSRIs), tricyclic antidepressants (TCAs), and monoamine oxidase inhibitors (MAOIs).[3] A common side effect for many of these medications is weight gain. Furthermore, it is felt that this weight gain may be a factor in noncompliance with treatment.[4] Other patients may also be at risk of weight gain from these medications—including patients who receive these medications for sleep disorders, fibromyalgia, or one of many other diagnoses. Recent reviews of the literature indicate that no specific mechanism has been identified to explain the observed weight gain. Some have proposed increased appetite, changes in metabolic rate, or hormonal changes, while others have suggested that the increased food intake is a result of effective treatment for the depressive disorder.[5] Medications most commonly associated with weight gain include:

Amitriptyline (Elavil®)	Venlafaxine (Effexor®)
Amoxapine (Asendin®)	Paroxetine (Paxil®)
Clomipramine (Anafranil®)	Luoxetine (Prozac®)
Desipramine (Norepramine®, Pertofrane®)	Sertraline (Zoloft®)
Doxepin (Adapin®, Sinequan®)	Mitrazapine (Remeron®)
Imipramine (Janimine®, Tofranil®)	Clozapine (Clozaril®)

Nortriptyline (Aventyl®, Pamelor®)	Olanzapine (Zyprexa®)
Protriptyline (Vivactil®)	Sertindole (Serlect®)
Trimipramine (Rhotramine®, Surmontil®)	Risperidone (Risperdal®)

Most individuals who begin treatment for depression are not treated in an inpatient hospital setting; therefore, providing nutrition education is not a common standard of practice with this diagnosis. Collaboration with mental health professionals is encouraged. Providing basic nutrition education and a means of referral to a registered dietitian are certainly important steps for the nutrition professional to take.[6] General suggestions for registered dietitians who work with patients on these medications include:

- Establish a baseline diet and weight history.
- Provide drug-nutrient information.
- Assist patients to develop a plan to monitor weight regularly.
- Develop a physical activity plan for each patient. Exercise assists in treatment of depression and will help maintain weight status.
- Reinforce the importance of continuing prescribed medications and conferring with the prescribing physician before making any changes.[7]

References
1. Psychiatric disorders in America, the epidemiologic catchment area study. New York: The Free Press; 1990.
2. National Institue of Mental Health Depression 2000. http://www.nimh.nih.gov/publicat/depression.cfm. 2005. Ref Type: Electronic Citation.
3. Kupfer DJ. The pharmacological management of depression. Dialogues Clin Neurosci. 2005;7:191–205.
4. Fava M. Weight gain and antidepressants. J Clin Psychiatry. 2000;61 Suppl 11:37–41.
5. Schwarz S, Leweling H. Multiple sclerosis and nutrition. Mult Scler. 2005;11:24–32.
6. Lacey JM, Houser RA. Dietetics and mental health counseling: time for partnership. J Am Diet Assoc. 2001;101:1313–14.
7. Kalarchian MA, Marcus MD, Levine MD et al. Behavioral treatment of obesity in patients taking antipsychotic medications. J Clin Psychiatry. 2005;66:1058–63.

glutamate binding to the NMDA receptor and has been shown to improve cognition and behavior in individuals with moderate-severe AD. Other treatments that have been investigated include the drug selegiline (Eldepryl®),[110] gingko biloba, estrogen replacement, statins, antioxidants, and anti-inflammatory drugs, but current evidence does not support use of these treatments.[121, 122] Nonpharmacological measures that have shown promise include reality orientation and cognitive stimulation.[122] Psychiatric medications may also be used to treat symptoms associated with AD, such as agitation and depression.[122] For more on the nutritional implications of depression and its treatment, see Box 20.3.

All other interventions are supportive in nature and provide for comfort and safety.

Nutrition Therapy for Dementia

Nutritional Implications Nutrition problems for AD are uniquely associated with altered brain function with regard to general cognition as well as feeding behavior, and well-planned and -executed interventions are necessary to prevent malnutrition. In the early stages of the disease, affected persons may find it challenging to shop for groceries, prepare foods, and remember whether or not they have eaten. As AD progresses, they may no longer remember what to do with food and utensils placed in front of them and require increasing assistance and encouragement from caregivers in order to consume sufficient nutrients. Chewing difficulty and dysphagia may also occur in the late stages of the disease. Confusion and agitation can compound feeding difficulties.[123]

These feeding challenges in AD often result in weight loss.[123] Weight loss and low BMI have been linked to cognitive decline, overall disease severity, morbidity, and mortality.[123] More rarely, some individuals with AD experience hyperphagia and consequent weight gain.[123]

Nutrition Diagnosis Common nutrition diagnoses consistent with AD include inadequate energy intake, inadequate oral food/beverage intake, inadequate fluid intake, evident protein-energy malnutrition, inadequate protein-energy intake, inadequate intakes of multiple nutrients (which may be specified as deemed appropriate), underweight, involuntary weight loss, swallowing difficulty, impaired ability to prepare foods/meals, and self-feeding difficulty.

Nutrition Intervention The goals for nutrition interventions in AD are to maximize nutritional intake in a manner that minimizes undesired weight loss and confusion while enhancing enjoyment and quality of life. Caregiver education has been identified as an important variable for community-dwelling individuals with AD and includes focus on providing companionship during meals and encouragement as well as assistance at mealtimes. Environmental factors such as better lighting and dinner music have also been shown to be helpful in improving behavior and increasing food intake. As AD progresses to the point where it becomes difficult to manipulate utensils, finger foods can be helpful in effectively increasing food intake. Modification of texture and thickening of liquids when necessary can reduce aspiration risk and improve intake of patients with chewing and swallowing difficulties. Dietary supplementation may entail liquid supplements as well as modifications of foods to increase energy density and enhance overall intake. Supplementation interventions have been shown to be least effective in individuals with the lowest BMIs; thus, it may be most helpful to institute interventions early when weight loss begins.[123] The impact of raising intakes of omega-3 fatty acids, B vitamins, and vitamins C and E in the early stages of AD is also being investigated in experimental models for potential to counteract neuronal degeneration, but current evidence does not support efficacy of these measures in treatment of existing AD.[116, 124] Table 20.10 summarizes nutrition therapy goals and interventions for dementia.

Monitoring and Evaluation Ongoing monitoring of individuals with AD must include assessment of ability to manage self care, as it involves food acquisition, preparation, and consumption. Additionally, adequacy of dietary intake should be regularly monitored along with body weight status such that changes in intake and weight may be detected early and appropriate interventions undertaken.

Neurotrauma and Spinal Cord Injury

Traumatic Brain Injury

Definition Traumatic brain injuries (TBI) are classified as either penetrating brain injuries or closed head injuries. A penetrating head injury involves actual penetration of the skull and direct damage to brain tissue.[125]

Table 20.10 Nutrition Goals and Interventions for Alzheimer's Disease and Other Forms of Dementia

- Maintain a reasonable weight (BMI 22 to 27).
- Provide sufficient kcal and nutrients to meet the needs of the individual. Use creative feeding techniques designed to address specific eating-related behavioral problems.
- Minimize confusion in the environment: offer one food at a time, use simple place settings; plate should be different color than table; add condiments prior to serving.
- Offer as many finger foods as possible.
- Strive to maintain independence in self feeding but provide assistance when patient is unable to maintain adequacy.
- Provide alternate opportunities to eat if the patient has difficulty sitting still—i.e., snack cart, finger foods.
- Minimize choking/aspiration. Adjust the texture of foods as the patient needs. Use verbal and physical cues to remind to chew and swallow.
- Allow adequate time for eating (30–45 minutes/meal).
- Maximize food intake during lunch when cognition is usually best.

Adapted from: Ham R. Dementia. *Nutrition management for older adults.* Washington, DC: Nutrition Screening Initiative (NSI); 2003:66–82.

Epidemiology About 1.4 million TBIs occur each year in the United States. Of these, about 1.1 million individuals are treated and released from emergency departments, 235,000 are hospitalized, and 50,000 die as a result of their injuries.[125, 126] About 60% of TBIs occur in males.[127] It has been estimated that approximately 3.17 million people, or 1.1% of the population, in the United States live with a long-term disability related to a TBI.[127]

Etiology Falls are the predominant cause of TBIs (28%) in the United States and are most common in adults over the age of 75 and children under the age of 5. Motor vehicle accidents account for 20% of TBIs with the highest rates among adolescents. Nineteen percent of TBIs are classified as "struck by/against events," which includes collision with a moving or stationary object. Many of these are related to participation in sports and recreational activities, are mild, and do not require medical care. Firearm use accounts for most TBIs due to assault (11%) and is the primary cause of fatal TBI. Bicycle accidents (3%) and other causes/unknowns (19%) make up the balance of TBI cases.[126] **Shaken baby syndrome** is a cause of TBI among infants.

Pathophysiology Injuries to the brain occurring due to TBI have been classified as primary and secondary.[125] The primary brain injury is that which results from the initial penetration or mechanical thrashing of the brain against the interior of the skull leading to lacerations and crushing of brain tissue. Secondary injuries are those occurring due to the physiological changes that occur during the aftermath of the initial injury. These often include cerebral edema, hemorrhage, hematoma, and infection. These changes can adversely affect brain function by leading to rising intracranial pressure (ICP) and impaired blood flow. Ultimately, it is the secondary injuries that can lead to destruction of neurons, glia, and axons.[125]

Clinical Manifestations Signs and symptoms can vary from mild to severe depending on the extent and location of the injury.[128] There may be no or only brief loss of consciousness (only a few seconds or minutes) with a mild TBI. Additional symptoms common with a mild TBI include headache, light-headedness, dizziness, blurred vision, fatigue, memory and concentration problems, and behavioral or mood changes, to name a few. A moderate to severe TBI may be characterized by a persistent and worsening headache, seizures, slurred speech, dilated pupils, nausea and vomiting, confusion, agitation, and loss of consciousness. The location and extent of injury may result in long-term consequences, including hemiparesis, changes in cognitive function, and other neurologic deficits such as changes in speech, movement, sensory function, emotions, and personality.[128]

Medical Diagnosis Diagnosis can incorporate a variety of neuroimaging techniques, but X-rays and CT are most common. Additionally, the Glasgow Coma Scale (GCS) shown in Table 20.11 is used to evaluate and rank the severity of TBI. Motor response, verbal response, and eye opening are ranked on a scale of 3 to 15, with the most severe level of coma correlating with the lowest number.[129] Long-term outcomes following TBI are best with higher GCS scores.

Treatment The acute treatment of TBI focuses on stabilization of the patient by maintaining oxygenation and blood flow to the brain while preventing shock. Interventions may include

Table 20.11 Glasgow Coma Scale

Eye Opening Response	Score
Spontaneous—open with blinking at baseline	4 points
To verbal stimuli, command, speech	3 points
To pain only (not applied to face)	2 points
No response	1 point
Verbal Response	**Score**
Oriented	5 points
Confused conversation, but able to answer questions	4 points
Inappropriate words	3 points
Incomprehensible speech	2 points
No response	1 point
Motor Response	**Score**
Obeys commands for movement	6 points
Purposeful movement to painful stimulus	5 points
Withdraws in response to pain	4 points
Flexion in response to pain (decorticate posturing)	3 points
Extension response in response to pain (decerebrate posturing)	2 points
No response	1 point

Categorization:

Coma: No eye opening, no ability to follow commands, no word verbalizations (3–8)

Head Injury Classification:

Severe Head Injury	= GCS score of 8 or less
Moderate Head Injury	= GCS score of 9 to 12
Mild Head Injury	= GCS score of 13 to 15

Sources: Teasdale G, Jennett B. Assessment of coma and impaired consciousness: A practical scale. *Lancet* 1974; 2(7872):81–84.

Teasdale G, Jennett B. Assessment and prognosis of coma after head injury. *Acta Neurochir* 1976; 34:45–55.

mechanical ventilation, blood pressure medications, and intravenous fluids. Depending on the level of injury, intracranial pressure and cerebral perfusion pressure will be monitored and treated as necessary. Many patients with severe head injuries will require surgery for removal or repair of hematomas and contusions.[128]

Chronic medical treatments focus on preventing further complications and promoting rehabilitation for functional recovery. As discussed previously in relation to treatment for stroke, rehabilitation and chronic care can occur in several different types of facilities, including specialized rehabilitation hospitals, outpatient, or in-home care settings. The program of care should utilize a multidisciplinary team involving psychiatry, physical therapy, occupational therapy, speech and language pathology, nutrition therapy, kinesiotherapy, and physical medicine.[44, 130]

Nutrition Therapy for Traumatic Brain Injury

Nutritional Implications Traumatic brain injury results in the systemic inflammatory response that is often described in burns and sepsis (see Chapter 22). This metabolic and inflammatory response results in hypermetabolism, hyperglycemia and insulin resistance, increased gluconeogenesis, lipolysis, and protein wasting and the magnitude of the metabolic response is directly proportional to the severity of the injury.[131] Providing timely exogenous substrate via enteral and/or parenteral nutrition is important to minimizing the catabolism of body lipids and protein as well as blunting the inflammatory response.[131] TBI patients who are not aggressively supported nutritionally have been estimated to lose as much as 15% of body weight in one week. Evidence indicates that not feeding the injured patient by the end of the first week post injury increases mortality rate.[132–135] Accordingly, management guidelines published by the Brain Trauma Foundation for severe TBI recommend attainment of full caloric requirements by day 7 post-injury.[136] Notably, catabolism, as evidenced by nitrogen excretion, appears to peak in the second week post injury and begins to slow after that time period. Research indicates that nitrogen balance is generally not achieved until after the third week.[132]

Nutrition Diagnosis Common nutrition diagnoses in patients with TBI include hypermetabolism, inadequate energy intake, inadequate oral food/beverage intake, inadequate intake from enteral/parenteral nutrition, inadequate fluid intake, evident protein-energy malnutrition, inadequate protein-energy intake, inadequate or excessive protein intake, swallowing difficulty, chewing (masticatory) difficulty, altered GI function, altered nutrition-related laboratory values, food-medication interaction, involuntary weight loss, impaired ability to prepare foods/meals, and self-feeding difficulty.

Nutrition Intervention Aggressive nutrition support is an important component of acute care for the TBI patient.[132, 136] Enteral nutrition is generally the most appropriate route unless the patient also sustained injuries to the gastrointestinal tract. However, TBI patients may have impaired gastric emptying related to vagal nerve damage, the stress response, or medications such as narcotics that slow peristalsis. Therefore, a nasoenteric feeding tube into the small bowel may be the best option. Parenteral nutrition is often undertaken if enteral access cannot be achieved within 72 hours of injury. As the acute phase subsides and the patient becomes more stable, a percutaneous endoscopic gastrostomy (PEG) provides the optimal nutritional route for patients unable to progress to an oral diet. Particularly in the early stages, slowed gastrointestinal motility due to narcotics, paralytic agents, etc., may require use of stool softeners and laxatives.[131] Progression to an oral diet as quickly as possible is certainly ideal from a quality of life perspective. However, oral diets are often delayed due to high prevalence of dysphagia related to both impaired cognition and physiological impairment of swallowing.[131] Dental and facial fractures and other complications may also delay the reintroduction of foods. Timing of an initial swallowing assessment depends on the severity of injury and patient progress, but typically one can be undertaken within 2–4 weeks of admission and diet consistency prescribed appropriately.[131] Most TBI patients achieve independent oral feeding within 6 months of injury.[131] During the transition to oral feeding, hydration status should be closely monitored as adequate fluid intake can be difficult to achieve in the presence of dysphagia and altered cognition.[131]

Energy requirements in individuals with TBI are highly variable. A recent review found that mean energy expenditures varied from 75% to 200% of that predicted using predictive equations. Use of paralytic agents, barbiturates, or sedatives reduce metabolic rate by 12%–32% while propanolol and morphine reduce metabolic rate somewhat less.[137] Due to the high degree of variability, measurement of actual expenditure with indirect calorimetry is most appropriate when available. In the absence of indirect calorimetry data, providing energy at 140% of estimated resting energy expenditure (REE) should be sufficient for most patients. Additional calories will be needed if there are additional injuries or stressors.[131] Providing excess protein has not been shown to offset protein catabolism and urinary nitrogen excretion, which appear to peak around 8–14 days post-injury. Current recommendations for protein are 1.5–2.0 g/kg body weight/day.[131]

Monitoring and Evaluation
Monitoring status changes is critical in TBI patients in order to progress nutritional care as appropriate. In the early stages, monitoring tolerance of enteral feedings is paramount with advancement to meet nutritional requirements in as timely a manner as possible. As mentioned above, slowed peristalsis can impair feeding tolerance as well as bowel function. Consequently, regular monitoring of these parameters in addition to appropriate preventive measures are important to detect problems early and institute necessary interventions. Since meeting fluid intake requirements can also be a challenge, close monitoring of fluid intake, urine output, skin turgor, and biochemical measures of hydration such as BUN and sodium is important to ensure sufficient hydration. Regardless of feeding route, energy and protein intake as well as body weight should be monitored closely to ensure optimal outcome for the patient with TBI. The patient may be at particular risk when transitioning to reliance on an oral diet.

Spinal Cord Injury

Definition
Injury to the spinal cord can involve fracture or compression of the vertebrae with consequent damage to the nerve cells. Spinal cord injury (SCI) can be further defined as either a complete or partial injury. A complete injury indicates that there is no function below the point of the spinal cord injury, and can result in **paraplegia** or **quadriplegia**. A partial injury indicates that the individual can experience sensation and may have voluntary movement. The level of residual function after a partial injury will vary from patient to patient.

Epidemiology
The National Spinal Cord Injury Statistical Center estimates that there are approximately 12,000 new cases of SCI in the United States each year.[138] The average age at injury is 40.2 years and 80.9% of SCIs in the United States occur among males.

Etiology
Most SCIs are a result of trauma and accidents, most commonly from automobile accidents. Other causes are falls, violence (most commonly gunshot wounds), and sports injuries.[138] Disease entities such as tumors, spina bifida, multiple sclerosis, and infectious disease can also result in SCI.

Pathophysiology
As described at the beginning of this chapter, the vertebrae and corresponding nerves of the spinal cord are named by the location: cervical, thoracic, lumbar, and sacral. These divisions also designate the effects of injuries to the spinal cord (see Figure 20.4). The level of the injury or disease will determine the overall signs and symptoms the patient will experience. For example, SCI at the level of the cervical nerves will result in damage to the nerves innervating the head, neck, diaphragm, arms, and hands, while SCI at the level of thoracic nerves affects chest and abdominal muscles. Lumbar nerves innervate leg muscles, and sacral nerves affect bowel, bladder, and sexual function.[139]

Complete or partial paralysis may result from SCI. Paraplegia is defined as paralysis involving the lower body below the umbilicus; quadriplegia or tetraplegia is paralysis involving all four limbs. Complications from paralysis may include pain, thrombosis or embolism, breathing difficulty, sexual dysfunction, **autonomic dysreflexia**, and **spasticity**. Nutrition therapy is an important component of treatment for other complications including bowel dysfunction, urinary tract problems, pressure ulcers (see Chapter 9 for a discussion of wound healing), and weight control.

Treatment
Acute spinal cord disease will be treated according to the underlying cause. Many inflammatory conditions of the spinal cord will be treated with corticosteroids, while acute trauma will focus first on stabilization of the patient and prevention of acute complications, and then on the disability and appropriate rehabilitation.

Nutrition Therapy for Spinal Cord Injury

Nutritional Implications
Nutritional needs for acute SCI are similar to those of traumatic brain injury in that trauma initiates an inflammatory and metabolic stress response (see Chapter 22). As the patient moves from acute care to rehabilitative care, nutritional implications shift to restorative and supportive goals. The primary long-term nutrition-related issues post SCI include obesity and cardiovascular disease due to immobility, neurogenic bowel involving fecal incontinence and difficulty with evacuation, and risk of pressure ulcers.[140]

Nutrition Diagnosis Nutrition diagnoses will vary from the acute care to rehabilitation settings as well as by individual status and complications. Those likely to be associated with spinal cord injury include hypometabolism, excessive energy intake, excessive fat intake, inappropriate intake of food fats, inadequate or excessive protein intake, inadequate fiber intake, inadequate vitamin and/or mineral intake, swallowing difficulty, altered GI function, altered nutrition-related laboratory values, overweight/obesity, impaired ability to prepare foods/meals, and self-feeding difficulty.

Nutrition Intervention Evidence suggests that early nutrition support is linked to better patient outcomes. Thus, a nutrition assessment should be conducted within 48 hours of injury to determine nutrient needs, identify potential for nutrition-related complications, and make recommendations for nutrition support.[140] In the acute phase of metabolic stress, energy requirements in patients with a SCI are estimated to be at least 10% below predicted needs due to reduced metabolic activity in denervated muscle. Consequently, measuring energy requirements with indirect calorimetry is recommended for accuracy. If indirect calorimetry is not available, a predictive equation may be used with admission weight, an injury factor of 1.2, and an activity factor of 1.1. As patients move into the rehabilitation phase, evidence suggests 22.7 kcal/kg for quadriplegic patients and 27.9 kcal/kg for those with paraplegia.[140] Protein requirements in the acute phase should be estimated at 2.0 g/kg of body weight per day. This declines to 0.8–1.0 g/kg/day during the rehabilitation and community living phases in the absence of pressure ulcers or infection.[140]

Rehabilitation presents its own unique nutritional concerns and interventions. Neurogenic bowel can be a problem when it results in slowed transit time and formation of hardened stool. The Consortium on Spinal Cord Medicine recommends 1 mL fluid/kcal consumed plus 500 mL *or* 40 mL/kg body weight plus 500 mL daily.[140] It is also suggested that the registered dietitian prescribe 15 g/day of fiber for patients with neurogenic bowel, slowly increasing to 30 g/day as tolerated. Patients deemed to be at risk for pressure ulcers based on biochemical, anthropometric, and lifestyle factors should begin nutrition support measures.[140] Biochemical parameters linked to pressure ulcer risk include albumin, prealbumin, zinc, vitamin A, and vitamin C. Pressure ulcer risk is lowest in individuals with a normal body weight, adequate nutrient intake, and no history of smoking or alcohol abuse, so these factors must be taken into consideration when determining risk. Prevention may include medical food supplements as well as enteral or parenteral nutrition. Once pressure ulcers are present, energy and protein needs are elevated and additional energy and protein are required for healing. Indirect calorimetry is preferred to measure energy requirements, but, in the absence of indirect calorimetry, these may be estimated at 30–40 kcal/kg of body weight per day or by using a predictive equation with a 1.2 stress factor for stage II ulcer or 1.5 stress factor for stage III and IV ulcers.[140] Optimal protein intake is estimated at 1.2–1.5 g/kg/day for stage II pressure ulcers and 1.5–2.0 g/kg/day for stage III and IV pressure ulcers.[140] There is currently insufficient evidence to establish a benefit of supplementation with arginine or glutamine. Several micronutrients are critical to wound healing, including vitamin A, vitamin C, vitamin E, zinc, and iron. Evidence analysis suggests a daily multivitamin and mineral supplement providing no more than 100% of the RDA for these nutrients unless there is a suspected or documented deficiency. Hydration status should also be monitored when pressure ulcers are present as additional losses may occur via wound drainage, evaporation from the site of the pressure ulcer, fever, or use of an air-fluidized bed.[140]

Preventing excessive weight gain is crucial, because SCI patients have a lower metabolic rate and, consequently, lower energy needs than noninjured patients.[140] Because of this, they are at higher risk for obesity-related comorbidities such as metabolic syndrome, cardiovascular disease, and diabetes.[140] When assessing body weight, however, standard BMI and skinfold measures should not be used as their development was based on a non-SCI population. Rather, it is recommended to use the Metropolitan Life Insurance tables (see online sources such as http://healthlinks.washington.edu/nutrition/section12.html) to determine ideal body weight. The listed weight for height in the tables should be adjusted by one of two methods: for an individual with quadriplegia, the listed table weight for height should be reduced by 10%–5% or 15–20 pounds; for an individual with paraplegia, the listed table weight for height should be reduced by 5%–10% or 10–15 pounds.[140] Others have alternatively suggested a BMI cut point of >22 kg/m² to identify individuals within the SCI population who are at high risk for obesity and obesity-related comorbidities.[141] Bioelectrical impedance analysis or dual-energy X-ray absorptiometry may be used to assess body composition and thereby determine degree of obesity. Individuals with SCI typically have higher fat mass and lower lean body mass than non-SCI individuals.[140] When determining overall energy requirements, type of wheelchair may play a role, with use of manual standard wheelchairs resulting in increased energy consumption compared to ultralight or power-assisted wheelchairs.[140]

Monitoring and Evaluation Ongoing nutrition monitoring of the person with SCI must focus on nutritional parameters associated with prevention of pressure ulcers, energy needs and weight management, and fiber-intake monitoring for optimal management of neurogenic bowel.[140]

Conclusion

Nutrition for neurological conditions may involve all aspects of the disease process and the subsequent medical treatment. When treating the numerous conditions affecting the nervous system—both acute and chronic—nutrition therapy should be at the forefront of patient assessment and intervention.

Beth Zupec-Kania, RD, CD *Clinical dietitian at Children's Hospital of Wisconsin and Consultant to The Charlie Foundation, an organization that trains hospitals to administer the ketogenic diet.*

How has working on a ketogenic diet (KD) medical team expanded your role as a dietitian?

Over the last century, researchers have consistently shown that the KD is an effective treatment for difficult-to-control epilepsy. In addition, it is rapidly emerging as a treatment for other brain disorders. It is rewarding to watch children's lives improve so drastically when they follow the meal plans that I designed for them. My role on the medical team has expanded from a discretionary provider of nutrition care to a primary provider of Medical Nutrition Therapy.

The ketogenic diet is very different from what is recommended in the *Dietary Guidelines for Americans*; can consuming this much fat be healthy?

Our culture has adopted the belief that fat is unhealthy. Certainly, *trans* fats are unhealthy and saturated fats should be limited, but an appropriate balance of polyunsaturated, monounsaturated, and saturated fats can be present in a healthy diet. The unusual aspect of this therapy is that the majority of energy is from fat. This high fat intake changes the metabolic performance of the brain—which naturally occurs for most people during illness and fasting. Blood lipid levels are evaluated at times prior to initiation of the diet and periodically during treatment. Our approach is to screen patients for a family history of heart disease on their intake document prior to initiating the diet so that we can tailor heart-healthy fat sources at the start. If lipids are elevated we modify the balance of dietary fats. Lowering the ratio (fat content) of the diet can also be trialed.

In your practice, have you seen some children become seizure free on this treatment?

Yes, very often. I recall one of the first children that I managed on the KD was a 2-year-old who had dozens of daily drop seizures requiring him to wear a helmet. He was antisocial, combative, and an overall unhappy toddler. His mother, a single parent, was exhausted from taking care of him and depressed over his lack of development. He gained seizure control within the first few weeks of being on the diet and his neurologist weaned him of off all of his seizure medications within two months. When he returned for his three-month visit in our clinic, no one could believe he was the same child. Not only were the seizures gone, he was affectionate and interactive and regaining developmental milestones. I saw him recently with his mother at a fund raiser. He is a healthy, active teenager who has remained seizure free.

How do you manage some of the most common side effects of the ketogenic diet?

One education session that I have with families is simply about ensuring that the child drinks adequate fluids during this therapy. A low-carbohydrate diet has a diuretic effect, which can compound constipation and nausea (especially while going into ketosis). Kidney stones have been reported by some centers as problematic, but in our experience, we've seen them only in children who were concurrently receiving seizure medications that are known to cause stones. Constipation is the main adverse effect of this therapy and preventative steps are recommended as follows:

Prevention:

- Adequate hydration
- High-fiber vegetables as well as 25 g of lettuce daily (free food)
- Avocado
- Medium-chain triglyceride oil (MCT)
- Bowel program—i.e., daily toilet routine following breakfast meal

Treatment:

- Carbohydrate-free products: Miralax, Milk of Magnesia, mineral oil, glycerin suppository, Colace 1% solution or suppository, Fleets Enema.

What are some of the biggest challenges in working with patients on the ketogenic diet?

The first challenge is that the diet is still not widely accepted within our health care system. Neurologists are still telling families that "it's too difficult" and there is also a lack of funding for most facilities to train and support a dietitian in this role.

Secondly, even in facilities that have programs, there needs to be a thorough screening process prior to initiating this therapy. Children who have difficult-to-control seizures may also have nutrition intake problems such as poor appetite, difficulty chewing and/or swallowing, and an overall history of poor growth. These problems may be due to their neurological impairment and/or a side-effect of their medication(s). The first step in utilizing this therapy is to screen the child for these problems and then determine if they can safely eat and drink the required foods and fluids. For many of the children, this may be the first time that

(continued)

feeding issues have been addressed. This may delay the initiation of the therapy but is essential. In addition, there are metabolic disorders for which the diet is contraindicated. Laboratory evaluations are required in children who are suspected of having these disorders.

A third challenge is that once a child is on the KD, he or she may experience drug toxicity even when the same dosage of medication is continued. I see this occur especially when a child is on two or more anti-epileptic medications or on very large doses of any one. This effect may occur despite no significant change in their blood level of the medication. An example of this is that the KD can exacerbate acidosis, which is an adverse effect of the same medications that can cause kidney stones (mentioned earlier).

Finally, although most people would expect that the diet itself is a great challenge, I have found that families overwhelmingly agree that the extra effort in using this therapy is worth the improvement in quality of life and is far easier than helplessly watching their child have seizures.

When you represent the Charlie Foundation and go into a hospital to train a medical team on the ketogenic diet, who are the members of the team and what are their roles?

- Neurologist: Medical director of KD therapy, selects candidates, orders laboratory assessments and supervises treatment.

- Dietitian: Pre-diet screening and evaluation, diet calculation, family education, initiation plan, follow-up care.
- Nurse
  - Outpatient Clinician: Assists in coordinating the pre-keto screening, admission and follow-up care.
  - Inpatient: Assists w/initiating diet, urine ketone testing.
- Pharmacist: Researches carbohydrate content of medications and supplements.
- Social worker:
  - Medical: assists family with financial services such as special need programs, which may cover nutritional supplementation.
  - Psychiatric Social Work: Supports family in understanding the expectations of KD therapy and adjusting to the limitations that this therapy may place on family lifestyle.

How important is the role of a dietitian when administering this treatment?

The dietitian's role is imperative; the ability to provide ketogenic diet therapy is dependent on the availability of a qualified dietitian.

Linda White Gray, RD, CDE *Senior Dietitian; Supervisor, Outpatient Dietitians; Coordinator, Nutrition Services and Dietetic Internship, UC San Francisco Medical Center*

How long have you been an RD? Which neurological disorders do you primarily see as an RD?

I have been a registered dietitian for 10 years and specialize in the nutritional management of amyotrophic lateral sclerosis (ALS), or Lou Gehrig's disease.

How do you nutritionally assess these patients? Is there anything unique that would be useful for future dietetic internship students to be aware of?

I take a systematic approach to nutrition assessment using the American Dietetic Association's nutrition care process (NCP). The five categories of nutrition assessment data include:

1. **Anthropometric Measurements.** I collect routine anthropometric measurements including height, weight, usual body weight (or baseline weight), percent weight change, and BMI.

2. **Biochemical Data, Medical Tests and Procedures.** An ALS patient's protein profile, such as albumin and prealbumin, is less utilized in the ambulatory setting. Resting metabolic rate can be measured using indirect calorimetry, but this method is often reserved for the research setting.

3. **Nutrition-Focused Physical Findings.** ALS can afflict multiple systems that directly impact nutritional status. I review respiratory function via pulmonary function tests, which aids in the determination of timing of PEG placement. Limb function is assessed to determine the patient's ability to self-feed and degree of ambulation. While ALS does not directly affect the muscles of the digestive tract, I review gastrointestinal symptoms such as constipation, gastroesophageal reflux disease, or feeding intolerances, which are more common in ALS patients. Importantly, I collect data on the patient's bulbar function, or ability to swallow, via formal swallow studies conducted by the speech and language pathologist.

4. **Food-/Nutrition-Related History.** In keeping with patient-centered care, I start by evaluating the patient's current nutrition knowledge of ALS and collect his or her questions and concerns. I assess appetite and nutrient adequacy by way of 24-hour diet recalls or food diaries, with focus on calorie, protein, fiber, and fluid intake. I probe for early indications of dysphagia or changes in the ability to eat, such as food or fluid consistency modifications and changes in meal duration. I review the patient's list of nutritionally-relevant

prescription and over-the-counter medications. Importantly, food access and eating environment should be addressed.

5. **Client History.** Client history can reveal information about both the patient's medical history and psychosocial status. I consider the patient's well-being, cultural and religious beliefs, support systems, and past medical history.

It is useful for practitioners to know that the role of nutrition in neuromuscular disease is being extensively researched. It has been well established that malnutrition causes rapid muscle wasting that leads to greater fatigue and weakness. Studies in ALS have found that nutrition is a prognostic factor in survival—those with malnutrition have at least a seven-fold increased risk for death.[1]

Are there common nutritional problems associated with neurological disorders?

Depending on the type of neuromuscular disorder and its affected muscle groups, nutrition problems can present because of any of the following reasons:

- Poor appetite—This is typically a function of reduced activity level. But psychosocial aspects may also play a role, including depression and anxiety.
- Bulbar dysfunction—When the bulbar muscles are affected by the disease, this leads to chewing and swallowing impairment. Also common is difficulty with hypersialorrhea (excessive saliva production) and thick, ropy oral secretions which can impair an individual's dietary intake. Individuals with bulbar dysfunction need mechanically altered diet consistencies and may need the help of a long-term feeding tube to prevent weight loss and aspiration.
- Limb dysfunction—When the upper limb muscles are affected by the disease, this can affect the individual's physical ability to eat (move the food from plate to mouth).
- Respiratory dysfunction—If respiratory muscles are impacted by the disease, individuals may experience fatigue from labored breathing, which can make eating more difficult.
- Gastrointestinal dysfunction—The gut, as an involuntary muscle, is not directly affected by ALS. But constipation is a common problem in neuromuscular disease due to inadequate intake of fiber and fluids, coupled with limited physical activity.

The most common nutritional problems associated with neuromuscular disease are the following:

1. Inadequate calorie and protein intake, which presents as weight loss and subsequent decrease in energy level and endurance. There is also growing evidence to support that ALS patients exhibit hypermetabolism, raising their daily calorie needs.[2]

2. Dysphagia

3. Dehydration

4. Constipation

5. Gastroesophageal reflux disease (GERD)

What is the biggest challenge of working with patients with neurological disorders?

ALS is a progressive disease. The rate of progression, and thereby the patient's nutrition needs, will change over the course of the disease. The key is in providing individualized care and frequent, serial follow-up of the patient to adjust nutrition care plans as appropriate.

Has the treatment changed much in the last ten years? How do you stay abreast of changes in this field?

Because of ongoing research on the role of nutrition in ALS, nutrition is emphasized as a key treatment strategy to help control symptoms of disease progression and to enhance an individual's quality of life. I stay abreast of changes in this field by regular networking with registered dietitians at certified ALS centers across the country, annual neurology conferences, and regular review of the published scientific literature. I also created (in 2007) and moderate an online listserv for RDs working in ALS.

Any advice for dietetic students about counseling clients with neurological disorders?

Respect patient autonomy and the patient's role in making care decisions. Neuromuscular disease, particularly ALS, is a condition in which patients lose aspects of control in many situations—such as the ability to perform activities of daily living and loss of independence. Nutrition and what a patient chooses to eat is one area where the patient can feel empowered to take control. Emphasize the positive, and involve the patient in setting nutrition goals.

References
1. Desport, JC, Preux, PM, Truong, TC, Vallat, JM, Sautereau, D, & Couratier, P. Nutritional status is a prognostic factor for survival in ALS patients. *Neurology.* 1999 Sep 22; 53 (5): 1059–63.
2. Desport, JC, Preux, PM, Magy, L, Boirie, Y, Vallat, JM, Beaufrere, B, & Couratier, P. Factors correlated with hypermetabolism in patients with amyotrophic lateral sclerosis. *Am J Clin Nutr.* 2001; 74:328–34.

Application of the Nutrition Care Process: Parkinson's Disease

Introduction:

78-year-old Caucasian female diagnosed with Parkinson's disease ten years ago—currently treated with Sinemet. She recently moved from her assisted living apartment to a skilled care facility. She states: "Every time I eat, it feels like it gets stuck in my throat. I am scared that I will choke." Patient was determined to have dysphagia secondary to her Parkinson's disease.

Rx: Sinemet: carbidopa/levodopa, 50/200 mg controlled release tablet BID

1. How is the pathophysiology of Parkinson's related to her development of dysphagia?

2. How does her medication, Sinemet, assist with control of her disease?

Nutrition Assessment:

Anthropometric Measurements:

Ht. 5'4" Wt. 119 lbs. UBW (1 year ago) 130 lbs.

Food-/Nutrition-Related History:

Previously enjoyed eating a varied diet with good appetite. Cannot prepare own meals and most recently has needed assistance with feeding.

Biochemical Data:

Albumin 3.4 mg/dL; Prealbumin 14 mg/dL

3. Identify at least two nutrition problems based on the nutrition assessment and medical history. Determine the diagnostic term for each nutrition problem. Next, identify the etiology of each nutrition problem. Finally, identify the signs and symptoms that support the evidence for these nutrition problems. An example for this case has been provided.

Diagnosis	Related to	Etiology	As Evidenced by	Sign/Symptoms
EXAMPLE: Involuntary weight loss	R/T	Difficulty swallowing	AEB	9% weight loss in one year.

Nutrition Intervention:

4. Identify the nutrition prescription for this patient by recommending the appropriate dietary modification for her diagnosis.

5. Recommend energy and protein requirements for this patient.

Nutrition Monitoring and Evaluation:

6. Determine nutrition criteria for monitoring and evaluation for each nutrition diagnosis that you identified.

EXAMPLE: Nutrition Care Outcome

Involuntary weight loss	Measure weight weekly.	Patient's weight will stabilize at 119#.

WEB LINKS

The Epilepsy Foundation of America National voluntary organization involved in advocacy, education, and research of epilepsy.
www.efa.org

The Charlie Foundation   An advocacy organization to promote awareness, research, and training to hospital teams on the ketogenic diet for childhood epilepsy.
www.charliefoundation.org

National Multiple Sclerosis Society   A foundation that provides information and support to those affected by MS, facilitates professional education, and funds MS research.
http://www.nationalmssociety.org

National Parkinson Foundation (NPF) NPF began to provide therapy for Parkinson's patients but now provides all levels of neurological care.
http://www.parkinson.org

The National Institute of Neurological Disorders and Stroke (NINDS) Government-sponsored research institute that is a part of the National Institutes of Health (NIH).
http://www.ninds.nih.gov

Alzheimer Research Forum   Nonprofit organization which creates resources for those involved in Alzheimer's research.
http://www.alzforum.org

The National Library of Medicine's Consumer Health Information Database   Extensive information from the National Institutes of Health and other trusted sources on over 500 diseases and conditions. Click on "Health Topics," and then "Brain and Nerves."
http://www.nlm.nih.gov/medlineplus

Spinal Cord Injury Information Network Federally funded website for resources to support those with spinal cord injury. From this site it is also possible to access data from the National Spinal Cord Injury Statistical Center (NSCISC), which maintains the national SCI database.
http://www.spinalcord.uab.edu

1. What is epilepsy? What are the signs and symptoms of epilepsy?

2. Describe the classic ketogenic diet that can be used in the treatment of epilepsy. List nutrition implications associated with the diet. What nutrients must be supplemented when on this treatment?

3. Which risk factors for stroke can be modified by diet? List the nutritional implications associated with stroke.

4. What is Parkinson's disease, and what are its signs and symptoms? Which neurotransmitter is progressively lost?

5. How can diet be modified to maximize the effectiveness of L-dopamine in the treatment of Parkinson's disease?

6. Describe the pathophysiology and clinical manifestations of multiple sclerosis (MS). What nutritional interventions are used for MS, and are they effective?

7. What is apolipoprotein E and what is its association with Alzheimer's disease? List the nutritional implications associated with Alzheimer's disease.

8. What are the nutritional implications associated with traumatic head injury?

Complementary and Alternative Medicine Remedies Used in Diseases of the Neurological System

Remedy	Scientific Name	CAM Use	Side Effects and/or Risks
Aconite, monkshood; fuzi	*Aconitum carmichaeli*	Treat pain	No forms of the plant are recommended for human consumption.
Amalaka, amla; Indian gooseberry	*Emblica officinalis*	Treat irritability, insomnia	Adverse reactions rare.
Ascorbic acid/ vitamin C		Treat depression	High intakes may cause GI upset or mild diarrhea. May increase activity of anticoagulants. May interfere with B_{12}, copper, chromium absorption and metabolism. Aspirin, corticosteroids, & indomethacin increase urinary vitamin C losses. High doses may reduce efficacy of select chemotherapeutic medications. Supplementation above DRI (or >70–90 mg/d) not recommended. May increase plasma estrogen levels in patients taking HRT or oral contraceptives. Large doses may interfere with anticoagulant medications.
Ashwagandha; winter cherry, Indian ginseng	*Withania somnifera*	Treat insomnia; treat Alzheimer's disease, memory loss	Lacks sufficient human research to recommend use. Possible hyperthyroidism, thyroid dysfunction. May interfere with thyroid and antithyroid medications.
Burdock; niu bang zi	*Arctium lappa*	Treat back pain	May increase hypoglycemic effects of insulin and oral hypoglycemic agents. Burdock products may be significantly contaminated with atropine at high enough levels to cause toxicity.
L-carnitine		Treat chronic fatigue syndrome; reduce memory loss	Appears to be safe. Might have antithyroid actions; possibly contraindicated in individuals with hypothyroidism.
Castor oil plant; eranda, vatari	*Ricinus communis*	Treat headache, back pain, sciatica	Avoid use if pregnant or breast-feeding. Can cause malabsorption of fat-soluble vitamins, fluid, & electrolytes. Prolonged use can result in laxative dependency.
Cat's claw; gambir; gao teng	*Uncaria rhynchophylla*	Treat tremors, seizures, convulsions; treat chronic pain	Should not be taken with antihypertensive or anticoagulant medications.
Coenzyme Q_{10}		Alleviate fibromyalgia symptoms; treat chronic fatigue syndrome, Parkinsonism	Anorexia, diarrhea, epigastric discomfort, mild nausea
Cobalamin/ vitamin B_{12}		Treat depression, psychosis, Alzheimer's disease, memory loss; assist with myelin regeneration; treat diabetic neuropathy	Folic acid supplementation may mask vitamin B_{12} deficiency. Supplemental vitamin C or iron may interfere with bioavailability of vitamin B_{12}. The following drugs may interfere with B_{12} absorption: zidovudine, antacids, clofibrate, colchicines, cycloserine, erythromycin, isoniazid, metformin, aldomet, neomycin, nitrous oxide, oral contraceptives, sulfonamides, tetracycline.

(continued)

Complementary and Alternative Medicine Remedies Used in Diseases of the Neurological System (*Continued*)

Remedy	Scientific Name	CAM Use	Side Effects and/or Risks
DHEA (dehydro-epiandrosterone)		Alleviate fibromyalgia symptoms; treat chronic fatigue syndrome	Elevated liver function test values, slightly decreased hemoglobin levels and RBC count, hirsutism, gynecomastia. Contraindicated in patients with benign prostatic hyperplasia, estrogen-responsive tumors, or prostate cancer. Women should not use while pregnant or breast-feeding. May interact with the following medications: alprazolam, calcium channel blockers, carbamazepine, dexamethasone, phenytoin, danazol, insulin, and insulin-sensitizing drugs.
Dong quai; angelica	*Angelica sinensis*	Treat headache	Should not be combined with blood-thinning medications (warfarin, heparin, aspirin) or supplements (garlic, ginkgo, vitamin E, fish oil).
Echinacea, purple coneflower	*Echinacea purpurea*, *E. angustifolia*	Treat chronic fatigue syndrome	Should not be used >8 weeks. May adversely influence fertility. Contraindicated in patients with severe illnesses, including autoimmune disease, collagen diseases, HIV infection, leukemia, multiple sclerosis, or tuberculosis. Should be avoided by pregnant or breast-feeding women. Tinctures may contain significant concentrations of alcohol.
Ephedrine; ephedra; ma huang	*Ephedra sinica*	Treat multiple sclerosis	FDA has issued warnings against use of ephedra as an appetite suppressant. Adverse reactions may include hypertension, palpitations, tachycardia, arrhythmias, cardiac arrest. Interacts with alcohol & caffeine, beta blockers, MAOIs, phenothiazines, theophylline.
Evening primrose oil	*Oenothera biennis*	Provide source of omega-3 fatty acids; treat attention deficit hyperactivity disorder, memory loss	Nausea. Increases risk of seizures if taken with phenothiazines; avoid administration with evening primrose oil.
Folic acid		Treat depression, insomnia, irritability, dementia	Supplementation $\geq 400\,\mu g$ can mask pernicious anemia caused by vitamin B_{12} deficiency; dosages ≥ 5 mg may cause abdominal cramps, diarrhea, and rash; doses ≥ 15 mg can alter sleep patterns or cause vivid dreaming; large doses exacerbate neuropathy in those with B_{12} deficiency; doses >1 mg/d interfere with anticonvulsant medications. The following medications can affect absorption or activity of folate: oral contraceptives, aspirin, indomethacin, methotrexate, famotidine, tetracycline, erythromycin, sulfonamides, & cholestyramine.
Gingko biloba; ying xing	*Gingko biloba*	Optimize brain function; treat headache, dizziness, depression, anxiety, Alzheimer's disease; reduce memory loss; treat diabetic neuropathy	No agreed-upon standards exist; different formulations will vary in actual content. Affects platelet activating factor; use with caution in patients taking anticoagulant or antiplatelet therapy. May cause elevated blood glucose levels in those taking insulin or oral hypoglycemic agents. Increases blood pressure when combined with thiazide diuretics. Do not use with trazodone.
Ginseng; ren shen; fivefinger; tartar root; redberry; sang	*Panax ginseng*, *P. quinquefolium*	Improve mental performance, reduce memory loss	Pregnant or breast-feeding women should not use. Should be used cautiously by patients with cardiovascular disease, diabetes, hypertension, or hypotension, and in those receiving steroid therapy.
Glucosamine		Treat back pain	No interactions reported. Monitor blood glucose levels in patients with diabetes.
Goldenseal; yellow root; yellow puccoon; yellow Indian paint; ground raspberry	*Hydrastis canadensis*	Treat chronic fatigue syndrome	CNS depression, abdominal cramps & pain, diarrhea, nausea, vomiting. Might inhibit CYP3A4.
Gotu kola; brahmi	*Centella asiatica*	Rejuvenate brain cells, nerves; treat nerve disorders, inc. convulsions, epilepsy, tetanus, dementia	Well tolerated. High doses may interfere with actions of antidiabetic medications & cholesterol-lowering drugs. Contraindicated in pregnancy and breast-feeding.
Grape seed extract	*Vitis vinifera*, *V. coignetiae*	Alleviate fibromyalgia symptoms	Essentially nontoxic. Might potentiate anticoagulant and antiplatelet agents.
Hawthorn	*Crategus oxyacantha*	Treat insomnia	Augments the effect of cardiac medications (digoxin) or blood pressure drugs.

Remedy	Scientific Name	CAM Use	Side Effects and/or Risks
Kava kava, 'awa	*Piper methysticum*	Treat headache, back pain; treat obsessive-compulsive disorder	Changes in motor reflexes and judgment, vision changes, hypertension. Alcohol increases kava toxicity. Administration with alprazolam may result in coma. Causes additive sedative effects when taken with CNS depressants. Large doses of kava exert digitalis-like effect. Avoid administration with levodopa as kava increases Parkinsonian symptoms. Do not administer with pentobarbital; may have additive effects.
Magnesium		Treat autism, attention deficit hyperactivity disorder	Wide margin of safety in individuals with healthy kidneys. Can mutually interfere with absorption of antibiotics in the tretracycline and fluoroquinolone families, as well as nitrofurantoin, penicillamine, ACE inhibitors, phenytoin, and H2 blockers. Might increase effectiveness of sulfonylureas, potentially creating risk of hypoglycemia.
Melatonin		Treat seasonal affective disorder, insomnia	May cause sedation and impair balance. Might impair insulin sensitivity and glucose tolerance. Benzodiazepine drugs may impair melatonin release.
Mistletoe	*Phorandendron leucarpum*	Ease anxiety, panic disorder	All plant parts of mistletoe are toxic. May increase hypotensive effects of antihypertensive medications. Increases sedative effects of CNS depressants. Causes cytotoxic & immunostimulant effects.
Niacin/vitamin B$_3$		Treat mania, depression, anxiety, dementia, memory loss	Dosages >100 mg daily frequently cause skin flushing, especially the face, which may be accompanied by stomach distress, itching, and headache. Slow-release niacin may induce hepatic inflammation. High doses may present risks in individuals with history of ulcer disease, gout, and excess alcohol consumption. May increase serum levels of anticonvulsant medications.
Peppermint	*Mentha piperita*	Treat headache, chronic pain	Enteric-coated peppermint oil believed to be reasonably safe; non-enteric-coated peppermint oil can cause heartburn.
Pyridoxine/ vitamin B$_6$		Treat depression, autism	Excessive intake (>2000 mg daily) can cause sensory neuropathy. May cause or worsen acne symptoms. Doses >5 mg may interfere with effects of levodopa when drug is taken alone.
Reishi; ling zhi	*Ganoderma lucidum*	Treat neuralgia	No known drug interactions. Widely believed to be safe.
Riboflavin/ vitamin B$_2$		Reduce migraine severity, frequency; treat depression	No safety issues. Oral contraceptives may reduce levels of riboflavin.
St. John's wort	*Hypericum perforatum*	Alleviate fibromyalgia symptoms, treat chronic fatigue syndrome; reduce memory loss	Avoid administration with alcohol, MAOIs, narcotics, OTC cold & flu medications, digoxin, drugs metabolized by CYP3A, indinavir, amphetamines, SSRIs, tricyclic antidepressants. May result in phototoxicity. May influence hepatic microsomal enzymes.
Schisandra; gomishi; wu wei zi	*Schizandra chinensis*	Treat insomnia	Side effects rare, but in large doses it may cause GI distress or appetite depression. May enhance elimination of hepatically metabolized drugs. Avoid use in pregnant or breast-feeding women.
Selfheal; all heal; xi ku cao	*Prunella vulgaris*	Treat headache, dizziness	None reported.
Thiamin/vitamin B$_1$		Treat depression, psychosis, irritability, Alzheimer's disease, memory loss	Appears to be safe. Loop diuretics may cause depletion of thiamin.
Turmeric; haridra	*Curcuma longa*	Treat chronic pain	GI ulceration with high doses or prolonged use. Avoid administration with anticoagulants. Decreases immunosuppressive effects. Avoid administration with NSAIDs, may inhibit platelet function and increase risk of bleeding.
Valerian	*Valerian* spp.	Treat headache; ease anxiety, panic disorder, obsessive-compulsive disorder	Exerts a mild sedative effect without hangover effects. May cause additive effects when taken with alcohol, CNS depressants. Many extract products may contain alcohol

(continued)

Complementary and Alternative Medicine Remedies Used in Diseases of the Neurological System (*Continued*)

Remedy	Scientific Name	CAM Use	Side Effects and/or Risks
Vitamin E		Treat depression; treat Alzheimer's disease, memory loss	Relatively safe, although high doses (>50 IU/day) may cause mild antiplatelet effects. May potentiate anticoagulant or antiplatelet medications. Could interact with herbs and supplements that possess anticoagulant or antiplatelet effects (garlic, policosanol, ginkgo).
Willow	*Salix* spp.	Treat headache, chronic pain	Increases risk of bleeding when taken with anticoagulants. May reduce effectiveness of diuretics. May increase risk of GI ulceration and bleeding when taken with NSAIDs. Contraindicated in patients with salicylate hypersensitivity. Should be avoided by pregnant and breast-feeding women.
Zinc		Treat autism	Long-term use significantly above nutritional requirements can cause numerous toxic effects (anemia, neutropenia, arrhythmias), increases LDL levels, decreases HDL levels, decreases glucose clearance, impairs immune function.

Sources: Fetrow CW, Avila JR. *Professional's Handbook of Complementary and Alternative Medicines*, 3rd ed. Philadelphia: Lippincott Williams & Wilkins, 2004.

Bratman S, Girman AM. *Mosby's Handbook of Herbs and Supplements and their Therapeutic Uses*. St. Louis: Mosby/Elsevier, 2003.

Fragakis AS, Thomson C. *The Health Professional's Guide to Popular Dietary Supplements*, 3rd ed. Chicago: American Dietetic Association, 2007.

REFERENCES

1. Sherwood L. Human physiology: from cells to systems. 6th ed. Belmont (CA): Brooks/Cole; 2009.

2. Epilepsy Foundation Website. Available at http://www.epilepsyfoundation.org/about/statistics.cfm. Updated: 2009 Jun 14.

3. Sirven JI. Classifying seizures and epilepsy: a synopsis. Semin Neurol. 2002;22:237–46.

4. The National Institute for Neurological Disorders and Stroke. Epilepsy: Hope Through Research. Available at http://www.ninds.nih.gov/disorders/epilepsy/detail_epilepsy.htm#131263109. Updated: 2009 Jul 23.

5. Marsh ED, Golden JA. Developing an animal model for infantile spasms: pathogenesis, problems and progress. Dis Model Mech. 2009;2:329–35.

6. Kossoff EH, McGrogan JR. Worldwide use of the ketogenic diet. Epilepsia. 2005;46:280–89.

7. Friedlander WJ. The rise and fall of bromide therapy in epilepsy. Arch Neurol. 2000;57:1782–85.

8. Malagon-Valdez J. [The new antiepileptic drugs: their indications and side effects]. Rev Neurol. 2004;39:57075.

9. Schmidt D, Loscher W. Drug resistance in epilepsy: putative neurobiologic and clinical mechanisms. Epilepsia. 2005;46:858–77.

10. Sisodiya SM. Genetics of drug resistance. Epilepsia. 2005;46 Suppl 10:33–38.

11. Ramani R. Vagus nerve stimulation therapy for seizures. J Neurosurg Anesthesiol. 2008;20:29–35.

12. Dietrich S, Smith J, Scherzinger C et al. [A novel transcutaneous vagus nerve stimulation leads to brainstem and cerebral activations measured by functional MRI]. Biomed Tech (Berl). 2008;53:104–11.

13. Cross JH. Epilepsy surgery in childhood. Epilepsia. 2002;43 Suppl 3:65–70.

14. Bailey EE, Pfeifer HH, Thiele EA. The use of diet in the treatment of epilepsy. Epilepsy Behav. 2005;6:4–8.

15. Wilder RM. The effects of ketonemia on the course of epilepsy. Mayo Clin Proc. 1921;2:307–308.

16. Neal EG, Chaffe H, Schwartz RH et al. The ketogenic diet for the treatment of childhood epilepsy: a randomised controlled trial. Lancet Neurol. 2008;7:500–506.

17. Sirven J, Whedon B, Caplan D et al. The ketogenic diet for intractable epilepsy in adults: preliminary results. Epilepsia. 1999;40:1721–26.

18. Kossoff EH, Rowley H, Sinha SR, Vining EP. A prospective study of the modified Atkins diet for intractable epilepsy in adults. Epilepsia. 2008;49:316–19.

19. Kossoff EH, Rho JM. Ketogenic diets: evidence for short- and long-term efficacy. Neurotherapeutics. 2009;6:406–14.

20. Pfeifer HH, Lyczkowski DA, Thiele EA. Low glycemic index treatment: implementation and new insights into efficacy. Epilepsia. 2008;49 Suppl 8:42–45.

21. Likhodii S, Nylen K, Burnham WM. Acetone as an anticonvulsant. Epilepsia. 2008;49 Suppl 8:83–86.

22. Fukao T, Lopaschuk GD, Mitchell GA. Pathways and control of ketone body metabolism: on the fringe of lipid biochemistry. Prostaglandins Leukot Essent Fatty Acids. 2004;70:243–51

23. Schwechter EM, Veliskova J, Velisek L. Correlation between extracellular glucose and seizure susceptibility in adult rats. Ann Neurol. 2003;53:91–101.

24. Yudkoff M, Daikhin Y, Nissim I, Lazarow A, Nissim I. Ketogenic diet, brain glutamate metabolism and seizure control. Prostaglandins Leukot Essent Fatty Acids. 2004;70:277–85.

25. Peterson SJ, Tangney CC, Pimentel-Zablah EM, Hjelmgren B, Booth G, Berry-Kravis E. Changes in growth and seizure reduction in children on the ketogenic diet as a treatment for intractable epilepsy. J Am Diet Assoc. 2005;105:718–25.

26. Zupec-Kania B, Zupanc ML. Long-term management of the ketogenic

diet: seizure monitoring, nutrition, and supplementation. Epilepsia. 2008;49 Suppl 8:23–26.

27. Kossoff EH, Zupec-Kania BA, Amark PE et al. Optimal clinical management of children receiving the ketogenic diet: recommendations of the International Ketogenic Diet Study Group. Epilepsia. 2009;50:304–17.

28. Berg MJ, Stumbo PJ, Chenard CA, Fincham RW, Schneider PJ, Schottelius DD. Folic acid improves phenytoin pharmacokinetics. J Am Diet Assoc. 1995;95:352–56.

29. Zupec-Kania B. KetoCalculator: a web-based calculator for the ketogenic diet. Epilepsia. 2008;49 Suppl 8:14–16.

30. Wirrell EC, Darwish HZ, Williams-Dyjur C, Blackman M, Lange V. Is a fast necessary when initiating the ketogenic diet? J Child Neurol. 2002;17:179–82.

31. Vaisleib II, Buchhalter JR, Zupanc ML. Ketogenic diet: outpatient initiation, without fluid, or caloric restrictions. Pediatr Neurol. 2004;31:198–202.

32. Stafstrom CE, Bough KJ. The ketogenic diet for the treatment of epilepsy: a challenge for nutritional neuroscientists. Nutr Neurosci. 2003;6:67–79.

33. Gilbert DL, Pyzik PL, Freeman JM. The ketogenic diet: seizure control correlates better with serum beta-hydroxybutyrate than with urine ketones. J Child Neurol. 2000;15:787–90.

34. The National Institute for Neurological Disorders and Stroke. Stroke: Challenges, Progress, and Promise. 2009. Bethesda (MD), National Institutes of Health. Ref Type: Pamphlet

35. Lloyd-Jones D, Adams R, Carnethon M et al. Heart disease and stroke statistics—2009 update: a report from the American Heart Association Statistics Committee and Stroke Statistics Subcommittee. Circulation. 2009;119:480–86.

36. Chiuve SE, Rexrode KM, Spiegelman D, Logroscino G, Manson JE, Rimm EB. Primary prevention of stroke by healthy lifestyle. Circulation. 2008;118:947–54.

37. Towfighi A, Saver JL, Engelhardt R, Ovbiagele B. Factors associated with the steep increase in late-midlife stroke occurrence among US men. J Stroke Cerebrovasc Dis. 2008;17:165–68.

38. Saposnik G, Ray JG, Sheridan P, McQueen M, Lonn E. Homocysteine-lowering therapy and stroke risk, severity, and disability: additional findings from the HOPE 2 trial. Stroke. 2009;40:1365–72.

39. Brott T, Adams HP, Jr., Olinger CP et al. Measurements of acute cerebral infarction: a clinical examination scale. Stroke. 1989;20:864–70.

40. Adjusted-dose warfarin versus low-intensity, fixed-dose warfarin plus aspirin for high-risk patients with atrial fibrillation: Stroke Prevention in Atrial Fibrillation III randomised clinical trial. Lancet. 1996;348:633–38.

41. Tissue plasminogen activator for acute ischemic stroke. The National Institute of Neurological Disorders and Stroke rt-PA Stroke Study Group. N Engl J Med. 1995;333:1581–87.

42. Fagan SC, Morgenstern LB, Petitta A et al. Cost-effectiveness of tissue plasminogen activator for acute ischemic stroke. NINDS rt-PA Stroke Study Group. Neurology. 1998;50:883–90.

43. The International Stroke Trial (IST): a randomised trial of aspirin, subcutaneous heparin, both, or neither among 19435 patients with acute ischaemic stroke. International Stroke Trial Collaborative Group. Lancet. 1997;349:1569–81.

44. Duncan PW, Zorowitz R, Bates B et al. Management of Adult Stroke Rehabilitation Care: a clinical practice guideline. Stroke. 2005;36:e100–e143.

45. Dennis MS, Lewis SC, Warlow C. Routine oral nutritional supplementation for stroke patients in hospital (FOOD): a multicentre randomised controlled trial. Lancet. 2005;365:755–63.

46. Gaenslen A, Gasser T, Berg D. Nutrition and the risk for Parkinson's disease: review of the literature. J Neural Transm. 2008;115:703–13.

47. Horowski R, Horowski L, Vogel S, Poewe W, Kielhorn FW. An essay on Wilhelm von Humboldt and the shaking palsy: first comprehensive description of Parkinson's disease by a patient. Neurology. 1995;45:565–68.

48. Hirtz D, Thurman DJ, Gwinn-Hardy K, Mohamed M, Chaudhuri AR, Zalutsky R. How common are the "common" neurologic disorders? Neurology. 2007;68:326–37.

49. Nussbaum RL, Polymeropoulos MH. Genetics of Parkinson's disease. Hum Mol Genet. 1997;6:1687–91.

50. Jankovic J. Parkinson's disease: clinical features and diagnosis. J Neurol Neurosurg Psychiatry. 2008;79:368–76.

51. Frank C, Pari G, Rossiter JP. Approach to diagnosis of Parkinson disease. Can Fam Physician. 2006;52:862–68.

52. Hoehn MM, Yahr MD. Parkinsonism: onset, progression and mortality. Neurology. 1967;17:427–42.

53. Biglan KM, Holloway RG. A review of pramipexole and its clinical utility in Parkinson's disease. Expert Opin Pharmacother. 2002;3:197–210.

54. Van CG, Flamez A, Cosyns B et al. Treatment of Parkinson's disease with pergolide and relation to restrictive valvular heart disease. Lancet. 2004;363:1179–83.

55. Kaduszkiewicz H, Zimmermann T, Beck-Bornholdt HP, van den BH. Cholinesterase inhibitors for patients with Alzheimer's disease: systematic review of randomised clinical trials. BMJ. 2005;331:321–27.

56. Delong MR, Juncos JL. Parkinson's disease and other movement disorders. In: Kasper DL, Braunwald E, Fauci AS et al, eds. Harrison's principles of internal medicine. 16th ed. New York: McGraw-Hill; 2006.

57. Gasior M, Rogawski MA, Hartman AL. Neuroprotective and disease-modifying effects of the ketogenic diet. Behav Pharmacol. 2006;17:431–39.

58. Vanitallie TB, Nonas C, Di RA, Boyar K, Hyams K, Heymsfield SB. Treatment of Parkinson disease with diet-induced hyperketonemia: a feasibility study. Neurology. 2005;64:728–30.

59. Palhagen S, Lorefalt B, Carlsson M et al. Does L-dopa treatment contribute to reduction in body weight in elderly patients with Parkinson's disease? Acta Neurol Scand. 2005;111:12–20.

60. Brannan T, Martinez-Tica J, Yahr MD. Effect of dietary protein on striatal dopamine formation following L-dopa administration: an in vivo study. Neuropharmacology. 1991;30:1125–27.

61. Stacy M. Pharmacotherapy for advanced Parkinson's disease. Pharmacotherapy. 2000;20:8S-16S.

62. Pare S, Barr SI, Ross SE. Effect of daytime protein restriction on nutrient intakes of free-living Parkinson's disease patients. Am J Clin Nutr. 1992;55:701–07.

63. Holden K and Remig VM. Parkinson's disease: assessing and managing unique nutrition needs. Chicago (IL): American Dietetic Association; 1999. Ref Type: Pamphlet.

64. Cooper MK, Brock DG, McDaniel CM. Interaction between levodopa and enteral nutrition. Ann Pharmacother. 2008;42:439–42.

65. Miller JW, Selhub J, Nadeau MR, Thomas CA, Feldman RG, Wolf PA. Effect of L-dopa on plasma homocysteine in PD patients: relationship to B-vitamin status. Neurology. 2003;60:1125–29.

66. Tan EK, Cheah SY, Fook-Chong S et al. Functional COMT variant predicts response to high dose pyridoxine in Parkinson's disease. Am J Med Genet B Neuropsychiatr Genet. 2005;137B:1–4.

67. Delikanaki-Skaribas E, Trail M, Wong WW, Lai EC. Daily energy expenditure, physical activity, and weight loss in Parkinson's disease patients. Mov Disord. 2009;24:667–71.

68. Pfeiffer RF. Gastrointestinal dysfunction in Parkinson's disease. Lancet Neurol. 2003;2:107–16.

69. Patrick A, Epstein O. Review article: gastroparesis. Aliment Pharmacol Ther. 2008;27:724–40.

70. Niv E, Fireman Z, Vaisman N. Postpyloric feeding. World J Gastroenterol. 2009;15:1281–88.

71. Alexopoulos GS, Katz IR, Reynolds CF, III, Ross RW. Depression in older adults. J Psychiatr Pract. 2001;7:441–46.

72. Wijesekera LC, Leigh PN. Amyotrophic lateral sclerosis. Orphanet J Rare Dis. 2009;4:3.

73. Vucic S, Kiernan MC. Pathophysiology of neurodegeneration in familial amyotrophic lateral sclerosis. Curr Mol Med. 2009;9:255–72.

74. Traynor BJ, Alexander M, Corr B, Frost E, Hardiman O. Effect of a multidisciplinary amyotrophic lateral sclerosis (ALS) clinic on ALS survival: a population based study, 1996–2000. J Neurol Neurosurg Psychiatry. 2003;74:1258–61.

75. Miller RG, Mitchell JD, Lyon M, Moore DH. Riluzole for amyotrophic lateral sclerosis (ALS)/motor neuron disease (MND). Cochrane Database Syst Rev. 2007;CD001447.

76. Van DP, Robberecht W. Recent advances in motor neuron disease. Curr Opin Neurol. 2009.

77. The National Institute for Neurological Disorders and Stroke. NINDS Guillain Barre' Syndrome Information Page. Available at http://www.ninds.nih.gov/disorders/gbs/gbs.htm. 6-23-2009. 7-31-0009. Ref Type: Electronic Citation

78. van Doorn PA, Ruts L, Jacobs BC. Clinical features, pathogenesis, and treatment of Guillain-Barre syndrome. Lancet Neurol. 2008;7:939–50.

79. Kuwabara S. Guillain-barre syndrome. Curr Neurol Neurosci Rep. 2007;7:57–62.

80. Newswanger DL, Warren CR. Guillain-Barre syndrome. Am Fam Physician. 2004;69:2405–10.

81. Drachman DB, Brodsky RA. High-dose therapy for autoimmune neurologic diseases. Curr Opin Oncol. 2005;17:83–88.

82. Alshekhlee A, Miles JD, Katirji B, Preston DC, Kaminski HJ. Incidence and mortality rates of myasthenia gravis and myasthenic crisis in US hospitals. Neurology. 2009;72:1548–54.

83. Gilhus NE. Autoimmune myasthenia gravis. Expert Rev Neurother. 2009;9:351–58.

84. Romi F, Gilhus NE, Aarli JA. Myasthenia gravis: clinical, immunological, and therapeutic advances. Acta Neurol Scand. 2005;111:134–41.

85. Meriggioli MN. Myasthenia gravis with anti-acetylcholine receptor antibodies. Front Neurol Neurosci. 2009;26:94–108.

86. Onodera H. The role of the thymus in the pathogenesis of myasthenia gravis. Tohoku J Exp Med. 2005;207:87–98.

87. Meriggioli MN. Use of immunoassays in neurological diagnosis and research. Neurol Res. 2005;27:734–40.

88. National Multiple Sclerosis Society. The Disease-Modifying Drugs: Newly Diagnosed. Available at http://www.nationalMSsociety.org 2009.

89. Multiple Sclerosis International Federation. About MS. a . 2009. 7-1-2009. Ref Type: Electronic Citation

90. Alonso A, Hernan MA. Temporal trends in the incidence of multiple sclerosis: a systematic review. Neurology. 2008;71:129–35.

91. Hernan MA, Olek MJ, Ascherio A. Geographic variation of MS incidence in two prospective studies of US women. Neurology. 1999;53:1711–18.

92. Cantorna MT. Vitamin D and multiple sclerosis: an update. Nutr Rev. 2008;66:S135-S138.

93. Ascherio A, Munger K. Epidemiology of multiple sclerosis: from risk factors to prevention. Semin Neurol. 2008;28:17–28.

94. Oksenberg JR, Baranzini SE, Sawcer S, Hauser SL. The genetics of multiple sclerosis: SNPs to pathways to pathogenesis. Nat Rev Genet. 2008;9:516–526.

95. Munger KL, Zhang SM, O'Reilly E et al. Vitamin D intake and incidence of multiple sclerosis. Neurology. 2004;62:60–65.

96. Ascherio A, Munger KL. Environmental risk factors for multiple sclerosis. Part I: the role of infection. Ann Neurol. 2007;61:288–99.

97. Fleming JO, Cook TD. Multiple sclerosis and the hygiene hypothesis. Neurology. 2006;67:2085–86.

98. Polman CH, Reingold SC, Edan G et al. Diagnostic criteria for multiple sclerosis: 2005 revisions to the "McDonald Criteria." Ann Neurol. 2005;58:840–46.

99. National Multiple Sclerosis Society. Available at http://www.nationalMSsociety.org 2009.

100. Schwarz S, Knorr C, Geiger H, Flachenecker P. Complementary and alternative medicine for multiple sclerosis. Mult Scler. 2008;14:1113–19.

101. Mehta LR, Dworkin RH, Schwid SR. Polyunsaturated fatty acids and their potential therapeutic role in multiple sclerosis. Nat Clin Pract Neurol. 2009;5:82–92.

102. Schwarz S, Leweling H. Multiple sclerosis and nutrition. Mult Scler. 2005;11:24–32.

103. Swank RL, Dugan BB. Effect of low saturated fat diet in early and late cases of multiple sclerosis. Lancet. 1990;336:37–39.

104. Gilgun-Sherki Y, Melamed E, Offen D. Oxidative stress induced-neurodegenerative diseases: the need for antioxidants that penetrate the blood brain barrier. Neuropharmacology. 2001;40:959–75.

105. White LJ, McCoy SC, Castellano V et al. Resistance training improves strength and functional capacity in persons with multiple sclerosis. Mult Scler. 2004;10:668–74.

106. 2009 Alzheimer's disease facts and figures. Alzheimers Dement. 2009;5:234–70.

107. Berrios GE. Alzheimer's disease: A conceptual history. Int J Geriatr Psych. 1990;5:355–65.

108. Bu G. Apolipoprotein E and its receptors in Alzheimer's disease: pathways, pathogenesis and therapy. Nat Rev Neurosci. 2009;10:333–44.

109. Williamson J, Goldman J, Marder KS. Genetic aspects of Alzheimer disease. Neurologist. 2009;15:80–86.

110. Alzheimers. Available at www.alz.org 2009.

111. Moreira PI, Duarte AI, Santos MS, Rego AC, Oliveira CR. An integrative view of the role of oxidative stress, mitochondria and insulin in Alzheimer's disease. J Alzheimers Dis. 2009;16:741–61.

112. Muhammad S, Bierhaus A, Schwaninger M. Reactive oxygen species in diabetes-induced vascular damage, stroke, and Alzheimer's disease. J Alzheimers Dis. 2009;16:775–85.

113. Purnell C, Gao S, Callahan CM, Hendrie HC. Cardiovascular risk factors and incident Alzheimer disease: a systematic review of the literature. Alzheimer Dis Assoc Disord. 2009;23:1–10.

114. Barberger-Gateau P, Raffaitin C, Letenneur L et al. Dietary patterns and risk of dementia: the Three-City cohort study. Neurology. 2007;69:1921–30.

115. Boudrault C, Bazinet RP, Ma DW. Experimental models and mechanisms underlying the protective effects of n-3 polyunsaturated fatty acids in Alzheimer's disease. J Nutr Biochem. 2009;20:1–10.

116. van der Beek EM, Kamphuis PJ. The potential role of nutritional components in the management of Alzheimer's Disease. Eur J Pharmacol. 2008;585:197–207.

117. The National Institute for Neurological Disorders and Stroke. NINDS Alzheimer's Disease Information Page. Available at http://www.ninds.nih.gov/disorders/alzheimersdisease/alzheimersdisease.htm. Updated: 2009 Jul 21.

118. American Psychiatric Association. Diagnostic and Statistical Manual of Mental Disorders. 4th ed, text revision. American Psychiatric Association; 2000.

119. McKhann G, Drachman D, Folstein M, Katzman R, Price D, Stadlan EM. Clinical diagnosis of Alzheimer's disease: report of the NINCDS-ADRDA Work Group under the auspices of Department of Health and Human Services Task Force on Alzheimer's Disease. Neurology. 1984;34:939–44.

120. Folstein MF, Folstein SE, McHugh PR. "Mini-mental state." A practical method for grading the cognitive state of patients for the clinician. J Psychiatr Res. 1975;12:189–98.

121. DeKosky ST, Williamson JD, Fitzpatrick AL et al. Ginkgo biloba for prevention of dementia: a randomized controlled trial. JAMA. 2008;300:2253–62.

122. Farlow MR, Miller ML, Pejovic V. Treatment options in Alzheimer's disease: maximizing benefit, managing expectations. Dement Geriatr Cogn Disord. 2008;25:408–22.

123. Smith KL, Greenwood CE. Weight loss and nutritional considerations in Alzheimer disease. J Nutr Elder. 2008;27:381–403.

124. Fotuhi M, Mohassel P, Yaffe K. Fish consumption, long-chain omega-3 fatty acids and risk of cognitive decline or Alzheimer disease: a complex association. Nat Clin Pract Neurol. 2009;5:140–52.

125. Greve MW, Zink BJ. Pathophysiology of traumatic brain injury. Mt Sinai J Med. 2009;76:97–104.

126. Traumatic Brain Injury. http://www.cdc.gov/TraumaticBrainInjury/tbi_concussion.html 2009.

127. Summers CR, Ivins B, Schwab KA. Traumatic brain injury in the United States: an epidemiologic overview. Mt Sinai J Med. 2009;76:105–10.

128. The National Institute for Neurological Disorders and Stroke. Traumatic Brain Injury: Hope Through Research. Bethesda (MD): National Institutes of Health; 2002. Ref Type: Pamphlet.

129. Teasdale G, Jennett B. Assessment of coma and impaired consciousness. A practical scale. Lancet. 1974;2:81–84.

130. Jumisko E, Lexell J, Soderberg S. The meaning of living with traumatic brain injury in people with moderate or severe traumatic brain injury. J Neurosci Nurs. 2005;37:42–50.

131. Cook AM, Peppard A, Magnuson B. Nutrition considerations in traumatic brain injury. Nutr Clin Pract. 2008;23:608–620.

132. Adelson PD, Bratton SL, Carney NA et al. Guidelines for the acute medical management of severe traumatic brain injury in infants, children, and adolescents. Pediatr Crit Care Med. 2003;4:S1-S491.

133. Bessey PQ. Metabolic Response to Critical Illness. ACS Surgery. 2004. 7-29-2009. Ref Type: Electronic Citation.

134. Perel P, Yanagawa T, Bunn F, Roberts I, Wentz R, Pierro A. Nutritional support for head-injured patients. Cochrane Database Syst Rev. 2006;CD001530.

135. Slone DS. Nutritional support of the critically ill and injured patient. Crit Care Clin. 2004;20:135–57.

136. Guidelines for the management of severe traumatic brain injury. J Neurotrauma. 2007;24 Suppl 1:S1–106.

137. Foley N, Marshall S, Pikul J, Salter K, Teasell R. Hypermetabolism following moderate to severe traumatic acute brain injury: a systematic review. J Neurotrauma. 2008;25:1415–31.

138. National Spinal Cord Injury Statistical Center. Available at http://www.spinalcord.uab.edu 2009.

139. Kasper DL, Braunwald E, Fauci AS et al. Diseases of the spinal cord. In: Hauser SL, Ropper AH, eds. Harrison's principles of internal medicine. 17th ed. New York: McGraw-Hill Medical Publishing Division; 2008.

140. American Dietetic Association. Spinal Cord Injury (SCI) Evidence-Based Nutrition Practice Guideline 2009. Available at https://www.adaevidencelibrary.com 2009.

141. Laughton GE, Buchholz AC, Martin Ginis KA, Goy RE. SHAPE SCI Research Group. Lowering body mass index cutoffs better identifies obese persons with spinal cord injury. Spinal Cord. 2009;47(10):757–62.

21

Diseases of the Respiratory System

Ethan A. Bergman, PhD, RD, CD, FADA
Central Washington University
Susan N. Hawk, PhD, RD
Central Washington University

CHAPTER OUTLINE

Introduction

Respiratory disease places a significant burden on health care systems throughout the world. Health care costs related to respiratory diseases in the United States are estimated at over $25 billion annually. One in every six deaths in America is related to lung disease, the third leading cause of death.[1] Asthma and chronic obstructive pulmonary disease (COPD) are among the ten leading chronic problems associated with reduction in activity.[2] Nutritional status and pulmonary function are interdependent. In a healthy individual the respiratory system receives oxygen for cellular metabolism and expires waste products. Metabolic fuels—carbohydrate, protein, and lipid—are metabolized, using oxygen and producing carbon dioxide. The type of fuel an individual receives can affect physiological conditions and interfere with normal respiration. Malnutrition can evolve

2,3-diphosphoglycerate (DPG)—an important regulator for the affinity of hemoglobin for oxygen. The synthesis of 2,3-biphosphoglycerate in red blood cells (RBC) is critical for controlling hemoglobin affinity for oxygen

acute lung injury (ALI)—term designating clinical and radiographic changes in lung function associated with critical illness. ARDS is the most severe form of acute lung injury

acute respiratory distress syndrome (ARDS)—respiratory failure (RF) resulting from an acute insult to the lungs that occurs when the respiratory system is no longer able to perform its normal functions

aerophagia—the swallowing of too much air resulting in gas and bloating

aspiration pneumonia—an inflammatory response in the lung that results from aspiration of inhaled materials (saliva, nasal secretions, bacteria, liquids, food, or gastric contents) into the airway below the level of the vocal cords

asthma—a chronic inflammatory disorder of the airway involving many cells and cellular elements such as mast cells, eosinophils, T lymphocytes, macrophages, neutrophils, and epithelial cells

auscultation—a technique used during physical examination in which a stethoscope is used to evaluate the sounds created in body organs

autosomal recessive—method of hereditary disease transmission in which the patient receives two chromosomes bearing the gene anomaly, one from each parent

bronchial hyperreactivity—tendency of the smooth muscle of the tracheobronchial tree to narrow in response to a stimulus; present in virtually all symptomatic patients with asthma

bronchitis—a condition characterized by inflammation and eventual scarring of the lining of the bronchial tubes accompanied by restricted airflow, excessive mucus production, and a persistent cough

bronchopulmonary dysplasia (BPD)—a chronic lung disorder that may affect infants who have been exposed to high levels of oxygen therapy and ventilator support

chest physiotherapy—physical therapy that includes a variety of techniques designed to reduce or prevent infection by clearing pooled secretions and/or infected materials from the lungs

chronic obstructive pulmonary disease (COPD)—a disease that limits airflow through either inflammation of the lining of the bronchial tubes or destruction of alveoli

clubbing—changes in fingers and toes due to hypoxemia; fingers and toes show a curve at a tip of the nail with flattening surface

cor pulmonale—an increase in size of the right ventricle of the heart caused by resistance to the passage of blood through the lungs; can lead to heart failure

cyanosis—blue-tinged mucous membranes and skin due to inadequate oxygen supply

cystic fibrosis (CF)—disease characterized by abnormally thick mucus secretions from the epithelial surfaces of various organ systems, including the respiratory tract, the gastrointestinal tract, the liver, the genitourinary system, and the sweat glands

dyspnea—shortness of breath or difficulty breathing

emphysema—a condition characterized by thinning and destruction of the alveoli, resulting in decreased oxygen transfer into the bloodstream and shortness of breath

esophageal stricture—a significant narrowing of the esophagus that may significantly interfere with swallowing

hyperinflation of the lungs—results from loss of elasticity of the alveoli, causing air to be trapped; often seen in emphysema

hypoxemia—condition in which there is an inadequate supply of oxygen in the blood

inhaled antibiotics—medications designed to reduce airway infection, particularly *Pseudomonas aeruginosa*, commonly seen in CF

inhaled anti-inflammatory agents—class of medications that often includes inhaled corticosteroids

inhaled bronchodilators—medications used to maximize airway size and improve clearance of mucus

inhaled mucolytics—class of medications, including pulmozyme™, which hydrolyzes the DNA in sputum of cystic fibrosis patients and reduces sputum viscosity

leukotrienes—powerful inflammatory mediators produced by the body that are important in inflammation and allergic reactions because of their ability to constrict blood vessels and attract a variety of types of immune cells

mechanical ventilation—artificial ventilation using a ventilator or respirator; performed with a piece of equipment designed to intermittently or continuously assist or control pulmonary ventilation

minute ventilation—the volume of air per unit time moved into or out of the lungs; measured by collecting expired volume for a fixed time

necrotizing enterocolitis (NEC)—a condition that occurs primarily in premature infants or sick newborns, in which intestinal tissue dies; the cause for this disorder is unknown, but it is thought to be due to decreased blood flow to the bowel, which keeps it from producing the normal protective mucus; if an infant is suspected of having necrotizing enterocolitis, feedings are stopped to allow the bowel to rest

orthopnea—difficulty breathing while lying down

pancreatic function tests—tests to measure pancreatic function, including serum amylase or lipase, a test for the amount of fat in the stool, and an X-ray of the anatomical features of the pancreas and common bile duct

percussion—a technique used during physical examination in which the hands are used to strike the body's surface, and the sounds that are transmitted from the underlying tissues and organs are evaluated

plasma prothrombin concentrations—a measure of blood clotting ability

pleural effusion—accumulation of fluid between the two outer membranes surrounding the lungs

pneumonia—inflammation of the lungs, usually caused by bacteria, viruses, or fungi

pulmonary consolidation—changes in tissue structure of the lungs; often visualized as opaque components on a chest X-ray

rales—bubbly sounds heard via stethoscope that may indicate pulmonary pathology

secondary polycythemia—condition in which an excessive number of red blood cells are produced; occurs in response to compensation for chronic hypoxemia

surfactant—substance secreted by the alveolar cells of the lung that serves to maintain the stability of pulmonary tissue by reducing the surface tension of fluids that coat the lung

sweat chloride test—a test to measure the amount of chloride in the sweat by stimulating the skin to produce a large amount of sweat that is then absorbed by a special filter paper and analyzed for chloride content

tracheostomy—a surgical opening placed in the trachea to assist breathing

tumor necrosis factor (TNF-α)—one type of cytokine which has been found to possess a wide range of proinflammatory actions

upper respiratory infection (URI)—a nonspecific term used to describe acute infections involving the nose, sinuses, pharynx, larynx, trachea, and bronchi; often referred to as the common cold

from pulmonary disorders and can contribute to declining pulmonary status. This chapter will review the normal anatomy and physiology of the respiratory system and discuss the major diagnoses that have a nutritional impact.

Normal Anatomy and Physiology of the Respiratory System

The pulmonary system includes two divisions: the upper and lower respiratory tracts, with their respective organs. The upper respiratory tract includes the nose, nasal cavity, frontal and maxillary sinuses, larynx, and trachea. The lower respiratory tract includes the lungs, bronchi, and alveoli. The lungs allow the body to obtain oxygen to support cellular metabolic functions and to remove carbon dioxide produced by this process. In the course of a day, over 8,000 liters of air meet 8,000–10,000 liters of blood pumped by the heart through the pulmonary artery.[3] The air enters the lungs through the trachea, which divides into the right and left bronchi, supplying the right and left lung (see Figure 21.1). The bronchi further divide again and again into smaller and smaller bronchioles. The bronchi-

oles end in small air sacs called alveoli, which are paper thin. Each alveoli is imbedded with millions of capillaries that are responsible for the exchange of oxygen and carbon dioxide. The skeleton and muscles surrounding the lungs, particularly the intercostal and diaphragmatic muscles, support the respiratory system.

The lungs undergo a period of growth and maturation during the first two decades of life. By 10–12 years of age the maximal number of alveoli is attained. Full maturation of the respiratory system is achieved by age 20 for females and 25 for males. Aging is associated with a progressive decrease in lung function; however, unless affected by disease, the lungs are capable of providing adequate gas exchange during the entire lifespan.[4]

Gas exchange occurs in the alveolar-capillary unit which, in the adult lung, covers approximately the area of a tennis court and contains more than 100 million capillaries (see Figure 21.2). The alveolar-capillary unit consists of the capillary endothelium and its basement membrane, the interstitial space, and the alveolar epithelium and its basement membrane. The interstitial space consists of the tissue layers between the lungs' air sacs (alveoli), and is very thin, which allows for efficient exchange of

Figure 21.1 Anatomy of the Pulmonary System

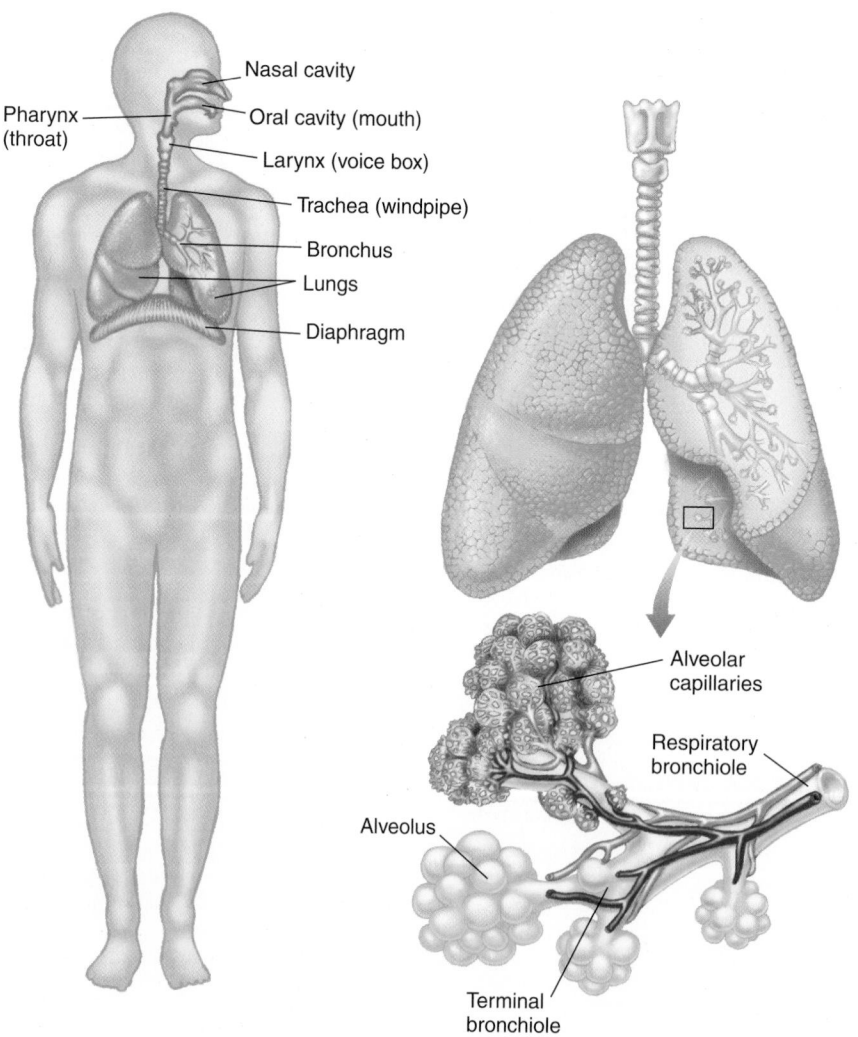

Source: R. Rhoades and R. Pflanzer, *Human physiology*, 4e, copyright © 2003, p. 633.

Figure 21.2 Gas Exchange between Alveoli and Capillaries

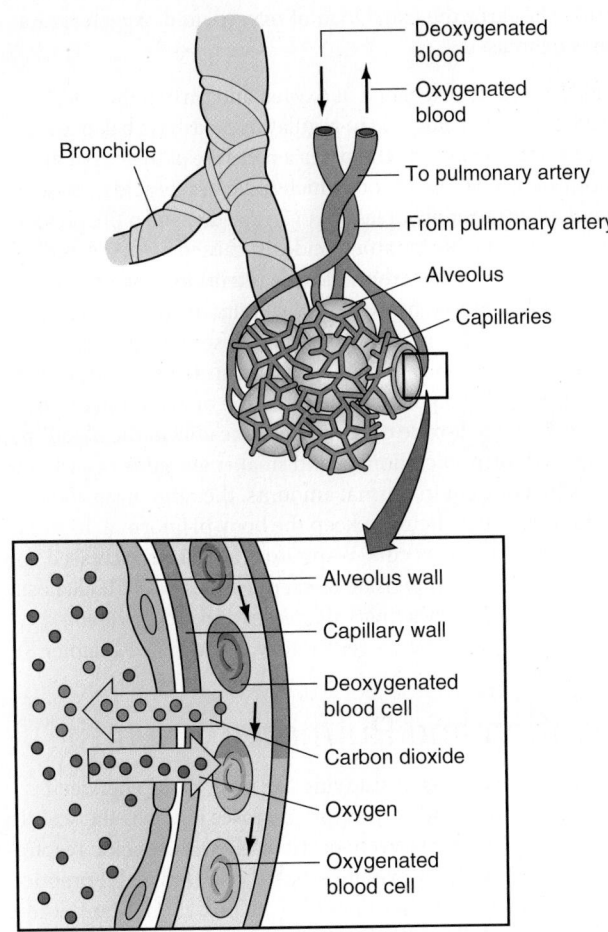

- Deoxygenated blood
- Oxygenated blood
- Bronchiole
- To pulmonary artery
- From pulmonary artery
- Alveolus
- Capillaries

- Alveolus wall
- Capillary wall
- Deoxygenated blood cell
- Carbon dioxide
- Oxygen
- Oxygenated blood cell

Source: From the Merck Manual of Medical Information—Home Edition, edited by Robert S. Porter. Copyright 2006 by Merck & Co., Inc., Whitehouse Station, NJ. Available at: http://www.merck.com/mmhe

gas if ventilation is adequate. The alveolar epithelium consists of two types of cells: type I and type II cells. Type I cells form the structure of the alveolar wall. Type II cells are responsible for producing **surfactant**, a fluid secreted by the cells of the alveoli that reduces the surface tension of pulmonary fluids and contributes to the elastic properties of pulmonary tissue. At rest, the alveolar ventilation is approximately equal to the cardiac output.

The lungs also play an important role in protecting the body against infection and harmful environmental toxins. Inhaled particles such as smoke, bacteria, and viruses pass through the nose and are trapped in the lungs by a sticky mucus substance, which serves to keep the airway moist. The cells that line the trachea, bronchi, and bronchioles contain tiny, hairlike cells called cilia, which beat with a rhythm fast and forceful enough to propel the mucus and unwanted cells upward toward the pharynx where they can be coughed out or swallowed (see Figure 21.3). Additionally, the epithelial surface of the alveoli contains macrophages or scavenger cells that engulf and destroy the inhaled bacteria.

Measures of Pulmonary Function

The initial evaluation of pulmonary function is generally accomplished with the physical examination and the tools of **percussion** and **auscultation**. Using these techniques for evaluating sounds, abnormalities in breathing and underlying organs may be detected. For example, dull or low-pitched sounds may suggest **pulmonary consolidation** or **pleural effusion**, which may occur with damaged cells during a disease process such as pneumonia. **Rales** are bubbly sounds created by a change in air flow that could occur when inflammation is present or when excessive mucus changes air flow.

Pulmonary function tests are used to detect lung diseases or to monitor the progression of a particular disease. In pulse oximetry, which is often used at bedside or in any outpatient

Figure 21.3 Mucociliary Transport System

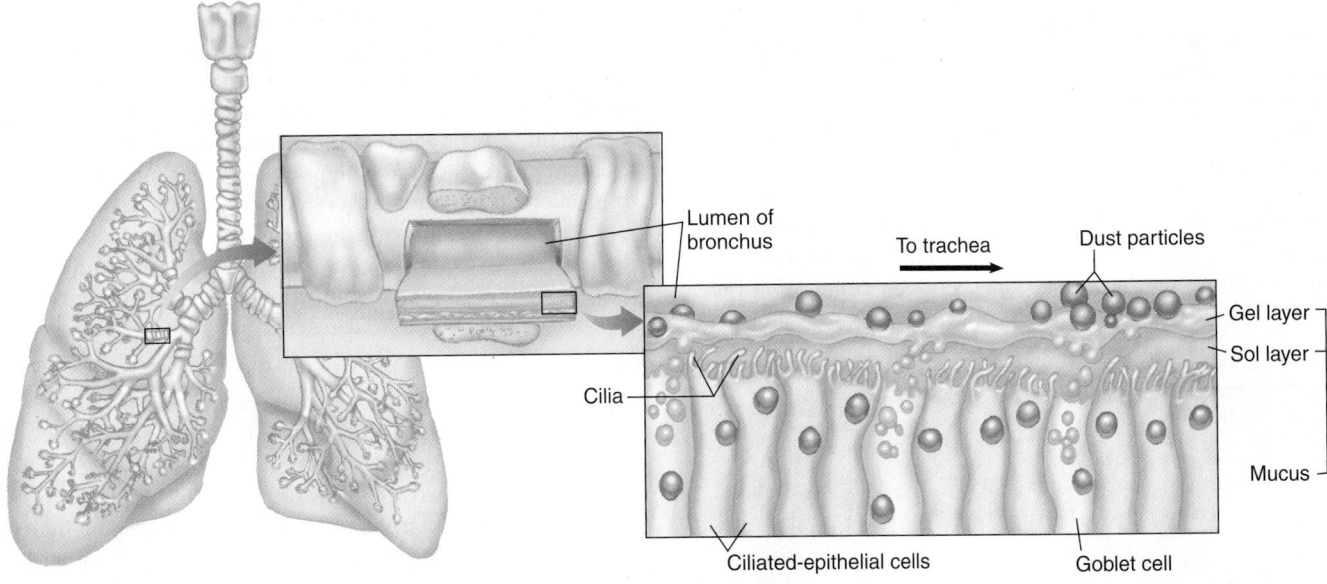

- Lumen of bronchus
- To trachea
- Dust particles
- Gel layer
- Sol layer
- Mucus
- Cilia
- Ciliated-epithelial cells
- Goblet cell

Source: R. Rhoades and R. Pflanzer, *Human physiology*, 4e, copyright © 2003, p. 635.

Figure 21.4 Pulse Oximeter

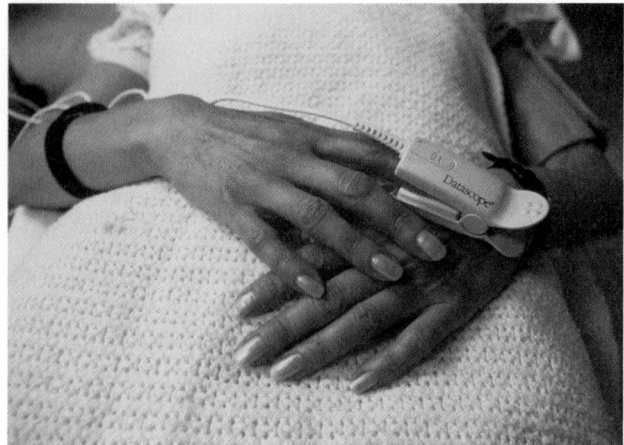

Source: Josh Sher/Photo Researchers, Inc.

setting, light waves measure the oxygenation of arterial blood. The pulse oximeter (shown in Figure 21.4) is able to detect the percentage of oxygen within the hemoglobin molecule based on the color of the blood. The finger or ear lobe is the most common site for using the pulse oximeter. The most common pulmonary function test is done with a machine called a spirometer.[5] During spirometry, the patient breathes into a tube attached to the machine, which calculates the amount of air the lungs can hold and the rate the air can be inhaled and exhaled. The results of the test are compared with those of healthy individuals of similar height and age, and of the same sex and race.

Common spirometry measurements include FVC (forced vital capacity), the total volume of air expired after a full inspiration, and FEV1 (forced expiratory volume in one second), the volume of air exhaled in the first second after a deep inhalation. In addition to spirometry, another test used to evaluate lung function is gas diffusion. Gas diffusion measures how well oxygen and other gases pass through the lung's air sacs and are absorbed by the blood. A reduced diffusing capacity could indicate pulmonary disease.

Evaluation of arterial blood gases (ABGs) determines the pH, oxygen content, and carbon dioxide content of the blood and can also be used to measure pulmonary function (see Table 21.1). Changes in pulmonary function are reflected by changes in the partial pressure of dissolved carbon dioxide ($PaCO_2$) and partial pressure of dissolved oxygen (PaO_2). Changes in the $PaCO_2$ measure how well carbon dioxide is able to move out of the blood into the airspaces of the lung, and then out with the exhaled air. Changes in PaO_2 measure how well oxygen is able to move from the air into the lungs. Oxygen saturation is the

measure of the amount of oxygen carried by the red blood cells and can be calculated using the partial pressure of dissolved oxygen (PaO_2). In patients with pulmonary disease, fewer red blood cells carry the usual load of oxygen, and oxygen saturation is decreased.

In addition to the exchange of oxygen and carbon dioxide, the lungs play a major role in regulating acid-base balance as discussed in Chapter 8. The pH is a measure of hydrogen ion concentration (H^+) in blood, which indicates its acid or base (alkaline) nature; a pH of less than 7 is acidic, and a pH greater than 7 is alkaline. Respiratory acidosis, caused by decreased ventilation, results in carbon dioxide retention, while respiratory alkalosis, caused by increased ventilation, results in loss of carbon dioxide. The respiratory system responds quickly to acid-base disturbances through changes in **minute ventilation** that alter blood pH by regulating the retention or excretion of carbon dioxide. Carbon dioxide dissolves more readily in the blood than oxygen and forms bicarbonate and smaller amounts of carbonic acid. When present in normal amounts, the ratio of carbonic acid to bicarbonate helps to keep the body pH normal. In situations where acidosis occurs, respiratory activity is increased and the lungs quickly compensate to excrete excess CO_2. If alkalosis is present, respiratory activity decreases and CO_2 is retained, producing a compensatory respiratory acidosis (see Chapter 8).

Nutrition and Pulmonary Health

One of the difficulties in studying the relationship between nutrition and diet and respiratory diseases involves the scarcity of evidenced-based research, particularly with specific respiratory diseases. Without evidence-based research, valid practice recommendations are not possible (see Box 21.1). Many studies looking particularly at food and nutrient intake related to the etiology of respiratory disease are based on retrospective population studies, rather than intervention trials. Unfortunately, population studies are often contradictory due to a variety of factors, including sampling errors, selection bias, and low statistical power. Even intervention studies sometimes show conflicting outcomes, resulting from differences in study design.

Malnutrition has been shown to have an adverse affect on clinical outcomes. The impact of protein-energy malnutrition on lung function has been examined in both clinical and animal studies. Also, the effects of weight loss on pulmonary function in individuals without lung disease have been described.[6] Malnutrition associated with poor intake appears to have an impact on the strength and endurance of respiratory muscles, particularly the diaphragm, and may also cause reductions in lung parenchyma (respiratory bronchioles, alveoli, and capillaries). With continued malnutrition, increased incidence of pulmonary infection may also occur, as a result of depressed immune function.[7]

There is mounting evidence correlating the role of dietary antioxidants such as vitamin C, vitamin E, β-carotene, and selenium with healthy lung function. A variety of antioxidants are present in the extracellular fluid (ECF) and appear to play an important role in protecting the lungs from oxidant injury as the result of the inflammatory process caused by the inhalation of cigarette smoke and other pollutants.[8] Results of a three-year population-based study looking at dietary intake of antioxi-

TABLE 21.1 Normal Blood Gas Values

Blood Gas	Normal Values
Partial pressure of oxygen (PaO_2)	75–100 mm Hg
Partial pressure of carbon dioxide ($PaCO_2$)	35–45 mm Hg
pH	7.35–7.45
Oxygen saturation (O_2Sat)	94%–100%
Bicarbonate—(HCO_3)	22–26 mEq/liter

BOX 21.1 RESEARCH TO PRACTICE

Serum Cholesterol and Pulmonary Health

The relationship between high serum cholesterol levels and the incidence of cardiovascular disease is well established. However, additional research has demonstrated an association between low serum cholesterol levels (<160 mg/dL) and increased risk for noncardiovascular mortality, including respiratory disease. Low serum cholesterol is thought to be an indicator of poor health and may be lowered in the wasting syndrome associated with an underlying disease process.[1, 2]

The relationship between low cholesterol levels and poor respiratory health is not clearly understood. Since cholesterol is a component of pulmonary surfactant, it has been suggested that low circulating cholesterol levels could result in impaired production of pulmonary surfactant.[3] Using data from the Third National Health and Nutrition Examination Survey, Cirillo et al. investigated the relationship between serum lipids and pulmonary health.[4, 5] Among young and middle-aged persons with low cholesterol levels, higher LDL components were associated with worse lung function, while higher HDL components were associated with better lung function, suggesting a possible association between the lipid sub-fractions and pulmonary health. However, in older persons with low cholesterol levels, lower LDL components were associated with a greater risk of respiratory disease. No association between HDL levels and lung function was seen in this age group. The authors concluded that the lipoprotein components may be markers for some aspect of diet or lifestyle that may affect lung function. Further research is needed to identify how cholesterol levels affect pulmonary health. Of specific interest is whether the use of drug treatment to lower cholesterol levels may affect pulmonary health in the elderly.

References

1. Jacobs D, Blackburn H, Higgins M, Reed D, Iso H, McMillan G, Neaton J, Nelson J, Potter J, Rifkind B, et al. Report of the Conference on Low Blood Cholesterol: Mortality Associations. Circulation. 1992;86:1046–60.
2. Brescianini S, Maggi S, Farchi G, Mariotti S, Carlo A, Baldereschi M, et al. Low cholesterol and increased risk of dying: Are low levels clinical warning signs in the elderly? Results from the Italian longitudinal study on aging. Am Geriatr Soc. 2003;51:991–96.
3. Volpato S, Zuliani G, Guralnik JM, Palmieri E, Fellin R. The inverse association between age and cholesterol level among older patients: the role of poor health status. Gerontology. 2001;47:36–45.
4. Cirillo D, Agrawai Y, Cassano P. Lipids and pulmonary function in the third national health and nutrition examination survey. Am J Epidemiol. 2002;155:842–48.
5. Iribarren C, Jacobs D, Sidney S, Claxton AJ, Gross M, Sadler M, et al. Serum total cholesterol and risk of hospitalization and death from respiratory disease. Int J Epidemiol. 1997;26:1191–1202.

dants and lung function found a positive association between pulmonary function and intakes of vitamin C, vitamin E, and carotenoids.[9, 10] Of all the carotenoids studied, lutein/zeaxanthin had the strongest relationship to pulmonary function as measured by FEV and FEV1. The strong relationship between antioxidants and respiratory health has prompted the use of antioxidants as a novel therapy for a reducing inflammation in a variety of respiratory illnesses.[11] This relationship has not held up within clinical trials for prevention of lung cancer and actually may be contraindicated.

Cigarette smoking is associated with reduced levels of antioxidants in various body fluids. Smokers have been shown to have depleted levels of serum ascorbate, α-tocopherol, β-carotene, and selenium.[8] A meta-analysis examining the relationship between cigarette smoking and nutrient intake showed that smokers had higher intakes of energy, total fat, saturated fat, cholesterol, and alcohol and lower intakes of antioxidants and fiber compared to nonsmokers.[12] The metabolic turnover for vitamin C is about 35 mg/day greater for smokers than non-smokers. Because of this, it is recommended that people who smoke consume an additional 35 mg/day of vitamin C beyond the DRI.[13]

Respiratory disease often includes a variety of symptoms that may affect dietary intake, including early satiety, anorexia, weight loss, cough, and **dyspnea** during eating. As the disease progresses, these symptoms may have a marked impact on nutritional status. For this reason, it is important that a nutritional assessment be initiated that includes an evaluation of weight history, nutrient intake, medication use, biochemical markers (albumin and prealbumin, lipid profile), and functional status.

Asthma

Definition

Asthma is a chronic inflammatory disorder of the airway involving many cells and cellular elements, such as mast cells, eosinophils, T lymphocytes, macrophages, neutrophils, and epithelial cells. Inflammation is the primary problem in asthma and is thought to be primarily immunoglobulin E (IgE) mediated (see Chapter 9). In susceptible individuals, this inflammation causes recurrent episodes of wheezing, breathlessness, chest tightness, and coughing, particularly at night or in the early morning. These episodes are usually associated with airflow obstruction that is often reversible either spontaneously or with treatment.

Epidemiology

The prevalence of asthma in western countries has increased since the early 1980s across all age, sex, and racial groups.[14] Asthma is the third leading cause of hospitalization and chronic illness among children under 15 years of age.[15]

Etiology

Asthma is usually divided into two types: allergic and nonallergic asthma.[14] Of the two, allergic asthma is the most common and is triggered predominantly by inhaled indoor allergens such as dust mite allergen, pet dander, pollen, and mold.[16] Nonallergic asthma is caused by other factors, such as anxiety, stress, exercise, cold air, dry air, hyperventilation, smoke, viruses, or other irritants. Patients with asthma are at greater risk for life-threatening allergic reactions to foods.[17] Persistent asthma has been associated with elevated IgE to egg and wheat, though food allergies are rarely a cause of asthma.

Epidemiological studies have demonstrated that there is an association between obesity and asthma. A systematic review of the literature indicated that weight loss resulted in the improvement of at least one asthma symptom.[18]

Pathophysiology

When asthma occurs, bronchi and bronchioles respond to stimuli by contraction of smooth muscle (bronchoconstriction). The mucosa is inflamed and edematous, with an increased production of mucus. This results in a partial or totally obstructed airway.

Clinical Manifestations

The initial symptoms the patient may experience include cough, dyspnea, and a tight feeling in the chest. Signs may include wheezing, increased respiratory rate, and labored breathing. Increased heart rate (tachycardia) and hypoxia may also be observed. Longer, prolonged episodes of asthma may result in respiratory alkalosis that can proceed to respiratory acidosis as well.

Treatment

An acute episode of asthma requires immediate attention to dilate airways and improve oxygenation. These interventions would include the use of a bronchodilator with a β-adrenergic agent such as ipratropium and theophylline (see Table 21.4). Chronic and long-term control of asthma is achieved using a variety of agents, including steroids and leukotriene receptor antagonists. Other treatments will include the use of environmental control of potential allergens and the development of controlled breathing techniques.

Treatment of asthma includes removal of any items that are known to be asthma triggers from the patient's environment. When these measures do not work, there are a variety of medications used to control symptoms. Generally, medications are divided into those that provide quick relief and those designed to provide long-term control. Quick-relief medicines, usually bronchodilators, are used to ease the wheezing, coughing, and tightness of the chest that occur during asthma episodes. Long-term medications include anti-inflammatory agents, such as oral steroids, that are designed to make the airways less sensitive and prevent them from reacting as easily to triggers.

Nutrition Therapy for Asthma

It is thought that diet and nutrition play a role in the development and treatment of asthma. Increases in asthma, particularly in developed countries, have paralleled increases in obesity. Data from NHANES I showed that adults with asthma had a 46% higher prevalence of obesity than those without asthma; this ratio remained relatively constant from NHANES I (1971–1975) through NHANES III (1988–1994).[19] This relationship has been consistently observed in women and men.[20] The strongest

association of BMI with asthma severity in women has been associated with early menarche.[21] In children, increased risk of new-onset asthma has been associated with increased BMI;[22] in fact, obese children are at a 50% greater risk for developing asthma.[23]

Reasons for the relationship between obesity and asthma are not clear. Some possible reasons include the direct effects of obesity on the mechanical functioning of the lungs, changes in the immune system or an inflammatory response related to obesity, hormonal influences, and the interrelationship between the genes responsible for asthma and obesity.[24]

A number of studies have demonstrated a protective effect of breast-feeding against the development and severity of asthma in children, while other studies have not shown this effect.[25–28] Differences in study design and sampling errors have been cited as factors that may contribute to these conflicting results. Using specific criteria to assess study design, Oddy et al. evaluated twenty-nine previous studies, looking at the relationship between breast-feeding and the development of asthma and allergic diseases.[22] The criteria included standards for assessing breast-feeding status and duration; the use of strict diagnostic criteria for asthma; controlling for other confounding factors such as the mother's education, mother's diet, socioeconomic status, smoking status, and housing type and allergen exposure; and the appropriate use of the statistics, including sufficient statistical power. Of the fifteen studies that met these criteria, all demonstrated a protective effect of breast-feeding; of the remaining twelve studies that did not meet the criteria, four showed a positive effect of breast-feeding, three showed a negative effect, and five showed no effect. Based on this review, these authors concluded that there is clear evidence to demonstrate that breast-feeding does protect against asthma and allergy in childhood. However, the protective effect of breast-feeding on asthma and allergy in adolescence and adulthood still needs to be confirmed since some researchers suggest it is dependent on the exclusivity or the time course of the breast-feeding.[29, 30]

Leukotrienes are chemical mediators produced by the body that contribute to the development of asthma. Leukotrienes, which are synthesized from arachidonic acid, modulate the inflammatory response resulting in tissue edema, mucus secretion, smooth-muscle proliferation, and powerful bronchoconstriction. Two types of leukotriene-based medications have been developed to combat asthma: leukotriene inhibitors that interfere with the actual synthesis of leukotrienes (Zyflo™), and leukotriene antagonists that block their action at the receptor level (Accolate™ and Singulair™).

One possible approach to preventing the synthesis of leukotrienes is through dietary modification. Normally, human inflammatory cells contain high amounts of the omega-6 fatty acid, arachidonic acid, and low amounts of omega-3 fatty acids.[31, 32] Because both omega-6 and omega-3 fatty acids are metabolized by a common pathway, an excess of omega-3 fatty acids interferes with the metabolism of the omega-6 fatty acids and reduces their incorporation into tissue lipids. Although there is some evidence that omega-3 fatty acid supplementation can decrease the production of inflammatory agents, primarily leukotrienes, in asthmatic patients, evidence that omega-3 fatty acid supplementation decreases the clinical severity of asthma in controlled trials has been inconsistent.[33–37] Supplementation of omega-3 fatty acids, vitamin C, and zinc either singly or in combination significantly improved asthma control and pulmonary function tests in addition to inflammatory markers in children.[38] However, a recent meta-analysis of 3,129 articles found no rationale for using omega-3 supplements to prevent or treat asthma.[39] Similarly, an earlier Cochrane review of nine randomized controlled studies of omega-3 fatty acid supplementation from fish or fish products in asthmatic patients found no consistent effects on clinical outcome measures of asthma, including pulmonary function tests, asthmatic symptoms, medication use, and **bronchial hyperreactivity**.[40] Moreover, levels of omega-3 fatty acids do not appear to be lower in asthmatic patients than in the general population.[37]

Interest has also been focused on the antioxidant vitamins—A, C, and E—and the carotenoids. Studies examining the protective effects of antioxidants and the consumption of fruits and vegetables on the development and treatment of asthma have not been conclusive.[41–44] However, low consumption of fruit and the nutrients manganese and vitamin C is associated with adult asthma.[45] Of all the antioxidant vitamins, the data linking vitamin C with asthma appears the strongest.[46] As an antioxidant, vitamin C may modify oxidative insults from inhaled or infectious agents and reduce cellular inflammation. However, a Cochrane review of nine randomized, controlled trials investigating the treatment of asthma using vitamin C supplementation concluded that there is insufficient data to either refute or confirm the role of vitamin C in the management of patients with asthma.[44]

Medications prescribed for treatment of asthma may have a number of nutritionally relevant side effects, including dry mouth, throat irritation, nausea, vomiting, and diarrhea. Long-term use of corticosteroids has been associated with increased serum glucose levels and sodium retention, as well as changes in bone mineral density. The short-term use of inhaled corticosteroids at conventional or usual doses (for two to three years) has not been associated with loss of bone mineral density (BMD) or fractures in adult patients with asthma or mild chronic obstructive pulmonary disease.[47] Higher doses of inhaled steroids have been associated with biochemical markers of increased bone turnover, but data on BMD and fractures are not available. There is a need for further randomized studies that look at the effects of long-term use of conventional and higher doses of inhaled corticosteroids on BMD.

Until the relationship between diet and the etiology of asthma is confirmed, it is most important to recommend a nutritionally adequate diet for individuals suffering from asthma. For overweight individuals, weight loss may result in reversibility of asthma.[48] Individuals who are leaner are also more responsive to corticosteroid treatment than people who are overweight or obese.[49] The dietitian can play a significant role in assessing nutrition status and making appropriate recommendations based on specific needs. Because of the potential benefits of breastfeeding, mothers should be encouraged to breast-feed whenever possible. In light of the positive relationship between BMI and the development asthma, assisting individuals and families with weight control is also important.

Bronchopulmonary Dysplasia

Definition

Bronchopulmonary dysplasia (BPD), also called chronic lung disease of prematurity, is characterized by pulmonary inflammation and impaired growth and development of the alveoli. The definition of BPD is based on the severity of the condition. All babies with BPD have received supplemental oxygen for at least 28 days and are less than 32 weeks gestational age. Babies characterized with mild BPD do not require supplemental oxygen at 36 weeks postmenstrual age (PMA) or discharge. Those with moderate BPD need <30% oxygen by 36 weeks PMA. In cases of severe BPD, babies still require ≥30% O_2 at 36 weeks PMA.[50] Factors associated with BPD include extreme prematurity (birth weight <1500 g), perinatal infection, and the presence of patent ductus arteriosus (PDA).[51, 52]

Etiology

The etiology of BPD is complex and multifactorial, with genetics playing a role in susceptibility to the condition.[53–55] Genetic factors contribute to 53% of the differences seen in patients with BPD.[56] Approximately 20%–40% of infants born prior to 28 weeks gestation and weighing less than 1,000 g develop BPD.[57] While the exact mechanism underlying BPD remains to be fully elucidated, it is linked to ventilator trauma. When born before 28 weeks, infants have immature lungs and require supplemental oxygen on a ventilator for extended periods of time. Prolonged exposure to high oxygen concentrations has been identified as one cause for BPD. Other causative agents may include proinflammatory proteins and reactive oxygen species.[58, 59] The premature infant has a poorly developed antioxidant system and therefore is highly susceptible to oxidative damage.

Poor vitamin A status is another possible factor in the development of BPD. Vitamin A is important in normal alveolar development and surfactant production, and supports the integrity and regeneration of respiratory epithelial cells. Vitamin A deficiency in animals results in epithelial lesions similar to those seen in BPD.[60] Most premature infants are born with low

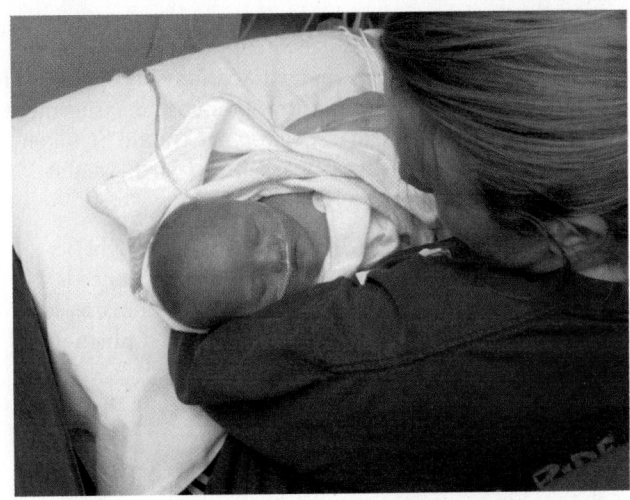

Infants with immature lungs receive supplemental oxygen via a nasal cannula.

Source: http://www.jimmcintosh.com
Courtesy of Jim McIntosh.

serum vitamin A levels and lower levels of vitamin A transport carrier protein, a retinol-binding protein. When supplemented with vitamin A, BPD infants have a significant drop in oxygen use and a reduced death rate.[61, 62]

Treatment

Infants with BPD often require prolonged, intensive hospitalization accompanied by **mechanical ventilation**. The supplemental oxygen helps stabilize breathing, and aides in the growth and weight gain of the infant.[62] Parenteral or enteral nutrition support or both are often required. The best method for preventing BPD is good prenatal care for the pregnant woman, including maintenance of good nutritional status, which helps ensure that infants are born full term. If a premature delivery is expected, use of prenatal steroids can help improve surfactant levels and antioxidant status, resulting in less neonatal morbidity and mortality.[63]

Nutrition Therapy for Bronchopulmonary Dysplasia

Nutrition Assessment and Diagnosis

The causes of growth failure include concomitant organ dysfunction resulting in congestive heart failure and renal insufficiency in some infants, and decreased nutrient intake, **hypoxemia**, and increased energy requirements in others. Poor nutrient intake is often related to a variety of factors, including poor swallowing function, oral aversion, reflux esophagitis, fatigue during feeding, and the need for fluid restriction resulting from fluid retention.[64] Increased oxygen consumption reflected by increased work of breathing is thought to be one factor related to the increased energy requirement seen in these infants. However, the resting metabolic rate is also higher and contributes to their increased energy and nutrient needs.[65] Infants with BPD are also at risk for delayed skeletal mineralization and osteopenia.

A complete nutrition assessment to evaluate specific nutrient needs is essential. Infants with BPD have difficulty maintaining weight gain and achieving development similar to normal healthy infants. Since infants with BPD are at nutritional risk and may have unique nutrition needs, nutrition screening should performed. Anthropometric measurements of length, weight, and head circumference should be taken routinely, using growth charts for very low-birth-weight children.[66]

Possible nutrition diagnoses in infants with BPD might be: Increased energy expenditure or increased nutrient needs, inadequate intake from enteral/parenteral nutrition, breastfeeding difficulty, swallowing difficulty, altered GI function, and/or underweight.

Nutrition Intervention

Because of immature swallowing function and oral aversion related to endotracheal and suctioning stimuli, nutrition support initially often includes nasogastric tube feedings. Nutrient requirements are often complicated by the need for fluid and sodium restrictions and the use of diuretics and other medications, such as steroids, that may cause nutrient losses and catabolism. When there is no longer risk of aspiration, and swallowing functions have matured, specially designed infant formulas to meet kcal, protein, calcium, phosphorus, vitamin, and mineral needs are used.

Energy and Macronutrient Needs The energy needs of infants with BPD are generally 15% to 25% higher than those of healthy, normal infants.[67] An estimated 120 to 130 kcal/kg are often required to provide for appropriate growth.[68] Energy intake and weight/length gains should be monitored closely, and energy goals should be adjusted to maintain proper growth. In some instances, energy intakes of 130 to 160 kcal/kg may be required due to increased metabolic demand. The metabolism of carbohydrate versus fat may increase the production of carbon dioxide (CO_2). Although high-carbohydrate feedings have been shown to increase CO_2 production in infants with BPD, they have not been associated with worsening of pulmonary function.[69, 70]

Protein catabolism induced by corticosteroid medications can be significant. Protein intakes of 3–4 g/kg/day have been recommended to meet needs.[71, 72] Intakes of greater than 4 g/kg/day should be avoided because of the risk of acidosis in preterm infants with immature kidneys. This occurs because the premature infant has a reduced ability to excrete both acid and the renal solutes (urea nitrogen and electrolytes). Elevations in blood urea nitrogen (BUN) and serum ammonia are used to monitor protein tolerance.

Vitamins and Minerals Lung injury in the preterm infant appears to occur within a short time after delivery, and oxidation appears to be a major contributor to this process. Unfortunately, attempts to deliver antioxidants directly to the lung have not been successful. As discussed earlier, preterm infants are born with reduced stores of vitamin A. Spears et al. found that low plasma retinol concentrations during the first month of life were significantly associated with increased risk for developing BPD in very low-birth-weight infants (<1250 g).[73] It has been determined that low plasma retinol concentration up to six months of age can produce a similar effect. The need for long-term oxygen support was significantly greater for infants with low vitamin A status versus those with higher vitamin A status. Supplementation of 1500 to 2800 IU/kg/day or 450 to 840 µg/kg/day of vitamin A appears to be safe and has led to decreased incidence of BPD, decreased days on mechanical ventilation and supplemental oxygen, and a decreased number of days in intensive care.[74, 75]

Electrolyte balance, particularly of sodium and chloride, is essential to maintain growth. The use of diuretics to treat pulmonary edema and fluid overload is common. Diuretics increase urinary losses of sodium, potassium, chloride, and calcium. Because preterm infants receiving diuretic therapy often exhibit signs of electrolyte depletion, close monitoring of electrolyte balance and correction of electrolyte abnormalities are necessary.

Infants are at significant risk for delayed skeletal maturation and osteopenia. Low body calcium stores are aggravated by the use of diuretics. The use of corticosteroids is also likely to reduce bone mineral deposition. Infant formulas specifically designed for the premature infant provide sufficient minerals to support bone growth (see Box 21.2). Table 21.2 summarizes the nutrient needs of infants with BPD.

Table 21.2 Increased Needs for Infants
with Bronchopulmonary Dysplasia

BOX 21.2 CLINICAL APPLICATIONS

Preterm Infant Formulas

- **Human milk** produced by mothers of infants born prematurely (or before term) is referred to as *preterm milk*. Preterm milk varies in nutrient composition from the milk of mothers of infants born at term. Preterm milk alone does not meet the increased nutrient needs of the premature infant and must be fortified.

- **Human milk fortifiers** are powdered or liquid supplements that are added to breast milk in order to increase the nutrient content to meet the needs of the premature infant.

- **Preterm infant formulas** are also available for premature infants whose mothers cannot provide human milk. Preterm infant formulas are available in 20 kcal/oz and 24 kcal/oz concentrations and contain nutrients to meet the special needs of the premature infant. Most premature infants require 24 kcal/oz for adequate growth. Infant formulas are available ready to feed, or as powders or concentrates; powders and concentrates require the addition of water to dilute them to the appropriate caloric concentration.

Table 21.2 Increased Needs for Infants with Bronchopulmonary Dysplasia

Energy	120–160 kcal/kg/day	May be 15% to 25% higher than normal
Protein	3–4 g/kg/day	Avoid providing over 4.
Vitamin A	1500–2800 IU/kg/day or 450–840 micrograms/kg/day	These levels appear to be safe and sustain adequate Vitamin A levels.
Electrolytes (sodium, potassium, and chloride)	Monitor and supplement when required	Diuretic therapy may increase urinary losses.
Calcium	Monitor and supplement when required	Diuretic therapy may increase urinary losses.

Feeding Practices Early, aggressive nutrition support may provide protection from the development of BPD and improve the nutritional status of infants with BPD. Once BPD has developed, nutrition plays a crucial role in preventing further damage and promoting healing. Initially very low-birth-weight infants may be placed on parenteral nutrition support. Gradual transition from parenteral to enteral feedings, using human milk, should be initiated as soon as possible; human milk is highly digestible and contains many antiinfective components that will help reduce the risk of infection.[76] Transition may take weeks or even months in some infants.[77] Enteral feeding advancement may be delayed due to poor peristalsis and medical and respiratory instability. Once initiated, continuous nasogastric feedings are often used, with transition to bolus feedings as respiratory status improves. Stimulation of oral-motor skills should also occur to prepare the infant for oral feeding.

Because these infants often have higher energy needs, they are often placed on high-kcal enteral formulas to meet these needs. These increased nutrient requirements can complicate the need for fluid restriction. Breast milk is the preferred feeding because it reduces sepsis and the **necrotizing enterocolitis (NEC)** often seen in the premature infant. The incidence of NEC is significantly reduced in very low-birth-weight infants if probiotics are fed enterally with breast milk.[78] However, breast milk alone cannot meet the high energy needs of the premature infant. Several human milk fortifiers are available to increase the nutrient content of breast milk. If human milk is not available, formulas for premature infants may be substituted. Premature infant formulas are available in 20 kcal/oz and 24 kcal/oz concentrations. (See Box 21.2.)

When it is necessary to increase the caloric density of infant formula beyond 24 kcal/oz, it is preferable that the formula be prepared with the addition of either powdered or concentrated preterm infant formula rather than carbohydrate or fat additives. While carbohydrate or fat additives can be used to increase the caloric density of breast milk or formula, they also dilute the protein and mineral concentrations of the feeding. Formulas fortified with carbohydrate and fat may contain inadequate amounts of calcium and phosphorus; supplementation of these minerals may be needed. Infants fed high-energy formulas supplemented with carbohydrate or fat have demonstrated increased gains in weight and fat mass, but not lean body mass.[79]

Infants often continue to have a number of feeding problems after hospital discharge (see Box 21.3). In a study designed to identify the nutritional risk factors seen in infants after hospital discharge, Johnson et al. found that parents expressed a number of concerns about feeding, including getting the infant to take enough food; the frequency of feeding; resistance to feeding; the occurrence of gagging, vomiting, and choking; and the need for knowledge about feeding techniques.[80] Infants with feeding problems often require several hours a day to feed. Dietitians can play an important role in helping families and caregivers manage the nutritional care of these infants. Comprehensive nutrition counseling and follow-up after discharge are necessary. Providing education and support to caregivers helps to ensure that these infants receive adequate nutrition to support appropriate growth.

Chronic Obstructive Pulmonary Disease

Definition

Chronic obstructive pulmonary disease (COPD) is a progressive disease that limits airflow through either inflammation of the lining of the bronchial tubes (**bronchitis**) or destruction of alveoli (**emphysema**). Frequently, both conditions coexist as part of this disorder.

Epidemiology

COPD is the fourth leading cause of death in America and the number of deaths of women from COPD exceeded those of men for the first time in 2002.[81] The primary risk factor for

Common Feeding Problems for Low-Birth-Weight Infants

"My baby falls asleep when I feed her. She doesn't seem to have the energy for feeding."

This is a fairly common problem in the first weeks after the baby comes home. The solution depends on the reason for the problem:

- The infant isn't able to maintain oxygen status during feeding and might benefit from supplemental oxygen during feedings, or from "pacing" the feeding to allow for feeding breaks to maintain adequate oxygen status.

- The infant is overwhelmed by too much adjacent activity while feeding. Decrease the exposure to light, noise, and movement while feeding.

- Small babies have limited stomach capacity and are not able to take much at each feeding. Provide smaller, more frequent feedings or concentrate the formula or breast milk to increase the caloric density.

- Some babies easily become fatigued and do not get enough total formula or breast milk in a 24-hour period. Provide smaller, more frequent feedings; concentrate the formula to increase kcal; or provide supplemental tube feedings.

"My baby gets really upset when I try to feed her. Feeding doesn't seem to bring her pleasure."

Some possible reasons for this include the following:

- Some infants may have gastro-esophageal reflux with or without aspiration; this should be explored and treated immediately.

- Some infants associate feeding with unpleasant things that happened to and around their mouths early in life. They should never be force-fed. A feeding therapist can help develop a more structured approach to feeding that will gradually desensitize the infant to oral feeding.

- The flow of milk from the bottle or the breast may be too rapid or too slow. Changing the bottle nipple type or size, or the hole in the nipple, may help. Some breast-fed infants do better if the breast milk "let down" is established before they are put to the breast; the initial volume of milk which follows "let down" overwhelms some babies.

"My baby coughs and gags when I feed her. Sometimes this even leads to spitting up."

It is important to find the cause for the coughing and gagging. Coughing or gagging that results in apnea or color changes is serious and should be evaluated promptly. Other causes might be the following:

- Infants have trouble coordinating suck-swallow-breathe. A feeding therapist can help establish a program to pace the feedings until the infant has developed the neurological maturity to overcome this problem.

- Infants might tire at the end of the feeding and lose ability to coordinate suck-swallow-breathe. Pacing the feedings so the infant doesn't get too tired (as above) or providing smaller, more frequent feedings may help.

- Milk may flow too rapidly from bottle or breast. See discussion above for possible solutions.

Adapted with permission from: Some Common Feeding Problems for Low Birth Weight Infants. Gaining and Growing: Ensuring Nutrition Care of Preterm Infants. Available at http://depts.washington.edu/growing/Feed/Oralprob.htm

---

the development of COPD is smoking; even after the cessation of smoking, the inflammatory stress continues to damage the lung tissue.[82] Other risk factors include air pollution, second-hand smoke, history of childhood infections, and occupational exposure to certain industrial pollutants. Although normal lung function gradually declines with age, individuals who are smokers have a more rapid decline—twice the rate of nonsmokers.[83] Low body weight has also been shown to be a risk factor for the development of COPD, even after adjusting for other potential risk factors including smoking and age.[84]

Etiology

Chronic bronchitis is one of the principal classifications of COPD and is defined using the criteria of a productive cough and shortness of breath that lasts about 3 months or more each year for 2 or more years in a row. As stated previously, cigarette smoking is the primary cause; the longer and more heavily a person smokes, the more likely it is that he or she will develop bronchitis. Second-hand smoke may also cause bronchitis. Chronic bronchitis is seen in people of all ages but is more common in individuals over the age of 45.[81] Females

are more than twice as likely to be diagnosed with chronic bronchitis as males.

Emphysema develops gradually over years, usually as a result of the exposure to cigarette smoke. Approximately 95% of Americans diagnosed with emphysema are 45 years of age or older.[81] In the past, more males than females suffered from emphysema; however, in the past few years the incidence in women has significantly increased so that the difference in the prevalence rates between the sexes has become statistically insignificant. Given that not all smokers develop severe emphysema, their susceptibility or resistance may be linked to genetic factors.[85]

Most cases of emphysema are caused by inflammation of the airways and lung tissue.[85] The inflammatory process can increase oxidative stress and contribute to the development of emphysema.[86–88] In rare cases emphysema is caused by the deficiency of a protein called alpha 1-antitrypsin (ATT) or alpha 1-protease inhibitor. ATT is produced by the liver and is released into the bloodstream, where it travels to the lungs to protect them from the destructive actions of common illnesses and exposures, particularly tobacco smoke. Only about 5% of emphysema in the United States is caused by ATT deficiency.[81] Unlike the common form of emphysema seen in otherwise healthy individuals who have smoked for many years, this ATT deficiency form of emphysema may occur at a younger age and after minimal exposure to tobacco smoke.

Pathophysiology: Chronic Bronchitis

In chronic bronchitis, repeated exposure to cigarette smoke and other pollutants results in a generalized inflammatory response. This includes decreased cilia function, increased phagocytosis, and suppressed amounts of immunoglobulin A (IgA). Chronic inflammation causes hyperplasia of the mucus-secreting cells, resulting in edema of the bronchioles. The walls of the airways thicken and mucus glands become hyperplastic. The damaged cilia are unable to clean mucus from the airways, and the patient is unable to increase the work of breathing enough to overcome the signs and symptoms of the disease. The thickened mucus provides an environment conducive to bacterial growth, and chronic respiratory infections are common.

Clinical Manifestations: Chronic Bronchitis

Chronic bronchitis is characterized by decreased air flow rates (↓FEV), dyspnea, hypoxemia, and hypercapnia. Signs of chronic hypoxemia include **cyanosis** (shown in Figure 21.5), **clubbing**, and **secondary polycythemia**.

The quality of life for persons suffering from COPD diminishes as the disease progresses, resulting in an inability to work and possibly limiting normal day-to-day physical exertion (see Table 21.3). Malnutrition is also common in individuals with COPD.[89] As the effort to breathe increases, the intake of food intake may decrease. Often, individuals with COPD eventually require supplemental oxygen, and they may have to rely on mechanical respiratory assistance. Chronic respiratory disease leads to increased work of the heart. Cardiac complications of COPD include **cor pulmonale** and left ventricular failure (see Chapter 13).

Pathophysiology: Emphysema

The destruction of actual lung tissue differentiates emphysema from chronic bronchitis, even though emphysema often develops as a late complication of chronic bronchitis. The loss of

Figure 21.5 Cyanosis

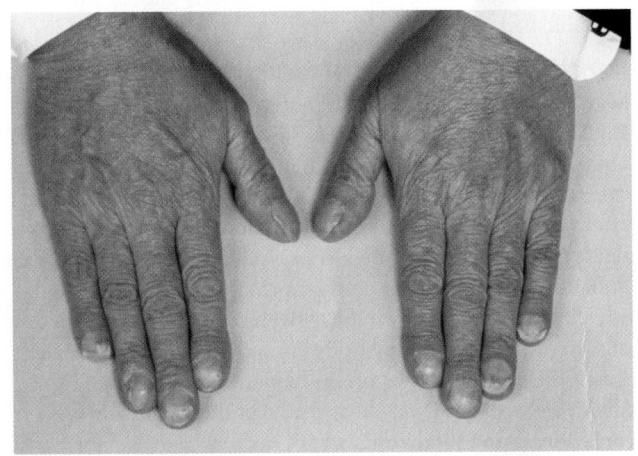

Source: John Radcliffe Hospital/Photo Researchers, Inc.

Table 21.3 Classifications of COPD by Severity

Stage 0 (at risk)
- Normal spirometry
- Chronic symptoms (cough, sputum production)

Stage 1 (mild COPD)
- FEV1/FVC <70%
- FEV1 >80% predicted
- With or without chronic symptoms

Stage II (moderate COPD)
- FEV1/FVC <70%
- FEV1 <80% predicted
- With or without chronic symptoms

Stage III (severe COPD)
- FEV1/FVC <70%
- FEV1 <50% predicted
- With or without chronic symptoms

Stage IV (very severe COPD)
- FEV1/FVC <70%
- FEV1 <30% predicted or
- FEV1 <50% predicted with symptoms of chronic respiratory failure
- FVC = forced vital capacity, FEV1 = forced expiratory volume in one second

Source: National Guideline Clearinghouse. Global strategy for the diagnosis, management, and prevention of chronic obstructive pulmonary disease. Available at http://www.guideline.gov/summary/summary.aspx?ss=14&doc_id=14175&string=

connective tissue results in a loss of surface area and decreased amounts of surfactant. Since the bronchioles lose their elasticity, they collapse during exhalation and trap air in the lungs.

Clinical Manifestations: Emphysema

Emphysema results in a decreased forced expiratory volume (FEV). Though inspiration is not impaired, expiration is, because air is trapped within the lungs. This inability to expire results in dyspnea and **orthopnea**, and causes hypercapnia and respiratory acidosis. The increased use of accessory muscles for expiration causes the development of a "barrel chest." Patients with emphysema do not experience hypoxemia until the last stages of the disease, when extreme fatigue and physical exhaustion prevent adequate oxygen intake.

Treatment

Medical treatment for individuals with COPD involves lifestyle changes, including smoking cessation, avoiding smoke and other air pollutants, exercising as tolerated, and good nutrition.[81] Treatment strategies are based on assessment of disease severity and response to various therapies. Pharmacologic treatment is used to prevent and control symptoms, improve health status, and improve exercise tolerance. Medications include bronchodilators, β-agonists to relax smooth muscle, and anticholinergics to decrease airway contraction and mucus production. Steroids are used to decrease swelling, and mucolytic agents work to decrease the viscosity of secretions (see Table 21.4).

Pulmonary rehabilitation programs provide a comprehensive, multidisciplinary approach to treatment that combines education

Table 21.4 Medications Used in Diseases of the Respiratory System

Type of Medication	Brand Names	Action	Common Side Effects
Bronchodilators *Types of bronchodilators:* β2-agonist Anticholinergics Theophyllines	*β2-Agonists:* Albuterol® Pirbuterol® Salbutamol® *Anticholinergics:* Ipatropium®, Tiotropium®	Used to open or relax the bronchial tubes and relieve shortness of breath. May be taken as an inhaler or pill.	Fast heartbeat, shakiness, and cramping of hands, legs, and feet; dry mouth, particularly with the anticholinergics. Severe nausea and vomiting with theophyllines.
Steroids (*corticosteroids*)	Predinisone®, Prednisolone®, Solu-Medrol®, Solu-Cortef®	Used to reduce inflammation in the bronchial tubes. May be used as an inhaler or taken orally.	Side effects depend on the dose, length of use, and whether taken orally or inhaled. For inhaled steroids, most common side effects include sore mouth, hoarse voice, and infections in throat and cough. Orally in high doses or low doses for a long period of time, the side effects include: altered fluid/electrolyte balance, hypertension, mood swings, increased appetite, weight gain, hyperglycemia, osteoporosis, hyperlipidemia, poor wound healing, growth retardation in children.
Antibiotics Doxycycline Hyclate Amoxicillin Macrolides Fluoroquinolones	Doryx® Vibramycin® Augmentin®	Treat respiratory infection.	Nausea, vomiting, diarrhea.
Leukotriene Inhibitors Zafirlukast Aileuton Montelukast	Accolate® Zyflo® Singulair®	Mediates inflammation of COPD.	Headache, nausea, diarrhea, infection. Singulair®—caution with grapefruit/grapefruit juice.
Mucolytics Acetylcysteine	Mucomyst®	Makes mucus in the lungs thinner and less sticky.	Nausea, vomiting, runny nose, drowsiness, clammy skin.
Immune System Modifier Xolair	Omalizumab®	Inhibits the binding of IgE and allergic response	Cold- or flu-like symptoms. Muscle or joint pain. Itching; fever.
Pancreatic Enzymes	Creon®, Pancrease®, Pancrease MT®, Pancrecarb®, Ultrase®, Viokase®, Ultrase MT®	Replace deficient pancreatic enzymes for patients with cystic fibrosis.	Take immediately before meals; do not take on empty stomach.

with therapeutic exercise. Team members may include nurses, occupational therapists, registered dietitians, respiratory therapists, physical therapists, and physicians. These programs help the patient understand and cope with these chronic conditions, with the ultimate goal of improving overall quality of life.

Nutrition Therapy for Chronic Obstructive Pulmonary Disease

Nutrition Assessment and Diagnosis

Malnutrition occurs in many patients with moderate to severe COPD. The incidence of malnutrition depends on the severity of the disease. Malnutrition due to COPD can weaken respiratory muscles, resulting in altered ventilation, poor muscle strength, and impaired immune function.[90, 91]

A complete nutrition assessment is necessary to identify patients who are at nutritional risk. As indicated in Table 21.5, this assessment should include an evaluation of anthropometric measurements and food/nutrition-related history, including a detailed nutrient intake evaluation, medication

and herbal supplement use, and physical activity and function. Measuring loss of fat-free body mass may be a better prognostic indicator of mortality than weight loss or BMI in malnourished patients.[92] Based on this assessment, nutrition problems can be identified and goals for the individual patient can then be established.

Anthropometric Measurements Weight loss and low BMI have been associated with increased mortality in patients with COPD, regardless of disease severity.[92] Weight loss occurs frequently, particularly in individuals with emphysema, and is associated with increased resting energy expenditure secondary to the work of breathing, reduced nutrient intake, and inefficient fuel metabolism.[92] In contrast, individuals with bronchitis frequently have normal or above-normal BMI. Losses of lean body mass (LBM), however, have been seen in both conditions. In a study examining body composition of individuals with COPD, Engelen et al. found depletion in LBM in 37% of individuals with emphysema and 12% of those with chronic bronchitis. Even in individuals with normal body weight, depletion of LBM was found in 16% of those with emphysema and 8% of those with chronic bronchitis.[93]

Table 21.5 Nutrition Assessment for the Respiratory System

Client History (personal, patient, family, medical and social)	• Medical diagnoses • Previous medical conditions or surgeries • Socioeconomic status • Support systems • Education—primary language • Psychological status—anxiety
Food-/Nutrition-Related History	• Medications: Careful attention to use of corticosteroids • Food security • Ability to consistently purchase adequate amounts of food on a daily basis • Ability to feed self • Ability to cook and prepare meals • Food allergies, preferences, or intolerances • Previous food restrictions • Ethnic, cultural, and religious influences • Use of alcohol, vitamin, mineral, herbal or other type of supplements • Previous nutrition education or nutrition therapy • Eating pattern: 24-hour recall, diet history, food frequency
Nutrition-Focused Physical Findings	• Ability to chew; use and fit of dentures • Problems swallowing • Nausea, vomiting • Constipation, diarrhea • Heartburn • Any other symptoms interfering with ability to ingest normal diet • Muscular strength and coordination • Physical assessment: temporal wasting; presence of edema
Anthropometric Measurements	• Height • Current weight • Weight history: highest adult weight; usual body weight • Reference weight (BMI)
Biochemical Data, Medical Tests and Procedures	• Visceral Protein Assessment: Standard • Immunocompetence: Standard • Hematological Assessment: Standard • Other specific labs: Electrolytes, pH, glucose, arterial blood gases

Respiratory System—Specific Biochemical Data, Medical Tests, and Procedures:

Vitamin D	Serum alkaline phosphatase provides indirect measure
Vitamin K	Prothrombin time
Vitamin A	Serum carotene, retinal binding protein
Vitamin E	Serum tocopherol, erythrocyte hemolysis
Zinc	Serum zinc

Excessive weight gain, particularly excessive body fat, may be deleterious by increasing the workload of an already compromised respiratory system.[94] Individuals who are morbidly obese have difficulty breathing caused by restrictions on the chest wall due to the accumulation of fat in and around the thoracic cage, diaphragm, and abdomen. This results in reduced lung volume accompanied by poor oxygen and carbon dioxide exchange.

Patients who are more than 40% above IBW should be evaluated individually to determine the most appropriate intervention that will provide long-term benefits.[94] The primary goal should be to prevent further weight gain and promote moderate weight loss, if appropriate. For patients whose health status is borderline, particularly patients who have a history of weight or appetite fluctuations, weight loss during exacerbations of the disease, or weight gain associated with prolonged steroid use, weight reduction is contraindicated. Weight reduction may exacerbate an existing risk for weight loss associated with the disease and lead to reduced pulmonary function.

Food/Nutrition-Related History Designing a nutrition care plan requires identification of the possible causes of reduced or inadequate intake. Oral intake in individuals with COPD is often inadequate because of a number of factors, including anorexia, early satiety, dyspnea, bloating, and fatigue.[95, 96]

Low dietary intake, weight loss, and cachexia occur in individuals with moderate to severe COPD because of symptoms of dyspnea, fatigue, dyspepsia, and early satiety. Taste perceptions may be altered with chronic mouth breathing and appetite may be further reduced as a result of depression. Nonetheless, even individuals with adequate dietary intake may lose weight, as elevations in both the resting energy expenditure (REE) and total energy expenditure independent of the REE are seen in individuals with COPD.[94, 97, 98]

As discussed earlier, a number of dietary factors, particularly the antioxidants, influence respiratory health and can play a key role in protecting against COPD.[99] Prospective studies of men and women show that a diet rich in fruits, vegetables, and fish may result in a low incidence of COPD. In contrast, a diet rich in refined grains, cured and red meats, desserts, and french fries may increase COPD risk.[100, 101] Numerous studies have shown that smokers have lower intakes of antioxidant vitamins, specifically vitamins C, A, and E, and β-carotene.[102–104] Individuals with COPD undergo oxidative damage during both exacerbations of the disease and stable periods. During periods of exacerbation, serum concentrations of vitamins A and E have been shown to decrease.[105] Vitamin E supplementation, however, does not appear to be correlated with improved lung function.[106] Similarly, omega-3 fatty acids do not appear to be protective against reductions in FEV1 in COPD patients.[107]

Phosphate is essential for the synthesis of adenosine triphosphate (ATP) and **2,3-diphosphoglycerate (DPG)**, both of which are critical for pulmonary function.[108] Respiratory and peripheral muscle stores of phosphate have been shown to be depleted in individuals with COPD.[109] Medical treatment with drugs commonly used for COPD, including corticosteroids, diuretics, and bronchodilators, is associated with hypophosphatemia and likely contributes to the depleted phosphate stores.[110] Serum phosphate levels need to be closely monitored in patients with pulmonary disease or respiratory failure to ensure adequate levels.

Osteoporosis, with resulting bone fractures, has been shown to be a significant problem in patients with advanced COPD.[111–113] A number of risk factors have been related to the pathophysiology of osteoporosis in these individuals. These include smoking, suppression of estrogen or testosterone levels, vitamin D deficiency, low BMI, low fat-free mass, and decreased mobility.[114, 115] Body weight is closely related to bone mineral density (BMD). Weight loss and malnutrition are likely involved in the pathogenesis of low BMD in individuals with COPD. Low serum 25-hydroxyvitamin D levels have also been documented in individuals with COPD, suggesting that vitamin D deficiency due to poor intake and decreased sun exposure may also play a role in the bone disease. The intake of calcium and vitamin D should be assessed, particularly in individuals with a reduced intake. According to the current nutritional recommendations, 1200 to 1500 mg/day of calcium and at least 400 IU of vitamin D should be provided.[111, 117]

Medication Use An evaluation of medication usage is also important in order to determine any resultant impacts on appetite and nutrition status. The use of glucocorticosteroids in the treatment of COPD has been shown to increase the incidence of osteoporosis.[118] Glucocorticosteroids decrease the intestinal absorption of calcium and increase urinary excretion, resulting in an increase in parathyroid hormone levels and bone resorption. Bone mineral density should be measured in individuals who have COPD, particularly in those receiving long-term glucocorticoid treatment (>7.5 mg prednisone/day), with follow-up testing every two years.[119, 120]

Physical Activity and Function A common complaint is that the basic activities of daily living, including eating, require effort. Individuals with COPD often complain that they tire easily when eating or experience dyspnea during eating and drinking. Fatigue resulting from dyspnea may also interfere with eating. Chewing and swallowing may be impaired since both activities change breathing patterns and reduce oxygen uptake. Chronic mouth breathing or certain medications may also cause changes in taste perceptions and/or xerostomia (dry mouth).

Patients with COPD often suffer from **hyperinflation of the lungs** with accompanying flattening of the diaphragm and reduced abdominal volume, leading to unnecessary fullness and bloating at mealtime.[114] If the individual eats too much at a meal, the increased positive pressure applied to the diaphragm by the stomach results in breathing difficulty. **Aerophagia** often occurs in COPD and may also cause gastric bloating. Additional factors that may contribute to poor nutrition include depression and difficulty in shopping and preparing foods.

Nutrition Intervention

Energy and Nutrient Needs In COPD, the respiratory muscles need to generate a large force to expand the thoracic rib cage. Though this increase in the energy cost of breathing is often cited as a major cause of the increased REE, this does not appear to be the only cause of the hypermetabolism. Increased systemic inflammation has also been cited as a cause of hypermetabolism evidenced by elevations in **tumor necrosis factor (TNF-α)** in these patients.[120] In addition, the thermogenic effects of medications used to treat COPD, including the bronchodilating drugs, may also play a role.

Maintaining optimal energy balance in the individual with COPD is essential in order to preserve body weight, lean body mass, and general well being. According to the ADA Evidence Analysis, the prevalence of a BMI <20 kg/m² may occur in as many as 30% of individuals with COPD (ADA evidence analysis COPD). Respiratory muscle function is severely affected by declining nutrition status and is closely linked to body weight and lean body mass. It is essential that the individual with COPD receive sufficient kcal and protein to maintain body weight, lean body mass, and adequate nutrition status. Thorsdottir and Gunnarsdottir found that energy intakes of 125% to 156% (average 140%) above basal energy expenditure and protein intakes of 1.2 to 1.7 grams/kg body weight (average 1.2 g/kg) were adequate to avoid protein losses in patients admitted to the hospital with exacerbation of their COPD.[121] Malnourished patients need additional energy and protein to provide for repletion. Because the predicted REE using the Harris-Benedict formula significantly underestimates the measured REE by 10% to 15% for patients with COPD, indirect calorimetry is the best method to assess kcal needs without overfeeding or underfeeding.[122] When indirect calorimetry is not available, providing 25–30 kcal/kg of body weight appears appropriate with approximately 20% of total kcal from protein (1.2–1.7 grams of protein/kg body weight), depending on the patient's individual needs and with special attention to the degree of inflammation and activity.[118, 121, 123] In the hospitalized patient with COPD, particularly patients who have compromised pulmonary function, it may be necessary to provide ventilatory support using mechanical ventilation. In these patients, overfeeding is of primary concern because it is associated with increased CO_2 production, which can further complicate ventilation. Although glucose and protein have been shown to stimulate ventilatory drive, excess glucose administration (0.5 mg/kg/minute) increases CO_2 production and makes it difficult to wean or remove patients from mechanical ventilation.[108] In spite of this, when total kcal are provided in moderate amounts (approximately 30% above basal needs), the macronutrient composition of the feeding has little effect on CO_2 production.[124] The production of excess CO_2 occurs when patients are overfed (>1.5 × REE).

Food and/or Nutrient Delivery Commercial enteral formulas that have been specifically designed for individuals with respiratory disease contain a lower carbohydrate content (30%) and higher lipid content (50%). Compared to the other macronutrients, and fat in particular, metabolized carbohydrate yields the greatest amount of CO_2. Controlled clinical trials using these modified formulas have demonstrated decreased CO_2 production when compared to standard formulas equal in kcal but higher in carbohydrate.[125, 126] However, improvement in clinical outcomes with the use of these formulas has not been consistently demonstrated.[7, 108, 127] One potential negative side effect of higher-fat meals or supplements is delayed gastric emptying, which may result in abdominal discomfort, bloating, or early satiety.[127]

Table 21.6 provides examples of nutrition interventions that can be used to treat and manage many of the problems associated with inadequate food and beverage intake common in patients with COPD. Particular attention needs to be paid to consuming foods that are not only good sources of both kcal and protein but are also nutrient dense. Studies show that COPD patients who receive dietary counseling and advice on food fortification consume more energy and protein and gain more weight than those who do not receive nutrition education.[128] Instructing individuals to rest before meals to avoid fatigue may be helpful. Eating smaller, more frequent meals

Table 21.6 Using the Evidence Analysis Library to Plan Nutrition Interventions for Individuals with COPD

Diet Composition	Use patient preference to guide food suggestions; limited evidence supports a specific macronutrient mix.
Medical Food Supplements	Use a variety of high-calorie, high-protein supplements to increase overall nutritional intake. These can include modular supplements added to foods or liquid supplements as an addition to the regular diet.
Vitamin/Mineral Supplements	Assess ability to meet nutritional needs from a variety of foods and then make recommendations for additional micronutrient supplementation. Specific attention should be given to calcium, vitamin D, and sources of antioxidants.

Source: Evidence-based Nutrition Practice Guideline on Chronic Obstructive Pulmonary Disease. Available at www.adaevidencelibrary.com

rather than three larger ones may help to alleviate the feeling of fullness and bloating. The use of nutrition supplements to provide additional kcal and protein has shown mixed results, suggesting that nutritional supplementation alone is not sufficient to improve physical and nutrition status.[129]

Coordination of Nutrition Care: Physical Exercise Physical exercise as part of the nutritional rehabilitation of patients with COPD is an equally important component of treatment for COPD. Nutrition support combined with exercise as part of a pulmonary rehabilitation program has been shown to have the best overall effect on increasing body weight, fat-free mass, and respiratory muscle strength in stable patients with COPD.[130] The type of exercise prescribed depends on the severity of COPD. As the disease progresses, patients often experience dyspnea, particularly on exertion, which limits their ability to do strenuous exercise. An evaluation of randomized controlled trials comparing different exercise protocols for patients with COPD concluded that strength/resistance training (versus endurance training) was associated with the greatest improvements in quality of life.[130] Upper extremity exercise training emphasizing the ability to perform activities of daily living has resulted in improvement in tests related to daily living and reduced fatigue with these activities.[131] Skeletal muscle dysfunction has been recognized as an indicator of the advanced stages of COPD; strength training may help to improve skeletal muscle function and overall well being.

Cystic Fibrosis

Definition

Cystic fibrosis (CF) is a disease characterized by abnormally thick mucus secretions from the epithelial surfaces of various organ systems, including the respiratory tract, the gastrointestinal tract, the liver, the genitourinary system, and the sweat glands.

Epidemiology

CF is the most common **autosomal recessive** disease in the United States, affecting approximately 30,000 children and adults.[132] One in 31 Americans are carriers of the defective gene that causes CF. To have CF, an individual must inherit two defective genes, one from each parent. Each time two carriers conceive, there is a 25% chance that their child will have CF; a 50% chance that the child will be a carrier of the CF gene; and a 25% chance that the child will be a noncarrier. Most individuals with CF are diagnosed by age 3; however, nearly 10% of new cases are diagnosed at age 18 or older. The median survival age for individuals with CF is now projected to be 37.4 years.[133]

Etiology

CF is caused by an abnormal mutation of the cystic fibrosis transmembrane conductance regulator (CFTR), which is a type of protein classified as an ATP-binding cassette (ABC) transporter.[134] Mutations of this protein prevent CFTR from functioning normally. A number of different mutations can occur in the CFTR, and can be linked to severity of the disease and the rate of decline of a patient.[135]

Pathophysiology

The CFTR proteins are responsible for the transport of sugars, peptides, inorganic phosphate, chloride, and metal cations across the cellular membrane. CFTR transports chloride ions (Cl^-) across the membranes of cells in the lungs, liver, pancreas, digestive tract, reproductive tract, and skin. In individuals with CF, CFTR's failure to function properly results in thick viscous secretions that eventually lead to obstruction of the glands and ducts in the affected organs.

Approximately 50% of individuals with CF have pulmonary symptoms that include a chronic cough and wheezing. Respiratory symptoms may begin during the first month of life. Pulmonary insufficiency leading to pulmonary failure is the major cause of death in individuals with CF. Because of this, early effective respiratory treatment is essential. A variety of aerosol therapies are used to increase airflow, reduce the thick mucus accumulation, and reduce infection. They include **inhaled bronchodilators, inhaled anti-inflammatory agents, inhaled mucolytics,** and **inhaled antibiotics** (see Table 21.4). **Chest physiotherapy,** designed to reduce airway obstruction by improving the clearance of secretions, is also an integral part of pulmonary care. Although CF affects many organ systems, the lungs are the most seriously involved, and respiratory failure and death result for more than 90% of patients.[135] Lung transplantation is an important aggressive treatment option for CF patients with end-stage respiratory failure and may involve either a bilateral lung transplant or a heart-lung transplant. (See discussion of lung transplantation later in this chapter.)

Pancreatic insufficiency secondary to blocked ducts occurs in CF patients and prevents enzymes from reaching the small intestine.[136] Symptoms of pancreatic involvement include frequent passage of bulky, foul-smelling, oily stools (steatorrhea); abdominal distension; and a poor growth pattern with decreased subcutaneous tissue and muscle mass despite a normal appetite. Enteric-coated digestive enzyme supplements are effective in enhancing absorption of fat and protein.[137] Cystic fibrosis-related diabetes with its subsequent clinical complications develops in approximately 10%–15% of adults.[136, 138] Other complications include cirrhosis of the liver with portal hypertension and varices (see Chapter 16 for a discussion of these problems). Children and adults with CF have an increased amount of sodium and chloride in their sweat. Because of this, excessive sweating in hot weather or fever may lead to increased risk of dehydration and electrolyte imbalances.

Medical Diagnosis

In addition to a complete medical history and physical examination, the diagnosis of CF is made using one or more of the following tests: a **sweat chloride test**, a blood test to confirm mutations of the CFTR gene, sputum cultures to test for infections typical in CF, **pancreatic function tests**, and pulmonary function tests. Fasting blood levels of carotenoids, vitamins A and E, essential fatty acids, and cholesterol are reduced in patients with steatorrhea (dietary fat malabsorption).

In May 2005, the FDA approved the first DNA-based test to detect CF. The test directly analyzes human DNA to find genetic variations indicative of the disease and will be used to diagnose both children with the disease and adults who are carriers of the CF gene.[139]

Nutrition Therapy for Cystic Fibrosis

A multidisciplinary approach to treatment of the individual with CF is essential in order to assist individuals and their families in meeting their complex medical and nutritional needs. The dietitian is responsible for assessment of nutritional status, including determining energy requirements; providing nutrition counseling; and helping to plan intervention strategies. Poor weight gain and poor growth occur due to poor intake, pancreatic insufficiency, and increased nutrient needs as evidenced by low BMI, poor bone mass, low serum nutrient status, and high infection rates. Early identification of CF is important in ensuring proper nutrition. The Cystic Fibrosis Foundation has developed ongoing consensus guidelines for treatment of CF, including nutrition treatment.

Nutritional Implications

The relationship between nutrition status and long-term survival of individuals with CF is well documented. Pancreatic insufficiency results in poor digestion, poor absorption of fat and fat-soluble vitamins, and loss of bile and bile salts. Chronic pulmonary infections and deteriorating pulmonary function may lead to anorexia, increased energy requirements, and malnutrition. The more malnourished the patient, the poorer their lung function.[140] Children with CF may have stunted growth and inadequate weight gain because of inadequate supplies of nutrients due to a combination of anorexia and poor nutrient absorption.[141] Lower-than-average height and weight are particularly pronounced in infants, adolescents, and individuals newly diagnosed with CF. Increased infection rates and inflammation may occur due to the effects of CF on fatty acid, cholesterol, and sphingolipid metabolism.[142] Individuals with CF are often diagnosed with osteopenia and osteoporosis occurring secondary to pancreatic insufficiency, which causes malabsorption of calcium, phosphorus, magnesium, vitamins D and K, and the chronic use of corticosteroid medications.[143] While the link between bone disease and cystic fibrosis is widely recognized, facilities are inconsistent in the levels of vitamin supplements they provide to their patients. One study found that only 18% of treatment centers prescribe vitamin K and only 61% assess bone health.[144] Infants and children with CF are at risk for developing hyponatremia (low serum sodium levels) because of salt loss through the skin. When sodium intake is inadequate, lethargy, vomiting, and dehydration may occur.

Nutrition Assessment

Anthropometric Measurements Early detection of poor growth allows for appropriate intervention and treatment. The type of CFTR mutation can influence height, weight, and transverse chest width in children.[145] There are three periods when special attention needs to be focused on growth and nutritional status: during the first 12 months after diagnosis of CF, from birth until 12 months of age for infants diagnosed at birth, and during the peripubertal growth period (9 to 16 years for girls and 12 to 18 years for boys).[146] Growth charts from the National Center for Health Statistics/Center for Disease Control and Prevention (NCHD/CDC) that plot weight, head circumference, length, and height should be used to monitor growth status (available at http://www.cdc.gov/growthcharts). Children are considered to be at nutritional risk if they are between the 10th and 25th weight-for-length percentile and are considered to have nutritional failure if they are at less than the 10th weight-for-length percentile.[146] Mid-arm circumference and triceps fatfold measurement provide information about adequacy of subcutaneous fat stores (energy) and lean body mass (muscle).

Biochemical Data and Medical Tests Patients with severe CF genotypes may develop essential fatty acid deficiency (low blood levels) in the absence of clinical symptoms.[147] Thus, the use of fatty acid supplementation in CF warrants consideration and is an area of active research. Vegetable oils such as canola, soy, and flaxseed are good sources of both energy and linolenic acids and should be encouraged in the diets of patients with CF.

Glucose intolerance and cystic fibrosis-related diabetes (CFRD) occur in approximately 10% to 15% of CF patients over the age of 20.[146] CFRD is rarely found in young children and occurs more frequently between 18 and 21 years.[148] The exact cause of diabetes mellitus in CF is not clearly understood but is thought to be related to inflammation and the accumulation of fibrous tissues in the pancreas, which interferes with normal insulin production. Individuals with CFRD have an increased morbidity and mortality. They are often underweight and have more advanced pulmonary involvement than those without CFRD. Malnutrition and low BMI are warning signs of the development of CFRD. Wilson et al. found that, at diagnosis of CFRD, 57% of teenagers and adults attending the Toronto Cystic Fibrosis Clinics had a BMI below the 10th percentile for age.[149]

Like type 2 diabetes (see Chapter 17), CFRD may be present for years before diagnosis. Symptoms of CFRD include polyuria (excessive urination) and polydipsia (excessive thirst), failure to gain or maintain weight despite aggressive nutrition intervention, poor growth velocity, an unexpected decline in pulmonary function, and a failure to progress normally through puberty.[146, 150] Diagnosis of CFRD is made using an oral glucose tolerance test (OGTT), a test of the body's ability to metabolize carbohydrate (see Chapter 17). Since a majority of patients do not have fasting hyperglycemia, using hemoglobin A1c (which measures the amount of glucose that has adhered to the hemoglobin cell) as a screening tool for CFRD is not appropriate; results of this test are normal in a majority of cases.

Vitamin A is important for vision, the integrity and proliferation of epithelial cells, and normal immunity. Studies examining the vitamin status in individuals with CF indicate that deficiency of vitamin A may be common.[146, 151] Since plasma vitamin A levels can be decreased in infection, levels measured during acute illness can be misleading. Beta-carotene, a precursor of vitamin A, also functions as an antioxidant. Serum β-carotene levels in patients with CF have been shown to be low.[146, 152]

The major function of vitamin D is to control calcium absorption. Low vitamin D intake has been reported among CF patients and studies have documented low serum 25-hydroxy vitamin D concentrations despite daily supplementation of vitamin D.[153] This is of particular importance because of the increased prevalence of osteoporosis and bone fractures seen in patients with CF.[154, 155] Serum vitamin D levels need to be carefully monitored to ensure that the patient is receiving an appropriate dosage of the vitamin (see Table 21.9). Vitamin D-binding protein may also be a valuable nutritional marker for CF patients.[156]

Vitamin K is necessary for the biosynthesis of normal clotting factors and plays an important regulatory role in bone formation and mineralization. Data show that vitamin K deficiency may be common in CF.[157, 158] Since measurement of serum vitamin K levels is not practical, **plasma prothrombin concentrations** are used to measure vitamin K status instead. Because colonic bacteria are also a source of vitamin K, and periods of antibiotic therapy can disrupt vitamin K synthesis in the colon, vitamin K status during prolonged antibiotic treatment needs to be evaluated.

Vitamin E is a powerful antioxidant. Deficiencies lead to hemolytic anemia, neuromuscular degeneration, and retinal and cognitive changes. Low vitamin E levels as well as symptomatic deficiency states have been reported in patients with CF, even those taking pancreatic enzymes and multivitamins.[146, 159] Current recommendations for supplementation of vitamin E may not be adequate, and require further study (see Table 21.9).

Children and adolescents with CF have also been shown to have iron and zinc deficiency.[146] Iron status should be monitored yearly by checking hemoglobin and hematocrit levels. Zinc deficiency may be present even though plasma zinc levels are normal.[160]

Individuals with CF have an increased incidence of osteopenia and osteoporosis, and an increased risk of fractures. Contributing factors include deficiencies of vitamin D, vitamin K, and calcium; use of corticosteroids; disease severity; inactivity and low body mass; hypogonadism; and malnutrition.[146] Children age 8 and older who have risk factors for bone disease should have their bone mass evaluated using dual-energy X-ray absorptiometry (DXA). Additionally, children at risk for poor bone health should have serum calcium, phosphorus, parathyroid hormone, and 25-hydroxyvitamin D levels measured.

Table 21.7 outlines the timeline for nutrition assessment, including growth status, for individuals with CF.

Table 21.7 Assessment of Nutrition Status for Individuals with Cystic Fibrosis

Nutrition Parameter	Frequency of Assessment
Anthropometric Measurements:	
Weight (to 0.1 kg)	Every 3 months
Height (length <2 years) (to 0.1 cm)	Every 3 months
Head circumference (to 0.1 cm)	Every 3 months
Mid-arm circumference (to 1.0 mm)	Annually
Triceps skinfold	Annually
Mid-arm muscle area, mm^2	Annually
Mid-arm fat area, mm^2	Annually
Food/Nutrition-Related History:	
Dietary intake (24-hour recall)	Annually
Nutritional supplement intake	Annually
Anticipatory dietary and feeding behavior guidance	As indicated
Biochemical Data:	
β-carotene	At physician's discretion
Vitamins, A, D, E, K	Annually
Essential fatty acids	Check in infants and those who are failure to thrive (FTT)
Calcium/bone status	>age 8 years if risk factors are present
Iron	Annually
Zinc	Annually
Sodium	If dehydration is suspected or exposed to heat stress
Albumin	Annually

Adapted from: Borowitz D, Baker RD, Stalling V. Consensus report on nutrition for pediatric patients with cystic fibrosis. *J Pediatr Gastroenterol.* 2002; 35:246–59.

Nutrition Diagnosis

Due to the complexity of the disease and numerous nutrition implications, it is possible for a patient to present with multiple nutrition diagnoses at any time. This requires careful and accurate nutrition assessment and timely interventions. Common nutrition diagnostic labels for cystic fibrosis include: Inadequate energy intake, inadequate oral food/beverages intake, increased nutrients needs, malnutrition, inappropriate intake of fats, inappropriate intake of types of carbohydrates, inconsistent carbohydrate intake, inadequate vitamin and mineral intakes, altered GI function, altered nutrition-related laboratory values, underweight, involuntary weight loss, self-monitoring deficit, limited adherence to nutrition-related recommendations, and poor nutrition quality of life.

Nutrition Intervention

Nutrition-Related Medication Management: Pancreatic Enzyme Therapy
Because a significant number of individuals with CF have pancreatic insufficiency, malabsorption of dietary fat, protein, fat-soluble vitamins, and other nutrients often occurs. Pancreatic insufficiency has a strong influence on nutrition status and is a predictor of long-term outcome. The thickened secretions obstruct the pancreatic ducts and prevent the secretions of lipase, amylase, proteases, and bicarbonate. When pancreatic insufficiency is present, individuals are treated with pancreatic enzyme

extracts. All enzyme products contain the various enzymes synthesized by the pancreas, including amylase, proteases, and lipase in varying amounts. Commercial enzyme products vary in lipase activity from 4,000 to 25,000 U lipase/capsule. They are available in powder form as tablets that are acid labile, or as enteric-coated microspheres—the enteric coating is designed to protect the enzyme from destruction by the acidic environment of the stomach.

Pancreatic enzymes are always taken when food or beverages are consumed. The dosage for enzymes is individualized based on the patient's diet, nutritional status, degree of pancreatic insufficiency, intestinal pH, and GI anatomy and physiology.[161] Infants may be given 2,000 to 4,000 lipase units per 120 mL of formula or breast-feeding. The recommended enzyme dose for children begins with 1,000 lipase units/kg per meal for children under 4 years and 500 lipase units/kg per meal for children over 4 years of age. As children grow older, they require less lipase per kilogram body weight because they tend to ingest less fat per kilogram of body weight. Usually, one half the standard dosage is given with snacks.

Because of inconsistencies in enzyme formulations, the FDA has issued a rule requiring manufacturers of pancreatic enzyme supplements to obtain approval for their products.[162] Prior to obtaining approval, manufacturers will need to test the enzymes in clinical trials and demonstrate that they are safe and effective. This rule means that the FDA now requires pancreatic enzymes to meet the same standards of testing as any other new drug.

Energy and Macronutrients
Adequate kcal to support normal growth and development are essential, especially in the presence of pancreatic insufficiency. Energy intake should be based on the patterns of weight gain and growth in children. Energy needs for children with CF without respiratory infection are comparable to that of healthy children (100% to 110% of Recommended Dietary Allowances [RDA]).[163] However, if an individual has significant growth deficits, lung disease, or malabsorption, energy requirements may be significantly increased (110% to 200% of the RDA).[161] The 2002 Nutrition Consensus Report states that there is no perfect method to estimate the kcal needs of a person with CF; instead, a steady rate of weight gain in growing children should be the goal.[146] For adults, the desired outcome is to maintain an acceptable weight in relation to height with optimal fat and muscle stores.

To obtain adequate kcal and compensate for any fat malabsorption, individuals with CF often require a greater fat intake (35%–40% of total kcal) than what is normally recommended for the general population (25%–35% of total kcal).[146] Fat restriction is not recommended, because fat is an important energy source, and pancreatic enzyme replacement therapy is used to aid its absorption. Medium-chain triglycerides (MCT) require less lipase activity than long-chain fatty acids and may be utilized as a better source of fat kcal. MCT have a fatty acid chain length between 6 and 12 carbons, making them short enough to be water soluble. They require less bile salt for solubilization and can be transported as free fatty acids through the portal system. Adequate protein intake (approximately 15%–20% of kcal) is essential to meet the needs of the growing child and maintain protein stores. Good nutrition also plays an important role in preparing the individual with cystic fibrosis for potential transplant later in life.

Table 21.8 Nutrition Management of Cystic Fibrosis-Related Diabetes

- Combine management of diet for both CF and diabetes.
- Aim for >100% RDA for energy.
- Provide 3 meals and 3 snacks per day.
- Fat should account for 35%–45% of total calories.
- No restrictions should be placed on total carbohydrate intake; carbohydrate calories should be distributed throughout the day.
- If insulin for diabetes management, carbohydrate counting should be taught.
- Allow flexibility in meal planning.

Sources: Reprinted from *Clinical Nutrition*, Vol. 19(2), Wilson, DC, Kalnins D, Stewart C, Hamilton N, Hanna K, Durie PR, et al, Challenges in the dietary treatment of cystic fibrosis related diabetes, pages 87–93, Copyright 2000, with permission from Elsevier.

Nutrition management is critical for the health and survival of patients with CFRD (see Table 21.8). Since a majority of these patients have difficulty maintaining weight, kcal restriction is never appropriate. For patients on insulin, carbohydrate counting offers a great degree of flexibility. Patients should be able to eat as they choose with appropriate insulin coverage.[164] Although carbohydrate is not restricted, patients should be taught to distribute carbohydrate kcal throughout the day and to avoid concentrated carbohydrate loads.

Vitamin and Mineral Supplements
Individuals with CF who are adequately treated with pancreatic enzymes may continue to have malabsorption of fat-soluble vitamins (A, D, E, and K).[146, 165] Patients with liver disease and accompanying disturbances of enterohepatic bile circulation are at significant risk for this malabsorption.[146] The combination of oral pancreatic enzyme therapy and supplementation of the fat-soluble vitamins will improve vitamin status. Fat-soluble vitamin supplementation needs to be individually adjusted based on age, dietary intake, and disease progression (see Table 21.9). These vitamins can be given in liquid, chewable, or pill form, depending on the age of the individual. Laboratory monitoring of each of these vitamins should be done at the time of diagnosis and at least yearly to ensure that the patient is receiving adequate amounts (see Table 21.7).

To prevent bone disease, patients should receive adequate nutrition—especially calcium, vitamin D, and vitamin K—participate in appropriate weight-bearing exercise, and take steps to minimize malabsorption, especially with adequate pancreatic enzyme replacement.

Individuals with CF should consume a higher-salt diet, particularly during the summer months or if they live in hot climates. The amounts of salt to add to the diet will vary based on sodium losses. Monitoring of sodium will dictate the amount needed. Breast-fed infants or individuals participating in sports activities are particularly susceptible to sodium depletion. For infants and small children, sodium chloride solutions available through pharmacies provide the most accurate method for supplying additional sodium. Adequate sodium intake for adolescents and adults is usually not a problem because of the amount of sodium available in the food supply, particularly in processed foods.

Table 21.9 Recommendations for Daily Vitamin Supplementation for Children and Adolescents with Cystic Fibrosis

Age	Vitamin A (IU)	Vitamin E (IU)	Vitamin D (IU)	Vitamin K (IU)
0–12 months	1,500	40–50	400	0.3–0.5*
1–3 years	5,000	80–150	400–800	0.3–0.5*
4–8 years	5,000–10,000	100–200	400–800	0.3–0.5*
>8 years	10,000	200–400	400–800	0.3–0.5*

*Currently commercially available products do not have ideal doses for supplementation.

Note: These fat-soluble vitamins are given in addition to an age-appropriate dose of non-fat-soluble vitamins.

Adapted from: Borowitz D, Baker RD, Stalling V. Consensus report on nutrition for pediatric patients with cystic fibrosis. *J Pediatr Gastroenterol.* 2002;35:246–59.

Zinc supplementation (~1 mg elemental zinc/kg/day) has been recommended for children who are failing to thrive or who have short stature. Supplementation for children with as much as 5 mg of zinc per day improves energy intake, increases forced expiratory volume, and decreases infections.[166]

Recommended Infant Feeding Breast-feeding is recommended for infants under the first year of age. Proprietary infant formulas may also be used. Caloric density greater than 20 kcal/ounce may be needed to support growth. This can be achieved by fortifying breast milk or by concentrating formula. Solid foods should be added at the appropriate age (see Table 21.10). If infants taking solid food are not achieving appropriate growth, additional kcal can be added to infant formula in the form of carbohydrate polymers (e.g., Polycose® from Abbott Nutrition) and/or fats such as vegetable oil or MCT oil. Families need to be counseled that when table food is introduced, the diet must be moderately high in fat and protein. When breast milk is discontinued, whole milk should be added. The diet should be supplemented with both fat-soluble and water-soluble vitamins in age-appropriate doses (see Table 21.9). Supplemental iron and fluoride may need to be given if the dietary intakes are inadequate. Supplementation of sodium chloride may be necessary for infants, particularly during the summer months. At all ages, appropriate doses of pancreatic enzymes must be given prior to meals and snacks to ensure nutrient absorption.[146] Participation in family mealtimes should be encouraged to increase energy intake.[167]

Children need to be monitored for appropriate growth. Additional kcal in the form of in-between meal snacks and nutrient-dense foods, particularly foods high in fat, need to be added to the diet to support growth. Appropriate feeding behaviors should be encouraged at each age. School-aged children (5–10 years) are at higher risk for decreased growth rate. Participation in activities increases energy expenditure and may also lead to limited time for consumption of snacks and taking of enzymes. Adolescence (11–18 years) is a time associated with accelerated growth and pubertal development; nutritional counseling directed specifically toward the adolescent may be necessary. Female adolescents are at greatest risk for poor nutrition because of their higher energy and nutrient requirements, poorer eating habits, and slower growth.[168] At this age, pulmonary infections,

Table 21.10 Developmental Approaches to Nutrition Counseling for Individuals with Cystic Fibrosis

Infants

- Breast-feeding is recommended for most infants as the primary source of nutrition during the first year of life. Human milk fortifiers may be used to increase the nutrient density of breast milk if needed.
- Iron-fortified infant formula may be used. The caloric density of the standard infant formula (20 kcal/ounce) may need to be increased by concentrating the formula or using fat and/or carbohydrate additives, as appropriate.
- Solid foods should be added at 4–6 months developmental age, according to recommendations of the American Academy of Pediatrics.
- Infant cereal should be prepared with breast milk or infant formula, not water or juice; additional fat or carbohydrate kcal may be added to infant cereal if needed to achieve expected rate of growth.
- Pancreatic enzymes should be given prior to each feeding.
- Vitamin supplements with additional fluoride and iron should be given.

Toddlers or Preschool Age (1–4 Years)

- Whole milk should be encouraged for the child with CF.
- Adding kcal to table food may help with maintenance of growth at this stage; avoid giving low-fat or low-kcal foods.
- Regular mealtimes and snack times should be encouraged.
- Dietitians should inquire about feeding behaviors to promote positive interactions and prevent negative behaviors; grazing behavior should be discouraged.
- Pancreatic enzymes and vitamins are continued.

School Age (5–10 Years)

- A normal, healthy diet with a variety of food should be the basis of the diet.
- This may be a high-risk period for decreased growth rate in children with CF; identify factors that may interfere with meeting nutritional needs, such as activities that may lead to limited time for snacks and enzyme adherence and the progression of the disease.
- Pancreatic enzymes and vitamins are continued.

Adolescence (11–18 Years)

- Associated with high nutrient requirements due to accelerated growth, pubertal development, and high levels of physical activity.
- Nutritional counseling will be more effective if directed toward the adolescent, not the parent.
- Adolescents may be more receptive to efforts to improve muscular strength and body image rather than stressing weight gain and improved disease status.
- Continue pancreatic enzymes and vitamins.

Adapted from: Borowitz D, Baker RD, Stalling V. Consensus report on nutrition for pediatric patients with cystic fibrosis. J Pediatr Gastroenterol. 2002;35:246–59.

CFRD, or liver disease may develop, making nutrition management more complex.

The use of nutritional supplements may be helpful in an effort to add additional kcal, protein, and other nutrients to the diet. Homemade foods high in kcal and protein, such as fortified beverages or puddings, may be beneficial. The addition of supplemental enteral feedings may be needed if adequate kcal intake cannot be achieved or growth is compromised. Nocturnal tube feedings are encouraged to promote normal eating behaviors during the day. Standard formulas (1.5–2.0 kcal/cc) containing complete protein and long-chain fatty acids are usually well tolerated.[146] Pancreatic enzymes should be taken before these feedings are given.

Pneumonia

Definition

Pneumonia is defined as an inflammation of the lungs, usually caused by bacteria, viruses, or fungi. Once the offending agent enters the lungs, it usually settles in the alveoli where it can grow rapidly. The infection causes deterioration of lung function resulting in fluid accumulation and breathing difficulty.[169]

Epidemiology

Prior to 1936, pneumonia was the leading cause of death in the United States, but with the use of antibiotics, the incidence of pneumonia has been substantially reduced. In 2006, pneumonia and influenza collectively ranked as the eighth leading cause of death.[170] Among the elderly, pneumonia is the fourth overall leading cause of death and the leading infectious cause of death in this subset of the population.[171]

Etiology

Although the microorganisms responsible for pneumonia are present in the environment and are inhaled into the lungs all the time, the cilia and microphages present in the lungs help prevent them from entering the alveoli. However, the normal defense mechanisms may be compromised in certain populations, including the elderly, infants and young children, and individuals with health problems such as COPD, diabetes mellitus, asthma, alcoholism, congestive heart failure, and sickle cell anemia. Individuals living with HIV/AIDS and those undergoing chemotherapy or organ transplants who are immunocompromised are also at substantial risk for developing pneumonia.

There are two major categories of pneumonia based on the cause: that which is community acquired and that which is hospital acquired. Community-acquired pneumonia (CAP) occurs when infected persons cough or sneeze and spread the bacteria, primarily *Streptococcus pneumoniae* or *Pneumococcal pneumonia*, to those around them.[169]

Hospital-acquired pneumonia (HAP), also known as nosocomial pneumonia or health-care associated pneumonia (HCAP), often affects patients who are in the intensive care unit (ICU) or are on a mechanical ventilator. Hospital-acquired pneumonia occurs fairly frequently and has a high rate of mortality. The incidence of hospital-acquired pneumonia is highest among the elderly, the very young, and those who are already debilitated by other diseases.[172]

Health care-associated pneumonia (HCAP) is acquired in other health care settings such as nursing homes, rehabilitation centers, and convalescent homes and may be caused by any of the same causative agents as pneumonia.[173]

Aspiration Pneumonia

Another common cause for the development of pneumonia is aspiration of inhaled materials (saliva, nasal secretions, bacteria, liquids, food, or gastric contents) into the airway below the level of the vocal cords. **Aspiration pneumonia** occurs when the aspirated material causes an inflammatory response in the lung. Aspiration is the leading cause of pneumonia in the ICU and contributes to the morbidity and mortality of patients.[173]

There are a number of normal defense mechanisms which help to prevent aspiration.[174] During the swallowing process the epiglottis, a thin cartilage structure located at the base of the tongue, folds over the top of the larynx to prevent food and liquid from entering the trachea (see Chapter 14). The lower esophageal sphincter (LES) also prevents the upward movement of gastric contents into the esophagus. In addition, food particles and fluids that may be aspirated into the lungs are entrapped in the mucus layer of the respiratory epithelium. The mechanical beating of cilia on the respiratory epithelium advances the mucus and entrapped particles upward so it can be cleared by the cough. Cough is an important mechanism that allows clearance of foreign material and secretions from the airway.

A number of risk factors may contribute to aspiration (see Table 21.11).[175] Patients with head injuries often have delayed gastric emptying and a decreased gag and cough reflex, and thus are at high risk for aspiration. Aspiration occurs more often in patients with high gastric residuals, although there is no relationship between the actual volume of the residuals and aspiration risk.[176] Individuals who have neurological impairments such as stroke and Parkinson's disease are a high risk for dysphagia (difficulty in swallowing) and subsequent aspiration (see Chapter 14). Hyperglycemia can result in disordered motility throughout the gastrointestinal tract. Patients with diabetes who have gastroparesis (delayed gastric emptying, vomiting, nausea, or bloating caused by stomach nerve or muscle damage) may aspirate, particularly if they receive gastric tube feedings. In addition, patients with abnormalities of the gastrointestinal tract including **esophageal stricture** and gastroesophageal reflux disease (GERD), and those who require mechanical ventilation, are also at high risk for developing aspiration.

Aspiration can be a serious side effect of enteral tube feeding, particularly with gastric feeding (see Chapter 5).[177] Patients fed using both a nasogastric feeding tube and a gastrostomy tube are at high risk for aspiration. The likely cause is the reflux of gastric contents from the stomach into the pharynx, where they are aspirated into the lungs. In addition, the presence of a nasogastric feeding tube may interfere with the effectiveness of the lower esophageal sphincter, resulting in gastrointestinal reflux.

Table 21.11 Risk Factors Associated with Aspiration

- Decreased level of consciousness
- Neuromuscular disease and structural abnormalities of the upper gastrointestinal tract
- Endotracheal intubation or mechanical ventilation
- Vomiting
- Delayed gastric emptying (seen in diabetes, hyperglycemia, electrolyte abnormalities, and with drugs known to reduce gastric emptying)
- Advanced age (60+ years)
- Poor oral care
- High gastric residual volumes
- Prolonged supine position (head of the bed flat)
- Presence of a nasogastric tube
- Noncontinuous or intermittent tube feeding
- Large diameter of feeding tube and/or malpositioned feeding tube

Source: McClave N. Risk factors for aspiration. *JPEN.* 2002;226:S26–S33.

Patients with Tracheostomies

A **tracheostomy** is a surgical opening made in the trachea to assist breathing. A tracheostomy tube is inserted through the surgical opening (stoma), as shown in Figure 21.6. A tracheostomy is usually performed for one of the following reasons: (1) to bypass an obstruction in the trachea, (2) to clean and remove secretions from the trachea and prevent them from entering into the lungs, or (3) to more easily and safely deliver oxygen to the lungs when the patient is not able to breathe without assistance. Sometimes children or adults require permanent tracheostomies to breathe. Patients who are unable to breathe on their own usually also require a mechanical ventilator in addition to the tracheostomy.

Nutritional Implications

There are many potential complications related to the presence of a tracheostomy tube, including the inability to speak or swallow normally.[178] Patients with tracheostomy tubes who are on mechanical ventilation are often at high risk for aspiration. Frequently these patients require tube feeding for nutritional support. Once they are weaned (removed) from the ventilator, the tracheostomy tube may remain in place to remove secretions from the trachea.

When it is safe for these patients to eat orally, the viscosity of the food often makes a difference.[178] The dietitian often works closely with the speech pathologist to determine whether the patient can safely swallow food without aspiration and, if so, the consistency of food that is appropriate. The speech pathologist is specially trained to evaluate swallowing and make recommendations about food consistencies. Because respiration is momentarily halted during swallowing, for a patient to eat orally, the act of swallowing needs to be combined successfully with respiration. Normally, exhalation of air both precedes and follows the swallow, ensuring that remnants of the food bolus are not aspirated. Successful feeding also requires the presence of an effective cough and intact upper airway reflexes.[179] Liquids that are thin in consistency are more frequently aspirated than those with more viscosity, because they move quickly into the pharynx, which requires rapid closure of the larynx[178] (see Chapter 14). Also, solid foods that generate thin liquids during

chewing (from oral secretions) may also present a problem, because the liquid can leak into the airway before the solid food bolus is swallowed. Thicker liquids with a honeylike texture can often be swallowed more efficiently, bypassing the incomplete laryngeal closure. Foods that require little or no chewing, such as pureed food, may be better tolerated by patients who aspirate thin liquids. Since aspiration risk may also be increased if a patient becomes fatigued during mealtime, smaller, more frequent meals may be beneficial by helping to reduce fatigue.

Tracheostomy tubes come in many varieties, including both cuffed and uncuffed tubes. A cuff is a soft balloon around the distal (far) end of the tube that can be inflated to prevent oral secretions from entering the lungs. The cuffs are inflated with air, foam, or sterile water. When the cuff is deflated, the tube allows air around it for vocalization. It is not completely clear whether the cuff should be inflated or deflated when feeding patients orally. A small study of 12 patients with tracheostomy tubes found that there was nearly a threefold increase in the aspiration rate when the cuff was inflated, versus deflated.[180]

Respiratory Failure

Respiratory failure (RF) occurs when the respiratory system is no longer able to perform its normal functions. It can result from long-standing chronic lung disease like COPD or cystic fibrosis, or as a result of an acute insult to the lung as seen with **acute lung injury (ALI)** and **acute respiratory distress syndrome (ARDS)**.

Acute lung injury encompasses a variety of changes in lung function associated with critical illness. The diagnostic criteria includes an acute onset, bilateral infiltrates in the lungs, hypoxemia without cardiac changes.[181] ARDS may result from direct damage to lung tissue as is seen with pneumonia, COPD, and aspiration or from an indirect insult such as sepsis, trauma, or burns (see Table 21.12). Ventilator-associated lung injury may contribute to a significant number of new cases of ARDS.[182] ARDS usually occurs within 24–72 hours of the predisposing factors.[182, 183] Although the actual mechanism for the lung injury seen in ARDS is not totally understood, it occurs in response to a variety of cytokines (small, hormonelike proteins) released from inflammatory cells that can trigger production of reactive oxygen species (ROS).[184] ARDS is characterized by dyspnea,

Figure 21.6 Tracheostomy Tube

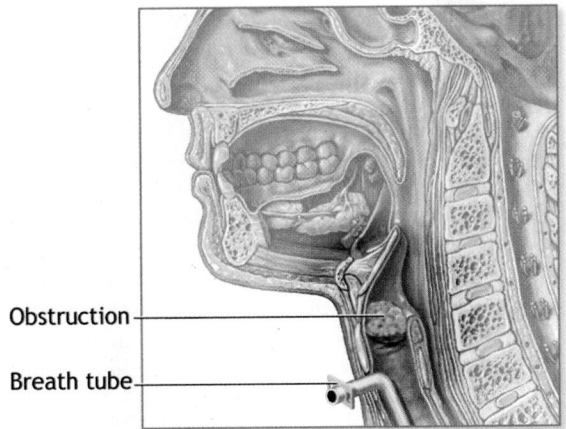

Obstruction

Breath tube

Source: Copyright © 2006 A.D.A.M., Inc.

Table 21.12 Conditions Associated with the Development of Acute Respiratory Distress Syndrome

Direct Injury	Indirect Injury
Pneumonia	Sepsis
Aspiration of gastric contents	Severe trauma
Inhalation injury	Acute pancreatitis
Near drowning	Cardiopulmonary bypass
Pulmonary contusion	Massive transfusions
Fat embolism	Drug overdose
Pulmonary edema	
Post-lung translation	

Adapted from: Mackay A, M Al-Hadded. Acute lung injury and acute respiratory distress syndrome. Cont Edu Anaesth Crit Care & Pain. 2009:9:152–156.

severe hypoxemia, decreased lung compliance, loss of surfactant, and leakage of a protein-rich fluid into the interstitium and alveolar lumen.[185–187] The use of moderate-dosage corticosteroids early on is effective in reducing the time spent on a ventilator and decreasing the length of stay in the ICU.[188]

Nutrition Therapy for Respiratory Failure

Nutritional Implications

Although patients with RF, particularly those with ALI and ARDS, may be initially managed with supplemental oxygen, progressive hypoxemia often requires intubation and mechanical ventilation (see Figure 21.7). Patients who require ventilator support are not able to consume foods orally and will require an alternative form of nutritional support. The route of nutritional support will be determined by the patient's underlying illness and gastrointestinal function; nonetheless, enteral nutrition is the preferred method of support due to its role in maintaining gastrointestinal function, the reduced risk of sepsis, and its lower cost. In cases where enteral nutrition support is not possible, parenteral nutrition is used.

Nutrition Assessment

Nutritional needs vary widely depending on the age of the patient, the patient's prior nutritional status, and the underlying disease process, which is often hypermetabolic. Often patients with RF are poorly nourished, particularly if they have a longstanding history of COPD or CF. They may have low plasma levels of vitamins A and E and elevated damage resulting from peroxidation and lipid peroxides. Lipid peroxides are the products of chemical damage done by oxygen free radicals to the polyunsaturated fatty acids of cell membranes.[189] A complete nutrition assessment to evaluate the patient's individual nutrition needs, including anthropometric and laboratory

Figure 21.7 Mechanical Ventilation

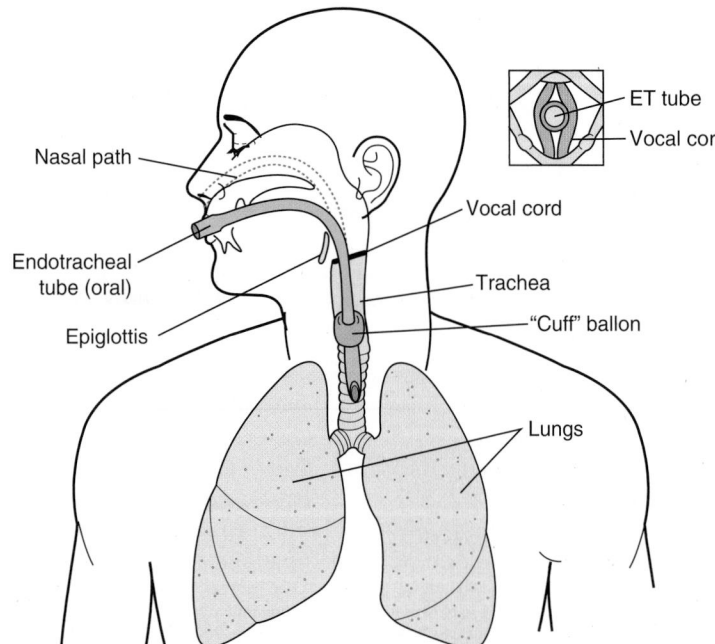

status (including antioxidants), should be completed. Poor nutritional status is often associated with fluid shifts. Because these patients often have fluid imbalances, careful interpretation of laboratory data is essential. For example, if the patient is protein deficient, fluid often shifts from the blood to interstitial spaces. This dehydration can result in serum values that are elevated due to the removal of blood volume.

Nutrient Requirements Total caloric requirements can either be estimated using a predictive equation or directly measured using indirect calorimetry. Although indirect calorimetry is the preferred method for assessing kcalorie needs, a number of predictive equations are also available (see Chapter 3). The provision of 25 kilocalories per kilogram of usual body weight (or 130% of basal energy expenditure) appears to be adequate for most patients.[108, 190] Once the kcalorie requirements have been estimated, the patient's pulmonary status, body weight, and fluid balance must be closely monitored to ensure that overfeeding does not occur.

Patients with RF often do not have the pulmonary reserve needed to clear excess carbon dioxide. Because overfeeding is associated with increased CO_2 production, it is important that a careful assessment of the patient's nutrient requirements be made, particularly if the patient requires mechanical ventilation. Overfeeding occurs most commonly in patients fed via multiple routes such as a tracheostomy and oral intake.[191] The increased ventilatory demand associated with overfeeding is related to both excess glucose administration (>5 mg/kg per minute) and excess energy intake.[108] Increases in carbon dioxide production from overfeeding may result in difficulty in weaning or removing patients from mechanical ventilation, which further complicates their medical care (see Chapter 5 for a discussion of overfeeding).

Patients with ALI/ARDS do have increased protein requirements resulting from hypermetabolism. A range of 1.2 to 1.5 g/kg/day of protein appears appropriate and should be adjusted to promote nitrogen retention, without being excessive.[190] An evaluation of urine urea nitrogen is one method for assessing adequacy of protein intake, though it does have its limitations (see Chapter 3).

Nutrition Intervention

The goals of nutrition care for patients with RF are to meet their nutritional needs; preserve and restore lean body mass, particularly respiratory muscle mass; maintain fluid balance; and facilitate weaning form mechanical ventilation (see Table 21.13).

Enteral and Parenteral Nutrition Several enteral products high in fat and low in carbohydrate have been developed specifically for patients with acute lung injury and respiratory failure, and have demonstrated decreases in both $PaCO_2$ and time on mechanical ventilation.[108] Because ARDS is also associated with the development of pulmonary edema, the use of fluid-restricted enteral formulations (1.5–2 kcal/cc) may be helpful in patients whose hemodynamic status requires fluid restriction.[108]

ARDS is also associated with the production of oxygen free radicals and inflammatory mediators derived from arachidonic acid. These mediators and free radicals have

Table 21.13 Nutrition Practice Guidelines for Patients with Respiratory Failure

- Patients should undergo nutrition screening to identify those who require nutrition assessment and development of a nutrition care plan. (B)
- Energy intake should be kept at or below estimated needs in patients with demonstrated carbon dioxide retention. (B)
- Modified enteral formulas consisting of omega-3 fatty acids may be beneficial in patients with early ARDS. (B)
- A fluid-restricted nutrient formulation should be used in patients with ARDS whose hemodynamic status necessitates fluid restriction. (B)
- Serum phosphate levels should be monitored closely. (B)

Note: (B) refers to strength of evidence to support the recommendations based on the following criteria: A = good research based evidence supported by prospective, randomized trials; B = fair evidence based on well designed studies without randomization; C = based on expert opinion.

Adapted from: Specific guidelines for disease—adults. *JPEN.* 2002;26(1):SA63–SA64.

been shown to cause lung inflammation, edema, alveolar damage, and lung collapse. Recent research has shown that the dietary fatty acids, particularly eicosapentaenoic acid (EPA), found in fish oil, and gamma-linolenic acid (GLA), found in borage oil, can reduce the severity of the inflammatory injury by altering the availability of arachidonic acid in tissue phospholipids.[192] Patients with ARDS receiving a high-fat enteral product supplemented with omega-3 fatty acids (fish oil and borage oil) and antioxidants (Oxepa™, Abbott Nutrition) spent less time on mechanical ventilation, had briefer stays in the ICU, and had a decreased incidence of organ failure.[192]

Vitamin and Mineral Supplementation Patients with ARDS have low levels of oxidant defense enzymes, high levels of ROS, and tissue oxidative damage.[193] Supplementation with α-tocopherol, β-carotene, and vitamin C at levels higher than the DRI has been associated with substantial increases in serum α-tocopherol and β-carotene, and appears to prevent further oxidative damage.[194]

Phosphate is essential for optimal pulmonary function and normal contractibility of the diaphragm. The length of hospital stay and dependence on mechanical ventilation have been shown to be increased in critically ill patients who have hypophosphatemia.[195] The phosphate balance of patients in respiratory failure needs to be closely monitored, particularly for those patients receiving parenteral nutrition support. Phosphate supplementation should be initiated whenever hypophosphatemia is present.

Transplantation

Definition and Epidemiology

Lung transplantation is a surgical procedure in which one or both lungs are replaced with healthy organs from a human donor. Lung transplantation is an accepted option for patients with end-stage lung disease. In 2005, approximately 3,500 people in the United States were waiting for a lung transplant, but only 25% received a transplant.[196] In some cases where the heart has also been weakened, both the heart and lungs will be replaced. Until 1989, combined heart-lung transplants were the most common form of lung transplantation, but more recently single and double lung transplants have become more common. The United Network for Organ Sharing indicates that there are

about 100,000 people waiting for organ donations, about 1,900 of whom are waiting for lungs; about 80 are waiting for heart and lungs.[197]

Pathophysiology

Transplant recipients are at high risk for rejection of the transplanted lung. The body's immune system considers the transplanted organ as an invader (similar to infection) and may attack it. Because of the rejection risk, patients must take immunosuppressive (anti-rejection) medications (see Chapter 9). Rejection occurs most often during the first three months after transplantation, but the immunosuppressive medication may need to be taken indefinitely.[198] Common immunosuppressive drugs used are cyclosporine, tacrolimus, mycophenolate mofetil, azathioprine, and prednisone. The nutritional side effects of these medications may include GI distress (nausea, vomiting, diarrhea, and/or constipation), increases in blood pressure, edema, and alterations in blood sugar levels. They also lower the body's immunity and increase the risk of infection.

Nutrition Therapy for Transplantation

Nutritional Implications

Patients with lung disease may be underweight, normal weight, or overweight. Overweight patients with lung disease, however, are often very sedentary and have significant increases in body fat mass rather than lean body mass (LBM). Lean body mass depletion has been associated with a higher rate of mortality in patients awaiting transplantation. Poor nutritional status and low LBM have a significant impact on the length of time postoperative transplant patients spend on ventilation, and can double the length of stay in the ICU.[199] Patients with a pretransplant BMI <17 kg/m² or >25 kg/m² have an increased risk of dying within 90 days posttransplant.[200] This risk is significantly higher in patients with a BMI >27 kg/m².

Nutrition Assessment

A comprehensive nutrition assessment of transplant recipients should include a physical assessment, dietary history, anthropometric measures, and laboratory values. Specific assessment issues are included in Table 21.5. Direct measurement of body composition using DEXA or dilution techniques may be more helpful than weight or BMI to assess changes in LBM. Although bioelectrical impedance analysis (BIA) has become widely available for assessing body composition, single tests using BIA may not be valid because of the fluid shifts that are often seen in these patients.

Nutrient Requirements Protein and energy requirements are affected by the stress of surgery, postoperative complications, episodes of rejection, and the use of immunosuppressant drugs, particularly corticosteroids. The use of indirect calorimetry to assess kcal needs is indicated when the patient's medical condition is complicated by posttransplant complications. When indirect calorimetry is not available, 130% to 150% of BEE (or 35 kcal/kg body weight) is usually adequate.[201] Adequate amounts of protein are required for wound healing and to prevent infection. Protein needs may be increased due to surgical stress and the use of corticosteroids. Nitrogen balance studies suggest protein requirements range from 1.5 to 2.0 g/kg per day.[201, 202] These requirements may decrease to 1 g/kg as the dose of corticosteroids is reduced to maintenance levels.[201]

Nutrition Intervention

The physical and nutrition status of patients awaiting transplant may decline. Therefore, nutrition support is very important during the pretransplant period. The goal is to optimize nutrition status as much as possible by increasing kcal and protein. Most patients waiting for an organ transplant are able to eat.[202] Small, frequent meals composed of nutrient-dense foods and supplements should be encouraged. Enteral feedings may be indicated if the patient is unable to eat adequate amounts.

The acute posttransplant period may be complicated by rejection, infection, and surgical complications.[202] The nutrition goals during this time are to provide adequate nutrients to promote wound healing, treat changes in electrolyte balance, and achieve optimal blood glucose control.[201, 202] Most transplant patients are allowed to eat three to five days after transplantation. Nutrition support should be considered when the patient is not able to eat or if oral feeding is delayed. When nutrition support is required, the use of enteral nutrition is preferred. Patients who are malnourished or at risk for extended NPO (nothing by mouth) status may benefit from immediate posttransplant tube feeding. When enteral nutrition support is not possible, parenteral nutrition is advocated. Parenteral nutrition should be administered cautiously because of postoperative hyperglycemia due to metabolic stress, infection, and the use of corticosteroids. Hyperglycemia has been shown to impair wound healing and increase the risk of infection.[201] When the oral diet is initiated, simple carbohydrates should be limited.

Upper Respiratory Infection

Definition and Epidemiology

Upper respiratory infection (URI), generally known as the common cold, is a nonspecific term used to describe acute infections which involve the nose, sinuses, pharynx, larynx, trachea, and bronchi.[203] URIs occur more frequently during the winter months. Adults develop an average of two to four colds per year, while children develop an average of six viral respiratory tract infections each year.[203, 204]

Pathophysiology

Transmission of the organisms which cause URIs occurs primarily by aerosol, droplet (sneezing or coughing), or direct hand-to-hand contact with infected secretions. Onset of symptoms occurs one to three days after exposure to the infectious agent.[203] Runny nose, nasal congestion, and sneezing are common symptoms of URIs, often accompanied by sore throat, cough, and headache. Treatment for URIs is usually directed towards minimizing symptoms. Rest and increased fluid intake are recommended. Hundreds of over-the-counter (OTC) medications are available; however, none have proven to be suitable in controlling the course of the infection, only effective in improving symptoms.[203]

Clinical Manifestations

The signs and symptoms associated with an upper respiratory infection are usually mild and brief in duration. These include the cardinal symptoms of inflammation (*rubor, calor, dolor,* and *functio laesa*; see Chapter 9). Fluid and mucus released in response to the inflammation obstruct upper airways. If the inflammation includes the larynx, trachea, and upper bronchi, the patient will experience labored breathing. These symptoms may be more difficult for infants and children due to their immature respiratory system and accessory muscles of the neck.

Nutritional Implications

Both vitamin C and zinc have been studied in preventing or decreasing the cold symptoms. Twenty years ago, Linus Pauling first stimulated public interest in the use of large doses of vitamin C (>1 gram/day) to prevent infections associated with the common cold.[205] Subsequent research has concluded that large doses of vitamin C do not have a significant effect on the incidence of the cold. However, a few studies have demonstrated that taking supplemental vitamin C (~2 grams) may decrease the duration and severity of the cold, an effect likely related to the antihistamine effects that occur with large vitamin C doses.[205–207] These doses represent intakes above the Tolerable Upper Intake Level (UL) for vitamin C and may result in gastrointestinal disturbances.

The use of zinc lozenges and nasal spray has been promoted to reduce the duration of the common cold. However, numerous controlled trials have found conflicting results regarding the effectiveness of zinc in reducing both its symptoms and duration.[208–210] Taking zinc lozenges at recommended levels (12.8 mg of zinc/lozenge) for every two to three hours while awake may result in a dietary intake of zinc above the UL of 40 mg/day. Short-term use of zinc lozenges (~5 days) has not resulted in serious side effects; however, long-term use (six to eight weeks) may result in copper deficiency.

A variety of other nutrients may be important in maintaining lung function and preventing respiratory infections. High serum levels of vitamin C, vitamin E, vitamin A, betacryptoxaanthin, selenium, calcium, and chloride are correlated with high FEV1.[211] The efficacy of vitamin A in preventing lower respiratory tract infections has also been studied in children. Findings show no relationship between vitamin A and decreased chances of lower respiratory infection rates in children under the age of 7.[212] However, there does appear to be a link between vitamin D and upper respiratory infections. Individuals with low 25(OH)D levels have a greater incidence of upper respiratory infections, especially patients who already suffer from asthma or COPD.[213]

Conclusion

This chapter has reviewed the anatomy and physiology of the respiratory tract along with the detailed changes that occur during the disease process. Many individuals with respiratory disease are at high nutritional risk and need specialized nutrition support. Nutritional care during respiratory disease functions both to support the normal function of the respiratory tract and to provide important interventions that serve as crucial components of medical care.

Julie Matel, MS, RD, *Clinical Dietitian at Lucille Packard Children's Hospital (http://www.lpch.org);*
Cystic Fibrosis Dietitian

Lucille Packard Children's Hospital, located in Stanford, CA, is a large teaching hospital with numerous outpatient clinics. The Cystic Fibrosis Center at LPCH is a center for excellence and cares for approximately 300 pediatric patients with CF. The RD sees each patient during quarterly clinic visits and manages the nutritional care for CF patients that are admitted to LPCH for acute respiratory exacerbations.

How do you nutritionally assess patients with cystic fibrosis?

The most useful assessment measures are current weight, weight trends, and BMI. The Cystic Fibrosis Foundation (CFF) has been instrumental in developing consensus guidelines for pediatric patients. One goal for infants and children with CF is for them to be at the 50th percentile wt/l or for BMI on the growth charts. In addition, I measure interval weight gain between visits. If they are below their goal BMI, their interval gain should be higher. I also assess whether their growth is age appropriate. Based on their anthropometrics, I assign a nutrition risk score. Many cystic fibrosis (CF) dietitians have developed a similar protocol for assessing nutrition risk. CFF consensus recommends that a detailed diet history be obtained once a year but I will perform this assessment more often if they have a higher nutrition risk score. Clinically, I ask about their stooling patterns (how many, consistency and/or if they are experiencing any pain or other related problems) and review enzyme therapy if they are on enzymes. Since fat-soluble vitamins are not as well absorbed in pancreatic-insufficient patients, these labs are checked annually. More recently, we have started screening 6-year-old CF clients for glucose intolerance since they are at increased risk of developing CF-related diabetes or abnormal glucose tolerance. It has been found that lung function is negatively affected by glucose intolerance/diabetes. Iron status is difficult to determine. CF clients with poor lung function tend to be anemic due to their chronic infections. Since iron supplements may support lung infections with pseudomonas bacteria, we

are very careful to determine the cause of their anemia before prescribing iron.

What is the biggest challenge working with patients with CF clients?

As with many chronic diseases, CF has a large treatment burden so motivating clients to improve their nutrition can be very difficult. What I have found is that most clients will routinely take their digestive enzymes. They are much less likely to adhere to their nutrition supplements and vitamins. When working to maximize calorie intake, I recommend families take the Ellyn Satter approach to feeding. Avoid forcing your child to eat. Instead, provide energy-dense food offered at regular meal and snack times. Try to teach families how to add calories to their diet without actually increasing the amount of food they eat. For example, by adding some avocado and an extra slice of cheese to a sandwich, calorie content can be increased by more than 200 kcal.

When CF clients develop a respiratory infection, their weight will drop quickly due to increased needs and decreased intake. The infection will cause increased sputum production and excessive coughing, which contributes to nausea and vomiting. However, with appropriate antibiotic therapy patients usually regain weight with extra nutrition support provided in the hospital setting.

Advice for future dietetic interns?

I have been a dietitian for over 10 years. While I do believe that when starting out in dietetics, diverse work experience is the way to go, after working in several areas of practice in both inpatient and outpatients settings, I have found specializing in one area gave me a greater sense of accomplishment and more time to become an expert. I continue to update my knowledge by participating in the National CF Conference annually and I have joined the CF practice group. This group has been extremely helpful in sharing research articles and standards of practice, which allow me to stay up to date, which can be a challenge in a busy work environment.

Application of the Nutrition Care Process: COPD

Introduction:

Stella is a 62-year-old female who was diagnosed with COPD five years ago. She has a medical history of bronchitis and upper respiratory infections spanning most of her adult life and admits to smoking one pack of cigarettes per day for the past forty-six years. She quit smoking during the past year.

Stella has currently been admitted to the hospital complaining of severe dyspnea and fatigue—"I'm hardly able to do anything for myself now. Even taking a bath or getting dressed makes me short of breath." Stella is diagnosed with acute exacerbation of COPD secondary to pneumonia. She will be discharged on oxygen therapy for the first time and referred to an outpatient pulmonary rehabilitation program.

(continued)

1. What risk factors for emphysema/COPD are present in Stella's history?

2. Describe the causes for the following symptoms present in Stella's history and physical exam: anorexia, dyspnea, fatigue, early satiety, bloating.

Nutrition Assessment:

Food/Nutrition-Related History:

When you talk to Stella about her food intake at home, she indicates that her appetite is poor. "Sometimes I feel bloated; I fill up so quickly—after a few bites of food." "Food doesn't taste good anymore." She tells you that her highest adult weight was 145–150 lbs (about 5 years ago). She complains that her dentures are fitting very loosely.

Usual dietary intake:

Coffee, juice, and dry cereal with a small amount of milk in the morning and one other larger meal during the day, usually at lunch—which consists of meat, vegetables, rice, potatoes, or pasta. She admits that she eats only small amounts. At night she often has a bowl of soup. She drinks Pepsi throughout the day (usually 3 12-oz. cans).

24-hour recall:

½ cup coffee, few sips of orange juice, ½ cup oatmeal with 1 tsp sugar and small amount of 1% milk

½ cup chicken noodle soup, 2 saltine crackers, ½ cup coffee

Drank Pepsi throughout the day (~32 ounces)

Vitamin/mineral supplements: None

3. What factors can you identify from her nutrition history which probably contribute to her poor food intake?

Anthropometric Measurements:

Height: 5'3", weight: 119 lbs.

Physical Exam:

Vital Signs: Temperature: 98.8°, blood pressure (BP): 130/88, respiratory rate (RR): 22

Extremities: 1+ bilateral pitting edema

Chest/lungs: Decreased breath sounds; prolonged expiration with wheezing; using accessory muscles at rest.

4. Evaluate Stella's current weight, usual body weight, ideal body weight, BMI. How does her bilateral pitting edema affect your evaluation of her weight?

Biochemical Data:

Albumin: 3.4 g/dL Sodium: 136 mEq/L Potassium: 3.7 mEq/L pH: 7.29 $PaCO_2$: 50.9 mmHg PaO_2: 77.7 mmHg O_2Sat: 92% HCO_3: 29.6 mEq/L

5. Evaluate Stella's biochemical data. What do they tell you about her nutritional status? Define each of the following blood gases and interpret her values: pH, $PaCO_2$, PaO_2, O_2Sat, HCO_3.

Nutrition Diagnosis:

6. Identify at least two nutrition problems based on the nutrition assessment and medical history. Determine the diagnostic term for each nutrition problem. Next, identify the etiology of each nutrition problem. Finally, identify the signs and symptoms that support the evidence for these nutrition problems.

Nutrition Intervention:

7. Identify the nutrition prescription for this patient by recommending the appropriate dietary modification for her diagnosis.

8. Calculate energy and protein requirements for Stella.

Nutrition Monitoring and Evaluation:

9. Determine nutrition criteria for monitoring and evaluation for each nutrition diagnosis that you identified.

WEB LINKS

American Academy of Allergy, Asthma, and Immunology This website provides many resources for patients and for health care professionals regarding allergies, asthma, and other associated diseases.
http://www.aaaai.org

American Lung Association The website for the American Lung Association contains excellent information about numerous respiratory-related disorders.
http://www.lungusa.org

American Thoracic Society The American Thoracic Society publishes many papers and guidelines for health professionals related to various respiratory diseases. They also publish three online respiratory-related journals (see "Relevant Online Journals").
http://www.thoracic.org

Centers for Disease Control and Prevention (CDC) A governmental agency whose mission is to promote health and quality of life by preventing and control-

ling disease, injury, and disability. The CDC maintains national health statistics, conducts research, and provides services concerning prevention of illness and injury.
http://www.cdc.gov

Cystic Fibrosis Foundation A nonprofit organization dedicated to improving the care of patients with cystic fibrosis and helping to find a cure for cystic fibrosis. The Cystic Fibrosis Foundation also supports research related to the treatment

of cystic fibrosis. This website contains excellent information for professionals and patients.
http://www.cff.org

National Heart, Lung and Blood Institute This website contains excellent information for health care professionals and patients about various diseases related to the heart, blood vessels, and lungs.
http://www.nhlbi.nih.gov

National Institute of Allergy and Infections Diseases Conducts and supports basic and applied research to better understand, treat, and ultimately prevent infectious, immunologic, and allergic diseases. The website contains good information about numerous infectious diseases.
http://www3.niaid.nih.gov/

United Network for Organ Sharing A nonprofit organization that collects and manages data related to organ transplants. They also assist in facilitating organ transplants.
http://www.unos.org

RELEVANT ONLINE JOURNALS

The following journals are online publications for health care professionals who treat patients with respiratory diseases:

American Journal of Respiratory and Critical Care Medicine An online journal published by the American Thoracic Society.
http://ajrccm.atsjournals.org

RT Online journal for health care professionals who treat respiratory related disorders. Contains articles on nutrition management of various pulmonary diseases.
http://www.rtmagazine.com

END-OF-CHAPTER QUESTIONS

1. Describe the three major functions of the respiratory system in human health. What methods are used to measure pulmonary function?

2. Describe the role of nutrition in pulmonary health. Which nutrients have been associated with normal pulmonary function? How does smoking affect vitamin C requirements?

3. Based on supportive evidence, what are the important nutrition factors to keep in mind when treating patients of various age groups with asthma?

4. Define bronchopulmonary dysplasia (BPD). Why does it occur? How is vitamin A related to BPD?

5. Define cystic fibrosis (CF). What organ systems are involved in the disease? How does this organ involvement affect nutrition status? You receive a nutrition referral from a physician for a 9-year-old male with CF. He is below the 10th percentile weight for height. Outline an appropriate nutrition protocol for someone his age.

6. Describe aspiration pneumonia. As a dietitian, what procedures or methods would you recommend to help to prevent the occurrence of aspiration pneumonia in a patient receiving tube feeding?

7. Define respiratory failure (RF). What are the goals of nutrition therapy for RF? Outline a nutrition protocol for a patient with RF. Which nutrients are of specific concern?

REFERENCES

1. American Lung Association. Epidemiology and disease trend reports. Available at: www.lungusa.org

2. U.S. Department of Health and Human Services. Office of Disease Prevention and Health Promotion. Healthy People 2010. Available from: www.healthypeople.gov

3. Sherwood L. Fluid and acid-base balance. In: Human physiology. from cells to systems. Belmont (CA): Brooks-Cole/Thomson Learning; 2010.

4. Janssens JP, Pache JC, Nicod JP. Physiological changes in respiratory function associated with aging. Eur Respir J. 1999; 13:197–205.

5. Petty TL. Spriometry Made Simple. Available at http://www.nlhep.org/resources/SpirometryMadeSimple.htm

6. Sahebjami H. Effects of nutritional depletion on lung parenchyma. Eur Respir J. 2003; 24:113–24.

7. Pingleton SK. Enteral nutrition in patients with respiratory disease. Eur Respir J. 1998; 9:364–70.

8. Young IS, Roxborough HE, Woodside JV. Antioxidants and respiratory disease. Antioxidants in Human Health. CAB International: 1999; 293–311.

9. Schunemann HJ, Bryden JB, Grant B, Freudenheim, J, Muti P, Browne R, et al. The relation of serum levels of antioxidants vitamins C and E, retinol and carotenoids with pulmonary function in the general population. An J Respir Crit Care Med. 2001; 163:1246–55.

10. Schunemann HJ, McCann S, Grant B, Trevisan M, Muti P, Freudenheim J. Lung function in relation to intake of carotenoid and other antioxidant

vitamins in a population-based study. Am J Epidemiol. 2002;155:463–71.

11. Braskett M, Riedl MA. Novel anti-oxidant approaches to the treatment of upper airway inflammation. Curr Opin Allergy Clin Immunol. 2010;10:34-41.

12. Dallongeville J, Marecaux N, Fruchart J, Amouyel P. Cigarette smoking is associated with unhealthy patterns of nutrient intake: a meta-analysis. J Nutr. 1998; 128(9):1450–57.

13. Food and Nutrition Board Institute of Medicine. Dietary Reference Intakes for Vitamin C, Vitamin E, Selenium, and Carotenoids. Washington (DC): National Academy Press; 2000, 152–53.

14. Asthma Overview. Asthma and Allergy Foundation of America. Available at http://www.aafa.org/display.cfm?id=8

15. <1>Asthma & Children Fact Sheet. American Lung Association. Available at http://www.lungusa.org/site/apps/nl/content3.asp?c=dvLUK9O0E&b=2058817&content_id={05C5FA0A-A953-4BB6-BB74-F07C2ECCABA9}&notoc=1

16. Gaffin JM, Phipatanakul W. The role of indoor allergens in the development of asthma. Curr Opin Allergy Clin Immunol. 2009; 9:128–35.

17. Spergel JM, Fiedler J. Food allergy and additives: triggers in asthma. Immunol Allergy Clin N Am. 2004;25:149–67.

18. Eneli IU, Skybo T, Camargo CA. Weight loss and asthma: a systematic review. Thorax. 2008;63:671–76.

19. Ford ES, Mannino DM. Time trends in obesity among adults with asthma in the United States: findings from three national surveys. J Asthma. 2005;42:91–95.

20. Beuther DA, Sutherland ER. Overweight, obesity and incident of asthma. Am J Respir Crit Care Med. 2007; 175:661–66.

21. Varraso, R, Siroux V, Maccario J, Pin I, Kauffmann F. Asthma severity is associated with body mass index and early menarche in women. Am J Resp Crit Care Med. 2005;171:334–39.

22. Oddy WH, Peet JK. Breastfeeding, asthma, and atopic disease: An epidemiological review of the literature. J Hum Lact. 2003;19:250–61.

23. Flaherman V, Rutherford GW. A meta-analysis of the effect of high weight on asthma. Arch Dis Child. 2006;91: 334–39.

24. Weiss ST. Obesity: insight into the origins of asthma. Nature Immunology. 2005;6:537–39.

25. Rust G, Thompson CJ, Minor P, Davis-Mitchell W, Holloway K, Murray V. Does breastfeeding protect children from asthma? Analysis of NHANES III survey data. J Nat Med Assoc. 2001;93:1139–48.

26. Peat J, Allen J, Oddy W Webb K. Breastfeeding and asthma: appraising the controversy. Pediatr Pulmon. 2003;35:331–34.

27. Sears M, Taylor DB, Poulton R. Breastfeeding and asthma: appraising the controversy—a rebuttal. Pediatr Pulmon. 2003;36:366–68.

28. Kemp A, Kakakious A. Asthma prevention: breast is best? J Paediatr Child Health. 2004;40:337–39.

29. Fredriksson P, Jaakkola N, Jaakkola J. Breastfeeding and childhood asthma: a six-year population-based cohort study. BMC Pediatrics. 2007;7:39.

30. Oddy WH, Sherriff JL, de Klerk NH, Kendall GE, Sly PD, Beilin LJ, Blake, KB, Landau, LI, Stanley FJ. The relation of breastfeeding and body mass index to asthma and atopy in children: a prospective cohort study to age 6 years. Am J Public Health. 2004;94:1531–37.

31. Simopoulos A. Omega-3 fatty acids in inflammation and autoimmune diseases. J Amer Coll Nutr. 2002;21(6):495–505.

32. Riediger ND, Othman RA, Suh M, Moghadasian MH. A systemic review of the roles of n-3 fatty acids in health and disease. J Am Diet Assoc. 2009;109: 668–79.

33. Dry J, Vincent D. Effect of fish oil diet on asthma: results of a 1-year double-blind study. Int Arch Allergy Appl Immunol. 1991;95(2–3):156–57.

34. Hodge L, Salome CM, Hughes JM, Liu-Brennan D, Rimmer J, Allman, M, et al. Effect of dietary intake of omega-3 and omega-6 fatty acids on severity of asthma in children. Eur Respir J. 1998;11:361–65.

35. Oddy, WH, de Klerk NH, Kendall GE, Mirhshahi S, Peat JK. Ratio of omega-6 to mega-3 fatty acids and childhood asthma. J Asthma. 2004;41:310–26.

36. Wong KW. Clinical Efficacy of n-3 fatty acid supplementation in patients with asthma. J Am Diet Assoc. 2005;105:98–105.

37. Woods RK, Raven JM, Walters EH, Abramson MJ, Thien FC. Fatty acid levels and risk of asthma in young adults. Thorax. 2004;59:105–10.

38. Biltagi MA, Baset AA, Bassiouny M, Kasrawi MA, Attia M. Omega-3 fatty acids, vitamin C and Zn supplementation in asthmatic children: a randomized self-controlled study. Acta Paediatr. 2009 Apr 98(4):737–42.

39. Anandan C, Nurmatov U, Sheikh A. Omega 3 and 6 oils for primary prevention of allergic disease: systemic review and meta-analysis. Allergy. 2009;64: 840–48.

40. Thien FK, Woods, R, De Luca S, Abramson MJ. Dietary marine fatty acids (fish oil) for asthma in adults and children. The Cochrane Database of Systematic Reviews. 2002; Issue 2, Art. No.:CD001283.DOI: 10.1002/14651858. CD001283.

41. Barros R, Moreira A, Fonseca J, Ferrza de Oliveira J, Delgado L, Castel-Branco MG, Haahtela T, Lopes C, Moreira P. Adherence to the Mediterranean diet and fresh fruit intake are associated with improved asthma control. Allergy. 2008:63:917–23.

42. Chatzi L, Apostolaki G, Bibakis I, Skypala I, Bibaki-Liakou V, Tzanakis N, Kogevinas M, Cullinan P. Protective effect of fruits, vegetables and the Mediterranean diet on asthma and allergies among children in Crete. Thorax. 2007:62: 677–83.

43. Gao J, Gao X, Li W, Zhu W, Thompson P. Observational studies on the effect of dietary antioxidants on asthma: A meta-analysis. Respirology. 2008;13: 528–36.

44. Kaur B, Rowe BH, Arnold E. Vitamin C supplementation for asthma. The Cochrane Database of Systemic Reviews. 2009; Issue 1. Art No.: CD000993.DOI: 10.1002/14651858.CD000993.

45. Patel BD, Welch AA, Bingham SA, Luben RN, Day NE, Khaw K-T, Lomas DA, Wareham NJ. Dietary antioxidants and asthma in adults. Thorax. 2006; 61:388-93.

46. Weiss, S. Diet as a risk factor for asthma. Ciba Found Symp. 1997:206: 244–51.

47. Jones A, Fay Jk, Burr M, Stone M, Hood K, Roberts G. Inhaled corticosteroid effects on bone metabolism in

asthma and mild chronic obstructive pulmonary disease. The Cochrane Database of Systematic Reviews 2002, Issue 1. Art No.: CD003537.DOI: 10.1002/14651858. CD003537.

48. Eneli IU, Skybo T, Camargo CA. Weight loss and asthma: a systematic review. Thorax. 2008;63:671–76.

49. Sutherland, ER, Lehman EB, Teodorescu, M, Wechsler ME. Body mass index and phenotype in subjects with mild-to-moderate persistent asthma. J Allergy Clin Immunol. 2009;123:1328–34.e1.

50. Ehrenkranz RA, Walsh MC, Vohr BR, et al. Validation of the National Institutes of Health consensus definition of bronchopulmonary dysplasia. Pediatrics. Dec 2005;116(6):1353–60.

51. American Thoracic Society Documents. Statement on the care of the child with chronic lung disease of infancy and childhood. Am J Crit Care Med. 2003;168:356–96.

52. Carlson SJ. Current nutrition management of infants with chronic lung disease. Nutr Clin Pract. 2004;19:581–86.

53. Cohen J, Van Marter LJ, Sun Y, Allred E, Leviton A, Kohane IS. Perturbation of gene expression of the chromatin remodeling pathway in premature newborns at risk for bronchopulmonary dysplasia. Genome Biology. 2007;8:R210.

54. Pavlovic J, Papagaroufalis C, Xanthou M, Liu W, Fan R, Thomas NJ, Apostolidou I, Papthoma E, Megaloyianni E, DiAngelo S, Floros J. Genetic variants of surfactant proteins A, B, C and D in bronshopulmonary dysplasia. Disease Markers. 2006;22:277–91.

55. Shetty A, Gruen JR, Bhandari V. Is there a genetic susceptibility to bronchopulmonary dysplasia? Current Respiratory Medicine Reviews. 2006;2:253–62.

56. Bhandari V, Bizzarro MJ, Shetty, Zhong X, Page GP, Zhang H, Ment LR, Gruen JR. Familial and Genetic Susceptibility to Major Neonatal Morbidities in Preterm Twins. Pediatrics. 2006;117:1901–06.

57. Smith VC, Zupancic JA, McCormick MC, Croen LA, Greene J, Escobar GJ, Richardson DK. Trends in severe bronchopulmonary dysplasia rates between 1994 and 2002. J Pediatr. 2005;146: 469–73.

58. Gitto E, Reiter RJ, Sabatino G, Buonocore G, Romeo C, Gitto P, Bugge C, Tri-marchi G, Ignazio B. Correlation among cytokines, bronchopulmonary dysplasia and modality of ventilation in preterm newborns: improvements with melatonin treatment. J Pineal Res. 2005;39:287–93.

59. Choi CW, Kim BI, Kim H, Park JD, Choi J, Son, DW. Increase of interleukin-6 in tracheal aspirate at birth: a predictor of subsequent bronchopulmonary dysplasia in preterm infants. Acta Paediatrica. 2006;95:38–43.

60. Zachman RD. Role of vitamin A in lung development. J Nutr. 1995;125: 1634S–1638S.

61. Darlow BA, Graham PJ. Vitamin A supplementation to prevent mortality and short and long-term morbidity in very low birthweight infants (review). The Cochrane Collaboration, Wiley Publishers. 2007;4.

62. Thomas W, Speer CP. Prevention and treatment of bronchopulmonary dysplasia: current status and future prospects. Journal of Perinatology. 2007;27:S26–S32.

63. Davis, JW and Sweet DG. Pathophysiology and prevention of bronchopulmonary dysplasia. Current Pediatric Reviews. 2008;4:2–14.

64. Carlson SJ. Current nutrition management of infants with chronic lung disease. Nutr Clin Pract. 2004;19:581–86.

65. Kurzner ST, Garg M, Bautista DB, Bader D, Merritt RJ, Warburton D, et al. Growth failure in infants with bronchopulmonary dysplasia: nutrition and elevated resting metabolic expenditure. Pediatrics. 1988;81:379–84.

66. Sherry B, Mei Z, Grummer-Strawn L, Dietz WH. Evaluation of and recommendations for growth references for very low birth weight (< or = 1500 grams) infants in the United States. Pediatrics. 2003;111:750–58.

67. Denne SC. Energy expenditure in infants with pulmonary insufficiency: is there evidence for increased energy needs? J Nutr. 2001;131:935S-937S.

68. Specific guidelines for disease— pediatrics. Pulmonary: bronchopulmonary dysplasia. JPEN. 2002;26:118SA–119SA.

69. Pereira GR, Baumgart S, Bennett MJ, Stalling VA, Georgieff MK, Hamosh M, et al. Use of high-fat formula for premature infants with bronchopulmonary dysplasia; metabolic, pulmonary, and nutritional needs. J Pediatr. 1994;124:605–11.

70. Chessex P, Belanger S, Piedboeuf B, Pineault M. Influence of energy substrates on respiratory gas exchange during conventional mechanical ventilation of preterm infants. J Pediatr. 1995;126:619–24.

71. American Thoracic Society Documents. Statement on the care of the child with chronic lung disease of infancy and childhood. Am J Crit Care Med. 2003;168:356–96.

72. Carlson SJ. Current nutrition management of infants with chronic lung disease. Nutr Clin Pract. 2004;19:581–86.

73. Spears K, Cheney C, Zerzan J. Low plasma retinol concentrations increase risk of developing bronchopulmonary dysplasia and long-term respiratory disability in very-low-birth-weight infants. Am J Clin Nutr. 2004;80:1589–94.

74. American Thoracic Society Documents. Statement on the care of the child with chronic lung disease of infancy and childhood. Am J Crit Care Med 2003; 168:356-96.

75. Darlow BA, Graham PJ. Vitamin A supplementation for preventing morbidity and mortality in very low birthweight infants. The Cochrane Database of Systematic Reviews 2002, Issue 4. Art. No.: CD000501. DOI:10.1002/14651858. CD000501.

76. Groh-Wargo S, Sapsford A. Enteral nutrition support of the preterm infant in the neonatal intensive care unit. Nutr Clin Pract. 2009;24:363–76.

77. Carlson SJ. Current nutrition management of infants with chronic lung disease. Nutr Clin Pract. 2004;19:581–86.

78. Lin H, Su B, Chen A, Lin TW, Tsai C, Yeh T, OH W. Oral probiotics reduce the incidence and severity of necrotizing enterocolitis in very low birth weight infants. Journal of Pediatrics. 2005; 115:1–4.

79. Romera G, Figueras J, Rodriguez-Miguelez JM, Ortega J, Jimenez R. Energy intake, metabolic balance and growth in preterm infants fed formulas with different nonprotein energy supplements. J Pediatr Gastroenterol Nutr. 2004;38:407–13.

80. Johnson DB, Cheney C, Monsen ER. Nutrition and feeding in infants with bronchopulmonary dysplasia after initial hospital discharge: risk factors for growth failure. J Am Diet Assoc. 1998;98:649–56.

81. Chronic Obstructive Pulmonary Disease Fact Sheet. America Lung

Association. Available at http://www.lungusa.org/site/apps/nlnet/content3.aspx?c=dvLUK9O0E&b=2058829&content_id={EE451F66-996B-4C23-874D-BF66586196FF}&notoc=1

82. MacNee W. AMC of chronic obstructive pulmonary disease pathology, pathogenesis, and pathophysiology. BMJ. 2006;332:1202–04.

83. Rennard S. COPD: overview of definitions, epideminology, and factors affecting its development. Chest. 1998;113:235S–241S.

84. Harik-Khan RI, Fleg JL, Wise RA. Body mass index and the risk of COPD. Chest. 2002;121:370–76.

85. Sharafkhaneh A, Hanania NA, Kim V. Pathogenesis of Emphysema. Proc Am Thorac Soc. 2008;5:475–77.

86. Ochs-Balcom HM, Grant BJB, Muti P, Sempos CT, Browne RW, McCann SE, Trevisan M, Cassano PA, Iacoviello L, Schunemann HJ. Antioxidants, oxidative stress, and pulmonary function in individuals diagnosed with asthma or COPD. European Journal of Clinical Nutrition. 2006;60:991–99.

87. MacNee W. Oxidants and COPD. Current Drug Targets—Inflammation and Allergy. 2005;4:627–41.

88. Psarras S, Caramori G, Papadopoulos N, Papi A. Oxidants in asthma and in chronic obstructive pulmonary disease (COPD). Current Pharmaceutical Design. 2005;11:2053–62.

89. Odencrants S, Ehnfors M, Ehrenberg A. Nutritional status and patient characteristics for hospitalized older patients with chronic obstructive pulmonary disease. Journal of Clinical Nursing. 2008;17:1771–78.

90. Engelen MP. Protein metabolism in chronic respiratory disease. Eur Respir Mon. 2003;24:25–33.

91. Hopkinson NS, Tennant RC, Dayer MJ, Swallow EB, Hansel TT, Moxham J, Polkey MI. A prospective study of decline in fat free mass and skeletal muscle strength in chronic obstructive pulmonary disease. Respiratory Research. 2007;8:25–32.

92. King DA, Cordova F, Scharf SM. Nutritional aspects of chronic obstructive pulmonary disease. Proc Am Thorac Soc. 2008;5:529–23.

93. Engelen MP, Schols AM, Lamers RJ, Wouters EF. Different patterns of chronic tissue wasting among patients with chronic pulmonary disease. Clin Nutr. 1999;18(5):275–80.

94. Shols AM, Fredric EW, Soeters PB, Westerterp KB, Wouter EF. Resting energy expenditure in patients with chronic obstructive pulmonary disease. Am J Clin Nutr. 1991;54:983–87.

95. Gronberg AM, Slinde F, Engstrom C-P, Hulthen L, Larsson S. Dietary problems in patients with severe chronic obstructive pulmonary disease. J Hum Nutr Dietet. 2005;18:445–52.

96. Martin-Harris B. Optimal patterns of care in patients with chronic obstructive pulmonary disease. Semin Speech Lang. 2000;21:311–19.

97. Congleton J. The pulmonary cachexia syndrome: aspects of energy balance. Proc Nutr Soc. 1999;58:321–28.

98. Gosker HR, Wouter EFM, van der Vusse GJ, Schols AM. Skeletal muscle dysfunction in chronic obstructive pulmonary disease and chronic heart failure: underlying mechanisms and therapy perspectives. Am J Clin Nutr. 2000;71:1033–47.

99. Aniwidyaningsih W, Varraso R, Cano N, Pison C. Impact of nutritional status on body functioning in chronic obstructive pulmonary disease and how to intervene. Curr Opin Clin Nutr Metab Care. 2008 Jul;11(4):435–42.

100. Varrasso R, Fung TT, Hu FB, Willett W, Camargo CA. Prospective study of dietary patterns and chronic obstructive pulmonary disease among US men. Thorax. 2007;62(9):786–91.

101. Varrasso R, Fung TT, Barr G, Hu FB, Willett W, Camargo Jr CA. Prospective study of dietary patterns and chronic obstructive pulmonary disease among US women. Am J Clin Nutr. 2007;86(2):488–95.

102. Dallongeville J, Marecau N, Fruchart JC, Amouyel P. Cigarette smoking is associated with unhealthy patterns of nutrient intake: a meta-analysis. J Nutr. 1998;128:1450–57.

103. Ma J, Hampl JS, Betts NM. Antioxidant intakes and smoking status: data from the continuing survey of food intakes by individuals 1994–1996. Am J Clin Nutr. 2000;71:774–80.

104. Watson L, Margetts B, Hawarth P, Dorward M, Thompson R, Little P. The association between diet and chronic obstructive pulmonary disease in subjects selected from general practice. Eur Respir J. 2002;20:313–18.

105. Tug T, Karatas F, Terzi SM. Antioxidant vitamins (A, C and E) and malondialdehyde levels in acute exacerbation and stable periods of patients with chronic obstructive pulmonary disease. Clin Invest Med. 2004;27:23–28.

106. Nadeem A, Raj HG, Chhabra SK. Effect of vitamin E supplementation with standard treatment on oxidant-antioxidant status in chronic obstructive pulmonary disease. Indian J Med Res. 2008;128:705–11.

107. McKeever TM, Lewis SA, Cassano PA, Ocka M, Burnery P, Britton J, Smit HA. The relation between dietary intake of individual fatty acids, FEV1 and respiratory disease in Dutch adults. Thorax. 2008 Mar;63(3):208–14.

108. Specific guidelines for disease—adults. Pulmonary disease. JPEN. 2002;26:SA63–SA64.

109. Fiaccadori E, Coffrini E, Fraccia C, Rampulla C, Montaga T, Borghetti A. Hypophosphatemia and phosphorus depletion in respiratory and peripheral muscles of patients with respiratory failure due to COPD. Chest. 1994;105:1392–98.

110. Fiaccadori E, Coffrini E, Ronda N, Vezzani A, Cacciani G, Fracchia C, et al. Hypophosphatemia in course of chronic obstructive pulmonary disease. Prevalence, mechanisms, and relationships with skeletal muscle phosphorus content. Chest. 1990;97:857–68.

111. Biskobing D. COPD and osteoporosis. Chest. 2002;121(2):609–20.

112. Ionescu AA, Schoon E. Osteoporosis in chronic obstructive pulmonary disease. Eur Respir J 2002; 22(Suppl 46):64s-75s.

113. Katsura H, Kida K. A comparison of bone mineral density in elderly female patients with COPD and bronchial asthma. Chest. 2002;122(6):1949–55.

114. Shols AM. Nutritional modulation as part of the integrated management of chronic obstructive pulmonary disease. Proc Nutr Soc. 2003;62:781–91.

115. Vrieze A, de Greef MHG, Wykstra PJ, Wempe JB. Low bone mineral density in COPD patients related to worse lung function, low weight and decreased fat-free mass. Osteoporos Int. 2007;18:1197–1202.

116. Ionescu AA, Schoon E. Osteoporosis in chronic obstructive pulmonary disease. Eur Respir J. 2002;22(Suppl 46):64s–75s.

117. Gluck O, Colice G. Recognizing and treating glucocorticoid-induced osteoporosis in patients with pulmonary diseases. Chest. 2005;125:1859–76.

118. American Dietetic Association. Evidence Analysis Library. Evidence-based Nutrition Practice Guideline on Chronic Obstructive Pulmonary Disease published at www.adaevidencelibrary.com

119. Biskobing D. COPD and osteoporosis. Chest. 2002;121:609–20.

120. Shols AM. Nutrition and respiratory disease. Clin Nutr. 2001;20(supplement 1):173–79.

121. Thorsdottir I, Gunnarsdottir I. Energy intake must be increased among recently hospitalized patients with chronic obstructive pulmonary disease to improve nutritional status. J Am Diet Assoc. 2002;102:247–49.

122. Mallampalli A. Nutritional management of the patient with chronic obstructive pulmonary disease. Nutr Clin Prac. 2004;19(6):550–56.

123. Felbinger, T, Suchner U, Peter K, Askanazi J. Nutrition support in respiratory disease. In: Payne-James J, Grimble G, Silk D, editors. Artificial nutrition support in clinical practice. Cambridge University Press; 2000, 537–52.

124. Talpers SS, Romberger DJ, Bunce SB, Pingleton SK. Nutritionally associated increased carbon dioxide production. Excess total calories vs high proportion of carbohydrate calories. Chest. 1992;102:551–55.

125. al-Saddy NM, Blackmore CM, Bennett ED. High fat, low carbohydrate, enteral feeding lowers $PaCO_2$ and reduces the period of ventilation in artificially ventilated patients. Intensive Care Med. 1989;15(5):290–95.

126. Kuo CD, Shiao GM, Lee JD. The effect of high-fat and high-carbohydrate diet loads on gas exchange and ventilation in COPD patients and normal subjects. Chest. 1993;104(1):189–96.

127. Akrabawi SS, Mobarhan S, Stolz RR, Ferguson PW. Gastric emptying, pulmonary function, gas exchange, and respiratory quotient after feeding a moderate versus high fat enteral formula meal in chronic obstructive pulmonary disease. Nutrition. 1996; 12:260–65.

128. Weekes CE, Emery PW, Elia M. Dietary counselling and food fortification in stable COPD: a randomised trial. Thorax. 2009;64:326–31.

129. Ferreira IM, Brooks, D, Lacasse Y, Goldstein RS, White J. Nutritional supplementation for stable obstructive pulmonary disease. The Cochrane Database of Systematic Reviews 2005, No.: CD000998.pub2. doi: 10.1002/14651858. CD000998.pub2

130. Puhan MA, Schunemann HJ, Frey M, Scharplatz M, Bachmann LM. How should COPD patients exercise during respiratory rehabilitation? Comparison of exercise modalities and intensities to treat skeletal muscle dysfunction. Thorax. 2005;60:367–75.

131. Costi S, Crisafulli E, Antoni FD, Beneventi C, Fabbri LM, Clini EM. Effects of unsupported upper extremity exercise training in patients with COPD: a randomized clinical trial. Chest. 2009;136:387–95.

132. Cystic Fibrosis Foundation. Available at http://www.cff.org/AboutCF/

133. Cystic Fibrosis Foundation, 2008. Annual Report. Available at http://www.cff.org/UploadedFiles/aboutCF Foundation/AnnualReport/2008-Annual-Report.pdf

134. Human Genome Project Information. CFTR: The Gene Associated with Cystic Fibrosis. Available from http://www.ornl.gov/sci/techresources/Human_Genome/posters/chromosome/cftr.shtml

135. Venuta F, Quattrucci S, Rendina EA, De Giacomo T, Mercadante E, Moretti M, et al. Improved results with lung transplantation for cystic fibrosis: a 6-year experience. Interactive Cardiovascular and Thoracic Surgery. 2001;3:21–24. Available at www.icvts.org

136. Stallings VA, Stark LJ, Robinson KA, Feranchak AP, Quinton H. Clinical Practice Guidelines on Growth and Nutrition Subcommittee; Ad Hoc Working Group. Evidence-Based Practice Recommendations for Nutrition-Related Management of Children and Adults with Cystic Fibrosis and Pancreatic Insufficiency: Results of a Systematic Review. J Am Diet Assoc. 2008; 108: 832–839.

137. Konstan MW, Stern RC, Trout JR, Sherman JM, Eigen H, Wagener JS, Duggan C, Wohl, MEB, Colin P. Ultrase MT12 and ultrase MT20 in the treatment of exocrine pancreatic insufficiency in cystic fibrosis: safety and efficacy. Aliment Pharmacol Ther. 2004;20:1365–71.

138. Dobson L, Sheldon C, Hattersley AT. Understanding cystic-fibrosis-related diabetes: best thought of as insulin deficiency? J R Soc Med. 2004;97(S44):26–35.

139. U.S. Food and Drug Administration. FDA News. FDA approves first DNA-based test to detect cystic fibrosis. Available from http://www.accessdata.fda.gov/scripts/cdrh/cfdocs/psn/transcript.cfm?show=41

140. Godzik J, Cofta S, Piorunek T, Batura-Gabryel H, Kosicki J. Relationship between nutritional status and pulmonary function in adult cystic and pulmonary function in adult cystic fibrosis patients. J Physiol Pharmacol. 2008;59 Suppl 6:253–60.

141. Hardin, DS. A review of the management of two common clinical problems found in patients with cystic fibrosis: cystic fibrosis-related diabetes and poor growth. Horm Res. 2007;68(S5):113–16.

142. Worgall TS. Lipid metabolism in cystic fibrosis. Curr Opin Clin Metab Care. 2009 Mar;12(2):105–99.

143. Hayek K. Medical nutrition therapy for cystic fibrosis: beyond pancreatic enzyme replacement therapy. J Am Diet Assoc. 2006;106(8):1186–88.

144. Urquhart DS, Fitzpatrick M, Cope J, Jaffe A. Vitamin K prescribing patterns and bone health surveillance in UK children in cystic fibrosis. J Hum Nutr Diet. 2007;20:605–10.

145. Umlawska W, Susanne C. Growth and nutritional status in children and adolescents with cystic fibrosis. Annals of Human Biology. 2008;35(2):145–53.

146. Borowitz D, Baker R, Stalling V. Consensus report on nutrition for pediatric patients with cystic fibrosis. J Pediatr Gastroenterol. 2002;35:246–59.

147. Van Biervliet S, Vanbillemont G, Van Biervliet J-P, Declercq D, Robberecht E, Christophe A. Relation between fatty acid composition and clinical status or genotype in cystic fibrosis patients. Ann Nutr Metab. 2007;51:541–49.

148. Moran A, Hardin D, Rodman D, Allen HF, Beall RJ, Borowitz C, et al. Diagnosis, screening and management of cystic fibrosis related diabetes mellitus: a consensus conference report. Diabetes Res Clin Pract. 1999;45:61–73.

149. Wilson, DC, Kalnins D, Stewart C, Hamilton N, Hanna K, Durie PR, et al. Challenges in the dietary treatment of cystic fibrosis related diabetes. Clin Nutr. 2000;19:87–93.

150. White H, Pollard K, Etherington C, Clifton 1, Morton AM, Owen D, Conway SP, Peckham DG. Nutritional decline in cystic fibrosis related diabetes: the effect of intensive nutritional intervention. J Cyst Fibros. 2009;8:179–85.

151. Mrugacz M, Tobolvzyk J, Minarowska A. Retinol binding protein status in relation to ocular surface changes in patients with cystic fibrosis treated with daily vitamin A supplements. Eur J Pediatr. 2005;164:202–06.

152. Rust P, Eichler I, Renner S, Elmadfa I. Effects of long-term oral beta-carotene supplementation on lipid peroxidation in patients with cystic fibrosis. Int J Vitam Nutr Res. 1998:68(2)83–87.

153. Gordon CM, Anderson EJ, Herlyn K, Hubbard JL, Pizzo A, Gelbard R, Lapey A, Merkel PA. Nutrient status of adults with cystic fibrosis. J Am Diet Assoc. 2007;107:2114–19.

154. Elkin SL, Fairney A, Burnett S, Kemp M, Kyd P, Burgess J, et al. Vertebral deformities and low bone mineral density in adults with cystic fibrosis: a cross-sectional study. Osteoporos Int. 2001;12:366–72.

155. Flohr F, Lutz A, App EM, Matthys H, Reincke M. Bone mineral density and quantitative ultrasound in adults with cystic fibrosis. Eur J Endo. 2002;146:531–36.

156. Speeckaert MM, Wehlou C, Vandewalle S, Taes YE, Robberecht E, Delanghe JR. Vitamin D binding protein, a new nutritional marker in cystic fibrosis patients. Clin Chem Lab Med. 2008;46:365–70.

157. Von Horn, JHL, Hendriks, JJE, Vermeer C, Forget P. Vitamin K supplementation in cystic fibrosis. Arch Dis Child. 2003;88:974–75.

158. Conway SP, Wolfe SP, Brownlee MB, White H, Oldroyd B, Truscott JG, et al. Vitamin K status among children with cystic fibrosis and its relationship to bone mineral density and bone turnover. Pediatrics. 2005;115:1325–31.

159. Lancellotti L, D'Orazio C, Mastella G, Mazzi G., Lippi U. Deficiency of Vitamins E and A in cystic fibrosis is independent of pancreatic function and current enzyme and vitamin supplementation. Eur J Pediatr. 1996;155:281–85.

160. Easley D, Krebs N, Jefferson M, Miller L, Erskine J, Accurso F, et al. Effect of pancreatic enzymes on zinc absorption in cystic fibrosis. JPGN. 1998;26:136–39.

161. Stallings VA, Stark L, Robinson KA, Feranchak, AP, Quinton H. Evidence-based practice recommendations for nutrition-related management of children and adults with cystic fibrosis and pancreatic insufficiency: results of a systematic review. J Am Diet Assoc. 2008;108:832–39.

162. U.S. Food and Drug Administration. Questions and answers on exocrine pancreatic insufficiency drug products. Available at http://www.fda.gov/Drugs/DrugSafety/PostmarketDrugSafetyInformationforPatientsandProviders/ucm149337.htm

163. Consensus conference. Management of patients with cystic fibrosis. Monday the 18th and Tuesday the 19th of November, 2002, Luxenbourg Palace, Paris. Observation, nutrition, gastroenterology and metabolism. Text of the recommendations (long version). Arch Pediatr. 2003;10(suppl 3):382s–397s.

164. Moran A. Cystic fibrosis-related diabetes: an approach to diagnosis and management. Pediatr Diabetes. 2000;1:41–48.

165. Feranchak AP, Sontag MK, Wagener JS, Hammond KB, Accurso FJ, Sokol RJ. Prospective, long-term study of fat soluble vitamin status in children with cystic fibrosis identified by newborn screen. J Pediatr. 1999;135(5):601–10.

166. Van Biervliet S, Vande Velde S., Van Biervliet JP, Robberecht E. The effect of zinc supplements in cystic fibrosis patients. Ann Nutr Metab. 2008;52:152–56.

167. Janicke DM, Mitchell MJ, Quittner AL, Piazza-Waggoner C, Stark LJ. The impact of behavioral intervention on family interactions at mealtime in pediatric cystic fibrosis. Children's Health Care. 2008;37:49–66.

168. Barclay A, Allen JR, Blyler E, Yap J, Gruca MA, Asperen PV, Cooper P, Gaskin KJ. Resting energy expenditure in females with cystic fibrosis: is it affected by puberty? European Journal of Clinical Nutrition. 2007;61:1207–12.

169. American Lung Association. Pneumonia Fact Sheet. Available at American Lung Association.

170. Center for Disease Control and Prevention. Available at http://www.cdc.gov/nchs/FASTATS/lcod.htm

171. Solh, AA. Pnemonia in the Elderly. Current Respiratory Medicine Reviews. 2006;2:75–87.

172. Beers MH, editor. The Merck Manual Medical Library. Hospital acquired. Available at http://merck.com/mmhe/print/sec04/ch042/ch042c.html

173. National Heart, Lung and Blood Institute: Disease Conditions Index. Available at http://www.nhlbi.nih.gov/health/dci/Diseases/pnu/pnu_types.html

174. McClave SA, Demen MT, DeLegge MH, DiSario JA, Heyland DK, Maloney JP, Metheny NA, Moore FA, Scolapio JS, Spain DA, and Zaloga GP. North American Summit on Aspiration in the Critically Ill Patient: Consensus Statement. J Parenteral and Enteral Nutrition. 2002;26:S80–S85.

175. Metheny NA. Risks for Aspiration. J Parenteral and Enteral Nutrition. 2008;26:S26–S33.

176. Metheny NA, Schallom L, Oliver DA, Clouse RE. Gastric residual volume and aspiration in critically ill patients receiving gastric feedings. Am J Crit Care. 2008 Nov;17(6):512–19.

177. Gomes GF, Pisani JC, Macedo ED Campos AC. The nasogastric feeding tube as a risk factor for aspiration and aspiration pneumonia. Curr Opin Clin Nutr Metab Care. 2003;6:327–33.

178. Murray KA, Brzozowski LA. Swallowing in patients with tracheotomies. AACN Clinical Issues: Advanced Practice in Acute and Critical Care 1998; volume 9, No.3. Available at www.aacn.org/AACN/jrnlci.nsf/GetArticle/ArticleTen93?openDocument=

179. Hughes T. Neurology of swallowing and oral feeding disorders: assessment and management. J Neurol Neurosurg Pyschiatry. 2003;70 (Suppl III): 48–52.

180. Davis, DG, Bears S, Barone JE, Corvo PR, Tucker JB. Swallowing with a tracheostomy tube in place: does cuff inflation matter? J Inten Care Med. 2002;17:132–35.

181. Mackay A, M Al-Hadded. Acute lung injury and acute respiratory distress

syndrome. Cont Edu Anaesth Crit Care & Pain. 2009:9:152–156.

182. Xiaoming J, Eng M, Malhotra A, Saeed M, Mark RG, Talmor D. Risk factors for acute respiratory distress syndrome in patients mechanically ventilated for greater than 48 hours. Chest. 2008;133(4):853–61.

183. Ubodi K, Childs E. Acute Respiratory Distress Syndrome. Am Fam Physician. 2003;67(2):315–22.

184. Bhatia M, Moochhala S. Role of inflammatory mediators in the pathophysiology of acute respiratory distress syndrome. J Pathol. 2004;202:145–56.

185. Zwischenberger JB, Alpard SK, Bidani A. Early complications, Respiratory failure. Chest Surg Clin N Am. 1999;9(3):543–64.

186. Ware LB, Matthay MA. The acute respiratory distress syndrome. New Eng J Med 2000. 342(18):1334–40.

187. Michaels AJ. Management of post traumatic respiratory failure. Crit Care Clin. 2004;20:83–99.

188. Bream-Rouwenhorst HR, Beltz EA, Ross MB, Moores KG. Recent developments in the management of acute respiratory distress syndrome in adults. Am J Health-Syst Pharm. 2008;65:29–36.

189. Singer P, Singer JA, Shapiro H, Lev S. Oxidative stress in the ICU. Current Nutrition & Food Science. 2007;3:209–15.

190. Cerra FB, Benitez MR, Blackburn GL, Irwin, RS, Jeejeebohy K, Katz DP, et al. Applied Nutrition in ICU patients. A consensus state of the American College of Chest Physicians. Chest. 1997;11:759–78.

191. Reid C. Frequency of under- and overfeeding in mechanically ventilated ICU patients: causes and possible consequence. J Hum Nutr Dietet. 2006;19:13–22.

192. Gadek J, DeMichele SJ, Karlstad MD, Pacht ER, Donahoe M, Albertson TE, et al. Effect of enteral feeding with eicosapentaenoic acid, γ-linolenic acid, and antioxidants in patients with acute respiratory distress syndrome. Crit Care Med. 1999;27:1409–20.

193. Ciencewicki J, Trivedi S, Kleeberger SR. Oxidants and the pathogenesis of lung diseases. J Allergy Clin Immunol. 2008;122:456–68.

194. Nelson J, DeMichele SJ, Pacht ER, Wennberg AK. Effect of enteral feeding with eicosapentaenoic acid, γ-linolenic acid, and antioxidants on antioxidant status in patients with acute respiratory distress syndrome. JPEN. 2003;27:98–104.

195. Marik PE, Bedigian MK. Refeeding hypophosphatemia in critically ill patients in an intensive care unit. Arch Surg. 1996;131:1043–47.

196. American Lung Association, Lung Transplants. Available at http://www .lungusa.org/site/c.dvLUK9OO0E/ b.23012/k.A039/Lung_Transplants.htm

197. United Network for Organ Sharing. Organ Procurement and Transplantation Network. Available at http://www.unos .org/data/default.asp?displayType=usData

198. Neuringer I, Chalermskulrat WJ, Aris RM. Special problems in long-term survivors of lung transplantation. Available at http://www.chestnet.org/ education/online/pccu/vol19/ lessons11_12/index.php

199. Schwebel C, Pin I, Barnoud D, Devouassoux G, Brichon PY, Chaffanjon PH, et al. Prevalence and consequences of nutritional depletion in lung transplant candidates. Eur Respir J. 2000;18: 1050–55.

200. Madill J, Gutierrez C, Grossman J, Allard J, Chan C, Hutcheon M, et al. Nutritional assessment of the lung transplant patient: body mass index as a predictor of 90-day mortality following transplantation. J Heart Lung Transplant. 2001;20(3):288–96.

201. Hasse JM. Nutrition assessment and support of organ transplant recipients. JPEN. 2001;25:120–31.

202. Beers MH, editor. The Merck Manual Medical Library. Respiratory tract infections. Available at www.merck .com/mmhe/print/sec23/ch273/ch273i .html

203. Specific guidelines for disease— adults. Solid organ transplantation. JPEN. 2002;26(1):SA74–SA75.

204. Mossad SB. Upper respiratory tract infections. The Cleveland Clinic Disease Management Project. Available at http://www.clevelandclinicmeded.com/ medicalpubs/diseasemanagement/ infectious-disease/upper-respiratory- tract-infection/

205. Douglas RM, Hemilä H, Chalker E, Treacy B. Vitamin C for preventing and treating the common cold. Cochrane Database Syst Rev. 2007 Jul 18;(3):CD000980.

206. Johnston CS, Martin LJ, Cai X. Antihistamine effect of supplemental ascorbic acid and neutrophil chemotaxix. J Am Coll Nutr. 1992;11(2):172–76.

207. Heimer KA, Hart AM, Martin LG, Rubio-Wallace S. Examining the evidence for the use of vitamin C in the prophylaxis and treatment of the common cold. J Am Acad Nurse Pract. 2009;21:295–300.

208. Jackson JL, Lesho E, Peterson C. Zinc and the common cold: a meta-analysis revisited. J Nutr. 2000;1512S– 1515S.

209. Prasad AS, Beck FW, Bao B, Snell D, Fitzgerald JT. Duration and severity of symptoms and levels of plasma interleukin-1 receptor antagonist, soluble tumor necrosis factor receptor, and adhesion molecules in patients with common cold treated with zinc acetate. J Infect Dis. 2008;197:795–802.

210. Caruso TJ, Prober CG, Gwaltney JM Jr. Treatment of naturally acquired common colds with zinc: a structured review. Clin Infect Dis. 2007;45:569–74.

211. McKeever TM, Lewis SA, Smit HA, Burney P, Cassano PA, Britton J. A multivariate analysis of serum nutrient levels and lung function. Respir Res. 2008 Sep 29;9:67.

212. Chen H, Zhuo Q, Yuan W, Wang J, Wu T. Vitamin A for preventing lower respiratory tract infections in children up to seven years of age. Indian Pediatr. 2009 May;46(5):403–04.

213. Ginde AA, Mansbach JM, Camargo CA. Association between serum 25-hydroxyvitamin D level and upper respiratory tract infection in the Third National Health Nutrition Examination Survey. Arch Intern Med. 2009 Feb 23;169(4):384–90.

22

Metabolic Stress and the Critically Ill

Marcia Nahikian Nelms, PhD, RD, LD
The Ohio State University

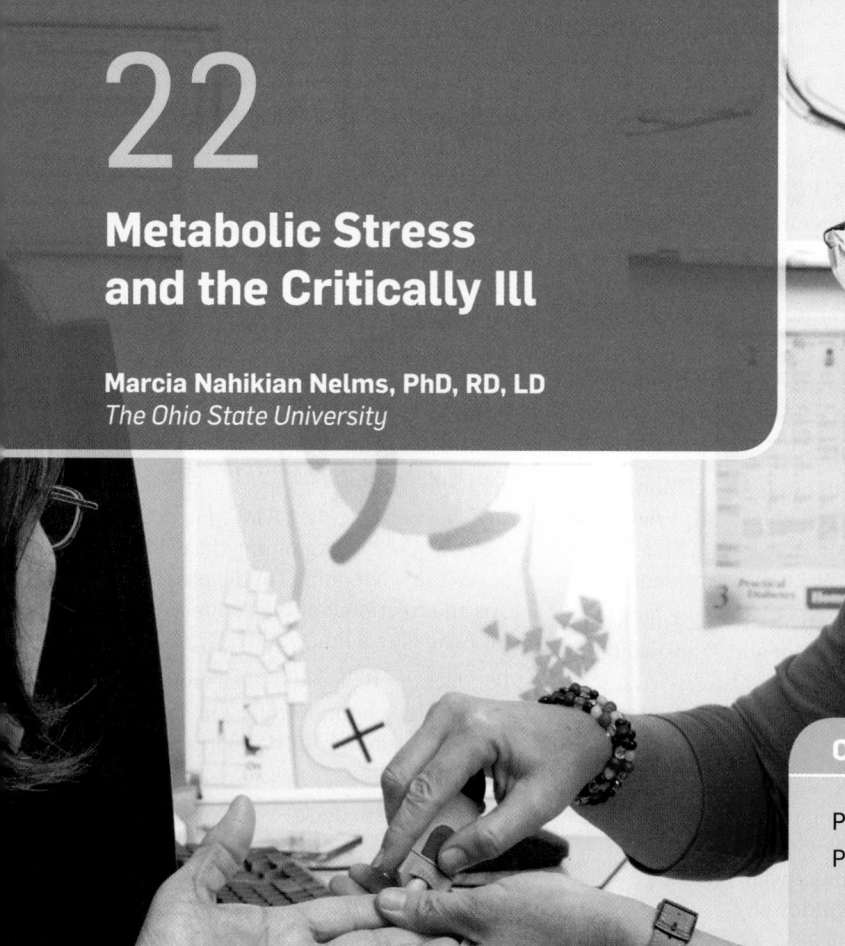

Introduction

Chapter 9 discussed the body's response to injury at the cellular level. The type of cellular response depends on the ability of the cell to react, adapt, and repair itself after exposure to injury. This response may be temporary and completely reversible, but in some situations, the cell is permanently damaged and is no longer functional. Cell responses may include inappropriate accumulation of substances within the cell; changes in size, number, or shape; and the inflammatory response.

In this chapter, the body's unique responses to specific types of stress and injury are explored. As described in the 2009 ASPEN guidelines for nutrition support therapy in the adult critically ill: "Traditionally nutrition support in the critically ill population was regarded as adjunctive care designed to provide exogenous fuels to support the patient during the stress response. This support had three main objectives: to preserve lean body mass, to maintain immune function, and to avert metabolic complications" (p. 278).[1] With increased research and experience in nutrition support, we understand that nutritional care in the critically ill can accomplish additional goals, including reduction of the stress, inflammatory, and immune responses as well as the prevention of cellular injury. Nutrition support is a crucial component of the medical care for the critically ill.[1]

With these goals in mind, this chapter seeks to first describe the physiological effects of metabolic stress. We know that the stress condition places the patient at the highest level of nutritional risk. In no other situation can nutritional deficits result in such dramatic and often severe consequences for a patient.

Research over the previous twenty years has led us to a clearer understanding of not only the body's need for nutrition support during critical illness but also the role of nutrition in mediating the stress response. Planning nutrition care goes beyond the simple decision to feed the patient during critical illness; it is imperative to understand who to feed, what to feed, when to feed, and how much to feed.

branched-chain amino acid—an amino acid that has a branched side chain; these include isoleucine, leucine, and valine

c-reactive protein—a protein released as a response to inflammation

ceruloplasmin—a protein used in copper transport

Curling's ulcer—ulceration of gastric or duodenal tissue as a result of burn or trauma

debride—to remove dead or injured tissue

epidural anesthesia—an anesthetic drug placed into the epidural space of the lumbar or sacral region of the spine, causing loss of sensation from the abdomen and pelvis to the lower limbs

esophagectomy—a surgical procedure resecting or removing the esophagus

fibronectin—an acute-phase glycoprotein involved in the regulation of cell growth and differentiation, wound healing, and vascular integrity

gangrene—tissue death due to lack of blood flow and oxygen

general anesthesia—total loss of sensation and consciousness as a result of an anesthetic drug

gluconeogenesis—the metabolic pathway through which glucose is formed from noncarbohydrate sources

glycogenolysis—the metabolic pathway through which glycogen is converted to glucose

hypoperfusion—reduced blood flow

hypotension—low blood pressure

hypoxic injury—cellular injury as a result of oxygen deprivation

local anesthesia—loss of sensation only in the area where an anesthetic drug is placed

menhaden oil—hydrogenated and partially hydrogenated oils from the menhaden fish (a small plankton-feeding fish)

MODS—multiorgan distress syndrome; a "[d]isease involving more than one of the vital organs, such as the heart, lungs, kidney, liver"[2]

MSOF—multisystem organ failure; a [d]isease involving more than one of the vital organs, such as the heart, lungs, kidney, liver"[2]

necrotizing fasciitis—inflammation of the connective tissue leading to necrosis of the tissue; may be caused by infection, injury, or an autoimmune reaction

sepsis—a systemic inflammatory response and immunosuppressive process that prevents an adequate response to infection or trauma; may result in organ dysfunction or hypoperfusion abnormalities

serum amyloid A—family of apolipoproteins associated with high-density lipoprotein (HDL) in plasma; considered to be an acute-phase protein released in response to inflammation

silver nitrate—colloidal silver; used as an antibacterial treatment in burns

silver sulfadiazine cream—a sulfa medication used to prevent and treat bacterial or fungal infections

Physiological Response to Starvation

Malnutrition occurs when there is an inadequate nutrient supply—that is, during starvation. But the body can also experience malnutrition when it is unable to utilize nutrients appropriately or when its nutritional needs are so high that current intake cannot meet those demands—that is, during metabolic stress. The body's reaction to these two situations is quite different.

The most important difference between the physiological response to starvation and the response to metabolic stress is the adaptation that occurs during starvation. It was not until completion of the hallmark studies of Dr. Ancel Keys and his subsequent publication of *The Biology of Human Starvation* (1950) that the manner in which the body responds to starvation through physiological adaptations was understood (see Box 22.1). Keys and his research team followed the effects of starvation in 36 conscientious objectors during World War II.[3]

One of the most important distinctions between starvation and metabolic stress is the difference in energy and fuel substrate requirements. During starvation, the body responds to a reduction in food intake by reducing its overall energy needs; the basal metabolic rate is reduced so that fewer kcal are needed. In contrast, energy requirements are increased during metabolic stress and injury. The next major difference between starvation and metabolic stress is the source of fuel that is used to meet energy requirements. Under normal circumstances, the body uses a mixture of fuels (primarily carbohydrate and lipid) for energy. But since humans have a limited ability to store carbohydrate, the primary source of fuel shifts from glucose to lipids during periods of starvation as glucose availability decreases. Lipolysis becomes preferential and the accumulated lipid stores serve as the primary energy source. This adaptation for the use of lipid as the primary fuel and the subsequent metabolism of

Table 22.1 Comparison of Metabolism during Normal Nutritional States versus Starvation

Normal Nutritional State	Starvation
Metabolic rate matches current physical activity requirements and body composition.	Decrease in metabolic rate to ensure conservation of energy.
Carbohydrate and lipid are efficiently metabolized sources of energy providing 55%–85% of energy requirements.	Decreased need for glucose utilization.
	Utilization of lipid as main source of energy.
Protein is used for maintenance of protein structures and to meet ongoing protein synthesis requirements.	Preservation of lean mass, minimizing protein loss.

ketones allow for preservation of muscle mass and prevent the complications of protein deficiency (infection and decreased transport protein synthesis).[4] Table 22.1 outlines both normal nutrient metabolism and the key adaptations that occur during starvation. Figure 22.1 summarizes the changes in metabolism during starvation.

Unfortunately, when the body is faced with an injury, infection, or disease causing metabolic stress, the normal adaptations that should occur do not occur.

Physiological Response to Stress

Definition

Metabolic stress is the hypermetabolic, catabolic response to acute injury or disease. Diagnoses that may lead to metabolic stress include trauma as seen in a gunshot wound or motor vehicle accident (MVA); closed head injury (see Chapter 20); burns; severe inflammation such as in pancreatitis; cancer; **sepsis**; **hypoxic injury** as seen in acute renal failure; and necrosis

Ancel Keys's Human Starvation Study

Taylor Nelms, MA, *Cambridge University UK*

In 1939, as the war in Europe intensified and the United States prepared to enter the conflict, Ancel Keys, a new member of the faculty of the University of Minnesota, was asked by the War Department to develop a food ration for paratroopers. Keys had just founded the Laboratory of Physiological Hygiene beneath the bleachers of Memorial Stadium. The solution Keys and his collaborators concocted in the lab—the infamous K-ration—was so successful that the U.S. military assigned it to *all* their troops. Partly because of this success, Keys received support from the U.S. government as the war came to a close to conduct a large-scale comprehensive study on the physiological and psychological effects of starvation.

Keys put 36 conscientious objectors through a diet and exercise program designed to recreate the living conditions of occupied Europe. His human subjects ate simple, starchy foods and root vegetables, and were required to walk at least 22 miles per week. After three months of this semi-starvation, the young men were re-fed and rehabilitated. The resulting two-volume publication, *The Biology of Human Starvation*, was an instant classic, not merely for its detailed investigation of the physical consequences of protein-energy malnutrition, but also for its description of the condition's psychological effects.[1] Deprived of food, the men became depressed and lost their motivation; later, they became obsessed with food, licking their plates, hoarding food, and even cheating.[2]

Ancel Keys was one of the twentieth century's most important physiologists and nutritionists. Besides inventing the K-ration and studying starvation, he was the first to uncover a relationship between a high intake of saturated fat, the level of cholesterol, and the development of cardiovascular disease. He was one of the first medical scientists to utilize mathematical regression in the study of human health.[3] Finally, and perhaps most importantly, Keys demonstrated the importance of cultural and economic factors in the health of human populations. Throughout his career in physiology, nutrition, and public health, Keys emphasized the mutability of the body and our ability to prevent disease through simple modifications in lifestyle.

References
1. Keys A, Brozek J, Henschel A, Mickelsen O, Taylor HL. The biology of human starvation. Minneapolis: University of Minnesota Press; 1950.
2. Kalm LM, Semba RD. They starved so that others are better fed: remembering Ancel Keys and the Minnesota experiment. J Nutr. 2005;135(6):1347–52.
3. VanItallie TB. Ancel Keys: a tribute. Nutr Metab. 2005;2:4.

of tissue such as in **gangrene** or after major surgery. The degree of metabolic stress generally correlates with the seriousness of the injury.[5, 6] In critical care medicine, rankings for severity of illness use scoring systems such as the Glasgow Coma Scale (see Table 20.11 in Chapter 20), the Acute Physiology and Chronic Health Evaluation (APACHE), the Injury Severity Score (ISS), or the Abdominal Trauma Index (ATI).

Epidemiology

Injuries caused by physical force (falls, gunshot wounds, stabbing, drowning, and other accidents) are classified as trauma. Trauma is a leading cause of death for young people, accounting for more than 70% of all deaths for those aged 15–24 years.[7]

Etiology

The metabolic consequences of injury and stress are a result of numerous factors including hormone release, acute-phase protein synthesis, hypermetabolism, increased reliance on gluconeogenesis and its subsequent production of glucose, and shifts in fluid balance and decreased urine output.[5, 8–10]

Clinical Manifestations

The stress response has been described as a progression through three phases: the ebb phase, the flow phase, and finally the recovery or resolution phase.[11, 12] The ebb phase encompasses the immediate period after injury (2–48 hours). This period is characterized by shock resulting in hypovolemia and decreased oxygen availability to tissues. The decrease in blood volume results in decreased cardiac output and urinary output. The goal of medical care during this acute period is to restore blood flow to organs, maintain oxygenation of all tissues, and stop all hemorrhaging. As the patient stabilizes hemodynamically, the acute period of the flow phase begins. This phase encompasses the classic signs and symptoms of metabolic stress: hypermetabolism, catabolism, and altered immune and hormonal responses. The final adaptation phase or recovery phase indicates a resolution of the stress with a return to anabolism and normal metabolic rate.

Pathophysiology

Hormones, acute-phase proteins, the immune system, and altered cellular metabolism direct the physiological changes that characterize metabolic stress. Stress and injury activate the hormones that direct a "flight or fight" response, including glucagon, cortisol, epinephrine, and norepinephrine. Their primary purpose is to mobilize nutrient stores to meet the immediate energy demand. Increased levels of glucagon serve to increase glucose production from amino acids (**gluconeogenesis**). Cortisol increases both gluconeogenesis and free fatty acid mobilization and decreases overall protein synthesis with an increased catabolism of skeletal muscle. The catecholamines (epinephrine, norepinephrine) increase energy availability by stimulating **glycogenolysis** and increasing the release of fatty acids.

Release of either glucagon or cortisol can result in hyperglycemia during the stress response. Even though insulin levels are increased during metabolic stress, insulin resistance diminishes

Figure 22.1 Changes in Metabolism during Starvation

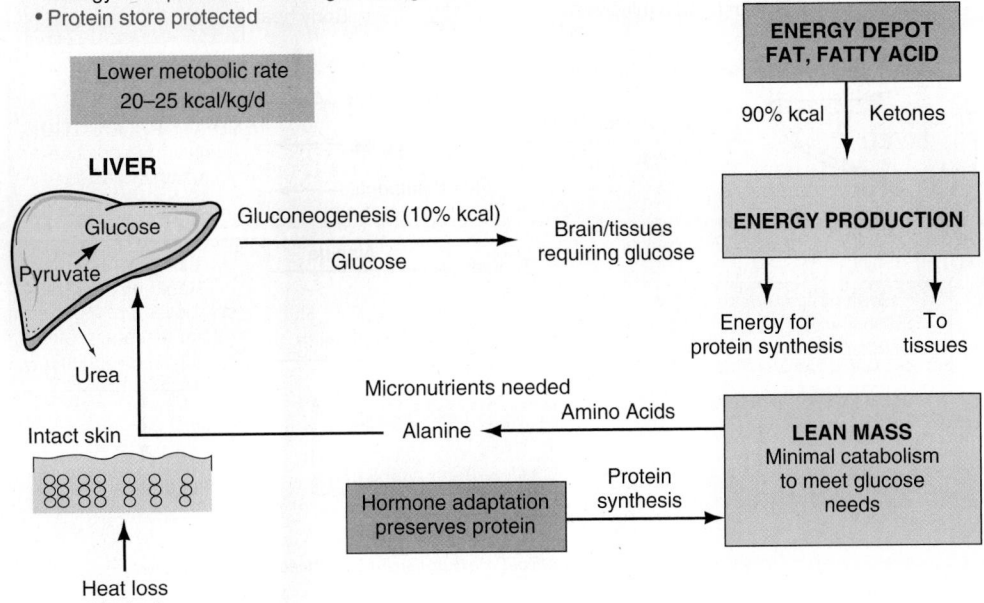

Metabolism Response to Starvation (Short Term)
No Injury or "Stress" (Protective Adaptation Occurs)

- Overall energy needs decrease
- Metabolic rate decreases 20-25 kcal/kg/d
- Energy from fat storage >90% of kcal
- Energy from protein <10% for gluconeogenesis
- Protein store protected

Source: Reprinted with permission from Medscape.com, 2009. Available at: http://cme.medscape.com/viewarticle/432384

the hormones' effectiveness.[13, 14] This contributes to the degree of hyperglycemia. The multicenter NICE-SUGAR trial evaluating the use of intensive insulin therapy to maintain normal blood glucose levels has produced mixed results, leaving the best protocol for controlling blood glucose levels in the critically ill undetermined. The original study published in 2001 demonstrated a reduction of morbidity and mortality for critically ill patients who received intensive insulin therapy, but this has not been consistently demonstrated in subsequent studies with the risk of hypoglycemia being a major concern.[15–17] It has been proposed that insulin therapy not only controls hyperglycemia seen in metabolic stress but may also reduce catabolism and inflammation, which improves the immune response but does contribute to risk of hypoglycemia.[14, 17]

The increased rate of gluconeogenesis creates reliance on protein as a source of glucose. The need for the amino acids alanine and glutamine is particularly increased. Since alanine is the primary substrate required for gluconeogenesis, there is an increased catabolism of skeletal muscle to make alanine available to the liver. Glutamine is a non-essential amino acid that is significant in both metabolic and immunologic pathways. For example, glutamine is the primary fuel for enterocytes within the gastrointestinal tract and for T-lymphocytes.[4] In injury and stress, the synthesis rate may be unable to accommodate the increased requirements. Negative nitrogen balance is a consistent marker during

metabolic stress. Figure 22.2 and Table 22.2 demonstrate the substantial nitrogen loss and complications that accompany skeletal muscle catabolism. (Figure 22.3 and Table 22.3 provide a summary of metabolic changes in stress and trauma.)

Positive acute-phase proteins are often used as markers of the stress response (see Chapter 3). These include **fibronectin, C-reactive protein, ceruloplasmin,** and **serum amyloid A.**

Figure 22.2 Increased Nitrogen Loss during Metabolic Stress

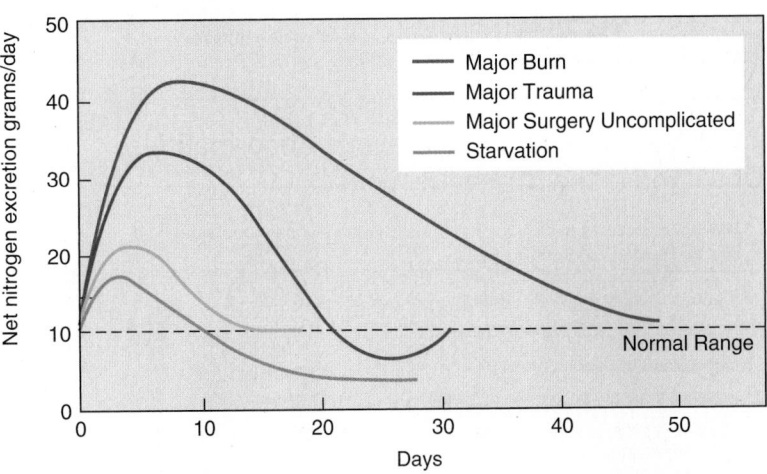

Source: JJ Diaz, R Pousman, G Jensen, Vanderbilt University Medical Center, *Critical Care Nutrition Practice Management Guidelines.*

Figure 22.3 Summary of Metabolic Changes in Stress and Trauma

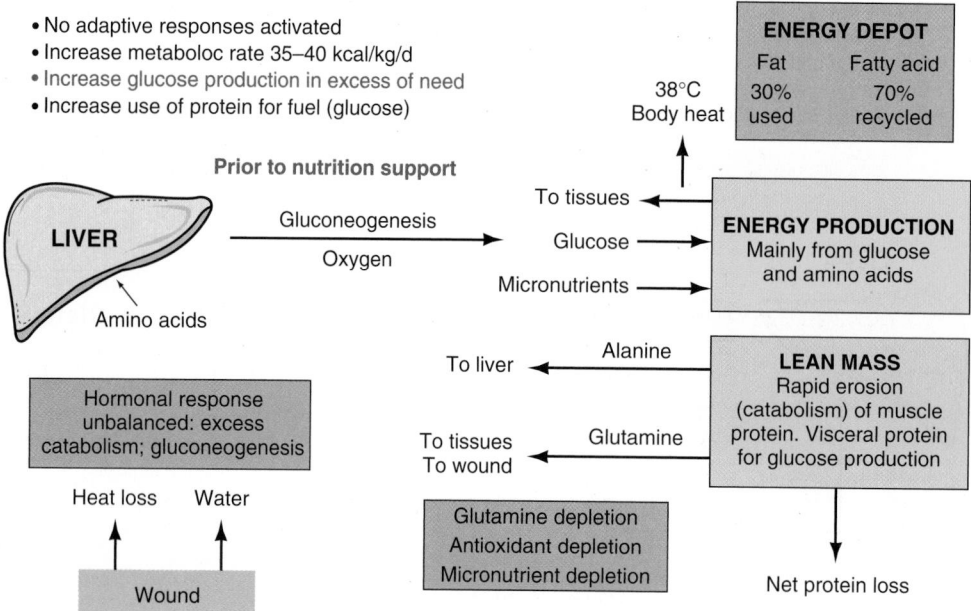

Catabolic Insult-Induced Protein-Energy Malnutrition
(Protein and Energy Production Abnormal)

- No adaptive responses activated
- Increase metaboloc rate 35–40 kcal/kg/d
- Increase glucose production in excess of need
- Increase use of protein for fuel (glucose)

Prior to nutrition support

LIVER

Gluconeogenesis
Oxygen
Amino acids

38°C
Body heat

ENERGY DEPOT
Fat	Fatty acid
30%	70%
used	recycled

To tissues
Glucose
Micronutrients

ENERGY PRODUCTION
Mainly from glucose
and amino acids

Hormonal response
unbalanced: excess
catabolism; gluconeogenesis

Heat loss   Water

Wound

To liver — Alanine

To tissues — Glutamine
To wound

LEAN MASS
Rapid erosion
(catabolism) of muscle
protein. Visceral protein
for glucose production

Glutamine depletion
Antioxidant depletion
Micronutrient depletion

Net protein loss

Source: JJ Diaz, R Pousman, G Jensen, Vanderbilt University Medical Center, *Critical Care Nutrition Practice Management Guidelines.*

Table 22.2 Complications Relative to Loss of Lean Body Mass[1]

Lean Body Mass (% Loss of Total)	Complications (Related to Lost Lean Mass)	Associated Mortality (%)
10	Impaired immunity, increased infection	10
20	Decreased healing, weakness, infection	30
30	Too weak to sit, pressure sores, pneumonia, no healing	50
40	Death, usually from pneumonia	100

[1]Assuming no preexisting loss.

Reprinted from: Demling R, DeSanti L. Effect of a catabolic state with involuntary weight loss on acute and chronic respiratory disease. Available at http://www.medscape.com/viewprogram/1816_index

Table 22.3 Summary of Metabolic Abnormalities Observed in the Stress Response

- Increased levels of glucagon, cortisol, epinephrine, norepinephrine
- Hyperglycemia and insulin resistance
- Increased basal metabolic rate
- Increased rate of gluconeogenesis
- Catabolism of skeletal muscle
- Increased urinary nitrogen excretion—negative nitrogen balance
- Increased synthesis of positive acute-phase proteins—CRP, fibronectin, ceruloplasmin
- Decreased synthesis of negative acute-phase proteins—albumin, prealbumin

An acute-phase protein is defined as "one whose plasma concentration increases (positive acute-phase proteins) or decreases (negative acute-phase proteins) by at least 25% during inflammatory disorders."[10, 11] The release of the acute-phase proteins is regulated by a variety of cytokines and other communication molecules within the immune system. Cytokines include interleukins (interleukin-1/IL-1, interleukin-6/IL-6), leukotrienes, tumor necrosis factor, and interferons (see Chapter 9). IL-6 directly affects protein metabolism by decreasing acute-phase proteins such as albumin and prealbumin and increasing other acute-phase proteins such as C-reactive protein.[5] Table 22.4 summarizes the metabolic effects of pro-inflammatory cytokines.

As discussed previously, cytokines are proteins that, in small amounts, affect behavior of other cells (see Chapter 9). The injury or stress induces cytokine production. Cytokines then act on target cells whose behavior can result in loss of appetite, fever, inflammation, and metabolic abnormalities such as hyperglycemia and catabolism.

Table 22.4 Proinflammatory Cytokines and their Metabolic Effects

Cytokines	Metabolic Effect
Tumor necrosis factor (TNF)	Altered metabolism: catabolism, hypermetabolism
Interleukin-1 (IL-1)	Increased body temperature
Interleukin-6 (IL-6)	Activation and release of cellular communication/mediators

Medical Diagnosis and Treatment

Treatment for metabolic stress will involve the interventions appropriate for the underlying injury or illness. Interventions may include lung-protective ventilation, broad-spectrum antibiotics, medications such as steroids for anti-inflammatory treatment, continuous renal replacement, intensive insulin therapy, and nutrition support.

Nutrition Therapy for Metabolic Stress

Nutrition therapy for metabolic stress must take into consideration the patient's primary medical diagnosis along with accommodations for the metabolic changes that have occurred. There is a delicate balance between prevention of protein-energy malnutrition (PEM) and prevention of the possible complications of nutrition support. As stated previously, critically ill patients are at significant nutritional risk. Nutritional status prior to the current illness is an important predictor of morbidity and mortality. Whereas from a healthy state it takes approximately two weeks for a patient to progress to PEM, in the presence of preexisting malnutrition, it takes only a few days. As already mentioned, the level of injury and the subsequent risks are often quantified using scoring systems; the Glasgow Coma Scale or the APACHE II score are common in critical care medicine. Use of a scoring system may give the clinician an idea of the level of metabolic stress that might be expected for the individual patient.

Nutrition Assessment

Table 22.5 outlines nutrition assessment parameters for critically ill patients. Many of the standard measures for nutritional assessment are neither valid nor reliable in this population.[1, 18] The clinician must take into consideration these challenges for nutrition assessment, acknowledge the factors that impact assessment in this population, and make recommendations based on the most reliable information available. Constant re-evaluation of data, patient status, and tolerance to nutrition support is a necessity in the critically ill population.

Information about prior diet history or weight status is not always available. Patients with this level of injury or disease are often on mechanical ventilation or are unable to communicate. Family members will be an important source of information.

Measured weight may not be reflective of actual weight due to changes in fluid balance secondary to fluid resuscitation, losses from wounds, and loss of blood. Various institutions have differing recommendations on the body weight that should be used with each of the resting energy expenditure (REE) equations.[19–21] Either actual body weight, ideal body weight, or adjusted body weight for obese patients could potentially be used to calculate energy requirements. For example, in the American College of Chest Physicians equation, BMI determines the weight that is used in the calculation (see Table 3.12). As discussed earlier in this chapter, lean body mass is the most metabolically active tissue and represents the largest proportion of the REE. Historically, the rationale for use of an adjusted body weight was based on the assumption that 20%–25% of fat mass is metabolically inactive. As Walker and Heuberger point

Table 22.5 Nutrition Assessment for Metabolic Stress

Medical/Social History	• Diagnoses
	• Medications
	• Previous medical conditions or surgeries
Food-/Nutrition-Related History	• Ability to chew; use and fit of dentures
	• Problems swallowing
	• Nausea, vomiting
	• Constipation, diarrhea
	• Any other symptoms interfering with ability to ingest normal diet
	• Ability to feed self
Anthropometric Measurements	• Height
	• Current weight
	• Weight history (if available): highest adult weight; usual body weight
	• Reference weight (BMI)
Biochemical Data, Medical Tests and Procedures	
Visceral Protein Assessment:	Albumin
	Prealbumin
Acute-Phase Proteins:	CRP
	Fibronectin
	Serum amyloid A
	Ceruloplasmin
Hematological Assessment:	Hemoglobin
	Hematocrit
	Platelet count
Other Laboratory Indices:	Electrolytes
	Glucose
	Lactate

out in their recent review, there is no research to support this calculation.[19] In general, most clinicians feel that the risk of initially overfeeding in the critically ill justifies the use of ideal body weight in overweight or obese patients. Otherwise, in normal-weight individuals, actual body weight is used in the calculations for REE.

Biochemical indices to assess visceral protein status (transport proteins) may be more reflective of the level of stress than of the patient's actual protein status, and may also be affected by fluid balance, wound losses, and the use of blood products needed for stabilization of the patient. When assessing protein status in metabolic stress, these laboratory values should be nutritionally interpreted with caution.

Indirect calorimety is the method of choice to establish energy and protein requirements for the critically ill.[1, 22] When this is unavailable, the clinician must estimate energy requirements with use of prediction equations such as those shown in Table 22.6. As discussed in Chapters 3 and 5, meta-analysis has attempted to establish the equation that is most consistent with measured energy requirements.[23–26] Walker and Heuberger examined the literature supporting seven different equations that are frequently used in the acute care population. Because each equation was established in a particular population, its accuracy changes significantly when it is applied to a different subset of patients.[19]

Table 22.6 Possible Prediction Equations

Estimation of total energy requirements for the critically ill patient when indirect calorimetry is unavailable:

A.S.P.E.N. Guidelines for the Provision and Assessment of Nutrition Support Therapy in the Critically Ill Patient

BMI 20–25: 22–25 kcal/kg actual body weight

BMI >30: 22–25 kcal/kg ideal body weight

Mifflin-St. Jeor Equation

REE for females: $10\,W + 6.25\,H - 5\,Age - 161$

REE for males: $10\,W + 6.25\,H - 5\,Age + 5$

[W = weight in kg; H = height in cm; Age = age in years]

American College of Chest Physicians Equation

REE = 25 × weight (kg)

[If BMI = 16–25, use usual body weight; if BMI >25, use ideal body weight; and if BMI <16, use existing body weight for first 7–10 days, then use ideal body weight.]

Penn State 2003 Equation

REE = (0.85 × value from Harris-Benedict equation) + (175 × T_{max}) + (33 × V_e) − 6,443

[V_e = minute volume (in L/min); T_{max} = maximum body temp in previous 24 hours]

Ireton-Jones 1997 Equation

REE = (5 × weight) − (11 × age) + (244 if male) + (239 if trauma present) + (840 if burns present) + 1,784

[W = weight in kg; H = height in cm; A = age in years]

Swinamer 1990 Equation

REE = (945 × body surface area) − (6.4 × age) + (108 × temperature) + (24.2 × respiratory rate) + (817 × V_T) − 4,349

[V_T = tidal volume in liters]

Permissive Underfeeding Guidelines:

18–22 kcal/kg IBW

1.5–2.5 g protein/kg IBW

Activity Factors	Average Injury Factors
Out of bed 1.2	Surgery 1.0–1.3
Confined to bed 1.1	Infection 1.0–1.4
	Skeletal trauma 1.2–1.4
	Head injury 1.5

Protein Requirements

RDA 0.8 g protein/kg

Minor surgery 1–1.1 g protein/kg

Major surgery 1.2–1.5 g protein/kg

Burn 1.5–2.0 g protein/kg

Sources: Boullata J, Williams J, Cottrell F, Hudson L, Compher C. Accurate determination of energy needs in hospitalized patients. J Am Diet Assoc. 2007; 107: 393–401. McClave SA, Martindale RG, Vanek VW, McCarthy M, Roberts P, Taylor B, Ochoa JB, Napolitano L, Cresci G; A.S.P.E.N. Board of Directors; American College of Critical Care Medicine; Society of Critical Care Medicine. Guidelines for the Provision and Assessment of Nutrition Support Therapy in the Adult Critically Ill Patient: Society of Critical Care Medicine (SCCM) and American Society for Parenteral and Enteral Nutrition (A.S.P.E.N.). J Parenter Enteral Nutr. 2009;33:277–316.

Pirat A, Tucker AM, Taylor KA, Jinnah R, Finch CG, Canada TD, Nates JL. Comparison of measured versus predicted energy requirements in critically ill cancer patients. Respir Care. 2009;54:487–94.

Malone AM. Permissive underfeeding: its appropriateness in patients with obesity, patients on parenteral nutrition, and non-obese patients receiving enteral nutrition. Curr Gastroenterol Rep. 2007;9:317–22.

Table 22.7 Potential Benefits of Permissive Underfeeding

- Lower omega-6 fatty acid intake provides reduced substrate for proinflammatory mediator synthesis.
- Limited carbohydrate intake may result in reduced hyperglycemia.
- Lower nutrient oxidation.
- Reduced DNA damage.
- Decreased hypermetabolism with resultant reduced carbon dioxide production.

Reprinted with permission Malone AM. Permissive underfeeding: its appropriateness in patients with obesity, patients on parenteral nutrition, and non-obese patients receiving enteral nutrition. Curr Gastroenterol Rep. 2007;9:317–22.

Factors that affect the accuracy include the presence of obesity, mechanical ventilation, stress, and trauma. Another retrospective chart analysis found that in the 395 patients evaluated, none of the equations accurately predicted the measured energy expenditure.[26] The ADA Evidence Analysis work group could not support the use of the Harris-Benedict, Ireton-Jones, or Fick equations in hospitalized, critically ill populations.[24] Their review also stated that there was not enough data to either support or reject the use of the Mifflin-St. Jeor equation in this population. In the most recent analysis of 202 mechanically ventilated patients, the Penn State equation had the highest accuracy when compared to measured energy expenditure.[25]

When providing nutrition support to the critically ill, it is crucial to avoid overfeeding and its subsequent metabolic complications, which include increased carbon dioxide production and hyperglycemia. Steps to prevent these complications include the concept of permissive underfeeding.[1, 27–29] The protective benefits of permissive underfeeding are summarized in Table 22.7.

A.S.P.E.N and ADA recommendations state that when the BMI is >30, the patient should receive approximately 22–25 kcal/kg of ideal body weight. Protein requirements should be estimated at approximately 2 g protein/kg/ideal body weight. For normal-weight individuals, 22–25 kcal/kg of actual body weight are recommended during critical illness.[1, 22]

Nutrition Diagnosis

Nutrition diagnoses during metabolic stress and critical illness may include: increased energy expenditure; increased nutrient needs; inadequate protein-energy intake; altered GI function; and impaired nutrient utilization.

Nutrition Intervention

For critically ill individuals, nutritional needs can rarely be met by the oral route, and alternate feeding routes, including both enteral (EN) and parenteral (PN) nutrition, are standard components of the medical and nutritional care. See Figure 5.3 in Chapter 5 for guidance in selecting a feeding route. A.S.P.E.N. 2009 guidelines document Grade B evidence that "enteral nutrition should be started early within the first 24–48 hours following admission" (p. 280).[1]

When compared to parenteral nutrition (PN), EN is more cost-effective and is associated with reduced infectious complications, fewer surgical interventions, and, in some studies, fewer hospital days.[1, 30–34] The first step in developing the nutrition prescription, once access to the GI tract has been established, is choosing the appropriate enteral formula. Figure 22.4 presents an example of a formula selection protocol. Decisions will be influenced by the ability to meet energy and protein needs within the fluid volume that can be best tolerated by the patient.

In addition to energy and protein needs, formula selection should address the specific types of nutrients that have increased requirements during metabolic stress. Nutrients that are altered during metabolic stress or that have been determined to assist in the body's response to stress are often included in specialized formulas (see Table 22.8) or are supplemented separately.[1, 35–37] The amino acid glutamine is recommended for all burn, trauma, and ICU patients. Even though glutamine is a nonessential amino acid, the body's synthesis rate cannot meet the increased needs during the stress of critical illness. Most specialized formulas provide 0.3–0.4 g/kg/day of glutamine, and glutamine can be added to an enteral formula that does not already contain it. Glutamine is the preferred fuel for the enterocytes and assists in maintaining intestinal membrane permeability. The supplementation of another nonessential amino acid, arginine, is controversial. Because negative outcomes were documented in patients with sepsis, arginine is not recommended for use in these individuals.[35]

Branched-chain amino acids include isoleucine, leucine, and valine. Interest in their role during starvation and metabolic stress is linked to their particular metabolism within the skeletal muscle.[38] For example, leucine is completely oxidized for energy within the muscle and provides more ATP than glucose. The increased metabolism of branched-chain amino acids within the skeletal muscle may spare other substrates that are required to meet the metabolic demands of stress.[4]

Other nutrients that should be considered include the antioxidants: vitamin C, vitamin E, selenium, copper, and zinc.[1, 4, 35] The protocol for supplementation proposed by the Inflammation and the Host Response to Injury Collaborative Research Program includes 100 mg IV vitamin C every 8 hours; 400 µg IV selenium daily; and 1500 IU vitamin E every 12 hours for

Figure 22.4 Enteral Nutrition Protocol

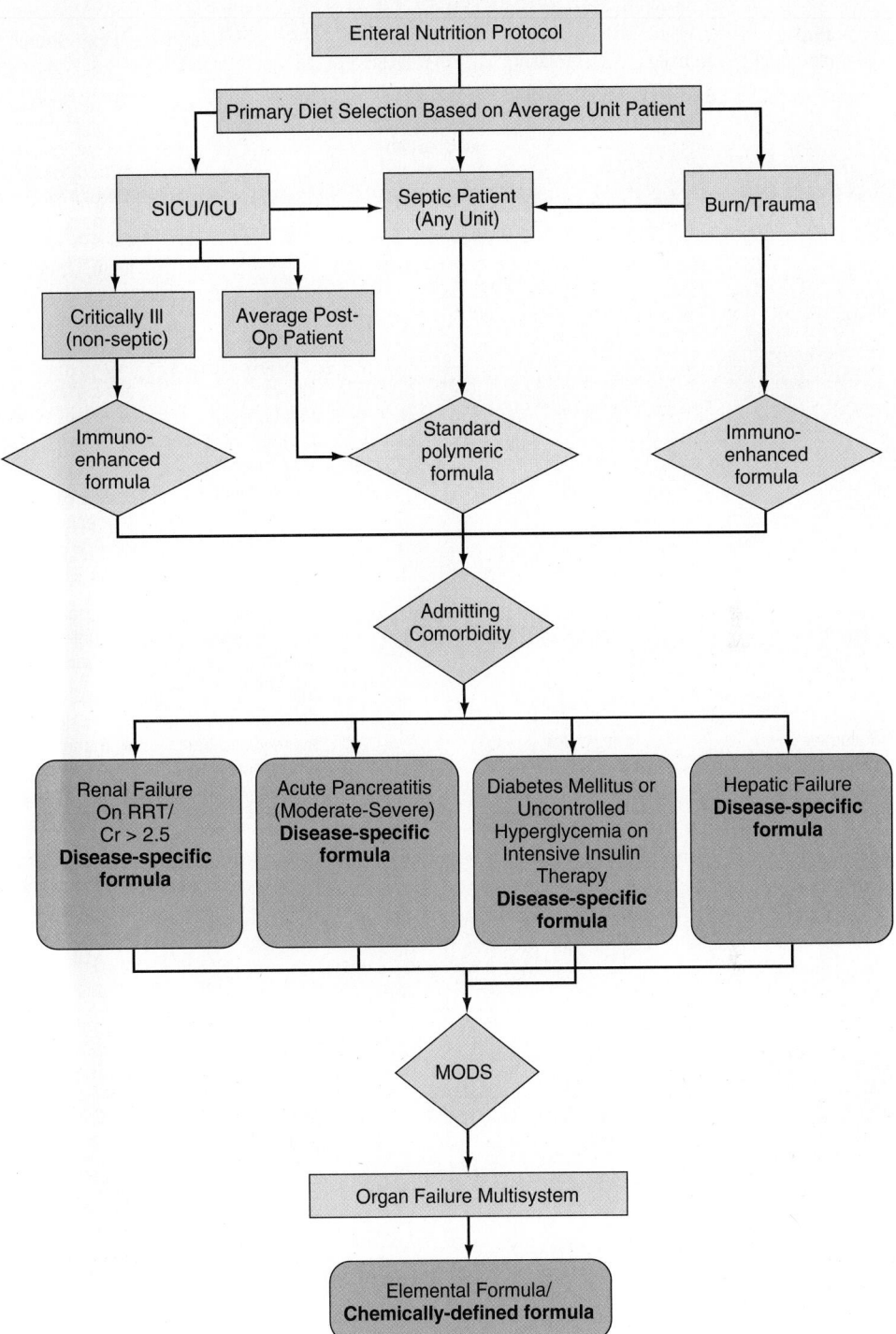

Source: JJ Diaz, R Pousman, G Jensen, Vanderbilt University Medical Center, *Critical Care Nutrition Practice Management Guidelines.*

7 days or until discharged from the ICU. Vitamin C and selenium are given intravenously for the first two days and then are changed to enteral dosages.[38] Specialized enteral formulas for metabolic stress do contain these added nutrients and may provide adequate amounts of antioxidants without supplementation of individual nutrients. Further research needs to be conducted to establish these recommendations.

The fatty acids eicosapentaenoic (EPA) and docosohexaenoic acid (DHA) have shown positive effects when used in critically

Table 22.8 Enteral Formulas in Metabolic Stress

Formula/ Manufacturer	Caloric Density	Osmolality	Carbohydrate Source (% of total kcal)	Protein Source (% of total kcal)	Lipid Source (% of total kcal)	Rationale for metabolic stress
Oxepa© (Ross)	1.5 kcal/mL	535 mOsm/kg	28.1% (sugar [sucrose], corn maltodextrin)	16.9%	55.2% Canola oil, medium-chain triglycerides, marine oil (may contain one or more of the following: anchovy, menhaden, salmon, sardine, tuna), borage oil	• Concentrated calories for fluid-restricted patients • Unique, patented oil blend—contains 4.6 g/L of EPA* and 4 g/L of GLA† • Elevated levels of antioxidants vitamin C, vitamin E, and beta-carotene
Pivot© (Ross)	1.5 kcal/mL	595 mOsm/kg water	46% (corn syrup solids)	25% (partially hydrolyzed sodium caseinate, whey protein hydrolysate)	Structured lipid (interesterified sardine oil and medium-chain triglycerides), soy oil, canola oil	• Arginine—13 g/L (3.5% of calories) • Glutamine (inherent)—6.5 g/L • Omega-3 fatty acids (EPA, 2.6 g/L; DHA, 1.3 g/L)
Crucial© (Nestle)	1.5 kcal/mL	375 mOsm/kg water	36% (maltodextrin)	25% (hydrolyzed casein with added amino acid fortification)	39% Marine oil, MCT and soybean oil MCT:LCT ratio: 50:50 n6:n3 ratio: 1.5:1	• EPA/DHA: 4.3 g/L • Supplemental L-arginine: 12 g/L
Impact© (Nestle)	1.0 kcal/mL	375 mOsm/kg water	53% (hydrolyzed cornstarch)	22% (sodium and calcium caseinates, L-arginine 12.5 g/L)	25% (palm kernel oil, sunflower oil, **menhaden oil**)	• EPA/DHA: 1.7 g/L • Supplemental L-arginine: 12.5 g/L • Dietary nucleotides: 1.2 g/L
Impact Glutamine© (Nestle)	1.3 kcal/mL	630 mOsm/kg water	46% (maltodextrin)	24% (wheat protein hydrolysate, free amino acids, sodium caseinate)	30% (palm kernel oil, menhaden oil, sunflower oil) MCT:LCT ratio 27:73 n6:n3 ratio: 1.4:1	• Glutamine: 15 g/L • L-arginine: 16.3 g/L • Dietary nucleotides: 1.6 g/L • n-6:n-3 ratio: 1.4:1.0
Impact with Fiber	1.0 kcal/mL	375 mOsm/kg water	53% (maltodextrin)	22% (sodium and calcium caseinates (milk), L-arginine)	25% (palm kernel oil, menhaden oil, sunflower oil) MCT:LCT ratio 27:73 n6:n3 ratio: 1.4:1	• Meets 100% RDI for 24 key micronutrients: 1500 mL • Fiber content (source) 10 g/L (BENEFIBER® soluble fiber, soy fiber) • EPA/DHA: 1.7 g/L • Supplemental L-arginine: 12.5 g/L • Dietary nucleotides: 1.2 g/L
Nutren Replete© (Nestle)	1 kcal/mL	300–350 mOsm/kg water	25%	45%, calcium-potassium caseinate	30%	• MCT:LCT ratio 25:75 • n6:n3 ratio 2.3:1
Periative© (Ross)	1.3 kcal/mL	460 mOsm/kg water	55% (corn maltodextrin)	20.5% (partially hydrolyzed sodium caseinate, hydrolyzed lactalbumin)	Canola oil, medium-chain triglycerides	• Includes 1.6 g of NutraFlora® scFOS®/8 fl oz (6.5 g/L and 9.8 g/1500 mL) • Contains added arginine (8 g/L total arginine at 2.5%)

†GLA = gamma-linolenic acid (from borage oil)

*EPA = eicosapentaenoic acid (from marine oil)

ill patients. Immunomodulating formulas such as Oxepa® and Pivot® are examples of these products.

Sources of fiber added to enteral feedings provide substrates that assist in maintenance of beneficial bacteria in the gastrointestinal tract (probiotics, prebiotics, and synbiotics) and may assist in preventing diarrhea.[39] As discussed in Chapter 15, substrates such as inulin, guar gum, and other soluble fibers are fermented to short-chain fatty acids and lactate. Enteral nutri-

tion formulas with added probiotics, prebiotics, and synbiotics may be important options to consider for the prevention of infectious complications associated with critical illness.[40]

Complications of enteral feeding may include metabolic complications (hyperglycemia and electrolyte imbalances), aspiration, or mechanical complications (clogging or misplacement of the tube). See Chapter 5 for details about these complications and their prevention.

Parenteral nutrition (PN) should be reserved for those cases of prolonged nothing by mouth (NPO) status lasting longer than 7 days; when the patient is malnourished and enteral access cannot be obtained; when enteral nutrition support cannot meet the patient's needs or is not tolerated; or when a major surgical procedure will prevent the patient from starting enteral nutrition and the patient will enter surgery in a malnourished state.[1, 22, 38] Chapter 5 provides details regarding the design of PN prescriptions. An important concern for PN in metabolic stress may be its contribution to hyperglycemia. Additionally, the use of parenteral lipids may be immunosuppressive. The A.S.P.E.N. guidelines recommend avoiding use of soy-based parenteral lipids for the first seven days of admission.[1] Other complications of PN may include catheter occlusion, catheter-related infection, hypertriglyceridemia, intestinal atrophy, electrolyte disturbances, and refeeding syndrome in previously malnourished individuals.

Sepsis, Systemic Inflammatory Response Syndrome (SIRS), and Multiorgan Distress Syndrome (MODS)/Multisystem Organ Failure (MSOF)

Definition

Sepsis has historically been defined as an uncontrolled inflammatory response to infection or trauma. It is now understood that sepsis is actually an immunosuppressive process that prevents an adequate response to infection.[8, 9, 41] The Surviving Sepsis Campaign was formed in 2002 and included the Society for Critical Care Medicine, along with eleven other international organizations. They defined severe sepsis as "infection-induced organ dysfunction or hypoperfusion abnormalities." Septic shock is defined as "severe sepsis with hypotension not reversed with fluid resuscitation and associated with organ dysfunction or hypoperfusion abnormalities."[8]

Systemic inflammatory response syndrome (SIRS) is an additional classification of this condition not necessarily caused by an infectious process. SIRS may occur after major surgery or trauma, or with other conditions such as myocardial infarction.[9]

Multiorgan distress syndrome (MODS) is also referred to as multisystem organ failure (MSOF). The terms are often used interchangeably. This condition results from the complications of sepsis and SIRS. MODS/MSOF may include dysfunction within cardiac, respiratory, and renal systems.

Epidemiology

Over 750,000 cases of sepsis were identified in 2008.[7] Durthaler states that sepsis accounts for a greater number of deaths than breast cancer, colon cancer, AIDS, and congestive heart failure combined.[41]

Etiology

As stated previously, sepsis was originally thought to be a result of an overwhelming systemic infection. Researchers are now beginning to realize the complexity of the cascade of events associated with sepsis and that the similar progression characteristic of SIRS can occur without infection. A combination of proinflammatory cytokine release, imbalance of coagulation factors, altered cellular metabolism, **hypoperfusion**, and **hypotension** direct the physiological changes that occur with sepsis, SIRS, and MODS (MSOF).[9]

Clinical Manifestations

The initial major signs of sepsis include increased white blood cell count (>12,000 mm^3), increased heart rate (>90 beats per minute) and respirations (>20 breaths/minute), and fever (>38 °C) or hypothermia (<36 °C) (8). C-reactive protein, fibrinogen, complement proteins, and other acute-phase proteins are elevated.[10] Other laboratory values may include increased serum lactate and serum glucose.

The diagnostic criteria for SIRS and MODS (MSOF) as defined by Dellinger and colleagues[8] are presented in Table 22.9 and Table 22.10, respectively.

Pathophysiology

Sepsis is initiated by an originating source of infection and/or trauma. Russell explains that when individual cell markers recognize microorganisms, the systemic response causes the release of inflammatory mediators including TNF, interferon, and interleukins[9] (see Chapter 9). Inflammation results in vascular permeability, which allows the shift of fluid into the lungs and other third spaces. High levels of nitric oxide contribute to this vascular permeability. There is also an imbalance of coagulation factors that can contribute to production of multiple thrombi.[9] As sepsis continues, there is a shift from an inflammatory response to an anti-inflammatory response with resulting energy (inability to mount an immune response) and organ dysfunction. The increased rate of gluconeogenesis

Table 22.9 Diagnostic Criteria for SIRS

SIRS is present with any two of the following conditions:
- Temperature >38.0°C or <36.0°C
- Heart rate >90 beats per minute
- Respiratory rate >20 breaths per minute
- Partial pressure of carbon dioxide (PCO_2) <32 mm Hg
- Leukocytosis (white blood cell [WBC] count >12,000 mcL^{-1})
- Leukopenia (WBC count <4000 mcL^{-1})
- Normal WBC count with >10% immature forms

Sources: Dellinger RP, Levy MM, Carlet JM, Bion J, Parker MM, Jaeschke R, Reinhart K, Angus DC, Brun-Buisson C, Beale R, Calandra T, Dhainaut JF, Gerlach H, Harvey M, Marini JJ, Marshall J, Ranieri M, Ramsay G, Sevransky J, Thompson BT, Townsend S, Vender JS, Zimmerman JL, Vincent JL; International Surviving Sepsis Campaign Guidelines Committee; American Association of Critical-Care Nurses; American College of Chest Physicians; American College of Emergency Physicians; Canadian Critical Care Society; European Society of Clinical Microbiology and Infectious Diseases; European Society of Intensive Care Medicine; European Respiratory Society; International Sepsis Forum; Japanese Association for Acute Medicine; Japanese Society of Intensive Care Medicine; Society of Critical Care Medicine; Society of Hospital Medicine; Surgical Infection Society; World Federation of Societies of Intensive and Critical Care Medicine. Surviving Sepsis Campaign: international guidelines for management of severe sepsis and septic shock: 2008. Crit Care Med. 2008 Jan;36(1):296–327. Erratum in: Crit Care Med. 2008 Apr;36(4):1394–96.

Note for permissions: These diagnostic criteria are also published on the Surviving Sepsis website (available from http://www.survivingsepsis.org/GUIDELINES/Pages/default.aspx).

Table 22.10 Diagnostic Criteria for MODS (MSOF)

In MODS, organ dysfunction is identified by:

- Arterial hypoxemia (PaO$_2$/fraction of inspired oxygen [FiO$_2$] ratio of <300 torr)
- Acute oliguria (urine output <0.5 mL · kg^{-1}·hour^1 or 45 mmol/L for at least 2 hours)
- Creatinine >2.0 mg/dL
- Coagulation abnormalities (international normalized ratio >1.5 or activated partial thromboplastin time >60 seconds)
- Thrombocytopenia (platelet count <100,000 mcL^{-1})
- Hyperbilirubinemia (plasma total bilirubin >2.0 mg/dL or 35 mmol/L)
- Tissue-perfusion variable: hyperlactatemia (>2 mmol/L)
- Hemodynamic variables:
  - arterial hypotension (systolic blood pressure [SBP] <90 mm Hg)
  - mean arterial pressure [MAP] <70 mm Hg
  - SBP decrease >40 mm Hg

References: Dellinger RP, Levy MM, Carlet JM, Bion J, Parker MM, Jaeschke R, Reinhart K, Angus DC, Brun-Buisson C, Beale R, Calandra T, Dhainaut JF, Gerlach H, Harvey M, Marini JJ, Marshall J, Ranieri M, Ramsay G, Sevransky J, Thompson BT, Townsend S, Vender JS, Zimmerman JL, Vincent JL; International Surviving Sepsis Campaign Guidelines Committee; American Association of Critical-Care Nurses; American College of Chest Physicians; American College of Emergency Physicians; Canadian Critical Care Society; European Society of Clinical Microbiology and Infectious Diseases; European Society of Intensive Care Medicine; European Respiratory Society; International Sepsis Forum; Japanese Association for Acute Medicine; Japanese Society of Intensive Care Medicine; Society of Critical Care Medicine; Society of Hospital Medicine; Surgical Infection Society; World Federation of Societies of Intensive and Critical Care Medicine. Surviving Sepsis Campaign: international guidelines for management of severe sepsis and septic shock: 2008. Crit Care Med. 2008 Jan;36(1):296–327. Erratum in: Crit Care Med. 2008 Apr;36(4):1394–96.

Note for permissions: These diagnostic criteria are also published on the Surviving Sepsis website (available from http://www.survivingsepsis.org/GUIDELINES/Pages/default.aspx).

results in significant catabolism of skeletal muscle mass. These metabolic abnormalities result in hyperglycemia and increased serum lactate (see Figure 22.5).

Treatment

Evidence-based medical protocols have been developed to address the immediate treatment of an individual diagnosed with sepsis. Treatment of sepsis will center on treating the source of infection or trauma and supporting the patient with lung-protective ventilation, with antibiotics, and with hemodynamic, renal, and metabolic support.[8] Intensive insulin therapy, antimicrobial agents, coagulation-modulating drugs [such as activated Protein C (drotrecogin alfa activated)], and nutrition support are all included in the medical protocols, and represent crucial steps in the effective treatment of sepsis.[8, 9, 41]

Nutrition Therapy for Sepsis

Nutrition support is a critical step in the treatment and resolution of sepsis. Meeting the needs of these critically ill patients presents many challenges: abnormalities of metabolism, difficulty estimating and/or measuring nutritional requirements, fluid/volume restrictions, and multi-system organ dysfunctions. Nutrition support plans will follow the standard procedures for nutrition therapy for metabolic stress previously discussed.

Burns

Definition

Burns are a result of tissue injury caused by exposure to heat, chemicals, radiation, or electricity. They may result from

Figure 22.5 Pathophysiology of Sepsis, SIRS, and MODS

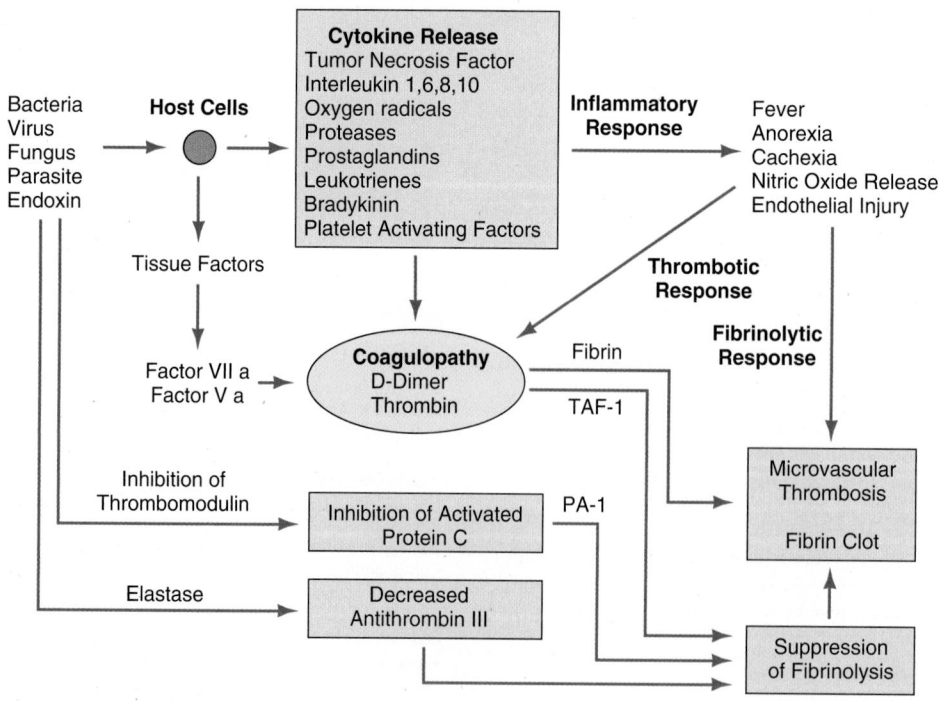

TAF-1: thrombin-activatable fibrinolysis inhibitor
PA-1: plasminogen-activator inhibitor-1

Source: Sat Sharma, MD, FRCPC, FCCP, Anand Kumar, MD, FRCPC, FCCP, FCCM, *Current Opinion in Pulmonary Medicine* 9(3):199–209, 203. © 2003 Lippincott Williams & Wilkins.

injury to the skin, but the damage may extend into muscle and bone. The depth of the wound and the percentage of the body surface area that is affected are used to classify burn injury. The wounds are described as superficial, superficial partial thickness, deep partial thickness, or full thickness.

Epidemiology

Medical treatment is required for more than 500,000 victims each year, with approximately 25,000 admissions to specialized burn centers. Approximately 4,500 deaths from burns were reported for the year 2007.[42]

Etiology

Burns may result from thermal exposure, which includes direct contact with a heat source such as hot water or flames. These are the most common type of burns and occur both in the home and the workplace. Burns can also be caused by chemical or electrical exposure. Damage to the body occurs when the electrical current moves through tissues or bone. The severity of the burn will be correlated with the amount of voltage, the location where it enters the body, and the amount of time that the exposure continues. Chemical burns occur when the body is exposed to an acid or alkali such as battery acid, drain cleaner, or other environmental sources.

Clinical Manifestations

Signs and symptoms that an individual experiences will be determined by the extent of the burn injury—by both depth and body surface area affected. Age, nutritional status, and other comorbidities will have an impact on the physiological response to the injury, treatment, and recovery.

Table 22.11 summarizes the characteristics of burn classifications. Superficial burns involve the top layer of epidermis, which will characteristically redden. An example of this type of burn may result from sunburn. Superficial burns are painful but typically heal easily and quickly. Superficial partial thickness burns produce open, weeping wounds that are very painful. Partial thickness deep burns involve destruction of the epidermis and dermis. Full thickness burns destroy all layers of the skin and can involve underlying muscle, organs, and bone.[43]

The Rule of "Nines" is one method that is used to make a rapid estimation of body surface area (BSA) that has been burned. In this method, the body is divided into portions with a value or derivative of nine. For example, an entire leg and foot is estimated to be approximately 9% of the total body surface area (see Figure 22.6). Estimation of the affected body surface area assists in assessment of the extent of the injury, and helps provide the basis for prescribing fluid and medications.

Pathophysiology

While superficial burns and some superficial partial thickness burns are generally treated on an outpatient basis, other levels of burn injury require treatment in specialized burn units. The initial burn shock is a result of the extensive inflammatory process and involves rapid fluid shifts and accumulation and, therefore, fluid loss from the wound. Initial treatment of a burn will focus on fluid resuscitation and stabilization of all organ systems.[43, 44]

The physiological response to burn injury progresses through the same phases of metabolic stress as described earlier in this chapter: hypermetabolism, catabolism, and altered immune and hormonal response. Respiratory complications are multifactorial. They may originate from inhalation of smoke and other toxic substances or occur as complications secondary to fluid resuscitation, pain, inflammation, and infection.

Treatment

Management of the burn wound involves application of topical agents such as **silver sulfadiazine cream** and **silver nitrate** to prevent infections. Other treatment involves the complex procedures used to clean, **debride**, and dress the wounds. Full thickness burns require skin grafting or the use of skin substitutes for acceptable closure of the wounds.

Table 22.11 Characteristics of Burns Based on Depth

Classification	Cause	Characteristics			
		Appearance	**Sensation**	**Healing Time**	**Scarring**
Superficial	Ultraviolet light, very short flash (flame exposure)	Dry and red; blanches with pressure	Painful	3 to 6 days	None
Superficial partial-thickness	Scald (spill or splash), short flash	Blisters; moist, red and weeping; blanches with pressure	Painful to air and temperature	7 to 20 days	Unusual; potential pigmentary changes
Deep partial-thickness	Scald (spill), flame, oil, grease	Blisters (easily unroofed); wet or waxy dry; variable color (patchy to cheesy white to red); does not blanch with pressure	Perceptive of pressure only	More than 21 days	Severe (hypertrophic) risk of contracture
Full-thickness burn	Scald (immersion), flame, steam, oil, grease, chemical, high-voltage electricity	Waxy white to leathery gray to charred and black; dry and inelastic; does not blanch with pressure	Deep pressure only	Never (if the burn affects more than 2% of the total surface area of the body)	Very severe risk of contracture

Adapted with permission of Robert Hugh Demling, M.D. (Editor, BurnSurgery.org) from http://www.burnsurgery.org/Modules/BurnWound/rationale/burn_injury/depth_assessment.htm

Figure 22.6 Rule of Nines to Estimate Body Surface Area

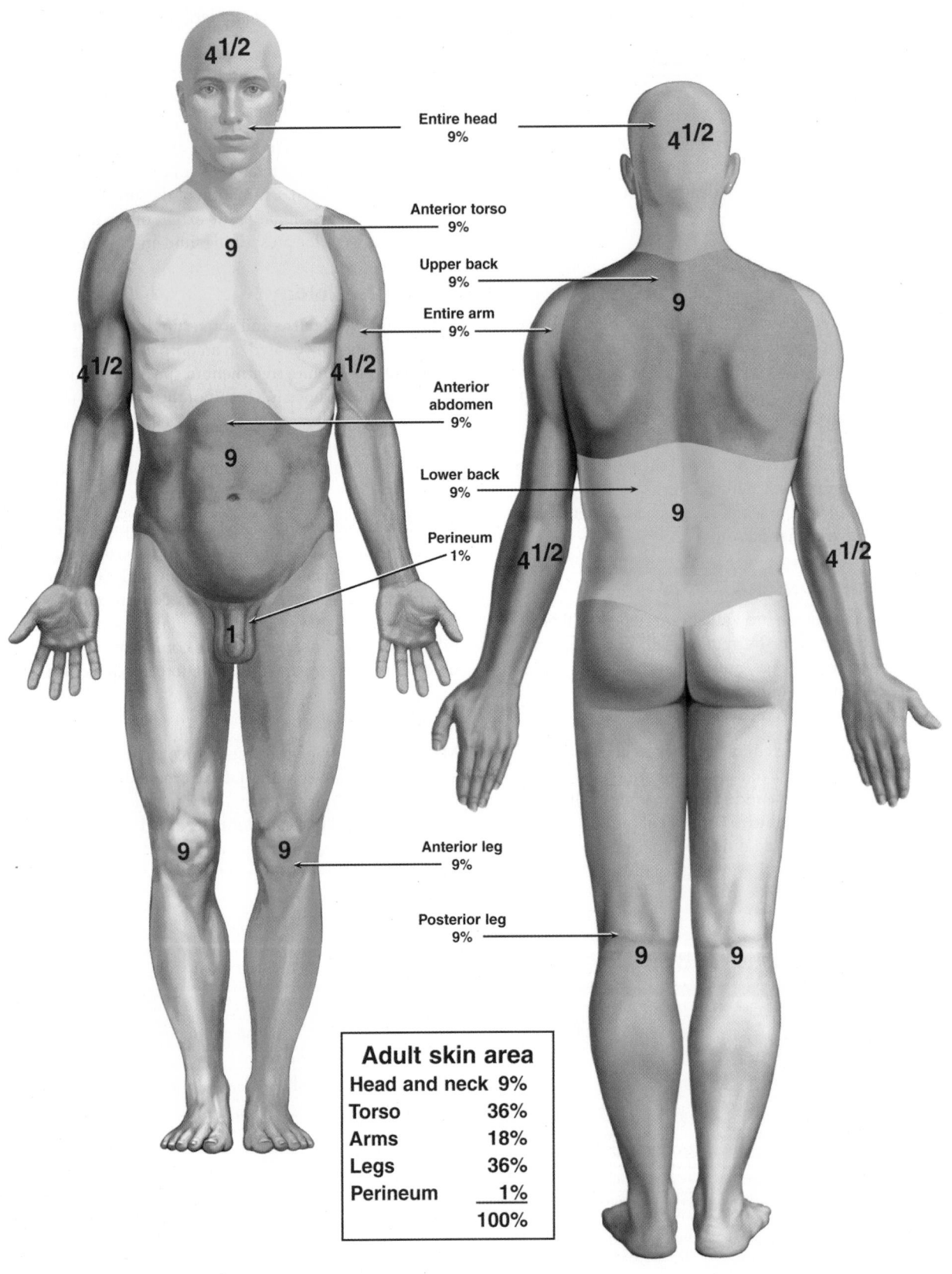

Entire head 9%

Anterior torso 9%

Upper back 9%

Entire arm 9%

Anterior abdomen 9%

Lower back 9%

Perineum 1%

Anterior leg 9%

Posterior leg 9%

Adult skin area

Head and neck	9%
Torso	36%
Arms	18%
Legs	36%
Perineum	1%
	100%

Nutrition Therapy for Burns

The patient with a burn injury is at significant nutritional risk due to the hypermetabolic, catabolic response that occurs after the injury. It is estimated that as much as 20% of body protein can be lost within the first two weeks of burn injury.[44] Fluid imbalance, pain, and immobility make it difficult for the patient with extensive burns to maintain his or her nutritional status. The patient also requires optimal nutrition therapy to support wound healing during the treatment and healing process.

Nutrition Assessment

Standard nutritional parameters are affected by both the nature of the burn injury and the metabolic stress response. Weight may fluctuate considerably due to fluid shifts and resuscitation. High protein losses from wounds and the overall acute inflammatory response can affect accurate interpretation of the protein markers albumin, prealbumin, or C-reactive protein in relationship to nutritional status.

Energy requirements are measured by indirect calorimetry or estimated using standard equations.[44] A specific equation that incorporates and accounts for the extent of the burn injury, called the Curreri equation, was developed for use in burn patients.[45] Research and practice, however, have indicated that this equation best estimates energy requirements at the peak of burn injury and that calculations do not necessarily accommodate the changes that occur from day to day in the patient. While the degree of hypermetabolism and catabolism correlate with the patient's level of injury, researchers agree that the level of hypermetabolism does not increase beyond that reached for a 50%–60% total body surface area burn.[46, 47] Other factors that contribute to increased energy needs include fever, infection, or the development of sepsis.

Protein requirements can be based on 1.5–2 g protein/kg. It is unlikely that the negative nitrogen balance that occurs during the catabolic phase of burn injury can be totally prevented, but the goal should be to minimize losses and promote wound healing. Monitoring daily calorie counts, wound closure, and acceptance of engraftment provides practical measurements of adequate nutritional support.

Nutrition Diagnosis

Common nutrition diagnoses that occur for the burn patient include: increased energy expenditure; increased nutrient needs; impaired nutrient utilization; altered GI function; involuntary weight loss; and altered nutrition-related laboratory values.

Nutrition Intervention

Most severe burn injuries require nutrition support in addition to or as a substitute for an oral diet. As previously discussed, early enteral feeding is recommended during metabolic stress and for those who are critically ill. In burn patients, EN that is initiated within 24 hours of injury has been associated with prevention of infections (in particular bacterial translocation), the prevention of **Curling's ulcer**, and the reduction of protein catabolism.[18, 44] Curling's ulcer is also treated prophylactically with medications such as H_2 antagonists.

The enteral feeding prescription for a burn patient is developed by following the same steps used with other diagnoses, but will need to accommodate the special metabolic requirements of this stressed state. In severe burn patients, ileus (general paralysis of the GI tract) is common during the burn shock period, but enteral feeding is generally tolerated when delivered to the small bowel. When choosing a formula, the clinician should focus on those with higher amounts of protein (20%–25% of kcal) and consider those formulas with supplemental glutamine and omega-3 fatty acids.[18, 44] More research is needed to determine the exact amounts or duration of therapy for glutamine, arginine, or omega-3 fatty acids for promotion of wound healing.[47]

When enteral feeding cannot meet nutritional needs, PN should be prescribed. This may be used in combination with enteral or oral feedings, or can provide all of the patient's nutritional needs. In a recent study examining nutrition support for patients with **necrotizing fasciitis**, 94% of patients required either EN or PN for an average of 24 days (range 1–68 days).[48]

Careful attention will be necessary to avoid overfeeding and control hyperglycemia. There is a tenuous line between providing the amount of energy needed to meet metabolic requirements and contributing to further metabolic complications.

Oxandrolone, an anabolic steroid, has been used to promote protein synthesis in burns affecting >15% BSA.[49] Additional vitamins, minerals, and trace elements are often supplemented for wound healing (see Chapter 9, Table 9.5). In burn patients, higher amounts of these nutrients are often prescribed to replace the large amounts lost via the wound exudates, but also to ensure adequate support for engraftment and overall wound healing. Supplementation with vitamin C, vitamin A, vitamin E, and zinc may be included for nutrition protocols in burn intensive care units (see Table 22.12). Depending on the route of nutrition support, the supplementation in the formula chosen may be adequate to meet the client needs. Identifying specific deficiencies and then providing supplementation for that nutrient may also be a route for providing these micronutrients.[50]

Table 22.12 Nutrient Supplementation for Burns for Children[a]

Micronutrient	Enteral Supplementation[b]	Parenteral Supplementation
Multivitamin with trace elements[c]	1 tablet/day	1 single dose vial/day
Zinc[d]	25 mg/day	50 µg/kg/day
Copper[d]	2.5 mg/day	20 µg/kg/day
Selenium	50–170 mg/day	2 µg/kg/day
Vitamin C	200 mg/day	200 µg/kg/day

[a]children >3 years of age

[b]children receiving adult formulas or specialty formulas for wound healing may not require additional supplementation of individual nutrients

[c]Vitamins A, E, iron, B complex are provided only as part of multivitamin/trace element preparation

[d]addition of a multivitamin supplement with trace elements may be sufficient for meeting requirements

Source: Reprinted from Burns by Prelack K, Dylewski M, Sheridan RL. Practical guidelines for nutritional management of burn injury and recovery, 2007;33:14–24. Copyright (2007), with permission from Elsevier.

Figure 22.7 Combination Feeding Protocol

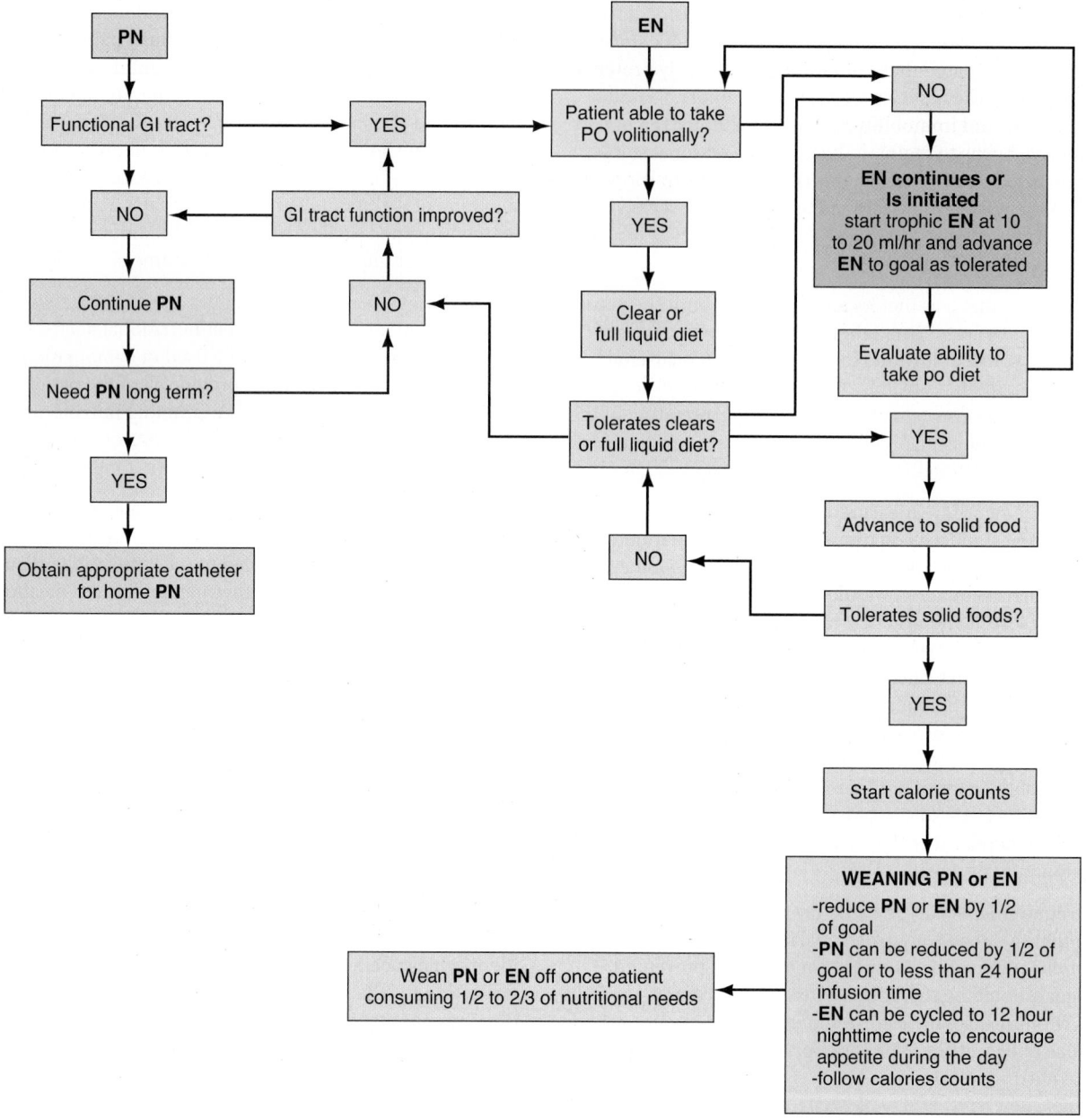

Source: JJ Diaz, R Pousman, G Jensen, Vanderbilt University Medical Center, Critical Care Nutrition Practice Management Guidelines.

As recovery proceeds and the patient is able, oral feedings can be initiated. Weaning from nutrition support is recommended when the patient is able to meet at least 60% of nutritional needs orally (see Figure 22.7). Nutritional requirements will need to be adjusted as the patient heals and the focus of therapy is rehabilitation.

Surgery

Definition

Surgery is defined as an operative procedure used to diagnose, repair, or treat an organ or tissue. Surgery can be further classified by the seriousness of the procedure (major or minor), the necessity (elective or emergency), or the specific purpose of the procedure (diagnostic, excision, palliative, reconstructive, or transplant).

Epidemiology

In 2006, over 53.3 million surgical procedures were performed in the United States.[51] These included a wide range of procedures such as cardiac catherizations, mobilization of fracture, and hysterectomy.

Etiology

Many surgical procedures do not pose nutritional risk. Nonetheless, if the patient enters surgery malnourished or overnourished, or if the surgical procedure will interrupt normal nutrition processes, the individual will be at nutritional risk postoperatively. Age and coexisting diagnoses will have an impact on the outcome and recovery from the surgical procedure. Malnutrition increases the risk of the most common postoperative complications, including wound dehiscence (improper wound opening after suture closure) and infections.[5]

Screening and prognostic tools have been developed to identify those patients most likely to be at nutritional risk (see Chapter 3). Preoperative changes in weight, albumin, and C-reactive protein have been quantified to predict outcome in a number of research studies. One of these, The National VA Surgical Risk Study, evaluated the relationship of numerous characteristics to complications and mortality rate in over 50,000 surgical patients.[52] These researchers found that preoperative albumin was a better predictor of complications and mortality than any of the other characteristics such as age, smoking, and other laboratory values.

Clinical Manifestations

The signs and symptoms experienced with surgery depend on the type of procedure. Patients are required to refrain from eating or drinking for at least twelve hours before surgery. The patient will receive **general**, **epidural**, or **local anesthesia**. Postoperatively, the patient may also have a nasogastric tube in place to remove gastric secretions and/or a urinary catheter in place to remove urine until normal voluntary control of urination has returned. Other general concerns for surgical patients include maintenance of respiratory function and circulation, prevention of infection, wound healing, and pain control.

General anesthesia may result in a postoperative ileus (lack of motility), a general paralysis of the gastrointestinal tract. Resolution of the ileus generally occurs within 24–48 hours, depending on the type of procedure. Traditionally, the patient was prevented from eating or drinking until the ileus was resolved, and the production of gas or bowel movement was a sign of resolution of ileus. Because it is difficult to ascertain when GI function has returned, however, the determination of when to actually begin postoperative feeding has been a topic of recent debate.[53–57] Many patients cannot withstand additional weight loss if they have entered surgery with the presence of nutritional deficits. Further weight loss may increase the chances of complications and lengthen hospital stay. Allowing a patient to eat as soon after surgery as it is possible and safe to do so is recommended.

Nutrition Therapy for Postsurgery

Major surgical procedures, including gastrointestinal procedures such as **esophagectomy** or pancreatic resections, result in the cycle of metabolic stress (as described throughout this chapter) postoperatively.[5] Furthermore, the ability to meet nutritional needs by an oral diet may be inhibited or precluded by the postoperative status of the GI tract, pain, and other systemic effects of a major surgery. Nutritional needs are increased in order to accommodate repair and healing postoperatively. Recovery of functional status requires adequate nutritional support.

Nutrition Assessment

Energy and protein requirements are established individually and should support postoperative healing. Resting energy expenditure (REE) with appropriate activity and injury factors is typically used to estimate energy requirements, but as stated earlier, this method is not without limitations. Protein requirements are elevated above the RDA of 0.8 g/kg but vary depending on the type of surgery and preoperative nutritional status. Table 22.6 provides guidelines for injury factors to consider when estimating energy and protein requirements for surgical patients.

For those patients who experience metabolic stress postoperatively, nutritional needs will increase to meet the demands of hypermetabolism and catabolism. Enteral and parenteral feeding may be initiated immediately after surgery if prolonged NPO status is anticipated. Examples include major bowel or pancreatic resections (see Chapters 15, 17).

Nutrition Diagnosis

Common nutrition diagnoses for individuals postoperatively include: increased energy expenditure; inadequate oral/food beverage intake; increased nutrient needs; altered GI function; and impaired nutrient utilization.

Nutrition Intervention

The surgeon and registered dietitian will recommend the progression for postoperative feeding on an individual basis. It is recommended, though, that the patient should be progressed from nil per os (NPO) to solid food as quickly as possible.[53] In a randomized trial of 96 patients undergoing major abdominal surgery, patients tolerated advancement to solid food after an initial trial of 500 mL of clear liquids.[57] More recent research and meta-analyses support the advancement to solid food after surgery without the traditional transition from clear liquids to slow addition of solid food.[54–56]

Conclusion

Metabolic stress represents the most challenging and demanding environment for nutritional care. In no other situation are the complications of inadequate nutritional support more evident. Research continues to provide practical application of nutrition support to optimize outcomes for the critically ill.

PRACTITIONER INTERVIEW

Kathryn Sikorski, RD, CDE, *Santa Clara Valley Medical Center, San Jose, CA*

How long have you been an RD?

Received a BS in Nutritional Science in 1987; registered since 1990; I have worked in the Burn Unit for the past five years.

Describe the population you would expect to see in a burn unit.

There are a number of situations and circumstances that seem to increase the possibility of sustaining a burn injury.

(continued)

The very young, the elderly, and people in general who are less able to protect themselves are at higher risk, especially for scalds, hot oil burns, accidental ignition of clothing, and so on. Some jobs carry an increased risk for burn injury: roofers (hot tar), restaurant workers (scalds, hot oil), electricians (electrical injury), plumbers, welders, laboratory workers (chemical burns, gas explosions). Some people engage in lifestyle practices that place them at higher risk; many burn injuries occur in the context of some sort of intoxication (alcohol or drugs). Sometimes people get burned because of lapses in judgment (e.g., smoking while working on a car engine, pouring gasoline on a fire, responding to a "dare" of some sort). Poverty and homelessness also increase the likelihood of a burn injury due to substandard housing, cooking or heating with open flames, faulty electrical wiring, and so on.

How do you determine the caloric and protein requirement of a burn patient?

Burn patients have higher energy and protein requirements than other patients due to the hypermetabolic nature of burn injuries. There are a number of formulas that have been developed to help estimate the energy needs of burn patients. None of them is perfect—some tend to over or underestimate actual energy requirements. I generally compare a number of formulas to arrive at a range for my energy goal and provide 20% of total kcalories as protein. Of course, it is possible to more accurately determine energy requirements by using indirect calorimetry. Some facilities use this technique routinely to set energy goals for their burn patients, but it hasn't proven to be a practical option at the burn center where I work. Performing an indirect calorimetry is costly, involving additional labor costs for the technician to run the test and the physician to evaluate the test results, as well as the cost of the equipment itself.

Are there common nutritional problems associated with burn patients?

Finding a way for a person with a burn injury to be able to eat more than his or her usual intake can be challenging. People with burn injuries usually don't feel well. The medications that they take for pain may make them groggy or nauseated. If the face is burned, eating can be painful. If the hands are burned, it's hard to manage with a knife and fork. For patients with severe injuries, enteral support is frequently necessary in order to meet the elevated kcalorie and protein needs. The staff on the burn unit is very aware of the importance of nutrition for these patients, and we are usually in agreement when it comes to being proactive about providing nutrition and being creative about helping our patients to meet their nutrition goals.

What do you look for in evaluating nutrition support?

There are a number of ways to evaluate the results of your intervention with burn patients. We monitor lab values that reflect nutritional status, such as albumin and sometimes prealbumin. For patients with severe injuries, nitrogen balance studies are used to assess the adequacy of protein intake. Calorie counts can be helpful, too, in determining if a patient is meeting his or her nutrition goals. Sometimes you can obtain a lot of information from relatively "low tech" sources, like monitoring serial weights or wound healing. It's always helpful to keep in mind that the outcome we are striving for is a healed patient.

What is the biggest challenge of working with patients with burns?

You need to be creative and resourceful in order to help patients meet their high nutrition goals. You need to be able to integrate a patient's nutrition therapy in with all the other therapies he or she may be receiving that could result in interruptions in feeding, such as surgery, dressing changes, and physical and occupational therapy. Pediatric patients can be especially challenging, because you can't really reason with a 2-year-old about why she needs to eat more. But I also find the burn center to be one of the most rewarding areas to work in, and it's usually a favorite with the student interns, too. The burn unit strongly promotes a multidisciplinary approach to patient care; you truly feel like you're part of a team there. And, because most patients stay on the unit until they are healed, you really get to see the results of your efforts on their behalf. You can see the cause and effect relationship between medical nutrition therapy and wound healing, and that's very satisfying.

Has the treatment changed much in the last ten years? How do you stay abreast of changes in this field?

Many of the advances in burn treatment have to do with wound management issues: skin replacement products, cultured cells, and specialized dressings. Since I have been involved in burn care, the use of an anabolic steroid, oxandrolone, has become more common to help promote lean body synthesis. There are enteral formulas that may have some particular benefits for burn patients ("immune-enhanced" formulas with a modified fatty acid profile and, possibly, arginine or glutamine), but we haven't seen the kind of evidence that would make us want to include these products on our formulary. I was fortunate a couple of years ago to be able to attend the annual meeting of the American Burn Association. There were many dietitians there from burn facilities all over the country, and it was great to be able to meet with them and compare notes about how we were managing our patients. I look forward to attending more of those meetings in the future.

Any advice for dietetic students about counseling clients with burns?

I think that flexibility and empathy are probably among the most important traits to have when counseling burn patients. Sometimes the smallest gesture, like bringing a patient a special treat from the cafeteria or walking across the street for a burger from McDonald's (they're high in protein!) for a teenager who's tired of hospital food, can make a big difference in how well your patients tolerate their diet. I try to treat my patients the way I would want to be treated.

Application of the Nutrition Care Process: Metabolic Stress

Introduction:

Lisa is a 19-year-old female admitted through the ER after a high-speed MVA, a head-on collision with a truck. She was a restrained front seat passenger. She was admitted with a closed head injury with GCS = 10 E4 V2 M4; 15% TBSA partial thickness burns to arms and upper trunk. Physician's orders include: vital signs: Q 1 hr; pulse oximeter; intake and output Q 4 hr; circulation checks Q 4 hr. Begin lactated Ringer's at 75 mL/hr. Other orders per burn unit protocol.

Nutrition Assessment:

Food/Nutrition-Related History:

Previously enjoyed eating a varied diet with good appetite. Family interview indicates no previous nutrition problems.

1. Identify the contributing factors from Lisa's diagnosis that may contribute to metabolic stress.

2. Describe the metabolic changes that may contribute to changes in Lisa's nutritional status.

Anthropometric Measurements:

Ht. 5'5" Wt. 120# (reported by family)

3. Calculate Lisa's BMI and IBW.

Biochemical Data:

Total Protein 5.5 g/dL, Albumin 2.9 g/dL, Prealbumin 22 mg/dL, Hgb 10.5 g/dL, Hct 35%

4. Which, if any, of these labs could be used to assess Lisa's nutritional status? Explain.

Nutrition Diagnosis:

5. Identify at least two nutrition problems based on the nutrition assessment and medical history. Determine the diagnostic term for each nutrition problem. Next, identify the etiology of each nutrition problem. Finally, identify the signs and symptoms that support the evidence for these nutrition problems.

Nutrition Intervention:

6. Determine Lisa's energy and protein requirements.

7. Lisa has had a nasoduodenal tube placed. Outline the specific nutrition prescription you would recommend, including formula choice, volume, and rate progression.

Nutrition Monitoring and Evaluation:

8. Determine nutrition criteria for monitoring and evaluation for each nutrition diagnosis that you identified.

WEB LINKS

Surviving Sepsis Campaign Combined efforts of European Society of Intensive Care Medicine, the International Sepsis Forum, and the Society of Critical Care Medicine to coordinate care and research for individuals with sepsis. *http://www.survivingsepsis.org/Pages/default.aspx*

Merck Online Medical Manual General medical information about burns. *http://www.merck.com/mmhe/sec24/ch289/ch289a.html*

National Library of Medicine Patient education information about burns with a multimedia presentation. Available at:

http://www.nlm.nih.gov/medlineplus/tutorials/burns/htm/_no_50_no_0.htm

Society for Critical Care Medicine Overview of pathophysiology and treatment of sepsis. *http://www.survivingsepsis.org/*

END-OF-CHAPTER QUESTIONS

1. List five conditions that can cause stress and result in a hypermetabolic state.

2. Describe the differences between the body's responses to starvation and to stress.

3. Describe the ebb and flow response to stress.

4. Describe the hormonal response to stress. How do these hormones affect protein and carbohydrate metabolism?

5. How do cytokines influence the metabolic response to stress?

6. Why might omega-3 fatty acids, branched-chain amino acids, glutamine, and arginine be used in nutrition support for a patient with a hypermetabolic condition?

7. Describe how burns are classified. What are the nutrition implications of burns? How does the burn classification affect the kilocalorie and protein requirements of a burn patient?

8. After surgery, when might a patient be advanced to solid food? Describe a situation in which advancement to solid food would not be recommended.

9. What is sepsis? How is the immune system involved? What are the nutrition implications of sepsis?

10. What is SIRS; what is MODS/MSOF? How does SIRS differ from sepsis?

1. McClave SA, Martindale RG, Vanek VW, McCarthy M, Roberts P, Taylor B, Ochoa JB, Napolitano L, Cresci G. A.S.P.E.N. Board of Directors; American College of Critical Care Medicine; Society of Critical Care Medicine. Guidelines for the Provision and Assessment of Nutrition Support Therapy in the Adult Critically Ill Patient: Society of Critical Care Medicine (SCCM) and American Society for Parenteral and Enteral Nutrition (A.S.P.E.N.). J Parenter Enteral Nutr. 2009;33:277–316.

2. Surviving Sepsis Campaign. Available at http://www.survivingsepsis.org/Pages/default.aspx

3. Kalm LM, Semba RD. They starved so that others are better fed: remembering Ancel Keys and the Minnesota experiment. J Nutr. 2005;135(6):1347–52.

4. Gropper S, Smith J, Groff JL. Advanced Nutrition and Human Metabolism. 5th ed. Belmont (CA):Brooks/Cole/Cengage Learning; 2009.

5. Kudsk and Sacks KA, Sacks GS. Nutrition in the care of the patient with surgery, trauma and sepsis. In: Modern nutrition in health and disease. 10th ed. Philadelphia (PA): Lippincott Williams & Wilkins; 2004.

6. Wooley JA, Btaiche IF, Good KL. Metabolic and nutritional aspects of acute renal failure in critically ill patients requiring continuous renal replacement therapy. Nutr Clin Pract. 2005;20(2):176–91.

7. Centers for Disease Control. National Center for Health Statistics. Available at http://www.cdc.gov/nchs/datawh/statab/unpubd/mortabs/hist290a.htm

8. Dellinger RP, Levy MM, Carlet JM, Bion J, Parker MM, Jaeschke R, Reinhart K, Angus DC, Brun-Buisson C, Beale R, Calandra T, Dhainaut JF, Gerlach H, Harvey M, Marini JJ, Marshall J, Ranieri M, Ramsay G, Sevransky J, Thompson BT, Townsend S, Vender JS, Zimmerman JL, Vincent JL. International Surviving Sepsis Campaign Guidelines Committee; American Association of Critical-Care Nurses; American College of Chest Physicians; American College of Emergency Physicians; Canadian Critical Care Society; European Society of Clinical Microbiology and Infectious Diseases; European Society of Intensive Care Medicine; European Respiratory Society; International Sepsis Forum; Japanese Association for Acute Medicine; Japanese Society of Intensive Care Medicine; Society of Critical Care Medicine; Society of Hospital Medicine; Surgical Infection Society; World Federation of Societies of Intensive and Critical Care Medicine. Surviving Sepsis Campaign: international guidelines for management of severe sepsis and septic shock: 2008. Crit Care Med. 2008 Jan;36(1):296-327. Erratum in: Crit Care Med. 2008 Apr;36(4):1394–96.

9. Russell JA. Management of Sepsis. N Eng J Med. 2006; 355:1699–713.

10. Gabay C, Kushner I. Acute phase proteins and other systemic responses to inflammation. NEJM. 1999;340(6):448–54.

11. Gariballa S, Forster S. Effects of acute-phase response on nutritional status and clinical outcome of hospitalized patients. Nutrition. 2006;22:750–57.

12. Cuthbertson DP. The metabolic response to injury and its nutritional implications: retrospect and prospect. J Parenter Enteral Nutr. 1979;3:108–14.

13. Van den Berghe G, Wouters P, Weekers F, Verwaest M, Brucynincick F, Schetz M, et al. Intensive insulin therapy protects the endothelium of critically ill patients. J Clin Invest. 2005 Aug;115(8):2277–86.

14. Langouche L, Vanhorebeek I, Van den Berghe G. The role of insulin therapy in critically ill patients. Treat. Endocrinol. 2005;4(6):353–60.

15. NICE-SUGAR Study Investigators. Intensive versus conventional glucose control in critically ill patients. N Engl J Med. 2009 Mar 26; 360(13):1283–97. Epub 2009 Mar 24.

16. Wiener RS, Wiener DC, Larson RJ. Benefits and risks of tight glucose control in critically ill adults: a meta-analysis [published correction appears in JAMA. 2009;300(8):936]. JAMA. 2008;300(8):933–44.

17. Bellomo R, Egi M, What is a NICE-SUGAR for patients in the intensive care unit? Mayo Clin Proc. May 2009;84(5):400–402.

18. Charney P, Malone A. ADA Pocket Guide to Enteral Nutrition. Chicago: American Dietetic Association; 2006.

19. Walker RN, Heuberger RA. Predictive equations for energy needs for the critically ill. Respr Care. 2009;54:509–21.

20. Krenitsky J. Adjusted body weight: pro: evidence to support use of adjusted body weight in calculating calorie requirements. Nutr Clin Pract. 2005;20:468–73.

21. Ireton Jones CS, Turner WW. Adjusted body weight, con: why adjust body weight in energy-expenditure calculations. Nutr Clin Pract. 2005;20:474–79.

22. American Dietetic Association Evidence Library. Effects of enteral versus parenteral nutrition. Available at http://www.adaevidencelibrary.com/topic.cfm?cat=1032

23. Frankenfield D, Roth-Yousey L, Compher C. Comparison of predictive equations for resting metabolic rate in healthy nonobese and obese adults: a systematic review. J Am Diet Assoc. 2005;105:775–89.

24. Frankenfield D, Hise M, Malone A, Russell M, Gradwell E, Compher C. Evidence Analysis Working Group. Prediction of resting metabolic rate in critically ill adult patients: Results of a systematic review of the evidence. J Am Diet Assoc. 2007;107:1552–61.

25. Frankenfield DC, Coleman A, Alam S, Cooney RN. Analysis of estimation methods for resting metabolic rate in critically ill adults. J Parenter Enteral Nutr. 2009;33:27–36.

26. Boullata J, Williams J, Cottrell F, Hudson L, Compher C. Accurate determination of energy needs in hospitalized patients. J Am Diet Assoc. 2007;107:393–401.

27. Malone AM. Permissive underfeeding: its appropriateness in patients with obesity, patients on parenteral nutrition, and non-obese patients receiving enteral nutrition. Curr Gastroenterol Rep. 2007;9:317–22.

28. Choban PS, Dickerson RN. Morbid obesity and nutrition support: is bigger different? Nutr Clin Prac. 2005;20: 480–87.

29. Dickerson RN. Hypocaloric feeding of obese patients in the intensive care unit. Curr Opin Clin Nutr Metabol Care. 2005;8:189–96.

30. Kang W, Kudsk KA. Is there evidence that the gut contributes to mucosal immunity in humans? JPEN. 2007;31: 246–58.

31. Farber MS, Moses J, Korn M. Reducing costs and patient morbidity in the enterally fed intensive care unit patient. JPEN. 2005;20(1 Suppl): S62–S69.

32. Jeejeebhoy KN. Enteral feeding. Curr Opin Gastroenterol. 2005;21(2):187–91.

33. Binnekade JM. Review: enteral nutrition reduces infections, need for surgical intervention, and length of hospital stay more than parenteral nutrition in acute Evid Based Nurs. 2005;8(1):19.

34. Marik PE, Zaloga GP. Immunonutrition in critically ill patients: a systematic review and analysis of the literature. Intensive Care Med. 2008;34:1980–90.

35. O'Keefe GE, Shelton M, Cuschieri J, Moore EE, Lowry SF, Harbrecht BG, Maier RV, and the Inflammation and the Host Response to Injury Collaborative Research Program. Inflammation and the host response to injury, a large scale collaborative project: patient-oriented research core—standard operating procedures for clinical care VIII—nutritional support of the trauma patient. J Trauma. 2008;65:1520–28.

36. Grimble RF. Immunonutrition. Curr Opin Gastroenterol. 2005;21(2):216–22.

37. DeSouza DA, Greene LJ. Intestinal permeability and systemic infection. Crit Care Med. 2005;233(5):1125–35.

38. Laviano A, Muscaritoli M, Cascino A, Preziosa I, Inui A, Mantovani G, et al. Branched chain amino acids: the best compromise to achieve anabolism? Curr Opin Clin Nutr Metab Care. 2005;8:408-14.

39. Rushdi TA, Pichard C, Khater YH. Control of diarrhea by fiber enriched diet in ICU patients on enteral nutrition: a prospective randomized controlled trial. Clin Nutr. 2004;23:1344–52.

40. Bengmark S. Bio-ecological control of acute pancreatitis: the role of enteral nutrition, pro and synbiotics. Curr Opin Clin Nutr Metab Care. 2005;8(5):557–61.

41. Durthaler JM, Ernst FR, Johnston JA. Managing severe sepsis: a national survey of current practices. Am J Health Syst Pharm. 2009;66:45–53.

42. American Burn Association. Burn Incidence and Treatment in the US: 2007 Fact Sheet Available at http://www.ameriburn.org/resources_factsheet.php

43. Wolf SE. Burns. In: Merck Manual. 18th ed. 2009. Available at http://www.merck.com/mmpe/sec21/ch315/ch315a.html?qt=burn&alt=sh

44. Chan MM, Chan GM. Nutritional therapy for burns in children and adults. Nutrition. 2009;25:261–69.

45. Curreri PW. Supportive therapy in burn care. Nutritional replacement modalities. J Trauma. 1979 Nov;19(11 Suppl):906–08.

46. Mayes T, Gottschlich MM. Burns and wound healing. In: The science and practice of nutrition support. Dubuque (IA): Kendall Hunt; 2001, 391–420.

47. Thompson C, Fuhrman MP. Nutrients and wound healing: still searching for the magic bullet. Nutr Clin Pract. 2005;20(3):331–47.

48. Graves C, Saffle J, Morris S, Stauffer T, Edelman L. Caloric requirements in patients with necrotizing fasciitis. Burns. 2005;31(1):55–59.

49. Demling RH, DeSanti L. The rate of restoration of body weight after burn injury, using the anabolic agent oxandrolone, is not age dependent. Burns. 2001;27(1):46–51.

50. Prelack K, Dylewski M, Sheridan RL. Practical Guidelines for nutritional management of burns. Burns. 2007;33:14–24.

51. Cullen KA, Hall M, Golosinskiy, A. Ambulatory Surgery in United States. National Health Statistics Report. 2009;11:1–28.

52. Gibbs J, Cull W, Henderson W, Daley J, Hur K, Khuri SF. Preoperative serum albumin level as a predictor of operative mortality and morbidity: Results from the National VA Surgical Risk Study. Arch Surg. 1999;134:36–42.

53. Schulman AS, Sawyer RG. Have you passed gas yet? Time for a new approach to feeding patients postoperatively. Practical Gastroenterology. 2005:10:82–88.

54. Lassen K, Kjaeve J, Fetveit T, Tranø G, Sigurdsson HK, Horn A, Revhaug A. Allowing normal food at will after major upper gastrointestinal surgery does not increase morbidity: a randomized multicenter trial. Ann Surg. 2008 May;247(5):721–29.

55. Charoenkwan K, Phillipson G, Vutyavanich T. Early versus delayed (traditional) oral fluids and food for reducing complications after major abdominal gynaecologic surgery. Cochrane Database Syst Rev. 2007 Oct 17;(4):CD004508.

56. Lewis SJ, Andersen HK, Thomas S. Early enteral nutrition within 24 h of intestinal surgery versus later commencement of feeding: a systematic review and meta-analysis. J Gastrointest Surg. 2009;13:569–75.

57. Steed HL, Capstick V, Flood C, Schepansky A, Schulz J, Mayes DC. A randomized controlled trial of early versus "traditional" postoperative oral intake after major abdominal gynecologic surgery. Am J Obstet Gynecol, 2002; May;186(5):861–65.

23

Neoplastic Disease

Deborah A. Cohen, DCN, RD
Assistant Professor, University of New Mexico

Introduction

When people speak about cancer, they may fail to understand that this term encompasses over one hundred different disease types. Each disease type has its own unique characteristics, and all types share some common characteristics. Therefore, this chapter will focus on the universal features of the disease process, consistent modes of therapy, and most importantly, the relationship to nutrition. Nutrition can be discussed as a factor in prevention, and compromised nutrition can be considered a complication of the disease or the treatment process. Some cancers, particularly lung and head and neck cancers, affect an individual's nutritional status even before the cancer is diagnosed. In addition, treatment of cancer, including surgery, chemotherapy, and radiation, can have a significant impact on an individual's nutritional status. Weight loss, **anorexia**, alterations in metabolism, and lean body mass wasting that often occur in cancer patients can have a profound impact on morbidity, mortality, the ability to withstand or tolerate treatment, and quality of life.

Definition

Cancer is a disorder of cell growth and regulation. These abnormal cells know no limits for cellular replication and produce cells that serve no purpose.

Epidemiology

Cancer is a major cause of mortality in the United States, second only to cardiovascular disease (see Figure 23.1). In 2008, 1.44 million new cases of cancer (745,180 men, 692,000 women) were diagnosed and 565,650 persons (294,120 men, 271,530 women) died from cancer. Cancer incidence and mortality trends indicate stabilization in rates for all cancer sites for men between 1995 and 2004 and for women between 1999 and 2004.[1] Most cases of cancer occur in older individuals. Statistics indicate that two-thirds of all cases were in those

absolute neutrophil count (ANC)—a measure of the number of neutrophil granulocytes (also known as polymorphonuclear cells, PMNs, polys, granulocytes, segmented neutrophils, or segs) present in the blood; neutrophils are a type of white blood cell that fights against infection

adjuvant—usually, treatment "in addition to" initial treatment. For example, one or more anticancer drugs may be used in combination with surgical therapy as part of the cancer treatment regimen

adjuvant chemotherapy—the use of drugs as additional treatment for patients with cancers that are thought to have spread outside their original sites

anorexia—lack of appetite

antiemetic—a pharmacologic agent that reduces nausea

Barrett's esophagus—pre-malignant condition that is considered a risk factor for esophageal adenocarcinoma; a complication of severe chronic GERD involving changes in the cells of the tissue that line the bottom of the esophagus

biomarker—a biological molecule used as a marker to measure or indicate the effects or progress of a disease or condition

brachytherapy—a type of radiation therapy in which radioactive materials are placed in direct contact with the tissue being treated

CA-125—a protein that is secreted into the blood by ovarian cells and is used to monitor progress in the treatment of ovarian cancer

cachexia—weight loss, wasting of muscle, loss of appetite, and general debility that can occur during a chronic disease

cancer—a class of diseases characterized by uncontrolled cell division and the ability of these cells to invade other tissues, either by direct growth into adjacent tissue (invasion) or by migration of cells to distant sites (metastasis)

carcinoembryonic antigen (CEA)—a glycoprotein present in fetal gastrointestinal tissue and in the cells or serum of adults having certain types of cancers; it is used clinically to monitor the effectiveness of a treatment, such as for colorectal cancer

carcinogen—substance that causes cancer

carcinogenesis—the multistep process (initiation, promotion, and progression) through which normal cells are transformed into cancer cells

combination chemotherapy—the use of two or more antineoplastic agents to achieve maximum destruction of malignant cells

combination therapy—the use of two or more therapeutic agents/processes for the treatment of a neoplasm

cytologic—refers to tests that help to determine the morphologic features of a cell

dumping syndrome—condition in which food moves too quickly from the stomach to the small intestine

dysgeusia—altered taste

dysphagia—difficulty swallowing

dysphonia—difficulty speaking

emetogenic—an agent that causes nausea and/or vomiting

fractionation—a method of radiation therapy that delivers the total prescribed dose of radiation into several smaller-dose treatments in order to reduce side effects

initiation—the first phase in cancer cell development; the exposure of cells to an appropriate dose of a carcinogen (initiator)

metastasis—spread of cancer from the primary site to nearby or distant areas through the blood or lymph

mucositis—inflammation of a mucous membrane (e.g., mouth sores)

nadir—the lowest point, usually in reference to the white blood cell count

neoadjuvant chemotherapy—refers to chemotherapy used prior to primary treatment, which is typically surgery

neoplasm—literally means "new growth"; an abnormal mass of tissue, the growth of which exceeds and is uncoordinated with that of normal tissue

neutropenia—low white blood cell count

odynophagia—painful swallowing

palliative—refers to a noncurative treatment which reduces symptoms such as pain

primary cancer—the location or organ/cells from which the cancer originated

prognosis—a prediction of the probable course and outcome of a disease

progression—the third phase in cancer cell development; the orderly transformation of a preneoplastic lesion to a tumor and, ultimately, invasive cancer

prokinetic—a pharmacologic agent that promotes gastric emptying

promotion—the second phase in cancer cell development; process induced in a normal cell that has been exposed to a carcinogen, resulting in transformation into a cancer cell (promoters are not necessarily carcinogenic)

salvage—additional treatment, used in hope of a cure or to prolong life, in a patient with recurrence of a malignancy following initial treatment

steatorrhea—fat malabsorption resulting in severe diarrhea

telomere—the end section of a human chromosome

terminal—refers to a condition or disease for which there is no cure

Tumor Node Metastases (TNM) Staging System—a systematic way of describing the size, location, and spread of a tumor; T describes the primary tumor according to its size, N applies to the lymph nodes and whether cancer cells have spread to them, and M refers to metastasis and whether the cancer has spread to distant sites

xerostomia—dry mouth, often the result of damage to the salivary glands

---

over age 65.[2] As of 2008, cancer of the lung and bronchus remains the number one killer in both men and women. Prostate and colorectal cancers are second and third leading causes of cancer cases and mortality in men, while breast and colorectal cancers are the second and third leading causes in women (see Figures 23.2 and 23.3).

Cancer rates do vary by ethnicity. For all cancer sites combined, African-American men have a 19% higher cancer incidence rate and a 37% higher cancer mortality rate than white men. African-American women have a 6% lower incidence rate than white females for all cancer sites combined, and yet they have a 17% higher mortality rate.[1] Possible explanations include differences in smoking prevalence, risk factors, access to high-quality regular screening (breast, cervical, and colorectal cancers), and timely, high-quality treatments (many cancers).[1]

Etiology: Role of Genetics and Nutrition

Carcinogenesis is a multistep process in which normal cells are transformed into cancer cells. Many factors play a role in carcinogenesis, including exposure to **carcinogens** such as chemicals, physical agents, radiation, and infectious microorganisms. Genetics and nutritional factors also play a role. Though only a small percentage of cancers are actually considered hereditary, all cancers involve genetics to a certain degree. Genetic research has explained the mechanisms of cancer development. Damage to a gene may occur as a result of exposure to chemicals, physical agents (ionizing radiation, ultraviolet radiation, asbestos), viral agents (Epstein-Barr virus, human papilloma virus), and bacterial agents (*Helicobacter pylori*). Genes may also

Figure 23.1 Change in U.S. Death Rates* from 1991 to 2006

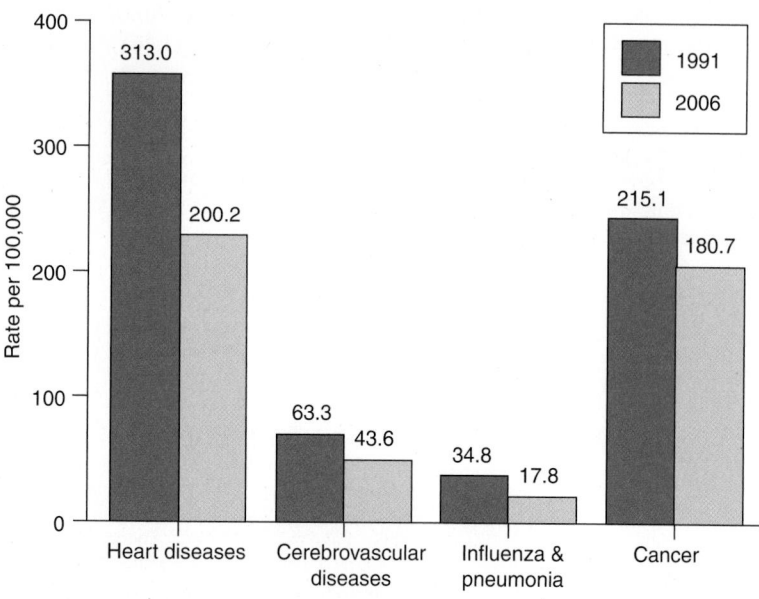

*Age-adjusted to 2000 US standatd population.

Sources: US Mortality Data, National Center for Health Statistics, Centers for Disease Control and Prevention, 2009.

Source: Reprinted by the permission of the American Cancer Society, Inc. from www.cancer.org. All rights reserved.

Figure 23.2 2009 Estimated U.S. Cancer Deaths

2009 Estimated US Cancer Deaths		Men 292,540	Women 269,800		
Lung and bronchus	30%			26%	Lung and bronchus
Prostate	9%			15%	Breast
Colon and rectum	9%			9%	Colon and rectum
Pancreas	6%			6%	Pancreas
Leukemia	4%			5%	Ovary
Liver and intrahepatic bile duct	4%			4%	Non-Hodgkin's lymphoma
Esophagus	4%			3%	Leukemia
Urinary bladder	3%			3%	Uterine corpus
Non-Hodgkin's lymphoma	3%			2%	Liver and intrahepatic bile duct
Kidney and renal pelvis	3%			2%	Brain/ONS
All other sites	25%			25%	All other sites

ONS = other nervous system.

Source: Reprinted by the permission of the American Cancer Society, Inc. from www.cancer.org. All rights reserved.

alter the way we apply nutrition therapy and cancer prevention strategies (see Chapter 10).

Evidence suggests that one-third of the more than 500,000 cancer deaths that occur in the United States each year can be attributed to diet and physical activity habits, with another third due to cigarette smoking.[3] Epidemiologic studies and laboratory experiments have provided substantial evidence for a relationship between environmental factors and carcinogenesis relating to cigarettes and lung cancer, and ultraviolet radiation and skin cancer. Establishing a strong link between diet and cancer has proven to be much more difficult due to the limitations of case-control studies, and timing and variations in nutrient intake.[4]

Biomarkers are distinctive biological or biologically derived indicators (such as a biochemical metabolite in the body) that may be used to identify nutrient exposure and help improve the precision of epidemiologic studies. In studies of dietary interventions, nutritional biomarkers can be used as a measure of internal dose, which is an indication of the amount of nutrient available to the tissues after absorption and metabolism. The marker can also be used as a measure of dietary change or compliance with a new dietary regimen.[5] Biomarkers for many nutrients are not always reliable, sensitive, or specific. Many factors can affect the use of nutritional biomarkers to validate nutrient intake, including physiology, absorption, nutrient-nutrient interactions in the body, cooking methods, and tissue and renal saturation levels. For example, vitamin C, which is often studied for its antioxidant properties in cancer research, is stored largely in white blood cells, and thus the amount of vitamin C in these cells does not correlate with vitamin C intake. Nonetheless, biomarkers have a significant potential for helping to establish the cause-and-effect relationship between diet and cancer in the future.

Information obtained from animal models is useful and does provide important evidence to substantiate certain relationships that are alluded to in epidemiologic studies. However, data from animals often cannot be directly applied to humans. Genetic technology will soon be useful in applying animal/laboratory research to human biology. The role of genetic research and nutrigenomics in determining the mechanisms and the role of nutrients in cancer formation will have a significant impact on detection, screening, and prevention of cancer.

Epidemiologic studies compare patterns of intake of nutrients between certain population groups having high and low incidence of a particular type of cancer. Most individuals consume thousands of compounds each day. Relating a nutrient

be affected by nutritional components such as antioxidants, soy protein, fat, calories, alcohol, and phytochemicals. Nutritional genomics (nutrigenomics), the study of genetic variations that cause different phenotypic responses to diet among humans, is a relatively new field that is already opening new and exciting opportunities for research in the field of nutrition and may

Figure 23.3 2009 Estimated U.S. New Cancer Cases*

2009 Estimated US Cancer Cases*				
		Men 766,130	**Women** 713,220	
Prostate	25%		27%	Breast
Lung and bronchus	15%		14%	Lung and bronchus
Colon and rectum	10%		10%	Colon and rectum
Urinary bladder	7%		6%	Uterine corpus
Melanoma of skin	5%		4%	Non-Hodgkin's lymphoma
Non-Hodgkin's lymphoma	5%		4%	Melanoma of skin
Kidney and renal pelvis	5%		4%	Thyroid
Leukemia	3%		3%	Kidney and renal pelvis
Oral cavity	3%		3%	Ovary
Pancreas	3%		3%	Pancreas
All other sites	19%		22%	All other sites

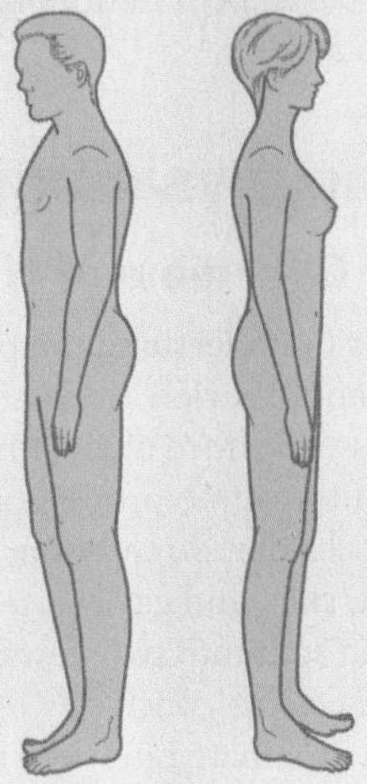

*Excludes basal and squamous cell skin cancers and in situ carcinomas except urinary bladder.

Source: Reprinted by the permission of the American Cancer Society, Inc. from www.cancer.org. All rights reserved.

to cancer has been exceedingly difficult due to compounding factors that include environmental exposures (cigarette smoking, for example), ethnicity, body mass, age, sex, socioeconomic status, genetics, physical activity levels, and numerous lifestyle and occupational hazards. The critical limiting feature of most human studies is the imprecision of quantifying nutrient intake. A number of tools are utilized, including food records/diaries, diet histories, food frequencies, and 24-hour recalls. All methodologies have inherent strengths and weaknesses. The method used must be appropriate for the population and nutrient(s) being studied (see Chapter 3).

Despite the limitations of the various research designs, epidemiologic studies currently suggest that heavy consumption of red meats, including meat that has been processed, increases cancer risk, especially for colorectal, bladder, prostate, breast, gastric, oral, and pancreatic cancers.[6] In addition, epidemiological research has demonstrated a cancer-preventative effect for the consumption of fruits and vegetables on the development of cancers of the mouth, pharynx, esophagus, stomach, colon, rectum, larynx, lung, ovary (vegetables only), bladder (fruits only), and kidney.[7] Research also supports inverse relationships between cancer risk and intakes of whole grains, fiber, vitamin D, saturated fat, and *trans* fat, and physical activity. Research supports direct relationships between cancer risk and intakes of total fat/certain types of fat (e.g., saturated fat) and alcohol; obesity (as measured by a high body mass index [BMI]); and certain food preparation methods such as smoking, salting, and pickling foods, and high-temperature cooking of meats.[8]

In a study of more than 900,000 U.S. adults, death rates from all cancers combined for individuals with a BMI of >40 were 52% higher for men and 62% higher for women than the death rates in men and women of normal weight. In both men and women, BMI of at least 40 was also significantly associated with higher rates of death due to cancer of the esophagus, colon and rectum, liver, gallbladder, pancreas, and kidney. On the basis of associations observed in that study, the authors estimated that current patterns of overweight and obesity in the United States could account for 14% of all deaths from cancer in men and 20% of those in women.[9] A few recent epidemiologic reports have associated metabolic syndrome with an increased incidence of certain types of cancer, specifically colorectal cancer, prostate cancer, and breast cancer.[10]

Cancer Prevention and Screening

Prevention of cancer can be addressed on two levels: primary and secondary. In primary prevention, specific factors are identified as part of the cancer process, and these factors are acted upon to decrease their potential activity as a carcinogen. For example, smoking cessation would be a primary prevention of cancer. Primary prevention of cancer refers to personal and communitywide efforts, whereas secondary prevention of cancer consists of measures for early detection and intervention. Screening for cancer risk is considered to be a secondary level of prevention. The American Cancer Society (ACS) publishes guidelines on nutrition and cancer prevention every five years. These guidelines, developed by a national panel of experts in cancer research, prevention, epidemiology, public health, and policy, represent the most current scientific evidence relating to dietary/physical activity patterns and cancer risk.[3] The ACS guidelines (see Table 23.1) are consistent with both guidelines from the American Heart Association for the prevention of coronary heart disease and recommendations for general health promotion, as defined by the Department of Health and Human Services's 2005 *Dietary Guidelines for Americans*.[11]

The 2009 ACS Cancer Screening Guidelines report summarizes the rationale for the guidelines and presents recent data and issues pertaining to early cancer detection. The ACS recommends screening for several forms of cancer:

- *Breast cancer:* Clinical breast examination for women in their 20s and 30s every three years, and for women aged 40+, annually; annual mammography for women aged 40 and over
- *Cervical cancer:* Papanicolaou test and pelvic examination at least every three years in women aged 21+; after age 30, if a woman has had three normal tests, Pap tests recommended every two to three years
- *Colorectal cancer:* Annual fecal occult blood test in adults aged 50+, colonoscopy every ten years starting at age 50
- *Prostate cancer:* Serum prostate-specific antigen and digital rectal examination for men aged 50+ yearly.

There are no general screening guidelines for lung, oral, endometrial, or ovarian cancers, but there are warning signs

Table 23.1 Cancer Prevention Guidelines

Recommendations for Individual Choices

Maintain a healthy weight throughout life.

- Balance calorie intake with physical activity.
- Avoid excessive weight gain throughout life.
- Achieve and maintain a healthy weight if currently overweight or obese.

Adopt a physically active lifestyle.

- **Adults:** Engage in at least 30 minutes of moderate to vigorous physical activity, above usual activities, on 5 or more days of the week; 45 to 60 minutes of intentional physical activity are preferable.
- **Children and adolescents:** Engage in at least 60 minutes per day of moderate to vigorous physical activity at least 5 days per week.

Eat a healthy diet, with an emphasis on plant sources.

- Choose foods and drinks in amounts that help achieve and maintain a healthy weight.
- Eat 5 or more servings of a variety of vegetables and fruits every day.
- Choose whole grains over processed (refined) grains.
- Limit intake of processed and red meats.

If you drink alcoholic beverages, limit your intake.

- Drink no more than 1 drink per day for women or 2 per day for men.

ACS Recommendation for Community Action

Public, private, and community organizations should work to create social and physical environments that help people adopt and maintain healthful nutrition and physical activity behaviors.

- Increase access to healthful foods in schools, worksites, and communities.
- Provide safe, enjoyable spaces for physical activity in schools.
- Provide for safe, physically active transportation (such as biking and walking) and recreation in communities.

Source: Reprinted by the permission of the American Cancer Society, Inc. from www.cancer.org. All rights reserved.

and symptoms that may aid with detection of all cancers. See Table 23.2 for a summary of these warning signs.

Another level of cancer prevention includes chemoprevention. Research in this area uses specific agents to "reverse, suppress, or prevent carcinogenesis before the development of invasive malignancy."[12] For example, the use of Tamoxifen to block hormonally driven breast cancers has been the target of several large clinical trials.[13] Nutritional factors and vaccines are currently being studied as potential chemopreventive agents. These include carotenoids, resveratrol, quercetin

Table 23.2 Signs and Symptoms of Cancer

- Unexplained weight loss, fever, fatigue, pain
- Skin changes (darker looking skin, reddened, excessive hair growth)
- Change in bowel habits or bladder function
- Sores that do not heal
- White patches on inside of mouth or tongue
- Unusual bleeding or discharge
- Thickening or lump in the breast or other parts of the body
- Indigestion or trouble swallowing
- Recent change in a wart or mole or any new skin change
- Nagging cough or hoarseness

Source: Reprinted by the permission of the American Cancer Society, Inc. from www.cancer.org. All rights reserved.

(a flavenoid), silymarin (a flavenoid), catechins (found in green and black tea), curcurmin, diallyldisulfide (garlic), and thymoquinone (black cumin).[6] In addition, several substances derived from various spices (capsaicin, gingerol, anethole, diogenin, eugenol) appear to have promising chemopreventive effects based on a limited number of preliminary animal and in vitro studies.[6]

Pathophysiology

Brief Overview of Normal Cell Growth

In order to understand the pathophysiology of cancer cells, it is important to review the basic principles of normal cell growth. All cells reproduce during the embryonic phase, but only some cells continue to reproduce after the first few months following birth. Cells that do reproduce, such as those of the liver, bone marrow, skin, and gastrointestinal tract, copy their DNA exactly and then split into two new daughter cells, which allows these types of cells to constantly regenerate. Cells that reproduce do so at an innate rate—the rate at which they are genetically programmed to reproduce—and this rate may be decreased or increased depending on genetic factors. In general, the cells of the bone marrow and the gastrointestinal tract have the fastest rates of replication.

Cells are classified as cycling cells, nondividing cells, and resting cells. Cycling cells divide continuously; the epithelial cells that line the gastrointestinal tract are an example. Nondividing cells divide before they differentiate (specialize), and then they do not divide again. Resting cells remain dormant initially, but certain conditions can stimulate their replication and growth.

Genetic controls for cellular division and growth include two basic sets of genes called oncogenes and tumor-suppressor genes. Oncogenes stimulate growth, and suppressor genes, as their name implies, suppress cellular growth. Examples of suppressor genes include the RB gene, which codes for the master "brake" of the cell cycle, and the P53 gene, which codes for the protein monitoring of cell health and the reliability of the cellular DNA. It is thought that cellular growth is also controlled by a counting system based on **telomeres**. Telomeres are end pieces of chromosomes that become shorter after each cell division. When the telomere shortens to a specific length, the cell will stop dividing.

Normal cellular reproduction is controlled by a combination of factors: genetic controls, hormones, and growth substances secreted by distant cells; local growth factors; and chemical cues from neighboring cells. Examples of hormones and systemic growth factors include epidermal growth factor, fibroblast growth factor, erythropoietin, insulin-like growth factors, and platelet-derived growth factor. Local growth factors include interleukins and cytokines. Cells also receive messages from neighboring cells that provide information about cellular type and the physical space available for cellular growth.

Cancer Cell Growth

Unlike a normal cell, whose growth is closely regulated, a cancer cell reproduces at an uncontrolled rate. The cancer cell becomes autonomous from the normal growth signals and genetic

Figure 23.4 Cancer Progression

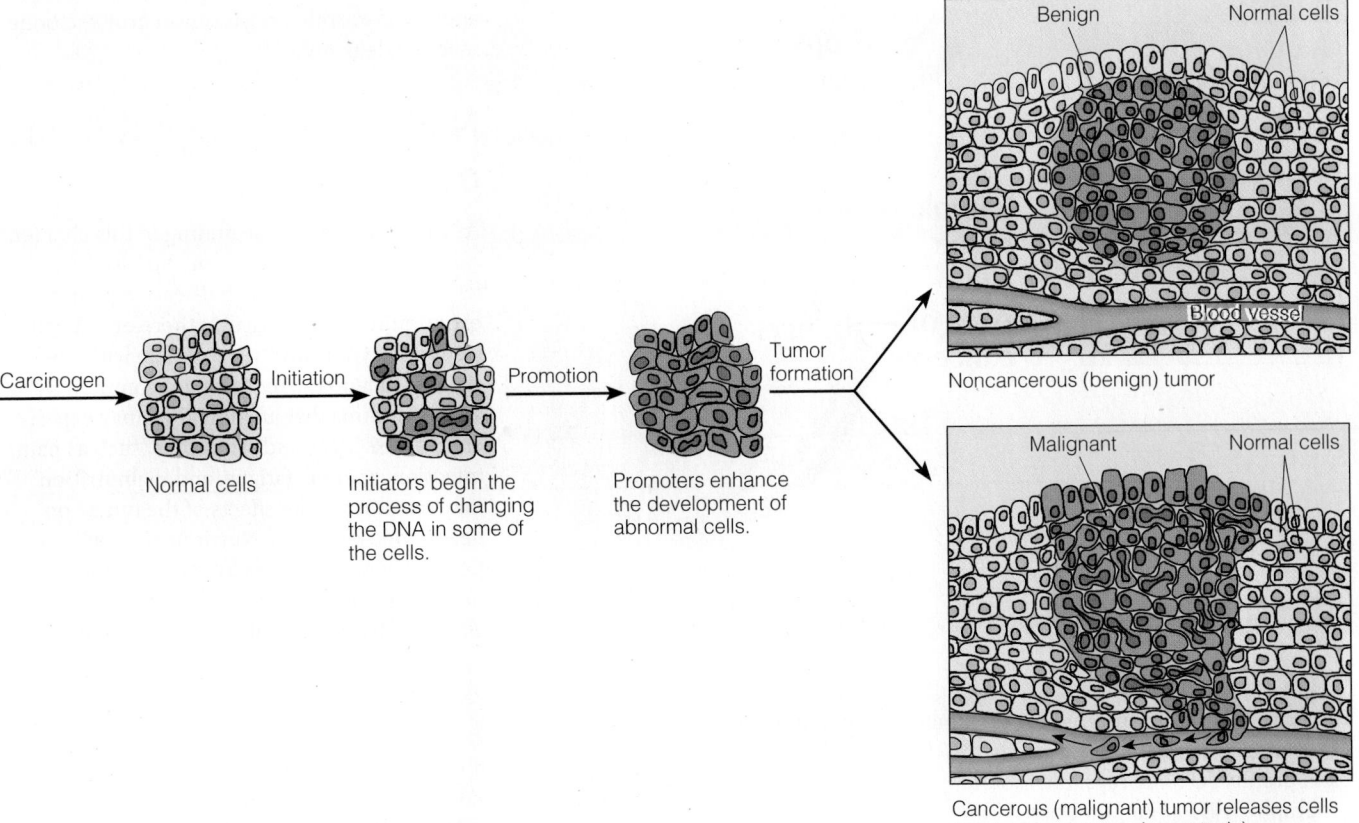

Carcinogen → Initiation → Promotion → Tumor formation

Normal cells

Initiators begin the process of changing the DNA in some of the cells.

Promoters enhance the development of abnormal cells.

Benign — Normal cells

Blood vessel

Noncancerous (benign) tumor

Malignant — Normal cells

Cancerous (malignant) tumor releases cells into the bloodstream (metastasis)

Source: S. Rolfes, K. Pinna, and E. Whitney, *Understanding Normal and Clinical Nutrition*, 7e, copyright © 2006 p. 880.

control, and may even secrete its own growth factor. In a cancer cell, an enzyme is secreted that destroys the telomere, leading to the loss of the cell's internal clock or counting mechanism—permitting uncontrolled cellular replication. The process of cell differentiation may change, and a specific cell type may take on other traits. The physical characteristics of the cancer cell are altered: the nucleus and cytoplasm may be enlarged or misshapen, the mitosis rate is usually higher, and there may be derangements in the chromosome sequence.

The change from a normal cell to a cancer cell theoretically involves several steps. These include **initiation, promotion**, and **progression**. Figure 23.4 outlines this process. It is difficult in some situations to distinguish between initiation and promotion, but in general, initiation occurs as a result of exposure to an initiating agent, such as tobacco.[14] An initiating agent predisposes the cell to genetic mutation. Factors that promote the cell's movement through the carcinogenic changes include some hormones such as estrogen or testosterone. These promoters require an activation of the carcinogen as well as a failure of natural immunity and cellular repair mechanisms. Conditions must be conducive for the **neoplasm** (tumor) to continue to grow. Tumor growth rate is dependent on characteristics of the host such as age, sex, overall health, nutritional status, and immune function.

Characteristics of the tumor also affect the cell's ability to grow and the growth rate. The original cell type (and its natural rate

of proliferation), as well as the availability of an adequate blood supply for the cancer cells, are crucial factors that determine how quickly the cancer cells will grow. Cancer cells may grow locally at the original (primary) site of cell transformation or spread to distant sites. This distant spread is called **metastasis**. Specific cancer types have typical routes for metastasis that include the lymphatic system, circulatory system, or nearby body cavities. For example, breast cancer typically metastasizes to brain and lung tissue through both the circulatory and lymphatic systems.

Medical Diagnosis

When cancer is diagnosed, a series of blood and physical tests, **cytologic** tests, imaging, and biochemical tests is performed. Biochemical analysis of blood, serum, urine, and other body fluids can detect tumor biomarkers and also help to determine if the cancer has metastasized. Tumor markers are also used to determine if a patient is responding to treatment. For example, the **carcinoembryonic antigen (CEA)** is often measured to monitor colon cancer while **CA-125** is useful in monitoring treatment for ovarian cancer.

Diagnostic procedures provide useful information regarding tumor size, localization of the tumor for biopsy or resection, and assessment of the anatomical extent of the disease. Tumor imaging techniques are valuable for visualization of the

Figure 23.5 Biopsy Procedure

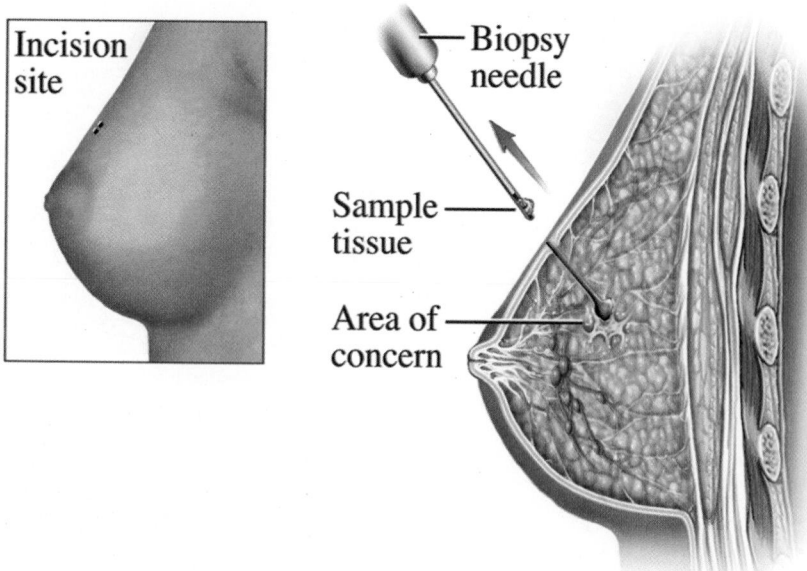

Incision site

Biopsy needle

Sample tissue

Area of concern

comas), lymphatic tissue (lymphomas), glial cells of the central nervous system (gliomas), and blood-forming organs, primarily the bone marrow (leukemias).[15]

Nutritional Manifestations of Cancer

As mentioned at the beginning of this chapter, cancer encompasses over one hundred different disease types. Each disease type has its own unique characteristics; therefore, specific signs and symptoms will correlate with the specific diagnosis. Yet there are common signs and symptoms that an individual may experience. These signs and symptoms, such as pain, infection, anemia, fatigue, and malnutrition, may result from the effects of the tumor on nearby body systems. Nutritional manifestations such as cancer **cachexia** are frequently present in patients with cancer. Nutritional effects of treatment will be discussed later in this chapter.

tumor in relation to internal organs. Imaging techniques include magnetic resonance imaging (MRI), computerized axial tomography (CT), X-rays, ultrasound, positron emission tomography (PET), mammograms, bone scans, and endoscopy. Invasive diagnostic techniques allow for direct visualization of the tumor and may include needle or excisional biopsy (see Figure 23.5), cytologic aspiration, and laparoscopy. All diagnostic procedures provide information that is clinically useful in determining the tissue type of the tumor, the primary site of the malignancy, the extent of disease in the body, and the tumor's potential to recur. This information comprises the critical first step in developing a treatment plan. Precisely which diagnostic technique is used depends on a number of factors, including the patient's presenting signs and symptoms, the clinical status of the patient, the anticipated goal of treatment when diagnosis is made, biologic characteristics of the suspected malignancy, diagnostic equipment available in the community, and insurance approval of diagnostic procedures.[15]

The tumor is then classified and assigned a "stage" on the basis of cell type, tissue of origin, whether it is benign or malignant, degree of differentiation, anatomic site, and function. The TNM Committee of the International Union Against Cancer (IUCC) and the American Joint Committee on Cancer (AJCC) have agreed on the **Tumor Node Metastases (TNM) Staging System** (T = depth of tumor invasion, surface spread and tumor size; N = absence or presence and extent of regional lymph nodes; M = absence or presence of distant metastases). (See Box 23.1.) Staging of tumors helps to assist with treatment planning, provides prognostic information, assists in treatment evaluation, and helps to identify individuals who may be eligible for clinical trials.[16]

Tumors are named according to the tissues from which they arise, and commonly use the suffix "-oma." Cancers include those of epithelial tissue (carcinomas), connective tissue (sar-

Cachexia

Cachexia is one of the most common causes of death among patients with cancer, and is present in 80% at death.[17] The word "cachexia" was used by a Roman physician to describe a state of "marked emaciation and severe malaise that frequently accompanies chronic, severe, and often lethal diseases."[18] As illustrated in Figure 23.6, cachexia is characterized by involuntary weight loss, tissue wasting (particularly lean body mass and adipose tissue), inability to perform daily activities, and metabolic alterations. These alterations in glucose, amino acid/ protein, and lipid metabolism can have an impact on the patient's nutritional and medical status with a subsequent impact on quality of life, morbidity, and mortality.

Though the pathophysiology of cancer cachexia is not completely understood, it seems to be attributable, at least in part, to metabolic alterations that lead to increased energy expenditure. These alterations are thought to be partially attributable to increased levels of circulating C-reactive protein, fibrinogen, white blood cells, and pro-inflammatory cytokines (e.g., IL-1, IL-6, TNF-α).[19] Chemical mediators involved in cachexia include cytokines, hormones, neurotransmitters, serotonin, interleukins, interferons, prostaglandins, tumor necrosis factor, neuropeptide Y, substance P, bradykinins, and glutamate.[20]

Figure 23.6 The Cachexia Journey

Normal	Mild cachexia	Moderate cachexia	Severe cachexia	Death

Weight loss	Below ideal body weight	Muscle wasting obvious

Not all patients progress along the entire pathway.

Tan BHL, Fearon KCH. Cachexia: prevalence and impact in medicine. *Curr Opin Clin Nutr Metab Care.* 2008;11:400–407.

Tumor Node Metastases (TNM) Staging System

The American Joint Committee on Cancer (AJCC) developed the *TNM classification system* as a tool for doctors to stage different types of cancer based on certain standards. It has replaced many of the older staging systems. In the TNM system, each cancer is assigned a T, N, and M category.

The **T** category describes the original (primary) tumor. The tumor size is usually measured in centimeters (2 and ½ centimeters is about 1 inch) or millimeters (10 millimeters = 1 centimeter).

- **TX** means the tumor can't be measured.
- **T0** means there is no evidence of primary tumor (it cannot be found).
- **Tis** means the cancer is in situ (the tumor has not started growing into the structures around it).
- The numbers **T1**, **T2**, **T3**, and **T4** describe the tumor size and/or level of invasion into nearby structures. The higher the T number, the larger the tumor and/or the more it has grown into nearby tissues.

The **N** category describes whether or not the cancer has spread into nearby lymph nodes.

- **NX** means the nearby lymph nodes cannot be evaluated.

- **N0** means nearby lymph nodes do not contain cancer.
- The numbers **N1**, **N2**, and **N3** describe the size, location, and/or the number of lymph nodes involved. The higher the N number, the more the lymph nodes are involved.

The **M** category tells whether there are distant metastases (spread of cancer to other parts of body).

- **MX** means metastasis can't be evaluated.
- **M0** means that no distant metastases were found.
- **M1** means that distant metastases were found (the cancer has spread to distant organs or tissues).

Each cancer type has its own version of this classification system, so letters and numbers don't always mean the same thing for every kind of cancer. For example, for some cancers, classifications may have subcategories, such as T3a and T3b, while others may not have an N3 category.

Stage Grouping

Once the T, N, and M have been learned, they are combined, and an overall "stage" of 0, I, II, III, or IV is assigned. (Sometimes these stages are subdivided as well, using letters such as IIIA and IIIB.)

For example, a T1, N0, M0 breast cancer would mean that the primary breast tumor is less than 2 cm across (T1), does not have lymph node involvement (N0), and has not spread to distant parts of the body (M0). This would make it a stage I cancer.

A T2, N1, M0 breast cancer would mean that the cancer is more than 2 cm but less than 5 cm across (T2), has reached only the lymph nodes in the underarm area (N1), and has not spread to distant parts of the body. This would make it a stage IIB cancer.

Stage 0 is *carcinoma in situ* for most cancers. This means the cancer is at a very early stage, is only in the area where it first developed, and has not spread. Not all cancers have a stage 0. Stage I cancers are the next least advanced and often have a good prognosis (outlook for survival). As the stage number goes up the cancers are more advanced (bigger and more widespread), but in many cases they can still be treated.

These "cachectic factors" are presumably tumor-specific since, for example, lung and gastrointestinal tumors including pancreatic cancer are well known for causing cachexia with much higher incidence than breast and hematopoietic tumors.[21] Changes in taste and smell perception, psychologic factors, uncontrolled pain, and therapy-induced side effects also play an important role in the severity of cachexia, but vary from one patient to another.

Cancer cachexia has been described as involving three phases: precachexia, moderate cachexia, and advanced cachexia.[22] Upon clinical examination of the patient, cachexia may be further classified into symptomatic and asymptomatic cachexia. An algorithm for the classification of cachexia has been proposed by the European Society for Clinical Nutrition and Metabolism (Figure 23.7). Standard methods of nutritional therapy, including enteral and parenteral nutrition support, may not be effective in improving the outcomes of cancer patients due to alterations in metabolism. In addition, cancer chemotherapy, radiation, and surgery can exacerbate an already abnormal metabolic milieu.

There are no standard criteria with which to diagnose cachexia. Diagnosis usually stems from the presenting signs and symptoms. These include weight loss, anorexia, muscle wasting, fatigue, and early satiety. Treatment of the anorexia-cachexia syndrome (discussed in the section Nutrition Intervention) may include nutrition therapy strategies.

Abnormalities in Carbohydrate, Protein, and Lipid Metabolism

The normal physiologic conservation mechanisms seen during periods of simple starvation do not occur in the presence of cancer. During periods of simple starvation, free fatty acids from adipose tissue supply energy to the liver and muscle. The free fatty acids are converted to ketone bodies that can then be utilized by most tissues in the body as a source of energy.

Figure 23.7 Classification of Cachexia and Precachexia

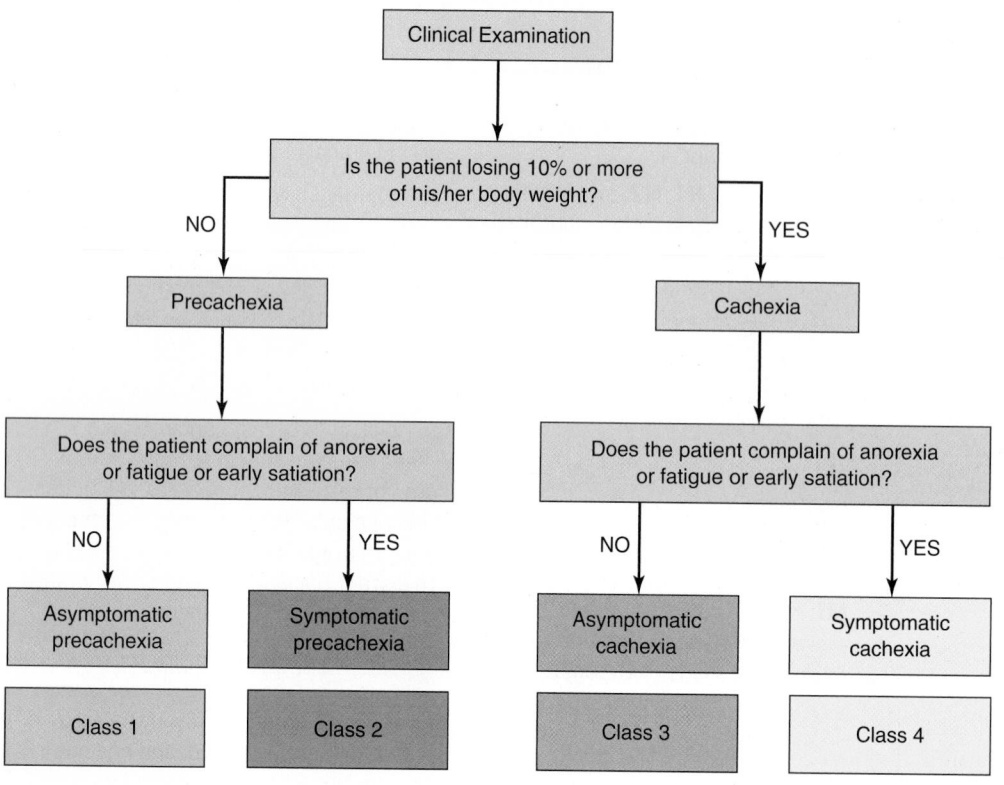

Source: Bozzetti F, Mariani L. Defining and classifying cancer cachexia: a proposal by the SCRINIO Working Group. *JPEN J Parenter Enteral Nutr* 2009;33:316–367. © Copyright 2009 SAGE Publications. Reprinted by Permission of SAGE Publications.

neous presence of both results in the most intense muscular atrophy. Simple anorexia alone cannot fully explain wasting of lean body mass and increased protein breakdown observed in cancer cachexia. In some animal models, cachexia develops in the complete absence of anorexia. It seems likely that cancer cachexia is provoked by chemical mediators originating from the host and/or the tumor, including insulin, insulin-like growth factor, growth hormone, glucagons, glucocorticosteroids, ketone bodies, arginine, β-adrenergic agonists, prostaglandins, interferon, interleukins, tumor necrosis factor, and proteolysis-inducing glycoprotein.[27]

Alterations in lipid metabolism also occur in the presence of malignancy. Fat is the body's primary fuel source, both under normal physiologic conditions and during simple starvation. Abnormalities that occur in the presence of cancer include increased lipid metabolism, decreased lipogenesis, and decreased activity of lipoprotein lipase (LPL), the enzyme responsible for triglyceride clearance from the plasma.[25] Mobilization of fatty acids from adipose tissue may occur before weight loss, suggesting the presence of a lipid-mobilizing factor (LMF) produced either by the tumor or host tissues.[26] Figure 23.8 summarizes the development of the nutritional manifestations of cancer.

Treatment: Focus on Nutritional Concerns

A patient may undergo one or more of the following treatments depending upon the cancer diagnosis: chemotherapy, radiation, and surgery. Because chemotherapy is given systemically—in other words, it is infused into the body's circulation—all cells of the body are exposed to the toxic effects of the drug, and almost all body systems will be affected. Chemotherapy affects rapidly dividing cells; hence, the rapidly dividing gastrointestinal cells will be significantly affected by most chemotherapeutic agents. This is one reason most patients undergoing chemotherapy experience nausea and other gastrointestinal (GI) problems during their course of treatment. Because hair follicles also contain rapidly dividing cells, alopecia (hair loss) is also a common side effect of chemotherapy. The effects of surgical treatment and radiation therapy depend on the particular diagnosis.

The type of treatment an individual may undergo depends on a number of factors, including location of the tumor, size of the tumor, and the health of the individual. In general, large bulky tumors are not curable with drugs. Drugs may not be able to

Ketone bodies inhibit glucose utilization and protein degradation from lean body mass. Therefore, protein is not used as a primary energy source. Serum insulin levels decline with increasing ketone body formation.

In malignancy, several biochemical changes occur. The most important carbohydrate abnormalities are insulin resistance, increased glucose synthesis, gluconeogenesis, increased Cori cycle activity, and decreased glucose tolerance and turnover.[23] Most solid tumors produce large amounts of lactate, which is converted back to glucose in the liver. This cyclic metabolic pathway, in which glucose is converted back to lactate by glycolysis and then reconverted to glucose in the liver, is known as the Cori cycle. Abnormal elevations in Cori cycle activity in the presence of malignancy, which have been noted in malnourished cancer patients, are reported to account for up to 300 kcal/d loss of energy.[24] Gluconeogenesis, or the production of glucose in the liver from non-glucose sources such as lactate, uses ATP molecules and is very inefficient for the host. This is known as futile cycling and may be responsible, at least in part, for the increased energy expenditure seen in many cancer patients.[25] A 40% increase in hepatic glucose production has been reported in weight-losing cancer patients, in contrast to the reduced production seen in patients with anorexia nervosa. Thus, changes in carbohydrate metabolism in cancer patients probably arise as a consequence of meeting the metabolic demands of the tumor, and may contribute to the development of the cachectic state.[26]

In cancer cachexia, amino acids are not spared as they are during simple starvation, and depletion of lean body mass occurs. Muscle wasting may be due to increased protein catabolism (hypercatabolism) or decreased protein synthesis; the simulta-

Figure 23.8 Factors Influencing Development of Anorexia and Cachexia

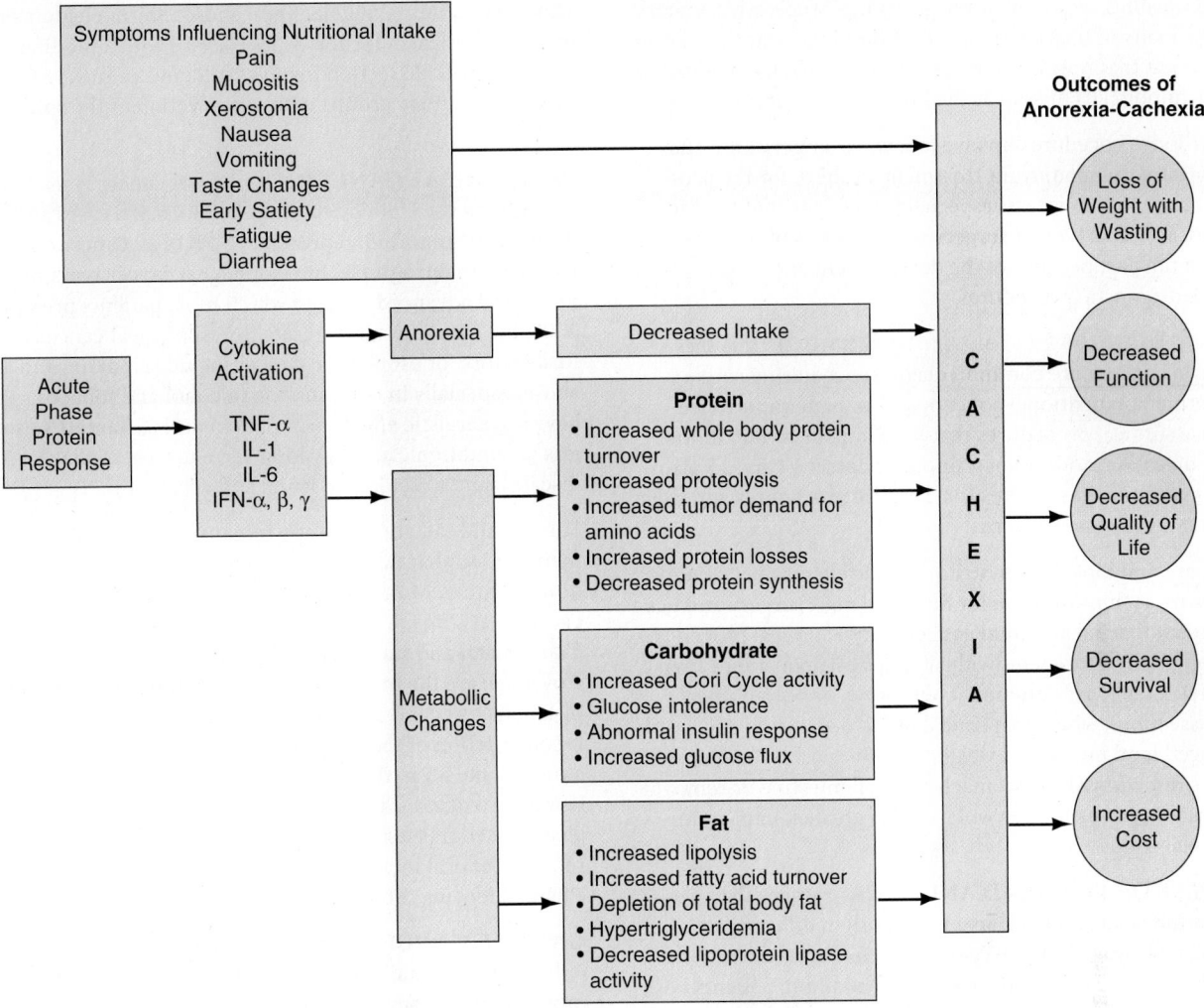

Source: C. Yarbro, M. Frogge, and M. Goodman, *Cancer Symptom Management*, 3e, copyright © 2004, Jones and Barlett.

penetrate into a solid tumor in amounts sufficient to kill the cells. Also, most cells in a bulky tumor may not be replicating at the time of treatment and thus survive to reestablish the tumor mass. The longer a tumor has been present, the greater the likelihood that it has already metastasized, resulting in ineffective drug therapy. Understanding the treatment pathways for cancer is crucial in planning nutrition therapy for the cancer patient. The extent and type of treatment plan will determine the potential adverse effects on the patient's nutritional status.

Surgery

Surgical treatment can be considered **primary**, **adjuvant**, **combination**, **salvage**, or **palliative**. Primary (definitive) treatment indicates that surgery will be the only therapy an individual will receive—for example, surgical removal of a small tumor. Adjuvant therapy includes chemoreductive therapy (debulking)—i.e., chemotherapy to reduce the size/ bulk of a tumor in addition to surgical removal of the mass. Combination therapy includes surgical resection followed by additional routes for treatment, such as radiation or chemotherapy. Salvage therapy involves use of extensive surgery to treat local recurrence after a less extensive primary approach has been implemented—for example, mastectomy after lumpectomy

and radiation.[28] Palliative surgery is used to ameliorate disease and/or treatment-related symptoms without attempting to cure the cancer. Surgical removal of a tumor that is causing a spinal cord compression is an example of palliative therapy. Surgical removal of a tumor may be undertaken if it has been established that the tumor can be excised without damaging too much of the surrounding area, if the tumor is blocking an important pathway (GI tract, esophagus, or trachea), or if the tumor is small enough to be successfully excised. Chemotherapy or radiation is often employed before surgical excision in order to shrink the tumor, making it easier to remove. Because surgical removal of any tumor can cause cancer cells to leak or metastasize, surgery may not be an option for all cancers. If a tumor has invaded a major organ, successful surgical resection may be impossible.

If a tumor is localized or has limited local-regional spread, the goal of surgery may be to cure the disease. This is particularly applicable to cancer of the bladder, breast, cervix, colon, endometrium, larynx, head and neck, kidney, lung, ovaries, and testes. Cure is achieved by the mechanical removal of all cancer cells. In circumstances in which tumor resection cannot be performed, multiple therapies with radiation, chemotherapy, or a combination of chemotherapy and radiation may reduce the size of the cancer, making it amenable to surgical resection for cure.[29] However, cancer cells are frequently found in operative wound washings or in the drainage from postoperative wounds. Because many of these patients never develop recurrent cancer, it is thought that host immune defenses are effective in destroying any tumor cells missed at the time of resection.[28]

Second-look procedures involve follow-up surgery after the original surgery or adjuvant treatment to check for the presence or absence of disease, especially in cancers that tend to recur locally. This type of surgery is not used as often today as it was in the past because of the development and use of less invasive diagnostic procedures.

Cancer Diagnoses Requiring Surgery This section reviews the more common diagnoses requiring surgery for treatment. Nutritional concerns are of high importance for those surgical procedures that have impact on nutritional intake, digestion, and/or absorption. Chapters 14 and 15 also address common gastrointestinal surgical procedures and their resulting nutritional concerns.

Significant and long-term nutritional side effects due to cancer surgery are seen most often for head and neck surgery and gastric surgery (partial and total gastrectomy). Surgical procedures may potentially alter an individual's physical appearance (mastectomy), alter organ function (colostomy), or both (radical neck dissection), which can contribute to depression and result in reduced food intake and weight loss. Surgical treatment that involves resection of the stomach or small intestine or removal of the esophagus or tongue will have the greatest impact on nutritional status.

CANCERS OF THE HEAD AND NECK Surgery that involves the head and neck area may result in difficulty in chewing and swallowing, **dysgeusia**, **xerostomia**, alterations in smell, and difficulty with speaking. In addition, patients with head and neck cancers typically have a history of alcohol, tobacco, and/or substance abuse and significant weight loss prior to diagnosis, and therefore often present with malnutrition. It is estimated that 60% of patients with head and neck cancers will present with malnutrition at the time of diagnosis.[30] Surgical procedures for head and neck cancers can significantly alter both appearance and function, depending on the surgical and reconstructive procedures employed. In addition to the primary surgery, the patient may also undergo lymph node dissection. The head and neck have approximately 300 lymph nodes, all of which have the potential to contain metastatic cancer cells that may need to be removed. A radical neck dissection (RND) involves removal of all lymph nodes on one side of the neck, internal and external veins, external carotid artery, the sternocleidomastoid muscle (one of the muscles that functions to flex the head), internal jugular (neck) vein, submandibular gland (one of the salivary glands), the hypoglossal nerve, portions of the vagus nerve, tail of the parotid gland, and the spinal accessory nerve (a nerve that helps control speech, swallowing, and certain movements of the head and neck). While there is a good probability that most, if not all, of the cancer has been removed, a RND can cause severe nutritional deficits. In anticipation of the long rehabilitation and healing that is required before oral intake can be resumed, placement of a gastrostomy or jejunostomy at the time of surgery is a standard procedure. The patient will also require speech therapy soon after surgery.

A modified RND involves removal of all lymph nodes on both sides of the neck with preservation of the spinal accessory nerve, internal jugular vein, and/or sternocleidomastoid muscle. Physical deformity is less severe with a modified RND. Selective neck dissection involves selective removal of one or more lymph node groups with preservation of the spinal accessory nerve.

ESOPHAGEAL CANCER Esophageal cancer is relatively uncommon in the United States. In 2008, 16,470 cases and 14,280 deaths were reported, representing 1% of all cancer cases.[1] However, the **prognosis** for esophageal cancer remains poor due to the advanced stage in which most patients present. In 2008, the five-year survival rate for esophageal cancer was 16%. Risk factors for esophageal cancer include smoking and alcohol abuse, especially in combination (alcohol and tobacco appear to have a synergistic effect on carcinogenesis), **Barrett's esophagus** (a condition caused by long-term gastroesophageal reflux disease), and a diet low in fruits and vegetables.

Surgery and radiation therapy remain the mainstays of treatment for esophageal cancer; nonetheless, the overall results are disappointing. Many patients with esophageal cancer develop cachexia at some point in the progression of their disease.[30] Trans-hiatal and trans-thoracic esophagectomy, while controversial, are the most common procedures for treatment of esophageal cancer. In most cases, the stomach is used for reconstruction of the esophagus. If the disease involves the stomach and a gastric resection is necessary, the esophagus can be reconstructed from part of the small or large intestine.[31] Surgical procedures for esophageal cancer can substantially delay recovery of oral intake, and often patients require placement of a jejunal feeding tube during surgery.

GASTRIC CANCER Despite a universal decrease in the incidence and mortality of gastric cancer, it remains the second most common cause of cancer-related death in the world.[32] Because most gastric cancers are diagnosed at an advanced stage, the overall five-year survival rate of gastric cancer in the United States is less than 25%.[1] One of the important predisposing factors for the development of gastric cancer is repeated infection with *H. pylori*. Surgical intervention is the only potentially curative therapy for the treatment of gastric cancer. Partial (subtotal) or total gastrectomy increases a patient's risk for vitamin B_{12} deficiency, due to removal of parietal cells and loss of intrinsic factor, which is necessary for the transport and absorption of B_{12}. Calcium and iron absorption will also be reduced due to a reduction in the secretion of hydrochloric acid, which is also produced by the parietal cells in the gastric mucosa; this may lead to deficiencies of these minerals. **Dumping syndrome**, delayed gastric emptying, early satiety, nausea, and vomiting may also occur as a result of a partial or total gastrectomy. (See Chapter 14.)

INTESTINAL CANCERS Cancers of the small intestine are relatively uncommon. Surgical excision of parts of the small bowel can have significant effects on the ability to digest and absorb nutrients. Malabsorption of nutrients (especially vitamin B_{12}) and **steatorrhea** are common following small bowel resection. The extent of the side effects depends in large part on the amount of small bowel that is resected, the ability of the remaining portion to adapt, and whether the ileocecal valve could be spared during the surgery.

BOX 23.2 **CLINICAL APPLICATIONS**

Lifestyle Risk Factors for Colorectal Cancer

Several lifestyle-related factors have been linked to colorectal cancer. In fact, the links between diet, weight, and exercise and colorectal cancer risk are some of the strongest for any type of cancer.

Certain Types of Diets

A diet that is high in red meats (beef, lamb, or liver) and processed meats (hot dogs and some luncheon meats) can increase colorectal cancer risk. Cooking meats at very high temperatures (frying, broiling, or grilling) creates chemicals that might increase cancer risk, although it's not clear how much this might contribute to an increase in colorectal cancer risk. Diets high in vegetables and fruits have been linked with a decreased risk of colorectal cancer. Whether other dietary components (fiber, certain types of fats, etc.) affect colorectal cancer risk is not clear.

Physical Inactivity

If you are not physically active, you have a greater chance of developing colorectal cancer. Increasing activity may help reduce your risk.

Obesity

If you are very overweight, your risk of developing and dying from colorectal cancer is increased. Although obesity raises the risk of colon cancer in both men and women, the link seems to be stronger in men.

Smoking

Long-term smokers are more likely than non-smokers to develop and die from colorectal cancer. Smoking is a well-known cause of lung cancer, but some of the cancer-causing substances are swallowed and can cause digestive system cancers, such as colorectal cancer.

Heavy Alcohol Use

Colorectal cancer has been linked to the heavy use of alcohol. At least some of this may be due to the fact that heavy alcohol users tend to have low levels of folic acid in the body. Still, alcohol use should be limited to no more than 2 drinks a day for men and 1 drink a day for women.

Type 2 Diabetes

People with type 2 diabetes have an increased risk of developing colorectal cancer. Both type 2 diabetes and colorectal cancer share some of the same risk factors (such as excess weight). But even after taking these into account, people with type 2 diabetes still have an increased risk. They also tend to have a less favorable prognosis after diagnosis.

Source: Reprinted by the permission of the American Cancer Society Inc. from www.cancer.org. All rights reserved.

Colorectal cancer remains the third most common cancer and the third leading cause of cancer deaths in men and women.[1] Cancers of the colon and rectum are treated primarily with surgery and chemotherapy. Risk factors for colorectal cancer include family history, diet (particularly red and processed meats; see Box 23.2), lack of physical activity, obesity, smoking, and history of inflammatory bowel disease. Surgeries performed for colorectal cancer are colon resection with reanastomosis, colostomy (temporary or permanent), and abdominal perianal resection.[33] A colostomy is created when a portion of the colon is resected, the distal part of the colon is brought through the abdominal wall, and an artificial opening is created through which waste material passes out of the body from the bowel (see Figure 23.9). When the descending colon is resected, the effects on nutrition are usually minimal because most water and electrolytes have already been reabsorbed. However, when the ascending or transcending portion of the colon is resected,

Figure 23.9 Colostomy and Ileostomy

Colostomy

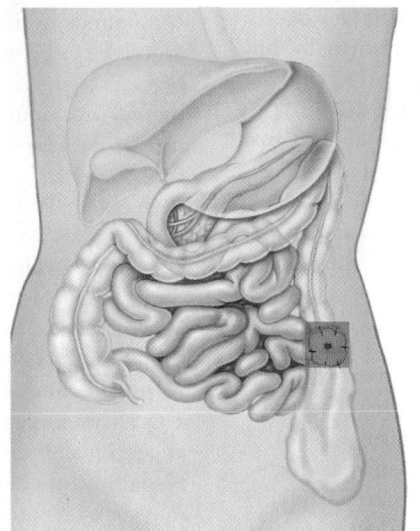

Ileostomy

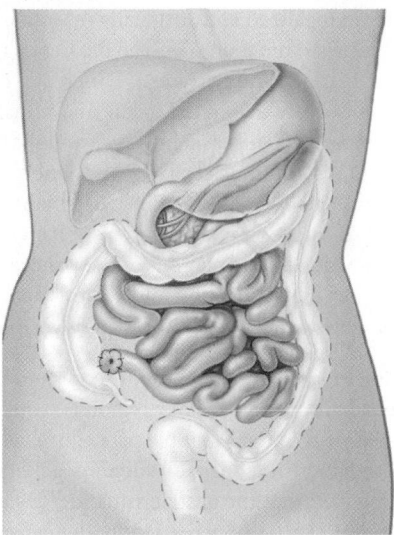

In a colostomy, the rectum and anus are removed, and the stoma is formed from the remaining colon.

In an ileostomy, the entire colon, rectum, and anus are removed, and the stoma is formed from the ileum.

Source: S. Rolfes, K. Pinna, and E. Whitney, *Understanding Normal and Clinical Nutrition,* 7e, copyright © 2006, p. 761.

Colon and Rectal Cancer Screening Guidelines

Beginning at age 50, both men and women at average risk for developing colorectal cancer should follow one of these five testing schedules:

Tests that find polyps and cancer:

- flexible sigmoidoscopy every 5 years*
- colonoscopy every 10 years
- double contrast barium enema every 5 years*
- CT colonography (virtual colonoscopy) every 5 years*

Tests that mainly find cancer:

- fecal occult blood test (FOBT) every year*,**

- fecal immunochemical test (FIT) every year*,**
- stool DNA test (sDNA), interval uncertain*

People should begin colorectal cancer screening earlier and/or undergo screening more often if they have any of the following colorectal cancer risk factors:

- a personal history of colorectal cancer or adenomatous polyps
- a personal history of chronic inflammatory bowel disease (Crohn's disease or ulcerative colitis)
- a strong family history of colorectal cancer or polyps (cancer or polyps

in a first-degree relative [parent, sibling, or child] younger than 60 or in 2 or more first-degree relatives of any age)

- a known family history of hereditary colorectal cancer syndromes such as familial adenomatous polyposis (FAP) or hereditary non-polyposis colon cancer (HNPCC)

*Colonoscopy should be done if test results are positive.

**For FOBT or FIT used as a screening test, the take-home multiple sample method should be used. A FOBT or FIT done during a digital rectal exam in the doctor's office is not adequate for screening.

Source: Reprinted by the permission of the American Cancer Society Inc. from www.cancer.org. All rights reserved.

---

there is a higher risk for electrolyte and fluid loss that result in electrolyte (especially potassium) abnormalities and dehydration. In general, most colostomies result in a stool that is more liquid in consistency. (See Chapter 15.)

Colon cancer, if detected and treated at an early stage, before it has metastasized, may be cured by surgical therapy. See Box 23.3 for colorectal cancer screening guidelines.

PANCREATIC CANCER Pancreatic cancer is the fourth leading cause of cancer-related death for both men and women, and is responsible for 6% of all cancer-related deaths.[1] Because of the frequent inability to diagnose pancreatic cancer when it is still localized and surgically resectable, and the lack of effective systemic therapies, the incidence rates are virtually the same as the mortality rates.[34] Pancreatic cancer is frequently diagnosed late due to the vague and nonspecific symptoms that accompany the disease; however, weight loss and anorexia are common symptoms at diagnosis. Ninety percent of patients present with weight loss, 75% with malnutrition, and 60% with anorexia at the time of diagnosis.[35] Pancreatic exocrine insufficiency due to obstruction of the pancreatic duct commonly results in malabsorption and steatorrhea. Most patients with pancreatic cancer also have hyperglycemia.

The standard surgical procedure for neoplasms of the pancreatic head and periampullary region is pancreaticoduodenectomy, also known as the Whipple procedure, which involves removal of the pancreatic head, duodenum, gallbladder, and bile duct with or without the gastric antrum.[34] Delayed gastric emptying is common after pancreaticoduodenectomy. The nutrition consequences of delayed gastric emptying are most significant in those patients with some degree of nutrition depletion preoperatively and in older patients with significant medical comorbid conditions.[34] There are numerous potential nutrition complica-

tions for this procedure and thus, the placement of a jejunostomy feeding tube is common practice. (See Chapter 16.)

Chemotherapy

Chemotherapy includes medications that interrupt different stages of cell cycle replication. Chemotherapeutic agents are most lethal to cells that are undergoing continual proliferation, which is logical since cells of many common tumor types are actively dividing.

Chemotherapy agents are classified into the following groups: alkylating, anti-metabolites (folate antagonists), purine/pyrimidine antagonists, anthracyclines, platinum antitumor compounds, antibiotics, nitrosureas, mitototic inhibitors, microtubule targeting agents, topoisomerase inhibitors, cytokines, biologic response modifiers, monoclonal antibodies, immunotherapy, hormones, and enzymes. Chemotherapeutic agents are rarely used as single agents. The use of combinations of drugs in cancer chemotherapy is a commonly employed practice and has been one of the major advances made in this field. Because different agents work to kill the cancer cell in different ways, using **combination chemotherapy** is advantageous. This approach decreases incidence of drug resistance, allows for an additive or synergistic effect of the drugs, and decreases the potential overall toxicity or at least the toxicity to any one organ system. **Adjuvant chemotherapy** refers to chemotherapy after surgery. This chemotherapy has the theoretical advantage of eliminating any residual or metastatic cells, thus improving patient survival. **Neoadjuvant chemotherapy** refers to chemotherapy that is administered before surgery; this chemotherapy is indicated when the tumor size is too large for an effective resection.

Because most of the normal cells in the body are in a resting stage, they are somewhat protected from the lethal effects of most chemotherapeutic agents. Normal, healthy cells in the human body that are susceptible to the effects of these lethal agents include those that are frequently dividing, such as the cells of the bone marrow (red blood cells, white blood cells, and platelets), the epithelial lining of the gastrointestinal tract, and the cells of the hair follicles. The most common side effects, therefore, are due to toxicity to these cells, and include **neutropenia**, thrombocytopenia, anemia, diarrhea, **mucositis** (discussed later in this chapter), and alopecia. Some chemotherapeutic agents are also known to cause cardiotoxicity, neurotoxicity, and nephrotoxicity. (See Table 23.3, which presents the side effects of chemotherapeutic agents.) Specific nutritional side effects of chemotherapy will also be discussed later in this chapter.

Radiation

Radiation therapy (RT) alone is the most common treatment for certain types of head and neck cancers, such as cancer of the nasopharynx, larynx, and oropharynx.[36, 37] Radiation may be used to cure the cancer, as in Hodgkin's disease, testicular seminomas, thyroid carcinomas, localized cancers of the head and

Table 23.3 Side Effects of Antineoplastic Agents

Drug Class and Examples	Mechanism of Action	Cell Cycle Specificity	Common Side Effects
Alkylating Agents			
busulfan, carboplatin, chlorambucil, cisplatin, cyclophosphamide, dacarbazine, hexamethyl melamine, ifosfamide, melphalan, nitrogen mustard, oxaliplatin, thiotepa	Alters DNA structure by misreading DNA code, initiating breaks in the DNA molecule, cross-linking DNA strands	Cell cycle-nonspecific	Bone marrow suppression, nausea, vomiting, cystitis (cyclophosphamide, ifosfamide), stomatitis, alopecia, gonadal suppression, renal toxicity (cisplatin)
Nitrosureas			
carmustine (BCNU), lomustine (CCNU), semustine (methyl CCNU), streptozocin	Similar to the alkylating agents; cross the blood-brain barrier	Cell cycle-nonspecific	Delayed and cumulative myelosuppression, especially thrombocytopenia; nausea, vomiting
Topoisomerase Inhibitors			
irinotecan, topotecan	Induce breaks in the DNA strand by binding to enzyme topoisomerase I, preventing cells from dividing	Cell cycle-specific	Bone marrow suppression, diarrhea, nausea, vomiting, hepatotoxicity
Antimetabolites			
5-azacytidine, capecitabine (Xeloda), cytarabine, edatrexate fludarabine, 5-fluorouracil (5-FU), FUDR, gemcitabine, hydroxyurea, leustatin, 6-mercaptopurine, methotrexate, pentostatin, 6-thioguanine	Interfere with the biosynthesis of metabolites or nucleic acids necessary for RNA and DNA synthesis	Cell cycle-specific (S phase)	Nausea, vomiting, diarrhea, bone marrow suppression, proctitis, stomatitis, renal toxicity(methotrexate), hepatotoxicity
Antitumor Antibiotics			
bleomycin, dactinomycin, daunorubicin, doxorubicin (Adriamycin), idarubicin, mitomycin, mitoxantrone, plicamycin	Interfere with DNA synthesis by binding DNA; prevent RNA synthesis	Cell cycle-nonspecific	Bone marrow suppression, nausea, vomiting, alopecia, anorexia, cardiac toxicity (daunorubicin, doxorubicin)
Mitotic Spindle Poisons			
Plant alkaloids: etoposide, irinateran, teniposide, vinblastine, vincristine (VCR), vindesine, vinorelbine.	Arrest metaphase by inhibiting mitotic tubular formation (spindle); inhibit DNA and protein synthesis.	Cell cycle-specific (M phase)	Bone marrow suppression (mild with VCR), neuropathies (VCR), stomatitis
Taxanes: paclitaxel, docetaxel	Arrest metaphase by inhibiting tubulin depolymerization	Cell cycle-specific (M phase)	Bradycardia, hypersensitivity reactions, bone marrow suppression, alopecia, neuropathies
Hormonal Agents			
androgens and antiandrogens, estrogens and antiestrogens, progestins and antiprogestins, aromatase inhibitors, luteinizing hormone- releasing hormone analogs, steroids	Bind to hormone receptor sites that alter cellular growth; block binding of estrogens to receptor sites (antiestrogens); inhibit RNA synthesis; suppress aromatase of P450 system, which decreases estrogen level	Cell cycle-nonspecific	Hypercalcemia, jaundice, increased appetite, masculinization, feminization, sodium, and fluid retention, nausea, vomiting, hot flashes, vaginal dryness
Miscellaneous Agents			
asparaginase, procarbazine	Unknown or too complex to categorize	Varies	Anorexia, nausea, vomiting, bone marrow suppression, hepatotoxicity, anaphylaxis, hypotension, altered glucose metabolism

Reprinted from: *Brunner and Suddarth's Textbook of Medical-Surgical Nursing.* 11th ed. Copyright © 2008 by Lippincott Williams & Wilkins.

neck, and cancers of the uterine cervix. RT may also be used to control malignant disease when a tumor cannot be removed surgically or when local nodal metastasis is present, or it can be used prophylactically to prevent leukemic infiltration of the brain or spinal cord.

RT is delivered with electromagnetic rays (gamma rays and X-rays) and charged particles (electrons). RT destroys cancer cells by altering cellular and nuclear material, especially DNA. The most harmful tissue disruption is due to the alteration of the DNA molecule within the cells of the tissue. Ionizing radiation breaks the strands of the DNA helix, leading to cell death. As with chemotherapy, those cells that are continually proliferating are the cells most sensitive to the effects of radiation. As discussed earlier, these include epithelial cells, bone marrow cells, lymph tissue, and hair cells. But, unlike chemotherapy's toxic effects, toxicity of RT is localized to the region being irradiated. Toxicity may be exacerbated when concomitant chemotherapy is administered.

The goal of treatment planning is to uniformly irradiate a specified target while minimizing the dose to surrounding normal tissues.[38, 39] Normal cells as well as the targeted cancer cells are susceptible to the toxicity of RT; therefore, custom-made lead blocks are developed for each patient to protect vital organs that may be in the RT field and which may be damaged during RT.

RT may be administered externally or internally, either alone or in combination. External beam RT is administered by linear accelerators on a daily basis (in an outpatient setting), usually five days per week, for a period of time (approximately five to six weeks) depending on the cancer being treated. **Brachytherapy** involves the placement of radioactive sources either within an existing body cavity (e.g., the vagina) in close proximity to the tumor (intracavity), or directly within a tumor (interstitial), potentially sparing surrounding tissues from damage due to external bean therapy.[38] Intracavity brachytherapy is used to treat cancers of the prostate and cervix, pituitary adenomas, and acoustic neuromas. Intravascular brachytherapy may be used to treat heptocellular carcinoma or liver metastases.[39]

A common use of RT is in combination with surgery and/or chemotherapy. When combined with surgery, RT may be given prior to (preoperative), following (postoperative), or during (intraoperative) surgery. Delayed wound healing is common when RT is used in combination with surgery, especially when RT is administered postoperatively. Chemotherapy may be administered prior to (neoadjuvant), during (concomitant), or following RT (maintenance).[40] RT may also be used as a palliative form of therapy; for instance, it can be used to control intracranial swelling due to metastatic disease in order to help alleviate pain. Whole-brain RT for metastatic disease is not used with the intention of curing the disease, but rather to control symptoms such as headaches and visual problems. Palliative therapy is often delivered at higher doses for shorter periods of time. A trend in recent years is the use of accelerated **fractionation**, in which the overall time of RT administration is reduced; however, the number of fractions, size, or total dose of the radiation is not reduced. Accelerated fractionation is an established and accepted mode of therapy for head and neck cancers.[39] Toxicities may be increased, particularly in those receiving concomitant chemotherapy. Common adverse effects

of RT to the head and neck area include fatigue, mucositis, dysgeusia, xerostomia secondary to salivary gland destruction, **dysphagia**, **odynophagia**, and severe esophagitis. Symptoms will manifest by approximately day ten of RT and continue until about two to three weeks after RT has ended. Mucositis, xerostomia, and odynophagia due to esophagitis may be severe and debilitating enough to warrant a temporary discontinuation of RT. The individual is at high risk of dehydration due to inadequate fluid intake, and may need intravenous fluids for hydration and electrolyte correction. Esophageal tissue may become extremely irritated and friable to the extent that oral intake is impossible. A surgically placed feeding tube may be indicated to provide nutritional support for these patients. Nasoenteric placed feeding tubes may be contraindicated due to the fragile nature of the esophagus and increased risk of bleeding during the placement. In addition, the esophagus may be totally or partially obstructed by the tumor mass, which prevents placement of a feeding tube in this manner.

RT to the abdominal and pelvic area can result in radiation enteritis, a severe, often debilitating disease that can take up to ten years after RT to manifest. Radiation enteritis is one of the most feared complications of abdominal and pelvic radiation. Bowel injuries that result in fistulas, strictures, and chronic malabsorption are potentially life-threatening complications, and have a significant impact on quality of life.[41] Symptoms include severe diarrhea and malabsorption. Medical therapies are primarily supportive. Total parenteral nutrition may be necessary to prevent weight loss and correct electrolyte abnormalities during periods when bowel rest is necessary.

Other Therapies
Hematopoietic Stem Cell Transplantation
Hematopoietic stem cell transplantation (HSCT) is a therapeutic option for those with hematalogic and some non-hematologic malignancies (see Chapter 19). Most transplants are now performed with peripheral blood stem cells as opposed to bone marrow. HSCT is a potentially curative treatment for hematologic malignancies, and is used primarily in the treatment of hematologic and lymphoid cancers (chronic and acute leukemia, lymphoma, Hodgkin's disease, multiple myeloma, myelodysplastic syndromes), ovarian cancer, and germ cell tumors. It may also be used in the treatment of other disorders including aplastic anemia, sickle cell disease, and severe combined immunodeficiency.[42]

Stem cells used in HSCT may be obtained from one of three sources: a donor (related or unrelated), a genetically identical twin, or the individual who will be undergoing HSCT. Donor transplantation is referred to as allogeneic HSCT. If the donor is a genetically identical twin, it is referred to as syngeneic HSCT. When patients donate their own stem cells, the procedure is referred to as autologous HSCT. Those patients undergoing an allogeneic HSCT are at highest risk for developing transplant-related morbidity and mortality due to complications secondary to graft-versus-host disease.

The stem cells are harvested (either surgically, if bone marrow cells are used, or by a procedure called apheresis if peripheral stem cells are used) from the donor or the patient. After a certain period of time, the patient undergoes high-dose chemotherapy and/or total body irradiation (TBI), which is referred to as the preparative or conditioning regimen. Stem cells are infused once the conditioning regimen is complete, usually after several days, and the day of stem cell infusion is referred to as day 0.

The conditioning regimen serves two purposes: (1) to provide sufficient immunosuppression to prevent rejection and allow engraftment of donor stem cells, and (2) to eradicate malignant cells. Major toxicities are associated with the conditioning regimen (high-dose chemotherapy and/or total body radiation). Patients may also experience toxicities that occur as a result of graft rejection, known as graft-versus-host disease (GVHD), and infectious complications associated with immunosuppression. GVHD only occurs in those individuals receiving an allogeneic HSCT. The most common toxicities that are associated with the high doses of chemotherapy (and TBI)—nausea, vomiting, mucositis, diarrhea, pancytopenia (including neutropenia), fevers, and infections—are typically seen immediately after the conditioning regimen is delivered (see Chapter 19). Severity of the toxicities depends on the intensity and type of agent used in the conditioning regimen. If oral intake is inadequate, parenteral nutrition support is initiated to meet nutritional needs. Tube feeding has not been successful in these patients due to the severity of nausea, vomiting, mucositis, delayed gastric emptying, and diarrhea. In addition, low platelet counts associated with the conditioning regimen significantly increase risk of bleeding during placement of a feeding tube.

Late complications are more commonly observed after an allogenic HSCT, and occur as a consequence of the effects of long-term damage to normal tissues, either from the conditioning regimen or from immunosuppressive agents and GVHD. Long-term complications include: growth retardation, infertility, endocrine failure, avascular joint necrosis, osteopenia, cataracts, renal insufficiency, restrictive pulmonary defects, neurocognitive defects, and secondary malignancies.[43] Delayed gastric emptying may persist for months following a bone marrow transplant.

GVHD, which occurs in a high percentage of allogeneic HSCT recipients, is associated with serious long-term medical problems, including immunosuppression, organ dysfunction, and infections. Acute GVHD is graft rejection that occurs in the first 100 days after HSCT infusion, while chronic GVHD is graft rejection that occurs after day 100. GVHD is thought to be mediated by donor T-lymphocytes that induce tissue damage in the skin, liver, and/or GI tract. Skin manifestations may include a mild to severe erythematous papular rash. Liver manifestations may include bile duct damage with cholestatic liver test abnormalities. The GI tract can be affected anywhere from the mouth to the anus. GI manifestations range from dry mouth and oral ulcerations to small intestinal GI mucosal damage leading to severe diarrhea (>1 liter per day) and malabsorption.

Anorexia, weight loss (especially loss of lean body mass), and nutritional deficiencies are common and often severe. Patients with moderate to severe GI GVHD usually require parenteral nutrition (PN). These patients have elevated energy needs. Protein needs increase as well, to support replacement GI losses (and dermal losses if skin GVHD is also present) and to meet amounts needed for protein synthesis (see Chapter 19). Protein needs are also elevated because these patients are given

high doses of glucocorticoids for immunosuppression. These drugs increase nitrogen losses and cause lean body mass/skeletal muscle loss. Enteral nutrition is usually contraindicated in patients with severe GI GVHD due to severe delayed gastric emptying and severely damaged GI mucosa, which causes malabsorption.

Primary treatment of either acute or chronic GVHD includes the administration of high doses of immunosuppressant medications (including cyclosporine, tacrolimus, methotrexate, and prednisone). Corticosteroids, such as prednisone, are administered at such high doses that the side effects of these medications, which have significant nutritional implications, can be seen within days after the drug is first given. Nutrition-related side effects include hyperglycemia (which is usually resistant to diet manipulation and requires insulin), nitrogen catabolism, hypertension, and sodium retention. Long-term complications of corticosteroids include diabetes, osteoporosis, Cushing's syndrome, weight gain, muscle wasting (noted especially in the extremities), and poor wound healing.

All patients undergoing autologous, syngeneic, or allogeneic HSCT should be considered at high nutritional risk and should undergo a complete nutritional assessment by a clinical dietitian prior to beginning the conditioning regimen. In addition, those undergoing allogeneic HSCT need to be monitored closely throughout their transplant course and in the outpatient setting after discharge for nutritional complications due to immunosuppression or infections, late complications due to the conditioning regimen, and acute and/or chronic GVHD.

Biological Response Modifiers Biological response modifiers (BRMs) are agents used to boost or restore the ability of the immune system to combat cancer. BRMs, which are also referred to as immunotherapy, biological therapy, or biotherapy, use the body's own immune system, either directly or indirectly, to eradicate cancer cells. Biological therapies use interferons, interleukins, monoclonal antibodies, growth factors, gene therapy, and nonspecific immunomodulating agents.

BRMs are used to stop, control, or suppress processes that permit cancer growth; make cancer cells more recognizable for destruction by the body's own immune system; enhance the killing power of the body's own immune system; enhance the body's ability to repair or replace normal cells damaged or destroyed by other forms of cancer treatment; and prevent cancer cells from spreading to other parts of the body.[44] While some BRMs are part of standard treatment, some are only available to patients enrolled in clinical trials.

Interferons and interleukins are cytokines that are produced by the human body and can be synthesized by drug companies. They are used to fight certain types of cancer, including metastatic renal cancer and metastatic melanoma. Colony-stimulating factors (CSFs) are given to increase bone marrow cell production, including red blood cells, white blood cells, and platelets. They are used to decrease the period of neutropenia or thrombocytopenia in those patients receiving chemotherapeutic agents that cause a reduction of these cells. The patient receiving CSFs is less prone to develop infections or anemia and less likely to bleed excessively. G-CSF (Filgrastim), GM-CSF (Sargramostim), and erythropoietin (Epogen) are commonly prescribed CSFs used in cancer therapy. Monoclonal antibodies react with certain types of cancers to make the body's immune response more effective in fighting the cancer. Sometimes monoclonal antibodies are "attached" to a chemotherapeutic agent to make destruction of the cancer cells more effective. Rituxan® (rituximab) for the treatment of non-Hodgkin's lymphoma, Erbitux (cetuximab) for the treatment of colorectal cancer, and Herceptin® (trastuzumab) for the treatment of breast cancer are monoclonal antibodies approved by the Food and Drug Administration (FDA). Side effects of these drugs include mild to moderate flulike symptoms.

Gene therapy is currently only available in experimental clinical trials. Gene therapy works by inserting a gene into a person's immune cell to enhance its ability to eradicate cancer cells. Nonspecific immunomodulating agents work to directly or indirectly stimulate the body's own immune system, especially immunoglobulins and cytokines, to destroy malignant cells.

Common side effects of BRMs include bone pain, fatigue, fever, anorexia, rashes at the site of injection, and flulike symptoms (nausea, vomiting, chills, fever, and anorexia).

Nutrition Therapy

As noted above, patients who are receiving treatment for cancer may experience nausea, vomiting, early satiety, dysgeusia, diarrhea, mucositis, xerostomia, constipation, weight loss, and anemia. All of these side effects have the potential to place the individual at nutritional risk, and, if not successfully treated, may result in malnutrition. The primary goal of nutrition therapy for the cancer patient is to prevent malnutrition, because reversing it may prove to be very difficult. The metabolic alterations discussed above (see Figure 23.8) can cause severe anorexia and cachexia that may not be amenable to common nutrition interventions. Nutritional treatment relies heavily on careful screening and assessment of patients who are at high risk for developing malnutrition, which may result from either the cancer itself or the medical treatment, including surgery, chemotherapy, and/or radiation. Clinical dietitians need to be aware of those cancer diagnoses and treatments that are most likely to cause malnutrition in this population. Pharmacologic agents, such as appetite stimulants, **prokinetics**, **antiemetics**, and anabolic agents, may be useful for control of symptomatic treatment in combination with nutrition therapy.

As discussed in Chapters 2–4, the nutrition care process provides a systematic approach to assess patients for nutritional imbalance and malnutrition. It allows for the early identification and naming of nutrition-related problems (nutrition diagnoses), and establishes a clear and logical pathway for nutrition interventions. Aggressive identification and treatment of nutrition-related side effects can stabilize or reverse weight loss in 50% to 88% of oncology patients.[45] Malnutrition is a common cause of morbidity and mortality in cancer patients, and is more prevalent in those who have been sick for longer periods of time, those who have had multiple treatments, and those with certain types of cancer including lung, pancreatic, GI cancers, head and neck cancers, and ovarian cancer. Data indicate that malnutrition reduces responsiveness to chemotherapy and RT, increases perioperative morbidity in cancer patients, worsens quality of life, and diminishes the likelihood of survival.[46–48]

Nutrition Assessment

Nutrition screening, as discussed in Chapter 3, is a reliable method for identifying those patients at nutritional risk. Subjective global assessment (SGA) is one such method that has been successfully used for the cancer patient population. A positive correlation has been demonstrated between the use of SGA and objective assessments such as anthropometry, albumin, total cholesterol, weight, and weight loss.[49]

The Patient Generated SGA (PG-SGA) and the Scored PG-SGA are recent modifications to the original SGA. The health care professional completes the SGA, while the patient completes the PG-SGA. Both are applicable in the inpatient and outpatient oncology settings. The Scored PG-SGA is a further adaptation of the PG-SGA and allows for triaging of specific nutrition interventions, as well as facilitating quantitative outcomes data collection.[50] (See Chapter 3.) The Scored PG-SGA has been shown to be accurate at distinguishing well-nourished patients from malnourished patients, and has a high sensitivity and specificity. It is a quick, valid, and reliable nutrition assessment tool that enables malnourished hospital patients with cancer to be identified and triaged for nutrition support.[51]

Because the factors that impact the nutritional status of patients with cancer are extensive and complex, it is important to conduct a comprehensive nutrition assessment examining data from nearly all of the assessment domains: Anthropometrics, Biochemical Data, Medical Tests and Procedures, Nutrition-Focused Physical Findings, Treatments/Therapy/Alternative Medicine, and Food/Nutrition-Related History.

Accurate height and weight measurements should be performed as part of an initial nutrition assessment. A detailed weight history obtained either from the patient or a significant other allows the clinical dietitian to determine if weight has changed significantly in the past six months. Usual or preillness weight should be obtained. If weight loss has occurred, a determination is made as to whether weight loss was voluntary or involuntary. If weight loss occurred involuntarily, the cause of the weight loss should be investigated. In the cancer patient, weight change may occur due to any one or a combination of the following: anorexia, depression, surgery, anxiety, nausea, vomiting, taste changes, xerostomia, diarrhea, or constipation. Once a weight history has been established, current weight needs to be documented. Weighing a patient can be achieved using a variety of standard scales available in the facility. A chair scale can be employed if a cancer patient is physically weak and lacks the strength to stand on a scale. Many outpatient clinics, as well as inpatient facilities, have chair scales available. If a patient has edema (sacral, pedal, ascites), this should be documented and taken into account when assessing current weight.

All weight changes should be considered within the context of time. The Scored PG-SGA includes criteria with which to evaluate weight changes. Weight loss as well as weight gain (which could signify edema) can have important nutritional implications. Lack of weight loss, however, may not necessarily indicate that lean body mass has not been lost. In patients with ovarian cancer, for example, a significant amount of lean body mass may have been lost due to anorexia and poor food intake due to early satiety (usually as a result of the growing tumor that pushes up against the stomach wall). However, ovarian cancer metastases to the liver may cause significant ascites, and thus weight gain due to fluid retention. The gain from ascites may negate the loss from lean body mass; therefore, it is possible that neither the patient nor the health care professional will detect a weight change.

Anthropometric measurements that are also useful include skinfold measurements (to measure subcutaneous fat) and mid-arm muscle circumference (to assess lean body mass). Serial measurements are useful when monitoring weight to determine if fat and/or lean body mass is being lost or accrued. These measurements, while providing useful information, need to be assessed cautiously in cancer patients, because the "norms" on which they are based represent healthy individuals and hence may not have direct applications to the cancer population.

Serum hepatic proteins, such as albumin, prealbumin, and transferrin, have been, historically, the most commonly monitored and assessed biochemical markers utilized in the nutrition assessment process. Because serum hepatic proteins have a long half-life and relatively large body pool, they are relatively insensitive to rapid changes in nutrition. Serum hepatic protein levels can assist the clinical dietitian in identifying patients who are the most ill and thus at risk for developing serious nutritional deficits. A patient with a decreased albumin, prealbumin, or transferrin level is less likely to meet energy and nutrient requirements volitionally and therefore will probably require aggressive nutrition therapies.[52]

Serum albumin levels are affected by many factors. These include changes in plasma volume (for example, intravenous fluids given in preparation for chemotherapy; dehydration secondary to nausea, vomiting, diarrhea, mucositis), GI bleeding, severe diarrhea, renal and liver disease, burns, massive trauma, blood losses (such as those that occur during surgery or with trauma), and chemotherapy. Because serum albumin is affected by so many different factors, some or all of which may be present in the cancer patient, it may not be the best biochemical tool with which to assess nutritional status in these patients.

Nonetheless, serum albumin has been shown to be an excellent predictor of survival in patients with cancer. Several studies have shown that a serum albumin level below normal can be used to predict disease outcomes in many groups of patients, including those with Hodgkin's disease and lung cancer.[53] Low serum hepatic protein concentration is not necessarily indicative of nutrient deficit, but is more likely a response to inflammatory processes.

C-reactive protein (CRP) is the most sensitive indicator of inflammation because it increases in serum concentration as much as 1,000-fold (see Chapter 3 for a full discussion of CRP). Testing for the serum concentration of CRP at baseline may identify a subset of patients for whom a decline in nutritional status is linked to the presence of an active inflammatory response, a recognized precursor of cachexia.[54]

Delayed cutaneous hypersensitivity (DCH) has been used to measure immunological competence and to indirectly measure nutritional status. Because DCH is significantly affected by a compromised immune system, and many cancer patients have depressed immune function secondary to the disease process, to chemotherapy, or to radiation, use of DCH as part of the nutrition assessment process is not appropriate for oncology patients.

A thorough assessment of the patient's clinical signs and symptoms, as well as previous therapies, is imperative before determining a nutrition care plan. A complete physical assessment should include observation for signs of edema, ascites, and bitemporal and other muscle wasting (especially in all four extremities). Gastrointestinal tract assessment includes determining whether there is any history of anorexia, change in appetite, nausea and vomiting, diarrhea, constipation, abdominal pain, early satiety, mouth sores (mucositis), taste changes, or dysphagia. A complete oral assessment should also be completed to determine the individual's ability to chew and to assess the condition of the teeth, gums, and tongue.

Upon initial assessment, a review of an individual's most recent dietary intake is important to determine what kinds of foods can be tolerated, and what, if any, special diets he or she is on, including the use of alternative diets, herbal therapy, vitamin supplements, or nutritional supplements. Many individuals seek out complementary and alternative medicine (CAM) therapies (see Chapter 11) after a cancer diagnosis and may or may not report this to their health care providers. It is important to carefully obtain detailed information regarding alternative therapies and special supplements, because some therapies may be nutritionally inadequate, interfere with conventional therapy, and be harmful to the individual. By developing a strong rapport with the patient, the health care professional can elicit information without making the patient feel as if he or she is "doing something wrong or dangerous." Patients are more likely to divulge information if they feel that they can trust the health care professional. The primary priority for the clinical dietitian is to assess whether the alternative therapy has the potential to cause harm. The CAM feature at the end of the chapter lists the numerous contemporary and alternative remedies associated with neoplastic disease, and Box 23.4 outlines the potentially harmful alternative treatments most commonly used by cancer patients. Finally, it is important to determine if the individual has tried any of the several liquid nutritional supplements available on the market and, if so, which one(s) they preferred and/or disliked.

Determining Nutrient Requirements Provision of adequate kcalories is essential to maintain current weight and/or prevent treatment- or disease-associated loss. The following energy needs equations may be used (in addition to the Harris-Benedict or Mifflin-St. Jeor equations) to determine kcalorie needs of cancer patients:[55]

- Obese patients: 21–25 kcalories/kg
- Non-ambulatory or sedentary adults: 25–30 kcalories/kg
- Slightly hypermetabolic patients or those patients who need to gain weight, or are anabolic: 30–35 kcalories/kg
- Hypermetabolic or severely stressed patients or those with malabsorption: 35 kcalories/kg or greater as needed

Meeting protein needs is important to prevent or reduce negative nitrogen balance and to meet protein synthesis needs. Protein needs are especially elevated in those patients with

BOX 23.4 CLINICAL APPLICATIONS

Popular Alternative Therapies Used by Cancer Patients

	Chemical Components	Clinical Efficacy as Cancer Therapy	Potential for Adverse Reactions	Caution
Pau D'Arco	Quinine compounds	No published clinical evidence	High—includes mild to moderate nausea, vomiting + anticoagulant effects	People taking anticoagulants and patients with thrombocytopenia
DHEA	Steroid precursor (pregnenolone)	May stimulate cancer growth in patients with hormone-sensitive cancers including prostate, breast and endometrial	High—aggressiveness, fatigue, headache, insomnia, elevated liver function tests, decreased Hgb and RBC	Patients with cancer-sensitive hormones should avoid; decreases HDL levels and may increase the risk for heart disease
Goldenseal	Alkaloids	No published clinical evidence	High—can produce significant changes in blood pressure + anticoagulant effects	Patients on antihypertensive medications (beta-blockers, calcium channel blockers, digoxin) should avoid
Mistletoe (Iscador)	Amines, including tyramine, histamine, acetylcholine, beta-phentolamine	Mistletoe does demonstrate immune system modulation and antineoplastic activity—studies too small and short duration to demonstrate evidence	High—may increase hypotensive effects of antihypertensive medications; CNS depressant	Mistletoe plant is toxic and should not be used as a home remedy
Kombucha Tea	Made by incubating the Kombucha mushroom in black tea	No published clinical evidence	Several deaths have been reported after ingestion	Mushrooms may be contaminated with potentially pathogenic bacteria
Astragalus	Betaine, beta-sitosterol, choline, glycosides, plant acids	Demonstrates immune enhancing properties; however, large clinical trials have not been conducted	Low—may increase hypotensive effects of antihypertensive medications	Those patients who are taking immunosuppressants and those with autoimmune disorders should avoid
714X	Nitrogen, camphor, ammonia, ethanol	No published clinical evidence	Moderate—flulike symptoms, inflammation at injection site	FDA has placed an import ban on this compound
PC-SPES	Combination of chrysanthemum, isatis, licorice, *Ganoderma lucidum, Panax pseudoginseng, Rabdosia rubescens,* saw palmetto, and skullcap	No published clinical evidence	High—similar to hormonal drugs: gynecomastia, dyspepsia, nausea, fatigue, leg cramps, diarrhea, angina, hot flashes, and thromboembolic effects	Recalled and removed from the market by the FDA after several batches were found to be contaminated with prescription drugs
Essiac	Usually prepared as a tea; contains burdock root, rhubarb root, sheep sorrel, and slippery elm bark	No published clinical evidence	Moderate—nausea, vomiting, increased urination, flulike symptoms	
Laetrile	Amygdalin, cyanide	No controlled clinical trials have ever been conducted	High—attributed to cyanide poisoning and include nausea, vomiting, headache, dizziness, low blood pressure, bluish discoloration of the skin due to lack of oxygen in the blood, liver damage, fever, mental confusion, coma, death	Not approved by the FDA as a cancer treatment in the US, preparations have been found to be contaminated by bacteria
Cartilage (bovine, shark)	Collagen, glycosaminoglycans	Inhibitors of angiogenesis have been found in cartilage; ongoing clinical trials in patients with advanced cancer; however, the cumulative data is inconclusive regarding its effectiveness as a cancer treatment	Mild to moderate—injectable form: inflammation at injection site, dysgeusia, fatigue, nausea, dyspepsia, and dizziness; powdered shark cartilage form: nausea, vomiting, abdominal cramping/bloating, constipation, hypotension, hyperglycemia, generalized weakness, and hypercalcemia	More published clinical evidence needed

	Chemical Components	Clinical Efficacy as Cancer Therapy	Potential for Adverse Reactions	Caution
Milk Thistle	Silymarin (a flavonoid mixture)	No published clinical trials in patients with cancer	Mild to moderate—mild laxative effect, allergic reaction	Decreases the activity of the cytochrome P450 enzyme system and may affect the clearance of certain chemotherapeutic agents
Hydrazine sulfate	Hydrazine sulfate	No evidence of anticancer activity in randomized clinical trials	Mild to moderate—nausea, vomiting, dizziness, sensory and motor neuropathies; highly toxic when taken with alcohol or barbiturates	Increases the incidence of lung, liver, and breast tumors in animals
Antineoplaston Therapy	Substances (amino acid derivatives, peptides, and essential amino acids) isolated from normal human blood and urine	No published clinical data	Mild to moderate—nausea, vomiting, flatulence, chills, rashes, fever, joint pain, changes in blood pressure, and body odor during therapy	Research needed

severe diarrhea and/or malabsorption. Protein needs may be calculated based on body weight (kg) using the following guidelines:[55]

- Normal or maintenance protein needs: 0.8–1.0 g/kg
- Nonstressed cancer patients: 1.0–1.5 g/kg
- Bone marrow transplant or HSCT patients: 1.5 g/kg
- Increased protein needs (protein-losing enteropathy, hypermetabolism, extreme wasting): 1.5–2.5 g/kg
- Hepatic or renal compromise including BUN approaching 100 mg/dL or elevated ammonia: 0.5–0.8 g/kg

Many cancer patients, especially those undergoing chemotherapy and/or radiation, can become dehydrated easily. Those patients receiving chemotherapeutic agents that damage the GI mucosa and cause diarrhea are at particularly high risk for developing dehydration. Patients undergoing radiation to the head and neck area are also at high risk for dehydration due to their inability to take adequate oral fluids secondary to pain and inflammation of the mouth, throat, and esophagus. High-risk patients need to be assessed frequently for signs and symptoms of dehydration (dark, concentrated urine, decreased urine output, dry mouth, acute weight loss). Fluid needs can be calculated using the same formulas used for most other patients without renal disease (30–35 mL/kg).

Vitamins and minerals act as cofactors for essential processes both in health and illness. Deficiencies of vitamins (especially folate, vitamin C, and retinol) and minerals (magnesium, zinc, copper, and iron) can occur in cancer patients due to the direct effects of the tumor, effects of cytokines, infectious processes, chemotherapy, radiation, or inadequate food intake. Micronutrient requirements have not been established for those individuals diagnosed with cancer, because the precise needs of these patients have not been well documented. Use of a daily multivitamin and mineral supplement that contains <150% of the DRI may be beneficial for most patients undergoing chemotherapy and/or radiation therapies.

Nutrition Diagnosis

Nutrition diagnoses associated with malignancy will vary considerably dependent on the primary tumor site and the extent of tumor burden. They may include involuntary weight loss; increased energy and protein needs; altered GI function; or inadequate oral food/beverage intake.

Nutrition Intervention

Nausea and Vomiting Nausea/vomiting is one of the most common side effects that occurs as a result of oncologic therapies and can be debilitating.

Nausea and vomiting have multifactorial etiologies (see Chapter 14). Causes of nausea and vomiting in cancer patients include chemotherapy, radiation, narcotic analgesics, odors (including food odors, perfumes), and delayed gastric emptying. Nausea and vomiting associated with chemotherapy can be classified as acute, delayed, or anticipatory.[56] Acute nausea and vomiting occur within 24 hours of administration of chemotherapy. The most **emetogenic** chemotherapeutic agents include cisplatin, methotrexate, doxorubicin, and cyclophosphamide (see Table 23.4). Delayed nausea and vomiting usually begin 24 hours after the chemotherapy has been administered and may last up to a week. Delayed nausea and vomiting are most commonly seen after the administration of cisplatin, carboplatin, cyclophosphamide, or doxorubicin.[57] Anticipatory nausea and vomiting most commonly occurs before the initiation of chemotherapy, but may also occur during or after the initiation of chemotherapy. This type of nausea and vomiting often results from inadequate prevention and/or poorly controlled nausea and vomiting during the first chemotherapy and is more commonly seen in pediatric patients. Nausea and vomiting related to RT are dependent on the field being irradiated. Almost 100% of patients undergoing total body irradiation (TBI) during bone marrow transplantation experience emesis, while radiation of the cranium only is considered low risk (about 10% to 30% of patients experience emesis).[57] Upper- and mid-abdominal RT can also result in nausea and vomiting starting one to two hours after treatment and persisting for several hours.[58]

A thorough assessment of the causes of nausea and vomiting will help with treatment. Patients who are experiencing nausea and vomiting due to certain odors are encouraged to take precautions in avoiding noxious odors. Nausea from cooking odors

Table 23.4 Relative Emetogenic Potential of Antineoplastic Drugs

Emetogenic Potential (% of patients)	Agent	Onset/Duration of Response (hours)
High (>90%)	Cisplatin	1.5–56
	Cyclophosphamide	9–28
	Dacarbazine	4–24
	Mechlorethamine	0.5–24
Moderately high (60%–90%)	Cisplatin	1.5–56
	Cyclophosphamide	9–28
	Methotrexate	4–12
	Carboplatin	6–46
	Doxorubicin	3.5–34
Moderate (30%–60%)	Cyclophosphamide	9–28
	Methotrexate	4–12
	Doxorubicin	3.5–34
Moderately low (10%–30%)	Methotrexate	4–12
	Fluorouracil	3–10
	Etoposide	3.5–34
Low (<10%)	Hydroxyurea	8–48
	Vinblastine	3.5–34
	Bleomycin	3.5–24
	Tamoxifen	12–36
	Chlorambucil	48–56

Source: From "Closing the Gap in Prophylactic antiemetic therapy: Patient Factors in Calculating the Emetogenic Potential of Chemotherapy," by K.M. Doherty, 1999, Clinical Journal of Oncology Nursing, 3, p. 114. Copyright 1999 by the Oncology Nursing Society. Adapted with permission.

can be minimized by using a microwave oven, opening windows when cooking, taking a walk when meals are being cooked, and avoiding frying of foods, which emits more odors than most other forms of cooking. Patients should ask friends and family members to avoid perfumes when they are visiting. In addition, a patient's medication list should be reviewed for potential causes of nausea and vomiting. A common cause of nausea and vomiting is the use of narcotic analgesics (morphine, codeine, fentanyl), which are prescribed for many cancer patients for chronic pain. Usually the nausea and vomiting that results from these agents occurs acutely at the beginning of therapy and resolves with chronic use. Other medications known to cause nausea and vomiting include antibiotics, digoxin, and anticholinergic agents. Assessing the patient for signs and symptoms of early satiety is also important. Delayed gastric emptying can result in nausea and vomiting. Small, frequent meals may be helpful, as well as the administration of prokinetics.

By far, the most common cause of nausea and vomiting in cancer patients is chemotherapy; this is referred to as chemotherapy-induced nausea and vomiting (CINV). A thorough review of the patient's chemotherapy regimen will assist the dietitian in determining the severity of the nausea and vomiting that can be anticipated. The patient should be advised to eat only a small, low-fat meal the morning of the first treatment and to avoid fried, greasy, and favorite foods for several days following the treatment. A clear liquid diet for the first few days after therapy may be indicated. To provide kcalories and maintain hydration, consumption of electrolyte-fortified beverages such as Gatorade;

nutritional fruit beverages such as Resource Breeze (Nestlé Nutrition) and Enlive (Abbott Nutrition) and non-acidic fruit drinks (apple and grape juice, nectars) should be encouraged. It is important for patients to avoid favorite foods at any time the chance for emesis is high, since once a favorite food has been vomited, the likelihood of its subsequent consumption is low. The same principle applies to the use of "creamy" liquid nutritional drinks. A patient who has vomited a nutritional beverage will associate vomiting with that beverage, even if told that their vomiting was probably caused by chemotherapy. This reduces the likelihood that the clinical dietitian will be able to successfully encourage the use of nutritional drinks later in therapy for weight gain.

Whenever possible, the clinical dietitian should play an active role in assisting the patient to avoid the onset or at least minimize the occurrence of nausea and vomiting. One important intervention is to encourage patients to take their antiemetics as instructed by their physician. To encourage adequate intake and maximal control of nausea and vomiting, antiemetics should be taken at least 30–45 minutes before a meal is consumed. Patients should be encouraged to take their antiemetics even if they do not feel nauseated at the time, especially while actively receiving treatment.

Effective and consistent use of antiemetic agents by the patient with nausea and vomiting is important for the control of both delayed and anticipatory CINV.[56] In addition, control of nausea and vomiting may assist in maintaining a patient's nutritional intake. Antiemetics are classified as $5\text{-}HT_3$ receptor antagonists, dopamine receptor antagonists, corticosteroids, cannabinoids, and benzodiazepines. The $5\text{-}HT_3$ receptor antagonists, developed in the mid-1980s, have become the accepted gold standard for the control of CINV. $5\text{-}HT_3$ receptor antagonists specifically block the binding of serotonin to the receptors on the vagal nerve that trigger the emetic response.[59] They include Granisetron, Ondansetron, Dolasetron, and Palonosetron. Corticosteroids, such as prednisone and dexamethasone, are occasionally used in combination with other antiemetics for short-term control of nausea and vomiting; however, their long-term use is contraindicated due to their significant adverse effects, including immunosuppression, hyperglycemia, osteoporosis, skeletal muscle wasting, GI irritation, and mood changes. Dronabinol (Marinol®) is a cannabinoid that contains delta-9-tetrahydrocannabinol and is FDA approved for the treatment of CINV. Cannabinoids, however, are only used in selected patients due to the cultural and societal constraints as well as their relatively low therapeutic index.[60]

Nonpharmacologic alternative methods that have been used with varying success to prevent or treat nausea and vomiting include acupressure, acupuncture, hypnosis, and guided imagery. The website for the National Center for Complementary and Alternative Medicine (http://nccam.nih.gov) is useful for finding complementary and alternative therapies in the treatment of nausea and vomiting, as well as for finding general information about cancer.

Early Satiety A common complaint expressed by cancer patients is "I just can't eat as much as I used to" or "I get full right after I start eating." This describes the symptom of early satiety, which is caused primarily by delayed gastric emptying. It is important for the patient with early satiety to eat small, frequent meals that are nutrient dense. Beverages should also contain nutrients and should be consumed between meals rather than with meals so as not to add to the feeling of fullness. Consumption of raw vegetables, such as salads, and other high-fiber foods should be avoided. Prokinetics, medications that increase gastric emptying, may be useful. Metoclopramide, for example, is a motility agent that selectively stimulates gastric emptying and may be useful for the patient with early satiety. A potential side effect of metoclopramide is diarrhea; therefore, it should not be used by patients that are already at risk for diarrhea.

Mucositis Mucositis, also known as stomatitis, is irritation and inflammation of the epithelial cells of the mucosal membranes lining the gastrointestinal tract that can occur at any point in the GI tract from the mouth to the anus. Its manifestations may range from generalized swelling and inflammation to obvious ulceration and hemorrhage.[61] Mucositis-associated pain is the main source of cancer treatment-related pain, which afflicts from 40% to 70% of patients receiving chemotherapy or radiotherapy. It has been proposed that mucositis is related to direct and indirect cytotoxicity, local tissue cytokine and immune activity, and bacterial colonization of the ulcerative lesion.[62] The growth factor filgrastim may also cause mucositis. Other causes of mucositis include viral, bacterial, and fungal infections, radiation, stem cell transplant therapy, and graft-versus-host disease.

The patient with oral mucositis should have a thorough and systematic assessment of the mouth. Alterations in the oral mucosa may include color changes of the tongue, lips, and gingiva; changes in moisture; and changes in integrity, including cracks, fissures, ulcers, blisters and lesions. The presence of white plaques is generally indicative of fungal infections such as candidiasis. The disruption of the mucosal barrier in the oral cavity increases the risk of infections.

Chemotherapy-induced mucositis commonly occurs five to seven days after chemotherapy is initiated and may continue until the patient recovers from the immunosuppression (referred to as the **nadir**) or until the **absolute neutrophil count** is >500. Symptoms will include pain and burning with chewing and swallowing. Mucositis may be severe enough to cause the patient to completely forgo any food or fluids, which can lead to dehydration and acute weight loss. Good oral hygiene is important for the patient with oral mucositis in order to prevent infection. Box 23.5 outlines the ACS recommendations for preventing and treating mouth sores. Oral glutamine has been utilized for the prevention and treatment of oral mucositis; however, studies have not shown any beneficial effects of glutamine on the prevention of mucositis in chemotherapy or radiation patients.[63]

Narcotic analgesics may be required for pain. Topical therapies including agents such as sucralfate, nystatin, and clotrimazole troches also help treat pain and infections, but may cause taste changes. Patients with oral mucositis may need nutrition education to provide guidelines for eating until the mucositis resolves. The patient should be encouraged to eat only soft, non-fibrous, non-acidic foods. Hot foods should be avoided as they can burn the already tender, fragile mucosa. Liquids should be encouraged to prevent dehydration; non-acidic juices such as nectars may be helpful. High-kcalorie, high-protein

milkshakes or nutritional supplements may be beneficial at this time (see Table 23.5).

Diarrhea Because antineoplastic agents target those cells that have the highest replication rate, they often cause diarrhea. In the GI tract, antineoplastic agents, especially antimetabolites (e.g., 5-fluorouracil), inhibit mitosis in rapidly proliferating crypt cells, leading to a disproportionate increase in the number of immature crypt cells.[64]

When mucositis is present in the oral mucosa, it can be assumed that it may also be present in the stomach and in the small and large intestine, resulting in diarrhea, which may at times become severe. Dehydration can occur rapidly. The patient with diarrhea should be encouraged to drink small amounts of fluid frequently throughout the day. Large amounts of fruit juices should be avoided as excessive fructose can exacerbate diarrhea. Gatorade®, Pedialyte®, clear liquid nutritional beverages, and other oral rehydration fluids are recommended. Patients should be encouraged to use the antidiarrheal medications as prescribed by their physicians. Antidiarrheals include loperamide (Imodium®) and diphenoxylate (Lomotil®). Instructing the patient to increase their intake of foods high in soluble fiber may help with the treatment of diarrhea; however, often these patients have a poor appetite and may have a difficult time increasing their intake of foods in general. (For further details on nutrition therapy for diarrhea, see Chapter 15.)

Dysgeusia "Meats have a bitter taste." "My morning coffee just doesn't taste the same." These common complaints may be related to another typical nutritional problem in cancer patients—dysgeusia. Dysgeusia, or alterations in taste, can have a profound effect on a patient's ability to ingest an adequate amount of nutrition. The presence of certain tumors can elicit taste changes even before a diagnosis is made. Many chemotherapeutic agents, specifically cisplatin, and radiation to the head and neck area cause dysgeusia. Taste changes that occur include a metallic taste (usually due to the chemotherapeutic agent cisplatin), no taste sensation (aguesia), a heightening of certain tastes (especially sweets), or aversions to foods the patient liked to eat in the past.

Patients who experience a metallic taste in their mouth should be advised to avoid metal utensils and instead use plastic utensils. If nutritional supplements are consumed, they should be poured into a glass first, as often the metal container may also be offensive. Meats are often not tolerated. To ensure an adequate protein intake, the patient should be encouraged to incorporate other high-protein foods into the diet, including peanut butter, cottage cheese, cheese, poultry, and soy meat substitutes. Patients with aguesia should be encouraged to use more highly spiced and flavorful foods, such as marinated foods. Sweet foods often taste too sweet to individuals undergoing cancer therapy. Many homemade drinks and nutritional beverages may be too sweet for these patients. Alternative options may be to have the patient try a nonsweet supplement such as Osmolite® (Ross Laboratories) or one that is juice or yogurt based.

Xerostomia Xerostomia, reduced saliva production, is a common side effect of head and neck radiation and chemotherapy (methotrexate, 5-fluorouracil, paclitaxel, carboplatin, cisplatin). Other causes of xerostomia include dehydration,

Guidelines for Preventing and Treating Mouth Sores

What the patient can do

- Check mouth twice a day using a small flashlight and a padded Popsicle stick. If you wear dentures, take them out before you inspect your mouth. Tell your doctor or nurse if your mouth looks or feels different or if you notice changes in how things taste.

- Follow the plan below for mouth care 30 minutes after eating and every 4 hours while awake, or at least twice a day unless your doctor or nurse gives you other instructions:

  Brush your teeth using a soft nylon bristle toothbrush. To soften the bristles even more, soak the brush in hot water before brushing and rinse brush with hot water during brushing. If the toothbrush hurts, use a Popsicle stick with gauze wrapped around it or a cotton swab instead. Or you can get soft foam mouth swabs from the drugstore.

  Rinse toothbrush well in hot water after use and store in a cool, dry place.

  Use a non-abrasive toothpaste that contains fluoride. Note that whitening toothpastes may contain hydrogen peroxide, which can irritate sore mouths.

  Remove and clean your dentures between meals on a regular time schedule. If you have sores under your dentures, leave your dentures out between meals and at night. Clean dentures well between uses, and store them in an anti-bacterial soak. If your dentures fit poorly, do not use them during treatment.

- Gently rinse your mouth before and after meals and at bedtime with one of the following solutions (stir or shake the solution well, then swish it around and gently gargle, then spit it out):
  1 teaspoon baking soda
  2 cups water
  Or
  1 teaspoon salt
  1 teaspoon baking soda
  1 quart water

- If you normally floss, keep flossing at least once a day unless you are told not to do so. Tell your doctor if this causes bleeding or other problems. If you do not usually floss, talk with your doctor before you start.

- Avoid store-bought mouthwashes, which often contain alcohol or other irritants.

- Keep lips moist with petroleum jelly, a mild lip balm, or cocoa butter.

- Drink at least 2 to 3 quarts of fluids each day, if your doctor approves.

- If mouth pain is severe or makes it hard to eat, ask your doctor about medicine that can be swished 15–20 minutes before meals or painted on a painful sore with a cotton swab before meals. If this does not work, you may need stronger pain medicines.

- To promote healing, ask your doctor about using Maalox® or Milk of Magnesia®. You can use these products to help sores by allowing them to settle and separate, pouring the liquid off the top of the solution, and then swabbing the pasty part onto the sore area with a cotton swab. Rinse your mouth with water after 15–20 minutes.

- Sip warm tea slowly.

- Eat chilled foods and fluids (for instance, Popsicles, ice cubes, frozen yogurt, sherbet, or ice cream).

- Eat soft foods that are moist and easy to swallow.

- Eat small, frequent meals of bland, moist, non-spicy foods. Avoid raw vegetable and fruits, and other hard, dry, or crusty foods, such as chips or pretzels.

- Avoid very salty or high-sugar foods.

- Avoid acidic fruits and juices, such as tomato, orange, grapefruit, lime, or lemon.

- Avoid fizzy drinks, alcohol, and tobacco.

- Create a pleasant mealtime atmosphere.

Source: Reprinted by the permission of the American Cancer Society Inc. from www.cancer.org. All rights reserved.

---

chronic graft-versus-host disease of the GI tract, medications (narcotic analgesics, antianxiety agents, antihistamines, beta blockers, antidepressants, diuretics), Sjögren's syndrome, and aging. The severity of xerostomia is correlated with the severity of oral discomfort, dysgeusia, dysphagia, and **dysphonia**. Drugs used to treat cancer can make saliva thicker, causing the mouth to feel dry.[61]

Treatment of xerostomia may include use of artificial saliva (saliva substitutes) and/or mouth moisturizers. There are several artificial salivas available on the market; however, patient compliance may be a problem due to their consistency, taste, and cost.[65] In addition, the duration of action is short as they are quickly removed from the mouth with swallowing. Mouth moisturizing lubricants come in the form of gels, lozenges, and mouthwashes. Sugar-free gum and sour-flavored sugar-free hard candies may help increase the flow of saliva in the mouth and are less expensive than artificial saliva. One study found chewing gum to be more effective than artificial saliva for the treatment of radiation-induced xerostomia.[66] Denture wearers may not be able to chew gum for the treatment of xerostomia.

Anorexia Lack of appetite, or anorexia, is a challenging problem for both patients and clinical dietitians. The prevalence of anorexia in cancer patients is estimated at approximately 50% of patients upon diagnosis. While the exact prevalence is unknown, it is generally acknowledged that anorexia and reduced food intake are frequent occurrences in cancer patients.[67] Anorexia has multiple etiologies in the cancer patient, including circulating cytokines, hormones, depression, therapy (surgery, radiation, chemotherapy), learned food aver-

Table 23.5 High-Kilocalorie, High-Protein Nutritional Beverages

Beverage	Manufacturer	Kcalories (per 240 mL)*	Protein (g per 240 mL)*
Ensure HP	Abbott Nutrition	230	12
Ensure Plus	Abbott Nutrition	360	13
Enlive	Abbott Nutrition	250	9
2 Cal HN	Abbott Nutrition	475	19.9
HI-CAL	Abbott Nutrition	475	19.8
Boost	Nestlé Nutrition	240	10
Boost High Protein	Nestlé Nutrition	240	15
Boost Plus	Nestlé Nutrition	360	14
Resource Breeze	Nestlé Nutrition	250	9
Resource Plus	Nestlé Nutrition	360	13
Resource 2.0	Nestlé Nutrition	480	20
Resource Healthshake	Nestlé Nutrition		
4 oz		200	6
6 oz		300	9
Resource Shake Plus	Nestlé Nutrition	480	15
Resource Benecalorie (1.5 oz serving)	Nestlé Nutrition	330	7
Carnation Instant Breakfast Essentials (powder made with 8 oz whole milk)	Nestlé Nutrition	220	13
Carnation Instant Breakfast Essentials—Ready To Drink	Nestlé Nutrition	250	14
Carnation Instant Breakfast Essentials VHC	Nestlé Nutrition	560	22.5
Carnation Instant Breakfast Essentials Lactose Free Plus	Nestlé Nutrition	375	13
Replete	Nestlé Nutrition	250	15.6
Nutren 1.5	Nestlé Nutrition	375	15
Nutren 2.0	Nestlé Nutrition	500	20
Scandishake (made with 8 oz whole milk)	Axcan Pharm	600	12

*Unless otherwise noted.

sions, fatigue, and certain medications. Chronic anorexia and reduced energy intake can lead to weight loss and exacerbate the development of cancer cachexia. In general, patients with hematologic malignancies and breast cancer seldom exhibit weight loss and cachexia; however, most other solid tumors, including those of the GI tract, head and neck cancer, and lung cancer, are associated with anorexia and cachexia.[25] While nutrition therapy for the treatment of anorexia (see Table 23.6) is helpful for some patients, improvement requires intensive counseling and motivation by the patient. Exercise may help to increase appetite, but many patients may be unable to increase their physical activity for a variety of reasons, including profound fatigue, severe thrombocytopenia (platelets <20,000 μ^3), severe immunosuppression, and side effects from therapy, such as nausea, vomiting, or diarrhea. Exercise, on the other hand, may actually relieve fatigue, prevent muscle wasting, and improve the ability to perform activities of daily living by improving endurance levels.

Pharmacologic interventions have been found to be relatively useful for stimulating appetite in cancer patients. To date, two agents have been recognized as useful for appetite stimulation: megestrol acetate and corticosteroid agents. Agents that have been found to be less effective or unproven include dronabinol, cyproheptadine, hydrazine, and metoclopramide.

Megestrol acetate is available in tablet form and as a liquid suspension. Patient compliance will be significantly enhanced when the liquid suspension is taken as opposed to tablets. The optimal dose is 800 mg/day, which can be provided in 20 mL liquid suspension, whereas the tablets contain only

Table 23.6 Nutrition Therapy for the Treatment of Anorexia

- Eat smaller, more frequent meals.
- Maximize your intake when appetite is most normal.
- Limit fluid with meals to avoid feeling of fullness.
- Keep favorite foods readily available at all times.
- Mild exercise, as tolerated (check with physician).
- Eat meals in a pleasant environment.
- A glass of wine before a meal may help to stimulate the appetite (check with the physician first).
- Avoid noxious odors; ventilate eating area.
- Find a liquid nutritional supplement that is appealing and drink only 2–4 ounces at a time (to avoid a feeling of fullness); keep unopened beverage in the refrigerator.
- Try relaxation exercises before mealtimes.
- Consider pharmacologic agents/appetite stimulants.

20 mg each, meaning the patient must take 40 tablets to achieve a full appetite stimulating dose. Appetite may improve in as little as 24 hours after initial administration. Adverse effects of these drugs include hyperglycemia, peripheral edema, increased risk of thromboembolic events, breakthrough uterine bleeding, hypertension, and Cushing's syndrome.[68]

Corticosteroids (dexamethasone, prednisone, methylprednisolone) are known appetite stimulants. Although several randomized, placebo-controlled studies demonstrated that corticosteroids induce a usually temporary (limited to a few weeks) effect on indicators such as appetite, food intake, sensation of well-being, and performance status, none of the studies showed a beneficial effect on body weight.[69–73] Corticosteroids have multiple adverse side effects, including sodium retention, fluid retention, hyperglycemia, lean body mass wasting, hypertension, Cushing's syndrome, osteoporosis, immunosuppression, and delayed wound healing. They should thus be used cautiously in cancer patients.

Pharmacologic agents used in the treatment of weight loss in cancer patients include growth hormone, insulin-like growth factor-1 (IGF-1), testosterone, dihydrotestosterone, and the testosterone analogues oxandrolone and nandrolone decanoate. Oxandrolone is the only one of these approved for use by the FDA for weight gain following disease-related weight loss. It has been used clinically for the treatment of cancer-induced wasting. Oxandrolone is a synthetic agent with a chemical structure similar to that of testosterone and is currently a schedule III controlled substance. It is administered orally and, in general, appears to be safe and well tolerated. Oxandrolone is associated with gains in weight and lean body mass as well as improvements in nitrogen balance and functional status in many patients. It is indicated as an adjuvant therapy to promote weight gain after weight loss following extensive surgery, chronic infection, or severe trauma, and in some patients who, without pathophysiologic reasons, fail to gain or maintain normal weight.[74] Oxandrolone has been studied in patients with human immunodeficiency virus (HIV), acquired immunodeficiency syndrome (AIDS), chronic obstructive pulmonary disease, and severe muscle wasting due to paraplegia or hemiplegia.

Physical activity is a key factor influencing the maintenance of lean body mass. Whenever appropriate, cancer patients should be advised to engage in some form of exercise. Prolonged inactivity due to bed rest, certain medications (corticosteroids), aging, and weight loss are associated with the loss of skeletal muscle tissue. Many cancer patients, however, may be unable to participate in regular physical activity, either due to their medical condition, fatigue, or advanced age. Physical therapists will be helpful in determining the appropriate activity for individual patients.

In cancer patients, feeding studies have shown generally disappointing results in altering the course of catabolic changes that occur in cancer cachexia.[75] Protein anabolism mechanisms are impaired; anabolic hormone levels such as insulin and IGF-1 are decreased, while levels of cortisol, a catabolic hormone, are elevated. In addition, the ability to utilize nutrients effectively appears to be altered in cancer patients.

Administration of omega-3 fatty acids or high-purity EPA capsules has been associated with weight stabilization in weight-losing patients with advanced pancreatic cancer.[76, 77] On the other hand, in a study on 200 patients with unresectable pancreatic cancer, those who were provided with an oral liquid supplement enriched with omega-3 fatty acids and antioxidants failed to show a therapeutic advantage compared with subjects provided with an identical supplement without the omega-3 fatty acids or antioxidants; both supplements were equally effective at arresting weight loss.[78] A recent systematic review found no significant clinical benefits for EPA or DHA in terms of weight, lean muscle mass, survival, or quality of life in cancer patients with cachexia or anorexia.[79] Further studies are needed to determine if omega-3 fatty acids are beneficial for the treatment of cancer cachexia. Patient compliance with nutritional supplements enriched with omega-3 fatty acids may be problematic due to the taste and cost of the product.

Pharmacologic agents are being investigated as potential therapy for the treatment of cancer cachexia. These include anticytokine therapies (pentoxifylline, thalidomide), non-steroidal anti-inflammatory drugs, melatonin, b_2 agonists, and anabolic agents (growth hormone, insulin-like growth factor, testosterone, dihydrotestosterone, and testosterone analogues such as oxandrolone, discussed previously).

Nutrition Support

The use of nutritional support (enteral and parenteral nutrition) has been a controversial area in the field of oncology nutrition. A thorough nutrition assessment is critical when determining whether or not a patient needs nutrition support. The principle "when the gut works, use it" applies to cancer patients. When oral intake continues to decline despite aggressive efforts at dietary and pharmacological interventions, or when the GI tract becomes non-functional, the RD and the health care team managing the patient should consider nutrition support options. While there is no evidence to support the routine use of enteral or parenteral nutrition in well-nourished cancer patients undergoing surgery, chemotherapy, or radiation, enteral nutrition is indicated for any malnourished cancer patient whose GI tract is functional, who is undergoing anticancer therapy, and who has a reasonable prognosis. Nutrition support is considered an aggressive form of therapy and should be utilized only when other aggressive medical approaches (i.e., chemotherapy, surgery, radiation) are also being used to treat the cancer. Nutrition support is inappropriate for most **terminal** cancer patients or for patients with a poor prognosis for whom all medical anticancer therapies have been exhausted.

The practice guidelines for nutrition support of adults with cancer of the American Society for Parenteral and Enteral Nutrition (ASPEN) include the following:

- Patients with cancer are nutritionally at risk and should undergo nutrition screening to identify those who require formal nutrition assessment with development of a nutrition care plan.
- Nutrition support therapy should not be used *routinely* in patients undergoing major cancer operations.
- Perioperative nutrition support therapy may be beneficial in moderately or severely malnourished patients if administered for 7–14 days preoperatively, but the potential benefits of nutrition support must be weighed against the potential

risks of the nutrition support therapy itself and of delaying the operation.

- Nutrition support therapy should not be used *routinely* as an adjunct to chemotherapy.

- Nutrition support therapy should not be used *routinely* in patients undergoing head and neck, abdominal, or pelvic irradiation.

- Nutrition support therapy is appropriate in patients receiving active anticancer treatment who are malnourished and who are anticipated to be unable to ingest and/or absorb adequate nutrients for a prolonged period of time.

- The palliative use of nutrition support therapy in terminally ill cancer patients is rarely indicated.

- Omega-3 fatty acid supplementation may help stabilize weight in cancer patients on oral diets experiencing progressive, unintentional weight loss.

- Patients should not use therapeutic diets to treat cancer.

- Immune-enhancing enteral formulas containing mixtures of arginine, nucleic acids, and essential fatty acids may be beneficial in malnourished patients undergoing major cancer operations.[80]

Enteral nutrition support may be beneficial in malnourished patients undergoing RT for head and neck cancers. Enteral nutrition may allow patients to complete their entire course of RT without interruption while minimizing weight loss. Enteral nutrition may also be indicated in patients with head and neck cancer undergoing extensive surgery, patients with esophageal cancer (many of whom suffer from dysphagia) following esophagectomy, and gastric cancer patients following total or subtotal gastrectomy. Many recommend that a feeding tube route be placed prior to initiation of treatment therapies.[30]

Providing parenteral nutrition support to cancer patients with advanced disease has significant ethical implications, increases the risk of metabolic and infectious complications, and is expensive. A patient's lack of appetite and food intake is often a greater source of concern to health care providers, family members, and caretakers than it is to the patient. Caretakers feel a sense of frustration and powerlessness, and feeding the patient gives them some sense of control over their loved one's care and well-being. Fear of death by starvation is common. Orrevall et al. studied patients with advanced cancer who had received home parenteral nutrition (HPN) and found that the desperate and chaotic situation in the family influenced a patient's willingness to accept HPN.[81] It is unknown whether HPN can improve a patient's quality of life. Quality of life is very difficult to measure, and few studies have documented quality of life in advanced cancer patients who are on HPN. Bozzetti et al. studied patients with advanced cancer who received HPN with respect to nutritional status, length of survival, and quality of life. Their study suggests that in malnourished, chronically obstructed patients with advanced cancer resistant to conventional curative therapy, HPN may help to prolong survival past seven months (in approximately one-third of patients), improve their quality of life (in 20% to 40% of them), or at least maintain it until two months prior to death.[82]

Before discharging a patient from the hospital, insurance (including Medicare) guidelines regarding HPN reimbursement need to be clarified. Medicare will not approve the use of HPN unless it is documented in the medical record that HPN will be required for at least three months and that enteral nutrition is not a viable means of feeding the patient.[83]

Conclusion

Research has consistently demonstrated that a healthy, nutritious diet that is low in fat, low in saturated fat, and high in fruits, vegetables, and whole grains is important for the prevention of many types of cancers. Physical activity also appears to play a role in cancer prevention. Paying close attention to diet and nutrition is important during the treatment of cancer, because many treatment options can significantly affect a person's nutritional status. Obtaining adequate kcalories, protein, fluids, vitamins, and minerals, and preventing weight loss, which occurs so frequently in cancer patients, may help to improve quality of life for many of those undergoing surgical and/or pharmacologic treatment of cancer. Research in the areas of nutrigenomics and the biologic basis of anorexia-cachexia promises to offer new insight into cancer prevention as well as new treatment options.

PRACTITIONER INTERVIEW

Robin Gaff, RD, LD, CSO *Clinical Outpatient Dietitian,* **James Cancer Hospital**—*The Ohio State University Medical Center in Columbus, Ohio*

Background. I began in food service in an extended care facility and worked my way up from dietary aide to Certified Food Service Manager. Moving around to different states as my husband's job transferred him gave me the opportunity to work at a variety of health care facilities, expand my education, and take time off to have a family. I eventually completed college and a dietetic internship and became a registered dietitian.

Specialization in this area of practice. A move to Texas was pivotal in the beginning of my oncology journey, both as a professional—as I was hired as a hospital's clinical manager and worked the Oncology Unit—and personally, as I was diagnosed with metastatic thyroid cancer. Both became a learning and an educational experience. After relocating once again to Columbus, Ohio, I was hired to do a clinical research study at The Ohio State Medical Center

(continued)

on proactive nutrition with an oral supplement for head and neck cancer patients. Then a wonderful job opportunity opened up for me to be hired on as the first outpatient clinical dietitian for the James Cancer Hospital and I built the outpatient system up from ground zero. I began in the Breast Cancer Clinic, but soon was spreading out to the other clinics to see multiple types of cancers, including GI, melanoma, pancreatic, prostate, and ENT. Many co-workers tell me I was very verbal and visible, which helped the growth of the outpatient clinic, but I still needed to continue to learn all I could about not only nutrition, but how it related to all of these cancers. When I found out about the ADA Board Certified Specialist in Oncology nutrition, I studied with oncology nurses, and took and passed the test. I also took the opportunity to attend the ADA seminar and obtained a certification in weight management, as this is an area of concern in oncology.

Role of dietitian in oncology care. There are many roles a dietitian can take on when in oncology. One is becoming an integral part of the patient support team. Nutrition assessment is important, but with cancer patients, there is more to the picture. As a specialist, I provide expertise in my field and promote the importance of oncology nutrition through education for the patient and caregivers, for medical staff, and throughout the community. It is important to take the opportunity for professional growth through attending lectures, writing articles, applying for grants, doing research, and attending conferences, not just on nutrition, but other oncology and medical subjects, when one is able. Educational growth of and on the subject of oncology nutrition can be promoted through various community events, cooking demonstrations, and lectures, to name a few possibilities.

How can a dietitian make a difference in the care of a patient?

First, by having care and compassion; next, passion for your job; and then the education (required and to continue learning) in order to provide the best care possible for the special and at times emotional and difficult patient population. Some specific interventions that improve a patient's quality of life include suggestions for treatment-related side effects that affect a patient's nutritional status. The patient's family and caregivers are often helped when given the task of cooking for their loved one who has just had surgery and is preparing for chemotherapy and radiation treatments. In taking control of the nutrition component, the dietitian allows other medical staff to focus on their specific roles.

What specialized training does the RD need?

Having a well-rounded nutrition background provides a good basis, but working in the field is helpful in learning all the aspects of oncology care as part of the patient's health care team. Professional networking, attending conferences, and joining the Oncology Dietetics Practice Group provide great resources. After being in the field for the required time, I suggest taking the ADA boards for oncology nutrition, which is a boost for being specialized.

Many new dietitians may be apprehensive about working with patients who may have a terminal illness. How have you handled this?

Having a passion for what I do and a compassionate and caring attitude helps in this area. Knowing that I am providing the tools for them to improve quality of life through nutrition interventions and that the patient and their caregivers are appreciative of the information I give helps me deal with this situation. Nutrition is one area that can give them some control in their life. Also, keep in mind that there are so many medical and technical advances that you shouldn't assume all cancer patients are terminal. I tend to view a cancer diagnosis as a "survivorship journey" and not a "battle."

How do you see the future role of dietitians in oncology nutrition?

I see it expanding as the need grows for this specialized component of the oncology health care team. I am hopeful to see a dietitian in all major cancer centers, for inpatient and outpatient services, and in research. Proactive interventions can make a difference in decreasing nutrition-related side effects, which can keep the patient's treatment on schedule and ultimately save the facility money. As insurance reimbursements for nutrition services improve, so will the number of dietitians in the field.

Application of the Nutrition Care Process: Cancer

Introduction:

Mr. J is a 70-year-old male with a history of lung cancer. He was first diagnosed with limited small cell lung cancer a little over one year ago. Mr. J initially received radiation therapy and has been followed for the previous year. He now presents with a new tumor on left upper lobe, multiple mediastinal lymph nodes, and a left supraclavicular node. He describes being short of breath and easily tired. Chemotherapy is now planned for Mr. J.

1. What type of cancer is small cell lung cancer? What is the usual cause of this type of lung cancer?

2. Describe the basic mechanisms for radiation and chemotherapy. How do these treatments interfere with cancer cell growth?

3. What are the common nutritional side effects of each of these treatment methods?

Nutrition Assessment:

Food/Nutrition-Related History:

Mr. J states that he is having difficulty eating.

Anthropometric Measurements:

Ht. 6'2" Wt. 165 lbs. Pre-illness wt (2 years ago) 205 lbs.

Biochemical Data:

Prealbumin: 17 mg/dL

Nutrition Diagnosis:

4. Identify at least two nutrition problems based on the nutrition assessment and medical history. Determine the diagnostic term for each nutrition problem. Next, identify the etiology of each nutrition problem. Finally, identify the signs and symptoms that support the evidence for these nutrition problems.

Nutrition Intervention:

5. Identify the nutrition prescription for this patient by recommending the appropriate dietary modification for his diagnosis.

6. Calculate energy and protein requirements for this patient.

Nutrition Monitoring and Evaluation:

7. Determine nutrition criteria for monitoring and evaluation for each nutrition diagnosis that you identified.

WEB LINKS

American Institute for Cancer Research   Nonprofit organization that supports both research on diet and cancer prevention and public health education. Numerous sources for both consumers and professionals.
http://www.AICR.org

American Cancer Society   This site provides specific information for patients, survivors, and professionals. Information for diagnosis and treatment—including nutrition—is available.
http://www.cancer.org

Cancer Facts   This online resource for cancer patients, their families, and caregivers focuses on providing background and treatment-related information about specific diagnoses.
http://www.cancerfacts.com

National Cancer Institute   This government-sponsored site provides links and information about specific diagnoses and current research. Links to nutrition education for therapy regimens as well as cancer prevention.
http://www.nci.nih.gov/cancerinfo

END-OF-CHAPTER QUESTIONS

1. Carcinogenesis is the process by which normal cells are transformed into cancer cells. How could a nutritional factor act as a carcinogen?

2. In general, how does the growth of a cancer cell differ from that of a normal, healthy cell?

3. Explain the following diagnosis in terms of how a malignancy is classified: Stage II diffuse large B-cell lymphoma.

4. A patient diagnosed with cancer may be treated with one or a combination of the following treatments: chemotherapy, radiation, and/or surgery. Describe the basic mechanism or rationale behind each method used to treat a malignancy.

5. How do radiation and chemotherapy affect healthy cells within the body? What are the cells that are primarily affected? How is this related to the side effects that are often experienced with treatment?

6. Another category of treatment includes biological response modifiers or immunotherapy. What is the general mechanism for these treatments?

7. What is the subjective global assessment? What are the primary nutrition assessment factors that this tool identifies in cancer patients and those undergoing cancer therapies?

8. Many cancer patients present with some nutritional symptoms such as weight loss or taste changes. How might a malignancy affect nutritional status even before it is diagnosed?

9. Name three common nutritional problems that a cancer patient might experience. Identify interventions for each.

10. Identify one complementary and alternative therapy that is commonly used by cancer patients. Explain the indications for this therapy and/or its risks.

Complementary and Alternative Medicine Remedies Used in Neoplastic Diseases

Remedy	Scientific Name	CAM Use	Side Effects and/or Risks
Aloe vera; kumari; sabila	*Aloe* spp.	Treat cervical, lung cancers	Causes active secretion of fluids and electrolytes (hypokalemia) in lumen, inhibits reabsorption of fluids from colon.
Amalaka, amla; Indian gooseberry	*Emblica officinalis*	Prevent/treat cancer	Adverse reactions rare.
Ascorbic acid/ vitamin C		Prevent/treat cancer; antioxidant	High intakes may cause GI upset or mild diarrhea. May increase activity of anticoagulants. May interfere with B_{12}, copper, chromium absorption and metabolism. Aspirin, corticosteroids, and indomethacin increase urinary vitamin C losses. High doses may reduce efficacy of select chemotherapeutic medications. Supplementation above DRI (or >70–90 mg/d) not recommended. May increase plasma estrogen levels in patients taking HRT or oral contraceptives. Large doses may interfere with anticoagulant medications.
Ashwagandha; winter cherry, Indian ginseng	*Withania somnifera*	Prevent/treat cancer	Lacks sufficient human research to recommend use. Possible hyperthyroidism, thyroid dysfunction. May interfere with thyroid and antithyroid medications.
Astragalus, milk vetch; huang qi	*Astragalus membranaceus*	Adjunct to radiation, chemotherapy; treat melanoma, bladder, bone, breast, cervical, colorectal, endometrial, kidney, liver, lung, ovarian cancers; see Box 23.4 for information on efficacy and potential for adverse reactions	None reported.
Beta-carotene		Reduce cancer risk, esp. cervical cancer	No adverse reactions reported.
Burdock; niu bang zi	*Arctium lappa*	Prevent/treat cancer	May increase hypoglycemic effects of insulin and oral hypoglycemic agents. Burdock products may be significantly contaminated with atropine at high enough levels to cause toxicity.
Chinese, Baikal skullcap; huang qin	*Scutellaria baicalensis*	Treat bone, liver cancers	Hepatotoxicity, pneumonitis. Stupor, confusion, seizures. Products have been found to be contaminated with a similar-looking plant known as germander (*Teucrium chamaedrys*) that can cause hepatitis.
Coenzyme Q_{10}		Reduce cancer risk; treat fibrocystic disease	Anorexia, diarrhea, epigastric discomfort, mild nausea.
Copper		Prevent/treat cancer	Nausea may occur with doses of 10 mg, and 60 mg may cause vomiting.
Echinacea, purple coneflower	*Echinacea purpurea, E. angustifolia*	Improve immune system during radiation, chemotherapy	Should not be used >8 weeks. May adversely influence fertility. Contraindicated in patients with severe illnesses, including autoimmune disease, collagen diseases, HIV infection, leukemia, multiple sclerosis, or tuberculosis. Should be avoided by pregnant or breast-feeding women. Tinctures may contain significant concentrations of alcohol.
Folic acid		Reduce risk of cervical, colon cancers	Supplementation ≥400 μg can mask pernicious anemia caused by vitamin B_{12} deficiency; dosages ≥5 mg may cause abdominal cramps, diarrhea, and rash; doses ≥15 mg can alter sleep patterns or cause vivid dreaming; large doses exacerbate neuropathy in those with B_{12} deficiency; doses >1 mg/d interfere with anticonvulsant medications. The following medications can affect absorption or activity of folate: oral contraceptives, aspirin, indomethacin, methotrexate, famotidine, tetracycline, erythromycin, folfonamides, and cholestyramine.
Dong quai; angelica	*Angelica sinensis*	Treat esophageal, liver cancers	Should not be combined with blood-thinning medications (warfarin, heparin, aspirine) or supplements (garlic, ginkgo, vitamin E, fish oil).
Garlic; ajo; lasuna	*Allium sativum*	Reduce risk of colon, esophageal, lung, stomach cancers	Headache, fatigue, altered platelet function with potential for bleeding, offensive odor, GI upset, diarrhea, sweating, changes in the intestinal flora, hypoglycemia. Discontinue use of garlic at least 7 days prior to surgery.

Complementary and Alternative Medicine Remedies Used in Neoplastic Diseases (*Continued*)

Remedy	Scientific Name	CAM Use	Side Effects and/or Risks
Goldenseal; yellow root; yellow puccoon; yellow Indian paint; ground raspberry	*Hydrasitis canadensis*	Inhibit liver cancer growth; see Box 23.4 for information on efficacy and potential for adverse reactions	CNS depression, abdominal cramps and pain, diarrhea, nausea, vomiting. Might inhibit CYP3A4.
Gotu kola; brahmi	*Centella asiatica*	Treat cancer	Well tolerated. High doses may interfere with actions of antidiabetic medications and cholesterol-lowering drugs. Contraindicated in pregnancy and breast-feeding.
Grape Seed Extract	*Vitis vinifera, V. coignetiae*	Prevent/treat cancer	Essentially nontoxic. Might potentiate anticoagulant and antiplatelet agents.
Hare's ear; chai hu	*Bupleurum chinense*	Treat bone cancer	Large doses can cause nausea and vomiting, facial and extremity edema, gastrointestinal distention and constipation.
Licorice root; mulethi; gan cao	*Glycyrrhiza glabra*	Treat kidney tumors	Licorice intake over a long period of time can lead to hypertension, salt and water retention, swelling, depletion of potassium, headache, and/or sluggishness; can worsen ascites; can interact with certain drugs (diuretics, digitalis, antiarrhythmic agents, and corticosteroids).
Magnesium		Reduce cancer risk	Wide margin of safety in individuals with healthy kidneys. Can mutually interfere with absorption of antibiotics in the tretracycline and fluoroquinolone families, as well as nitrofurantoin, penicilamine, ACE inhibitors, phenytoin, and H_2 blockers. Might increase effectiveness of sulfonylureas, potentially creating risk of hypoglycemia.
Mistletoe	*Phorandendron leucarpum*	Treat cancer, esp. breast, ovarian, prostate; see Box 23.4 for information on efficacy and potential for adverse reactions	All plant parts of mistletoe are toxic. May increase hypotensive effects of antihypertensive medications. Increases sedative effects of CNS depressants. Causes cytotoxic and immunostimulant effects.
Neem; nimb	*Azadirachta india*	Prevent/treat cancer	None known; no full scientific evaluation of toxicity and side effects has been conducted.
Melatonin		Treat cancer, adjunct to chemotherapy, radiation	May cause sedation and impair balance. Might impair insulin sensitivity and glucose tolerance. Benzodiazepine drugs may impair melatonin release.
Noni, Indian mulberry, ashyulka	*Morinda citirola*	Prevent/treat cancer	No known side effects, but no safety studies have been conducted. Use by patients with severe liver disease not recommended. May interact with potassium-sparing diuretics due to high potassium content.
Poke, inkberry; fiolaca	*Phytolacca Americana*	Treat cancer	FDA has classified pokeweed as a herb of undefined safety due to narcotic-like effects. Most all parts of pokeweed appear to be toxic.
Pyridoxine/ vitamin B_6		Reduce risk of cervical cancer	Excessive intake (>2000 mg daily) can cause sensory neuropathy. May cause or worsen acne symptoms. Doses >5 mg may interfere with effects of levodopa when drug is taken alone.
Reishi; ling zhi	*Ganoderma lucidum*	Prevent/treat cancer, esp. colorectal, kidney, liver cancers; treat fibrocystic disease	No known drug interactions. Widely believed to be safe.
St. John's wort	*Hypericum perforatum*	Prevent infiltration of chest wall in breast cancer	Avoid administration with alcohol, MAOIs, narcotics, OTC cold and flu medications, digoxin, drugs metabolized by CYP3A, indinavir, amphetamines, SSRIs, tricyclic antidepressants. May result in phototoxicity. May influence hepatic microsomal enzymes.
Saw palmetto	*Serenoa repens, Sabal serrulata*	Treat prostate cancer	May cause intraoperative hemorrhage, GI complaints, nausea, vomiting, and diarrhea. May also have additive anticoagulant effects and prolong bleeding time.
Selenium		Prevent/treat cancer	May reduce efficacy of statin medications (HMG-CoA-reductase inhibitors).

(continued)

Complementary and Alternative Medicine Remedies Used in Neoplastic Diseases (*Continued*)

Remedy	Scientific Name	CAM Use	Side Effects and/or Risks
Self heal; all heal; xi ku cao	*Prunella vulgaris*	Treat cancer	None reported.
Turmeric; haridra	*Curcuma longa*	Treat breast, cervical, colorectal, lung, prostate cancers	GI ulceration with high doses or prolonged use. Avoid administration with anticoagulants. Decreases immunosuppressive effects. Avoid administration with NSAIDs; may inhibit platelet function and increase risk of bleeding.
Vitamin A		Prevent/treat cancer	Toxic in doses 10 times RDA (900 mcg/d for men; 700 mcg/d for women). HMG-CoA-reductase inhibitors and oral contraceptives are associated with increased blood levels of vitamin A. Bile acid sequestrants, mineral oil, orlistat, and neomycin may reduce vitamin A absorption. Antacids may interfere with vitamin A and beta-carotene absorption.
Vitamin E		Reduce colon cancer risk; antioxidant	Relatively safe, although high doses (>50 IU/day) may cause mild antiplatelet effects. May potentiate anticoagulant or antiplatelet medications. Could interact with herbs and supplements that possess anticoagulant or antiplatelet effects (garlic, policosnol ginkgo).
Zinc		Prevent/treat cancer, esp. prostate cancer	Long-term use significantly above nutritional requirements can cause numerous toxic effects (anemia, neutropenia, arrhythmias, increased LDL levels, decreased HDL levels, decreased glucose clearance, impaired immune function).

Sources: Fetrow CW, Avila JR. *Professional's Handbook of Complementary and Alternative Medicines*, ed 3. Philadelphia: Lippincott Williams & Wilkins, 2004.

Bratman S, Girman AM. *Mosby's Handbook of Herbs and Supplements and their Therapeutic Uses*. St. Louis: Mosby/Elsevier, 2003.

Fragakis AS, Thomson C. *The Health Professional's Guide to Popular Dietary Supplements*, ed 3, Chicago: American Dietetic Association, 2007.

Memorial Sloan-Kettering Cancer Center. About Herbs, Botanicals & Other Products. Accessed from http://www.mskcc.org/mskcc/html/11570.cfm on December 15, 2009.

REFERENCES

1. Jemal A, Siegel R, Ward E, Hao Y, Xu J, Murray T, Thun MJ. Cancer statistics. CA Cancer J Clin. 2008;58(2):71–96.

2. American Cancer Society. Statistics for 2008. Available at http://www.cancer.org/docroot/STT/stt_0_2008.asp?sitearea=STT&level=1. Updated: 2009.

3. Kushi LH, Byers T, Doyle C, Bandera EV, McCullough M, Gansler T, Andrews KS, Thun MJ, and The American Cancer Society 2006 Nutrition and Physical Activity Guidelines Advisory Committee. American Cancer Society Guidelines on Nutrition and Physical Activity for Cancer Prevention: Reducing the Risk of Cancer with Healthy Food Choices and Physical Activity. CA Cancer J Clin. 2006;56(5):254–81.

4. Gonzalez CA, Riboli E. Diet and cancer prevention: where we are, where are we going. Nutr Cancer. 2006;56(2):225–31.

5. Le Marchand L, Hankin JH, Carter FS, Essling C, Luffey D, Franke AA, et al. A pilot study on the use of plasma carotenoids and ascorbic acid as markers of compliance to a high fruit and vegetable dietary intervention. Cancer Epidemio Biomarkers Prev. 1994;3(3):245–51.

6. Anand P, Kunnumakara AB, Sundaram C, Harikumar KB, Tharakan ST, Lai OS, Sung B, Aggarwal BB. Cancer is a preventable disease that requires major lifestyle changes. Pharm Res. 2008;25(9):2097–2116.

7. Vainio H, Weiderpass E. Fruits and vegetables in cancer prevention. Nutr Cancer. 2006;54(1):111–42.

8. Greenwald P, Clifford CK, Milner JA. Diet and cancer prevention. Eur J Cancer. 2006;37(8):948–65.

9. Calle EE, Rodriguez C, Walker-Thurmond K, Thun MJ. Overweight, obesity and mortality from cancer in a prospectively studied cohort of U.S. adults. NEJM.2003; 348(17):1625–38.

10. Zhou JR, Blackburn GL, Walker WA. Symposium introduction: metabolic syndrome and the onset of cancer. Am J Clin Nutr. 2007;86(suppl):817S–819S.

11. U.S. Food and Drug Administration. United States Department of Agriculture and U.S. Department of Health and Human Services. Dietary Guidelines for Americans 2005. 6th ed. Home and Garden Bulletin no. 232. Washington (DC): 2005. Available at http://www.usda.gov/cnpp/DG2005/index.html

12. Brawley OW, Kramer BS. Prevention and early detection of cancer. In: Kasper DL, Braunwald E, Fauci AS, Hauser SL, Longo DL, Jameson L, Isselbacher KJ, editors. Harrison's principles of internal medicine. 16th ed. [Internet]. Boston (MA): McGraw-Hill; 2005. Available at http://www.accessmedicine.com/resourceTOC.aspx?resourceID=4

13. Fabian CJ, Kimler BF. Selective estrogen-receptor modulators for primary prevention of breast cancer. J Clin Oncol. 2005;23(8):1644–55.

14. Longo DL. Approach to the patient with cancer In: Kasper DL, Braunwald E,

Fauci AS, Hauser SL, Longo DL, Jameson L, Isselbacher KJ, editors. Harrison's principles of internal medicine. 16th ed [Internet]. Boston (MA): McGraw-Hill; 2005. Available at http://www.accessmedicine.com/resourceTOC.aspx?resourceID=4

15. McCance KL, Roberts LK. Biology of cancer. In: McCance KL, Huether SE, Editors. Pathophysiology: The biologic basis for disease in adult and children. 4th ed. Philadelphia (PA): Mosby; 2002.

16. Beahrs OH, Henson DE, Hutter DVP, et al. American Joint Committee on Cancer: Manual for staging of cancer. 5th ed. Philadelphia (PA): Lippincott, 1997.

17. Nelson KA. The cancer anorexia-cachexia syndrome. Semin Onco. 2000;27:64–68.

18. Bozzetti F, Mariani L. Defining and classifying cancer cachexia: a proposal by the SCRINIO Working Group. J Parenter Enteral Nutr. 2009;33:361–67.

19. Durham WJ, Dillon EL, Sheefield-Moore M. Inflammatory burden and amino acid metabolism in cancer cachexia. Curr Opin Clin Nutr Metab Care. 2009;12:72–77.

20. Sanchez O. Insights into novel biological mediators of clinical manifestations in cancer: AACN Clin Issues Adv Pract Acute Crit Care. 2004;15:112–18.

21. Capra S, Ferguson M, Ried K. Cancer: Impact of nutrition intervention outcome—nutrition issues for patients. Nutrition. 2000;17:769–72.

22. Tan BHL, Fearon KCH. Cachexia: prevalence and impact in medicine. Curr Opin Clin Nutr Metab Care. 2008;11:400–07.

23. Mantovani G, Maccio A, Massa E, Madeddu C. Managing cancer related anorexia/cachexia. Drugs. 2001;61: 499–514.

24. Eden D, Edstrom S, Bennegard K, Schersten T, Lundholm K. Glucose flux in relation to energy expenditure with and without cancer during periods of fasting and feeding. Cancer Res. 1984;44: 1718–24.

25. Inui A. Cancer anorexia-cachexia syndrome: Current issues in research and management. CA Cancer J Clin. 2002;52:72–91.

26. Tisdale MJ. Metabolic abnormalities in cachexia and anorexia. Nutrition. 2000;16:1013–14.

27. Baracos VE. Regulation of skeletal-muscle-protein turnover in cancer-associated cachexia. Nutrition. 2000;16(10):1015–18.

28. Pfeifer K. Surgery. In SE Otto (Ed.), Oncology nursing. 4th ed. Philadelphia (PA): Mosby; 2000.

29. Chabner BA, Thompson EC. Modalities of Cancer Therapy. In: The Merck Manual. Available at http://www.merck.com/mmpe/sec11/ch149/ch149b.html#S11_CH149_T002. Updated: 2009 Jul.

30. Bower MR, Martin RCG. Nutritional management during neoadjuvant therapy for esophageal cancer. Journal of Surgical Oncology. 2009;100:82–87.

31. Vaporciyan AA, Swisher SG. Esophageal cancer. In: Feig BW, Berger DH, Furhman DH (Eds.). The MD Anderson surgical oncology handbook. 4th ed. Philadelphia (PA): Lippincott Williams & Wilkins; 2006.

32. Chan AO, Wong BC, Lam SK. Gastric cancer: Past, present and future. Can J Gastroenterol. 2001;15: 469–74.

33. Murphy ME. Colorectal cancers. In: Langhorn M, Fulton J, Otto SE (Ed.), Oncology Nursing. 5th ed. Philadelphia (PA): Mosby, 2007.

34. Wolff RA, Abbruzzese J, Evans DB. Neoplasms of the exocrine pancreas. In Holland Frei. (Eds.), Cancer medicine. 5th ed. Ontario: BC Decker, Inc.; 2000.

35. Spitz FR, Bouvet M, Fuhrman GM, Berger DH. Pancreatic adenocarcinoma. In: Feig BW, Berger DH, Furhman DH (Eds.). The MD Anderson surgical oncology handbook. 4th ed. Philadelphia (PA): Lippincott Williams & Wilkins; 2006.

36. Kajiyama Y. Ann Thorac Cardiovasc Surg. The reality and the reliability of esophageal cancer treatment: which will you choose for yourself? 2009;15;211–12.

37. National Cancer Institute. Esophageal Cancer Treatment. (PDQ) Available at http://www.cancer.gov/cancertopics/pdq/treatment/esophageal/healthprofessional

38. Mundt AJ, Roeske JC, Chung TD, Weichselbaum RR. Principles of Radiation Oncology. In Kufe DW, Pollock RE, Weichselbaum RR, Bast RC, Gansler TS, Holland JF, Frei E, editors. Holland-Frei: Cancer Medicine. 7th ed. Hamilton: BC Becker Inc.; 2006.

39. Sharma RA, Vallis KA, McKenna WG. Basics of radiation therapy. In: Abeloff's clinical oncology. 4th ed. Churchill Livingstone; 2008.

40. Chu E, DeVita VT. Principles of medical oncology. In: DeVita, Hellman, and Rosenberg's cancer: principles & practice of oncology. 8th ed. Philadelphia (PA): Lippincott Williams & Wilkins; 2007.

41. Abayomi J, Kirwan J, Hackett A. The prevalence of chronic radiation enteritis following radiotherapy for cervical or endometrial cancer and its impact on quality of life. Eur J Oncol Nurs. 2009;13:262–67.

42. Copelan EA. Hematopoietic stem-cell transplantation. NEJM. 2006;354:1813–26.

43. Childs RW. Allogeneic hematopoietic stem cell transplantation. In: DeVita, Hellman, and Rosenberg's cancer: principles & practice of oncology. 8th ed. Philadelphia (PA): Lippincott Williams & Wilkins; 2007.

44. National Cancer Institute. Biological Therapies for Cancer: Questions and Answers. Available at http://www.cancer.gov/cancertopics/factsheet/Therapy/biological. Updated: 2006 Jun.

45. Ottery FD, Kasenic S, DeBolt S, Rodgers K. Volunteer network accrues >1900 patients in 6 months to validate standardized nutritional triage. Proceedings of ASCO 1998, 17, abstract 282.

46. Andreyev H, Norman A, Oates J, Cunningham D. Why do patients with weight loss have a worse outcome when undergoing chemotherapy for gastrointestinal malignancies? Eur J Cancer. 1998;34:503–09.

47. Van Bokhorst-de van der Schuer, Van Leeuwen PA, Kulk DJ, Klop WM, Sauerwein HP, Snow GB, Quak JJ. The impact of nutritional status on the prognoses of patients with advanced head and neck cancer. Cancer. 1999;86:519–27.

48. Bosaeus I, Daneryd P, Lundholm K. Dietary intake, resting energy expenditure, weight loss, and survival in cancer patients. J Nutr. 2002;132(Supp 11):3465S–3466S.

49. Sungurtekin H, Sungurtekin U, Oner O, Okke D. Nutrition assessment in critically ill patients. Nutr Clin Pract. 2009;23(6):635–41.

50. McCallum PD. Patient-generated subjective global assessment. In: The clinical guide to oncology nutrition.

Chicago:American Dietetic Association; 2000.

51. Bauer J, Capra S, Ferguson M. Use of the scored patient-generated subjective global assessment (PG-SGA) as a nutrition assessment tool in patients with cancer. Eur J Clin Nutr. 2002;56:779–85.

52. Fuhrman MP, Charney P, Mueller CM. Hepatic proteins and nutritional assessment. J Am Diet Assoc. 2004;104:1258–64.

53. Tayek JA. Nutritional and biochemical aspects of the cancer patient. In Heber D, Blackburn GL, editors, Nutritional oncology. San Diego: Academic Press; 1999.

54. Slaviero KA, Read JA, Clarke SJ, Rivory LP. Baseline nutritional assessment in advanced cancer patients receiving palliative chemotherapy. Nutrition & Cancer. 2003;46:148–57.

55. Martin C. Calorie, protein, fluid, and micronutrient requirements. In MacCallum PD, Polisena CG, editors. The clinical guide to oncology nutrition. Chicago (IL): American Dietetic Association; 2000.

56. Schnell FM. Chemotherapy-induced nausea and vomiting: the importance of acute antiemetic control. The Oncologist. 2003;8:187–98.

57. Tipton JM, McDaniel RW, Barbour L, Johnston MP, Kayne M, LeRoy P, Ripple ML. Putting evidence into practice: evidence-based interventions to prevent, manage, and treat chemotherapy-induced nausea and vomiting. Clin J Oncol Nurs. 2007;11(1):69–78. Review. Erratum in: Clin J Oncol Nurs. 2007;11(2):186.

58. Harding RK, Young RW, Anno GH. Radiotherapy-induced emesis. In Andrews PLR, Sanger GJ, editors, Emesis in anticancer therapy: Mechanisms and treatment. London: Chapman & Hall Medical; 1993.

59. Doherty KM. Closing the gap in prophylactic antiemetic therapy: patient factors in calculating the emetogenic potential of chemotherapy. Clin J Oncol Nurs. 1999;3:113–19.

60. Walsh D, Nelson KA, Mahmoud FA. Established and potential therapeutic applications of cannabinoids in oncology. Support Care Cancer. 2003;11:137–43.

61. Kwong K. Prevention and treatment of oropharyngeal mucositis following cancer therapy: Are there new approaches? Cancer Nurs. 2004;27:183–205.

62. Haas ML. Oral mucositis in radiation/chemotherapy: treatment similarities. Oncology. 2009;23(8 Suppl):23–26.

63. Stokman MA, Spijkervet FKL, Boezen HM, Schouten JP, Roodenburg JLN, de Vries EGE. Preventive intervention possibilities in radiotherapy and chemotherapy-induced oral mucositis: results of meta-analyses. J Dent Res. 2006;85(8):690–700.

64. Cherry J. Evaluation and management of treatment-related diarrhea in patients with advanced cancer: a review. J Pain Symptom Manage. 2008;36(4):413–23. Review.

65. Porter SR, Scully C, Hegarty AM. An update of the etiology and management of xerostomia. Oral Surgery Oral Medicine Oral Pathology Oral Radiology & Endodontic. 2004;97:28–46.

66. Davies AN. A comparison of artificial saliva and chewing gum in the management of xerostomia in patients with advanced cancer. Palliat Med. 2000;14:197–203.

67. Laviano A, Meguid M, Rossi-Fanelli F. Improving food intake in anorectic cancer patients. Curr Opin Clin Nutr Metab Care. 2003;6:421–26.

68. Jatoi A, Windschitl HE, Loprinzi CL, Sloan JA, Dakhil SR, Mailliard JA., et al. Dronabinol versus megestrol acetate versus combination therapy for cancer-associated anorexia: a North Central Cancer Treatment Group study. Journal of Clin Oncol. 2002;20:567–73.

69. Moertel C, Scutt AG, Reiteneier RJ, et al. Corticosteroid therapy of pre-terminal gastrointestinal cancer. Cancer. 1974;33:1607–09.

70. Willox J, Corr J, Shaw J, et al. Prednisolone as an appetite stimulant in patients with cancer. BMJ.1984;288:27.

71. Bruera E, Roca E, Cedaro L, et al. Action of oral methlyprednisolone in terminal cancer patients: a prospective randomized double-blind study. Cancer Treat Rep. 1985;69:751–54.

72. Robustelli Della Cuna G, Pellagrini A, Piazzi M. Effect of methylprednisolone sodium succinate on quality of life in pretreatment cancer patients: a placebo-controlled multi-cancer study. Eur J Cancer Clin Oncol. 1989;25:1817–21.

73. Mantovani G, Maccio A, Massa E, Madeddu C. Managing cancer related anorexia/cachexia. Drugs. 2001;61:499–514.

74. Langer CJ, Hoffman JP, Ottery FD. Clinical significance of weight loss in cancer patients: rationale for the use of anabolic agents in the treatment of cancer-related cachexia. Nutrition, 2001;17 (suppl 1), S1–S18.

75. Baracos VE. Management of muscle wasting in cancer-associated cachexia: understanding gained from experimental studies. Cancer. 2001; 92:1669–77.

76. Wigmore SJ, Ross JA, Falconer JS, et al. The effect of polyunsaturated fatty acids on the progress of cachexia in patients with pancreatic cancer. Nutrition. 1996;12 (suppl):27–30.

77. Wigmore SJ, Barber MD, Ross JA, et al. Effect of oral eicosapentaenoic acid on weight loss in patients with pancreatic cancer. Nutr Cancer. 2000;36:177–84.

78. Fearon KCH, von Meyenfeldt MF, Moses AGW, et al. Effect of a protein and energy dense n-3 fatty acid enriched oral supplement on loss of weight and lean tissue in cancer cachexia: a randomized double blind trial. Gut. 2003;52:1479–86.

79. Mazzotta P, Jeney CM. Anorexia-cachexia syndrome: a systematic review of the role of dietary polyunsaturated fatty acids in the management of symptoms, survival, and quality of life. J Pain Symptom Manage. 2008;37(6):1069–74.

80. August DA, Huhmann MB; American Society for Parenteral and Enteral Nutrition (A.S.P.E.N.) Board of Directors. A.S.P.E.N. clinical guidelines: nutrition support therapy during adult anticancer treatment and in hematopoietic cell transplantation. J Parenter Enteral Nutr. 2009;33(5):472–500.

81. Orrevall Y, Tishelman C, Herrington MK. The path from oral nutrition to home parenteral nutrition: a qualitative interview study of the experiences of advanced cancer patients and their families. Clin Nutr. 2004;23:1280–87.

82. Bozzetti F, Cozzagliio L, Biganzloi E, Chiavenna G, De Cicco M, Donati D, Gilli G, Percolla S, Pironi L. Quality of life and length of survival in advanced cancer patients on home parenteral nutrition. Clin Nutr. 2002;21(4),281–88.

83. Ireton-Jones C. Home enteral nutrition from the provider's perspective. J Parenter Enteral Nutr 2002;26:S8–S9.

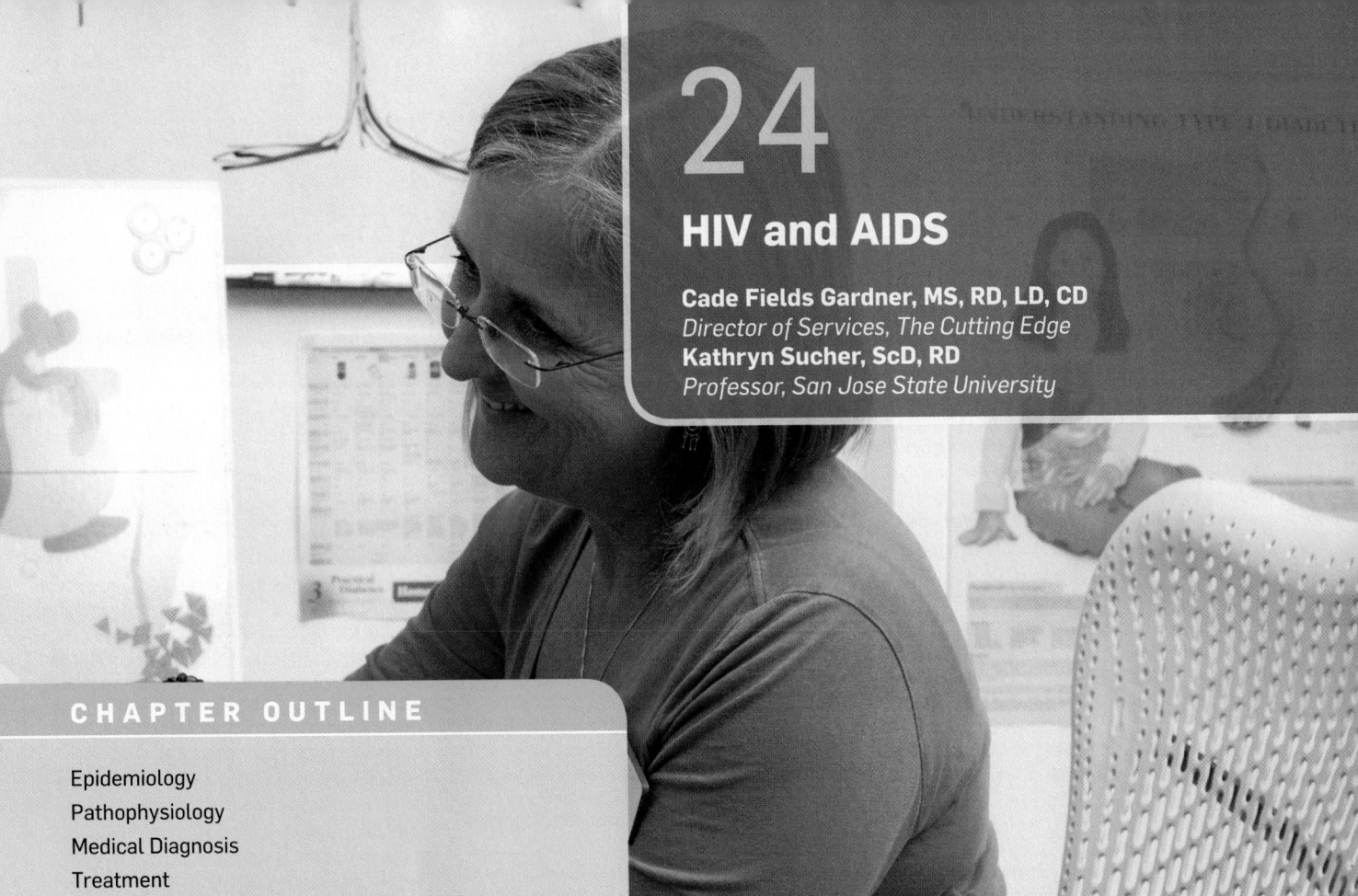

24

HIV and AIDS

Cade Fields Gardner, MS, RD, LD, CD
Director of Services, The Cutting Edge
Kathryn Sucher, ScD, RD
Professor, San Jose State University

CHAPTER OUTLINE

Epidemiology

Pathophysiology

Medical Diagnosis

Treatment

    Anti-HIV Therapies · Prevention and Treatment of
    Opportunistic Disease · Prevention and Treatment of
    Wasting · Supplements and Herbal Therapies

Nutrition Therapy

    Nutritional Implications of Disease and Treatment ·
    Nutrition Assessment · Nutrition Diagnosis · Nutrition
    Intervention · Monitoring and Evaluation

Introduction

The **human immunodeficiency virus (HIV)** is a virus that targets host immune cells and turns them into viral factories for HIV reproduction. Progression to the symptomatic condition of **acquired immunodeficiency syndrome (AIDS)** has the potential to make the infected individual vulnerable to opportunistic infections that can cause a range of disabilities or death. However, the progression from HIV infection to AIDS and from AIDS to death is no longer seen as an inevitable outcome if effective medical and nutritional interventions are initiated.

HIV is transmitted from person to person through infected body fluids. HIV is a relatively weak virus, however, and does not survive well outside the body. Thus, transmission through casual contact and the environment has been deemed unlikely. A review of the early history of HIV infection suggests that although HIV is not as easily transmitted as other infectious diseases, it has become one of the more important health threats in last four decades.

HIV and AIDS present a wide variety of challenges for maintenance of nutrition status. Changes in nutritional status can result from HIV infection, disease complications and co-infections, and disease treatments. Social, economic, and clinical issues all interact with nutritional status. Populations already at risk for nutritional compromise because of lack of health care access, lifestyle choices (such as smoking, alcohol, and drug abuse), **food/nutrition insecurity** (the lack of consistent access to an adequate and appropriate food supply), and comorbidities such as hepatitis, diabetes, or other conditions may find their health status worsened with HIV infection and AIDS. Alterations in nutrient intake, absorption, metabolism, and excretion have been documented throughout the disease spectrum. While these interactions are complex and multifactorial in nature, this chapter will concentrate on clinical aspects of the disease and potential interventions.

Epidemiology

By the end of 2007, there were an estimated 33 million people with HIV infection in the world; most resided in southern and eastern African countries.[1] Estimates of new infections in 2007 were 2.7 million people, including approximately 370,000 children. More than 2 million people may have died from complications related to HIV infection in 2007. According to the World Health Organization, the HIV epidemic has leveled off, but globally, AIDS is still considered to be among the leading causes of deaths, especially in Africa. At the end of 2006, an estimated 1.1 million persons in the United States were living with diagnosed or undiagnosed HIV/AIDS.[2] In 2007, 42,655 new cases of HIV/AIDS in adults, adolescents, and children were diagnosed in the 33 states with long-term, confidential

acquired immune deficiency syndrome or *acquired immunodeficiency syndrome (AIDS)*—immune dysfunction characterized by the destruction of immune cells, leaving the body open to infection

acute-phase proteins—proteins that increase or decrease in the blood during an inflammatory process, such as C-reactive protein and fibrinogen

AIDS-related wasting syndrome (AWS)—defined by the Centers for Disease Control and Prevention (CDC) as a 10% weight loss without an identifiable cause that is accompanied by fever or diarrhea for 30 days or more

antiretroviral (ARV)—refers to medications targeted to interrupt the retrovirus life cycle

antiretroviral therapy (ART)—refers to the combination of medications that are typically used for controlling and reducing viral load

avidity—measure of the strength of antigen-antibody binding

body cell mass (BCM)—kcalorie-using protein stores in the body; primarily muscle and organ tissues

CD4 cell—immune cell that is one of the primary targets of HIV for infection

food insecurity—lack of adequate access by all people, at all times, to sufficient food for an active and healthy life, including at a minimum a readily available supply of nutritionally adequate and safe foods and an assured ability to acquire acceptable foods in a socially acceptable way

fusion inhibitors—medications that interrupt the viral replication cycle by inhibiting fusion of the HIV virus to the target cell; also known as *entry inhibitors*

genotype—an organism's genetic makeup

HAART—highly active antiretroviral therapy; a combination of ARVs that is able to fully suppress the virus

human immunodeficiency virus (HIV)—a retrovirus that targets many cells in the body, particularly CD4 T helper cells

inflammatory response—the body's response to infection or injury that is cortisol driven and allows for the breakdown of labile body protein to increase the amino acid pool for the purpose of synthesizing protective and healing proteins

integrase inhibitors—medications that interrupt the viral replication cycle by inhibiting integrase enzymes that allow the transcribed viral DNA to integrate into the host cell DNA

lactic acidosis—an accumulation of lactic acid in the body characterized by abdominal pain, vomiting, and rapid breathing; this condition occurs in diabetes and as a potential side effect of medications

lipodystrophy syndrome—loss or absence of fat, or the abnormal distribution of fat in the body, in HIV infection; these changes are likely hormonally mediated; subcutaneous fat loss is most apparent in peripheral limbs and facial areas; fat deposits are most commonly central, located in the dorsocervical area, breast area, and abdominal region

nutrition insecurity—nutritionally inadequate diets as the result of diminished purchasing power, nutrition knowledge, etc.

oxidative stress—the imbalance of prooxidant production and the body's antioxidant supplies that yields cell damage

phenotype— physical or biochemical characteristics of an organism, as determined by both the genetic makeup and environmental influences

PLHA—people living with HIV and AIDS; other acronyms include PLWHA (people living with HIV/AIDS) and PWA (people with AIDS)

primary HIV infection—the time of the initial seroconversion to HIV infection; usually involves a spike in the level of the virus and sometimes is accompanied by a flu like syndrome

protease inhibitors—medications that interrupt the viral replication cycle by inhibiting protease enzymes that allow the viral proteins to be cleaved for reassembly into viral cores

protein turnover rate—a combination of the rates of catabolism and anabolism of protein stores in the body

retrovirus—a virus that carries RNA rather than DNA; RNA must be transcribed prior to integrating into the host cell DNA to reproduce

reverse transcriptase inhibitors—medications that interrupt the viral replication cycle by inhibiting reverse transcriptase enzymes that allow the viral RNA to be transcribed to DNA before being integrated into the host cell DNA

seroconversion—the change in a person's antibody status from negative to positive

viral load—the level of virus or viral markers measured in the blood

viron—a complete viral particle, consisting of RNA or DNA surrounded by a protein shell and constituting the infective form of a virus

---

name-based HIV reporting.[3] The majority of new infections are in minority populations, women, and youth who may have less than optimal access to health care.

Pathophysiology

HIV is a **retrovirus** approximately 0.1 microns in diameter, about 1/70th of the diameter of the **CD4** immune cell that it particularly targets. It contains nine genes, six of which are essential to penetrate and infect the target cells and produce copies of the virus. The other three genes are used to provide the necessary information to produce new viral particles in the host cell. HIV is most typically transmitted via blood through sexual contact (generally penile anal or penile vaginal intercourse), blood transfusion, intravenous needle sharing, and perinatally (from mother to child) through blood or breast milk. As with other infections, there is both an exposure and dose requirement for the body to become infected, and it is possible to be exposed without **seroconversion**, such as in the case of accidental needle sticks with very low doses of HIV transferred.[4]

This retrovirus targets many cells in the body, including gastrointestinal cells, organ cells, and immune cells, among others (see Box 24.2 for more detail). The resulting immunodeficiency syndrome is most closely related to the infection of activated CD4 (T helper cells), which become a viral factory, as shown in Figure 24.1. The virus identifies the target CD4 cell and fuses to the surface. HIV injects RNA, enzymes, and other substances that assist in viral integration and replication. Using its own kit of injected substances, HIV RNA is transcribed to DNA particles using reverse transcriptase enzymes. The DNA is carried to the nucleus and integrated into the host DNA using the HIV's integrase enzymes. At this point, the integrated viral materials can remain dormant until they are activated, at which time they command the cell to become a viral factory that manufactures viral components. Protease enzymes cleave the viral proteins for assembly into viral cores. Once fully assembled, the virus is ready to bud out of the infected host cell. As the host CD4 cell manufactures, assembles, and releases viruses, it is incapacitated and destroyed. In addition, macrophages harboring HIV are rendered dysfunctional. It is through this process that the immune system is compromised and HIV disease progresses. Table 24.1 summarizes related infections and other complications along with selected nutritional implications.

Primary HIV infection is often accompanied by flu-like symptoms and a reduction in CD4 cell counts. As CD4 and other cells are damaged and rendered dysfunctional, the body's defenses against infection and malignancy may decline. There is

HISTORICAL PERSPECTIVES

History of the HIV/AIDS Epidemic

HIV probably started its spread before 1970 and evolved into the current pandemic by the latter half of the 1970s.[1] In the late 1970s and early 1980s, there were conditions called "Slim Disease" in Zaire (now the Democratic Republic of Congo), Uganda, and Tanzania later attributed to HIV infection. HIV may have affected at least five continents, including North and South America, Europe, Africa, and Australia, by the early 1980s.[2] There may have been between 100,000 to 300,000 persons infected by the time the description of AIDS was documented in 1981.[3, 4] Rare cases of *Pneumocystis carinii* pneumonia (abbreviated PCP, though now referred to as *Pneumocystis jiroveci* in the form that is infectious to humans) were documented and on the rise in both California and New York when the Centers for Disease Control and Prevention (CDC) took notice in April of 1981 and published a report noting an unidentified cause for five cases of PCP in the Los Angeles area.

In 1982, the disease syndrome was dubbed with several names, including the gay-related immune deficiency (GRID) and the community-acquired immune dysfunction. By July 1982, there were a total of 452 cases from twenty-three states reported to CDC, including cases in Haitians and patients with hemophilia. The disease was dubbed the acquired immune deficiency syndrome at that time, suggesting the immune deficiency was acquired and the manifestation was multifactorial. When a child who had received multiple blood transfusions was diagnosed and the first cases of possible mother-to-child transmission were reported in December of 1982, it became apparent that the disease was caused by an infectious agent. By mid-1983, a virus was isolated as an etiologic agent for AIDS and was named the lymphadenopathy-associated virus (LAV). Later that year, the first globally oriented meeting to discuss an "AIDS epidemic" was convened, and surveillance was initiated. At that time, there were 3,064 reported cases and 1,292 reported deaths attributed to AIDS in the United States. In 1984, isolation of the human T lymphotropic virus type 3 (HTLV III) was announced, along with the hope that a vaccine for testing would be developed within two years. By 1986, the name human immunodeficiency virus (HIV) was adopted, and the director of the WHO suggested that there may have been as many as 10 million people with HIV infection worldwide. By the end of that year, 38,401 cases had been reported, 31,741 of which were in the Americas.

During the 1980s, the first anti-HIV treatment, azidothymidine (AZT), was tested using a double-blind methodology. Studies were discontinued early due to a significant difference in survival noted in the first six months. Many more developments in care and treatment were achieved over the subsequent decades, including medication combinations effective in reducing viral burden and progression of the disease.

References
1. Mann J. AIDS: a worldwide pandemic in Current Topics in AIDS, volume 2. Gottlieb MS, Jeffries DJ, Mildvan D, Pinching AJ, Quinn TC, editors. John Wiley & Sons; 1989.
2. AVERT. The History of AIDS 1981–1986 [monograph on the Internet]. West Sussex (UK): AVERT; 2006. Available at http://www.avert.org/his81_86.htm
3. Hymes KB, Greene JB, Marcus A. Kaposi's sarcoma in homosexual men: a report of eight cases. Lancet. 1981;2:598–600.
4. Osmond DH. Epidemiology of HIV/AIDS in the United States. HIV InSite Knowledge Base. 2003. Available at http://hivinsite.ucsf.edu?InSite?pages=kb-01-03#S1X

a strong relationship between CD4+ cell counts, HIV **viral load** (in copies per milliliter) and progression to a diagnosis of AIDS.[5] The higher the viral burden of HIV, the more CD4 cells are infected, rendered dysfunctional, and destroyed, leaving the body open to opportunistic infections and cancers. Lower viral load set points are associated with prolonged survival and reduced numbers of AIDS-defining diagnoses, as outlined by the CDC for surveillance of cases in the United States. However, the level of destruction of CD4 cells appears to be more precisely predictive of survival than viral load.[6] The reduction in numbers and function of CD4 cells is associated with an increase in incidence of opportunistic disease. These opportunistic infections and

CLINICAL APPLICATIONS

Review of the Immune System

This section provides a brief review of immune functions targeted by HIV. Immunity can be classified as: (1) primary or secondary, according to whether it is an initial or subsequent contact with the antigen; (2) humoral or cellular, depending on the activities of B cells or T cells, respectively; and (3) active or passive, depending on whether it is acquired through contact with an antigen or provided by a transfer of presensitized or activated immune cells (see Chapter 9). B cells can neutralize and destroy invaders that are not incorporated into host cells, and T cells will kill a host cell based on the expression of a foreign antigen on its surface. B cells can also go on to produce antibodies. Helper T cells (also called CD4 cells) can secrete cytokines, which stimulate proliferation of T cells. Cytotoxic/suppressor T cells kill infected host cells, modulate B and T cell responses, and activate macrophages.

Figure 24.1 HIV Life Cycle

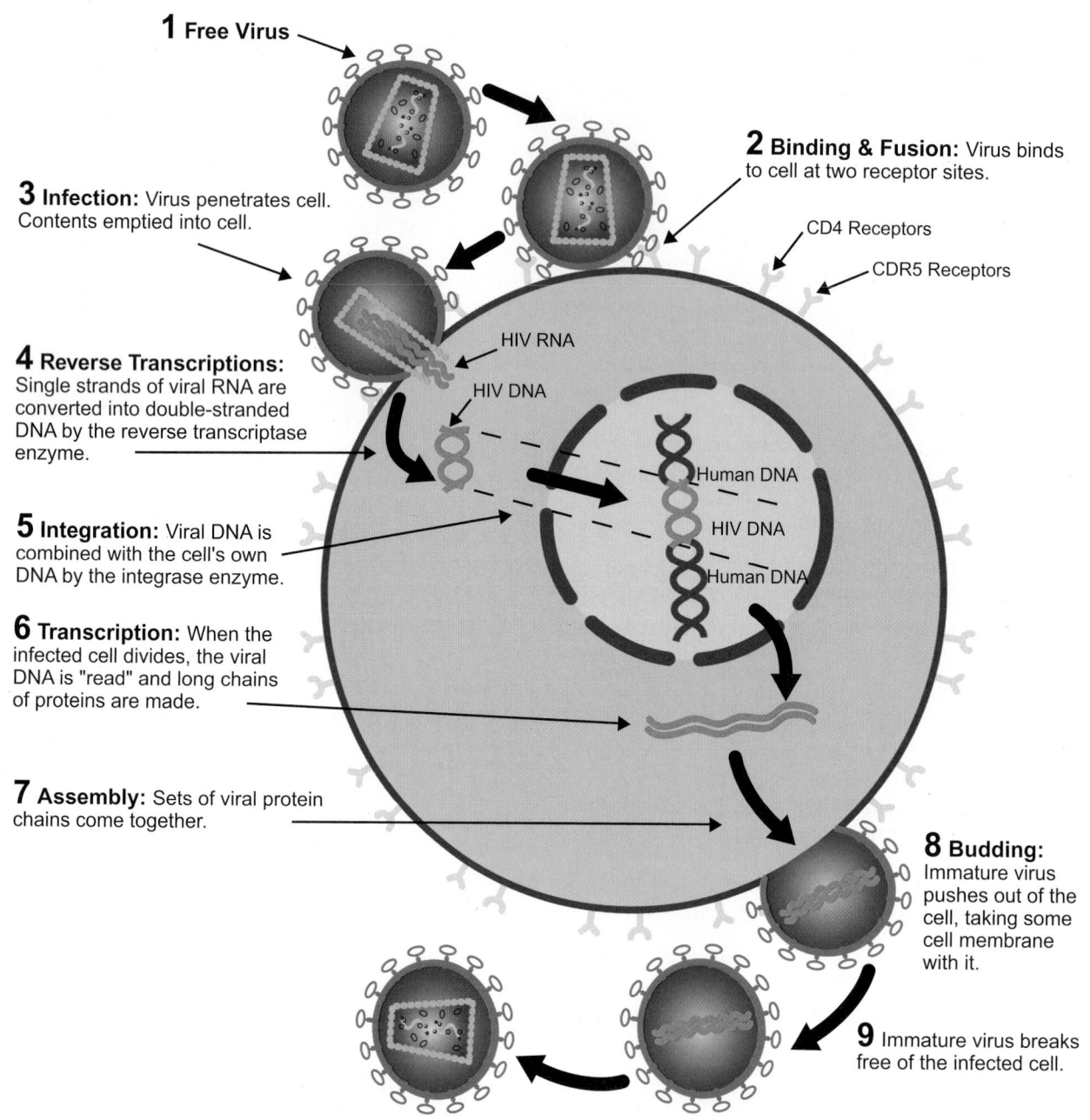

1 Free Virus

2 Binding & Fusion: Virus binds to cell at two receptor sites.

CD4 Receptors

CDR5 Receptors

3 Infection: Virus penetrates cell. Contents emptied into cell.

HIV RNA

HIV DNA

Human DNA

HIV DNA

Human DNA

4 Reverse Transcriptions: Single strands of viral RNA are converted into double-stranded DNA by the reverse transcriptase enzyme.

5 Integration: Viral DNA is combined with the cell's own DNA by the integrase enzyme.

6 Transcription: When the infected cell divides, the viral DNA is "read" and long chains of proteins are made.

7 Assembly: Sets of viral protein chains come together.

8 Budding: Immature virus pushes out of the cell, taking some cell membrane with it.

9 Immature virus breaks free of the infected cell.

10 Maturation: The protein chains in the new viral particle are cut by the protease enzyme into individual proteins that combine to make working virus.

Sources: http://www.aidsinfonet.org/factsheet_detail.php?fsnumber=106 (AIDS infonet)

malignancies (see Table 24.2) further assault the body's normal functions and are associated with nutritional decline and mortality.

The response of the intestinal tract includes the rapid turnover of gut tissue and activation of immune cells on the intestinal surface.[7,8] HIV infection is associated with depletion of CD4 cells in the GI tract, which is where more than 60% of the body's T lymphocytes reside. This provides a large reservoir of HIV-infected cells in the gut and increases risk for the malab-

sorption of nutrients.[9–11] As malnutrition progresses, malabsorption worsens, further contributing to health decline.[12] Oral manifestations can be caused by fungal infection, viral infection, bacterial conditions, neoplastic problems, salivary gland disease, and other problems. Oral lesions can lead to mouth itching, pain, a burning sensation (especially when eating spicy or acidic foods), and taste changes.

Neurologic disorders, such as neuropathy and dementia, are common in HIV infection and treatment. Neuropathy is often

Table 24.1 Selected Clinical Manifestations of HIV Infection and Treatment

System	Examples	Nutritional Implications
Cardiac	• Pericarditis/endocarditis • Pulmonary hypertension • Coronary artery disease	Patients at risk for congestive heart failure and myocardial infarction may benefit from dietary modulation.
Neurologic	• Neuropathy • Dementia/altered mental status	Pain, weakness, and hypersensitivity to touch may impair capacity to exercise and appetite; impairment in cognitive function can interfere with self-care, including food-related activities.
Gastrointestinal Tract and Symptoms	• Rapid intestinal cell turnover; infection of intestinal immune cells • Nausea/vomiting • Abdominal pain • Diarrhea	Immature enterocytes, reduction in intestinal enzyme production, and blockage of intestinal surface can lead to malabsorption with or without diarrhea. Often the side effect of medication therapies, nausea, vomiting, abdominal pain, and diarrhea may lead to inadequate nutrient intake or nutrient losses and require both dietary and medication interventions.
Hematologic	Anemias	Anemias and related fatigue can impair physical capacity and the ability to maintain body composition.
Hepatic	• Hepatitis • Biliary tract disorders	Common co-infections and disorders with HIV may lead to dietary restrictions consistent with hepatic disease; intolerance to fat and deficiency of pancreatic hormones can result in malabsorption.
Immune System	• Infiltration and destruction of immune cell function and numbers • Increased threat of opportunistic events	The body's response to infection is dependent on the severity of the infection or trauma; repeated opportunistic events can lead to progressive wasting, particularly of body cell mass, and greater risk for morbidities and mortality.
Musculoskeletal	Myopathy	Fatigue and muscle weakness limit the capacity to exercise and maintain body composition.
Oral	• Candidiasis • Herpes • Periodontal disease • Salivary gland disease	Oral pain, taste changes, and dry mouth can lead to a reduction in food intake and increased risk for weight loss.
Pulmonary	• Bacterial pneumonia • Kaposi's sarcoma • Fungal pneumonia	Fever, fatigue, difficulty breathing, and persistent coughing may impair adequate food intake; pulmonary involvement may reduce exercise capacity and limit the ability to maintain nutritional status through exercise.
Renal	HIV-associated nephropathy (HIVAN)	Progression to end-stage renal disease is less common with ART, but can lead to significant dietary restrictions.
Systemic	Inflammatory process as the body fights HIV infection	Catabolism of muscle stores and fluid shifts lead to subclinical wasting with or without weight losses; hormonal changes that occur with chronic inflammation can lead to alterations in nutrient metabolism.

peripheral and associated with pain, numbness, tingling, and burning sensations. This type of neuropathy may start with the feet and can spread. Peripheral neuropathy has been related to a number of the antiretroviral therapies. Dementia and altered cognitive functions can be related to HIV infection, other infections, and nutrient deficiencies.

Pulmonary disorders may occur both related to and unrelated to HIV infection. The level of CD4+ cell count is related to the risk for the development of several types of pulmonary diseases. Pulmonary hypertension has been demonstrated in patients with HIV infection without AIDS-defining diagnoses. Some of the many symptoms associated with pulmonary problems, including persistent coughing, chest pain, fatigue, fever, and shortness of breath, may make it more difficult for the patient to maintain adequate food intake. Fortunately, **antiretroviral (ARV)** medication has significantly reduced AIDS-related pulmonary disorders.

Cardiac manifestations may include infections and inflammation, cardiomyopathy, pulmonary hypertension, and coronary artery disease. Cardiomyopathy may be related to infectious

agents, inflammation processes, and pharmacologic therapies. For instance, the antiretroviral medication zidovudine has been associated with skeletal muscle myopathy. Much of the coronary artery disease risk is associated with medication interactions that increase blood lipid levels. However, overall it appears that there may be a protective effect of treatment medication on the mortality associated with myocardial disease.[13, 14] Chronic inflammation, immune dysfunction, concomitant conditions (such as insulin resistance, diabetes, renal involvement, and hypertension), and opportunistic conditions place a patient infected with HIV at a higher risk for cardiac disease. However, in addition to HIV infection, there are many other cofactors to consider such as smoking, drinking, and other risk factors.

Anemias are a common occurrence in chronic HIV infection, with higher prevalence in more symptomatic phases of the disease.[15] Anemia can be related to chronic disease, hormonal alterations, infections, and medications.

Renal failure in the form of HIV-associated nephropathy (HIVAN) is associated with risk factors such as infection, certain medications, male gender, and black African ancestry. Tubular

Table 24.2 Opportunistic Infections and Preventive Medications

Pathogen	Indication	First choice	Alternative
Pneumocystis pneumonia (PCP)	CD4+ count <200 cells/μL (AI) or oropharyngeal candidiasis (AII) CD4+ <14% or history of AIDS-defining illness (BII) CD4+ count >200 but <250 cells/μL if monitoring CD4+ count every 1–3 months is not possible (BII)	Trimethoprim-sulfamethoxazole (TMP-SMX), 1 DF PO daily (AI); or 1 SS daily (AI)	• TMP-SMX 1 DS PO tiw (BI); or • Dapsone 100 mg PO daily or 50 mg PO bid (BI); or • Dapsone 50 mg PO daily + pyrimethamine 50 mg PO weekly + leucovorin 25 mg PO weekly (BI); or • Aerosolized pentamidine 300 mg via Respigard II™ nebulizer every month (BI); or • Atovaquone 1500 mg PO daily (BI); or • Atovaquone 1500 mg + pyrimethamine 25 mg + leucovorin 10 mg PO daily (CIII)
Toxoplasma gondii encephalitis	Toxoplasma IgG positive patients with CD4+ count <100 cells/μL (AII) Seronegative patients receiving PCP prophylaxis not active against toxoplasmosis should have toxoplasma serology retested if CD4+ count decline to <100 cells/μL (CII) Prophylaxis should be initiated if seroconversion occurred (AII)	TMP-SMX, 1 DS PO daily (AII)	• TMP-SMX 1 DS PO tiw (BIII); or • TMP-SMX 1 SS PO daily (BIII); • Dapsone 50 mg PO daily + pyrimethamine 50 mg PO weekly + leucovorin 25 mg PO weekly (BI); or • (Dapsone 200 mg + pyrimethamine 75 mg + leucovorin 25 mg) PO weekly (BI); • (Atovaquone 1500 mg +/– pyrimethamine 25 mg + leucovorin 10 mg) PO daily (CIII)
Mycobacterium tuberculosis infection (TB) (Treatment of latent TB infection of LTBI)	(+) diagnostic test for LTBI, no evidence of active TB, and no prior history of treatment for active or latent TB (AI); (−) diagnostic test for LTBI, but close contact with a person with infectious pulmonary TB and no evidence of active TB (AII); A history of untreated or inadequately treated healed TB (i.e., old fibrotic lesions) regardless of diagnostic tests for LTBI and no evidence of active TB (AII)	Isoniazid (INH) 300 mg PO daily (AII) or 900 mg PO b/w (BII) for 9 months—both plus pyridoxine 50 mg PO daily (BIII); or For persons exposed to drug-resistent TB, selection of drugs after consultation with public health authorities (AII)	• Rifampin (RIF) 600 mg PO daily × 4 months (BIII); or • Rifabutin (RGB) (dose adjusted bases on concomitant ART) × 4 months (BIII)
Disseminated *Mycobacterium avium* complex (MAC) disease	CD4+ count <50 cells/μL – after ruling out active MAC infection (AI)	Azithromycin 1200 mg PO once weekly (AI); or Clarithromycin 500 mg PO bid (AI); or Azithromycin 600 mg PO twice weekly (BIII)	• RFB 300 mg PO daily (BI) (dosage adjustment based on drug-drug interactions with ART; rule out active TB before starting RFB
Streptococcus pneumonize infection	CD4+ count >200 cells/μL and no receipt of pneumococcal vaccine in the past 5 years (AII) CD4+ count <200 cells/μL – vaccination can be offered (CIII) In patients who received polysaccharide pneumococcal vaccination (PPV) when CD4+ count <200 cells/μL but has increased to >200 cells/μL in response to ART (CIII)	23-valent PPV 0.5 mL IM × 1 (BII) Revaccination every 5 years may be considered (CIII)	
Influenza A and B virus infection	All HIV-infected patients (AIII)	Inactivated influenza vaccine 0.5 mL IM annually (AIII)	
Histoplasma capsulatum infection	CD4+ count ≤150 cells μ/L and at high risk because of occupational exposure or live in a community with a hyperendemic rate of histoplasmosis (>10 cases/100 patient-years) (CI)	Itraconazole 200 mg PO daily (CI)	

Table 24.2 Opportunistic Infections and Preventive Medications (*Continued*)

Pathogen	Indication	First choice	Alternative
Coccidioidomycosis	Positive IgM or IgG serologic test in a patient from a disease-endemic area; and CD4+ count <250 cells/μL (CIII)	Fluconazole 400 mg PO daily (CIII) Itraconazole 200 mg PO bid (CIII)	
Varicella-zoster virus (VZV) infection	Pre-exposure prevention: Patients with CD4+ count ≥200 cells/μL who have not been vaccinated, have no history of vericella or herpes zoster, or who are seronegative for VZV (CIII) Note: routine VZV serologic testing in HIV-infected adults is not recommended Post-exposure—close contact with a person who has active caricella or herpes zoster For susceptible patients (those who have no history of vaccination or of either condition, or are known to be VZV seronegative) (AIII)	Pre-exposure prevention: Primary varicella vaccination (Varivax™), 2 doses (0.5 mL SQ) administered 3 months apart (CIII) If vaccination results in disease because of vaccine virus, treatment with acyclovir is recommended (AIII) Post-exposure therapy: Varicella-zoster immune globulin (VariZIG™) 125 IU per 10 kg (maximum of 625 IU) IM, administered within 96 hours after exposure to a person with active varicella or herpes zoster (AIII) Note: As of June 2007, VariZIG can be obtained only under a treatment IND (1-800-843-7477, FFF Enterprises)	• VZV-susceptible household contacts of susceptible HIV-infected persons should be vaccinated to prevent potential transmission of VZV to their HIV-infected contracts (BIII) Alternative post-exposure therapy: • Post exposure varicella vaccine (Varivax) 0.5 mL SQ × 2 doses, 3 months apart if CD4+ count >200 cells/μL (CIII); or • Pre-emptive acyclovir 800 mg PO 5x/day for 5 days (CIII) • These two alternatives have not been studied in the HIV population
Human Papillomavirus (HPV) infection	Women aged 15–25 yrs (CIII)	HPV quadravalent vaccine 0.5 mL IM months 0, 2, and 6 (CIII)	
Hepatitis A virus (HAV) infection	HAV-susceptible patients with chronic liver disease, or who are injection-drug users, or men who have sex with men (AII). Certain specialists might delay vaccination until CD4+ count >200 cells/μL (CIII)	Hepatitis A vaccine 1 mL IM × 2 doses - at 0 & 6–12 months (AII) IgG antibody response should be assesses 1 month after vaccination; non-responders should be revaccinated (BIII)	
Hepatitis B virus (HBV) infection	All HIV patients without evidence of prior exposure to HBV should be vaccinated with HBV vaccine, including patients with CD4+ count <200 cells/μL (AII) *Patients with isolated anti-HBc:* (BII) (consider screening for HBV DNA before vaccination to rule out occult chronic HBV infection) *Vaccine non-responders:* Defined as anti-HBs <10 IU/mL 1 month after a vaccination series For patients with low CD4+ count at the time of first vaccination series, certain specialists might delay revaccination until after a sustained increase in CD4+ count with ART.	Hepatitis B vaccine IM (Engerix-B® 20 μg/mL or Recombivax HB® 10 μg/mL) at 0, 1, and 6 months (AII) Anti-HBs should be obtained one month after completion of the vaccine series (BIII) Revaccinate with a second vaccine series (BIII)	Some experts recommend vaccinating with 40 μg doses of either vaccine (CIII) Some experts recommend revaccinating with 40 μg doses of either vaccine (CIII)
Malaria	Travel to disease-endemic area	Recommendations are the same for HIV-infected and -uninfected patients. One of the following three drugs is usually recommended depending on location: atovaquone/proguanil, doxycycline, or mefloquine. Refer to the following website for the most recent recommendations based on region and drug susceptibility. http://www.cdc.gov/malaria/ (AII)	

Definitions of abbreviations: DS = double strength; PO = by mouth; SS = single strength; bid = twice daily; tiw = 3 times weekly; SQ = subcutaneous; IM = intramuscular

Reprinted from: 34. Kaplan JE, et al Benson C, Holmes KH, Brooks JT, Pau A, Masur H; Centers for Disease Control and Prevention (CDC); National Institutes of Health; HIV Medicine Association of the Infectious Diseases Society of America. Guidelines for prevention and treatment of opportunistic infections in HIV-infected adults and adolescents: recommendations from CDC, the National Institutes of Health, and the HIV Medicine Association of the Infectious Diseases Society of America. MMWR Recomm Rep. 2009 Apr 10;58(RR-4):1–207; quiz CE1–4.

necrosis is associated with several medications commonly used in HIV infected patients, such as acyclovir, tenofovir, adefovir, and cidofovir, among other nephrotoxic medications. Nephrolithiasis (formation of kidney stones) has been associated with indinavir, and adequate hydration is a key recommendation for its use.[16] Volume depletion, which can occur in patients experiencing diarrhea related to infections or medication intolerance, increases the risk for renal involvement and hyponatremia, hypokalemia, and other imbalances.

The **AIDS-related wasting syndrome (AWS)** is an AIDS-defining diagnosis that was added into the CDC definition in 1987.[17] The definition states that weight loss of 10% without any known cause accompanied by fever or diarrhea for more than a month is AWS and qualifies for the diagnosis of AIDS. The limitations of this definition are many. Significant weight loss often occurs during opportunistic infection or other events. With the advent of combination antiretroviral therapies that can achieve low and undetectable viral loads, wasting should be less prevalent. Yet in studies on the effect of antiretrovirals on nutritional status, the results continue to be mixed, with AWS continuing to define nearly 20% of AIDS cases in the United States.[18–21] The etiology of wasting may be related to hormonal deficiencies (testosterone or thyroid), the cytokine dysregulation often associated with chronic inflammation/infection, and metabolic demands of medications.[20]

Medical Diagnosis

HIV type 1 (HIV 1) is the most common subtype of the virus found outside of the western region of Africa. There are several options for diagnosis of HIV 1 infection and AIDS. There is a "window period" of primary infection during which immune responses mount an effort to prevent HIV seroconversion. During this period, HIV antigen expression varies widely and may initially remain below the level detectable by many of the screening methodologies. Screening tests that are licensed by the Food and Drug Administration (FDA) include the testing of serum or plasma samples using laboratory measures with a high sensitivity to HIV antibodies. These screening tests are usually an Enzyme Linked Immunoabsorbent Assay (ELISA) or "rapid test" that can anticipate the possibility of HIV infection, though they are more sensitive than specific. Duplicate screening can be conducted when a test result appears reactive, and is followed by a confirmatory test such as an ELISA and Enzyme Immunoassay (EIA) to determine the concentration of antibody present.

Comprehensive testing recommendations include testing for early infection, antibody, antigen, and viral RNA levels. Available assays to identify early infection (within two weeks) evaluate p24 antigen levels and titers of viral RNA or DNA, which determine viral load. Antibody levels may increase progressively for about three to five months post exposure, and then tend to retain a set point level. As the infection transitions from recent to established, there is an increase in the aggressiveness and affinity of HIV antibodies. Testing modifications have allowed for the differentiation in antibody titer and **avidity** to allow an estimation of whether an infection is established or recent. Combination ELISA testing for both antigen and antibodies allows for earlier diagnosis of HIV infection. Though early detection may be most important for blood donor agencies, it may also provide some clinical management advantages. Confirmation tests include Western blot, modified Western blot, indirect immunofluorescent antibody assay (IFA), and line immunoassay (LIA) (see Box 24.3 for descriptions of these tests).

While HIV 1 is the primary subtype of HIV infection seen in the Americas, HIV 2 is endemic in Western Africa. Testing for HIV 2 includes a similar battery of tests and combination HIV 1 and 2 tests have been developed. Once HIV is diagnosed, differentiation between the two strains requires very sensitive testing with specific ELISA, Western blot, radio immunoprecipitation (RIPA), or polymerase chain reaction (PCR) tests. Definitive testing for newborns is difficult before the age of six months, because of the presence of maternal antibodies and the limited ability of the child's immature immune system to produce anti-

---

BOX 24.3 CLINICAL APPLICATIONS

Tests/Methods Used to Diagnose HIV Infection

1. *ELISA (enzyme-linked immunosorbent assay)*—HIV antigens interact with antibodies to the HIV virus in a sample of blood. An enzyme linked to an antibody is then added and reacts with the HIV antigen/antibody. A catalyst is added and the enzyme-linked complex changes color and can be detected.

2. *Western Blot/Modified Western Blot*—Detects specific HIV proteins in a given sample. It uses gel electrophoresis to separate the proteins, which are then transferred to a membrane where they are identified using antibodies specific to the target protein.

3. *Indirect Immunofluorescent Antibody Assay (IFA)*—HIV antibodies in the sample will interact with HIV-infected cells. The test sample is then modified to enable the bonded HIV antibody to fluoresce under UV light.

4. *Line Immunoassay (LIA)*—Measures antibodies to five HIV-1 antigens.

5. *Radio Immunoprecipitation (RIPA)*—Based on culturing HIV-infected cells in the presence of radiolabelled amino acids, allowing their incorporation into the viral proteins. This method allows for detecting low levels of antibodies.

6. *Polymerase Chain Reaction (PCR)*—Technique to amplify a single or few copies of a piece of DNA across several orders of magnitude, generating thousands to millions of copies of a particular DNA sequence.

bodies. Generally, PCR and antigen assays are most appropriate to diagnose the child who may have been infected during birth or through breast feeding with maternal HIV. Several alternative diagnostic tools are also available, including saliva tests, urine tests, and finger stick tests. Inconsistent results and other problems that may yield inconclusive, false positive, or false negative results have been reported, suggesting that careful testing procedures and confirmatory testing are important to an accurate diagnosis.

In addition to adequate testing for the presence and levels of HIV in the blood, testing for viral mutations through **genotype** and **phenotype** assays can assist the process of choosing the most potentially effective therapies. A genotype assay is used to identify points in the DNA sequence of the virus genome that may be mutated and cause resistance to treatment. Phenotype assays are used to determine viral susceptibility to a particular drug. In some cases, genotype predictions of drug resistance and phenotype testing for resistance may show different results or discordance. The prediction of virus susceptibility based on genotype and phenotype testing may assist in the effective use of medications and help to limit the exposure to medications and their associated toxicities when there is likely to be resistance.

The diagnosis of advanced HIV infection as AIDS, according to the case definition set forth by the CDC, is a reportable disease in the U.S. and is broken down into classifications that provide a schematic for management of the disease (see Table 24.3). There are three criteria for the official diagnosis of AIDS, including: a CD4+ cell count of less than 200 cells per microliter, a CD4+ cell count that comprises less than 14% of lymphocytes present, and/or an AIDS-defining illness. CD4+ cell counts and a range of complications related to immune dysfunction figure into the classification of the disease. CD4+ cell counts that range from 600–1,200 cells per microliter generally provide adequate immune defense against opportunistic disease. AIDS-defining illnesses include opportunistic infections (e.g., candidiasis, cytomegalovirus, *Mycobacterium* spp., *Pneumocystis jiroveci*), opportunistic malignancies (e.g., Kaposi's sarcoma, Burkitt's lymphoma, invasive cervical cancer), or other conditions (e.g., recurrent pneumonias, recurrent bacterial infections, wasting syndrome).[22]

In addition to AIDS-defining diagnoses, the CDC has outlined CD4 level categories and clinical categories of HIV infection. The CD4+ cell count categories are based on the lowest value ever tested in a person, or "nadir" result (see Table 24.3). Category 1 is defined as a nadir CD4+ cell count of 500 per microliter or more. Category 2 shows higher risk for some immune deficit-related conditions with 200–499 CD4+ cells per microliter. Category 3 is the highest risk category with criteria of less than 200 CD4+ cells per microliter. There are three clinical categories that describe potential clinical manifestations of immune dysfunction in adults and adolescents above 13 years old. Clinical Category A includes primary HIV infection, apparently asymptomatic HIV infection, and persistent generalized lymphadenopathy. Clinical Category B includes conditions that may be attributable to HIV infection or a defect in cell-mediated immunity other than those listed in Category C. Category C lists a series of the AIDS-defining diagnoses.

The CDC case definition for children less than 13 years old is similar to the adult schematic, featuring both immunity decline

Table 24.3 CDC Clinical and Immune Cell Categories of HIV Infection

Classification/ Categories*	Sample Criteria
CD4+ Cell Count Categories	
Category 1	≥500 cells/μL
Category 2	200–499 cells/μL
Category 3+	<200 cells/μL
Clinical Categories	
Category A	No symptoms other than persistent generalized lymphadenopathy or those associated with primary HIV infection.
Category B	Symptomatic conditions attributed to HIV infection or defect in cell-mediated immunity or require management that is complicated by HIV infection. Examples include: bacillary angiomatosis, oral or vaginal candidiasis, cervical dysplasia, oral hairy leukoplakia, idiopathic thrombocytopenia, listeriosis, peripheral neuropathy, persistent diarrhea, persistent fever.
Category C+	Includes diseases that are opportunistic and define AIDS. Examples are: disseminated candidiasis, invasive cervical cancer, cryptosporidiosis, cytomegalovirus, HIV related encephalopathy, disseminated or extrapulmonary histo plasmosis, Kaposi's sarcoma, several types of lymphoma, disseminated or extrapulmonary *Mycobacterium* spp., *Pneumocystis jeroveci*, recurrent *Salmonella* septicemia, wasting syndrome.

*Disease diagnosis remains at the most advanced stage diagnosed regardless of current condition. For example, if a patient has ever had a diagnosis in Clinical Category C, that diagnosis of AIDS remains regardless of the patient returning to an asymptomatic condition.
+CD4 Cell Category 3 and Clinical Category C are criteria for the diagnosis of AIDS.
Source: Centers for Disease Control and Prevention. 1993 Revised Classification System for HIV Infection and Expanded Surveillance Case Definition for AIDS Among Adolescents and Adults (1992).

and symptoms. In the children's categories, the severity of symptoms is considered. The 1994 revised pediatric categories for immune suppression (listed in Box 24.4) are age based.[23] Clinical categories include N (no symptoms), A (mild symptoms), B (moderate symptoms), and C (severe symptoms).

These categories are not a dynamic way to classify a person's current condition; the classifications for individuals appear to represent a linear, progressively forward development of HIV infection when this is not necessarily the case. For instance, if a person is diagnosed with an AIDS-defining condition, they will remain categorized with the AIDS diagnosis even when they recover or reverse the condition and have no current AIDS-defining condition. A person who has been diagnosed with a condition listed in Category B will remain in that category even when they are apparently asymptomatic. However, a person previously diagnosed as Category B who develops a Category C condition is reclassified as Category C from that point forward. It will be important to know not only the category of conditions from a patient's medical history, but also his or her current condition in order to make a determination about clinical management strategies. In reality, a person can progress to the diagnosis of AIDS without a continued progression to an early death.

BOX 24.4 | Life Cycle Perspectives

Classifications of Pediatric HIV Infection

In 1994, the CDC revised its classifications for pediatric HIV infection. As with the adult categories, pediatric classification relies on both clinical categories and immunologic status. The expected CD4 cell counts differ according to age, as shown in the following table.

CD4 Cell Count by Age

Immune Suppression	Less than 1 year old	1–5 years old	6–12 years old
Normal	≥1500/μL	≥1000/μL	≥500/μL
Moderate	750–1499/μL	500–999/μL	200–499/μL
Severe	<750/μL	<500/μL	<200/μL

This information is then considered in a second step that includes the Clinical Categories. These categories are:

- *Category N:* No symptoms or one of the conditions noted in the Adult Classification Category A.

- *Category A:* Mild symptoms, including two or more symptoms of dermatitis, lymphadenopathy, hepatomegaly, parotitis, recurrent or persistent upper respiratory infection, or splenomegaly.

- *Category B:* Moderate symptoms that may be attributed to HIV infection, including anemias (low hemoglobin, neutropenia, thrombocytopenia that persists), a single episode of bacterial infection, candidiasis, cardiomyopathy, hepatitis, herpes simplex (recurrent), lymphoid interstitial pneumonia, nephropathy, persistent fever or diarrhea, and others.

- *Category C:* Severe symptoms, including all the conditions in Adult Category C.

The combination of the immune suppression categories and clinical categories yields the results shown in the following table.

Immunologic Category	Clinical Category N	Clinical Category A	Clinical Category B	Category C
Normal	N1	A1	B1	C1
Moderate	N2	A2	B2	C2
Severe	N3	A3	B3	C3

Table 24.4 WHO Clinical and Immune Cell Categories of HIV Infection

Categories	Sample Criteria
Primary HIV Infection	Acute retroviral syndrome, but no complicating opportunistic infection or immune dysfunction
Clinical Stage 1	Primarily asymptomatic as above, possible persistent generalized lymphadenopathy
Clinical Stage 2	Weight losses that are <10% of body weight, herpes zoster, minor mucocutaneous manifestations, recurrent bacterial upper respiratory tract infections, fungal infections of fingers, papular pruritic eruptions, or seborrhoeic dermatitis
Clinical Stage 3	Weight loss of >10% of body weight, persistent constitutional symptoms (fever, diarrhea), oral candidiasis or hairy leukoplakia, acute necrotizing ulcerative gingivitis or necrotizing ulcerative periodontitis, pulmonary tuberculosis, severe bacterial infections, unexplained anemia, neutropenia, and/or thrombocytopenia for more than a month (confirmatory testing is required for anemias)
Clinical Stage 4	HIV wasting syndrome (>10% weight loss with chronic diarrhea, weakness, fever), opportunistic events as described in Clinical Category 3, as well as chronic herpes simplex virus, recurrent severe or radiological bacterial pneumonia, pneumocystis pneumonia, Kaposi's sarcoma, CMV, cryptosporidiosis, isosporiasis, any disseminated mycosis (e.g., coccidiomycosis, histoplasmosis, penicilliosis), recurrent non-typhoidal salmonella septicaemia, and/or lymphoma (cerebral or B cell non-Hodgkin)

Source: World Health Organization. Interim WHO Clinical Staging of HIV/AIDS and HIV/AIDS Case Definitions for Surveillance (2005).

With adequate treatment, a return to and maintenance of an asymptomatic state is achievable for extended periods of time.

The WHO has also developed a clinical staging system, which was tested to improve it for international use in 1993.[24] This system attempts to categorize patients into four clinical stages according to the severity of disease manifestations (see Table 24.4). The current set of staging guidelines has separated primary infection from three clinical stages, similar to the CDC classifications.[25] Additional diagnoses that may be recognized in the WHO set of guidelines include fungal nail infection and dermatitis for Clinical Stage 2 and unexplained but confirmed anemias for Clinical Stage 3.

Treatment

Treatment for chronic HIV infection includes antiretroviral (ARV) medications, prevention and treatment for opportunistic events, modulation of altered hormonal milieu, and maintenance and restoration of nutritional status. The progression of HIV infection to AIDS and the rate of mortality have dropped significantly since the introduction of combinations of therapies. The advent of effective combinations of antiretroviral drugs has reduced the reported cases of HIV's symptomatic manifestation as AIDS by nearly one third, and reported deaths by nearly half, since 1995.[26] Prior to the common use of combination medication treatments (which became widespread in 1996), HIV infection was considered a progressive disease with little chance for a return to health because of recurrent infections and other illnesses that led inevitably to death. Wasting and nutritional decline were commonly reported, with the anticipation that though approximately 20% of the initial AIDS diagnoses were made based on AIDS-related wasting, nearly all patients would experience this type of decline prior to their death. In 1989 it was reported that wasting was a strong predictor of the timing of death.[27] Since then, despite the success of HIV treatment in improving survival of infected patients in the developed world, weight loss remains an independent predictor of mortality.[20]

Anti-HIV Therapies

Antiretroviral (ARV) medications are used to lower viral load, and the goal is to achieve and maintain an undetectable level of less than 50 copies/mL in serial tests. There are currently five classes of antiretroviral medications, including **entry/ fusion inhibitors**, nucleoside/nucleotide **reverse transcriptase inhibitors** and non-nucleoside reverse transcriptase inhibitors, **integrase inhibitors**, and **protease inhibitors** . A summary of ARV treatments is shown in Table 24.5. Combinations of ARV medications that inhibit the various segments of the life cycle of HIV infection effectively have been referred to as "highly active antiretroviral therapy" or **HAART**, and generally include the use of three or more medications. Recommendations for the use of **antiretroviral therapies (ART)** in adults and children have been developed by the CDC and are generally followed by most practitioners.[28–31]

The categories of HIV infection used in guidelines to determine treatment strategies include the degree of immunosuppression as well as opportunistic events.[28–31] Anti-HIV treatments are aimed at interrupting the viral life cycle at one or more points. However, because the virus can reproduce rapidly, generating between a billion and a trillion **virons** per day, the potential for alterations in genetic structure (or "mutation"), and hence drug resistance, is high. Several strains may exist in a single person, and depending on the suppression of the particular strains, some strains may still replicate, leading to immune destruction despite treatment. Combination medications that attack different parts of the viral life cycle or that can be used with strains that are resistant to some of the drugs in a class can be used to prevent disease progression and further health decline.

The transmission of a mutated virus from one person to another can also limit the number and types of medications that can successfully suppress viral burden and prevent disease progression. A virus that has not undergone mutations due to the use of ART is referred to as a "Wild Type" virus and is generally susceptible to most ARV medications. If the presence of ARVs causes the virus to mutate, then drug resistance can follow. Once mutations occur, ART can "fail" as viral load increases and CD4+ cell destruction accelerates. ARVs may be introduced and discontinued based on observation of these effects or through direct and quicker genotype and phenotype testing for specific types of mutations (see the "Diagnosis" section).

Adherence to therapy regimens is a major determinant of ART success. Successful ART requires nearly perfect adherence at 95% or better. However, actual adherence rates are estimated to be much lower, largely because of fear and experience of the side effects and symptoms of these potent chemotherapy combinations.[32] The most common challenges to adherence include pill burden, complexity of regimen, understanding of appropriate use of the medications (including diet interactions), and po-

tential or experienced adverse effects. With adequate preparation, screening for the most appropriate therapy regimens, and careful management of adverse effects of therapy, adherence can be improved. Treatment is only one aspect of the medical management of HIV infection. Interestingly, even in so-called asymptomatic states where viral load is fully suppressed to undetectable levels, the effects of HIV infection continue to exist with evidence of continuing **inflammatory responses** that can alter body functions and maintenance and in part be responsible for the wasting seen in AIDS.

Prevention and Treatment of Opportunistic Disease

Opportunistic infections and other diseases that occur as a result of immune dysfunction are associated with further nutritional and health decline. Opportunistic infections may initiate a reduction in food intake and lead to episodic wasting as well as activate CD4+ cells, making them active viral factories and a target for further spread of the infection in the body.[33] Prophylaxis and early treatment for opportunistic events can help to prevent nutritional decline and disease progression. Guidelines for the prevention and treatment of opportunistic infections and other events in people living with AIDS (**PLHA**) have been published by the CDC.[34] Prevention efforts with medications are recommended for varying CD4+ levels according to the risk or history of infection. Table 24.2 includes a summary of selected treatments for opportunistic events.

Prevention and Treatment of Wasting

Pharmacologic interventions targeted to reduce risk and reverse wasting and weight loss include appetite stimulants, anti-catabolic and anabolic medications, and hormone replacement therapies. A summary of selected medications is shown in Table 24.6.

Two of the primary medications used for treatment of anorexia in HIV infection include megestrol acetate (Megace) and dronabinol (Marinol). Megestrol acetate is an antineoplastic and progestational agent that has been used to successfully improve appetite and weight gain in cancer-related and HIV-related wasting. Recent pharmacokinetic information on megestrol acetate suggests that the original formulation (Megace) had a diminished absorption when taken with food. However, a new formulation utilizing nanocrystal technology overcomes this effect and decreases the required dose volume in the product Megace ES as compared to Megace.[35] Dronabinol (Marinol), a cannabinoid, has been used in cases of anorexia, nausea, and pain. Some concern has been expressed about the potential for abuse and immunosuppression of cannabinoid drugs; however, research has explored and refuted these issues as serious concerns.[36–39] There has also been some exploration of "medical marijuana" as a means to improve appetite in chronic HIV infection. While both oral and smoked forms of the drug appear to have some effect on appetite, the smoked form may show localized immunosuppressive effects in the lungs.[36] In addition, there is always a concern about the interaction between a medication and the primary treatments for HIV. One study examined the potential for smoked and oral cannabinoids to interact with nelfinavir and indinavir. While the cannabinoids appeared to exert an effect in decreasing concentrations of the antiretrovirals, the authors stated that it

Table 24.5 Food- and Nutrition-Related Medication Interactions for Selected Antiretroviral Therapies

Drug/Abbreviation	Class	Brand Name	Diet Requirements	Diarrhea	Nausea/ Vomiting	Appetite Loss	Abdominal Pain	Taste Change	Lipid Alterations*	Glucose Intolerance	Lipo- dystrophy
Abacavir/ABC	NRTI	Ziagen	No food restrictions	X	**X**	X	X				
Amprenavir/APV	PI	Agenerase	Avoid taking with high-fat meal	X	**X**		**X**	**X**	**X**	**X**	
Atazanavir/ATV	PI	Reyataz	Take with food	X	X			X		X	X
Combivir(AZT & 3TC)	NRTI		No food restrictions	X	X	X	X			X	
Darunavir/TMC114/r	PI	Prezista	Take with food	X	X		X		X	X	X
Delavirdine/DLV	NNRTI	Rescriptor	No food restrictions; take with acidic beverage	X	X	X					
Didanosine/ddI	NRTI	Videx	Take without food	**X**	**X**	**X**	X	X	X		
Efavirenz/EFV	NNRTI	Sustiva	Avoid taking with high-fat meal	**X**	X	X	X		X		
Emtricitabine/FTC	NRTI	Emtriva	No food restrictions	**X**	X	X			X		
Enfuvirtide/T 20	FI	Fuzeon	No food restrictions			X		X	X		
Epzicom (abacavir & 3TC)	NRTI		No food restrictions	X	X	**X**	**X**	X	X	**X**	
Efavirenz	NNRTI	Sustiva	Take on an empty stomach before going to sleep	X			X				X
Fosamprenavir/FPC	PI	Lexiva	No food restrictions	X	X		X				
Indinar	PI	(Crixivan)	Take with lots of water, on empty stomach or with low-fat snack	X	**X**		X				X

Drug	Class	Brand	Food restrictions								
Lamivudine/3TC	NRTI	Epivir	No food restrictions	X	X	X	X			X	X
Lopinavir/LPV/r	PI	Kaletra	Take with food (no food restrictions)	**X**	**X**	**X**	**X**		X	X	X
Maraviroc	FI	Selzentry, Celsentri	No food restrictions	**X**	X	X	X				X
Nelfinavir/NFV	PI	Viracept	Take with food	X	X	X	X			X	X
Nevirapine/NVP	NNRTI	Viramune	No food restrictions		**X**	X	X				
Raltegravir	II	Isentress	No food restrictions	**X**	**X**	X	X				X
Ritonavir/RTV	PI	Norvir	Take with food	**X**	**X**	X	X	X	X	X	X
Saquinavir/SQV	PI	Fortovase or Invirase	Take with food	**X**	X	**X**	X	X	X		X
Stavudine/d4T	NRTI	Zerit	No food restrictions	**X**	**X**	X	X		X		X
Tenofovir/TDF	NNRTI	Viread	No food restrictions (high-fat meal increases bioavailability)	**X**	**X**	X	X				X
Tipranavir/TPV	PI	Aptivus	Take with a full meal, preferably high-fat	**X**	X	X	X		X		X
Trizivir TZV (3TC, ACT & AZT)	NRTI		No food restrictions	X	**X**	**X**	**X**			**X**	
Truvada (tenofovir & emtricitabine)	NRTI		No food restrictions	X	**X**	**X**	X				**X**
Zidovudine/AZT or ZVD	NRTI	Retrovir	No food restrictions	**X**	**X**	**X**	X				

Bolded X suggests increased incidence or severity.

*Protease inhibitors are often associated with elevated lipids; complete information on the effect of darunavir on lipids was unavailable, however elevated total cholesterol was reported by Grinsztejn B et al. 2005.

Sources: Drug Facts and Comparisons, 2006; New Mexico AIDS Education and Training Center. Current Antiretroviral Drugs (Fact Sheet 401). Revised August 28, 2008. Available at http://www.aidsinfonet.org/fact_sheets/view/401#anchor50580; Pronsky, ZA. Food Medication Interaction, 15th ed. 2008.

Table 24.6 Selected Non-Nutrient Therapies for Wasting

Therapy	Description
Androgens	• *Nandrolone:* injection, occasional liver function and alkaline phosphatase increase, enhanced effect on improved body cell mass with exercise regimen • *Oxandrolone:* occasional nausea and vomiting, potential for hepatotoxicity, enhanced improvement of body cell mass when used with exercise; monitor liver function and blood lipids • *Oxymetholone:* possible edema and potential for hepatoxicity, glucose intolerance, enhanced improvement of body cell mass when used with exercise; monitor liver function and blood lipids • *Testosterone:* injection, potential for hepatoxicity, elevated blood lipids, enhanced improvement of body cell mass when used with exercise; monitor liver function and blood lipids
Appetite Stimulants	• *Dronabinol:* some CNS effects; modest weight gains • *Megestrol acetate:* hyperglycemia, hypogonadism (with long-term use), adrenal insufficiency (especially in children); megestrol acetate ES (nano crystal form) eliminates food effect; may be used with testosterone therapy
Exercise	Both resistance and aerobic exercises are important to maintain body composition and normalize metabolism, especially in conditions such as wasting, insulin resistance, cardiovascular risk, and lipodystrophy
Growth Hormone	Used for both wasting and lipodystrophy (fat accumulation); in lower doses, may cause muscle discomfort, potential for fluid retention, hypertension, fat atrophy, insulin resistance
Insulin Sensitizing Agents	• *Metformin:* reduces hepatic insulin resistance and may be used to reduce the effect of fat accumulation; caution with renal or hepatic dysfunction; may cause nausea, vomiting, fullness, diarrhea, flatulence; monitor for lactic acidosis • *Glitazones:* reduces peripheral insulin resistance and may be used to reduce the effect of fat atrophy; caution with liver dysfunction; may cause increases in blood lipids

therapies, exercise, and adequate diet.[47, 48] Oral agents include oxandrolone and oxymetholone. Injection and patch agents are testosterone based and the injectable nandrolone decanoate is a synthetic hormone. Caution should be taken for patients with active liver disease, and liver function should be carefully monitored at higher doses.[49] Growth hormone and related treatments have demonstrated effects in both improving **body cell mass (BCM)** and reducing central fat accumulation in **lipodystrophy syndrome**.[50, 51] The dose used for reversal of wasting is substantially higher than the dose recommended for use in reducing central fat accumulation.

Supplements and Herbal Therapies

Supplements should be considered as pharmaceutical interventions with a potential for interactions with prescribed therapies.[52] For instance, garlic has been recommended as a natural anti-fungal agent and large doses have been recommended to assist in treatment of infections. However, garlic supplements can potentially interact with ARVs or other therapies because of their effects on the liver's enzyme processing.[53] In some cases, therapies can increase the serum level of the medications, causing a risk for toxicity; in others, they can reduce the medication levels, causing a reduction in effectiveness of the drug.[54]

Supplements and herbal medications are commonly used by PLHA (Table 24.7). While some supplemental herbal and nutrient preparations have the potential to be beneficial, potential medication interactions have been suggested for some of the supplements.[55] There are a few herbal medications that have known toxic effects that are particularly important to avoid in diseases such as chronic HIV infection. Liver toxicities are associated with borage, coltsfoot, and germander. Renal toxicity is associated with calamus. Hypertension is associated with ephedra and glycerin. Vaso-occlusive disease is associated with comfrey and life root. Complete information regarding contents and potency may not be available for such supplements, since they are not regulated like traditional medications are. Additional studies are required to determine the safety and efficacy of complementary and alternative medicines when used in conjunction with ARVs and other medications.

Nutrition Therapy

Support for the maintenance and restoration of nutritional status is essential to the management of chronic HIV infection. Poor clinical outcomes have been documented in patients who are malnourished compared to those who are not. Intervention with macronutrient and micronutrient therapies to support immune function could play an important role in the course of HIV disease.[19, 56]

HIV infection itself can present a formidable challenge to the maintenance of nutritional status and health. Because there is no cure for HIV infection, treatment strategies are aimed at preventing and slowing immune compromise and containment of complications of immune suppression, side effects of medication, opportunistic disease, development of chronic disease, and nutritional compromise. In order to provide integrated nutritional care, the clinician should have a good working knowledge of the disease, common complications and co-diseases, treatments, and potential for interactions and adverse events.[57–59]

is unlikely that the short-term clinical effects of the reduction of medication levels would be a concern.[37] Even so, it may be of concern that cannabinoids delivered through smoking may lead to some destruction of lung immunity and slowed ability to heal lung tissues.

Several medications have been used to specifically support and restore lean tissues, normalize hormonal balances and function, and improve health and quality of life. Low sex hormone levels have been treated with hormone replacement therapy.[40] Both men and women experiencing hypogonadism may lose lean and fat tissues, including bone mineral density, and experience related anemias.[41–43] Testosterone replacement is available in injection, patch, and gel forms, and acceptability and efficacy may vary among patients.[44] Physiologic dosing of androgens has been used in women less commonly, but may provide some benefit, including support for the maintenance of lean tissues.[45, 46] Improved muscle function may be enhanced with the combination use of hormone replacement, anabolic

Table 24.7 Effects of Selected Herbal and Other Complementary Therapies in HIV Disease

Therapy	Description, Claims, and Comments
Aloe vera; kumari, sabila	Used to inhibit HIV; may reduce absorption of some medications
Alpha-lipoic acid	Antioxidant; possible improvement of neuropathy; slowed in vitro replication of HIV
Anemarrhena, zhi mu	Used for potential anti-HIV and anti-cancer properties; may increase insulin stimulation
Ascorbic acid (vitamin C)	Improves antioxidant capacity of body and immune system; increases iron absorption from non-heme sources; upper limit is 2,000 in adults and less in children; increases urinary losses of oxalate and calcium
Ashwagandha; winter cherry, Indian ginseng	Used for antidementia, antibacterial, anticancer, and antioxidant; reduces gastric acidity; possible increase in serum T4 (thyroid); can potentiate barbiturate drugs
Astragalus, milk vetch; huang qi	In combination with other herbal medications for anti-HIV (in vitro information) as immune enhancer
Atractylodes; bai zhu; cang zhu	Anti-HIV activity effect on enzymes; potential for increased insulin and hypoglycemia or loss of blood glucose control
Burdock; niu bang zi	Immune modulator; increases insulin and may cause hypoglycemia or loss of blood glucose control
Cat's claw; gambir; gao teng	Used as antifungal, antiherpes; may inhibit platelet aggregation (anticoagulant)
Chromium	Used to improve insulin sensitivity and improve lipid profile; speculated as potential treatment for lipodystrophy/metabolic syndrome in HIV; in vitro and animal studies suggest cellular DNA damage associated with long-term intake; picolinate form may have increased absorption rate; suggested competition for iron binding to proteins
Cobalamin (vitamin B$_{12}$)	Used to reverse low levels of B$_{12}$; in vitro inhibition of HIV 1; suggested to reduce effects of fatigue, anemia, neuropathy, cognitive impairment
Coenzyme Q10	Decreased blood levels are seen in HIV infection; suggested to improve immunity, energy; some gastric distress is associated with supplementation
Dehydro epiandrosterone (DHEA)	Adrenal hormone supplement used to prevent and treat muscle wasting, improve resistance exercise performance; suggested to increase testosterone levels; possible small improvement in immune function; improved mood; caution is suggested for cancer risk
Dong quai	Anticoagulant; may change effectiveness of estrogen replacement; may work synergistically with calcium channel blockers
Echinacea, purple coneflower	Anticancer; immunostimulatory; may inhibit metabolism of drugs using the cytochrome P 450 enzyme pathway
Garlic; ajo; lasuna	Antifungal and antimicrobial properties; hypoglycemic agent—should be used with caution in patients with altered blood glucose control; may work synergistically with anticoagulants, antihypertensives, lipid-lowering agents; competes with liver processing of ARVs; inhibits uptake of iodine by thyroid gland
Ginkgo biloba	Antioxidant; memory enhancer; anticoagulant; may interact adversely with ARVs
Glutamine	Immune enhancing; antiwasting; treatment in diarrhea
Goldenseal; yellow root; yellow puccoon; yellow Indian paint; ground raspberry	Antibiotic properties; long-term use can decrease absorption and utilization of B vitamins; may exacerbate hypertension; high doses can also cause nausea, vomiting, and tingling in the hands and feet (known as neuropathy)
Gotukola; brahmi	Immunomodulatory; possible anticancer; suggested to have antioxidant effects; hypoglycemic and may reduce blood glucose control
Grape seed extract	Antioxidant, anticancer, anti-inflammatory; anti-HIV effects hypothesized
L-carnitine	Large doses required; suggested for antioxidant (decrease TNF alpha cytokine), immunomodulation (increase CD4 cells), cognitive enhancement, and other problems; treatment for vomiting and lactic acidosis associated with NRTI ARVs; IV form used for neuropathy treatment related to NRTI ARVs, oral form shows promise in some neuropathy improvement; speculated as potential treatment for lipoatrophy; some nausea, gastritis, cramps, and diarrhea associated with oral doses; diet high in carbohydrate and low in fat is recommended as adjunctive treatment for carnitine deficiency; slight improvements in lipid profile may be seen
Licorice root; glycerrhizin; mulethi; gan cao	Antiviral, antibiotic; suggested to reduce HIV infection by increasing cell fluidity; suggested to reduce liver damage related to infection and medications; interacts with potassium-losing diuretics to increase risk of hypokalemia; may cause hyperglycemia, electrolyte imbalance, and increase risk for side effects with MAOI medications; oral contraceptives may have reduced effectiveness; may increase interferon and isoniazid drugs
N acetyl cysteine, NAC	Glutathione precursor; antioxidant; anti-inflammatory; decreased viral load and increased CD4 cell count
Selenium	Immune enhancement (lymphocyte function), antioxidant cofactor; high doses are associated with nausea, diarrhea, peripheral neuropathy, fatigue, and immune dysfunction
St. John's wort; hypericum	Antidepressant, antianxiety, anti-HIV; contraindicated with the use of medications processed by the CYP3A4 and P glycoprotein pathways, including protease inhibitors and NNRTIs; reduced effectiveness of oral contraceptives; antagonistic to antihypertensive medications
Zinc	Immune enhancement; antifungal; competes with copper and long-term high doses are associated with anemias and immunotoxicity; adverse effect on LDL:HDL; high dose can cause nausea, vomiting, taste changes

Source: NIH, National Center for Complementary and Alternative Medicine. Available at http://nccam.nih.gov/

Nutritional Implications of Disease and Treatment

After exposure, seroconversion to an HIV positive status starts the lifelong inflammatory process and, as previously mentioned, the development of AIDS-related wasting syndrome (AWS). This process is similar to other infections in that it is the severity of infection that determines the level of inflammatory response. A simplified overview from a nutritional response standpoint includes a continued cortisol response to HIV infection, leading to breakdown of labile body protein stores to feed the inflammatory response. This response includes alterations in nutrient intake and utilization along with changes in the sensitivity and levels of hormones that regulate nutritional status.[60, 61] Even during "asymptomatic" phases of HIV infection, body composition testing shows alterations reflecting the inflammatory process and suggesting that subclinical wasting can continuously and progressively occur.[62] However, ongoing cohort studies suggest that the lack of changes in prevalence of wasting may reflect nutritional insecurity more than the disease process when compared to the pre-HAART era.[63]

Initial infection with HIV is handled by the body similarly to other infections. The body breaks down labile protein stores, primarily muscle, to enhance the amino acid pool and allow the synthesis of proteins that initiate and maintain immune and healing responses.[64, 65] **Protein turnover rates** and the level of immune system activity are elevated throughout the course of HIV infection and disease, while anabolic rates are more directly regulated by nutrient substrate availability. HIV has been associated with additional abnormal protein metabolism comparable with tuberculosis.[66] Deranged protein metabolism has also been observed in HIV-infected children, suggesting that protein catabolism may not be downregulated even during accelerated protein synthesis.[67]

Changes in body composition during weight loss in HIV infection appear to fit the profile of starvation or marasmus.[68] This suggests that, though weight loss episodes are generally matched to an opportunistic event, a combination of increased nutrient needs, decreased nutrient intake, and malabsorption may play a role in precipitating weight loss episodes. Patients with advanced HIV infection may even consume more kilocalories than their asymptomatic and HIV negative counterparts, while still losing ground.[69] Patients who lose weight during opportunistic infection may not fully recover their weight or normalize their body composition.[34] Conflicting evidence suggests that wasting may be more complicated than simple starvation, and weight loss might not always be significant while a wasting process occurs. A longitudinal evaluation of 172 HIV-infected men revealed a loss of lean tissues without significant changes in body weight. This suggests that body composition changes may be a common occurrence in chronic HIV infection, and be driven by cytokine mediators rather than inadequate dietary intake or by altered androgen levels.[70]

Alternate definitions of wasting have been offered to include both weight losses and changes in body composition that increase risk for functional impairment.[71] Weight loss and wasting can accompany opportunistic infection. Evidence suggests that 10% weight loss for any reason and over any period of time is a strong predictor of death in HIV-infected patients, that even less than 5% weight loss may be a risk factor for mortal-

Table 24.8 Suggested Criteria for the Diagnosis of Wasting in HIV Disease

Parameter	Criteria
Weight loss	10% loss over 12 months or 7.5% loss over 6 months
Body mass index (BMI)	<20
Body cell mass	5% loss over 6 months or <35% of weight if BMI is less than 27 in men, <23% of weight if BMI is less than 27 in women

ity, and that there may be an increase at the 5% level in the post-HAART era.[72–74] A body mass index (BMI) of less than 20 suggests an increased risk for mortality in patients with HIV infection.[75] BMIs of less than 18 are markers of mild to moderate malnutrition and are associated with a significantly reduced survival.[76] Patients with higher BMI levels, on the other hand, have a decreased risk for HIV disease progression.[77]

In addition to weight loss, body composition changes without weight loss can compromise clinical and nutritional status. Additional wasting criteria may include a loss of 5% BCM over a period of six months and a BMI of <20 (see Table 24.8).

In addition, other factors influence the wasting process. Some medications may have the effect of reducing muscular protein synthesis, making it difficult to recover lost protein stores as HIV viral burden and associated inflammation are decreased.[78] Treatments for wasting and for events that can trigger a wasting process, such as opportunistic infections, are both important to the restoration of patient health and quality of life.

In addition to weight loss and wasting, there are many other nutrition implications of HIV disease.[79] Because of the inflammatory nature of chronic HIV infection, evident even when viral load is fully suppressed, nutrient metabolism can be expected to change. Positive and negative **acute-phase proteins** are continually produced, and exert an effect on both macronutrient and micronutrient status. Documented micronutrient changes include lower serum levels of selenium, zinc, magnesium, calcium, iron, manganese, copper, carotene, choline, glutathione, and vitamins A, B_6, B_{12}, and E.[80–84] Elevated levels of folate, niacin, and carnitine have been documented as well.

Of specific interest are the associations between micronutrient levels and complications of immune impairment, HIV infection process and progress, and treatment interactions.[85] Low levels of vitamin B_{12} are associated with neurologic changes, bone marrow toxicity in patients taking zidovudine, and accelerated progression of HIV disease. **Oxidative stress** conditions have been associated with lower vitamin E and C levels.[86] While there are clearly associations between micronutrient status and disease management outcomes, the ability to reverse these complications remains unclear.[87]

Treatments for HIV and Opportunistic Infection

Suppression of viral load by antiretrovirals has shown varying effects on nutritional status. Weight gains are commonly seen in patients on successful ARV therapy.[18] When changes in body shape and composition were noted after the introduction of HAART, the effects of these medications were explored.[88, 89] Varying speculations suggested that alterations in body shape and metabolic indices are multifactorial and may be related to many risk factors, including ART, other medications, HIV as a chronic inflammation, and lifestyle behaviors, such as smoking, drug and alcohol abuse, and diet and exercise.

In one study, the type of medication regimen did not appear to affect triglyceride or insulin responses, while others showed a specific impact on these, hormonal balances, or other indicators of nutritional risk.[90–93] However, it appears to be a combination of successful ARV therapy, pretherapy viral load, nadir CD4 count, and improvement of CD4 counts that are predictive of body composition and other changes. For instance, baseline CD4 count, change in CD4 count, higher baseline viral load, and zidovudine use were associated with changes in fat deposition. Protease inhibitor and zidovudine use were predictive of bone mineral losses. Interestingly, this study suggested that HAART was not associated with fat mass changes, despite the popular opinion that there is a connection.[94] It has been speculated that a reduction in inflammatory processes can be accomplished with effective ARV therapy, improving markers of health risk that are commonly associated with inflammation.[95] However, contradictory studies have suggested that the potential side effects of ARV therapy may overcome the positive effects of weight improvement.[96]

In addition to ARV therapy, the prevention and early treatment of opportunistic infection is an important strategy to maintain and restore nutritional status. Weight loss episodes and health decline are triggered by both acute and chronic co-infection. Nonetheless, the practitioner should be aware that some treatments for opportunistic diseases have nutritional implications of their own (see Table 24.9). Medications, vaccinations, and other treatment strategies are employed in conjunction with nutrition-related therapies to preserve normal body composition and functions as much as possible.

Nutrition Assessment

Nutritional risk screening should be conducted on all patients with HIV infection, with follow-up assessment for identified risk factors. Table 24.10 illustrates various aspects of nutrition screening and assessment. Initial screening may include items from a self screener that can identify general risk factors, such as access to food, consumption of food groups, symptoms, and others. Full assessment can then be completed on patients who are at risk, or if a baseline of measurements is desired. Assessment factors include physical factors, biochemical factors, and nutrition-related behaviors and social factors. Physical evaluation should include at least the minimum data set of height, weight, and body mass index (BMI). In addition, physical measures such as anthropometry, body composition, and examination for clinical signs of overnutrition and undernutrition can help to provide an overview of the patient's nutritional status and identify issues to address. Biochemical measures will include disease-related measures, nutrition-related measures, and selected measures of concomitant disease factors such as diabetes, cardiovascular disease, and others. Patient food access and food choices should be explored with consideration to psychosocial and economic factors.

Physical Assessment Weight loss continues to be a complication of HIV infection, despite HAART use.[97] A history of weight changes will help to establish the level of

Table 24.9 Nutritional Implications of Selected Treatments for Opportunistic Events

OI	Medications	Nutritional Implications
Candidiasis	Fluconazole	Take with/without food; may cause nausea, vomiting, abdominal pain, taste changes
Cytomegalovirus	Valganciclovir Ganciclovir Foscarnet:	Valganciclovir/ganciclovir: take with food; may cause nausea, diarrhea, anemia; Foscarnet may cause anorexia, nausea, vomiting, abdominal pain, diarrhea, taste changes, anemia, renal abnormalities
Hepatitis C	Peg interferon	Ensure hydration; may cause anorexia, nausea, vomiting, abdominal pain, diarrhea, dry mouth and taste changes, anemia
Herpes Simplex	Acyclovir	Ensure hydration; may cause nausea, vomiting, abdominal pain, diarrhea, renal abnormalities
Mycobacterium Avium Complex (MAC)	Azithromycin or clarithromycin	Azithromycin: take without food; may cause nausea, diarrhea; Clarithromycin: take with/without food; may cause dyspepsia, abnormal taste, abdominal pain, diarrhea
P. jeroveci (PCP)	Trimethoprim sulfamethoxazole (TMP SMX)	Take with food, ensure hydration; long-term use may require supplemental folic acid; may cause nausea, vomiting, anorexia

Sources: Bartlett JG, *Pocket Guide to Adult HIV/AIDS Treatment;* January 2005. Pronsky ZM, *HIV Medications. Food Interactions,* 15th ed; 2008.

nutritional risk. Weight losses of 10% are strong predictors of a fourfold to sixfold increase in mortality in HIV infection.[75] Smaller weight losses of 3% or 5% between six-month intervals may also be predictive of mortality. BMI is a strong predictor of survival and may also be a reasonable surrogate marker for changes in CD4 counts for the initiation of ART in resource-limited settings.[77, 98]

The causes of weight-loss episodes should be considered, because the health risk may vary according to the type of weight lost. Starvation, including dieting efforts and malabsorption, can yield lower losses of body cell mass (BCM) than the kcal imbalances that result from infections. Weight losses that are related to inadequate food intake and malabsorption are likely to be mostly fat. Weight losses related to infections and injuries can be mostly body cell mass, a more detrimental condition. Patterns of weight loss in patients with HIV infection suggest that full reconstitution of weight and body composition is not always achieved when weight loss is associated with opportunistic infection.[34, 99] Careful monitoring to catch the problem of weight loss early may help to prevent this gradual decline due to a series of weight-loss events.[100]

Both total volumes of body compartments and patterning of fat tissues will be useful in monitoring nutritional and clinical status in HIV-infected patients, as well as evaluating this risk factor for mortality.[101] Body composition evaluation can be

Table 24.10 Nutritional Screening and Assessment Factors in HIV Infection

Factor	Description	Comments
Dietary Evaluation	Access to food, food consumption	• Self-screeners can contain questions on food group intake, meals per day, and food access • Assessments may include food resources and intake analysis • Food intake can be compared to estimated requirements to determine counseling needs
Physical Assessment	Weight, body mass index, physical examination for clinical signs of deficiency and excess of macronutrients and micronutrients; anthropometry to characterize body composition and patterning (including body shape changes associated with lipodystrophy); body composition analysis	• Body mass index of <20, weight loss of 5%–10%, or body cell mass loss of ≤5% over any time period is associated with risk for morbidity and mortality • Three-compartment body composition analysis using equations validated in HIV can be used to evaluate wasting and inflammatory responses associated with infection
Biochemical Assessment	Immunologic profile, hematologic profile, lipid profile, liver function, renal function, electrolytes, glucose and insulin levels; inflammatory markers	• Biochemical testing is based on the need for routine evaluation for medication and disease cofactors • Additional testing can be done if problems are expected or anticipated according to an individual patient's medical history
Medical History	Past and current information on diagnoses and symptoms; family history of diabetes, cardiovascular diseases, cancers, and renal disease; history of smoking, alcohol, and drug use; medication history and current profile (including herbal, supplement, and other complementary therapies)	• In conjunction with other assessment criteria, risk for nutrition-related problems can be anticipated

achieved in a number of ways, including DEXA, MRI, CT, total body potassium counting, deuterium hydroxide, underwater weighing, and others (see Chapter 3). In the clinical setting, bioelectrical impedance analysis (BIA) is a convenient, inexpensive, and relatively common method for evaluating body composition (see Box 24.5).[102] Tetrapolar BIA provides resistance and reactance readings that can be used in equations that have been validated for use in fluid-shifted populations, such as people living with chronic HIV infection.[103] Such evaluation has allowed the observation that during the wasting of protein stores, muscle tissues are likely reduced, while the organ tissue component of body cell mass is relatively preserved, as demonstrated by increased resting energy expenditure per kilogram of body cell mass without indications of hypermetabolism at a cellular level.[104]

While anthropometric evaluations, such as abdominal and peripheral (mid-upper arm, thigh, and calf) circumferences and fatfolds, are helpful in estimating body composition, it is likely that their most appropriate use in chronic HIV infection will be to monitor fat patterns and changes over time. Changes seen in fat patterns include losses of subcutaneous fat stores (lipoatrophy) and gains in central fat stores. Fat losses are most apparent in the peripheral limbs and in the facial area. Fat gains tend to be centralized, concentrated around the dorsocervical area (as shown in Figure 24.2), in upper back and breast areas, and in the belly region as visceral fat.

In addition to anthropometry, measures using dual energy x-ray absorptiometry (DEXA or DXA) and computed tomography (CT) scans or magnetic resonance imaging (MRI) scans have been used to identify and differentiate changes in subcutaneous and visceral fat stores.[105] Additional anthropometric measures that are specific to fat changes seen in chronic HIV infection include facial fat measures, dimension measures of breast and dorsocervical fat pad areas, and abdominal measures to differentiate subcutaneous fat gain from potential visceral fat gains.[106] Losses in subcutaneous fat and/or gains in abdominal fat stores can both result from and lead to problems of insulin resistance, which can be addressed (at least in part) through dietary and exercise modulation.

In addition to the more quickly lethal problems of weight loss and body composition changes, weight gain and obesity have become important aspects of nutrition assessment and therapy. Weight gain and obesity should be categorized according

---

BOX 24.5 CLINICAL APPLICATIONS

Bioelectrical Impedance Analysis (BIA) in Adults with HIV Infection

Bioelectrical impedance analysis (BIA) testing has been used to estimate fat compartments in primarily healthy people. The linear equations developed to estimate fat volume were typically based on the gold standard of underwater weighing. In populations with fluid shifts, usually seen in injury and infection, such equations were not considered valid, limiting the use of BIA in the clinical setting. More recent investigation has provided the clinician with nonlinear equations to estimate fat-free mass (FFM) and the FFM subcompartment of body cell mass (BCM). These equations have been validated in healthy populations, obese populations, and people living with HIV infection.[1] This capability of BIA technology can be used to anticipate wasting and monitor the severity and resolution of injury and infection according to the recommended criteria shown in Table 24.10. Thus, BIA utilization in chronic inflammatory diseases (such as chronic HIV infection) assists the clinician to monitor the status of infections and effectiveness of treatments that affect nutritional status.

The body can be compartmentalized according to body functions. A two-compartment model may consist of FFM and fat mass. Using the nonlinear equations validated for use in HIV infection, a three-compartment model helps to differentiate two functional masses within fat-free mass: BCM and extracellular mass (ECM). Descriptions of these tissues are shown here:

- BCM: the fat-free mass body compartment that contains muscle and organ tissues, the most volatile of which is muscle; responsible for nearly all kcal use in the adult body; related to health maintenance and survival.

- ECM: the fat-free mass body compartment that contains bone, collagen, and extracellular fluids, the most volatile of which are extracellular fluids; provides structure and transport in the body.

- Fat: all tissue weight that remains after the calculations of fat-free mass; fat mass by BIA includes both stored and essential fat. It should be noted that because the fat component is not directly estimated, but rather calculated by subtracting FFM from total body weight, the accuracy of this compartment estimate may be reduced. In addition, linear equations developed with the gold standard of underwater weight may not consider the essential fat compartment as a component of fat as it is in the case of the HIV-validated nonlinear equations.

Normal levels of each compartment are important to maintain in order to preserve normal body processes. Expected levels of BCM vary according to sex and height; of ECM, according to sex and weight. For purposes of this discussion, the BIA predictive equations that are validated for HIV infec-

tion will be the foundation for expected values and results, and expected values may vary if other equations are used. For men, the functional level of BCM starts at approximately 40% of ideal body weight by height (using Hamwi equations). For women, the functional level of BCM begins at approximately 30% of ideal weight. Expected ranges of ECM are approximately 40% to 45% of current weight for men and 37% to 45% of current weight for women. Fat ranges are functional at between 11% and 22% of current weight for men and between 20% and 32% of current weight for women.

Movement of compartments can give clues about changes in body function and metabolic processes such as infection. Multiple types of and the potential reasons for nutritional compromise can be identified using the nonlinear equations. BIA results can be used to determine the level and source of a problem that affects nutritional status. Serial measures can assist in monitoring both the effects of nutritional therapies and the nutritional response to acute events. Following is an overview of how observed changes in body compartments should be interpreted in fluid-shifted patients with chronic inflammatory conditions, such as HIV-infected patients.

- **Decreased BCM:** Rapid and more severe losses are usually associated with acute phases of infectious disease or injury. Slower and less severe losses are associated with a starvation process (including inadequate intake, malabsorption, and increased losses of nutrients) or reduced physical activity.

- **Increased BCM:** Indicates a resolution of starvation and/or acute events, exercise/activity, and anabolic medications. The inability to improve BCM after the apparent resolution of an acute injury or infection suggests a non-responsive condition or anabolic block that may occur with chronic HIV infection.

- **Decreased ECM:** A drop below expected values indicates dehydration. A return to expected levels indicates the resolution of an acute event that caused extracellular fluid increases

(fluid shifts). Note that increases in the fat compartment (which is approximately 11% fluid compared to fat-free mass, which is more than 75% fluid) may reduce the percentage of weight that is ECM.

- **Increased ECM:** An increase to expected levels indicates rehydration. Increases to levels above expected values indicate fluid shifts related to acute events, such as injury or infection. Note that a decrease in fat volume can increase relative ECM levels. Note that BIA values for ECM should be carefully interpreted with additional information on hydration and infection or injury; this is because only total ECM volume is reflected in the levels estimated by BIA. Mixed events can include both the intravascular dehydration at the same time as extravascular fluid shifts (edema). The shifts in each compartment (BCM and ECT) can add up to a total volume within the expected range.

- **Decreased fat:** A decrease in fat volume indicates a loss of fat and imbalance in kcal related to starvation-related problems or exercise/activity. Fat volume can also indicate metabolic alterations (such as lipoatrophy). The evaluation of the fat compartment should include additional indicators to differentiate types of fat losses.

- **Increased fat:** An increase after a previous loss indicates a resolution of starvation or caloric imbalance. Increases in fat without concurrent increases in BCM or weight can indicate metabolic alterations such as fat accumulation and non-response to nutrient-based therapies or anabolic block.

Source: AMERICAN JOURNAL OF CLINICAL NUTRITION: A JOURNAL REPORTING THE PRACTICAL APPLICATION OF OUR WORLD-WIDE KNOWLEDGE OF NUTRITION by Kotler, Burastero, Wang, Pierson. Copyright 1996 by AMERICAN SOCIETY FOR NUTRITION. Reproduced with permission of AMERICAN SOCIETY FOR NUTRITION in the format Textbook via Copyright Clearance Center.

Reference
1. Kotler DP, Burastero S, Wang J, Pierson RN Jr. Prediction of body cell mass, fat-free mass, and total body water with bioelectrical impedance analysis: effects of race, sex, and disease. Am J Clin Nutr. 1996;64(3 Suppl):489S–497S.

Figure 24.2 Dorsocervical Fatpad

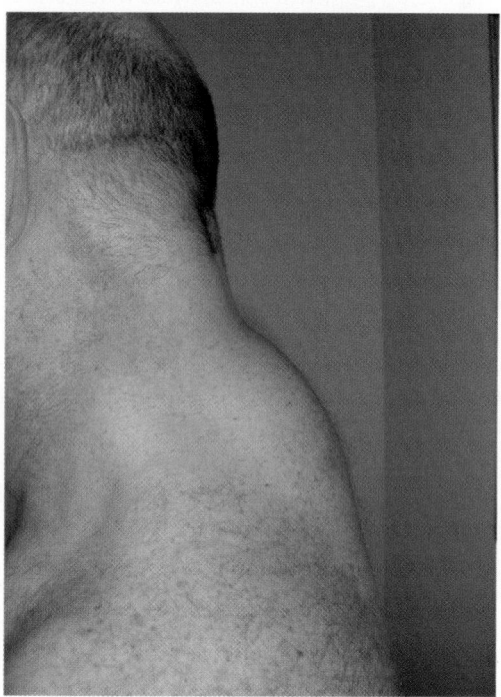

Source: © 2007 Cade Fields-Gardner.

to the type of additional weight that is seen. With efforts to restore body cell mass, weight gain to a high body mass index may be a result of muscle hypertrophy. However, fat gain can be differentiated as normal, subcutaneous fat deposition or abnormal, visceral fat accumulation. Box 24.6 describes selected lipodystrophy-specific anthropometry techniques.

Biochemical Assessment Biochemical measures that assist in nutrition-related assessment include measures of disease progression (e.g., viral load, CD4 count), current inflammatory process (e.g., C-reactive protein), and general nutritional status, with inflammatory processes in mind (e.g., albumin, transferrin, others). Selected measures are shown in Table 24.11. Because viral load and CD4 counts have been associated with changes in weight, a high viral load or a low CD4 count should be considered a nutritional risk factor.[77, 107] Other traditional biochemical measures of nutritional status such as albumin can be used, but the clinician should take into account the potential for changes in interpretation of results in the presence of chronic inflammatory processes.

Tests to evaluate co-diagnoses and side effects of medication that may effect diet and nutrition-related therapies include liver function tests, renal function tests, insulin and blood sugar testing, testosterone and other hormone levels, and hematologic labs for identification of anemias (relatively common in chronic HIV infection). A finding of altered levels of micro-

BOX 24.6 CLINICAL APPLICATIONS

Anthropometry for Fat Patterning in Lipodystrophy

In addition to standard anthropometric measures, there are some additional circumferences and skinfolds that can help to characterize fat patterning and changes associated with lipodystrophy. This box describes the anthropometric measures and methods that may be used to evaluate abdominal fat deposition in order to differentiate between normal gains and deep tissue gains of fat, to track changes in dorsocervical fat deposition, and to track peripheral fat losses in the facial areas.

Abdominal fat evaluation requires a circumference at the abdominal level and four skinfold measures at the same level, including abdominal, right side, left side, and back. Methods for circumferences are the same as for standardized methods. The abdominal circumference should be measured at the level of the navel so that it can be repeated in the same place over time (see Figure 24.3).

Abdominal skinfolds are taken from the front approximately one inch to the right of the navel; on the right and left sides at the mid-axillary line, and on the back about one inch from the spinal column (see Figures 24.4, 24.5, and 24.6).

The abdominal circumference and skinfolds are then entered into the following equation in order to monitor differences over time and estimate changes in both subcutaneous and deep tissue fat accumulation. It should be noted that deep or visceral fat accumulation should be differentiated from ascites in patients at risk for ascites.

Figure 24.3 Abdominal Circumference

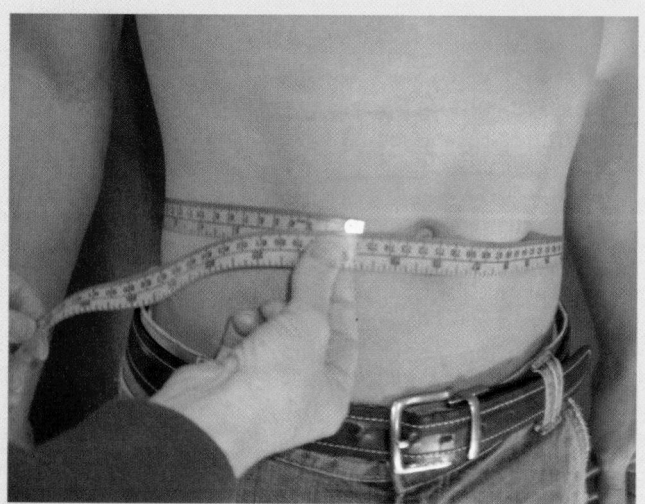

Source: © 2007 Cade Fields-Gardner.

Total abdominal area (TAA) cm² $= \pi \times$ ((abdominal circumference in cm/2) $\times \pi)^2$

Total visceral area (TVA) cm² $= \pi \times$ ((((abdominal circumference in cm/2) $\times \pi)^2) -$ (((abdominal skinfold + right side skinfold + back skinfold + left side skinfold)/8) $\times 10))^2$

Figure 24.4 Abdominal Skinfold

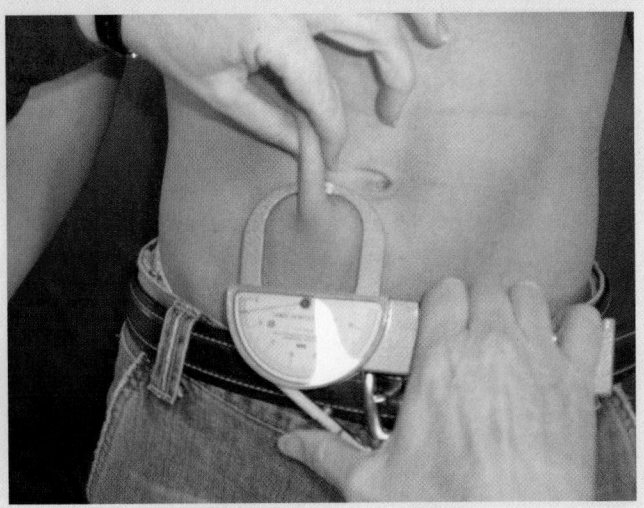

Source: © 2007 Cade Fields-Gardner.

Figure 24.6 Back Skinfold

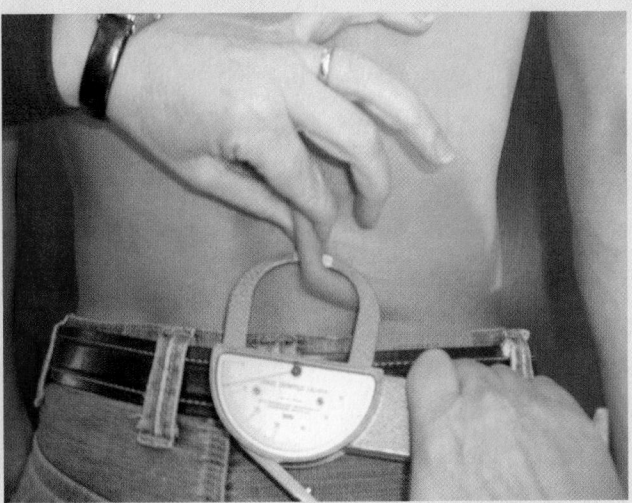

Source: © 2007 Cade Fields-Gardner.

Figure 24.5 Right Side Skinfold

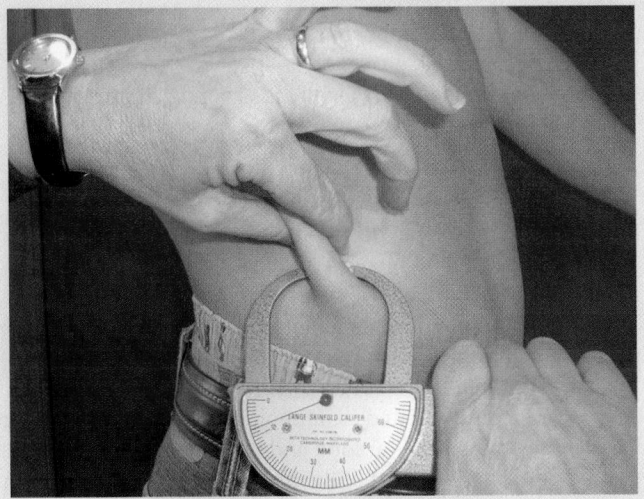

Source: © 2007 Cade Fields-Gardner.

Figure 24.7 Infraorbital Skinfold

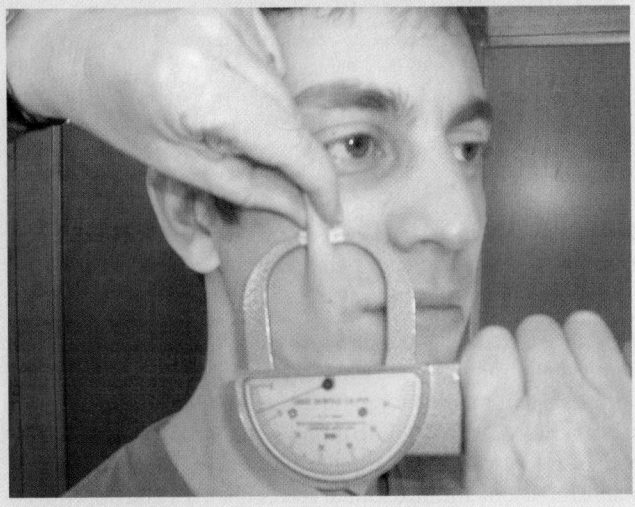

Source: © 2007 Cade Fields-Gardner.

When the abdominal circumference increases and the ratio of total visceral area (TVA) to total abdominal area (TAA) increases, deep fat accumulation may be occurring. Conversely, when increased abdominal circumference shows a decreased TVA:TAA, fat may be accumulated in the subcutaneous area and would be considered a normal fat gain. As abdominal circumferences decrease, noting which compartment decreases will help to differentiate between the loss of fat in the subcutaneous area and deep fat losses.

To effectively monitor peripheral lipoatrophy, it is best to have a baseline measure to compare with follow-up measures. Changes over time in both circumferences and

skinfolds can help to differentiate muscle from fat wasting in the arm, thigh, and calf. Facial fat changes are often the most distressing to patients and can be tracked through the facial skinfold measures shown in Figures 24.7, 24.8, and 24.9.

The infraorbital skinfold can be measured on the zygomatic process under the eye (see Figure 24.7). This measure should be done very carefully because it may be somewhat painful to some who have lost fat padding in this area. The buccal skinfold can be measured just to the right of the corner of the right side of the mouth (see Figure 24.8). The sub-mandible skinfold is measured at the midpoint between the bottom middle of the chin and the back curve point of

(continued)

Figure 24.8 Buccal Skinfold

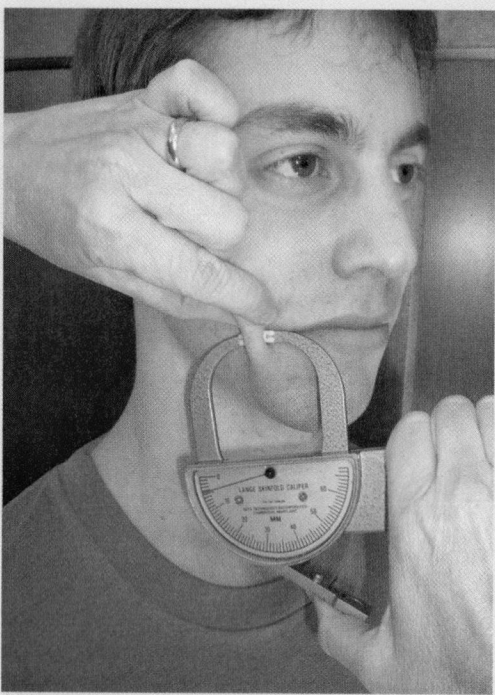

Source: © 2007 Cade Fields-Gardner.

Figure 24.9 Sub-Mandible Skinfold

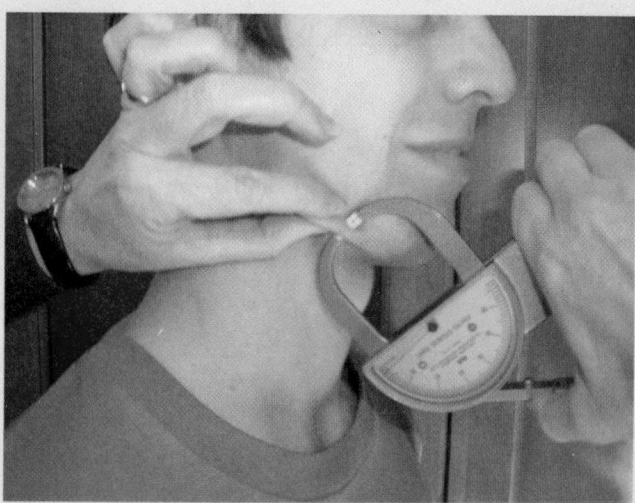

Source: © 2007 Cade Fields-Gardner.

the mandible. The thumb and index finger hold the skinfold at the edge of the bone (see Figure 24.9).

While there are currently no "normal levels" in the literature for each of these measures, comparison of serial measures will give clues about both the process of and treatment efforts toward reversing facial lipoatrophy.

nutrients in HIV infection is not universal. There should be a reason to suspect deficiency before routinely checking micronutrient biochemical status markers.[108]

Medical History Assessment In addition to biochemical and physical assessment, current and past medical conditions and therapies should be included in a full nutrition assessment. A history of nutrition-related problems, such as past episodes of wasting or bouts of diarrhea, can also help to establish reasons for nutritional status changes and appropriate choices for interventions. Current uses and histories of medications, including antiretrovirals, antibiotics, and other medications, will assist in establishing the potential for nutritional risk and the need for educational interventions (see Tables 24.5 and 24.9). Current and past history of acute or chronic infection, disease, or injury will also assist to determine the causes of and risks for altered nutritional status. Concomitant diseases, such as diabetes, insulin resistance, hepatitis, renal dysfunction, pancreatic dysfunction, cardiovascular disease, osteoporosis, cancers, **lactic acidosis**, and others, should be prioritized along with HIV infection for nutrition-related intervention.

Over-the-counter medications and herbal or other non-nutrient supplements should be included in the medical history. It is especially important to assess the use of such supplements and to discuss and anticipate any adverse effects or interactions wherever possible. (See Table 24.7.)

Dietary Evaluation Food intake, food-related behaviors, and food/nutrition insecurity should be evaluated. An assess-

ment of food intake can be accomplished in many ways, including a food recall, food records, and food frequencies. Dietary intake can be compared with estimated needs for fluids, kcal, protein, and micronutrients to determine the most appropriate diet-related interventions. Factors that may affect food intake such as fatigue and/or inability to purchase or prepare food, excessive coughing related to pulmonary disorders, etc. should be identified. The need for food and nutrient supplements should be evaluated on an individual basis, with consideration given to medication profiles and the potential for interactions or toxicities.

Nutrition Diagnosis

Common nutritional problems associated with HIV/AIDS include: increased energy expenditure; inadequate oral food/beverage intake; excessive bioactive substances; increased nutrient needs; malnutrition; inadequate protein-energy intake; swallowing difficulty; altered GI function; impaired nutrient utilization; altered nutrition-related laboratory values; food-medication interaction; underweight; involuntary weight loss; impaired ability to prepare foods/meals; physical inactivity; and intake of unsafe food.

Nutrition Intervention

Intervention plans are implemented according to the setting and wishes of the patient and his or her care provider. The health care team, including the patient and/or appropriate care providers, should agree to and be capable of implement-

Table 24.11 Selected Biochemical Measures in HIV Disease

Measure	Criteria or Expected Values	Evaluation in HIV Infection
Immunologic		
CD4 cell count	398–1535/μL	<200/μL defines AIDS; decreased levels are prognostic for opportunistic disease and often associated with body mass index levels
Viral load (PCR)	Undetectable	Elevated levels are prognostic for immune deficits
Hematologic		
Hemoglobin	F: 12.1–15.6 g/dL; M: 14.6–17.5 g/dL	Decreased in anemia; elevated in dehydration, chronic testosterone replacement
Hematocrit	F: 34%–45%; M: 41%–51%	Decreased in anemia; elevated in dehydration, chronic testosterone replacement
Mean corpuscular volume	78–93 cubic microns/RBC	Increased in folate- or vitamin B_{12}-deficiency anemia, associated with zidovudine; decreased in iron-deficiency anemia
Ferritin	F: 12–150 ng/mL; M: 30–320 ng/mL	Elevated in inflammation; decreased in iron-deficiency anemia
Transferrin	212–360 mg/dL	Elevated in iron deficiency; decreased in malnutrition
Albumin	3.5–5.0 mg/dL	Decreased in malnutrition; rapid decrease with acute inflammation
Prealbumin (transthyretin)	18–38 mg/dL	Decreased in acute catabolism, inflammation, malnutrition
Organ Function		
AST	M: 10–37 U/L; F: 10–31 U/L	Elevated in hepatitis or due to medication interactions
ALT	M: 4–40 U/L; F: 4–31 U/L	
BUN	8–23 mg/dL	Elevated in diabetes; low in malnutrition
Creatinine	Adult: 0.4–1.2 mg/dL	Elevated in renal disease, wasting
Endocrine		
Glucose	Fasting: 70–99 mg/dL	Elevated in diabetes, pancreatitis, chronic malnutrition
Insulin	Fasting: 4 27 uIU/mL	Elevated in metabolic syndrome, type 2 diabetes
Glycated hemoglobin A1c	4%–6%	Elevated in diabetes, iron deficiency
Testosterone	350–1080 ng/dL	Decreased in hypogonadism, AIDS wasting
Cardiovascular		
Total cholesterol	120–199 mg/dL	Elevated in hyperlipidemia, diabetes, obesity, infection
HDL	40–60 mg/dL	Decreased in starvation, obesity, diabetes, smoking, liver disease, AIDS
LDL	<100 mg/dL	Elevated in hyperlipidemia, lower in advanced AIDS
Triglycerides	Fasting: <150 mg/dL	Elevated in hyperlipidemia, AIDS
C-reactive protein (CRP)	Regular: <0.8 mg/dL	High-sensitivity CRP will provide risk as low ($\leq$1 mg/L), average (1–3 mg/L), or high ($\leq$3 mg/L)
Electrolytes		
Sodium	136–144 mEq/L	Decreased in diarrhea, vomiting, AIDS
Potassium	3.5–5.5 mEq/L	Decreased in diarrhea, vomiting, chronic stress/fever

ing the nutrition care plan. Goals for the nutrition care plan may include prevention of adverse events related to therapies, restoration of adequate nutritional status, management of co-conditions (e.g., diabetes, liver disease, renal dysfunction), and others. Research strongly supports the effectiveness of nutrition intervention for prevention weight loss, improved caloric intake and/or symptoms, and improved CD4 count and quality of life.[109] Implementation may include the coordination of services and products to target maintenance and improvement of nutritional status as well as direct patient care. All aspects of patient care should be considered in order to reduce the possibility of adverse events and effects of therapy, and to enhance the efficacy of and adherence to recommended therapies.

With the advent of HAART, there are many more opportunities to implement plans in the outpatient setting. Routine appointments should be scheduled according to the nutrition care plan and as often as necessary to monitor and adjust the plan.

Because the coordination with the overall health care plan is an important feature of the nutrition care plan in chronic HIV infection, communication and documentation should be shared between key members of the health care team.

Nutrition therapy can include the restorative or modulating effects of macronutrients, micronutrients, or other therapies aimed at improving nutritional status. Because chronic HIV infection results in a heterogeneous set of complications, it is likely that each HIV-infected person will require a different set of dietary and other recommendations according to their assessment.[110] While there are occasional reports of toxicities and potential for adverse interactions with medication therapies, food and supplement levels of nutrients can be compared to recommended intakes as well as to upper limits that have been documented in the literature. Some components of foods or supplements may also have the potential for interactions, so careful monitoring should be established to identify any adverse effects.[111, 112]

Macronutrient Therapy Alterations in macronutrients for therapeutic purposes can address disease and treatment complications, such as symptom management, weight maintenance (including gain and loss according to needs), and management of concomitant disease processes. A summary of nutrient interventions is shown in Table 24.12 and an overview of dietary recommendations for symptom relief is shown in Table 24.13.

Fluid recommendations are based on needs for hydration maintenance with any limitations or enhanced needs, such as renal dysfunction or fever and sweating. Kcal recommendations are based on the balance needed to maintain a healthy weight. This may include the need for additional kcal to gain weight, or for fewer kcal in order to lose weight.[113] Although

Table 24.12 Nutrient-Based Therapy Recommendations

Nutrient	Recommendations
Fluids	• Hydration maintenance is the goal, and standard fluid intake recommendations can apply • Additional fluids are recommended in cases of dehydration, fluid losses through diarrhea or sweating • Restrictions are recommended in cases of renal insufficiency
Kcal	• Weight maintenance is the goal with additional kcal typically recommended if weight gain is desired and a mild restriction in kcal to achieve desired weight losses • Additional energy may be required during bouts of opportunistic conditions that increase metabolic rate • Increased kcal requirements during pregnancy and lactation should be incorporated into recommendations
Carbohydrate	• The amount and types of carbohydrate recommended are based on both energy needs and carbohydrate tolerance • Insulin resistance and diabetes may require dietary modification to modulate glucose and insulin levels
Protein	• The amount and types of protein recommended are based on the need for protein stores maintenance • Additional protein is likely to be needed in cases of inflammation, fever, and during pregnancy • Any protein losses should be restored with increased protein intake and activity to promote protein stores maintenance • Renal disease or other conditions may require protein restriction or other changes in protein recommendations
Fat	• The amount and types of fat recommended are based on energy needs, cardiovascular risk, and inflammatory conditions • Weight maintenance may require increases or decreases in fat kcal • Cardiovascular risk may require lower fat intake and a higher ratio of unsaturated fats • Omega-3 fatty acid sources may be recommended to help reduce inflammation effects as well as improve lipid profiles

Table 24.12 Nutrient-Based Therapy Recommendations (*Continued*)

Nutrient	Recommendations
Vitamins and Minerals	• Recommendations are based on individual needs; for instance, during bouts of diarrhea, the replacement of electrolytes and any potential losses of vitamins and minerals (such as fat-soluble vitamins during steatorrhea and zinc during larger volume losses of fluids) are essential to balance • Upper limits of toxicity, the potential for nutrient interactions with medications and disease, and balance in micronutrient intake should be considered in recommendations (for instance, iron supplementation is controversial due to the potential for increasing the risk of opportunistic infection without overcoming the inflammatory-mediated drop in iron availability) • Specific conditions that alter micronutrient requirements, such as pregnancy and lactation or child growth and development, should be considered in recommendations
Fiber	• Fiber recommendations are similar to those in healthy individuals • Fiber has been suggested to improve glucose tolerance, affect glycemic response to foods, and reduce the potential for cardiovascular risk and altered fat deposition seen in lipodystrophy

Table 24.13 Dietary Recommendations for Selected Symptom Management

Symptom	Education and Counseling Recommendations
Appetite Loss	• Determine the cause of anorexia and pursue treatment • Offer tips: 　• Eat favorite foods often in relaxed settings 　• Add flavors and have foods of various colors for more interest 　• Keep a supply of snacks handy for when appetite increases 　• Fix and eat foods that don't take as much energy or enjoy takeout food • If significant weight loss occurs, consider appetite stimulants • Refer to resources for food if food insecurity is a problem • Monitor nutrient intake, weight, and body composition
Diarrhea	• Determine the cause of diarrhea and pursue treatment • Offer tips 　• Replace fluids and electrolytes with juices, sports drinks, gelatin, broths 　• Eat bland foods that are lower in fiber and residue 　• Avoid fatty and gassy foods 　• If lactose is a problem, find substitutes for lactose-containing products • Anti-diarrheal medications should be considered for significant acute and chronic diarrhea • Monitor fluid intake, hydration, weight, and body composition

Table 24.13 Dietary Recommendations for Selected Symptom Management (Continued)

Symptom	Education and Counseling Recommendations
Heartburn/ Reflux	• Determine any cause of heartburn and reflux and pursue treatment • Offer tips: • Eat small amounts of food more often throughout the day • Keep the upper body elevated for at least an hour after eating • Avoid alcohol, caffeinated beverages, and very spicy and fatty foods • Stimulate saliva production by chewing sugarless gum • If overweight, lose weight and avoid tight-fitting clothes • If reflux is chronic, consider medications to control the symptoms • Monitor food intake, weight
Nausea/ Vomiting	• Determine the cause of nausea and vomiting and pursue treatment • Offer tips: • Replace any lost fluids and electrolytes • Try bland, non-odorous foods • Drink beverages between meals and not with meals • Eat smaller, more frequent meals • Keep upper body elevated for at least an hour after meals • Reduce fatty foods if early satiety is a problem • If vomiting leads to chronic reduced food intake, evaluate for antiemetic medications • Monitor food and food intake, fluid status, weight
Oral Lesions	• Determine and pursue appropriate treatment for oral lesions • Offer tips: • Eat moist, soft foods and finely diced foods and keep mouth moist between meals • Avoid irritating spicy or acid-containing foods • Eat foods at room temperature or cooler temperatures • Avoid tough foods or foods that require chewing • Consider topical medications to ease pain in eating and advise on good oral hygiene • Monitor food intake, weight

recovery of fat before recovery of lean tissues is a normal process during weight regain, emphasis on lean tissue gain can be achieved with aggressive efforts to introduce kcal, protein, and muscle-supporting activities. In some cases, weight loss may be appropriate to improve health. When body composition has been altered, however, weight loss through dieting may be risky, especially if there is no buffer in the volume of body cell mass to lose when lean tissue is lost during dieting. Kcal balance and other nutrient-based therapies may be best used as one of several therapies aimed at improving overall health.

Protein recommendations usually include extra protein to buffer the catabolism of protein stores that accompanies the inflammatory response to viral and other infections.[114] It is believed that additional protein helps to improve immune and other inflammation-mediated responses, but additional research needs to be conducted on protein requirements and intake. Diets are generally high in protein in developed countries, allowing this recommendation to be made without any particular diet change. However, in cases where baseline levels of protein intake are low or food insecurity limits the ability to obtain adequate high-quality protein, enhancing protein intake may be required. For patients with renal dysfunction, bone mineral losses, and long-term diabetes, the quality of protein and the best sources may be more specifically tailored. In such instances, a balance of animal and plant-based protein sources will be especially important. Overdoses of protein pose additional burden and risk in patients with potential for renal or bone problems. Special care should be taken to tailor protein recommendations to individual needs and limitations.

Carbohydrate and fat kcal may be modulated according to assessed risk for insulin resistance, cardiovascular disease, and other conditions where treatment is related to dietary interventions. Limited evidence supports the use of the glycemic index, fiber, or specific fatty acids in individuals with HIV.[115, 116] Because HIV infection is associated with chronic inflammatory response, it may be prudent to provide education on balancing daily dietary intake to minimize problems and reduce associated risks.

Nutritional supplements have been recommended for patients who require a boost in macronutrient intake. Patients may use recipes to create their own nutritional supplements and/or obtain ready-to-use commercial supplements in the form of beverages and bars. While some food-based kcal may be displaced by such therapy, kcal-containing supplements have been shown to yield an increase in energy intake and may assist in adherence to a weight gain regimen.[97]

Micronutrient Therapy Altered micronutrient levels have been documented in both pediatric and adult HIV disease, in research that suggests a worsening associated with disease progression. Alterations in micronutrients may exacerbate immune compromise, disease progression, damage due to oxidative stress, morbidity, and even risk for HIV transmission independent of CD4 cell counts,[87, 117] While there have been many speculations about the benefit of reversing a low level of selected micronutrients, evidence of the impact on HIV disease to date has been equivocal.[86, 118–121] Table 24.14 provides information on selected micronutrients that have been targeted for supplementation in PLHA.

Reversal of many true nutrient deficiencies is possible with supplementation.[122] Micronutrient supplementation remains controversial in most areas, but is generally recommended when nutrient intake may be inadequate.[123] The use of multivitamin/mineral supplementation may have a supportive role in slowing disease progression, particularly in patients without access to antiretroviral therapies.

Additional micronutrient therapies may include those targeted to treat anemias, diarrhea, and potential dietary deficiencies. Special care should be taken in each instance to consider the effect of chronic inflammation in HIV infection. While there is some hope for health benefits and reversal of nutrient deficits or otherwise low serum levels, in the case of diarrhea and wasting,

Table 24.14 Selected Micronutrients in HIV Infection

Nutrient	DRI/UL*	Comments
Vitamin A	700–900 µg/3000 µg	Vitamin A and beta-carotene have been associated with immune function, pregnancy outcomes, and growth and development of children (especially with diarrhea); supplementation may not have an impact beyond normalizing serum values; maternal supplementation of vitamin A is associated with greater risk of mother-to-child transmission of HIV infection; however, supplementation along with zinc in undernourished children may assist in reducing diarrhea and opportunistic infections
Riboflavin	1.1–1.3 mg/no UL	Supplementation has been suggested along with thiamin in cases of lactic acidosis
Folic Acid	400 mg/1000 mg	Folate supplementation is generally recommended in cases of pregnancy
Pyridoxine	1.5–1.7 mg/100 mg	Has been touted to reduce peripheral neuropathy; however, excesses can contribute to peripheral neuropathy
Cyanocobalamin	2.4 mcg/no UL	Vitamin B_{12} may be malabsorbed due to changes in stomach pH; it is related to cognitive function and has been commonly supplemented as a part of oral multivitamin/mineral supplements, through sublingual, nasal, and intravenous doses
Ascorbic Acid	75–90 mg/2000 mg	Supplemented to improve antioxidant status and reduce oxidative stress; high doses, but as little as 1000 mg per day, can interact with ARVs
Alpha-Tocopherol	15 mg/1000 mg	Supplementation has the potential for immunostimulation; should be avoided when taking the ARV amprenavir
Iron	8–18 mg/350 mg	Red blood cell iron is affected by inflammatory processes and the shunting to storage forms; iron supplementation is not generally recommended unless the threat of iron-deficiency anemia-related mortality is imminent because of the potential to increase the risk of opportunistic infection and progression of disease
Selenium	55 mg/400 mg	Low values are an independent predictor of survival; low selenium values may be related to inflammatory process; selenium is a cofactor in antioxidant protection, and low values are related to immune dysfunction; excess selenium is associated with immune dysfunction
Zinc	8–11 mg/40 mg	Low zinc values are seen in HIV infection and associated with inflammatory processes and an increased mortality; zinc supplementation has been associated with fewer opportunistic infections and slower disease progression; high zinc intake has been associated with disease progression; zinc restoration is associated with reversal of weight losses and amelioration of diarrhea; effective ART does not appear to reverse low levels of zinc

*DRI/UL Dietary Reference Intake/Upper Level

Sources: Abrams B, Duncan D, Hertz Piddiotto. A prospective study of dietary intake and acquired immune deficiency syndrome in HIV seropositive homosexual men. *J Acquir Immune Defic Syndr.* 1993;6(8):949–58.

Baum MK. Role of micronutrients in HIV infected intravenous drug users. *J Acquir Immune Defic Syndr.* 2000;25 suppl 1: S49–S52.

Baum MK, Campa A, Lai S, Lai H, Page JB. Zinc status in human immunodeficiency virus type 1 infection and illicit drug use. *Clin Infect Dis.* 2003;37 suppl 2:S117–S123.

Baum MK, Shor Posner G. Micronutrient status in relationship to mortality in HIV 1 disease. *Nutr Rev.* 1998;56:S135–S139.

Bowers JM, Bert Moreno A. Treatment of HAART induced lactic acidosis with B vitamin supplements. *Nutr Clin Pract.* 2004;19(4):375–78.

Butensky E, Kennedy CM, Lee MM, Harmatz P, Miaskowski C. Potential mechanisms for altered iron metabolism in human immunodeficiency virus disease. *J Assoc Nurses AIDS Care.* 2004;15(6):31–45.

Duggan C, Fawzi W. Micronutrients and child health: studies in international nutrition and HIV infection. *Nutr Rev.* 2001;59:358–69.

Fawzi W. Nutritional factors and vertical transmission of HIV 1. Epidemiology and potential mechanisms. *Ann NY Acad Sci.* 2000;918:99–114.

Slain D, Amsden JR, Khakoo RA, Fisher MA, Lalka D, Hobbs GR. Effect of high dose vitamin C on the steady state pharmacokinetics of the protease inhibitor indinavir in healthy volunteers. *Pharmacotherapy.* 2005;25(2):165–70.

Wellinghausen N, Kern WV, Jochle W, Kern P. Zinc serum level in human immunodeficiency virus infected patients in relation to immunological status. *Biol Trace Elem Res.* 2000;73:139–49.

a randomized placebo-controlled trial showed no significant benefit for short-term oral supplementation of vitamins A, C, and E, along with zinc and selenium, in reducing the risk for mortality or reversing hematologic parameters.[121, 124]

Children who are perinatally infected with HIV often show multiple nutritional problems, including micronutrient deficiencies, which hinder normal growth and development as well as immune function.[125] However, infants and children of HIV-infected pregnant women who had been supplemented with multivitamins prenatally showed improved nutritional status.[121] Box 24.7 provides more information about supplementation in pregnant women and infants.

In general, studies are equivocal and have not provided an adequate basis to make generalized recommendations. Indi-vidualized assessment, care plans, and counseling are likely to be most appropriate for micronutrient supplementation, in addition to diet-related strategies.

Exercise Exercise has been recommended to balance activity with kcal intake, improve muscle volume and function, and normalize lipid and energy metabolism.[126] Exercise has the potential to mitigate the loss of muscle mass in wasting conditions and to improve the recovery from bouts of protein wasting.[127] The combination of diet interventions and exercise is considered first line therapy for cardiovascular disease and altered fat metabolism and deposition in chronic HIV infection.[128, 129] General recommendations include a tailored, routine exercise program that includes both aerobic and resistance components at least three times per week. As with any exercise

Pregnancy Outcomes and Infant Feeding

Much research has concentrated on the role of nutrients in pregnancy outcome and prevention of mother-to-child transmission of HIV infection. The supplementation of multivitamins/ minerals shows some promise in improving pregnancy outcomes in women who are malnourished. Improvements in birth weight and a reduced number of preterm births have been observed.[1] The mother's nutritional status does not always have bearing on transmission rates of HIV. For instance, a low level of vitamin A has been associated with greater transmission risk between mother and child, yet supplementation with vitamin A has shown disappointing results in curbing this problem.[2] In fact, single-nutrient supplementation of vitamin A may be related to an increase in mother-to-child transmission of HIV, whereas multiple-vitamin supplementation showed promising results in the reduction of both transmission through breast-feeding and child mortality.[3] Maternal micronutrient status may have less effect on child mortality than the child's nutritional status, and should be considered regardless of breast-feeding.[4]

General recommendations for feeding newborns of HIV-infected mothers in developed countries include the preferred use of breast milk substitutes and formulas that will meet nutritional needs of the infant and eliminate the exposure to breast milk virus.[5] In developing countries, this recommendation may be less feasible, affordable, and sustainable. For these instances, the WHO has outlined breast-feeding recommendations and emphasized the need to provide an unbiased representation of risks and benefits so that mothers can make educated and personalized decisions about breast-feeding their infants.[6,7]

References
1. Fawzi WW, Msamanga GI, Spiegelman D, Urassa EJ, McGrath N, Mwakagile D, et al. Randomised trial of effects of vitamin supplements on pregnancy outcomes and T cell counts in HIV-1-infected women in Tanzania. Lancet. 1998;351:1477–82.
2. Dreyfuss ML, Fawzi WW. Micronutrients and vertical transmission of HIV-1. Am J Clin Nutr. 2002;75:959–70.
3. Fawzi WW, Msamanga GI, Hunter D, Renjifo B, Antelman G, Bang H, et al. Randomized trial of vitamin supplements in relation to transmission of HIV-1 through breast-feeding and early child mortality. AIDS. 2002;16;1935–44.
4. Fawzi W. Nutritional factors and vertical transmission of HIV-1. Epidemiology and potential mechanisms. Ann NY Acad Sci. 2000;918:99–114.
5. Jackson DJ, Chopra M, Witten C, Sengwana MJ. HIV and infant feeding: issues in developed and developing countries. J Obstet Gynecol Neonatal Nurs. 2003;32:117–27.
6. Gupta A, Mathur GP, Sobti JC. World Health Assembly recommends exclusive breast-feeding for first six months. J Indian Med Assoc. 2002;100(8):510–11, 515.
7. Habicht JP; WHO Expert Consultation. Expert consultation on the optimal duration of exclusive breastfeeding: the process, recommendations, and challenges for the future. Adv Exp Med Biol. 2004;554;79–87.

program, adherence is an important factor in realizing benefits, and may present a challenge to the health care team designing and tailoring such interventions. Exercise and diet are both cost-effective therapies to improve nutritional status, and may be combined with medications or other therapeutic strategies to improve effectiveness.[130, 131]

For metabolic alterations, including fat accumulation, exercise may prove to be an essential and effective therapy. Aerobic training, in particular, may improve lipid metabolism, lower blood lipids, and reduce visceral fat accumulation.[129, 130, 132] Table 24.15 summarizes the effects of exercise in this population.

Table 24.15 Effects of Exercise

Type	Effects	Uses in HIV Infection
Progressive Resistance	Preserves, improves muscle mass; improves strength, improves effect of anabolic therapies	Restoration of BCM during recovery from weight loss and wasting, prevention of BCM volume and function loss, improvement of muscle function
Aerobic	Improves insulin sensitivity, increases HDL C, improves endurance, improves effect of anabolic therapies	Restoration of normal activity level after illness or debilitation, reduced effect of insulin resistance on body shape and fat changes, prevention and treatment adjunct for cardiovascular health efforts

Education and Counseling Education on relevant nutrition-related topics and counseling on nutritional and other therapies is a key feature of all interventions. Baseline education on nutrition principles, including balanced diet and food safety, should be provided to all patients (see Box 24.8).[133, 134] Additional education on topics such as drug-nutrient interactions or symptom management can be provided in individual or group sessions. Individualized counseling should include all facets of the nutrition care plan, and how other aspects of the overall health care plan may affect nutritional status and well being. In addition, recommendations for food intake can be customized to a prescription level in counseling sessions.[135] Referrals should be made for other aspects of care that affect nutritional status and health, including smoking cessation, alcohol or drug rehabilitation, HIV and other infectious disease transmission, and psychosocial economic factors that can affect food access, choices, and metabolism.[136] Counseling activities may be essential to the successful use of nutritional supplements to achieve weight gain objectives.[137] Medication interactions with food and nutrients are an important feature of education and counseling for patients taking ARVs and other therapies (see Tables 24.5 and 24.9). Lastly, since HIV-infected clients are now living longer, they may develop nutrition-related chronic diseases as seen among other adults in developed countries. Therefore, lifestyle and dietary recommendations should be provided to prevent or manage these diseases.

Symptom Management Side effects of medications and disease-related symptoms require adequate management

Counseling Concepts in HIV Infection

Clinicians should include nutrition-related counseling in their patient care plans. Basic education on nutrition-related concepts and principles will help the patient to make diet and health-related decisions. Tailored counseling can address specific problems and conditions that may benefit from nutrients or nutrition-related care.

Nutrition Education Basics Checklist:

- Nutrition priorities: water and fluids, kilocalories, protein, and micronutrients
- Weight maintenance and kilocalorie balance

- Role of exercise and lifestyle choices, including alcohol use, drug abuse, and smoking
- Food and water safety
- Food shopping and storage
- Food preparation and dining out

Nutrition Counseling Checklist:

- Food interactions with medication regimens
- Symptom management
- Nausea and vomiting
- Diarrhea and constipation
- Appetite loss

- Heartburn and bloating
- Weight losses
- Weight gains: overweight and obesity
- Additional conditions
- Lipodystrophy: fat gain, fat loss
- Insulin resistance and diabetes
- Hyperlipidemias/altered blood fats
- Hepatic conditions
- Gastrointestinal conditions
- Renal conditions

---

to support medication adherence, prevent additional disease progression, and improve quality of life.[138] Common symptoms related to the complications of disease and treatments include those that can interfere with nutritional maintenance and the apparent acceleration of chronic diseases such as cardiovascular disease, diabetes, osteoporosis, anemias, and others. The most commonly reported symptoms include fatigue and diarrhea.

Symptom management often includes a nutrition component, but may also include non-nutrient therapies. For instance, diarrhea can result from opportunistic infection, medications, and organ dysfunction. If adequately evaluated, the appropriate therapy may include treatment of opportunistic infections identified, alteration of medication or doses, additional medication therapy to reduce diarrhea, and dietary management. It has been reported that use of medium-chain fatty acids and vitamin A and beta-carotene supplementation increased fat absorption and decreased steatorrhea. Additional research is needed to determine if pancreatic and/or lactase enzymes, L-glutamine supplements, probiotics, calcium carbonate, the BRAT diet, or amino acid-based elemental diets are effective. Some of the medications used to manage diarrhea may include anti-diarrheals.[139] (Table 24.16 outlines selected medication strategies for symptom management.)

Monitoring and Evaluation

Chronic HIV infection is a multifactorial disease that presents in a variety of ways that may differ between patients. Problems, associated interventions, and monitoring and evaluation of appropriate outcomes are all important features of the nutrition care plan, and evaluations should be conducted routinely. Interventions should be monitored at points in time when measurable achievements are expected. For instance, if the goal is to reduce weight and maintain a buffer of body cell mass,

Table 24.16 Selected Medication Therapies for Symptom Management

Symptom	Medication Therapies: Generic (Brand Name)
Anorexia	• Dronabinol (Marinol): synthetic THC/cannabinoid; may cause dry mouth, avoid alcohol. • Megestrol acetate (Megace, Megace ES): progestational agent; may cause edema, potential for hypogonadism; take with food if needed to prevent dyspepsia.
Constipation	Cisapride (Propulsid) Note: This drug can interact unfavorably with protease inhibitors and non-nucleoside reverse transcriptase inhibitors, increasing risk of potentially fatal changes in heart rhythms.
Diarrhea	Loperimide (Immodium), diphenolxylate/atropine (Lomotil), Kaopectate, tincture of opium, Pepto Bismol, psyllium (Metamucil), pancreatic enzymes (Ultrase, Pancrease)
Gastroesophageal Reflux Disease (GERD)	• *Proton Pump Inhibitors:* rabeprazole (AcipHex), esomeprazole (Nexium), lansoprazole (Prevacid), omeprazole (Prilosec), pantoprazole (Protonix) Note: These drugs reduce acid and may affect ARV absorption, particularly atazanavir; in some cases, ARVs may speed the metabolism of these medications, rendering them less effective. • *H2 Blockers:* nizatidine (Axid), famotidine (Pepcid), cimetidine (Tagamet), ranitidine (Zantac) Note: These drugs reduce acid and may affect ARV absorption; some interactions with ARVs have the potential to increase toxicity (e.g., cimetidine increases protease inhibitor levels).

Symptom	Medication Therapies: Generic (Brand Name)
Heartburn	Antacids (Alka Selzer, Bromo Selzer, Maalox, Mylanta, Rolaids, Tums) Note: Caution should be taken when administered with ARVs or other medications that require an acid environment for absorption, such as agenerase, atazanavir, tipranavir, delavirdine; Alka Selzer contains aspirin and Bromo Selzer contains acetaminophen.
Nausea and/or Vomiting	Procholorperazine (Compazine), cola syrup (Emetrol), metoclopramide (Reglan), chlorpromazine (Thorazine), thiethylperzaine (Roecan), ondansetron (Zofran) Note: Many of these medications can interact with ARVs, causing a need for dose adjustments.

measurable changes may be achieved in three months and an appointment for follow-up should be made to confirm that progress is being made and to adjust the plan appropriately to support and improve outcomes. If a short-term plan is put into place in a hospital setting, follow-up should be scheduled prior to discharge or should be a part of the discharge plan and communication with outpatient service providers.

Optimal outcomes should be matched to the problem list based on assessment, as with any other disease state. Prevention of nutritional compromise, restoration of nutritional well being, and support for the overall health care plan are important general features of desired outcomes. Improved knowledge, therapy adherence, management of disease and co-diseases, and support for physical and psychosocial/economic well being related to nutrition and food are more specific goals that can be tailored to each patient's needs. Clear documentation is necessary to make the case for the supportive role of nutrition therapy in HIV infection and disease management.

In addition to individualized outcomes, monitoring and research on the benefits of providing support to prevent malnutrition and improve nutritional status will be important to demonstrate the benefits that nutrition therapy can have in patients living with chronic HIV infection.[140]

Conclusion

HIV infection creates a number of challenges to the maintenance of nutritional status. While weight loss and wasting are not an inevitable part of the natural history of HIV infection,[27] chronic inflammation due to HIV infections, co-conditions, and complications assault nutritional status continuously.

With long-term survival, it will be an important practice to prioritize health issues, including conditions directly and indirectly related to nutritional status. Treatments for HIV infection can both support and inhibit nutritional status maintenance and improvement. There is an ongoing need for education, counseling, and a variety of other interventions to mitigate nutritional decline and disease progression. Successful nutrition-related therapies will help to stabilize nutritional and other types of clinical status markers. Patient-centered care through inclusion of the patient in the health care team that specializes in HIV-related care may be best suited for providing ongoing care to patients living with this complex and lifelong disease.[141]

PRACTITIONER INTERVIEW

Melody O'Donnell, R.D., Clinical Dietitian, *University of California, San Francisco Medical Center*

I have worked as a registered dietitian for two years and ever since becoming an RD I have spent about 50% of my time in direct care to HIV patients. I work in the HIV 360 Positive Care Center at UCSF and do video conferences with patients all over San Francisco that may have access to an HIV clinic but not an HIV dietitian. I normally see patients ages 30–60 because many of the younger clients go to clinics that focus on providing care to younger adults. I would estimate that 50% of the clients I see have been infected with HIV for over ten years.

I nutritionally assess my patients not only for HIV/AIDS nutritional markers but also for chronic co-morbid diseases. I look for family medical history; the patient's weight history (highest and lowest weight since diagnosis); history of opportunistic infections; current medications; appetite; symptoms of the entire gastrointestinal tract; dentition, allergies and/or food intolerances; dietary supplements; any other co-morbidities such as type 2 diabetes mellitus; hyperlipidemia, hepatitis C, tuberculosis; and monitor current lab values—HIV viral load, CD4 count, renal function tests (BUN and creatinine), liver function tests, lipid panel, and glucose.

In the past, there was more concern about protein (muscle) wasting, but now with the newer categories of medication for HIV, chronic long-term effects of HIV are more evident. There is a much greater focus on the conditions associated with aging. In addition, for many of my patients, food security is an ongoing concern. It is a significant issue that many practitioners overlook, especially if the patient looks well nourished. Many of my patients do not even have a kitchen, so I need to have resources available for them so they can have access to healthy food.

(continued)

The most common nutrition problems I see in my clients are hyperlipidemia, hypertension, metabolic syndrome, type 2 diabetes, and acute weight loss related to opportunistic infections. In the last two years we have also started testing for vitamin D deficiency, as many of my patients have steatorrhea and subsequently have low serum vitamin D levels. Many clients who have lived with HIV for fifteen to twenty years or longer still think that they are going to die from AIDS and thus may continue to smoke and/or practice other poor health habits. In reality, though, they should be concerned with developing chronic diseases that will more likely cause their death than HIV.

Some of my biggest nutrition challenges working with patients with HIV/AIDS are keeping up with all the medications they take and their side effects, treating nutrition problems associated with chronic conditions, and treatments such as those associated with hepatitis C, which often causes acute weight loss and depression. There have been tremendous changes in the treatment of HIV/AIDS in the last twenty years. The upside of working with HIV/AIDS in the San Francisco Bay area is that it is a hub for community-based HIV research. Working with other health professions involved in research helps keep me abreast of the field, and they know that nutrition is a critical part of HIV treatment—I am appreciated and considered part of the HIV health care team.

When I started my dietetic education I never thought that I would eventually be working in a public health setting, specifically in infectious diseases, but I was open to all the opportunities provided by my internship. HIV turned out to be to be my passion. My advice to dietetic students seeing HIV patients is to not get overwhelmed by all the numbers and medications; listen and learn from your clients. If you ask them about their HIV history, they will tell you their stories, and from both a medical and physiological standpoint, it is one of the most fascinating areas of medicine to work in. Working in the HIV/AIDS community, you are able to utilize your clinical skills and MNT knowledge, and because it has become a chronic disease, I have developed many long-term relationships with clients, which are not common in the acute care setting.

Application of the Nutrition Care Process: HIV and Lipodystrophy

Introduction:

RE is a 49-year-old Caucasian male who has been HIV positive for approximately 14 years. He presented to his physician with complaints of abdominal discomfort and heartburn related to his enlarged abdomen. His doctor previously diagnosed Mr. E with lipodystrophy syndrome based on physical examination and insulin resistance. He was referred to the dietitian for recommendations for nutrition-related interventions.

1. Describe the pathophysiology of HIV infection.

2. How is HIV infection treated?

3. Describe the lipodystrophy syndrome associated with HIV infection.

4. What may be the etiology of insulin resistance in an individual being treated for HIV?

Nutrition Assessment:

Food/Nutrition-Related History:

The registered dietitian's interview indicates that the patient describes gaining weight quickly over the last six months, and most appears to be in the abdomen. Heartburn began in the last few weeks and he is concerned about his body shape. He stopped exercising regularly about a year ago after he moved to the suburbs and away from his usual gym.

Dietary Intake: AM: scrambled eggs with sausage and toast, orange juice, coffee; Mid-day: meatloaf sandwich with cheese or gravy-dipped Italian beef with cola; PM: casserole with bread or rolls and water. Snacks throughout the day are typically potato chips or crackers. Drinks 3 large mugs of coffee and 1–2 colas throughout the day. Partner cooks large meals for family and friends on weekends.

Medications: Currently taking efavirenz and combovir (lamivudine and zidovudine), previously taking lopinavir/ritonavir and combovir, just prescribed rosiglitazone/metformin combination

Anthropometrics:

Ht. 5'11" Wt. 205 lbs. UBW 175 lbs.

Abdominal circumference: 104 cm, abdominal fatfold: 3 mm

Biochemical Data:

Viral load undetectable (below 50 copies per mL), CD4+ count: 660 per mL

5. What is REs body mass index? What does his anthropometric data indicate about his body composition?

6. Which assessment data support the diagnosis of lipodystrophy?

7. Which medications may contribute to lipodystrophy and/or insulin resistance?

Nutrition Diagnosis:

8. Identify at least two nutrition problems suggested by the nutrition assessment and medical history. Determine the diagnostic term for each nutrition problem. Next, identify the etiology of each nutrition diagnosis. Finally, identify the signs and symptoms that support the evidence for the diagnoses.

Diagnosis	Related to	Etiology	As Evidenced by	Sign/Symptoms
EXAMPLE: Involuntary weight loss	R/T	poor dietary intake secondary abdominal pain	AEB	25% unplanned weight loss in 6 weeks

Nutrition Intervention:

9. Based on your PES statements, what would be your nutrition Rx and goals for intervention?

10. What non-nutrient recommendations would you suggest that this patient, his physician, and his health care team explore in order to improve lipodystrophy-related discomfort and heartburn?

Monitoring and Evaluation:

11. Based on your intervention, what criteria would you monitor in this patient for improvement?

WEB LINKS

Research and Reviews: Adult AIDS Clinical Trials Group Includes information and summaries; research links.
http://www.aactg.org

Medscape HIV/AIDS Page Covers conference and journal reviews and provides online access to AIDS Reader articles.
http://www.medscape.com/hiv

HIV InSite Clinician and Patient Materials and Information: Online Textbook on HIV/AIDS through University of California, San Francisco School of Medicine.
http://hivinsite.ucsf.edu

Health Resources and Services Administration (HRSA) Guide to Nutrition in HIV/AIDS Includes clinician guidelines/algorithms and patient handouts.
http://www.aidsetc.org/pdf/p02-et/et-30-20-01/nutr_guide_0602.pdf

Johns Hopkins AIDS Service
http://www.hopkins-aids.edu/

Pocket Guide for ARVs Literature reviews.
http://www.hopkins-hivguide.org/literature_review/index.html?categoryId=9351&siteId=7151

The AIDS InfoNet Fact sheets and guides for patients and health care providers.
http://www.aidsinfonet.org

National Institutes of Health AIDS Information Patient and Provider Fact Sheets on topics related to HIV/AIDS.
http://www.aidsinfo.nih.gov/other/factsheet.aspx

GUIDELINES AND OTHER RESOURCE LINKS

AIDS Resource List Links for information resources.
http://www.specialweb.com/aids

Food and Nutrition Technical Assistance/Academy of Educational Development Funded by United States Agency for International Development, FANTA provides services and monographs appropriate for use in resource-limited settings and developing countries.
http://www.fantaproject.org

AVERT Provides information on HIV/AIDS related care in resource-limited settings.
http://www.avert.org/hivcare.htm

HIV/AIDS Dietetic Practice Group of the American Dietetic Association Nutrition and HIV/AIDS information, including continuing education.
http://www.hivaidsdpg.org

Project Inform Information on treatment, community programs, and research.
http://www.projinf.org

1. What kind of virus is the human immunodeficiency virus (HIV)? Which types of cells does it target?

2. List two factors that are predictive of disease progression. List one nutrition-related factor that is predictive of survival.

3. List two examples of how HIV or opportunistic infection compromises gastrointestinal function.

4. List two cofactor diseases or conditions that are related to HIV infection and its treatment and may require nutritional therapies.

5. Define *AIDS-related wasting syndrome (AWS)*.

6. Describe the inflammatory process that is seen in early and asymptomatic HIV infection. How can it trigger a wasting process?

7. At which levels of weight loss and body cell mass loss is there an increased risk for morbidity and mortality in HIV infection?

8. Which antiretroviral (ARV) medications should be taken on an empty stomach? Which should be taken with food?

9. Describe potential side effects of two medications used to treat HIV infection that can alter nutritional status.

10. Describe three types of dietary interventions that may be required in treated HIV infection.

11. List four types of non-nutrient interventions that are used to improve nutritional status.

12. Describe how HIV infection and its treatment may contribute to insulin resistance and diabetes.

13. Describe the characteristics of lipodystrophy syndrome and the potential uses of nutrition-related therapies to reduce its effects.

REFERENCES

1. 2008 Report on the Global AIDS Epidemic | HIV and AIDS Estimates and Data, 2007 and 2001© Joint United Nations Programme on HIV/AIDS (UNAIDS) 2008. Available at http://www.who.int/hiv/data/2008_global_summary_AIDS_ep.png

2. CDC. HIV Prevalence Estimates—United States, 2006. MMWR. 2008; 57(39):1073–76.

3. CDC. HIV/AIDS Surveillance Report, 2007. Vol. 19. Atlanta (GA): U.S. Department of Health and Human Services, CDC: 2009.

4. Quinn TC, Wawer MJ, Sewankambo N, Serwadda D, Li C, Wabwire-Mangen F, et al. Viral load and heterosexual transmission of human immunodeficiency virus type 1. N Engl J Med. 2000;342: 921–29.

5. Mellors JW, Munoz A, Giorgi JV, Margolick JB, Tassoni CJ, Gupta P, et al. Plasma viral load and CD4+ lymphocytes as prognostic markers of HIV-1 infection. Ann Intern Med. 1997;126:946–54.

6. MacArthur RD, Perez G, Walmsley S, Baxter JD, Mullin CM, Neaton JD. Terry Beirn Community Programs for Clinical Research on AIDS (CPCRA) 042/045; Canadian HIV Trials Network (CTN) 102 Protocol Teams. Comparison of prognostic importance of latest CD4+ cell count

and HIV RNA levels in patients with advanced HIV infection on highly active antiretroviral therapy. HIV Clin Trials. 2005;6:127–35.

7. Schneider T, Ullrich R, Zeitz M. The immunologic aspects of human immunodeficiency virus infection in the gastrointestinal tract. Semin Gastrointest Dis. 1996;7:19–29.

8. Castello-Branco LR, Lewis DJ, Ortigõ-de-Sampaio MB, Griffin GE Gastrointestinal immune responses in HIV infected subjects. Mem Inst Oswaldo Cruz. 1996 May–Jun;91(3):363–66.

9. Smith PD, Mai UE. Immunopathophysiology of gastrointestinal disease in HIV infection. Gastroenterol Clin North Am. 1992;21:331–35.

10. Smith PD, Meng G, Salazar-Gonzalez JF, Shaw GM. Macrophage HIV-1 infection and the gastrointestinal reservoir. J Leukoc Biol. 2003;74:642–49.

11. Hill A, Balkin A. Risk factors for gastrointestinal adverse events in HIV treated and untreated patients. AIDS Review 2009 11(1):30–38.

12. Ott M, Wegner A, Caspary WF, Lembcke B. Intestinal absorption and malnutrition in patients with the acquired immunodeficiency syndrome. Z Gastroenterol. 1993;31:661–65.

13. Bijl M, Dieleman JP, Simoons M, van der Ende ME. Low prevalence of cardiac abnormalities in an HIV-seropositive population on antiretroviral combination therapy. J Acquir Immune Defic Syndr. 2001;27:318–20.

14. Kuritzkes DR, Currier J. Cardiovascular risk factors and antiretroviral therapy. New Engl J Med 2003;348(8):679–80.

15. Moyle G. Anaemia in persons with HIV infection: prognostic marker and contributor to morbidity. AIDS Rev. 2002;4:13–20.

16. Szczech LA. HIV-related renal disease and the utility of empiric therapy—not everyone needs to be biopsied. Nature Clinical Practice Nephrology. 2008;5:20–21.

17. Centers for Disease Control and Prevention. Revision of the CDC surveillance case definition for acquired immunodeficiency syndrome. MMWR. 1987;36 (Suppl 1S).

18. Carbonnel F, Maslo C, Beaugerie L, Carrat F, Wirbel E, Aussel C, Gobert JG, Girard PM, Gendre JP, Cosnes J, Rozenbaum W. Effect of indinavir on HIV-related wasting. AIDS. 1998;12(14):1777–84.

19. Moore RD, Chaisson RE. Natural history of HIV infection in the era of combination antiretroviral therapy. AIDS. 1999;13:1933–42.

20. Tang AM, Forrester J, Spiegelman D, et al. Weight loss and survival in HIV-positive patients in the era of highly active antiretroviral therapy. J Acquir Immune Defic Syndr. 2002;31:230–36.

21. Mangili A, Murman DH, Zampini AM, Wanke CA. HIV/AIDS: Nutrition and HIV Infection: Review of Weight Loss and Wasting in the Era of Highly Active Antiretroviral Therapy from the Nutrition for Healthy Living Cohort. Clin Infect Dis. 2006 Mar 15;42(6):836–42.

22. Centers for Disease Control and Prevention. 1993 revised classification system for HIV infection and expanded surveillance case definition for AIDS among adolescents and adults. MMWR. 1992;41(RR-17):1–19.

23. Centers for Disease Control and Prevention. Revised classification system for HIV infection in children less than 13 years of age. MMWR. 1994;43 (RR-12):1–10.

24. World Health Organization. WHO International Collaborating Group for the Study of the WHO Staging System. AIDS. 1993;7:711–18.

25. World Health Organization. WHO Case Definitions of HIV for Surveillance and Revised Clinical Staging and Immunological Classification of HIV-Related Disease in Adults and Children. 2007. Available at www.who.int/hiv/pub/guidelines/HIVstaging150307.pdf

26. Centers for Disease Control and Prevention. HIV Mortality Slides: 1987–2002. Rockville (MD): CDC; c2002. Available at http://www.cdc.gov/hiv/graphics/images/l285/l285.pdf

27. Kotler DP, Tierney AR, Wang J, Pierson RN Jr. Magnitude of body-cell-mass depletion and the timing of death from wasting in AIDS. Am J Clin Nutr. 1989;50:444–47.

28. Guidelines for the Use of Antiretroviral Agents in HIV-1-Infected Adults and Adolescents - November 3, 2008. Available at http://aidsinfo.nih.gov/Guidelines/GuidelineDetail.aspx?MenuItem=Guidelines&Search=Off&GuidelineID=7&ClassID=1

29. A Guide To Primary Care For People With HIV/AIDS, 2004 edition. Available at http://www.hab.hrsa.gov/tools/primarycareguide/PCGchap1.htm

30. Department of Health and Human Services (DHHS). Guidelines for the use of antiretroviral agents in HIV-1-infected adults and adolescents. November 3, 2008. Available at http://aidsinfo.nih.gov/contentfiles/adultandAdolescentGL.pdf

31. Health Resources and Services Administration (HRSA). Guidelines for the use of antiretroviral agents in pediatric HIV infection. http://www.aidsinfo.nih.gov/Guidelines/GuidelineDetail.aspx?GuidelineID=8

32. World Health Organization. Adherence to long-term therapies: evidence for action. Geneva: World Health Organization; 2003.

33. Macallan DC, Noble C, Baldwin C, Foskett M, McManus T, Griffin GE. Prospective analysis of patterns of weight change in stage IV human immunodeficiency virus infection. Am J Clin Nutr. 1993;58:417–24.

34. Kaplan JE, et al. Benson C, Holmes KH, Brooks JT, Pau A, Masur H. Centers for Disease Control and Prevention (CDC); National Institutes of Health; HIV Medicine Association of the Infectious Diseases Society of America. Guidelines for prevention and treatment of opportunistic infections in HIV-infected adults and adolescents: recommendations from CDC, the National Institutes of Health, and the HIV Medicine Association of the Infectious Diseases Society of America. MMWR Recomm Rep. 2009 Apr 10;58(RR-4):1–207; quiz CE1–4.

35. AIDS Info US Dept of Health and Human Services, HIV/AIDS. Available at http://www.aidsinfo.nih.gov/DrugsNew/DrugDetailNT.aspx?MenuItem=Drugs&Search=On&int_id=63. Updated: 2007 Apr 9

36. Kraft B, Kress HG. Cannabinoids and the immune system. Of men, mice and cells. Schmerz. 2004;18:203–10.

37. Cabral GA, Staab A. Effects on the immune system. Handb Exp Pharmacol. 2005;(168):385–423.

38. Haney M. Effects of smoked marijuana in healthy and HIV+ marijuana smokers. J Clin Phamacol. 2002;42(11 Suppl):34A–40S.

39. Kosel BW, Aweeka FT, Benowitz NL, Shade SB, Hilton JF, Lizak PS, Abrams DI. The effects of cannabinoids on the pharmacokinetics of indinavir and nelfinavir. AIDS. 2002;16:543–50.

40. Crum NF, Furtek KJ, Olson PE, Amling CL, Wallace MR. A review of hypogonadism and erectile dysfunction among HIV-infected men during the pre- and post-HAART eras: diagnosis, pathogenesis, and management. AIDS Patient Care STDS. 2005;19(10):655–71.

41. Clay PG, Lam AI. Testosterone replacement therapy for bone loss prevention in HIV-infected males. Ann Pharmacother. 2003;37(4):582–85.

42. Behler C, Shade S, Gregory K, Abrams D, Volberding P. Anemia and HIV in the antiretroviral era: potential significance of testosterone. AIDS Res Hum Retroviruses. 2005;21:200–206.

43. Miller MG, Mulligan T. Human immunodeficiency virus and hypogonadal bone disease. Pharmacotherapy. 2005;25:632–34.

44. Clay PG. Program savings associated with switching testosterone intramuscular injections to topical gel in HIV-infected males. Curr Med Res Opin. 2004;20(4):461–68.

45. Mylonakis E, Koutkia P, Grinspoon S. Diagnosis and treatment of androgen deficiency in human immunodeficiency virus-infected men and women. Clin Infect Dis. 2001;33:857–64.

46. Mazer NA, Shifren JL. Transdermal testosterone for women: a new physiological approach for androgen therapy. Obstet Gynecol Surv. 2003;58:489–500.

47. Strawford A, Barbieri T, Van Loan M, Parks E, Catlin D, Barton N, et al. Resistance exercise and supraphysiologic androgen therapy in eugonadal men with HIV-related weight loss: a randomized controlled trial. JAMA. 1999;281: 1282–90.

48. Johns K, Beddall MJ, Corrin RC. Anabolic steroids for the treatment of weight loss in HIV-infected individuals. Cochrane Database Syst Rev. 2005 Oct 19;(4):CD005483

49. Orr R, Fiatarone Singh M. The anabolic androgenic steroid oxandrolone in the treatment of wasting and catabolic disorders: review of efficacy and safety. Drugs. 2004;64:725–50.

50. Falutz J, Allas S, Kotler D, Thompson M, Koutkia P, Albu J, et al. A placebo-controlled, dose-ranging study of a growth hormone releasing factor in HIV infected patients with abdominal fat accumulation. AIDS. 2005;19:1279–87.

51. Haugaard SB, Andersen O, Flyvbjerg A, Orskov H, Madsvad S, Iversen J. Growth factors, glucose and insulin kinetics after low dose growth hormone therapy in HIV-lipodystrophy. J Infect. 2006 Jun;52(6):389–98.

52. Powers HJ. Riboflavin (vitamin B-2) and health. Am J Clin Nutr. 2003;77:1352–60.

53. Mills E, Montori V, Perri D, Phillips E, Koren G. Natural health product-HIV drug interactions: a systematic review. Int J STD AIDS. 2005;16:181–86.

54. Mills E, Foster BC, van Heeswijk R, Phillips E, Wilson K, Leonard B, et al. Impact of African herbal medicines on antiretroviral metabolism. AIDS. 2005;19:95–97.

55. Collins RA, Ng TB, Fong WP, Wan CC, Yeung HW. A comparison of human immunodeficiency virus type 1 inhibition by partially purified aqueous extracts of Chinese medicinal herbs. Lif Sci. 1997; 60(23):PL345–51.

56. Macallan DC. Nutrition and immune function in human immunodeficiency virus infection. Proc Nutr Soc. 1999;58:743–48.

57. Gerbert B, Caspers N, Moe J, Clanon K, Abercrombie P, Herzig K. The mysteries and demands of HIV care: qualitative analyses of HIV specialists' views on their expertise. AIDS Care. 2004;16:363–76.

58. Heslin KC, Andersen RM, Ettner SL, Kominski GF, Belen TR, Margenstern H, et al. Do specialist self-referral insurance policies improve access to HIV-experienced physicians as a regular source of care? Med Care Res Rev. 2005;62:583–600.

59. Wilson IB, Landon BE, Ding L, Zaslavsky AM, Shapiro MF, Bozzette SA, Cleary PD. A national study of the relationship of care site HIV specialization to early adoption of highly active antiretroviral therapy. Med Care. 2005;43:12–20.

60. Fraker P. Nutritional immunology: methodological considerations. J Nutr Immunol. 1994;2:87–92.

61. Semba RD. Micronutrients and the pathogenesis of human immunodeficiency virus infection. In: Fitzpatrick DW, Anderson JE, L'Abbe ML, eds. Proceedings of the 16th International Congress of Nutrition. Ottawa: Canadian Federation of Biological Societies. 1998:349–51.

62. Ott M, Lembcke B, Fischer H, Jager R, Polat H, Geier H, et al. Early changes of body composition in human immunodeficiency virus-infected patients: tetrapolar body impedance analysis indicates significant malnutrition. Am J Clin Nutr. 1993;57:15–19.

63. Wanke C. Pathogenesis and consequences of HIV-associated wasting. J Acquir Immune Defic Syndr. 2004;27(Suppl4):S277–S279.

64. Scrimshaw NS, SanGiovanni JP. Synergism of nutrition, infection, and immunity: an overview. Am J Clin Nutr. 1997;66(suppl):464S–477S.

65. Campbell IT. Limitations of nutrient intake. The effect of stressors: trauma, sepsis and multiple organ failure. Eur J Clin Nutr. 1999;53(Suppl 1):S143–S147.

66. Paton NI, Ng YM, Chee CB, Persaud C, Jackson AA. Effects of tuberculosis and HIV infection on whole-body protein metabolism during feeding, measured by the [15N]glycine method. Am J Clin Nutr. 2003;78:319–25.

67. Jahoor F, Abramson S, Heird WC. The protein metabolic response to HIV infection in young children. Am J Clin Nutr. 2003;78:182–89.

68. Forrester JE, Spiegelman D, Woods M, Knox TA, Fauntleroy JM, Gorbach SL. Weight and body composition in a cohort of HIV-positive men and women. Public Health Nutr. 2001;4:743–47.

69. Grunfeld C, Pang M, Shimizu L, Shigenaga JK, Jensen P, Feingold KR. Resting energy expenditure, caloric intake, and short-term weight changes in human immunodeficiency virus infection and the acquired immunodeficiency syndrome. Am J Clin Nutr. 1992;55:455–60.

70. Roubenoff R, Grinspoon S, Skolnik PR, Tchetgen E, Abad L, Spiegelman D, et al. Role of cytokines and testosterone in regulating lean body mass and resting energy expenditure in HIV-infected men. Am J Physiol Endrocrinol Metab. 2002;283:E138–E145.

71. Polsky B, Kotler DP, Steinhart C. HIV-associated wasting in the HAART era: guidelines for assessment, diagnosis, and treatment. AIDS Patient CARE STDS. 2001;15:411–23.

72. Ferrando SJ, Rabkin JG, Lin SH, McElhiney M. Increase in body cell mass and decrease in wasting are associated wtih increasing potency of antiretroviral therapy for HIV infection. AIDS Patient Care STDS. 2005;19(4):216–23.

73. Wheeler DA. Weight loss and disease progression in HIV infection. AIDS Read. 1999;9:347–53.

74. Tang AM, Forrester J, Spiegelman D, Knox TA, Tchetgen E, Gorbach SL. Weight loss and survival in HIV-positive patients in the era of highly active antiretroviral therapy. J Acquir Immune Defic Syndr. 2002;31:230–36.

75. Wanke CA, Silva M, Ganda A, Fauntleroy J, Spiegelman D, Knox TA, et al. Role of acquired immune deficiency syndrome-defining conditions in human immunodeficiency virus-associated wasting. Clin Infect Dis. 2003;37(Suppl 2):S81–S84.

76. van der Sande MA, Schim van der Loeff MF, Aveika AA, Sabally S, Togun T, Sarge-Nije R, et al. Body mass index at time of HIV diagnosis: a strong and independent predictor of survival. J Acquir Immune Defic Sydr. 2004;37:1288–94.

77. Jones CY, Hogan JW, Snyder B, Klein RS, Rompalo A, Schuman P, Carpenter CC, HIV Epidemiology Research Study Group. Overweight and human immunodeficiency virus (HIV) progression in women: association HIV disease progression and changes in body mass index in women in the HIV epidemiology research study cohort. Clin Infect Dis. 2003;37(Suppl 2):S69–S80.

78. Hong-Brown LQ, Brown DR, Lang CH. HIV antiretroviral agents inhibit protein synthesis and decrease ribosomal protein S6 and 4EBP1 phosphorylation in C2C12 myocytes. AIDS Res Hum Retroviruses. 2005;21:854–62.

79. Ambrus JL Sr, Ambrus JL Jr. Nutrition and infectious diseases in developing countries and problems of acquired immunodeficiency syndrome. Exp Biol Med. 2004;229:464–72.

80. Semba RD, Tang AM. Micronutrients and the pathogenesis of human immunodeficiency virus infection. Br J Nutr. 1999;81:181–89.

81. Skurnick JH, Bogden JD, Baker H, Kemp FW, Sheffet A, Quattrone G, et al. Micronutrient profiles in HIV-1-infected heterosexual adults. J Acquir Immune Defic Syndr Hum Retrovirol. 1996; 12:75–83.

82. Bogden JD, Baker H, Frank O, Perez G, Kemp F, Bruening K, Louria D. Micronutrient status and human immunodeficiency virus (HIV) infection. Ann NY Acad Sci. 1990;587:189–95.

83. Bogden JD, Kemp FW, Han S, Li W, Bruening K, Denny T, Oleske JM, Lloyd J, Baker H, Perez G, Kloser P, Skurnick J, Louria DB. Status of selected nutrients and progression of human immunodeficiency virus type 1 infection. Am J Clin Nutr. 2000;72(3):809–15.

84. Jimenez-Exposito MJ, Bullo Bonet M, Alonso-Villaverde C, Serrano P, Garcia-Lorda P, Garcia-Luna PP, Masana L, Salas-Salvado J. Micronutrients in HIV-infection and the relationship with the inflammatory response. Med Clin. 2002;119(20):765–69.

85. Tang AM, Smit E. Selected vitamins in HIV infection: a review. AIDS Patient Care STDS. 1998;12:249–50.

86. Evans P, Halliwell B. Micronutrients: oxidant/antioxidant status. Br J Nutr. 2001;85 suppl 2:S67–S74.

87. Lanzillotti JS, Tang AM. Micronutrients and HIV disease: a review pre- and post-HAART. Nutr Clin Care. 2005; 8:16–23.

88. Silva M, Skolnik PR, Gorbach SL, Spiegelman D, Wilson IB, Fernandez-DiFranco MG, et al. The effect of protease inhibitors on weight and body composition in HIV-infected patients. AIDS. 1998;12:1645–51.

89. Lee GA, Rao MN, Grunfeld C. The effect of HIV protease inhibitors on carbohydrate and lipid metabolism. Curr HIV/AIDS Rep. 2005;2:39–50.

90. Thomas-Geevarghese A, Raghavan S, Minolfo R, Holleran S, Ramakrishnan R, Ormsby B, et al. Postprandial response to a physiologic caloric load in HIV-positive patients receiving protease inhibitor-based or nonnucleoside reverse transcriptase inhibitor-based antiretroviral therapy. Am J Clin Nutr. 2005;82:146–54.

91. Schutt M, Zhou J, Meier M, Klein HH. Long-term effects of HIV-1 protease inhibitors on insulin secretion and insulin signaling in INS-1 beta cells. J Endocrinol. 2004;183:445–54.

92. Hong-Brown LQ, Brown DR, Lang CH. HIV antiretroviral agents inhibit protein synthesis and decrease ribosomal protein S6 and 4EBP1 phosphorylation in C2C12 myocytes. AIDS Res Hum Retroviruses. 2005;21:854–62.

93. Hong-Brown LQ, Pruznak AM, Frost RA, Vary TC, Lang CH. Indinavir alters regulators of protein anabolism and catabolism in skeletal muscle. Am J Physiol Endocrinol Metab. 2005;289:E382–E390.

94. McDermott AY, Terrin N, Wanke C, Skinner S, Tchetgen E, Shevitz AH. CD4+ cell count, viral load, and highly active antiretroviral therapy use are independent predictors of body composition alterations in HIV-infected adults: a longitudinal study. Clin Infect Dis. 2005;41:1662–70.

95. Young EM, Considine RV, Sattler FR, Deeg MA, Buchanan TA, Degawa-Yamauchi M, et al. Changes in thrombolytic and inflammatory markers after initiation of indinavir- or amprenavir-based antiretroviral therapy. Cardiovasc Toxicol. 2004;4:179–86.

96. Schwenk A, Steuck H, Kremer G. Oral supplements as adjunctive treatment to nutritional counseling in malnourished HIV-infected patients: randomized controlled trial. Clin Nutr. 1999;18:371–74.

97. Tang AM. Weight loss, wasting and survival in HIV-positive patients: current strategies. AIDS Read. 2003;13(12 Suppl): S23–S27.

98. Zachariah R, Teck R, Ascurra O, Humblet P, Harries AD. Targeting CD4 testing to a clinical subgroup of patients could limit unnecessary CD4 measurements, premature antiretroviral treatment and costs in Thyolo Distric, Malawi. Trans R Soc Trop Med Hyg. 2006;100: 24–31.

99. Sheehan LA, Macallan DC. Determinants of energy intake and energy expenditure in HIV and AIDS. Nutrition. 2000;16:101–06.

100. American Dietetic Association. Evidence to support the assessment of body composition for people with HIV infection. Available at https://www.adaevidencelibrary.com/conclusion.cfm?conclusion_statement_id=250968

101. Ott M, Fischer H, Polat H, Helm EB, Frenz M, Caspary WF, et al. Bioelectrical impedance analysis as a predictor of survival in patients with human immunodeficiency virus infection. J Acquir Immune Defic Syndr Hum Retrovirol. 1995;9:20–25.

102. American Dietetic Association. Evidence to support certain methodologies in the measurement of body composition of people with HIV infection. Available at https://www.adaevidencelibrary.com/conclusion.cfm?conclusion_statement_id=250998

103. Kotler DP, Burastero S, Wang J, Pierson RN Jr. Prediction of body cell mass, fat-free mass, and total body water with bioelectrical impedance analysis: effects of race, sex, and disease. Am J Clin Nutr. 1996;64(3 Suppl):489S–497S.

104. Schwenk A, Hoffer-Belitz E, Jung B, Kremer G, Burger B, Salzberger B, et al. Resting energy expenditure, weight loss, and altered body composition in HIV infection. Nutrition. 1996;12:595–601.

105. Schwenk A. Methods of assessing body shape and composition in HIV-associated lipodystrophy. Curr Opin Infect Dis. 2002;15:9–16.

106. Fields-Gardner C. Anthropometry Measures. Cary (IL): Hi-R-Ed; 2001. Available at http://www.Hi-R-Ed.org

107. Batterham MJ, Garsia R, Greenop P. Prevalence and predictors of HIV-associated weight loss in the era of highly active antiretroviral therapy. Int J STD AIDS. 2002;13:744–47.

108. Henderson RA, Talusan K, Hutton N, Yolken RH, Caballero B. Serum and plasma markers of nutritional status in children infected with the human immunodeficiency virus. J Am Diet Assoc. 1997;97:1377–81.

109. American Dietetic Association. Evidence to support Medical Nutrition Therapy for people with HIV infection? Conclusion. Available at https://www.adaevidencelibrary.com/conclusion.cfm?conclusion_statement_id=250707

110. Coyne-Meyers K, Trombley LE. A review of nutrition in human immunodeficiency virus infection in the era of highly active antiretroviral therapy. Nutr Clin Pract. 2004;19(4):340–55.

111. Piscitelli SC, Berstein AH, Welden N, Gallicano KD, Falloon J. The effect of garlic supplements on the pharmacokinetics of saquinavir. Clin Infect Dis. 2002;34:234–38.

112. Kupferschmidt HH, Fattinger KE, Ha HR, Follath F, Krahenbuhl S. Grapefruit juice enhances the bioavailability of the HIV protease inhibitor saquinavir in

man. Br J Clin Pharmacol. 1998;45: 355–59.

113. American Dietetic Association. Evidence to support a particular dietary intake of energy for people with HIV infection. Available at https://www.adaevidencelibrary.com/conclusion.cfm?conclusion_statement_id=250676

114. American Dietetic Association. Evidence to support a particular dietary intake of protein for people with HIV infection. Available at https://www.adaevidencelibrary.com/conclusion.cfm?conclusion_statement_id=250705

115. American Dietetic Association. Evidence to support a particular dietary intake of carbohydrate for people with HIV infection. Available at https://www.adaevidencelibrary.com/conclusion.cfm?conclusion_statement_id=250706

116. American Dietetic Association. Evidence to support the consumption of specific fatty acids for people with HIV infection. Available at https://www.adaevidencelibrary.com/conclusion.cfm?conclusion_statement_id=250677

117. Mehendale SM, Shepherd ME, Brookmeyer RS, Semba RD, Divekar AD, Gangakhedkar RR, et al. Low carotenoid concentration and the risk of HIV seroconversion in Pune, India. J Acquir Immune Defic Syndr. 2001;26:352–59.

118. Hegde HR, Woodman RC, Sankaran K. Nutrients as modulators of energy in acquired immune deficiency syndrome. J Assoc Physicians India. 1999;47:318–25.

119. Patrick L. Nutrients and HIV: part one – beta carotene and selenium. Altern Med Rev. 1999;4:403–13.

120. Patrick L. Nutrients and HIV: part two – vitamins A and E, zinc, B-vitamins, and magnesium. Altern Med Rev. 2000;5:39–51.

121. American Dietetic Association. Evidence to support micronutrient supplementation for people with HIV infection. Available at https://www.adaevidencelibrary.com/conclusion.cfm?conclusion_statement_id=250924

122. Coodley G. Update on vitamins, minerals, and the carotenoids. J Physicians Assoc AIDS Care. 1995;2(1):24–29.

123. Buys H, Hendricks M, Eley B, Hussey G. The role of nutrition and micronutrients in pediatric HIV infection. SADJ. 2002;57:454–56.

124. Kelly P, Musonda R, Kafwembe E, Kaetano L, Keane E, Farthing M. Micronutrient supplementation in the AIDS diarrhea-wasting syndrome in Zambia: a randomized controlled trial. AIDS. 1999;13:495–500.

125. Cunningham-Rundles S, Ahrn S, Abuav-Nussbaum R, Dnistrian A. Development of immunocompetence: role of micronutrients and microorganisms. Nutr Rev. 2002;60:S68-S72.

126. O'Brien K, Nixon S, Glazier RH, Tynan AM. Progressive resistive exercise interventions for adults living with HIV/AIDS. Cochrane Database Syst Rev. 2004;4:CD004248.

127. Zinna EM, Yarasheski KE. Exercise treatment to counteract protein wasting of chronic diseases. Curr Opin Clin Nutr Metab Care. 2003;6(1):87–93.

128. Scevola D, DiMatteo A, Lanzarini P, Uberti F, Scevola S, Bernini V, et al. Effect of exercise and strength training on cardiovascular status in HIV-infected patients receiving highly active antiretroviral therapy. AIDS. 2003;17(Suppl 1): S123–S129.

129. Thoni GJ, Fedou C, Brun JF, Fabre J, Renard E, Reynes J, et al. Reduction of fat accumulation and lipid disorders by individualized light aerobic training in human immunodeficiency virus infected patients with lipodystrophy and/or dyslipidemia. Diabetes Metab. 2002;28:397–404.

130. Shevitz AH, Wilson IB, McDermott AY, Spiegelman D, Skinner SC, Antonsson K, et al. A comparison of the clinical and cost-effectiveness of 3 intervention strategies for AIDS wasting. J Acquir Immune Defic Syndr. 2005;38:399–406.

131. American Dietetic Association. Evidence to support physical activity for people with HIV infection. Available at https://www.adaevidencelibrary.com/conclusion.cfm?conclusion_statement_id=250967

132. Jones SP, Doran DA, Leatt PB, Maher B, Pirmohamed M. Short-term exercise training improves body composition and hyperlipidaemia in HIV-positive individuals with lipodystrophy. AIDS. 2001;15:2049–51.

133. American Dietetic Association. Evidence to support education on foodborne illness for people with HIV infection and their caregivers. Available at https://www.adaevidencelibrary.com/topic.cfm?cat=3772

134. U.S. Department of Agriculture. Food Safety and Inspection Service. Food Safety for People with HIV/AIDS. Available at www.fsis.usda.gov/PDF/Food_Safety_for_People_with_hiv_Text.pdf; Sept 2006.

135. McDermott AY, Shevitz A, Must A, Harris S, Roubenoff R, Gorbach S. Nutrition treatment for HIV wasting: a prescription for food as medicine. Nutr Clin Pract. 2003;18:86–94.

136. Normen L, Chan K, Braitstein P, Anema A, Bondy G, Montaner JS, et al. Food insecurity and hunger are prevalent among HIV-positive individuals in British Columbia, Canada. J Nutr. 2005;135: 820–25.

137. Rabeneck L, Palmer A, Knowles JB, Seidehamel RJ, Harris CL, Merkel KL, et al. A randomized controlled trial evaluating nutrition counseling with or without oral supplementation in malnourished HIV-infected patients. J Am Diet Assoc. 1998;98:434–38.

138. Spirig R, Moody K, Battegay M, DeGeest S. Symptom management in HIV/AIDS. Advancing the conceptualization. Adv Nurs Sci. 2005;28:333–34.

139. American Dietetic Association. Evidence to support dietary treatment of diarrhea/malabsorption in people with HIV infection. Available at https://www.adaevidencelibrary.com/topic.cfm?cat=3778

140. Young JS. HIV and medical nutrition therapy. J Am Diet Assoc. 1997;97 (10 Suppl 2):S161–S166.

141. American Dietetic Association. Position of the American Dietetic Association and the Dietitans of Canada: nutrition intervention in the care of persons with human immunodeficiency virus infection. J Am Diet Assoc. 2004;104:1425–41.

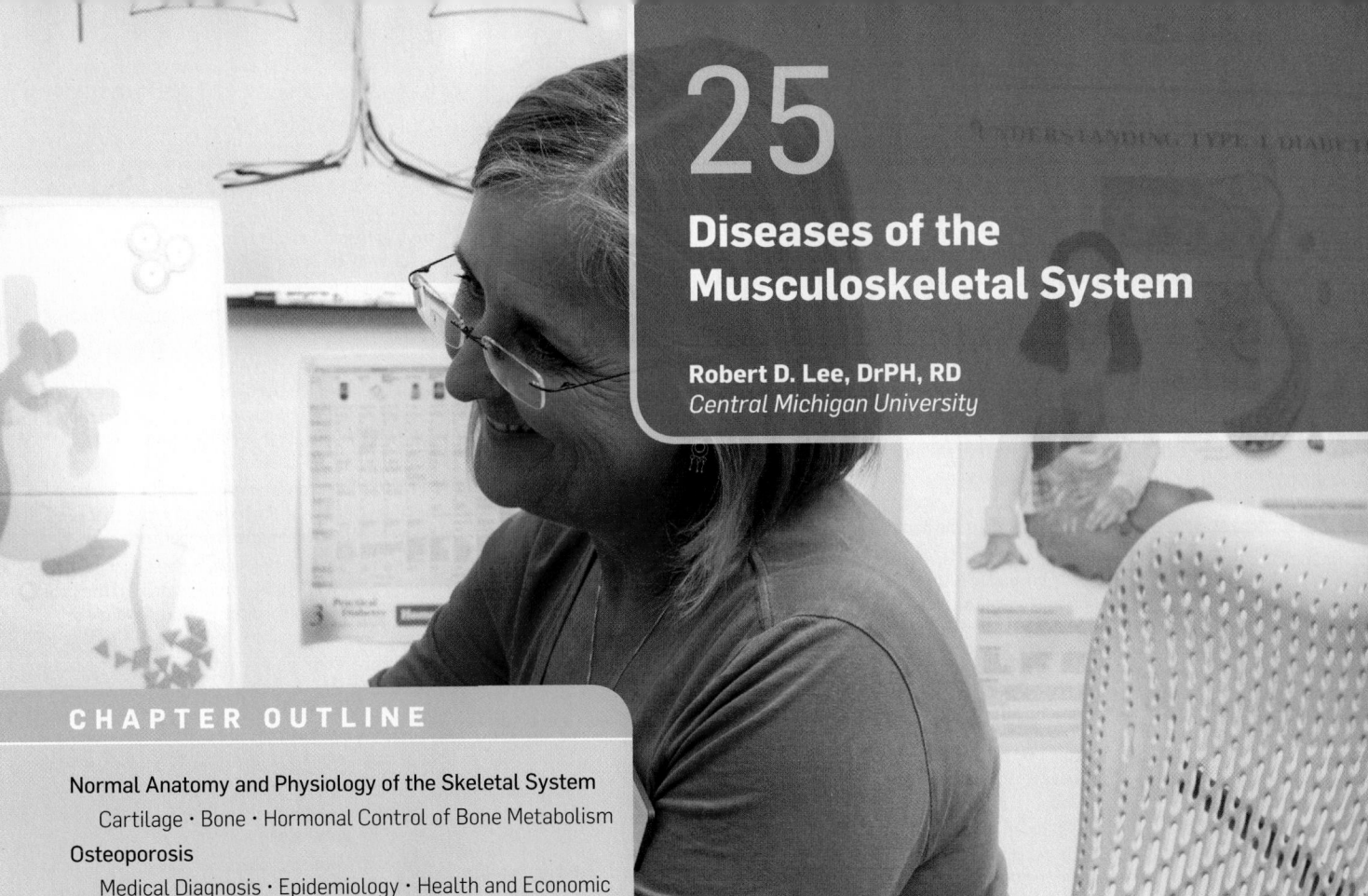

25

Diseases of the Musculoskeletal System

Robert D. Lee, DrPH, RD
Central Michigan University

to the body, and protects and maintains the position of soft tissues. The bones also serve as a reservoir for calcium and phosphorus, so that the levels of these important minerals in bodily fluids can be maintained within certain physiologically acceptable ranges.[2] Consequently, diseases of this system can severely limit mobility and the performance of the activities of daily life, can cause considerable pain and disability, and have the potential to deform the body, which, in turn, can adversely affect respiration, digestion, nervous system function, and nutritional status.

This chapter addresses the more common diseases of the musculoskeletal system that have the potential to be prevented or managed by nutrition therapy. These include osteoporosis, Paget disease, rickets, osteomalacia, osteoarthritis, rheumatoid arthritis, gout, fibromyalgia, systemic lupus erythematosus, and scleroderma.

Introduction

The musculoskeletal system comprises the bones and cartilage of the skeleton, as well as the muscles, tendons, and ligaments attached to the bones. The skeleton is divided into the axial and appendicular skeleton. The axial skeleton forms the axis of the body and includes the bones of the skull, vertebral column, and thorax. The appendicular skeleton consists of the bones of the upper extremities, the lower extremities, the shoulder, and the hip.[1] The musculoskeletal system makes movement in the external environment possible, gives shape and stability

Normal Anatomy and Physiology of the Skeletal System

Composed of cartilage, ligaments, tendons, and bones, the skeletal system forms a strong, tightly bound but flexible framework for the body. Far from being an inert scaffold for the body, the skeletal system is composed of numerous metabolically active cells and tissues that interact physiologically with other organ systems of the body and are in a continual state of change throughout the life cycle.

GLOSSARY

amenorrhea—the absence of menstrual cycles when they would be expected to occur

balneotherapy—treating disease by bathing; it can involve hot water, steam, or application of hot packs to the body; cold water; contrasting hot and cold baths or showers; a steam bath followed by immersion in cold water; or immersion in mineral springs or a hot bath to which various medicinal herbs or minerals have been added

calcitonin—a polypeptide hormone secreted by the "C" cells of the thyroid gland when the blood calcium concentration is high; it lowers blood calcium by inhibiting bone resorption, promoting bone formation, and reducing renal reabsorption of calcium and phosphorus

carpopedal—referring or pertaining to the hand and foot

cholecalciferol—a naturally occurring form of vitamin D produced in humans when the precursor molecule 7-dehydrocholesterol present in the skin is exposed to sunlight or to ultraviolet radiation; also known as vitamin D_3

chondroblasts—cells that are actively forming cartilage

chondrocytes—cells surrounded by cartilage and located inside small cavities known as lacunae

chondroitin—a nutritional supplement used by some to treat osteoarthritis; it is a specific glycosaminoglycan found in the proteoglycans of articular cartilage

chondroitin sulfate—the most common polysaccharide found in the proteoglycan molecules of cartilage

cortical bone—dense bone that forms the external surfaces of all bones, the shafts of the long bones, and a shell that caps the ends of the long bones

ergocalciferol—a form of vitamin D produced by exposing the plant steroid ergosterol to ultraviolet irradiation; also known as vitamin D_2

fibromyalgia—a chronic musculoskeletal disorder characterized by widespread muscle pain, joint stiffness, disturbed sleep, fatigue, headache, cognitive and memory problems, paresthesias, and numerous tender points. The word comes from the Latin term for fibrous tissue (*fibro*) and the Greek terms for muscle (*myo*) and pain (*algia*)

glucosamine—a nutritional supplement used by some as a treatment for osteoarthritis; it is an amino sugar and a raw material for synthesizing glycosaminoglycans and proteoglycans, which are important constituents of articular cartilage

gout—swelling, redness, heat, pain, and stiffness in a joint due to the formation of uric acid crystals in the synovial fluid, resulting in inflammation within the joint and in the surrounding tissues

hydroxyapatite—a crystallized calcium phosphate salt that gives bones their stiffness

hypercalciuria—excessive calcium in the urine

hypochondriasis—a somatoform disorder (i.e., a physical ailment stemming from a psychological problem) characterized by an unfounded belief that one is suffering from a serious illness

oligomenorrhea—abnormally infrequent menstrual cycles

osseous tissue—the group of cells and cell products that collectively form bone; bone tissue

osteoarthritis—a condition involving progressive loss of articular cartilage and inflammation of the tissues composing the joint, resulting in joint pain, stiffness, and limited joint movement

osteoblasts—cells that synthesize, deposit, and then orient the fibrous proteins of the organic matter of the bone matrix

osteoclasts—bone-removing cells that dissolve the mineral component of the bone matrix, playing a major role in bone resorption

osteocytes—mature osteoblasts surrounded and entrapped by the matrix they have synthesized and then calcified

osteogenic cells—stem cells capable of developing into osteoblasts

osteomalacia—a condition in which the organic matrix of the bones of adults is inadequately mineralized, resulting in muscular weakness, bone pain, and, in advanced cases, deformities of the ribs, pelvis, and bones of the legs

osteopenia—a term used to describe a bone mineral density that is low but not low enough to meet the diagnostic criterion for osteoporosis

osteophyte—a bony outgrowth near the joint affected by osteoarthritis; also referred to as a bone spur

osteoporosis—a disease resulting from a decreased amount of bone mineral and organic matrix which weakens bones, making them more susceptible to fracture

Paget disease—a localized disorder of bone remodeling resulting from rapid bone resorption followed by rapid formation of new bone that is structurally disorganized and more susceptible to deformities and fractures

pannus—an abnormal, destructive tissue that develops on the synovial membrane of patients with advanced rheumatoid arthritis; inflammatory cells in pannus secrete enzymes that are destructive to articular cartilage and subchondral bone

paresthesia—an abnormal sensation in the skin that may be described as burning, pricking, or like ants crawling on the skin

periarticular muscles—those muscles located near a joint

rheumatoid arthritis—a chronic inflammatory disease in which the synovial membrane of the joint becomes inflamed, resulting in swelling, stiffness, pain, limited range of motion, joint deformity, and disability

rheumatologist—a medical doctor specializing in diseases of the muscles and joints that are classified as rheumatic diseases

rickets—a condition characterized by inadequate mineralization of the organic matrix in the bones of children usually caused by a deficiency of vitamin D and resulting in bowing of the legs and skeletal deformity of the rib cage

scleroderma—a chronic autoimmune disease of unknown etiology affecting multiple organ systems including the skin, lungs, gastrointestinal tract, heart, kidneys, and, to a lesser extent, the joints and tendons

subchondral bone—bone located beneath the articular cartilage of a joint

synovial fluid—a protein-rich, slippery fluid contained inside a fibrous capsule that lubricates and nourishes the cartilage covering the ends of bones at their joints

synovial membrane—a membrane lining the capsule that encloses synovial joints and secretes synovial fluid, which lubricates and nourishes the cartilage at the end of bones

systemic lupus erythematosus (SLE; lupus)—an autoimmune, chronic inflammatory disease that affects the connective tissue; affects skin, joints, kidneys, central nervous system, and mucous membranes and eventually spreads to all tissues, invoking a systemic reaction with pain, fever, sensitivity to light, and skin lesions

trabecular bone—loosely organized bone that has a sponge-like appearance and is found at the ends of long bones

T-score—the number of standard deviations that the patient's BMD is either above or below the mean BMD for healthy young adults of the same sex and race; measure that compares a patient's bone mineral density (BMD) to a standard, healthy BMD, which is set at the mean BMD of healthy young adults of the same sex and race as the patient

Cartilage

Cartilage is a flexible yet firm connective tissue consisting of cells and collagen fibers surrounded by an amorphous gel-like matrix. Cartilage is formed by cells called **chondroblasts** that actively secrete and surround themselves with a gel-like matrix until they become trapped in little cavities known as lacunae. Once the chondroblasts are enclosed in lacunae, they are called **chondrocytes**. The matrix is composed of collagen fibers, protein-carbohydrate complexes known as proteoglycans, and water.[1,3] Collagen is a fibrous protein that serves as a major constituent of cartilage, tendons, ligaments, bone, and skin, and is the most common protein found in the body. Collagen fibers are tough and flexible, resist stretching, and give cartilage its form and tensile strength. Proteoglycans are large macromolecules shaped like a test-tube brush, with a central protein core to which are attached numerous polysaccharides that project out much like bristles do. The most common polysaccharide found in the proteoglycans of cartilage is **chondroitin sulfate**.[3]

The proteoglycans attract and hold water, which gives cartilage its elasticity, stiffness, and ability to resist compression. It is estimated that 65% to 80% of the wet weight of cartilage is water.[1]

Cartilage gives shape to such structures as the external ear, the tip of the nose, and the larynx. It is the precursor of the axial and appendicular skeleton during embryonic development, the growth zone of the long bones of children, and an important component of various types of joints in the mature skeleton.[3] The costal cartilage of the rib cage firmly binds the ribs and clavicles to the sternum, while the fibrocartilage of the intervertebral discs joins the bodies of the vertebral bones and allows limited movement between adjacent vertebrae. The articular cartilage of the synovial joints allows free movement of the bones of the hands, feet, arms, legs, and so on. The articular cartilage covering the tips of the bones at the joints provides a remarkably smooth surface, resulting in extremely low friction between the two bones during movement of the joint. It increases the surface area between the two bones, helps to dissipate mechanical stresses applied to bones, and aids in transmitting the load down the bone.[3, 4]

Cartilage is free of blood vessels except for the period when it is changing into bone during the skeletal development of childhood. Consequently, gasses, nutrients, and wastes must travel by solute diffusion between the blood vessels outside the cartilage and the chondrocytes embedded within the matrix of the cartilage. Although this diffusion is facilitated by the high water content of cartilage, it is a relatively slow process. As a result, chondrocytes have low rates of metabolism and cell division, and when injured, cartilage heals slowly.[1, 3] Cartilage undergoes constant turnover as its worn-out matrix components are degraded by enzymes produced by the chondrocytes, which then secrete new matrix to replace the old. Thus, the integrity of joints is dependent upon healthy cartilage, which is maintained by properly functioning chondrocytes.[5] However, in the latter decades of life, the secretion of new matrix fails to keep up with losses, and the articular cartilage gradually thins, or in the case of osteoarthritis, is absent.[3]

Bone

There are two ways in which the term *bone* is used in this chapter. It is used to refer to specific body structures such as the pelvis, femur, and mandible, which are organs composed of multiple tissue types including nerves, blood vessels, adipose tissue, bone marrow, cartilage, and fibrous tissue. The term *bone* is also used to refer to bone tissue or **osseous tissue**, which is the major component of bone and makes up most of the mass of individual bones.[3] Osseous tissue is a connective tissue having an *organic* component that is mineralized or calcified by being impregnated with an *inorganic* component. The organic or protein component includes bone cells embedded in a matrix of collagen fibers and proteoglycans. The protein collagen comprises 90% of this organic matrix.[2] The organic component of osseous tissue contributes approximately one-third of the dry weight of bones and gives bone a degree of flexibility. Without this component, bones would be brittle and likely to shatter when stressed. The inorganic or mineral component consists of approximately 85% **hydroxyapatite** (a crystallized calcium phosphate salt), about 10% calcium carbonate, and smaller amounts of magnesium, sodium, potassium, fluoride, sulfate, and carbonate.[1, 2, 3] The inorganic component, responsible for

about two-thirds of the dry weight of bones, gives stiffness to bones and allows them to easily support the weight of the body without bending. When bones lack this mineral component, they are soft and can easily bend, as is the case with the childhood disease rickets and the condition known as osteomalacia, which is seen in adults.[3, 5]

In addition to serving as the body's framework, bones are a ready source of calcium and phosphorus for maintaining physiologic concentrations of these minerals in the extracellular fluids. The calcium content of the adult human body ranges from 1100 to 2000 g, with 99% of this in the bones. The bones provide two calcium reserves: a massive (99% of total available calcium) but less readily accessible one in the form of hydroxyapatite (crystallized calcium phosphate salt) and a much smaller pool (1% of total available calcium) that can be quickly released into the extracellular fluid. The average adult body contains about 500 to 800 g of phosphorus and 85% to 90% of this is in the bones.[2, 3, 6] In health, the body maintains tight control of these minerals in the serum. The normal serum concentrations of calcium and phosphorus are 9.0 to 10.5 mg/dL and 3.0 to 4.5 mg/dL, respectively.[6] Of the two minerals, abnormalities of serum calcium concentrations are the most critical. Hypocalcemia, an abnormally low serum calcium concentration, can cause excessive excitability of the nervous system and lead to facial grimacing, muscle spasms, and tetany (an inability of the muscle to relax). **Carpopedal** spasm results from tetany occurring in the hands and feet and can be a sign of hypocalcemia. In extreme hypocalcemia, laryngeal spasm, respiratory arrest, and convulsions can occur. Hypercalcemia, an abnormally high serum calcium concentration, can result in fatigue, depression, mental confusion, anorexia, nausea, vomiting, and constipation.

The Cells of Osseous Tissue Four principal types of cells are found in osseous tissue:

1. **Osteogenic cells** are stem cells capable of differentiating (developing a more specialized form or function) into *osteoblasts*, and are the only source of new osteoblasts. Osteogenic cells are active during normal growth of the skeletal system during childhood and adolescence, and in adulthood are activated in response to bone injury such as a fracture and in response to the stress placed on bones during weight-bearing exercise.

2. **Osteoblasts**, or bone-building cells, synthesize, deposit, and then orient the fibrous proteins of the organic matter of the bone matrix (collagen, proteoglycans, and other proteins) and then participate in the calcification or mineralization of the bone matrix, in a process known as bone formation or mineral deposition.[2, 3] Active osteoblasts are found on the surface of newly forming bone. They remove ions of calcium, phosphate, and other minerals from the blood plasma and deposit them within the bone matrix, thus hardening it. In response to bone fracture or the stress of weight-bearing exercise, the osteogenic cells multiply more rapidly and then differentiate to become osteoblasts.[1, 3]

3. **Osteocytes** are mature osteoblasts surrounded and entrapped by the matrix they have synthesized and calcified, and represent the vast majority of cells in bone.[1, 3] Osteocytes reside in the tiny, fluid-filled cavities within the

calcified matrix called lacunae, which are interconnected by narrow channels called canaliculi. Delicate cytoplasmic processes extend from the osteocytes and pass through the canaliculi. These processes connect with those of other osteocytes, thus allowing the osteocytes to chemically signal each other. The fluid-filled canaliculi also serve as channels for the passage of nutrients and metabolites between the osteocytes and nearby blood vessels. Although the osteocytes neither deposit nor remove bone, they are actively involved in maintaining the bony matrix by monitoring the amount of strain (bending) a bone experiences when it is mechanically loaded (for example, by weight-bearing exercise) and then communicating this information to osteoblasts on the bone surface.[2, 3, 6] The osteoblasts can then build up and strengthen the bone where needed in response to the stress.

4. **Osteoclasts** are bone-removing cells that secrete hydrochloric acid (pH of about 4) to dissolve the mineral component of bone matrix and an enzyme called acid phosphatase that digests the collagen and other protein components of the bone matrix. This process is known as bone resorption or mineral resorption.[3] The dissolved minerals are released into the blood and made available for other uses.

Skeletal Growth and Development Throughout life, the bones of the skeleton are in a continual state of change. Linear and circumferential growth of the long bones are the most prominent observable features of skeletal development during childhood and adolescence. In females, mature height is reached at about age 16 to 18 years, and in males, at about age 18 to 20 years. In addition to linear and circumferential skeletal growth during childhood and adolescence, bones are continually changing their size and shape in response to changes in forces applied to the skeleton in a process known as remodeling.[7] For example, as children begin to walk and then become increasingly physically active, bones develop ridges, spines, and bumps on their surfaces in response to stresses placed on bones by the exercising muscles. Remodeling also involves the repair of microscopic bone damage resulting from excessive or accumulated stresses and the maintenance of serum calcium levels by moving calcium from the bones into the blood.[3, 7] Compared to sedentary people, the bones of those regularly engaged in physical activity or heavy manual labor tend to be stronger and have a greater density and mass. Even within the same person, the bones of the dominant arm (e.g., the right arm for a person who is right-handed) are stronger than the bones in the nondominant arm.[2] In remodeling, the osteoclasts remove bone from low-stress areas where it is not needed, while the osteoblasts lay down new bone in high-stress areas where it is needed.[1, 3]

Cortical and Trabecular Bone The bones of the skeleton are made from two different types of osseous tissue: **cortical bone** and **trabecular bone**. Cortical bone (also known as compact bone) is dense and has no open spaces visible to the naked eye. It forms the external surfaces of all bones, the shafts of the long bones (e.g., the femur and humerus), and a shell that caps the ends of the long bones. Trabecular bone (also known as cancellous bone), on the other hand, is loosely organized with a sponge-like appearance. Trabecular bone is found at the ends (or "heads") of the long bones. The vertebrae, pelvis, sternum, and scapulae are primarily trabecular bone with a thin covering of cortical bone.[2, 3] Figure 25.1, which shows the

Figure 25.1 Cortical and Trabecular Bone in a Human Femur

In this photo of the proximal end of a human femur cut longitudinally, cortical bone forms a dense outer shell around the sponge-like trabecular bone. At the top or head of the bone, trabecular bone predominates and the shell of cortical bone is thin, unlike the shaft where thick cortical bone provides strength and rigidity.

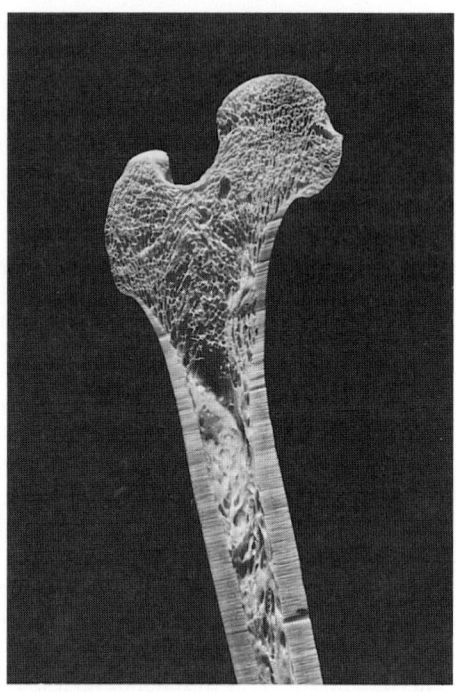

Source: S. Rolfes, K. Pinna, and E. Whitney, Understanding Normal and Clinical Nutrition, 7e, copyright © 2006, p. 429.

Courtesy of Hologic Inc.

proximal end of the femur cut longitudinally, illustrates some of the differences between cortical and trabecular bone. The shaft or middle part of the bone has a dense, thick wall composed of cortical bone enclosing a space inside the shaft called the medullary cavity, which contains bone marrow. The end of the bone has a spongy appearance due to the trabecular bone that predominates there. The trabecular bone at the end is capped with a thin layer of cortical bone.[3]

Cortical and trabecular bone are considered lamellar bone because the osseous tissue of both is organized in layers known as lamellae. However, the arrangement of the lamellae differs between the two types of bone, as shown in Figure 25.2, which represents the microscopic appearance of a biopsy specimen removed from the shaft of the long bone. In the cortical bone, the lamellae are arranged concentrically, in onion-like layers, around a central canal that runs parallel to the long axis of the bone. Each concentric lamella is connected to its adjacent lamellae by canaliculi. A central canal and its surrounding lamellae constitute an osteon, which is the basic structural unit of cortical bone. The central canals are joined by traverse or diagonal canals. These canals provide passages through which blood vessels and nerves pass.[1, 3]

Trabecular bone consists of a framework of crossed and interconnected plates, rods, and spicules called trabeculae that form

Figure 25.2 Organization of a Long Bone

(a) A long bone cut longitudinally shows the dense cortical bone covering the sponge-like trabecular bone. Enclosed within the shaft is the medullary cavity, which contains bone marrow. (b) A small piece removed from the long bone and enlarged illustrates the organization of cortical and trabecular bone. In cortical bone, the lamellae are arranged concentrically around a central canal running parallel to the long axis of the bone. A central canal and its surrounding lamellae constitute an osteon. (c) A magnified spicule of trabecular bone shows that the lamellae do not surround a central canal because the osteocytes have a nearby blood supply. Consequently, in trabecular bone no osteons are formed. (d) A magnification of lamellae from cortical bone shows the osteocytes and each concentric lamella connected to its adjacent lamellae by canaliculi.

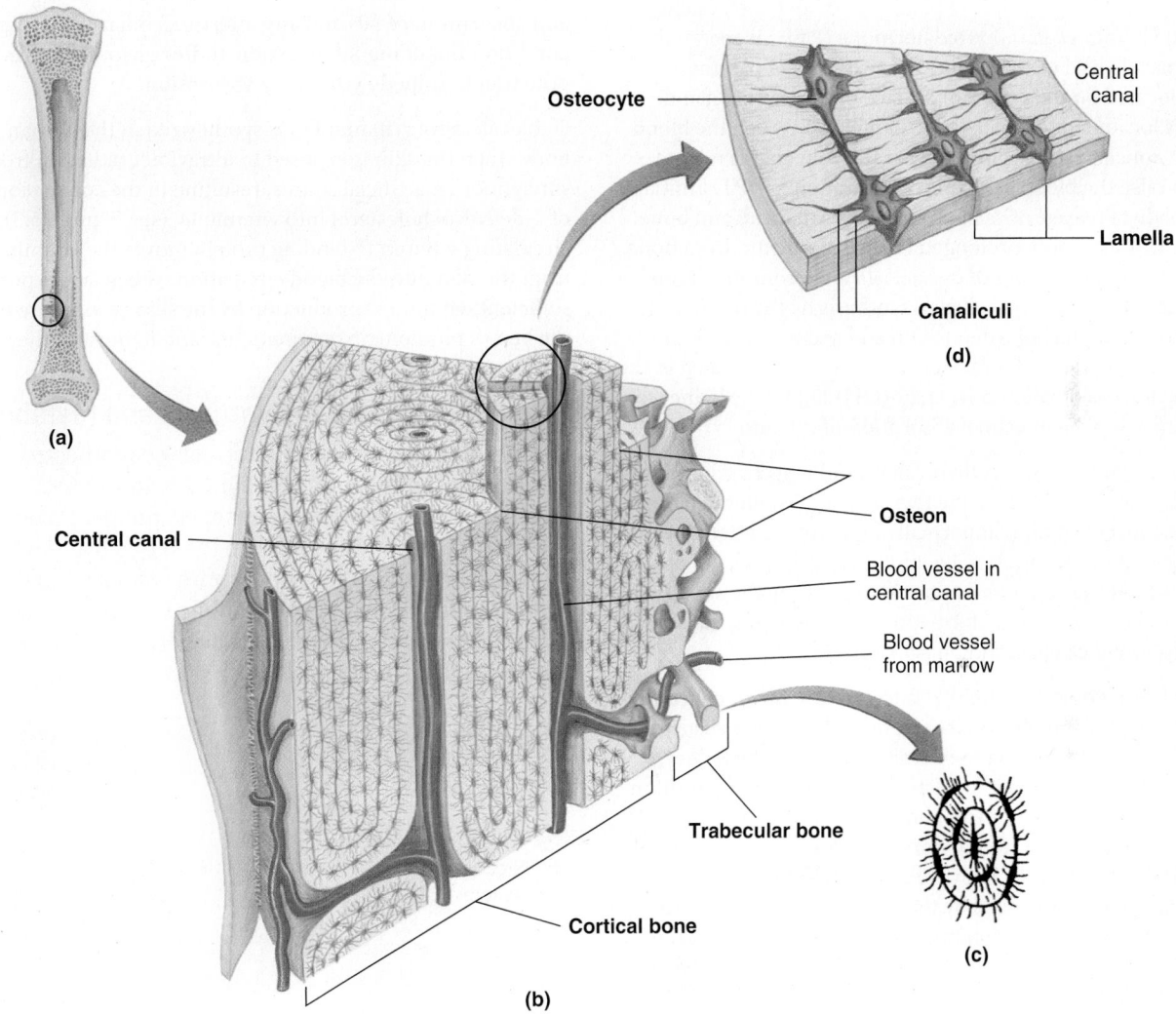

Source: L. Sherwood, *Human Physiology: From Cells to Systems,* 5e, copyright © 2004, p. 738.

a lattice-like pattern.[2, 3] The space between the rods and plates is filled with bone marrow. The matrix of the rods and plates is arranged in lamellae but not in layers surrounding a central canal. The central canal is not needed because no osteocyte is very far from a source of blood. Consequently, no osteons are formed in trabecular bone. The arrangement of the rods and plates occurs in response to the stresses placed on the bone and is not random. As mentioned earlier, the osteocytes act as strain sensors and signal the osteoblasts to form bone where it is needed in response to the mechanical stresses bones receive, for example, from weight-bearing exercise.[3, 6] The unique structure of trabecular bone imparts considerable tensile strength and weight-bearing properties while simultaneously being rela-

tively light. The densely packed calcified matrix of cortical bone makes it much more rigid than trabecular bone.[1, 3]

Although roughly 75% of the weight of the skeleton is composed of cortical bone and about 25% is composed of trabecular bone, the relative quantity of cortical and trabecular bone varies considerably in different types of bones and even within the same bone depending on the need for lightness and strength.[1, 3, 6] Through the process of remodeling, bones are able to adapt and respond to changing physical demands and stresses imposed on them. Studies indicate that each year as much as 18% of the total mineral content of the skeleton is deposited and removed and that the rate of turnover in trabecular bone is much greater than in cortical bone. This observation has an impact on the prevention and treatment of osteoporosis, as discussed later in this chapter.[5, 6]

Hormonal Control of Bone Metabolism

Maintaining calcium and phosphorus homeostasis is a complex process requiring the body to balance the dietary intake of these minerals, their fecal and urinary losses, and their flux in and out of bone. Several hormones participate in this process, including cortisol, growth hormone, and the thyroid hormones, but the primary regulators are parathyroid hormone, calcitonin, and vitamin D. Parathyroid hormone (PTH) is secreted by the parathyroid glands. There are two pairs of parathyroid glands located on the posterior surface of the thyroid gland, which is located in the neck (see Chapter 17). When the blood calcium concentration is low, the parathyroid glands release PTH to raise the blood calcium concentration.[1, 3, 6] PTH initiates an immediate release of calcium from the canaliculi and bone cells, as well as a more prolonged release of calcium from bone, by increasing the number of osteoclasts and promoting bone resorption. PTH inhibits collagen synthesis by the osteoblasts, which then inhibits bone deposition and promotes calcium reabsorption by the kidneys. PTH promotes the final step in the body's synthesis of vitamin D_3 [1,25-$(OH)_2D_3$] by the kidneys, thus enhancing the intestinal absorption of calcium.[1, 3]

Calcitonin (also known as thyrocalcitonin) is secreted by the parafollicular or "C" cells of the thyroid gland when the blood calcium concentration is abnormally high. Its secretion lowers blood calcium concentration by inhibiting the activity of osteoclasts (and thus bone resorption), stimulating the activity of osteoblasts (and thus bone formation), and reducing the renal reabsorption of calcium and phosphate.[1, 3, 6]

The major function of vitamin D is to increase blood concentrations of calcium and phosphorus by promoting their absorption by the GI tract, promoting their reabsorption by the kidney, and stimulating osteoclast formation and thus bone resorption and the release of calcium and phosphorus from bone.[1, 3, 6] Vitamin D is actually a steroid hormone given the fact that it is synthesized within the body, and, like other hormones, is a chemical messenger carried by the blood from one organ to another.[3, 8, 9] Because it is synthesized by the body, it is not an essential nutrient (i.e., one that must be obtained from the diet) and therefore, technically speaking, is not a vitamin.

There are two major physiologically relevant forms of vitamin D: **ergocalciferol** (vitamin D_2) and **cholecalciferol** (vitamin D_3). The two forms have identical biological activity and differ only slightly in their molecular structure. Vitamin D without a subscript represents either D_2 or D_3. Ergocalciferol is produced from the ultraviolet irradiation of the plant steroid ergosterol. Cholecalciferol is the naturally occurring form and is produced photochemically when the precursor molecule 7-dehydrocholesterol is exposed to ultraviolet light from the sun or from an ultraviolet-emitting lamp. Both ergocalciferol and cholecalciferol can be obtained from the diet, while cholecalciferol is the form produced in humans and higher animals when the skin is exposed to sunlight. The few foods that are naturally good sources of vitamin D include fish liver oils, the flesh of fatty fish, the liver and fat of aquatic mammals such as polar bears and seals, and the eggs of hens

fed vitamin D.[8] In North America, most dietary vitamin D is obtained from fortified milk and milk products and other fortified foods such as breakfast cereals, soy milk, and margarine. In Canada and the United States, all commercially available milk, regardless of its fat content, must be fortified with 385 IU/liter or 400 IU/quart. However, several studies have shown that the actual vitamin D content of fortified milk varies considerably and that as much as 70% of milk sampled did not contain the required amount of added vitamin D.[8] Because it is fat soluble, vitamin D absorption is dependent on many of the same mechanisms involved in the digestion and absorption of fat, and any intestinal, biliary, or lymphatic condition impeding fat digestion and/or absorption has the potential to impede vitamin D absorption.

Cholecalciferol (vitamin D_3) is synthesized in the human body when the skin is exposed to ultraviolet radiation from sunlight or an artificial source, resulting in the conversion of 7-dehydrocholesterol into vitamin D_3 (see Figure 25.3). A circulating vitamin D-binding protein moves the vitamin D_3 from the skin into the blood circulation. When sun exposure is sufficient, vitamin D production by the skin is adequate to meet the body's physiologic requirements, and in most of the world,

Figure 25.3 Vitamin D Synthesis and Metabolism

Vitamin D can be obtained in the diet or synthesized when skin is exposed to ultraviolet light. Cholecalciferol and ergocalciferol must then be changed into their biologically active forms by having added to them two hydroxyl (OR) groups, in a process known as hydroxylation. The first hydroxylation occurs in the liver and the second occurs in the kidneys, resulting in 1,25-dihydroxycholecalciferol, the active form of vitamin D.

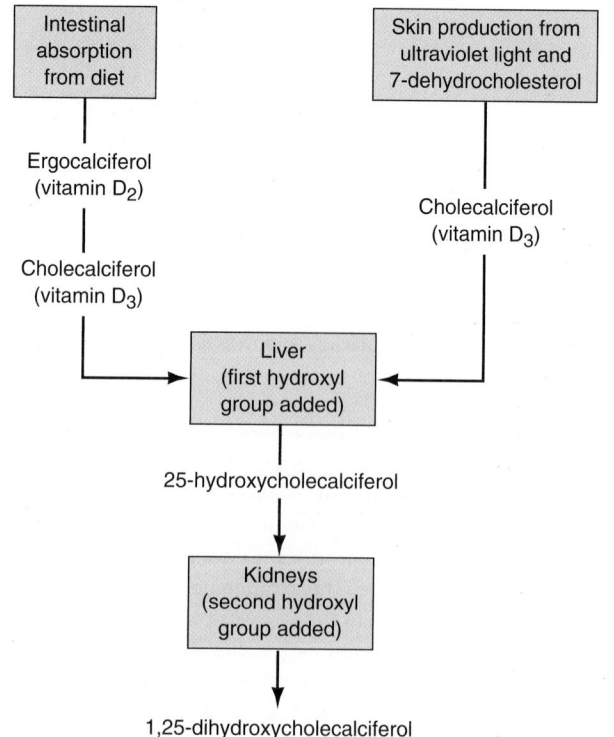

Source: CM Porth, Pathophysiology: Concepts of Altered Health Status, 7e fig. 56-4 pg. 1362.

this is the major source of the vitamin for humans. However, several factors can limit vitamin D synthesis in the skin, including the topical application of a sunscreen, clothing that covers the skin and limits sun exposure, increased melanin pigmentation in the skin, limited time outside in the sun due to being housebound or institutionalized, and advanced age. Cutaneous vitamin D synthesis is decreased by as much as fourfold in persons aged 65 years compared to those aged 20 to 30 years. The season of the year, time of day, and latitude can also impact cutaneous vitamin D production. Persons living above 40° north latitude or above 40° south latitude may have no cutaneous vitamin D production during three to four months of winter. Persons living in far northern or southern latitudes may not experience vitamin D production for up to six months.[8, 10] As shown in Figure 25.4, much of the United States and all of Canada lie above 40° north latitude. Research suggests that a large proportion of North Americans and Europeans, for example, have suboptimal sun exposure and, thus, inadequate vitamin D synthesis. Consequently, for these individuals, adequate dietary intake of vitamin D is necessary to meet their requirements, and in this situation vitamin D becomes a true vitamin.[9]

Whether obtained from the diet or synthesized in the skin, ergocalciferol and cholecalciferol are both biologically inactive until they are modified by the liver and then by the kidney to form the biologically active form of vitamin D: 1,25-dihydroxyvitamin D $(1,25\text{-}(OH)_2D)$.[8, 9] The first step in this activation process occurs in the liver, where a hydroxyl group (OH) is added to the molecule to form 25-hydroxyvitamin D (see Figure 25.3). From the liver, 25-hydroxyvitamin D is transported to the kidney where a second hydroxyl group is added to form the biologically active 1,25-dihydroxyvitamin D. Because kidney disease can interfere with this final step, patients with renal failure will be vitamin D-deficient unless they receive supplements of the vitamin in its biologically active form. Calcitriol is a form of 1,25-dihydroxyvitamin D available by prescription that can be administered orally or by intravenous injection. Over-the-counter vitamin D supplements contain either cholecalciferol or ergocalciferol and are not suitable for renal failure patients because both forms require activation by the liver and kidney before becoming biologically active.

Osteoporosis

Osteoporosis, the most common bone disease in humans, is characterized by the loss of bone mass and deterioration of bone microarchitecture, compromised bone strength, and an increased susceptibility to fracture and painful morbidity.[7, 11, 12] The scanning electron micrographs shown in Figure 25.5 dramatically illustrate the differences between normal and osteoporotic trabecular bone in terms of bone mass and microarchitecture. The U.S. National Institutes of Health (NIH) defines osteoporosis as "a skeletal disorder characterized by compromised bone strength predisposing a person to an increased risk of fracture. Bone strength reflects the integration of two main

Figure 25.4 Sunlight Exposure and Vitamin D Deficiency Risk

Much of the United States and all of Canada lie above 40° north latitude. Research indicates that people residing in this area have insufficient duration and intensity of sunlight exposure for optimal synthesis of vitamin D. Unless vitamin D is obtained from fortified foods or from supplements, many North Americans are at risk of deficiency of this important vitamin.

Source: S. Rolfes, K. Pinna, and E. Whitney, Understanding Normal and Clinical Nutrition, 7e, copyright © 2006, p. 378.

features: bone density and bone quality."[13] Bone quality is influenced by factors such as bone architecture, bone turnover, mineralization, and the accumulation of damage to the bone (e.g., microfractures). Because bone quality is difficult to quantify, clinicians rely on a diagnosis or history of a fragility fracture as the only clinically applicable index of bone quality.[14] A fragility fracture is one that occurs in the absence of trauma or following minimal trauma.[7, 11] For example, it is not uncommon for a fracture of an osteoporotic hip to *precede* and result in a fall

Figure 25.5 Scanning Electron Micrographs

(a) Healthy trabecular bone, and (b) Osteoporotic trabecular bone

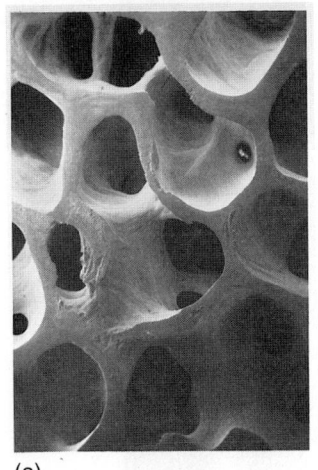 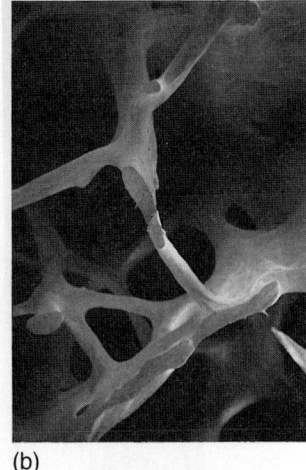

(a)　　　　　　　　　　　(b)

Source: S. Rolfes, K. Pinna, and E. Whitney, *Understanding Normal and Clinical Nutrition,* 7e, copyright © 2006, p. 429.

(note that the fracture occurs *before* the fall) or for a fracture to occur after bumping into a table or kitchen counter top.

Medical Diagnosis

The most widely used method for diagnosing osteoporosis is measurement of bone mineral density.[15, 16] Bone mineral density (BMD) accounts for approximately 70% of bone strength, and because there is currently no accurate measure of overall bone strength, BMD is frequently used as a surrogate measure.[13] BMD is expressed in grams of mineral per area (g/cm²) and is best measured using dual-energy X-ray absorptiometry (DXA).[17] In DXA, X-ray beams of two different energy levels are projected through the body and received by a detector opposite the X-ray source (see Chapter 3). As the X-ray beam passes through the body, some of its energy is absorbed by the body's tissues, particularly dense tissue such as bone; this reduction in energy is known as attenuation. Because bone and soft tissue have different densities, they attenuate or absorb X-ray energy differently, and the differences in attenuation between soft tissues and bone at the two different energy levels are used to calculate BMD.[18, 19] Figure 25.6 shows a DXA scan of the hip (see Figure 3.12 Chapter 3 for a photo of a DXA instrument).

Osteoporosis is generally diagnosed by comparing a patient's BMD (determined by DXA) with the mean normal BMD in a population of healthy young adults of the same sex and race by using what is referred to as a "T-score."[7, 16, 19, 20] The **T-score** compares a patient's BMD with the mean in a healthy young reference population, which is considered the standard for peak bone mass. The T-score is the number of standard deviations above or below the mean BMD for normal young adults of the same sex and race as the patient. The most widely accepted diagnostic criterion for osteoporosis is that developed by the World Health Organization (WHO), which defines osteoporosis as a T-score at or below 22.5. For example, a diagnosis of osteoporosis can be made in the case of a 60-year-old white female if her BMD is 2.5 standard deviations or more below the mean BMD of young adult white females. When the T-score is between 1.0 and 2.5 standard deviations below the mean

BMD of the young adult mean value, a condition known as **osteopenia** is present. Osteopenia is not considered a diagnosis but rather a term used to describe a bone mineral density that is somewhat low but not so low as to warrant a diagnosis of osteoporosis. Risk of fracture is increased as well, but not to the extent seen in osteoporosis.[5, 11, 18, 21]

Measurements of BMD at any skeletal site have value in predicting fracture risk, and the lower the BMD, the greater the fracture risk.[11] It is estimated that fracture risk increases 1.5 to 3.0 times for each standard deviation decrease in bone density.[18, 22] Clinical determinations of BMD are usually made at the lumbar spine and hip, but hip BMD is the best predictor of risk of hip fracture and is useful for predicting fractures at other sites.[7, 11] Table 25.1 summarizes the WHO criteria for evaluating BMD.

Quantitative ultrasound of the calcaneus (the bone of the heel) is gaining recognition as a useful screening method for identifying persons at high risk of osteoporosis who might then be referred for additional evaluation of BMD using DXA.[23] Quantitative ultrasound is radiation free, relatively simple to use, portable, and inexpensive, and measurements take one or two minutes to perform.[17, 24] When used in conjunction with clinical risk assessment, quantitative ultrasound is particularly useful at identifying people at high risk of fracture who warrant additional assessment of BMD with DXA.[23] A commercially available quantitative ultrasonography instrument is shown in Figure 25.7.

Many North Americans are unaware that their bone health is in jeopardy because osteoporosis is frequently not diagnosed.[25] Data from the third National Health and Nutrition Examination Survey show that, among adults aged 65 years and older, only 25% of men and about 42% of women who have osteoporosis are aware of their condition.[21] This is largely due to physicians and other health care providers failing to discuss osteoporosis with patients or to recommend screening for persons at risk. In one study of 1,500 women aged 40 to 69 years who participated in a primary care health plan, only 49% reported that a health care provider had discussed the topic of osteoporosis with them.[26] In another study of hip fracture patients in four Midwestern health care systems, fewer than 25% received bone mineral density testing. Prescription drug therapy for osteoporosis was prescribed in only 3% to 26% of these hip fracture patients, depending on the site of hospitalization.[27] Despite the relative ease of screening for osteoporosis and availability of effective measures for preventing and treating the condition, it all too often is under-recognized and under-treated.[25]

Figure 25.6 A DXA Scan of the Left Hip

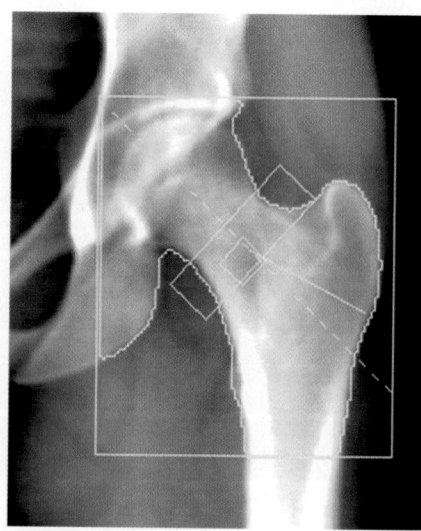

Source: Photo courtesy of Hologic, Inc.

Table 25.1 World Health Organization T-Score Criteria for Classifying Normal Bone Mineral Density (BMD), Osteopenia, and Osteoporosis

Classification	T-Score
Normal BMD	−1.0 or greater
Osteopenia	Between −1.0 and −2.5
Osteoporosis	−2.5 or less
Severe Osteoporosis	−2.5 or less and fragility fracture

Source: Adapted from the World *Health* Organization and the International Society for Clinical Densitometry.

Figure 25.7 Quantitative Ultrasound of the Calcaneus

This compact and lightweight instrument measures the transmission of high-frequency sound waves (ultrasound) through the calcaneus (heel bone). Quantitative ultrasonography (QUS) is a clinically useful and cost-effective screening approach for identifying persons at risk for osteoporosis who would benefit from more definitive evaluation of bone mineral density using DXA.

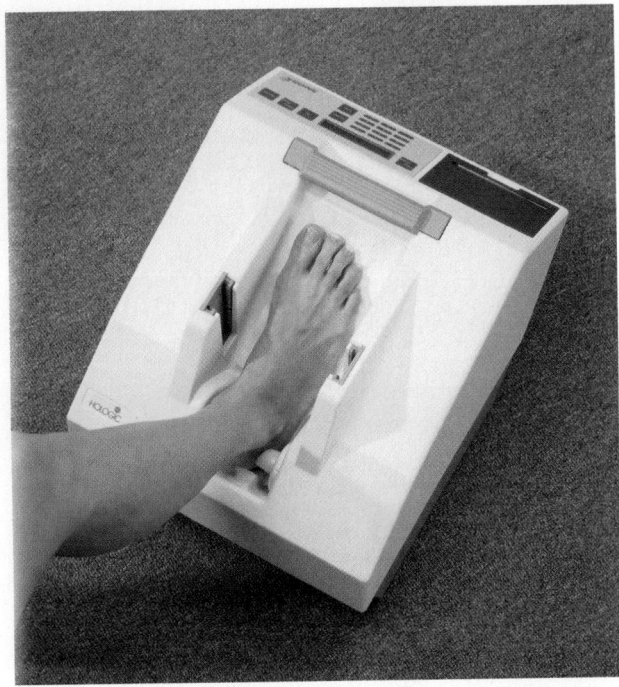

Source: Photo courtesy of Hologic, Inc.

Epidemiology

Osteoporosis is a major global public health problem, particularly among postmenopausal white women, but its prevalence is increasing across all groups, including males and nonwhite persons, as life expectancy increases and the world population expands. In the United States and Canada, approximately 25% of postmenopausal white women have osteoporosis and an additional 50% of postmenopausal white women have low BMD, and thus are at increased risk of fracture and future development of osteoporosis.[11, 14] In the United States, as many as 8 million women and 2 million men have osteoporosis and an additional 18 million persons have a low bone mineral mass that puts them at increased risk of developing osteoporosis.[7] Although osteoporosis is often considered a disease of women, it is a major health care problem in men. According to World Health Organization (WHO) estimates, the risk of suffering an osteoporotic fracture over the course of life is about 40% for women and about 13% for men.[14]

Bone mineral density rapidly increases during the growth spurt of adolescence (11 to 14 years of age in females and 13 to 17 years of age in males), after which it continues to increase at a much slower rate until maximum bone mineral density is reached in the late 20s or 30s.[7, 18] As early as adolescence, differences in BMD between the sexes begin to appear, with females having a lower BMD than males. Adult females begin losing bone mineral earlier and at a faster rate than adult males. The rate of bone mineral loss in females averages about 1% per year as menopause approaches, increases to 2% to 6% per year for one to five years after the onset of menopause, and then by the tenth year after the onset of menopause returns to about 1% per year and remains at that level throughout the remainder of life.[18] On average, U.S. females live seven to eight years longer than U.S. males, which contributes to a female's greater lifetime loss of bone mineral and increased incidence of osteoporosis and risk of fracture. By age 70 years, the average white woman has lost 30% of her bone mass while some white women can lose as much as 50%. Between the age of 50 and 60 years, bone mineral loss in males averages about 1% per year, and continues at this rate throughout the remainder of life, with total losses rarely exceeding 25%.[18] Other factors responsible for the lower prevalence of osteoporosis among males compared to females is their greater body mass, greater bone size, absence of a decrease in endogenous sex hormone production analogous to menopause, and shorter average life-span.[7, 28]

Race and ethnicity influence BMD and risk of fracture. Beginning in early childhood and on through the remainder of the life cycle, blacks tend to have higher BMD values than whites and Asians.[7, 18] Hip fractures are much more frequent among whites than nonwhites, with U.S. white women reporting three times as many hip fractures as African-American women.[20] Risk of osteoporosis and fracture increases with age, and as the number of older people increases worldwide, osteoporosis and fracture will become even greater concerns. The current global estimate of 330 million persons aged 65 years or over is expected to increase nearly fivefold to 1.6 billion persons by the year 2050, in large part due to increased life expectancy in Asia and Latin America. The WHO estimates that the number of hip fractures will rise from 1.7 million in 1990 to as many as 6.3 million by 2050. In 1990, approximately 50% of hip fractures in the elderly occurred in North America and Europe, but by 2050 it is estimated that this number will drop to 25% and that more than half of all hip fractures will occur in Asia.[20, 29]

Health and Economic Impact of Fractures

The most common fracture sites are the vertebrae (spine), proximal femur (hip), and distal forearm (wrist).[7] Fractures of the spine occur two to three times as frequently as hip fractures, but in contrast to hip fractures, are not as easily diagnosed and are often asymptomatic (without symptoms).[29, 30] Spine fractures are a major source of chronic back pain, disfigurement, low self-esteem, and depression, and result in a slight increase in mortality. In the elderly, multiple compression fractures of the vertebral bodies can cause a loss in stature of several inches and an unnatural curvature of the back known as kyphosis. Postural and height changes associated with kyphosis can restrict the activities of daily life such as bending and reaching. Severe kyphosis is associated with restrictive lung disease and such abdominal symptoms as pain, distension, loss of appetite, early satiety, gastroesophageal reflex disease, and constipation.[7, 11] Fractures of the wrist are the least debilitating and only 20% of these require hospitalization.[21]

Hip fractures are the most severe in terms of impact on morbidity and mortality and are associated with a high incidence of deep vein thrombosis and pulmonary embolism, every year resulting in about 300,000 hospitalizations in the United States and more than 30,000 in Canada.[7, 21, 30] Of these hip fracture patients, as many as 20% die within the first year of a fracture, 20% end up in a nursing home within a year, and many become isolated, depressed, or afraid to leave home because of fear of another fall. So devastating are the consequences of a hip fracture that in a study of women 75 years and older, 80% preferred death to a severe hip fracture resulting in nursing home placement.[21] Although the overall prevalence of hip fractures is greater among females, the likelihood of death from hip fractures is considerably greater among males. In some studies, males hospitalized for hip fracture were nearly twice as likely to die than were hospitalized females.[28] Despite the serious nature of fractures, including those of the hip, few people die as a direct result of fractures. Death is generally the indirect result of complications from the fracture that either trigger or hasten a downward spiral in health. This is particularly the case among frail, elderly persons with underlying health conditions or those who are already living in a nursing home and who might have died even in the absence of a hip fracture.[21]

Morbidity resulting from osteoporosis is expensive in terms of both indirect and direct costs. Indirect costs include reduced productivity due to disability and premature death and reduced earnings because of workdays lost by patients and caregivers. Indirect costs are difficult to estimate but are considered sizeable.[21] Direct costs include those related to inpatient and outpatient health care, nursing home care, home health care, durable medical equipment, and pharmaceuticals. The annual direct costs of osteoporotic fractures in the United States in 2002 were estimated at $12 to $18 billion dollars. A majority of these costs are due to the inpatient treatment of fractures and their sequelae (diseases or illnesses occurring as a consequence of another condition or event).[13, 14, 21] Fractures of the hip are the most expensive to treat. The initial cost of hospitalization following a hip fracture can vary from $30,000 to $44,000, with an additional $15,000 in costs for follow-up and outpatient care during the first year following the fracture, and over a lifetime costs can exceed $81,000.[21, 31] The annual costs related to osteoporosis could more than triple by the year 2040 due to the rising cost of health care and the increasing number of persons likely to experience osteoporosis and osteoporosis-related fractures.[11]

Etiology

Osteoporosis can be categorized as either primary or secondary. In primary osteoporosis, no specific cause of the condition can be identified, whereas secondary osteoporosis results from or is "secondary to" a specifically identifiable cause, such as a disease or the use of certain drugs. Primary osteoporosis is more common than secondary and is generally a disease of the elderly, resulting from the cumulative impact of bone mineral loss and deterioration of bone structure that occur with aging.[21] Primary osteoporosis is sometimes referred to as "age-related osteoporosis" or as "postmenopausal osteoporosis," because it is often diagnosed in elderly, postmenopausal women. Table 25.2 outlines factors associated with increased risk of osteoporosis fracture. Risk factors which are modifiable are considered prime targets for interventions used in osteoporosis prevention or nonpharmacologic treatment. Secondary osteoporosis is more commonly seen in premenopausal women than in men. Dis-

Table 25.2 Risk Factors for Osteoporosis Fracture

Major Risk Factors
- Low bone mineral density
- Personal history of fracture as an adult
- History of fracture in a parent or sibling
- Female sex
- Age 65 years or older
- Caucasian race
- Menopause before age 45 years
- Premenopausal amenorrhea for 1 year or more
- Glucocorticoid therapy for >3 months
- Recurrent falls

Minor Risk Factors
- Impaired vision
- Dementia
- Alcoholism
- Low lifelong calcium and vitamin D intake
- Physical inactivity
- Poor health/frailty
- Current cigarette smoking
- Body weight <127 lb (<58 kg)

Sources: Adapted from: National Osteoporosis Foundation. *Physician's guide to prevention and treatment of osteoporosis.* Washington (DC): National Osteoporosis Foundation, 2003; Lindsay R, Cosman F. Osteoporosis. In Fauci AS, Braunwald E, Kasper DL, Hauser SL, Longo DL, Jameson JL, Loscalzo J (eds.), *Harrison's Principles of Internal Medicine,* 17th ed. New York: McGraw-Hill, 2008, 2397–408; Brown JP, Josse RG. 2002 clinical practice guidelines for the diagnosis and management of osteoporosis in Canada. *CMAJ.* 2002;167(10 suppl):S1–S34.

eases associated with increased risk of osteoporosis are listed in Table 25.3, and drugs associated with increased risk of osteoporosis are listed in Table 25.4. Secondary osteoporosis accounts for approximately 10% to 30% of cases of osteoporosis in postmenopausal women and as many as two-thirds of cases in men.[12, 21]

Risk of osteoporosis (regardless of whether it is primary or secondary) and osteoporosis-related fractures is influenced by a number of factors, as outlined in Table 25.2. Risk factors having a particularly strong bearing on the development of peak BMD from childhood through early adulthood include genetic susceptibility and family history, female sex, Caucasian race, premenopausal amenorrhea, physical inactivity, and low lifetime calcium and vitamin D intakes. Low lifetime intakes of fluoride, magnesium, and zinc may also have a bearing.[22, 32] Genetics is a particularly important determinant of peak bone mass and of subsequent fracture risk.[7, 32]

Factors increasing the rate and/or degree of bone demineralization once peak BMD has been achieved include female sex, premenopausal amenorrhea, menopause before age 45 years, advanced age, glucocorticoid therapy, cigarette smoking, physical inactivity, and low intakes of calcium and vitamin D. Low intakes of fluoride, magnesium, zinc, and vitamin K may also have a bearing on bone demineralization.

It is important to note that low bone mass is only one of several risk factors for osteoporotic fracture, and that consideration

Table 25.3 Diseases Associated with Increased Risk of Osteoporosis

Nutritional Disorders	Hypogonadal States
• Biliary cirrhosis	• Amenorrhea
• Celiac disease	• Anorexia nervosa
• Gastrectomy	• Hyperprolactinemia
• Inflammatory bowel disease	• Klinefelter syndrome
• Malabsorption syndromes	• Turner syndrome
• Malnutrition	**Miscellaneous**
• Parenteral nutrition	• Alcoholism
Genetic Disorders	• Ankylosing spondylitis
• Ehlers-Danlos syndrome	• Immobilization
• Glycogen storage diseases	• Organ transplantation
• Hemochromatosis	• Acromegaly
• Homocystinuria	• Cushing's syndrome
• Marfan syndrome	• Renal failure
• Menkes' syndrome	• Rheumatoid arthritis
• Osteogenesis imperfecta	
Endocrine Disorders	
• Hyperparathyroidism	
• Thyrotoxicosis	
• Type 1 diabetes mellitus	

Sources: Adapted from: Lindsay R, Cosman F. Osteoporosis. In Fauci AS, Braunwald E, Kasper DL, Hauser SL, Longo DL, Jameson JL, Loscalzo J (eds.), *Harrison's Principles of Internal Medicine*, 17th ed. New York: McGraw-Hill, 2008, 2397–408; U.S. Department of Health and Human Services. Bone Health and Osteoporosis: A Report of the Surgeon General. Rockville (MD): U.S. Department of Health and Human Services, Office of the Surgeon General, 2004; Kirk D, Fish SA. Medical management of osteoporosis. *Am J Manag Care.* 2004;10:445–55.

should be given to any factor increasing the likelihood of falling. Among these are impaired vision, dementia, alcoholism, physical inactivity, poor health, and frailty. In addition, the use of any medication that might impair balance or alertness should be considered, especially among the elderly who often simultaneously take several prescription and over-the-counter drugs (a practice known as polypharmacy), some of which may affect balance and alertness when used in combination. Unsafe conditions in the physical environment that might increase the

Table 25.4 Drugs Associated with Increased Risk of Osteoporosis

- Glucocorticoids
- Cyclosporine A and Tacrolimus
- Cytotoxic drugs
- Anticonvulsants
- Excessive alcohol
- Excessive thyroxine
- Gonadotropin-releasing hormone agonists
- Heparin
- Lithium

Sources: Adapted from: Lindsay R, Cosman F. Osteoporosis. In Fauci AS, Braunwald E, Kasper DL, Hauser SL, Longo DL, Jameson JL, Loscalzo J (eds.), *Harrison's Principles of Internal Medicine*, 17th ed. New York: McGraw-Hill, 2008, 2397–408; U.S. Department of Health and Human Services. Bone Health and Osteoporosis: A Report of the Surgeon General. Rockville (MD): U.S. Department of Health and Human Services, Office of the Surgeon General, 2004; Kirk D, Fish SA. Medical management of osteoporosis. *Am J Manag Care.* 2004;10:445–55.

chance of falls should also be considered; these include uneven floors, slick floors, poorly illuminated halls and stairways, extension cords stretched across walkways, the upturned edges or corners of rugs or carpets, and inadequate railings or handholds in stairways and bathrooms.

Prevention

Preventing osteoporosis involves strategies to encourage individuals to change certain behaviors during childhood, adolescence, and early adulthood in order to reduce their risk of developing osteoporosis in the latter decades of life. The goal is to promote an individual's genetically determined peak BMD, slow the rate of bone mineral loss once demineralization begins in middle age, and reduce overall fracture incidence. Preventive strategies are generally recommended for the entire population and must be shown to be safe and reasonably effective. It should be noted, however, that the results of research on the efficacy of some preventive strategies are sometimes inconclusive and perhaps even contradictory. This is largely due to the difficulty of establishing cause-effect relationships between disease outcome and diet and other lifestyle factors, particularly when those relationships are weak or are influenced by factors not included in the experimental design. Despite the sometimes inconclusive nature of the research upon which preventive strategies are based, the scientific community has reached a consensus on a set of key recommendations for reducing the risk of osteoporosis and fractures. These key recommendations include adequate calcium and vitamin D intake, weight-bearing and muscle-strengthening exercise, fall prevention, smoking cessation, and avoiding excessive alcohol intake.

Calcium Calcium is a primary bone-forming mineral required in adequate amounts throughout the entire life cycle to achieve peak bone mass, maintain bone mass, minimize bone mineral loss, and reduce the incidence of osteoporosis-related fracture. Inadequate calcium intake during growth will result in suboptimal development of peak bone mass and increased risk of osteoporosis in the latter decades of life.[2, 18] During adulthood, a calcium intake insufficient to maintain serum calcium levels leads to increased secretion of parathyroid hormone which, in turn, stimulates the resorption of bone in order to raise serum calcium levels.[7] The Adequate Intakes (AI) for calcium and vitamin D for various life stage and gender groups as established by the Institute of Medicine of the National Academy of Sciences are shown in Table 25.5. These recommended intake levels are based on the best available scientific data and appear sufficient to achieve peak bone mass during growth and to minimize bone mineral loss in adulthood.[8] The mean (average) estimated daily calcium intakes for American females and males of different age groups are shown in Table 25.6. These data, from the 1999–2000 National Health and Nutrition Examination Survey (NHANES 1999–2000), indicate that the average calcium intakes of Americans are, in most instances, well below the recommended levels.

There is convincing evidence that for women and men living in a Western-style environment where the risk of osteoporosis tends to be high (e.g., North America and Europe), calcium intakes of less than 400 mg/day have an adverse impact on bone mineral density and are associated with increased risk of osteoporosis and fracture.[7, 22] However, there is less certainty about osteoporosis and fracture risk among these women and men

Table 25.5 Adequate Intakes (AI) for Calcium and Vitamin D by Life Stage and Gender Group

Life Stage/Gender Group	AI for Calcium	AI for Vitamin D*
Children, 1–3 years old	500 mg/d	5 μg/d
Children, 4–8 years old	800 mg/d	5 μg/d
Adolescents, 9–18 years old	1300 mg/d	5 μg/d
Females & males, 19–50 years old	1000 mg/d	5 μg/d
Females & males, 51–70 years old	1200 mg/d	10 μg/d
Females & males, >70 years old	1200 mg/d	15 μg/d
Pregnancy and Lactation		
≤18 years old	1300 mg/d	5 μg/d
19–50 years old	1000 mg/d	5 μg/d

*As cholecalciferol. 1 μg of cholecalciferol = 40 international units (IU) of vitamin D.

Sources: Adapted from: Standing Committee on the Scientific Evaluation of Dietary Reference Intakes, Food and Nutrition Board, Institute of Medicine. Dietary Reference Intakes for Calcium, Phosphorus, Magnesium, Vitamin D, and Fluoride. Washington, DC: National Academy Press, 1997.

Table 25.6 Mean Estimated Daily Calcium Intakes of Americans by Age and Gender*

Age	Females	Males
All ages	765 mg/d	966 mg/d
Under 6 years	785 mg/d	916 mg/d
6–11 years	860 mg/d	915 mg/d
12–19 years	793 mg/d	1081 mg/d
20–39 years	797 mg/d	1025 mg/d
40–59 years	744 mg/d	969 mg/d
60 years and over	660 mg/d	797 mg/d

*Data are from the 1999–2000 National Health and Nutrition Examination Survey (NHANES 1999–2000).

Source: U.S. Department of Health and Human Services, Centers for Disease Control and Prevention, National Center for Health Statistics.

when their calcium intakes are in the range of 600 to 800 mg/day, intakes that are typical of women living in North America and Europe. Calcium requirements vary considerably among humans, depending on each individual person's biology, diet, lifestyle, and environment. Furthermore, risk of osteoporosis is influenced by numerous factors other than calcium intake. Paradoxically, in many developing countries where average calcium intake tends to be lower than in developed nations, hip fracture incidence also tends to be lower than in developed nations. Although research to explain this paradox is lacking, it is assumed that the explanation lies with certain dietary and lifestyle practices of people living in less-industrialized countries, where calcium intakes are low. Among these practices are lower intakes of animal proteins, sodium, and caffeine; increased consumption of fruits, vegetables, legumes, and whole-grain products; and higher levels of physical activity and sun exposure, all of which have the potential to reduce calcium losses (and thus calcium requirements) and to enhance bone mineral density.[8, 22]

Because calcium requirements vary among individuals, calcium intakes less than those recommended by the AI may be sufficient to maintain calcium nutriture in some persons. However, given the impracticality of determining one's calcium requirement and the relative ease of achieving an adequate calcium intake, North Americans are advised to maintain calcium intakes at the levels suggested by the AI (shown in Table 25.5), preferably by consuming a variety of calcium-rich foods. These include milk (liquid and powdered), milk products (e.g., cheese, yogurt, and kefir), dark green vegetables (mustard and turnip greens, kale, and broccoli), some nuts (e.g., almonds), some seeds (e.g., sesame), tofu (manufactured using calcium sulfate as opposed to magnesium chloride), corn tortillas, and a variety of calcium-fortified foods such as citrus juice and soy beverages. Persons who are lactose intolerant typically avoid dairy products because they have insufficient production of lactase, the gastrointestinal enzyme necessary to digest lactose, the disaccharide present in milk. Approximately 75% of the world's population and about 20% of North Americans are lactose intolerant. This condition is more common in persons who are of African, Hispanic, Asian, and Native North American ancestry. Persons with lactose intolerance should be advised to consume a variety of calcium-rich, lactose-free foods. Lactose-free milk and tablets containing the enzyme lactase are available for those who wish to consume dairy products. Examples of dairy and nondairy food sources of calcium are shown in Tables 25.7 and 25.8, respectively.

Calcium supplements may be required for persons having difficulty achieving the recommended intake of calcium from foods alone.[2, 18] The most common and least expensive calcium supplements are those containing calcium carbonate, which should be taken with meals because calcium carbonate requires acid to make the calcium more soluble and absorbable. Supplements containing calcium citrate may be taken anytime. Calcium carbonate interferes with the absorption of iron and should not be taken at the same time as an iron supplement; however, calcium citrate does not interfere with the absorption of iron from supplements. Persons taking prescription or over-the-counter medications should check for any incompatibility with calcium supplements prior to their use. The amount of elemental calcium provided by a supplement is listed on the Supplement Facts label, and should be noted. Calcium absorption is enhanced when smaller doses (500 mg) are taken two or more times a day. The Tolerable Upper Intake Level (UL) for calcium for all persons 1 year of age and older, including pregnant or lactating females, is 2500 mg/day, and total calcium intake (calcium obtained from food plus any taken in supplemental form) should not routinely exceed this level. There is no evidence that calcium intake in excess of the AI is beneficial. A patient's usual calcium intake should be evaluated and, if low, the patient should be encouraged to increase the consumption of calcium-rich foods to achieve the recommended calcium intake. A simple approach for arriving at a rough estimate of a patient's calcium intake is shown in Box 25.1. If necessary, a commercially available calcium supplement with vitamin D should be recommended at a dosage sufficient to meet the AI without routinely exceeding the UL. Calcium supplements containing dolomite and bonemeal should not be used because they may be contaminated with lead.[2, 18]

Table 25.7 Dairy Food Sources of Calcium[1]

Food, Standard Amount	Calcium (mg)	Kcalories
Plain yogurt, nonfat (13 g protein/8 oz), 8-oz container	452	127
Romano cheese, 1.5 oz	452	165
Pasteurized process Swiss cheese, 2 oz	438	190
Plain yogurt, low-fat (12 g protein/8 oz), 8-oz container	415	143
Fruit yogurt, low-fat (10 g protein/8 oz), 8-oz container	345	232
Swiss cheese, 1.5 oz	336	162
Ricotta cheese, part skim, ½ cup	335	170
Pasteurized process American cheese food, 2 oz	323	188
Provolone cheese, 1.5 oz	321	150
Mozzarella cheese, part-skim, 1.5 oz	311	129
Cheddar cheese, 1.5 oz	307	171
Fat-free (skim) milk, 1 cup	306	83
Muenster cheese, 1.5 oz	305	156
1% low-fat milk, 1 cup	290	102
Low-fat chocolate milk (1%), 1 cup	288	158
2% reduced fat milk, 1 cup	285	122
Reduced fat chocolate milk (2%), 1 cup	285	180
Buttermilk, low-fat, 1 cup	284	98
Chocolate milk, 1 cup	280	208
Whole milk, 1 cup	276	146
Ricotta cheese, whole milk, ½ cup	255	214
Blue cheese, 1.5 oz	225	150
Mozzarella cheese, whole milk, 1.5 oz	215	128
Feta cheese, 1.5 oz	210	113

[1]Foods are ranked by milligrams of calcium per standard amount. The number of kcalories is per standard amount.

Source: Agricultural Research Service (ARS) Nutrient Database for Standard Reference, Release 17.

Table 25.8 Nondairy Food Sources of Calcium[1]

Food, Standard Amount	Calcium (mg)	Kcalories
Fortified ready-to-eat cereals (various), 1 oz	236–1043	88–106
Soy beverage, calcium fortified, 1 cup	368	98
Sardines, Atlantic, in oil, drained, 3 oz	325	177
Tofu, firm, prepared with calcium sulfate, ½ cup	253	88
Pink salmon, canned, with bone, 3 oz	181	118
Collards, cooked from frozen, ½ cup	178	31
Molasses, blackstrap, 1 Tbsp	172	47
Spinach, cooked from frozen, ½ cup	146	30
Soybeans, green, cooked, ½ cup	130	127
Turnip greens, cooked from frozen, ½ cup	124	24
Ocean perch, Atlantic, cooked, 3 oz	116	103
Oatmeal, instant, fortified, 1 packet prepared	99–110	97–157
Cowpeas, cooked, ½ cup	106	80
White beans, canned, ½ cup	96	153
Kale, cooked from frozen, ½ cup	90	20
Okra, cooked from frozen, ½ cup	88	26
Soybeans, mature, cooked, ½ cup	88	149
Blue crab, canned, 3 oz	86	84
Beet greens, cooked from fresh, ½ cup	82	19
Pak-choi, Chinese cabbage, cooked from fresh, ½cup	79	10
Clams, canned, 3 oz	78	126
Dandelion greens, cooked from fresh, ½ cup	74	17
Rainbow trout, farmed, cooked, 3 oz	73	144

[1]Foods are ranked by milligrams of calcium per standard amount. The number of kcalories is per standard amount. The bioavailability of calcium may vary. Both calcium content and bioavailability should be considered when selecting dietary sources of calcium. Some plant foods have calcium that is well absorbed, but the large quantity of plant foods that would be needed to provide as much calcium as in a glass of milk may be unachievable for many. Many other calcium-fortified foods are available, but the percentage of calcium that can be absorbed is unavailable for many of them.

Source: Agricultural Research Service (ARS) Nutrient Database for Standard Reference, Release 17.

As long as total calcium intake is kept at or below the UL, calcium supplementation is not harmful unless contraindicated by some medical condition or by a potential adverse drug or nutrient interaction. To avoid **hypercalciuria**, patients with a history of kidney stones should have a 24-hour urine calcium determination before beginning calcium supplementation.[7] Most authorities support supplementation if dietary sources of calcium do not provide sufficient calcium to reach the AI on most days of the week. The best sources of calcium are foods which, in addition to calcium, provide other key nutrients needed for optimal bone development and overall good health. Supplemental calcium will not protect a person against bone loss caused by poor diet, estrogen deficiency, physical inactivity, smoking, alcohol abuse, or various medical disorders or treatments.[7, 22]

Vitamin D As previously mentioned, the major function of vitamin D is to increase blood concentrations of calcium and phosphorus by (1) promoting their absorption by the GI tract, (2) promoting their reabsorption by the kidney, and (3) stimulating osteoclast formation, and thus bone resorption and the release of calcium and phosphorus from bone.[1, 3, 6] An overt deficiency of vitamin D is linked to rickets in children and osteomalacia in adults, both of which are discussed in detail later in this chapter. Although the fortification of milk with vitamin D (introduced in the 1930s) has nearly eliminated rickets in North America, vitamin D insufficiency continues to be common throughout North America, particularly among those living above 40° North latitude who have inadequate sun exposure during the winter months.[8, 10] The best indicator of vitamin D status is the concentration of serum 25-hydroxyvitamin D, because it represents both dietary vitamin D intake and vitamin D synthesized by the body. Vitamin D status is considered

A Simple Approach for Estimating Dietary Calcium Intake

Step 1: For each of the following dairy or calcium-fortified foods, enter the number of servings consumed per day. Multiply the number of servings for each food by the calcium content per serving to arrive at the total calcium from the food per day.

Food	Servings of food/day	Calcium/serving of food	Total calcium from food
Milk, fat-free or 1% low-fat, 1 cup	_____	× 300 mg	= _____ mg
Soy beverage, calcium-fortified, 1 cup	_____	× 350 mg	= _____ mg
Yogurt, nonfat or low-fat, 8-oz container	_____	× 400 mg	= _____ mg
Cheese, 1.5 oz	_____	× 300 mg	= _____ mg
Orange juice, calcium-fortified, 1 cup	_____	× 350 mg	= _____ mg
Breakfast cereals, calcium-fortified, 1 oz	_____	× 200–1000 mg[a]	= _____ mg

Step 2: Add together the amounts of calcium from the above dairy and calcium-fortified foods. _____ mg

Step 3: Add an additional 300 mg to account for calcium obtained from food sources other than the above dairy and calcium-fortified foods to arrive at a rough estimate of total calcium intake per day from all foods. _____ mg

[a] The calcium content of fortified ready-to-eat breakfast cereals will vary. Check the Nutrition Facts label for calcium content.

Source: Adapted from: *Physician's Guide to Prevention and Treatment of Osteoporosis.* Washington, DC: National Osteoporosis Foundation, 2003.

adequate when the serum 25-hydroxyvitamin D concentration is consistently >50 µmol/L (20 ng/mL).[7, 10] Research shows that the prevalence of vitamin D insufficiency is particularly high in persons living year-round in northern latitudes who have dark skin and who are older. The high prevalence of vitamin D insufficiency observed in these groups indicates that vitamin D fortification of milk is not an effective strategy for ensuring adequate vitamin D nutriture.[10] Because of its central role in maintaining calcium homeostasis and promoting bone health, vitamin D supplementation should be considered for people who have limited sunlight exposure for cultural or medical reasons, people who are dark skinned and live outside the tropics, and older individuals.[22]

Vitamin D insufficiency in the elderly is linked to age-related bone loss and increased risk of fracture, and vitamin D supplementation is an effective strategy for preventing fracture in the frail elderly.[18] Research has shown that supplementation with calcium and vitamin D slows bone loss in elderly males and females, and reduces fracture risk, including fractures of the hip, by as much as 20% to 30%.[7, 18, 22, 28] Providing an adequate intake of vitamin D and calcium is a safe, effective, and inexpensive way to reduce risk of fracture.[11] Vitamin D and calcium supplementation is now advocated as the basic minimum for treating osteoporosis and reducing risk of fracture in older females and males.[22, 28] The potential risk of vitamin D toxicity is generally overstated. Although the tolerable upper intake level (UL) for vitamin D for children and adults has been set at 2000 IU/day, scientific evidence indicates that toxicity rarely occurs until intakes exceed 10,000 IU/day.[33]

Physical Activity Bone mineral density increases in response to the stress bones receive from weight-bearing or impact-type physical activities such as physically demanding occupations, walking, jogging, jumping rope, climbing stairs, high-impact aerobics, weight/resistance training, and a variety

of sports such as tennis, soccer, basketball, and gymnastics. Persons living in rural communities and in countries where physical activity is maintained into adult life have a lower fracture risk compared to those living in more urban, sedentary societies.[22] Research has shown that athletes have a higher BMD than those in the general population and that significant bone mineral loss can result from prolonged bed rest, paralysis, and periods of weightlessness.[7, 14] Although improvements in BMD are most notable when impact exercise begins during growth and before the age of puberty, benefits have been observed in adult males and premenopausal and postmenopausal females. Research into the effects of physical activity on BMD and fracture risk in adults is complicated by small sample sizes, the short-term nature of most studies (less than two years), poor subject compliance, and high dropout rates. The results show that regular, moderate-intensity exercise initiated during adult life increases BMD by 1% to 2% and that impact-type activities result in greater improvements in BMD than do nonimpact activities (e.g., stretching, yoga, weight lifting, and resistance training). Subjects demonstrating a higher degree of compliance with recommendations to engage in regular, moderate physical activity experienced a greater degree of improvement in BMD. Very high levels of physical activity can be detrimental to bone health, particularly in premenopausal women who experience **oligomenorrhea** or **amenorrhea**. In adult male runners, the greatest improvements in BMD are seen in those averaging fifteen to twenty miles per week, whereas longer weekly distances result in little additional benefit or, in some instances, an actual reduction in BMD.[14]

In postmenopausal females, impact activities such as walking, dancing, and jumping have been shown to slow or prevent bone loss and reduce risk of hip fracture.[12, 14, 18] In a prospective cohort study of more than 61,000 postmenopausal women, those who walked for at least four hours per week had a 41%

lower risk of hip fracture compared to women who walked less than one hour per week.[32] Although the benefits of nonimpact activities on BMD in postmenopausal females are inconsistent, activities leading to improvements in flexibility, balance, agility, and muscle strength are associated with fewer falls and a significant reduction in fracture risk. This is particularly the case when coupled with an assessment and modification of fall hazards in the home, reviewing prescription medications for side effects that may affect stability and balance, and checking and correcting hearing and vision.[11, 14]

Cigarette Smoking
Cigarette smoking has been shown to be causally related to lower bone density, increased bone mineral loss, and increased risk of fracture in males and females.[7, 12, 18, 28, 34] There are several mechanisms that place smokers at increased risk of poor bone health. Nicotine and cadmium in tobacco smoke are toxic to osteoblasts. Tobacco smoke appears to reduce intestinal calcium absorption, and smokers generally have lower intakes of vitamin D and lower serum levels of 25-hydroxyvitamin D compared with nonsmokers.[34] Smokers are more likely to consume excessive amounts of alcohol than are nonsmokers, which can adversely impact bone health. Smokers tend to have lower body weights, are less physically active, are more frail, and experience higher rates of chronic disease. Some of the medications used to treat these chronic diseases (e.g., glucocorticoids used to treat lung disease) can indirectly impact bone health and increase fracture risk.[7, 34] Smoking accelerates the metabolism of estrogen and may reduce the benefits of hormone replacement therapy in females. On average, female cigarette smokers reach menopause one to two years earlier than nonsmokers, which extends the postmenopausal period during which the rate of bone loss is accelerated.[12, 34]

Tobacco smoking is the leading cause of preventable death in the United States and in most developed nations. It has been shown to have negative health impacts on people at all stages of life and to harm nearly every organ of the body, causing many diseases and damaging the overall health of smokers. Quitting smoking has immediate and long-term benefits, reducing risks for diseases caused by smoking and improving health.[34] Members of the health care team have a professional obligation to model good health behaviors to their patients by not smoking, to encourage their patients not to smoke, and to support the efforts of their patients who wish to quit smoking.

Alcohol
There is no consistent evidence that moderate alcohol consumption has an adverse impact on bone health and fracture risk.[18, 22] The *Dietary Guidelines for Americans* defines moderate alcohol consumption as no more than one drink per day for women and no more than two drinks per day for men.[35] Some research suggests that moderate alcohol consumption may increase BMD and reduce bone mineral loss, although this beneficial effect may only be seen in women.[18, 22, 34] However, heavy alcohol use is associated with decreased BMD, reduced bone formation, and increased risk of fracture. Chronic alcoholism is considered a major risk factor for osteoporosis. Heavy alcohol use increases calcium and magnesium losses from the body and adversely impacts vitamin D and overall nutritional status.[34] Even moderate alcohol consumption increases the risk of falling and other types of skeletal trauma, thus increasing the risk of fracture.[34]

Other Nutrients and Food Components
The role of several other nutrients and food components in the prevention of osteoporosis has been studied. Among these are phosphorus, protein, fruits and vegetables, sodium, caffeine, fluoride, and trace minerals. Phosphorus is an essential bone-forming mineral, and an adequate supply is necessary for optimal bone health throughout life. According to data from NHANES 1999–2000, mean intakes of phosphorus are well above the RDAs for both sexes at all age levels except for males 6 to 11 years of age and females 9 to 18 years of age. In these two groups, mean intakes are slightly less than the RDA. Apart from these two age categories, it appears that the phosphorus intakes of Americans are adequate except for instances of malnutrition, intestinal malabsorption, and excessive use of phosphorus-binding antacids.[18, 22] Concerns have been raised about diets high in phosphorus and low in calcium resulting in a high phosphorus:calcium ratio, particularly in relation to the consumption of carbonated soft-drinks. The ratio of phosphorus to calcium in the diet can vary widely with no detectable effect on the absorption and retention of either mineral or on BMD. Phosphorus-rich carbonated soft-drinks appear to only have a negligible effect on calcium excretion.[18, 22]

The observation that hip fractures are common in countries where meat and dairy food consumption is high has led to interest in the relationship between high-protein diets and risk of osteoporosis. As dietary protein intake increases, so does urinary calcium excretion. This has led some to believe that excess protein intake in industrialized nations, particularly from meat and dairy products, is causally linked to increased risk of osteoporosis despite the relatively high calcium intakes typically seen in these Western nations. However, the association between protein intake (both total and animal protein) and risk of osteoporosis and fracture is not clear. Dietary protein is an essential component of the organic matrix of bone and is necessary to maintain production of hormones and growth factors that modulate bone synthesis (36). A low protein intake in the elderly appears to increase risk of osteoporotic fracture, and elderly patients who have protein-energy malnutrition are more likely to be frail and to fall. Hip-fracture patients receiving dietary protein supplements during hospitalization experience lower rates of complications and mortality following surgery and have shorter hospitalizations than similar patients who do not receive protein supplements.[18, 21, 22]

Because animal proteins are rich in sulphur-containing amino acids, they contribute to an acidic environment within the body (see Chapter 8), which places a demand on bone as a source of skeletal salts. It is thought that calcium is removed from bone in order to neutralize the acid generated from a high-meat diet.[22, 36, 37] However, diets providing ample fruits and vegetables and vegetable proteins result in a more alkaline environment within the body that does not necessitate the removal of calcium from bone to maintain an appropriate acid-base homeostasis. Research shows that diets providing potassium, magnesium, fruits, vegetables, and vegetable proteins are associated with higher BMD.[22, 37] Patients should be encouraged to maintain good nutritional status and adequate protein intakes, and to consume plant foods from various groups consistent with the recommendations of the *Dietary Guidelines for Americans*.[35]

As dietary sodium intake increases, so does urinary calcium excretion. Thus, high-sodium diets are associated with increased excretion of calcium in the urine.[18, 35, 36] Although there is convincing evidence that reducing sodium intake results in a lowering of urinary calcium excretion, research has not yet demonstrated whether a low-sodium diet is an effective approach for preventing osteoporosis and reducing risk of fracture. However, considering the beneficial effects on blood pressure of diets providing ample amounts of vegetables and fruits, adequate intakes of nonfat dairy products, and moderate amounts of lean meats, there is a strong rationale for patients to follow such a dietary pattern.[22, 35]

Medical Management

Once a patient is diagnosed with osteoporosis or found to have low BMD, the patient should be evaluated to determine whether this condition is primary or is secondary to some other disease (see Table 25.3) or to drug use (see Table 25.4). This evaluation should begin with a thorough history and physical examination as well as several basic laboratory tests (see Table 25.9), which will rule out the most common causes of secondary osteoporosis. In patients with secondary osteoporosis, treatment of the underlying causes should be addressed before initiating osteoporosis-specific treatment. Patients with primary osteoporosis or those with secondary osteoporosis in which the underlying cause is being addressed should then undergo risk factor modification, dietary treatment, and drug therapy as needed. Risk factor modification includes efforts to encourage smoking cessation, control of excessive alcohol use, and moderate physical activity, while dietary treatment would include adequate intakes of calcium and vitamin D and promotion of overall good nutrition. If necessary, pharmacologic agents can be used to treat the osteoporosis or, in patients with low BMD, to prevent further loss of BMD in order to prevent osteoporosis.

Pharmacologic Prevention and Treatment

Several drugs have been approved by the U.S. Food and Drug Administration (FDA) and Health Canada's Therapeutic Products Directorate (TPD) for the prevention and treatment of osteoporosis.[7, 11, 14] Included among these are estrogens, selective estrogen receptor modulators, bisphosphonates, and teriparatide, a synthetic form of parathyroid hormone. Even when drugs are used in the prevention and treatment of osteoporosis, it is essential to ensure adequate intakes of calcium and vitamin D, as well as to advise patients to refrain from smoking, engage in regular weight-bearing exercise, and use alcohol in modera-

Table 25.9 Basic Laboratory Tests to Determine Secondary Osteoporosis

Laboratory Test	Normal Range for Adults	Evaluation in Osteoporosis
Albumin, serum	3.5–5.0 g/dL (35–50 g/L)	Increased in dehydration. Decreased in malnutrition, anorexia nervosa, malabsorption syndromes, liver disease.
Alkaline phosphatase, serum	30–120 U/L (0.5–2.0 μKat/L)	Increased during active bone growth including pathologic new bone growth (e.g., Paget disease) and metastatic bone cancer. Increased during fracture healing, rheumatoid arthritis, and hyperparathyroidism. Decreased in malnutrition.
Calcium, total, serum	9.0–10.5 mg/dL (2.25–2.75 mmol/L)	Increased in hyperparathyroidism, hyperthyroidism, metastatic bone cancer, Paget disease, and vitamin D toxicity. Decreased in hypoparathyroidism, malabsorption syndromes, rickets, osteomalacia, and vitamin D deficiency.
Creatinine, serum Female Male	0.5–1.1 mg/dL (44–97 μmol/L) 0.6–1.2 mg/dL (53–106 μmol/L)	Elevated in renal failure and diseases that decrease renal function. Increased in acromegaly.
Free testosterone, serum Female Male	<1 ng/mL (<3.5 nmol/L) 3–10 ng/mL (10–35 nmol/L)	Decreased testosterone in a male is associated with increased risk of osteoporosis.
Parathyroid hormone, serum	10–65 pg/mL (10–65 ng/L)	Increased in hyperparathyroidism, hypocalcemia, malabsorption syndromes, rickets, vitamin D deficiency. Decreased in hypoparathyroidism, hypercalcemia, and metastatic bone cancer.
Phosphorus, serum	3.0–4.5 mg/dL (0.97–1.45 mmol/L)	Increased in hypoparathyroidism, acromegaly, metastatic bone cancer, and hypocalcemia. Decreased in hyperparathyroidism, hypercalcemia, vitamin D deficiency, and rarely in malnutrition and inadequate dietary phosphorus intake.
Thyroid-stimulating hormone, serum	2–10 μU/mL (2–10 mU/L)	Increased in primary hypothyroidism (i.e., thyroid dysfunction). Decreased in hyperthyroidism and secondary hypothyroidism (e.g., due to pituitary or hypothalamus dysfunction).
Total protein, serum	6.4–8.3 g/dL (64–83 g/L)	Increased in dehydration. Decreased in malnutrition, anorexia nervosa, malabsorption syndromes, liver disease.

tion, if at all. All drugs have the potential for adverse side effects or drug-nutrient interactions (see Table 25.12) and in many instances they are expensive. Consequently, the therapeutic benefits of drug treatment must be weighed against their potential adverse effects and their cost, which is sometimes high.

Estrogen is a generic term for a group of female sex hormones that promote fertility in female mammals. They include estradiol, estriol, and estrone, and are marketed under a variety of names and forms. Estrogens are approved by the U.S. FDA and Health Canada's TPD for the prevention and treatment of osteoporosis in postmenopausal women. Estrogens reduce bone turnover, prevent bone loss, increase BMD by 5% to 10% in the hip, spine, and total body, significantly reduce fracture risk, and are effective in relieving the hot flashes and night sweats that often accompany menopause.[7, 38] Although several randomized controlled trials have shown that use of estrogens to treat women with postmenopausal osteoporosis reduces vertebral and hip fractures by about 33% and osteoporotic fractures at other sites by about 23%, there are potential serious side effects to such treatment, including irregular vaginal bleeding, breast tenderness, increased risk of breast cancer, and increased risk of thrombus (a stationery blood clot attached to the wall of a blood vessel that obstructs blood flow) and embolus (a blood clot carried by blood flow from the site of formation to a smaller blood vessel). In one study, women receiving combined estrogen-progestin treatment experienced an increased risk of breast cancer (26% increased risk), coronary heart disease (29%), stroke (41%), and thromboembolism (obstruction of a blood vessel by a thrombus) (111%) compared to women receiving a placebo.[39] For the past several decades, hormone therapy has been the primary pharmacologic approach for preventing and treating postmenopausal osteoporosis, but the potential risks of hormone treatment and the availability of other effective drugs is casting doubt on the use of hormones as the preferred therapeutic option.

Selective estrogen receptor modulators (SERMs) are non-hormonal agents that have tissue-specific effects in estrogen-responsive target tissues. In the estrogen-responsive tissues of the skeletal and cardiovascular system, the effects of SERMs are similar to those of estrogen. In breast and uterine tissue, however, SERMs have no estrogen-like effects.[7, 20, 38, 40] Raloxifene is a SERM approved by the FDA and TPD for the prevention and treatment of osteoporosis in postmenopausal women. When raloxifene binds to estrogen receptors in bone, it has an estrogen-like effect that increases BMD and reduces spine fracture risk by 30%.[11, 38] In contrast, raloxifene has no estrogen-like effects when it binds to estrogen receptors in breast and uterine tissue, and consequently it does not cause breast tenderness or vaginal bleeding, and it does not increase the risk of breast or uterine cancer. Thus, SERMs have several advantages over the use of estrogens for the prevention and treatment of postmenopausal osteoporosis. However, raloxifene use is associated with increased risk of thromboembolism, is not effective in treating perimenopausal symptoms (e.g., hot flashes and night sweats), and may actually worsen hot flashes.[20, 38]

Bisphosphonates are potent nonhormonal drugs that reduce bone resorption. They act exclusively in bone by binding to hydroxyapatite (the primary mineral component of bone) at sites of bone resorption and impairing osteoclast function, reducing osteoclast number, and promoting the death of osteoclasts.[7, 14, 20, 28, 41] Bisphosphonates approved by the FDA and TPD for preventing and treating postmenopausal osteoporosis, osteoporosis in men, and glucocorticoid-induced osteoporosis in both sexes are alendronate, risedronate, and ibandronate. Bisphosphonates are considered the first choice for the pharmacological treatment of osteoporosis in men.[28] Bisphosphonates have been shown to decrease bone turnover and increase bone mass in the lumbar spine and hip by as much as 8% and 6%, respectively, and to reduce spine and hip fracture risk by as much as 50%.[7, 28, 38] The bisphosphonates differ from each other in terms of their potency, ability to inhibit bone resorption, toxicity, and dosing regimen.

Bisphosphonates are poorly absorbed by the gastrointestinal tract and should only be taken first thing in the morning on an empty stomach with at least one cup of plain water. They are poorly absorbed even when taken on an empty stomach with plain water, and their absorption is markedly reduced if taken with food or with any beverage other than water.[7, 14, 38] Although generally well tolerated by patients, they may cause esophageal and gastric irritation. Consequently, after taking bisphosphonates, patients should remain in an upright position (sitting or standing) and avoid bending over for at least 30 to 60 minutes.

Bisphosphonates are as effective at treating osteoporosis when taken once weekly as they are when taken once daily, and the once-weekly dosing schedule has the advantage of improved patient convenience and adherence, and reduced risk of gastrointestinal complications. There is ongoing research on the safety and efficacy of once-monthly oral and once-yearly intravenous administration of bisphosphonates.[7]

Parathyroid hormone (PTH) is a peptide hormone largely responsible for calcium homeostasis within the body.[3, 7] Teriparatide is a synthetic version of PTH produced using recombinant DNA technology. Structurally similar to human PTH and having similar effects within the human body, teriparatide acts on the osteoblasts (the bone-building cells) to stimulate new bone growth and increase BMD; thus, it has an anabolic (or tissue building) effect on the skeleton. In this respect, teriparatide's mode of action is different than the previously mentioned drugs, which primarily reduce osteoclast activity and thus are considered antiresorptive agents. Teriparatide has been approved by the FDA and the TPD for treatment of osteoporosis in both females and males who have a high risk of fracture.[7, 14] Because it is a peptide hormone, it must be administered by injection, typically by daily subcutaneous injections. Research has shown that teriparatide stimulates new bone formation and lowers risk of spinal fractures by 65% and risk of fracture at other sites (wrist, ribs, hip, ankle, and foot) by 53%.

Paget Disease

Paget disease is a localized, progressive, often crippling disorder of bone remodeling resulting from overactive osteoclasts that cause rapid bone resorption followed by rapid formation of new bone by osteoblasts. The structure of the new bone is haphazard, disorganized, and structurally inferior, leaving the diseased bone more subject to bowing, deformity, fracture, and poor healing following a fracture.[21, 40] The bones most often affected by Paget disease are the upper femur, pelvis, vertebral bodies, skull, and tibia.[40] While pain (including headaches) is the most common presenting symptom, approximately 70% of patients with Paget disease are asymptomatic. Other clinical manifestations include bowing of the long bones with resulting gait abnormalities, nerve paralysis, hearing loss, facial deformity, tooth loss, and cardiovascular disease.[5, 40]

After osteoporosis, Paget disease is the second most common bone disorder and is diagnosed in about 3% of persons over age 40 years. The prevalence is greater in males and increases with age, and there are wide geographic variations in its frequency. The disease typically begins insidiously, and then slowly progresses over many years. Although the etiology of Paget disease is unknown, there is evidence suggesting that genetic and viral factors play a role. Diagnosis is based on x-rays showing characteristic bone changes and deformities. Biochemical markers of bone formation and resorption, such as serum alkaline phosphatase and urinary hydroxyproline, are used to confirm the diagnosis, evaluate the severity of the disease, and determine the patient's response to treatment.[5, 40] Treatment of Paget disease involves use of nonsteroidal anti-inflammatory agents for pain and use of bisphosphonates to suppress osteoclast activity and decrease bone resorption and formation.[40] Persons with Paget disease should also maintain an adequate intake of calcium and vitamin D.

Rickets and Osteomalacia

Rickets and **osteomalacia** are related diseases in which there is insufficient mineralization of the organic matrix of bone, in most cases due to vitamin D deficiency. As discussed earlier in this chapter, approximately two-thirds of bone is mineral and the remaining one-third is an organic matrix. In rickets and osteomalacia, the organic matrix is present but it is not sufficiently mineralized, unlike in osteoporosis, where there is a loss of both the organic matrix and bone mineral content. Rickets is a disease of childhood and osteomalacia is seen in adults.

Rickets

Epidemiology, Etiology, and Clinical Manifestations
Rickets is characterized by inadequate maturation and mineralization of the cartilaginous growth plate and inadequate mineralization of the organic matrix within the bones of children.[2, 5] The most common cause of rickets is vitamin D deficiency due to inadequate vitamin D intake and/or inadequate sunlight exposure, which leads to decreased intestinal calcium absorption, inadequate bone mineralization, and a low serum calcium concentration. Risk factors for rickets are outlined in Table 25.10. A low serum calcium concentration stimulates secretion of parathyroid hormone (PTH), which in turn mobilizes calcium and phosphorus from bone in order to maintain an acceptable blood calcium concentration. Other causes of rickets include calcium deficiency, disorders of vitamin D metabolism, and hypophosphatemia (low serum phosphate concentration).[6, 42] The symptoms of rickets are generally seen between 6 and 36 months of age and include lethargy, weakness, growth stunting, enlargement of the ends of the long bones and ribs, an abnormally shaped thorax, and bowing of the legs.[5] Figure 25.8 illustrates the bowing of legs due to rickets. Although rickets is primarily seen in developing countries and among immigrants in developed countries, cases are reported in the United States and Canada.[43] Inadequate sunlight exposure to provide optimum vitamin D synthesis is likely to occur in those living in temperate climates above the 40th parallel, and in persons living closer to

Table 25.10 Risk Factors for Rickets

- Maternal vitamin D deficiency
- Prolonged breast-feeding without vitamin D supplementation
- Living in a temperate climate
- Lack of sunlight exposure
- Dark skin pigmentation
- Calcium deficiency
- Intake of phytates from diets high in unrefined grains

the equator who have limited sun exposure because of social or religious customs.

Prevention At birth, most infants have adequate stores of vitamin D to cover their needs for the first few months of life. The American Academy of Pediatrics and the Canadian Paediatric Society regard breast-feeding as the optimal method for feeding infants.[44, 45] Despite the nutritional and immunological superiority of breast-feeding compared to feeding commercially available infant formulas or cow's milk, the vitamin D content of human milk is normally low and is insufficient to meet the infant's needs for vitamin D. Commercially available infant formulas sold in the United States and Canada are required to contain vitamin D at levels sufficient to meet the needs of infants.[46] Vitamin D-fortified cow's milk is also a suitable source of vitamin D and can be recommended for children when they reach an appropriate age, which is at about one year. Commercially available soy beverages (except soy-based infant formu-

Figure 25.8 Rickets

Rickets often results in bowing of the legs. A deficiency of vitamin D and/or calcium leads to inadequate mineralization of the bones, which then lack the strength to support the weight of the upper body once a child begins to stand and walk.

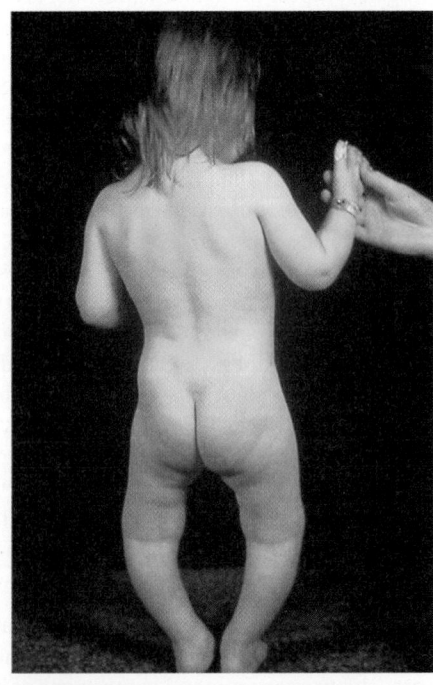

Source: S. Rolfes, K. Pinna, and E. Whitney, Understanding Normal and Clinical Nutrition, 7e, copyright © 2006, p. 377.

Courtesy of Hologic Inc.

las), rice beverages, or other vegetarian beverages, regardless of whether they are fortified, are inappropriate alternatives to breast milk, infant formulas, or pasteurized whole cow's milk in the first two years of life.[44]

Although an infant's vitamin D requirements can be met through sunlight exposure, determining what is adequate sunlight exposure for any given infant or child is difficult. Sunlight exposure varies considerably due to such factors as the season of the year, cloud cover, pollution, time spent in the shade, and the amount of body surface area covered by clothing when outdoors. Furthermore, vitamin D synthesis in response to sunlight exposure will be less for individuals who have darker skin pigmentation and for those who use sunscreens to limit exposure to ultraviolet light in order to reduce their risk of skin cancer.[46] Consequently, the American Academy of Pediatrics recommends that beginning at two months of age, exclusively breast-fed infants be given a multivitamin supplement designed for infants containing 5 mg (200 IU) of vitamin D. Infants receiving less than 500 mL per day of infant formula should also be given a multivitamin supplement designed for infants. Children and adolescents who do not receive regular sunlight exposure and who do not consume at least 500 mL (2 cups) per day of vitamin D-fortified milk should also take an age-appropriate multivitamin supplement containing at least 5 mg (200 IU) of vitamin D.[45]

Treatment Rickets is treated by a balanced diet that is age appropriate and that provides adequate intake of vitamin D, calcium, and phosphorus. Skeletal deformities can be prevented by maintaining good posture, body positioning, and bracing. After the disease is controlled, skeletal deformities may require surgical treatment.[5, 6]

Osteomalacia

Etiology and Clinical Manifestations Osteomalacia is a generalized bone condition affecting adults in which the organic matrix of bone is inadequately mineralized. Sometimes regarded as the adult form of rickets, the primary clinical manifestations of osteomalacia are muscular weakness, bone pain, and in advanced cases, deformities of the ribs, pelvis, and bones of the legs.[5] The most common causes of osteomalacia are vitamin D deficiency or impaired vitamin D action, calcium deficiency, and hypophosphatemia. Vitamin D deficiency can result from inadequate sun exposure, low dietary intake, or malabsorption of vitamin D due to biliary tract or intestinal diseases that impair the absorption of fat and fat-soluble vitamins.[5] Drugs such as phenytoin, phenobarbital, and rifampin stimulate hepatic breakdown of vitamin D and accelerate its loss from the body.[6] Diseases of the liver and kidney can impair the body's ability to convert vitamin D absorbed from the diet or synthesized in the skin into its biologically active form. A major consequence of long-standing vitamin D deficiency is reduced intestinal calcium absorption leading to hypocalcemia which, in turn, results in increased secretion of PTH. Increased secretion of PTH stimulates the removal of mineral from the organic matrix of bone in order to raise the serum calcium concentration but also results in a weakening of bone. Hypophosphatemia can result from excessive renal phosphate losses seen in renal tubular acidosis or from inadequate absorption due to long-term use of antacids that bind dietary phosphate in the GI tract and prevent its absorption.[5]

Treatment Elderly persons are at increased risk of osteomalacia due to diets low in calcium and vitamin D, a lack of sun exposure, decreased efficiency in synthesizing vitamin D when exposed to sunlight, and increased incidence of intestinal malabsorption problems that accompany aging.[5] The treatment of osteomalacia should address the underlying causes. Effective treatment of vitamin D deficiency may require a multivitamin providing as much as 20 mg of vitamin D, which is somewhat greater than the AI shown in Table 25.5 but still well below the Tolerable Upper Intake Level of 50 mg. Calcium intake should be adequate and may require use of an appropriate supplement providing as much as 1500 to 2000 mg of elemental calcium per day.[6] Patients with severe, long-standing vitamin D deficiency may initially require pharmacologic doses of up to 1250 mg per week for 3 to 12 weeks, followed by maintenance therapy of 20 mg per day. Patients taking drugs that accelerate hepatic breakdown of vitamin D (e.g., phenytoin, phenobarbital, and rifampin) and those with intestinal malabsorption problems will require vitamin D in doses much greater than the AI.[6] Patients with liver or kidney diseases that prevent the steps necessary to activate vitamin D into its biologically active form (1,25-dihydroxyvitamin D) will require a form of vitamin D that is already biologically active, such as calcitriol.[5, 6]

Arthritic Conditions

Definition and Epidemiology

Arthritic conditions encompass more than 100 different diseases and conditions affecting the joints, the tissues surrounding the joints, and the connective tissues. Often referred to as "arthritis," arthritic conditions are among the most common diseases in the world and include osteoarthritis, rheumatoid arthritis, and gout, which are addressed in this section.[47] Arthritic conditions affect nearly one in six North Americans and are the leading cause of disability among Americans 18 years of age and older.[48] Among Americans 65 years of age and older, arthritic symptoms are second only to hypertension as the most commonly reported chronic condition. It is estimated that by the year 2020, approximately 60 million Americans will be affected by arthritic conditions.

Common misconceptions about arthritic conditions are that they only affect older persons, that they are an inevitable consequence of aging, and that there are limited options for managing the symptoms of arthritis. Although arthritic conditions affect one out of every two people age 65 years and older, they are diagnosed in people of all ages, including children and teens. In fact, most people with arthritic conditions are younger than 65 years. Juvenile rheumatoid arthritis affects 70,000 to 100,000 children in the United States and is one of the most common chronic conditions of childhood.[47]

There are several factors known to increase the risk of arthritic conditions, three of which are modifiable: overweight, joint injuries, and infections. Overweight and obesity increase the risk of several arthritic conditions, particularly osteoarthritis of the knee in females and gout in males. Precautions should be taken in the workplace to avoid repetitive joint use which increases risk of certain arthritic conditions. Sport-related injuries to joints and connective tissues can be prevented by following injury prevention strategies such as using protective equipment, exercising to strengthen muscles, and warming up before exercising and cooling down afterwards. Because infectious diseases (e.g., Lyme disease) can be associated with arthritic conditions, their prevention and treatment in a timely and effective manner are important strategies. Nonmodifiable risk factors include female sex, age, and family history. Compared with males, arthritic conditions are diagnosed more frequently in females, who account for 60% of cases in persons aged 15 years and older. Risk increases with age and in persons with a family history.[4, 47]

Osteoarthritis

Epidemiology, Etiology, and Clinical Manifestations

Osteoarthritis (OA) is the most common arthritic condition and is a leading cause of physical disability, increased health care costs, and impaired quality of life.[47, 49] It is estimated that 12% of the U.S. population 25 years of age and older have clinical signs and symptoms of OA.[49] OA is not a specific entity but a disease process involving all the structures of the joint, including the articular cartilage, the **subchondral bone**, the **synovial fluid** and membranes, the ligaments, and the nerves and muscles supporting the joint. However, the most striking changes seen in OA are those within the load-bearing articular cartilage.[4] The progressive loss of articular cartilage and the inflammation of the other tissues composing the joint result in joint pain, stiffness, limited joint movement, wasting of **periarticular muscles**, and in some instances joint instability and deformity.[47] The joints of the fingers, feet, lumbar and cervical vertebrae, hips, and knees are those most commonly affected by OA.[3, 47]

Figure 25.9 illustrates the basic structures of the normal joint and a joint with OA. In the early stages of OA, the articular cartilage thickens, but as the condition progresses the cartilage thins and softens. Surface cracks develop in the articular cartilage and it loses its smooth surface. Eventually these cracks extend completely through the articular cartilage down to the bone, portions of the articular cartilage become completely eroded, and the exposed surface of the subchondral bone becomes thickened and polished. Cysts form within the bone as synovial fluid leaks through the cracked and eroded articular cartilage. Dislodged fragments of cartilage and bone may float freely within the joint cavity. As the OA progresses, growth of cartilage and bone at the joint margins leads to the formation of abnormal bony outgrowths called **osteophytes** or bone spurs.[4, 47]

Major risk factors for OA include age (considered the most powerful risk factor), female sex, family history, major trauma to a joint or to soft tissues surrounding a joint, repetitive joint stress related to occupation, and obesity. Although OA is common in the joints typically stressed by ballet dancers, baseball pitchers, and prizefighters, there is no evidence that long-distance running or jogging increases risk of OA in the joints of the lower extremities. In addition to the added stress on the joints of the lower extremities resulting from obesity, excess body fat may also have a direct metabolic effect on the cartilage of joints.[47] Weight loss has great potential for reducing risk of OA of the knee. A loss of only 5 kg (11 lb) can reduce the risk of symptomatic knee OA by 50%.[4]

Treatment Treatment of OA is focused on reducing joint inflammation, reducing pain, maintaining joint mobility, and minimizing disability. Nonpharmacologic treatments include improving body posture, proper footwear, weight reduction as indicated, periodic rest of the affected joint (but rarely is complete immobilization advised), and application of heat to the affected joint. In patients with OA, physical activity has been shown to decrease joint pain and disability while improving function, quality of life, cardiovascular fitness, and muscular strength.[4] Periarticular muscle strength is a major factor protecting the articular cartilage from stress and injury. Therapeutic exercises to strengthen the periarticular muscles of the knee can decrease pain to an extent similar to that achieved by the use of nonprescription analgesics. Therapeutic exercise also can reduce disability, anxiety, and depression, and improve functional status.

Because no pharmacologic agent has been shown to prevent OA, slow its progression, or reverse it, the aim of drug therapy is pain relief as an adjunct to nonpharmacologic treatment.[4] The mainstays of drug therapy are the nonsteroidal anti-inflammatory drugs (NSAIDs) which include aspirin, acetaminophen, indomethacin, ibuprofen, naproxen, and the cyclooxygenase-2 (COX-2) inhibitors celecoxib and valdecoxib. NSAIDs provide effective relief from mild to moderate pain associated with OA and other inflammatory conditions such as rheumatoid arthritis, gout, and toothache.[41] Except for the COX-2 inhibitors, all the above NSAIDs are available at low cost without a prescription. Notwithstanding consumer familiarity with NSAIDs and their nonprescription status, they are associated with gastrointestinal (GI) blood loss and ulceration,

Figure 25.9 Pathophysiology of Osteoarthritis

A normal joint (left) and a joint affected by osteoarthritis (right) are compared. Early osteoarthritic changes shown on the left side of the affected joint include thickening of the articlar cartilage, narrowing of the joint space, and development of surface cracks in the cartilage. Late changes shown on the right side of the affected joint include erosion of cartilage, development of bone cysts as synovial fluid comes into contact with bone, and formation of osteophytes or bone spurs.

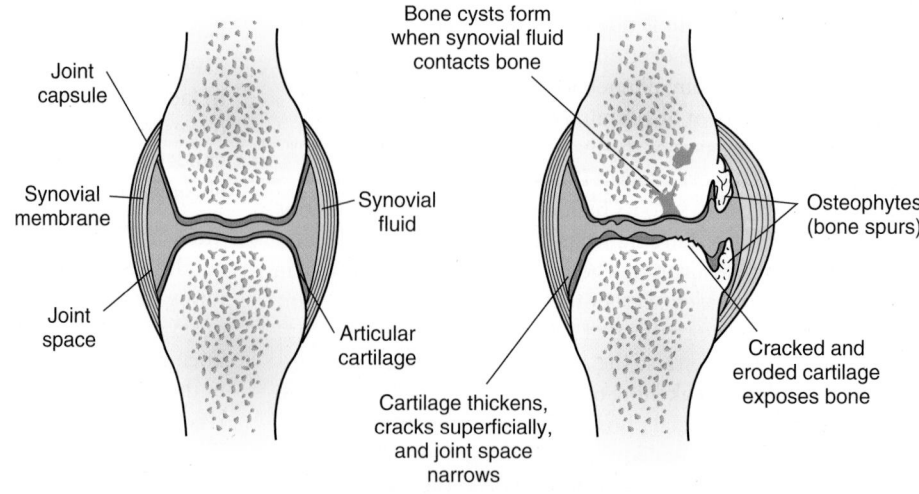

Source: CM Porth, Pathophysiology: Concepts of Altered Health Status 7e.[47]

reduced platelet aggregation, prolonged bleeding time, and greater risk of death, particularly among the elderly and those with a history of peptic ulcer disease or upper GI bleeding. Use of NSAIDs has been linked to as many as 30% of all hospitalizations and deaths related to peptic ulcer disease in persons aged 65 years and older.[4] Although the COX-2 inhibitors are much more expensive and no more effective than the other NSAIDs in reducing inflammation and pain, they have been aggressively marketed as being associated with a lower incidence of peptic ulcer disease. However, use of COX-2 inhibitors significantly increases risk of stroke, heart attack, and death from coronary heart disease, and there is uncertainty whether celecoxib and valdecoxib protect against GI complications.[50] Despite concerns about physical and psychological dependence associated with their use, opioids such as codeine can be useful for some OA patients with chronic pain. Side effects associated with opioids include nausea, vomiting, constipation, urinary retention, mental confusion, drowsiness, and respiratory depression.[4, 41]

In recent years, glucosamine and chondroitin have attracted considerable attention in the public and medical literature as dietary supplements that are potentially useful in treating OA.[49, 51-53] **Glucosamine** and **chondroitin** have been used for decades in Europe as dietary supplements for treating OA. In the United States and Canada, they are aggressively marketed for treatment of OA and widely sold in supermarkets, pharmacies, and health food stores. Glucosamine and chondroitin are naturally occurring components of articular cartilage. Glucosamine is an amino sugar and a raw material for synthesizing glycosaminoglycans and proteoglycans, which are important constituents of articular cartilage. Chondroitin is a specific glycosaminoglycan found in the proteoglycans of articular cartilage. They are extracted from animal tissue; glucosamine comes from the shells of crab, lobster, and shrimp, and chondroitin comes from the cartilage of calves, cows, steers, whales, and sharks. Between the years 1997 and 2000, glucosamine supplements were ranked as the third best-selling nutritional product, with more than $300 million in sales in 2000.[52]

Research on cell cultures and animals shows that both compounds have anti-inflammatory activity, favorably affect cartilage metabolism, and have anti-arthritic effects.[49] Preliminary human research indicates that glucosamine and chondroitin can favorably modify the progression of OA and that both compounds may be effective therapies in the symptomatic treatment of OA.[49, 51, 52] However, the only long-term randomized controlled trial on the safety and efficacy of glucosamine and chondroitin is the Glucosamine/Chondroitin Arthritis Intervention Trial (GAIT), conducted by National Center for Complementary and Alternative Medicine in collaboration with the National Institute of Arthritis and Musculoskeletal and Skin Diseases. The GAIT, which is described in Box 25.2, studied the effectiveness of glucosamine alone, chondroitin alone, and the combination of glucosamine and chondroitin in treating knee pain from OA compared to a placebo. Additional complementary and alternative therapies used for musculoskeletal disorders are listed in the CAM feature at the end of this chapter.

Rheumatoid Arthritis

Epidemiology, Etiology, and Clinical Manifestations
Rheumatoid arthritis (RA) is a chronic inflammatory disease in which the synovial membrane of the joint becomes inflamed, resulting in swelling, stiffness, pain, limited range of motion, joint deformity, and disability.[47, 54] RA affects approximately 0.8% of the population (range 0.3% to 2.1%), is more common in older persons, and affects females three times more often than males, although this difference between the sexes diminishes in older groups.[54, 55] Although RA primarily affects the joints, it also can affect other tissues, resulting in anorexia, weight loss, fatigue, and generalized aching and stiffness.

The clinical course of RA is variable and characterized by periods of exacerbation and remission. In some patients the joint inflammation results in mild to moderate pain and stiffness lasting for short periods of time, while in others it can progress to debilitating, irreversible joint deformity and destruction. In most instances, RA begins insidiously with patients experiencing fatigue, anorexia, generalized weakness, and vague musculoskeletal symptoms for weeks or months, during which time diagnosis is difficult. Eventually, inflammation of the **synovial membrane** in the joints of the hands, wrists, knees, and feet results in warmth, redness, swelling, stiffness, and pain around these joints. As the inflammation progresses, rapid division and growth of cells in the synovial membrane cause an abnormal thickening of the synovial membrane known as **pannus**, which can eventually fill the synovial cavity and invade the joint margin, as illustrated in Figure 25.10. With further progression, inflamed cells in the pannus release enzymes that digest the adjacent bone and cartilage, causing joint deformity, severe pain, and consolidation and immobility of the joint known as ankylosis. This leads to muscle atrophy from disuse, stretching of ligaments, and changes in tendons. Although treatment can slow the progression of the disease, many of these destructive changes are permanent once they occur.[47, 54]

The body's immune system plays an important role in RA and it is thought that the disease results from an aberrant immune response (see Chapter 9) in which a person's immune system mistakenly regards the body's own healthy joint tissues as foreign, and produces antibodies which then damage one's own healthy joint tissues. These self-damaging antibodies are referred to as autoantibodies. The diseases in which these play a causative role are known as autoimmune diseases and include RA, as well as two other musculoskeletal conditions, systemic lupus erythematosus (discussed later) and **scleroderma**. It is not known what causes RA or even whether RA is a single disease or several different diseases having common features. Research suggests that the autoimmune response is triggered by an infectious agent in a genetically susceptible person.[47, 54]

Treatment The goals of RA treatment are to reduce pain and inflammation, protect the joint from destruction, maintain the function of the joint and surrounding structures, and control any systemic manifestations.[54] Treatment is multidisciplinary and involves physicians, physical therapists, occupational therapists, clinical psychologists, and dietitians. A number of pharmacologic agents, such as NSAIDs (discussed earlier in this chapter), glucocorticoids, immunosupressive drugs, and what are referred to as disease-modifying antirheumatic drugs (DMARDs), are used. The glucocorticoid drug prednisone is commonly used to treat RA, but its use is associated with side effects such as increased risk of osteoporosis, peptic ulcer disease, esophagitis, and glucose intolerance, increased appetite, weight gain, and retention of sodium and fluids.[41] When prednisone is used systemically at relatively low

BOX 25.2 **RESEARCH TO PRACTICE**

The Glucosamine/Chondroitin Arthritis Intervention Trial (GAIT)

Glucosamine and chondroitin sulfate are the two most widely used dietary supplements marketed for osteoarthritis (OA), the most common form of arthritis in North America.[1, 2, 3] The pain, stiffness, and limited mobility associated with OA of the knees are among the most common reasons that people seek alternative medical treatments. In 2001, more than 5 million Americans had used either glucosamine or chondroitin, and in 2004, Americans spent an estimated $734 million on these supplements, making them among the most widely used dietary supplements sold in the United States.[4] Despite the popularity of these supplements, there have been unanswered questions about the safety and efficacy of glucosamine and chondroitin as a treatment of knee pain for OA patients. A meta-analysis of 17 double-blind, randomized placebo-controlled clinical trials indicated that glucosamine and chondroitin are likely beneficial in treating the symptoms of OA, but the degree of benefit was unclear due to methodological flaws in some of the trials.[5] Those trials that were publicly funded tended to show little, if any, benefit from the supplements, while those that were funded by the supplement makers showed positive effects.[4]

To provide a more definitive answer to this question, the National Center for Complementary and Alternative Medicine and the National Institute of Arthritis and Musculoskeletal and Skin Diseases funded the Glucosamine/Chondroitin Arthritis Intervention Trial (GAIT), a double-blind, randomized, placebo-controlled trial in which 1,583 patients

with symptomatic knee OA were randomly assigned to one of five different groups. Each group received one of the following treatments: 1,500 mg per day of glucosamine hydrochloride, 1,200 mg per day of chondroitin sulfate, both glucosamine hydrochloride and chondroitin sulfate, 200 mg of celecoxib per day, or a placebo.[2] The primary outcome measure was a 20% reduction in knee pain between the beginning and ending of the 24-week study. The results of the GAIT were that the glucosamine hydrochloride and chondroitin sulfate alone or in combination were not significantly better than the placebo in reducing knee pain by 20%. There was some evidence suggesting that, in OA patients with moderate to severe knee pain, the combination of glucosamine and chondroitin sulfate is effective in reducing pain. However, because GAIT was not specifically designed to address this patient subgroup, further research is needed to conclusively establish a benefit in these patients.

References
1. Towheed TE, Anastassiades TP. Glucosamine and chondroitin for treating symptoms of osteoarthritis, evidence is widely touted by incomplete. JAMA. 2000;283:1483–84.
2. Clegg DO, Reda DJ, Harris CL, Klein MA, O'Dell JR, Hooper MM, et. al. Glucosamine, chrondroitin sulfate, and the two in combination for painful knee osteoarthritis. N Eng J Med. 2006;354:795–808.
3. Hochberg MC. Nutritional supplements for knee osteoarthritis—still no resolution. N Eng J Med. 2006;354:858–60.
4. Kolata G. Supplements fail to stop arthritis pain, study says. New York Times, February 23, 2006, page A23.
5. McAlindon TE, LaValley MP, Gulin JP, Felson DT. Glucosamine and chondroitin for treatment of osteoarthritis: a systematic quality assessment and meta-analysis. JAMA, 2000;283:1469–75.

Figure 25.10 Pathophysiology of Rheumatoid Arthritis

A normal joint (left) and a joint affected by rheumatoid arthritis (right) are compared. Early changes shown on the left side of the affected joint include inflammation and thickening of the synovial membrane. Late changes shown on the right side of the affected joint include development of pannus, erosion of articular cartilage and bone, and filling of the joint space by pannus.

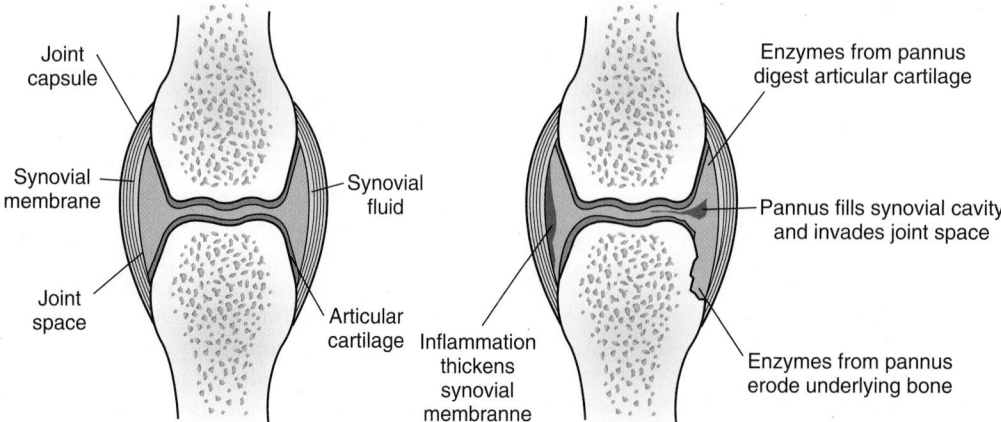

Source: CM Porth, Pathophysiology: Concepts of Altered Health Status 7e.

doses or injected directly into a joint, these side effects can be minimized. DMARDs are a chemically and pharmacologically disparate group of drugs that modify the course of RA but are also toxic and have potentially serious side effects. The most commonly used DMARD is the folic acid antagonist methotrexate, which can cause gastrointestinal upset, mouth ulcers, and liver toxicity. The simultaneous administration of folic acid can reduce the severity of side effects without diminishing treatment effectiveness.[41, 54]

Nutrition and Rheumatoid Arthritis

The role of diet and nutritional supplements in the etiology and treatment of RA has been a topic of considerable interest and speculation for decades. However, studying the relationship between diet and RA is problematic given the multiple risk factors for RA, the episodic nature of RA symptoms, the expense of conducting long-term randomized controlled trials, the difficulty of measuring diet, and the difficulty of assessing subjects' compliance to dietary change.[56]

There is evidence that lower intakes of vegetables, fruits, and dietary vitamin C are associated with increased risk of developing RA. It has been observed that there is a lower risk of developing RA in countries such as Italy and Greece, where oil-rich fish, olive oil, vegetables, and fruits are consumed in greater amounts compared to other countries that have a higher risk of developing RA.[55, 57, 58] Several randomized, double-blind, placebo-controlled trials have shown that dietary fish oil supplementation is effective in reducing the symptoms associated with chronic RA.[54, 57, 59] Fish oils contain two important omega-3 long-chain polyunsaturated fatty acids: eicosapentaenoic acid (EPA) and docosahexaenoic acid (DHA). Consumption of EPA and DHA reduces the synthesis of chemicals known to stimulate joint inflammation and cartilage degradation, and RA patients taking EPA and DHA report a reduction of morning stiffness, decreased joint tenderness, and less need to take NSAIDs to relieve pain.[59, 60] Consumption of omega-6 polyunsaturated fatty acids, on the other hand, tends to increase the synthesis of proinflammatory chemicals.[57, 61, 62] When used in conjunction with drugs to treat RA, EPA and DHA can potentially reduce the severity of side effects of NSAIDs, glucocorticoids, immunosupressive drugs, and DMARDs.[59] Current research suggests that the effective dose of fish oils is between 2.6 g/d and 7.1 g/d, but it generally takes two to three months before the anti-inflammatory effects of fish oils are noticed. Because most commercially available fish oil preparations provide approximately 300 mg of EPA or DHA per capsule, more than eight capsules per day would be necessary to achieve a dose of 2.6 g/d. Because of the high cost of fish oil capsules, a more economical approach for some patients might be to purchase bottled fish oils.[59] Considering the observation that high doses of fish oil can suppress immunity in animals, further research is needed to study the long-term safety of fish oil supplementation in humans.[57]

There is considerable speculation about the benefits of excluding foods that are thought to aggravate the symptoms of RA from the diet, and many patients report an improvement in symptoms after excluding such foods as red meats, dairy products, cereals, and wheat gluten.[56, 57, 62] Some of this improvement is likely due to the episodic nature of RA: the fact that it is characterized by periods of remission and exacerbation. Although controlled studies in which these foods have been eliminated have yielded inconsistent results, a better experimental design is to give patients capsules containing the alleged problem food antigen without the patient knowing what is in the capsule. This technique is referred to as a blind challenge test, because the patient does not know what he or she is taking and is therefore "blinded" to the treatment. Most patients who report an improvement in symptoms during elimination diets when aware of what they are eating notice no change in symptoms during blind challenge tests (when unaware of what they are eating). Because a small number of patients do report changes in symptoms in response to the elimination or addition of certain foods, further evaluation and testing to confirm a food allergy may be warranted (see Chapter 9). However, the prevalence of immune sensitivity to specific foods by patients with RA is similar to that found in the general population.[56, 57, 62] In some RA patients, symptoms improve after eliminating meat from the diet. However, it is not known whether the improvement is due to eliminating meat or a consequence of consuming more foods rich in antioxidants such as vegetables and fruits.[56, 57] Because poor nutritional status is common in patients with RA, emphasis must be given to maintaining optimal nutritional status.[57, 62]

Gout

Epidemiology and Etiology **Gout** is an inflammatory disease resulting in swelling, redness, heat, pain, and stiffness in the affected joint. Gout occurs when the serum concentration of uric acid becomes elevated to the point that uric acid crystals begin to precipitate in the synovial fluid, initiating an inflammatory response within the joint and surrounding tissues. Uric acid is the end product of the metabolism of the purines adenine and guanine from DNA and RNA, which are part of all human tissue and are found in many foods. In most cases of gout, the elevation of serum uric acid (hyperuricemia) results from overproduction of uric acid, inadequate elimination of uric acid by the kidney, or a combination of both.[47, 63]

Gout is one of the most painful arthritic conditions, is much more common in males than in females, and is the most common cause of inflammatory arthritis in males >40 years of age.[63] Risk factors associated with gout include genetics, male sex, older age, overweight, excessive alcohol consumption (three or more drinks per day), a diet rich in purines (see Table 25.11),

Table 25.11 Foods That Are Very High or Moderately High in Purines

Very-High-Purine Foods	Moderately-High-Purine Foods
• Anchovies and sardines	• Asparagus
• Fish roe	• Cauliflower
• Gravies, meat-based	• Legumes, dried
• Meat extracts	• Meat (beef, pork, fish, and poultry)
• Kidneys, beef	• Meat soups or broth
• Liver, beef or calf	• Mushrooms
• Sweetbreads[1]	• Oatmeal
• Nutritional or brewer's yeast	• Spinach
	• Wheat germ or bran

[1] Sweetbreads are animal pancreas and thymus, usually beef.

Table 25.12 Drug–Nutrient Interactions for Medications Used to Treat Diseases of the Musculoskeletal System

Classification	Mechanism	Generic	Brand Names	Possible Food–Drug Interactions
Anti-gout	Xanthine oxidase inhibitor	Allopurinol	Aloprim, Zyloprim	Drink 2.5–3 L fluids/day; avoid large doses of vit C; N/V, gastritis, abdominal pain, diarrhea; limit alcohol
Anti-gout, anti-inflammatory	Inhibits mitosis	Colchicine	Colchicine	↓ purine diet during acute attack; ↓ kcal if wt loss needed; anorexia, ↓ wt.; may ↓ absorption of B₁₂; N/V, diarrhea, abdominal pain; avoid alcohol
Disease-Modifying Antirheumatic Drugs (DMARDs) Antineoplastic, antipsoriatic, antiarthritic (rheumatoid)	Dihydrofolate inductase inhibitor	Methotrexate	Methotrexate, Rheumatrex	Encourage ↑ fluid intake to ↑ urine output; food delays absorption and ↓ peak concentration & bioavailability; ↓ folate absorption; anorexia, ↓ wt., dehydration, altered taste, N/V, diarrhea; avoid alcohol
Glucocorticoids Anti-inflammatory, immunosuppressant, hormone	Mimics the action of cortisol	Prednisone	Deltasone	Caution with DM—↑ glucose; highly protein bound; may need ↑ K, PO₄, Ca, and ↑ vits A, C, D, ↑ protein, and ↓ dietary Na; avoid alcohol
Anti-osteoporosis	Selective estrogen receptor modulator	Raloxifene	Evista	Adequate Ca & vit D intake essential; ↑ wt.; limit alcohol
Parathyroid hormone	PTH acts on bone-building cells called osteoblasts to stimulate new bone growth and improve bone density	Teriparatide	Forteo	Nausea, constipation
Hormone replacement therapy	Increases calcium abs, decreases osteoclast activity in bone	Estrogen, estrogen/progesterone	Estrogen: Cenestin, Estrace, Estradiol, Ogen, Premarin, Climara, Estraderm Estrogen/Progesterone: Activella, Femhrt, Prempro, Premphase, Combipatch	↑ foods high in Ca, vit D, Mg, folate, B₆; may need Ca, vit C suppl >1 g/d; appetite changes, ↑ Ca absorption, N/V, diarrhea; limit alcohol; ↓ glucose tolerance
Bisphosphonates	Bind permanently to the surfaces of the bones and slow down the osteoclasts	Alendronate, etidronate risedronate	Fosamax, Didrocal, Actonel	Nausea, abdominal pain, loose bowel movements, no Ca supplement or MVI for 2 hrs before or after
Opioids	Bind to specific opioid receptors in the central nervous system and in other tissues	Morphine, codeine, oxycodone, hydrocodone, loperamide, heroin	Oxycontin, Roxicodone, Imodium, Roxinal	N/V, constipation, dry mouth
Steroids	Mimics the action of cortisol	Prednisone	Deltasone	Caution with DM—↑ glucose; highly protein bound; may need ↑ K, PO₄, Ca, and ↑ vits A, C, D, ↑ protein, and ↑ dietary Na; avoid alcohol
Immunosuppressant, antineoplastic	Attacks the white blood cells	Cylclophosphamide	Cytoxan	↑ fluid needs; N/V, abdominal pain, dry mouth, diarrhea
Immunosuppressant	Attacks the white blood cells	Mycophenolate mofetil (MMF)	CellCept	Take on an empty stomach; take Mg supplement separately; N/V, diarrhea
Immunosuppressant	Attacks the white blood cells	Azathioprine	Imuran	Diarrhea, vomiting, anorexia, steatorrhea
Antineoplastic	Blocks the metabolism of cells	Methotrexate	Rheumatrex, Trexall	Mouth ulcers, diarrhea; ↑ fluid needs; food delays absorption and ↓ bioavailability; folate or MVI may ↓ response of drug; may ↓ absorption of fat, vit B₁₂, Ca, and folate; anorexia, ↓ wt.; avoid alcohol
Immunosuppressant	Blocks the production of folic acid	Cotrimoxazole	Bactrim, Septra	May need folate supplement; anorexia, N/V, diarrhea; adequate fluid
Immunosuppressant	Attacks the white blood cells	Ciclosporin, tacrolimus	Neoral, Sandimmune	No K supplement or salt substitute, caution with grapefruit; anorexia, N/V, diarrhea; increases glucose

exposure to lead, and use of certain drugs such as aspirin, diuretics, nicotinic acid, cyclosporine, and levodopa. It rarely occurs in children, young adults, or premenopausal females.

Clinical Manifestations Compared with plasma, synovial fluid is a poor solvent for uric acid, particularly at temperatures less than 37°C, which helps explain why the joint most commonly affected by gout is the great toe. Other sites commonly affected by gout are joints in the periphery of the body such as the instep, ankles, heels, knees, wrists, fingers, and elbows.[47, 63, 64] The symptoms of gout usually occur rapidly, sometimes overnight, resulting in sudden, severe joint pain and swelling; shiny, red skin around the joint; and extreme tenderness around the joint. These symptoms typically go away within five to ten days, even without treatment, and the patient may not have another attack for months or years. An acute attack of gout can be precipitated by excessive exercise, certain medications, purine-rich foods, excessive alcohol consumption, or crash dieting.[47, 64, 65] Over a period of years, chronic hyperuricemia can result in the formation of uric acid crystals in the joints and the soft tissues surrounding the joints, which form large deposits called tophi that resemble lumps just below the skin. If left untreated, recurrent attacks of gout will result in persistent swelling, stiffness, mild to moderate joint pain, and eventually permanent joint damage and disability.[47, 63, 64]

Treatment Initial treatment of gout involves use of nonsteroidal anti-inflammatory drugs (NSAIDs) to relieve pain and reduce inflammation. Glucocorticoids and colchicine are also used to reduce inflammation. Once the pain and inflammation of the acute attack are controlled, the hyperuricemia is treated using the drug allopurinol or one of several other drugs that increase urinary excretion of uric acid. Lifestyle modifications helpful in managing gout include moderate alcohol consumption, avoiding purine-rich foods, and achieving and maintaining a healthy body weight. Crash dieting resulting in overly rapid weight loss will reduce urinary excretion of uric acid and thus possibly precipitate an attack of gout.[64] As shown in Table 25.11, purine-rich foods include sardines, fish roe, anchovies, meat-based gravies, meat extracts, food yeast, and several animal organs used for food including the liver, kidneys, pancreas, and thymus, the latter two euphemistically known as "sweetbreads."

Systemic Lupus Erythematosus

Definition and Epidemiology

Systemic lupus erythematosus (SLE or known simply as **lupus)** is a chronic inflammatory disease affecting practically every organ system, but especially the musculoskeletal system. It can be clinically manifest in a wide variety of ways that may mimic other diseases (such as RA), which makes diagnosis of SLE difficult.[47] It is a major autoimmune disease, with a prevalence ranging from 15 to 50 cases per 100,000 persons in North America. Ninety percent of cases of SLE occur in women of child-bearing age, and it is seen more commonly in individuals of African, Asian, and Hispanic descent than in those of European descent. The prevalence is greatest among those of African descent. It is estimated that approximately 500,000 persons in the United States have SLE.[66]

Medical Diagnosis and Clinical Manifestations

SLE is diagnosed based on the presence in the blood of a variety of antibodies capable of damaging one's own tissues (autoantibodies) and certain clinical features that are uncovered during a thorough physical exam and medical history. The most common early symptoms of SLE are pain and arthritis in multiple joints (similar to that seen in RA), which can occur in as many as 90% of patients.[47] Also common are fatigue and muscle pain. The onset of SLE can be rapid or very slow and the course of the disease is chronic and is characterized by a pattern of acute flare-ups (exacerbations) followed by periods of remissions. However, a complete, permanent remission (the disappearance of symptoms in the absence of treatment) is rare.

Treatment

There is no cure for SLE and the mainstay of treatment is the pharmacologic management of acute, periodic flare-ups and management of the less severe, chronic forms of the disease. Drugs used to treat the musculoskeletal manifestations of SLE range from over-the-counter nonsteroidal anti-inflammatory drugs (e.g., such as aspirin, ibuprofen, and naproxen) and antimalarial drugs (e.g., hydroxychloroquine) during the chronic periods of remission to relatively high-dose corticosteroids during instances of severe flare-ups.[47, 66]

Nutrition and Systemic Lupus Erythematosus

Identification of the potential relationships between diet and SLE has been complicated by the difficulty of separating the effects of the many other factors known to be associated with SLE—including genetic, environmental, hormonal, viral, and psychoneurological factors—from the impact of diet.[67] Most research on diet and SLE has involved studying the impact of various diets, foods, and nutrients on SLE in animals, particularly strains of mice that are prone to developing lupus. While animal models are often useful in uncovering the etiology and pathogenesis of SLE, no animal model perfectly mimics the disease in humans. Our understanding of the role of diet in SLE in humans is limited because there have been no large-scale clinical trials examining this role, and the relatively few human studies that have been conducted have had a small number of subjects and lacked a control group.[67, 68] Another important factor complicating research into the effects of diet on SLE is the disorder's exacerbation and remission pattern, which makes it difficult to know whether a subject's improvement during a diet trial was due to the experimental treatment or to an unexpected period of remission.

Despite the limits of our knowledge, there is some evidence that diet can affect the course of SLE as assessed using clinical, immunological, and biochemical measures. One of the most consistent findings is that restriction of energy, protein, and fat (especially saturated and omega-6 polyunsaturated fatty acids) significantly reduces the severity of SLE in mice and prolongs the animal's life span. Consumption of fish oils rich in long-chain omega-3 polyunsaturated fatty acids such as eicosapentaenoic acid (EPA) and docosahexaenoic acid (DHA) has also been shown to significantly improve SLE in animal and human trials. Some patients with SLE seem to benefit from eating foods that are good sources of vitamin E, beta-carotene, selenium, and calcium and from avoiding supplements containing protein, zinc, and iron.[67, 68]

Fibromyalgia

Definition and Epidemiology

Fibromyalgia is a chronic musculoskeletal disorder characterized by widespread muscle pain, joint stiffness, disturbed sleep, fatigue, headache, cognitive and memory problems (sometimes referred to as "fibro fog"), **paresthesias**, and numerous tender points, which are specific muscle-tendon sites throughout the body that are painful when pressed.[69, 70] Fibromyalgia—also referred to as fibromyalgia syndrome or FMS—is the second most common musculoskeletal condition encountered by **rheumatologists**, affecting about 3 to 6 million Americans, or about 1 in 50.[69] In the United States, the prevalence is estimated to be about 2% overall, with 3.4% of females and 0.5% of males affected.[74] Approximately 80% to 90% of cases of fibromyalgia are diagnosed in females, and the condition is more common in middle-aged and older persons, with the prevalence increasing with age.[70] The condition is found in all types of climates, among most ethnic groups, and in most countries.[70] Unlike osteoarthritis or rheumatoid arthritis, fibromyalgia is not crippling, deforming, or disabling.[70]

Etiology

The etiology of fibromyalgia is unknown, but it can be associated with a traumatic or emotionally stressful event such as a motor vehicle crash or with injuries due to repetitive motion. Fibromyalgia is more often seen in persons diagnosed with autoimmune diseases such as rheumatoid arthritis or systemic lupus erythematosis. Other factors linked to an increased risk of a fibromyalgia diagnosis include sleep disturbances, low levels of serotonin in the brain, reduced levels of growth hormone, and such psychological abnormalities as depression, anxiety, somatoform disorders (physical ailments stemming from psychological problems), and **hypochondriasis**.[70]

Medical Diagnosis and Clinical Manifestations

Establishing a diagnosis of fibromyalgia is complicated by the lack of any objective diagnostic tests specific for the condition, and the fact that the symptoms of fibromyalgia are common to many other illnesses. Consequently, clinicians generally rule out other potential causes of symptoms before diagnosing fibromyalgia, and patients typically see several physicians before ultimately being diagnosed. The American College of Rheumatology (ACR) has established the following diagnostic criteria for fibromyalgia: (1) a history of pain that is widespread (i.e., on both sides of the body as well as above and below the waist) and that has lasted for at least three months and (2) excessive tenderness or pain when pressure is applied to at least eleven of eighteen different tender points that the ACR has identified in the neck, shoulders, arms, back, and legs.[69–71] The eighteen

Figure 25.11 Fibromyalgia Tender Points

Location of the 18 different tender points designated by the American College of Rheumatology in its criteria for diagnosing fibromyalgia

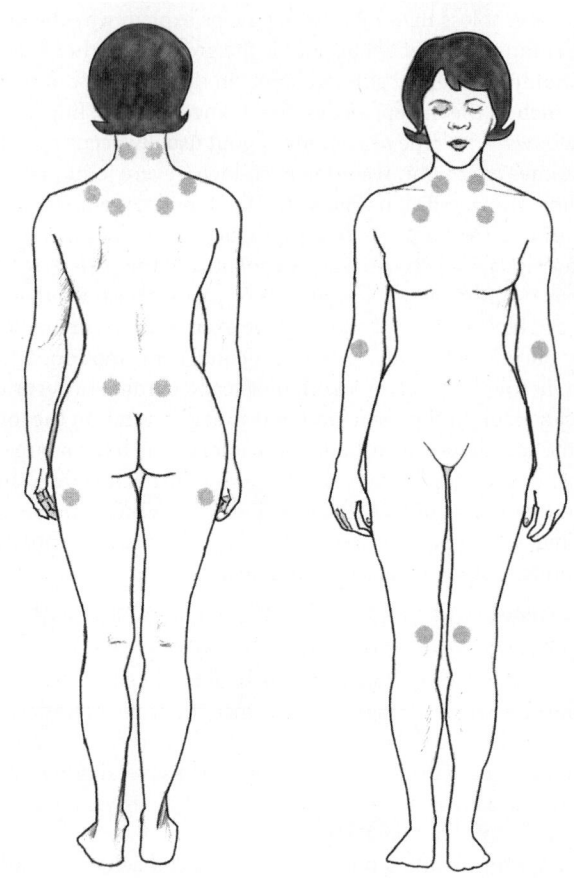

Source: Courtesy of National Institute of Arthritis and Musculoskeletal and Skin Diseases. Questions and Answers about Fibromyalgia. Bethesda, MD: National Institutes of Health, 2004.

different tender points designated by the ACR are shown in Figure 25.11.

Treatment

Treating fibromyalgia involves a multidisciplinary approach aimed at improving the quality of sleep; treating depression, anxiety, and pain; increasing physical activity; and using various approaches to help patients better cope with stressful events and improve their ability to relax.[69, 71, 72] The initial aim of therapy is improving the quality of restorative sleep by educating patients about lifestyle practices that can improve the quality of sleep and through the use of antidepressant drugs such as tricyclic antidepressants and cyclobenzaprine. Depression and anxiety are treated using psychiatric counseling and, if necessary, pharmacologic agents such as serotonin/norephineprine reuptake inhibitors and selective serotonin reuptake inhibitors.[70] Some fibromyalgia patients may periodically require analgesics for pain control.[70, 71] Regular physical activity involving aerobic exercise, muscle strengthening, and flexibility training have all been shown to improve aerobic fitness, overall well-being, and sleep quality, and to reduce pain, fatigue, and depression.[71] Low-impact activities such as water aerobics, bicycling, walking, and aerobic dance are particularly helpful. Initially, patients should begin exercising at a low level, and then gradually increase the duration and intensity of activity until they are exercising 20 to 30 minutes 3 to 4 days per week.[70, 71]

Psychological counseling and cognitive behavioral therapy have been shown to decrease the severity of pain and fatigue, assist in alleviating depression and anxiety, and improve the ability to cope with stressful life events. Fibromyalgia patients have reported improvement in symptoms from instruction in meditation and relaxation techniques and participation in stress management workshops.[70, 71] Some patients report improvement of symptoms from acupuncture, chiropractic spinal manipulation, massage, hypnotherapy, biofeedback, ultrasound treatments, and **balneotherapy**.[71, 73]

Intensive patient education about fibromyalgia and the various available treatment approaches has been shown effective at improving the quality of sleep, increasing adherence to a regular exercise program, reducing pain, reducing depression and anxiety, and improving coping skills and the overall quality of life.[71, 74, 75] An effective approach to patient education has been a group format involving lectures, distribution of educational literature, group discussions, and demonstrations.[71] Participation in a multidisciplinary treatment program involving group exercise sessions, group pain and stress management lectures, dietary education lectures, and massage therapy sessions resulted in improvements in pain, depression, and self-perceived health status compared to fibromyalgia patients who did not participate in the program.[72]

Nutrition and Fibromyalgia

Although some fibromyalgia patients report symptom improvement after eating or avoiding certain foods, there is no consensus within the scientific community that the condition is significantly improved by any specific dietary practice, in large part due to the absence of rigorous randomized controlled trials on the effect of diet on fibromyalgia.[69] In one relatively small, short-term observational study, 18 fibromyalgia patients who followed a low-sodium, uncooked vegan diet for three months reported an improvement in pain, sleep quality, joint stiffness, and self-perceived health status compared to a group of 15 other fibromyalgia patients who followed their conventional diet.[76] Uncooked vegan diets include liberal quantities of berries, fruits, vegetables, nuts, germinated grains, and sprouts, all of which are sources of numerous vitamins, minerals, dietary

fibers, and phytochemicals.[77] Because this diet has a very low caloric density, it promotes weight loss in overweight persons, which could explain some of the reported improvements in symptoms.[76]

As discussed earlier in this chapter, studying the relationship between diet and fibromyalgia and other musculoskeletal conditions is complicated by the cyclic nature of the symptoms patients experience; in other words, patients experience periods of remission followed by periods of exacerbation. When an improvement in symptoms coincides with a dietary change, one must ask whether the change in symptoms is due to the dietary change or to the patient experiencing a transient remission. The psychosomatic nature of fibromyalgia and the placebo effect must be considered as well. Often, patients will report an improvement in symptoms because they believe or hope that a dietary change will be helpful. Many of the prevalent ideas about the influence of diet on fibromyalgia and other musculoskeletal conditions are based on anecdotal evidence instead of rigorously designed, randomized controlled trials.[73] The common notion that monosodium glutamate (MSG) in the diet can exacerbate pain in persons with fibromyalgia is based on anecdotal evidence and is not supported by sound scientific research.[78]

Our current knowledge of the role of diet in treating fibromyalgia and other musculoskeletal conditions is limited, and there is much that is yet to be learned in this area. It is important to keep an open mind and to actively research these questions. However, the dietitian should avoid giving a patient hope of relief from symptoms or hope of a cure by recommending some dietary regimen if there is no sound scientific evidence supporting both the safety and efficacy of that dietary approach.

Conclusion

Historically, the role of nutrition has been long recognized in the development of both rickets and osteomalacia. Today, there is increasing concern regarding the impact of osteoporosis on both health care expenditures and overall quality of life within the population of the United States and Canada. This chapter has focused on the role of nutrition therapy in both treatment and prevention of this most common bone disease. New technology and treatment for osteoporosis provide further opportunities for the registered dietitian in impacting the effects of this disease. As research continues, additional roles for nutrition may be identified to assist in both treatment and prevention of other musculoskeletal diseases, including arthritis.

Nancy Duhaime, MS, RD, LD *Clinical Dietitian Bartlett Regional Hospital (BRH), Juneau, Alaska, and Wild-flower Court, a long-term care hospital that contracts with BRH for 20 hours/week*

How long have you been an RD? How long have you worked in long-term care?

I've been an RD for twenty-five years and have worked in long-term care consistently for the past five years. Previously, I ran an elderly nutrition program in Connecticut for three years, and I was program director for an Ameri-Corps program (in Vermont) that helped match community volunteers with elders in their own homes. Although this was not a "nutrition" job, the volunteers played a vital role in getting meals and groceries delivered, cleaning out refrigerators, and even helping with some food preparation. It was a very rewarding job.

Roughly what percentage of your residents have arthritis or osteoporosis? How does this affect their nutritional status?

Approximately 35% of the residents have some type of degenerative joint disease, including gout and osteoarthritis. Many have other comorbidities that may impact on their nutritional status as well. I pay attention to their ability to hold a utensil and feed themselves. To gain insight into how arthritis, especially in the hands, can impact nutrition, "buddy" tape your thumb to the index finger so you cannot use it. Now try eating a meal using utensils, especially soup! We had a resident who had arthritis along with COPD. When he first came to live at the home, he would come to the common dining area, but he wouldn't eat the food put in front of him. He complained of the smell or look. As the RD, I was called in to talk with him regarding meal preferences. As I spoke with him he disclosed that he didn't like people to see him eat because he had a difficult time holding the utensils and made a mess all over his clothing. We put together a care plan that allowed him to eat in his room, with assistance as needed, as well as eat in the common dining room when "finger foods" were on the menu, for example a sandwich (hamburger) that could be cut up and held in his hand. His food intake improved and he still had times when he came out for some of the social aspects of eating.

Are there common nutritional problems associated with arthritis?

The effects of medications on the stomach and appetite are the biggest common problems I have seen. NSAIDS for pain relief are hard on the GI tract. Also, some of the treatments for osteoporosis are hard on the gut. Fosamax, for example, requires one to sit upright for at least 30 minutes after taking it and to take it 30 minutes before eating (not to mention the risk for esophageal erosion if not done properly). It's quite a "dance" between nurse's aides who get folks up and out of bed in the morning and medication nurses who are on a time schedule to provide prescribed medications; and then the cooks have a scheduled time for breakfast—if Mrs. Smith doesn't get up early so that the medication nurse can give her the Fosamax in a timely manner, she gets a cold breakfast when everyone else gets a hot one. Also, if posture is compromised, as it usually is with osteoporosis, the slow compression of the spine leads to more pressure on the internal organs, especially the stomach, causing early satiety. Additionally, it can impact on regular bowel movements, which can also decrease food intake.

So many people are involved in one person's care; it takes a great deal of teamwork for it to all fall into place. I always try to find something positive to say to the nurse's aides, nurses, or food service staff to let them know they are all part of a very important team. If they all work together, it's fewer nutritional problems for the resident, and in the long run, less work for them.

What is the biggest challenge working with patients with arthritis?

Because they move slowly and are in pain, they usually need some assistance walking or getting to the toilet. It's imperative that all staff working with these folks have patience and understanding. If you have to ask someone to take you to the bathroom every time you feel the urge and that person is busy and speaks in harsh tones while telling you "Just a minute!" you don't want to eat or drink because it's too uncomfortable to get help to use the toilet. As an RD, I have a lot of things to do, and I have to keep reminding myself that I must be available when the resident is available mentally and physically.

Any advice for dietetic students about counseling clients with arthritis?

Taking time to build rapport is so important. Constant pain is not pleasant, and many of these residents have been in constant pain for years. They learn to live with it, but usually upon the first meeting, they can be grouchy, mean, insulting, and not much fun to visit. However, I have come to understand that because of their constant pain, they strike out but don't intend to be mean or insulting. Take a gentle, caring approach and validate that they are in pain. Ask if it would be better to come back at another time. Doing that often softens them to a point where their "kind" self shines through. We all have bad days and we very likely snap at someone ourselves sometimes. I give the resident a chance to know that they have snapped because of their pain. Then I offer them an opportunity to start over again. It's worked every time.

Any advice for dietetic students?

Well, it's been twenty-five years since my internship, but I remember feeling overwhelmed—yet I've felt that more recently as well. I was out of the field of dietetics for over twelve years, and five years ago I took the plunge to return to clinical dietetics. I was very scared. How well would I remember all of the information I had learned in school some twenty-five years ago—and what would I do about the fact that some things had changed since that time? Well, I'm happy to say I survived, just like the students will survive their first experiences. What I found helpful was being able to communicate with people in a friendly way, making sure they knew that I was interested in what they were saying, and when I didn't have an answer to their questions, I would reassure them that I would get back to them. (I can't tell you how many times I had to run to the nearest diet manual to look things up just to be sure I was remembering them correctly.)

I was even scared of the doctors, and my worst fears came true one day (but my good training came through for me!). One day, in front of the patient, a doctor asked me a very detailed question about the patient's insulin dose and carbohydrates intake. I knew it was something I should have had an immediate answer for, but I didn't. I had to admit to that I did not know the answer, but quickly followed with the statement, "I need to check a reference to be sure I'm accurate." They were fine with this, and I was able to look up the information and provide the answer some time later. I keep in mind that if I were the patient in the bed, I'd feel better knowing that someone was checking their own accuracy rather than trying to look like they know it all.

Application of the Nutrition Care Process: Musculoskeletal Disease

Introduction:

CC: Presents to emergency room after fall on ice—diagnosed with fracture of her right wrist. LL is a 61-year-old white female. She is a sixth-grade teacher, married with three college-aged children. This patient was referred for a DXA (dual energy x-ray absorptiometry) scan to evaluate her bone mineral density by the orthopedic surgeon who treated her wrist fracture. She was diagnosed with osteopenia as her BMD is 2.0 standard deviations below the mean.

Past Med Hx: Has not had menstrual cycle for past year; no other contributory history.

Family History: Grandmother and 2 great aunts have osteoporosis. Her mother died at an early age.

1. What is osteopenia?

2. What risk factors does this patient present with?

3. What additional questions would be important for the medical team to establish in evaluating her diagnosis?

Nutrition Assessment:

Food/Nutrition-Related History:

Reports good appetite; eats a variety of foods but states that she has routinely tried to lose weight, which means she has restricted kcal off and on for previous five years. Calcium intake is from yogurt 3–4 times per week and the amount of milk on her breakfast cereal. She states that she has purposefully stopped eating cheese in order to help her lose weight.

Anthropometric Measurements:

Ht: 5'6" Wt: 172 lbs. Usual body weight: 160–180 lbs. for past five years.

Biochemical Data:

5 (OH) D Level 11 ng/ml; 21 Mol/L All other labs WNL.

4. Why would vitamin D be measured? What form of vitamin D is recommended to be evaluated? Why?

Nutrition Diagnosis:

5. Identify at least two nutrition problems based on the nutrition assessment and medical history. Determine the diagnostic term for each nutrition problem. Next, identify the etiology of each nutrition problem. Finally, identify the signs and symptoms that support the evidence for these nutrition problems.

Nutrition Intervention:

6. Identify the nutrition prescription for this patient by recommending the appropriate dietary modification for her diagnosis.

7. Recommend energy and protein requirements for LL.

8. What level of calcium supplementation should be initiated for this patient?

9. Discuss the importance of vitamin D to calcium absorption. Should she begin vitamin D supplementation also?

Nutrition Monitoring and Evaluation:

10. Determine nutrition criteria for monitoring and evaluation for each nutrition diagnosis that you identified.

11. What other medications might her physician consider to prevent osteoporosis in this patient?

National Osteoporosis Foundation (NOF) The NOF is a voluntary health organization based in Washington, DC, that promotes education, awareness, training, and research on osteoporosis and bone health. It is a resource for information on osteoporosis for the general public and health professionals.
http://www.nof.org

National Institute of Arthritis and Musculoskeletal and Skin Diseases (NIAMS) The NIAMS, a division of the National Institutes of Health, supports research into the causes, treatment, and prevention of arthritis and musculoskeletal and skin diseases. It is a valuable source of authoritative and scientifically accurate information on various musculoskeletal diseases.
http://www.niams.nih.gov

Osteoporosis Canada Osteoporosis Canada is a not-for-profit foundation serving people who are at risk of or who have osteoporosis and working with individuals and communities in the prevention and treatment of osteoporosis. It is a source of scientifically sound information on osteoporosis for the general public and for health professionals.
http://www.osteoporosis.ca

International Society for Clinical Densitometry (ISCD) The ISCD is a multidisciplinary, not-for-profit organization that provides a central resource for a number of scientific disciplines with an interest in bone mass measurement. It is a source of scientifically sound information on bone mass measurement.
http://www.iscd.org

International Osteoporosis Foundation (IOF) The IOF is an international nongovernmental organization based in Berne, Switzerland, whose mission is to advance the understanding of osteoporosis and to promote prevention, diagnosis, and treatment of the disease worldwide. It is a source of scientifically sound information on osteoporosis for the general public and for health professionals, presented from an international perspective.
http://www.iofbonehealth.org/

The Mayo Clinic Fibromyalgia. A medical site providing an excellent overview of fibromyalgia and current research.
http://www.mayoclinic.com/health/fibromyalgia/DS00079

National Fibromyalgia Association A nonprofit site that provides information for the layperson and links to current research.
http://www.fmaware.org/site/PageServer.

END-OF-CHAPTER QUESTIONS

1. Describe osseous tissue and its composition.

2. List the four principal types of cells found in osseous tissue and describe their functions.

3. Briefly describe the hormonal control of bone metabolism.

4. What is osteoporosis? List risk factors for this disorder. Describe the diagnostic measurement of bone mineral density (BMD). Is it an appropriate measure of osteoporosis?

5. Describe the dietary and pharmacologic prevention and treatment of osteoporosis.

6. Describe Paget disease and its treatment.

7. What are the similarities and differences between osteoarthritis, rheumatoid arthritis, and gout? Describe the dietary and pharmacologic treatments for these disorders.

Complementary and Alternative Medicine Remedies Used in Diseases of the Musculoskeletal System

Remedy	Scientific Name	CAM Use	Side Effects and/or Risks
Amalaki, amla; Indian gooseberry	*Emblica officinalis*	Treat osteoporosis	Adverse reactions rare.
Ashwagandha; winter cherry, Indian ginseng	*Withania somnifera*	Treat arthritis, rheumatism, paralysis	May induce abortion. May potentiate the sedative effect of barbiturates.
Beta-carotene		Reduce pain in osteoarthritis	No adverse reactions reported.
Bloodroot, red root, red puccoon	*Sanguinaria Canadensis*	Treat rheumatism	Avoid use in those with glaucoma. Dizziness, vertigo, nausea, vomiting, skin irritation, esophageal burning, burning of the gums, systemic burning, oral leukoplakia.
Boron		Improve bone density	Theoretical increase in serum estrogen when taken with estrogenic drugs.
Cat's claw; gambir; gao teng	*Uncaria rhynchophylla*	Treat arthritis	May cause diarrhea and lower blood pressure.

Complementary and Alternative Medicine Remedies Used in Diseases of the Musculoskeletal System (*Continued*)

Remedy	Scientific Name	CAM Use	Side Effects and/or Risks
Castor oil plant; eranda, vatari	*Ricinus communis*	Treat arthritis, rheumatism, back pain	Avoid use if pregnant or breast-feeding. Can cause malabsorption of fat-soluble vitamins, fluid, & electrolytes. Prolonged use can result in laxative dependency.
Chondroitin		Treat osteoarthritis	No reported interactions.
Chromium		Increase lean muscle	Long-term effects of high doses and increased cellular concentrations not known.
Coenzyme Q$_{10}$		Alleviate fibromylagia symptoms; treat Parkinsonism	Anorexia, diarrhea, epigastric discomfort, mild nausea
DHEA (dehydro-epiandrosterone)		Alleviate fibromylagia symptoms	Elevated liver test values, slightly decreased hemoglobin levels and RBC count, hirsutism, gynecomastia. Contraindicated in patients with benign prostatic hyperplasia, estrogen-responsive tumors, or prostate cancer. Women should not use while pregnant or breast-feeding. May interact with the following medications: alprazolam, calcium channel blockers, carbamazepine, dexamethasone, phenytoin, danazol, insulin, and insulin-sensitizing drugs.
Dong quai; angelica	*Angelica sinensis*	Treat arthritis	Should not be combined with blood-thinning medications (warfarin, heparin, aspirin) or supplements (garlic, ginkgo, vitamin E, fish oil).
Evening primrose oil	*Oenothera biennis*	Treat osteoarthritis	Should not be combined with blood-thinning medications (warfarin, heparin, aspirin) or supplements (garlic, ginkgo, vitamin E, fish oil).
Glucosamine		Treat osteoarthritis; alleviate back, joint pain	No reported interactions.
Guggula; bedellium	*Commiphora mukal*	Treat arthritis, rheumatism, gout	Headache, mild nausea, eructation, hiccough, and loose stools. Hypersensitivity rash.
Grape seed extract	*Vitis vinifera, V. coignetiae*	Alleviate fibromylagia symptoms	None.
Kava kava, 'awa	*Piper methysticum*	Treat back pain	Changes in motor reflexes and judgment, vision changes, hypertension. Alcohol increases kava toxicity. Administration with alprazolam may result in coma. Causes additive sedative effects when taken with CNS depressants. Large doses of kava exert digitalis-like effect. Avoid administration with levodopa as kava increases Parkinsonian symptoms. Do not administer with pentobarbital, may have additive effects.
Malabar nut: adhosa, vasaka	*Adhatoda vasika*	Treat rheumatism	Inadequate information available regarding potential side effects.
Neem; nimb	*Azadirachta india*	Treat arthritis, rheumatism	None known; no full scientific evaluation of toxicity and side effects has been conducted.
Poke, inkberry; fiolaca	*Phytolacca Americana*	Treat arthritis, rheumatism, bursitis	Nausea, vomiting, stomach cramps, diarrhea, weakness, hematemesis, hypotension, tachycardia. No study supports the use of pokeweed for any proposed claim.
Potassium		Prevent muscle cramps	May produce hyperkalemia when taken in combination with ACE inhibitors (aptopril, enalapril). Do not use with potassium-sparing diuretics.
Selenium		Treat osteoarthritis	May reduce efficacy of statin medications (HMG-CoA-reductase inhibitors).
St. John's wort	*Hypericum perforatum*	Alleviate fibromylagia symptoms	Avoid administration with alcohol, MAOIs, narcotics, OTC cold & flu medications, digoxin, drugs metabolized by CYP3A, indinavir, amphetamines, SSRIs, tricyclic antidepressants. May result in phototoxicity. May influence hepatic microsomal enzymes.

(continued)

Complementary and Alternative Medicine Remedies Used in Diseases of the Musculoskeletal System (*Continued*)

Remedy	Scientific Name	CAM Use	Side Effects and/or Risks
Turmeric; haridra	*Curcuma longa*	Treat arthritis	Avoid administration with alcohol, MAOIs, narcotics, OTC cold & flu medications, digoxin, drugs metabolized by CYP3A, indinavir, amphetamines, SSRIs, tricyclic antidepressants. May result in phototoxicity. May influence hepatic microsomal enzymes.
Vitamin A		Treat osteoarthritis	Toxic in doses 10 times RDA (900 mcg/d for men; 700 mcg/d for women). HMG-CoA-reductase inhibitors and oral contraceptives are associated with increased blood levels of vitamin A. Bile acid sequestrants, mineral oil, orlistat, and neomycin may reduce vitamin A absorption. Antacids may interfere with vitamin A and beta-carotene absorption.
Willow	*Salix* spp.	Treat osteoarthritis, rheumatism	Increases risk of bleeding when taken with anticoagulants. May reduce effectiveness of diuretics. May increase risk of GI ulceration and bleeding when taken with NSAIDs. Contraindicated in patients with salicylate hypersensitivity. Should be avoided by pregnant and breast-feeding women.
Zinc		Treat osteoarthritis	Supplements may interfere with absorption of copper and iron. Zinc and penicillin bind, interfering with absorption of both. May bind with warfarin. May increase toxicity of cisplatin.

Sources: Fetrow CW, Avila JR. *Professional's Handbook of Complementary and Alternative Medicines*, 3rd ed. Philadelphia: Lippincott Williams & Wilkins, 2004.
Bratman S, Girman AM. *Mosby's Handbook of Herbs and Supplements and their Therapeutic Uses.* St. Louis: Mosby/Elsevier, 2003.
Fragakis AS, Thomson C. *The Health Professional's Guide to Popular Dietary Supplements*, 3rd ed. Chicago: American Dietetic Association, 2007.
Memorial Sloan-Kettering Cancer Center. About Herbs, Botanicals & Other Products. Available at http://www.mskcc.org/mskcc/html/11570.cfm

REFERENCES

1. Porth CM. Structure and function of the musculoskeletal system. In Porth CM, Matfin G (eds.). Pathophysiology: concepts of altered health status. 8th ed. Philadelphia (PA): Lippincott Williams & Wilkins; 2009, 1454–64.

2. Heaney RP. Bone biology in health and disease. In Shils ME, Shike M, Ross AC, Cabellero B, Cousins RJ (eds.). Modern nutrition in health and disease. 10th ed. Philadelphia (PA): Lippincott Williams & Wilkins; 2006, 1314–25.

3. Saladin KS. Anatomy and physiology: the unity of form and function. 4th ed. Boston (MA): McGraw-Hill; 2007.

4. Felson DT. Osteoarthritis. In Fauci AS, Braunwald E, Kasper DL, Hauser SL, Longo DL, Jameson JL, Loscalzo J (eds.). Harrison's principles of internal medicine. 17th ed. New York: McGraw-Hill; 2008, 2158–65.

5. Gunta KE. Disorders of musculoskeletal function: developmental and metabolic disorders. In Porth CM, Matfin G. (eds.). Pathophysiology: Concepts of Altered Health States. 8th ed. Philadelphia (PA): Lippincott Williams & Wilkins; 2009, 1493–1518.

6. Bringhurst FR, Demay MB, Krane SM, Kronenberg HM. Bone and mineral metabolism in health and disease. In Fauci AS, Braunwald E, Kasper DL, Hauser SL, Longo DL, Jameson JL, Loscalzo J (eds.). Harrison's Principles of Internal Medicine, 17th ed. New York: McGraw-Hill; 2008, 2365–77.

7. Lindsay R, Cosman F. Osteoporosis. In Fauci AS, Braunwald E, Kasper DL, Hauser SL, Longo DL, Jameson JL, Loscalzo J (eds.), Harrison's principles of internal medicine. 17th ed. New York: McGraw-Hill; 2008, 2397–408.

8. Standing Committee on the Scientific Evaluation of Dietary Reference Intakes, Food and Nutrition Board, Institute of Medicine. Dietary Reference Intakes for Calcium, Phosphorus, Magnesium, Vitamin D, and Fluoride. Washington (DC): National Academy Press, 1997.

9. Norman AW, Henry HH. Vitamin D. In Bowman BA, Russell RM (eds.). Present knowledge in nutrition. 9th ed. Washington (DC): International Life Sciences Institute Press, 2006, 198–210.

10. Calvo MS, Whiting SJ. Prevalence of vitamin D insufficiency in Canada and the United States: importance to health status and efficacy of current food fortification and dietary supplement use. Nutr Rev. 2003;61:107–13.

11. National Osteoporosis Foundation. Physician's guide to prevention and treatment of osteoporosis. Washington (DC): National Osteoporosis Foundation; 2003.

12. Kirk D, Fish SA. Medical management of osteoporosis. Am J Manag Care. 2004;10:445–55.

13. National Institutes of Health. Osteoporosis Prevention, Diagnosis, and Therapy. NIH consensus statements. 2000;17(1):1–52. Available at http://consensus.nih.gov/cons/111/111_statement.pdf

14. Brown JP, Josse RG. 2002 clinical practice guidelines for the diagnosis and management of osteoporosis in Canada. CMAJ. 2002;167(10 suppl):S1–S34.

15. Writing Group for the ISCD Position Development Conference. Diagnosis of osteoporosis in men, premenopausal women, and children. J Clin Densitom. 2004;7:17–26.

16. Writing Group for the ISCD Position Development Conference. Nomenclature and decimal places in bone densitometry. J Clin Densitom. 2004;7:45–49.

17. Theodorou SJ, Theodorou DJ, Sartoris DJ. Evaluation of osteoporosis in orthopedic practice: a review of current diagnostic modalities. Am J Orthop. 2003;32:178–88.

18. Dawson-Hughes B. Osteoporosis. In Shils ME, Shike M, Ross AC, Cabellero B, Cousins RJ (eds.), Modern nutrition in health and disease. 10th ed. Philadelphia (PA): Lippincott Williams & Wilkins; 2006,1339–52.

19. Richmond B. DXA scanning to diagnose osteoporosis: do you know what the results mean? Cleve Clin J Med. 2003;70:353–60.

20. Brown SA, Rosen CJ. Osteoporosis. Med Clin N Am. 2003;87:1039–63.

21. U.S. Department of Health and Human Services. Bone Health and Osteoporosis: A Report of the Surgeon General. Rockville (MD): U.S. Department of Health and Human Services, Office of the Surgeon General; 2004.

22. Prentice A. Diet, nutrition and the prevention of osteoporosis. Public Health Nutr. 2004;7:227–43.

23. Dargent-Molina P, Piault S, Breart G. A comparison of different screening strategies to identify elderly women at high risk of hip fracture: results of the EPIDOS prospective study. Osteoporosis Int. 2003;14:969–77.

24. Rothenberg RJ, Boyd JL, Holcomb JP. Quantitative ultrasound of the calcaneus as a screening tool to detect osteoporosis. J Clin Densitom. 2004;7:101–10.

25. Mazanec D. Osteoporosis screening: time to take responsibility. Arch Intern Med. 2004;164:1047–48.

26. Gallagher TC, Geline O, Comite F. Missed opportunities for prevention of osteoporotic fracture. Arch Intern Med. 2002;162:450–56.

27. Harrington JT, Broy SB, Derosa AM, Licata AA, Shewmon DA. Hip fracture patients are not treated for osteoporosis: a call to action. Arthritis Rheum. 2002;47:651–54.

28. Olszynski WP, Davison KS, Adachi JD, Brown JP, Cummings SR, et. al. Osteoporosis in men: epidemiology, diagnosis, prevention and treatment. Clin Ther. 2004;26:15–28.

29. Black DM, Steinbuch M, Palermo L, Dargent-Molina P, et al. An assessment tool for predicting fracture risk in post-menopausal women. Osteoporosis Int. 2001;12:519–528.

30. Cheung AM, Feig DS, Kapral M, Diaz-Grandos N, Dodin S. Prevention of osteoporosis and osteoporotic fractures in postmenopausal women: recommendation statement from the Canadian Task Force on Preventive Health Care. CMAJ. 2004;170:1665–67.

31. Braithwaite RS, Col NF, Wong JB. Estimating hip fracture morbidity, mortality, and costs. J Am Geriatr Soc. 2003;51:364–70.

32. Feskanich D, Willett W, Colditz G. Walking and leisure-time activity and risk of hip fracture in postmenopausal women. JAMA. 2002;288:2300–06.

33. Holick MF. Vitamin D. In Shils ME, Shike M, Ross AC, Cabellero B, Cousins RI, editors. Modern nutrition in health and disease. 10th ed. Philadelphia (PA): Lippincott Williams & Wilkins; 2006, 376–95.

34. U.S. Department of Health and Human Services. The Health Consequences of Smoking: A Report of the Surgeon General. Rockville (MD): U.S. Department of Health and Human Services, Office of the Surgeon General; 2004.

35. U.S. Department of Health and Human Services and U.S. Department of Agriculture. Dietary Guidelines for Americans, 2005. 6th ed. Washington (DC): U.S. Government Printing Office; January 2005.

36. Atkinson SA, Ward WE. Clinical nutrition 2: the role of nutrition in the prevention and treatment of adult osteoporosis. CMAJ. 2001;165:1511–14.

37. Tucker KL, Hannan MT, Chen H, Cupples LA, Wilson PWF, Kiel DP. Potassium, magnesium, and fruit and vegetable intakes are associated with greater bone mineral density in elderly men and women. Am J Clin Nutr. 1999;69:727–36.

38. Reginster JY. Prevention of post-menopausal osteoporosis with pharmacological therapy: practice and possibilities. J Intern Med. 2004;255:615–28.

39. Rossouw JE, Anderson GL, Prentice RL, et al. Risks and benefits of estrogen plus progestin in healthy postmenopausal women: principal results from the Women Health Initiative randomized controlled trial. JAMA. 2002;288:321–33.

40. Favus MJ, Vokes TJ. Paget disease and other dysplasias of bone. In Fauci AS, Braunwald E, Kasper DL, Hauser SL, Longo DL, Jameson JL, Loscalzo J (eds.), Harrison's Principles of Internal Medicine. 17th ed. New York: McGraw-Hill; 2008, 2408–16.

41. Page C, Curtis M, Sutter M, Walker M, Hoffman B. Integrated Pharmacology. 2nd ed. Edinburgh, Scotland: Mosby International; 2002.

42. Pettifor JM. Nutritional rickets: deficiency of vitamin D, calcium, or both? Am J Clin Nutr. 2004;80(suppl):1725S–1729S.

43. Weisberg P, Scanlon KS, Li R, Cogswell ME. Nutritional rickets among children in the United States: review of cases reported between 1986 and 2003. Am J Clin Nutr. 2000;80(suppl):1697S–1705S.

44. Canadian Paediatric Society, Dietitians of Canada and Health Canada. Nutrition for Healthy Term Infants. Minister of Public Works and Government Services, Ottawa, 1998.

45. Gartner LM, Eidelman AI. American Academy of Pediatrics policy statement: breastfeeding and the use of human milk. Pediatr. 2005;115:496–506.

46. Gartner LM, Greer FR. Prevention of rickets and vitamin D deficiency: new guidelines for vitamin D intake. Pediatr. 2003;111:908–10.

47. Rizzo DB. Disorders of musculoskeletal function: rheumatic disorders. In Porth CM, Matfin G. (eds.) Pathophysiology: Concepts of Altered Health States. 8th ed. Philadelphia (PA): Lippincott Williams & Wilkins, 2009, 1519–43.

48. U.S. Department of Health and Human Services. Prevalence of disabilities and associated health conditions among adults—United States, 1999. Morb Mortal Wkly Rep. 2001;50:120–25.

49. Towheed TE, Anastassiades TP. Glucosamine and chondroitin for treating symptoms of osteoarthritis, evidence is widely touted by incomplete. JAMA. 2000;283:1483–84.

50. Topol EJ. Arthritis medicines and cardiovascular events house of coxibs. JAMA. 2005;293:366–68.

51. McAlindon TE, LaValley MP, Gulin JP, Felson DT. Glucosamine and chondroitin for treatment of osteoarhritis: a systematic quality assessment and meta-analysis. JAMA, 2000;283:1469–75.

52. Biggee BA, McAlindon T. Glucosamine for osteoarthritis: part I, review of the clinical evidence. Med Health R I. 2004;87:176–79.

53. Biggee BA, McAlindon T. Glucosamine for osteoarthritis: part II, biological and metabolic controversies. Med Health R I. 2004;87:180–81.

54. Lipsky PE. Rheumatoid arthritis. In Kasper DL, Braunwald E, Fauci AS, Hauser SL, Longo DL, Jameson JL (eds.). Harrison's principles of internal medicine. 16th ed. New York: McGraw-Hill; 2005, 1968–77.

55. Pattison DJ, Symmons DPM, Young A. Does diet have a role in the aetiology of rheumatoid arthritis? Proc Nutr Soc. 2004;63:137–43.

56. Martin RH. The role of nutrition and diet in rheumatoid arthritis. Proc Nutr Soc. 1998;57:231–34.

57. Rennie KL, Hughes J, Lang R, Jebb SA. Nutritional management of rheumatoid arthritis: a review of the evidence. J Hum Nutr Diett. 2003;16:97–109.

58. McCann K. Nutrition and rheumatoid arthritis. Explore. 2007;3:616–18.

59. Cleland LG, James MJ, Proudman SM. The role of fish oils in the treatment of rheumatoid arthritis. Drugs. 2003;63:845–53.

60. Volker D, Fitzgerald P, Major G, Garg M. Efficacy of fish oil concentrate in the treatment of rheumatoid arthritis. J Rheumatol. 2000;27:2343–346.

61. Adam O, Beringer C, Kless T, Lemmen C, Adam A, Wiseman M, Adam P, Klimmek R, Forth W. Anti-inflammatory effects of a low arachidonic acid diet and fish oil in patients with rheumatoid arthritis. Rheumatol Int. 2003;23:27–36.

62. Calder PC. Session 3: Joint Nutrition Society and Irish Nutrition and Dietetic Institute Symposium on 'Nutrition and autoimmune disease' PUFA, inflammatory processes and rheumatoid arthritis. Proc Nutr Soc. 2008;67:409–18.

63. Kim KY, Schumacher HR, Hunsche E, Wertheimer AI, Kong SX. A literature review of the epidemiology and treatment of acute gout. Clin Ther. 2003;25:1593–617.

64. Schumacher HR, Chen LX. Gout and other crystal-associated arthropathies. In Fauci AS, Braunwald E, Kasper DL, Hauser SL, Longo DL, Jameson JL, Loscalzo J (eds.), Harrison's principles of internal medicine. 17th ed. New York: McGraw-Hill; 2008, 2165–69.

65. Morgan SL, Baggott JE. Nutrition and diet in rheumatic diseases. In Shils ME, Shike M, Ross AC, Cabellero B, Cousins RJ (eds.), Modern Nutrition in Health and Disease, 10th ed. Philadelphia (PA): Lippincott Williams & Wilkins, 2006, 1326–338.

66. Hahn BH. Systemic Lupus Erythematosis. In Fauci AS, Braunwald E, Kasper DL, Hauser SL, Longo DL, Jameson JL, Loscalzo J (eds.), Harrison's principles of internal medicine. 17th ed. New York:McGraw-Hill; 2008, 2075–83.

67. Brown AC. Lupus erythematosus and nutrition: a review of the literature. J Ren Nutr. 2000;10:170–83.

68. Leiba A, Amital H, Gershwin ME, Shoenfeld Y. Diet and lupus. Lupus. 2001;10:246–48.

69. National Institute of Arthritis and Musculoskeletal and Skin Diseases. Questions and Answers about Fibromyalgia. Bethesda (MD): National Institutes of Health; 2004. Available at http://www.niams.nih.gov/hi/topics/fibromyalgia/fibrofs.htm

70. Langford CA, Gilliland BC. Fibromyalgia. In Fauci AS, Braunwald E, Kasper DL, Hauser SL, Longo DL, Jameson JL, Loscalzo J (eds.), Harrison's principles of internal medicine. 17th ed. New York: McGraw-Hill; 2008, 2175–77.

71. Goldenberg DL, Burckhardt C, Crofford L. Management of fibromyalgia syndrome. JAMA. 2004;292:2388–95.

72. Lemstra M, Olszynski WP. The effectiveness of multidisciplinary rehabilitation in the treatment of fibromyalgia: a randomized controlled trial. Clin J Pain. 2005;21:166–74.

73. Holdcraft LC, Assefi N, Buchwald D. Complementary and alternative medicine in fibromyalgia and related syndromes. Best Pract Res Clin Rheumatol. 2003;17:667–83.

74. Nicassio PM, Radojevic V, Weisman MH, Schuman C, Kim J, Schoenfeld-Smith K, Krall T. A comparison of behavioral and educational interventions for fibromyalgia. J Rheumatol. 1997;24:2000–2007.

75. Alamo MM, Moral RR, Pérula de Torres LA. Evaluation of a patient-centered approach in generalized musculoskeletal chronic pain/fibromyalgia patients in primary care. Patient Educ Couns. 2002;48:23–31.

76. Kaartinen K, Lammi K, Hypen M, Nenonen M, Hanninen O, Rauma AL. Vegan diet alleviates fibromyalgia symptoms. Scand J Rheumatol. 2000;29:308–13.

77. Hanninen O, Kaartinen K, Rauma AL, Nenonen M, Torronen R, Kakkinen AS, Adlercreutz H, Laakso J. Antioxidants in vegan diets and rheumatic disorders. Toxicology. 2000;155:45–53.

78. Geenen R, Janssens EL, Jacobs JW, van Staveren W. Hypothesis: dietary glutamate will not affect pain in fibromyalgia. J Rheumatol. 2004;31:785–87.

26

Metabolic Disorders

Elaina Jurecki, MS, RD
Nutritional Specialist
BioMarin Pharmaceutical Inc.
Joyce Wong, MS, RD
Regional Metabolic Nutrition Coordinator
Kaiser Permanente Medical Center, Northern California

CHAPTER OUTLINE

Introduction

Inborn errors of metabolism are a group of diseases that affect a wide variety of metabolic processes. Disease results when there is defective processing or transport of small molecules such as amino acids, fatty acids, metals, or sugars in the body. For some of these small molecules, cells in the body possess enzymes that can catalyze synthesis of a given substrate when it is needed. The classical inborn error of metabolism is caused by a defect in the activity of one of these enzymes. The resulting disorder may lead not only to the accumulation of compounds with harmful effects, but also to deficiencies of substances that are essential for normal growth and development.[1]

Sir Archibald Garrod first postulated that metabolic disorders were inherited biochemical blocks in normal metabolic pathways in 1908. Garrod used the term "inborn errors of metabolism" to describe these disorders of lifetime duration. He hypothesized that specific enzymes were synthesized under the direction of genes. This one-gene-one-enzyme relationship was later proven in 1940 by La Du et al. when they were able to identify the defective or missing enzyme resulting in an abnormal metabolic process in four different disorders. This led to the concept that genes controlled metabolism and that disease states were created by blocks in this metabolic flow that yielded accumulated precursors and deficient products. Garrod's observations were the beginning of human biochemical

alanine shunt—a process which allows for the production of glucose from protein sources by the conversion of the amino acid alanine to pyruvate to glucose in the liver; the glucose produced can then be used in the muscle to generate additional alanine

amylopectin—the insoluble component of starch

ataxia—a nonspecific clinical manifestation implying dysfunction of parts of the nervous system that coordinate movement

autosomal recessive inheritance—inheritance of a trait as the result of inheriting a recessive gene for a particular trait from each parent

binding site—active site of the enzyme that binds to and acts on a particular substrate

carnitine—a substrate needed for the normal metabolism of fat for energy; it is mostly found in beef and lamb but can be synthesized endogenously from the amino acids L-lysine and L-methionine

chaperoning effect—refers to the effect of proteins ("chaperones") that assist the noncovalent folding/unfolding and the assembly/disassembly of other structures, such as enzymes, to help to improve their normal biological functions

cofactors—vitamins or other nutrients needed for the proper function of certain enzymes

dephosphorylate—to remove a phosphate group from an organic molecule

docosahexanoic acid (DHA)—a fatty acid that can be produced from the essential omega-3 fatty acid linolenic acid, and that can also be found in many types of fish; DHA has been found to be beneficial for proper brain and vision development as well as in the management of hypertriglyceridemia

dysphagia—difficulty with swallowing

encephalopathy—a degenerative disease of the brain

euglycemia—maintenance of normal blood sugar levels

feedback inhibition—regulatory mechanism to limit production of certain substrates, which involves the down-regulation of the enzyme involved in the production of the substrate by that substrate when the substrate accumulates a certain level

fibroblasts—connective tissue cells found in the skin

free radicals—compounds formed when the breakage of the electron bonds of a molecule leaves an exposed electron; this electron will attack another molecule for its electron, and this process can continue, eventually leading to cell damage; antioxidants such as vitamins C and E can be helpful in stopping this destructive process

genotype—the mutant or absent gene responsible for a metabolic defect

glucagon—a hormone secreted by the islets of Langerhans (of the pancreas) in response to hypoglycemia, which acts to stimulate glycogenolysis and is used to treat hypoglycemia

gluconeogenisis—the process of synthesizing glucose from fatty acids or glycerol

glycogenolysis—the process of breaking down glycogen to produce glucose

glycolysis—the anaerobic enzymatic conversion of glucose to lactate or pyruvate, which results in the production of energy in the form of ATP

hepatocytes—the cells of the liver

hepatomegaly—an enlarged liver

histological—pertaining to the minute structure, composition, and function of the tissues

inorganic phosphate—a phosphate compound not derived from an organic origin

intralipid—an intravenous fat emulsion used to prevent or correct deficiency of essential fatty acids and to provide energy

ketone bodies—the substances acetone, acetoacetic acid, and beta-hydroxybutyric acid, which are normal metabolic products of lipid metabolism within the liver and are oxidized in the muscle

ketosis—an abnormally elevated concentration of ketone bodies in the body tissues and fluids

medium-chain triglyceride (MCT) oil—synthetic oil containing fatty acids that are 6 to 8 carbons in chain length

metabolic decompensation—an inability to maintain metabolic balance leading to derangements in biochemical and clinical parameters

neuropathy—a general term denoting functional disturbances and/or pathological changes in the peripheral nervous system

oligosaccharide—a carbohydrate which, upon hydrolysis, yields a small number of monosaccharides

osteopenia—a decrease in bone mass to below-normal levels

parenchymal—refers to the essential elements of an organ

periorifacial acrodermatitis—a disease of the skin surrounding the mouth area

phenotype—the presentation (i.e., anatomical, physiological) associated with a specific disorder

pyruvate complex disorders—disorders involving dysfunction in the metabolism of pyruvate, the end product of glycosis, via either the Kreb cycle or gluconeogenesis, resulting in the production of lactic acid

raffinose—a trisaccharide composed of galactose, fructose, and glucose that is found in grains, vegetables, and legumes; humans lack the enzyme to digest this compound to individual sugar molecules

retinopathy—a noninflammatory disease in the retina of the eye

rhabdomyolysis—the disintegration of muscle, associated with excretion of myoglobin in the urine

stachyose—a tetrasaccharide consisting of 2 galactose, 1 fructose, and 1 glucose units

steatosis—excessive amounts of fat found in the stool

substrates—substances that an enzyme acts on to make a product

tandem mass spectroscopy—the methodology used to detect a large number of organic acid compounds on a filter paper blood spot for diagnosing an inborn error of metabolism

tetrahydrobiopterin—a cofactor needed to stabilize the enzyme phenylalanine hydroxylase

---

genetics, and he defined the principle of genetically determined biochemical individuality, noting that "no two individuals of a species are absolutely identical in bodily structure, neither are their chemical processes carried out on exactly the same line."[2] Consequently, there is considerable human variation in the structure and activity of these proteins, and in their role of synthesizing and breaking down compounds, though only a few individuals are so impaired that ingestion of the Dietary Reference Intakes (DRI) of nutrients will create severe disease.[3, 4]

Epidemiology and Inheritance

Over 200 genetic disorders in which there is toxicity, deficiency, or overproduction of normally occurring **substrates** and/or products of metabolic flow have been reported. Most inborn errors of metabolism are inherited through an autosomal recessive transmission of the affected gene, meaning that both parents, who usually have no clinical symptoms, must each pass the defective gene to the offspring for the disease to be manifested (see Chapter 10). With **autosomal recessive inheritance**, carrier parents have a 25% chance of having an affected child with each pregnancy. A few disorders are X-linked, meaning that the defective enzyme is produced from an abnormal gene found on the X chromosome. In these disorders, males are symptomatic since they have only one X chromosome, but females are rarely affected because they have a functioning gene on their other X chromosome.

The mutated or absent gene responsible for the metabolic defect is present from fertilization. This is often referred to as the **genotype** of the disorder. Although present in utero, the effects of the disorder may not manifest immediately. Two individuals can be affected with the same metabolic disorder, but experience different degrees of severity and time of onset. A less severe enzymatic defect may allow an individual to function adequately for years until a stressor such as infection, dehydration, increased protein intake, or a growth spurt results in an accumulation of toxic levels of the substance. The variation in the presentation of the disorder is often referred to in terms of its **phenotype**. A particular disorder may have a phenotypical variation that ranges from a severe neonatal presentation to a diet- or vitamin-responsive form to an adult onset form. In some situations, the phenotype of the disorder is closely associated with the genotype, meaning that certain mutations are known to result in a typical presentation. In other disorders, there is little genotype-phenotype correlation resulting in a wide variation of presentations in affected individuals.

Although the incidence of each individual metabolic disorder is very rare, cumulatively it is reported that they occur as frequently as one in every 2,500 births.[5] With current diagnostic methods and mass screening programs, practitioners are finding many of these disorders to be significantly more common than previously believed. The diagnosis of an inborn error of metabolism is devastating to a family, for it represents not only an inherited disorder ("something I gave to my child"), but also a chronic disorder, often associated with a shortened life span. Despite major diagnostic achievements, patients and their families are often disappointed by the absence of specific therapies and by the investigational nature of many of the treatment options. Although diet modifications can alleviate the manifestations of many of these disorders, many patients still suffer irreversible damage before a diagnosis is made. Optimal management of these disorders depends on identifying affected subjects while they are presymptomatic or before irreversible disease has occurred.[6]

Pathophysiology of Impaired Metabolism

Impaired metabolism of nutrients can occur as a direct result of deficient or absent enzyme activity. It can also be caused by a defective gene that results in a change to the **binding site** of **cofactors** that are needed for a specific enzyme to function properly. Figure 26.1 represents a typical enzymatic pathway, where compound A is converted to D, with intermediates B and C. The arrow leading from D back to A signifies that an accumulation of D results in inhibition of the pathway. The X indicates that enzyme 3, one of three enzymes required in this pathway, is absent or not fully functional. As a result of the defective enzyme 3, the precursors, both immediate (B) and

distant (A), are unable to be metabolized and can build up. The precursors can accumulate as a direct result of the block or as a result of impaired **feedback inhibition** related to the inability to produce the end substrate. Toxic metabolites can be produced from the precursors that are built up as a result of the blocked pathway (enzyme 3).

In other disorders, the enzymatic block can result in a deficiency of much-needed end products. In Figure 26.1, the absence of enzyme 3 leads to the inability to produce substrate D. The inability to produce vital nutrients such as glucose, essential amino acids, or **ketone bodies** can lead to serious consequences such as hypoglycemia, coma, growth failure, kidney dysfunction, and even death.

Dietary interventions for treating many of these disorders consist of rigid restrictions of protein, fat, or carbohydrates that would in turn limit intake of many micronutrients. Therefore, a secondary consequence of these metabolic defects can be nutritional deficiencies. Nutrition therapy for these disorders is aimed at not only treating the underlying disorder, but also at providing sufficient nutrients to promote adequate growth and development. Frequent assessment of anthropometric, biochemical, and clinical indices is required to ensure optimal care of these patients.[6]

Medical Diagnosis/Newborn Screening

Newborn babies are required to perform myriad complex physiologic tasks. They must assume independent control of body temperature, blood glucose levels, disposal of toxic metabolites, intake and assimilation of nutrients, and many other physical processes. Any significant illness in a newborn impedes these adaptive processes. Metabolic diseases can greatly complicate the clinical picture for ill neonates, because symptoms of one disease may closely resemble those of others that have quite different etiologic and pathogenic mechanisms. Waiting until sepsis and other more common causes of illness are ruled out before initiating a specific diagnostic evaluation is inadvisable, as is indiscriminate study of all ill newborns for metabolic disorders.[7]

The detection of inborn errors of metabolism is complicated by the fact that many symptoms are nonspecific and can be similar from one inborn error of metabolism to the next. Some of the symptoms of these disorders include failure to feed, vomiting, hypotonia, hypertonia, seizures, and lethargy progressing to coma. These symptoms can also be seen in neonates suffering from infection, cardiopulmonary dysfunction, intracranial hemorrhage, congenital structural abnormalities of the brain, or trauma. Because most neonates with an inborn error of metabolism look normal at birth and have no gross physical anomalies, detection of these disorders can be elusive. Screening for these metabolic disorders involves testing for specific metabolites. Elevations or deficiencies of certain substrates can point to the location of the enzymatic block and thus the specific disorder. Nonselective neonatal screening, or newborn screening (NBS), is associated with the screening of all newborns for a limited number of the more common inborn errors of metabolism. The benefit of testing all infants in a population is that treatment can start before the infant has suffered any

Figure 26.1 Schematic of Enzymatic Pathway

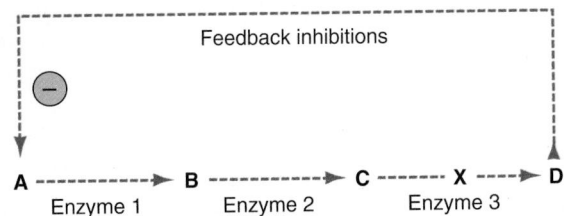

damage as a result of the disease. Selective neonatal screening is the testing of an individual known to be at increased risk for a genetic disorder. Testing of siblings in a family with a child diagnosed with an inborn error of metabolism or confirmatory testing of a positive neonatal screening test are examples of selective screening.[7, 8]

In the United States, all states have screening programs that test newborns for various metabolic disorders between 24 and 48 hours of age. A new testing methodology, **tandem mass spectroscopy**, allows clinicians to screen for over 30 metabolic disorders by analyzing metabolites in a blood spot collected on a filter paper from a simple heel prick on the newborn. This technique is relatively inexpensive and provides the ability to screen for a large number of metabolic conditions pre-symptomatically. Currently, all states routinely screen for the 22 core conditions listed in Table 26.1, but screening for secondary disorders in addition to these varies from state to state.[9] Tandem mass spectroscopy has not only proven successful in identifying many affected individuals prior to the development of symptoms, but it is helping us to learn more about the true incidences of many disorders and to better understand certain conditions.

Establishment of metabolite levels that are diagnostic of a metabolic disease in a newborn can be problematic. The goal is to attempt to identify the majority of infants with an inborn error while minimizing the number of false positives, which can potentially label an infant with a disease that he or she turns out not to have. When cut-off levels are established, these values are used to suggest the presence of an abnormal amount of a metabolite, and when they are set too high, there is the risk of missing an affected infant. On the other hand, since expanded NBS has been implemented in many states, the identification of metabolic diseases that are so mild that the individual may have potentially been completely asymptomatic and led a disease-free life has become more common. This has added an additional burden on the clinicians, as they need to determine which diseases do not warrant prompt

diagnosis and treatment due to their mild nature. An example of this occurred after implementation of NBS to detect methylcrotonylglycinuria, a disorder in leucine (a branched-chain amino acid) metabolism. Many newborns identified with this metabolic disorder were found to have several family members that were unaffected and leading normal lives without treatment. Many states are currently no longer screening for this condition.[10]

Clinical Manifestations of Inborn Errors of Metabolism

Although the use of tandem mass spectroscopy screening permits the identification of many more metabolic disorders than in the past, it is still necessary to be aware of other clues that may indicate the presence of a metabolic defect. Not all disorders are universally screened for, and presentation of these disorders can occur prior to the arrival of the screening test results.

The major and less common clinical manifestations of inborn errors of metabolism are summarized in Table 26.2. These symptoms usually appear 24 hours or more after birth and may be attributed to ingestion of the precursor substrate of the defective enzyme. It is more likely, however, that the primary source of the accumulated precursor is endogenous, resulting from the normal catabolism that accompanies transfer from the intrauterine to the extrauterine environment. This response can also occur later in life as a result of episodes causing endogenous catabolism such as infection, fasting, or fever.

Table 26.1 Metabolic Disorders Routinely Screened for in All 50 States

Amino acid disorders:	Urea cycle disorders:
• Homocystinuria	• Argininosuccinate aciduria
• Maple syrup urine disease	• Citrullinemia type 1
• Phenylketonuria	Biotinidase deficiency
Organic acid disorders:	Galactosemia
• Glutaric acidemia type 1	Fatty acid disorders:
• 3-hydroxy-3-methylglutaric aciduria	• Carnitine update defects
• Isovaleric aciduria	• Long-chain 3-hydroxy acyl CoA dehydrogenase
• 3-methylcrotonyl carboxylase	
• Methylmalonic acidemia	• Medium-chain acyl CoA dehydrogenase
• Cobalamin A and B defects	
• Beta ketothiolase	• Trifunctional protein deficiency
• Propionic acidemia	• Very long-chain acyl CoA dehydrogenase
• Multiple carboxylase	

Source: Adapted from the National Newborn Screening and Genetics Resource Center at the http://genes-r-us.uthscsa.edu/resources/newborn/newborn_menu.htm

Table 26.2 Major Clinical Manifestations of Inborn Errors of Metabolism: Common Signs and Symptoms

Neurologic Signs	Gastrointestinal Signs
• Poor suck	• Poor feeding
• Lethargy (progressing to coma)	• Vomiting
	• Diarrhea
• Abnormalities of tone (hypertonia and hypotonia)	• Reflux
• Loss of reflexes	
• Seizures	

Respiratory Signs	Organ Dysfunction
• Hyperpnea	• Hepatomegaly
• Respiratory failure	• Hepatic dysfunction
• Tachypnea	• Cardiomegaly
• Apnea	• Cardiomyopathy

Rarer Findings	
• Abnormal smell	• Renal stones
• Abnormal hair	• Ectopia lentis (eye lens displacement)
• Self mutilation	• Corneal clouding
• Blood in urine	• Arthritis
• Ataxia	• Dystonia
• Myopathy	• Cirrhosis
• Dysmorphic features	

Sources: Adapted from: Burton BK. Inborn errors of metabolism in infancy: a guide to diagnosis. 1998. *Pediatr.* 102(6):1–15, and Cederbaum SD. Diagnosing metabolic disease in the neonate. 1989. *Metabolic Currents.* 2(1):1–7.

Any neonate or infant who presents with a history of postnatal onset of neurologic dysfunction after an interval of good health should raise the suspicion of a metabolic disorder. Neonates with a metabolic disorder generally show acute central nervous system symptoms, including generalized or partial seizures. Infants and older children with metabolic disorders present with neurologic symptoms that are characteristically intermittent in their severity. The child may show poor growth or failure to thrive, significant developmental delay, and/or specific neurologic deficits such as **ataxia** or visual dysfunction.[8]

Although the symptoms of inborn errors of metabolism are nonspecific and sometimes subtle, a few of these disorders will have characteristic signs. There can be a striking, unusual odor present in the diaper such as burnt sugar, sweaty feet, or a musty odor. These odors can also present in the ear wax and the sweat of the older infant and child. Even without other indications, a mother's observation that she thinks the infant "smells funny" should be a clue to further investigate the potential for an inborn error of metabolism.[4]

The lack of specificity in the clinical manifestations of inborn errors of metabolism puts greater weight on the laboratory evaluation in making the correct diagnosis. Laboratory studies, as identified in Table 26.3, fall into two categories: routine studies available in many clinical laboratories, and specialized studies only done at laboratories set up to do these tests. The routine studies can identify the child with hypoglycemia, a disturbance of acid-base balance, hyperammonemia, or **ketosis**. A sequence of selected screening tests, performed in consultation with a biochemical geneticist, follows the finding of abnormal routine chemistries.

The specialized assays can be done on one or more specimens of serum, plasma, cerebral spinal fluid, and/or urine. A urine organic acid analysis, a plasma and urine amino acid screen, a plasma carnitine profile, and an acylcarnitine profile are sent to laboratories specializing in biochemical genetic disorders. In addition to the biochemical tests, fibroblast and DNA studies can provide additional information about the diagnosis. **Fibroblasts** (skin cells) grown from a skin biopsy or peripheral blood leukocytes can be used to determine the extent of enzymatic

activity present. DNA studies can provide information about the specific mutation(s) involved. Knowledge about the extent of enzymatic activity in addition to any information about the phenotype associated with a particular mutation can be useful in counseling and in managing the treatment of an individual. If the death of a child is imminent, and an inborn error of metabolism is the suspected cause, efforts should be made to obtain post-mortem biological samples for diagnostic purposes. If an inborn error of metabolism is in fact the cause, the ability to make a post-mortem diagnosis can be significant, especially for genetic counseling purposes.[8, 11]

Medical Approaches to Treatment

Acute Medical Therapy

Though some metabolic disorders may appear to be benign, many others cause a variety of severe problems beginning early in life unless an effective treatment is quickly initiated. Any metabolic stress that results in endogenous catabolism, such as an infection or surgery, can lead to the accumulation of toxic metabolites; these metabolites will lead to rapid illness and require acute management. The correction of acid-base status and hydration are essential and of immediate importance. Removal of the accumulated toxic metabolites is of utmost importance to help minimize the deleterious consequences of these compounds. This can be achieved by flushing them out of the body by administering high volumes of intravenous fluids, using medications that can bind and detoxify them as described below, or in more extreme circumstances, initiating dialysis for rapid removal. Patients will also need medications to control fever and antibiotics for management of infection. Those patients who are more acutely ill may also need pressor support (medications to maintain blood pressure), medical assistance for organ impairment, and correction of electrolyte imbalances.

Use of Scavenger Drugs to Remove Toxic By-Products

Pharmaceutical products such as sodium benzoate and sodium phenylbutyrate are needed to remove the excess nitrogen that is produced as a result of a block in the urea cycle. **Carnitine** is used in many organic acid disorders to remove and detoxify waste products that accumulate in disorders affecting the metabolism of proteins, carbohydrates, and fats. (See Table 26.4 for drug-nutrient interaction information for these and other medications/supplements used in the treatment of metabolic disorders.)

Nutrition Therapy for Inborn Errors of Metabolism: General Guidelines

Acute Nutrition Therapy

Maintenance of an adequate caloric intake is necessary for prevention of tissue catabolism and must be achieved by parenteral or oral routes of administration. Offending metabolites, such as protein for urea cycle defects and aminoacidopathies, should be restricted. Restriction of the essential nutrients, such as proteins, can not take place over a prolonged period as the body will become catabolic if deprived for an extended period of

Table 26.3 Routine and Specialized Medical Diagnostic Laboratory Studies for Inborn Errors of Metabolism

Routine Studies	Specialized Studies
• Complete blood count with differential	• Plasma quantitative amino acids
• Urinalysis	• Urine quantitative amino acids
• Blood gases	• Plasma carnitine profile
• Serum electrolytes and blood pH	• Blood acylcarnitine profile
• Blood glucose	• Urine organic acids
• Plasma ammonia	
• Urine reducing substances	
• Urine ketones	
• Blood lactate and pyruvate	

Sources: Adapted from: Burton BK. Inborn errors of metabolism in infancy: a guide to diagnosis. 1998. *Pediatr.* 102(6):1–15, and Cederbaum SD. Diagnosing metabolic disease in the neonate. 1989. *Metabolic Currents.* 2(1):1–7.

Table 26.4 Drug–Nutrient Interactions for Medications/Supplements Used in Metabolic Disorders

Generic	Brand Names	Classification	Mechanism	Possible Food–Drug Interactions
Arginine	None	Nutrient supplement	Anti atherogenic, antioxidant, and immunomodulatory actions; it may also have wound repair activity	Nausea, abdominal cramps, and diarrhea
Biotin	None	Nutrient supplement	Replaces biotin in metabolic conditions resulting in increased excretion	None
Citrulline	None	Nutrient supplement	Provides alternate metabolic pathway by correcting deficiency of urea cycle intermediates	None
Cystadane	Betaine	Antihomocystinuric	Binds with homocysteine to form the amino acid methionine	Diarrhea, nausea, stomach upset
Metronidazole	Flagyl	Bowel disease, inflammatory, suppressant	Decreases bacterial synthesis of propionates in the intestine to aid with maintaining metabolic control	Diarrhea; loss of appetite; nausea or vomiting; stomach pain or cramps; change in taste sensation; dryness of mouth; unpleasant or sharp metallic taste
Levocarnitine	Carnitor	Nutrient supplement	Carnitine binds with toxic organic acids to allow excretion from the body	Abdominal or stomach cramps; diarrhea; headache; nausea or vomiting
Glycine	None	Nutrient supplement	Provides alternate metabolic pathway in urea cycle disorders and in disorders of leucine catabolism	None
Riboflavin	None	Nutrient supplement	Used for cofactor therapy to stimulate residual enzymatic activity	None
Sodium benzoate	Ammonul	Antihyperammonemic	Decreases ammonia production by providing alternative pathway for nitrogen excretion	Dry mouth; increased thirst; loss of appetite; N and V; unusual tiredness or weakness; increased hunger
Sodium phenylbutyrate	Buphenyl	Antihyperammonemic	Decreases ammonia production by providing alternative pathway for nitrogen excretion	Nausea or vomiting; stomach pain; unpleasant taste; changes in taste; decreased appetite; unusual tiredness or weakness
Thiamin (B₁)	None	Nutrient supplement	Used for cofactor therapy to stimulate residual enzymatic activity	↑ kcal intake requires ↑ thiamin supplementation
Alpha-lipoic acid	None	Supplement	Antioxidant and cofactor for respiratory chain	None reported

time. In general, restriction of the offending metabolite should not be extended beyond a 48-hour period. At that point, gradual introduction of the offending essential nutrient should begin to take place in a step wise fashion, i.e., in 10% increments until the final feeding goal is achieved. This does not apply when the offending nutrient is not essential to the individual, such as with fructose in hereditary fructose metabolism disorders, as described in a subsequent section. Long-term management of many inborn errors of metabolism is quite distinctive from the acute management.[3, 12]

The Nutrition Care Process for Inborn Errors of Metabolism

Implementation of the four steps of the nutrition care process (NCP) in the management of inborn errors of metabolism (IEMs) helps the dietitian to identify a standardized approach for nutritional delivery in these conditions. The dietitian plays a key role in the primary treatment for these disorders. Knowing how to apply the International Dietetics and Nutritional Terminology to these diseases can help to provide a uniform method

of documenting the nutritional services critical in the management of these patients. Since most of these conditions are rare, standard treatment guidelines do not exist.

In the first step of the NCP, the nutritional assessment for IEMs will include a food/nutrition-related history, anthropometric measurements, biochemical data including the metabolic labs as well as other medical tests and procedures used, client history, and any nutrition-focused physical findings that can also be a clue to their metabolic stability (see Table 26.5).

The nutrition diagnoses for IEMs will be related to their clinical diagnosis and presentation, because the identity of the enzyme involved and degree of severity of its deficiency directly impact the type of dietary treatment implemented. Food and nutrient intake can play an important part in making the diagnosis; for instance, if a patient has a protein aversion, this could be a clue that the patient has a disorder of protein metabolism. The patient's behavior/environment can provide important information on the type of educational needs that the patient will have and what type of treatment plan will work best in that circumstance.

Table 26.5 Nutrition Assessment for Inborn Errors of Metabolism

Medical/Social History	• Diagnoses/date of diagnosis • Family history/genetic history • Birth/prenatal history • Medications • Previous medical conditions or surgeries • Socioeconomic status/food security • Support systems • Education level—primary language
Food-/Nutrition-Related History	• Ability to suck in infant or chew in older child • Problems swallowing, feeding aversions • Nausea, vomiting, reflux • Behavioral management • Compliance to dietary restrictions • Constipation, diarrhea • Feeding pattern: 24-hour recall and/or 3-day diet record • Nutrition support (i.e., tube feedings, TPN)
Physical Assessment	• Signs and symptoms of nutrient deficiencies or toxicities • Tone • Physical activity level • Signs and symptoms associated with metabolic condition
Anthropometric Measurements	• Height/length • Current weight/birth weight • Weight history if adult: highest adult weight; usual body weight • Body mass index
Biochemical Data, Medical Tests and Procedures	• Albumin • Prealbumin • Complete blood count • General chemistries (i.e., electrolytes, LFTs, BUN, creatinine, etc.) • Newborn screening results • Labs pertinent to metabolic condition (e.g., phenylalanine level, plasma amino acids, ammonia, carnitine profile)

Nutrition intervention, the third step of the NCP, is typically the primary treatment for the IEM, whether it is a protein restriction for an inborn error in amino acid metabolism or frequent carbohydrate feedings in a glycogen storage disorder. This will be described in more detail in subsequent sections. Nutrition education and counseling are critical to the outcome of the patient; hence, close coordination of the nutritional care is essential.

The ongoing monitoring and evaluation of an IEM reveals the effectiveness of the treatment plan in the management of that disease, which ultimately impacts the overall outcome of the patient. Paying close attention to food/nutrition-related history, anthropometric measurements, biochemical data, and ongoing nutrition-focused physical findings is essential in these patients' management. Use of the NCP in IEMs will help the dietitian to provide comprehensive care to these patients. The goals of the NCP help to provide a standard system of management, which

is essential given the rarity of each of these IEMs and the lack of established treatment guidelines to follow.[13]

Chronic Nutrition Therapy

Approaches to the chronic management of metabolic disorders are based on one or more of the therapy options described in this section.[6]

Restriction of Precursors This treatment is aimed at limiting the substrate or substrates prior to the block. In Figure 26.1, the block in enzyme 3 requires the restriction of the precursors A and B. This usually results in restriction of one of the macronutrients. In the case of amino acid disorders, this treatment involves the restriction of dietary protein intake, since proteins found in natural foods contain varying amounts of each amino acid. In order to limit the individual offending amino acid(s), a dietary protein restriction is typically implemented. In disorders of fat metabolism, total fat intake is generally restricted, because all fats will eventually be oxidized to the point where the enzymatic block occurs in the body. In certain disorders of carbohydrate metabolism, sources of foods containing particular monosaccharides, such as fructose, must be restricted.

Replacement of the End Products In Figure 26.1, with the block of enzyme 3, substrate D needs to be replaced. Substrates that should have been formed from the enzymatic reactions are still needed for energy production, protein turnover, feedback inhibition of metabolic pathways, and/or production of other substrates. In the case of metabolic blocks in these enzymatic reactions, an alternative means of achieving the end goal of the reaction must be provided. In glycogen storage disease type I, for example, the enzymatic block results in an inability to produce glucose. Consequently, providing a continuous source of glucose through frequent feedings is the therapy for this disorder. Similarly, in the amino acid disorder homocystinuria, where the amino acid homocysteine is unable to be converted to cysteine, cysteine consequently needs to be supplemented in the diet.

Providing Alternate Substrates for Metabolism In the case of long-chain fatty acid oxidation defects, medium-chain triglycerides (MCT)—fatty acids that are 6 to 8 carbons in chain length—are used as a supplemental energy source because they are able to bypass the metabolic defect and be metabolized for energy in place of the longer-chain fats found in the food supply. In glycogen storage disease type III, protein is used as an alternative source of glucose production, via the **alanine shunt**, since the use of carbohydrates is compromised due to the enzymatic block.

Supplementation of Vitamins or Other Cofactors Use of certain vitamins, at pharmacologic doses, can in certain instances increase the enzymatic activity or facilitate biochemical processes in various inborn errors of metabolism. When an increased amount of a cofactor to the enzyme is present, this may help stimulate residual activity of the enzyme if the mutation or defect is found in the binding site of the cofactor to the enzyme. Additional amounts of a cofactor can also provide for a **chaperoning effect** where the cofactor helps to stabilize the enzyme complex, allowing for an increase in enzyme activity. The body will destroy proteins that look foreign to it, which in many cases include the defective enzyme in an inborn error. The cofactor can help the enzyme protein unfold to look more like the native form, and less foreign to the body, which helps to minimize the amount of enzyme that gets degraded. This ultimately allows for greater enzyme activity as well. This type of chaperoning effect has been suggested as the mode of action for increasing enzyme activity in phenylketonuria (PKU) with use of the cofactor tetrahydrobiopterin.[14] There are forms of methylmalonic acidemia and maple syrup urine disease (MSUD) that are considered to be "vitamin responsive" to the appropriate vitamin cofactors—vitamin B_{12} and thiamin, respectively. Thus, the vitamin is administered in pharmacologic amounts to increase residual enzymatic activity. These disorders are discussed further in subsequent sections.

Several examples of more common inborn errors of metabolism will be discussed in the remainder of the chapter; references are available with more detailed listings of less common metabolic disorders.[15] Examples of disorders to be discussed are amino acid disorders; organic acidemias; urea cycle disorders; mitochondrial disorders; disorders related to vitamin metabolism; disorders of carbohydrate metabolism, including galactosemia; hereditary fructose intolerance; glycogen storage diseases; and disorders of fat metabolism.

Amino Acid Disorders

Epidemiology, Etiology, and Clinical Manifestations

Amino acid disorders include conditions affecting the metabolism of a single amino acid, such as phenylketonuria (PKU) or isovaleric acidemia (IVA), as well as disorders involving multiple amino acids, such as maple syrup urine disease (MSUD). Table 26.6 lists some of the more common disorders of amino acid metabolism and their affected amino acids. Discovery of several of these conditions has been relatively recent, as most disorders were found less than 50 years ago. Incidence of these disorders can vary significantly, with the most common amino acid disorder, PKU, having an incidence of 1 in 10,000 births. The rarer disorders, isovaleric academia or homocystinuria, however, have incidences of less than 1 in 100,000 births. Incidence of these disorders can vary based on ethnicity. Those populations that tend to be more homogenous have been found to have higher rates of certain disorders. For example, MSUD has been detected at an incidence of 1 in 290,000 births in state screening programs. In the Mennonite Community, a homogeneous group in Pennsylvania, there is a much higher prevalence for this condition at 1 in every 790 births.[16]

Treatment recommendations have changed over time and can still vary quite a bit among different parts of the world. This is attributed to the rarity of these conditions limiting patient populations and making it difficult to develop consensus treatment recommendations. With implementation of newborn screening programs, treatments are instituted earlier, with improvement in outcomes being achieved. Many of these patients can now live into adulthood, and it is even possible for some affected women to have successful pregnancies. However, those patients with late treated or untreated amino acid disorders can still have serious complications, including mental retardation, coma, and even death. In general, amino acid disorders are usually characterized by infants who are initially poor feeders and hypotonic. Some of these patients may require long-term gastrostomy feedings due to their poor feeding skills or compromised neurologic status. Many of the milder forms of these disorders may not present in individuals until after the introduction of infant formula or cow's milk, both of which have higher protein content compared to breast milk.

Phenylketonuria

Phenylketonuria (PKU) is the most common amino acid disorder, and the first recognized disorder of metabolism. The defect in phenylalanine hydroxylase results from mutations in the gene coding for this enzyme, which lead to a reduction in or an absence of activity of the enzyme. The deficiency or absence of the phenylalanine hydroxylase enzyme (as shown in Figure 26.2) leads to the inability to convert the essential amino acid phenylalanine (Phe) into the amino acid tyrosine. Consequently, tyrosine becomes a conditionally essential amino acid for individuals with PKU. As a result of the buildup of Phe, phenylacetic and phenylpyruvic acids, which are phenylketones, will also increase—hence the name of this condition, phenylketonuria.

An untreated individual with PKU will likely become mentally retarded with severe behavioral problems, resulting in the need for institutionalization for the majority of her or his life. Untreated PKU can lead to additional neurologic abnormalities, seizures, and eczema. Affected individuals can be found to have a musty or mousy odor because of the accumulation of phenylketones in their urine. Individuals with PKU have been reported to be fairer than other family members due to decreased pigmentation related to the inhibition of the tyrosinase enzyme, needed to make melanin for pigmentation, by phenylalanine. Hyperphenylalaninemia has been found to be toxic to the brain, and demyelination of white matter has been seen on MRI (magnetic resonance imaging) studies of affected individuals. An MRI brain imaging study of 33 early-treated children as compared to age-matched controls found that severe white matter abnormalities were related to the most significant cognitive impairment.[17]

The decreased amounts of serotonin, epinephrine, norepinephrine, and dopamine reported in individuals with PKU are presumably caused by a decrease in their synthesis as a result of the accumulation of Phe, which inhibits the enzymes needed to make these neurotransmitters. Deficiencies of these neurotransmitters have been implicated as the cause of the cognitive, behavioral, and emotional problems found in individuals with this disorder.[18]

Previously, health care providers allowed for the discontinuation of dietary treatment when a child reached school age,

Table 26.6 Amino Acid Disorders and Affected Amino Acid(s)

Metabolic Condition	Affected Amino Acid(s)
Homocystinuria	Methionine
Maple syrup urine disease	Leucine, isoleucine, valine
Phenylketonuria	Phenylalanine
Tyrosinemia	Tyrosine, phenylalanine

Figure 26.2 Enzymatic Defect in Phenylketonuria

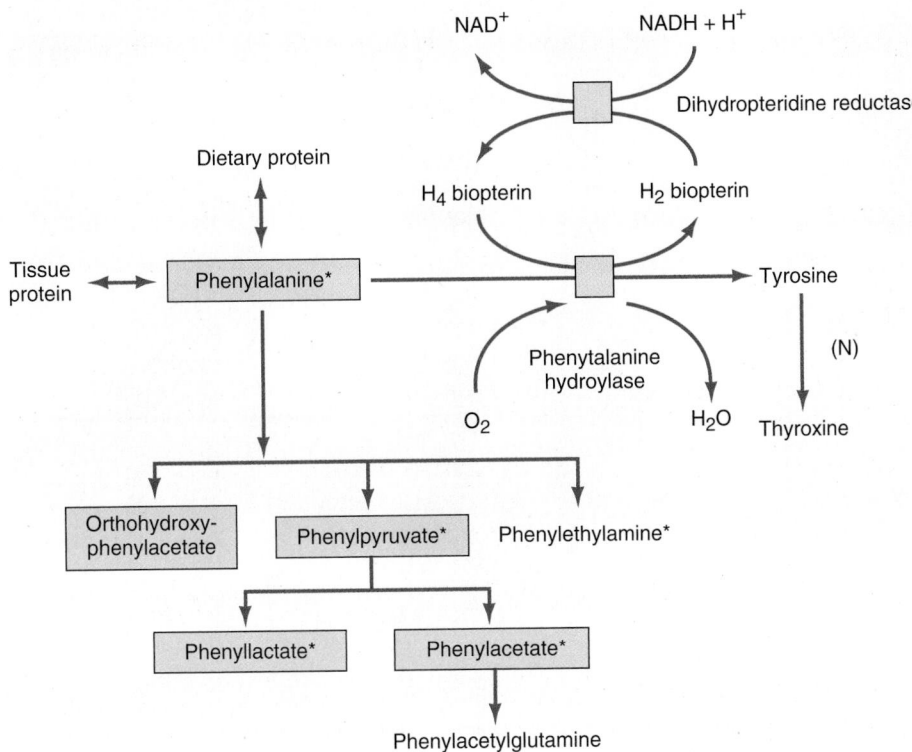

* Accumulates in untreated PKU

(N) = several steps

☐ Indicates sites of possible enzyme defects

Source: Modification of Figure A, p. 1 of Acosta PA, Yannicellis S, *Nutrition Support Protocols*, 4e. Ross Metabolic Formula System; © 2001.

since the brain was believed to have been fully developed, and hence not impacted by the accumulation of Phe. However, a meta-analysis evaluating the relation between blood Phe levels and IQ in 3,361 PKU patients showed that every 100 umol/L increase in blood Phe above the targeted range predicted an average reduction of 1.3 to 3.1 IQ points over the individual's lifetime.[19] A U.S. collaborative follow-up study of 73 early-treated adult PKU patients demonstrated that those that discontinued their diets had lower IQ scores as compared to those who never went off the diet.[20] The higher incidence of neurocognitive dysfunction in children with PKU increases their risk for learning problems in school.[21] Today, most health care providers recommend dietary treatment for life in order to optimize the cognitive and psychosocial abilities of PKU patients, which leads to an improvement in their overall quality of life. Individuals with PKU who are well treated can expect to lead normal and complete lives without a shortened life span.

Saproterin hydrochloride is an oral formulation of the biologically active **tetrahydrobiopterin,** cofactor to the phenylalanine hydroxylase enzyme. Administering sapropterin has been shown to be effective in lowering blood Phe levels in some patients with PKU, by activating residual activity of the PAH enzyme.[22–25] This medication has recently been approved by the Food and Drug Administration for use in the management of PKU. Saproterin has been found to improve tolerance to dietary Phe, thus allowing a greater intake of natural protein.[26, 27] This medication may be a suitable tool to use in addition to the

Phe-restricted diet in the treatment of PKU patients responsive to this form of therapy. Dietary guidelines for the use of sapropterin in the management of PKU have been recently published.[28]

Nutrition Therapy for Amino Acid Disorders

Nutrition Intervention Treatment of amino acid disorders is based on the restriction of dietary protein, which is the source of the offending amino acid(s). Consequently, protein needs must be met by using a synthetic formula containing all the essential amino acids except for the specific offending amino acid(s). The main principle of dietary treatment is to allow the maximum amount of natural protein in the diet, in order to promote growth, while still maintaining laboratory levels of the specific amino acid(s) in the blood within the desired, targeted range. The targeted blood Phe levels for PKU can vary between medical centers, but the National Institute of Health published guidelines in 2000, which recommended targeted blood levels of 120–360 umol/L (2–6 mg/dL) for children less than 12 years of age, and 120–900 umol/L (2–15 mg/dL) for those over 12 years of age.[29]

These guidelines now report that "this statement is more than five years old and is provided solely for historical purposes" as there is increasing evidence that demonstrates cognitive and social-emotional problems developing in early and continuously diet-treated individuals whose blood Phe levels have been allowed to relax later in life. These treatment guidelines need to be revisited to encourage stricter blood Phe control in order to promote more optimal outcomes.[30]

A diet prescription is developed by initially assessing the protein and kcal needs of the patient. Although there are guidelines available for protein requirements by age (see Table 26.7), the amount of protein allowed in the diet is ultimately based on achieving targeted blood amino acid levels. This tends to be based on the level of enzymatic activity determined by the specific gene mutation. Null mutations can lead to no enzyme activity and therefore require a very strict dietary protein restriction, whereas other mutations can lead to more, milder forms of the disorder that allow patients to have a greater protein intake. Frequent monitoring of blood levels after the diagnosis is made is the best way to determine the amount of the specific amino acid or protein to include in the diet. In PKU, blood Phe levels are closely monitored to help maintain metabolic control.

> **Sample PES Statement:** Excessive bioactive substance intake related to an average daily intake of 345 mg phenylalanine (125% of recommended intake) as evidenced by increasing blood levels of phenylalanine and dietary record.

Table 26.7 Protein, Kcal, Phenylalanine, and Tyrosine Recommendations for Children and Adults with PKU

Age	PHE[1,2] (mg/kg)	TYR[1] (mg/kg)	Protein (g/kg)	Energy (kcal/kg)	Fluid[3] (mL/kg)
Infants					
0 to <3 mo	25–70	300–350	3.50–3.00	120 (145–95)	160–135
3 to <6 mo	20–45	300–350	3.50–3.00	120 (145–95)	160–130
6 to <9 mo	15–35	250–300	3.00–2.50	110 (135–80)	145–125
9 to <12 mo	10–35	250–300	3.00–2.50	105 (135–80)	135–120
Age	**(mg/day)**	**(g/day)**	**(g/day)**	**(kcal/day)**	**(mL/day)**
Children					
1 to <4 yr	200–400	1.72–3.00	≥30	1300 (900–1800)	900–1800
4 to <7 yr	210–450	2.25–3.50	≥35	1700 (1300–2300)	1300–2300
7 to <11 yr	220–500	2.55–4.00	≥40	2400 (1650–3300)	1650–3300
Women					
11 to <15 yr	250–750	3.45–5.00	≥50	2200 (1500–3000)	1500–3000
15 to <19 yr	230–700	3.45–5.00	≥55	2100 (1200–3000)	1200–3000
≥19 yr	220–700	3.75–5.00	≥60	2100 (1400–2500)	2100–2500
Men					
11 to <15 yr	225–900	3.38–5.50	≥55	2700 (2000–3700)	2000–3700
15 to <19 yr	295–1100	4.42–6.50	≥65	2800 (2100–3900)	2100–3900
≥19 yr	290–1200	4.35–6.50	≥70	2900 (2000–3300)	2000–3300

[1]Modify prescription based on frequently obtained blood and/or plasma values and growth in infants and children and frequently obtained plasma values and weight maintenance in adults.

[2]PHE requirements of premature infants may be greater than highest value noted.

[3]Under normal circumstances, offer minimum of 1.5 mL fluid to neonates and 1.0 mL to children and adults for each kcal ingested.

Source: Reprinted with permission from: Table 11, p. 12. Acosta PB, Yannicelli S. *Nutrition support protocols.* 4th ed. Columbus (OH): Ross Products Division of Abbott Laboratories; 2001.

Allowing as much protein as possible in the diet from natural sources is important for the promotion of adequate growth and to allow for the greatest variety in a very restrictive diet. Dietary protein is often calculated by closely keeping track of the specific amino acid of concern. In PKU, for example, counting milligrams of Phe is more precise than counting grams of dietary protein. There are many food tables and nutrition software applications available that contain listings of the amino acid content in foods (see Table 26.8). The prescribed amount of amino acid(s) allowed in the diet for children and adults typically comes from fruits, vegetables, and a limited amount of grain products (due to the higher protein content) such as rice, cereals, or crackers. In infants, the allowed amounts of the specific amino acid(s) are provided by standard infant formula and/or breast milk. Although the protein content of breast milk can vary quite a bit between women, and the amount ingested during nursing may be hard to determine, many clinics still encourage mothers of PKU infants to breast feed because of the known health benefits. For more precise quantification, mothers can use expressed breast milk in order to provide an accurate measurement of the breast milk (and subsequent Phe) intake.

After determination of the amount of amino acid(s) (protein) to be supplied by the diet, the balance of the calculated protein needs is provided by a specifically designed metabolic formula or food. The metabolic formula is usually a powder that contains all of the amino acids except for the one(s) that are not able to be metabolized appropriately. The powder formula can contain a carbohydrate source, a fat source that includes essential fatty acids, and vitamins and minerals.

Metabolic formulas and foods containing little carbohydrate or fat have recently been developed in an attempt to minimize the caloric contribution of these products. It is important to determine whether extra kcal are needed to promote satiety and to decrease the intake of foods containing the offending amino acid(s). On the other hand, as more individuals are successfully treated and reaching adulthood, lower caloric needs are indicated due to decreasing rates of growth. Higher-kcal formulas can result in excessive weight gain leading to overweight and obesity in these individuals. Newer formulas and foods that contain little or no vitamins and minerals have also been developed in order to improve the palatability of these products. These products need to be used in conjunction with a vitamin/mineral supplement. Additional calcium may be needed if using an unfortified product, given the restricted use of dairy products and inadequate amount of calcium in a general multiple vitamin/mineral supplement. Metabolic formula bars, gels, drink packs, and tablets are now available for certain disorders in order to meet the demands of the more active, older patients who do not want to prepare a liquid formula each day.

Intake of small amounts of metabolic formula throughout the day provides for the best utilization of the synthetic amino acids. Studies have shown increased nitrogen loss when the formula is consumed 1 to 2 times per day versus 4 to 5 times per day.[31] Although an essential part of the treatment, the coverage of metabolic formulas by various health insurance plans can vary considerably among states.

Calculated protein needs are determined by weight in a growing child and are usually 25% to 30% higher than the standard

Table 26.8 Phenylalanine Content of Selected Foods

Food	Weight (g)	Approximate Measure	PHE (mg)	TYR (mg)	Protein (g)	Energy (kcal)
Fruits						
Orange, fresh whole, peeled	184	1 large	57	29	1.7	86
Peach, fresh whole	175	1 large	33	25	1.6	68
Pear, whole	230	1 large	25	5	0.9	133
Pears, canned juice pack	76	½ with liquid	7	2	0.3	38
Pear, dried halves	18	1 half	9	3	0.3	47
Grapes, American	92	1 cup	12	10	0.6	62
Vegetables						
Carrots, raw	61	1 medium	37	26	0.6	25
Carrots, canned slices	146	1 cup	61	42	0.9	36
Carrots, frozen prepared	146	1 cup	55	38	0.9	54
Broccoli, cooked	78	½ cup	90	47	1.9	27
Cauliflower, cooked	62	½ cup	41	25	1.1	14
Green beans, cooked	125	1 cup	86	55	2.4	44
Breads, Crackers, and Cereals						
Bread, white	25	1 slice	93	55	1.9	66
Saltine crackers	15	5 crackers	68	27	1.4	63
Cream of wheat, regular, prepared with water	33	1 svg (3 tbsp)	187	110	3.5	122
Cheerios®	28	1 cup	165	96	3.2	103
Cinnamon Toast Crunch®	31	¾ cup	73	41	1.6	134

Source: Nutrient data obtained from the USDA National Nutrient Database for Standard Reference at http://www.nal.usda.gov/fnic/foodcomp/search

DRIs for age due to the synthetic nature of the protein in the formula.[32] Although guidelines are available to help with determining protein and amino acid requirements, nutrient needs are best determined by frequent monitoring of growth and laboratory indices.[33, 34] Protein and amino acid requirements taper per unit body weight as the child ages, due to decreased growth velocity. The kcal supplied by the metabolic foods or formulas and the permitted amount of dietary protein are subtracted from the calculated caloric needs of the individual. Caloric needs vary by disorder and are sometimes estimated higher to minimize catabolism. Calculated caloric requirements may also be lower than the DRI for age if the disease results in neurologic compromise causing hypotonia and/or inactivity, leading to decreased energy use.[35] The balance of the kcal needs are provided by a protein-free kcal supplement and/or by the inclusion of specially designed low-protein foods.

Low-protein foods are those that have been designed to provide less than one gram of protein per serving. In addition to providing extra kcal, these foods are used to provide for a greater variety and quantity of foods allowed in the diet. For example, there are special low-protein breads, pasta, cheese, and rice that contain much less protein than their regular counterparts. Low-protein meat and meat substitutes such as low-protein hamburger patties or chicken nuggets have also been recently developed and can add more normalcy to a previously very limited diet. Many companies now supply these foods. These special food items must be obtained via the Internet and by mail, as they cannot be found at local grocery stores. Consequently, they will cost much more than the regular foods they are replacing. Similar to the metabolic formulas, coverage of low-protein foods by health insurance plans and state agencies varies considerably by state.

During illness or metabolic crisis, the main goals of therapy are to (1) decrease intake of offending amino acid(s) by eliminating or reducing the intake of natural proteins, (2) provide sufficient kcal to prevent catabolism, and (3) provide enough fluids to flush out any toxic metabolites. Aggressive measures such as temporary nasogastric feedings, parenteral nutrition, and/or intravenous hydration may be necessary to achieve these goals. It is advisable to start some metabolic formula feedings (not containing the offending amino acids) as soon as tolerated, and after two days of illness the offending amino acid(s) need to be gradually reintroduced to prevent catabolism due to the development of an essential amino acid deficiency. There are several companies throughout the United States that specialize in the preparation of custom parenteral amino acid solutions individualized for the various aminoacidopathies. Treatment for any associated illness that leads to **metabolic decompensation** must be started immediately. Many patients are provided with emergency care protocols to help them receive prompt treatment and prevent further metabolic decompensation.

Monitoring and Evaluation Given the restricted nature of the diet, patients with amino acid disorders are at risk for nutritional deficiencies despite the fact that the majority of the metabolic formulas used are fortified with adequate amounts of vitamins and minerals. Most patients who are drinking the prescribed amount of metabolic formula should have their nutritional needs met. However, since the bulk of their diet comes from synthetic sources, there can still be problems associated with the absorption and utilization of these nutrients. For example, there have been several reports of trace metal deficiencies, including zinc, selenium, and iron, in PKU patients on their prescribed diet.[36–38]

Nutritional deficiencies become a concern for those patients taking a suboptimal amount of metabolic formula. They are not only receiving insufficient protein for growth, but may also show signs of other nutritional deficiencies such as anemia, **neuropathy**, or **osteopenia** associated with inadequate intakes of iron, vitamin B_{12}, vitamin D, and calcium. Growth retardation, associated with the diet restriction and metabolic imbalances, has been noted in patients with PKU and other inborn errors of metabolism.[39–41] The development of newer, more palatable formulas that are incomplete in nutrient composition necessitates frequent monitoring to ensure that patients are receiving adequate intakes of kcal, protein, essential fats, vitamins, and minerals. Patients should have their laboratory studies evaluated periodically for nutritional indices such as albumin, total protein, complete blood count, transthyretin, essential fatty acids, calcium, magnesium, and zinc. Box 26.1 presents an example of the NCP, including monitoring and evaluation, for a patient with PKU.

Bone mineral status can be a major concern in certain amino acid disorders. Results from several studies have shown decreased bone mineral density in PKU patients, despite adequate intakes of calcium and magnesium.[42–44] The explanation for this observation is currently unclear. Studies have indicated that compliance could play a part in the decreased bone mineralization, because it was noted that decreased densities were more significant in patients greater than 10 years of age with compromised metabolic control.[43, 45]

Another nutritional concern is the possibility of developing amino acid deficiencies. It is important to provide enough essential amino acids in the diet for protein synthesis and tissue turnover. The diet prescription aims to provide just enough amino acids to meet this need. Over-restriction of the diet may occur because of the parents' desire to achieve metabolic stability, resulting in low blood levels rather than achieving levels in the recommended range. This over-restriction can lead to anorexia and/or increased catabolism, both of which can compromise metabolic control. Over-restriction of amino acids may also occur secondary to insufficient advancement of the diet to account for growth. This may happen if patients fail to keep appointments with the clinic or to get regular blood testing.

In some amino acid disorders, low levels of certain offending amino acids can develop in response to the patient's attempts to achieve the desired level of one particular amino acid. In maple syrup urine disease (MSUD), for instance, there is a metabolic block affecting the metabolism of the essential branched-chain amino acids leucine, isoleucine, and valine. When leucine is restricted to maintain normal plasma levels, then intakes of the other two amino acids (valine and isoleucine) will also be limited and plasma levels will be subsequently low. **Periorificial acrodermatitis** may develop as a result of specific amino acid deficiencies.[46, 47] Supplementation with specific amino acids is required to normalize their levels and needs to be continued as part of the patient's diet therapy for life.

BOX 26.1 CLINICAL APPLICATIONS

Case Example for Phenylketonuria

Mary is a 4-year-old girl with PKU. Weight = 17 kg, Height = 100 cm. She is currently taking 150 grams of a metabolic formula powder, which is mixed with water to make 32 oz. of liquid formula. This provides her with 22.5 g of (Phe-free) protein and 720 kcal. The remainder of her protein and kcal are provided by her diet. Mom reports that Mary is often not hungry because her formula fills her up. She also reports difficulty getting Mary to eat the prescribed amount of Phe.

Flow Sheet:

Month	Phe Level	Caloric Intake	Protein Intake	Phe Intake
January	246 umol/L	1400	30 g	325 mg
February	228 umol/L	1450	29.5 g	315 mg
March	318 umol/L	1350	29.5 g	315 mg
April	432 umol/L	1400	29.5 g	320 mg

Estimated needs (based on Table 26.7):

- 30–35 g protein
- 1400–1700 kcalories
- 310–320 mg phenylalanine

Nutrition Assessment:
Biochemical Data, Medical Tests and Procedures

Labs: Blood Phe exceeding the upper normal limit for blood drawn in April.

Anthropometric Measurements

Mary's weight is between the 50th and 75th percentile; her height is at the 25th–50th percentile growth curves.

Food/Nutrition-Related History

Mary's dietary Phe intake from natural proteins has been generally within the prescribed range, but blood Phe levels have increased over the last few months. Kcal and protein intakes are at the lower end of the recommended range.

Client/Patient History

As stated above.

Comparative Standards

CDC growth chart, targeted blood Phe range of 120–360 umol/L as recommended by NIH guidelines, estimated nutrient needs per recommendations in Table 26.7.

Nutrition Diagnosis:

Problem: Inadequate nutrient intake—Mary's protein needs have been increasing due to growth. Her recorded protein intake over the past three months did not meet the recommended amount. The increase in blood Phe could be related to an inadequate protein intake.

Nutrition Intervention:

Modify concentration/composition: Mary would benefit from an increased protein intake, but if she consumed more protein from her diet, she would also receive more Phe. Consuming more metabolic formula would likely further decrease her appetite and daily food intake. This could lead to a further compromise in metabolic control since Phe is an essential amino acid, and a lower than required intake could lead to catabolism resulting in an increase in her blood level. Changing Mary's metabolic formula to one that is more concentrated in protein and provides fewer kcal will greatly help her to meet her nutrient needs. This type of formula is designed for older children and adults, and also contains more vitamins and minerals to better meet her nutrient needs.

Mary will transition to the concentrated formula to provide her more protein in less volume. The new metabolic formula provides 25 grams of protein with only 350 kcal in 85 grams of powder. Mixed with water, this amount of powder will make 16 oz of formula. If Mary keeps her dietary protein intake restricted at 7 grams, she will now be receiving a total of 32 grams of protein each day. A decrease in her Phe intake to 300–310 mg each day will also help to decrease blood Phe levels. The decreased volume and kcal from formula will help to stimulate Mary's appetite, but since she has to also slightly decrease her dietary Phe intake, she will have to include more special low-protein foods in her diet. These specially designed foods often are fairly high in kcal and will help ensure she meets her caloric needs while not exceeding her Phe prescription. See the adjusted meal plan shown in the table.

Nutrition Monitoring and Evaluation:

Food intake will be reassessed to ensure that Mary is meeting her dietary kcal, protein, and Phe prescribed amounts. Fluid/beverage intake will be assessed to determine how well she is transitioning over to the new formula. Her biochemical labs will be evaluated to reassess her metabolic control, checking that her blood Phe level returns to within the targeted range.

Original Meal Plan			Adjusted Meal Plan		
Food/Beverage	Phe (mg)	Kcal	Food/Beverage	Phe (mg)	Kcal
Breakfast			**Breakfast**		
¾ cup Froot Loops	56	82	½ cup Froot Loops	37	55
1 banana	43	105	½ cup low-protein cereal loops	1	52
8 oz formula	0	180	1 banana	43	105
			4 oz formula	0	88
Lunch			**Lunch**		
1 slice low-protein bread	15	100	2 slices low-protein bread	30	200
1 slice low-protein cheese	30	60	1 slice low-protein cheese	30	60
20 goldfish crackers	36	52	20 low-protein pretzels	34	112
½ cup canned peaches	17	29	½ cup canned peaches	17	29
1 tsp mayonnaise	2	40	1 tsp mayonnaise	2	40
8 oz formula	0	180	4 oz formula	0	88
			Juice box	0	100
Dinner			**Dinner**		
1 cup low-protein pasta	8	150	1 cup low-protein pasta	8	150
¼ cup spaghetti sauce	27	25	¼ cup spaghetti sauce	27	25
¾ cup broccoli	49	16	½ cup broccoli	33	11
8 oz formula	0	180	4 oz applesauce	6	97
			4 oz formula	0	88
Snack			**Snack**		
8 oz formula	0	180	4 oz formula	0	88
1 cup popcorn	35	35	1 cup popcorn	35	35
Total	318	1414	Total	303	1418

Management of Metabolic Disorders during Pregnancy

As previously noted, many women with inborn errors of metabolism have been reported to have successful pregnancies. But, pregnancy adds additional nutrition concerns for these women. Ensuring adequate intakes of protein, kcal, vitamins, and minerals to promote the growth and development of the fetus needs to be balanced with the dietary restrictions needed to maintain metabolic balance and optimize the health of the mother. It is important to maintain good metabolic control because elevations of certain metabolites are known to be teratogenic to the developing fetus. This is the case in maternal PKU, where elevated blood Phe levels in the mother with PKU have been found to result in microcephaly, cardiac defects, and mental retardation in the offspring. The International Maternal PKU Study has developed guidelines for the management of PKU during pregnancy.[1] There are additional case reports of successfully managed pregnancies in other amino acid disorders. It is usually recommended that blood levels be within the goal range prior to conception and throughout the pregnancy. Tolerance to dietary protein is much lower during the first trimester, and will increase in the second and third trimester due to increasing needs of the fetus. This allows for a greater dietary protein intake for the mother. Pregnancies in women with PKU are considered high-risk pregnancies, and thus need to be followed closely in the metabolic clinics as well as by the obstetrician. This will help ensure the most optimal outcome for mother and baby.[2]

References
1. Matalon K, Acosta PB, Castiglioni L, Austin V, Rohr F, Wenz E, et al. Protocol for Nutrition Support of Maternal PKU. Maternal PKU Collaborative Study funded by the National Institute of Child Health and Human Development. 1998.
2. Van Calcar SC, Harding CO, Davidson SR, Barness LA, Wolff JA. Case reports of successful pregnancy in women with maple syrup urine disease and propionic academia. Am J Med Genet. 1992;44(5):641–46.

Organic Acidemias

Organic acidemias (OAs) are a subset of amino acid disorders that are different in that they do not involve a dysfunction in the metabolism of an amino acid, but rather an intermediary product of metabolism (see Table 26.9 for examples). The amino acid will typically have the nitrogen or amino group removed, and then the rest of this compound will be further processed to generate energy for the body. When there is a block in the pathway generating energy from these compounds, accumulation of compounds known as organic acids will occur. Toxic manifestations of these disorders result from the accumulation of these organic acids in the blood, cerebrospinal fluid, or urine. The accumulation of the organic acids not only causes acidosis, but is usually associated with neurologic disease. The primary enzyme deficiencies responsible for the generation of these compounds cause significant derangements of intracellular and mitochondrial metabolism. There are over fifty different OAs known today, including disorders in protein, carbohydrate, and lipid metabolism, and several of the more common disorders are listed in Table 26.9.[48]

Similar to those with amino acid disorders, individuals with OAs can present at almost any age. Many develop symptoms in the first week of life, but others can have a late-onset form. More recently, the application of tandem mass spectrometry for newborn screening of inborn errors of metabolism has led to earlier diagnosis and improvement in outcomes for many of these conditions. The OA propionic acidemia will now be described in order to illustrate the nature of these disorders.[49]

Epidemiology, Etiology, and Clinical Manifestations of Propionic Acidemia

Propionic acidemia (PPA) is a disorder caused by mutations in the propionyl-CoA carboxylase gene. This leads to a deficiency in the propionyl-CoA carboxylase enzyme activity that is involved in the catabolism of the amino acids valine, isoleucine, methionine, and threonine as well as cholesterol and odd-carbon-numbered fatty acids. As a consequence of this defect, there is an accumulation of propionic acid and an interruption in the formation of intermediates for the Krebs cycle for energy production. The exact incidence of PPA has not been established, but the disease is thought to be quite rare, at less than 1 in 100,000. Newborn screening is now possible for this disorder.

The neonate can present with acute neurological deterioration after an initial symptom-free period, ranging from hours to days after birth. Typically the progression of symptoms moves from feeding refusal and vomiting to progressive weight loss, generalized hypotonia, and abnormal posturing and movements, and then on to lethargy, seizures, and coma leading to severe brain damage or death within a few days if it is not promptly treated. The degree of illness depends on the severity of the enzyme impairment caused by the genetic lesion.[48, 50]

Nutrition Therapy for Propionic Acidemia

Nutrition Intervention The mainstay of nutrition therapy is to restrict the offending dietary precursor amino acids (isoleu-

Table 26.9 Organic Acidemias and Affected Amino Acid(s)

Metabolic Condition	Affected Amino Acid(s)
Glutaric acidemia	Lysine, tryptophan
3-hydroxyisobutyric aciduria	Valine
Isovaleric acidemia	Leucine
3-methylcrotonyl glycinuria	Leucine
Methylmalonic acidemia	Valine, isoleucine, methionine, threonine
Propionic acidemia	Valine, isoleucine, methionine, threonine

cine, methionine, threonine, and valine) in order to reduce elevated concentrations of the toxic organic acids. Fasting should be avoided to prevent catabolism of muscle protein, which leads to the build-up of the organic acids as well. The foundation of nutritional therapy is a moderate-protein, high-energy diet that include dietary protein as a source of the offending precursor amino acids, a metabolic formula providing a protein source that is free of or low in the precursor amino acids, and additional energy sources (carbohydrates and fat) to promote anabolism. The exact amount of precursor amino acids is determined by the child's age, biochemical data, growth parameters, and health status. The diet must be sufficient in all nutrients including minerals and vitamins to prevent deficiencies. Essential fatty acid deficiency has been described in patients with OAs and should be monitored for periodically.[51, 52] Feeding problems are extremely common in these patients, and in order to maintain a good nutritional status it is almost invariably necessary to give feedings via nasogastric tube or gastrostomy at some stage. The use of tube feeding makes it easier to ensure that feedings are nutritionally complete and balanced.[51, 53]

The process to calculate the diet prescription is similar to that used in PKU. After kcal and protein needs are determined, along with the allowed amounts of offending amino acids provided by natural protein, the dietary prescription can be developed. Since a larger percentage of natural proteins contain valine, as compared to the other offending amino acids (isoleucine, threonine, and methionine), this is the amino acid that is monitored when determining natural protein intake. Patients with greater enzyme activity who are allowed a greater natural protein intake may be able to more simply count grams of dietary protein instead. The remaining protein prescribed to meet total protein needs will be provided from a special metabolic formula. There are also special protein-free formulas and low-protein foods that can be used to ensure an adequate caloric intake. Since individuals with PPA do not tolerate fasting well, frequent feedings are recommended.

Most patients with severe PPA have recurrent episodes of metabolic decompensation, most commonly precipitated by simultaneous illnesses with fever or endogenous catabolism due to inadequate kcal and/or protein intakes. Acute intercurrent episodes are prevented or minimized by detecting these situations early and making prompt dietary changes. Patients should have detailed instructions for a sick day diet, in which natural protein intakes are reduced by half and total kcal are increased through protein-free sources. During acute illnesses, an emergency diet should be provided where all sources of natural protein are stopped. In both, energy supply is augmented using carbohydrates and lipids such as solutions consisting of a protein-free formula base powder or a mixture of glucose polymer and lipids diluted in an oral rehydration solution.[54]

Monitoring and Evaluation Similar parameters are monitored as in PKU. Nutritional deficiencies are also a significant concern with these disorders. There are additional biochemical markers that will be monitored including plasma ammonia, blood gases, lactate, glucose, uric acid, and urinary ketones, as these values indicate how stable the patients are metabolically. Regular amino acid analysis is important, and in some institutions, levels of propionic acid in the blood can be monitored. The measurement of carnitine and acylcarnitines in blood may be a useful determinant of metabolic control.[50, 54]

Adjunct Medical and Nutritional Therapies

There are many additional pharmacologic and dietary treatments used to manage amino acid disorders and organic acidemias. Certain antibiotics have been used to decrease intestinal bacterial production of toxic metabolites, such as propionate and ammonia.[54] Other medications have been utilized to detoxify the accumulated metabolites. Carnitine is used to bind a variety of organic acids, and sodium benzoate and sodium phenylbutyrate can bind with excess nitrogen to decrease ammonia production.[55] The use of these therapies in conjunction with protein restriction is needed to optimize the outcomes of these patients.

Urea Cycle Disorders

Epidemiology, Etiology, and Clinical Manifestations

Urea cycle disorders result in an impaired capacity of the body to excrete nitrogen in the form of urea. Ammonia is normally converted to urea in the liver by means of several biochemical steps and eventually excreted as urea (see Figure 26.3). The urea cycle resides primarily in the **hepatocytes** (cells of the liver), but this process can also take place to a lesser extent in the kidneys and small intestine. It is an essential biochemical pathway for excretion of waste nitrogen extracted from the amino acids in the body. A disruption in any one of the eight biochemical pathways involved results in disorders of the urea cycle. These pathways are governed by enzymes (see Table 26.10) that undergo a cascade of enzymatic transformations to convert the toxic ammonia molecule to nontoxic, water-soluble urea, which contains 2 nitrogen groups and is eliminated in the urine. A block in the urea cycle can result from an enzyme deficiency (carbamyl phosphate synthetase I, ornithine transcarbamylase, argininosuccinic acid synthetase, argininosuccinic acid lyase, arginase) in the urea cycle pathway. Alternatively, in other disorders a transport defect in the intestine and/or the kidneys (hyperornithinemia hyperammonemia homocitrullinuria syndrome, and lysinuric protein intolerance) can result in the depletion of an amino acid essential to the normal function of the cycle.

Disorders of the urea cycle have cumulatively been reported to occur in approximately 1 out of 50,000 births.[5] This figure may not have accounted for those individuals with late-onset

Table 26.10 Medical Diagnosis: Enzymatic Defects of the Urea Cycle

Enzyme Defects in the Urea Cycle Pathway:

CPS—carbamyl phosphate synthetase I

OTC—ornithine transcarbamylase

AS—argininosuccinate synthetase

AL—argininosuccinate lyase

ARG—arginase

Enzyme Defects Outside of the Urea Cycle Pathway Leading to Deficiencies of Amino Acids Needed for the Urea Cycle Pathway:

HHH—hyperornithinemia, hyperammonemia, homocitrullinemia

LPI—lysinuric protein intolerance

NAGS—N acetylglutamate synthase

Figure 26.3 The Urea Cycle Pathway in Urea Cycle Disorders

Illustrated here is the complete urea cycle pathway, labeled with enzymes that may be defective in urea cycle disorders. Pharmacologic pathways used for nitrogen removal are also shown. The compounds in the shaded boxes are formed to remove nitrogen: urea, by the body in a normal individual; and the other compounds, through the use of medications such as benzoate or phenylbutyrate.

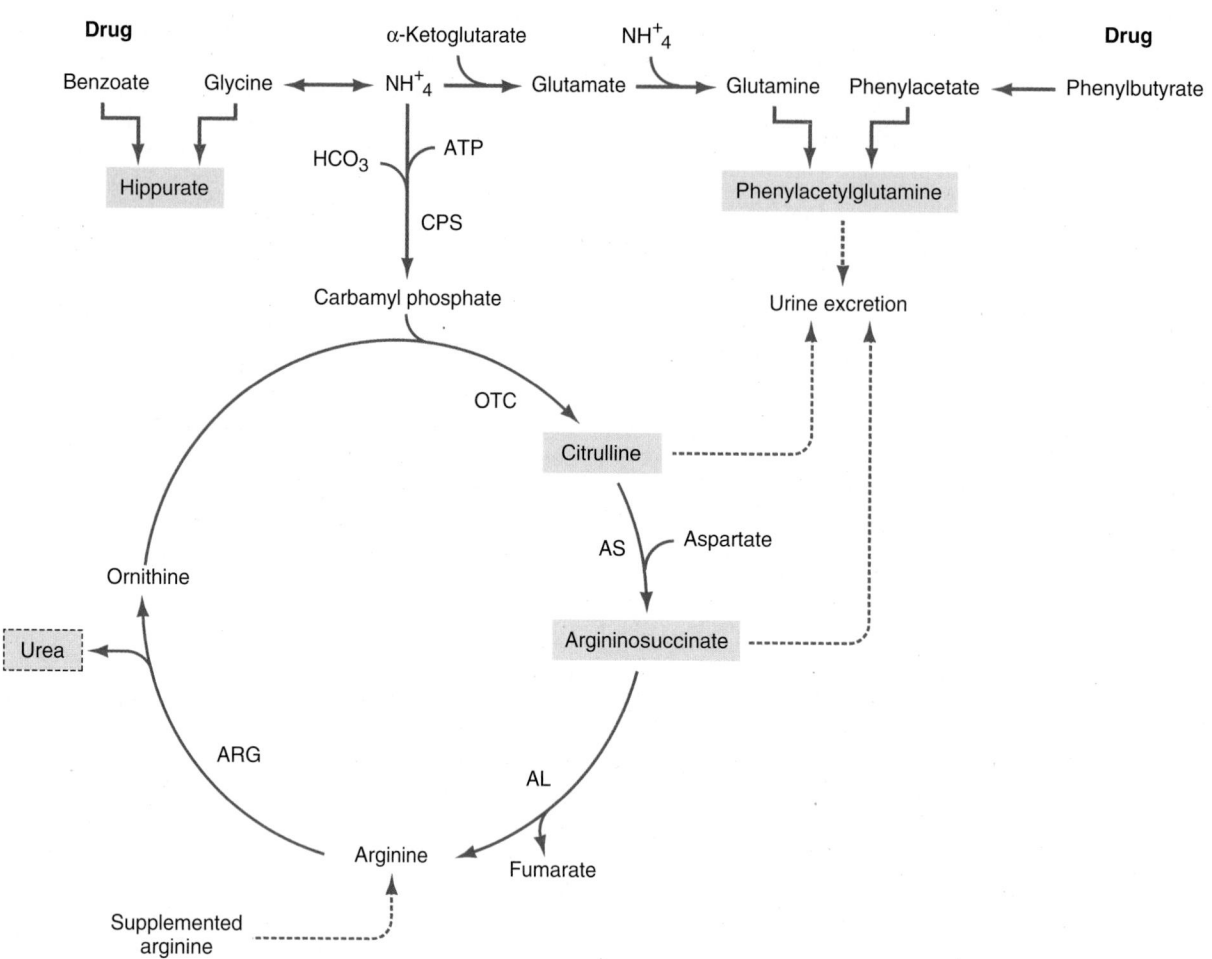

Source: Reprinted from Proceedings of a consensus conference for the management of patients with urea cycle disorders, J Pediatrics. vol. 138:57, Summas M, Tuckman M. Copyright 2001, with permission of Elsevier.

disease. Expanded newborn screening has made the presymptomatic identification of individuals affected with urea cycle disorders possible.

With the exception of ornithine transcarbamylase (OTC) deficiency, all have an autosomal recessive mode of inheritance. OTC deficiency is inherited as an X-linked dominant trait that is usually lethal in males. Urea cycle disorders are a common cause of inherited hyperammonemia, but because of the severe consequences of these disorders for the patient, they should be distinguished from other inborn errors of metabolism with secondary hyperammonemia, such as fatty acid oxidation disorders and organic acidemias. In general, the earlier in the process of converting nitrogen atoms into urea that the defect in the biochemical pathway occurs, the more severe and resistant to treatment the hyperammonemia will be (e.g., CPS and OTC are most severe). There is considerable heterogeneity in the magnitude of hyperammonemia and in the age of initial presentation. This is based on both the position of the block within the urea cycle and the degree of the enzyme deficiency. The most severe cases have no enzyme activity and present

with hyperammonemic coma in the first week of life, with a very poor rate of survival. Patients with the milder forms have some residual enzyme activity and their clinical presentation occurs later in life (ranging from infancy to adulthood) with recurrent episodes of hyperammonemia.[56] If diagnosed early and aggressively treated, these individuals can expect to live to adulthood with minimal neurologic damage.

Infants with complete enzyme deficiencies are usually born at term with no prenatal complications, because the maternal circulation detoxifies the accumulating ammonia. The typical initial symptoms of a child with hyperammonemia include poor feeding and somnolence, which may not be recognized by new parents. As a result, advice and care are sought later when the child's illness has progressed and has become more severe. When medical care for their infant is obtained, his or her plasma ammonia levels are usually quite elevated. The hyperammonemia is associated with clinical symptoms of lethargy, confusion, and vomiting, and even in coma in extreme situations. A loss of thermoregulation, with a low core temperature and feeding disruption, correlates with the somnolence.

Abnormal posturing and **encephalopathy** are often related to the degree of central nervous system swelling and increased pressure on the brain stem. Seizures are seen in approximately 50% of severely hyperammonemic neonates. Hyperventilation caused by cerebral edema leads to respiratory alkalosis, a common symptom in the early stages of the hyperammonemic attack. This can progress to respiratory arrest as pressure on the brain stem increases.[57]

In patients with partial enzyme deficiencies, the first recognized clinical episode may be delayed for months or years. Presentation may occur when infants are weaned from formula to cow's milk, which contains a greater protein concentration. Older children may develop viral illnesses with subsequent endogenous catabolism leading to hyperammonemia. A woman with a partial enzyme deficiency could lead an asymptomatic life until childbirth. The protein load created by the involution of the uterus after pregnancy, along with the increased stress of childbirth, can exceed the metabolic threshold in that individual and send her into a hyperammonemia crisis. The hyperammonemia is typically less severe, and the symptoms more subtle, than in patients with complete enzyme deficiencies.

In addition to the common symptoms of hyperammonemia, patients with partial enzyme defects can present with a wide spectrum of other symptoms such as learning disorders, hyperactivity, or self-injurious behaviors.[58]

Results of therapy in infants with complete or near complete enzyme deficiencies have been less than optimal, with outcomes such as below-normal development and/or death. If the individual has suffered a neurologic insult as a result of a severe hyperammonemic crisis, he/she will usually be severely impaired and require significant assistance with activities of daily living. If serious brain swelling and coma are prevented in the neonatal period, or if onset of the hyperammonemic crisis is delayed, physical growth and mental development are more nearly normal with nutrition and pharmacological support. If diagnosis is anticipated and treatment is begun during the neonatal period in affected siblings, a relatively normal outcome is observed, even with severe enzyme defects.[59]

Acute Medical Treatment

The most effective treatment of the acute hyperammonemic crisis that occurs in the newborn period is dialysis. Initially, these patients will be dialyzed in order to detoxify them by rapidly removing the ammonia. Once the dialysis phase is complete, the medications sodium benzoate and/or sodium phenylacetate are used to scavenge excess ammonia from the bloodstream. Usually, intravenous preparations of these medications are used in acute situations. The pharmacologic approach (see Figure 26.3), rather than dialysis, can be used as the sole

method to remove excess nitrogen in cases of milder elevations in blood ammonia. Sodium benzoate can effectively conjugate with the amino acid glycine to form hippurate, which is efficiently excreted in the urine; similarly, sodium phenylacetate is conjugated to form phenylacetylglutamine, and this compound is excreted in the urine. Both provide pathways for disposing of nitrogen that cannot be excreted as urea and would otherwise accumulate as ammonia. Intravenous citrulline and/or arginine is also provided to optimize the urea cycle and further promote nitrogen removal.

Chronic Medical Treatment

Oral preparations of sodium phenylbutyrate, sodium benzoate, citrulline, and arginine are used routinely for maintaining ammonia homeostasis. Liver transplantation for patients with urea cycle defects has been used, typically in the more severe forms of the disease.

Nutrition Therapy for Urea Cycle Disorders

Acute Nutrition Intervention Initial treatment includes administration of intravenous fluids, because many infants are dehydrated at presentation due to anorexia and poor oral intake. Overhydration should be avoided, however, because most patients with urea cycle defects have some degree of cerebral swelling. Caloric supplementation should be maximized in an effort to reverse catabolism and nitrogen turnover. In addition to glucose, **intralipid** administration can provide more kcal. Since feedings of all protein should be halted temporarily, kcal are provided as carbohydrates and fat. Complete protein restriction is recommended only for a 24- to 48-hour period in order to avoid depletion of essential amino acids, which would result in further protein catabolism and nitrogen release.[60, 61] If enteral feedings cannot be tolerated, TPN should be started. When protein can be reintroduced into the diet, a standard TPN solution with a low percentage of amino acids may be used if TPN is only anticipated to be needed for a short duration. If enteral feedings are not going to be resumed quickly, however, then the use of a special TPN solution rich in the essential branched-chain amino acids should be considered. A comparison of these products is presented in Table 26.11.

Chronic Nutrition Intervention Diet is one of the mainstays of the treatment for patients with urea cycle disorders. The protein intake should be adjusted to account for the severity of the enzymatic defect, along with the patient's age, growth rate, and individual preferences. Some children will have an aversion to protein, whereas others do not. Most patients, except for those with arginase deficiency, will need supplemental arginine. In normal children and adults, arginine is not an essential amino acid, because it can be synthesized

Table 26.11 Branched-Chain Amino Acid Contents of Parenteral Solutions (mg amino acid/100 mL)

Amino Acid	10% Travasol	15% Clinisol	10% TrophAmine	20% Prosol	HepatAmine	Vamin 18 EF	Vaminolact
Isoleucine	600	749	820	1080	900	560	310
Leucine	730	1040	1400	1080	1100	790	700
Valine	580	960	780	1440	840	730	360

Source: With kind permission from Springer Science+Business Media: J Inherit Metab Dis. 2007;10.1007/s10545-007-0718-4. Nutritional management of patients with urea cycle disorders, Singh RH.

by the urea cycle. However, in many urea cycle disorders, this amino acid becomes essential because of the enzymatic block and hence needs to be supplemented back into the diet.

Ideally, protein should be given in the exact amount needed for growth and maintenance without any excess. In reality, however, this has proven to be impossible. Tissue protein is constantly being synthesized and degraded. During growth, a net protein synthesis occurs, but during fasting or illnesses, with an increased rate of degradation exceeding the protein synthesis rate, the outcome is a negative nitrogen balance. Negative nitrogen balance results in an increased flux through the urea cycle and ultimately an elevated ammonia level. During adulthood, protein needs are decreased, because there is less need for growth and a decreased rate of tissue turnover. The protein-restricted diets prescribed need to be maintained throughout life. Table 26.12 shows the difference in protein recommendations for age for the general population versus those affected with a urea cycle disorder.[62]

For patients with the more severe variants of the disorder, some of the natural protein that is provided from foods, which includes a mixture of essential and nonessential amino acids, may be replaced with an essential amino acid mixture to ensure that the individual receives a sufficient amount of these proteins. When one of these mixtures is ingested, the body's surplus nitrogen is used to synthesize the nonessential amino acids, thereby reducing the load on the urea cycle. In general, it is suggested that about 25% to 50% of total protein intake be provided as essential amino acids. There are special metabolic formulas designed for treatment of urea cycle disorders that are complete and provide essential amino acids along with fatty acids, carbohydrates, vitamins, and minerals as well as low-volume formula preparations that do not include fat or carbohydrate sources.[62]

Monitoring and Evaluation As with any low-protein diet, care must be taken to ensure the diet is nutritionally complete. The amino acid intake needs to be balanced, and the essential amino acid intake adequate. Risk of micronutrient deficiency is equally as great as risk of an essential amino acid deficiency, particularly for iron and zinc; hence, supplementation of these nutrients needs to be considered. It is also important to ensure an adequate energy intake. Many patients may be anorexic secondary to elevated ammonia levels, and may not only have a low protein intake but an energy deficit as well. Enteral nutrition support should be considered as a means to ensure a sufficient intake in these patients.

Clinical, biochemical, and nutritional monitoring should be continuous. The lower the protein intake is, the more meticulous the monitoring must be. Any changes to the diet are made in an incremental manner by no more than 10% at any one time. Weights need to be assessed frequently, as poor weight gain in infants/children may indicate catabolism and the potential for hyperammonemia. In addition to weight, other parameters should be monitored to confirm that the patient's nutrition status is adequate. These include clinical features such as the linear growth of the patient and the appearance of the hair, skin, and nails. Biochemical tests include the plasma concentrations of ammonia and essential amino acids. Other monitoring tests may include hemoglobin, hematocrit, albumin, prealbumin, transferrin, and total protein. Given the restricted nature of the diet, it is important to monitor the status of the branched-chain amino acids since use of the nitrogen scavenging medications can further exasperate deficiencies of these nutrients. Patients deficient in branched-chain amino acids have been found to display a peri-oral dermatitis.[63] More recent work has suggested that patients may benefit from supplementation of branched-chain amino acids in addition to the amounts found in the essential amino acid supplement. Branched-chain amino acids are essential for adequate protein synthesis, and inadequate intakes of these specific amino acids despite an adequate overall protein intake can still contribute to catabolism and compromise metabolic control.[64]

A detailed flow sheet can help document trends that relate dietary protein intake to certain laboratory indices. Figure 26.4 shows an example of a clinical flow sheet. Medications used in the management of these conditions should be noted, since many are prescribed on a per kilogram body weight basis and need to be adjusted to account for growth in relation to laboratory results. Including all of these parameters on the flow sheet helps the clinician to make the necessary adjustments to treatments promptly and accurately.

All patients are at risk of decompensation with metabolic stress, including fasting and intercurrent illness, where a rapid increase in blood ammonia can occur. It is important to change the diet to increase energy content and reduce protein intake orally while it is still tolerated. If a child is unable to tolerate the diet orally or is decompensating, then hospitalization is required for the administration of intravenous medications, fluids, and energy sources. An emergency care protocol should be developed for the patient so that when he or she arrives at the hospital, appropriate and urgent care can be administered.

Table 26.12 Comparison of Daily Protein Recommendations for the General Population and Those Affected with a Urea Cycle Disorder

Age	FAO/WHO/UNU Safe Protein Recommendations (mean + 2 SD) (1985)	Protein Recommendations for Individuals with UCDs
0–3 months	2.25 g/kg	1.25–2.20 g/kg
3–6 months	1.86 g/kg	1.8–2.0 g/kg
6–9 months	1.65 g/kg	1.6–1.8 g/kg
9–12 months	1.48 g/kg	1.4–1.6 g/kg
1–4 years	1.09–1.26 g/kg	8–12 g/day
4–7 years	1.01–1.06 g/kg	12–15 g/day
7–11 years	1.0 g/kg	14–17 g/day
11- to 15-year-old girls	0.9–1.0 g/kg	20–23 g/day
11- to 15-year-old boys	0.96–0.98 g/kg	20–23 g/day
15- to 19-year-old girls	0. 8–0.9 g/kg	20–23 g/day
15- to 19-year-old boys	0.86–0.92 g/kg	21–24 g/day

Sources: Leonard JV. The nutritional management of urea cycle disorders. *J Pediatr.* 2001;138:S40–S45. Acosta PB, Yannicelli S. *Nutrition support protocols.* 4th ed. Columbus (OH): Ross Products Division of Abbott Laboratories; 2001.

Figure 26.4 Sample Flow Sheet Used for the Monitoring of Patients with Urea Cycle Defects

Patient Name: _____   Date of Birth: _____

Patient Medical Record #: _____   Diagnosis: _____

Date	Age	Wt/Ht	OFC	Medications	Plasma Citrulline	Plasma Intamine	NH₃/alb	H/H	Formula	Protein Intake	Kcal

Mitochondrial Disorders

Etiology and Clinical Manifestations

Mitochondria are essential for the production of energy in all types of tissues. They are found in all cells in the body (except mature red blood cells) and are considered the "power houses" or energy producers of the cell. Mitochondrial disorders result either from defects of the respiratory chain that produces energy or from defects affecting the overall number and function of the mitochondria. A defect in function of the mitochondrial system results in diseases affecting multiple organs with a wide range of symptoms, including cardiomyopathy, blindness, deafness, endocrine problems, and muscle disorders. Individuals presenting with a mitochondrial disorder can range from infants, as in the case of Leigh's disease, to adults with progressive neurologic or muscle disorders. Table 26.13 lists the spectrum of conditions associated with mitochondrial mutations leading to disruption in function. Mitochondrial disorders are renowned for their variability in clinical features and genetic causes, making it difficult to determine their true prevalence.[65, 66]

Medical Diagnosis

Some of the most common mitochondrial disorders such as MELAS (mitochondrial encephalomyopathy with lactic acidosis and stroke-like episodes) and NARP (neurogenic muscular weakness, ataxia, and retinitis pigmentosa) result from identi-fied genetic mutations and thus can be tested for through a blood test. Skin and muscle tissue can be assayed for further **histological** and biochemical analyses in order to make a diagnosis.

In a healthy person, energy is produced through several metabolic processes involving the use of fatty acids and carbohydrates within the mitochondria. Disorders of the mitochondria include fatty acid transport disorders, fatty acid oxidation defects, **pyruvate complex disorders**, and respiratory chain

Table 26.13 Spectrum of Medical Conditions Related to Mitochondrial Dysfunction and Used to Make the Diagnosis

• Ataxia	• Hypoglycemia
• Autism	• Hypotonia
• Blindness	• Language delays
• Cardiomyopathy	• Liver failure
• Deafness	• Neuropathy
• Dementia	• Pancreatitis
• Developmental delay	• Renal failure
• Diabetes	• Seizures
• Epilepsy	• Strokes
• Gastrointestinal motility problems (diarrhea, dysmotility)	• Weakness

defects. Because the respiratory chain defects are the predominant types of mitochondrial disorders, the remainder of this section will focus on these defects. Disorders of fatty acid transport and oxidation will be discussed in a subsequent section.

Respiratory Chain Defects

The respiratory chain (see Figure 26.5) is made up of five complexes that undergo changes in their oxidative state to produce ATP (adenosine triphosphate). Carbohydrates are eventually metabolized to pyruvate, which will then enter into the Krebs cycle. Electrons generated from the Krebs cycle and from beta-oxidation of fatty acids are used in the production of energy via the complexes of the respiratory chain. Defects affecting several of the individual complexes have been identified. These defects lead to decreased energy production in various tissues, and subsequently to clinical symptoms, such as hypotonia, developmental delay, and failure to thrive. Attempts to facilitate the function of the respiratory chain through the administration of pharmacological doses of several vitamins and nutrients have resulted in limited success thus far.[66]

Nutrition Therapy for Mitochondrial Disorders

Nutrition Intervention There is no definite treatment for mitochondrial disorders. The use of dietary intervention may help alleviate some of the symptoms and/or delay progression of the disease, but will not prevent the debilitating effects of the disorder. Therapy for defects of the respiratory chain entails the use of vitamin cofactors in pharmacological amounts equal to approximately 100 to 1000 times the DRI for age to enhance the activity of the various complexes. Riboflavin and thiamin serve as cofactors, while vitamin E and lipoic acid are used to protect

Table 26.14 Recommended Cofactor Doses for Mitochondrial Disorders

Cofactor	Suggested Dosing Range
Coenzyme Q_{10}	5–15 mg/kg/day
Vitamin K	40–80 mg/day
Vitamin C	0.25–4 g/day
Vitamin E	400–1200 IU/day
Selenium	50–100 μg/day
Thiamine	25 mg/day
Riboflavin	25 mg/day
Pantothenate	25 mg/day
Carnitine	50 mg/kg/day

Source: Marriage B, Clandinin MT, Macdonald IM, Glerum DM. Cofactor treatment improves ATP synthetic capacity in patients with oxidative phosphorylation disorders. *Mol Genet Metab.* 2004 Apr:81(4):263–72.

against **free radicals** that are produced from ongoing aberrant reactions (see Chapter 13 for a discussion of free radicals). Vitamins C and K and coenzyme Q_{10} are used as artificial electron receptors and transporters.[67] Prescription doses of these cofactors vary widely between clinics, but Table 26.14 shows the ranges prescribed.[68] The reported efficacy of vitamin therapy can vary significantly, but given the benign nature of most of the vitamins prescribed, many clinics will recommend a trial of one or more of these supplements. Since fasting produces stress on the respiratory chain, frequent feedings and a nighttime snack are recommended. The use of uncooked cornstarch at night has been suggested as a therapeutic option for some of these conditions.

Figure 26.5 Respiratory Chain Pathway

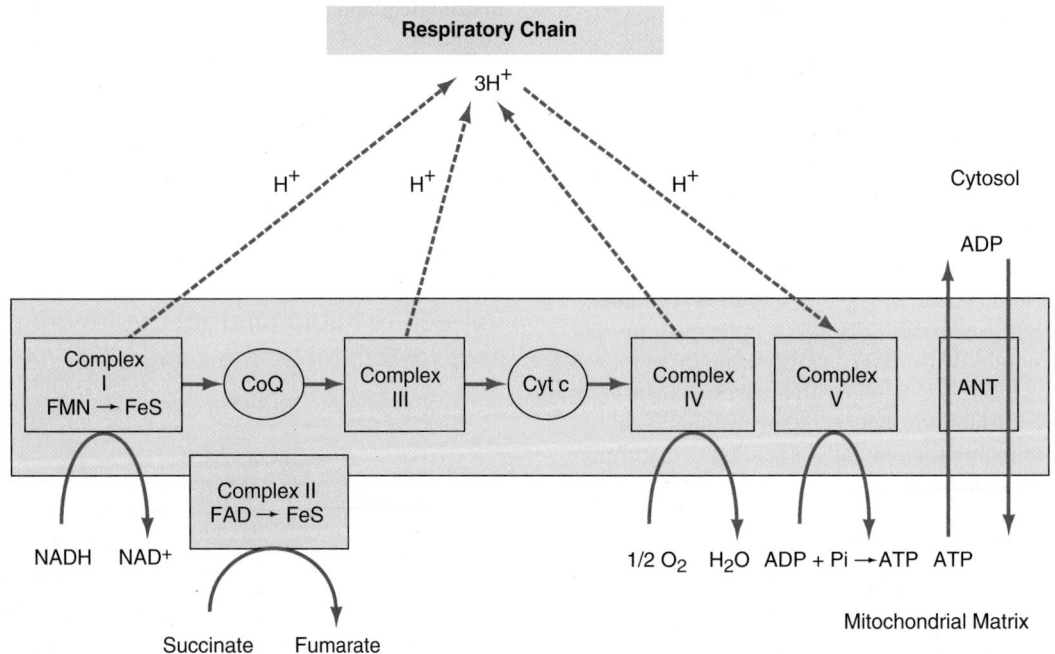

Source: From slide presentation by Marriage B, "Cofactor treatment in oxidative phosphorylation disorders: is it effective?" Presented at 9th Ross Metabolic Conference: Advances in Management of Inherited Metabolic Disorders, April 4–6, 2003.

Monitoring and Evaluation Nutrition monitoring of these patients is dependent on their symptoms as well as the degree of intervention. Some individuals can have issues with poor weight gain and failure to thrive if they have **dysphagia** or diarrhea, while others may be at risk for obesity with the frequent feedings and addition of cornstarch at night. The results of vitamin/cofactor therapy need to be assessed routinely to determine the effectiveness of the treatment and to identify any problems that may be related to the supplements (e.g., diarrhea associated with carnitine therapy). If the supplements are not noted to be beneficial, then they should be discontinued, as their cost can be high and is usually not covered by medical insurance.

Adjunct Medical Therapies

Carnitine and glycine have been used in the treatment of several of these disorders. They act by conjugating with toxic metabolites, removing them from the body.[67]

Disorders Related to Vitamin Metabolism and Vitamin-Responsive Metabolic Disorders

Etiology

Sometimes a pharmacologic dose of a vitamin is all that is necessary to maintain normal enzymatic function and eliminates the need for other treatment modalities, such as a restrictive diet. This is a possibility not only in disorders of vitamin metabolism, but also in other types of metabolic disorders, which may present in vitamin-responsive forms.

Vitamins are needed for various functions. They can be used as cofactors for various enzymatic reactions, antioxidants, or electron receptors. In many cases, vitamins that are used as cofactors for enzymes function by helping to activate the enzyme. A defect in one of these enzymes could result in an inability of the vitamin to bind to the active site of the enzyme to activate its function. In other circumstances, the vitamin can stabilize the enzyme through its chaperoning effect, as described previously. Whether there is a defect in the vitamin's ability to bind or stabilize the enzyme, providing large doses of the vitamin can sometimes help to overcome the defect and allow greater enzyme function. This can occur in certain amino acid disorders, such as maple syrup urine disease, where there is a thiamin-responsive form.

There are other disorders directly affecting the metabolism of the vitamin rather than an enzyme with which the vitamin interacts. An example is biotinidase deficiency, where there is a defect in biotin recycling that causes an excessive amount of this vitamin to be excreted from the body. This results in a biotin deficiency that can be corrected by supplementing with pharmacological amounts of this vitamin.

In general, the vitamin-responsive form of a metabolic disorder tends to be less common than the non-responsive form of the same condition (see "Supplementation of Vitamins or Other Cofactors").

Nutrition Therapy for Vitamin-Responsive Metabolic Disorders

Nutrition Intervention There are some amino acid disorders, including methylmalonic acidemia and MSUD, that have a vitamin-responsive form. A pharmacologic dose of vitamin B_{12} will be given to a newly diagnosed methylmalonic acidemia patient to determine whether the enzymatic defect is vitamin responsive.[39] Only a small percentage of patients with methylmalonic acidemia have been found to be responsive to vitamin B_{12}, but those who are do not require additional treatment and need not adhere to a restrictive diet. Vitamin B_{12} is a cofactor for the methylmalonyl CoA mutase enzyme (see Figure 26.6); the adeno form of this vitamin is required in order for this enzyme to function properly. Consequently, there have been reports of patients diagnosed with methylmalonic acidemia with a malfunctioning enzyme due to a vitamin B_{12} deficiency, and not because of a genetic mutation. These individuals will have the classical clinical presentation of methylmalonic acidemia, including failure to thrive, developmental delay, and hypotonia. This can be caused by following a strict vegan diet, or may even develop in an infant who is breast feeding from a vitamin B_{12}-deficient mother.[69] Supplementation with B_{12} leads to normalization of methylmalonic acid levels.

Administration of a vitamin can be the primary treatment modality in some metabolic conditions. In holocarboxylase synthetase deficiency and biotinidase deficiency, for example, biotin in doses from 100 to 1000 times the DRI for age is the only treatment required and will prevent serious complications. These disorders result from an inability to regenerate biotin from endogenous sources. Individuals with one of these conditions typically present in infancy with feeding and respiratory difficulties, vomiting, seizures, lethargy, and hypotonia related to the lactic acidosis, ketosis, and hyperammonemia created by the accumulation of the intermediate compounds due to the metabolic block. Frequently infants with these conditions have an erythematous skin rash (see Figure 26.7) and varying degrees of alopecia associated with a biotin deficiency created by the metabolic block. If left untreated, individuals can progress to dehydration, coma, and even death. Once treatment with pharmacologic doses of biotin begins, individuals are expected to lead a healthy and normal life, without a compromised life expectancy. Treatment with this vitamin needs to be continued throughout life. These are extremely rare disorders, and the incidence has not yet been determined.

Monitoring and Evaluation It needs to be emphasized that the vitamins administered in pharmacologic amounts are being utilized as drugs to treat specifically responsive metabolic disorders. Even in those disorders in which the precise biochemical defect may not have been elucidated, cofactor administration is based on reasonable and focused hypotheses concerning underlying etiologies. Such administration is in sharp contrast to the use of "megavitamin" supplements in a random fashion in conditions for which a response has not been documented, or in patients with ill-defined dysfunctions presenting no clinically valid justifications for therapeutic trials.[3]

Most of the vitamins used in the treatment of metabolic disorders are water soluble, so issues of toxicity are minimized.

Figure 26.6 Vitamin B$_{12}$ (adenosyl cobalamin) and Methylmalonic Acidemia (MMA)

Adenosyl cobalamin (AdoCbl) is necessary for the conversion of L-methylmalonyl CoA to succinyl CoA by the methylmalonyl CoA mutase enzyme.

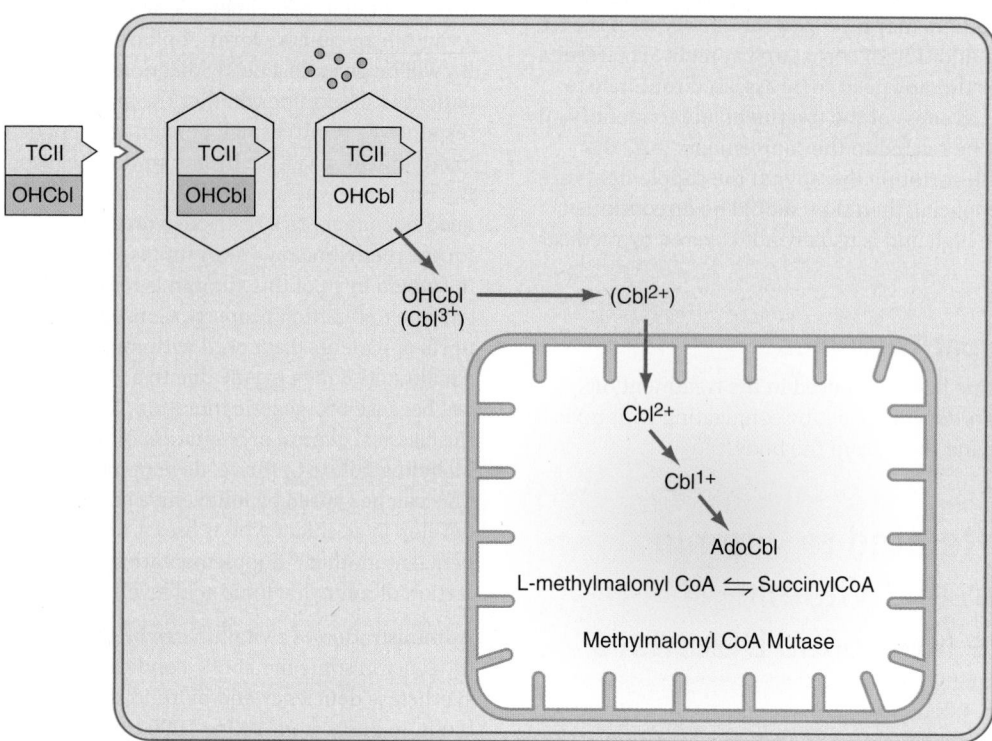

TCII = transcobalamin II

OHCbl = hydroxycobalamin

Source: Modification of Figure 1, p. 620 of Shevell MI, Matiaszuk N, Ledley F, Rosenblatt DS, "Varying neurological phenotypes among muto & mut- patients with methylmalonyl CoA mutase deficiency," *Am J Med Genetics*; 1993.

Figure 26.7 Infant with Erythematous Skin Rash Associated with Biotin Deficiency Secondary to Holocarboxylase Synthetase Deficiency

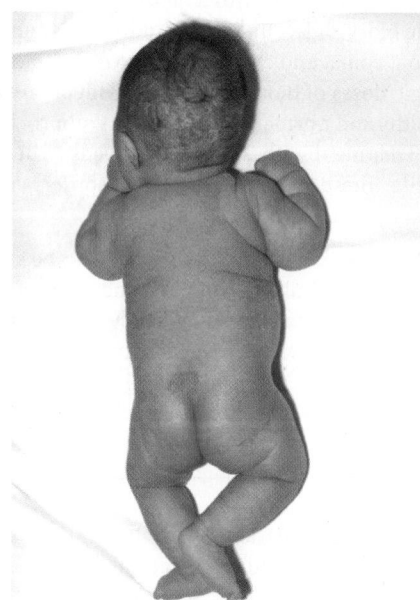

Source: Figure 4.3, p. 28 of Nyhan WL and Ozard PT, "Multiple Carboxylase Deficiency," *Atlas of Metabolic Diseases* 1st ed. Chapman and Hall Medical; 1998.

Courtesy of New Zealand Dermatological Society Incorporated. Published online at: www.dermnetnz.org

When fat-soluble vitamins (E and K) are used, such as in the treatment of mitochondrial disorders, toxicity can be of concern. Because these vitamins can affect clotting factors, laboratory studies assessing clotting times should be assessed periodically. Compliance with vitamin supplementation can be a concern, especially when a large amount and number of vitamins and/or cofactors are prescribed, as in the case of mitochondrial disorders. The cost associated with purchasing these supplements is of concern, as many health insurance companies will not cover them. Further research documenting the effectiveness of these nutrients in the treatment of metabolic disorders is needed in order to justify advocacy of insurance coverage for these products that is similar to current coverage for medications.

Disorders of Carbohydrate Metabolism

Disorders of carbohydrate metabolism include problems with processing the simple sugars galactose and fructose in the body, and a group of disorders called glycogen storage diseases that result in various defects in synthesizing and releasing glucose. Table 26.15 summarizes the carbohydrate disorders to be discussed here, along with their main clinical symptoms and available treatments. Prompt diagnosis and dietary treatment significantly help to improve the overall outcome for patients affected with one of these types of conditions.

Table 26.15 A Summary of Carbohydrate Disorders, Associated Symptoms, and Treatments Used

Carbohydrate Disorder	Main Clinical Symptoms	Treatment
Galactosemia	Failure to thrive, hepatomegaly jaundice, vomiting, and sepsis	Diet restricted in lactose/galactose
Hereditary fructose intolerance	Vomiting, failure to thrive, diarrhea, hypoglycemia, liver dysfunction, and aversion to sweets	Diet restricted in fructose, sucrose, and sorbitol
Glycogen storage disease type 1: glucose-6-phosphatase deficiency	Hypoglycemia, poor growth, lactic acidosis, hyperlipidemia, elevated uric acid levels, and osteoporosis	Frequent feedings of foods high in carbohydrate and low in fat content, continuous nocturnal tube feedings, and intermittent cornstarch feedings

Galactosemia

Classical galactosemia is an enzyme defect in galactose metabolism leading to failure to thrive, **hepatomegaly** (enlarged liver), and life-threatening sepsis in the newborn period. Vomiting and jaundice may develop as early as a few days after milk feedings are begun. Anorexia, failure to gain weight or to grow, or even weight loss ensues. In the absence of treatment, **parenchymal** damage to the liver leads to the development of cirrhosis. Patients may present with edema, ascites, bleeding problems, and an enlarged spleen. Because galactose is a monocsaccharide found in milk products, if milk feedings are continued, the disease may be rapidly fatal. Other complications associated with the continued ingestion of galactose include cataract formation, mental retardation, and renal tubular dysfunction. All states now screen for galactosemia in the newborn period before symptoms are present or complications of the disease have taken place.[70] Galactosemia has been reported to occur in approximately one in every 60,000 to 80,000 births in the United States.

In patients with galactosemia, the enzyme defect occurs in uridyltransferase, which normally converts galactose, in the form of galactose-1-phosphate, to glucose-1-phosphate (see Figure 26.8). Glucose-1-phosphate can then produce free glucose or be utilized to produce energy via **glycolysis**. The other enzymes involved in galactose metabolism are normal in these patients. Galactose-1-phosphate accumulates in tissues and may be responsible for many of the clinical manifestations of the disease. There are several known mutations that result in several specific types of galactosemia. The classic form of galactosemia results in the absence of enzymatic activity, whereas the Duarte and Black variants have enzymatic activity ranging from 10% to 50%, resulting in the need for a less stringent diet.

Despite early diagnosis and nutrition intervention, there are still problems associated with the disorder. Although IQ levels have been reported to be normal in these patients,[16] many affected individuals with speech concerns and ovarian dysfunction (in females) are still identified.[71] Several theories have been proposed to explain the poor outcomes in these individuals, including fetal damage in utero through exposure to galactose via maternal circulation, damage before nutrition intervention, de novo galactose synthesis, and an overrestriction of dietary galactose and a subsequent deficiency of uridine diphosphate galactose. More questions than answers regarding optimal treatment to alleviate the clinical manifestations associated with galactosemia still abound and further research may lead to additional interventions that will optimize the outcomes of these patients.

Nutrition Therapy for Galactosemia

Nutrition Intervention The treatment for galactosemia is exclusion of galactose from the diet. Treatment with a lactose/galactose-free diet leads to immediate reversal of symptoms. Lactose, the principal sugar of mammalian milks, is the predominant dietary source of galactose. It is a disaccharide in which glucose and galactose are linked together. The mainstay of the diet for an infant is the substitution of a soy-based formula for a milk-based formula. The use of soy formulas has been questioned because of the presence of sugars containing galactose, such as **raffinose** and **stachyose**. However, it is believed that these galactose **oligosaccharides** are not hydrolyzed to their component sugars by human intestinal mucosa.[6]

Education of the family is directed at helping them recognize the potential dairy and non-dairy sources of galactose and lactose. Table 26.16 lists some of these potentially galactose containing ingredients. Many processed foods, such as cakes, cookies, muffins, and French fries, contain milk products, and consequently galactose. Several fruits and vegetables, such as persimmons, dried figs, and tomatoes, contain galactose. Some

Figure 26.8 Enzymatic Pathway for Galactose Metabolism and Site of Enzymatic Defect, Galactose-1-Phosphate Uridyltransferase, in Galactosemia

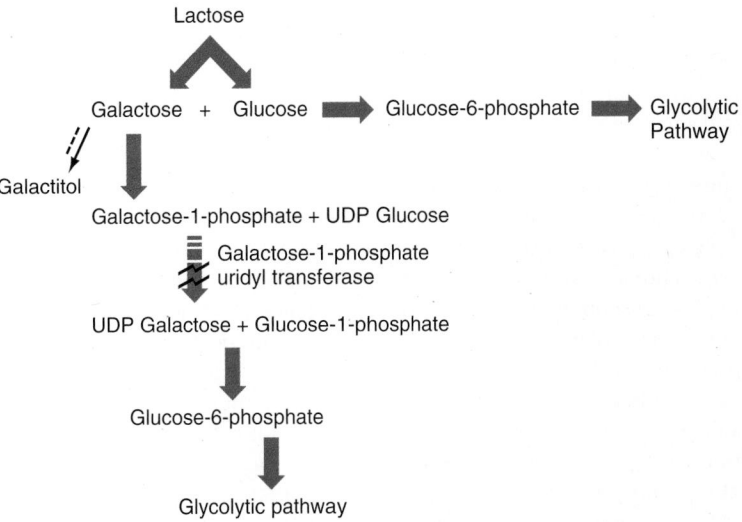

Table 26.16 Ingredients That Contain Galactose

- Butter
- Calcium caseinate
- Nonfat milk
- Dry milk and milk protein
- Hydrolyzed protein made from casein or whey (avoid if unspecified)
- Lactalbumin
- Milk and milk solids
- Organ meats
- Sodium caseinate

- Whey and whey solids
- Buttermilk and solids
- Casein
- Cream
- Lactose
- Milk chocolate
- Cheese
- Sour cream
- Yogurt

Source: Parents of Galactosemic Children, Inc. [homepage on the Internet]. Gauiter (MS): Parents of Galactosemic Children, Inc.; 2006. Available at http://www. galactosemia.org/

Sample PES Statement: Limited adherence to nutrition-related recommendations as evidenced by diet record documenting lactose-containing products related to lack of knowledge for identification of galactose-containing foods.

of the galactose in fruits and vegetables may be found as the galactose oligosaccharides, raffinose and stachyose, and hence is most likely unavailable for absorption. Medications should be checked for galactose content because lactose can be used as a sweetener or filler. The determination of the galactose-1-phosphate level in the red blood cells is helpful in monitoring adherence to the diet.

Treatment for the milder forms of galactosemia has been a controversial subject in recent years. Given the level of enzymatic activity, some metabolic centers are choosing not to restrict the diet of those individuals with certain variant forms of galactosemia. Other centers are only restricting the diet during the first year of life when the intake of milk represents a significant percentage of the infant's diet.

Monitoring and Evaluation Maintaining a galactose-restricted diet is not harmful as long as alternative sources of those nutrients abundant in dairy foods, such as calcium and vitamins A and D, are provided. Calcium and vitamin D supplementation is often required by children and adults with galactosemia. Kcal, protein, vitamin, and mineral needs are similar to those of other individuals unaffected by this disease.

Hereditary Fructose Intolerance

Fructose is predominantly metabolized in the liver, kidney, and small intestine in a specialized pathway composed of three enzymes—fructokinase, aldolase type B, and triokinase—that convert fructose into intermediates of the glycolytic gluconeogenic pathway. Hereditary fructose intolerance is caused by a deficiency in activity of the fructose 1-phosphate aldolase enzyme (aldolase B). A deficiency of the fructoaldolase B enzyme results in accumulation of fructose 1-phosphate in tissues that possess fructokinase, causing depletion of **inorganic phosphate** and ATP. Liver and renal impairment is thought to be due to either the direct toxic effect of fructose 1-phosphate or a deficiency in tissue ATP content. Fructose-induced hypoglycemia results from inhibition of both **gluconeogenesis** and **glycogenolysis**. Symptoms appear only when fructose, sucrose,

or sorbitol is introduced into the diet, and they are not specific, so diagnosis may be overlooked. Symptoms may occur early in life and can be severe; later they are less apparent, because the child develops an aversion to sweets. Vomiting is such a constant finding that its absence in a subject ingesting fructose argues against the diagnosis. Poor feeding, diarrhea, and later failure to thrive are less frequent. Some manifestations reflect liver impairment: hepatomegaly, bleeding tendency, jaundice, edema, or ascites. The diagnosis of hereditary fructose intolerance should be strongly suspected in an infant with liver failure who has consumed sucrose or fructose. This can readily be seen in an infant ingesting a sucrose-based formula such as Isomil, or who has been given fruit juice or prescribed a sucrose-containing medication.

As this disorder is currently not routinely screened for, some infants with severe liver failure may die if the diagnosis is overlooked. In the majority of diagnosed cases, outcome is excellent. With a fructose-free diet, vomiting stops almost immediately. In addition, all the clinical and laboratory findings return to normal within one to two weeks, with the exception of hepatomegaly, which may persist for many years. Hepatomegaly and **steatosis** disappear with the fructose-free diet between the ages of 5 and 10 years, but diet recommendations are continued for life.[72, 73] Normal growth parameters are usually achieved within two to three years after diagnosis and intervention.

Nutrition Therapy for Hereditary Fructose Intolerance

Nutrition Intervention Fructose is a widely distributed natural compound. As the free monosaccharide, it is found in honey and in numerous vegetables and fruits, where it can account for up to 40% of the dry weight. As the disaccharide sucrose, which consists of one molecule of fructose attached to a molecule of glucose, it is found in many more nutrients and constitutes an important source of dietary carbohydrate. Listings of the fructose content of various foods are available, but families need to be vigilant about reading labels because product ingredients are always changing. Fructose (especially in the form of high-fructose corn syrup) is now extensively used in food processing as a sweetening additive in foods, medications, and even infant formulas. Average intake of this sugar has steadily increased since the beginning of this century to amounts exceeding 100 grams per day in adults. Successful treatment requires strict avoidance of all dietary fructose and sucrose. Calorie, protein, vitamin, and mineral requirements are similar to those of normal individuals.[68]

Monitoring and Evaluation Given the omission of most fruits and some vegetables from the diet, a vitamin and mineral supplement is usually indicated. One must ensure that the vitamin preparation is free of fructose, as there are many that contain this sugar, especially among the liquid preparations made for children.

It is very challenging for individuals to eat outside the home while following the necessary dietary restriction. New foods need to be closely scrutinized before ingestion to ensure that they are fructose free.

It is typical for individuals with this disorder to acquire an aversion to sweet foods. A positive aspect of this disease has been the lack of dental caries among patients due to a self-imposed sucrose-free diet.

Glycogen Storage Diseases

Glycogen storage diseases are caused by deficiencies of enzymes that regulate the synthesis or degradation of glycogen. Glycogen, a polysaccharide composed of glucose units, is the main carbohydrate reserve in the body. This polysaccharide is assembled through a process of chain elongation and branching by the sequential addition of glucose units, achieved by glycogen synthetase and branching enzymes. Catabolism of the polysaccharide is achieved by the opposite process, through phosphorylase and debranching enzymes (see Figure 26.9). Glycogen is abundant in the liver, where it is used predominantly to form blood glucose, and in muscle, where it is a fuel for muscle contraction. The brain, on the other hand, although utilizing glucose preferentially for its metabolic needs, does not glycogen to any significant extent. Hence, the brain is dependent on receiving glucose from the blood supply.[74]

The syndromes of glycogen storage disease can be divided into at least eight different types with respect to clinical and chemical manifestations and according to the enzymatic deficiency. Most of the various types affect the liver and are a consequence of deficient activity of an enzyme directly involved in degradation of glycogen to glucose-6-phosphate, or rarely, the synthesis of glycogen from glucose-6-phosphate. The enzymes noted by an asterisk in Figure 26.9 are those whose absences are known to lead to the accumulation of an abnormal amount of glycogen. The common clinical manifestation is abnormal glycogen deposition, primarily in liver, muscle, or both. In some of these defects, the accumulated glycogen is of normal structure. On the other hand, rare defects—that is, type IV—result in glycogen of an abnormal structure, more resembling **amylopectin**, which is the component of starch that is the storage product of the plant world. It is probably treated as a foreign substance despite the fact that the body produces it, and such a hostile reaction may be the basis of the early cirrhosis that is characteristic of this disease. Table 26.17 summarizes the types of glycogen storage diseases that are now recognized, the main tissues affected, and the main clinical symptoms.[75] There is a wide range of incidence in the occurrence of glycogen storage diseases. The more common glycogen storage disease, type I, occurs about once in every 100,000 births, while frequency of the less common forms such as type II has not yet been determined.

Glycogen Storage Disease Type I Glycogen storage disease type I (GSD I), the most commonly diagnosed type, actually occurs despite correct functioning in the enzymes required for both the synthesis of glycogen and its degradation to glucose-6-phosphate. As a consequence, many of its clinical and chemical features are unique among the types of glycogen storage diseases. GSD I is characterized by a deficiency of the enzyme glucose-6-phosphatase. The inability to **dephosphorylate** glucose-6-phosphate results in hypoglycemia and its metabolic consequences. Figure 26.10 illustrates the metabolic pathway involved in GSD I. Since patients with GSD I are at risk of death or hypoglycemic damage to the brain in early infancy, prompt diagnosis, the avoidance of fast-

Figure 26.9 Summary of the Enzymes Involved in the Synthesis and Degradation of Glycogen in Tissue

Those enzymes carrying an asterisk are associated with a specific type of glycogen storage disease.

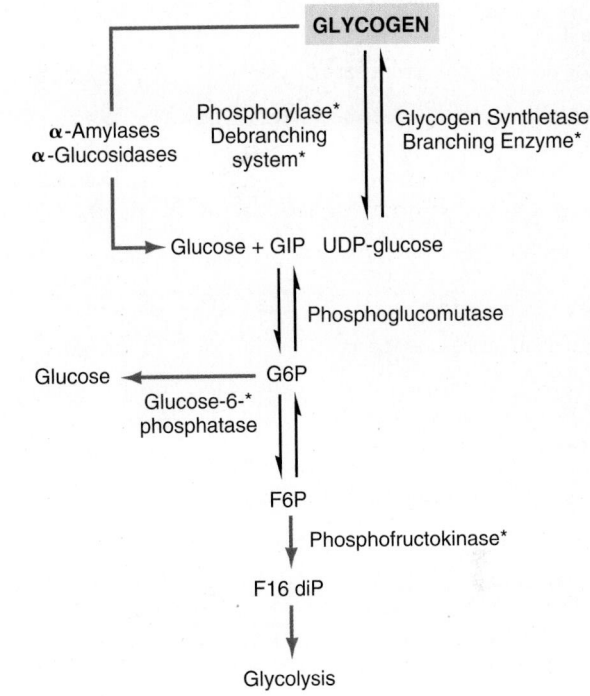

Source: Figure 3, p. 107 of Ryman BE, "The glycogen storage diseases," *J Clin Pathology,* vol. 27, suppl. 8, 1974.

ing, and the provision of free glucose are important in getting the patient through this critical period.[76]

The most constant and life-threatening feature of this disease is the low blood glucose levels which result from relatively short periods of fasting. Fasting for as little as 2 to 4 hours is almost always associated with a decrease in blood glucose to less than 70 mg/dL, and it is not uncommon to observe 6- to 8-hour fasting levels of 5 to 10 mg/dL. In normal individuals, blood

Figure 26.10 Metabolic Pathway Affected in Glycogen Storage Disease Type I

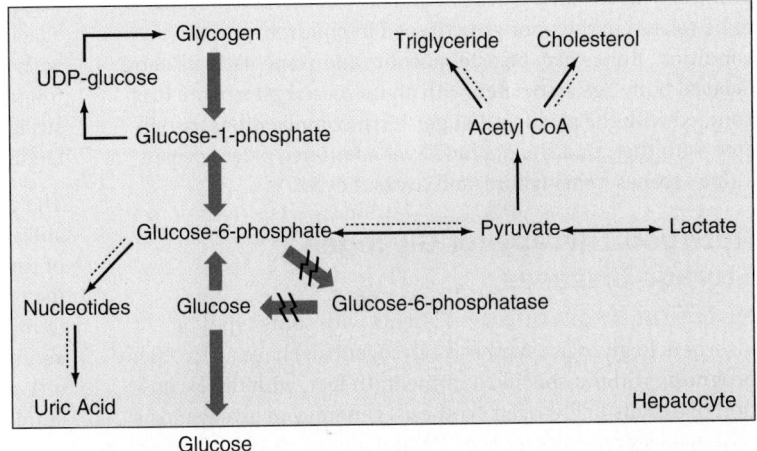

Table 26.17 Listing of the More Common Glycogen Storage Diseases, Tissues Affected, and Clinical Symptoms

Glycogen Storage Disease and Enzyme Defect	Tissue(s) Affected	Symptoms
Type I Glucose-6-phosphatase deficiency	Liver, kidney, small intestine	Hypoglycemia, hepatomegaly, lactic acidosis, hyperlipidemia, hyperuricemia, poor growth, and osteoporosis
Type II Lysosomal acid alpha-glucosidase deficiency	Most tissues	Cardiomegaly in infants, hypotonia
Type III Glycogen debranching enzyme	Liver and muscle	Hepatomegaly, variable hypoglycemia; similar to type I
Type IV Branching enzyme	Liver, leukocytes, and fibroblasts	Progressive cirrhosis with hepatosplenomegaly, failure to thrive, and ascites
Type V Muscle phosphorylase deficiency	Muscle	Easy fatigability, painful cramps after strenuous exercise, myoglobinuria
Type VI Liver phosphorylase defect	Liver, leukocytes	Mild hepatomegaly, mild hypoglycemia and hyperlipidemia, growth retardation
Type VII Muscle phosphofructokinase deficiency	Muscle	Easy fatigability, painful cramps after strenuous exercise, myoglobinuria, degenerative brain disease
Type IX Defect in 1 of the 4 subunits of phosphorylase kinase	Liver and muscle	Mild hepatomegaly, mild hypoglycemia and hyperlipidemia, growth retardation

glucose levels are maintained within a relatively narrow range by hormones excreted by the liver, such as **glucagon**, which releases glucose from stored glycogen, as well as through gluconeogenesis, the process of synthesizing glucose from amino acids. In patients with GSD I, it is possible to degrade glycogen to glucose-6-phosphate, but, in the absence of the glucose-6-phosphatase enzyme, no glucose is released and blood glucose levels continue to fall. The liver is further stimulated, which leads to continued glycogen degradation resulting in secondary manifestations of GSD I that cause the characteristic elevations in blood lipids, lactate, and uric acid. Hepatomegaly is often present at birth and can progress to huge enlargement of the liver if the disease is poorly controlled. This enlargement is due to the accumulation of glycogen as the metabolic block compromises the ability to release free glucose. The chronic lactic acidosis with elevated glucagon and low insulin levels seems to be related to the poor growth seen in children with this condition. Bones may be osteoporotic, and some patients show delayed bone age associated with an increased phosphate loss coupled with the acidosis. Longer-term complications associated with this condition include liver adenoma, osteoporosis, kidney stones, renal failure, and ovarian cysts.[76]

Nutrition Therapy for Glycogen Storage Diseases

Nutrition Intervention Patients with some types of disorders in glycogen synthesis (glycogenosis) have an excellent prognosis without specific treatment. In fact, with the exception of defects in glycogen synthesis, generalized glycogenosis (glycogen storage disease type II), and glucose 6-phosphatase

deficiency, most patients with hepatic glycogenosis have a favorable prognosis and are successfully managed with some attention to the frequency of feeding. This, however, has not been true of most patients with GSD I. Dietary manipulation has evolved as the preferred method of management for this condition. Frequent oral feedings, high in carbohydrate, are recommended to maintain blood glucose levels above 70 mg/dL. Total parenteral nutrition or continuous nasogastric infusions of glucose correct most of the metabolic abnormalities associated with GSD I. The impracticality of these forms of management, however, led to the development of a dietary treatment plan that includes frequent daytime meals followed by continuous drip nocturnal enteral feedings. Hypoglycemia and death have been reported following malfunctions of the pump or dislodging of the tube.[77, 78] There are unanswered questions related to compensatory mechanisms that can develop to compensate for a lack of glucose production in individuals with GSD 1, as many with this condition have demonstrated improvements in fasting tolerance as they get older.[79] Nonetheless, the current consensus is that dietary treatment, consisting of frequent glucose feedings, needs to be maintained for life.

The use of uncooked cornstarch has been introduced due to its ability to maintain blood glucose levels over an extended period of time. The cornstarch is slowly digested by the body, allowing for the gradual release of glucose. This helps the individual to maintain a fed state for an extended period of time, from about 3 to 6 hours. The typical regimen of cornstarch consists of 1 to 2 grams cornstarch per kilogram body weight per dose, administered every 3 to 6 hours. Some individuals have even been able to forgo nocturnal feedings by ingesting cornstarch at bedtime and once in the middle of the night. Cornstarch should not be mixed in beverages containing citric acid, nor should it be heated, because that will result in a breakdown in the starch molecules that allows more rapid digestion and interferes with the release of glucose over an extended time period.[80]

The goals of nutritional therapy for GSD I are to prevent hypoglycemia, correct metabolic derangement, and provide optimal nutrition to support growth and development. Carbohydrates from complex sources should provide about 60% to 70% of the total caloric intake. A portion of the carbohydrate may be provided by the uncooked cornstarch. Protein should come from lean sources, providing 10% to 15% of total kcal. Fat should provide less than 30% of total kcal, and dietary saturated fat should be limited. Two-thirds of the kcal will be given during daytime feedings to allow for carbohydrates to be distributed frequently and evenly over the 24-hour period. One-third of kcal are administered overnight, primarily from a carbohydrate source via cornstarch or tube feedings. Meals high in starch, without free glucose, tend to provide a more sustained increase in blood glucose levels and, except during symptoms suggestive of hypoglycemia, glucose solutions alone are not recommended.[77, 78]

Monitoring and Evaluation Compliance with dietary treatment reduces the risk of developing hypoglycemia. Daytime schedules must be flexible enough to include ingestions of feedings as needed. School-age children need to carry high-carbohydrate snacks with them at all times. Infants and toddlers are given a formula concentrated in glucose polymers for symptomatic hypoglycemia. Refusal by toddlers of breakfast can cause potential problems. In such cases, mothers are encouraged to use a nasogastric tube for overnight feedings and to keep it in place until after the first meal in the morning. Since illness, particularly vomiting and diarrhea, can be life threatening to GSD I patients, sick day guidelines must be provided. If sick children are not eating in adequate amounts, the tube should be left in place to permit continuous infusion of carbohydrate. If this is not tolerated, then the child must be hospitalized to receive intravenous glucose. (See Box 26.3 for a clinician's perspective on management of GSD I in children.)

BOX 26.3 CLINICAL APPLICATIONS

Case Report for Glycogen Storage Disease Type 1

Nutrition Assessment:

Johnny is a 2-year-old boy who was diagnosed with glycogen storage disease type 1 (GSD 1) by 4 months of age, after presenting with hypoglycemia due to an extended fasting period of 6 hours. A gastrostomy tube was placed during the initial hospitalization to accommodate overnight feedings. Mother also administers tube feedings when Johnny refuses to eat meals during the day. Since then, he has been maintained on a regimen that consists of daytime feedings administered every 3 hours and continuous overnight feedings. Daytime meals consist of high-complex carbohydrate, low-fat foods with limited intakes of dairy and fruit. When Johnny refuses to eat, he is administered the formula that is used for overnight feedings. This is a low-fat formula (containing only 10% of total kcal from fat) with an added modular consisting of glucose polymers as the carbohydrate source. It contains ample amounts of protein (up to 20% of total calories) which are provided by a protein modular that also contains vitamins and minerals. The carbohydrate is administered at a rate of 6 to 8 mg glucose per kg body weight per minute. Mother checks Johnny's blood glucose level whenever he seems cranky or sleepy, and if it is less than 70 mg/dL, she will administer a mixture of glucose polymers and water via the gastrostomy tube followed by some crackers or cereal taken orally.

Johnny has maintained fairly good metabolic control with recent labs of lactate of 2.7 mmol/L (normal is less than 2.2), triglyceride of 200 (normal is less than 150), and uric acid of 6.0 mg% (normal is less than 8). He has been growing appropriately, following the 25th to 50th percentile growth curve for height and 50th to 75th percentile for weight. Biochemical parameters used to monitor nutritional status, which include complete blood cell count, albumin, total protein, and transthyretin, have been within normal limits. Intake of total kcal and protein adequately meet estimated nutrient needs based on age and weight. He takes an additional vitamin/mineral supplement to ensure that his DRI for age are met.

During a recent clinic visit to the Metabolic Center, his mother indicated that Johnny was refusing more daytime feedings and, consequently, she needed to use the tube to administer over one third of his meals. There was a tendency to overfeed Johnny via gastrostomy tube feedings as his mother was attempting to compensate for a decreased oral intake. His weight velocity has also increased, from the 50th to 75th percentile growth curve.

Nutrition Diagnosis:

A decreased oral intake is a common problem encountered by children with this disease. This is because parents, desiring to maintain euglycemia, tend to overcompensate with increased frequency of feedings. The ability for the child to experience the sensation of hunger is significantly blunted. There is also a tendency to overfeed the child, leading to excessive weight gain.

Nutrition Intervention:

Cornstarch should be considered since it can help to reduce frequency of feedings. Since it is a concentrated source of carbohydrate, which is free of fat, it can help reduce the amount of kcal administered.

Cornstarch has been added to Johnny's daytime feedings to allow for more extended periods between meals. Johnny

(continued)

now receives 2 tbsp cornstarch, providing approximately 2 g cornstarch/kg body weight/dose, at the completion of nocturnal tube feedings at 9 a.m. He is then allowed to go for a 4- to 5-hour period without eating or receiving a gastrostomy tube feeding to allow him an opportunity to become hungry and eat an afternoon meal. Johnny receives another bolus of cornstarch at 2 p.m. and again at 7 p.m. in between meals. Since the institution of cornstarch, Johnny has been eating more meals and has actually been expressing hunger prior to mealtime. Blood glucose control has remained fairly stable, with levels remaining between 70 and 140 mg/dL. He did experience a moderately low level one day, around 5 p.m., after having a very active day. Mother was encouraged to ensure that Johnny is eating more snacks or glucose-containing beverages, such as sports drinks, during these times. Mother is happy with the new feeding regimen as Johnny has started eating a greater variety of foods and he has become less dependent on the tube for daytime feedings.

Monitoring and Evaluation:

Johnny's growth, dietary intake, feeding frequency and history, and blood glucose levels will be evaluated on an ongoing basis to ensure that the current nutritional intervention is effective in optimizing his outcome.

Another significant issue with dietary management is the patient's adjustment to a decreased oral intake, since a significant amount of their caloric needs must be administered during overnight feedings. The omission of fructose and galactose (fruits, juice, milk, dairy products, and sweets that contain sucrose) from the GSD I patient's diet further limits the already restricted diet. This omission is necessary, because these sugars cannot be converted into glucose and are subsequently metabolized via alternative pathways leading to an increased production of lactic acid and lipids. Consequently, it is recommended that infants receive a formula that contains glucose polymers as the carbohydrate source.

A multivitamin/mineral supplement is usually indicated, since a high percentage of the kcal may be coming from "empty" kcal sources, such as cornstarch, and because of the limitation of food groups, including fruits and dairy, in an attempt to limit ingestion of fructose and galactose. Calcium supplementation is usually needed due to a limited intake of dairy products. Additional iron supplementation is frequently warranted for patients relying on cornstarch for night feedings, because the cornstarch can chelate and inhibit the absorption of iron. Dietary compliance is necessary to achieve the full growth potential of GSD I children and for the prevention of secondary complications of the disease.[68, 77]

Prognosis for patients with GSD I is still unclear. Prognosis during infancy has dramatically improved with the administration of dietary management. Biochemical and clinical aberrations can be substantially improved by treatment aimed at decreasing the hepatic stimulus for glycogenolysis through maintenance of blood glucose concentration between 70 and 150 mg/dL. This treatment has been very effective in providing the necessary milieu for normal growth and development. The fact that a number of patients have reached adulthood suggests that this type of dietary management is sufficient for some patients. There are many long-term complications reported in individuals with this disease, including nephropathy, hepatic adenomas, hepatocellular carcinoma, osteoporosis, anemia, pulmonary hypertension, acute pancreatitis, and polycystic ovaries. Transplantation of the liver provides a definitive cure of this otherwise life-time disease; however, the magnitude of the procedure would indicate its reservation for a small number of patients with refractory disease or hepatic malignancy.[79]

Disorders of Fat Metabolism

Etiology and Clinical Manifestations

Fatty acids are transported into the mitochondria for oxidation via a complex system.[81, 82] The fatty acids must be taken into the mitochondria using the carnitine transport system (see Figure 26.11). Carnitine palmitoyltransferase I (CPT 1) is an enzyme that allows a carnitine molecule to bind to a fatty acid molecule outside of the mitochondrion while releasing the coenzyme A segment. This acylcarnitine compound is then taken into the mitochondrion via the carnitine acylcarnitine translocase enzyme. In the mitochondrion, another enzyme, carnitine palmitoyltransferase II (CPT 2), attaches the acylated fatty acid released from the carnitine with the coenzyme A segment. The carnitine is then released to allow for the transport of more fatty acids into the mitochondria.

Once they are inside the mitochondria, the process of beta-oxidation is used to convert the fatty acids to ketone bodies, which can be used as an energy substrate in various tissues. Each time a fatty acid enters the beta-oxidation pathway, a two-carbon segment of the fatty acid is cleaved off; this segment can subsequently be used to produce ATP via the Krebs cycle. Electrons produced in these reactions are transferred to the electron transport chain. The resulting shorter fatty acid then reenters the pathway to produce additional two-carbon segments. The process of cleaving off a two-carbon segment requires four separate enzymes that are specific to the fatty acid, based on its carbon chain length. There are enzymes specific for short-, medium-, and long-chain fatty acids. Given the complexity of the many enzymes involved in the metabolism of fat, a number of disorders are known to affect this macronutrient's metabolism[81–85] (see Table 26.18).

The presentation of these disorders in patients can vary considerably. An affected individual can present in the neonatal period when breast milk or an infant formula, both of which are high in long-chain fatty acids, is introduced. In the case of medium-chain acyl-CoA dehydrogenase deficiency (MCADD), presentation does not usually occur until the infant or child is sleeping through the night. The fasting period during the night leads to the release of fatty acids that cannot be properly utilized for energy and results in hypoglycemia and/or death.

Figure 26.11 Fatty Acid Transport System

The enzyme CPT 1 is needed to bring the fatty acyl CoA across the outer mitochondria. Inside the mitochondria, CPT 2 is needed to release the fatty acyl CoA from the carnitine so that it can be oxidized.

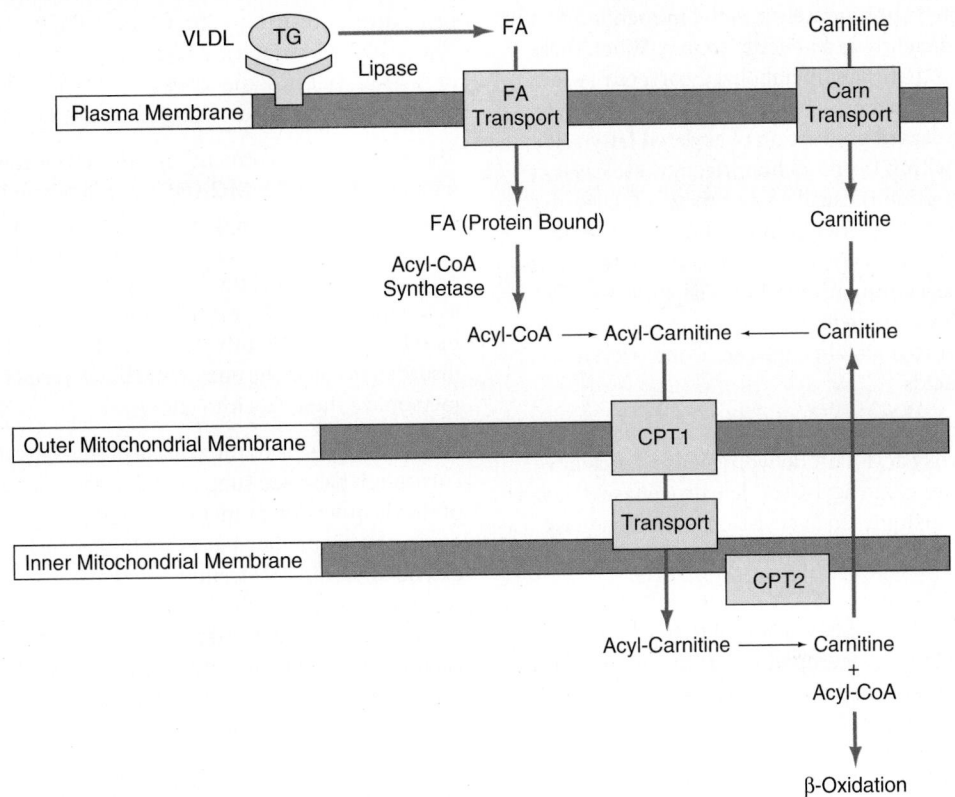

Source: Figure 1, p. 1 of Bennett MU, Ross, *Metabolic Currents*, Vol 9, #1.

Many individuals, however, can go undiagnosed for years until a metabolic stressor such as a viral illness or prolonged fast occurs. Disorders of carnitine metabolism have also been found. These disorders can present at various ages with symptoms including hypertrophic cardiomyopathy, muscle weakness, liver enlargement, or **rhabdomyolysis**. Secondary carnitine deficiency can occur with some of the fatty acid defects because the carnitine is needed to remove the free fatty acids that can not be metabolized. The implementation of newborn screening programs have now identified many individuals affected by these disorders and the incidence of LCHADD, previously believed to be quite rare, is now estimated to be 1 in 110,000 births.[86]

Table 26.18 Disorders of Fat Metabolism

- Carnitine acylcarnitine translocase deficiency
- Carnitine palmotyl transferase (CPT) I deficiency
- Carnitine palmotyl transferase (CPT) II deficiency
- Very-long-chain acyl-CoA dehydrogenase (VLCAD) deficiency
- Tri-functional protein (TFP) deficiency
- Long-chain 3 hydroxy acyl-CoA dehydrogenase (LCHAD) deficiency
- Medium-chain acyl-CoA dehydrogenase (MCAD) deficiency
- Short-chain 3 hydroxy acyl-CoA dehydrogenase (SCHAD) deficiency
- Short chain acyl-CoA dehydrogenase (SCAD) deficiency

Nutrition Therapy for Disorders of Fat Metabolism

Acute Nutrition Intervention The metabolic crises associated with disorders of fatty acid metabolism are usually precipitated by febrile illness, fasting, and/or excess energy expenditure. Acute treatment is aimed at stopping catabolism and the subsequent release of free fatty acids. Use of antipyretic and antiemetic medications to reduce fever and vomiting is helpful. Intravenous dextrose is used to provide a constant source of glucose. The high-glucose feedings used in order to prevent or recover from hypoglycemia may necessitate the use of insulin to achieve **euglycemia**. The dextrose should not be discontinued until the patient is able to maintain his/her blood sugar and to tolerate enteral feedings. The use of medium-chain triglycerides in treatment of disorders besides MCAD is an effective way to provide additional calories without increasing the load on the enzymatic block. Fluids should be provided at a rate of 1.5–2 times maintenance requirements in order to flush out the metabolites. Carnitine can be given either intravenously or by mouth to help conjugate the excess fatty acids. An emergency treatment protocol should be developed for each patient in order to ensure prompt and appropriate treatment.

Chronic Nutrition Intervention The main goals of therapy involve prevention of fasting, limiting intake of fatty acids, and/or providing an alternate substrate for metabolism.[84]

Fasting guidelines vary as glycogen stores increase with age. Table 26.19 reflects these guidelines.[86] Dietary intake of fatty acids may be mildly restricted (<30% of kcal) in some disorders such as carnitine transport disorders, since all types of fatty acids are not transported appropriately into the mitochondria and cannot be used as efficiently as an energy source. When there is a specific fatty acid that is not metabolized correctly (which typically takes place in medium-chain fatty acid oxidation defects), there is an increased production of acylated fatty acids, which can build up within the mitochondria; in these cases, a more severe fat restriction is usually recommended. Disorders such as LCHADD require a strict fat restriction because the accumulated fat metabolites are believed to contribute to clinical symptoms such as **retinopathy**. Additional guidelines for a diet high in complex carbohydrates and low in simple sugars are provided in order to maintain euglycemia and prevent the release of free fatty acids.

The dietary treatment recommendations for long-chain fats in long-chain 3 hydroxy acyl-CoA dehydrogenase deficiency (LCHADD), very-long-chain acyl-CoA dehydrogenase deficiency (VLCADD), and medium-chain acyl-CoA dehydrogenase deficiency (MCADD) are quite different and will be further delineated.

Patients with LCHADD have been reported to have problems ranging from retinopathy to neuropathy. For this reason, diet recommendations for fat restriction are stricter than in other disorders. Total fat is restricted to no more than 25% to 30% of total kcal consumed, distributed as 20% to 25% MCTs and only 5% to 10% long-chain fatty acids. In this disorder, medium-chain triglycerides (MCT), containing fatty acids of 6 to 10 carbons in length, can be used as an alternative substrate since they can bypass the enzymatic block and be used for energy production. It is recommended that the MCT be administered three to four times throughout the day in order to provide a steady source of energy.[87] Infants diagnosed with LCHADD are usually started on an infant formula containing a high percentage of **MCT oil**. The inclusion of small amounts of expressed breast milk can be calculated into the feeding plan in order to provide the infant with the immunologic benefits of maternal breast milk.

The dietary recommendations for VLCADD are similar to those for LCHADD, but are not as restrictive. With experience from newborn screening programs, many individuals with mild phenotypes are now identified. Fat recommendations are less restrictive, with 15% to 20% of the calories coming from long-chain fatty acid sources. MCT are recommended at 10% to 15% of calories, but in some cases are only provided during periods of higher energy needs such as exercise or fever.[86]

The mainstay of treatment for MCADD is the avoidance of fasting. Infants with MCADD are initially fed every 3 hours. Parents are instructed to wake the infants for the feeding. The parents are also instructed on monitoring blood glucose levels. As previously noted, the fasting period can usually be extended as the infant gets older if the baby is able to maintain his or her blood sugar consistently. MCT oil should not be used in this disorder, because the enzymatic block results in the inability to metabolize these fats into energy.

Since fasting leads to the release of free fatty acids, uncooked cornstarch has been suggested as a therapeutic option in many of the disorders of fat metabolism. It provides a steady release of glucose over a long period of time, which is beneficial during extended periods of strenuous activity or before bedtime. The use of continuous enteral feeding is another method of providing a steady source of energy substrate in order to minimize catabolism. Supplementation of the omega-3 fatty acid **docosahexaenoic acid (DHA)** to prevent the progression of retinal degeneration in LCHADD has been studied, but results thus far have shown little benefit.[88]

Monitoring and Evaluation Due to the fat restriction and the current availability of fat-free food products, over-restriction of fat intake is a concern. Fat is essential for fat-soluble vitamin absorption and is needed to supply the body with essential fatty acids and energy. Essential fatty acid deficiency can occur as a result of the dietary restriction; therefore, the prescribed diet should provide at least 3% of its calories from linoleic acid and 2% of its calories from linolenic acid. These long-chain fatty acids should be counted as part of the patient's prescribed amount of long-chain fats. Red blood cell levels of essential fatty acids need to be monitored periodically to ensure intake and absorption of these fats is sufficient.[88, 89] Table 26.20 lists the compositions of various vegetable fats.

Another concern, especially with LCHAD, can be excessive weight gain—because of the supplementation of the diet with a significant amount of MCT oil, essential plant oils, and pos-

Table 26.19 Fasting Guidelines for Disorders of Fat Metabolism

The following are suggested maximum fasting periods (when stable—not ill).

Age	Fasting Hours during the Day	Fasting Hours Overnight
Neonate	3 hours	3 hours
<6 months	4 hours	4 hours
6–12 months	4 hours	6–8 hours
12–36 months	4 hours	10–12 hours

Source: With kind permission from Springer Science+Business Media: J Inherit Metab Disease 2009:32(4):498–505. Treatment recommendations in long-chain fatty acid oxidation defects: consensus from a workshop, Spiekerkotter U, Linder M, Santer R, Grotzke M, Baumgartner MR, Boehles H, Das A, Haase C, Hennermann JB, Karall D, de Klerk H, Knerr I, Koch HG, Plecko B, Roschinger W, Schwab KO, Scheible D, Wijburg FA, Zschocke J, Mayatepek E, Wendel U.

Table 26.20 Fatty Acid Composition of Various Fats (Per 1 tsp/5 cc)

Source	Linoleic (mg)	Linolenic (mg)
Canola	990	495
Corn	2600	Trace
Flaxseed	575	2425
Peanut	1440	Trace
Safflower	3350	Trace
Walnut	2380	470

Case Report of Medium-Chain Acyl-CoA Dehydrogenase Deficiency

Case Presentation:

James is a 6 month old with medium-chain acyl-CoA dehydrogenase deficiency (MCADD). He is receiving 4–6 oz of formula every 4 hours, and has been demonstrating appropriate growth. His early morning blood sugars have been in the 70–90 mg/dL range in the past month. James just started on rice cereal and some jarred baby foods, and typically eats 2 tbsp of these foods twice a day. James is also starting to teethe, and his mom noted a low-grade fever before she put him to bed. He had only half of his 4 a.m. feeding, and his blood sugar at 7:30 in the morning was 70 mg/dL. He only had half a bottle at 8 a.m. At 9 a.m., his mother tried to feed him some cereal, but he only ate 1 tbsp and vomited a half-hour later. At 11 a.m., his mother tried to give James another bottle due to his poor early morning intake, but he only took 1 oz of formula and fell asleep.

Nutrition Assessment:

Given James's decreased intake and vomiting, his caloric intake over the last 12 hours was poor.

Nutrition Diagnosis:

In addition to poor caloric intake, James also has a low-grade fever creating added stress and increasing catabolism. Given his low-normal blood sugar first thing in the morning, James is at risk for hypoglycemia.

Nutrition Intervention:

James's mother contacted the Metabolic Clinic and was advised to take him with his emergency care protocol to the urgent care clinic for IV hydration with glucose. The glucose will provide kcal to stop the catabolic process, while the IV fluids will help to flush out any toxic metabolites that have accumulated. He will also be given IV carnitine to conjugate the toxic metabolites, detoxifying them so that they are excreted via urine output. James will receive an antipyretic to reduce his fever, in order to prevent further catabolism and help promote oral intake. The health care team will evaluate James to ensure that he does not have an infection or other condition necessitating further treatment.

Monitoring and Evaluation:

James's oral intake will be monitored to ensure that he is able to resume his usual feeding schedule before discharge.

---

sibly cornstarch and because of the desire to minimize fasting in these patients. The calories provided by these supplements need to be accounted for and subtracted from estimated energy needs.

Adjunct Medical Therapies

The use of carnitine in these disorders is controversial: some believe that it can be harmful because it can promote a greater influx of fatty acids into the mitochondria, disrupting their function, while others believe it is helpful because it acts as a scavenger, removing the acylated fatty acids. Carnitine is usually supplemented if the amount of free carnitine is low and/or the ratio of esterified to free carnitine is high. Overall benefit of carnitine supplementation in these disorders is still to be determined.

Conclusion

Major advances have been made in identifying the specific enzymatic defects in more than 200 disorders caused by inborn errors of metabolism, allowing for accurate diagnosis. However, despite these major diagnostic achievements, patients and their families have been disappointed by the absence of specific therapies for some of the more debilitating disorders. Therapeutic endeavors have primarily involved attempts to alter the disease course by manipulations at the level of the metabolic or biochemical defects. Restriction of the accumulated precursor by dietary management or administration of appropriate metabolic inhibitors and supplementation of the deficient metabolic products are the main treatment strategies. The uses of alternative metabolic pathways for excretion of toxic metabolites, and the use of cofactors to stimulate residual enzymatic function, have been employed in various disorders. Several medications and hormones have been used in attempts to correct for the metabolic imbalance. Growth hormone, in its function of stimulating anabolism, has helped to stabilize patients and ultimately has led to improvement in metabolic control. Transplantation of organs capable of producing the normal enzyme has been utilized when other treatment options result in a less than optimal outcome. The applicability of this approach is restricted by the significant surgical risk as well as the limited number of organs available for transplantation.

Nutrition management of patients with inborn errors of metabolism in the clinical setting has been changing. With the implementation of expanded newborn screening, there is more information about the phenotypes of specific disorders and nutrition therapy can be targeted to address the specific needs of that patient and in some situations even be eliminated. The ability to target therapies or lessen diet restrictions can mean a more normalized diet, which can enhance compliance and reduce the risk for nutrient deficiencies. Because patients are now surviving longer, clinicians are challenged to integrate common adult nutrition issues such as diabetes or hypercholesterolemia into their metabolic therapies. Pregnancy is another area that requires a great deal of coordination of care. There are a myriad of nutritional concerns that are added to the care of a woman with an inborn error of metabolism. Metabolic status must be optimal even prior to pregnancies as some metabolites are toxic to the fetus. Increased protein needs for the growing fetus must be carefully balanced with the mother's ability to metabolize the added protein. Metabolic clinics need to work with patients to transition them to adult medical programs that can not only manage their general health needs, but address their metabolic needs as well. Transition of care should optimally start as early as elementary school. The dietitian needs to provide age-appropriate nutrition education directly to the patient so she/he will ultimately be able to follow the prescribed treatment plan while living independently.

Development of adjunct therapies will help to improve the outcomes of patients with inborn errors of metabolism, but nutrition therapy will remain the cornerstone of treatment of inborn errors.

PRACTITIONER INTERVIEW

Kathleen Huntington, MS, RD, LD *Metabolic Clinic, Child Development & Rehabilitation Center (CDRC), Oregon Health & Science University, Portland*

How long have you been an RD?

I've been an RD for over 20 years and I've worked in the Metabolic Clinic at CDRC since March 1987. The most common metabolic disorder we see is phenylketonuria (PKU), but we see all of them. For PKU our program implements an assessment for growth velocity, essential fatty acids, serum methylmalonic acid (MMA) levels (to check vitamin B_{12} level), CBC, ferritin, vitamin D 25 OH, and phenylalanine/tyrosine (PHE/TYR) levels when they come to the clinic. We review historical PHE levels previous to the clinic visit going back either six or twelve months and average the lab results. In addition we use a database that reports out an ordering profile for medical foods for the clients. This database is unique to our program and is not available in other metabolic centers. This ordering profile tells us immediately whether there may be a potential problem involving nutritional adequacy if the amount of medical protein that is ordered does not match or exceed the amount of medical protein prescribed. For other metabolic conditions, we conduct a biochemical review that is specific to the abnormal chemistry associated with the enzyme block along with a growth assessment and a review of their medical food ordering profile.

What are the common nutrition problems that you see in your clients?

Common nutrition problems for clients diagnosed with amino acidopathies, organic acidemias, and urea cycle defects involve compromised protein intake because of their protein-restricted diets compounded by inconsistent use of their medical protein formulas. Many patients demonstrate low vitamin D levels, suggesting that the vitamin D in medical formulas should be increased. Geographic location may contribute to this effect as well because of the low intensity of sunlight a significant part of the year. Working with this population, my biggest challenges are: (1) parents who are unable to organize themselves to meet the challenge of day-to-day planning and implementing medical nutritional therapy in the home, plus (2) health plan providers who deny coverage for medical foods and dietitian services. This results in additional costs for the parents to shoulder that can impact on treatment outcome in so far as the medical foods are concerned.

How has the science of dietetics changed for you in your practice? What do you do to make sure that you correlate your practice with the latest science? What advice can you give to newly practicing dietitians and students?

There have been considerable advances in the knowledge and understanding of genetics and nutritional genomics. Metabolic genetic diseases are not necessarily specific to the pediatric population and can be seen across the lifespan. To keep up with changes in the field, I read journal articles and go to conferences. My advice to students is to focus on biochemistry and metabolism; take advantage of courses and/or training in genetics, nutritional genomics, counseling for behavior change; and expand your skills in desktop publishing for development of patient education materials.

Application of the Nutrition Care Process: PKU

Introduction:

Nine-day-old female who had a positive newborn screen for PKU at 2 days of age. 40 weeks gestation. Physical exam indicates normal infant reflexes and all other assessments WNL.

1. What is the screening procedure for PKU?

2. Why is it important to accomplish screening as soon after birth as possible?

Nutrition Assessment:

Food/Nutrition-Related History:

Soy formula—4 oz. every four hours.

Recommended Treatment Plan:

Initial nutrition therapy goal to reduce phenylalanine levels to treatment range of 2–6 mg/dL. Prescription included 52 grams of Phenex-1 and 15 grams of Similac powder with iron providing 25 mg/kg of phenylalanine, 316 mg/kg of tyrosine, 3.5 g protein/kg, and 122 kcal/kg. 16 oz of formula yields 20 kcal/oz.

3. What is the goal of the PKU diet?

4. What is Phenex? How is it different from Similac or other standard infant formulas?

5. What would be the possible long-term complications of untreated PKU?

6. How long will the infant have to remain on a special diet?

7. In general, how will her diet change as she grows?

Anthropometric Measurements:

5.5 lb birthweight; 18.5 inches length.

Biochemical Data:

Day 2: phenylalanine 15 mg/dL; Day 8: phenylalanine 34.4 mg/dL and tyrosine 1.2 mg/dL.

Nutrition Diagnosis:

8. Identify at least two nutrition problems based on the nutrition assessment and medical history. Determine the diagnostic term for each nutrition problem. Next, identify the etiology of each nutrition problem. Finally, identify the signs and symptoms that support the evidence for these nutrition problems.

Nutrition Intervention:

9. Compare the prescribed diet to your calculation of energy and protein requirements.

Nutrition Monitoring and Evaluation:

10. Determine nutrition criteria for monitoring and evaluation for each nutrition diagnosis that you identified.

WEB LINKS

United Mitochondrial Disease Foundation Information about vitamin therapy for mitochondrial diseases.
http://www.umdf.org

Parents of Galactosemic Children Support for parents regarding dietary management of galactosemia.
http://www.galactosemia.org

PKU News Latest research on PKU and information on new products.
http://www.pkunews.org

National Organization for Rare Disorders Information about many rare disorders.
http://www.rarediseases.org

Fatty Acid Oxidation Disorders Family Support Group Parent support organization with latest research on fatty acid oxida-tion disorders, family stories, and new products.
http://www.fodsupport.org

Organic Acidemia Association Parent support organization with latest research on organic acid disorders, family stories, and new products.
http://www.oaanews.org

MSUD Family Support Group Parent support organization with latest research on maple syrup urine disease, family stories, and new products.
http://www.msud-support.org

Genetic Metabolic Dietitians, International Professional organization for dietitians and other health practitioners working in the field of metabolic disorders.
http://www.gmdi.org

Society for Inherited Metabolic Disorders Professional organization for dietitians and other health practitioners working in the field of metabolic disorders.
http://www.simd.org

National Urea Cycle Disorders Foundation Parent support organization with latest research on urea cycle disorders, family stories, and new products.
http://www.nucdf.org

Online Mendelian Inheritance in Man (OMIM) Detailed description of inherited metabolic disorders.
http://www.ncbi.nlm.nih.gov/omim/

1. What is the difference between selective and nonselective screening for metabolic disorders?

2. What amino acids are restricted in maple syrup urine disease?

3. What amino acid becomes essential secondary to the enzymatic block in PKU?

4. What is periorfacial acrodermatitis, and when does it occur?

5. What is the role of carnitine?

6. How does fasting affect disorders of fat metabolism?

7. Which vitamins are used as cofactors in mitochondrial disorders?

8. What are some clinical signs/symptoms that would indicate the need for a further metabolic work-up?

9. What vitamin deficiency can present as methylmalonic acidemia?

10. What vitamins are used for their antioxidant properties in the treatment of mitochondrial disorders?

11. What sugars are restricted in GSD I?

12. What percentage of protein intake should come from essential amino acids in urea cycle disorders?

13. Which food groups are restricted in galactosemia, and why?

14. What is the mainstay of dietary management for GSD I?

15. What becomes significantly elevated in the blood of a person with a urea cycle disorder prior to treatment?

16. How is the diet prescription different in an individual with a more severe enzymatic defect vs. a less severe defect?

17. What are the initial symptoms of a child with a UCD?

18. How much protein should be given for an individual with a UCD?

19. What parameters should be monitored in an individual with a UCD?

20. In GSD, there are deficiencies of enzymes involved in synthesizing and degrading which polysaccharide?

21. Describe the diet composition on an individual with GSD I.

22. What typically happens when a person with HFI ingests fructose?

23. What are some of the complications that an individual with HFI will encounter after ingestion of fructose?

24. Describe successful dietary treatment of HFI.

25. What is monitored in the blood to measure dietary compliance in galactosemia?

26. What are some possible theories on why some patients with galactosemia experience poor outcomes?

27. What types of formula are used to treat infants with galactosemia?

28. Why is MCT oil used in disorders of long chain fat metabolism? Why can't it be used in MCADD?

29. In what conditions can you breast feed a child with a metabolic disorder? When would it be contraindicated?

30. What are the three main goals for the acute treatment of an amino acid disorder?

REFERENCES

1. Collins, JE, Leonard JV. The dietary management of inborn errors of metabolism. Human Nutrition: Applied Nutrition. 1985;39A:255–72.

2. Garrod AE, Oxon MD, Lond FRCP. The Croonian Lectures on inborn errors of metabolism. Lancet. 1908:172(4427); 1–7.

3. Packman, S. Nutritional therapy in inborn errors of metabolism. Perinatol Neonatol. 1986;10:33–45.

4. Schmidt K. A primer to the inborn errors of metabolism for perinatal and neonatal nurses. J Perinat Neonatal Nurs. 1989;2(4):60–71.

5. Applegarth DA, Toone JR, Lowry RB. Incidence of inborn errors of metabolism in British Columbia, 1969–1996. Pediatrics. 2000 Jan;105(1):e10.

6. Elsas LJ, Acosta PB: Nutrition support of inherited metabolic disease. In Shils ME, Olson, JA, Shike M. Ross AC (Editors): Modern Nutrition in Health and Disease. Williams & Wilkins, A Waverly Company; 1999, 1003–56.

7. Cederbaum SD. Diagnosing metabolic disease in the neonate. Metabolic Currents. 1989;2(1):1–8.

8. Burton, BK: Inborn errors of metabolism in Infancy: a guide to diagnosis. Pediatrics. 1998;102(6):e69.

9. National Newborn Screening and Genetics Resource Center. National Newborn Screening Status Report. Austin: National Newborn Screening and Genetics Resource Center; 2006. Available at http://genes-r-us.uthscsa.edu/resources/genetics/reports.htm

10. Ensenauer R, Vockley J, Willard JM, Huey JC, Sass JO, Edland SD, Burton BK, Berry SA, Santer R, Grunert S, Koch HG, Marquardt I, Rinaldo P, Hahn S, Matern D. A common mutation is associated with a mild, potentially asymptomatic phenotype in patients with isovaleric acidemia diagnosed by newborn screening. Am J Hum Genet. 2004;75(6):1136–42.

11. Saudebray JM, Nassogne MC, deLanley P, Towati G. Clinical approach to inherited metabolic disorders in neonates: an overview. Semin. Neonatal. 2002;7(1):17–26.

12. Ogier de Bawny H. Management and emergency treatments of neonates with a suspicion of inborn errors of metabolism. Semin Neonatal. 2002;7(1):17–26.

13. Nevin-Folino N, Hanson C. Using the Nutrition Care Process in the neonatal intensive care unit. ICAN 2009;1(4):190–204.

14. Scriver CR. The PAH Gene, phenylketonuria, and a paradigm shift. Hum Mutat 2007;28(9):831–45.

15. Scriver C, Beaudet A, Sly W, and Valle D. (Editors). The metabolic basis of inherited diseases. 8th ed. New York: McGraw-Hill; 2001.

16. Nyhan WL, Pinar TO. Atlas of metabolic diseases. London: Chapman & Hall Medical; 1998.

17. Anderson PJ, Wood SJ, Francis DE, Coleman L, Anderson V, Boneh A. Are neuropsychological impairments in children with early-treated phenylketonuria related to white matter abnormalities or elevated phenylalanine levels? Dev Neuropsych. 2007;32:645–68.

18. Weglage J, Grenzeback M, Pietsch M, Feldmann R, Linnenbank R, Deneche J, Koch HG. Behavioural and emotional problems in early-treated adolescents with phenylketonuria in comparison with diabetic patients and healthy controls. J Inherit Metab Dis. 2000;23:487–96.

19. Waisbren SE, Noel K, Fahrbach K, Cella C, Frame D, Dorenbaum A. Phenylalanine blood levels and clinical outcomes in phenylketonuria: a systematic literature review and meta-analysis. Mol Genet Metab. 2007;92:63–70.

20. Koch, R, Burton, B, Hoganson G, et al. Phenylketonuria in adulthood: a collaborative study. JIMD. 2002;25:333–46.

21. Gassio R, Fuste E, Lopez-Sala A, Artuch R, Vilaseca MA, Campistol J. School performance in early and continuously treated phenylketonuria. Pediatr Neurol 2005;33:267–71.

22. Levy HL, Milanowski A, Chakrapani A, et al. Efficacy of sapropterin dihydrochloride (tetrahydrobiopterin, 6R-BH4) for reduction of phenylalanine concentration in patients with phenylketonuria: a phase III randomized placebo-controlled study. Lancet. 2007;370:504–10.

23. Burton BK, Grange DK, Milanowski A, et al. The response of patients with phenylketonuria and elevated serum phenylalanine to treatment with oral sapropterin dihydrochloride (6R-tetrahydrobiopterin): a phase II, multicentre, open-label, screening study. JIMD. 2007;30:700–77.

24. Matalon R, Koch R, Michals-Matalon K, et al. Biopterin responsive phenylalanine hydroxylase deficiency. Genet Med 2004;6:27–32.

25. Muntau AC, Roschinger W, Habich M, et al. Tetrahydrobiopterin as an alternative treatment for mild phenylketonuria. N Engl J Med. 2002;347:2122–32.

26. Hennermann JB, Buhrer C, Blau N, Vetter B, Monch E. Long-term treatment with tetrahydrobiopterin increases phenylalanine tolerance in children with severe phenotype of phenylketonuria. Mol Genet Metab. 2005;86(Suppl1):S148–S152.

27. Trefz FK, Burton BK, Longo N, et al. Efficacy of sapropterin dihydrochloride in increasing phenylalanine tolerance in children with phenylketonuria: a phase III randomized, double-blind, placebo-controlled study. J Pediatr. 2009;154:700–707.

28. Singh R, Jurecki E, Rohr F. Recommendations for personalized dietary adjustments based on patient response to tetrahydrobiopterin (BH4) in phenylketonuria. Top Clin Nutr. 2008;23:149–57.

29. Phenylketonuria: Screening and Management. National Institutes of Health, Consensus Development Conference Statement, National Institute of Child Health and Development, 2000 Oct 16–18;17(3):1–27.

30. Van Spronsen FJ, Burgard P. The truth of treating patients with phenylketonuria after childhood: the need for a new guideline. JIMD. 2008; doi 10.1007/s10545-008-0918-6

31. MacDonald A, Rylance G, Davies P, Asplin D, Hall SK, Booth IW. Administration of protein substitute and quality of control in phenylketonuria: a randomized study. J Inherit Metab Disease. 2003;26:319–26.

32. Kindt E, Halvorsen S. The need of essential amino acids in children: an evaluation based on the intake of phenylalanine, tyrosine, leucine, isoleucine and valine in children with phenylketonuria, tyrosine amino transferase defect and maple syrup urine disease. Am J Clin Nutr. 1980;33:279–86.

33. Ney D, Bay C, Saudubray JM, Kelts DG, Kulovich S, Sweetman L, Nyhan WL. An evaluation of protein requirements in methylmalonic academia. J Inherit Meta Disease. 1985;8:132–42.

34. Parsons HG, Carter RJ, Unrath M, Snyder FF. Evaluation of branched-chain amino acid intake in children with maple syrup urine disease and methylmalonic aciduria. J Inherit Metab Disease. 1990;13:125–36.

35. Feillet F, Bodamer AF, Dixon, MA, Sequeira S, Leonard JV. Resting energy expenditure in disorders of propionate metabolism. J Peds. 2000;136(5):659–63.

36. Acosta PB, Yannicelli S, Singh R, Elsas LJ, Mofidi S, Steiner RD. Iron status of children with phenylketonuria undergoing nutrition therapy assessed by transferring receptors. Genet Med. 2004;6:96–101.

37. Greeves LG, Carson DJ, Craig BG, McMaster D. Potentially life-threatening cardiac dysrhythmia in a child with selenium deficiency and phenylketonuria. Acta Paediatr Scand. 1990;79(12):1259–62.

38. Gropper SS, Acosta PB, Clarke-Sheehan N. Trace element status of children with PKU and normal children. JADA. 1988; 88:459–65.

39. Wolf, B, Hsia YE, Sweetman, L, Gravel R, Harris DJ, Nyhan WL. Propionic Acidemia: a clinical update. J. Peds. 1981;99(6):835–46.

40. Dobbelaere D, Michaud L, Debrabander A, Vanderbecken S, Gottrand F, Turck D, Farriaux JP. Evaluation of nutritional status and pathophysiology of growth retardation in patients with phenylketonuria. J Inherit Metab Disease. 2003;26:1–11.

41. Acosta PB, Yannicelli S, Singh R, et al. Nutrient intakes and physical growth of children with phenylketonuria undergoing nutrition therapy. JADA. 2003;103(9):1167–73.

42. Allen, JR, Humphries IR, Waters DL, Roberts DC, Lipson AH, Howman-Giles RG, Gaskin KJ. Decreased bone mineral density in children with phenylketonuria. Am J Clin Nutr. 1994;59:419–22.

43. Al-Qadresh A, Schulpis KH, Athanasopoulou H, Mengreli C, Skarpalezou A, Voskaki I. Bone mineral status in children with phenylketonuria under treatment. Acta Paediatr. 1998 Nov;87(11):1162–66.

44. Modan-Moses D, Vered I, Schwartz G, et al. Peak bone mass in patients with phenylketonuria. JIMD. 2007;300:202–08.

45. Yanicelli S, Medeiros DM. Elevated plasma phenylalanine concentrations may adversely affect bone status of phenylketonuric mice. J Inherit Metab Disease. 2002;25(5):347–61.

46. Giacoia GP, Berry G. Acrodermatitis Enteropathica-like syndrome secondary to isoleucine deficiency during treatment of maple syrup urine disease. AJDC. 1993;147:954–56.

47. De Raeve L, De Meirleir L, Ramet J, Vandenplas Y, Gerlo E. Acrodermatitis enteropathica-like cutaneous lesions in organic aciduria. J Peds. 1994;124(3): 416–20.

48. Nyhan WL, Ozand PT, Barshop B. Propionic academia. In: Atlas of metabolic diseases. 2nd ed. London: Hodder Arnold; 2005, 8–17.

49. Dionisi-Vici C, Deodato F, Roschinger W, Rhead W, Wilcken B. Classical organic acidurias, propionic aciduria, methylmalonic aciduria and isovaleric aciduria: long-term outcome and effects of expanded newborn screening using tandem mass spectrometry. JIMD. 2006;29:383–89.

50. Surtees RAH, Matthew EE, Leonard JV. Neurologic outcome of propionic academia. Pediatr Neurol. 1992;8:333–37.

51. Yannicelli, S. Nutrition therapy of organic acidemias with amino acid-based formulas: Emphasis on methylmalonic and propionic academia. JIMD. 2006; 29:281–87.

52. Wendel U, de Baulny O. (2006) Branched-chain organic acidurias/acidemias. In: Fernandes J, Saudubray JM, van den Berghe G, Walter JH. (eds.). Inborn metabolic diseases, diagnosis and treatment. 4th rev. ed. Germany: Springer Medizin Verlag, 247–62.

53. North KN, Korson MS, Gopai YR, Rohr FJ, Brazelton TB, Waisbren SE, Warman ML. Neonatal-onset propionic academia: neurologic and development profiles, and implications for management. J Pediatr 1995;126:916–22.

54. Bain MD, Borriello SP, Tracey BM, Jones M, Reed PJ, Chalmers RA, et al. Contribution of gut bacterial metabolism to human metabolic disease. Lancet. 1998;1078–79.

55. Walter JH. L-Carnitine in inborn errors of metabolism: What is the evidence? J. Inherit Metab Dis. 2003;26:181–88.

56. Summar M, Tuckman M. Proceedings of a consensus conference for the management of patients with urea cycle disorders. J Pediatr. 2001;138:S6–S10.

57. Leonard JV, Morris AA. Urea cycle disorders. Semin. Neonatal. 2002;7(1): 27–35.

58. Gropman AL, Summar M, Leonard JV. Neurological implications of urea cycle disorder. J. Inherit Metab Dis. 2007;30:865–69.

59. Brusilow SW, Horwich AL. Urea cycle enzymes. In Scriver C, Beaudet A, Sly W, Valle D, editors: The metabolic basis of inherited diseases. New York: McGraw-Hill; 2001, 1909–63.

60. Feillet F, Leonard JV. Alternative pathway therapy for urea cycle disorders. J Inherit Metab Dis. 1998;21(1):101–11.

61. Summar M. Current strategies for the management of neonatal urea cycle disorders. J Pediatr. 2001;138:S30–S39.

62. Leonard JV. The nutritional management of urea cycle disorders. J Pediatr 2001;138:S40–S45.

63. Singh RH. Nutritional management of patients with urea cycle disorders. J Inherit Metab Dis. 2007;30:880–887

64. Buse MB, Reid SS. Leucine: A possible regulator of protein turnover in muscle. JClin Invest. 1975:56(5):1250–61.

65. Wallace DC. Common causes of complex disorders. Exceptional Parent. 1997; June:39–52.

66. Smeitink JAM. Mitochondrial disorders: Clinical presentation and diagnostic dilemmas. J Inherit Metab Disease 2003;26:199–207.

67. Przyrember H. Therapy of Mitochondrial Disorders. J Inherit Metab Dis. 1987;10:129–46.

68. Acosta PA, Yannicelli S. Nutrition support protocols. 4th ed. Ross Metabolic Formula System. 2001.

69. Schrock-Kelley S, Abbott M, Jurecki E, Packman S. Acquired methylmalonic academia in a breast fed infant: issues for the genetic counselor. National Society of Genetic Counselors, 17th Annual Education Conference; 1998.

70. Holton JB, Walker JH, Tyfield LA. Galactosemia. In: Scriver C, Beaudet A, Sly W, Valle D, editors: The metabolic basis of inherited diseases. Vol. 1. New York: McGraw-Hill; 2001, 1553–87.

71. Potter NL, Lazarus J-AC, Johnson JM, Steiner RD, Shriberg LD. Correlates of language impairment in children with galactosemia. J Inherit Metab Dis. 2008;31:524–32.

72. Morrow, G. Diagnosis and Discussion. Hereditary Fructose Intolerance. Arch Pediatr Adolesc Med. 1997;151:1166.

73. Steinmann B, Gitzelmann R, Van den Berghe G. Disorders in fructose metabolism. In: Scriver C, Beaudet A, Sly W, and Valle D, editors: The metabolic basis of inherited diseases. Vol. 1. New York: McGraw-Hill; 2001, 1489–520.

74. Ryman, BE. The glycogen storage diseases. J. Clin. Path. 1974;(Suppl 8):106–21.

75. Chen, YT, Burchell A. 1995. Glycogen storage diseases. In: Scriver C, Beaudet A, Sly W, Valle D, editors. The metabolic and molecular basis of inherited disease. Vol. 1. New York: McGraw-Hill; 2001, 935–65.

76. Greene HL, Slonim AE, Burr IM. Type I glycogen storage disease: A metabolic basis for advanced in treatment. Adv. Pediatr. 1979;26:63–92.

77. Daeschel IE, Janick LS, Kramish MJ, Coleman RA. Diet and growth of children with glycogen storage dieases types I and III. JADA. 1983;83:135–41.

78. Folk CC. Greene HL. Dietary management of type I glycogen storage disease. JADA. 1984;84:293–301.

79. Moses SW. Historical highlights and unsolved problems in glycogen storage disease type 1. Eur J Pediatr. 2002;161(Suppl 1):s2–9.

80. Chen YT, Buzzarre CH, Lee MM, Sidbury JB, Coleman RA. Type 1 glycogen storage disease: nine years of management with cornstarch. Eur J Pediatr. 1993;152 (Suppl 1):s56–s59.

81. Pollitt RJ. Disorders of mitochondrial long-chain fatty acid oxidation. J. Inherit Metab Dis. 1995;18:473–90.

82. Bennett MJ. Mitochondrial fatty acid oxidation defects: biochemistry, diagnosis and therapy. Metabolic Currents (Ross). 1998;9(1):1–6.

83. Przyrembel H, Jakobs C, Ijlst L, De Klerk JBC, Wanders RJA. Long-chain 3-hydroxyacyl-CoA dehydrogenase

deficiency. J Inherit Metab Disease. 1991;14:674–80.

84. Hale DE, Bennett MJ. Fatty acid oxidation disorders: a new class of metabolic diseases. J Peds. 1992;121(1):1–11.

85. Tein I. Carnitine transport: Pathophysiology and metabolism of known molecular defects. J Inherit Metab Disease. 2003;26:147–69.

86. Spiekerkotter U, Linder M, Santer R, Grotzke M, Baumgartner MR, Boehles H, Das A, Haase C, Hennermann JB, Karall D, de Klerk H, Knerr I, Koch HG, Plecko B, Roschinger W, Schwab KO, Scheible D, Wijburg FA, Zschocke J, Mayatepek E, Wendel U. Treatment recommendations in long-chain fatty acid oxidation defects: consensus from a workshop. J Inherit Metab Disease. 2009:32(4):498–505.

87. Lund AM, Dixon MA, Vreken P, Leonard JV, Morris AAM. What is the role of medium-chain triglycerides in the management of long-chain 3-hydroxyacyl-CoA dehydrogenase deficiency? J Inherit Metab Disease. 2003;26:353–60.

88. Lund AM, Dixon MA, Vreken P, Leonard JV, Morris AAM. Plasma and erythrocyte fatty acid concentrations in long-chain 3-hydroxyacyl-CoA dehydrogenase deficiency. J Inherit Metab Disease. 2003;26:410–412.

89. Ruiz-Sanz JI, Aldamiz-Echevarria L, Arrizabalaga J, Aquino L, Jimeno P, Perez-nanclares G, et al. Polyunsaturated fatty acid deficiency during dietary treatment of very long-chain acyl-CoA dehydrogenase deficiency. Rescue with soybean oil. J Inherit Metab Disease. 2001;24:493–503.

Chapter 1

1. *Identify members of the clinical nutrition care team. What are the major tasks performed by the clinical dietitian and the chief clinical manager?*

   The clinical nutrition care team is made up a director of nutrition services, chief clinical manager or clinical nutrition manager, clinical dietitians, registered dietetic technicians (DTR), and dietetic assistants or dietary clerks. The clinical dietitian uses the nutrition care process to perform nutritional assessment and diagnosis, design appropriate nutritional interventions, and evaluate and monitor patients. The chief clinical manager oversees the clinical dietitians, DTRs, and dietetic assistants/clerks. They are involved in the processes of hiring, evaluating, and training employees, reviewing productivity reports, writing job descriptions, scheduling employees, developing policies and procedures, designing performance standards, and developing and implementing goals and objectives of the department.

2. *What are the five components needed for critical thinking skills? Why is supervised practice a requirement for becoming a RD?*

   The five components of critical thinking include specific knowledge base, experience, competence, attitudes, and standards. Supervised practice is a requirement because it is real-life experience that encourages intellectual curiosity and information retention. While textbook and classroom learning promote specific knowledge base competencies, it is through working with actual patients that a student can gain confidence and the experience necessary to fully develop critical thinking skills.

3. *Why is continuing education necessary for the practice of dietetics?*

   Continuing education is important for the dietetic professional so that they may apply the most current information to their practice. Nutritional science is a field that is continually changing and advancing, and as such, requires the nutritional professional to stay current on evolving topics.

4. *What are the components of medical problem solving? How does evidence-based dietetics practice contribute to critical thinking skills?*

   Problem solving involves identifying the problem and using the gathered information to solve the problem. Components of problem solving include making decisions and using diagnostic reasoning. Decision making involves several activities, including identifying and defining the problem, assessing options for solutions to the problem, weighing each option against a set of criteria, testing possible options, evaluating the consequences of the decision, and finally, making a decision. The result of this process is an informed decision supported by evidence and reasoning. Diagnostic reasoning involves clinical judgments that result in an informal judgment or a formal diagnosis. A dietitian uses diagnostic reasoning to determine the nutrition diagnosis and monitor a patient's progress in response to nutrition therapy. To make effective nutritional care decisions, the dietitian must use critical thinking skills, which include five components—specific knowledge base, experience, competence, attitudes, and standards. Evidence-based dietetics practice combines professional expertise and judgment with client, customer, and community values, and evaluates results. Research is constantly evolving in the field of nutrition, so the dietitian must keep current on research findings in order to make informed decisions.

5. *A new friend finds out you are a nutrition major and asks for your advice about overeating late in the day. She tells you that she has no time to eat lunch and wants to save money, but then she eats too much when she gets home. Suggest three possible solutions. What are the possible consequences (both positive and negative) of each solution? How could your attitude affect each solution?*

   Answers will vary but here are some possible solutions/consequences:

   - Make sure to eat a balanced breakfast—will keep you full throughout the morning; requires time and possibly planning on friend's part

   - Pack snacks that are affordable and nutritious to snack on throughout the day, like apples, protein snack bars and nuts—some snack items are "empty calories"; fruits out of season can be very expensive and can spoil quickly

   - Perhaps a "meal on the go" option, like a smoothie/possibly high in calories but also could be good source of fruit and vitamins

   - Pack lunch the night before school so that it is on hand when you need it during the day; depending on the lunch, it could be expensive and time consuming but will definitely help prevent overeating between lunch and dinner

   Our attitude reflects our values, so we need to be sensitive when offering solutions, as our values and beliefs may not match our patient's.

6. *Why is outcomes research necessary for the advancement of dietetics practice?*

   Outcomes research is the evaluation of care that focuses on the status of participants after receiving care. Therefore, as we evaluate the success or failure of a particular method of care, we can improve the level of care we provide and share our results with other dietitians so that other health care professionals can also benefit from our experience. In this regard, each dietitian's experience can play a role in improving the nutritional care provided by other dietitians.

Chapter 2

1. *List the internal and external factors that influence a person's ability to maintain optimal nutritional health.*

   Internal Factors

   Human Biological Factors (determines nutrient requirements):

   • Biological Factors (age, sex, genetics)

   • Physiological Stages (growth, pregnancy, lactation, aging)

   • Pathological Factors (disease, trauma, altered organ function or metabolism)

   Lifestyle Factors (determines food, physical activity, and related choices):

   • Attitudes/beliefs

   • Knowledge

   • Behaviors

   Food and Nutrient Factors (determines the type and amount of nutrients available for use by the body):

   • Intake/composition

   • Quantity

   • Quality

   • Feeding route and form

   External Factors

   Environmental Factors (impact consumption and lifestyle):

   • Social (cultural food practices/beliefs, parenting, peer influences)

   • Economic (household finances, economy of the community/country)

   • Food Safety and Sanitation (contaminated or unwholesome food, unsafe food handling)

   • Food availability/access

   System Factors (impact delivery and services):

   • Health care system

   • Educational system

   • Food supply system (industry, agriculture, institutions)

2. *What is the purpose of providing nutrition care?*

   The purpose of providing nutrition care is to restore a state of nutritional balance by influencing whatever factors are contributing to the imbalance or altered state of nutritional status.

3. *List the four steps of ADA's Standardized Nutrition Care Process (NCP). Briefly describe each step.*

   Nutrition Assessment: a systematic process of obtaining, verifying, and interpreting data in order to make decisions about the nature and cause of nutrition-related problems.

   Nutrition Diagnosis: the identification and descriptive labeling of an actual occurrence of a nutrition problem that dietetics practitioners are responsible for treating independently. PES (problem, etiology, signs and symptoms) statements are used in this step. It is the format used in the NCP to write a nutrition diagnosis.

   Nutrition Intervention: a specific set of activities and associated materials used to address a (nutrition-related) problem. It is a client-driven process and should be logically linked to the cause of the problems. The intent of the intervention is to change a nutrition-related behavior, risk factor, environmental condition, or aspect of health status for an individual or larger group. This includes the establishment of ideal goals and expected outcomes.

   Nutrition Monitoring and Evaluation: an active commitment to measuring and recording the appropriate outcome indicators relevant to a nutrition diagnosis in order to determine the degree to which progress is being made and whether or not the client's goals are being met. Progress is monitored, measured, and evaluated on a planned schedule.

4. *Why is it important to have standardized nutrition diagnostic terminology in the practice of dietetics?*

   It is important to have standardized nutrition diagnostic terminology because it allows for a way to classify, measure, and report outcomes of nutrition interventions in order to demonstrate the effectiveness of nutrition care. It also allows for a more logical and systematic process.

5. *List and briefly describe the domains that are part of the assessment terminology.*

   • **Food/Nutrition-Related History (FH):** food and nutrient intake, medication/herbal supplement intake, knowledge/beliefs/attitudes and behaviors, food and supply availability, physical activity, and nutrition quality of life

   • **Biochemical Data, Medical Tests, and Procedures (BD):** lab data (electrolytes, glucose) and tests (gastric emptying time, resting metabolic rate)

   • **Anthropometric Measurements (AD):** height, weight, BMI, growth pattern indices/percentile ranks, and weight history

   • **Nutrition-Focused Physical Findings (PD):** physical appearance, muscle and fat wasting, swallow function, appetite

   • **Client History (CH):** personal history, medical/health/family history, treatments and complementary/alternative medicine use, and social history

6. *List and briefly describe the three domains of nutrition diagnostic terms.*

   • **Intake Domain:** nutrition problems related to intake of energy, nutrients, fluids, and bioactive substances through oral diet or nutrition support

- **Clinical Domain:** nutrition problems that are related to medical or physical conditions (includes problems with swallowing, chewing, digestion, absorption, and maintaining appropriate weight)

- **Behavioral-Environmental Domain:** problems that are related to knowledge, attitudes/beliefs, physical environment, or access to food and food safety

7. *Explain what P, E, and S are in the nutrition diagnosis.*

- The problem (P) describes in a general way an alteration in the client's nutritional status. Words such as *excessive*, *inadequate*, and *inappropriate* are frequently used.

- The etiology (E) or related factors are those factors that contribute to the cause or existence of a particular problem.

- The signs and symptoms (S) are the defining characteristics obtained from the subjective and objective nutrition assessment data. These provide evidence of the problem and describe the severity of the problem.

A PES statement is generally written in the following format: the problem (P) related to the etiology (E) as evidenced by the signs and symptoms (S).

8. *Write an example of a PES nutrition diagnosis.*

"Inadequate energy intake (P) related to changes in taste and appetite (E) as evidenced by average daily kcal intake 50% less than estimated recommendations (S)."

9. *How does the nutrition diagnosis relate to the other steps in the nutrition care process?*

The nutrition diagnosis is the missing link between the nutrition assessment and the nutrition intervention. An accurate nutrition diagnosis is generated from a focused nutrition assessment and sets the stage for the next two steps of the NCP (Intervention and Monitoring/Evaluation). Once a nutrition diagnosis is made, then the specific nutrition intervention can be determined.

10. *How does a nutrition diagnosis differ from a medical diagnosis?*

A nutrition diagnosis consists of the identification and descriptive labeling of an actual occurrence of a nutrition problem that dietetics practitioners are responsible for treating independently. It may also address behaviors that impact food choices. In contrast, a medical diagnosis distinguishes one disease from another and describes the nature of that disease.

11. *Describe the criteria used to evaluate the quality of PES statements.*

The criteria used are a series of questions specific to the problem (P), etiology (E), and signs and symptoms (S).

- Problem: can a dietetics practitioner impact, improve, or resolve the nutrition problem?

- Etiology: is the etiology truly the root cause; and is there an intervention that will address the root cause, or at least lessen the significance of the signs and symptoms?

- Signs and Symptoms: are the signs and symptoms that are used to describe the problem specific enough to be measured; and will measuring the signs and symptoms indicate if the problem is resolved or improved?

In general, the criteria aim to find out if the problems are clearly stated and if the assessment data used to identify the nutrition diagnosis supports and links to the diagnostic statement, etiology, and signs and symptoms.

12. *What are the domains in the standardized language for nutrition intervention?*

- **Food and/or Nutrient Delivery (ND):** meals, snacks, enteral/parenteral feeding, supplements, assistance, environment, and medication management

- **Nutrition Education (E):** impart knowledge to help patients/clients to voluntarily manage/modify food choices and eating behavior to maintain and improve health

- **Nutrition Counseling (C):** collaborative, supportive process that fosters responsibility for self-care to treat an existing condition and promote health

- **Coordination of Nutrition Care (RC):** consultation with, referral to, or coordination of nutrition care with other health care providers that assists in treating nutrition-related problems

13. *List and briefly describe the four types of outcome measures that can be monitored and evaluated in the NCP.*

- **Direct Nutrition Outcomes:** knowledge gained, behavior changes, food or nutrient intake changes, and improved nutritional status

- **Clinical and Health Status Outcomes:** laboratory values, anthropometry and body composition, blood pressure, and risk factor profile

- **Patient/Client-Centered Outcomes:** quality of life, satisfaction, self-efficacy, and self-management

- **Health Care Utilization and Cost Outcomes:** medication changes, special procedures, planned/unplanned health care visits

14. *What is meant by "outcomes management system"? Why is it important in dietetic practice?*

It is a system that evaluates the effectiveness and efficiency of the entire NCP: assessment, diagnosis, implementation, cost, and other factors; it links care processes and resource utilization with outcomes when nutrition care is provided to a number of patients. It is important in dietetic practice because this system can measure and evaluate aggregate data, which can then be combined with data from other nutrition care providers and be part of evidenced-based research that demonstrates the benefit and effectiveness of nutrition care.

Chapter 3

1. *What is the difference between nutritional status and nutritional risk?*

Nutritional status involves evaluating indices that reflect the body's nutrient stores. It is altered when stores of energy, protein, water, vitamins, or minerals fluctuate as a result of increased need, increased utilization, altered intake, or altered utilization.

Nutritional risk assessment involves the ability of the clinician to predict potential nutritional problems based on the client's current health status. Certain factors may increase or decrease a client's nutritional risk. Knowing the pathophysiology, treatment, and clinical course of a disease or diagnosis allows one to identify nutrition problems for an individual and ultimately determine the nutritional diagnosis.

2. *How is nutritional screening different from nutritional assessment?*

Nutritional screening is defined as the "process of identifying patients, clients, or groups who may have a nutrition diagnosis and benefit from a nutrition assessment and intervention by a registered dietitian (RD). The screening can be performed by a dietetic technician or other trained personnel. It is required that all patients receive nutrition screening within 48 hours of their admission to a hospital."

Nutritional assessment consists of gathering data, which may be both subjective and objective in nature. The collected data is analyzed so that the current and potential nutritional problems can be identified. The assessment must be performed by a registered dietitian (RD). In contrast with nutritional screening, it is not possible or necessary to complete a full nutritional assessment of every patient admitted to a clinic or hospital.

3. *Describe the difference between subjective data and objective data that are collected for a nutritional assessment. List three pieces of objective information and three pieces of subjective information that could be collected for nutritional assessment.*

Subjective data include information gathered in interviews with the patient, family members, or significant others. They may include the client's perception of his/her medical condition, dietary intake, lifestyle conditions, current medication or supplement intake, and family medical history. Subjective data also include the interviewer's observations. Three pieces of subjective data that could be collected for nutritional assessment are reported food intake in the last 24 hours; reported intake of current medications; patient reporting that he/she has access only to a microwave to prepare meals.

Objective data include information obtained from verifiable sources, such as the current medical record and previous medical histories. Objective data can include anthropometric measurements, biochemical data, and medical tests and procedures. Three pieces of objective data that could be collected for nutritional assessment are weight, BMI, and albumin levels.

4. *Name and briefly describe four methods used to collect dietary assessment data. List the advantages and disadvantages of each method.*

- **24-Hour Recall:** the clinician guides the client through a recall of all food and drink consumed within the previous 24 hours.

   Advantages: it takes a short amount of time, little cost, and no risk to the client

   Disadvantages: it may be inaccurate as it is based on memory; does not always show typical eating patterns; clients may report data they think the clinician wants to hear (over or under-report intake)

- **Food Record/Food Diary:** the client documents his/her intake as it occurs over a specified period of time (typically 3–5 days) and should include a sampling of weekdays and weekends. Clients may estimate or weigh their food intake.

   Advantages: it is not completely reliant on memory; may be more representative of their actual intake

   Disadvantages: under-reporting is common; client may change food habits during recording period; there is more of a burden on the client

- **Food Frequency:** the client retrospectively identifies how often and in what quantities he/she consumes a specific food or food groups.

   Advantages: inexpensive; quick to administer

   Disadvantages: response rates are lower because it is self-administered; the questionnaire may not include ethnic or child-appropriate foods or quantities that are realistic

- **Observation of Food Intake/"Calorie Count":** actual food intake is observed and recorded when a kcal or a kcal-protein count is ordered in an acute or long-term care setting

   Advantages: it can be quite accurate; no burden on patient

   Disadvantage: it can be time consuming for personnel involved

5. *Describe two methods that are used to analyze dietary intake.*

- Analysis based on **USDA's MyPyramid** quantifies food consumed from each of the groups of the food pyramid. It gives the clinician an overview of adequacy, variety, moderation, and balance. It does not, however, quantify macro- or micronutrients. Using MyPyramid with macronutrient data allows the overall quality of the diet to be assessed.

- Analysis based on **exchanges/carbohydrate counting** provides a quick, rough estimate of protein, kcal, carbohydrate, and fat in the diet. Carbohydrate counting is based on an estimation of carbohydrate only and is used primarily by individuals with diabetes who are balancing their insulin dosages with dietary intake of carbohydrates.

- **Specific nutrient analysis** uses food composition data from the USDA and other published data on food to calculate dietary intake of both macronutrients and various micronutrients.

- **Computerized dietary analysis** uses computerized data analysis programs to analyze dietary intake. It is similar to specific nutrient analysis with the exception that food composition data is stored in a database and analyzed with the use of an electronic device.

6. *Which anthropometric measurements are collected for nutritional assessment? Briefly describe each measurement and explain the accuracy of each in determination of body composition and/or health status.*

Height/stature is the measurement of supine or standing height. It can also be estimated with arm span or knee height measurements. Height is necessary for monitoring growth in children, interpreting weight in adults, calculating energy and protein requirements, and to calculate creatinine-height index. Height alone, however, is not an accurate determination of body composition or health status.

Weight is measured using a variety of scales and is the most common measure of anthropometrics. It is a gross measurement of all body compartments and does not distinguish body composition or fluid shifts. Because of its relationship to growth, development, and health, it is a vital component of nutrition assessment.

Body mass index (BMI) can be calculated once the height and weight are known. BMI has been correlated with overall mortality and nutritional risk. It does not estimate body composition, but can indicate obesity better than height and weight alone.

Skinfold measurements involve measuring a double fold of skin and subcutaneous adipose tissue while excluding muscle tissue. Measurement sites include the chest, triceps, subscapular, midaxillary, suprailiac, abdomen, thigh, and calf. The most commonly used site is the tricep. These measurements are used to estimate energy reserves—both fat and somatic protein—in subcutaneous tissue. Skinfold measurements do provide an estimate of body composition; however, more accurate methods do exist.

Bioelectrical impedance analysis (BIA) involves sending a low-frequency electrical current throughout the body and measures the impedance of that current, which can then estimate components of body composition such as body cell mass, fat-free mass, fat mass, and total body water. This method is considered to be a precise method of body composition assessment.

7. *List four blood proteins used in nutrition assessment. Describe the effectiveness of each as markers in measuring nutritional status.*

Albumin is the most abundant serum protein and the most well-known measure of visceral protein status, although it is not the best. Albumin has a long half-life (17–21 days), which decreases its sensitivity to short-term changes in protein status or short-term interventions. Albumin levels are affected by trauma, surgery, metabolic stress, and hydration status. It is important to note that albumin changes may reflect illness, not necessarily nutritional status.

Prealbumin appears to be a consistent indicator of risk for malnutrition. Because of its short half-life (2–3 days), prealbumin levels reflect short-term modifications in nutritional intake and interventions. Prealbumin levels rise with renal disease and Hodgkin's disease, and decrease with hepatitis, cirrhosis, malabsorption, and hyperthyroidism. Like albu-

min, prealbumin levels may decrease with illness and stress, and not necessarily due to malnutrition.

Transferrin transports iron throughout the body. Because of its short half-life (8–10 days), it can serve as an indicator of protein status. However, its concentration is directly affected by iron status. When iron stores are low, transferrin will increase. Hepatic and renal disease, inflammation, and congestive heart failure can also affect transferrin levels.

Retinol binding protein (RBP) transports vitamin A (retinol) throughout the body and is considered to be one of the more sensitive indicators of protein status. Because of its extremely short half-life (10–12 hours), it will reflect short-term changes in nutrition support interventions. It is important to note that RBP levels are affected by renal failure, hyperthyroidism, cystic fibrosis, liver failure, vitamin A deficiency, zinc deficiency, and metabolic stress.

8. *Describe how energy requirements can be determined or estimated. How is the energy requirement affected by stress?*

Energy requirements can be estimated by using a variety of established equations such as Harris-Benedict, Mifflin-St. Jeor, the Food and Agriculture Organization of the United Nations, and the World Health Organization. These equations use information on weight, height, age, and activity level in order to estimate energy requirements. In some situations, it may be possible to measure energy needs by indirect calorimetry. This type of measurement analyzes the amounts of oxygen and carbon dioxide in both inspired and expired air, as well as the volume of gas exchanged. However, it is more common in the clinical environment to calculate the estimated energy requirements.

Disease, infection, and trauma can affect an individual's energy requirements. Because patients with these conditions can become hypermetabolic, it is important to add a stress factor when using the equations.

9. *List the hematological measurements collected for nutritional assessment.*

Hemoglobin (Hgb), hematocrit (Hct), mean corpuscular volume (MCV), mean corpuscular hemoglobin (MCH), mean corpuscular hemoglobin concentration (MCHC), ferritin, transferrin saturation, protoporphoryn, serum folate, serum B_{12}.

Chapter 4

1. *Briefly describe the four types of intervention strategies in the nutrition care process.*

Intervention strategies are selected based on the nutrition diagnoses and their etiologies and are intended to change (a) nutritional intake, (b) nutrition-related knowledge or behavior, (c) environmental conditions, or d) access to supportive care and services.

2. *Describe two ways that the house or regular diet can be modified to accommodate patients' needs.*

The house or regular diet may be modified for patients with impaired chewing ability so that softer foods are served. Soft diets contain foods that are easy to chew and usually omit raw fruits and vegetables.

Another means to modify foods is to add single nutrients such as protein through the use of "modular" products. Protein modulars, such as ProMod® or Beneprotein®, can be added to both foods and beverages but will need to be mixed well. There is a slight change in taste and consistency.

3. *What are the similarities and differences between a clear and full liquid diet? When are they used, and what are their limitations?*

For very short periods (two or three meals), liquid diets (clear or full) consisting of broth, juice, cream soups, and milk may be served to patients who are beginning to eat after a long period without food. Both diets consist of foods that are liquid at room temperature (juice, tea, coffee, gelatin, broth, popsicles), but the full liquid diet begins the introduction of milk products (milk, ice cream, pudding). A clear liquid diet is intended to provide fluid and energy in a form that requires minimal digestion and contributes to limited residue in the gastrointestinal tract. It may be used during acute gastrointestinal distress, during gastrointestinal medical testing (such as a colonoscopy), or prior to surgery. Clear liquid diets are inadequate in kilocalories (kcal), protein, vitamins, and minerals, so they should be used only when medically necessary. Historically, the clear liquid diet has been used as a progression toward solid food after a surgical procedure or when the GI tract required minimal stimulation, but more recent opinions question this necessity.

A full liquid diet also has been used as a transitional diet between liquids and solid foods. Because this diet includes milk and milk products, it may present a problem due to the large amounts of lactose. Table 4.2 outlines the basic principles of these liquid diets.

4. *Describe two nutrition interventions that could be suggested to increase nutrient density.*

For calories, add butter or margarine to cooked cereals, soups, vegetables, or casseroles. For protein, add liquid egg substitutes to shakes, soups, vegetables, or casseroles.

5. *What are bioactive substance supplements? Provide an example and why one might be prescribed.*

Bioactive substances are defined as food substances added to a food product, or taken as a supplement, that have a specific intended health purpose. For example, a patient with hyperlipidemia may be instructed to consume stanol esters as a supplement. The registered dietitian, using the American Heart Association's guideline for the National Cholesterol Education Program, would instruct the patient on the amount of stanol esters recommended for the intended reduction of lipid levels.

6. *What are the skills and resources needed for an RD to provide effective nutrition education?*

Nutrition education is defined as "a formal process to instruct or train a patient/client in a skill or to impart knowledge to help patients/clients voluntarily manage or modify food choices and eating behavior to maintain or improve health."

The skills and resources needed by the RD include effective communication, use of terms that can be understood by patient, reading materials appropriate to the client, visual aids to support the verbal information provided, and sensitive listening skills.

7. *How does nutrition education differ from nutrition counseling? List the characteristics of the role of the RD as a counselor. Pick one and provide its description and a possible situation where it might be used.*

Nutrition education is defined as "a formal process to instruct or train a patient/client in a skill or to impart knowledge to help patients/clients voluntarily manage or modify food choices and eating behavior to maintain or improve health"; whereas nutrition counseling is defined as "a supportive process, characterized by a collaborative counselor-patient/client relationship, to set priorities, establish goals, and create individualized action plans that acknowledge and foster responsibility for self-care to treat an existing condition and promote health." The difference is that nutrition education primarily involves the transfer of knowledge and/or skill building.

The role of the RD as counselor has evolved from a clinician who mainly provides information on what and how to eat to one who is able to evaluate and take into consideration the complex social and physiological factors that influence food and lifestyle choices. The RD as counselor has to be supportive, needs to be involved and dedicated to the process, and should collaborate with the patient while making decisions, provide individualized counseling, develop a relationship with the patient, and make sure the patient understands the skills needed to take care of him-/herself.

Individualized: Unlike nutrition information that describes healthy eating guidelines for a group or population, such as the dietary guidelines, nutrition counseling can take on many different shapes and forms. It is based on well-known and researched counseling theories and strategies that are tailored to an individual's needs and environment.

8. *What is the difference between a closed-ended and open-ended question? What are the disadvantages and advantages of using these two types of questions in a nutrition counseling situation?*

Closed-ended questions are designed to obtain either a yes or no response or very brief answer. It is generally recommended that use of this type of question be limited, as a closed-ended question tends to prompt a response thought be correct or preferred.

Open-ended questions are valuable in that they allow the responder to provide a great deal of information and do not limit a response to a short answer; they do require that the counselor listen very carefully to what is being said. Answers may become lengthy and it may be necessary to redirect the client. The major advantage to using open-ended questions is that clients are less likely to feel threatened and more likely to provide honest responses. They may also provide information that guides appropriate secondary and probing questions.

Chapter 5

1. *Describe two ways that the house or regular diet can be modified to accommodate patient needs.*

Modification of the regular diet for texture can be used to accommodate patient needs. This is important for patients who are unable to chew food properly. The texture can be mechanically softer in order to make foods easier to chew and/or swallow.

A test diet is used to prepare patients for specific tests that will be administered. For example, a dietitian may prescribe a high-fat diet for 2–3 days prior to a stool fat test, which is used to determine the intestine's ability to absorb fat.

2. *What is the difference between clear and full liquid diets? When are they used, and what are their limitations?*

A clear liquid diet provides kcal in a form that requires minimal digestion, and therefore does not irritate the GI tract. It may be used during acute GI distress or medical testing, such as a colonoscopy, or prior to surgery. It consists of clear fluids and foods that are clear liquid at body temperature and leave minimal residue. Examples include clear fruit juices, bouillon, consommé, clear broth, gelatin, fruit ice, plain hard candy, sugar, and honey. The major limitation to these diets is that they are inadequate in kcal, protein, vitamins, and minerals. Therefore, they are only used when absolutely necessary and can be administered only for a couple of days. Furthermore, the clear liquid diet has been used as a progression toward solid food after a surgical procedure, but again, only if necessary. A full liquid diet also has been used as a transitional diet between liquids and solid foods. Unlike the clear liquid diet, it includes milk and milk products; therefore, a limitation is that it may present a problem due to the large amounts of lactose, especially if a patient is lactose intolerant. Full liquid diets consist of clear liquid foods and cream soups, milk, ice cream, puddings, and yogurt. Commercially available liquid supplements can also be used as full liquid diets.

3. *What are the advantages and disadvantages of enteral and parenteral nutrition support?*

Advantages of enteral nutrition support include cost effectiveness, reduced rate of infectious complications in critically ill patients, improved wound healing, reduced surgical interventions, and maintenance of GI function. However, the disadvantages of enteral feeding include administration difficulty, poor tolerance, and difficulty meeting the nutritional requirements of some patients. These disadvantages can be minimized by carefully choosing which patients will receive enteral feeding, thorough physical and nutritional assessments, and use of the standardized protocols.

In general, disadvantages of parenteral nutrition are decreased gut function, infections, metabolic complication in critically ill patients, and cost. Disadvantages of peripheral parenteral nutritional support (PPN) include irritation to small veins, the difficulty of maintaining peripheral access for more than a few days, and that it may not provide sufficient kcal for many patients. Therefore, its use is declining, and most hospitals use central line catheters (TPN) instead.

4. *Describe three ways enteral and two ways parenteral nutrition can be delivered to the patient.*

Enteral nutrition can be delivered to the patient through (1) nasogastric feeding, (2) nasointestinal feeding (nasointestinal feeding tubes enter the gastrointestinal tract through the nose and reside in the duodenum or jejunum), used for short-term nutrition administration, or (3) gastrostomy, used for more long-term feeding. In nasogastric feeding the tube extends from the nose into the stomach. Feeding tubes can be placed through the skin without a surgical incision, which is referred to as percutaneous gastrostomy. If this is done using an endoscope, then the procedure is called a percutaneous endoscopic gastrostomy, or PEG. Nasogastric feeding is the most common, the easiest to achieve and maintain, and the least expensive, and is acceptable in many circumstances. In nasointestinal feeding the tubes enter the GI tract through the nose and extend into the duodenum or jejunum.

Parenteral nutrition can be delivered through (1) a central venous catheter, (2) a peripherally inserted central catheter, or (3) peripheral venous catheters. The most common parenteral access is a central venous catheter (CVC), or central line, inserted percutaneously (through the skin) at the bedside while the patient is under local anesthesia. On the other hand, peripheral venous catheters are inserted into an arm vein and are only meant for short-term use. The peripherally inserted central catheter (PICC) is frequently used. Unlike the central catheter, which requires a bedside surgical procedure by an MD for insertion, the PICC can be inserted by specially trained nurses; this increases the availability of the procedure and decreases costs.

5. *List five factors that might influence selection of an enteral formula (e.g., viscosity). Explain why each factor is important when choosing a formula.*

The five factors that might influence selection of an enteral formula include nutrition density, osmolality/osmolarity, vitamin/mineral content, protein content, and cost.

Nutrient density is monitored because many patients are unable to tolerate large volumes of fluid. The nutrient density of an enteral formula is measured in kcal per mL, and usually, the difference in these formulas is the amount of water added, depending upon what the patient can handle.

Osmolality/osmolarity is considered because an iso-osmolar enteral formula can minimize the development of dumping syndrome (diarrhea that results from rapid movement of fluids into the GI tract due to hyperosmolar or concentrated fluids).

Vitamin and mineral content is a factor in selection of an enteral formula. Some patients have special needs, such as those under stress and with healing wounds; therefore, enteral formulas may need to be adjusted according to the patient's current needs.

Protein content is an important factor also. Some formulas contain protein from peptides, used for patients with enzyme deficiency or other conditions resulting in malabsorption. Also, formulas with specialized amino acid profiles are available for patients with renal failure, hepatic failure, stress, and inborn errors of metabolism. Some formulas are supplemented with additional amounts of specific amino acids such as glutamine or arginine.

Finally, cost of the formula is considered as well. Traditionally, enteral products have been inexpensive, but newer products and those for specialized indications are increasingly expensive. Patients who purchase formula from a retail grocery store typically pay much more than large institutions that purchases thousands of cases.

6. *What are medium-chain triglycerides (MCT), and why are they added to some enteral products? What is the most common source that is currently used?*

Medium-chain triglycerides are 8- and 10-carbon chain fatty acids, extracted from coconut oil and then reesterfied to glycerol. MCTs are different in that they don't require pancreatic lipase or emulsification for digestion; therefore, they are used clinically to supply kcal to patients with a variety of pancreatic and GI disorders. However, medium-chain fatty acid oxidation results in an increased production of ketone bodies, which can be utilized for the conversion of fatty acids for energy by those who have complications with the liver and pancreases. Coconut oil is a common source of MCTs.

7. *List four complications that might occur when feeding a patient enterally. Describe and provide the rationale for three factors that should be monitored.*

Four complications that might occur when feeding a patient enterally are

1. dehydration/tube feeding syndrome: hyperosmolar, non-ketotic dehydration can develop due to inadequate free water administration;

2. electrolyte imbalances: depending on the electrolyte content of the enteral solutions, changes in electrolyte concentrations may develop;

3. hyperglycemia: during periods of physiological stress, such as those caused by severe illness or severe infection (sepsis), hyperglycemia can appear even in patients with no previous history of diabetes; and

4. under/overfeeding: overfeeding may result in hyperglycemia, hypertriglyceridemia, and hepatic steatosis (fatty liver), while underfeeding may contribute to the development of malnutrition and increased length of hospital stay.

Factors to be monitored include (1) intake/output and weight, which allow the medical team to monitor the sufficiency of the nutrient intake and utilization of nutrients by the body, (2) blood glucose concentration, and (3) I and Os for edema or dehydration.

8. *Calculate the caloric content and protein amount in one liter of parenteral solution composed of 25% dextrose and 4.25% amino acids. If 250 milliliters of a 20% fat emulsion were added, how many more kcal would be provided?*

One liter would contain 1020 kcal and 42.5 g of protein. If 250 mL of a 20% fat emulsion were added, this would provide an additional 500 kcal.

Chapter 6

1. *What is the primary purpose of the medical record in the clinical setting? What other functions does it serve?*

The primary purpose of the medical record in the clinical setting is to provide legal documentation of medical care that the client has received. The record serves as a description of exactly what happened during the medical care. Other functions it serves include communication between members of the health care team, evaluation of medical care for that client, funding and resource management, continuous quality improvement, third-party reimbursement, accreditation, and research.

2. *Why should standardized language and abbreviations be used in the medical record?*

The use of standardized language and medical abbreviations ensures accuracy of the medical record. Health professionals with access to the records will have a uniform understanding of what the information written down means for the patient and his or her care. Clear, concise wording in the medical record, using terminology consistent with the prospective payment system, will also facilitate reimbursement for services. Authors of one study predicted that the adoption of standardized language terms will improve accountability and reimbursement and enhance patient care overall.

3. *Describe each of the four sections that constitute a SOAP note.*

The four sections of the SOAP note include: Subjective data (S), Objective data (O), Assessment (A) and Plan (P).

Subjective data include patient information or data collected from the patient or caregiver. This information can be placed into four major categories: diet-related information, lifestyle/psychosocial or emotional information, medical history information, and learning/motivation information. This section may include symptoms expressed by the patient; descriptions by the patient of her or his pain, discomfort, and/or dysfunction; dietary history; or the presence of symptoms that interfere with the ability to eat. See Table 6.3.

Objective data include empirical information which is drawn from physical tests and medical staff observations that are of consequence to the patient's nutritional status. This information can come from physical examinations (temperature, pulse, blood pressure or respiratory rate), X-ray examinations, other imaging techniques, or biochemical tests that are of nutritional relevance. Examples of objective data would be the patient's age, gender, and anthropometric information. Medical diagnosis and current medical care are noted here. Objective data can also include sensory information noted by the medical staff member such as smells, how an organ feels during a physical exam, or the visual recording of patient skin coloration. See Table 6.4.

Assessment is the nutrition diagnosis or interpretation of the patient's nutrition problems. Assessment should include the nutrition problems with their supporting data, stated as nutrition diagnoses in the PES format. In the assessment section of a SOAP note, conclusions are drawn from the subjective and objective data in order to support the nutrition diagnosis. See Table 6.5.

Plan will include an outline of interventions necessary to treat each nutrition problem. The plan may include requests for additional information needed to address the patient's nutrition problems. Specific nutrition therapy recommendations are stated here. Finally, goals and objectives may be included with a specific measure and timeline for evaluation of the intervention. See Table 6.6.

4. *How does medical record documentation fit within the nutrition care process?*

In medical record documentation, the subject matter is the nutrition care process for a patient. The purposes may include not only the legal documentation of care, but also communication with other health care providers or collection of research data.

5. *How does charting by exception differ from the other charting methods?*

Charting by exception (CBE) is an abbreviated approach to medical charting that involves recording *only* unusual or out-of-the-ordinary events. The CBE format includes a standardized nutritional care plan. After the initial charting of the care plan, only significant data and/or unanticipated responses to the proposed plan should be included in the record. In most CBE formats, a flowchart is used to document assessments and interventions. An asterisk (*) indicates an abnormal finding on an assessment or an abnormal response to an intervention. The findings are explained in the comments section of the form. The progress notes are used to document revisions in the plan of care and specific interventions.

6. *Why might a dietitian keep a personal notebook in the clinical setting?*

The goal of medical charting is to create complete, concise documentation—for legal reasons, and, more importantly, for medical reasons. With this in mind, health professionals, especially novice health professionals, should consider keeping a personal medical notebook where they can chart everything involved with each patient's treatment, where they can brainstorm problems, and, perhaps most importantly, where they can include their subjective experiences and express their own emotional and intellectual responses to their experiences.

7. *Why must information in a chart note be kept confidential?*

All medical record information is confidential. A confidential communication is given by one person to another with the trust and confidence that such information will not be disclosed. The federal U.S. law that assures patients of the confidentiality of their medical information is the Health Insurance Portability and Accountability Act of 1996, or HIPAA. HIPAA protects information about clients that is gathered by examination, observation, conversation, or treatment. A dietitian cannot discuss a client's status with other clients or staff who are uninvolved in the client's care. A legal suit can be brought against a dietitian who has disclosed information about clients without their consent.

Chapter 7

1. *What are electrolytes? What are anions and cations?*

Electrolytes are substances that separate into charged particles (ions) when dissolved in water or other solvents

and thus become capable of conducting an electric current. Anions are ions with negative charges while cations have positive charges.

2. *List the electrolytes which are primarily found in extracellular fluid and intracellular fluid. What is the normal range of concentration for these electrolytes in the serum?*

Sodium, potassium, and chloride are the primary electrolytes found in intracellular and extracellular fluid.

The normal serum values are as follows:
Sodium: 135–142 mEq/L, Potassium: 3.8–5.0 mEq/L, Chloride: 95–102 mEq/L.

3. *What is the difference between osmolality and osmolarity?*

Osmolarity is defined as the number of osmols (standard unit of osmotic pressure) per liter of solution (mOsm/L), while osmolality is defined as the number of osmols per kilogram of solvent (water) (mOsm/Kg).

4. *Describe three factors that influence the movement of solutes through semi-permeable membranes.*

1. Molecular size: Smaller molecules transport across more easily than larger molecules.

2. Method of transport: Solutes transported across the membrane by active transport move more easily than those transported by facilitated diffusion or simple diffusion.

3. Electrical charge of a solute can dictate its affinity for a specific active transport.

5. *List three mechanisms by which the body regulates the movement of fluid and solutes to ensure homeostasis.*

Hypothalmic thirst mechanism, renal function, and hormonal control are factors that regulate movement of fluids and solutes.

6. *Explain how the renin-angiotensin-aldosterone system can affect blood volume.*

A decrease in blood volume will decrease hydrostatic pressure, thus stimulating the release of the hormone renin from the kidney. Renin stimulates conversion of angiotensinogen to angiotensin I, and a second activation converts angiotensin I to angiotensin II. Increasing amounts of angiotensin II stimulate release of aldosterone from the adrenal cortex. Aldosterone directly influences the kidney to retain Na^+. When Na^+ levels increase, increased osmotic pressure will pull fluid back into the blood; thus, blood volume will increase back to its normal range

7. *Explain how aldosterone and arginine vasopressin can affect urine volume.*

Aldosterone directly influences the kidney to retain Na^+. Subsequently, osmotic pressure increases and urine volume decreases.

Arginine vasopressin causes fluid to be reabsorbed in the tubules of the kidney. This decreases urine volume, resulting in increased blood volume.

8. *Discuss how calcium and phosphate balance are maintained in the body.*

Calcium and phosphorus balance is maintained in the body by hormonal influence of intestinal absorption, exchange between extracellular fluid and bone, and renal excretion of these minerals. When plasma levels of calcium are low, parathyroid hormone (PTH) is secreted from the parathyroid glands. PTH works to raise serum calcium levels by pulling calcium from the bone and decreasing renal excretion of calcium. PTH also stimulates absorption of calcium in the small intestine by activation of vitamin D. When necessary, PTH also acts to increase phosphorus excretion. Calcitonin, another hormone, originates from the thyroid gland. It is capable of decreasing serum calcium levels acts by inhibiting osteoclast activity.

9. *Physiologically, what does hyper- or hypovolemia describe? What are the common causes of hyper- and hypovolemia?*

Hyper- or hypovolemia describe conditions involving abnormal volume of circulating blood.

Hypovolemia is an extracellular fluid deficit and is almost always related to renal or extrarenal loss of fluids. Extrarenal losses include any excess loss of fluid outside of renal excretion, including gastrointestinal losses such as in vomiting or diarrhea. Losses through the skin occur during exposure to heat such as increasing body temperature (fever) or increased environmental heat. Excess loss through the skin can also occur through burns or draining wounds.

The most common cause of hypervolemia is a decrease in urinary output as seen in acute renal failure. Excess intravenous fluids or the failure of the kidney to accommodate a rapid ingestion of fluids quickly enough may also cause hypervolemia.

10. *Is there a difference between hypervolemia and hyponatremia? Explain your answer.*

Hypervolemia is excess extracellular fluid and is usually due to a decrease in urinary output. Hyponatremia can be caused by either a decrease in the amount of sodium, an increase in the amount of water in the ECF, or a combination of both. It can also occur through a combination of a sodium restriction used for nutrition therapy and use of diuretics.

11. *Do laboratory values of serum Na^+ >145 mEq/L; serum osmolality >295; and urine osmolality >800 mOsm/kg indicate hypernatremia or hyponatremia? List three signs/ symptoms that accompany this condition.*

Hypernatremia. Cellular dehydration results in an increasing severity of neurological symptoms ranging from lethargy and agitation to seizures and coma. Body temperature can be elevated, skin is flushed, and mucous membranes are dry.

12. *Describe three common conditions that can result in hypokalemia. What are common signs and symptoms of hypokalemia? Hyperkalemia?*

Hypokalemia can result from inadequate nutritional intake of potassium, increased renal loss of potassium (use of loop diuretics), or increased loss from the gastrointestinal tract (vomiting, nasogastric suction, and diarrhea). Hypokalemia also results from a shift of potassium from the ECF to the ICF. Signs and symptoms of hypokalemia include muscle weakness, diminished deep tendon reflexes, and cardiovascular dysrhythmias which can lead to cardiac arrest. The most common cause of hyperkalemia is inadequate excretion of potassium, commonly found with acute renal failure and chronic kidney disease. Shifts in potassium from the ICF to the ECF can also result in hyperkalemia. Signs and symptoms are a result of the neuromuscular effects of altered potassium levels—muscle weakness, paralysis, paresthesias, and cardiac dysrhythmias, which can lead to cardiac arrest.

13. *How do changes in blood pH affect blood potassium levels?*

Hydrogen ions are excreted to correct acidosis, potassium ions are retained, and hyperkalemia may develop.

Chapter 8

1. *Define the following terms: pH, volatile and nonvolatile (fixed) acid, and buffer.*

The measure of H ions in solution is expressed as pH. The pH scale is from 1 to 14, with 7 being neutral; less than 7, acidic; and more than 7, alkaline (basic).

Volatile acids can be converted to a gaseous form and eliminated by the lungs. Nonvolatile acids occur through metabolism of carbohydrate, protein, and lipid, and cannot be exhaled.

A buffer is a protein or chemical that decreases the change in pH when an acid or base is added to the solution.

2. *What organ controls the level of pCO_2 in the blood? What organ controls $[HCO_3^-]$ in the blood?*

The lungs have the ability to change respiratory rate and depth of breathing to control either release or retention of pCO_2 in the blood.

The kidneys reabsorb the majority of HCO_3^- in the blood to maintain a normal physiological pH.

3. *What is the basic problem in respiratory acidosis? Respiratory alkalosis?*

Respiratory acidosis occurs when there is an inability of the lungs to expire CO_2 adequately. The retention of carbon dioxide causes an excess of acid in relation to base.

Respiratory alkalosis generally occurs as a result of hyperventilation. Rapid breathing results in a decreased pCO_2 causing an excess of base in relationship to acid.

4. *What is the most important difference between metabolic acid-base disorders and those of a respiratory origin?*

Respiratory acid-base disorders are caused by abnormal levels of CO_2, which is a volatile acid. Metabolic acid-base disorders are a result of excessive loss/gain of base (HCO_3) or an excessive loss/gain of fixed (nonvolatile) acids. Changes in blood pH can occur rapidly with changes in respiratory rate. Metabolic acid-base disorders occur more slowly.

5. *Name some conditions that might result in respiratory acidosis.*

Sleep apnea, cardiac arrest, myasthenia gravis, extreme obesity, injury or trauma to the chest wall, chronic obstructive pulmonary disease, and pneumonia can result in respiratory acidosis.

6. *Name some conditions that might result in metabolic acidosis.*

   Diarrhea, fistula drainage, end-stage renal failure, ketoacidosis (due to diabetes mellitus, alcoholism, or starvation), and lactic acidosis (due to diabetes mellitus or salicylate overdose) can result in metabolic acidosis.

7. *What is an anion gap?*

   The anion gap is considered a very useful tool to distinguish between the main types of metabolic acidosis. It is calculated by subtracting the chloride and bicarbonate levels from the sodium plus potassium levels.

   $$\text{Anion gap} = ([Na^+] + [K^+]) - ([Cl^-] + [HCO_3^-])$$

   As sodium and potassium are the main extracellular cations, and chloride and bicarbonate are the main anions, the result should reflect the remaining anions. Normally, this concentration is about 8–16 mmol/l. An elevated anion gap (i.e., >16 mmol/L) can indicate particular types of metabolic acidosis, particularly certain poisons, lactate acidosis, and ketoacidosis.

8. *How can respiratory mechanisms compensate for a metabolic alkalosis? Are there any major limitations to this compensation?*

   Respiratory mechanisms such as respiratory rate, tidal volume, ventilation, and rate of CO_2 removal decrease to compensate for metabolic alkalosis. Rate of carbonic acid formation and rate of H^+ generation from CO_2 increase to re-establish acid-base balance. The need for oxygenation of blood is the major limitation of respiratory compensation.

Chapter 9

1. *When researchers study the prevalence of atherosclerosis in developing countries and compare this to the prevalence in industrialized nations, this is an example of:*

   B. epidemiology.

2. *Infectious disease is an example of which etiological category of disease?*

   B. acquired

   For the other two answers that you did not choose, give an example of that category of disease.

   Examples of multifactorial disease: atherosclerosis, osteoporosis, and diabetes mellitus

   Examples of genetic disease: cystic fibrosis, sickle cell anemia, and hemophilia

3. *Mrs. J is meeting with her physician to discuss her recent diagnosis of breast cancer. As her physician outlines the probable response to therapy and how she expects Mrs. J to respond, the physician is actually discussing:*

   B. prognosis.

4. *Mrs. J's physician also states that the initial goal of her treatment is to find no indication of disease after five years postchemotherapy. In this discussion, the MD is outlining what we could call the _____ of her disease.*

   A. remission

5. *When Mark twists his ankle in practice, the trainer immediately places cold packs and elevates the ankle. What symptoms do these two actions prevent?*

   Cold packs and elevation will help to reduce inflammation and the pain that is associated with it. Cold packs will result in vasoconstriction to the injured area. This will limit blood flow and reduce swelling and pain associated with vasodilation and increased vascular permeability. Elevation decreases blood flow as well and will assist in reducing swelling.

6. *How do nonsteroidal anti-inflammatory drugs (NSAIDs) treat the acute inflammatory process? Give an example of an NSAID.*

   Nonsteroidal anti-inflammatory medications (NSAIDs) treat acute inflammation by blocking this process in one or more steps. Salicylates (aspirin) inhibit prostaglandin production. Other NSAIDs further prevent synthesis of prostaglandins by blocking cyclooxygenase enzyme 2. Ibuprofen is an example of an NSAID.

7. *List and describe factors that can influence an individual's susceptibility to infectious disease.*

   - **Gender.** Although gender sometimes plays a role, in many cases the underlying mechanism is differential exposure due to occupational and recreational activities.

   - **Age.** The immune system takes time to develop; therefore, the young do not have the full spectrum of immunological defenses available to the adult. As humans grow older, several immune mechanisms decrease, including secretion of mucous and sebaceous glands and the production of cytokines, including an interferon. However, natural killer cells that attack infected cells and tumor cells increase.

   - **Nutritional status.** Malnutrition is a major cause of immunodeficiency. Research has suggested that certain nutrients may play a role in supporting the immune response.

   - **Physical activity.** Intensive physical training may lead to periods of depressed immune response and increased risk of upper respiratory tract infections.

   - **Hormones.** Levels of various hormones play a role in an individual's susceptibility to infectious disease. Individuals with diabetes have an increased risk of fungal and staphylococcal infections, while women with low estrogen have a higher vaginal pH and thus are more susceptible to vaginal infections.

   - **Stress.** Stress activates the fight-or-flight response, resulting in several physiological changes that impact the immune response.

8. *Describe an example of natural resistance.*

   Intact skin and mucous membranes provide not only a physical barrier against infection and injury but the first chemical response. Intact epithelial surfaces such as skin and the lining of the body's tubular structures such as the gastrointestinal, respiratory, and genitourinary tracts are excellent barriers to most pathogens. In addition to providing a physical barrier, skin and mucous membrane components produce chemical barriers, such as the lysozyme produced by sweat glands. This enzyme damages peptidoglycan, a critical component of bacterial cell walls.

9. *What are the differences between antigens, haptens, and immunogens?*

An antigen is a structure that can combine with a cell of the immune system or an antibody but does not necessarily induce activation of the cell or formation of an antibody. An antigen is generally a large protein or polysaccharide attached to foreign substances. Membranes of neoplastic cells can be antigens and the body's own cell membranes can also serve as self-antigens. An immunogen is an antigen that can induce an immune response. The difference between an antigen and an immunogen is that the immunogen is foreign to the host whose immune system interacts with it. Hapten refers to an incomplete antigen which when combined with a larger molecule can elicit an immune response. An example of this occurs in an allergic response.

10. *Describe humoral and cellular immunity, specific and nonspecific immunity, and active and passive immunity.*

Humoral and cellular immunity are the two arms of the immune system. The humoral arm refers to the antibodies that appear in serum and B cells that become plasma cells, which produce antibodies. The cellular part consists of the T cells, macrophages, monocytes, and polymorphonuclear leukocytes that interact with pathogens at the cellular level.

Specific and nonspecific immunity are two branches of the immune system. Cells of the nonspecific immune system (macrophages, monocytes, natural killer cells, and polymorphonuclear leukocytes) can react immediately with any antigen. This initial reaction is often sufficient to eliminate the pathogen or reduce its numbers significantly enough to prevent initiation of the disease process. This response does not increase or improve with repeated exposures. Inflammation is a nonspecific immune response. In the specific immune response, each B and T cell is programmed to attack one specific antigen, but these cells can interact with other antigens that are closely related or very similar. The specific immune system takes time to respond initially, but it improves with additional exposures and responds more rapidly on subsequent encounters with the organism. Thus, it normally protects the human from reinfection.

The response to a pathogen normally involves an initial contact with the nonspecific immune system, which often is capable of eliminating the organism by itself. The nonspecific immune system then stimulates the specific immune system to seek out and target any remaining pathogens. In some cases, the specific immune system can eliminate the pathogen; in others, it merely tags it or alters it in such a way that it becomes more susceptible to the cells of the nonspecific immune system. The two systems thus work together and are interdependent.

Active and passive immunity are types of specific immunity. In active immunity, individuals synthesize their own antibodies or activate immune cells. In passive immunity, they receive antibodies or activated cells produced by another individual. Both active and passive immunity can be described as either natural (occurring without human intervention) or artificial (resulting from human intervention). The four types of immunity are:

- *Active natural immunity:* Mounting an immune response to an infectious organism.

- *Active artificial immunity:* Mounting an immune response to vaccination.

- *Natural passive immunity:* An antibody from the mother goes to the fetus across the placenta. Both regular breast milk and colostrum contain antibodies, but the concentration is higher in colostrum.

- *Passive artificial immunity:* Transferring antibodies or immune cells produced in one organism to another organism to prevent the action of a virus or toxin before it does damage.

11. *Briefly describe the function of each of the three groups of white blood cells: macrophages/monocytes, microphages/granulocytes/polymorphonuclear leukocytes, and lymphocytes and natural killer cells.*

Monocytes and macrophages are highly specialized; they ingest and destroy bacteria, aged cells, and neoplastic cells in a process called phagocytosis. They are major antigen-presenting cells that can break down antigens into small pieces that they "present" on their cell surface. This function is very important for initiating an immune response. Polymorphonuclear leukocytes (PMN), also called granulocytes, are a second group of cells involved in the nonspecific immune response. PMN move easily between blood and tissues, so their number in the blood increases or decreases in infections depending on the type of organism involved. A large pool of PMN is available in bone marrow for rapid response to an infection. The lymphoid stem cell produces the cells of the specific immune system—T cells (T lymphocytes) and B cells (B lymphocytes). Natural killer cells are also produced from the lymphoid stem cell; however, they work as a component of the nonspecific or innate immune response.

12. *How are mast cells involved in the symptoms of allergies?*

When histamines, concentrated in mast cells, are released, they are the cause of many of the symptoms experienced during the allergic reaction.

13. *Briefly describe the functions of T helper cells, Th1 and Th2 cells, and cytotoxic T cells. What are CD4 and CD8 cells?*

T helper cells determine how the immune system will respond to various antigens. They stimulate cell and cytokine action, and they interact with other cells of the immune system, causing them to become more immunologically active and to proliferate. There are two subsets of helper T cells: Th1 and Th2. T helper cells trigger B cells to make antibodies, activate macrophages, and promote the differentiation of other T cells.

Th1 activate the cellular immune system. Th2 increase the production of antibodies. Cytotoxic T cells (also called CD8 cells) kill cells infected by viruses or transformed by cancer CD4 is a marker found predominantly on helper T cells that interacts with MHC class II molecules on antigen-presenting cells; CD8 is a marker found predominantly on cytotoxic T cells that interacts with MHC class I molecules on target cells.

14. *Briefly describe the functions of B cells—plasma cells, memory B cells, and antibody-producing cells. What are the immune functions for each of the five antibodies produced by B cells?*

Plasma cells facilitate efficient production of protein antibodies. Memory B cells produce a rapid antibody response the next time the person is exposed to the antigen, and they normally block infection and prevent symptoms. Antibodies assist in the destruction or neutralization of antigens.

- **IgA (immunoglobulin A)** is the predominant immunoglobulin in secretions and stops pathogens at the point of entry (e.g., gastrointestinal tract).

- **IgD (immunoglobulin D)** is present on the surfaces of B cells.

- **IgE (immunoglobulin E)** is the immunoglobulin class that is involved in allergic responses to food and respiratory allergens, such as animal dander and pollens, and in countering parasitic infections.

- **IgG (immunoglobulin G)** is the predominant immunoglobulin class produced during secondary immune responses; the most prevalent immunoglobulin in the blood. It is the second antibody made during an initial infection, where it functions to bring the ongoing infection under control, and the first made in subsequent infections, where it usually prevents reinfection.

- **IgM (immunoglobulin M)** is the predominant immunoglobulin class expressed by virgin B lymphocytes and secreted during primary immune responses. It is the first antibody found in new infections, where its 10 binding sites help ensure that once it binds to the antigen it will stay attached.

15. *What is meant by "antigen-presenting cell," and which cells in the body can serve this function?*

When a pathogen or foreign material invades and is engulfed by a cell, the MHC molecule selects fragments of it, transports them to the cell surface, and displays them for the T cells. The cell that picked up the invader has now become an antigen-presenting cell (APC). The term APC refers to a cell that presents the antigen alerting the B and T cells to the presence of a "non-self" substance. Monocytes, macrophages, and dendritic cells are types of antigen-presenting cells.

16. *What is the immune function of the lymphatic system?*

The major functions of the lymphatic system are to concentrate antigens from all over the body into a few lymphoid organs, to circulate lymphocytes through lymphoid organs in order to allow antigens to interact with antigen-specific cells, and to carry antibodies and other immune cells to the bloodstream and tissues.

17. *There are several soluble mediators of the immune system. List and briefly describe their function.*

Soluble mediators are the complement proteins and cytokines. Complement proteins are activated by other elements of the immune response and are involved in destroying infected cells and some pathogens. Complement can also kill beneficial neighboring cells and participates in transplant rejection. Cytokines mediate communication among the cells of the immune system as well as between the cells of the immune system and other body systems, including the nervous system.

18. *How do major histocompatibility complexes I and II aid the immune system in distinguishing between self and non-self?*

Major histocompatibility complexes I and II are cell surface molecules that help the organism identify pathogens as foreign or self; they are important in antigen presentation to T cells, play a role in transplantation rejection, and influence the susceptibility to certain autoimmune diseases.

19. *How can T helper cells be activated? After activation, what is their response?*

Activation of T cells requires that an antigen be "presented" to the T cells. The T cell is not activated unless its receptor recognizes both the antigen and the MHC molecule it is attached to. Once T helper cells are activated they can act in one of two ways: T1 activate the cellular immune system, while Th2 cells increase the production of antibodies.

20. *Why is it critical to match MHC antigens for tissues used in transplantation? What might happen if they are not matched?*

MHC antigens play an important role in transplant rejection. If the MHC antigen on the transplanted organ or tissue is different from the MHC antigens on the recipient's tissues, it signals the presence of the transplanted tissue as foreign (non-self) and initiates an immune response. The immune system attacks the transplanted cells that display MHC antigens that are different from those found on the recipient's cells, leading to rejection of the transplanted tissue.

21. *What is the difference between active and passive immunization?*

In passive immunization, antibodies are administered to either prevent the disease or decrease the severity of the symptoms. Immunity is fast acting but the antibodies are short lived, and no immunological memory is induced. The major risk is serum sickness.

In active immunization, the individual is exposed to an antigen in a harmless form, and produces antibodies as well as activated T and B cells and memory cells. The memory immune response, initiated upon exposure to the pathogen or a booster, prevents infection and provides additional antibodies and memory cells. In some cases, such as influenza, the residual antibodies are more important than the memory response due to the short incubation period of the disease, and more frequent vaccinations are often required to maintain the antibody level.

22. *Describe one way that malnutrition can compromise immunity.*

In addition to the example in the current text: Adolescents who were prenatally and/or are currently undernourished produce a significantly lower antibody response to vaccination.

23. *List common food allergies. What are the roles of nutrition intervention in the diagnosis and treatment of food allergies?*

Common food allergies: peanuts, tree nuts, fish, shellfish, cow's milk, eggs, wheat, and soy nutrition intervention can be used to identify the food causing the allergic reaction. The double-blind, placebo-controlled food challenge has long been the standard procedure in diagnosing food allergies. Once the allergen is identified, nutrition therapy consists of avoidance of all sources of that particular food. Education will include development of skills for food identification, including reading food labels, steps to contact food manufacturers, and methods to avoid cross-contamination. Finally, nutritional equivalents for the food that is being eliminated need to be identified, and the importance of including them in the diet reinforced, so that adequate nutritional status is maintained.

Chapter 10

1. *What is a genome? How does knowledge of its content possibly affect dietary recommendations for individuals?*

A genome is the entire set of genes of a given organism. The human genome is the blueprint for approximately 20,000–25,000 different proteins. Knowledge of an individual's genetic background and how that may interact with diet may lead to the creation of individual dietary recommendations.

2. *What are the differences between genotype, haplotype, epigenotype, and phenotype?*

Genotype is the specific variants of a gene present in the two alleles in an individual that can result in specific traits or disorders. Haplotype is a group of gene variants that occur together and these variants may work together to produce a specific phenotype. Phenotype is the actual expression of inherited genes. Epigenotype is an individual's unique pattern of gene expression due to patterns of DNA methylation and histone modification. It displays greater variability than the genotype and may be more responsive to environmental influences.

3. *Define the following terms:* autosomal dominant; autosomal recessive; X-linked dominant; X-linked recessive; Y-linked; heterozygous alleles; *and* homozygous alleles. *Name one autosomal recessive disorder, one autosomal dominant disorder, and an X-linked recessive disorder.*

autosomal dominant: an inheritance pattern of a dominant allele on an autosome

autosomal recessive: an inheritance pattern of a recessive allele on an autosome

X-linked dominant: an inheritance pattern of a dominant allele on the X chromosome; disorders are rare

X- linked recessive: an inheritance pattern of a recessive allele on the X chromosome; related disorders are more common in males who carry only one X chromosome

Y-linked: inheritance based on Y chromosome; disorders are very rare and occur only in males

heterozygous alleles: having two different alleles or variants of a given gene

homozygous alleles: having two identical alleles or variants of a given gene

Autosomal recessive disorders: phenylketonuria and cystic fibrosis

Autosomal dominant disorder: familial hypercholesterolemia

X-linked recessive disorders: red-green colorblindness, hemophilia

4. *What is the difference between a monogenic disorder and a polygenic disorder?*

A monogenic disorder is caused by a single gene. Polygenic disorders are caused by multiple genes interacting with each other.

5. *Define single nucleotide polymorphisms. How are they identified? Give an example of one and explain what it means.*

Single nucleotide polymorphisms (SNPs) are defined as those genetic variants or polymorphisms in which a single nucleotide has been exchanged for another. SNPs are generally identified by the gene name, the location of the affected nucleotide within the gene sequence, the common nucleotide in that position, and an arrow indicating that a less common nucleotide is present.

Example: *MTHFR* 667C→T (ala→val) indicates that there is a SNP at nucleotide number 667 in the methylene tetrahydrofolate reductase gene characterized by a thymine in place of the more common cytosine.

6. *What is meant by epigenetic regulation? How could the nutrients folate, choline, methionine, and vitamin B_{12} affect gene expression?*

Epigenetic regulation relates to the pattern of gene expression regulated by modifications to DNA. Gene expression is regulated in many ways, including DNA methylation, histone methylation, acetylation, or phosphorylation, and transcription factors. All of these regulatory mechanisms can be influenced by early programming in response to nutrition and other environmental factors in fetal life or infancy as well as throughout the life span. Epigenetic patterns may also be passed from one generation to the next; thus, individual patterns may reflect environmental exposures of previous generations.

Dietary folate, vitamin B_{12}, choline, and methionine are of particular interest since these are primary sources of methyl groups, and dietary adequacy may influence DNA methylation patterns and thus genomic stability and gene expression. A deficiency of methyl groups related to lack of the above-listed nutrients means that as cells divide, methylation may be reduced, and some of that transcriptional regulation is lost. Impaired methylation of DNA is related strongly to impaired fetal development and cancer.

7. *For each of the following disorders, list at least one gene that is linked to its occurrence: obesity, type 2 diabetes, and colon cancer. For each gene listed, describe its possible role in the development of the disorder.*

Obesity: fat mass and obesity-associated gene (*FTO*). The *FTO* variant may play a role in control of food intake and possibly a preference for energy-dense foods.

Type 2 diabetes: gene encoding for peroxisome proliferator-activated receptor gamma (PPARγ). PPARγ is a receptor on the cell nucleus that plays a central role in adipocyte development and function. PUFAs are natural ligands for this receptor—meaning that they bind to and activate it—but thiazolidinedione drugs (i.e., rosiglitazone) treat diabetes also by interacting with PPARγ to enhance insulin sensitization. Thus, PPARγ activation is associated with greater insulin sensitivity. Because fatty acids with longer chain length and greater desaturation have a higher affinity for PPARγ, diets high in saturated fat are likely to have little effect on PPARγ and have been associated with insulin resistance.

Colon cancer: genes encoding for *N*-acetyl transferase 2 (NAT2) and cytochrome P450 1a2 (CYP1A2) enzymes in the liver. Both enzymes are involved in biotransformation of incoming xenobiotics into harmless substances for excretion. Individuals exhibiting different phenotypes of these enzymes metabolize xenobiotics at different rates and are thus often classified as slow, intermediate, or rapid acetylators. NAT2 and CYP1A3 transform some xenobiotics into genotoxic substances. Both are integral to the biotransformation of heterocyclic amines from cooked meat into genotoxic substances, which by definition have the potential to cause cancer. Epidemiological research has long linked cooked meats to increased risk of colon cancer.

8. *Describe an example of "developmental origins of adult disease."*

The "developmental (or fetal) origins of adult disease" relates metabolic status and fetal adaptation in the womb to disease risk in later life. Nutrient deprivation *in utero* leads to fetal adaptation to a deprived environment by an increased efficiency in use of nutrients. This has been termed a "predictive adaptive response" in which the fetus predicts the postnatal environment based on fetal nutritional conditions and adapts in order to maximize ability to survive postnatal life. This adaptation is epigenetically regulated and is referred to as metabolic imprinting or metabolic programming. It may manifest itself through alterations in appetite regulation, decreased physical activity, altered adipocyte metabolism, and altered mitochondrial function. As a consequence of the previously-listed adaptations, nutritional deprivation in both the fetal environment and early childhood has been linked to a predisposition for metabolic syndrome, obesity, diabetes, and cardiovascular disease in later life.

Chapter 11

1. *Match the following examples to these routes of administration. (Choose from: oral, sublingual, buccal, rectal, intramuscular, intravenous, and inhalation.)*

   a. Dissolved under the tongue = Sublingual

   b. Insulin given into the muscle = Intramuscular

   c. Dextrose given into a peripheral vein = Intravenous

   d. Asthma medication that is delivered by puffs through a breathing device = Buccal

2. *What factors could affect the dissolution of a medicine?*

   Substances added to formulations of medications that decrease dissolution are binders, lubricants, and coating agents. Also, the tablet formulation of the medication; hard, round, and large tablets dissolve more slowly. Disintegrants (ingredients that dissolve readily in water) increase dissolution. Coloring and flavoring agents have varying effects on dissolution and a generic medication may have different dissolution rates than the original medication.

3. *Distribution of the drug is defined as:*

   The movement of a drug throughout the body to the target site where it can act and perform its function.

   What is the major physiological factor that can affect this?

   b. Blood flow

4. *What organ is primarily involved in the metabolism of a drug?*

   The liver is the primary site of metabolic activity for drugs.

5. *Name three factors that can affect metabolism of a drug.*

   Three factors that can affect the metabolism of a drug are inducers, inhibitors, and dosage. An inducer works to stimulate synthesis of the enzymes, increasing action potential. Inhibitors decrease metabolism and generally lead to increased drug effect, whereas inducers will increase metabolism and generally lead to decreased drug effect. The dosage range with therapeutic efficacy is referred to as the "therapeutic window." Levels below the "therapeutic window" dosage range of a medication may not be effective, and those above may result in toxicity.

6. *How are drugs excreted? Give an example of how disease (affecting an organ function) affects drug excretion.*

   Depending on the chemical nature of the drug, most are removed by either urinary or biliary excretion, but some can be excreted via the lungs or bowel. Urinary excretion of drugs can occur in all three stages of urinary filtration and can be altered by changes in urinary flow rates or kidney function. This may occur as a result of another medication, or as a result of disease or injury. For example, changes in creatinine clearance significantly alter the effectiveness of medications. If an individual has renal insufficiency from any etiology, drug levels must be adjusted to ensure therapeutic levels.

7. *Polypharmacy means:*

   The administration of excessive drugs at one time or concurrent use of a large number of drugs, which increases the risk of interactions. Other features of polypharmacy may include the use of medications without a reason; the use of multiple medications for the same condition; the use of medications that interact with one another; the use of inappropriate dosages; the use of additional drugs to treat side effects of medications; and overall improvement when medications are discontinued.

Who is at risk?

The elderly are at risk due to failing health, which can necessitate multiple medications; altered GI function, which can affect absorption; and decreased mental capacity (memory), which can cause accidental misuse of medicine.

Determine how each of the following could be considered a drug-nutrient interaction:

a. *When the use of methotrexate causes a change in taste*

Appetite and subsequent food ingestion can be affected by taste, smell, and saliva production. Some medications may cause a decrease in saliva production. Since adequate solution is necessary for taste, patients on medications that decrease saliva production express difficulty eating, decreased appetite, or anorexia, ultimately due to dry mouth. Other medications may actually result in a perceived abnormal taste. Patients have reported experiencing metallic, salty, sweet, and simply foul tastes after taking some medications.

b. *When antacids bind phosphorus*

Adequate and efficient nutrient absorption requires exposure to enzymes in the appropriate metabolic environment, adequate transit time, sufficient GI tract surface area, and any transporters necessary for absorption. Any medication that speeds gastric emptying or affects the pH of gastric juices could therefore interfere with nutrient absorption. For example, since calcium supplements are absorbed best in an acidic environment, the chronic use of proton pump inhibitors may affect calcium absorption by decreasing stomach acidity.[1] Calcium, which is found in many antacids, and phosphate will precipitate if mixed together, causing both calcium and phosphate to not be absorbed.

c. *When Lasix increases the renal excretion of potassium*

In the kidney, the loop diuretics reduce the reabsorption of NaCl and also diminish the lumen-positive potential that derives from K^+ recycling. Use of the diuretic Lasix (furosemide) or any other medications in this class can result in hypokalemia (low serum potassium).

d. *When phenobarbital decreases folate metabolism*

Drugs can interfere with macronutrient, vitamin, and mineral metabolism. For example, phenytoin (Dilantin), used for treatment of seizures, inhibits both vitamin D and folate metabolism. Long-term use may result in megaloblastic anemia secondary to folate deficiency.

Chapter 12

1. *How is the energy content of food determined today? What is a kilojoule? How does it differ from a kilocalorie?*

The energy content of a food or beverage is generally determined by first measuring the amount of carbohydrate, protein, fat, and alcohol it contains, using relatively simple laboratory techniques, and then multiplying the number of grams of carbohydrate, protein, fat, and alcohol in the food or beverage by the energy values for each of the macronutrients.

A kilojoule (kjoule or kJ) is a unit of measurement for energy. It is the amount of work required to move 1 kilogram for 1 meter with the force of 1 newton. A kilocalorie (kcalorie or kcal) is the amount of heat required to raise 1000 mL (1 liter) of water 1° Celsius; 1 kcal = 4.2 kJ. The kilojoule is the most commonly used measure of food energy; however, the kilocalorie is the measure of food energy most commonly used in the United States.

2. *Describe the three main components of energy expenditure. What is the difference between basal energy expenditure and resting energy expenditure?*

Total energy expenditure can be divided into three major components: resting energy expenditure, the thermic effect of food, and physical activity-related energy expenditure. Resting energy expenditure (REE) is the energy necessary to sustain life and to keep such vital organs as the heart, lungs, brain, liver, and kidneys functioning. The thermic effect of food (TEF) is a measurable increase in energy expenditure over and above resting energy expenditure that can be measured for several hours following a meal. The thermic effect of food is the energy required to digest, absorb, metabolize, and store the nutrients contained in foods that are consumed and to eliminate the resulting by-products and wastes. Physical activity-related energy expenditure is the amount of energy expended doing physical activity. It is the most highly variable component of 24-hour energy expenditure and can vary due to the person's body weight, the number of muscle groups used in the activity, and the intensity, duration, and frequency of the activity.

Basal energy expenditure (BEE) is the lowest rate of energy expenditure of an individual. It is measured in the morning when a subject is in a postabsorptive state (no food consumed during the previous 12 to 14 hours) and is lying motionless in a thermally neutral environment (a room temperature that is perceived as neither hot nor cold). These strict conditions often make obtaining a true BEE impractical in the clinical setting. REE, on the other hand, can be measured at any time of day after a subject has quietly rested for the previous 30 minutes. Basal energy expenditure is generally 10% to 20% less than resting energy expenditure.

3. *Describe three methods that can be used to estimate or determine a person's energy requirement.*

Equations: In most instances, an individual's energy requirements are estimated using one of several empirically derived equations. Equations usually use variables such as sex, age, weight, height, and physical activity level.

Indirect calorimetery: The most commonly used approach for measuring energy requirements in critically ill patients and in human metabolic research. When the differences in oxygen and carbon dioxide in inspired and expired air are known and the volume of air moving through a subject's lungs is measured, the body's energy expenditure can be calculated. A mask or hood is placed over the subject's face and the amount of air flow through the lungs per minute (known as minute ventilation) is measured by various types

---

[1] O'Connell MB, Madden DM, Murray AM, Heanery RP, Kerzner LJ. Effects of proton pump inhibitors on calcium carbonate absorption in women: a randomized crossover trial. Am J Med. 2005;118:778–81.

of instruments that are built into the mask. Gas analyzers in the monitor measure the oxygen and carbon dioxide content of both inspired and expired air.

Direct calorimetry: This is a very impractical approach for measuring energy expenditure in the critically ill or those afraid of enclosed spaces. It involves using a highly sophisticated chamber that is capable of determining a subject's total energy expenditure by measuring the amount of heat given off by the subject's body through evaporation, convection, and radiation. Once inside the calorimeter, the subject's activity is monitored, and the subject's response to clinically prepared meals can be studied. If necessary, samples of urine and feces can be collected for analysis. In some instances, subjects may remain inside the calorimeter for up to 24 hours or longer.

4. *List five substances produced in the body that can affect appetite. Pick two of them and describe them in more detail (source, function, and effect).*

Appetite is influenced by a number of signals to the brain that are primarily orchestrated by the hypothalamus region and the secretion of pancreatic and gastrointestinal hormones such as (1) insulin, (2) glucagon, (3) amylin, (4) cholecystokinin, and (5) ghrelin. Insulin is produced in the beta-cells of the pancreas in response to rising plasma glucose levels. Insulin stimulates cellular uptake of glucose. Insulin will decrease appetite and food intake. Ghrelin is a peptide hormone that is produced mainly by the stomach and stimulates appetite. Ghrelin levels are normally increased during fasting, but immediately following food intake, ghrelin levels decline. Decreased ghrelin levels decrease appetite and food intake.

5. *What is the clinical implication of excess adipose tissue in specific body locations? How should it be measured—BMI, waist circumference, or waist-to-hip ratio (WHR)? Why?*

Body fat distribution can be divided into two clinically significant categories: (1) abdominal or central body fat distribution, and (2) lower body fat distribution. Excessive adipose tissue located deep within the abdomen and surrounding the intestines and liver is associated with increased risk of type 2 diabetes, hypertension, dyslipidemia, coronary heart disease, and metabolic syndrome, even when BMI is within the healthy weight range. An increased amount of adipose tissue located within the hips and thighs is not associated with these increased risks; in fact, there is research suggesting that adipose tissue located within the lower body is inversely related to (i.e., lowers) risk of these conditions. Some clinicians prefer using the waist-hip ratio instead of the waist circumference, citing research suggesting that the WHR is somewhat better at predicting risk of coronary heart disease than is waist circumference alone. WHR takes into account the protective effects of larger hip circumferences, which may result from increased adipose tissue or from increased muscle mass in the hips and thighs, both of which are associated with a lower risk of type 2 diabetes, hypertension, dyslipidemia, coronary heart disease, and metabolic syndrome. However, both waist circumference and WHR have been shown useful in assessing body fat distribution and evaluating disease risk. Because BMI does not distinguish between lean tissue and adipose tissue or

indicate how fat is distributed, it cannot predict disease risk when used alone.

6. *List at least five factors that have contributed to the growing problem of obesity. Pick one factor and explain how you would address this concern with a client who needs to lose weight.*

Among the key factors contributing to obesity are 1 and 2) specific medical and psychiatric disorders or their treatment, 3) genetics, 4) disordered eating patterns, and 5) an obesigenic environment that promotes a high energy intake and discourages physical activity.

In terms of weight loss I would address the obesigenic environment. Clients would be encouraged to increase energy expenditure by walking whenever possible, for example, taking the stairs instead of the elevator. Fast foods are often chosen for their convenience, but they can be energy dense due to high amounts of fat and sugar. Clients could be educated in identifying less energy-dense foods on fast-food menus.

7. *What are the key elements of nutrition therapy for weight loss? What additional benefits are associated with increasing physical activity?*

The cornerstone of weight reduction therapy is an individually planned low-kcal diet (LCD) that reduces energy intake by 500 to 1000 kcal/day and achieves a slow but progressive weight loss of 1 to 2 pounds per week. Fat should make up 30% or less of total calories. Protein should make up approximately 15% of total calories. Carbohydrates should make up 55% or more of total calories. Fiber intake should be 20–30 g per day.

Increased physical activity appears to be crucial for maintaining weight loss. Physical activity has the added benefit of minimizing loss of lean body mass, reducing LDL-cholesterol levels, increasing HDL-cholesterol levels, improving insulin sensitivity, and improving fitness.

8. *What medications are currently approved for weight loss? Pick one, describe its mechanism of action, and list possible side effects. Describe one surgery that is performed for weight reduction. List possible complications associated with the surgery.*

The U.S. Food and Drug Administration has approved two drugs for long-term use for weight loss and the maintenance of weight loss: sibutramine and orlistat. Several have been approved for short-term treatment (6–12 weeks), including mazindol, diethylpropion, benzphetamine, phendimetrazine, and phentermine.

Sibutramine is available only by prescription and is marketed under the trade name Meridia. It is a serotonin-norepinephrine reuptake inhibitor that acts on receptors in the hypothalamus to suppress appetite. It can result in a 7% reduction in body weight after one year, but when used in conjunction with an intensive program of diet, exercise, and behavior therapy, losses can increase to 10% to 15% of body weight. It should not be used by patients with CVD or uncontrolled hypertension because it increases heart rate (4–5 beats/minute) and blood pressure (1–2 mm Hg).

Adjustable gastric banding (AGB) involves an inflatable silicone ring or band that is laparoscopically introduced into the abdominal cavity and secured around the upper part of the stomach to create a small pouch with a narrow opening or stoma at the bottom of the pouch through which food passes into the rest of the stomach. The pouch restricts the amount of food that can be consumed at one time and initially has a capacity of about 30 mL, with the potential to stretch somewhat to eventually hold up to 90 mL. The band is connected by a tube to an access port that is placed under the skin of the abdomen and is inflated when saline is injected into the access port using a syringe. The inflation of the band can be adjusted by the amount of saline injected into or removed from the access port. As the band is inflated, the stoma becomes narrowed, thus delaying the emptying of the pouch and giving the patient a greater sense of fullness and further restricting the patient's food intake. If it is necessary to permit greater food intake, the band can be deflated to enlarge the stoma, allowing faster emptying of the pouch. AGB is the most commonly performed restrictive procedure. It is also the simplest and least invasive bariatric surgery. Consequently, hospitalization and postoperative recovery are shorter. The band can be adjusted to suit the patient's needs and the procedure is fully reversible.

Postoperative complications include pulmonary embolism, anastomotic leaks, injury to the spleen, wound infection, and, rarely, death. Some patients may experience nausea and vomiting when too much food is consumed at one time and others may experience "dumping syndrome," which is characterized by nausea, flushing, bloating, and diarrhea after eating a food or beverage that is high in refined carbohydrates. For patients undergoing AGB and VSG, there are no intestinal malabsorption problems, only a reduction in gastric volume and rate of gastric emptying. In these two procedures, nutrient deficiencies are uncommon unless the diet is unbalanced and has a low nutrient density.

Chapter 13

1. *Define the following terms: systolic, diastolic, stroke volume, and cardiac output.*

   Systolic: The phase of blood circulation in which the heart's pumping chambers (ventricles) are actively pumping blood. The ventricles are squeezing (contracting) forcefully, and the pressure against the walls of the arteries is at its highest.

   Diastolic: The phase of blood circulation in which the heart's pumping chambers (ventricles) are being filled with blood. During this phase, the ventricles are at their most relaxed, and the pressure against the walls of the arteries is at its lowest.

   Stroke Volume: The volume of blood ejected with each contraction of the left ventricle.

   Cardiac Output: Amount of blood pumped per minute; it is the product of heart rate multiplied by stroke volume.

2. *Describe the factors that will influence stroke volume and mean arterial pressure (MAP).*

   Stroke volume is regulated by end-diastolic volume (EDV), mean aortic blood pressure, and strength of ventricular contraction. EDV, often referred to as the preload, is the amount of blood in the ventricles at the end of diastole,

which is mostly determined by venous return. Greater EDV means the ventricles will be increasingly stretched. The stretching of the ventricles increases their force of contraction, leading to a greater amount of blood that will be ejected. The fraction of EDV that is ejected from the heart by contraction of the ventricles is termed the ejection fraction (EF). The mean aortic blood pressure or mean arterial pressure (MAP), termed afterload, also affects stroke volume. It is the average force exerted against the walls of the arteries by the blood, over a cardiac cycle. MAP represents the resistance the ventricles must contract against in order to eject blood; therefore, greater resistance means the ventricles will eject less blood. Finally, the strength of ventricular contraction is affected by the hormones epinephrine and norepinephrine, and the sympathetic stimulation of the heart. These mechanisms increase contractility by increasing the amount of calcium available to the myocardial cells.

Mean arterial pressure (MAP) is determined through a combination of cardiac output and total peripheral resistance. The MAP must be high enough to force blood through systemic circulation without causing vascular damage. The regulation of MAP involves the sympathetic nervous system, the renin-angiotensin system, and renal function. All three affect blood pressure. Heart rate is dependent upon the balance between parasympathetic activity, which decreases heart rate, and sympathetic activity, which increases heart rate, and both are regulated by the cardiovascular control center. Total peripheral resistance is dependent upon the radium of arterioles (most importantly) and blood viscosity (primarily determined by the number of red blood cells). The radius is controlled by local metabolic controls in skeletal muscle, which may cause vasodilation and increase blood flow to muscles in order to meet metabolic needs. This would decrease resistance by increasing the radius of the vessel. Vasoconstriction will decrease vessel radius and increase resistance, and is caused by sympathetic activity and epinephrine. The hormones vasopressin and angiotensin II also cause vasoconstriction. They affect blood pressure in other ways, as well. When vasopressin (antidiuretic hormone) is released, there is an increase in the reabsorption of water, which increases blood volume, thus increasing blood pressure. Angiotensin II causes aldosterone to be secreted by the adrenal cortex, which causes an increase in sodium and chloride reabsorption, leading to retention of water and increased blood pressure. Additionally, the circulatory system contains baroreceptors, which constantly monitor blood pressure through both short-term and long-term adjustments. The carotid sinus baroreceptor and aortic arch baroreceptor monitor MAP and pulse pressure. If either cardiac output or peripheral resistance increases without a compensating fall in the other, blood pressure increases.

3. *What is the definition of* hypertension? *Explain the pathophysiology for the lifestyle factors known to contribute to the development of hypertension.*

   Hypertension refers to a chronic elevation of blood pressure. A systolic pressure/diastolic pressure measure greater than or equal to 140/90 mmHg is considered to be hypertensive. Lifestyle factors that contribute to the development of HTN include diet, exercise, smoking, stress, and weight management. Because development of HTN has such a strong genetic component, poor lifestyle choices may simply exacerbate the problem.

Cigarette smoking interferes with the action of nitrous oxide, thus impairing endothelial relaxation and vasodilation because of its interference with the action of nitrous oxide. Hypertension related to the endocrine system may occur with adrenal disorders that cause excessive secretion of epinephrine and norepinephrine. As previously discussed, this will increase cardiac output and peripheral resistance by vasoconstriction, both of which increase BP. Hyperinsulinemia as seen in type 2 diabetes and metabolic syndrome (often associated with increased weight) may also play a role in the development of hypertension in some individuals, though the relationships remain unclear and controversial.

4. *List the major classifications of medications used to treat hypertension. Describe their mechanism of effect. Describe the DASH diet.*

Loop diuretics: act by inhibiting sodium, chloride, and potassium reabsorption in the loop of Henle; increase prostaglandins, resulting in vasodilation

Thiazides: inhibit the reabsorption of sodium, chloride, and potassium; primarily act in the distal tubule and the ascending loop of Henle

Carbonic anhydrase inhibitors: block carbonic anhydrase, preventing the exchange of hydrogen ions with sodium and water

Potassium-sparing diuretics: prevent sodium-potassium exchange within the collecting and convoluted tubules; reduce aldosterone stimulation

ACE inhibitors: competitively block the conversion of angiotensin I into angiotensin II; results in vasodilation, decrease in vasopressin release; increase kinin levels that lead to vasodilations; increase release of fibrinolytic substances

Beta-adrenergic blocking agents: block β-receptors in the heart to decrease rate and cardiac output

Alpha-receptor antagonists: block the vascular muscle action that normally responds to sympathetic stimulation; this reduces stroke volume.

Calcium channel blocking agents: affect the movement of calcium, causing the blood vessels to relax, reducing vasoconstriction

Aldosterone antagonists: suppress the actions of aldosterone

The DASH diet focuses on the use of a variety of foods that not only reduce sodium intake but increase potassium, magnesium, calcium, and fiber within a moderate energy intake. The eating plan is low in saturated fat, cholesterol, and total fat, and emphasizes fruits, vegetables, and low-fat dairy foods. It also includes whole-grain products, fish, poultry, and nuts. It is limited in red meat, sweets, and sugar-containing beverages, and rich in magnesium, potassium, calcium, protein, and fiber.

5. *List the risk factors associated with development of atherosclerosis. Which risk factors are alterable? List the dietary changes in the ATP III guidelines.*

The risk factors for development of atherosclerosis are family history, age, sex, obesity, dyslipidemia, hypertension, diabetes, physical inactivity, and cigarette smoking. Those that are alterable include obesity, dyslipidemia, hypertension, physical inactivity, atherogenic diet, and cigarette smoking. Dietary changes in the ATP III guidelines include dietary fat intake within 25%–30% of total caloric intake; reduce saturated fat intake to <7% of total kcal or less than 16 grams for an individual on a 2000-kcal a day diet; reduce cholesterol to <200 mg/day; consider increased viscous (soluble) fiber (10–25 g/day) and plant stanols/sterols (2 g/day); eat for weight management; if TG levels are = 500 mg/dL, then reduce fat intake to = 15% of total kcal.

6. *What are the four mechanisms that can initiate a myocardial infarction (MI)? How does the rupture of an atheromatous plaque result in an MI?*

The four following mechanisms can initiate an MI: (1) sudden blockage of a coronary artery, (2) hemorrhage into an atheromatous plaque, (3) arterial spasm, and (4) increase in myocardial oxygen demand. The rupture or tearing of the fibrous cap (which separates the core and the plaque) exposes the plaque to the blood, allowing blood to push into the fissure. Therefore, hemorrhaging occurs, and clotting ensues. The result may be thrombotic occlusion of the artery. In some cases, occlusion will occur because blood seeping into the plaque causes enlargement, which may obstruct the coronary artery.

7. *Compare peripheral arterial disease (PAD) to atherosclerosis—how are they similar and how do they differ? Describe several complications associated with PAD.*

Peripheral arterial disease (PAD) is used to describe occlusion of blood flow in noncoronary arteries, especially in the lower extremities (pelvis and legs). PAD has been used by some to describe all noncoronary atherosclerotic disease including involvement in the carotid arteries. An atherosclerotic plaque in the leg can result in PAD.

Similarities and differences are found among the risk factors for these diseases. Risk factors for the development of atherosclerosis are important to the development of PAD. However, the impact of the individual risk factors on atherosclerotic disease development and progression in the periphery are not the same as in the coronary arteries. The most influential independent risk factors for PAD development and progression are cigarette smoking and diabetes mellitus. While dyslipidemia and hypertension are important and do increase risk, they do not appear to be as influential as diabetes and cigarette smoking. Also, elevated LDL, low HDL, and elevated serum triglycerides do impact risk of PAD, similar to AS. Finally, the effect of hypertension on PAD appears to be much more subdued than the effect in the cerebral or coronary arteries.

PAD is a major risk factor for lower extremity amputation because ischemic conditions can lead to gangrene. One symptom (complication) associated with the ischemic conditions of PAD is intermittent claudication, which is a cramp-like pain, most common in the calf, associated with activity. Strenuous activity leads to earlier onset and greater pain intensity.

8. *What are the primary causes of heart failure? Describe the clinical progression of heart failure. What is meant by "cardiac remodeling" and what is its etiology?*

The underlying causes of heart failure can be either structural or functional in nature. The primary causes are ischemic heart disease, hypertension, and dilated cardiomyopathy. Heart failure is described as a progressive disorder because even if there is no additional injury to the myocardium there will be a continued deterioration in function. As compensation for the impairment in function, the renin-angiotensin-aldosterone system initiates changes in blood pressure that exacerbate the dysfunction. As a result, the heart becomes weakened and dilated, myocardial fibrosis limits the ability of the walls to respond to stress, and oxidative damage further impairs contractility. Therefore, the overall structure of the heart does not allow for proper functioning. The change is referred to as *cardiac remodeling*. It will result in a hypertrophied and/or dilated left ventricular chamber. Cardiac remodeling is mediated to some extent by neurohormonal systems. Heart failure patients typically have elevated levels of norepinephrine, angiotensin II, aldosterone, endothelin, vasopressin, and cytokines, which can have adverse effects on cardiac structure. Sodium retention and peripheral vasoconstriction result in increases in arterial blood pressure, which increase myocardial workload. Other substances mediate oxidative stress, causing myocardial cell damage and myocardial fibrosis and further altering the structure of the heart, and therefore cardiac remodeling.

9. *What are the nutritional implications associated with heart failure and cardiac cachexia?*

Individuals with heart failure have difficulty eating and many experience a syndrome of malnutrition called cardiac cachexia, characterized by extreme skeletal muscle wasting, fatigue, and anorexia. Further complications from heart failure contributing to nutrition problems include decreased blood flow to the GI tract causing slowed peristalsis, early satiety, and impaired nutrient absorption, and side effects from drugs such as nausea, vomiting, and anorexia. Nutrient deficiencies are also common side effects from the use of diuretics and other medications. Nutritional care during congestive heart failure is difficult. A diet restricted in both sodium and fluid is crucial to control acute symptoms and may help reduce the cardiac workload.

Chapter 14

1. *Define and describe the four basic functions of the GI tract as discussed in this chapter.*

Motility is the movement of the food consumed along the GI tract. Both propulsive contractions and mixing movements serve not only to move foodstuffs toward sites of digestion and absorption, but to mix foods with digestive secretions and maximize potential absorption.

Secretions of the GI tract include water, electrolytes, enzymes, bile salts, and mucus.

Digestion is the process by which complex molecules are converted to their simplest form. Carbohydrates are digested from their most complex form of polysaccharides to the monosaccharides: glucose, fructose, and galactose. Proteins are converted from polypeptides to single amino acids, di-, and tri-peptides. Lipids are digested to their simplest forms: free fatty acids, monoglycerides, glycerol, phospholipids, and cholesterol.

Absorption is the final step occurring after digestion. Basic molecules are absorbed along with water, electrolytes, vitamins, and minerals to provide essential nutrients to every cell.

2. *Considering the basic functions of saliva, what are the possible consequences of xerostomia?*

Xerostomia is dry mouth, which can result from damage to the salivary glands. Reduced saliva production can result in (1) decreased moistening and lubricating of food, making swallowing difficult; (2) delayed digestion of carbohydrate; (3) bacterial proliferation due to decreased lysozyme production and food remaining in the oral cavity; (4) decreasing taste because food reacts with taste buds when in solution; (5) increased dental caries because saliva isn't present to act as a buffer to neutralize acids; (6) poor oral hygiene due to undissolved food, dead cells, and foreign substances; and (7) difficult speech due to less free movement of the lips and tongue.

3. *An imbalance of pressure at the lower esophageal sphincter (LES) may result in the symptoms associated with gastroesophageal reflux disease. What factors may affect LES pressure?*

Factors that can lower LES pressure and thus contribute to LES incompetence include (1) increased secretion of the hormones gastrin, estrogen, and progesterone; (2) presence of other medical conditions, such as hiatal hernia, scleroderma, or obesity; (3) cigarette smoking; (4) use of medications, including dopamine, morphine, and theophylline; and (5) specific foods such as those high in fat, chocolate, spearmint, peppermint, alcohol, and caffeine.

4. *Explain the potential nutritional and metabolic consequences of prolonged vomiting.*

Prolonged nausea and vomiting can have negative consequences. Forceful vomiting can either rupture the esophagus (Boerhaave's syndrome) or tear the lower esophageal sphincter (Mallory-Weiss tear). Bleeding or hematemesis is a serious outcome of these injuries. Continued vomiting also can result in dehydration and acid-base imbalances. Malnutrition can be a long-term consequence for the patient if he or she is not able to ingest an adequate diet for a prolonged amount of time. If gastric contents are aspirated into the lungs, aspiration pneumonia is a likely result. Prolonged vomiting may lead to learned food aversions. This is similar to anticipatory nausea and vomiting. When a negative consequence is linked to a particular food, most people choose to avoid eating that food.

5. *Peptic ulcer disease (PUD), in many cases, is linked to an infection. What is the origin of this infection, and how is it treated?*

Helibacter pylori (H. pylori) is the bacteria now believed to cause 92% of duodenal ulcers and 70% of gastric ulcers. *H. pylori* lives under the mucous layer of the stomach and attaches to mucus-secreting cells lining the stomach. The organism breaks down urea to produce ammonia, which helps neutralize acid in the immediate vicinity of these bacteria and enhances their survival. The *H. pylori* organisms subsequently produce various proteins that damage mucosal cells, attracting lymphocytes and causing persistent inflammation. By-products released by the organism result in damage to the epithelium and impair the mucous barrier within the stomach.

Treatment of peptic ulcer disease associated with *H. pylori* infections includes regimens of three to four medications (triple/quadruple therapy). The recommended therapy involves a 7- to 14-day course of two antibiotics with bismuth and one of the proton pump inhibitors.

6. *Identify three major goals for nutrition interventions to assist in the control of symptoms associated with PUD.*

Currently, goals for nutrition therapy include supporting medical treatment, maintaining or improving nutritional status, and providing a diet that minimizes symptoms of PUD. Nutrition interventions should restrict foods that may increase acid secretion or cause direct irritation to gastric mucosa. Additionally, it is recommended that patients avoid any foods they do not individually tolerate. Other components of NT will include timing and size of meals. Patients should not lie down after eating and avoid eating large meals close to bedtime. Smaller, more frequent meals may be better tolerated, but there is some controversy regarding whether this might increase the overall amounts of acid that are secreted.

7. *Complications of PUD may result in surgical resection. What are the physiological consequences of gastric resection?*

Physiological consequences of gastric resection include changes in gastric emptying and transit time when the normal pathway for digestion and absorption is interrupted. This can lead to dumping syndrome. Additionally, when portions of the stomach are resected, valuable components of digestion may be altered or lost. These issues combine to place the patient at significant nutritional risk due to decreased oral intake, maldigestion, and/or malabsorption.

8. *What are the potential nutritional complications of gastric resection?*

Several vitamin and mineral deficiencies can occur due to gastric resection. Due to the changes in gastric anatomy, there may be a reduction in secretion of intrinsic factor. This would prevent normal B_{12} absorption and lead to a subsequent deficiency. If left untreated this deficiency may lead to cardiovascular, hematologic, or neurologic abnormalities, or pernicious anemia. Iron deficiency is also common. The cause is multifactorial, including a decrease in HCl, decreased dietary intake, and possible malabsorption. Risk of osteoporosis is also increased due to decreased absorption of calcium. It is recommended that both calcium and vitamin D supplements be prescribed for these patients.

9. *Explain the symptomatic and etiological differences between early and late dumping syndrome.*

Early dumping syndrome usually occurs 10–20 minutes after eating. Symptoms of early dumping syndrome include cramping, abdominal pain, hypermotility, diarrhea, dizziness, weakness, and tachycardia. Early dumping syndrome occurs as a result of an increased osmolar load quickly entering the small intestine from the stomach. The chyme entering the small intestine is hyperosmolar, causing fluid to be drawn into the small intestine in an attempt to dilute intestinal contents. This leads to the symptoms associated with early dumping syndrome.

Late dumping syndrome can occur anywhere from 1 to 3 hours after eating, and is especially common after consuming simple carbohydrates. In this situation, rapid absorption in the small intestine stimulates insulin release. After quick movement and absorption of food through the small intestine, there is no longer any substrate for the insulin to act upon. This results in hypoglycemia and its symptoms of shakiness, sweating, confusion, and weakness.

Chapter 15

1. *What are pre- and probiotics? How do they affect the health of the GI tract?*

Prebiotics are substances in food that stimulate the beneficial flora of the large intestine. Examples of such foods are inulin, which is a fructooligosaccharide derived from chicory, and oligosaccharides.

Probiotics are products containing microorganisms that are manufactured and sold as food products such as yogurt with live culture and supplements. Research is currently underway to determine their role in the promotion of a beneficial environment for the health of the colon and in prevention and treatment of disease.

Lactate and short-chain fatty acids that result from fermentation by these microbes can be absorbed from the colon and utilized elsewhere in the body or utilized by the colon for support of its own tissue growth.

2. *Describe the types of diarrhea and compare/contrast their possible etiologies. Are there nutritional consequences of diarrhea? Describe dietary measures that are commonly recommended for diarrhea.*

Diarrhea can be classified in several different ways. First, diarrhea can be either acute or chronic in origin. Acute diarrhea is short term, whereas diarrhea lasting several weeks is considered chronic, and is usually associated with more health concerns such as electrolyte imbalances, malabsorption, dehydration, and malnutrition. Nutrition implications of diarrhea are dependent on the volume of gastrointestinal losses and length of the disease course. Large-volume losses can quickly lead to dehydration and electrolyte and acid-base imbalances.

Hyponatremia and hypokalemia are both common with diarrhea. Metabolic acidosis may occur due to excessive loss of bicarbonate ions in stool output.

Diarrhea can also be classified as either osmotic or secretory. When there is an increase in osmotically active particles in the intestine, the body reacts by pulling water into the lumen in an attempt to normalize osmolality; this may result in diarrhea. Osmotic diarrhea can be caused by maldigestion of nutrients, excessive sorbitol or fructose intake, enteral feeding, and some laxatives. In general, when the causative agent is removed, osmotic diarrhea will cease. This is one of the major differentiations between osmotic and secretory diarrhea. Secretory diarrhea also results from excessive fluid and electrolyte secretions into the intestine. The difference here is that the underlying disease is what causes excessive secretions. Furthermore, secretory diarrhea does not resolve when the patient is made NPO. Bacterial infections often produce enterotoxins that result in this type of diarrhea. Protozoa, viruses, and other infections can also cause secretory diarrhea.

The dietary measures that are most commonly recommended are: (1) Replace fluids and electrolytes; (2) Include foods rich in soluble fiber; (3) Limit fat and foods high in insoluble fibers; (4) Avoid caffeine; (5) Avoid lactose as it may not be well tolerated; and (6) Use foods with pro- or prebiotics.

3. *Describe the pathophysiology of irritable bowel syndrome (IBS) and its recommended medical treatment. What is the role of NT in the treatment of IBS?*

The pathophysiology of IBS is complex and not completely understood. In IBS, abnormal motility is considered to be one of the major factors involved in symptoms of abdominal pain and altered bowel habits. Since etiology of IBS is currently unknown, treatment for IBS is guided by the patient's symptoms. For those patients with diarrhea, antidiarrheal agents can be used. These medications assist by decreasing motility and increasing consistency of the stool. Tricyclic antidepressants and selective serotonin reuptake inhibitors (SSRIs) are also used to treat IBS, and can commonly assist in the control of chronic pain. Medications used to treat constipation-predominant IBS should include bulking agents and osmotic laxatives. Bulking agents are supplements or medications that add psyllium, bran, or other sources of fiber to the diet.

Nutrition therapy goals for IBS will focus on decreasing anxiety, normalizing dietary patterns, ensuring adequate nutritional intake including sufficient fiber, and taking necessary steps to reduce gas production. Nutrition therapy should be initiated after a careful diet history. Food diaries or diet history should focus on dietary components that the patient has associated with any increase in GI symptoms. Both prebiotics and probiotics have received attention for their potential use in IBS. Adding these foods and supplements may be beneficial in the overall MNT plan. Due to problems with gas and flatulence, providing recommendations to relieve these symptoms will also be beneficial.

4. *Describe the pathophysiology of inflammatory bowel disease (IBD) by comparing Crohn's disease and ulcerative colitis. Medically, what is recommended for the treatment of IBD? What are the potential nutritional consequences of IBD? Describe common NT recommendations for IBD.*

Inflammatory bowel disease (IBD) is characterized as an autoimmune, chronic inflammatory condition of the gastrointestinal tract. IBD is actually the term designating a syndrome consisting of two diagnoses: ulcerative colitis (UC) and Crohn's disease (CD). UC is primarily located in the colon and rectum. Damage to intestinal mucosa in UC usually involves only the first two layers of tissue (mucosa and superficial submucosa). Crohn's disease can affect any portion of the gastrointestinal tract from mouth to anus but most commonly affects the ileum and colon. Crohn's disease can damage all layers of gastrointestinal mucosa and can lead to strictures and bowel obstruction.

Treatments for both UC and Crohn's disease include antibiotics, immunosuppressive medications, immunomodulators, and biologic therapies as well as surgical intervention.

The symptoms of IBD can result in decreased oral intake and increased protein and energy needs. Many patients electively restrict eating during acute exacerbations of the disease in order to minimize symptoms. Pain commonly causes generalized anorexia, which further decreases dietary intake. When infection is present or when the patient is febrile, energy needs are increased. Protein needs are increased, in some cases up to 150% of normal requirements. Micronutrients, especially iron, zinc, magnesium, and electrolytes, are at risk for deficiency due to their losses in blood and diarrhea. Medication and surgery may also affect the nutritional status for certain nutrients.

The common nutritional recommendation during flare-ups of IBD is to provide adequate energy and protein, enterally or parenterally. The diet should be advanced as tolerated, but should be individualized to limit problematic foods (gas-forming, spicy, or fried foods). Patients should take a multivitamin and minerals to ensure adequate intake. Patients with IBD are often deficient in B_{12} and iron. When the disease is in remission, adequate protein and calorie intakes with physical activity will help build muscle. Foods high in antioxidants as well as pre- or probiotics may decrease inflammation and reduce symptoms.

5. *How can diet help prevent and treat diverticulosis? Describe the pathophysiology of diverticulitis.*

Diverticulosis is defined as abnormal presence of outpockets or pouches on the surface of the small intestine or colon. Evidence suggests development of diverticulosis is related to low fiber intake, history of constipation, and the resulting long-term increased colonic pressure. Factors that may increase risk for development of diverticulosis include obesity, decreased physical activity, steroids, alcohol and caffeine intake, and cigarette smoking. Treatment for diverticulosis involves only nutrition therapy, with specific focus on increased fiber intake, and use of pro- and prebiotic supplementation.

Diverticulitis is acute inflammation of the diverticula. Foodstuffs and bacteria collect in diverticula and they become infected. Further complications can include development of bleeding, abscess, obstruction, fistula, or perforation.

6. *What is short bowel syndrome, and what factors increase its incidence after surgery? Describe the role of NT in the treatment of short bowel syndrome.*

Short bowel syndrome (SBS) results from a large resection of the small intestine. The American Gastroenterological Association defines short bowel syndrome as occurring when, after surgery or congenitally, a patient is left with less than 200 cm of functional small intestine. Those patients who also experience resections of the large intestine will have additional symptoms that contribute to complications of their SBS. Several factors determine the prognosis for this condition: extent of remaining small intestine, presence of the colon, presence of the ileocecal valve, health of the remaining gastrointestinal tract, and any comorbid conditions the individual may have. Though each case is highly individualized, most research agrees a resection of more than 70% of the GI tract will result in severe nutritional and metabolic complications.

Maintenance of nutritional and hydration status is critical for individuals with SBS. Aggressive nutrition support and careful progression to an oral diet require careful attention by the health care team. Immediately postoperatively patients should receive total parenteral nutrition. As diarrhea decreases, oral diets can begin. Diet is then progressed slowly to a low-residue, low-fat, lactose-free, low-oxalate diet. Caffeine and alcohol should not be initially consumed. Insoluble fiber is generally not tolerated initially, but sources of soluble fiber may actually assist in promoting mucosal health. Soluble fiber, like other sources of prebiotics, assists in production of short-chain fatty acids that are a primary fuel for the colon. As tolerated, one food item at a time is added to the diet.

7. *Describe the primary nutrition-related concerns of people who have undergone colostomies and ileostomies.*

When a certain part of the intestinal tract is removed, normal physiology and function of that portion is lost. This loss of function will produce changes in motility, absorption, and the way waste products are handled—all of which potentially can impact nutritional status. Larger resection will result in the most nutritional complications. Resections of the terminal ileum and loss of the ileocecal valve tend to result in significant fluid, electrolyte, vitamin, and mineral deficiencies. The ileocecal valve controls the rate of movement from the small intestine to the large; hence, when it is absent, motility is much faster, which interrupts normal absorption and causes dehydration.

Goals for nutrition therapy include decrease risk of obstruction; maintain normal fluid and electrolyte balance; reduce excessive fecal output; and minimize gas and flatulence (to reduce odor and inflation of the appliance).

Chapter 16

1. *List the major functions of the liver, pancreas, and gallbladder.*

The liver synthesizes bile salts and helps metabolize protein, carbohydrates, fat, and vitamins. It also detoxifies substances in the body and monitors the nutrient-rich blood delivered via the portal vein. The liver is important in making blood cells and blood clotting as well. There are over 500 known functions of the liver. The gallbladder has three functions: to concentrate, store, and secrete (into the duodenum) bile. The pancreas has endocrine and exocrine functions. As an exocrine gland, the pancreas functions to produce the enzymes and bicarbonate needed for digestion. Its endocrine functions include producing hormones (primarily insulin and glucagon) that regulate the usage of body fuels, mainly glucose.

2. *What is jaundice? What is the difference between conjugated and unconjugated bilirubin? What disorders could cause elevated unconjugated jaundice, and what disorders could cause elevated conjugated jaundice?*

Jaundice is the result of elevated concentrations of bilirubin in extracellular fluid, which cause a yellowing of tissue. There are three primary causes of jaundice: rapid destruction of RBCs with release of bilirubin into the blood, decreased liver function, or obstructed bile ducts preventing secretion of bilirubin into the GI tract. Unconjugated (indirect) bilirubin is formed in macrophages of the spleen and marrow. This is accomplished through a number of reactions, where heme is converted to biliverdin, and then to unconjugated bilirubin during hemolysis. It is indirect reacting, water insoluble, and not excretable. Conjugated (direct) bilirubin is formed in the liver. It is direct reacting, water soluble, and excretable. Elevated unconjugated jaundice could be caused from hyperbilirubinemia, often seen in newborns, because their bilirubin enzyme transport systems are immature; and in liver disease, since the bilirubin cannot be conjugated. Conjugated jaundice can be caused by obstructions to the bile duct (gallstones, for example). There are also certain cancers that can cause this type of jaundice.

3. *What is portal hypertension (include causes, signs, and symptoms)? Why would portal hypertension cause ascites and esophageal varices? What is the medical and nutrition treatment for portal hypertension?*

The portal vein in the abdominal cavity drains blood from the GI tract and spleen. When blood pressure is elevated in the portal vein (usually due to liver disease), it is called portal hypertension. Symptoms include ascites, GI bleeding from varices, and encephalopathy (reduced mental capacity and unconsciousness). Factors that contribute to ascites include reduced osmotic pressure due to failure of the liver to produce serum proteins, like albumin, and increased sodium retention due to decreased renal blood flow and renal filtration rate. Factors that contribute to hepatic encephalopathy include degree of liver failure, diversion of portal blood, bleeding from varices, and sepsis. Signs and symptoms include changes in mental status and personality and neuromuscular changes (asterixis, or "flaps"). We do not know for certain what causes hepatic encephalopathy but it is believed to be caused by the liver's inability to metabolize substances toxic to the brain, including ammonia. For ascites, medical treatment includes sodium restriction and/or diuretics. Fluid can be withdrawn from the abdominal cavity by insertion of a catheter, a process called paracentesis. However, because large amounts of protein are also lost with the removal of fluid, protein intake must compensate for such losses. TIPS, a procedure that uses a shunt to reroute blood flow to the liver, thus reducing portal hypertension, is also used to treat ascites. Treatment of hepatic encephalopathy aims to reduce blood ammonia levels. A common medication, lactulose, is a non-absorbable disaccharide that functions by limiting ammonia production in the intestines and also stimulates the transfer of ammonia from body tissues to the gut lumen. Antibiotics are also prescribed to limit the ammonia-producing bacteria in the intestines.

4. *List the types of viral hepatitis and their modes of transmission. How is alcoholic hepatitis different from viral hepatitis?*

Hepatitis A: transmitted via oral-fecal route, which includes contaminated drinking water, food, and sewage

Hepatitis B: transmitted via blood transfusions, improperly sterilized medical instruments, dental drills, or contaminated tattooing needles

Hepatitis C: transmitted through contaminated blood or body fluids. Most commonly transmitted by sharing contaminated needles, but can also occur if a person is recipient of blood clotting factors before 1987. Infants born to HCV-positive mothers are at an increased risk, as are hemodialysis patients.

Hepatitis D: transmitted through contact with contaminated blood. Interestingly, person must be contaminated with HBV to become infected with HDV.

Hepatitis E: transmitted similarly to HAV

With viral hepatitis, the liver becomes inflamed due to the virus, whereas with alcoholic hepatitis, the liver is injured from chronic consumption of alcohol.

5. *What is cirrhosis? List some of the causes of this liver disorder. What are the common complications? How does cirrhosis cause hypoglycemia and hyperglycemia? What parameters can you use to nutritionally assess a patient with cirrhosis?*

Cirrhosis is end-stage liver disease caused by a wide variety of chronic liver diseases. Scar tissue replaces healthy liver tissue, obstructing blood flow, and ultimately resulting in loss of liver function. Complications of cirrhosis include encephalopathy, portal hypertension (ascites), hepatorenal syndrome, and esophageal varices. In cirrhotic patients, the liver cells, called hepatocytes, are severely damaged. There is a lack of glycogen stored in the liver so hypoglycemia is common, particularly after an overnight fast. Hyperglycemia is also a common condition with cirrhotic patients, as 40%–50% also have diabetes mellitus. Parameters commonly used to nutritionally assess a patient with cirrhosis include low serum protein levels and weight, though weight may not be accurate due to ascites and liver damage. Assessment should include food/liquid intake and intolerances; lifestyle factors, including alcohol intake; lab values related to liver function and vitamin/mineral status; and current and past medical diagnoses and treatments. Many of the methods often used for anthropometric and biochemical assessment of nutritional status are not reliable for individuals with cirrhosis. Current recommendations include Subjective Global Assessment (SGA) and mid-arm circumference of triceps, skin fold thickness, and indirect calorimetry.

6. *What are the possible biochemical causes of hepatic encephalopathy? What is the amount and type of protein used in MNT for hepatic encephalopathy?*

There are three possible causes of hepatic encephalopathy. The first theory is called the ammonia hypothesis, which suggests that ammonia accumulates because of decreased levels of glutamate, necessary for the formation of urea via the urea cycle in the liver. If not enough glutamate is present, ammonia toxicity can occur. The second theory is the neurotoxin hypothesis, which proposes that intestinal bacteria produce neurotoxins, such as ammonia, tyramine, GABA, and manganese. The neuronal GABA receptor complex contains a binding site for neurosteroids, which may be responsible for hepatic encephalopathy. The third theory states that hyperglycemia leads to an increase in skeletal uptake of branched-chain amino acids, resulting in an abnormal ratio of branched-chain amino acids to aromatic amino acids in the plasma, and the accumulation of aromatic amino acids in the brain. In turn, the brain has higher serotonin levels and the neurotransmitter DOPA is inhibited. Intake of up to 1.6 g/kg/day is recommended with lactovegetarian (dairy and vegetable) sources preferred. Lower amounts of protein lower in aromatic and higher in branched-chain amino acids should be given only if encephalopathy does not respond to treatment.

7. *What is recommended for a postoperative cholecystectomy diet?*

After removal of the gallbladder, in most cases oral intake can be resumed immediately. The diet is advanced to normal foods as tolerated, with a possible increase in fiber to help manage diarrhea.

8. *Describe the difference between acute and chronic pancreatitis. List the pertinent labs for each type. What are common nutritional problems associated with chronic pancreatitis?*

Chronic pancreatitis develops over a period of many years with many acute attacks. Acute pancreatitis is marked by acute attacks with the presentation of such GI disturbances as nausea, vomiting, abdominal distension, and steatorrhea. Acute pancreatitis labs include elevated serum amylase and/or lipase and BUN levels. Chronic pancreatitis labs should include blood glucose, triglycerides, hematocrit, hemoglobin, and white blood cells and pancreatic enzymes. Common nutritional problems associated with chronic pancreatitis include steatorrhea, malabsorption of vitamin B_{12} and fat-soluble vitamins, and glucose intolerance.

Chapter 17

1. *List the three chemical classes of endocrine hormones. For each class, pick one hormone, name its production site, and briefly describe its function.*

There are three chemical classes of hormones:

1. Peptides and proteins

2. Amines

3. Steroids

Class of Endocrine Hormone	Hormone	Production Site	Function
Peptide and Protein Hormone	Insulin	Pancreas	Insulin is responsible for cellular glucose uptake and fed-state metabolism.
	Glucagon	Pancreas	Glucagon is responsible for fasted-state metabolism.
	Leptin	Adipose tissue	Leptin influences food intake and metabolic rate.
	GI hormones: gastrin, secretin, cholecystekinin, motilin	GI tract	They control the GI tract, liver, pancreas, and gallbladder.
	Parathyroid hormone (PTH)	Parathyroid gland	PTH controls the plasma calcium and phosphate concentration. It also stimulates vitamin D activation.
Amine Hormones	Thyroid hormones: thyroxine, triiodothyronine, calcitonin	Thyroid gland	They control metabolic rate, growth, brain function, and plasma calcium level.
	Epinephrine and norepinephrine	Adrenal medulla	They control organic metabolism, cardiovascular function, and response to stress.
	Dopamine	Hypothalamus	Dopamine prohibits prolactin secretion.
Steroid Hormones	Sex hormones: estrogen, progestin, testosterone	Gonads	They influence growth and reproductive development.
	Cortisol	Adrenal cortex	Cortisol influences the stress response, the immune system, and metabolism.
	Androgens	Adrenal cortex	Androgens are responsible for the pubertal growth spurt and also control the sex drive in women.
	Aldosterone	Adrenal cortex	Aldosterone controls sodium and potassium excretion by the kidneys.

2. *Describe the action of insulin on carbohydrate, lipid, and protein metabolism.*

Action of Insulin on Carbohydrate, Lipid, and Protein Metabolism		
Carbohydrates	Lipids	Protein
• Insulin stimulates the glucose uptake and glycogenesis. • Insulin inhibits glycogenesis and gluconeogenesis.	• Insulin stimulates the triglyceride synthesis. • Insulin inhibits lipolysis.	• Insulin stimulates protein synthesis. • Insulin inhibits protein degradation.

3. *What is the definition of diabetes mellitus (DM)? List the classifications for DM and briefly explain similarities and differences for their epidemiology, etiology, pathophysiology, and clinical manifestations. Describe three ways diabetes can be diagnosed.*

Diabetes Mellitus (DM): DM is not a single disease but it is a diverse group of disorders that differ in origin and severity. All forms of DM share one common character-istic: hyperglycemia, which results from defects in insulin production, insulin action, or both.

DM can be classified under 3 major categories.

- **Type 1 DM (T1DM):** It results from the failure of the body to produce insulin. All type 1 diabetics are insulin dependent; therefore, this category is also termed insulin-dependent diabetes mellitus (IDDM).

- **Type 2 DM (T2DM):** It results from insulin resistance. It is a condition in which cells fail to use insulin properly; this results in increased need for insulin and eventually the pancreas loses the ability to produce insulin. Type 2 diabetics can be classified as IDDM or non-insulin-dependent diabetes mellitus (NIDDM).

- **Gestational DM (GDM):** It occurs in pregnant women who have never had diabetes before but have high blood sugar (glucose) levels during pregnancy; it may precede development of type 2 (or rarely type 1) DM.

	Type 1 DM	Type 2 DM	Gestational Diabetes

Epidemiology	• 5%–10% of all cases of DM.
	• Develops most frequently in children and adolescents.
	• But it is affecting older adults increasingly.
	• Gender distribution is equal.
	• Idiopathic DM is more common in Asians and those with African ancestry.
		• 90%–95% of all cases of DM.
		• Develops most frequently in adults.
		• But it is affecting children and adolescents increasingly.
		• The elderly and people of color are disproportionately affected.
		• It is more prevalent in American Indians, non-Hispanic black, and Hispanic Americans.
		• It is estimated that about ⅓ of all the individuals with T2DM are undiagnosed.
			• 7% of all pregnancies are complicated by GDM.
			• Women at risk for GDM have following characteristics:
			  • Obesity
			  • Personal history of GDM
			  • Glycosuria
			  • Strong family history of DM
			  • Prior poor obstetrical outcome
			  • High-risk ethnic group

(continued)

	Type 1 DM	Type 2 DM	Gestational Diabetes
Etiology	• There are two types: immune-mediated and idiopathic. • Immune-mediated results from cellular-mediated autoimmune destruction of beta-cells of the pancreas. • Idiopathic refers to having no known cause; individuals with idiopathic diabetes produce no insulin and have no evidence of autoimmunity.	• In some cases heredity may be a factor. • Obesity, especially central obesity, seems to increase the degree of insulin resistance. • Physical inactivity can increase risk of developing T2DM unrelated to body weight. • High-birth-weight infants and low-birth-weight infants are at an increased risk of developing T2DM.	• Metabolic alterations develop to meet the maternal and fetal demands for energy and nutrients during the second and third trimester. • These alterations affect insulin secretion, which in turn affect glucose, amino acid, and lipid metabolism.
Pathophysiology	• T1DM is characterized by an absolute deficiency of insulin due to destruction of pancreatic beta-cells, resulting in the inability of cells to use glucose for energy. • When glucose cannot enter the cells, it results in hyperglycemia and cell starvation. • This results in polyphagia, polydipsia, and polyuria.	• T2DM is characterized by insulin resistance and relative insulin deficiency; both defects must be present for T2DM to manifest. • Insulin resistance is caused by a cell-receptor defect resulting in the body's inability to use insulin. • Defective insulin secretory response results in excess production of glucose from the liver. • The need for insulin is increased so the pancreas increases the production of insulin and eventually loses the ability to produce insulin.	• It is pathophysiologically similar to T2DM. • It affects the fetus; maternal hyperglycemia results in fetal hyperglycemia, which results in fetal hyperinsulinemia and macrosomia.

Clinical manifestations	• It is sudden in onset.
	• It can result in ketoacidosis.
	• Untreated ketoacidosis can result in hypovolemic shock.
	• Long-term complications of hyperglycemia are neuropathy, nephropathy, retinopathy, and cardiovascular disease (CVD).
		• The onset is insidious; individuals may be asymptomatic for 6–10 years but present with complications associated with DM.
		• Uncontrolled T2DM can lead to hyperglycemic hyperosmolar nonketotic syndrome (HHNS).
		• Long-term complications of hyperglycemia are neuropathy, nephropathy, retinopathy, and cardiovascular disease (CVD).
			• Complications associated with GDM include:
			  • Hypertension (preeclampsia)
			  • Polyhydramnios
			  • Difficult birth
			  • Preterm delivery
			  • Higher rate of cesarean sections
			• Fetal and neonatal complications include:
			  • Macrosomia
			  • Hypoglycemia
			  • Respiratory distress syndrome
			  • Hypocalcemia
			  • Hyperbilirubinemia
			  • Polycythemia

The three different ways of diagnosing diabetes are:

1. Casual plasma glucose ≥200 mg/dL (≥11.1 mmol/L) in addition to some symptoms like unexpected weight loss, polydipsia, polyuria.

2. Fasting plasma glucose ≥126 mg/dL (≥7.0 mmol/L).

3. 2-hour postprandial glucose ≥200 mg/dL (≥11.1 mmol/L) during an oral glucose tolerance test (OGTT).

4. *What is meant by glycemic control, and why is it important? Describe the physiological consequences of poor glycemic control. Which laboratory measurements are indicators of short- and long-term glycemic control? How often are they checked?*

Glycemic control refers to the body's ability to maintain glucose homeostasis.

Poor glycemic control results in acute complications such as hypoglycemia, hyperglycemia, diabetic ketoacidosis, hyperglycemic hyperosmolar nonketotic state, Somogyi effect, and dawn phenomenon.

Long-term chronic complications include macrovascular diseases such as dyslipidemia and hypertension; and microvascular diseases such as nephropathy, retinopathy, and neuropathy.

The best measure of long-term glycemic control is the hemoglobin A_1C test, which measures glucose associated with blood proteins that have a 120-day turnover. Fructosamine is also an indicator of long-term glycemic control.

Short-term glycemic control is best monitored using the self-monitoring of blood glucose (SMBG) method in which the individual can use a portable monitor to test up to several times a day. Urine glucose and urine ketones are also short-term indicators of glycemic control.

5. *List and describe the types of insulin that are now available. How is insulin dosage determined, and how can insulin be administered? What are the differences in nutrition therapy recommendations for a person with diabetes who is using insulin on the conventional plan versus a person using intensive insulin therapy?*

The different types of insulin available are as follows:

1. Rapid-acting insulin

- It begins to work about 5 minutes after injection.

- It peaks in about 1 hour.

- It continues to work for 2–4 hours.

2. **Regular or short-acting insulin (human)**
   - It usually reaches the bloodstream within 30 minutes after injection.
   - It peaks anywhere from 2 to 3 hours after the injections.
   - It is effective for approximately 3–6 hours.

3. **Intermediate-acting insulin (human)**
   - It generally reaches the bloodstream about 2–4 hours after injection.
   - It peaks 4–12 hours later.
   - It is effective for 12–18 hours.

4. **Long-acting insulin**
   - It reaches the bloodstream 6–10 hours after injection.
   - It is usually effective for 20–24 hours.

5. **Very long-acting insulin**
   - It starts working within 1 hour after injection.
   - It keeps working evenly for 24 hours after injection.

6. **Mixed insulin**
   - It consists of combinations of different insulins, which can meet the needs of different patients.

Insulin dose can be determined by using mathematical formulas based on body weight and is further adjusted based on blood glucose levels.

There are three basic types of insulin administration regimens:
- Fixed (conventional or standard therapy)
- Flexible (intensive insulin therapy)
- Continuous subcutaneous insulin infusion

Nutritional therapy recommendations in conventional plan versus intensive insulin therapy:
- **Conventional (standard) insulin therapy** consists of a constant dose of basal insulin combined with short- or rapid-acting insulin. Individuals using conventional therapy must synchronize administration of insulin and food intake to avoid hypoglycemia, which requires strict adhesion to meal plans and patterns.

- **Flexible (intensive) insulin therapy** requires multiple daily injections of bolus insulin before meals in addition to basal insulin once or twice daily. Insulin can be adjusted to correspond to food intake, thereby mimicking endogenous insulin secretion in a person without diabetes and allowing for more flexibility regarding meal planning, exercise, and response to hyperglycemia.

6. *Briefly describe several meal planning approaches that are used with individuals with diabetes mellitus. Select a meal plan you would use for a 70-year-old man with an eighth-grade education (type 2 DM), a 13-year-old female athlete (type 2 DM), and a 32-year-old pregnant woman (gestational DM), and justify your answers.*

Some of the meal planning approaches that are used for individuals with diabetes mellitus.
- **Healthful Eating Guidelines:** These guidelines teach making healthy food choices the primary goal. These guidelines are especially useful for people newly diagnosed with diabetes and for those who may be less educated or have low math skills, which are important for more advanced methods such as exchange lists and carbohydrate counting.

- **Plate method:** The plate method teaches portion control, consistency, and basic food categories. This method is also best for newly diagnosed patients and those who are not willing or able to use a more complex system.

- **Menus:** The menus teach the portion control and meal spacing. Menu plans are individualized for those who may have special needs. The main drawback of this method is that self-sufficiency is not developed.

- **Carbohydrate counting:** In this method the individual plans the meals based on the amount of carbohydrate consumed. Food carbohydrate sources are starches, fruits, milk/yogurt, and sweets. Nonstarch vegetables do not need to be counted unless eaten in servings containing >15 g of carbohydrate.

There are two ways of counting carbohydrate:
- The amount of food containing 15 g of carbohydrate counts as one carbohydrate choice.
- The total amount (grams) of carbohydrate in a meal or snack can be counted.

The amount of carbohydrate tolerated varies among individuals and can also be matched to insulin dosage or physical activity.
- **Exchange System:** In this system, foods are grouped together in lists of equal macronutrient value so that they may easily be exchanged or substituted for easy meal planning. Some individuals may find the exchange list to be conceptually difficult to understand and may not be able to utilize it effectively.

Selection of meal plan for different individuals:
- **70-year-old man:** Individualized menus, healthful eating guidelines, and plate method. I chose these methods because of the old age, some elderly develop dementia, and may have a low education level (eighth grade).

- **13-year-old female athlete:** Carbohydrate counting and healthful eating guidelines. I chose these methods because of her age, educational level, and physical activity.

- **32-year-old pregnant woman:** Healthful eating guidelines and carbohydrate counting. I chose these methods because of the short duration of therapy.

7. *List the seven classes of diabetes medications. Briefly describe their effects and mechanisms of action.*

The seven classes of oral diabetes medications are:

1. **Alpha-glycosidase inhibitors (AGIs):** They help in delaying intestinal absorption of glucose.

2. **Amylin analogs:** They delay gastric emptying, decrease postprandial glucagon release, suppress appetite.

3. **Biguanides:** They decrease hepatic glucose production, increase insulin uptake in muscles.

4. **Incretin mimetics:** They mimic glucose-dependent insulin secretion, suppress elevated glucagon secretion, and also delay gastric emptying.

5. **Meglitinides:** They stimulate insulin secretion in presence of glucose.

6. **Sufonylurea agents:** They stimulate insulin secretion.

7. **Thiazolidinediones:** They decrease insulin resistance.

8. *For individuals with type 2 diabetes, why is weight management often included as a component of nutrition therapy? Why is it important for the treatment of type 2 diabetes?*

   Overweight and obesity are risk factors for T2DM. Obesity is also an independent risk factor for hypertension, dyslipidimia, and CVD, the major cause of death in those with diabetes. Moderate weight loss improves glycemic control and reduces the risk for CVD; therefore, weight loss is recommended for individuals with BMI >25.0 kg/m^2.

Chapter 18

1. *List the top three diseases and four risk factors for developing kidney disease.*

   The top three diseases for developing kidney disease are diabetes, hypertension, and glomerulonephritis. The four factors that contribute to developing kidney disease include:

   - Ethnicity: African-Americans are nearly four times as likely to develop kidney failure as white Americans; Native Americans are nearly two times as likely; and Hispanic Americans have nearly twice the risk of non-Hispanic whites.

   - Family history: CKD runs in families, so one's risk is greater if a family member has kidney failure. Hereditary factors such as polycystic kidney disease (PKD) may increase risk.

   - A direct and forceful blow to the kidneys.

   - Prolonged consumption of over-the-counter painkillers that combine aspirin, acetaminophen, and other medicines such as ibuprofen.

2. *How does development of diabetes lead to kidney disease?*

   Diabetic nephropathy is the most common cause of CKD in the United States. People with either type 1 or type 2 diabetes are at increased risk. The risk is greater if blood sugars are not controlled. The earliest detectable change in the course of diabetic nephropathy is a thickening in the glomerulus, perhaps caused by hyperglycemia and a change in the basement membrane of the tissue. Since the glomerulus is responsible for filtering the blood and the fluid that eventually forms urine, as these glomeruler changes occur, the kidney may start allowing more protein (albumin) than normal to be excreted in the urine. As diabetic nephropathy progresses, increasing numbers of glomeruli are destroyed

and increasing amounts of albumin are excreted, which can be detected by a urinalysis. As the number of functioning nephrons declines, each remaining nephron must clear an increasing solute load. Eventually, the limit to the amount of solute that can be cleared is achieved and the concentration in body fluids increases, leading to azotemia and uremia. Because the progression is slow (microalbuminuria can continue up to 5–10 years before other symptoms develop), the body can partially adapt to the changes.

3. *How is kidney function assessed? Explain the formula used for this assessment.*

   Kidney function is assessed based on the glomerular filtration rate (GFR), which is reflected in clearance tests that measure the rate at which substances are cleared from the plasma by the glomeruli. The onset of renal failure is not usually apparent until 50% to 70% of renal function is lost. The National Kidney Disease Education Program (NKDEP) has defined CKD as having a GFR of less than 60 mL/min/1.73 m^2 for three months or longer and/or albuminuria of more than 30 mg of urinary albumin per gram of urinary creatinine.

4. *Symptomatically, how does acute renal failure (ARF) differ from chronic kidney disease (CKD)? Which conditions most commonly lead to ARF?*

   ARF is a disorder characterized by abrupt cessation or reduction in GFR and accumulation of nitrogenous wastes. Normal urine output is 1 to 1.5 L per day. During the period when GFR declines and reaches its lowest level, ARF patients may produce <500 mL of urine. ARF patients are likely to develop fluid and electrolyte disorders, azotemia, and wasting, particularly if they are both oliguric and hypercatabolic (common complications of ARF). With CKD, reduction of GFR is more gradual and individuals may not realize they have kidney disease until they have lost 90% of kidney function.

   The prevalence of ARF is estimated at 1% for all hospitalized patients, 3% to 5% for general medical-surgical patients, 5% to 25% for those in intensive care units, 5% to 20% for open-heart surgery patients, 10% to 30% for those receiving aminoglycoside therapy (a group of antibiotics used to treat Gram-negative bacteria), 20% to 60% for those with severe burns, 20% to 30% for those with rhabdomyolysis (destruction of muscle tissue accompanied by the release of myoglobin into the bloodstream), and 15% to 25% for those treated with cisplatinum, bleomycin, and vinblastine (chemotherapeutic agents).

5. *What is nephrotic syndrome? How are its signs and symptoms like protein-energy malnutrition? What is the cause of the edema? Should you recommend a high level of dietary protein for patients with nephrotic syndrome? Why or why not?*

   Nephrotic syndrome (NS) is an abnormal condition that is marked by a deficiency of albumin in the blood and its excretion in the urine due to altered permeability of the glomerular basement membranes.

   The signs and symptoms of nephrotic syndrome are like protein-energy malnutrition: NS resembles kwashiorkor. In both NS and kwashiorkor, albumin levels are low, plasma volume is expanded, and albumin pools shift from the extravascular space to the vascular space. Muscle wasting is common in those patients with massive and continual proteinuria and can often be masked by edema. Edema occurs when the amount of fluid flowing into the interstitium exceeds maximal lymph flow.

Energy and protein needs should be calculated using usual or ideal body weight since anthropometrics will be skewed due to the presence of edema. Current protein recommendations are 0.8 to 1.0 g/kg/day. This level of intake is believed to decrease proteinuria without reducing serum albumin.

6. *Describe the five stages of chronic kidney disease and include GFR in your explanation. For each stage, list the recommended frequency of nutrition assessment and components of assessment.*

Stage	Description	GFR (mL/min/1.73 m²)	Action
—	At increased risk	≥60 (with CKD risk factors)	Screening CKD risk reduction
1	Kidney damage with normal or increased GFR	≥90	Diagnosis and treatment Treatment of comorbid conditions Slowing progression CVD risk reduction
2	Kidney damage with mild decrease in GFR	60–89	Estimating progression
3	Moderate decrease in GFR	30–59	Evaluating and treating complications
4	Severe decrease in GFR	15–29	Preparation for kidney replacement therapy
5	Kidney failure	<15 (or dialysis)	Replacement (if uremia present)

7. *Describe the two renal replacement treatments used in kidney failure. How do they differ with respect to access to the circulatory system, type of dialysate, length of dialysis, frequency of dialysis, and kcal obtained from glucose during dialysis? At what stage of kidney disease is dialysis initiated?*

The two major types of renal replacement therapy used for patients with CKD Stage 5 are hemodialysis (HD) and peritoneal dialysis (PD). The most common method is hemodialysis. Dialysis is a renal replacement procedure that removes excessive and toxic by-products of metabolism from the blood, thus replacing the filtering function of healthy kidneys. It can maintain life once CKD progresses to the end stage (Stage 5), even though endocrine and metabolic functions of the kidney are not totally replaced.

In hemodialysis, the selective membrane is a man-made dialyzer sometimes referred to as an artificial kidney. The most common types are the hollow fiber and parallel-plate dialyzers. Patients receiving hemodialysis first need to undergo a procedure that allows continual access to the bloodstream. The preferred permanent access site is an arteriovenous fistula (AVF), created surgically by fashioning in the forearm a subcutaneous joining of the radial

artery and the cephalic vein. If the patient's veins are not adequate for this procedure, an arteriovenous graft (AVG) can be created with polytetrafluoroethylene (Teflon). The AV fistula requires 4–6 weeks to become fully functional. Hemodialysis treatments are typically prescribed three times a week for an average of 4 hours per treatment. Although most hemodialysis treatments are done at a dialysis center, home treatments can be an option for some patients. Daily home hemodialysis (DHHD) is conducted 5 to 7 days per week for 2 to 3 hours at a time, and nocturnal home hemodialysis (NHHD) is performed 3 to 6 nights per week during sleep.

In peritoneal dialysis, the lining of the patient's peritoneal wall serves as the selective membrane. Access to the patient's blood supply is gained via a catheter of silicone rubber or polyurethane, placed surgically into the peritoneal cavity. The usual dwell time is 4 to 6 hours, followed by the draining of used dialysate and its replacement with fresh solution requiring an additional 30–40 minutes. Generally, in PD patients with normal peritoneal transport capacity, approximately 60% of the daily dialysate glucose load is absorbed—this results in a glucose absorption of approximately 100 to 200 g of glucose per 24 hours.

8. *What are the protein and energy recommendations for patients undergoing HD or PD? Should body weight be adjusted to determine these amounts? Why or why not? Which formula is used to determine edema-free body weight?*

Since the protein requirement in CKD patients on HD and PD is generally higher than in healthy individuals, it is important to accurately assess protein status in these patients. During steady-state healthy conditions (where nitrogen balance is neutral) the net amount of protein catabolized is nearly the same as the amount of protein ingested. The sum of all nitrogen losses plus the change in body nonprotein nitrogen (mainly urea nitrogen) is expressed as total nitrogen appearance (TNA). The TNA reflects the net breakdown of protein, which again in the steady state should be nearly equal to the intake of nitrogen, of which protein is the major source.

For patients on HD and PD, additional losses of nitrogen in the dialysate must also be considered; therefore, the TNA in CKD patients is equal to the sum of dialysate, urine, and fecal nitrogen losses, and the post-dialysis increment in body urea-nitrogen content. PNA (protein equivalent to total nitrogen appearance) is expressed in terms of grams of protein per day (g/day). PNA may be affected by protein intake and anabolic and catabolic factors.

Protein needs are determined by edema-free and fat-free body mass. PNA is typically normalized (nPNA) to some function of body weight, is a measure of protein nitrogen appearance, and is expressed in terms of grams of protein per kilogram per day (g/kg/day). PNA is calculated automatically as part of routine urea kinetic modeling results. nPNA requires a 24-hour urine collection.

According to NKF K/DOQI Nutrition Guidelines, it is recommended that the adjusted edema-free body weight (aBW$_{ef}$) should be based on the National Health Nutrition Evaluation Survey (NHANES) II data.

9. *Why must dietary potassium, sodium, phosphates, and fluid be restricted during HD? Which vitamins are supplemented during HD and which ones must not be taken in excess?*

The potassium restriction for HD patients varies, depending on the degree of kidney function, serum potassium levels, modality, and drug therapy. For the most part, a diet that allows 50 to 70 mmol/day or approximately 2 to 3 g/day of potassium is commonly prescribed. Those who are oliguric or anuric are at an increased risk for hyperkalemia and should have a more stringent dietary restriction. Fluid and sodium allowances are highly individualized and based primarily on residual urine output and dialysis modality. Other considerations include blood pressure control, interdialytic weight gains in HD patients, presence of edema, and congestive heart failure. The interdialytic weight gain goal in HD patients should not exceed 5% of body weight. Higher fluid gains can lead to sudden changes in blood volume and hypotension during the hemodialysis treatment. Excessive Na intake causes thirst and may increase the interdialytic weight gain leading to increased dialysis time.

Supplementation of water-soluble vitamins due to increased losses during dialysis, anorexia, and poor dietary intake is typically indicated for HD patients. Other reasons include a renal diet that is low in fresh fruits/vegetables, whole grains, and dairy products; altered metabolism; impaired synthesis; resistance to the actions of some vitamins; and decreased intestinal absorption. In general, the "renal" vitamin contains B vitamins, folic acid, and vitamin C. Fat-soluble vitamins and minerals are not included. Preparations containing vitamin A or high doses of vitamin C should be avoided.

10. *What are the risk factors for cardiovascular disease in CKD? What are the current dietary recommendations used to help prevent cardiovascular disease in patients with CKD?*

It has been found that end-stage renal disease patients have higher hyperhomo-cysteinemia levels when compared to the general population. Hyperhomocysteinemia has been linked to cardiovascular disease (CVD) in both the dialysis and general populations. Some studies have suggested that hyperhomocysteinemia is an independent risk factor for CVD. However, a causal relationship between hyperhomocysteinemia and CVD has yet to be clarified in CKD because both inflammation and malnutrition also impact hyperhomocysteinemia levels in this population. Since hyperhomocysteinemia can occur as a result of folic acid, riboflavin, and pyridoxine deficiencies, the recommendation is to be sure dialysis patients are prescribed and routinely take a vitamin supplement that meets the current recommendations.

11. *What is osteitis fibrosa? How are calcium, phosphate, PTH, and vitamin D involved in its pathophysiology? What are the dietary recommendations to help control its development?*

Osteitis fibrosa cystica is a form of bone disease characterized by rapid bone turnover with an excess of collagen production and inadequate mineralization. Prolonged exposure to elevations of PTH causes the development of osteitis fibrosa. Patients are often resistant to vitamin D therapy and require a surgical parathyroidectomy. Most HD and PD patients will require management of SHPT (secondary hyperparathyroid) through the use of dietary phosphorus restriction, phosphate binders, and oral or IV supplementation of vitamin D or vitamin D analogs. Serum PTH, calcium, phosphorus, and Ca × P product should be closely monitored for those on vitamin D therapy.

12. *Anemia is very common in patients with CKD. Why does it occur, and how is it treated?*

CKD patients are not able to synthesize adequate amounts of endogenous erythropoietin, a hormone that is made by renal tubular cells, which leads to decreased red cell production in the bone marrow and low hemoglobin (Hgb) levels. Unlike PD patients, HD patients also incur increased blood loss from dialysis procedures. The discovery of recombinant human erythropoietin (rHuEPO) was a major breakthrough and has significantly improved the quality of life in CKD patients. Treatment with rHuEPO and iron is common practice today.

13. *List the long-term nutrition therapy goals for post-renal transplant patients.*

The goals of the transplant diet in the long term include the management of obesity, blood pressure, insulin resistance, diabetes, and hyperlipidemia; maintenance of electrolyte balance; and maximized bone health.

Nutrient	Chronic Phase (after 8 weeks)
Protein	1.0 g/kg; limit with chronic graft dysfunction
Calories	Maintain desirable weight
Carbohydrates	50%–60% of total kcal; emphasis on complex CHO and 20–30 g dietary fiber (5–10 g per day soluble fiber)
Fats	25%–35% of total kcal with saturated fat <7% of total kcal; up to 10% of kcal from PUFA, and up to 20% of kcal from MUFA
Cholesterol	<200 mg per day; consider plant stanols/sterols, 2 g per day
Potassium	No restriction unless hyperkalemia exists
Sodium	2000–4000 mg with hypertension
Calcium	1200–1500 mg
Phosphorus	1200–1500 mg (supplements may be needed)
Vitamins/Minerals/Trace Elements	Dietary reference intake; may need additional vitamin D
Fluids	No restriction unless graft not functioning

14. *What are the common causes of kidney stone formation?*

Kidney stones (renal lithiasis or nephrolithiasis) form as a result of abnormal crystallization of calcium, oxalate, struvite, cystine, hydroxyapatite, or uric acid that is unable to be excreted normally in the urine.

Chapter 19

1. *Describe the abnormal laboratory values and clinical signs and symptoms indicative of:*

   a. *Iron-deficiency anemia:* Low hematocrit and hemoglobin, small red blood cells, low serum ferritin, low serum iron level, high iron binding capacity (TIBC) in the blood, low MCH, low MCHC. **Signs/Symptoms:** Pale skin, fatigue, irritability, weakness, SOB, sore tongue, brittle nails, pica, decreased appetite, headache.

b. *Pernicious anemia:* Low hematocrit and hemoglobin with elevated MCV (low red blood cell count with large-sized red blood cells), CBC showing low white blood count and low platelets. **Signs/symptoms:** Shortness of breath, fatigue, loss of appetite, diarrhea, tingling/numbness in limbs, sore mouth, unsteady gait, tongue problems, impaired sense of smell, bleeding gums, pallor, rapid heart rate.

c. *Hemolytic anemia:* Elevated indirect bilirubin levels, low serum haptoglobin, hemoglobin in the urine, hemosiderin in the urine, increased urine and fecal urobilinogen, elevated absolute reticulocyte count, low red blood cell count (RBC) and hemoglobin, elevated serum LDH. **Signs/Symptoms:** Chills, fatigue, pale skin, SOB, rapid heart rate, yellow skin (jaundice), dark urine, and enlarged spleen.

d. *Sickle cell anemia:* Hemoglobin decreased, elevated bilirubin, high white blood cell count, elevated serum potassium, elevated serum creatinine, blood oxygen saturation may be decreased. **Signs/Symptoms:** Paleness, yellow eyes/skin, fatigue, breathlessness, rapid heart rate, delayed growth and puberty, susceptibility to infections, ulcers, jaundice, and bone pain

2. *Describe the abnormal laboratory values and clinical signs and symptoms of:*

a. *Leukemia:* Increased WBCs. **Signs/symptoms:** Mucositis, ulceration, chronic fatigue, malaise, poor nutritional status delays healing, lethargy, nutrient deficiencies.

b. *Aplastic anemia:* Low WBC, low RBC, and low platelets. **Signs/symptoms:** Headache, dizziness, nausea, shortness of breath, bruising, lack of energy or tiring easily (fatigue), abnormal paleness or lack of color of the skin, blood in stool.

c. *Thalassemia:* Decreased hemoglobin and low RBC counts. **Signs/symptoms:** In severe thalassemia: fatigue, weakness, pale skin or jaundice, protruding abdomen with enlarged spleen and liver, dark urine, abnormal facial bones, and poor growth.

3. *Describe the homeostatic mechanisms involved in hematopoiesis.*

Erythropoietin, a hormone produced by the kidneys, regulates red blood cell production. Its synthesis is regulated by the kidney's ability to monitor blood volume and its production is stimulated by low oxygen levels in the blood. Iron availability dictates the rate of heme synthesis. With an iron deficiency, circulating RBCs die off without adequate replacement, the oxygen tension falls, and erythropoietin production is increased, but the homeostatic mechanisms eventually fail to compensate, resulting in anemia.

4. *List dietary strategies that can be used to increase intake of folate, cyanocobalamin, iron, zinc, and vitamin A. Why are these nutrients especially important in hematopoiesis?*

Patients may require supplementation; education on foods/cereal fortified with folate, iron, and cyanocobalamin; instructions to limit carbonated beverages; and handouts with lists of foods high in these particular vitamins. Dietary iron, zinc, copper, all the B vitamins, and vitamins A, K, and E are closely tied to normal hematopoiesis. They are involved with WBC production, RBC production, and cell division.

5. *The elderly, pregnant women, and children are at increased risk for anemia. Explain the different contributing factors that heighten risk for these groups.*

With pregnant women, hemodilution in addition to the demands of the fetus results in an increased utilization of iron and depletion of liver stores. Children may have poor Fe absorption, poor iron density, and the presence of increased amounts of calcium, which is another divalent cation that can alter iron uptake from the gastrointestinal tract. Additionally, growing children have an increased need for iron, which may not be met by typical dietary iron intake. Obese children may also be at higher risk for anemia due to the increase of the hormone leptin, which alters the transport, absorption, and storage of iron from food. In elderly people, a decreased immune system may put them at increased risk of anemia. In addition, iron-deficiency anemia is often present in association with various chronic diseases.

Chapter 20

1. *What is epilepsy? What are the signs and symptoms of epilepsy?*

Epilepsy is a neurological disorder characterized by recurrent seizures, generally more than two unprovoked seizures. A seizure can occur in either a portion of or the entire brain and typically presents as sudden deviations in the way the person feels or acts. For example, the patient may experience a change in smell, vision, or hearing. Possible signs include staring, muscle jerking, twitching, head drops, falls, incontinence, salivation, cyanosis, and/or loss of consciousness.

2. *Describe the classic ketogenic diet that can be used in the treatment of epilepsy. List nutrition implications associated with the diet. What nutrients must be supplemented when on this treatment?*

The classic ketogenic diet consists of a high-fat (70%–90% of energy) and low-carbohydrate diet. The ratio of the macronutrients is 4 g of fat to every 1 g of protein + carbohydrate. This diet results in the production of ketones and a shift of energy utilization in the brain so that instead of metabolizing glucose for energy it uses ketones instead. Although the exact mechanism of its effectiveness is not fully understood, it is thought that ketosis is responsible for the diet's anticonvulsant effect. A recent meta-analysis reported that the ketogenic diet reduced seizures by 90% in a third and by 50% in half of the patients. Recent class I and II data confirmed clear anti-epileptic effects of these diets, with efficacy rates similar to main-line anti-epilepsy drugs. Ketogenic diets are considered valid epilepsy treatment options.

The major variations of ketogenic diets in use today include the classic ketogenic diet; medium-chain triglyceride

(MCT) diet; low-glycemic-index treatment (LGIT); and modified Atkins diet (MAD). The classic ketogenic diet is the most widely used and requires adherence to a strict high-fat dietary regimen.

Individuals with epilepsy and seizure disorders can be at nutritional risk. For infants, children, and adolescents, impaired ability to consume adequate nutrients due to limited food choices may interfere with the ability to achieve optimal growth and development. The ketogenic diet is low in fiber and calcium and, in addition to these two nutrients, a multivitamin and mineral supplement is recommended.

3. *Which risk factors for stroke can be modified by diet? List the nutritional implications associated with stroke.*

Multiple factors are thought to place an individual at risk for stroke. Those can be categorized as modifiable or unmodifiable. Unmodifiable risk factors include age, gender, and race. Modifiable stroke risk factors include hypertension, cardiovascular disease, diabetes mellitus, dyslipidemia, asymptomatic carotid stenosis, atrial fibrillation, cigarette smoking, physical inactivity, and obesity. Although dietary factors are not specifically identified, diet does play a critical role in prevention and management of hypertension, cardiovascular disease, diabetes, dyslipidemias, and obesity and, therefore, prevention of stroke.

4. *What is Parkinson's disease, and what are its signs and symptoms? Which neurotransmitter is progressively lost?*

Parkinson's disease is a neuromuscular, neurodegenerative disease caused by the loss of dopamine-producing cells in the substantia nigra portion of the brain, resulting in resting tremor, rigidity, slowed movement, stooped posture and postural instability, mask-like facial features, and a shuffling gait. Other symptoms such as depression, anxiety, sleep disturbances, sensory abnormalities, and pain are often experienced prior to development of the motor symptoms.

5. *How can diet be modified to maximize the effectiveness of L-dopamine in the treatment of Parkinson's disease?*

Two of the most important drug-nutrient interactions involve the medication levodopa (L-dopa). L-dopa is taken orally and transported from the small intestine to the blood-brain barrier; at both of these sites, protein carriers transport the medication. Because amino acids also utilize these same carriers, they compete with L-dopa for transport. Therefore, it has been proposed that high protein intake interferes with optimal levels of L-dopa. Manipulation of protein intake may promote increased therapeutic effects of the medication. The specific method of prescribing and scheduling protein in the diet has varied throughout the literature, but it is generally understood that the benefit of protein manipulation should be apparent within 7–10 days. One strategy is to limit overall protein intake to 0.5–1 g protein/kg with an even distribution of protein throughout all meals. The second strategy limits protein intake during the waking hours with increased amounts of protein at the evening meal and before bed. Ketogenic diets are now being investigated for multiple neurodegenerative diseases, including Parkinson's disease. These diets are thought to be neuro-protective based on some preliminary data that show relief of Parkinson's disease symptoms; however,

more research is needed in this area before these diets are considered for this disease state. Pyridoxine (vitamin B_6) is required as a cofactor for dopa decarboxylase (DDC)—one of the enzymes needed for the conversion of L-dopa to dopamine. Vitamin supplements and foods high in pyridoxine may expedite this conversion before the L-dopa reaches the brain and reduce the amount of dopamine actually transported to the brain.

6. *Describe the pathophysiology and clinical manifestations of multiple sclerosis (MS). What nutritional interventions are used for MS, and are they effective?*

The term "multiple sclerosis" refers to the many areas of scarring (sclerosis) that result from destruction of the tissues that wrap around nerves (myelin sheath). This destruction is called demyelination. Sometimes the nerve fibers that send messages (axons) are also damaged. Over time, the brain may shrink in size because axons are destroyed. When these two components of the nerves are damaged, the communications between neurons is altered. Any nerve can be affected in MS; therefore, symptoms vary widely. Numbness, tingling (paresthesia), uncoordination (ataxia), and weakness are all common symptoms. Some individuals experience visual problems from optic neuritis such as double vision, blurred vision, or blindness. Other symptoms can include difficulty swallowing, constipation, and bladder dysfunction.

Nutrition interventions employed as adjunct medical therapies include supplementation with omega-3 fatty acids and restriction of saturated fat. Nutrition therapy for MS began with the Swank diet, which restricted saturated fat while supplementing with omega-3 fatty acids. Though anecdotally the symptoms of MS appeared to be reduced by this diet, these studies were not experimentally controlled and have not been replicated. Overall, studies published to date have not definitively demonstrated clear benefit of omega-3 fatty acids in treatment of MS, but it is possible that an anti-inflammatory effect of omega-3 fatty acids could be beneficial in MS management. Further research in this area is warranted. Other nutrition interventions that are often proposed include supplementation with antioxidants, including beta-carotene, vitamin C, vitamin E, and selenium. To date there is no research to support their use. People with MS should strive to meet the recommended daily intake of these nutrients by eating a well-balanced diet. Supplementation with vitamin D and calcium appears to be warranted in those patients who are at increased risk of osteoporosis because of long-term steroid use.

7. *What is apolipoprotein E and what is its association with Alzheimer's disease? List the nutritional implications associated with Alzheimer's disease (AD).*

Lipoproteins are responsible for carrying lipids (including cholesterol) in the bloodstream. Apolipoprotein E is associated with several types of lipoproteins. Lipoproteins are also produced by cells in the brain, where they deliver cholesterol and lipids to neurons. The *APOE* ε4 variant is found in approximately 15% of the general population, but in 40% of patients with AD. Individuals with one copy of the ε4 variant have a 3–4 times greater risk of AD than those without ε4.

Nutritional implications for AD are uniquely associated with altered brain function with regard to general cognition as well as feeding behavior.

- Amnesia: In the early stages of the disease, affected persons may find it challenging to shop for groceries, prepare foods, and remember whether or not they have eaten. As AD progresses, they may no longer remember what to do with food and utensils placed in front of them and require increasing assistance and encouragement from caregivers in order to consume sufficient nutrients.

- Apraxia: Chewing difficulty and dysphagia may also occur in the late stages of the disease. Confusion and agitation can compound feeding difficulties.

- Anorexia: Lack of desire to eat, possible psychological basis.

These feeding challenges in AD often result in weight loss. Weight loss and low BMI have been linked to cognitive decline, overall disease severity, morbidity, and mortality. More rarely, some individuals with AD experience hyperphagia and consequent weight gain.

8. *What are the nutritional implications associated with traumatic brain injury?*

Traumatic brain injury (TBI) results in a systemic inflammatory response, evidenced by hypermetabolism, hyperglycemia and insulin resistance, increased gluconeogenesis, lipolysis, and protein wasting. The magnitude of the metabolic response is directly proportional to the severity of the injury. Providing timely exogenous substrate via enteral and/or parenteral nutrition has been shown to minimize the catabolism of body lipids and protein as well as blunt the inflammatory response. TBI patients who are not aggressively supported nutritionally have been estimated to lose as much as 15% of body weight in one week. Evidence indicates that not feeding the injured patient by the end of the first week postinjury increases mortality rate, but even with nutrition support nitrogen balance is generally not achieved until after the third week.

Chapter 21

1. *Describe the three major functions of the respiratory system in human health. What methods are used to measure pulmonary function?*

- Exchange of oxygen and carbon dioxide

- Protection of the body against infection

- Acid-base regulation

Methods used to measure pulmonary function include:

- The initial evaluation of pulmonary function is generally accomplished with the physical examination and the tools of percussion and auscultation. Using these techniques for evaluating sounds, abnormalities in breathing and underlying organs may be detected.

- In pulse oximetry, which is often used at bedside or in any outpatient setting, light waves measure the oxygenation of arterial blood. The pulse oximeter is able to detect the percentage of oxygen within the hemoglobin molecule based on the color of the blood.

- The most common pulmonary function test is done with a machine called a spirometer. During spirometry, the patient breathes into a tube attached to the machine, which calculates the amount of air the lungs can hold and the rate the air can be inhaled and exhaled. The results of the test are compared with those of healthy individuals of similar height and age, and of the same sex and race. Common spirometry measurements include FVC (forced vital capacity), the total volume of air expired after a full inspiration, and FEV1 (forced expiratory volume in one second), the volume of air exhaled in the first second after a deep inhalation. In addition to spirometry, another test used to evaluate lung function is gas diffusion. Gas diffusion measures how well oxygen and other gases pass through the lung's air sacs and are absorbed by the blood. A reduced diffusing capacity could indicate pulmonary disease.

- Evaluation of arterial blood gases (ABGs) determines the pH (acidity), oxygen content, and carbon dioxide content of the blood and can also be used to measure pulmonary function. Changes in pulmonary function are reflected by changes in the partial pressure of dissolved carbon dioxide ($PaCO_2$) and partial pressure of dissolved oxygen (PaO_2). Changes in the $PaCO_2$ measure how well carbon dioxide is able to move out of the blood into the airspaces of the lung, and then out with the exhaled air. Changes in PaO_2 measure how well oxygen is able to move from the air into the lungs. Oxygen saturation is the measure of the amount of oxygen carried by the red blood cells and can be calculated using the partial pressure of dissolved oxygen (PaO_2). In patients with pulmonary disease, fewer red blood cells carry the usual load of oxygen, and oxygen saturation is decreased.

2. *Describe the role of nutrition in pulmonary health. Which nutrients have been associated with normal pulmonary function? How does smoking affect vitamin C requirements?*

Malnutrition has been shown to have an adverse effect on clinical outcomes. The impact of protein-energy malnutrition on lung function has been examined in both clinical and animal studies. Also, the effects of weight loss on pulmonary function in individuals without lung disease have been described. Malnutrition associated with poor intake appears to have an impact on the strength and endurance of respiratory muscles, particularly the diaphragm, and may also cause reductions in lung parenchyma (respiratory bronchioles, alveoli, and capillaries). With continued malnutrition, increased incidence of pulmonary infection may also occur, as a result of depressed immune function.

There is mounting evidence correlating the role of dietary antioxidants such as vitamin C, vitamin E, β-carotene, and selenium with healthy lung function. A variety of antioxidants are present in the extracellular fluid (ECF) and appear to play an important role in protecting the lungs from oxidant injury as the result of the inflammatory process caused by the inhalation of cigarette smoke and other pollutants.

3. *Based on supportive evidence, what are the important nutrition factors to keep in mind when treating patients of various age groups with asthma?*

BMI, breast-feeding, and leukotrienes are important to keep in mind when treating patients with asthma.

- Increases in asthma, particularly in developed countries, have paralleled increases in obesity. Data from NHANES I showed that adults with asthma had a 46% higher prevalence of obesity than those without asthma and this relationship has been consistently observed in women and men. The strongest association of BMI with asthma severity in women has been associated with early menarche. In children, increased risk of new-onset asthma has been associated with increased BMI; in fact, obese children are at a 50% greater risk for developing asthma.

- A number of studies have demonstrated a protective effect of breast-feeding against the development and severity of asthma in children.

- Leukotrienes are chemical mediators produced by the body that contribute to the development of asthma. Leukotrienes, which are synthesized from arachidonic acid, modulate the inflammatory response resulting in tissue edema, mucus secretion, smooth-muscle proliferation, and powerful bronchoconstriction. One possible approach to preventing the synthesis of leukotrienes is through dietary modification. Normally, human inflammatory cells contain high ratios of omega-6 fatty acid (arachidonic acid) to omega-3 fatty acids. Because a common pathway metabolizes both of these fatty acids, an excess of omega-3 fatty acids interferes with the metabolism of the omega-6 fatty acids and reduces their incorporation into tissue lipids.

4. *Define* bronchopulmonary dysplasia (BPD). *Why does it occur? How is vitamin A related to BPD?*

Bronchopulmonary dysplasia (BPD) is characterized by pulmonary inflammation and impaired growth and development of the alveoli. The definition of BPD is based on the severity of the condition and ranges from mild to severe depending on requirements from supplemental oxygen at 36 weeks. However, all babies with BPD have received supplemental oxygen for at least 28 days and are less than 32 weeks gestational age. Factors associated with BPD include extreme prematurity (birth weight <1500 g), perinatal infection, and the presence of patent ductus arteriosus (PDA).

The etiology of BPD is complex and multifactorial, with genetics playing a role in susceptibility to the condition. Approximately 20%–40% of infants born prior to 28 weeks gestation and weighing less than 1000 g develop BPD. Ventilator trauma is also linked to BPD. When born before 28 weeks, infants have immature lungs and require supplemental oxygen on a ventilator for extended periods of time. Prolonged exposure to high oxygen concentrations has been identified as one cause for BPD. Other causative agents may include pro-inflammatory proteins and reactive oxygen species. The premature infant has a poorly developed antioxidant system and therefore is highly susceptible to oxidative damage.

Poor vitamin A status is another possible factor in the development of BPD. Vitamin A is important in normal alveolar development and surfactant production, and supports the integrity and regeneration of respiratory epithelial cells. Most premature infants are born with low serum vitamin A levels and lower levels of vitamin A transport carrier protein, a retinol-binding protein. When supplemented with vitamin A, BPD infants have a significant drop in oxygen use and a reduced death rate.

5. *Define* cystic fibrosis (CF). *What organ systems are involved in the disease? How does this organ involvement affect nutrition status? You receive a nutrition referral from a physician for a 9-year-old male with CF. He is below the 10th percentile weight for height. Outline an appropriate nutrition protocol for someone his age.*

Cystic fibrosis (CF) is a disease characterized by abnormally thick mucus secretions from the epithelial surfaces of various organ systems, including the respiratory tract, the gastrointestinal tract, the liver, the genitourinary system, and the sweat glands.

- Pancreatic insufficiency results in poor digestion, poor absorption of fat and fat-soluble vitamins, and loss of bile and bile salts. As a result, pancreatic enzyme supplements might be necessary.

- CF and deteriorating pulmonary function may lead to anorexia, increased energy requirements, and malnutrition.

- Individuals with CF are at risk for osteopenia and osteoporosis because of pancreatic insufficiency, vitamin and mineral malabsorption (Ca, P), and the chronic use of corticosteroid medications.

- Adequate kcal are essential to support normal growth and development. Energy needs for children with CF without respiratory infection are comparable to that of healthy children (100%–110% of RDA). However, if an individual has significant lung disease or malabsorption, energy requirements may be significantly increased (120%–150% of the RDA).

Nutrition protocol for a patient with CF:

A steady rate of weight gain in children should be the goal, and it is necessary to monitor for appropriate growth. Additional kcal are required in the diet to support growth. These kcal can be added in the form of between-meal snacks and nutrient-dense foods (particularly those high in fat), whereas low-fat and low-kcal foods should be avoided. Adequate protein intake is also essential to meet growth needs and maintain protein stores. The use of nutritional supplements may be helpful in an effort to increase kcal, protein, and other nutrients to the diet. Foods high in energy and protein, such as fortified beverages or puddings, may be beneficial. The addition of supplemental enteral feedings may be needed if adequate kcal intake cannot be achieved or growth is compromised. Nocturnal tube feedings are encouraged as a means to promote normal eating behaviors during the day.

Deficiency in several vitamins and minerals might be a problem. These include vitamin A, vitamin D, vitamin K, vitamin E, iron, and zinc. Vitamin supplementation needs to be individually adjusted based on age, dietary intake, and disease progression. Laboratory monitoring of each of these vitamins should be done at the time of diagnosis and at least yearly to ensure that the patient is receiving adequate amounts. It is recommended that individuals with CF should consume a high-salt diet, as infants and children with CF are at risk for developing hyponatremia.

School-aged children (5–10 years) are at higher risk for decreased growth rate and therefore should be monitored appropriately. Certain activities may need to be limited to decrease energy expenditure and increase time for consumption of snacks and taking of enzymes.

6. *Describe aspiration pneumonia. As a dietitian, what procedures or methods would you recommend to help prevent the occurrence of aspiration pneumonia in a patient receiving tube feeding?*

Aspiration pneumonia occurs when aspirated materials (saliva, nasal secretions, bacteria, liquids, food, or gastric contents) cause an inflammatory response in the lung. Patients fed using both a nasogastric feeding tube and a gastrostomy tube are at high risk for aspiration. The likely cause is the reflux of gastric contents from the stomach into the pharynx, where they are aspirated into the lungs. In addition, the presence of a nasogastric feeding tube may interfere with the effectiveness of the lower esophageal sphincter, resulting in gastrointestinal reflux.

Prevention of aspiration associated with tube feeding may include elevating the head of the patient's bed 30°–45° to prevent gastroesophageal reflux and regularly checking the placement of the tube feeding to ensure proper placement and that it does not move upward beyond the pyloric sphincter. The following procedures have also been proposed to prevent aspiration: use small-bowel feedings, rather than gastric feedings, to prevent gastric reflux; and a continuous method versus an intermittent or bolus method. If the patient has diabetes, maintaining rigid blood glucose control may be helpful.

7. *Define* respiratory failure (RF). *What are the goals of nutrition therapy for RF? Outline a nutrition protocol for a patient with RF. Which nutrients are of specific concern?*

Respiratory failure (RF) occurs when the respiratory system is no longer able to perform its normal functions. It can result from long-standing chronic lung disease like COPD or cystic fibrosis, or as a result of an acute insult to the lung. The goals for nutrition care for patients with RF are to meet their nutrition needs, preserve and restore lean body mass, and maintain fluid balance.

Nutrition protocol for a patient with RF:

- A complete nutrition assessment to evaluate the patient's individual nutrition needs, including anthropometric and laboratory status, should be completed. Caloric requirements can be determined using a predictive equation or directly measured using indirect calorimetry. Once the kcal requirements have been estimated, the patient's pulmonary status, body weight, and fluid balance must be closely monitored to ensure that overfeeding does not occur. Patients with ARDS have an increased protein requirement due to hypermetabolism.

- A range of 1.2 to 1.5 g/kg/day of protein is appropriate and should be adjusted to promote nitrogen retention, without being excessive. An evaluation of urine urea nitrogen is one method for assessing adequacy of protein intake.

- Patients with ARDS have low levels of oxidant defense enzymes, high levels of ROS, and tissue oxidative damage; therefore, nutrients of specific concern include α-tocopherol, β-carotene, and vitamin C. Supplementation of α-tocopherol, β-carotene, and vitamin C at levels higher than the DRI has been associated with substantial increases in serum α-tocopherol and β-carotene and appears to prevent further oxidative damage.

- Phosphate is an additional nutrient of concern as it is essential for optimal pulmonary function and normal contractibility of the diaphragm. The length of hospital stay and dependence on mechanical ventilation have been shown to be increased in critically ill patients who have hypophosphatemia. The phosphate balance of patients in respiratory failure needs to be closely monitored, particularly for those patients receiving parenteral nutrition support. Phosphate supplementation should be initiated whenever hypophosphatemia is present.

Chapter 22

1. *List five conditions that can cause stress and result in a hypermetabolic state.*

Five different conditions that may cause stress and lead to a hypermetabolic state include (1) a closed head injury, (2) a burn, (3) severe inflammation, (4) cancer, or (5) sepsis/SIRS.

2. *Describe the differences between the body's responses to starvation and to stress.*

When the body is responding to starvation, it will increasingly rely on body tissues such as lipid stores for energy, and it will reduce the basal metabolic rate so as to require less energy overall. In metabolic stress, the response is different. Energy requirements are increased and glucose is used as an energy fuel.

3. *Describe the ebb and flow response to stress.*

The ebb phase in the response to stress begins shortly after the injury occurs. During the ebb phase, shock brings about hypovolemia, thus providing less oxygen to tissues. Lower blood volume causes decreased cardiac and urinary output. At this point the goal of treatment is to restore blood flow, get oxygen to tissues, and stop bleeding.

Once the blood pressure stabilizes, the flow phase begins. During this phase metabolic stress occurs and leads to hypermetabolism, catabolism, and altered immune and hormonal responses. This phase does not end until the metabolic stress is resolved.

4. *Describe the hormonal response to stress. How do these hormones affect protein and carbohydrate metabolism?*

During stress the body goes into a "flight or fight" response, releasing stress hormones such as glucagon, cortisol, epinephrine, and norepinephrine. These hormones work to mobilize energy stores so that the immediate energy demand can be met. Specifically, glucagon increases glucose production from amino acids, while cortisol increases levels of glucose, AA, and FFA in the blood and decreases the amount of protein synthesis. Epinephrine and norepinephrine stimulate the breakdown of glycogen and also increase FFA release. Insulin is also released, but the body is insulin resistant at this point.

Overall, these hormones work to increase the breakdown of body stores of carbohydrate and protein in order to use them as an energy source.

5. *How do cytokines influence the metabolic response to stress?*

Cytokines regulate the release of acute-phase proteins. During stress, cytokines such as interleukins, leukotrienes, tumor necrosis factor, and interferons are released and act on target cells to produce stress reactions such as anorexia, fever, inflammation, and possibly hyperglycemia or catabolism.

6. *Why might omega-3 fatty acids, branched-chain amino acids, glutamine, and arginine be used in nutrition support for a patient with a hypermetabolic condition?*

Unlike most fats that have an immunosuppressive role, omega-3 fatty acids seem to play an important role in improving the body's ability to respond to stress.

Branched-chain amino acids may play a role in providing energy within the skeletal muscle. Using them might spare other substances that are needed for energy by other tissues during stress.

Although arginine and glutamine are not essential AA, the body cannot synthesize them at a rate to meet the increased needs. Arginine in particular may help inhibit immunosuppression during stress and improve nitrogen balance, but use of it has also been shown to have a negative outcome. Glutamine supplementation might lower infection rate and prevent translocation of bacteria from the GI tract and is often given orally.

7. *Describe how burns are classified. What are the nutrition implications of burns? How does the burn classification affect the kilocalorie and protein requirements of a burn patient?*

Burns are classified by the depth of the wound and the percentage of body surface that is affected. Wounds are described as superficial, superficial partial thickness, deep partial thickness, or full thickness.

After a burn injury the body's response is hypermetabolic and catabolic, meaning that a burn victim is at significant nutritional risk. Body protein may be lost and wound healing requires intense nutrition therapy; however, fluid loss, pain, and immobility may complicate intake and make it difficult to maintain nutritional status. As the % of body surface area increases, the kcal and protein needs increase, especially at the peak of injury. Protein requirements are estimated at 1.5–2.0 g/kg/day to help minimize negative nitrogen balance. Calorie requirements are estimated to be 22–25 kcal/kg of actual body weight.

8. *After surgery, when might a patient be advanced to solid food? Describe a situation in which advancement to solid food would not be recommended.*

After surgery a patient should be advanced to solid food as quickly as possible (and as tolerated) to help maintain their nutritional status and to help maintain the GI tract.

Advancement to solid food may not be advised when the surgery interrupts normal nutrition processes, such as with bariatric surgery.

9. *What is sepsis? How is the immune system involved? What are the nutrition implications of sepsis?*

Sepsis is presently thought to be an immunosuppressive process due to an infection that prevents an adequate response to that infection. An infection or trauma triggers sepsis and causes the immune system to release inflammatory mediators such as TNF, interferon, and interleukins. Since the body is not able to mount a proper immune response, the reaction becomes anti-inflammatory.

During sepsis there are increased nutritional risks and complications due to abnormalities of metabolism, difficulty determining nutritional needs, fluid shifts, and multi-system organ dysfunctions.

10. *What is SIRS; what is MODS/MSOF? How does SIRS differ from sepsis?*

SIRS (systemic inflammatory response syndrome) is another classification of sepsis, which may not be caused by an infection, but may occur after a major surgery or trauma, or with other conditions such as myocardial infarction.

MODS/MSOF is multiorgan distress syndrome/multisystem organ failure (two terms are used for the same condition); a "disease involving more than one of the vital organs, such as the heart, lungs, kidney, liver."

Sepsis, unlike SIRS, must be due to an infection, whereas SIRS may be brought on by infection or trauma.

Chapter 23

1. *Carcinogenesis is the process by which normal cells are transformed into cancer cells. How could a nutritional factor act as a carcinogen?*

Diet and nutrition are viewed more appropriately as modifiers, rather than initiators, of carcinogenesis. Caloric intake, type and amount of fat, protein, amino acids, vitamins, minerals, fiber, and other dietary constituents have been studied in regard to their influence on the development of neoplasms. Genes may also be affected by nutritional components such as antioxidants, soy protein, fat, calories, alcohol, and phytochemicals. Nutritional genomics (nutrigenomics), the study of genetic variations that cause different phenotypic responses to diet among humans, is a relatively new field that is already opening new and exciting opportunities for research in the field of nutrition and may alter the way we apply nutrition therapy and cancer prevention strategies.

2. *In general, how does the growth of a cancer cell differ from that of a normal, healthy cell?*

Unlike a normal cell, whose growth is closely regulated, a cancer cell reproduces at an uncontrolled rate. The cancer cell becomes autonomous from the normal growth signals and genetic control, and may even secrete its own growth factor. In some cancer cells, an enzyme is secreted that destroys the telomere structure, leading to the loss of the cell's internal clock or counting mechanism—permitting uncontrolled cellular replication.

3. *Explain the following diagnosis in terms of how a malignancy is classified: Stage II diffuse large B-cell lymphoma.*

Tumors are classified and assigned a "stage" on the basis of cell type, tissue of origin, whether it is benign or malignant, and degree of differentiation, anatomic site, and function. The Tumor Node Metastases (TNM) Staging System is a descriptor (usually numbers I to IV) of how much the cancer has spread (T = depth of tumor invasion, surface spread and tumor size; N = absence or presence and extent of regional lymph nodes; M = absence or presence of distant metastases). Large B cell is a non-Hodgkin's lymphoma. The cells are fairly large when viewed under the microscope.

4. *A patient diagnosed with cancer may be treated with one or a combination of the following treatments: chemotherapy, radiation, and/or surgery. Describe the basic mechanism or rationale behind each method used to treat a malignancy.*

Chemotherapy is a drug therapy that prevents cancer cells from dividing. However, chemotherapy is given systemically so it also affects healthy cells resulting in side effects. The kinds of side effects that the patient would have depend on the type and dose of chemotherapy. Healthy cells usually recover after chemotherapy.

Radiation therapy (RT) destroys cancer cells by altering cellular and nuclear material, especially DNA. The radiation's effect is localized to the target area, but can affect surrounding cells as well.

Surgical treatment can be considered primary, adjuvant, combination, salvage, or palliative. Primary (definitive) treatment indicates that surgery will be the only therapy an individual will receive; for example, surgical removal of a small tumor. Adjuvant therapy could also include chemoreductive therapy, chemotherapy to reduce the size/bulk of a tumor, in addition to surgical removal of the mass. Combination therapy includes surgical resection followed by additional treatments, such as radiation and/or chemotherapy. Salvage therapy involves use of extensive surgery to treat local recurrence after a less extensive primary approach has been implemented—for example, mastectomy after lumpectomy and radiation. Palliative surgery is used to ameliorate disease and/or treatment-related symptoms without attempting to cure the cancer. Chemotherapy or radiation is often employed before surgical excision in order to shrink the tumor, making it easier to remove. Because surgical removal of any tumor can cause cancer cells to leak or metastasize, surgery may not be an option for all cancers. If a tumor has invaded a major organ, successful surgical resection may be impossible. So, it is used only if it is determined that the tumor can be excised without too much damage to surrounding areas or if there is a slim chance of the cells metastasizing.

5. *How do radiation and chemotherapy affect healthy cells within the body? What are the cells that are primarily affected? How is this related to the side effects that are often experienced with treatment?*

Radiation and chemotherapy affect rapidly dividing cells. The rapidly dividing gastrointestinal cells will be significantly affected by most chemotherapeutic agents. This is the reason most patients undergoing chemotherapy experience nausea and other gastrointestinal (GI) problems such as diarrhea or mucositis during their course of treatment. Because hair follicles also contain rapidly dividing cells, alopecia is also a common side effect of chemotherapy.

6. *Another category of treatment includes biological response modifiers or immunotherapy. What is the general mechanism for these treatments?*

Biological response modifiers (BRMs) are agents used to boost or restore the ability of the immune system to combat cancer. BRMs, which are also referred to as immunotherapy, biological therapy, or biotherapy, use the body's own immune system, either directly or indirectly, to eradicate cancer cells. Biological therapies use interferons, interleukins, monoclonal antibodies, growth factors, gene therapy, and nonspecific immunomodulating agents. BRMs are used to stop, control, or suppress processes that permit cancer growth; make cancer cells more recognizable for destruction by the body's own immune system; enhance the killing power of the body's own immune system; enhance the body's ability to repair or replace normal cells damaged or destroyed by other forms of cancer treatment; and prevent cancer cells from spreading to other parts of the body.

7. *What is the subjective global assessment? What are the primary nutrition assessment factors that this tool identifies in cancer patients and those undergoing cancer therapies?*

Subjective global assessment (SGA) is a validated form of nutritional assessment based on medical history of weight change, changes in dietary intake, and gastrointestinal symptoms. It also includes a physical examination. For the cancer patient population, a positive correlation has been demonstrated between the use of SGA and objective assessments such as anthropometry, albumin, total cholesterol, weight, and weight loss.

8. *Many cancer patients present with some nutritional symptoms such as weight loss or taste changes. How might a malignancy affect nutritional status even before it is diagnosed?*

In malignancy, several biochemical changes occur. The most important carbohydrate abnormalities are insulin resistance, increased glucose synthesis, gluconeogenesis, increased Cori cycle activity, and decreased glucose tolerance and turnover. In cancer cachexia, amino acids are not spared as they are during simple starvation, and depletion of lean body mass occurs. Muscle wasting may be due to increased protein catabolism (hypercatabolism) or decreased protein synthesis; the simultaneous presence of both results in the most intense muscular atrophy. Alterations in lipid metabolism also occur in the presence of malignancy. Abnormalities that occur in the presence of cancer include increased lipid metabolism, decreased lipogenesis, and decreased activity of lipoprotein lipase (LPL). Mobilization of fatty acids from adipose tissue may occur before weight loss, suggesting the presence of a lipid-mobilizing factor (LMF) produced either by the tumor or host tissues.

9. *Name three common nutritional problems that a cancer patient might experience. Identify interventions for each.*

1. Nausea and Vomiting: to assess the causes of nausea and vomiting will help with treatment. Patients should avoid noxious odors. Nausea from cooking odors can be minimized by using a microwave oven, opening windows when cooking, taking a walk when meals are being cooked, and avoiding frying of foods, which emits more odors than most other forms of cooking. For vomiting the patient should be advised to eat only a small, low-fat meal the morning of the first treatment and to avoid fried, greasy, and favorite foods for several days following the treatment. A clear liquid diet for the first few days after therapy may be indicated. To provide calories and maintain hydration, consumption of electrolyte-fortified beverages such as Gatorade should be encouraged. It is important for patients to avoid favorite foods at any time the chance for emesis is high, since once a favorite food has been vomited, the likelihood of its subsequent consumption is low. The same principle applies to the use of "creamy" liquid nutritional drinks.

2. Early Satiety: is caused primarily by delayed gastric emptying. It is important for the patient with early satiety to eat small, frequent meals that are nutrient dense. Beverages should also contain nutrients and should be consumed between meals rather than with meals so as not to add to the feeling of fullness. Consumption of raw vegetables, such as salads, and other high-fiber foods should be avoided. Prokinetics, medications that increase gastric emptying, may be useful.

3. Mucositis: (also known as stomatitis) may be eased by the consumption of soft, nonfibrous, nonacidic foods. Liquids at room temperature should be frequently consumed to avoid dehydration. The use of nutritional supplements is recommended.

10. *Identify one complementary and alternative therapy that is commonly used by cancer patients. Explain the indications for this therapy and/or its risks.*

Milk thistle (active ingredient silymarin): Laboratory studies demonstrate that silymarin functions as a potent antioxidant, stabilizes cellular membranes, stimulates detoxification pathways, stimulates regeneration of liver tissue, inhibits the growth of certain cancer cell lines, exerts direct cytotoxic activity toward certain cancer cell lines, and may increase the efficacy of certain chemotherapy agents. No clinical trials in individuals with cancer have been published. Few adverse side effects have been reported for milk thistle, but little information about interactions with anticancer medications or other drugs is available.

Chapter 24

1. *What kind of virus is the human immunodeficiency virus (HIV)? Which types of cells does it target?*

HIV is a retrovirus that targets CD4 immune cells. It contains nine genes, six of which are essential to penetrate and infect the target cells and produce copies of the virus. The other three genes are used to provide the necessary information to produce new viral particles within the host cell. This retrovirus targets many cells in the body, including gastrointestinal cells, organ cells, and immune cells, among others. The resulting immunodeficiency syndrome is most closely related to the infection of activated CD4 (T helper cells), which become a viral factory. The virus identifies the target CD4 cell and fuses to the surface. HIV injects RNA, enzymes, and other substances that assist in viral integration and replication.

2. *List two factors that are predictive of disease progression. List one nutrition-related factor that is predictive of survival.*

Two factors that are predictive of disease progression are:

- Diagnoses of AIDS-defining illnesses, including opportunistic infections (e.g., candidiasis, cytomegalovirus, *Mycobacterium* spp., *Pneumocystis jiroveci*), opportunistic malignancies (e.g., Kaposi's sarcoma, Burkitt's lymphoma, invasive cervical cancer), or other conditions (e.g., recurrent pneumonias, recurrent bacterial infections, wasting syndrome).

- A CD4+ cell count of less than 200 cells per microliter, a CD4+ cell count that comprises less than 14% of lymphocytes present, and/or an AIDS-defining illness. CD4+ cell counts and a range of complications related to immune dysfunction figure into the classification of the disease. CD4+ cell counts that range from 600 to 1200 cells per microliter generally provide adequate immune defense against opportunistic disease.

One nutrition-related factor that is predictive of survival is a BMI >18. Wasting and nutritional decline are strong predictors of the timing of death since, despite the success of HIV treatment in improving survival of infected patients in the developed world, weight loss remains an independent predictor of mortality.

3. *List two examples of how HIV or opportunistic infection compromises gastrointestinal function.*

HIV may be transmitted through the gastrointestinal tract, which is an important barrier to infection and immune organ in the body. The response of the intestinal tract includes the rapid turnover of gut tissue and activation of immune cells on the intestinal surface. HIV infection is associated with depletion of CD4 cells in the GI tract, which is where more than 60% of the body's T lymphocytes resides, providing a large reservoir of HIV-infected cells in the gut and increased risk for the malabsorption of nutrients.[1] As malnutrition progresses, malabsorption worsens, and health further deteriorates. Fungal infection, viral infection, bacterial conditions, neoplastic problems, salivary gland disease, and other conditions can cause oral manifestations. Oral lesions can lead to mouth itching, pain, a burning sensation (especially when eating spicy or acidic foods), and taste changes.

4. *List two cofactor diseases or conditions that are related to HIV infection and its treatment and may require nutritional therapies.*

---

[1]Smith PD, Mai UE. Immunopathophysiology of gastrointestinal disease in HIV infection. Gastroenterol Clin North Am. 1992;21:331–35.

- Pulmonary disorders may occur both related to and unrelated to HIV infection. The level of CD4+ cell count is related to the risk for the development of several types of pulmonary diseases. With some of the many symptoms associated with pulmonary problems, including persistent coughing, chest pain, fatigue, fever, and shortness of breath, it may become more difficult for the patient to maintain adequate food intake. Pulmonary hypertension has been demonstrated in patients with HIV infection without AIDS-defining diagnoses. Antiretroviral (ARV) medication has significantly reduced AIDS-related pulmonary disorders.

- Cardiac manifestations may include infections and inflammation, cardiomyopathy, pulmonary hypertension, and coronary artery disease. Cardiomyopathy may be related to infectious agents, inflammation processes, and medication therapies. For instance, the antiretroviral (ARV) medication zidovudine has been associated with skeletal muscle myopathy. Much of the coronary artery disease risk is associated with medication interactions that increase blood lipid levels. However, overall it appears that there may be a protective effect of treatment medication on the mortality associated with myocardial disease. Chronic inflammation, immune dysfunction, concomitant conditions (such as insulin resistance, diabetes, renal involvement, and hypertension) and opportunistic conditions place a patient infected with HIV at a higher risk for cardiac disease.

5. *Define* AIDS-related wasting syndrome (AWS).

   AIDS-related wasting syndrome (AWS) is defined by the Centers for Disease Control and Prevention (CDC) as a 10% weight loss without an identifiable cause that is accompanied by fever or diarrhea for 30 days or more.

6. *Describe the inflammatory process that is seen in early and asymptomatic HIV infection. How can it trigger a wasting process?*

   An HIV positive status starts the lifelong inflammatory process that is similar to other infections in that it is the severity of infection that determines the level of inflammatory response. A simplified overview from a nutritional response standpoint includes a continued cortisol response to HIV infection, leading to breakdown of labile body protein stores to feed the inflammatory response. This response includes alterations in nutrient intake and utilization along with changes in hormone sensitivity and levels that regulate nutritional status. Even during "asymptomatic" phases of HIV infection, body composition testing shows alterations reflecting the inflammatory process and suggesting that subclinical wasting can continuously and progressively occur.

   Because of the inflammatory nature of chronic HIV infection, evident even when viral load is fully suppressed, nutrient metabolism can be expected to change. Positive and negative acute-phase proteins are continually produced, and exert an effect on both macronutrient and micronutrient status. Documented micronutrient changes include lower serum levels of selenium, zinc, magnesium, calcium, iron, manganese, copper, carotene, choline, glutathione, and vitamins A, B_6, B_{12}, and E. Elevated levels of folate, niacin, and carnitine have been documented as well.

   HIV causes the body to break down labile protein stores, primarily muscle, to enhance the amino acid pool and allow the synthesis of proteins that initiate and maintain immune and healing responses. Protein turnover rates and the level of immune system activity are elevated throughout the course of HIV infection and disease, while anabolic rates are more directly regulated by nutrient substrate availability.

7. *At which levels of weight loss and body cell mass loss is there an increased risk for morbidity and mortality in HIV infection?*

   Alternate definitions of wasting have been offered to include both weight losses and changes in body composition that increase risk for functional impairment. Evidence suggests that 10% weight loss for any reason and over any period of time is a strong predictor of death in HIV-infected patients, and that even less than 5% weight loss may be a risk factor for mortality. A body mass index (BMI) of less than 20 suggests an increased risk for mortality in patients with HIV infection. BMIs of less than 18 are markers of mild to moderate malnutrition and are associated with a significantly reduced survival. Patients with higher BMI levels, on the other hand, have a decreased risk for HIV disease progression.

   In addition to weight loss, body composition changes without weight loss can compromise clinical and nutritional status. Additional wasting criteria may include a loss of 5% BCM over a period of six months and a BMI of <20.

8. *Which antiretroviral (ARV) medications should be taken on an empty stomach? Which should be taken with food?*

   - Reyataz, Prezista, Kaletra, Viracept, Norvir, Fortovas—take with food

   - Videx—take without food

   - Agenerase, Sustiva—avoid taking with high-fat meal

9. *Describe potential side effects of two medications used to treat HIV infection that can alter nutritional status.*

   The zidovudine in combovir has been known to cause or exacerbate lipodystrophy. The protease inhibitor combination of lopinavir/ritonavir has been known to cause insulin resistance as well as body fat redistribution.

10. *Describe three types of dietary interventions that may be required in treating HIV infection.*

    Goals for the nutrition care plan may include:

    - **Protein recommendations:** Usually include extra protein to buffer the catabolism of protein stores that accompanies the inflammatory response to viral and other infections. It is believed that additional protein helps to improve immune and other inflammation-mediated responses, but additional research needs to be conducted on protein requirements and intake. For patients with renal dysfunction, bone mineral losses, and long-term diabetes, the quality of protein and the best sources may be more specifically tailored. In such instances, a balance of animal- and plant-based protein sources will be especially important. Overdoses of protein pose an additional burden and risk in patients with potential for renal or bone problems. Special care should be taken to tailor protein recommendations to individual needs and limitations.

- **Micronutrient therapy:** Altered micronutrient levels have been documented in both pediatric and adult HIV disease, in research that suggests a worsening associated with disease progression. Alterations in micronutrients may exacerbate immune compromise, disease progression, damage due to oxidative stress, morbidity, and even risk for HIV transmission independent of CD4 cell counts.

The use of multivitamin/mineral supplementation may have a supportive role in slowing disease progression, particularly in patients without access to antiretroviral therapies. Additional micronutrient therapies may include those targeted to treat anemias, diarrhea, and potential dietary deficiencies. Special care should be taken in each instance to consider the effect of chronic inflammation in HIV infection.

- **Macronutrient therapy:** Alterations in macronutrients for therapeutic purposes can address disease and treatment complications, such as symptom management, weight maintenance (including gain and loss according to needs), and management of concomitant disease processes. Nutritional supplements have been recommended for patients who require a boost in macronutrient intake. Patients may use recipes to create their own nutritional supplements and/or obtain ready-to-use commercial supplements in the form of beverages and bars. While some food-based kcal may be displaced by such therapy, kcal-containing supplements have been shown to yield an increase in energy intake and may assist in adherence to a weight gain regimen.

11. *List four types of non-nutrient interventions that are used to improve nutritional status.*

   1. **Exercise:** Exercise has been recommended to balance activity with kcal intake, improve muscle volume and function, and normalize lipid and energy metabolism.

   2. **Education/counseling:** Referrals should be made for other aspects of care that affect nutritional status and health, including smoking cessation, alcohol or drug rehabilitation, HIV and other infectious disease transmission, and psychosocial or economic factors that can affect food access, choices, and metabolism. Counseling activities may be essential to the successful use of nutritional supplements to achieve weight gain objectives. Medication interactions with food and nutrients are an important feature of education and counseling for patients taking ARVs and other therapies.

   3. **Symptom management:** Symptom management often includes a nutrition component, but may also include non-nutrient therapies. For instance, diarrhea can result from opportunistic infection, medications, and organ dysfunction. If adequately evaluated, the appropriate therapy may include treatment of opportunistic infections identified, alteration of medication or doses,

additional medication therapy to reduce diarrhea, and dietary management.

   4. **Prevention and treatment for wasting:** Medication interventions targeted to reduce risk and reverse wasting and weight loss include appetite stimulants, anti-catabolic and anabolic medications, and hormone replacement therapies.

12. *Describe how HIV infection and its treatment may contribute to insulin resistance and diabetes.*

Medication regimens for HIV treatment may affect triglycerides, hormonal balances, or insulin responses in patients. Suppression of viral load by antiretrovirals has shown varying effects on nutritional status. Weight gains are commonly seen in patients on successful ARV therapy. Because of the inflammatory nature of chronic HIV infection, evident even when viral load is fully suppressed, nutrient metabolism can be expected to change.

Losses in subcutaneous fat and/or gains in abdominal fat stores can both result from and lead to problems of insulin resistance, which can eventually lead to diabetes if not addressed (at least in part) through dietary and exercise modulation.

13. *Describe the characteristics of lipodystrophy syndrome and the potential uses of nutrition-related therapies to reduce its effects.*

Lipodystrophy is associated with a loss or absence of fat, or the abnormal distribution of fat in the body. In HIV infection, these changes are likely hormonally mediated; subcutaneous fat loss is most apparent in peripheral limbs and facial areas; fat deposits are most commonly central, located in the dorsocervical area, breast area, and abdominal region.

Potential uses of nutrition therapies to reduce its effects are kcal recommendations based on the balance needed to maintain a healthy weight. This may include the need for additional kcal to gain weight, or for fewer kcal in order to lose weight. Although recovery of fat before recovery of lean tissues is a normal process during weight regain, emphasis on lean tissue gain can be achieved with aggressive efforts to introduce kcal, protein, and muscle-supporting activities. In some cases, weight loss may be appropriate to improve health. When body composition has been altered, however, weight loss through dieting may be risky, especially if there is no buffer in the volume of body cell mass to lose when lean tissue is lost during dieting. Kcal balance and other nutrient-based therapies may be best used as one of several therapies aimed at improving overall health.

Carbohydrate and fat kcal may be modulated according to assessed risk for insulin resistance, cardiovascular disease, and other conditions where treatment is related to dietary interventions. Limited evidence supports the use of the glycemic index, fiber, or specific fatty acids in individuals with HIV. Because HIV infection is associated with a chronic inflammatory response, it may be prudent to provide education on balancing daily dietary intake to minimize problems and reduce associated risks.

For metabolic alterations, including fat accumulation, exercise may prove to be an essential and effective therapy. Aerobic training, in particular, may improve lipid metabolism, lower blood lipids, and reduce visceral fat accumulation.

Chapter 25

1. *Describe osseous tissue and its composition.*

   Osseous tissue is a major structural and supportive connective tissue that forms bones for the skeletal system. It assists in movement, protects vital organs, and stores calcium phosphate. Three primary components of osseous tissue are osteoblasts, osteocytes, and osteoclasts. Osseous matrix is composed of two components: osteocollagenous fibers (organic) and calcium phosphate crystals (inorganic). The inorganic components of the bone matrix mineralize and calcify the organic component. These osteocollagenous fibers help with the flexibility of the bone, while the calcium phosphate crystals provide the rigidity, so the body can be supported.

2. *List the four principal types of cells found in osseous tissue and describe their functions.*

   The four types of cells found in osseous tissue and their respective functions are as follows:

   1. **Osteogenic cells** are cells capable of mitotic division and differentiating into osteoblasts, and are the only source of new osteoblasts. Osteogenic cells are active during normal growth of the skeletal system during childhood and adolescence, and in adulthood are activated in response to bone injury such as a fracture and in response to the stress placed on bones during weight-bearing exercise.

   2. **Osteoblasts** synthesize, deposit, and then orient the fibrous proteins of the organic matter of the bone matrix (collagen, proteoglycans, and other proteins) and then participate in the calcification or mineralization of the bone matrix, in a process known as bone formation or mineral deposition. Active osteoblasts are found on the surface of newly forming bone. They remove ions of calcium, phosphate, and other minerals from the blood plasma and deposit them within the bone matrix, thus hardening it. In response to bone fracture or the stress of weight-bearing exercise, the osteogenic cells multiply more rapidly and then differentiate to become osteoblasts.

   3. **Osteocytes** are mature osteoblasts surrounded and entrapped by the matrix they have synthesized and calcified, and represent the vast majority of cells in bone. Osteocytes reside in the tiny, fluid-filled cavities within the calcified matrix called lacunae, which are interconnected by narrow channels called canaliculi. Delicate cytoplasmic processes extend from the osteocytes and pass through the canaliculi. These processes connect with those of other osteocytes, thus allowing the osteocytes to chemically signal each other. The fluid-filled canaliculi also serve as channels for the passage of nutrients and metabolites between the osteocytes and nearby blood vessels. Although the osteocytes neither deposit nor remove bone, they are actively involved in maintaining the bony matrix by monitoring the amount of strain (bending) a bone experiences when it is mechanically loaded (for example, by weight-bearing exercise) and then communicating this information to osteoblasts on the bone surface. The osteoblasts can then build up and strengthen the bone where needed in response to the stress.

   4. **Osteoclasts** are bone-removing cells that secrete hydrochloric acid (pH of about 4) to dissolve the mineral component of bone matrix and an enzyme called acid phosphatase that digests the collagen and other protein components of the bone matrix. This process is known as bone resorption or mineral resorption. The dissolved minerals are released into the blood and made available for other uses.

3. *Briefly describe the hormonal control of bone metabolism.*

   Several hormones participate in the control of bone metabolism, including cortisol, growth hormone, and the thyroid hormones, but the primary regulators are parathyroid hormone, calcitonin, and vitamin D. Parathyroid hormone (PTH) is secreted by the parathyroid glands. There are two pairs of parathyroid glands located on the posterior surface of the thyroid gland. When the blood calcium concentration is low, the parathyroid glands release PTH to raise the blood calcium concentration. PTH initiates an immediate release of calcium from the canaliculi and bone cells, as well as a more prolonged release of calcium from bone, by increasing the number of osteoclasts and promoting bone resorption. PTH inhibits collagen synthesis by the osteoblasts, which then inhibits bone deposition and promotes calcium reabsorption by the kidneys. PTH promotes the final step in the body's synthesis of vitamin D_3 [$1,25\text{-}(OH)_2D_3$] by the kidneys, thus enhancing the intestinal absorption of calcium. Calcitonin is secreted by the parafollicular (or "C") cells of the thyroid gland when the blood calcium concentration is abnormally high. Its secretion lowers blood calcium concentration by inhibiting the activity of osteoclasts (bone resorption), stimulating the activity of osteoblasts (bone formation), and reducing the renal reabsorption of calcium and phosphate. The major function of vitamin D is to increase blood concentrations of calcium and phosphorus by promoting their absorption by the GI tract. This promotes their reabsorption by the kidney, and stimulates osteoclast formation (and bone resorption) and the release of calcium and phosphorus from bone.

4. *What is osteoporosis? List risk factors for this disorder. Describe the diagnostic measurement of bone mineral density (BMD). Is it an appropriate measure of osteoporosis?*

   Osteoporosis is a disease resulting from a decreased amount of bone mineral and organic matrix, which weakens bones, making them more susceptible to fracture. It is the most common bone disease in humans and is characterized by the loss of bone mass and deterioration of bone microarchitecture, compromised bone strength, and an increased susceptibility to fracture and painful morbidity. The major risk factors for this disorder include:

   - Age ≥65 years

   - Female sex

- Being of a certain race/ethnicity
- Premenopausal amenorrhea
- Malabsorption syndromes
- Family history
- Physical inactivity
- Low lifetime calcium and vitamin D intake

Bone mineral density (BMD) accounts for approximately 70% of bone strength, and because there is currently no accurate measure of overall bone strength, BMD is frequently used as a surrogate measure. BMD is expressed in grams of mineral per area (g/cm^2) and is best measured using dual-energy X-ray absorptiometry (DXA). In DXA, X-ray beams of two different energy levels are projected through the body and received by a detector opposite the X-ray source. As the X-ray beam passes through the body, the body's tissues, particularly dense tissue such as bone, absorb some of its energy; this reduction in energy is known as attenuation. Because bone and soft tissue have different densities, they attenuate or absorb x-ray energy differently, and the differences in attenuation between soft tissues and bone at the two different energy levels are used to calculate BMD.

Osteoporosis is generally diagnosed by comparing a patient's BMD (determined by DXA) with the mean normal BMD in a population of healthy young adults (considered the standard for peak bone mass) by using what is referred to as a "T-score." The T-score represents the number of standard deviations above or below the mean normal BMD for normal young adults of the same sex and race as the patient. Therefore, BMD is an appropriate measure of osteoporosis because determining a patient's T-score can indicate whether bone loss has occurred.

5. *Describe the dietary and pharmacologic prevention and treatment of osteoporosis.*

Preventing osteoporosis involves strategies to encourage individuals to change certain behaviors during childhood, adolescence, and early adulthood in order to reduce their risk of developing osteoporosis in the latter decades of life. The goal is to promote an individual's genetically determined peak BMD, slow the rate of bone mineral loss once demineralization begins in middle age, and reduce overall fracture incidence. Key preventive strategies include adequate calcium and vitamin D intake, weight-bearing and muscle-strengthening exercise, fall prevention, smoking cessation, and avoiding excessive alcohol intake.

Calcium and vitamin D are two dietary elements that are vital for bone health. Prevention of osteoporosis can be achieved by consuming the recommended adequate Intake of calcium, which is 1000 mg/d for adults 19–50 years, and 1200 mg/d for adults over 50 years. Sources of calcium-rich foods are milk (liquid and powdered), milk products (e.g., cheese, yogurt, and kefir), dark green vegetables (mustard and turnip greens, kale, and broccoli), some nuts (e.g., almonds), some seeds (e.g., sesame), tofu (manufactured using calcium sulfate as opposed to magnesium chloride), corn tortillas, and a variety of calcium-fortified foods such

as citrus juice and soy beverages. Persons who are lactose intolerant can consume calcium-rich, lactose-free foods such as soymilk, lactose-free milk, and tablets. Milk products are fortified with vitamin D and exposure to the sun will provide adequate vitamin D status as well.

Other nutrients involved in the treatment of osteoporosis are phosphorus, vitamin K, protein, fruit and vegetable intake, sodium, caffeine, fluoride, and trace minerals. Phosphorus is an essential bone-forming mineral and throughout life an adequate supply is necessary for optimal bone health. Vitamin K is also essential for calcium regulation and bone formation. Rich sources of vitamin K are dark green, leafy vegetables such as broccoli, brussels sprouts, dark green lettuce, collard greens, or kale. Eating one or more servings of these should be enough to meet the daily-recommended target of 120 micrograms/day for men and 90 micrograms/day for women. For those individuals who have poor calcium absorption from the diet or vitamin D insufficiency, supplementation may be required. The most common and least expensive calcium supplements are those containing calcium carbonate, which should be taken with meals because calcium carbonate requires acid to make the calcium more soluble and absorbable. Supplements containing calcium citrate may be taken at any time in reference to meals. Calcium carbonate interferes with the absorption of iron and should not be taken at the same time as an iron supplement; however, calcium citrate does not interfere with the absorption of iron from supplements. Vitamin D and calcium supplementation is now advocated as the basic minimum for treating osteoporosis and reducing risk of fracture in older females and males.

Pharmacological treatments of osteoporosis are estrogens, selective estrogen receptor modulators, bisphosphonates, and teriparatide, a synthetic form of parathyroid hormone.

- **Estrogens (female sex hormones)** reduce bone turnover, prevent bone loss, increase BMD by 5% to 10% in the hip, spine, and total body, significantly reduce fracture risk, and are effective in relieving the hot flushes and night sweats that often accompany menopause.

- **Selective estrogen receptor modulators (SERMs)** are nonhormonal agents that have tissue-specific effects in estrogen-responsive target tissues. In the estrogen-responsive tissues of the skeletal and cardiovascular system, the effects of SERMs are similar to those of estrogen while in breast and uterine tissue SERMs have no estrogen-like effects. Raloxifene is a SERM that binds to estrogen receptors in bone. It has an estrogen-like effect that increases BMD and reduces spine fracture risk by 30%.

- **Bisphosphonates** are potent nonhormonal drugs that reduce bone resorption. They act exclusively in bone by binding to hydroxyapatite (the primary mineral component of bone) at sites of bone resorption and impairing osteoclast function, reducing osteoclast number, and promoting the death of osteoclasts.

- **Teriparatide** is a synthetic version of PTH that is structurally similar to human PTH and has similar effects within the human body. Teriparatide acts on the osteoblasts (the bone-building cells) to stimulate new bone growth and increase BMD; thus, it has an anabolic (or tissue building) effect on the skeleton.

6. *Describe Paget disease and its treatment.*

Paget disease is a localized, progressive, often crippling disorder of bone remodeling resulting from overactive osteoclasts that cause rapid bone resorption followed by rapid formation of new bone by osteoblasts. The structure of the new bone is structurally inferior, leaving the diseased bone more subject to bowing, deformity, fracture, and poor healing following a fracture. The bones most often affected by Paget disease are the upper femur, pelvis, vertebral bodies, skull, and tibia. While pain (including headaches) is the most common presenting symptom, approximately 70% of patients with Paget disease are asymptomatic. Other clinical manifestations include bowing of the long bones with resulting gait abnormalities, nerve paralysis, hearing loss, facial deformity, tooth loss, and cardiovascular disease.

Treatment of Paget disease involves use of nonsteroidal anti-inflammatory agents for pain and use of bisphosphonates to suppress osteoclast activity and decrease bone resorption and formation. Persons with Paget disease should also maintain an adequate intake of calcium and vitamin D.

7. *What are the similarities and differences between osteoarthritis, rheumatoid arthritis, and gout? Describe the dietary and pharmacologic treatments for these disorders.*

The similarities between osteoarthritis, rheumatoid arthritis and gout include that they are all categorized as inflammatory diseases resulting in swelling, pain, heat, redness, stiffness, and eventual deformity in the affected joints.

The differences among these disorders are:

- Osteoarthritis (OA) is a degenerative disease of the cartilage. It is the progressive loss of articular cartilage, which is the load-bearing cartilage, due to mechanical stresses. The major risk factors for OA are age, female sex, family history, major trauma to a joint or to soft tissues surrounding a joint, repetitive joint stress related to occupation, and obesity.

- Rheumatoid arthritis (RA) is different from the other two in that it is considered an autoimmune disease. The body's own immune system attacks the healthy joint tissue. The inflammation of the synovial membrane in the joints leads to abnormal thickening and further deformity and immobility of the joint as the disease progresses. Rheumatoid arthritis is common in the elderly and three times more likely to occur among females than in males. RA may also affect other tissues in the body, resulting in anorexia, weight loss, fatigue, and generalized aching and stiffness.

- Gout usually occurs in men over 40 years of age. It is characterized by excessive build-up of uric acid in the synovial fluid. Risk factors associated with gout include genetics, male sex, older age, overweight, excessive alcohol consumption, eating foods rich in purines, exposure to lead, and use of certain drugs such as aspirin, diuretics, nicotinic acid, cyclosporine, and levodopa.

The treatments for these disorders include:

Because no pharmacologic agent has been shown to prevent OA, slow its progression, or reverse it, the aim of drug therapy is pain relief as an adjunct to nonpharmacologic treatment. The mainstays of drug therapy are the nonsteroidal anti-inflammatory drugs (NSAIDs), which include aspirin, acetaminophen, indomethacin, ibuprofen, naproxen, and the cyclooxygenase-2 (COX-2) inhibitors celecoxib, and valdecoxib. Dietary supplements that have been marketed for preventing osteoarthritis are glucosamine and chondritin.

The treatment of rheumatoid arthritis includes glucocorticoids and immunosuppressive drugs in addition to NSAIDs. These are commonly known as the disease-modifying antirheumatic drugs (DMARDs). A common glucocorticoid drug commonly used to treat RA is Prednisone. Prednisone has major side effects such as osteoporosis, peptic ulcer, etc., but if it is used at low doses or injected directly into the joints, these side effects can be greatly reduced. Dietary treatment for RA is to improve nutritional status by consuming a diet high in fruits and vegetables, because they are rich in antioxidants. Dietary supplementation of fish oils (EPA and DHA) is also beneficial because they reduce the synthesis of chemicals that stimulate joint inflammation and cartilage degradation.

Hyperuricemia in gout is treated with a drug known as allopurinol, which increases the urinary secretion of uric acid. Dietary treatment of gout includes moderate alcohol consumption, avoiding purine-rich foods, and maintaining a healthy body weight. Purine-rich foods include: sardines, fish roe, anchovies, meat-based gravies, meat extracts, food yeast, and several animal organs used for food, including the liver, kidneys, pancreas, and thymus.

Chapter 26

1. *What is the difference between selective and nonselective screening for metabolic disorders?*

Nonselective neonatal screening, or newborn screening (NBS), is associated with the screening of all newborns for a limited number of the more common inborn errors of metabolism. On the other hand, selective neonatal screening is the testing of an individual known to be at increased risk for a genetic disorder. If a child is diagnosed with an inborn error of metabolism, then selective screening may be performed on the siblings in the family.

2. *What amino acids are restricted in maple syrup urine disease?*

The restricted amino acids in maple syrup urine disease are the branched-chain amino acids—leucine, isoleucine, and valine.

3. *What amino acid becomes essential secondary to the enzymatic block in PKU?*

Tyrosine becomes essential secondary to the enzymatic block in PKU.

4. *What is periorfacial acrodermatitis, and when does it occur?*

Periorfacial acrodermatitis is a disease of the skin surrounding the mouth area. It may occur as a result of specific amino acid deficiencies. For instance, it may occur in individuals with maple syrup urine disease.

5. *What is the role of carnitine?*

Carnitine is used in many organic acid disorders to remove and detoxify waste products that accumulate in disorders affecting the metabolism of proteins, carbohydrates, and fats.

6. *How does fasting affect disorders of fat metabolism?*

Normally, fasting triggers the release of fatty acids. In the case of medium-chain acyl-CoA dehydrogenase deficiency (MCADD), the infant's or child's fasting period occurs during the night and leads to the release of fatty acids that cannot be properly utilized for energy, which results in hypoglycemia and/or death. There are fasting guidelines for disorders of fat metabolism that provide the suggested maximum fasting periods at a given age.

7. *Which vitamins are used as cofactors in mitochondrial disorders?*

Riboflavin and thiamin serve as cofactors in mitochondrial disorders.

8. *What are some clinical signs/symptoms that would indicate the need for a further metabolic work-up?*

A defect in function of the mitochondrial system results in diseases affecting multiple organs with a wide range of symptoms, including cardiomyopathy, blindness, deafness, endocrine problems, and muscle disorders. The most common mitochondrial disorders such as MELAS (mitochondrial encephalomyopathy with lactic acidosis and stroke-like episodes) and NARP (neurogenic muscular weakness, ataxia, and retinitis pigmentosa) result from identified genetic mutations and thus can be tested for through a blood test. Skin and muscle tissue can be assayed for further histological and biochemical analyses in order to make a diagnosis.

9. *What vitamin deficiency can present as methylmalonic acidemia?*

Vitamin B_{12} is a cofactor for the methylmalonyl CoA mutase enzyme. The adeno form of this vitamin is required in order for the proper function of this enzyme. There have been reports of patients diagnosed with methylmalonic acidemia with a malfunctioning enzyme due to a vitamin B_{12} deficiency.

10. *What vitamins are used for their antioxidant properties in the treatment of mitochondrial disorders?*

Vitamin E and lipoic acid are used to protect against free radicals in the treatment of mitochondrial disorders.

11. *What sugars are restricted in GSD I?*

Both fructose and galactose (fruits, juice, milk, dairy products, and sweets that contain sucrose) are restricted in glycogen storage disease type I (GSD I). These sugars are restricted because they cannot be converted into glucose and are metabolized through alternative pathways, which leads to an increased production of lactic acid and lipids.

12. *What percentage of protein intake should come from essential amino acids in urea cycle disorders?*

It is suggested that 25%–50% of protein intake should come from essential amino acids in urea cycle disorders.

13. *Which food groups are restricted in galactosemia, and why?*

The food groups that are restricted in galactosemia are dairy products and nondairy sources that contain galactose or lactose. Many processed foods, such as cakes, cookies, muffins, and French fries, contain milk products, and consequently galactose. Several fruits and vegetables, such as persimmons, dried figs, and tomatoes, contain galactose. Some of the galactose in fruits and vegetables may be found as the galactose oligosaccharides, raffinose and stachyose. Lactose in dairy products is a disaccharide of glucose and galactose. In patients with galactosemia, galactose cannot be metabolized due to a defect of an enzyme, uridyltransferase, which converts galactose to glucose. Because galactose is a monosaccharide found in milk products, if milk feedings are continued, the disease may be rapidly fatal. Other complications associated with the continued ingestion of galactose include cataract formation, mental retardation, and renal tubular dysfunction.

14. *What is the mainstay of dietary management for GSD I?*

The mainstay of dietary management for GSD I is frequent oral feedings, high in carbohydrate, in order to maintain blood glucose levels above 70 mg/dL. Total parenteral nutrition or continuous nasogastric infusions of glucose correct most of the metabolic abnormalities associated with GSD I. The impracticality of these forms of management, however, led to the development of a dietary treatment plan that includes frequent daytime meals followed by continuous drip nocturnal enteral feedings.

15. *What becomes significantly elevated in the blood of a person with a urea cycle disorder prior to treatment?*

Ammonia levels become significantly elevated in the blood of a person with a urea cycle disorder prior to treatment.

16. *How is the diet prescription different in an individual with a more severe enzymatic defect vs. a less severe defect?*

For patients with the more severe variants of urea cycle disorders, some of the natural protein that is provided from foods, which includes a mixture of essential and nonessential amino acids, may be replaced with an essential amino acid mixture to ensure that the individual receives a sufficient amount of these proteins. When one of these mixtures is ingested, the body's surplus nitrogen is used to synthesize the nonessential amino acids, thereby reducing the load on the urea cycle. Individuals with less severe defects have more choices for food selection for protein ingestion.

17. *What are the initial symptoms of a child with a UCD?*

The initial symptoms of a child with a UCD include poor feeding and somnolence.

18. *How much protein should be given for an individual with a UCD?*

The protein intake should be adjusted to account for the severity of the enzymatic defect, along with the patient's age, growth rate, and individual preferences. Some children will have an aversion to protein, whereas others do not. During growth, younger individuals require more protein for growth and maintenance, whereas during adulthood, protein needs are decreased because there is less need for growth and decreased rate of tissue turnover.

19. *What parameters should be monitored in an individual with a UCD?*

Clinical, biochemical, and nutritional parameters should be monitored in an individual with a UCD. Clinical parameters include the linear growth of the patient and the appearance of the hair, skin, and nails. Biochemical tests include the plasma concentrations of ammonia and essential amino acids. Other monitoring tests may include hemoglobin, hematocrit, albumin, prealbumin, transferrin, and total protein. Given the restricted nature of the diet, it is important to monitor the status of the branched-chain amino acids since use of the nitrogen-scavenging medications can further worsen deficiencies of these nutrients.

20. *In GSD, there are deficiencies of enzymes involved in synthesizing and degrading which polysaccharide?*

In GSD, deficiencies of enzymes are involved in synthesizing and degrading glycogen.

21. *Describe the diet composition for an individual with GSD I.*

The diet composition for an individual with GSD I should include the following: 60% to 70% of the total caloric intake from carbohydrates from complex sources and 10% to 15% of total caloric intake from protein from lean sources. A portion of the carbohydrate may be provided by the uncooked cornstarch. Less than 30% of total caloric intake should come from fat, and dietary saturated fat should be limited. Two-thirds of the kcal should be given during daytime feedings to allow for carbohydrates to be distributed frequently and evenly over the 24-hour period. One-third of kcal should be administered overnight, primarily from a carbohydrate source via cornstarch or tube feedings. Meals high in starch, without free glucose, tend to provide a more sustained increase in blood glucose levels and, except during symptoms suggestive of hypoglycemia, glucose solutions alone are not recommended.

22. *What typically happens when a person with HFI ingests fructose?*

Ingestion of fructose may cause direct toxic effects on the liver causing inhibition of both gluconeogenesis and glycogenolysis. Thus, fructose-induced hypoglycemia results from inhibition of both gluconeogenesis and glycogenolysis. Symptoms include constant vomiting.

23. *What are some of the complications that an individual with HFI will encounter after ingestion of fructose?*

Some of the complications that an individual with HFI will encounter after ingestion of fructose include hepatomegaly, bleeding tendency, jaundice, edema, and/or ascites.

24. *Describe successful dietary treatment of HFI.*

Successful dietary treatment of HFI requires strict avoidance of all dietary fructose and sucrose. Most fruits and some vegetables that contain fructose and sucrose are to be avoided, and therefore a vitamin and mineral supplement is needed. One must ensure that the vitamin preparation is free of fructose, as there are many that contain this sugar, especially among the liquid preparations made for children. It is very challenging for individuals with HFI to eat outside the home while following the necessary dietary restriction. New foods need to be closely scrutinized before ingestion to ensure that they are fructose free. Listings of the fructose content of various foods are available, but families need to be vigilant about reading labels because product ingredients are always changing.

25. *What is monitored in the blood to measure dietary compliance in galactosemia?*

The determination of the galactose-1-phosphate level in the red blood cells is helpful in monitoring dietary compliance in galactosemia.

26. *What are some possible theories on why some patients with galactosemia experience poor outcomes?*

Some possible theories that have been proposed to explain the poor outcomes in patients with galactosemia include fetal damage in utero through exposure to galactose via maternal circulation, damage before nutrition intervention, de novo galactose synthesis, and an overrestriction of dietary galactose and a subsequent deficiency of uridine diphosphate galactose.

27. *What types of formula are used to treat infants with galactosemia?*

Soy-based formulas are used to treat infants with galactosemia.

28. *Why is MCT oil used in disorders of long-chain fat metabolism? Why can't it be used in MCADD?*

In long-chain fat metabolism disorders, medium-chain triglycerides (MCT), containing fatty acids of 6 to 10 carbons in length, can be used as an alternative substrate since they can bypass the enzymatic block and be used for energy production. MCT oil should not be used in MCADD because the enzymatic block also causes medium-chain fatty acids not to be metabolized into energy.

29. *In what conditions can you breast-feed a child with a metabolic disorder? When would it be contraindicated?*

Breast-feeding a child with a metabolic disorder depends on the degree of severity of enzymatic activity determined by the specific gene mutation present in the infant. If the level of enzymatic activity is mild, the breast milk can be tolerated. However, if the level of enzymatic activity is severe, then breast milk cannot be tolerated, as in the case of the multiple amino acid disorder propionic acidemia.

30. *What are the three main goals for the acute treatment of an amino acid disorder?*

The three main goals of therapy are the following: (1) decrease intake of offending amino acid(s) by eliminating or reducing the intake of natural proteins, (2) provide sufficient kcal to prevent catabolism, and (3) provide enough fluids to flush out any toxic metabolites.

Appendix B—Answers to Application of the Nutrition Care Process Questions

Chapter 5

1. *Identify at least two nutrition problems based on the nutrition assessment and medical history. Determine the diagnostic term for each nutrition problem. Next, identify the etiology of each nutrition problem. Finally, identify the signs and symptoms that support the evidence for these nutrition problems.*

   (P) Altered nutrition related labs (chloride) (E) related to excessive vomiting (S) as evidenced by serum chloride level of 90 mmol/L

   (P) Malnutrition (E) related to insufficient food and beverage intake (S) as evidenced by 15 kg wt loss/6 mo (15%); albumin/prealbumin = 2.6 g/dL and 17 mg/dL, respectively

2. *Determine the amount of energy (kcal) and protein provided by the initial PN solution.*

   The 1300 mL of PN solution contains 990 kcal and 80 g of protein.

3. *Calculate the grams of carbohydrate, protein, and lipid provided by this prescription. How many kcal/kg and grams of protein/kg does it provide? Calculate the patient's nutritional needs. Compare the two.*

   The 1392 mL of PN solution contains 1060 kcal with 86 g of protein, 32 g of fat, and 107 g of carbohydrates. It provides 1060 kcal/66 kg = 16 kcal/kg; 86 g protein/66kg = 1.3 g/kg.

   Patient's BMI is 24.2, within normal range (BMI 16–24.9). Pt's nutritional needs = approx. 1750 kcal (25 kcal × 70 kg; adjusted for dehydration); protein = approx. 105 g (1.5 g. × 70 kg). Her PN Rx does not meet her calculated energy or protein requirements.

4. *Is this patient at risk for refeeding syndrome? Why? What can be done to prevent it?*

   Since she is receiving less than her required protein/calories she is not at risk for overfeeding. For monitoring refeeding syndrome, serum levels of potassium, phosphorus, and magnesium need to be assessed. In order to prevent refeeding syndrome, it is suggested that any enteral or parenteral feeding be started at small amounts and advanced as tolerated.

5. *What clue in the patient's admission history gives support to the patient's low chloride level at the initiation of PN?*

   The patient's admission history reports excessive vomiting, which can lead to decreased chloride levels.

6. *On 1/10, the RD recommended that NaCl be increased by 30 mEq, and then by 20 mEq of Na Acetate on 1/12. Why?*

   The chloride and acetate are used to control the acid-base balance in the body. The RD added the NaCl to increase chloride levels and prevent acidosis, and then the Na Acetate was added 2 days later to balance the bicarbonate with an acid, as there was a significant rise in chloride levels.

7. *Determine nutrition criteria for monitoring and evaluation for each nutrition diagnosis that you identified.*

   Nutrition-related lab—Monitoring for: reduced N&V symptoms, serum electrolyte level, pH, and albumin and prealbumin

   Outcome: electrolyte concentration and pH will be within normal range and albumin/prealbumin will increase to within normal range

Chapter 6

1. *Outline the subjective information that should be included in a chart note for this session.*

   In this chart note, the subjective information should include the diet-related information, lifestyle/psychosocial/emotional reasons for the visit, medically-related reasons, and learning and motivation-related reasons. An example follows:

   Mr. J is visiting the outpatient clinic for counseling on weight reduction after being referred by his physician. Mr. J expresses interest in losing weight because of his family history of diabetes. He reports disliking sweets, rarely eating vegetables, and drinking about 8 cups of fruit juice daily. He also consumes tea and coffee throughout the day with added sugar. Mr. J reports eating two meals a day and never eating breakfast. He classifies himself as fairly inactive.

2. *List all the objective information.*

   Anthropometrics: 52 YOM, Ht/wt: 67"/195#, BMI: 30.5

   Biochemical: Not Applicable

   Clinical: Not Applicable

   Dietary Assessment: Usual dietary intake: 2800 kcal

3. *Write a PES statement for one other nutrition problem.*

   (P) Overweight/obesity related to (E) excessive caloric intake and physical inactivity as evidenced by (S) BMI 30.5

4. *Write your assessment.*

   Patient is obese, consumes excessive fruit juice, is physically inactive, and has a family history of diabetes. Nutritional assessment confirms excessive caloric intake.

5. *Determine appropriate interventions for the nutrition problem you identified.*

   One goal is to decrease total daily caloric intake (i.e., excess fruit juices and sugar-sweetened beverages) and increase physical activity level so that weight loss can be achieved and a healthy BMI range maintained. This can be done through counseling with a dietitian/weight loss counselor and educating the patient on his dietary intake and lifestyle factors that are contributing to his obesity. Patient will be recommended to keep a record (food journal) and encouraged to consume whole fruits and vegetables as well as whole grains. Patient will also be advised to eat breakfast. Regarding physical activity, the patient will be educated on the benefits of exercise and recommended to participate in 30–60 minutes of moderate to vigorous activity each day.

6. *Design your evaluation/outcome measures for this problem.*

Client will make appointment in one month to meet with RD to review food journal kept by patient and discuss progress and barriers associated with the patient's new diet and exercise routine. Weight will be assessed.

Chapter 7

1. *What signs and symptoms in the physical assessment provide evidence for the diagnosis of dehydration?*

Fever, fatigue, 5–10 episodes of diarrhea each day, sunken eyes, reduced capillary refill (approximately 2 seconds), 7-pound deficit of body weight when compared to usual BW, Na and Cl serum levels are abnormally high.

2. *How might Max's laboratory values be affected by his hydration status?*

Concentrations may appear higher than normal due to lowered blood volume.

3. *What factors should be identified in a urinalysis that may also be consistent with dehydration?*

Increased color and clarity of urine and high specific gravity.

4. *Identify at least two nutrition problems revealed by the nutrition assessment and medical history. Next, identify the etiology of each nutrition problem. Finally, identify the signs and symptoms that support the evidence for these nutrition problems.*

(P) Inadequate fluid intake related to (E) increased losses of fluids and fever as evidenced by (S) diarrhea, hypernatremia, and elevated serum chloride ion and a 7-lb weight loss in 2 weeks.

(P) Altered nutrition-related lab values ($\uparrow$ serum Na & Cl) related to (E) inadequate fluid intake & fever as evidenced by (S) diarrhea, elevated serum chloride, and hypernatremia and a 7-lb weight loss in 2 weeks.

5. *Calculate Max's fluid needs.*

Max's fluid requirements based on his body weight: 35–40 mL/kg

Current body weight: 178 lbs; 80.9 kg

For current body weight, his fluid requirements are 35–40 mL/kg $\times$ 80.9 kg = 2831.5–3236 mL

6. *Outline the nutrition therapy recommendations you would make for Max as he begins to try oral intake. Address both liquids and progression of his diet.*

- In order to restore normal fluid, electrolyte, and acid-base balance, Max should try using oral rehydration solutions such as Pedialyte. He should avoid beverages with caffeine.

- As stool begins to form, the introduction of solid foods is advised, one food at a time and as tolerated by the patient. Bananas, rice, applesauce, and toast are some foods known to increase stool consistency.

- Max should stick to the recommended foods suggested by the dietitian, and avoid foods with fiber, fat, lactose, high-fructose corn syrup, and sorbitol.

- It is recommended to eat small, frequent meals every 3–4 hours and to stop consuming a food if it makes the diarrhea worse.

- He should avoid spicy foods as well as high-fiber and gas-producing foods.

- Foods with prebiotics and probiotics are encouraged.

7. *Determine nutrition criteria for monitoring and evaluation for each nutrition diagnosis that you identified.*

- Weight

- Record food and fluid intake as well as tolerance

- Monitor occurrence and amount of diarrhea

- Monitor serum electrolyte labs

Chapter 8

1. *Outline the items from Mr. N's case that fall into each of the following categories:*

A. *Signs*

Blood pressure abnormal, high respiratory rate, high heart rate

B. *Symptoms:*

Pain, dizziness, difficulty breathing, confusion

C. *Laboratory abnormalities*

Low hemoglobin, elevated white blood cell count

2. *Upon examination of his medical record, you find that Mr. N has the following diagnoses: renal insufficiency; chronic obstructive pulmonary disease; and a history of coronary heart disease. Which of these might interfere with his ability to maintain a normal acid-base balance?*

Renal insufficiency may interfere with the excretion of HCO_3^-, increasing blood pH while obstructive pulmonary disease may decrease the ability of the patient to expire CO_2. The retention of CO_2 in the blood can lead to respiratory acidosis.

3. *The physician has ordered arterial blood gases. The values that you note as abnormal are as follows: pH 7.47; pCO$_2$ 46 mmHg; pO$_2$ 83 mmHg; HCO$_3^-$ 32 mEq/L.*

A. *Classify the pH.*

A pH of 7.47 is higher than normal range.

B. *Assess pCO₂.*

   A pCO_2 of 46 mmHg is higher than normal range.

C. *Assess HCO₃⁻.*

   A serum HCO_3^- level of 32 mEq/L is higher than normal range.

D. *Do you see any indication of compensation? Why or why not?*

   Yes; respiratory compensation can help correct the elevated blood pH.

E. *Identify the primary acid-base disorder.*

   Metabolic alkalosis

F. *How do his diagnoses relate to this acid-base imbalance?*

   Renal insufficiency can result in low serum bicarbonate levels, thus causing metabolic alkalosis. Obstructive pulmonary disease can decrease the patient's ability to expire CO_2, subsequently leading to respiratory acidosis. It is not related to this acid-base imbalance of metabolic alkalosis.

Chapter 12

1. *Using her diet history, estimate her daily energy intake. Compare this with her estimated energy and protein requirements.*

   Daily energy intake using exchange list

8 oz skim milk	90 kcal
1 bagel	320 kcal
1 tsp. peanut butter	30 kcal
1 C ice milk	240 kcal
1 banana	60 kcal
20 pretzels	80 kcal
3 oz skinless baked chicken	105 kcal
1 C vegetables	50 kcal
2 C salad	50 kcal
8 oz skim milk	90 kcal
Total	**1115 kcal**

   EER based on Table 12.2 Female 9–18 yrs: EER w/PA of 1.56 = 2682 kcal/day

   Protein requirement: 33 g/day

   The patient is not meeting her energy needs. On average her daily caloric intake is deficient by 1567 kcal.

2. *Calculate NL's BMI. Calculate NL's percent usual body weight and the % weight loss.*
   $$BMI = 41 \text{ kg}/(1.62 \text{ m})^2 = 15$$
   $$\% \text{ UBW} = 91/105 \times 100 = 87\%$$
   $$\% \text{ weight loss} = (105 - 91)/105 \times 100 = 13\%$$

3. *Compare her body fat percentage, triceps skinfold, and arm muscle area to population standards.*

   A BMI value of 15 is considered underweight. Being at 87% of her usual body weight, she would be considered to

have mild malnutrition; however, since she has lost 13% of her body weight in 9 months, it can be considered a severe weight loss. Population standards are located in the appendix. Her tricep skinfold is 10th percentile, which is considered below average, but her arm muscle area is 28 cm², which is considered average.

4. *Evaluate her biochemical data. Which are not within a normal range?*

   Albumin, prealbumin, Hgb, Hct, and ferritin are within normal range. Sodium is still within the normal range but is borderline low. Potassium is just below the normal range. Chloride is within the normal range but is borderline high. These abnormal values may indicate an electrolyte imbalance, which may be due to bulimia. Her glucose levels are low, which may be due to her inadequate calorie intake.

5. *Identify at least two nutrition problems that can be identified as a result of the nutrition assessment and medical history. Determine the diagnostic term for each nutrition problem. Next, identify the etiology of each nutrition problem. Finally, identify the signs and symptoms that support the evidence for these nutrition problems.*

   - Inadequate energy intake related to skipping meals and eating low-fat meals as evidenced by average daily intake of approximately 1100 kcal compared to recommended intake of 2700 kcal.

   - Disordered eating pattern related to avoidance of eating calorically dense food, skipping meals, and excessive physical activity as evidenced by eating in private, 13% weight loss in 9 months, and a BMI of 15 (underweight).

6. *Why is this patient at age 16 still premenarchal? Does this place her at any medical risk?*

   The patient may still be premenarchal due to her low body fat. If she has anorexia nervosa, this can result in decreased secretion of luteinizing hormone and follicle-stimulating hormone from the anterior pituitary; this results in decreased estrogen, which causes the amenorrhea. Prolonged amenorrhea can put her at risk for premature cessation of linear bone growth, failure to achieve expected adult height, and reduced bone mineral density.

7. *Using the diagnostic criteria for eating disorders and your nutritional assessment data, identify her risk for an eating disorder. What are your conclusions?*

   The patient's BMI is below the 5th percentile for her age. The patient is at risk for eating disorders because of her distorted perception of her weight and its effects on her physical performance. Her slightly abnormal sodium, potassium, and chloride lab values as well as her excessive exercising meet the criteria for bulimia. The abnormal lab values may be indicative of purging behaviors. The patient does not eat meals with the rest of her family, which may indicate that she is trying to hide a disordered eating pattern. She meets the criteria for the restricting subtype of anorexia nervosa because of weight loss through dieting, fasting, or excessive exercise.

Chapter 13

1. *What are the criteria for diagnosis with Stage 2 essential HTN? What factors allow for that diagnosis for this patient?*

   The criteria for diagnosis with Stage 2 essential HTN is a systolic BP ≥160 or diastolic BP ≥100. Patient's BP was 160/100 and he is a smoker. Furthermore, he is African American and both parents had high BP and heart disease.

2. *What would be the complications of untreated hypertension?*

   Hypertension can cause congestive heart failure, kidney failure, myocardial infarction, stroke, and aneurysms if left untreated. Vision problems may occur due to blood vessels bursting or bleeding within the eyes. Hypertension may also cause ventricular arrhythmias and sudden cardiac death.

3. *What are the mechanisms through which thiazide diuretics and ACE inhibitors treat high blood pressure?*

   In order to change BP, either cardiac output or peripheral resistance must be altered. Thiazides (hydrochlorothiazide) decrease blood volume by increasing urinary output; they inhibit renal sodium and water reabsorption. ACE inhibitors (Captopril) are vasodilators that reduce BP by decreasing peripheral vascular resistance by interfering with production of angiotensin II from angiotensin I and inhibiting degradation of bradykinin.

4. *Assess the client's usual intake using appropriate nutrition criteria.*

   Diet appears to be low in potassium, magnesium, fiber, and calcium, and high in total fat, cholesterol, and sodium. Weight indicates that the diet is likely too high in total kcal as well.

5. *Evaluate the patient's weight history. Is this a risk factor? What other possible risk factors does this patient present with?*

   Patient BMI = 31.5; BMI ≥30 or greater is classified as obese. This is considered as one of the risk factors for HTN and various other health problems. Other possible risk factors this patient presents with are high blood pressure 160/100, African American ancestry, family history of heart disease, cigarette smoking, and abnormal cholesterol profile.

6. *Identify at least two nutrition problems that can be identified as a result of the nutrition assessment and medical history. Determine the diagnostic term for each nutrition problem. Next, identify the etiology of each nutrition problem. Finally, identify the signs and symptoms that support the evidence for these nutrition problems.*

   a. P: Excessive energy intake

   E: Related to excessive snacking on calorically dense foods

   S: As evidenced by consumption of donuts & peanut butter and a BMI 31.5.

   b. P: Excessive Na intake

   E: Related to high-salt food and snack choices

   S: As evidenced by consumption of processed sandwich meats, crackers, and popcorn; BP of 160/100

7. *Identify the nutrition prescription for GG by recommending the appropriate nutrition therapy for his diagnosis.*

   Decrease caloric intake by 500 kcal and increase physical activity. Consume foods that adhere to the DASH recommendations and decrease Na intake to ~2.3 grams.

8. *Recommend energy and protein requirements for GG.*

   Mifflin-St Jeor equations for male:

   $$10 \times \text{WEIGHT (Kg)} + 6.25 \times \text{HEIGHT (cm)} - 5 \times \text{AGE} + 5$$
   $$= 10(111.4) + 6.25(188) - 5(49) + 5$$
   $$= 2049 \text{ kcal/day}$$

   Activity level (sedentary) = 0.3(BMR) = 615 kcal/day

   EER = 2049 + 615 = 2664 kcal/day

   Protein requirement: $0.8 \times 111.4 = 89$ g/day

9. *List the data that should be collected for monitoring and evaluation for each nutrition diagnosis that you identified.*

   The following data can be collected for monitoring and evaluation for each nutrition diagnosis identified:

   - Assess weight and amount of physical activity and monitor changes in weight

   - Assess 24-hr diet intake and evaluate kcal and DASH nutrient intake

   - Assess BP and monitor changes

Chapter 14

1. *Mrs. Flores's endoscopy indicated that her biopsy was positive for Helicobacter pylori. What is this and how is it related to her duodenal ulcer?*

   Helicobacter pylori is a spiral-shaped, flagellated, Gram-negative rod that lives under the mucous layer of the stomach and attaches to mucus-secreting cells lining the stomach. These organisms break down urea to produce ammonia, which helps neutralize acid in the immediate vicinity of these bacteria and enhances their survival. The *H. pylori* organisms subsequently produce various proteins that damage mucosal cells, attracting lymphocytes and causing persistent inflammation. By-products released by the organism result in damage to the epithelium and impair the mucous barrier within the stomach.

2. *This patient was prescribed three different medications. How does each of these work? What are the current recommendations for treatment of Helicobacter pylori infection? Are there any drug-nutrient interactions that need to be addressed?*

   Metronidazole is an antibiotic that eliminates bacteria and other microorganisms that cause infections of the reproductive system, gastrointestinal tract, skin, vagina, and other areas of the body. It selectively blocks some of the cell functions in these microorganisms, resulting in their demise.

   Omeprazole is a proton-pump inhibitor, and provides not only symptom relief, but also symptom resolution in most cases, including those involving more significant ulcers and/or damage to the esophagus. Because proton-pump inhibitors offer the most effective means of impeding acid production, they are useful in treating serious ulcer conditions.

Clarithromycin is used to treat infections caused by bacteria. Infections treated by Clarithromycin include pneumonia, bronchitis, and infections of the ears, sinuses, skin, and throat. It also is used to treat and prevent disseminated Mycobacterium avium complex (MAC) infection. It is used in combination with other medications to eliminate *H. pylori*. It is in a class of medications called macrolide antibiotics. It works by stopping the growth of bacteria.

Treatment of peptic ulcer disease associated with *H. pylori* infections includes regimens of three to four medications (triple/quadruple therapy). The recommended therapy involves a 7- to 14-day course of two antibiotics with bismuth and one of the proton-pump inhibitors. Eradication rates associated with triple/quadruple therapy range from 86% to 98% if patients comply with triple/quadruple therapy treatment regimens. However, frequently occurring adverse effects such as nausea, vomiting, and abdominal pain associated with these regimens significantly hinder patient compliance and most often the 7-day treatment is recommended.

Drug-nutrient interactions: With Metronidazole, alcohol should be avoided because flushing, fast heartbeats, nausea, and vomiting may occur. Omeprazole is a proton-pump inhibitor and may therefore interfere with vitamin B_{12} absorption from food by slowing the release of gastric acid into the stomach. Clarithromycin can cause diarrhea, nausea, heartburn, abnormal taste, stomach pain, and headaches.

3. *What admission laboratory values are abnormal? Interpret their significance in relationship to both her diagnosis and nutritional status.*

Protein: 5.9 g/dL (normal 6.4–8.3 g/dL); albumin: 3.4 g/dL (normal 3.5–5 g/dL)

Low protein and albumin are indicative of an infection. Low protein could be due to tissue damage that has occurred to the gastric mucosa. The low albumin is due to her recent weight loss since she is eating less to avoid pain.

Hgb: 11.5 g/dL (normal 12–16 g/dL); Hct: 36% (normal 35%–45%)

Hgb is below the normal range and Hct is low but still within normal limits. Both would seem to indicate bleeding and possible anemia.

4. *Identify at least two nutrition problems that can be found as a result of the nutrition assessment and medical history. Next, identify the etiology of each nutrition problem. Finally, identify the signs and symptoms that support the evidence for these nutrition problems.*

- Inadequate protein-energy intake related to pain associated with eating and decreased food and beverage intake as evidenced by a 24% weight loss in the prior 6-week period.

- Altered nutrition-related albumin and hematocrit/hemoglobin labs related to decreased food and beverage intake and possible GI bleeding as evidenced by 24% weight loss in the prior 6 weeks; albumin 3.4 g/dL, Hgb 11.5 g/dL, Hct 36%.

Chapter 15

1. *What is a 72-hour fecal fat test? Interpret her result of 11.5 g/24 hours.*

This test evaluates digestion of fats by determining excessive excretion of lipids. Stool is collected for 3 days after ingesting 100 grams fat/day. If more than 6 grams of fat are present in the stool after 24 hours, the diagnosis of steatorrhea can be made. Her result indicates fat malabsorption and steatorrhea.

2. *What do the results of her small bowel biopsy tell you about the change in the anatomy of the small intestine? How is celiac disease related to this change?*

When the small intestinal mucosa is exposed to certain sequences of amino acids found in the prolamin fraction of wheat (gliadin), rye (secalin), and barley (hordein), there appear to be both a toxic and an inflammatory response. This response damages villi; height is reduced, and they are flattened in appearance. Lack of surface area and reduction of enzyme production cause both malabsorption and maldigestion.

3. *What is the etiology of celiac disease? How do AGA and EMA antibodies assist in this diagnosis?*

It is well understood that damage to the intestinal mucosa occurs when the small intestine is exposed to the prolamin fraction—α-gliadin and other protein components of gluten. Damage to the intestinal mucosa is accompanied by an infiltration of white blood cells into the mucosa. This inflammatory response is also reflected in production of IgA antigliadin and antiendomysial antibodies. These antibodies to gliadin (AGA) and endomysium (EMA), respectively, are now used to diagnose CD and reflect the autoimmune nature of this disease.

4. *Identify at least two nutrition problems that can be found as a result of the nutrition assessment and medical history. Next, identify the etiology of each nutrition problem. Finally, identify the signs and symptoms that support the evidence for these nutrition problems.*

- (P) Inadequate oral food/beverage intake

(S) related to avoiding eating secondary to concern for symptoms of diarrhea

(E) as evidenced by small amounts of soup, toast, and crackers at two meals only

- (P) Altered GI function

(S) related to intolerance to wheat, rye, oats, and barley secondary to celiac disease

(E) as evidenced by diarrhea following eating

5. *Identify the nutrition prescription for this patient by recommending the appropriate dietary modification for her diagnosis.*

Ideal Goal: Minimize/eliminate symptoms of diarrhea following eating

Interventions: Comprehensive nutrition education to provide purpose of gluten-free dietary modification to avoid wheat, oats, rye, and barley. Education will also include more advanced topics related to label reading for targeted ingredients.

Ideal Goal: Goal is for patient to tolerate regular meals and snacks and increase her caloric and protein intakes to meet recommended amounts after adhering to a gluten-free diet.

6. *Recommend energy and protein requirements for KM.*

   1400–1500 kcal and 55–65 grams protein per day

7. *Determine nutrition criteria for monitoring and evaluation for each nutrition diagnosis that you identified.*

   The following should be monitored and evaluated in 6–8 weeks:

   - Dietary intake to assess gluten-free food knowledge and intake and amount of food and beverage intake

   - Assess tolerance to the diet and symptoms of diarrhea.

Chapter 16

1. *Describe the pathophysiology of acute pancreatitis. What signs and symptoms support the diagnosis?*

   The pathophysiology of acute pancreatitis is not fully understood but it seems to be related to premature activation of trypsin inside the pancreas leading to autodigestion of pancreatic cells. The destroyed pancreatic cells then release enzymes into the bloodstream, which results in elevated amylase and lipase levels. Acute pancreatitis can be asymptomatic but most cases present with upper abdominal pain that radiates towards the back, nausea, vomiting, abdominal distention, steatorrhea, and fever (indicating an infection in the body). In severe cases, low blood pressure and dehydration can occur. Mr. Cowan presents with upper abdominal pain that radiates toward his back, nausea, vomiting, and fever. His amylase and lipase levels are elevated.

2. *What are the common causes of acute pancreatitis?*

   The three most common causes of acute pancreatitis are bile duct obstruction caused from a gallstone, alcohol, or hypertryglyceridemia.

3. *Why is Mr. Cowan not NPO?*

   Current research supports the immediate refeeding of acute pancreatitis patients because there has been no difference found between fed and fasted patients in regard to pain, lab values, and GI symptoms. In fact, immediate refeeding reduced hospital stay duration. He may have been NPO until the presence of a gallstone was determined.

4. *What admission laboratory values are abnormal? Interpret their significance in relationship to both his diagnosis and his nutritional status.*

   Mr. Cowan's value for total protein was 3.6 g/dL. Normal TP is between 6.0 and 8.3 g/dL. A lower-than-normal lab value for TP indicates such conditions as liver disease, malnutrition (weight loss), malabsorption, and protein-losing enteropathy. We also see that Mr. Cowan has lost about 10 pounds recently, which would help to explain his low TP value. Alb was 3.5 g/dL, which falls WNL of 3.4–5.4 g/dL. Lipase appears in the blood when the pancreas is damaged. A normal lab value would fall between 0 and 160 U/L. Mr. Cowan's lipase was 521 U/L. The patient's amylase was 925 U/L and normal range is 23–85 U/L, which obviously supports the diagnosis.

5. *What factor in this client's history is consistent with risk for acute pancreatitis?*

   Alcohol usage is a risk factor for acute pancreatitis. Obesity is also considered a risk factor, but Mr. Cowan has a BMI of 26 (overweight) currently and was as high as 28.5 when he weighed 210 pounds. His history of overweight could play a role in the development of pancreatitis. Biliary stones are also a risk factor for acute pancreatitis.

6. *Identify at least two nutrition problems suggested by the nutrition assessment and medical history. Determine the diagnostic term for each nutrition problem. Next, identify the etiology of each nutrition diagnosis. Finally, identify the signs and symptoms that support the diagnoses.*

   1. (P) Malnutrition related to (E) nausea and vomiting as evidenced by (S) weight loss of 8.5% (18 pounds).

   2. (P) Inadequate oral intake related to (E) altered GI function as evidenced by (S) NPO.

Chapter 17

1. *What is the difference between type 1 and type 2 DM? Why is it assumed that this patient has type 2?*

   - **Type 1 DM (T1DM):** It results from the failure of the body to produce insulin. Little or no insulin is produced by pancreatic beta cells. Type 1 diabetics are dependent on exogenous insulin; therefore, this category is also termed insulin-dependent diabetes mellitus (IDDM). The onset of type 1 DM is usually abrupt and before the age of 30.

   - **Type 2 DM (T2DM):** It results from insulin resistance. It is a condition in which cells fail to use insulin properly; this results in increased need for insulin and eventually the pancreas loses the ability to produce insulin. Type 2 diabetics can be classified as IDDM or non-insulin-dependent diabetes mellitus (NIDDM). Patients normally do not require exogenous insulin. The onset is gradual, usually after age 30, and associated with weight gain and obesity. It develops most frequently in adults, but it is affecting children and adolescents increasingly. Metabolic syndrome is common.

   It is assumed that the patient has T2DM because

   - She is obese.

   - She has a family history of T2DM (sister has T2DM).

   - She has metabolic syndrome.

2. *What risk factors does this patient present with? What symptoms may indicate that she has complications of type 2 DM?*

   This patient's risk factors for T2DM include:

   - Older age

   - Obesity

   - Family history of T2DM

   - Increased susceptibility to infections (frequent bladder infections)

- Peripheral neuropathy indicated by slow-healing wounds in the extremities, and tingling and numbness in the extremities

3. *Evaluate this patient's weight. Calculate and interpret her BMI.*

   The patient weighs 155 lb and she is 5'0" tall. This patient's BMI is 30.3, which categorizes her as Class I obese.

4. *What admission laboratory values are abnormal? Interpret their significance in relationship to both her diagnosis and nutritional status.*

Para-meters	Normal Value	Patient's Value	Relationship to Diagnosis and Nutrition Status
BUN	8–18 mg/dL	26 mg/dL	This lab value is high and is indicative of decreased renal function associated with diabetes.
Cr	0.6–1.2 mg/dL	1.2 mg/dL	High level of creatinine indicates possible chronic kidney disease.
Choles-terol	120–199 mg/dL	300 mg/dL	High value is indicative of dyslipidemia associated with diabetes.
HDL	>55 mg/dL (women)	35 mg/dL	Low value is indicative of dyslipidemia associated with diabetes.
LDL	<130 mg/dL	140 mg/dL	High value is indicative of dyslipidemia associated with diabetes.
Glucose	70–110 mg/dL	325 mg/dL	High blood glucose is due to decreased insulin sensitivity.
Hb_{A1C}	3.9%–5.2%	8.5%	High value indicates poor long-term glycemic control.

5. *Identify at least two nutrition problems suggested by the nutrition assessment and medical history. Determine the diagnostic term for each nutrition problem. Next, identify the etiology of each nutrition diagnosis. Finally, identify the signs and symptoms that support the evidence for the diagnoses.*

   The possible PES (Problem, Etiology, Signs and symptoms) statements for this patient are:

   - **PES 1:** (P) Inappropriate carbohydrate intake related to (E) physical inactivity and food and nutrition-related knowledge deficit as evidenced by (S) belief that carbohydrates are found only in foods with sugar; BS of 325 mg/dL; and H_{A1c} of 8.5%

   - **PES 2:** (P) Excessive energy intake related to (E) excessive carbohydrate intake and physical inactivity as evidenced by (S) BMI of 30.3

Chapter 18

1. *Define each of the patient's current symptoms. Explain how they are related to chronic kidney disease.*

Symptoms associated with inadequate kidney function are directly related to the kidney's inability to perform its normal homeostatic control functions.

The patient's current symptoms include anorexia (lack of appetite resulting in decreased dietary intake), N/V (nausea & vomiting), and pruritus (an intense feeling of itchiness). These symptoms are directly related to her inability to excrete urea and uric acid. As the serum levels of urea/uric acid increase, these are very common side effects. The inability to urinate is due to her Stage 5 CKD, and the systemic increase in fluid is demonstrated by her recent 4 kg weight gain, edema, shortness of breath, and increased blood pressure.

2. *How are hypertension and type 2 diabetes mellitus related to her kidney disease?*

   These two diagnoses represent the major causes of renal disease.

   It is proposed that hyperglycemia damages the glomerulus, and this results in a loss of protein into the urine. As diabetic nephropathy progresses, increasing numbers of glomeruli are destroyed and increasing amounts of albumin are excreted. As the number of functioning nephrons declines, each remaining nephron must clear an increasing solute load. Eventually, the limit to the amount of solute that can be cleared is achieved and the concentration in body fluids increases, leading to hypertension, azotemia, and uremia.

   Kidney disease and hypertension can be related in several ways. Kidney disease can result in poorly controlled blood pressure and hypertension can also result in kidney disease. As kidney disease progresses, there is a loss of renin-angiotensin control of blood pressure and fluid balance. Furthermore, damage to vessels of the kidney by hypertension and cardiovascular disease can affect the overall function of the kidney.

3. *Identify each of her medications. What is the rationale for each?*

   Captopril: Antihypertensive; angiotensin-converting enzyme inhibitor, which assists with control of hypertension

   Erythropoietin: Synthetic hormone that stimulates synthesis of RBC; kidney failure results in inability to synthesize erythropoietin, causing chronic anemia

   Vitamin/mineral supplement: To supplement a variety of micronutrients and replace those lost in treatment (dialysis)

   Calcitriol: 25 OH-vitamin D replacement; kidney failure results in inability to activate vitamin D and due to chronic low levels of parathyroid hormone, this vitamin replacement will increase calcium absorption

   Glucophage: Oral hypoglycemic to assist with control of type 2 diabetes mellitus

4. *Assess her dietary intake and compare it to her recommended nutritional needs.*

   Mrs. J's intake versus her recommended nutritional needs:

Nutrient	Mrs. J's Intake	Hemodialysis Recommendations
Protein	30 g (low)	1.2 g/kg/day, at least 50% high biological value Mrs. J: 160/2.2 = 1.2 × 72.7 = 87 g (85–90 g)
Energy	850 kcal (low)	$aBW_{ef} = BW_{ef} - [(SBW - BW_{ef}) \times 0.25]$ 160 + [(100–160) × 0.25] = 145 145/2.2 = 65.6 kg 35 kcal/kg if less than 60 years of age Mrs. J: 65.9 kg × 35 = 2300–2400 kcal
Fat	20 g (low)	25%–35% of total kcal 2300 × 0.25 = 575 kcal 575 kcal/9 kcal/g − 840 kcal = 63–93 g
Cholesterol	45 mg (low)	<200 mg/day
Fiber	6 g (low)	20–30 g/day
Potassium	826 mg (low)	2–3 g/day
Sodium	2010 mg (w/i range)	2–3 g/day
Fluid	<1 L (low)	1 L plus urine output
Calcium	300 mg (low)	Not to exceed 1500 mg/day from phosphate binders and no more than 2000 mg/day total, including dietary sources
Phosphorus	240 mg (low)	1000 mg (17 mg × 72.7) = 1200 mg

5. *Identify at least two nutrition problems based on the nutrition assessment and medical history. Determine the diagnostic term for each nutrition problem. Next, identify the etiology of each nutrition problem. Finally, identify the signs and symptoms that support evidence for these nutrition problems.*

Inadequate protein-energy intake related to recent poor appetite and nausea and vomiting secondary to CKD as evidenced by typical recent daily intake of 850 kcal and 30 g protein compared to recommended 2300–2400 kcal and 90 g protein (hemodialysis)

Altered nutrition-related laboratory values related to inappropriate intake of carbohydrates as evidenced by BG of 200 and A1c 8.9%

6. *Identify the nutrition prescription for this patient by recommending the appropriate dietary modification for her diagnosis.*

Nutrition Rx: 2300–2400 kcal; 90 g protein, with 45 g protein from high-biological value sources; sodium and potassium 2–3 g, phosphorus 1.2 g; fluid <1 L plus urine output; calcium <1500 mg.

Dialysis should improve her appetite. Pt needs to consume enough calories and protein (50% HBV). This can be achieved by consuming foods that are calorically dense. Pt's only HBV protein was from processed deli meat (ham) and American cheese, which are high in sodium. Although her Na intake is within recommended range, as she increases

her dietary intake, salty deli meats, canned tomato sauce, and processed cheese should be replaced with lower-Na alternatives such as baked chicken for the deli meat and a low-Na tomato sauce or a garlic, oil, and spice sauce for the canned tomato sauce. Reducing Na intake will also help by decreasing thirst, helping to avoid consumption of excess fluids and fluid weight gain between dialysis treatments.

American cheese also contains phosphorus, as does Pepsi. Pepsi is also a concentrated source of sugar. Mrs. J should d/c consumption of soda and replace it with lemon-flavored water. Liquids should be consumed only when she feels thirsty. In order to consume adequate protein and calories while limiting intake of sodium, fluid, carbohydrates, and phosphate, Mrs. J may benefit from trying renal nutrition support supplements.

7. *Determine nutrition criteria for monitoring and evaluation for each nutrition diagnosis that you identified.*

Patient will be evaluated for diet compliance by patient's self-reporting of a 24-hour recall. Her weight will be monitored before and after dialysis treatments, along with nutrition-related lab values related to her nutritional status.

Chapter 19

1. *The patient was diagnosed with hypochromic microcytic anemia, most likely secondary to iron deficiency. Define the terms* hypochromic *and* microcytic. *How are they related to his medical diagnosis?*

Hypochromic anemia is characterized by a decrease in hemoglobin in the erythrocytes, so that patients are abnormally pale. Microcytic type anemia, smaller-than-normal red blood cells, is caused by impaired heme synthesis, as a result of inability to absorb, transport, store, or utilize iron, or impaired synthetic abilities from deficiencies of protein, iron, ascorbate, vitamin A, pyridoxine, copper, or manganese. An iron deficiency, where iron is (a) deficient in the diet, (b) competing with another element (as in lead poisoning), or (c) unavailable due to being bound to another compound in food (as in millet), will cause microcytic and hypochromic anemia.

The child has been diagnosed with frequent upper respiratory infections as well as multiple ear infections; this is important because iron status and immunity are interrelated.

2. *What are signs and symptoms in the patient's history that are consistent with this diagnosis?*

Fatigue (excessive tiredness), loss of appetite, low consumption of meat, impaired immune deficiency as seen by his frequent respiratory and ear infections, and poor performance and behavioral issues at school.

3. *What additional dietary history information will be important for the RD to determine?*

A food frequency questionnaire would be useful in determining the child's and family's food culture and to help the RD suggest dietary modifications to improve their nutrient intake.

4. *Evaluate his anthropometric information. Are there any particular concerns?*

Based on his growth pattern, he is short and underweight. This may be genetic and it may indicate that the patient suffers iron-deficiency anemia and other nutrient deficiencies.

5. *Define each of his laboratory values.*

   MCV (mean corpuscular volume) indicates the size of the red blood cell (RBC). MCHC (mean corpuscular hemoglobin concentration) is the amount of hemoglobin relative to the size of the cell. A low MCHC is indicative of a hypochromic microcytic anemia. Hgb (hemoglobin) is the protein that carries oxygen in the RBC; hematocrit is the percentage of whole blood that is composed of red blood cells.

6. *Identify at least two nutrition problems based on the nutrition assessment and medical history. Determine the diagnostic term for each nutrition problem. Next, identify the etiology of each nutrition problem. Finally, identify the signs and symptoms that support the evidence for these nutrition problems.*

   - (P) Inadequate mineral intake (iron) (E) related to food and nutrient knowledge deficit (S) as evidenced by a diet low in heme iron and 24–48 oz of milk per day; MCV 65; MCHC 27; Hgb 10.5; Hematocrit 33%

   - (P) Limited access to food (E) related to lack of financial resources to purchase sufficient quantity of food (S) as evidenced by the fact that the mother is employed, but with four children, and appears to have inadequate access to food.

7. *How will his iron-deficiency anemia be treated?*

   His iron deficiency should be treated by the incorporation of nutrient-dense foods high in iron (especially heme iron) and/or a multiple vitamin and mineral supplement with iron. Diet should be sufficient in protein and calories.

8. *Determine at least one nutrition intervention that you would recommend.*

   The family should be provided with comprehensive education regarding which foods are high in iron, especially those provided by WIC, and cooking techniques that increase the iron content of the food.

9. *Determine nutrition criteria for monitoring and evaluation for each nutrition diagnosis that you identified.*

   The patient's food history should also be taken so that his eating habits can be monitored, and to seek improvement. Because the patient is small for his age, it is important that both his stature and weight be remeasured after intervention. If labs are ordered, his Hg and HCT values as well as MCV and MCHC should be monitored.

Chapter 20

1. *How is the pathophysiology of Parkinson's related to her development of dysphagia?*

   Parkinson's disease is a progressive disorder of the central nervous system. Motor impairment of the muscles in the throat not only impairs swallowing (leading to malnourishment) but also increases the risk for aspiration pneumonia.

2. *How does her medication, Sinemet, assist with control of her disease?*

The medication Sinemet is used for treating symptoms associated with Parkinson's disease and parkinsonism-like symptoms caused by other conditions. It is an antidyskinetic combination of two drugs, Carbidopa and levodopa. Levodopa is transformed by the brain into a substance that helps decrease tremors and other symptoms of Parkinson's disease. Carbidopa helps levodopa enter the brain.

3. *Identify at least two nutrition problems based on the nutrition assessment and medical history. Determine the diagnostic term for each nutrition problem. Next, identify the etiology of each nutrition problem. Finally, identify the signs and symptoms that support the evidence for these nutrition problems.*

   According to the information presented, the patient has been experiencing loss of appetite in addition to difficulty swallowing plus impaired ability to prepare food or meals and now needs assistance to eat. This results in:

   - (P) Inadequate protein-energy intake related to (E) difficulty eating and in swallowing as evidenced by (S) 8.5 % weight loss in 1 year and an albumin of 3.4 mg/dL.

   - (P) Self-feeding difficulty related to (E) impaired ability to prepare food/meals and need for assistance at meals as evidence by (S) a diagnosis of Parkinson's and dysphagia.

4. *Identify the nutrition prescription for this patient by recommending the appropriate dietary modification for her diagnosis.*

   Depending on the degree of dysphagia, her diet will need to be modified for texture, consistency, and thickness. If severe, she may need to be fed via a feeding tube. Patients receiving enteral feeding should be placed on a bolus feeding schedule as opposed to continuous feeding, as L-dopa dosing can then be administered between the bolus feeds, thus increasing the likelihood of achieving optimal therapeutic levels.

5. *Recommend energy and protein requirements for this patient.*

   35 kcal/kg current weight (~2000 kcal), 0.8–0.9 g of protein/kg (50 g) or 10% of the energy requirement, 30% of kcal from fat and 60% from carbohydrates. Manipulation of protein intake (less during the day and more at the evening meal) may promote increased therapeutic effects of the medication.

6. *Determine nutrition criteria for monitoring and evaluation for each nutrition diagnosis that you identified.*

   Measure weight and assess nutrient intake weekly. Monitor prealbumin and albumin blood levels monthly.

Chapter 21

1. *What risk factors for emphysema/COPD are present in Stella's history?*

   Included in Stella's history is smoking, the primary risk factor for the development of COPD. Other risk factors are a history of bronchitis and respiratory infections. In addition, females are more than twice as likely to be diagnosed with chronic bronchitis as males.

2. *Describe the causes for the following symptoms present in Stella's history and physical exam: anorexia, dyspnea, fatigue, early satiety, bloating.*

- Anorexia and dyspnea: Severe dyspnea results in difficulty eating. Chewing and swallowing may be impaired, since both activities alter breathing patterns and reduce oxygen uptake. Chronic mouth breathing or certain medications may also cause changes in taste perceptions. Stella reports that food doesn't taste good to her, she has a reduced appetite, and her dentures no longer fit correctly related to gradual weight loss over 5 years.

- Fatigue: Stella's body is not getting enough oxygen and this would cause fatigue. She is also consuming too few calories, which could also cause fatigue.

- Early satiety and bloating: Patients with COPD often experience hyperinflation of the lungs with accompanying flattening of the diaphragm and reduced abdominal volume, leading to unnecessary fullness and bloating at mealtime. Aerophagia (the swallowing of too much air resulting in gas and bloating) is often seen in COPD.

3. *Look at Stella's laboratory data. What do they tell you about her nutrition status? Define each of the following blood gases and interpret her values: pH, $PaCo_2$, PaO_2, O_2Sat, HCO_3.*

Stella has a low albumin count, which indicates malnutrition. Sodium is a bit low, but still in normal range. Her potassium is also within normal range.

Blood Gas	Stella's Value	Normal Values
Partial pressure of oxygen (PaO_2)	77.7 mm Hg	75–100 mmHg
Partial pressure of carbon dioxide ($PaCO_2$)	50.9 mm Hg	35–45 mmHg
pH	7.29	7.35–7.45
Oxygen saturation (O_2Sat)	92%	94%–100%
Bicarbonate (HCO_3)	29.6 mEq/L	22–26 mEq/L

Changes in pulmonary function are reflected by changes in the partial pressure of dissolved carbon dioxide ($PaCO_2$), and by changes of partial pressure of dissolved oxygen (PaO_2). Changes in the $PaCO_2$ measure how well carbon dioxide is able to move out of the blood into the airspaces of the lung, and out in the exhaled air. Changes in PaO_2 measure how well oxygen is able to move from the air into the lungs. Oxygen saturation is the measure of the amount of oxygen carried by the red blood cells and can be calculated using the partial pressure of dissolved oxygen (PaO_2). In patients with pulmonary disease, fewer red blood cells carry the usual load of oxygen, and oxygen saturation is decreased.

The pH is a measure of hydrogen ion concentration (H+) in blood. Respiratory acidosis, caused by decreased ventilation, results in carbon dioxide retention, while respiratory alkalosis, caused by increased ventilation, results in loss of carbon dioxide. Carbon dioxide dissolves more readily in the blood than oxygen and forms bicarbonate and smaller amounts of carbonic acid. When present in normal amounts, the ratio of carbonic acid to bicarbonate helps to keep the body pH normal. In situations where acidosis occurs, respiratory activity is increased and the lungs quickly compensate to excrete excess CO_2. If alkalosis is present, respiratory activity automatically decreases and CO_2 is retained, producing a compensatory respiratory acidosis. Stella's increased $PaCO_2$ caused respiratory acidosis, which is indicated by her low pH. Her increased HCO_3, which causes alkalosis, is inconsistent with her acidosis.

4. *Evaluate Stella's current weight, usual body weight, ideal body weight, BMI. How does her 1+ bilateral pitting edema affect your evaluation of her weight?*

UBW: 145–150 lbs.

IBW: 115 lbs.

BMI: 21

Although Stella's current weight is close to her ideal body weight and her BMI is normal, her 1+ pitting edema indicates that it is not her "dry" weight and that she is retaining fluid. She is probably underweight, which is consistent with her reporting weight loss in the past 5 years.

5. *What factors can you identify from her nutrition history that probably contribute to her poor food intake?*

She states that appetite is poor, she feels bloated, she fills up quickly, and food doesn't taste good.

6. *Based on the information presented previously, what recommendations would you make to enhance Stella's nutrition status? Be specific.*

For her unintentional weight loss, Stella should eat calorie-dense foods. Suggestions for increasing the calorie content of her meals would be to add butter, sauces, or gravies. For early satiety, I would suggest she eat her high-calorie foods first and limit fluids with meals. For mealtime dyspnea and fatigue, I would suggest she eat more slowly, and rest before meals. To relieve bloating, she should eat small, frequent meals and avoid gas-producing foods. Stella should eat about 5–6 small meals per day to compensate for some of her satiety and bloating problems and ensure she is getting sufficient calories. She should try to increase her consumption of fruits and vegetables, dairy products, and whole grains. Stella should also begin to take a vitamin and mineral supplement if tolerated. It is important that Stella have properly fitting dentures, so I would refer her to a dentist as well. Although the carbonated soda provides calories, it may be contributing to bloating. She should drink it only with meals. Between meals should try to consume noncarbonated drinks such as orange juice with added glucose polymers, which would provide additional calories plus other nutrients.

Energy needs:

Based on 25–30 kcal/kg and caloric needs of patients with COPD = ~140% above REE, Stella's caloric requirements would be between 1350 and 1600 kcal/day based on her current weight. Even though the 1350–1600 kcal requirement was not based on her dry weight, improving her nutritional status and weight is a goal.

Protein needs:

To halt wasting, patients are recommended to eat 1.2–1.7 g protein/kg body weight, so Stella's protein needs range from 65 to 90 g/day.

Chapter 22

1. *Identify the contributing factors from Lisa's diagnosis that may contribute to metabolic stress.*

   Lisa's diagnosis of a moderate closed head injury indicates that she is in metabolic stress and that her nutritional needs are increased both for energy and protein. In addition, she has suffered from burns on 15% of her body surface area, indicating again that her needs are increased.

2. *Describe the metabolic changes that may contribute to metabolic stress.*

   The metabolic consequences of injury and stress are a result of numerous factors, including hormone release, acute-phase protein synthesis, hypermetabolism, increased reliance on gluconeogenesis and its subsequent production of glucose, and shifts in fluid balance and decreased urine output. Hormones such as glucagon, cortisol, epinephrine, and norepinephrine are secreted during the stress response in order to mobilize nutrient stores to meet the immediate energy demand. Gluconeogenesis, glycogenolysis, and lipolysis occur as the body attempts to increase available glucose. The increased rate of gluconeogenesis creates reliance on protein as a source of glucose, which causes increased catabolism of skeletal muscle.

3. *Calculate Lisa's BMI and IBW.*

   BMI = 20; IBW = 114–130 lbs

4. *Which, if any, of these labs could be used to assess Lisa's nutrition status? Explain.*

   High protein losses from burn wounds and the overall acute inflammatory response can affect accurate interpretation of the protein markers total protein, albumin, prealbumin, or C-reactive protein in relationship to nutritional status. Shifts in fluid balance will influence other hematological markers.

5. *Identify at least two nutrition problems.*

   • Inadequate intake of protein/energy related to increased nutrient (protein/energy) requirements d/t increased metabolic response to stress as evidence by 15% TSA burns with moderate closed head injury and GCS score of 10.

   • Self-feeding difficulty related to patient's disorientation and decreased responsiveness as evidenced by a GCS score of 10.

6. *Determine Lisa's energy and protein requirements.*

   Energy: 22–25 kcal/kg of actual body weight = 1200–1360 kcal/day

   Protein: 54.5 × 2 g/kg = 109 g of protein

7. *Lisa has had a nasoduodenal tube placed. Outline the specific nutrition prescription you would recommend, including formula choice, volume, and rate progression.*

Formula given: Novartis (1.0 kcal/mL, 53% CHO, 22% protein, 25% lipid). Given a volume of 1200 mL/day in feedings/day initiated at 10–24 mL/hr. The rate is advanced in increments of 10–25 mL/hr every 4–8 hours until goal rate is established (50 mL/hr).

8. *Determine nutrition criteria for monitoring and evaluation for each nutrition diagnosis that you identified.*

   • Assess amount of formula received and monitor GI tolerance, hydration, and blood labs (BS, TGs, electrolytes).

   • Monitor wound healing to determine if protein/kcal intake is adequate.

Chapter 23

1. *What type of cancer is small cell lung cancer? What is the usual cause of this type of lung cancer?*

   Small cell lung cancer is a fast-growing type of lung cancer. It tends to spread much more quickly than non-small cell lung cancer. There are three different types of small cell lung cancer: small cell carcinoma, mixed small cell/large cell carcinoma, and combined small cell carcinoma. Smoking almost always causes small cell lung cancer. This type of lung cancer is rare in those who have never smoked. Small cell is the most aggressive form of lung cancer.

2. *Describe the basic mechanisms for radiation and chemotherapy. How do these treatments interfere with cancer cell growth?*

   Chemotherapy is drug therapy that prevents cancer cells from dividing. However, chemotherapy is given systemically so it also affects healthy cells, resulting in side effects. Radiation therapy (RT) destroys cancer cells by altering cellular and nuclear material, especially DNA.

3. *What are the common nutritional side effects of each of these treatment methods?*

   Radiation targets cancerous cells; the radiation's effect is localized to the target area but can affect surrounding cells as well. Side effects of RT to the head and neck area include fatigue, mucositis, dysgeusia, xerostomia secondary to salivary gland destruction, dysphagia, odynophagia, and severe esophagitis. RT to the abdominal and pelvic area can result in radiation enteritis. Symptoms include severe diarrhea and malabsorption. Chemotherapy affects rapidly dividing cells. The rapidly dividing gastrointestinal cells will be significantly affected by most chemotherapeutic agents. This is the reason most patients undergoing chemotherapy experience nausea and other gastrointestinal (GI) problems such as diarrhea or mucositis during their course of treatment. Specialized nutritional support may be needed in both treatments.

4. *Identify at least two nutrition problems based on the nutrition assessment and medical history. Determine the diagnostic term for each nutrition problem. Next, identify the etiology of each nutrition problem. Finally, identify the signs and symptoms that support the evidence for these nutrition problems.*

   a. (P) Inadequate oral food/beverage intake related to (E) self-feeding difficulty—short of breath and tires easily as evidence by (S) 15% weight loss (30#) over 2 years.

b. (P) Increased energy expenditure related to (E) increased resting metabolic rate as evidenced by (S) diagnosis of lung cancer that has metastasized.

5. *Identify the nutrition prescription for this patient by recommending the appropriate dietary modification for his diagnosis.*

Consume small, frequent, calorically dense meals/supplements. Limit fluids with meals. Keep favorite foods readily available at all times. Eat earlier in the day or when less tired. Avoid noxious odors and ventilate eating area. Consider pharmacologic agents/appetite stimulants.

6. *Calculate energy and protein requirements for this patient.*

Energy requirement: 35 kcal/kg of weight = 2600 kcal, 1.5 g/kg of protein (112.5 g) or 17%; 30% from fat, 53% from carbohydrates.

7. *Determine nutrition criteria for monitoring and evaluation for each nutrition diagnosis that you identified.*

See patient in 1 week. Patient will document food intake per daily food diary. Measure weight and monitor prealbumin blood levels, if available.

Chapter 24

1. *Describe the pathophysiology of HIV infection.*

HIV is most typically transmitted via blood through sexual contact (generally penile anal or penile vaginal intercourse), blood transfusion, intravenous needle sharing, and prenatally (from mother to child) through blood or breast milk.

HIV is a retrovirus, which targets many cells in the body, including gastrointestinal cells, organ cells, and immune cells. The resulting immunodeficiency syndrome is most closely related to the infection of activated CD4 (T helper cells), which become a viral factory. The virus identifies the target CD4 cell and fuses to the surface. HIV injects RNA, enzymes, and other substances that assist in viral integration and replication. Using its own kit of injected substances, HIV RNA is transcribed to DNA particles using reverse transcriptase enzymes. The DNA is carried to the nucleus and integrated into the host DNA using viral integrase enzymes. At this point, the integrated viral materials can remain dormant until they are activated, at which time they command the cell to become a viral factory that manufactures viral components. Protease enzymes cleave the viral proteins for assembly into viral cores. Once fully assembled, the virus is ready to bud out of the infected host cell. As the host CD4 cell manufactures, assembles, and releases viruses, it is incapacitated and destroyed. In addition, macrophages harboring HIV are rendered dysfunctional. It is through this process that the immune system is compromised and HIV disease progresses. Many infections and other complications along with selected nutritional implications are stimulated by HIV infection.

2. *How is HIV infection treated?*

Treatment for chronic HIV infection includes antiretroviral (ARV) medications, prevention and treatment for opportunistic events, modulation of the altered hormonal milieu, and maintenance and restoration of nutritional status.

3. *Describe the lipodystrophy syndrome associated with HIV infection.*

Lipodystrophy syndrome is a disturbance of lipid (fat) metabolism that involves the partial or total absence of fat and often the abnormal deposition and distribution of fat in the body. Lipodystrophy syndrome appears to be associated with the protease inhibitor drugs used in the treatment of AIDS.

In this lipodystrophy syndrome, the face, arms, and legs become thin due to loss of subcutaneous fat. They are characterized by abnormalities in fatty (adipose) tissue associated with total or partial loss of body fat, abnormalities of carbohydrate and lipid metabolism, severe resistance to naturally occurring and synthetic insulin, and immune system dysfunction. These disorders are differentiated by degrees of severity, and by areas or systems of the body affected.

4. *What may be the etiology of insulin resistance in an individual being treated for HIV?*

The medication regimen for HIV treatment may affect triglycerides, hormonal balances, or insulin responses in patients. Suppression of viral load by antiretrovirals has shown varying effects on nutritional status. Weight gains are commonly seen in patients on successful ARV therapy. Because of the inflammatory nature of chronic HIV infection, evident even when viral load is fully suppressed, nutrient metabolism can be expected to change.

Losses in subcutaneous fat and/or gains in abdominal fat stores can both result from and lead to problems of insulin resistance, which can be addressed (at least in part) through dietary and exercise modulation.

5. *What is RE's body mass index? What do his anthropometric data indicate about his body composition?*

BMI = 28.6, which puts him in the overweight body category, and his abdominal circumference and fatfold indicate deep fat accumulation.

6. *Which assessment data support the diagnosis of lipodystrophy?*

Recent weight gain, mostly in the abdomen, and use of zidovudine, which is associated with changes in fat deposition; abdominal circumference: 104 cm; abdominal fatfold: 3 mm.

7. *Which medications may contribute to lipodystrophy and/or insulin resistance?*

Protease inhibitors and zidovudine are medications that may contribute to lipodystrophy and/or insulin resistance. Protease inhibitor and zidovudine use are also predictive of bone mineral losses. The antiretroviral (ARV) medication zidovudine has been associated with skeletal muscle myopathy and fat deposition. Much of the coronary artery disease risk for HIV infection is associated with medication interactions that increase blood lipid levels.

8. *Identify at least two nutrition problems suggested by the nutrition assessment and medical history. Determine the diagnostic term for each nutrition problem. Next, identify*

the etiology of each nutrition diagnosis. Finally, identify the signs and symptoms that support the evidence for the diagnoses.

(P) Undesirable food choices (E) related to food and nutrition knowledge deficit (S) as evidenced by daily intake of caffeinated beverages (3 large mugs of coffee and 1–2 colas per day) and high-fat meals and snacks consisting of potato chips, sausage, cheeses, and gravies.

(P) Overweight (E) related to inactivity and excessive caloric intake (S) as evidenced by 30 lb weight gain in last 6 months, BMI 28.6, cessation of regular exercise, and 24-hour recall indicating intake of 4000 kcal compared to recommended 2300 kcal.

9. *Based on your PES statements what would be your nutrition Rx and goals for intervention?*

Rx: Nutrient recommendation: 2300 kcal/day; d/c caffeine; and reduce foods associated with GERD. Incorporate physical activity into lifestyle daily.

Goals: weight loss though decreased daily caloric intake and increased physical activity; decrease symptoms of heartburn

Intervention: Increase patient's knowledge of foods/beverages/eating habits that may trigger heartburn; education on energy content of foods/beverage and energy expenditure of physical activity.

- Education on how to read nutrition labels for fat and calorie content of purchased food; provide handout on fat and caloric content of foods.

- Instruct client on suitable substitutes for high-fat food items (lean meat products such as chicken breast, turkey breast, and egg whites instead of sausage, beef, whole eggs, and meatloaf; grilled or baked options instead of fried; and limiting the use of gravy and sauces).

- Educate patient on foods and food habits associated with GERD such as caffeine (coffee/tea/cola) and suitable substitutions (water, water with Crystal Light, diet lemonade, decaf green tea, decaf chai tea, ginger tea, and light cranberry juice).

- Provide tips on methods to limit exposure to caloric temptation, such as packing a homemade lunch every day in a cooler; this should be a balanced meal containing whole grains, lean meat, and at least two fruits or vegetables.

- Refer patient to trainer two times/week to begin new exercise program.

10. *What non-nutrient recommendations would you suggest that this patient, his physician, and his health care team explore in order to improve lipodystrophy-related discomfort and heartburn?*

Recommendations for patient that are non-nutrient related: To relieve symptoms of heartburn, start patient on proton pump inhibitor therapy, such as omeprazole 20 mg/per day.

11. *Based on your intervention, what criteria would you monitor in this patient for improvement?*

Meet with client in 1 month to measure/evaluate his weight, 24-hour recall, and occurrence of heartburn.

Chapter 25

1. *What is osteopenia?*

Osteopenia is a condition in which the bone mineral density is low but not low enough to be diagnosed as osteoporosis. The T-score for osteopenia is between 1.0 and 2.5 standard deviations below the mean BMD of the young adult mean value. If osteopenia is not treated, it may result in osteoporosis.

2. *What risk factors does this patient present with? What additional questions would be important for the medical team to establish in evaluating her diagnosis?*

The risk factors that this female presents are that she is postmenopausal, is over age 50, has poor dietary intake of calcium-rich foods, and has a fracture. She also has a family history of osteoporosis.

More information is needed about her diet history, which includes levels of caffeine intake; level of physical activity, including any weight-bearing exercises; any past successes in weight loss; alcohol consumption; any history of smoking; serum 25-hydroxyvitamin D concentration to detect vitamin D deficiency; and current thyroid status.

3. *Why would vitamin D be measured? What form of vitamin D is recommended to be evaluated? Why?*

The major function of vitamin D is to increase blood concentrations of calcium and phosphorus by (1) promoting their absorption by the GI tract, (2) promoting their reabsorption by the kidney, and (3) stimulating osteoclast formation, and thus bone resorption and the release of calcium and phosphorus from bone. Therefore, a deficiency in vitamin D could indicate a risk factor for osteoporosis.

The best indicator of vitamin D status is the concentration of serum 25-hydroxyvitamin D, because it represents both dietary vitamin D intake and vitamin D synthesized by the body. Adequate vitamin D status is reflected by a serum 25-hydroxyvitamin D concentration that is consistently >50 μmol/L (20 ng/mL).

4. *Identify at least two nutrition problems based on the nutrition assessment and medical history. Determine the diagnostic term for each nutrition problem. Next, identify the etiology of each nutrition problem. Finally, identify the signs and symptoms that support the evidence for these nutrition problems.*

Diagnosis	Etiology	Sign/Symptoms
Inadequate calcium intake	Related to food- and nutrition-related deficit concerning food and supplemental sources of calcium.	As evidenced by an estimated mineral intake from diet that is less than the recommended intake and a BMD of 2.0 standard deviations below the mean.
Increased calcium and vitamin D needs	Related to increased requirements for calcium & vitamin D with onset of menopause and family history of osteoporosis	As evidenced by an estimated intake of foods/supplements containing needed nutrient less than estimated requirements, a BMD of 2.0 standard deviations below the mean, and a 25 (OH) D level of 11 ng/mL; 21 Mol/L.

5. *Identify the nutrition prescription for this patient by recommending the appropriate dietary modification for her diagnosis.*

Nutrition education/counseling should be provided to provide the patient with information on the benefits of adequate calcium and vitamin D for bone health, good sources of calcium, and the specific daily intake values for the patient. A calcium supplement may be necessary to meet the 1500 mg/day requirement. A vitamin D supplement is recommended—10 μg/day.

From the given information, it is unclear as to the patient's physical activity level, but a bone-healthful lifestyle should include a variety of weight-bearing, balance, and flexibility exercises. The presence of osteopenia will need to be considered when creating an exercise plan.

6. *Recommend energy and protein requirements for LL.*

   a. Based on basal energy expenditure, with the addition of a stress factor (skeletal trauma), thermic effect of food, and physical activity, the energy requirements for the patient are: ~1800 kcal.

   RMR: $10 (78 \text{ kg}) + 6.25 (167.6 \text{ cm}) - 5(61) - 161 = 1364$

   TEF: $1364 (0.1) = 136$

   PA: $1500 (1.0) = 1,500$

   Stress factor: $1,500 \times 1.2 = 1,800$

   b. The protein requirements for the patient are: $78 \text{ kg} \times 0.8 \text{ g} = 62.4 \text{ g}$.

7. *What level of calcium supplementation should be initiated for this patient?*

After menopause, the calcium recommendation is 1000 mg to 1500 mg/day, which is the Adequate Intake. Exceeding the AI will not have any benefits for the treatment of the condition. Therefore, in order to reach her AI the patient should be encouraged to consume calcium-rich foods along with calcium supplements. This patient should take Ca in small doses, which is 500 mg, two or more times per day with meals because calcium absorption is enhanced at smaller doses and when it is spread throughout the day. In addition, she should not exceed the UL of 2500 mg/day (from foods and supplements). Calcium carbonate and calcium citrate are two supplements that provide more elemental calcium than a regular multivitamin. One thousand IU of vitamin D should be taken daily.

8. *Discuss the importance of vitamin D to calcium absorption. Should she begin vitamin D supplementation also?*

Vitamin D plays an important role in maintaining calcium homeostasis and promoting bone health. Vitamin D facilitates the absorption of calcium from the small intestine, as well as increases the blood concentrations of calcium and phosphorus by promoting their reabsorption by the kidney. Without vitamin D, bones can become thin and brittle and prone to fractures, leading to rickets in children and ostomalacia in adults. Vitamin D deficiency can result from inadequate sun exposure, low dietary intake, or malabsorption of vitamin D due to biliary tract or intestinal diseases that impair the absorption of fat and fat-soluble vitamins.

Vitamin D supplements should be recommended since studies have shown that dietary intake of vitamin D does not elevate serum levels adequately.

9. *Determine nutrition criteria for monitoring and evaluation for each nutrition diagnosis you identified.*

Progress will be evaluated in a follow-up appointment by way of a 24-hour dietary recall, and the calcium intake of patient will be compared to the recommended intake of 1500 mg/day. In addition, the patient's food journal may be evaluated for adequate intake of sodium, protein, fruits, and vegetables. Vitamin D serum levels should be monitored.

10. *What other medications might her physician consider to prevent osteoporosis in this patient?*

This patient's physician might have also considered estrogen (hormone replacement therapy), and selective estrogen receptor modulators (SERMS). Estrogens reduce bone turnover, prevent bone loss, increase BMD by 5% to 10% in the hip, spine, and total body, significantly reduce fracture risk, and are effective in relieving the hot flushes and night sweats that often accompany menopause. However, studies have shown that they may increase the risk of developing breast cancer.

(SERMs) are nonhormonal agents that have tissue-specific effects in estrogen-responsive target tissues. In the estrogen-responsive tissues of the skeletal and cardiovascular system, the effects of SERMs are similar to those of estrogen while in breast and uterine tissue SERMs have no estrogen-like effects. Raloxifene is a SERM approved by the FDA and TPD for the prevention and treatment of osteoporosis in postmenopausal women. When raloxifene binds to estrogen receptors in bone, it has an estrogen-like effect that increases BMD and reduces spine fracture risk by 30%. Biophosphonates and PTH hormone are recommended if the patient has osteoporosis.

Chapter 26

1. *What is the screening procedure for PKU?*

A blood sample from the infant is taken, usually from the baby's heel, within 24 hours after birth. A follow-up test is usually done 7–10 days later. In the United States, newborn PKU screening is performed with the Guthrie inhibition assay or the McCamon-Robins fluorometric test with blood spotted on filter paper. Treatment is not needed for infants who have phenylalanine blood concentrations of less than 10 mg/dL.

2. *Why is it important to accomplish screening as soon after birth as possible?*

Newborn screening allows for early detection/diagnosis and treatment for PKU. Accumulation of phenylketones rises soon after birth when there is absence of phenylalanine hydroxylase enzyme to metabolize phenylalanine. It is important to screen for PKU soon after birth because untreated PKU or late-detected PKU, by contributing to high blood concentrations of phenylalanine, will result in irreversible neurological defects, coma, and even death. The symptoms of PKU are gradual, progressive, and often overlooked until it has reached a severe stage.

3. *What is the goal of the PKU diet?*

The main goal of the PKU diet is to allow the maximum amount of dietary protein in order to promote growth, while still maintaining the levels of phenylalanine in the desired range of 2–6 mg/dL for children ages 12 years and younger, as per the 2000 guidelines from the National Institutes of Health consensus guidelines. For individuals above 12 years the diet becomes more relaxed, with recommendations of 2–15 mg/dL. The amount of protein that is allowed in an individual's diet is based on the level of activity of the deficient enzyme.

4. *What is Phenex? How is it different from Similac or other standard infant formulas?*

Phenex-1 is a powder formula that contains all amino acids except for phenylalanine and is specially designed for infants with PKU. Phenex contains a carbohydrate source, a fat source including essential fatty acids, and vitamins and minerals. It is also fortified with tyrosine, an essential amino acid for individuals with PKU, and taurine. Phenex also contains carnitine (found in breast milk). Similac and other standard infant formulas contain significant amounts of all amino acids required for the development of a healthy infant; therefore, it contains phenylalanine. Since tyrosine is not an essential amino acid for normal infants, it is not present in standard formulas.

5. *What would be the possible long-term complications of untreated PKU?*

Untreated PKU results in the accumulation of phenylketones including phenylpyruvate, phenyllactic acid, phenylacetic acid, and phenylacetylglutamine, which are highly toxic to the brain. As a result, individuals will develop severe mental retardation and seizures that would require lifelong institutionalization. As children with PKU grow older they will have an abnormally small head—microcephaly—and will have stunted growth as well as delayed mental and social skills. Other complications include musty odor due to high levels of phenylketones in the skin and urine and decreased pigmentation of skin and hair due to the retardation of the tyrosinase enzyme activity by phenylalanine.

6. *How long will the infant have to remain on a special diet?*

Individuals with PKU have to maintain the PKU diet throughout their lives in order to ensure normal cognitive development in childhood and maintain psychosocial abilities in adulthood. Research studies indicate that adults who do not maintain adequate diet control have degenerated white matter in the brain as shown in brain scans. It has also been found that individuals who do not consume adequate amounts of tyrosine fail to produce dopamine, an important neurotransmitter. These individuals therefore have impaired cognitive function and may suffer from depression.

7. *In general, how will her diet change as she grows?*

As an infant and growing child, her diet would be based on strict phenylalanine restriction in order to avoid harmful defects brought about by hyperphenylalanine. This would mean counting the specific milligrams of phenylalanine allowed in her diet according to age. Her treatment plan indicates that she has been prescribed Phenex-1 as well as Similac, so that her high phenylalanine levels are normalized from 34.4 mg/dL to 2–6 mg/dL. These two formulas would provide her with the maximum amount of proteins for proper growth. Similac will maintain her intake of phenylalanine from 2–6 mg/dL, whereas Phenex-1 would provide all the essential amino acids. Her protein intake will be calculated by weight and higher than the standard recommended dietary intake for age. This is because of the higher protein needs for PKU patients.

As she weans from formulas, most foods rich in protein, such as dairy products, eggs, meat, poultry, beans, and nuts, are avoided. However, foods low in protein content such as fruits, vegetables, and some grains such as rice cereals and crackers are prescribed to adequately maintain phenylalanine levels, and these will be the basis of her diet. As a growing child with PKU, she would also be kept on continued amino acid supplementation to provide her with essential amino acids without adding additional phenylalanine into her diet. Her protein intake, kcal intake, and growth rate should be monitored. It is important to change her caloric intake if she is experiencing poor growth rate. In addition, phenylalanine levels have to be constantly monitored because they change with age, diet, and health. Protein and kcal requirements increase during an illness due to increased catabolism, and therefore a strict diet regimen should be followed to avoid intake of high-Phe protein and its associated complications.

As she reaches adulthood and with successful growth, she will require a lower caloric intake because of decreased rate of growth. Her PKU diet could also be relaxed as regards the phenylalanine restriction (2–15 mg/dL). However, adults who keep on the PKU diet have better mental coordination and psychosocial skills. Adults with PKU who follow suboptimal or even optimal PKU diets will experience certain complications such as abnormal nutrient utilization and absorption, resulting in selenium deficiency, and decreased bone mineralization. These would lead to bone loss and anemia. Therefore, this patient would have to have her laboratory indices checked periodically to monitor her albumin, total protein, transthyritin, complete blood count, essential fatty acids, calcium, and magnesium.

8. *Identify at least two nutrition problems based on the nutrition assessment and medical history. Determine the diagnostic term for each nutrition problem. Next, identify the etiology of each nutrition problem. Finally, identify the signs and symptoms that support the evidence for these nutrition problems.*

Two nutrition problems based on the nutrition assessment and medical history:

1. Phenylalanine levels are too high (Day 2: phenylalanine 15 mg/dL; Day 8: phenylalanine 34.4 mg/dL) when recommended level of phenylalanine is 2–6 mg/dL for children 12 and under.

2. Food/nutrition-related history of soy formula 4 oz every four hours when soy is a food that is not recommended for patients with PKU.

P: Excessive bioactive substance (phenylalanine) intake

E: Related to food- and nutrition-related knowledge deficit concerning recommended bioactive substance (phenylalanine) intake and frequent intake of foods (soy formula) containing phenylalanine

S: As evidenced by increasing blood levels of phenylalanine; Day 2: phenylalanine 15 mg/dL; Day 8: phenylalanine 34.4 mg/dL

P: Impaired nutrient utilization

E: Unable to metabolize phenylalanine through normal biochemical pathways

S: As evidenced by elevated blood levels of phenylalanine: Day 2: phenylalanine 15 mg/dL; Day 8: phenylalanine 34.4 mg/dL

9. *Compare the prescribed diet to your calculation of energy and protein requirements.*

   Calculation

   For a 0 to <3 month infant recommended energy is 120 (124–95) kcal/kg and protein is 3.50–3.00 g/kg.

   Infant weighs 5.5 lb or 2.75 kg

   Calculated energy needs: 2.75 kg $\times$ 120 kcal/kg = 330 kcal

   or: 2.75 kg $\times$ the range (95–124 kcal/kg) = 261.25–341 kcal

   Calculated protein needs: 2.75 kg $\times$ 3.00–3.50 g/kg = 8.25–9.625 g protein

Prescription included 52 grams of Phenex-1 and 15 grams of Similac powder with iron providing 25 mg/kg of phenylalanine, 316 mg/kg of tyrosine, 3.5 g protein/kg, and 122 kcal/kg. 16 oz. of formula yields 20 kcal/oz.

Energy: 2.75 kg $\times$ 122 kcal/kg = 335.50 kcal

Protein: 2.75 kg $\times$ 3.5 g/kg= 9.625 g protein

The prescribed diet (336 kcal) is within range for the calculated energy requirement (261–341 kcal). In addition, the protein in the prescribed diet (9.625 g) is within the range for the calculated protein requirement (8.25–9.625 g).

10. *Determine nutrition criteria for monitoring and evaluation for each nutrition diagnosis that you identified.*

    Goal: Maintain phenylalanine blood levels with normal range but provide adequate nutrients for normal growth and development.

    Monitor and evaluate monthly for:

    - Phe blood levels and quantity and type of formula consumed

    - Adequate intakes of kcal, protein, essential fats, vitamins, and minerals. In addition, patients should have their laboratory studies evaluated periodically for nutritional indices such as albumin, total protein, complete blood count, transthyretin, essential fatty acids, calcium, magnesium, and zinc.

This is a combined list of Nutrition Assessment and Monitoring and Evaluation terms. Indicators that are shaded are used ONLY for nutrition assessment. The rest of the indicators are used for assessment and monitoring and evaluation.

FOOD/NUTRITION-RELATED HISTORY (FH)

Food and nutrient intake, medication/herbal supplement intake, knowledge/beliefs/attitudes and behavior, food and supply availability, physical activity, nutrition quality of life.

Food and Nutrient Intake (1)

Composition and adequacy of food and nutrient intake, meal and snack patterns, current and previous diets and/or food modifications, and eating environment.

Diet History (1.1)

Description of food and drink regularly provided or consumed, past diets followed or prescribed and counseling received, and the eating environment.

Diet order (1.1.1)

- ❑ General, healthful diet — FH-1.1.1.1
- ❑ Modified diet (specify) _____ — FH-1.1.1.2
- ❑ Enteral nutrition order (specify) _____ — FH-1.1.1.3
- ❑ Parenteral nutrition order (specify) _____ — FH-1.1.1.4

Diet experience (1.1.2)

- ❑ Previously prescribed diets — FH-1.1.2.1
- ❑ Previous diet/nutrition education/counseling — FH-1.1.2.2
- ❑ Self-selected diet/s followed — FH-1.1.2.3
- ❑ Dieting attempts — FH-1.1.2.4

Eating environment (1.1.3)

- ❑ Location — FH-1.1.3.1
- ❑ Atmosphere — FH-1.1.3.2
- ❑ Caregiver/companion — FH-1.1.3.3
- ❑ Appropriate breastfeeding accommodations/facility — FH-1.1.3.4

Energy Intake (1.2)

Total energy intake from all sources, including food, beverages, supplements, and via enteral and parenteral routes.

Energy intake (1.2.1)

- ❑ Total energy intake — FH-1.2.1.1

Food and Beverage Intake (1.3)

Type, amount, and pattern of intake of foods and food groups, indices of diet quality, intake of fluids, breast milk and infant formula

Fluid/Beverage intake (1.3.1)

- ❑ Oral fluids amounts — FH-1.3.1.1
- ❑ Food-derived fluids — FH-1.3.1.2
- ❑ Liquid meal replacement or supplement — FH-1.3.1.3

Food intake (1.3.2)

- ❑ Amount of food — FH-1.3.2.1
- ❑ Types of food/meals — FH-1.3.2.2
- ❑ Meal/snack pattern — FH-1.3.2.3
- ❑ Diet quality index — FH-1.3.2.4
- ❑ Food variety — FH-1.3.2.5

Breast milk/infant formula intake (1.3.3)

- ❑ Breast milk intake — FH-1.3.3.1
- ❑ Infant formula intake — FH-1.3.3.2

Enteral and Parenteral Nutrition Intake (1.4)

Specialized nutrition support intake from all sources, e.g., enteral and parenteral routes.

Enteral and Parenteral Nutrition Intake (1.4.1)

- ❑ Access — FH-1.4.1.1
- ❑ Formula/solution — FH-1.4.1.2
- ❑ Discontinuation — FH-1.4.1.3
- ❑ Initiation — FH-1.4.1.4
- ❑ Rate/schedule — FH-1.4.1.5

Bioactive Substance Intake (1.5)

Alcohol, plant stanol and sterol esters, soy protein, psyllium and β-glucan, and caffeine intake from all sources, e.g., food, beverages, supplements, and via enteral and parenteral routes.

Alcohol intake (1.5.1)

- ❑ Drink size/volume — FH-1.5.1.1
- ❑ Frequency — FH-1.5.1.2
- ❑ Pattern of alcohol consumption — FH-1.5.1.3

Bioactive substance intake (1.5.2)

- ❑ Plant sterol and stanol esters — FH-1.5.2.1
- ❑ Soy protein — FH-1.5.2.2
- ❑ Psyllium and β-glucan — FH-1.5.2.3

Caffeine intake (1.5.3)

- ❑ Total caffeine — FH-1.5.3.1

Macronutrient Intake (1.6)

Fat and cholesterol, protein, carbohydrate, and fiber intake from all sources including food, beverages, supplements, and via enteral and parenteral routes.

Fat and cholesterol intake (1.6.1)

- ❑ Total fat — FH-1.6.1.1
- ❑ Saturated fat — FH-1.6.1.2
- ❑ Trans fatty acids — FH-1.6.1.3
- ❑ Polyunsaturated fat — FH-1.6.1.4
- ❑ Monounsaturated fat — FH-1.6.1.5
- ❑ Omega-3 fatty acids — FH-1.6.1.6
- ❑ Dietary cholesterol — FH-1.6.1.7
- ❑ Essential fatty acids — FH-1.6.1.8

Protein intake (1.6.2)

- ❑ Total protein — FH-1.6.2.1
- ❑ High biological value protein — FH-1.6.2.2
- ❑ Casein — FH-1.6.2.3
- ❑ Whey — FH-1.6.2.4
- ❑ Amino acids — FH-1.6.2.5
- ❑ Essential amino acids — FH-1.6.2.6

Carbohydrate intake (1.6.3)

- ❑ Total carbohydrate — FH-1.6.3.1
- ❑ Sugar — FH-1.6.3.2
- ❑ Starch — FH-1.6.3.3
- ❑ Glycemic index — FH-1.6.3.4
- ❑ Glycemic load — FH-1.6.3.5
- ❑ Source of carbohydrate — FH-1.6.3.6

Fiber intake (1.6.4)

- ❑ Total fiber — FH-1.6.4.1
- ❑ Soluble fiber — FH-1.6.4.2
- ❑ Insoluble fiber — FH-1.6.4.3

Micronutrient Intake (1.7)

Vitamin and mineral intake from all sources, e.g., food, beverages, supplements, and via enteral and parenteral routes.

Vitamin intake (1.7.1)

- ❑ A (1)
- ❑ C (2)
- ❑ D (3)
- ❑ E (4)
- ❑ K (5)
- ❑ Thiamin (6)
- ❑ Riboflavin (7)
- ❑ Niacin (8)
- ❑ Folate (9)
- ❑ B6 (10)
- ❑ B12 (11)
- ❑ Multivitamin (12)
- ❑ Other (specify) _____ (13)

Mineral/element intake (1.7.2)

- ❑ Calcium (1)
- ❑ Chloride (2)
- ❑ Iron (3)
- ❑ Magnesium (4)
- ❑ Multi-mineral (9)
- ❑ Multi-trace element (10)
- ❑ Potassium (5)
- ❑ Phosphorus (6)
- ❑ Sodium (7)
- ❑ Zinc (8)
- ❑ Other, (specify) _____ (11)

Medication and herbal supplement use (2)

Prescription and over-the-counter medications, including herbal preparations and complementary medicine products used.

Medication and herbal supplements (2.1)

- ❑ Medications, specify prescription or OTC — FH-2.1.1
- ❑ Herbal/complementary products (specify) — FH-2.1.2
- ❑ Misuse of medication (specify) — FH-2.1.3

Knowledge/Beliefs/Attitudes (3)

Understanding of nutrition-related concepts and conviction of the truth and feelings/emotions toward some nutrition-related statement or phenomenon, along with readiness to change nutrition-related behaviors.

Food and nutrition knowledge (3.1)

- ❑ Area(s) and level of knowledge — FH-3.1.1
- ❑ Diagnosis specific or global nutrition-related knowledge score — FH-3.1.2

Beliefs and attitudes (3.2)

- ❑ Conflict with personal/family value system — FH-3.2.1
- ❑ Distorted body image — FH-3.2.2
- ❑ End-of-life decisions — FH-3.2.3
- ❑ Motivation — FH-3.2.4
- ❑ Preoccupation with food — FH-3.2.5
- ❑ Preoccupation with weight — FH-3.2.6
- ❑ Readiness to change nutrition-related behaviors — FH-3.2.7
- ❑ Self-efficacy — FH-3.2.8
- ❑ Self-talk/cognitions — FH-3.2.9
- ❑ Unrealistic nutrition-related goals — FH-3.2.10
- ❑ Unscientific beliefs/attitudes — FH-3.2.11

Behavior (4)

Patient/client activities and actions, which influence achievement of nutrition-related goals.

Adherence (4.1)

- ❑ Self-reported adherence score — FH-4.1.1
- ❑ Nutrition visit attendance — FH-4.1.2
- ❑ Ability to recall nutrition goals — FH-4.1.3
- ❑ Self-monitoring at agreed upon rate — FH-4.1.4
- ❑ Self-management as agreed upon — FH-4.1.5

Avoidance behavior (4.2)

- ❑ Avoidance — FH-4.2.1
- ❑ Restrictive eating — FH-4.2.2
- ❑ Cause of avoidance behavior — FH-4.2.3

Bingeing and purging behavior (4.3)

- ❑ Binge eating behavior — FH-4.3.1
- ❑ Purging behavior — FH-4.3.2

Mealtime behavior (4.4)

- ❑ Meal duration — FH-4.4.1
- ❑ Percent of meal time spent eating — FH-4.4.2
- ❑ Preference to drink rather than eat — FH-4.4.3
- ❑ Refusal to eat/chew — FH-4.4.4
- ❑ Spitting food out — FH-4.4.5
- ❑ Rumination — FH-4.4.6
- ❑ Patient/client/caregiver fatigue during feeding process resulting in inadequate intake — FH-4.4.7
- ❑ Willingness to try new foods — FH-4.4.8
- ❑ Limited number of accepted foods — FH-4.4.9
- ❑ Rigid sensory preferences — FH-4.4.10

Social network (4.5)

- ❑ Ability to build and utilize social network — FH-4.5.1

Edition: 2009

Factors Affecting Access to Food and Food/Nutrition-Related Supplies (5)

Factors that affect intake and availability of a sufficient quantity of safe, healthful food as well as food/nutrition-related supplies.

Food/nutrition program participation (5.1)

❑ Eligibility for government programs	FH-5.1.1
❑ Participation in government programs	FH-5.1.2
❑ Eligibility for community programs	FH-5.1.3
❑ Participation in community programs	FH-5.1.4

Safe food/meal availability (5.2)

❑ Availability of shopping facilities	FH-5.2.1
❑ Procurement, identification of safe food	FH-5.2.2
❑ Appropriate meal preparation facilities	FH-5.2.3
❑ Availability of safe food storage	FH-5.2.4
❑ Appropriate storage technique	FH-5.2.5

Safe water availability (5.3)

❑ Availability of potable water	FH-5.3.1
❑ Appropriate water decontamination	FH-5.3.2

Food and nutrition-related supplies availability (5.4)

❑ Access to food and nutrition-related supplies	FH-5.4.1
❑ Access to assistive eating devices	FH-5.4.2
❑ Access to assistive food preparation devices	FH 5.4.3

Physical Activity and Function (6)

Physical activity, cognitive and physical ability to engage in specific tasks, e.g., breastfeeding, self-feeding.

Breastfeeding (6.1)

❑ Initiation of breastfeeding	FH-6.1.1
❑ Duration of breastfeeding	FH-6.1.2
❑ Exclusive breastfeeding	FH-6.1.3
❑ Breastfeeding problems	FH-6.1.4

Nutrition-related ADLs and IADLs (6.2)

❑ Physical ability to complete tasks for meal preparation	FH-6.2.1
❑ Physical ability to self-feed	FH-6.2.2
❑ Ability to position self in relation to plate	FH-6.2.3
❑ Receives assistance with intake	FH 6.2.4
❑ Ability to use adaptive eating devices	FH 6.2.5
❑ Cognitive ability to complete tasks for meal preparation	FH-6.2.6
❑ Remembers to eat, recalls eating	FH-6.2.7
❑ Mini Mental State Examination Score	FH-6.2.8
❑ Nutrition-related activities of daily living (ADL) score	FH-6.2.9
❑ Nutrition-related instrumental activities of daily living (IADL) score	FH-6.2.10

Physical activity (6.3)

❑ Physical activity history	FH-6.3.1
❑ Consistency	FH-6.3.2
❑ Frequency	FH-6.3.3
❑ Duration	FH-6.3.4
❑ Intensity	FH-6.3.5
❑ Type of physical activity	FH-6.3.6
❑ Strength	FH-6.3.7
❑ TV/screen time	FH-6.3.8
❑ Other sedentary activity time	FH-6.3.9
❑ Involuntary physical movement	FH-6.3.10

Nutrition-Related Patient/Client-Centered Measures (7)

Patient/client's perception of his/her nutrition intervention and its impact on life.

Nutrition quality of life (7.1)

❑ Nutrition quality of life responses	FH-7.1.1

ANTHROPOMETRIC MEASUREMENTS (AD)

Height, weight, body mass index (BMI), growth pattern indices/percentile ranks, and weight history.

Body composition/growth/weight history (1.1)

❑ Height/length	AD-1.1.1
❑ Weight	AD-1.1.2
❑ Frame size	AD-1.1.3
❑ Weight change	AD-1.1.4
❑ Body mass index	AD-1.1.5
❑ Growth pattern indices/percentile ranks	AD-1.1.6
❑ Body compartment estimates	AD-1.1.7

BIOCHEMICAL DATA, MEDICAL TESTS AND PROCEDURES (BD)

Laboratory data, (e.g., electrolytes, glucose, and lipid panel) and tests (e.g., gastric emptying time, resting metabolic rate).

Acid-base balance (1.1)

❑ Arterial pH	BD-1.1.1
❑ Arterial bicarbonate	BD-1.1.2
❑ Partial pressure of carbon dioxide in arterial blood, $PaCO_2$	BD-1.1.3
❑ Partial pressure of oxygen in arterial blood, PaO_2	BD-1.1.4
❑ Venous pH	BD-1.1.5
❑ Venous bicarbonate	BD-1.1.6

Electrolyte and renal profile (1.2)

❑ BUN	BD-1.2.1
❑ Creatinine	BD-1.2.2
❑ BUN:creatinine ratio	BD-1.2.3
❑ Glomerular filtration rate	BD-1.2.4
❑ Sodium	BD-1.2.5
❑ Chloride	BD-1.2.6
❑ Potassium	BD-1.2.7
❑ Magnesium	BD-1.2.8
❑ Calcium, serum	BD-1.2.9
❑ Calcium, ionized	BD-1.2.10
❑ Phosphorus	BD-1.2.11
❑ Serum osmolality	BD-1.2.12
❑ Parathyroid hormone	BD-1.2.13

Essential fatty acid profile (1.3)

❑ Triene:Tetraene ratio	BD-1.3.1

Gastrointestinal profile (1.4)

❑ Alkaline phophatase	BD-1.4.1
❑ Alanine aminotransferase, ALT	BD-1.4.2
❑ Aspartate aminotransferase, AST	BD-1.4.3
❑ Gamma glutamyl transferase, GGT	BD-1.4.4
❑ Gastric residual volume	BD-1.4.5
❑ Bilirubin, total	BD-1.4.6
❑ Ammonia, serum	BD-1.4.7
❑ Toxicology report, including alcohol	BD-1.4.8
❑ Prothrombin time, PT	BD-1.4.9
❑ Partial thromboplastin time, PTT	BD-1.4.10
❑ INR (ratio)	BD-1.4.11
❑ Fecal fat	BD-1.4.12
❑ Amylase	BD-1.4.13
❑ Lipase	BD-1.4.14

Gastrointestinal profile, cont'd (1.4)

❑ Other digestive enzymes (*specify*)	BD-1.4.15
❑ D-xylose	BD-1.4.16
❑ Hydrogen breath test	BD-1.4.17
❑ Intestinal biopsy	BD-1.4.18
❑ Stool culture	BD-1.4.19
❑ Gastric emptying time	BD-1.4.20
❑ Small bowel transit time	BD-1.4.21
❑ Abdominal films	BD-1.4.22
❑ Swallow study	BD-1.4.23

Glucose/endocrine profile (1.5)

❑ Glucose, fasting	BD-1.5.1
❑ Glucose, casual	BD-1.5.2
❑ HgbA1c	BD-1.5.3
❑ Preprandial capillary plasma glucose	BD-1.5.4
❑ Peak postprandial capillary plasma glucose	BD-1.5.5
❑ Glucose tolerance test	BD-1.5.6
❑ Cortisol level	BD-1.5.7
❑ IGF-binding protein	BD-1.5.8
❑ Thyroid function tests (TSH, T4, T3)	BD-1.5.9

Inflammatory profile (1.6)

❑ C-reactive protein	BD-1.6.1

Lipid profile (1.7)

❑ Cholesterol, serum	BD-1.7.1
❑ Cholesterol, HDL	BD-1.7.2
❑ Cholesterol, LDL	BD-1.7.3
❑ Cholesterol, non-HDL	BD-1.7.4
❑ Total cholesterol:HDL cholesterol	BD-1.7.5
❑ LDL:HDL	BD-1.7.6
❑ Triglycerides, serum	BD-1.7.7

Metabolic rate profile (1.8)

❑ Resting metabolic rate, measured	BD-1.8.1
❑ RQ	BD-1.8.2

Mineral profile (1.9)

❑ Copper, serum or plasma	BD-1.9.1
❑ Iodine, urinary excretion	BD-1.9.2
❑ Zinc, serum or plasma	BD-1.9.3
❑ Other	BD-1.9.4

Nutritional anemia profile (1.10)

❑ Hemoglobin	BD-1.10.1
❑ Hematocrit	BD-1.10.2
❑ Mean corpuscular volume	BD-1.10.3
❑ Red blood cell folate	BD-1.10.4
❑ Red cell distribution width	BD-1.10.5
❑ B12, serum	BD-1.10.6
❑ Methylmalonic acid, serum	BD-1.10.7
❑ Folate, serum	BD-1.10.8
❑ Homocysteine, serum	BD-1.10.9
❑ Ferritin, serum	BD-1.10.10
❑ Iron, serum	BD-1.10.11
❑ Total iron-binding capacity	BD-1.10.12
❑ Transferrin saturation	BD-1.10.13

Protein profile (1.11)

❑ Albumin	BD-1.11.1
❑ Prealbumin	BD-1.11.2
❑ Transferrin	BD-1.11.3
❑ Phenylalanine, plasma	BD-1.11.4
❑ Tyrosine, plasma	BD-1.11.5
❑ Amino acid, other, specify	BD-1.11.6

Edition: 2009

Urine profile (1.12)
- ❑ Urine color — BD-1.12.1
- ❑ Urine osmolality — BD-1.12.2
- ❑ Urine specific gravity — BD-1.12.3
- ❑ Urine test, specify — BD-1.12.4
- ❑ Urine volume — BD-1.12.5

Vitamin profile (1.13)
- ❑ Vitamin A, serum or plasma retinol — BD-1.13.1
- ❑ Vitamin C, plasma or serum — BD-1.13.2
- ❑ Vitamin D, 25-hydroxy — BD-1.13.3
- ❑ Vitamin E, plasma alpha-tocopherol — BD-1.13.4
- ❑ Thiamin, activity coefficient for erythrocyte transketolase activity — BD-1.13.5
- ❑ Riboflavin, activity coefficient for erythrocyte glutathione reductase activity — BD-1.13.6
- ❑ Niacin, urinary N'methyl-nicotinamide concentration — BD-1.13.7
- ❑ Vitamin B6, plasma or serum pyridoxal 5'phosphate concentration — BD-1.13.8
- ❑ Other — BD-1.13.9

NUTRITION-FOCUSED PHYSICAL FINDINGS (PD)

Findings from an evaluation of body systems, muscle and subcutaneous fat wasting, oral health, suck/swallow/breathe ability, appetite, and affect.

Nutrition-focused physical findings (1.1)
- ❑ Overall appearance — PD-1.1.1
  (*specify*) _____
- ❑ Body language — PD-1.1.2
  (*specify*) _____
- ❑ Cardiovascular-pulmonary — PD-1.1.3
  (*specify*) _____
- ❑ Extremities, muscles and bones — PD-1.1.4
  (*specify*) _____
- ❑ Digestive system (mouth to rectum) — PD-1.1.5
  (*specify*) _____
- ❑ Head and eyes — PD-1.1.6
  (*specify*) _____
- ❑ Nerves and cognition — PD-1.1.7
  (*specify*) _____
- ❑ Skin — PD-1.1.8
  (*specify*) _____
- ❑ Vital signs — PD-1.1.9
  (*specify*) _____

CLIENT HISTORY (CH)

Current and past information related to personal, medical, family, and social history.

Personal History (1)

General patient/client information such as age, gender, race/ethnicity, language, education, and role in family.

Personal data (1.1)
- ❑ Age — CH-1.1.1
- ❑ Gender — CH-1.1.2
- ❑ Race/Ethnicity — CH-1.1.3
- ❑ Language — CH-1.1.4
- ❑ Literacy factors — CH-1.1.5
- ❑ Education — CH-1.1.6
- ❑ Role in family — CH-1.1.7
- ❑ Tobacco use — CH-1.1.8
- ❑ Physical disability — CH-1.1.9
- ❑ Mobility — CH-1.1.10

Patient/Client/Family Medical/Health History (2)

Patient/client or family disease states, conditions, and illnesses that may have nutritional impact.

Patient/client OR family nutrition-oriented medical/health history (2.1)

Specify issue(s) and whether it is patient/client history (P) or family history (F)
- ❑ Patient/client chief nutrition complaint (*specify*) _____ — CH-2.1.1
- ❑ Cardiovascular (*specify*) _____ P or F — CH-2.1.2
- ❑ Endocrine/metabolism (*specify*) _____ P or F — CH-2.1.3
- ❑ Excretory (*specify*) _____ P or F — CH-2.1.4
- ❑ Gastrointestinal (*specify*) _____ P or F — CH-2.1.5
- ❑ Gynecological (*specify*) _____ P or F — CH-2.1.6
- ❑ Hematology/oncology (*specify*) _____ P or F — CH-2.1.7
- ❑ Immune (e.g., food allergies) (*specify*) _____ P or F — CH-2.1.8
- ❑ Integumentary (*specify*) _____ P or F — CH-2.1.9
- ❑ Musculo-skeletal (*specify*) _____ P or F — CH-2.1.10
- ❑ Neurological (*specify*) _____ P or F — CH-2.1.11
- ❑ Psychological (*specify*) _____ P or F — CH-2.1.12
- ❑ Respiratory (*specify*) _____ P or F — CH-2.1.13

Treatments/therapy/alternative medicine (2.2)

Documented medical or surgical treatments, complementary and alternative medicine that may impact nutritional status of the patient
- ❑ Medical treatment/therapy (*specify*) _____ — CH-2.2.1
- ❑ Surgical treatment (*specify*) _____ — CH-2.2.2
- ❑ Complementary/alternative medicine (*specify*) _____ — CH-2.2.3

Social History (3)

Patient/client socioeconomic status, housing situation, medical care support and involvement in social groups.

Social history (3.1)
- ❑ Socioeconomic factors (*specify*) _____ — CH-3.1.1
- ❑ Living/housing situation (*specify*) _____ — CH-3.1.2
- ❑ Domestic issues (*specify*) _____ — CH-3.1.3
- ❑ Social and medical support (*specify*) _____ — CH-3.1.4
- ❑ Geographic location of home (*specify*) _____ — CH-3.1.5
- ❑ Occupation (*specify*) _____ — CH-3.1.6
- ❑ Religion (*specify*) _____ — CH-3.1.7
- ❑ History of recent crisis (*specify*) _____ — CH-3.1.8
- ❑ Daily stress level — CH-3.1.9

COMPARATIVE STANDARDS (CS)

Energy Needs (1)

Estimated energy needs (1.1)
- ❑ Total energy estimated needs — CS-1.1.1
- ❑ Method for estimating needs — CS-1.1.2

Macronutrient Needs (2)

Estimated fat needs (2.1)
- ❑ Total fat estimated needs — CS-2.1.1
- ❑ Type of fat needed — CS-2.1.2
- ❑ Method for estimating needs — CS-2.1.3

Estimated protein needs (2.2)
- ❑ Total protein estimated needs — CS-2.2.1
- ❑ Type of protein needed — CS-2.2.2
- ❑ Method for estimating needs — CS-2.2.3

Estimated carbohydrate needs (2.3)
- ❑ Total carbohydrate estimated needs — CS-2.3.1
- ❑ Type of carbohydrate needed — CS-2.3.2
- ❑ Method for estimating needs — CS-2.3.3

Estimated fiber needs (2.4)
- ❑ Total fiber estimated needs — CS-2.4.1
- ❑ Type of fiber needed — CS-2.4.2
- ❑ Method for estimating needs — CS-2.4.3

Fluid Needs (3)

Estimated fluid needs (3.1)
- ❑ Total fluid estimated needs — CS-3.1.1
- ❑ Method for estimating needs — CS-3.1.2

Micronutrient Needs (4)

Estimated vitamin needs (4.1)
- ❑ A (1)
- ❑ C (2)
- ❑ D (3)
- ❑ E (4)
- ❑ K (5)
- ❑ Thiamin (6)
- ❑ Riboflavin (7)
- ❑ Niacin (8)
- ❑ Folate (9)
- ❑ B6 (10)
- ❑ B12 (11)
- ❑ Other (*specify*) (12)
- ❑ Method for estimating needs (13)

Estimated mineral needs (4.2)
- ❑ Calcium (1)
- ❑ Chloride (2)
- ❑ Iron (3)
- ❑ Magnesium (4)
- ❑ Potassium (5)
- ❑ Phosphorus (6)
- ❑ Sodium (7)
- ❑ Zinc (8)
- ❑ Other (*specify*) (9)
- ❑ Method for estimating needs (10)

Weight and Growth Recommendation (5)

Recommended body weight/body mass index/growth (5.1)
- ❑ Ideal/reference body weight (IBW) — CS-5.1.1
- ❑ Recommended body mass index (BMI) — CS-5.1.2
- ❑ Desired growth pattern — CS-5.1.3

Edition: 2009

American Dietetic Association, International Dietetic & Nutrition Terminology (IDNT) Reference Manuel: Standardized Language for the Nutrition Care Process. © American Dietetic Association. Reprinted with permission.

INTAKE NI

Defined as "actual problems related to intake of energy, nutrients, fluids, bioactive substances through oral diet or nutrition support"

Energy Balance (1)

Defined as "actual or estimated changes in energy (kcal) balance"

❑ Unused	NI-1.1
❑ Increased energy expenditure	NI-1.2
❑ Unused	NI-1.3
❑ Inadequate energy intake	NI-1.4
❑ Excessive energy intake	NI-1.5

Oral or Nutrition Support Intake (2)

Defined as "actual or estimated food and beverage intake from oral diet or nutrition support compared with patient goal"

❑ Inadequate oral food/ beverage intake	NI-2.1
❑ Excessive oral food/beverage intake	NI-2.2
❑ Inadequate intake from enteral/parenteral nutrition	NI-2.3
❑ Excessive intake from enteral/parenteral nutrition	NI-2.4
❑ Inappropriate infusion of enteral/parenteral nutrition (use with caution)	NI-2.5

Fluid Intake (3)

Defined as "actual or estimated fluid intake compared with patient goal"

❑ Inadequate fluid intake	NI-3.1
❑ Excessive fluid intake	NI-3.2

Bioactive Substances (4)

Defined as "actual or observed intake of bioactive substances, including single or multiple functional food components, ingredients, dietary supplements, alcohol"

❑ Inadequate bioactive substance intake	NI-4.1
❑ Excessive bioactive substance intake	NI-4.2
❑ Excessive alcohol intake	NI-4.3

Nutrient (5)

Defined as "actual or estimated intake of specific nutrient groups or single nutrients as compared with desired levels"

❑ Increased nutrient needs (specify) _____	NI-5.1
❑ Malnutrition	NI-5.2
❑ Inadequate protein-energy intake	NI-5.3
❑ Decreased nutrient needs (specify) _____	NI-5.4
❑ Imbalance of nutrients	NI-5.5

Fat and Cholesterol (5.6)

❑ Inadequate fat intake	NI-5.6.1
❑ Excessive fat intake	NI-5.6.2
❑ Inappropriate intake of fats (specify) _____	NI-5.6.3

Protein (5.7)

❑ Inadequate protein intake	NI-5.7.1
❑ Excessive protein intake	NI-5.7.2
❑ Inappropriate intake of amino acids (specify) _____	NI-5.7.3

Carbohydrate and Fiber (5.8)

❑ Inadequate carbohydrate intake	NI-5.8.1
❑ Excessive carbohydrate intake	NI-5.8.2
❑ Inappropriate intake of types of carbohydrate (specify) _____	NI-5.8.3
❑ Inconsistent carbohydrate intake	NI-5.8.4
❑ Inadequate fiber intake	NI-5.8.5
❑ Excessive fiber intake	NI-5.8.6

Vitamin (5.9)

❑ Inadequate vitamin intake (specify) _____ NI-5.9.1

❑ A (1)	❑ Riboflavin (7)
❑ C (2)	❑ Niacin (8)
❑ D (3)	❑ Folate (9)
❑ E (4)	❑ B6 (10)
❑ K (5)	❑ B12 (11)
❑ Thiamin (6)	
❑ Other (specify) _____ (12)	

❑ Excessive vitamin intake (specify) _____ NI-5.9.2

❑ A (1)	❑ Riboflavin (7)
❑ C (2)	❑ Niacin (8)
❑ D (3)	❑ Folate (9)
❑ E (4)	❑ B6 (10)
❑ K (5)	❑ B12 (11)
❑ Thiamin (6)	
❑ Other (specify) _____ (12)	

Mineral (5.10)

❑ Inadequate mineral intake (specify) _____ NI-5.10.1

❑ Calcium (1)	❑ Potassium (5)
❑ Chloride (2)	❑ Phosphorus (6)
❑ Iron (3)	❑ Sodium (7)
❑ Magnesium (4)	❑ Zinc (8)
❑ Other (specify) _____ (9)	

❑ Excessive mineral intake (specify) _____ NI-5.10.2

❑ Calcium (1)	❑ Potassium (5)
❑ Chloride (2)	❑ Phosphorus (6)
❑ Iron (3)	❑ Sodium (7)
❑ Magnesium (4)	❑ Zinc (8)
❑ Other (specify) _____ (9)	

CLINICAL NC

Defined as "nutritional findings/problems identified that relate to medical or physical conditions"

Functional (1)

Defined as "change in physical or mechanical functioning that interferes with or prevents desired nutritional consequences"

❑ Swallowing difficulty	NC-1.1
❑ Biting/Chewing (masticatory) difficulty	NC-1.2
❑ Breastfeeding difficulty	NC-1.3
❑ Altered GI function	NC-1.4

Biochemical (2)

Defined as "change in capacity to metabolize nutrients as a result of medications, surgery, or as indicated by altered lab values"

❑ Impaired nutrient utilization	NC-2.1
❑ Altered nutrition-related laboratory values (specify) _____	NC-2.2
❑ Food-medication interaction	NC-2.3

Weight (3)

Defined as "chronic weight or changed weight status when compared with usual or desired body weight"

❑ Underweight	NC-3.1
❑ Involuntary weight loss	NC-3.2
❑ Overweight/obesity	NC-3.3
❑ Involuntary weight gain	NC-3.4

BEHAVIORAL-ENVIRONMENTAL NB

Defined as "nutritional findings/problems identified that relate to knowledge, attitudes/beliefs, physical environment, access to food, or food safety"

Knowledge and Beliefs (1)

Defined as "actual knowledge and beliefs as related, observed or documented"

❑ Food- and nutrition-related knowledge deficit	NB-1.1
❑ Harmful beliefs/attitudes about food- or nutrition-related topics (use with caution)	NB-1.2
❑ Not ready for diet/lifestyle change	NB-1.3
❑ Self-monitoring deficit	NB-1.4
❑ Disordered eating pattern	NB-1.5
❑ Limited adherence to nutrition-related recommendations	NB-1.6
❑ Undesirable food choices	NB-1.7

Physical Activity and Function (2)

Defined as "actual physical activity, self-care, and quality-of-life problems as reported, observed, or documented"

❑ Physical inactivity	NB-2.1
❑ Excessive physical activity	NB-2.2
❑ Inability or lack of desire to manage self-care	NB-2.3
❑ Impaired ability to prepare foods/meals	NB-2.4
❑ Poor nutrition quality of life	NB-2.5
❑ Self-feeding difficulty	NB-2.6

Food Safety and Access (3)

Defined as "actual problems with food access or food safety"

❑ Intake of unsafe food	NB-3.1
❑ Limited access to food	NB-3.2

Date Identified	Date Resolved

#1 Problem _____
Etiology _____
Signs/Symptoms _____

#2 Problem _____
Etiology _____
Signs/Symptoms _____

#3 Problem _____
Etiology _____
Signs/Symptoms _____

Edition: 2009

Problem _____

Etiology _____

Signs/Symptoms _____

Nutrition Prescription

The patient's/client's individualized recommended dietary intake of energy and/or selected foods or nutrients based on current reference standards and dietary guidelines and the patient's/client's health condition and nutrition diagnosis (*specify*).

_____

_____

_____

Intervention #1 _____

Goal(s) _____

_____

Intervention #2 _____

Goal(s) _____

_____

Intervention #3 _____

Goal(s) _____

FOOD AND/OR NUTRIENT DELIVERY   ND

Meal and Snacks (1)
Regular eating event (meal); food served between regular meals (snack).
- General/healthful diet   ND-1.1
- Modify distribution, type, or amount of food and nutrients within meals or at specified time   ND-1.2
- Specific foods/beverages or groups   ND-1.3
- Other   ND-1.4
  (*specify*) _____

Enteral and Parenteral Nutrition (2)
Nutrition provided through the GI tract via tube, catheter, or stoma (enteral) or intravenously (centrally or peripherally) (parenteral).
- Initiate EN or PN   ND-2.1
- Modify rate, concentration, composition or schedule   ND-2.2
- Discontinue EN or PN   ND-2.3
- Insert enteral feeding tube   ND-2.4
- Site care   ND-2.5
- Other   ND-2.6
  (*specify*) _____

Supplements (3)

Medical Food Supplements (3.1)
Commercial or prepared foods or beverages that supplement energy, protein, carbohydrate, fiber, fat intake.
Type
- Commercial beverage   ND-3.1.1
- Commercial food   ND-3.1.2
- Modified beverage   ND-3.1.3
- Modified food   ND-3.1.4
- Purpose   ND-3.1.5
  (*specify*) _____

Vitamin and Mineral Supplements (3.2)
Supplemental vitamins or minerals.
- Multivitamin/mineral   ND-3.2.1
- Multi-trace elements   ND-3.2.2
- Vitamin   ND-3.2.3
  - A (1)   Riboflavin (7)
  - C (2)   Niacin (8)
  - D (3)   Folate (9)
  - E (4)   B6 (10)
  - K (5)   B12 (11)
  - Thiamin (6)
  - Other (*specify*) _____ (12)
- Mineral   ND-3.2.4
  - Calcium (1)   Potassium (5)
  - Chloride (2)   Phosphorus (6)
  - Iron (3)   Sodium (7)
  - Magnesium (4)   Zinc (8)
  - Other (*specify*) _____ (9)

Bioactive Substance Supplement (3.3)
Supplemental bioactive substances.
- Initiate   ND-3.3.1
- Dose change   ND-3.3.2
- Form change   ND-3.3.3
- Route change   ND-3.3.4
- Administration schedule   ND-3.3.5
- Discontinue   ND-3.3.6
  (*specify*) _____

Feeding Assistance (4)
Accommodation or assistance in eating.
- Adaptive equipment   ND-4.1
- Feeding position   ND-4.2
- Meal set-up   ND-4.3
- Mouth care   ND-4.4
- Other   ND-4.5
  (*specify*) _____

Feeding Environment (5)
Adjustment of the factors where food is served that impact food consumption.
- Lighting   ND-5.1
- Odors   ND-5.2
- Distractions   ND-5.3
- Table height   ND-5.4
- Table service/set up   ND-5.5
- Room temperature   ND-5.6
- Other   ND-5.7
  (*specify*) _____

Nutrition-Related Medication Management (6)
Modification of a drug or herbal to optimize patient/client nutritional or health status.
- Initiate   ND-6.1
- Dose change   ND-6.2
- Form change   ND-6.3
- Route change   ND-6.4
- Administration schedule   ND-6.5
- Discontinue   ND-6.6
  (*specify*) _____

NUTRITION EDUCATION   E

Initial/Brief Nutrition Education (1)
Build or reinforce basic or essential nutrition-related knowledge.
- Purpose of the nutrition education   E-1.1
- Priority modifications   E-1.2
- Survival information   E-1.3
- Other   E-1.4
  (*specify*) _____

Comprehensive Nutrition Education (2)
Instruction or training leading to in-depth nutrition-related knowledge or skills.
- Purpose of the nutrition education   E-2.1
- Recommended modifications   E-2.2

Comprehensive Nutrition Education (2), cont'd
- Advanced or related topics   E-2.3
- Result interpretation   E-2.4
- Skill development   E-2.5
- Other   E-2.6
  (*specify*) _____

NUTRITION COUNSELING   C

Theoretical Basis/Approach (1)
The theories or models used to design and implement an intervention.
- Cognitive-Behavioral Theory   C-1.1
- Health Belief Model   C-1.2
- Social Learning Theory   C-1.3
- Transtheoretical Model/Stages of Change   C-1.4
- Other   C-1.5
  (*specify*) _____

Strategies (2)
Selectively applied evidence-based methods or plans of action designed to achieve a particular goal.
- Motivational interviewing   C-2.1
- Goal setting   C-2.2
- Self-monitoring   C-2.3
- Problem solving   C-2.4
- Social support   C-2.5
- Stress management   C-2.6
- Stimulus control   C-2.7
- Cognitive restructuring   C-2.8
- Relapse prevention   C-2.9
- Rewards/contingency management   C-2.10
- Other   C-2.11
  (*specify*) _____

COORDINATION OF NUTRITION CARE   RC

Coordination of Other Care During Nutrition Care (1)
Facilitating services with other professionals, institutions, or agencies during nutrition care.
- Team meeting   RC-1.1
- Referral to RD with different expertise   RC-1.2
- Collaboration/referral to other providers   RC-1.3
- Referral to community agencies/programs (*specify*)   RC-1.4

Discharge and Transfer of Nutrition Care to New Setting or Provider (2)
Discharge planning and transfer of nutrition care from one level or location of care to another.
- Collaboration/referral to other providers   RC-2.1
- Referral to community agencies/programs (*specify*)   RC-2.2

Edition: 2009

PG-SGA Scoring Guide

Note: PG-SGA is also available in several languages for non–English-speaking clients and caregivers.

The Patient-Generated Subjective Global Assessment (PG-SGA) provides a comprehensive evaluation of nutritional status level, which can then be used to determine the level of medical nutrition therapy required. The tool includes prognostic components of client history (amount and pattern of weight loss, qualitative assessment of nutritional intake, and standard performance status scales) and clinical history (nutrition impact symptoms, disease process, metabolic stress, and physical examination). Serial assessments using the PG-SGA are necessary in cancer patients to monitor any changes in nutritional status, as there is high risk for nutrition deterioration in this population. The PG-SGA scoring is based on the following parameters.

The first four boxes of the scored PG-SGA are filled out by the client, who provides a current history of weight change, food intake, symptoms, and functional capacity. The check-off format enables clients to be more forthcoming about symptoms that adversely impact intake and quality of life and that are not often thought of in a nutritional context by clinicians. After the client completes the first four boxes, the dietetics professional, doctor, nurse, or other therapist trained in PG-SGA completes the lower section.

Scoring is based on a scale from 0 to 4 points, ranging from no nutritional impact to mild, moderate, severe, and potentially life threatening. The points are determined by adding the checked off points in parentheses on the form, as well as from Boxes 1–4.

Box 1:	the point score for the weight loss during the past month if available (or the past 6 months if this is the only information available) plus the points for what happened to the weight during the past 2 weeks
Box 2:	the highest point category checked off by the client
Box 3:	the additive score, for all symptoms checked off by the client
Box 4:	the highest point category checked off by the client
Disease section:	one point for each diagnosis identified in Box 2
Metabolic section:	a score based on metabolic stressors identified in Box 3
Physical section:	a score based on the physical assessment; refer to Box 4

Once each of these evaluations is made, the trained clinician proficient in nutrition physical assessment determines a global physical scoring (well-nourished or moderately or severely malnourished) using criteria outlined in Box 5. Triaging nutrition intervention is then determined using the information provided in Box 6.

Source: Reprinted with permission from Ottery FD, Kasenic S, DeBolt S, Roger K. Volunteer network accrues >1900 patients in 6 months to validate standardized nutritional triage. Abstract 282. Meeting of the American Society of Clinical Oncology, 1998. Reprinted with permission from the American Society of Clinical Oncology.

Scored Patient-Generated Subjective Global Assessment (PG-SGA)

Patient ID Information

History

1. Weight:

In summary of my current and recent weight:

I currently weigh about _____ pounds
I am about _____ feet _____ inches tall

One month ago I weighed about _____ pounds
Six months ago I weighed about _____ pounds

During the past two weeks my weight has:
☐ decreased   ☐ not changed   ☐ increased

[ ]

2. Food Intake: As compared to my normal, I would rate my food intake during the past month as:
☐ unchanged
☐ more than usual
☐ less than usual
   I am now taking:
   ☐ normal food but less than normal
   ☐ little solid food
   ☐ only liquids
   ☐ only nutritional supplements
   ☐ very little of anything
   ☐ only tube feedings or only nutrition by vein

[ ]

3. Symptoms: I have had the following problems that have kept me from eating enough during the past two weeks (check all that apply):
☐ no problem eating
☐ no appetite, just did not feel like eating
☐ nausea   ☐ vomiting
☐ constipation   ☐ diarrhea
☐ mouth sores   ☐ dry mouth
☐ things taste funny or have no taste ☐ smells bother me
☐ problems swallowing ☐ feel full quickly
☐ pain; where? _____
☐ other*

*Examples: depression, money, or dental problems

[ ]

4. Activities and Function: Over the past month, I would generally rate my activity as:
☐ normal with no limitations
☐ not my normal self, but able to be up and about with fairly normal activities
☐ not feeling up to most things, but in bed or chair less than half the day
☐ able to do little activity and spend most of the day in bed or chair
☐ pretty much bedridden, rarely out of bed

[ ]

Additive Score of the Boxes 1–4 [ ] A

The remainder of this form will be completed by your doctor, nurse, or therapist. Thank you.

5. Disease and its relation to nutritional requirements

All relevant diagnoses (specify) _____

Primary disease stage (circle if known or appropriate)   I   II   III   IV   Other _____

Age _____

6. Metabolic demand
☐ no stress   ☐ low stress   ☐ moderate stress   ☐ high stress

7. Physical

Numerical score from Box 2 [ ]

Numerical score from Box 3 [ ]

Numerical score from Box 4 [ ]

Global Assessment
☐ Well-nourished or anabolic (SGA-A)
☐ Moderate or suspected malnutrition (SGA-B)
☐ Severely malnourished (SGA-C)

Total numerical score of Boxes A+B+C+D [ ]
(See triage recommendations below)

Clinician Signature _____ RD RN PA MD DO Other _____ Date _____

Nutritional Triage Recommendations: Additive score is used to define specific nutritional interventions including patient and family education, symptom management including pharmacologic intervention, and appropriate nutrient intervention (food, nutritional supplements, enteral, or parenteral triage). First line nutrition intervention includes optimal symptom management.

0–1   No intervention required at this time. Reassessment on routine and regular basis during treatment.

2–3   Patient and family education by dietitian, nurse, or other clinician with pharmacologic intervention as indicated by symptom survey (Box 3) and laboratory values as appropriate.

4–8   Requires intervention by dietitian, in conjunction with nurse or physician as indicated by symptoms survey (Box 3).

≥9   Indicates a critical need for improved symptom management and/or nutrient intervention options.

Source: © 2000, American Dietetic Association. "The Clinical Guide to Oncology Nutrition." Used with permission.

TABLE D.1

Criteria for Scoring Weight Loss

Weight Loss in 1 Month	Weight Loss in 6 Months	Points
10% or greater	20% or greater	4
5–9.9%	10–19.9%	3
3–4.9%	6–9.9%	2
2–2.9%	2–5.9%	1
0–1.9%	0–1.9%	0

TABLE D.2

Scoring Criteria for Diseases or Conditions

Category	Points
Cancer	1
AIDS	1
Pulmonary or cardiac cachexia	1
Presence of decubitus, open wound, or fistula	1
Presence of trauma	1
Age greater than 65 years	1

TABLE D.3

Scoring of Metabolic Stressors

Stressor	None (0)	Low (1)	Moderate (2)	High (3)
Fever (°F)	no fever	>99 and <101	≥101 and ≤102	≥102
Fever duration	no fever	<72 hr		>72 hr
Steroids	no steroids	low-dose steroids (<10 mg prednisone equivalents/day)	moderate steroids (≥10, <30 mg prednisone equivalents/day)	high-dose steroids (≥30 mg prednisone equivalents/day)

TABLE D.4

Components of Quick Physical Examination (none to +++)

Fat status	Muscle Status	Fluid Status
eyes	temples	skin & skin turgor
triceps fat pinch	shoulders	eyes
anterior lower ribs	clavicle	ankles
	scapula	sacrum
	thumb/index press	abdomen for ascites
	thigh and calf	

TABLE D.5

PG-SGA Staging Guide

	Stage A	Stage B	Stage C
Category	Well-nourished	Moderately malnourished or suspected of being malnourished	Severely malnourished
Weight	No weight loss or recent non-fluid weight gain	A. Approximately 5% weight loss within 1 month (or 10% in 6 months) B. No weight stabilization or continued weight loss	A. >5% loss in 1 month (or >10% loss in 6 months) B. No weight stabilization or weight gain
Intake	No deficit or significant recent improvement	Definite decrease in intake	Severe deficit in intake
Nutrition impact symptoms	None or significant recent improvement allowing adequate intake	Presence of nutrition impact symptoms (Box 3 of PG-SGA)	Presence of nutrition impact symptoms (Box 3 of PG-SGA)
Functionality	No deficit or significant recent improvement	Moderate functional deficit or recent functional deterioration	Severe functional deficit or recent significant functional deterioration
Physical exam	No deficit or chronic deficit in the face of recent improvement in all history categories listed above	Evidence of mild to moderate loss of subcutaneous fat and/or muscle mass and/or muscle tone on palpation	Obvious signs of malnutrition (eg, severe loss of subcutaneous tissues, possible edema)

TABLE D.6

Triaging Nutritional Intervention

Additive scores are used to define specific nutritional intervention pathways, including education and/or symptom management, aggressive oral nutrition, and enteral/parenteral triage.	
Additive score of 0–1	Indicates that no intervention is required at this time. While these examples include only the client section of the form, the total additive scores include addition of both the client and clinician scores.
Additive score of 2–3	Indicates a need for client education by a dietitian or nurse, with pharmacologic triage by the nurse or physician as indicated by the symptom survey.
Additive score of 4–8	Requires the intervention of the dietitian, working in conjunction with the nurse or physician as indicated by the symptom check-off for pharmacologic management.
Additive score of >9	Indicates a critical need for symptom management and/or nutritional intervention. These clients require an interdisciplinary discussion to address all the aspects that are impacting the nutritional status, as well as the potential need for non-oral nutritional options, including enteral and parenteral nutrition. This decision should be dictated by the presence or absence of GI function.

Reprinted with permission from Ottery FD, Kasenic S, DeBolt S, Rogers K. Volunteer network accrues >1900 patients in 6 months to validate standardized nutritional triage. Abstract 282. Meeting of the American Society of Clinical Oncology, 1998.

Figure E1.1

Weight-for-Age Percentiles: Boys, Birth to 36 Months

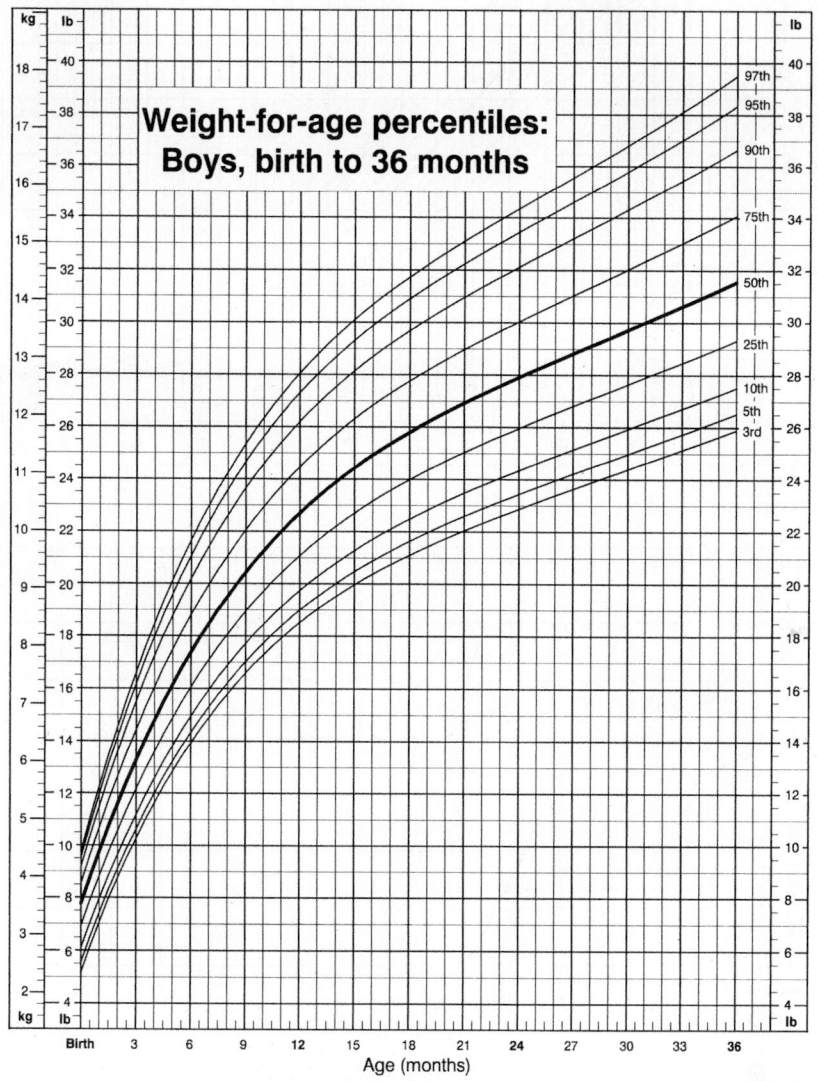

Weight-for-age percentiles: Boys, birth to 36 months

SOURCE: Developed by the National Center for Health Statistics in collaboration with
the National Center for Chronic Disease Prevention and Health Promotion (2000).

CDC
CENTERS FOR DISEASE CONTROL
AND PREVENTION

Figure E1.3
Length-for-Age Percentiles: Boys, Birth to 36 Months

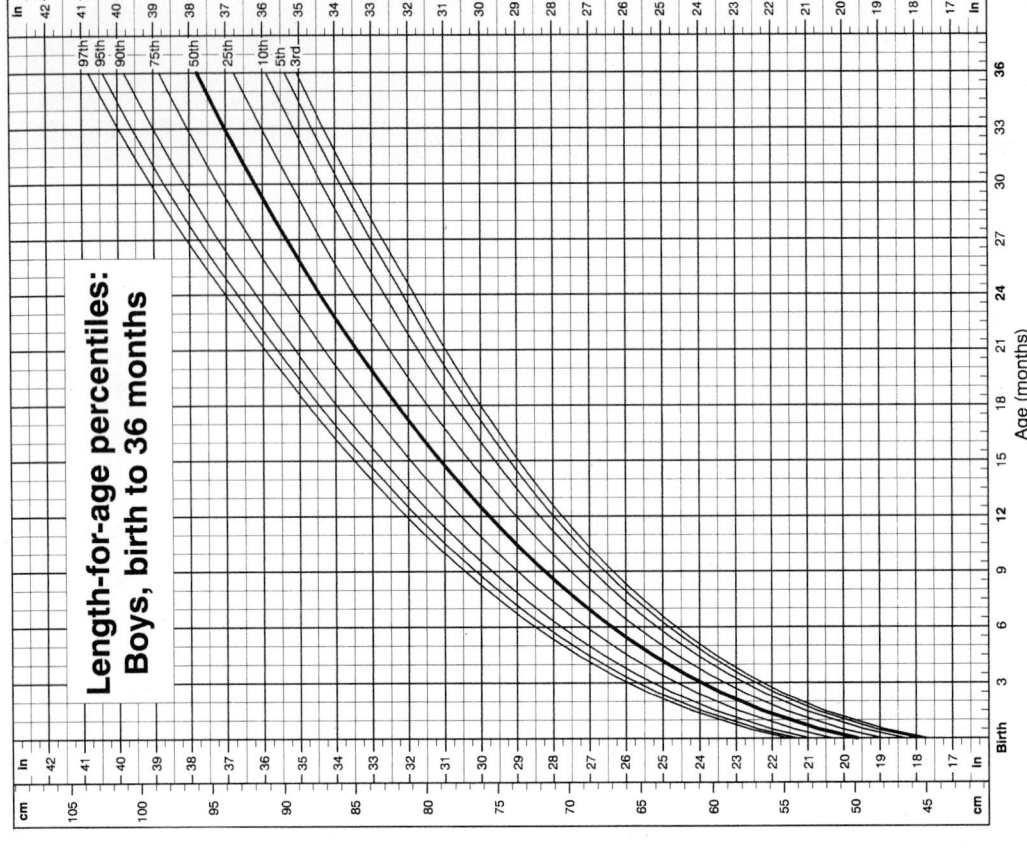

Length-for-age percentiles:
Boys, birth to 36 months

SOURCE: Developed by the National Center for Health Statistics in collaboration with the National Center for Chronic Disease Prevention and Health Promotion (2000).

Figure E1.2
Weight-for-Age Percentiles: Girls, Birth to 36 Months

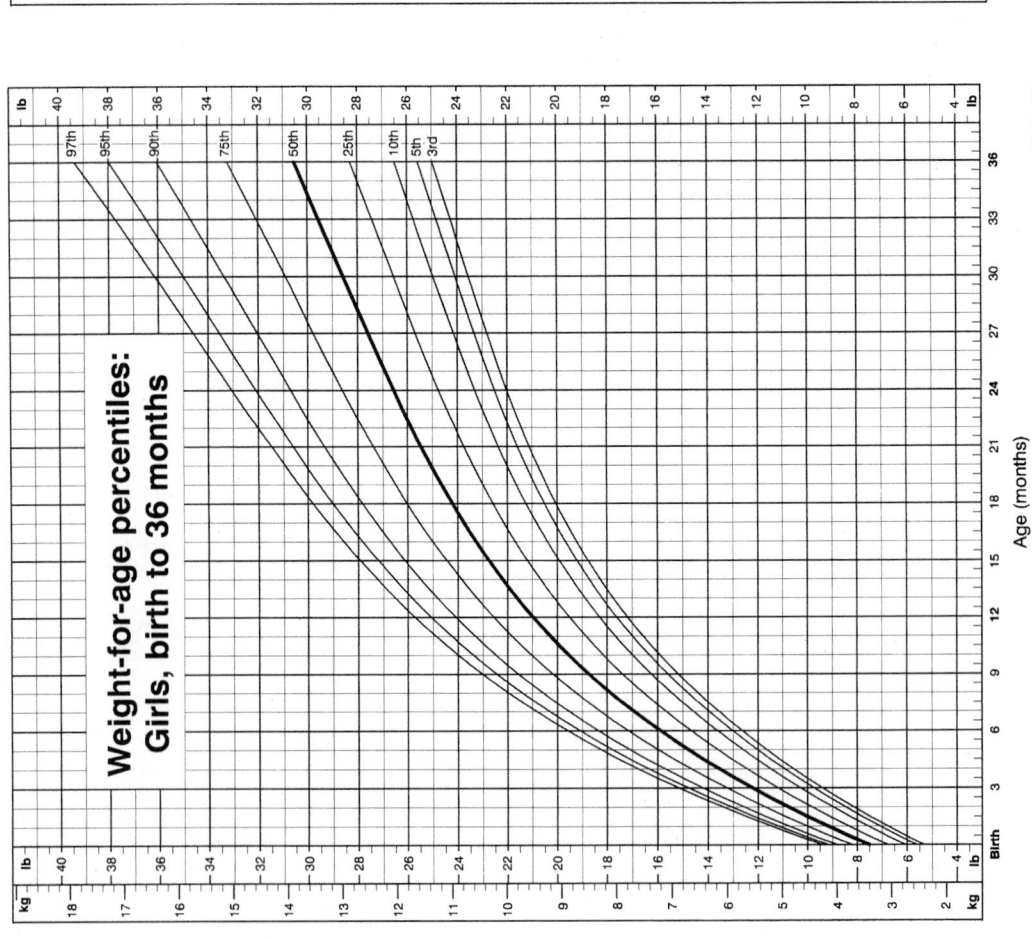

Weight-for-age percentiles:
Girls, birth to 36 months

SOURCE: Developed by the National Center for Health Statistics in collaboration with the National Center for Chronic Disease Prevention and Health Promotion (2000).

Figure E1.4
Length-for-Age Percentiles: Girls, Birth to 36 Months

Figure E1.5
Weight-for-Length Percentiles: Boys, Birth to 36 Months

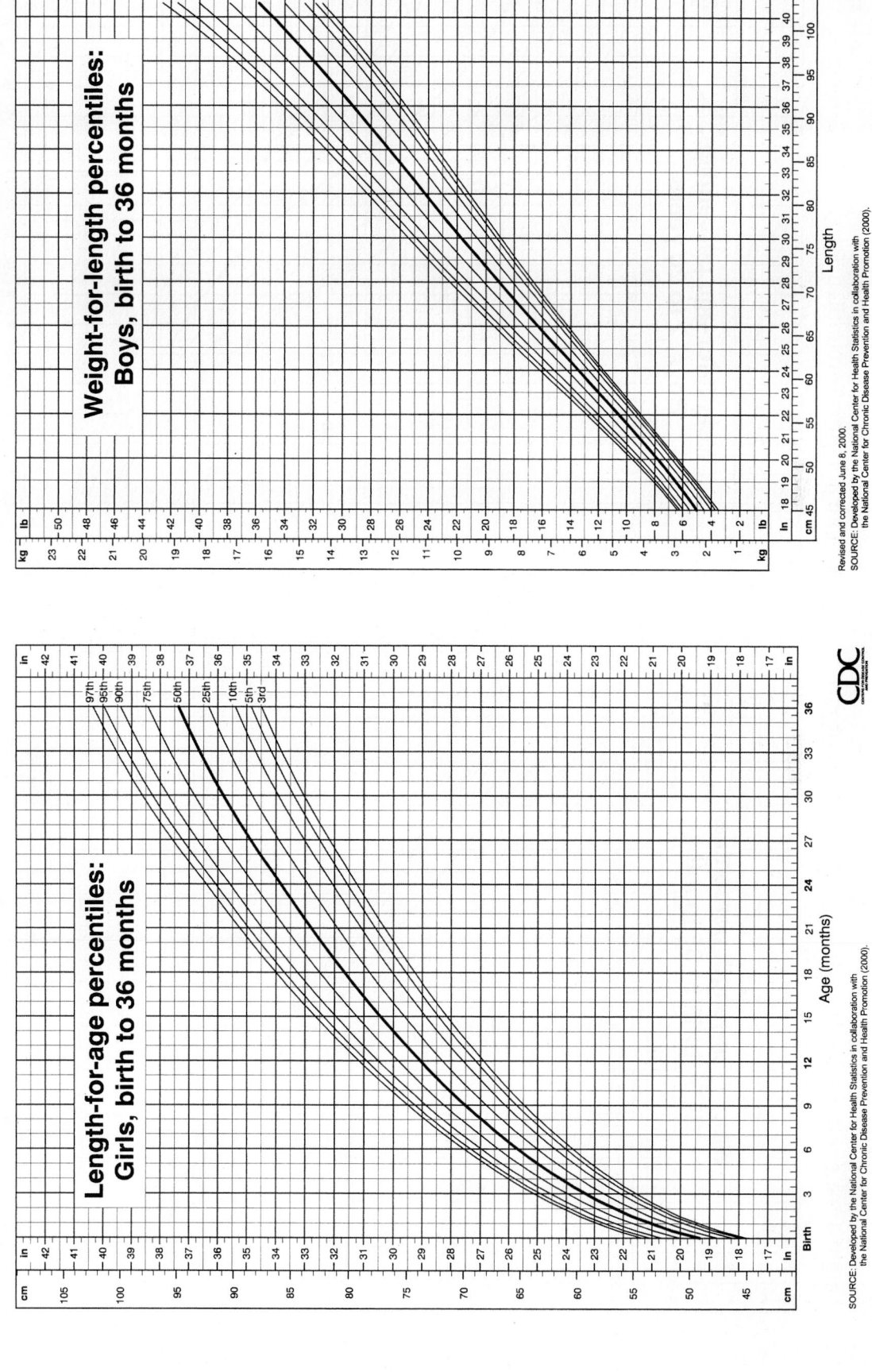

Length-for-age percentiles: Girls, birth to 36 months

Weight-for-length percentiles: Boys, birth to 36 months

SOURCE: Developed by the National Center for Health Statistics in collaboration with the National Center for Chronic Disease Prevention and Health Promotion (2000).

Revised and corrected June 8, 2000.
SOURCE: Developed by the National Center for Health Statistics in collaboration with the National Center for Chronic Disease Prevention and Health Promotion (2000).

Figure E1.7

Weight-for-Age Percentiles: Boys, 2 to 20 Years

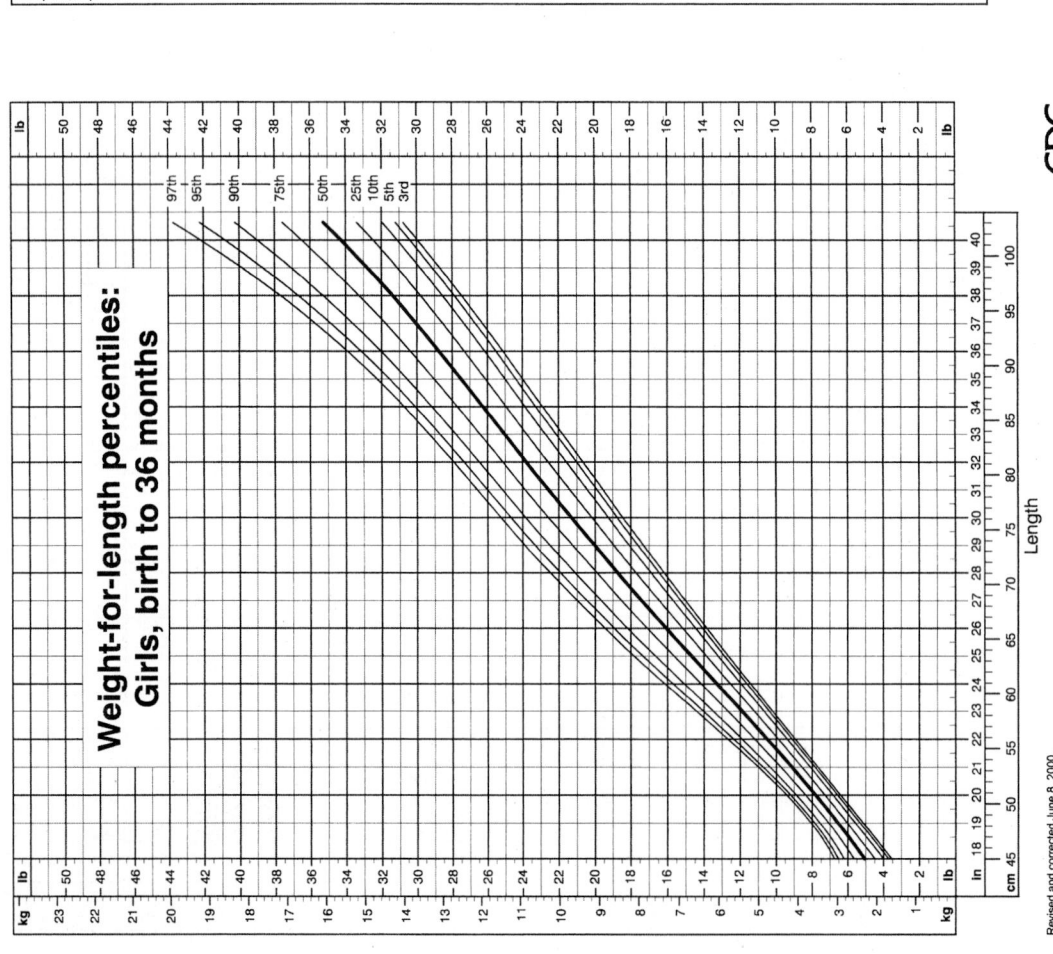

Weight-for-age percentiles:
Boys, 2 to 20 years

SOURCE: Developed by the National Center for Health Statistics in collaboration with
the National Center for Chronic Disease Prevention and Health Promotion (2000).

Figure E1.6

Weight-for-Length Percentiles: Girls, Birth to 36 Months

Weight-for-length percentiles:
Girls, birth to 36 months

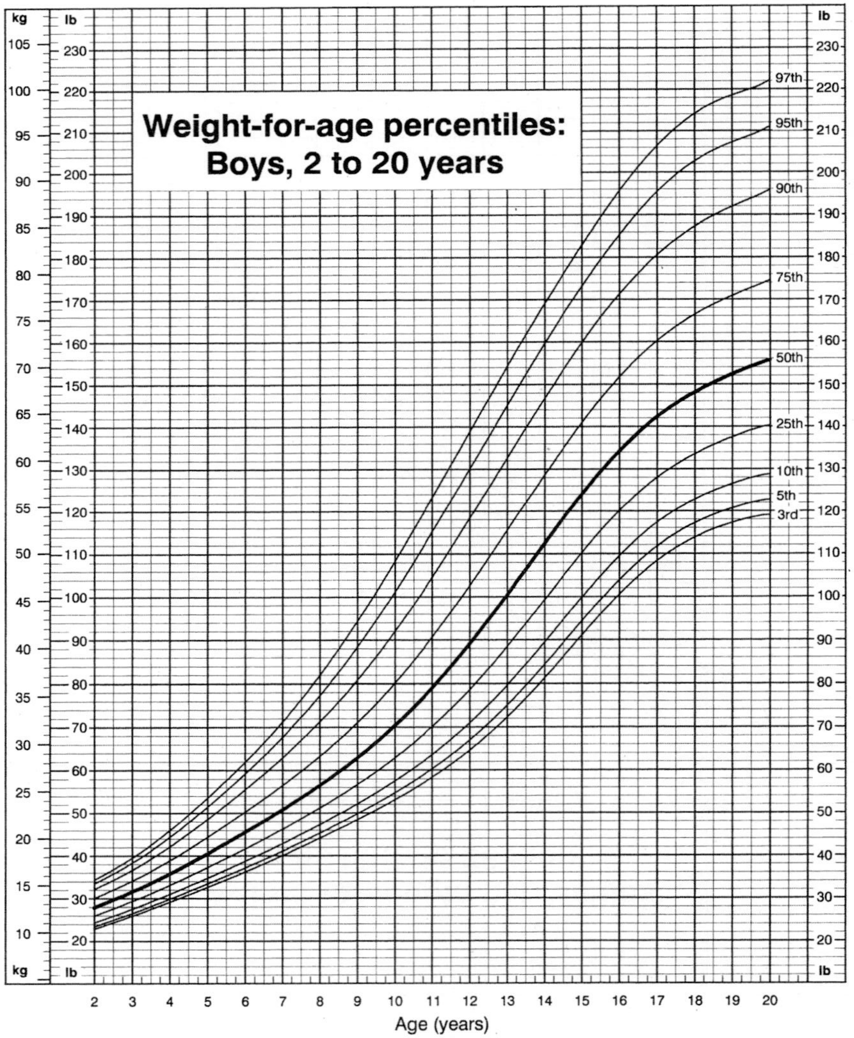

Revised and corrected June 8, 2000.
SOURCE: Developed by the National Center for Health Statistics in collaboration with
the National Center for Chronic Disease Prevention and Health Promotion (2000).

Figure E1.9
Stature-for-Age Percentiles: Boys, 2 to 20 Years

Figure E1.8
Weight-for-Age Percentiles: Girls, 2 to 20 Years

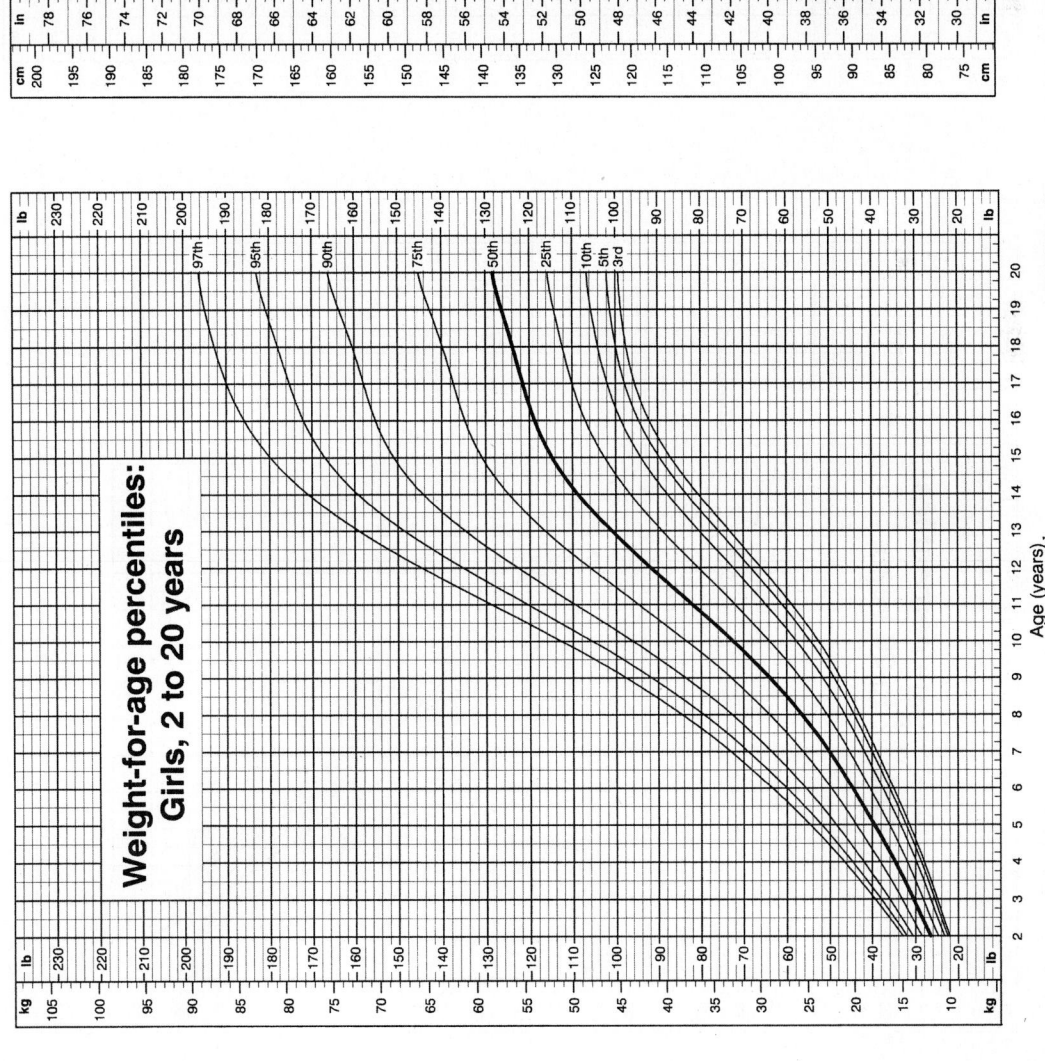

**Stature-for-age percentiles:
Boys, 2 to 20 years**

97th
95th
90th
75th
50th
25th
10th
5th
3rd

Age (years)

SOURCE: Developed by the National Center for Health Statistics in collaboration with
the National Center for Chronic Disease Prevention and Health Promotion (2000).

**Weight-for-age percentiles:
Girls, 2 to 20 years**

97th
95th
90th
75th
50th
25th
10th
5th
3rd

Age (years).

SOURCE: Developed by the National Center for Health Statistics in collaboration with
the National Center for Chronic Disease Prevention and Health Promotion (2000).

Figure E1.11
Weight-for-Stature Percentiles: Boys, 2 to 20 Years

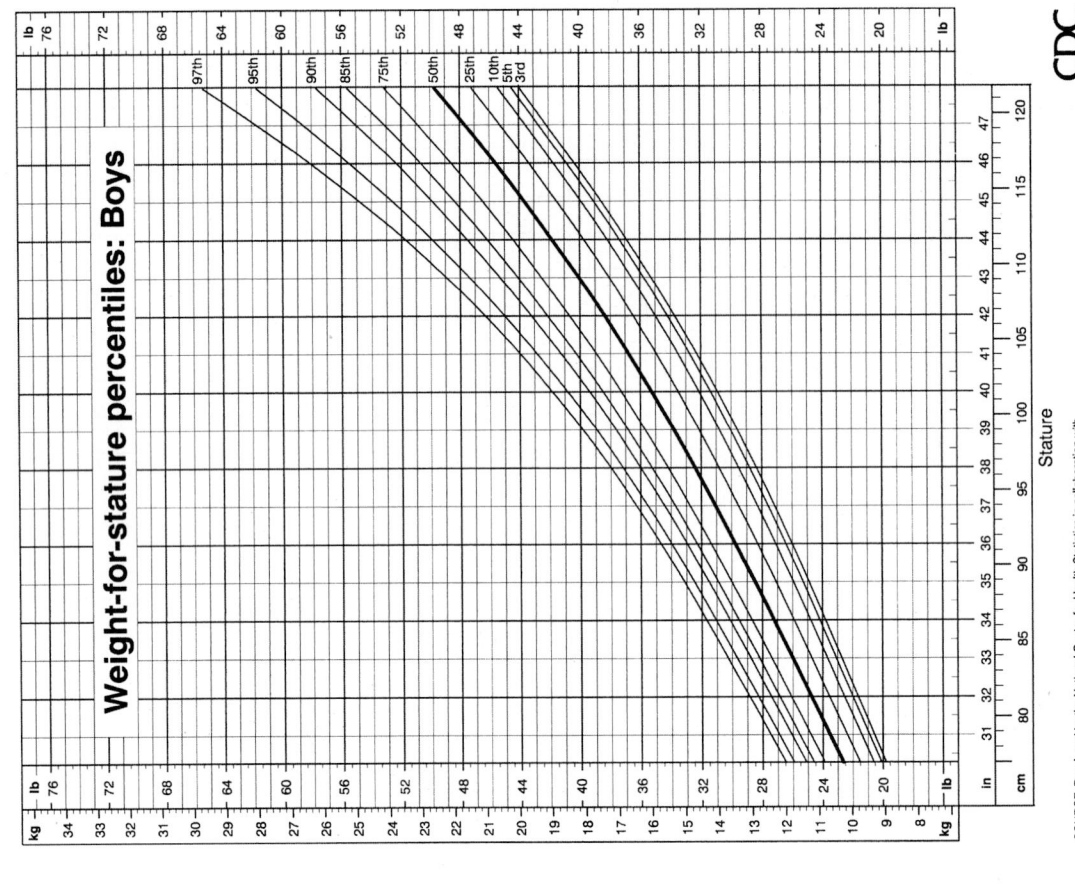

Weight-for-stature percentiles: Boys

SOURCE: Developed by the National Center for Health Statistics in collaboration with
the National Center for Chronic Disease Prevention and Health Promotion (2000).

Figure E1.10
Stature-for-Age Percentiles: Girls, 2 to 20 Years

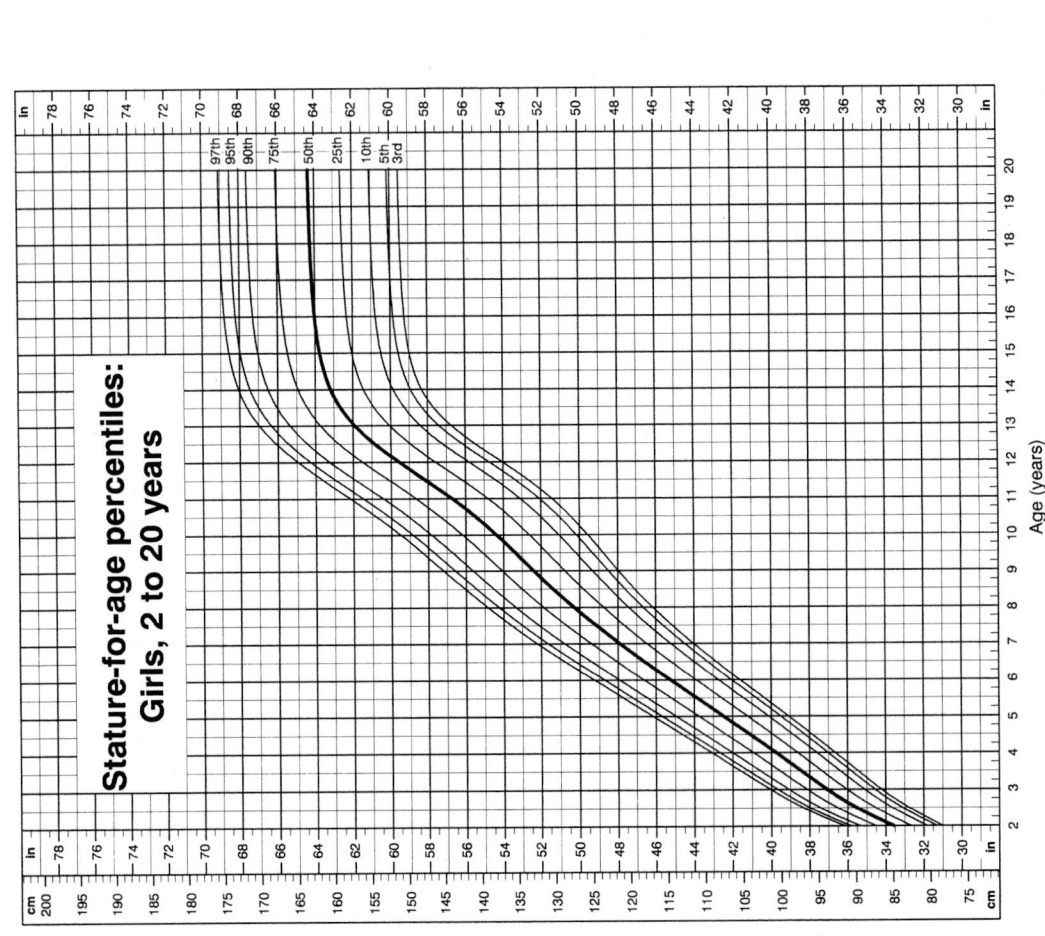

**Stature-for-age percentiles:
Girls, 2 to 20 years**

SOURCE: Developed by the National Center for Health Statistics in collaboration with
the National Center for Chronic Disease Prevention and Health Promotion (2000).

Weight-for-Stature Percentiles: Girls, 2 to 20 Years

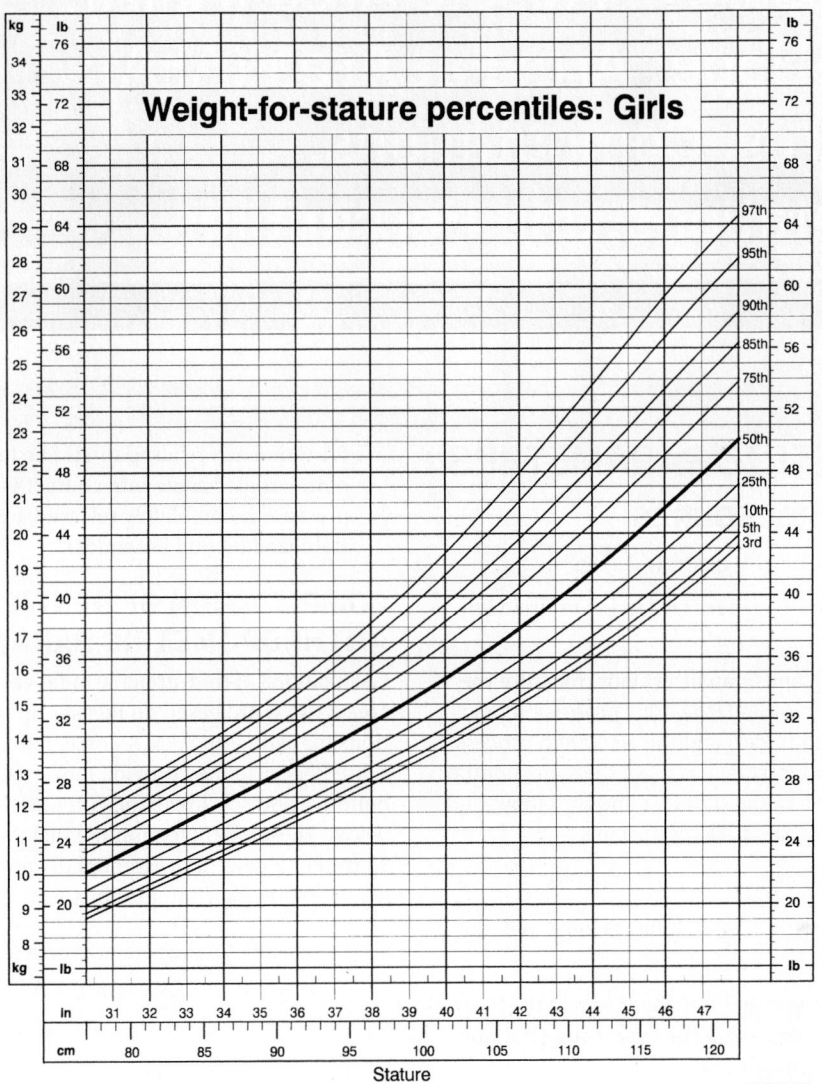

Weight-for-stature percentiles: Girls

Stature

SOURCE: Developed by the National Center for Health Statistics in collaboration with
the National Center for Chronic Disease Prevention and Health Promotion (2000).

Appendix E2—Triceps Skinfold Thickness, Arm Muscle Area (AMA), and Mid-Upper Arm Fat Area (AFA): Procedures, Computations, Interpretations, and Percentiles

ANTHROPOMETRIC PARAMETERS

MEASUREMENT PROCEDURES

Mid-Upper Arm Circumference (AC)

1. Keeping the subject's right arm parallel to the body, bend the elbow 90 degrees.

2. Using either a metallic tape or an insert tape, measure the distance between the acromion (the bony protrusion on the back of the upper shoulder) and the olecranon process (tip) of the elbow. If an insert tape is used, the same number should appear at the top of the shoulder and the elbow, and the midpoint is given by the mark on the tape.

3. Mark the midpoint between these two landmarks.

4. Ask the subject to relax the arm, so it hangs loose and parallel to the body.

5. Position the metric tape around the upper arm at the marked midpoint. Make sure that the tape is snug but not so tight as to indent or pinch the skin.

6. Record the AC to the nearest 0.1 cm.

Triceps Skinfold Thickness (TSF)

1. Locate the previously marked midpoint on the posterior or back side of the right upper arm.

2. With the subject's arm hanging down loosely at the side, palpate the measurement site at the midpoint to become familiar with distinguishing muscle from adipose soft tissue.

3. From 1 cm above the midpoint, grasp a vertical pinch of skin and only the subcutaneous fat layer between the thumb and index finger. The skinfold should be gently pulled away from the underlying muscle.

4. Place the skinfold calipers at the midpoint and release the jaw pressure slowly while maintaining a grasp of the skinfold. Three readings should be taken in quick succession, and the average of the three recorded to the nearest

0.1 mm. Each reading should be taken as soon as the jaws of the caliper come into contact with the skin after the pressure is completely released and the dial reading has stabilized (about 4 seconds).

COMPUTATION OF DERIVED ANTHROPOMETRIC PARAMETERS

Calculations of mid-upper arm fat area (AFA) and arm muscle area (AMA) are based on measurements of AC of TSF. Equations for estimating AMA corrected for bone area are presented because they provide more accurate assessments of bone-free muscle area. The computational steps are outlined below. These examples assume that AC is 30 cm and TSF is 35 mm (2.5 cm). (It is essential to convert TSF to centimeters for the following computations.)

1. Determine total upper arm area (TAA):

$$\text{TAA (cm}^2) = \text{AC}^2/4 \times \pi$$
$$\text{TAA} = 30^2/12.57 = 71.6 \text{ cm}^2$$

2. Determine Uncorrected AMA:

$$\text{AMA (cm}^2) = [\text{AC} - (\text{TSF} \times \pi)]^2/4 \times \pi$$
$$\text{AMA} = [30 - (2.5 \times 3.1416)]^2/12.57$$
$$\text{AMA} = 490.44/12.57$$
$$\text{AMA} = 39 \text{ cm}^2$$

3. Determine corrected AMA (AMA_C):

$$\text{Males} = \text{AMA}_C - 10 \text{ cm}^2 = 29.0 \text{ cm}^2$$
$$\text{Females} = \text{AMA}_C - 6.5 \text{ cm}^2 = 32.5 \text{ cm}^2$$

4. Determine the AFA as the difference between TAA and *uncorrected* AMA:

$$\text{AFA (cm}^2) = \text{TAA} - \text{AMA}$$
$$\text{AFA} = 71.6 - 39.0 = 32.6 \text{ cm}^2$$

Note: Arm fat index (AFI) or percent fat area (%FA) can be determined as follows:

$$\text{AFI or \%FA} = (\text{AFA}/\text{TAA}) \times 100$$
$$\text{AFI or \%FA} = (32.6/71.6) \times 100 = 45.5\%$$

INTERPRETATION OF ANTHROPOMETRIC PARAMETERS

Comparing the percentile ranking of a specific individual on the various anthropometric measurements with a classification scheme is the basis for interpreting these values. Reference data in percentiles for TSF, AMA, and AFA appear in the appendices. Because the reference data are those compiled by Frisancho from the NHANES I and II data, it is appropriate to use the classification categories derived statistically from these data. The table below displays the percentile categories and their interpretation for arm muscle and arm fat areas as well as total body weight.

Percentile Rank	AMA	AFA	Total Body Weight
<5	Muscle deficit	Fat deficit	Total body wasting
5.1–15	Below average	Below average	Below average
15.1–85	Average	Average	Average
>85	Above average musculature	Excess fat	Excess total body weight

FRAME SIZE

MEASUREMENT PROCEDURE

Elbow breadth is most accurately measured using either sliding or spreading calipers and the following procedure:

1. Raise the subject's right arm so that the forearm is parallel to the body and flexed to a 90-degree angle.

2. Facing the subject, palpate the lateral and medial epicondyles of the humerus (the two prominent bones on either side of the elbow) and place the caliper jaws parallel or slightly at a slant to these two sites.

3. Measure the greatest bony width across the elbow joint twice to the nearest 0.1 cm; take the average of the two measurements.

Note: If sliding calipers are not available, elbow breadth can be estimated by placing the thumb and index finger of one hand parallel to the body on the epicondyles. The distance between the tips of the thumb and index finger is measured as the elbow breadth.

MEASUREMENT INTERPETATION

The National Health and Nutrition Examination Survey (NHANES) classification of frame size as small, medium, or large is based on the Frame Index 2 value, which is derived thus:

$$\text{Frame index 2} = \text{elbow breadth (mm)/stature (cm)} \times 100$$

This index accommodates age-related changes in weight and stature. The computation is made after converting the units of elbow breadth from centimeters to millimeters (by multiplying by 10). The derived Frame Index 2 is compared with the reference values for small, medium, and large frames, as defined below for the appropriate age and gender of the subject.

Age (yrs)	Male			Female		
	Small	Medium	Large	Small	Medium	Large
18.0–24.9	<38.4	38.4 to 41.6	>41.6	<35.2	35.2 to 38.6	>38.6
25.0–29.9	<38.6	38.6 to 41.8	>41.8	<35.7	35.7 to 38.7	>38.7
30.0–34.9	<38.6	38.6 to 42.1	>42.1	<35.7	35.7 to 39.0	>39.0
35.0–39.9	<39.1	39.1 to 42.4	>42.4	<36.2	36.2 to 39.8	>39.8
40.0–44.9	<39.3	39.3 to 42.5	>42.5	<36.7	36.7 to 40.2	>40.2
45.0–49.9	<39.6	39.6 to 43.0	>43.0	<37.2	37.2 to 40.7	>40.7
50.0–54.9	<39.9	39.9 to 43.3	>43.3	<37.2	37.2 to 41.6	>41.6
55.0–59.9	<40.2	40.2 to 43.8	>43.8	<37.8	37.8 to 41.9	>41.9
60.0–64.9	<40.2	40.2 to 43.6	>43.6	<38.2	38.2 to 41.8	>41.8
65.0–69.9	<40.2	40.2 to 43.6	>43.6	<38.2	38.2 to 41.8	>41.8
70.0–74.9	<40.2	40.2 to 43.6	>43.6	<38.2	38.2 to 41.8	>41.8

Adapted from Frisancho AR: *Anthropometric standards for the assessment of growth and nutritional status*, Ann Arbor, MI, 1990, The University of Michigan Press, p. 28.

Means, Standard Deviations, and Percentiles of Triceps Skinfold Thickness (mm) by Age for Males and Females of 1 to 74 Years

| Age (yr) | N | Mean | SD | Percentiles | | | | | | | | |
				5	10	15	25	50	75	85	90	95
MALES												
1.0–1.9	681	10.4	2.9	6.5	7.0	7.5	8.0	10.0	12.0	13.0	14.0	15.5
2.0–2.9	677	10.0	2.9	6.0	6.5	7.0	8.0	10.0	12.0	13.0	14.0	15.0
3.0–3.9	717	9.9	2.7	6.0	7.0	7.0	8.0	9.5	11.5	12.5	13.5	15.0
4.0–4.9	708	9.2	2.7	5.5	6.5	7.0	7.5	9.0	11.0	12.0	12.5	14.0
5.0–5.9	677	8.9	3.1	5.0	6.0	6.0	7.0	8.0	10.0	11.5	13.0	14.5
6.0–6.9	298	8.9	3.8	5.0	5.5	6.0	6.5	8.0	10.0	12.0	13.0	16.0
7.0–7.9	312	9.0	4.0	4.5	5.0	6.0	6.0	8.0	10.5	12.5	14.0	16.0
8.0–8.9	296	9.6	4.4	5.0	5.5	6.0	7.0	8.5	11.0	13.0	16.0	19.0
9.0–9.9	322	10.2	5.1	5.0	5.5	6.0	6.5	9.0	12.5	15.5	17.0	20.0
10.0–10.9	334	11.5	5.7	5.0	6.0	6.0	7.5	10.0	14.0	17.0	20.0	24.0
11.0–11.9	324	12.5	7.0	5.0	6.0	6.5	7.5	10.0	16.0	19.5	23.0	27.0
12.0–12.9	348	12.2	6.8	4.5	6.0	6.0	7.5	10.5	14.5	18.0	22.5	27.5
13.0–13.9	350	11.0	6.7	4.5	5.0	5.5	7.0	9.0	13.0	17.0	20.5	25.0
14.0–14.9	358	10.4	6.5	4.0	5.0	5.0	6.0	8.5	12.5	15.0	18.0	23.5
15.0–15.9	356	9.8	6.5	5.0	5.0	5.0	6.0	7.5	11.0	15.0	18.0	23.5
16.0–16.9	350	10.0	5.9	4.0	5.0	5.1	6.0	8.0	12.0	14.0	17.0	23.0
17.0–17.9	337	9.1	5.3	4.0	5.0	5.0	6.0	7.0	11.0	13.5	16.0	19.5
18.0–24.9	1752	11.3	6.4	4.0	5.0	5.5	6.5	10.0	14.5	17.5	20.0	23.5
25.0–29.9	1251	12.2	6.7	4.0	5.0	6.0	7.0	11.0	15.5	19.0	21.5	25.0
30.0–34.9	941	13.1	6.7	4.5	6.0	6.5	8.0	12.0	16.5	20.0	22.0	25.0
35.0–39.9	832	12.9	6.2	4.5	6.0	7.0	8.5	12.0	16.0	18.5	20.5	24.5
40.0–44.9	828	13.0	6.6	5.0	6.0	6.9	8.0	12.0	16.0	19.0	21.5	26.0
45.0–49.9	867	12.9	6.4	5.0	6.0	7.0	8.0	12.0	16.0	19.0	21.0	25.0
50.0–54.9	879	12.6	6.1	5.0	6.0	7.0	8.0	11.5	15.0	18.5	20.8	25.0
55.0–59.9	807	12.4	6.0	5.0	6.0	6.5	8.0	11.5	15.0	18.0	20.5	25.0
60.0–64.9	1259	12.5	6.0	5.0	6.0	7.0	8.0	11.5	15.5	18.5	20.5	24.0
65.0–69.9	1774	12.1	5.9	4.5	5.0	6.5	8.0	11.0	15.0	18.0	20.0	23.5
70.0–74.9	1251	12.0	5.8	4.5	6.0	6.5	8.0	11.0	15.0	17.0	19.0	23.0

Means, Standard Deviations, and Percentiles of Triceps Skinfold Thickness (mm) by Age for Males and Females of 1 to 74 Years (*Continued*)

Age (yr)	N	Mean	SD	Percentiles								
				5	10	15	25	50	75	85	90	95
FEMALES												
1.0–1.9	622	10.4	3.1	6.0	7.0	7.0	8.0	10.0	12.0	13.0	14.0	16.0
2.0–2.9	614	10.5	2.9	6.0	7.0	7.5	8.5	10.0	12.0	13.5	14.5	16.0
3.0–3.9	652	10.4	2.9	6.0	7.0	7.5	8.5	10.0	12.0	13.0	14.0	16.0
4.0–4.9	681	10.3	3.0	6.0	7.0	7.5	8.0	10.0	12.0	13.0	14.0	15.5
5.0–5.9	673	10.4	3.5	5.5	7.0	7.0	8.0	10.0	12.0	13.5	15.0	17.0
6.0–6.9	296	10.4	3.7	6.0	6.5	7.0	8.0	10.0	12.0	13.0	15.0	17.0
7.0–7.9	330	11.1	4.2	6.0	7.0	7.0	8.0	10.5	12.5	15.0	16.0	19.0
8.0–8.9	276	21.1	5.4	6.0	7.0	7.5	8.5	11.0	14.5	17.0	18.0	22.0
9.0–9.9	322	13.4	5.9	6.5	7.0	8.0	9.0	12.0	16.0	19.0	21.0	25.0
10.0–10.9	329	13.9	6.1	7.0	8.0	8.0	9.0	12.5	17.5	20.0	22.5	27.0
11.0–11.9	302	15.0	6.8	7.0	8.0	8.5	10.0	13.0	18.0	21.5	24.0	29.0
12.0–12.9	323	15.1	6.3	7.0	8.0	9.0	11.0	14.0	18.5	21.5	24.0	27.5
13.0–13.9	360	16.4	7.4	7.0	8.0	9.0	11.0	15.0	20.0	24.0	25.0	30.0
14.0–14.9	370	17.1	7.3	8.0	9.0	10.0	11.5	16.0	21.0	23.5	26.5	32.0
15.0–15.9	309	17.3	7.4	8.0	9.5	10.5	12.0	16.5	20.5	23.0	26.0	32.5
16.0–16.9	343	19.2	7.0	10.5	11.5	12.0	14.0	18.0	23.0	26.0	29.0	32.5
17.0–17.9	291	19.1	8.0	9.0	10.0	12.0	13.0	18.0	24.0	26.5	29.0	34.5
18.0–24.9	2588	20.0	8.2	9.0	1.1	12.0	14.0	18.5	24.5	28.5	31.0	36.0
25.0–29.9	1921	21.7	8.8	10.0	12.0	13.0	15.0	20.0	26.5	31.0	34.0	38.0
30.0–34.9	1619	23.7	9.2	10.5	13.0	10.5	17.0	22.5	29.5	33.0	35.5	41.5
35.0–39.9	1453	24.7	9.3	11.0	13.0	15.5	18.0	23.5	30.0	35.0	37.0	41.0
40.0–44.9	1391	25.1	9.0	12.0	14.0	16.0	19.0	24.5	30.5	35.0	37.0	41.0
45.0–49.9	962	26.1	9.3	12.0	14.5	16.5	19.5	25.5	32.0	35.5	38.0	42.5
50.0–54.9	1006	26.5	9.0	12.0	15.0	17.5	20.5	25.5	32.0	36.0	38.5	42.0
55.0–59.9	880	26.6	9.4	12.0	15.0	17.0	20.5	26.0	32.0	36.0	39.0	42.5
60.0–64.9	1389	26.6	8.8	12.5	16.0	17.5	20.5	26.0	32.0	35.5	38.0	42.5
65.0–69.9	1946	25.1	8.5	12.0	14.5	16.0	19.0	25.0	30.0	33.5	36.0	40.0
70.0–74.9	1462	42.0	8.5	11.0	13.5	15.5	18.0	24.0	29.5	32.0	35.0	38.5

From Frisancho AR: *Anthropometric standards for the assessment of growth and nutritional status*, Ann Arbor, 1990, The University of Michigan Press, p. 54.

Means, Standard Deviations, and Percentiles of Upper Arm Muscle Area (cm²) by Height (cm) for Boys and Girls of 2 to 17 years

| Height (cm) | N | Mean | SD | Percentiles | | | | | | | | |
				5	10	15	25	50	75	85	90	95
BOYS: 2 TO 11 YR												
87–092	94	12.9	2.2	9.3	10.4	10.6	11.2	12.9	14.2	15.0	15.8	16.5
93–098	373	13.7	2.4	10.2	10.9	11.2	12.1	13.5	15.3	15.9	16.5	17.0
99–104	587	14.6	3.1	10.9	11.7	12.2	13.0	14.5	15.9	16.5	17.1	18.4
105–110	587	15.7	3.1	12.0	12.8	13.3	14.1	15.4	17.0	17.8	18.6	19.8
111–116	588	16.7	2.9	12.6	13.6	14.3	15.0	16.6	18.1	18.9	19.6	20.7
117–122	496	18.1	3.5	14.1	14.5	15.0	16.1	17.7	19.7	20.7	21.6	23.4
123–128	376	19.5	3.6	15.0	15.9	16.3	17.4	19.2	21.2	22.3	23.2	24.2
129–134	359	21.6	4.3	16.1	17.3	18.4	19.3	21.1	23.2	34.7	25.3	27.9
135–140	354	22.9	4.2	17.2	18.1	18.9	20.4	22.6	24.9	26.1	27.2	30.2
141–146	325	25.1	5.2	19.3	20.1	20.8	21.9	24.0	27.2	29.0	30.5	34.0
147–152	266	27.5	4.8	21.2	22.4	23.2	24.8	27.0	29.8	31.8	32.9	34.4
153–158	150	29.8	7.2	22.3	23.2	24.3	25.4	28.4	32.2	34.8	36.9	40.1
159–164	65	32.5	6.5	23.7	24.5	25.3	27.5	31.9	35.6	39.7	41.4	44.5
BOYS: 12 TO 17 YR												
141–146	31	26.8	4.7	20.7	21.4	22.7	24.1	25.6	30.3	32.8	33.9	36.3
147–152	90	28.2	4.1	22.4	23.4	24.1	25.6	27.5	30.2	33.1	34.2	36.1
153–158	181	31.4	6.4	22.7	24.9	26.1	27.5	30.4	34.1	36.4	39.1	41.5
159–164	218	35.0	7.7	23.7	26.7	27.8	30.2	34.1	38.6	41.5	44.3	48.4
165–170	323	40.8	9.3	28.1	29.7	31.5	34.2	40.0	45.6	49.0	52.9	58.9
171–176	431	46.6	9.9	32.8	35.2	36.6	39.5	45.8	52.6	56.0	59.1	66.0
177–182	431	50.3	9.3	36.1	38.7	40.8	43.4	49.5	56.4	59.4	62.6	65.9
183–188	269	53.4	11.2	38.3	41.3	42.8	46.1	52.6	57.8	63.0	67.5	74.3
189–194	99	55.4	9.9	41.4	44.2	45.7	48.9	53.9	60.3	65.0	68.5	74.0

TABLE E2.2

Means, Standard Deviations, and Percentiles of Upper Arm Muscle Area (cm²) by Height (cm) for Boys and Girls of 2 to 17 years (*Continued*)

| Height (cm) | N | Mean | SD | Percentiles | | | | | | | | |
				5	10	15	25	50	75	85	90	95
GIRLS: 2 TO 10 YR												
87–092	154	12.6	2.1	9.5	10.1	10.5	11.0	12.6	14.2	14.8	15.5	16.2
93–098	384	13.2	2.1	10.1	10.7	11.0	11.8	13.2	14.4	15.3	15.8	16.9
99–104	533	14.1	2.3	10.6	11.2	11.7	12.5	14.0	15.5	16.4	16.9	18.0
105–110	550	14.8	2.4	11.3	11.9	12.4	13.2	14.6	16.3	17.3	17.9	18.9
111–116	543	15.9	2.8	12.3	13.0	13.5	14.2	15.7	17.4	18.4	19.1	20.3
117–122	465	17.0	2.8	13.0	13.9	14.4	15.2	16.7	18.5	19.6	20.3	21.4
123–128	372	18.2	2.8	14.2	15.0	15.5	16.2	17.9	19.6	20.8	21.6	22.9
129–134	333	20.1	4.6	15.3	16.1	16.8	17.6	19.7	21.7	22.9	23.8	25.4
135–140	303	21.6	4.2	16.1	17.4	18.1	19.2	21.1	23.8	24.8	26.3	27.9
141–146	258	23.3	4.0	17.6	18.5	19.5	20.5	23.0	25.8	27.9	28.8	30.6
147–152	161	25.2	5.2	18.5	20.0	20.7	21.7	24.4	27.8	30.0	31.2	32.9
153–158	66	26.7	6.7	19.4	20.1	22.4	23.0	25.4	29.2	31.8	34.0	38.2
GIRLS: 11 TO 17 YR												
141–146	53	23.8	4.4	17.1	19.3	19.5	21.0	23.4	25.5	27.9	28.6	33.4
147–152	119	25.2	4.6	18.5	19.7	20.9	22.0	24.3	28.0	29.8	30.3	34.4
153–158	305	29.1	6.4	20.8	22.0	23.0	24.7	28.3	32.9	35.0	37.5	39.2
159–164	587	32.2	7.1	23.3	24.8	26.0	27.7	31.2	35.7	38.0	40.1	43.5
165–170	715	34.2	7.4	25.0	26.6	27.8	29.5	33.2	37.6	40.2	42.8	46.9
171–176	367	34.9	8.0	25.9	27.1	28.0	29.9	33.7	38.0	41.2	43.5	47.6
177–182	113	37.8	8.4	28.6	29.5	30.5	31.7	35.9	41.1	45.9	47.8	58.2

From Frisancho AR: *Anthropometric standards for the assessment of growth and nutritional status*, Ann Arbor, 1990, The University of Michigan Press, p. 51.

Means, Standard Deviations, and Percentiles of Muscle Area (cm²) by Age for Adult Females of Small, Medium, and Large Frames

Age (yr)	N	Mean	SD	5	10	15	25	50	75	85	90	95
FEMALES WITH SMALL FRAMES												
18.0–24.9	651	26.2	6.0	18.2	19.6	20.7	22.5	25.5	29.2	31.2	32.8	36.2
25.0–29.9	486	27.8	7.4	19.5	20.6	21.6	23.2	26.9	30.8	33.3	35.2	38.1
30.0–34.9	413	28.6	7.8	19.1	21.6	22.4	24.5	27.8	31.4	33.7	36.2	38.8
35.0–39.9	368	20.6	10.1	19.7	21.4	22.9	24.4	28.8	32.5	35.4	37.5	42.2
40.0–44.9	350	29.8	6.6	20.9	22.1	23.4	25.7	28.9	33.2	36.0	37.9	41.8
45.0–49.9	241	29.2	7.4	19.1	21.5	22.6	24.3	28.3	33.3	36.1	38.7	41.2
50.0–54.9	256	30.3	7.1	20.8	22.1	23.9	25.5	29.1	33.4	36.7	38.5	41.3
55.0–59.9	223	30.9	1.6	20.4	22.3	23.6	25.8	30.2	34.8	37.8	41.3	45.1
60.0–64.9	351	31.9	8.7	20.9	22.4	23.6	25.8	31.2	36.4	39.1	41.1	16.2
65.0–69.9	491	31.3	8.1	19.4	22.1	23.7	25.7	30.6	35.4	39.8	41.8	45.7
70.0–74.9	367	32.0	9.9	20.3	22.5	24.1	25.9	30.3	36.1	39.8	42.6	47.3
FEMALES WITH MEDIUM FRAMES												
18.0–24.9	1296	29.3	7.0	19.8	21.9	23.2	24.9	28.4	32.8	35.2	37.2	40.7
25.0–29.9	964	30.0	7.2	20.7	22.1	23.3	25.0	29.0	33.9	36.8	39.0	43.3
25.0–29.9	964	30.0	7.2	20.7	22.1	23.3	25.0	29.0	33.9	36.8	39.0	43.3
30.0–34.9	814	32.0	9.1	21.4	23.1	24.2	26.3	30.8	36.1	39.4	41.8	46.4
35.0–39.9	728	32.7	8.4	21.4	23.6	24.9	27.3	31.4	37.3	40.8	43.0	47.0
40.0–44.9	696	33.7	12.1	21.2	23.2	25.1	27.2	31.6	37.7	43.1	47.1	52.3
45.0–49.9	484	33.8	8.8	22.2	23.6	25.5	27.9	32.2	37.9	72.5	45.4	49.6
50.0–54.9	502	35.0	9.7	22.5	25.2	26.2	28.5	33.7	40.0	43.5	46.7	51.4
55.0–59.9	442	36.3	11.5	23.7	25.3	26.6	28.7	34.5	41.5	44.9	49.2	53.4
60.0–64.9	695	35.1	9.1	23.0	25.3	26.5	29.2	33.9	39.9	43.7	46.1	49.4
65.0–69.9	971	35.7	10.0	22.4	24.8	26.4	29.1	34.6	40.7	44.5	48.1	51.9
70.0–74.9	731	35.3	9.7	22.2	24.3	26.1	28.9	34.0	40.0	44.4	46.7	51.3
FEMALES WITH LARGE FRAMES												
18.0–24.9	641	34.4	10.7	21.9	23.8	25.3	27.3	31.9	38.7	43.9	47.5	55.8
25.0–29.9	471	36.7	11.5	22.2	25.4	26.8	29.3	34.5	42.0	46.8	50.3	60.1
30.0–34.9	392	38.8	12.3	24.0	25.8	27.3	30.1	36.3	45.1	50.7	55.1	61.2
35.0–39.9	357	41.6	14.4	23.9	27.4	29.1	32.2	39.1	47.2	53.7	61.0	72.1
40.0–44.9	344	43.5	16.6	26.2	28.8	30.5	32.9	40.3	49.5	54.4	58.7	71.6
45.0–49.9	236	43.0	15.8	25.0	28.0	29.4	32.5	39.7	49.0	58.3	62.8	69.9
50.0–54.9	146	42.4	13.1	25.1	28.4	30.1	33.4	39.6	49.5	54.8	59.7	68.4
55.0–59.9	213	45.2	16.9	27.0	30.0	32.4	35.8	42.0	51.0	58.5	62.2	65.7
60.0–64.9	341	43.1	14.2	26.6	29.1	31.2	33.9	40.7	49.8	54.8	57.5	67.6
65.0–69.9	482	42.5	13.4	26.4	28.4	30.6	33.5	40.0	48.7	55.3	58.7	66.5
70.0–74.9	363	41.5	11.6	25.7	28.8	30.2	32.8	40.1	48.7	51.4	54.8	60.3

From Frisancho AR: *Anthropometric standards for the assessment of growth and nutritional status*, Ann Arbor, 1990, The University of Michigan Press, p. 63.

Note: Values for females aged 18 years and older have been adjusted for bone area by subtracting 6.5 cm² from the calculated mid upper arm muscle area.

Means, Standard Deviations, and Percentiles of Muscle Area (cm²) by Age for Adult Males of Small, Medium, and Large Frames

Age (yr)	N	Mean	SD	Percentiles 5	10	15	25	50	75	85	90	95
MALES WITH SMALL FRAMES												
18.0–24.9	443	45.6	10.6	30.8	33.8	35.8	38.7	44.6	51.3	55.2	58.1	63.2
25.0–29.9	318	48.2	9.8	33.5	36.8	39.2	41.8	47.6	53.5	57.7	61.2	63.7
30.0–34.9	237	49.6	10.2	35.0	37.5	38.9	42.0	48.8	56.4	60.0	62.7	66.9
35.0–39.9	212	51.2	10.4	34.7	38.7	40.9	44.1	50.7	57.5	61.7	63.8	70.0
40.0–44.9	210	51.5	10.1	34.9	38.1	40.6	44.2	51.6	58.2	61.6	64.5	66.9
45.0–49.9	220	49.7	10.8	32.8	36.5	38.9	42.9	49.1	55.7	59.5	63.3	68.8
50.0–54.9	225	49.1	11.2	33.8	36.0	38.2	41.5	47.6	55.5	60.7	63.8	69.3
55.0–59.9	204	47.9	10.1	31.2	35.4	37.8	41.7	47.8	54.3	58.8	61.4	64.2
60.0–64.9	318	48.7	11.2	32.5	36.3	38.7	41.4	48.0	54.6	59.6	62.2	68.0
65.0–699	446	45.1	10.7	26.7	31.5	34.7	37.6	44.7	52.5	56.1	58.5	62.7
70.0–74.9	314	43.5	10.3	27.7	30.8	32.9	36.1	43.4	49.6	53.4	56.6	59.9
MALES WITH MEDIUM FRAMES												
18.0–24.9	875	50.5	10.5	35.5	38.2	40.8	43.6	49.5	56.5	60.8	63.2	69.3
25.0–29.9	626	54.0	11.3	37.0	40.1	42.9	46.8	53.2	60.9	65.6	67.7	73.0
30.0–34.9	472	55.0	10.4	38.5	42.2	44.8	48.0	54.3	61.8	65.7	68.6	72.7
35.0–39.9	416	56.7	11.7	39.9	43.1	45.2	48.8	55.9	46.0	69.0	71.6	75.6
40.0–44.9	413	56.7	11.0	39.2	42.6	45.8	49.2	56.3	64.0	68.0	71.1	74.4
45.0–49.9	433	56.6	11.2	39.0	42.6	45.6	49.4	55.9	63.7	69.6	72.8	76.2
50.0–54.9	440	55.3	11.7	37.6	41.8	44.5	47.7	54.2	62.5	65.9	69.6	74.1
55.0–59.9	403	55.1	10.8	39.2	42.5	44.4	48.5	54.8	62.2	66.7	69.5	75.0
60.0–64.9	627	52.3	10.8	34.5	38.3	41.6	45.0	52.1	69.2	63.3	66.3	70.4
65.0–69.9	886	49.8	10.5	33.4	37.2	39.6	43.0	49.2	56.7	60.1	62.4	68.1
70.0–74.9	626	47.8	10.8	30.8	34.6	36.9	40.6	74.5	54.4	59.1	62.0	66.8
MALES WITH LARGE FRAMES												
18.0–24.9	431	55.7	12.2	37.6	40.8	43.0	47.3	54.6	62.5	67.0	71.6	76.7
25.0–29.9	305	60.3	12.0	42.6	45.7	48.4	52.6	60.4	67.3	72.8	75.8	81.2
30.0–34.9	230	62.8	13.4	44.2	46.9	49.2	53.3	62.6	70.6	75.3	78.8	84.0
35.0–39.9	203	61.6	13.3	43.2	46.0	48.9	51.8	59.9	70.3	76.6	79.4	82.8
40.0–44.9	204	61.8	12.3	44.9	47.4	49.6	53.2	60.0	69.8	74.4	79.4	83.7
45.0–49.9	214	61.1	13.0	42.9	46.3	48.1	52.4	59.6	67.5	71.1	74.9	86.4
50.0–54.9	214	60.5	12.8	41.8	46.0	47.8	51.6	59.4	67.6	72.5	77.6	85.4
55.0–59.9	198	60.2	12.0	42.3	45.0	47.9	52.9	59.8	66.9	71.8	75.3	83.8
60.0–64.9	311	57.9	12.1	38.9	43.9	46.8	50.1	57.5	65.8	59.0	71.8	77.4
65.0–69.9	439	54.5	12.7	35.6	39.4	41.7	46.0	53.7	62.7	66.9	70.7	75.6
70.0–74.9	310	52.0	12.4	33.2	38.3	40.3	43.6	51.6	59.0	63.8	67.2	72.2

From Frisancho AR: *Anthropometric standards for the assessment of growth and nutritional status*, Ann Arbor, 1990, The University of Michigan Press, p. 62.

Note: Values for males aged 18 years and older have been adjusted for bone area by subtracting 10.0 cm² from the calculated mid upper arm muscle area.

TABLE E2.4

Means, Standard Deviations, and Percentiles of Upper Arm Fat Area (cm²) by Age for Males and Females 1 to 74 Years

| Age (yr) | N | Mean | SD | Percentiles | | | | | | | | |
				5	10	15	25	50	75	85	90	95
MALES												
1.0–1.9	681	7.5	2.2	4.5	4.9	5.3	5.9	7.4	8.9	9.6	10.3	11.7
2.0–2.9	672	7.4	2.3	4.2	4.8	5.1	5.8	7.3	8.6	9.7	10.6	11.6
3.0–3.9	715	7.6	2.4	4.5	5.0	5.4	5.9	7.2	8.6	9.7	10.6	11.8
4.0–4.9	707	7.3	2.5	4.1	4.7	5.2	5.7	6.9	8.5	9.3	10.0	11.4
5.0–5.9	676	7.4	3.1	4.0	4.5	4.9	5.5	6.7	8.3	9.8	10.9	12.7
6.0–6.9	298	7.7	4.1	3.7	4.3	4.6	5.2	6.7	8.6	10.3	11.2	15.2
7.0–7.9	312	8.1	4.2	3.8	4.3	4.7	5.4	7.1	9.6	11.6	12.8	15.5
8.0–8.9	296	8.9	5.0	4.1	4.8	5.1	5.8	7.6	10.4	12.4	15.6	18.6
9.0–9.9	322	10.1	6.2	4.2	4.8	5.4	6.1	8.3	11.8	15.8	18.2	21.7
10.0–10.9	333	12.0	7.3	4.7	5.3	5.7	6.9	9.8	14.7	18.3	21.5	37.0
11.0–11.9	324	13.6	9.4	4.9	5.5	6.2	7.3	10.4	16.9	22.3	26.0	32.5
12.0–12.9	348	13.9	9.6	4.7	5.6	6.3	7.6	11.3	15.8	21.1	27.3	35.0
13.0–13.9	350	13.0	9.2	4.7	5.7	6.3	7.6	10.1	14.9	21.2	25.4	32.1
14.0–14.9	358	13.3	10.2	4.6	5.6	6.3	7.4	10.1	15.9	19.5	25.5	31.8
15.0–15.9	356	12.8	9.0	5.6	6.1	6.5	7.3	9.6	14.6	20.2	24.5	31.3
16.0–16.9	350	13.9	9.5	5.6	6.1	6.9	8.3	10.5	16.6	20.6	24.8	33.5
17.0–17.9	337	12.9	8.9	5.4	6.1	6.7	7.4	9.9	15.6	19.7	23.7	28.9
18.0–24.9	1752	16.9	10.8	5.5	6.9	7.7	9.2	13.9	21.5	26.8	30.7	37.2
25.0–29.9	1250	18.8	11.6	6.0	7.3	8.4	10.2	16.3	23.9	29.7	33.3	40.4
30.0–34.9	940	20.4	11.4	6.2	8.4	9.7	11.9	18.4	25.6	31.6	34.8	41.9
35.0–39.9	832	20.1	10.5	6.5	8.1	9.6	12.8	18.8	25.2	29.6	33.4	39.4
40.0–44.9	828	20.4	11.2	7.1	8.7	9.9	12.4	18.0	25.3	30.1	35.3	42.1
45.0–49.9	867	20.1	11.0	7.4	9.0	10.2	12.3	18.1	24.9	29.7	33.7	40.4
50.0–54.9	879	19.4	10.3	7.0	8.6	10.1	12.3	17.3	23.9	29.0	32.4	40.0
55.0–59.9	807	19.2	10.2	6.4	8.2	9.7	12.3	17.4	23.8	28.4	33.3	39.1
60.0–64.9	1259	19.1	10.2	6.9	8.7	9.9	12.1	17.0	23.5	28.3	31.8	38.7
65.0–69.9	1773	18.0	9.8	5.8	7.4	8.5	10.9	16.5	22.8	27.2	30.7	36.3
70.0–74.9	1250	17.5	9.4	6.0	7.5	8.9	11.0	15.9	22.0	25.7	29.1	34.9

TABLE E2.4

Means, Standard Deviations, and Percentiles of Upper Arm Fat Area (cm²) by Age for Males and Females 1 to 74 Years (*Continued*)

| Age (yr) | N | Mean | SD | Percentiles | | | | | | | | |
				5	10	15	25	50	75	85	90	95
FEMALES												
1.0–1.9	622	7.3	2.3	4.1	4.6	5.0	5.6	7.1	8.6	9.5	10.4	11.7
2.0–2.9	614	7.7	2.3	4.4	5.0	5.4	6.1	7.5	9.0	10.0	10.8	12.0
3.0–3.9	651	7.8	2.5	4.3	5.0	5.4	6.1	7.6	9.2	10.2	10.8	12.2
4.0–4.9	680	8.0	2.6	4.3	4.9	5.4	6.2	7.7	9.3	10.4	11.3	12.8
5.0–5.9	672	8.5	3.4	4.4	5.0	5.4	6.3	7.8	9.8	11.3	12.5	14.5
6.0–6.9	296	8.7	3.9	4.5	5.0	5.6	6.2	8.1	10.0	11.2	13.3	16.5
7.0–7.9	329	9.8	4.5	4.8	5.5	6.0	7.0	8.8	11.0	13.2	14.7	19.0
8.0–8.9	275	11.3	6.5	5.2	5.7	6.4	7.2	9.8	13.3	15.8	18.0	23.7
9.0–9.9	321	13.1	7.3	5.4	6.2	6.8	8.1	11.5	15.6	18.8	22.0	27.5
10.0–10.9	329	14.1	7.7	6.1	6.9	7.2	8.4	11.9	18.0	21.5	25.3	29.9
11.0–11.9	302	16.3	9.7	6.6	7.5	8.2	9.8	13.1	19.9	24.4	28.2	36.8
12.0–12.9	323	16.9	8.9	6.7	8.0	8.8	10.8	14.8	20.8	24.8	29.4	34.0
13.0–13.9	360	19.1	11.0	6.7	7.7	9.4	11.6	16.5	23.7	28.7	32.7	40.8
14.0–14.9	370	20.4	11.0	8.3	9.6	10.9	12.4	17.7	25.1	29.5	34.6	41.2
15.0–15.9	309	20.7	11.4	8.6	10.0	11.4	12.8	18.2	24.4	29.2	32.9	44.3
16.0–16.9	343	23.5	10.9	11.3	12.8	13.7	15.9	20.5	28.0	32.7	37.0	46.0
17.0–17.9	291	23.9	13.0	9.5	11.7	13.0	14.6	21.0	29.5	33.5	38.0	51.6
18.0–24.9	2588	25.2	13.4	10.0	12.0	13.5	16.1	21.9	30.6	37.2	42.0	51.6
25.0–29.9	1921	28.1	14.7	11.0	13.3	15.1	17.7	24.5	34.8	42.1	47.1	57.5
30.0–34.9	1619	31.6	16.1	12.2	14.8	17.2	20.4	28.2	39.0	46.8	52.3	64.5
35.0–39.9	1453	33.6	16.8	13.0	15.8	18.0	21.8	29.7	41.7	49.2	55.5	64.9
40.0–44.9	1390	34.3	16.2	13.8	16.7	19.2	23.0	31.3	42.6	51.0	56.3	64.5
45.0–49.9	961	36.0	17.2	13.6	17.1	19.8	24.3	33.0	44.4	52.3	58.4	68.8
50.0–54.9	1004	36.7	15.9	14.3	18.3	21.4	25.7	34.1	45.6	53.9	57.7	65.7
55.0–59.9	879	37.6	17.7	13.7	18.2	20.7	26.0	34.5	46.4	53.9	59.1	69.7
60.0–64.9	1389	37.1	16.0	15.3	19.1	21.9	26.0	34.8	45.7	51.7	58.3	68.3
65.0–69.9	1946	34.7	15.1	13.9	17.6	20.0	24.1	32.7	42.7	49.2	53.6	62.4
70.0–74.9	1463	32.9	14.6	13.0	16.2	18.8	22.7	31.2	41.0	46.4	51.4	57.7

From Frisancho AR: *Anthropometric standards for the assessment of growth and nutritional status*, Ann Arbor, 1990, The University of Michigan Press, p. 52.

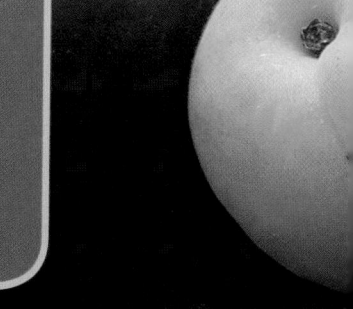

Arm Anthropometry Nomogram for Children

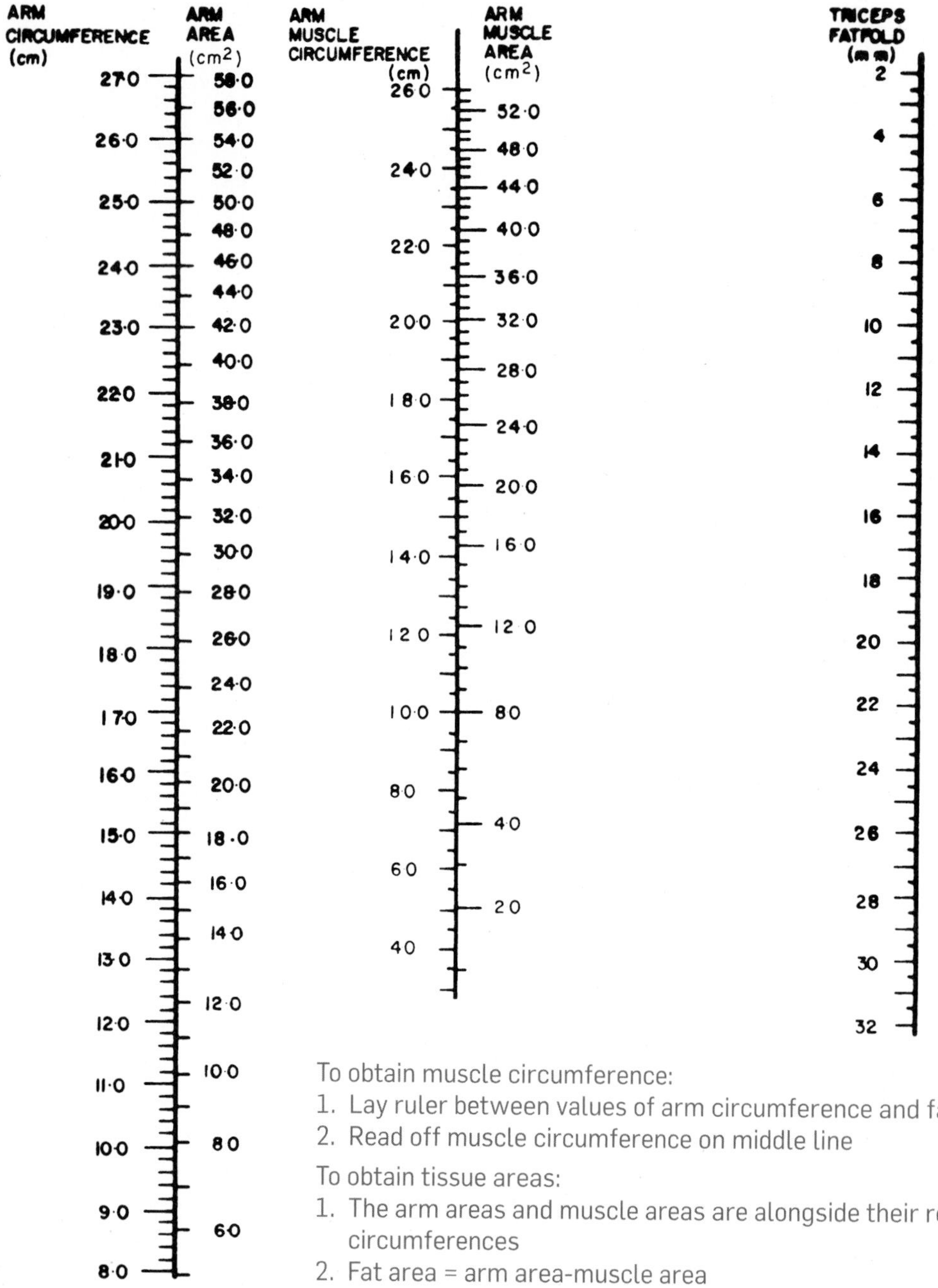

ARM CIRCUMFERENCE (cm)	ARM AREA (cm²)	ARM MUSCLE CIRCUMFERENCE (cm)	ARM MUSCLE AREA (cm²)	TRICEPS FATFOLD (mm)
27·0	58·0	26·0		2
	56·0		52·0	
26·0	54·0		48·0	4
	52·0	24·0	44·0	
25·0	50·0			6
	48·0		40·0	
24·0	46·0	22·0		8
	44·0		36·0	
23·0	42·0	20·0	32·0	10
	40·0		28·0	
22·0	38·0	18·0		12
	36·0		24·0	
21·0	34·0	16·0	20·0	14
20·0	32·0			16
	30·0	14·0	16·0	
19·0	28·0			18
18·0	26·0	12·0	12·0	20
	24·0			
17·0	22·0	10·0	8·0	22
16·0	20·0			24
		8·0	4·0	
15·0	18·0			26
	16·0	6·0		
14·0	14·0		2·0	28
13·0	12·0	4·0		30
12·0				32
11·0	10·0			
10·0	8·0			
9·0	6·0			
8·0				

To obtain muscle circumference:
1. Lay ruler between values of arm circumference and fatfold
2. Read off muscle circumference on middle line

To obtain tissue areas:
1. The arm areas and muscle areas are alongside their respective circumferences
2. Fat area = arm area-muscle area

Arm Anthropometry Nomogram for Adults

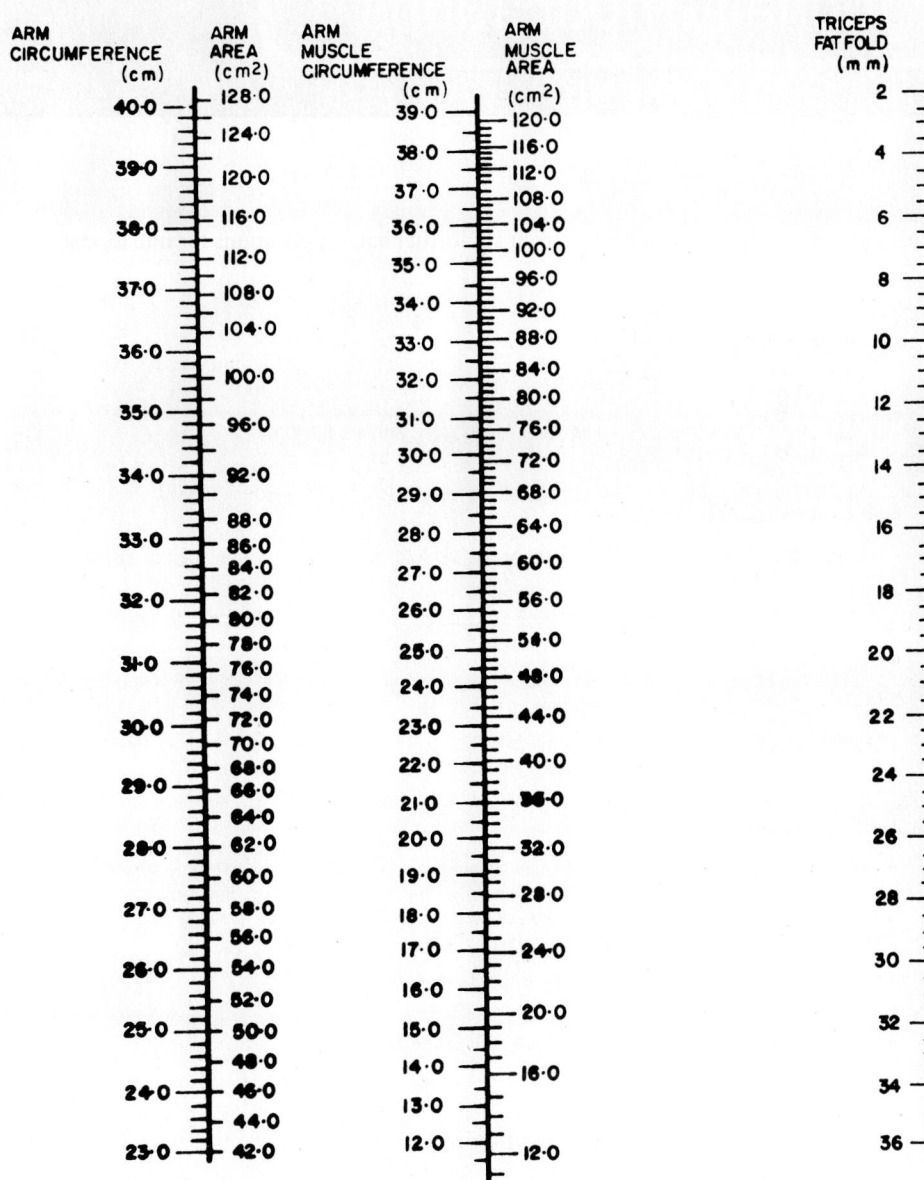

To obtain muscle circumference:
1. Lay ruler between value of arm circumference and fatfold
2. Read off muscle circumference on middle line

To obtain tissue areas:
1. The arm area and muscle area are alongside their respective circumferences
2. Fat area = arm area-muscle area

Appendix F—Routine Laboratory Tests With Nutritional Implications

This table presents a partial listing of some uses of commonly performed lab tests in adults that have implications for nutritional problems.

Laboratory Test Acceptable Range Description

Hematology		
Red blood cell (RBC) count	Male: 4.7–6.1 million/L Female: 4.2–5.4 million/L	Number of RBC; aids anemia diagnosis.
Hemoglobin (Hb)	Male: 13.5–17.5 g/dL Female: 12.0–16.0 g/dL	Hemoglobin content of RBC; aids anemia diagnosis.
Hematocrit (Hct)	Male: 42%–52% Female: 37%–47%	Percentage RBC in total blood volume; aids anemia diagnosis.
Mean corpuscular volume (MCV)	80–95 fL	RBC size; helps to distinguish between microcytic and macrocytic anemias.
Mean corpuscular hemoglobin concentration (MCHC)	32%–36% Hb/cell	Hb concentration within RBCs; helps to distinguish iron-deficiency anemia.
White blood cell (WBC) count	5000–10,000/mm³	Number of WBC; general assessment of immunity; infection Creatine kinase (CK)/creatine phosphokinase (CPK)
Blood Chemistry **Serum Proteins**		
• Total protein	6.4–8.3 g/dL	Protein levels are not specific to disease or highly sensitive; they can reflect poor protein intake, illness or infections, changes in hydration or metabolism, pregnancy, or medications.
• Albumin	3.5–5.0 g/dL	May reflect illness or PEM; slow to respond to improvement or worsening of disease.
• Transferrin	215–365 mg/dL; 60 yr: 180–380 mg/dL	May reflect illness, PEM, or iron deficiency; slightly more sensitive to changes than albumin.
• Prealbumin (transthyretin)	13–36 mg/dL	May reflect illness or PEM; more responsive to health status changes than albumin or transferrin.
• C-reactive protein	<1 mg/dL	Indicator of inflammation or disease.
Serum Enzymes		
• Creatine kinase (CK)/creatine phosphokinase (CPK)	Male: 55–170 U/L Female: 30–135 U/L	Different forms of CK are found in muscle, brain, and heart. High levels in blood may indicate heart attack, brain tissue damage, or skeletal muscle injury.
• Lactic dehydrogenase (LDH)	100–190 U/L	LDH is found in many tissues. Specific types may be elevated after heart attack, lung damage, or liver disease.
• Alkaline phosphatase	30–120 U/L	Found in many tissues; often measured to evaluate liver function.
• Aspartate aminotransferase (AST, formerly SGOT)	0–35 U/L	Usually monitored to assess liver damage; elevated in most liver diseases. Levels are somewhat increased after muscle injury.
• Alanine aminotransferase (ALT, formerly SGPT)	4–36 U/L	Usually monitored to assess liver damage; elevated in most liver diseases. Levels are somewhat increased after muscle injury.

Laboratory Test Acceptable Range Description

Serum Electrolytes		
• Sodium	136–145 mEq/L	Helps to evaluate hydration status or neuromuscular, kidney, and adrenal functions.
• Potassium	3.5–5.0 mEq/L	Helps to evaluate acid-base balance and kidney function; can detect potassium imbalances.
• Chloride	98–106 mEq/L	Helps to evaluate hydration status and detect acid-base and electrolyte imbalances.
Other		
• Glucose	74–106 mg/dL	Detects risk of glucose intolerance, diabetes mellitus, and hypoglycemia; helps to monitor diabetes treatment.
• Glycosylated hemoglobin (Hb A1c)	4.0%–5.9% of Hb	Used to monitor long-term blood glucose control (approximately 1 to 3 months prior).
• Blood urea nitrogen (BUN)	10–20 mg/dL	Primarily used to monitor kidney function; value is altered by liver failure, dehydration, or shock.
• Uric acid	Male: 4.0–8.5 mg/dL Female: 2.7–7.3 mg/dL	Used for detecting gout or changes in kidney function; levels affected by age and diet; varies among different ethnic groups.
• Creatinine (serum or plasma)	Male: 0.6–1.2 mg/dL Female: 0.6–1.2 mg/dL	Used to monitor renal function.

Note: L = microliter; dL = deciliter; fL = femtoliter; ng = nanogram; U/L = units per liter; mEq = milliequivalents.

Source: KD Pagana and TJ Pagana. Manual of Diagnostic and Laboratory Tests, 4th ed. (St Louis: Mosby, 2010).

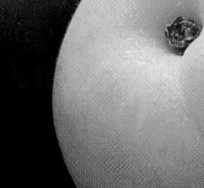

Appendix G—Normal Values for Physical Examination

Vital Signs

Temperature
Rectal: C = 37.6°/F = 99.6°
Oral: C = 37°/F = 98.6° (±10)
Axilla: C = 37.4°/F = 97.6°

Blood Pressure: average 120/80 mmHg

Heart Rate (beats per minute)

Age	At Rest Awake	At Rest Asleep	Exercise or Fever
Newborn	100–180	80–160	<220
1 week–3 months	100–220	80–200	<220
3 months–2 years	80–150	70–120	<200
2–10 years	70–110	60–90	<200
11 Years–Adult	55–90	50–90	<200

Respiratory Rate (breaths per minute)

Age	Respirations
Newborn	35
1–11 months	30
1–2 years	25
3–4 years	23
5–6 years	21
7–8 years	20
9–11 years	19
12–13 years	19
14–15 years	18
16–17 years	17
17–18 years	16–18
Adult	12–20

Cardiac Exam: carotid pulses equal in rate, rhythm, and strength; normal heart sounds; no murmurs present

HEENT Exam (head, eyes, ears, nose, throat)
Mouth: pink, moist, symmetrical; mucosa pink, soft, moist, smooth
Gums: pink, smooth, moist; may have patchy pigmentation
Teeth: smooth, white, shiny
Tongue: medium red or pink, smooth with free mobility, top surface slightly rough

Eyes: pupils equal, round, reactive to light and accommodation
Ears: tympanic membrane taut, translucent, pearly gray; auricle smooth without lesions; meatus not swollen or occluded; cerumen dry (tan/light yellow) or moist (dark yellow/brown)
Nose: external nose symmetrical, nontender without discharge; mucosa pink; septum at the midline
Pharynx: mucosa pink and smooth
Neck: thyroid gland, lymph nodes not easily palpable or enlarged

Lungs: chest contour symmetrical; spine straight without lateral deviation; no bulging or active movement within the intercostal spaces during breathing; respirations clear to auscultation and percussion

Peripheral Vascular: normal pulse graded at 3+, which indicates that pulse is easy to palpate and not easily obliterated; pulses equal bilaterally and symmetrically

Neurological: normal orientation to people, place, time, with appropriate response and concentration

Skin: warm and dry to touch; should lift easily and return back to original position indicating normal turgor and elasticity

Abdomen: umbilicus flat or concave positioned midway between xyphoid process and symphysis pubis; bowel motility notes normal air and fluid movement every 5–15 seconds; graded as normal, audible, absent, hyperactive, or hypoactive

Appendix H—Nutritional Deficiencies Revealed by Physical Examination

TABLE H.1

Nutrition Deficiencies Revealed by Physical Examination[3,8,9,12,18]

Deficient Nutrient	Findings
General Survey	
Protein, calories	Loss of weight, muscle, or fat stores; growth retardation, infection
Protein, thiamine	Edema (ankles and feet) (rule out sodium and water retention, pregnancy, protein-losing enteropathy)
Obesity	Excessive fat stores
Vitamin A	Poor growth
Iron	Anemia, fatigue
Skin	
Protein, vitamin C, zinc	Poor wound healing, pressure ulcers
Fat, vitamin A	Xerosis (rule out environmental, lack of hygiene, aging, uremia, hypothyroidism)
Vitamin C	Follicular hyperkeratosis
Niacin	Mosaic dermatitis (plaques of skin in center, peeling at periphery on shins)
Zinc	Slow wound healing
Vitamin K or C	Red, swollen skin lesions
Dehydration (fluid)	Delayed wound healing, acneiform rash, skin lesions, hair loss
	Excessive bleeding, petechiae, ecchymoses; small red, purple, black or blue, hemorrhagic spots
	Poor skin turgor
Nails	
Iron	Koilonychia (rule out cardiopulmonary disease)
Protein deficiency	Dull, lusterless with transverse ridging across nail plate
Vitamin A, C	Pale, poor blanching, irregular, mottled
Protein, calories	Bruising, bleeding
Vitamin C	Splinter hemorrhages
Hair	
Protein	Hair lacks shine, luster (cause may be environmental or chemical)
	Thin, sparse (fine, silky, and sparse with wide gaps between hairs)
Protein, copper	Dyspigmentation (lightening of normal hair color; consider if hair is bleached or dyed)
	Flag sign (alternating bands of light and dark hair in young children): rare
	Easily plucked
Copper	Corkscrew hair (Menkes syndrome)
Face	
Protein	Diffuse depigmentation, swelling
	Moon face (rounded cheeks with pursed mouth, seen in preschoolers)
Calcium	Facial paresthesias
Eyes	
Iron, folate, or vitamin B$_{12}$	Pale conjunctivae (anemia)
Vitamin A	Bitot's spots (more common in children)
Pyridoxine, niacin, riboflavin	Corneal xerosis
Hyperlipidemia	Keratomalacia
	Angular palpebritis
	Corneal arcus, xanthelasma
Nose	
Riboflavin, niacin, pyridoxine	Seborrhea on nasolabial area, nose bridge, eyebrows, and backs of ears (rule out poor hygiene)

(continued)

Nutrition Deficiencies Revealed by Physical Examination[3,8,9,12,18] *(Continued)*

Deficient Nutrient	Findings
Lips and mouth	
Niacin, riboflavin	Cheilosis
	Angular scars
Riboflavin, pyridoxine, niacin, iron	Angular stomatitis
Tongue	
Niacin, riboflavin, folic acid, iron, B₁₂	Atrophic filiform papillae
	Glossitis
Zinc	Taste atrophy
Riboflavin	Magenta tongue
Teeth	
Excess sugar, vitamin C	Edentia, caries
Fluorosis	Mottled
Gums	
Vitamin C	Spongy, bleeding, receding
Neck	
Iodine	Enlarged thyroid
Protein, bulimia	Enlarged parotids (bilateral)
Excess fluid	Venous distention, pulsations
Thorax	
Protein, calories	Decreased muscle mass and strength, shortness of breath, fatigue; decreased pulmonary function
Cardiac system	
Thiamine	Heart failure
Gastrointestinal system	
Protein, calories, zinc, vitamin C	Poor wound healing
Protein	Hepatomegaly
Urinary tract	
Dehydration	Dark, concentrated urine
Overhydration	Light, dilute urine
Musculoskeletal system	
Vitamin D, calcium	Rickets, osteomalacia
Vitamin D	Persistently open anterior fontanel (after age 18 months), craniotabes (softening of skull across back and sides before age 1 year)
	Epiphyseal enlargement (painless) at wrist, knees, and ankles
	Pigeon chest and Harrison's sulcus (horizontal depression on lower chest border)
Protein	Emaciation, muscle wasting, swelling, pain, pale hair patches
Vitamin C	Swollen, painful joints
Thiamine	Pain in thighs, calves
Nervous system	
Protein	Psychomotor changes (listless, apathetic)
	Mental confusion
Thiamin, B₆	Weakness, confusion, depressed reflexes, paresthesias, sensory loss, calf tenderness
Niacin, vitamin B₁₂	Dementia
Calcium, magnesium	Tetany

Appendix I—Evidenced-Based Analysis Process

Evidenced-based practice depends on the practitioner's ability to analyze the research literature on relevant topics. The following five steps were created by the ADA to help carry out the process of evidence analysis. Some if not all of these steps can be used by individuals to help critique research that has been published in journals or presented at conferences. The steps and several of the rubrics that are used in the process are presented below.

Step 1. Ask Good Questions

This first step is choosing the topics to investigate. The ***Anticipated Patient Outcomes*** of the nutrition care process serve as the context for the formulation of questions for evidence analysis. The general question posed by evidence analysis is: What evidence suggests that there is some association between an intervention or assessment method and some expected outcomes?

Step 2. Conduct Literature Review for the Question

Once the question has been chosen, a systematic search of the literature is conducted to find evidence related to the question. The research studies and reports are classified by type of evidence. Reports are classified as primary sources (and by their type of study design) or secondary sources, which are systematic reviews based on original research.

Step 3. Critically Appraise Each Report

Each report is reviewed for relevance to the question and critiqued for its scientific validity. The following rubrics are used to organize the information.

Evidence Worksheet[1]

Citation:	
Study Design:	
Class:	Based on classes of evidence reports
Quality Rating:	+, 0, − Based on quality criteria checklist*
Research Purpose:	
Inclusion Criteria:	
Exclusion Criteria:	
Description of Study Protocol:	**Recruitment** **Design** **Blinding used (if applicable)** **Intervention (if applicable)** **Statistical Analysis**
Data Collection Summary:	**Timing of Measurements** **Dependent Variables** • Variable 1: brief description (how measured?) • Variable 2: brief description (how measured?) • etc. **Independent Variables** **Control Variables**
Description of Actual Data Sample:	**Initial N:** (e.g., 731 (298 males, 433 females)) **Attrition (final N):** **Age:** **Ethnicity:** **Other relevant demographics:** **Anthropometrics** (e.g., were groups same or different on important measures) **Location:**

[1]*Source* for Evidence Worksheet: Reprinted from Appendix 7: Evidence Abstract Worksheet Template, pp. 75–76, of *Evidence Analysis Manual: Steps in the ADA Evidence Analysis Process*, ISBN 978-0-88091-429-1, published by the American Dietetic Association.

Summary of Results:	Variables	Treatment Group Measures and Confidence Intervals	Control Group Measures and Confidence Intervals	Statistical Significance of Group Difference
	Dep var 1	Mean, CI. e.g., 4.5 ± 2.2	Mean, CI. e.g., 1.5 ± 2.0	Stat signif difference between groups e.g., p = 0.002
	Dep var 2			
	Etc.			
	Other Findings			
Author Conclusion:				
Review Comments:				

*Relevance Questions rubric

Relevance & Validity Questions[2]

Possible responses for each question include: "Yes," "No," "Unclear," and "NA."

Relevance Questions

1. Would implementing the studied intervention or procedure (if found successful) result in improved outcomes for the patients/clients/population group? (NA for some Epi studies)

2. Did the authors study an outcome (dependent variable) or topic that the patients/clients/population group would care about?

3. Is the focus of the intervention or procedure (independent variable) or topic of study a common issue of concern to dietetics practice?

4. Is the intervention or procedure feasible? (NA for some epidemiological studies)

If the answers to all of the above relevance questions are "Yes," the report is eligible for designation with a plus (+) on the Evidence Quality Worksheet, depending on answers to the following validity questions.

Validity Questions

1. **Was the *research question* clearly stated?**

   a. Was the specific intervention(s) or procedure (independent variable(s)) identified?

   b. Was the outcome(s) (dependent variable(s)) clearly indicated?

   c. Were the target population and setting specified?

2. **Was the *selection* of study subjects/patients free from bias?**

   a. Were inclusion/exclusion criteria specified (e.g., risk, point in disease progression, diagnostic or prognosis criteria), and with sufficient detail and without omitting criteria critical to the study?

   b. Were criteria applied equally to all study groups?r

   c. Were health, demographics, and other characteristics of subjects described?

   d. Were the subjects/patients a representative sample of the relevant population?

3. **Were *study groups comparable*?**

   a. Was the method of assigning subjects/patients to groups described and unbiased? (Method of randomization identified if RCT.)

   b. Were distribution of disease status, prognostic factors, and other factors (e.g., demographics) similar across study groups at baseline?

   c. Were concurrent controls used? (Concurrent preferred over historical controls.)

   d. If cohort study or cross-sectional study, were groups comparable on important confounding factors and/or were preexisting differences accounted for by using appropriate adjustments in statistical analysis?

   e. If case control study, were potential confounding factors comparable for cases and controls?

   f. If case series or trial with subjects serving as own control, this criterion is not applicable. Criterion may not be applicable in some cross-sectional studies.

   g. If diagnostic test, was there an independent blind comparison with an appropriate reference standard (e.g., "gold standard")?

4. **Was method of handling *withdrawals* described?**

   a. Were follow-up methods described and the same for all groups?

   b. Was the number, characteristics of withdrawals (i.e., dropouts, lost to follow up, attrition rate) and/or response rate (cross-sectional studies) described for each group? (Follow-up goal for a strong study is 80%.)

   c. Were all enrolled subjects/patients (in the original sample) accounted for?

   d. Were reasons for withdrawals similar across groups?

   e. If diagnostic test, was decision to perform reference test not dependent on results of test under study?

[2]*Source* for Relevance & Validity Questions: Reprinted from Appendix 8: Quality Criteria Checklist: Primary Research, pp. 77–79, of *Evidence Analysis Manual: Steps in the ADA Evidence Analysis Process*, ISBN 978-0-88091-429-1, published by the American Dietetic Association.

5. **Was *blinding* used to prevent introduction of bias?**

   a. In intervention study, were subjects, clinicians/practitioners, and investigators blinded to treatment group, as appropriate?

   b. Were data collectors blinded for outcomes assessment? (If outcome is measured using an objective test, such as a lab value, this criterion is assumed to be met.)

   c. In cohort study or cross-sectional study, were measurements of outcomes and risk factors blinded?

   d. In case control study, was case definition explicit and case ascertainment not influenced by exposure status?

   e. In diagnostic study, were test results blinded to patient history and other test results?

6. **Were *intervention*/therapeutic regimens/exposure factor or procedure and any comparison(s) described in detail? Were *intervening factors* described?**

   a. In RCT or other intervention trial, were protocols described for all regimens studied?

   b. In observational study, were interventions, study settings, and clinicians/provider described?

   c. Were the intensity and duration of the intervention or exposure factor sufficient to produce a meaningful effect?

   d. Were the amount of exposure and, if relevant, subject/patient compliance measured?

   e. Were co-interventions (e.g., ancillary treatments, other therapies) described?

   f. Were extra or unplanned treatments described?

   g. Was the information for 6d, 6e, and 6f assessed the same way for all groups?

   h. In diagnostic study, were details of test administration and replication sufficient?

7. **Were *outcomes* clearly defined and the *measurements valid and reliable*?**

   a. Were primary and secondary endpoints described and relevant to the question?

   b. Were nutrition measures appropriate to questions and outcomes of concern?

   c. Was the period of follow-up long enough for important outcome(s) to occur?

   d. Were the observations and measurements based on standard, valid, and reliable data collection instruments/tests/procedures?

   e. Was the measurement of effect at an appropriate level of precision?

   f. Were other factors accounted for (measured) that could affect outcomes?

   g. Were the measurements conducted consistently across groups?

8. **Was the *statistical analysis* appropriate for the study design and type of outcome indicators?**

   a. Were statistical analyses adequately described and the results reported appropriately?

   b. Were correct statistical tests used and assumptions of test not violated?

   c. Were statistics reported with levels of significance and/or confidence intervals?

   d. Was "intent to treat" analysis of outcomes done (and as appropriate, was there an analysis of outcomes for those maximally exposed or a dose-response analysis)?

   e. Were adequate adjustments made for effects of confounding factors that might have affected the outcomes (e.g., multivariate analyses)?

   f. Was clinical significance as well as statistical significance reported?

   g. If negative findings, was a power calculation reported to address type 2 error?

9. **Are conclusions supported by results with biases and limitations taken into consideration?**

   a. Is there a discussion of findings?

   b. Are biases and study limitations identified and discussed?

10. **Is bias due to study's <u>funding or sponsorship</u> unlikely?**

    a. Were sources of funding and investigators' affiliations described?

    b. Was there no apparent conflict of interest?

Minus/Negative (−)
If most (six or more) of the answers to the above validity questions are "No," the report should be designated with a minus (−) symbol on the Evidence Quality Worksheet.

Neutral (~)
If the answers to validity criteria questions 2, 3, 6, and 7 do not indicate that the study is exceptionally strong, the report should be designated with a neutral (~) symbol on the Evidence Quality Worksheet.

Plus/Positive (+)
If most of the answers to the above validity questions are "Yes" (including criteria 2, 3, 6, 7 and at least one additional "Yes"), the report should be designated with a plus symbol (+) on the Evidence Quality Worksheet.

Step 4. Writing the Evidence Summary
The reports are synthesized into a table summarizing the research relevant to the question.

Step 5: Grading the Conclusion Statement
A concise conclusion statement (the answer to the question) is developed and assigned a grade to indicate the overall strength or weakness of evidence informing the conclusion statement.

Strength of Evidence Elements	Grades				
	I Good/Strong	II Fair	III Limited/Weak	IV Expert Opinion Only	V Grade Not Assignable
Quality Scientific rigor/validity Considers design and execution	Studies of strong design for question Free from design flaws, bias, and execution problems	Studies of strong design for question with minor methodological concerns, OR Only studies of weaker study design for question	Studies of weak design for answering the question OR Inconclusive findings due to design flaws, bias, or execution problems	No studies available Conclusion based on usual practice, expert consensus, clinical experience, opinion, or extrapolation from basic research	No evidence that pertains to question being addressed
Consistency Of findings across studies	Findings generally consistent in direction and size of effect or degree of association, and statistical significance with minor exceptions at most	Inconsistency among results of studies with strong design, OR Consistency with minor exceptions across studies of weaker design	Unexplained inconsistency among results from different studies OR single study unconfirmed by other studies	Conclusion supported solely by statements of informed nutrition or medical commentators	NA
Quantity Number of studies Number of subjects in studies	One to several good quality studies Large number of subjects studied Studies with negative results have sufficiently large sample size for adequate statistical power	Several studies by independent investigators Doubts about adequacy of sample size to avoid Type I and Type II error	Limited number of studies Low number of subjects studied AND/OR Inadequate sample size within studies	Unsubstantiated by published research studies	Relevant studies have not been done
Clinical Impact Importance of studied outcomes Magnitude of effect	Studied outcome relates directly to the question Size of effect is clinically meaningful Significant (statistical) difference is large	Some doubt about the statistical or clinical significance of the effect	Studied outcome is an intermediate outcome or surrogate for the true outcome of interest OR Size of effect is small or lacks statistical and/or clinical significance	Objective data unavailable	Indicates area for future research
Generalizability To population of interest	Studied population, intervention, and outcomes are free from serious doubts about generalizability	Minor doubts about generalizability	Serious doubts about generalizability due to narrow or different study population, intervention, or outcomes studied	Generalizability limited to scope of experience	NA

[3]*Source* for Conclusion Grading Table: Reprinted from Appendix 17: Conclusion Grading Table, p. 88, of *Evidence Analysis Manual: Steps in the ADA Evidence Analysis Process*, ISBN 978-0-88091-429-1, published by the American Dietetic Association.

Appendix J—Enteral Formulas

The large number of enteral formulas available allows patients to meet a wide variety of medical needs. The first step in narrowing the choice of formulas is to determine the patient's ability to digest and absorb nutrients. Table J.1 lists examples of standard formulas for patients who can adequately digest and absorb nutrients, and Table J.2 provides examples of chemically defined/hydrolyzed formulas for patients with a limited ability to digest or absorb nutrients. Each formula is listed only once, although a formula may have more than one use. A high-protein formula, for example, may also be a fiber-containing formula. Tables J.3 and J.4 list modules that can be used to prepare modular formulas or enhance enteral formulas.

The information shown in this appendix reflects the literature provided by manufacturers and does not suggest endorsement by the authors. Manufacturers frequently add new formulas, discontinue old ones, and change formula composition. Consult the manufacturers' literature and websites for updates and additional examples of enteral formulas.[1] The following products are listed in this appendix:

❏ *Abbott Nutrition:* Glucerna 1.0 Cal, Jevity 1 Cal, Jevity 1.5 Cal, Nepro with Carb Steady, Optimental, Osmolite 1 Cal, Oxepa, Pivot 1.5 Cal, Polycose Powder, Promote, Promote with Fiber, Pulmocare, Suplena with Carb Steady

❏ *Nestlé Nutrition:* Compleat, Compleat Pediatric, Crucial, Diabetisource AC, Fibersource HN, Impact, Impact 1.5, Impact Glutamine, Isosource HN, MCT Oil, Microlipid, Novasource Renal, Nutren 1.0, Nutren 1.0 Fiber, Nutren 1.5, Nutren 2.0, Nutren Glytrol, Nutren Junior, Nutren Pulmonary, Nutren Replete, Nutren Replete Fiber, NutriHep, Peptamen, Peptamen Junior, Resource Beneprotein Instant Protein Powder, Vivonex Pediatric, Vivonex T.E.N.

[1]Sources for the information in this appendix: Abbott Nutrition, **www.abbottnutrition.com**, Nestlé Nutrition, **www.nestle-nutrition.com**.

TABLE J.1
Standard Formulas

Product	Volume to Meet 100% RDI[a] (mL)	Energy (kcal/mL)	Protein or Amino Acids (g/L)	Carbohydrate (g/L)	Fat (g/L)	Notes
Lactose-Free, Standard Formulas						
Compleat	1313	1.07	48	128	40	Blenderized formula, 6 g fiber/L
Nutren 1.0	1500	1.00	40	127	38	25% fat from MCT
Osmolite 1 Cal	1321	1.06	44	144	35	20% fat from MCT
Lactose-Free, Fiber-Enhanced Formulas						
Jevity 1 Cal	1321	1.06	44	155	35	14 g fiber/L
Nutren 1.0 Fiber	1500	1.00	40	127	38	14 g fiber/L
Promote with Fiber	1000	1.00	63	138	28	14 g fiber/L
Lactose-Free, High-Kcalorie Formulas						
Jevity 1.5 Cal	1500	1.50	64	216	50	22 g fiber/L
Nutren 1.5	1000	1.50	60	169	68	50% fat from MCT
Nutren 2.0	750	2.00	80	196	104	75% fat from MCT
Lactose-Free, High-Protein Formulas						
Fibersource HN	1165	1.20	53	160	39	20% fat from MCT, 10 g fiber/L
Isosource HN	1165	1.20	53	160	39	20% fat from MCT, low residue
Promote	1000	1.00	63	130	26	20% fat from MCT, low residue
Specialized Formulas: Pediatric (1 to 10 Years)						
Compleat Pediatric	Varies[b]	1.00	38	130	39	Blenderized formula, 7 g fiber/L
Nutren Junior	Varies[b]	1.00	30	110	50	21% fat from MCT

[a] RDI = Reference Daily Intakes, which are labeling standards for vitamins, minerals, and protein. Consuming 100 percent of the RDI will meet the nutrient needs of most people using the product.

[b] Depends on age of child

(continued)

TABLE J.1

Standard Formulas (*Continued*)

Product	Volume to Meet 100% RDI[a] (mL)	Energy (kcal/mL)	Protein or Amino Acids (g/L)	Carbohydrate (g/L)	Fat (g/L)	Notes
Specialized Formulas: Glucose Intolerance						
Diabetisource AC	1250	1.20	60	100	59	36% kcal from carbohydrate, 15 g fiber/L
Glucerna 1.0 Cal	1420	1.00	42	96	54	34% kcal from carbohydrate, 14 g fiber/L
Nutren Glytrol	1400	1.00	45	100	48	40% kcal from carbohydrate, 15 g fiber/L
Specialized Formulas: Immune System Support						
Impact	1500	1.00	56	130	28	Enriched with arginine, nucleic acids, and omega-3 fatty acids
Impact 1.5	1250	1.50	84	140	69	Same as above
Impact Glutamine	1000	1.30	78	150	43	Same as above, and enriched with glutamine
Specialized Formulas: Chronic Kidney Disease (CKD)						
Nepro with Carb Steady	948	1.80	81	167	96	Low potassium, low phosphorus; to be used after dialysis has been instituted
Novasource Renal	1000	2.00	74	200	100	Low in electrolytes; to be used after dialysis has been instituted
Suplena with Carb Steady	948	1.80	45	205	96	Low in protein and electrolytes; for patients with CKD (stage 3 or 4)
Specialized Formulas: Respiratory Insufficiency						
Nutren Pulmonary	1000	1.50	68	100	95	55% kcal from fat, 40% fat from MCT
Oxepa	946	1.50	63	105	94	Enriched with omega-3 fatty acids and antioxidants; for mechanically ventilated patients
Pulmocare	947	1.50	63	106	93	55% kcal from fat, 20% fat from MCT, enriched with antioxidant nutrients
Specialized Formulas: Wound Healing						
Nutren Replete	1000	1.00	62	113	34	Enhanced with vitamins and minerals; for patients recovering from surgery, burns, and pressure ulcers
Nutren Replete Fiber	1000	1.00	62	113	34	Same as above; 14 g fiber/L

Note: MCT = Medium-chain triglycerides

[a] RDI = Reference Daily Intakes, which are labeling standards for vitamins, minerals, and protein. Consuming 100 percent of the RDI will meet the nutrient needs of most people using the product.

[b] Depends on age of child

TABLE J.2

Chemically Defined/Hydrolyzed Formulas

Product	Volume to Meet 100% RDI[a] (mL)	Energy (kcal/mL)	Protein or Amino Acids (g/L)	Carbohydrate (g/L)	Fat (g/L)	Notes
Specialized Chemically Defined Formula: Hepatic Insufficiency						
NutriHep	1000	1.50	40	290	21	Free amino acids, high in branched-chain amino acids and low in aromatic amino acids
Specialized Chemically Defined Formulas: Immune System Support						
Crucial	1000	1.50	94	134	68	Enriched with arginine, omega-3 fatty acids, antioxidant nutrients, and zinc
Pivot 1.5 Cal	1500	1.50	94	172	51	Enriched with arginine, glutamine, omega-3 fatty acids, and antioxidant nutrients
Specialized Chemically Defined Formulas: Malabsorption						
Optimental	1422	1.00	51	139	28	Contains MCT and arginine, enriched with antioxidants and omega-3 fatty acids
Peptamen	1500	1.00	40	127	39	70% fat from MCT
Vivonex T.E.N.	2000	1.00	38	210	3	Powder form, 100% free amino acids, enriched with glutamine
Specialized Chemically Defined Formulas: Pediatric (1 to 10 Years)						
Peptamen Junior	Varies[b]	1.00	30	138	38	60% fat from MCT
Vivonex Pediatric	Varies[b]	0.80	24	130	24	Powder form, 100% free amino acids

Note: MCT = Medium-chain triglycerides

[a] RDI = Reference Daily Intakes, which are labeling standards for vitamins, minerals, and protein. Consuming 100 percent of the RDI will meet the nutrient needs of most people using the product.

[b] Depends on age of child

TABLE J.3

Protein and Carbohydrate Modules

Product	Major Ingredient	Energy (kcal/g)	Nutrient Content (g/100 g)
Resource Beneprotein Instant Protein Powder	Whey protein	3.6	86 g protein
Polycose Powder	Hydrolyzed cornstarch	3.8	94 g carbohydrate

TABLE J.4

Fat Modules

Product	Major Ingredient	Energy (kcal/g)	Fat Content (g/100 mL)
MCT Oil	Coconut oil	7.7	93
Microlipid	Safflower oil	4.5	50

Product	Oil (%)		Fatty Acid Content (%)					Egg Yolk Phospho-lipids	Glycerin (%)	Kcal/mL	Osmolality (mOsm/L)
	Safflower	Soybean	Linoleic	Oleic	Palmitic	Linolenic	Stearic				
Intravenous Fat Emulsion											
Intralipid 10%		10	50	26	26	9	305	1.2	2.25	1.1	260
Intralipid 20%		20	50	26	10	9	3.5	1.2	2.25	2	260
Liposyn II 10%	5	5	65.8	17.7	8.8	4.2	3.4	1.2	2.5	1.1	276
Liposyn II 20%	10	10	65.8	17.7	8.8	4.2	3.4	1.2	2.5	2.0	258
Liposyn III 10%		10	54.5	22.4	10.5	8.3	4.2	1.2	2.5	1.1	284
Liposyn III 20%		20	54.5	22.4	10.5	8.3	4.2	1.2	2.5	2.0	292
Liposyn III 30%		30	54.5	22.4	10.5	8.3	4.2	1.2	2.5	2.9	293

	Aminosyn 3.5% (HOSPIRA)	Aminosyn B 3.5% (HOSPIRA)	Aminosyn 5% (HOSPIRA)	Aminosyn B.5% (HOSPIRA)	Travasol 5.5% (Baxter)	TrophAmine 6% (B. Braun)
Crystalline Amino Acid Infusions						
Amino Acid Concentration	3.5%	3.5%	5%	5%	5.5%	6%
Nitrogen (g/100 mL)	0.55	0.54	0.79	0.77	0.925	0.93
Amino Acids (Essential) (mg/100 mL)						
Isoleucine	252	231	360	300	263	490
Leucine	329	350	470	500	340	840
Lysine	252	368	360	525	318	490
Methionine	140	60	200	86	318	200
Phenylalanine	154	104	220	149	340	290
Threonine	182	140	260	200	230	250
Tryptophan	56	70	80	100	99	129
Valine	290	175	400	250	252	470
Amino Acids (Nonessential) (mg/100 mL)						
Alanine	448	348	640	497	1140	320
Arginine	343	356	490	509	570	730
Histidine[a]	105	105	150	150	241	290
Proline	300	253	430	361	230	410
Serine	147	186	210	265		230
Taurine						15
Tyrosine	31	95	44	135	22	140
Aminoacetic Acid (Glycine)	448	175	640	250	1140	220
Glutamic Acid		258		369		300
Aspartic Acid		245		350		190
Cysteine						<14

	Aminosyn 3.5% (HOSPIRA)	Aminosyn B 3.5% (HOSPIRA)	Aminosyn 5% (HOSPIRA)	Aminosyn B.5% (HOSPIRA)	Travasol 5.5% (Baxter)	TrophAmine 6% (B. Braun)
Electrolytes (mEq/L)						
Sodium	7	16.3		19.3		5
Potassium			5.4			
Chloride					22	<3
Acetate Phosphate (mM/L)	46	25.2	86	35.9	48	56
Osmolarity (mOsm/L)	357	308	500	438	575	526
Supplied in (mL)	1000[b]	1000[c]	500[d]	500[c]	500[e]	500
			1000[d]	1000[c]	1000[e]	
					2000[e]	
Labeled Indications						
Peripheral Parenteral Nutrition	Yes	Yes	Yes	Yes	Yes	Yes
Central PN	No	No	Yes	Yes	Yes	Yes
Protein Sparing	Yes	Yes	Yes	Yes	Yes	No

	Aminosyn 7% (HOSPIRA)	Aminosyn-OF 7% (HOSPIRA)	Aminosyn B 7% (HOSPIRA)	Aminosyn 8.5% (HOSPIRA)
Amino Acid Concentration	7%	7%	7%	8.5%
Nitrogen (g/100 mL)	1.1	1.07	1.07	1.34
Amino Acids (Essential) (mg/100 mL)				
Isoleucine	510	534	462	620
Leucine	660	831	700	810
Lysine	510	475	735	624
Methionine	280	125	120	340
Phenylalanine	310	300	209	380
Threonine	370	360	280	460
Tryptophan	120	125	140	150
Valine	560	452	350	680
Amino Acids (Nonessential) (mg/100 mL)				
Alanine	900	490	695	110
Arginine	690	861	713	850
Histidine[a]	210	220	210	260
Proline	610	570	505	750
Serine	300	347	371	370
Taurine		50		
Tyrosine	44	44	189	44
Aminoacetic Acid (Glycine)	900	270	350	110
Glutamic Acid		576	517	
Aspartic Acid		370	490	
Cysteine				
Electrolytes (mEq/L)				
Sodium		3.4	31.3	
Potassium	5.4			5.4

	Aminosyn 7% (HOSPIRA)	Aminosyn-OF 7% (HOSPIRA)	Aminosyn B 7% (HOSPIRA)	Aminosyn 8.5% (HOSPIRA)
Chloride				35
Acetate Phosphate (mM/L)	105	32.5	50.3	90
Osmolarity (mOsm/L)	700	586	612	860
Supplied in (mL)	500[d]	250[g]	500[c]	500[d]
		500[g]		1000[d]
Labeled Indications				
Peripheral Parenteral Nutrition	Yes	Yes	Yes	Yes
Central PN	Yes	Yes	Yes	Yes
Protein Sparing	Yes	No	Yes	Yes

	Aminosyn II 8.5% (HOSPIRA)	Travasol 8.5% without electrolytes (Baxter)	FreAmine III 8.5% (B. Braun)
Amino Acid Concentration	8.5%	8.5%	8.5%
Nitrogen (g/100 mL)	1.3	1.43	
Amino Acids (Essential) (mg/100 mL)			
Isoleucine	561	406	590
Leucine	850	526	770
Lysine	893	492	620
Methionine	146	492	450
Phenylalanine	253	526	480
Threonine	340	356	340
Tryptophan	170	152	130
Valine	425	390	560
Amino Acids (Nonessential) (mg/100 mL)			
Alanine	844	1760	600
Arginine	865	880	810
Histidine[a]	255	372	240
Proline	614	356	950
Serine	450		500
Taurine			
Tyrosine	230	34	
Aminoacetic Acid (Glycine)	425	1760	1190
Glutamic Acid	627		
Aspartic Acid	595		
Cysteine			<20
Electrolytes (mEq/L)			
Sodium	33.3	10	
Potassium			
Chloride		34	<3
Acetate	61.1	73	72
Phosphate (mM/L)			10
Osmolarity (mOsm/L)	742	890	810

	Aminosyn II 8.5% (HOSPIRA)	Travasol 8.5% without electrolytes (Baxter)	FreAmine III 8.5% (B. Braun)
Supplied in (mL)	500[c]	500[h]	500[i]
	1000[c]	1000[h]	1000[i]
		2000[h]	
Labeled Indications			
Peripheral Parenteral Nutrition	Yes	Yes	Yes
Central PN	Yes	Yes	Yes
Protein Sparing	Yes	No	Yes

	TrophAmine 10% (B. Braun)	Aminosyn 10% (HOSPIRA)	Aminosyn-OF 10% (HOSPIRA)	Aminosyn II 10% (HOSPIRA)
Amino Acid Concentration	10%	10%	10%	10%
Nitrogen (g/100 mL)	1.55	1.57	1.52	1.53
Amino Acids (Essential) (mg/100 mL)				
Isoleucine	820	720	760	660
Leucine	1400	940	1200	1000
Lysine	820	720	677	1050
Methionine	340	400	180	172
Phenylalanine	480	440	427	298
Threonine	420	520	512	400
Tryptophan	200	160	180	200
Valine	780	800	673	500
Amino Acids (Nonessential) (mg/100 mL)				
Alanine	540	1280	698	993
Arginine	1200	980	1227	1018
Histidine[a]	480	300	312	300
Proline	680	860	812	722
Serine	380	420	495	530
Taurine	25		70	
Tyrosine	240	44	40	270
Aminoacetic Acid (Glycine)	360	1280	385	500
Glutamic Acid	500		620	738
Aspartic Acid	320		527	700
Cysteine	<16			
Electrolytes (mEq/L)				
Sodium	5		3.4	45.3
Potassium		5.4		
Chloride	<3			
Acetate Phosphate (mM/L)	97	148	46.3	71.8
Osmolarity (mOsm/L)	875	1000	829	873
Supplied in (mL)	500[f]	500[d]	1000[i]	500[c]
		1000[d]		1000[c]
Labeled Indications				
Peripheral Parenteral Nutrition	Yes	Yes	Yes	Yes
Central PN	Yes	Yes	Yes	Yes
Protein Sparing	No	Yes	No	Yes

	Travasol 10% (Baxter)	FreAmine III 10% (B. Braun)	Novamine (HOSPIRA)	Aminosyn 15% (HOSPIRA)	Aminosyn II15% (HOSPIRA)
Amino Acid Concentration	10%	10%	11.4%	15%	15%
Nitrogen (g/100 mL)	1.65	1.53	1.8	2.37	2.3
Amino Acids (Essential) (mg/100 mL)					
Isoleucine	600	690	570	749	990
Leucine	730	910	790	1040	1500
Lysine	580	730	900	1180	1575
Methionine	400	530	570	749	258
Phenylalanine	560	560	790	1040	447
Threonine	420	400	570	749	600
Tryptophan	180	150	190	250	300
Valine	580	660	730	960	750
Amino Acids (Nonessential) (mg/100 mL)					
Alanine	2070	710	1650	2170	1490
Arginine	1150	950	1120	1470	1527
Histidine[a]	480	280	680	894	450
Proline	680	1120	680	894	1083
Serine	500	590	450	592	795
Taurine					
Tyrosine	40		30	39	405
Aminoacetic Acid (Glycine)	1030	1400	790	1040	750
Glutamic Acid			570	749	1107
Aspartic Acid			660	434	1050
Cysteine		<24			
Electrolytes (mEq/L)					
Sodium		10			62.7
Potassium					
Chloride	40	, 3			
Acetate	87	≈<89	114	151	107.6
Phosphate (mM/L)		10			
Osmolarity (mOsm/L)	1000	≈<	1057	1388	1300
	950				
Supplied in (mL)	250[l,m]	500[i]	500[n]	500[n]	2000[o]
	500[l,m]	1000[i]	1000[n]	1000[n]	
	1000[l,m]				
	2000[l,m]				
Labeled Indications					
Peripheral Parenteral Nutrition	Yes	Yes	Yes	Yes	Yes
Central PN	Yes	Yes	Yes	Yes	Yes
Protein Sparing	Yes	No	Yes	No	No

[a] Histidine is considered an essential amino acid in infants and in renal failure.
[b] With 7 mEq/L sodium from the antioxidant sodium hydrosulfite.
[c] Includes 20 mg/dL sodium hydrosulfite.
[d] Includes 5.4 mEq/L potassium from the antioxidant potassium metabisulfite.
[e] With ≈ 3 mEq/L sodium bisulfite.
[f] With <50 mg sodium metabisulfite per 100 mL.
[g] From the antioxidant sodium hydrosulfite.
[h] With 3 mEq/L sodium bisulfite.

[i] With <0.1 g sodium bisulfite per 100 mL.
[j] With 230 mg sodium hydrosulfite per 100 mL.
[k] Potassium derived from the antioxidant potassium metabisulfite.
[l] Acetate in Viaflex container ≈ 60 mEq/L; osmolarity is 970 mOsm/L.
[m] Sizes also come in Viaflex containers.
[n] With 30 mg sodium metabisulfite.
[o] With 60 mg sodium hydrosulfite per 100 mL.

	Aminosyn-RF 5.2% (HOSPIRA)	Aminess 5.2% (Baxter)	5.4% NephrAmine (B. Braun)	RenAmin (Baxter)
Amino Acid Formulations for Renal Failure				
Amino Acid Concentration	5.2%	5.2%	5.4%	6.5%
Nitrogen (g/100 mL)	0.79	0.66	0.65	1
Amino Acids (Essential) (mg/100 mL)				
Isoleucine	462	525	560	500
Leucine	726	825	880	600
Lysine	535	600	640	450
Methionine	726	825	880	500
Phenylalanine	726	825	880	490
Threonine	330	375	400	380
Tryptophan	165	188	200	160
Valine	528	600	640	820
Histidine	429	412	250	420
Amino Acids (Nonessential) (mg/100 mL)				
Cysteine			<20	
Arginine	600			630
Alanine				560
Proline				350
Glycine				300
Serine				300
Tyrosine				40
Electrolytes (mEq/L)				
Sodium			5	
Acetate	≈<105	50	≈<44	60
Potassium	5.4			
Chloride			<3	31
Osmolarity (mOsm/L)	475	416	435	600 (mOsm/L)
Supplied in (mL)	300[a]	400[b]	250[c]	250[d]
				500[d]

[a]With 60 mg potassium metabisulfite per 100 mL.
[b]In 500 mL bottle.
[c]With <; 0.05 g sodium bisulfite per 100 mL.
[d]With <3 mEq sodium bisulfite.

	STRESS FORMULATION		HEPATIC FORMULATION	
	4% BranchAmin (Baxter)	FreAmine HBC 6.9% (B. Braun)	AminosynHBC 7% (HOSPIRA)	HepatAmine (B. Braun)
Amino Acid Formulation for High Metabolic Stress and in Hepatic Failure/Hepatic Encephalopathy				
Amino Acid Concentration	4%	6.9%	7%	8%
Nitrogen (g/100 mL)	0.443	0.97	1.12	1.2
Amino Acids (Essential) (mg/100 mL)				
Isoleucine	1380	760	789	900
Leucine	1380	1370	1576	1100
Lysine		410	265	610
Methionine		250	206	100
Phenylalanine		320	228	100
Threonine		200	272	450
Tryptophan		90	88	66
Valine	1240	880	789	840
Amino Acids (Nonessential) (mg/100 mL)				
Alanine		400	660	770
Arginine		580	507	600
Histidine[a]		160	154	240
Proline		630	448	800
Serine		330	221	500
Tyrosine			33	
Glycine		330	660	900
Cysteine		<20		<20
Electrolytes (mEq/L)				
Sodium		10	7[b]	10
Chloride	≈	<3		<3
Acetate		≈ 57	72	≈ 62
Phosphate (mM/L)				10
Osmolarity (mOsm/L)	316	620	665	785
Supplied in (mL)	500	750[c,d]	500[b]	500[c]
			1000[b]	
Labeled Indications				
Peripheral Parenteral Nutrition	Yes[e]	Yes	Yes	Yes
Central PN	Yes[e]	Yes	Yes	Yes

[a] Histidine is considered an essential amino acid in infants and in renal failure.

[b] With 60 mg sodium hydrosulfite.

[c] With <100 mg sodium bisulfite/100 mL.

[d] In 100 mL bottles.

[e] Must be admixed with a complete amino acid injection.

Appendix L1—Choose Your Foods: Exchange Lists for Diabetes

The Exchange System

The exchange system sorts foods into groups by their proportions of carbohydrate, fat, and protein (Table L1.1). These groups may be organized into several exchange lists of foods (Tables L1.2 through L1.12). For example, the carbohydrate group includes these exchange lists:

❑ Starch

❑ Fruits

❑ Milk (fat-free, reduced-fat, and whole)

❑ Sweets, Desserts, and Other Carbohydrates

❑ Nonstarchy Vegetables

Then any food on a list can be "exchanged" for any other on that same list. Another group for alcohol has been included as a reminder that these beverages often deliver substantial carbohydrate and kcalories, and therefore warrant their own list.

Serving Sizes

The serving sizes have been carefully adjusted and defined so that a serving of any food on a given list provides roughly the same amount of carbohydrate, fat, and protein, and, therefore, total energy. Any food on a list can thus be exchanged, or traded, for any other food on the same list without significantly affecting the diet's energy-nutrient balance or total kcalories. For example, a person may select 17 small grapes or ½ large grapefruit as one fruit exchange, and either choice would provide roughly 15 grams of carbohydrate and 60 kcalories. A whole grapefruit, however, would count as 2 fruit exchanges.

To apply the system successfully, users must become familiar with the specified serving sizes. A convenient way to remember the serving sizes and energy values is to keep in mind a typical item from each list (review Table L1.1).

TABLE L1.1

The Food Lists

Lists	Typical Item/Portion Size	Carbohydrate (g)	Protein (g)	Fat (g)	Energy[a] (kcal)
Carbohydrates					
Starch[b]	1 slice bread	15	0–3	0–1	80
Fruits	1 small apple	15	—	—	60
Milk					
Fat-free, low-fat, 1%	1 c fat-free milk	12	8	0–3	100
Reduced-fat, 2%	1 c reduced-fat milk	12	8	5	120
Whole	1 c whole milk	12	8	8	160
Sweets, desserts, and other carbohydrates[c]	2 small cookies	15	varies	varies	varies
Nonstarchy vegetables	½ c cooked carrots	5	2	—	25
Meat and Meat Substitutes					
Lean	1 oz chicken (no skin)	—	7	0–3	45
Medium-fat	1 oz ground beef	—	7	4–7	75
High-fat	1 oz pork sausage	—	7	8+	100
Plant-based proteins	½ c tofu	varies	7	varies	varies
Fats	1 tsp butter	—	—	5	45
Alcohol	12 oz beer	varies	—	—	100

[a]The energy value for each exchange list represents an approximate average for the group and does not reflect the precise number of grams of carbohydrate, protein, and fat. For example, a slice of bread contains 15 grams of carbohydrate (60 kcalories), 3 grams protein (12 kcalories), and a little fat—rounded to 80 kcalories for ease in calculating. A ½ cup of vegetables (not including starchy vegetables) contains 5 grams carbohydrate (20 kcalories) and 2 grams protein (8 more), which has been rounded down to 25 kcalories.

[b]The Starch list includes cereals, grains, breads, crackers, snacks, starchy vegetables (such as corn, peas, and potatoes), and legumes (dried beans, peas, and lentils).

[c]The Sweets, Desserts, and Other Carbohydrates list includes foods that contain added sugars and fats such as sodas, candy, cakes, cookies, doughnuts, ice cream, pudding, syrup, and frozen yogurt.

Appendix D from Nutritional Sciences: From Fundamentals to Food by McGuire & Beerman; copyright holders are American Dietetic Association & American Diabetes Association.

The Foods on the Lists

Foods do not always appear on the exchange list where you might first expect to find them. They are grouped according to their energy-nutrient contents rather than by their source (such as milks), their outward appearance, or their vitamin and mineral contents. For example, cheeses are grouped with meats (not milk) because, like meats, cheeses contribute energy from protein and fat but provide negligible carbohydrate.

For similar reasons, starchy vegetables such as corn, green peas, and potatoes are found on the Starch list with breads and cereals, not with the vegetables. Likewise, bacon is grouped with the fats and oils, not with the meats.

Diet planners learn to view mixtures of foods, such as casseroles and soups, as combinations of foods from different exchange lists. They also learn to interpret food labels with the exchange system in mind.

Controlling Energy, Fat, and Sodium

The exchange lists help people control their energy intakes by paying close attention to serving sizes. People wanting to lose weight can limit foods from the Sweets, Desserts, and Other Carbohydrates and Fats lists, and they might choose to avoid the Alcohol list altogether. The Free Foods list provide low-kcalorie choices.

By assigning items like bacon to the Fats list, the exchange lists alert consumers to foods that are unexpectedly high in fat. Even the Starch list specifies which grain products contain added fat (such as biscuits, cornbread, and waffles) by marking them with a symbol to indicate added fat (the symbols are explained in the table keys). In addition, the exchange lists encourage users to think of fat-free milk as milk and of whole milk as milk with added fat, and to think of lean meats as meats and of medium-fat and high-fat meats as meats with added fat. To that end, foods on the milk and meat lists are separated into categories based on their fat contents (review Table L1.1). The Milk list is subdivided for fat-free, reduced fat, and whole; the meat list is subdivided for lean, medium fat, and high fat. The meat list also includes plant-based proteins, which tend to be rich in fiber. Notice that many of these foods bear the symbol for "high fiber."

People wanting to control the sodium in their diets can begin by eliminating any foods bearing the "high sodium" symbol. In most cases, the symbol identifies foods that, in one serving, provide 480 milligrams or more of sodium. Foods on the Combination Foods or Fast Foods lists that bear the symbol provide more than 600 milligrams of sodium. Other foods may also contribute substantially to sodium.

Planning a Healthy Diet

To obtain a daily variety of foods that provide healthful amounts of carbohydrate, protein, and fat, as well as vitamins, minerals, and fiber, the meal plan for adults and teenagers should include at least:

- ❏ Two to three servings of nonstarchy vegetables
- ❏ Two servings of fruits
- ❏ Six servings of grains (at least three of whole grains), beans, and starchy vegetables
- ❏ Two servings of low-fat or fat-free milk
- ❏ About 6 ounces of meat or meat substitutes
- ❏ *Small* amounts of fat and sugar

The actual amounts are determined by age, gender, activity levels, and other factors that influence energy needs. Refer to Chapter 17 as you read through these sections to get an idea of how exchange lists can be useful in planning a diet.

TABLE L1.2

Starch

The Starch list includes bread, cereals and grains, starchy vegetables, crackers and snacks, and legumes (dried beans, peas, and lentils). 1 starch choice = 15 grams carbohydrate, 0–3 grams protein, 0–1 grams fat, and 80 kcalories.

Note: In general, one starch exchange is ½ cup cooked cereal, grain, or starchy vegetable; ⅓ cup cooked rice or pasta; 1 ounce of bread product; ¾ ounce to 1 ounce of most snack foods.

Food	Serving Size
Bread	
Bagel, large (about 4 oz)	¼ (1 oz)
⚠ Biscuit, 2½ inches across	1
Bread	
☺ reduced-kcalorie	2 slices (1½ oz)
white, whole-grain, pumpernickel, rye, unfrosted raisin	1 slice (1 oz)
Chapatti, small, 6 inches across	1
⚠ Cornbread, 1¾ inch cube	1 (1½ oz)
English muffin	½
Hot dog bun or hamburger bun	½ (1 oz)
Naan, 8 inches by 2 inches	¼
Pancake, 4 inches across, ¼ inch thick	1
Pita, 6 inches across	½

Food	Serving Size
Roll, plain, small	1 (1 oz)
⚠ Stuffing, bread	⅓ cup
⚠ Taco shell, 5 inches across	2
Tortilla, corn, 6 inches across	1
Tortilla, flour, 6 inches across	1
Tortilla, flour, 10 inches across	⅓
⚠ Waffle, 4-inch square or 4 inches across	1
Cereals and Grains	
Barley, cooked	⅓ cup
Bran, dry	
☺ oat	¼ cup
☺ wheat	½ cup
☺ Bulgur (cooked)	½ cup

Starch (*Continued*)

Food	Serving Size
Cereals	
☺ bran	½ cup
cooked (oats, oatmeal)	½ cup
puffed	1½ cups
shredded wheat, plain	½ cup
sugar-coated	½ cup
unsweetened, ready-to-eat	¾ cup
Couscous	⅓ cup
Granola	
low-fat	¼ cup
⚠ regular	¼ cup
Grits, cooked	½ cup
Kasha	½ cup
Millet, cooked	⅓ cup
Muesli	¼ cup
Pasta, cooked	⅓ cup
Polenta, cooked	⅓ cup
Quinoa, cooked	⅓ cup
Rice, white or brown, cooked	⅓ cup
Tabbouleh (tabouli), prepared	½ cup
Wheat germ, dry	3 Tbsp
Wild rice, cooked	½ cup
Starchy Vegetables	
Cassava	⅓ cup
Corn	½ cup
on cob, large	½ cob (5 oz)
☺ Hominy, canned	¾ cup
☺ Mixed vegetables with corn, peas, or pasta	1 cup
☺ Parsnips	½ cup
☺ Peas, green	½ cup
Plantain, ripe	⅓ cup
Potato	
baked with skin	¼ large (3 oz)
boiled, all kinds	½ cup or ½ medium (3 oz)
⚠ mashed, with milk and fat	½ cup
french fried (oven-baked)[a]	1 cup (2 oz)

Food	Serving Size
☺ Pumpkin, canned, no sugar added	1 cup
Spaghetti/pasta sauce	½ cup
☺ Squash, winter (acorn, butternut)	1 cup
☺ Succotash	½ cup
Yam, sweet potato, plain	½ cup
Crackers and Snacks[b]	
Animal crackers	8
Crackers	
⚠ round-butter type	6
saltine-type	6
⚠ sandwich-style, cheese or peanut butter filling	3
⚠ whole-wheat regular	2–5 (¾ oz)
☺ whole-wheat lower fat or crispbreads	2–5 (¾ oz)
Graham cracker, 2½-inch square	3
Matzoh	¾ oz
Melba toast, about 2-inch by 4-inch piece	4
Oyster crackers	20
Popcorn	3 cups
⚠ ☺ with butter	3 cups
☺ no fat added	3 cups
☺ lower fat	3 cups
Pretzels	¾ oz
Rice cakes, 4 inches across	2
Snack chips	
fat-free or baked (tortilla, potato), baked pita chips	15–20 (¾ oz)
⚠ regular (tortilla, potato)	9–13 (¾ oz)
Beans, Peas, and Lentils[c]	
The choices on this list count as 1 starch + 1 lean meat.	
☺ Baked beans	⅓ cup
☺ Beans, cooked (black, garbanzo, kidney, lima, navy, pinto, white)	½ cup
☺ Lentils, cooked (brown, green, yellow)	½ cup
☺ Peas, cooked (black-eyed, split)	½ cup
🧂 ☺ Refried beans, canned	½ cup

KEY

☺ = More than 3 grams of dietary fiber per serving.

⚠ = Extra fat, or prepared with added fat. (Count as 1 starch + 1 fat.)

🧂 = 480 milligrams or more of sodium per serving.

[a]Restaurant-style french fries are on the Fast Foods list.

[b]For other snacks, see the Sweets, Desserts, and Other Carbohydrates list. For a quick estimate of serving size, an open handful is equal to about 1 cup or 1 to 2 ounces of snack food.

[c]Beans, peas, and lentils are also found on the Meat and Meat Substitutes list.

Fruits

The Fruits list includes fresh, frozen, canned, and dried fruits and fruit juices. 1 fruit choice = 15 grams carbohydrate, 0 grams protein, 0 grams fat, and 60 kcalories.

Note: In general, one fruit exchange is ½ cup canned or fresh fruit or unsweetened fruit juice; 1 small fresh fruit (4 ounces); 2 tablespoons dried fruit.

Food	Serving Size
Fruit[a]	
Apple, unpeeled, small	1 (4 oz)
Apples, dried	4 rings
Applesauce, unsweetened	½ cup
Apricots	
canned	½ cup
dried	8 halves
☺ fresh	4 whole (5½ oz)
Banana, extra small	1 (4 oz)
☺ Blackberries	¾ cup
Blueberries	¾ cup
Cantaloupe, small	⅓ melon or 1 cup cubed (11 oz)
Cherries	
sweet, canned	½ cup
sweet fresh	12 (3 oz)
Dates	3
Dried fruits (blueberries, cherries, cranberries, mixed fruit, raisins)	2 Tbsp
Figs	
dried	1½
☺ fresh	1½ large or 2 medium (3½ oz)
Fruit cocktail	½ cup
Grapefruit	
large	½ (11 oz)
sections, canned	¾ cup
Grapes, small	17 (3 oz)
Honeydew melon	1 slice or 1 cup cubed (10 oz)
☺ Kiwi	1 (3½ oz)
Mandarin oranges, canned	¾ cup

Food	Serving Size
Mango, small	½ (5½ oz) or ½ cup
Nectarine, small	1 (5 oz)
☺ Orange, small	1 (6½ oz)
Papaya	½ or 1 cup cubed (8 oz)
Peaches	
canned	½ cup
fresh, medium	1 (6 oz)
Pears	
canned	½ cup
fresh, large	½ (4 oz)
Pineapple	
canned	½ cup
fresh	¾ cup
Plums	
canned	½ cup
dried (prunes)	3
small	2 (5 oz)
☺ Raspberries	1 cup
☺ Strawberries	1¼ cup whole berries
☺ Tangerines, small	2 (8 oz)
Watermelon	1 slice or 1¼ cups cubes (13½ oz)
Fruit Juice	
Apple juice/cider	½ cup
Fruit juice blends, 100% juice	⅓ cup
Grape juice	⅓ cup
Grapefruit juice	½ cup
Orange juice	½ cup
Pineapple juice	½ cup
Prune juice	⅓ cup

KEY

☺ = More than 3 grams of dietary fiber per serving.

△ = Extra fat, or prepared with added fat. (Count as 1 starch + 1 fat.)

🧂 = 480 milligrams or more of sodium per serving.

[a]The weight listed includes skin, core, seeds, and rind.

TABLE L1.4

Milk

The Milk list groups milks and yogurts based on the amount of fat they have (fat-free/low fat, reduced fat, and whole). Cheeses are found on the Meat and Meat Substitutes list and cream and other dairy fats are found on the Fats list.

Note: In general, one milk choice is 1 cup (8 fluid ounces or ½ pint) milk or yogurt.

Food	Serving Size
Milk and Yogurts	
Fat-free or low-fat (1%)	
1 fat-free/low-fat milk choice = 12 g carbohydrate, 8 g protein, 0–3 g fat, and 100 kcal.	
Milk, buttermilk, acidophilus milk, Lactaid	1 cup
Evaporated milk	½ cup
Yogurt, plain or flavored with an artificial sweetener	⅔ cup (6 oz)
Reduced-fat (2%)	
1 reduced-fat milk choice = 12 g carbohydrate, 8 g protein, 5 g fat, and 120 kcal.	
Milk, acidophilus milk, kefir, Lactaid	1 cup
Yogurt, plain	⅔ cup (6 oz)
Whole	
1 whole milk choice = 12 g carbohydrate, 8 g protein, 8 g fat, and 160 kcal.	
Milk, buttermilk, goat's milk	1 cup
Evaporated milk	½ cup
Yogurt, plain	8 oz

Food	Serving Size	Count as
Dairy-Like Foods		
Chocolate milk		
fat-free	1 cup	1 fat-free milk + 1 carbohydrate
whole	1 cup	1 whole milk + 1 carbohydrate
Eggnog, whole milk	½ cup	1 carbohydrate + 2 fats
Rice drink		
flavored, low fat	1 cup	2 carbohydrates
plain, fat-free	1 cup	1 carbohydrate
Smoothies, flavored, regular	10 oz	1 fat-free milk + 2½ carbohydrates
Soy milk		
light	1 cup	1 carbohydrate + ½ fat
regular, plain	1 cup	1 carbohydrate + 1 fat
Yogurt		
and juice blends	1 cup	1 fat-free milk + 1 carbohydrate
low carbohydrate (less than 6 grams carbohydrate per choice)	⅔ cup (6 oz)	½ fat-free milk
with fruit, low-fat	⅔ cup (6 oz)	1 fat-free milk + 1 carbohydrate

Sweets, Desserts, and Other Carbohydrates

1 other carbohydrate choice = 15 grams carbohydrate, variable grams protein, variable grams fat, and variable kcalories.

Note: In general, one choice from this list can substitute for foods on the Starch, Fruits, or Milk lists.

Food	Serving Size	Count as
Beverages, Soda, and Energy/Sports Drinks		
Cranberry juice cocktail	½ cup	1 carbohydrate
Energy drink	1 can (8.3 oz)	2 carbohydrates
Fruit drink or lemonade	1 cup (8 oz)	2 carbohydrates
Hot chocolate		
regular	1 envelope added to 8 oz water	1 carbohydrate + 1 fat
sugar-free or light	1 envelope added to 8 oz water	1 carbohydrate
Soft drink (soda), regular	1 can (12 oz)	2½ carbohydrates
Sports drink	1 cup (8 oz)	1 carbohydrate
Brownies, Cake, Cookies, Gelatin, Pie, and Pudding		
Brownie, small, unfrosted	1¼-inch square, ⅞ inch high (about 1 oz)	1 carbohydrate + 1 fat
Cake		
angel food, unfrosted	1/12 of cake (about 2 oz)	2 carbohydrates
frosted	2-inch square (about 2 oz)	2 carbohydrates + 1 fat
unfrosted	2-inch square (about 2 oz)	1 carbohydrate + 1 fat
Cookies		
chocolate chip	2 cookies (2¼ inches across)	1 carbohydrate + 2 fats
gingersnap	3 cookies	1 carbohydrate
sandwich, with creme filling	2 small (about ⅔ oz)	1 carbohydrate + 1 fat
sugar-free	3 small or 1 large (¾–1 oz)	1 carbohydrate + 1–2 fats
vanilla wafer	5 cookies	1 carbohydrate + 1 fat
Cupcake, frosted	1 small (about 1¾ oz)	2 carbohydrates + 1–1½ fats
Fruit cobbler	½ cup (3½ oz)	3 carbohydrates + 1 fat
Gelatin, regular	½ cup	1 carbohydrate
Pie		
commercially prepared fruit, 2 crusts	⅙ of 8-inch pie	3 carbohydrates + 2 fats
pumpkin or custard	⅛ of 8-inch pie	1½ carbohydrates + 1½ fats
Pudding		
regular (made with reduced-fat milk)	½ cup	2 carbohydrates
sugar-free or sugar- and fat-free (made with fat-free milk)	½ cup	1 carbohydrate
Candy, Spreads, Sweets, Sweeteners, Syrups, and Toppings		
Candy bar, chocolate/peanut	2 "fun size" bars (1 oz)	1½ carbohydrates + 1½ fats
Candy, hard	3 pieces	1 carbohydrate
Chocolate "kisses"	5 pieces	1 carbohydrate + 1 fat
Coffee creamer		
dry, flavored	4 tsp	½ carbohydrate + ½ fat
liquid, flavored	2 Tbsp	1 carbohydrate
Fruit snacks, chewy (pureed fruit concentrate)	1 roll (¾ oz)	1 carbohydrate
Fruit spreads, 100% fruit	1½ Tbsp	1 carbohydrate
Honey	1 Tbsp	1 carbohydrate

Sweets, Desserts, and Other Carbohydrates (*Continued*)

Food	Serving Size	Count as
Jam or jelly, regular	1 Tbsp	1 carbohydrate
Sugar	1 Tbsp	1 carbohydrate
Syrup		
chocolate	2 Tbsp	2 carbohydrates
light (pancake type)	2 Tbsp	1 carbohydrate
regular (pancake type)	1 Tbsp	1 carbohydrate
Condiments and Sauces[a]		
Barbeque sauce	3 Tbsp	1 carbohydrate
Cranberry sauce, jellied	¼ cup	1½ carbohydrates
Gravy, canned or bottled	½ cup	½ carbohydrate + ½ fat
Salad dressing, fat-free, low-fat, cream-based	3 Tbsp	1 carbohydrate
Sweet and sour sauce	3 Tbsp	1 carbohydrate
Doughnuts, Muffins, Pastries, and Sweet Breads		
Banana nut bread	1-inch slice (1 oz)	2 carbohydrates + 1 fat
Doughnut		
cake, plain	1 medium (1½ oz)	1½ carbohydrates + 2 fats
yeast type, glazed	3¾ inches across (2 oz)	2 carbohydrates + 2 fats
Muffin (4 oz)	¼ muffin (1 oz)	1 carbohydrate + ½ fat
Sweet roll or Danish	1 (2½ oz)	2½ carbohydrates + 2 fats
Frozen Bars, Frozen Desserts, Frozen Yogurt, and Ice Cream		
Frozen pops	1	½ carbohydrate
Fruit juice bars, frozen, 100% juice	1 bar (3 oz)	1 carbohydrate
Ice cream		
fat-free	½ cup	1½ carbohydrates
light	½ cup	1 carbohydrate + 1 fat
no sugar added	½ cup	1 carbohydrate + 1 fat
regular	½ cup	1 carbohydrate + 2 fats
Sherbet, sorbet	½ cup	2 carbohydrates
Yogurt, frozen		
fat-free	⅓ cup	1 carbohydrate
regular	½ cup	1 carbohydrate + 0–1 fat
Granola Bars, Meal Replacement Bars/Shakes, and Trail Mix		
Granola or snack bar, regular or low-fat	1 bar (1 oz)	1½ carbohydrates
Meal replacement bar	1 bar (1⅓ oz)	1½ carbohydrates + 0–1 fat
Meal replacement bar	1 bar (2 oz)	2 carbohydrates + 1 fat
Meal replacement shake, reduced kcalorie	1 can (10–11 oz)	1½ carbohydrates + 0–1 fat
Trail mix		
candy/nut-based	1 oz	1 carbohydrate + 2 fats
dried fruit-based	1 oz	1 carbohydrate + 1 fat

KEY

= 480 milligrams or more of sodium per serving.

[a] You can also check the Fats list and Free Foods list for other condiments.

Nonstarchy Vegetables

The Nonstarchy Vegetables list includes vegetables that have few grams of carbohydrates or kcalories; starchy vegetables are found on the Starch list. 1 nonstarchy vegetable choice = 5 grams carbohydrate, 2 grams protein, 0 grams fat, and 25 kcalories.

Note: In general, one nonstarchy vegetable choice is ½ cup cooked vegetables or vegetable juice or 1 cup raw vegetables. Count 3 cups of raw vegetables or 1½ cups of cooked vegetables as one carbohydrate choice.

Nonstarchy Vegetables[a]	
Amaranth or Chinese spinach	Kohlrabi
Artichoke	Leeks
Artichoke hearts	Mixed vegetables (without corn, peas, or pasta)
Asparagus	Mung bean sprouts
Baby corn	Mushrooms, all kinds, fresh
Bamboo shoots	Okra
Beans (green, wax, Italian)	Onions
Bean sprouts	Oriental radish or daikon
Beets	Pea pods
⌂ Borscht	☺ Peppers (all varieties)
Broccoli	Radishes
☺ Brussels sprouts	Rutabaga
Cabbage (green, bok choy, Chinese)	⌂ Sauerkraut
☺ Carrots	Soybean sprouts
Cauliflower	Spinach
Celery	Squash (summer, crookneck, zucchini)
☺ Chayote	Sugar pea snaps
Coleslaw, packaged, no dressing	☺ Swiss chard
Cucumber	Tomato
Eggplant	Tomatoes, canned
Gourds (bitter, bottle, luffa, bitter melon)	⌂ Tomato sauce
Green onions or scallions	⌂ Tomato/vegetable juice
Greens (collard, kale, mustard, turnip)	Turnips
Hearts of palm	Water chestnuts
Jicama	Yard-long beans

KEY

☺ = More than 3 grams of dietary fiber per serving.

⌂ = 480 milligrams or more of sodium per serving.

[a]Salad greens (like chicory, endive, escarole, lettuce, romaine, spinach, arugula, radicchio, watercress) are on the Free Foods list.

Meat and Meat Substitutes

The Meat and Meat Substitutes list groups foods based on the amount of fat they have (lean meat, medium-fat meat, high-fat meat, and plant-based proteins).

Food	Amount
Lean Meats and Meat Substitutes	
1 lean meat choice = 0 grams carbohydrate, 7 grams protein, 0–3 grams fat, and 100 kcalories.	
Beef: Select or Choice grades trimmed of fat: ground round, roast (chuck, rib, rump), round, sirloin, steak (cubed, flank, porterhouse, T-bone), tenderloin	1 oz
Beef jerky	1 oz
Cheeses with 3 grams of fat or less per oz	1 oz
Cottage cheese	¼ cup
Egg substitutes, plain	¼ cup
Egg whites	2
Fish, fresh or frozen, plain: catfish, cod, flounder, haddock, halibut, orange roughy, salmon, tilapia, trout, tuna	1 oz
Fish, smoked: herring or salmon (lox)	1 oz
Game: buffalo, ostrich, rabbit, venison	1 oz
Hot dog with 3 grams of fat or less per oz (8 dogs per 14 oz package) *Note: May be high in carbohydrate.*	1
Lamb: chop, leg, or roast	1 oz
Organ meats: heart, kidney, liver *Note: May be high in cholesterol.*	1 oz
Oysters, fresh or frozen	6 medium
Pork, lean	
Canadian bacon	1 oz
rib or loin chop/roast, ham, tenderloin	1 oz
Poultry, without skin: Cornish hen, chicken, domestic duck or goose (well-drained of fat), turkey	1 oz
Processed sandwich meats with 3 grams of fat or less per oz: chipped beef, deli thin-sliced meats, turkey ham, turkey kielbasa, turkey pastrami	1 oz
Salmon, canned	1 oz
Sardines, canned	2 medium
Sausage with 3 grams of fat or less per oz	1 oz
Shellfish: clams, crab, imitation shellfish, lobster, scallops, shrimp	1 oz
Tuna, canned in water or oil, drained	1 oz
Veal, lean chop, roast	1 oz

Food	Amount
Medium-Fat Meat and Meat Substitutes	
1 medium-fat meat choice = 0 grams carbohydrate, 7 grams protein, 4–7 grams fat, and 130 kcalories.	
Beef: corned beef, ground beef, meatloaf, Prime grades trimmed of fat (prime rib), short ribs, tongue	1 oz
Cheeses with 4–7 grams of fat per oz: feta, mozzarella, pasteurized processed cheese spread, reduced-fat cheeses, string	1 oz
Egg	1
Note: High in cholesterol, so limit to 3 per week.	
Fish, any fried product	1 oz
Lamb: ground, rib roast	1 oz
Pork: cutlet, shoulder roast	1 oz
Poultry: chicken with skin; dove, pheasant, wild duck, or goose; fried chicken; ground turkey	1 oz
Ricotta cheese	2 oz or ¼ cup
Sausage with 4–7 grams of fat per oz	1 oz
Veal, cutlet (no breading)	1 oz
High-Fat Meat and Meat Substitutes	
1 high-fat meat choice = 0 grams carbohydrate, 7 grams protein, 8 grams fat, and 150 kcalories. These foods are high in saturated fat, cholesterol, and kcalories and may raise blood cholesterol levels if eaten on a regular basis. Try to eat 3 or fewer servings from this group per week.	
Bacon	
pork	2 slices (16 slices per lb or 1 oz each, before cooking)
turkey	3 slices (½ oz each before cooking)
Cheese, regular: American, bleu, brie, cheddar, hard goat, Monterey jack, queso, and Swiss	1 oz
⚠ Hot dog: beef, pork, or combination (10 per lb-sized package)	1
Hot dog: turkey or chicken (10 per lb-sized package)	1
Pork: ground, sausage, spareribs	1 oz
Processed sandwich meats with 8 grams of fat or more per oz: bologna, pastrami, hard salami	1 oz
Sausage with 8 grams fat or more per oz: bratwurst, chorizo, Italian, knockwurst, Polish, smoked, summer	1 oz

KEY

☺ = More than 3 grams of dietary fiber per serving.

⚠ = Extra fat, or prepared with added fat. (Add an additional fat choice to this food.)

🧂 = 480 milligrams or more of sodium per serving (based on the sodium content of a typical 3-oz serving of meat, unless 1 or 2 oz is the normal serving size).

Meat and Meat Substitutes (*Continued*)

Food	Serving Size	Count as
Plant-Based Proteins		
1 plant-based protein choice = variable grams carbohydrate, 7 grams protein, variable grams fat, and variable kcalories. Because carbohydrate content varies among plant-based proteins, you should read the food label.		
"Bacon" strips, soy-based	3 strips	1 medium-fat meat
☺ Baked beans	⅓ cup	1 starch + 1 lean meat
☺ Beans, cooked: black, garbanzo, kidney, lima, navy, pinto, whiteª	½ cup	1 starch + 1 lean meat
☺ "Beef" or "sausage" crumbles, soy-based	2 oz	½ carbohydrate + 1 lean meat
"Chicken" nuggets, soy-based	2 nuggets (1½ oz)	½ carbohydrate + 1 medium-fat meat
☺ Edamame	½ cup	½ carbohydrate + 1 lean meat
Falafel (spiced chickpea and wheat patties)	3 patties (about 2 inches across)	1 carbohydrate + 1 high-fat meat
Hot dog, soy-based	1 (1½ oz)	½ carbohydrate + 1 lean meat
☺ Hummus	⅓ cup	1 carbohydrate + 1 high-fat meat
☺ Lentils, brown, green, or yellow	½ cup	1 carbohydrate + 1 lean meat
☺ Meatless burger, soy-based	3 oz	½ carbohydrate + 2 lean meats
☺ Meatless burger, vegetable- and starch-based	1 patty (about 2½ oz)	1 carbohydrate + 2 lean meats
Nut spreads: almond butter, cashew butter, peanut butter, soy nut butter	1 Tbsp	1 high-fat meat
☺ Peas, cooked: black-eyed and split peas	½ cup	1 starch + 1 lean meat
⌕ ☺ Refried beans, canned	½ cup	1 starch + 1 lean meat
"Sausage" patties, soy-based	1 (1½ oz)	1 medium-fat meat
Soy nuts, unsalted	¾ oz	½ carbohydrate + 1 medium-fat meat
Tempeh	¼ cup	1 medium-fat meat
Tofu	4 oz (½ cup)	1 medium-fat meat
Tofu, light	4 oz (½ cup) 1 lean meat	

KEY

☺ = More than 3 grams of dietary fiber per serving.

⚠ = Extra fat, or prepared with added fat. (Add an additional fat choice to this food.)

⌕ = 480 milligrams or more of sodium per serving (based on the sodium content of a typical 3-oz serving of meat, unless 1 or 2 oz is the normal serving size).

ªBeans, peas, and lentils are also found on the Starch list; nut butters in smaller amounts are found in the Fats list.

TABLE L1.8

Fats

Fats and oils have mixtures of unsaturated (polyunsaturated and monounsaturated) and saturated fats. Foods on the Fats list are grouped together based on the major type of fat they contain. 1 fat choice = 0 grams carbohydrate, 0 grams protein, 5 grams fat, and 45 kcalories.

Note: In general, one fat exchange is 1 teaspoon of regular margarine, vegetable oil, or butter; 1 tablespoon of regular salad dressing. When used in large amounts, bacon and peanut butter are counted as high-fat meat choices (see Meat and Meat Substitutes list). Fat-free salad dressings are found on the Sweets, Desserts, and Other Carbohydrates list. Fat-free products such as margarines, salad dressings, mayonnaise, sour cream, and cream cheese are found on the Free Foods list.

Food	Serving Size
Monounsaturated Fats	
Avocado, medium	2 Tbsp (1 oz)
Nut butters (*trans* fat-free): almond butter, cashew butter, peanut butter (smooth or crunchy)	1½ tsp
Nuts	
almonds	6 nuts
Brazil	2 nuts
cashews	6 nuts
filberts (hazelnuts)	5 nuts
macadamia	3 nuts
mixed (50% peanuts)	6 nuts
peanuts	10 nuts
pecans	4 halves
pistachios	16 nuts
Oil: canola, olive, peanut	1 tsp
Olives	
black (ripe)	8 large
green, stuffed	10 large
Polyunsaturated Fats	
Margarine: lower-fat spread (30%–50% vegetable oil, *trans* fat-free)	1 Tbsp
Margarine: stick, tub (*trans* fat-free) or squeeze (*trans* fat-free)	1 tsp
Mayonnaise	
reduced-fat	1 Tbsp
regular	1 tsp
Mayonnaise-style salad dressing	
reduced-fat	1 Tbsp
regular	2 tsp
Nuts	
Pignolia (pine nuts)	1 Tbsp
walnuts, English	4 halves
Oil: corn, cottonseed, flaxseed, grape seed, safflower, soybean, sunflower	1 tsp
Oil: made from soybean and canola oil—Enova	1 tsp
Plant stanol esters	
light	1 Tbsp
regular	2 tsp

Food	Serving Size
Salad dressing	
reduced-fat 🧂	2 Tbsp
Note: May be high in carbohydrate.	
regular 🧂	1 Tbsp
Seeds	
flaxseed, whole	1 Tbsp
pumpkin, sunflower	1 Tbsp
sesame seeds	1 Tbsp
Tahini or sesame paste	2 tsp
Saturated Fats	
Bacon, cooked, regular or turkey	1 slice
Butter	
reduced-fat	1 Tbsp
stick	1 tsp
whipped	2 tsp
Butter blends made with oil	
reduced-fat or light	1 Tbsp
regular	1½ tsp
Chitterlings, boiled	2 Tbsp (½ oz)
Coconut, sweetened, shredded	2 Tbsp
Coconut milk	
light	⅓ cup
regular	1½ Tbsp
Cream	
half and half	2 Tbsp
heavy	1 Tbsp
light	1½ Tbsp
whipped	2 Tbsp
whipped, pressurized	¼ cup
Cream cheese	
reduced-fat	1½ Tbsp (¾ oz)
regular	1 Tbsp (½ oz)
Lard	1 tsp
Oil: coconut, palm, palm kernel	1 tsp
Salt pork	¼ oz
Shortening, solid	1 tsp
Sour cream	
reduced-fat or light	3 Tbsp
regular	2 Tbsp

KEY

🧂 = 480 milligrams or more of sodium per serving.

Free Foods

A "free" food is any food or drink choice that has less than 20 kcalories and 5 grams or less of carbohydrate per serving.
- Most foods on this list should be limited to 3 servings (as listed here) per day. Spread out the servings throughout the day. If you eat all 3 servings at once, it could raise your blood glucose level.
- Food and drink choices listed here without a serving size can be eaten whenever you like.

Food	Serving Size
Low Carbohydrate Foods	
Cabbage, raw	½ cup
Candy, hard (regular or sugar-free)	1 piece
Carrots, cauliflower, or green beans, cooked	¼ cup
Cranberries, sweetened with sugar substitute	½ cup
Cucumber, sliced	½ cup
Gelatin	
dessert, sugar-free	
unflavored	
Gum	
Jam or jelly, light or no sugar added	2 tsp
Rhubarb, sweetened with sugar substitute	½ cup
Salad greens	
Sugar substitutes (artificial sweeteners)	
Syrup, sugar-free	2 Tbsp
Modified Fat Foods with Carbohydrate	
Cream cheese, fat-free	1 Tbsp (½ oz)
Creamers	
nondairy, liquid	1 Tbsp
nondairy, powdered	2 tsp
Margarine spread	
fat-free	1 Tbsp
reduced-fat	1 tsp
Mayonnaise	
fat-free	1 Tbsp
reduced-fat	1 tsp
Mayonnaise-style salad dressing	
fat-free	1 Tbsp
reduced-fat	1 tsp
Salad dressing	
fat-free or low-fat	1 Tbsp
fat-free, Italian	2 Tbsp
Sour cream, fat-free or reduced-fat	1 Tbsp
Whipped topping	
light or fat-free	2 Tbsp
regular	1 Tbsp
Condiments	
Barbecue sauce	2 tsp
Catsup (ketchup)	1 Tbsp
Honey mustard	1 Tbsp
Horseradish	

Food	Serving Size
Lemon juice	
Miso	1½ tsp
Mustard	
Parmesan cheese, freshly grated	1 Tbsp
Pickle relish	1 Tbsp
Pickles	
🧂 dill	1½ medium
sweet, bread and butter	2 slices
sweet, gherkin	¾ oz
Salsa	¼ cup
🧂 Soy sauce, light or regular	1 Tbsp
Sweet and sour sauce	2 tsp
Sweet chili sauce	2 tsp
Taco sauce	1 Tbsp
Vinegar	
Yogurt, any type	2 Tbsp
Drinks/Mixes	
Any food on the list—without a serving size listed—can be consumed in any moderate amount.	
🧂 Bouillon, broth, consommé	
Bouillon or broth, low-sodium	
Carbonated or mineral water	
Club soda	
Cocoa powder, unsweetened (1 Tbsp)	
Coffee, unsweetened or with sugar substitute	
Diet soft drinks, sugar-free	
Drink mixes, sugar-free	
Tea, unsweetened or with sugar substitute	
Tonic water, diet	
Water	
Water, flavored, carbohydrate free	
Seasonings	
Any food on this list can be consumed in any moderate amount.	
Flavoring extracts (for example, vanilla, almond, peppermint)	
Garlic	
Herbs, fresh or dried	
Nonstick cooking spray	
Pimento	
Spices	
Hot pepper sauce	
Wine, used in cooking	
Worcestershire sauce	

KEY

🧂 = 480 milligrams or more of sodium per serving.

Combination Foods

Many foods are eaten in various combinations, such as casseroles. Because "combination" foods do not fit into any one choice list, this list of choices provides some typical combination foods.

Food	Serving Size	Count as
Entrees		
Casserole type (tuna noodle, lasagna, spaghetti with meatballs, chili with beans, macaroni and cheese)	1 cup (8 oz)	2 carbohydrates + 2 medium-fat meats
Stews (beef/other meats and vegetables)	1 cup (8 oz)	1 carbohydrate + 1 medium-fat meat + 0–3 fats
Tuna salad or chicken salad	½ cup (3½ oz)	½ carbohydrate + 2 lean meats + 1 fat
Frozen Meals/Entrees		
☺ Burrito (beef and bean)	1 (5 oz)	3 carbohydrates + 1 lean meat + 2 fats
Dinner-type meal	generally 14–17 oz	3 carbohydrates + 3 medium-fat meats + 3 fats
Entrée or meal with less than 340 kcalories	about 8–11 oz	2–3 carbohydrates + 1–2 lean meats
Pizza		
cheese/vegetarian, thin crust	¼ of a 12 inch (4½–5 oz)	2 carbohydrates + 2 medium-fat meats
meat topping, thin crust	¼ of a 12 inch (5 oz)	2 carbohydrates + 2 medium-fat meats + 1½ fats
Pocket sandwich	1 (4½ oz)	3 carbohydrates + 1 lean meat + 1–2 fats
Pot pie	1 (7 oz)	2½ carbohydrates + 1 medium-fat meat + 3 fats
Salads (Deli-Style)		
Coleslaw	½ cup	1 carbohydrate + 1½ fats
Macaroni/pasta salad	½ cup	2 carbohydrates + 3 fats
Potato salad	½ cup	1½–2 carbohydrates + 1–2 fats
Soups		
Bean, lentil, or split pea	1 cup	1 carbohydrate + 1 lean meat
Chowder (made with milk)	1 cup (8 oz)	1 carbohydrate + 1 lean meat + 1½ fats
Cream (made with water)	1 cup (8 oz)	1 carbohydrate + 1 fat
Instant	6 oz prepared	1 carbohydrate
with beans or lentils	8 oz prepared	2½ carbohydrates + 1 lean meat
Miso soup	1 cup	½ carbohydrate + 1 fat
Oriental noodle	1 cup	2 carbohydrates + 2 fats
Rice (congee)	1 cup	1 carbohydrate
Tomato (made with water)	1 cup (8 oz)	1 carbohydrate
Vegetable beef, chicken noodle, or other broth-type	1 cup (8 oz)	1 carbohydrate

KEY

☺ = More than 3 grams of dietary fiber per serving.

△ = Extra fat, or prepared with added fat.

🧂 = 600 milligrams or more of sodium per serving (for combination food main dishes/meals).

Fast Foods

The choices in the Fast Foods list are not specific fast-food meals or items, but are estimates based on popular foods. Ask the restaurant or check its website for nutrition information about your favorite fast foods.

Food	Serving Size	Count as
Breakfast Sandwiches		
Egg, cheese, meat, English muffin	1 sandwich	2 carbohydrates + 2 medium-fat meats
Sausage biscuit sandwich	1 sandwich	2 carbohydrates + 2 high-fat meats + 3½ fats
Main Dishes/Entrees		
☺ Burrito (beef and beans)	1 (about 8 oz)	3 carbohydrates + 3 medium-fat meats + 3 fats
Chicken breast, breaded and fried	1 (about 5 oz)	1 carbohydrate + 4 medium-fat meats
Chicken drumstick, breaded and fried	1 (about 2 oz)	2 medium-fat meats
Chicken nuggets	6 (about 3½ oz)	1 carbohydrate + 2 medium-fat meats + 1 fat
Chicken thigh, breaded and fried	1 (about 4 oz)	½ carbohydrate + 3 medium-fat meats + 1½ fats
Chicken wings, hot	6 (5 oz)	5 medium-fat meats + 1½ fats
Oriental		
Beef/chicken/shrimp with vegetables in sauce	1 cup (about 5 oz)	1 carbohydrate + 1 lean meat + 1 fat
Egg roll, meat	1 (about 3 oz)	1 carbohydrate + 1 lean meat + 1 fat
Fried rice, meatless	½ cup	1½ carbohydrates + 1½ fats
Meat and sweet sauce (orange chicken)	1 cup	3 carbohydrates + 3 medium-fat meats + 2 fats
☺ Noodles and vegetables in sauce (chow mein, lo mein)	1 cup	2 carbohydrates + 1 fat
Pizza		
Pizza		
cheese, pepperoni, regular crust	⅛ of a 14 inch (about 4 oz)	2½ carbohydrates + 1 medium-fat meat + 1½ fats
cheese/vegetarian, thin crust	¼ of a 12 inch (about 6 oz)	2½ carbohydrates + 2 medium-fat meats + 1½ fats
Sandwiches		
Chicken sandwich, grilled	1	3 carbohydrates + 4 lean meats
Chicken sandwich, crispy	1	3½ carbohydrates + 3 medium-fat meats + 1 fat
Fish sandwich with tartar sauce	1	2½ carbohydrates + 2 medium-fat meats + 2 fats
Hamburger		
large with cheese	1	2½ carbohydrates + 4 medium-fat meats + 1 fat
regular	1	2 carbohydrates + 1 medium-fat meat + 1 fat
Hot dog with bun	1	1 carbohydrate + 1 high-fat meat + 1 fat
Submarine sandwich		
less than 6 grams fat	6-inch sub	3 carbohydrates + 2 lean meats
regular	6-inch sub	3½ carbohydrates + 2 medium-fat meats + 1 fat
Taco, hard or soft shell (meat and cheese)	1 small	1 carbohydrate + 1 medium-fat meat + 1½ fats
☺ Salad, main dish (grilled chicken type, no dressing or croutons)		1 carbohydrate + 4 lean meats
Salad, side, no dressing or cheese	Small (about 5 oz)	1 vegetable
Sides/Appetizers		
△ French fries, restaurant style	small	3 carbohydrates + 3 fats
	medium	4 carbohydrates + 4 fats
	large	5 carbohydrates + 6 fats
Nachos with cheese	small (about 4½ oz)	2½ carbohydrates + 4 fats
Onion rings	1 serving (about 3 oz)	2½ carbohydrates + 3 fats
Desserts		
Milkshake, any flavor	12 oz	6 carbohydrates + 2 fats
Soft-serve ice cream cone	1 small	2½ carbohydrates + 1 fat

KEY

☺ = More than 3 grams of dietary fiber per serving.

△ = Extra fat, or prepared with added fat.

⌂ = 600 milligrams or more of sodium per serving (for fast-food main dishes/meals).

TABLE L1.12

Alcohol

1 alcohol equivalent = variable grams carbohydrate, 0 grams protein, 0 grams fat, and 100 kcalories.

Note: In general, one alcohol choice (½ ounce absolute alcohol) has about 100 kcalories. For those who choose to drink alcohol, guidelines suggest limiting alcohol intake to 1 drink or less per day for women, and 2 drinks or less per day for men. To reduce your risk of low blood glucose (hypoglycemia), especially if you take insulin or a diabetes pill that increases insulin, always drink alcohol with food. While alcohol, by itself, does not directly affect blood glucose, be aware of the carbohydrate (for example, in mixed drinks, beer, and wine) that may raise your blood glucose.

Alcoholic Beverage	Serving Size	Count as
Beer		
light (4.2%)	12 fl oz	1 alcohol equivalent + ½ carbohydrate
regular (4.9%)	12 fl oz	1 alcohol equivalent + 1 carbohydrate
Distilled spirits: vodka, rum, gin, whiskey, 80 or 86 proof	1½ fl oz	1 alcohol equivalent
Liqueur, coffee (53 proof)	1 fl oz	1 alcohol equivalent + 1 carbohydrate
Sake	1 fl oz	½ alcohol equivalent
Wine		
dessert (sherry)	3½ fl oz	1 alcohol equivalent + 1 carbohydrate
dry, red or white (10%)	5 fl oz	1 alcohol equivalent

Appendix L2—Exchange Calculation Form

Nutrition Rx: _____ KC als _____   % CHO (_____ g CHO)   _____% Protein   (_____ g Pro)   _____% Fat   (_____ g Fat)

	Exchange	No. of Ex	g CHO	Total g CHO	g Prot	Total g Protein	g Fat	Total g Fat	Cal/Exch	Total Exchange Calories
	FF, LF MILK		12		8		1		100	
	RF MILK		12		8		5		120	
	WH MILK		12		8		8		160	
	NONSTARCHY VEG		5		2		0		25	
	FRUIT		15		0		0		60	
Rx g CHO less Total g CHO so far ÷ 15 = Starch Ex			☐							
	STARCH		15		3		1		80	
Rx g Prot less Total g Prot so far ÷ 7 = Meat Ex					☐					
	L MEAT		0		7		3		45	
	MF MEAT		0		7		5.2		75	
	HF MEAT		0		7		8		100	
Rx g Fat less Total g Fat so far ÷ 5 = Fat Ex							☐			
	FAT		0		0		5		45	
TOTALS										

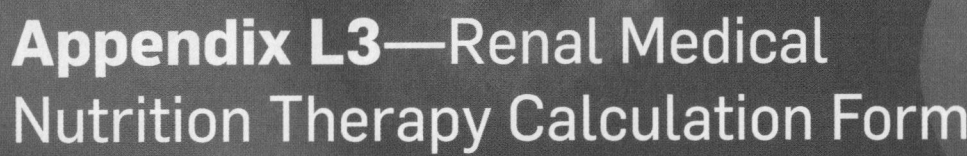

Patient's name: _____   _____ grams protein

Date: _____   _____ calories

Your dietitian is: _____   _____ milligrams phosphorus

Telephone number: _____   _____ milligrams sodium

Your Daily Meal Plan

Breakfast   **Sample Menu**

Milk   _____ choices   _____

Nondairy milk substitute   _____ choices   _____

Meat   _____ choices   _____

Starch   _____ choices   _____

Fruit   _____ choices   _____

Fat   _____ choices   _____

High-calorie   _____ choices   _____

Salt   _____ choices   _____

Snack

                                _____ choices   _____

                                _____ choices   _____

Lunch

Milk   _____ choices   _____

Nondairy milk substitute   _____ choices   _____

Meat   _____ choices   _____

Starch   _____ choices   _____

Vegetable   _____ choices   _____

Fruit   _____ choices   _____

Fat   _____ choices   _____

High-calorie   _____ choices   _____

Salt   _____ choices   _____

Snack

                                _____ choices   _____

                                _____ choices   _____

Dinner

Milk   _____ choices   _____

Nondairy milk substitute   _____ choices   _____

Meat   _____ choices   _____

Starch   _____ choices   _____

Vegetable   _____ choices   _____

Fruit   _____ choices   _____

Fat   _____ choices   _____

High-calorie   _____ choices   _____

Salt   _____ choices   _____

Snack

                                _____ choices   _____

                                _____ choices   _____

Appendix M1—Saturated Fat, Total Fat, Cholesterol, and Omega-3 Content of Meat, Fish, and Poultry (3 oz. Portions)

Saturated Fat, Total Fat, Cholesterol, and Omega-3 Content of Meat, Fish, and Poultry in 3-Ounce Portions Cooked Without Added Fat

Source	Saturated Fat g/3 oz	Total Fat g/3 oz	Cholesterol mg/3 oz	Omega-3 mg/3 oz
Lean Red Meats				
Beef (rump roast, shank, bottom round, sirloin)	2.1	7	66	–
Lamb (shank roast, sirloin roast, shoulder roast, loin chops, sirloin chops, center leg chop)	9.1	21	81	–
Pork (sirloin cutlet, loin roast, sirloin roast, center roast, butterfly chops, loin chops)	3.4	9.2	69	–
Veal (blade roast, sirloin chops, shoulder roast, loin chops, rump roast, shank)	2	6	97	–
Organ Meats				
Liver				
Beef	1.6	4.2	324	–
Calf	2.2	5.9	477	–
Chicken	1.8	6	479	–
Sweetbread	7.3	21.3	250	–
Kidney	0.9	2.9	329	–
Brains	2.5	10.7	1,747	–
Heart	1.4	4.8	164	–
Poultry				
Chicken (without skin)				
Light (roasted)	.9	3	64	–
Dark (roasted)	2.1	7	64	–
Turkey (without skin)				
Light (roasted)	0.9	3	59	–
Dark (roasted)	2.1	6.1	72	–
Fish				
Haddock	0.1	0.08	33	0.22
Flounder	1.1	6	44	0.47
Salmon	1.2	7.0	40	1.88
Tuna, light, canned in water	0.13	<1	10	0.24
Shellfish				
Crustaceans				
Lobster	0.1	0.5	61	0.07
Crab meat				
Alaskan King Crab	0.2	1	65	0.38
Blue Crab	0.2	2	85	0.45
Shrimp	0.25	1	166	0.28
Mollusks				
Abalone	0.25	1	143	0.15
Clams	0.2	1.7	57	0.33
Mussels	0.7	3.8	48	0.70
Oysters	1.3	4.2	93	1.06
Scallops	0.31	1	40	0.36
Squid	0.35	1	227	0.84

Food	Milligrams per serving	% DV*
Selected Food Sources of Heme Iron		
Chicken liver, cooked, 3½ ounces	12.8	70
Oysters, breaded and fried, 6 pieces	4.5	25
Beef, chuck, lean only, braised, 3 ounces	3.2	20
Clams, breaded, fried, ¾ cup	3.0	15
Beef, tenderloin, roasted, 3 ounces	3.0	15
Turkey, dark meat, roasted, 3½ ounces	2.3	10
Beef, eye of round, roasted, 3 ounces	2.2	10
Turkey, light meat, roasted, 3½ ounces	1.6	8
Chicken, leg, meat only, roasted, 3½ ounces	1.3	6
Tuna, fresh bluefin, cooked, dry heat, 3 ounces	1.1	6
Chicken, breast, roasted, 3 ounces	1.1	6
Halibut, cooked, dry heat, 3 ounces	0.9	6
Crab, blue crab, cooked, moist heat, 3 ounces	0.8	4
Pork, loin, broiled, 3 ounces	0.8	4
Tuna, white, canned in water, 3 ounces	0.8	4
Shrimp, mixed species, cooked, moist heat, 4 large	0.7	4
Egg, 1 whole	0.6	4
Selected Food Sources of Nonheme Iron		
Ready-to-eat cereal, 100% iron fortified, ¾ cup	18.0	100
Oatmeal, instant, fortified, prepared with water, 1 cup	10.0	60
Soybeans, mature, boiled, 1 cup	8.8	50
Lentils, boiled, 1 cup	6.6	35
Beans, kidney, mature, boiled, 1 cup	5.2	25
Beans, lima, large, mature, boiled, 1 cup	4.5	25
Beans, navy, mature, boiled, 1 cup	4.5	25
Ready-to-eat cereal, 25% iron fortified, ¾ cup	4.5	25
Beans, black, mature, boiled, 1 cup	3.6	20
Beans, pinto, mature, boiled, 1 cup	3.6	20
Molasses, blackstrap, 1 tablespoon	3.5	20
Tofu, raw, firm, ½ cup	3.4	20
Spinach, boiled, drained, ½ cup	3.2	20
Spinach, canned, drained solids, ½ cup	2.5	10
Black-eyed peas (cowpeas), boiled, 1 cup	1.8	10
Spinach, frozen, chopped, boiled, ½ cup	1.9	10
Grits, white, enriched, quick, prepared with water, 1 cup	1.5	8
Raisins, seedless, packed, ½ cup	1.5	8
Whole wheat bread, 1 slice	0.9	6
White bread, enriched, 1 slice	0.9	6
Balance (Energy) Bar, 1 bar	4.5	25
Clif Bar, 1 bar	4.5	25
Luna Bar	6.3	35

*DV 5 Daily Value. DVs are reference numbers developed by the Food and Drug Administration (FDA) to help consumers determine if a food contains a lot or a little of a specific nutrient. The FDA requires all food labels to include the percent DV (%DV) for iron. The percent DV tells you what percent of the DV is provided in one serving. The DV for iron is 18 milligrams (mg). A food providing 5% of the DV or less is a low source while a food that provides 10%–19% of the DV is a good source. A food that provides 20% or more of the DV is high in that nutrient. It is important to remember that foods that provide lower percentages of the DV also contribute to a healthful diet. For foods not listed in this table, please refer to the U.S. Department of Agriculture's Nutrient Database Web site: www.nal.usda.gov/fnic/cgi-bin/nut_search.pl.

Sources: U.S. Department of Agriculture, Agricultural Research Service. 2003. USDA Nutrient Database for Standard Reference, Release 16. Nutrient Data Laboratory Home Page, http://www.nal.usda.gov/fnic/foodcomp. Office of Dietary Supplements, National Institutes of Health. Dietary Supplement Fact Sheet: Iron [monograph on the Internet]. Bethesda (MD): Office of Dietary Supplements; 2005. Available at http://ods.od.nih.gov/factsheets/iron.asp#en10.

Dietary and Supplemental Calcium Guide

Food	Serving Size	Approximate Calcium Content (mg)*
Calcium fortified orange juice	1 cup	290–300
Milk	1 cup	285–300
Dry Powdered Milk	½ cup	400
Evaporated Skim Milk	½ cup	400
Sardines	3 oz	300 mg
Yogurt	1 cup	275–450
Salmon, canned, with bone	3 oz	205
Cheese	1 oz	175–275
Tofu, firm	½ cup	155–260
Ice cream	½ cup	90–135
Frozen yogurt	½ cup	105
Nonfat cream cheese	1 oz	100
Almonds	1 oz	100
Greens	½ cup	50–100
White beans	½ cup	100
Dried figs	3	80
Broccoli	½ cup	45
Soy milk**	1 cup	Varies widely by brand (check label)
Fortified cereals	1 serving	Varies widely by brand (check label)
Multivitamin with minerals	1 dose	0–210

*Calcium content of foods may vary; read labels to determine the actual calcium content of a certain food.

**The nutrient content of soy milk varies greatly depending on the manufacturing process and whether the product is fortified with nutrients such as calcium.

Reading Labels

The food label on most products will list its calcium content. The amount of calcium in a product is expressed as a percentage (%). This percentage is based on the recommended calcium intake for many adults of 1000 mg per day.

Example: A product label says the food contains 30% of the daily need for calcium:

30% of 1000 mg per day = 300 mg of calcium

These percentages are meant to be used as a guide; however, since individuals may have different calcium needs, they may not be exact for all adults.

Calcium Needs (mg)	Calcium Needed Per Day Based on Food Labels
1000	100%
1200	120%
1500	150%

Calcium Supplements

If you are unable to take in enough calcium through the diet, calcium supplements are available to help meet calcium needs. The most common types of calcium supplements are calcium carbonate and calcium citrate.

Calcium is best absorbed in doses of 500 mg or less. If you need to take more than 500 mg of calcium supplements per day, consider taking several smaller doses throughout the day.

Commonly Available Calcium Supplements				
Brand	Calcium (mg) Per Tablet	Vitamin D (IU) Per Tablet	Approximate Cost Per Tablet	Calcium Source
Tums®	200	0	$0.02	Calcium carbonate
Extra Strength Tums®	300	0	$0.04	Calcium carbonate
Oscal® 500	500	0	$0.10	Calcium carbonate
Oscal® 500 + D	500	200	$0.11	Calcium carbonate
Caltrate® 600	600	0	$0.09	Calcium carbonate
Caltrate® 600 + D	600	200	$0.09	Calcium carbonate
Caltrate® 600 Plus®*	600	200	$0.11	Calcium carbonate
Viactiv®*	500	100	$0.11	Calcium carbonate
Citrical®	200	0	$0.07	Calcium citrate
Citrical® + D	315	200	$0.11	Calcium citrate

*Also contain additional vitamins &/or minerals; Viactiv® contains 20 calories per piece.

Reprinted (with modifications to table of food sources) from: University of Virginia Health System Digestive Health Center of Excellence. Calcium and Vitamin D [monograph on the Internet]. Charlottesville (VA): University of Virginia; 2005. Available at http://www.healthsystem.virginia.edu/internet/digestive-health/nutrition/calcium1.cfm.

Appendix M4—Resistant Starch Guide

Types of Resistant Starch

(http://www.resistantstarch.com/NR/rdonlyres/
1CAF6D0F-5648-464E-954E-0E341B995F3A/0/
RSConsumptionFactSheet.pdf)

RS1: Physically inaccessible starch • e.g., partly milled grains and seeds and legumes

RS2: Granular starch/Native starch granule • e.g., native un-cooked potato starch and green banana

RS3: Retrograded starch, mainly retrograded amylase • e.g., cooled-cooked potato, bread, corn flakes

Examples of Naturally-Occurring Resistant Starch*		
Food	Resistant Starch (Amount in gms)	Types of Resistant Starch
Navy beans ½ cup cooked	9.8	RS1–RS3
Banana, raw 1 medium, peeled	4.7	RS2
Cold potato ½" diameter	3.2	RS3
Lentils ½ cup cooked	2.5	RS1–RS3
Cold pasta 1 cup	1.9	RS1
Pearl barley ½ cup cooked	1.6	RS3
Oatmeal 1 cup cooked	0.7	RS1–RS3
Wholegrain bread 2 slices	0.5	RS3

*Amounts may vary

Appendix N—Common Medical Abbreviations

AAL	anterior axillary line
ab lib	at pleasure; as desired (ab libitum)
ac	before meals
ACTH	adrenocorticotropic hormone
AD	Alzheimer's disease
ad lib	as desired (ad libitum)
ADA	American Dietetic Association, American Diabetes Association
ADL	activities of daily living
AGA	antigliadin antibody
AIDS	acquired immunodeficiency syndrome
ALP (Alk phos)	alkaline phosphatase
ALS	amyotrophic lateral sclerosis
ALT	alanine aminotransferase
amp	ampule
ANC	absolute neutrophil count
ANCA	antisaccharomyces antibodies
AP	anterior posterior
ARDS	adult respiratory distress syndrome
ARF	acute renal failure, acute respiratory failure
ASA	acetylsalicylic acid, aspirin
ASCA	antineutrophil cytoplasmic antibodies
ASHD	arteriosclerotic heart disease
AV	arteriovenous
BANDS	neutrophils
BCAA	branched-chain amino acids
BE	barium enema
BEE	basal energy expenditure
BG	blood glucose
bid	twice a day (bis in die)
bili	bilirubin
BM	bowel movement
BMI	body mass index
BMR	basal metabolic rate
BMT	bone marrow transplant
BP (B/P)	blood pressure
BPD	bronchopulmonary dysplasia
BPH	benign prostate hypertrophy
bpm	beats per minute, breaths per minute
BS	bowel sounds, breath sounds, or blood sugar
BSA	body surface area

BUN	blood urea nitrogen
c	cup
c	with
C	centigrade
C.C.E.	clubbing, cyanosis, or edema
c/o	complains of
CA	cancer; carcinoma
CABG	coronary artery bypass graft
CAD	coronary artery disease
CAPD	continuous ambulatory peritoneal dialysis
cath	catheter, catheterize
CAVH	continuous arteriovenous hemofiltration
CBC	complete blood count
cc	cubic centimeter
CCK	cholecystokinin
CCU	coronary care unit
CDAI	Crohn's disease activity index
CDC	Centers for Disease Control and Prevention
CHD	coronary heart disease
CHF	congestive heart failure
CHI	closed head injury
CHO	carbohydrate
CHOL	cholesterol
CKD	chronic kidney disease
cm	centimeter
CNS	central nervous system
COPD	chronic obstructive pulmonary disease
CPK	creatinine phosphokinase
Cr	creatinine
CR	complete remission
CSF	cerebrospinal fluid
CT	computed tomography
CVA	cerebrovascular accident
CVD	cardiovascular disease
CVP	central venous pressure
CXR	chest X-ray
d/c	discharge
D/C	discontinue
D_5NS	dextrose, 5% in normal saline
D_5W	dextrose, 5% in water

(continued)

DASH	Dietary Approaches to Stop Hypertension
DBW	desirable body weight
DCCT	Diabetes Control and Complications Trial
DKA	diabetic ketoacidosis
dL	deciliter
DM	diabetes mellitus
DRI	Dietary Reference Intake
DTR	deep tendon reflex
DTs	delirium tremens
DVT	deep vein thrombosis
Dx	diagnosis
e.g.	for example
ECF	extracellular fluid
ECG/EKG	electrocardiogram
EEG	electroencephalogram
EGD	esophagogastroduodenoscopy
ELISA	enzyme-linked immunosorbent assay
EMA	antiendomysial antibody
EMG	electromyography
EOMI	extra-ocular muscles intact
ER	emergency room
ERT	estrogen replacement therapy
ESR	erythrocyte sedimentation rate
F	Fahrenheit
FACSM	Fellow, American College of Sports Medicine
FBG	fasting blood glucose
FBS	fasting blood sugar
FDA	Food and Drug Administration
FEF	forced expiratory flow
FEV	forced expiratory volume
FFA	free fatty acid
FH	family history
FTT	failure to thrive
FUO	fever of unknown origin
FVC	forced vital capacity
FX	fracture
g	gram
g/dL	grams per deciliter
GB	gallbladder
GERD	gastroesophageal reflux disease
GFR	glomerular filtration rate
GI	gastrointestinal
GM-CSF	granulocyte/macrophage colony stimulating factor
GTF	glucose tolerance factor
GTT	glucose tolerance test
GVHD	graft versus host disease
h	hour
H & P (HPI)	history and physical

HAV	hepatitis A virus
HbA$_{1c}$	glycosylated hemoglobin
HBV	hepatitis B virus
HC	head circumference
Hct	hematocrit
HCV	hepatitis C virus
Hgb	hemoglobin
HDL	high-density lipoprotein
HEENT	head, eyes, ears, nose, throat
Hg	mercury
Hgb	hemoglobin
HHNS	hyperosmolar hyperglycemic nonketotic (syndrome)
HIV	human immunodeficiency virus
HLA	human leukocyte antigen
HOB	head of bed
HR	heart rate
HS or h.s.	hours of sleep
HTN	hypertension
Hx	history
I & O (I/O)	intake and output
i.e.	that is
IBD	inflammatory bowel disease
IBS	irritable bowel syndrome
IBW	ideal body weight
ICF	intracranial fluid
ICP	intracranial pressure
ICS	intercostal space
ICU	intensive care unit
IGT	impaired glucose tolerance
IM	intramuscularly
inc	incontinent
IU	international unit
IV	intravenous
J	joule
K	potassium
kcal	kilocalorie
KCl	potassium chloride
kg	kilogram
KS	Kaposi's sarcoma
KUB	kidney, ureter, bladder
L	liter
lb	pounds
LBM	lean body mass
LCT	long-chain triglyceride
LDH	lactic dehydrogenase
LES	lower esophageal sphincter
LFT	liver function test
LIGS	low intermittent gastric suction

LLD	left lateral decubitus position
LLQ	lower left quadrant
LMP	last menstrual period
LOC	level of consciousness
LP	lumbar puncture
LUQ	left upper quadrant
lytes	electrolytes
MAC	midarm circumference
MAMC	midarm muscle circumference
MAOI	monoamine oxidase inhibitor
MCHC	mean corpuscular hemoglobin concentration
MCL	midclavicular line
MCT	medium-chain triglyceride
MCV	mean corpuscular volume
mEq	milliequivalent
mg	milligram
Mg	magnesium
MI	myocardial infarction
mm	millimeter
mmHg	millimeters of mercury
MNT	medical nutrition therapy
MODY	maturity onset diabetes of the young
MOM	Milk of Magnesia
mOsm	milliosmol
MR	mitral regurgitation
MRI	magnetic resonance imaging
MS	multiple sclerosis, morphine sulfate
MVA	motor vehicle accident
MVI	multiple vitamin infusion
N	nitrogen
N/V	nausea and vomiting
NG	nasogastric
NH_3	ammonia
NICU	neurointensive care unit, neonatal intensive care unit
NKA	no known allergies
NKDA	no known drug allergies
NPH	neutral protamine Hagedorn insulin
NPO	nothing by mouth
NSAID	nonsteroidal antiinflammatory drug
NTG	nitroglycerin
O_2	oxygen
OA	osteoarthritis
OC	oral contraceptive
OHA	oral hypoglycemic agent
OR	operating room
ORIF	open reduction internal fixation
OT	occupational therapist
OTC	over the counter

$paCO_2$	partial pressure of dissolved carbon dioxide in arterial blood
paO_2	partial pressure of dissolved oxygen in arterial blood
pc	after meals (post cibum)
PCM	protein-calorie malnutrition
PD	Parkinson's disease
PE	pulmonary embolus
PED	percutaneous endoscopic duodenostomy
PEEP	positive end expiratory pressure
PEG	percutaneous endoscopic gastrostomy
PEM	protein-energy malnutrition
PERRLA	pupils equal, round, and reactive to light and accommodation
pH	hydrogen ion concentration
PKU	phenylketonuria
PMI	point of maximum impulse
PMN	polymorphonuclear
PN	parenteral nutrition
PO	by mouth (per os)
PPD	packs per day
PPN	peripheral parenteral nutrition
prn	may be repeated as necessary (pro re nata)
pt	patient
PT	patient, physical therapy, prothrombin time
PTA	prior to admission
PTT	prothromboplastin time
PUD	peptic ulcer disease
PVC	premature ventricular contraction
PVD	peripheral vascular disease
q	every (quaque)
qd	every day (quaque die)
qh	every hour (quaque hora)
qid	four times daily (quater in die)
qns	quantity not sufficient (quantum non sufficiat)
qod	every other day
R/O	rule out
RA	rheumatoid arthritis
RBC	red blood cell
RBW	reference body weight
RD	registered dietitian
RDA	Recommended Dietary Allowance
RDS	respiratory distress syndrome
REE	resting energy expenditure
RLL	right lower lobe
RLQ	right lower quadrant
ROM	range of motion
ROS	review of systems
RQ	respiratory quotient
RR	respiratory rate

(continued)

RUL	right upper lobe	tid	three times daily (ter in die)
RUQ	right upper quadrant	TKO	to keep open
Rx	take, prescribe, or treat	TLC	total lymphocyte count
s̄	without	TNM	tumor, node, metastasis
S/P	status post	TPN	total parenteral nutrition
SBGM	self blood glucose monitoring	TSF	triceps skinfold
SBO	small bowel obstruction	TSH	thyroid stimulating hormone
SBS	short bowel syndrome	TURP	transurethral resection of the prostate
SGOT	serum glutamic oxaloacetic transaminase	U	unit
SGPT	serum glutamic pyruvic transaminase	UA	urinalysis
SOB	shortness of breath	UBW	usual body weight
SQ	subcutaneous	UL	Tolerable Upper Intake Level
ss	half	URI	upper respiratory intake
stat	immediately	UTI	urinary tract infection
susp	suspension	UUN	urine urea nitrogen
T	temperature	VLCD	very-low-calorie diet
T & A	tonsillectomy and adenoidectomy	VOD	venous occlusive disease
T, tbsp	tablespoon	VS	vital signs
t, tsp	teaspoon	w.a.	while awake
T_3	triiodothyronine	WBC	white blood cell
T_4	thyroxine	WNL	within normal limits
TB	tuberculosis	wt	weight
TEE	total energy expenditure	WW	whole wheat
TF	tube feeding	yo	year old
TG	triglyceride		
TIA	transient ischemic attack		
TIBC	total iron binding capacity		

Note: Abbreviations can vary from institution to institution. Although the student will find many of the accepted variations listed in this appendix, other references may be needed to supplement this list.

Appendix O—Conversion Factors: Milliequivalents/Milligrams of Electrolytes

Milliequivalents to Milligrams

Cations		Anions	
Milliequivalents	Milligrams	Milliequivalents	Milligrams
1 mEq Potassium (K^+)	39 mg	1 mEq Chloride (Cl^-)	35.5 mg
1 mEq Sodium (Na^{2+})	23 mg	1 mEq Bicarbonate (HCO_3^-)	61 mg
1 mEq Calcium (Ca^{2+})	20 mg	1 mEq Potassium (PO_4^{3-})	31.67 mg
1 mEq Magnesium (Mg^{2+})	12.2 mg		

The equivalent weight of an electrolyte is its molecular weight divided by its valence. Therefore, because the molecular weight of K^+ is 39 and its valance is one, 39/1 is 39 grams. Milliequivalents would be 1/1000 of the equivalents or 39 milligrams. One milliequivalent of Na^+ is 23 milligrams [(23 grams/1) divided by 1000].

Glossary

2,3-diphosphoglycerate (DPG)—an important regulator for the affinity of hemoglobin for oxygen. The synthesis of 2,3-biphosphoglycerate in red blood cells (RBC) is critical for controlling hemoglobin affinity for oxygen

24-hour energy expenditure—the total amount of energy expended by a human in a 24-hour period, made up of three main components: resting energy expenditure, thermic effect of food, and physical activity-related energy expenditure

24-hour recall—dietary assessment method in which the clinician interviews the client to obtain a list of all foods/beverages consumed in the previous 24 hours

absolute neutrophil count (ANC)—a measure of the number of neutrophil granulocytes (also known as polymorphonuclear cells, PMNs, polys, granulocytes, segmented neutrophils, or segs) present in the blood; neutrophils are a type of white blood cell that fights against infection

acanthrosis nigricans—diffuse, velvety-thickening hyperpigmenation of the skin; it may be present at the nape of the neck, axillae, area beneath the breasts, intertriginous areas (where skin touches or rubs together), and exposed areas (elbows, knuckles); thought to be the result of insulin resistance

acarbose—also known as Precose™; a medication—an alpha glucoside inhibitor—that slows the digestion of starch; used in diabetes treatment and to prevent dumping syndrome

acetylation—in genomics, modification of histones by attachment of acetyl groups

acetylcholine—excitatory neurotransmitter involved in stimulation of parietal cells

achalasia—motility disorder characterized by an absence or weakened peristalsis within the esophagus

achlorhydria—lack of gastric hydrochloric acid secretions

acid-base balance—maintenance of homeostasis between acidity and alkalinity within body systems

acidemia—condition of excess acid in the blood consistent with a pH <7.35

acidosis—conditions that produce excess acid in the blood

acquired immune deficiency syndrome or acquired immunodeficiency syndrome (AIDS)—immune dysfunction characterized by the destruction of immune cells, leaving the body open to infection

acquired immunity—immune response that results from exposure to an antigen or immunoglobulin

active immunity—immunity produced due to exposure to an antigen (e.g., infection or vaccination)

acute coronary syndrome—condition characterized by an episode of acute unstable angina

acute (fulminant) hepatic/liver failure—the severe impairment of hepatic functions in the absence of preexisting liver disease

acute renal failure (ARF)—refers to kidney dysfunction of short duration or any sudden, severe impairment of kidney function

acute respiratory distress syndrome (ARDS)—respiratory failure (RF) resulting from an acute insult to the lungs that occurs when the respiratory system is no longer able to perform its normal functions

acute tubular necrosis (ATN)—a common cause of acute renal failure when cells of the renal tubules die. Often associated with trauma and other serious illnesses.

acute-phase proteins—proteins that increase or decrease in the blood during an inflammatory process, such as C-reactive protein and fibrinogen

ADA's Evidence-Based Guides for Practice—specific nutrition practice guidelines available on ADA's Evidence Analysis Library

adaptive thermogenesis—energy expenditure above and beyond the thermic effect of food and resting energy expenditure that is seen in response to overfeeding, traumatic injury, changes in hormonal status, and exposure to a cold environment

Addison's disease—also known as *chronic adrenal insufficiency*; a rare endocrine disorder in which the adrenal glands do not produce enough steroid hormones; some common symptoms include fatigue, dizziness, muscle weakness, weight loss, difficulty in standing up, vomiting, anxiety, diarrhea, headache, sweating, changes in mood and personality, and joint and muscle pains

adhesion—scar tissue that forms between two body surfaces, usually as a result of surgery or injury

adjuvant—usually, treatment "in addition to" initial treatment; for example, one or more anticancer drugs may be used in combination with surgical therapy as part of the cancer treatment regimen

adjuvant chemotherapy—the use of drugs as additional treatment for patients with cancers that are thought to have spread outside their original sites

adrenocortical hormones—hormones secreted by the adrenal gland which include glucocorticoids, androgens, estrogens, etc.

aerophagia—the swallowing of air resulting in gas and bloating

afferent—carrying blood to the designated site; for example, the afferent arteriole carries blood to the glomerulus

ageusia—inability to taste

AIDS-related wasting syndrome (AWS)—defined by the Centers for Disease Control and Prevention (CDC) as a 10% weight loss without an identifiable cause that is accompanied by fever or diarrhea for 30 days or more

akinesia—loss of voluntary movement

alanine shunt—a process that allows for the production of glucose from protein sources by the conversion of the amino acid alanine to pyruvate to glucose in the liver; the glucose produced can then be used in the muscle to generate additional alanine

alcoholic liver disease (ALD)—liver disease associated with alcoholism

algorithm—a finite set of well-defined instructions for accomplishing a task; given an initial state, an algorithm will terminate in a corresponding recognizable end-point

alkalemia—condition of excess base in the blood consistent with a pH >7.45

alkalosis—conditions that produce excess base in the blood

allele—a copy of a specific gene situated in a given locus on a chromosome

allergen—an antigen that triggers an allergic response

allergy—an inappropriate and harmful immune reaction to a harmless nonpathogenic substance; also called *hypersensitivity*

allogeneic—having a different genetic composition; in bone marrow transplant, refers to receipt of bone marrow from a donor of different genetic composition.

allograft—a tissue/organ graft between two genetically different individuals from the same species

alpha-glucosidase—a digestive enzyme found in the brush border cells of the small intestine that cleaves more complex carbohydrates into sugars

alternative medicine—unconventional therapeutic systems used by clients in place of or parallel to conventional biomedicine; typically administered by trained practitioner

Alzheimer's disease—the most common form of dementia, characterized by formation of amyloid plaques in the brain and neurofibrillary tangles within neurons

amenorrhea—the absence of menstrual cycles when they would be expected to occur

amylin—a hormone synthesized by pancreatic ß-cells that contributes to glucose control during the postprandial period

amyloid—a starch-like substance present in diseased tissues

amyloid plaques—cellular deposits found between nerve cells

amyloid precursor protein (APP)—protein from which beta-amyloid is formed

amylopectin—the insoluble component of starch

amyotrophic lateral sclerosis—a progressive neurological disease that causes destruction of the motor neurons of the nervous system, resulting in muscle weakness, twitching, and atrophy; also known as *Lou Gehrig's disease*

anabolic—refers to building up or synthesis of larger organic molecules from smaller organic molecular subunits

anaphylactic shock—a life-threatening IgE-mediated allergic reaction; in humans, symptoms include swelling (especially of the lips and face), vomiting, diarrhea, difficulty in breathing, and a sudden drop in blood pressure; also called *anaphylaxis*

anasarca—generalized edema with accumulation of serum in the connective tissue

anastomosis—the surgical connection of body parts, especially hollow tubular parts like those of the GI tract

anemia—abnormal blood constituents resulting from various etiologies; anemia is a symptom and is often a result of the decrement in blood constituents, although some forms of elevated blood components that are nonfunctional may be referred to as an "anemia"

anergy—antigen-specific nonresponsiveness by a T or B cell in which the cell is present but cannot respond

aneurysm—a weakened portion of a blood vessel wall

angina—chest pain caused by oxygen deficit to the heart; two forms are stable and unstable

angiogenesis—the formation of new blood vessels and expanded systems for nutrient delivery and waste removal; a result of chemokines and hormonal messages that up-regulate the formation of these processes; cancer cells can produce various messengers that trigger this up-regulation, thus allowing the rapidly dividing abnormal cells to acquire materials for growth and spread

anion gap (AG)—the difference between unmeasured anions and cations; anion gap (AG) = (serum Na^+) – (serum Cl^- + HCO_3^-); normal AG = 12 to 14 mEq/L

ankle brachial index (ABI)—ratio of Doppler-recorded systolic blood pressures between upper and lower extremities; a measure of peripheral vascular disease

anorexia—lack of appetite

anorexigenic—appetite inhibiting

anovulation—lack of ovulation during the menstrual cycle

anti-emetic—a pharmacologic agent that reduces nausea

antibody—a protein molecule found in serum and tissues that is secreted by B cells in response to a specific antigen that can bind to that antigen and neutralize or help destroy it; also called *immunoglobulin*

anticariogenic—describes foods or conditions that assist in prevention of dental caries

anticodons—tRNA coding sequences; these sequences are complementary to the codons in mRNA and thus serve as anticodons

antidiuresis—inhibition of water losses through the kidney's reaction to hormones and abnormal cell signals, which reduce tubular losses

antigen—a substance that is specifically bound by an antibody or lymphocytes; used by the immune system to recognize pathogens and altered cells; see *immunogen*

antigen-presenting cell (APC)—a cell capable of displaying fragments of antigens from pathogens and altered cells joined to major histocompatability molecules on its surface in a manner that can be recognized by T cells

antiretroviral (ARV)—refers to medications targeted to interrupt the retrovirus life cycle

antiretroviral therapy (ART)—refers to the combination of medications that are typically used for controlling and reducing viral load

antisense strand—the noncoding strand of DNA

antiseptics—agents that kill microbes within living tissue

antitoxin—an antibody to an exotoxin

apheresis—removal of harmful blood components or substitution of more desirable constituents in blood

aplastic anemia—idiopathic anemia from abnormal, deficient, or absent red cell production due to bone marrow disorders

apolipoprotein—protein portion of the lipoprotein; provides cellular stability and allows for cellular recognition and binding

apolipoprotein E—a protein that carries cholesterol in blood and that appears to play a role in brain function

apoptosis—genetically programmed cell death

arginine vasopressin (AVP)—previously known as *anti-diuretic hormone*; a hormone that acts on the renal tubules to reduce urine output in response to dehydration and hyperosmolality

aromatic amino acids (AAA)—amino acids containing an aromatic side chain (phenylalanine, tyrosine, and tryptophan)

arteriosclerosis—a general term for thickening of the walls of the blood vessels with a resulting loss of vascular elasticity and narrowed lumen

arteriovenous fistula (AVF)—a connection of an artery and vein to provide circulatory access for hemodialysis

arteriovenous graft (AVG)—the planting of an artificial vessel connecting an artery and vein; used when patients' blood vessels are fragile and a fistula is not feasible

ascites—accumulation or retention of free fluid within the peritoneal cavity

aspiration—inspiration of foreign matter into the lung

aspiration pneumonia—an inflammatory response in the lung that results from aspiration of inhaled materials (saliva, nasal secretions, bacteria, liquids, food, or gastric contents) into the airway below the level of the vocal cords

asterixis—abnormal involuntary movements that primarily affect the extremities

asthma—a chronic inflammatory disorder of the airway involving many cells and cellular elements such as mast cells, eosinophils, T lymphocytes, macrophages, neutrophils, and epithelial cells; it is triggered by either an IgE allergic reaction or nonallergic factors and results in inflammation of the airway and reversible airway obstruction

asymmetric muscle weakness—muscle weakness occurring unequally in different parts or sides of the body

ataxia—a nonspecific clinical manifestation implying dysfunction of parts of the nervous system that coordinate movement

atherosclerosis (AS)—thickening of the blood vessel walls specifically caused by the presence of plaque

atopic—refers to a milder IgE-mediated allergic response

atrophic gastritis—atrophy of the lining of the stomach, which contains the parietal cells that produce intrinsic factor, proteases, and hydrochloric acid; this form of inflammation is often accompanied by bacterial overgrowth due to elevation of the gastric pH

atrophy—reduction in size of muscle cells; wasting of body tissue that occurs from disuse, disease, or malnutrition

attenuated—refers to an antigen rendered less virulent but still capable of eliciting an immune response

auscultation—a technique used during physical examination in which a stethoscope is used to evaluate the sounds created in body organs

autoantibody—an antibody to self-antigens

autocrine—a type of communication between hormones and other chemical messengers that is released from a cell at a distance from the target cell

autograft—a tissue graft from one area to another on the same individual

autoimmunity—an immune response to one's own tissues

autologous—transplant of one's own body tissue; in bone marrow transplant, refers to treatment through receipt of one's own bone marrow

autonomic division—components of the nervous system that control involuntary functions of the body

autonomic dysreflexia—a complication of spinal cord injury and paralysis; combination of stimuli that result in sudden increase in blood pressure and changes in heart function

autosomal dominant—an inheritance pattern of a dominant allele on an autosome

autosomal recessive inheritance—inheritance of a trait as the result of inheriting a recessive gene for a particular trait from each parent

autosomal recessive—an inheritance pattern of a recessive allele on an autosome; method of hereditary disease transmission in which the patient receives two chromosomes bearing the gene anomaly, one from each parent

autosomes—non-sex-determining chromosomes; a human has 22 autosomes

avidity—measure of the strength of antigen-antibody binding

axon—part of a neuron that transmits outgoing signals to other neurons

axon terminals—structure at the end of an axon that releases neurotransmitters

azotemia—a buildup of nitrogenous waste products such as urea in the blood and body fluids

B cell—a lymphocyte derived from the bone marrow, which differentiates into a plasma cell that makes an antibody

bacterial overgrowth syndrome—malabsorption and malnutrition that result from cross-contamination of bacteria from the colon to the small intestine

balneotherapy—treating disease by bathing; it can involve hot water, steam, or application of hot packs to the body; cold water; contrasting hot and cold baths or showers; a steam bath followed by immersion in cold water; or immersion in mineral springs or a hot bath to which various medicinal herbs or minerals have been added

BALT (bronchial-associated lymphatic tissue)—secondary lymphoid organs of the bronchial tree

bariatric—referring to medical treatment of morbid obesity

baroreceptor—in general, any sensor of pressure changes

Barrett's esophagus—a complication of severe chronic GERD involving changes in the cells of the tissue that line the bottom of the esophagus; these esophageal cells become irritated when the contents of the stomach back up, and there is a small but definite increased risk of cancer of the esophagus; a pre-malignant condition that is considered a risk factor for esophageal adenocarcinoma

basal energy expenditure—the minimum level of energy expended by the body to sustain life; it is measured in the morning when a subject is in a postabsorptive state, comfortably lying motionless in a supine position, and in a thermally neutral environment

basophils—polymorphonuclear leukocytes containing granules that stain with basic dyes; they have much in common with mast cells, including the release of histamine and leukotrienes, which contribute to allergic responses and inflammation

beta-amyloid—a part of the amyloid precursor protein found in the insoluble deposits outside neurons, which forms the core of plaques

beta-hydroxybutyrate (BHB)—ketone measured in serum as a marker for patients on the ketogenic diet

betadine—a povidone-iodine containing solution that is used topically to destroy microorganisms

bile—an emulsifying agent produced in the liver and eventually secreted into the duodenum

biliary cirrhosis—liver cirrhosis in which there is interference with intrahepatic bile flow

biliary sludge—a mixture of particulate matter and mucus that forms in bile

biliary stasis—intrahepatic impairment of bile flow

biliary tract (tree)—the common anatomical term for the path by which bile that has been secreted by the liver travels on its way to the small intestine

bilirubin—the yellow breakdown product of hemoglobin molecules produced by normal heme metabolism; it is normally excreted from the body via bile secretions

binding site—active site of the enzyme that binds to and acts on a particular substrate

biomarker—a biological molecule used as a marker to measure or indicate the effects or progress of a disease or condition

biotransformation—modification of a drug through metabolism

body cell mass (BCM)—kcalorie-using protein stores in the body; primarily muscle and organ tissues

body mass index (BMI)—weight in kilograms divided by height in meters squared ($BMI = kg \div m^2$); although technically not a body composition assessment technique, it correlates well with estimates of body composition derived from skinfold measurements and underwater weighing (hydrodensitometry), and can easily be calculated from weight and height; it is also known as Quetelet's index, named after its developer, Adolphe Quetelet (1796–1874), a Belgian statistician, astronomer, mathematician, and sociologist; the formula for calculating body mass index is: $BMI = (\text{weight in kg}) \div (\text{height in m})^2$

bolus feedings—rapid administration of 250 to 500 mL of formula several times daily

bone marrow—soft tissue in the cavities of bones where stem cells become red and white blood cells

bone resorption—a process whereby osteoclasts destroy an area of bone as the first step in bone remodeling

borborygmas—stomach "growling"

brachytherapy—a type of radiation therapy in which radioactive materials are placed in direct contact with the tissue being treated

bradykinesia—delayed or slowed body movements

brain stem—the part of the brain that connects the brain to the spinal cord and controls autonomic body functions

branched-chain amino acid—an amino acid that has a branched side chain; these include isoleucine, leucine, and valine

bronchial hyperreactivity—tendency of the smooth muscle of the tracheobronchial tree to narrow in response to a stimulus; present in virtually all symptomatic patients with asthma

bronchitis—a condition characterized by inflammation and eventual scarring of the lining of the bronchial tubes accompanied by restricted airflow, excessive mucus production, and a persistent cough

bronchopulmonary dysplasia (BPD)—a chronic lung disorder that may affect infants who have been exposed to high levels of oxygen therapy and ventilator support

buccal—refers to placement of a drug in the cheek

C-reactive protein—a protein released as a response to inflammation

CA-125—a protein that is secreted into the blood by ovarian cells and is used to monitor progress in the treatment of ovarian cancer

cachexia—weight loss, wasting of muscle, loss of appetite, and general debility that can occur during a chronic disease

calcitonin—a polypeptide hormone secreted by the "C" cells of the thyroid gland when the blood calcium concentration is high; it lowers blood calcium by inhibiting bone resorption, promoting bone formation, and reducing renal reabsorption of calcium and phosphorus

calculus—calcified deposits that have formed around the teeth

cancer—a class of diseases characterized by uncontrolled cell division and the ability of these cells to invade other tissues, either by direct growth into adjacent tissue (invasion) or by migration of cells to distant sites (metastasis)

carcinoembryonic antigen (CEA)—a glycoprotein present in fetal gastrointestinal tissue and in the cells or serum of adults having certain types of cancers; it is used clinically to monitor the effectiveness of a treatment, such as for colorectal cancer

carcinogen—substance that causes cancer

carcinogenesis—the multistep process (initiation, promotion, and progression) through which normal cells are transformed into cancer cells

cardiac cachexia—CVD-associated malnutrition/wasting syndrome characterized by extreme skeletal muscle wasting, fatigue, and anorexia

cardiac output—the volume of blood ejected from the left ventricle each minute; mathematically defined as heart rate × stroke volume

cariogenic—describes foods or conditions that contribute to dental caries

cariostatic—describes foods or conditions that neither contribute to nor prevent dental caries

carnitine—a substrate needed for the normal metabolism of fat for energy; it is mostly found in beef and lamb but can be synthesized endogenously from the amino acids L-lysine and L-methionine

carpopedal—referring or pertaining to the hand and foot

catecholamines—the chemical classification of adrenomedullary hormones

CD—"cluster designation"; an international nomenclature system of leukocyte cell surface molecules (CD number)

CD4—a marker found predominantly on helper T cells that interacts with MHC class II molecules on antigen-presenting cells

CD4 cell—immune cell that is one of the primary targets of HIV for infection

CD8—a marker found predominantly on cytotoxic T cells that interacts with MHC class I molecules on target cells

celiac disease (CD)—inflammation of the small intestine caused by gluten found in various grains, including wheat

cellular immunity—immune protection provided by the action of immune cells, especially T cells, polymorphonuclear leukocytes, and macrophages

central nervous system—the brain, spinal cord, and the associated nerves

central venous catheter (CVC)—intravenous access device inserted into large veins such as the subclavian, jugular, or femoral veins in the center of the body

cerebellum—the part of the brain that is responsible for maintaining the body's balance and coordination

cerebral cortex—the outer layer of nerve tissue surrounding the cerebral hemispheres

cerebral hemisphere—one side of the cerebrum; each side contains four lobes (frontal, parietal, occipital, and temporal)

ceruloplasmin—a protein used in copper transport

Chagas disease—a parasitic disease caused by *Trypanosoma cruzi*

chaperoning effect—refers to the effect of proteins ("chaperones") that assist the non-covalent folding/unfolding and the assembly/disassembly of other structures, such as enzymes, to help to improve their normal biological functions

cheilosis—fissures that develop at the edges of the mouth

chest physiotherapy—physical therapy that includes a variety of techniques designed to reduce or prevent infection by clearing pooled secretions and/or infected materials from the lungs

cholecalciferol—a naturally occurring form of vitamin D produced in humans when the precursor molecule 7-dehydrocholesterol present in the skin is exposed to sunlight or to ultraviolet radiation; also known as vitamin D_3

cholecystectomy—surgical removal of the gallbladder

cholecystitis—inflammation of the gallbladder

choledocholithiasis—gallstones that are present in the common bile duct but are usually formed in the gallbladder

cholelithiasis—the presence or formation of gallstones

cholestasis—condition in which the flow of bile from the liver is blocked

cholinergic—resembling acetylcholine; stimulated by or releasing acetylcholine or a related compound

chondroblasts—cells that are actively forming cartilage

chondrocytes—cells surrounded by cartilage and located inside small cavities known as *lacunae*

chondroitin—a nutritional supplement used by some to treat osteoarthritis; it is a specific glycosaminoglycan found in the proteoglycans of articular cartilage

chondroitin sulfate—the most common polysaccharide found in the proteoglycan molecules of cartilage

chromatin—the entire complement of DNA plus the histone proteins with which it is associated

chromosomes—units of the genome, each consisting of a long molecule of DNA that encodes numerous genes plus histone proteins; there are 22 autosomes and 2 sex chromosomes located within the nucleus of a human cell

chronic kidney disease (CKD)—kidney damage or GFR <60 mL/min/1.73m² for >3 months; kidney damage is defined as pathologic abnormalities or markers of damages

chronic myeloproliferative disease—long-term hyperplasia of hematological tissues, with concomitant overproduction of abnormal cells, growth factors, chemokines, cytokines, and hormones involved in hematopoiesis

chronic obstructive pulmonary disease (COPD)—a disease that limits airflow through either inflammation of the lining of the bronchial tubes or destruction of alveoli

chyme—partially digested food in a semifluid state

cicatrix—scar tissue formation with calcification or hardening of the connective tissue used in repair of tissue damage

cirrhosis—(1) any pathological condition where fibrous connective tissue invades an organ, usually as a consequence of inflammation or other injury; (2) end-stage liver disease characterized by damage to hepatic parenchymal cells with modular regeneration and fibrosis, associated with failure of hepatic cell function, interference with hepatic blood flow, frequently jaundice, portal hypertension, ascites, and ultimately hepatic failure

Class I MHC antigen—glycoproteins found on nucleated cells and encoded by the A, B, and C locus of the major histocompatibility complex; they present antigens to cytotoxic (CD8 +) T cells

Class II MHC antigen—glycoproteins found on nucleated cells and encoded by the Dr, Dq, or DP locus of the major histocompatibility complex; they present antigens to helper (CD4 +) T cells

claudication—pain in arms and legs due to inadequate blood flow to those muscles

clear liquid diet—diet consisting of liquids that contribute minimal residue to the gastrointestinal tract; includes fruit juices without pulp, carbonated sodas, broth, tea, coffee, water, popsicles, fruit ice, Jell-O (gelatin), and liquid nutritional supplements (e.g., Resource Breeze®)

clinical manifestations—signs and symptoms associated with a particular condition

clubbing—changes in fingers and toes due to hypoxemia; fingers and toes show a curve at a tip of the nail with flattening surface

codon—a series of three nucleotides in mRNA that encodes a specific amino acid

cofactors—vitamins or other nutrients needed for the proper function of certain enzymes

collagen—a fibrous protein found in connective tissue

colloid osmotic pressure (oncotic pressure)—the osmotic pressure attributed to proteins and other macromolecules

colonocyte—epithelial cell of the large intestine or colon

colostomy—a procedure in which the rectum only is surgically removed, and the end of the colon is attached to the stoma

combination chemotherapy—the use of two or more antineoplastic agents to achieve maximum destruction of malignant cells

combination therapy—the use of two or more therapeutic agents/processes for the treatment of a neoplasm

compassionate use—the use of an investigational drug outside of a clinical trial to treat a patient with a serious or immediately life-threatening disease or condition who has no comparable or satisfactory alternative treatment options; also called expanded access (Food and Drug Administration, http://www.fda.gov/ForConsumers/ByAudience/ForPatientAdvocates/AccesstoInvestigationalDrugs/ucm176098.htm)

complement—a group of serum proteins activated in a cascade that produces compounds that lyse cells and mediate immune reactions

complementary and alternative medicine (CAM)—a collective term for:
- **complementary medicine**—unconventional modalities used by clients in addition to conventional biomedicine; may involve practitioner, but often self-prescribed
- **alternative medicine**—unconventional therapyeutic systems used by clients in place of or parallel to conventional biomedicine; typically administered by trained practitioner

congestive heart failure (CHF) or heart failure (HF)—impairment of the ventricles' capacity to eject blood from the heart or to fill with blood

constipation—a decrease in frequency of bowel movements with straining with defecation and/or hard stools

continuous feedings—administration of formula for 10 to 24 hours daily, using a pump to control the feeding rate

continuous renal replacement therapy (CRRT)—type of renal replacement therapy used to treat patients in acute renal failure, particularly those with multiple organ failure; the types of patients treated tend to be hemodynamically unstable, have poor cardiac output, and be unable to tolerate hemodialysis

contracture—shortening of muscle tissue resulting in immobility

cor pulmonale—an increase in size of the right ventricle of the heart caused by resistance to the passage of blood through the lungs; can lead to heart failure

coronary artery disease (CAD)—general term for all causes of heart disease characterized by narrowing of vessels supplying blood to the heart

cortical bone—dense bone that forms the external surfaces of all bones, the shafts of the long bones, and a shell that caps the ends of the long bones

creatinine clearance—rate at which creatinine is filtered through the kidney; often used as a measure of kidney function

Crohn's disease—a chronic inflammatory bowel disease (IBD) that can affect the entire gastrointestinal tract but most commonly affects the ileum and colon

Curling's ulcer—ulceration of gastric or duodenal tissue as a result of burn or trauma

Current Procedural Terminology (CPT) codes—numeric codes used to describe a medical service; these codes were developed by American Medical Association with the Health Care Financing Administration

Cushing's syndrome—a disorder resulting from prolonged exposure to high levels of glucocorticoid hormones; symptoms include muscle weakness, thinning of the skin, moon-shaped face, weight gain, and diabetes mellitus

cyanosis—blue-tinged mucous membranes and skin due to inadequate oxygen supply

cyclosporine—immunosuppressant medication that is often prescribed after organ transplant

CYP 3A4—a specific cytochrome enzyme involved in drug metabolism

cystic fibrosis (CF)—disease characterized by abnormally thick mucus secretions from the epithelial surfaces of various organ systems, including the respiratory tract, the gastrointestinal tract, the liver, the genitourinary system, and the sweat glands

cystine—a sulfur-containing amino acid, which is produced by the actions of acids on proteins that contain this compound

cytochrome P-450 isoenzymes (CP450)—family of enzyme systems responsible for drug metabolism

cytokines—soluble substances (glycoproteins) secreted by one cell that cause it or other cells to proliferate, differentiate, migrate, or become activated (i.e., that act as local messengers or regional hormones); hundreds of different cytokines have been discovered and categorized, all with very distinct functions in the homeostatic and genetic controls over cellular function

cytologic—refers to tests that help to determine the morphologic features of a cell

cytotoxic T cells (Tc)—T lymphocytes that kill cells infected by viruses or transformed by cancer

dawn phenomenon—an increase in blood glucose in the early morning, most likely due to increased glucose production in the liver after an overnight fast

debride—to remove dead or injured tissue

dehiscence—separation of wound edges

dehydration—a deficit of water in the body

delayed hypersensitivity—a cell-mediated inflammatory allergic reaction in the skin (e.g., poison ivy) that takes 24 to 48 hours to appear

dendrite—branches extending out from the neuron that assist in transmission of impulses

dendritic cells—antigen-trapping and antigen-presenting white blood cells with nerve-like processes (e.g., Langerhans cells and interdigitating cells)

dental caries—decay of the teeth that begins when acid dissolves the enamel that covers the tooth

dentin—the hard tissue of the tooth surrounding the central core of nerves and blood vessels

dephosphorylate—to remove a phosphate group from an organic molecule

developed nation—a nation that is generally regarded as one with a high standard of living, a high per capita income, a well-developed infrastructure (e.g., public utilities and systems for transport, public health, and public education), high literacy, long life expectancy, and so on, when compared to the global average

developing nation—a nation that is generally regarded as one with a low standard of living, a low per capita income, a relatively poorly developed infrastructure (e.g., public utilities and systems for transport, public health, and public education), low literacy rates, low life-expectancy, and so on, when compared to the global norm

dextrans—glucose polymer used to expand plasma volume and chelate iron in solution

diabetes insipidus—chronic excretion of very large amounts of pale urine of low specific gravity

diabetes mellitus—a diverse group of disorders that share the primary symptom of hyperglycemia resulting from defective insulin production, insulin action, or both

dialysate—fluid used by the dialysis procedure to assist in removal of metabolic byproducts, wastes, and toxins; composition is determined by individual patient requirements

dialysis—renal replacement procedure that removes excessive and toxic byproducts of metabolism from the blood, thus replacing the filtering function of healthy kidneys

diarrhea—frequent or unusually liquid bowel movements

diastole—relaxation phase of the cardiac cycle; during this phase, ventricles empty and blood fills the atria

diastolic blood pressure—pressure that occurs as ventricles relax (diastole phase of the cardiac cycle)

dietary fiber—the component of plant matrix that serves as the plant cell wall or intercellular structure and is not digestible

diffusion—passage of particles through a semipermeable membrane

digoxin—cardiac glycoside that is prescribed to alter the contractions of the heart

dinucleotides—paired nucleotide sequences

direct calorimetry—a technique to determine energy expenditure using a highly sophisticated chamber capable of measuring the amount of heat released by a subject's body through evaporation, convection, and radiation

disinfectants—agents that kill microbes on inanimate objects or surfaces

dissolution—dissolving of a medication

diuresis—the production of excessive amounts of urine

diverticulitis—an acute inflammation of the diverticula

diverticulosis—an abnormal presence of outpockets or pouches (diverticula) on the surface of the small intestine or colon

DMT-1—transmembrane protein divalent metal transporter which traffics iron

docosehexanoic acid (DHA)—a fatty acid that can be produced from the essential omega-3 fatty acid linolenic acid, and that can also be found in many types of fish; DHA has been found to be beneficial for proper brain and vision development as well as in the management of hypertriglyceridemia

doubly labeled water—a technique to determine energy expenditure in which subjects drink a known amount of water containing two different stable isotopic forms of water: $H_2^{18}O$ and ^2H_2O; the rate that this water disappears from the subject's body is used to calculate the subject's energy expenditure

dumping syndrome—a group of symptoms that occurs with rapid passage of large amounts of food from the stomach into the small intestine; symptoms include dizziness, sweating, decreased blood pressure, and diarrhea

dysgeusia—abnormalities in or reduced ability to taste

dyspepsia—vague upper abdominal symptoms that may include upper abdominal pain, bloating, early satiety, nausea, or belching

dysphagia—difficulty swallowing

dysphonia—difficulty speaking

dysplasia—abnormal cell growth

dyspnea—shortness of breath or difficulty breathing

edema—the accumulation of excess fluid in cells, tissue, or a cavity, resulting in swelling

edentulous—without any teeth

efferent—carrying blood away from the designated site; for example, the efferent arteriole carries blood away from the glomerulus

ejection fraction—the percentage of the LVEDV that is ejected in the systolic phase; in normal, apparently healthy adults, the typical ejection fraction is 50% to 60%; defined mathematically as stroke volume ÷ LVEDV

electrolytes—those substances that bear an electrical charge (ions)

electroneutrality—state in which the sum of the charges of the anions equals the sum of the charges of the cations

embolus—blood clot that breaks from the cellular surface and freely moves through the circulation

emetogenic—an agent that causes nausea and/or vomiting

emphysema—a condition characterized by thinning and destruction of the alveoli, resulting in decreased oxygen transfer into the bloodstream and shortness of breath

enamel—hard outer layer of teeth consisting of hydroxyapatite; this mineral is composed of calcium, phosphorous, fluoride, chloride, sodium, and magnesium

encephalopathy—a degenerative disease of the brain

end-stage renal disease (ESRD)—kidney disease in which kidney function declines to 10% to 15% of normal; the term *CKD* is now preferred over this older term; ESRD is equivalent to CKD Stage 5 or when a patient requires renal replacement therapy

endorphins—neuropeptides that assist with pain control

endoscopy—examination of the interior of a canal by means of an endoscope

endotoxins—toxins found in bacteria, often as part of the cell wall, that stimulate an immune response

enteral nutrition (EN)—feeding through the gastrointestinal tract using a tube, catheter, or stoma that delivers nutrients distal to (or beyond) the oral cavity

environmental factors—social and economic factors (wages, transportation, etc.) that impact both lifestyle choices and the consumption of food and nutrients; other external factors such as food safety and sanitation determine the quality of food that is consumed, and food availability/access contributes to the amount and type of food consumed

eosinophil—a polymorphonuclear leukocyte containing granules that produce substances that damage parasites and decrease inflammation; these granules stain with acid dyes

eosinophilic esophagitis—abnormal infiltration of eosinophils into the esophagus; may be associated with food allergy as well as other conditions such as scleroderma, gastroesophageal reflux disease, and infection

epidemiology—the study of the rates of disease within a given population

epidural—refers to placement of a drug into the spinal fluid

epidural anesthesia—an anesthetic drug placed into the epidural space of the lumbar or sacral region of the spine, causing loss of sensation from the abdomen and pelvis to the lower limbs

epigastric—referring to the upper abdominal region

epigenetics—inheritance of information based on gene expression levels rather than gene sequence; regulated by genomic modifications such as DNA methylation and histone methylation or acetylation

epilepsy—a neurological disorder characterized by recurrent seizures, generally more than two unprovoked seizures

epinephrine—a chemical made by the adrenal gland that relaxes smooth muscles, constricts blood vessels, regulates arterial blood pressure, and prepares body for "fight or flight" responses; when it is used to treat severe allergic reactions, it is sometimes referred to as *adrenaline* (an older term)

ergocalciferol—a form of vitamin D produced by exposing the plant steroid ergosterol to ultraviolet irradiation; also known as vitamin D$_2$

eructation—belch or burp

eryptosis—suicidal death of erythrocytes characterized by cell shrinkage, membrane loss, cell disintegration, and engulfing by macrophages

erythroblastosis fetalis—an antigen-induced hemolytic anemia of the newborn or premature infant, as a result of incompatibility of maternal Rh factors with the neonate

erythropoiesis—production of erythrocytes or red blood cells

erythropoietin—the hormone produced in the kidney that regulates marrow production of red blood cells

esophageal phase of swallowing—esophageal peristalsis carries the bolus through the esophagus and LES and into the stomach

esophageal stricture—a significant narrowing of the esophagus that may significantly interfere with swallowing

esophagectomy—a surgical procedure resecting or removing the esophagus

Estimated Energy Requirement (EER)—the average dietary energy intake that is predicted to maintain energy balance in a healthy adult of a defined age, gender, weight, height, and level of physical activity, consistent with good health; in children and pregnant and lactating women, the EER includes the needs associated with the deposition of tissues or the secretion of milk at rates consistent with good health

etiology—the cause of disease

euglycemia—maintenance of normal blood sugar levels

evidence-based dietetics practice—dietetics practice in which systematically reviewed scientific evidence is used to make food and nutrition practice decisions

evidence-based practice guides—a series of guiding statements and treatment algorithms that are developed using a systematic review process of identifying, analyzing, and synthesizing scientific evidence

excipients—those substances added to formulations of medications, such as color or coating agents

excitatory—in the context of the neurological system, refers to a stimulus that results in neural response

exercise-induced allergic syndrome—an allergic reaction that occurs when a food allergen is consumed in combination with physical activity

exocrine pancreas—part of the pancreas that secretes digestive enzymes and bicarbonate into the duodenal lumen

exons—expressed sequences in mRNA; sequences that are translated into the final protein product

exotoxins—toxins produced by bacteria

expected outcomes—the desired change(s) to be achieved over time as a result of nutrition intervention

extracellular fluid (ECF)—the interstitial fluid and the plasma, constituting about 20% of the weight of the body; sometimes used to mean all fluid outside of cells, usually excluding transcellular fluid

extracorporeal shockwave lithotripsy (ESWL)—a common procedure used to treat kidney stones whereby shock waves are used to break down the stones into smaller pieces

exudate—fluid and cellular debris that seep from blood vessels, usually as a result of inflammation

facultative urine—excess water that is excreted through urination

fasciculations—involuntary twitching or movement of muscle

fasting hypoglycemia—also called postabsorptive hypoglycemia; low blood glucose that is often related to an underlying disease and can be diagnosed from a blood glucose level below 50 mg/dL after an overnight fast, between meals, or after physical activity; symptoms are similar to those of diabetes-related hypoglycemia and may include hunger, sweating, shakiness, dizziness, light-headedness, sleepiness, confusion, difficulty speaking, anxiety, and weakness

fatty liver—yellow discoloration of the liver due to fatty degeneration of liver parenchymal cells

feedback inhibition—regulatory mechanism to limit production of certain substrates, which involves the down-regulation of the enzyme involved in the production of the substrate by that substrate when the substrate accumulates a certain level

ferritin—the storage protein for iron

ferroportin—transmembrane iron export regulating protein

fibrin—a filamentous protein; for blood clotting to occur, fibrinogen must be converted to fibrin

fibroblasts—connective tissue cells capable of forming collagen fibers and found in the skin

fibromyalgia—a chronic musculoskeletal disorder characterized by widespread muscle pain, joint stiffness, disturbed sleep, fatigue, headache, cognitive and memory problems, paresthesias, and numerous tender points. The word comes from the Latin term for fibrous tissue (*fibro*) and the Greek terms for muscle (*myo*) and pain (*algia*)

fibronectin—an acute-phase glycoprotein involved in the regulation of cell growth and differentiation, wound healing, and vascular integrity

first-set reaction—rejection of a foreign tissue graft due to antibodies and activated cells formed in response to the graft; usually occurs one to two weeks after the tissue is transplanted

fistula—an abnormal opening or passage between two internal organs or from an internal organ to the surface of the body

flatulence—perceived excess gas in the intestinal tract

foam cells—macrophage cells containing lipid; found within the fatty streaks in the development of atherosclerosis

focal segmental glomerulosclerosis (FSGS)—describes scarring in scattered regions of the kidney, typically limited to one part of the glomerulus and to a minority of glomeruli in the affected region; FSGS may result from a systemic disorder, or it may develop as an idiopathic kidney disease, without a known cause

food and nutrient factors—the amount and type of foods and nutrients that are consumed and therefore made available to the body

food frequency—dietary assessment method in which the client describes the frequency and quantity of his or her consumption of certain foods/food groups

food-induced eczema—atopic dermatitis; a chronic skin disorder that usually begins in infancy and causes itching; many children with this diagnosis also suffer from other allergic conditions such as asthma and rhinitis

food insecurity—lack of adequate access by all people, at all times, to sufficient food for an active and healthy life, including at a minimum a readily available supply of nutritionally adequate and safe foods and an assured ability to acquire acceptable foods in a socially acceptable way

fractionation—a method of radiation therapy that delivers the total prescribed dose of radiation into several smaller-dose treatments in order to reduce side-effects

free radicals—compounds formed when the breakage of the electron bonds of a molecule leaves an exposed electron; this electron will attack another molecule for its electron, and this process can continue, eventually leading to cell damage; antioxidants such as vitamins C and E can be helpful in stopping this destructive process

frontal lobe—a division of the cerebrum that is responsible for voluntary movement, speech, and complex thought

fructose—a disaccharide absorbed by a facilitated transport mechanism but not against a concentration gradient; when the concentration of fructose in the small intestine is greater than that of glucose, its rate of absorption slows and the unabsorbed fructose is fermented in the colon, causing diarrhea; osmotic diarrhea has been reported in persons who have overconsumed sodas sweetened with high-fructose corn syrup or fruit juices

full liquid diet—diet consisting of all beverages allowed on clear liquid diets with addition of milk, ice cream, yogurt, and liquid nutritional supplements (e.g., Ensure®, Boost®)

functional fiber—undigested plant components that have an established physiological function for humans

fundoplication—a surgical technique used to suture the fundus of the stomach around the esophagus to prevent reflux

fusion inhibitors—medications that interrupt the viral replication cycle by inhibiting fusion of the HIV virus to the target cell; also known as *entry inhibitors*

GALT (gut-associated lymphatic tissue)—lymphoid tissue including Peyer's patches, the appendix, and solitary lymph nodes in the submucosa

gamma globulins—a group of serum proteins, including most antibody molecules, that migrate fastest toward the cathode during electrophoresis

gangrene—tissue death due to lack of blood flow and oxygen

gastrectomy—surgery to resect a portion of or the entire stomach

gastrin—primary hormone released to stimulate digestion and production of hydrocholric acid in the stomach

gastritis—inflammation of the gastric mucosa

gastroesophageal reflux disease (GERD)—chronic or recurrent gastric pain due to reflux of gastric secretions into the lower esophagus

gastroparesis—delayed emptying of the stomach

gastrostomy—an opening into the stomach

gene expression—the level of activity of a specific gene in producing mRNA and, subsequently, protein; expression can be regulated by many variables, including diet

general anesthesia—total loss of sensation and consciousness as a result of an anesthetic drug

genetic imprinting—expression of specific genes, which depends on the parent of origin; some genes are expressed only from the maternal allele and others are expressed only from the paternal allele

genome—the entire set of genes of a given organism

genotype—the specific variants of a gene present in the two alleles in an individual that can result in specific traits or disorders

gentamycin—an antibiotic

gingiva—the gums

glomerular filtration rate (GFR)—the filtration ability of the glomerulus; used as an index of kidney function; normal value is approximately 125 mL/min

glomerulonephritis—nephritis marked by inflammation of the capillaries of the renal glomeruli and membrane tissue that serves as a filter

glomerulosclerosis—development of scar tissue within the glomerulus

glomerulus—a network of thin-walled capillaries closely surrounded by a pear-shaped epithelial membrane called the Bowman's capsule (within the kidney)

glossitis—inflammation of the tongue

glucagon—a hormone produced by the α cells of the Islets of Langerhans in the pancreas that works in concert with insulin to maintain blood glucose levels; it promotes glycogenolysis, gluconeogenesis, and lipolysis, while inhibiting glycogenesis and triglyceride synthesis

gluconeogenesis—the metabolic pathway through which glucose is formed from noncarbohydrate sources (e.g., amino acids or glycerol)

glucosamine—a nutritional supplement used by some as a treatment for osteoarthritis; it is an amino sugar and a raw material for synthesizing glycosaminoglycans and proteoglycans, which are important constituents of articular cartilage

GLUT-4—glucose transporter that transports glucose between blood and cells; it is the only glucose transporter responsive to insulin

glycemic control—control of blood glucose

glycogenolysis—the metabolic pathway through which glycogen is converted to glucose

glycolysis—the anaerobic enzymatic conversion of glucose to lactate or pyruvate, which results in the production of energy in the form of ATP

glycosuria—the presence of glucose in the urine

gout—swelling, redness, heat, pain, and stiffness in a joint due to the formation of uric acid crystals in the synovial fluid, resulting in inflammation within the joint and in the surrounding tissues

graft-versus-host (GVH) rejection—a life-threatening reaction in which transplanted immunocompetent cells, usually T cells, attack the tissues of the immunocompromised recipient

Guillain Barré syndrome—an acute peripheral nervous system disease characterized by progressive paralysis

H$_2$ blockers—medications that interrupt the production of acid in the stomach

HAART—highly active antiretroviral therapy; a combination of ARVs that is able to fully suppress the virus

haplotype—a group of gene variants that occur together

hapten—a nonimmunogenic, low-molecular weight molecule that can be recognized by an antibody; it can initiate an immune response if it is conjugated to a "carrier" molecule

health insurance—financial protection against health care costs associated with treatment of disease or accidental injury

helper T cells (TH)—a subset of T cells that triggers B cells to make antibodies, activates macrophages, and promotes the differentiation of other T cells

hematemesis—the vomiting of blood

hematocrit—packed red blood cell volume expressed as a percentage of whole blood upon centrifugation

hematopoiesis—production of blood cells

hematopoietic stem cell—an undifferentiated bone marrow cell that is a precursor for multiple cell types; also called pluripotential stem cells

hematuria—the presence of blood in the urine

heme—iron-containing, non-protein portion of the hemoglobin molecule that contains iron in the ferrous (+3) state

hemochromatosis—iron overload; elevated levels of iron that can cause tissue damage, especially in the liver

hemoconcentration—the decrease in free water circulating in the blood supply, causing increased levels of proteins, electrolytes, wastes, and nutrients per deciliter of blood; elevations of several laboratory values are present, and dehydration signs and symptoms may be present

hemodialysis—a type of renal replacement therapy whereby wastes or uremic toxins are filtered from the blood by a semi-permeable membrane and removed by the dialysis fluid

hemoglobin—the four-pyrrole ring compound in red blood cells that contains iron centers and is responsible for the transport of oxygen

hemoglobinemia—excess free hemoglobin build-up in circulation

hemoglobinuria—excessive free hemoglobin spillage into the urine

hemojuvelin—protein sensor for iron levels in tissues

hemolytic anemia—an anemia brought on by the rapid, premature destruction of red blood cells in circulation, which may be precipitated by vitamin E deficiency

hemophilia—an inherited disorder of blood clotting, with pronounced bleeding upon tissue injury

hemorrhage—bleeding

hemorrhagic stroke—stroke caused by rupture of a blood vessel (e.g., aneurysm)

hemostasis—normal blood flow and blood clotting

Henderson-Hasselback equation—pH = pK$_a$ + [H$_2$CO$_3$]/[HCO$_3^-$]

hepatic encephalopathy—a syndrome characterized by central nervous system dysfunction in association with liver failure

hepatitis—inflammation of the liver and liver disease involving degenerative or necrotic alterations of hepatocytes

hepatocytes—the cells of the liver

hepatomegaly—enlargement of the liver

hepatosteatosis—accumulation of fat in the interstitial tissue of the liver

hepcidin—hormone regulating iron homeostasis

hephestin—membrane-bound, copper-containing oxidizing protein that incorporates iron into transferrin

heterozygous—having two different alleles or variants of a given gene

hiatal hernia—protrusion of part of the stomach through the diaphragm into the space normally occupied by the esophagus, heart, and lungs

high-fiber diet—a diet high in fiber (6 to 10 g above the usual recommendation of 20 to 35 g/day)

histamine—paracrine released from parietal cells involved in production of hydrochloric acid; also released from mast cells and basophils as a component of inflammatory and immune responses; it contributes to inflammation and IgE-mediated allergic reaction by causing the dilation of local blood vessels and smooth muscle contraction; histamine release produces some of the symptoms of immediate hypersensitivity reactions

histological—pertaining to the minute structure, composition, and function of the tissues

histone—class of simple proteins found in the cell's nucleus, around which the DNA is wrapped

hives—an itchy skin condition with raised red lumps, often due to an allergic reaction; also called *urticaria*

HLA antigens—antigens specific to the individual that cause a rejection reaction in a host receiving transplantation or foreign cells

holotranscobalamin—the fraction of metabolically active B_{12} that is composed of cobalamin linked to transcobalamin in circulation

homozygous—having two identical alleles or variants of a given gene

hormones—blood-borne chemical messengers that act on target cells located in a different part of the body from the endocrine gland that produces them

human biological factors—conditions that determine a person's nutrient requirements; one's age, gender, and stage of growth and development are used to estimate kcal and nutrient needs; illnesses that alter organ function or metabolism influence not only the amount of nutrients required but also the form of nutrients that the body needs and can tolerate

human immunodeficiency virus (HIV)—a retrovirus that targets many cells in the body, particularly CD4 T helper cells

humoral immunity—immunity due to soluble factors such as antibodies circulating in the body's fluids, mainly serum and lymph; "humors" is an old term for body fluids

hyaline—a histological term used to describe tissue injury that has a glassy, pink appearance

hydrogenation—the addition of hydrogen atoms; in the food industry, the addition of hydrogens to unsaturated fatty acids in order to increase their degree of saturation; results in the formation of *trans* fatty acids (with hydrogens on opposite sides of the C-C double bond)

hydrophilic—water loving, or attracting water

hydroxyapatite—a crystallized calcium phosphate salt that gives bones their stiffness

hypercalcemia—high serum calcium

hypercalciuria—an excess of calcium in the urine

hypercapnia—the term used to describe an excess of the blood gas carbon dioxide (CO_2)

hyperemia—increased blood flow to a body tissue

hyperinflation of the lungs—results from loss of elasticity of the alveoli, causing air to be trapped; often seen in emphysema

hyperkalemia—high serum potassium

hypermagnesemia—high serum magnesium levels

hypernatremia—abnormally high levels of serum sodium

hyperosmolar hyperglycemic nonketotic syndrome—a complication of type 2 diabetes mellitus that usually develops after a period of hyperglycemia combined with inadequate fluid intake

hyperosmolar—having a higher osmolality than body fluids (>300 mOsm/kg)

hyperoxaluria—an excess of oxalate in the urine

hyperphosphatemia—high serum phosphorus

hyperplasia—increased number of cells

hyperreflexia—overresponse or exaggeration of response to a neural stimulus (e.g., twitching)

hypersensitivity—an inappropriate and harmful immune reaction to a harmless, nonpathogenic substance; also called *allergy*

hypertension—condition of chronically elevated blood pressure

hyperthyroidism—excess thyroid secretion

hypertrophic cardiomyopathy—a genetic disorder causing abnormal thickening of the left ventricular wall

hypertrophy—increase in cell size

hyperuricosuria—a disorder of uric acid metabolism

hypervolemia—increased blood volume

hypocalcemia—low serum calcium

hypochondriasis—a somatoform disorder (i.e., a physical ailment stemming from a psychological problem) characterized by an unfounded belief that one is suffering from a serious illness

hypochromic—abnormally pale in color upon visual inspection under a microscope

hypoglycemia—a low serum glucose; generally considered to be <70 mg/dL

hypokalemia—low serum potassium

hypomagnesemia—low serum magnesium levels

hyponatremia—abnormally low concentrations of sodium ions in the circulating blood

hypoperfusion—reduced blood flow

hypophosphatemia—low serum phosphorus

hyporesponsiveness—hormone resistance on the part of target cells/tissues

hypotension—low blood pressure

hypothyroidism—deficient thyroid secretion

hypovolemia—decreased blood volume

hypoxemia—condition in which there is an inadequate supply of oxygen in the blood

hypoxic injury—cellular injury as a result of oxygen deprivation

iatrogenic—an adverse condition in a patient resulting from treatment, usually by a physician; iatrogenic literally means "brought forth by a physician"

iatrogenic, iatrogenicity—harm caused by treatment or procedures performed by healthcare personnel or sustained through hospitalization, medical intervention, or prescription

IgA (immunoglobulin A)—the predominant immunoglobulin in secretions

IgA nephropathy—a form of glomerular disease that results when immunoglobulin A (IgA) forms deposits in the glomeruli, where it creates inflammation

IgD (immunoglobulin D)—an immunoglobulin present on the surfaces of B cells

IgE (immunoglobulin E)—the immunoglobulin class that is the predominant mediator of immediate hypersensitivity reactions (allergies)

IgG (immunoglobulin G)—the predominant immunoglobulin class produced during secondary immune responses; the most prevalent immunoglobulin in the blood

IgM (immunoglobulin M)—the predominant immunoglobulin class expressed by virgin B lymphocytes and secreted during primary immune responses

IL-2—interleukin-2; a lymphokine required by activated T cells for growth

ileostomy—a procedure in which the colon and rectum are surgically removed, and the end of the ileum is attached to the stoma

ileus—decreased or absent motility of the bowel and forward movement of bowel contents

immediate hypersensitivity—a hypersensitivity reaction that appears within minutes after the exposure to the allergen

immune complex—a cluster of antibodies bound to antigens

immunodeficiency—decrease in or lack of an immune response due to absence or defect of one or more components of the immune system

immunogen—an antigen capable of inducing an immune response because it is foreign to the host

implantable port—intravenous access device that is completely under the skin, is placed in the vein on the upper chest wall, and exits the body near the xyphoid process, axilla, or abdominal wall

indirect calorimetry—an approach to determine energy expenditure by measuring a subject's oxygen consumption, carbon dioxide production, and minute ventilation (the amount of air a subject breathes in one minute)

infarct—cellular necrosis as a result of lack of oxygen

inflammatory bowel disease (IBD)—an autoimmune, chronic inflammatory condition of the gastrointestinal tract; IBD is actually the term designating a syndrome consisting of two diagnoses: *ulcerative colitis* and *Crohn's disease*

inflammatory response—the body's response to infection or injury that is cortisol driven and allows for the breakdown of labile body protein to increase the amino acid pool for the purpose of synthesizing protective and healing proteins

inhalation—refers to placement of a drug so that it is breathed into the respiratory system

inhaled anti-inflammatory agents—class of medications that often includes inhaled corticosteroids

inhaled antibiotics—medications designed to reduce airway infection, particularly *Pseudomonas aeruginosa*, commonly seen in CF

inhaled bronchodilators—medications used to maximize airway size and improve clearance of mucus

inhaled mucolytics—class of medications, including pulmozyme™, which hydrolyzes the DNA in sputum of cystic fibrosis patients and reduces sputum viscosity

inhibitory—in the context of the neurological system, refers to a stimulus that results in a decreased neural response

initiation—the first phase in cancer cell development; the exposure of cells to an appropriate dose of a carcinogen (initiator)

innate immunity—immune response resulting from natural barriers and resistance that are present at birth

inorganic phosphate—a phosphate compound not derived from an organic origin

insensible losses—fluid loss that cannot be easily measured (usually refers to fluid lost via sweat and respirations)

insulin—a hormone produced by the ß cells of the Islets of Langerhans in the pancreas to regulate blood glucose; it promotes uptake, utilization, and storage of nutrients

insulin resistance—resistance of body cells to the action of insulin

integrase inhibitors—medications that interrupt the viral replication cycle by inhibiting integrase enzymes that allow the transcribed viral DNA to integrate into the host cell DNA

interferon (INF)—a group of cytokines that regulate the immune system and protect cells from viruses

interleukin—now used primarily as a naming convention for cytokines/lymphokines/chemokines/growth factors (IL-number)

intermittent feedings—administration of formula several times daily, over 20 to 30 minutes

intracellular fluid (ICF)—the fluid within the tissue cells

intractable— resistant to treatment

intradermal—refers to injection under the outermost layer of skin

intralipid—an intravenous fat emulsion used to prevent or correct deficiency of essential fatty acids and to provide energy

intramuscular—refers to injection into the muscle

intraperitoneal—refers to injection into the body's peritoneal cavity

intrathecal—refers to injection of a drug into the membrane surrounding the central nervous system

intravenous pyelogram (IVP)—radiographic imaging of the kidneys, ureter, and bladder using X-ray and contrast dye that is injected intravenously

intravenous (IV)—refers to injection directly into a vein

intravenously (IV)—by vein, in reference to administration of drugs or nutrients

intrinsic acute renal failure—acute renal failure associated with damage to the renal anatomy

intrinsic factor (IF)—the protein produced by the parietal cells in the stomach lining that is responsible for the pick-up of cyanocobalamin from protein in foods

introns—intervening sequences in mRNA that are enzymatically excised during posttranscriptional processing prior to translation into a protein

inulin—a fructooligosaccharide derived from chicory; intravenous inulin is used as a diagnostic test for kidney function since it is not utilized by the body and is excreted in the urine

ionization—process of producing negatively or positively charged ions

irritable bowel syndrome (IBS)—a bowel disorder characterized by abdominal pain with diarrhea and/or constipation

ischemia—inadequate supply of oxygen

ischemic heart disease (IHD)—heart disease characterized by inadequate blood supply to the heart

ischemic stroke—stroke caused by an interruption of blood flow to the tissue

isograft—tissue transplanted between two genetically identical individuals; also called *syngraft*

isotonic—having the same osmolality as body fluids (approximately 300 mOsm/kg)

jaundice—a clinical manifestation of hyperbilirubinemia, consisting of deposition of bile pigments in the skin, resulting in a yellowish staining of the skin and mucous membranes

jejunostomy—an opening into the jejunum

karyotype—a chart that displays chromosome pairs in order according to size

Kayexalate—a medication used to reduce high serum potassium; exchanges sodium for potassium in the intestine

kernicterus—infiltration of excessive amounts of bilirubin into the neurons of the spinal cord and brain

ketoacidosis—an acid-base imbalance caused by an increase in concentration of ketones in the blood

ketogenic diets—nutrition therapy characterized by diets high in fat and restrictive in carbohydrates to produce a therapeutic ketosis (increase in ketones in serum and urine)

ketone bodies—the substances acetone, acetoacetic acid, and beta-hydroxybutyric acid, which are normal metabolic products of lipid metabolism within the liver and are oxidized in the muscle

ketosis—an abnormally elevated concentration of ketone bodies in the body tissues and fluids

kilocalorie (kcalorie or kcal)—the amount of heat required to raise 1000 mL (1 liter) of water 1° Celsius

kilojoule (kjoule or kJ)—the SI (Système International d'Unités or International System of Units) unit of measurement for energy; the amount of work required to move 1 kilogram for 1 meter with the force of 1 newton; 1 kcal = 4.2 kJ (to convert kcal to kJ, multiply kcal by 4.2)

Korsakoff's psychosis—condition characterized by amnesia, confabulation (false memories), and hallucinations

Kupffer cells—specialized phagocytic cells of the reticuloendothelial system found on the luminal surface of the hepatic sinusoids; they filter bacteria and small foreign proteins out of the blood and dispose of worn-out red blood cells

Kussmaul breathing (Kussmaul respirations)—rapid, deep, and labored breathing commonly seen in people who have ketoacidosis or who are in a diabetic coma; Kussmaul breathing is named for Adolph Kussmaul, the nineteenth century German doctor who first noted it

lactic acidosis—an accumulation of lactic acid in the body characterized by abdominal pain, vomiting, and rapid breathing; this condition occurs in diabetes and as a potential side effect of medications

lactoferrin—a protein in plasma and secretions (milk, mucus, bile), secreted by leukocytes, that can bind iron; it helps prevent infection by depriving bacteria of the iron necessary for their growth

Langerhans cell—dendritic cell that traps and processes antigens in the epidermal layer of the skin and then migrates through lymphatics to lymph nodes where it presents the antigen to T cells

laparoscopically—describes the process of using a laparoscopic procedure through which an instrument is used to see structures within the abdomen and pelvis; in this way, a number of surgical procedures can be performed without the need for a large surgical incision

latent autoimmune diabetes of adulthood (LADA)—sometimes called T1.5DM, a slowly progressive form of T1DM; individuals are often diagnosed as T2DM, but have positive pancreatic islet antibodies, especially to glutamic acid decarboxylase (GADA)

learned food aversion—avoidance of certain foods due to association with unpleasant GI symptoms

left ventricular end diastolic volume (LVEDV)—the amount of blood in the left ventricle at the end of the diastolic phase and immediately prior to systolic ejection of blood

left ventricular end systolic volume (LVESV)—the amount of blood that remains in the left ventricle at the conclusion of the systolic phase

left ventricular hypertrophy (LV hypertrophy)—enlargement of the left ventricle; most commonly related to hypertension and/or congestive heart failure

leukocytes—white blood cells (WBC); a generic term for several types of WBC that arise from the same parent cell in the bone marrow

leukocytopoiesis—the production of all categories of white blood cells from the pluripotent (able to differentiate into multiple cell types) stem cells found in the bone marrow

leukocytosis—high white blood cell count

leukotrienes—powerful inflammatory mediators produced by the body (as metabolic products of arachidonic acid) that are important in inflammation and allergic reactions because of their ability to constrict blood vessels and attract a variety of types of immune cells

lifestyle factors—a person's knowledge, attitudes/beliefs, and behavior patterns directly impact the choices that are made regarding food and physical activity; assessment of these factors provides information about a person's ability and/or readiness to make behavior changes

limbic lobe—component of the brain involved in control of emotions

lipoatrophy—an immune response related to source and purity of insulin resulting in thinning of subcutaneous fat at the injection site, which causes concaving or pitting of fatty tissue

lipodystrophy syndrome—loss or absence of fat, or the abnormal distribution of fat in the body, in HIV infection; these changes are likely hormonally mediated; subcutaneous fat loss is most apparent in peripheral limbs and facial areas; fat deposits are most commonly central, located in the dorsocervical area, breast area, and abdominal region

lipogenesis—the synthesis of triglyceride from carbohydrates and proteins

lipohypertrophy—thickening of subcutaneous fat at an insulin injection site

local anesthesia—loss of sensation only in the area where an anesthetic drug is placed

lock and key model—description of communication between two cells; action between two substances within the body; in order for action to occur, the two cells must fit together as a lock and key might

low-residue diet—a diet low in fiber and other food constituents that may contribute to bulk in the large intestine

lower esophageal sphincter (LES)—the junction between the esophagus and the stomach

lymph—extracellular fluid containing white blood cells (mostly lymphocytes) and antibodies that bathe tissues

lymph nodes—small organs of the immune system where mature B and T lymphocytes respond to an antigen; they are distributed widely throughout the body and linked by lymphatic vessels that bring in antigens from surrounding tissue

lymphatic system—a system of vessels through which lymph travels, consisting of lymphatic vessels and lymph nodes at the intersection of vessels

lymphocyte—a small mononuclear cell with a thin rim of cytoplasm that has antigen-specific receptors

lymphokine—a soluble molecule used for communication between lymphocytes and other cells

lysosomes—cytoplasmic granules that contain hydrolytic enzymes and are involved in the digestion of phagocytosed material

macrocytic—refers to abnormally large cell size

macrophage—a large, mononuclear phagocytic antigen-presenting cell derived from the blood monocyte and found in tissues

macrosomia—refers to the condition of abnormally large infants whose mothers have diabetes

major histocompatibility complex (MHC)—a cluster of genes encoding polymorphic cell-surface molecules (MHC class I and class II) that help the organism identify pathogens as foreign; they are important in antigen presentation to T cells, play a role in transplantation rejection, and influence the susceptibility to certain autoimmune diseases; MHC antigens are also called *HLA antigens*

MALT (mucosa-associated lymphatic tissue)—lymphoid tissue found in the surface mucosa of the respiratory, gastrointestinal, and genitourinary tracts

mast cell—a tissue cell found primarily in mucosal and connective tissue that is similar to the basophil (which is found in blood)

mechanical ventilation—artificial ventilation using a ventilator or respirator; performed with a piece of equipment designed to intermittently or continuously assist or control pulmonary ventilation

medical doctor—a health professional who has earned a post-bachelor degree of doctor of medicine or doctor of osteopathy and who has passed a licensing examination

medical foods—foods administered under the supervision of a physician and intended for the specific dietary management of a disease for which distinctive nutritional requirements are established

medium-chain triglyceride (MCT) oil—synthetic oil containing fatty acids that are 6 to 8 carbons in chain length

medium-chain triglycerides (MCTs)—triglycerides composed of fatty acids with 8 carbons (octanoic and decanoic fatty acids)

megaloblastic—refers to an immature, large red blood cell that is oval in shape and abnormal

meiosis—cell division to produce gametes (sperm and ova) that results in the production of cells with half the complement of chromosomes

membrane attack complex—the final product of the complement cascade that forms a pore on the surface of the target cell, which results in lysis of the cell

membranous nephropathy—disease diagnosed when a kidney biopsy reveals unusual deposits of immunoglobulin G and complement C3, substances created by the body's immune system; 75% of cases are idiopathic

memory cells—lymphocytes produced on the first encounter with an antigen that produce a rapid, more vigorous response upon subsequent exposures, which often prevents reinfection

menhaden oil—hydrogenated and partially hydrogenated oils from the menhaden fish (a small plankton-feeding fish)

metabolic acidosis—condition resulting from either loss of bicarbonate or retention of nonvolatile acid

metabolic alkalosis—condition resulting from either retention of bicarbonate or loss of nonvolatile acid

metabolic decompensation—an inability to maintain metabolic balance leading to derangements in biochemical and clinical parameters

metabolic water—water that is produced through nutrient metabolism

metabolomics—study of the collection of all metabolites present in a living organism

metaplasia—replacement of one cell type with another

metastasis—spread of cancer from the primary site to nearby or distant areas through the blood or lymph

methylation—the addition of methyl (-CH_3) groups; DNA methylation patterns can be inherited and impact patterns of gene expression

microalbuminuria—the leaking of small amounts of albumin into the urine by the kidneys

microarray—technology used to measure expression of thousands of genes simultaneously

microcytic—refers to abnormally small cell size

migrating motility complex (MMC)—weak contractions of the gastrointestinal tract that serve to assist in clearing waste

minor histocompatibility antigens—cell surface-processed peptides not encoded by the MHC that can contribute to graft rejection

minute ventilation—the volume of air per unit time moved into or out of the lungs; measured by collecting expired volume for a fixed time

mitosis—cell division that produces two cells that are genetically identical to the progenitor cell

MODS—multiorgan distress syndrome; a disease involving two or more vital organs

monoamine oxidase (MAO) inhibitors—group of medications that block the enzyme system that inactivates some neurotransmitters

monoclonal antibody—an antibody produced by an immortal B cell line that reacts with a single antigenic determinant (a specific part of an immunogen that stimulates a specific immune response)

monocyte—a large, mononuclear, phagocytic white blood cell that develops into a macrophage when it enters tissue

monogenic—arising from a single gene

monounsaturated fats—sources of fat that have a predominant amount of fatty acids with one carbon-carbon double bond within their chemical structures

morbidity—the state of being diseased

mortality—the incidence of death in a population

MSOF—multisystem organ failure; a disease involving two or more vital organs

mucositis—inflammation of a mucous membrane (e.g., mouth sores)

multiple sclerosis—a disorder characterized by demyelination of cells within the CNS, inflammation, and development of scar tissue, causing numbness, tingling, incoordination, weakness, and varying degrees of blindness

myasthenia gravis—a neuromuscular disorder that affects the skeletal muscles and causes muscle weakness, particularly of the face, eyes, arms, and legs

myelin—the covering or insulation of the axon that ensures proper communication between neurons

myeloma—a tumor composed of cells derived from hemopoietic tissues of the bone marrow; a plasma cell tumor

myocardial cells—cells found in the myocardium

myocardial infarction (MI)—necrosis of the myocardial cells as a result of oxygen deprivation.

myxedematous—refers to nonpitting edema; noun form is *myxedema*

Na$^+$/K$^+$ pump—the enzyme-based mechanism that moves potassium ions into and sodium ions out of a cell by active transport

nadir—the lowest point, usually in reference to the white blood cell count

nasogastric feeding tube—a tube that is inserted nasally (through the nose) into the stomach

nasointestinal feeding tube—a tube that is inserted nasally (through the nose) past the stomach into the intestine

natural killer cells (NK cells)—large granular lymphocyte cells that attack tumors and virally infected cells but do not exhibit antigenic specificity; also called *killer cells (K cells)* and *null cells*

necrosis—general term referring to cell death

necrotizing enterocolitis (NEC)—a condition that occurs primarily in premature infants or sick newborns, in which intestinal tissue dies; the cause for this disorder is unknown, but it is thought to be due to decreased blood flow to the bowel, which keeps it from producing the normal protective mucus; if an infant is suspected of having necrotizing enterocolitis, feedings are stopped to allow the bowel to rest

necrotizing fasciitis—inflammation of the connective tissue leading to necrosis of the tissue; may be caused by infection, injury, or an autoimmune reaction

negative feedback—a regulatory mechanism in which a change in a controlled variable triggers a response that opposes the change, thus maintaining a relatively steady state for the regulated factor

negative nitrogen balance—net loss of protein in the body

negative selection—the process in which B and T cells that react to self molecules are deleted or functionally inactivated during their development

neoadjuvant chemotherapy—refers to chemotherapy used prior to primary treatment, which is typically surgery

neoplasm—literally means "new growth"; an abnormal mass of tissue, the growth of which exceeds and is uncoordinated with that of normal tissue

nephritic syndrome—a condition of inflammation of the glomerulus, resulting in hematuria, proteinuria, and oliguria

nephrolithiasis—kidney stones, a common disorder in the United States

nephron—basic functioning unit of the normal kidney; each nephron has two main parts: the glomerulus and the tubule

nephropathy—renal disease that results from damage to blood vessels from hyperglycemia

nephrotic syndrome—a clinical condition consisting of losses of protein in the urine exceeding 3.5 g/day, hyperlipidemia, and low albumin levels (< 3.5 g/dL) with edema; also characterized by increased permeability of the glomerular capillary basement membranes, often caused by diabetes-induced glomerulosclerosis, systemic lupus erythematosus, renal vein thrombosis, or hypersensitivity to toxic agents

neurofibrillary tangles—collections of twisted tau found in the cell bodies of neurons in Alzheimer's disease

neuroglycopenia—inadequate glucose supply to the brain

neuromodulator—substance released that will increase or decrease the activity of specific neurotransmitters

neuron—a nerve cell in the central nervous system

neuropathy—a general term denoting functional disturbances and/or pathological changes in the peripheral nervous system

neuropeptide—protein messengers within the brain and nervous system that assist in communication between neurons

neurotransmitter—a chemical messenger that communicates between neurons

neutropenia—low white blood cell count

neutrophil—the most numerous type of polymorphonuclear leukocyte, with granules that stain with acid and basic dyes; it is phagocytic and enters tissues early in inflammation

non-coding DNA—sequences of DNA that lie within or between expressed genes and whose function is largely unknown; over 95% of DNA in humans is made up of non-coding DNA; sometimes referred to as "junk DNA"

nonalcoholic fatty liver disease (NAFLD)—a wide spectrum of non-alcohol-related liver diseases ranging from fatty liver (steatosis), to nonalcoholic steatohepatitis (NASH), to cirrhosis

nonalcoholic steatohepatitis (NASH)—non-alcohol-related liver inflammation caused by a buildup of fat in the liver

nonexercise activity thermogenesis (NEAT)—the energy expended through physical activity involved in performing the ordinary activities of daily life; it excludes energy expended in activities to obtain physical exercise or involving sports-like activity

nonspecific immune system—all aspects of immunity not directly mediated by antigen-specific lymphocytes

norepinephrine—a neurotransmitter released from sympathetic postganglionic fibers; formerly referred to as noradrenaline; also a stress hormone that affects parts of the brain where attention and responding actions are controlled

normalized protein equivalent of nitrogen appearance (nPNA)—an assessment of protein catabolic rate

NPO—*nil per os*, which is Latin meaning "nothing per mouth"

nucleotide—the building block of a nucleic acid, consisting of a ribose sugar, a phosphate group, and a nitrogenous base

nurse—a health care professional who has earned at least an associate's degree in nursing, has been licensed by the state, and assists patients in activities related to maintaining or recovering health

nutrigenetics—the interaction between an individual's genetic profile, i.e., genotypes of specific genes and the function of proteins encoded by those genes, with nutrients and other bioactive food components

nutrigenomics—the study of mechanisms by which nutrients and other food-derived bioactive substances interact with the genome to influence gene expression

nutrition assessment—a systematic method for obtaining, verifying, and interpreting data needed to identify nutrition-related problems, their causes, and significance

nutrition care process (NCP)—a systematic problem-solving method developed by the ADA that dietetics professionals use to think critically, make decisions addressing nutrition-related problems, and provide safe, effective, high-quality nutrition care

nutrition diagnosis—the identification and descriptive labeling of an actual occurrence of a nutrition problem that dietetics practitioners are responsible for treating independently

nutrition insecurity—nutritionally inadequate diets as the result of diminished purchasing power, nutrition knowledge, etc.

nutrition intervention—a specific set of activities and associated materials used to address a (nutrition-related) problem

nutrition monitoring and evaluation—an active commitment to measuring and recording the appropriate outcome indicators relevant to a nutrition diagnosis in order to determine the degree to which progress is being made and whether or not the client's goals are being met

nutrition screening—process of identifying patients, clients, or groups who may have a nutrition diagnosis and benefit from nutrition assessment and intervention by a registered dietitian

nutritional genomics—a field of study that describes the application of genetic technology to food and nutrition and includes nutrigenetics and nutrigenomics

obesigenic—promoting or encouraging the development of obesity; an obesigenic environment is one that promotes weight gain and the development of obesity by encouraging consumption of energy and discouraging physical activity

obesity—an excess of body fat or adipose tissue; obesity can be defined as a proportion of body weight that is adipose tissue (percent body fat) that is greater than some standard; because it is often impractical in the clinical setting to measure the percent of body fat using body composition analysis, obesity is generally defined as a BMI = 30.0 kg/m² for adults; for children and adolescents, obesity is defined as a BMI-for-age at or above the 95th percentile using the CDC growth charts; the term *obesity* comes from the Latin *obesus*, meaning "one who has become plump through eating"

obligatory urine—the amount of fluid necessary for the body to excrete waste products and solutes (approximately 500 mL)

obstruction—blockage

obstructive acute renal failure—acute renal failure related to the obstruction of urine flow

occipital lobe—portion of the cerebral cortex controlling vision

occupational therapist—a health professional who has obtained a master's degree and passed a national registration exam, who helps individuals with mentally, physically, developmentally, or emotionally disabling conditions improve their ability to perform tasks in their daily living and working environments

octreotide—medication that mimics the action of somatostatin

odynophagia—painful swallowing

oligomenorrhea—abnormally infrequent menstrual cycles

oligosaccharide—a carbohydrate which, upon hydrolysis, yields a small number of monosaccharides

oliguria—urine output less than 400 mL, which is the minimum amount of normal urine that can carry away the daily load of metabolic waste products

omeprazole—a type of proton pump inhibitor used to treat GERD and peptic ulcer disease

oncotic pressure—pressure exerted by large protein molecules in blood plasma, which usually do not cross the capillaries; these molecules decrease the fluid that can leak out of the capillaries into the tissue

ophthalmic—refers to placement of a drug into the eye

oral allergy syndrome—food allergy symptoms of the mouth and pharynx, which usually occur within minutes of contact between the allergen and the oral mucosa

oral glucose tolerance test (OGTT)—timed glucose challenge to examine efficiency of the body in metabolism of glucose

oral preparatory phase—tongue, teeth, and mandible involved in chewing of food and preparation of bolus; food is mixed with saliva, pressed against the hard palate, and formed into a bolus

oral transit phase of swallowing—tongue moves bolus to back of throat

orexigenic—appetite stimulating

orogastric feeding tube—a tube that is inserted orally (through the mouth) into the stomach

orthopnea—shortness of breath associated with lying in the supine position

osmolality—number of water-attracting particles per weight of water in kilograms (expressed as mOsm/kg)

osmolarity—number of millimoles of liquid or solid in a liter of solution; the number of osmols (standard unit of osmotic pressure) per liter of solution (mOsm/L)

osmosis—movement of fluid across a semipermeable membrane from a lower concentration of solutes to a higher concentration of solutes

osmotic pressure—the pressure that must be applied to a solution to prevent the passage of solvent into it when solution and pure solvent are separated by a membrane permeable only to the solvent

osseous tissue—the group of cells and cell products that collectively form bone; bone tissue

osteitis fibrosa cystica—a form of high-turnover bone disease caused by overproduction of parathyroid hormone (PTH), which increases the rate of bone turnover

osteoarthritis—a condition involving progressive loss of articular cartilage and inflammation of the tissues composing the joint, resulting in joint pain, stiffness, and limited joint movement

osteoblasts—cells that synthesize, deposit, and then orient the fibrous proteins of the organic matter of the bone matrix

osteoclasts—bone-removing cells that dissolve the mineral component of the bone matrix, playing a major role in bone resorption

osteocytes—mature osteoblasts surrounded and entrapped by the matrix they have synthesized and then calcified

osteogenic cells—stem cells capable of developing into osteoblasts

osteomalacia—a condition in which the organic matrix of the bones of adults is inadequately mineralized, resulting in muscular weakness, bone pain, and, in advanced cases, deformities of the ribs, pelvis, and bones of the legs

osteopenia—a term used to describe a bone mineral density that is low but not low enough to meet the diagnostic criterion for osteoporosis

osteopetrosis—death of bone cells through excessive calcification

osteophyte—a bony outgrowth near the joint affected by osteoarthritis; also referred to as a bone spur

osteoporosis—a disease resulting from a decreased amount of bone mineral and organic matrix that weakens bones, making them more susceptible to fracture

ostomy—an artificial opening created by surgical procedure

otic—refers to placement of a drug into the ear

outcome—the measurable consequence of disease

outcome measures—data used to evaluate the success of interventions; includes direct nutrition outcomes, clinical and health status outcomes, patient/client-centered outcomes, and health care utilization and cost outcomes

outcomes management system—a system that evaluates the effectiveness and efficiency of the entire NCP: assessment, diagnosis, implementation, cost, and other factors; it links care processes and resource utilization with outcomes

outcomes research—evaluation of care that focuses on the status of participants after receiving care

overweight—an excess of body weight in relationship to height; for adults, overweight is generally defined as a body mass index or BMI of 25.0 kg/m² to 29.9 kg/m²; for children and adolescents, overweight can be defined as a BMI-for-age-and-sex at or above the 85th percentile but less than the 95th percentile using the CDC growth charts

oxalate—a salt of oxalic acid produced by the body's metabolism and excreted in the urine

oxidative stress—a disturbance in the pro-oxidant—antioxidant balance in favor of the former, leading to potential cell damage; indicators include damaged DNA bases, protein oxidation products, and lipid peroxidation products

oxytocin—a hormone that stimulates contraction of the uterus during childbirth, and promotes ejection of milk from mammary glands during breast-feeding

Paget disease—a localized disorder of bone remodeling resulting from rapid bone resorption followed by rapid formation of new bone that is structurally disorganized and more susceptible to deformities and fractures

palliative—refers to a noncurative treatment that reduces symptoms such as pain

pancreatic function tests—tests to measure pancreatic function, including serum amylase or lipase, a test for the amount of fat in the stool, and an X-ray of the anatomical features of the pancreas and common bile duct

pancreatic polypeptide—a polypeptide produced by the F cells of the Islets of Langerhans in the pancreas; its function is not yet known

pancreatitis—inflammation of the pancreas

pancytopenia—a reduction in the numbers of all the blood elements—white, red, other cells, and proteins

pannus—an abnormal, destructive tissue that develops on the synovial membrane of patients with advanced rheumatoid arthritis; inflammatory cells in pannus secrete enzymes that are destructive to articular cartilage and subchondral bone

paracentesis—a procedure in which fluid is withdrawn from a body cavity via a trocar and cannula, needle, or other hollow instrument

paracrine—a name for a neurotransmitter that is released from a cell that is close to the target cell

paraplegia—paralysis involving the lower body below the umbilicus

parasympathetic branch—division of the autonomic nervous system that is involved in control of gastrointestinal, cardiac, and respiratory systems

parenchymal—refers to the essential elements of an organ

parenteral—refers to injection into the body's circulatory system through a blood vessel

parenteral nutrition (PN)—administration of nutrition directly into the circulatory system (also known as total parenteral nutrition [TPN], central venous nutrition [CVN], or intravenous hyperalimentation [IVH])

paresthesia—an abnormal sensation in the skin that may be described as burning, pricking, or like ants crawling on the skin; often consistent with electrolyte imbalance

parietal cell—one of the gastric gland cells that lies on the basement membrane covered by chief cells, and secretes hydrochloric acid

parietal lobe—portion of the cerebral cortex responsible for the sensations of pain, touch, taste, temperature, and pressure; related to mathematical and logical thinking

Parkinson's disease—a neuromuscular, neurodegenerative disease resulting in the loss of dopamine-producing cells in the substantia nigra portion of the brain, leading to resting tremor, rigidity, slowed movement, stooped posture and postural instability, mask-like facial features, and a shuffling gait

passive immunity—immunity due to the transfer of antibodies or activated T cells produced by another individual

pathogenesis—the clinical course of disease

pathophysiology—(1) alterations from normal anatomy and physiology that occur as a result of disease or injury; (2) the study of disease

patient-centered care—care that considers patients' cultural traditions, personal preferences and values, family situations, and lifestyles

pelvic floor—refers to the pelvic diaphragm, the sphincter mechanism of the lower urinary tract, the upper and lower vaginal supports, and the internal and external anal sphincters; it is a network of muscles, ligaments, and other tissues that hold up the pelvic organs (vagina, rectum, uterus and bladder)

pelvic floor dysfunction—weakening of the pelvic floor that can cause the organs to shift, bulge, and push outward against each other, resulting in urinary or fecal incontinence or obstruction, vaginal prolapse or pain, sexual dysfunction, and other problems.

peptic ulcer disease—ulceration or perforation in the lining of the stomach, duodenum, or esophagus

percent weight for height—percentage used to evaluate a child's growth pattern relative to population standards

percussion—a technique used during physical examination in which the hands are used to strike the body's surface, and the sounds that are transmitted from the underlying tissues and organs are evaluated

percutaneous endoscopic gastrostomy (PEG)—a procedure used by a physician to insert a feeding tube through the skin and into the stomach using an endoscope

percutaneous nephrolithotomy—a surgical procedure in which a surgeon makes an incision in the back and creates a tunnel to the kidney to remove a kidney stone

perforation—a break in the integrity of the tissue

periarticular muscles—those muscles located near a joint

periodontal disease—a bacterial infection that destroys the attachment fibers and supporting bone that hold the teeth in the mouth

periorifacial acrodermatitis—a disease of the skin surrounding the mouth area

peripheral arterial disease (PAD)—atherosclerotic heart disease of all vessels except specific coronary vessels; term used interchangeably with peripheral vascular disease

peripheral nervous system—all components of the nervous system except for the brain and spinal cord (central nervous system)

peripheral parenteral nutrition (PPN)—administration of nutrition into a vein in the arm or back of the hand (also known as peripheral venous nutrition [PVN])

peripheral vascular disease (PVD)—atherosclerotic heart disease of all vessels except specific coronary vessels

peripherally inserted central catheter (PICC)—intravenous access device inserted into the arm and threaded into the subclavian vein to the vena cava

peritoneal dialysis—a type of renal replacement therapy during which the peritoneal cavity serves as the reservoir for the dialysate and the peritoneum acts as the semipermeable membrane across which excess body fluid and solutes are removed

peritonitis—an inflammation of the peritoneum membrane

pernicious anemia—the anemia associated with B_{12} deficiency that is slow, aggressive, and potentially life threatening; it is specific to gastrointestinal dysfunction, namely, to gastric enterocytic atrophy, with diminished availability of intrinsic factor, HCl, and enzymes; neuropathy (especially peripheral) results from prolonged deficiency; the nervous system has a decreased ability to regenerate as well as regain function and feeling in the affected areas

PES—problem, etiology, and signs and symptoms; the format used in the NCP to write a nutrition diagnosis; it clarifies a specific nutrition problem and logically links the nutrition diagnosis to nutrition intervention and to monitoring and evaluation

Peyer's patches—distinct lymphoid nodules in the intestine that are part of the gut-associated lymphoid tissue (GALT)

phagocytosis—the engulfment of a particle or a microorganism by leukocytes such as macrophages and neutrophils, normally followed by destruction of the particle

pharmacist—a licensed health professional with a doctorate of pharmacy who compounds and dispenses medications, checks laboratory results for therapeutic drug levels, and reviews risk for drug interactions

pharmacogenomics—the interaction between drugs and an individual's genome that can impact drug efficacy and toxicity

pharmacokinetics—study of drug absorption, distribution, metabolism, and excretion

pharmacology—study of drugs, their properties, and their effects

pharmacotherapy—use of drugs for treatment of disease and health maintenance

pharyngeal phase of swallowing—the involuntary swallowing reflux begins, and the bolus is carried through the pharynx to the top of the esophagus; the entrance to the trachea (larynx) closes, and the soft palate lifts and closes off entrance to the nose

phase angle—calculates a mathematical relationship between resistance and reactance; for use with bioelectrical impedance to calculate body composition; higher values for phase angle appear to be consistent with greater body muscle mass and lower risk for morbidity and mortality; values range from 3 to 12

phenotype—the expressed physical or biochemical characteristics of an organism, as determined by both the genetic makeup and environmental influences

phlebotomy—blood removal through a venous puncture; blood draw

physical activity-related energy expenditure—energy expended in voluntary body movement resulting from the daily activities of life, physical exercise, sports, and play, and nonvoluntary behaviors such as spontaneous muscle contractions, maintenance of posture, and fidgeting; it is the most variable component of 24-hour energy expenditure, depending on how physically active a person is

pica—eating of abnormal items, or non-nutritive substances, such as laundry starch, clay, ice, dirt, paint chips, etc.

pK—the constant degree of dissociation (the ability of an acid to release its hydrogen ions) for a given solution; this is a constant amount for any given solution

plaque—the noncalcified accumulation of oral microorganisms and their by-products that adhere to the teeth

plasma—the portion of the blood in which blood constituents are dissolved or suspended; it contains water, proteins, electrolytes, gases, non-proteinaceous compounds, wastes, and nutrients

plasma cells—large antibody-producing cells that develop from activated B cells; also called *antigen-forming cells (AFC)*

plasma prothrombin concentrations—a measure of blood clotting ability

plasmapheresis—treatment that removes blood from the body, separates out certain cells from the plasma, and then returns the blood back to the body

pleural effusion—accumulation of fluid between the two outer membranes surrounding the lungs

PLHA—people living with HIV and AIDS; other acronyms include PLWHA (people living with HIV/AIDS) and PWA (people with AIDS)

pneumonia—inflammation of the lungs, usually caused by bacteria, viruses, or fungi

polydipsia—excessive thirst

polygenic—arising from multiple genes interacting with each other

polyhydramnios—excessive accumulation of amniotic fluid

polymorphisms—DNA sequences of specific genes that vary among individuals

polymorphonuclear leukocytes (PMN)—leukocytes with a multilobed nucleus and cytoplasmic granules that take up acid and basic dyes; also known as *granulocytes*, *PMNs*, and *polys*

polyphagia—excessive hunger

polyunsaturated fats—sources of fat that have a predominant amount of fatty acids that contain more than one double bond in their chemical structures

polyuria—frequent urination

porphyria—a cluster of blood-related disorders characterized by abnormal porphyrin synthesis or metabolism; these disorders are hereditary and vary greatly depending upon which enzyme in the cascade of reactions is affected

portal hypertension—abnormally increased pressure in the portal venous system; frequently seen in cirrhosis of the liver and in other conditions that cause obstruction of the portal vein

positive nitrogen balance—net accumulation of protein in the body

positive selection—the rescue from apoptosis of T cells in the thymus that can recognize self-MHC molecules

posttranscriptional processing—the processing of newly transcribed RNA to excise introns, thus creating the final mRNA product prior to translation of mRNA into a protein

posttranslational modification—modification of a newly synthesized protein to its active form through changes such as phosphorylation or cleavage of specific sections

pre-diabetes mellitus—blood glucose levels that are higher than normal but not yet high enough to be diagnosed as diabetes

prebiotics—substances in food that stimulate the beneficial flora of the large intestine

preeclampsia—development of hypertension, with symptoms of proteinuria and edema, during pregnancy

prerenal azotemia—uremic symptoms associated with acute renal failure

pressor agents—substances that cause blood pressure to increase

primary cancer—the location or organ/cells from which the cancer originated

primary HIV infection—the time of the initial seroconversion to HIV infection; usually involves a spike in the level of the virus and sometimes is accompanied by a flulike syndrome

primary immune response—the immune response that occurs when the naive lymphocyte first encounters its antigen

privileged sites—nonvascularized locations in the body where foreign grafts are not rejected

probiotics—products containing microorganisms manufactured and sold as food products and supplements

prognosis—a prediction of the probable course and outcome of a disease, including expected response to treatment

Prognostic Inflammatory Nutrition Index (PINI)—a combination of serum C-reactive protein (CRP), alpha 1-acid glycoprotein (AAG), prealbumin (PA), and albumin measurements that is scored as a measurement of nutritional risk

progression—the third phase in cancer cell development; the orderly transformation of a preneoplastic lesion to a tumor and, ultimately, invasive cancer

prohepcidin—precursor to hepcidin; a marker for hemostasis

prokinetic agents—medications that cause the lower esophageal sphincter to close tightly, preventing gastric reflux; they also act to decrease transit time (i.e., they increase peristalsis) of stomach contents

promoter region—regulatory sequence in a gene to which molecules, such as fatty acids, can bind in order to induce expression of that specific gene; molecules can also bind to the promoter region to suppress transcription of a specific gene

promotion—the second phase in cancer cell development; process induced in a normal cell that has been exposed to a carcinogen, resulting in transformation into a cancer cell (promoters are not necessarily carcinogenic)

prophylaxis—preventative administration of a compound to avoid consequences of a disease state

prospective payment system—a system developed by U.S. government to reimburse health care providers for inpatient health services at a predetermined rate for a particular diagnosis and level of care

prospectively—refers to collecting data as it occurs or happens

protease inhibitor—a medication that prevents protein replication; a common class of drug that is used to prevent human immunodeficiency virus replication: it interrupts the viral replication cycle by inhibiting protease enzymes that allow the viral proteins to be cleaved for reassembly into viral cores

protein-losing enteropathy—increased fecal loss of serum protein, especially albumin, causing hypoproteinemia

protein turnover rate—a combination of the rates of catabolism and anabolism of protein stores in the body

proteinuria—the presence of too much protein in the urine

proteomics—study of the complement of proteins produced by a living organism

proton pump inhibitors—class of medications that block the H^+,K^+-ATPase enzyme, a component in HCl production, and hence reduce acid secretion in the stomach

protoporphyrin—the derivative of hemoglobin containing four pyrrole rings without the iron centers

pulmonary consolidation—changes in tissue structure of the lungs; often visualized as opaque components on a chest X-ray

pyelonephritis—inflammation of both the parenchyma of a kidney and the lining of its renal pelvis, especially due to bacterial infection

pyloroplasty—enlarging the pyloric sphincter

pyruvate complex disorders—disorders involving dysfunction in the metabolism of pyruvate, the end product of glycosis, via either the Kreb cycle or gluconeogenesis, resulting in the production of lactic acid

quadriplegia—paralysis involving all arms and legs; also known as tetraplegia

raffinose—a trisaccharide composed of galactose, fructose, and glucose that is found in grains, vegetables, and legumes; humans lack the enzyme to digest this compound to individual sugar molecules

rales—abnormal ("bubbly") respiratory sounds made when air flows through liquid present in the airways, and that may indicate pulmonary pathology

reactive hypoglycemia—low blood glucose levels that occur within 4 hours after a meal; also called postprandial hypoglycemia; symptoms are similar to those of diabetes-related hypoglycemia and may include hunger, sweating, shakiness, dizziness, light-headedness, sleepiness, confusion, difficulty speaking, anxiety, and weakness; a blood glucose level below 70 mg/dL at the time of symptoms and relief after eating will confirm the diagnosis

reduction and fixation of fracture—a method to surgically repair a bone fracture

refeeding syndrome—metabolic alterations that may occur during nutritional repletion of starved patients

refractory celiac disease—initial or subsequent failure of a strict gluten-free diet to restore normal intestinal architecture and function in patients who have celiac-like enteropathy

renal osteodystrophy—a general term that refers to bone disease related to CKD, caused by over- or underproduction of PTH or by exposure to aluminum

renal threshold—a concentration level of glucose in the blood above which the kidneys pass it through into the urine

resistant starch—indigestible starch that can be found naturally in foods such as beans and peas; produced during food processing or from chemical modification

respiratory acidosis—condition resulting from excess acid in the blood secondary to carbon dioxide retention

respiratory alkalosis—condition resulting from excess base in the blood secondary to increased carbon dioxide expiration

resting energy expenditure—energy expended by the body at rest to keep vital organ systems functioning, including the heart, kidneys, brain, liver, and lungs; it accounts for approximately 60% to 75% of 24-hour energy expenditure and is roughly 1 kcal/kg body weight/hour

reticulocytes—immature red blood cells; normal ranges for circulating erythrocytes exist, and levels reflect the ability of the bone marrow to produce precursor cells in normal amounts

retinopathy—a noninflammatory disorder of the retina, the tissue layer within the back of the eye that senses light and transmits sensory information to the brain

retroperitoneal—lying behind the peritoneum (lining of the abdominal cavity)

retrospectively—refers to collecting data from events that have already happened

retrovirus—a virus that carries RNA rather than DNA; RNA must be transcribed prior to integrating into the host cell DNA to reproduce

reverse transcriptase inhibitors—medications that interrupt the viral replication cycle by inhibiting reverse transcriptase enzymes that allow the viral RNA to be transcribed to DNA before being integrated into the host cell DNA

rhabdomyolysis—an acute condition of skeletal muscle destruction; the disintegration of muscle, associated with excretion of myoglobin in the urine

rheumatoid arthritis—a chronic inflammatory disease in which the synovial membrane of the joint becomes inflamed, resulting in swelling, stiffness, pain, limited range of motion, joint deformity, and disability

rheumatologist—a medical doctor specializing in diseases of the muscles and joints that are classified as rheumatic diseases

rickets—a condition characterized by inadequate mineralization of the organic matrix in the bones of children usually caused by a deficiency of vitamin D and resulting in bowing of the legs and skeletal deformity of the rib cage

salt resistant—describes an individual whose body presents resistance to change in blood pressure as a result of salt intake

salt sensitive—describes an individual who experiences an increase in blood pressure as a result of salt intake

salvage—additional treatment, used in hope of a cure or to prolong life, in a patient with recurrence of a malignancy following initial treatment

saturated fats—sources of fat that have a predominant amount of fatty acids that contain all single bonds within their chemical structures

scleroderma—a chronic autoimmune disease of unknown etiology affecting multiple organ systems including the skin, lungs, gastrointestinal tract, heart, kidneys, and, to a lesser extent, the joints and tendons

screening and referral system—a supportive system to the Nutrition Care Process and Model that helps identify those persons who would benefit from nutrition care

second-set rejection—accelerated rejection of an allograft due to previous exposure to some of the antigens on the graft

secondary hyperparathyroidism—high levels of PTH in the circulation that stimulate bone turnover, which may be accompanied by hyperplasia of the parathyroid glands

secondary immune response—rapid, more vigorous immunologic response by memory lymphocytes after the first encounter with an antigen; produced upon subsequent exposures to the antigen; often prevents reinfection

secondary polycythemia—condition in which an excessive number of red blood cells are produced; occurs in response to compensation for chronic hypoxemia

secretagogues—medications that increase secretion of insulin

seizure—episode of spontaneous, uncontrolled electrical activity in the brain

sense strand—the coding strand of DNA that is transcribed into RNA

sensible losses—fluid loss that can be measured (usually refers to fluid lost via urine excretion)

sensitivity—the likelihood that an individual with a disease or condition will be correctly identified when administered a test designed to detect that particular disease or condition

sepsis—a systemic inflammatory response and immunosuppressive process that prevents an adequate response to infection or trauma; may result in organ dysfunction or hypoperfusion abnormalities

seroconversion—the change in a person's antibody status from negative to positive

serum—fluid that is obtained after whole blood is separated into solid and liquid components; the fraction of blood containing water after the removal of cellular components; this should be differentiated from plasma, which contains serum, proteins, and clotting factors.

serum amyloid A—family of apolipoproteins associated with high-density lipoprotein (HDL) in plasma; considered to be an acute-phase protein released in response to inflammation

serum osmolality—a measure of the concentration of solute molecules in the blood

serum sickness—a Type III hypersensitivity response following the administration of a passive antibody in foreign serum

severe combined immune deficiency (SCID)—disease due to several mechanisms that produce an early block in differentiation pathways of both B and T lymphocytes, resulting in infants who are born lacking all major immune defenses

shaken baby syndrome—signs and symptoms that occur as a result of brain injury caused by violently shaking or impacting the head of an infant or small child

short bowel syndrome (SBS)—decreased digestion and absorption that result from a large resection of the small intestine

sickle cell anemia—a hereditary disease of genetically altered red blood cells that have a sickled shape, carry abnormally formed hemoglobin, and have abnormal transport capabilities for oxygen; the disease is thought to confer protection against malaria

sideroblastic anemia—a form of anemia characterized by the appearance of sideroblasts, immature ferritin-containing blast marrow cells in circulation

signs—observable phenomena such as heart or respiratory rate

silver nitrate—colloidal silver; used as an antibacterial treatment in burns

silver sulfadiazine cream—a sulfa medication used to prevent and treat bacterial or fungal infections

single nucleotide polymorphisms (SNPs)—situations in which one nucleotide is replaced by another in a gene, potentially leading to altered function

Sjögren's syndrome—a chronic, systematic, autoimmune inflammatory disorder, etiology unknown, characterized by dryness of mucous membranes and often associated with lupus or rheumatoid arthritis

social worker—a professional with at least a bachelor's degree in social work who provides persons, families, or vulnerable populations with psychosocial support, advises family caregivers, counsels patients, and helps plan for patients' needs after discharge

soma—major body portion of the neuron

somatic division—portion of the peripheral nervous system that carries messages from the body back to the brain

somatization—the physical manifestation of stress

somatostatin—a hormone and neurotransmitter that inhibits release of peptide hormones in several tissues; e.g., it is produced by the Δ cells of the Islets of Langerhans in the pancreas to control secretion of growth hormone from the anterior pituitary gland

sorbitol—a sugar alcohol; it is used as a sugar substitute

spasticity—involuntary muscle contraction that results in rigidity

specific gravity—the weight of a solution (e.g., urine) in comparison to an equal amount of distilled water; this is used to measure the concentrating ability of the kidney

specific immune system—body system responsible for immunity mediated by antigen-specific lymphocytes

specificity—the likelihood that an individual who does not have a particular condition or disease will be correctly excluded when administered a test designed to detect that condition or disease

speech-language pathologist—a health professional who has earned a master's degree and passed a national examination, who assesses, diagnoses, treats, and helps to prevent speech, language, cognitive, communication, voice, swallowing, fluency, and other related disorders

sphincter—a circular muscle that prevents movement or passage through the circle when contracted; sphincter muscles are located throughout the GI tract and are crucial control factors for peristalsis

spleen—a lymphoid organ in the abdominal cavity that filters blood

splenomegaly—enlargement of the spleen

stable angina—chest pain associated with increased oxygen demand such as occurs with physical exertion

stachyose—a tetrasaccharide consisting of 2 galactose, 1 fructose, and 1 glucose units

stadiometer—a calibrated device used to measure stature

standardized language—a uniform terminology that is used to describe practice

statin—a type of medication that is used to treat hyperlipidemias

stearic acid—an 18-carbon saturated fatty acid

steatorrhea—excess fat in the stool (>6 g/24 hrs), resulting from fat malabsorption and causing diarrhea

steatosis—excessive amounts of fat found in the stool

stem cells—nondifferentiated, primitive cells that have the ability both to multiply and to differentiate into more specialized cells that display unique functions

sterilization—a process that destroys all living organisms

stoma—an opening; e.g., a surgically created artificial opening into the abdomen

stomatitis—inflammation of the membrane in the mouth

stop codon (nonsense codon)—the codon in mRNA that signals completion of translation

stroke volume—the volume of blood that is ejected from the left ventricle with each systolic phase; defined mathematically as LVEDV – LVESV

struvite—a form of kidney stones composed of ammonium and magnesium phosphate; they resemble hard crystals

stylet—wire guide within the enteral tube that assists with insertion

subchondral bone—bone located beneath the articular cartilage of a joint

subcutaneous—refers to injection into the body under the skin

sublingual—refers to placement of a drug under the tongue

substrates—substances that an enzyme acts on to make a product

suppressor T cell—a T lymphocyte that suppresses (turns off) specific immune responses; this may or may not be a separate subclass of T cells

surfactant—substance secreted by the alveolar cells of the lung that serves to maintain the stability of pulmonary tissue by reducing the surface tension of fluids that coat the lung

surgical gastrostomy—an opening into the stomach that requires a surgical procedure

sweat chloride test—a test to measure the amount of chloride in the sweat by stimulating the skin to produce a large amount of sweat that is then absorbed by a special filter paper and analyzed for chloride content

sympathetic branch—portion of the peripheral nervous system that prepares the body for action; controls flight or fight response

symptoms—complaints experienced/verbalized by a patient

synapse—space or gap between nerve cells across which neurotransmitters pass

synbiotics—products that contain both prebiotics and probiotics

syncope—temporary loss of consciousness; fainting

synovial fluid—a protein-rich, slippery fluid contained inside a fibrous capsule that lubricates and nourishes the cartilage covering the ends of bones at their joints

synovial membrane—a membrane lining the capsule that encloses synovial joints and secretes synovial fluid, which lubricates and nourishes the cartilage at the end of bones

system factors—external factors (health care, education, and food supply systems) that influence the type of services that are available to individuals and how these services are delivered

systemic lupus erythematosus (SLE; lupus)—an autoimmune, chronic inflammatory disease that affects the connective tissue; affects skin, joints, kidneys, central nervous system, and mucous membranes and eventually spreads to all tissues, invoking a systemic reaction with pain, fever, sensitivity to light, and skin lesions

systole—contraction phase of the cardiac cycle; during this phase blood is ejected from the ventricles into the aorta and pulmonary artery

systolic blood pressure—pressure exerted when ejected from the ventricles (systole phase of the cardiac cycle)

T cells—lymphocytes that differentiate in the thymus

T-score—the number of standard deviations that the patient's BMD is either above or below the mean BMD for healthy young adults of the same sex and race; measure that compares a patient's bone mineral density (BMD) to a standard, healthy BMD, which is set at the mean BMD of healthy young adults of the same sex and race as the patient

tandem mass spectroscopy—the methodology used to detect a large number of organic acid compounds on a filter paper blood spot for diagnosing an inborn error of metabolism

tau—a protein that is a principal component of the paired helical filaments in neurofibrillary tangles; helps to maintain the structure of microtubules in normal nerve cells

telomere—the end section of a human chromosome

temporal lobe—portion of the cerebral cortex responsible for hearing and memory

temporomandibular joint (TMJ) syndrome—a condition of facial pain in the joints of the lower jaw

terminal—refers to a condition or disease for which there is no cure

tetrahydrobiopterin—a cofactor needed to stabilize the enzyme phenylalanine hydroxylase

Th1—a subset of the T helper cells that secretes cytokines, which trigger cell-mediated immune responses that promote inflammation and antiviral responses

Th2—helper T cells that predominate in the response to allergens and parasites and that make cytokines that promote antibody responses

thalassemia—a group of related blood disorders involving abnormal globin subunits in the hemoglobin molecule; these are hereditary and are most common in persons of Mediterranean or southeastern Asian descent

thermic effect of food—energy expended by the body to digest, absorb, and metabolize food; it accounts for about 10% of 24-hour energy expenditure

"third space" fluid—shift of fluid from the intravascular space to a nonfunctional space

thrombocytes—platelets; essential to blood clotting, these pieces of larger immature cells contribute to the formation of a thrombus (clot) by aggregating (coalescing) upon chemical activation after endothelial wall (blood vessel) tissue injury

thrombocytosis—low number of platelets

thrombopoietin—a stimulatory protein in red bone marrow that responds to the need for more platelets post-injury; causes an increase in the production of new platelets and also signals other systems to speed up the maturation and activation of the new platelets; up-regulates the complex mechanisms in hemostasis under conditions of injury and trauma

thrombus—blood clot

thymus—a primary lymphoid organ located in the chest, where T lymphocytes differentiate, proliferate, and are positively and negatively selected

topically—refers to placement of a drug on the skin

total iron-binding capacity (TIBC)—the capacity for the binding of iron by blood constituents; a surrogate measure for transferrin, since it binds the majority of iron

toxic megacolon—a very inflated colon with abdominal distention, and sometimes fever, abdominal pain, or shock

trabecular bone—loosely organized bone that has a sponge-like appearance and is found at the ends of long bones

tracheostomy—a surgical opening placed in the trachea to assist breathing

transcobalamin I–III—a group of proteins that are responsible for the transfer of vitamin B_{12}

transcription—the manufacture of RNA from DNA

transcription factor—a protein that activates transcription of a gene or genes by interacting with RNA polymerase in a gene promoter region

transcriptome—the complement of transcripts (mRNA) produced during gene expression

transferrin—the protein responsible for the transport of iron

transferrin saturation—the saturation of the carrier protein for iron, which is a sensitive indicator of iron status and stages of anemia

transient ischemic attacks (TIAs)—"mini-strokes"; an episode of ischemia where blood flow is quickly restored and symptoms resolve within 24 hours

translation—the assembly of a polypeptide chain based on the sequence of mRNA

transplantation—grafting an organ (e.g., kidney or heart) or cells (e.g., bone marrow) from one individual to another

tropic hormone—a hormone that regulates secretion of another hormone

tubules—component of the nephron responsible for reabsorption and secretion (within the kidney); designated as the proximal convoluted tubule, the loop of Henle, and the distal convoluted tubule

tumor necrosis factor (TNF-α)—one type of cytokine which has been found to possess a wide range of proinflammatory actions; it induces programmed cell death, primarily in tumor cells but for any cell with a receptor; also involved in immunoregulation

Tumor Node Metastases (TNM) Staging System—a systematic way of describing the size, location, and spread of a tumor; T describes the primary tumor according to its size, N applies to the lymph nodes and whether cancer cells have spread to them, and M refers to metastases and whether the cancer has spread to distant sites

tunneled catheter—intravenous access device that is placed in the vein on the upper chest wall and exits the body near the xyphoid process, axilla, or abdominal wall

U-100—refers to the units of insulin in each mL; U-100 is equivalent to 100 units per mL

ulceration—nonhealing break in skin or tissue surface

ulcerative colitis (UC)—a chronic inflammatory bowel disease (IBD) primarily located in the colon and rectum

ultrafiltrate—referring to the initial filtration of metabolic byproducts from the filtered blood within the tubule

ultrafiltration—a form of filtration that provides additional pressure to achieve more concentrated filtration

unstable angina—chest pain that occurs at rest

upper respiratory infection (URI)—a nonspecific term used to describe acute infections involving the nose, sinuses, pharynx, larynx, trachea, and bronchi; often referred to as the common cold

urea kinetic modeling—a quantitative method by which an individualized hemodialysis treatment prescription can be developed; requires measures of pre and post dialysis BUN, weights and treatment times

uremia (uremic syndrome)—a general term used to encompass a cluster of symptoms resulting from disordered biochemical processes as chronic kidney disease progresses; early symptoms include fatigue, delayed thinking, and pruritis

ureterorenoscopy—a nonsurgical procedure where a surgeon uses a fiberoptic instrument called a ureteroscope to remove a stone lodged in the ureter

uric acid—a crystalline acid occurring as an end product of purine metabolism; a common constituent of renal calculi

vaccine—a substance made from the whole organism or parts that contain critical antigenic components or genes for those components; it stimulates a primary immune response that produces antibodies and memory cells that protect against subsequent infection by that organism

vagotomy—severing of the vagus nerve; often a component of gastric surgery

vagus nerve—tenth cranial nerve; one of its major functions is to coordinate the autonomic nervous system communication between organs of digestion

validity—the quality of producing desired results

vasomotor—referring to nerves that innervate smooth muscles in the walls of arteries and veins and can cause their constriction or dilation

vasopressin—the primary endocrine factor that regulates urinary H_2O loss and overall H_2O balance; regulates blood pressure via this hormone's pressor effects on blood vessels; also known as antidiuretic hormone (ADH)

ventricular fibrillation—uncontrolled contractions of the ventricle; often associated with myocardial infarction

ventricular tachycardia—rapid heartbeat originating from the ventricle

viral load—the level of virus or viral markers measured in the blood

viron—a complete viral particle, consisting of RNA or DNA surrounded by a protein shell and constituting the infective form of a virus

viscosity—thickness of a liquid

volvulus—the twisting of the bowel causing obstruction

von Willebrand factor—a protein on the platelet membrane surface that is sensitive to the chemical signals of an injured cell and causes the platelet to become sticky and adhere to other platelets and blood constituents

water intoxication—uncontrolled, excessive water consumption resulting in dilutional complications

Wernicke-Korsakoff syndrome—manifestation of thiamin deficiency usually seen in individuals suffering from alcoholism

Wernicke's encephalopathy—condition characterized by confusion, nystagmus (involuntary eye movement), anisocoria (unequal size of the pupils), ataxia, and sluggishness

X-linked dominant—an inheritance pattern of a dominant allele on the X chromosome; such disorders are relatively rare

X-linked recessive—an inheritance pattern of a recessive allele on the X chromosome; related disorders are more common in males, who carry only one X chromosome

xenobiotics—chemicals that are found in an organism but are not produced by it or expected to be there, such as drugs or pollutants

xenograft—tissue transplantation between individuals from different species

xerostomia—dry mouth, often the result of damage to the salivary glands and decreased salivary production

Y-linked—inheritance based on the Y chromosome; disorders are extremely rare and occur only in males

Index

Dietary Reference Intakes (DRI)

The Dietary Reference Intakes (DRI) include two sets of values that serve as goals for nutrient intake—Recommended Dietary Allowances (RDA) and Adequate Intakes (AI). The RDA reflect the average daily amount of a nutrient considered adequate to meet the needs of most healthy people. If there is insufficient evidence to determine an RDA, an AI is set. AI are more tentative than RDA, but both may be used as goals for nutrient intakes.

In addition to the values that serve as goals for nutrient intakes (presented in the tables on these two pages), the DRI include a set of values called Tolerable Upper Intake Levels (UL). The UL represent the maximum amount of a nutrient that appears safe for most healthy people to consume on a regular basis. Turn the page for a listing of the UL for selected vitamins and minerals.

Estimated Energy Requirements (EER), Recommended Dietary Allowances (RDA), and Adequate Intakes (AI) for Water, Energy, and the Energy Nutrients

Age(yr)	Reference BMI (kg/m²)	Reference height, cm (in)	Reference weight, kg (lb)	Water[a] AI (L/day)	Energy EER[b] (kcal/day)	Carbohydrate RDA (g/day)	Total fiber AI (g/day)	Total fat AI (g/day)	Linoleic acid AI (g/day)	Linolenic acid[c] AI (g/day)	Protein RDA (g/day)[d]	Protein RDA (g/kg/day)
Males												
0–0.5	—	62 (24)	6 (13)	0.7[e]	570	60	—	31	4.4	0.5	9.1	1.52
0.5–1	—	71 (28)	9 (20)	0.8[f]	743	95	—	30	4.6	0.5	13.5	1.5
1–3[g]	—	86 (34)	12 (27)	1.3	1046	130	19	—	7	0.7	13	1.1
4–8[g]	15.3	115 (45)	20 (44)	1.7	1742	130	25	—	10	0.9	19	0.95
9–13	17.2	144 (57)	36 (79)	2.4	2279	130	31	—	12	1.2	34	0.95
14–18	20.5	174 (68)	61 (134)	3.3	3152[h]	130	38	—	16	1.6	52	0.85
19–30	22.5	177 (70)	70 (154)	3.7	3067[h]	130	38	—	17	1.6	56	0.8
31–50				3.7	3067[h]	130	38	—	17	1.6	56	0.8
>50				3.7	3067[h]	130	30	—	14	1.6	56	0.8
Females												
0–0.5	—	62 (24)	6 (13)	0.7[e]	520	60	—	31	4.4	0.5	9.1	1.52
0.5–1	—	71 (28)	9 (20)	0.8[f]	676	95	—	30	4.6	0.5	13.5	1.5
1–3[g]	—	86 (34)	12 (27)	1.3	992	130	19	—	7	0.7	13	1.1
4–8[g]	15.3	115 (45)	20 (44)	1.7	1642	130	25	—	10	0.9	19	0.95
9–13	17.4	144 (57)	37 (81)	2.1	2071	130	26	—	10	1.0	34	0.95
14–18	20.4	163 (64)	54 (119)	2.3	2368	130	26	—	11	1.1	46	0.85
19–30	21.5	163 (64)	57 (126)	2.7	2403[i]	130	25	—	12	1.1	46	0.8
31–50				2.7	2403[i]	130	25	—	12	1.1	46	0.8
>50				2.7	2403[i]	130	21	—	11	1.1	46	0.8
Pregnancy												
1st trimester				3.0	+0	175	28	—	13	1.4	+25	1.1
2nd trimester				3.0	+340	175	28	—	13	1.4	+25	1.1
3rd trimester				3.0	+452	175	28	—	13	1.4	+25	1.1
Lactation												
1st 6 months				3.8	+330	210	29	—	13	1.3	+25	1.1
2nd 6 months				3.8	+400	210	29	—	13	1.3	+25	1.1

NOTE: For all nutrients, values for infants are AI. Dashes indicate that values have not been determined.

[a] The water AI includes drinking water, water in beverages, and water in foods; in general, drinking water and other beverages contribute about 70 to 80 percent, and foods, the remainder. Conversion factors: 1 L = 33.8 fluid oz; 1 L = 1.06 qt; 1 cup = 8 fluid oz.

[b] The Estimated Energy Requirement (EER) represents the average dietary energy intake that will maintain energy balance in a healthy person of a given gender, age, weight, height, and physical activity level. The values listed are based on an "active" person at the reference height and weight and at the midpoint ages for each group until age 19. Chapter 12 provides equations and tables to determine estimated energy requirements.

[c] The linolenic acid referred to in this table and text is the omega-3 fatty acid known as alpha-linolenic acid.

[d] The values listed are based on reference body weights.

[e] Assumed to be from human milk.

[f] Assumed to be from human milk and complementary foods and beverages. This includes approximately 0.6 L (~3 cups) as total fluid including formula, juices, and drinking water.

[g] For energy, the age groups for young children are 1–2 years and 3–8 years.

[h] For males, subtract 10 kcalories per day for each year of age above 19.

[i] For females, subtract 7 kcalories per day for each year of age above 19.

SOURCE: Reprinted with permission from the *Dietary Reference Intakes* series, 1997, 1998, 2000, 2001, 2002, 2005 by the National Academy of Sciences, Courtesy of the National Academies Press, Washington, D.C.

Recommended Dietary Allowances (RDA) and Adequate Intakes (AI) for Vitamins

Age (yr)	Thiamin RDA (mg/day)	Riboflavin RDA (mg/day)	Niacin RDA (mg/day)[a]	Biotin AI (µg/day)	Pantothenic acid AI (mg/day)	Vitamin B$_6$ RDA (mg/day)	Folate RDA (µg/day)[b]	Vitamin B$_{12}$ RDA (µg/day)	Choline AI (mg/day)	Vitamin C RDA (mg/day)	Vitamin A RDA (µg/day)[c]	Vitamin D AI (µg/day)[d]	Vitamin E RDA (mg/day)[e]	Vitamin K AI (µg/day)
Infants														
0–0.5	0.2	0.3	2	5	1.7	0.1	65	0.4	125	40	400	5	4	2.0
0.5–1	0.3	0.4	4	6	1.8	0.3	80	0.5	150	50	500	5	5	2.5
Children														
1–3	0.5	0.5	6	8	2	0.5	150	0.9	200	15	300	5	6	30
4–8	0.6	0.6	8	12	3	0.6	200	1.2	250	25	400	5	7	55
Males														
9–13	0.9	0.9	12	20	4	1.0	300	1.8	375	45	600	5	11	60
14–18	1.2	1.3	16	25	5	1.3	400	2.4	550	75	900	5	15	75
19–30	1.2	1.3	16	30	5	1.3	400	2.4	550	90	900	5	15	120
31–50	1.2	1.3	16	30	5	1.3	400	2.4	550	90	900	5	15	120
51–70	1.2	1.3	16	30	5	1.7	400	2.4	550	90	900	10	15	120
>70	1.2	1.3	16	30	5	1.7	400	2.4	550	90	900	15	15	120
Females														
9–13	0.9	0.9	12	20	4	1.0	300	1.8	375	45	600	5	11	60
14–18	1.0	1.0	14	25	5	1.2	400	2.4	400	65	700	5	15	75
19–30	1.1	1.1	14	30	5	1.3	400	2.4	425	75	700	5	15	90
31–50	1.1	1.1	14	30	5	1.3	400	2.4	425	75	700	5	15	90
51–70	1.1	1.1	14	30	5	1.5	400	2.4	425	75	700	10	15	90
>70	1.1	1.1	14	30	5	1.5	400	2.4	425	75	700	15	15	90
Pregnancy														
≤18	1.4	1.4	18	30	6	1.9	600	2.6	450	80	750	5	15	75
19–30	1.4	1.4	18	30	6	1.9	600	2.6	450	85	770	5	15	90
31–50	1.4	1.4	18	30	6	1.9	600	2.6	450	85	770	5	15	90
Lactation														
≤18	1.4	1.6	17	35	7	2.0	500	2.8	550	115	1200	5	19	75
19–30	1.4	1.6	17	35	7	2.0	500	2.8	550	120	1300	5	19	90
31–50	1.4	1.6	17	35	7	2.0	500	2.8	550	120	1300	5	19	90

NOTE: For all nutrients, values for infants are AI. The glossary on the inside back cover defines units of nutrient measure.

[a] Niacin recommendations are expressed as niacin equivalents (NE), except for recommendations for infants younger than 6 months, which are expressed as preformed niacin.

[b] Folate recommendations are expressed as dietary folate equivalents (DFE).

[c] Vitamin A recommendations are expressed as retinol activity equivalents (RAE).

[d] Vitamin D recommendations are expressed as cholecalciferol and assume an absence of adequate exposure to sunlight.

[e] Vitamin E recommendations are expressed as α-tocopherol.

Recommended Dietary Allowances (RDA) and Adequate Intakes (AI) for Minerals

Age (yr)	Sodium AI (mg/day)	Chloride AI (mg/day)	Potassium AI (mg/day)	Calcium AI (mg/day)	Phosphorus RDA (mg/day)	Magnesium RDA (mg/day)	Iron RDA (mg/day)	Zinc RDA (mg/day)	Iodine RDA (µg/day)	Selenium RDA (µg/day)	Copper RDA (µg/day)	Manganese AI (mg/day)	Fluoride AI (mg/day)	Chromium AI (µg/day)	Molybdenum RDA (µg/day)
Infants															
0–0.5	120	180	400	210	100	30	0.27	2	110	15	200	0.003	0.01	0.2	2
0.5–1	370	570	700	270	275	75	11	3	130	20	220	0.6	0.5	5.5	3
Children															
1–3	1000	1500	3000	500	460	80	7	3	90	20	340	1.2	0.7	11	17
4–8	1200	1900	3800	800	500	130	10	5	90	30	440	1.5	1.0	15	22
Males															
9–13	1500	2300	4500	1300	1250	240	8	8	120	40	700	1.9	2	25	34
14–18	1500	2300	4700	1300	1250	410	11	11	150	55	890	2.2	3	35	43
19–30	1500	2300	4700	1000	700	400	8	11	150	55	900	2.3	4	35	45
31–50	1500	2300	4700	1000	700	420	8	11	150	55	900	2.3	4	35	45
51–70	1300	2000	4700	1200	700	420	8	11	150	55	900	2.3	4	30	45
>70	1200	1800	4700	1200	700	420	8	11	150	55	900	2.3	4	30	45
Females															
9–13	1500	2300	4500	1300	1250	240	8	8	120	40	700	1.6	2	21	34
14–18	1500	2300	4700	1300	1250	360	15	9	150	55	890	1.6	3	24	43
19–30	1500	2300	4700	1000	700	310	18	8	150	55	900	1.8	3	25	45
31–50	1500	2300	4700	1000	700	320	18	8	150	55	900	1.8	3	25	45
51–70	1300	2000	4700	1200	700	320	8	8	150	55	900	1.8	3	20	45
>70	1200	1800	4700	1200	700	320	8	8	150	55	900	1.8	3	20	45
Pregnancy															
≤18	1500	2300	4700	1300	1250	400	27	12	220	60	1000	2.0	3	29	50
19–30	1500	2300	4700	1000	700	350	27	11	220	60	1000	2.0	3	30	50
31–50	1500	2300	4700	1000	700	360	27	11	220	60	1000	2.0	3	30	50
Lactation															
≤18	1500	2300	5100	1300	1250	360	10	14	290	70	1300	2.6	3	44	50
19–30	1500	2300	5100	1000	700	310	9	12	290	70	1300	2.6	3	45	50
31–50	1500	2300	5100	1000	700	320	9	12	290	70	1300	2.6	3	45	50